Garden Seed Inventory
Sixth Edition

An Inventory of Seed Catalogs
Listing All Non-Hybrid Vegetable Seeds
Available in the United States and Canada

Originally Compiled and Edited by Kent Whealy
Subsequent Inventories Updated by Joanne Thuente
Programming, Statistics and Graphs by John Francis
Front Cover Art Copyright © by Judith Ann Griffith

Garden Seed Inventory (Sixth Edition)

Copyright © 2004 by Seed Savers Exchange, Inc.
All Rights Reserved

Seed Savers Exchange
3094 North Winn Road
Decorah, Iowa 52101
(Office) 563-382-5990
(Fax)   563-382-5872
www.seedsavers.org

ISBN 1-882424-60-3 (Softcover)
ISBN 1-882424-59-X (Hardcover)
Library of Congress Catalog Card Number 2004095357

Seed Savers, Seed Savers Exchange, Seed Savers Exchange "logo" (cupped hands sharing seeds), Heritage Farm, and The Flower and Herb Exchange are federally registered trademarks of Seed Savers Exchange, Inc. Any unauthorized use is strictly prohibited.

Printed on Recycled Paper with Soy Ink
in the United States of America

Contents

Introduction	4
How to Use This Book	12
How to <u>Really</u> Use This Book	12
Seed Companies and Source Codes	13
Taxonomy	44
Amaranth	47
Artichoke, Globe	51
Asparagus	52
Bean / Bush / Dry	54
Bean / Bush / Snap	64
Bean / Bush / Wax	77
Bean / Pole / Dry	80
Bean / Pole / Snap	84
Bean / Pole / Wax	91
Beet / Garden	92
Beet / Sugar	97
Broccoli	98
Broccoli Raab	101
Brussels Sprouts	102
Cabbage	104
Cabbage / Red	108
Cabbage / Savoy	110
Cardoon	111
Carrot	112
Cauliflower	120
Celeriac	124
Celery	125
Chicory	129
Chinese Cabbage	133
Collard	139
Corn / Dent	140
Corn / Flint	144
Corn / Flour	147
Corn / Pop	151
Corn / Sweet	154
Corn / Other Species	158
Cornsalad	159
Cowpea	160
Cress	166
Cucumber	168
Eggplant	175
Endive and Escarole	180
Fava Bean	183
Florence Fennel	186
Garlic	187
Gourd	198
Ground Cherry	203
Jerusalem Artichoke	204
Kale	205
Kohlrabi	209
Leek	210
Lettuce / Head	214
Lettuce / Leaf	225
Lettuce / Romaine	233
Lima Bean / Bush	239
Lima Bean / Pole	241
Melon / Honeydew and Casaba	244
Melon / Muskmelon	246
Melon / Other	253
Mushroom	256
Mustard Greens	261
Okra	263
Onion / Bunching	266
Onion / Common / Red-Purple	269
Onion / Common / White	271
Onion / Common / Yellow-Brown	273
Onion / Multiplier / Root	277
Onion / Multiplier / Topset	279
Onion / Other Species	280
Parsley	281
Parsnip	285
Pea / Garden	287
Pea / Edible Pod	294
Pea / Soup	298
Peanut	300
Pepper / Hot	302
Pepper / Sweet / Bell	316
Pepper / Sweet / Non-Bell	322
Pepper / Other Species	328
Potato	334
Quinoa	341
Radish	343
Runner Bean	351
Rutabaga (Swede Turnip)	353
Salsify	355
Scorzonera (Black Salsify)	356
Sorghum	357
Soybean	359
Spinach	362
Squash (*C. maxima*)	364
Squash (*C. mixta*)	372
Squash (*C. moschata*)	374
Squash (*C. pepo*)	378
Squash / Other Species	390
Sunflower / Ornamental	391
Sunflower / Edible	394
Sweet Potato	396
Swiss Chard	400
Tepary Bean	403
Tomatillo	405
Tomato / Orange-Yellow	406
Tomato / Pink-Purple	416
Tomato / Red	426
Tomato / Other Colors	462
Tomato / Other Species	471
Turnip	472
Watermelon	475
Miscellaneous	484
Miscellaneous Cucurbitaceae	494
Miscellaneous Leguminosae	496
Miscellaneous Solanaceae	500
Disease Abbreviations	502
Symbols and Other Abbreviations	503

Introduction

Seed Savers Exchange (SSE) is pleased to offer serious gardeners our *Sixth Edition* of the *Garden Seed Inventory*. This unique and comprehensive catalog of catalogs provides access to all non-hybrid vegetable varieties available from every mail-order seed company in the U.S. and Canada. Today's gardeners are blessed with access to a vast array of the finest vegetables ever developed, but this rich garden heritage continues to undergo rapid changes. As examples, 57% of the nearly 5,000 non-hybrid vegetable varieties offered in 1984 catalogs had been dropped by 2004, but 2,559 entirely new varieties have been introduced just during the last six years! The destructive forces transforming the seed industry and threatening this irreplaceable genetic diversity include takeovers and consolidation within the mail-order garden seed industry, the profit-motivated hybrid bias of most seed companies, and plant breeding for mechanical harvest and cross-country shipping. Counter-balancing these negative trends, an increasing number of specialty companies and individuals are introducing a vast array of excellent and fascinating vegetable varieties, including heirlooms that have never been offered before commercially and an increasing number of foreign varieties available for the first time in the West.

For more than two decades, the garden seed industry has undergone extensive consolidation, but those buyouts and mergers appear to mostly have stabilized. The greatest losses occurred from 1984 to 1987, when 23.5% of the mail-order seed companies in the U.S. and Canada went out of business or were bought out. Transnational agrichemical corporations went on a buying spree, purchasing small seed companies and replacing their regionally-adapted collections with more profitable hybrids and patented varieties. Many of the collections that were dropped represented the life's work of several generations of seedsmen, and were well adapted to regional climates and resistant to local diseases and pests. New corporate owners usually switched to generalized varieties that grew reasonably well throughout the U.S., thus assuring the greatest sales in the company's new nationwide market. Irreplaceable genetic resources were thoughtlessly destroyed by marketing decisions to maximize the short-term profits of corporations that often ended up not even owning those seed companies the following year.

By the late 1970s many of SSE's members were very concerned, because their favorite varieties were disappearing from seed catalogs without warning. Those losses seemed to be escalating, but there was no overall view of the garden seed industry to determine what was being lost or how quickly. In 1981 I decided to compile and publish a comprehensive inventory of every mail-order vegetable seed catalog in the U.S. and Canada. That rather crude *First Edition* of the *Garden Seed Inventory* (*GSI-1*) took three years to compile on a tiny, ancient Apple computer and was finally published in December 1984. The book was an inventory of 230 mail-order catalogs, contained descriptions of 4,949 non-hybrid vegetable varieties available at that time, and a coded list of every company offering each variety. The ongoing editions of this unique sourcebook have proven to be unequaled preservation tools that clearly identify the varieties about to be dropped, which has allowed preservation projects and gardeners to buy up endangered seeds while sources still exist.

Any inventory is out-of-date before it is published and becomes less effective with each passing year. In 1987 SSE decided to publish an updated, greatly improved *Garden Seed Inventory: Second Edition* (*GSI-2*). SSE's office staff obtained every 1987 mail-order catalog in the U.S. and Canada, which were compared to printouts of their 1984 offerings, then company codes were either added or deleted, and all of the varietal descriptions were also updated. Immediately it became apparent that the garden seed industry was in an almost violent state of flux. Out of 230 seed companies inventoried in 1984, 54 went out of business or had been bought out by 1987. Sadly, but predictably, the majority were smaller companies that had been rich sources of unique varieties. *GSI-2* also included 39 new companies that emerged during the same three-year period, which brought the total back up to 215, but the 39 new companies were offering only 21% as many unique varieties as the number of unique varieties that disappeared when the 54 former companies were lost.

The following chart shows the number of companies that went out of business between each of SSE's inventories, and also the companies that emerged during the same period. The figures in parentheses are the percentages of losses and gains when compared to the total number of companies in the previous inventory. (Remember that the years between SSE's inventories are not equal: three years elapsed between *GSI-1* and *GSI-2*, four years between *GSI-2* and *GSI-3*, three years between *GSI-3* and *GSI-4*, four years between *GSI-4* and *GSI-5*, and six years between *GSI-5* and *GSI-6*.) So those percentages have also been divided by the number of years since the previous inventory, which shows that annual losses of companies have slowed from a high of 7.8% per year (1984-1987) down to 4.2% recently (1998-2004). Total gains surpassed total losses about 1990 and have continued to increase ever since.

Mail-Order Vegetable Seed Companies in the U.S. and Canada

	GSI-1 (1984)	GSI-2 (1987)	GSI-3 (1991)	GSI-4 (1994)	GSI-5 (1998)	GSI-6 (2004)
Companies Lost		-54	-41	-26	-32	-64
(% of previous total)		(-23.4)	(-19.1%)	(-11.7%)	(-13.1%)	(-25.1%)
(% lost annually)		(-7.8%)	(-4.8%)	(-3.9%)	(-3.3%)	(-4.2%)
Companies Gained		+39	+49	+47	+43	+83
(% of previous total)		(+16.9%)	(+22.8%)	(+21.1%)	(+17.6%)	(+32.5%)
(% gained annually)		(+5.6%)	(+5.7%)	(+7.0%)	(+4.4%)	(+5.4%)
Total Companies	230	215	223	244	255	274

The next chart focuses on "Relative Availability" (the number of companies that were offering each variety), and those statistics clearly show that about half of all non-hybrid vegetable varieties have always been right on the edge of being lost. For example, in 1984 the 230 companies (in the chart above) were offering 4,949 total varieties (below). The next-to-last line of the chart below shows that 13 of the most common varieties were available from more than 100 sources (were being offered by more than 100 different companies out of 230). But that same year 2,660 of the varieties (53.7%) were "one source varieties" (and that single source could have been any one of the 230 companies inventoried for GSI-1). Those disturbing percentages for "one source varieties" have barely changed since 1984.

Relative Availability (Sources per Variety)

Available from:	GSI-1 (1984) # of Varieties (percentage)	GSI-2 (1987) # of Varieties (percentage)	GSI-3 (1991) # of Varieties (percentage)	GSI-4 (1994) # of Varieties (percentage)	GSI-5 (1998) # of Varieties (percentage)	GSI-6 (2004) # of Varieties (percentage)
1 source	2,660 (53.7%)	2,841 (53.5%)	3,002 (51.8%)	3,525 (54.4%)	3,839 (52.5%)	4,226 (49.8%)
2 sources	659 (13.3%)	711 (13.4%)	883 (15.2%)	941 (14.5%)	1,076 (14.7%)	1,345 (15.8%)
3 sources	304 (6.1%)	363 (6.8%)	431 (7.4%)	459 (7.0%)	568 (7.8%)	646 (7.6%)
4 sources	183 (3.7%)	206 (3.9%)	243 (4.2%)	269 (4.1%)	337 (4.6%)	390 (4.6%)
5 sources	147 (3.0%)	157 (3.0%)	172 (3.0%)	217 (3.3%)	211 (2.9%)	290 (3.4%)
6-10	360 (7.3%)	395 (7.4%)	454 (7.8%)	449 (6.9%)	558 (7.6%)	689 (8.1%)
11-20	280 (5.7%)	286 (5.4%)	299 (5.2%)	321 (4.9%)	366 (5.0%)	435 (5.1%)
21-50	254 (5.1%)	242 (4.6%)	228 (3.9%)	248 (3.8%)	255 (3.5%)	356 (4.2%)
51-100	89 (1.8%)	92 (1.7%)	77 (1.3%)	81 (1.2%)	97 (1.3%)	109 (1.3%)
101+	13 (0.3%)	14 (0.3%)	8 (0.1%)	6 (0.1%)	6 (0.1%)	7 (0.1%)
Totals	4,949 (100%)	5,307 (100%)	5,797 (100%)	6,483 (100%)	7,313 (100%)	8,494 (100%)

The total number of non-hybrid varieties being offered (in the previous chart) has steadily increased since 1984. At first glance these increasing totals would appear to indicate that losses are no longer occurring, but a more detailed examination of those statistics (see the following chart) reveals that the original varieties available in 1984 are still being lost at a steady rate. What's actually happening is that the continuing losses of original varieties in *GSI-1* are being obscured by even greater increases of newly introduced varieties.

The following statistics for "Total Availability" reach back to 1981 when U. S. and Canadian seed catalogs were first gathered at the beginning of this study. (Although the total number of varieties available in the 1981 and 1984 catalogs can be determined, significant losses occurred during the three years of compilation before *GSI-1* was published in December 1984.) The first set of figures (below) gives the number of "Varieties Dropped" since the previous inventory, followed by "percent of previous total" (the varieties dropped divided by the total varieties in the previous inventory), and that percentage is then divided by the number of years since the previous inventory to determine the "percent dropped annually." The second set of figures shows all of the new "Varieties Added," followed by the percent of increase based on the previous inventory's total, and then the percent that were added annually. The third set of figures are the "Total Varieties" in each inventory (including the beginning total from the 1981 catalogs), followed by percentages of losses and gains based on that 1981 total. Apparently the total number of varieties was in decline in 1981, bottomed out about 1984 and has increased steadily ever since.

Total Availability (Losses and Gains Between Inventories)

	1981	*GSI-1* (1984)	*GSI-2* (1987)	*GSI-3* (1991)	*GSI-4* (1994)	*GSI-5* (1998)	*GSI-6* (2004)
Varieties Dropped:			-943	-1,263	-1,108	-1,059	-1,791
(% of previous total)			(-19.1%)	(-23.8%)	(-19.1%)	(-16.3%)	(-24.5%)
(% dropped annually)			(-6.4%)	(-6.0%)	(-6.4%)	(-4.1%)	(-4.1%)
Varieties Added:			+1,301	+1,753	+1,794	+1,889	+2,657
(% of previous total)			(+26.3%)	(+33.0%)	(+30.9%)	(+29.1%)	(+36.3%)
(% added annually)			(+8.8%)	(+8.3%)	(+10.3%)	(+7.3%)	(+6.1%)
Total Varieties	5,534	4,949	5,307	5,797	6,483	7,313	8,494
(% of the 1981 total)	(100%)	(81.1%)	(95.9%)	(104.8%)	(117.1%)	(132.1%)	(153.4%)

A closer examination of the 4,226 "one source varieties" for 2004 in the "Relative Availability" chart (on the previous page) reveals that just a handful of companies are responsible for increasing the genetic diversity available to North American gardeners. Only 10% of the 274 companies in *GSI-6* are responsible for 58.1% of the "one source varieties" offered in 2004. The following chart shows the top 27 companies that offered the most one source varieties in *GSI-6*, followed by the number of unique varieties in their 2004 catalogs, and then their percentage of the unique varieties within these top 27 companies.

Companies Offering the Most "One Source Varieties" in *GSI-6*

Top 27 Companies (top 10 percent)	**Code**	**Unique Varieties**	**Percent**
Sand Hill Preservation Center	Sa9	345	8.1 %
Horus Botanicals	Ho13	326	7.7 %
Native Seeds/SEARCH	Na2	253	6.0 %
Seed Dreams	Se17	145	3.4 %

INTRODUCTION

TomatoFest® Homegrown Seeds	To9	98	2.3 %
Tanager Song Farm	Ta5	96	2.3 %
Thompson & Morgan, Inc.	Tho	91	2.2 %
Marianna's Heirloom Seeds	Ma18	75	1.8 %
Seeds from Italy	Se24	72	1.7 %
Sourcepoint Organic Seeds	So12	72	1.7 %
Redwood City Seed Co.	Red	69	1.6 %
Florida Mycology Research Center	Fl3	67	1.6 %
Synergy Seeds	Syn	66	1.6 %
Baker Creek Heirloom Seeds	Ba8	64	1.5 %
Territorial Seed Co.	Ter	62	1.5 %
Eastern Native Seed Conservancy	Ea4	61	1.4 %
Prairie Garden Seeds	Pr3	58	1.4 %
Western Hybrid Seeds, Inc.	WE10	56	1.3 %
Meyer Seed International	ME9	51	1.2 %
Salt Spring Seeds	Sa5	48	1.1 %
Evergreen Y.H. Enterprises	Ev2	43	1.0 %
Southern Exposure Seed Exchange	So1	42	1.0 %
Seed Savers Catalog	Se16	41	1.0 %
Seeds of Change	Se7	39	1.0 %
Berton Seeds Co. Ltd.	Ber	38	1.0 %
Stokes Seeds, Inc.	Sto	38	1.0 %
Terra Edibles	Te4	38	1.0 %
One Source Varieties		**2,454**	**58.1 %**

Many of these same companies have also introduced the most "New Unique Varieties." In the 2004 catalogs there were 2,657 "new unique varieties" (defined as zero sources from 1981-1998, and one source in 2004). The following chart of new unique varieties introduced since *GSI-5* reveals that only 10% of the 274 companies in *GSI-6* introduced 60.6% of the new varieties during the last six years.

Companies Introducing the Most "New Unique Varieties" since *GSI-5*

Top 27 Companies (top 10 percent)	**Code**	**Unique Varieties**	**Percent**
Sand Hill Preservation Center	Sa9	160	6.3 %
Native Seeds/SEARCH	Na2	142	5.5 %
Seed Dreams	Se17	129	5.0 %
Horus Botanicals	Ho13	116	4.5 %
TomatoFest® Homegrown Seeds	To9	98	3.8 %
Tanager Song Farm	Ta5	96	3.8 %
Marianna's Heirloom Seeds	Ma18	72	2.8 %
Seeds from Italy	Se24	72	2.8 %
Thompson & Morgan, Inc.	Tho	68	2.7 %
Baker Creek Heirloom Seeds	Ba8	63	2.5 %
Eastern Native Seed Conservancy	Ea4	61	2.4 %
Meyer Seed International	ME9	51	2.0 %
Sourcepoint Organic Seeds	So12	42	1.6 %
Synergy Seeds	Syn	39	1.5 %

Redwood City Seed Co.	Red	38	1.5 %
Florida Mycology Research Center	Fl3	37	1.4 %
Seed Savers Catalog	Se16	36	1.4 %
Territorial Seed Co.	Ter	35	1.4 %
Salt Spring Seeds	Sa5	33	1.3 %
J. L. Hudson, Seedsman	Hud	31	1.2 %
Kitazawa Seed Co.	Kit	30	1.2 %
Terra Time & Tide	Te8	30	1.2 %
Bayou Traders Peppermania	Ba16	29	1.1 %
Corona Seeds WorldWide	CO30	26	1.0 %
Turtle Tree Seed Farm	Tu6	26	1.0 %
Heirloom Tomatoes	He18	25	1.0%
Terra Edibles	Te4	25	1.0%
New Unique Varieties		**1,610**	**60.6%**

Gardeners should be extremely pleased to learn that 2,657 new unique varieties have been introduced within the past six years, but these gains are extremely fragile and could obviously be wiped out by the loss of just a few companies. The following chart reveals that substantial increases (200% or more from 1981 to 2004) have occurred for certain types of plants. In the column on the right are the codes of companies in *GSI-6* that offered a substantial portion of the unique varieties in these plant types (and in parentheses is the total number of the unique varieties that each company offered). Again, these substantial gains are the result of outstanding efforts by only a few remarkable companies.

Plant Types That Increased More Than 200% Since 1981

Plant Type	1981	1984	1987	1991	1994	1998	2004	Company Codes (unique varieties)
Bean/Pole/Dry	25	19	29	25	30	35	85	Na2 (43)
Bean/Pole/Snap	68	62	80	91	90	111	139	Se17 (9)
Corn/Flour	26	22	41	62	61	67	74	Na2 (31)
Garlic	18	18	18	51	110	165	274	Fi6 (29), Fi7 (16)
Gourd	19	19	23	23	22	24	88	Sa9 (18)
Jerusalem Artichoke	4	4	9	14	19	12	16	Ho13 (7), Moo (3), Ma13 (2)
Lettuce/Leaf	51	45	61	78	95	134	184	OSB (9), Me9 (9)
Lettuce/Romaine	28	24	38	53	67	77	109	Me9 (14)
Miscellaneous Cucurbitaceae	11	10	15	17	17	18	26	Ev2 (3), Sa9 (2), Hud (2), Ho13 (2)
Miscellaenous Leguminosae	25	25	37	49	47	65	80	Ho13 (11), Kit (9), Hud (5)
Miscellaneous Solanaceae	2	2	3	4	3	18	27	Hud (5), Ho13 (5)
Okra	23	22	27	30	33	43	51	Ho13 (13)
Onion Multiplier/Topset	2	2	4	5	5	5	9	Ho13 (6)
Pea/Soup	4	3	4	6	14	36	47	Ho13 (19), Na2 (9)
Pepper/Hot	106	98	115	166	238	264	325	Pep (30), Na2 (17), Te8 (16)
Pepper/Other Species	16	15	21	33	72	108	143	Red (21), Te8 (14), Ba16 (14)
Potato	61	55	73	193	230	163	129	Ron (18), Moo (7)
Quinoa	2	2	7	28	28	21	22	So12 (5), Hud (2)
Sorghum	10	10	12	18	29	38	48	Na2 (5), Sa9 (4), Syn (4), Ho13 (3)
Squash/Mixta	10	9	17	46	40	29	35	Na2 (8)
Squash/Moschata	28	25	36	42	39	41	68	Na2 (11), Sa9 (7), Ba8 (5)

	1981	1984	1987	1991	1994	1998	2004	
Sweet Potato	23	20	24	21	28	70	79	Sa9 (41), Al6 (6)
Tepary Bean	10	4	10	27	35	36	36	Na2 (17)
Tomatillo	4	4	7	12	14	10	18	So25 (2)
Tomato/Other Colors	9	8	12	20	41	116	190	To9 (16), Sa9 (11)
Tomato/Orange-Yellow	33	28	37	57	82	156	208	To9 (26), Sa9 (11), Ma18 (10)
Tomato/Pink-Purple	40	34	39	54	68	144	245	Sa9 (30), Ma18 (20)

After *GSI-1* was published, it became readily apparent that the ongoing losses could be at least partially offset if alternative seed companies continued to emerge and offer unique collections. That movement has rapidly gained momentum, which is a bright ray of hope in an otherwise bleak situation. SSE is pleased that our various editions of the *Garden Seed Inventory* have focused attention on the admirable efforts of these specialized seed companies striving to offer unique varieties to the gardening public. Gardeners can encourage and support this promising trend by patronizing these exceptional companies, which will help to ensure their economic survival.

Again, the highly successful efforts by these companies tend to mask the statistical evidence of the steadily continuing losses of traditional varieties. Certain types of vegetables have been decimated, especially the biennials and many of the food crops that are easily hybridized. For example, most gardeners tend to just purchase a six-pack of various *Brassicas* (members of the cabbage family) and a bundle of onion plants at their local greenhouse or grocery store, which strongly encourages the widespread use of only a handful of the most common varieties. To make matters worse, endangered *Brassicas* and onions are some of the most difficult plant types for small seed companies to permanently maintain and continue to reoffer.

The following chart lists the plant types that have suffered losses of 25% or more since 1981.

Plant Types That Lost 25% or More Since 1981

Plant Type	**1981**	**1984**	**1987**	**1991**	**1994**	**1998**	**2004**	**(% of 1981)**
Asparagus	22	20	23	19	19	17	14	(-36.4 %)
Beet/Sugar	24	20	13	9	10	9	11	(-54.2 %)
Broccoli	50	46	38	33	37	32	32	(-36.0 %)
Brussels Sprouts	28	21	15	14	14	13	14	(-50.0 %)
Cabbage	134	117	104	77	60	57	47	(-64.9 %)
Cabbage/Red	35	31	28	22	19	21	16	(-54.3 %)
Cabbage/Savoy	30	26	24	16	30	30	18	(-40.0 %)
Cauliflower	152	131	118	100	94	81	55	(-63.8 %)
Celeriac	21	16	17	16	16	19	14	(-33.3 %)
Corn/Other Species	17	15	20	14	17	14	8	(-52.9 %)
Cornsalad	25	24	27	27	24	22	17	(-32.0 %)
Onion/Common/White	60	55	44	48	44	40	30	(-50.0 %)
Onion/Common/Yellow-Brown	95	83	72	58	55	55	47	(-50.5 %)
Pea/Garden	182	160	148	127	123	124	135	(-25.4 %)
Radish	183	162	165	153	149	155	138	(-25.0 %)
Salsify	11	9	12	11	5	3	3	(-73.7 %)
Spinach	56	49	51	39	54	46	31	(-46.6 %)
Squash/Other Species	3	3	3	3	3	2	1	(-66.7 %)
Turnip	64	57	54	38	39	39	38	(-40.6 %)

One of the main purposes of each *Garden Seed Inventory* is to reveal the ongoing gains and losses of commercially available, non-hybrid vegetable varieties. The varietal listings within each plant type's section show which varieties are being offered in the current catalogs, but that assemblage is only a snapshot at the time of that inventory. To better illustrate the ongoing changes within the seed trade, we have developed graphs for each plant type that are designed to show the flow of losses and gains that have occurred across time. Also, those graphs are followed by three statistical statements that describe the status of that plant type. Finally, at the end of each plant type's section there is a list of all varieties dropped since 1981. The information provided by these three elements (graphs, statistical statements and lists of dropped varieties) gives an extremely clear picture of the ongoing changes that have occurred within the mail-order garden seed industry.

The graphs in the following example visually illustrate the flow of gains and losses since this study began in 1981. The height of the columns and the numbers (in parenthesis) below each column show the number of varieties of that plant type offered in each inventory. The steep downward slope of the columns makes it quite clear that open-pollinated cauliflowers are steadily disappearing at an alarming rate. The columns are divided into sections with various shadings, designed to show all of the gains and subsequent losses of varieties in each inventory. For example, black portion of each column shows the number of varieties available in 1981 catalogs at the beginning of this study (far left column), and then tracks how many of those remained available in each subsequent inventory. Looking only at those black sections: 152 cauliflower varieties were available in 1981, but those varieties had declined to 131 by 1984 (when *GSI-1* was finally published). Further losses reduced those varieties to 95 in 1987, 61 in 1991 and 36 in 1994, but then the remaining 1981 varieties increased to 42 in 1998 (six of the previously dropped varieties became available again), and finally dropped to 29 in 2004. The dark grey section at the top of the 1987 column shows that 23 new varieties were introduced between 1984 and 1987 and then tracks the subsequent losses of those varieties. Likewise, the differently shaded section at the top of the 1991, 1994, 1998 and 2004 columns show the new introductions and then track subsequent losses.

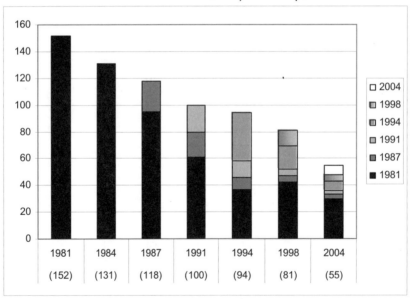

Cauliflower
Brassica oleracea (Botrytis Group)

Number of varieties listed in 1981 and still available in 2004: 152 to 30 (20%).
Overall change in number of varieties from 1981 to 2004: 152 to 55 (-64%).
Number of 2004 varieties available from 1 or 2 sources: 34 out of 55 (62%).

Three statistical statements, located immediately below each set of graphs, further characterize the status of that plant type. In the example, above, the first statement reveals that only 20% of the cauliflower varieties available in 1981 (30 out of the original 152) are still being offered in 2004. The second statement explains that even when all of the new varieties introduced between 1981 and 2004 are included, the number of open-pollinated cauliflowers has still decreased by 64%. The last statement indicates the degree to which the 2004 varieties are imperiled (their relative rarity), because 62% of the open-pollinated cauliflower varieties being sold in 2004 are available from only one or two sources (this is the criteria that SSE uses when buying samples of varieties in danger of being dropped). Finally, following all of the plant descriptions and source codes in the Cauliflower section of this inventory, there is an alphabetical list of all varieties dropped since this study began in 1981 (and the year each was dropped). This *Sixth Edition* of the *Garden Seed Inventory* took two years to research and compile – normally it takes one year – because seed sales on the internet have absolutely exploded! Although we have extensively researched the varieties available on the internet, many (if not most) of those websites are relatively minor and obscure. Since this ongoing inventory and the validity of its more than two decades of statistics depend on being able to re-inventory each source several years from now, whether or not these websites will still be around is of major concern to us. For example, think about the table of "Mail-Order Vegetable Seed Companies in the U.S. and Canada" at the beginning of this introduction. There is a tremendous difference between the disastrous fire that destroyed Abundant Life Seed Foundation and their hundreds of offerings, compared to an individual who decides to sell a couple of varieties on the internet for only a year or two. Therefore, we have only inventoried and included those websites that appear to be both substantial and stable. Although we fully realize those are arbitrary decisions, the alternative was to not inventory any new website-only sources, which would have deprived our readers of access to a substantial number of unique varieties. Many companies are now offering their customers the choice of ordering from their mail-order catalogs or from new websites, while some companies that previously sent out mail-order catalogs have now switched entirely to website sales. Because of these rapid changes, which are bound to increase and accelerate, there is the distinct possibility that researching and compiling a *Seventh Edition* of the *Garden Seed Inventory* may become virtually impossible. That decision, however, won't have to be made for a few years.

During the first half of the previous century, substantial losses were the norm while commercially available vegetables were rapidly being superseded by superior varieties developed in dynamic breeding programs at State Universities and Agricultural Experiment Stations. But public sector plant breeding programs are in rapid decline, and the garden seeds currently being dropped from mail-order catalogs are the best home garden varieties we will ever see. Far from being obsolete or inferior, these are the cream of our vegetable crops. If our vegetable heritage is allowed to die out, home gardeners will become increasingly more dependent on transnational seed companies and the generic and hybrid and patented varieties they choose to offer. And that means giving up our right to determine the quality of the food our families grow and consume, and losing the ability of gardeners and farmers to save their own seeds which is the reason that all of this incredible diversity exists in the first place.

Ever since *GSI-1* was published in 1984, SSE has used revenue generated by sales of these inventories to purchase seeds of endangered varieties (which are then permanently maintained at Heritage Farm, SSE's headquarters near Decorah, Iowa). SSE's members also have constantly been encouraged to buy up these endangered varieties, so we hope that our combined efforts have been in time to rescue a substantial portion of our endangered vegetable heritage. It seems ironic that today's gardeners have access to such a vast array of the best vegetable varieties ever developed, and yet so many are in immediate danger of being lost forever. This rapidly disappearing garden heritage is the end result of millions of years of natural selection, 12,000 years of human selection, and (sometimes) decades of costly plant breeding. Backyard gardeners are emerging as the most vitally concerned stewards of this irreplaceable genetic wealth, and we must quickly accept our responsibility. Try to imagine what it would cost – in terms of time and energy and money – to develop this many outstanding varieties. But they already exist. All we have to do is save them.

– Kent Whealy
Executive Director
Seed Savers Exchange

(November 2004)

How To Use This Book

Each variety in this inventory has a separate record which consists of: **Variety Name**, (Possible Synonyms), maturity, description, *Source history*, and **Source codes**. Let's briefly examine each part:

Variety Name - also known as the plant's "common name." Some names have been reversed for better groupings (example: Chicago Warted Hubbard is listed as Hubbard, Chicago Warted). This groups all Hubbard squash together and, within that group, the Hubbard varieties are arranged alphabetically.

Synonyms - other names for the same variety. If synonyms occur, they are listed in parentheses following the Variety Name. A synonym is listed even if mentioned by only one company. Almost no combining has been done on the basis of synonyms, because companies sometimes drop a variety and then attempt to switch their customers to a similar variety by claiming it is a synonym.

Days to maturity - a range from shortest to longest maturities listed by all sources. The minimum maturity will be of value to gardeners in short season areas. The maximum maturity can be misleading, especially for commercial varieties bred for winter growing areas or those tested where cool temperatures delay maturation.

Plant description - a composite pieced together from sometimes dozens of catalog descriptions. All sizes, therefore, are *ranges* of everything being offered. "Bushy det. upright 14-26 in. plants" does not mean that bean plants within a population will vary from 14-26 inches in height. Instead, it either means that various companies may be offering slightly different strains or that the variety was tested under different conditions that affected its growth.

Source history - the number of companies that offered the variety in each inventory. These statistics show the flow of gains and losses in availability during each inventory, and gives a clear picture of which varieties are about to be lost.

Source codes - 3-letter codes for the 274 mail-order seed companies in this inventory. Each code is the first three letters in that company's name. When that would give two or more companies the same code, a number replaces the third letter. Wholesale companies, which only sell to other companies, have codes in all capital letters. Codes for retail companies (and companies with both wholesale and retail sales) have only the first letter capitalized. This lets gardeners tell at a glance which companies sell to individuals. All codes are listed under "Seed Companies" on pages 13-43 with addresses, prices of catalogs, and descriptions of specialties.

To order seed of any variety in this inventory: look up one of its "retail codes" (pages 13-43), send the correct amount for that company's catalog or price list, and buy the seed. Many people mistakenly think that companies only sell seeds in the spring, but SSE has good luck buying seeds any time of year. When looking up a company with a wholesale code, you will see that all sales are "Wholesale only." Do not request those catalogs, because such companies only sell to other companies. Sometimes, however, you can have a local business (greenhouse or garden seed store) order the seeds for you.

How To *Really* Use This Book

Anyone can figure out *how* to buy seeds using this inventory. For those interested in genetic preservation and working with other gardeners to save our vegetable heritage, let's discuss *which* seeds you should buy.

Garden Seed Inventory was designed as a preservation tool to save endangered commercial varieties about to be dropped from seed catalogs. After each variety's description are statistics showing how many companies offered that variety in each inventory. This "Source History" is printed in italics and shows the number of companies in 1981, 1984, 1987, 1991, 1994, 1998 and 2004. The flow of gains and losses is perfectly clear. You can easily see which varieties are going down fast. Buy them!

The "Source History" also highlights several other dangerous situations. Hundreds of varieties listed in this inventory have only been available from a single source every year since 1981. Buy them! Even varieties with several sources may actually be a grouping of one wholesale supplier and several of its retailers. If that wholesale company drops varieties or goes out of business, there can still be a grace period of a couple years before the retailers sell out their remaining stocks of seeds.

There is a large amount of excellent Oriental and European varieties currently being offered. The Peoples Republic of China is a botanical treasure chest that is just beginning to open. Often business contacts in the East are shaky and many Oriental varieties are only available for a single year. Buy them! Outstanding European varieties are also being offered by a few specialty companies, but these windows to foreign varieties can close quickly.

Each plant type in this *Sixth Edition* ends with lists of all dropped varieties. If you are maintaining any of those varieties, SSE would appreciate being sent a list. We will get back to you if our purchasing efforts have missed any of the varieties that you are keeping. SSE would also be pleased to have you offer such seeds to other gardeners through a listing in *Seed Savers Yearbook*.

Seed Savers Exchange is a grassroots network of gardeners and plant collectors who maintain and distribute rare varieties of food crops. *Seed Savers 2004 Yearbook* lists names and addresses of 800 SSE members who are offering nearly 12,000 "heirloom" vegetable and fruit varieties. Since SSE was founded in 1975, our members have distributed an estimated one million samples of garden seeds often on the verge of extinction. SSE's projects will link you with thousands of other dedicated gardeners who are determined that our vegetable heritage will not die out. Membership information is in the center of each Seed Savers Catalog (Se16) or at www.seedsavers.org.

Seed Companies (and Source Codes)

Ab2 **Abbe Hills Seed,** L. J. Krouse, 825 Abbe Hills Road, Mt. Vernon, IA 52314. Phone: 319-895-6924. E-mail: lkrouse@cornellcollege.edu. Send SASE for information letter. Sells one variety of Reid's Yellow Dent-type open-pollinated untreated corn in one-bushel bags, for livestock feed or silage. Retail and wholesale. Offered 1 corn variety in 2004.

ABB **Abbott and Cobb, Inc.,** PO Box 307, Feasterville, PA 19053-0307. Phone: 215-245-6666. Web Site: www.acseed.com. Strictly wholesale. Specializes in sweet corn, watermelon, peppers and squash. Offered 55 non-hybrid vegetable varieties in 2004 including: 2 beet, 2 collard, 2 corn, 2 cress, 2 kale, 2 melon, 1 mustard greens, 1 okra, 1 onion other, 3 parsley, 4 pepper, 2 radish, 1 rutabaga, 2 spinach, 15 squash, 3 Swiss chard, 2 turnip, 4 watermelon and 4 miscellaneous.

Ada **Adams-Briscoe Seed Co.,** 325 E. Second St., P.O. Box 19, Jackson, GA 30233-0019. Phone: 770-775-7826. Fax: 770-775-7122. E-mail: abseed@juno.com. Web Site: www.abseed.com. Free catalog. Specializes in farm, lawn and wildlife seed that is grown in the southeastern U.S. Retail and wholesale. Offered 125 non-hybrid vegetable varieties in 2004 including: 16 bean, 1 beet, 1 broccoli, 4 cabbage, 1 Chinese cabbage, 3 collard, 9 corn, 19 cowpea, 3 cucumber, 1 eggplant, 2 kale, 4 lettuce, 10 lima, 4 melon, 2 mustard greens, 4 okra, 6 pea, 6 pepper, 3 radish, 1 spinach, 9 squash, 1 sunflower ornamental, 1 sunflower edible, 4 tomato, 4 turnip, 1 miscellaneous leguminosae and 5 miscellaneous.

Ag7 **Agrestal Organic Heritage Seed Co.,** P.O. Box 646, Gormley, ON L0H 1G0, Canada. Fax: 905-888-0094. E-mail: agrestal@sympatico.ca. Free catalog. Small, family-owned and operated Canadian seed company offering certified organic open-pollinated heirloom seed. Retail and wholesale. Offered 171 non-hybrid vegetable varieties in 2004 including: 14 bean, 4 beet, 7 carrot, 1 Chinese cabbage, 17 corn, 8 cucumber, 3 eggplant, 3 endive, 2 kale, 27 lettuce, 4 melon, 1 mustard greens, 3 okra, 2 onion other, 2 parsley, 9 pea, 10 pepper, 5 radish, 1 spinach, 12 squash, 3 sunflower ornamental, 3 sunflower edible, 2 Swiss chard, 1 tepary, 22 tomato, 1 miscellaneous leguminosae and 4 miscellaneous.

Al6 **Ken Allan,** 61 S. Bartlett St., Kingston, ON K7K 1X3, Canada. E-mail: allan@kingston.net. Send SASE for catalog (in Canada), or 50-cent Canadian stamp or $1 (in U.S.). Specializes in organically grown vegetable seeds, particularly sweet potatoes and tall peas. Sweet potatoes for Canada only. Retail only. Offered 28 non-hybrid vegetable varieties in 2004 including: 2 bean, 1 corn, 3 pea, 3 pepper, 14 sweet potato and 5 tomato.

Al25 **Along the Garden Path,** A Keenan Family Renaissance, General Delivery, Springbrook, ON K0K 3C0, Canada. Web Site: www.eagle.ca/~akeenan/Gardenpath.html. On-line catalog. Commited to preserving heritage vegetable varieties. Offered 36 non-hybrid vegetable varieties in 2004 including: 2 bean, 3 cucumber, 1 eggplant, 1 ground cherry, 4 lettuce, 1 lima, 1 pea, 3 pepper, 4 squash and 16 tomato.

Alb **Alberta Nurseries and Seeds Ltd.,** Box 20, Bowden, AB T0M 0K0, Canada. Phone: 403-224-3544. Fax: 403-224-2455. Free catalog in Canada only; $2 to U. S. Specializes in early-maturing seeds for short-season areas. Retail and internet sales. Offered 131 non-hybrid vegetable varieties in 2004 including: 1 asparagus, 17 bean, 5 beet, 1 broccoli, 1 brussels sprout, 5 cabbage, 6 carrot, 2 cauliflower, 1 celeriac, 2 celery, 1 Chinese cabbage, 2 cress, 5 cucumber, 1 endive, 1 fava bean, 1 fennel, 1 garlic, 1 kale, 2 kohlrabi, 1 leek, 9 lettuce, 1 onion bunching, 3 onion common, 1 onion multiplier, 2 onion other, 2 parsley, 3 parsnip, 11 pea, 4 pepper, 2 potato, 9 radish, 1 runner bean, 2 rutabaga, 1 salsify, 10 squash, 1 sunflower edible, 2 Swiss chard, 5 tomato, 1 turnip, 1 watermelon, 1 miscellaneous leguminosae and 2 miscellaneous.

All **Allen, Sterling and Lothrop,** 191 U.S. Rt. 1, Falmouth, ME 04105. Fax: 207-781-4143. $1 for catalog, refundable. Specializes in vegetable seeds adapted to northern New England. Company founded in 1911. Retail only. Offered 143 non-hybrid vegetable varieties in 2004 including: 15 bean, 3 beet, 2 broccoli, 1 brussels sprout, 4 cabbage, 4 carrot, 1 cauliflower, 1 celery, 2 Chinese cabbage, 2 corn, 1 cress, 8 cucumber, 1 eggplant, 2 endive, 1 fava bean, 1 fennel, 1 gourd, 1 kale, 1 kohlrabi, 1 leek, 10 lettuce, 2 lima, 2 melon, 1 okra, 1 onion bunching, 1 onion common, 1 onion other, 2 parsley, 1 parsnip, 13 pea, 4 pepper, 6 radish, 1 runner bean, 1 rutabaga, 1 salsify, 1 spinach, 18 squash, 2 sunflower ornamental, 2 sunflower edible, 4 Swiss chard, 10 tomato, 1 turnip, 1 watermelon and 4 miscellaneous.

Ami **Amishland Heirloom Seeds,** c/o Lisa Von Saunder, Box 365, Reamstown, PA 17567-0365. Web Site: www.amishlandseeds.com. On-line catalog. Specializes in rare and colorful heirloom vegetables and old-fashioned flower seed and rare locally grown varieties that were grown for generations by Pennsylvania German farmers, commonly known as Pennsylvania Dutch. Offered 89 non-hybrid vegetable varieties in 2004 including: 7 bean, 4 eggplant, 1 ground cherry, 2 kale, 1 lettuce, 3 lima, 3 pea, 11 pepper, 1 salsify, 1 tomatillo, 48 tomato, 1 miscellaneous cucurbitaceae, 1 miscellaneous leguminosae, 2 miscellaneous solanaceae and 3 miscellaneous.

Ap6 **Appalachian Seeds,** P.O. Box 248, Flat Rock, NC 28731. E-mail: seeds@appalachianseeds.com. Web Site: www.appalachianseeds.com. On-line catalog. Offers organically grown heirloom tomato seeds. Offered 67 tomato varieties in 2004.

Au2 **Aurora Farm,** 3492 Phillips Road, Creston, BC V0B 1G2, Canada. Phone: 250-428-4404. Fax: 250-428-4404. E-mail: aurora@kootenay.com. Web Site: www.kootenay.com/~aurora. U.S. address is PO Box 697, Porthill, ID 83853. Sells open-pollinated and biodynamic heritage seeds only. Retail and wholesale prices are listed in the same catalog. Offered 22 non-hybrid vegetable varieties in 2004 including: 1 carrot, 1 kale, 1 leek, 5 lettuce, 1 parsley, 1 spinach, 1 squash, 1 Swiss chard, 5 tomato and 5 miscellaneous.

Ba6 **Barney's Ginseng Patch,** 433 SSE Highway B, Montgomery City, MO 63361. Phone: 573-564-2575 (after 5 p.m.). $3 for booklet entitled "Grow Ginseng for Profit" which includes an updated price list. Specializes in ginseng and goldenseal seed and planting roots. Retail and wholesale. Offered 1 miscellaneous variety in 2004.

Ba8 **Baker Creek Heirloom Seeds,** 2278 Baker Creek Road, Mansfield, MO 65704. Phone: 417-924-8917. Fax: 417-924-8887. E-mail: seeds@rareseeds.com. Web Site: www.rareseeds.com. Free catalog. Specializes in heirloom seeds including many rare types and Oriental and European varieties. Offers over 725 kinds of vegetables, herbs and flowers. Retail, wholesale and internet sales. Offered 585 non-hybrid vegetable varieties in 2004 including: 4 amaranth, 2 artichoke, 2 asparagus, 4 bean, 7 beet, 3 broccoli, 1 broccoli raab, 3 brussels sprout, 6 cabbage, 7 carrot, 3 cauliflower, 1 celeriac, 2 celery, 6 Chinese cabbage, 10 chicory, 1 collard, 16 corn, 2 cornsalad, 26 cowpea, 1 cress, 17 cucumber, 29 eggplant, 3 endive, 2 fava bean, 1 fennel, 8 garlic, 9 gourd, 2 ground cherry, 2 kale, 1 kohlrabi, 4 leek, 26 lettuce, 5 lima, 59 melon, 3 mustard greens, 8 okra, 2 onion bunching, 6 onion common, 2 onion other, 2 parsley, 1 parsnip, 5 pea, 25 pepper, 15 radish, 3 runner bean, 1 rutabaga, 1 salsify, 2 sorghum, 1 soybean, 1 spinach, 80 squash, 8 sunflower ornamental, 1 sunflower edible, 4 Swiss chard, 2 tomatillo, 72 tomato, 6 turnip, 36 watermelon, 6 miscellaneous cucurbitaceae, 2 miscellaneous leguminosae, 6 miscellaneous solanaceae and 9 miscellaneous.

Ba16 **Bayou Traders Peppermania,** Attn: Beth Boyd, 12999 Murphy Road, Suite K, Stafford, TX 77477. Web Site: www.peppermania.com. On-line catalog. Specializes in organically grown pepper seeds. Offered 54 pepper varieties in 2004.

BAL **Ball Seed Company,** 622 Town Road, West Chicago, IL 60185-2698. Web Site: www.ballseed.com. Wholesale only. Offered 52 non-hybrid vegetable varieties in 2004 including: 1 artichoke, 1 cabbage, 2 cauliflower, 4 cucumber, 3 eggplant, 1 leek, 6 lettuce, 1 melon, 3 onion common, 2 parsley, 17 pepper, 7 squash, 1 Swiss chard and 3 tomato.

Ban **The Banana Tree,** 715 Northampton St., Easton, PA 18042. Phone: 610-253-9589. E-mail: faban@enter.net. Web Site: www.banana-tree.com. "Strange and wonderful collection" of exotics and ornamentals, mostly tropicals. Providing tropical plants and seeds to botanical gardens, professional growers and individuals around the world since 1955.

Offered 42 non-hybrid vegetable varieties in 2004 including: 1 amaranth, 2 Chinese cabbage, 1 cress, 2 eggplant, 1 fennel, 2 gourd, 1 ground cherry, 1 kale, 1 melon, 1 onion common, 1 onion other, 1 pea, 6 pepper, 3 radish, 4 miscellaneous cucurbitaceae, 5 miscellaneous leguminosae, 2 miscellaneous solanaceae and 7 miscellaneous.

Be4 **Berlin Seeds,** 3628 State Route 39, Millersburg, OH 44654. Phone: 330-893-2091. Free catalog. Specializes in quality seeds for the home gardener at economical prices. Offers "many tried and true varieties of vegetable and flower seeds," both open-pollinated and hybrid. Mostly retail, with wholesale on some items. Offered 189 non-hybrid vegetable varieties in 2004 including: 28 bean, 5 beet, 3 broccoli, 1 brussels sprout, 2 cabbage, 8 carrot, 2 cauliflower, 4 celery, 1 Chinese cabbage, 4 corn, 1 cowpea, 8 cucumber, 1 eggplant, 1 endive, 1 gourd, 1 ground cherry, 2 kohlrabi, 1 leek, 20 lettuce, 5 lima, 5 melon, 1 mustard greens, 1 okra, 1 onion common, 1 onion other, 1 parsley, 1 parsnip, 15 pea, 1 peanut, 7 pepper, 3 potato, 11 radish, 1 salsify, 1 sorghum, 1 soybean, 1 spinach, 16 squash, 1 sunflower edible, 2 Swiss chard, 10 tomato, 6 watermelon, 2 miscellaneous leguminosae and 1 miscellaneous.

BE6 **Bejo Seeds, Inc.,** 1972 Silver Spur Place, Oceano, CA 93455. Web Site: www.bejoseeds.com. Strictly wholesale. Offered 20 non-hybrid vegetable varieties in 2004 including: 1 beet, 1 carrot, 2 cauliflower, 5 chicory, 4 endive, 1 fennel, 1 leek, 1 onion bunching, 3 parsley and 1 radish.

Bec **Becker's Seed Potatoes,** R.R. 1, Trout Creek, ON P0H 2L0, Canada. Phone: 705-724-2305. E-mail: beckers@vianet.ca. Free catalog. Canadian orders only. Not shipping to the U.S. Small family-operated business "offering the largest selection of seed potato varieties in Canada to home gardeners." Retail only. Offered 21 potato varieties in 2004.

Ber **Berton Seeds Co. Ltd.,** 4260 Weston Road, Weston, ON M9L 1W9, Canada. Fax: 416-745-3954. E-mail: info@bertonseeds.com. Web Site: www.bertonseeds.com. On-line catalog; printed catalog $12. Sales to Canada and the U.S. Specializes in European chicories. Retail, wholesale and internet sales. Offered 105 non-hybrid vegetable varieties in 2004 including: 1 artichoke, 1 asparagus, 14 bean, 2 beet, 1 broccoli, 2 cabbage, 1 cardoon, 2 carrot, 1 cauliflower, 1 celeriac, 1 celery, 1 Chinese cabbage, 16 chicory, 1 cornsalad, 1 cress, 2 cucumber, 3 eggplant, 5 endive, 1 fava bean, 2 fennel, 2 kohlrabi, 8 lettuce, 1 lima, 3 melon, 2 onion common, 1 onion other, 2 parsley, 3 pea, 5 pepper, 2 radish, 1 rutabaga, 2 spinach, 4 squash, 2 Swiss chard, 4 tomato, 2 watermelon and 2 miscellaneous.

Bi3 **Big Creek Ginseng and Garlic Farm,** R. R. #1, LaSalette, ON N0E 1H0, Canada. E-mail: darrell@bigcreekgarlic.com. Web Site: www.bigcreekgarlic.com. On-line catalog. Canada's largest grower of garlic. Grows more than 30 acres of garlic with varieties from all over the world. Offered 117 non-hybrid vegetable varieties in 2004 including: 110 garlic, 6 onion multiplier and 1 miscellaneous.

Bl7 **Blue Moon Farm,** 3584 Poosey Ridge, Richmond, KY 40475-9780. Web Site: http://bluemoongarlic.hypermart.net. $2 for catalog, refundable with order from that catalog. Small family-owned company launched in 1988. Specializes in garlic, all planted, hand-tended, harvested, cured and mailed to customers by the two owners. Retail only. Offered 16 garlic varieties in 2004.

Bo9 **Boone's Native Seed Company,** PO Box 10363, Raleigh, NC 27605. E-mail: rboonenc@earthlink.net. Free catalog. Specializes in open-pollinated traditional and heirloom tomatoes and select chili peppers. Untreated seed. Retail only. Offered 35 non-hybrid vegetable varieties in 2004 including: 10 pepper, 1 sunflower edible, 1 tomatillo and 23 tomato.

Bo17 **Botanical Interests, Inc.,** 660 Compton St., Broomfield, CO 80020. On-line catalog. Retail catalog at www.gardentrails.com. Wholesale at www.gardenguides.com. Family owned business offers a full line of untreated flower, vegetable and herb seeds as well as a Certified Organic seed line. Exclusive USA distributor of Mr. Fothergill's seeds from England. Heirloom seeds are packed in beautiful informative seed packets. Retail and wholesale catalogs/price lists are separate but contain identical offerings. Offered 156 non-hybrid vegetable varieties in 2004 including: 1 amaranth, 8 bean, 2 beet, 2 broccoli, 1 broccoli raab, 6 carrot, 1 cauliflower, 2 celery, 4 Chinese cabbage, 1 chicory, 1 corn, 1 cornsalad, 1 cowpea, 1 cress, 3 cucumber, 2 eggplant, 2 endive, 1 fava bean, 1 fennel, 3 gourd, 3 kale, 1 leek, 11 lettuce, 1 lima, 6 melon, 2 mustard greens, 1 okra, 1 onion bunching, 2 onion common, 2 onion other, 2 parsley, 4 pea, 12 pepper, 5 radish, 1 runner bean, 1 soybean, 1 spinach, 18 squash, 3 sunflower ornamental, 1 sunflower edible, 3 Swiss chard, 1 tomatillo, 13 tomato, 2 turnip, 3 watermelon, 2 miscellaneous cucurbitaceae, 1 miscellaneous leguminosae and 9 miscellaneous.

Bo18 **Borghese Gardens,** 1010 Somerset W, Ottawa, Canada. E-mail: Borghese@Onebox.com. Web Site: www.borghesegardens.com. Canada only. On-line catalog. Includes a wide variety of rare and unusual seeds. Offered 66 non-hybrid vegetable varieties in 2004 including: 1 asparagus, 1 beet, 4 cabbage, 2 carrot, 2 cauliflower, 1 celery, 6 corn, 1 cress, 2 cucumber, 2 eggplant, 1 kale, 1 leek, 2 melon, 1 okra, 2 onion common, 1 parsnip, 7 pepper, 1 radish, 1 salsify, 3 sorghum, 6 squash, 12 sunflower ornamental, 4 sunflower edible, 1 miscellaneous cucurbitaceae and 1 miscellaneous.

Bo19 **Botanikka Seeds,** P.O. Box 182, Iron Ridge, WI 53035. E-mail: mark@botanikka.com. Web Site: www.botanikka.com. On-line catalog. Specializes in heirloom and rare specialty seeds with an emphasis on herbs and vegetables. Grows, imports and packages garden seeds for sale through retail stores, farm markets and on their web site. Retail sales; bulk and special request information available. Offered 241 non-hybrid vegetable varieties in 2004 including: 1 artichoke, 1 asparagus, 10 bean, 8 beet, 4 broccoli, 2 broccoli raab, 2 brussels sprout, 4 cabbage, 1 cardoon, 6 carrot, 4 cauliflower, 1 celeriac, 1 celery, 3 Chinese cabbage, 1 chicory, 1 collard, 4 corn, 1 cornsalad, 3 cress, 5 cucumber, 6 eggplant, 2 endive, 1 fennel, 9 gourd, 3 kale, 3 kohlrabi, 2 leek, 11 lettuce, 1 lima, 9 melon, 1 mustard greens, 1 okra, 1 onion bunching, 3 onion common, 2 onion other, 4 parsley, 2 parsnip, 5 pea, 29 pepper, 4 radish, 2 rutabaga, 1 salsify, 1 spinach, 27 squash, 3 Swiss chard, 1 tomatillo, 37 tomato, 2 miscellaneous cucurbitaceae, 1 miscellaneous leguminosae, 1 miscellaneous solanaceae and 3 miscellaneous.

Bo20 **Bobba-Mike's Garlic Farm,** PO Box 261, Orrville, OH 44667. E-mail: garlic@garlicfarm.com. Web Site: www.garlicfarm.com. On-line catalog. Specializes in organically grown gourmet garlic. Wholesale prices available for German White and Music varieties. Offered 15 non-hybrid vegetable varieties in 2004 including: 14 garlic and 1 leek.

Bo21 **Boundary Garlic,** Sonia Stairs/Henry Caron, Box 273, Midway, BC V0H 1M0, Canada. E-mail: sonia@garlicfarm.ca. Web Site: www.garlicfarm.ca. On-line catalog. Specializes in organic garlic seed and sells by mail order across Canada. Offered 13 garlic varieties in 2004.

Bor **Leonard Borries,** 16293 E. 1400th Ave., Teutopolis, IL 62467. Phone: 217-857-3377. Free price list. Dealers in open-pollinated field corn since 1969. Pound and bushel amounts available. Sells Boone County White, Henry Moore Yellow, Reid's Yellow Dent and Krug's Yellow. Separate retail and wholesale catalogs each containing identical offerings. Offered 4 corn varieties in 2004.

Bou **Bountiful Gardens,** 18001 Shafer Ranch Road, Willits, CA 95490. Phone: 707-459-6410. Fax: 707-459-1925. E-mail: bountiful@sonic.net. Web Site: www.bountifulgardens.org. Free retail catalog in U.S.; $2 Canada/Mexico; $4 international. Offers open-pollinated, untreated heirloom seeds that are adaptable to varied conditions. Offered 264 non-hybrid vegetable varieties in 2004 including: 2 amaranth, 1 artichoke, 2 asparagus, 16 bean, 10 beet, 5 broccoli, 3 brussels sprout, 6 cabbage, 1 cardoon, 8 carrot, 7 cauliflower, 2 celery, 6 Chinese cabbage, 2 chicory, 1 collard, 9 corn, 1 cornsalad, 1 cowpea, 1 cress, 3 cucumber, 2 eggplant, 1 endive, 3 fava bean, 1 fennel, 1 gourd, 3 kale, 1 kohlrabi, 3 leek, 18 lettuce, 1 lima, 6 melon, 2 mustard greens, 2 okra, 1 onion bunching, 8 onion common, 2 onion other, 3 parsley, 2 parsnip, 12 pea, 6 pepper, 3 quinoa, 5 radish, 4 runner bean, 1 rutabaga, 1 scorzonera, 5 sorghum, 3 soybean, 5 spinach, 13 squash, 1 sunflower edible, 3 Swiss chard, 30 tomato, 3 turnip, 3 watermelon, 4 miscellaneous leguminosae and 15 miscellaneous.

Br12 **Brown's Omaha Plant Farms, Inc.,** 110 McLean Avenue, P.O. Box 787, Omaha, TX 75571. E-mail: mail@bopf.com. Web Site: www.bopf.com. Free catalog. Family based business which has been providing home gardeners with high quality onion plants and other vegetable plants for over 60 years. Included here for sweet potato plants. Offered 4 sweet potato varieties in 2004.

Bu1 Burgess Seed and Plant Co., 905 Four Seasons Road, Bloomington, IL 61701. E-mail: CustomerService@eburgess.com. Web Site: www.eburgess.com. Free catalog. Unique and popular vegetable seeds and plants since 1913. Retail only. Offered 95 non-hybrid vegetable varieties in 2004 including: 2 asparagus, 12 bean, 2 beet, 1 cabbage, 1 carrot, 1 cauliflower, 2 corn, 1 cowpea, 4 cucumber, 1 eggplant, 3 gourd, 1 ground cherry, 1 Jerusalem artichoke, 1 leek, 1 lettuce, 1 lima, 5 melon, 2 okra, 2 onion common, 1 onion multiplier, 7 pea, 3 pepper, 7 potato, 4 radish, 1 soybean, 8 squash, 1 sunflower edible, 1 sweet potato, 9 tomato, 1 turnip, 5 watermelon, 2 miscellaneous solanaceae and 1 miscellaneous.

Bu2 W. Atlee Burpee and Co., 300 Park Ave., Warminster, PA 18991. Phone: 800-333-5808. Fax: 800-487-5530. Web Site: www.burpee.com. Call 800-888-1447 to place an order. Free catalog. Separate wholesale and retail catalogs; retail catalog contains additional varieties. Extensive selection including many original introductions. Offered 295 non-hybrid vegetable varieties in 2004 including: 2 amaranth, 1 asparagus, 28 bean, 6 beet, 1 broccoli, 1 broccoli raab, 1 brussels sprout, 2 cabbage, 7 carrot, 1 celery, 4 Chinese cabbage, 2 chicory, 1 collard, 4 corn, 1 cornsalad, 1 cowpea, 1 cress, 7 cucumber, 2 eggplant, 2 endive, 1 fennel, 3 gourd, 2 kale, 1 kohlrabi, 3 leek, 27 lettuce, 6 lima, 7 melon, 1 mustard greens, 1 okra, 2 onion bunching, 3 onion common, 1 onion multiplier, 2 onion other, 3 parsley, 1 parsnip, 18 pea, 1 peanut, 17 pepper, 6 potato, 9 radish, 3 runner bean, 1 rutabaga, 1 salsify, 1 spinach, 37 squash, 7 sunflower ornamental, 2 sunflower edible, 5 Swiss chard, 5 sweet potato, 1 tomatillo, 25 tomato, 1 turnip, 7 watermelon, 1 miscellaneous cucurbitaceae, 3 miscellaneous leguminosae and 5 miscellaneous.

Bu3 D. V. Burrell Seed Growers Co., P.O. Box 150, Rocky Ford, CO 81067. Phone: 719-254-3318. Fax: 719-254-3319. Free catalog. Family-owned business founded in 1900. Fine collection of cantaloupe, watermelon, vegetable, flower and herb seeds. Retail and wholesale catalogs. Offered 215 non-hybrid vegetable varieties in 2004 including: 1 artichoke, 1 asparagus, 14 bean, 4 beet, 1 broccoli, 1 brussels sprout, 6 cabbage, 4 carrot, 2 cauliflower, 2 celery, 1 Chinese cabbage, 1 collard, 7 corn, 1 cowpea, 5 cucumber, 2 eggplant, 1 endive, 1 fava bean, 4 gourd, 1 kale, 1 kohlrabi, 1 leek, 10 lettuce, 1 lima, 17 melon, 2 mustard greens, 2 okra, 8 onion common, 2 onion other, 3 parsley, 1 parsnip, 12 pea, 25 pepper, 5 radish, 1 rutabaga, 1 salsify, 2 spinach, 30 squash, 1 sunflower edible, 3 Swiss chard, 11 tomato, 1 turnip, 13 watermelon, 1 miscellaneous cucurbitaceae and 1 miscellaneous.

Bu8 Bunton Seed Co., 939 East Jefferson St., Louisville, KY 40206. E-mail: info@buntonseed.com. Web Site: www.buntonseed.com. Catalog $1. Vegetable, herb and flower seeds. Retail and wholesale prices listed in same catalog. Offered 179 non-hybrid vegetable varieties in 2004 including: 1 asparagus, 26 bean, 2 beet, 1 broccoli, 1 brussels sprout, 6 cabbage, 5 carrot, 1 cauliflower, 1 celery, 2 Chinese cabbage, 2 collard, 8 corn, 3 cowpea, 4 cucumber, 1 eggplant, 1 endive, 3 gourd, 2 kale, 2 kohlrabi, 1 leek, 11 lettuce, 9 lima, 6 melon, 2 mustard greens, 3 okra, 3 onion common, 1 parsley, 1 parsnip, 9 pea, 1 peanut, 5 pepper, 10 radish, 1 rutabaga, 1 salsify, 1 spinach, 23 squash, 1 sunflower edible, 6 tomato, 3 turnip, 7 watermelon, 1 miscellaneous cucurbitaceae and 1 miscellaneous.

But Butterbrooke Farm, 78 Barry Road, Oxford, CT 06478. Phone: 203-888-2000. Seed co-op. Specializes in open-pollinated, chemically untreated, old-type hardy vegetable strains. Send SASE for price list and membership information. Offered 56 non-hybrid vegetable varieties in 2004 including: 1 asparagus, 5 bean, 1 beet, 1 broccoli, 2 cabbage, 1 carrot, 1 cauliflower, 1 celery, 1 collard, 1 corn, 1 eggplant, 1 endive, 1 kale, 1 kohlrabi, 6 lettuce, 1 lima, 1 melon, 1 mustard greens, 1 okra, 2 parsley, 1 parsnip, 3 pea, 1 pepper, 2 radish, 1 rutabaga, 1 spinach, 7 squash, 1 sunflower edible, 1 Swiss chard, 2 tomato, 1 turnip, 1 watermelon and 3 miscellaneous.

CA25 Carolina Seeds/New England Seed Co., 3580 Main St., Bldg. 10, Hartford, CT 06120. Phone: 800-825-5477. Fax: 877-229-8487. E-mail: newseedco@aol.com. Web Site: www.neseed.com. Free catalog. Merged with New England Seed Co. Carries a full line of vegetable seeds and a collection of seeds from Semiorto-Sementi, an importer and seed producer in Salerno, Italy. Retail and wholesale catalog and internet sales. Offered 206 non-hybrid vegetable varieties in 2004 including: 4 amaranth, 1 asparagus, 7 bean, 1 beet, 1 broccoli, 2 broccoli raab, 1 brussels sprout, 7 cabbage, 3 carrot, 3 cauliflower, 2 celery, 1 Chinese cabbage, 4 chicory, 2 collard, 6 corn, 7 cucumber, 1 eggplant, 1 endive, 1 fennel, 6 gourd, 1 kohlrabi, 1 leek, 18 lettuce, 1 lima, 1 mustard greens, 1 okra, 3 onion common, 2 onion other, 5 parsley, 5 pea, 20 pepper, 2 radish, 1 sorghum, 1 spinach, 27 squash, 6 sunflower ornamental, 2 sunflower edible, 1 Swiss chard, 1 tomatillo, 34 tomato, 1 turnip, 7 watermelon, 1 miscellaneous cucurbitaceae, 1 miscellaneous leguminosae and 2 miscellaneous.

Ca26 Calypso Farms, 4225 Escondido Canyon Road, Acton, CA 93510. Web Site: www.calypsofarms.com. Family owned business offering vegetable and flower seeds. Offered 14 non-hybrid vegetable varieties in 2004 including: 1 onion common, 1 radish, 6 squash, 5 sunflower ornamental, 1 sunflower edible.

CH14 Chesmore Seed Co., 5030 Highway 36 East, Saint Joseph, MO 64507. E-mail: chesmore@chesmore.com. Web Site: www.chesmore.com. Wholesale only. Wholesale seedsman since 1878. Offered 314 non-hybrid vegetable varieties in 2004 including: 1 artichoke, 1 asparagus, 44 bean, 4 beet, 1 broccoli, 1 brussels sprout, 5 cabbage, 6 carrot, 1 cauliflower, 2 celery, 2 Chinese cabbage, 2 collard, 13 corn, 7 cowpea, 8 cucumber, 2 eggplant, 2 endive, 5 gourd, 2 kale, 2 kohlrabi, 1 leek, 13 lettuce, 6 lima, 11 melon, 3 mustard greens, 3 okra, 2 onion bunching, 8 onion common, 1 onion other, 4 parsley, 3 parsnip, 19 pea, 2 peanut, 30 pepper, 11 radish, 2 rutabaga, 1 salsify, 2 spinach, 42 squash, 1 sunflower edible, 4 Swiss chard, 12 tomato, 3 turnip, 16 watermelon and 3 miscellaneous.

Ch15 **Charley's Farm,** Charley and Ginny Hein, 54 E. Stutler Road, Spokane, WA 99224. E-mail: charleysfarm @att.net. Web Site: www.charleysfarm.com. On-line catalog. Certified organic grower offering heirloom garlic varieties. Offered 19 garlic varieties in 2004.

Cl3 **Clifton Seed Company,** 2586 NC 403 West, PO Box 206, Faison, NC 28341. Phone: 910-267-2690. E-mail: carolync@cliftonseed.com. Web Site: www.cliftonseed.com. Free catalog. Family business founded in 1928. Separate but identical retail and wholesale catalogs. Offered 117 non-hybrid vegetable varieties in 2004 including: 21 bean, 2 beet, 2 broccoli raab, 3 cabbage, 2 carrot, 1 cauliflower, 1 Chinese cabbage, 4 collard, 9 cowpea, 2 cucumber, 3 gourd, 4 kale, 5 lima, 3 mustard greens, 1 okra, 1 onion bunching, 10 pea, 2 pepper, 2 radish, 2 rutabaga, 27 squash, 2 turnip, 6 watermelon and 2 miscellaneous.

CO23 **Condor Seed Production, Inc.,** PO Box 6485, Yuma, AZ 85366. Phone: 928-627-8803. E-mail: info@condorseed.com. Web Site: www.condorseed.com. Strictly wholesale. Offered 180 non-hybrid vegetable varieties in 2004 including: 3 artichoke, 1 asparagus, 5 beet, 2 broccoli, 2 broccoli raab, 1 brussels sprout, 5 cabbage, 5 carrot, 4 cauliflower, 2 celery, 6 Chinese cabbage, 3 chicory, 2 cress, 7 cucumber, 2 eggplant, 2 endive, 1 fennel, 5 kale, 2 kohlrabi, 1 leek, 14 lettuce, 5 melon, 4 mustard greens, 5 okra, 2 onion bunching, 16 onion common, 4 parsley, 2 parsnip, 10 pepper, 7 radish, 1 salsify, 2 spinach, 16 squash, 9 Swiss chard, 1 tomatillo, 3 tomato, 1 turnip, 8 watermelon, 2 miscellaneous cucurbitaceae and 7 miscellaneous.

CO30 **Corona Seeds WorldWide,** 590-F Constitution Ave., Camarillo, CA 93012. E-mail: coronaseed@aol.com. Web Site: www.coronaseeds.com. Strictly wholesale. Vegetable and herb seed for commercial growers. Offered 334 non-hybrid vegetable varieties in 2004 including: 3 amaranth, 4 artichoke, 1 asparagus, 12 beet, 2 broccoli, 2 broccoli raab, 6 carrot, 4 cauliflower, 1 celery, 3 Chinese cabbage, 3 chicory, 2 cornsalad, 4 cress, 3 cucumber, 5 eggplant, 7 endive, 1 fennel, 5 gourd, 1 ground cherry, 4 kale, 3 leek, 46 lettuce, 15 melon, 3 mustard greens, 5 okra, 4 onion bunching, 2 onion common, 2 onion other, 5 parsley, 1 parsnip, 3 pea, 15 pepper, 8 radish, 1 rutabaga, 1 salsify, 5 spinach, 21 squash, 5 Swiss chard, 2 tomatillo, 91 tomato, 2 turnip, 6 watermelon, 3 miscellaneous cucurbitaceae and 12 miscellaneous.

Co31 **Cooper Seeds,** 131 Eaton St., Lawrenceville, GA 30045. Web Site: www.cooperseeds.com. On-line catalog. Serving home, farm and garden needs since 1890. Offered 163 non-hybrid vegetable varieties in 2004 including: 17 bean, 1 beet, 1 cabbage, 2 carrot, 1 cauliflower, 2 collard, 7 corn, 12 cowpea, 4 cucumber, 1 eggplant, 5 gourd, 1 kale, 6 lettuce, 11 lima, 5 melon, 2 mustard greens, 3 okra, 1 parsley, 1 parsnip, 5 pea, 1 peanut, 17 pepper, 3 potato, 6 radish, 1 rutabaga, 1 soybean, 1 spinach, 13 squash, 3 Swiss chard, 15 tomato, 3 turnip, 7 watermelon, 2 miscellaneous cucurbitaceae and 1 miscellaneous.

Co32 **The Cottage Gardener,** 4199 Gilmore Road, RR 1, Newtonville, ON L0A 1J0, Canada. E-mail: heirlooms @cottagegardener.com. Web Site: www.cottagegardener .com. Sales within Canada only. Paper catalog $2. Heirloom seed and plant nursery in southern Ontario specializing in seeds of heirloom perennials, annuals, herbs and vegetables. Offered 138 non-hybrid vegetable varieties in 2004 including: 2 amaranth, 14 bean, 6 beet, 2 broccoli, 1 brussels sprout, 2 cabbage, 5 carrot, 4 cucumber, 1 eggplant, 3 kale, 1 leek, 9 lettuce, 6 melon, 2 onion common, 1 onion other, 7 pea, 11 pepper, 3 radish, 3 runner bean, 1 spinach, 11 squash, 1 sunflower ornamental, 1 sunflower edible, 3 Swiss chard, 37 tomato and 1 miscellaneous.

Com **Comstock, Ferre and Co.,** 263 Main St., Wethersfield, CT 06109. Phone: 860-571-6590. E-mail: comstock@tiac.net. Web Site: www.comstockferre.com. Free catalog. Established in 1820, incorporated in 1853. "From packets to pounds." Offers vegetable varieties, numerous heirlooms, annuals, herbs and perennials. Retail and wholesale prices are listed in the same catalog. Offered 264 non-hybrid vegetable varieties in 2004 including: 2 amaranth, 1 artichoke, 1 asparagus, 18 bean, 7 beet, 1 broccoli, 2 broccoli raab, 1 brussels sprout, 5 cabbage, 1 cardoon, 5 carrot, 3 cauliflower, 1 celeriac, 3 celery, 4 Chinese cabbage, 7 chicory, 1 collard, 4 corn, 1 cornsalad, 2 cowpea, 3 cress, 6 cucumber, 2 eggplant, 3 endive, 1 fava bean, 1 fennel, 6 gourd, 1 ground cherry, 2 kale, 2 kohlrabi, 1 leek, 15 lettuce, 4 lima, 3 melon, 4 mustard greens, 1 okra, 2 onion bunching, 4 onion common, 2 onion other, 3 parsley, 1 parsnip, 7 pea, 25 pepper, 7 radish, 3 runner bean, 1 rutabaga, 1 salsify, 1 scorzonera, 1 soybean, 1 spinach, 26 squash, 5 sunflower ornamental, 1 sunflower edible, 6 Swiss chard, 1 tomatillo, 15 tomato, 4 turnip, 3 watermelon, 2 miscellaneous cucurbitaceae, 2 miscellaneous leguminosae and 15 miscellaneous.

CON **Conida Seed Co.,** 515 East Main St., PO Box 129, Hazelton, ID 83335. Strictly wholesale. Offered 65 non-hybrid vegetable varieties in 2004 including: 47 bean and 18 pea.

Coo **The Cook's Garden,** PO Box 1889, Southampton, PA 18966-0895. Phone: 800-457-9703. E-mail: catalog@cooks garden.com. Web Site: www.cooksgarden.com. Free catalog, which lists heirloom culinary vegetable seeds from around the world, many are organic. Retail only. Also many herbs, cut flowers and annual vines plus some seed saving books and supplies. Offered 281 non-hybrid vegetable varieties in 2004 including: 2 artichoke, 23 bean, 7 beet, 2 broccoli, 1 broccoli raab, 1 brussels sprout, 2 cabbage, 2 cardoon, 5 carrot, 2 celeriac, 2 celery, 4 Chinese cabbage, 15 chicory, 6 corn, 3 cornsalad, 4 cress, 3 cucumber, 6 eggplant, 3 endive, 1 fennel, 2 gourd, 1 kale, 5 kohlrabi, 2 leek, 45 lettuce, 2 melon, 1 mustard greens, 1 onion bunching, 6 onion common, 1 onion multiplier, 2 onion other, 4 parsley, 9 pea, 10 pepper, 4 potato, 6 radish, 3 runner bean, 1 sorghum, 1 soybean, 14 squash, 7 sunflower ornamental, 1 sunflower edible, 5 Swiss chard, 2 tomatillo, 34 tomato, 1 turnip, 2 watermelon, 1 miscellaneous leguminosae and 14 miscellaneous.

Cor CORNS, Carl L. and Karen D. Barnes, Rt. 1, Box 32, Turpin, OK 73950. Phone: 580-778-3615. SASE for price list and information letter. Sells open-pollinated organically grown dent corn. Retail only. Offered 1 corn variety in 2004.

Cr1 Crosman Seed Corp., PO Box 110, East Rochester, NY 14445. Phone: 585-586-1928. Fax: 585-586-6093. Web Site: www.crosmanseed.com. Free catalog. Specializes in "tried and true" varieties for the home gardener. Retail and wholesale. Offered 120 non-hybrid vegetable varieties in 2004 including: 1 amaranth, 1 asparagus, 9 bean, 2 beet, 1 broccoli, 1 brussels sprout, 5 cabbage, 1 cardoon, 3 carrot, 1 cauliflower, 1 celery, 1 Chinese cabbage, 1 chicory, 2 collard, 1 corn, 1 cress, 5 cucumber, 2 eggplant, 2 endive, 1 kale, 1 kohlrabi, 10 lettuce, 2 lima, 2 melon, 2 mustard greens, 2 okra, 4 onion common, 1 onion other, 2 parsley, 1 parsnip, 6 pea, 6 pepper, 4 radish, 1 runner bean, 1 rutabaga, 2 spinach, 11 squash, 1 sunflower ornamental, 1 sunflower edible, 1 Swiss chard, 11 tomato, 2 turnip, 2 watermelon and 2 miscellaneous.

Da6 Dakota Garlic, 16790 Happy Landing Way, Monument, CO 80132. E-mail: sales@dakotagarlic.com. Web Site: www.dakotagarlic.com. On-line sales. Garlic is grown on the family farm in North Dakota. Retail only. Offered 10 garlic varieties in 2004.

DAl Daisy Farms, 28355 M-152, 9995 SW 66 St., Dowagiac, MI 49047. Phone: 269-782-6321. E-mail: daisyfarms@beanstalk.net. Web Site: www.daisyfarms.net. Free price list. Retail, wholesale and internet sales. Offered 3 non-hybrid vegetable varieties in 2004 including: 2 asparagus and 1 miscellaneous.

Dam William Dam Seeds Ltd., Box 8400, Dundas, ON L9H 6M1, Canada. Phone: 905-628-6641. Fax: 905-627-1729. E-mail: willdam@damseeds.com. Web Site: www.damseeds.com. Sales to Canada only. Catalog free within Canada. Small family owned and operated company. Many European varieties. Quality untreated seeds produced by world renowned breeders. Unable to ship to the United States due to new U.S. import regulations. Offered 298 non-hybrid vegetable varieties in 2004 including: 1 amaranth, 1 artichoke, 1 asparagus, 32 bean, 7 beet, 5 broccoli, 1 broccoli raab, 2 brussels sprout, 7 cabbage, 10 carrot, 3 cauliflower, 2 celeriac, 3 celery, 8 Chinese cabbage, 5 chicory, 1 collard, 5 corn, 1 cornsalad, 1 cress, 6 cucumber, 1 eggplant, 4 endive, 1 fava bean, 1 fennel, 1 garlic, 1 gourd, 1 ground cherry, 4 kale, 3 kohlrabi, 2 leek, 32 lettuce, 1 lima, 1 melon, 3 mustard greens, 1 onion bunching, 9 onion common, 2 onion multiplier, 2 onion other, 5 parsley, 18 pea, 7 pepper, 14 potato, 11 radish, 2 rutabaga, 1 salsify, 1 scorzonera, 2 spinach, 21 squash, 1 sunflower edible, 5 Swiss chard, 1 tomatillo, 17 tomato, 4 turnip, 1 watermelon, 1 miscellaneous cucurbitaceae, 3 miscellaneous leguminosae and 11 miscellaneous.

De6 De Bruyn Seed Co., 101 E. Washington, Zeeland, MI 49464. Phone: 616-772-2316. Fax: 616-772-0011. E-mail: debruynseed@egl.net. Web Site: www.debruynseed.com. Free price list. Specializes in sweet corn and pumpkins. Separate but identical retail and wholesale price lists/catalog. Offered 184 non-hybrid vegetable varieties in 2004 including: 1 asparagus, 22 bean, 4 beet, 1 broccoli, 1 brussels sprout, 7 cabbage, 5 carrot, 2 cauliflower, 1 Chinese cabbage, 1 collard, 5 corn, 2 cowpea, 6 cucumber, 1 eggplant, 1 fava bean, 2 gourd, 1 kale, 1 kohlrabi, 1 leek, 8 lettuce, 6 lima, 4 melon, 2 mustard greens, 2 okra, 3 onion common, 1 onion other, 1 parsley, 1 parsnip, 12 pea, 6 pepper, 13 potato, 6 radish, 1 runner bean, 1 rutabaga, 1 sorghum, 1 soybean, 1 spinach, 28 squash, 5 sunflower ornamental, 3 sunflower edible, 1 Swiss chard, 4 tomato, 2 turnip, 3 watermelon and 3 miscellaneous.

Dgs Dan's Garden Shop, P.O. Box 67, Adamstown, MD 21710-0067. E-mail: dan@dansgardenshop.com. Web Site: www.dansgardenshop.com. On-line catalog. Offered 84 non-hybrid vegetable varieties in 2004 including: 8 bean, 1 beet, 1 broccoli raab, 3 carrot, 1 cauliflower, 2 Chinese cabbage, 1 collard, 1 cucumber, 2 eggplant, 1 endive, 1 fava bean, 1 fennel, 2 kohlrabi, 2 lettuce, 1 lima, 2 melon, 1 mustard greens, 1 okra, 3 onion common, 2 onion other, 2 parsley, 2 parsnip, 6 pea, 5 pepper, 7 radish, 1 salsify, 1 spinach, 11 squash, 1 sunflower ornamental, 1 sunflower edible, 1 Swiss chard, 4 tomato, 1 turnip, 1 watermelon, 1 miscellaneous cucurbitaceae and 2 miscellaneous.

Dil Howard Dill Enterprises, R.R. #1, 400 College Road, Windsor, NS B0N 2T0, Canada. Phone: 902-798-2728. Web Site: www.howarddill.com. Free price list. Specializes in the biggest and the smallest pumpkins. Formerly known as Dill's Garden Giant. Retail only. Offered 12 squash varieties in 2004.

Dom Dominion Seed House, PO Box 2500, Georgetown, ON L7G 5L6, Canada. Web Site: www.dominion-seed-house.com. Free catalog. Canadian orders only. In business since 1928. Offered 52 non-hybrid vegetable varieties in 2004 including: 1 artichoke, 6 bean, 2 beet, 1 carrot, 1 cucumber, 1 endive, 1 garlic, 1 gourd, 7 lettuce, 1 mushroom, 2 onion multiplier, 3 parsley, 2 pea, 3 pepper, 9 potato, 1 radish, 2 squash, 4 sunflower ornamental, 1 sunflower edible and 3 tomato.

DOR Dorsing Seeds, Inc., PO Box 2552, Nyssa, OR 97913. Phone: 208-674-1020. E-mail: doug@dorsingseeds.com. Strictly wholesale. Offered 299 non-hybrid vegetable varieties in 2004 including: 2 artichoke, 1 asparagus, 29 bean, 6 beet, 2 broccoli, 1 brussels sprout, 5 cabbage, 10 carrot, 1 cauliflower, 1 celery, 4 Chinese cabbage, 2 chicory,

4 collard, 3 cress, 14 cucumber, 2 eggplant, 2 endive, 1 fava bean, 1 fennel, 1 gourd, 3 kale, 2 kohlrabi, 1 leek, 34 lettuce, 4 lima, 8 melon, 5 mustard greens, 4 okra, 6 onion bunching, 18 onion common, 2 onion other, 3 parsley, 3 parsnip, 23 pea, 9 pepper, 19 radish, 1 rutabaga, 3 spinach, 19 squash, 8 Swiss chard, 10 tomato, 5 turnip, 10 watermelon, 1 miscellaneous cucurbitaceae, 1 miscellaneous leguminosae and 5 miscellaneous.

Dow **Down on the Farm Seed,** PO Box 184, Hiram, OH 44234. $1 for catalog, refundable with order. Specializes in heirloom, open-pollinated, untreated seeds. Retail only. Offered 23 non-hybrid vegetable varieties in 2004 including: 2 bean, 1 beet, 1 carrot, 1 cucumber, 1 lettuce, 1 melon, 1 okra, 1 onion common, 2 parsley, 1 pea, 1 runner bean, 1 spinach, 6 squash, 1 sunflower ornamental, 1 watermelon and 1 miscellaneous.

Du7 **Dutton Berry Farm,** Paul Dutton, Rt. 30, Newfane, VT 05345. Bought Gilfeather® Turnip Seed business in 2002 from Elysian Hills Tree Farm. Gilfeather® Turnip seed is available at the store for $2.25 per pack or by special order through the mail. Offered 1 turnip in 2004.

Ea4 **Eastern Native Seed Conservancy,** P.O. Box 451, Great Barrington, MA 01230. E-mail: natseeds@aol.com. Web Site: www.enscseeds.org. On-line seed listing. Non-profit organization endeavoring to preserve rare genetic seed stock. Offered 165 non-hybrid vegetable varieties in 2004 including: 2 bean, 1 corn, 7 cucumber, 1 eggplant, 1 ground cherry, 1 kale, 3 lettuce, 4 melon, 2 pea, 9 pepper, 1 radish, 1 sorghum, 12 squash, 118 tomato and 2 watermelon.

Ear **Early's Farm and Garden Centre Inc.,** 2615 Lorne Ave., Saskatoon, SK S7J 0S5, Canada. Phone: 306-931-1982. Fax: 306-931-7110. E-mail: sales@earlysgarden.com. Web Site: www.earlysgarden.com. $2 for catalog to U.S. addresses, free within Canada. Founded in 1907. Specializes in vegetables for cool/northern climates. "Saskatchewan's only locally-owned company publishing a mail-order garden seed catalog." Retail price list, with quantity prices available on request. Offered 190 non-hybrid vegetable varieties in 2004 including: 2 amaranth, 1 artichoke, 1 asparagus, 14 bean, 6 beet, 2 broccoli, 2 brussels sprout, 4 cabbage, 9 carrot, 3 cauliflower, 1 celeriac, 2 celery, 3 Chinese cabbage, 2 corn, 2 cress, 8 cucumber, 1 eggplant, 1 endive, 1 fava bean, 1 gourd, 1 kale, 2 kohlrabi, 2 leek, 11 lettuce, 1 lima, 4 melon, 1 okra, 2 onion bunching, 7 onion common, 2 onion other, 3 parsley, 2 parsnip, 13 pea, 1 peanut, 7 pepper, 12 potato, 10 radish, 1 runner bean, 1 rutabaga, 1 spinach, 12 squash, 2 sunflower ornamental, 1 sunflower edible, 5 Swiss chard, 9 tomato, 3 turnip, 3 watermelon, 1 miscellaneous cucurbitaceae and 4 miscellaneous.

Ec2 **Echo Valley Herb Farm,** 3717 Old Hwy. 2 North, Troy, MT 59935-9736. Phone: 406-295-5862. E-mail: echovalleyherb@yahoo.com. Free price list. Specializes in Red German garlic. Former supplier to Silver Springs Nursery, which has turned over its seed garlic production and sales to Echo Valley. Retail price list. Offered 1 garlic in 2004.

Ech **ECHO Seed Sales,** 17391 Durrance Road, North Fort Myers, FL 33917. Phone: 239-543-3246. E-mail: gju@echonet.org. Web Site: www.echonet.org. Free catalog. Retail only. Non-profit organization specializing in tropical or subtropical fruits and vegetables and those underutilized in underdeveloped nations. Seed sale profits are used to cover some of the cost for free packets of seeds sent to small farmers in the Third World. Offered 45 non-hybrid vegetable varieties in 2004 including: 4 amaranth, 2 bean, 2 carrot, 2 corn, 1 eggplant, 4 gourd, 4 lettuce, 3 okra, 1 pea, 2 sorghum, 1 soybean, 2 squash, 4 sunflower edible, 1 tepary, 2 tomato, 2 miscellaneous cucurbitaceae, 3 miscellaneous leguminosae and 5 miscellaneous.

ECK **Eckroat Seed Co.,** PO Box 17610, 1106 Martin Luther King Road, Oklahoma City, OK 73136. Strictly wholesale. Offered 6 non-hybrid vegetable varieties in 2004 including: 4 cowpea, 1 miscellaneous leguminosae and 1 miscellaneous.

Eco **Ecogenesis Inc.,** 1267-2384 Yonge St., Toronto, ON M4P 3E5, Canada. Phone: 416-485-8333. E-mail: ecogenesis @hotmail.com. Web Site: www.ecogenesis.ca. $5 for catalog. Specializes in early, untreated insect- and disease-resistant vegetable varieties for organic gardens. Separate retail and wholesale catalogs; retail catalog contains additional varieties. Offered 47 non-hybrid vegetable varieties in 2004 including: 8 bean, 3 beet, 2 broccoli, 1 cabbage, 3 carrot, 1 Chinese cabbage, 3 corn, 3 cucumber, 3 lettuce, 1 melon, 1 onion common, 1 onion other, 1 pea, 1 pepper, 1 radish, 1 spinach, 3 squash, 1 sunflower edible, 8 tomato and 1 watermelon.

Ed3 **Jack Edelman, Ph.D.,** PO Box 20799, Brooklyn, NY 11202-0799. E-mail: thechromosomekid@37.com. Send SASE for price list. Offers only tubers of Chayote squash (*Sechium edule*). Retail and wholesale. Offered 1 miscellaneous cucurbitaceae in 2004.

Enc **Enchanted Seeds,** PO Box 6087, Las Cruces, NM 88006. Phone: 505-523-6058. E-mail: enchantedseeds @aol.com. Web Site: www.enchantedseeds.com. Free catalog. Specializes in chili peppers. Retail and wholesale. Offered 68 non-hybrid vegetable varieties in 2004 including: 1 onion common, 66 pepper and 1 tomatillo.

Eo2 **E.O.N.S., Inc.,** Eden Organic Nursery Services, Inc., P.O. Box 4604, Hallandale, FL 33008. Phone: 954-455-0229. E-mail: info@eonseed.com. Web Site: www.eonseed.com. Free price list. Offers garden seeds, gardening supplies and safe organic products for home and garden pest control. Retail and wholesale. Offered 81 non-hybrid vegetable varieties in 2004 including: 2 amaranth, 1 artichoke, 1 asparagus, 3 bean, 1 beet, 1 cabbage, 2 carrot, 1 celery, 3 Chinese cabbage, 1 chicory, 1 collard, 7 corn, 1 cowpea, 1 cucumber, 1 endive, 1 fennel, 2 gourd, 1 ground cherry, 1 leek, 3 lettuce, 1 lima, 3 melon, 1 okra, 2 onion bunching, 2 onion other, 1 parsley, 1 pea, 12 pepper, 1 radish, 1 salsify, 1 scorzonera, 1 soybean, 3 squash, 1 sunflower edible, 1 Swiss chard, 6 tomato, 1 miscellaneous cucurbitaceae, 1 miscellaneous leguminosae and 6 miscellaneous.

Ers E & R Seed, 1356 E 200 S, Monroe, IN 46772. Phone: 260-692-6071. Free catalog. Offering over 1,000 varieties of vegetable and flower seeds including many organically grown, open-pollinated and heirloom varieties. Retail and wholesale. Offered 509 non-hybrid vegetable varieties in 2004 including: 1 amaranth, 1 artichoke, 1 asparagus, 49 bean, 11 beet, 5 broccoli, 1 brussels sprout, 11 cabbage, 9 carrot, 3 cauliflower, 1 celeriac, 4 celery, 4 Chinese cabbage, 1 chicory, 3 collard, 26 corn, 1 cornsalad, 3 cowpea, 3 cress, 12 cucumber, 1 eggplant, 2 endive, 1 fennel, 1 garlic, 8 gourd, 1 ground cherry, 3 kale, 3 kohlrabi, 3 leek, 32 lettuce, 13 lima, 8 melon, 3 mustard greens, 2 okra, 1 onion bunching, 5 onion common, 1 onion multiplier, 2 onion other, 5 parsley, 3 parsnip, 30 pea, 1 peanut, 18 pepper, 10 potato, 20 radish, 1 runner bean, 1 rutabaga, 1 salsify, 3 spinach, 51 squash, 7 sunflower ornamental, 3 sunflower edible, 5 Swiss chard, 8 sweet potato, 1 tomatillo, 70 tomato, 4 turnip, 21 watermelon, 1 miscellaneous cucurbitaceae, 1 miscellaneous leguminosae and 4 miscellaneous.

Ev2 Evergreen Y.H. Enterprises, PO Box 17538, Anaheim, CA 92817. Phone: 714-637-5769. Fax: 714-637-5769. E-mail: eeseedsyh@aol.com. Web Site: www.evergreenseeds.com. On-line catalog only. Specializes in Oriental vegetable seeds. Retail and wholesale. Offered 141 non-hybrid vegetable varieties in 2004 including: 5 amaranth, 1 bean, 4 carrot, 1 celery, 20 Chinese cabbage, 1 corn, 1 cress, 3 cucumber, 6 eggplant, 1 fava bean, 7 gourd, 1 kale, 2 kohlrabi, 4 lettuce, 2 melon, 2 mustard greens, 6 onion bunching, 4 onion other, 5 pea, 6 pepper, 8 radish, 7 soybean, 3 squash, 1 tomato, 3 turnip, 8 miscellaneous cucurbitaceae, 13 miscellaneous leguminosae and 16 miscellaneous.

Fa1 Farmer Seed and Nursery Co., Division of Plantron, Inc., 818 NW 4th St., Faribault, MN 55021. Free catalog. Retail only. Offered 105 non-hybrid vegetable varieties in 2004 including: 2 asparagus, 13 bean, 4 beet, 1 cabbage, 6 carrot, 1 cauliflower, 1 corn, 3 cucumber, 1 garlic, 2 gourd, 1 Jerusalem artichoke, 1 kohlrabi, 1 leek, 6 lettuce, 2 lima, 1 melon, 4 onion common, 1 onion multiplier, 1 onion other, 1 parsley, 12 pea, 1 pepper, 6 potato, 5 radish, 2 rutabaga, 1 spinach, 11 squash, 1 sunflower edible, 2 Swiss chard, 3 sweet potato, 1 tomato, 3 watermelon, 1 miscellaneous solanaceae and 3 miscellaneous.

Fe5 Fedco Seeds, PO Box 520, Waterville, ME 04903. Phone: 207-873-7333. Web Site: www.fedcoseeds.com. $2 for catalog. Specializes in selections for cold climates and short growing seasons. Offers vegetable, herb, flower, cover crop and green manure seeds. All seeds are untreated. Retail and wholesale. Offered 389 non-hybrid vegetable varieties in 2004 including: 3 amaranth, 1 artichoke, 40 bean, 7 beet, 2 broccoli, 1 broccoli raab, 1 brussels sprout, 4 cabbage, 8 carrot, 1 cauliflower, 3 celery, 7 Chinese cabbage, 5 chicory, 1 collard, 12 corn, 2 cornsalad, 3 cowpea, 3 cress, 7 cucumber, 2 eggplant, 3 endive, 1 fava bean, 1 fennel, 7 gourd, 4 kale, 2 kohlrabi, 4 leek, 46 lettuce, 1 lima, 5 melon, 1 mustard greens, 1 okra, 1 onion bunching, 5 onion common, 2 onion other, 4 parsley, 2 parsnip, 22 pea, 25 pepper, 9 radish, 1 runner bean, 1 rutabaga, 1 scorzonera, 2 sorghum, 1 soybean, 2 spinach, 39 squash, 6 sunflower ornamental, 2 sunflower edible, 6 Swiss chard, 1 tepary, 1 tomatillo, 38 tomato, 5 turnip, 9 watermelon, 1 miscellaneous cucurbitaceae, 1 miscellaneous leguminosae and 13 miscellaneous.

Fi1 Henry Field's Seed and Nursery Co., Order Processing Center, P.O. Box 397, Aurora, IN 47001-0397. Phone: 513-354-1494. E-mail: service@henryfields.com. Web Site: www.henryfields.com. Free catalog. Founded in 1893. Offered 195 non-hybrid vegetable varieties in 2004 including: 1 artichoke, 2 asparagus, 25 bean, 7 beet, 2 broccoli, 1 brussels sprout, 1 cabbage, 5 carrot, 1 cauliflower, 1 Chinese cabbage, 2 collard, 4 corn, 4 cowpea, 6 cucumber, 1 fennel, 3 garlic, 2 gourd, 1 Jerusalem artichoke, 1 kale, 2 kohlrabi, 2 leek, 6 lettuce, 5 lima, 3 melon, 1 mustard greens, 2 okra, 1 onion bunching, 4 onion common, 1 onion multiplier, 2 onion other, 1 parsley, 14 pea, 1 peanut, 5 pepper, 10 potato, 9 radish, 1 sorghum, 2 spinach, 18 squash, 1 sunflower ornamental, 3 sunflower edible, 2 Swiss chard, 5 sweet potato, 13 tomato, 1 turnip, 9 watermelon and 1 miscellaneous.

Fi2 Field and Forest Products Inc., N3296 Kozuzek Road, Peshtigo, WI 54157. Phone: 715-582-4997. Fax: 715-582-0181. E-mail: ffp@mari.net. Web Site: www.fieldforest.net. Free catalog. Specializes in mushroom spawn for the home gardener and commercial operations. Retail and wholesale. Offered 27 mushroom varieties in 2004.

Fi6 Filaree Farm, 182 Conconully Hwy., Okanogan, WA 98840. Phone: 509-422-6940. E-mail: filaree@northcascades.net. Web Site: www.filareefarm.com. $2 for 40-page catalog. Specializes in certified organic garlic and green manure seeds. Currently testing over 400 garlic strains and offering more than 100 for sale, "the widest selection of garlics in the U.S." Retail and wholesale. Offered 137 non-hybrid vegetable varieties in 2004 including: 136 garlic and 1 miscellaneous.

Fi7 Fish Lake Garlic Man, RR 2, Demorestville, ON K0K 1W0, Canada. Phone: 613-476-8030. Within Canada, send SASE for price list and information on growing garlic; $3 plus self-addressed envelope for requests from other countries. Specializes in growing better quality and higher yielding garlic. Promotes garlic self-sufficiency and use as a healing and preventative medicine. Strictly organic. Retail only. Offered 18 non-hybrid vegetable varieties in 2004 including: 17 garlic and 1 leek.

Fis Fisher's Garden Store, 20750 E. Frontage Road, PO Box 236, Belgrade, MT 59714. Phone: 406-388-6052. Free catalog. Breeders and growers of vegetable and flower seeds for high altitudes and short growing seasons including many varieties originated by them. Established in Montana in 1923. Offered 124 non-hybrid vegetable varieties in 2004 including: 1 asparagus, 11 bean, 4 beet, 1 broccoli, 1 brussels sprout, 6 cabbage, 4 carrot, 2 cauliflower, 1 celery, 1 Chinese cabbage, 1 collard, 4 corn, 1 cress, 5 cucumber, 1 kale, 1 kohlrabi, 9 lettuce, 1 melon, 1 onion bunching, 5 onion common, 1 onion other, 2 parsley, 1 parsnip, 8 pea, 3 pepper, 3 potato, 7 radish, 1 runner bean, 1 rutabaga, 1 salsify, 1 spinach, 16 squash, 1 sunflower ornamental, 1 sunflower edible, 3 Swiss chard, 9 tomato, 2 turnip and 2 miscellaneous.

Fl3 **Florida Mycology Research Center,** P.O. Box 18105, Pensacola, FL 32523-8105. Phone: 850-327-4378. E-mail: FloridaMycology@cs.com Web Site: www.mushroomsfmrc.com. $10 for their main catalog (free if you mention reading about them in this *Garden Seed Inventory*), containing the world's largest selection of mushroom spores and live cultures, growing supplies and publications since 1972. Annual subscription to their quarterly color mushroom journal is $20. Retail and wholesale. Offered 74 mushroom varieties in 2004.

Fo4 **Four Seasons Nursery,** Division of Plantron, Inc., 1706 Morrissey Drive, Bloomington, IL 61704. Phone: 309-663-9551. E-mail: customerservice@4seasonsnurseries.com. Web Site: www.4seasonsnurseries.com. Free catalog. Retail only. Offered 1 leek in 2004.

Fo7 **Fox Hollow Seed Company,** 204 Arch Street, Kittan-

~~~~ OUT OF BUSINESS ~~~~

ning, PA 16201. E-mail: seeds@alltel.net. Web Site: www.foxhollowseed.com. Free catalog. Specializes in open-pollinated, heirloom, untreated varieties including many certified organic vegetable seeds. Retail only. Offered 327 non-hybrid vegetable varieties in 2004 including: 42 bean, 3 beet, 2 broccoli, 1 brussels sprout, 8 cabbage, 1 cardoon, 8 carrot, 2 cauliflower, 1 celeriac, 1 celery, 3 Chinese cabbage, 4 chicory, 2 collard, 10 corn, 1 cornsalad, 3 cress, 7 cucumber, 2 eggplant, 2 endive, 1 fava bean, 1 fennel, 2 kale, 1 kohlrabi, 1 leek, 23 lettuce, 7 lima, 7 melon, 1 mustard greens, 1 okra, 9 onion common, 2 onion other, 4 parsley, 1 parsnip, 12 pea, 37 pepper, 9 radish, 2 runner bean, 1 rutabaga, 1 salsify, 2 sorghum, 2 spinach, 13 squash, 1 sunflower edible, 2 Swiss chard, 1 tomatillo, 60 tomato, 4 turnip, 7 watermelon, 1 miscellaneous cucurbitaceae and 8 miscellaneous.

Fo13 **Forget Me Not Heritage Seed,** 729 Erbsville Road, Waterloo, ON N2J 3Z4, Canada. Web Site: www.forgetmenotseeds.com. On-line catalog. Will ship small quantities of seed to the U.S. for home garden use. Offers untreated, open-pollinated seed grown with respect for the earth. Offered 58 non-hybrid vegetable varieties in 2004 including: 7 bean, 3 beet, 1 broccoli, 2 cabbage, 2 carrot, 3 cucumber, 4 lettuce, 1 mustard greens, 3 onion common, 1 onion other, 3 pea, 5 pepper, 1 radish, 2 runner bean, 1 spinach, 4 squash, 1 Swiss chard and 14 tomato.

Fre **Fred's Plant Farm,** 4589 Ralston Road, Martin, TN 38237. Phone: 800-550-2575. Free catalog. Extensive collection of sweet potatoes and yams, including some oldtimers. Retail and wholesale. Offered 9 sweet potato varieties in 2004.

Fun **Fungi Perfecti,** PO Box 7634, Olympia, WA 98507. Phone: 360-426-9292. Fax: 360-426-9377. E-mail: mycomedia@aol.com. Web Site: www.fungi.com. Free brochure. $4.95 for 80-page catalog. Family-owned business specializing in high-quality gourmet and medicinal mushroom kits, spawn, plug spawn, along with equipment and techniques for their cultivation. Retail and wholesale. Offered 16 mushroom varieties in 2004.

Ga1 **Irish Eyes and Garden City Seeds,** P.O. Box 307, Thorp, WA 98946. Phone: 509-964-7000. E-mail: potatoes@irish-eyes.com. Web Site: www.irish-eyes.com. Free catalog. Specializes in vegetables that are trialed and tasted to ensure quick maturity, good flavor and vigor. Retail and wholesale. Offered 392 non-hybrid vegetable varieties in 2004 including: 2 amaranth, 1 asparagus, 15 bean, 6 beet, 3 broccoli, 3 cabbage, 14 carrot, 1 cauliflower, 1 celery, 4 Chinese cabbage, 1 chicory, 7 corn, 1 cornsalad, 2 cress, 6 cucumber, 2 eggplant, 2 endive, 1 fava bean, 20 garlic, 1 gourd, 6 kale, 3 kohlrabi, 2 leek, 23 lettuce, 3 melon, 3 onion bunching, 12 onion common, 6 onion multiplier, 2 onion other, 2 parsley, 1 parsnip, 12 pea, 15 pepper, 53 potato, 8 radish, 1 runner bean, 1 rutabaga, 1 spinach, 22 squash, 3 sunflower ornamental, 1 sunflower edible, 3 Swiss chard, 6 sweet potato, 1 tomatillo, 97 tomato, 1 turnip, 4 watermelon, 1 miscellaneous leguminosae and 6 miscellaneous.

Ga7 **Gardenimport Inc.,** 135 West Beaver Creek, P.O. Box 760, Richmond Hill, ON L4B 1C6, Canada. Phone: 905-731-1950. E-mail: flower@gardenimport.com. Web Site: www.gardenimport.com. On-line catalog; $5 for two-year catalog, refundable with order. Specializes in flower and vegetable seed from Suttons in England, tested for suitability to Ontario growing conditions. Retail only. Offered 22 non-hybrid vegetable varieties in 2004 including: 2 bean, 1 cardoon, 1 cucumber, 1 ground cherry, 5 lettuce, 1 onion multiplier, 2 pea, 1 runner bean, 1 squash, 6 tomato and 1 miscellaneous.

Ga17 **Garlicsmiths,** 967 Mingo Mountain Road, Kettle Falls, WA 99141. Web Site: www.garlicsmiths.com. Free catalog. Specializes in gourmet garlic. Small family-owned business since 1989, certified organic by Washington State Department of Agriculture. Retail and wholesale prices are listed in the same catalog. Offered 38 non-hybrid vegetable varieties in 2004 including: 37 garlic and 1 leek.

Ga19 **TheGarlicStore.com,** Yucca Ridge Farm, 46050 Weld County Road 13, Fort Collins, CO 80524. Phone: 970-568-7664. E-mail: TheChiefClove@TheGarlicStore.com. Web Site: www.TheGarlicStore.com. Phone (business hours only--10 a.m. to 6 p.m. Mountain Time, Monday through Friday): 800-854-7219. Catalog is available on-line only, but price list and order form may be requested by phone or mail.

Specializes in certified organic garlic and also offers books and other information on how to grow garlic. Retail only. Offered 61 non-hybrid vegetable varieties in 2004 including: 60 garlic and 1 leek.

Ga21 **Garden Medicinals and Culinaries,** P.O. Box 320, Earlysville, VA 22936. Phone: 434-964-9113. Web Site: www.gardenmedicinals.com. Specializes in medicinal and culinary herbs plus a small collection of open-pollinated vegetables, primarily tomatoes and peppers. Retail and wholesale. Offered 65 non-hybrid vegetable varieties in 2004 including: 1 amaranth, 1 cardoon, 1 collard, 4 cress, 1 fennel, 19 garlic, 1 gourd, 1 Jerusalem artichoke, 1 mustard greens, 4 onion multiplier, 2 onion other, 2 parsley, 5 pepper, 6 tomato, 1 miscellaneous cucurbitaceae and 15 miscellaneous.

Ga22 **The Garden Path Nursery,** 395 Conway Road, Victoria, BC V9E 2B9, Canada. E-mail: thegardenpath @shaw.ca. Web Site: www.earthfuture.com/gardenpath. Catalog $2. Offering Seeds of Victoria vegetable and flower seeds. Organically grown and locally harvested, open-pollinated seeds. Offered 64 non-hybrid vegetable varieties in 2004 including: 1 bean, 2 beet, 1 broccoli, 1 cabbage, 1 celeriac, 1 Chinese cabbage, 1 cornsalad, 1 cress, 1 cucumber, 1 endive, 4 fava bean, 1 ground cherry, 4 kale, 1 leek, 5 lettuce, 1 onion bunching, 1 onion other, 2 parsley, 1 pea, 1 radish, 1 rutabaga, 2 Swiss chard, 1 tomatillo, 23 tomato and 5 miscellaneous.

Ge2 **Germania Seed Co.,** 5978 N. Northwest Hwy., P.O. Box 31787, Chicago, IL 60631. Phone: 773-631-6631. Web Site: www.germaniaseed.com. $15 for catalog, refundable with initial order of $50 or more. Offers over 4,100 varieties of seed and over 3,500 varieties of plug and plant material. Many items are heirlooms. Offered 211 non-hybrid vegetable varieties in 2004 including: 3 amaranth, 1 artichoke, 1 asparagus, 4 bean, 3 beet, 1 broccoli, 1 brussels sprout, 7 cabbage, 1 cardoon, 3 carrot, 5 cauliflower, 1 celeriac, 2 celery, 2 chicory, 1 collard, 2 corn, 1 cowpea, 1 cress, 7 cucumber, 2 eggplant, 2 endive, 1 fava bean, 1 fennel, 1 gourd, 1 ground cherry, 1 kale, 3 kohlrabi, 1 leek, 12 lettuce, 1 lima, 3 melon, 1 mustard greens, 1 okra, 1 onion bunching, 5 onion common, 2 onion other, 4 parsley, 1 parsnip, 3 pea, 24 pepper, 2 radish, 1 runner bean, 5 sorghum, 1 spinach, 16 squash, 12 sunflower ornamental, 3 sunflower edible, 2 Swiss chard, 1 tomatillo, 38 tomato, 1 turnip, 3 watermelon, 1 miscellaneous cucurbitaceae, 1 miscellaneous leguminosae and 6 miscellaneous.

Gl2 **Glendale Enterprises,** 297 Railroad Avenue, De Funiak Springs, FL 32433. Phone: 850-859-2141. Fax: 850-859-2181. Free brochure. Small family-run business selling Chufa and Velvet Beans. Offered 2 non-hybrid vegetable varieties in 2004 including: 1 miscellaneous leguminosae and 1 miscellaneous.

Gl4 **Glacier Gourmet Garlic,** Gene and Helen Gray, 1488 Helena Flats Road, Kalispell, MT 59901. Web Site: www.garlicgourmet.com. Free brochure. Family owned and operated farm specializing in fine quality, organically grown garlic for gourmet cooking and gardening seed. Offered 39 garlic varieties in 2004.

GLO **Fred C. Gloeckner and Co. Inc.,** 600 Mamaroneck Ave., Harrison, NY 105281631. E-mail: info@fred gloeckner.com. Web Site: www.fredgloeckner.com. Strictly wholesale. Offered 95 non-hybrid vegetable varieties in 2004 including: 5 amaranth, 2 broccoli, 6 cabbage, 1 cauliflower, 2 celery, 1 chicory, 7 corn, 2 cucumber, 1 eggplant, 1 fennel, 11 lettuce, 3 onion common, 2 onion other, 2 parsley, 14 pepper, 3 sorghum, 4 squash, 15 sunflower ornamental, 2 sunflower edible, 2 Swiss chard, 7 tomato and 2 miscellaneous.

Go6 **Gourmet Garlic Gardens,** 12300 FM 1176, Bangs, TX 76823. Phone: 325-348-3049. E-mail: bob@web-access.net. Web Site: www.gourmetgarlicgardens.com. Free catalog. Catalog also available on-line. Small family-owned business specializing in gourmet garlic, grown without toxic chemical pesticides or herbicides. Retail only. Offered 30 non-hybrid vegetable varieties in 2004 including: 29 garlic and 1 leek.

Go8 **Golden Harvest Organics, LLC,** 404 N. Impala Drive, Fort Collins, CO 80521. Web Site: www. ghorganics.com. On-line catalog. Organically grown tomato seeds. Offered 41 tomato varieties in 2004.

Go9 **Gourdgeous Farm, LLC,** 5174 County Road 675 E, Bradenton, FL 34211. Web Site: www.gourdgeousfarm.com. On-line seed list. Florida-based family business providing high quality farm-grown gourds and gourd seeds to customers throughout the world. Offered 17 non-hybrid vegetable varieties in 2004 including: 12 gourd, 3 squash and 2 miscellaneous cucurbitaceae.

Goo **Good Seed Co.,** 195 Bolster Creek Road, Oroville, WA 98844. Phone: 509-485-2281. E-mail: moonmt@ televar.com. Web Site: www.goodseedco.net. On-line catalog. Specializes in open-pollinated, heirloom and homestead seeds adapted to northern gardens. Retail only. Offered 82 non-hybrid vegetable varieties in 2004 including: 4 bean, 3 beet, 2 broccoli, 1 brussels sprout, 3 cabbage, 1 carrot, 1 cauliflower, 1 celeriac, 3 Chinese cabbage, 1 chicory, 1 collard, 5 corn, 1 endive, 1 fava bean, 3 garlic, 2 kale, 7 lettuce, 2 melon, 2 onion other, 1 parsley, 3 pea, 6 pepper, 1 quinoa, 1 radish, 1 runner bean, 1 soybean, 1 spinach, 8 squash, 1 sunflower edible, 1 Swiss chard, 1 tomatillo, 9 tomato and 3 miscellaneous.

Gou **Gourmet Gardener,** 12287 117th Drive, Live Oak, FL 32060. Fax: 407-650-2691. E-mail: information@gourmet gardener.com. Web Site: www.gourmetgardener.com. On-line catalog. $5 for complete catalog along with planting calendar and coupon for free seeds. Formerly Herb Gathering. Offers unique, exceptional seeds, many from leading importers. Specializes in gourmet vegetables. Retail only. Offered 57 non-hybrid vegetable varieties in 2004 including: 2 artichoke, 7 bean, 4 beet, 1 broccoli raab, 1 brussels sprout, 2 cabbage, 1 cardoon, 2 carrot, 1 celeriac, 2 celery, 2 eggplant, 2 leek, 1 melon, 1 mustard greens, 1 okra, 1 onion bunching, 2 onion common, 1 onion multiplier, 1 parsnip, 4 pea, 2 pepper, 2 radish, 1 salsify, 5 squash, 2 Swiss chard, 3 tomato, 1 turnip and 2 miscellaneous.

Gr27 Green Thumb Seeds, 17011 W. 280th St., Bethany, MO 64424. Free catalog. Offers heirloom seeds for the home gardener. Retail and wholesale. Formerly Yoder Greenhouse (Yod). Offered 179 non-hybrid vegetable varieties in 2004 including: 20 bean, 5 beet, 7 cabbage, 7 carrot, 2 cauliflower, 3 celery, 1 Chinese cabbage, 1 corn, 1 cress, 6 cucumber, 1 eggplant, 1 ground cherry, 1 kale, 3 kohlrabi, 16 lettuce, 5 lima, 6 melon, 2 okra, 3 onion common, 1 parsley, 1 parsnip, 12 pea, 1 peanut, 5 pepper, 13 radish, 1 rutabaga, 1 salsify, 2 spinach, 19 squash, 1 sunflower edible, 1 Swiss chard, 17 tomato, 1 turnip, 10 watermelon, 1 miscellaneous solanaceae and 1 miscellaneous.

Gr28 Greta's Organic Gardens, Box 352, St-Isidore, ON K0C 2B0, Canada. Web Site: www.perc.ca/PEN/1997-03/s-kryger.html. Sales within Canada only. On-line seed listing. Specializes in organically grown, open-pollinated and unusual vegetable, herb and flower seeds. Offered 147 non-hybrid vegetable varieties in 2004 including: 1 amaranth, 8 bean, 2 corn, 1 cress, 1 cucumber, 1 ground cherry, 1 mustard greens, 2 onion other, 2 parsley, 23 pepper, 2 radish, 1 salsify, 1 sorghum, 1 spinach, 14 squash, 3 sunflower ornamental, 1 sunflower edible, 2 tomatillo, 73 tomato and 7 miscellaneous.

Gr29 GreenDealer Exotic Seeds, PO Box 37328, Louisville, KY 40233-7328. Web Site: www.greendealer-exotic-seeds.com. On-line catalog. Retail and bulk. Offered 199 non-hybrid vegetable varieties in 2004 including: 1 asparagus, 1 beet, 2 cabbage, 1 carrot, 2 cauliflower, 1 celery, 2 Chinese cabbage, 10 corn, 2 cress, 2 cucumber, 2 eggplant, 1 gourd, 1 kale, 1 leek, 6 melon, 2 onion common, 1 parsnip, 1 pea, 116 pepper, 1 radish, 1 salsify, 4 sorghum, 1 soybean, 7 squash, 7 sunflower ornamental, 1 sunflower edible, 1 tomatillo, 17 tomato, 1 watermelon, 2 miscellaneous cucurbitaceae and 1 miscellaneous.

GRI G. S. Grimes Seeds, 11335 Concord-Hambden Road, Concord, OH 44077. Phone: 440-352-6650. E-mail: bwatson @grimesseeds.com. Strictly wholesale. Formerly H. G. German Seeds, Inc. Offered 85 non-hybrid vegetable varieties in 2004 including: 2 amaranth, 1 asparagus, 2 bean, 6 cabbage, 1 cauliflower, 1 celery, 1 collard, 1 cress, 3 cucumber, 1 eggplant, 1 fennel, 1 leek, 3 lettuce, 1 lima, 1 melon, 2 onion other, 4 parsley, 1 pea, 15 pepper, 7 squash, 10 sunflower ornamental, 2 sunflower edible, 1 Swiss chard, 1 tomatillo, 10 tomato, 1 watermelon, 1 miscellaneous leguminosae and 4 miscellaneous.

Gur Gurney's Seed and Nursery Co., PO Box 4178, Greendale, IN 47025-4178. Web Site: www.gurneys.com. Free catalog. Retail only. Specializes in garden seed and nursery items. Offered 164 non-hybrid vegetable varieties in 2004 including: 1 artichoke, 2 asparagus, 17 bean, 4 beet, 1 cabbage, 4 carrot, 1 cauliflower, 2 Chinese cabbage, 3 corn, 2 cowpea, 2 cucumber, 1 eggplant, 4 garlic, 4 gourd, 1 Jerusalem artichoke, 1 kohlrabi, 2 leek, 7 lettuce, 4 lima, 1 melon, 1 okra, 1 onion bunching, 5 onion common, 1 onion multiplier, 2 onion other, 2 parsley, 1 parsnip, 10 pea, 2 peanut, 6 pepper, 8 potato, 6 radish, 1 rutabaga, 1 sorghum, 2 spinach, 16 squash, 5 sunflower ornamental, 4 sunflower edible, 2 Swiss chard, 4 sweet potato, 9 tomato, 1 turnip, 6 watermelon and 4 miscellaneous.

HA3 Chas. C. Hart Seed Co., PO Box 290169, 304 Main St., Wethersfield, CT 06129-0169. Phone: 860-529-2537. E-mail: hartseed@earthlink.net. Web Site: www.hartseed.com. Strictly wholesale. Specializes in conventionally produced and open-pollinated varieties. Offered 209 non-hybrid vegetable varieties in 2004 including: 1 artichoke, 1 asparagus, 20 bean, 3 beet, 3 broccoli, 2 broccoli raab, 1 brussels sprout, 8 cabbage, 1 cardoon, 5 carrot, 2 cauliflower, 2 celery, 3 Chinese cabbage, 2 chicory, 2 collard, 5 corn, 1 cornsalad, 1 cowpea, 1 cress, 8 cucumber, 1 eggplant, 2 endive, 1 fava bean, 1 fennel, 2 gourd, 1 kale, 2 kohlrabi, 1 leek, 12 lettuce, 5 lima, 2 melon, 2 mustard greens, 3 okra, 1 onion bunching, 5 onion common, 2 onion other, 3 parsley, 1 parsnip, 18 pea, 12 pepper, 9 radish, 1 runner bean, 1 rutabaga, 1 salsify, 1 spinach, 23 squash, 2 sunflower ornamental, 2 sunflower edible, 3 Swiss chard, 6 tomato, 3 turnip, 2 watermelon, 1 miscellaneous cucurbitaceae, 1 miscellaneous leguminosae and 3 miscellaneous.

Ha5 Harris Seeds, 355 Paul Road, Rochester, NY 14624-0966. Phone: 800-514-4441. Fax: 877-892-9197. E-mail: kmcguire@harrisseeds.com. Web Site: www.harrisseeds.com. Free catalog. Specializes in treated, untreated and organic seeds. Separate retail and wholesale catalogs, each containing varieties not in the other. Offered 140 non-hybrid vegetable varieties in 2004 including: 14 bean, 2 beet, 2 carrot, 1 collard, 9 corn, 1 cucumber, 1 eggplant, 3 endive, 1 fava bean, 3 gourd, 3 kale, 1 kohlrabi, 1 leek, 16 lettuce, 3 lima, 1 mustard greens, 1 okra, 5 onion common, 1 onion other, 2 parsley, 1 parsnip, 11 pea, 10 pepper, 4 radish, 1 rutabaga, 1 sorghum, 20 squash, 6 sunflower ornamental, 2 sunflower edible, 3 Swiss chard, 3 tomato, 1 turnip, 2 watermelon, 2 miscellaneous leguminosae and 2 miscellaneous.

Ha14 Hardscrabble Enterprises, Inc., PO Box 1124, Franklin, WV 26807-1124. Phone: 304-358-2921. E-mail: hardscrabble@mountain.net. Price list available for SASE. Specializes in Shiitake mushroom spawn. Separate but identical retail and wholesale price lists. Offered 1 mushroom variety in 2004.

HA20 Harris Moran Seed Company, PO Box 4938, Modesto, CA 95352-4938. Web Site: www.harrismoran.com. Strictly wholesale. Formerly "Ferry-Morse Seed Co." Offered 61 non-hybrid vegetable varieties in 2004 including: 20 bean, 6 carrot, 3 cauliflower, 1 leek, 13 lettuce, 2 pepper, 2 radish, 9 squash and 5 tomato.

Ha1 Halifax Seed Company Ltd., 5860 Kane St., PO Box 8026, Stn. A, Halifax, NS B3K 5L8, Canada. Web Site: www.halifaxseed.ca. Sales to Canada only. Catalog is free within Atlantic Canada, $1 elsewhere in Canada. "Serving The Maritime Gardener Since 1866." Offers traditional varieties that perform well in cool climates and short seasons. Retail and wholesale prices are listed in the same catalog. Offered 107 non-hybrid vegetable varieties in 2004 including: 1 asparagus, 10 bean, 5 beet, 4 cabbage, 5 carrot, 1 cauliflower, 1 celery, 1 Chinese cabbage, 1 chicory, 2 cress, 3 cucumber, 1 endive, 1 fava bean, 1 fennel, 1 garlic, 1 kale, 2 kohlrabi, 1 leek, 6 lettuce, 1 lima, 1 onion bunching, 1 onion common, 1 onion multiplier, 1 onion other, 2 parsley, 1 parsnip, 10 pea, 1 pepper, 5 radish, 1 runner bean, 1 rutabaga, 1 spinach, 13 squash, 3 sunflower ornamental, 2 sunflower edible, 4 Swiss chard, 4 tomato, 2 turnip, 1 watermelon and 3 miscellaneous.

He8 Heirloom Seeds, PO Box 245, West Elizabeth, PA 15088-0245. Phone: 412-384-0852. Fax: 412-384-0852. E-mail: mail@heirloomseeds.com. Web Site: www.heirloomseeds.com or www.heirloomtomatoes.com. $1 for catalog, refundable with order. Small family-run seed house selling only open-pollinated vegetable and flower seeds, many first introduced in the 1700s and 1800s. Offered 474 non-hybrid vegetable varieties in 2004 including: 1 amaranth, 1 artichoke, 30 bean, 10 beet, 5 broccoli, 2 brussels sprout, 12 cabbage, 1 cardoon, 12 carrot, 3 cauliflower, 2 celery, 1 Chinese cabbage, 2 chicory, 3 collard, 9 corn, 1 cornsalad, 2 cowpea, 1 cress, 15 cucumber, 4 eggplant, 3 endive, 1 fava bean, 4 gourd, 4 kale, 2 kohlrabi, 3 leek, 35 lettuce, 6 lima, 26 melon, 2 mustard greens, 5 okra, 3 onion bunching, 6 onion common, 2 onion other, 2 parsley, 1 parsnip, 14 pea, 49 pepper, 12 radish, 1 runner bean, 2 rutabaga, 1 salsify, 1 sorghum, 1 soybean, 5 spinach, 26 squash, 6 sunflower ornamental, 4 sunflower edible, 5 Swiss chard, 1 tomatillo, 97 tomato, 5 turnip, 15 watermelon, 2 miscellaneous cucurbitaceae and 5 miscellaneous.

He17 Heirloom Acres, PO Box 194, New Bloomfield, MO 65063. Web Site: www.heirloomacres.net. On-line catalog only. Family-owned and operated business offering the finest in heirloom, open-pollinated and other selected vegetable, herb and flower seeds. Offered 296 non-hybrid vegetable varieties in 2004 including: 2 asparagus, 29 bean, 7 beet, 3 broccoli, 1 brussels sprout, 9 cabbage, 5 carrot, 2 cauliflower, 2 celery, 6 Chinese cabbage, 2 collard, 19 corn, 4 cowpea, 11 cucumber, 3 eggplant, 1 endive, 1 fava bean, 5 gourd, 1 ground cherry, 3 kale, 2 kohlrabi, 1 leek, 11 lettuce, 8 lima, 7 melon, 4 mustard greens, 5 okra, 1 onion bunching, 4 onion common, 2 parsnip, 12 pea, 2 peanut, 17 pepper, 12 radish, 1 salsify, 1 sorghum, 2 spinach, 20 squash, 5 Swiss chard, 1 tomatillo, 41 tomato, 3 turnip, 12 watermelon, 2 miscellaneous cucurbitaceae, 1 miscellaneous solanaceae and 3 miscellaneous.

He18 Heirloom Tomatoes, Donna Meinschein, 5423 Princess Drive, Rosedale, MD 21237. Web Site: www.heirloomtomatoes.net. On-line seed listing. Website founded and created by the late Chuck Wyatt and now maintained by Donna Meinschein. Specializes in heirloom tomatoes. Offered 226 tomato varieties in 2004.

Hi6 High Mowing Seeds, 813 Brook Road, Wolcott, VT 05680-4223. Phone: 802-888-1800. Web Site: www.highmowingseeds.com. Free catalog. Formerly "The Good Seed Company of Vermont." Specializes in rare New England heirlooms, 100% certified organically grown by a network of growers in the Vermont area and especially suited to the Northeast U.S. Retail and wholesale. Offered 157 non-hybrid vegetable varieties in 2004 including: 22 bean, 4 beet, 1 broccoli, 2 carrot, 4 Chinese cabbage, 4 corn, 5 cucumber, 1 eggplant, 1 fennel, 9 garlic, 3 kale, 12 lettuce, 5 melon, 1 onion bunching, 1 onion common, 2 onion other, 2 parsley, 9 pea, 6 pepper, 14 potato, 2 radish, 2 soybean, 1 spinach, 14 squash, 1 sunflower ornamental, 1 sunflower edible, 2 Swiss chard, 1 tomatillo, 20 tomato, 2 watermelon and 3 miscellaneous.

Hi13 Hirt's Greenhouse and Flowers, 13867 Pearl Road, Strongsville, OH 44136. Web Site: www.hirts.com. On-line catalog. One of Ohio's oldest horticultural establishments specializing in unusual vegetables, perennials, flowers, houseplants and bulbs. Offered 178 non-hybrid vegetable varieties in 2004 including: 14 bean, 4 beet, 1 broccoli, 1 broccoli raab, 1 brussels sprout, 6 carrot, 2 Chinese cabbage, 1 chicory, 1 collard, 3 corn, 1 cowpea, 6 cucumber, 2 eggplant, 1 endive, 1 fava bean, 5 gourd, 1 ground cherry, 2 kale, 1 kohlrabi, 2 leek, 10 lettuce, 2 lima, 5 melon, 1 mustard greens, 1 okra, 3 onion bunching, 1 onion common, 1 parsnip, 11 pea, 5 pepper, 7 radish, 3 runner bean, 1 salsify, 1 spinach, 19 squash, 7 sunflower ornamental, 2 sunflower edible, 5 Swiss chard, 2 tomatillo, 19 tomato, 2 turnip, 6 watermelon, 1 miscellaneous cucurbitaceae, 2 miscellaneous leguminosae, 2 miscellaneous solanaceae and 3 miscellaneous.

Hig Seeds Trust/High Altitude Gardens, 4150 B Black Oak Drive, Hailey, ID 83333-8447. Phone: 208-788-4363. E-mail: support@seedstrust.com. Web Site: www.seedstrust.com. Catalog $2. Small regional seed company specializing in short-season, open-pollinated, hardy seeds adapted to cold high mountain climates. Separate retail and wholesale catalogs, each containing additional varieties not in the other. Offered 154 non-hybrid vegetable varieties in 2004 including: 1 asparagus, 9 bean, 3 beet, 2 broccoli, 1 brussels sprout, 6 cabbage, 6 carrot, 1 cauliflower, 1 celery, 3 Chinese cabbage, 1 chicory, 3 corn, 1 cornsalad, 2 cress, 4 cucumber, 2 endive, 1 fava bean, 2 kale, 1 leek, 13 lettuce, 1 melon, 1 onion bunching, 2 onion common, 2 onion other, 2 parsley, 1 parsnip, 9 pea, 8 pepper, 3 radish, 1 spinach, 10 squash, 1 sunflower edible, 1 Swiss chard, 44 tomato, 2 turnip, 1 watermelon and 2 miscellaneous.

HO1 Hollar and Company, Inc., PO Box 106, Rocky Ford, CO 81067. Phone: 719-254-7411. E-mail: bob@hollerseeds.com. Web Site: www.hollarseeds.com. Strictly wholesale. On-line price list. Specializes in seeds of vine crops including watermelon, cantaloupe, cucumber, gourds, summer squash, winter squash and pumpkins. Offered 65 non-hybrid vegetable varieties in 2004 including: 9 cucumber, 5 gourd, 9 melon, 1 pepper, 32 squash and 9 watermelon.

Ho2 Holmes Seed Co., 2125 46th St. NW, Canton, OH 44709. Phone: 800-435-6077. Web Site: www.holmesseed.com. Strictly wholesale. In business for over 105 years. Offering an extensive catalog of commercial vegetable seed varieties. Specializes in sweet corn, beans, pumpkins and squash. Offered 450 non-hybrid vegetable varieties in 2004 including: 1 asparagus, 67 bean, 11 beet, 2 broccoli, 1 broccoli raab, 1 brussels sprout, 9 cabbage, 8 carrot, 4 cauliflower, 2 celery, 4 Chinese cabbage, 1 chicory, 3 collard, 17 corn, 9 cucumber, 1 eggplant, 3 endive, 1 fava bean, 1 fennel, 13 gourd, 1 kale, 3 kohlrabi, 3 leek, 30 lettuce, 10 lima, 5 melon, 3 mustard greens, 4 okra, 1 onion bunching, 6 onion common, 1 onion other, 4 parsley, 2 parsnip, 29 pea, 28 pepper, 17 radish, 2 rutabaga, 1 salsify, 3 sorghum, 2 spinach, 77 squash, 5 Swiss chard, 37 tomato, 5 turnip, 8 watermelon, 1 miscellaneous cucurbitaceae and 3 miscellaneous.

Ho8 Horizon Herbs, PO Box 69, Williams, OR 97544. Phone: 541-846-6704. Fax: 541-846-6233. E-mail: herbseed @horizonherbs.com. Web Site: www.horizonherbs.com. $2 for *Strictly Medicinal®* catalog and growing guide. Retail and wholesale. Specializes in common and rare medicinal herb seeds and books on medicinal herb cultivation by Richo Cech. Offered 29 non-hybrid vegetable varieties in 2004 including: 1 artichoke, 2 cress, 1 fennel, 5 gourd, 1 parsley, 2 pepper, 1 squash, 1 sunflower edible, 3 miscellaneous cucurbitaceae and 12 miscellaneous.

Ho12 Howe Sound Seeds, PO Box 109, Bowen Island, BC V0N 1G0, Canada. Phone: 604-947-0943. Fax: 604-947-0945. $1 for catalog. Canada only. Specializes in open-pollinated late-Victorian vegetable varieties at affordable prices. Retail only. Offered 29 non-hybrid vegetable varieties in 2004 including: 1 artichoke, 4 bean, 1 beet, 1 cardoon, 1 carrot, 1 cauliflower, 1 chicory, 1 endive, 2 kale, 2 lettuce, 1 lima, 2 melon, 1 parsley, 4 pea, 3 radish, 2 squash and 1 tomato.

Ho13 Horus Botanicals, 341 Mulberry, Salem, AR 72576. $3 for catalog. Specializes in open-pollinated strains, especially heirlooms and ethnic types. Retail only. Offered 1096 non-hybrid vegetable varieties in 2004 including: 16 amaranth, 105 bean, 13 beet, 8 broccoli, 1 broccoli raab, 6 cabbage, 2 cardoon, 12 carrot, 1 cauliflower, 1 celeriac, 7 celery, 38 Chinese cabbage, 3 chicory, 3 collard, 48 corn, 39 cowpea, 2 cress, 11 cucumber, 17 eggplant, 9 fava bean, 41 garlic, 9 gourd, 4 ground cherry, 7 Jerusalem artichoke, 16 kale, 3 kohlrabi, 5 leek, 49 lettuce, 18 lima, 16 melon, 15 mustard greens, 26 okra, 4 onion bunching, 4 onion common, 21 onion multiplier, 2 onion other, 3 parsnip, 49 pea, 8 peanut, 108 pepper, 4 potato, 2 quinoa, 19 radish, 6 runner bean, 5 rutabaga, 10 sorghum, 13 soybean, 1 spinach, 35 squash, 1 sunflower ornamental, 1 sunflower edible, 7 Swiss chard, 5 sweet potato, 11 tepary, 4 tomatillo, 116 tomato, 13 turnip, 13 watermelon, 3 miscellaneous cucurbitaceae, 28 miscellaneous leguminosae, 10 miscellaneous solanaceae and 39 miscellaneous.

HPS Horticultural Products and Services Div., 334 W. Stroud Street, Randolph, WI 53956. Phone: 800-322-7288. Web Site: www.hpsseed.com. R. H. Shumway's commercial catalog. Includes both new and heirloom seeds. Offered 91 non-hybrid vegetable varieties in 2004 including: 2 amaranth, 5 cabbage, 2 carrot, 2 cauliflower, 1 celery, 1 cowpea, 6 cucumber, 2 eggplant, 1 lettuce, 1 melon, 1 okra, 2 onion other, 2 parsley, 23 pepper, 8 squash, 6 sunflower ornamental, 19 tomato, 5 watermelon and 2 miscellaneous.

Hud J. L. Hudson, Seedsman, Star Route 2 Box 337, LaHonda, CA 94020. E-mail: inquiry@jlhudsonseeds.net. Web Site: www.jlhudsonseeds.net. $1 for catalog. Specializes in seeds of rare and unusual plants, Zapotec varieties and seeds of useful wild and cultivated plants. Established in 1911, successors to Harry Saier. Stresses open-pollinated, unpatented, non-hybrid seeds and preservation of biological and cultural diversity since 1973. Offered 252 non-hybrid vegetable varieties in 2004 including: 7 amaranth, 7 bean, 5 beet, 1 broccoli, 2 broccoli raab, 1 brussels sprout, 4 cabbage, 3 carrot, 1 cauliflower, 1 celery, 4 chicory, 2 collard, 12 corn, 3 cornsalad, 4 cowpea, 2 cress, 4 cucumber, 4 eggplant, 6 gourd, 3 ground cherry, 3 kale, 1 kohlrabi, 1 leek, 10 lettuce, 1 lima, 10 melon, 2 mustard greens, 3 okra, 2 onion bunching, 2 onion common, 2 onion other, 2 parsley, 3 pea, 15 pepper, 3 quinoa, 6 radish, 1 sorghum, 2 spinach, 18 squash, 7 sunflower ornamental, 3 sunflower edible, 3 Swiss chard, 2 tomatillo, 26 tomato, 3 turnip, 4 watermelon, 5 miscellaneous cucurbitaceae, 10 miscellaneous leguminosae, 8 miscellaneous solanaceae and 18 miscellaneous.

Hum Ed Hume Seeds, PO Box 73160, Puyallup, WA 98373. E-mail: jim@humeseeds.com. Web Site: www.humeseeds.com. Free price list. Specializes in seeds for cool and short-season climates, with special selections for Alaska and for autumn planting. Retail catalog and wholesale rack sales in Pacific Northwest. Offered 135 non-hybrid vegetable varieties in 2004 including: 1 artichoke, 11 bean, 4 beet, 2 broccoli, 1 broccoli raab, 1 brussels sprout, 4 cabbage, 7 carrot, 1 cauliflower, 1 celery, 2 Chinese cabbage, 1 chicory, 1 collard, 1 cornsalad, 1 cress, 4 cucumber, 1 endive, 1 fava bean, 1 fennel, 1 ground cherry, 2 kale, 2 kohlrabi, 1 leek, 7 lettuce, 1 melon, 1 okra, 1 onion bunching, 3 onion common, 2 onion other, 2 parsley, 1 parsnip, 13 pea, 2 pepper, 7 radish, 1 runner bean, 1 rutabaga, 17 squash, 4 sunflower ornamental, 1 sunflower edible, 4 Swiss chard, 1 tomatillo, 9 tomato, 1 turnip, 1 watermelon and 3 miscellaneous.

In8 Independent Seed, PO Box 106, Port Townsend, WA 98368. E-mail: mail@independentseed.com. Web Site: www.independentseed.com. Previous staff, growers and volunteers of Abundant Life Seed Foundation have joined together to form Independent Seed and continue to work toward a healthy seed system. Specializing in quality, open-pollinated, organically grown seed from independent seed farms. Retail only. Offered 138 non-hybrid vegetable varieties in 2004 including: 3 amaranth, 9 bean, 5 beet, 2 broccoli, 3 cabbage, 5 carrot, 1 cauliflower, 2 celery, 4 Chinese cabbage, 4 corn, 1 cornsalad, 5 cucumber, 2 fava bean, 3 kale, 4 leek, 11 lettuce, 2 onion bunching, 3 onion common, 5 parsley, 2 parsnip, 7 pea, 1 pepper, 4 quinoa, 5 radish, 1 runner bean, 1 rutabaga, 2 spinach, 11 squash, 2 sunflower ornamental, 4 sunflower edible, 6 Swiss chard, 12 tomato, 1 turnip and 5 miscellaneous.

Jo1 **Johnny's Selected Seeds,** 955 Benton Avenue, Winslow, ME 04901. Phone: 207-861-3901. Fax: 800-437-4290. E-mail: staff@johnnyseeds.com. Web Site: www.johnnyseeds.com. Free catalog. Vegetable, flower, herb and farm seed and garden accessories. Many heirlooms and new introductions. Extensive trial grounds throughout North America. Satisfaction 100% guaranteed. Retail and wholesale. Offered 284 non-hybrid vegetable varieties in 2004 including: 4 amaranth, 1 artichoke, 26 bean, 7 beet, 1 broccoli, 2 broccoli raab, 2 cabbage, 3 carrot, 1 cauliflower, 1 celeriac, 2 celery, 5 Chinese cabbage, 2 chicory, 1 collard, 8 corn, 1 cornsalad, 3 cress, 4 cucumber, 3 eggplant, 4 endive, 1 fava bean, 1 fennel, 3 garlic, 7 gourd, 1 Jerusalem artichoke, 2 kale, 4 leek, 26 lettuce, 1 lima, 2 melon, 2 mustard greens, 2 onion bunching, 6 onion common, 1 onion multiplier, 4 onion other, 4 parsley, 1 parsnip, 8 pea, 20 pepper, 8 potato, 5 radish, 2 runner bean, 1 rutabaga, 1 salsify, 1 scorzonera, 3 soybean, 27 squash, 11 sunflower ornamental, 2 sunflower edible, 4 Swiss chard, 3 tomatillo, 18 tomato, 2 turnip, 1 watermelon, 1 miscellaneous leguminosae and 17 miscellaneous.

Jo6 **Johnny Pepperseed,** 6407 Market St., Charlotte, NC 28215-4221. Web Site: www.johnnypepperseed.com. On-line only. Offering fine quality pepper seeds. Offered 38 non-hybrid vegetable varieties in 2004 including: 1 parsley, 29 pepper, 2 tomatillo and 6 tomato.

Jor **Jordan Seeds Inc.,** 6400 Upper Afton Road, Woodbury, MN 55125-1146. Phone: 651-738-3422. E-mail: seeds@jordanseeds.com. Web Site: www.jordanseeds.com. Free catalog. Specializes in untreated seed for organic growers. Retail and wholesale. Offered 383 non-hybrid vegetable varieties in 2004 including: 2 asparagus, 52 bean, 5 beet, 2 broccoli, 1 brussels sprout, 7 cabbage, 8 carrot, 2 cauliflower, 2 celery, 6 Chinese cabbage, 2 collard, 20 corn, 1 cowpea, 11 cucumber, 1 eggplant, 1 endive, 1 fennel, 12 gourd, 1 ground cherry, 3 kale, 2 kohlrabi, 2 leek, 14 lettuce, 2 lima, 8 melon, 4 mustard greens, 2 okra, 4 onion bunching, 5 onion common, 3 onion other, 3 parsley, 2 parsnip, 30 pea, 24 pepper, 15 radish, 2 rutabaga, 4 sorghum, 1 spinach, 66 squash, 9 sunflower ornamental, 4 sunflower edible, 3 Swiss chard, 2 tomatillo, 10 tomato, 4 turnip, 10 watermelon, 3 miscellaneous cucurbitaceae, 1 miscellaneous leguminosae and 4 miscellaneous.

Jun **J. W. Jung Seed Co.,** 335 S. High St., Randolph, WI 53957-0001. Phone: 800-297-3123. E-mail: info@jungseed.com. Web Site: www.jungseed.com. Free retail catalog. Quality seeds since 1907. Specializes in seeds and nursery stock suitable for northern climates. Offered 224 non-hybrid vegetable varieties in 2004 including: 2 asparagus, 29 bean, 7 beet, 3 carrot, 1 cauliflower, 1 celeriac, 1 celery, 1 Chinese cabbage, 1 chicory, 9 corn, 8 cucumber, 1 endive, 1 fennel, 3 garlic, 4 gourd, 1 ground cherry, 1 Jerusalem artichoke, 1 kale, 2 kohlrabi, 2 leek, 14 lettuce, 4 lima, 1 melon, 1 mustard greens, 1 onion bunching, 3 onion common, 3 onion multiplier, 2 onion other, 3 parsley, 2 parsnip, 17 pea, 1 peanut, 6 pepper, 13 potato, 10 radish, 1 rutabaga, 1 salsify, 1 sorghum, 1 spinach, 22 squash, 6 sunflower ornamental, 1 sunflower edible, 3 Swiss chard, 5 sweet potato, 1 tomatillo, 14 tomato, 1 turnip, 3 watermelon and 4 miscellaneous.

Kes **Kester's Wild Game Food Nurseries Inc.,** PO Box 516, Omro, WI 54963. Phone: 920-685-2929. Fax: 920-685-6727. Web Site: www.kestersnursery.com. Free price list. Mostly food plants for waterfowl and erosion control. Offers seed of Giant Wild Rice. Retail only. Offered 4 miscellaneous varieties in 2004.

Ki4 **John Scheepers Kitchen Garden Seeds,** 23 Tulip Drive, P.O. Box 638, Bantam, CT 06750-5323. Phone: 860-567-6086. E-mail: lance@johnscheepers.com. Web Site: www.kitchengardenseeds.com. Free catalog. One of the oldest and most prestigious seed importers offering the finest heirlooms, cottage garden flowers, gourmet vegetables and aromatic herbs from around the world. Retail only. Offered 111 non-hybrid vegetable varieties in 2004 including: 1 artichoke, 3 bean, 4 beet, 3 carrot, 1 cauliflower, 1 celeriac, 2 Chinese cabbage, 2 chicory, 1 cornsalad, 4 cress, 2 cucumber, 2 eggplant, 1 endive, 1 fennel, 2 garlic, 3 kale, 1 leek, 20 lettuce, 1 lima, 1 mustard greens, 1 okra, 1 onion bunching, 2 onion common, 1 onion multiplier, 1 onion other, 2 parsley, 3 pea, 8 pepper, 5 potato, 6 radish, 8 squash, 4 Swiss chard, 1 tomatillo, 7 tomato, 1 turnip, 1 watermelon and 3 miscellaneous.

Kil **Kilgore Seed Co.,** P.O. Box 2082, Lake City, FL 32056-2082. Free catalog. Specializes in vegetables that are for the Gulf Coast States and tropical and subtropical areas. Retail and bulk prices. Offered 167 non-hybrid vegetable varieties in 2004 including: 11 bean, 3 beet, 1 broccoli, 1 broccoli raab, 1 brussels sprout, 6 cabbage, 3 carrot, 2 cauliflower, 1 celery, 3 Chinese cabbage, 3 collard, 2 corn, 1 cornsalad, 9 cowpea, 3 cucumber, 2 eggplant, 2 endive, 1 fennel, 1 gourd, 1 kale, 1 kohlrabi, 1 leek, 10 lettuce, 8 lima, 5 melon, 2 mustard greens, 4 okra, 1 onion bunching, 4 onion common, 1 onion other, 2 parsley, 5 pea, 15 pepper, 5 radish, 1 rutabaga, 1 spinach, 15 squash, 2 Swiss chard, 11 tomato, 3 turnip, 8 watermelon, 1 miscellaneous cucurbitaceae, 1 miscellaneous leguminosae and 3 miscellaneous.

Kit **Kitazawa Seed Co.,** P.O. Box 13220, Oakland, CA 94661-3220. Phone: 510-595-1188. E-mail: kitaseed@pacbell.net. Web Site: www.kitazawaseed.com. Free catalog. Product specialty is vegetable seeds from Japan and other parts of Asia for Japanese, Chinese, Thai, Korean, Indian, Vietnamese and Philippine cuisines. Supplying seeds to commercial growers, wholesalers and home gardeners since 1917. Retail and wholesale. Offered 152 non-hybrid vegetable varieties in 2004 including: 2 amaranth, 3 bean, 2 broccoli raab, 3 carrot, 2 celery, 20 Chinese cabbage, 3 cress, 2 cucumber, 3 eggplant, 1 fava bean, 4 gourd, 1 ground cherry, 1 kale, 2 lettuce, 4 melon, 10 mustard greens, 7 onion bunching, 2 onion common, 2 onion other, 1 parsley, 7 pea, 5 pepper, 17 radish, 5 soybean, 4 squash, 5 turnip, 2 miscellaneous cucurbitaceae, 12 miscellaneous leguminosae and 20 miscellaneous.

Kus **The KUSA Seed Research Foundation,** PO Box 761, Ojai, CA 93024. $2.50 plus long SASE for seed and literature catalog. Specializes in heirloom edible seed crops, especially cereal grains, for the garden and mini-farm. All proceeds go to further the work of non-profit KUSA Seed Research Foundation. Retail only. Offered 6 non-hybrid vegetable varieties in 2004 including: 1 amaranth, 4 miscellaneous leguminosae and 1 miscellaneous.

La1 **D. Landreth Seed Co.,** 650 N. North Point Road, Baltimore, MD 21237. Phone: 410-325-2045. Fax: 410-325-2046. Toll-free telephone is 800-654-2407. Web Site: www.landrethseeds.com. Founded in 1784, making it the oldest seedhouse in the United States. Specializes in rare and heirloom vegetable and flower seeds. Retail and wholesale. Offered 456 non-hybrid vegetable varieties in 2004 including: 2 artichoke, 1 asparagus, 48 bean, 15 beet, 8 broccoli, 2 broccoli raab, 2 brussels sprout, 14 cabbage, 1 cardoon, 13 carrot, 5 cauliflower, 2 celery, 3 Chinese cabbage, 3 chicory, 3 collard, 10 corn, 1 cornsalad, 2 cowpea, 1 cress, 20 cucumber, 17 eggplant, 4 endive, 2 fava bean, 1 fennel, 2 garlic, 6 gourd, 4 kale, 2 kohlrabi, 1 leek, 36 lettuce, 10 lima, 28 melon, 5 mustard greens, 4 okra, 9 onion common, 1 onion other, 5 parsley, 2 parsnip, 15 pea, 1 peanut, 20 pepper, 14 radish, 1 runner bean, 1 rutabaga, 1 salsify, 1 soybean, 3 spinach, 36 squash, 3 sunflower ornamental, 4 sunflower edible, 5 Swiss chard, 1 tomatillo, 25 tomato, 5 turnip, 11 watermelon, 3 miscellaneous cucurbitaceae, 2 miscellaneous leguminosae, 1 miscellaneous solanaceae and 8 miscellaneous.

La8 **Ladybug Herbs of Vermont,** 943 Richard Woolcutt Road, Wolcott, VT 05680-4176. Phone: 802-888-5940. E-mail: organic@ladybugherbsofvermont.com. Web Site: www.Ladybugherbsofvermont.com. Catalog $2. Acquired "Herbs, Naturally" in the summer of 1997 from Tinmouth Channel Farm. Sells culinary, medicinal and ornamental herb seeds, herb plants and herbal products. Retail only. Offered 14 non-hybrid vegetable varieties in 2004 including: 2 cress, 1 fennel, 2 garlic, 1 onion bunching, 3 onion other, 2 parsley and 3 miscellaneous.

Lan **Landis Valley Heirloom Seed Project,** Landis Valley Museum, 2451 Kissel Hill Road, Lancaster, PA 17601. Phone: 717-569-0401. Fax: 717-560-2147. E-mail: bikers2@nbn.net. Web Site: www.landisvalleymuseum.org. $3 for catalog. Specializes in open-pollinated vegetable varieties with Pennsylvania German origin from 1740 to 1940. Retail and wholesale catalog. Offered 77 non-hybrid vegetable varieties in 2004 including: 13 bean, 3 beet, 4 cabbage, 1 carrot, 4 corn, 1 cucumber, 3 gourd, 1 ground cherry, 1 kale, 1 leek, 3 lettuce, 4 lima, 2 melon, 2 onion other, 1 parsley, 4 pea, 1 radish, 1 runner bean, 1 scorzonera, 1 spinach, 1 squash, 16 tomato, 1 turnip, 2 watermelon, 1 miscellaneous cucurbitaceae and 4 miscellaneous.

Lej **Le Jardin du Gourmet,** PO Box 75, St. Johnsbury Center, VT 05863-0075. Phone: 802-748-1446. Fax: 802-748-1446. E-mail: orderdesk@artisticgardens.com. Web Site: www. artisticgardens.com. Free catalog. Specializes in shallots and gourmet vegetable seeds. Also offers herb and flower seeds. Retail only. Offered 182 non-hybrid vegetable varieties in 2004 including: 1 amaranth, 1 artichoke, 1 asparagus, 6 bean, 4 beet, 2 broccoli, 1 brussels sprout, 1 cardoon, 8 carrot, 2 cauliflower, 1 celeriac, 2 Chinese cabbage, 4 chicory, 1 collard, 4 corn, 3 cornsalad, 2 cress, 4 cucumber, 2 eggplant, 3 endive, 1 fennel, 2 garlic, 1 gourd, 1 Jerusalem artichoke, 2 kohlrabi, 4 leek, 12 lettuce, 4 melon, 3 okra, 1 onion bunching, 2 onion common, 4 onion multiplier, 2 onion other, 3 parsley, 1 parsnip, 4 pea, 10 pepper, 11 radish, 1 scorzonera, 2 spinach, 19 squash, 3 sunflower ornamental, 4 Swiss chard, 2 tomatillo, 9 tomato, 6 turnip, 1 miscellaneous leguminosae, 1 miscellaneous solanaceae and 13 miscellaneous.

Lin **Lindenberg Seeds Limited,** 803 Princess Ave., Brandon, MB R7A 0P5, Canada. Web Site: www.lindenbergseeds.mb.ca. Canada only. Specializes in seeds selected for the short sunny prairie growing season. Offered 193 non-hybrid vegetable varieties in 2004 including: 1 asparagus, 16 bean, 8 beet, 1 broccoli, 1 brussels sprout, 7 cabbage, 1 cardoon, 8 carrot, 2 cauliflower, 1 celeriac, 1 celery, 3 Chinese cabbage, 1 chicory, 3 corn, 1 cress, 8 cucumber, 1 eggplant, 1 endive, 1 fava bean, 1 fennel, 2 kale, 3 kohlrabi, 1 leek, 18 lettuce, 1 lima, 1 melon, 1 okra, 1 onion bunching, 5 onion common, 2 onion multiplier, 2 onion other, 4 parsley, 1 parsnip, 15 pea, 4 pepper, 11 potato, 11 radish, 1 runner bean, 2 rutabaga, 1 spinach, 12 squash, 1 sunflower edible, 5 Swiss chard, 1 tomatillo, 11 tomato, 1 turnip, 3 watermelon and 5 miscellaneous.

LO8 **Lonestar Seed Company,** P.O. Box 831553, San Antonio, TX 78283. Web Site: www.lonestarseed.com. On-line catalog. Wholesale only. Selling quality seeds for four generations. Offered 152 non-hybrid vegetable varieties in 2004 including: 1 amaranth, 1 artichoke, 1 asparagus, 10 bean, 3 beet, 1 broccoli, 1 brussels sprout, 3 cabbage, 6 carrot, 2 cauliflower, 1 celery, 2 Chinese cabbage, 1 collard, 1 corn, 1 cowpea, 4 cucumber, 2 eggplant, 1 endive, 1 gourd, 1 kale, 2 kohlrabi, 1 leek, 12 lettuce, 5 lima, 5 melon, 2 mustard greens, 4 okra, 7 onion common, 1 onion other, 2 parsley, 4 pea, 13 pepper, 10 radish, 1 runner bean, 1 rutabaga, 1 spinach, 15 squash, 2 sunflower ornamental, 3 Swiss chard, 1 tomatillo, 11 tomato, 3 turnip, 1 miscellaneous cucurbitaceae and 1 miscellaneous.

Loc **Lockhart Seeds, Inc.,** PO Box 1361, 3 North Wilson Way, Stockton, CA 95205. Phone: 209-466-4401. Fax: 209-466-9766. Retail price list upon request. Main emphasis is commercial growers, fruit stands and large market gardeners. Offered 281 non-hybrid vegetable varieties in 2004 including: 3 artichoke, 15 bean, 4 beet, 2 broccoli, 1 broccoli raab, 1 brussels sprout, 5 cabbage, 1 cardoon, 5 carrot, 1 cauliflower, 1 celeriac, 1 celery, 10 Chinese cabbage, 4 chicory, 1 collard, 5 corn, 3 cowpea, 3 cucumber, 3 eggplant, 3 endive, 2 fava bean, 1 fennel, 6 gourd, 3 kale, 2 kohlrabi, 1 leek, 13 lettuce, 3 lima, 13 melon, 3 mustard greens, 2 okra, 13 onion common, 1 onion other, 3 parsley, 1 parsnip, 8 pea, 22 pepper, 9 radish, 1 rutabaga, 35 squash, 1 sunflower ornamental, 1 sunflower edible, 3 Swiss chard, 1 tomatillo, 38 tomato, 4 turnip, 8 watermelon, 3 miscellaneous cucurbitaceae, 3 miscellaneous leguminosae and 5 miscellaneous.

Ma13 **Mapple Farm,** 129 Beech Hill Road, Weldon, NB E4H 4N5, Canada. Phone: 506-734-3361. E-mail: wingate @nbnet.nb.ca. Send SASE for catalog within Canada, $1 to U.S. Specializes in certified organic seeds and unusual and/or not commonly offered varieties, all grown by them. Retail and wholesale. Offered 39 non-hybrid vegetable varieties in 2004 including: 1 bean, 1 corn, 1 cucumber, 1 garlic, 3 Jerusalem artichoke, 4 onion multiplier, 1 pea, 2 soybean, 1 squash, 8 sweet potato, 1 tomatillo, 12 tomato, 1 miscellaneous solanaceae and 2 miscellaneous.

Ma18 **Marianna's Heirloom Seeds,** 1955 CCC Road, Dickson, TN 37055. Phone: 615-446-9191. $1 for catalog. Specializes in Italian cooking tomatoes, peppers and eggplant. Retail only. Formerly known as Pomodori di Marianna. Offered 363 non-hybrid vegetable varieties in 2004 including: 18 eggplant, 38 pepper, 306 tomato and 1 miscellaneous solanaceae.

Ma19 **Main Street Seed and Supply Co.,** 401 Main St., Bay City, MI 48706. E-mail: service@mainstreetseedand supply.com. Web Site: www. mainstreetseedandsupply.com. On-line catalog. Family owned and operated business since 1981. Offers a variety of vegetable seed for the farmer and home gardener alike. Small quantities and bulk amounts available. Offered 143 non-hybrid vegetable varieties in 2004 including: 1 asparagus, 12 bean, 4 beet, 1 broccoli, 1 brussels sprout, 4 cabbage, 5 carrot, 2 cauliflower, 1 Chinese cabbage, 1 collard, 4 corn, 3 cowpea, 5 cucumber, 1 eggplant, 1 endive, 1 fava bean, 3 gourd, 1 ground cherry, 1 kale, 2 kohlrabi, 1 leek, 7 lettuce, 4 lima, 4 melon, 2 mustard greens, 2 okra, 1 onion bunching, 2 onion common, 1 onion other, 2 parsley, 1 parsnip, 8 pea, 3 pepper, 8 radish, 1 rutabaga, 1 salsify, 1 sorghum, 1 soybean, 1 spinach, 27 squash, 1 Swiss chard, 3 tomato, 2 turnip, 3 watermelon and 2 miscellaneous.

MAY **Earl May Seed and Nursery,** 208 N. Elm Street, Shenandoah, IA 51603. Phone: 712-246-1020. Fax: 712-246-1760. Web Site: www.earlmay.com. Bulk seed only. In business since 1919. Garden centers located in Iowa, Nebraska, Kansas and Missouri. Offered 152 non-hybrid vegetable varieties in 2004 including: 1 asparagus, 16 bean, 4 beet, 1 broccoli, 2 cabbage, 4 carrot, 1 cauliflower, 1 celery, 1 Chinese cabbage, 4 corn, 5 cucumber, 1 eggplant, 1 endive, 1 gourd, 1 ground cherry, 1 kale, 2 kohlrabi, 1 leek, 9 lettuce, 4 lima, 4 melon, 1 mustard greens, 2 okra, 1 onion bunching, 2 onion common, 1 onion other, 1 parsley, 1 parsnip, 11 pea, 8 pepper, 9 radish, 1 rutabaga, 1 salsify, 1 sorghum, 1 spinach, 18 squash, 7 sunflower ornamental, 2 sunflower edible, 2 Swiss chard, 1 tomatillo, 5 tomato, 1 turnip, 5 watermelon, 1 miscellaneous cucurbitaceae, 1 miscellaneous leguminosae and 3 miscellaneous.

Mcf **McFayden Seed Co. Ltd.,** 30 9th St., Brandon, MB R7A 6N4, Canada. Web Site: www.mcfayden.com. Canada only. Specializes in vegetable and flower seeds, hardy nursery stock and kitchen and garden accessories. Retail only. Offered 84 non-hybrid vegetable varieties in 2004 including: 9 bean, 4 beet, 3 carrot, 1 cauliflower, 4 cucumber, 1 fava bean, 2 garlic, 1 Jerusalem artichoke, 2 kohlrabi, 1 leek, 9 lettuce, 2 melon, 1 mushroom, 2 onion multiplier, 1 parsnip, 12 pea, 1 pepper, 10 potato, 4 radish, 1 runner bean, 1 rutabaga, 3 squash, 3 sunflower ornamental, 2 sunflower edible, 1 Swiss chard, 2 tomato and 1 watermelon.

Me7 **Melissa's Seeds,** PO Box 242, Hastings, MN 55033. Phone: 715-556-1398. Send two first-class stamps for catalog. Offers open-pollinated seeds selected for flavor and reliability in short-season home gardens. Formerly "Noel's Seeds." Catalog and seed requests received after August 1 will be processed the following January-February. Retail only. Offered 59 non-hybrid vegetable varieties in 2004 including: 3 bean, 2 beet, 1 broccoli, 3 cabbage, 2 carrot, 1 Chinese cabbage, 3 corn, 3 cucumber, 3 lettuce, 1 okra, 4 onion common, 1 onion other, 1 parsnip, 2 pea, 2 pepper, 1 radish, 2 rutabaga, 2 salsify, 2 soybean, 1 spinach, 5 squash, 1 sunflower edible, 2 Swiss chard, 7 tomato, 1 turnip, 2 watermelon and 1 miscellaneous.

ME9 **Meyer Seed International,** 4321 Fitch Avenue, Baltimore, MD 21236. Phone: 888-503-7333. Wholesale only. Offered 756 non-hybrid vegetable varieties in 2004 including: 1 amaranth, 1 asparagus, 64 bean, 12 beet, 3 broccoli, 2 broccoli raab, 1 brussels sprout, 4 cabbage, 1 cardoon, 11 carrot, 1 cauliflower, 3 celery, 9 Chinese cabbage, 3 chicory, 3 collard, 24 corn, 1 cornsalad, 7 cowpea, 4 cress, 5 cucumber, 11 eggplant, 7 endive, 2 fava bean, 1 fennel, 15 gourd, 7 kale, 2 kohlrabi, 5 leek, 82 lettuce, 13 lima, 8 mustard greens, 4 okra, 8 onion bunching, 5 onion common, 2 onion other, 4 parsley, 3 parsnip, 33 pea, 89 pepper, 10 radish, 2 rutabaga, 1 salsify, 2 sorghum, 2 spinach, 79 squash, 9 sunflower ornamental, 4 sunflower edible, 16 Swiss chard, 3 tomatillo, 139 tomato, 3 turnip, 11 watermelon, 1 miscellaneous cucurbitaceae, 1 miscellaneous solanaceae and 12 miscellaneous.

Mel **Mellinger's, Inc.,** 2310 W. South Range Road, P.O. Box 157, North Lima, OH 44452. Phone: 330-549-9861. E-mail: mellgarden@aol.com. Web Site: www.mellingers.com.

Free catalog within the U.S., containing items for country living including seeds, bulbs, perennials, trees, shrubs and lawn/garden supplies. Retail and wholesale. Offered 168 non-hybrid vegetable varieties in 2004 including: 1 artichoke, 2 asparagus, 9 bean, 3 beet, 1 broccoli, 1 brussels sprout, 5 cabbage, 4 carrot, 1 cauliflower, 1 celeriac, 1 celery, 1 Chinese cabbage, 1 collard, 8 corn, 1 cress, 5 cucumber, 1 eggplant, 2 endive, 1 fava bean, 1 garlic, 6 gourd, 1 ground cherry, 1 Jerusalem artichoke, 1 kale, 1 kohlrabi, 2 leek, 9 lettuce, 4 lima, 2 melon, 1 mustard greens, 2 okra, 1 onion other, 3 parsley, 1 parsnip, 9 pea, 1 peanut, 9 pepper, 3 radish, 1 rutabaga, 1 sorghum, 1 spinach, 19 squash, 1 sunflower edible, 3 Swiss chard, 9 sweet potato, 1 tomatillo, 13 tomato, 2 turnip, 2 watermelon, 1 miscellaneous cucurbitaceae, 1 miscellaneous leguminosae and 6 miscellaneous.

Mey **Meyer Seed Co. of Baltimore,** 600 S. Caroline St., Baltimore, MD 21231. Phone: 410-342-4224. Free catalog. Retail and wholesale. Quality seeds since 1911. Offered 250 non-hybrid vegetable varieties in 2004 including: 1 amaranth, 1 asparagus, 29 bean, 6 beet, 3 broccoli, 2 broccoli raab, 1 brussels sprout, 7 cabbage, 4 carrot, 1 cauliflower, 2 celery, 3 Chinese cabbage, 2 chicory, 4 collard, 6 corn, 4 cowpea, 2 cress, 6 cucumber, 2 eggplant, 2 endive, 1 fava bean, 1 fennel, 1 garlic, 1 gourd, 3 kale, 1 kohlrabi, 1 leek, 11 lettuce, 12 lima, 5 melon, 3 mustard greens, 2 okra, 1 onion bunching, 3 onion common, 1 onion multiplier, 1 onion other, 4 parsley, 1 parsnip, 14 pea, 1 peanut, 14 pepper, 8 potato, 9 radish, 1 rutabaga, 1 salsify, 1 soybean, 1 spinach, 21 squash, 1 sunflower ornamental, 2 sunflower edible, 5 Swiss chard, 5 sweet potato, 8 tomato, 3 turnip, 7 watermelon, 1 miscellaneous leguminosae and 6 miscellaneous.

Mi8 **Milk Ranch Specialty Potatoes, LLC,** 20094 Hwy. 149, Powderhorn, CO 81243. Phone: 970-641-5634. E-mail: craig@milkranch.com. Web Site: www.milkranch.com. Free catalog. Specializes in certified seed potatoes, both new and heirloom. Retail only. Offered 40 potato varieties in 2004.

Mi12 **Mikamoki Seeds,** 66 South Orchard St., Logan, OH 43138. Web Site: www.mikamoki.safeshopper.com. On-line catalog. Offered 34 non-hybrid vegetable varieties in 2004 including: 3 amaranth, 1 okra, 2 parsley, 10 pepper, 1 runner bean, 2 sunflower ornamental, 1 sunflower edible, 11 tomato, 1 miscellaneous cucurbitaceae, 1 miscellaneous leguminosae and 1 miscellaneous.

MIC **Henry F. Michell Co.,** PO Box 60160, King of Prussia, PA 19406. Strictly wholesale. Offered 99 non-hybrid vegetable varieties in 2004 including: 1 artichoke, 1 bean, 1 broccoli, 1 brussels sprout, 2 cabbage, 3 cauliflower, 3 celery, 1 collard, 5 cucumber, 2 eggplant, 1 kale, 1 kohlrabi, 11 lettuce, 2 melon, 2 okra, 4 onion common, 21 pepper, 6 squash, 1 sunflower edible, 1 Swiss chard, 1 tomatillo, 23 tomato, 1 turnip and 4 watermelon.

Mo13 **Morgan County Seeds,** Norman and Vera Kilmer, Owners, 18761 Kelsay Road, Barnett, MO 65011-3009. Phone: 573-378-2655. Free catalog. Offers a full line of vegetable seeds, with a large selection of untreated and open-pollinated varieties. Retail and wholesale. Offered 268 non-hybrid vegetable varieties in 2004 including: 1 asparagus, 28 bean, 6 beet, 1 broccoli, 1 brussels sprout, 8 cabbage, 5 carrot, 3 cauliflower, 2 celery, 5 Chinese cabbage, 2 collard, 15 corn, 3 cowpea, 7 cucumber, 1 eggplant, 1 endive, 1 fava bean, 1 fennel, 2 garlic, 6 gourd, 1 ground cherry, 3 kale, 2 kohlrabi, 1 leek, 15 lettuce, 9 lima, 7 melon, 4 mustard greens, 3 okra, 1 onion bunching, 3 onion common, 2 onion other, 2 parsley, 2 parsnip, 15 pea, 2 peanut, 14 pepper, 11 radish, 1 rutabaga, 1 salsify, 1 sorghum, 1 spinach, 35 squash, 1 sunflower edible, 3 Swiss chard, 13 tomato, 2 turnip, 8 watermelon, 1 miscellaneous cucurbitaceae, 1 miscellaneous solanaceae and 4 miscellaneous.

Mo20 **Mountain Meadow Seeds,** Dennis and Christy Vickery, Rt. 1 Box 23-2, Augusta, WV 26704-9703. E-mail: MountainMeadowSeeds@frontier.net. Web Site: www.MountainMeadowSeeds.com. On-line catalog. Small family nursery offering seeds of flowering vines, herbs, vegetables, annuals and perennials. Bulk quantities available. Offered 24 non-hybrid vegetable varieties in 2004 including: 1 broccoli, 1 cabbage, 1 carrot, 3 cucumber, 1 melon, 9 pepper, 2 sunflower ornamental, 1 sunflower edible, 2 tomato, 1 watermelon and 2 miscellaneous.

Moo **Moose Tubers,** PO Box 520, Waterville, ME 04903. Phone: 207-873-7333. Web Site: www.fedcoseeds.com. $2 for catalog. Specializes in untreated seed potatoes for northern climates and lists over 40 varieties. A division of Fedco Seeds and listed in Fedco Seeds catalog under Moose Tubers. Retail and wholesale prices are listed in the same catalog. Offered 57 non-hybrid vegetable varieties in 2004 including: 4 Jerusalem artichoke, 1 onion common, 2 onion multiplier and 50 potato.

MOU **Mountain Valley Seed Inc.,** 1800 South West Temple #600, Salt Lake City, UT 84115. Phone: 801-486-0480. E-mail: info@mvseeds.com. Web Site: www.mv-seeds.com. Strictly wholesale. Specializes in varieties for colder climates with extensive sales of AAS Winners. Offered 184 non-hybrid vegetable varieties in 2004 including: 1 asparagus, 17 bean, 4 beet, 1 broccoli, 1 brussels sprout, 5 cabbage, 4 carrot, 2 cauliflower, 1 celery, 1 Chinese cabbage, 1 collard, 5 corn, 6 cucumber, 1 eggplant, 1 endive, 1 fava bean, 1 fennel, 1 gourd, 1 kale, 2 kohlrabi, 1 leek, 11 lettuce, 1 lima, 6 melon, 1 okra, 1 onion bunching, 5 onion common, 2 onion other, 2 parsley, 1 parsnip, 14 pea, 21 pepper, 6 radish, 1 rutabaga, 1 spinach, 24 squash, 3 sunflower ornamental, 1 sunflower edible, 3 Swiss chard, 1 tomatillo, 9 tomato, 1 turnip, 5 watermelon, 1 miscellaneous cucurbitaceae, 3 miscellaneous leguminosae and 2 miscellaneous.

Mus **Mushroompeople,** 560 Farm Road, P.O. Box 220, Summertown, TN 38483. Phone: 931-964-2200. E-mail: mushroom@thefarm.org. Web Site: www.mushroompeople.com. Free catalog. Specializes in spawn for Shiitake and other mushrooms, plus plugs, sawdust and grain. Wide range of mushroom books and videos. For home or commercial mushroom growers. Offered 9 mushroom varieties in 2004.

Na2 **Native Seeds/SEARCH,** 526 N. 4th Ave., Tucson, AZ 85705. Phone: 520-622-5561. E-mail: info@nativeseeds.org. Web Site: www.nativeseeds.org. $1 for seed list, $25 for quarterly newsletter (*The Seedhead News*) and annual seed list that also includes packaged desert foods, crafts from indigenous farming areas, ethnobotany and related books, gift baskets and other unique items. Conserving Southwest native crops and their wild relatives, specializing in vegetables that have sustained native peoples throughout the southwestern U.S. and northern Mexico. First catalog in 1984. Offered 326 non-hybrid vegetable varieties in 2004 including: 10 amaranth, 74 bean, 60 corn, 12 cowpea, 2 fava bean, 9 gourd, 1 lima, 7 melon, 1 okra, 10 pea, 29 pepper, 10 sorghum, 37 squash, 3 sunflower ornamental, 26 tepary, 1 tomatillo, 2 tomato, 14 watermelon, 3 miscellaneous leguminosae and 15 miscellaneous.

Na6 **Natural Gardening Company,** PO Box 750776, Petaluma, CA 94975-0776. E-mail: info@naturalgardening.com. Web Site: www.naturalgardening.com. Free catalog. Started the nation's first certified organic nursery. Offers untreated vegetable seeds, many certified organic and certified organic seedlings. Specializes in tomatoes. Offered 151 non-hybrid vegetable varieties in 2004 including: 1 amaranth, 1 artichoke, 13 bean, 3 beet, 1 broccoli, 1 broccoli raab, 1 cabbage, 1 cardoon, 3 carrot, 1 cauliflower, 1 celery, 3 Chinese cabbage, 1 chicory, 1 collard, 1 cornsalad, 3 cucumber, 5 eggplant, 2 endive, 1 fava bean, 1 fennel, 1 gourd, 2 kale, 1 leek, 16 lettuce, 1 lima, 5 melon, 1 mustard greens, 3 onion common, 2 onion other, 2 parsley, 1 parsnip, 4 pea, 9 pepper, 4 potato, 2 radish, 1 runner bean, 2 soybean, 1 spinach, 13 squash, 1 sunflower ornamental, 1 sunflower edible, 3 Swiss chard, 1 tomatillo, 23 tomato, 1 turnip, 1 watermelon and 4 miscellaneous.

Ni1 **Nichols Garden Nursery,** 1190 Old Salem Road NE, Albany, OR 97321-4580. Phone: 541-928-9280. Fax: 800-231-5306. E-mail: nichols@gardennursery.com. Web Site: www.nicholsgardennursery.com. Free catalog. Unusual varieties for the gardener cook, especially Asian and European specialties as well as quality garden varieties, herbs and elephant garlic. Separate retail and wholesale catalogs. Offered 303 non-hybrid vegetable varieties in 2004 including: 3 amaranth, 2 artichoke, 17 bean, 5 beet, 4 broccoli, 1 broccoli raab, 1 cabbage, 1 cardoon, 4 carrot, 1 cauliflower, 1 celeriac, 2 celery, 11 Chinese cabbage, 5 chicory, 1 collard, 8 corn, 2 cornsalad, 5 cress, 4 cucumber, 2 endive, 1 fava bean, 1 fennel, 11 garlic, 13 gourd, 1 ground cherry, 8 kale, 2 kohlrabi, 3 leek, 22 lettuce, 1 lima, 3 melon, 2 mustard greens, 1 okra, 2 onion bunching, 4 onion common, 3 onion multiplier, 4 onion other, 6 parsley, 1 parsnip, 10 pea, 15 pepper, 12 radish, 2 runner bean, 1 rutabaga, 1 salsify, 1 sorghum, 1 soybean, 1 spinach, 21 squash, 1 sunflower ornamental, 2 sunflower edible, 5 Swiss chard, 2 tomatillo, 24 tomato, 4 turnip, 1 watermelon, 2 miscellaneous cucurbitaceae, 1 miscellaneous leguminosae and 27 miscellaneous.

No12 **Northwest Mycological Consultants,** 702 NW 4th St., Corvallis, OR 97330. Phone: 541-753-8198. Fax: 541-752-3401. $2 for catalog. Specializes in top quality mushroom spawn (Shiitake, Oyster, etc.) and provides supplies and consulting for the mushroom grower. Retail and wholesale prices are listed in the same catalog. Offered 7 mushroom varieties in 2004.

Ol2 **Old Sturbridge Village,** One Old Sturbridge Village Road, Sturbridge, MA 01566. Phone: 508-347-0270. Web Site: www.osvgifts.org. On-line catalog. Specializes in vegetables, flowers and herbs that are appropriate for gardeners re-creating early 19th-century gardens. Offered 41 non-hybrid vegetable varieties in 2004 including: 2 bean, 2 beet, 2 cabbage, 1 carrot, 1 cucumber, 1 fennel, 1 lettuce, 2 onion common, 2 onion other, 2 parsley, 1 parsnip, 1 pea, 1 pepper, 2 radish, 1 runner bean, 1 rutabaga, 1 salsify, 1 sorghum, 8 squash, 2 tomato, 1 turnip, 1 watermelon, 1 miscellaneous cucurbitaceae, 1 miscellaneous leguminosae and 2 miscellaneous.

OLD **Olds Seed Solutions,** PO Box 7790, Madison, WI 53707. Web Site: www.seedsolutions.com. Strictly wholesale. Midwest seed supplier since 1888. Specializes in open-pollinated, untreated seed that is easy to grow when sown directly in the garden. Packet size and bulk for retailers. No catalog available for home gardeners. Formerly L. L. Olds Seed Company. Offered 205 non-hybrid vegetable varieties in 2004 including: 1 asparagus, 22 bean, 2 beet, 1 broccoli, 1 brussels sprout, 4 cabbage, 7 carrot, 1 cauliflower, 2 Chinese cabbage, 2 collard, 6 corn, 3 cowpea, 5 cucumber, 1 eggplant, 2 endive, 1 fava bean, 2 gourd, 1 ground cherry, 3 kale, 3 kohlrabi, 2 leek, 13 lettuce, 9 lima, 5 melon, 2 mustard greens, 4 okra, 1 onion bunching, 2 onion common, 4 parsley, 1 parsnip, 11 pea, 1 peanut, 7 pepper, 12 radish, 1 runner bean, 1 rutabaga, 1 salsify, 1 sorghum, 1 soybean, 1 spinach, 34 squash, 2 sunflower ornamental, 2 sunflower edible, 4 Swiss chard, 2 tomato, 3 turnip, 7 watermelon and 1 miscellaneous.

Ont **Ontario Seed Company,** PO Box 7, Waterloo, ON N2J 3Z6, Canada. Phone: 519-886-0557. E-mail: seeds@oscseeds.com. Web Site: www.oscseeds.com. Catalog and sales to Canada only. Family-owned business since 1913. Specializes in untreated, heirloom, open-pollinated seeds. Retail and bulk prices listed in the same catalog. Retail store open to the public. Offered 188 non-hybrid vegetable varieties in 2004 including: 1 amaranth, 1 artichoke, 1 asparagus, 20 bean, 1 beet, 1 broccoli, 1 broccoli raab, 1 brussels sprout, 7 cabbage, 5 carrot, 2 cauliflower, 1 celeriac, 2 celery, 2 Chinese cabbage, 4 chicory, 1 collard, 4 corn, 1 cornsalad, 2 cress, 6 cucumber, 2 eggplant, 2 endive, 2 fava bean, 1 fennel, 1 kale, 2 kohlrabi, 1 leek, 12 lettuce, 2 lima, 2 melon, 1 mustard greens, 6 onion common, 1 onion multiplier, 2 onion other, 3 parsley, 1 parsnip, 12 pea, 1 peanut, 10 pepper, 9 radish, 1 runner bean, 2 rutabaga, 1 salsify, 1 spinach, 16 squash, 5 sunflower ornamental, 2 sunflower edible, 3 Swiss chard, 11 tomato, 1 turnip, 3 watermelon, 2 miscellaneous leguminosae and 3 miscellaneous.

Or10 **Organica Seed Co.,** PO Box 611, Wilbraham, MA 01095. Web Site: www.organicaseed.com. On-line catalog. Offers untreated, organic vegetable, flower and herb seeds. Offered 125 non-hybrid vegetable varieties in 2004 including: 1 artichoke, 1 asparagus, 12 bean, 4 beet, 1 broccoli, 1 brussels sprout, 4 cabbage, 4 carrot, 1 cauliflower, 1 celery, 1 collard, 4 corn, 4 cucumber, 3 eggplant, 2 kale, 1 kohlrabi, 1 leek, 10 lettuce, 2 lima, 1 melon, 2 mustard greens, 1 okra, 2 onion common, 1 parsnip, 6 pea, 9 pepper, 5 radish, 1 soybean, 1 spinach, 14 squash, 4 Swiss chard, 1 tomatillo, 11 tomato, 3 turnip, 2 watermelon, 1 miscellaneous cucurbitaceae, 1 miscellaneous leguminosae and 1 miscellaneous.

Orn **Ornamental Edibles,** Specialty Seeds by Mail, 3272 Fleur De Lis Court, San Jose, CA 95132. Phone: 408-929-7333. Web Site: www.ornamentaledibles.com. Free catalog. Specializes in an international selection of seeds for both "edible landscaping" and "specialty market" growers. Retail only. Offered 178 non-hybrid vegetable varieties in 2004 including: 2 amaranth, 2 artichoke, 4 bean, 7 beet, 1 broccoli, 1 broccoli raab, 3 carrot, 2 celery, 10 Chinese cabbage, 4 chicory, 2 cornsalad, 3 cress, 1 cucumber, 4 eggplant, 4 endive, 1 fennel, 3 kale, 1 kohlrabi, 2 leek, 45 lettuce, 1 melon, 3 mustard greens, 1 onion bunching, 5 onion common, 1 onion other, 2 parsley, 1 pea, 11 pepper, 4 radish, 1 spinach, 8 squash, 1 sunflower ornamental, 4 Swiss chard, 1 tomatillo, 20 tomato, 1 turnip and 11 miscellaneous.

OSB **Osborne International Seed Co.,** 2428 Old Highway 99 S Road, Mount Vernon, WA 98273. E-mail: cosborne @osborneseed.com. Web Site: www.osborneseed.com. Strictly wholesale. Specializes in vegetables that are suitable for the Pacific Northwest. Offered 243 non-hybrid vegetable varieties in 2004 including: 2 artichoke, 1 asparagus, 20 bean, 7 beet, 2 broccoli raab, 3 carrot, 1 cauliflower, 1 celeriac, 8 Chinese cabbage, 1 collard, 6 corn, 2 cornsalad, 3 cress, 2 cucumber, 2 eggplant, 5 endive, 3 fava bean, 2 fennel, 6 gourd, 3 kale, 6 leek, 39 lettuce, 4 onion bunching, 3 onion common, 2 onion other, 3 parsley, 1 parsnip, 13 pea, 16 pepper, 10 radish, 1 rutabaga, 4 soybean, 30 squash, 8 sunflower ornamental, 3 sunflower edible, 4 Swiss chard, 2 tomatillo, 4 tomato, 1 turnip and 9 miscellaneous.

PAG **Page Seed Co.,** 1 A Green Street, PO Box 158, Greene, NY 13778. Phone: 607-656-4107. E-mail: pageseed @aol.com. Strictly wholesale. Specializes primarily in home garden varieties. Offered 168 non-hybrid vegetable varieties in 2004 including: 1 asparagus, 15 bean, 5 beet, 1 broccoli, 1 broccoli raab, 4 cabbage, 5 carrot, 1 cauliflower, 2 Chinese cabbage, 1 chicory, 2 collard, 1 corn, 1 cowpea, 1 cress, 8 cucumber, 1 eggplant, 2 endive, 1 fava bean, 1 gourd, 1 kale, 1 kohlrabi, 1 leek, 16 lettuce, 3 lima, 3 melon, 2 mustard greens, 1 okra, 4 onion common, 1 onion other, 3 parsley, 1 parsnip, 16 pea, 8 pepper, 9 radish, 2 rutabaga, 1 spinach, 21 squash, 3 sunflower ornamental, 1 sunflower edible, 3 Swiss chard, 9 tomato, 2 turnip and 2 miscellaneous.

Par **Park Seed Co.,** 1 Parkton Avenue, Greenwood, SC 29647-0001. Phone: 800-845-3369. E-mail: info@parkseed .com. Web Site: www.parkseed.com. Free catalog. Separate retail and wholesale catalogs; retail catalog contains additional varieties. Flower seed specialists since 1868. Also offers a full line of vegetables and a number of small fruit varieties. Offered 112 non-hybrid vegetable varieties in 2004 including: 1 amaranth, 1 artichoke, 1 asparagus, 14 bean, 2 beet, 3 carrot, 1 cauliflower, 1 celeriac, 1 celery, 1 corn, 2 cowpea, 1 fennel, 3 garlic, 4 gourd, 2 leek, 8 lettuce, 2 lima, 1 mushroom, 2 okra, 1 onion bunching, 1 onion common, 1 onion multiplier, 2 onion other, 1 parsley, 9 pea, 2 peanut, 5 pepper, 4 potato, 2 radish, 1 runner bean, 2 soybean, 5 squash, 4 sunflower ornamental, 2 sunflower edible, 1 Swiss chard, 7 sweet potato, 1 tomatillo, 1 tomato, 2 watermelon, 1 miscellaneous cucurbitaceae, 1 miscellaneous leguminosae and 5 miscellaneous.

Pe1 **Peace Seeds,** 2385 SE Thompson St., Corvallis, OR 97333. $1 for Annual Seed List. Specializes in organically grown vegetable, herb and medicinal seeds; also offers seeds of rare, endangered and genetically important wild plants. Peace Seeds Resource Journal, Vol. 9, 2001, 76 pages, $20 + $2 postage. Offered 75 non-hybrid vegetable varieties in 2004 including: 3 amaranth, 3 bean, 1 broccoli, 1 cabbage, 1 collard, 6 corn, 1 fava bean, 1 garlic, 1 ground cherry, 6 kale, 2 leek, 14 lettuce, 2 melon, 3 parsley, 1 parsnip, 4 pea, 1 pepper, 2 runner bean, 6 soybean, 4 squash, 1 sunflower ornamental, 1 tepary, 5 tomato, 1 miscellaneous cucurbitaceae and 4 miscellaneous.

Pe2 **Peaceful Valley Farm Supply,** PO Box 2209, Grass Valley, CA 95945. Phone: 888-784-1722. Fax: 530-272-4794. E-mail: contact@groworganic.com. Web Site: www. groworganic.com. Tools and supplies for organic gardeners and farmers since 1976. Free 140-page catalog with over 2,000 items including fertilizers, organic and/or open-pollinated vegetable and cover crop seeds, weed and pest controls, beneficial insects, irrigation and (in Fall) bulbs, garlic, onions, potatoes and fruit trees. Offered 343 non-hybrid vegetable varieties in 2004 including: 3 amaranth, 1 artichoke, 1 asparagus, 15 bean, 8 beet, 3 broccoli, 1 broccoli raab, 1 brussels sprout, 3 cabbage, 9 carrot, 1 cauliflower, 1 celeriac, 1 celery, 6 Chinese cabbage, 1 chicory, 15 corn, 2 cress, 8 cucumber, 2 eggplant, 2 endive, 2 fava bean, 2 fennel, 9 garlic, 4 gourd, 5 kale, 4 leek, 32 lettuce, 1 lima, 10 melon, 2 mustard greens, 2 okra, 2 onion bunching, 7 onion common, 2 onion multiplier, 2 onion other, 5 parsley, 1 parsnip, 10 pea, 21 pepper, 8 potato, 7 radish, 2 runner bean, 1 sorghum, 1 soybean, 3 spinach, 21 squash, 7 sunflower ornamental, 5 sunflower edible, 6 Swiss chard, 2 tomatillo, 53 tomato, 2 turnip, 4 watermelon, 2 miscellaneous leguminosae and 12 miscellaneous.

Pe6 **Peters Seed and Research,** PO Box 1472, Myrtle Creek, OR 97457-0137. Phone: 541-874-2615. Fax: 541-874-3426. E-mail: psr@pioneer-net.com. Web Site: www.

pioneer-net.com/psr/. $2 for catalog. Specializes in open-pollinated, uncommon vegetables, many bred at their research facility. Offers many Northern varieties and a unique tomato collection. Retail, with some wholesale on contract-grown items. Offered 129 non-hybrid vegetable varieties in 2004 including: 1 amaranth, 3 artichoke, 1 asparagus, 2 bean, 6 beet, 2 broccoli, 1 brussels sprout, 2 cabbage, 3 carrot, 2 Chinese cabbage, 4 corn, 2 cress, 4 cucumber, 2 eggplant, 1 kohlrabi, 8 lettuce, 3 melon, 4 mustard greens, 1 okra, 1 onion bunching, 1 onion other, 1 parsley, 1 parsnip, 3 pea, 3 pepper, 3 radish, 1 rutabaga, 4 sorghum, 1 spinach, 13 squash, 2 sunflower edible, 6 Swiss chard, 26 tomato, 1 turnip, 4 watermelon and 6 miscellaneous.

Pe7 Pepper Joe's, Inc., 7 Tyburn Court, Timonium, MD 21093. Send SASE for catalog. Specializes in hot peppers. All seeds are grown organically with no pesticides. Retail only. Offered 33 non-hybrid vegetable varieties in 2004 including: 27 pepper and 6 tomato.

Pep Pepper Gal, PO Box 23006, Ft. Lauderdale, FL 33307. Phone: 954-537-5540. Fax: 954-566-2208. E-mail: peppergal@mindspring.com. Web Site: www.peppergal.com. Free catalog. Offers seed for over 300 hot, sweet and ornamental peppers, plus tomatoes and gourds. Retail and wholesale. Offered 267 non-hybrid vegetable varieties in 2004 including: 1 fennel, 11 gourd, 2 onion other, 3 parsley, 228 pepper, 9 squash, 2 tomatillo, 8 tomato, 1 miscellaneous cucurbitaceae and 2 miscellaneous.

Pin Pinetree Garden Seeds, Box 300, New Gloucester, ME 04260. Phone: 207-926-3400. Fax: 888-527-3337. E-mail: pinetree@superseeds.com. Web Site: www.superseeds.com. Free catalog. Specializes in flavorful varieties for home gardeners and offers some unique material. "Smaller packets and lower prices." Over 300 gardening books plus tools and bulbs. Retail only. Offered 296 non-hybrid vegetable varieties in 2004 including: 4 amaranth, 2 artichoke, 1 asparagus, 27 bean, 9 beet, 2 broccoli, 1 broccoli raab, 2 cabbage, 6 carrot, 1 cauliflower, 1 celeriac, 3 celery, 4 Chinese cabbage, 2 chicory, 1 collard, 6 corn, 1 cornsalad, 1 cress, 7 cucumber, 4 eggplant, 2 endive, 1 fava bean, 1 fennel, 1 garlic, 5 gourd, 1 ground cherry, 4 kale, 2 kohlrabi, 2 leek, 20 lettuce, 2 lima, 2 melon, 2 mustard greens, 2 okra, 2 onion bunching, 7 onion common, 2 onion multiplier, 2 onion other, 4 parsley, 1 parsnip, 9 pea, 17 pepper, 10 potato, 8 radish, 1 runner bean, 2 rutabaga, 1 salsify, 3 sorghum, 1 soybean, 1 spinach, 26 squash, 9 sunflower ornamental, 1 sunflower edible, 5 Swiss chard, 2 tomatillo, 24 tomato, 2 turnip, 3 watermelon, 3 miscellaneous cucurbitaceae, 6 miscellaneous leguminosae and 12 miscellaneous.

Pl3 Ned W. Place, 21600 Conant Road, Wapakoneta, OH 45895. Web Site: www.greenfieldfarms.org. SASE for free catalog. Specializes in open-pollinated corn varieties. Retail and wholesale. Offered 4 total corn varieties in 2004.

Pla Plants of the Southwest, 3095 Agua FriaRoad, Santa Fe, NM 87507-5411. Phone: 800-788-7333. Fax: 505-438-8800. E-mail: contact@plantsofthesouthwest.com. Web Site: www.plantsofthesouthwest.com. $3.50 for 85-page color retail catalog; call for wholesale prices. Specializes in heirloom and traditional Southwestern native plants including berry- and nut-producing trees and shrubs. Catalog also includes native grasses, wildflowers, ancient and drought-tolerant vegetables including little-known Native American crops. Offered 105 non-hybrid vegetable varieties in 2004 including: 2 amaranth, 10 bean, 2 beet, 1 broccoli, 2 cabbage, 3 carrot, 1 cauliflower, 8 corn, 1 cucumber, 1 fava bean, 1 kale, 4 lettuce, 1 lima, 2 melon, 1 onion other, 2 parsley, 3 pea, 35 pepper, 1 quinoa, 1 radish, 2 runner bean, 1 spinach, 3 squash, 1 sunflower ornamental, 1 sunflower edible, 1 Swiss chard, 1 tepary, 1 tomatillo, 5 tomato, 2 watermelon and 5 miscellaneous.

Pp2 P & P Seed Co., 56 East Union Street, Hamburg, NY 14075-5007. Phone: 800-449-5681. Send long SASE for price list. Specializes in giant varieties, such as 300-pound watermelons and 800-pound pumpkins. Support company for World Pumpkin Confederation. Retail only. Offered 20 non-hybrid vegetable varieties in 2004 including: 1 beet, 1 cabbage, 1 carrot, 1 corn, 1 cucumber, 1 fava bean, 1 kohlrabi, 1 melon, 1 onion common, 1 radish, 1 rutabaga, 3 squash, 2 sunflower edible, 2 tomato, 1 watermelon and 1 miscellaneous leguminosae.

Pr3 Prairie Garden Seeds, Box 118, Cochin, SK S0M 0L0, Canada. Phone: 306-386-2737. E-mail: prairie.seeds@sk.sympatico.ca. Web Site: www.prseeds.ca. Within Canada, send $2 for catalog; all other countries use on-line catalogue. Specializes in open-pollinated vegetable and flower varieties grown without agricultural chemicals and seeds for dryland, short season growing. Offers many heritage seeds with some historical information. Retail only. Offered 307 non-hybrid vegetable varieties in 2004 including: 7 amaranth, 57 bean, 6 beet, 1 broccoli, 1 broccoli raab, 1 brussels sprout, 1 cabbage, 5 carrot, 1 Chinese cabbage, 1 chicory, 5 corn, 1 cress, 10 cucumber, 4 eggplant, 9 fava bean, 4 ground cherry, 2 kale, 1 kohlrabi, 1 leek, 11 lettuce, 7 melon, 1 mustard greens, 2 onion bunching, 2 onion other, 3 parsley, 1 parsnip, 27 pea, 9 pepper, 8 radish, 3 runner bean, 2 rutabaga, 3 soybean, 1 spinach, 19 squash, 1 sunflower ornamental, 1 sunflower edible, 2 Swiss chard, 3 tepary, 2 tomatillo, 59 tomato, 3 turnip, 4 watermelon, 1 miscellaneous cucurbitaceae, 7 miscellaneous leguminosae, 2 miscellaneous solanaceae and 5 miscellaneous.

Pr9 Prizeseeds.com, Web Site: www.prizeseeds.com. Offered 103 non-hybrid vegetable varieties in 2004 including: 1 beet, 2 carrot, 3 cucumber, 8 eggplant, 1 gourd, 2 melon, 25 pepper, 2 radish, 19 squash, 37 tomato, 1 turnip and 2 watermelon.

Ra3 Rainforest Mushroom Spawn, Box 1793, Gibsons, BC V0N 1V0, Canada. $2 for catalog, refunded on first order. Retail only. Offered 14 mushroom varieties in 2004.

Ra5 Rancid Sawdust, Jim Johnson Jr., 3421 Bream St., Gautier, MS 39553. Fax: 228-497-5488. Web Site: www.rancidsawdust.com. On-line catalog. Offered 342 non-hybrid vegetable varieties in 2004 including: 25 bean, 1 beet, 1 broccoli raab, 2 cabbage, 2 carrot, 3 cauliflower, 1 celery, 2 Chinese cabbage, 1 chicory, 1 collard, 6 corn, 3 cucumber, 3 eggplant, 1 fava bean, 27 gourd, 1 kale, 1 kohlrabi, 1 leek, 2 lettuce, 3 lima, 3 melon, 1 onion bunching, 2 onion common, 4 pea, 14 pepper, 3 radish, 1 runner bean, 1 salsify, 1 sorghum, 1 soybean, 1 spinach, 31 squash, 1 Swiss chard, 2 tomatillo, 159 tomato, 22 watermelon, 3 miscellaneous cucurbitaceae, 1 miscellaneous leguminosae and 4 miscellaneous.

Ra6 Rachel's Tomato Seed Supply, 3421 Bream St., Gautier, MS 39553. E-mail: rachel@seedman.com. Web Site: www.rachelssupply.com. Online catalog. Offering vegetable varieties from around the world. Offered 327 non-hybrid vegetable varieties in 2004 including: 29 gourd, 3 mushroom, 90 pepper, 13 squash, 2 tomatillo, 188 tomato and 2 miscellaneous cucurbitaceae.

Rai Raintree Nursery, 391 Butts Road, Morton, WA 98356. Phone: 360-496-6400. Fax: 888-770-8358. E-mail: sam@raintreenursery.com. Web Site: www.raintreenursery.com. Free 90-page catalog and backyard growers guidebook. Fruits, berries, nuts and bamboo for the edible landscape, most tested in the Pacific Northwest for over 20 years. Included here for its mushroom spawn listings. Retail only. Offered 3 mushroom varieties in 2004.

Re6 Renee's Garden, 7389 W. Zayante Road, Felton, CA 95018. Web Site: www.reneesgarden.com. On-line catalog. Renee Shepherd offers seeds of heirloom and cottage garden flowers, aromatic herbs and gourmet vegetables from around the world. Seeds available on-line or through garden centers and nurseries. Retail and wholesale. Offered 38 non-hybrid vegetable varieties in 2004 including: 2 bean, 1 beet, 2 carrot, 1 cress, 1 gourd, 1 kale, 1 leek, 1 onion common, 2 onion other, 2 parsley, 3 pea, 2 runner bean, 5 squash, 3 sunflower ornamental, 2 sunflower edible, 3 Swiss chard, 1 tomatillo, 1 tomato and 4 miscellaneous.

Re8 Rex's Seed Co., 5308 51st Avenue North, Crystal, MN 55429-3612. Web Site: www.rexseedco.com. On-line catalog. Family-owned and operated business serving the home gardener, truck farmer, hobbyist and greenhouse grower. Offered 207 non-hybrid vegetable varieties in 2004 including: 1 beet, 1 cardoon, 2 carrot, 1 cornsalad, 1 cress, 1 fennel, 1 gourd, 1 lettuce, 1 onion bunching, 2 onion other, 1 parsley, 1 parsnip, 56 pepper, 3 squash, 1 tomatillo, 124 tomato, 1 miscellaneous cucurbitaceae, 1 miscellaneous leguminosae and 7 miscellaneous.

Red Redwood City Seed Co., PO Box 361, Redwood City, CA 94064. Phone: 650-325-7333. Web Site: www.ecoseeds.com. Free catalog to US, Mexico and Canada; $2 overseas. Specializes in seeds of endangered traditional varieties (all open-pollinated or non-hybrid), plus pamphlets and books for the gardener. Two annual catalog supplements on the internet, listing seasonal or rare items where they have only a dozen packets available. All seeds offered are listed on the internet. (Whenever "Red" is the only source for a variety, that description is copyrighted by Redwood City Seed Co.) Retail and wholesale with wholesale catalog only on the internet. Offered 217 non-hybrid vegetable varieties in 2004 including: 3 amaranth, 6 bean, 2 broccoli, 2 carrot, 2 Chinese cabbage, 1 collard, 17 corn, 5 cucumber, 6 eggplant, 1 fennel, 2 kale, 1 leek, 7 lettuce, 3 lima, 4 melon, 1 mustard greens, 2 okra, 1 onion bunching, 2 onion common, 3 parsley, 3 pea, 1 peanut, 87 pepper, 1 radish, 1 runner bean, 1 soybean, 1 spinach, 13 squash, 2 sunflower ornamental, 2 sunflower edible, 2 Swiss chard, 1 tepary, 1 tomatillo, 8 tomato, 4 watermelon, 1 miscellaneous cucurbitaceae, 3 miscellaneous leguminosae and 14 miscellaneous.

REE Reed's Seeds, 3334 N.Y.S. Rt. 215, Cortland, NY 13045-9433. Strictly wholesale. Specializes in cole crops. Deals only with commercial growers. Offered 3 non-hybrid vegetable varieties in 2004 including: 1 cabbage and 2 cauliflower.

Ren Renaissance Acres Organic Herb Farm, 4450 Valentine Road, Whitmore Lake, MI 48189. Phone: 734-449-8336. E-mail: raohf@provide.net. Web Site: www.provide.net/~raohf. $3.00 for catalog. Propagates over 300 varieties of culinary, medicinal, scented, native, dye and ceremonial herb plants and seeds using the strictest organic growing methods. Retail only. Offered 15 non-hybrid vegetable varieties in 2004 including: 1 cardoon, 1 cress, 1 fennel, 2 onion other, 2 parsley, 1 sorghum, 1 tomatillo and 6 miscellaneous.

Rev Revolution Seeds, 204 North Waverly St., Homer, IL 61849. Phone: 217-896-3267. E-mail: bodhi@prairienet.org. Web Site: www.walkinplace.org/seeds. $1.00 for catalog. Specializes in rare, heirloom and historic open-pollinated varieties from the around the world, especially squash and other *cucurbits*. Retail only. Offered 80 non-hybrid vegetable varieties in 2004 including: 1 bean, 4 corn, 1 eggplant, 15 garlic, 5 gourd, 1 melon, 1 mustard greens, 1 okra, 1 onion multiplier, 1 pea, 1 pepper, 2 runner bean, 45 squash and 1 miscellaneous cucurbitaceae.

Ri2 Richters, 357 Hwy. 47, Goodwood, ON L0C 1A0, Canada. Phone: 905-640-6677. Fax: 905-640-6641. E-mail: orderdesk@richters.com. Web Site: www.richters.com. Free catalog. Family-owned company offering 800 types of herbs, plus unusual gourmet vegetables list. Offers seed of Siberian, Korean and American ginseng. Retail and wholesale. Offered 177 non-hybrid vegetable varieties in 2004 including: 2 amaranth, 1 artichoke, 5 bean, 3 beet, 2 broccoli, 1 broccoli raab, 1 cabbage, 1 cardoon, 4 carrot, 1 cauliflower, 1 celery, 4 Chinese cabbage, 1 chicory, 1 cornsalad, 3 cress, 4 cucumber, 1 eggplant, 2 endive, 1 fennel, 4 garlic, 1 gourd, 1 Jerusalem artichoke, 3 kale, 3 leek, 6 lettuce, 1 lima, 1 melon, 1 mustard greens, 1 okra, 2 onion bunching, 1 onion common, 5 onion multiplier, 6 onion other, 7 parsley, 2 pea, 15 pepper, 6 potato, 4 radish, 1 spinach, 9 squash, 1 sunflower edible, 1 Swiss chard, 1 tomatillo, 8 tomato, 1 watermelon, 4 miscellaneous cucurbitaceae, 2 miscellaneous leguminosae and 40 miscellaneous.

Ri12 Rich Farm Garden Supply, 985 W. State Road 32, Winchester, IN 47394. Web Site: www.richfarmgarden.com. Complete on-line catalog; occasionally mails partial lists and product information upon request. Full line of heirloom vegetable, herb and flower seeds, rare and heirloom nursery stock. Offered 380 non-hybrid vegetable varieties in 2004 including: 4 amaranth, 1 artichoke, 1 asparagus, 39 bean, 5 beet, 2 broccoli, 2 brussels sprout, 4 cabbage, 1 cardoon, 4 carrot, 2 cauliflower, 1 celeriac, 2 celery, 5 Chinese cabbage, 3 chicory, 2 collard, 25 corn, 1 cowpea, 1 cress, 9 cucumber, 2 eggplant, 2 endive, 3 fava bean, 5 garlic, 7 gourd, 1 ground cherry, 5 kale, 2 kohlrabi, 3 leek, 20 lettuce, 3 lima, 14 melon, 2 mustard greens, 3 okra, 7 onion common, 2 onion other, 1 parsley, 1 parsnip, 8 pea, 1 peanut, 33 pepper, 9 radish, 3 runner bean, 1 rutabaga, 1 scorzonera, 1 soybean, 2 spinach, 35 squash, 6 sunflower ornamental, 2 sunflower edible, 2 Swiss chard, 3 sweet potato, 2 tomatillo, 61 tomato, 3 turnip, 3 miscellaneous cucurbitaceae, 1 miscellaneous leguminosae and 6 miscellaneous.

RIS Rispens Seeds, Inc., PO Box 310, Beecher, IL 60401-0310. Strictly wholesale. Deals only with commercial growers. Offered 209 non-hybrid vegetable varieties in 2004 including: 25 bean, 2 beet, 3 cabbage, 2 carrot, 2 cauliflower, 5 celery, 2 Chinese cabbage, 1 chicory, 2 collard, 5 corn, 4 cowpea, 2 cucumber, 1 eggplant, 9 gourd, 1 kale, 1 kohlrabi, 2 leek, 18 lettuce, 4 lima, 2 melon, 3 mustard greens, 1 okra, 3 onion bunching, 3 onion common, 1 onion other, 3 parsley, 2 parsnip, 5 pea, 19 pepper, 3 radish, 1 sorghum, 1 spinach, 47 squash, 1 sunflower edible, 2 Swiss chard, 2 tomatillo, 9 tomato, 2 turnip, 6 watermelon and 2 miscellaneous.

Roh P. L. Rohrer and Bro. Inc., PO Box 250, Smoketown, PA 17576. Phone: 717-299-2571. Fax: 800-468-4944. E-mail: info@rohrerseeds.com. Web Site: www.rohrerseeds.com. Free catalog. Quality farm and garden seeds since 1918. Locally grown by Amish and Mennonite gardeners for their own use as well as organic and heirloom seeds. Retail and wholesale catalogs; wholesale catalog contains varieties not in the other. Offered 357 non-hybrid vegetable varieties in 2004 including: 1 amaranth, 1 artichoke, 34 bean, 8 beet, 3 broccoli, 1 broccoli raab, 1 brussels sprout, 4 cabbage, 1 cardoon, 8 carrot, 2 cauliflower, 3 celery, 6 Chinese cabbage, 2 chicory, 1 collard, 14 corn, 1 cornsalad, 1 cowpea, 2 cress, 11 cucumber, 4 eggplant, 3 endive, 1 fava bean, 1 garlic, 9 gourd, 1 ground cherry, 3 kale, 2 kohlrabi, 2 leek, 27 lettuce, 9 lima, 6 melon, 3 mustard greens, 2 okra, 1 onion bunching, 7 onion common, 1 onion multiplier, 2 onion other, 4 parsley, 1 parsnip, 19 pea, 1 peanut, 12 pepper, 8 potato, 12 radish, 3 runner bean, 1 rutabaga, 1 salsify, 2 sorghum, 3 soybean, 1 spinach, 31 squash, 12 sunflower ornamental, 3 sunflower edible, 4 Swiss chard, 2 tomatillo, 20 tomato, 4 turnip, 6 watermelon, 3 miscellaneous cucurbitaceae, 3 miscellaneous leguminosae, 1 miscellaneous solanaceae and 11 miscellaneous.

Ron Ronnigers Potato Farm, Star Route, Moyie Springs, ID 83845. Fax: 208-267-3265. E-mail: smallpotatoes @ronnigers.com. Web Site: www.ronnigers.com. Free information-packed "Catalog and Growers Guide." Small family farm offering the largest selection of certified organically grown seed potato varieties in the United States to home gardeners, small farmers and hobbyists. Retail and bulk. Offered 84 non-hybrid vegetable varieties in 2004 including: 11 garlic, 1 onion bunching, 3 onion common, 3 onion multiplier, 1 pea, 63 potato and 2 miscellaneous.

Ros Roswell Seed Co., 115 S. Main St., PO Box 725, Roswell, NM 88202-0725. Phone: 505-622-7701. Fax: 505-623-2885. Free catalog. Established in 1900. Specializes in vegetable, flower, field and lawn seeds adapted to the Southwest. Retail and wholesale. Offered 162 non-hybrid vegetable varieties in 2004 including: 1 asparagus, 9 bean, 2 beet, 1 broccoli, 1 brussels sprout, 5 cabbage, 4 carrot, 1 cauliflower, 3 Chinese cabbage, 1 collard, 4 corn, 1 cowpea, 5 cucumber, 2 eggplant, 1 kale, 1 kohlrabi, 1 leek, 7 lettuce, 5 lima, 10 melon, 2 mustard greens, 3 okra, 1 onion bunching, 6 onion common, 1 onion other, 1 parsley, 3 pea, 20 pepper, 6 radish, 1 rutabaga, 2 spinach, 23 squash, 1 sunflower edible, 2 Swiss chard, 9 tomato, 2 turnip, 10 watermelon, 1 miscellaneous cucurbitaceae, 1 miscellaneous leguminosae and 2 miscellaneous.

RUP Rupp Seeds, Inc., 17919 Co. Rd. B, Wauseon, OH 43567. Phone: 419-337-1841. Fax: 419-337-5491. Wholesale only. Specializes in vegetables for the commercial grower or market gardener. Offered 506 non-hybrid vegetable varieties in 2004 including: 1 asparagus, 72 bean, 10 beet, 2 broccoli, 1 brussels sprout, 10 cabbage, 6 carrot, 4 cauliflower, 1 celeriac, 1 celery, 6 Chinese cabbage, 3 collard, 16 corn, 1 cornsalad, 4 cowpea, 2 cress, 6 cucumber, 1 eggplant, 6 endive, 1 fennel, 14 gourd, 1 ground cherry, 4 kale, 2 kohlrabi, 3 leek, 31 lettuce, 9 lima, 9 melon, 3 mustard greens, 3 okra, 3 onion bunching, 5 onion common, 2 onion other, 5 parsley, 3 parsnip, 30 pea, 43 pepper, 17 radish, 2 rutabaga, 1 salsify, 4 sorghum, 1 soybean, 2 spinach, 73 squash, 2 sunflower ornamental, 2 sunflower edible, 5 Swiss chard, 2 tomatillo, 46 tomato, 5 turnip, 12 watermelon, 3 miscellaneous cucurbitaceae and 5 miscellaneous.

Sa5 Salt Spring Seeds, Box 444, Ganges P.O., Salt Spring Island, BC V8K 2W1, Canada. Phone: 250-537-5269. Web Site: www.saltspringseeds.com. $2 for catalog for first-time customers. Open-pollinated, organically grown seed, all adapted to northern climates. Retail only. Offered 186 non-hybrid vegetable varieties in 2004 including: 2 amaranth, 32 bean, 2 beet, 2 carrot, 3 celery, 2 Chinese cabbage, 1 chicory, 1 corn, 1 cornsalad, 2 cress, 2 fava bean, 13 garlic, 1 Jerusalem artichoke, 3 kale, 20 lettuce, 2 onion bunching, 1 parsley, 5 pea, 14 pepper, 3 radish, 6 soybean, 4 squash, 1 Swiss chard, 1 tepary, 1 tomatillo, 51 tomato, 1 miscellaneous leguminosae and 9 miscellaneous.

Sa6 The Sandy Mush Herb Nursery, 316 Surrett Cove Road, Leicester, NC 28748. Phone: 828-683-2014. $4 for catalog. Extensive herb, perennial, native plant and ivy selections, plus a number of gourmet vegetable varieties. Retail only. Offered 21 non-hybrid vegetable varieties in 2004 including: 1 cardoon, 1 Chinese cabbage, 1 cornsalad, 1 cress, 1 fennel, 1 lettuce, 1 onion multiplier, 2 onion other, 1 parsley, 1 pepper, 1 miscellaneous leguminosae and 9 miscellaneous.

Sa9 **Sand Hill Preservation Center,** Heirloom Seeds and Poultry, 1878 230th St., Calamus, IA 52729. Phone: 563-246-2299. E-mail: sandhill@fbcom.net. Web Site: www.sandhillpreservation.com. Free catalog. All varieties are open-pollinated and grown on site. Retail only. Offered 913 non-hybrid vegetable varieties in 2004 including: 6 amaranth, 24 bean, 3 beet, 1 broccoli, 2 brussels sprout, 7 cabbage, 1 cardoon, 6 carrot, 2 cauliflower, 1 celeriac, 1 celery, 6 Chinese cabbage, 1 collard, 49 corn, 1 cornsalad, 6 cowpea, 19 cucumber, 21 eggplant, 1 endive, 2 fennel, 1 garlic, 41 gourd, 2 Jerusalem artichoke, 2 kale, 2 kohlrabi, 1 leek, 8 lettuce, 3 lima, 17 melon, 2 mustard greens, 7 okra, 2 onion common, 1 onion multiplier, 2 parsley, 2 parsnip, 6 pea, 32 pepper, 1 quinoa, 10 radish, 2 rutabaga, 1 salsify, 1 scorzonera, 7 sorghum, 3 soybean, 1 spinach, 76 squash, 5 sunflower ornamental, 2 sunflower edible, 4 Swiss chard, 61 sweet potato, 4 tomatillo, 384 tomato, 4 turnip, 28 watermelon, 6 miscellaneous cucurbitaceae, 4 miscellaneous leguminosae, 5 miscellaneous solanaceae and 13 miscellaneous.

Sa12 **Sand Mountain Herbs,** 321 County Road 18, Fyffe, AL 35971. Web Site: www.sandmountainherbs.com. On-line catalog. Specializes in unique, high quality herb seeds. Offered 16 non-hybrid vegetable varieties in 2004 including: 1 artichoke, 1 cardoon, 1 celery, 1 cress, 2 onion other, 1 parsley, 1 miscellaneous cucurbitaceae and 8 miscellaneous.

SAK **Sakata Seed America, Inc.,** 18095 Serene Drive, Morgan Hill, CA 95038. Strictly wholesale. Offered 20 non-hybrid vegetable varieties in 2004 including: 3 Chinese cabbage, 1 fava bean, 7 lettuce, 1 onion bunching, 1 parsley, 1 pea, 3 soybean, 2 squash and 1 miscellaneous leguminosae.

Sau **Saunders Seed Co., Inc.,** 101 W. Broadway, Tipp City, OH 45371. Phone: 937-667-2313. Send SASE for price list. Separate wholesale and retail price lists, each containing varieties not in the other. Retail price list offers quantities from 1 oz. to 1 lb. Offered 189 non-hybrid vegetable varieties in 2004 including: 1 asparagus, 24 bean, 4 beet, 1 broccoli, 1 brussels sprout, 9 cabbage, 4 carrot, 1 cauliflower, 1 celery, 2 Chinese cabbage, 1 collard, 7 corn, 3 cowpea, 6 cucumber, 1 eggplant, 2 endive, 2 kale, 2 kohlrabi, 1 leek, 10 lettuce, 6 lima, 7 melon, 2 mustard greens, 2 okra, 5 onion common, 1 onion other, 2 parsley, 1 parsnip, 10 pea, 1 peanut, 7 pepper, 4 potato, 11 radish, 1 rutabaga, 1 salsify, 1 spinach, 20 squash, 1 sunflower edible, 2 Swiss chard, 8 tomato, 3 turnip, 8 watermelon and 2 miscellaneous.

Scf **South Carolina Foundation Seed Assn.,** 1162 Cherry Road, Box 349952, Clemson, SC 29634. Phone: 864-656-2520. Send SASE for price list. Specializes in vegetable varieties developed by state agricultural colleges and USDA plant breeders. Retail only. Offered 124 non-hybrid vegetable varieties in 2004 including: 17 bean, 1 carrot, 1 Chinese cabbage, 3 corn, 15 cowpea, 3 cucumber, 2 gourd, 1 leek, 4 lettuce, 9 lima, 5 melon, 1 mustard greens, 2 okra, 1 onion other, 2 parsley, 2 pea, 3 peanut, 13 pepper, 2 radish, 1 runner bean, 1 soybean, 1 spinach, 11 squash, 9 sweet potato, 1 tomatillo, 7 tomato, 2 turnip, 1 miscellaneous cucurbitaceae, 1 miscellaneous leguminosae and 2 miscellaneous.

SE4 **Seedway, Inc.,** 1225 Zeager Road, Elizabethtown, PA 17022-9427. Phone: 800-952-7333. Fax: 717-367-0387. E-mail: info@seedway.com. Web Site: www.seedway.com. Wholesale only. Specializes in seeds for commercial growers. Offered 257 non-hybrid vegetable varieties in 2004 including: 52 bean, 2 beet, 2 broccoli raab, 1 brussels sprout, 2 cabbage, 4 carrot, 2 cauliflower, 2 celeriac, 1 celery, 3 chicory, 3 collard, 10 corn, 8 cowpea, 3 cucumber, 2 eggplant, 4 endive, 1 fava bean, 1 fennel, 8 gourd, 1 kale, 2 kohlrabi, 1 leek, 26 lettuce, 7 lima, 1 melon, 2 mustard greens, 1 okra, 3 onion bunching, 2 onion other, 4 parsley, 3 parsnip, 22 pea, 4 pepper, 8 radish, 1 runner bean, 1 rutabaga, 1 sorghum, 1 spinach, 39 squash, 1 sunflower edible, 4 Swiss chard, 1 tomatillo, 1 tomato, 2 turnip, 3 watermelon and 4 miscellaneous.

Se7 **Seeds of Change,** PO Box 15700, Santa Fe, NM 87592. Phone: 888-762-7333. Web Site: www.seedsofchange.com. Free catalog. Specializes in vegetables, flowers and herbs that are all 100% certified organic, open-pollinated, public domain varieties. Wide selection of traditional and heirloom varieties. Separate retail and wholesale catalogs, each containing varieties not in the other. Offered 415 non-hybrid vegetable varieties in 2004 including: 9 amaranth, 26 bean, 4 beet, 4 broccoli, 1 broccoli raab, 1 brussels sprout, 3 cabbage, 1 cardoon, 10 carrot, 1 cauliflower, 5 celery, 7 Chinese cabbage, 2 chicory, 2 collard, 18 corn, 1 cornsalad, 1 cress, 9 cucumber, 7 eggplant, 1 endive, 4 fava bean, 1 fennel, 12 garlic, 4 gourd, 1 Jerusalem artichoke, 4 kale, 6 leek, 36 lettuce, 13 melon, 3 mustard greens, 4 okra, 8 onion common, 2 onion other, 3 parsley, 1 parsnip, 6 pea, 28 pepper, 10 potato, 4 quinoa, 8 radish, 4 runner bean, 1 rutabaga, 1 sorghum, 1 soybean, 3 spinach, 32 squash, 11 sunflower ornamental, 5 sunflower edible, 4 Swiss chard, 3 tepary, 2 tomatillo, 46 tomato, 2 turnip, 9 watermelon, 4 miscellaneous leguminosae, 1 miscellaneous solanaceae and 15 miscellaneous.

Se8 **Seeds West Garden Seeds,** 317 14th St. NW, Albuquerque, NM 87125. Phone: 505-843-9713. E-mail: seeds@nmia.com. Web Site: www.seedswestgardenseeds.com. $2 for catalog. Specializes in open-pollinated heirloom vegetable seeds and traditional cottage garden annuals selected for performance in hot dry short-season growing conditions of the West and Southwest. All untreated and many organic and grown regionally. Retail and wholesale. Offered 170 non-hybrid vegetable varieties in 2004 including: 12 bean, 5 beet, 3 broccoli, 5 cabbage, 7 carrot, 1 cauliflower, 3 Chinese cabbage, 2 chicory, 7 corn, 1 cornsalad, 1 cress, 4 cucumber, 2 eggplant, 2 endive, 2 gourd, 3 kale, 2 leek, 16 lettuce, 3 melon, 1 mustard greens, 1 onion bunching, 1 onion other, 1 parsley, 6 pea, 20 pepper, 7 radish, 1 runner bean, 3 spinach, 17 squash, 1 sunflower edible, 2 Swiss chard, 1 tepary, 1 tomatillo, 16 tomato, 4 watermelon, 1 miscellaneous cucurbitaceae and 5 miscellaneous.

SE14 **Seeds by Design, Inc.,** PO Box 602, Maxwell, CA 95955. Phone: 530-438-2126. Fax: 530-438-2171. E-mail: info@seedsbydesign.com. Web Site: www.seedsbydesign.com. Strictly wholesale. Specializes in heritage vegetables and herbs. Offered 517 non-hybrid vegetable varieties in 2004 including: 3 asparagus, 7 beet, 3 broccoli, 2 brussels sprout, 14 cabbage, 8 carrot, 2 cauliflower, 2 celery, 2 Chinese cabbage, 3 chicory, 5 corn, 1 cornsalad, 5 cress, 15 cucumber, 10 eggplant, 1 fennel, 11 gourd, 1 ground cherry, 2 kale, 34 lettuce, 23 melon, 1 mustard greens, 5 okra, 3 onion bunching, 12 onion common, 2 onion other, 5 parsley, 83 pepper, 11 radish, 1 sorghum, 1 spinach, 64 squash, 8 sunflower ornamental, 2 sunflower edible, 7 Swiss chard, 4 tomatillo, 124 tomato, 3 turnip, 20 watermelon, 2 miscellaneous cucurbitaceae, 1 miscellaneous solanaceae and 4 miscellaneous.

Se16 **Seed Savers Catalog,** 3094 North Winn Road, Decorah, IA 52101. Phone: 563-382-5990. Fax: 563-382-5872. Web Site: www.seedsavers.org. Free 92-page color catalog offers a truly unique selection of outstanding vegetables, flowers and herbs including many family heirlooms from the Seed Savers Exchange and traditional varieties from Eastern Europe and the former Soviet Union. Project-related revenue from seed sales is being used to permanently maintain Heritage Farm's vast seed collections of 24,000 rare varieties. Retail catalog (626 varieties) also contains bulk prices (613 varieties) for specialty growers, truck gardeners, CSA growers and other seed companies. (About one-quarter of the catalog's varieties are Certified Organic and that number will steadily increase). Wholesale price list available on request. Offered 408 non-hybrid vegetable varieties in 2004 including: 30 bean, 8 beet, 3 broccoli, 1 brussels sprout, 5 cabbage, 5 carrot, 2 cauliflower, 11 corn, 12 cucumber, 9 eggplant, 1 fennel, 15 garlic, 4 gourd, 1 ground cherry, 3 kale, 3 leek, 40 lettuce, 3 lima, 27 melon, 4 okra, 5 onion common, 2 parsley, 9 pea, 36 pepper, 8 potato, 7 radish, 3 runner bean, 2 rutabaga, 1 sorghum, 3 soybean, 2 spinach, 32 squash, 13 sunflower ornamental, 4 sunflower edible, 3 Swiss chard, 2 tomatillo, 62 tomato, 2 turnip, 16 watermelon, 1 miscellaneous cucurbitaceae, 1 miscellaneous leguminosae, 2 miscellaneous solanaceae and 5 miscellaneous.

Se17 **Seed Dreams,** PO Box 106, Port Townsend, WA 98368. E-mail: gowantoseed@yahoo.com. Free catalog. Specializes in the preservation of rare, heirloom, open-pollinated and Native American varieties. Seeds are distributed through Independent Seed, a group of folks in Port Townsend, Washington, working to make open-pollinated, organic seed grown by independent seed farmers available to home gardeners and market farmers. Retail and wholesale. Offered 486 non-hybrid vegetable varieties in 2004 including: 8 amaranth, 94 bean, 5 beet, 2 broccoli, 1 cabbage, 1 carrot, 4 celery, 1 Chinese cabbage, 3 collard, 28 corn, 16 cowpea, 11 cucumber, 4 eggplant, 5 fava bean, 9 gourd, 2 kale, 2 leek, 32 lettuce, 5 lima, 8 melon, 4 okra, 1 onion bunching, 5 onion common, 1 onion other, 3 parsley, 3 parsnip, 18 pea, 2 peanut, 22 pepper, 5 quinoa, 6 radish, 5 runner bean, 1 rutabaga, 6 sorghum, 23 soybean, 44 squash, 4 sunflower ornamental, 5 sunflower edible, 4 Swiss chard, 3 tomatillo, 66 tomato, 6 watermelon, 1 miscellaneous cucurbitaceae, 4 miscellaneous leguminosae and 3 miscellaneous.

Se24 **Seeds from Italy,** PO Box 149, Winchester, MA 01890. E-mail: seeds@growitalian.com. Web Site: www.growitalian.com. Free catalog. Offers more than 300 varieties of traditional Italian vegetable, herb and flower seeds along with growing instructions, tips and Italian recipes. Retail only. Offered 173 non-hybrid vegetable varieties in 2004 including: 1 artichoke, 17 bean, 1 beet, 2 broccoli, 5 broccoli raab, 3 cabbage, 2 cardoon, 2 carrot, 2 cauliflower, 1 celeriac, 2 celery, 26 chicory, 2 cornsalad, 3 cucumber, 3 eggplant, 6 endive, 5 fennel, 1 gourd, 1 kale, 13 lettuce, 9 melon, 4 mushroom, 6 onion common, 1 onion other, 2 parsley, 2 pea, 10 pepper, 3 radish, 2 spinach, 14 squash, 1 sunflower edible, 3 Swiss chard, 11 tomato, 2 miscellaneous leguminosae and 5 miscellaneous.

Se25 **Seeds for Survival,** PO Box 111, Waterport, NY 14571. E-mail: customercare@seedsforsurvival.com. Web Site: www.seedsforsurvival.com. Online catalog. Specializes in preparing non-hybrid vegetable seeds for long-term storage so they remain viable and vigorous well beyond their normal life span. Retail only. Offered 151 non-hybrid vegetable varieties in 2004 including: 1 asparagus, 12 bean, 4 beet, 5 broccoli, 6 cabbage, 4 carrot, 3 cauliflower, 1 celeriac, 1 Chinese cabbage, 1 chicory, 1 collard, 7 corn, 5 cucumber, 2 eggplant, 4 kale, 3 kohlrabi, 7 lettuce, 1 lima, 5 melon, 4 okra, 5 onion common, 1 parsley, 2 parsnip, 7 pea, 14 pepper, 2 radish, 2 rutabaga, 1 spinach, 14 squash, 2 Swiss chard, 1 tomatillo, 9 tomato, 8 watermelon, 1 miscellaneous cucurbitaceae, 3 miscellaneous leguminosae and 2 miscellaneous.

Se26 **Seeds,** 410 Whaley Pond Road, Graniteville, SC 29829. E-mail: seedsout@mindspring.com. Web Site: www.vegetableseedwarehouse.com. Online catalog. Small family-owned business offering heirloom vegetable and herb varieties, many certified organic. Retail only. Offered 521 non-hybrid vegetable varieties in 2004 including: 1 amaranth, 2 artichoke, 1 asparagus, 34 bean, 11 beet, 5 broccoli, 1 broccoli raab, 3 brussels sprout, 10 cabbage, 6 carrot, 5 cauliflower, 1 celery, 5 Chinese cabbage, 5 chicory, 3 collard, 11 corn, 2 cornsalad, 14 cowpea, 1 cress, 16 cucumber, 4 eggplant, 6 endive, 1 fava bean, 7 gourd, 5 kale, 1 kohlrabi, 2 leek, 29 lettuce, 11 lima, 18 melon, 3 mustard greens, 4 okra, 2 onion bunching, 7 onion common, 2 onion other, 5 parsley, 2 parsnip, 9 pea, 58 pepper, 14 radish, 1 runner bean, 1 rutabaga, 1 salsify, 1 soybean, 2 spinach, 31 squash, 3 sunflower ornamental, 7 Swiss chard, 3 tomatillo, 120 tomato, 3 turnip, 11 watermelon, 2 miscellaneous cucurbitaceae, 1 miscellaneous leguminosae and 7 miscellaneous.

Se27 **Seed Movement,** P.O. Box 1988, Port Townsend, WA 98368. E-mail: seedmovement@earthlink.net. Free brochure-pricelist available by e-mail. Mailing address may change. Offering 100% certified organically grown seeds. All varieties are bred, selected and evaluated under less than ideal "real life" conditions, thus singling out only the workhorse varieties. Seed Movement hopes to establish a truly sustainable loop between the eaters, the farmers, the seed growers and the breeders. Bulk quantities available for most varieties. Offered 17 non-hybrid vegetable varieties in 2004 including: 1 beet, 3 carrot, 1 chicory, 1 endive, 1 kale, 2 lettuce, 1 spinach, 2 Swiss chard, 4 tomato and 1 miscellaneous.

Se28 **Seeds Etc.,** 2616 Turk Drive, Marysville, WA 98271. E-mail: info@seedsetc.com. Web Site: www.seedsetc.com. On-line catalog. Offers seed from well established and proven seed collectors from around the world. Retail only. Offered 99 non-hybrid vegetable varieties in 2004 including: 1 asparagus, 2 beet, 2 cabbage, 3 cauliflower, 2 Chinese cabbage, 8 corn, 2 cucumber, 2 eggplant, 2 gourd, 1 kale, 1 leek, 1 melon, 1 okra, 1 onion bunching, 2 onion common, 1 parsnip, 1 pea, 17 pepper, 2 radish, 1 salsify, 4 sorghum, 1 soybean, 9 squash, 7 sunflower ornamental, 3 sunflower edible, 1 tomatillo, 18 tomato, 2 miscellaneous cucurbitaceae and 1 miscellaneous.

Sey **Seymour's Selected Seeds,** 334 W. Stroud St., Randolph, WI 53956. Phone: 803-853-9516. Fax: 888-739-6687. Web Site: www.seymourseedusa.com. Free catalog. Specializes in English cottage garden varieties. Retail only. Offered 6 non-hybrid vegetable varieties in 2004 including: 1 amaranth, 5 sunflower ornamental.

SGF **S & G Flowers,** 5300 Katrine Ave., Downers Grove, IL 60515. Strictly wholesale. Distributes to large commercial dealers only. Purchased Novartis Seeds, Vaughan's Seed Company and Rogers Northrup King Seed Company. Offered 51 non-hybrid vegetable varieties in 2004 including: 3 cabbage, 1 cauliflower, 1 celery, 1 cress, 3 cucumber, 1 eggplant, 1 fennel, 5 lettuce, 1 okra, 5 onion common, 2 onion other, 2 parsley, 16 pepper, 3 squash, 4 tomato, 1 watermelon and 1 miscellaneous.

Sh7 **Sharp Bros. Seed Co.,** PO Box 140, Healy, KS 67850. Free price list. Specializes in native grasses. Included here for Maximillian sunflower. Retail and wholesale. Offered 1 edible sunflower in 2004.

Sh9 **Shaffer Seed and Supply Co.,** 1203 E. Tuscarawas St., Canton, OH 44707. Phone: 330-452-8866. $2 for catalog. Retail and wholesale. Offered 322 non-hybrid vegetable varieties in 2004 including: 1 asparagus, 30 bean, 7 beet, 2 broccoli, 1 broccoli raab, 1 brussels sprout, 11 cabbage, 13 carrot, 5 cauliflower, 5 celery, 2 Chinese cabbage, 1 chicory, 2 collard, 12 corn, 3 cowpea, 1 cress, 14 cucumber, 1 eggplant, 4 endive, 1 fava bean, 1 garlic, 4 gourd, 3 kale, 2 kohlrabi, 1 leek, 17 lettuce, 7 lima, 10 melon, 3 mustard greens, 4 okra, 5 onion common, 1 onion other, 4 parsley, 2 parsnip, 16 pea, 1 peanut, 18 pepper, 10 radish, 1 rutabaga, 1 salsify, 1 sorghum, 1 soybean, 1 spinach, 41 squash, 4 sunflower ornamental, 1 sunflower edible, 3 Swiss chard, 25 tomato, 4 turnip, 7 watermelon, 2 miscellaneous cucurbitaceae and 4 miscellaneous.

Sh12 **Shoulder to Shoulder Farm,** Wild Garden Seed, PO Box 1509, Philomath, OR 97370. Phone: 541-929-4068. Web Site: www.wildgardenseed.com. Free catalog. Specializes in unusual salad greens, beneficial insectary plants and native Northwest species useful to developing stable farm ecosystems. Retail and bulk prices are listed in the same catalog. Offered 64 non-hybrid vegetable varieties in 2004 including: 2 amaranth, 1 beet, 1 broccoli raab, 2 celery, 6 Chinese cabbage, 1 chicory, 2 cress, 1 fennel, 8 kale, 17 lettuce, 4 mustard greens, 3 parsley, 2 quinoa, 1 squash, 2 Swiss chard, 1 turnip and 10 miscellaneous.

Sh13 **Shawnee Seed Company,** Linda Parker, 944 Cedar Creek Road, Makanda, IL 62958. Phone: 618-457-0114. E-mail: shawneeseedco@hotmail.com. Free seed list. Specializes in select heirloom varieties that are hardy for USDA Zone 6 without irrigation. Retail only. Offered 107 non-hybrid vegetable varieties in 2004 including: 29 bean, 1 corn, 3 cowpea, 2 cucumber, 1 gourd, 7 lima, 3 melon, 2 okra, 1 pea, 2 sunflower ornamental, 49 tomato, 4 watermelon and 1 miscellaneous leguminosae.

Sh14 **Sheffield's Seed Company, Inc.,** 269 Auburn Road, Route 34, Locke, NY 13092. E-mail: seed@sheffields.com. Web Site: www.sheffields.com. Free catalog. Specializes in woody plants and herbaceous seeds. Retail and wholesale prices are listed in the same catalog. Offered 23 non-hybrid vegetable varieties in 2004 including: 2 carrot, 1 cornsalad, 4 lettuce, 1 onion bunching, 1 onion common, 2 onion other, 1 parsley, 3 pepper, 2 radish, 1 squash, 1 sunflower ornamental, 1 miscellaneous cucurbitaceae and 3 miscellaneous.

Sh15 **Shady Oaks Ginseng Farm,** Chris and Leslie Burdette, Rt. 1 Box 209, Poca, WV 25159. Phone: 888-304-5638. E-mail: shadyoaksfarm@aol.com. Web Site: www.shadyoaksginseng.net. Free price list or view on-line. Specializes in ginseng and goldenseal roots and ginseng seed. Offered 1 miscellaneous variety in 2004.

Shu **R. H. Shumway, Seedsman,** 334 W. Stroud St., Randolph, WI 53956-1274. Phone: 800-342-9461. Fax: 888-437-2733. Web Site: www.rhshumway.com. Free catalog. Specializes in heirloom, heritage and open-pollinated seeds. Separate retail and wholesale catalogs, each containing varieties not in the other. Offered 415 non-hybrid vegetable varieties in 2004 including: 2 amaranth, 1 artichoke, 3 asparagus, 46 bean, 12 beet, 1 broccoli, 1 brussels sprout, 11 cabbage, 9 carrot, 3 cauliflower, 1 celeriac, 3 celery, 3

Chinese cabbage, 3 collard, 29 corn, 1 cornsalad, 8 cowpea, 1 cress, 16 cucumber, 3 eggplant, 1 endive, 1 fava bean, 1 fennel, 6 garlic, 6 gourd, 1 ground cherry, 1 Jerusalem artichoke, 3 kale, 2 kohlrabi, 3 leek, 13 lettuce, 10 lima, 14 melon, 1 mustard greens, 4 okra, 1 onion bunching, 3 onion common, 1 onion multiplier, 1 onion other, 2 parsley, 2 parsnip, 19 pea, 2 peanut, 16 pepper, 9 potato, 12 radish, 1 runner bean, 1 rutabaga, 1 salsify, 3 sorghum, 2 spinach, 35 squash, 1 sunflower edible, 2 Swiss chard, 7 sweet potato, 41 tomato, 4 turnip, 15 watermelon, 3 miscellaneous cucurbitaceae, 1 miscellaneous leguminosae and 6 miscellaneous.

Si5 Silver Creek Supply, RD 1 Box 70, Port Trevorton, PA 17864. Phone: 570-374-8010. E-mail: silcreek@uplink.net. Web Site: www.silvercreeksupply.net. On-line catalog. Retail only. Offered 104 non-hybrid vegetable varieties in 2004 including: 16 bean, 3 beet, 1 broccoli, 3 cabbage, 3 carrot, 2 cauliflower, 1 celery, 4 corn, 1 cucumber, 1 eggplant, 2 endive, 2 gourd, 1 kale, 1 kohlrabi, 7 lettuce, 3 lima, 1 okra, 2 onion common, 3 parsley, 1 parsnip, 9 pea, 8 pepper, 6 radish, 1 salsify, 1 soybean, 1 spinach, 13 squash, 1 sunflower edible, 1 tomato, 1 turnip and 4 watermelon.

Sk2 Skyfire Garden Seeds, 1313 23rd Road, Kanopolis, KS 67454. E-mail: seedsaver@myvine.com. Web Site: www.grapevine.net/~mctaylor. Free catalog. Specializes in open-pollinated, heirloom, non-GMO treated seeds. Retail only. Offered 242 non-hybrid vegetable varieties in 2004 including: 14 bean, 5 beet, 1 broccoli, 4 cabbage, 31 carrot, 3 cauliflower, 5 corn, 9 cucumber, 4 eggplant, 1 kale, 2 kohlrabi, 16 lettuce, 2 lima, 14 melon, 2 okra, 1 onion other, 1 parsley, 1 parsnip, 6 pea, 22 pepper, 12 radish, 3 runner bean, 1 spinach, 29 squash, 4 sunflower ornamental, 1 Swiss chard, 44 tomato, 1 turnip, 1 watermelon and 2 miscellaneous.

So1 Southern Exposure Seed Exchange, PO Box 460, Mineral, VA 23117. Phone: 540-894-9480. Fax: 540-894-9481. E-mail: gardens@southernexposure.com. Web Site: www.southernexposure.com. $2 for catalog. Offers over 500 varieties of open-pollinated heirloom and traditional vegetables, flowers and herbs. Many are suited to hot, humid and disease-prone areas. Seed is untreated and much of it is grown organically. Retail and bulk prices are listed in the same catalog. Offered 426 non-hybrid vegetable varieties in 2004 including: 2 amaranth, 25 bean, 5 beet, 4 broccoli, 1 brussels sprout, 5 cabbage, 1 cardoon, 7 carrot, 2 cauliflower, 1 celeriac, 1 celery, 5 Chinese cabbage, 3 chicory, 7 collard, 20 corn, 6 cowpea, 3 cress, 12 cucumber, 10 eggplant, 1 endive, 1 fava bean, 1 fennel, 9 garlic, 3 gourd, 2 ground cherry, 4 kale, 2 kohlrabi, 2 leek, 31 lettuce, 9 lima, 9 melon, 2 mustard greens, 8 okra, 3 onion bunching, 2 onion common, 2 onion multiplier, 2 onion other, 2 parsley, 1 parsnip, 6 pea, 3 peanut, 38 pepper, 5 radish, 1 runner bean, 1 rutabaga, 1 salsify, 4 sorghum, 2 spinach, 29 squash, 8 sunflower ornamental, 3 sunflower edible, 3 Swiss chard, 2 tomatillo, 75 tomato, 4 turnip, 8 watermelon, 1 miscellaneous cucurbitaceae, 2 miscellaneous leguminosae, 1 miscellaneous solanaceae and 13 miscellaneous.

So9 Sow Organic, P.O. Box 527, Williams, OR 97544. Phone: 541-846-7173. Web Site: www.organicseed.com. Free price list. Specializes in open-pollinated varieties grown and processed by the company and certified organic by the Oregon Tilth Certified program. Formerly "Southern Oregon Organics." Separate retail and wholesale price lists. Offered 132 non-hybrid vegetable varieties in 2004: 2 amaranth, 13 bean, 2 beet, 2 broccoli, 2 cabbage, 2 carrot, 1 celery, 4 Chinese cabbage, 6 corn, 1 cress, 4 cucumber, 3 kale, 3 leek, 16 lettuce, 4 melon, 1 mustard greens, 3 onion common, 1 onion multiplier, 1 onion other, 2 parsley, 1 pea, 3 pepper, 5 radish, 2 spinach, 18 squash, 2 sunflower ornamental, 1 sunflower edible, 2 Swiss chard, 15 tomato, 1 turnip, 3 watermelon, 1 miscellaneous leguminosae and 5 miscellaneous.

So12 Sourcepoint Organic Seeds, 1220 2640 Road, Hotchkiss, CO 81419-9475. $3 for catalog. Offers only organically grown or wild-crafted seeds. Retail only. Offered 276 non-hybrid vegetable varieties in 2004 including: 10 amaranth, 1 asparagus, 28 bean, 2 beet, 1 broccoli raab, 5 carrot, 2 celeriac, 2 celery, 9 Chinese cabbage, 1 chicory, 1 collard, 19 corn, 2 cowpea, 2 cress, 3 cucumber, 2 fava bean, 1 fennel, 3 gourd, 1 ground cherry, 1 Jerusalem artichoke, 1 kale, 2 leek, 12 lettuce, 2 lima, 8 melon, 1 mustard greens, 1 okra, 2 onion bunching, 1 onion multiplier, 1 onion other, 3 parsley, 4 parsnip, 9 pea, 1 peanut, 9 pepper, 9 quinoa, 9 radish, 1 runner bean, 1 salsify, 1 scorzonera, 6 sorghum, 5 soybean, 16 squash, 1 sunflower ornamental, 4 sunflower edible, 1 Swiss chard, 6 tepary, 1 tomatillo, 6 tomato, 2 turnip, 3 watermelon, 10 miscellaneous leguminosae, 1 miscellaneous solanaceae and 40 miscellaneous.

So25 Solana Seeds, 17 Place Leger, Repentigny, QC J6A 5N7, Canada. Web Site: www.solanaseeds.netfirms.com. On-line catalog. Small seed company in Quebec, Canada offering a variety of vegetable, flower and exotic plant seeds including many rare and unusual heirlooms. Retail only. Offered 237 non-hybrid vegetable varieties in 2004 including: 4 bean, 1 broccoli raab, 1 carrot, 3 Chinese cabbage, 2 corn, 8 cucumber, 4 eggplant, 2 ground cherry, 1 lettuce, 14 melon, 1 mustard greens, 1 okra, 1 onion common, 2 onion other, 23 pepper, 2 radish, 5 squash, 2 sunflower ornamental, 1 sunflower edible, 5 tomatillo, 135 tomato, 9 watermelon, 2 miscellaneous cucurbitaceae, 2 miscellaneous solanaceae and 6 miscellaneous.

SOU Southern States Cooperative, 6606 West Broad Street, Richmond, VA 23230. Web Site: www.southernstates.com. Free catalog. Mainly wholesale, with retail sales through its stores in six mid-Atlantic states. Separate retail and wholesale catalogs; wholesale catalog contains additional varieties. Offered 167 non-hybrid vegetable varieties in 2004 including: 31 bean, 4 beet, 4 cabbage, 4 carrot, 1 cauliflower, 1 Chinese cabbage, 4 collard, 3 corn, 7 cowpea, 1 cress, 5 cucumber, 1 eggplant, 1 gourd, 3 kale, 8 lettuce, 12 lima, 5 melon, 3 mustard greens, 3 okra, 1 parsley, 1 parsnip, 11 pea, 7 pepper, 5 radish, 1 rutabaga, 1 salsify, 1 spinach, 15 squash, 3 Swiss chard, 7 tomato, 3 turnip, 9 watermelon and 1 miscellaneous.

St18 **Stellar Seeds,** 3801 40th St. NE, Salmon Arm, BC V1E 1Z6, Canada. Web Site: www.stellarseeds.com. $2.00 for catalog. Certified organic farm offering many heritage varieties along with unique cultivars. Offered 59 non-hybrid vegetable varieties in 2004 including: 6 bean, 1 beet, 2 carrot, 1 celery, 1 Chinese cabbage, 1 chicory, 1 cress, 1 cucumber, 1 endive, 1 ground cherry, 1 kale, 1 leek, 13 lettuce, 1 mustard greens, 1 onion common, 1 parsley, 2 parsnip, 4 pea, 1 spinach, 2 squash, 1 Swiss chard, 1 tomatillo, 12 tomato and 2 miscellaneous.

Ste **Steele Plant Co.,** 202 Collins St., Gleason, TN 38229. Phone: 731-648-5476. Send SASE for catalog. Specializes in sweet potato, cabbage and onion plants. Separate but identical retail and wholesale catalogs. Offered 11 sweet potato varieties in 2004.

Sto **Stokes Seeds, Inc.,** Box 548, Buffalo, NY 14240-0548. Phone: 800-396-9238. Fax: 888-834-3334. E-mail: stokes @stokeseeds.com. Web Site: www.stokeseeds.com. Free catalog. Specializes in short-season vegetables. Retail and wholesale prices are listed in the same catalog. Offered 323 non-hybrid vegetable varieties in 2004 including: 2 amaranth, 1 asparagus, 39 bean, 7 beet, 1 broccoli, 2 broccoli raab, 2 cabbage, 6 carrot, 1 cauliflower, 1 celeriac, 7 celery, 4 Chinese cabbage, 5 chicory, 1 collard, 7 corn, 2 cornsalad, 2 cress, 4 cucumber, 3 eggplant, 3 endive, 1 fava bean, 1 fennel, 11 gourd, 3 kale, 2 kohlrabi, 5 leek, 31 lettuce, 3 lima, 3 mustard greens, 1 okra, 7 onion bunching, 8 onion common, 2 onion other, 7 parsley, 2 parsnip, 16 pea, 26 pepper, 14 radish, 1 runner bean, 2 rutabaga, 1 salsify, 1 sorghum, 2 soybean, 1 spinach, 31 squash, 4 sunflower ornamental, 2 sunflower edible, 5 Swiss chard, 1 tomatillo, 13 tomato, 2 turnip, 2 watermelon, 2 miscellaneous cucurbitaceae, 1 miscellaneous leguminosae, 1 miscellaneous solanaceae and 8 miscellaneous.

Sw8 **Swan View Farm, LLC,** 345 Rocky Woods Lane, Bigfork, MT 59911-6324. Phone: 406-837-4438. E-mail: orderinfo@thepowerofgarlic.com. Web Site: www.thepower ofgarlic.com. Free catalog. Specializes in stiffneck garlic varieties. Offered 11 garlic varieties in 2004.

Sw9 **Swallowtail Garden Seeds,** 122 Calistoga Road, #178, Santa Rosa, CA 95409. E-mail: info@swallowtailgarden seeds.com. Web Site: www.swallowtailgardenseeds.com. On-line catalog. Offering over 750 varieties of vegetable, flower and herb seeds. Offered 101 non-hybrid vegetable varieties in 2004 including: 2 artichoke, 6 bean, 4 beet, 1 broccoli, 1 cardoon, 1 carrot, 3 corn, 2 cucumber, 15 lettuce, 1 melon, 3 pea, 14 pepper, 2 radish, 2 runner bean, 5 squash, 1 Swiss chard, 37 tomato and 1 watermelon.

Syn **Synergy Seeds,** Box 323, Orleans, CA 95556. Web Site: www.synergyseeds.com. On-line catalog. Specializes in experimental breeding projects and home-grown family favorites relied on year after year, all grown with sustainable methods. Does custom and contract seed growing, consultations and seed-banking workshops. Retail and wholesale. Offered 285 non-hybrid vegetable varieties in 2004 including: 7 amaranth, 23 bean, 3 beet, 1 broccoli raab, 3 carrot, 2 celery, 4 Chinese cabbage, 1 chicory, 6 corn, 1 cornsalad, 7 cowpea, 4 cress, 6 cucumber, 1 fennel, 2 ground cherry, 1 kale, 1 leek, 24 lettuce, 13 melon, 2 onion bunching, 1 onion common, 1 onion other, 2 parsley, 1 parsnip, 1 pea, 4 quinoa, 4 radish, 1 rutabaga, 6 sorghum, 34 soybean, 1 spinach, 14 squash, 4 sunflower ornamental, 5 sunflower edible, 4 tepary, 2 tomatillo, 44 tomato, 3 turnip, 11 watermelon, 1 miscellaneous cucurbitaceae, 7 miscellaneous leguminosae, 1 miscellaneous solanaceae and 21 miscellaneous.

Ta5 **Tanager Song Farm,** PO Box 2143, Toccoa, GA 30577. E-mail: tanagersongfarm@aol.com. Web Site: www. tanagersongfarm.com. Online catalog. Specializes in open-pollinated, unique tomatoes, beans, peas, peppers and lettuces with a "history". Retail only. Offered 421 non-hybrid vegetable varieties in 2004 including: 7 bean, 1 gourd, 31 lettuce, 1 lima, 12 pea, 25 pepper, 344 tomato.

Te4 **Terra Edibles,** Box 164, 535 Ashley St., Foxboro, ON K0K 2B0, Canada. Phone: 613-961-0654. Fax: 613-968-6369. E-mail: karyn@magma.ca. Web Site: www.terra edibles.ca. Free catalog. Specializes in organically grown heirlooms, edible landscaping plants and varieties that require little space or are extra-nutritious. Retail only. Offered 213 non-hybrid vegetable varieties in 2004 including: 56 bean, 3 beet, 2 broccoli, 2 carrot, 2 Chinese cabbage, 1 corn, 4 cucumber, 1 eggplant, 1 ground cherry, 1 kale, 1 leek, 13 lettuce, 1 lima, 5 melon, 1 onion bunching, 1 onion common, 2 onion other, 2 parsley, 5 pea, 4 pepper, 3 radish, 1 runner bean, 3 soybean, 10 squash, 1 Swiss chard, 1 tomatillo, 76 tomato, 1 turnip, 3 watermelon, 1 miscellaneous cucurbitaceae and 5 miscellaneous.

TE7 **Terra Organics, LLC,** PO Box 171, Maxwell, CA 95955. Free price list. The first wholesale seed company to offer a full line of multiple species of organic seeds. Wholesale only. Offered 221 non-hybrid vegetable varieties in 2004 including: 15 bean, 7 beet, 2 broccoli, 1 broccoli raab, 2 cabbage, 5 carrot, 1 cauliflower, 1 celery, 4 Chinese cabbage, 1 chicory, 3 corn, 1 cowpea, 5 cucumber, 2 eggplant, 2 endive, 1 fennel, 1 gourd, 5 kale, 1 leek, 20 lettuce, 1 lima, 5 melon, 2 mustard greens, 2 okra, 1 onion bunching, 2 onion common, 2 onion other, 3 parsley, 4 pea, 29 pepper, 6 radish, 1 spinach, 17 squash, 2 sunflower edible, 5 Swiss chard, 2 tomatillo, 45 tomato, 3 turnip, 3 watermelon and 6 miscellaneous.

Te8 **Terra Time & Tide,** 590 E. 59th St., Jacksonville, FL 32208. E-mail: LTDTerra@aol.com. Web Site: www.pepper hot.com. Free catalog. Specializes in hot pepper seeds. Retail only. Offered 71 pepper varieties in 2004.

Ter **Territorial Seed Co.,** PO Box 158, Cottage Grove, OR 97424-0061. Phone: 541-942-9547. Fax: 888-657-3131. E-mail: joshk@territorial-seed.com. Web Site: www.territorial seed.com. Free catalog. Specializes in varieties for Maritime climate (west of the Cascades). Selected for year-round, fresh-from-the-garden food, all grown without chemical pesticides and using only organic fertilizers. Retail and wholesale. Offered 408 non-hybrid vegetable varieties in 2004 including: 1 amaranth, 3 artichoke, 1 asparagus, 26 bean, 6 beet, 5 broccoli, 2 broccoli raab, 1 brussels sprout, 4 cabbage, 1 cardoon, 9 carrot, 6 cauliflower, 1 celeriac, 2 celery, 5 Chinese cabbage, 2 chicory, 1 collard, 8 corn, 1 cornsalad, 1 cowpea, 3 cress, 8 cucumber, 2 eggplant, 2 endive, 5 fava bean, 2 fennel, 24 garlic, 4 gourd, 1 ground cherry, 5 kale, 2 kohlrabi, 3 leek, 40 lettuce, 1 lima, 1 melon, 4 mustard greens, 5 onion bunching, 5 onion common, 5 onion multiplier, 3 onion other, 4 parsley, 1 parsnip, 15 pea, 24 pepper, 11 potato, 7 radish, 3 runner bean, 1 rutabaga, 1 sorghum, 5 soybean, 4 spinach, 29 squash, 10 sunflower ornamental, 2 sunflower edible, 4 Swiss chard, 1 sweet potato, 2 tomatillo, 50 tomato, 2 turnip, 4 watermelon, 5 miscellaneous leguminosae and 12 miscellaneous.

Th3 Thomas Jefferson Center for Historic Plants, Monticello, PO Box 316, Charlottesville, VA 22902. Phone: 434-984-9816. Fax: 434-984-0358. E-mail: pcornett@ monticello.org. Web Site: www.monticello.org. $2 for annual journal and catalog. Specializes in vegetables and flowers that were grown by Thomas Jefferson. Retail only. Offered 44 non-hybrid vegetable varieties in 2004 including: 2 amaranth, 1 artichoke, 3 bean, 1 beet, 1 cabbage, 1 cardoon, 1 cowpea, 1 eggplant, 1 fennel, 1 kale, 4 lettuce, 2 lima, 1 melon, 1 onion common, 1 parsley, 3 pea, 2 pepper, 2 runner bean, 1 spinach, 1 squash, 4 tomato, 1 watermelon, 2 miscellaneous cucurbitaceae, 2 miscellaneous leguminosae and 4 miscellaneous.

Tho Thompson & Morgan, Inc., PO Box 1308, Jackson, NJ 08527-0308. Phone: 800-274-7333. Web Site: www.thompson-morgan.com. Free full-color catalog. Offers rare and exotic varieties, many unobtainable elsewhere. Retail only. Offered 247 non-hybrid vegetable varieties in 2004 including: 3 amaranth, 1 artichoke, 15 bean, 9 beet, 6 broccoli, 1 broccoli raab, 2 brussels sprout, 1 cabbage, 5 carrot, 4 cauliflower, 1 celeriac, 2 celery, 1 Chinese cabbage, 3 chicory, 1 corn, 1 cornsalad, 4 cress, 2 cucumber, 2 eggplant, 2 endive, 7 fava bean, 2 fennel, 1 ground cherry, 1 Jerusalem artichoke, 2 kale, 1 kohlrabi, 7 leek, 34 lettuce, 1 lima, 2 melon, 2 mustard greens, 1 okra, 4 onion bunching, 2 onion other, 5 parsley, 3 parsnip, 21 pea, 5 pepper, 7 radish, 12 runner bean, 2 rutabaga, 1 salsify, 1 scorzonera, 1 soybean, 1 spinach, 6 squash, 10 sunflower ornamental, 2 sunflower edible, 5 Swiss chard, 1 tomatillo, 18 tomato, 2 turnip, 1 miscellaneous cucurbitaceae, 3 miscellaneous leguminosae and 7 miscellaneous.

Thy The Thyme Garden, 20546 Alsea Highway, Alsea, OR 97324. Phone: 541-487-8671. E-mail: herbs@thymegarden .com. Web Site: www.thymegarden.com. $2.00 for catalog. Specializes in organic herb seeds and plants. Offered 32 non-hybrid vegetable varieties in 2004 including: 3 amaranth, 1 cardoon, 2 cress, 1 fennel, 3 onion other, 2 parsley, 9 pepper, 1 miscellaneous cucurbitaceae and 10 miscellaneous.

To1 Tomato Growers Supply Company, PO Box 2237, Fort Myers, FL 33902. Web Site: www.tomatogrowers.com. Free catalog. Specializes in tomatoes and peppers. Offering a large selection of carefully chosen heirloom and open-pollinated seed. Retail only. Offered 363 non-hybrid vegetable varieties in 2004 including: 9 eggplant, 1 ground cherry, 87 pepper, 5 tomatillo, 261 tomato.

To3 Totally Tomatoes, 334 West Stroud Street, Randolph, WI 53956. Phone: 800-345-5977. Fax: 888-477-7333. Web Site: www.totallytomato.com. Free catalog. Offers hundreds of varieties of tomatoes and peppers, about half are open-pollinated. Retail and wholesale. Offered 193 non-hybrid vegetable varieties in 2004 including: 1 ground cherry, 57 pepper, 135 tomato.

To9 TomatoFest® Homegrown Seeds, P.O. Box W-1, Carmel, CA 93921. Web Site: www.tomatofest.com. On-line catalog. Certified organic, open-pollinated tomato seeds from Gary Ibsen with photos and information on several hundred rare tomato varieties. Offered 355 tomato varieties in 2004.

To10 Tomato Bob's Heirloom Tomatoes, Robert Price, 5764 Saucony Drive, Hilliard, OH 43026. E-mail: questions @tomatobob.com. Web Site: www.tomatobob.com. On-line seed list. Offers over 100 heirloom tomato varieties. Offered 115 non-hybrid vegetable varieties in 2004 including: 9 pepper, 106 tomato.

Tt2 T & T Seeds, Ltd., Box 1710, Winnipeg, MB R3C 3P6, Canada. $3 for catalog. Canada only. Offers seeds of early-short season hardy varieties. Retail only. Offered 120 non-hybrid vegetable varieties in 2004 including: 2 amaranth, 1 artichoke, 1 asparagus, 12 bean, 4 beet, 3 cabbage, 5 carrot, 1 cauliflower, 1 celery, 2 corn, 1 cress, 6 cucumber, 1 eggplant, 1 fava bean, 1 garlic, 1 gourd, 2 kohlrabi, 1 leek, 7 lettuce, 1 melon, 1 onion bunching, 3 onion common, 2 onion other, 3 parsley, 1 parsnip, 9 pea, 1 pepper, 13 potato, 3 radish, 1 runner bean, 1 rutabaga, 5 squash, 1 sunflower ornamental, 3 sunflower edible, 1 Swiss chard, 11 tomato, 1 watermelon, 1 miscellaneous leguminosae and 5 miscellaneous.

Tu6 Turtle Tree Seed Farm, Camphill Village, Copake, NY 12516. Phone: 518-329-3038. E-mail: turtle@ taconic.net. Free catalog. Specializes in bio-dynamically grown, open-pollinated vegetables home-grown by Turtle Tree or by other experienced bio-dynamic growers. Retail only. Offered 229 non-hybrid vegetable varieties in 2004 including: 1 amaranth, 16 bean, 4 beet, 2 broccoli, 1 broccoli raab, 7 cabbage, 7 carrot, 1 celeriac, 1 celery, 9 Chinese cabbage, 8 corn, 1 cornsalad, 1 cowpea, 2 cress, 4 cucumber, 2 eggplant, 2 fennel, 3 kale, 2 kohlrabi, 4 leek, 37 lettuce, 5 melon, 1 mustard greens, 2 onion bunching, 8 onion common, 2 onion other, 3 parsley, 3 parsnip, 8 pea, 9 pepper, 6 radish, 2 runner bean, 2 sorghum, 1 soybean, 4 spinach, 13 squash, 1 sunflower ornamental, 2 sunflower edible, 4 Swiss chard, 27 tomato, 4 watermelon and 7 miscellaneous.

Twi **Otis Twilley Seed Co.,** PO Box 4000, Hodges, SC 29653. Phone: 800-622-7333. Fax: 864-227-5108. Web Site: www.twilleyseed.com. Free catalog. Specializes in sweet corn and watermelons. Retail and wholesale. Offered 177 non-hybrid vegetable varieties in 2004 including: 2 artichoke, 18 bean, 6 beet, 1 broccoli raab, 1 cardoon, 3 carrot, 1 cauliflower, 2 celery, 1 Chinese cabbage, 3 collard, 8 corn, 1 cornsalad, 2 cowpea, 3 cress, 1 cucumber, 1 eggplant, 2 endive, 1 fennel, 4 gourd, 2 kale, 1 leek, 10 lettuce, 3 lima, 3 mustard greens, 1 okra, 2 onion bunching, 4 onion common, 2 onion other, 4 parsley, 11 pea, 16 pepper, 3 radish, 1 rutabaga, 1 sorghum, 26 squash, 4 sunflower ornamental, 2 sunflower edible, 4 Swiss chard, 1 tomatillo, 2 tomato, 2 turnip, 5 watermelon, 1 miscellaneous cucurbitaceae and 5 miscellaneous.

Und **Underwood Gardens,** 1414 Zimmerman Road, Woodstock, IL 60098. Fax: 888-382-7041. Web Site: www.grandmasgarden.com. $3 for catalog. Retail sales, with bulk seed available depending on harvest. Specializes in hard-to-find, untreated, open-pollinated and heirloom seeds, many of which are grown organically. Offered 127 non-hybrid vegetable varieties in 2004 including: 3 amaranth, 5 bean, 2 beet, 1 broccoli, 1 brussels sprout, 1 cabbage, 2 carrot, 1 cauliflower, 1 celery, 1 Chinese cabbage, 4 corn, 1 cornsalad, 3 cress, 2 cucumber, 1 eggplant, 2 garlic, 1 gourd, 1 ground cherry, 1 kale, 9 lettuce, 1 lima, 4 melon, 2 okra, 2 onion common, 2 onion multiplier, 1 parsley, 4 pea, 1 peanut, 13 pepper, 1 radish, 2 runner bean, 1 sorghum, 1 spinach, 8 squash, 1 sunflower ornamental, 1 sunflower edible, 3 Swiss chard, 1 tomatillo, 23 tomato, 3 watermelon, 1 miscellaneous cucurbitaceae, 1 miscellaneous leguminosae, 1 miscellaneous solanaceae and 6 miscellaneous.

Uni **Seed Program - University of Hawaii,** Ag. Diagnostic Service Ctr., 1910 East-West Road, Room 108, Honolulu, HI 96822. Phone: 808-956-7890. Fax: 808- 956-3894. E-mail: seed@ctahr.hawaii.edu. Web Site: www2.ctahr.hawaii.edu/seed. Free price list. Specializes in vegetables developed by University of Hawaii horticulturists that do well in tropical climates. Retail only. Offered 16 non-hybrid vegetable varieties in 2004 including: 3 bean, 1 cauliflower, 1 corn, 1 eggplant, 1 lettuce, 1 onion bunching, 1 onion common, 1 pea, 2 pepper, 1 soybean and 3 tomato.

Up2 **Upper Canada Seeds,** 8 Royal Doulton Drive, Don Mills, ON M3A 1N4, Canada. Phone: 416-447-5321. E-mail: uppercanadaseeds@rogers.com. Free catalog. Offers top quality, untreated organically grown seeds that are suitable for the climatic conditions of Eastern Canada and the Northeastern United States. All seeds are open-pollinated and many are heirlooms that have disappeared from most seed catalogues long ago. Retail only. Offered 160 non-hybrid vegetable varieties in 2004 including: 5 bean, 3 beet, 1 broccoli, 1 brussels sprout, 2 cabbage, 3 carrot, 1 corn, 3 cucumber, 1 eggplant, 1 kohlrabi, 1 leek, 5 lettuce, 3 melon, 2 onion common, 1 parsnip, 2 pea, 5 pepper, 2 radish, 1 runner bean, 1 rutabaga, 1 spinach, 6 squash, 108 tomato and 1 watermelon.

Ver **Vermont Bean Seed Co.,** 334 W. Stroud St., Randolph, WI 53956. Phone: 800-349-1071. Fax: 888-500-7333. E-mail: info@vermontbean.com. Web Site: www.vermontbean.com. Free catalog. Known for having a large selection of beans, plus gourmet specialities and rare and unusual flowers. All seeds are untreated. Retail only. Offered 266 non-hybrid vegetable varieties in 2004 including: 2 asparagus, 82 bean, 4 beet, 1 broccoli, 8 carrot, 2 cauliflower, 1 celeriac, 1 celery, 3 Chinese cabbage, 1 chicory, 7 corn, 3 cowpea, 5 cucumber, 1 endive, 1 fava bean, 6 garlic, 3 gourd, 1 Jerusalem artichoke, 2 kohlrabi, 2 leek, 10 lettuce, 10 lima, 2 melon, 1 mushroom, 1 mustard greens, 1 onion bunching, 3 onion common, 2 onion multiplier, 2 onion other, 3 parsley, 1 parsnip, 18 pea, 7 pepper, 7 potato, 5 radish, 5 runner bean, 1 rutabaga, 1 sorghum, 1 soybean, 1 spinach, 12 squash, 3 sunflower ornamental, 3 Swiss chard, 7 sweet potato, 10 tomato, 2 turnip, 2 watermelon, 1 miscellaneous cucurbitaceae, 4 miscellaneous leguminosae and 4 miscellaneous.

Ves **Vesey's Seeds Ltd.,** PO Box 9000, Calais, ME 04619-6102. Phone: 800-363-7333. Fax: 800-686-0329. E-mail: Veseys@Veseys.com. Web Site: www.veseys.com. Canadian address is Vesey's Seeds Ltd., PO Box 9000, Charlottetown, PE C1A 8K6, Canada. Free catalog. Specializes in quality vegetable and flower seeds for short growing season areas. Retail catalog, but offers many varieties to commercial customers that are not in retail catalog. Offered 160 non-hybrid vegetable varieties in 2004 including: 2 amaranth, 1 asparagus, 14 bean, 5 beet, 4 carrot, 2 cauliflower, 1 celery, 2 Chinese cabbage, 1 chicory, 2 corn, 1 cornsalad, 2 cucumber, 1 fava bean, 1 fennel, 1 gourd, 1 kale, 2 kohlrabi, 2 leek, 24 lettuce, 1 lima, 3 onion bunching, 2 onion common, 2 onion other, 2 parsley, 2 parsnip, 17 pea, 3 pepper, 6 radish, 1 runner bean, 2 rutabaga, 1 sorghum, 1 spinach, 16 squash, 10 sunflower ornamental, 2 sunflower edible, 6 Swiss chard, 1 tomatillo, 9 tomato, 1 turnip, 1 watermelon and 2 miscellaneous.

Vl3 **Vilmorin, Inc.,** 2551 N. Dragoon St., #131, Tucson, AZ 85745-1454. Wholesale only. Offered 40 non-hybrid vegetable varieties in 2004 including: 9 bean, 1 chicory, 3 cornsalad, 7 endive, 1 fava bean, 3 leek, 11 lettuce, 1 melon, 2 radish and 2 squash.

Vi4 **Victory Seed Company,** P.O. Box 192, Molalla, OR 97038. Phone: 503-829-3126. E-mail: info@VictorySeeds.com. Web Site: www.VictorySeeds.com. $2 for catalog (refunded with order). Family-owned and operated retail packet seed company selling only open-pollinated varieties. Offered 321 non-hybrid vegetable varieties in 2004 including: 1 artichoke, 1 asparagus, 15 bean, 9 beet, 3 broccoli, 1 brussels sprout, 9 cabbage, 1 cardoon, 8 carrot, 2 cauliflower, 1 celeriac, 1 celery, 2 Chinese cabbage, 1 chicory, 3 collard, 6 corn, 5 cowpea, 8 cucumber, 1 eggplant, 3 endive, 1 fava bean, 1 fennel, 5 gourd, 2 ground cherry, 3 kale, 4 kohlrabi, 2 leek, 18 lettuce, 6 lima, 9 melon, 2 mustard greens, 4 okra, 1 onion bunching, 5 onion common, 2 onion other, 5 parsley, 2 parsnip, 11 pea, 14 pepper, 9 radish, 1 runner bean, 2 rutabaga, 1 salsify, 1 sorghum, 2 spinach, 25 squash, 1 sunflower ornamental, 2 sunflower edible, 4 Swiss chard, 1 tomatillo, 74 tomato, 4 turnip, 10 watermelon, 2 miscellaneous cucurbitaceae, 1 miscellaneous solanaceae and 3 miscellaneous.

Vi5 **Virtual Seeds,** 92934 Coyote Drive, Astoria, OR 97103. Web Site: www.virtualseeds.com. On-line catalog. No orders shipped outside the U.S. Retail only. Offered 138 non-hybrid vegetable varieties in 2004 including: 1 amaranth, 1 beet, 2 broccoli, 2 cabbage, 1 cardoon, 3 carrot, 3 cauliflower, 1 celery, 3 Chinese cabbage, 5 corn, 1 cornsalad, 1 cress, 1 cucumber, 1 eggplant, 1 fennel, 18 gourd, 1 leek, 1 lettuce, 1 melon, 2 onion bunching, 1 onion common, 1 onion other, 1 parsnip, 1 pea, 11 pepper, 1 radish, 1 salsify, 4 sorghum, 30 squash, 14 sunflower ornamental, 2 tomatillo, 12 tomato, 3 miscellaneous cucurbitaceae and 6 miscellaneous.

We2 **Western Biologicals,** PO Box 283, Aldergrove, BC V4W 2T8, Canada. Phone: 604-856-3339. E-mail: western @prismnet.bc.ca. $3 for 40-page catalog which includes detailed cultivation information, mushroom kits, live cultures and spawn. Free two-page Mushroom Information flyer. Retail and wholesale prices are listed in the same catalog. Offered 19 mushroom varieties in 2004.

WE10 **Western Hybrid Seeds, Inc.,** PO Box 1169, Hamilton City, CA 95951. Phone: 530-342-3410. Strictly wholesale. Specializes in heritage tomato and pepper varieties and unusual vegetable varieties for home gardeners, as well as old European varieties. Offered 598 non-hybrid vegetable varieties in 2004 including: 1 artichoke, 2 asparagus, 9 bean, 11 beet, 6 broccoli, 2 broccoli raab, 3 brussels sprout, 14 cabbage, 1 cardoon, 14 carrot, 8 cauliflower, 1 celeriac, 4 celery, 6 Chinese cabbage, 4 collard, 4 corn, 1 cornsalad, 3 cress, 23 cucumber, 10 eggplant, 2 endive, 1 fava bean, 1 fennel, 11 gourd, 1 ground cherry, 3 kale, 4 kohlrabi, 3 leek, 61 lettuce, 33 melon, 3 mustard greens, 4 okra, 7 onion bunching, 16 onion common, 2 onion other, 5 parsley, 3 parsnip, 6 pea, 70 pepper, 15 radish, 4 runner bean, 2 rutabaga, 1 salsify, 1 sorghum, 4 spinach, 60 squash, 8 Swiss chard, 1 tomatillo, 94 tomato, 4 turnip, 24 watermelon, 3 miscellaneous cucurbitaceae, 1 miscellaneous leguminosae, 1 miscellaneous solanaceae and 12 miscellaneous.

We11 **Christopher E. Weeks Peppers,** PO Box 3207, Kill Devil Hills, NC 27948. Phone: 252-335-7525. E-mail: peppers@pinn.net. Free price list. Specializes in rare hot and sweet pepper varieties. Retail only. Offered 36 pepper varieties in 2004.

We14 **Weavers Seed of Oregon,** P.O. Box 67, Crabtree, OR 97335. Phone: 541-924-9701. Free price list. Specializes in garlic, grown in state-inspected fields. Retail and wholesale prices on the same price list. Offered 3 non-hybrid vegetable varieties in 2004 including: 1 garlic, 1 leek, 1 onion multiplier.

We19 **West Coast Seeds,** 3925 64th St., R.R. #1, Delta, BC V4K 3N2. Phone: 604-952-8820. Fax: 877-482-8828. E-mail: info@westcoastseeds.com. Web Site: www.westcoast seeds.com. Free catalog. All prices are in Canadian dollars. Specializes in seeds for organic growing. Retail and wholesale. Offered 220 non-hybrid vegetable varieties in 2004 including: 3 amaranth, 2 artichoke, 18 bean, 5 beet, 4 broccoli, 1 broccoli raab, 1 brussels sprout, 4 cabbage, 1 cardoon, 7 carrot, 3 cauliflower, 1 celeriac, 1 celery, 7 Chinese cabbage, 4 chicory, 1 collard, 3 corn, 1 cornsalad, 3 cress, 3 cucumber, 2 endive, 3 fava bean, 2 garlic, 1 gourd, 1 ground cherry, 3 kale, 1 kohlrabi, 3 leek, 15 lettuce, 1 melon, 3 mustard greens, 2 onion bunching, 2 onion common, 2 onion other, 2 parsley, 1 parsnip, 12 pea, 10 pepper, 5 radish, 3 runner bean, 1 rutabaga, 1 soybean, 1 spinach, 24 squash, 7 sunflower ornamental, 3 sunflower edible, 4 Swiss chard, 1 tomatillo, 19 tomato, 2 turnip, 1 watermelon and 9 miscellaneous.

We20 **Weed Farm Herbs,** 613 Quaker St., Lincoln, VT 05443. Free catalog. Specializes in certified organic herb seeds appropriate for northern gardens. Retail only. Offered 15 non-hybrid vegetable varieties in 2004 including: 1 fennel, 3 garlic, 2 onion other, 2 parsley and 7 miscellaneous.

Wet **Wetsel Seed Co., Inc.,** 128 West Market St., Harrisonburg, VA 22801. E-mail: hortus@rica.net. Free catalog. Separate retail and wholesale catalogs; wholesale catalog contains additional varieties. Offered 234 non-hybrid vegetable varieties in 2004 including: 1 artichoke, 1 asparagus, 30 bean, 5 beet, 2 broccoli, 1 brussels sprout, 10 cabbage, 4 carrot, 1 cauliflower, 1 celeriac, 1 celery, 3 Chinese cabbage, 4 collard, 4 corn, 4 cowpea, 1 cress, 11 cucumber, 1 eggplant, 2 endive, 2 gourd, 5 kale, 1 kohlrabi, 1 leek, 11 lettuce, 7 lima, 6 melon, 3 mustard greens, 4 okra, 1 onion bunching, 4 onion common, 2 onion other, 2 parsley, 1 parsnip, 13 pea, 10 pepper, 4 potato, 9 radish, 1 runner bean, 1 rutabaga, 1 salsify, 1 spinach, 23 squash, 4 sunflower ornamental, 1 sunflower edible, 2 Swiss chard, 10 tomato, 3 turnip, 10 watermelon, 1 miscellaneous cucurbitaceae and 3 miscellaneous.

Wi1 **Wildlife Nurseries, Inc.,** PO Box 2724, Oshkosh, WI 54903-2724. Phone: 920-231-3780. Free price list; $3 for informative color catalog. Specializes in waterfowl and wildlife food plants. Retail and wholesale. Offered 5 non-hybrid vegetable varieties in 2004 including: 1 sunflower edible and 4 miscellaneous.

Wi2 **Willhite Seed Inc.,** PO Box 23, Poolville, TX 76487. Phone: 800-828-1840. Fax: 817-599-5843. Web Site: www.willhiteseed.com. Free color catalog. Family-owned seed company providing quality seed to gardeners and commercial growers. Offering over 450 varieties of vegetable, flower and herb seed with an extensive expertise in watermelon crop production. Separate but identical retail and wholesale catalogs. Offered 267 non-hybrid vegetable varieties in 2004 including: 1 asparagus, 23 bean, 3 beet, 1 broccoli, 1 broccoli raab, 1 brussels sprout, 4 cabbage, 6 carrot, 1 cauliflower, 1 celery, 5 Chinese cabbage, 2 collard, 8 corn, 1 cornsalad, 18 cowpea, 10 cucumber, 1 eggplant, 1 endive, 5 gourd, 3 kale, 1 kohlrabi, 1 leek, 12 lettuce, 7 lima, 19 melon, 2 mustard greens, 6 okra, 1 onion bunching, 5 onion common, 1 onion other, 1 parsley, 7 pea, 13 pepper, 7 radish, 1 rutabaga, 1 soybean, 1 spinach, 26 squash, 2 sunflower ornamental, 2 sunflower edible, 3 Swiss chard, 1 tomatillo, 12 tomato, 3 turnip, 32 watermelon, 1 miscellaneous cucurbitaceae, 1 miscellaneous leguminosae and 2 miscellaneous.

Wi15 **Wildseed Farms,** 425 Wildflower Hills, PO Box 3000, Fredericksburg, TX 78624-3000. Phone: 800-848-0078. Fax: 830-990-8090. Web Site: www.wildseedfarms.com. Free award-winning catalog that contains excellent how-to-plant information. Offers over 80 species of wildflowers, garden variety flowers and culinary herbs. Offered 15 non-hybrid vegetable varieties in 2004 including: 1 fennel, 1 onion other, 2 parsley, 8 sunflower ornamental, 2 sunflower edible and 1 miscellaneous.

Wl23 **Wild West Seed Inc.,** P.O. Box 327, Albany, OR 97321. Phone: 541-928-7100. Fax: 541-928-7101. Web Site: www.wildwestseed.com. Strictly wholesale. Specializes in vegetable, herb, flower and wildflower seeds. Offered 211 non-hybrid vegetable varieties in 2004 including: 1 amaranth, 1 asparagus, 6 beet, 2 broccoli, 1 brussels sprout, 8 cabbage, 5 carrot, 1 cauliflower, 1 celery, 5 Chinese cabbage, 4 collard, 3 cress, 11 cucumber, 1 eggplant, 2 endive, 1 fennel, 3 gourd, 4 kale, 2 kohlrabi, 1 leek, 17 lettuce, 7 melon, 4 mustard greens, 4 okra, 2 onion bunching, 7 onion common, 2 onion other, 5 parsley, 3 parsnip, 13 pepper, 10 radish, 1 rutabaga, 1 salsify, 1 spinach, 31 squash, 3 sunflower ornamental, 1 sunflower edible, 4 Swiss chard, 1 tomatillo, 12 tomato, 4 turnip, 11 watermelon, 1 miscellaneous cucurbitaceae and 3 miscellaneous.

Wo6 **Wood Prairie Farm,** Jim and Megan Gerritsen, 49 Kinney Road, Bridgewater, ME 04735. Phone: 800-631-8027. Fax: 800-300-6494. E-mail: orders@woodprairie.com. Web Site: www.woodprairie.com. Free catalog. Specializes in seed potatoes that are certified by both the State of Maine and the Maine Organic Farmer and Gardener Association. Retail and wholesale. Offered 17 non-hybrid vegetable varieties in 2004 including: 1 onion multiplier and 16 potato.

Wo8 **World Wide Exotic Seed Company,** PO Box 1488, Friendswood, TX 77549-1488. Web Site: www.seedman.com. On-line catalog. Extensive collection of rare, exotic and unusual seeds from around the world. Offered 94 non-hybrid vegetable varieties in 2004 including: 1 asparagus, 1 beet, 3 cabbage, 3 carrot, 2 cauliflower, 1 celery, 1 Chinese cabbage, 10 corn, 1 cress, 2 cucumber, 3 eggplant, 1 gourd, 1 kale, 1 leek, 2 melon, 1 onion bunching, 2 onion common, 1 parsnip, 1 pea, 10 pepper, 2 radish, 1 salsify, 4 sorghum, 1 soybean, 7 squash, 7 sunflower ornamental, 3 sunflower edible, 1 tomatillo, 17 tomato, 2 miscellaneous cucurbitaceae and 1 miscellaneous.

Yu2 **Yuko's Open-Pollinated Seeds,** Yuko Horiuchi, 405 Lake Avenue East, Carleton Place, ON K7C 3R9, Canada. Web Site: www.comnet.ca/~yuko/seeds. On-line catalog. Organically grown vegetable seeds. Offered 45 non-hybrid vegetable varieties in 2004 including: 1 amaranth, 5 bean, 4 Chinese cabbage, 1 cucumber, 1 eggplant, 1 Jerusalem artichoke, 2 lettuce, 4 melon, 2 mustard greens, 1 onion multiplier, 1 onion other, 1 pea, 2 pepper, 1 squash, 16 tomato and 2 miscellaneous leguminosae.

Taxonomy
Based on *Hortus Third (1976 Revision)*, compiled by Liberty Hyde Bailey

Amaranth	*Amaranthus spp.*
Artichoke, Globe	*Cynara scolymus*
Asparagus	*Asparagus officinalis*
Bean / Bush / Dry	*Phaseolus vulgaris*
Bean / Pole / Dry	*Phaseolus vulgaris*
Bean / Bush / Snap	*Phaseolus vulgaris*
Bean / Pole / Snap	*Phaseolus vulgaris*
Bean / Bush / Wax	*Phaseolus vulgaris*
Bean / Pole / Wax	*Phaseolus vulgaris*
Beet / Garden	*Beta vulgaris (Crassa Group)*
Beet / Sugar and Mangel	*Beta vulgaris (Crassa Group)*
Broccoli	
Heading	*Brassica oleracea (Botrytis Group)*
Sprouting	*Brassica oleracea (Italica Group)*
Broccoli Raab	*Brassica rapa (Ruvo Group)*
Brussels Sprouts	*Brassica oleracea (Gemmifera Group)*
Cabbage	*Brassica oleracea (Capitata Group)*
Cabbage / Red	*Brassica oleracea (Capitata Group)*
Cabbage / Savoy	*Brassica oleracea (Capitata Group)*
Cardoon	*Cynara cardunculus*
Carrot	*Daucus carota var. sativus*
Cauliflower	*Brassica oleracea (Botrytis Group)*
Celeriac	*Apium graveolens var. rapaceum*
Celery	*Apium graveolens var. dulce*
Chicory	*Cicorium intybus*
Chinese Cabbage	
Non-Heading Types	*Brassica rapa (Chinensis Group)*
Heading & Semi-Heading Types	*Brassica rapa (Pekinensis Group)*
Collard	*Brassica oleracea (Acephala Group)*
Corn / Dent	*Zea mays*
Corn / Flint	*Zea mays*
Corn / Flour	*Zea mays*
Corn / Pop	*Zea mays*
Corn / Sweet	*Zea mays*
Corn / Other Species	*Zea spp.*
Cornsalad	
Common (Smooth Leaved)	*Valerianella locusta*
Italian (Hairy Leaved)	*Valerianella eriocarpa*
Cowpea	*Vigna unguiculata subsp. unguiculata*
Cress	
Peppergrass	*Lepidium sativum*
Upland Cress	*Barbarea verna*
Watercress	*Nasturtium officinale*
Cucumber	*Cucumis sativus*
Eggplant	*Solanum melongena var. esculentum*
Endive (Escarole)	*Cicorium endiva*
Fava Bean	*Vicia faba*
Florence Fennel	*Foeniculum vulgare var. azoricum*
Garlic	*Allium sativum*

Gourd
- White Flowered Gourds .... *Lagenaria siceraria*
- Wax Gourds .... *Benincasa hispida*

Ground Cherry .... *Physalis spp.*
Jerusalem Artichoke .... *Helianthus tuberosus*
Kale
- Common .... *Brassica oleracea (Acephala Group)*
- Chinese .... *Brassica oleracea (Alboglabra Group)*
- Siberian .... *Brassica napus (Pabularia Group)*

Kohlrabi .... *Brassica oleracea (Gongylodes Group)*
Leek .... *Allium ampeloprasum (Porrum Group)*
Lettuce / Head .... *Lactuca sativa*
Lettuce / Leaf .... *Lactuca sativa*
Lettuce / Romaine .... *Lactuca sativa*
Lettuce / Unknown .... *Lactuca sativa*
Lima Bean / Bush .... *Phaseolus lunatus*
Lima Bean / Pole .... *Phaseolus lunatus*
Melon / Muskmelon
- Muskmelon .... *Cucumis melo (Reticulatus Group)*
  (netted and slips from vine)
- European Cantaloupe .... *Cucumis melo (Cantalupensis Group)*
  (no net and does not slip)

Melon / Honeydew and Casaba .... *Cucumis melo (Inodorus Group)*
Melon / Other
- Oriental Pickling Melon .... *Cucumis melo (Conomon Group)*
- Vine Peach .... *Cucumis melo (Chito Group)*
- Vine Pomegranate .... *Cucumis melo (Dudaim Group)*
- Serpent (Armenian) Cucumber .... *Cucumis melo (Flexuosus Group)*

Mustard Greens .... *Brassica juncea*
Okra .... *Abelmoschus esculentus*
Onion / Bunching .... *Allium fistulosum*
Onion / Common / Red-Purple .... *Allium cepa (Cepa Group)*
Onion / Common / White .... *Allium cepa (Cepa Group)*
Onion / Common / Yellow-Brown .... *Allium cepa (Cepa Group)*
Onion / Multiplier / Root .... *Allium cepa (Aggregatum Group)*
Onion / Multiplier / Topset .... *Allium cepa (Proliferum Group)*
Onion / Other Species .... *Allium spp.*
Parsley .... *Petroselinum crispum*
Parsnip .... *Pastinaca sativa*
Pea / Garden .... *Pisum sativum var. sativum*
Pea / Edible Pod .... *Pisum sativum var. macrocarpon*
Pea / Soup .... *Pisum sativum var. sativum*
Peanut .... *Arachis hypogaea*
Pepper / Hot .... *Capsicum annuum*
Pepper / Sweet / Bell .... *Capsicum annuum*
Pepper / Sweet / Non-Bell .... *Capsicum annuum*
Pepper / Other Species .... *Capsicum spp.*
Potato .... *Solanum tuberosum*
Quinoa .... *Chenopodium quinoa*
Radish .... *Raphanus sativus*
Runner Bean .... *Phaseolus coccineus*
Rutabaga (Swede Turnip) .... *Brassica napus (Napobrassica Group)*
Salsify .... *Tragopogon porrifolius*

Scozonera or Black Salsify	*Scorzonera hispanica*
Sorghum	*Sorghum bicolo*
Soybean	*Glycine max*
Spinach	*Spinacia oleracea*
Squash / Maxima	*Cucurbita maxima*
Squash / Mixta	*Cucurbita mixta*
Squash / Moschata	*Cucurbita moschata*
Squash / Pepo	*Cucurbita pepo*
Squash / Other Species	*Cucurbita spp.*
Sunflower	*Helianthus annuus*
Sweet Potato	*Ipomoea batatas*
Swiss Chard	*Beta vulgaris (Cicla Group)*
Tepary Bean	*Phaseolus acutifolius (Latifolius Group)*
Tomatillo	*Physalis ixocarpa*
Tomato / Orange-Yellow	*Lycopersicon lycopersicum*
Tomato / Pink-Purple	*Lycopersicon lycopersicum*
Tomato / Red	*Lycopersicon lycopersicum*
Tomato / Other Colors	*Lycopersicon lycopersicum*
Tomato / Other Species	*Lycopersicon spp.*
Turnip	*Brassica rapa (Rapifera Group)*
Watermelon	*Citrullus lanatus*

Amaranth
Amaranthus spp.

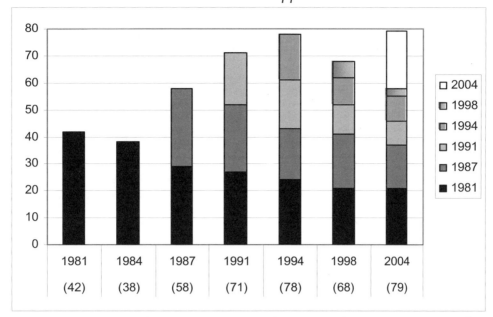

Number of varieties listed in 1981 and still available in 2004: 42 to 21 (50%).
Overall change in number of varieties from 1981 to 2004: 42 to 79 (188%).
Number of 2004 varieties available from 1 or 2 sources: 49 out of 79 (62%).

A. caudatus - (Love Lies Bleeding, Caudatus Red) - 60 days - Ropes of red blooms trail from main stems, 2-3 ft. plants, extensively cultivated in India for its nutritious seeds, leaves are boiled and eaten like spinach, pre-1700. *Source History: 1 in 1981; 1 in 1984; 7 in 1987; 13 in 1991; 15 in 1994; 19 in 1998; 28 in 2004.* **Sources: Ba8, Co32, Com, GLO, Gr28, He8, Ho13, Hud, Jo1, Na6, Ni1, Ont, Pe2, Pin, Pr3, Ri12, Se7, Shu, So9, Sto, Ter, Th3, Tho, Thy, Tt2, Und, Ves, We19.**

A. viridus - Bush form plant, from India, rare. *Source History: 1 in 2004.* **Sources: Syn.**

African Spinach - (Sahel) - 45 days - High quality of leaves is retained with frequent pickings, heavy producer. *Source History: 1 in 2004.* **Sources: Sa9.**

Alegria - 110-115 days - *A. cruentus*, produces blond seed typically used for a traditional confection called alegria in Central Mexico which is made with popped seed and honey. *Source History: 1 in 1991; 2 in 1994; 1 in 1998; 3 in 2004.* **Sources: Ho13, Na2, So12.**

Amaranth - (Hinn Choy) - 90 days - Stoutly branched 3-8 ft. annual, high protein seeds ground or popped or sprouted or cooked whole, needs long growing season, stands hot weather well. *Source History: 6 in 1981; 5 in 1984; 3 in 1987; 1 in 1991; 2 in 1994; 1 in 2004.* **Sources: So12.**

Bayam - 30 days - Rare southeast Asian cultivar, resembles and cooks like spinach, continues producing even in extreme hot weather, tender, tasty, easy to grow, sow late spring, Zone 8. *Source History: 1 in 1987; 2 in 2004.* **Sources: Ev2, Syn.**

Black Leaved - Dark purple-maroon plant grows to 6 ft., white seed. *Source History: 1 in 2004.* **Sources: Ho13.**

Burgundy - 90-100 days - *A. hypochondriacus*, 5-6 ft. plant, smaller tassels, deep-red seedheads, white seeds, resembles a giant Love-Lies-Bleeding, harvest young leaves for salads and grain for baking or hot cereal. *Source History: 3 in 1987; 6 in 1991; 5 in 1994; 8 in 1998; 8 in 2004.* **Sources: Ho13, Orn, Pe2, Pr3, Se7, Sh12, So12, Tu6.**

Calilu - Vigorous plant grows to 7 ft., does not require long days for good growth. *Source History: 1 in 1994; 1 in 1998; 1 in 2004.* **Sources: Ech.**

Dreadicus - (Mercado Dreadicus, San Martin) - *A. hypochondriacus*, a rare variety from Mexico, 8-10 ft. plants, thick matted seedheads, golden seeds, productive popping type. *Source History: 1 in 1987; 2 in 1991; 2 in 1994; 3 in 1998; 2 in 2004.* **Sources: Ho13, Syn.**

Early Splendor - *A. tricolor*, early maturing version of Molten Fire with large brilliant crimson and purple and bronze leaves, edible as greens, 3.5 ft. tall plants. *Source History: 1 in 1981; 1 in 1984; 15 in 1987; 16 in 1991; 11 in 1994; 10 in 1998; 8 in 2004.* **Sources: CA25, Cr1, Ge2, GLO, GRI, HPS, Hud, Tt2.**

Elephant Head - (Elephant's Trunk) - 65-80 days - *A. gangeticus*, brought to the U.S. in the 1880s from Germany, so named because the bloom ends up at the bottom of the plant and turns into a large structure with a long protuberance that looks like an elephant's trunk. *Source History: 2 in 1991; 5 in 1994; 8 in 1998; 9 in 2004.* **Sources: Co32, Fe5, Ho13, Hud, Pe1, Pe2, Pe6, Se7, Sh12.**

Golden - Has a few side branches, yellow-brown seed heads. *Source History: 1 in 1991; 1 in 1994; 1 in 2004.* **Sources: Pr3.**

Golden Giant - 98-110 days - *A. cruentus*, late season

grain amaranth, med-tall 6 ft. plants, edible green leaves with golden-brown veins and stalks and flowers, large seeds are mixed light and dark golden, can produce over 1 lb. of seed per plant. *Source History: 1 in 1981; 1 in 1984; 3 in 1987; 5 in 1991; 6 in 1994; 8 in 1998; 3 in 2004.* **Sources: Bou, Ho13, Se7.**

Golden Grain - 100-150 days - *A. hypochondriacus*, earliest maturing strain, yellow seedheads, annual, 6 ft. tall, still grown for grain in Southwest and Mexico, ground into flour or sprouted or popped, nutritious, harvest any time after mature, even after frost. *Source History: 1 in 1981; 1 in 1984; 2 in 1987; 3 in 1991; 3 in 1994; 3 in 1998; 5 in 2004.* **Sources: Pla, Red, Ri12, So1, So9.**

Grain Amaranth - (White Grain Amaranth, White-Seeded) - *A. hypochondriacus*, 4-6 ft. tall, white or tan nutrient-rich seeds are ground for flour or popped or sprouted or added to breads, good edible leaves, may have to hill up as seedheads get heavy. *Source History: 4 in 1981; 3 in 1984; 2 in 1987; 4 in 1991; 2 in 1994; 3 in 1998; 4 in 2004.* **Sources: Jo1, Kus, Shu, Thy.**

Greek - 55-70 days - *A. gangeticus*, 6-7 ft. plant, light-green leaves used as a steamed vegetable in Greece, gold multi-headed flower plumes yield an abundance of dark-purple edible seeds, rare. *Source History: 1 in 1994; 1 in 2004.* **Sources: Se7.**

Green Leaf - (Green Leaved) - 40-50 days - Oval to heart shaped, yellow-green leaves are large tender and delicious, excellent for cooking and salads, very easy to grow, adaptive to a wide range of growing conditions, known as Calaloo in the Caribbean. *Source History: 1 in 1987; 1 in 1991; 3 in 1994; 3 in 1998; 3 in 2004.* **Sources: CO30, Eo2, Ev2.**

Green Thumb - 80-100 days - *A. hypochondriacus*, dwarf upright plant, 6-8 in. high, 8-10 in. spread, produces one major vivid-green flower spike, long-lasting, garden performance rated excellent by H.G. German. *Source History: 2 in 1987; 4 in 1991; 5 in 1994; 4 in 1998; 3 in 2004.* **Sources: Ge2, GLO, Pin.**

Guarijio Indian Grain Amaranth - *A. hypochondriacus x A. hybridus*, white seeded grain used for tamales, pinole or popping, from the Rio Mayo in Sonora, Mexico, for planting during the summer rains. *Source History: 2 in 1981; 2 in 1984; 1 in 1987; 1 in 1991; 1 in 1994; 1 in 1998; 1 in 2004.* **Sources: Na2.**

Hartman's Giant - *A. gangeticus*, 6-8 ft. purple plants, purple seeds, used for flour, cereal and cover crops. *Source History: 1 in 1987; 1 in 1991; 1 in 1994; 1 in 1998; 1 in 2004.* **Sources: Pe1.**

Himalayan Hands - 105-110 days - *A. cruentus*, 5-6 ft. tall, attractive, lime-green, long-fingered seedheads, yellow seeds, originally from 6000' Kulu Valley of Himachal Pradesh in India. *Source History: 1 in 2004.* **Sources: So12.**

Hinn Choy - (Tampala, Hsien Shu, Giensok, Edible Amaranth Spinach, Chinese Spinach) - 40-60 days - *A. tricolor*, grows spring until fall, 6-8 ft. tall, small green paddle-shaped leaves, fast maturing, unique taste, heat res., tender. *Source History: 9 in 1981; 9 in 1984; 7 in 1987; 7 in 1991; 5 in 1994; 2 in 1998; 4 in 2004.* **Sources: CO30, Ni1, Ri2, Vi5.**

Hopi Red Dye - (Komo) - 75 days - *A. cruentus x A. powelli*, 5-6 ft. tall, entire plant is deep reddish-purple, used by Hopis as a ceremonial food dye, in Hopi land this readily crosses with wild *A. powelli*, edible black seeds, young plants can be eaten as greens, from high cool desert, spring planting. *Source History: 2 in 1981; 2 in 1984; 3 in 1987; 7 in 1991; 8 in 1994; 6 in 1998; 5 in 2004.* **Sources: Fe5, Pin, Se7, Thy, Und.**

Illumination - A Japanese innovation, earlier and more uniform than *Tricolor splendens*, heavy stalks of pendant foliage form a wide 3-5 ft. column the upper third of which is luminescent crimson topped with gold, lower leaves are green and chocolate. *Source History: 6 in 1987; 4 in 1991; 5 in 1994; 2 in 1998; 3 in 2004.* **Sources: Bu2, Ge2, Par.**

Inchoi - Tender purple leaves harvested when young, needs heat. *Source History: 1 in 2004.* **Sources: Yu2.**

Intense Purple - 120-130 days - *A. hypochondriacus*, tall version of Burgundy, stocky 5-8 ft. plant, dark-purple with white seeds, resists drought. *Source History: 2 in 1987; 1 in 1991; 1 in 1994; 1 in 1998; 1 in 2004.* **Sources: Se17.**

Joseph's Coat - (Summer Poinsettia, Chinese Spinach) - *A. gangeticus*, 36 in. plant, used primarily for greens, tricolored. *Source History: 1 in 1981; 1 in 1984; 7 in 1987; 6 in 1991; 6 in 1994; 7 in 1998; 8 in 2004.* **Sources: Bu2, CA25, CO30, Ers, Hud, LO8, Tho, Und.**

K432 - 95 days - Shorter, high yielding var. from Rodale Research, 3-5 ft. height depending on fertility and moisture availability, green plants, pale-pink seed heads, light-tan seeds, good quality grain for cooking and baking. *Source History: 2 in 1991; 4 in 1994; 4 in 1998; 1 in 2004.* **Sources: Pr3.**

Kahlalu Leaf - 40-50 days - *A. gangeticus*, 6-8 ft. tall, edible leaves used like spinach, deep green leaf color, purple veins, adapts to most locations, from Jamaica. *Source History: 1 in 1994; 1 in 1998; 2 in 2004.* **Sources: Ho13, Ri12.**

Kinnauri Dhankar - 90-100 days - *A. cruentus*, old landrace from the Himalayas, burgundy and pale green leaves on 8 ft. plant, white seeds. *Source History: 2 in 1998; 3 in 2004.* **Sources: In8, Se17, So12.**

Lotus Purple - *A. gangeticus*, 6-8 ft. thick-topped plants, purple seeds, used for grain, vegetable or salad. *Source History: 1 in 1987; 1 in 1991; 3 in 1994; 1 in 1998; 1 in 2004.* **Sources: Ho13.**

Love-Lies-Bleeding, Green - *A. caudatus*, same as red but with yellow-green plumes, 3 ft. tall. *Source History: 1 in 1998; 2 in 2004.* **Sources: GLO, Ni1.**

Manali - Grows 5-6 ft. tall, lime-green seed heads with an abundance of light yellow seed, from Kulu Valley in Himalayas, India. *Source History: 1 in 2004.* **Sources: So12.**

Manna - (Manna de Montana) - 98-110 days - *A. hypochondriacus*, grain amaranth, 4-6 ft. tall, orange-gold seed similar to sesame, also ground into flour or cooked grain and used for popping, rare. *Source History: 1 in 1981; 1 in 1984; 2 in 1987; 2 in 1991; 2 in 1994; 2 in 1998; 3 in 2004.* **Sources: Ri2, Se7, Se17.**

Mano de Obispo - (Cockscomb) - *C. cristata*, vivid

magenta flowers, edible seed. *Source History: 1 in 2004.* **Sources: Na2.**

Mayo Indian Grain Amaranth - 90 days - *A. cruentus*, leaves used as quelites (greens), black seeds are used for esquite (parched), pinole and atole, from Sonora, Mexico, for planting during summer rains. *Source History: 1 in 1981; 1 in 1984; 1 in 1987; 1 in 1991; 1 in 1994; 2 in 1998; 4 in 2004.* **Sources: Ga21, Na2, Sa9, So1.**

Merah - (Coleus Leaf Merah) - 65-80 days - *A. tricolor*, round, brilliant magenta leaves, delicious in salads. *Source History: 3 in 1991; 4 in 1994; 3 in 1998; 3 in 2004.* **Sources: Ech, Ga1, Red.**

Mercado - (Mercado Dreadicus) - 125 days - *A. hypochondriacus*, plants 7-9 ft., very high protein grain from southern Mexico, can be cooked whole, milled into flour, popped like popcorn or added like seeds to salads and baked goods, likes sandy somewhat fertile soil, drought resistant. *Source History: 2 in 1991; 3 in 1994; 3 in 1998; 3 in 2004.* **Sources: Se7, Se17, So12.**

Mexican Grain - *A. cruentus*, green plants and flowers, blond seed, originally from a gardener in Hobbs, NM. *Source History: 1 in 2004.* **Sources: Na2.**

Molten Fire - (Summer Poinsettia) - *A. gangeticus*, used primarily for greens, large scarlet and green leaves, dark-red seedhead, 4 ft. tall plants. *Source History: 2 in 1981; 2 in 1984; 12 in 1987; 9 in 1991; 6 in 1994; 3 in 1998; 1 in 2004.* **Sources: Ear.**

Mora - Tall plant, red leaves and plumes, black seeds. *Source History: 1 in 2004.* **Sources: Sa5.**

Multicolor - *A. cruentus*, seedheads vary in color from red to yellow to green, slightly later maturity than *A. hypochondriacus*, yields are usually greater. *Source History: 1 in 1987; 1 in 1991; 3 in 1994; 4 in 1998; 1 in 2004.* **Sources: Ho13.**

Nepalese - 100-110 days - *A. cruentus*, fairly late-maturing red-headed strain, pale yellow-white seeds, 6-7 ft. tall, for higher elevations, high yielding. *Source History: 1 in 1987; 1 in 1991; 3 in 1994; 1 in 1998; 3 in 2004.* **Sources: Ho13, So12, Syn.**

New Mexico - *A. hypochondriacus*, beautiful pink and white inflorescences yield golden seeds, from a dooryard garden near Rinconada. *Source History: 1 in 1991; 1 in 1994; 1 in 2004.* **Sources: Na2.**

Opopeo - 100 days - *A. cruentus*, vigorous plants grow over 7 ft., good for early season greens, white seed is tasty popped or ground for flour, huge maroon-red central plume is a nice addition to dried arrangements, from Opopeo, Mexico. *Source History: 1 in 1994; 3 in 1998; 5 in 2004.* **Sources: Ba8, Ga1, Ho13, Jo1, Mi12.**

Orange Giant - 100 days - *A. cruentus*, beautiful burnt orange seed heads on 7-8 ft. plants, white seeds, high protein content. *Source History: 3 in 2004.* **Sources: In8, Sa9, Se17.**

Oscar Blanco - *A. cruentus*, grain amaranth with 10-12 ft. tall plants, dull pink flower heads up to 24 in. long, beige-white seeds, from Bolivia. *Source History: 2 in 2004.* **Sources: In8, Se17.**

Paiute - *A. cruentus*, edible leaves and seeds, from a garden on the Kaibab Southern Paiute Reservation in Utah. *Source History: 1 in 1991; 1 in 1994; 1 in 2004.* **Sources: Na2.**

Pendant - (Love-Lies-Bleeding, Pendant Feather) - Plants 4-5 ft. tall, pink racemes, pink seed. *Source History: 1 in 1981; 1 in 1984; 4 in 1987; 3 in 1991; 3 in 1994; 2 in 1998; 3 in 2004.* **Sources: Ear, Ho13, Pe1.**

Perfecta - *A. tricolor*, to 36 in., branching plant, red, yellow and green colors. *Source History: 3 in 2004.* **Sources: CA25, Ech, HPS.**

Plainsman - Red flowers, excellent grain producer, developed by Rodale Research Center. *Source History: 1 in 2004.* **Sources: Ba8.**

Polish - Plant resembles Burgundy but with deep-purple seeds, slightly better flavored salad leaf, originally from Poland. *Source History: 1 in 1994; 1 in 1998; 2 in 2004.* **Sources: Ho13, Hud.**

Popping - *A. cruentus*, 6 ft. plant, high-yielding multicolor seedheads, may be popped in a frying pan or wok, rare old world variety. *Source History: 1 in 1987; 2 in 1991; 2 in 1994; 4 in 1998; 4 in 2004.* **Sources: Ho13, Pr3, Se17, Syn.**

Purple - (Prince's Feather) - 110 days - *A. cruentus*, ornamental grain amaranth, annual, 6 ft. tall, red and green seed heads, green to red variegated foliage, harvest any time in fall once mature. *Source History: 1 in 1981; 1 in 1984; 1 in 1987; 3 in 1991; 4 in 1994; 2 in 1998; 2 in 2004.* **Sources: Pla, Sa5.**

Puteh - (Besar, Putah) - 80 days - Most tender variety with light green spade-shaped leaves, 5-6 in. long by 5 in. wide on 12-18 in. tall plants, mild flavor, good for salads or cooked like spinach, from Malaysia. *Source History: 2 in 1991; 1 in 1994; 1 in 1998; 1 in 2004.* **Sources: Red.**

Pygmy Torch - 80-100 days - *A. hypochondriacus*, grain variety, 4 ft. plants, deep-maroon erect spikes, black seeds. *Source History: 2 in 1987; 5 in 1991; 6 in 1994; 7 in 1998; 3 in 2004.* **Sources: Hud, Pin, Pr3.**

Rama's Finger - *A. cruentus*, Garwhali cultivar from northern India, pale yellow-green columnar seedheads with yellow seeds, 6-7 ft. tall. *Source History: 1 in 2004.* **Sources: So12.**

Red - Whole plant is very dark-red, broad leaves, very decorative, grows to 3 ft. high, some of the red color bleeds out when cooked, used as a food dye, India. *Source History: 1 in 1981; 1 in 1984; 1 in 1998; 1 in 2004.* **Sources: Sa9.**

Red Amaranth R158 - 95 days - Re-selected R158, most uniform and refined grain amaranth available, mostly white seed, 6-7 ft. plants, red stalks and leaves and seedheads, matures in the North. *Source History: 1 in 1981; 1 in 1984; 1 in 1987; 2 in 1991; 1 in 1998; 1 in 2004.* **Sources: Pr3.**

Red Garnet - 50 days - Spoon shaped, oval leaves, medium green with red center and veins. *Source History: 3 in 2004.* **Sources: ME9, Orn, WI23.**

Red Leaf - (Red Leaved, Vegetable/Leaf) - 40-50 days - Leaf variety popular in southern Asia, Africa and the West Indies, 12-18 in. multi-stalked plant, oval-heart shaped green leaves overlaid with burgundy red (coleus-like), heat tol., plant thinly in warm soil (70 deg. F/21 deg. C), a tasty colorful salad and cooking green. *Source History: 2 in 1987; 4 in 1991; 2 in 1994; 3 in 1998; 11 in 2004.* **Sources: Bo17, Bou, Dam, Eo2, Ev2, Jo1, Kit, Lej, Roh, Se26, We19.**

Red Leaf Grain - Red foliage and flower heads. *Source History: 1 in 2004.* **Sources: Sa9.**

Red Stripe Leaf - 28 days - *A. gangeticus*, edible amaranth, leaves with some red color, very easy to grow, sow seeds in spring and summer, harvest when plants are young, beautiful. *Source History: 2 in 1981; 2 in 1984; 2 in 1987; 3 in 1991; 3 in 1994; 4 in 1998; 2 in 2004.* **Sources: Ho13, Hud.**

Rio San Lorenzo - *A. hypochondriacus*, blond seed used as a grain, from the southernmost part of Durango, Mexico. *Source History: 1 in 1991; 1 in 1994; 1 in 1998; 1 in 2004.* **Sources: Na2.**

Ruby Red - *A. gangeticus*, 8 ft. plants, produces up to 1 lb. of purple seed per plant, Earthstar Botanical strain. *Source History: 1 in 1994; 1 in 1998; 1 in 2004.* **Sources: Syn.**

Southern Tepehuan Baute - *A. cruentus*, leaves are used for greens, blond seed, from Durango. *Source History: 1 in 1998; 1 in 2004.* **Sources: Na2.**

Tarahumara Okite - Grows 4-6 ft. tall, brilliant red stems and flowers, seeds and young leaves edible, collected from a ranch above Batopilas, an old silver mining town located along the Rio Batopilas at the bottom of Barranca del Cobre. *Source History: 1 in 2004.* **Sources: Na2.**

Tender Leaf - Light green, tender stems and leaves, fast growing, Taiwan. *Source History: 1 in 2004.* **Sources: Ev2.**

Tiger Eye - 50 days - Name refers to red splotch in middle of broad green leaf, very branchy 3 ft. plant, very productive in good soil, makes many shoots for a number of cuttings, Taiwan. *Source History: 1 in 1981; 1 in 1984; 1 in 2004.* **Sources: Sa9.**

Tiger Leaf - *A. mangostana*, herb-like vegetable, cooked like spinach, green and red striped, wide leaves, very small seeds, warm season crop. *Source History: 2 in 1994; 2 in 1998; 2 in 2004.* **Sources: Ban, Ech.**

Tricolor - (Joseph's Coat) - *A. tricolor*, leaves are splashed with red and yellow and green, 3 ft. tall plants, leaves are highly regarded as a spinach in India and China, stem is sliced and eaten in salads, tops served like asparagus. *Source History: 3 in 1981; 2 in 1984; 9 in 1987; 11 in 1991; 11 in 1994; 8 in 1998; 11 in 2004.* **Sources: CA25, Com, Fe5, GLO, GRI, Mey, Mi12, Sey, Sto, Th3, Ves.**

Vietnamese Red - 50 days - Rare variety from Vietnam, 6 ft. tall plant with beautiful red flower head, flavorful red-green leaves, black seeds, used in Oriental dishes. *Source History: 1 in 2004.* **Sources: Ba8.**

Viridis - Long graceful tassels looking very like millet except for the rich electric green coloring. *Source History: 1 in 1987; 1 in 1991; 1 in 1994; 1 in 1998; 1 in 2004.* **Sources: Tho.**

Vleta - *A. gangeticus*, traditional green and grain producer, almost vanished, Greece. *Source History: 1 in 1994; 1 in 1998; 2 in 2004.* **Sources: Se17, Syn.**

Warihio - 105 days - *A. hypochondriacus*, grows 6-8 ft. tall, red foliage, burgundy heads with dark purple seeds, native staple of Sonora area, Mexican ceremonial, rare. *Source History: 3 in 2004.* **Sources: Mi12, Ri12, Se7.**

White - 97 days - *A. cruentus*, chosen for heavier yielding seedheads and greater uniformity in height. *Source History: 1 in 1994; 1 in 1998; 1 in 2004.* **Sources: So12.**

White Leaf - 50-55 days - Large, tender, light green leaves and stems with high iron and calcium content, tolerates hot, dry and moist areas, harvest young leaves for salads, leaves and stems used in stir fry. *Source History: 2 in 2004.* **Sources: Kit, We19.**

White Round Leaf - 30 days - *A. gangeticus*, edible amaranth with light-green leaves. *Source History: 1 in 1981; 1 in 1984; 1 in 1987; 3 in 1991; 4 in 1994; 2 in 1998; 1 in 2004.* **Sources: Ev2.**

Varieties dropped since 1981 (and the year dropped):

A. gangeticus (1998), *A. hypochondriacus* (1998), *All Red* (2004), *Amaranth R153* (1983), *Amont* (1998), *Bicolor* (1994), *Black Seeded* (1998), *Bolivia 153* (2004), *Chihuahuan Ornamental* (2004), *Chinese* (1987), *Chinese Spinach* (1987), *Edible Amaranth* (1982), *Erect Amaranths* (1987), *Flaming Fountain* (2004), *Fote Te* (2004), *Green* (1987), *Green Purple White* (1991), *Green Top* (2004), *Guegui (or Ramdana)* (1994), *Headdress* (1998), *Hijau* (2004), *Indian Spinach* (2004), *K112-R* (1994), *Long Season* (1998), *Magic Fountains* (1994), *Medium Season* (1998), *Mirah* (2004), *Mountain Pima Green* (1998), *MT-3* (2004), *Oeschberg* (1994), *Paniculatus* (1998), *Pinang* (2004), *Prima Nepal* (2004), *Purple Giant* (1998), *Quintonil* (1994), *R1011* (1994), *R1017* (1987), *R158, Reselected* (1998), *Ramdana* (1998), *Red Striped Hinn Choy* (1991), *Rodale Multiflora* (1998), *Rose Beauty* (2004), *San Martin* (1998), *Scarlet Torch* (2004), *Short Season Amaranth* (1984), *Sunrise* (2004), *Sunset Orange* (2004), *Taiwan Salad* (1987), *Tampala* (2004), *Vermillion* (1998), *Yen Chao* (1994).

Artichoke, Globe
Cynara scolymus

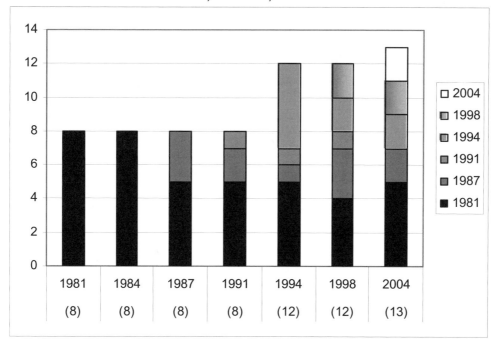

Number of varieties listed in 1981 and still available in 2004: 8 to 5 (63%).
Overall change in number of varieties from 1981 to 2004: 8 to 13 (163%).
Number of 2004 varieties available from 1 or 2 sources: 6 out of 13 (46%).

Early Green Provence - (Vert de Provence) - Perennial, Early, med-sized plants, small long heads with narrow green scales, excel. flavor. *Source History: 1 in 1981; 1 in 1984; 1 in 2004.* **Sources: Gou.**

Emerald - 170 days - Thornless var. with attractive glossy, green globes, uniform shape, exceptionally high seed germination, has shown hardiness to 0 degrees F, PVP 1992 Tom Kurapas Farming. *Source History: 1 in 1998; 4 in 2004.* **Sources: CO30, Loc, Pe6, Se26.**

Globe Artichoke - Perennial, Silver-green 3-4 ft. plants, frost sensitive, produces large thick-fleshed scaled flower buds late the first summer, good pickled or boiled and serve with butter. *Source History: 7 in 1981; 6 in 1984; 5 in 1987; 7 in 1991; 9 in 1994; 10 in 1998; 7 in 2004.* **Sources: Fi1, Gur, Ho8, Lej, Ri2, Sa12, Th3.**

Green Globe - (Early Green Globe) - Tender perennial, grown commercially on Gulf and Calif. coasts, 3-6 ft. plants, 3-4 in. heads, thick scales, needs mild damp long season and rich soil, produce in 18 mo., large flower heads, edible flower buds made up of thick fleshy scales and solid centers, not hardy, should be harvested when young. *Source History: 28 in 1981; 23 in 1984; 26 in 1987; 23 in 1991; 21 in 1994; 20 in 1998; 34 in 2004.* **Sources: Ba8, BAL, Bo19, Bu3, CH14, CO23, CO30, Com, DOR, Ear, Ers, Ge2, HA3, He8, Ho12, La1, LO8, Loc, Mel, MIC, Na6, Ni1, Or10, Orn, OSB, Ri12, Roh, Sw9, Ter, Tt2, Twi, Vi4, We19, Wet.**

Green Globe Improved - Perennial, Vastly improved, deep-green globe-shaped buds without purple tinge, sharp spines almost eliminated, vigorous prolific 5 ft. plants, produces first year from seed. *Source History: 4 in 1981; 4 in 1984; 6 in 1987; 8 in 1991; 8 in 1994; 7 in 1998; 8 in 2004.* **Sources: Hum, Ont, Pe2, Pin, Se26, Shu, Tho, WE10.**

Green Globe, Cal - Deep green globe-shaped buds, no purple tinge, very few sharp spines, vigorous and prolific, often produces the first year from seed. *Source History: 1 in 1994; 1 in 1998; 1 in 2004.* **Sources: Bou.**

Green Globe, French - Perennial, Globe flower heads with thick scales picked before flowering, thistle type plant, also used for dried flowers. *Source History: 1 in 1981; 1 in 1984; 1 in 1987; 1 in 1991; 1 in 1994; 1 in 1998; 1 in 2004.* **Sources: Dam.**

Imperial Star - 90-150 days - Developed to be grown as an annual, more productive, sweeter milder flavor, thornless 4.5 in. flower buds are round and slow to spread open when mature, tends to be more tender than other Green Globe strains, from the U of California, PVP 1991 Regents of U of CA. *Source History: 5 in 1994; 15 in 1998; 17 in 2004.* **Sources: CO23, CO30, Coo, Dom, DOR, Eo2, Fe5, Gou, Jo1, Ki4, Loc, Ni1, Orn, OSB, Par, Pe6, Ter.**

Kiss of Burgundy - Upright plant to 6 ft., semi-thornless, light burgundy color, heat, drought and cold tolerant. *Source History: 2 in 2004.* **Sources: CO23, CO30.**

Northern Star - 300 days - First hardy strain bred for over wintering and spring harvest, has survived sub-zero temperatures without protection, for trial only, from Peters Seed and Research Breeding Program. *Source History: 1 in*

1998; 1 in 2004. **Sources: Pe6.**

Romanesco - Italian heirloom, touches of bronze and purple on stems and heads, edible and ornamental. *Source History: 1 in 2004.* **Sources: Twi.**

Violetto - 110-150 days - Purple artichoke heads look like flowers on these plants, fruit is more elongated than the green globe type, start indoors in late winter for a midsummer crop even in Maine, in Italy which includes Zones 8-10 the plants are started in spring and transplanted in autumn. *Source History: 1 in 1987; 2 in 1991; 3 in 1994; 3 in 1998; 6 in 2004.* **Sources: Coo, Pin, Se24, Sw9, Ter, We19.**

Violetto di Romagna - (Purple of Romagna) - Large round purple head, more tender than green types, requires a mild climate, popular in Italy for hundreds of years. *Source History: 1 in 1987; 1 in 1991; 3 in 2004.* **Sources: Ba8, Ber, La1.**

Varieties dropped since 1981 (and the year dropped):

Carciofo (1987), *Grande Beurre* (1998), *Purple* (1998), *Purple Sicilian* (2004), *Violet* (2004), *Violetto di Jesi* (2004).

Asparagus
Asparagus officinalis

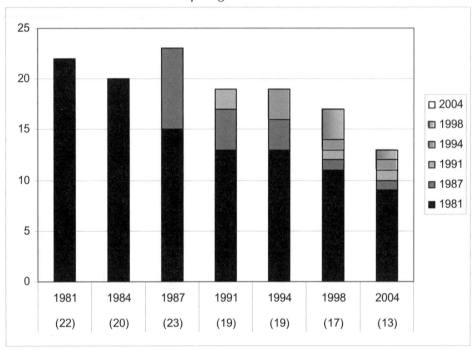

Number of varieties listed in 1981 and still available in 2004: 22 to 9 (41%).
Overall change in number of varieties from 1981 to 2004: 22 to 13 (-59%).
Number of 2004 varieties available from 1 or 2 sources: 7 out of 13 (54%).

Argenteuil Early - (D'Argenteuil Hative, Argenteuil, Precoce d'Argenteuil) - Perennial, Early variety with thick delicious white stems and purple tips, highly esteemed in Europe, heirloom from France. *Source History: 1 in 1981; 2 in 1987; 3 in 1991; 3 in 1994; 5 in 1998; 3 in 2004.* **Sources: Ba8, Ber, Lej.**

Asparagus - Open pollinated variety with stalk form ranging from tall, round and slender to short, flattened and thick, most have purple bracts and tight purple tips, suitable for garden culture or naturalizing. *Source History: 1 in 1994; 1 in 1998; 1 in 2004.* **Sources: So12.**

Barr's Mammoth - Perennial, Weather tolerant, rust and FW tolerant. *Source History: 1 in 1998; 1 in 2004.* **Sources: Shu.**

California 500 - Perennial, Light-green thick spears with tight tips, res. to some rusts, adapted to Pacific coast, UC/Davis, early Mary Washington str., no purple tips or elongated nodes, 1953. *Source History: 6 in 1981; 5 in 1984; 3 in 1987; 2 in 1991; 1 in 1994; 1 in 1998; 1 in 2004.* **Sources: Alb.**

Conover's Colossal - (Connover's Colossal) - Old English variety with stumpy, 1.5 in. spears, lacks disease resistance of modern varieties, developed from an unknown European variety by S. B. Conover in 1868, who was a produce commissioner merchant in New York's old West Washington Market, 15-40 spears per crown according to 1890's literature. *Source History: 2 in 1981; 2 in 1984; 1 in 1987; 1 in 1991; 1 in 1994; 1 in 1998; 2 in 2004.* **Sources: Bou, He17.**

Martha Washington - Superior strain dev. by George and Roy Brooks of New Jersey, tightly budded and uniform, high quality stalks, rust res., FW tol., an outstanding private stock. *Source History: 2 in 1981; 2 in 1984; 9 in 1987; 4 in 1991; 4 in 1994; 3 in 1998; 4 in 2004.* **Sources: Bo18, Gr29, Se28, Wo8.**

Mary Washington - Perennial, Most popular U.S. variety, 60 day cutting season, early long straight spears with tight tips, long-standing, standard commercial strain, resistant to some rusts and blight. *Source History: 91 in 1981; 82 in 1984; 80 in 1987; 61 in 1991; 56 in 1994; 52 in 1998; 55 in 2004.* **Sources: Ba8, Bo19, Bu1, Bu2, Bu8, But, CA25, CH14, CO23, Com, Cr1, DAI, De6, DOR, Eo2, Ers, Fa1, Fis, Ga1, Ge2, GRI, Gur, HA3, Hal, He17, Hig, HO2, Jun, La1, Lin, LO8, Ma19, MAY, ME9, Mel, Mey, Mo13, MOU, OLD, Ont, Or10, PAG, Ri12, RUP, Sau, SE14, Se26, Sh9, Shu, Tt2, Ver, Vi4, WE10, Wet, WI23.**

Mary Washington Improved - Medium green, early, vigorous, highest quality, uniform, heavy yield, excellent for freezing, Twilley's strain selected for tighter tips and larger spurs, winter hardy and somewhat rust res., thrives almost anywhere in Canada and U.S. *Source History: 2 in 1987; 1 in 1991; 2 in 1994; 2 in 1998; 3 in 2004.* **Sources: Jor, Se25, Ves.**

Mary Washington Rust Res. - (MW Giant Rust-Proof, Giant MW, Giant Wash. Rust Res., Giant Wash., Improved MW) - Perennial, Apparently immune to rust, larger thicker stalks, less stringy, heavy yields, high quality, used by many commercial growers, fine home garden variety. *Source History: 5 in 1981; 5 in 1984; 4 in 1987; 3 in 1991; 2 in 1994; 2 in 1998; 1 in 2004.* **Sources: Fi1.**

Purple Passion - Derived from a 400 year-old purple heirloom asparagus of the small Italian-French valley of Aldinga, from asparagus breeder, Brian Benson, large, sweet, deep-burgundy spears are more tender than green asparagus, mild, nutty flavor. *Source History: 7 in 1998; 12 in 2004.* **Sources: Bou, DAI, Fa1, Gur, Jun, Mel, OSB, Par, Pe6, Pin, Shu, Ter.**

Sweet Purple - Deep burgundy spears, pleasant nutty flavor, 20% higher sugar content, dev. for specialty and fresh market, resembles Purple Passion. *Source History: 2 in 1998; 6 in 2004.* **Sources: Bu1, CO30, Fi1, Pe2, SE14, Ver.**

UC 72 - (Mary Washington UC 72) - Perennial, University of California, sel. from UC 711 which it outyields, tol. to FW and rust, higher yields than 500 str. but not as uniform or early, large light-green tight spears. *Source History: 8 in 1981; 7 in 1984; 13 in 1987; 10 in 1991; 11 in 1994; 11 in 1998; 6 in 2004.* **Sources: Bu3, Jor, Ros, SE14, WE10, Wi2.**

Viking - (Mary Washington Improved, Vineland No. 35, Mary Washington V35) - Perennial, Imp. Mary Wash. type, very vigorous growth, large spears, tightly folded buds, slightly paler green than Mary Wash., heavy yields, considerable res. to rust, good home or market var., dev. at Vineland Hort. Exp. Station in Ontario in 1945. *Source History: 15 in 1981; 12 in 1984; 11 in 1987; 8 in 1991; 4 in 1994; 3 in 1998; 2 in 2004.* **Sources: Dam, Ear.**

Viking KB3 - Perennial, Selected from Viking 2K for a more uniform and taller spear, better hardiness and fusarium tolerance, seed hand selected in isolated growing areas. *Source History: 1 in 1981; 1 in 1984; 2 in 1987; 3 in 1991; 3 in 1994; 2 in 1998; 1 in 2004.* **Sources: Sto.**

Varieties dropped since 1981 (and the year dropped):

California 309 (1994), *Centennial* (1991), *D'Argenteuil Hative* (1987), *Giant Washington* (1991), *Glen Smith* (1998), *Lorella* (1991), *Mary Washington 500* (1994), *Mary Washington, Giant* (1998), *Paradise* (1983), *Purple Argenteuil* (1987), *Roberts Improved* (2004), *Roberts Strain* (2004), *Rutgers Fe. Syn #2* (1987), *UC 157* (2004), *UC 309* (1987), *UC 800* (1998), *Viking 2G* (1998), *Waltham* (1998), *Washington Pedigreed* (1991).

Bean / Bush / Dry
Phaseolus vulgaris

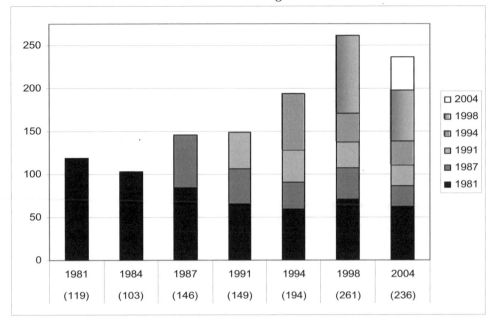

Number of varieties listed in 1981 and still available in 2004: 119 to 62 (52%).
Overall change in number of varieties from 1981 to 2004: 119 to 236 (198%).
Number of 2004 varieties available from 1 or 2 sources: 180 out of 236 (76%).

Adventist - Small yellow bean from the mountains of northern Idaho, excel. in soup or stews. *Source History: 1 in 1981; 1 in 1984; 1 in 1987; 1 in 1991; 2 in 1994; 4 in 1998; 1 in 2004.* **Sources: Syn.**

African Premier - Extremely productive in hot, dry conditions, pretty seeds are rusty brown with pink overtones and lilac stripes. *Source History: 1 in 1998; 1 in 2004.* **Sources: Ho13.**

Algarrobo - 90-95 days - Compact 2 ft. bush, white seed with maroon thunderbird design, Columbia. *Source History: 2 in 1998; 1 in 2004.* **Sources: Se17.**

Alluvias - Large white seeds, from Chihuahua, Mexico. *Source History: 1 in 1994; 1 in 2004.* **Sources: Na2.**

Anasazi - (Anasazi-Analog) - 90-95 days - Maroon and white Jacob's Cattle bean identified as one of the few cultivated crops grown by the Anasazi cliff dwellers (Anasazi means "The Ancient Ones" in Navajo), sweet flavor, meaty texture. *Source History: 2 in 1987; 5 in 1991; 6 in 1994; 9 in 1998; 13 in 2004.* **Sources: Bou, Bu3, Eco, HO2, Ho13, Pla, Ri12, Se8, Se17, So9, So12, Te4, Vi4.**

Andrew Kent - 100-105 days - Upright bush, large buff kidney beans with red streaks, fresh shelling and drying, heirloom from Scotland via Leonard Alexander from North Haven Island, Maine. *Source History: 3 in 1998; 4 in 2004.* **Sources: Jo1, So12, Ta5, Te4.**

Appaloosa - 85-110 days - Compact 18-24 in. plant, some short runners, 5 in. long flat pods, med-size gold and white seed, flavorful and firm when cooked, very attractive type from Vermont. *Source History: 2 in 1981; 2 in 1984; 1 in 1987; 1 in 1991; 3 in 1994; 7 in 1998; 9 in 2004.* **Sources: Fo7, Na2, Pe2, Pr3, Ra5, Se17, Se25, Shu, TE7.**

Appaloosa (New Mexico) - 90-110 days - White bean with maroon and black mottling, from Velarde in northern New Mexico, bush habit, used for dry beans, beautiful seed. *Source History: 1 in 1981; 1 in 1984; 1 in 1987; 1 in 1991; 1 in 1994; 2 in 1998; 2 in 2004.* **Sources: Ho13, Pla.**

Appaloosa, Vermont - 80-110 days - Compact, 18-24 in. plant with short runners, half of the slender seed is cream colored, the other half is maroon and splashed with cream, a distinctive ragged boundary diagonally encircles the seed. *Source History: 1 in 1994; 4 in 1998; 4 in 2004.* **Sources: Hi6, HO2, Ho13, Ver.**

Arikara Yellow - 85 days - Slim round pods with 4-6 light yellow-brown seeds with red-brown eye ring, drought tol., named for the Dakota Arikara tribe Lewis and Clark encountered, one of the bean varieties that sustained the members of the expedition through the arduous Fort Mandan winter, grown by the native people of the Dakotas, 1800s. *Source History: 2 in 1994; 5 in 1998; 5 in 2004.* **Sources: Co32, Pr3, Se16, Se17, Th3.**

Arkansas Bush - 90 days - Tan and dark brown striped seed, very productive. *Source History: 1 in 1998; 1 in 2004.* **Sources: Sa9.**

Aunt Emma's - Mennonite heirloom, large round white seeds are quicker cooking than others, used in baked beans and salads. *Source History: 1 in 1998; 1 in 2004.* **Sources: Te4.**

Auntie Brown - Excel. baker, buff-green with brown eye ring, 100 years old, from Missouri. *Source History: 1 in*

1998; 1 in 2004. **Sources: Ho13.**

Azores - Oval bright yellow seed, reliable and prolific, cooks quickly, disease free. *Source History: 3 in 1998; 2 in 2004.* **Sources: Ho13, Te4.**

Aztec Half Runner - 60-75 days - No need to stake 3 ft. runners, 8 in. pods, plump white seeds, flavor resembles limas, drought res., potato bean cultivated by Anasazi Indians and probably Aztecs. *Source History: 3 in 1994; 2 in 1998; 5 in 2004.* **Sources: Co32, Ri12, Roh, Se8, So12.**

Baja Azufrados - Bright, sulfur colored seeds, from Todos Santos in Baja, California, Mexico. *Source History: 1 in 1998; 1 in 2004.* **Sources: Na2.**

Baker - *Source History: 1 in 1998; 1 in 2004.* **Sources: Sh13.**

Bean Soup Special - 90 days - Kidney type, large red dappled seed, reliable yields, cooks quickly, good soup bean. *Source History: 1 in 2004.* **Sources: Te4.**

Beautiful - 88-100 days - Unusual heirloom bean, creamy-white seeds with bright maroon splashes, clean spicy flavor, remarkably easy to digest--little gas upset. *Source History: 1 in 1987; 1 in 1991; 1 in 1994; 3 in 1998; 3 in 2004.* **Sources: Ge2, Ho13, Sa9.**

Bert Goodwin's - 75 days - Long vigorous plant with some runners, pods 7 in. long, large seeds mottled with tan and brown. *Source History: 1 in 1991; 2 in 1994; 3 in 1998; 1 in 2004.* **Sources: Sh13.**

Bjlovar Beauty - Hort. type, round buff-pink seeds, from Croatia. *Source History: 1 in 1998; 1 in 2004.* **Sources: Ho13.**

Black Coco - 60-95 days - First pods form early and make fine snap beans, then fast-developing shell beans, finally plump black quick-cooking dry beans, 6 in. pods on 14-16 in. plants, heirloom. *Source History: 1 in 1987; 3 in 1991; 5 in 1994; 7 in 1998; 8 in 2004.* **Sources: Ers, Ho13, Ri12, Sa5, Se7, Ter, Ver, We19.**

Black Mexican - (Frijol Negro) - Familiar black bean used in Mexican cookery, later maturity than pinto, 85%. *Source History: 1 in 1981; 1 in 1984; 2 in 1987; 2 in 1991; 2 in 1994; 2 in 1998; 1 in 2004.* **Sources: Red.**

Black Turtle - (Black Turtle Soup, Black Bean Soup, Black Mexican) - 85-105 days - Popular for black bean soup, widely grown from Southwest to Cuba and down into South America, half-runner, disease and heat res., hardy, used in Black Turtle soup for over 150 years. *Source History: 12 in 1981; 9 in 1984; 11 in 1987; 16 in 1991; 23 in 1994; 25 in 1998; 28 in 2004.* **Sources: Bu2, Co32, Com, CON, Eco, Ers, Fe5, Gr28, He8, He17, Hi6, HO2, Jor, Jun, Na6, Or10, PAG, Pe2, Pin, Pr3, Ra5, Sa9, Se25, Se26, Sh13, So9, Tu6, Ver.**

Black Turtle Soup, Midnight - 104 days - Upright growing black bean strain, tall bush does not sprawl like the old var., pods retain dark-purple color as they dry, small black beans, dev. by Cornell, excel. for soups and stews and refrying. *Source History: 1 in 1981; 1 in 1984; 2 in 1987; 3 in 1991; 3 in 1994; 4 in 1998; 4 in 2004.* **Sources: Ami, Fo13, Ga1, Jo1.**

Blacked Twiner - Old Australian var., climbs to 3 ft., long pods are nearly stringless, small black seeds with pink specks, good home garden type. *Source History: 1 in 1998; 1 in 2004.* **Sources: Ho13.**

Bolas Maycoba - *Source History: 1 in 1994; 1 in 1998; 1 in 2004.* **Sources: Na2.**

Boleta - (Bolita) - 90-95 days - Dev. from Pinto by Spanish people in northern New Mexico, smaller and lighter brown than Pinto, low bush plant with short runners does not require staking, used in burritos or refried beans. *Source History: 1 in 1981; 1 in 1984; 1 in 1987; 2 in 1991; 3 in 1994; 3 in 1998; 2 in 2004.* **Sources: Pla, So12.**

Boston Beauty - Pretty maroon and white seed, heirloom. *Source History: 2 in 1998; 1 in 2004.* **Sources: Hi6.**

Boston Favorite - 95 days - Very old hort. type, kidney shaped seeds, pinkish with red streaks, family heirloom. *Source History: 1 in 1998; 1 in 2004.* **Sources: Sa9.**

Brown Beauty - Shiny, oval, brown seeds, productive soup or baking bean. *Source History: 1 in 1998; 1 in 2004.* **Sources: Ho13.**

Buckskin - 75-85 days - Early, excellent soup or dry bean, appearance and eating quality similar to Pinto when cooked, heirloom from eastern Oregon, PVP 2000 Novartis Seeds, Inc. *Source History: 1 in 1981; 1 in 1984; 1 in 1987; 2 in 1991; 2 in 1994; 2 in 1998; 1 in 2004.* **Sources: Tu6.**

Bumblebee - 95-98 days - Large white seeds with black splotch, high vitamin and mineral content, so named for its large size, Maine heirloom. *Source History: 1 in 1987; 1 in 1998; 1 in 2004.* **Sources: Ver.**

Burlescyna - Pinto type, deep purple-brown seeds, rare. *Source History: 1 in 2004.* **Sources: Syn.**

Calypso - (Yin Yang) - 70-90 days - Good baking bean with texture similar to Yellow Eye, strong 15 in. plant, round blue and white seeds dries to black and white with contrasting dot, 4-5 seeds per pod, Caribbean. *Source History: 1 in 1994; 8 in 1998; 8 in 2004.* **Sources: Fe5, Ho13, Jun, Ra5, Rev, Se25, Sh13, Ver.**

Canadian Wild Goose - Small white seeds spotted with blue, productive. *Source History: 1 in 1991; 1 in 1994; 2 in 1998; 1 in 2004.* **Sources: Syn.**

Candy - *Source History: 1 in 1998; 1 in 2004.* **Sources: Sa5.**

Cannellini - (Cannellone, Cannelino) - 75-85 days - Italian bean, white kidney-shape seeds, dry beans make delicious Italian-style baked beans, shell beans are main ingredient for minestrone, easy to grow, first appeared in America in the early 1800s. *Source History: 3 in 1991; 6 in 1994; 10 in 1998; 17 in 2004.* **Sources: Coo, Fe5, Hi6, Jo1, Lin, Na6, Ni1, Pin, Ri12, Sa5, Se7, Se24, Se25, Syn, Ter, Tu6, Ver.**

Carmine - Outstanding yielder developed from Red Mexican bean, small red seed, same uses as kidney beans. *Source History: 1 in 1998; 1 in 2004.* **Sources: Te4.**

Chickashaw - Prolific plant produces many runners, lime-green seed with red ring at hilum, matures medium-late to late, good for soups, stews, shell and baking, southern U.S. variety. *Source History: 1 in 2004.* **Sources: Te4.**

Chinese Yellow - 75-80 days - Marrowfat bean that is genetically close to Mennonite and Hutterite Soup Bean, the stout bushy plant bears good crops of pale yellow-green seed, excel. disease resistance. *Source History: 1 in 2004.* **Sources: Te4.**

Clem and Sarah's Big Bean - 80-90 days - Large, round, white seed is twice the size of Navy bean, delicious, Harmon family heirloom. *Source History: 2 in 1998; 1 in 2004.* **Sources: Se17.**

Coco Prague - (Coco Rose de Prauge) - European horticultural bean, fiery red pod mottled with white, used fresh or dried, traditional fresh bean for a French vegetable soup called soupe au pistou. *Source History: 1 in 1981; 1 in 1984; 1 in 1987; 1 in 2004.* **Sources: Lej.**

Coco Rubico - 60-63 days - French horticultural bean, flat speckled pods are 5-6 in. long, kidney shaped seeds are buff with purple markings, fresh, shelled or dried, good yields, introduced in 1976. *Source History: 3 in 1994; 8 in 1998; 7 in 2004.* **Sources: HO2, Jun, Pr3, Se8, Twi, Ver, VI3.**

Cranberry - (Vermont Cranberry) - 65-75 days - Early, 6-7 in. pods turn striped red when mature, plump round seeds are rusty brown with pale mottles, easy to shell, used for shelling green or dry beans. *Source History: 1 in 1981; 1 in 1984; 2 in 1987; 4 in 1991; 3 in 1994; 8 in 1998; 6 in 2004.* **Sources: Fo7, Ma19, Pe2, Ri12, Te4, TE7.**

Cranberry, Iroquois - *Source History: 1 in 2004.* **Sources: Se17.**

Cranberry, Minnesota - 85 days - Semi-twining plant, pinto type seed, hardy and productive. *Source History: 1 in 2004.* **Sources: Sa9.**

Cubanos - 85 days - Semi-vining bush produces 4 in. flat pods with small black seeds, purple blossoms, excel. drought resistance, high yields. *Source History: 1 in 2004.* **Sources: Te4.**

Cuetzalan - 100 days - *Source History: 1 in 2004.* **Sources: Se17.**

Dalmation - (Dalmatian) - Bush plant to 24 in. tall, pretty white seed with patterns of red and orange, shell or dry, heirloom. *Source History: 1 in 1998; 2 in 2004.* **Sources: Ho13, Ra5.**

Del Norte - 80-85 days - Hardy bushes do not require staking, small brown-yellow seed, nutty flavor, no seasoning required, from the village of Cerrillos. *Source History: 1 in 2004.* **Sources: Pla.**

Delgado Black Bean - Small black beans, very ancient bush type, withstands considerable cold and drought, used to dye the black shawls of the women in Oaxaca, Zapotec seed. *Source History: 1 in 1981; 1 in 1984; 1 in 1987; 1 in 1998; 1 in 2004.* **Sources: Se17.**

Dog Bean - 70-85 days - Beautiful, purple spotted, white seed, rare. *Source History: 2 in 1998; 2 in 2004.* **Sources: Se17, Syn.**

Dolores de Hidalgo Canario - 90 days - Large tubular bean, cafe au lait color, from central Mexico. *Source History: 1 in 1987; 1 in 1991; 1 in 1994; 1 in 1998; 1 in 2004.* **Sources: Na2.**

Drabo - 85 days - Large white oblong seed, cooks quickly, rich flavor, high quality, heirloom from Switzerland. *Source History: 1 in 1994; 1 in 1998; 4 in 2004.* **Sources: Ag7, Fe5, Ho13, Tu6.**

Duane Baptiste's Potato Bean - Old Six Nations bean traditionally used as a thickener in Native Canadian corn soup, large white seeds, germinates well, consistently produces high yields even in cool weather, highly adaptable, excellent flavor baked but do not soak as this toughens the skin. *Source History: 3 in 1998; 5 in 2004.* **Sources: Ea4, Ho13, Sa5, Syn, Te4.**

Dutch Brown - (Bruine Bonen) - 83-90 days - Compact plants are 15-20 in. tall, produces heavy yields of brown beans for dry use, flavor of dry bean is nut-like, also used for early green beans, popular old Pennsylvania Dutch variety for baking, boiling, soups and stews. *Source History: 1 in 1981; 1 in 1984; 1 in 1987; 4 in 1991; 5 in 1994; 6 in 1998; 2 in 2004.* **Sources: Dam, Ver.**

Etna - 60-70 days - Similar to French Horticultural with earlier maturity and more disease res., 6 in. flat pods, buff-red seeds, BCMV, PVP 1992 Seminis Veg Seeds. *Source History: 1 in 1994; 3 in 1998; 5 in 2004.* **Sources: Bu1, CH14, ME9, Ter, Ver.**

Family - 90 days - *Source History: 1 in 2004.* **Sources: Se17.**

Fisher Bean - Origin traced to the Iroquois in eastern North America, mentioned in Rhineland in 1820, possibly the same as Pennsylvania Dutch Egg bean (Oibuhne) or All-in-One-Bean (Eenbuhne), snap or dry, round, light-tan seed with deep-maroon circle around the eye, good yields. *Source History: 1 in 1994; 2 in 1998; 1 in 2004.* **Sources: Lan.**

Flageolet - 75-100 days - Grown extensively in Europe as a gourmet green shell bean, pure-white seed with white eye, vigorous 24 in. bush, dry beans seem to produce own sauce when cooked. *Source History: 3 in 1981; 2 in 1984; 2 in 1987; 4 in 1991; 6 in 1994; 6 in 1998; 6 in 2004.* **Sources: Com, Na6, Pe2, Pr3, Ri2, TE7.**

Flageolet Chevrier, Green - Most valued shell bean in France, a gourmet item, often served with lamb, use fresh or dried, high yields. *Source History: 1 in 1981; 1 in 1984; 2 in 1987; 1 in 1991; 1 in 1994; 1 in 1998; 1 in 2004.* **Sources: Coo.**

Flambeau - 76-80 days - French flageolet, slender long pods packed with 8-10 small vivid mint-green seeds, tender but firm, flavor reminiscent of fresh limas, prolific, easily shelled, excellent frozen. *Source History: 1 in 1987; 3 in 1991; 2 in 1994; 1 in 1998; 2 in 2004.* **Sources: Jo1, Te4.**

Flor de Mayo - 100 days - Mexican heirloom, seed color ranges from lilac to purple to tan, described by Santa Fe chefs as having "a mild smoky flavor". *Source History: 1 in 1994; 1 in 1998; 1 in 2004.* **Sources: Se17.**

Four Corners Gold - Quick cooking, white seed mottled with gold, from the Four Corners area. *Source History: 1 in 1998; 1 in 2004.* **Sources: Ho13.**

Fradinho - Brazilian variety, plant produces some runners, brown seed with darker brown stripes, high yields, reliable, heat tolerant, preferred over the black bean in northeast Brazil. *Source History: 1 in 1998; 1 in 2004.* **Sources: Te4.**

Frijol de Cerocahui - Small, beige/tan seed, originally collected in Cerocahui in the Barranca del Cobre from a farmer who left the area due to drug trafficking violence. *Source History: 1 in 2004.* **Sources: Na2.**

Frijol Gringo - Medium size white seed, matures late, originally collected in 1983 from within the Barranca del Cobre (Copper Canyon). *Source History: 1 in 2004.* **Sources: Na2.**

Frijol Rojo - (Red Mexican Chili) - See catalog for description. *Source History: 1 in 1991; 1 in 1994; 1 in 1998; 1 in 2004.* **Sources: Red.**

Garcia Bolita - Light tan-brown beans from Garcia, Colorado, Bolita is Spanish for "little ball", which refers to the shape of the bean. *Source History: 1 in 1998; 1 in 2004.* **Sources: Na2.**

Gaucho - 55-75 days - Rich tan bean, early sheller, good as green snap, shell or dry soup and baking, Argentina. *Source History: 1 in 1987; 1 in 1991; 1 in 1994; 2 in 1998; 2 in 2004.* **Sources: Ho13, Tu6.**

Gerald's White - Great Northern type, some plants will sprawl, flattened white seed, very tasty, midseason. *Source History: 1 in 2004.* **Sources: Te4.**

Gipsy - 70 days - Borlotto bean, flat mottled green-red pods to 6.5 in., scarlet at maturity, good soup bean. *Source History: 1 in 2004.* **Sources: OSB.**

Golondrinas - (Cinco Minutos) - Gold and white Jacob's Cattle bean, flowers in low desert but is not heat tolerant, from Aguacaliente, Chihuahua, Mexico. *Source History: 1 in 1994; 2 in 2004.* **Sources: Na2, Se17.**

Grandma Turner's - 80-90 days - *Source History: 1 in 2004.* **Sources: Se17.**

Granny Bean - Old Kentucky bean, small, cream colored seed, productive and reliable. *Source History: 1 in 1998; 1 in 2004.* **Sources: Ho13.**

Great Northern - (Montana White, White Marrow) - 65-90 days - Hardy dark-green 24-26 in. plant, straight flat 5 in. pods with 5-6 long thin-skinned white fine flavored seeds, fine flavored, excel. for baking or soups, large navy type, cooks 1/3 more quickly, heavy yields, grows well in the North, heirloom bean grown by the Mandan Indians, grown in gardens after 1907. *Source History: 29 in 1981; 26 in 1984; 34 in 1987; 29 in 1991; 29 in 1994; 31 in 1998; 22 in 2004.* **Sources: CH14, CON, De6, Ers, Gr27, He17, HO2, Jun, Mey, Mo13, OLD, Pr3, Ra5, Ri12, Roh, RUP, Se25, Se26, Sh9, Shu, Te4, Ver.**

Hidatsa Red - (Hidatsa Indian Red) - 80-100 days - Twining bushes, dark red seeds, very productive, from the former Oscar Will Seed Co. of Bismarck ND via the Seed Savers Exchange, originated with the Hidatsa Indians on the Missouri River, 1881. *Source History: 1 in 1994; 5 in 1998; 3 in 2004.* **Sources: Sa9, Se17, Te4.**

Hidatsa White - *Source History: 1 in 2004.* **Sources: Se17.**

Hopi Black - 105-100 days - Desert variety from the American Southwest, small plants, small black seed, can be used for dye, extremely drought resistant, dries quickly, does not do as well in high humidity. *Source History: 2 in 1994; 5 in 1998; 5 in 2004.* **Sources: Co32, Na2, Se17, So12, Te4.**

Hopi Gold - 90 days - Good yields of med. size, oblong shaped, golden brown seed. *Source History: 1 in 2004.* **Sources: So12.**

Hopi Pink - 85-90 days - Large flattened seeds, bright pink when fresh, productive, collected from dry-farm fields near Hotevilla. *Source History: 1 in 1998; 3 in 2004.* **Sources: Ho13, Na2, So12.**

Hopi Purple String - Beautiful purple seed with black crescent moon-shaped stripes, from Hopi farmers in village of Hotevilla. *Source History: 1 in 1994; 2 in 1998; 1 in 2004.* **Sources: Na2.**

Hopi Red - Bushy low pole plant, red pods, large mottled seed, light maroon over red/beige, original seed from farmers near Hotevilla on the Hopi reservation. *Source History: 1 in 2004.* **Sources: Na2.**

Hopi String Bean - 65-90 days - Bushy plant, big pods, large mottled beige seed, fast-growing, prolific. *Source History: 1 in 1987; 1 in 1998; 1 in 2004.* **Sources: So12.**

Hopi White - 100 days - Climbs to 5 ft. with support, small white seed, good as snap or dry, good production in arid conditions. *Source History: 1 in 1998; 2 in 2004.* **Sources: Se17, So12.**

Horticultural, Dwarf - (Wren's Egg, Cranberry Bean, Longpod, Kievits, Roman Bean, Improved Pinto) - 53-66 days - Mainly used for green shells, but also good as snaps and dry, vigorous 15-17 in. plant, thick flat 4.5-6 x .75 in. pods become splashed with maroon when mature, oval pinkish-buff seeds striped with maroon, good for cool climates and Atlantic Coast. *Source History: 33 in 1981; 26 in 1984; 29 in 1987; 24 in 1991; 19 in 1994; 14 in 1998; 10 in 2004.* **Sources: Be4, Bu8, Dam, Fi1, Fis, Gur, Sau, Sh9, Shu, Sk2.**

Horticultural, French - (October Bean, French Dwarf Horticultural, Horticultural French's Dwarf) - 63-68 days - Vigorous semi-runner, flat oval 6-8 x .5 in. straight pods, yellow splashed with red when dry, purple seed, high yields, uniform, excel. green or shell or dry, good freezer, quite long on heavy soils, halo blight res., heirloom, some list it as a pole bean. *Source History: 20 in 1981; 17 in 1984; 18 in 1987; 20 in 1991; 22 in 1994; 26 in 1998; 19 in 2004.* **Sources: All, Bu8, CA25, Cl3, HA3, Ha5, HO2, Lin, ME9, Ont, PAG, RIS, RUP, SE4, Shu, Si5, SOU, Sto, Ver.**

Horticultural, Robin's Egg - English shell bean, half-runner twines to 3 ft., best when staked, blue-purple and white seeds, great as snap, shelly or dry bean. *Source History: 1 in 1981; 1 in 1984; 3 in 1998; 3 in 2004.* **Sources: Ho13, In8, Se17.**

Horticultural, Taylor Dwarf - (October, Taylor Long Podded Dwarf Hort., Improved Pinto, Shelley Bean, Taylor Hort. Improved) - 52-68 days - Thick flat oval cream and red 6 in. pods, red splashed on buff seeds, used early as green snaps but primarily used for green shell beans, 14-18 in. semi-runner plants, does well in cool climates, strain of Dwarf Hort., dates back to the early 1800s when it may have been brought to the U.S. from Italy. *Source History: 31 in 1981; 24 in 1984; 32 in 1987; 38 in 1991; 43 in 1994; 51 in 1998; 37 in 2004.* **Sources: Ada, All, Bou, CH14, Cl3, Co31, Com, CON, De6, DOR, Ers, Fe5, Gr27, HA3, He17, HO2, Jo1, Jor, Kil, La1, Loc, ME9, Mey, Mo13, OLD, Ont, PAG, Ra5, RIS, Roh, RUP, Si5, So1, SOU, Vi4, Wet, Wi2.**

Horticultural, Volcano - 66-68 days - Bush plants bear broad flat 6 in. pods, splashed with red, green and yellow, buff/red seed, resembles French Horticultural but earlier and shorter pod length, BCMV tolerant. *Source History: 3 in 2004.* **Sources: ME9, OSB, Sto.**

Horticultural, White - Large, white seeds with purple streaks. *Source History: 1 in 1998; 1 in 2004.* **Sources: Pr3.**

Hutterite Soup - 60-85 days - Plump light-green distinctive seeds, rapidly dissolve into a creamy white soup, Hutterites are a sect from Austria following the religious teaching of Jacob Hutter, they moved to Canada in the 1750s, this is one of their heirloom seeds. *Source History: 3 in 1991; 5 in 1994; 14 in 1998; 11 in 2004.* **Sources: Fe5, Fo7, Ho13, Lan, Pe1, Roh, Se7, Se17, Sh13, So9, So12.**

Indian Woman Yellow - 60-90 days - Short plant to 20-24 in., yellow-buff seeds, nutty flavor, early maturing dry bean used for soup, good for northern dry areas. *Source History: 1 in 1991; 2 in 1994; 2 in 1998; 4 in 2004.* **Sources: Ga1, Goo, Se7, Te4.**

Ireland Creek Annie's - 70-75 days - Heirloom introduced into Canada from England in the 1920s, very early, 5 in. pods, 5 seeds per pod, fine flavor, makes its own thick sauce, disease res. *Source History: 1 in 1987; 2 in 1991; 3 in 1994; 2 in 1998; 6 in 2004.* **Sources: Ho13, Sa5, Sa9, Se16, Syn, Ta5.**

Jacob's Cattle - (Trout, Coach dog, Dalmation, Anasazi) - 80-100 days - Beautiful bean of ancient origin, 24 in. bush, kidney-shaped white seeds speckled with deep-maroon, very popular in New England and other cool short season areas, good yields of dry beans for baking or soup, possibly dev. in Virginia by Jacob Trout. *Source History: 15 in 1981; 13 in 1984; 11 in 1987; 16 in 1991; 27 in 1994; 31 in 1998; 36 in 2004.* **Sources: Co32, Eco, Fe5, Fi1, Fo7, Ga1, Gr28, HA3, Hal, He8, He17, Hi6, Ho13, Jo1, Jun, Lan, Mo13, Ol2, Pe2, Pin, Pr3, Ra5, Ri12, Roh, Se7, Se25, Sh13, Shu, Sk2, So1, Syn, Te4, Tu6, Up2, Ver, Yu2.**

Jacob's Cattle Gasless - 80-85 days - Trout type, prolific bushes, pale pink flowers, green pods ripen to yellow, kidney shaped white seeds splashed with maroon, bred by Dr. Sumner and Dr. Radcliff Pike (UNH, late 1950s) who crossed Jacob's Cattle x Mexican Black Turtle, primarily a baking bean, although edible briefly as green snap (not stringless), according to Dr. Pike, low flatulence was not a breeding goal, but accidental. *Source History: 2 in 1994; 2 in 1998; 1 in 2004.* **Sources: Se16.**

Jacob's Gold - Vigorous bush, large white seeds with gold eye and spots. *Source History: 2 in 1998; 1 in 2004.* **Sources: Ho13.**

Jelly - *Source History: 1 in 2004.* **Sources: Se17.**

John's Old Bean - 75 days - Upright plant to 1.5 ft., 6 in. flat green pods mature to deep yellow with red streaks, 4-5 large seeds per pod, tan mottled with purple, eaten as snap when young, green shell or dried. *Source History: 2 in 2004.* **Sources: Se17, Te4.**

Kahl, Yellow-Green - Seeds are shaped like a navy bean but twice the size, fast cooking bean. *Source History: 1 in 1991; 2 in 1994; 2 in 1998; 1 in 2004.* **Sources: Pr3.**

Kidney, Aztec Red - 65-80 days - Vibrant red seed, delicious eating at all stages, high yields. *Source History: 1 in 1998; 1 in 2004.* **Sources: Sa5.**

Kidney, Brick Red - Traditional kidney bean size. *Source History: 1 in 1991; 1 in 2004.* **Sources: Pr3.**

Kidney, California Red - 100 days - Large red kidney-shaped beans, quite hardy, very high yielding. *Source History: 1 in 1981; 1 in 1984; 1 in 1987; 1 in 1991; 1 in 1994; 1 in 1998; 1 in 2004.* **Sources: Sto.**

Kidney, Charlevoix Dark Red - 89-105 days - Attractive med-sized beans provide a tasty thick broth in soups and stews and chili, week earlier than other dark red types, AN res., 6 in. pods, 20-24 in. bush, USDA/MI State, 1963. *Source History: 2 in 1981; 1 in 1984; 1 in 1987; 1 in 2004.* **Sources: Se16.**

Kidney, Dark Red - 90-100 days - Stands autumn rains well, traditional chili bean, hardy 22 in. dark-green vigorous plants, heavy crops of inedible 6 in. pods that contain 5 seeds, for baking or soup, mold res. *Source History: 18 in 1981; 17 in 1984; 28 in 1987; 27 in 1991; 25 in 1994; 25 in 1998; 21 in 2004.* **Sources: Bo19, Bu1, CH14, CON, De6, Ers, Gr27, He8, He17, HO2, Jor, Jun, Mo13, OLD, Ra5, Ri2, RUP, Se25, Sh9, Shu, TE7.**

Kidney, Large Red - *Source History: 1 in 1991; 2 in 1994; 1 in 1998; 1 in 2004.* **Sources: Pr3.**

Kidney, Light Red - 109 days - Pink-mahogany beans, larger than Charlevoix, excel. for baking or soups or stews. *Source History: 5 in 1981; 5 in 1984; 4 in 1987; 3 in 1991; 3 in 1994; 4 in 1998; 1 in 2004.* **Sources: CON.**

Kidney, Mull - Bush plant, excel. yields of deep maroon seeds, from Lancaster County, PA. *Source History: 1 in 1998; 1 in 2004.* **Sources: Lan.**

Kidney, Pink - Compact plants, seeds slightly smaller than red kidneys, flavorful, hold their form when cooked. *Source History: 1 in 1994; 1 in 2004.* **Sources: Pr3.**

Kidney, Purple - Large, maroon-purple seeds, productive. *Source History: 1 in 1991; 1 in 1998; 1 in 2004.* **Sources: Ho13.**

Kidney, Ralph Dutcher White - 80-85 days - Highly praised for its flavor, cooks down into a rich gravy. *Source History: 1 in 1998; 1 in 2004.* **Sources: Se16.**

Kidney, Red - (Chili Bean, Mahogany) - 90-110 days - Long oval red/brown seeds, 5.5-6 x .5 in. inedible pods, 22-28 in. plant, old dry bean for baking or soups or Mexican dishes, strains from pink to dark mahogany-red, possibly dates back to 7000 B.C. *Source History: 37 in 1981; 33 in 1984; 35 in 1987; 29 in 1991; 25 in 1994; 27 in 1998; 19 in 2004.* **Sources: Be4, But, Com, Dam, Fi1, Fo7, Gur, Hi6, La1, ME9, Mey, Or10, Ri12, Sau, Se26, Si5, TE7, Ver, Wet.**

Kidney, Redkloud - (Light Red Kidney) - 80-100 days - Light red kidney type baking bean dev. at Cornell, 15 days earlier, can be grown into Nova Scotia, res. to HB and AN and BCMV, good yields, fine quality. *Source History: 7 in 1981; 6 in 1984; 6 in 1987; 3 in 1991; 2 in 1994; 2 in 1998; 2 in 2004.* **Sources: Fe5, SE4.**

Kidney, Troomly's Dark Red - Large seeds, early maturity, heirloom. *Source History: 1 in 1998; 1 in 2004.* **Sources: Ho13.**

Kidney, White - 88-100 days - Inedible large 5.5-6 in.

pods contain 5 or 6 large white beans, milder flavor than red kidneys, strong bushy hardy 24-28 in. vigorous prolific plants, excel. baker. *Source History: 12 in 1981; 9 in 1984; 8 in 1987; 6 in 1991; 6 in 1994; 9 in 1998; 1 in 2004.* **Sources: Si5.**

Kilham Goose - Bush growth habit, shiny round seeds, purple on white. *Source History: 1 in 1994; 2 in 1998; 2 in 2004.* **Sources: Ho13, In8.**

King of the Early - (King of the Earlies) - 65-88 days - Mottled red seed, hearty flavor, ripens early, Maine family heirloom. *Source History: 2 in 1994; 5 in 1998; 3 in 2004.* **Sources: Fe5, Fo7, Se25.**

Kintoki - Exceptionally flavored, plump red bean, Japan. *Source History: 1 in 2004.* **Sources: Sa5.**

Kiva - (Taos Kiva) - *Source History: 1 in 1994; 1 in 1998; 1 in 2004.* **Sources: Sh13.**

Lightning - *Source History: 1 in 1998; 2 in 2004.* **Sources: Se17, Sh13.**

Limelight - (Dwarf Italian) - 38-45 days - Very early maturing dwarf Italian flat pod type, tender snap beans when under 5 in., also excel. shell beans, compact 12-15 in. plants, thick fleshy fiberless pods, lima-like flat white seed, dev. in Canada. *Source History: 4 in 1981; 4 in 1984; 5 in 1987; 7 in 1991; 6 in 1994; 5 in 1998; 1 in 2004.* **Sources: Fe5.**

Lina Cisco's Bird Egg - 85 days - Hort. type used as a dry bean, large tan seed with maroon markings, brought to Missouri by covered wagon in the 1880s by Lina Cisco's grandmother, Lina was one of the six original members of the Seed Savers Exchange, which started in 1975. *Source History: 1 in 2004.* **Sources: Se16.**

Low's Champion - (Dwarf Red Cranberry) - 65-68 days - New England heirloom, broad flat 4-5 in. pods, for snaps when young or later for green shell or dry, mahogany-brown seeds, old as the hills, named by the Aaron Low Seed Co. in 1884. *Source History: 2 in 1981; 1 in 1984; 2 in 1987; 1 in 1991; 5 in 1994; 11 in 1998; 2 in 2004.* **Sources: Ho13, Ver.**

Maine Sunset - 80-85 days - Dependable production of plump, oval, ivory seed with irregular splotches of maroon and red-orange around the hilum, high yields, rare. *Source History: 4 in 1998; 3 in 2004.* **Sources: Fe5, Ho13, Ri12.**

Manitoba Black - Small black seed, good soup type from Canada. *Source History: 1 in 1998; 1 in 2004.* **Sources: Ho13.**

Marble, White - Small round white seed, strong bean flavor, productive. *Source History: 1 in 1994; 2 in 1998; 1 in 2004.* **Sources: Ho13.**

Marfax - 80-85 days - Heirloom gold-colored soup bean from New England, resembles Swedish Brown but is earlier and higher yielding. *Source History: 1 in 1981; 1 in 1984; 1 in 1987; 2 in 1991; 3 in 1994; 5 in 1998; 5 in 2004.* **Sources: Fe5, Hi6, Ho13, Sa9, Ta5.**

Mary Ison's Little Brown Bunch - Short vine or bush, small light brown seeds with darker brown markings, very productive. *Source History: 1 in 1998; 1 in 2004.* **Sources: Ho13.**

Mayocoba - Vayo type, veined beige seed, from Rio Mayo valley, adapted to low hot desert, traditionally planted during the summer rains. *Source History: 1 in 1981; 1 in 1984; 1 in 1987; 1 in 2004.* **Sources: Na2.**

Mimbres Cave - 100-110 days - Sprawling vines, high yields of white seeds mottled with maroon-brown, fresh or dry use, from ancient Pueblos of southwest New Mexico. *Source History: 1 in 2004.* **Sources: So12.**

Mitla Black - 70-110 days - Pre-Columbian variety from the Mitla Valley, Oaxaca, Mexico, twining plant habit, high-yielding, delicious and nutritious in soups, good results from Southwest to the Canadian border. *Source History: 1 in 1987; 3 in 1991; 3 in 1994; 2 in 1998; 2 in 2004.* **Sources: Na2, Se17.**

Money - 80-95 days - Attractive red-speckled variety from England, good yield, excellent for baked beans and soup. *Source History: 1 in 1981; 1 in 1987; 2 in 1991; 2 in 1994; 3 in 1998; 3 in 2004.* **Sources: Ho13, Ta5, Ver.**

Monos Negros - (Black Monkeys) - 90 days - Related to Black Turtle lineage, rich earthy flavor, early, very productive, originated in San Salvador. *Source History: 1 in 1994; 1 in 1998; 2 in 2004.* **Sources: Ag7, Fe5.**

Montcalm - (Montcalm Dark Red Kidney) - 105 days - Early str. of Red Kidney produced on upright non-sprawling plants, excel. flavor, also good as a green shelled bean. *Source History: 1 in 1994; 2 in 1998; 4 in 2004.* **Sources: Roh, Ter, Tu6, We19.**

Montezuma Red - (Montezuma Red Man) - 85-100 days - Dry bush bean being grown by Indians when Cortez arrived in Mexico, found in 3,000 year old pre-historic tombs, strain obtained from Guatemalan Indians, compact plants, with too much water will require support, dark mahogany seeds, excel. in chili, baked or refried beans, adapts to a wide range of soils and growing conditions, good disease res. *Source History: 2 in 1981; 1 in 1984; 2 in 1987; 4 in 1991; 4 in 1994; 6 in 1998; 7 in 2004.* **Sources: Ag7, Ni1, Pr3, Ri12, Sa5, Se17, So12.**

Mrociumere - Bush grows 14 in. tall, pretty tan seeds, dusted with purple, from Kenya. *Source History: 2 in 1998; 3 in 2004.* **Sources: Hi6, Ho13, Sa5.**

Murdock - *Source History: 1 in 2004.* **Sources: Se17.**

Native Whispering - Traditional native variety. *Source History: 1 in 2004.* **Sources: So9.**

Navajo - *Source History: 1 in 2004.* **Sources: Se17.**

Navy - (Boston Pea, White Navy, Small White Navy, White Bean, Navy Pea Bean) - 85-100 days - Plants 16-24 in. tall, 4 in. pods with small white beans, will not mush up when cooked, firm skins stand reheating well, excel. for baked beans or soup, high yields. *Source History: 16 in 1981; 11 in 1984; 19 in 1987; 13 in 1991; 19 in 1994; 17 in 1998; 17 in 2004.* **Sources: Bo19, CH14, De6, Gr27, He8, He17, HO2, Jor, Mey, Mo13, OLD, Pr3, Ra5, RUP, Sau, Se26, Ver.**

Navy, Fleetwood - 80-88 days - Small white seed, very well adapted, dependable. *Source History: 1 in 1981; 1 in 1984; 1 in 1987; 2 in 1991; 2 in 1994; 2 in 1998; 2 in 2004.* **Sources: CON, Ers.**

Navy, Sanilac - 90-95 days - Standard of excellence, highly improved White Navy strain, true bush, pods held high for easy harvest, uniform maturity, heavy yield, glossy white beans, MI/AES/1956. *Source History: 6 in 1981; 6 in 1984; 5 in 1987; 2 in 1991; 2 in 1994; 2 in 1998; 1 in 2004.* **Sources: Be4.**

Navy, Seafarer - 90-100 days - Early dry navy bean for soup or baking, small round shiny white seed, tol. to BCMV and HB and AN, heavy yields, vigorous bushes, the one used in pork and beans. *Source History: 8 in 1981; 6 in 1984; 3 in 1987; 2 in 1991; 2 in 1994; 1 in 1998; 1 in 2004.* **Sources: Se17.**

New Mexico Bolitas - Veined beige-tan bean, staple, Rio Grande irrigated, Hispanic variety. *Source History: 1 in 1987; 1 in 1991; 1 in 1994; 1 in 1998; 1 in 2004.* **Sources: Na2.**

Nez Perce - 80-90 days - Similar to Squaw Yellow, half-climber bush, fibrous pods, small, yellow-brown seeds, rich taste, do not get mushy when cooked, from the Nez Perce in the 1930s. *Source History: 1 in 1991; 1 in 1994; 3 in 1998; 3 in 2004.* **Sources: Ri12, Sa5, Se17.**

Nicaraguan Bush - Usually a bush type plant, small red Mexican type seed, high yields, early. *Source History: 1 in 2004.* **Sources: Te4.**

Norwegian Brown - 80-85 days - Similar to Swedish Brown and Dutch Brown, nutty flavor, from Norway around 1900. *Source History: 1 in 1991; 2 in 1994; 3 in 1998; 3 in 2004.* **Sources: Ga1, Pr3, Sa5.**

Odawa Indian - 80-100 days - Rare heirloom from Odawa tribe in Harbor Springs, MI, via the Wright family for three generations, round white seed with tan shield figure. *Source History: 3 in 1998; 2 in 2004.* **Sources: Ri12, Se17.**

Odawa Soup - Broad pods contain oval seeds, half white, half pinto colored, grown by Native Americans. *Source History: 1 in 1998; 2 in 2004.* **Sources: Pr3, Sa5.**

Ojito Bolita - Light tan-beige seed, from Ojito, NM. *Source History: 1 in 2004.* **Sources: Na2.**

O'odham Pink - Sprawling plants, pink seed, produces in early spring or late fall in low desert, not heat tolerant, from desert borderlands of Sonora and Arizona. *Source History: 1 in 2004.* **Sources: Na2.**

Orca - 80 days - Seed has beautiful black and white yin yang pattern. *Source History: 1 in 1998; 1 in 2004.* **Sources: Te4.**

Paint - 100-110 days - Offshot of the Yellow Eye dry soup bean, white background with yellow eye marking, drought res., early, from eastern Washington. *Source History: 1 in 1981; 1 in 1984; 1 in 1987; 1 in 1991; 2 in 1994; 1 in 1998; 1 in 2004.* **Sources: Se7.**

Painted Pony - Flageolet type, small seed with unusual coloring, brown over white with a distinctive white eye shadowed in black, exceptional taste. *Source History: 1 in 1994; 2 in 1998; 2 in 2004.* **Sources: Ho13, Sh13.**

Palomino - 100 days - Closely related to Jacob's Cattle, this trout type kidney bean is yellow-brown with white instead of red. *Source History: 1 in 2004.* **Sources: Te4.**

Pawnee - 90-100 days - Reliable and productive, white and brown spotted seed, retains pattern after cooking, nice baking bean. *Source History: 2 in 1994; 5 in 1998; 2 in 2004.* **Sources: Ho13, Se7.**

Pea Bean - 90 days - Many small pods produced on small compact plants, high yields of small white plump navy beans, from central Canada in the 1930s. *Source History: 1 in 1991; 1 in 1994; 1 in 2004.* **Sources: Te4.**

Pepa de Zapallo - 85 days - Semi-climber, large flat gold seed with maroon swirls, does well in dry conditions, beautiful, from Chile. *Source History: 3 in 1998; 3 in 2004.* **Sources: Ho13, Se17, Ta5.**

Peregion - 95 days - Similar to cornfield but much earlier, upright bush will climb to 4 ft, 5 in. flat green pods, small brown seeds streaked with tan, disease resistant, from eastern Oregon. *Source History: 1 in 1981; 1 in 1984; 1 in 1987; 1 in 1991; 1 in 1994; 1 in 1998; 2 in 2004.* **Sources: Shu, Ver.**

PI 194582 (Guatemala) - Half runner plant, small black seeds. *Source History: 1 in 2004.* **Sources: Ho13.**

Pink - 85-95 days - True bush, quite similar to Red Mexican, ideal chili or soup bean, holds shape when cooked, long narrow pods, small smooth med-pink sweet tender beans. *Source History: 3 in 1981; 2 in 1984; 2 in 1987; 2 in 1991; 2 in 1994; 2 in 1998; 1 in 2004.* **Sources: Sh13.**

Pinto - 85-95 days - 85 days to dry shell stage, 4.5-6.5 in. pods borne near crowns of 20 in. half-runner plants, good for Southwestern dryland conditions, snap/green shell/dry/refried, short broad oval pods with 5-6 broad oval light-buff seeds speckled brown. *Source History: 29 in 1981; 26 in 1984; 32 in 1987; 32 in 1991; 28 in 1994; 30 in 1998; 26 in 2004.* **Sources: Be4, Bu2, CH14, CON, De6, Ers, Fi1, He8, He17, HO2, Jor, Mey, Mo13, OLD, Pe2, Ra5, Ri12, RUP, Sau, Se25, Se26, Sh9, Shu, So9, Ver, Wet.**

Pinto III - 95 days - Bush field bean, very popular dry shell bean for winter use, best adapted soup bean for the South. *Source History: 3 in 1981; 3 in 1984; 1 in 1987; 1 in 1991; 1 in 1994; 1 in 1998; 1 in 2004.* **Sources: Wi2.**

Pinto Improved - (Dwarf Horticultural) - 62 days - Heavy producer, 5 in. long, stringless pods at snap stage, extra full at shell stage, also excel. for dry beans, large plump seeds when mature. *Source History: 1 in 1981; 1 in 1984; 1 in 1987; 1 in 1991; 1 in 1998; 1 in 2004.* **Sources: Ros.**

Pinto, Agassiz - 92 days - Extra early maturing pinto, erect plant, similar to Agate but matures about 5 days sooner, thin skinned, buff and brown streaked seeds, cooks up firm yet tender, PVP 1992 Novartis Seeds, Inc. *Source History: 1 in 1994; 3 in 1998; 2 in 2004.* **Sources: Pr3, Te4.**

Pinto, Agate - 92-98 days - Bush plants without usual semi-runners and sprawling habit, rectangular med-sized buff mottled beans, spicy flavor, relatively short cooking time, PVP expired 2002. *Source History: 1 in 1981; 1 in 1984; 1 in 1987; 4 in 1991; 4 in 1994; 3 in 1998; 2 in 2004.* **Sources: Fe5, Sa5.**

Pinto, Blue - Bush plant with a few runners, blue fresh seeds dry to black. *Source History: 1 in 1998; 1 in 2004.* **Sources: Ho13.**

Pinto, Frijol en Seco - 90 days - Bush plants, early high yielder with excel. flavor, drought tolerant, collected in Estancia, New Mexico. *Source History: 2 in 1991; 2 in 1994; 3 in 1998; 2 in 2004.* **Sources: Na2, Pla.**

Pinto, Hopi - Classic brown and beige pinto bean, dry farmed by Hopi farmers in northeastern Arizona. *Source*

History: 1 in 1994; 1 in 2004. **Sources: Na2.**

Pinto, Hopi Black - 95-100 days - Beige seeds with black mottling, high yields, dry farmed in the Hopi fields of northeastern Arizona. *Source History: 1 in 1994; 2 in 2004.* **Sources: Na2, So12.**

Pinto, Mexican - Sprawling growth habit, early and heavy yields. *Source History: 1 in 1991; 1 in 1994; 1 in 1998; 1 in 2004.* **Sources: Pr3.**

Pinto, Nodak - 85 days - Outstanding yields, matures a week or two earlier than most pintos, split res., ND AES/USDA- ARS/WA AES. *Source History: 2 in 1998; 2 in 2004.* **Sources: Ga1, Te4.**

Pinto, Sonoran - Brown-speckled, light beige seed, high yields, matures late, irrigated winter crop in frost-free desert. *Source History: 1 in 1987; 1 in 2004.* **Sources: Na2.**

Pinto, Spanish Tolosana - Maroon and white seeds, quick cooker. *Source History: 1 in 1994; 3 in 1998; 2 in 2004.* **Sources: Gr28, Ho13.**

Piros-Feher - 70-90 days - Semi-vining plant to 24-30 in., green pods, white seed with striking deep red markings, Hungarian heirloom. *Source History: 1 in 2004.* **Sources: Se17.**

Pretzel - Bush plant, purple flowers are attractive to beneficial insects, pods twist like a pretzel, abundant yields, good soup bean, pre-1900. *Source History: 1 in 2004.* **Sources: Lan.**

Pueblo Black - Med. size black seed, very productive, from the Indians of New Mexico. *Source History: 1 in 1998; 1 in 2004.* **Sources: Ho13.**

Rajman Himalayan - 90-95 days - Staple among mountain villages near Tibet, ovoid, off-white seed mottled with red, productive, 9,000-10,000 ft., heirloom from Kinnaur in Himachal Pradesh, India. *Source History: 1 in 2004.* **Sources: So12.**

Raven - 90-95 days - Compact bush bears many pale purple blossoms, yellow and green pods, glossy black seed, PVP 1998 Michigan State University. *Source History: 1 in 2004.* **Sources: Se17.**

Red Mexican - (Montezuma's Red) - 85-100 days - True bush 14 in. plants, red-brown seeds, popular with commercial canners but almost unknown to gardeners, excel. baker, will not become soggy or fall apart, heirloom bean prized in California since 1855, rare. *Source History: 2 in 1981; 2 in 1984; 2 in 1987; 3 in 1991; 3 in 1994; 2 in 1998; 3 in 2004.* **Sources: Se7, TE7, Ver.**

Red Peanut Bean - 50-56 days - True bush 14 in. plants, 4 in. pods grow green turning red at maturity, suggested for dry climates, used for both green shells and dry, plant often for continuous yield. *Source History: 1 in 1981; 1 in 1984; 1 in 1987; 2 in 1991; 2 in 1994; 3 in 1998; 1 in 2004.* **Sources: Ers.**

Rice - *Source History: 1 in 1998; 1 in 2004.* **Sources: Se17.**

Rojo de Seda - (Red Silk) - Grown by Lencas Indians in Honduras and El Salvador, plant loses its leaves early making for easy picking, pretty red seed, disease resistant. *Source History: 1 in 1994; 1 in 1998; 2 in 2004.* **Sources: Se17, Te4.**

Ruckle - (Walcherse) - Large white beans, milder flavor than red kidneys, excel. for soups and baking, grown by the Ruckle family of Salt Spring Island since the turn of the century. *Source History: 1 in 1991; 1 in 1994; 1 in 1998; 1 in 2004.* **Sources: Sa5.**

Russian Soup - 80 days - Vigorous small plant bears 3-4 in. pods with 4-6 light colored seeds, productive. *Source History: 1 in 2004.* **Sources: Te4.**

Sacrament Bean - 80 days - Medium size white seed with maroon marking around the hilum which resembles a monstrance, good producer, from Germany. *Source History: 1 in 1994; 1 in 1998; 1 in 2004.* **Sources: Lan.**

Sangre de Toro - Small red bean used for commercial production in El Salvador, name translates "bull's blood". *Source History: 1 in 1994; 3 in 1998; 1 in 2004.* **Sources: Te4.**

Santa Maria Pinquito - (Pinquito, Bush Pinquito) - 75-90 days - Vigorous plant, slender pods with 7-10 tiny pink dry beans, fantastic flavor, low starch, do not break up when cooked, from early California Spaniards. *Source History: 6 in 1981; 4 in 1984; 3 in 1987; 6 in 1991; 4 in 1994; 6 in 1998; 4 in 2004.* **Sources: Ho13, Ni1, Red, So12.**

Scandinavian - 105 days - Vigorous plant produces flavorful large brown seed. *Source History: 1 in 2004.* **Sources: Te4.**

Scarlet Beauty - 90 days - Sprawling plant bears long heavy pods, red seed with beige stripes, great for chili, developed by New Hampshire plant breeder, Elwyn Meader, 1954. *Source History: 2 in 2004.* **Sources: Pr3, Ver.**

Scarlet Beauty Elite - Compact bush, early, flavorful, scarlet and pink bean marbled with tan, red striped pods. *Source History: 4 in 1998; 2 in 2004.* **Sources: Ho13, Sh13.**

Scott's Choice - 85-100 days - Short plant, short flat pods, round golden seed, cooks into a flavorful soup, Scott family heirloom, pre-1880. *Source History: 1 in 1998; 1 in 2004.* **Sources: Hi6.**

Seneca - *Source History: 1 in 2004.* **Sources: Se17.**

Serene - Tasty shell bean holds its shape when cooked, reddish pods, large, egg shaped seeds that are violet-gray with darker striations and yellow around the hilum. *Source History: 2 in 1998; 1 in 2004.* **Sources: Sa5.**

Shortcut Query - Heavy set of 4 in. maroon pods, small shiny black seeds, drought res. *Source History: 1 in 1998; 2 in 2004.* **Sources: Ho13, Syn.**

Six Nation - (Iroquois) - 80 days - Plants to 14 in. tall, red seeds speckled with white, excel. quality. *Source History: 1 in 1994; 4 in 1998; 2 in 2004.* **Sources: In8, Se17.**

Small Red - 95 days - Distinctive spicy flavor, best bean for chili, smaller than kidney beans, high yielding. *Source History: 2 in 1981; 1 in 1984; 2 in 1987; 1 in 1991; 1 in 1994; 3 in 1998; 2 in 2004.* **Sources: CON, Ers.**

Smith River Super Speckle - Large red seeds with white speckles on one end, good soup bean, productive. *Source History: 1 in 1998; 1 in 2004.* **Sources: Ho13.**

Soldier - (Johnson Bean, Human) - 75-100 days - Popular New England heirloom, hardy drought res. 18 in. plants, large white bean with red/brown eye figure, good in cooler climates, excel. flavor, soup or bake, pre-1800. *Source*

History: 13 in 1981; 12 in 1984; 16 in 1987; 19 in 1991; 17 in 1994; 21 in 1998; 17 in 2004. **Sources: CON, Ers, Fe5, Fo7, He8, Hi6, Jo1, Jun, Me7, Ol2, Ont, Ra5, Ri12, Se25, Shu, Up2, Ver.**

Spanish Black - Bush plant with short runners, clusters of 3-4 pods, small black seed. *Source History: 1 in 1998; 1 in 2004.* **Sources: Ho13.**

Speckled Bays - (Bayo, Speckled Bale, Taylor's Horticultural) - 95-100 days - Plump cream seed with red speckles, pinto shape, cooks quickly, very tender and hardy, chill tol., prolific, Washington and Oregon coastal heirloom. *Source History: 1 in 1981; 1 in 1987; 1 in 1991; 2 in 1994; 4 in 1998; 3 in 2004.* **Sources: Ho13, Ter, We19.**

Squaw Yellow - (Indian Woman) - 60-75 days - Small golden-yellow beans about the size of navy beans, pods too fibrous for use as snaps. *Source History: 1 in 1981; 1 in 1984; 2 in 1987; 2 in 1991; 1 in 1994; 2 in 1998; 2 in 2004.* **Sources: Fis, Syn.**

Stevenson Blue Eye - Pretty oval white seed with blue-black pattern around the eye. *Source History: 2 in 1998; 1 in 2004.* **Sources: Ho13.**

Stockbridge Indian - *Source History: 1 in 2004.* **Sources: Se17.**

Summerfelt - Productive golden yellow soup bean from Germany. *Source History: 1 in 1998; 1 in 2004.* **Sources: Ho13.**

Swedish Brown - 82-95 days - Popular Scandinavian bean from the 1890s, nutty flavor, bake or soup, compact 15 in. bushes, many pods per plant, 5 to 7 beans each, brown seed with white eye, good for the North. *Source History: 5 in 1981; 5 in 1984; 7 in 1987; 9 in 1991; 11 in 1994; 12 in 1998; 13 in 2004.* **Sources: Ers, Fo7, Ho13, Jun, Ri12, Se7, Se16, Se25, Sh13, So9, Ta5, Ver, Vi4.**

Taos Red - One of the Pueblo's oldest crops, light-red to maroon seeds with darker red mottling, similar to Hopi Red, grown with irrigation in northern New Mexico. *Source History: 1 in 1994; 2 in 1998; 2 in 2004.* **Sources: Na2, Se17.**

Tarahumara Azufrado - (Sulphur Bean) - Egg-shaped, yellow and beige seed, tasty staple, unique, adapted to low hot desert areas of Sierra Madre in Mexico, planted during summer rains, Tarahumara Indian seed. *Source History: 1 in 1981; 1 in 1984; 1 in 1987; 1 in 1991; 1 in 1994; 1 in 1998; 2 in 2004.* **Sources: Na2, Se17.**

Tarahumara Bakamina - Small semi-bush plant, long pods, tiny long shiny dark-red bean, excellent green beans, rare. *Source History: 1 in 1987; 1 in 1991; 1 in 2004.* **Sources: Na2.**

Tarahumara Canario - 90+ days, Long med-size veined pale-yellow/beige bean, from various locations in Tarahumara country. *Source History: 1 in 1987; 1 in 1991; 1 in 1994; 1 in 2004.* **Sources: Na2.**

Tarahumara Capirame - 90-95 days - Large long beige bean with maroon stripes and flecks, from Tarahumara people of the Sierra Madre of northern Mexico. *Source History: 1 in 1987; 1 in 1991; 1 in 1994; 1 in 1998; 2 in 2004.* **Sources: Se17, So12.**

Tarahumara Kakamira - High yielding small red bean, vigorous and drought res., for soup and refries. *Source History: 1 in 1998; 1 in 2004.* **Sources: Te4.**

Tarahumara Norteno - "Northern", elongated beige bean with dark brown ring around the hilum, creamy texture, originally collected in 1984 from Kirare, Chihuahua, Mexico. *Source History: 1 in 1987; 1 in 2004.* **Sources: Na2.**

Tarahumara Reds - Med. size, red-purple seeds, long season. *Source History: 1 in 1987; 1 in 1991; 1 in 1998; 1 in 2004.* **Sources: Ho13.**

Tierra del Fuego - 65-80 days - Originally from the tip of Patagonia, cream seed with red stripes, shell or dry use. *Source History: 1 in 1991; 1 in 1994; 1 in 2004.* **Sources: Bou.**

Tigre - Productive, tall bush, pretty pink seeds with dark maroon streaks. *Source History: 1 in 1998; 2 in 2004.* **Sources: Ho13, Se17.**

Tonawanda, Seneca - *Source History: 1 in 2004.* **Sources: Se17.**

Tongue of Fire - (Horto Tongues of Fire, Borlotto Lingua di Fuoco) - 70-80 days - Horticultural bean from Tierra del Fuego, flat 6-7 in. ivory-tan pods with reddish streaks and spots, large round seeds, superior texture and flavor, good fresh, frozen, canned, pods dry very early. *Source History: 3 in 1987; 8 in 1991; 9 in 1994; 17 in 1998; 13 in 2004.* **Sources: Bou, Fo7, Ho13, Pr3, Sa5, Se17, Se24, Sh13, So25, Sw9, Te4, Ver, Yu2.**

Trout, Black and White - 80-100 days - Productive kidney resembles Jacob's Cattle, black and white seed, for cooler, short season. *Source History: 2 in 1998; 1 in 2004.* **Sources: Ho13.**

Tutelo - Old native variety. *Source History: 1 in 2004.* **Sources: Eco.**

Tzutuhil Red - Slim pods contain small squarish, dark burgundy seed, originally collected in Guatemala where they are planted on the sides of volcanoes. *Source History: 1 in 2004.* **Sources: Se17.**

Ukrainian - Large round white marrowfat type seed with excellent flavor. *Source History: 1 in 1998; 1 in 2004.* **Sources: Te4.**

Uncle Willie's - 90 days - Bush cranberry type, cream colored seeds with crimson streaks, early, first obtained in San Juan Islands off the Washington coast. *Source History: 1 in 1981; 1 in 1984; 1 in 1987; 1 in 1991; 1 in 1994; 1 in 1998; 1 in 2004.* **Sources: Se17.**

Vayo Blanco - Large pale beige seed with orange ring around hilum, high yields, from the high cool desert near Durango in Sierra Madre of Mexico, traditionally planted during the summer rains. *Source History: 1 in 1981; 1 in 1984; 1 in 1987; 1 in 1991; 1 in 1994; 1 in 2004.* **Sources: Na2.**

Vermont Cranberry - 75/shell, 98/dry days, Popular New England heirloom bean, color and shape of a cranberry, 5-6 seeds per pod, green shell or dry, unique sweet taste, fine quality, reliable, hardy, easy to shell, pre-1876. *Source History: 4 in 1981; 4 in 1984; 5 in 1987; 4 in 1991; 8 in 1994; 9 in 1998; 9 in 2004.* **Sources: Fe5, Fo13, Goo, Jo1, Pin, Pr3, Shu, Te4, Ver.**

Walcherse - (Walcherse White) - White dry beans,

much larger than navy beans, outstanding flavor for soups and baking. *Source History: 1 in 1981; 1 in 1984; 1 in 1987; 1 in 1991; 1 in 1994; 1 in 1998; 1 in 2004.* **Sources: Dam.**

White Marrow - 100 days - Pods contain 5 or 6 med-sized beans, excel. for soups and baking, 100 days to dry shell stage, flat straight 5 in. pods, larger than navy, easier to shell. *Source History: 2 in 1981; 2 in 1984; 1 in 1987; 1 in 1991; 1 in 1994; 1 in 1998; 1 in 2004.* **Sources: Sto.**

White Marrowfat - (White Marrow, White Egg) - 65-100 days - Prolific half runner plants, straight flat 4.5-5 in. inedible pods, larger than regular navy and better for baking, 5 or 6 white egg shaped seeds per pod, 100 days to dry shell stage, was a major market bean and staple for Civil War soldiers in the 1860s. *Source History: 6 in 1981; 3 in 1984; 2 in 1987; 3 in 1991; 5 in 1994; 6 in 1998; 7 in 2004.* **Sources: CON, Fo7, Ho13, La1, Ont, RUP, Syn.**

Wild Goose - 70 days - Small, bushy plants produce heavy yields of pods with 6-10 seeds each, dark seeds are mottled with tan, one of many beans reputed to have been found in a wild goose's craw, early season maturity. *Source History: 1 in 1998; 1 in 2004.* **Sources: Te4.**

Wild Pigeon Bean - 90-95 days - Heirloom Iroquois bean from Grand River Reserve, Ontario, Canada, twining plant benefits from support, small round pea-bean, grey seed speckled with black, looks like a bird's egg. *Source History: 1 in 2004.* **Sources: Se17.**

Yellow Eye - (Dot Eye, Molasses Face, Yellow Eyed China Bean, Steuben) - 85-90 days - White with a yellow eye, excel. for soup and baking, 18 in. plants, hardy and prolific, very popular New England field bean, from Maine. *Source History: 5 in 1981; 3 in 1984; 4 in 1987; 5 in 1991; 5 in 1994; 10 in 1998; 3 in 2004.* **Sources: Fo7, Ho13, Jun.**

Yellow Eye, Maine - 92 days - Most popular dry baking bean in Maine, cooked quality superior to other Yellow Eye strains, somewhat later than Soldier or Trout, plump, oval, medium size seed. *Source History: 1 in 1981; 1 in 1984; 1 in 1987; 4 in 1991; 6 in 1994; 7 in 1998; 5 in 2004.* **Sources: Fe5, Hi6, Jo1, Tu6, Ver.**

Yellow Eye, Steuben - (Dot Eye Bean, Molasses Face, Yellow-Eyed China Bean) - 85-95 days - Early-maturing heirloom, 18 in. bush plant, white seeds with prominent large yellow eye, large mild beans absorb the flavors of soups and stews, modest yields, good germ. under adverse conditions, disease res. *Source History: 1 in 1991; 3 in 1994; 2 in 1998; 2 in 2004.* **Sources: Gr28, Ver.**

Yellow Squaw - (Indian Woman) - 60-75 days - Old sulphur-yellow soup bean, short maturity, good for short season areas, rare. *Source History: 1 in 1981; 1 in 1984; 1 in 1987; 1 in 1991; 1 in 1994; 1 in 2004.* **Sources: Ri12.**

Zuni Shalako - 70-80 days - Gold and white Jacob's Cattle bean given to Zuni farmers by the giant Shalako Kachina at the winter ceremony, plant during summer rains in low desert, bushy compact plants produce 6 in. pods. *Source History: 1 in 1991; 1 in 1994; 3 in 1998; 3 in 2004.* **Sources: Ag7, Se17, So12.**

Varieties dropped since 1981 (and the year dropped): *Alamos Black* (1998), *Appaloosa, Black and White* (1998), *Appaloosa, New Mexico Black* (2004), *Appaloosa, New Mexico Red* (2004), *Aruba* (1998), *Aurora* (2004), *Aztec* (2004), *Beka Brown* (2004), *Black* (2004), *Black Jason* (2004), *Blower* (2004), *Boy* (2004), *Bullot* (1991), *Canary* (2004), *Cayuga* (2004), *Centralia* (2004), *Coco Nain Blanc* (1991), *Colored Half Runner* (1987), *Cranberry, Goose* (2004), *Cuban Black Bean* (1982), *Dr. Wyche's Russian* (1998), *Dutch Shell Yellow* (1987), *Early White Dwarf* (1991), *Eastern Tarahumara Amarillo* (1994), *Esther's Swedish* (2004), *Flageolet Chelinex* (1987), *Flamata* (1987), *Flor de Rio* (2004), *Forty Bushel* (1994), *Frijol Amarillo Oro* (2004), *Frijol Blanco* (2004), *Frijol Canario* (1987), *Geril's* (2004), *Gold Nugget* (1998), *Golden Heirloom* (2004), *Gotlands Speckled* (2004), *Granny Hagler Half-Runner* (2004), *Green Seeded Flageolet* (1994), *Hignell's Italian* (2004), *Hopi White Lima* (2004), *Horsehead* (1998), *Horticultural* (1998), *Horticultural No. 4, Dwarf* (1998), *Horticultural, Cranberry* (2004), *Horticultural, Dwarf Longpod* (1994), *Horticultural, Scarlet Beauty Elite* (2004), *Humason's Best* (2004), *Idaho Pinto* (1984), *Immigrant* (2004), *Italian Zebra* (2004), *Japanese* (2004), *Kahl, White* (2004), *Kenearly Baking Bean* (1998), *Kidney, Dove Creek* (1991), *Kidney, Large Royal* (2004), *Kidney, Wine Red* (1994), *Kingston Black Haybeans* (1987), *Krol* (2004), *Kutzner's Russian Soup* (2004), *Large White Marrow* (1984), *Larry Locke's* (2004), *Lemon Yellow* (1991), *Lucas Navy* (2004), *Maccarone* (2004), *Madeira* (1998), *Mansell Magic* (2004), *Mayo Azufrado* (1987), *Mecosta Light Red Kidney* (1984), *Mennonite* (2004), *Missouri Bill's* (2004), *Mojave* (2004), *Molasses Face* (2004), *Montezuma Black* (1994), *Moreno* (2004), *Mortgage Lifter* (1994), *Mrs. Kramer's* (2004), *Mt. Pima Azufrado* (1998), *Mt. Pima Frijol Canelo* (1991), *Navy, French* (2004), *Navy, Gratiot* (1994), *Navy, Michilite* (2004), *Navy, Seaway* (1998), *Negro Taxumal* (2004), *Nombre de Dios Vayo Blanco* (1994), *Old Green* (1983), *Old Rumanian* (2004), *Oregon White Light Lima* (1998), *Othello* (2004), *Peruvian Goose* (2004), *Pinto (UI #111)* (2004), *Pinto (WY #166)* (1982), *Pinto II* (1987), *Pinto, Agate (Black)* (1994), *Pinto, Mayo Colima* (2004), *Pinto, San Juan* (1998), *Pinto, Tarahumara Black* (2004), *Plata* (2004), *Poroto Granada* (2004), *Puerto Rican Pinto* (1987), *Raquel* (1998), *Red Bean* (1991), *Red Beans* (1994), *Red Eye* (2004), *Red Jason* (2004), *Red Seeded Flageollet* (1987), *Red Speckled Bayou* (1998), *Rhodenizer's* (2004), *Rockwell* (2004), *Rowley* (2004), *Ruby Dwarf Horticultural* (1991), *Sacramento* (1994), *Salad* (1991), *San Luis Potosi Black Bean* (1991), *San Luis Potosi Flor de Mayo* (1991), *Sarah Ross Black Bean* (2004), *Sinaloa Azufrado* (1998), *Soissons Nain Blanc* (1998), *Soissons Nain Hatif* (1984), *Soldier, European* (2004), *Sonoran Canario* (1994), *Sonoran Vayo* (1987), *Spaulding* (2004), *Stinger* (2004), *Suppenbohne Wybelsum* (2004), *Tarahumara Azufrado (No. 2)* (1987), *Tarahumara Black Large* (1991), *Tarahumara Black Medium Size* (1998), *Tarahumara Chiba Busira Mix* (1991), *Tarahumara Cinco Minutos* (1998), *Tarahumara Frijol Cuerenteno* (1998), *Tarahumara Gauapata con Laja* (1998), *Tarahumara Laja and Azufrado* (1998), *Tarahumara Multis*

(1998), *Texas Pink* (2004), *Tohono O'Odham Red* (1994), *Vert Suma* (1987), *Viva Pink* (1983), *Wade Bush* (1982), *Warihio Bolitas* (1991), *Western Tarahumara Amarillo* (1991), *White Hailstones Navy* (2004), *White Improved Navy* (1998), *White Pea Navy* (1983), *White Wonder* (1987), *Wild Rice Bean* (1998), *World War* (2004), *Yaqui Ojo de Cabra* (1991), *Yellow Eye Improved* (1984), *Yellow Eye, Kenearly* (2004), *Yellow Eye, Maine Improved* (2004), *Yellow Seeded Flageolet* (2004), *Zert* (2004).

Bean / Bush / Snap
Phaseolus vulgaris

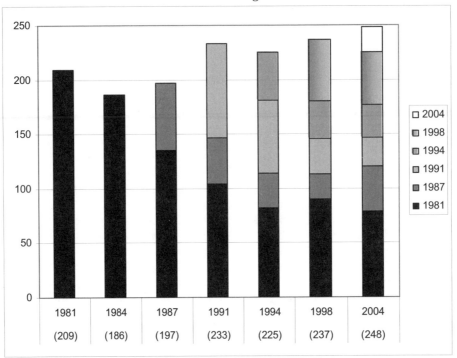

Number of varieties listed in 1981 and still available in 2004: 209 to 79 (38%).
Overall change in number of varieties from 1981 to 2004: 209 to 248 (119%).
Number of 2004 varieties available from 1 or 2 sources: 144 out of 248 (58%).

Admires - 50 days - Early, long stringless 1 in. wide pods, large white seeds, tasty, BCMV and AN res., Europe. *Source History: 1 in 1987; 1 in 1991; 1 in 1994; 1 in 1998; 2 in 2004.* **Sources: Dam, Pr3.**

Alicante - 59 days - Straight, slender dark green pods, 5.5-6 in., very slow seed development, best for summer to early fall harvest, tolerates AN, BCMV. *Source History: 1 in 2004.* **Sources: OSB.**

Ambra - (HMX 0104) - 52-58 days - Upright plants, smooth medium green, straight 6 in. pods held high on the plant, holds well in the field, tolerates BCMV and CT. *Source History: 7 in 2004.* **Sources: CH14, Cl3, HA20, ME9, RIS, SE4, Sto.**

Amigo - PVP expired 1992. *Source History: 1 in 2004.* **Sources: Se17.**

Aramis - 53-70 days - Stringless filet var. adapted to mech. harv., round 5.5 in. med-green pods with some purple streaks, vigorous erect 14-16 in. plants, res. to BCMV and AN, 1984. *Source History: 1 in 1981; 1 in 1984; 1 in 1987; 2 in 1991; 3 in 1994; 2 in 1998; 1 in 2004.* **Sources: Pr3.**

Astro - 50-53 days - Nearly round straight smooth med- dark green 6-6.25 x .4 in. pods, white seed, BCMV and NY15 res., 18- 21 in. vigorous plant, mainly for shipping, widely adapted, 1966. *Source History: 15 in 1981; 13 in 1984; 11 in 1987; 5 in 1991; 2 in 1994; 2 in 1998; 2 in 2004.* **Sources: CON, La1.**

Aunt Alley - Early, thick tasty pods, heirloom. *Source History: 1 in 1987; 1 in 1998; 1 in 2004.* **Sources: Te4.**

Aunt Lara's - Ozark Mountain heirloom, produces many small tan seeds splotched with black, snap and shell. *Source History: 1 in 1998; 1 in 2004.* **Sources: Ho13.**

Baby Bop - 51 days - Gourmet bean with delicate flavor, compact plant, tiny 4 in. pods held high on the plant, cooks quickly, high yields, good for canning and freezing, PVP 2002 Seminis Vegetable Seeds, Inc. *Source History: 3 in 2004.* **Sources: Ers, MAY, Ver.**

Baccicia - (Bachicha) - 52-60 days - Bush habit, tender 6.5 in. pods, dull red and white speckled seed, distinct flavor, not recommended for freezing, Italian heirloom. *Source History: 3 in 1981; 2 in 1984; 2 in 1987; 2 in 1991; 2 in 1994; 2 in 1998; 3 in 2004.* **Sources: Loc, Se7, Se17.**

Baroma - 58 days - Bush Romano grows 18 in. tall,

medium green, stringless, flat pods, best harvested at 6 in. length. *Source History: 1 in 2004.* **Sources: Ni1.**

BBL 274 - 60 days - Large, spreading bush, 5.5-6 in. pods, fine flavor and texture, small white seeds dev. slowly. *Source History: 1 in 1998; 1 in 2004.* **Sources: OSB.**

Benchmark - 53-55 days - Erect plant, medium dark green 6 in. pods held high on the plant, compares to Bronco, concentrated maturity, BCMV NY15, PVP 2001 Novartis Seeds, Inc. *Source History: 7 in 1998; 15 in 2004.* **Sources: CH14, Fe5, HO2, Jor, Jun, La1, Loc, Mcf, ME9, MOU, RIS, RUP, SE4, Ter, Ver.**

Benton - (Southern Belle) - 59 days - Straight 6 in. dark-green pods in the Tendercrop class, excel. color, tolerant to bacterial brown spot and NY15, good heat tol., PVP expired 2004. *Source History: 3 in 1991; 4 in 1994; 3 in 1998; 1 in 2004.* **Sources: Bu3.**

Blue Bloom - 90 days - Old Tennessee hills semi-runner with 3 in. pods, brown-black seeds with violet tinge at shelly stage. *Source History: 1 in 1998; 2 in 2004.* **Sources: Ho13, Se17.**

Blue Lagoon - 55-60 days - The best of the Blue Lake beans, high quality, tender, 5-6 in. pods set on 18-24 in. erect plants, sweet flavor, great for canning, freezing or fresh use, not recommended for shipping. *Source History: 12 in 1998; 6 in 2004.* **Sources: Fa1, Fi1, HO2, MIC, Ni1, Twi.**

Blue Lake Bush - 50-65 days - AAS/1961, vigorous spreading 15-18 in. bush, round straight stringless tender 6-7 in. pods, good sweet flavor, seeds and fiber form slowly, holds in good condition longer, prolific, heavy yields, white seed, excel. freezer, res. to BCMV and NY15. *Source History: 54 in 1981; 49 in 1984; 51 in 1987; 43 in 1991; 37 in 1994; 32 in 1998; 31 in 2004.* **Sources: Ag7, Alb, All, Be4, Com, Cr1, De6, Ear, Ers, Fis, Fo7, Gr27, He8, He17, Hig, Hum, Lin, LO8, Ma19, Mo13, Ont, Or10, Pr3, Ra5, Ri12, Ros, Sau, Se7, Shu, Tu6, Up2.**

Blue Lake Bush 274 - 54-61 days - Dev. from Blue Lake Pole, dark-green round/creaseback 5.5-6.5 in. pods, slow seed dev., long season, 12-22 in. plants, res. to BCMV and NY15, white seed, heavy yields, processor type, mech. harv., fine texture and quality, seed and fiber slow to develop, widely adapted, 1961. *Source History: 38 in 1981; 38 in 1984; 43 in 1987; 51 in 1991; 50 in 1994; 61 in 1998; 56 in 2004.* **Sources: Ada, Bo17, Bu1, Bu2, Bu8, CA25, CH14, CON, Coo, Dam, Dgs, DOR, Eco, Fa1, Fe5, Fi1, Ga1, GRI, Gur, HA3, Ha5, Hi13, HO2, Jor, Jun, Kil, La1, Loc, MAY, ME9, Mel, Mey, Mo13, MOU, OLD, PAG, Par, Pin, RIS, Roh, RUP, SE4, Se26, Sh9, Si5, Sk2, So1, SOU, Sto, TE7, Twi, Ver, Vi4, WE10, Wet, Wi2.**

Blue Lake Select - Prolific, produces well over a long season, used for pickling, freezing and fresh eating. *Source History: 1 in 1994; 1 in 1998; 1 in 2004.* **Sources: Syn.**

Blue Wonder - 55 days - Medium green 5.5 in. pods held high on the plant, good flavor, high yields, resists BCMV. *Source History: 1 in 2004.* **Sources: Par.**

Boby Bianco - Thin crisp pods, sweet flavor. *Source History: 1 in 2004.* **Sources: Se24.**

Borlotto - (Borlotti Bush) - 68-73 days - Bright rosy-red and cream colored pods, grow like regular bush beans and harvest when beans swell in the pods, boil or steam lightly until tender, also used as dry bean, Italian heirloom. *Source History: 1 in 1991; 1 in 1994; 4 in 1998; 7 in 2004.* **Sources: Coo, Fo7, Ho13, Na6, Se8, Shu, So25.**

Borlotto Lingua di Fuoco Nano - 70 days - Cream colored pods speckled with red, 5-7 seeds per pod, creamy white with red splotches, fresh shell or dry bean. *Source History: 1 in 1987; 1 in 1991; 1 in 1994; 1 in 1998; 2 in 2004.* **Sources: Ber, Se24.**

Borlotto of Vigevano - 70 days - Classic borlotto for fresh cooked shell beans and soups. *Source History: 1 in 2004.* **Sources: Se24.**

Bountiful - (Early Six Weeks, Bountiful Stringless, Early Bountiful) - 42-51 days - Straight flat broad stringless 6-7 in. pods, heavy yielding, 16-18 in. plants, good for freezing, light-tan seeds, yellowish-green foliage, home garden or ship, resists rust, mildew and beetles, intro. in Genesee County NY in 1898. *Source History: 43 in 1981; 35 in 1984; 30 in 1987; 25 in 1991; 22 in 1994; 27 in 1998; 26 in 2004.* **Sources: All, Co32, Com, CON, Dow, Eco, Fo7, HA3, He8, Ho12, La1, ME9, Mey, Or10, Pin, Ra5, Roh, Se7, Se16, Se26, Sh13, Shu, So9, Ver, Wet, Wi2.**

Brio - 54 days - Fresh market variety with medium dark-green slim 5 in. pods, high yields, excellent freezer, BCMV, PVP 1992 Seminis Vegetable Seeds, Inc. *Source History: 3 in 1994; 6 in 1998; 7 in 2004.* **Sources: HO2, Jor, ME9, RUP, SE4, Sto, Ver.**

Bronco - 53-58 days - Asgrow variety resembling Strike except the pods are med. dark-green and shiny and mature slightly earlier, white seeds, high yields, BCMV res., PVP 1990 Seminis Vegetable Seeds, Inc. *Source History: 1 in 1987; 3 in 1991; 3 in 1994; 11 in 1998; 13 in 2004.* **Sources: CH14, HO2, Jor, La1, ME9, OSB, RIS, Roh, RUP, SE4, Si5, Sto, We19.**

Buckskin Girl - 80-90 days - Long, narrow pods contain seeds in many shades of slate green-grey with a darker spot on the hilum, delicate, sweet flavor. *Source History: 1 in 1998; 1 in 2004.* **Sources: Se17.**

Canadian Wonder - Vigorous hardy prolific bushy plants, broad flat 7-8 in. pods, dark-red seeds, hard to beat for general use, popular variety from England. *Source History: 2 in 1981; 2 in 1984; 1 in 1987; 1 in 1991; 1 in 1994; 2 in 1998; 1 in 2004.* **Sources: Ho13.**

Capitole - 52-75 days - Huge cropper produces round pods perfect for serving whole as a gourmet baby bean, BCMV res., heat tol. *Source History: 1 in 1998; 1 in 2004.* **Sources: Tho.**

Capricorn - 59 days - Straight 6 in. round pods held high off the ground, hand pick or mechanical harvest, white seeds, resists BCMV, NY15, NL8 and CTV, PVP 2002 Syngenta Seeds, Inc. - Vegetables. *Source History: 5 in 2004.* **Sources: Cl3, Ers, RIS, RUP, SE4.**

Carlo - 55 days - Improved Strike variety, 5 in. dark green pods held high on the upright plant, BCMV, PVP 2001 Seminis Vegetable Seeds, Inc. *Source History: 3 in 1998; 3 in 2004.* **Sources: Jor, RUP, SE4.**

Carr Virginia - 75-85 days - Used as snap or dry bean, good for high altitude areas, brought to Colorado from

Virginia in the late 1800s. *Source History: 1 in 2004.* **Sources: Ri12.**

Caruso - Filet bean, 8 in. round green pods, excellent flavor, productive, resists BCMV and AN. *Source History: 1 in 2004.* **Sources: Coo.**

Castano - 55 days - Dark green, 5.5 in. pods, good disease resistance, PVP 2002 Syngenta Seeds, Inc. - Vegetables. *Source History: 1 in 2004.* **Sources: RUP.**

Charon - 58 days - High yields of 6 in. long straight pods, white seeds, tender and flavorful, fresh, canning and freezing, productive even when stressed, adaptable, resists BCMV, NL8, CTV and HB, Novartis var., PVP 2002 Syngenta Seeds, Inc. - Vegetables. *Source History: 9 in 2004.* **Sources: CH14, Cl3, Ers, HO2, Jun, ME9, RUP, SE4, Ver.**

Chevrier Vert - 68 days - Most valued shell bean in France, very productive 24 in. plants, each pod contains 6 or 7 seeds, used as young green beans or green shells or dried beans, originated in 1878. *Source History: 1 in 1981; 1 in 1984; 1 in 1987; 2 in 1991; 2 in 1994; 3 in 1998; 3 in 2004.* **Sources: Gou, Se8, So12.**

Cloudburst - 55 days - Upright plants, 6 in. round, medium light green pods held high, white seeds, harvests and ships well, BCMV, PVP 2001 Seminis Vegetable Seeds, Inc. *Source History: 2 in 2004.* **Sources: Jor, SE4.**

Coco Nain Blanc Precoce - 60 days - Early standard variety, white seeded, for fresh use or shell beans, from Holland. *Source History: 1 in 1981; 1 in 1984; 2 in 1998; 2 in 2004.* **Sources: Gou, Ho13.**

Commodore - (Bush Kentucky Wonder, Dwarf Kentucky Wonder, Commander Freezer) - 58-63 days - AAS/1938, large upright 15-21 in. plants, heavy yields, fleshy round 7.5-8.5 x .5 in. dark-green stringless pods, cooks in half the time, very tasty and tender. *Source History: 30 in 1981; 27 in 1984; 29 in 1987; 24 in 1991; 22 in 1994; 23 in 1998; 15 in 2004.* **Sources: Ada, Be4, Bu8, CH14, CON, DOR, He8, ME9, Mey, MOU, Ros, RUP, Sau, Se26, Sh9.**

Commodore Improved - (Bush Kentucky Wonder) - 58-65 days - AAS/1945, vigorous upright open 16 in. plants, dark-green round stringless 6.5-8 in. slightly curved pods, reddish-purple long round seeds, excel. flavor and quality. *Source History: 20 in 1981; 15 in 1984; 12 in 1987; 13 in 1991; 10 in 1994; 9 in 1998; 5 in 2004.* **Sources: Co31, Dgs, Kil, Wet, Wi2.**

Comrades - (Ukrainian Comrades) - 50-55 days - Rare Ukrainian bean, lavender blooms, yellow pods with shiny black or gold seeds, some green pods, early ripening. *Source History: 2 in 2004.* **Sources: Se17, Ta5.**

Concesa - 55 days - Large, upright plant bears heavy yields of slender, glossy dark green round pods, 5.5-6 in., tolerates R, BCMV, CalMV, AN. *Source History: 3 in 2004.* **Sources: HA20, OSB, SE4.**

Contender - (Buff Valentine, Early Contender, Contender Stringless) - 40-55 days - Dark-green 12-20 in. plants, med-green round/oval stringless 6-8 x .5 in. pods, buff mottled seeds, res. to BCMV NY15 GP and heat, S.E. Veg. Breeding Lab/SC, 1949, possibly derived from pre-1855 'Early Valentine'. *Source History: 79 in 1981; 72 in 1984; 69 in 1987; 56 in 1991; 53 in 1994; 48 in 1998; 56 in 2004.* **Sources: Ada, Alb, Ba8, Be4, Bo17, Bo19, Bou, Bu2, Bu8, CH14, Co31, CON, Dgs, DOR, Ech, Ers, Fa1, Fi1, Fis, Fo7, Goo, Gr27, Gur, HA3, He8, He17, Hi6, Hi13, Hum, Jor, Kil, La1, LO8, MAY, ME9, Mey, Mo13, OLD, Pe2, Red, Ri12, Ros, RUP, Sa9, Scf, Se26, Shu, Sk2, So1, SOU, TE7, Tu6, Ver, WE10, Wet, Wi2.**

Corbett Refugee - *Source History: 1 in 2004.* **Sources: Se17.**

Daisy - (Daisy Haricot Nain) - 55 days - Sets meaty 5-7 in. pods above foliage, good flavor and quality, seeds dev. slowly for longer season, heavy yields, from Germany, BCMV res., white seed, can or freeze. *Source History: 19 in 1981; 16 in 1984; 15 in 1987; 9 in 1991; 3 in 1994; 1 in 1998; 1 in 2004.* **Sources: Lej.**

Dandy - 54 days - Upright compact bush, slender straight dark-green 4 in. pod, high yields, low in fiber, small seeds dev. slowly, gourmet-type small haricot verts, PVP expired 2001. *Source History: 3 in 1981; 3 in 1984; 7 in 1987; 8 in 1991; 9 in 1994; 7 in 1998; 2 in 2004.* **Sources: Tt2, Twi.**

Dart - Fresh market var., upright plant, smooth straight pods set high on the plant, early maturity. *Source History: 1 in 2004.* **Sources: HA20.**

Daytona - 54-58 days - Fresh market and shipping var. with superior performance during hot weather, excel. for late season harvest, 6-7 in. pods, white seed, BV1 and rust res., dev. by Ferry-Morse, PVP application abandoned. *Source History: 4 in 1998; 3 in 2004.* **Sources: Se26, Sto, Twi.**

Derby - (FM-175) - 57 days - AAS/1990, strong upright weather-tolerant disease-resistant plants, pods up to 7 in. long with slow seed development for prolonged harvest, resistant to lodging and BCMV, introduced in 1990, PVP 1992 HA20. *Source History: 45 in 1991; 46 in 1994; 51 in 1998; 40 in 2004.* **Sources: Be4, Bou, Bu8, CA25, CH14, Cl3, Co31, Com, Dam, Ear, Fi1, Gr27, GRI, Gur, HA3, HA20, HO2, Jor, Jun, La1, MAY, ME9, Mey, Mo13, Ni1, Or10, PAG, RIS, RUP, Sau, Sh9, Shu, Si5, SOU, Sto, Twi, Ver, Wet, Wi2, Yu2.**

Dieul Fin Precoce - (Petit Gris) - The old favorite Petit Gris, long, large purple streaked pods, large purple speckled seeds, crops well, very fine flavor. *Source History: 1 in 1987; 1 in 1991; 1 in 1994; 2 in 1998; 2 in 2004.* **Sources: Bou, Pr3.**

Distinction - 58 days - Medium dark green pods to 6 in., white seeds, good shipper. *Source History: 1 in 2004.* **Sources: RUP.**

Dona Bobolink - Beautiful plump seeds, white with an irregular burgundy pattern. *Source History: 1 in 2004.* **Sources: Se17.**

Dusky - 52-57 days - Straight, dark green round pods, 6 in. long, white seeds, fresh market and shipping, BCMV, NY15, NL8, BS and HB, PVP 2003 Syngenta Seeds, Inc. - Vegetables. *Source History: 6 in 2004.* **Sources: CH14, Cl3, RIS, RUP, SE4, Sto.**

Eagle - 54-57 days - Round med-green straight slender 6 in. pods, widely adapted Southern canner, high yields, sensitive to brown spot, res. to BCMV and NY15, PVP expired 1991. *Source History: 3 in 1981; 3 in 1984; 3 in 1987; 3 in 1991; 2 in 1994; 5 in 1998; 1 in 2004.* **Sources: HO2.**

Earliserve - (Early Serve) - 41-56 days - Of Blue Lake Bush and Slenderette heritage, very early, sturdy upright plants 18 in. high x 18 in. wide, straight slender 4 in. pods set high on the plant for easy picking, true Blue Lake color, prolific, because of earliness and slenderness beans should be picked more often than other varieties, good flavor, BCMV tol., home garden or fresh market or processing, PVP application abandoned 1988. *Source History: 6 in 1987; 15 in 1991; 15 in 1994; 12 in 1998; 10 in 2004.* **Sources: Bu3, Ers, Fo7, Hig, HO2, La1, Me7, Sk2, Vi4, Wet.**

Empress - (Experimental Bean 121) - 55 days - Straight dark-green stringless meaty 5.75-6 in. pods, white seed, fine quality, for fresh or frozen, originally introduced by Gurney's Seed Co. as Experimental Bean 121, re-named Empress in 1979, PVP expired 2000. *Source History: 3 in 1981; 2 in 1984; 4 in 1987; 2 in 1991; 1 in 2004.* **Sources: Se16.**

Envy - Blue Lake type, uniform dark green pods, 4-5 sieve range, slow seed development allows for long harvest period, fresh market, PVP 1993 HA20. *Source History: 1 in 2004.* **Sources: HA20.**

Espada - 56 days - Large vigorous bush plants, dark-green round 6 in. pods, for fresh market or processing, res. to BCMV NY15 HB2 and AN, PVP 1992 HA20 *Source History: 1 in 1991; 2 in 1994; 1 in 1998; 3 in 2004.* **Sources: Ha5, HA20, Wi2.**

Eva's Chow Chow Beans - Prolific, small bush plant, white seed with burgundy eye, used as green beans, also good dried, resist bugs and disease, 150 year-old Pennsylvania heirloom from 87 year-old Eva, grown by her Pennsylvania German family for 5 generations. *Source History: 1 in 2004.* **Sources: Ami.**

E-Z Pick - 55-75 days - A Blue Lake type stringless bean on a new type of plant, 24 in. plants have thick stiff upright stems which do not flop, this keeps beans high and clean and protects them from mold in damp weather, straight round 6-6.5 in. pods hang freely, bred-in easy release makes for easy manual and mech. harvesting with less plant damage, white seed, good texture, for fresh use or freezing or canning, PVP 1987 Novartis Seeds, Inc. *Source History: 3 in 1987; 7 in 1991; 11 in 1994; 2 in 1998; 3 in 2004.* **Sources: Fa1, Jo1, Pr3.**

Fandango - 56-62 days - Erect plant with small leaves, medium green 5-6 in. round pods, white seeds develop slowly, BCMV AN, fresh market and processing. *Source History: 3 in 1998; 2 in 2004.* **Sources: OSB, SE4.**

Festina - 56 days - Blue Lake type, erect plants with very straight 6 in. pods, white seeds, excellent flavor, for cooking, freezing and canning, excellent shipper, mechanical harvest, withstands heat and drought, BCMV-1, PVP 2001 Seminis Veg. Seeds, Inc. *Source History: 6 in 2004.* **Sources: HO2, Mcf, OSB, RUP, SE4, Ver.**

Fin de Bagnols - (Shoestring Bean) - 50-60 days - French string bean, long fine pods, best when picked every 2 or 3 days when very young before strings appear, old European gourmet variety first grown over one hundred years ago. *Source History: 6 in 1981; 5 in 1984; 6 in 1987; 3 in 1991; 4 in 1994; 4 in 1998; 7 in 2004.* **Sources: Coo, Gou, Na6, Pr3, Se16, So12, Te4.**

Finaud - 70 days - Perfectly straight slender pods are borne very early on short sturdy plants, holds much longer at the ideal 4-6 in. long stage than other vars., two pickings 2-3 weeks apart give best results, introduced in 1988. *Source History: 2 in 1991; 2 in 1994; 1 in 1998; 1 in 2004.* **Sources: Se8.**

Firstmark - 49 days - Med. green 6 in. pods set high on the plant, white seeds, good disease tol., mechanical or hand harvest, similar to Bronco in fiber, PVP applied for. *Source History: 2 in 2004.* **Sources: DOR, ME9.**

Flageolet Rouge - Flageolets can be eaten fresh or dry, for dry flageolets pick up the whole plant when the pods are full and hang the bundles upside down in the shade away from the sun and rain. *Source History: 1 in 1987; 1 in 1991; 1 in 1994; 1 in 1998; 1 in 2004.* **Sources: Lej.**

Flaro - 67 days - Flageolet best eaten as green shell bean, also dried, rich in iron, thiamine and protein. *Source History: 3 in 2004.* **Sources: Ers, La1, Ver.**

Florence - 49-56 days - Flavorful, early, high yielding variety with good heat tolerance, erect plants for easy harvest, 6.5 in. dark-green pods, white seeds, sets well in hot weather, BCMV, AN and white mold res., can or freeze. *Source History: 3 in 1994; 16 in 1998; 10 in 2004.* **Sources: Ers, Fi1, Ha5, Par, Pin, Pr3, SE4, Sh9, Ver, We19.**

Fordhook Standard - 57 days - Small leafed plant, sweet fleshy pods, easy harvest, trademarked. *Source History: 1 in 2004.* **Sources: Bu2.**

Fowler - 51-70 days - Compact plants, 16 in. tall, 5 in., slender, round, stringless pods, 7 dark brown seeds per pod, withstands heat and drought, large leaves suppress weeds, tender and flavorful, Oregon heirloom from Don Fowler. *Source History: 1 in 1998; 1 in 2004.* **Sources: So1.**

French Filet - 50-56 days - Straight pods, best at 5-7 in. length, unusual nutty flavor. *Source History: 3 in 2004.* **Sources: Bo17, Bu2, Hi13.**

French Flageolet - 90 days - Compact bush plants, green pods with 5 or 6 seeds, for snaps/green shell/dry, dry beans seem to produce own sauce, famous gourmet bean from south of France. *Source History: 1 in 1981; 1 in 1984; 2 in 1987; 1 in 1991; 1 in 1994; 1 in 1998; 1 in 2004.* **Sources: Ni1.**

Frenchy - (Frenchie) - 43-53 days - Gourmet bean, 12 in. plants, dark-green round slender 3.75 in. pods, extremely tender, much milder and sweeter, BCMV tol., finest quality, PVP expired 1999. *Source History: 2 in 1981; 2 in 1984; 2 in 1987; 3 in 1991; 2 in 1994; 2 in 1998; 1 in 2004.* **Sources: Ra5.**

Full Measure - *Source History: 1 in 2004.* **Sources: Se17.**

Gaia - 55 days - Unique romano type, 18-20 in. bush, 5-6 in. round green pods with large brown seeds which make a savory soup, good flavor. *Source History: 1 in 1998; 1 in 2004.* **Sources: Pe1.**

Gator Green - 53-55 days - Round 6-7 in. pods, holds

firmness and flavor after picking, retains color and flavor when frozen, yields well in poor conditions, slim pods, white seed, market. *Source History: 9 in 1981; 6 in 1984; 6 in 1987; 5 in 1991; 6 in 1994; 6 in 1998; 2 in 2004.* **Sources: HO2, Sh9.**

Gator Green 15 - 53 days - Upright plants, slender med-green round/oval 6-7 in. pods, high yields, matures all at once in midseason, res. to BCMV, for fresh market, white seed, PVP expired 1991. *Source History: 6 in 1981; 5 in 1984; 5 in 1987; 9 in 1991; 7 in 1994; 8 in 1998; 4 in 2004.* **Sources: Mey, RUP, Se26, Twi.**

Giant Stringless - 53-58 days - Earlier str. of Burpee Stringless, vigorous semi-spreading 16-18 in. plants, good yields of round light-green straight stringless meaty pods up to 12 in. long, good freezer. *Source History: 9 in 1981; 8 in 1984; 6 in 1987; 5 in 1991; 5 in 1994; 6 in 1998; 4 in 2004.* **Sources: CON, Fo7, La1, Wet.**

Gina - 55 days - Bush Romano, strong upright plants, thick 5-6 in. long pods, sweet, beany flavor, crisp texture, good yields, excellent disease resistance, PVP expired 1991. *Source History: 1 in 1998; 2 in 2004.* **Sources: Ver, We19.**

Grandma Stober's Chow Chow - 90 days - Similar to New England's Soldier bean, white seeds with red eye, good producer, excel. in chow-chows and relishes, from Lancaster County Pennsylvania farm family. *Source History: 1 in 1994; 2 in 1998; 2 in 2004.* **Sources: Lan, Roh.**

Granny Easley's - *Source History: 1 in 2004.* **Sources: Se17.**

Green Bean - Round 8 in. pods, crisp, sweet, disease res., pick for cooking use when young and stringless, high yielding. *Source History: 1 in 1987; 1 in 1991; 1 in 1994; 1 in 1998; 1 in 2004.* **Sources: Ev2.**

Green Isle - 55 days - Round green pod snap bean, straight 6-8 in. pods held well above ground, early and heavy and long yield, good freezer and canner, white seed, tol. to BCMV, 1967. *Source History: 11 in 1981; 7 in 1984; 5 in 1987; 4 in 1991; 6 in 1994; 3 in 1998; 1 in 2004.* **Sources: CON.**

Green Ruler - 51 days - Flat med-green pods, low fiber and string, spreading upright 19 in. plants, res. to BCMV, Dr. Honma at Michigan State University, an improved Bush Romano, earlier with larger pods. *Source History: 6 in 1981; 4 in 1984; 5 in 1987; 4 in 1991; 5 in 1994; 4 in 1998; 2 in 2004.* **Sources: CON, HO2.**

Greencrop - (Bush Kentucky Wonder) - 45-55 days - AAS/1957, upright 18-20 in. plants, flat-oval stringless med-green 6.5-8 x .5 in. dia. pods, white seed, long picking stage, bred for the North by NH/AES and USDA, 1956, tol. to mosaic, crop holds up well, matures at one time, widely adapted. *Source History: 29 in 1981; 25 in 1984; 31 in 1987; 32 in 1991; 32 in 1994; 26 in 1998; 23 in 2004.* **Sources: Bu2, Bu3, Cl3, Co31, CON, Dam, Dgs, DOR, Fe5, Fo7, Hi13, Jor, Kil, ME9, Mey, Pr3, RUP, SE4, Se26, Sk2, Sto, Vi4, Wi2.**

Greencrop, Asgrow Strain - 55 days - Elite strain of Greencrop with same size, appearance and disease tolerance, white seed. *Source History: 1 in 2004.* **Sources: Sto.**

Greensleeves - 56 days - Dark-green straight round pods held well off the ground, res. to BCMV and NY15, excellent for freezing, sturdy plants, PVP expired 1993. *Source History: 3 in 1981; 2 in 1984; 3 in 1987; 3 in 1991; 1 in 1994; 1 in 1998; 2 in 2004.* **Sources: Bu2, Hi13.**

Grenoble - 52-62 days - Bronco/Matador type, long round 5.5-6 in. pods, straight, white seeds, fresh use, freezing, BCMV, A, HB, Royal Sluis variety. *Source History: 6 in 2004.* **Sources: Jo1, Jor, RIS, RUP, SE4, Sto.**

Haricot Vert, Cupidon - 55 days - French filet bean, 18-22 in. plant, 6-8 in. stringless, medium green pods, flavorful, productive over long season, brown seeds. *Source History: 1 in 2004.* **Sources: Ter.**

Haricot Vert, Delinel - 51-64 days - French bean, vigorous, 18 in. plants, violet flowers, straight pencil-thin 6.5-7 in. pods, unlike many haricot verts it stays truly stringless, gather every 2-3 days, BCMV and AN res. *Source History: 2 in 1981; 2 in 1984; 4 in 1987; 3 in 1991; 6 in 1994; 13 in 1998; 7 in 2004.* **Sources: Com, Dam, Fo7, Mcf, Pr3, Tho, Ver.**

Haricot Vert, Maxibel - 51-65 days - The first full-size, high-quality filet on the market, sturdy bush plants bear straight, dark green 7-8 in. pods, stringless, excellent flavor, brown/tan seeds, favored by specialty market growers. *Source History: 15 in 1998; 25 in 2004.* **Sources: Be4, Dam, Fe5, Gou, Hi6, HO2, Jo1, Jor, Jun, Lej, ME9, OSB, Par, Pin, RUP, SE4, Se7, Se8, Shu, St18, Ter, Ver, Ves, VI3, We19.**

Haricot Vert, Nickel - 52-60 days - Replaces Astrel as the premier French filet baby bean on the market, stronger plant with more erect branches, better temperature tolerance, stronger resistance to foliar disease and root rot, bears stringless, 4 in. pods high off the ground, mostly 4 sieve size, excellent flavor, PVP 1997 Vilmorin, S.A. *Source History: 13 in 1998; 16 in 2004.* **Sources: Coo, HO2, Jor, ME9, Na6, Orn, RUP, SE4, Sto, Sw9, Ter, Und, Ver, VI3, We19, Wi2.**

Haricot Vert, Rolande - 55 days - Refined variety with extra crisp texture, long, extra slim, deep green 6 in. pods, sturdy, disease res. plants, fine French delicacy. *Source History: 2 in 2004.* **Sources: Pe2, Re6.**

Harvester - 50-60 days - Mech. harvest type, pods set high on hardy upright 21 in. plants, round med-green straight stringless 5-6 x .4 in. pods, BCMV RR NY15 and rust res., Northern marketer, tender smooth pods, widely adapted, 1957. *Source History: 36 in 1981; 31 in 1984; 24 in 1987; 20 in 1991; 16 in 1994; 12 in 1998; 7 in 2004.* **Sources: Ber, CON, DOR, Mey, MOU, WE10, Wi2.**

Heavyweight II - 53 days - Stocky 18 in. plants, sweet tender 8 in. pods, disease resistant, productive. *Source History: 1 in 2004.* **Sources: Bu2.**

Hialeah - 53 days - Gator Green type, round 5.5 to 6.5 in. pods set high on sturdy, upright plants, white seeds, high yield and recovery, good for mechanical harvest, BCMV, PVP 1992 HA20. *Source History: 5 in 1994; 13 in 1998; 9 in 2004.* **Sources: CH14, Cl3, HA20, HO2, ME9, RIS, RUP, SE4, Sto.**

HMX 5991 - 55 days - *Source History: 1 in 2004.* **Sources: SE4.**

Horto - 55-72 days - Imp. hort. bean with stringless, flat, 6 in. pods, red speckled seed, use as snap, shell or dry bean. *Source History: 3 in 1998; 2 in 2004.* **Sources: Hum, SE4.**

Hurricane - 48 days - Excellent early shipping var., light green 5.5 in. round pods, white seeds, BCMV and bean rust tolerant, PVP 2002 Seminis Veg. Seeds, Inc. *Source History: 1 in 2004.* **Sources: SE4.**

Hystyle - (Petit Gris) - 53 days - Medium early, tender round medium dark-green pods, good flavor, white/green seed, BCMV NY15 and BS res., tol. to lodging, heavy yields, local market or processing, PVP 1988 HA20. *Source History: 1 in 1987; 1 in 1991; 1 in 1994; 4 in 2004.* **Sources: Dam, HA20, RIS, Se26.**

Italian, Early Bush - 50 days - Erect, 16-18 in. plants, 5.5 in. pods, robust bean flavor, stringless, good for freezing. *Source History: 1 in 1998; 1 in 2004.* **Sources: Bu2.**

Jade - (GB45-1-3-1) - 53-60 days - Strong, upright bush plants, long round straight 5-7 in. pods with rich traditional bean flavor, easy to pick, produces quality pods later in the season than other beans, BCMV, curly top and rust res., PVP 1992 Novartis Seeds, Inc. *Source History: 13 in 1994; 26 in 1998; 27 in 2004.* **Sources: Be4, Bu3, CH14, Dam, De6, Eo2, Ers, Fe5, Gr27, HO2, Jo1, Jor, Jun, Loc, Mey, MOU, OSB, Pin, Pr3, Ri12, RIS, Roh, Sa5, SOU, Ter, Ver, We19.**

Jo Bean - 50-75 days - Kentucky heirloom, 3 in. pods contain small brown and tan streaked seeds. *Source History: 2 in 1998; 1 in 2004.* **Sources: Ho13.**

Joy - (Everbearing Joy) - 54-60 days - Improved Top Crop variety, sturdy 20 in. plants, straight round 5.5 in. med-dark green pods held high, dark-brown seed slow to dev., extended harvest, CT and BCMV res., vigorous hardy and flavorful. *Source History: 2 in 1981; 2 in 1984; 1 in 1987; 2 in 1994; 2 in 1998; 2 in 2004.* **Sources: Fo7, La1.**

Jumbo - (New Jumbo, Romano Jumbo, Jumbo Pod) - 54-70 days - Bush Romano type resulting from a cross of Romano x Kentucky Wonder, very large flat pods, bright dark-green, strong beany flavor, up to 12 in. long x 1 in. wide, best at 6-8 in., large brown striped seeds, gourmet quality, vigorous 15-16 in. bush plants may flop over with weight of pods, still stringless at 10 in., home and market, BCMV, PVP expired 2000. *Source History: 20 in 1981; 18 in 1984; 25 in 1987; 26 in 1991; 23 in 1994; 18 in 1998; 4 in 2004.* **Sources: Fis, Pe6, Pr3, Syn.**

Kebarika - 70-80 days - Beautiful late maturing bean, violet blossoms, wide 6 in. long pods, dried seeds are royal purple splashed with white, heat and drought tol., from Kenya, East Africa. *Source History: 3 in 1994; 2 in 1998; 1 in 2004.* **Sources: Syn.**

Kentucky Dreamer - 54-58 days - Imp. Bush Kentucky 125 type, upright plant, concentrated pod set, 6 in. pods, white seed, resists BCMV and NY 15, PVP applied for. *Source History: 4 in 2004.* **Sources: CH14, DOR, ME9, OLD.**

Kentucky Wonder No. 125 - (Bush Kentucky No. 125, Kentucky Wonder Exp. No. 125) - 51-59 days - Sturdy 20 in. plants, flat/oval 6-7 x .5 in. med-green pods borne high, low in fiber, distinctive flavor and heavy yields of KY Wonder in improved new bush form, good holding ability, fine color and shape, white seed, garden, market, shipping, BCMV NY15 res., PVP expired 2003. *Source History: 14 in 1981; 14 in 1984; 19 in 1987; 23 in 1991; 28 in 1994; 26 in 1998; 22 in 2004.* **Sources: Ada, Bu2, Bu3, CH14, Ers, Ha5, HO2, La1, Loc, MAY, ME9, Mo13, Par, RIS, Roh, RUP, SE4, Se26, Sh9, SOU, Twi, Wet.**

Kentucky Wonder, Bush - (Improved Commodore, Dwarf Kentucky Wonder, Kentucky Wonder Green Bush) - 52-65 days - Sel. from Kentucky Wonder pole, tender stringless round fleshy 6-8 in. pods, fine quality and flavor, heavy yielder for very long period, vigorous, brilliant carmine seeds, introduced in the late 1800s. *Source History: 13 in 1981; 13 in 1984; 11 in 1987; 13 in 1991; 8 in 1994; 13 in 1998; 14 in 2004.* **Sources: Bo19, CON, De6, Fi1, Gur, He8, He17, Jor, La1, Ma19, Ra5, Ri12, Se16, Shu.**

La France - 56 days - Filet type, holds well on the plant, no strings. *Source History: 1 in 1998; 1 in 2004.* **Sources: Bu2.**

Labrador - 56 days - Slim, dark green 5.5 in. pods, low fiber, high pod set, high yields, A, BCMV, PVP 1986 Seminis Vegetable Seeds, Inc. *Source History: 1 in 1981; 1 in 1984; 1 in 1987; 3 in 1991; 3 in 1994; 2 in 1998; 6 in 2004.* **Sources: Alb, Hal, HO2, Jor, ME9, RUP.**

Lake Largo - 58 days - Blue Lake type, round straight dark-green 6.5 in. pods, holds well, flavorful, very productive, excel. freezer, market growers and processing, PVP expired 1991. *Source History: 4 in 1981; 4 in 1984; 10 in 1987; 5 in 1991; 1 in 1994; 1 in 1998; 1 in 2004.* **Sources: Lin.**

Landmark - 52-55 days - Round med-green pod, 7 x 3/8 in., good yield potential, good North and South, BCMV and NY15 res., white seed, fresh market and shipping. *Source History: 1 in 1987; 7 in 1991; 9 in 1994; 7 in 1998; 3 in 2004.* **Sources: Bu3, DOR, Jor.**

Landstar - 52 days - Plant to 18 in., dark green round pods, 6.5-7 in., white seeds, excellent flavor, slow to become fibrous, high pod set suitable for machine or hand harvest, compare with Bronco. *Source History: 3 in 2004.* **Sources: DOR, ME9, OLD.**

Leon - 55 days - Medium light green 5.5 in. round pods, slow seed development allows for excellent field holding ability, resists BCMV, NY15 and NL8, PVP 2003 Syngenta Seeds, Inc. - Vegetables. *Source History: 2 in 2004.* **Sources: Cl3, SE4.**

Lizzie Miller - 60-70 days - White seed with tan pattern on hilum that looks like a woman in a shawl. *Source History: 1 in 2004.* **Sources: Se17.**

Lodi - PVP 1997 Seminis Veg. Seeds, Inc. *Source History: 1 in 1998; 1 in 2004.* **Sources: ME9.**

Macedonia Snap - *Source History: 1 in 2004.* **Sources: Se17.**

Magnum - (XP B211) - 51-55 days - Kentucky Wonder Bush type with significantly higher yields and straighter pods, flat med-green 7 in. pods with brown seeds, used for fresh market and shipping, PVP 1992 Seminis Vegetable Seeds, Inc. *Source History: 2 in 1991; 3 in 1994; 6 in 1998; 5 in 2004.* **Sources: Ga1, ME9, RUP, SE4, Ver.**

Magpie - (Superlative) - 60-65 days - Strong plants grow 24-28 in. tall, 7 in. pod, black and white seeds resemble the plumage of the magpie bird, French var. from the early 1900s. *Source History: 1 in 1991; 4 in 1994; 5 in 1998; 5 in 2004.* **Sources: Pr3, Ri12, Se7, Syn, Te4.**

Mammy - Plants grow 3 ft. with short runners, 4 in. pods, small white seed. *Source History: 1 in 1998; 1 in 2004.* **Sources: Ho13.**

Marconi Nano - 55-60 days - Green Roma type, bush plant. *Source History: 1 in 2004.* **Sources: Se24.**

Marconi Nano, Black - Medium size, green, flat, wide, stringless pods, black seeds. *Source History: 1 in 1987; 1 in 1991; 1 in 1994; 1 in 1998; 1 in 2004.* **Sources: Ber.**

Marconi Nano, White - Medium size, green, flat, wide, stringless pods, white seeds. *Source History: 1 in 1991; 1 in 1994; 1 in 1998; 1 in 2004.* **Sources: Ber.**

Marseilles - 52 days - Small French bean, 4.5 in. long, dark green, straight pods, HB tol., MV and AN res. *Source History: 3 in 1998; 2 in 2004.* **Sources: Fi1, Hum.**

Masai - 47-60 days - Short plant to 12 in., slim pods, 4.5 in. long, early, high yields, pod set concentrated on upper half of bush, good disease and cold tolerance, good for container growing and late summer planting, PVP 1997 Rogers Seed Co. *Source History: 1 in 1994; 1 in 1998; 3 in 2004.* **Sources: De6, Jun, Pin.**

Matador - 56-60 days - Dark green, 5.5-7 in. round pods, excellent standability, high pod set, hand pick, very high yield potential, good heat tolerance, white seeds, AN BCMV res., PVP 1996 Seminis Veg. Seeds, Inc. *Source History: 9 in 1998; 9 in 2004.* **Sources: HO2, Jor, Lin, ME9, RUP, SE4, Sto, Ver, Ves.**

Medinah - 53 days - Flavorful, tender green filet bean, 5 in. straight pods held high on upright plant, resists BCMV, AN, HB, 5 sieve size, PVP 2001 Novartis Seeds, Inc. *Source History: 2 in 2004.* **Sources: Eo2, HO2.**

Mercury - 54 days - Upright bush, round pods, white seeds, produces good concentrated pod set under high temperatures in southern trials, resists BCMV, R, fresh market, PVP 2002 Syngenta Seeds, Inc. - Vegetables. *Source History: 2 in 2004.* **Sources: SE4, Twi.**

Minuette - Blue Lake type, glossy dark green pods set high on the plant, fresh market, PVP 1999 HA20. *Source History: 1 in 2004.* **Sources: HA20.**

Mirada - 55 days - Imp. Blue Lake type, upright bush habit, 6-6.5 in. straight pods, BV1 and NY14 res., PVP 1996 Rogers Seed Company. *Source History: 3 in 1994; 7 in 1998; 4 in 2004.* **Sources: Bu8, Cl3, SE4, Twi.**

Modus - Bushy plants bear 6 in. long pods, resists BCMV, AN, HB, PVP 1992 Nunza B.V. *Source History: 1 in 2004.* **Sources: Tho.**

Montana Green - 50 days - Round med-green 6-7 in. pods, entirely stringless, crisp and tender, light-buff seeds. *Source History: 1 in 1981; 1 in 1984; 2 in 1987; 2 in 1991; 2 in 1994; 2 in 1998; 1 in 2004.* **Sources: Fis.**

Montano - 50 days - Improved sel. of Dutch Princess, vigorous erect plant habit, concentrated set of dark-green round very straight pods, slow seed dev., res. to most bean diseases, very productive, suited for canning and freezing. *Source History: 1 in 1991; 1 in 1994; 1 in 1998; 1 in 2004.* **Sources: Dam.**

Montpellier - 56-61 days - French filet, upright bush plants bear high quality pods can grow up to 7.5 in. and remain tender and slim, delicate flavor, easily harvested pods borne high on the plant, harvest before .25 in. diameter for best quality, slow to dev. seed, BCMV and AN tolerant. *Source History: 7 in 2004.* **Sources: Ag7, Ha5, Kit, La1, Ont, Red, Ver.**

Morgane - Finest Haricot Vert for commercial production, up to five harvests from a single planting, 7-7.5 in. long pods, brown seed, nice flavor, BCMV and AN res. *Source History: 1 in 1991; 1 in 1994; 1 in 2004.* **Sources: Coo.**

Mountaineer Half Runner - (Old Dutch Half Runner) - 56-66 days - Heavy yields, BCMV res., oval/round light-green 4-5 x .4 in. pods eventually become stringy, vigorous bush plants with 36 in. non-climbing runners, excel. for baking. *Source History: 14 in 1981; 13 in 1984; 14 in 1987; 12 in 1991; 13 in 1994; 13 in 1998; 11 in 2004.* **Sources: Co31, CON, DOR, Fo7, HO2, ME9, Par, RUP, Se26, SOU, Wet.**

Mrs. Neidigh's Six-Week Bean - White seeded variety with strings, tender 4-5 in. pods, excel. as snaps, can also dry, from a Lancaster County, PA farm. *Source History: 2 in 1994; 1 in 1998; 1 in 2004.* **Sources: Lan.**

Nomad - 75 days - Stringless, straight pods, 5-5.5 in. long, resists BCMV and AN, introduced by Thompson and Morgan. *Source History: 1 in 2004.* **Sources: Tho.**

Norma - *Source History: 1 in 2004.* **Sources: Se17.**

Normandie - 54 days - High quality baby bean, tall, upright plant, straight, tender, round 4.5 in. pods, BV-1A res., HB tol., can be machine harvested, for fresh market, specialty and home garden, 2 sieve size. *Source History: 1 in 1998; 3 in 2004.* **Sources: HO2, Jo1, Twi.**

Old Dutch Half Runner - 60 days - Bush plants with 3 ft. non-climbing runners, 5 in. long light-green pods, pick before pods fill out, stringless when young, can also be used for dry baking, old favorite. *Source History: 1 in 1991; 1 in 1994; 1 in 1998; 1 in 2004.* **Sources: He8.**

Opera - 75 days - Kenyan type, vigorous plants with upright growth habit, dark green shiny stringless pods, 4-5 in. long, resists HB, BCMV and AN. *Source History: 1 in 2004.* **Sources: Tho.**

Opus - (XP B223) - 53 days - High yielding Asgrow var., 5.5 in. round med-green pods, white seeds, BCMV, moderate rust res., good for commercial use, PVP 1992 Seminis Vegetable Seeds, Inc. *Source History: 1 in 1991; 3 in 1994; 5 in 1998; 5 in 2004.* **Sources: HO2, Jor, ME9, RUP, SE4.**

Oregon Giant Bush - Bush plant, huge flat 5-6 in. pods stay snappy when mature, delicious fresh or canned. *Source History: 1 in 1991; 1 in 1994; 1 in 1998; 1 in 2004.* **Sources: So9.**

Oregon Trail - 55 days - Vigorous, productive, large tender bean with Blue Lake flavor, color and texture, Dr. Baggett at OSU, extensive pre-release home garden trials as OSU 4911, 1984. *Source History: 1 in 1981; 1 in 1984; 2 in 1987; 2 in 1991; 1 in 1994; 1 in 1998; 1 in 2004.* **Sources: Sa5.**

OSU 91-G - Processing Blue Lake type, 5 sieve, high yields. *Source History: 1 in 2004.* **Sources: HA20.**

Ozark Bunch - Half runner to 2 ft., green pods with pink streaks, strings, small cream colored seeds with black speckles. *Source History: 1 in 1998; 1 in 2004.* **Sources: Ho13.**

Parson - *Source History: 1 in 2004.* **Sources: Se17.**

Pink Eye Half Runner - 50 days - Bushes 12-14 in. with short runners, dark-green 3-3.5 in. pods ripen to red, immature pink seed dries to ox-blood red in about 85 days - excel. as green beans or dried, home garden. *Source History: 1 in 1981; 1 in 1984; 1 in 1987; 1 in 1991; 1 in 1994; 1 in 1998; 1 in 2004.* **Sources: SOU.**

Pink Half Runner - (Red Peanut) - 50-60 days - Dual purpose bean for snaps or baking, bright-red pods mature to 4.5 in., pink seeds. *Source History: 3 in 1981; 3 in 1984; 3 in 1987; 2 in 1991; 4 in 1994; 5 in 1998; 6 in 2004.* **Sources: Bu8, Co31, HO2, Ho13, Sau, Se26.**

Podsquad - 45-53 days - A potentially high-yielding fresh market bean with fiber to withstand long-distance shipping and extensive handling, long (6 in.) round slim med-green pods, white seed, BCMV tol., PVP 1991 Seminis Vegetable Seeds, Inc. *Source History: 1 in 1987; 4 in 1991; 4 in 1994; 6 in 1998; 1 in 2004.* **Sources: RUP.**

Potato Patch - Semi-vining plant, 3 in. pods contain white seeds, flavorful, also good as shell bean. *Source History: 1 in 1998; 1 in 2004.* **Sources: Ho13.**

Primo - (FM-168, Rio Plata, Slim Romano) - 53-59 days - Excel. processing quality with more narrow pods, BV1 PSV R res., PVP 1992 HA20. *Source History: 1 in 1991; 2 in 1994; 2 in 1998; 1 in 2004.* **Sources: HA20.**

Prince, The - 55 days - Plants 14-20 in. tall, long flat pods, red-purple seeds marked with gold, immense crops, excel. for freezing, widely grown, highly recommended by the Royal Horticultural Society. *Source History: 2 in 1981; 2 in 1984; 2 in 1987; 1 in 1991; 2 in 1994; 3 in 1998; 4 in 2004.* **Sources: Bou, Hi13, Ho13, Tho.**

Probe - High quality fresh market variety, med. green 6.5 in. pods set high on the plant, white seeds, holds pod color, good shipper, excellent flavor and disease tolerance. *Source History: 2 in 2004.* **Sources: HA20, ME9.**

Prosperity - 53 days - Upright bush plants with concentrated pod set for easy picking and mechanical harvesting, smooth med-green pods, BCMV and rust res., PVP 1995 HA20. *Source History: 1 in 1994; 3 in 1998; 2 in 2004.* **Sources: Cl3, ME9.**

Provider - 48-54 days - Early USDA shipper, 16-18 in. vines, straight med-green 5-8 in. round/creaseback pods in clusters, low fiber, purple seed, widely adapted, slow to wilt, heavy early yields, BCMV NY15 PMV and rust res., from Dr. Hoffman of South Carolina, 1965. *Source History: 47 in 1981; 40 in 1984; 47 in 1987; 50 in 1991; 49 in 1994; 58 in 1998; 46 in 2004.* **Sources: All, Be4, CH14, CON, Dam, DOR, Ers, Fe5, Ga1, Gr27, Gr28, HA3, Ha5, Hal, He8, He17, Hi6, Hig, HO2, Jo1, Jor, ME9, Mey, Mo13, Na6, Pe2, Pin, Ra5, RIS, Roh, RUP, Sa5, Scf, SE4, Se7, Sh9, Shu, Si5, So12, Sto, Syn, TE7, Tu6, Ver, Ves, WE10.**

Purple Pod - 51-52 days - Round stringless 5 x .5 in. purple pods that turn dark-green when cooked, dark-green plants, buff/cream colored seed. *Source History: 7 in 1981; 5 in 1984; 6 in 1987; 6 in 1991; 5 in 1994; 3 in 1998; 2 in 2004.* **Sources: But, Pla.**

Purple Queen - 52-56 days - Glossy purple pods turn dark-green when boiled, round 7.5 in. pods, light-brown seeds, vigorous disease res. plants, excel. flavor, an improved Royalty strain. *Source History: 1 in 1981; 1 in 1984; 2 in 1987; 7 in 1991; 8 in 1994; 19 in 1998; 17 in 2004.* **Sources: Bu2, DOR, Fi1, He17, Hum, Jor, La1, Loc, ME9, Mo13, Na6, Ni1, Orn, Pr3, Sa5, Und, Ver.**

Purple Teepee - 51-55 days - Compact plants, deep purple 5-6 in. pods, tender smooth round and slightly curved, heavy yielding, easy to spot for picking, excellent flavor, good fresh canned or frozen, color turns a deep forest-green when cooked. *Source History: 3 in 1987; 7 in 1991; 6 in 1994; 5 in 1998; 6 in 2004.* **Sources: Bo17, Dom, Hig, Pr3, Se26, Tho.**

Radar - 50-57 days - French string bean, can stay on vine longer without becoming stringy than other French gourmet beans, fine quality, highly productive. *Source History: 1 in 1981; 1 in 1987; 2 in 1991; 2 in 1994; 1 in 1998; 1 in 2004.* **Sources: Hi13.**

Rapids - 48-50 days - Early maturing, low fiber var. with exceptional cold soil vigor, 5.5-6 in. dark green pods, white seeds, BV1 R res., PVP 1992 HA20. *Source History: 3 in 1998; 3 in 2004.* **Sources: HA20, HO2, RUP.**

Record - Dutch Princess type produces an abundance of 4 in. green pods, mostly stringless, excellent flavor and freezing qualities. *Source History: 1 in 2004.* **Sources: Dam.**

Regalfin - Haricot vert, long, narrow pods splashed with purple, slender white seeds streaked with black-purple, France. *Source History: 1 in 1987; 2 in 1991; 2 in 1994; 2 in 1998; 1 in 2004.* **Sources: So12.**

Remus - 40-50 days - Unique plant habit, clusters of long tender pods borne above tops of plants, up to 10 in. long, easy to harvest, fiberless, excel. flavor, heavy yields, BCMV, European. *Source History: 3 in 1981; 2 in 1984; 2 in 1987; 4 in 1991; 2 in 1994; 1 in 1998; 1 in 2004.* **Sources: Pr3.**

Rhapsody - 54 days - Medium dark green 5.5 in. round pods, white seed, good flavor, slow fiber development, good heat tolerance, resists several races of rust, BCMV, PVP 2001 HA20. *Source History: 1 in 2004.* **Sources: SE4.**

Roma - (Bush Romano, Italian Flat, Italian) - 53-70 days Bush Romano type, spreading upright plant, broad flat med-green 4.5-5.5 x .75 in. pods, white seed and fiber dev. slowly, can be mech. harv., snaps or dry beans, distinctive flavor, excel. fresh or frozen, PVP expired 1991. *Source History: 42 in 1981; 35 in 1984; 26 in 1987; 15 in 1991; 11 in 1994; 7 in 1998; 5 in 2004.* **Sources: All, Be4, Ma19, Ri12, Sh9.**

Roma II - (Roma Bush) - 50-60 days - Improved bush Romano type, med-green 4.5-5 x .6-.7 in. smooth wide pods, slow to dev. seed and fiber, for snaps or French cut or dry horticultural beans, very flavorful, NY15 BCMV and rust res., PVP expired 1997. *Source History: 16 in 1981; 16 in 1984; 32 in 1987; 42 in 1991; 41 in 1994; 51 in 1998; 52 in 2004.* **Sources: Alb, Bu2, Bu3, Bu8, CH14, Cl3, Co31, Com,**

Dam, De6, DOR, Ers, Fe5, Gur, Hal, He17, Hi13, HO2, Hum, Jor, Jun, Kil, La1, Lin, Loc, ME9, Mel, Mey, Mo13, MOU, OLD, Or10, OSB, PAG, Par, RIS, Roh, RUP, Sau, SE4, Se26, Shu, SOU, Sto, Ter, Tt2, Twi, Ver, Ves, Vi4, Wet, Wi2.

Romanette - (New Romanette, Improved Romanette) - 55-60 days - Early Romano type bean, combines flavor and green flat pod of Romano Pole with a strong bush habit, plant is 16-18 in. tall x 20 in. wide, pods are 5-6 in. long x .75 in. dia., straighter than most Romano types, BCMV and NY15 res., white seed. *Source History: 2 in 1987; 4 in 1991; 5 in 1994; 6 in 1998; 8 in 2004.* **Sources: Cr1, Ear, Jo1, La1, MAY, OLD, SOU, Wet.**

Romano 14 - 45-56 days - Low spreading plants with pods held high, can mech. harv., med-green broad flat 5.5-6.5 x .75 in. pods, freezing and home garden, buff and white seed, tol. to some rusts. *Source History: 9 in 1981; 7 in 1984; 4 in 1987; 9 in 1991; 8 in 1994; 8 in 1998; 7 in 2004.* **Sources: CON, Fis, Jor, La1, Ont, So1, Wi2.**

Romano 26 - 60 days - Harris Seeds sturdy variety with pods that resemble traditional Italian Pole Romano but grow on an erect bush, med. green flat pods with the Romano flavor, easy to pick, BCMV res. *Source History: 2 in 1994; 1 in 1998; 1 in 2004.* **Sources: Se26.**

Romano 942 - 57 days - Improved Romano 26 type with 6-6.5 in. pods held high off the ground on upright plants, flavorful, resists BV1 and NY15 strains of BCMV. *Source History: 2 in 2004.* **Sources: Ha5, HA20.**

Romano Gold - 50-56 days - Italian novelty, bushy, dark green plants bear pods high, 4-5.5 in. long, high quality bean, some disease resistance, popular for roadside market and PYO farms, PVP application pending. *Source History: 3 in 2004.* **Sources: Dom, Sto, Ver.**

Romano Purpiat - 75 days - Purple Romano bush type, flat broad stringless glossy purple pods, 5 in., excellent flavor, purple color turns bright green when cooked, more tolerant of cool weather sowing conditions. *Source History: 1 in 2004.* **Sources: Tho.**

Romano, Bush - (Italian Bush) - 50-70 days - Wide flat med-green stringless 5-6 x 1 in. pods, good freezer and home garden bean, snap or green shell, BCMV and NY15 res., 15-20 in. slow-spreading plant, Romano flavor. *Source History: 27 in 1981; 23 in 1984; 18 in 1987; 17 in 1991; 13 in 1994; 14 in 1998; 12 in 2004.* **Sources: Ag7, Bo19, But, Fi1, Fo7, Gr27, Gur, HA3, He8, HO2, Pin, So9.**

Royal Burgundy - (Purple Burgundy, Royal Purple Burgundy, Purple Queen Bush, Purple Pod Bush) - 50-60 days - Round stringless 5-6 in. purple pods turn dark green when cooked, vigorous erect 15-20 in. bushes, buff seeds, high yields, good in colder soils, bean beetle res., U of NH. *Source History: 44 in 1981; 38 in 1984; 41 in 1987; 42 in 1991; 49 in 1994; 54 in 1998; 57 in 2004.* **Sources: Alb, Be4, Bu8, Co32, Com, CON, Coo, Dam, De6, Ear, Ers, Fe5, Fo7, Gr27, Gr28, Gur, He8, Hi6, HO2, Jo1, Jun, La1, Lin, Ma19, MAY, Mcf, Me7, ME9, Mel, Mey, MOU, OLD, Ont, Or10, OSB, Pin, Ra5, Roh, RUP, SE4, Se7, Se26, Shu, Si5, Sk2, So25, SOU, St18, Sto, Te4, Ter, Tu6, Twi, Ver, Ves, We19, Wet.**

Royal Duke - 50 days - Round pods like Bush Blue Lake except purple and turn green when cooked, more res. to Chlorosis (yellow leaves) caused by alkaline soil. *Source History: 1 in 1991; 1 in 1994; 1 in 2004.* **Sources: Fis.**

Royalnel - (Fin de Fin) - *Source History: 1 in 1987; 1 in 1991; 1 in 1994; 1 in 1998; 1 in 2004.* **Sources: Lej.**

Royalty Purple Pod - (Royalty) - 50-60 days - Purple bushes with short runners and purple flowers, bright-purple stringless slightly curved round 5-5.5 in. pods, cook to green, Mexican bean beetles avoid it, buff seeds germ. in cold wet soil, very tender, home gardens, BCMV res., bred by E.M. Meader, U of NH, intro. by Billy Hepler Seed Co. in 1957. *Source History: 37 in 1981; 31 in 1984; 32 in 1987; 34 in 1991; 27 in 1994; 31 in 1998; 29 in 2004.* **Sources: Ag7, All, Ba8, Bo19, Bou, Bu1, CH14, CON, Fa1, Ga1, HA3, He8, Hig, Hud, Jor, La1, Pe2, Pr3, RUP, Sa9, Sau, Se7, Se16, Sh9, So1, So12, TE7, Tt2, Wi2.**

Rushmore - 49-52 days - Asgrow variety, erect plant, 5-6 in. pods, brown seeds, does well in cold soil, good shipper, PVP 1997 Seminis Veg. Seeds, Inc. *Source History: 2 in 1994; 8 in 1998; 2 in 2004.* **Sources: HO2, Jor.**

Sable - 58 days - Medium size bush, 5 in. round dark green pods, good lodge resistance, resists BCMV and hard blight, good for mechanical harvest. *Source History: 2 in 2004.* **Sources: ME9, SOU.**

Safari - French filet bean with dark green, slender, round 5 in. pods, 24 in. plant. *Source History: 1 in 1994; 1 in 1998; 1 in 2004.* **Sources: Ga7.**

Saratoga - 48 days - Compact plants bear medium-dark green, 5.5 in. round fleshy pods, excellent blanched quality, can be frozen, very early, BCMV res. *Source History: 5 in 2004.* **Sources: Bu1, Ers, La1, MOU, Ver.**

Savannah - 55 days - Dark green 5.5-6 in. pods set high on upright plants, excellent gourmet flavor, quality is retained over extended harvest period, tolerant to BCMV CT BYM and AN. *Source History: 5 in 2004.* **Sources: Ha5, HA20, Loc, ME9, OSB.**

Saxa - Dwarf plants produce large crops of round, med-thick, 5-6 in. tender pods, golden tan seeds, excel. for canning or freezing, best flavor when young, pick every 2-3 days, Germany. *Source History: 1 in 1981; 1 in 1984; 1 in 1994; 2 in 1998; 1 in 2004.* **Sources: Ho13.**

SB 4223 - 57 days - Dark green, round 6 in. pods held high on upright bush, good shipper, plant resists lodging, BCMV, NY15, NL8, HB and BBS resistant. *Source History: 1 in 2004.* **Sources: Cl3.**

SB 4243 - 57 days - Medium green, round 5.25 in. pods, resists Golden Mosaic Virus so does well for late winter and spring planting in Florida and Mexico, BCMV, NY15, NL8 and BGMV. *Source History: 1 in 2004.* **Sources: Cl3.**

Sentry - 55 days - Fiberless 5.5 in. pods, PVP 1989 Seminis Vegetable Seeds, Inc. *Source History: 2 in 1991; 2 in 1994; 3 in 1998; 2 in 2004.* **Sources: CH14, Jor.**

Sequoia - (Purple Pod Bush Romano) - 53-70 days - Deep-purple plants and pods make harvest easier, 5 in. pods, great Romano flavor and meatiness, beige beans are good for canning, tolerant of cool weather conditions. *Source History:*

3 in 1991; 2 in 1994; 10 in 1998; 4 in 2004. **Sources: Bu2, Coo, La1, Te4.**

Seville - 52-56 days - Round 6 in. med-green pods with white seeds, used for fresh market and shipping, an elegant looking bean borne high on a straight vine, holds sieve size very well, BCMV NY15, PVP 1992 Novartis Seeds, Inc. *Source History: 1 in 1991; 13 in 1994; 15 in 1998; 7 in 2004.* **Sources: Bu1, HO2, Jor, ME9, RUP, Sh9, Twi.**

Shade - 54-56 days - Straight, slender, dark green round pods, 5.5 in., good flavor, excellent eating quality, white seed, tolerates BV-1, CTV, heat tolerant. *Source History: 5 in 2004.* **Sources: Cl3, HA20, ME9, SE4, Sto.**

Slenderette - (Slankette) - 53-56 days - Slender dark-green stringless 5 in. pods, small white seeds dev. slowly, multiple disease res., vigorous 14-20 in. vines, pods on upper half of plant, freeze or can, BCMV NY15 PMV and CT res., PVP expired 1992. *Source History: 16 in 1981; 16 in 1984; 15 in 1987; 20 in 1991; 26 in 1994; 28 in 1998; 26 in 2004.* **Sources: Ag7, Be4, Bou, Co31, Coo, Dom, DOR, Ers, Gr27, He17, Jor, Lin, MAY, Mo13, MOU, OLD, Ont, PAG, Pla, Roh, RUP, Sa9, Shu, SOU, Ver, Wet.**

Soleil - 54-59 days - Upright plant bears concentrated set of 4-5 in. round yellow pods, 2 sieve size, crisp and tender, buttery flavor, steam briefly to retain rich gold color, hand pick and mechanical harvest, PVP 1999 Vilmorin, S.A. *Source History: 5 in 2004.* **Sources: HO2, ME9, RUP, Ver, VI3.**

Sonata - (FMX 602) - Med. size bush produces dark green, straight, smooth pods, early midseason, good disease res., grows well from South Florida to New Jersey. *Source History: 1 in 1998; 1 in 2004.* **Sources: ME9.**

Spartan Arrow - 42-52 days - Good Northern var., oval straight stringless med-green 5.5-6.5 in. pods, light buff seeds, BCMV and NY15 res., conc. set, upright 22-24 in. plants, holds quality well, MI/AES, 1963. *Source History: 17 in 1981; 13 in 1984; 12 in 1987; 8 in 1991; 5 in 1994; 5 in 1998; 4 in 2004.* **Sources: CON, HO2, Jor, RIS.**

Spartan Half Runner - (Striped Half Runner) - 53 days - Plant type similar to White Half Runner, light-green oval 4.5-5 in. pods, tan seed with dark-brown stripes, for snaps or dry. *Source History: 6 in 1981; 5 in 1984; 5 in 1987; 2 in 1991; 3 in 1994; 1 in 2004.* **Sources: Se26.**

Speculator - 48-55 days - Sturdy upright plant, long straight dark-green tender pods held well above soil, good for market or canning and also dark enough for freezing, easy to pick, BCMV tolerant. *Source History: 1 in 1981; 1 in 1984; 1 in 1987; 1 in 1991; 1 in 1994; 1 in 1998; 1 in 2004.* **Sources: Sto.**

Spezia - 59 days - Early Romano type, PVP applied for. *Source History: 2 in 2004.* **Sources: DOR, ME9.**

Sprite - 50-54 days - Pods set high and mature together for mech. harv., round straight stringless med-green 5-5.5 x .4 in. pods, 16-18 in. upright plants, small seed dev. at picking stage. *Source History: 12 in 1981; 9 in 1984; 7 in 1987; 4 in 1991; 6 in 1994; 2 in 1998; 1 in 2004.* **Sources: DOR.**

St. Andreas mit Faden - *Source History: 1 in 1998; 1 in 2004.* **Sources: Sh13.**

Stallion - 53 days - Round, slender 5.5 in. pods, medium dark green, brown seed, shipper, tolerates BCMV and HB, PVP 2001 Seminis Vegetable Seeds, Inc. *Source History: 2 in 2004.* **Sources: RUP, Sto.**

State Half Runner - (State White Half Runner) - 52-60 days - Vigorous bushes, 3 ft. non-climbing runners, silver-green flat/oval slightly curved 4 in. pods, stringless when young, BCMV nem. heat and drought res., pleasing strong flavor, excel. high yielding home garden variety, needs no staking. *Source History: 11 in 1981; 10 in 1984; 14 in 1987; 12 in 1991; 12 in 1994; 16 in 1998; 17 in 2004.* **Sources: Bu8, CA25, CH14, Cl3, CON, DOR, Ers, HO2, La1, ME9, RUP, Sau, SE4, Se26, Sh9, Shu, SOU.**

Stoltzfus String - *Source History: 1 in 2004.* **Sources: Se17.**

Storm - 51 days - Straight, medium green 5.5 in. pods, brown seed, good shipper, BCMV and rust tolerant, PVP 2001 Seminis Vegetable Seeds, Inc. *Source History: 5 in 2004.* **Sources: CH14, RIS, RUP, SE4, Sto.**

Straight 'N Narrow - 53-57 days - High quality French filet bean, 5 in. pods, upright 18 in. bush plants, served whole because of its smaller size, BCMV and AN res. *Source History: 8 in 1998; 11 in 2004.* **Sources: Bu1, Dam, Ear, Ers, Fa1, Ha5, La1, Ni1, Pin, Tt2, Ver.**

Strike - 45-55 days - Vigorous upright plants, round med-green straight stringless slender 5.5 in. very smooth pods, conc. for easy picking, BCMV and NY15 res., good shipper, PVP expired 1994. *Source History: 6 in 1981; 6 in 1984; 3 in 1987; 14 in 1991; 16 in 1994; 22 in 1998; 24 in 2004.* **Sources: Be4, CH14, Cl3, DOR, Ers, Fe5, Gr27, He17, HO2, Jor, La1, Loc, ME9, Mo13, RIS, Roh, RUP, SE4, Se26, Si5, Sto, Tt2, Wet, Wi2.**

Stringless Green Pod - (Burpee's Stringless Green Pod, Landreth's Stringless Green Pod) - 50-54 days - Stringless fleshy round curved 5-6 in. med-green heart-shaped pods, tol. to heat and drought, light coffee-brown seed, snap or dry, high yields, upright 20 in. plants, Keeney introduction sold to Burpee's in 1894. *Source History: 41 in 1981; 37 in 1984; 33 in 1987; 25 in 1991; 20 in 1994; 19 in 1998; 16 in 2004.* **Sources: Alb, Bu2, Bu8, Cr1, De6, Ear, Ers, HA3, Ho12, Lin, LO8, Ont, PAG, Sau, Se26, SOU.**

Stringless Green Pod Improved - (Improved Burpee Stringless Green Pod) - 52-53 days - Nearly-round slightly-curved fiberless 6 in. green pods, outstanding quality green bean in the Pecos Valley. *Source History: 3 in 1981; 3 in 1984; 3 in 1987; 2 in 1991; 1 in 1994; 1 in 1998; 1 in 2004.* **Sources: Ros.**

Stringless Green Pod, Burpee's - (Burpee's Dwarf Green Stringless, Landreth's Stringless Green Pod, Improved Stringless Green Pod) - 50-54 days - Claimed to be the only stringless green podded bean when it was introduced in 1894 by W. Atlee Burpee, stringless fiberless waxy deep-green 6.5 in. slightly curved pods in clusters, heart-shaped cross section, buff to dark seed, heavy yields, canning or freezing, original stock seed from N. B. Kenney, dev. by Calvin Keeney, LeRoy, NY, 1894. *Source History: 32 in 1981; 28 in 1984; 27 in 1987; 27 in 1991; 26 in 1994; 26 in 1998; 20 in 2004.* **Sources: All, Be4, CH14, Co31, CON, Fo7, Hi13,**

HO2, La1, MAY, ME9, Mey, MOU, Roh, RUP, Se16, Sh9, Shu, Wet, Wi2.

Stringless Green Pod, Landreth's - 52-55 days - Slightly curved round med-green 5.5 in. pods, thick meaty juicy brittle tender stringless fiberless and tasty, excel. green podded bush bean to 20 in., for home or market or canning, dev. in 1885 for Landreth's Seeds, the oldest seed house in the country which was established in 1784. *Source History: 2 in 1981; 2 in 1984; 3 in 1987; 3 in 1991; 6 in 1994; 7 in 1998; 6 in 2004.* **Sources: CON, Dgs, DOR, Fo7, Jor, La1.**

Striped Half Runner - (Spartan) - 55-60 days - Vigorous half runner, often planted in the corn, slender light-green 5 x .4 in. pods, stringy when mature, for snap beans early and shell beans later. *Source History: 6 in 1981; 5 in 1984; 5 in 1987; 2 in 1991; 1 in 1994; 1 in 1998; 1 in 2004.* **Sources: Bu8.**

Sulphur - (Brimstone, Golden Cranberry, China Yellow) - 52-63 days - Round light-green 4-5 in. pods, Southern bean known for its short cooking time, distinctive flavor and ability to form a thick gravy when baked or boiled, 16 in. bushes, yellow seed with faint brown eye ring, home garden and fresh market, introduced about 1839, re-introduced in 1893 by the Ford Seed Company. *Source History: 9 in 1981; 6 in 1984; 9 in 1987; 6 in 1991; 7 in 1994; 6 in 1998; 7 in 2004.* **Sources: Ho13, La1, Ri12, Shu, So1, Te4, Ver.**

Tanya's Pink Pod - Striking snap bean, also good dry, tan seed. *Source History: 1 in 2004.* **Sources: Sa5.**

Tarahumara Ejotero Negro - Dark lilac flowers, black seeds, early maturing, originally collected from the southern edge of Tarahumara country in Chihuahua, Mexico. *Source History: 1 in 2004.* **Sources: Na2.**

Tasman - 57 days - *Source History: 1 in 2004.* **Sources: SE4.**

Tema - (XP B201) - 52-55 days - Round fiberless 5 in. dark-green pods with brown seeds, gourmet flavor, exceptional eating quality, germinates in cold soil, very high yields, used for fresh market and shipping, BCMV res., PVP 1992 Seminis Vegetable Seeds, Inc. *Source History: 4 in 1991; 5 in 1994; 9 in 1998; 16 in 2004.* **Sources: CH14, Ers, Fa1, He17, HO2, Jor, Mcf, ME9, Mo13, RIS, Roh, RUP, SE4, Sto, Tt2, Ver.**

Tendercrop - 46-61 days - Topcrop x Tenderpod, upright 18-21 in. plants, round straight slender 5-7 in. dark-green pods in clusters, pointed long tips, res. to BCMV NY15 and PMV, purple seed, USDA/Beltsville, 1958, heavy yields, good home garden var., also suited for mech. harv., holds well, adapted to North Midwest and West. *Source History: 40 in 1981; 36 in 1984; 33 in 1987; 31 in 1991; 27 in 1994; 22 in 1998; 21 in 2004.* **Sources: Al6, Ber, Bo19, CH14, CON, Ers, Fi1, Gr27, He17, HO2, Hum, La1, Mo13, MOU, Ni1, Par, Ra5, Ri12, RUP, Se26, Shu.**

Tenderette - 55-56 days - Vigorous, straight dark-green round/slight creaseback 6 in. pods, BCMV and NY15 res., bears during hot weather, pods borne high on 18-20 in. upright plants, suited for mech. harv., excel. quality pods, white seed, resembles Tendercrop but with Blue Lake pods, stringless and fiberless, widely adapted, 1962. *Source History: 25 in 1981; 23 in 1984; 23 in 1987; 23 in 1991; 23 in 1994; 26 in 1998; 26 in 2004.* **Sources: Be4, Bu8, CA25, CH14, Co31, CON, DOR, Ers, Fi1, Gr27, Gur, Ha5, He8, HO2, La1, ME9, Mel, Mey, Ri12, RUP, Sau, Se26, Sh9, Shu, SOU, Wet.**

Tendergreen - (Tendergreen Stringless) - 45-57 days - AAS/1933, strong erect vigorous leafy dark-green 18-20 in. plants, round meaty stringless dark-green 5.5-7 x .4 in. pods, BCMV res., brownish-purple seeds, extended season, heavy yields even in hot weather, all purpose, home or market, thought to have been dev. by Calvin Keeney, introduced in 1922. *Source History: 48 in 1981; 43 in 1984; 45 in 1987; 34 in 1991; 28 in 1994; 27 in 1998; 28 in 2004.* **Sources: Ada, Alb, All, Be4, Bou, Bu8, But, Com, CON, Gr27, La1, Lin, Mey, MOU, Ont, Pe2, Ra5, Ri2, Sa5, Sa9, Sau, Sh9, Sk2, SOU, TE7, Tt2, WE10, Yu2.**

Tendergreen Improved - (Tendergreen Long) - 45-56 days - AAS/1933, large upright 16-20 in. plant, straight round med-dark green 5.5-7 x .4 in. pods, longer and smoother, very meaty, purple seed mottled with tan, heavy yields, fine quality, for canning or freezing or shipping. *Source History: 52 in 1981; 50 in 1984; 45 in 1987; 38 in 1991; 37 in 1994; 39 in 1998; 31 in 2004.* **Sources: Bu3, CH14, Cr1, De6, Dgs, DOR, Ear, Ers, Fa1, Fi1, Ge2, HA3, Hal, He8, He17, Jor, Jun, Kil, MAY, Mcf, Mel, Mo13, OLD, PAG, Pin, Roh, RUP, SE4, Shu, Wet, Wi2.**

Tenderlake - (E 6207) - 51-53 days - Upright plant, med-dark green 6 in. pods, round to slightly creaseback, outstanding processing characteristics, good carpel adherence, BCMV res., white seed, PVP expired 2000. *Source History: 1 in 1981; 1 in 1984; 2 in 1987; 5 in 1991; 5 in 1994; 3 in 1998; 6 in 2004.* **Sources: HA20, HO2, Jo1, Ni1, Par, SOU.**

Tenderpick, Burpee's - 54 days - Vigorous plants, tender 5.5 in. dark green pods with curved tips, white seeds, good for fresh eating, canning and freezing, super germination. *Source History: 1 in 1994; 1 in 1998; 1 in 2004.* **Sources: Bu2.**

Tenderpod - (Burpee's Tenderpod) - 48-50 days - AAS/1941, brittle meaty deep-green 4.5-5.5 in. pods with curved tips, stringless and fiberless, for canning freezing dry shell, early. *Source History: 7 in 1981; 6 in 1984; 10 in 1987; 8 in 1991; 6 in 1994; 4 in 1998; 8 in 2004.* **Sources: Bu2, CH14, CON, Ers, He8, He17, La1, Mo13.**

Tennessee Green Pod - (Brown Bunch, Case Knife) - 48-62 days - Flat broad stringless 7 x .75 in. dark-green meaty pods, high yields, good only in the South, yellow-brown seed becomes tough when over-mature, for home gardens. *Source History: 24 in 1981; 20 in 1984; 21 in 1987; 14 in 1991; 14 in 1994; 11 in 1998; 9 in 2004.* **Sources: Ada, CON, Ers, LO8, Se26, Sh13, Shu, SOU, Wet.**

Teseo - 58 days - High quality, fiberless, dark green 5-6 in. pods, produces high yields under a wide range of conditions, BV1 A res., PVP 1997 Rogers Seed Co. *Source History: 2 in 1998; 2 in 2004.* **Sources: RUP, SE4.**

Thibodeau du Comte Beauce - 40-60 days - Beautiful deep rose-pink blossoms, straight green pods striped with purple-red, maroon stripes and spots on white seed, heirloom from Beauce County, Quebec. *Source*

History: 1 in 2004. **Sources: Te4.**

Thousand-to-One - (Refugee) - 75 days - Prolific heirloom bush plant produces hundreds of 5 in. pods, each containing 5-6 tri-color seeds, white, black or gray mottled, excel. flavor, great for canning and freezing, pre-1920. *Source History: 2 in 1994; 5 in 1998; 3 in 2004.* **Sources: Ho13, Te4, Th3.**

Topcrop - 45-53 days - AAS/1952, strong upright 18-24 in. dark-green plants, straight med-green stringless round meaty 6-7 in. pods, BCMV NY15 and PMV res., heavy crops, concentrated pickings, oblong brown seeds mottled with buff, USDA/Beltsville, 1950. *Source History: 83 in 1981; 75 in 1984; 73 in 1987; 61 in 1991; 48 in 1994; 54 in 1998; 45 in 2004.* **Sources: Ada, Alb, Be4, Bu1, Bu3, Bu8, CH14, Co31, CON, Cr1, De6, DOR, Ers, Fa1, Fi1, Fis, Gr27, Gur, He8, He17, Jor, Jun, La1, LO8, Ma19, MAY, ME9, Mel, Mey, Mo13, MOU, OLD, Ri12, Ros, RUP, Sau, Sh9, Shu, Sk2, So1, SOU, Ver, Vi4, Wet, Wi2.**

Totem - 63-65 days - French filet bean, 36 in. plants, 8-10 in. long, slender pods, retains tenderness, popular with fine restaurant market growers, white seed. *Source History: 1 in 1994; 1 in 1998; 1 in 2004.* **Sources: Ter.**

Triomphe de Farcy - 48-59 days - Traditional French string bean for early planting, long thin straight 6-7 in. pods are medium green with purple stripes, dark brown-purple mottled seed, must harvest at a very early stage before strings appear, gourmet variety. *Source History: 2 in 1981; 2 in 1984; 5 in 1987; 6 in 1991; 6 in 1994; 8 in 1998; 4 in 2004.* **Sources: Bu2, Coo, Gou, So12.**

Trueblue - 54 days - Widely adapted Blue Lake type with medium dark green pods, 5-6 in. long, good heat tolerance, BCMV res., PVP 1991 HA20. *Source History: 1 in 1991; 1 in 1994; 3 in 1998; 4 in 2004.* **Sources: HA20, ME9, Orn, SOU.**

Tweed Wonder - Large productive plants, dark purple seeds, good flavor, late maturing bean to be grown in areas with a long growing season, from Australia. *Source History: 1 in 1998; 1 in 2004.* **Sources: Te4.**

Valentine, Stringless Black - (Black Valentine) - 48-70 days - Straight slender dark-green 6-6.5 in. nearly round pods, stringless all stages, 16-18 in. plants, hardy, good for early plantings, good shipper, very old heirloom, pre-1850, introduced by seedsman Peter Henderson in 1897. *Source History: 14 in 1981; 12 in 1984; 12 in 1987; 14 in 1991; 17 in 1994; 22 in 1998; 25 in 2004.* **Sources: Ba8, Co32, Com, CON, DOR, Fe5, Fo7, Fo13, He8, Hi6, Ho13, Hud, Mey, Pla, Ra5, Sa9, Se8, Se16, Se26, So1, Te4, TE7, Und, Ver, Vi4.**

Valentine, Stringless Red - 47 days - Hearty erect heavy yielding plants, large clusters of 5 in. med-green pods, curved cylindrical and creasebacked, crisp tender flesh, extra early, pre-1850. *Source History: 2 in 1981; 2 in 1984; 1 in 1987; 2 in 1991; 1 in 1994; 3 in 1998; 1 in 2004.* **Sources: Ho13.**

Venito - Bush plant with glossy leaves, long straight pods. *Source History: 1 in 2004.* **Sources: So25.**

Venture - 48-55 days - Blue Lake heritage, as early as Provider, excellent quality in this maturity, cold soil tol., stiff erect stems hold pods well off the ground for easy harvest, dark-green 6.5 in. long pods somewhat lumpy with a firm but tender texture, sweet flavor, productive, white seed, fresh use or freezing, BCMV res., PVP 1989 Novartis Seeds, Inc. *Source History: 4 in 1987; 13 in 1991; 19 in 1994; 20 in 1998; 10 in 2004.* **Sources: Alb, Ga1, Hig, HO2, Jor, Pr3, RUP, Ter, Twi, We19.**

Vernel - 60-70 days - Best flageolet bean say French garden magazines, heavy crops of slim green pods, 5-6 green colored seeds, harvest at any stage, use fresh or dry, BCMV and AN res. *Source History: 1 in 1981; 1 in 1984; 3 in 1991; 4 in 1994; 3 in 1998; 1 in 2004.* **Sources: Lej.**

Volunteer Half Runner - 60 days - Strain of Half Runner, light green, 4-6 in. pods, white seeds, continuous set for 3 or 4 hand harvests, BCMV and rust res., dev. at the U of TN by USDA plant breeder, Dr. Mullins. *Source History: 3 in 2004.* **Sources: HO2, RUP, Shu.**

Wade - 52-55 days - Straight round stringless dark-green slender 6-6.5 in. pods, BCMV NY15 and PMV res., reddish-brown buttery flavored seeds, heavy yields, pods wilt slowly, USDA, 1952. *Source History: 10 in 1981; 8 in 1984; 6 in 1987; 4 in 1991; 5 in 1994; 5 in 1998; 1 in 2004.* **Sources: Fo7.**

White Half Runner - (Mountaineer, Mississippi Skip Bean, Early White Half Runner) - 50-60 days - High yield, med-light green round/oval 4-4.5 x .4 in. pods, good for canning, freezing fresh or shell, dries white, Southern semi-runner, stringless when young, heat and drought res., BCMV res., not recommended for the North. *Source History: 40 in 1981; 38 in 1984; 35 in 1987; 33 in 1991; 32 in 1994; 38 in 1998; 32 in 2004.* **Sources: Ada, Be4, Bu2, Bu8, CA25, CH14, CON, De6, Ers, Fi1, He17, HO2, Kil, La1, Ma19, ME9, Mel, Mey, Mo13, Pr3, RIS, Roh, RUP, Sau, SE4, Sh9, Shu, So1, SOU, Ver, Wet, Wi2.**

White Hull Bunch Bean - Creamy yellow to white pods develop a rich pink color along the midrib at maturity, frequent picking stimulates additional fruit set, heirloom. *Source History: 1 in 1998; 1 in 2004.* **Sources: Scf.**

Xera - 54-65 days - Upright bush to 21 in., shiny, dark green, 5.5-6 in. pods, good yields over a long period, A and BCMV-B res. *Source History: 1 in 1998; 1 in 2004.* **Sources: Eo2.**

Yer Fasulyasi - 45 days - Turkish heirloom, small bushes, sweet stringless pods, round brown seed, very early, high yields. *Source History: 2 in 2004.* **Sources: In8, Se17.**

Yugoslavian Blush - 85 days - Blush pink seed, good as snap or dry but not good for freezing or canning, declared an endangered var. by the U of WA Ag Dept., brought into Washington state by immigrants about 1875. *Source History: 1 in 2004.* **Sources: Ri12.**

Zane - 60-70 days - Virginia heirloom maintained by the Counts family, small white seed, for fresh or frozen use. *Source History: 1 in 2004.* **Sources: Ri12.**

Zodiac - 55 days - High yielding shipping variety with superior flavor, high pod set, light green pods, white seeds, BCMV res. *Source History: 3 in 1998; 3 in 2004.* **Sources: HO2, RUP, SE4.**

Varieties dropped since 1981 (and the year dropped):

Acclaim (1998), *Ahron* (1987), *Aiguille Vert* (2004), *Aiguillon* (1994), *Ajax Broad Stringless* (1991), *Allure* (2004), *Alpental* (1994), *Amateur* (1982), *Apennine* (1987), *Applause* (1998), *Aristocrop* (1987), *Astrel* (2004), *Atlanta* (2004), *Atlantic* (1998), *Atlas* (2004), *Avalanche* (1991), *Baby French* (1994), *Baffin* (2004), *Bahalores* (2004), *Ballack* (1991), *BE 205* (1998), *Beeline* (1994), *Belami* (1984), *Berthold Everbearing* (1991), *Bertina Climbing Bush* (1994), *Bestcrop* (1982), *Black Resistant Valentine* (1987), *Blazer* (1994), *Blue Dawn (E 5202)* (1987), *Blue Jay* (2004), *Blue Knight* (1994), *Blue Lake Bush 47* (2004), *Blue Lake Bush 109* (2004), *Blue Lake Bush 141* (1994), *Blue Lake Bush (Oregon 1604)* (1991), *Blue Lake Bush 2117* (2004), *Blue Lake Bush 290* (1991), *Blue Lake Bush, Oregon* (2004), *Blue Lake Bush, Salem* (1994), *Bluecrop* (1991), *Bounty* (2004), *Brezobel Stringless* (1994), *Broad* (1984), *Broker* (2004), *Broker's Choice* (1994), *Brown Seeded Romano 14* (1987), *Burly* (1998), *Burpee's Richgreen* (1983), *Bush Baby Bean Mini Green* (1998), *Buttergreen* (1998), *BX 156-2-3-6* (1998), *BX207* (2004), *Caesar* (1994), *Calgreen* (1998), *Camile* (1998), *Canadian Filet* (2004), *Canberra* (2004), *Canyon* (1998), *Cascade* (1994), *Castel* (2004), *Celina* (1987), *Charlie Murphy* (2004), *Checkmate* (1991), *Cheverbel* (1991), *Clyde* (1998), *Coktel* (1994), *Columbus* (1998), *Conquest* (1987), *Cordoba* (2004), *Couch's* (2004), *Countess Gourmet* (1994), *Crossville* (1987), *Cumberland* (1987), *Cypress Stringless* (1998), *Cyrus* (1998), *Daria* (1987), *Davis Shell* (2004), *Decibel* (2004), *Del Rey* (1987), *Demeter* (1998), *Dubresco* (1987), *Duchess* (2004), *Dufrix* (1983), *Dutch Princess* (2004), *Dutch Stringless Green Pod* (1998), *Early Gallatin* (2004), *Early Harvest* (1991), *Endurance* (2004), *English Long Green* (2004), *Etalon* (1991), *Executive* (1991), *Exp. 116* (1987), *Expressway* (1991), *Extender* (1998), *Fanion* (1987), *Fantastico* (1994), *Favornel* (1987), *Fetiche* (1984), *Feyenoord Stringless* (1991), *Flair* (1983), *Flatbush* (1998), *Flaveol* (1991), *Flevoro* (2004), *Flits* (1983), *Flo* (1991), *FM-103* (1994), *FM-208* (1991), *FM-216* (1994), *Forum* (1998), *Frazier's Choice* (2004), *Gallatin* (1994), *Gallatin Valley 50* (1991), *Gardengreen* (1994), *Gator Green Improved* (1998), *Gentry* (1998), *Giant Stringless Green Pod* (2004), *Gill's Relic* (2004), *Gitana* (2004), *Goodwin's Bush Butterbean* (2004), *Gourmet* (1991), *Green Beans* (1991), *Green Genes* (1998), *Green Lantern* (1998), *Green Needle* (1984), *Green Pak* (1998), *Green Perfection* (1982), *Green Run* (1994), *Greenway* (1991), *Groffy* (1998), *H 732A* (2004), *Hallord Select* (1994), *Haricot Vert, Label* (1998), *Haricot Vert, Lobith* (1998), *Haricot Vert, Monel* (1998), *Hawkesbury Wonder* (2004), *Homestyle* (2004), *Idaho Refugee* (2004), *Italian Colombo* (1994), *Janus* (1987), *Javelin* (2004), *Kenya* (1991), *Lake Erie* (1987), *Lake Geneva* (1987), *Lake Seneca* (1987), *Lake Shasta* (1987), *Lake Superior* (1994), *Lakeland* (1987), *Lancer* (1998), *Larma* (1994), *Lasso Gourmet* (1994), *Laureat* (1998), *Lavra* (1987), *Legacy* (1991), *Limbourg* (1991), *Loch Ness* (1987), *Marbel* (1998), *Masterpiece* (1994), *McHarvest* (1994), *Miami* (1991), *Mikado* (1994), *Milano* (2004), *Miry* (1994), *Mon Petit Cheri* (2004), *Morinaka* (2004), *Mr. Fearn's* (2004), *Mt. Hood (E 6211)* (1987), *Mustang* (2004), *Myriad* (1991), *Narbonne* (2004), *Nathanael Greene* (1991), *Nerina* (2004), *Nervina* (1994), *Nirda* (1984), *Non Plus Ultra* (1994), *Noordster* (1998), *Olympia* (1987), *Oregon 83 (OSU 4883)* (1987), *OSU 1604-B* (1987), *Parfaco* (1984), *Parisian String* (2004), *Parker Half Runner* (2004), *Paymaster* (1987), *Peak* (1998), *Perfect Improved* (1994), *Perla* (1984), *Pirate* (1987), *Plano* (1991), *Plentiful Stringless* (1998), *Portico* (1991), *Prelude* (2004), *Prestige Salad* (2004), *Primel* (1987), *Processor* (2004), *Producer* (1991), *Profit Maker* (1994), *Provider, White Seeded* (2004), *Purple King* (1998), *Purple Rouge* (1998), *Purple Royalty Improved* (1998), *R 104* (1987), *Raider* (1994), *Rainier* (1994), *Rapid Green* (2004), *Rapier* (2004), *Rebel* (1994), *Regal Salad* (2004), *Remora* (1987), *Resistant Valentine* (1987), *Resisto* (1994), *Rio Belges* (1983), *Romano 71 Bush* (1987), *Romano 123* (1991), *Romano 635* (1998), *Romano II* (1991), *Romano Improved* (2004), *Romano, Mrs. Marotti's* (2004), *Royal Purple* (1991), *Royalette* (1994), *RS 1384* (2004), *RS 5402* (2004), *Rudolpho* (1991), *Seminole* (2004), *Settler* (1998), *Settler II Green* (2004), *Settler IV Purple* (2004), *SG 251* (1987), *Shamrock* (2004), *Shore* (2004), *Slenderella* (1994), *Slenderwhite* (1994), *Slimgreen* (1994), *Spring Green* (1998), *Sprout* (1994), *Spurt* (1998), *Stiletto* (1994), *Stretch* (1987), *Stringless Green Pod, Giant* (2004), *Sujinashi Green* (1987), *Summergreen* (1994), *Sunbird* (1994), *Sunray* (2004), *Sunray, Early* (1998), *Sunray, Super* (1998), *Tavera* (2004), *Teepee, Cropper* (1994), *Teepee, Harvester* (1998), *Teepee, Santos* (1998), *Tempo* (1982), *Tenderlong* (1987), *Tidal Wave* (1994), *Tobacco Patch* (1984), *Topcrop 32* (1998), *Topper, Burpee's* (1998), *Topset* (1991), *Tornado* (1991), *Torrent* (1991), *Triumph* (1998), *Valentine, Resistant Asgrow* (2004), *Vanderpool Six Week* (2004), *Vernandon* (2004), *Vilbel* (2004), *Volare* (1991), *Wando* (1994), *White Early Dwarf Coco* (1983), *White Settler Filet* (2004), *Widuco* (1998), *Win* (1991), *Wondergreen* (1984), *Wrangler* (2004).

Bean / Bush / Wax
Phaseolus vulgaris

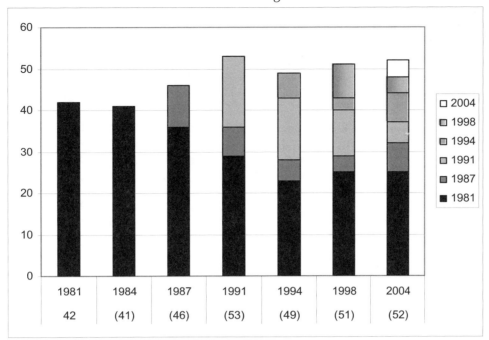

Number of varieties listed in 1981 and still available in 2004: 42 to 25 (60%).
Overall change in number of varieties from 1981 to 2004: 42 to 52 (124%).
Number of 2004 varieties available from 1 or 2 sources: 22 out of 52 (42%).

Berggold - 75 days - Erect bush with compact growth habit, 5 in. long straight stringless pods held high on the plant, excellent flavor, bears heavily, BCMV and AN resistant. *Source History: 1 in 2004.* **Sources: Tho.**

Beste Von Allen - 60-75 days - Long yellow pods, white seed with a black eye, production decreases in extemely hot weather, original seed from the Seed Savers Exchange. *Source History: 1 in 1994; 1 in 1998; 1 in 2004.* **Sources: Lan.**

Beurre de Rocquencourt - 48-70 days - Old French variety, straight yellow juicy 5-6 in. pods, thinner and slightly longer than Brittle Beurre, keep picked for maximum yields, good where night temps are cool, black seed, productive in cool weather. *Source History: 7 in 1981; 5 in 1984; 6 in 1987; 5 in 1991; 4 in 1994; 6 in 1998; 8 in 2004.* **Sources: Co32, Gou, Ri12, Se7, Se8, Se17, Se24, St18.**

Brittle Wax - (Round Pod Kidney Wax, Brittle Beurre) - 45-58 days - Light golden-yellow slightly curved round fleshy 6-7 in. pods, stringless and fiberless, high yields, all purpose, long harvest, white seed with black eye, good yielding bushy plants, also a fine shell bean, intro. in 1901 by Johnson and Stokes of Philadelphia. *Source History: 17 in 1981; 13 in 1984; 15 in 1987; 12 in 1991; 15 in 1994; 18 in 1998; 16 in 2004.* **Sources: Alb, Co32, CON, Ers, Fa1, Fo7, He8, Hi13, Ho12, Hud, La1, ME9, Ont, Pr3, Se16, Se24.**

Buttercrisp - 59 days - Bush plant bears 5.5 in. bright yellow pods, fine flavor, prolific yielder, resists curly top, rust and BCMV. *Source History: 1 in 1994; 3 in 1998; 2 in 2004.* **Sources: Ech, Fe5.**

Butterwax, Eastern - 44-58 days - High yields, round/oval meaty 6-7 in. slightly curved pods, retains golden color, no green tips, sturdy 16-18 in. bush, good for Atlantic Canada, rich flavor, 1958. *Source History: 13 in 1981; 10 in 1984; 8 in 1987; 8 in 1991; 5 in 1994; 2 in 1998; 1 in 2004.* **Sources: Ves.**

Butterwax, Golden - (Butterwax, Yellow) - 60 days - Sturdy upright plants have excel. seedling vigor, high yield of straight 6-6.5 in. pods, seed development in the pod is slow, color and fresh appearance after harvest holds longer, will not scuff up when machine harvested, sets from late July until frost, ozone damage and rust res., dev. by Agway. *Source History: 2 in 1981; 2 in 1984; 2 in 1987; 5 in 1991; 4 in 1994; 2 in 1998; 3 in 2004.* **Sources: Fi1, HO2, SE4.**

Century Gold - 56-62 days - High yielding white-seeded wax bean, round straight fleshy 6 in. pods, very golden-yellow when mature, very long-holding quality, slow seed dev., PVP expired 1993. *Source History: 4 in 1981; 4 in 1984; 2 in 1987; 1 in 1991; 2 in 2004.* **Sources: In8, Se17.**

Cherokee Wax - (Surecrop Stringless Wax, Bountiful Wax, Valentine Wax, Cherokee Yellow Pod) - 43-58 days - AAS, vigorous hardy erect 16-18 in. vines, straight flat deep-golden yellow oval 5-6.5 x .4 in. pods, excel. color, BCMV and NY15 res., stringless all stages, slight fiber, black seed, USDA and Clemson/SC, 1946. *Source History: 41 in 1981; 35 in 1984; 32 in 1987; 30 in 1991; 28 in 1994; 29 in 1998; 27 in 2004.* **Sources: CH14, Com, Cr1, De6, Ers, Fo13, Ge2, Gur, He17, Hi6, Jor, La1, MAY, ME9, Mey, Mo13, OLD, Or10, Ra5, Ri12, RIS, RUP, Se7, Sh13, Shu, Sk2, Ver.**

Cherokee Wax, Resistant - 50-56 days - Standard wax bean shipping var., oval bright-yellow straight 5.5-6.5 in. slightly curved pods, res. to BCMV and NY15, large vigorous erect plant, heavy yields. *Source History: 11 in 1981; 11 in 1984; 7 in 1987; 6 in 1991; 5 in 1994; 5 in 1998; 6 in 2004.* **Sources: CON, DOR, HO2, Kil, Sh9, Si5.**

Dandy Gold - 54 days - Early, erect compact bush with a very concentrated pod set, smooth round 4 in. pods high on the bush, ideal for French cut/canning/freezing/dill beans. *Source History: 1 in 1987; 1 in 1991; 1 in 1994; 1 in 1998; 1 in 2004.* **Sources: Tt2.**

Dorabel - 55-62 days - Popular wax bean in France, colors up early and stays slim, stringless pods 4.5-5 in. long, ideal for a yellow filet bean, exceptional flavor when large or small, freezes well, BCMV res. *Source History: 1 in 1991; 14 in 1994; 13 in 1998; 8 in 2004.* **Sources: Be4, Coo, HO2, ME9, Pin, Se8, VI3, Wi2.**

Dragon Langerie - (Dragon's Tongue Bean, Horticultural Wax, Merveille de Piemonte) - 57-95 days - Unique Dutch wax bean, use for snaps when pods are creamy-yellow covered with thin purple stripes, pick as a shell bean when stripes change from purple to red, use for dry bean when fully mature, remarkably good flavor and yields, delicious conversation piece. *Source History: 1 in 1981; 1 in 1984; 3 in 1987; 8 in 1991; 12 in 1994; 21 in 1998; 24 in 2004.* **Sources: Al25, Ami, Bou, Bu2, Coo, Fe5, Ga1, Ho13, In8, Jo1, Jor, Pin, Pr3, Sa5, Se16, Se17, Shu, Sk2, So12, St18, Te4, Ter, Und, Ver.**

Eureka - 56-60 days - Tender slightly white 5.5-6.5 in. pods, heavy concentrated set, adaptable for mechanical harvest, good yields, disease res., PVP 1989 Seminis Vegetable Seeds, Inc. *Source History: 2 in 1991; 6 in 1994; 8 in 1998; 9 in 2004.* **Sources: CH14, HO2, Jor, ME9, RIS, Roh, RUP, SE4, Si5.**

EZ Gold - 54 days - Sturdy, upright 18 in. plants bear heavy yields of medium thick, uniform round 5.5 in. long yellow pods set high on the plant for easy picking, slow seed development, resists BCMV, AN, HB. *Source History: 6 in 2004.* **Sources: Bu1, Ers, Fa1, Hal, Jun, Ver.**

Fortin's Family - Handed down by the Fortin family from Quebec for many years, productive and flavorful, appears to be the same as Quebec Coffee. *Source History: 2 in 1998; 1 in 2004.* **Sources: Te4.**

Gold Mine - 47-55 days - Bright yellow Gold Rush type with less fiber, 5 in. pods, PYO, fresh market and mech. harv., BCMV, brown spot and some halo blight tolerance, PVP 1992 Seminis Vegetable Seeds, Inc. *Source History: 1 in 1991; 2 in 1994; 9 in 1998; 9 in 2004.* **Sources: Bu2, HO2, Jor, ME9, RUP, SE4, Sto, Tt2, Ves.**

Goldcrop Wax - (Golden Bountiful Wax) - 45-65 days - AAS/1974, straight round bright-yellow 5-6.5 in. pods, white seed, from USDA in WA, res. to BCMV and CT and heat, no strings or fiber, small compact upright plants, fine quality, sets in hot weather, pods held well off ground, suited for home gardens or mech. harv., resists hot weather blossom drop. *Source History: 33 in 1981; 32 in 1984; 31 in 1987; 27 in 1991; 17 in 1994; 16 in 1998; 14 in 2004.* **Sources: Bo19, CH14, CON, Ers, Gr27, HA3, He17, HO2, Jor, La1, ME9, Mo13, Sa9, Si5.**

Golden Rocky - (Rocquencourt) - 44-65 days - Long straight stringless bright-yellow 6 in. pods, excel. for freezing and canning, long harvest period, good cold tolerance, less susceptible to insect damage, 1984. *Source History: 2 in 1981; 2 in 1984; 4 in 1987; 6 in 1991; 7 in 1994; 8 in 1998; 3 in 2004.* **Sources: Hi6, Hig, Tu6.**

Golden Rod - 52-58 days - Upright vigorous 14-16 in. plants, plump 5-6 in. pods, brittle and meaty, fine rich flavor, BCMV and NY15 tol., white seeds dev. slowly, fine quality and appearance, PVP expired 1995. *Source History: 3 in 1981; 3 in 1984; 7 in 1987; 8 in 1991; 5 in 1994; 10 in 1998; 9 in 2004.* **Sources: Cl3, HA20, HO2, Jun, Pr3, SE4, Twi, Ver, We19.**

Golden Wax - (Unrivalled Wax, Early Golden Wax, Stringless Golden Wax, Yellow Golden Wax) - 45-60 days - Fine flavor, entirely stringless 4.5-5.5 in. golden-yellow pods, ideally suited for Northern climates, res. to BCMV, freezes well, white seed with purple/brown eye, early dependable cropper, for fresh or canning or freezing, buttery flavor. *Source History: 21 in 1981; 17 in 1984; 15 in 1987; 13 in 1991; 17 in 1994; 16 in 1998; 14 in 2004.* **Sources: Ada, Alb, De6, DOR, LO8, Ma19, ME9, Ont, Ri2, Te4, TE7, Ves, Vi4, Wi2.**

Golden Wax Improved - (Topnotch Strain, Early Golden Wax Improved, Golden Wax Dual Purpose) - 50-52 days - Longer podded str., straight stringless thick flat creamy-yellow 4-6 in. pods, res. to BCMV and rust, productive, brown-eyed white seed, compact hardy 16-19 in. plants, for fresh or canning or freezing or dry. *Source History: 28 in 1981; 24 in 1984; 25 in 1987; 18 in 1991; 15 in 1994; 15 in 1998; 10 in 2004.* **Sources: All, Bu8, Ear, Ers, Gur, Lin, Mel, Ros, RUP, Shu.**

Golden Wax, Rustproof - 50 days - Broad flat/oval stringless gold-yellow 5.5-6 in. pods held high, erect vines, oval shiny white seeds with red-brown eye mark, very rust res., hardy, high quality. *Source History: 4 in 1981; 3 in 1984; 3 in 1987; 2 in 1991; 2 in 1994; 2 in 1998; 1 in 2004.* **Sources: Fi1.**

Golden Wax, Topnotch - (Topnotch Strain, Topnotch Golden Wax Improved) - 50-55 days - Imp. Golden Wax type, stringless straight flat broad light-yellow 5.5 x .6 in. pods, 15-18 in. vigorous upright plant, white seed with brown eye, disease res., seed resembles Golden Wax but has an inch longer pod, less fiber, very productive. *Source History: 24 in 1981; 17 in 1984; 20 in 1987; 19 in 1991; 15 in 1994; 22 in 1998; 15 in 2004.* **Sources: Be4, Bu3, CH14, CON, Fo7, HA3, Hal, Jor, La1, Mey, PAG, Sau, Sh9, Sk2, Wet.**

Goldito - 52 days - Semi-upright plant, straight, smooth 4.5-5 in. pods distributed evenly on the plant for easy mechanical or hand harvest, white seed, BCMV tolerant. *Source History: 4 in 2004.* **Sources: HO2, Jor, RUP, Sto.**

Goldkist - 55-59 days - Long slender golden-yellow 5.5-6 in. pods, superb eating quality, tender and flavorful, for best flavor and maximum productivity harvest pods when young, BV1 NY15 SS and rust res., PVP 1990 Novartis Seeds, Inc. *Source History: 1 in 1987; 8 in 1991; 15 in 1994;*

27 in 1998; 5 in 2004. **Sources: Alb, Fis, ME9, Sa5, Tt2.**

Goldrush - 54 days - Round golden-yellow 5.7 in. straight pods hang in clusters around mainstem, retains prime condition a long time, BCMV and NY15 res., improved Brittle Wax type, PVP expired 1994. *Source History: 2 in 1981; 2 in 1984; 2 in 1987; 4 in 1991; 5 in 1994; 6 in 1998; 6 in 2004.* **Sources: CH14, Ear, Loc, OSB, Sto, Ves.**

Hildora - Round yellow 6-8 in. pods held high on sturdy, erect plants, vigorous, disease tolerant, European origin. *Source History: 1 in 2004.* **Sources: Coo.**

Ice - 50-65 days - Old Appalachian family heirloom, abundant tiny light-green pods on every bush, small white seed, fresh or canned. *Source History: 2 in 1991; 3 in 2004.* **Sources: Hi6, Syn, Te4.**

Indy Gold - 47-59 days - Gold Rush type with green-tipped golden, round, 6 in. pods held high on the plant for easy picking, preferred yellow wax bean for growers, shippers and consumers, adaptable throughout most of the U.S., BCMV BV1 NY15 res., PVP 2001 Novartis Seeds, Inc. *Source History: 5 in 1998; 22 in 2004.* **Sources: Be4, Bo17, Bu3, Bu8, CH14, Cl3, Dam, De6, Dom, Ers, Fe5, HO2, Jo1, Jor, Mcf, ME9, Mey, Pin, Roh, RUP, SE4, Ves.**

Keygold - 51-56 days - Compact plants, round firm 6 in. straight slender pods, fine yellow color, very early, very hardy, BCMV and NY15 res., good under stress, wax bean for North, PVP expired 1997. *Source History: 4 in 1981; 4 in 1984; 5 in 1987; 4 in 1991; 2 in 1994; 2 in 1998; 1 in 2004.* **Sources: Pr3.**

Kinghorn Wax - (Kinghorn Wax Resistant, White Seeded Brittle Wax) - 45-56 days - Vigorous 14-24 in. plants, slender golden-yellow 5.5-6 x .4 in. slightly curved pods in clusters, round to slightly oval, stringless, disease res., pure-white seed, heavy yields, good home canner and freezer, popular commercial canner. *Source History: 38 in 1981; 29 in 1984; 20 in 1987; 18 in 1991; 18 in 1994; 16 in 1998; 8 in 2004.* **Sources: CON, Cr1, HA3, MOU, Ont, PAG, Sh9, Up2.**

Kishwaukee Yellow - Seed color may be black, tan or mottled, good flavor, from a tribe in Michigan. *Source History: 1 in 1998; 1 in 2004.* **Sources: Ho13.**

Klondyke - 60 days - Robust, compact plants, high yields of straight, 6 in. pods, fresh market or home garden. *Source History: 2 in 1998; 2 in 2004.* **Sources: ME9, RUP.**

La Victoire - French type, crisp pods, great bean flavor. *Source History: 1 in 1981; 1 in 1984; 1 in 1987; 1 in 2004.* **Sources: Se24.**

Lois Monihan's Candy Wax - Half-climber vine, flattened round seeds in shades of light to medium brown, tender and sweet, productive. *Source History: 1 in 2004.* **Sources: Se17.**

Major - 63 days - Yellow round 5.5-6 in. yellow pods, very slow to develop seeds, full-flavored, highly productive, black seeds. *Source History: 1 in 1987; 1 in 2004.* **Sources: Ter.**

Mont D'Or - (Golden Butter) - 57 days - Heavy yielding French variety, straight oval true-yellow pods, delicate light flavor, res. to BCMV and AN. *Source History: 1 in 1981; 1 in 1984; 1 in 1987; 1 in 1991; 1 in 1994; 1 in 1998; 1 in 2004.* **Sources: Bou.**

Nugget - 52-56 days - Sturdy plant grows 15-20 in. tall, bears 5-6 in. round, bright yellow pods retain color, rich buttery taste, a real standout in producer trials all across the U.S. and Canada, BCMV res., PVP 1999 HA20 *Source History: 7 in 1998; 7 in 2004.* **Sources: Cl3, HA20, ME9, SE4, Sto, Ter, We19.**

Pencil Pod Black Wax - (Butter Bean, Eastern Black Wax) - 50-65 days - Dev. from a cross of Improved Black Wax x Black Eyed Wax, round-oval curved golden-yellow 5-7 x .4 in. pods, black seeds, rust and mosaic res., stringless, excel. quality, vigorous bushy 14-21 in. plants, an all-purpose old-time favorite, introduced in 1900. *Source History: 75 in 1981; 66 in 1984; 62 in 1987; 53 in 1991; 51 in 1994; 52 in 1998; 49 in 2004.* **Sources: Alb, All, Be4, Ber, CH14, Com, CON, Cr1, Dam, De6, Dgs, DOR, Dow, Ear, Ers, Fo7, Gr27, Gr28, HA3, Hal, He8, He17, Ho12, Hum, Jor, La1, Lin, LO8, Ma19, Mo13, Ont, Or10, PAG, Pin, Pr3, Ra5, Roh, RUP, Sa9, Sau, Se17, Sh9, Shu, So1, St18, Te4, Tt2, Up2, Ver.**

Pisarecka Zlutoluske - 55-65 days - Bushes to 2 ft., white flowers, slim golden pods with long red-brown seeds, flavorful both raw and cooked, very productive. *Source History: 3 in 2004.* **Sources: Se17, Sh13, St18.**

Puregold Wax - 58-62 days - AAS/1948, 13-16 in. bush, heavy crops of round slim brittle fiberless stringless 5.5 in. pods, white seeds with brown eyes, res. to BCMV. *Source History: 6 in 1981; 5 in 1984; 5 in 1987; 4 in 1991; 5 in 1994; 1 in 1998; 1 in 2004.* **Sources: Alb.**

Roc D'Or - (Rocdor, Racdor) - 50-60 days - French variety with true yellow color, upright plants, straight tender crisp pods 6-6.5 in. long x .3 in. dia., stringless, early to color, hold their slenderness and color well, superior texture, delicate light and buttery flavor, small black seed, good cool soil germination, BCMV and AN res., very productive, make a second planting for the freezer. *Source History: 5 in 1987; 9 in 1991; 13 in 1994; 19 in 1998; 19 in 2004.* **Sources: Dam, Dom, Ga1, Gr28, Ha5, HO2, Jo1, Jun, ME9, Na6, Ni1, Orn, OSB, Par, Pr3, Roh, Se7, Sto, Ver.**

Roma Gold - 55 days - Long fleshy tender stringless pods, good even when pods are quite large and plump, very productive. *Source History: 1 in 1991; 1 in 1994; 2 in 1998; 1 in 2004.* **Sources: Alb.**

Round Pod Kidney Wax - (Brittle Wax) - 52-58 days - White kidney-shaped seed with black eye, excel. home garden var., good freezer, round thick slightly curved med-yellow 5.5-6.5 x .4 in. pods. *Source History: 12 in 1981; 8 in 1984; 7 in 1987; 4 in 1991; 3 in 1994; 3 in 1998; 3 in 2004.* **Sources: Ear, Lin, Sh9.**

Slenderwax - 56-58 days - Newly developed and refined at Cornell, carries the instant wax gene released by NY State which turns the bean bright-yellow earlier, straighter more uniform pods 5 in. long x .4 in. dia., med-round, tender, no fiber, borne high in huge clusters on vigorous 16-20 in. plants, white seed, BCMV and NY15 res., tol. to heat and white mold fungi, garden and market, excel. frozen or canned, PVP 1988 Musser Seed Co. *Source History: 8 in 1987; 15 in 1991; 15 in 1994; 9 in 1998; 6 in*

2004. **Sources: Be4, Ha5, Lin, MAY, OLD, SOU.**

Sunburst - 51 days - Smooth yellow pods, 5.5 in., white seed, tolerates BCMV, some HB resistance. *Source History: 3 in 2004.* **Sources: Dom, RUP, Sto.**

Sungold - (Sungold Wax) - 47-56 days - Round straight golden-yellow stringless 5.5-6 in. pods in clusters, large 18-24 in. bushes, fine flavor, Geneva NY/AES, res. to BCMV and NY15, widely adapted. *Source History: 15 in 1981; 11 in 1984; 9 in 1987; 10 in 1991; 10 in 1994; 4 in 1998; 1 in 2004.* **Sources: CON.**

Sunrise - 55 days - Strong vigorous plant, waxy yellow 6-6.5 in. pods, excel. color and flavor and texture, slow seed development, plant spring or summer, PVP expired 1991. *Source History: 2 in 1981; 1 in 2004.* **Sources: Se17.**

Unidor - 54-58 days - Sturdy upright bush, pods borne high, white seed, slow seed development, excel. processing bean, BCMV, AN, HB res. *Source History: 2 in 1998; 4 in 2004.* **Sources: OSB, RUP, Sto, Ves.**

Vald'or - 52 days - Larger version of Nugget, upright bushy plants bear 5.5 in. straight bright yellow pods, very tender, excel. flavor, white seed, BV-1A tol., PVP 2001 HA20. *Source History: 1 in 1998; 2 in 2004.* **Sources: Dam, Sto.**

Wax Romano - 59 days - Compact bush plant, wide flat med-yellow 6 in. pods res. to BV1 and NY15 str. of bean mosaic virus. *Source History: 1 in 1991; 4 in 1994; 6 in 1998; 2 in 2004.* **Sources: Roh, Se17.**

Varieties dropped since 1981 (and the year dropped):

Anniversary Gold (1991), *Belidor* (1994), *Black African* (1994), *Bonanza Wax* (1991), *Brittle Wax, Burpee's* (1998), *Brittle Wax, Faribo* (1994), *Buerre de Rocquencourt* (1994), *Bush Baby Bean Mini Yellow* (2004), *Constanza* (1991), *Earliwax* (2004), *Echo* (1987), *Galagold* (2004), *Golden Sands* (1998), *Golden Sinclair* (2004), *Goldie* (2004), *Goldimmens* (1987), *Honey Gold* (1994), *Majestic* (1994), *Midas* (1987), *Minidor* (1998), *Moongold* (1994), *Orlinel* (2004), *Rapier Wax* (2004), *Romano 264* (2004), *Romano, Sunny Yellow* (1998), *Sally's Sunshine* (1998), *Settler III Yellow* (2004), *Sundial* (1994), *Sunkist* (1998), *Sunrae* (2004), *Surecrop Stringless Wax* (2004), *Topcrop Stringless Bush Wax* (1991), *Wax-216* (1991), *Waxidor* (1998), *Yellow Podded Bountiful* (1987).

Bean / Pole / Dry
Phaseolus vulgaris

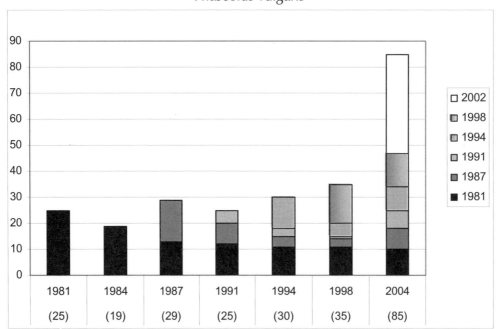

Number of varieties listed in 1981 and still available in 2004: 25 to 10 (40%).
Overall change in number of varieties from 1981 to 2004: 25 to 85 (340%).
Number of 2004 varieties available from 1 or 2 sources: 75 out of 85 (88%).

1500 Year Old Cave Bean - 80-90 days - Ancestor of the Anasazi bean, vigorous vines to 10 ft., tasty slender green pods, white seeds with maroon speckles resemble Jacob's Cattle beans but have richer flavor, longer cooking time required, original seed found in a cave in New Mexico, carbon dated to 1500 years. *Source History: 2 in 1991; 2 in 1994; 1 in 1998; 2 in 2004.* **Sources: Ri12, Se17.**

Amarillo del Norte - Rare, beautiful golden bean from Vadito, New Mexico, 8000', looks similar to Tarahumara Frijol Amarillo and Hopi Yellow. *Source History: 1 in 1998; 1 in 2004.* **Sources: Na2.**

Amish Knuttle - (Corn Hill Bean) - 100 days -

Vigorous 8 ft. vines, 3-4 in. long pods with 4-5 seeds with a round-square shape, purple-grey speckled with bright garnet-red, flavorful, need to be trellised or grown among corn, from Amish farmers of southeastern Pennsylvania. *Source History: 1 in 2004.* **Sources: Ver.**

Aunt Jean's Bean - 90 days fresh/107-110 dry, beautiful half red and half white bean that is almost round, 6 seeds per pod, can be used as a green bean, also good as a dry bean, from Canada. *Source History: 1 in 1991; 2 in 1994; 2 in 1998; 3 in 2004.* **Sources: Bou, Pr3, Sa5.**

Bald Hornet - Round purple and white seed. *Source History: 1 in 2004.* **Sources: Ho13.**

Blanche McGhee - Strong vines, round, maroon seeds, productive, from Indiana. *Source History: 1 in 1998; 1 in 2004.* **Sources: Ho13.**

Brita's Foot Long - Long pods, white seeds, flavorful. *Source History: 1 in 1998; 1 in 2004.* **Sources: Sa5.**

Chihuahuan Ojo de Cabra - (Eye of the Goat) - May be Pinto precursor, adapted to the plains near Casas Grandes in the Sierra Madre of Mexico, traditionally planted during the summer rains. *Source History: 1 in 1981; 1 in 1984; 1 in 2004.* **Sources: Na2.**

Colorado Bolitas - Reddish-beige heirloom, high yields, dry farmed at 7,000 ft. *Source History: 1 in 1987; 1 in 2004.* **Sources: Na2.**

Davis Black - Misleading name since equal numbers of three different colors of seeds are produced, including white, black or brown. *Source History: 1 in 2004.* **Sources: Scf.**

Dolloff - Vermont heirloom, large flattened seeds are pink with gold swirls, pre-1920. *Source History: 1 in 1998; 1 in 2004.* **Sources: Ho13.**

Dolores Hidalgo Frijol Negro - 105 days - Mexican black bean, dark lilac flowers, purple mottled pods, sweet flavor, snap or dry. *Source History: 1 in 1987; 1 in 1991; 1 in 1994; 1 in 2004.* **Sources: Na2.**

Dove - Flattened, round seeds are blue-gray when fresh, gray tone when dry. *Source History: 1 in 1998; 1 in 2004.* **Sources: Ho13.**

Fagy Utan - Large round cream colored seed with blue-purple streaks, from Hungary. *Source History: 1 in 1998; 1 in 2004.* **Sources: Ho13.**

Flagg - Large flattened seeds are buff with dark purple or black streaks, very productive, from the Northeast. *Source History: 2 in 1998; 1 in 2004.* **Sources: Ho13.**

Frijol Colorado - Large flat dark red beans, kidney shape, good as green beans, originally collected in the central and southern Sierra Tarahumara region in Chihuahua, Mexico. *Source History: 1 in 2004.* **Sources: Na2.**

Frijol de Sinaloa - Large beige-yellow seed, dry farmed, high yields, Sonora, Mexico. *Source History: 1 in 1994; 1 in 2004.* **Sources: Na2.**

Frijol Manzano - Medium size round beans with slight lavender tinge, matures late, high yields, from Gomez Farias, Chihuahua, Mexico. *Source History: 1 in 2004.* **Sources: Na2.**

Golden Lima - 85-90 days - Not a true lima, med-large flattened pink bean with gold overlay, most beautiful, heavy cropper, originally offered by Abundant Life Seed Fdn. in the 1980s. *Source History: 1 in 1981; 1 in 1984; 1 in 1987; 1 in 1991; 1 in 2004.* **Sources: Se16.**

Gramma Walters - 65-80 days - Heirloom cranberry bean from eastern Washington, early, bears heavily. *Source History: 1 in 1981; 1 in 1984; 1 in 1987; 2 in 1991; 2 in 1994; 2 in 1998; 1 in 2004.* **Sources: Sa5.**

Guadalupe y Calvo Negro - Purple stemmed plant, dark lilac flowers, round shiny black seeds, from southern Sierra Madre, Chihuahua, Mexico. *Source History: 1 in 2004.* **Sources: Na2.**

Hidatsa Shield Figure - 90-100 days - Green pods, white seed with brown-red shield marking along the face, interplanted with corn where it climbs up the stalks as in the garden of Buffalo Bird Woman, Hidatsa Indians were experts at raising crops of corn, squash, beans and sunflowers in the Missouri River Valley of North Dakota, 1917. *Source History: 3 in 2004.* **Sources: Ag7, Se16, Se17.**

Hopi Beige - 85-90 days - Good production in arid conditions. *Source History: 1 in 2004.* **Sources: So12.**

Hopi Light Yellow - (Seka Mori, Grease Bean) - Large veined ochre bean, colorful mottled pods, high yields, dry farm staple, grown near Moenkopi. *Source History: 1 in 1987; 1 in 1991; 1 in 1994; 1 in 1998; 1 in 2004.* **Sources: Na2.**

Hopi Yellow - 'Sikya mori', large bronze seeds, good as green bean, dry farmed or irrigated, from Hopi country. *Source History: 1 in 2004.* **Sources: Na2.**

Horticultural, Bird's Egg - 110 days - Very large, striped, 6 in. pods, 6 seeds per pod, robust flavor, good as a shell bean, holds its shape when cooked. *Source History: 2 in 1987; 2 in 1991; 2 in 1994; 3 in 1998; 1 in 2004.* **Sources: Pr3.**

Horticultural, Brockton - 85 days - Beautiful red striped pods, wonderful flavor, used only as a dry bean, intro. in 1885 by the Aaron Low Seed Co. after securing seed from a vendor in Brockton, MA. *Source History: 1 in 2004.* **Sources: Se16.**

Horticultural, Cranberry - 75 days - Largest podded and seeded horticultural type, 5-6 ft. plant, thick oval 6-7 in. dark-green pods turn yellow-green tinged with purple at maturity, light buff seeds with dark red splashes and a deep-orange eye ring. *Source History: 1 in 1987; 1 in 1991; 1 in 1994; 1 in 1998; 3 in 2004.* **Sources: Loc, Se16, Se17.**

Indian Mound - *Source History: 1 in 2004.* **Sources: Se17.**

Jack and the Beanstalk - 80-85 days - Long vines to 20 ft., hence the name, best flavor when immature, Polish heirloom. *Source History: 1 in 2004.* **Sources: Ri12.**

Lazy Wife - (Hoffer Lazy Wife, Lazy House Wife) - 80 days - Probably introduced by German immigrants, may have been grown here as early as 1810 as White Cranberry Pole, named Lazy Wife because they were the first snap beans that did not need de-stringing, vines spread at the base, after a slow start they bear heavily and continuously until fall frost, straight stringless pods 5-6 in. long, distinctive shiny white seeds crossed with faint gray patterns. *Source History: 1 in 1981; 1 in 1984; 2 in 1987; 1 in 1991; 1 in 1994; 2 in 1998; 4 in 2004.* **Sources: He8, Lan, Se16, Shu.**

Mayflower - (Amish Knuttle) - 100 days - Vigorous 8 ft. vines, round to pyramid shaped seeds in soft grays with bright red speckles, exceptional flavor, given to the PA Amish by Native Americans, said to have been brought over on the Mayflower in 1620 by Ann Hutchinson. *Source History: 1 in 1998; 2 in 2004.* **Sources: Ag7, Se16.**

Mimbres - *Source History: 1 in 1998; 1 in 2004.* **Sources: Sh13.**

Mostoller Wild Goose - 60 days - Plant with short runners, large seeds are buff with a brown patch around the hilum, grown by the Mostoller family over one hundred years, Pennsylvania legend has it that the seed was taken from the craw of a Canadian goose shot in Somerset County in 1864 by John Mostoller, possibly grown by the Cornplanter Indians along the Upper Allegheny River. *Source History: 3 in 1994; 3 in 1998; 3 in 2004.* **Sources: Lan, Pr3, Te4.**

Mt. Pima Burro and Caballito - Gray/brown on white, high yields, collected in Mayocoba, Sonora, 1984. *Source History: 1 in 2004.* **Sources: Na2.**

Mt. Pima Frijol Blanco - White seed, collected in the Yecora and Mayocoba region of eastern Sonora, Mexico. *Source History: 1 in 2004.* **Sources: Na2.**

Mt. Pima Ojo de Cabra - 'Eye of the Mountain Goat', large brown on beige bean, collected near Mayocoba, Sonora, Mexico. *Source History: 1 in 1987; 1 in 1998; 1 in 2004.* **Sources: Na2.**

Mt. Pima Pintados - Jacob's Cattle type with beautiful plum and white seeds, high yields, collected in Yecora, Sonora, Mexico. *Source History: 1 in 2004.* **Sources: Na2.**

Mt. Pima Vapai Bavi - Small veined yellow-beige bean, also sulfur and light pink seeds, dry farmed, from Mayocoba, Sonora, Mexico. *Source History: 1 in 1987; 1 in 1991; 1 in 1994; 1 in 1998; 1 in 2004.* **Sources: Na2.**

Neabel's - Good for shelling and dry bean. *Source History: 1 in 2004.* **Sources: Sa5.**

Nicaraguan - Productive 6 ft. vines, mix of black, red and tan seeds. *Source History: 1 in 2004.* **Sources: Ho13.**

October - 90-122 days - Vines grow to over 7 ft., 5.5-7 in. pods are streaked with fuchsia and pale yellow at shell stage, maturing to purple with tan streaks when dry, 6-7 tan/maroon seeds per pod, shelled beans have tender skin, good flavor. *Source History: 1 in 2004.* **Sources: Te4.**

O'odham Vayo - Climbing vayo type, mixed gold and light tan seeds, sweet mild flavor, creamy texture, also good green bean. *Source History: 1 in 2004.* **Sources: Na2.**

Or du Rhin - 85-95 days - French type, plump, shiny black, oval seeds, high yields. *Source History: 1 in 2004.* **Sources: So12.**

Papa de Rola - (Dove Breast) - 85-100 days - Outstanding heirloom soup bean from Portugal, white flowers, 4 in. plump pods, oval seeds are half white and half beige, streaked and speckled with red, brown and purple. *Source History: 1 in 2004.* **Sources: Se17.**

Pinto, Juan Gonzales - Old str. from New Mexico, productive 6-7 ft. vines, flattened seeds, buff with brown markings. *Source History: 1 in 2004.* **Sources: Ho13.**

Pinto, Pinacate - Brown speckles on beige-gray, originally from arid runoff farm in Mexico in the Sierra El Pinacate Protected Zone. *Source History: 1 in 2004.* **Sources: Na2.**

Pinto, Rio Bavispe - Beige-gray seed with brown speckles, matures early, from the Rio Bavispe watershed in the foothills of northeastern Sonora. *Source History: 1 in 2004.* **Sources: Na2.**

San Luis Potosi Ojo de Cabra - Colorful flowers and pods, large brown goat's eye seed, matures late, good green bean, from San Luis Potosi, central Mexico. *Source History: 1 in 2004.* **Sources: Na2.**

San Luis Potosi Vayo Blanco - Large beige seeds, irregular shape, late maturing, from San Luis de la Paz, central Mexico. *Source History: 1 in 2004.* **Sources: Na2.**

Snowcap - 75-110 days - Large flat pods contain large flattened seeds with markings similar to Mostoller Wild Goose, half the seed is white, the other half is beige with brown markings which remain when cooked. *Source History: 3 in 1994; 6 in 1998; 7 in 2004.* **Sources: Fo7, Ho13, Pr3, Ra5, Sh13, Ver, Vi4.**

Sonoran Azufrado - Round, sulfur-colored seed, good green bean, matures very late, collected in lowland Sonora in 1985. *Source History: 1 in 2004.* **Sources: Na2.**

Succotash - Named after its resemblance to purple corn kernels, produces best in the fall. *Source History: 2 in 2004.* **Sources: Ho13, Se17.**

Superba - Red-streaked pods, huge round seeds, buff with red splotches. *Source History: 1 in 1998; 1 in 2004.* **Sources: Ho13.**

Tarahumara Black and Blue - Dark lilac flowers, mature pods purple, medium-large black, blue and purple seeds, matures late, from Sierra Tarahumara. *Source History: 1 in 2004.* **Sources: Na2.**

Tarahumara Caballito and Burro - Gray/brown and white Jacob's Cattle bean, high yields, collected in 1984 from Cerocahui, Chihuahua, Mexico. *Source History: 1 in 1987; 1 in 2004.* **Sources: Na2.**

Tarahumara Café - Large seeds resemble unroasted coffee beans, high yields, collected in Kirare, Chihuahua, Mexico. *Source History: 1 in 2004.* **Sources: Na2.**

Tarahumara Carpinteros - Black and white Jacob's Cattle bean, prolific with a little shade in Tucson, collected in central and southern Tarahumara country, Chihuahua, Mexico. *Source History: 1 in 1987; 1 in 2004.* **Sources: Na2.**

Tarahumara Chokame - Lilac flowers, colorful mottled pods, small round shiny black bean with earthy flavor, from the bottom of Batopilas Canyon in the Sierra Madres of Chihuahua, Mexico. *Source History: 1 in 1987; 1 in 1998; 1 in 2004.* **Sources: Na2.**

Tarahumara Choliwame - Purple bean mottled gray, from Sierra Tarahumara, Chihuahua, Mexico. *Source History: 1 in 1987; 1 in 1991; 1 in 1994; 1 in 2004.* **Sources: Na2.**

Tarahumara Flor de Mayo - (May Flower) - Purple star with white rays, matures late, from Creel, Chihuahua, Mexico. *Source History: 1 in 1987; 1 in 1991; 2 in 1994; 1 in 2004.* **Sources: Na2.**

Tarahumara Frijol Amarillo - Dark-yellow seed,

adapted to high cool desert regions of the Sierra Madre in Mexico, traditionally planted during the summer rains, Tarahumara Indian seed. *Source History: 1 in 1981; 1 in 1984; 1 in 1987; 1 in 1991; 1 in 1994; 1 in 2004.* **Sources: Na2.**

Tarahumara Frijol Negro - Small leaves and pods resemble that of tepary bean, tiny black seeds cook quickly, originally collected in 1984 from Kirare, Chihuahua, Mexico. *Source History: 1 in 2004.* **Sources: Na2.**

Tarahumara Mantequilla - (Frijol Burrito) - 'Butter', American pinto type, dark brown and light beige seed, matures late, from thorn forest and high elevation within Sierra Tarahumara of Chihuahua, Mexico. *Source History: 1 in 1987; 1 in 1991; 1 in 1994; 1 in 1998; 1 in 2004.* **Sources: Na2.**

Tarahumara Ojo de Cabra - (Eye of the Goat) - 100 days - Primitive pinto-like bean, seed has dark stripes and specks on beige resembling a goat's eye, adapted to high cool desert areas of the Sierra Madre in Mexico, traditionally planted during summer rains. *Source History: 1 in 1981; 1 in 1984; 1 in 1987; 1 in 1991; 1 in 1994; 5 in 1998; 3 in 2004.* **Sources: Fo7, Ho13, Na2.**

Tarahumara Purple - Large, shiny, deep purple seeds with sweet taste, smooth texture, original seeds from central and eastern Tarahumara country, Chihuahua, Mexico. *Source History: 1 in 2004.* **Sources: Na2.**

Tarahumara Purple Ojos - Seed colors range from deep, dark purple stripes on tan to solid purple and tan with speckles, collected from remote area of Huetozacachi, Chihuahua, Mexico. *Source History: 1 in 1998; 1 in 2004.* **Sources: Na2.**

Tarahumara Purple Star - Large white seed with purple radiating from the hilum, very late maturity, from central and southern Tarahumara country, Chihuahua, Mexico. *Source History: 1 in 2004.* **Sources: Na2.**

Tarahumara Sitakame - Royal purple and related shades of beige, matures late, from Sierra Madre in southern Chihuahua, Mexico. *Source History: 1 in 1987; 1 in 1991; 1 in 1998; 1 in 2004.* **Sources: Na2.**

Tarahumara Star - 120 days - Dark lilac flowers, large black and white rayed seed, definitely daylength sensitive, plant in late summer with rains in the low desert only, originally collected from the heart of the Sierra Tarahumara in Chihuahua, Mexico. *Source History: 1 in 1987; 1 in 1991; 1 in 2004.* **Sources: Na2.**

Tarahumara Vayito - Large beige bean, gold ring around hilum, staple from arid eastern Sierra Madre. *Source History: 1 in 1987; 1 in 2004.* **Sources: Na2.**

Tarahumara Vayo - Mixed beige, medium size, Copper Canyon, grown with other types. *Source History: 1 in 1987; 1 in 1991; 1 in 1994; 1 in 1998; 1 in 2004.* **Sources: Na2.**

Tender Frost - Tender pods when picked young, white seeds, handed down in family for 100 years. *Source History: 1 in 1998; 1 in 2004.* **Sources: Scf.**

Tohono O'odham Vayo Amarillo - Large golden seeds with sweet flavor, creamy texture, Santa Rosa, Arizona. *Source History: 1 in 2004.* **Sources: Na2.**

Toni's Red - Small seeds are half white and half rose-red, resemble seed of Lazy Wife bean but are smaller, grown by a Mr. Bailey's great, great grandfather before the Civil War. *Source History: 1 in 1998; 1 in 2004.* **Sources: Scf.**

True Cranberry - 95 days - Vines grow 6 ft., round and entirely deep-purple seed, really looks just like a large berry, rich flavor, extremely rare, from Vermont. *Source History: 1 in 1981; 1 in 1984; 1 in 1987; 1 in 1991; 1 in 1994; 3 in 1998; 5 in 2004.* **Sources: Co32, Fe5, Hi6, Ho13, Se16.**

Vadito Bolita - Light to tan-beige seed, grown in Vadito, New Mexico at 8000 ft. *Source History: 1 in 1998; 1 in 2004.* **Sources: Na2.**

Vayo de Sonora Sur - Medium size, light tan seed, some light and pink included, excellent flavor, from Alamos, Sonora region, 5000 ft. elevation. *Source History: 1 in 2004.* **Sources: Na2.**

Vermont Cranberry Pole - 60 days - Climbing form of Vermont Bush Cranberry, suggested for use as shell beans but excel. dried for baking, abundant yields, does well in all climates, flavorful, very old heirloom. *Source History: 1 in 1981; 1 in 1984; 2 in 1987; 3 in 1991; 2 in 1994; 3 in 1998; 1 in 2004.* **Sources: Sh13.**

Wren's Egg - (Speckled Cranberry, Mammoth Podded Horticultural, Dwarf Horticultural) - 65-69 days - Heirloom hort., large oval buff seeds with red stripes, slender stringless 5 in. pods, all purpose, bears all season, from England in 1825. *Source History: 4 in 1981; 4 in 1984; 3 in 1987; 5 in 1991; 6 in 1994; 6 in 1998; 7 in 2004.* **Sources: Fo7, He8, Lan, Ra5, Roh, Se17, Shu.**

Wren's Egg Improved - (Speckled Cranberry, King Mammoth Podded Horticultural) - 65 days - New highly imp. disease res. strain, med-dark 5 in. pods, very productive, excel. shelling qualities. *Source History: 1 in 1981; 1 in 1984; 1 in 1987; 1 in 1991; 1 in 1994; 1 in 1998; 1 in 2004.* **Sources: Ver.**

Yin Yang - 100 days - Rare heirloom, half white and half red, yellow or black with distinctive sharp boundaries that curve around the seed, good yields. *Source History: 1 in 1994; 2 in 1998; 3 in 2004.* **Sources: Fo7, Sa5, Te4.**

Yoeme Ojo de Cabra - Small goat's eye bean, beige background with gold-brown lines, green or dry, matures late, from Vicam, Sonora, traditional Yoeme village. *Source History: 1 in 2004.* **Sources: Na2.**

Yoeme Vayo - Beige seeds, matures early, high yields, from Yoeme village outside of Ures, Sonora. *Source History: 1 in 2004.* **Sources: Na2.**

Zapotec - *Source History: 1 in 2004.* **Sources: Se17.**

Varieties dropped since 1981 (and the year dropped):

Barnett (2004), *Bertha* (1991), *Bogen* (1991), *Bolito* (1983), *Box* (1998), *Brown Goose* (1998), *Burt's Bean* (1991), *Cannellini, Large* (2004), *Chapman's Climbing Bumblebee* (2004), *Cornfield* (2004), *Ellen's Black* (1998), *Ellen's Brown Striped* (1998), *Fred Taylor's* (2004), *Friederich Dry Pole* (1998), *Frijol Canario* (1987), *Good Mother Stallard*

(2004), *Goose Bean* (2004), *Guatemalan Highlands* (2004), *Horticultural Pole* (1998), *Horticultural, Cranberry* (2004), *Horticultural, King* (2004), *Horticultural, Worcester* (2004), *Horticulture, Tall* (1994), *Inge Hanle* (1998), *Italian Spotted* (1987), *Jerusalem* (1987), *London Horticultural* (1983), *Mexican Climbing* (1994), *Milpa Black* (1983), *Montezuma* (2004), *Musick Pinto* (1998), *Nunas Popping* (1994), *Oceana's Summer* (2004), *Old Cornfield* (1983), *Pinto, Pioneer* (1998), *Pole Cherry Horticultural* (1998), *Purple Buddha* (1991), *Red Chili* (2004), *Red Lazy Wife* (1991), *Scotch* (2004), *Smooth Brown* (1984), *Tarahumara Estacas* (1991), *Tarahumara Pink Green Bean* (1991), *Yaqui String* (1998), *Zebra Horticultural* (1984).

Bean / Pole / Snap
Phaseolus vulgaris

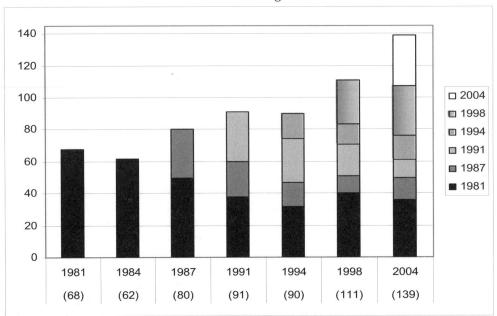

Number of varieties listed in 1981 and still available in 2004: 68 to 36 (53%).
Overall change in number of varieties from 1981 to 2004: 68 to 139 (207%).
Number of 2004 varieties available from 1 or 2 sources: 104 out of 139 (75%).

Alice Whitis - 70-90 days - Vigorous vines bear tasty green pods, bright deep rose seeds, Alice Whitis, a stern woman, lived in Acorn, KY, in the 1930s, lived to almost 100 years old, was known to carry her shoes to church and put them on when she arrived. *Source History: 1 in 2004.* **Sources: Se17.**

Anderson's Purple King - *Source History: 1 in 2004.* **Sources: Se17.**

Aunt Ada's Italian Pole - Unusual heirloom from Italy, brought to the U.S. by the Botanelli family around 1900, loses quality when over mature. *Source History: 1 in 1998; 1 in 2004.* **Sources: Tu6.**

Aunt Jean's - 90 days - White seed with maroon eye, sweet nutty flavor. *Source History: 1 in 2004.* **Sources: Fe5.**

Berta Talaska - Tall vigorous plants, stringless flat pods used as snaps, plump pods can be shelled and eaten as a green shell bean and mature pods are shelled and used as kidney bean substitute, brought from Portugal 50 years ago by Berta Talaska. *Source History: 1 in 2004.* **Sources: Al6.**

Big Mama - *Source History: 1 in 2004.* **Sources: Se17.**

Blauhilde - Stringless blue pole bean, 10-12 in. long fleshy pods, higher yielding and healthier than other blue pole types, turns green when cooked. *Source History: 1 in 1987; 1 in 1991; 1 in 1994; 1 in 1998; 2 in 2004.* **Sources: Coo, Dam.**

Blue Coco - 55-60 days - Name refers to the purple (blue) color of the pods and the chocolate (coco) color of the seeds, vines climb to 8 ft., rose-colored blooms, curved pods are 5-6 in. long, nice meaty flavor, very early, will produce under hot dry conditions, French heirloom, 1775. *Source History: 1 in 1987; 1 in 1991; 1 in 1994; 1 in 1998; 3 in 2004.* **Sources: In8, Se17, So1.**

Blue Lake FM-1 - (Fillbasket, White Creaseback) - 55-65 days - Round smooth dark-green stringless 6-7 in. meaty pods, heavy yields, vigorous 5.5 ft. mosaic res. plant, excel. for canning, freezing, soups and baking. *Source History: 9 in 1981; 9 in 1984; 7 in 1987; 8 in 1991; 10 in 1994; 8 in 1998; 11 in 2004.* **Sources: Bu3, DOR, Jun, MAY, Ont, OSB, Par, Roh, Shu, SOU, TE7.**

Blue Lake FM-1K - (Stringless Blue Lake FM-1K) - 66 days - Dark-green round 5.5-6 in. pods, white seed, also good as shell bean, starts producing 5 days later than Prime Pak but finishes with it, can/freeze/home, BCMV res. *Source History: 2 in 1981; 2 in 1984; 4 in 1987; 5 in 1991; 5 in*

1994; 5 in 1998; 5 in 2004. **Sources: CH14, Ga1, Jor, Mo13, WE10.**

Blue Lake Pole - 62-75 days - Straight stringless bright dark-green round/oval 5.5-7 x .4 in. pods, good climber, sets pods from base to top, not mech. harv., good fresh, canned, frozen, baked, dev. in the Willamette Valley of Oregon, good for Canadian prairies and Pacific Northwest. *Source History: 50 in 1981; 43 in 1984; 44 in 1987; 40 in 1991; 33 in 1994; 31 in 1998; 30 in 2004.* **Sources: Ada, Alb, All, Bo17, Bou, Bu8, Com, Ers, Fi1, Fis, Gur, HA3, He17, Hum, Ki4, Kit, LO8, Loc, ME9, Mey, MOU, Na6, Or10, PAG, Sau, Sto, Sw9, Ter, Ves, We19.**

Blue Lake S-7, Stringless - 60 days - Improved Blue Lake Pole variety with sparse foliage, vines grow to 7 ft., dark green 6.5 in. long round pods, tender, meaty and stringless, slow seed development, exellent flavor, grows in hot or cool weather, up to 10 days earlier and produces bigger yields than other strains. *Source History: 4 in 2004.* **Sources: Bu1, Dam, Jun, Ver.**

Blue Lake Stringless - (FM-1) - 60-70 days - Vigorous 5-6.5 ft. plants, straight dark-green round slightly curved 6 x .4 in. pods, stringless and fiberless all stages, heavy yields until frost if kept picked clean, holds flavor through processing, good fresh/green shelled/dry, BCMV res. *Source History: 22 in 1981; 18 in 1984; 16 in 1987; 15 in 1991; 12 in 1994; 14 in 1998; 11 in 2004.* **Sources: Be4, Ear, Fo7, Hi6, Hud, La1, OLD, Ri12, Ros, Ver, Wet.**

Blue Lake Stringless S-7 - 59-62 days - Vines to 7 ft., thick flat med-green 6 in. pods borne on spurs extending out from the vine, white seed, easy harvest, for local fresh market or shipping, res. to rust. *Source History: 1 in 1981; 1 in 1984; 1 in 1987; 1 in 1991; 1 in 1994; 1 in 2004.* **Sources: Shu.**

Blue Lake, Black Seeded - 55-63 days - Heavy yielding 6 ft. plants, round straight stringless 6-7.5 x .4 in. fleshy pods, med- green leaves, black seed, for freezing or canning. *Source History: 3 in 1981; 1 in 1984; 1 in 1987; 1 in 1991; 1 in 1994; 4 in 1998; 4 in 2004.* **Sources: Pe2, Roh, Se7, So12.**

Blue Lake, Oregon - 60 days - Round/oval stringless 6.5 in. pods, small white seeds excel. for dried shell use, very popular, basis for an entire industry in Oregon. *Source History: 1 in 1981; 1 in 1984; 1 in 1987; 1 in 1991; 1 in 1994; 1 in 1998; 1 in 2004.* **Sources: Ni1.**

Blue Lake, White Seeded - (Blue Lake Stringless White) - 60-66 days - Round/oval med-green straight smooth stringless 6- 7 in. pods, white seeds make excel. dry beans for baking, pods good fresh or canned or frozen, res. to BCMV. *Source History: 10 in 1981; 8 in 1984; 7 in 1987; 7 in 1991; 6 in 1994; 5 in 1998; 4 in 2004.* **Sources: Bu2, Hi13, RUP, Sh9.**

Blue Ribbon - 69 days - Excel. yielding creaseback type, 5 ft. plants, known for excel. flavor, will do well in drought areas, produces continuously so pick early and often, 1940. *Source History: 1 in 1981; 1 in 2004.* **Sources: Sa9.**

Blue Tip - White seeds, round pods, no blue on it, handed down in a family for generations. *Source History: 1 in 1998; 1 in 2004.* **Sources: Scf.**

Bobis a Grano Bianco - Medium sized, green, stringless pod. *Source History: 1 in 1991; 1 in 1994; 1 in 1998; 1 in 2004.* **Sources: Ber.**

Borlotti - Classic Italian pole bean with medium to large, tan seeds splashed with red-black to magenta streaks, use fresh or dry for storage, very popular in Italian and Portuguese cuisine. *Source History: 1 in 1994; 2 in 1998; 1 in 2004.* **Sources: Coo.**

Borlotto Fiamma Rampicante - *Source History: 1 in 1987; 1 in 1991; 1 in 1994; 1 in 1998; 1 in 2004.* **Sources: Ber.**

Borlotto Lamon - 75-80 days - White seeds mottled with rose, from Venice where it is used in pasta fagiolo. *Source History: 1 in 2004.* **Sources: Se24.**

Borlotto Sanguigno - Red and white spotted seed. *Source History: 1 in 1987; 1 in 1991; 1 in 1994; 1 in 1998; 1 in 2004.* **Sources: Ber.**

Brejo - 65-75 days - Vigorous 8 ft. vines with violet flowers, stringless 6-7 in. pods are somewhat contorted, green pods sprayed with purple, unique, flattened tan colored seeds with black markings, valuable old Indian strain from Willamette Valley. *Source History: 1 in 1981; 1 in 1984; 1 in 1987; 1 in 1991; 1 in 1994; 4 in 1998; 3 in 2004.* **Sources: Ho13, In8, Se17.**

Cascade Giant - 58-80 days - Stringless version of Oregon Giant with 8-10 in. pods, 6 ft. tall plants, seeds are dark-green mottled with purple, flavor and color identical to its parent, adapts to cool, damp weather and tolerant of drought, good canner, fairly drought tol., adaptable to cool damp conditions, dev. at Oregon State U. *Source History: 4 in 1994; 7 in 1998; 2 in 2004.* **Sources: Se7, Ter.**

Case Knife - 60-65 days - Heirloom, one of the oldest green beans in America, noted in literature in 1820, strong climber, flat stringless 8 in. pods, white seed, lots of good eating. *Source History: 1 in 1981; 1 in 1984; 1 in 1987; 2 in 1998; 6 in 2004.* **Sources: Ri12, Sh13, Shu, Te4, Th3, Ver.**

Celina's Wide Green - Excellent for fresh snap and dry use. *Source History: 1 in 2004.* **Sources: Sa5.**

Cherokee Cornfield - 65-90+ days, Generations-old mix of over 20 different variations, has always been grown with cornstalks, will not produce as abundantly if the variations are separated, plant 2 packets to insure maximum variation, fun to shell, originally from the Cherokees in TN. *Source History: 3 in 1998; 5 in 2004.* **Sources: Fo13, Ho13, Ri12, Se17, So1.**

Cherokee Trail of Tears - (Cherokee Black) - 65-95 days - Cultivated by the Cherokee Nation and carried with them when displaced by white settlers in the 1800s, vines to 8 ft., 6 in. purple pods, small shiny black seed, interplant with corn and squash, prolific, snap or dry soup bean. *Source History: 2 in 1987; 1 in 1991; 5 in 1994; 10 in 1998; 11 in 2004.* **Sources: Co32, Eco, Fo13, He8, Ho13, Pr3, Ri12, Se16, Se17, Sh13, Yu2.**

Chinese Beige - Productive bean used as snap, shell and dry, rare. *Source History: 1 in 2004.* **Sources: Syn.**

Climbing French - (Climbing Canadian Wonder, True and Tender, Veitch's Climbing French) - 65-75 days - Beautiful lilac flowers, 4-7 in. stringless pods, shiny dark purple seeds, excel. fresh eating qualities, good disease res., popular in the Northeast in the 1930s. *Source History: 2 in 2004.* **Sources: Ag7, Se16.**

Cobra - 75 days - Decorative mauve flowers, smooth, round stringless 7-8 in. long pods, heavy yields. *Source History: 2 in 2004.* **Sources: Ga7, Tho.**

Cornfield, Black - Does well when grown in a cornfield to vine on the corn stalks, shiny black seeds, believed to have originated in Germany. *Source History: 1 in 1998; 1 in 2004.* **Sources: Scf.**

Cornfield, Striped - Striped beans, used for fresh, canning, freezing and crafting, often grown among corn as a natural trellis. *Source History: 1 in 2004.* **Sources: Fe5.**

Cornfield, Susie's - *Source History: 1 in 2004.* **Sources: Se17.**

Country Woman - (Signora Della Campagna) - Good producer of fresh shell and dried beans, green and red curved pods, cream colored seeds with red splotches, flavorful. *Source History: 1 in 2004.* **Sources: Se24.**

Cutshort, Greasy - (White Cutshort) - 85 days - So named because the seeds grow so closely together in the pods that the seed ends are flattened or "cut short", slick or "greasy" pod, excel. for fresh eating, formerly used for drying as "leather breeches beans" which were strung like peppers on a string and hung on the porch to dry, heirloom. *Source History: 2 in 1998; 3 in 2004.* **Sources: Ho13, Scf, So1.**

Cutshort, Ohio - 80 days - High yields of 3-4 in. tender pods. *Source History: 1 in 2004.* **Sources: Sa9.**

Cutshort, Old Time - 75 days - Pods to 3-4 in. long with speckled seeds, high yields. *Source History: 1 in 2004.* **Sources: Sa9.**

Dade - 55-62 days - Thick flat oval med-dark green 7.5 in. pods, 5-6.5 ft. plants, multiple disease res., high yields, for hot humid South, stringless when young, vigorous, FL/AES, 1962. *Source History: 13 in 1981; 8 in 1984; 8 in 1987; 7 in 1991; 5 in 1994; 7 in 1998; 2 in 2004.* **Sources: Kil, Scf.**

Domatsu Snap Vine - Long thin pods borne in clusters, 7-8 in. *Source History: 1 in 2004.* **Sources: Pe1.**

Dow Purple Pod - 75-80 days - Vigorous climber, 7-8 in., flat purple pods, tan seeds, germinates in cool wet soil, hand-me-down, originally from Illinois, very scarce seed. *Source History: 1 in 1981; 1 in 1984; 1 in 1987; 4 in 1991; 4 in 1994; 2 in 1998; 1 in 2004.* **Sources: Se7.**

Early Riser - 40-55 days - European pole bean, dark-green 8-10 in. pods, white seed, good seedling vigor and yield, BV1A res. *Source History: 1 in 1987; 2 in 1991; 1 in 1994; 2 in 1998; 2 in 2004.* **Sources: Pe2, Tu6.**

Emerite - 50-70 days - Straight slender 7-9 in. pods in heavy clusters, sweet beany flavor, good for freezing, early, pick when 4-5 in. long for tender green beans or when mature at 7 in. long for a crisp brittle snap, BCMV res., obtained from Vilmorin, one of the oldest and best seed houses in France. *Source History: 6 in 1991; 10 in 1994; 11 in 1998; 13 in 2004.* **Sources: Coo, Dam, Gou, Ki4, Mcf, Na6, Pin, Se8, Sto, Syn, Ver, Vl3, Wi2.**

Fagiolo Meraviglia Venezia - 75-80 days - Very productive yellow pole roma type, great taste, needs staking but requires less room than bush beans, from C. Faraone Mennela of Italy. *Source History: 1 in 1981; 1 in 1984; 1 in 2004.* **Sources: Se24.**

Fasold - 90 days - Semi-climbing vines, mauve blooms, long round fleshy stringless pods, flavorful, slow to develop seeds, long harvest period, huge yields, BCMV, less vigorous habit makes it a good variety for growing under glass for earlier harvest. *Source History: 1 in 2004.* **Sources: Tho.**

Fawcett's - Stringless snap, speckled purple pods change to bright green when cooked, stays crunchy as it matures, very prolific. *Source History: 1 in 2004.* **Sources: Sa5.**

Fortex - 60-70 days - Extra-long round stringless French pole bean, pods grow to over 11 in., may be picked at 7 in. for slender filet beans, walnut-brown seed, for fresh use and freezing, introduced in 1988. *Source History: 2 in 1991; 4 in 1994; 5 in 1998; 11 in 2004.* **Sources: Dam, Fe5, Jo1, Jun, Mcf, Sw9, Ter, Tu6, Ver, Vl3, We19.**

Garden of Eden - 65 days - Medium green pods retain sweet tender flavor, brown seeds with dark brown stripes, heirloom from New Jersey via Spain or Portugal. *Source History: 1 in 1998; 1 in 2004.* **Sources: Jo1.**

Genuine Cornfield - (Cutshort, Creaseback) - 72-90 days - Introduced before 1835, was commonly planted in corn patches, very good fresh shell bean before maturity, 5-7 in. pods turn purple as they mature, buff colored seed striped with brown. *Source History: 1 in 1981; 1 in 1984; 3 in 1994; 4 in 1998; 6 in 2004.* **Sources: Ada, Bu8, Co32, Fo7, He8, He17, Ho13, Sau, Se25, Shu, So1, Syn, Wet.**

Georgia Black - Long, fat pods, strings, heirloom. *Source History: 1 in 2004.* **Sources: Syn.**

Gila - (Gila River) - 90 days - Heirloom, 6 ft. vines, 6 in. pods, 6 seeds per pod, color and markings similar to New Mexico Cave bean. *Source History: 1 in 1987; 1 in 1998; 1 in 2004.* **Sources: Se17.**

Gnuttle Amish - (Amish Gnuttle, Amish Knuttle, Amish Gnuddel) - 90-100 days - Vines 5-8 ft., 5 in. pods with cream-pink dotted purple, known as cutshort since square ended seeds are packed tightly in the pods, prolific, unusual, dates back to the 1840s. *Source History: 1 in 1987; 3 in 1991; 4 in 1994; 4 in 1998; 4 in 2004.* **Sources: Ami, Ho13, Lan, Roh.**

Golden of Bacau - 60-70 days - Flat golden romano type, 6-10 in. long pods, 1 in. wide, excel. flavor, stringless at all stages, tender and sweet, from northern Romania. *Source History: 1 in 2004.* **Sources: Se16.**

Grandma Nellie's - Flat pods with dark brown flat seeds, brought from Russia to Western Canada early in the 20th century. *Source History: 1 in 2004.* **Sources: Pr3.**

Grandma Nespeca's - *Source History: 1 in 2004.* **Sources: Se17.**

Granny - *Source History: 1 in 2004.* **Sources: Se17.**

Granny Richmond - Small tan and brown striped seed, from Wilbraham, MA family where it had been grown since at least the 1870s. *Source History: 1 in 2004.* **Sources: Ea4.**

Grape Bean - Kentucky heirloom with 5 in., flat green pods, round purple seeds. *Source History: 1 in 1998; 1 in 2004.* **Sources: Ho13.**

Greasy, Brown Speckled - 73 days - Cutshort type, 3.5 in pods, round, brown-speckled seeds, 8 per pod. *Source History: 1 in 2004.* **Sources: Sa9.**

Greasy, Tennessee - 70 days - Pole snap bean with 3 in. pods, 6-8 tan striped seeds. *Source History: 1 in 1998; 1 in 2004.* **Sources: Sa9.**

Green Anellino - 85 days - Italian heirloom, small crescent-shaped bean with rich Romano flavor, pods are curved so each forms a half circle near tip, pick when young before strings develop, in great demand by chefs for its flavor and unique shape. *Source History: 1 in 1981; 1 in 1984; 1 in 1987; 1 in 1991; 1 in 1994; 1 in 1998; 2 in 2004.* **Sources: Coo, Pin.**

Griggs Black - Vigorous vines benefit from trellising, black seeds, green snap or dry use. *Source History: 1 in 2004.* **Sources: Scf.**

Hawaiian Wonder - Rust resistant. *Source History: 1 in 1987; 2 in 1991; 1 in 1994; 1 in 1998; 1 in 2004.* **Sources: Uni.**

Heirloom Pole - Horticultural bean, vines to 6-7 ft. long, pale lilac flowers, 6-8 in. pods mature to bright fuchsia pink stripe, used as snap, shelly or dry bean, originated in Lancaster County, PA, rare. *Source History: 1 in 2004.* **Sources: Ami.**

Helda - 58-65 days - Long 9 in. broad light-green stringless pods .75 in. wide, disease res. climbing plant, vigorous, for outdoors and especially greenhouse culture, bred in southern Europe. *Source History: 1 in 1987; 1 in 1991; 3 in 1994; 1 in 1998; 5 in 2004.* **Sources: Jun, Par, Ter, Twi, Ver.**

Hilda - Flat green pods grow 12 in. long and retain flavor and tenderness, disease tolerant. *Source History: 1 in 2004.* **Sources: Coo.**

Hillbilly Purple - 60-90 days - Prolific vines, striking, deep purple pods, light beige, flattened kidney shaped seed. *Source History: 1 in 1998; 1 in 2004.* **Sources: So12.**

Honey - Medium-large, flattened seeds are honey colored, long pods tend to twist, from Czechoslovakia. *Source History: 1 in 1998; 1 in 2004.* **Sources: Ho13.**

Horticultural, New Hampshire - 50-70 days - Beautiful, deep purple pods contain large, white and purple seeds which are larger with darker markings than other hort. strains. *Source History: 2 in 1998; 1 in 2004.* **Sources: Ho13.**

Hunter Stringless - 55-68 days - Earlier more disease res. strain, healthier growth and higher yields, long flat 8 in. pods, also suited for greenhouse growing, European sword type slicing bean. *Source History: 1 in 1981; 1 in 1984; 1 in 1987; 1 in 1991; 2 in 1994; 1 in 1998; 1 in 2004.* **Sources: Dam.**

Ideal Market - 65-70 days - Originally introduced in 1914 as Black Creaseback by Van Antwerp's Seed Store of Mobile, AL, later renamed in the seed trade by Chris Reuter Seed Company as Reuter's Ideal Market, 5-6 in. pods, fleshy, brittle, stringless, fine texture, vigorous and hardy. *Source History: 1 in 2004.* **Sources: Se16.**

Italian Pole - 60-70 days - Considered by many to be the finest flavored snap bean, 5-6 ft. vines, very tender fleshy flat pods, 5-7 in. long, light-mauve seeds with black stripes and speckles. *Source History: 1 in 1981; 1 in 1984; 1 in 1987; 1 in 1991; 3 in 1994; 4 in 1998; 3 in 2004.* **Sources: All, Se7, Sh13.**

Jeminez - 67-75 days - Vigorous vines, 7-8 in. flat-oval straight pods are uniquely colored with purple-red tinting and streaking over dark green, BCMV resistant. *Source History: 2 in 1991; 3 in 1994; 4 in 1998; 2 in 2004.* **Sources: Sa9, Tho.**

Juanita Smith Beans - Cornfield bean, tender round pods, black and white appaloosa speckled seed, heirloom originally from Juanita Smith. *Source History: 1 in 1998; 1 in 2004.* **Sources: Scf.**

Kahnawake Mohawk - (Cherokee Shellout, Steak) - 110 days - Cornfield bean from the Powhatan Indians of North Carolina, climbs to over 9 ft., 5 in. pods, tan seeds with dark brown markings, reliable yields in all weather types, used fresh or dry. *Source History: 1 in 2004.* **Sources: Ag7.**

Kentucky Blue - 51-73 days - AAS/1991, Kentucky Wonder x Blue Lake, flavor similar to Kentucky Wonder with traces of Blue Lake's sweetness, disease res. vines need support, straight smooth pods, best flavor and tenderness at 6-7 in., BCMV res., bred by Dr. Calvin Lamborn, PVP 1992 Novartis Seeds, Inc. *Source History: 41 in 1991; 51 in 1994; 47 in 1998; 46 in 2004.* **Sources: Be4, Bo17, Bou, Bu2, Bu8, CH14, Co31, De6, Dgs, Ers, Fa1, Fi1, Gr27, Gur, Ha5, Hal, He17, HO2, Hum, Jor, Jun, La1, Lin, Ma19, MAY, Mey, Mo13, MOU, Ont, OSB, Par, Pe6, Pin, RIS, Roh, RUP, Se26, Sh9, Si5, SOU, Ter, Tt2, Twi, Ver, Ves, Wet.**

Kentucky King - 57 days - Upright plants, 7 in. long pods borne in clusters, easy harvest, an Italian type bean with that classic Kentucky Wonder flavor. *Source History: 2 in 2004.* **Sources: Bu2, Hi13.**

Kentucky Wonder - (Old Homestead, Texas Pole, Egg Harbour, Kentucky Wonder Green Pod, Improved Kentucky Wonder) - 58-72 days - Vigorous 5-7 ft. plant, straight med-green flat/oval 7-10 in. pods in clusters, rust res., white or brown seeded, high yields, extended season, distinctive flavor, good firmness, introduced before 1864 in Kentucky, formerly known as Texas Pole, given its present name by seedsman James J.H. Gregory in 1877. *Source History: 110 in 1981; 97 in 1984; 98 in 1987; 85 in 1991; 75 in 1994; 72 in 1998; 72 in 2004.* **Sources: Alb, All, Ba8, Be4, Bo19, Bu1, Bu2, Bu3, But, CA25, CH14, Co32, Com, CON, Cr1, De6, Ear, Ers, Fe5, Fi1, Fo13, Ga1, Ge2, Goo, Gur, HA3, Hal, He17, Hi13, HO2, Hum, Jor, Kit, Lin, LO8, Loc, Ma19, MAY, ME9, Mel, Mey, Mo13, MOU, Na6, Ni1, OLD, Ont, Or10, OSB, PAG, Pe2, Pin, Ra5, Ri2, Ri12, RIS, Roh, Ros, RUP, Sa9, Sau, Se7, Se16, Se26, Si5, Sol, So9, SOU, Sto, TE7, Vi4, WE10.**

Kentucky Wonder 191 - (White Kentucky 191) - 62-68 days - Thick oval silvery-green 6-6.5 x .5 in. pods, white seed, med- green 5-6 ft. plants, good where rust is a problem, for snaps or dry shell. *Source History: 10 in 1981; 9 in 1984; 9 in 1987; 8 in 1991; 7 in 1994; 5 in 1998; 6 in 2004.* **Sources: Ada, Co31, SE4, Se26, SOU, Wet.**

Kentucky Wonder Brown Seeded - (Old Homestead Brown Seeded) - 58-75 days - Brown seeded strain, large clusters of round curved 7-10 x .5 in. silvery-green fleshy pods, almost stringless, distinct flavor, excel. freezer, also for shells, introduced before 1864. *Source History: 18 in 1981; 16 in 1984; 18 in 1987; 15 in 1991; 15 in 1994; 19 in 1998; 13 in 2004.* **Sources: Ada, Bu8, Co31, Dam, DOR, Fa1, La1, Sh9, Shu, Sk2, Ter, Ver, Wi2.**

Kentucky Wonder Rust Res. - (Buff/Brown Seed, White Seed) - 63-67 days - Round/oval med-green stringless 6-7 in. pods, good quality when young, popular pole bean, for home garden and market. *Source History: 4 in 1981; 4 in 1984; 5 in 1987; 2 in 1991; 2 in 1994; 3 in 1998; 1 in 2004.* **Sources: Kil.**

Kentucky Wonder White Seeded - (Kentucky Wonder 191, No. 191, White Kentucky Wonder 191) - 64-65 days - Vigorous and prolific, thick oval 6-9 x .5 in. slivery green pods, climbs 4-5 ft., well adapted to the Pacific coast and the Southeast states, rust res., almost stringless, first introduced in the 1850s. *Source History: 15 in 1981; 14 in 1984; 11 in 1987; 9 in 1991; 7 in 1994; 6 in 1998; 3 in 2004.* **Sources: Bu8, La1, Sau.**

Kentucky Wonder, Black Seeded - 84 days - Kentucky Wonder type with long, large, stringless, fiberless, fleshy pods, 6-8 in. long, 8-10 seeds per pod, good flavor and texture, heirloom from central Ohio, original seed from Tom Knoche's Aunt Marge who maintained this var. for 60 years. *Source History: 1 in 1998; 1 in 2004.* **Sources: So1.**

Kwintus - 55-90 days - European climbing bean selected for greenhouse use but does equally well outdoors, covered with long, flat green 12 in. pods that stay tender and flavorful regardless of size, bears early, both the first and last bean to be picked, formerly Early Riser. *Source History: 1 in 1994; 2 in 1998; 5 in 2004.* **Sources: Coo, Ki4, OSB, Par, VI3.**

Landfrauen - 55 days - Swiss heirloom with unequaled bean flavor and aroma, tender stringless light-green 5-6 in. pods with purple mottling, excel. fresh, frozen or dried. *Source History: 1 in 1994; 1 in 1998; 1 in 2004.* **Sources: Ami.**

Lazy Wife, Hoffer - Used as green or snap, stringless, so named because of the ease with which it grows and cooks, German origin. *Source History: 1 in 1998; 1 in 2004.* **Sources: Roh.**

Leather Britches - 70 days - Stringless 3 in. pods, no need for snapping, great for canning or freezing whole, stir fry and salads, pods grow two by two in clusters for easy harvest, dried pods reminded pioneer women of leather britches hanging out to dry, thus the name, discovered in the Ohio River Valley by early settlers. *Source History: 1 in 1994; 1 in 1998; 1 in 2004.* **Sources: Sh13.**

Louisiana Purple Pod - 60-90 days - Heirloom strain, fast growing and prolific, 7.5 in. purple pods turn green when cooked, practically stringless when young. *Source History: 3 in 1981; 2 in 1984; 3 in 1987; 3 in 1991; 3 in 1994; 4 in 1998; 2 in 2004.* **Sources: Ho13, So1.**

Manoa Wonder - Nematode resistant. *Source History: 1 in 1987; 2 in 1991; 1 in 1994; 1 in 1998; 1 in 2004.* **Sources: Uni.**

Mantra - 90 days - Medium green stringless pods up to 10 in. long, sweet flavor, early prolific cropper, suitable for growing under glass or outdoors, BCMV resistant. *Source History: 1 in 2004.* **Sources: Tho.**

Marsingill Black - Often grown among corn as a natural trellis, black seeds, resists beetle damage, possibly from Europe. *Source History: 1 in 1998; 1 in 2004.* **Sources: Scf.**

Mazlenk Visoki Zeleni - Seeds are either blue or tan, 8 in. pods, from Eastern Europe. *Source History: 1 in 1998; 1 in 2004.* **Sources: Ho13.**

McCaslan - 61-66 days - Southern favorite for snaps and shelling, 7-8 in. long flat med-green slightly curved pods, prolific vigorous vines, everbearing, excel. quality, always stringless, grown before 1900 by the McCaslan family in Georgia, introduced in 1912. *Source History: 17 in 1981; 11 in 1984; 15 in 1987; 18 in 1991; 16 in 1994; 18 in 1998; 14 in 2004.* **Sources: Ada, Bu8, Co31, Fo7, HO2, RUP, Sau, Se26, Sh9, Shu, So1, SOU, Ver, Wet.**

McCaslan 42 - 61-65 days - Broad flat curved med-dark green 7.5-8 x .6 in. pods, well adapted to Calif. heat, 5-6.5 ft. plants, sets from ground up, long season, stringless, ivory seed, 1962. *Source History: 5 in 1981; 5 in 1984; 4 in 1987; 2 in 1991; 2 in 1994; 5 in 1998; 1 in 2004.* **Sources: De6.**

McCormick's Wonder - Flattened, square, pink-tan seeds with pinto markings, good yield. *Source History: 1 in 2004.* **Sources: Pr3.**

Metro o Stringa Rampicante - Long, thin, round, green, stringless pods. *Source History: 1 in 1987; 1 in 1991; 1 in 1994; 1 in 1998; 1 in 2004.* **Sources: Ber.**

Missouri Wonder - (Nancy Hall) - 60-75 days - For snaps and green shell, thick round/oval med-green slightly curved fleshy 6-8 in. pods, home garden, stringless when young, prolific sure cropper. *Source History: 12 in 1981; 12 in 1984; 11 in 1987; 11 in 1991; 10 in 1994; 12 in 1998; 10 in 2004.* **Sources: Bu8, Ers, Fo7, He17, Ho13, OLD, Sau, Sh13, Shu, SOU.**

Musica - 55-67 days - Early Romano type produces massive crop of huge tender broad flat 9-10 in. pods over a long season, pods have slightly scalloped edges, delicious true bean flavor, matures a week ahead of Blue Lake Pole, BCMV res. *Source History: 2 in 1991; 2 in 1994; 1 in 1998; 6 in 2004.* **Sources: Pe2, Re6, Ter, Ver, VI3, We19.**

Neckarkonigin - 60 days - Improved Phenomeen type, long slim firm 9 in. stringless pods, very fleshy, excel. flavor, often used for French-style slicing, stok spercie bonen from Germany. *Source History: 2 in 1981; 1 in 1984; 1 in 1987; 1 in 1991; 1 in 1994; 1 in 1998; 1 in 2004.* **Sources: Dam.**

Neopolitan Pole - 60 days - Improved Romano, large sweet pods. *Source History: 1 in 2004.* **Sources: Bu2.**

New Mexico Cave - 90-100 days - Found originally at an archaeological dig in New Mexico in a pitch-sealed clay jar, white seed with black blotches, interplant with corn and squash, snap or dry. *Source History: 1 in 1987; 5 in 1998; 1 in 2004.* **Sources: Se17.**

Northeaster - (Kwintus, Quintus) - 55 days - Extremely early pole bean, flat med-green 8 in. pods, tender and stringless, rich fine flavor, vigorous strong vines, outdoor or off-season low-light greenhouse. *Source History:*

1 in 1981; 1 in 1984; 2 in 1987; 5 in 1991; 5 in 1994; 5 in 1998; 5 in 2004. **Sources: Fe5, Ga1, Hig, Jo1, Pr3.**

Ohio Pole - 85 days - Large dark-maroon seeds frosted with white on one end, 6-8 seeds per pod, heirloom. *Source History: 1 in 1991; 1 in 1994; 2 in 1998; 2 in 2004.* **Sources: Ho13, Sh13.**

Old Homestead - 65-75 days - Pods 9-10 in., almost stringless, meaty flavor, freezes well, introduced in the 1850s. *Source History: 2 in 1991; 2 in 1994; 2 in 1998; 2 in 2004.* **Sources: Fo7, He8.**

Old Time Golden Stick - Early maturing variety, good for canning, grown for many generations in Fentriss County, TN. *Source History: 1 in 2004.* **Sources: Fe5.**

Old Timey - Hardy climbing vines produce many 4-5 in. round pods filled with a colorful array of seeds, used as snaps when young or dried for soup. *Source History: 1 in 1998; 1 in 2004.* **Sources: Scf.**

Oregon Giant - (Paul Bunyan) - 63-70 days - Distinctive mottled thick broad 7-9 in. slightly curved pods, light-green splashed purplish-blue, Northwest favorite, stringless, long everbearing vines, very scarce, 1931. *Source History: 10 in 1981; 9 in 1984; 7 in 1987; 7 in 1991; 8 in 1994; 5 in 1998; 2 in 2004.* **Sources: Pr3, Se17.**

Ozark Heirloom - Scrambling climber does not require support, both smooth and wrinkled pod types, used as snap bean when young, shelled or for soups, stews and baking when dry, seed is spotted like a pinto, prolific, plant often has both mature pods and blossoms at the same time. *Source History: 1 in 2004.* **Sources: Te4.**

Paterge Head - Markings on the seed coat resemble that of a partridge head, good canner, grown in Fentriss County, TN since the early 1800s. *Source History: 1 in 2004.* **Sources: Fe5.**

Poamoho - 66 days - Stringless, nematode and rust resistant, well adapted to tropical conditions, dev. by the Univ. of Hawaii. *Source History: 1 in 1987; 3 in 1991; 2 in 1994; 1 in 1998; 1 in 2004.* **Sources: Uni.**

Potomac - 67 days - Vigorous climbing vines, slightly curved 6.5 in. stringless pods, 8 seeds per pod, prior to 1860 on the Virginia side of the Potomac River, after the Civil War it was carried west by the Barley family to Tehama County, California where it has been grown for over 125 years, introduced in 1990 by CV So1. *Source History: 1 in 1991; 1 in 1994; 1 in 1998; 1 in 2004.* **Sources: So1.**

Priscilla Polish - Heirloom, 7 in. pods, plump white maroon and black streaked seeds, 7 per pod, excellent as a snap or soup bean, exceptional flavor, late producer. *Source History: 1 in 1987; 2 in 1991; 2 in 1994; 1 in 1998; 1 in 2004.* **Sources: So9.**

Purple Peacock - 60-70 days - Quick growing vines covered with purple and lilac flowers, long purple pods filled with tender, round seeds, heavy yields, turns green during cooking, good producer under a wide range of conditions, good disease res. *Source History: 1 in 1991; 1 in 1994; 3 in 1998; 6 in 2004.* **Sources: Ami, Ga22, Hig, Ma13, Ni1, We19.**

Purple Pod Pole - (Purple Peacock) - 63-65 days - European heirloom, found by Henry Field in an Ozark garden in the 1930s, climbs over 6 ft., good yields and quality, large meaty stringless reddish-purple pods blanch light-green. *Source History: 8 in 1981; 5 in 1984; 4 in 1987; 6 in 1991; 7 in 1994; 8 in 1998; 7 in 2004.* **Sources: Fo7, Gur, Pla, Pr3, Se16, Shu, Ver.**

Purple Prize - Excellent for fresh snap and cooked dry beans. *Source History: 1 in 2004.* **Sources: Sa5.**

Ram's Horn - Plant grows 12 ft. tall, unique silver and black seeds. *Source History: 1 in 1991; 1 in 1994; 1 in 1998; 1 in 2004.* **Sources: Sh13.**

Rattlesnake - (Preacher Bean) - 60-90 days - Vine grows vigorously to 10 ft., distinctive dark-green pods streaked with purple, ripen to yellow, 7-8 in. long, light buff seeds splashed with dark-brown resembles a rattlesnake's color, fine flavor, good res. to drought. *Source History: 5 in 1981; 2 in 1984; 6 in 1987; 6 in 1991; 16 in 1994; 21 in 1998; 25 in 2004.* **Sources: Ada, Ag7, Bo19, Co31, CON, Fe5, Fo7, He8, Hi6, Ho13, Hud, La1, Pin, Pla, Pr3, Sa9, Scf, Se16, Se26, Sk2, So1, Te4, Und, Ver, Vi4.**

Romano Pole - (Italian Green Pod Pole, Italian Flat, Italian Pole) - 60-70 days - Buff/brown seed with white eye, high yields, meaty thick flat med-green stringless 5-6.75 x .75 in. pods edible when young, BCMV res., 72 in. plants, unique flavor, popular in Europe. *Source History: 66 in 1981; 60 in 1984; 52 in 1987; 49 in 1991; 48 in 1994; 40 in 1998; 35 in 2004.* **Sources: Bo17, Bu1, Bu2, Com, Fi1, Fo7, HA3, He8, Hi13, Hud, Jor, La1, LO8, Loc, Na6, Ni1, OLD, Ont, PAG, Pin, Red, Ri12, Roh, Se7, Se8, Se17, Se26, Sh9, Shu, Si5, So9, Tho, Ver, Vi4, WE10.**

Ruth Bible - 52 days - Heirloom saved by the Buoy family in Kentucky since at least 1832, 12-18 ft. vines bear heavily, 3.5 in. pods, brownish-tan seeds, best picked when small and tender, introduced in 1984 by CV So1. *Source History: 1 in 1981; 1 in 1984; 1 in 1987; 2 in 1991; 2 in 1994; 2 in 1998; 2 in 2004.* **Sources: Ho13, So1.**

Santa Anna - 75-80 days - Good producer of long slim pods with great texture and flavor, pick young. *Source History: 1 in 2004.* **Sources: Se24.**

Socks Bean - 85-90 days - Vines grow 6-8 ft. tall, 12 different seed types including brown, white, black, gray and speckled, original seed from a Tennessee gardener whose dog Socks came trotting down the road with a bag of beans in her mouth, heirloom. *Source History: 1 in 2004.* **Sources: Se17.**

Stregonta - Early maturing pole type for fresh shell and dry beans, bright red and white long pods, cream colored seeds with red veins, high yields, popular Italian variety. *Source History: 1 in 2004.* **Sources: Se24.**

Striped Creaseback - (Nancy Davis) - 69 days - Planted in corn for green shelled beans, used as snaps when very young, 4.5-5 ft. tall vines, good climber, pods 4.5-5 in. long, stringless when young, oval mottled buff seed with brown stripes. *Source History: 1 in 1981; 1 in 1984; 1 in 1987; 2 in 1991; 1 in 1994; 1 in 1998; 1 in 2004.* **Sources: SOU.**

Sunward - 66 days - Dark-green pods, 5.5-6 in. long, good raw, canned, frozen or cooked, BCMV res. *Source History: 1 in 1994; 1 in 1998; 1 in 2004.* **Sources: Eco.**

Supermarconi Rampicante, Black - Stringless pole type Roma, large, green, flat, wide pods, pods remain crisp and tender with no strings when 10 in. long, great taste. *Source History: 1 in 1987; 1 in 1991; 1 in 1994; 1 in 1998; 2 in 2004.* **Sources: Ber, Se24.**

Supermarconi Rampicante, White - Large, green, flat, wide, stringless pod. *Source History: 1 in 1991; 1 in 1994; 1 in 1998; 1 in 2004.* **Sources: Ber.**

Trionfo Violetto - (Violet Snap, Trionfo Purple Pole) - 64-75 days - Italian heirloom, vigorous vines, dark-green purple veined leaves, deep lavender flowers, long deep-purple stringless 7 in. pods turn green in cooking, excellent flavor. *Source History: 2 in 1991; 5 in 1994; 7 in 1998; 9 in 2004.* **Sources: Al25, Coo, Jo1, Pin, Se17, Se24, So9, Sw9, Syn.**

Tsing Tou - High yields of black seeds, rare. *Source History: 1 in 2004.* **Sources: Syn.**

Turkey Craw - (Turkey Gizzard) - Southeastern heirloom from Virginia, North Carolina and Tennessee, according to folklore the original seed was obtained by a hunter from the craw of a turkey he shot, often used as a cornfield bean, stringless 3.5-4 in. pods cling to the vine, excellent fresh flavor, buff seeds frosted with brown on one end, good canned frozen or dried. *Source History: 2 in 1987; 1 in 1991; 1 in 1994; 3 in 1998; 4 in 2004.* **Sources: Ho13, Hud, Scf, So1.**

Ura - 60 days - Plants grow to over 8 ft., 6-7 in. dark green pods, exceptional flavor, white seeds. *Source History: 1 in 2004.* **Sources: Ter.**

Valena Italian - 75-90 days - Flat pods used when young as green beans, mature pods are tan with maroon streaks, large, egg-shaped tan seeds with dark brown streaks, family heirloom from Victory Seed Company, still grown in Italy. *Source History: 1 in 2004.* **Sources: Vi4.**

Violet Podded Stringless - 70-80 days - Vines grow 7-10 ft. tall, like purple beans in general it sprouts when soil is cool, flowers first, handles cool conditions well, highly refined pods retain good flavor and tenderness even when they grow long, sets abundantly, snap or dry. *Source History: 1 in 1987; 1 in 1991; 1 in 1994; 1 in 1998; 1 in 2004.* **Sources: Ter.**

Violetto Rampicante - Long, round, purple, stringless pods. *Source History: 1 in 1987; 1 in 1991; 1 in 1994; 1 in 1998; 1 in 2004.* **Sources: Ber.**

Westland - European type stringless double, 5-7 in. long, good yields, excel. freezer, stok spercie bonen. *Source History: 1 in 1981; 1 in 1984; 1 in 1987; 1 in 1991; 1 in 1994; 1 in 1998; 1 in 2004.* **Sources: Dam.**

White and Green Hull - Produces both white to pale yellow pods and green pods on the same stem, used fresh and canned. *Source History: 1 in 1998; 1 in 2004.* **Sources: Scf.**

White Greasy, Mullins Strain - 50 days - Reselected superior strain of North Carolina white greasy beans, 5.5 in. flat oval medium green pods, white seeds, Dr. Mullins, U of TN. *Source History: 1 in 2004.* **Sources: RUP.**

Yoeme Purple String - Seeds are purple on beige, heat tolerant plants, snap or shell. *Source History: 1 in 1998; 1 in 2004.* **Sources: Na2.**

Varieties dropped since 1981 (and the year dropped):

Alabama No. 1 Purple Pod (1998), *Anelli* (1991), *Berner Landfrauen* (2004), *Blue Lake (Ponderosa Special)* (1991), *Blue Lake No. 7* (1987), *Champagne* (1987), *Cockstone* (1987), *Cornfield, White* (2004), *Crystal (Greenhouse)* (1991), *Cutshort, Honeycutt Pioneer* (1998), *Davis Purple* (1998), *Earlipol* (2004), *Fabulon* (1994), *Foot Long Snapper* (1991), *Garafal Oro* (2004), *Garrafal de Encarnada* (1998), *Garrafal Oro* (1991), *Giant Kentucky* (1994), *Glastada* (2004), *Gold Buddha* (1991), *Grandma Black's Pole* (2004), *Hickman's* (2004), *Idaho Wonder* (2004), *Jembo Polish* (2004), *JPB 4396* (2004), *Kalbo Veense* (1984), *Kapral* (2004), *Kentucky North* (1991), *Kentucky Pioneer Cutshort* (1991), *Kentucky Wonder, Mr. Waters* (2004), *Kolba* (1987), *Large Early Greasy* (1994), *Logan Giant* (1998), *London Horticultural Cranberry* (1983), *Malibu* (2004), *Meralda* (2004), *Miss Kelly* (1987), *Morse's Pole No. 191* (2004), *Non Plus Ultra* (1994), *Pardola* (1994), *Phenomeen* (1991), *Pioneer Pinto* (1984), *Portugal* (1991), *Prime Pak* (2004), *Princesse Race Pevir* (1994), *Promo* (1991), *Red Speckled Fall* (1998), *Rentegevers* (1994), *Robison Purple Pod* (2004), *Romano, Garafal Oro* (1994), *Romano, Spanish Gigantic* (1998), *Selka Improved* (1994), *Selma Star* (1998), *Selma Zebra* (2004), *Serbo* (1994), *SG 255* (1987), *Speckled Italian Bean* (1987), *Sultans Emerald Moon* (2004), *Sultans Golden Moon* (1998), *White Creaseback* (1987), *White Frost* (2004), *Wonder Lake* (2004).

Bean / Pole / Wax
Phaseolus vulgaris

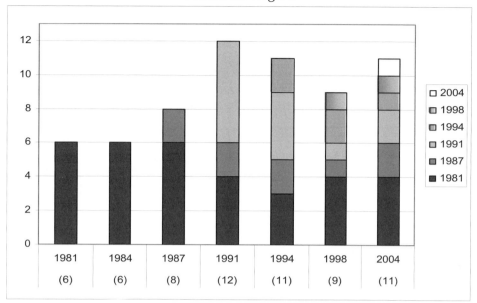

Number of varieties listed in 1981 and still available in 2004: 6 to 4 (67%).
Overall change in number of varieties from 1981 to 2004: 6 to 11 (183%).
Number of 2004 varieties available from 1 or 2 sources: 8 out of 11 (73%).

Appaloosa Goose - Golden pods, distinctive grey-white seeds, very productive, same lineage as Gold Buddha. *Source History: 1 in 1991; 1 in 1994; 1 in 1998; 1 in 2004.* **Sources: Sh13.**

Goldfield - 80 days - Flat, broad, bright yellow stringless pods, 10 in. long, 1 in. wide, early and vigorous, BCMV resistant. *Source History: 1 in 2004.* **Sources: Tho.**

Goldmarie - (Gold Marie Wax) - 55-67 days - Productive 7-9 ft. vines bear delicious, tender, golden 8-12 in. pods, white seeds, grows well in the greenhouse. *Source History: 5 in 1994; 7 in 1998; 5 in 2004.* **Sources: Dam, Ho13, In8, Pr3, We19.**

Grandma Nellie's Yellow Mushroom - Produces an abundance of light yellow pods, tender at 5 in. length, cooked bean flavor resembles that of mushrooms, CV So1 introduction, seed from Marge Mozelisky via her grandmother. *Source History: 1 in 2004.* **Sources: So1.**

Kentucky Wonder Wax - (Kentucky Wonder Yellow Wax, Golden Podded Climbing Bean, Kentucky Wonder Brown Seed Wax) - 61-73 days - An heirloom, straight slightly curved flattened oval golden-yellow 6-9 x .6 in. pods, almost stringless, always everbearing, oval light-chocolate seeds, prolific yields especially in cooler climates, all-purpose. *Source History: 34 in 1981; 28 in 1984; 28 in 1987; 27 in 1991; 21 in 1994; 20 in 1998; 15 in 2004.* **Sources: All, Com, Ear, Fi1, HA3, Jun, Ma19, OLD, Ont, Or10, Sa9, Sto, Ver, WE10, Wet.**

Meraviglia di Venezia - (Marvel of Venice) - 75 days - Most wax beans emerge green, these emerge butter-yellow and ripen deeper yellow, broad flat finger-length stringless Italian slicing bean, 8-9 in. long .75 in. wide, sweet juicy flavor, white seeds, snaps or green shells, identical to Goldmarie, Italian heirloom. *Source History: 2 in 1981; 2 in 1984; 2 in 1987; 1 in 1991; 1 in 1994; 1 in 1998; 4 in 2004.* **Sources: Ber, Jo1, Ni1, Sw9.**

Neckargold - 60 days - Very long stringless pole bean, similar to Neckarkonigin but bears heavy crop of long oval deep-yellow pods, do not plant before June in Canada, white seed. *Source History: 1 in 1981; 1 in 1984; 1 in 1987; 1 in 1991; 1 in 1994; 1 in 1998; 1 in 2004.* **Sources: Dam.**

Ramdor - 45-50 days - Early maturing pole bean, heavy clusters of straight, 7 in. yellow pods, crisp texture, rich flavor. *Source History: 1 in 2004.* **Sources: Ves.**

Romano, Burro D' Ingegnoli - 74 days - Vigorous vines to 8 ft., 5 in. flat creamy yellow pods, good flavor, popular in Italy. *Source History: 1 in 2004.* **Sources: Pin.**

White Hull Runner - Twining plant bears pale yellow pods tipped with pink which contrasts with green foliage color for easy picking, requires trellis, canning and freezing, *Source History: 1 in 2004.* **Sources: Fe5.**

Yellow Anellino - Heirloom, cousin of Green Anellino, bright pale-yellow pods are curved so that each forms a half circle near tip, pick when young before strings develop. *Source History: 1 in 1981; 1 in 1984; 1 in 1987; 2 in 1991; 2 in 1994; 2 in 1998; 2 in 2004.* **Sources: Coo, Se24.**

Varieties dropped since 1981 (and the year dropped):

B.B. Wax (2004), *Frima* (1998), *Gold Straw* (1994), *Golden, Burpee's* (1994), *Goldstraw* (2004), *Lazer* (1994), *Paille D'Or* (1998), *Wonder of Venice Wax* (1998), *Yellow Wonder* (2004).

Beet / Garden
Beta vulgaris (Crassa Group)

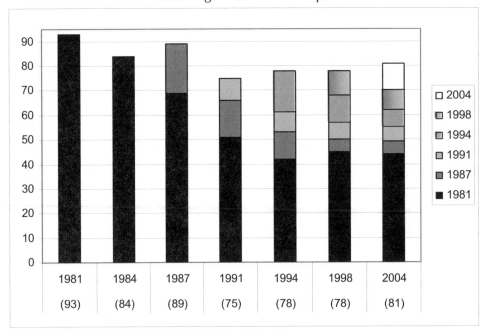

Number of varieties listed in 1981 and still available in 2004: 93 to 44 (47%).
Overall change in number of varieties from 1981 to 2004: 93 to 81 (87%).
Number of 2004 varieties available from 1 or 2 sources: 50 out of 81 (62%).

Albina Vereduna - (Snowhite, Albina) - 50-65 days - Large ice-white globe, thick durable skin, stores well in ground, white very sweet flesh, potato-like texture, curled wavy greens, root crown can turn greenish if not hilled up with soil, Holland. *Source History: 2 in 1981; 2 in 1984; 3 in 1987; 4 in 1991; 5 in 1994; 6 in 1998; 7 in 2004.* **Sources: Ho13, Ki4, Orn, Pe6, Pr3, Sw9, Tho.**

Albino - (Albino White) - 50 days - Round white beet, delicious sweet flavor, fairly smooth, plain green tops, does not bleed like red beets, makes unusual sliced beets and pickles. *Source History: 6 in 1981; 6 in 1984; 3 in 1987; 4 in 1991; 3 in 1994; 3 in 1998; 7 in 2004.* **Sources: Co32, Gou, Ho13, Lej, Se16, Se26, Se28.**

Always Tender - 75 days - Late maturing, unsuitable for summer use but stores extremely well, remains tender and sweet until late spring. *Source History: 1 in 1981; 1 in 1984; 2 in 1987; 2 in 1991; 2 in 1994; 2 in 1998; 1 in 2004.* **Sources: Ves.**

Baby Ball - 50 days - Tender, sweet, ball shaped red roots, uniform size, healthy greens, pick when baby size or mature. *Source History: 3 in 2004.* **Sources: Pe2, Re6, Roh.**

Big Top - 55 days - Extra-tall, bright green, large tops, deep red roots, bunched greens. *Source History: 1 in 2004.* **Sources: Jo1.**

Black Knight - (Detroit Dark Red) - *Source History: 1 in 1991; 1 in 1994; 1 in 1998; 1 in 2004.* **Sources: La1.**

Boltardy - (Bolthardy) - 57-60 days - Highly bolt res. Detroit type, withstands early spring cold during emergence, deep-red ringless flesh, smooth skin, high yields, stores well, from Holland. *Source History: 5 in 1981; 4 in 1984; 4 in 1987; 3 in 1991; 2 in 1994; 4 in 1998; 5 in 2004.* **Sources: Bou, He8, Se26, Tho, WE10.**

Bordo - Tender, sweet, excellent quality roots, deep purple color, top shape, tender greens, from the former Soviet Union. *Source History: 2 in 2004.* **Sources: In8, Se17.**

Bull's Blood - 50-60 days - Matures 35 days to baby leaf tops, 55 days for beet root, exceptionally beautiful dark red-purple tops with remarkable flavor, as beautiful as Radicchio but easier and faster to grow with a sweeter flavor, heat resistant, sel. for the darkest colored leaves by seedsman Kees Sahin in the Netherlands from the French variety Crapaudine. *Source History: 3 in 1998; 54 in 2004.* **Sources: Ag7, Ba8, Bo19, Bou, Bu2, CO23, CO30, Co32, Com, Coo, Dam, Dom, DOR, Eco, Ers, Fe5, Fo13, Ga1, He8, He17, Hi6, HO2, Hud, In8, Jo1, Ki4, La1, ME9, Mo13, Na6, Ni1, Orn, OSB, Pe6, Pin, Pr3, Ri12, RUP, SE14, Se16, Se26, Se27, Sh12, Shu, Sk2, St18, Te4, TE7, Ter, Ves, Vi4, WE10, We19, WI23.**

Carillon - 58 days - Sweet, dark roots ideal for making pickles and relishes, peel easily, more yields per square foot. *Source History: 1 in 2004.* **Sources: Coo.**

Cheltenham Green Top - Slicing type with long, deep red roots, slightly rough texture, excel. flavor. *Source History: 1 in 1998; 1 in 2004.* **Sources: Bou.**

Chioggia - (Dolce di Chioggia, Di Chioggia, Candystripe Beet, Bull's Eye Beet, Bassano) - 52-65 days - Heirloom named after a fishing town in Italy, 2.5 in. roots, alternating red and white concentric rings make it a real specialty item, mild flavored greens used raw or cooked, pre-1840s. *Source History: 3 in 1987; 11 in 1991; 17 in 1994; 31 in 1998; 62 in 2004.* **Sources: Ag7, Ba8, Ber, Bo19, Bou, Bu2, CO30, Co32, Com, Coo, Dam, Dom, Ers, Fe5, Gal, Ge2, He8, Hi6, Hi13, HO2, Ho13, Hud, Jo1, Jun, Ki4, La1, Loc, Mcf, ME9, Na6, Ni1, Orn, OSB, Pe2, Pe6, Pin, Ri2, Ri12, Roh, RUP, Se7, Se8, SE14, Se16, Se24, Se26, Shu, Sol, So12, Sw9, Syn, TE7, Ter, Tho, Tt2, Und, Up2, Ver, Ves, Vi4, WE10, We19.**

Crimson Globe - 65 days - Med-sized globe-shaped roots, deep-crimson, slightly zoned, average 3 in. dia., excel. flavor, tender, grows quickly for early use, med-short tops, home and market. *Source History: 3 in 1981; 2 in 1984; 1 in 1987; 2 in 1991; 1 in 1994; 2 in 1998; 2 in 2004.* **Sources: Bou, WE10.**

Crimson King - (Improved Crimson Globe) - Medium sized globe shaped 3 in. roots, deep crimson, slightly zoned, excel. flavor, quick grower. *Source History: 1 in 1994; 1 in 1998; 1 in 2004.* **Sources: Bou.**

Crosby's Egyptian - (Early Wonder, Crosby's Early Egyptian, Crosby's Extra Early Egyptian) - 50-60 days - Standard early bunching beet for table or market, flattened heart-shaped roots, dark-red flesh with lighter zones, 17 in. med-green erect tops, uniform shape, even maturity, smooth skin, largest early var., holds well, keeps shape until fall, parent strain of the Egyptian beet, introduced from Germany in 1865, commercially available in 1880 after seedsman James Gregory purchased this beet from market gardener Josiah Crosby. *Source History: 36 in 1981; 31 in 1984; 30 in 1987; 25 in 1991; 28 in 1994; 28 in 1998; 23 in 2004.* **Sources: Ba8, Bo19, CO30, DOR, Ers, Fo7, Gou, He8, La1, Lej, Lin, LO8, Mey, Ri12, Se8, Se26, Sh9, Shu, Sol, SOU, Vi4, WE10, Wet.**

Crosby's Green Top - 60 days - Special Harris sel., bright glossy green tops in spring or summer or fall, dark-red flattened globes, fine taproots, quick growing, fine quality, uniform. *Source History: 1 in 1981; 1 in 1984; 2 in 1987; 2 in 1991; 2 in 1994; 2 in 1998; 1 in 2004.* **Sources: ME9.**

Cylinder Long Red - (Cylindra Long Red, Long Red Cylandra) - 55-60 days - Cylindrical 6-8 in. long dark-red roots, no zoning, uniform slices, med-sized rich green tops, popular for canning and freezing. *Source History: 1 in 1981; 1 in 1984; 2 in 1987; 3 in 1991; 1 in 1994; 2 in 1998; 6 in 2004.* **Sources: Bo18, Gr29, Sau, Se28, Sh9, Wo8.**

Cylinder, Deep - 55 days - Blocky cylinder-shaped beet, brilliant red, fine textured tasty flesh, can be sliced uniformly. *Source History: 1 in 1981; 1 in 1984; 1 in 1987; 2 in 1991; 2 in 1994; 2 in 1998; 1 in 2004.* **Sources: Tt2.**

Cylindra - (Cylindra Long, Formanova, Butter Slicer, Tendersweet Cylindra, Great Cylindra) - 45-80 days - Long smooth and cylindrical, excel. for slicing, half-long 6-8 x 1.5-2 in. dia. dark-red roots, good dark-red interior color, small reddish-green tops, uniform Danish garden variety, grows almost entirely underground, introduced into the U.S. from Europe in 1892. *Source History: 43 in 1981; 38 in 1984; 49 in 1987; 51 in 1991; 48 in 1994; 57 in 1998; 57 in 2004.* **Sources: Be4, Bo19, Bu1, Bu2, Bu3, CH14, Com, Dam, DOR, Ers, Fa1, Fe5, Fi1, Fis, Ga1, Goo, Gr27, He8,** He17, HO2, Hum, Jor, Jun, La1, Lin, Ma19, MAY, Mcf, ME9, Mel, Mey, Mo13, MOU, Ni1, OLD, Or10, PAG, Pe2, Pe6, Pin, Ra5, Roh, RUP, SE14, Se16, Se26, Shu, Sk2, SOU, TE7, Tho, Twi, Ver, Vi4, WE10, Wet, WI23.

Deacon Dan - 60-100 days - Large red beet measures 6-8 in. across at the shoulder, striped red and white flesh is all red after cooking, often weighs several pounds, good keeper, retains sweetness in storage, favorite of Deacon Dan Burkholder who headed a Mennonite community in central PA in the early 1800s. *Source History: 2 in 1994; 3 in 1998; 2 in 2004.* **Sources: Ho13, Lan.**

Detroit 6, Rubidus - 53 days - Firm smooth skin, deep-red flesh, totally fiberless even when 5 in. in dia., matures one week sooner than Boltardy, excel. vigor and yields. *Source History: 1 in 1994; 1 in 1998; 1 in 2004.* **Sources: Tho.**

Detroit Crimson Globe - 60-70 days - Improved Detroit type, rich-maroon globe-shaped beets, indistinct lighter zones, very uniform, ideal for successional sowing, excel. freezer. *Source History: 1 in 1981; 1 in 1984; 1 in 1987; 1 in 1994; 1 in 2004.* **Sources: Tho.**

Detroit Dark Red - 45-70 days - Dev. from Early Blood turnip, nearly globe blood-red 2.5-3 in. dia. roots, little zoning, main crop canner for home gardens, all purpose, prolific, DM res., solid root, good keeper, the standard for beets, heirloom from 1892. *Source History: 79 in 1981; 70 in 1984; 87 in 1987; 87 in 1991; 79 in 1994; 81 in 1998; 89 in 2004.* **Sources: Ada, Ag7, Alb, All, Be4, Bo17, Bo19, Bu1, Bu3, Bu8, CA25, CH14, Cl3, CO23, CO30, Co31, Co32, Com, Cr1, Dam, De6, Dow, Ear, Eco, Ers, Fa1, Fi1, Fis, Fo7, Fo13, Ga1, Ge2, Goo, Gr27, Gur, HA3, Hal, He8, He17, Hi6, Hi13, Hig, HO2, Jor, Kil, La1, Lej, Lin, LO8, Ma19, MAY, Mcf, ME9, Mel, Mo13, MOU, OLD, Or10, Orn, OSB, Pe2, Pin, Pr3, Pr9, Ri12, Roh, RUP, Sa5, Sau, SE4, Se7, Se16, Se26, Si5, Sk2, Sol, So9, SOU, TE7, Tt2, Tu6, Twi, Up2, Ver, Ves, Vi4, Wet, Wi2, WI23.**

Detroit Dark Red Improved - 60-65 days - A leading main-crop beet for home and market gardens, smooth globe, deep-red inside and out, indistinct zoning, sweet tender fine-grained flesh, tops make excellent boiling greens, sow at 2-week intervals. *Source History: 2 in 1987; 2 in 1991; 2 in 1994; 2 in 1998; 2 in 2004.* **Sources: Jun, Shu.**

Detroit Dark Red, Medium Top - 58-68 days - Perfectly round, deep-red skin, dark-red flesh, little zoning, 2.5-3 in. diameter, small collar and taproot, DM res., 13-16 in. dark-green tops sometimes are red-tinged. *Source History: 13 in 1981; 11 in 1984; 12 in 1987; 11 in 1991; 13 in 1994; 11 in 1998; 7 in 2004.* **Sources: ABB, Bu2, DOR, Fe5, Loc, Sh9, Sto.**

Detroit Dark Red, Perfected - (Perfected Detroit) - 55-58 days - Reselected str., superior shape, darker color inside and out, very early and productive, large 16-18 in. dark-green tops ringed with deep-red, no zones, globe roots flattened at base, holds shape if crowded, all purpose, canneries or wholesale. *Source History: 22 in 1981; 19 in 1984; 10 in 1987; 7 in 1991; 5 in 1994; 4 in 1998; 3 in 2004.* **Sources: Fi1, Gur, Ros.**

Detroit Dark Red, Short Top - 55-60 days - Excel.

during hot weather, plant in late spring or late summer, small 12-13 in. dark-green tops turning red, round dark-red roots, no zones, excellent canner, widely adapted, for home or market, very tender, no side roots, holds well. *Source History: 21 in 1981; 19 in 1984; 17 in 1987; 12 in 1991; 10 in 1994; 11 in 1998; 8 in 2004.* **Sources: Fi1, Ha5, La1, Mey, PAG, Par, SE14, WE10.**

Detroit Supreme - 60-65 days - An improved Detroit Medium Top with mildew resistance, smooth-skinned uniform globes with blood-red flesh, little zoning even in hot weather, small taproot, glossy dark-green tops with typical red veining supported by a small crown, processing and fresh market, dev. by Alf Christianson Seed Co. *Source History: 4 in 1987; 9 in 1991; 16 in 1994; 14 in 1998; 8 in 2004.* **Sources: HO2, Hum, La1, Lin, OSB, Pe6, SOU, Sto.**

Dewing's Early Blood Turnip - Leaves are almost black, violet-red skin, sweet red flesh with paler red rings, blood turnip is an old term for garden beet, especially in the U.K. *Source History: 2 in 2004.* **Sources: In8, Se17.**

Drum - 48-60 days - Tall tops, dark green tinged with maroon, excel. for greens, dark purple-red 3 in. flattened globes, bunching beet used as a first early var. or for midsummer sowing and storage. *Source History: 1 in 1998; 1 in 2004.* **Sources: Eco.**

Dwergina - (Baby Beets) - 58 days - Small round short-topped beet in the Detroit Dark Red class, intense color, tender dense smooth texture, completely free from zoning, tiny taproot, best picked as baby beets although still relatively small at maturity, tops are ready to add to salads in 2 weeks, Holland. *Source History: 3 in 1987; 6 in 1991; 2 in 1994; 2 in 1998; 2 in 2004.* **Sources: Bou, Pr3.**

Early Blood Turnip - (Early Dark Blood Turnip, Blood Turnip) - 48-68 days - Med-small fairly coarse tops, turnip-shaped dark red roots, sweet dark flesh with pink zones, good all-purpose home garden variety, dates back to 1825. *Source History: 13 in 1981; 10 in 1984; 8 in 1987; 5 in 1991; 7 in 1994; 5 in 1998; 4 in 2004.* **Sources: Ol2, Se16, Th3, Wet.**

Early Wonder - (Asgrow Wonder, Greentop, Green Top Bunching, Model) - 40-60 days - Half flat with round bottom, smooth 3 in. roots, vivid-red flesh with light zones, bright-green 16-18 in. tops, good for greens, quick growing, good flavor, for home garden bunching or for shipping, introduced in 1911. *Source History: 73 in 1981; 67 in 1984; 68 in 1987; 62 in 1991; 54 in 1994; 51 in 1998; 44 in 2004.* **Sources: Alb, All, Ba8, Bo17, Bu8, But, CH14, Co32, Com, Dam, De6, DOR, Ear, Ers, Fis, Fo13, Goo, Gr27, HA3, Hal, He8, He17, Ho12, La1, Lej, Lin, LO8, MAY, Mel, Mo13, MOU, Or10, PAG, Pin, RUP, Sa9, Sau, Se26, Sh9, Shu, So1, Sto, Wi2, WI23.**

Early Wonder Green Top - (Early Wonder Green Top Bunching) - 53-58 days - Resel. for longer lasting tops, slightly flattened globe, dark-red flesh with some zoning, glossy 17 in. tops, garden or market. *Source History: 6 in 1981; 5 in 1984; 5 in 1987; 5 in 1991; 3 in 1994; 3 in 1998; 1 in 2004.* **Sources: Twi.**

Early Wonder Smooth Leaf - 54 days - Early maturing variety for both beet greens and bunching, leaves to 18 in., flesh well colored and fiberless, for fresh market or local sales. *Source History: 1 in 1981; 1 in 1984; 1 in 1987; 1 in 1991; 1 in 1994; 1 in 1998; 1 in 2004.* **Sources: SE4.**

Early Wonder Stays Green - 52-53 days - Stays green in cool weather longer, erect glossy longer lasting 16-18 in. tops, slightly flattened globe, dark-red flesh, some zones, garden or processing, 1961. *Source History: 8 in 1981; 6 in 1984; 7 in 1987; 3 in 1991; 1 in 1994; 1 in 1998; 2 in 2004.* **Sources: ME9, Mey.**

Early Wonder Tall Top - 48-60 days - Tall vigorous dark-green 16-18 in. tops tinged with maroon, excel. for greens, dark purplish-red 3 in. flattened globes, bunching beet for home or market gardens, used as a first early variety or for midsummer sowing and storage. *Source History: 14 in 1981; 12 in 1984; 11 in 1987; 16 in 1991; 15 in 1994; 17 in 1998; 23 in 2004.* **Sources: Bo19, Bu3, CO23, CO30, Dgs, Fe5, Ga1, Hi6, Hig, Ho13, Jo1, Jor, Kil, Orn, Pe2, Pla, Roh, SE14, TE7, Ter, Vi4, WE10, We19.**

Edmund's Blood - (Edmund's Early Blood Turnip) - Red-purple roots, purple leaves, good as an early variety. *Source History: 1 in 1981; 1 in 1984; 1 in 1987; 1 in 1991; 1 in 1994; 3 in 1998; 1 in 2004.* **Sources: Ho13.**

Excaliber - *Source History: 1 in 2004.* **Sources: CO30.**

Feuer Kugel - 65-75 days - Translates Fire Globe, smooth skin, remains sweet and tender to a large size, flavorful raw and cooked, good storage beet, from Switzerland, rare. *Source History: 2 in 1998; 5 in 2004.* **Sources: In8, Pe2, Se17, So12, Tu6.**

First Crop - 45 days - Matures very quickly and uniformly, attractive deep-red globes, sweet tender flesh, excel. variety for both market growers and home gardeners. *Source History: 1 in 1981; 1 in 1984; 1 in 1987; 1 in 1991; 1 in 1994; 2 in 1998; 1 in 2004.* **Sources: Ves.**

Flat Egyptian - (Rouge Noir Plate d'Egypte, Chata de Egipto, Extra Early Flat Egyptian, Egyptian Early Flat) - 35-50 days - Flattened globe, 3-5 in. across, deep purple-red, short tops, used widely to force or transplant. *Source History: 9 in 1981; 7 in 1984; 5 in 1987; 4 in 1991; 2 in 1994; 1 in 2004.* **Sources: Par.**

Formanova - (Cylindra, Improved Cylindra, Cook's Delight, Formanova LD) - 45-60 days - Unique cylindrical shape, 6-8 x 2-3 in., smooth and tapered, dark-red flesh, no zones, popular slicer, freezes and stores well, from Germany or Denmark (?), hill to prevent sunscald, very uniform, home canning or commercial processing, 1964. *Source History: 21 in 1981; 20 in 1984; 19 in 1987; 20 in 1991; 18 in 1994; 11 in 1998; 4 in 2004.* **Sources: Alb, Ear, Hal, Me7.**

Forona - 54-70 days - Danish import, smooth cylindrical ruby-red root, 2 x 8 in. long, very sweet, short tops, high quality, very productive, use when young and tender, gets coarse if allowed to grow, developed especially for canning. *Source History: 3 in 1987; 2 in 1991; 3 in 1994; 6 in 1998; 6 in 2004.* **Sources: Bou, Eo2, Gur, Jo1, Se26, Sw9.**

Garnet - (Garnet Short Top, Ruby Queen) - 55-58 days - Smooth deep-red globes, no zones or side roots, short dark-green tops tinged with red, sel. from Detroit Short Top, bunch or can, uniform. *Source History: 10 in 1981; 9 in 1984; 5 in 1987; 2 in 1991; 1 in 1994; 1 in 1998; 1 in 2004.*

Sources: Lin.

German Lutz - 65 days - Sweet tender roots are good winter keepers, big tender glossy green tops with midribs, the best to grow for greens. *Source History: 1 in 1981; 1 in 1984; 1 in 1987; 1 in 1991; 1 in 1994; 1 in 1998; 1 in 2004.* **Sources: Ni1.**

Gladiator - 45-60 days - Early baby canner, dark-red roots, very little zoning, med-green 12-13 in. tops, holds round shape well when crowded, small taproot, for processing or market. *Source History: 6 in 1981; 6 in 1984; 5 in 1987; 2 in 1991; 1 in 1994; 1 in 1998; 1 in 2004.* **Sources: Fi1.**

Golden Beet - 50-60 days - Round orange roots turn gold-yellow when cooked, does not bleed, roots grow fast, retains sweet flavor when quite large, excellent for pickles, very tender and mild, well suited to California climate, pre-1820s. *Source History: 7 in 1981; 7 in 1984; 15 in 1987; 17 in 1991; 14 in 1994; 19 in 1998; 16 in 2004.* **Sources: Ba8, Bou, CO30, Coo, Fis, Hig, Ho13, Jo1, Na6, Ni1, Or10, Orn, Pe6, Se8, Sw9, Ter.**

Golden Detroit - (Yellow Detroit) - 50-55 days - Golden orange roots, sweetest when young but do not become fibrous when larger, tops used as cooked greens, roots are pickled, grilled or used in salads. *Source History: 8 in 2004.* **Sources: HO2, ME9, OSB, Ri2, TE7, Tt2, WE10, WI23.**

Golden Tankard - 110 days - Huge, smooth roots with a tankard shape that tapers quickly at the bottom, firm creamy yellow flesh is very sweet, keeps all winter if stored properly, excel. table quality or good stock food. *Source History: 1 in 1994; 2 in 1998; 2 in 2004.* **Sources: Ho13, Shu.**

Golden Wonder - 55 days - Golden 2.5 in. roots that do not bleed. *Source History: 1 in 1991; 1 in 1994; 1 in 1998; 1 in 2004.* **Sources: Ge2.**

Golden, Burpee's - (Burpee's Golden Delight, Burpee's Golden Globe) - 50-55 days - Dual purpose beet for greens and roots, will not bleed like red beets, best when small but stay sweet and will not become fibrous when larger, bright-golden, smooth, very early, tops are quite tender, commercially available since 1828. *Source History: 29 in 1981; 26 in 1984; 24 in 1987; 21 in 1991; 19 in 1994; 27 in 1998; 36 in 2004.* **Sources: Ag7, Alb, Be4, Bo19, Co32, Com, Dam, Ers, Fa1, Fe5, Ga1, Gou, He17, Hi13, HO2, Hum, Jor, Jun, La1, Lin, Loc, Ma19, ME9, Pe2, Pin, Roh, RUP, Se16, Sh9, Shu, Sk2, Sto, Syn, Tho, Twi, Ver.**

Green Top Bunching - (Letherman's Green Top, Early Tender Bunching) - 52-60 days - Bright-red slightly flattened globes, slight zones, tender 15-18 in. tops hold bright-green color, excel. bunching and greens beet, for home or market, 1940. *Source History: 21 in 1981; 20 in 1984; 17 in 1987; 15 in 1991; 15 in 1994; 19 in 1998; 13 in 2004.* **Sources: All, CO23, Ers, He8, HO2, La1, Loc, ME9, RIS, Sh9, Sto, TE7, Vi4.**

Iride - 58 days - Selection of Detroit Dark Red with smooth red globes with deep rich color, maroon-tinged tops, excellent used fresh or for canning. *Source History: 1 in 2004.* **Sources: Coo.**

JBT 1002 White Beet - (Blankoma) - 55 days - Highly refined strain of the heirloom Albina Vereduna, tall green tops, small sweet roots, faint green root crowns, white flesh, table use, will size up for fall storage. *Source History: 1 in 1998; 7 in 2004.* **Sources: CO30, Coo, Ear, Jo1, OSB, Ter, We19.**

Little Ball - (Burpee's Little Ball, Little Mini Ball, Little Ball Feet, Mini Ball, Kleine Bol) - 42-58 days - Gourmet baby beet, quick forming smooth dark-red 1 in. roots, no zones, suited for late summer or spring planting, never get oversized, fresh or can or pickle. *Source History: 20 in 1981; 18 in 1984; 14 in 1987; 16 in 1991; 18 in 1994; 9 in 1998; 1 in 2004.* **Sources: La1.**

Little Mini Ball - 54 days - One of the first true baby beets, round silver dollar sized roots at maturity, used for whole pak pickled beets, short green tops excel. for greens. *Source History: 1 in 1981; 1 in 1984; 1 in 1987; 2 in 1991; 2 in 1994; 1 in 1998; 1 in 2004.* **Sources: Ear.**

Long Season - (Winter Keeper, Lutz) - 75-80 days - Very large rough beet, deep-red throughout, holds excel. quality, stays sweet and tender, light-green tops, grows slowly, late home garden harvest, heirloom. *Source History: 3 in 1981; 3 in 1984; 4 in 1987; 8 in 1991; 9 in 1994; 7 in 1998; 6 in 2004.* **Sources: Ha5, He8, Hud, Me7, Pp2, Se25.**

Lutz Green Leaf - (New Century, Winter Keeper, Lutz Green, Lutz Salad, Lutz Green Top) - 60-80 days - Smooth purple-red top-shaped beet, 2.25-3 in. dia., lighter zones, half-long taproot, long glossy 14-18 in. tops with pink midribs, good for greens, excel. keeper, grows large without getting woody, good fresh, for winter and fall use. *Source History: 30 in 1981; 26 in 1984; 27 in 1987; 28 in 1991; 27 in 1994; 33 in 1998; 38 in 2004.* **Sources: Bu2, Coo, Dam, Ers, Fe5, Fi1, Ga22, Ga22, Gr27, HA3, Hi13, HO2, Hud, Hum, Jun, La1, Lan, ME9, Mey, PAG, Pe2, Pin, RUP, Sa5, Sa9, Se7, Se25, Se26, Shu, Si5, Sol, So9, Te4, Twi, Up2, Vi5, WE10, Wet.**

MacGregor's Favorite - (Dracena Beet, McGregor's Red Bedding) - 60-90 days - Scottish heirloom, long dark roots send up a profusion of spear-shaped metallic-purple leaves, both root and leaves edible, very ornamental as well as fine tasting. *Source History: 2 in 1987; 4 in 1991; 3 in 1994; 3 in 1998; 5 in 2004.* **Sources: Ho13, In8, Mcf, Se16, Se17.**

Mona Lisa - 75 days - Smooth round roots, good for baby beet production, glossy green tops, 10-12 in., red veins, small tap root, monogerm variety, holds well into late fall. *Source History: 2 in 2004.* **Sources: CO30, OSB.**

Monogram - 60-65 days - Dark-red globe with smooth skin, good flavor and vigor, each seed produces one seedling for easy thinning. *Source History: 1 in 1994; 2 in 1998; 1 in 2004.* **Sources: Tho.**

New Century - (Lutz's Green Leaf) - 80 days - Long season, long tapered root, very tender, can be used when small or grown to large size for winter use. *Source History: 1 in 1981; 1 in 1984; 1 in 1987; 1 in 1991; 1 in 1994; 1 in 1998; 1 in 2004.* **Sources: Roh.**

Paonazza d'Egitto - 45 days - Early beets are bright purple with concentric lighter circles, very sweet flavor.

Source History: 1 in 2004. **Sources: Pin.**

Pronto - 50-58 days - Smooth crowned roots with uniform size, shape and color, will not become fibrous when left in the ground but should be harvested at 1 in. diameter. *Source History: 1 in 2004.* **Sources: Ki4.**

Punakera - *Source History: 1 in 2004.* **Sources: Se17.**

Red Ball, Burpee's - 60 days - Dark-red fiberless flesh, almost no lighter zones, retains color when cooked, uniform globe-shaped 3 in. roots, erect med-sized plants, smooth deep-red skin. *Source History: 6 in 1981; 5 in 1984; 6 in 1987; 4 in 1991; 2 in 1994; 1 in 1998; 1 in 2004.* **Sources: Bu2.**

Red Crapaudine - 53 days - Long well-rooted variety with rough black skin and dark-red flesh, very popular for its fine quality and sweet taste, roots do not push out of the ground, possibly dates back 100 years ago to the time of Charlemagne. *Source History: 1 in 1981; 1 in 1998; 2 in 2004.* **Sources: Gou, Pr3.**

Red Success - *Source History: 1 in 2004.* **Sources: CO30.**

Red Titan - *Source History: 1 in 2004.* **Sources: CO30.**

Rocket - 100 days - Straight, cylindrical, dark red root, productive, for processing and slicing. *Source History: 1 in 2004.* **Sources: BE6.**

Rossa Quarantina d'Egitto - 70-90 days - Entirely red root. *Source History: 1 in 1987; 1 in 1991; 1 in 1994; 1 in 1998; 1 in 2004.* **Sources: Ber.**

Rote Kugel - (Red Globe) - 60-65 days - Dark red, round smooth root, strong tops, bunching and storage, Switzerland. *Source History: 1 in 2004.* **Sources: Tu6.**

Round Purple - Large deep-purple roots, edible purple leaves, old strain. *Source History: 1 in 1987; 1 in 1991; 1 in 1994; 1 in 1998; 1 in 2004.* **Sources: Ho13.**

Ruby Queen - 45-70 days - AAS/1957, smooth med-dark red globes, dark-red flesh, no zones, med-dull green 10-12 in. tops tinged with maroon, home gardens and processing, holds shape well when crowded, good on poor soils, popular in the Northeast, fine quality, uniform. *Source History: 84 in 1981; 80 in 1984; 73 in 1987; 62 in 1991; 63 in 1994; 63 in 1998; 47 in 2004.* **Sources: Alb, Be4, Bu3, CH14, Cl3, CO23, CO30, Cr1, De6, DOR, Ear, Ers, Fa1, Fi1, Fo7, Gr27, Gur, Hal, He8, He17, HO2, Jor, Jun, La1, MAY, ME9, Mey, Mo13, MOU, Orn, PAG, Pin, RIS, RUP, Sau, SE14, Se26, Sh9, Si5, Sk2, Sto, Twi, Und, Vi4, WE10, Wi2, WI23.**

Ruby Queen Improved - 53 days - Globe to round med-bright red roots with little zoning, 13 in. med-green tops, for processing or market. *Source History: 1 in 1981; 1 in 1984; 1 in 1987; 1 in 1991; 1 in 1994; 1 in 1998; 1 in 2004.* **Sources: ABB.**

Sangria - 50-56 days - Improved Detroit type, globe-shaped roots, deep blood-red flesh, sweeter flavor than most, dark-green maroon-tinged 12-14 in. tops, small collar, extra-slow bolting, excel. processor, also good marketer. *Source History: 3 in 1981; 3 in 1984; 2 in 1987; 1 in 1991; 11 in 1994; 13 in 1998; 4 in 2004.* **Sources: Kil, Lin, Se8, Vi4.**

Sweet Supra - 55 days - Deep red, perfectly round roots remain tender and sweet even under difficult growing conditions, excel. for greens, from Holland. *Source History: 1 in 1994; 1 in 1998; 1 in 2004.* **Sources: Se8.**

Sweetheart - 58 days - Detroit Dark Red x sugar beet, round, not smooth or uniform, great flavor, tender, twice the sugar, red throughout, med-tall red-tinged green tops, U of NH/1958. *Source History: 1 in 1981; 1 in 1984; 1 in 1987; 3 in 1991; 2 in 1994; 3 in 1998; 2 in 2004.* **Sources: Ho13, Pr3.**

Tardel - 65 days - Slow to bolt, uniform globe-shaped roots, dark sweet tender flesh, ideal for pickling or freezing, kroten from the Elite Zaden Seedhouse of Holland. *Source History: 1 in 1981; 1 in 1984; 1 in 2004.* **Sources: Tho.**

Winter Keeper - (Long Season, Green Leaf, Lutz Green Leaf, Longkeeper) - 60-80 days - Produces greens in cold weather, very large, almost mangel type, rather rough, 6-8 in., deep red, slightly tapering root, harvest tops separately, keeps well, for gardens. *Source History: 5 in 1981; 5 in 1984; 6 in 1987; 7 in 1991; 6 in 1994; 6 in 1998; 6 in 2004.* **Sources: Be4, Pla, Sto, Te4, Ter, We19.**

Varieties dropped since 1981 (and the year dropped):

Action (1998), *Baby Golden* (1994), *Badger Baby* (1994), *Best of All* (1987), *Bikores* (1998), *Boldet* (1991), *Burgundy* (1984), *Cardenal* (1994), *Cavalier* (1991), *Cherry Beet* (1987), *Crimson Beauty* (1991), *Crimson Tide* (1987), *Cut and Come Again* (1991), *Cyndor* (2004), *Dark Red Canner* (1998), *Dark Red Globe* (1998), *D'Egypte* (1984), *Detroit 243* (2004), *Detroit Dark Red (Short Top 12)* (1991), *Detroit Dark Red, Ferry Strain* (1994), *Detroit Dark Red, Morse Strain* (1998), *Detroit New Globe* (1984), *Dwergina* (99999), *Early Blood Turnip Improved* (1991), *Early Red Ball* (1991), *Early Wonder Dark Strain* (2004), *Early Wonder Improved* (1994), *Excalibur* (2004), *Firechief* (1991), *Greenleaf* (2004), *Half Long Red* (1991), *Holmes Fireball* (1991), *Honey Red* (1991), *Indian* (2004), *King Red* (1998), *Little Egypt* (1994), *Lola* (2004), *Long Blood Red* (1987), *Luxor* (1987), *Matchless Green Top* (1991), *Miniature* (1998), *Mobile* (1994), *Mona* (2004), *Mono Germ (Burgundy Strain)* (1991), *Monodet* (1991), *Mono-King Burgundy* (1987), *Mono-King Explorer* (1987), *Monopoly* (2004), *Nero* (1994), *New Early* (1984), *New Globe* (1991), *Oblong Red* (1984), *Perfecta* (1991), *Red Baron* (1994), *Red Brigadier* (1994), *Red Pack* (1998), *Red Sun* (1991), *Red Velvet* (2004), *Redheart* (1982), *Regala* (1983), *Replata* (1998), *Rote Ruben* (1987), *Ruby Red* (2004), *Smoothie* (1994), *Snow White* (2004), *Snowflesh* (2004), *Special Early, Stokes* (2004), *Spinel Baby Beet* (2004), *Spring Red* (1984), *Staysgreen* (1991), *Turnip Rooted* (1987), *Uniball* (1987), *Vermilion* (1991), *Waltan* (1991), *Wonder Green* (1987).

Beet / Sugar
Beta vulgaris (Crassa Group)

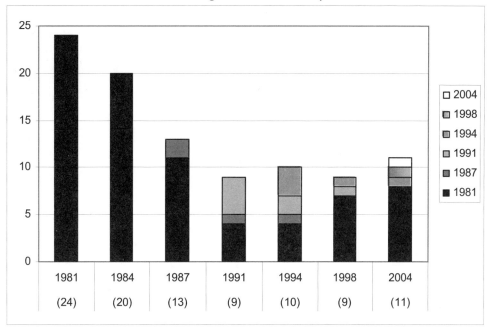

Number of varieties listed in 1981 and still available in 2004: 24 to 8 (33%).
Overall change in number of varieties from 1981 to 2004: 24 to 11 (46%).
Number of 2004 varieties available from 1 or 2 sources: 6 out of 11 (55%).

Colossal Long Red Mangel - 90-100 days - Roots up to 24 in. long and 15 lbs., bright-red skin, white flesh tinted with rose, roots grow 2/3 above ground so easily harvested, low cost feed for cattle. *Source History: 1 in 1981; 1 in 1984; 1 in 1987; 1 in 1991; 1 in 1994; 4 in 1998; 5 in 2004.* **Sources: Ho13, Re8, Sa9, Shu, Tu6.**

Giant Half Sugar - 120 days - Long oval roots, white skin with green shoulder, white flesh, high yielding (Half Sugar and Intermediate types may be mangel x sugar beet crosses). *Source History: 1 in 1981; 1 in 1984; 1 in 1998; 1 in 2004.* **Sources: Shu.**

Giant Red Sugar Beet - Large red roots, sow 4 to 6 pounds of seed per acre. *Source History: 1 in 1981; 1 in 1984; 2 in 1987; 1 in 1991; 1 in 1994; 1 in 2004.* **Sources: Ont.**

Giant Western Sugar Beet - (Great Western) - 80 days - Grown for sugar and stock feed, heavy yields, 18-20% sugar, sow at 5 lbs./acre, can grow to 10 lbs. in rich loamy deep soils, yellow flesh, 20-40 tons per acre. *Source History: 9 in 1981; 7 in 1984; 8 in 1987; 4 in 1991; 5 in 1994; 4 in 1998; 4 in 2004.* **Sources: HO2, Ho13, Se25, Shu.**

Klein Wanzleben Sugar Beet - Top U.S. str., ideal for sugar manufacture, hardy white sugar beet with long tapering roots, excel. for sheep forage, good winter greens for human consumption. *Source History: 5 in 1981; 3 in 1984; 3 in 1987; 2 in 1991; 1 in 1994; 1 in 1998; 1 in 2004.* **Sources: Ri2.**

Mammoth Long Red Mangel - 95-120 days - Large heavy yielding mangel, dull-red above ground, clear-red below, white flesh, 25-60 lbs., valuable winter feed for cattle and sheep and poultry, oval to spindle-shaped roots, 15-20 lbs., adapted to alkaline soils, tremendous food value per acre. *Source History: 32 in 1981; 27 in 1984; 19 in 1987; 15 in 1991; 13 in 1994; 13 in 1998; 12 in 2004.* **Sources: Ba8, De6, He17, Hud, Jun, La1, Ma19, Mo13, Pr3, Ros, RUP, Se25.**

Mangel Wurzel - (Stock Beet, Red Mangel Wurtzel) - 70-100 days - Large pink roots with red skins grow up to 2 ft. long, widely used for livestock feed, mangel greens can be used as an alternative to beet greens or Swiss chard. *Source History: 2 in 1987; 3 in 1991; 4 in 1994; 7 in 1998; 5 in 2004.* **Sources: Hal, Lan, Ol2, Ri12, Roh.**

Sugar Beet - 85-110 days - Heavy yields, widely adapted, long and thick, white flesh, 2-3 lb., home gardeners can use a few slices to add sweetness to canned beets, also for molasses or wine. *Source History: 4 in 1981; 3 in 1984; 1 in 1994; 3 in 1998; 6 in 2004.* **Sources: Bo19, Com, Ers, ME9, RUP, SE14.**

Yellow Cylindrical - Rare European heirloom, large oblong golden yellow mangel beets, small roots are sweet and tasty, mature roots used as high quality stock feed. *Source History: 1 in 2004.* **Sources: Ba8.**

Yellow Intermediate - (Giant Yellow Intermediate) - 110 days - Large half-long oval roots, grey-green shoulder well above ground, orange base, solid white flesh, 12-18 in. tops, easy to dig and top, dev. by Vilmorin in the 1800s. *Source History: 4 in 1981; 3 in 1984; 2 in 1987; 1 in 1991; 3 in 1994; 3 in 1998; 2 in 2004.* **Sources: Bou, Se7.**

Yellow Mangel - 70 days - Heirloom, orange-yellow beet that grows to 10 lbs. without losing delicious flavor, chard-like greens are great steamed, overwinters well. *Source History: 2 in 1991; 1 in 1994; 1 in 1998; 1 in 2004.* **Sources: Syn.**

Varieties dropped since 1981 (and the year dropped):

Danish Sludstrup Mangel (1984), *Giant Half Sugar Red Top* (1994), *Giant Sugar Beet* (1987), *Giant White Sugar Beet* (1991), *Giant White Sugar Rose* (1987), *Giant Yellow Intermediate* (1987), *Giant Yellow Vauriac Mangel* (1984), *Half Sugar Rose* (1987), *Half Sugar White* (1991), *Juane d'Eckendorf* (1991), *Our Ideal* (1982), *Prize Winner Yellow Glove* (1991), *Prizewinner Mangel* (1998), *Red Otofte* (1991), *Red Top Giant* (1987), *Sludstrup Yellow Intermediate* (1987), *USH-21* (1991), *White Sugar Beet Improved* (1983).

Broccoli
Brassica oleracea (Botrytis Group)

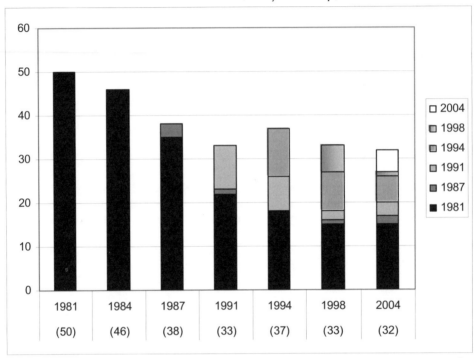

Number of varieties listed in 1981 and still available in 2004: 50 to 15 (30%).
Overall change in number of varieties from 1981 to 2004: 50 to 32 (64%).
Number of 2004 varieties available from 1 or 2 sources: 13 out of 32 (41%).

Atlantic - 55-70 days - Fast growing, excel. uniformity and quality, well rounded compact solid med-sized bluish heads, low growing compact plants, heavy crops of side shoots, 1960. *Source History: 6 in 1981; 4 in 1984; 4 in 1987; 5 in 1991; 4 in 1994; 5 in 1998; 3 in 2004.* **Sources: Fi1, Se25, WE10.**

Calabrese - (Early Green Sprouting Calabrese, Early Italian Green Sprouting, El Centro) - 58-90 days - Dark-green 18-30 in. plant, close-beaded 3-8 in. bluish-green central head followed by abundant side shoots, bears until frost, sow early March under glass or May-June for fall harvest after 90 day maturity, home or market, can or freeze, brought to this country by Italian immigrants in the late 1880s. *Source History: 34 in 1981; 29 in 1984; 23 in 1987; 25 in 1991; 26 in 1994; 24 in 1998; 37 in 2004.* **Sources: Alb, Ba8, Ber, CO23, CO30, Co32, DOR, Ear, Fo7, Fo13, HA3, He8, Ho13, Hud, Jor, La1, Ma19, ME9, Mey, PAG, Pin, Pr3, Ri2, Ri12, Se7, Se8, Se16, Se24, Se25, Si5, So1, Te4, TE7, Und, Vi4, Wet, WI23.**

Cavolo Broccolo Ramosa Calabrese Grande - Multiple branched 2 ft. plants, continuous production of 2 in. green heads, fine flavor. *Source History: 2 in 2004.* **Sources: In8, Se17.**

Christmas Purple Sprouting - 220 days - Turns emerald-green when cooked, excel. quality, juicy and full-flavored, good for freezing, sow in July or early August for late January harvest in Pacific Northwest. *Source History: 2 in 1981; 2 in 1984; 1 in 1987; 1 in 2004.* **Sources: Tho.**

De Cicco - (De Cicco Italian Green Sprouting) - 48-85 days - Old reliable European var., compact 2-3 ft. light-green plants, 3-4 in. bluish-green central head, then lots of med-sized side shoots, non-uniform in maturity, more variable and more productive than hybrids, excel. quality, freezer, intro. 1890. *Source History: 33 in 1981; 28 in 1984; 24 in 1987; 25 in 1991; 24 in 1994; 26 in 1998; 35 in 2004.* **Sources: All, Bo19, Bou, Dam, Fo7, Ga1, Goo, HA3, He8,**

He17, Hi6, Hig, HO2, Jo1, La1, ME9, MIC, Mo20, Na6, Orn, Pe2, Pla, Ri2, Ri12, Se8, SE14, Se16, Se25, Se26, So1, So9, Te4, TE7, Vi4, WE10.

De Cicco, Early - 55 days - Week earlier than regular str., close-beaded flat 4 in. heads, vigorous 24 in. plants, many side shoots, good freezer, fine quality, attractive, good home garden var. *Source History: 5 in 1981; 5 in 1984; 4 in 1987; 1 in 1991; 1 in 1994; 2 in 1998; 2 in 2004.* **Sources: Cr1, Mey.**

Early Green - 70 days - Spring var., large green heads, many side shoots, grows to 14 in. tall. *Source History: 2 in 1991; 3 in 1994; 6 in 1998; 4 in 2004.* **Sources: Eco, Pe2, Se7, So9.**

Enchantment - 75 days - Large blue-green heads, good side shoot production. *Source History: 1 in 1994; 1 in 1998; 1 in 2004.* **Sources: Eco.**

Gai Lan - (Chinese Broccoli, Flower Broccoli, Oriental Broccoli) - 60-70 days - *B. oleracea* var. *alboglabra*, stems and leaves have broccoli-like flavor, harvest thickened stalk with two sets of leaves just as white flower buds form, right before they open, enjoy several harvests, best for spring and midsummer planting. *Source History: 5 in 2004.* **Sources: Bo17, Roh, Se26, Tu6, We19.**

Green Goliath - (Burpee's Green Goliath, Emerald Spring) - 55-70 days - Large tight-budded blue-green central heads, matures over a three week period, many side shoots after head is cut, dev. for home gardens, good fresh or frozen. *Source History: 5 in 1981; 3 in 1984; 6 in 1987; 18 in 1991; 12 in 1994; 14 in 1998; 14 in 2004.* **Sources: Be4, Bu2, Dam, Ers, Fi1, Fis, Ga1, Hi13, Ho13, La1, Me7, Se25, So1, We19.**

Green Mountain - 60-85 days - Dark-green sprouting broccoli, large solid dark blue-green fine-budded heads, med-large laterals produce over long season, home or market, good freezer, 1953. *Source History: 4 in 1981; 1 in 1984; 1 in 1987; 1 in 1991; 1 in 1994; 1 in 1998; 2 in 2004.* **Sources: Ho13, Lej.**

Green Sprouting - (Early Calabrese, Green Sprouting Early) - 55-80 days - Compact dark-green 5-6 in. central head, uniform maturity, lateral heads over a long season, slow bolting European var. for short season areas with cold nights. *Source History: 18 in 1981; 16 in 1984; 17 in 1987; 12 in 1991; 17 in 1994; 16 in 1998; 17 in 2004.* **Sources: Be4, Bu8, CH14, Ers, Ge2, He17, La1, Loc, Mo13, OLD, Ont, Red, RUP, SE14, Sk2, WE10, Wi2.**

Health Plus - 70 days - Med. size solid heads, also good for sprouting, excellent quality. *Source History: 1 in 1998; 3 in 2004.* **Sources: Ers, Hum, Se17.**

Italian Green Sprouting - (Calabrese, De Cicco, Early Italian Green Sprouting, Italian Sprouting, Calambria, Waltham 29) - 60-90 days - Standard Calabrese type, large blue-green cauliflower-like central heads, good side shoot production, for home gardens and bunching, early heavy producer, intro. to the seed trade before 1920. *Source History: 33 in 1981; 29 in 1984; 27 in 1987; 18 in 1991; 16 in 1994; 13 in 1998; 8 in 2004.* **Sources: Bo17, Dam, GLO, Lin, Ni1, Sh9, Sto, Up2.**

Minaret - 75-85 days - Smaller more uniform version of Romanesco with spiraled conical lime-green heads, mild flavor much like cauliflower, good for small gardens, for spring and fall crops and winter harvest in mild areas. *Source History: 1 in 1991; 4 in 1994; 2 in 1998; 3 in 2004.* **Sources: Coo, Sw9, We19.**

Munchkin - 60 days - Open-pollinated version of Small Miracle, extra large blue-green heads produced on space-saving plants, long harvest time, secondary shoots develop after main head is removed. *Source History: 6 in 2004.* **Sources: Ers, GLO, Ni1, Se8, Se26, Ver.**

Nutri-Bud - 60-80 days - Vigorous plants to 24 in. tall, 4-6 in. head, med. size side shoots, high in free glutamine which is one of the building blocks of protein, dev. by Alan Kapuler. *Source History: 1 in 1998; 4 in 2004.* **Sources: In8, Pe1, Pe2, Se7.**

Purple Sprouting - (Late Purple Sprouting) - 220 days - Not same var. as spring planted Purple Sprouting, English seed bred for overwintering, 24 to 36 in. bush with ample purple spring flowers, longer flowering strain. *Source History: 1 in 1981; 1 in 1984; 4 in 1987; 5 in 1991; 7 in 1994; 4 in 1998; 9 in 2004.* **Sources: Bou, Coo, Ga22, Ho13, Roh, Ter, Tho, Vi5, We19.**

Purple Sprouting, Early - (Purple Sprouting) - 120-220 days - Very frost hardy, bushy plant grows slowly through winter, sprouts out prolifically in spring, for spring harvest. *Source History: 7 in 1981; 6 in 1984; 5 in 1987; 4 in 1991; 3 in 1994; 3 in 1998; 9 in 2004.* **Sources: Ba8, Bo19, Bou, He8, Ho13, La1, Se26, Tho, WE10.**

Redhead - 220 days - Improved purple sprouting type with more vigor and larger primary heads, good size secondary shoots picked over 8 week period, good flavor. *Source History: 1 in 2004.* **Sources: Tho.**

Romanesco - (Cavalo Broccola Gialastro, Roman Broccoli) - 75-100 days - Unique type, beautiful spiraling apple-green head on cauliflower-type plant, widely grown in northern Italy, better taste and texture than finest broccoli, different flavor, snap off individual spears or together, good in Northern areas, does best in the fall. *Source History: 4 in 1981; 3 in 1984; 7 in 1987; 14 in 1991; 18 in 1994; 15 in 1998; 15 in 2004.* **Sources: Ba8, Bo19, Bou, CO30, Co32, Dam, Fe5, He8, La1, Lej, Ni1, Se16, Se24, Tho, WE10.**

Rosalind - 60-65 days - Large purple-headed variety with better color and finer texture than other purples, color changes to rich dark green when cooked, withstands summer heat but does better during cooler fall weather. *Source History: 1 in 1991; 2 in 1994; 1 in 1998; 1 in 2004.* **Sources: Ter.**

Rudolph - 150 days - Early winter sprouting var., produces an abundance of purple spears by mid-December when planted in mid-July, *Source History: 2 in 2004.* **Sources: Ho13, Ter.**

Second Wave - Compact, high dome with fine beads. *Source History: 1 in 2004.* **Sources: La1.**

Spigariello - (Minestra Nera) - 65 days - Non-heading early form of broccoli, leaves and tops have delicious broccoli flavor, cutting encourages new growth, excellent for soups, stews, stir fry and salads, sow in July, harvest until hard frost. *Source History: 3 in 2004.* **Sources: CA25, Ni1,**

Red.

Spring Royalty - New breed of purple broccoli with larger central heads and many nice side shoots, sweet and tender when raw, dark green when cooked, 210 days from August 1 sowing, PSR Breeding Program. *Source History: 1 in 2004.* **Sources: Pe6.**

Thompson - 70-92 days - Late season variety with medium to large heads, fine flavor, matures two weeks later than most, uneven maturity means an extended harvest, good for fall planting. *Source History: 1 in 1994; 2 in 1998; 2 in 2004.* **Sources: Fe5, Ga1.**

Thompson 92 - Slow-maturing variety produces uniform, large heads, good side shoot production, heat and disease tolerant, excellent stress resistance, good for early spring plantings, P36 introduction. *Source History: 1 in 1994; 1 in 1998; 1 in 2004.* **Sources: Pe6.**

Umpqua - (Umpquah) - 50-75 days - Large vigorous plant, 6-7 in. head, good side shoot production, hollow stem, dev. by Tim Peters, named after the Umpquah River Valley in western Oregon, introduced in 1990. *Source History: 2 in 1991; 5 in 1994; 3 in 1998; 3 in 2004.* **Sources: Sa9, Ter, Tu6.**

Waltham 29 - (Early Green Waltham 29, Waltham, Italian Green Sprouting) - 60-95 days - Compact stocky 20 in. plants, dark blue-green solid med-green heads, large crops of side shoots, not for spring, late summer and fall harvest, can survive dry spells, dev. to withstand the increasing cold of fall maturity, MA/AES, 1951. *Source History: 51 in 1981; 44 in 1984; 46 in 1987; 51 in 1991; 51 in 1994; 52 in 1998; 48 in 2004.* **Sources: Ada, All, Be4, Bo19, Bu3, But, CO23, Com, Dam, De6, DOR, Ear, Ers, Goo, HA3, He8, He17, Hig, HO2, Ho13, Hum, Jor, Kil, La1, LO8, Loc, MAY, ME9, Mel, Mey, MOU, Or10, Pin, Roh, Ros, RUP, Sau, Se7, SE14, Se25, Se26, Sh9, Shu, So1, Vi4, WE10, Wet, WI23.**

White Sprouting - 75-105 days - A unique broccoli, produces white heads that look like small cauliflowers, very mild flavor, good cauliflower substitute in areas where it is difficult to grow. *Source History: 3 in 1981; 3 in 1984; 5 in 1987; 3 in 1991; 2 in 1994; 2 in 1998; 4 in 2004.* **Sources: Bou, Ho13, Tho, Vi5.**

White Sprouting, Late - 250 days - Late type of overwintering broccoli, has some cauliflower in its genes, sow in May or June, mulch over winter, produces many golden-white heads in late spring. *Source History: 2 in 1981; 2 in 1984; 1 in 1987; 1 in 1991; 1 in 1994; 1 in 1998; 1 in 2004.* **Sources: Ter.**

Varieties dropped since 1981 (and the year dropped):

Albert (2004), *Autumn Spear* (1982), *Boram 92* (2004), *Bronzino* (1994), *Burpee's Greenbud de Cicco* (1987), *CR-1* (2004), *Dandy Early* (2004), *Early Compact* (1987), *Early Green Sprouting* (1987), *Early One* (1991), *Early One - Green Sprouting* (1987), *Early White Pearl* (1998), *Early White Sprouting* (1998), *El Centro* (1987), *Emerald Spring* (1998), *Fat Shan* (1998), *Fiorentino* (1994), *Grande* (1991), *Green Sprouting Improved* (1998), *Italian Early Sprouting* (1998), *King Purple* (1998), *King Robert Purple* (1994), *Late Crop* (1987), *Little Guy* (2004), *Medium Late* (1998), *Medium Late 143* (1984), *Medium Late 143-A* (1987), *Medium Late 145* (1991), *Medium Late 423* (1998), *Morse's 4638* (1991), *Nine Star Perennial* (1994), *Northwest 29* (1991), *Overwintering* (1987), *Pacifica* (1991), *Packer* (1987), *Precoce Romanesco* (1987), *Pronto* (2004), *Red Arrow* (2004), *Romanesco Improved* (2004), *Spartan Early* (1994), *Spartan Early K Strain* (1994), *Special Early One* (1984), *Spiral Point* (1998), *St. Valentine* (1991), *Toor 92* (2004), *Topper* (1991), *Vietnamese* (1998).

Broccoli Raab
Brassica rapa (Ruvo Group)

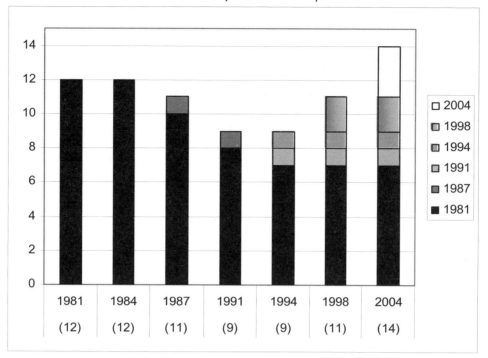

Number of varieties listed in 1981 and still available in 2004: 12 to 7 (58%).
Overall change in number of varieties from 1981 to 2004: 12 to 14 (117%).
Number of 2004 varieties available from 1 or 2 sources: 6 out of 14 (43%).

Broccoli Raab - (Rapini, Rapa, Di Rapa, Sparachetti, Italian Turnip, Ruvo Kale, Crime di Rapa) - 40-70 days - Italian non-heading broccoli which is a type of turnip grown for asparagus-like early spring shoots, 12 in. plants bolt readily. *Source History: 10 in 1981; 8 in 1984; 14 in 1987; 12 in 1991; 10 in 1994; 9 in 1998; 15 in 2004.* **Sources: Ba8, Bo17, CO30, Coo, Hi13, HO2, Hum, Na6, Ni1, Ont, Ra5, Se7, Se26, Tho, Tu6.**

Di Rapa - (De Rapa) - 85 days - Non-heading var. from Europe, produces an abundant crop of turnip-like leaves and slender flower stalks and small flower heads that are eaten as greens or cooked, quick-growing vegetable for early spring or fall sowing, pick just as flower buds start to open, favorite winter greens in Italy. *Source History: 1 in 1981; 1 in 1984; 2 in 1987; 4 in 1991; 2 in 1994; 4 in 1998; 4 in 2004.* **Sources: Dgs, Gou, Hud, Orn.**

Di Rapa, Maceratese - 50-55 days - Non-heading rape grown for its numerous large, succulent leaves, tender stems, can be grown through the summer but does best in spring and fall, harvest before flowers open, outstanding with pasta. *Source History: 1 in 2004.* **Sources: Se24.**

Di Rapa, Novantina - 30 days - Non-heading broccoli, 14-16 in. tall, top 6-8 in. of plant is harvested, pungent flavor, steamed instead of eaten raw, excel. cooked with pasta. *Source History: 1 in 1998; 3 in 2004.* **Sources: CA25, Pin, Se24.**

Fall Broccoli Raab - (Italian Turnip, Turnip Broccoli, Broccoli Headed Turnip, Sparachetti, Rapa) - Winter annual sown in late fall, winters over for very early spring turnip-like leafy shoots, tops and tender flower shoots used as greens, 18-22 in. tall plants, med-green strap leaves, greens similar to turnip tops. *Source History: 3 in 1981; 3 in 1984; 2 in 1987; 2 in 1991; 3 in 1994; 7 in 1998; 9 in 2004.* **Sources: Bo19, CO23, HA3, Hud, Kit, La1, Mey, WE10, Wi2.**

Late Rappone - (Rappone, Fall Raab, Rapa, Raab, Turnip Broccoli) - 60 days - Non-heading, tops and tender flower shoots are used as spicy greens, large heavy plant with dark-green leaves, develops small clusters of buds on short stems, for late summer or early fall harvest. *Source History: 3 in 1981; 1 in 1984; 1 in 1987; 1 in 1991; 2 in 1994; 2 in 1998; 1 in 2004.* **Sources: Com.**

Quarantino - 40 days - Medium sized plants, non-heading, more pungent flavor than broccoli, very early. *Source History: 1 in 1981; 1 in 1984; 1 in 1998; 2 in 2004.* **Sources: Fe5, Se24.**

Rapa da Foglia Senza Testa - 30-40 days - Non-heading rape, jagged leaves, tasty in a fresh salad or cooked, sow anytime. *Source History: 1 in 2004.* **Sources: Se24.**

Salad Rappone - (Rappone, Salad, Fall Raab) - 40-50 days - Can winter over, earlier but less long-standing, dark-green plants, 14 x 14 in. dia., slender flowering shoots used for spicy greens. *Source History: 3 in 1981; 2 in 1984; 1 in 1987; 1 in 1991; 4 in 1994; 5 in 1998; 1 in 2004.* **Sources:**

Loc.

<u>Sessantina Grossa</u> - 35-60 days - Thick, tender shoots and buds, strong plant, similar to Quarantina but larger. *Source History: 1 in 1998; 3 in 2004.* **Sources: CA25, Jo1, Se24.**

<u>Sorrento</u> - 40-55 days - Fast maturing variety with large, uniform, dark green florets, blue-green foliage, grows to 30 in. tall, leaves, stems and unopened flower buds also eaten. *Source History: 3 in 1998; 7 in 2004.* **Sources: Cl3, ME9, OSB, Pr3, SE4, Sto, Ter.**

<u>Spring Rapini</u> - (Spring Rapa, Spring Raab, Spring Annual, Spring, Rapine, Sparachetti, Annual Spring Sprouting, Broccoli Headed Turnip) - 60 days - Plant early in spring, harvest before hot weather, no central head but many strap-leaves and side shoots, looks like mustard with many dime-sized green buds, early European branching var. for tops and tender flower shoots, will not winter over. *Source History: 12 in 1981; 11 in 1984; 15 in 1987; 15 in 1991; 18 in 1994; 31 in 1998; 27 in 2004.* **Sources: Bo19, Bu2, CO23, CO30, Com, Dam, HA3, Ho13, Jo1, Kil, Kit, La1, ME9, Mey, OSB, PAG, Ri2, Roh, Sh9, Sh12, So12, So25, Syn, TE7, Twi, WE10, We19.**

<u>Super Rapini</u> - 60 days - Mediterranean specialty, delicious greens, piquant flavor, Italian heirloom. *Source History: 1 in 2004.* **Sources: Pe2.**

<u>Zamboni</u> - 45-70 days - Turnip-leaf raab, 22 in. tall erect blue-green plant, larger bud than traditional Rapine or Spring Raab, more uniform bud set for fewer harvests, higher yields, Italian heirloom. *Source History: 4 in 2004.* **Sources: Cl3, SE4, Sto, Ter.**

Varieties dropped since 1981 (and the year dropped):

De Brocoletto (1998), *Di Napoli* (1987), *Early Pugliese* (1998), *Italian Turnip* (2004), *Raab 7 Top* (1991), *Salad* (1991).

Brussels Sprouts
Brassica oleracea (Gemmifera Group)

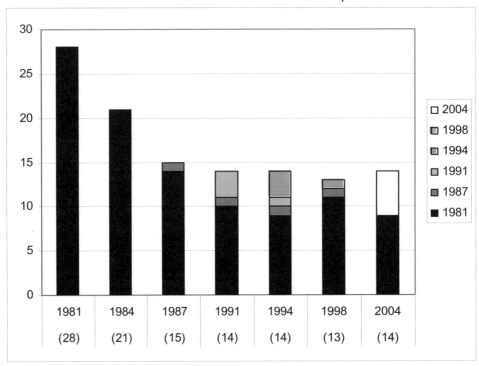

Number of varieties listed in 1981 and still available in 2004: 28 to 9 (32%).
Overall change in number of varieties from 1981 to 2004: 28 to 14 (50%).
Number of 2004 varieties available from 1 or 2 sources: 9 out of 14 (64%).

<u>Bedford Fillbasket</u> - 85-95 days - Robust 3-4 ft. plants produce largest most solid sprouts and heaviest yields, produces continuous crop from early autumn until Christmas. *Source History: 1 in 1981; 1 in 1984; 2 in 1987; 3 in 1991; 2 in 1994; 2 in 1998; 1 in 2004.* **Sources: Tho.**

<u>Catskill</u> - (Long Island Improved) - 85-110 days - Dwarf or semi-dwarf, heavy yields, large dark-green firm early 1.25-1.75 in. sprouts, compact 20-24 in. plants, med-green leaves, used mainly for fall crops, ship or freeze, hardy and uniform, dev. by Arthur White of Arkport, NY, 1941. *Source History: 19 in 1981; 17 in 1984; 15 in 1987; 10 in 1991; 12 in 1994; 12 in 1998; 15 in 2004.* **Sources: Ba8, Bo19, Co32, Cr1, Dam, He8, HO2, Kil, La1, Ri12, SE4, SE14, Se26, So1, WE10.**

<u>Catskill Improved</u> - 85-90 days - Dwarf selection, early, heavy yields, will survive several frosts, large compact

sprouts, remove leaves between sprouts and they will develop all the way up the stem. *Source History: 3 in 1981; 3 in 1984; 3 in 1987; 3 in 1991; 3 in 1994; 1 in 1998; 2 in 2004.* **Sources: Alb, Ear.**

Dark Beauty - Nice looking, med. size dark green sprouts, sweet flavor, good winter hardiness and disease res. in the Pacific NW, no other o.p. variety like it, from PSR breeding program. *Source History: 1 in 2004.* **Sources: Pe6.**

Early Half Tall - 80-90 days - Early, heavy yields, harvest mid-to-late autumn. *Source History: 1 in 1981; 1 in 1984; 1 in 1987; 1 in 1991; 1 in 1994; 1 in 1998; 1 in 2004.* **Sources: Bou.**

Evesham - (Evesham Special) - Large sprouts with fine flavor, excellent yields, old-fashioned English variety. *Source History: 3 in 2004.* **Sources: Ba8, Se26, WE10.**

Falstaff - 200 days - Ornamental purple strain, tall plants with sprouts the color of red cabbage, as tasty as green sprouts with a milder, nuttier flavor, color intensifies after a hard frost, retained when cooked. *Source History: 2 in 2004.* **Sources: Coo, Tho.**

Groninger - 104 days - Green sprouts, fair yields, good for home gardens. *Source History: 1 in 2004.* **Sources: Fe5.**

Long Island - 90-115 days - Strong 24 in. plants, large thick close-jointed stalks, small 1.5 in. cabbage-like heads, fall frosts improve flavor, same culture as late cabbage, for winter use. *Source History: 11 in 1981; 9 in 1984; 9 in 1987; 7 in 1991; 6 in 1994; 9 in 1998; 8 in 2004.* **Sources: Bo19, Bu2, Lin, Ma19, Pe2, Ros, Se7, Se26.**

Long Island Improved - (Paris Market, Half Dwarf Improved, Long Island Improved Catskill Strain) - 80-115 days - Chief commercial Calif. sprout until the more uniform hybrids, 50-100 dark-green tight 1.25-1.5 in. heads, compact 20-24 in. plants, semi-upright, med-green leaves, heavy set of firm sprouts over an extended season, good freezer, heirloom from the 1890s. *Source History: 95 in 1981; 86 in 1984; 79 in 1987; 71 in 1991; 65 in 1994; 56 in 1998; 55 in 2004.* **Sources: All, Ba8, Be4, Bu3, Bu8, CA25, CH14, CO23, Com, De6, DOR, Ear, Ers, Fi1, Fis, Fo7, Ge2, Goo, HA3, He8, He17, Hi13, Hig, Hud, Hum, Jor, La1, Lej, LO8, Loc, ME9, Mel, Mey, MIC, Mo13, MOU, OLD, Ont, Or10, Pr3, Ri12, Roh, RUP, Sa9, Sau, SE14, Se16, Sh9, Shu, Up2, Vi4, WE10, Wet, Wi2, WI23.**

Mezzo Nano - 110 days - Open-pollinated variety that equals any hybrid, firm sprouts on 3 ft. stalks. *Source History: 1 in 2004.* **Sources: Sa9.**

Paris Market - Cold resistant semi-dwarf variety with many tight heads, straight from the Paris markets. *Source History: 1 in 1981; 1 in 1994; 1 in 1998; 1 in 2004.* **Sources: Und.**

Rubine Red - 80-100 days - Plants have red foliage and red sprouts, distinct flavor, beautiful garden plant, matures late, remove terminal when buds begin to form, hardy to 20 degrees F. *Source History: 3 in 1981; 2 in 1984; 2 in 1987; 15 in 1991; 9 in 1994; 7 in 1998; 5 in 2004.* **Sources: Bou, Dam, Gou, Ter, We19.**

Seven Hills - 100-133 days - Late, hardy, harvest in early winter or let stand until mid-winter. *Source History: 1 in 1981; 1 in 1984; 1 in 1987; 1 in 1991; 1 in 1994; 1 in 1998; 1 in 2004.* **Sources: Bou.**

Varieties dropped since 1981 (and the year dropped):

Anagor (1991), *Bedford Marsters Special* (1987), *British Allrounder* (1982), *Cambridge No. 5* (2004), *Catskill Large Isle* (1982), *Darkmar 21* (2004), *Early Button* (1991), *Early Dutch Improved* (1987), *Early Dwarf Danish* (2004), *Early Morn* (1982), *Early Pick* (1987), *Green Pearl* (1991), *Half Dwarf Improved* (1983), *Harola* (1987), *Lindo* (1994), *Medium Dutch Improved* (1987), *Monitor* (1987), *Noisette* (2004), *Red* (1998), *Roodnerf Late Supreme* (1998), *Semi-dwarf Paris Market* (1983), *Steikema* (1994), *Stiekema No. 1 Original* (1994), *Tall French* (1984).

Cabbage
Brassica oleracea (Capitata Group)

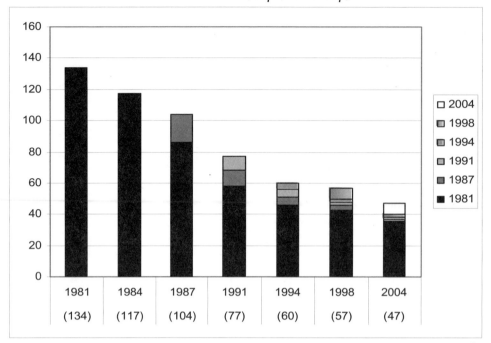

Number of varieties listed in 1981 and still available in 2004: 134 to 35 (26%).
Overall change in number of varieties from 1981 to 2004: 134 to 47 (35%).
Number of 2004 varieties available from 1 or 2 sources: 27 out of 47 (57%).

All Seasons - (Vandegaw, Succession, Wisconsin Yellows Resistant) - 85-95 days - Round and flattened on top, 8 in. deep x 10 in. dia., 9-14 lbs., heat res., fine flavor, strong grower, sure header, hard, med-short stem. *Source History: 12 in 1981; 12 in 1984; 12 in 1987; 8 in 1991; 6 in 1994; 9 in 1998; 12 in 2004.* **Sources: Bou, Ers, He8, La1, Mel, SE14, Se26, Sh9, Sk2, TE7, Und, WE10.**

Amager Green Storage - (Amager) - 120 days - Late maturing, large solid green heads, slightly flattened on top, 6-10 lbs., dev. for long term storage, properly stored keeps until spring and retains color. *Source History: 2 in 1981; 2 in 1984; 1 in 1987; 2 in 1991; 2 in 1994; 1 in 2004.* **Sources: REE.**

April - 170-250 days - Very hardy cold tol. spring cabbage, small heart, tight green pointed head, good flavor, produces heads very early in the spring, traditionally overwintered. *Source History: 2 in 1981; 1 in 1984; 2 in 1987; 1 in 1994; 1 in 2004.* **Sources: Tho.**

Brunswick - 85-90 days - Early Flat Dutch type with uniform flat drumhead, 6-7 in. deep x 9-12 in. dia., 6-9 lbs., large-ribbed plants with short stems, dependable yields even in cold harsh summers, storage and kraut. *Source History: 7 in 1981; 6 in 1984; 6 in 1987; 6 in 1991; 9 in 1994; 13 in 1998; 12 in 2004.* **Sources: Ba8, De6, DOR, Ers, HO2, Hud, La1, OLD, RUP, SE14, WE10, WI23.**

Cabbage-Sprouts - 70 days - Own cross of Golden Acre x brussels sprouts, most of them make good solid heads like cabbage, some produce heads along the stem like sprouts. *Source History: 1 in 1981; 1 in 1984; 1 in 1987; 1 in 1991; 1 in 1994; 1 in 1998; 1 in 2004.* **Sources: Fis.**

Charleston Wakefield - 70-74 days - Dark-green uniform very solid 6 x 8 in. conical heads, 4-6 lbs., yellows res., for Southern winter shipping to Northern markets, also home garden or trucking, slow to medium core, Southern market favorite, USDA and U of WI. *Source History: 31 in 1981; 24 in 1984; 22 in 1987; 15 in 1991; 16 in 1994; 17 in 1998; 9 in 2004.* **Sources: Ada, Kil, LO8, Mey, SE14, Se26, SOU, WE10, Wet.**

Charleston Wakefield, Henderson's - Larger and slightly later than Early Jersey Wakefield, 4-6 lbs., developed for the South by Peterson Henderson and Co. in 1892. *Source History: 1 in 2004.* **Sources: Ba8.**

Charleston Wakefield, Large - (Large Wakefield) - 74-105 days - Fine qualities of Jersey Wakefield and half again larger heads, less pointed, 5 lbs., very popular with Southern market gardeners and shippers, uniform. *Source History: 3 in 1981; 2 in 1984; 2 in 1987; 1 in 1991; 1 in 1994; 1 in 1998; 1 in 2004.* **Sources: La1.**

Coeur de Boeuf Moyen de la Halle - 60-70 days - Fine European heirloom, large pointed head, smooth green leaves. *Source History: 1 in 1998; 1 in 2004.* **Sources: Gou.**

Copenhagen Market - (Copenhagen Early Market, Early Copenhagen Market, Copenhagen) - 63-100 days - Round solid heads, 6-8 in. dia., 3-4 lbs., med-green small plants, yellows susc., good wrapper leaves, not inclined to burst, uniform, keeps well, short stem, largest early roundhead, very uniform in maturity, popular for market and shipping, intro. by H. Hartman and Company in 1909. *Source History: 83 in 1981; 73 in 1984; 74 in 1987; 65 in 1991; 60 in 1994; 59 in 1998; 52 in 2004.* **Sources: Alb, Ber, Bu3, Bu8, CA25, CH14, CO23, Com, Cr1, Ear, Ers, Fis, Fo7, Ge2, GLO, Gr27, GRI, HA3, Hal, He8, He17, HO2, Jor, Kil, La1, Lan, Lin, LO8, Loc, MAY, Mey, Mo13, Mo20, MOU, Ont, Or10, PAG, Ri12, RUP, Sa9, Sau, Se8, SE14, Se16, Se26, Sh9, Shu, Sto, Vi4, WE10, Wi2, WI23.**

Copenhagen Market Early Imp. - *Source History: 1 in 1994; 1 in 1998; 1 in 2004.* **Sources: DOR.**

Couve Tronchuda - (Couves, Portuguese Cabbage, CouveTronchuda de Valhascos) - Ball-shaped Portuguese cabbage, thick fleshy light-green leaves, widely used in Portugal and Brazil. *Source History: 2 in 1981; 2 in 1984; 2 in 1987; 2 in 1991; 1 in 1994; 1 in 1998; 2 in 2004.* **Sources: Com, Ho13.**

Cuor Di Bue - Old Italian variety with heart shaped, 3-5 lb. heads, name translates bull's heart. *Source History: 1 in 2004.* **Sources: Hud.**

Danish Ballhead - (True Hollander, Pennstate, Amager Short-Stemmed) - 85-110 days - Standard storage type, round blue-green 7-8 in. dia. heads, 5-7 lbs., adapted to the Northeast, does well in mountain areas, excel. yields, not yellows res., keeps until late spring, resists bolting and splitting, dependable, intro. by Burpee in 1887. *Source History: 73 in 1981; 66 in 1984; 60 in 1987; 52 in 1991; 51 in 1994; 42 in 1998; 38 in 2004.* **Sources: Alb, All, Bu3, Cr1, De6, Ers, Fa1, Fe5, Fo7, Goo, HA3, Hal, He8, He17, HO2, Hum, La1, Lin, Ma19, Me7, MOU, Ont, Or10, Pe6, Pin, RUP, Sa9, Sau, SE14, Se25, Sh9, Shu, Ter, Tt2, Up2, Vi4, We19, Wet.**

Danish Ballhead Short Stem - 100-108 days - Very hard fine-grained heads weigh 25% more than heads of equal size, 8 in. dia., 7 lbs., excel. keeper and shipper, med-dark green, round with slightly flat top. *Source History: 12 in 1981; 12 in 1984; 10 in 1987; 9 in 1991; 5 in 1994; 4 in 1998; 3 in 2004.* **Sources: Ge2, GLO, Jor.**

Danish Roundhead - (Burpees Danish Roundhead, Short Stem Ballhead) - 105 days - All-purpose fall or winter cabbage, round 7-8 in. heads, 5-7 lbs., uniformly solid and heavy, kraut and storage. *Source History: 3 in 1981; 3 in 1984; 3 in 1987; 3 in 1991; 2 in 1994; 3 in 1998; 1 in 2004.* **Sources: Com.**

Derby Day - 58-65 days - Golden Acre type, early cabbage with the handsome dark color usually associated with later varieties, compact plants, 2.5-4 lb. heads, holds well without splitting. *Source History: 1 in 1987; 2 in 1991; 3 in 1994; 3 in 1998; 2 in 2004.* **Sources: Ter, We19.**

Duor di Bue Grosso - (Large Oxheart) - 65-75 days - Early, medium size, conical, firm head, flavorful, sow spring/fall. *Source History: 1 in 2004.* **Sources: Se24.**

Dutch, Early Flat - (Early Dwarf Flat Dutch, Steins Drumhead) - 70-95 days - Solid flat heads, 7 in. deep x 9-11 in. dia., 6-10 lbs., few outside leaves so set close in rows, resists heat, popular in the South, sure header, medium stem, short core, withstands heat, fine quality, kraut or market, pre-1875, possibly pre-1855. *Source History: 42 in 1981; 37 in 1984; 38 in 1987; 29 in 1991; 29 in 1994; 20 in 1998; 14 in 2004.* **Sources: Ada, Bu3, CA25, Ge2, GLO, GRI, HO2, La1, RUP, Sau, Se26, Se28, So1, Wet.**

Dutch, Early Round - (Round Dutch) - 70-85 days - Solid uniform oval 6-7 in. dia. heads, 4-5 lbs., small compact short-stemmed plant, short core, slow bolt, good wrapper leaves, Southern favorite, excel. flavor, between drumhead and ballhead types, crinkled dark blue-green cup-shaped leaves. *Source History: 28 in 1981; 19 in 1984; 17 in 1987; 13 in 1991; 15 in 1994; 16 in 1998; 12 in 2004.* **Sources: Ada, Bu8, CA25, Fo7, He17, La1, MIC, Mo13, SE14, Se26, Sh9, Wet.**

Dutch, Late Flat - (Large Late Flat Dutch) - 100-110 days - Large firm flattened oval heads, 7 in. deep x 12-14 in. dia., 2-4 lbs., white interior, excel. keeper and shipper, makes good after dry spells, intro. by first European settlers, one of the best late fall and winter cabbages, known as Premium Late Flat Dutch prior to 1840, described by Burr in 1863, grown in England since the 1700s. *Source History: 46 in 1981; 41 in 1984; 40 in 1987; 36 in 1991; 40 in 1994; 35 in 1998; 39 in 2004.* **Sources: Bo18, Bu8, But, CA25, CH14, Cl3, Co31, De6, Ers, Fis, Ge2, GLO, Gr27, Gr29, GRI, Gur, He8, He17, HO2, La1, LO8, Ma19, MIC, Mo13, MOU, Ont, Ra5, Ri12, RIS, Roh, Ros, RUP, Sa9, SE4, SGF, Sh9, SOU, WI23, Wo8.**

Dutch, Premium Late Flat - (Premium Flat Dutch, Premium Late Dutch, Drumhead, Surehead) - 98-105 days - Large spreading short-stemmed plants, solid flat blue-green heads, 7-8 in. deep x 10-14 in. dia., 10-15 lbs., white interior, grows slow, sure heading, good keeper, high quality, for late fall and winter use, good shipper, leaves were used as a source of cut greens while the heads were forming by Virginia gardeners in the early 1900s, intro. by German immigrants about 1840, listed in 1924 catalog of D. M. Ferry and Company. *Source History: 36 in 1981; 33 in 1984; 26 in 1987; 31 in 1991; 25 in 1994; 26 in 1998; 22 in 2004.* **Sources: Bu1, Cr1, Fi1, HA3, He8, HPS, Jor, La1, Lin, ME9, Mel, Mey, Ol2, Sau, SE14, Se16, Se25, Shu, Si5, So1, WE10, Wet.**

Dutch, Stein's Early Flat - (Stein's Flat Dutch) - 83-97 days - Best and largest early flat cabbage in the South, 8 in. dia. or more, uniform, free from coarseness, excel. for kraut, keeps well, vigorous. *Source History: 10 in 1981; 9 in 1984; 9 in 1987; 6 in 1991; 7 in 1994; 7 in 1998; 5 in 2004.* **Sources: HPS, Loc, Sh9, Shu, Ter.**

Earliana - 60 days - Very early Golden Acre type, round compact 4.5-5 in. head, 2-2.5 lbs., med-green into interior, small frame, uniform size and maturity, well folded. *Source History: 5 in 1981; 5 in 1984; 4 in 1987; 7 in 1991; 4 in 1994; 4 in 1998; 3 in 2004.* **Sources: Be4, Bu2, Shu.**

Early Mountain Wakefield - *Source History: 1 in 2004.* **Sources: Hig.**

Emerald Acre - 60 days - Finest tasting early cabbage, very solid 4 in. dia. perfectly round med-green heads, almost hybrid-like uniformity, 3-3.5 lbs., good shipper, not yellows tol. *Source History: 1 in 1981; 1 in 1984; 2 in 1987; 2 in 1991; 2 in 1994; 2 in 1998; 1 in 2004.* **Sources: Me7.**

Erfurter Zwerg - Sturdy compact plants, medium size

heads, white curds, from Erfurt, Germany. *Source History: 1 in 2004.* **Sources: Tu6.**

First Early Market - 220 days - Overwintering English variety, loose head, tender sweet leaves, produces more heads when the stalk is cut at ground level. *Source History: 1 in 2004.* **Sources: Ga22.**

Glory of Enkhuizen - 75-100 days - Dark blue-green 7-9 in. round head, 8-10 lbs., few outer leaves permits close planting, med-framed spreading vigorous plant, good keeper and shipper, market or kraut, very desirable early variety, ideal for storage, dev. in the 1800s by Royal Dutch Sluis of Holland, an old seed house. *Source History: 23 in 1981; 17 in 1984; 16 in 1987; 14 in 1991; 10 in 1994; 9 in 1998; 8 in 2004.* **Sources: Ba8, Dam, Fo7, He8, Sk2, Vi4, WE10, WI23.**

Golden Acre - (Yellows Resistant Golden Acre, Resistant Detroit) - 50-70 days - Standard early ballhead, short-stemmed compact erect plants, solid 5.5-7 in. round gray-green heads, 3.5-5 lbs., yellows res., uniform, high yields, white interior, tightly folded, not long-standing, early garden use. *Source History: 103 in 1981; 94 in 1984; 87 in 1987; 77 in 1991; 69 in 1994; 69 in 1998; 64 in 2004.* **Sources: Alb, All, Bo18, Bo19, Bu8, CA25, CH14, CO23, Cr1, Dam, DOR, Ear, Ers, Fis, Fo7, Ga1, Ge2, GLO, Gr27, GRI, HA3, Hal, He8, He17, Hig, HO2, HPS, Hum, Jor, La1, Lin, Ma19, MAY, ME9, Mel, Mey, Mo13, MOU, OLD, Ont, PAG, Pin, RIS, Roh, Ros, RUP, Sa9, Sau, Se8, SE14, Se25, Se26, SGF, Sh9, SOU, Sto, Tt2, Vi4, Vi5, WE10, Wet, Wi2, WI23, Wo8.**

Golden Acre, Early - 62 days - Extra early Copenhagen type, small round solid 3 lb. head. *Source History: 1 in 1981; 1 in 1984; 1 in 1987; 1 in 1991; 1 in 1994; 2 in 1998; 1 in 2004.* **Sources: De6.**

Golden Acre, Wisconsin - (Detroit) - 60-67 days - Same head size and earliness as Golden Acre 84, slightly bluer and more yellows res., rounded slightly oval compact heads, 3-3.5 lbs., best quality production in spring. *Source History: 7 in 1981; 7 in 1984; 6 in 1987; 3 in 1991; 1 in 1994; 1 in 1998; 1 in 2004.* **Sources: SE4.**

Golden Acre, Yellows Resistant - (Golden Acre Resistant) - 60-70 days - Round solid uniform gray-green heads, 6 in. dia. x 5.5 in. deep, 3-4 lbs., res. to yellows and earlier, can plant close, small plant, med-core, good wrapper leaves. *Source History: 12 in 1981; 11 in 1984; 11 in 1987; 8 in 1991; 7 in 1994; 6 in 1998; 4 in 2004.* **Sources: BAL, Bu3, Kil, Si5.**

Greyhound - 64-73 days - Popular var. in Europe, similar to Jersey Wakefield but much earlier maturity, pointed head, small heart, very mild flavor, excel. quality, sow in early spring. *Source History: 4 in 1981; 1 in 1984; 3 in 1987; 2 in 1991; 2 in 1994; 2 in 1998; 1 in 2004.* **Sources: Bou.**

Jersey Wakefield - 62-65 days - Relatively small compact plants, light-green leaves, small compact 6-7 in. pointed heads, 2.5-3 lbs., not too tightly formed, for early spring shipping locally, earliest pointed head cabbage, few outer leaves, matures all at one time. *Source History: 14 in 1981; 13 in 1984; 12 in 1987; 11 in 1991; 9 in 1994; 5 in 1998; 5 in 2004.* **Sources: Bo19, Eco, Lin, SE14, So9.**

Jersey Wakefield, Early - (True American) - 60-75 days - Conical solid tightly folded heads, 5-7 in. dia. x 10-15 in. tall, 2-4 lbs., yellows res., can be planted close, smooth dark-green thick leaves, can be overwintered, resists splitting, first grown in the U.S. by Francis Brill of Jersey City, New Jersey in 1840. *Source History: 98 in 1981; 87 in 1984; 82 in 1987; 76 in 1991; 74 in 1994; 77 in 1998; 74 in 2004.* **Sources: Ada, Alb, Ba8, Bu2, Bu3, Bu8, But, CA25, CH14, Cl3, CO23, Co32, Coo, Cr1, Dam, De6, DOR, Ear, Ers, Fe5, Fis, Fo7, Fo13, Ga1, Ge2, GLO, Goo, Gr27, GRI, HA3, He8, He17, HO2, HPS, Hud, Hum, Jor, Kil, La1, Lan, Me7, ME9, Mey, Mo13, Na6, Ni1, Ol2, OLD, Ont, Pla, Pr3, Ri12, Ros, RUP, Sa9, Sau, Se7, Se8, Se16, Se25, Se26, Sh9, Shu, Sk2, So1, SOU, Th3, Up2, Vi4, WE10, We19, Wet, Wi2, WI23.**

Marion Market - 70-90 days - Mid-season yellows res. strain of Copenhagen Market, 10 days later but larger, round solid 6-7 in. heads, 5-7 lbs., U of WI, good early sort for diseased soils, med-large short-stemmed plant, bolt res., heavy yields, ships well. *Source History: 34 in 1981; 29 in 1984; 25 in 1987; 20 in 1991; 13 in 1994; 6 in 1998; 1 in 2004.* **Sources: Sau.**

Marner Allfroh - 60-70 days - Small garden cabbage, spring sow for small-to-medium sized very solid heads, earlier than similar types, sow thinly in May-June for tennis ball size heads, stands well without splitting, German variety, very expensive and rare. *Source History: 1 in 1981; 1 in 1991; 1 in 1994; 1 in 1998; 3 in 2004.* **Sources: Bou, Pe2, Tu6.**

Northern Giant - (McFayden Giant) - 105 days - Huge firm solid 12-15 in. drumheads, crisp greenish-white leaves, excel. for kraut, largest in Canada. *Source History: 2 in 1981; 2 in 1984; 1 in 1998; 1 in 2004.* **Sources: Pp2.**

Penn State Ballhead - (Penn State Danish Ballhead) - 90-110 days - AAS, high yielding Danish ballhead, large spreading plant, uniform flattened globe, 6-7 in. deep x 7-9 in. dia., 6-8 lbs., short stem, medium core, noted for very hard head and uniformity, kraut and storage, Prof. Meyers at Penn State. *Source History: 34 in 1981; 28 in 1984; 24 in 1987; 9 in 1991; 5 in 1994; 3 in 1998; 1 in 2004.* **Sources: Tu6.**

Primax - 60-63 days - High quality uniform European Golden Acre str., shorter stems and cores, darker-green wrapper leaves, round 3-4 lb. heads, fast growing, holds in field 2 weeks. *Source History: 3 in 1981; 3 in 1984; 3 in 1987; 4 in 1991; 2 in 1994; 2 in 1998; 3 in 2004.* **Sources: Eo2, Hig, Jo1.**

Primo - 52-65 days - Early compact ballheaded cabbage, 6-7 in., large outer leaves, 20-22 in. spread, excel. quality, good for successional sowings from February or March until mid-July. *Source History: 2 in 1981; 2 in 1984; 2 in 1987; 1 in 1991; 1 in 1994; 2 in 1998; 1 in 2004.* **Sources: Bou.**

Storage Marner - (Marner Lager Weiss) - Long storage variety, solid large heads. *Source History: 1 in 2004.* **Sources: Tu6.**

Sugar Loaf - *Source History: 1 in 1998; 1 in 2004.* **Sources: WE10.**

Winningstadt - (Winnigstadt) - Old German variety, compact, grey-green heads and fluted leaves, short stem, stores well, first listed in America by J. J. H. Gregory and Sons of Marblehead, MA in 1866. *Source History: 3 in 1994; 3 in 1998; 2 in 2004.* **Sources: Lan, Se16.**

Wisconsin All Seasons - (Wisconsin All Seasons Yellows Resistant) - 76-95 days - Yellows res. drumhead type, large hard flattened globes, 10-12 in. dia., 10 lbs., withstands adverse weather, spreading plants, plant summer or fall, good keeper, res. to yellows and wilt and heat, good for kraut, hardiest of all drumhead vars. *Source History: 23 in 1981; 19 in 1984; 20 in 1987; 16 in 1991; 15 in 1994; 12 in 1998; 5 in 2004.* **Sources: Gr27, Sau, Se25, Shu, Wet.**

Wisconsin Hollander No. 8 - (Wisconsin Hollander, Wisconsin No. 8, Hollander No. 8) - 78-115 days - Sel. from Hollander (Dutch Winter), 100% immune to yellows, deep blue-green, 7-8 in. dia and 7-9 lbs., ballhead, keeps until spring if properly stored, prolific. *Source History: 18 in 1981; 13 in 1984; 11 in 1987; 8 in 1991; 8 in 1994; 5 in 1998; 1 in 2004.* **Sources: Ros.**

Varieties dropped since 1981 (and the year dropped):

All Head Early (1991), *All Seasons Drumhead* (1991), *April Green* (2004), *April Monarch* (1984), *Autumn King* (1987), *Badger Babyhead* (1991), *Badger Market* (1991), *Bergkabis* (2004), *Biro* (1983), *Bola Verde* (1994), *Bola Verde 79* (1987), *Bonanza* (1994), *Braunschweiger* (1987), *Burpee's Surehead* (1991), *Cape Horn* (2004), *Choux de Bruxelles* (1998), *Christmas Drumhead* (2004), *Col de Machachi* (1991), *Copenhagen Market (Norseman)* (1991), *Copenhagen Market 53* (1987), *Copenhagen Market No. 6* (1994), *Copenhagen Market Select* (1987), *Copenhagen Market, Burpee's* (1994), *Copenhagen Resistant* (1987), *Copenhagen Resistant Imp.* (1987), *Copenhagen, Precoce di* (1994), *Custodian* (1987), *Decema* (1998), *Decema Extra* (1994), *Deen 523* (1991), *Delicatesse* (2004), *Desserts No. 126* (1987), *Detroit* (1983), *Durham Early* (1991), *Dutch, Ferry's Round* (1998), *Dutch, Ferry's Round* (1998), *Dutchy 522* (1991), *Early Ditmarsh* (1991), *Early Greenball* (1991), *Early Marvel* (2004), *Early Wonder* (1994), *Eastern Ballhead* (1987), *Evergreen Ballhead Medium Stem* (1991), *Express* (1994), *First Early Market 218* (2004), *Flowering* (1998), *Fuji Early* (1983), *Gigante Andino* (1994), *Globe 62M-TBR / Res Glory* (1994), *Goldack* (1987), *Golden Acre 84* (1987), *Golden Acre K Strain* (1991), *Golden Acre Primo* (1983), *Golden Acre Special 0212* (1991), *Golden Acre Supreme* (1994), *Golden Acre, Early Str. No.126* (1994), *Golden Acre, Elite* (2004), *Golden Self-Blanching* (1991), *Green Acre* (1987), *Green Storing* (1994), *Green Winter* (1991), *Greenback* (1998), *Greenhead* (1998), *Greensleaves* (1991), *Heads Up II Green* (2004), *Holland Select* (1994), *Holland Winter E. 40* (1991), *Houston Evergreen* (2004), *Jersey Queen* (1998), *Jersey Wakefield Pewa* (2004), *Jersey Wakefield Yellows Res.* (1994), *Jumbo* (2004), *Junior* (1987), *Langedijker Late W.K. Decema* (1998), *Langedijker Late White* (1994), *Lariat* (1991), *Late Danish L.D.* (1987), *Libra* (1987), *Lightning Express* (1982), *Lilliput Midget* (1987), *Little Rock Y.R.* (1991), *Morden Dwarf* (1987), *Myatts Offenham Compacta* (2004), *Noblesse* (1991), *Offenham* (2004), *Offenham, Hardy* (2004), *Offenheim III* (2004), *Oxheart Large* (1991), *Oxheart Medium* (1991), *Pixie* (2004), *Pride of the Market* (1982), *Quick-Green Storage* (1998), *Quintal Alsace* (2004), *Sakatas No. 1 Succession* (1987), *Shamrock* (1991), *Short Stem Hollander* (1991), *Starski* (1994), *Storage Green (Short Stem)* (1987), *Success* (1984), *Succession* (1998), *Superior Danish* (1994), *Superslaw* (1987), *Tastie* (1982), *TBR Globe* (1987), *TBR Globe 62 M* (1983), *Testa di Negro* (1994), *Tres Hatif de la St. Jean* (1998), *Treta* (1991), *Ultra Green (Short Stem)* (1984), *Vaugirard Winter* (1983), *Viking Extra Early Strain* (1994), *Viking Golden Acre, Stokes* (1994), *Wakker* (1987), *Wiam* (1983), *Widi* (1984), *Winter Green* (1991), *Wisconsin Ballhead Improved* (1984), *Wisconsin Copenhagen Y.R.* (1983), *Yellow Leaved Succession* (1987), *Yellows Resistant Globe* (1991), *Zwaans Jumbo* (1991).

Cabbage / Red
Brassica oleracea (Capitata Group)

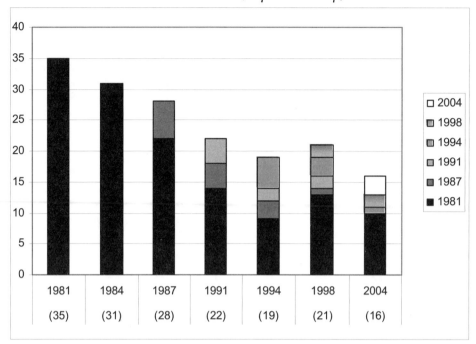

Number of varieties listed in 1981 and still available in 2004: 35 to 10 (29%).
Overall change in number of varieties from 1981 to 2004: 35 to 16 (46%).
Number of 2004 varieties available from 1 or 2 sources: 11 out of 16 (69%).

Cabeza Negra - Italian variety, 5-7 lb., solid heads, deep red color, good keeper, good for home gardens. *Source History: 1 in 2004.* **Sources: Hud.**

Early Red - 65 days - Compact frame with few outer leaves, 3-4 lbs., stores well. *Source History: 1 in 2004.* **Sources: Ga1.**

Langedijker Winter Keeper Red - 100 days - Original, the best red keeper, firm oblong dark-red heads, best planted somewhat earlier than other late reds to obtain maximum size heads. *Source History: 1 in 1981; 1 in 1984; 1 in 1987; 1 in 1991; 1 in 1994; 1 in 1998; 1 in 2004.* **Sources: Dam.**

Lasso - (Langedijker Sommer Laso) - 75 days - Beautiful bright-red round heads, uniform and solid, early heading, head size is largely dependent upon spacing, keeps in the field several weeks, can be stored. *Source History: 5 in 1981; 5 in 1984; 6 in 1987; 7 in 1991; 7 in 1994; 7 in 1998; 1 in 2004.* **Sources: Hig.**

Mammoth Red Rock - (Red Danish) - 78-105 days - Solid round flattened purple-red 7-10 in. dia. heads, 6-8 lbs., red to the center, excel. keeper, sure header, hard and tight, large-framed plant, medium core, for pickling and boiling, fine quality, heirloom from 1889. *Source History: 51 in 1981; 43 in 1984; 44 in 1987; 39 in 1991; 36 in 1994; 42 in 1998; 32 in 2004.* **Sources: Ba8, CA25, CO23, Co32, De6, DOR, Ear, Ers, Fe5, Fo13, Ge2, HA3, He17, Jor, Kil, La1, Mo13, MOU, OLD, Ont, Ri12, RUP, Sa9, SE14, Se16, Se25, Se28, Sh9, Shu, Vi4, WE10, Wi2.**

Pozky Koppus - Croatian var., small red, semi-heading type resembles a small collard, fair cold tolerance. *Source History: 1 in 2004.* **Sources: Ho13.**

Red Acre - (Resistant Red Acre, Improved Red Acre) - 74-100 days - Hard round deep-red/purple 5-7 in. heads, 2-4 lbs., small compact short-stemmed plants, solid even in early stages, sure heading, res. to yellows and splitting, uniform color, stands well, best red for storage. *Source History: 52 in 1981; 48 in 1984; 43 in 1987; 50 in 1991; 45 in 1994; 41 in 1998; 47 in 2004.* **Sources: All, Be4, Bo19, Bu3, Bu8, CH14, CO23, Com, Ers, Fis, Goo, Gr27, GRI, HA3, Hal, He8, He17, Hig, HO2, HPS, Hum, La1, Loc, Ma19, ME9, Mel, Mo13, Or10, PAG, Pe6, Pla, RIS, Roh, Ros, RUP, Se8, SE14, Se26, Sh9, Shu, Si5, Sk2, So1, Vi4, WE10, Wet, WI23.**

Red Acre Early - 75-76 days - Med-sized globe heads, deep purplish-red leaves, about 4 in. dia. and 2 lbs., best short season red cabbage available. *Source History: 3 in 1981; 3 in 1984; 3 in 1987; 2 in 1991; 2 in 1998; 1 in 2004.* **Sources: SGF.**

Red Danish - (Red Danish Ballhead) - 92-100 days - Oval purple-red 6-8 in. heads, 4-6 lbs., yellows tol., for late summer and fall, med-size plant, medium core, good wrapper leaves, stores well. *Source History: 9 in 1981; 7 in 1984; 8 in 1987; 1 in 1991; 1 in 1994; 1 in 1998; 1 in 2004.* **Sources: He8.**

Red Drumhead - 74-95 days - Round to slightly flattened heads, deep purplish-red, 7 in. dia., about 7 lbs., remarkably sweet, excel. winter keeper, very hardy, widely adapted. *Source History: 3 in 1981; 1 in 1984; 3 in 1987; 3 in 1991; 3 in 1994; 4 in 1998; 4 in 2004.* **Sources: Bou, In8, Lan, Se7.**

Red Express - 55-75 days - The first super early o.p. red cabbage to be released in years, small plant frame, dense, solid, oval heads, dark red color, 2-4 lbs., good flavor, great for salads, bred for Canada and northern U.S. *Source History: 1 in 1998; 10 in 2004.* **Sources: Bo18, Ers, Jo1, Lin, Pe2, Ri2, SE14, Shu, TE7, Tt2.**

Red Meteor - 75-80 days - Uniform deep-globe dark-red 6-6.5 in. dia. heads, 2-3 lbs., small plants with red leaves, very uniform color, plant spring or summer or fall, for home or shipping. *Source History: 7 in 1981; 6 in 1984; 2 in 1987; 3 in 1991; 2 in 1994; 5 in 1998; 1 in 2004.* **Sources: Loc.**

Red Rock - 98-103 days - Solid round heads, 8-9 in., up to 7 lbs., red throughout, vigorous and uniform, small to medium core, short stem, sure cropper, fine flavor, good keeper, solid and dense, standard red for home and market, excel. for cooking or salads or pickling. *Source History: 15 in 1981; 13 in 1984; 10 in 1987; 4 in 1991; 4 in 1994; 4 in 1998; 7 in 2004.* **Sources: Bo18, Bo19, Gr29, Mey, Ra5, Sau, Wo8.**

Red Verona Savoy - 100 days - Most unusual variety, pale red-maroon color, adds beauty and interest to late season garden, cooking with a little vinegar preserves its color. *Source History: 2 in 1981; 1 in 1998; 1 in 2004.* **Sources: Sa9.**

Rodynda - 100 days - Red storage cabbage with tight, 6-8 in. heads, excellent flavor, bio-dynamic variety developed by Dietrich Bauer at Dottenfelderhof, Germany, name stands for Rot Kohl-Dynamischen-Dauer, which means Red Bio-dynamic Storage Cabbage. *Source History: 1 in 1994; 1 in 1998; 2 in 2004.* **Sources: Pe2, Tu6.**

Soro - Medium large, early red cabbage. *Source History: 1 in 1998; 1 in 2004.* **Sources: Tu6.**

Varieties dropped since 1981 (and the year dropped):

Baby Early (1991), *Baby Late* (1987), *Black Head* (2004), *Dark-Red Winter* (1994), *Debut* (2004), *Earliest of All* (1994), *Early Red Baby (Original)* (1991), *Early Red Negro* (1983), *Heads Up II Red* (2004), *Langedijk Extra Early Red* (1987), *Langedijker Dark-Red* (1994), *Langedijker Earliest Red* (1998), *Langendijk* (2004), *Large Red Dutch Storage* (1987), *Late Red Baby (Original)* (1991), *Late Storing Red* (1994), *Mammoth German Red* (1987), *Meteor* (2004), *Preko* (1994), *Purple* (1998), *Red Danish, Reed's* (1994), *Red Debut* (2004), *Red Dutch* (1991), *Red Dwarf* (1987), *Red Earliest of All* (1994), *Red Hollander* (1991), *Red Rodan* (2004), *Red Savoy San Michele Italian* (1991), *Red Storage No .4004 (Med-Stem)* (1987), *Rode Kogel* (1991), *Rouge* (1998), *Rougette* (2004), *Round Red Dutch* (1987), *Storage Red (Short Stem)* (1991), *Wisconsin Red Acre* (1983).

Cabbage / Savoy
Brassica oleracea (Capitata Group)

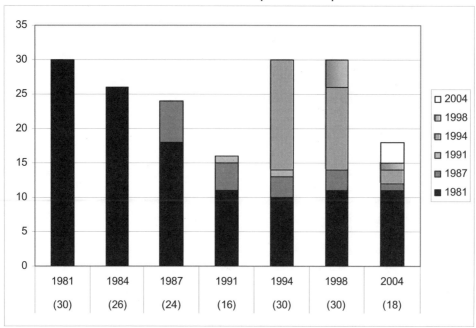

Number of varieties listed in 1981 and still available in 2004: 30 to 11 (37%).
Overall change in number of varieties from 1981 to 2004: 30 to 18 (62%).
Number of 2004 varieties available from 1 or 2 sources: 11 out of 18 (61%).

Aubervilliers - (D'Aubervilliers) - 75 days - Old French str., flat-round, blue-green head, early. *Source History: 1 in 1994; 2 in 1998; 2 in 2004.* **Sources: Gou, Ho13.**

Chieftain Savoy - (Chieftain Drumhead Savoy, Savoy Chief) - 80-105 days - AAS/1938, round flattened solid drumhead, 6-7 in. deep x 7-10 in. dia., 6-8 lbs., dark blue-green finely curled leaves, good wrapper leaves, white inside, stands well, stands fall frost, introduced in 1938. *Source History: 61 in 1981; 54 in 1984; 54 in 1987; 46 in 1991; 35 in 1994; 27 in 1998; 18 in 2004.* **Sources: Alb, Cl3, Com, Ers, Gr27, HA3, He8, Jor, Kil, Lin, Ont, Or10, PAG, Se26, Shu, So1, WE10, Wet.**

Des Vertus - 75-95 days - French savoy cabbage, 3 lb. heads with deep blue-green crinkled leaves, mild sweet flavor, very adaptable, stands heat well, can overwinter in mild areas. *Source History: 1 in 1981; 1 in 1984; 1 in 1987; 2 in 1991; 4 in 2004.* **Sources: Fe5, Ho13, In8, Se17.**

Drumhead Savoy - 85-90 days - Large coarsely crumpled heads, excel. color, very sure-heading, well blanched, very firm, crisp and tender, superior quality, pre-1885. *Source History: 5 in 1981; 4 in 1984; 3 in 1987; 3 in 1991; 5 in 1994; 6 in 1998; 7 in 2004.* **Sources: All, Ba8, Fo7, He8, Se8, Vi4, WE10.**

Holland White - (Holland Late Winter) - Solid white heads used for sauerkraut, coleslaw and salads, 4-7 lbs., stores well, plant May-June for Oct.-Dec. harvest. *Source History: 1 in 1994; 1 in 1998; 1 in 2004.* **Sources: Pe1.**

January King - 100-160 days - Frost res. English semi-savoy type, sow May-July, harvest Nov.-Jan., solid flat light-green heads, purple markings on wrapper leaves, 3-5 lbs., stands well, resists splitting, fine quality. *Source History: 8 in 1981; 7 in 1984; 7 in 1987; 8 in 1991; 9 in 1994; 6 in 1998; 8 in 2004.* **Sources: Bou, In8, Se7, Se26, So9, Ter, Vi5, We19.**

Langedijker Early - 65-70 days - Large round heads, very firm, remains in good marketable condition almost without splitting, a recommended market variety. *Source History: 1 in 1981; 1 in 1984; 1 in 1987; 2 in 1991; 2 in 1994; 2 in 1998; 1 in 2004.* **Sources: Fo7.**

Langedijker W.K. Savoy - 80 days - Original, 'W.K.' in name stands for Winter Keeper, very firm light-green heads, excel. storage quality, a recommended market variety. *Source History: 2 in 1981; 2 in 1984; 2 in 1987; 2 in 1991; 1 in 1994; 1 in 1998; 1 in 2004.* **Sources: Dam.**

Lisboa Savoy - 90 days - Large framed plant, crumpled green leaves, from Portugal. *Source History: 1 in 1998; 1 in 2004.* **Sources: Ho13.**

Ormskirk Late - 220-250 days - Huge framed savoy cabbage, large solid heart, fine quality, hardy to 10 degrees F., needs deeply dug soil, spring sown for early January harvest in the Northwest, Hurst, 1899. *Source History: 2 in 1981; 2 in 1984; 1 in 1998; 1 in 2004.* **Sources: Ho13.**

Perfection - 90 days - Dark-green deeply crumpled leaves, excel. keeper, fine flavor is even better after touched by frost, very tender. *Source History: 1 in 1981; 1 in 1984; 1 in 1998; 9 in 2004.* **Sources: De6, He17, Hig, Loc, Mo13, Roh, RUP, SE14, WI23.**

Perfection Drumhead Savoy - 90-95 days - Short-stemmed compact plants, solid nearly round heads with curled leaves, 6-8 lbs., mild flavor when touched by frost. *Source History: 1 in 1981; 1 in 1984; 1 in 1987; 1 in 1991; 2 in 1994; 8 in 1998; 6 in 2004.* **Sources: HO2, La1, Mel, Mey, SE14, Sh9.**

Testa di Ferro - Med-large head on short stemmed plant. *Source History: 1 in 1987; 1 in 1991; 1 in 1994; 1 in 1998; 1 in 2004.* **Sources: Ber.**

Vertus - 80 days - Large mid-season savoy type, med-framed plants with good vigor, green flattened-globe 8 in. heads weigh about 5 lbs. *Source History: 1 in 1981; 1 in 1984; 2 in 1994; 4 in 1998; 4 in 2004.* **Sources: Coo, Dam, Vi4, WE10.**

Verza di Verona - (San Michelle) - 80-90 days - Green savoy type, red tinge on top of large firm head, sow spring/fall, cold res., holds well in ground in fall, good storage. *Source History: 1 in 2004.* **Sources: Se24.**

Verza Montovano - 80-90 days - Sow spring or fall, better for fall crops, cold res., holds well in ground in fall, good storage. *Source History: 1 in 2004.* **Sources: Se24.**

Winter King - 80 days - Original, dark-green finely crumpled leaves, uniform heads on short stems, some frost res., overwintered in mild areas, excel. quality, special selection. *Source History: 2 in 1981; 2 in 1984; 1 in 1987; 1 in 1991; 1 in 1994; 1 in 1998; 1 in 2004.* **Sources: Dam.**

Winterfurst - Extremely late savoy, 5-7 lb. heads, allow 24 in. between plants, sow early, harvest late fall, very hardy, stores well, overwinters in mild climates for harvest in early spring. *Source History: 1 in 2004.* **Sources: Tu6.**

Varieties dropped since 1981 (and the year dropped):

Alexander No. 1 (1991), *Amager Tall Grami* (2004), *Amager Tall Winta* (2004), *American Savoy Improved* (1987), *Best of All* (2004), *Best of All* (2004), *Blue Max* (2004), *Bredase (Westlandse Putjes)* (1991), *Ditmarsch Ega 409* (2004), *Ditmarsch Midi 84* (2004), *Ditmarsch Special 0212* (2004), *Drumhead, Perfection* (1998), *Early Drumhead* (1991), *Early Perfection* (1987), *Elams Early* (2004), *Erstling Linga* (2004), *Estibal* (1987), *Green Curled Savoy* (1991), *Green Italian Savoy de Vertus* (1991), *January King 3* (2004), *Kappertjes* (1991), *Langendijk Dauer Dural* (2004), *Late Drumhead* (1991), *Lombarda Portuguesa* (1998), *Long Island Savoy* (1987), *Netted Savoy* (1984), *Novum* (1998), *Ormskirk Early* (2004), *Ostara* (1983), *Ruhm von Enkhuizen 386* (2004), *Savoy Pontoise* (1987), *Spivoy* (1991), *Tropic Giant* (1998), *Tundra* (2004), *Ukrainian Winter* (1998), *U-Neek* (1982), *Unira* (2004), *Vanguard Savoy* (1994), *Winter Keeper* (1991), *Wintergreen* (2004).

Cardoon
Cynara cardunculus

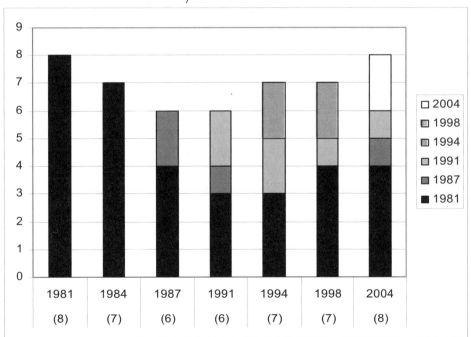

Number of varieties listed in 1981 and still available in 2004: 8 to 4 (50%).
Overall change in number of varieties from 1981 to 2004: 8 to 8 (100%).
Number of 2004 varieties available from 1 or 2 sources: 6 out of 8 (75%).

Bianco Avorio - Straight plant with some spines, thin ribs, beautiful ornamental. *Source History: 1 in 1987; 1 in 1991; 1 in 2004.* **Sources: Se24.**

Cardon d'Algiers - Algerian var. with huge leaves, pink-red stems, spiny. *Source History: 1 in 2004.* **Sources: Ho13.**

Cardoon - 60-110 days - Celery-like plant but artichoke family, piquant flavor, edible root and leafstalk, feed and water well or leafstalks get pithy, wrap in paper or burlap and mound up dirt in fall, 4 weeks to blanch, parboil stalks and hearts. *Source History: 5 in 1981; 5 in 1984; 13 in 1987; 17 in 1991; 21 in 1994; 23 in 1998; 34 in 2004.* **Sources: Ber, Bo19, Bou, Com, Fo7, Ga7, Ga21, HA3, He8, Ho12, La1, Lej, Lin, ME9, Na6, Ni1, Re8, Ren, Ri2, Ri12, Roh, Sa6, Sa12, Se7, So1, Sw9, Ter, Th3, Thy, Twi, Vi4, Vi5, WE10, We19.**

Gigante - 110-150 days - Striking 4-5 ft. tall plant with long-toothed silver-gray leaves and spineless stems, blanch in fall by tying up the leaves and wrapping a piece of cardboard around the plant, just before hard frost cut the stems an inch above the ground and mulch well for winter. *Source History: 1 in 1991; 1 in 1994; 1 in 1998; 1 in 2004.* **Sources: Coo.**

Gobbo di Nizza - 110 days - Hunchback of Nice, broader stemmed variety, can grow to 6 ft., will produce for 3-4 years. *Source History: 2 in 2004.* **Sources: Coo, Se24.**

Large Smooth - Artichoke family but more robust growing, plants 6 ft. tall, edible roots, smooth thickened leafstalks are blanched and then steamed or boiled. *Source History: 7 in 1981; 6 in 1984; 3 in 1987; 2 in 1991; 4 in 1994; 3 in 1998; 1 in 2004.* **Sources: Ho13.**

Plein Blanc Inorme - French strain, large plant with thick fleshy scales, tie up long stalks to blanch choice inner heart, similar to celery heart, plant in May, harvest in December. *Source History: 1 in 1981; 1 in 1984; 2 in 1987; 1 in 1998; 1 in 2004.* **Sources: Gou.**

Tenderheart - 120-150 days - Closely related to globe artichoke, thistles and deep-cut leaves, produces sidebranches and heavy flower head with purple bristles, grown for edible leafstalks and tender inner leaves, tied with heavy brown paper to blanch. *Source History: 4 in 1981; 3 in 1984; 3 in 1987; 3 in 1991; 2 in 1994; 3 in 1998; 4 in 2004.* **Sources: Cr1, Ge2, Loc, Sa9.**

Varieties dropped since 1981 (and the year dropped):

Bianco Pieno Inerme (2004), *Blanc Ameliore* (1998), *De Giorgi Strain* (1987), *Italian Dwarf* (1991), *Ivory Coast* (2004), *Ivory White Smooth* (1987), *Spineless* (1987), *White Improved* (1983).

Carrot
Daucus carota var. sativus

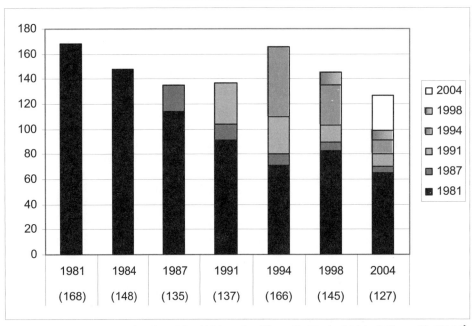

Number of varieties listed in 1981 and still available in 2004: 168 to 65 (39%).
Overall change in number of varieties from 1981 to 2004: 168 to 127 (76%).
Number of 2004 varieties available from 1 or 2 sources: 74 out of 127 (58%).

24 Karat - Long, thin, pointed root, flavorful, good fresh market variety, strong tops work for mechanical harvest, CV Harris Moran introduction. *Source History: 2 in 2004.* **Sources: HA20, Orn.**

Afghan Purple - 65 days - Bicolored roots have yellow core surrounded by purple, 9 in long, pleasant wild carrot flavor, texture is similar to celeriac or turnip root, not as sweet or as crunchy as garden carrots. *Source History: 1 in 1991; 1 in 1998; 1 in 2004.* **Sources: Ho13.**

Amarillo - (Amarillo Yellow) - 75 days - Tapered yellow root with large shoulders, distinctive sweet flavor, strong tops withstand dry conditions. *Source History: 1 in 1994; 1 in 1998; 4 in 2004.* **Sources: Bo19, He8, Sk2, WE10.**

Amsterdam Coreless - 55 days - Blunt-pointed bright

deep-orange roots, small short tops, extra early, excel. for greenhouse forcing or frame culture or in the garden, fine color and flavor. *Source History: 1 in 1981; 1 in 1984; 1 in 1987; 2 in 1991; 1 in 1994; 1 in 1998; 2 in 2004.* **Sources: La1, Sa9.**

Amsterdam Forcing - (Amsterdam) - 55-72 days - Baby carrot for greenhouse forcing and early outdoor use, 4-6 x 1 in., stump-rooted, deep-orange, very small core, almost coreless, slim and cylindrical. *Source History: 11 in 1981; 7 in 1984; 7 in 1987; 8 in 1991; 11 in 1994; 10 in 1998; 5 in 2004.* **Sources: Dam, DOR, Fo7, Fo13, Gou.**

Amsterdam Forcing, Bak Strain - (Amsterdam Coreless Bak Strain, Sweet Bak Strain) - 55 days - Quick growing strain, early coloring, 6-7 in. long, stump-rooted, very sweet and coreless, reddish-orange flesh. *Source History: 2 in 1981; 2 in 1984; 3 in 1987; 1 in 1991; 1 in 1994; 1 in 1998; 1 in 2004.* **Sources: Sto.**

Amsterdam Minicor - (Jung's First of All, Minicor) - 54-69 days - Cylindrical, stump rooted, 6-7 in., sweet tender fine-grained deep-orange flesh, never coarse or woody, uniform size and shape even when young, fine Dutch canning str. of Amsterdam, used in early stages by commercial packers of baby carrots, fresh or can or pickle. *Source History: 9 in 1981; 8 in 1984; 11 in 1987; 17 in 1991; 13 in 1994; 19 in 1998; 15 in 2004.* **Sources: Coo, Fe5, Hal, He8, Hig, Jor, OLD, Pe2, RUP, So1, Sto, Tu6, Ver, Ves, We19.**

Atomic Red - Red root. *Source History: 2 in 2004.* **Sources: ME9, Pe6.**

Autumn King 2 - 90 days - Large roots, 10-12 in. long, rich red color, heavy yields, no splitting when overwintered. *Source History: 1 in 2004.* **Sources: Tho.**

Autumn King Improved - (Autumn King) - 70 days - Imperator type, stump-rooted 7-9 in. carrot, good main crop var., large and well shaped, well colored crisp flesh, excel. for winter storage and freezing, crops late summer to autumn. *Source History: 1 in 1981; 1 in 1984; 3 in 1987; 3 in 1991; 4 in 1994; 4 in 1998; 9 in 2004.* **Sources: Bou, He8, Ho13, Se25, Shu, Ter, Vi5, WE10, We19.**

Babette - 55-85 days - Amsterdam type mini carrot, slender long roots with extra sweet crisp flesh, used as a "baby" carrot or up to 6-7 in. long, strong tops, from France. *Source History: 1 in 1994; 5 in 1998; 8 in 2004.* **Sources: Bo17, HO2, Orn, Re6, Roh, Se8, Sk2, Sw9.**

Baby Finger - (Baby Carrot, Little Finger) - Early 4 in. baby finger carrot, very small core, colors up quick, full sweetness and flavor at 3-4 in., very tender, used whole for table or canning. *Source History: 4 in 1981; 4 in 1984; 3 in 1987; 1 in 1991; 1 in 1994; 1 in 1998; 6 in 2004.* **Sources: Gr29, Lej, Ra5, Re8, Vi5, Wo8.**

Baby Spike - 55 days - Miniature Amsterdam type, holds 4 x .5 in. size well past maturity, sweet flavor, excel. color, crisp texture, good for processing, market growing and home gardening. *Source History: 9 in 1991; 12 in 1994; 8 in 1998; 2 in 2004.* **Sources: PAG, Tt2.**

Bambina - 60 days - Short, dark green foliage, very sweet flavor, long slender, almost coreless root, high yields. *Source History: 2 in 1998; 2 in 2004.* **Sources: Eo2, Sk2.**

Belgium White - (White Belgian) - 60-75 days - Old European variety, 8-10 in. long, pure-white roots with green tops, very mild flavor, practically coreless, delicious and unique, very productive and vigorous, dig before frost, not hardy, very scarce, 1885. *Source History: 3 in 1981; 2 in 1984; 2 in 1987; 2 in 1991; 1 in 1994; 2 in 1998; 5 in 2004.* **Sources: Bou, Ho13, La1, Se8, Sk2.**

Berlicummer - (Berlikum, Bercoro) - 65-75 days - Improved Nantes strain, up to 10 in. long, blunt-ended, coreless, orange color, no cracking, excellent keeper, exceptional taste, one of the best late varieties, good market type. *Source History: 1 in 1981; 1 in 1984; 2 in 1987; 2 in 1991; 1 in 1994; 2 in 1998; 4 in 2004.* **Sources: Ag7, Dam, Mcf, Sk2.**

Beta III - High carotene content, exceptional taste. *Source History: 1 in 1994; 2 in 1998; 1 in 2004.* **Sources: Ech.**

Bolina - (Bolina Round) - 65 days - Paris Market type, deep orange roots, 1.5-2.5 in. dia., sweet and flavorful, processing and home gardens. *Source History: 3 in 1998; 4 in 2004.* **Sources: CO30, HO2, Se8, Sk2.**

Brazilia - 70-75 days - Widely adapted with fair bolt tol. *Source History: 1 in 1994; 1 in 1998; 1 in 2004.* **Sources: Sk2.**

Brilliance - Nantes type with blunt end, deep magenta red/orange roots are smooth, cylindrical and coreless, very high beta carotene content, small, dark-green tops. *Source History: 1 in 1994; 1 in 1998; 2 in 2004.* **Sources: Sk2, WE10.**

Camberley - 74 days - Excellent combination of Berlicum and Danvers types, 7-9 in. roots are uniform deep-orange throughout, will grow in heavy soils and can be overwintered. *Source History: 1 in 1991; 2 in 1994; 2 in 1998; 1 in 2004.* **Sources: Ga1.**

Carentan - *Source History: 1 in 1981; 3 in 2004.* **Sources: Bo19, He8, Vi4.**

Chantenay - (Rouge Demi-Longue de Chantenay, Model) - 69-72 days - Deep-orange, 5-6 x 2.25 in. at shoulder, gradually tapers to a stump root, distinct core, yields heavily, keeps well, all purpose, for home or market or shipping. *Source History: 28 in 1981; 23 in 1984; 21 in 1987; 19 in 1991; 18 in 1994; 16 in 1998; 18 in 2004.* **Sources: All, Be4, Bou, Bu8, CH14, Co31, Hal, Lej, Pe2, Pla, Sau, Sh14, Sk2, So9, SOU, Te4, TE7, Wet.**

Chantenay, Early - (Early Chantenay Red Cored) - 70-72 days - Smooth, 5-6 in. long, half-long (abruptly stump-rooted or square at tip), small core, red-orange throughout, retains color when boiled, medium-sized tops, bunch after half grown. *Source History: 3 in 1981; 3 in 1984; 3 in 1987; 2 in 1991; 1 in 1994; 1 in 1998; 1 in 2004.* **Sources: Shu.**

Chantenay, Kinko 4 Inch - 50-90 days - Bright scarlet-orange, 4 in. long and broad at shoulder, stump-rooted, fine erect foliage, good uniformity, seldom cracks, slender core, grown all year, AL res., Johnny's introduction. *Source History: 3 in 1981; 2 in 1984; 3 in 1987; 2 in 1991; 1 in 1994; 1 in 1998; 3 in 2004.* **Sources: Jo1, Pr3, Se8.**

Chantenay, Kinko 6 Inch - 55-110 days - Erect tops, small conical stump-rooted carrot, deep red-orange, distinct core, sweet, slow bolt, strong and vigorous, crack and alternaria res., harvest when young. *Source History: 5 in 1981; 5 in 1984; 2 in 1987; 3 in 1991; 3 in 1994; 5 in 1998; 3 in 2004.* **Sources: Ga1, Hig, Syn.**

Chantenay, Kuroda - 70 days - Fall crop carrot, best if not planted before mid-June, med-sized smooth roots, 6 in. long x 2 in. dia. at crown, better eating quality than other Chantenay strains. *Source History: 2 in 1981; 2 in 1984; 4 in 1987; 3 in 1991; 2 in 1994; 2 in 1998; 4 in 2004.* **Sources: Ga1, Hig, Ri2, Se7.**

Chantenay, Long - (Chantenay, Long Supreme) - 68-74 days - Tapered deep-orange half-long 7 x 2 in. carrot, vigorous 18 in. dark-green tops, good quality and flavor, suitable for bunching, early market type, 1.5 in. at bottom. *Source History: 18 in 1981; 16 in 1984; 10 in 1987; 9 in 1991; 7 in 1994; 6 in 1998; 2 in 2004.* **Sources: La1, Sh9.**

Chantenay, Red - 70 days - Dev. from a special selection of Chantenay, good flavor, slightly earlier and shorter with rich red- orange color, used extensively for bunching by commercial growers. *Source History: 1 in 1981; 1 in 1984; 1 in 1987; 1 in 1991; 1 in 2004.* **Sources: Up2.**

Chantenay, Red Cored - (Chantenay Half Long Red Cored, Supreme Long, Burpee's Goldinhart, Chantenay Red Cored Supreme) - 60-75 days - Large, stump-rooted, deep red-orange to center, 5-7 x 2.5 in., 15-20 in. abundant erect tops, no light core, suitable for heavier soils, heavy producer, fine-grained, smooth refined shape, wide shoulders, longer Chantenay type for bunching, sweetens in storage, French introduction from the late 1800s, introduced in 1929. *Source History: 76 in 1981; 71 in 1984; 68 in 1987; 60 in 1991; 60 in 1994; 65 in 1998; 57 in 2004.* **Sources: Ag7, Alb, Bo18, Bo19, Bu2, CO23, CO30, Co32, De6, DOR, Dow, Ear, Eco, Ers, Fe5, Fis, Fo7, Fo13, Ga1, Gr27, HA3, HA20, He17, Hi13, Jun, La1, Lin, LO8, Loc, Ma19, ME9, Mel, Mey, Mo13, MOU, Ni1, OLD, Ont, OSB, PAG, Pe2, Pin, Ros, RUP, Se7, SE14, Se25, Sh9, So1, So12, Und, Ves, Vi4, WE10, Wi2, WI23, Wo8.**

Chantenay, Royal - 60-75 days - Standard garden carrot for heavy or shallow soils, cylindrical, stump-rooted, 5-7 x 2.5-3.5 in., sturdy 15-20 in. tops, red-orange inside and out, sweet fine-grained flesh, dependable heavy yields, bunch or store or home can or freeze or market, 1952. *Source History: 49 in 1981; 44 in 1984; 35 in 1987; 40 in 1991; 34 in 1994; 35 in 1998; 37 in 2004.* **Sources: Be4, Bo17, Bu2, Bu3, CO30, Dam, Dgs, DOR, Ear, Ga1, Ha5, HA20, He8, HO2, Hum, Jor, Ki4, MAY, Me7, ME9, Orn, Pe2, Ri2, Roh, Sa5, Sa9, SE4, Se7, SE14, Se26, So12, Sto, Ter, Tt2, Vi4, WE10, We19.**

Cosmic Purple - *Source History: 1 in 2004.* **Sources: ME9.**

Danvers - 65-75 days - Heat tol., 7-7.5 x 2 in. at shoulder, slightly tapered, well stumped, deep-orange throughout, home garden freezer and canner, adapted to all soils, very productive. *Source History: 10 in 1981; 7 in 1984; 6 in 1987; 4 in 1991; 7 in 1994; 5 in 1998; 10 in 2004.* **Sources: Bu8, CO23, DOR, Ga1, Pe2, Sk2, Ter, Tt2, We19, Wo8.**

Danvers 126 - (Half Long, Danvers Red Core #126) - 70-80 days - Adaptable to most soils, very productive, 7-8.5 x 1.75-2.25 in. roots, broad shoulders taper to slightly stumped end, deep-orange cortex and core without halo, uniform size and shape, large heavy 16-20 in. top, heat res., processing type, adaptable to most soils, 1947. *Source History: 32 in 1981; 31 in 1984; 35 in 1987; 38 in 1991; 40 in 1994; 41 in 1998; 40 in 2004.* **Sources: Ag7, Ba8, Bo17, Bo18, Bu3, CA25, Cl3, CO30, Cr1, Dam, Dgs, Ers, Fe5, Gr27, HA20, Hi6, HO2, Jor, Kil, Loc, ME9, Mel, Mo13, MOU, OLD, RIS, RUP, SE4, Se7, SE14, Se26, Sh9, So1, Sto, Syn, TE7, Vi4, WE10, Wi2, WI23.**

Danvers 126 Improved - 75 days - Heavier yields than Danvers, smooth orange-red 6-8 in. roots with well-colored cores, heat res. tops. *Source History: 1 in 1991; 1 in 1994; 1 in 1998; 1 in 2004.* **Sources: Twi.**

Danvers Half Long - 65-87 days - High yields in clay or heavy soils, uniform 6-8 x 2-2.5 in. roots, tapers to semi-blunt end, smooth red-orange skin, deep bright-orange flesh, nearly coreless, 16-20 in. tops, leading main crop var., excel. for fall crop, stores well, 1871. *Source History: 79 in 1981; 72 in 1984; 68 in 1987; 58 in 1991; 51 in 1994; 48 in 1998; 45 in 2004.* **Sources: Alb, All, Be4, Bu2, But, CH14, Co32, Com, De6, Ear, Fa1, Fi1, Fis, Fo7, Ge2, Gur, HA3, He8, He17, Hi13, Hud, Hum, La1, Lej, Lin, LO8, Ma19, MAY, Mey, Ont, Or10, PAG, Pin, Roh, Ros, Sau, Se16, Se25, Sh9, Shu, Si5, So1, SOU, Up2, Wet.**

Dragon - (Purple Dragon) - 60-90 days - Purple skin, solid orange to orange-yellow to bright yellow interior, sweet, spicy orange flesh, dev. by John Navazio. *Source History: 2 in 1998; 9 in 2004.* **Sources: Co32, Ga1, Ho13, In8, Ki4, Sa9, Se16, Se27, St18.**

Duke - 80-90 days - Very smooth bright-red, true European Nantes, 7-8 in. long, excel. crisp sweet flesh, grows vigorously on many different types of soil, from Denmark. *Source History: 1 in 1981; 1 in 1984; 1 in 1994; 1 in 1998; 2 in 2004.* **Sources: In8, Se27.**

Early Market - 68 days - Baby carrot, 2-6 in. long, excellent flavor, used fresh, for canning, freezing and pickling, greenhouse and outdoors, early, good summer crop carrot. *Source History: 1 in 2004.* **Sources: Bou.**

Early Scarlet Horn - 60 days - One of the oldest orange carrots still in cultivation, short, stumpy roots with very good flavor, use fresh or frozen, suited to shallow soils, early, pre-1700. *Source History: 2 in 1994; 2 in 1998; 1 in 2004.* **Sources: Lan.**

Flakkee - (Autumn King, Flakkee Giant Processor, Vita Longa, Long Flacoro) - 70-85 days - Chantenay type, up to 12 in. long x 2.5 in. dia., strong 14-16 in. tops, popular European variety used primarily for processing. *Source History: 3 in 1981; 3 in 1984; 3 in 1987; 6 in 1991; 6 in 1994; 7 in 1998; 3 in 2004.* **Sources: Ag7, DOR, WE10.**

Flakkee Long Red Giant - Strong very large carrot, up to 24 in. long x 4 in. dia., red-cored, stump-rooted, high yielding, excel. for storage, good-tasting, for human and animal use. *Source History: 1 in 1981; 1 in 1984; 1 in 1987; 1 in 1991; 1 in 1994; 2 in 1998; 3 in 2004.* **Sources: Dam, Pp2, Sk2.**

Flakkee Trofeo - 90 days - Beautiful dark green tops,

uniform yields of 5-6 in. long roots, flavorful, good for juicing, tops used for garnish. *Source History: 1 in 2004.* **Sources: Ter.**

Gold King - (Gold King Select) - 68-70 days - Slightly tapered, stump-rooted, 6 x 2.5 in., uniform reddish-orange color, smooth exterior, 15-20 in. tops, excel. on heavy soils, can and freeze. *Source History: 8 in 1981; 8 in 1984; 10 in 1987; 8 in 1991; 7 in 1994; 12 in 1998; 8 in 2004.* **Sources: Bo19, CH14, DOR, Fi1, Gr27, Jor, Se25, Sk2.**

Gold Pak - (Improved Imperator Strain) - 70-80 days - AAS/1956, smooth slender slightly tapered cylinder, 8.5-10 x 1.5 in., red-orange flesh, almost coreless, 13-16 in. bushy tops, good keeper and shipper, needs deeply worked loamy soil, not good on heavy clay, also for bunching and freezing. *Source History: 49 in 1981; 39 in 1984; 31 in 1987; 23 in 1991; 21 in 1994; 10 in 1998; 9 in 2004.* **Sources: Be4, Com, He8, HO2, Or10, Sh9, Sh14, Shu, Ver.**

Imperator - (Tendersweet) - 68-77 days - AAS, tapers to blunt end, 8-9 x 1.5 in., deep red-orange smooth skin, deep-orange flesh, coreless and fiberless, brittle but tender, used extensively for bunching, needs light and sandy or deeply worked soil, home or freeze or ship, introduced in 1928. *Source History: 51 in 1981; 47 in 1984; 47 in 1987; 31 in 1991; 26 in 1994; 22 in 1998; 18 in 2004.* **Sources: Fa1, Fo7, Gr27, HA3, He17, Hum, Lin, Ma19, Mey, Ont, PAG, Ri12, Ros, Sau, Shu, Si5, Ver, Wet.**

Imperator 58 - (Long Imperator 58) - 75-77 days - Slender 8-12 x 1.5-2 in. roots, slight taper, deep-orange flesh, small core, strong 13-20 in. tops, likes deep loam soil, excel. for fall crops in North, hybrid-like uniformity, fine quality and flavor, leading cello-pak marketer and shipper. *Source History: 36 in 1981; 33 in 1984; 28 in 1987; 36 in 1991; 36 in 1994; 34 in 1998; 27 in 2004.* **Sources: Alb, Bu3, CH14, CO23, Cr1, Dam, De6, DOR, Ers, Hal, Jor, Kil, La1, Loc, ME9, Mel, Mo13, MOU, Na6, OLD, Roh, SE14, Sh9, Vi4, WE10, Wi2, WI23.**

Imperator 58 Improved - 75-77 days - Dark-orange high-quality carrot, uniform 8 x 1.5 in. dia. roots, slight taper from top to bottom, excel. sweetness and flavor, tender crisp and coreless. *Source History: 2 in 1981; 2 in 1984; 6 in 1987; 3 in 1991; 3 in 1994; 2 in 1998; 2 in 2004.* **Sources: Be4, RUP.**

Imperator, Long - (Imperator Long Strain Bunching) - 75-79 days - Improved interior color, long 9-12 x 1.5 in. dia. rich orange- red roots, tapers to pointed tip, indistinct core, strong 16-20 in. tops, uniform, adapted only to deep sandy or loamy soils, deep color, longer strain for bunching. *Source History: 16 in 1981; 11 in 1984; 12 in 1987; 9 in 1991; 8 in 1994; 5 in 1998; 4 in 2004.* **Sources: All, Ear, La1, LO8.**

James' Scarlet Intermediate - 80 days - Fine main crop carrot, symmetrical roots taper to a point, high carotene content, excel. flavor, keeps well in the ground, fine freezer. *Source History: 2 in 1981; 1 in 1984; 1 in 1987; 1 in 1991; 1 in 1994; 1 in 1998; 1 in 2004.* **Sources: Bou.**

Japanese Imperial Long - 90-100 days - Roots 9-12 in. long, holds world record for longest carrot, grows up to 3 ft., long, thin roots are very dark orange, almost red, exceptionally fine flavor, best used for cooking. *Source History: 1 in 1994; 1 in 1998; 2 in 2004.* **Sources: Se7, Sk2.**

Japanese Long - 110-130 days - Deep orange-red roots can grow up to 2 ft., excel. cooked. *Source History: 1 in 1981; 1 in 1984; 1 in 1998; 1 in 2004.* **Sources: So12.**

Juwarot - (Juwarot Double Vitamin A) - 70-75 days - Highest vitamin A content ever recorded in carrots - 249 mg/kilo, twice normal, heavy yields, crisp sweet juicy roots, up to 8 in. long, excel. for juice, also freeze or can. *Source History: 1 in 1981; 1 in 1984; 2 in 1987; 1 in 1991; 3 in 1994; 2 in 1998; 1 in 2004.* **Sources: Com.**

Kintoki, Early Strain - Imported from Japan, true red color with no orange cast, 8 x 1.5 in. dia. roots, tender, sweet flavor, early spring strain. *Source History: 2 in 1981; 2 in 1984; 1 in 1987; 1 in 1991; 1 in 1994; 1 in 1998; 1 in 2004.* **Sources: Ho13.**

Kintoki, Kyoto Red - 120 days - Japanese var., glossy, deep scarlet, 12 in. long tapered roots, tender and sweet, excel. for juice, fresh or steamed. *Source History: 1 in 2004.* **Sources: Kit.**

Kono Shinkuroda Gosun - Japanese variety with sweet, deep red-orange roots, medium length. *Source History: 1 in 2004.* **Sources: Ba8.**

Kundulus - 50-68 days - Round forcing type carrot, deep-orange, quick maturing, grows well even in shallow or heavy soils, flats or frames or pots or windowsills or garden, freeze or can. *Source History: 2 in 1981; 2 in 1984; 2 in 1987; 3 in 1991; 2 in 1994; 1 in 1998; 1 in 2004.* **Sources: Hi13.**

Kuroda Nova - (New Kuroda, Kuroda) - 68-110 days - Chantenay type, deep red-orange, tender, sweet tapered roots, 7-8 in. long, excel. juicing carrot, stores well, does best in a mild climate. *Source History: 2 in 1981; 1 in 1984; 2 in 1987; 3 in 1991; 7 in 1994; 6 in 1998; 13 in 2004.* **Sources: Ag7, Alb, Bo17, Coo, Ers, Ev2, Fi1, Jun, Kit, Pe2, SE14, TE7, Wi2.**

Kuroda Shinn - 68-75 days - Popular Oriental market variety, 8-9 in. long root with stubby end, bright orange, tender and sweet, does well in a wide range of soil and climate conditions. *Source History: 2 in 1991; 12 in 1994; 12 in 1998; 12 in 2004.* **Sources: Ba8, CO23, DOR, Fo7, Lin, Pr3, Roh, Se8, Se26, Sk2, Ver, WE10.**

Kuttiger - 70 days - European delicacy from the early 1700s, white roots with green shoulders, 6-8 in. long, mild flavor, long storage. *Source History: 3 in 2004.* **Sources: Ri12, Sk2, Ter.**

Lady Finger - (Lady Finger Tiny Sweet, Amsterdam, Little Finger) - 65-95 days - Smooth cylindrical rich-orange roots, almost coreless, 3.5 x .6 in., small tops, sweet gourmet cooking carrot for serving whole, good in heavy soil, colors quick. *Source History: 9 in 1981; 8 in 1984; 11 in 1987; 12 in 1991; 15 in 1994; 13 in 1998; 7 in 2004.* **Sources: CH14, Fa1, Fo7, Jor, Lin, Pla, Sh9.**

Lange Rote Stumpfe 2 Zino - 85 days - Huge well shaped roots, 8.5-12 in., largest carrot. *Source History: 1 in 2004.* **Sources: Tho.**

Laval - Consistent production of tapered 7-10 in. long roots with blunt tip, withstands mechanical harvest. *Source History: 1 in 2004.* **Sources: HA20.**

Little Finger - (Lady Finger, Nantes Little Finger, Sweet Midget - Little Finger) - 50-68 days - Baby gourmet carrot, color develops quickly, can pull early, 3-3.5 x .5-.7 in. cylindrical blunt roots, very small core, smooth skin, deep-orange color, small weak tops, dev. in France for canning and pickling whole, must be harvested early. *Source History: 25 in 1981; 22 in 1984; 27 in 1987; 36 in 1991; 32 in 1994; 37 in 1998; 56 in 2004.* **Sources: Ag7, Alb, Ba8, Be4, Bo19, Bu2, Bu8, CA25, Cl3, Co31, Com, De6, Ear, Eo2, Ers, Ga1, Gr27, Gur, HA3, He8, He17, HO2, HPS, Hud, Hum, La1, LO8, Loc, Ma19, MAY, ME9, Mel, Mo13, Na6, OLD, Ont, Or10, Par, Pe2, Pin, Ri2, Ri12, Roh, RUP, Se7, SE14, Shu, Sk2, SOU, TE7, Und, Ver, WE10, Wet, Wi2, WI23.**

Long Orange - 70-86 days - Large scarlet-orange 12 x 2 in. roots, fairly smooth, tapers to pointed tip, strong deep-green tops, good for market while tender or stock feed when too mature, dates back to 1870. *Source History: 9 in 1981; 8 in 1984; 4 in 1987; 1 in 1991; 1 in 1994; 1 in 2004.* **Sources: SOU.**

Long Orange Improved - 85-88 days - Deep red-orange roots, 11-12 x 2.75 in. dia., tapers to point, heavy yields, likes light loose soil, lighter interior with distinct core, fine quality, good keeper, Dutch, 1620. *Source History: 5 in 1981; 4 in 1984; 3 in 1987; 5 in 1991; 5 in 1994; 3 in 1998; 1 in 2004.* **Sources: So1.**

Long Red Surrey - Long, tapering, yellow roots may exceed 1 ft. in length, yellow core, high yields, keeps well in the ground, will overwinter if mulched, long roots make it more drought tol. in sandy soils, introduced in 1834. *Source History: 1 in 1994; 2 in 1998; 1 in 2004.* **Sources: Ho13.**

Lubiana - 75 days - Long pointed roots with big shoulders, tall strong tops, sweet, spicy flavor that improves in storage as the starches turn into sugar, excellent storage carrot, ALS res. *Source History: 3 in 2004.* **Sources: Ga1, Ho13, Se17.**

Lunar White - *Source History: 1 in 2004.* **Sources: ME9.**

Lunga di San Valerio - *Source History: 1 in 1987; 1 in 1991; 1 in 1994; 1 in 1998; 1 in 2004.* **Sources: Ber.**

Manchester Table - 80-130 days - Roots grow 10 in. long by 2 in. wide, deep orange color, good crack res., strong tops resist blights, excellent juicer or for soups and stews, "table" in the name refers to the old days in England when carrots were divided into table use carrots or fodder type for livestock. *Source History: 1 in 2004.* **Sources: In8.**

Merida - 85 days - Bolt res. overwintering Nantes type for fall planting, also does well for spring and summer plantings, sweet flavor, 7 to 8 in. long roots, good juicing carrot. *Source History: 1 in 1994; 2 in 1998; 2 in 2004.* **Sources: Ho13, Sa9.**

Mignon - 50-68 days - Baby carrot, keeps its small size in the ground, unlike other baby types, good color and sweetness. *Source History: 3 in 1998; 3 in 2004.* **Sources: Dam, OSB, Sk2.**

Mini Sweet - 60 days - Early Japanese var., small slender root, 4 in., very smooth and almost coreless, sweet and tender, high quality. *Source History: 1 in 1991; 2 in 1994; 1 in 1998; 2 in 2004.* **Sources: Ev2, Kit.**

Nantaise - 72 days - Nantes selection with 7-8 in. roots, small core, deep orange color, sweet flavor, fall and spring harvest variety in Europe. *Source History: 1 in 2004.* **Sources: Ter.**

Nantaise Gem - Short root, from France. *Source History: 1 in 1987; 1 in 1991; 1 in 1994; 1 in 1998; 1 in 2004.* **Sources: Lej.**

Nantes - (Early Coreless) - 62-70 days - Small 10-12 in. tops, 6-7 x 1-1.5 in. dia., bright red-orange, blunt-ended, leave in ground until fall, good keeper, small core, fiberless, smooth skin, half long var. for forcing or home and early or late market, good fresh or frozen. *Source History: 23 in 1981; 22 in 1984; 25 in 1987; 27 in 1991; 26 in 1994; 19 in 1998; 18 in 2004.* **Sources: Be4, Bu3, Bu8, Ear, Fo7, Goo, He8, Jun, La1, LO8, Mey, Pin, Ri2, Sa5, Scf, Sk2, So12, TE7.**

Nantes 616, Special - 62 days - High quality well colored str. of Nantes, roots average 7 in. on muck and slightly shorter on high land, slightly tapered, med-sized tops, for bunching and storage. *Source History: 1 in 1981; 1 in 1984; 1 in 1987; 1 in 1991; 1 in 1994; 1 in 1998; 2 in 2004.* **Sources: Ear, Sto.**

Nantes di Chioggia - 75 days - Tender, crunchy, heartless Nantes type from near Venice. *Source History: 1 in 2004.* **Sources: Se24.**

Nantes Improved - 62 days - Cylindrical blunt-ended roots, 6 in. long x 1 in. dia., bright orange-red flesh, almost no core, well shaped. *Source History: 2 in 1981; 2 in 1984; 3 in 1987; 3 in 1991; 6 in 1994; 4 in 1998; 3 in 2004.* **Sources: Dom, Red, Vi4.**

Nantes, Armstrong - 65 days - Extra-long, 7-8 in., slender roots retain sweetness, flavor and tender crunch over a long period of time in the garden or field. *Source History: 2 in 1994; 2 in 1998; 2 in 2004.* **Sources: Sk2, Ter.**

Nantes, Coreless - (Scarlet Nantes, Touchon, Half-Long Coreless Nantes) - 68-72 days - Bright-orange throughout, coreless, fine-grained, 6-8 x 1.25-1.5 in., taper to a stump end, very small tops, smooth bright-orange skin, freezer type, cylindrical, excel. for forcing or for home or market garden, almost coreless. *Source History: 25 in 1981; 22 in 1984; 14 in 1987; 20 in 1991; 23 in 1994; 18 in 1998; 14 in 2004.* **Sources: All, Cr1, DOR, Fa1, Fi1, Fis, Gur, Ont, PAG, Sau, Se7, Sh9, St18, Tt2.**

Nantes, Early - (Nantes Tip Top) - 65 days - Very early, long and cylindrical, medium stump-rooted, 7 in., almost no core, coldframes or greenhouses or outdoors, good even in heavy soils, excel. freezer, England. *Source History: 2 in 1981; 2 in 1984; 2 in 1987; 2 in 1991; 2 in 1994; 1 in 1998; 2 in 2004.* **Sources: Bou, Tho.**

Nantes, Fancy - 65-68 days - Sturdy Nantes var. with unmatched color, smooth cylindrical 5-8 in. roots, 1.75 in. diameter, strong tops, uniformity not often found in open-pollinated vars., holds well in field, good keeper. *Source History: 1 in 1981; 1 in 1984; 1 in 1987; 2 in 1991; 3 in 1994; 3 in 1998; 3 in 2004.* **Sources: Fe5, Sk2, Tu6.**

Nantes, Forto - 65-70 days - Imp. European sel. of standard Nantes, smooth cylindrical 6.5 in. stump roots, well colored, brittle, good quality, strong grower even on heavier soils, maintains quality later in season, good keeper. *Source History: 1 in 1981; 1 in 1991; 1 in 1994; 2 in 1998; 2 in 2004.* **Sources: Dam, Jor.**

Nantes, Half Long - (Scarlet Nantes Half Long, Nantes Stump Rooted, Nantaise Amelioree Demi-Longue, Scarlet, Mezza Lunga) - 60-70 days - Uniform, smooth, completely cylindrical, 6-7 x 1-1.5 in., red-orange, fine-grained, almost coreless, blunt tip, small dark-green tops, satisfactory on muck soils, very popular for early market and home use, excel. quality, introduced in 1870. *Source History: 15 in 1981; 15 in 1984; 14 in 1987; 10 in 1991; 8 in 1994; 9 in 1998; 10 in 2004.* **Sources: Ber, Bu2, Com, Hi13, La1, Lin, Or10, Ros, Si5, So1.**

Nantes, Scarlet - (Early Coreless, Scarlet Nantes Coreless) - 65-75 days - Cylindrical, half-long, 7 x 1.25 in., bright red-orange flesh, fine-grained, nearly coreless, fine flavor, sweet and brittle, colors up early for baby carrots, for bunching or freezing or storage, excel. for juice. *Source History: 42 in 1981; 37 in 1984; 47 in 1987; 48 in 1991; 48 in 1994; 57 in 1998; 70 in 2004.* **Sources: Ag7, Au2, Bo17, Bo19, CA25, CH14, CO23, CO30, Co32, Dam, De6, Dgs, DOR, Eco, Ers, Fe5, Ga1, Gr27, HA3, Ha5, Hal, He8, He17, Hi6, Hig, HO2, Ho12, Hud, Hum, Jor, La1, Loc, Ma19, MAY, Mcf, ME9, Mo13, MOU, Na6, OLD, Par, Pe2, Pe6, Pr3, Pr9, Ri12, RIS, Roh, RUP, Sa9, SE4, Se7, SE14, Se16, Se26, Se27, Sh9, Shu, So9, So25, Tu6, Twi, Up2, Ver, Ves, Vi4, WE10, We19, Wi2, WI23.**

Nantes, Strong Top - 68-70 days - Half-long, cylindrical and blunt, 6-7 x 1.2-1.5 in., red-orange throughout, coreless, non-brittle tops will not break when pulled, favorite garden var., fine quality. *Source History: 10 in 1981; 9 in 1984; 8 in 1987; 6 in 1991; 4 in 1994; 3 in 1998; 3 in 2004.* **Sources: Alb, Kil, Sh9.**

Nantes, Tip Top - (Nantaise Tip Top) - 60-70 days - Longer European Nantes Touchon sel., early superior color, extremely straight, almost cylindrical, blunt tip, 7-7.5 in., coreless, smooth skin, do not crowd it. *Source History: 8 in 1981; 5 in 1984; 3 in 1987; 4 in 1991; 2 in 1994; 3 in 1998; 2 in 2004.* **Sources: Lej, Tu6.**

Nantes, Titan - Uniform, bright orange roots resist cracking, high carotene content. *Source History: 1 in 2004.* **Sources: Sk2.**

Nantesa Superior - Early, true Nantes type, very smooth uniform roots, well stumped, 5-6 x 1.2-1.5 in. dia., med-sized sparse tops, excel. external color, good internal color. *Source History: 1 in 1981; 1 in 1984; 1 in 1987; 1 in 1991; 1 in 1994; 1 in 1998; 1 in 2004.* **Sources: HA20.**

Nutri-Red - 70-80 days - Imperator type with high lycopene content, deep red roots, 8-9 in. long, not meant for eating raw, cooking enhances both flavor and color. *Source History: 14 in 2004.* **Sources: Ers, Fa1, Ga1, Ge2, Hum, Ki4, Mcf, Ni1, Par, Pin, Sk2, Tt2, Ver, We19.**

Oranza - 80 days - Main crop coreless type, cylindrical shape, 8-9 in., slicer for the processing industry, early packaging type for the fresh market grower, stores well. *Source History: 1 in 1987; 1 in 1991; 2 in 1994; 1 in 1998; 3 in 2004.* **Sources: BE6, SE4, Sk2.**

Orbit - 50 days - Smooth round 1 in. dia. baby carrots, sweet, do not split or crack or turn yellow when harvested past maturity. *Source History: 21 in 1991; 15 in 1994; 4 in 1998; 2 in 2004.* **Sources: HPS, LO8.**

Oxheart - (Guernade) - 72-80 days - Unique shape, short very thick roots, 6 in. x 4-5 in. dia., grows rapidly to over 1 lb., easy to pull and will not corkscrew, 16-18 in. tops, med-orange, excel. quality, stores well without weight loss, scarce, 1884. *Source History: 16 in 1981; 12 in 1984; 9 in 1987; 6 in 1991; 5 in 1994; 10 in 1998; 12 in 2004.* **Sources: Co32, Fis, Ga1, Ho13, Ol2, Pr9, Sa9, Se7, Se16, Shu, Sk2, Te4.**

Parabel - 60-65 days - Small round red-orange roots, 1-1.5 in., sweet and tender, very juicy, used fresh for eating whole, good for greenhouse, early. *Source History: 1 in 2004.* **Sources: Bou.**

Paris Market - (Parisian Market) - 50-68 days - Early, round as a radish, small and uniform, very sweet, red inside and out, does not need deep worked soil, gourmet restaurants and canning industry. *Source History: 6 in 1981; 5 in 1984; 2 in 1987; 2 in 1991; 3 in 1994; 1 in 1998; 1 in 2004.* **Sources: Hum.**

Pariser Market 4 - 55-60 days - Tender, crunchy, small round specialty type. *Source History: 1 in 2004.* **Sources: Se24.**

Parisian Ball - (Parisian Rondo, Round French, Round Paris Market, Round Carrot) - 60-65 days - Nineteenth century French heirloom, deep-orange 1-1.5 in. round roots, holds a long time without bursting or getting woody, excel. flavor. *Source History: 6 in 1981; 5 in 1984; 3 in 1987; 1 in 1991; 3 in 1994; 3 in 1998; 1 in 2004.* **Sources: Lej.**

Parmex - (Round Carrot) - 50-70 days - Paris Market type, improvement over the standard Planet, round roots 1-1.5 in. dia., may mature as early as peas (often prepared together in Germany and Holland), develop excellent flavor and bright color while still young, perfect for growing in heavy soil, for early home canning or freezing or specialty markets. *Source History: 2 in 1987; 6 in 1991; 1 in 1994; 4 in 1998; 5 in 2004.* **Sources: Coo, Hig, Jo1, OSB, Tho.**

Phalzer - Austrian heirloom, long, bright yellow to light orange roots that taper to a blunt end with a little point or tail, very productive, matures late, excellent storage. *Source History: 1 in 2004.* **Sources: Pr3.**

Red Elephant - Red-orange skin, very large, flavorful, received from England where it may no longer be grown due to new plant patenting laws, a first class variety. *Source History: 1 in 1981; 1 in 1984; 1 in 1987; 1 in 1991; 1 in 1994; 1 in 1998; 1 in 2004.* **Sources: Pr3.**

Red Intermediate, New - 90 days - Maincrop variety, slender tapering roots, 10-12 in. long, excellent flavor. *Source History: 1 in 2004.* **Sources: Tho.**

Red Muscade - 72 days - Sweet half-long European carrot, common in North Africa, good in hot climates, does not become bitter with the heat, sow in late summer for fall or spring harvest. *Source History: 1 in 1981; 1 in 1984; 1 in 1987; 1 in 1998; 1 in 2004.* **Sources: Syn.**

Robila - Berlicum type from Europe, deep orange roots, 6-8 in. long, sweet flavor, cooked or fresh. *Source History: 1 in 2004.* **Sources: Sk2.**

Rodelika - 72 days - Imp. Rothild, 6-8 in. long roots, sweet flavor, great for juice, good storage variety, does well in heavy soil, further bred and selected by breeder Dieter Bauer, Germany. *Source History: 2 in 2004.* **Sources: Ter, Tu6.**

Romeo - Sweet, crunchy, smooth round orange roots, do not require peeling. *Source History: 1 in 2004.* **Sources: Re6.**

Rondo - Parisian type, round, well colored roots, early. *Source History: 1 in 1998; 1 in 2004.* **Sources: CO30.**

Rothild - Dense tops, large, sweet storage carrot grows 8-10 in. long, deep red-orange color, almost coreless, high beta carotene content. *Source History: 1 in 1998; 1 in 2004.* **Sources: Coo.**

Rumba - (Tapered Nantes) - 72 days - Full flavored Nantes type with 6-7 in. roots, deep-orange color, excel. for fall in-ground storage. *Source History: 2 in 1991; 2 in 1994; 2 in 2004.* **Sources: In8, Jo1.**

Saint Valery - 70 days - Smooth, uniform root to 10-12 in. long, 2-3 in. at shoulder, thick, sweet, tender flesh with little core, sparse tops, highly productive. *Source History: 1 in 1994; 1 in 1998; 1 in 2004.* **Sources: Bou.**

Scarlet Horn, Early - (Early Horn, Early Scarlet Horn - Little Fingers) - 68 days - Stump-rooted 2-6 in. baby carrot for greenhouse forcing or outdoors, useful for successive sowings from March onward, early to late summer crops, excel. flavor, used for pickling or canning or eating fresh or freezing, from England. *Source History: 5 in 1981; 4 in 1984; 2 in 1987; 1 in 1991; 1 in 1994; 3 in 1998; 1 in 2004.* **Sources: Ho13.**

Scarlet Keeper - 85 days - Only for fall harvest and winter storage, non-bitter even after prolonged storage, cylindrical 7-10 in. dark-orange roots, red core, large strong tops, heavy yields. *Source History: 4 in 1981; 4 in 1984; 3 in 1987; 2 in 1991; 2 in 1994; 4 in 1998; 5 in 2004.* **Sources: Eco, Fe5, Ga1, In8, So12.**

Scarlet Wonder - 75-110 days - One of the main varieties in Japan, tapering deep-scarlet 12-15 in. roots, suitable only for summer sowing for fall or winter harvest, very sweet flavor, different. *Source History: 1 in 1981; 1 in 1984; 2 in 1987; 2 in 1991; 4 in 1994; 1 in 1998; 1 in 2004.* **Sources: Ev2.**

Shin Kuroda 5 Inch - 75 days - Flavorful Japanese variety with 3-5 in. tender roots, adapts to a wide range of soil and weather conditions, refined Chantenay type. *Source History: 3 in 2004.* **Sources: Ba8, Fe5, WE10.**

Solar Yellow - Yellow root. *Source History: 2 in 2004.* **Sources: ME9, Pe6.**

St. Valery - (James Scarlet) - 80-90 days - Smooth, bright red-orange roots can grow 10-12 in. long in light, sandy soil, sweet and tender, mentioned in 1885 by Vilmorin's of France as having been grown a "long time", rare. *Source History: 4 in 1981; 2 in 1984; 1 in 1987; 1 in 1991; 1 in 1994; 3 in 1998; 9 in 2004.* **Sources: Ba8, He8, Ho13, Pe2, Red, Se7, Se16, Sk2, WE10.**

Sweet 'n Short - (Short 'n Sweet) - 55-68 days - Shorter Goldinhart type, 3.5-4 x 2 in. dia., tapered, good in heavy or shallow soils, early summer or autumn crops, home garden, can or freeze. *Source History: 7 in 1981; 6 in 1984; 3 in 1987; 5 in 1991; 4 in 1994; 1 in 1998; 2 in 2004.* **Sources: Bu2, Hi13.**

Tendersweet - (Imperator 58) - 69-80 days - AAS, uniform deep red-orange, smooth, nearly coreless, 8-10 x 1.5-2 in., tapers to point, very sweet, 16-20 in. tops, leaf stems tinged purple, fine-grained crisp flesh, similar to Imperator but more useful home var., holds color for freezing. *Source History: 18 in 1981; 13 in 1984; 7 in 1987; 7 in 1991; 6 in 1994; 11 in 1998; 17 in 2004.* **Sources: Be4, Bu1, Ers, Fa1, Fi1, Gr27, Gur, He8, HO2, ME9, Mo20, SE14, Se26, Sh9, Sk2, So1, WE10.**

Thumbelina - 50-77 days - AAS/1992, 1 in. round gourmet carrot needs no peeling, stays crisp and sweet longer than other planet-type carrots, use raw, baked, steamed or boiled, produces in heavy soils. *Source History: 4 in 1991; 48 in 1994; 44 in 1998; 30 in 2004.* **Sources: Bo17, Bu8, Dam, Ear, Ers, Fe5, Ga1, Ge2, Hal, Hig, La1, Lin, Me7, Ni1, OLD, Pin, Ra5, Re8, Roh, Se8, Se26, Sh9, Shu, Sk2, Ter, Twi, Ver, Vi4, Vi5, We19.**

Tokita's Scarlet - 60-100 days - Top quality 7 in. carrot with broad shoulders, small core, heat res., dev. by Tokita Seeds Co., imported from Japan. *Source History: 2 in 1991; 3 in 1994; 2 in 1998; 1 in 2004.* **Sources: Ev2.**

Tonda di Parigi - 65 days - Uniform, deep orange, round roots harvested when 1-2 in. long, excellent sweet flavor, tender, popular market variety, 19th century heirloom from Paris. *Source History: 3 in 2004.* **Sources: Ba8, La1, Sk2.**

Topweight - 80 days - Large chunky bright-orange carrot, 10-12 x 3 in. at shoulder, received from England when outlawed by plant patenting laws, good grower in deep soils, keeps well in the soil, winter over type, yields under adverse conditions, fine quality. *Source History: 3 in 1981; 3 in 1984; 2 in 1987; 3 in 1991; 5 in 1994; 1 in 1998; 3 in 2004.* **Sources: Ho13, Sk2, Tu6.**

Touchon - (Nantes Special Long, French/Nantes/Scarlet Nantes/Super Touchon, Improved Coreless) - 60-75 days - Refined French var., deep-orange, 6.5-8 x 1-1.5 in., cylindrical, 10-12 in. tops, Nantes type but more intense color and longer, needs good soil, coreless, fine texture and quality, very disease res., can be crowded, plant spring or late summer, good in the North. *Source History: 26 in 1981; 26 in 1984; 27 in 1987; 22 in 1991; 26 in 1994; 26 in 1998; 15 in 2004.* **Sources: Bu2, Coo, Ear, Fo7, Gou, Hi13, Lej, Lin, Ni1, Pla, Se8, Sh9, Sk2, Tu6, Ves.**

Touchon Deluxe - 58 days - Sel. from Nantes Touchon for earliness uniformity extra length and exceptional color, very smooth cylindrical 7 in. stumped roots with rounded shoulders, early. *Source History: 1 in 1981; 1 in 1984; 1 in 1987; 1 in 1991; 1 in 1994; 1 in 1998; 1 in 2004.* **Sources: Sto.**

Uberlandia - Dev. by a Brazilian scientist, blooms soon after the roots are formed with no chilling or special attention required, this method works in Ft. Myers, FL and in the tropics, length of growing season to allow seed to mature farther north is not known. *Source History: 1 in 1994; 1 in 1998; 1 in 2004.* **Sources: Ech.**

White French - Vigorous hardy plants produce heavy crop, grows entirely below ground, practically coreless, delicious unique very mild flavor, strictly a gourmet table carrot. *Source History: 1 in 1991; 1 in 1994; 1 in 1998; 1 in 2004.* **Sources: Lej.**

Varieties dropped since 1981 (and the year dropped):

All Seasons (1987), *Amca* (1998), *Amini* (2004), *Amstel* (2004), *Amsterdam A.B.K.* (2004), *Amsterdam Forcing, Armstrong* (1998), *Amsterdam Long* (1987), *Amsterdam Rouge* (1984), *Amtou* (2004), *Amtou Original* (1983), *Baby Long* (2004), *Baby Orange* (1998), *Beacon* (1991), *Black Afghan* (1984), *Bunny Bite* (1982), *Caramba* (1998), *Chantenay (Kinko 8 Inch)* (1987), *Chantenay Advance* (1987), *Chantenay, Gold King Royal* (2004), *Chantenay, Half Long* (2004), *Chantenay, Imperial* (2004), *Chantenay, Improved* (1998), *Chantenay, Red Cored No. 5* (1998), *Chantenay, Red Cored No. 503* (1994), *Chrisma* (1991), *Danro* (2004), *Danvers Supreme* (1987), *Danvers, Red Cored* (2004), *Dominator* (1991), *Dutch Early Scarlet Horn* (1991), *Early Champion Scarlet Horn* (1983), *Early French Forcing* (1994), *El Presidente* (1994), *Empire* (2004), *Express* (2004), *Express 39 Nantes* (1987), *Extra Early Coreless* (1987), *Fakkel Mix* (2004), *Falcon* (1998), *Falcon II* (2004), *Fancy* (2004), *Fina* (2004), *Fincor* (1991), *First Choice* (1991), *Flak* (1998), *Flakko* (2004), *Foot Long* (1984), *Frantes* (1994), *French Forcing* (1987), *Frubund Nantes* (1991), *Giant Stock* (1991), *Glowing Ball* (1987), *Gokuwase 3-Sun* (1987), *Gold Nugget* (1987), *Gold Pak 28* (2004), *Gold Pak 263* (1994), *Gold Pak Elite* (1982), *Golden Ball* (1998), *Golden Beauty Nantes* (1987), *Goldilocks* (1982), *Goldinhart* (1998), *Greater Chantenay* (1991), *Grosa* (2004), *Hicolor 9* (1994), *Imperator 406* (1987), *Imperator 408* (1998), *Indian Long Red* (1998), *Indu* (2004), *Japanese Imperial* (1987), *Joba* (1994), *Jones Coreless* (1982), *King Imperator* (1991), *Kintoki (Regular Strain)* (1991), *Kokubu - Japanese Long* (1994), *Kurna* (2004), *Landa* (2004), *Lange Rote Stumpfe Ohne Herz* (1998), *Little Baby* (1998), *Little Finger, Improved* (2004), *Long - Duke Nantes* (1987), *Maravilla Andina* (1991), *Maravilla Bonanza* (1994), *Mini Express* (1994), *Morse's Bunching* (1983), *Naba* (1998), *Nangro* (1991), *Nantaise Improved* (2004), *Nantes 77* (1987), *Nantes (Strong Top No. 3)* (1998), *Nantes 1003 Duke* (2004), *Nantes Special* (1991), *Nantes, A and C* (1998), *Nantes, Amelioree* (1998), *Nantes, Baby* (1998), *Nantes, Baby Finger* (1994), *Nantes, Coreless Improved* (2004), *Nantes, Douceur* (2004), *Nantes, Express* (1998), *Nantes, Fruhbund Fast Crop* (1998), *Nantes, Long* (2004), *Nantes, Mexican Strain* (2004), *Nantes, Mexican Strong* (2004), *Nantes, Nancy* (1998), *Nantes, Romosa* (1998), *Nantes, Scarlet (K Strain)* (1994), *Nantes, Selma* (1994), *Nantes, Slendero* (1998), *Nantesa Superior Improved* (1998), *New King 8 Inch Chantenay* (1987), *Nobo* (2004), *Normu* (2004), *Nugget* (1991), *Orange Delight* (1994), *Orange Improved* (1991), *Oregon Chantenay* (1987), *Parisienne* (2004), *Petite Mignon* (2004), *Phil Barber's Carrot* (2004), *Planet* (1998), *Primenantes* (1987), *PSR 462* (2004), *Red Flakee* (1982), *Red Intermediate* (1998), *Redca* (2004), *Redco* (2004), *Regol* (2004), *Rojo* (1998), *Rola* (2004), *Rona* (2004), *Rondo-Vivace* (1983), *Royal Nantes* (1991), *Scarlet Imperial Long* (1998), *Senko* (1987), *Senta* (1983), *Short Fat Huge* (1994), *Spartan Bonus* (1998), *St. Valery Improved* (2004), *Sucram* (2004), *Suko* (2004), *Super Nantes* (1991), *Superator* (1987), *Superior Improved Nantes* (1991), *Sweet Cherry Ball* (1998), *Sweetheart* (2004), *Tall Top Nantes* (1987), *Tamahata* (1987), *Texas Gold Spike* (1998), *Tim Tom* (2004), *Tiny Sweet* (1983), *Top Pak* (2004), *Toponova Kurado 8 Inch* (1987), *Topscore* (2004), *Touching Earth II* (2004), *Ultra Pak* (1987), *Vertou* (1983), *Vita Longa* (1994), *Vriends Glorie* (1987), *Waltham Hicolor* (1991), *White Parsley* (2004), *Winter Hardy Chantenay* (1991), *Winter Scarlet* (1987), *Zino* (1998).

Cauliflower
Brassica oleracea (Botrytis Group)

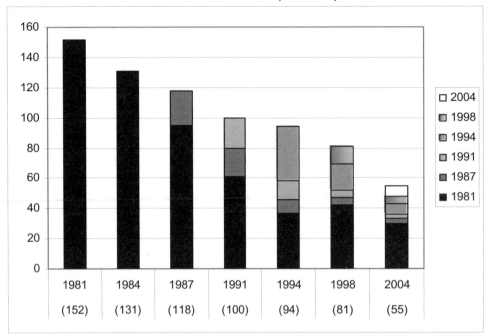

Number of varieties listed in 1981 and still available in 2004: 152 to 30 (20%).
Overall change in number of varieties from 1981 to 2004: 152 to 55 (-64%).
Number of 2004 varieties available from 1 or 2 sources: 34 out of 55 (62%).

Aalsmeer - A Walcheren winter cauliflower, 4-6 in. heads, large plants can withstand temps from 16 to -5 degrees F, depending on wind and snow cover, harvest late April to early May, developed in England. *Source History: 1 in 2004.* **Sources: We19.**

Absolute - Newly advanced commercial variety with superior curd density, excel. wrap, uniform maturity, widely adaptable, fresh market. *Source History: 1 in 2004.* **Sources: HA20.**

Alaska - Newly advanced commercial variety, smooth head is wrapped tightly, dense curd, holds well in the field, processing and fresh market. *Source History: 1 in 2004.* **Sources: HA20.**

All-The-Year-Round - 70-100 days - Large tight head surrounded by dark-green leaves, keeps in good condition a long time, ideal for frame culture and successional sowings all year, excel. freezer. *Source History: 1 in 1981; 1 in 1984; 2 in 1987; 3 in 1991; 5 in 1994; 3 in 1998; 8 in 2004.* **Sources: Bo19, Bou, In8, Se7, Se26, Sk2, Tho, WE10.**

Alpha Fortados - 65 days - Self-blanching var. grown commercially in Europe, med. size heads, does well in all areas of the U. S. *Source History: 1 in 1991; 2 in 1994; 2 in 1998; 1 in 2004.* **Sources: Fo7.**

Alverda - 80-90 days - Green-headed cauliflower on medium-size plants, heads have bright yellowish green curds, nice taste raw or cooked, easy to grow, no tying needed, PVP 1988 Sunglow and Rijk Zwaan Zaadteelten Zaadhandel B.V. *Source History: 1 in 1991; 4 in 1994; 4 in 1998; 3 in 2004.* **Sources: CO30, REE, Ves.**

Amazing - 60-85 days - Self-blanching, 7-10 in. domed heads, retains white color, upright leaves protect the solid curds, for late summer and fall harvest, heat and cold tolerant, excellent holding ability. *Source History: 2 in 1994; 4 in 1998; 7 in 2004.* **Sources: BE6, Dam, Jo1, Jun, Par, SE4, Ver.**

Andes - 67-72 days - Fine mid-summer var., impressive in numerous Northern trials, dense heavy well domed heads, strong blue-green foliage, quite uniform, good early or late, PVP expired 2001. *Source History: 2 in 1981; 2 in 1984; 5 in 1987; 10 in 1991; 9 in 1994; 9 in 1998; 3 in 2004.* **Sources: RIS, RUP, Ves.**

Armado - (Walcheren Winter 3) - 300 days - Solid, heavy heads, heavy crops, winter hardy, spring harvest. *Source History: 1 in 2004.* **Sources: Tho.**

Armado April - 240 days - Walcherin overwintering type, sow early August for April harvest, freezes out at 12 degrees F., good in the Pacific Northwest, especially around Portland. *Source History: 2 in 1981; 1 in 1984; 2 in 1987; 1 in 1991; 1 in 1998; 1 in 2004.* **Sources: Ter.**

Armado May - Walcherin overwintering cauliflower that heads in late April, an individual Armado strain for the small commercial grower. *Source History: 1 in 1981; 1 in 1984; 1 in 1987; 1 in 1991; 1 in 1994; 2 in 1998; 1 in 2004.* **Sources: Ter.**

Armado Tardo - (Walcherin Winter Armado Tardo) - 240 days - Walcherin overwintering cauliflower that heads in early May, an individual Armado strain for the small commercial grower, exceptional curds, robust and reliable. *Source History: 1 in 1981; 1 in 1984; 1 in 1987; 2 in 1991; 1 in 1998; 1 in 2004.* **Sources: Ter.**

Autumn Giant - 120 days - Largest and finest of the late cauliflowers, heads grow to enormous size, saleable when early varieties are finished. *Source History: 1 in 1981; 1 in 1984; 1 in 1987; 1 in 1991; 2 in 1994; 1 in 1998; 1 in 2004.* **Sources: Tho.**

Brocoverde - 62-75 days - Cauliflower x broccoli cross, sturdy upright plants, semi-domed heads, 14 oz., tastes like sweet cauliflower, twice the vitamin C and folic acid content than either parent. *Source History: 9 in 1998; 28 in 2004.* **Sources: Alb, Bo17, Bo18, Bo19, Com, Dam, Ers, Ge2, Gr29, HO2, HPS, Kil, La1, Lin, LO8, Ma19, MIC, Ont, Ra5, Ri12, Roh, Se26, Se28, Shu, Twi, Ver, Vi5, Wo8.**

Burpeeana - 58 days - Burpee str. of Super Snowball, upright vigorous plants, well rounded compact pure-white heads with smooth curds, 2 lbs., well protected by many long jacket leaves. *Source History: 5 in 1981; 5 in 1984; 3 in 1987; 3 in 1991; 4 in 1994; 2 in 1998; 1 in 2004.* **Sources: Ge2.**

Cauliflower of Macerata - 70-80 days - Compact green head, med. size, great taste, spring/fall sow. *Source History: 1 in 2004.* **Sources: Se24.**

Dominant - 60-78 days - Outstanding autumn cauliflower, large solid heavy head, inside leaves cover the heads for good sun protection, thick smooth white curds, heavy root system offers some dry weather tol., recommended for fall harvest. *Source History: 7 in 1981; 5 in 1984; 6 in 1987; 6 in 1991; 6 in 1994; 7 in 1998; 2 in 2004.* **Sources: Fe5, HO2.**

Early Pearl - 90 days - Largely used in Calif. coastal areas, well protected large pure-white heads, quite distinct, maturity follows that of Veitch Autumn Giant. *Source History: 1 in 1981; 1 in 1984; 1 in 1991; 1 in 1994; 1 in 1998; 2 in 2004.* **Sources: He8, WE10.**

Erfurt - *Source History: 1 in 1991; 1 in 1994; 1 in 1998; 1 in 2004.* **Sources: WE10.**

Galleon - Walcherin str., high quality, fine-curded head, matures in mid-May. *Source History: 1 in 1998; 1 in 2004.* **Sources: Ter.**

Igloo - 70 days - Late stove-pipe type for freezing or processing, heavy 7 in. solid white curds are shielded from frost by dense upright foliage, Snowball type head. *Source History: 2 in 1981; 2 in 1984; 2 in 1987; 1 in 1991; 1 in 1994; 1 in 1998; 1 in 2004.* **Sources: He8.**

Lateman - 90-98 days - Improved White Rock var., pure white color, 7-8 in. deep dome, self-wrapping, very uniform, heat res., suitable for close spacing. *Source History: 2 in 1991; 2 in 1994; 3 in 1998; 3 in 2004.* **Sources: BE6, SE4, Se26.**

Maystar - Latest maturing in the Walcherin series, harvest in late May. *Source History: 1 in 1998; 1 in 2004.* **Sources: Ter.**

Mechelse Early - *Source History: 1 in 1998; 1 in 2004.* **Sources: WE10.**

Palla di Neve - Medium size, compact white head. *Source History: 1 in 1987; 1 in 1991; 1 in 1994; 1 in 1998; 1 in 2004.* **Sources: Ber.**

Puakea - *Source History: 2 in 1991; 1 in 1994; 1 in 1998; 1 in 2004.* **Sources: Uni.**

Purple Cape - 85 days - Rich purple heads, excellent flavor, winter-heading type ready in late winter or early spring, hardy to Zone 6, popular in Europe but relatively unknown in the U. S., intro. from South Africa in 1808. *Source History: 2 in 1987; 3 in 1991; 4 in 1994; 4 in 1998; 5 in 2004.* **Sources: Bou, Se8, Se16, Ter, We19.**

Purple Head - 80-85 days - No need for tying, large deep-purple heads, turns green when cooked, mild broccoli-like flavor, easier to grow and holds longer in the garden than the whites. *Source History: 11 in 1981; 9 in 1984; 8 in 1987; 8 in 1991; 4 in 1994; 2 in 1998; 7 in 2004.* **Sources: Ge2, Gr29, Ra5, Se28, Sh9, Vi5, Wo8.**

Purple Queen - 70 days - Deep-purple 6 in. head with good dome, curds turn bright green after boiling, used for fresh market and home garden, best for fall. *Source History: 1 in 1991; 1 in 2004.* **Sources: CO30.**

Rosalind - 60-90 days - Early maturing purple cauliflower, better color and finer texture than other purples, heat adaptable, for planting through spring and summer. *Source History: 1 in 1991; 1 in 2004.* **Sources: We19.**

Self-Blanche - (White Self-Blanching) - 68-100 days - Wrapper leaves curl over head in cool fall weather, no tying until 6-8 in. dia., stops growing in hot weather instead of producing inferior heads, large plant, deep and smooth, fine-grained, pure-white, Dr. Homna, MSU. *Source History: 47 in 1981; 44 in 1984; 43 in 1987; 33 in 1991; 24 in 1994; 24 in 1998; 24 in 2004.* **Sources: BAL, Be4, CA25, De6, Ear, Fis, GRI, Gur, HA3, HO2, HPS, La1, Ma19, MIC, MOU, Ont, REE, RUP, Sa9, Se26, Sh9, Shu, Si5, Vi4.**

Shannon - 80-90 days - Romanesco type with beautiful pointed 6-7 in. heads with small pointed florets, best for fall harvest. *Source History: 2 in 2004.* **Sources: Ki4, OSB.**

Sicilian Purple - (Violetta di Sicilia) - 61-85 days - Huge deep-purple heads, cooks to emerald-green, high mineral content accounts for color, very mild flavor, insect res., easier to grow than white cauliflower. *Source History: 1 in 1981; 1 in 1984; 2 in 1987; 2 in 1991; 2 in 1994; 4 in 1998; 3 in 2004.* **Sources: Ba8, CA25, Se24.**

Snowball - 52-70 days - Large solid snow-white 6-7 in. dia. heads, well protected by large outer leaves, reliable sure-heading str., heavy yielding, DM tol., smooth white curds, uniform. *Source History: 18 in 1981; 16 in 1984; 14 in 1987; 7 in 1991; 5 in 1994; 6 in 1998; 9 in 2004.* **Sources: All, Bou, Bu8, De6, Lej, Pe2, Pla, Ri2, TE7.**

Snowball 16 - 70 days - Round deep solid compact 7-9 in. heads, pale-green protective leaves, good str. for fall crops, dwarf plants, will stand more severe frost late in the season. *Source History: 7 in 1981; 5 in 1984; 3 in 1987; 2 in 1991; 2 in 1994; 1 in 1998; 2 in 2004.* **Sources: CO23, Ear.**

Snowball A - 55 days - Deep solid ivory-white heads, medium to large, well protected by inner leaves, very reliable early strain, concentrated maturity. *Source History: 7 in 1981; 6 in 1984; 7 in 1987; 6 in 1991; 3 in 1994; 2 in 1998; 3 in 2004.* **Sources: CO23, Se26, Wet.**

Snowball A Improved - 60 days - Best when planted for fall crop, large white heads with good taste and freezing quality, recommended for market. *Source History: 1 in 1981; 1 in 1984; 1 in 1987; 1 in 1991; 1 in 1994; 1 in 1998; 2 in*

2004. **Sources: Com, Dam.**

Snowball A, Early - (Super Snowball, Super Snowball A, Super Snowball Early A**)** - 52-60 days - Smooth med-deep round 6- 7 in. pure-white heads, 1.5-2 lbs., uniform maturity requires prompt harvest and handling, for summer or fall harvest in the North, likes alkaline soil, uniform size and shape. *Source History: 25 in 1981; 22 in 1984; 18 in 1987; 19 in 1991; 13 in 1994; 12 in 1998; 8 in 2004.* **Sources: BAL, Bu3, CA25, CO30, Hal, HO2, Kil, Mo13.**

Snowball Cross - 74 days - Reliable yields of med. size, compact heads which are held higher than others for easy cultivation and harvest, suitable for warmer, drier climates. *Source History: 1 in 1998; 1 in 2004.* **Sources: Fo7.**

Snowball X - (Early Snowball X, Giant Snowball X, Snowdrift) - 60-70 days - Deep well rounded 6-7 in. heads, small curds, sure header, matures gradually, long strong protective outer leaves, large solid and pure-white, very adaptable, grows well in a wide range of conditions, old favorite home garden variety. *Source History: 20 in 1981; 19 in 1984; 14 in 1987; 13 in 1991; 12 in 1994; 6 in 1998; 4 in 2004.* **Sources: Alb, Bou, Fa1, WE10.**

Snowball Y - 65-75 days - Excel. Snowball type, smooth tight curds, deep 6-7 in. heads, large plants, erect outer leaves offer easy typing and good protection, vigorous grower, dwarf plants, for home or commercial use, can be harvested over a long period, recommended for fall crops, intro. in 1947 by Ferry Morse. *Source History: 17 in 1981; 13 in 1984; 9 in 1987; 8 in 1991; 10 in 1994; 8 in 1998; 6 in 2004.* **Sources: Bo19, Ers, Gr27, Mey, MIC, Mo13.**

Snowball Y Improved - 65-75 days - Self-blanching str., needs no tying, does well in California, deep well rounded smooth white 6-6.5 in. heads, for fall and winter harvest, reliable, heavy yielding, large plant, smooth curds, for market or freezing. *Source History: 19 in 1981; 17 in 1984; 17 in 1987; 20 in 1991; 22 in 1994; 22 in 1998; 24 in 2004.* **Sources: Bou, CH14, Cl3, CO23, CO30, Co31, Dgs, DOR, Jor, La1, Loc, ME9, Na6, OLD, Or10, RIS, Ros, RUP, SE14, Se25, SOU, WE10, Wi2, WI23.**

Snowball, Early - (Select Snowball Early, Super Select Snowball, Snowball X, Snowdrift, Dwarf Erfurt, Extra Early) - 50-85 days - Standard main crop Snowball type, deep smooth uniform solid 6 in. heads, dwarf compact plants, well adapted to warm and milder regions, matures over 2 to 3 weeks, used mainly as a fall harvested crop in short season areas, intro. by Peter Henderson and Company in 1888. *Source History: 55 in 1981; 49 in 1984; 49 in 1987; 39 in 1991; 35 in 1994; 36 in 1998; 34 in 2004.* **Sources: Be4, Bu1, But, CO23, Cr1, Ge2, GLO, Goo, HA3, He8, He17, Hig, Ho12, Hud, Hum, La1, Lin, LO8, MAY, Mcf, Mel, MOU, PAG, Pin, Ri12, Sa9, Sau, Se16, Se25, Sh9, Shu, Si5, So1, Vi4.**

Snowball, Self-Blanching - 60-70 days - Pure-white smooth heads, 6.5-8 in. dia., med-large plants, vigorous curled upright leaves will self-wrap in cool weather but not in warm weather for early crops, dev. by Dr. Homna, MSU. *Source History: 6 in 1981; 6 in 1984; 9 in 1987; 11 in 1991; 13 in 1994; 14 in 1998; 16 in 2004.* **Sources: Ba8, Bu3, Ers, Ge2, Gr27, He17, Jor, Mo13, Ni1, Roh, SE14, Se25, Sk2, So1, Und, WE10.**

Snowball, Super - (Early Super Snowball) - 52-60 days - Extra early Snowball str., all plants head and mature at same time, compact 6.5 in. deep heads, plants more dwarf and less upright than Snowball, 1.75-2.5 lbs., good wrapper leaves, for home or market gardens, excel. freezer, largest of the earlies, De Giorgi offers originator's seed. *Source History: 30 in 1981; 26 in 1984; 20 in 1987; 14 in 1991; 10 in 1994; 8 in 1998; 8 in 2004.* **Sources: Bo18, Ear, Fi1, Fis, Ra5, Se28, Tt2, Vi5.**

Snowcone - 55 days - Excellent quality, med. size, semi-domed heads, self-wrapping leaves, PVP 1993 HA20. *Source History: 1 in 1994; 1 in 2004.* **Sources: Ga1.**

Snowdrift - (Snowball X, Snowball Early, New Snowball) - 55-70 days - Pure-white variety of excel. quality, large deep solid heads, well protected by folding inner leaves, heavy yields. *Source History: 6 in 1981; 6 in 1984; 6 in 1987; 3 in 1991; 3 in 1994; 2 in 1998; 1 in 2004.* **Sources: Sh9.**

Snowman - 71 days - Med-late, upright plants with large dark-green leaves, large pure-white heads, well covered and protected by inner jacket leaves, high domed 2 lb. rounded heads. *Source History: 1 in 1981; 1 in 1984; 1 in 1994; 1 in 2004.* **Sources: HA20.**

Snowpak - (Snow Pak) - 70-92 days - Upright med-sized plants, excel. wrapper leaves, 6-8 in. heads, deep firm smooth curds, tested in South for winter production, fresh market and process. *Source History: 7 in 1981; 7 in 1984; 10 in 1987; 8 in 1991; 10 in 1994; 1 in 1998; 1 in 2004.* **Sources: Sh9.**

Snow's Winter White - (Snow's Overwintering White) - 110 days - Popular winter cauliflower, heads can grow from 10-16 in. in dia., cold tolerant, easier to grow than spring varieties, excel. freezer, hardy to Zone 6. *Source History: 1 in 1981; 1 in 1984; 2 in 1987; 1 in 1991; 3 in 1994; 3 in 1998; 2 in 2004.* **Sources: Bou, Ho13.**

Veitch's Self Protecting - 80 days - Firm white heads thoroughly protected by the leaves, for cutting in November and December, excel. freezer, 1885. *Source History: 1 in 1981; 1 in 1984; 1 in 1987; 2 in 1991; 1 in 2004.* **Sources: Bou.**

Violet - (Purple Broccoli) - Sometimes called purple broccoli because the heads are so irregular, curds are more like individual florets, tends to bolt quickly in warm weather, turns green when cooked. *Source History: 1 in 1987; 1 in 2004.* **Sources: Lej.**

Violetta Italia - 85 days - Large, vigorous plants produce large purple heads that blanch to green when cooked, mild broccoli-like flavor, attractive and easy to grow, Italy. *Source History: 1 in 1994; 1 in 1998; 6 in 2004.* **Sources: Ba8, Bo19, Com, La1, Sk2, WE10.**

Walcheren Winter Pilgrim - 300 days - Compact creamy white curds, excellent quality, winter hardy, spring harvest. *Source History: 1 in 2004.* **Sources: Tho.**

White Rock - (SG 112) - 69-77 days - Medium late maturity, medium sized plants, deep white firm 6-6.5 in. heads, excel. self-wrapping ability, Snowball type, vigorous,

PVP expired 2002. *Source History: 4 in 1981; 4 in 1984; 7 in 1987; 12 in 1991; 12 in 1994; 12 in 1998; 3 in 2004.*
Sources: RUP, SGF, Sto.

Varieties dropped since 1981 (and the year dropped):

Abundantia (1987), *Abuntia* (1991), *Alert* (2004), *Alpha* (1994), *Alpha Balanza* (1991), *Alpha Begum* (2004), *Alpha Paloma* (1994), *April* (1994), *Armado Clio* (1994), *Armado Quick* (1994), *Armado Spring (Blend)* (1991), *Autumn Giant 3* (1998), *Avalanche* (2004), *Batsman* (1994), *Blizzard* (1998), *Boshu Late* (1983), *Boshu Mid-Season* (1983), *Brio* (2004), *Bronze Leaved Italian Purple* (1998), *Cargill* (1998), *Casablanca* (2004), *Cervina* (1994), *Chartreuse* (1987), *Cloud Nine* (1994), *Corvilia* (2004), *Crystal* (1998), *Dania* (2004), *Danish Giant* (1991), *Danish Perfection* (2004), *Danova* (2004), *Delira* (1987), *Dok Elgon* (1998), *Dry Weather* (1987), *Dwarf* (1987), *Dwarf Erfurt* (1991), *Early Abundance* (1991), *Early Dominant* (1991), *Early March* (1994), *Early Purple Head* (1987), *Early Purple Sicilian* (2004), *Early Snowflake* (1991), *Early Tropical* (1991), *Elgon* (1983), *English Winter* (1994), *Erfu* (2004), *Erfurt, Holland Improved* (1994), *February* (1994), *Firstman* (1994), *Fleurly* (2004), *Flora Blanca 18* (2004), *Fluerly* (1994), *Garant* (1998), *Garant Dwarf* (1987), *Giant Vetus* (1983), *Green Ball Chartreuse* (1991), *Green Broccoli Type* (1998), *Grodan* (2004), *Harvester* (2004), *Harvester S* (2004), *Headman* (2004), *Hormade* (1998), *Ice Cap* (2004), *Idol* (2004), *Idol Original* (1987), *Igloory* (2004), *Imperial* (1987), *Imperial 10-6* (2004), *Imperial 10-6 Improved* (1987), *Imperial Special* (1987), *Inca* (2004), *Jura* (1994), *King* (2004), *Late Algiers* (1983), *Late Queen* (1991), *Lecerf* (1998), *Linas* (1994), *Maravilla Andina* (1991), *Master* (2004), *Matra* (1998), *Metropole* (1982), *Mini Snow (Trop. Snow 55)* (1984), *Monarch* (1994), *Monarch 73* (1991), *Mt. Hood* (1991), *New Early* (2004), *Newton Seale* (1991), *Nimba* (1984), *November-December* (1991), *Nozaki Early* (1983), *Olympus* (1994), *Panda* (1983), *Parnas* (1987), *Perron Special Early* (1987), *Pinnacle* (1994), *Pioneer* (1994), *Predominant* (1987), *Purple Giant* (2004), *Pybas No. 184* (1998), *Raket* (1991), *Royal Purple* (1991), *Selandia Osena* (1991), *Self-Blanche Improved* (1991), *Self-Blanche, Early* (1994), *SG 111* (1991), *SG 125* (1991), *SG 741* (1991), *Sierra Nevada* (2004), *Snow Giant* (1998), *Snow Peak* (1987), *Snow Star* (1994), *Snow White* (1987), *Snowball 15 (Mid Season)* (1983), *Snowball 16X* (1987), *Snowball 25* (1991), *Snowball 34* (1998), *Snowball 42* (1991), *Snowball 57 Improved* (1984), *Snowball 73* (1991), *Snowball 76* (1994), *Snowball 90* (2004), *Snowball 99* (1991), *Snowball 123* (2004), *Snowball 212* (1987), *Snowball 222* (1987), *Snowball 421* (1984), *Snowball 741* (1991), *Snowball A (Special Strain)* (1994), *Snowball D* (2004), *Snowball E* (1998), *Snowball M* (1987), *Snowball Monarch* (1987), *Snowball T-2* (2004), *Snowball T-2 Early* (1991), *Snowball T-3* (1998), *Snowball T-3 Early* (1994), *Snowball T-4* (1991), *Snowball T-6* (1987), *Snowball Y-76* (1994), *Snowball Z (E-0222)* (1994), *Snowball, Extra Early* (2004), *Snowball, Henderson's Early* (1994), *Snowball, Imperial 10-6* (1994), *Snowball, Pybas 184* (2004), *Snowball, Select Early* (1998), *Snowbird* (1982), *Snowflower* (2004), *Snowmound* (1991), *Solide* (1998), *St. Mark* (1987), *Starla* (2004), *Stovepipe* (1991), *Super Snowball Y650* (1983), *Suprimax* (1998), *T-2 Early* (2004), *Taipan* (1998), *Terezopolis* (1991), *Tornado* (1991), *Tropical 45 Days* (1987), *Tropical 55 Days* (1987), *Tropical Snow 55* (1983), *Veitch's Autumn Giant* (1994), *Verde Marchigano Smeraldo* (1987), *Vernon* (1998), *Vision* (1987), *Wallaby* (1998), *White Cloud* (1998), *White Diamond* (1994), *White Empress* (1987), *White Fox* (1998), *White Mountain* (1987), *White Pearl* (1991), *White Satin* (1998), *White Summer* (1998), *White Top* (2004), *Whitehorse* (1991), *Winner* (2004), *Zero* (2004).

Celeriac
Apium graveolens var. rapaceum

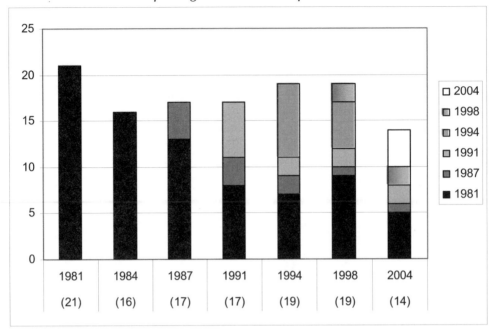

Number of varieties listed in 1981 and still available in 2004: 21 to 5 (24%).
Overall change in number of varieties from 1981 to 2004: 21 to 14 (67%).
Number of 2004 varieties available from 1 or 2 sources: 9 out of 14 (64%).

Alabaster - 120 days - Close relative of celery, forms fine-flavored bulbs, harvest at 2-4 in., white flesh, may be stored, discolors less when cooked, slow germination. *Source History: 11 in 1981; 9 in 1984; 5 in 1987; 4 in 1991; 5 in 1994; 5 in 1998; 2 in 2004.* **Sources: Loc, RUP.**

Brilliant - 100-110 days - Round smooth European str., white flesh, sweet, nutty flavor, less warted, keeps well, recommended for commercial purposes. *Source History: 2 in 1991; 5 in 1994; 7 in 1998; 9 in 2004.* **Sources: Ga22, Goo, OSB, Pe2, SE4, So12, Ter, Tu6, Ver.**

Celeriac - (Knob Celery, Turnip-Rooted Celery, Root Celery) - 112 days - Dev. large turnip-like roots, mound dirt around it, needs constant ample moisture, base for French cream of celery soup. *Source History: 12 in 1981; 11 in 1984; 10 in 1987; 9 in 1991; 11 in 1994; 8 in 1998; 3 in 2004.* **Sources: Ge2, Mel, Wet.**

Diamant - 110-120 days - No hollow crown, no internal browning, few off shoots, roots stand above ground and are easily harvested and cleaned. *Source History: 1 in 1991; 2 in 1994; 1 in 1998; 6 in 2004.* **Sources: Coo, Jo1, Ki4, Par, Ri12, SE4.**

Dolvi - 120-150 days - Large nearly round roots, white interior, fine texture and flavor either cooked or raw, multiple disease res. and vigor, excel. quality, stores well, versatile. *Source History: 2 in 1981; 1 in 1984; 1 in 1987; 1 in 1991; 1 in 1994; 1 in 1998; 1 in 2004.* **Sources: So12.**

Giant Prague - (Giant Smooth Prague, New Giant Prague, Fruilano) - 110-120 days - Turnip-rooted celery, culture and taste much like ordinary celery, large thick pure-white roots store well for winter use, hollow stalks are not used, for fresh use harvest at about 2 in., 2-4 in. roots are stored, used in soups and stews or boiled or sliced cold for salads, intro. in 1871. *Source History: 22 in 1981; 18 in 1984; 13 in 1987; 9 in 1991; 8 in 1994; 8 in 1998; 11 in 2004.* **Sources: Alb, Ba8, Ber, Bo19, Com, Dam, Ers, Fo7, Ont, Vi4, WE10.**

Large Smooth Prague - (Giant Smooth Prague, Large Prague, Smooth Prague, Turnip Rooted Celery) - 110-120 days - Large round evenly shaped roots about 4 in. across, turnip-shaped roots have flavor of celery, few rootlets, good winter keeper, roots eaten raw or cooked, pre-1870. *Source History: 15 in 1981; 13 in 1984; 14 in 1987; 11 in 1991; 10 in 1994; 11 in 1998; 9 in 2004.* **Sources: Ear, Gou, Jun, Lej, Pin, Se25, Shu, So1, Sto.**

Mentor - (Turnip Rooted Celery) - 110-115 days - Roots form an edible globe the size of a softball, nutty flavor, does not require mulching or blanching like celery. *Source History: 2 in 1998; 2 in 2004.* **Sources: Lin, We19.**

Monarch - 100-160 days - Improved Giant Prague type, autumn/early winter vegetable, very large roots with firm white flesh, do not discolor when cut or blanched, scab res. *Source History: 1 in 1987; 3 in 1991; 3 in 1994; 3 in 1998; 2 in 2004.* **Sources: Dam, Tho.**

Monostopalyi - Rare Hungarian variety, lumpy root is used in soups or grated raw, mild sweet celery flavor, dried for powder, requires constant moisture during growing season. *Source History: 1 in 2004.* **Sources: Ho13.**

Monstorpolgi - 100 days - Few side shoots, firm, large round roots. *Source History: 1 in 2004.* **Sources: Sa9.**

President - 110 days - Popular in Central European

kitchen gardens, trouble-free, slow grower. *Source History: 1 in 2004.* **Sources: Coo.**

Sedano R. Bianco Veneto - 90-95 days - White of Venice, grown for fiberless root, celery flavor, 3-4 in. dia., same storage as carrots. *Source History: 1 in 2004.* **Sources: Se24.**

Snow White - 100 days - Danish var. with well shaped, large, smooth, solid roots, 4-5 lbs., solid white interior, bolt res. *Source History: 1 in 1998; 1 in 2004.* **Sources: Ni1.**

Varieties dropped since 1981 (and the year dropped):

Alba (2004), *Arvi* (1994), *Balder* (2004), *Balder L.D.* (1982), *Bali* (2004), *Blanco* (2004), *Celery Root* (1991), *De Rueil* (1998), *Giant Prague HS III* (1983), *Globus* (1998), *Iram* (1991), *Jose* (2004), *Juvel* (2004), *Magdenburger* (1998), *Marble Ball* (1994), *Market Growers* (1991), *Monostorpalyi* (1994), *Paris Improved* (1991), *Prague* (2004), *Prague Model* (2004), *Snehvide* (2004), *Zwijndrechtse Zwindra* (1987).

Celery
Apium graveolens var. dulce

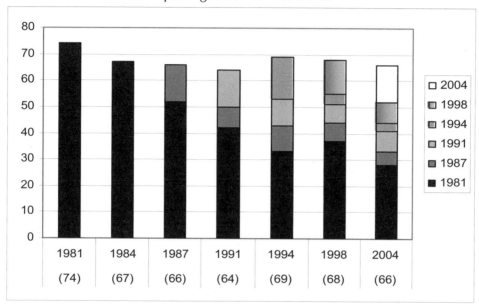

Number of varieties listed in 1981 and still available in 2004: 74 to 28 (38%).
Overall change in number of varieties from 1981 to 2004: 74 to 66 (89%).
Number of 2004 varieties available from 1 or 2 sources: 50 out of 66 (76%).

Afina - 60 days - Dutch cutting celery noted for its sturdy dark green foliage and rich aroma and taste, regenerates quickly after cutting, both the thin hollow stems and leaves are used as flavoring, leaves may be dried. *Source History: 1 in 1994; 1 in 1998; 1 in 2004.* **Sources: Fe5.**

Brydon's Prize Red - Heirloom gourmet cutting celery, dark red color from the crown, fine foliage, rare. *Source History: 1 in 2004.* **Sources: Syn.**

Celebration - 100-110 days - Grows 18-22 in. tall, dark green, broad stems, flavorful, hardy. *Source History: 1 in 2004.* **Sources: Se7.**

Celeri a Couper - Delicate leaves and stems used like parsley for soup bunches or salad, sweet flavor, known as Amsterdamse Donkeh G'roene in Holland. *Source History: 2 in 2004.* **Sources: In8, Se17.**

Celery Leaf - 90 days - Grown for its tasty fragrant leaves, has a small stalk, shorter and bushier than common celery, cut and come again if you leave some leaves. *Source History: 1 in 1981; 1 in 1984; 1 in 1987; 1 in 1998; 1 in 2004.* **Sources: Gou.**

Chinese Celery, Golden - 60 days - Thin long stalks, small leaves, very tasty, excellent for cooking, soup and salad, suitable for planting all year round. *Source History: 1 in 1987; 1 in 1991; 2 in 1994; 2 in 1998; 2 in 2004.* **Sources: Ev2, Ho13.**

Chinese Green Giant - 95 days - Medium-tall plant with many upright leaves, fine flavor, blight res. *Source History: 1 in 1991; 1 in 1994; 1 in 1998; 1 in 2004.* **Sources: La1.**

Cutting Leaf Celery - (Leaf Celery, Soup Celery, Zwolsche Krul, Kintsai, Chinese Celery) - 80 days - Almost wild small European celery, hollow brittle stalks, stronger and more pungent, leaves flavor soups and stews, keeps producing after cutting. *Source History: 3 in 1981; 3 in 1984; 6 in 1987; 7 in 1991; 5 in 1994; 10 in 1998; 12 in 2004.* **Sources: Com, Coo, Dam, Ho13, Jo1, Jun, Orn, Pin, Ri2, Sa5, Sa12, Ver.**

EA Special Strain - Tall Utah type developed from Ventura, upright glossy bright green stalks over 1 ft. long, better developed hearts, high yields, good disease res., slow to bolt. *Source History: 1 in 1994; 1 in 1998; 1 in 2004.* **Sources: Bou.**

Early Dell - 85-110 days - Grows 16-20 in., sweet flavor, good disease resistance, slow bolting. *Source History: 1 in 2004.* **Sources: Se7.**

E-Z Leaf - Resembles curly leaf parsley, 12 in. plant, an easy to add celery flavor when cooking. *Source History: 1 in 1998; 1 in 2004.* **Sources: Par.**

Florida 683 - (Marvelous) - 90-120 days - For East Coast and South, resembles Utah 52-70 but darker green, fairly smooth ribs, res. to RR and yellow and mosaic, boron deficiency tol., cylindrical, in Florida plants are 22-26 in., 10.5-11 in. ribs from butt to joint, fine heart formation, less tendency toward transverse cracking, fall and winter crops, FL/AES, 1964. *Source History: 14 in 1981; 14 in 1984; 14 in 1987; 10 in 1991; 8 in 1994; 6 in 1998; 2 in 2004.* **Sources: HO2, Sto.**

Florida 683 K Strain - 120 days - Sel. from Florida 683 for tol. to fusarium, compact cylindrical med-deep green 22 in. plant, 9.5 in. rib length. *Source History: 2 in 1981; 2 in 1984; 2 in 1987; 3 in 1991; 2 in 1994; 4 in 1998; 1 in 2004.* **Sources: RIS.**

Florida 683 Utah Type - Resembles Utah 52-70 but is somewhat smoother and has a fuller heart. *Source History: 1 in 1998; 1 in 2004.* **Sources: Sh9.**

Fordhook - (Emperor, Utah, Hauser) - 125-130 days - Fine large winter var., dwarf stocky dark green plants, smooth solid broad stalks blanch easily, string-free, disease and insect res. *Source History: 10 in 1981; 8 in 1984; 5 in 1987; 2 in 1991; 2 in 1994; 3 in 1998; 1 in 2004.* **Sources: Sh9.**

Fordhook, Burpee's - 130 days - Fall and winter green celery, stocky compact 15-18 in. plants, stalks blanch to silvery-white with large full tightly-folded hearts, fall use and winter storage. *Source History: 1 in 1981; 1 in 1984; 3 in 1987; 1 in 1991; 1 in 1994; 1 in 1998; 1 in 2004.* **Sources: Be4.**

French Dinant - (Chinese Celery) - 150 days - Unique cutting celery, forms a dense clump of narrow thin green stalks, stronger flavor for seasoning, res. to light frosts, dry for winter use. *Source History: 4 in 1981; 4 in 1984; 3 in 1987; 2 in 1991; 4 in 1994; 3 in 1998; 1 in 2004.* **Sources: Ni1.**

Galaxy - (Lathom Self Blanching) - 120 days - Thicker stems stay stringless longer, can be sown earlier without bolting to seed, stands ready to harvest for a long time. *Source History: 1 in 1991; 1 in 1994; 1 in 2004.* **Sources: Tho.**

Giant Pascal - 90-140 days - Standard late green celery, large heart blanches to creamy-white, thick round 8 in. stalks, large stocky dark-green 24 in. plants, for fall use or winter storage, improved Utah str., blight res., tender brittle good quality hearts, not a good shipper because of tender stalks. *Source History: 36 in 1981; 36 in 1984; 28 in 1987; 20 in 1991; 14 in 1994; 14 in 1998; 12 in 2004.* **Sources: Bo19, Com, Ers, Ge2, GLO, Gou, HA3, MIC, Sa9, SGF, Shu, Tt2.**

Giant Pascal Select Strain - Tall strain grows 28-30 in., good heart development, better rib strength than other Utah strains, highly res. to "brown check". *Source History: 1 in 1998; 1 in 2004.* **Sources: Sh9.**

Giant Red - 120 days - Very cold hardy, light-green tinged with purple, if earthed up as it grows will bleach a lovely shell-pink, hardy to Zone 3. *Source History: 1 in 1981; 1 in 1984; 2 in 1987; 1 in 1994; 3 in 1998; 1 in 2004.* **Sources: Ter.**

Giant Red Reselection - Large red stalks, beautiful yellow-pink hearts excel. for salads, good soup celery, reselected by CV Sh12 for better color, cold hardiness and disease res. *Source History: 2 in 2004.* **Sources: In8, Sh12.**

Golden Pascal - 114 days - Pale yellow-green stalks to 20 in. *Source History: 1 in 1991; 1 in 1994; 3 in 1998; 3 in 2004.* **Sources: Ba8, Mey, SE14.**

Golden Plume - (Wonderful, Stoke's Golden Plume) - 85-90 days - Early yellow market celery, full compact hearts blanch easily to clear-yellow, good sized stalks, fine quality, also home gardens, intro. by Ferry Morse in 1937. *Source History: 4 in 1981; 3 in 1984; 2 in 1987; 1 in 1991; 2 in 1994; 2 in 1998; 1 in 2004.* **Sources: Sto.**

Golden Self-Blanching - (American or Fall Dwarf Strain, Burpee's Golden Self-Blanching, Gigante Dorato) - 80-120 days - Fall dwarf str., 20-28 x 4.5 in. plants, full-hearted, pale-yellow/green, blanches clear and waxy, thick heavy stringless 9 in. ribs, extra large, nutty tasting base, no trenching needed, disease res., for home and market, intro. 1886. *Source History: 51 in 1981; 40 in 1984; 51 in 1987; 45 in 1991; 42 in 1994; 39 in 1998; 46 in 2004.* **Sources: Alb, All, Be4, Ber, Bo18, Bou, Bu3, Bu8, But, CA25, Com, Cr1, Dam, Ear, Eo2, Ers, Fo7, Ge2, Gr27, Gr29, GRI, HA3, He8, He17, HPS, Jor, LO8, MAY, ME9, MIC, Mo13, Ni1, Or10, Pin, Ra5, Se7, Se24, Se26, Sh9, So1, Und, Vi4, Vi5, WE10, Wet, Wo8.**

Golden Self-Blanching, Tall - (Shumway's Golden Self Blanching) - 85 days - Popular early market type, 20-30 in. plants, med-thick stalks, large stringless close set ribs, blanches readily to golden-yellow, delicate flavor. *Source History: 5 in 1981; 5 in 1984; 3 in 1987; 1 in 1991; 1 in 1994; 1 in 1998; 1 in 2004.* **Sources: Shu.**

Golden Yellow Self-Blanching - (Golden Yellow, Golden Self-Blanching Tall, Tall Golden Self-Blanching Perron's Strain) - 82-85 days - Med-thick solid 7 in. stalks blanch to yellow, 20 in. plants, home gardens. *Source History: 5 in 1981; 5 in 1984; 4 in 1987; 5 in 1991; 4 in 1994; 4 in 1998; 2 in 2004.* **Sources: GLO, Mel.**

Green Bay - 95 days - Compact plant, resists fusarium. *Source History: 1 in 1998; 1 in 2004.* **Sources: RIS.**

Green Pascal - 80-85 days - Dark green pascal type with outstanding flavor. *Source History: 1 in 1998; 3 in 2004.* **Sources: Gr27, Se24, WE10.**

Hauser - Excellent winter variety, compact green stalks, fine flavor, tender and crisp. *Source History: 1 in 2004.* **Sources: Ont.**

Kan Tsai - 60 days - Strong flavored Chinese celery, Asiatic origin, small tough plants with slender dark-green stems, grows quickly, easy to grow, widely adaptable. *Source History: 2 in 1981; 2 in 1984; 1 in 1987; 2 in 1991; 2 in*

1994; 2 in 1998; 2 in 2004. **Sources: Hig, Roh.**

Kintsai - (Kintsai Oriental Wild, Japanese Serina, Kin Tsai) - 60 days - Slender dark-green stems and lacy celery-like leaves, strong sharp celery flavor, quick and easy to grow, widely used in Oriental cooking, good for salads or soups or pickling or with meat, popular for its unique aroma and vivid green color when cooked. *Source History: 1 in 1981; 1 in 1984; 2 in 1987; 2 in 1991; 1 in 1994; 2 in 1998; 3 in 2004.* **Sources: Bo17, Kit, ME9.**

Par-Cel - 72 days - Unique Dutch celery with more delicate flavor than other varieties, looks like triple curled parsley, withstands heat and drought, cold tolerant, 18th century heirloom. *Source History: 1 in 1991; 4 in 1994; 3 in 1998; 4 in 2004.* **Sources: Coo, Fe5, Sa5, Tho.**

Paris Golden - (Yellow Self-Blanching) - 115 days - Stalks grow 18-20 in. tall, blanches easily to golden-yellow, both stem and heart are tender and delicious and free from string. *Source History: 1 in 1981; 1 in 1984; 2 in 1987; 1 in 1991; 1 in 1994; 1 in 1998; 1 in 2004.* **Sources: Ont.**

Pascal Giant - *Source History: 1 in 2004.* **Sources: Se17.**

Peto 285 - 98 days - Used for main season crop, dark green color, slightly taller than Utah 52-70, PVP 1994 Seminis Vegetable Seeds, Inc. *Source History: 1 in 1998; 1 in 2004.* **Sources: Sto.**

Picador - (Exp. A-865) - 95 days - Fairly ribbed, tol. to pithiness, res. to fusarium yellows, susceptible to CMV. *Source History: 2 in 1991; 3 in 1994; 5 in 1998; 1 in 2004.* **Sources: Sto.**

Pink - Imported from England, color remains pink even after cooking, blanches fairly easily and quickly, very hardy, will withstand some late frosts, 1894. *Source History: 2 in 1981; 2 in 1984; 2 in 1987; 2 in 1991; 3 in 1994; 3 in 1998; 1 in 2004.* **Sources: Ho13.**

Pink Plume - 110 days - English heirloom, large stems are tinged with a delicate pink, softening into ivory-white, very crisp, sweet and tender with a rich, nutty flavor. *Source History: 2 in 1998; 2 in 2004.* **Sources: Ho13, Shu.**

Promise - PVP 1994 The Regents of the U of CA. *Source History: 2 in 1998; 1 in 2004.* **Sources: CO23.**

PYC 9112 - 95 days - Improved XP-85 type, uniform size, high yields, resists fusarium. *Source History: 1 in 2004.* **Sources: RIS.**

Red - Retains red color when cooked, needs frost for color, flavorful, heirloom. *Source History: 1 in 1981; 1 in 1984; 1 in 1987; 1 in 1991; 2 in 1994; 3 in 1998; 2 in 2004.* **Sources: Hud, Sa5.**

Red Stalk - 80-85 days - Old English cultivar that used to be limited to the gardens of the wealthy, 14-25 in. biennial gives abundant yields of spicy celery seed, stalks stay red when cooked. *Source History: 1 in 1987; 3 in 1991; 2 in 1994; 5 in 1998; 5 in 2004.* **Sources: Ho13, Roh, Se7, So9, So12.**

Red Stalk, Giant - 90-100 days - Stalks to 10-15 in., slight red blush on inside of stalks, crisp and juicy, long-standing, heirloom. *Source History: 1 in 2004.* **Sources: Se7.**

Redventure - Giant Red x Ventura cross dev. by CV Sh12, the most adaptive and resilient green celery available to the fresh market grower, dark to brilliant red stalks, emerald green leaves. *Source History: 1 in 2004.* **Sources: Sh12.**

Sabroso - 95-100 days - Improved Picador type, thick plants, long straight petioles, medium smooth, fresh market or processing, yellows tolerant, PVP. *Source History: 2 in 2004.* **Sources: RIS, Sto.**

Solid White - Plants grows to 16-20 in., crisp solid tender stems, fine flavor, known before 1877. *Source History: 1 in 1981; 1 in 1984; 1 in 1987; 1 in 1991; 1 in 1994; 1 in 1998; 1 in 2004.* **Sources: Ho13.**

Starlet - (Starlet RS) - 98-120 days - Special disease tol. strain of Tall Utah 52-70R, res. to both races of fusarium, plants 25-27 in. tall, few basal buds, smooth, med. dark green ribs. *Source History: 2 in 1991; 1 in 1994; 1 in 1998; 1 in 2004.* **Sources: Twi.**

Summer Pascal - (Tall Fordhook) - 110-130 days - Summer Pascal str. for the East, dark-green, upright, 24 in., full heart, bolt and blight res., 9 in. to first joint, U of MA, for upland or muck. *Source History: 11 in 1981; 9 in 1984; 7 in 1987; 4 in 1991; 3 in 1994; 1 in 1998; 1 in 2004.* **Sources: Sh9.**

Tehama Sunrise - 85-90 days - Heat tolerant strain adapted to Sacramento Valley climate, robust, hardy deep green stalks. *Source History: 1 in 1994; 2 in 1998; 3 in 2004.* **Sources: Se17, So12, Syn.**

Tendercrisp - 90-105 days - Giant Pascal type, 24-26 in. plant, compact massive head, smooth tall thick dark-green stalks, 11-12 in. to first joint, high yields, good home garden variety. *Source History: 7 in 1981; 7 in 1984; 7 in 1987; 5 in 1991; 2 in 1994; 3 in 1998; 4 in 2004.* **Sources: Ba8, Ers, He8, WE10.**

Utah - (Utah Golden Crisp, Golden Crisp, Tall Utah, Utah Tall Green, Utah Green) - 70-125 days - Stringless green celery, tightly folded full hearts, broad thick well-rounded solid stalks, blanches pure-white and crisp, grows vigorous without getting punky, popular tall green celery, for late fall use and winter storage. *Source History: 18 in 1981; 15 in 1984; 15 in 1987; 13 in 1991; 12 in 1994; 9 in 1998; 8 in 2004.* **Sources: Alb, Be4, Ear, Fis, La1, Pe2, Ri12, TE7.**

Utah 15, Salt Lake - 115 days - Excel. green variety for fall use, well rounded thick stems, rich nutty flavor, relatively early. *Source History: 1 in 1981; 1 in 1984; 1 in 1987; 1 in 1991; 1 in 1994; 1 in 1998; 1 in 2004.* **Sources: Lin.**

Utah 52-70 - (Tall Utah 52-70, Utah Giant 52-70, Utah Improved 5270) - 90-130 days - Tall med-dark green stalk and foliage, upright compact habit, green Pascal str., 30 in. overall height, 10-11 in. to first joint, good heart development, strong roots, will bolt in cold weather, can absorb more magnesium from the soil, home garden or commercial, 1953. *Source History: 30 in 1981; 29 in 1984; 24 in 1987; 18 in 1991; 15 in 1994; 17 in 1998; 14 in 2004.* **Sources: Bo17, CA25, CH14, CO23, DOR, Hal, Hum, Kil, Mey, MIC, MOU, Pin, Sau, Ves.**

Utah 52-70 Improved - (Tall Utah 52-70 Improved) - 98-120 days - Vigorous strain of Utah often used by commercial celery growers near Oxnard, CA, tall dark-green plants, thick smooth stalks 10-12 in. long to the first joint if well grown, sweet without blanching, brown check res., seems to be tol. to some strains of Western celery mosaic. *Source History: 3 in 1987; 3 in 1991; 1 in 1994; 2 in 1998; 3 in 2004.* **Sources: HO2, Twi, WI23.**

Utah 52-70 Improved, Stokes - 98 days - Improved strain sel. from Utah 52-70, earlier and darker-green and more uniform, stalks form a more compact group, pithiness or punking out has been reduced. *Source History: 1 in 1981; 1 in 1984; 1 in 1987; 1 in 1991; 1 in 1994; 1 in 1998; 1 in 2004.* **Sources: Sto.**

Utah 52-70 Triumph - Dark-green var. with sturdy foliage and long thick fleshy stalks, holds color in the field. *Source History: 1 in 1991; 1 in 1994; 1 in 1998; 1 in 2004.* **Sources: SE4.**

Utah 52-70H - (Utah 52-70H Improved, Utah 52-70H - Tall Green, Tall Utah 52-70H) - 90-120 days - Improved strain of Tall Utah, more res. to celery diseases than other strains, long thick stalks often 10-11 in. to first joint, good heart development, strong roots, will bolt in cold weather. *Source History: 4 in 1981; 3 in 1984; 2 in 1987; 2 in 1991; 1 in 1994; 1 in 1998; 1 in 2004.* **Sources: Mo13.**

Utah 52-70R - Compact plants with broad thick dark-green stalks that can be used green or blanched to white by tying up and covering with soil up to leaves, vigorous transplants, maintains good quality late in harvest season. *Source History: 1 in 1987; 3 in 1991; 3 in 1994; 3 in 1998; 4 in 2004.* **Sources: Dam, Ers, Si5, Wi2.**

Utah 52-70R Improved - (Tall Utah 52-70R Improved) - 82-150 days - Green Pascal type for muck and mineral soils, taller 24 in. plants, longer ribs and higher rib count, bolts in cold in early stages, res. to brown cheek and Western celery mosaic, dark-green glossy 11-14 in. stalks, not ordinarily blanched. *Source History: 18 in 1981; 16 in 1984; 17 in 1987; 18 in 1991; 18 in 1994; 18 in 1998; 12 in 2004.* **Sources: Bu2, Bu3, CH14, CO30, Gr27, Jor, Loc, ME9, Roh, RUP, SE14, WE10.**

Utah Tall Green - 120-125 days - Large 20-24 in. plants, erect and full hearted, excel. disease res., recommended for Northern areas. *Source History: 1 in 1981; 1 in 1984; 1 in 1987; 1 in 1991; 2 in 1994; 2 in 1998; 1 in 2004.* **Sources: He17.**

Ventura - 52-100 days - Early Tall Utah type with more upright growth and better developed hearts, 28-31 in. plant, glossy bright-green stalks 12-13 in. long, extra heavy trimmed weight, high-yielding, widely adapted, tol. to some strains of Fusarium yellows, bolt res., PVP expired 2003. *Source History: 6 in 1987; 8 in 1991; 12 in 1994; 11 in 1998; 10 in 2004.* **Sources: Be4, Fe5, Ga1, Jo1, Na6, St18, Sto, Ter, Tu6, We19.**

White Queen - 60-65 days - Long white stalks, jagged green leaves, small seed, excellent quality for salads and cooking. *Source History: 2 in 2004.* **Sources: Kit, Orn.**

Wild Irish - Known as smallage in Ireland where it still grows wild in the country, small dark green plants grow to 12 in. tall, very strong celery flavor, good in soups. *Source History: 1 in 2004.* **Sources: Ho13.**

XP 266 - 95 days - Compact plant, resists fusarium. *Source History: 1 in 2004.* **Sources: RIS.**

Zwollsche Krul - Old Dutch variety, curly leaf strain with extra dark-green leaves, distinct celery flavor, used for flavoring and garnishes. *Source History: 1 in 1991; 2 in 2004.* **Sources: Ri12, Se17.**

Varieties dropped since 1981 (and the year dropped):

American (1984), *Amsterdam Fine* (2004), *Bishop* (1991), *Calmiro* (1991), *Celeri Dore* (1987), *Celeri Vert* (1983), *Chinese Celery, Shicun* (2004), *Clean-Cut* (1987), *Conquistador* (2004), *Cornell 619* (1991), *Cornell No. 19* (1991), *Curly* (1991), *Earlibelle* (1987), *Emerson Pascal* (1983), *Empire* (1998), *Florida 2-13* (1987), *Florida 693* (1987), *Florigreen* (1984), *Florimart 19* (1987), *FM-1218* (1991), *Fordhook Emperor* (1991), *Fordhook Giant* (2004), *Giant Pascal (Dwarf Strain)* (1987), *Giant Pink* (1998), *Giant White* (1998), *Golden Detroit* (1998), *Golden Detroit Self-Blanching* (1991), *Golden Medium* (2004), *Golden Self-Blanching Extra* (1987), *Golden Self-Blanching, Dwarf* (2004), *Golden Self-Blanching, Select* (1998), *Golden Spartan* (1998), *Grande* (1994), *Greensleeves* (1998), *Greensnap* (1983), *Hercules* (1998), *Heung Kunn* (1994), *Hopkins Fenlander* (2004), *Ivory Tower* (2004), *June Belle* (2004), *Lathom Self-Blanching* (1983), *M-9* (2004), *Martha Warde's* (2004), *Matador* (2004), *Monarch Golden Self-Blanch. 94* (1987), *Ponderosa* (1998), *Prize Pink* (1987), *PS 285* (2004), *PS 28588* (1998), *Smallage* (1998), *Smooth-Leaved* (1987), *Solid Red* (1998), *Starlet RS* (1998), *Strain 2-14* (1991), *Summertime* (2004), *Summit* (2004), *Superdora* (1998), *Supreme Golden* (1987), *Surepak* (1991), *Tall Fordhook* (1987), *Tall Green* (1998), *Triumph* (1994), *Utah 15 (Green)* (1991), *Utah 52-70H Improved* (1987), *Utah 52-70HK Improved* (2004), *Utah 52-70HK Strain, Tall* (2004), *Utah 52-75, Tall* (2004), *Utah Early Green* (1998), *Utah Improved* (1994), *Utah Salt Lake Select* (1991), *Verde a Costa Piena* (1994), *Waltham Summer Pascal* (1987), *Wensleydale Solid Giant White* (1982).

Chicory
Cichorium intybus

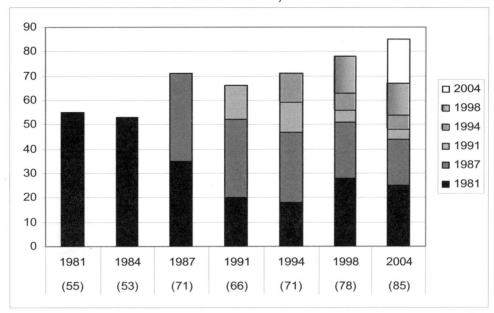

Number of varieties listed in 1981 and still available in 2004: 55 to 25 (45%).
Overall change in number of varieties from 1981 to 2004: 55 to 85 (155%).
Number of 2004 varieties available from 1 or 2 sources: 62 out of 85 (73%).

Asparagus Chicory - (Italian Dandelion) - 65-75 days - Plant sends up many leafy thick shoots, used in salads or cooked as an asparagus substitute, excel. flavor, some prefer it to asparagus, rich in vitamins and iron. *Source History: 2 in 1981; 2 in 1984; 1 in 1987; 2 in 1991; 1 in 1994; 1 in 1998; 1 in 2004.* **Sources: Com.**

Augusto - 70-80 days - Deep burgundy-red head, frost tol., bolt res., plant mid to late summer. *Source History: 1 in 1987; 5 in 1991; 4 in 1994; 1 in 1998; 1 in 2004.* **Sources: Se26.**

Barba di Capuccino - (Barbe de Capucin, Monk's Beard) - A forcing chicory, long leaves deeply cut like a dandelion, for salads or cooking, best sown in fall, also works in spring. *Source History: 2 in 1987; 1 in 1991; 1 in 1994; 1 in 1998; 2 in 2004.* **Sources: Lej, Se24.**

Baxter's Special - 70 days - Reselected San Pasquale type, dev. in the Texas market, heavy thick darker colored leaves, plants more uniform and vigorous, quite tasty, bunching and greens. *Source History: 2 in 1981; 2 in 1984; 2 in 1987; 1 in 1994; 1 in 1998; 2 in 2004.* **Sources: CO23, RIS.**

Bianca a Bergamo sel. Franchi - Nice heads formed by light green leaves, white stems, tasty, Franchi Special Sel. *Source History: 1 in 2004.* **Sources: Se24.**

Bianca di Milano - 75 days - Large dark-green leaved plant forces a large tuft of creamy-white leaves from which it gets its name Bianca or White of Milan, excellent radicchio for salads and seafoods, sow in May for autumn and winter crops. *Source History: 2 in 1987; 1 in 1991; 1 in 1994; 2 in 1998; 2 in 2004.* **Sources: Ber, We19.**

Bionda a Foglie Larghe - Light green cutting chicory, tall rounded leaves with smooth edge, nice mild flavor, for salads, sow spring, summer or fall. *Source History: 1 in 2004.* **Sources: Se24.**

Brindisina - Grown for the stems like puntarelle with thicker stems, white bulb at the base, great in salads. *Source History: 1 in 2004.* **Sources: Se24.**

Carmen - 75-110 days - Improved Marina or Chioggia type, crimson heads, white veins, 5.5 in. dia. *Source History: 1 in 1994; 3 in 1998; 3 in 2004.* **Sources: Ga1, Hig, Sto.**

Castelfranco - (Radicchio Castelfranco) - 85-95 days - Italian heirloom rarely seen in the markets, marbled red and white coloring is quite striking in salads, heads do not need cutting back to produce well. *Source History: 3 in 1981; 2 in 1984; 4 in 1987; 2 in 1991; 1 in 1994; 3 in 1998; 2 in 2004.* **Sources: Orn, Se24.**

Castelfranco Libra - Improved selection, cream colored round heads splashed with wine-red, popular Italian variety. *Source History: 1 in 2004.* **Sources: Ba8.**

Castelfranco Variegata - (Variegata di Castelfranco) - 85 days - Striped chicory from northern Italy, variegated green and red and white and yellow crumpled foliage, tender slightly bitter rounded leaves form loose heads, blanch like celery, needs protection from sun, for late autumn harvest. *Source History: 4 in 1981; 1 in 1984; 4 in 1987; 2 in 1991; 1 in 1994; 2 in 1998; 4 in 2004.* **Sources: Ber, Coo, Ho13, Se26.**

Catalogna - (Radichetta, Italian Dandelion, Cicoria Catalogna, Asparagus Chicory, Catalogna a Foglia Liscia) - 52-83 days - Leaves and seed stalks used for early greens, asparagus-like flavor, deep-cut long broad dandelion-like leaves, rapid grower, produces tender spring stalks from seed sown the previous summer. *Source History: 20 in 1981; 18 in 1984; 19 in 1987; 17 in 1991; 15 in 1994; 18 in 1998; 27 in 2004.* **Sources: Ba8, Bou, CA25, CO23, Com, DOR, Ers, Fo7, HA3, He8, Hud, La1, Loc, ME9, Mey, Ont, Ri12, Roh, Se8, SE14, Se26, So1, St18, Sto, Syn, Tho, We19.**

Catalogna a Foglie Frastagliata - 65 days - Straight, upright, long, white ribbed leaves with deeply indented edges, grows 12 in. tall, forms heavy bunches. *Source History: 1 in 1987; 1 in 1991; 1 in 1994; 2 in 1998; 3 in 2004.* **Sources: Ber, Orn, Ter.**

Catalogna Pugliese - Tall open plant, green leaves with long white stems, fair amount of serrated leaves, fresh or salad, best for fall. *Source History: 1 in 2004.* **Sources: Se24.**

Catalogna Special - 40-48 days - Can be harvested 3-4 weeks after transplanting or left to grow into heavy 18 in. tall bunches, long deep-green slender deeply-cut leaves with white midribs, slow bolting. *Source History: 1 in 1991; 1 in 1994; 1 in 1998; 2 in 2004.* **Sources: Dam, Jo1.**

Ceriolo - (Grumulo) - 120 days - Green heading chicory, outer leaves curl outwards, inner leaves form upright rosette, tapering leaf bases are very light-green to white, attractive, maturity is from fall planting. *Source History: 1 in 1981; 1 in 1984; 2 in 1987; 1 in 1991; 1 in 1994; 1 in 1998; 1 in 2004.* **Sources: Coo.**

Chicory - (Italian Dandelion, Common, Coffee Rooted) - 110 days - Flowers and young leaves used as greens, coffee substitute from dried ground-up roots, tangy leaves flavor soups and sauces and salads. *Source History: 6 in 1981; 5 in 1984; 8 in 1987; 7 in 1991; 5 in 1994; 8 in 1998; 8 in 2004.* **Sources: Cr1, Eo2, Fo7, Lej, Pe2, Se7, Sh12, Vi4.**

Chioggia - (Variegata di Chioggia) - 85 days - Heading type, variegated red and white, dev. from Castelfranco, tighter head, needs no forcing period to stimulate head formation, plant late summer. *Source History: 2 in 1981; 2 in 1984; 6 in 1987; 2 in 1991; 2 in 1994; 3 in 1998; 5 in 2004.* **Sources: Ber, Coo, Ont, Pin, Se24.**

Chioggia, Precose - 60-70 days - Forms compact crunchy purple-red heads without being cut back. *Source History: 1 in 1991; 1 in 1994; 1 in 1998; 1 in 2004.* **Sources: Jo1.**

Crystal Hat - 70 days - Non-forcing chicory, long oval heads look like romaine lettuce, easy culture, high yields, withstands summer heat and fall frosts, sweet tangy flavor, can store. *Source History: 3 in 1981; 2 in 1984; 2 in 1987; 1 in 1991; 1 in 1994; 1 in 1998; 1 in 2004.* **Sources: Ni1.**

Cuor d'Oro, Golden Heart - Frisee type with very frilly leaves, early, full head, large plant, use fresh and cooked, to blanche the heart, 10-12 days before harvest tie leaves up with twine or cover with a pail or pan. *Source History: 1 in 2004.* **Sources: Se24.**

Dandy Red - 35-55 days - Italian dandelion with serrated leaves with deep dark red midribs, higher percentage of darker reds, less pink and white, fewer leaf hairs and less toothiness than common red varieties from Italy, from John Navazio of Seed Movement. *Source History: 4 in 2004.* **Sources: Ba8, Fe5, Se27, So1.**

Della Catalogna - (Radichetta) - 49 days - Italian dandelion with bluish serrated leaves, heavy texture, bitter taste, grown as a fall crop, Italians chop the leaves, steam and saute in olive oil. *Source History: 1 in 1994; 1 in 1998; 1 in 2004.* **Sources: Fe5.**

Dentarella - (Asparagus Chicory, Catalogna Chicory) - 65 days - Italian stem chicory similar to Puntarella but with a straighter stem. *Source History: 1 in 1987; 1 in 1991; 1 in 1994; 1 in 1998; 1 in 2004.* **Sources: Coo.**

Di Bruxelles - *Source History: 1 in 1987; 1 in 1991; 1 in 1994; 1 in 1998; 1 in 2004.* **Sources: Ber.**

Early Treviso - 80 days - Selected from Rossa di Treviso for more precocious growth, can be seeded in mid-July after the days have begun to shorten noticeably when less likely to bolt, if heads do not form by Labor Day it should be cut back to an inch above the crown at the beginning of cool weather. *Source History: 1 in 1991; 1 in 1994; 1 in 1998; 1 in 2004.* **Sources: Coo.**

Frisee - 75 days - Beautiful, finely cut, frilly leaves form a nice head. *Source History: 1 in 1998; 1 in 2004.* **Sources: Tho.**

Gallatina - 55-65 days - The pre-flowering stalks are eaten like broccoli. *Source History: 1 in 1998; 1 in 2004.* **Sources: So12.**

Greenlof (Sugar Hat Chicory) - Combines the qualities of endive and witloof, can be used raw in salads or cooked as endive, sow in May or June for heads in the fall. *Source History: 1 in 1981; 1 in 1984; 1 in 1987; 1 in 1991; 1 in 1994; 1 in 1998; 1 in 2004.* **Sources: Dam.**

Grumolo Bionda - Forms a round light green cluster of leaves, for fall growing. *Source History: 1 in 1987; 1 in 1991; 1 in 1994; 1 in 1998; 1 in 2004.* **Sources: Se24.**

Grumolo Biondo Golden - Popular baby salad item in Italy, small rounded rosette head with thick, gold-green leaves, heirloom from the Piedmont region. *Source History: 1 in 2004.* **Sources: Ba8.**

Grumolo Verde - (Grumolo Dark Green) - Extremely cold-hardy chicory from the Piedmont region of Italy, green rounded leaves, sow mid to late summer, harvest late autumn to spring when heads form ground-hugging rosettes, plants will resprout if heads are cut rather than dug. *Source History: 2 in 1987; 4 in 1991; 3 in 1994; 3 in 1998; 4 in 2004.* **Sources: Ber, Hud, Ni1, Se24.**

Italian Dandelion, Red - 65 days - Deep green leaves with bright red stems, baby leaves add a tangy flavor to salads or good as a cooked green. *Source History: 3 in 2004.* **Sources: Dam, TE7, We19.**

Italico Rosso - 65 days - Rare red form of Dentralla chicory, mistakenly known as Red Italian Dandelion. *Source History: 1 in 2004.* **Sources: Coo.**

Magdeburg - (Coffee Chicory, Cicoria Siciliana, Large Rooted Magdeburg, Magdeburg) - 100-110 days - Med-green 15 in. tall plants, tangy dandelion-like foliage used as greens or for flavoring, 12-16 x 2 in. tapered roots are roasted and ground for a coffee substitute. *Source History: 18 in 1981; 15 in 1984; 13 in 1987; 9 in 1991; 7 in 1994; 7 in 1998; 4 in 2004.* **Sources: Loc, Ri2, Se25, Sto.**

Mantovana - Medium size compact head with green leaves, resembles a small iceberg lettuce, best grown as a fall crop. *Source History: 1 in 1987; 1 in 1991; 1 in 1994; 1 in 1998; 1 in 2004.* **Sources: Coo.**

Marina - (Chioggia Race Marina) - 90-110 days - Compact red heads 8-12 oz., sharp flavor, sow May-June, harvest in fall before frost, may bolt if sown too early. *Source History: 2 in 1987; 3 in 1991; 2 in 1994; 1 in 1998; 1 in 2004.* **Sources: Roh.**

Mechelse - Witloof or French endive, early, for forcing inside, after hard frost cut off growth to within 1 in. of base, bring inside and cover with 6 in. of sand or soil, plant roots upright in soil of pit or box, keep dark, forcing time depends on temperature and variety used, ready when tips show through soil. *Source History: 1 in 1981; 1 in 1984; 1 in 2004.* **Sources: Dam.**

Montmagny - Refined version of wild chicory, medium-green finely cut leaves, best for overwintering. *Source History: 1 in 1994; 1 in 1998; 2 in 2004.* **Sources: CO30, VI3.**

Palla di Fuoco Rossa - 65-80 days - Chioggia type, round heading radicchio from Italy, deep burgundy heads. *Source History: 1 in 1981; 1 in 1984; 1 in 1987; 1 in 1991; 1 in 1998; 1 in 2004.* **Sources: Fe5.**

Palla Rossa - 90-100 days - Heads have dark-green leaves outside and red inside, pure-white ribs, plant during summer, but susc. to intense heat, no forcing needed, easy to grow. *Source History: 1 in 1981; 1 in 1984; 3 in 1987; 2 in 1991; 5 in 1994; 6 in 1998; 6 in 2004.* **Sources: Coo, Goo, Hud, Ri12, Se7, So1.**

Palla Rossa Precoce - (Early Palla Rossa) - Head is surrounded by beautiful maroon spotted dark-green leaves, inner leaves form wine-red head with pure-white ribs. *Source History: 1 in 1987; 1 in 1991; 2 in 1994; 2 in 1998; 2 in 2004.* **Sources: Ber, CA25.**

Palla Rossa Special - 85 days - Large uniform heads with gorgeous dark red leaves, distinctive tangy taste without bitterness, resists bolting, good for both summer and fall harvest. *Source History: 1 in 1998; 2 in 2004.* **Sources: Ter, We19.**

Palla Rossa Tardiva - (de Treviso Tardiva) - Overlapping white ribbed leaves with red head, from Italy. *Source History: 1 in 1987; 1 in 1991; 2 in 1994; 1 in 1998; 1 in 2004.* **Sources: Ber.**

Pan di Zucchero - (Sugar Loaf) - 80-89 days - Elongated green heads, 16 in. high, white inner leaves, very tasty, best as a fall crop, stores well for three months, popular throughout Italy. *Source History: 2 in 1981; 2 in 1984; 6 in 1987; 4 in 1991; 2 in 1994; 3 in 1998; 3 in 2004.* **Sources: Ber, Fe5, Se24.**

Pan Zucchero sel. Borca - Light green sugarloaf type with large tight head, fall plant, reliable, Franchi Special Sel. *Source History: 1 in 2004.* **Sources: Se24.**

Puntarella - 120 days - A Italian winter chicory, grown for its twisted succulent stems, harvested in spring, will overwinter in all but the coldest areas. *Source History: 1 in 1987; 1 in 1991; 2 in 1994; 2 in 1998; 2 in 2004.* **Sources: Coo, Se24.**

Radicchio - 85-110 days - Italian red chicory, very fine leaved romaine-like head, unique tangy delicate flavor, color dev. best in cool weather, popular in fine Italian restaurants. *Source History: 1 in 1981; 1 in 1984; 1 in 1987; 6 in 1991; 4 in 1994; 7 in 1998; 5 in 2004.* **Sources: Bo17, Bu2, Ge2, Lin, Sa5.**

Radicchio, Beacon - 70-74 days - Outstanding color and size, 4 in. leaf size, well suited for summer production and harvest through the fall. *Source History: 1 in 1998; 1 in 2004.* **Sources: SE4.**

Radicchio, Firebird - 74 days - Vibrant red color, excel. choice for summer production. *Source History: 1 in 1998; 1 in 2004.* **Sources: BE6.**

Radicchio, Fossa Verona Tardiva - Grumolo type, non-heading, open plant with solid red tender leaves, decorative plant. *Source History: 1 in 2004.* **Sources: Se24.**

Radicchio, Indigo - 68-75 days - Solid, deep red heads with bright white veins, few outer leaves, excellent yields and outstanding uniformity. *Source History: 1 in 1998; 3 in 2004.* **Sources: Coo, Ki4, SE4.**

Radicchio, Inferno - 80 days - Deep red color, 4 in. head, TB tolerant, harvest into late fall and early winter. *Source History: 1 in 1998; 1 in 2004.* **Sources: BE6.**

Radicchio, Melrose - 67 days - Red and green outer leaves, medium red 5 in. head, productive, bolt tolerant. *Source History: 2 in 2004.* **Sources: BE6, SE4.**

Radicchio, Milan - 75-110 days - Slow growing, round, 4-6 in. red head, use leaves until fall, taste is slightly bitter and tart, very reliable. *Source History: 1 in 1994; 4 in 1998; 7 in 2004.* **Sources: GLO, Hal, Jun, Ni1, Se26, Ver, Ves.**

Radicchio, Palla Rossa 3 - 70-80 days - Nice heads, dark red leaves with small white stems, can be forced, spring/fall. *Source History: 1 in 2004.* **Sources: Se24.**

Radicchio, Palla Rossa sel. Agena - Attractive blood-red leaves, white stem color spreads into the red leaves, fall culture, Franchi Special Sel. *Source History: 1 in 2004.* **Sources: Se24.**

Radicchio, Prima Rossa - 65 days - Head size 5-8 in. across, dark red in the center with some white ribs, bitterness is reduced by soaking in cold water for 10 minutes. *Source History: 1 in 1998; 1 in 2004.* **Sources: CO30.**

Radicchio, Rossa Treviso sel. Svelta - Treviso type with deep maroon leaves, large white stem, stunning appearance, Franchi Sepcial sel. *Source History: 1 in 2004.* **Sources: Se24.**

Radicchio, Rossa Verona sel. Arca - Deep maroon leaf with thick white stem, forms a nice elongated head, Franchi Special Sel. *Source History: 1 in 2004.* **Sources: Se24.**

Radicchio, Rubello - 76-82 days - Vigorous grower with good bolting tolerance, excel. res. against leaf browning. *Source History: 1 in 1998; 1 in 2004.* **Sources: BE6.**

Radicchio, Rubicon - 74 days - Imp. Rubello type with good bolting tolerance, suitable for spring, summer and fall harvest. *Source History: 1 in 1998; 1 in 2004.* **Sources: BE6.**

Radicchio, Versuvio - 85-92 days - Taller and more vigorous than Fiero, very deep red color, suitable for year-round production. *Source History: 1 in 1998; 1 in 2004.* **Sources: Tho.**

Radichetta - (Italian Chicory, Ciccoria Catalogna, Asparagus Chicory, Small Rooted) - 55-75 days - Dark-green deeply- notched 18 x 3 in. leaves, light-green petioles, asparagus type for spring planting, rapid growing annual. *Source History: 11 in 1981; 9 in 1984; 7 in 1987; 3 in 1991; 2 in 1994; 1 in 1998; 4 in 2004.* **Sources: Loc, Ri12, Se24, Sh9.**

Red Orchid - (Orchidea Rossa) - 65-70 days - Early red chicory with attractive, deep red, rosette shaped head, easy to grow, heads up nicely, reliable, best for fall planting, sow anytime for baby leaves. *Source History: 1 in 1994; 1 in 1998; 1 in 2004.* **Sources: Se24.**

Red Ribbed - Italian dandelion with scarlet midrib contrasting against the dark green, serrated, spade shape leaves and colorful stems. *Source History: 1 in 1998; 1 in 2004.* **Sources: Orn.**

Rossa di Treviso - (Red Treviso, Treviso, Rouge de Trevise) - 80-110 days - Famous radicchio from Treviso Italy, long slender green leaves in summer, cool weather turns it deep-red with white veins, non-heading, tart and slightly bitter, crisp but tender, can be overwintered for spring harvest, must be cut back to form the oblong heads from winter forcing. *Source History: 8 in 1981; 8 in 1984; 12 in 1987; 13 in 1991; 14 in 1994; 18 in 1998; 14 in 2004.* **Sources: Ba8, Ber, CA25, Coo, Fe5, Fo7, HO2, Ho13, Lej, Na6, Ni1, Ont, Orn, Se24.**

Rossa di Verona - (Verona Red, Rossa de Verone, Red Chicory, Radicchio) - 85-100 days - Later and hardier than Treviso, round compact head, heart-shaped deep-red leaves, solid heart, sharp flavor is best used sparingly in green salads, can direct seed or transplant in May or June or July for fall harvest, cut off all leaves above the crown in early fall and new growth in cool weather produces the characteristic small cabbage-like head, or overwinter for spring harvest, or develops in 10-12 days if spring transplanted inside and covered with wet burlap. *Source History: 12 in 1981; 12 in 1984; 15 in 1987; 15 in 1991; 20 in 1994; 21 in 1998; 18 in 2004.* **Sources: Bou, Bu2, CO30, Com, Coo, Fo7, Hi13, Ho13, Hum, La1, Lej, Ont, Pin, Pr3, Se8, SE14, Se24, Se26.**

Rossa di Verona a Foglia Larga - Overlapping leaves become red in colder months. *Source History: 1 in 1987; 1 in 1991; 1 in 1994; 1 in 1998; 1 in 2004.* **Sources: Ber.**

Rossa di Verona a Palla - (Red Ball of Verona) - Famous forcing radicchio, forms tight round heads of dark-red leaves with broad white ribs after the green summer growth has been cut back, needs freezing weather to perform well, heads up quickly. *Source History: 1 in 1987; 2 in 1991; 2 in 1994; 1 in 1998; 2 in 2004.* **Sources: Ba8, Ber.**

Rouge de Verone - 85 days - Strong rugged plant, heavy red/green leaves, if cut foliage back in spring will produce clusters of apple-size heads, trim away heads in fall to winter force. *Source History: 3 in 1981; 3 in 1984; 3 in 1987; 4 in 1991; 6 in 1994; 5 in 1998; 5 in 2004.* **Sources: Ba8, He8, Loc, Ni1, PAG.**

Rubico - Chioggia type from Italy, round red and white heads form in about 70% of plants, should be direct seeded in garden not before June to mid-July, transplants will produce lower percentage of heads, earlier sowing will cause bolting, matures Sept.-Oct. *Source History: 1 in 1991; 1 in 1994; 1 in 1998; 1 in 2004.* **Sources: Dam.**

San Pasquale - (Italian Dandelion, All Seasons) - 70 days - Large very early plant, broader deeper-cut smooth thick meaty light-green leaves, tightly bunched blanching in center. *Source History: 10 in 1981; 9 in 1984; 10 in 1987; 10 in 1991; 12 in 1994; 13 in 1998; 6 in 2004.* **Sources: Ba8, CO23, DOR, ME9, Mey, Sto.**

Selvatica di Campo - (Wild of the Fields) - Wild chicory with long thin leaves, sharp flavor, salads, cooked green or stuffings. *Source History: 1 in 1987; 1 in 2004.* **Sources: Se24.**

Soncino - Root chicory with off-white, long thick roots, smooth and bitter. *Source History: 1 in 1987; 1 in 1991; 1 in 1994; 1 in 1998; 2 in 2004.* **Sources: Ber, Se24.**

Sottomarina - (Variegata di Sottomarina) - Large loose heads variegated green and red, use individual leaves or entire heads, plant early spring or late summer. *Source History: 2 in 1987; 1 in 1991; 1 in 1994; 1 in 1998; 1 in 2004.* **Sources: Ber.**

Spadona - (Lingua di Cane, Dog's Tongue Chicory, Spadona Verde) - 35 days - Plant spring or fall for cutting at about 6 in. tall, very cold hardy, fast grower, sow anytime. *Source History: 1 in 1981; 1 in 1984; 2 in 1987; 1 in 1991; 1 in 1994; 1 in 1998; 4 in 2004.* **Sources: Com, Coo, Hud, Se24.**

Sugar Loaf - (Pain de Sucre, Sugarlof) - 80-90 days - Large cos-type heading chicory for summer or fall culture, mild flavor, self blanching. *Source History: 1 in 1981; 1 in 1984; 1 in 1987; 2 in 1991; 5 in 1994; 5 in 1998; 6 in 2004.* **Sources: Ba8, Bo19, Com, Coo, ME9, SE14.**

Trieste Sweet - 40 days - Fast growing, smooth pale green leaves. *Source History: 1 in 2004.* **Sources: Coo.**

Triestina da Taglio - Large, light green straight leaves. *Source History: 1 in 1987; 1 in 1991; 1 in 1994; 1 in 1998; 1 in 2004.* **Sources: Ber.**

Variegata di Lusia - Large overlapping variegated cream and red leaves. *Source History: 1 in 1987; 1 in 1991; 1 in 1994; 1 in 1998; 1 in 2004.* **Sources: Ber.**

Witloof - (Belgian Endive, French Endive, Brussels Witloof, Large Brussels, White Endive) - 55-150 days - Med-dark green long leaves, inner leaves and hearts for salads, delightful mildly acid flavor, 15-18 in. tall plants, roots dried and roasted for a coffee substitute, trim tops of mature roots and transplant inside for winter forcing, resulting second growth crown is self-blanching and a delicate addition to winter salads, witloof is Flemish for "white leaf", discovered in Belgium in the mid-1800s. *Source History: 46 in 1981; 41 in 1984; 32 in 1987; 21 in 1991; 20 in 1994; 17 in 1998; 9 in 2004.* **Sources: Ba8, CA25, Com, Ge2, HA3, Ho12, Ki4, La1, Ra5.**

Witloof Improved - 140 days - Delicacy produced in two stages: obtain healthy roots at end of growing season, then force these roots inside during winter to get crisp compact heads. *Source History: 3 in 1981; 3 in 1984; 4 in 1987; 2 in 1991; 1 in 1994; 1 in 1998; 1 in 2004.* **Sources: Sto.**

Zuccherina di Trieste - 35 days - Cutting chicory, small rosette of smooth, tender thick green leaves, mild flavor, cut at 3-4 in. long, regrows rapidly, spring or fall. *Source History: 2 in 2004.* **Sources: Com, Se24.**

Varieties dropped since 1981 (and the year dropped):

Abruzzese o di Galatina (2004), *Adria* (1998), *Alba* (1987), *Alouette Radicchio* (2004), *Alto* (1998), *Cesare* (2004), *Di Chiavari* (1991), *Di Trieste* (1991), *Dolce Bianca* (1991), *Flash* (1994), *Frastigliata* (1991), *French Endive* (1991), *Giant Magdeburg* (1991), *Guilio* (2004), *Italian Rooted* (1994), *Large Leaf* (2004), *Large Rooted* (2004), *Liber L.O.* (1987), *Liber M.O.* (1987), *Little Cornet* (1991), *Luxor* (1987), *Maxor* (1987), *Maxor Coffee Improved* (1984), *Meilof* (1994), *Mideka* (1987), *Milanese* (2004), *Mitado* (1998), *Normato* (1983), *Omega Rouge de Verone* (1987), *Pain de Sucre* (1998), *Pain de Sucre Amelioree Elmo* (1987), *Pall Rossa, Medium* (1998), *Parolus* (1987), *Pont de Pierre* (1991), *Productiva* (1991), *R 'C'* (2004), *Radicchio, Fiero* (2004), *Radicchio, Livrette* (2004), *Radicchio, Wildfire* (2004), *Red Belgium* (1998), *Ronette* (1991), *Rossa di Precoce* (1998), *Semi-Early Evere* (1987), *Sicilian* (1987), *Silla Radicchio* (1998), *Snowflake* (2004), *Spadona da Taglio* (2004), *Striped* (1987), *Sugar Loaf French* (2004), *Sugar Loaf Improved* (1991), *Sugarhat* (1991), *Trilof* (1998), *Vatters Zucherhat* (1987), *Wild Improved* (1987), *Zoom (Inra)* (1987).

Chinese Cabbage
Brassica rapa

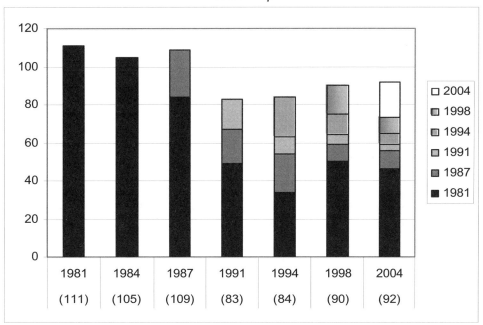

Number of varieties listed in 1981 and still available in 2004: 111 to 46 (41%).
Overall change in number of varieties from 1981 to 2004: 111 to 92 (83%).
Number of 2004 varieties available from 1 or 2 sources: 65 out of 92 (71%).

Aichi Hakusai - *B. rapa*, (Pekinensis Group), heading Chinese Cabbage is a staple in Asian cuisine, large oval heads weigh 10-12 lbs., young leaves used in salad, the main ingredient in Kimchee, the spicy national dish of Korea, which is made by pickling cabbage, garlic, red peppers and ginger, plant in July or August. *Source History: 1 in 1981; 1 in 1984; 1 in 1987; 1 in 1991; 1 in 1994; 1 in 1998; 2 in 2004.* **Sources: Ho13, Kit.**

Bau Sin - (Bao Sin) - 55-75 days - *B. juncea* var. *involutus*, a.k.a. Wrapped Heart Mustard Cabbage, rather dwarf plants with short leaves that fold in to form a tender blanched heart, 12 in. high x 18 in. wide, thick flat short wide petals, large tender crisp heads, mildly pungent, very tasty, for frying cooking and pickling *Source History: 1 in 1981; 1 in 1984; 1 in 1987; 3 in 1991; 3 in 1994; 3 in 1998; 6 in 2004.* **Sources: Ev2, Ho13, Loc, Sh12, So12, We19.**

Beka Maru - Japanese looseleaf type, tender light green stalks and leaves, fast grower, spring/fall sow. *Source History: 1 in 2004.* **Sources: Ho13.**

Big Stem Mustard - *B. rapa* (*Chinensis* Group), Chinese non-heading leaf type, mild and tender, small leaves but heavy stems, heat tol., quick growing, heavy stems are highly valued in Chinese cooking. *Source History: 2 in 1981; 2 in 1984; 1 in 1987; 1 in 2004.* **Sources: Ev2.**

Bok Choy - (Pak Choi, Bok Choi) - 45-60 days - *B. rapa*, (Chinensis Group), Chinese non-heading leaf type, bulbous plants with large white midribs, slight mustard taste, cool weather plant, does best in early spring or fall, long standing, used like celery or for greens, common throughout most of China. *Source History: 22 in 1981; 13 in 1984; 18 in 1987; 19 in 1991; 18 in 1994; 16 in 1998; 15 in 2004.* **Sources: Ban, Bo17, Com, Eo2, Fo7, Gur, In8, Ni1, Pr3, Ri2, Roh, Ros, Se26, So1, So12.**

Bok Choy, Purple Stalk - Attractive violet stems and veins, succulent and flavorful. *Source History: 1 in 2004.* **Sources: So12.**

Celery Mustard - 40 days - Stalks of this mustard resemble celery but are smooth and thick, excel. flavor, fine either boiled or eaten raw like celery. *Source History: 1 in 1981; 1 in 1984; 1 in 1987; 1 in 1991; 2 in 1994; 1 in 1998; 1 in 2004.* **Sources: Twi.**

Che-Foo - (Chihfu) - 65-75 days - *B. rapa*, (Pekinensis Group), heading Chinese cabbage, perfectly round very firm 5-6 lb. heads, vigorous, easy to grow, good shipper, old favorite that is still widely grown. *Source History: 3 in 1981; 2 in 1984; 1 in 1987; 1 in 1998; 1 in 2004.* **Sources: Ho13.**

Chihili - (Chihli) - 60-75 days - *B. rapa*, (Pekinensis Group), heading Chinese cabbage, cylindrical 14-22 x 4 in. heads, white midribs, broad well fringed leaves, mild flavor, seed directly as plants bolt if transplanted, popular in East. *Source History: 13 in 1981; 12 in 1984; 10 in 1987; 5 in 1991; 2 in 1994; 1 in 1998; 1 in 2004.* **Sources: Ear.**

Chihili, Long - *B. rapa*, (Pekinensis Group), heading Chinese cabbage. *Source History: 1 in 1987; 1 in 1991; 1 in 1994; 1 in 1998; 1 in 2004.* **Sources: Sau.**

China Choy - 70 days - Similar to Bok Choy but with a looser rosette of leaves, does well in fall plantings. *Source History: 1 in 1994; 1 in 1998; 2 in 2004.* **Sources: Se7, So9.**

Chinese - (Chinese Stem Mustard) - 70 days - Tightly packed heads with luscious white hearts, mild sweet flavor, 18-20 in. tall. *Source History: 1 in 1987; 2 in 1991; 3 in 1994; 2 in 1998; 1 in 2004.* **Sources: Lin.**

Chinese Broad Leaf - (Elephant Ear) - 45-52 days - Med-large 18-24 in. dia. plant, large broad med-green oval leaves with crumpled and scalloped edges, fairly tender, mild flavor. *Source History: 6 in 1981; 6 in 1984; 4 in 1987; 5 in 1991; 3 in 1994; 3 in 1998; 1 in 2004.* **Sources: Ev2.**

Chinese Flat Cabbage - 30-35 days - *B. rapa*, (Chinensis Group), Chinese non-heading leaf type, round dark-green leaves, tender pure-white leaf stalks, plants grow vertical in warm weather and horizontal in cold weather. *Source History: 1 in 1981; 1 in 1984; 1 in 1987; 1 in 1991; 1 in 1994; 2 in 1998; 1 in 2004.* **Sources: So12.**

Ching Chiang, Short Green Petiole - (Baby Pak Choy, Shanghai Pok Choi) - 30-45 days - Popular small 8 in. cabbage, smooth round leaves with thick green petioles, very tender and crisp, few strings, harvest when 6 in. tall, tolerates heat, rain, cold and dampness, grows best in a mild climate. *Source History: 1 in 1987; 1 in 1991; 2 in 1994; 4 in 1998; 4 in 2004.* **Sources: Ev2, Ter, Tu6, We19.**

Chirimen - 50-55 days - Chinese cabbage or Pe-tsai, loose heading type, heavily fringed leaves, heads weigh about 2.5 lbs., highly resistant to cold weather, excel. for stir fry. *Source History: 1 in 1981; 1 in 2004.* **Sources: Kit.**

Choy Sum, Long White Petiole - (Choi Sum) - 60 days - *B. rapa*, (Chinensis Group), Chinese non-heading leaf type, 12-14 in. flowering Bok Choy type with smaller sweet leaves and larger heart, Sum means heart in Chinese, early spring or late fall. *Source History: 5 in 1981; 4 in 1984; 5 in 1987; 5 in 1991; 3 in 1994; 2 in 1998; 2 in 2004.* **Sources: Ev2, Ho13.**

Choy Sum, Short White Petiole - 60 days - Tender plants harvested when 3-8 in. tall, tasty, excel. for boiling, stir-fry and soup, heat res. *Source History: 2 in 1994; 5 in 1998; 2 in 2004.* **Sources: Ev2, Ho13.**

Entsai - (Kwang Con) - 40-60 days - Long pointed leaves, very mild and tender, keep well watered, start harvesting when 12 in. tall to cut and come again, used in salads and stir-fried, not frost hardy, easy to grow. *Source History: 1 in 1987; 1 in 2004.* **Sources: Orn.**

Grelas de Santiago - Cold hardy var. from Chile, large green stemmed plant, long leaves with scalloped edges, cold hardy. *Source History: 1 in 2004.* **Sources: Ho13.**

Heavy Purple - Emerald leaves with toothed lobes, very dark purple stems, best used when young for cutting or baby rosettes, stem color is intensified by cool temps and bright light, muted by high nitrogen, bolting plants produce mild sweet purple salad shoots. *Source History: 1 in 2004.* **Sources: Sh12.**

Hiroshimana - 45-75 days - *B. rapa*, (Japonica Group), Japanese non-heading leaf type, wide rounded smooth dark-green leaves with wide pale green midribs, resembles Swiss chard, unique mild flavor, popular around Hiroshima. *Source History: 1 in 1981; 1 in 1984; 1 in 1987; 2 in 2004.* **Sources: Ho13, Kit.**

Hon Tsai Tai - (Yellow Flowering Purple Pak Choi) - 37-50 days - *B. rapa*, (Chinensis Group), Chinese non-heading leaf type, vigorous cool weather plant grown for tasty flower stalks, deep-cut dark-green leaves with purple-red veins, must keep moist during all stages, side dress with fertilizer just prior to bolting, up to 40 edible flowering stalks, from Japan. *Source History: 2 in 1981; 2 in 1984; 2 in 1987; 2 in 1991; 5 in 1994; 8 in 1998; 10 in 2004.* **Sources: Ev2, Jo1, Lej, Orn, OSB, Pin, Ri12, Sa9, Tu6, Und.**

Houshu - 70 days - Compact barrel shaped dark green heads, 9 in., matures a week later than Nozaki Early, introduced by Japanese ag. students at Rudolf Steiner College. *Source History: 3 in 2004.* **Sources: Ho13, Pe2, Tu6.**

Japanese White Celery Mustard - 45 days - *B. rapa*, (Chinensis Group), Chinese non-heading leaf type, Japanese var. of Pak Choi, originated in China, bulbous plants with broad pure-white stalks, deep-green spoon-shaped leaves are quite thick, grown extensively in northern Japan, res. to cold weather. *Source History: 3 in 1981; 2 in 1984; 1 in 1987; 2 in 1991; 2 in 1994; 1 in 1998; 1 in 2004.* **Sources: Ho13.**

Joi Choy - 50-55 days - An improved Lei Choi, thick white clasping stems, dark-green leaves, non-heading, expect about 20% off types because this is an open-pollinated line that is still being selected. *Source History: 1 in 1991; 1 in 1994; 2 in 1998; 4 in 2004.* **Sources: Bu2, Loc, Ri12, So12.**

Kaisin Hakusai - 70 days - Loose head type Chinese cabbage with fluffy top, frilly, light green outer leaves surround the yellow core leaves, young leaves used in salad, more mature leaves are steamed, pickled or used in stir fry. *Source History: 1 in 2004.* **Sources: Kit.**

Kate's Mustard Lettuce - Fine addition to a salad, a mustard and not a lettuce, bright green leaves with frilled edges, mild and tasty, sow in early spring. *Source History: 1 in 1998; 1 in 2004.* **Sources: Ho13.**

Kogane - 80 days - Non-heading type, bright green outer leaves, creamy yellow interior, open v-shape, 5-6 lbs. *Source History: 1 in 2004.* **Sources: OSB.**

Komatsuna - (Spinach Mustard) - 35-55 days - *B. rapa*, (Chinensis Group), Chinese non-heading leaf type, combines the flavors of mustard and spinach, quick growing tender glossy dark-green leaves, used in salads, lacks pungency, excel. bolt resistance, tolerates cold, can be grown year-round. *Source History: 4 in 1981; 3 in 1984; 4 in 1987; 6 in 1991; 8 in 1994; 13 in 1998; 22 in 2004.* **Sources: Ag7, Ba8, Dam, Ev2, He17, Ho13, Kit, ME9, Mo13, Ni1, Orn, OSB, Pe2, Pin, Se7, So12, So25, Sto, Tu6, Ver, We19, Yu2.**

Komatsuna, Torasan - 35 days - Unusual mustard or Oriental spinach is best used as salad greens when young. *Source History: 1 in 2004.* **Sources: Tho.**

Kosaitai - 50 days - Purple leaf stalks and leaf veins, yellow flower, entire plant can be harvested young and will regrow, young flowering shoots add a mild mustard flavor to fresh salads, excel. cooked flavor. *Source History: 1 in 2004.* **Sources: Kit.**

Kyona - 40 days - *B. rapa* (*Japonica* Group), Japanese non-heading leaf type, deeply-cut leaves, numerous white stems, produces all season, survival food for high altitude gardens, ground squirrels will not touch, frost only improves flavor. *Source History: 1 in 1981; 1 in 1984; 1 in 1987; 1 in 1991; 1 in 2004.* **Sources: Hig.**

Kyoto No. 3 - (Kayoto No. 3) - 80 days - *B. rapa*, (Pekinensis Group), heading Chinese cabbage, large late Japanese storage var., 5-15 lb. heads, sow July for October harvest, good in almost any climate, high yields, market, good keeper. *Source History: 5 in 1981; 5 in 1984; 3 in 1987; 2 in 1991; 1 in 1994; 2 in 1998; 2 in 2004.* **Sources: Dam, Kit.**

Lei Choi - (Pak Choi) - 47 days - *B. rapa*, (Chinensis Group), Chinese non-heading leaf type, spicy pure-white celery-like stalks, topped by small dark-green leaves, 10 to 14 erect 8-10 in. stalks, slow bolting, use raw or cooked, easy to grow, PVP expired 1995. *Source History: 7 in 1981; 7 in 1984; 8 in 1987; 9 in 1991; 11 in 1994; 5 in 1998; 1 in 2004.* **Sources: Roh.**

Market Pride - 70 days - Chinese cabbage or Petsai, standard Japanese autumn market type, light-green heavily savoyed outer leaves, cylindrical heads, 4-4.5 lbs., pure-white heart. *Source History: 2 in 1981; 1 in 1984; 1 in 1998; 1 in 2004.* **Sources: Se7.**

Maruba Santo - 30-50 days - *B. rapa*, (Japonica Group), Japanese non-heading leaf type, 18 in. plants with light-green leaves and large white stalks, excel. heat res., does well in semi-tropics of Southeast Asia, from Japan. *Source History: 2 in 1981; 2 in 1984; 2 in 1987; 1 in 1998; 1 in 2004.* **Sources: Ho13.**

Matsushima - 80-85 days - *B. rapa*, (Pekinensis Group), heading Chinese cabbage, grows fast, very dark-green outer leaves, light-green round 5-6 lb. heads with folded over leaves, cold hardy, Northern main crop type, limited winter storage. *Source History: 2 in 1981; 2 in 1984; 1 in 1987; 1 in 1998; 1 in 2004.* **Sources: Kit.**

Matsushima No. 2 - Cool weather, semi-heading type with med. green leaves, grows fast, from Japan. *Source History: 1 in 2004.* **Sources: Ho13.**

Mibuna - 30-40 days - *B. rapa*, (Japonica Group), Japanese non-heading leaf type, somewhat similar to Mizuna but with round leaves and greenish stems, for late summer planting, easy to grow, good resistance to heat and cold. *Source History: 1 in 1981; 1 in 1984; 1 in 1987; 4 in 1998; 6 in 2004.* **Sources: Dam, Ho13, Kit, ME9, Orn, We19.**

Michihli - (Michihili, Michihili Chihili, Slobolt, Chinese Celery Cabbage, Petsai) - 55-100 days - *B. rapa*, (Pekinensis Group), heading Chinese cabbage, solid cylindrical 14-24 x 4 in. dia. upright head, dark-green leaves, well blanched interior, sure heading, an improved Chihli, excel. fall crop, sow in July, very well known, accounts for 99% of all Chinese cabbage grown for market in the United States, 1948. *Source History: 97 in 1981; 91 in 1984; 85 in 1987; 71 in 1991; 72 in 1994; 65 in 1998; 57 in 2004.* **Sources: All, Ba8, Ban, Be4, Ber, Bo17, Bo19, Bu3, Bu8, CH14, CO23, Com, Cr1, Dam, DOR, Ear, Eo2, Ers, Ev2, Fo7, HA3, Hal, He17, HO2, Jor, Jun, Kil, La1, LO8, Loc, Ma19, Me7, ME9, Mey, Mo13, OLD, Ont, PAG, Ri2, Ri12, RIS, Roh, Ros, RUP, Sa9, Se8, Se25, Se26, Sh9, Shu, So1, Syn, Vi4, WE10, Wet, Wi2, WI23.**

Ming Choi - 49 days - *B. rapa*, (Pekinensis Group), Chinese Mustard with growth somewhat similar to Swiss Chard, large dark-green smooth leaves and clusters of thick white stems, leaves and stalks have endless uses in cooking and are a main ingredient in chow mein, delicious, easy to grow, PVP abandoned 1996. *Source History: 1 in 1991; 7 in 1994; 6 in 1998; 1 in 2004.* **Sources: Lin.**

Mizspoona Mix - 50-60 days - Gene pool from Mizuna x Tatsoi cross, thick tasty salad leaves in many shapes and shades of green, extremely cold hardy, CV Sh12 intro. *Source History: 2 in 2004.* **Sources: Se7, Sh12.**

Mizspoona Salad Select - 40 days - Careful reselection of Mizspoona Mix with better leaf quality, variably shaped, medium dark green leaves, some round, some pointed, juicy with a slight zing, from Frank Morton, Oregon. *Source History: 2 in 2004.* **Sources: Fe5, Sh12.**

Mizuna - (Kyona, Shuichoi, Japanese Potherb Mustard, Mizuna Early, Chinese Celery) - 40-65 days - *B. rapa*, (Japonica Group), Japanese non-heading leaf type, vigorous 12-24 in. plants, 2-4 lbs., many slender white .25 in. stalks, narrow dark-green 12 in. feathery deeply-cut leaves, grows fast, cold res., can overwinter, heat res. if watered, slightly hairy leaves and stalks, very mild when young. *Source History: 16 in 1981; 16 in 1984; 19 in 1987; 18 in 1991; 24 in 1994; 46 in 1998; 56 in 2004.* **Sources: Ba8, Bo17, Bou, CO23, CO30, Com, Coo, Dam, Fe5, Ga1, Goo, Gr29, He17, Hi6, Hi13, HO2, Ho13, Jo1, Kit, Loc, ME9, Mo13, Na6, Ni1, Orn, OSB, Pe2, Pe6, Pin, Ra5, Red, Ri2, Roh, RUP, Sa5, Sa6, Sa9, Se7, Se8, SE14, Se26, Se28, Shu, So1, So9, Sto, Syn, TE7, Ter, Tu6, Ver, Ves, Vi5, WE10, Wi2, WI23.**

Mizuna, Kyoto - 35-45 days - Deeply serrated, fringed narrow leaves form attractive rosettes, mild flavor, cold tolerant, can be cut several times, grows under low light conditions, bolts in April if overwintered. *Source History: 2 in 2004.* **Sources: Ki4, We19.**

Mizuna, Pink Petiole Mix - 40-60 days - Blond to green leaves with pink to purple leaf stems or petioles, color range is caused by outdoor temperature variation, warm days and cool nights, green stems when greenhouse grown, dev. by Frank Morton. *Source History: 1 in 2004.* **Sources: In8.**

Mizuna, Purple - Stable line of mild flavored mustards with leaves like Mizuna but with light purple stems resembling Purple Flowering Pac Choi, very mild flavor when young, cold hardy. *Source History: 1 in 1998; 3 in 2004.* **Sources: Ho13, In8, Sh12.**

Mizuna, Sakata - Japanese mustard, slender stalks, narrow leaves are deeply cut and fringed at the edges, long shelf life. *Source History: 1 in 1998; 1 in 2004.* **Sources: Orn.**

Molokheiya - (Egyptian Spinach, Moulukeyeh) - 70 days - Asian green originating in the Middle East for planting in warmer temperatures, holds well in the field, can be cut several times when 6-8 in. tall, exceptionally nutritious, used in Mesclun, as part of a braising mix and in soups, also used dried and powdered for improving the nutrient value of cakes and cookies. *Source History: 2 in 1998; 3 in 2004.* **Sources: Bou, Kit, Orn.**

Nagoda - Japanese winter str., semi-heading type to 12 in. across, med. green leaves, white stalks, tolerates cold. *Source History: 1 in 2004.* **Sources: Ho13.**

Napa - (Napa Cabbage) - 70 days - *B. rapa*, (Pekinensis Group), heading Chinese cabbage, light-green leaves form barrel-shaped head, 6-9 lbs., slow to bolt, seed from Japan. *Source History: 3 in 1981; 2 in 1984; 1 in 1987; 1 in 1991; 1 in 1994; 3 in 1998; 1 in 2004.* **Sources: Ev2.**

Nozaki Early - 60-65 days - *B. rapa*, (Pekinensis Group), heading Chinese cabbage, tall barrel-shaped 3-6 lb. heads, light-green leaves with broad white midribs, heads up extra early, good for spring and summer sowing, dependable, compact plant. *Source History: 6 in 1981; 5 in 1984; 2 in 1987; 2 in 1994; 4 in 1998; 2 in 2004.* **Sources: Ho13, Tu6.**

Orient Express - 43 days - Small, solid, 6 in. oblong heads of well-blanched, crisp leaves, pleasant peppery flavor, early, heat res. *Source History: 1 in 1994; 1 in 1998; 2 in 2004.* **Sources: Bu2, Roh.**

Osaka Purple - (Osaka Purple Leaf) - 70-80 days - Very large round dark-purple leaves with bright-white veins, harvest at 4-5 in. for salad, very mild flavor, good for overwintering, from Japan. *Source History: 6 in 1981; 6 in 1984; 4 in 1987; 7 in 1991; 8 in 1994; 13 in 1998; 16 in 2004.* **Sources: Bu2, CO23, CO30, Coo, DOR, Hi6, Ho13, Jo1, Orn, Pe2, Ri12, Sh12, So9, So12, Syn, Tu6.**

Pak Choi - (Pac Choy, Bok Choy, Bak Toi, Lei Choi, White Stalk Cabbage, White Celery Mustard, Japanese White) - 45-60 days - *B. rapa*, (Chinensis Group), Chinese non-heading leaf type, clusters of 8-12 wide white celery-like stalks, large nearly round smooth glossy green leaves, mildly pungent, hardy cool weather plant, grows fast, slow to bolt, long standing, very popular in cool climate areas. *Source History: 42 in 1981; 39 in 1984; 43 in 1987; 37 in 1991; 37 in 1994; 30 in 1998; 43 in 2004.* **Sources: Alb, Ba8, Bo19, CO23, Dgs, Ear, Eco, Ers, Ev2, Ga1, Goo, HA3, He17, Hi6, Hi13, Hig, Ho13, Hum, Jo1, Kit, La1, Lej, Lin, Loc, Mey, Na6, Ont, Pe2, Ri12, Roh, RUP, Sa5, Sa9, So25, Te4, TE7, Ter, Tu6, WE10, Wet, Wi2, WI23, Yu2.**

Pak Choi, Canton - (Canton Dwarf) - 42 days - *B. rapa*, (Chinensis Group), Chinese non-heading leaf type, round glossy dark-green leaves with wide thick white stalks, grown year-round in tropics and sub-tropics, bolts instantly if cool. *Source History: 1 in 1981; 1 in 1984; 2 in 1987; 7 in 1991; 5 in 1994; 3 in 1998; 3 in 2004.* **Sources: Ho13, Loc, SAK.**

Pak Choi, Canton Dwarf - 42 days - Baby bok choy with thick white stem, glossy, dark green leaves, slightly savoyed with white petioles, heat tolerant. *Source History: 1 in 1998; 4 in 2004.* **Sources: Jor, Kit, OSB, SAK.**

Pak Choi, Flowering - (Chinese Tsai Shim, Choi Sum) - 50 days - *B. rapa*, (Chinensis Group), Chinese non-heading leaf type, unusual because flower stalks are used instead of leaves, harvest when flowering begins, successive stalks will grow from leaf axil. *Source History: 1 in 1981; 1 in 1984; 1 in 1987; 2 in 1991; 2 in 1994; 2 in 1998; 2 in 2004.* **Sources: Ni1, Syn.**

Pak Choi, Flowering Purple - (Hon Tsai Tai, Chinese Tsaishim) - 40-60 days - *B. rapa*, (Chinensis Group), Chinese non-heading leaf type, cool weather plant, harvest purplish-red flower stalks at 8-12 in. just as buds start to open, quick to bolt, needs constantly moist soil. *Source History: 3 in 1981; 3 in 1984; 4 in 1987; 5 in 1991; 3 in 1994; 2 in 1998; 1 in 2004.* **Sources: Ni1.**

Pak Choi, Green Stalk - (Green Stemmed Pak Choy) - 55-70 days - *B. rapa*, (Chinensis Group), Chinese non-heading leaf type, cluster of green stalks topped by round dark-green leaves, smaller than regular Pak Choi, perfect to harvest young. *Source History: 2 in 1987; 1 in 1991; 3 in 1994; 5 in 1998; 2 in 2004.* **Sources: Ho13, Jor.**

Pak Choi, Hy Sawi Manis - 30 days - Malaysian var., tender plant to 12 in. tall, spoon-shaped dark green leaves with light green ribs, delicate flavor. *Source History: 1 in 1994; 1 in 1998; 1 in 2004.* **Sources: Ho13.**

Pak Choi, Shanghai - (Green Petiole Pak Choi, Shanghai Bok Choy, Green Stalk Bok Choy) - 40-50 days - *B. rapa*, (Chinensis Group), Chinese non-heading leaf type, thick tender light-green petioles, sturdy heat res. plant grows rapidly, can run instantly to seed in cool weather, firmer, more crisp, smaller, leaves are slightly darker, broad at the base tapering into the leaves. *Source History: 6 in 1981; 5 in 1984; 7 in 1987; 10 in 1991; 7 in 1994; 12 in 1998; 7 in 2004.* **Sources: Ev2, Ho13, Kil, Kit, Ni1, Orn, SAK.**

Pak Choi, Short White Petiole - White stem

(Pekinensis Group), dark green glossy leaves with thick, short white stalks, used in Cantonese cooking, heat res., from China. *Source History: 1 in 1994; 6 in 1998; 4 in 2004.* **Sources: Bou, Dam, Ev2, Ho13.**

Pak Choi, Spoon - (Bak Choi) - 60 days - *B. rapa*, (Chinensis Group), Chinese non-heading leaf type, large white spoon-shaped leafstalks form dense clump at base, dark-green rounded leaves, mild, likes warm weather, from southern China. *Source History: 2 in 1981; 1 in 1984; 1 in 1987; 1 in 1991; 1 in 1994; 1 in 2004.* **Sources: Ho13.**

Pak Choi, Takii Green - 40-50 days - Tender petioles picked when young for baby bok choy, sturdy, heat tolerant. *Source History: 1 in 2004.* **Sources: Loc.**

Peacock Tail - Tender delicious 12 in. leaves, excel. for stir-fry and pickling, young plants are very tender and less pungent. *Source History: 1 in 1991; 1 in 1994; 1 in 1998; 1 in 2004.* **Sources: Ev2.**

Prize Choi - 50 days - *B. rapa*, (Pekinensis Group), heading Chinese cabbage, large tender dark-green spoon-shaped leaves, thick rounded solid-white stems, celery-like base, vase-shaped 15-18 in. heads, compact and heavy, good quality. *Source History: 4 in 1981; 3 in 1984; 2 in 1987; 4 in 1991; 3 in 1994; 3 in 1998; 2 in 2004.* **Sources: Fe5, Ho13.**

Radish-Rooted - Tender, long edible leaves, mild flavor, also forms small 2-3 in. purple topped turnip. *Source History: 1 in 2004.* **Sources: Ho13.**

Red Giant - 23-45 days - Large dark purple-red savoy leaves with white midrib, sturdy, tender, mildly pungent, slow to bolt, winter hardy, very showy for the edible landscape, from Japan. *Source History: 1 in 1987; 7 in 1991; 8 in 1994; 14 in 1998; 12 in 2004.* **Sources: CO30, Coo, DOR, Ev2, Fe5, Ga1, HO2, La1, Loc, Ni1, OSB, RUP.**

Santo - (Santoh) - 35-60 days - *B. rapa*, (Pekinensis Group), semi-heading Chinese cabbage, open Pak Choi type, leaf extends farther down midrib, needs warmth and lots of moisture, suffers in hot summer sun, plant spring or fall. *Source History: 2 in 1981; 2 in 1984; 2 in 1987; 1 in 1991; 2 in 1994; 2 in 1998; 3 in 2004.* **Sources: Fe5, Kit, Yu2.**

Santo, Round Leaved - 50 days - *B. rapa*, (Pekinensis Group), semi-heading Chinese cabbage, round med-green leaves with white ribs, grows erect, res. to cold weather, excel. quality, easy to grow. *Source History: 2 in 1981; 2 in 1984; 1 in 1987; 1 in 1991; 1 in 1994; 1 in 2004.* **Sources: ME9.**

Seoul Cabbage - 35-70 days - Non-heading Chinese cabbage, can be used 30 days after sowing or matures to 2 lbs., very slow bolting, can be planted in spring or summer or fall, popular pickling type in China and Japan. *Source History: 1 in 1981; 1 in 1984; 2 in 1987; 2 in 1998; 7 in 2004.* **Sources: Gr29, Ho13, Kit, Ra5, Se28, Vi5, Wo8.**

Shantung - 60 days - *B. rapa*, (Pekinensis Group), semi-heading Chinese cabbage, smooth tender light-green leaves, spreading habit becomes upright with maturity, 4-6 lb. head, dense inner leaves blanch well. *Source History: 2 in 1981; 2 in 1984; 1 in 1987; 1 in 1991; 1 in 1994; 1 in 1998; 1 in 2004.* **Sources: Ho13.**

Shantung 4S - Harvest at 30 days when small and tender, light green leaves are delicious for stir fry. *Source History: 1 in 2004.* **Sources: Ho13.**

Shirona - 30 days - Large, mildly flavored Japanese vegetable, leaves grow 2 ft. tall with 2 in. wide white midribs, 2 lbs., cooked like Swiss chard which it resembles, but with thinner, more tender leaves with no bitterness, tolerates both cold and heat. *Source History: 1 in 1981; 1 in 1984; 1 in 1987; 1 in 1991; 1 in 1994; 1 in 1998; 3 in 2004.* **Sources: Ho13, Kit, Red.**

Small Cabbage Round Leaved - (Pe Tsai) - 30 days - Fast-growing looseleaf variety with light green round leaves and white flat petioles, 10-12 in. tall, easy to grow from spring through fall, matures in 3 to 4 weeks, excellent for stir-fry and salad. *Source History: 1 in 1987; 1 in 1991; 1 in 1994; 2 in 1998; 2 in 2004.* **Sources: Ev2, Ho13.**

Small Chinese Cabbage - (Tokyo Bekana) - 30 days - Very fast growing, yellowish-green leaves and white petioles, very delicious, excellent for stir-fry or soup. *Source History: 1 in 1987; 1 in 1991; 1 in 1994; 1 in 1998; 1 in 2004.* **Sources: Ev2.**

Snow Cabbage - (Shia Li Hon) - 50 days - So called because even in the coldest winters in northern China it can be seen peeking through the snow-covered ground, flat cabbage with very green puckered leaves, plant late summer and early fall, can withstand temperatures to 20 deg. F., widely used for pickling in the Orient. *Source History: 2 in 1987; 1 in 1991; 1 in 1994; 1 in 1998; 1 in 2004.* **Sources: Ev2.**

Swollen Stem Mustard - (Horned Mustard) - 45 days - Med-large loose heads of tangy frilly foliage, base of each leaf resembles a small celery stalk, very tender, sweet mustard flavor. *Source History: 2 in 1987; 3 in 1991; 2 in 1994; 2 in 1998; 1 in 2004.* **Sources: So12.**

Tah Tsai - (Tatsoi, Tai Sai, Pe Tsai, Spoon Mustard) - 40-55 days - *B. rapa*, (Chinensis Group), Chinese non-heading mustard, round spoon-shaped green leaves, thick broad tender crisp white stalks, 12-18 in. tall plants with large bulbous celery-like base, tangy mustard flavor. *Source History: 4 in 1981; 4 in 1984; 5 in 1987; 18 in 1991; 22 in 1994; 19 in 1998; 25 in 2004.* **Sources: Bo17, Bou, Coo, Ev2, Fe5, Ga1, Ga22, Hig, Ho13, In8, Jo1, Jor, Ki4, Kit, Ni1, Orn, OSB, Pe2, Se7, Se17, Sto, Te4, Ter, Ver, We19.**

Tah Tsoi - (Spinach Mustard, Spoon Cabbage) - 45-50 days - Easy to grow deep forest green cooking or salad green, grows in a compact, upright rosette, flavor almost as mild as spinach, free of oxalic acid. *Source History: 16 in 1998; 29 in 2004.* **Sources: Ba8, Bo19, CO23, Dam, Goo, He17, Hi6, Kit, Loc, ME9, Mo13, Na6, Pe6, RUP, Sa9, Se8, SE14, Se26, So1, So9, So12, So25, St18, TE7, Tu6, Ves, WE10, Wi2, WI23.**

Tai Sai, Giant - *B. rapa*, (Chinensis Group), Chinese non-heading leaf type, very refined version of Pak Choi with longer whiter stalks and more tender texture, thin to 8 in., bolts about May 10 in Lorane, Oregon, also excel. in fall. *Source History: 1 in 1981; 1 in 1984; 1 in 1987; 1 in 1991; 2 in 1994; 1 in 1998; 2 in 2004.* **Sources: ME9, Se7.**

Tendergreen - (Komatsuma, Mustard Spinach, Japanese Greens) - 21-50 days - *B. rapa*, (Chinensis Group), Chinese non-heading leaf type, 10 in. plants with 16-22 in.

spread, flat oblong smooth glossy dark-green leaves, pale-green midribs, slightly pungent, grows fast, cold and heat and drought res., slow bolt, flavor combines mustard and spinach, Japan. *Source History: 76 in 1981; 69 in 1984; 73 in 1987; 62 in 1991; 54 in 1994; 59 in 1998; 55 in 2004.* **Sources: Ada, All, Bu2, Bu8, CA25, CH14, Cl3, CO23, Com, De6, Dgs, DOR, Eo2, Ers, Fi1, Fis, Fo7, Gr27, Gur, HA3, He8, He17, HO2, Hum, Jor, Kil, LO8, MAY, ME9, Mel, Mey, Mo13, MOU, Ni1, OLD, PAG, Ri2, RIS, Ros, RUP, Sa9, Sau, Scf, Se26, Sh9, Shu, SOU, TE7, Vi4, Vi5, WE10, We19, Wet, Wi2, WI23.**

Tokyo Bekana - Japanese version of Pe Tsai, tender, wavy, yellow-green leaves, flat white stalks. *Source History: 1 in 1998; 1 in 2004.* **Sources: Ho13.**

Toraziroh - 45 days - *B. alboglabra*, Japanese str., 18 in. long, dark green leaves, strong distinct flavor with slight mustard tang, edible stems with flavor resembling pak choi, grows fast, relatively slow to bolt, very productive. *Source History: 2 in 1998; 2 in 2004.* **Sources: Fe5, Ho13.**

Tsai Shim - (Flowering Pak Choi) - 40 days - *B. rapa*, (Chinensis Group), Chinese non-heading leaf type of Tsoi Sum or rape, grown for its fine-flavored flower stalks, dark-green leaves with lighter stalk, highly heat res., harvest right after plant bolts but before flowering, leave 3 or 4 leaves on plant. *Source History: 5 in 1981; 5 in 1984; 3 in 1987; 2 in 1991; 2 in 1994; 2 in 1998; 1 in 2004.* **Sources: Loc.**

Tsoi Sim - 40-60 days - Quick growing bright green leaves on fleshy stems, entire plant is edible including the flowers, Chinese origin. *Source History: 1 in 1981; 1 in 1984; 1 in 1987; 1 in 1991; 3 in 1998; 5 in 2004.* **Sources: Dam, Jor, ME9, OSB, Sto.**

Tyfon - 40 days - *B. rapa*, cross of non-heading Chinese cabbage x fodder turnip, for greens or green manure, mild mustard flavor, amazingly rapid growth, strong taproots break up heavy soil, hardy to 10 deg. F., Holland. *Source History: 1 in 1981; 1 in 1984; 5 in 1987; 6 in 1991; 2 in 1994; 2 in 1998; 2 in 2004.* **Sources: Bou, Ter.**

Tyfon Holland Greens - 40 days - Stubble turnip x Chinese cabbage cross, tender mild greens, very hardy, cut back then re-harvest every 30-40 days - salads or cooked greens or fodder crop. *Source History: 1 in 1981; 1 in 1984; 1 in 1987; 2 in 1991; 2 in 1994; 2 in 1998; 2 in 2004.* **Sources: Ni1, Pin.**

Vitaminna - 45 days - Thick, crinkled, dark green leaves high in vitamin A, tolerates cool temps, slow bolting, Japan. *Source History: 2 in 2004.* **Sources: Kit, Yu2.**

Wong Bok - (Won Bok, Wong Bok Mandarin) - 70-85 days - *B. rapa*, (Pekinensis Group), heading Chinese cabbage, short oval broad head, 8-10 x 6-7 in. dia., some 10 lbs., light-green leaves, mild mustard-like flavor, productive old Mandarin var., heat and drought res., for summer and fall plantings in warm Southern areas, white heart, very crisp and tender, shorter and thicker than Michihli. *Source History: 28 in 1981; 23 in 1984; 27 in 1987; 20 in 1991; 15 in 1994; 18 in 1998; 8 in 2004.* **Sources: Ba8, Bou, Ers, Ev2, Ho13, Ni1, So1, WE10.**

Varieties dropped since 1981 (and the year dropped):

Aka Chirimen (1991), *Bok Choy, Baby* (2004), *Bok Choy, Tropical* (2004), *Celery Chinese Cabbage* (2004), *Chai Tai* (1991), *Challenger* (2004), *Chang Puh Extra Early* (1987), *Chao Chow* (1998), *Chin* (1991), *Chin Kang Cabbage* (1991), *Chin Tau* (1994), *China King* (1998), *Chinese - C.T.I.* (1994), *Chinese Heading Mustard* (1998), *Chinese Kale* (2004), *Chinese Mustard Green* (1998), *Chinese Mustard Red* (1998), *Chinese Smooth Leaf* (2004), *Chinese WR60* (1991), *Dynasty* (1998), *Fong Sun* (1994), *Giant Asian Red* (1994), *Granat* (2004), *Green Vegetable* (2004), *Heavy Pak Choi* (1987), *Hiroshima Green* (2004), *Hiroshima Spring* (1991), *Hong Kong Yow Choy* (1994), *Hung Chin Cabbage* (2004), *Japanese Giant Celery Mustard* (1987), *Japanese Mustard* (1983), *Japanese White Celery* (1991), *Japanese White Celery Cabbage* (1991), *Kai Choy, Waianae Strain* (1994), *Kailaan White Flowered* (1987), *Karashina* (1998), *Kashin* (1991), *Kireba Santo* (1987), *Komatsuna Late* (1991), *Korean Spring and Fall* (2004), *Kwan-Hoo Choi* (1994), *Kyo Mizuna* (1991), *Large Chinese Smooth Leaf* (1991), *Le Choi* (1998), *Light Pak Choi* (1987), *Liu Choy* (1991), *Mamang* (1994), *Michihli Tall* (1994), *Minato Santo* (2004), *Mizuna, Kyona* (2004), *Monument* (1991), *Nabana* (1998), *Nagaoka No. 2* (1991), *Nagaoka No. 50* (1991), *Nan Foong* (1994), *Napa Cabbage, Natsugumo* (1998), *Napa One* (2004), *Nikanme* (1991), *Nozaki No. 2* (1987), *Numfong Loose Leaved* (1987), *Ohgon* (1987), *Osaka Chusei Large* (1991), *Osaka Large Leaf Latest* (1994), *Osome* (1998), *Paak Tsoi Green Petiole* (1987), *Paak Tsoi Horse Ear* (1987), *Paak Tsoi Sum* (1987), *Pai Taisai White Stalk* (1982), *Paotoulien* (1987), *Pe Tsai* (2004), *Pot-Herb Mustard* (2004), *Rape-Mustard* (1991), *Round Chinese Pak Choi* (1991), *Round Leaved Santung* (2004), *Santo (Serrated Leaved)* (1991), *Santo, Large Semi-Heading* (1994), *Seppaku* (1991), *Seppaku Taina* (1983), *Shaho Tsai* (1991), *Shuho Napa* (1991), *Siew Choy* (1998), *Slow Bolt China Bok Choy* (1991), *Slow Bolt Usa Bok Choy* (1991), *South China Earliest* (1998), *South China Heading* (1994), *Spoon Cabbage* (2004), *Spring Giant* (1987), *Treasure Island* (1984), *Tse Tai, Purple* (1994), *Tsoi Sum White Stem* (1994), *Victor WR* (1991), *W-R 55 Days* (1998), *W-R 70 Days* (1998), *Yayoi Komatsuna* (1991), *Yellow Leaf Pe Tsai* (1991)

Collard
Brassica oleracea (Acephala Group)

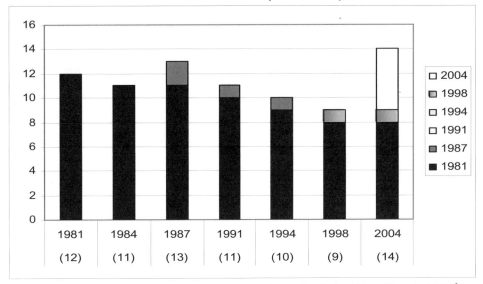

Number of varieties listed in 1981 and still available in 2004: 12 to 8 (67%).
Overall change in number of varieties from 1981 to 2004: 12 to 14 (117%).
Number of 2004 varieties available from 1 or 2 sources: 9 out of 14 (64%).

Afghanistan - *Source History: 1 in 2004.* **Sources: Se17.**

Cabbage Collards - 60-80 days - True heading collard with cabbage characteristics, combines both flavors, very hard compact dark-green heads on short stems, light-green ribs and veins. *Source History: 2 in 1981; 1 in 1984; 3 in 1987; 3 in 1991; 2 in 1994; 3 in 1998; 2 in 2004.* **Sources: DOR, SOU.**

Champion - 60-80 days - Longer standing Vates type, uniform vigorous darker blue-green 34 in. plant, high yields, holds longer without bolting, good hardiness, widely adapted, VA/AES. *Source History: 8 in 1981; 8 in 1984; 14 in 1987; 18 in 1991; 23 in 1994; 24 in 1998; 22 in 2004.* **Sources: ABB, Cl3, Ers, Fe5, Ha5, HO2, Jo1, La1, ME9, Mey, OSB, RIS, RUP, SE4, So1, SOU, Ter, Twi, WE10, We19, Wet, WI23.**

Georgia - (Georgia Southern, True Southern, Southern, True Georgia, Creole, Georgia Green) - 60-80 days - Old standard, erect spreading 36 in. plants, 2-3 in. loose clusters, huge cabbage-like blue-green slightly crumpled juicy leaves, tol. to heat and poor soil, slow bolt, non-heading type, a light freeze improves its mild cabbage-like flavor, pre-1880. *Source History: 74 in 1981; 66 in 1984; 66 in 1987; 61 in 1991; 57 in 1994; 62 in 1998; 60 in 2004.* **Sources: Ada, Ba8, Bou, Bu2, Bu8, But, CH14, Cl3, Co31, Com, Cr1, De6, DOR, Eo2, Ers, Fi1, Fis, Fo7, Ge2, GRI, He8, He17, Hi13, Jor, Kil, La1, Lej, LO8, Ma19, ME9, Mel, Mey, Mo13, Ni1, OLD, Or10, PAG, Ra5, Red, Ri12, Roh, Ros, RUP, Sa9, Sau, SE4, Se7, Se25, Se26, Sh9, Shu, So1, So12, SOU, Twi, Vi4, WE10, Wet, Wi2, WI23.**

Georgia Blue Stem - 60 days - Old popular var., very tall and long-stemmed with leaves far apart, leaves can be cropped and eaten as the plant will form new leaves again. *Source History: 2 in 1981; 2 in 1984; 3 in 1987; 3 in 1991; 1 in 1994; 2 in 1998; 1 in 2004.* **Sources: Ho13.**

Georgia Green - 70 days - Upright 30-36 in. plant, large juicy leaves, light frost improves flavor, tolerates drought and cold. *Source History: 1 in 1981; 1 in 1984; 1 in 1987; 1 in 1991; 1 in 1998; 2 in 2004.* **Sources: HA3, Hud.**

Giant Purple Flat Poll - *Source History: 1 in 2004.* **Sources: Se17.**

Green Glaze - 73-79 days - Upright plants 30-34 in. tall, bright-green sheen gives the surface a greasy look that many Southerners refer to as Greasy Collards, excel. res. to cabbage worm and cabbage looper, intro. in 1820 by David Landreth. *Source History: 2 in 1981; 2 in 1984; 3 in 1987; 3 in 1991; 4 in 1994; 6 in 1998; 6 in 2004.* **Sources: Ho13, Hud, Pe1, Se7, Shu, So1.**

Green Glaze, McCormack's - 75 days - Result of eight years of selecting for improved cold tolerance and uniform bright green glossy leaves, otherwise similar to Green Glaze, shows superior res. to cabbage worm and cabbage looper, has survived 0 degree temperature, highly recommended for southern and warm coastal areas, CV So1 introduction, 2000, Dr. Jeff McCormack. *Source History: 2 in 2004.* **Sources: Ga21, So1.**

Mesic Zero Degrees - Plants are 3 ft. tall, ruffled blue-green leaves with purple tinge, sweet flavor, cold hardy, from South Carolina. *Source History: 1 in 1998; 1 in 2004.* **Sources: Ho13.**

Morris Heading - (Morris Improved Heading, Carolina, Cabbage Collard) - 52-85 days - Old favorite variety, low growing 18-24 in. plants, loose heavy heads with short stems, smooth dark-green leaves with light-green veins, very slow to bolt. *Source History: 19 in 1981; 17 in 1984; 18 in 1987; 16 in 1991; 13 in 1994; 15 in 1998; 19 in 2004.* **Sources: Ada, CA25, Cl3, Co31, DOR, He8, HO2, Kil, La1, Mey, Ri12, Se26, Shu, So1, SOU, Vi4, WE10, Wet, WI23.**

<u>Syrian</u> - *Source History: 1 in 2004.* **Sources: Se17.**

<u>Variegated</u> - 80 days - Florida family heirloom, unusual collard with some of the plants showing leaf variegation when in flower with slower growth at this time, green color returns as seeds mature, plant life of up to five years is not uncommon in areas where winter temperatures remain above 20 degrees F, with over 3 in. stem dia., CV So1 introduction, 1999 via family friend of Walt Childs. *Source History: 1 in 2004.* **Sources: So1.**

<u>Vates</u> - (Blue Stem, Vates Non-Heading) - 55-90 days - Non-heading, long standing, bolt and frost res., winter hardy in South and Middle Atlantic, dark-green slick crumpled leaves, spreading vigorous plants, known for lack of purpling in veins and leaves, upright 24-32 in. plants with large cabbage-like leaves, can survive frost, VA/AES. *Source History: 85 in 1981; 68 in 1984; 66 in 1987; 56 in 1991; 55 in 1994; 53 in 1998; 47 in 2004.* **Sources: ABB, Ada, Bo19, Bu3, Bu8, CA25, CH14, Cl3, Cr1, Dam, Dgs, DOR, Ers, Fi1, Fo7, Goo, HA3, He8, He17, HO2, Hum, Jor, Kil, Loc, ME9, Mey, MIC, Mo13, MOU, Na6, OLD, Ont, PAG, Pin, RIS, RUP, SE4, Se26, Sh9, So1, Sto, Twi, Vi4, WE10, Wet, Wi2, WI23.**

Varieties dropped since 1981 (and the year dropped):

Carolina (1991), *Florida* (1991), *Georgia Long Standing* (1998), *Morris Heading Improved* (2004), *Tait's Heading Collard* (1983).

Corn / Dent
Zea mays

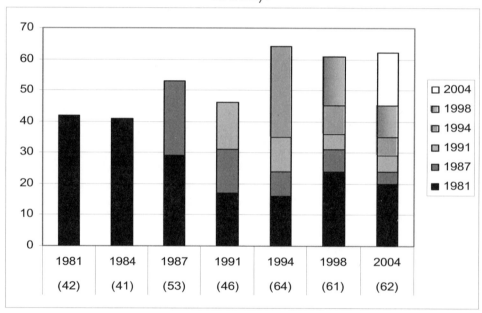

Number of varieties listed in 1981 and still available in 2004: 42 to 20 (48%).
Overall change in number of varieties from 1981 to 2004: 42 to 62 (148%).
Number of 2004 varieties available from 1 or 2 sources: 47 out of 62 (76%).

<u>Abbe Hills Yellow Dent</u> - (Neal's Yellow Dent) - 110 days - High protein feed corn, 12 in. ears, kernels in paired rows, selected and improved by the Neal family in Iowa since 1903 from Reid's Yellow Dent, dries down well, especially good for chopped silage. *Source History: 1 in 1981; 1 in 1984; 1 in 1987; 2 in 1991; 1 in 1994; 1 in 1998; 1 in 2004.* **Sources: Ab2.**

<u>Adams Early</u> - (Adams Extra Early) - 65-75 days - Early roasting ear corn, 5.5-6.5 ft. plants, cylindrical 7-8 x 2 in. ears with 12-14 rows, fairly tender and sweet when young, high yields. *Source History: 2 in 1981; 1 in 2004.* **Sources: Ho13.**

<u>All Purpose</u> - 135 days - Stalks to 11 ft., multi-colored kernels. *Source History: 1 in 1994; 1 in 1998; 1 in 2004.* **Sources: Sa9.**

<u>Arkansas Red and White</u> - Beautiful red or white 8 in. ears, 10-12 ft. stalks, makes fine cornmeal or grits, rare heirloom grown by the Byers family in Arkansas. *Source History: 1 in 2004.* **Sources: Ba8.**

<u>Aztec Red</u> - 160 days - Plants to 9 ft., huge red kernels used for hominy, soups and flour, pre-Columbian. *Source History: 1 in 1991; 1 in 1994; 2 in 1998; 2 in 2004.* **Sources: Ho13, Red.**

<u>Beasley's Red Dent</u> - 105-115 days - Heirloom discovered in Whitley County, Indiana in the early 1970s by Glenn Beasley, a yellow dent selected for red color, 9 ft.

stalks, 9-12 in. ears, 16-20 rows of red kernels or on occasion yellow and orange kernels, excellent blight and drought resistance. *Source History: 1 in 1987; 1 in 1991; 1 in 1994; 1 in 1998; 1 in 2004.* **Sources: Sa9.**

Big Chief - 120 days - Plants grow 10-14 ft. tall, 8-10 in. ears, kernel color includes red, blue, yellow, white and bicolors. *Source History: 2 in 2004.* **Sources: HO2, SE4.**

Bloody Butcher - 100-120 days - Grown in the U. S. since 1845, 10-12 ft. stalks, 2-6 ears per stalk, pink or red cobs, red kernels striped with darker red, an occasional white ear will appear, fine flavor, good for flour, cornmeal or corn-on-the-cob when young, originally from Virginia. *Source History: 2 in 1987; 3 in 1991; 5 in 1994; 21 in 1998; 37 in 2004.* **Sources: Ag7, Ba8, Bo18, Bo19, CA25, Ea4, Ers, GLO, Gr29, Ha5, He8, He17, HO2, Ho13, Hud, Jor, Jun, MAY, ME9, Mel, Mo13, Pin, RUP, SE14, Se16, Se26, Se28, Shu, Sk2, So1, Ter, Twi, Ver, Vi4, Vi5, WE10, Wo8.**

Bloody Mary - 100-110 days - Plants grow 6-12 ft. tall, 2 ears per plant, 10 in., 12-16 rows, huge kernels, brown to blood-red with ivory tips, common grinding corn from before the 1940s. *Source History: 1 in 2004.* **Sources: Se17.**

Blue Claredge - (Ohio Blue Clarage) - 100-113 days - Stalks grow 9-10 ft. tall, 14-16 rows of purple and white kernels on 7-8 in. ears., excel. for grinding or eating as sweet corn when young, dev. west of the Appalachian Mountains in the Ohio and West Virginia areas between 1830 and 1850. *Source History: 2 in 1991; 1 in 1994; 1 in 1998; 2 in 2004.* **Sources: Sa9, So1.**

Blue Dent - (Hopi Blue Dent) - 100-110 days - Hopi Indian corn, 5 ft. stalks, 8 rows, 8 in. long ears, dries to royal blue. *Source History: 2 in 1998; 5 in 2004.* **Sources: Ga1, Ho13, Pe2, Se26, Shu.**

Boone County White - 110-120 days - Very deep flat white slightly rough kernels, uniform, cylindrical 9-11 in. ears, 18-22 rows, white cobs, heavily leaved 9.5 ft. stalks, 7.9% protein, silage, stock feed or meal, dev. in Boone County in northwest Indiana by James Riley from a selection made in 1876 from White Mastadon. *Source History: 5 in 1981; 5 in 1984; 4 in 1987; 4 in 1991; 5 in 1994; 3 in 1998; 5 in 2004.* **Sources: Ba8, Bor, Ers, He17, Shu.**

Buck Lantz Dent - 126 days - Sturdy 11 ft. stalks, huge 12 in. long cobs, 16-22 rows of yellow shaded kernels. *Source History: 1 in 2004.* **Sources: Sa9.**

Carolina Gourdseed - Tall stalks, white kernels, used for corn meal or grits, one of the oldest corn varieties grown today. *Source History: 1 in 2004.* **Sources: Fe5.**

Cherokee Blue and White - 95-110 days - Pre-Columbian Native American corn, 7-10 ft. plants, 12-18 rows, good all-purpose type with large ears, beautiful white and blue-purple kernels. *Source History: 2 in 1998; 2 in 2004.* **Sources: Cor, Ho13.**

Cherokee White Eagle - *Source History: 1 in 2004.* **Sources: Se17.**

Dia de San Juan - Planted on June 24 which is the Dia de San Juan, when Southwestern people traditionally celebrate the coming of the summer rains, all-purpose white corn from north of Alamos, Sonora. *Source History: 1 in 2004.* **Sources: Na2.**

Duncan Dent - Family heirloom from Duncan, Arizona area, used for cornmeal and chicken feed, stalks 8-10 ft., thick ears, dates back to the 1900s. *Source History: 1 in 2004.* **Sources: Na2.**

Earth Tones - (Indian Summer) - 85-95 days - Dent flour corn, 8-12 ft. stalks, 8-10 in. ears, kernels are beautiful muted colors of gold, bronze, mauve-pink, green and brown, ground into flour, birdseed, decorative. *Source History: 14 in 2004.* **Sources: GLO, HO2, Hud, Jor, Jun, ME9, Mel, OLD, Par, Roh, Shu, Sto, Twi, Ver.**

Ernest Strubbe's Blue - 98 days - Beautiful corn, kernels range in shades of blue from turquoise to darker blue, occasional off-colored red ears will appear, dev. by Minnesota farmer, Mr. Strubbe, now deceased. *Source History: 1 in 1998; 1 in 2004.* **Sources: Sa9.**

Ernest Strubbe's Brown - 105 days - Stalks grow 7 ft. tall, kernels are beautiful shades of brown with occasional off colors. *Source History: 1 in 2004.* **Sources: Sa9.**

Ernest Strubbe's Orange - 97 days - Kernels colors are a gorgeous shade of bright orange, as a result of the genetics of this corn, a few ears will be red, some yellow as well as shades of orange, dev. by Mr. Strubbe, Minnesota farmer, now deceased. *Source History: 1 in 1998; 1 in 2004.* **Sources: Sa9.**

Ernest Strubbe's Pink - 100 days - Stalks to 7 ft., 8 in. ears with beautiful pink kernels, developed by Mr. Strubbe, a Minnesota farmer, now deceased. *Source History: 1 in 1998; 1 in 2004.* **Sources: Sa9.**

Ernest Strubbe's Purple - 107 days - Kernels are a nice shade of purple, named in honor of the developer, now deceased. *Source History: 1 in 1998; 1 in 2004.* **Sources: Sa9.**

Eureka Ensilage - 120 days - Stalks grow 11-13 ft. tall, large ears, white kernels. *Source History: 1 in 1994; 2 in 1998; 2 in 2004.* **Sources: Ers, Shu.**

Goliath Silo, Shumway's - (Ensilage Seed Corn) - 110-120 days - Produces an abundance of excel. silage, long broad leaves cover entire length of 12-15 ft. stalks, wind and drought res., solely a silo filler, one acre produces more ton of feed than 5 acres of ordinary corn, not recommended for areas beyond northern IL, 1934. *Source History: 1 in 1981; 1 in 1984; 1 in 1987; 1 in 1991; 1 in 1994; 1 in 1998; 3 in 2004.* **Sources: Ra5, Sa9, Shu.**

Green and Gold Dent - 95 days - Offshoot of Oaxacan Green dent, bright green and yellow kernels, large ears. *Source History: 1 in 2004.* **Sources: HO2.**

Greenfield - 113 days - Stalks to 9-10 ft. tall, 10 in. long ears with 15-20 rows of cream to yellow kernels, no lodging even after 60 mph winds on two occasions, respectable yields. *Source History: 1 in 2004.* **Sources: Sa9.**

Greenfield 114 - 114 days - Short-stalked corn with excellent standability, good yields, preferred by Amish farmers, high yields, reports of up to 190 bushels per acre, dev. by Ned W. Place. *Source History: 1 in 1994; 1 in 1998; 2 in 2004.* **Sources: Ers, Pl3.**

Henry Moore Yellow - 110 days - Large flat kernels, 9-10 in. ears, U of MO tests: 11.3% protein and .38 lysine, used for feeding, very tall heavily leaved stalks also excel.

for silage. *Source History: 2 in 1981; 2 in 1984; 1 in 1987; 2 in 1991; 3 in 1994; 1 in 1998; 2 in 2004.* **Sources: Bor, Ers.**

Hickory King, White - 75-130 days - Large wide deep white kernels, small white cob, grows on any soil, 8 ft. plant, 8-9 in. ear, 10-12 rows, excel. for hominy grits, cornnuts or cornmeal, used when young as roasting ears, also freezes well, good silage corn because of high bulk, an old favorite in the South where it is used extensively for roasting, north of the Mason-Dixon line they preferred Yellow Hickory King, pre-1850. *Source History: 24 in 1981; 21 in 1984; 25 in 1987; 20 in 1991; 20 in 1994; 20 in 1998; 18 in 2004.* **Sources: Ada, Ba8, Bou, Bu8, Co31, Ers, He8, He17, Ho13, Kil, Mel, Mo13, Ri12, Sau, Se25, Shu, So1, So12.**

Hickory King, Yellow - 85-90 days - Excel. for roasting ears or hominy or grits or cornmeal, also good silage corn due to high bulk, 12 ft. stalks produce 2 ears with large yellow kernels, tight husks, SNA has been holding field selections annually since 1907, from Appalachia, late 1800s. *Source History: 6 in 1981; 5 in 1984; 3 in 1987; 2 in 1991; 4 in 1994; 4 in 1998; 2 in 2004.* **Sources: Fo7, Ho13.**

John Haulk - Hardy stalks to 15 ft. tall, excel. for corn meal or animal feed, resists insect and mold damage, plants should be spaced 2 ft. apart and hilled to prevent lodging during high winds, heirloom from the foothills of South Carolina. *Source History: 1 in 1998; 1 in 2004.* **Sources: Scf.**

King White - Stalks to 8 ft., 7-8 in. ears, 12 rows of white and blue kernels, excel. for roasting or corn meal. *Source History: 1 in 2004.* **Sources: Scf.**

Krug's Yellow - (Krug's 90 Day) - 90 days - Medium flat seed, popular silage corn, U of Missouri tests: 10% protein and .30 lysine, 8-10 in. ears, high yields, proven in short season areas. *Source History: 3 in 1981; 3 in 1984; 2 in 1987; 3 in 1991; 5 in 1994; 2 in 1998; 3 in 2004.* **Sources: Bor, Ers, Pl3.**

Lancaster Surecrop - 110-120 days - Stalks grow 10-12 ft., ears 10-12 in. long, deep yellow kernels, strong root system, dev. around the turn of the century by Isaac Hershey, a Mennonite farmer from Lancaster County, Pennsylvania. *Source History: 2 in 1991; 3 in 1994; 6 in 1998; 3 in 2004.* **Sources: Ers, Lan, Shu.**

Manzano Yellow - Yellow to orange kernels are used for meal or animal feed, dry farmed in Manzano Mts. in New Mexico. *Source History: 1 in 1994; 1 in 1998; 1 in 2004.* **Sources: Na2.**

Mayo Tuxpeno - Large, fat ears on 10-12 ft. plants, yellow, blue and yellow or pink kernels, recent growout of a 1985 collection from Saneal, Sonora. *Source History: 1 in 1998; 1 in 2004.* **Sources: Na2.**

McCormack's Blue Giant - 85/100 days - First blue dent corn to be developed since before the turn of the century, dev. by Jeff McCormack from a cross between Hickory King and an unnamed heirloom blue dent, especially suited for clay soils and drought prone areas, 10-12' stalks not recommended for loose soils or high wind areas, two 7-9 in. ears per stalk, large wide smoky blue kernels, makes a light-blue flour, great for blue tortillas and blue corn chips, also can be used fresh as roasting ear corn, good leaf blight res., intro. in 1994 by CV So1. *Source History: 1 in 1994; 1 in 1998; 2 in 2004.* **Sources: Ho13, So1.**

Mesquakie - Productive, 7 ft. stalks, fat, 6-7 in. ears with either yellow and purple or light and dark red kernels which are tall and tooth-like with crimped tops, good meal corn, grown for over 100 years in the Midwest. *Source History: 1 in 1998; 1 in 2004.* **Sources: Ho13.**

Mexican June (Blue and White) - 90-110 days - Strong tall dependable Indian field corn, heat and drought res., tight heavy husk, 8-9 in. ears, usually blue and white kernels, for late plantings and roasting and silage, Southwest. *Source History: 6 in 1981; 6 in 1984; 8 in 1987; 5 in 1991; 4 in 1994; 4 in 1998; 1 in 2004.* **Sources: Co31.**

Mexican June (White) - 100-120 days - Favorite roasting variety throughout the South, 9-10 ft. tall, 8-9 in. ears have white kernels with an occasional blue kernel, heat, drought, lodging and ear worm resistance. *Source History: 1 in 1981; 1 in 1984; 1 in 1987; 1 in 1991; 2 in 1994; 3 in 1998; 5 in 2004.* **Sources: Ada, Fo7, Pp2, Ros, Sa9.**

Nokomis Gold Field Corn - Indian corn x corn-belt dent crosses, good standability, flint-dent type seeds with high test weight, bred by Michael Fields Ag. Institute. *Source History: 1 in 2004.* **Sources: Tu6.**

Norfolk Market - *Source History: 1 in 1981; 1 in 1984; 1 in 1994; 1 in 1998; 1 in 2004.* **Sources: Shu.**

Northwestern Red Dent - 90-100 days - Rare Oscar Will heirloom, 6-8 ft., sturdy stalks are not prone to lodging, tapered ears, 8-10 in. long, 12-16 rows of red kernels with white caps, dev. in the Midwestern U.S. *Source History: 1 in 1987; 2 in 1991; 1 in 1994; 2 in 1998; 2 in 2004.* **Sources: Ag7, Sa9.**

Nothstine Dent - 90-100 days - Glossy yellow kernels with white caps, 7-8 in. ears, 7 ft. stalks, dries early in the field, not high yield but excel. quality, northern Michigan heirloom, sweet meal and flour. *Source History: 1 in 1981; 1 in 1984; 1 in 1987; 1 in 1991; 1 in 1994; 2 in 1998; 1 in 2004.* **Sources: Jo1.**

Oaxacan Green - 70-105 days - Stalks 5-6 ft. tall, 6 in. ears with some smooth kernels in shades of green ranging from bronze to pea-green to emerald-green, grown for centuries by the Zapotec Indians of southern Mexico to make green flour tamales, traditionally grown with squash and beans which twine up the corn stalks, highly ornamental. *Source History: 1 in 1991; 1 in 1994; 5 in 1998; 16 in 2004.* **Sources: Ag7, Ech, GLO, Ha5, HO2, Ho13, Jor, Jun, ME9, Rev, Ri12, Sa9, Se7, Se16, Shu, So12.**

Ohio Calico - 115 days - Stalks grow 9-10 ft. tall with 10-12 in. ears, most kernels will be red striped with white, some solid red, others solid white. *Source History: 1 in 2004.* **Sources: Sa9.**

Pencil Cob Corn - 76-100 days - Old Shoepeg-type dry corn with unusual med-sized ears, deep grains, almost no cob, very sweet at milk stage, 6 ft. tall sturdy plants. *Source History: 4 in 1981; 4 in 1984; 8 in 1987; 7 in 1991; 5 in 1994; 4 in 1998; 8 in 2004.* **Sources: Ada, Co31, Ho13, Hud, Ri12, Se26, Shu, Wi2.**

Pride of Saline White - 120 days - Very stalky, 10 ft. plants, large 12 in. ears, 16 rows per cob, white kernels. *Source History: 1 in 1998; 1 in 2004.* **Sources: Sa9.**

Reid's Yellow Dent - 85-110 days - Old-timer, well adapted to Southern heat and soils, vigorous 6-7 ft. plant, 9-10 in. double well filled ears, 16 rows, deep close-set med-flat seed, 9.9 % protein.31 lysine, dev. by James L. Reid in northern Illinois from a Gordon Hopkins cross his father brought from Brown County Ohio in 1846, this late large reddish corn was crossed with an earlier yellow dent to create the modern Reid's Yellow Dent. *Source History: 14 in 1981; 14 in 1984; 10 in 1987; 10 in 1991; 10 in 1994; 11 in 1998; 13 in 2004.* **Sources: Ada, Ba8, Bor, CH14, Eo2, Ers, He17, La1, Pl3, Ri12, Shu, So1, Wi2.**

Silver King - (Wisconsin No. 7) - 100-110 days - Best white dent corn for Wisconsin/northern Illinois/Iowa, uniform creamy white kernels with a slight roughness, glistening white cobs 8.5-9.5 in. long, A.J. Goddard of Ft. Atkinson, Iowa, brought a bushel of this corn from Indiana to Fayette County, Iowa in 1862, foundation stock was secured by the Wisconsin Experimental Station, 1904. *Source History: 1 in 1981; 1 in 1984; 1 in 1987; 1 in 1991; 1 in 1994; 3 in 1998; 2 in 2004.* **Sources: Ers, Shu.**

Silvermine - 82 days - Extremely high yields, large heavy 11-12 in. roasting ears, creamy white kernels, 14-18 rows, res. to worm damage, excel. shipping var., heavy dark-green husk holds its color well, does well on poor soils, originated by J.A. Beagley of Sibly, IL, 1890. *Source History: 3 in 1981; 3 in 1984; 5 in 1987; 7 in 1991; 4 in 1994; 6 in 1998; 4 in 2004.* **Sources: Ada, Co31, Ers, Shu.**

Tennessee Red Cob - 100-120 days - Famous field corn, dependable yields of large deep-grained ears set on red cobs, used for either roasting or as grain for feeding, pre-1900. *Source History: 1 in 1981; 1 in 1984; 1 in 1987; 1 in 1998; 2 in 2004.* **Sources: Sa9, So1.**

Texas Gourdseed - 120 days - Reintroduced by Southern Exposure, originally brought to south Texas by German farmers in late 19th century, closely approximates orig. gourdseed characteristics, 8 ft. stalks, 2 ears per stalk, 18-22 rows of creamy-white narrow tightly compacted kernels, very easy to shell by hand, good disease (except for smut) and drought res., does well in clay soil, in south Texas considered best for tortillas. *Source History: 1 in 1987; 1 in 1991; 1 in 1994; 1 in 1998; 1 in 2004.* **Sources: So1.**

Tohono O'odham June - 90 days - Stalks to 8 ft., clear white-yellow kernels with a hint of pink on 6 in. ears, soft floury centers, floodwater farmed in midsummer with the desert rains. *Source History: 1 in 1991; 1 in 1994; 1 in 1998; 1 in 2004.* **Sources: Na2.**

Trucker's Favorite White - (Trucker's White) - 95-105 days - White dent field corn, also used for roasting ears at milk stage, hardy type for the South, 8-9 ft. stalks, 8-10 in. med-tapered ears, 14-18 rows, packs and ships well, stays in top condition for a long period, high yields with good soil and fertilization, excel. foliage for early fodder. *Source History: 27 in 1981; 22 in 1984; 20 in 1987; 15 in 1991; 13 in 1994; 12 in 1998; 14 in 2004.* **Sources: Ada, Bu8, CH14, Co31, Eo2, Ers, He8, He17, Kil, La1, Ra5, Se26, Shu, Wi2.**

Trucker's Favorite Yellow - (Trucker's Yellow) - 70-105 days.- Yellow strain of the early-maturing Trucker's Favorite, 8-12 in. ears with 12-14 rows, red cob, use for early quick crop or for late planting. *Source History: 7 in 1981; 6 in 1984; 4 in 1987; 7 in 1991; 8 in 1994; 7 in 1998; 7 in 2004.* **Sources: CH14, Co31, Ers, Fo7, He17, Ra5, Shu.**

Virginia Gourdseed - 120 days - Sturdy 8 ft. plants with fat stout ears, 20-26 rows of "horse tooth" white kernels, 2-3 ears per plant. *Source History: 1 in 2004.* **Sources: Sa9.**

White Dent, Tait's - Stalks 6 ft. tall, ears 8-10 in. long, adapted from Virginia southward to Georgia, rare heirloom. *Source History: 1 in 1981; 1 in 1984; 1 in 1987; 2 in 1991; 3 in 1994; 1 in 1998; 1 in 2004.* **Sources: Shu.**

White Eagle - 115-120 days - Used young as an eating/roasting corn, when mature as a dent, field or flour corn, 12 ft. stalks, large ears, most have blue kernels, some with a majority of white kernels, well adapted to the needs of America's early farmers, the high ear placement facilitated hand picking, stalks provided fodder for winter livestock feeding, this Cherokee Blue and White corn was brought over the Trail of Tears by the Cherokee, from Carl Barnes of Oklahoma. *Source History: 1 in 2004.* **Sources: Und.**

Yuman - Mature kernels are yellow with red chinmark, originally grown along the lower Colorado River. *Source History: 1 in 1994; 1 in 1998; 1 in 2004.* **Sources: Na2.**

Varieties dropped since 1981 (and the year dropped):

Adams Early Improved (2004), *Alabama White Dent* (2004), *Big Mixed* (1991), *Big Red* (1991), *Bloody Butcher, Southern* (1998), *Cherokee Princess* (1998), *Chickasaw Dent* (2004), *Conico Norteno Pepitillo* (1991), *Dark Red Dent* (1998), *Dwarf* (1998), *Funk's REB-87* (1991), *German Pencil Cob* (1987), *Gila River* (1998), *Gold Crown Red Dent* (1998), *Greenfield 120* (2004), *Hastings Prolific* (1998), *Holcomb's Prolific* (1991), *Hopi* (2004), *Hopi Magenta* (1998), *Kester's Early Dwarf* (1998), *Kirkland 89* (1998), *Leaming's Yellow* (1998), *Legg's Prolific White* (1994), *Leming* (1998), *Louis Miller Heirloom Dent* (1998), *Mayo Batchi* (2004), *Minnesota No. 13 (Mellum Str.)* (2004), *Mortgage Lifter* (2004), *Mosby's* (2004), *Mt. Pima Maiz Blanco* (1991), *Murdock* (2004), *Mustang* (1998), *Old Hickory King Composite* (2004), *Old Hickory King White* (1994), *Pencil Cob, Yellow* (1998), *Posole Maiz Blanco* (1994), *PSC 114 (Yellow)* (1991), *Rotten Clarage* (1998), *Snowflake* (1987), *So. Conico Norteno Pepitillo* (1991), *Southern Maiz Negro* (1994), *Tait's Norfolk Market* (1991), *Tarahumara Bachiachi* (1991), *Tarahumara Chapo (or Chato)* (1994), *Tarahumara Cherry/Blue* (1987), *Tarahumara Maiz Blanco* (1987), *Tarahumara Maiz Chato* (1998), *Tarahumara Maiz Colorado* (1987), *Tarahumara Maiz Colorado No. 2* (1987), *Tarahumara Maiz de Cerocahui* (1998), *Tarahumara Maiz Pinto* (1987), *Tarahumara Pepitillo* (2004), *Texas 17W* (1991), *Texas 28A* (1991), *Texas 30A* (1991), *Thompson's Prolific* (2004), *Thunderbolt Red Dent* (1998), *Trucker's Delight* (1991), *U.S. No. 13* (1994), *Urubama White* (1994), *Virginia White Gourdseed* (1991), *Wapsie Valley Dent* (1991), *White Cap* (1994), *White Dent* (1991), *White Surecropper* (2004), *Yaqui* (1998), *Yaqui June Corn* (1998), *Yellow Dent* (2004), *Yellow Surecropper* (2004), *Yoeme Vatchi* (2004).

Corn / Flint
Zea mays

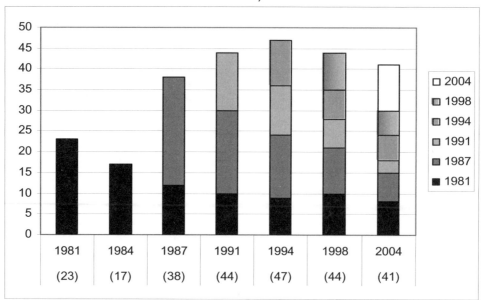

Number of varieties listed in 1981 and still available in 2004: 23 to 8 (35%).
Overall change in number of varieties from 1981 to 2004: 23 to 41 (178%).
Number of 2004 varieties available from 1 or 2 sources: 28 out of 41 (68%).

Alabama Coschatta - 110-120 days - Heirloom var. maintained and recommended by Glenn Drowns as one of the best multi-colored ornamental corns, 8 ft. plants, ears 8-12 in. long. *Source History: 2 in 1998; 1 in 2004.* **Sources: Sa9.**

Apache Yellow Flint - 105-110 days - Stalks grow 5-6 ft. tall, 8-9 in. ears, bright yellow, hard kernels, originally from Gila of southwest New Mexico. *Source History: 1 in 2004.* **Sources: So12.**

Bear Island Chippewa - 85-93 days - Originally collected from the Chippewa, 4-5 ft. plants, 5-8 in. ears, 8-12 rows, kernel colors include yellow, pink, white, red, blue and striped, grinding corn, old Oscar Will variety, 1925. *Source History: 3 in 2004.* **Sources: In8, Ri12, Se17.**

Blue - Flint/flour, native, Southwest type. *Source History: 1 in 1998; 1 in 2004.* **Sources: Pl3.**

Byron Yellow Flint - Northern golden yellow flint, 6-8 in. ears, 8 rows, matures early. *Source History: 1 in 2004.* **Sources: So12.**

Cheyenne Agency Striped - 81 days - Stalks 5-6 ft. tall, 8-12 rows of multi-colored flinty hard kernels per ear, resists drought. *Source History: 1 in 2004.* **Sources: Sa9.**

Chinook - 85-90 days - Flint/flour type, stalks from 5 to 5.5 ft. tall, two 5-6 in. ears on the main stalk with two tillers bearing one smaller but useable ear each, 10-14 rows, kernel colors range from dark-maroon through tan with shades of red and bronzy orange, Baggett/OSU. *Source History: 2 in 1994; 2 in 1998; 2 in 2004.* **Sources: RUP, Ter.**

Early Indian - 75-80 days - Extra early dry corn selected for high percentage of whites, but produces a beautiful range of colors, 4.5-5 ft. plants, 1 to 2 ears per stalk, 7-9 in. long, 8 rows of kernels, good for flour or meal, good choice for coastal or short-season areas, a northern corn. *Source History: 1 in 1994; 1 in 1998; 2 in 2004.* **Sources: ME9, Pe6.**

Guarijio Maiz Amarillo - Yellow semi-flint corn, some white kernels, 8 ft. tall plants, used for tamales, atole, pinole and as elote, dry-farmed. *Source History: 1 in 2004.* **Sources: Na2.**

Inca Rainbow - 85-90 days - Tall, vigorous stalks, 8-10 in. cobs on 7-9 ft. plants, multi-hued, for raw/roasting or ground corn meals. *Source History: 1 in 1998; 1 in 2004.* **Sources: So12.**

Indian Fingers - 100-110 days - Miniature ornamental corn, 6-7 ft. stalks, delicate 2.5-4.5 x .75 in. ears, small shiny kernels colored deep mahogany and yellow and red and purple and orange stripe on yellow, very attractive. *Source History: 1 in 1987; 5 in 1991; 2 in 1994; 5 in 1998; 5 in 2004.* **Sources: Ha5, ME9, RIS, Roh, RUP.**

Indian Flint - 105-110 days - Hard flint corn, 9-11 in. ears, ideal for grinding into corn meal which shows little or no color despite the multi-colored ears. *Source History: 1 in 1994; 1 in 1998; 1 in 2004.* **Sources: Ni1.**

Indian Ornamental - (Indian Corn, Indian Rainbow Corn, Ornamental Indian Corn, Rainbow, Indian) - 95-110 days - Vigorous and very productive, strong 6-8 ft. stalks, large decorative ears, excel. for roadside sales and fall decorations and winter arrangements, easy to grow, a rainbow of color combinations. *Source History: 33 in 1981; 28 in 1984; 32 in 1987; 32 in 1991; 32 in 1994; 33 in 1998; 28 in 2004.* **Sources: Be4, Bu3, Bu8, CH14, Com, Fi1, Gr27, Ha5, He8, Hi13, Jor, Jun, La1, ME9, Mel, Mey, Mo13,**

MOU, OLD, Ont, RUP, Sau, Si5, So25, SOU, Twi, Ver, Wi2.

Kaladia Ornamental Flint - 110 days - Stalks grow 9 ft. tall, thick 8-10 in. ears, beautiful colorful kernels with exquisite markings, dev. by Peter Kopcinski. *Source History: 1 in 2004.* **Sources: Fe5.**

King Philip, Improved - 90-110 days - Stalks to 6-7 ft. tall, some lodging, copper-red kernels, 8 rows, 8-12 in. long ears, red or white cobs, hardy in the north, commercially released in 1853, improved by John Brown of New Hampshire, originally from the Wampanoag Indians with King Philip as the chief. *Source History: 1 in 2004.* **Sources: Se17.**

Lenni Lenape, Delaware - Ears grow 8-10 in. long, dark blue kernels. *Source History: 1 in 2004.* **Sources: So12.**

Lenore - 100-108 days - Stalks to 6-7 ft., 75%+ kernels are shades of red and red-brown, the remainder are a colorful ornamental corn, dev. by Glenn Drowns in the early 1980s as part of a college project. *Source History: 1 in 1994; 1 in 2004.* **Sources: Sa9.**

Little Indian - 100-105 days - Ears average 2.5-3 in. long, ears remain small when not fertilized and isolated from other corns, used in stir fries, pickles and as finger food. *Source History: 4 in 1991; 6 in 1994; 8 in 1998; 5 in 2004.* **Sources: Bu3, Fo7, RUP, Sh9, Tt2.**

Longfellow Flint - 112-117 days - Standard 8-row Northern flint, late maturing, 10 ft. plants with 10 in. ears, orange kernels, makes a sweet cornmeal, highest yielding flint where it can be grown. *Source History: 2 in 1981; 1 in 1984; 1 in 1987; 1 in 1991; 1 in 1998; 1 in 2004.* **Sources: So12.**

Mandan Black - Short stalked, purple-black selection of Mandan corn, raised with great success in northeast Washington. *Source History: 1 in 1987; 1 in 2004.* **Sources: Ho13.**

Mandan Clay Red - 81 days - Stalks to 4 ft., clay pink-red kernels, excellent drought tolerance. *Source History: 1 in 2004.* **Sources: Sa9.**

Mini Seneca Indian - 103 days - Miniature ears of Indian corn in both dark and light colors, colorful 3-5 in. ears but can vary under certain growing conditions, do not fertilize to keep ears small, use in dried arrangements. *Source History: 1 in 1987; 14 in 1991; 12 in 1994; 15 in 1998; 16 in 2004.* **Sources: Bu3, CH14, Goo, HA3, HO2, Jor, Loc, Ont, OSB, RUP, Sau, Se26, So25, SOU, Wet, Wi2.**

Miniature Indian - 103 days - Multicolored flint corn mixture, 3-4 small 4-6 in. ears on 5-6 ft. plants, excellent for dried arrangements. *Source History: 1 in 1987; 8 in 1991; 9 in 1994; 10 in 1998; 5 in 2004.* **Sources: CA25, He17, Mo13, Shu, Si5.**

Mt. Pima Cristalino de Chihuahua - Med-large ears, hard shiny clear white kernels. *Source History: 1 in 1987; 1 in 1991; 1 in 1994; 1 in 1998; 1 in 2004.* **Sources: Na2.**

Nambe White - Long, slender ears with white kernels, from Nambe Pueblo, NM, grown at 6000 ft. elevation. *Source History: 1 in 1998; 1 in 2004.* **Sources: Na2.**

Ornamental - (Ornamental Flint, Ornamental Indian) - 100-110 days - Large decorative 7-9 in. ears, kernels in endless combinations of red and purple and orange and yellow and white and blue, strong stalks, used for Halloween and Thanksgiving decorations. *Source History: 17 in 1981; 16 in 1984; 15 in 1987; 11 in 1991; 12 in 1994; 11 in 1998; 9 in 2004.* **Sources: All, De6, Ear, Ers, Goo, HA3, Pin, RIS, Wet.**

Osage Red - 118 days - Rare many-purpose corn, 8 ft. stalks with 7-9 in. ears, 8-14 rows, kernels have varying shades of pink and purple. *Source History: 1 in 1987; 1 in 1991; 1 in 2004.* **Sources: Sa9.**

Pipestone - 81 days - Short stalks to 4 ft., 8 in. ears with red and yellow kernels. *Source History: 1 in 2004.* **Sources: Sa9.**

Rainbow - (Rainbow Flint, Rainbow Mixed, Rainbow Field, Rainbow Starch, Ornamental Rainbow) - 90-112 days - Large long smooth glossy multi-colored ears, sells well at farmers markets, used for fall decorations or poultry and livestock feed, endless color combinations, white and yellow and orange and red and purple and blue and striped kernels all on the same ear. *Source History: 18 in 1981; 16 in 1984; 20 in 1987; 21 in 1991; 16 in 1994; 16 in 1998; 19 in 2004.* **Sources: ABB, Bu1, Bu2, CA25, Fo7, Ge2, Gur, La1, MAY, ME9, Pla, Ri12, Roh, Ros, Sa9, SE14, Sh9, Vi4, WE10.**

Roy's Calais - (Abenaki Calais Flint) - 80-90 days - Vermont heirloom from the mid 1800s, 6-7 ft. plants, 7-9 in. cobs, 8-10 rows, has both golden yellow and dark maroon ears, does well in cold weather, good meal corn, withstood a killing frost in Calais, VT, on July 9, 1816, grown by local Abenaki Native American tribe for generations and given to settlers, in Roy Fair's family for many years. *Source History: 3 in 1998; 4 in 2004.* **Sources: Ag7, Fe5, Hi6, Ri12.**

Seedway Elite - 105-107 days - New strain of ornamental corn with 15% purple husks and stalks, colorful ears borne on strong sturdy plants, easy to grow. *Source History: 3 in 1981; 2 in 1984; 2 in 1987; 3 in 1991; 1 in 1994; 2 in 1998; 2 in 2004.* **Sources: Jor, SE4.**

Seneca Blue Bear Dance - Stalks 4-5 ft. tall, 6 in. ears have 8 rows of pearlescent, lavender kernels with blue-purple shades and cream kernels. *Source History: 1 in 1998; 2 in 2004.* **Sources: Ho13, Se17.**

Seneca Indian Corn - 110 days - Squaw corn, 7-8 ft. stalks, 8-9 in. ears, very good mix of colors, used mainly for autumn decorations, plant at least four rows to produce good ears, 10% purple stalks. *Source History: 2 in 1981; 1 in 1984; 3 in 1991; 2 in 1994; 5 in 1998; 5 in 2004.* **Sources: CH14, Dam, Ers, HO2, Ves.**

Seneca Red Stalker - 105-110 days - Ornamental 9-10 ft. decorative purple and red stalks, beautifully multi-colored ears, kernel colors include purple, burgundy, cream and gold. *Source History: 9 in 1991; 16 in 1994; 24 in 1998; 24 in 2004.* **Sources: Bo18, CH14, Coo, Dam, Ers, Gr29, He17, HO2, Jo1, Jor, ME9, Mo13, Ni1, Ont, OSB, RIS, Roh, RUP, SE4, Sh9, Si5, Twi, Vi5, Wo8.**

Squaw Corn - (Rainbow, Indian Flint, Calico Indian) - 100-110 days - Ornamental flint corn, large multi-colored ears, strong stalks, good for cornmeal which shows almost no trace of color. *Source History: 10 in 1981; 9 in 1984; 12 in 1987; 9 in 1991; 7 in 1994; 7 in 1998; 6 in 2004.* **Sources:**

He8, Lin, PAG, Se26, Shu, Ver.

Tarahumara Apachito - Small rose and clear kernels, farmer's favorite. *Source History: 1 in 1987; 1 in 1991; 1 in 1994; 1 in 1998; 1 in 2004.* **Sources: Na2.**

Tarahumara Golden Cristalino - Long slender 12-row cobs, pearly gold kernels. *Source History: 1 in 1987; 1 in 1991; 1 in 1994; 1 in 1998; 1 in 2004.* **Sources: Na2.**

Tarahumara Maiz Azul - Thin flint, used to make corn beer which is called tesguino, some color mix, for spring planting in the high cool desert of the Sierra Madre of Mexico. *Source History: 1 in 1981; 1 in 1984; 1 in 1987; 1 in 1991; 1 in 1994; 1 in 1998; 1 in 2004.* **Sources: Na2.**

Tarahumara Serape - Beautiful long, slender ears with pearly white, red and striped kernels, a Cristalino de Chihuahua landrace. *Source History: 1 in 1994; 1 in 1998; 1 in 2004.* **Sources: Na2.**

Vermont Yellow Flint - 82 days - Eight-row northern flint, 8 in. ears with deep golden kernels, nutty, rich flavor, ground for meal, also used for roasting at milk stage, matures in short season areas, possible Iroquois origin. *Source History: 2 in 2004.* **Sources: Ga1, Pe2.**

Wampum - 85-90 days - Earlier version of a Carousel type ornamental with ears 4-5 in. long, 1.25 in.dia., 16-20 rows of small to very small kernels, colors include many shades of yellow gold white pink red blue black and purple, much color variation on a single ear, husk color is both maroon and white, resembles Carousel and Mini Indian but matures 2 weeks earlier, dev. by Dr. James Baggett, OR State U. *Source History: 5 in 1994; 10 in 1998; 12 in 2004.* **Sources: Be4, Coo, Ers, HO2, Jo1, Mel, Ni1, RUP, SE14, Sh9, Ter, Ves.**

Varieties dropped since 1981 (and the year dropped):

Amber Flint Maize (1998), *Big Timber Flint* (1998), *Calico Flint* (1998), *Casados Blue* (2004), *Cherokee Red* (1998), *Colorama Mixed* (1984), *Dwarf Multicolored* (2004), *Eastern Tarahumara Pepitillo* (2004), *Fort Kent Golden* (2004), *Garland Flint* (1998), *Hernandez Blue* (2004), *Inca* (1983), *Jones Ornamental Flint* (1982), *Jung's Little Indian* (2004), *Little Indian, Jung's Improved* (2004), *Little Jewels* (2004), *Longfellow Yellow Flint* (1987), *Mayo Corriente* (2004), *Mohawk White Hominy* (1991), *Mt. Pima Ancho* (1994), *Mt. Pima Maiz Azul* (1994), *Mt. Pima Quamun* (1994), *O'odham Flint* (1991), *Purple Husk* (2004), *Rhode Island White Cap* (1987), *Santo Domingo Mix* (2004), *Saskatchewan White* (1994), *Saskatoon White* (2004), *Selected Ornamentals* (1983), *Squaw Improved* (1987), *Symphonie* (1991), *Tarahumara Blanco* (1994), *Tarahumara Bola* (1987), *Tarahumara Chiquita* (2004), *Tarahumara Conico Norteno* (1991), *Tarahumara Copper Flint* (1991), *Tarahumara Maiz Caliente* (1998), *Tarahumara Maiz Color de Rosa* (2004), *Tarahumara Maiz Pinto* (1987), *Tarahumara Onaveno* (1991), *Tarahumara Pink Bola* (1998), *Tarahumara Rosavewali* (1998), *Tarahumara Sitakame* (1994), *Tarahumara/Tepehuan Serrano* (2004), *Tepehuan Maiz Colorado* (2004), *Tepehuan Maize Rosero* (1998), *Vermont Yellow* (1998), *Western Tarahumara Amarillo* (1998), *Western Tarahumara Gileno* (1998).

Corn / Flour
Zea mays

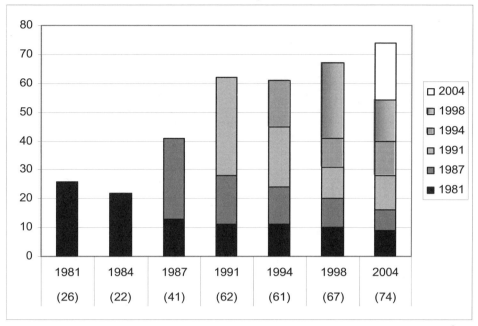

Number of varieties listed in 1981 and still available in 2004: 26 to 9 (35%).
Overall change in number of varieties from 1981 to 2004: 26 to 74 (285%).
Number of 2004 varieties available from 1 or 2 sources: 62 out of 74 (84%).

Acoma Blue - Plants 7-8 ft. tall, long ears with deep blue, small kernels, roasted and ground for hot porridge called "atole", grown under irrigation on the high mesas in northern New Mexico from May to October. *Source History: 1 in 1994; 3 in 1998; 2 in 2004.* **Sources: Na2, Se17.**

Acomita Blue - From the village of Acomita, below the Pueblo of Acoma. *Source History: 1 in 2004.* **Sources: Na2.**

Amarillo Cancha - *Source History: 1 in 2004.* **Sources: Red.**

Anasazi - 100 days - Beautiful heirloom from the Anasazi (ancient ones) tribe in the Southwest, stalk size varies from 6-9 ft. tall, multi-eared and multi-colored, excel. for grinding into flour, may be an ancestor of many of the Southwest corn races. *Source History: 1 in 1991; 1 in 1994; 4 in 1998; 4 in 2004.* **Sources: Ho13, Ri12, Se7, Se17.**

Apache - (Apache Red) - 110 days - Stalks 6 ft. tall, black to blackish red kernels, traditional corn of the San Carlos Apache of Arizona, red corns were viewed with disdain in the process of "Christianizing the heathens" during the Spanish hegemony in the Southwest, so apparently they selected the black colored corn. *Source History: 1 in 1991; 2 in 1994; 3 in 1998; 5 in 2004.* **Sources: Ho13, Na2, Ri12, Se7, So12.**

Aztec Giant White - 160 days - Plants to 8 ft. tall, sometimes 2-3 ears per node, ears 10 in. long, 4 rows, huge white kernels up to 1 in., used for parching, flour or made into soup. *Source History: 1 in 2004.* **Sources: Red.**

Big Mountain Blue - 110-120 days - Stalks to 7 ft., 1-2 large, 8 in. ears per stalk, blue kernels with occasional white, from Big Mountain, rare. *Source History: 2 in 1998; 1 in 2004.* **Sources: Se17.**

Blanco Para Pozole - Huge white kernels are up to twice the size of Hickory King, popular in Mexico for making hominy and roasting ears, unique and rare, seed collected at markets in Mexico, introduced by Baker Creek Heirloom Seeds. *Source History: 1 in 2004.* **Sources: Ba8.**

Cherokee Village - Flour corn, 10 ft. stalks with some purple, multicolored large flat kernels. *Source History: 1 in 2004.* **Sources: Ho13.**

Cherokee White - 110 days - Old, traditional corn from an eastern band of Cherokees, sturdy stalks grow 10-15 ft. tall, good for interplanting with beans, ears 6-9 in. long, 8 rows of creamy white kernels, makes very good flour. *Source History: 2 in 1998; 3 in 2004.* **Sources: Ho13, Se17, So1.**

Cheyenne Pencil - Tall stalks, white kernels, used as a flour corn, grown by Southwest Indians for many years. *Source History: 1 in 2004.* **Sources: Fe5.**

Chulpi Cancha - Plants 8-12 ft. tall, ears 8 in. long, 10-14 rows, can also be grown for corn-on-the-cob. *Source History: 1 in 2004.* **Sources: Red.**

Colorado Blue - 110 days - Large ears, dark blue kernels, for drying and making blue corn flour. *Source History: 1 in 1998; 1 in 2004.* **Sources: Ros.**

Cuzco Cancha - Plants 15 ft. tall, ears 6-8 in. long, 10 rows, large white variety, favorite in the ancient capitol of the Incan Empire. *Source History: 1 in 2004.* **Sources: Red.**

Encinal Blue - Grown in northern New Mexico at 7200 ft., harvested in 90 days for sweet corn, 120 days for grinding or seed. *Source History: 1 in 1994; 1 in 2004.* **Sources: Na2.**

Escondida Blue - Medium size ears, dark to light blue kernels, from Escondida in south-central New Mexico. *Source History: 1 in 2004.* **Sources: Na2.**

Hopi Blue - (Hopi Blue Starch) - 75-110 days - Ancient flint corn, bushy 5 ft. plants, 8-10 in. ears, smooth blue kernels, traditional staple of the Hopi Indians in northern Arizona. *Source History: 3 in 1981; 2 in 1984; 6 in 1987; 8 in 1991; 9 in 1994; 10 in 1998; 23 in 2004.* **Sources: Ba8, Bo19, Bu3, Ers, Fe5, He8, He17, HO2, Ho13, Hud, Jor, Lej, ME9, Mo13, Red, RUP, Se8, SE14, Se17, Se25, Sk2, Tu6, Vi4.**

Hopi Greasy Hair - (Wiekte) - Plum-colored kernels on 10-12 in. ears, often planted early at Hopi so the harvest can be used for the Home Dance ceremony in July. *Source History: 1 in 1991; 1 in 1994; 1 in 1998; 1 in 2004.* **Sources: Na2.**

Hopi Hominy - Long cobs with large white kernels, used for posole (stew) and grinding. *Source History: 1 in 2004.* **Sources: Na2.**

Hopi Kokoma - 95-105 days - Deep-red to burgundy kernels on long slender ears, traditionally used as a dye for baskets and kilts, dry farmed. *Source History: 1 in 1994; 1 in 1998; 1 in 2004.* **Sources: So12.**

Hopi Orange Red - 100 days - Not a true flint, soft, orange-red kernels, stalks to 9 ft., 1-2 large 12 in. cobs, grinds easily for meal and grits, tolerates alkaline soil and drought conditions, given to the Hopi by Kokopelli, the flute player. *Source History: 1 in 1994; 1 in 2004.* **Sources: Ag7.**

Hopi Pink - 70 days - Bushy 4-4.5 ft. plants, 12-14 rows of small beautiful pastel pink kernels on 8 in. slender ears, traditional, used for meal and parched corn, consistently does best in all drought tol. tests due in part to a long embryonic taproot, usually dry farmed. *Source History: 1 in 1987; 4 in 1991; 3 in 1994; 6 in 1998; 6 in 2004.* **Sources: Ho13, Na2, Ri12, Se7, Se17, So12.**

Hopi Purple - 80-105 days - Sturdy 6-8 ft. plants, 2 ears per stalk, 8 in. long, 10 rows, soft dried kernels are ground into flour of excel. flavor and quality, also an attractive ornamental, pre-Columbian. *Source History: 1 in 1981; 1 in 1991; 4 in 1994; 3 in 1998; 2 in 2004.* **Sources: Ho13, Se7.**

Hopi Red - 85-90 days - Ancient traditional flour corn with striking deep-red and an occasional blue kernel, two 7-8 in. ears on 7 ft. red-veined stalks, very adaptable, fairly drought and alkaline soil tol., second of the four sacred colors given by Kokopelli, the humpbacked flute player. *Source History: 1 in 1981; 1 in 1984; 1 in 1987; 3 in 1991; 3 in 1994; 1 in 1998; 3 in 2004.* **Sources: Na2, Red, Se17.**

Hopi Red, White and Blue - Plants 4-5 ft. tall, ears 5-9 in. long, 10-16 rows, each ear contains red, white and blue kernels. *Source History: 1 in 2004.* **Sources: Red.**

Hopi Speckled - White kernels with blue speckles intermixed with solid blue and white, dry farmed at Hopi. *Source History: 1 in 1994; 1 in 1998; 1 in 2004.* **Sources: Na2.**

Hopi White - 80-105 days - Ears 6-8 in. long with soft white kernels, very drought tolerant, usually grown in sand dunes, does best when planted in mid-spring in moist sandy ground. *Source History: 1 in 1981; 1 in 1984; 2 in 1987; 3 in 1991; 4 in 1994; 3 in 1998; 2 in 2004.* **Sources: Red, So12.**

Hopi Yellow - 85-95 days - Small 5-6 in. ears on 5-5.5 ft. stalks, drought tol., especially if planted 4-6 in. deep in moist sandy ground in mid-spring, for spring planting in the high cool desert, Hopi Indian seed. *Source History: 2 in 1981; 2 in 1984; 2 in 1987; 4 in 1991; 3 in 1994; 5 in 1998; 3 in 2004.* **Sources: Na2, Se17, So12.**

Incan White, Giant - Plants 12 ft. tall, 6-8 in. long ears, 8 rows of huge seeds used for parching, corn-nuts or hominy. *Source History: 1 in 2004.* **Sources: Red.**

Iroquois Calico - 120 days - Stalks to 6 ft. tall with 1-2 cobs, 6-9 in. long, 8 rows, kernel colors include pastel shades of mauve, red, yellow, blue and white, colors not as bright as calico vars. from the U.S., withstands cool soils, some drought and excessive heat and overnight temps near freezing, ceremonial corn, originally from Donnie "Miles" Hill's collection on the Six Nations Reserve, Ohsweken, Ontario, Canada. *Source History: 1 in 2004.* **Sources: Ag7.**

Iroquois White - (Six Nations Soup and Flour Corn) - 120 days - Heirloom sweet/flour corn traditionally used by the Iroquois Indians, 8 ft. plants, 12 in. ears, 8 rows, white rounded kernels, some 6- or 10-row cobs may appear, for corn soup and bread, tolerates some drought, excessive heat, cool soils and near freezing overnight temps, originally from Six Nations Reserve in Canada, collection of Donnie "Miles" Hill. *Source History: 1 in 1998; 2 in 2004.* **Sources: Ag7, Hud.**

Isleta Blue - 120 days - Tall stalks to 12 ft., very long purple-blue ears, from New Mexico Pueblo, for spring planting in the high cool desert. *Source History: 1 in 1981; 1 in 1984; 1 in 1987; 1 in 1991; 2 in 1994; 1 in 1998; 2 in 2004.* **Sources: Bou, Na2.**

Isleta White - Ears to 11 in., white kernels (and an occasional red), grown in the New Mexico Pueblo south of Albuquerque. *Source History: 1 in 1998; 1 in 2004.* **Sources: Na2.**

Jemez Blue - Large ears with blue, red and purple kernels, grown with irrigation at the Jemez Pueblo. *Source History: 1 in 1991; 1 in 1994; 1 in 1998; 1 in 2004.* **Sources: Na2.**

Jemez White - Medium to large ears, used for hominy and grinding, from Jemez Pueblo northern New Mexico. *Source History: 1 in 1991; 1 in 1994; 1 in 2004.* **Sources: Na2.**

Maiz Blando - White Mexican flour corn with very soft large flat kernels, makes excel. flour for tortillas and tamales, fairly long season. *Source History: 1 in 1991; 1 in 1994; 1 in 1998; 1 in 2004.* **Sources: Na2.**

Maiz Morado - Tall vigorous stalks, unusual 8-rowed ears with deep purple-red kernels, from Mexico. *Source History: 1 in 1991; 1 in 1994; 1 in 1998; 2 in 2004.* **Sources: Hud, Red.**

Mandan Bride - 89-98 days - Flour corn from the Mandan tribe in what is now North Dakota, distinct from flint because nearly whole kernel is soft, white, starchy and easy to grind, very early, 5-6 ft. plants, 8-12 rows of kernels, striking array of colors: purple, white, variegated, red, yellow, some with a beautiful translucent rosy effect, not recommended for areas below 40 degrees latitude. *Source History: 4 in 1981; 3 in 1984; 3 in 1987; 6 in 1991; 5 in 1994; 5 in 1998; 6 in 2004.* **Sources: Fe5, Ho13, Jo1, Pe2, Sa9, Tu6.**

Mandan Red - 70-98 days - Reddish-black selection of the Mandan strain, 5-6 in. long ears with 8-12 rows, pale yellow kernels can be eaten as sweet corn, starchy soft white kernels are easy to grind, good for parching, developed in northeastern Washington State. *Source History: 1 in 1991; 4 in 1994; 3 in 1998; 3 in 2004.* **Sources: Ho13, Ri12, Se7.**

Mayo Tosabatchi - 90 days - Soft white kernels, 70-75 days to elote, 90 to maturity, ground to make a soft flour/meal used for cookies, Blando de Sonora landrace from Sinaloa, Mexico. *Source History: 1 in 1987; 1 in 1991; 1 in 1994; 1 in 2004.* **Sources: Na2.**

Mexican Morado - Nutty tasting, huge pink-red kernels, superb for making hominy, grits, corn meal and roasting, seed collected at markets in Mexico, Baker Creek introduction. *Source History: 1 in 2004.* **Sources: Ba8.**

Mojave - Similar to Papago 60-day but faster, from Colorado River Indians, for planting during the summer rains in low hot desert. *Source History: 1 in 1981; 1 in 1984; 1 in 1987; 1 in 1991; 1 in 1994; 1 in 1998; 1 in 2004.* **Sources: Na2.**

Navajo Blue - (Alamo Navajo Blue) - 90-110 days - Stalks 7 ft. tall, short slender 6.5 in. ears, large blue-black kernels, used in making blue corn tortillas and piki bread, can be eaten fresh when young, also ornamental, dry farmed, northern Arizona. *Source History: 1 in 1987; 4 in 1991; 2 in 1994; 4 in 1998; 6 in 2004.* **Sources: Gr29, Pla, Ra5, Se8, Se28, Wo8.**

Navajo Gold - Yellow kernels with some white red and blue, grown near Gallup, New Mexico. *Source History: 1 in 1994; 1 in 1998; 1 in 2004.* **Sources: Na2.**

Navajo Robin's Egg - 114 days - Plants to 5 ft., long leaves, 8 in. ears, speckled blue, white and red kernels, good flavor, dry farmed. *Source History: 1 in 1987; 1 in 1991; 2 in 1994; 2 in 1998; 1 in 2004.* **Sources: Na2.**

Navajo White - Smaller kernels and more slender cobs than Acoma, dry farmed. *Source History: 1 in 1987; 2 in 1991; 1 in 1994; 1 in 1998; 1 in 2004.* **Sources: Na2.**

Osage Brown Flour - Old sacred flour corn of the Osage tribe, 6-7 ft. tall plants bear 4-6 in. long ears, kernels are various shades of brown, occasional dent and sweet kernels. *Source History: 1 in 2004.* **Sources: Ho13.**

Painted Mountain - 70-90 days - Cold and drought tolerant variety developed in Montana over the past 30 years from over 1000 native and commercial strains from around the world, multi-colored kernels and tassels, great for cornbread and muffins, plants grow to 5 ft., 7 in. ears, 13% protein content. *Source History: 1 in 1994; 3 in 1998; 17 in 2004.* **Sources: Ag7, Ba8, Bou, Ech, Fe5, Ga1, Ho13, Jo1, La1, Ni1, OSB, Pe1, Pe2, Pe6, Ri12, Sto, Ter.**

Parch, Supai Red - 110 days - Hopi flour corn, one of the best for parching, 6-7 ft. stalks with 1-2 large ears per stalk, white kernels with intense red blazes and starburst patterns. *Source History: 1 in 1998; 1 in 2004.* **Sources: Se7.**

Peruvian Morado - (Inca Purple) - 180 days - Plants to 15 ft., deep purple cobs, short, plump ears, 2.5-6 in. long, 8-12 rows of deep violet, round kernels, used in making "chicha", an ancient corn drink, ancient Incan variety originating in Peru. *Source History: 1 in 1998; 2 in 2004.* **Sources: Ag7, Hud.**

Pink and White - 90 days - Deeply rooted stalks grow 5-8 ft. tall, some tillering, can lodge, 8-14 in. ears, 8 rows, kernel color ranges from white to all shades of pink to almost garnet, kernel color is the same on each ear, excellent flour corn. *Source History: 1 in 2004.* **Sources: Se17.**

Posole - 100 days - Traditional variety of dry corn for making posole--the hominy of the Southwest, vigorous 12-14 ft. plants, large plump ears, solid white plump kernels. *Source History: 2 in 1987; 1 in 1991; 1 in 1994; 1 in 1998; 1 in 2004.* **Sources: Pla.**

Quapaw - Thin, tall 6-10 ft. plants, 6 in. ears, magenta-purple kernels, the Quapaws lived in the Ozarks. *Source History: 1 in 2004.* **Sources: Ho13.**

Red Midget - Plants grow 2.5-3.5 ft. tall with 3 ears per plant, some smaller ears on side shoots, 3-5 in. long, dark red kernels, rare. *Source History: 2 in 1998; 1 in 2004.* **Sources: Ho13.**

Rio Lucio Concho - Pearl white kernels used for chicos, posole or flour, grown at 8000 ft. in the Spanish village of Rio Lucio. *Source History: 1 in 1998; 1 in 2004.* **Sources: Na2.**

Sahuarita - Parching corn, family heirloom from the 1920s from Sahuarita, south of Tucson, AZ. *Source History: 1 in 2004.* **Sources: Na2.**

San Felipe Pueblo Blue - Small kernels on long slender ears, grown with irrigation in the Rio Grande Valley of New Mexico. *Source History: 1 in 1991; 1 in 1994; 1 in 1998; 1 in 2004.* **Sources: Na2.**

San Felipe Pueblo White - Ground for meal, whole kernels used in stews, from the northern New Mexico pueblo. *Source History: 1 in 1994; 1 in 1998; 1 in 2004.* **Sources: Na2.**

San Luis Concho - Small pearly kernels, short, cool season corn grown at 8500 ft. in San Luis, CO. *Source History: 1 in 1998; 1 in 2004.* **Sources: Na2.**

Santa Ana Pueblo Blue - 110-120 days - Stalks to 9 ft. tall with 1-2 ears, 8-9 in. long, beautiful dark blue kernels, flint/flour, originally cultivated in the Magdalena River basin in northern Mexico. *Source History: 1 in 1991; 1 in 2004.* **Sources: Ag7.**

Santo Domingo Blue - Large ears up to 15 in., large kernels of deep-blue, from Santo Domingo Pueblo of New Mexico. *Source History: 1 in 1991; 1 in 1994; 1 in 2004.* **Sources: Na2.**

Santo Domingo White - Staple corn grown in the Pueblo of northern New Mexico. *Source History: 1 in 1994; 1 in 1998; 1 in 2004.* **Sources: Na2.**

Taos Pueblo Blue Corn - (Taos Blue) - 100-125 days - Ancient traditional corn, grown on the highest mountain in New Mexico, irrigated from sacred Blue Lake, slim nearly black ears, many 12 in. and longer, drought res., has been grown successfully in all states in the U.S. except Alaska. *Source History: 4 in 1981; 4 in 1984; 3 in 1987; 5 in 1991; 2 in 1994; 2 in 1998; 3 in 2004.* **Sources: Na2, Ri12, Se17.**

Tarahumara Blando de Sonora - Large ears, soft white kernels, red cobs, used in milk stage for making tamales, one of the mainstays of Tarahumara corn production. *Source History: 1 in 1987; 1 in 1991; 1 in 1994; 1 in 1998; 1 in 2004.* **Sources: Na2.**

Tarahumara Harinoso de Ocho - Large ears, large flat kernels, grown at low elevations in the Barranca del Cobre, Chihuahua, Mexico. *Source History: 1 in 1998; 1 in 2004.* **Sources: Na2.**

Tarahumara Maiz Rojo - Color variant of Tarahumara Maiz Azul with kernels ranging in color from light red to blood-red to chocolate-red. *Source History: 1 in 1998; 1 in 2004.* **Sources: Na2.**

Tehama Dawn - Heavy crops of full 16-row ears, gold to fiery red kernels, Sunrise Farm heirloom. *Source History: 1 in 1994; 1 in 1998; 1 in 2004.* **Sources: Syn.**

Texas Shoepeg - Bright yellow elongated cone-shaped kernels with dented beaks, short red cobs have up to 26 rows, excel. for meal or boiling, kernels are sweet in milk stage, grown in Tucson, AZ. *Source History: 1 in 1991; 2 in 1994; 2 in 1998; 1 in 2004.* **Sources: Na2.**

Tohono O'odham 60-Day - Short stalks, 6-10 in. ears with white kernels, traditionally grown by the Tohono O'odham with the summer rains in floodwater fields. *Source History: 1 in 1994; 1 in 1998; 1 in 2004.* **Sources: Na2.**

Truchas Lumbroso - White kernels, some tinted pink, from Truchas, NM. *Source History: 1 in 2004.* **Sources: Na2.**

Tuscarora White - 90-115 days - Grows 6-8 ft. tall, long cobs with large flat kernels, 8 rows, good for growing with cornhill beans, original seed from a corn braid made on the Tuscarora Nation Reservation in New York, heirloom. *Source History: 1 in 2004.* **Sources: Se17.**

Vadito Blue - Pale to dark-blue kernels, used for grinding, grown at 8000 ft. in northern New Mexico. *Source History: 1 in 1994; 1 in 1998; 1 in 2004.* **Sources: Na2.**

Vadito Concho - Pearly, flour-flint corn for short, cool season areas at high elevation (8000 ft.), from Vadito, NM. *Source History: 1 in 1998; 1 in 2004.* **Sources: Na2.**

Vadito White - Matures early, posole or flour, from Vadito, NM. *Source History: 1 in 2004.* **Sources: Na2.**

Varieties dropped since 1981 (and the year dropped):

Acoma Cherry Blue (1994), *Acoma White* (2004), *Acoma Yellow* (1994), *African Zulu Maize* (1991), *Amarillo Terciopelo* (1994), *American Maize* (1991), *Assiniboine* (2004), *Bear Creek Flour* (1987), *Big Timber Flour* (1998), *Black Vernadeno* (1984), *Blue Corn* (2004), *Blue Tortilla Corn* (1991), *Caspecio Blue* (1994), *Cheyenne Red* (2004), *Chiapas Blue* (1984), *Chichito* (1991), *Cochiti Pueblo Blue* (2004), *Cochiti Pueblo White* (2004), *Concho White* (1994), *Gaspe* (2004), *Guadalajara All Purpose* (2004), *Havasupai* (1994), *Hopi Blue (#2)* (1998), *Hopi Blue Tall* (1994), *Hopi Blue, FPS Strain* (1998), *Hopi Blue, Sekwakae* (1994), *Hopi Blue/Black* (1987), *Hopi Cherry Blue* (1998), *Hopi Chin Mark* (2004), *Hopi Coconino* (1991), *Hopi Purple/Blue* (1987), *Hopi Turquoise* (1998), *Hopi White Hominy* (1998), *Jicarilla Apache White* (1998), *Maiz Negro* (1994), *Mandan Lavender* (2004), *Mandan Yellow* (1994), *Mandan, Bear Creek Strain* (1998), *Mayo Maiz Blando* (2004), *Mt. Pima Tuk Huna* (1991), *Navajo* (1998), *Navajo Whitish Flour* (1994), *Navajo Yellow* (2004), *Nichol's Hominy Corn* (1982), *O'odham 60 Day* (1994), *Papago* (1998), *Pawnee Blue* (2004), *Ponca Blue Flour* (2004), *Pueblo Blue* (1987), *Pueblo White* (1994), *Red Corn* (1991), *Red Field* (1991), *Rio Lucio Blue* (2004), *San Juan Pueblo Mix* (1994), *Santo Domingo Yellow/Blue* (2004), *Seibel's Red* (2004), *Seneca White* (1994), *Smoky Blue* (1998), *Tablita* (1987), *Taos Pueblo White Corn* (1991), *Taos White* (1994), *Tarahumara Bofo* (1998), *Tarahumara Gordo* (1998), *Tarahumara Maiz Azul Blanco* (1991), *Tarahumara Maiz Gordo "Rosari"* (2004), *Tepehuan Harinoso de Ocho* (1998), *Velarde Azul y Blanco* (2004), *Velarde Blue* (2004), *White American Maize* (1991), *White Posole Corn* (1991), *White Vernadeno* (1987), *Yaqui Blue* (1998), *Yoeme Blue* (2004), *Zulu Maize* (1987).

Corn / Pop
Zea mays

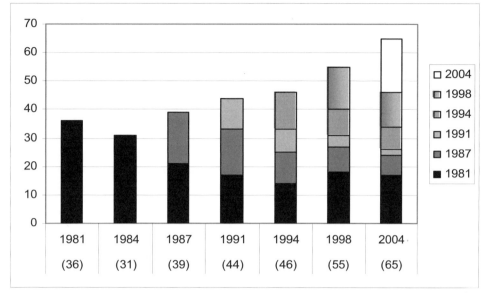

Number of varieties listed in 1981 and still available in 2004: 36 to 17 (47%).
Overall change in number of varieties from 1981 to 2004: 36 to 65 (181%).
Number of 2004 varieties available from 1 or 2 sources: 47 out of 65 (72%).

Amethyst - 115 days - Stalks over 8 ft. tall, 4-6 in. ears, 2-4 per stalk, pretty violet/light purple kernels, blond husk. *Source History: 2 in 2004.* **Sources: HO2, ME9.**

Baby Rice - 70-100 days - Old-time Wisconsin favorite, 4-5 ft. stalks, 2-6 small ears per plant, hulless, small white kernels pop into tender popcorn, over 100 years old. *Source History: 1 in 1981; 1 in 1984; 1 in 1987; 2 in 1991; 3 in 1994; 3 in 1998; 1 in 2004.* **Sources: Jun.**

Baby Rice, Hulless - 85-108 days - Stalks 4-7 ft., 3-6 small ears per stalk, very tender. *Source History: 1 in 1981; 1 in 1984; 1 in 1987; 1 in 1991; 2 in 1994; 1 in 1998; 1 in 2004.* **Sources: Sa9.**

Bearpaw - 90-110 days - New England heirloom, name comes from the somewhat flattened silk end of its ears, white-seeded, 4-5 in. blocky ears, 5-6 ft. stalks, dates to the early 1900s. *Source History: 1 in 1987; 2 in 1991; 2 in 1994; 2 in 1998; 1 in 2004.* **Sources: Pr3.**

Black - 80 days - Mennonite heirloom, 6 ft. stalks bear 2 ears each, 4-5 in. ears with black and dark maroon kernels, pops to bright white with black accents. *Source History: 1 in 2004.* **Sources: So1.**

Black Aztec - 75-100 days - Unusual black popcorn from South America, 6 ft. stalks, 7 in. ears, large blue-black kernels pop white and tender, pops half a bushel from 1 lb. of kernels, fine flavor, also makes decorative centerpieces. *Source History: 2 in 1981; 2 in 1984; 2 in 1987; 2 in 1991; 4 in 1994; 4 in 1998; 2 in 2004.* **Sources: He17, Jor.**

Blue Popcorn - Specialty food from the Southwestern U.S., late maturity, may not ripen in the North. *Source History: 1 in 1981; 1 in 1991; 1 in 1998; 1 in 2004.* **Sources: De6.**

Calico - 90-105 days - Brilliantly colored 6-8 in. ears, pops well for white popcorn, kernel colors are reds yellows mahogany brown white purple blue and lots of stripes and patterns, heirloom. *Source History: 2 in 1981; 2 in 1984; 2 in 1987; 3 in 1991; 2 in 1994; 8 in 1998; 11 in 2004.* **Sources: Fe5, GLO, Gr28, Jor, Loc, Mel, Shu, Sw9, Ter, Und, We19.**

Carousel - 104-110 days - Highly decorative popcorn, 7 ft. plants average 4-6 ears per plant plus tiller ears, 4-5 in. ears are uniform in size, wildly variable in color, delicate red and yellow and white and blue and purple kernels, dries beautifully, tiny seeds - 8,000 per lb. average, plant 8-10 in. apart. *Source History: 5 in 1987; 21 in 1991; 21 in 1994; 24 in 1998; 23 in 2004.* **Sources: ABB, Bu8, Com, Dam, De6, Ers, He17, HO2, Jor, Lej, Loc, Ma19, MAY, ME9, Mey, Mo13, MOU, Ni1, OLD, RIS, Sau, SE4, Twi.**

Chapalote - (Pinole Maiz) - 90-150 days - Ancient, slender-eared pinole maize, 10 ft. stalks, multiple tillers, ears at 6-8 ft. on the stalk, ears 10 in. long, small brown kernels-- the only brown corn, used by Indians to make a drink called pinole, originally collected in Sinaloa, Mexico. *Source History: 1 in 1987; 1 in 1991; 3 in 1994; 1 in 1998; 4 in 2004.* **Sources: Ag7, Na2, Ri12, Se17.**

Chapalote, Mayo Yellow - Flinty yellow corn, makes flavorful pinole, from Piedras Verdes area. *Source History: 1 in 2004.* **Sources: Na2.**

Charleston Black - Color is very dark burgundy-red, kernels progress from white to red, color darkens in storage. *Source History: 1 in 1998; 1 in 2004.* **Sources: Al6.**

Cherokee - 100-110 days - Stalks to 9 ft., 2-4 ears at shoulder height on the plant, shows considerable tolerance to heat and drought, good vigor, makes great popcorn, also used decoratively, blend of Cherokee popcorns brought over the Trail of Tears, seed from Carl Barnes of Oklahoma. *Source History: 1 in 2004.* **Sources: Und.**

Cherokee Long Ear - Ornamental corn with small kernels that pop large, colors include red, blue, orange, white and yellow. *Source History: 1 in 2004.* **Sources: So1.**

Chinese Baby - Small ears used for popping. *Source History: 1 in 1994; 1 in 2004.* **Sources: Bo19.**

Chires - (Chires Baby Corn) - 75-85 days - Stalks to 4-5 ft., produce 8-12 miniature ears, 4 in. long, popular in Chinese dishes, dried kernels can be popped. *Source History: 5 in 2004.* **Sources: Ho13, Pe1, Ri12, Roh, Se7.**

Chocolate Cherry - 120 days - Selected from Chocolate Pop, produces a more uniform, deep red cherry-brown color, bright pink silks, sturdy stalks grow 5.5-6 ft., two 6 in. ears per stalk, 16-18 rows of kernels, has the flavor of a good tasting yellow popcorn with more tender hulls, good res. to drought and earworms, intro. in 1998 by CV So1. *Source History: 1 in 1998; 1 in 2004.* **Sources: So1.**

Cochiti Pueblo - 90-110 days - Rare old var., stalks to 6-7 ft. tall, 2-3 small colored ears per stalk, rainbow of colors, rich, nutty flavor, has drought resistant gene. *Source History: 3 in 2004.* **Sources: Ho13, Sa9, Se17.**

Cranston Hulless - 120 days - Plants to 6 ft., up to 6 ears each, very tiny kernels on 2-3 in. long ears, can be used as miniature corn in Chinese dishes when picked young. *Source History: 1 in 2004.* **Sources: Sa9.**

Cutie Blues - 100 days - Dark-blue kernels on 4 in. ears. *Source History: 1 in 1994; 1 in 1998; 1 in 2004.* **Sources: Sto.**

Cutie Pink - 100 days - Slim ears 4 in. long, rose-pink kernels. *Source History: 1 in 1998; 1 in 2004.* **Sources: Sto.**

Cutie Pops - 100 days - Miniature popcorn, 4 in. ears, multicolored kernels, use as popcorn or ornamental corn in dried bouquets. *Source History: 1 in 1987; 2 in 1991; 3 in 1994; 2 in 1998; 1 in 2004.* **Sources: Sto.**

Golden - 85 days - Plants 4-5 ft. tall, 4-6 in. ears, 14 rows, golden kernels. *Source History: 1 in 1981; 1 in 1984; 1 in 1987; 1 in 1991; 1 in 2004.* **Sources: Red.**

Golden Queen - 90 days - *Source History: 1 in 1981; 1 in 1984; 1 in 1987; 1 in 2004.* **Sources: LO8.**

Indian Berries - Strawberry-shaped 2-3 in. ears, kernel colors include yellow, blue, white and red, very ornamental. *Source History: 1 in 2004.* **Sources: Ho13.**

Indian Popping Corn - (Little Indian Ornamental Popcorn) - 100 days - Small 4 in. ears, yellow and red kernels, 4-5 ft. plants, can be popped or used for decorations. *Source History: 1 in 1981; 1 in 1984; 2 in 1987; 3 in 1991; 3 in 1994; 2 in 1998; 3 in 2004.* **Sources: Ada, Co31, Pla.**

Japanese Hulless - (White Rice, Early White, Tom Thumb, Japanese White Hulless, Bumblebee) - 83-110 days - Hulless fiber-free and tender, 4-5 ft. vigorous plants, 4 in. ears, irregular rows, tender small slender white kernels, excel. popper that pops pure-white with no hard centers. *Source History: 10 in 1981; 10 in 1984; 16 in 1987; 18 in 1991; 23 in 1994; 22 in 1998; 26 in 2004.* **Sources: Ada, Be4, Bou, Bu3, Bu8, Eo2, Ers, Fi1, Gur, He17, Jor, Jun, La1, Me7, Mel, Mo13, Ri12, RUP, Sau, Se7, Se25, Se26, Sh9, Shu, So12, Wi2.**

Kauders - 105 days - Plants to 7-8 ft. tall, 6 in. ears with both yellow and white kernels, pops large. *Source History: 1 in 2004.* **Sources: Sa9.**

Ken - Plants grow 6-8 ft. tall, 8 in. ears with kernels in colors of red, yellow, white, rose and purple, some entirely purple, dev. by Ken Asmus of Oikos Tree Crops from a cross of two heirloom multi-colored popcorns, has grown them out for over 10 years, selecting for large size ears. *Source History: 1 in 2004.* **Sources: Hud.**

Lady Finger - (Lady Finger Rainbow) - Heirloom from the Amish in Ohio and Indiana, short stalky plants bear multiple 6-7 in. long slender ears, mostly deep-yellow kernels with some red and purple mixed in, excellent flavor, pops well, pretty and unusual in dried arrangements. *Source History: 2 in 1987; 2 in 1991; 3 in 1994; 3 in 1998; 5 in 2004.* **Sources: Ba8, Be4, Ers, Ho13, Hud.**

Laser - 90-95 days - Small ears 4-6 in. long, pointed kernels with calico stripes, 7.5-8 ft. stalks stand well, PVP 1996 Larry and Lucyle Eckler. *Source History: 2 in 1998; 2 in 2004.* **Sources: Jor, ME9.**

Little Bow Peep - 100 days - Pink-purple ornamental used for decorating and floral arrangements. *Source History: 2 in 1994; 1 in 1998; 1 in 2004.* **Sources: RUP.**

Little Boy Blue - 100-110 days - Dark-blue kernels on 4 in. ears. *Source History: 3 in 1994; 2 in 1998; 1 in 2004.* **Sources: RUP.**

Mahogany - 95 days - Tender gourmet type popping corn, 4-5 in. cobs, 2-3 per stalk, dark black/brown kernels. *Source History: 2 in 2004.* **Sources: HO2, ME9.**

Miniature - 100-110 days - Stalks to 6 ft., narrow cobs 3-5 in. shiny kernels in wide range of colors. *Source History: 1 in 1987; 2 in 1991; 4 in 1994; 4 in 1998; 7 in 2004.* **Sources: Eo2, Fa1, Gr29, SE4, Se28, Ver, Wo8.**

Miniature Blue - (Country Blue, Baby Blue) - 100-115 days - Stalks 5-6 ft. tall, 5 in. cobs, medium to deep blue kernels, excel. popping quality, ideal for fall decorations and dried arrangements. *Source History: 2 in 1991; 16 in 1994; 20 in 1998; 23 in 2004.* **Sources: Bo18, CA25, CH14, GLO, Gr29, Ha5, He17, HO2, Jor, Jun, Lej, Ma19, ME9, Mo13, OSB, Pin, Roh, SE4, Se28, Sh9, Shu, Twi, Wo8.**

Miniature Pink - (Country Pink, Pink Bo Peep, Little Miss Muffet) - 100-105 days - Each 5-6 ft. stalk bears 2-3 ears, 3-5 in. long, kernel color varies from pink to mauve to light-purple, used for dried arrangements or popping. *Source History: 11 in 1994; 15 in 1998; 19 in 2004.* **Sources: Bo18, Bu2, CA25, CH14, GLO, Gr29, He17, HO2, Jor, Jun, ME9, Mo13, Pin, Roh, SE4, Se28, Sh9, Twi, Wo8.**

Mountain View Farm - 90-95 days - Ears 5-6 in. long, pale yellow kernels pop white. *Source History: 2 in 2004.* **Sources: Pe2, Tu6.**

Neon Pink - 110 days - Small ears, 3-4.5 in., bright pink to dark pink kernels. *Source History: 1 in 2004.* **Sources: Jor.**

Onaveno - Ancient grinding corn used for pinole with flinty, cream colored kernels, from along the Rio Mayo on Sonora. *Source History: 1 in 1998; 1 in 2004.* **Sources: Na2.**

Pennsylvania Butter Flavored - 102 days - Stalks to

8 ft., 4-6 in. ears contain 26-28 rows of kernels, pre-1885 heirloom maintained by the Pennsylvania Dutch, CV So1 introduction, 1988. *Source History: 2 in 1991; 1 in 1994; 1 in 1998; 1 in 2004.* **Sources: So1.**

Pennsylvania Dutch Butter Flavor - 105 days - Pennsylvania Dutch heirloom from the late 1800s, stalks bear two 4-6 in. ears with small cream colored kernels, original seed from the Seed Savers Exchange. *Source History: 1 in 1994; 1 in 1998; 1 in 2004.* **Sources: Lan.**

Pink Beauty - 100 days - Plants to 5 ft. tall, 3-4 in. ears, kernels are bright shades of pink. *Source History: 1 in 1998; 1 in 2004.* **Sources: Sa9.**

Pink, Early - 85-90 days - Decorative pink to mauve colored kernels on 5-6 in. cobs, pops to fluffy white, 5 ft. stalks bear two ears, rich flavor. *Source History: 4 in 1998; 6 in 2004.* **Sources: Hud, ME9, OSB, Pin, Ter, We19.**

Popwhite - 105-112 days - White kerneled popcorn for the North, 4 in. ears, 3 to 4 per plant, good popping quality, early enough for most growing seasons. *Source History: 1 in 1981; 1 in 1984; 1 in 1987; 1 in 2004.* **Sources: Sa9.**

Pretty Pops - 95 days - Ornamental popcorn, 6 ft. plants, 5 in. ears, red and blue and black and yellow and purple and orange kernels turn white when popped, superb flavor, crunchy, nutty, never tough. *Source History: 1 in 1987; 3 in 1991; 2 in 1994; 2 in 1998; 1 in 2004.* **Sources: SE4.**

Purple - Lovely purple kernels, used for popping or as an ornamental corn, popular with the Amish. *Source History: 1 in 2004.* **Sources: Ba8.**

Red Beauty - 126 days - Plants to 6 ft., red kernels. *Source History: 1 in 1998; 1 in 2004.* **Sources: Sa9.**

Red Sunrise - 105 days - Ears 3-6 in., brown and yellow earth tone kernel color, over 50% purple husks. *Source History: 1 in 2004.* **Sources: ME9.**

Reventador - Old-fashioned pinole corn with translucent white kernels, once grown in Arizona with irrigation. *Source History: 1 in 1994; 1 in 1998; 1 in 2004.* **Sources: Na2.**

Shades of Brown - 100 days - *Source History: 1 in 2004.* **Sources: ME9.**

Smoke Signals - 105-115 days - Stalks average 8 ft., abundant production of slender 4-7 in. ears with kernels in shades of blue and pink, mahogany, white and yellow, ornamental and delicious, tender when popped. *Source History: 2 in 2004.* **Sources: Ha5, ME9.**

South American - (South American Giant) - 68-110 days - Late yellow popcorn, 76 in. plants, 7-8 in. ears with 12-14 rows, good bacterial wilt res. *Source History: 10 in 1981; 10 in 1984; 9 in 1987; 7 in 1991; 9 in 1994; 5 in 1998; 4 in 2004.* **Sources: Ada, Ers, He17, Se26.**

South American Yellow - (Dynamite) - 103-115 days - Large ears, large yellow-orange kernels, high yielding, a long-time favorite, 5 ft. stalks, 6 in. ears, good quality, popular with commercial poppers. *Source History: 7 in 1981; 7 in 1984; 7 in 1987; 8 in 1991; 10 in 1994; 7 in 1998; 6 in 2004.* **Sources: Bu8, Jor, Mo13, RUP, Se25, Wet.**

South American Yellow Giant - 110-115 days - Ears 6 in. long, yellow kernels. *Source History: 1 in 1981; 1 in 1984; 1 in 1987; 1 in 1991; 1 in 1994; 2 in 1998; 3 in 2004.* **Sources: ME9, Mey, Sh9.**

Spectrum - (Red Husk Spectrum) - 100-105 days - Sturdy stalk to 8.5-9.5 ft. tall, very colorful, 4.5 in. long ears, high percentage of red husks. *Source History: 2 in 1998; 2 in 2004.* **Sources: Jor, ME9.**

Strawberry - (Dwarf Strawberry, Red Strawberry) - 80-114 days - Very ornamental variety, 4 ft. stalks, 1.5 x 2 in. long ears, 2-4 per stalk, irregular rows of mahogany-red kernels, pops well. *Source History: 42 in 1981; 37 in 1984; 50 in 1987; 61 in 1991; 61 in 1994; 64 in 1998; 64 in 2004.* **Sources: Bo18, Bu1, Bu3, Bu8, CA25, CH14, Com, Coo, Dam, De6, Eo2, Ers, Fe5, Fo7, Ge2, Goo, Gr29, HA3, Ha5, He17, HO2, Hud, Jor, Jun, Lej, Lin, Loc, Ma19, MAY, ME9, Mel, Mey, Mo13, MOU, Ni1, OLD, Or10, OSB, Pla, Ri12, RIS, Roh, RUP, Sau, SE4, Se8, SE14, Se16, Se26, Se28, Sh9, Shu, Si5, So1, Sto, Tho, Twi, Ver, Vi4, Vi5, WE10, Wet, Wi2, Wo8.**

Thanksgiving Bouquet - 110 days - Plants to 7 ft., 2 ears per stalk, 6-7 in. long, most of the kernels in shades of brown, the others are colorful as well, beautiful ornamental selected by Glenn Drowns. *Source History: 1 in 2004.* **Sources: Sa9.**

Tom Thumb - (Hulless, Dwarf Rice, Squirrel Tooth) - 80-90 days - Very dwarf, stubby 3.5 in. ears, long narrow pointed white kernels, hulless, no hard centers, fine eating quality, good for coastal and short-season areas, dev. from a New Hampshire heirloom by the U of New Hampshire. *Source History: 4 in 1981; 3 in 1984; 3 in 1987; 6 in 1991; 8 in 1994; 7 in 1998; 9 in 2004.* **Sources: Ag7, Eco, Ga1, Hi6, Or10, Pe2, RUP, Sa9, Se16.**

Tom Thumb Yellow - 85-105 days - Suited for the far North, dwarf 3.5 ft. plants bear two to four 3-4 in. ears, refined from a New Hampshire heirloom by Prof. Meader, good popping quality, very tender. *Source History: 3 in 1981; 3 in 1984; 1 in 1987; 1 in 1991; 1 in 1994; 1 in 1998; 1 in 2004.* **Sources: Jo1.**

Tom Thumb, Ornamental - 85 days - Bred from yellow Tom Thumb, stalks 3.5 ft. tall, kernels of all colors. *Source History: 1 in 1998; 1 in 2004.* **Sources: Fis.**

White Rice - 110 days - Pearly white kernels, hulless. *Source History: 2 in 1981; 2 in 1984; 1 in 1987; 2 in 1991; 2 in 1994; 1 in 1998; 1 in 2004.* **Sources: Mey.**

Wilda's Pride - 110 days - Tall, sturdy plants produce 8-11 in. ears ranging from yellow and white, deep maroon-red, solid yellow, black-red, and a wide range of multi-colored ears, beautiful ornamental. *Source History: 1 in 1998; 3 in 2004.* **Sources: Ha5, Loc, WE10.**

Wisconsin Black - Ears 3-5 in. long, pointed shiny black kernels, delicious. *Source History: 1 in 1994; 1 in 1998; 1 in 2004.* **Sources: So9.**

Yellow Popcorn - 110 days - Small tender hulless popcorn, short strong stalks, slightly tapering 7 in. ears, outstanding Iowa heirloom. *Source History: 1 in 1981; 1 in 1984; 1 in 1991; 3 in 1998; 2 in 2004.* **Sources: MOU, Pr3.**

Varieties dropped since 1981 (and the year dropped):

Blue Boy (2004), *Brown* (1998), *Chocolate Pop* (2004), *Cochiti* (2004), *Colored Popcorns* (1982), *Cutie Brown* (2004), *Dwarf Cob Ornamental* (1998), *Dynamite* (2004), *Earligold* (1987), *Early White Hulless* (1987), *Eastern Sunburst* (1982), *Faribo Hulless White* (1991), *Golden* (2004), *Golden Crisp* (1987), *Golden Pearl* (1994), *Hulless* (2004), *Indian Calico* (1987), *Indian Dark* (1994), *Indian Light* (1994), *Indian Pink and Purple* (1994), *Indian Red* (1994), *Japanese Yellow Hulless* (2004), *Large Seed* (1994), *Licorice Bean* (1987), *McSmith* (1998), *Midget Pop* (1987), *New England White Pop* (1998), *New York Red Robin* (2004), *Ornamental, Barnes* (2004), *Papoose* (2004), *Pretty Pops, Shades of Brown* (2004), *Pueblo Multicolored* (1994), *Queens Golden* (1991), *Rainbow Baby* (1998), *Red* (1998), *Red Popcorn* (1984), *Rhodes' Yellow Pop* (1984), *Sonoran Reventador* (1994), *South American Dynamite* (2004), *Striped* (1998), *Tiny Tender Black* (1987), *Tom Thumb Blue* (1987), *White Hulless* (2004), *White Pearl Hulless Rice* (1987).

Corn / Sweet
Zea mays

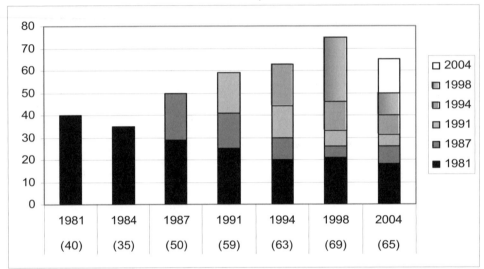

Number of varieties listed in 1981 and still available in 2004: 40 to 18 (45%).
Overall change in number of varieties from 1981 to 2004: 40 to 65 (163%).
Number of 2004 varieties available from 1 or 2 sources: 47 out of 65 (72%).

Anasazi Sweet - Plants grow 5-9 ft., 4-7 in. ears, possibly the most colorful sweet corn, very flavorful, ancient corn, reportedly found by archeologists in the Anasazi ruins. *Source History: 2 in 1998; 1 in 2004.* **Sources: Ho13.**

Art Verrell's - 60-70 days - Heirloom from southwestern Oregon, 4.5 ft. plant, 7 in. ears with 10 rows of uniform, white, tender kernels, tolerates cool weather, selected for over 30 years by Art Verrell for cold-soil germination, earliness, uniformity and fine eating quality, a northern corn, discovered and introduced by PSR. *Source History: 1 in 1991; 2 in 1994; 2 in 1998; 2 in 2004.* **Sources: Bou, Pe6.**

Ashworth - (Rat Selected) - 69-85 days - Sel. from composite of early vars., 4-5.5 ft. plants, fat well-filled 6-7 in. ears, 12 bright-yellow rows, matures with first early Canadian hybrids, excel. flavor, does not hold in milk stage long, New York heirloom from Fred Ashworth, who in the selection process planted the corn from the storage bins most preferred by rats, first commercially available in 1978. *Source History: 6 in 1981; 5 in 1984; 2 in 1987; 5 in 1991; 5 in 1994; 11 in 1998; 12 in 2004.* **Sources: Ag7, Ba8, Fe5, Ga1, Hi6, Ho13, Me7, Pe2, Pla, Se8, So1, Tu6.**

Aunt Mary's - 90-105 days - Old Ohio heirloom, two large white-kerneled ears per 6 ft. stalk, 8-10 rows, discovered in the 1930s by Lee Bonnewitz at his Aunt Mary's dinner table. *Source History: 1 in 1981; 1 in 1987; 1 in 1991; 1 in 1994; 1 in 1998; 1 in 2004.* **Sources: Se17.**

Bellingham Blue - 75-80 days - Short ears, kernels dry to navy, preserved by a Bellingham, WA railroad worker, rare heirloom. *Source History: 1 in 2004.* **Sources: Ri12.**

Black Aztec - 70-100 days - Grown by the Aztecs 2,000 years ago, vigorous 6 ft. plants, ears to 18 in. long, for white roasting ears in milk stage, jet black when dry, ground for an excel. blue meal or used for parching, drought tol., intro. into the seed trade in the 1860s. *Source History: 6 in 1981; 4 in 1984; 4 in 1987; 9 in 1991; 10 in 1994; 11 in 1998; 18 in 2004.* **Sources: Bou, Gr29, Hi6, Ho13, ME9, Ra5, Rev, Ri12, Roh, Se7, Se8, Se16, Se17, Se28, So12, Syn, Ver, Wo8.**

Black Mexican - (Black Aztec, Mexican Sweet, Black Sweet, Black Iroquois) - 62-86 days - Time-honored New England sweet corn, appears to have originated in upper New

York State, possibly derived from Iroquois "black puckers" which was a northern black flint corn, the mutation from the starchy to sugary condition may have caused a puckered appearance, thus the name "black puckers", possibly renamed by a seed company wanting to enhance its seed listing, which was not uncommon in the late 1800s, kernels white at table stage, blue-black at dry maturity, slender 7-8 in. ears, 8 rows, 5 ft. stalks, very hardy, stands all kinds of weather, intro. 1864. *Source History: 6 in 1981; 3 in 1984; 3 in 1987; 2 in 1991; 3 in 1994; 4 in 1998; 4 in 2004.* **Sources: Red, Sa9, Shu, So1.**

Blue Jade - 70-80 days - Plants to 2-3 ft. tall bear 3-6 small ears, steel-blue kernels turn jade-blue when boiled, one of the only sweet corns good for container culture. *Source History: 1 in 2004.* **Sources: Se16.**

Buhl - 81 days - Two ears per 6-7 ft. plant, tender sweet golden kernels, uniform plants, often confused for a hybrid, from Sandhill Preservation Center. *Source History: 1 in 1998; 3 in 2004.* **Sources: Ag7, Sa9, Syn.**

Candy Mountain - 68-73 days - Open-pollinated var. of extra sweet corn, 8 in. ears, golden yellow tender kernels, dev. from Kandy Korn, earlier and has medium-size ears. *Source History: 2 in 1991; 2 in 1994; 1 in 1998; 1 in 2004.* **Sources: Hig.**

Clem Bennett - *Source History: 1 in 2004.* **Sources: Se17.**

Cocopah - 80-100 days - Originally collected in 1868-69 and saved by 3 generations of Arizona prospectors, med-size ears, large multicolored kernels, white at milk stage, matures to shades of pink and red, occasional blue, white and yellow, fast- growing. *Source History: 1 in 1987; 1 in 1991; 1 in 1994; 2 in 2004.* **Sources: Ri12, Se17.**

Country Gentleman - (Shoepeg Corn) - 83-100 days - Standard late white corn, deep narrow small non-rowing kernels, 7-8 in. tapered ears, often 2 ears on each dark-green 7-8 ft. stalk, res. to Stewart's wilt, heavy yields, thin narrow shoepeg kernels, good home garden var., for fresh use or canning, intro. in 1890 by S. D. Woodruff and Sons of Orange, CT, and one year later by Peter Henderson and Co. *Source History: 43 in 1981; 36 in 1984; 30 in 1987; 24 in 1991; 23 in 1994; 19 in 1998; 36 in 2004.* **Sources: Ag7, Ba8, Bo17, Bou, Bu2, CH14, Coo, Ers, Fo7, He8, He17, Hi13, Ho13, La1, Mey, Mo13, Or10, Pe2, Red, Rev, Ri12, Roh, Sa9, Scf, Se16, Se25, Se26, Sh9, Sh13, Shu, Sk2, So1, SOU, Sw9, TE7, Ter.**

Double Play - 80 days - Open-pollinated bi-color variety, sweet tender kernels, 14-16 rows, mid to late season. *Source History: 1 in 2004.* **Sources: Te4.**

Double Red Sweet - Stalks to 6 ft., intense, anthocyanin purple color, 2 ears per plant. *Source History: 1 in 2004.* **Sources: Pe1.**

Double Standard - 73 days - First open-pollinated bi-color sweet corn, early maturing with unusually good germination in cool soil, somewhat variable 7 in. ears, 12-14 rows of yellow and white kernels--some ears with yellow kernels only, yellow and white seed unlike hybrid bi-color seed which is all yellow, better than average flavor and tenderness. *Source History: 1 in 1987; 2 in 1991; 2 in 1994; 1 in 1998; 1 in 2004.* **Sources: Jo1.**

Early Pearl - 84-86 days - Once offered by Chas. C. Hart Seed Co. this variety is being reintroduced after over 20 years of commercial neglect, uniform 6 ft. plants, 6-7 in. ears--2nd ear often larger and later, 12-14 rows of white kernels, extended harvest period, most delicious early at the plump milk stage. *Source History: 1 in 1987; 2 in 1991; 1 in 1994; 1 in 1998; 1 in 2004.* **Sources: Goo.**

Fisher's Earliest - 60 days - Very earliest open-pollinated sweet corn, 5-6 in. ears, 8 to 12 rows, golden-yellow kernels, thin cob, very sweet and tender, good for freezing and canning, dwarf plant. *Source History: 1 in 1981; 1 in 1984; 2 in 1987; 2 in 1991; 2 in 1994; 2 in 1998; 4 in 2004.* **Sources: Fis, Hig, So12, Syn.**

Golden Bantam - (Burpee's Golden Bantam, Golden Bantam 8 Row, Yellow Bantam) - 70-85 days - The standard yellow, 5-6 ft. stalk, slender 5.5-7 in. ears, 8 rows of med-deep broad golden kernels, 2 or more ears per stalk, high yields, excel. flavor, early main crop home garden var., good for freezing on cob, grown by a farmer named William Chambers of Greenfield, Massachusetts, intro. by W. Atlee Burpee in 1902. *Source History: 69 in 1981; 61 in 1984; 60 in 1987; 44 in 1991; 43 in 1994; 35 in 1998; 55 in 2004.* **Sources: Ba8, Bo19, Bou, Bu2, Bu3, Bu8, CH14, Com, Coo, Cr1, De6, Eco, Eo2, Ers, Fe5, Fi1, Fo7, Goo, Gur, HA3, He8, He17, Hi13, Ho13, Hud, La1, Lin, Ma19, Me7, Mo13, Ni1, Ont, Or10, Pe2, Pin, Pla, Rev, Ri12, Roh, Sa5, Se7, Se8, Se16, Se25, Se26, Shu, Sk2, So1, So9, Sw9, Ter, Tt2, Up2, Vi4, We19.**

Golden Bantam Improved - (Twelve Row Golden Bantam Improved, Golden Bantam 12 Row, Bantam Improved) - 75-82 days - Selection of Golden Bantam, small slender golden-yellow 7 in. ears, 10-14 rows, improved to stay tender longer, slightly longer ears, often 2 ears per stalk, broad evenly set sweet kernels, tolerates tight spacing and dry conditions, Burpee Seeds first introduced this var. into the North American seed trade about 1902. *Source History: 21 in 1981; 16 in 1984; 16 in 1987; 12 in 1991; 8 in 1994; 9 in 1998; 11 in 2004.* **Sources: Ag7, HA3, La1, MOU, OLD, Red, Ros, Sa9, Sh9, Und, Wi2.**

Golden Bantam, Early - (Extra Early Golden Bantam, Early Golden Bantam 8 Row) - 68-80 days - Old time favorite, excel. quality and flavor, for table use or freezing, 6.5-7 in. 8-rowed ears, sweet broad tender hulls, 5-6 ft. stalks. *Source History: 15 in 1981; 12 in 1984; 10 in 1987; 8 in 1991; 6 in 1994; 6 in 1998; 6 in 2004.* **Sources: All, But, Dam, Ear, OLD, Pr3.**

Golden Bantam, Extra Early - (Extra Early Bantam) - 69-80 days - Dwarf stalks average 5 ft., can be planted close together, 5-6 in. ears, bright golden-yellow kernels, exceptionally rich sugary flavor, vigorous, very hardy, heirloom. *Source History: 3 in 1987; 3 in 1991; 3 in 1994; 3 in 1998; 2 in 2004.* **Sources: Ga1, Hig.**

Golden Midget - (Golden Miniature, Dorinny) - 55-75 days - Very sweet and tender, 20-48 in. plants bear 3 to 5 small 3-5 in. butter-yellow 8-12 row ears, used fresh or frozen or cooked whole in soup or stew, very vigorous dwarf plants. *Source History: 20 in 1981; 19 in 1984; 18 in 1987; 5

in 1991; 5 in 1994; 7 in 1998; 2 in 2004. **Sources: Ho13, Pla.**

Guarijio Red - Red sweet corn, 8-10 ft. stalks, slender cobs, flavorful in the milk stage, creamy white kernels mature to red-orange, long season, exotic, wind tolerant, for planting during summer rains in low hot desert areas of the Sierra Madre in Mexico, Maiz Dulce landrace. *Source History: 1 in 1981; 1 in 1984; 1 in 1987; 1 in 1991; 1 in 1994; 2 in 1998; 1 in 2004.* **Sources: Ho13.**

Guarijio Sweet - Plants 6-8 ft. tall, yellow or burnt-orange kernels. *Source History: 1 in 1998; 1 in 2004.* **Sources: Na2.**

Hartline's 8-Row - 85 days - Plants grow 8 ft. tall, 8 in. long, 8 rows of kernels. *Source History: 1 in 2004.* **Sources: Sa9.**

Hawaiian - 97 days - Stalks grow 8 ft. tall, ears 10-12 in. long, super sweet kernels, will be tough if grown near other corn pollinating at the same time. *Source History: 1 in 1998; 1 in 2004.* **Sources: Sa9.**

Hawaiian Supersweet - 80 days - Bred in Hawaii, specifically for warmer climates, tighter husks to resist borer entrance, res. to corn stripe mosaic, retains sweet flavor 7-10 days after harvest without refrigeration. *Source History: 1 in 1987; 2 in 1991; 1 in 1994; 1 in 2004.* **Sources: Ho13.**

Hawaiian Supersweet #9 - *Source History: 1 in 1991; 1 in 1994; 1 in 1998; 1 in 2004.* **Sources: Uni.**

Honey Cream - 60-86 days - Very early, dwarf stalks suitable for small gardens, 6-7 in. ears with 10-12 rows, silvery-white very tender kernels, practially no hull, fresh or freeze or can. *Source History: 2 in 1981; 2 in 1984; 1 in 1987; 1 in 1991; 1 in 1994; 2 in 2004.* **Sources: Pe2, TE7.**

Honey Rainbow - Experimental, most are white bicolor sprinkled with blue, red and purple. *Source History: 1 in 2004.* **Sources: Syn.**

Hooker's Sweet Indian - 70-100 days - Grown by Ira Hooker near Olympia, Washington for 50 years, stalks 4-4.5 ft. tall with 2 or 3 thin 5-7 in. ears, finest tasting white sweet corn, matures blue-black, works well for small spaces. *Source History: 1 in 1981; 1 in 1984; 1 in 1987; 4 in 1991; 3 in 1994; 5 in 1998; 4 in 2004.* **Sources: Ho13, Se7, So9, Tu6.**

Hopi Massiqa - 90 days - Plants 5-6.5 ft., 8-10 in. long ears, kernels pastel to darker blue. *Source History: 1 in 1998; 1 in 2004.* **Sources: So12.**

Hopi Sweet Corn - 90 days - Bantam type, acclimatized by the Hopi Indians, for spring planting in the high cool desert. *Source History: 1 in 1981; 1 in 1984; 1 in 1998; 2 in 2004.* **Sources: Na2, So12.**

Howling Mob - 85 days - Stalks grow 6-7 ft., 14-16 even rows of pale yellow kernels, sweet flavor is similar to Silver Queen, introduced by C. D. Keller who named it after a trip to the market where his wagon was surrounded by buyers making loud demands for the ears of corn, introduced in 1905. *Source History: 2 in 1994; 2 in 1998; 1 in 2004.* **Sources: Shu.**

Jerde's Red - Plants grow 6-7 ft. tall, 6 in. ears, dries down to a unique plum color, Midwestern variety. *Source History: 1 in 1998; 1 in 2004.* **Sources: Ho13.**

Jubilee - 80 days - Open-pollinated Jubilee selection, 8 in. ears, 16-20 rows of tender yellow kernels, holds sweet flavor, easy to grow, good production, resists lodging. *Source History: 3 in 2004.* **Sources: Ba8, Pe2, TE7.**

Lindsey Meyer Blue - Possibly a str. of Black Mexican, 4-6 ft. tall plants, lots of tillering, 5-6 in. long ears, 8-12 rows, white kernels mature to blue-black, early maturity, maintained for over 60 years in the family of Mrs. Meyer of Pewaukee, WI. *Source History: 1 in 2004.* **Sources: In8.**

Little Giant - Sturdy 4-5 ft. tall plants, red stems, green leaves, 6-7 in. long husky ears, 8-12 rows of sweet yellow kernels, plant bears 2 good size ears and 2-3 smaller ears on the tillers. *Source History: 1 in 2004.* **Sources: In8.**

Luther Hill - 65-85 days - Developed in 1902 by Luther Hill of Andover Township, Sussex County, NJ, 5.5 ft. stalks produce two 6 in. ears and some suckers, excellent flavor, lacking in vigor, adapted to the Appalachian foothills, can be grown as far north as southern Ontario, offered only for seed-savers and gardeners on an experimental basis. *Source History: 1 in 1987; 4 in 1991; 5 in 1994; 12 in 1998; 7 in 2004.* **Sources: Fe5, Ho13, Pe2, Se17, So1, So9, Tu6.**

Martian Jewels - Plants to 6 ft., 2 ears per stalk. *Source History: 1 in 2004.* **Sources: Pe1.**

Midnight Snack - 80-85 days - Blue-black kernels at dry stage, sweet and pale-yellow at eating stage, sturdy 5.5 ft. plants, 7-7.5 in. ears with 14-16 rows, dev. by Prof. Meader at U of NH. *Source History: 1 in 1981; 1 in 1984; 1 in 1987; 1 in 1991; 1 in 1994; 1 in 1998; 1 in 2004.* **Sources: In8.**

Mini Purple - 80 days - Plants grow 4 ft. tall with 6 in. ears, sweet and flavorful. *Source History: 1 in 2004.* **Sources: Eo2.**

Montana Bantam - 65 days - Extra early strain of Golden Bantam, 6-7 in. ears with 8 rows of deep golden kernels, delicious flavor, good quality. *Source History: 1 in 1987; 1 in 1991; 1 in 1994; 1 in 1998; 1 in 2004.* **Sources: Fis.**

Mt. Pima Yellow - Dulcillo del Noroeste landrace from western Chihuahua, Mexico at 5,000 ft., grow with midsummer rains in the low desert due to daylength requirements. *Source History: 1 in 1991; 1 in 1994; 1 in 2004.* **Sources: Na2.**

Multicolor - 60 days - Flavorful sweet corn with blue, red, gold and orange kernels, 4-5 ft. stalks, 6 in. ears, original seed from an 80 year-old gentleman in upstate New York who has maintained it for over 50 years. *Source History: 1 in 1987; 1 in 1998; 1 in 2004.* **Sources: Ho13.**

Nuetta - 57 days - Grown in the Midwest for many years, 5 ft. stalks with 8 in. ears, 8 rows of bronze-orange kernels, not overly sweet flavor. *Source History: 1 in 2004.* **Sources: Sa9.**

Orchard Baby - 55-70 days - Small plants 3-5 ft. tall, 4-6 in. ears, 2-3 per stalk, old Oscar Will variety. *Source History: 1 in 1994; 3 in 1998; 4 in 2004.* **Sources: Ho13, Ma13, Pr3, Sa9.**

Pease Crosby - 67 days - Stalks 5-6 ft., average 1 ear with 12-14 rows of white kernels on 6 in. ears, 1860. *Source History: 1 in 1998; 1 in 2004.* **Sources: Sa9.**

Seneca Snow Prince - 71 days - Ears 7-8 in. long,

sweet, creamy flavor, good germination in cold soil. *Source History: 1 in 2004.* **Sources: Fis.**

Shaffer Eight Row - Small cobs with 8 rows of large kernels, good cold soil tolerance, over 100 years old. *Source History: 2 in 1998; 2 in 2004.* **Sources: Lan, Roh.**

Simonet - 58-75 days - Yellow full-size variety developed by Bob Simonet in Alberta, Canada, 6 in. ears with 12 rows. *Source History: 1 in 1981; 1 in 1984; 1 in 1987; 3 in 1991; 1 in 1994; 3 in 1998; 1 in 2004.* **Sources: Pr3.**

Six Shooter - 80 days - Oddity, 6 to 10 ears on each very tall vigorous stalk, heavy yields, very solid sweet meaty tender small white kernels, 10 rows, slender cob, roast or freeze. *Source History: 2 in 1981; 2 in 1984; 3 in 1987; 4 in 1991; 4 in 1994; 4 in 1998; 3 in 2004.* **Sources: Fo7, Se25, Shu.**

Stowell's Evergreen - 80-100 days - Leading white var. for home garden and market, needs long season, 8-9 x 2.5 in. med- tapered ears, 14-20 rows, clear-white kernels, 2-3 ears per 8-10 ft. stalk, holds quality, long harvest, oldest named var. still available, dev. from a cross between Menomony soft corn and northern sugar corn by Nathaniel Newman Stowell of New Jersey in 1848, after years of refining the str., he sold two ears of seed for $4 to a friend who agreed to use it only for his private use, his 'friend' then turned around and sold the seed for $20,000, intro. to the seed trade by Thoburn and Co. in 1856. *Source History: 38 in 1981; 30 in 1984; 25 in 1987; 20 in 1991; 20 in 1994; 24 in 1998; 25 in 2004.* **Sources: Ag7, Ba8, Ers, Fi1, Fo7, He8, He17, Ho13, La1, Lan, Mo13, Red, Ri12, Roh, Sa9, Sau, Se7, Se16, Se17, Sh9, Shu, Sk2, So1, So12, Vi4.**

Sunseeker - 75 days - Ears 6.5-7 in., good fresh or frozen, heirloom. *Source History: 1 in 1994; 1 in 1998; 1 in 2004.* **Sources: Eco.**

Sweet Baby Blue - Short plants only 2 ft. tall, very sweet, yellow-white kernels which turn purple-blue when dried, 3.5 in. ears. *Source History: 1 in 1998; 1 in 2004.* **Sources: Ho13.**

Texas Honey June - 97 days - Heirloom that does well in both northern and southern growing areas, 8 ft. stalks bear 2-3 ears, 5.5 in. long, 10-20 rows of white kernels, extremely tight husks deter insects. *Source History: 1 in 1994; 2 in 1998; 2 in 2004.* **Sources: Sa9, So1.**

Triple Play - 60-85 days - Short season corn selected from Hooker's Sweet Indian, produces 6 ft. stalks with tri-colored multiple cobs, yellow blue and white, good flavor, does best in cool soil, dev. in Oregon. *Source History: 1 in 1994; 1 in 1998; 3 in 2004.* **Sources: Ho13, Ri12, Se7.**

True Gold - 80 days - Selected for many years from parents of hybrid ancestry, 6-8 ft. plants, 8.5 in. cobs, golden yellow kernels, fine flavor, dependable and resists lodging. *Source History: 3 in 1994; 4 in 1998; 5 in 2004.* **Sources: Pe1, Pe2, Se7, Se17, So9.**

True Platinum - 78-84 days - Open-pollinated white sweet corn sel. from hybrid ancestors, 5-7 ft. stalks bear 8-9 in. ears, 10-14 rows, good flavor, an excel. parent for introducing the crinkle seed gene into any of the many fine maize starch corns, dev. by Dr. Alan Kapular. *Source History: 3 in 1994; 4 in 1998; 3 in 2004.* **Sources: Se7, Se17, So12.**

Whipple's White - 85 days - White mid-season sweet corn with 6 ft. plants, 14 rows of white kernels on flat, 7-9 in. ears, dates to 1919. *Source History: 1 in 1981; 1 in 1998; 1 in 2004.* **Sources: Sa9.**

Whipple's Yellow - 85 days - Plants 6 ft. tall, 14-16 rows of deep grained yellow kernels on fat, 7-9 in. ears, dates to 1921. *Source History: 1 in 1998; 1 in 2004.* **Sources: Sa9.**

White Sugar - 80 days - The first open-pollinated super sweet white corn, 7-7.5 ft. plant produces 1 to 2 fat 7-8 in. ears, must isolate from regular sweet corn, grows north or south, from PSR. *Source History: 1 in 1994; 1 in 1998; 2 in 2004.* **Sources: Bou, Pe6.**

Yukon Chief - 55-58 days - Dwarf stalks grow to only 3 ft. tall, 5-6 in. ears with 10-12 rows of golden yellow kernels, outstanding flavor for a normal sugary corn, bred for the north in the mid-1960s at U of AK AES. *Source History: 1 in 1991; 2 in 1998; 2 in 2004.* **Sources: Ga1, Pe2.**

Yukon Supreme - 53 days - Extremely early and productive sweet corn. *Source History: 1 in 2004.* **Sources: Sa9.**

Varieties dropped since 1981 (and the year dropped):

Altagold (1998), *Atkinson* (1998), *Bantam Evergreen* (1991), *Bi-Color* (1998), *Bronze-Orange* (1998), *Buhrow's White Desert* (1994), *Campbell* (1994), *Chemehuevi* (1998), *Dolce da Tavola* (1994), *Dorinny* (2004), *Dorinny, Kerr's* (1994), *Earliest White* (2004), *Early Alberta* (2004), *Early Golden Market* (1991), *Early Golden Sweet* (1984), *Early Large Adam* (2004), *Early Large Adams* (1991), *Early Spring Gold* (1998), *Evergreen Early* (2004), *Faribo Golden Midget* (1983), *Gill's Early Market* (1994), *Glacier* (1998), *Golden* (2004), *Golden Beauty* (1994), *Golden Early Market* (1998), *Golden Giant* (1998), *Golden Jubilee* (1998), *Golden Market* (1994), *Golden Sunshine* (1994), *Guatemalan Purple* (2004), *Hopi Bantam* (2004), *Hopi Early* (1991), *Ideal* (1987), *June White* (1994), *Kalakoa* (2004), *Kievit* (2004), *Kodiac* (2004), *Lorrie* (2004), *Malcombs* (2004), *Mandan Red Sweet* (2004), *Maricopa* (2004), *Mason's Midget* (1987), *Mayo Cacabatchi* (2004), *Neal's Paymaster* (1987), *New Zealander* (1998), *Paiute* (1994), *Pow Wow* (2004), *Puget Gold* (2004), *Queen Anne* (2004), *Rainbow Sweet* (1998), *Rangold* (2004), *Skyscraper* (1998), *Sweet Nothings* (2004), *Sweet Tooth* (1998), *White Midget* (1987), *White Sunglow* (1998), *Yuman Yellow* (2004)**.**

Corn / Other
Zea spp.

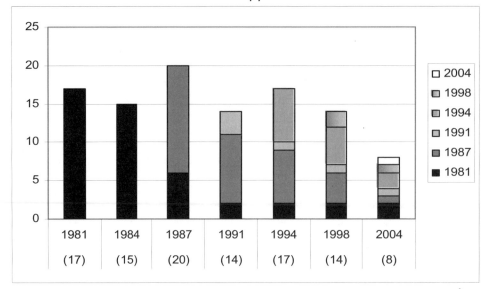

Number of varieties listed in 1981 and still available in 2004: 17 to 2 (12%).
Overall change in number of varieties from 1981 to 2004: 17 to 8 (47%).
Number of 2004 varieties available from 1 or 2 sources: 5 out of 8 (63%).

Baby - 65 days - Tender finger-like ears, delicately flavored, entirely edible, best harvested within 5 days of the appearance of silks, makes delicious hors d'oeuvres stir fries and pickles, excellent for freezing, long popular in the Orient and Germany. *Source History: 1 in 1987; 1 in 1991; 5 in 1994; 5 in 1998; 8 in 2004.* **Sources: Coo, Ev2, Gr28, Gr29, Ni1, Ra5, Vi5, Wo8.**

Japonica Striped Maize - Beautiful ornamental corn from Japan, grows 5-6 ft. tall, variegated leaves striped with green, white, yellow and pink, dark purple tassels, burgundy kernels, better color development when plants are widely spaced, listed in the 1890s as Striped-Leafed Japanese Maize. *Source History: 2 in 2004.* **Sources: Ho13, Se16.**

Jicarilla Apache Concho - (75-80) - Stalks to 3-5 ft., 6-8 in. ears with pearl-white kernels, tolerant of cool high elevations. *Source History: 1 in 1994; 1 in 1998; 1 in 2004.* **Sources: Na2.**

Parch, Magenta - 110 days - Stalks grow 5 ft. tall, 8 in. cobs with purple kernels, color does not fade when parched, dev. by independent plant breeders Carol Deppe, Ph.D. and Alan Kapuler, Director of Research at Seeds of Change. *Source History: 3 in 1998; 2 in 2004.* **Sources: Roh, Se7.**

Pod Corn - (Indian Ceremonial) - 100-110 days - Ears range in size from 5-14 in., each kernel is covered by a husk or pod called a glume which varies in color and size and shape, glume color on each ear can be either white, cream, red, brown or purple, same growing culture as corn, excel. for fall decorations *Source History: 7 in 1991; 12 in 1994; 10 in 1998; 17 in 2004.* **Sources: Bo18, CH14, GLO, Gr29, Ha5, HO2, Jo1, Jor, ME9, Ri12, RUP, SE4, Se16, Se28, Sto, Vi5, Wo8.**

Rainbow Inca - 85-130 days - Oregon grown composite of White Inca crossed with several Missouri Basin strains, 8-10 ft. stalk, 8 rows of multicolored seeds on a narrow cob, drought tol., maturity shows days for sweet corn and for mature dry seed. *Source History: 1 in 1981; 1 in 1984; 5 in 1994; 9 in 1998; 9 in 2004.* **Sources: Ho13, Pe1, Pe2, Se7, Se8, Se17, So9, Syn, Tu6.**

Santo Domingo Posole - Large white kernels used to make hominy, from northern New Mexico. *Source History: 1 in 1994; 1 in 1998; 1 in 2004.* **Sources: Na2.**

Teosinte, Annual - 100-125 days - *Zea mays spp. mexicana*, the wild ancestor of corn and its closest relative, annual to 10 ft. with 1-2 in. wide leaves and a tiny "ear" only 2-3 in. long, the "ear" consists of two rows of kernels back to back and has the power to disperse its seed--a trait which corn has lost, from Mexico. *Source History: 2 in 1981; 2 in 1984; 3 in 1987; 2 in 1991; 3 in 1994; 5 in 1998; 2 in 2004.* **Sources: Red, Se17.**

Varieties dropped since 1981 (and the year dropped):

Amarillo (1987), Bland's Extra Early (1984), Chishe (1987), Eastern Tarahumara Conico (1991), Eastern Tarahumara Tajawe (2004), Flor del Rio (2004), Gila Pima (2004), Gracillis (1987), Hispanic Pueblo Red (2004), Las Trumpas Concho (2004), Mayo Maiz Temporaldero (1994), Mayo Mix (1991), Mt. Pima Blanco (1998), Mt. Pima Hunu (1991), Multicolor Inca (1991), Northern Tepehuan Maizillo (1991), Perennial Teosinte (1987), Polyear (1987), Southern Maiz Colorado (1998), Striped Quadricolor (1991), Tarahumara Blue/Orange (1987), Tarahumara Cholii (1991), Tarahumara Chopo (1998), Tarahumara la Bufa Yellow

(1987), *Tarahumara Maiz Azul Negro* (1994), *Tarahumara Maiz Chomo* (2004), *Tarahumara Maiz de Cerocahui* (1998), *Tarahumara Multi-Colored* (1987), *Tarahumara Prieto/Colorado* (1991), *Tarahumara Tabloncillo Perla* (1991), *Teosinte, Diploid Perennial* (1998), *Teosinte, Northern Tepehuan* (2004), *Teosinte, Perennial* (1994), *Teosinte, Tarahumara Wild* (1994), *White American Maize* (1991).

Cornsalad
Valerianella locusta or V. eriocarpa

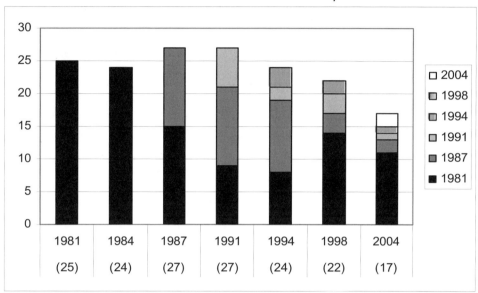

Number of varieties listed in 1981 and still available in 2004: 25 to 11 (44%).
Overall change in number of varieties from 1981 to 2004: 25 to 17 (68%).
Number of 2004 varieties available from 1 or 2 sources: 11 out of 17 (65%).

A Grosse Graine - (Mache a Grosse Graine, Big Seed) - 45 days - Widely esteemed in France and Europe as a salad green, early variety with large leaves, good color contrast in salads. *Source History: 3 in 1981; 3 in 1984; 3 in 1987; 5 in 1991; 6 in 1994; 6 in 1998; 2 in 2004.* **Sources: Lej, Orn.**

Big Seeded - (Large Seeded) - 60 days - Large seeded mache does much better in warmer weather than small seeded types, also has larger leaves. *Source History: 1 in 1991; 2 in 1994; 3 in 1998; 6 in 2004.* **Sources: Bo17, CO30, OSB, Pin, Se26, VI3.**

Broad Leaved - 48-60 days - Large light-seeded var., dark-green leaves, very hardy, often harvested right out of early winter snows, plant spring or fall, gourmet European salad plant. *Source History: 7 in 1981; 6 in 1984; 5 in 1987; 2 in 1991; 1 in 1994; 2 in 1998; 2 in 2004.* **Sources: Com, Sa9.**

Broadleaf Dutch - (Dutch) - Unique piquant flavor, grows very quickly, can be used when it has only 3 or 4 leaves, make successive sowings, popular salad plant in France and Germany. *Source History: 1 in 1981; 1 in 1984; 1 in 1987; 2 in 1991; 1 in 1994; 1 in 1998; 2 in 2004.* **Sources: Ers, Ri2.**

Cavallo - 45-50 days - Beloved Feldsalat of Germany, very early salad green, crisp green leaves, tangy taste. *Source History: 1 in 1981; 1 in 1984; 1 in 1987; 1 in 1991; 1 in 1994; 1 in 1998; 1 in 2004.* **Sources: Tho.**

Coquille de Louviers - 45-80 days - Deep-green spoon-shaped leaves, mild flavor, frost hardy, good fall crop in the North, may possibly be overwintered without protection in Southern areas. *Source History: 2 in 1981; 2 in 1984; 3 in 1987; 5 in 1991; 3 in 1994; 2 in 1998; 1 in 2004.* **Sources: Coo.**

Corn Salad - (Mache (French for "corn salad"), Lamb's Lettuce, Nut Lettuce, Fetticus) - 45-60 days - Small glossy plant, large smooth broad green leaves, quick growing, for early spring or late fall and winter use, culture like spinach, very high in vitamin C, very hardy, used as a salad plant in France or as a cold tol. winter green in England where it is fall planted. *Source History: 13 in 1981; 12 in 1984; 20 in 1987; 17 in 1991; 20 in 1994; 26 in 1998; 29 in 2004.* **Sources: Ba8, Ber, Fo7, HA3, He8, Hud, Hum, In8, Kil, ME9, Ont, Re8, Roh, RUP, Sa5, Sa6, Se7, Se26, Sh14, Shu, Sto, Syn, Tu6, Twi, Und, Ves, Vi5, WE10, Wi2.**

D'Etampes - 45-60 days - Heritage var. in cultivation over 100 years, narrow fleshy very dark-green leaves, res. to cold, from France. *Source History: 1 in 1981; 1 in 1984; 1 in 1987; 2 in 1991; 2 in 1994; 1 in 1998; 3 in 2004.* **Sources: Bu2, Coo, Hud.**

D'Orlanda - Larger leaves than Cambrai, great taste, for winter and spring salad, good cold res. *Source History: 1 in 2004.* **Sources: Se24.**

Large Dutch - (Large Seeded Dutch, Dutch) - 45-50 days - Dutch strain that produces an abundance of large leaves in winter and fall from a fall planting, large round dark-green leaves, mild nutty flavor, an old favorite for salad use. *Source History: 1 in 1981; 1 in 1984; 2 in 1987; 2 in 1991; 1 in 1994; 2 in 1998; 4 in 2004.* **Sources: Ba8, Bo19, Ni1, SE14.**

Large Leaved - 45 days - Tasty salad plant with oval light-green leaves, sow early in spring, ready in 6 weeks when the leaves should be pulled, not cut. *Source History: 2 in 1981; 2 in 1984; 2 in 1987; 2 in 1991; 2 in 1994; 3 in 1998; 1 in 2004.* **Sources: La1.**

Large Round Leaved - (Lamb's Lettuce, Field Salad, Fetticus) - Cool weather crop, useful to extend the salad season, mild pleasing flavor, sow in August for late fall greens, winters over with light mulch for spring harvest. *Source History: 1 in 1981; 1 in 1984; 3 in 1987; 2 in 1991; 2 in 1994; 2 in 1998; 2 in 2004.* **Sources: Dam, Fe5.**

Medallion - 40-45 days - Swiss sel., dark green, shiny leaves, mild, rich flavor. *Source History: 1 in 2004.* **Sources: Ter.**

Ronde Maraichere - Early productive French variety, round med-green leaves. *Source History: 1 in 1981; 1 in 1984; 2 in 1987; 1 in 1991; 1 in 1994; 1 in 1998; 1 in 2004.* **Sources: Lej.**

Valgros - 45 days - Large Dutch var. with elongated, dark-green leaves, sow in late August to mid-September for winter and spring harvest. *Source History: 1 in 1994; 1 in 1998; 2 in 2004.* **Sources: Ga22, VI3.**

Verte de Cambrai - (Lamb's Lettuce) - 45-75 days - Dark-green oval leaves, mild flavor, delicate texture, harvest early for baby rosettes, very cold tol., for overwintering, standard variety in France and Germany, excellent combined with more tart greens. *Source History: 3 in 1987; 6 in 1991; 8 in 1994; 14 in 1998; 16 in 2004.* **Sources: Bou, CO30, Fe5, Ga1, Hig, Hud, Ki4, Lej, Na6, Ni1, Orn, OSB, Se8, Se24, Sto, VI3.**

Vit - 50 days - Long glossy tender green leaves form a heavy bunch, mild minty flavor, most vigorous variety for spring and fall crops and overwintering, mildew tol. *Source History: 1 in 1987; 2 in 1991; 1 in 1994; 1 in 1998; 3 in 2004.* **Sources: Coo, Jo1, We19.**

Varieties dropped since 1981 (and the year dropped):

`Baval (2004), Blonde Shell Leaved (2004), Broad Leaf, Improved (1994), Deutscher (1987), Dunkelgruner Vollherziger (1987), Elan (2004), English (1987), Fetticus (1998), Feuilles Rondes Ameliorees (1991), Gayla (2004), Green Cabbaging Large Seeded (1998), Green de Cambria (2004), Green Full Heart (1994), Grote Noorde (1991), Grote Noordhollandse (1991), Lamb's Lettuce (2004), Large Green Cabbaging (1991), Large Leaved Italian (1987), Olanda Seme Grosso (1994), Ronde (1987), Slobolt (1994), Valerianella (2004), Verella (1991), Verte a Coeur Plein (1994), Verte d'Etampes (1998).*

Cowpea
Vigna unguiculata subsp. unguiculata

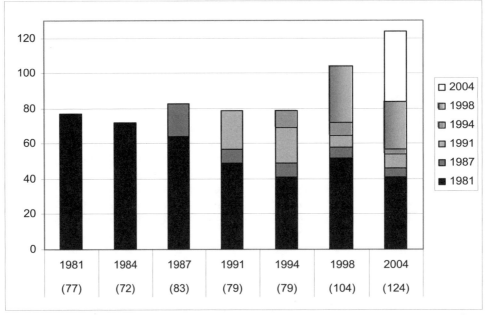

Number of varieties listed in 1981 and still available in 2004: 77 to 41 (53%).
Overall change in number of varieties from 1981 to 2004: 77 to 124 (161%).
Number of 2004 varieties available from 1 or 2 sources: 88 out of 124 (71%).

African - Southern cowpea thought to have come to the U.S. with some of the early slaves, tiny seed makes excel. wild game food, harvested as dried peas for human consumption since they are too tedious to shell fresh, seeds show an abundance of color patterns, grown along the coastal barrier islands since before the Civil War. *Source History: 1 in 1998; 1 in 2004.* **Sources: Scf.**

Alston Purple Hull - Mississippi heirloom, tall vines, tan, speckled or black seed in long pods. *Source History: 1 in 2004.* **Sources: Ho13.**

Bettersnap - Cream type pea with edible pods, 8-10 in. long, 12-16 seeds per pod, USDA, Charleston, SC. *Source History: 1 in 1998; 1 in 2004.* **Sources: Scf.**

Big Boy - 65-75 days - Heavy yielding green-hulled field pea, bunch type, 3-3.5 in. erect bushes with 8-9 in. pods, cream colored peas with light-brown eyes, wilt and nem. res., pods borne high on vines, easy to pick and shell, for fresh use or freezing or canning. *Source History: 14 in 1981; 12 in 1984; 13 in 1987; 11 in 1991; 7 in 1994; 6 in 1998; 6 in 2004.* **Sources: Ada, CH14, Co31, Kil, Shu, Wi2.**

Bisbee Black - Early bloomer, solid black seeds, from James Cowan of Missouri who got them from a Bisbee trucker who got them from an Arizona Native American, did okay in Tucson. *Source History: 1 in 1987; 1 in 1991; 2 in 1994; 2 in 1998; 2 in 2004.* **Sources: Ho13, Na2.**

Bisbee Red - Same source as Bisbee Black, larger seed is a dark-red color, more prolific because it blooms until frost, 15 peas to some pods, does well in the low desert. *Source History: 1 in 1987; 1 in 1991; 1 in 1994; 1 in 2004.* **Sources: Na2.**

Black Crowder - 63-70 days - Unusually long pods borne profusely, bunch type, deep-purple cast when shelled green, turn black as they dry, very prolific, green podded, easy to shell. *Source History: 8 in 1981; 6 in 1984; 5 in 1987; 3 in 1991; 3 in 1994; 4 in 1998; 5 in 2004.* **Sources: Ba8, Co31, Ho13, Sa9, Wi2.**

Blackeye Pea - (Southern Blackeye Pea, Blackeye Bean, Blackeye Crowder, Blackeye) - 60-78 days - Vigorous high yielding 24-40 in. plants, 7-9 in. pods, 10-12 peas per pod, wilt res., excel. as green shell or dry like winter beans, good producer even into Illinois, snaps at 45 days - dry at 60, easy to grow, cooks quickly. *Source History: 31 in 1981; 24 in 1984; 25 in 1987; 23 in 1991; 18 in 1994; 11 in 1998; 9 in 2004.* **Sources: Bo17, ECK, He17, Mey, Mo13, Ros, Sh9, Ver, Vi4.**

Blackeye, Botswana - *Source History: 1 in 2004.* **Sources: Se17.**

Blackeye, California - 55-75 days - Large table pea, either snap or shell, bunch type, cream with dark eye, res. to wilt and nematodes, dwarf prolific plants, 7-8 in. well filled pods, heavy yielding. *Source History: 12 in 1981; 10 in 1984; 14 in 1987; 13 in 1991; 15 in 1994; 14 in 1998; 18 in 2004.* **Sources: Ba8, Bou, Bu8, CH14, Co31, Ge2, Gur, HA3, He8, La1, Ma19, PAG, Ri12, Sau, Se26, Shu, TE7, Wet.**

Blackeye, California No. 5 - (Blackeye Cowpea No. 5) - 60-95 days - Improved Giant Ramshorn str., large smooth seed, cream with black eye, semi-spreading erect high-yielding plants, pod set concentrated at crown, 8-12 in. pods, Southern home garden favorite, fresh or frozen or dried, wilt and nem. res. *Source History: 20 in 1981; 19 in 1984; 21 in 1987; 19 in 1991; 17 in 1994; 17 in 1998; 19 in 2004.* **Sources: Ada, Bu1, Bu2, Bu3, Cl3, Com, Eo2, Fi1, Hi13, Jor, Kil, LO8, Loc, ME9, RIS, Roh, SE4, Syn, Wi2.**

Blackeye, California No. 46 - 55-65 days - Plants to 20 in., more compact than California Blackeye No. 5 with a higher crown set which increases yields dramatically, 6-8 in. pods, slightly smaller seed, RKN 1 and 3 res. *Source History: 1 in 1991; 1 in 1994; 6 in 1998; 3 in 2004.* **Sources: Ers, Ter, Wi2.**

Blue Goose - (Gray Crowder, Taylor) - 80 days - Vigorous 36 in. vines, 15-20 large speckled purple-gray peas per 10 in. pod, good flavor, table use or hay or soil building cover crop, rare southern heirloom, pre-1860. *Source History: 11 in 1981; 10 in 1984; 11 in 1987; 6 in 1991; 5 in 1994; 5 in 1998; 5 in 2004.* **Sources: Ada, Ba8, Ho13, ME9, Se26.**

Bohemian - *Source History: 1 in 2004.* **Sources: Se17.**

Brown Crowder - (Brown Crowder Mississippi Cowpea) - 54-90 days - Old time fine-flavored Southern table peas, dry seed is light-brown with darker eyes, 7-8 in. well filled pods, vigorous 28 in. vine, very productive, good market pea, used as either green shells or dried. *Source History: 23 in 1981; 19 in 1984; 22 in 1987; 17 in 1991; 17 in 1994; 12 in 1998; 9 in 2004.* **Sources: Ada, Ba8, Co31, Fi1, Ho13, Kil, La1, ME9, Sau.**

Brown Crowder, CT Dimpled - 66-70 days - Bush type, 12-15 peas per pod, prolific. *Source History: 1 in 1998; 1 in 2004.* **Sources: Wi2.**

Brown Crowder, Dimpled - 85 days - Runner type, spreading vines, brown seeds with tan eye, for canning, freezing or drying. *Source History: 1 in 2004.* **Sources: SOU.**

Burma - Small seeds are speckled purple and gray, young pods good as snaps, very productive 2 ft. plants. *Source History: 1 in 1998; 1 in 2004.* **Sources: Ho13.**

Calico Crowder - (Herford Peas, Polecat Peas, Calico Cowpea) - 65-89 days - Excel. mild flavor, 14 in. pods, med-large white peas with maroon-red splotches, running vine, prolific. *Source History: 10 in 1981; 9 in 1984; 7 in 1987; 5 in 1991; 4 in 1994; 4 in 1998; 3 in 2004.* **Sources: Ada, Ho13, So1.**

Cardinal - Robust vines bear very large peas, up to 20 red per pod, heavy producer, easy to grow. *Source History: 1 in 1998; 2 in 2004.* **Sources: Ba8, Ho13.**

Charcoal Gray - Abundance of small speckled gray seeds, rare. *Source History: 1 in 2004.* **Sources: Syn.**

Charleston Nemagreen - 64-72 days - Root-knot nematode resistant cream type, dev. by USDA Veg. Lab, Charleston, SC *Source History: 1 in 2004.* **Sources: Scf.**

Chinese Red Bean - (Chinese Red) - 85 days - Small sprawling bush type cowpea with long narrow brown pods, high in protein, widely adapted, Southern favorite. *Source History: 2 in 1981; 1 in 1984; 1 in 1987; 1 in 1991; 1 in 1994; 1 in 2004.* **Sources: ECK.**

Chori, Cream - Dried beans used like lentils, fresh green pods eaten like snap beans, small cream-tan seeds, heat tolerant, from India. *Source History: 1 in 2004.* **Sources: Hud.**

Chori, Red - Heat tolerant variety from India where fresh green pods used like snap beans, dried seeds cooked like lentils, small red-brown seeds. *Source History: 1 in 2004.* **Sources: Hud.**

Chosi, Brown Seeded - Bush plant, very small seeds, grown in India. *Source History: 1 in 1998; 1 in 2004.* **Sources: Ho13.**

Chosi, White Seeded - Small buff colored seeds, ripens earlier than the brown seeded var., steamed baby pods are a delicacy in India. *Source History: 1 in 1998; 1 in 2004.* **Sources: Ho13.**

Clay - Small bush with elongated, tan seeds, important staple of Southern soldiers during the Civil War, dates back to the 1860s. *Source History: 1 in 1998; 2 in 2004.* **Sources: Ba8, Ho13.**

Clemson Purple - 66 days - Thick stemmed plant grows 21 in. high, large purple pods contain 16 peas, easy picking and shelling, FW res., PVP 1993 South Carolina AES. *Source History: 1 in 1991; 5 in 1994; 4 in 1998; 3 in 2004.* **Sources: Scf, SOU, Wi2.**

Colossus - 58-65 days - Extra large light-brown crowder, straw-colored red-tinged 7-9 in. pod, excel. production, easy to pick and shell, good flavor, bushy plants, Clemson SC/AES. *Source History: 12 in 1981; 11 in 1984; 13 in 1987; 10 in 1991; 7 in 1994; 5 in 1998; 8 in 2004.* **Sources: Ada, Cl3, Co31, Scf, Se26, Shu, SOU, Wet.**

Corrientes - Dark-red seeds, excel. as green beans or shelled, extremely hardy and prolific, collected in Nayarit, Mexico. *Source History: 1 in 1994; 1 in 1998; 2 in 2004.* **Sources: Ba8, Se17.**

Cream - (Cream Crowder) - Med-large plants, slender pods, very sweet distinctive flavor, tasty. *Source History: 2 in 1981; 1 in 1984; 1 in 1987; 1 in 1991; 1 in 2004.* **Sources: HPS.**

Cream 8 - 64 days - Bush variety, produces long slender pods with kidney-shaped white peas, yields well, a delectable mix of the blackeye pea and long pod cream pea. *Source History: 5 in 1981; 5 in 1984; 6 in 1987; 4 in 1991; 3 in 1994; 3 in 1998; 2 in 2004.* **Sources: Vi4, Wi2.**

Cream 12 - 60 days - Bush type, almost round pea, heavy yielder, med-sized pod with smooth skin. *Source History: 4 in 1981; 4 in 1984; 5 in 1987; 2 in 1991; 2 in 1994; 5 in 1998; 3 in 2004.* **Sources: CH14, ECK, Ho13.**

Cream 40 - (Texas Cream 40) - 60 days - Distinct variety, early prolific semi-bush type, med-sized 6-8 in. slightly curved pods, small kidney-shaped white pea with an orange eye. *Source History: 4 in 1981; 4 in 1984; 6 in 1987; 4 in 1991; 3 in 1994; 4 in 1998; 4 in 2004.* **Sources: Ada, CH14, Co31, Wi2.**

Crudup - Large vining plants, long pods, red seeds. *Source History: 1 in 1998; 1 in 2004.* **Sources: Ho13.**

Dimpled Brown - 80 days - Pods grow 6.5 in. long, seeds have dimpled, flattened appearance, fine flavor, good producer. *Source History: 1 in 2004.* **Sources: Bu8.**

Dixie Lee - 60-65 days - Bunch to semi-bunch depending on soil fertility, mature light-yellow/green 8 in. pods, med-small brown peas, two heavy crops per season, nematode res., high yields. *Source History: 9 in 1981; 8 in 1984; 11 in 1987; 6 in 1991; 4 in 1994; 4 in 1998; 4 in 2004.* **Sources: Cl3, Scf, Se26, SOU.**

Double Green Delight - 65-71 days - High bush plant, mature pods are green with purple shading, dark straw color when dry, green seed color retained after blanching, dev. by USDA Veg. Lab, Charleston, SC. *Source History: 1 in 2004.* **Sources: Scf.**

Early Lady - 60-70 days - Tiny white peas, good yield. *Source History: 1 in 1991; 1 in 2004.* **Sources: Sa9.**

Ejotero - Long pods used as green beans, light-beige seeds, grown by Mayo Indians along the Rio Fuerte in Mexico. *Source History: 1 in 1994; 1 in 1998; 1 in 2004.* **Sources: Na2.**

Elite - 75 days - Bush type, perhaps most productive cream type, pods 7 in. long, bunched slightly above foliage level, easily shelled, peas are small to medium size. *Source History: 2 in 1981; 2 in 1984; 2 in 1987; 2 in 1991; 3 in 1994; 4 in 1998; 2 in 2004.* **Sources: Ho13, Wi2.**

Georgia Long - Runner type with long vines, med. size, black-eyed peas. *Source History: 1 in 1998; 1 in 2004.* **Sources: Ho13.**

Gray Speckled Palapye - Large pods contain gray speckled peas, flavorful, very early, perfect for the North, rare variety from a market in Palapye, Botswana. *Source History: 2 in 2004.* **Sources: Ba8, Se17.**

Green Dixie - 71 days - First blackeye southern pea that does not lose fresh green color at dry stage, concentrated pod set on high bushy plant, smooth seed with black eye, for fresh use, freezing and dry pack storage, dev. by USDA Veg. Lab, Charleston, SC *Source History: 1 in 2004.* **Sources: Scf.**

Green Pixie - 76 days - Concentrated set of pods held above the foliage on high bushy plants, light green pods mature to purple for fresh harvest, light straw color when dry, green color after blanching, dev. by USDA Vegetable Lab at Charleston, SC. *Source History: 1 in 2004.* **Sources: Scf.**

Guarijio Muni Cafe - Small white bean with dark eye, from the Rio Mayo watershed in Sonora, Mexico. *Source History: 1 in 1991; 1 in 1994; 1 in 1998; 1 in 2004.* **Sources: Na2.**

Haricot Rouge - (Haricot Rouge du Burkina Faso, Burkina Faso Red) - Bushy 4 ft. plants, solid dark red seeds, does well in heat, from Burkina, Faso, West Africa. *Source History: 1 in 1998; 4 in 2004.* **Sources: Ba8, Ho13, Se17, Syn.**

Hercules - 60-75 days - Late season variety but produces for four weeks, large pods and seeds, easy to pick and shell, fresh market and home garden use, very drought res., Clemson U. *Source History: 2 in 1981; 2 in 1984; 3 in 1987; 2 in 1991; 3 in 1994; 3 in 1998; 4 in 2004.* **Sources: Ada, Cl3, Co31, Scf.**

Holstein - Attractive seeds are mottled black and white, like the dairy cow, rare. *Source History: 1 in 1998; 2 in 2004.* **Sources: Ba8, Ho13.**

Italian Cream - Long thin pods contain 15-20 seeds, very good flavor, can be used for snaps when small, withstands hot weather, collected in the 1950s from a Cajun in southern Louisiana, origin unknown. *Source History: 1 in 1998; 1 in 2004.* **Sources: Ho13.**

Jet Black - Long vines, black peas are excellent in rice

dishes, old Southern variety, pre-1850. *Source History: 2 in 2004.* **Sources: Ba8, Se17.**

Kentucky Heirloom - *Source History: 1 in 2004.* **Sources: Sh13.**

Knuckle Purple Hull - (Knucklehull Purple Hull, Purple Hull Knucklehull, Purple Hull Brown Crowder) - 56-80 days - Brown sugar crowder with purple hull when mature, strong erect bush plant, not vining, big plump dark-brown dry peas form knuckles along the pod, fine flavor. *Source History: 16 in 1981; 13 in 1984; 16 in 1987; 14 in 1991; 11 in 1994; 13 in 1998; 10 in 2004.* **Sources: Ada, Bu8, Cl3, Co31, Ho13, Kil, Mey, SE4, Shu, Wi2.**

Knucklehull-VNR - 68-74 days - Crowder type, resists BICMV, cowpea stunt and root knot nematodes, resembles Knuckle Purple Hull, dev. by USDA Veg. Lab, Charleston, SC. *Source History: 1 in 2004.* **Sources: Scf.**

Lady - (Lady Finger) - 60-100 days - Tiny fine-flavored peas, bunch type, creamy white seeds, very prolific, used either green or frozen or dried, tender and delicious, heat tolerant, southern type. *Source History: 8 in 1981; 6 in 1984; 6 in 1987; 7 in 1991; 6 in 1994; 6 in 1998; 7 in 2004.* **Sources: Ada, Ba8, Co31, Ho13, Hud, Se26, Wi2.**

Magnolia - 65-70 days - Small sized peas, high yielding, excel. quality, disease res., good for canning and freezing, PVP expired 1993. *Source History: 3 in 1981; 2 in 1984; 2 in 1987; 3 in 1998; 3 in 2004.* **Sources: OLD, RUP, SE4.**

Mayo Colima - Seeds are shades of beige to orange with white eyes, from Sinaloa, Mexico. *Source History: 1 in 1998; 1 in 2004.* **Sources: Na2.**

Mayo Colima Pinto - Mottled cream brown and gray seeds, dry farmed staple in Los Capomos, Sinaloa, Mexico. *Source History: 1 in 2004.* **Sources: Na2.**

Mayo Speckled - Colima variety with pinto bean mottling over chocolate colored seeds, from Los Capomos, Sinaloa. *Source History: 1 in 2004.* **Sources: Na2.**

Minnesota 13 - (MN 13) - Short, bushy plants with pods held high on the plant, buff seeds with black mottling, best var. for northern climates, dev. at the U of Minnesota by Prof. Davis. *Source History: 1 in 1998; 2 in 2004.* **Sources: Ho13, Se17.**

Minnesota 157 - Resembles Minnesota 13 except the pods are purple and the seeds are buff with brown splotches, U of MN, Prof. Davis. *Source History: 1 in 1998; 1 in 2004.* **Sources: Ho13.**

Mississippi Cream - 70 days - New type tol. to RKN and most viruses, 7 in. pods are green to nearly white at green shell stage and straw colored when dry, green peas turn light-cream when dry, peas shell much easier if left overnight after picking, high yields, fresh or frozen or canned, PVP expired 2002. *Source History: 3 in 1981; 3 in 1984; 4 in 1987; 2 in 1991; 2 in 1994; 2 in 1998; 1 in 2004.* **Sources: Se26.**

Mississippi Pinkeye - 60-70 days - Young pods change from green to purple as they mature and then dark-purple when they dry, FW and RKN res., superior yields, PVP 1989 MS AES. *Source History: 2 in 1991; 1 in 1994; 3 in 1998; 5 in 2004.* **Sources: Be4, OLD, RUP, SE4, Se26.**

Mississippi Purple - 65-70 days - Mississippi Silver-Knuckle Purple Hull cross, 7 in. pods, green when young, turn purple at tip and along suture when ready to shell, bright-purple when mature, highly res. to disease, PVP expired 1993. *Source History: 8 in 1981; 7 in 1984; 7 in 1987; 7 in 1991; 5 in 1994; 6 in 1998; 7 in 2004.* **Sources: Ada, Co31, Ers, RIS, SE4, Se26, Wi2.**

Mississippi Purple Hull - 70 days - Crowder pea with bright-purple hull, large tan seeds, uniform maturity, easily shelled, res. to fusarium wilts and root knot nematodes. *Source History: 2 in 1981; 2 in 1984; 2 in 1987; 2 in 1998; 1 in 2004.* **Sources: Ho13.**

Mississippi Running White Conch - Very productive climbing conch type, cream-white seeds, tasty. *Source History: 1 in 2004.* **Sources: Ho13.**

Mississippi Shipper - 65-73 days - Purple pods, brown seed with dark eye, disease resistant, PVP 1988 Mississippi Ag. and Forestry Exp. Station. *Source History: 1 in 1994; 3 in 1998; 3 in 2004.* **Sources: ME9, OLD, RUP.**

Mississippi Silver - (Mississippi Silverskin, Mississippi Silver Crowder, Mississippi Silverhull) - 60-90 days - Early and more uniform maturing, easy shelling pods borne above upright plants may allow mech. harv., conc. set, 30 in. plants, some insect res., bunch type, smooth silvery 7 in. mature pods with some streaks and spots of rose, for hot humid climates in the South and East, large light-green and cream dry seed, MS/AES. *Source History: 28 in 1981; 25 in 1984; 25 in 1987; 27 in 1991; 23 in 1994; 22 in 1998; 19 in 2004.* **Sources: Ada, CH14, Cl3, Co31, De6, ECK, He17, Kil, Loc, Ma19, ME9, Mey, RIS, RUP, Scf, Sh9, So1, Ver, Wi2.**

Mississippi Silver Brown - (Mississippi Silverskin Brown) - 56-80 days - Mississippi Silver Brown Crowder, thick-stemmed bush var., brown seeded all-purpose field pea, for eating fresh or canning of freezing. *Source History: 3 in 1981; 3 in 1984; 6 in 1987; 3 in 1991; 1 in 1994; 1 in 1998; 2 in 2004.* **Sources: Se26, Shu.**

Missouri Heirloom - *Source History: 1 in 2004.* **Sources: Sh13.**

Mt. Pima Yori Muni - Small cream-colored seed with brown eye, from a Mountain Pima rancheria near Maicoba. *Source History: 1 in 2004.* **Sources: Na2.**

Nigerian - *Source History: 1 in 2004.* **Sources: Se17.**

Nyarit - 95 days - Pale white-yellow flowers, long slender pods, small grey and blue speckled seeds, good yields. *Source History: 1 in 2004.* **Sources: So12.**

October Pea - Requires an extremely long season, nearly 5 months, good as a long season cover crop to smother weeds, round, buff colored seeds. *Source History: 1 in 1998; 1 in 2004.* **Sources: Ho13.**

Ozark Razorback - Productive bush plants bear many tasty small peas, mottled half white and half red, very pretty, whippoorwill and cream pea cross developed by Horus Botanicals of Salem, Arkansas, rare. *Source History: 2 in 2004.* **Sources: Ba8, Ho13.**

Papago Cowpea - 83 days - Non-vining plant, black and white speckled seed, rare, for planting during the summer rains in low hot desert regions. *Source History: 1 in 1981; 1 in 1984; 2 in 1991; 3 in 1994; 3 in 1998; 4 in 2004.*

Sources: Ho13, Sa9, Se17, Syn.

Papago Surprise - Sel. from Papago, multi-color patterned seed, gold, brown and black, rare. *Source History: 1 in 2004.* **Sources: Syn.**

Paw's Old Gray Pea - Heirloom grown in Washington Parish, LA since 1900, originally from Missouri. *Source History: 1 in 2004.* **Sources: Fe5.**

Penny Rile - 80 days - Heavy yields of medium size, khaki-tan color peas, good for soup, originally grown by the Martin family for livestock feed in the winter and as their food source as supplies got low. *Source History: 2 in 2004.* **Sources: Ba8, Sa9.**

PI 189374 (Nigeria) - Bush plant bears many small pods with pink-cream seeds. *Source History: 1 in 2004.* **Sources: Ho13.**

PI 225901 (East Africa) - Tan seeds with pink tint speckled with dark gray, productive. *Source History: 1 in 2004.* **Sources: Ho13.**

PI 257463 (South Africa) - Productive, short plants, small buff seeds are mottled with red, insect and drought res. *Source History: 1 in 1998; 1 in 2004.* **Sources: Ho13.**

PI 339619 (Tanzania) - Semi-bush plant bears thin purple streaked pods, small seeds mottled with black, cream and grey. *Source History: 1 in 2004.* **Sources: Ho13.**

Pigott Family Heirloom - Brown speckled seed, flavorful, has been in the Pigott Family of Washington Parish, Louisiana, since the 1850s. *Source History: 3 in 2004.* **Sources: Ba8, Fe5, Se17.**

Pima Bajo - Small white beans with black and brown eyes, excel. green or dried, originally collected from the Pima Bajo living near the Rio Yaqui in Onavas, Sonora. *Source History: 1 in 1991; 1 in 1994; 1 in 1998; 1 in 2004.* **Sources: Na2.**

Pinkeye - (Pinkeye Cream) - Long pods with cream colored seeds with pink eye marking, good shell type. *Source History: 1 in 1981; 1 in 1984; 1 in 1987; 1 in 1991; 1 in 1998; 1 in 2004.* **Sources: Ho13.**

Pinkeye Purple Hull BVR - 63-75 days - Very similar to Pinkeye Purple Hull, same plant and pod characteristics, purple pods contain cream color seeds with maroon eyes, shells easily, BVR and Cowpea Mosaic Virus res., good for home gardening and commercial production, U of GA. *Source History: 2 in 1991; 4 in 1994; 9 in 1998; 7 in 2004.* **Sources: Ada, Cl3, Com, Sh9, So1, Vi4, Wi2.**

Purple Hull - 50-78 days - Popular Southern purple-hulled browneye crowder, strong vigorous vines, excel. for cooking green or freezing, white pea with small purple eye, can get two crops. *Source History: 10 in 1981; 10 in 1984; 9 in 1987; 7 in 1991; 5 in 1994; 1 in 1998; 1 in 2004.* **Sources: Sau.**

Purple Hull Pinkeye - (Purple Hull White Bunch Pinkeye, Pinkeye Purplehull) - 49-100 days - Delicious heavy yielding field pea, purple-hulled bush type, good disease res., young elongated white peas with pink or purple eyes, good soil improver, can produce two crops per season with favorable weather, compact plants. *Source History: 23 in 1981; 20 in 1984; 29 in 1987; 22 in 1991; 27 in 1994; 21 in 1998; 23 in 2004.* **Sources: Ba8, CH14, Co31, De6, Ers, Fi1, Gur, He8, He17, Kil, Loc, Ma19, ME9, Mo13, Par, RIS, Se26, Shu, SOU, Twi, Ver, Wet, Wi2.**

Purple Hull Vining Calhoun - 63 days - Pods are speckled to purplish, delicious green or dried, VW and rust res. *Source History: 1 in 1987; 1 in 1991; 1 in 1994; 2 in 1998; 1 in 2004.* **Sources: Syn.**

Purple Hull, Speckled - 62 days - Semi-crowder with curved pods grow to 7 in., pods are purple in shell and dry stage, oblong dark-brown pea with lighter speckles. *Source History: 4 in 1981; 4 in 1984; 6 in 1987; 4 in 1991; 3 in 1994; 3 in 1998; 2 in 2004.* **Sources: Ada, Shu.**

Queen Anne - (Queen Anne Blackeye) - 56-68 days - Early blackeye from Virginia Truck Experimental Station, compact 26 in. bush plant, no runners, excel. yields, 8-12 seeds per 7-9 in. pod, smaller than a blackeye, mech. harvest. *Source History: 10 in 1981; 9 in 1984; 10 in 1987; 7 in 1991; 7 in 1994; 7 in 1998; 7 in 2004.* **Sources: ME9, Mey, SE4, So1, SOU, Twi, Wet.**

Quickpick Pinkeye - 56-60 days - Upright, non-vining bush type plant, excellent yield, immune to Georgia isolate of blackeye cowpea mosaic virus, dev. by Louisiana AES, PVP 2001 Louisiana AES. *Source History: 1 in 2004.* **Sources: Scf.**

Rattlesnake - Tender, flavorful brown and tan seeds, eaten fresh or dried. *Source History: 1 in 2004.* **Sources: Hud.**

Red and Black - *Source History: 1 in 2004.* **Sources: Se17.**

Red Ripper - (Big Red Ripper, Mandy) - 70-85 days - Improved prolific red-seeded running vine pea, 12-14 in. pods with 18- 21 peas per pod, delicious flavor either green shelled or dry, heat res. plants. *Source History: 3 in 1981; 2 in 1984; 4 in 1987; 3 in 1991; 3 in 1994; 4 in 1998; 6 in 2004.* **Sources: Ada, Ba8, Sa9, Se26, So1, Tu6.**

Rice Pea - Southern cow pea, bush plants bear tiny white seeds that are slightly larger than rice, very tasty, cooks in 40 minutes, pre-1860. *Source History: 2 in 2004.* **Sources: Ba8, Se17.**

Rouge et Noir - Large red peas mature to black color, old southern variety from the 1880s from Roy Blunt, Washington Parish, Louisiana. *Source History: 1 in 2004.* **Sources: Ba8.**

Running Conch - 90 days - Non-climbing long vines, original from which others have been developed, harder to shell than modern varieties, valued for its ability to resist insects and weeds, late 1800s. *Source History: 1 in 1991; 1 in 1994; 1 in 1998; 2 in 2004.* **Sources: Ba8, Ho13.**

Sa-Dandy - (Sadandy) - 70-75 days - Similar to Texas Cream 40, peas slightly smaller, a bush type conch that thrives in hot Southern weather, good producer. *Source History: 6 in 1981; 6 in 1984; 6 in 1987; 5 in 1991; 3 in 1994; 3 in 1998; 2 in 2004.* **Sources: Ada, Se26.**

Saidy - *Source History: 1 in 2004.* **Sources: Se17.**

Silver Tip - Rare crowder type with large grey mottled seeds with a distinctive white blotch on one end, sprawling plants with 4-6 ft. runners, good for interplanting with corn, tasty. *Source History: 1 in 1998; 1 in 2004.* **Sources: Ho13.**

Six Week Browneye - (Brown-Eyed Six Weeks) - 42-

65 days - Popular small brown-eyed white peas, bunch type, very early. *Source History: 3 in 1981; 3 in 1984; 4 in 1987; 1 in 1991; 1 in 1994; 2 in 1998; 1 in 2004.* **Sources: Cl3.**

Sonoran Yori Muni - Small, brown-eyed, adapted from early Spanish seeds, for planting during the summer rains in low hot desert regions. *Source History: 1 in 1981; 1 in 1984; 1 in 1987; 1 in 1991; 1 in 1994; 1 in 1998; 1 in 2004.* **Sources: Na2.**

Stick Up - Once popular in French Fork, Louisiana, where people say it got them through the Great Depression, hardy, produces over a long season, small, red-brown seed. *Source History: 1 in 1998; 1 in 2004.* **Sources: Ho13.**

Tender Cream - Bushy plants with concentrated pod set, pod color is green when mature, green with purple shading at green shell stage, straw color when dry, moderate res. to cowpea curculio and RKN, susc. to bacterial blight, USDA, Charleston, SC. *Source History: 1 in 1998; 1 in 2004.* **Sources: Scf.**

Tennessee White Crowder - 65-100 days - Large round off-white seed with light-brown eye, well-filled pods are dark-green at green maturity, vining, long blossom period, comes close to cream pea flavor. *Source History: 8 in 1981; 7 in 1984; 7 in 1987; 5 in 1991; 5 in 1994; 2 in 1998; 2 in 2004.* **Sources: Ba8, SOU.**

Tetapeche Gray Mottled - Pea size with white eye, look like wild beans, Sonora. *Source History: 1 in 1987; 1 in 1991; 1 in 1994; 1 in 1998; 1 in 2004.* **Sources: Na2.**

Texas Cream 8 - 72 days - Green-tan pods held above foliage of bushy plants, small white peas, fresh market. *Source History: 1 in 2004.* **Sources: SE4.**

Texas Cream 40 - (Texas Cream 40 Improved Conch) - 62-75 days - An extra early blackeye x a mid-season cream pea, long cream-colored pods dev. above foliage, light-green peas, bush, two crops. *Source History: 8 in 1981; 8 in 1984; 4 in 1987; 5 in 1991; 3 in 1994; 1 in 1998; 3 in 2004.* **Sources: Kil, Se26, Vi4.**

Texas Pinkeye Purple Hull - 60 days - Tall, erect bush plants, no runners, green and purple pods when immature, purple when dry, kidney shaped peas with bright pink eye, average insect and disease tolerance, PVP 1995 Texas AES. *Source History: 2 in 1998; 1 in 2004.* **Sources: Wi2.**

Thai Black - Robust upright producer, some shattering, rare. *Source History: 1 in 2004.* **Sources: Syn.**

Tohono O'odham Cowpea - 75 days - White eyes with black splotches, excel. for green beans, prolific, rarely grown on the reservations, originally from Africa. *Source History: 1 in 1987; 2 in 1991; 1 in 1994; 3 in 1998; 3 in 2004.* **Sources: Na2, Se17, So12.**

Turkey Craw Crowder - Tall vines, long pods contain tan and red-brown seeds, productive. *Source History: 1 in 1998; 1 in 2004.* **Sources: Ho13.**

Turtle Peas - No trellis required, mature black seeds are used primarily for the Cuban dish, "black beans and rice", from the Pinar del Rio Province in western Cuba to Kentucky with a Mr. Hernandez when he immigrated about 75 years ago, has been in his family for over 100 years. *Source History: 1 in 1998; 1 in 2004.* **Sources: Scf.**

Washday - Half runner, small yellow-tan peas, cooks fast, cooked by women on wash day, from the 1800s. *Source History: 1 in 1998; 3 in 2004.* **Sources: Ba8, Ho13, Se17.**

Whippoorwill - 75-90 days - Good general purpose old standard var., very prolific tall plants, 10 in. pods held high, easily picked, smooth speckled buff-brown peas, bears over a long period. *Source History: 11 in 1981; 11 in 1984; 11 in 1987; 8 in 1991; 6 in 1994; 7 in 1998; 10 in 2004.* **Sources: Ada, Ba8, Fe5, Ho13, Sa9, Se17, Sh13, Th3, Vi4, Wi2.**

Whippoorwill, Purple Hull - Purple pods, flavorful tan and brown spotted peas, very hardy, produces under adverse conditions, very rare. *Source History: 1 in 1998; 2 in 2004.* **Sources: Ba8, Ho13.**

Whippoorwill, White - Large yields of creamy white peas, young pods can used in stir fry, Southern heirloom. *Source History: 1 in 2004.* **Sources: Ba8.**

White Acre - 65-80 days - Creamy-white small seeds, green pods, large bush type plants, matures early, bears over a long period, excel. fresh eating quality, mid-Atlantic heirloom. *Source History: 6 in 1981; 6 in 1984; 5 in 1987; 4 in 1991; 3 in 1994; 3 in 1998; 5 in 2004.* **Sources: Ada, Kil, Par, Se26, So1.**

Yori Cahui - Low desert green bean used like Asapargus beans, speckled brown dried peas used in soup, long vines bear 13 in. long pods, collected from the village of Ahome, near Los Mochis in Sinaloa. *Source History: 1 in 1998; 1 in 2004.* **Sources: Ba8.**

Zipper - (Zipper Cream) - 70-75 days - Named Zipper because so easy to shell, curculio res., thick pod walls stop stink bug and weevil damage, large seed and pods, bushy compact plants, high yields. *Source History: 8 in 1981; 8 in 1984; 5 in 1987; 10 in 1991; 7 in 1994; 7 in 1998; 6 in 2004.* **Sources: Ada, Fi1, He17, Kil, Shu, Wi2.**

Zipper Cream Crowder - 75 days - An improved strain of the old popular Cream Crowder, bushy compact plant, mild flavor, good producer, fresh market. *Source History: 1 in 1981; 5 in 1987; 3 in 1991; 1 in 1994; 4 in 1998; 4 in 2004.* **Sources: CH14, Cl3, Mo13, SE4.**

Varieties dropped since 1981 (and the year dropped):

Arkansas Crowder (2004), *Banquet Cream Pea* (1998), *Bettergreen* (2004), *Bettergro Blackeye* (2004), *Big Boy Purple Hull* (2004), *Blackeye White Crowder* (1998), *Blackeye WR5* (1991), *Blackeye, Extra Early* (1994), *Brown Sugar Crowder* (2004), *California Blackeye Ramshorn* (1991), *California Pea* (2004), *Champion* (2004), *Chinese Red Pea* (1991), *Coronet* (1991), *Cream Champion* (1991), *Cream Combine* (2004), *Cream Elite* (1991), *Crimson* (1991), *Crimson Brown Crowder* (1983), *Freezegreen* (1991), *Frijol Riata* (1998), *Grandma's Black Crowder* (2004), *Grey Adzuki* (2004), *Guarijio Frijol Gamuza* (2004), *Iron and Clay Mixed* (2004), *Jones Blackeye* (1983), *Lady Cream* (2004), *Minnesota* (2004), *Mississippi Brown Crowder* (1994), *Mississippi Silver (CA Grown)* (1998), *Oklahoma Blackeye* (1987), *Pinkeye Purple Hull, Louisiana* (2004), *Purple Hull 49* (2004), *Purple Hull Browneye Crowder* (1994), *Purple Hull Bush* (1998), *Purple Hull*

Pinkeye (CA Grown) (2004), *Purple Hull Pinkeye (TX Grown)* (1991), *Purple Hull Vining* (2004), *Purple Hull White Crowder* (1994), *Purple Tip Crowder* (1998), *Ramshorn* (1994), *Ramshorn, Early* (1998), *Royal Black Eye* (2004), *Royal Cream* (1991), *Royal Cream Crowder* (1991), *Royal Pink Eye* (1991), *Running Acre* (1991), *Snap-Pea* (1998), *Sunapee* (2004), *Suzanne* (1998), *Texas Big Boy* (2004), *Texas Purple Hull 49* (2004), *Thai* (1994), *Thailand Long* (1998), *Two Crop Brown* (2004), *Warihio Black Eyes* (1991), *White Lady* (1983), *Worthmore* (2004), *Zipper White Crowder* (1987).

Cress
Various genera

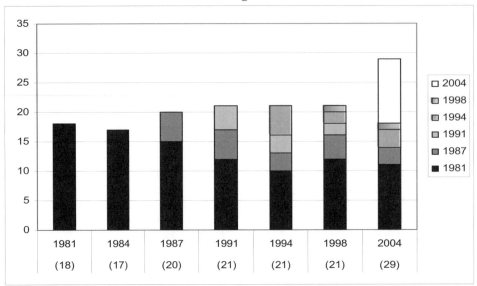

Number of varieties listed in 1981 and still available in 2004: 18 to 11 (61%).
Overall change in number of varieties from 1981 to 2004: 18 to 29 (161%).
Number of 2004 varieties available from 1 or 2 sources: 20 out of 29 (69%).

Broadleaf Cress - (Mega Cress) - 10-32 days - A distinct variety with much broader leaves, peppery flavor, harvest when leaves are 3-4 in. long, heat increases with the temperature. *Source History:1 in 1987; 2 in 1991; 3 in 1994; 2 in 1998; 5 in 2004.* **Sources: CO30, Coo, Ki4, Se8, St18.**

Cressida - 20-30 days - Very popular broadleaf cress type in Europe, bright green 4 x 2 in. frilly-edged leaves grow upright, can be grown indoors in flats or in the garden, suitable for 'cut and come again' scissor harvest, seeds can be sprouted. *Source History:1 in 1994; 1 in 2004.* **Sources: Ki4.**

Crinkly - Upright broadleaf cress with pretty ruffled and crinkled green leaves, 4-5 in., adds a peppery zing to salads, soups, sandwiches, fast grower. *Source History: 2 in 2004.* **Sources: Pe2, Re6.**

Curled Cress - (Curlicress, Peppergrass, Cresson, Fine Curled, Burpee's Curlycress) - 10-50 days - *Lepidium sativum*, garden cress with curled leaves and cotyledons, dwarf 8 in. tall plants, pungent flavor, dark-green finely cut curled leaves, slow to bolt, sow every two weeks year round, can harvest 10 days after seeding when 3 in. tall. *Source History: 30 in 1981; 27 in 1984; 22 in 1987; 18 in 1991; 13 in 1994; 20 in 1998; 21 in 2004.* **Sources: All, CO23, CO30, Coo, DOR, Fis, Ga1, HA3, He8, Hig, La8, ME9, Mel, OSB, Pe6, Ri2, SE14, So12, Twi, WE10, WI23.**

Curly Moss Cress - *Barbarea rupicola*, native to California, very dense and low growing, curly leaves, flavor is quite crisp, good salad green, easily grown in rock gardens. *Source History: 1 in 1981; 1 in 1984; 1 in 1987; 3 in 1991; 2 in 1994; 2 in 1998; 2 in 2004.* **Sources: Fe5, Orn.**

Garden Cress - (Land Cress, Peppergrass, Cresson de Jardin, American Cress) - 10-28 days - *Lepidium sativum*, tastes like watercress, likes moisture and some shade, extra curled, from Europe and Asia minor, good in cool greenhouse. *Source History: 8 in 1981; 7 in 1984; 13 in 1987; 13 in 1991; 13 in 1994; 13 in 1998; 16 in 2004.* **Sources: Ba8, Ber, Bo19, Bou, Com, Dam, Ers, Kit, Lej, Ni1, PAG, Pr3, SE14, So1, Ter, Tu6.**

Gerard's - Succulent, broad leaves with peppery flavor. *Source History: 1 in 1998; 1 in 2004.* **Sources: Sa5.**

Greek Cress - Fast grower with an unusual peppery taste, frilly leaves, good addition to salads. *Source History: 1 in 1994; 1 in 1998; 2 in 2004.* **Sources: Sa5, Tho.**

Land Cress - 35 days - Tasty water cress substitute, sow early spring onwards, overwinters too. *Source History:1 in 1987; 1 in 1991; 2 in 1994; 1 in 1998; 1 in 2004.* **Sources: Tho.**

Nigerian - Taller sel. from special seed collection, rare. *Source History: 1 in 2004.* **Sources: Syn.**

Peppercress - 40-45 days - Sharp tangy flavor similar to water cress, grows very fast and is ready for tasting in 10 days - sow in early spring until weather warms, start again in fall, cut and come again, good window sill or container plant, will tolerate shade. *Source History: 2 in 1987; 1 in 1991; 1 in 1994; 1 in 1998; 3 in 2004.* **Sources: Gr28, SE14, Tu6.**

Peppergrass - (Pepper Cress, Fine Curled, Extra Fine Triple Curled, Curly Cress, Extra Curled) - 10-45 days - *Lepidium sativum*, dark-green finely cut curly leaves, pungent flavor for salads or sandwiches, pressed seeds yield edible oil, also a garnish, tastes similar to watercress, grow early or late in frames or inside for winter. *Source History: 27 in 1981; 26 in 1984; 20 in 1987; 19 in 1991; 20 in 1994; 20 in 1998; 20 in 2004.* **Sources: Alb, Bo18, Cr1, Fo7, Ga21, Gr29, Hal, Lin, Ni1, Ont, Re8, Ren, RUP, Sto, Syn, Ter, Und, Vi5, We19, Wo8.**

Persian - Spicy salad plant with various leaf types, delicious peppery flavor, plant spring or fall. *Source History: 2 in 2004.* **Sources: Ho13, So9.**

Persian Broadleaf - Dark green strap-like leaves 2-6 in. long, mild cress flavor, nutritious. *Source History: 1 in 2004.* **Sources: Sh12.**

Persian Garden - 20-45 days - Early spring-sown green, rich in vitamins and minerals, native to Iran. *Source History: 1 in 2004.* **Sources: Se7.**

Plain Cress - (Plain Leaf, Cresson Alenois) - 40 days - *Lepidium sativum*, indented deep-green pungent leaves, sow thickly at any time in any soil, will even grow on sand, sow indoors on paper napkins. *Source History: 4 in 1981; 4 in 1984; 2 in 1987; 1 in 1991; 3 in 1994; 2 in 1998; 2 in 2004.* **Sources: Coo, Roh.**

Rishad - Rare Mideast heirloom. *Source History: 1 in 2004.* **Sources: Syn.**

Shallot - 45 days - *Lepidium sativum* var. *Mahantongo*, plant resembles cornsalad, spoon-shaped leaves, white flowers in April-May, flavor resembles garlic chives, best used in late fall as a salad green, seed requires cold stratification, old Pennsylvania Dutch heirloom from Pitman, Pennsylvania. *Source History: 1 in 2004.* **Sources: Und.**

Triple Curled - (Extra Triple Curled, Extra Curled, Double Curled) - 05-45 days - Finely cut and deep-green, grows quickly in pots or garden, nutritious year round salad crop, often forced. *Source History: 6 in 1981; 3 in 1984; 4 in 1987; 3 in 1991; 3 in 1994; 3 in 1998; 2 in 2004.* **Sources: Ear, Tho.**

Upland Cress - (Land Cress, Curled Upland, Creasy Greens, Winter Cress) - 45-60 days - *Barbarea verna*, dwarf plant with slender stalks, 4-6 in. tall x 10-12 in. wide, small oval notched 2 in. long leaves, very mild flavor for salads, grows quite densely, easily grown on high dry land. *Source History: 21 in 1981; 18 in 1984; 25 in 1987; 30 in 1991; 35 in 1994; 37 in 1998; 39 in 2004.* **Sources: ABB, Bo19, Bu2, CO23, CO30, Com, DOR, Ers, Fo7, Ga21, Ga22, Gr27, Ho13, Hud, Hum, Jo1, Kit, La1, La8, ME9, Mey, Ni1, Orn, OSB, Ri2, RUP, Sa6, SE14, Sh9, Shu, So1, SOU, Syn, Thy, Twi, WE10, We19, Wet, WI23.**

Upland Special - Water cress substitute, tends to be perennial, useful for wildlife feeding and the home garden, fall sowing for winter/spring harvest produces the largest and most tender plants, stronger flavor similar to arugula, PSR selection. *Source History: 1 in 2004.* **Sources: Pe6.**

Victoria - 25 days - Fast growing, broad leaved cress with spicy, peppery flavor, similar to nasturtium leaves but hotter. *Source History:2 in 1994; 1 in 1998; 1 in 2004.* **Sources: Pin.**

Watercress - (Cresson de Fontaine) - 50-180 days - *Nasturtium officinale*, hardy perennial, popular peppery tasting salad herb, plant near a spring or on the margin of a pond or stream, pungent, high in calcium and vitamins, easily grown if you have running water or even a leaky faucet. *Source History: 41 in 1981; 38 in 1984; 38 in 1987; 43 in 1991; 41 in 1994; 42 in 1998; 40 in 2004.* **Sources: ABB, Alb, Ban, Bo17, Bo19, Com, Coo, Ear, Ers, Ev2, Fe5, Fo7, Ga1, Ga21, Gr29, GRI, Hig, Ho8, Hud, Jo1, Ki4, Lej, ME9, Ni1, Orn, OSB, Pe2, Ri2, Ri12, Roh, Sa12, Se26, SGF, So1, So12, Sto, Tho, Thy, Tt2, We19.**

Watercress, Broad Leaved Imp. - (Large Leaf, Broad Leaf) - 50-60 days - *Nasturtium officinale*, special large-leaved strain, superior to common watercress, extensively grown in greenhouses for market in winter, needs very moist soil, harvest in cool temps. *Source History: 5 in 1981; 5 in 1984; 5 in 1987; 3 in 1991; 2 in 1994; 3 in 1998; 2 in 2004.* **Sources: CO30, Hal.**

Watercress, True - 45-180 days - Hardy aquatic plant, press seeds into mud in early spring or fall, creeping habit, healthful greens with small round leaves and slightly pungent spicy flavor. *Source History: 11 in 1981; 11 in 1984; 11 in 1987; 8 in 1991; 9 in 1994; 11 in 1998; 9 in 2004.* **Sources: DOR, Ge2, HA3, Kit, Mey, Ont, SE14, Twi, WE10.**

Wild Winter - *Barbarea vulgaris*, glossy, dark green leaves form a basal rosette, excel. addition to salads or as a cooked green. *Source History: 1 in 2004.* **Sources: Ga21.**

Winter Cress - *Barbarea vulgaris*, plants are short and full, fast growing and very frost hardy, easy to grow, will provide fresh greens well into the winter, also good window plant. *Source History: 1 in 1981; 1 in 1984; 1 in 1987; 2 in 1991; 2 in 1994; 2 in 1998; 1 in 2004.* **Sources: Ho8.**

Winter Cress, Variegated - *Barbarea vulgaris* var. *variegata*, white and pale green leaves used like watercress with similar flavor, stores under refrigeration for several weeks, good vitamin C source in winter, can be harvested in winter, known as scurvy grass in Philadelphia and the South. *Source History: 1 in 2004.* **Sources: Und.**

Wrinkled Crinkled Crumpled - 20-35 days - Spoon shaped, bright green leaves, blistered surface, ruffled edges, resists bolting longer than others, sweet, spicy addition to salads, dev. from Curly and Persian Cress by Frank Morton. *Source History: 9 in 2004.* **Sources: Fe5, Jo1, Ki4, ME9, Ni1, Sh12, Ter, Und, WI23.**

Varieties dropped since 1981 (and the year dropped):

Armada (1994), *Belle-Isle Cress* (1998), *Broad Leaved French* (1994), *Common* (1987), *Early Curled* (1994), *Mega* (1991), *Moss Curled* (2004), *Mustard Cress* (1984), *Pink Cress* (1991), *Reform Broadleaf* (2004), *Spring Cress* (2004), *Turkish* (2004), *Yellow Meadow Cress* (1994).

Cucumber
Cucumis sativus

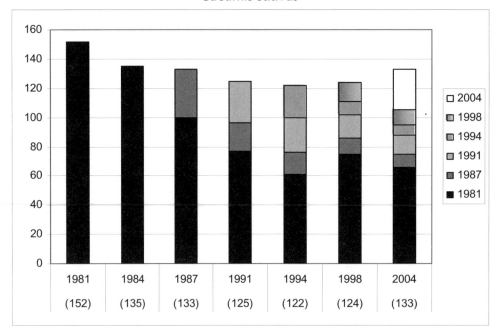

Number of varieties listed in 1981 and still available in 2004: 152 to 66 (43%).
Overall change in number of varieties from 1981 to 2004: 152 to 133 (88%).
Number of 2004 varieties available from 1 or 2 sources: 68 out of 133 (51%).

A and C Pickling - (Ace) - 50-55 days - Deep green fruit, 8-10 in. long, for salads or dill pickles, excel. home or market var., Abbott and Cobb introduction, 1928. *Source History: 4 in 2004.* **Sources: Ba8, Hud, Pr3, Se16.**

Addis - 56 days - Dark-green, straight, tends to taper to blossom end, white spine, res. to DM PM CMV and ALS, some res. to AN, vigorous vines, dev. for the South at North Carolina State University. *Source History: 2 in 1981; 2 in 1984; 2 in 1987; 2 in 1991; 3 in 1994; 1 in 1998; 1 in 2004.* **Sources: WE10.**

Apple Green - Rare Australian heirloom, large oval fruit, light green color, excel. sweet flavor, juicy. *Source History: 1 in 2004.* **Sources: Ga22.**

Ashley - (Ashley Long Green) - 53-70 days - Dark-green, 7-8 in., slight taper at stem end, white spines, ALS DM PM res., popular on Mid-Atlantic coast and in Southeast, med-sized vigorous vines, heavy producer, good shipper, resembles Marketer, Clemson SC/AES, 1956. *Source History: 42 in 1981; 33 in 1984; 28 in 1987; 20 in 1991; 21 in 1994; 19 in 1998; 15 in 2004.* **Sources: Ada, Be4, Cr1, DOR, Hud, Jun, Kil, La1, SE14, Se26, Sh9, So1, WE10, Wi2, WI23.**

Athens - 75 days - Large, dark green slicer, excellent quality. *Source History: 1 in 2004.* **Sources: Sa9.**

Aunt Rita's Monastery - Old Russian pickling cucumber, seed obtained from Aunt Rita, originated in a monastery. *Source History: 1 in 2004.* **Sources: Ea4.**

Bedfordshire Prize - 65 days - English slicing cucumber, 10 in. long, high yields over a long period. *Source History: 1 in 2004.* **Sources: Sa9.**

Beit Alpha CMMR - 56 days - Straight thick tender fruit, medium green skin with fine black spines, medium vines, CMV tol., for pickling or salads. *Source History: 1 in 1981; 1 in 1984; 2 in 1987; 2 in 1991; 2 in 1994; 2 in 1998; 2 in 2004.* **Sources: DOR, WE10.**

Beit Alpha MR - 55-57 days - Pickling or salad cucumber, med-green, straight, round ends, slight taper at blossom end, 5-6 x 1.25-2 in., CMV res., black spines, med-sized vines, Mideastern, rare in U.S. *Source History: 7 in 1981; 5 in 1984; 3 in 1987; 2 in 1991; 1 in 1994; 1 in 1998; 5 in 2004.* **Sources: Bou, CO23, Com, Or10, Se24.**

Bianco Lungo di Parigi - 62 days - Knobby cukes but otherwise similar to a regular slicer, yields over long period, even in early October. *Source History: 1 in 1991; 1 in 1994; 2 in 1998; 2 in 2004.* **Sources: Pin, Pr3.**

Bianco Primaticcio - White fruits grow 4-5 in., early, productive over long season, shows some res. to diseases transmitted by cucumber beetles, does well trellised. *Source History: 1 in 2004.* **Sources: Se24.**

Black Diamond - 53-67 days - Produces heavy crops of uniform cylindrical fruits, 7-9 x 2.25-3 in. dia., holds dark-green color, good keeper and slicer, similar to Early Fortune, introduced in 1920. *Source History: 4 in 1981; 2 in 1984; 3 in 1987; 1 in 1991; 1 in 2004.* **Sources: Sa9.**

Boothby's Blonde - 63 days - Heirloom from the Boothby family of Livermore, Maine, where it has been grown for several generations, 6-8 in. long slicer with pleasant flavor and texture, yellow-cream color with contrasting black spines. *Source History: 2 in 1994; 3 in 1998; 11 in 2004.* **Sources: Fe5, Hi6, Ho13, In8, La1, Pin, Se16, Se17, Sh13, So25, Tu6.**

Boston Pickling - (Green Prolific) - 50-60 days - Smooth, bright-green, 5.5-6 x 2.5-3 in., blunt ended, seldom too large for pickles, slight taper, black spine, very high yields, bears continually if kept picked, popular old reliable small cucumber for pickling, listed by D. M. Ferry and Co. in 1880. *Source History: 34 in 1981; 26 in 1984; 24 in 1987; 18 in 1991; 23 in 1994; 26 in 1998; 21 in 2004.* **Sources: All, Ba8, Co32, Com, DOR, Ers, Fo13, Gr27, HA3, He17, La1, Mo13, MOU, Or10, Pr9, Ri12, Sa9, Sh9, Shu, Wet, WI23.**

Boston Pickling Improved - 52 days - Blocky shape, bright green skin, great pickler, bears continually with frequent pickings, CMV and scab res. *Source History: 2 in 1991; 2 in 1994; 2 in 1998; 4 in 2004.* **Sources: He8, SE14, Sk2, WE10.**

Burpless - (Burpless Long Green, Burpless Delight) - 55-65 days - Long thin dark-green 8-10 in. fruits, res. to DM and PM, practically acid-free or burpless, non-bitter, smooth skin. *Source History: 8 in 1981; 7 in 1984; 9 in 1987; 6 in 1991; 6 in 1994; 6 in 1998; 3 in 2004.* **Sources: De6, Eo2, Ge2.**

Burpless Muncher - 65 days - Open-pollinated salad cucumber, bitter-free at all slicing stages, med-green 7 in. fruits, mosaic res. *Source History: 1 in 1981; 1 in 1984; 1 in 1987; 1 in 1991; 1 in 1994; 1 in 1998; 1 in 2004.* **Sources: All.**

Burpless No. 26 - 62 days - Smooth dark-green fruits with small white spines, harvest when 9 x 1 in. *Source History: 1 in 1987; 1 in 1991; 1 in 1994; 1 in 1998; 1 in 2004.* **Sources: SGF.**

Bush Champion - (Burpee's Bush Champion) - 60-80 days - Long season, produces both male and female flowers over an extended period, short compact productive mosaic res. vines, 9-11 in. bright-green straight slender fruits. *Source History: 4 in 1981; 3 in 1984; 6 in 1987; 9 in 1991; 4 in 1994; 8 in 1998; 5 in 2004.* **Sources: Bu2, Ge2, Hi13, Roh, Se7.**

Bush Crop - 48-62 days - Compact bushy 18-36 in. dwarf vines with no runners, 6-8 in. fruits, ideal for small gardens and containers. *Source History: 18 in 1981; 17 in 1984; 19 in 1987; 17 in 1991; 16 in 1994; 18 in 1998; 19 in 2004.* **Sources: CA25, CH14, De6, DOR, GRI, HPS, Jor, Jun, Ma19, MAY, Mey, MIC, MOU, Pin, Pr3, Roh, RUP, SGF, Wet.**

Bush Cucumber - 54 days - Heavy yields of 6 in. cucumbers on short runners. *Source History: 2 in 1981; 2 in 1984; 1 in 1987; 1 in 1998; 1 in 2004.* **Sources: De6.**

Bush Pickle - 45-55 days - Plants average 24-36 in. dia., good quantities of straight cylindrical 4-5 in. fruits, shorter production period than vining types, bred for small gardens and containers. *Source History: 15 in 1981; 15 in 1984; 21 in 1987; 19 in 1991; 20 in 1994; 20 in 1998; 23 in 2004.* **Sources: BAL, CH14, CO23, DOR, GLO, GRI, He17, HO2, HPS, Jor, Jun, Lin, Ma19, MAY, Mey, MIC, Mo13, Pr3, Roh, SGF, Sto, Ver, Wet.**

Bushy - 46-49 days - Well known popular old variety from southern Russia, very short vines, fruits 4.5-6 in. long with medium size seeds, some bumps with black points, high yields, extremely early. *Source History: 1 in 1994; 1 in 1998; 3 in 2004.* **Sources: In8, Se16, Se17.**

Caroline's Pickler - Consistent production of spineless fruit, crisp and firm, started out as a hybrid. *Source History: 1 in 2004.* **Sources: Al25.**

Chicago Pickling - (Boston Pickling, National Pickling, B.S. Westerfield's Pickling) - 55-60 days - Old timer for home gardens, thick square-ended med-green thin-skinned fruits, 6-7 x 2.5 in., high yielding, well warted, black spines, disease res., prolific, most widely used pickling variety, early, fine quality, intro. by D. M. Ferry, 1888. *Source History: 34 in 1981; 29 in 1984; 24 in 1987; 16 in 1991; 13 in 1994; 14 in 1998; 13 in 2004.* **Sources: Fa1, Fi1, Ge2, La1, Lin, Ont, PAG, Ri12, Sa9, Se8, Se25, Sh9, Up2.**

Chinese Yellow - Direct from mainland China, 10-15 in. long, 4-5 in. dia., dark yellow skin color, weighs 1-2 lbs., mild flavor, rare Chinese heirloom. *Source History: 1 in 1991; 1 in 1998; 2 in 2004.* **Sources: Ba8, Ho13.**

Cirrus - 66 days - Vigorous, medium size vines, dark-green fruits are slightly tapered at stem end, white spines, 7-8 in. long, disease res. *Source History: 1 in 1994; 1 in 1998; 1 in 2004.* **Sources: Eco.**

Clinton - 55 days - Vines grow 4-5 ft., 5-6 in. long dark green cylindrical fruit, few seeds, excel. firmness and brining quality, DM, PM, CMV, A, ALS, S res. *Source History: 1 in 1981; 1 in 1984; 1 in 1987; 1 in 1991; 1 in 2004.* **Sources: Ter.**

Cornichon - 53 days - Wonderful made into sour pickles, also good as sweet pickles, pick often to extend harvest. *Source History: 1 in 2004.* **Sources: Ag7.**

Cornichon de Bourbonne - 60 days - Used to make tiny sour cornichon pickles featured in French gourmet restaurants, harvest when the size of your little finger. *Source History: 1 in 1981; 1 in 1984; 2 in 1987; 1 in 1991; 1 in 1994; 1 in 1998; 1 in 2004.* **Sources: Sa9.**

Cornichon, Vert de Massy - 53 days - French var. dating back to the 1800s, black spine, may be picked for finger-sized tiny pickles or at 4 in. long for pickling whole or slicing, scab tol., excel. for specialty markets and home gardens. *Source History: 8 in 1991; 14 in 1994; 16 in 1998; 7 in 2004.* **Sources: Ho13, Ni1, Se8, Se26, Shu, So12, Wi2.**

Crystal Apple - (Lemon, Apple Crystal White) - 65 days - Imported from New Zealand, thin tender creamy-white skin, white spine, apple-shaped when ripe, up to 3 in. dia., mild flavor, very prolific, similar to Lemon cucumber, from the early 1930s. *Source History: 5 in 1981; 4 in 1984; 5 in 1987; 7 in 1991; 2 in 1994; 4 in 1998; 3 in 2004.* **Sources: Ho13, Pr3, So25.**

De Bouenil - White cucumber, spreading plants, standard size, in Europe these are the choice for larger pickles because the white color allows the cook to choose what color the relish will be, can also be used for whole pickles if harvested young. *Source History: 1 in 1991; 1 in 1994; 1 in 1998; 1 in 2004.* **Sources: Al25.**

Delikatess - 60 days - Med-sized well rounded 6-10 in.

fruits, pale-green with small warts, superior taste, no bitterness, bears abundantly, for pickles when small, for slicing when larger, from Germany. *Source History: 1 in 1981; 1 in 1984; 2 in 1991; 1 in 1994; 1 in 1998; 8 in 2004.* **Sources: Ba8, He8, In8, Sa9, Se17, Sk2, Vi4, WE10.**

Della's White - 50-60 days - Blocky fruit, 3 x 6 in., tender white skin and flesh, retains sweetness as it matures, productive vines, grown in Kansas for over 100 years. *Source History: 2 in 2004.* **Sources: Ho13, Se17.**

Double Yield Pickling - 50-60 days - Excel. var. for gherkins and dills, slender fruits, 5-6 in. long by 2 in. dia., retains dark-green color longer than most, high percent of sets produce two fruits at each leaf joint, susc. to black spot and scab, intro. by Joseph Harris and Company of Coldwater, NY, 1924. *Source History: 1 in 1981; 1 in 1984; 1 in 1987; 2 in 1991; 2 in 1994; 3 in 2004.* **Sources: Co32, Me7, Se16.**

Early Cluster - (Early Green Cluster, Russian) - 52-58 days - Blocky med-green fruits frequently borne in clusters, squared ends, 5.5 x 2.5 in., black spines, prolific old timer for pickling or slicing, good for trellis culture, good keeping quality, lacks disease res., pre-1860. *Source History: 12 in 1981; 11 in 1984; 8 in 1987; 9 in 1991; 7 in 1994; 4 in 1998; 4 in 2004.* **Sources: Ea4, Roh, Sa9, Shu.**

Early Fortune - 58-70 days - Dark olive-green, smooth, slightly tapering, 8-9 x 2.5 in. dia., white spine, few seeds, disease res., strong growing, holds color and freshness, table or shipping, selected by George Starr of Royal Oak, MI from a single plant found in a crop of Davis Perfect, intro. by J. B. Rice Seed Company in 1906. *Source History: 11 in 1981; 10 in 1984; 6 in 1987; 1 in 1991; 1 in 1994; 1 in 1998; 3 in 2004.* **Sources: Ba8, Se16, Te4.**

Early Ochiai - Extra early, green 7-8 in. fruits, black spines, produces abundant female flowers and fruits on every node, heavy yields, will not stand extreme heat, must be staked for straight fruit, will coil into serpent shapes if left to sprawl on the ground. *Source History: 2 in 1981; 1 in 1987; 1 in 2004.* **Sources: In8.**

Early Prolific Short Green - (Green Prolific, Early Frame) - Fairly plump, 6-8 in. long, does not get bitter, prolific, excellent for pickles, included in McMahon's 1806 Gardener's Catalog. *Source History: 1 in 2004.* **Sources: Ea4.**

Early Russian - (Russian, Cluster, Early Cluster) - 52-58 days - Short oval med-dark green 5 x 2 in. fruits, very early and hardy, for Northern short season areas, very mild, never bitter, uniform, productive, will bear all season, small fruits, amazingly productive, used primarily for pickling, very early variety, from Europe, presumably Russia, 1860. *Source History: 15 in 1981; 14 in 1984; 14 in 1987; 11 in 1991; 8 in 1994; 4 in 1998; 5 in 2004.* **Sources: Ea4, Pr3, Ri12, Se7, Syn.**

Earnest Family - 60-70 days - Mildly flavored, angular blocky fruit, 3-6 in. long, heirloom from Louise Earnest family in Kansas. *Source History: 1 in 2004.* **Sources: Se17.**

Edmonson - 70 days - Family heirloom from Kansas since 1913, whitish-green 4 in. fruits, good crisp flavor even when past prime, very hardy, prolific, good resistance to disease insects and drought, best for pickles but used as a slicer as well, CV So1 introduction, 1987. *Source History: 2 in 1987; 1 in 1991; 1 in 1994; 2 in 1998; 3 in 2004.* **Sources: Ho13, So1, Tu6.**

Everbearing - 55 days - Early, will bear entire season if kept closely picked, 5 in. fruits, best for pickling, disease res., U of WI. *Source History: 1 in 1987; 1 in 1994; 2 in 1998; 4 in 2004.* **Sources: Fi1, Gur, Und, Wet.**

Fancy Pickling, Shumway's - (National Pickling) - 50 days - Very productive, dark-green 5 in. fruits, solid thick texture, excel. flavor, black spines, widely used for pickles, also good for slicing. *Source History: 1 in 1981; 1 in 1984; 1 in 1987; 1 in 1991; 1 in 1994; 1 in 1998; 1 in 2004.* **Sources: Shu.**

Fine Meaux - (Fin de Meaux) - Fruits are longer and darker than Vert de Paris or Small Paris. *Source History: 1 in 1981; 1 in 1984; 3 in 1987; 1 in 1991; 1 in 1994; 1 in 1998; 1 in 2004.* **Sources: Lej.**

Heiwa Green Prolific - 65 days - Burpless, up to 18 in. long, prolific, no bitterness, also good greenhouse variety. *Source History: 2 in 1981; 2 in 1984; 1 in 1987; 3 in 1991; 1 in 1994; 2 in 1998; 1 in 2004.* **Sources: Ri12.**

Hmong Red - Blocky green fruit matures to gold, 3 x 8 in., productive, good keeper. *Source History: 1 in 1998; 2 in 2004.* **Sources: Ho13, Se17.**

Homemade Pickles - 55-60 days - Bush type plants, medium green, 5-6 in. fruit, small white spines, crisp, solid core, good pickling var., can be harvested at 1.5 in. length for baby sweets, pick regularly to maintain production, good disease res. *Source History: 16 in 1994; 32 in 1998; 48 in 2004.* **Sources: All, BAL, Be4, Bo17, Bo18, Bu1, CA25, Coo, Ear, Ers, Fa1, Fi1, Ga1, Ge2, Gr27, Gr29, He8, He17, HO2, HPS, Jor, Jun, La1, Lin, Me7, MIC, Mo13, OLD, PAG, Pin, Ra5, SE14, Se26, Se28, Sh9, Shu, Sk2, So1, So25, Sw9, Ter, Tt2, Ver, Vi4, Vi5, We19, Wi2, Wo8.**

Japanese Climbing - (Japanese Longfellow Climbing) - 58-65 days - Vigorous healthy climbing vines with strong tendrils, easily trained on trellis or fence, 7-9 x 3 in. dia., black spines, bears all season if kept picked, slicer, listed by Thorburn in 1892, Japanese introduction. *Source History: 4 in 1981; 4 in 1984; 2 in 1987; 2 in 1991; 1 in 1998; 3 in 2004.* **Sources: Ea4, Sa9, Se16.**

Japanese Long Pickling - 60-75 days - Very crisp and mild, easy to digest, 12-18 x 1.5 in. dia., shortest of the Long cucumbers, very small seeds, firm flesh, best at 12 in., dark-green, burpless. *Source History: 2 in 1981; 2 in 1984; 1 in 1987; 1 in 1991; 1 in 1994; 1 in 1998; 2 in 2004.* **Sources: Ba8, WE10.**

Katsura Giant Pickling - 70 days - Imported from Japan, old favorite, oblong, 12.5 x 4.5 in. dia., light-green skin turns almost white when fully ripe, very thick white flesh, for pickling. *Source History: 1 in 1981; 1 in 1984; 1 in 1987; 1 in 1991; 1 in 1994; 1 in 2004.* **Sources: Kit.**

Landis White - Large fruits which turn yellow when fully ripe. *Source History: 1 in 1998; 1 in 2004.* **Sources: Pr3.**

Langelang Giant - (Giant of Langeland) - 70 days - Large smooth fruit, 12 in. long, 4 in. dia., white flesh with excel. texture, small core. *Source History: 1 in 1991; 2 in*

1994; 2 in 1998; 2 in 2004. **Sources: He8, WE10.**

Lemon - (True Lemon, White Lemon, Lemon Apple, Apple Shaped, Apple, Crystal Apple, Lemon Crystal) - 58-70 days - Ripens light-yellow, 3-3.25 x 2.25-2.5 in., looks like a lemon, white flesh, good slicer and pickler, easy to digest, rust and drought res., stem end fairly flat, brown-flecked on blossom end, early maturing, introduced to the U.S. from Australian markets by Samuel Wilson, Mechanicsville, Pennsylvania, 1894. *Source History: 45 in 1981; 40 in 1984; 44 in 1987; 44 in 1991; 49 in 1994; 55 in 1998; 69 in 2004.* **Sources: Ba8, BAL, Bo17, Bo19, Bou, Bu2, Co32, Com, Coo, DOR, Ers, Fe5, Fi1, Fis, Fo7, Fo13, Ga1, Ge2, Gr28, He8, He17, Hi6, Hi13, HO2, Ho13, Hud, Hum, Jor, Ki4, La1, Loc, ME9, MOU, Na6, Ni1, Ont, Orn, OSB, Pe2, Pe6, Pla, Pr3, Red, Ri2, Ri12, Roh, Ros, Se7, Se8, SE14, Se16, Se26, Sh13, Shu, Sk2, So1, So9, So25, Sw9, Syn, Te4, TE7, Ter, Tt2, Und, Up2, WE10, We19, WI23.**

Little Leaf - (Arkansas Little Leaf H-19) - 55-72 days - Compact, highly branched vines, small 2 in. leaves, 4 in. fruit, white spine, excel. yields, good fruit set under a wide range of growing conditions, disease res., can be grown in the greenhouse, dev. and released by the U of Arkansas in 1991, PVP 1993 AR AES. *Source History: 2 in 1991; 21 in 1994; 13 in 1998; 11 in 2004.* **Sources: Be4, Co31, Ge2, GLO, Jo1, Jun, Lin, Pe6, So1, Syn, Ver.**

Long Green - (Windermoor Wonder) - 60-70 days - Med-dark green, 9-12 x 2.5 in., rounded ends, slight taper, no disease res., very prolific, fine flavor but poor appearance keeps it at home. *Source History: 13 in 1981; 10 in 1984; 6 in 1987; 2 in 1991; 4 in 1994; 4 in 1998; 4 in 2004.* **Sources: Alb, All, CA25, Wi2.**

Long Green Improved - (Long Green) - 60-72 days - Vigorous prolific dependable home garden var., med-green, 10-12 x 2.5-3 in., fine flavor, tapered ends, warted, black spine, can be used for pickling or slicing, sow all season, heavy yields, few seeds, home or market, introduced in 1842. *Source History: 51 in 1981; 39 in 1984; 41 in 1987; 31 in 1991; 30 in 1994; 26 in 1998; 18 in 2004.* **Sources: CH14, DOR, Ear, Fo7, HA3, He8, La1, Lin, Mel, Sa9, Sau, SE14, Se26, Sh9, Shu, WE10, Wet, WI23.**

Long White - 35-70 days - Large smooth white fruit can grow to 6 lbs., very crisp and mild, excellent for fresh eating, rare heirloom. *Source History: 3 in 2004.* **Sources: Ba8, Hi13, La1.**

Longfellow - (Vaughan) - 62-80 days - Long slender cylindrical dark-green fruits, 12-15 x 2.5 in., med-heavy yields, for home or market garden, shipping or greenhouse, holds color and crispness, introduced by the Rice Seed Company, 1927. *Source History: 13 in 1981; 8 in 1984; 5 in 1987; 2 in 1991; 2 in 1994; 3 in 1998; 5 in 2004.* **Sources: Ba8, Ri12, Sa9, Se16, Shu.**

Mandurian - 60-70 days - Non-vining compact plants, round fruit, light and dark green variegated skin, sweet, juicy flesh is crisp, generally eaten when the size of a baseball, can be left to grow to melon size, Asia. *Source History: 1 in 2004.* **Sources: Ag7.**

Marketer - (Long Marketer, Early Green, Marketer Long Green) - 55-70 days - AAS/1943, slender smooth tapered and dark-green, 8-9 x 2-2.5 in., stands intense heat of late Southern spring, uniform prolific heavy yields, susc. to DM, white spine, abrupt taper, some dark stippling, long harvest, vigorous. *Source History: 53 in 1981; 40 in 1984; 36 in 1987; 25 in 1991; 22 in 1994; 21 in 1998; 8 in 2004.* **Sources: DOR, La1, Lej, Mel, SE14, Sh9, WE10, WI23.**

Marketmore - 63-70 days - Vines to 4-6 ft., straight but tends to curl in young stages, 8-10 x 2-2.5 in., mosaic and scab res., dark-green, often pointed at blossom end, high yields, produces through hot or cold. *Source History: 7 in 1981; 7 in 1984; 9 in 1987; 12 in 1991; 9 in 1994; 7 in 1998; 14 in 2004.* **Sources: Be4, Ber, CO30, Com, Ga7, Lin, Ma19, Na6, Or10, Ri12, Roh, Se7, So25, St18.**

Marketmore 70 - 60-72 days - Popular Northeast var. dev. by Dr. Munger of Cornell, uniform dark-green glossy non-fading and straight, 8-9 x 2.5 in., tapered ends, res. to scab and CMV, white spine, for fall use, bears over a long period, fine quality, very little stippling, monoecious, well shaped, good variety for Northern areas. *Source History: 33 in 1981; 27 in 1984; 12 in 1987; 6 in 1991; 2 in 1994; 1 in 1998; 2 in 2004.* **Sources: Coo, Dom.**

Marketmore 76 - 58-75 days - Marketmore 70 str. dev. by Dr. Munger of Cornell, 8-9 x 2.25-2.5 in., dark green even in hot weather, sweet mild flavor, res. to scab CMV PM DM ALS and AN, tapers at both ends, blocky, white spines, no stippling, for home gardens or market in cool climate areas. *Source History: 45 in 1981; 43 in 1984; 55 in 1987; 63 in 1991; 67 in 1994; 66 in 1998; 66 in 2004.* **Sources: Ag7, All, Ba8, BAL, Bo19, Bu3, CA25, CH14, Cl3, CO23, Cr1, Dam, De6, DOR, Dow, Ers, Fe5, Fo7, Ga1, Gr27, GRI, HA3, Ha5, He8, He17, Hi6, HO1, HO2, Hum, Jo1, Jor, La1, LO8, ME9, Mey, MIC, Mo13, Mo20, MOU, Ont, OSB, PAG, Pe2, Pr9, RIS, Ros, RUP, Sau, Scf, SE4, SE14, Se25, Se26, Shu, Si5, Sk2, Sto, TE7, Tu6, Twi, Ver, Ves, Vi4, WE10, Wi2, WI23.**

Marketmore 80 - 61-68 days - Improved Marketmore 76, bitter-free even when large, blocky dark-green fruits, 8-9 x 2.25-2.5 in., PM DM ALS AN CMV and scab res., Dr. Henry Munger/Cornell. *Source History: 3 in 1981; 3 in 1984; 4 in 1987; 13 in 1991; 4 in 1994; 3 in 1998; 2 in 2004.* **Sources: Pe2, So1.**

Marketmore 86 - 56-68 days - Improved Marketmore 76 type, small vine or semi-bush growth habit, dark-green 8-9 in. long fruits are sweet with no bitterness under the skin, disease res., use for main season or late slicers, dev. by Dr. Henry Munger of Cornell U. *Source History: 11 in 1991; 20 in 1994; 21 in 1998; 18 in 2004.* **Sources: Co31, Fe5, HA3, Hal, Hig, HO1, HO2, Jor, Jun, La1, Me7, OLD, Roh, RUP, Sh9, SOU, Ter, We19.**

Marketmore 97 - 55 days - Non-bitter, straight, 9-11 in. long, dark green slicer, white spines, PM DM BCMV, scab and leaf spot res., excellent Marketmore variety developed by Dr. Henry Munger at Cornell University. *Source History: 1 in 1998; 2 in 2004.* **Sources: HA3, Ter.**

Masterpiece - 75 days - English heirloom, good outdoor variety with the length and quality of most frame types, dark skin, slightly spined, 8 in. fruit, crisp white flesh. *Source History: 1 in 1987; 1 in 1991; 1 in 1994; 1 in 1998; 1 in 2004.* **Sources: Tho.**

Mideast Prolific - 70-85 days - Vines 3-5 ft., several crops of 6-8 in, non-bitter fruit with smooth green thin skin, great for pickles and fresh eating. *Source History: 1 in 1994; 3 in 1998; 3 in 2004.* **Sources: Pe2, Se7, Syn.**

Midget Cucumber - (Mincu) - 55 days - Extra early, compact 24 in. vines yield lots of finger length cucumbers, ideal for pickling, very prolific, space saver. *Source History: 1 in 1981; 1 in 1984; 1 in 1987; 1 in 1991; 1 in 1994; 1 in 1998; 1 in 2004.* **Sources: Shu.**

Mincu - (Baby Mincu, Mincu Extra Early) - 48-53 days - Vigorous compact bushy 24 in. vines, clusters of 8-10 fruits near base of plant, 3-5 x 2 in., oval and uniform, crispy whole pickles early or slice later, early heavy cropper, good for the North, ideal for small gardens or containers, everbearing, MN/AES, 1937. *Source History: 15 in 1981; 14 in 1984; 11 in 1987; 5 in 1991; 5 in 1994; 5 in 1998; 5 in 2004.* **Sources: Alb, Ear, Fis, Hig, Mcf.**

Miniature White - 50-55 days - Productive vines rarely exceed 3 ft., yellow-white cucumber best eaten raw when under 3 in. long, no need to peel, sweet flavor. *Source History: 1 in 2004.* **Sources: Se16.**

Mirella - 61-67 days - Smooth skinned slicer from Middle East, shorter plumper fruits, solid green, no ridges, unusual sweet flavor, some like it and some do not, CMV res., susc. to scab, 1927. *Source History: 1 in 1981; 1 in 1984; 3 in 1998; 3 in 2004.* **Sources: Sa9, Se17, So25.**

Monastic - 65 days - Short, fat pickling type also used as a slicer, cream colored when young. *Source History: 1 in 1998; 1 in 2004.* **Sources: Sa9.**

Morden Early - 45-50 days - For short season areas, short vines, med-size fruits turn yellow early, drought res., pickle or slice, Morden Exp. Farm/Manitoba, 1956. *Source History: 6 in 1981; 6 in 1984; 5 in 1987; 5 in 1991; 5 in 1994; 7 in 1998; 7 in 2004.* **Sources: Alb, Ba8, Lin, Mcf, Sa9, Tt2, WE10.**

Mountain Pickling - 52 days - Selected from Boston Pickling, smooth, bright green fruits, black spines, 3 in. long, high yields, continued production with frequent pickings. *Source History: 1 in 2004.* **Sources: Hig.**

Muncher - 59-65 days - Smooth med-green 9 x 3 in. fruits, strong vigorous vines, mosaic res., prolific, tender, eat like an apple, not bitter or tough in any slicer stages, burpless. *Source History: 4 in 1981; 3 in 1984; 2 in 1987; 4 in 1991; 4 in 1994; 7 in 1998; 14 in 2004.* **Sources: Be4, Bu3, Bu8, He8, He17, HO1, La1, OLD, Ri2, SE14, Se26, TE7, Vi4, WE10.**

National Pickling - (Ohio MR17) - 50-58 days - Short thick and blunt-ended when small, smooth and cylindrical when larger, 6-7 x 2.5 in., dark-green, black spines, heavy yields, med-large vines, excel. pickler, dev. by the National Pickle Packers Association. *Source History: 52 in 1981; 43 in 1984; 39 in 1987; 35 in 1991; 33 in 1994; 38 in 1998; 36 in 2004.* **Sources: Ada, Alb, Bo19, Bu2, CA25, CH14, CO23, Co31, Dam, DOR, Ear, Ers, Fe5, Fis, Fo7, HA3, He8, He17, Hi6, HO1, La1, Lej, Mcf, Mel, Mo13, Pin, Red, Sau, Se26, Sh9, SOU, Syn, Tt2, Wet, Wi2, WI23.**

National Pickling Improved - 52 days - Slightly tapered, medium-green fruits, black spines. *Source History: 3 in 1991; 3 in 1994; 3 in 1998; 3 in 2004.* **Sources: Jor, SE14, WE10.**

North Carolina Heirloom Pickling - 60 days - Excellent pickler, 2-3 in. long creamy white cucumber, retains crispness when pickled. *Source History: 1 in 2004.* **Sources: Sa9.**

Northern Pickling - 48 days - Very early, med-green, black spines, sets heavily, short vines, best flavor when small, scab res., for home gardens and early market in short season areas, 1955. *Source History: 5 in 1981; 4 in 1984; 4 in 1987; 8 in 1991; 5 in 1994; 8 in 1998; 8 in 2004.* **Sources: Eco, Hig, Jo1, Pe2, Pr3, Roh, Se7, So9.**

Oriental Burpless - Dark-green skin and crisp flesh, truly burpless, delicious, good pickler when small, good slicer when mature. *Source History: 1 in 1987; 1 in 1998; 1 in 2004.* **Sources: Ri2.**

Parade - 35-40 days - Russian var., heavy set of uniform fruits that mature at relatively the same time, 2 x 5 in. long fruit, resistant to weather extremes, good for processing. *Source History: 1 in 1998; 2 in 2004.* **Sources: Ma13, Se16.**

Parisian Pickling - 60 days - Fine French heirloom gherkin or cornichon pickler, makes excel. pickles when picked small, also good slicer, introduced into the U.S. around 1892. *Source History: 1 in 1981; 1 in 2004.* **Sources: La1.**

Per Sotto Aceti - (Per Sott'aceto) - Small, light green, elongated fruit. *Source History: 1 in 1987; 2 in 1991; 1 in 1994; 1 in 1998; 1 in 2004.* **Sources: Ber.**

Perfection - Fine outdoor English cucumber, excel. 8-10 in. long fruits, bears well into the fall. *Source History: 1 in 1981; 1 in 1984; 1 in 1987; 2 in 1991; 2 in 1994; 1 in 1998; 2 in 2004.* **Sources: Eco, So9.**

Persian - Considered the finest for pickling or fresh eating, sweet and crunchy with almost no skin, the favorite var. from the Middle East. *Source History: 1 in 1991; 1 in 2004.* **Sources: Red.**

Piccolo di Parigi - Early green pickling type, small seeds even at 7 in. length, excel. taste, holds very well on the vine, productive over a long season. *Source History: 1 in 2004.* **Sources: Se24.**

Picklebush - 52 days - New pickler from Burpee, very compact productive vines 20-24 in. long, blocky white-spined light-green moderately warted fruits, 4.5 x 1.5 in. at maturity, PM and CMV tol., usable at any stage from small sweets to large dills, PVP 1987 Bu2. *Source History: 1 in 1987; 1 in 1991; 1 in 1994; 1 in 1998; 1 in 2004.* **Sources: Bu2.**

Pickler, Burpee's - (Pickler) - 53 days - Vigorous early-maturing black-spined pickler, good CMV tol., heavy yields over a long period, usable in all stages, med-green warty blocky and blunt tipped, 1957. *Source History: 5 in 1981; 3 in 1984; 3 in 1987; 5 in 1991; 3 in 1994; 4 in 1998; 3 in 2004.* **Sources: Bu2, Ev2, Hi13.**

Poinsett - 58-70 days - Widely adapted but most common in the South, dark-green and cylindrical, rounded ends, res. to DM PM AN1 AN2 and ALS, very hardy, good med-early market variety, blocky, good color, white spines, heavy yields, dependable, Dr. Barnes/Clemson, 1966. *Source History: 39 in 1981; 36 in 1984; 22 in 1987; 12 in 1991; 10 in 1994; 9 in 1998; 5 in 2004.* **Sources: Ada, La1, Mo20, Sau, Sh9.**

Poinsett 76 - 63-75 days - Improved Poinsett, adds scab res. and darker green fruit color, longer blocky straight fruits, 7.5-8.5 x 2.25-2.75 in., EM PM AN1 AN2 ALS and Scab res., not recommended for New York growers because susc. to CMV, cooperative release by Dr. Munger of Cornell and Clemson VA/AES. *Source History: 37 in 1981; 32 in 1984; 38 in 1987; 34 in 1991; 39 in 1994; 35 in 1998; 29 in 2004.* **Sources: Bu3, Bu8, Cl3, CO23, CO30, Dgs, DOR, Ers, He8, HO1, HO2, Kil, LO8, Loc, ME9, Mey, PAG, Ros, Scf, SE14, Se26, Shu, So1, SOU, Vi4, WE10, Wet, Wi2, WI23.**

Poona Kheera - 55 days - Imported from India, very unusual, smooth-skinned greenish-white small fruits, tender crisp and delicious, may be eaten at any stage skin and all. *Source History: 1 in 1981; 1 in 1984; 1 in 1987; 2 in 1991; 2 in 1994; 2 in 1998; 3 in 2004.* **Sources: Ba8, Sa9, Se17.**

Qualitas - *Source History: 1 in 2004.* **Sources: WE10.**

Rhinish Pickle - 52-55 days - Old German pickling var., black spines, small bumps. *Source History: 2 in 2004.* **Sources: He8, Sk2.**

Richmond Green Apple - Medium green, round, 3 in. fruit, mild flavor, from Australia. *Source History: 1 in 1998; 2 in 2004.* **Sources: Ba8, Ho13.**

Russian - 52 days - Fruits 6 in. long grow in clusters of 2 or 3, excel. for small pickles, does well when grown on a trellis, introduced in 1850. *Source History: 2 in 1991; 2 in 1994; 3 in 1998; 4 in 2004.* **Sources: Fo7, He8, Ho13, Ra5.**

Shintokiwa - 65-70 days - Harvest slim fruit for best flavor, trellis to avoid crooked fruit, 9-16 in., burpless, no bitterness, from Japanese ag students at Rudolf Steiner College, Turtle Tree Seed introduction. *Source History: 3 in 2004.* **Sources: Pe2, So12, Tu6.**

Smart Pickle - 45-55 days - Productive vines grow 3-6 ft., fine quality 4-7 in. fruit, good pickler, disease res. *Source History: 1 in 1991; 2 in 1994; 2 in 1998; 2 in 2004.* **Sources: Se7, So12.**

SMR 58 - (Everbearing) - 55-58 days - Everbearing, vigorous vines produce heavy yields all season, nearly cylincrical, top quality for processing, also good slicer, good disease res., University of Wisconsin. *Source History: 6 in 1981; 6 in 1984; 19 in 1987; 25 in 1991; 27 in 1994; 29 in 1998; 27 in 2004.* **Sources: Bu3, Bu8, CH14, CO23, De6, DOR, Ers, Gr27, HO1, Hum, Jor, La1, Loc, Ma19, MAY, ME9, Ni1, RIS, Ros, RUP, Sau, Se25, Se26, Sh9, Ter, Wet, WI23.**

Snow's Fancy Pickling - 50-60 days - Dark green, short slender fruit, great pickling variety, 5-6 in. long, selected from Chicago Pickling by J. C. Snow of the Snow Pickle Farm of Rockford, IL, intro. by Vaughan's of Chicago in 1905. *Source History: 3 in 2004.* **Sources: Ba8, Se16, Te4.**

Spacemaster - (Burpee's Spacemaster, Spacemaster Bush) - 58-60 days - Dark-green smooth slender cylindrical uniform 7- 8 in. fruits, dwarf 36 in. vines with no runners, CMV and scab res., adaptable to a wide range of climates, keep well picked to avoid misshapen fruits late in the season and to prolong the bearing period, Cornell. *Source History: 32 in 1981; 30 in 1984; 36 in 1987; 39 in 1991; 33 in 1994; 26 in 1998; 25 in 2004.* **Sources: All, Be4, Bo19, Bu2, Dam, DOR, Ear, Fis, Gr27, He17, Hi13, HPS, Hum, La1, Mel, Mo13, Mo20, Pin, Ros, Se26, Sh9, Shu, So1, So25, Yu2.**

Spacemaster 80 - 57-64 days - Dwarf 18-24 in. bush vines, large uniform 7-9 in. fruits, good flavor--never bitter, heavy yields, mildew and scab and mosaic res., thrives in full sun. *Source History: 4 in 1987; 7 in 1991; 5 in 1994; 13 in 1998; 14 in 2004.* **Sources: CH14, Cr1, Ers, Ga1, HO1, Jor, OLD, PAG, RUP, SE14, Se25, Vi4, WE10, WI23.**

Spring of Water - Large, tasty pickling cucumber from Leningrad, Russia, may be some slicing types, seed was slightly crossed in Russia, rare. *Source History: 1 in 2004.* **Sources: Ea4.**

Straight Eight - 52-75 days - AAS/1935, smooth straight cylindrical and deep-green with rounded blunt ends, tol. to mosaic, 8 x 2-2.5 in., early, prolific, for home use, vigorous, excel. quality, free from stippling, 1935. *Source History: 95 in 1981; 85 in 1984; 84 in 1987; 71 in 1991; 65 in 1994; 69 in 1998; 83 in 2004.* **Sources: Ag7, Alb, All, Be4, Bo17, Bou, Bu1, Bu2, Bu3, CA25, CH14, CO23, Co31, Com, Cr1, Dam, De6, DOR, Ear, Ers, Fa1, Fi1, Fis, Fo7, Fo13, Ga1, Ge2, Gr27, Gur, HA3, Hal, He8, He17, Hi6, Hi13, HO1, HO2, HPS, Jor, La1, Lin, LO8, Ma19, MAY, Mcf, ME9, Mel, Mey, MIC, Mo13, MOU, Ont, Or10, PAG, Pe2, Pe6, Pr3, Pr9, Ri2, Ri12, Roh, RUP, Sau, SE4, Se7, SE14, Se25, Se26, Sh9, Shu, Sk2, So1, SOU, Te4, TE7, Tt2, Up2, Ves, Vi4, WE10, Wet, Wi2, WI23.**

Straight Nine - 58-66 days - Imp. Straight Eight, straight fruits 9 in. long x 1.5 in. dia., prolific, vigorous, retains green color in heat and drought, DM and PM tol., widely adapted. *Source History: 1 in 1981; 1 in 1984; 7 in 1987; 8 in 1991; 10 in 1994; 10 in 1998; 8 in 2004.* **Sources: HO1, La1, MAY, OLD, Pin, Roh, Sk2, Wi2.**

Sumter - (Sumpter) - 50-55 days - Blocky, slightly tapered, med-green, white spines, for home or market, North or South, res. to PM DM ALS and scab, tol. to CMV, some res. to AN, Clemson. *Source History: 6 in 1981; 5 in 1984; 3 in 1987; 3 in 1991; 4 in 1994; 3 in 1998; 7 in 2004.* **Sources: Ag7, Dam, Scf, Se26, Sk2, TE7, WE10.**

Suyo Long - (Soo Yoo Long, Soo Yoh, Soo Yow, Suhyo Long) - 60-70 days - Cucumber from northern China, ribbed dark- green skin with heavy white spines, 10-18 x 1.5 in., PM res., almost seedless, requires staking, non-bitter, burpless, crisp and tender, prolific, heat res., does well in the interior of southern California. *Source History: 13 in 1981; 13 in 1984; 12 in 1987; 13 in 1991; 12 in 1994; 17 in 1998; 14 in 2004.* **Sources: Ag7, Al25, Ev2, Fe5, Ga1, Jo1, Na6, Pe2, Pe6, Red, Se7, So1, So9, Syn.**

Telegraph Improved - 60-62 days - Improved strain, shorter neck, keeps crisp and fresh longer, prolific cropper, all around variety, equally good for summer or winter, greenhouse or outdoors, introduced around 1897. *Source History: 2 in 1981; 2 in 1984; 4 in 1987; 2 in 1991; 2 in 1994; 4 in 1998; 7 in 2004.* **Sources: Ba8, Dam, Fe5, SE14, Se26, Ter, WE10.**

Telegraph, English - 75 days - Imported from Europe, dark-green slim straight and symmetrical, 15-17 in. average, non-bitter, excel. for slicing and eating, sel. from Rollinson's Telegraph. *Source History: 5 in 1981; 3 in 1984; 5 in 1987; 7 in 1991; 6 in 1994; 4 in 1998; 3 in 2004.* **Sources: Ear, Ont, Sto.**

Tendergreen Burpless - 55-65 days - Open-pollinated burpless cucumber, light green with white spines, 7-12 in. long, slight blocky shape, sweet flesh is tender, ideal slicer or pickler, good disease resistance, over 80 years old. *Source History: 3 in 2004.* **Sources: Ers, He8, Shu.**

Thai Green - Medium green fruit, 2 x 7 in. long, perfect for hot humid climates, from a seed dealer in Bangkok, hard to find outside of Thailand, Baker Creek Heirloom Seed introduction. *Source History: 1 in 2004.* **Sources: Ba8.**

Timun Hijan - (Tempaton Jenis) - Thin skinned fruit, 4-6 in. long, 1-2 in. across, mild flavor, good slicer, does well in southern gardens. *Source History: 1 in 1991; 1 in 1994; 1 in 1998; 1 in 2004.* **Sources: Red.**

Tokiwa - 70 days - Uniform, large slicer, good flavor, slow seed development. *Source History: 1 in 2004.* **Sources: Sa9.**

Tokiwa Jihai - 45 days - Dark-green tender sweet fruits 1.5 x 12 in., white spines, well known in Japan where it is succession sown from April to mid-July, does best when trained on a trellis. *Source History: 1 in 1981; 1 in 1984; 1 in 1991; 1 in 1994; 1 in 1998; 1 in 2004.* **Sources: Ev2.**

Ukrainian Pickling - 60 days - Eastern European type, light green fruits turn a russet color when ripe, remains solid for several weeks in storage. *Source History: 1 in 1998; 1 in 2004.* **Sources: Sa9.**

Uzbekski - 45-60 days - Light green skin turns golden brown, 6-8 in. long by 2.5-3.5 in. wide, retains crispness and flavor for up to 2 months after harvest. *Source History: 2 in 1998; 4 in 2004.* **Sources: In8, Sa9, Se17, So25.**

Vert Petit de Paris - (Parisian Cornichon) - 50-55 days - European cucumber used especially for making cornichons (tiny very sour French pickles), pick when as thick as your finger, very popular with French gardeners, originated in France circa 1870. *Source History: 4 in 1981; 4 in 1984; 3 in 1987; 2 in 1991; 1 in 1994; 2 in 1998; 2 in 2004.* **Sources: Ag7, Lej.**

Vorgebirstauben - Type uncertain. *Source History: 1 in 1991; 1 in 1994; 1 in 1998; 1 in 2004.* **Sources: WE10.**

Vyaznikosky - 60 days - Old Russian heirloom, prolific compact vines, blocky white fruit matures to yellow, good for cold climates. *Source History: 1 in 2004.* **Sources: Se17.**

Wautoma - 60 days - Consistently produces huge crops of high quality picklers, BCMV PM DM AN FW and leaf spot res., bitter-free, USDA. *Source History: 1 in 1998; 1 in 2004.* **Sources: Ter.**

White - 60-70 days - Old variety, plump shape, rough whitish skin, turns yellow-orange when very ripe, white flesh. *Source History: 2 in 1998; 3 in 2004.* **Sources: Bo19, Roh, Se17.**

White Cucumber - Very old variety, small, easy to grow, can be grown in the corn without any spraying. *Source History: 1 in 1981; 1 in 1984; 1 in 1991; 2 in 1994; 2 in 1998; 1 in 2004.* **Sources: Lan.**

White Spine Improved - (Fordhook White Spine) - 60-67 days - Early, med-green fruits, 7.5-8 x 2.5-2.75 in., long slender taper, white spines, very productive, uniform in size, popular old timer. *Source History: 10 in 1981; 8 in 1984; 7 in 1987; 4 in 1991; 2 in 1994; 2 in 1998; 1 in 2004.* **Sources: Wet.**

White Spine, Early - (Short Prickly/White Spine) - Pickling cucumber, 8-10 in. long, good taste, high yields, listed by McMahon in 1806 and Thorburn in 1824. *Source History: 1 in 2004.* **Sources: Ea4.**

White Wonder - (Long White) - 35-60 days - Ivory-white fruits even when mature, cylindrical, rounded ends, 7-9 x 2.5-3 in., excel. eating quality, ideal for pickles or slicing, highly productive even in hot weather, crisp firm mild flesh, vigorous vines, W. Atlee Burpee introduction in 1893. *Source History: 22 in 1981; 18 in 1984; 17 in 1987; 17 in 1991; 24 in 1994; 27 in 1998; 37 in 2004.* **Sources: Ag7, Ba8, Be4, Bo18, Bu1, Bu8, CA25, Co32, Com, Ers, Fo7, Gr29, He8, He17, Ho13, Hud, Jor, Ki4, La1, Ol2, Ra5, Ri12, Se8, SE14, Se16, Se26, Se28, Sh9, Shu, So1, SOU, Tho, Vi4, WE10, Wet, Wi2, Wo8.**

Wisconsin SMR - 56 days - Husky vigorous vines, blocky dark-green fruits, heavy yields. *Source History: 1 in 1987; 1 in 1991; 1 in 1994; 2 in 2004.* **Sources: LO8, SE4.**

Wisconsin SMR 18 - (SMR-18) - 53-60 days - Strong vigorous vines, 6 x 2.5 in. dark-green fruits--straight blocky well warted, black spines, small seed cavity, DM PM ALS AN1 and 2 res., CMV tol. in favorable conditions, heavy yields, for the North, Dr. Walker/U of WI. *Source History: 29 in 1981; 25 in 1984; 20 in 1987; 17 in 1991; 10 in 1994; 6 in 1998; 4 in 2004.* **Sources: All, Fi1, Kil, Sh9.**

Wisconsin SMR 58 - 53-64 days - Mid-season med-green straight blocky and slightly tapered, 6.5 x 2.5 in., blockier than Wisconsin SMR 18, CMV scab and spot rot res., black spines, uniform size and shape, everbearing if kept picked, very vigorous, heavy yields, long harvest, high quality, home or market in the North, Dr. Walker/U of WI, 1959. *Source History: 30 in 1981; 25 in 1984; 18 in 1987; 15 in 1991; 12 in 1994; 15 in 1998; 13 in 2004.* **Sources: Cr1, HA3, Hal, He17, HO2, Jun, Mey, MOU, Ont, PAG, SE14, Sto, WE10.**

Yamato, Extra Long - 60-75 days - Unusual Japanese var., dark-green straight fruits up to 15-24 in., smooth skin, must stake, excel. slicer even when large, thick firm flesh, fine special flavor. *Source History: 5 in 1981; 5 in 1984; 3 in 1987; 3 in 1991; 1 in 1994; 1 in 1998; 6 in 2004.* **Sources: CO30, Kit, Ni1, PAG, Se26, So1.**

Yellow Submarine - 58-60 days - Large fruit to 8 in. long by 3 in. wide, small seed cavity, light green > light yellow, pickling type, also used fresh, resists CMV, PM and scab. *Source History: 9 in 2004.* **Sources: Bu1, Ear, Ers, HO2, HPS, Jun, Shu, Tt2, Ver.**

Zeppelin - 67 days - Holds world record weights of 11 lb. 6 oz. indoors and 8 lb. 4 oz. outdoors, firm and juicy, can be thinly sliced or diced or chopped, for the Southern states. *Source History: 1 in 1981; 1 in 1984; 1 in 1987; 1 in 1994; 2 in 1998; 1 in 2004.* **Sources: Pp2.**

Varieties dropped since 1981 (and the year dropped):

A and C (1987), *Ace* (1987), *Amcogreen* (1982), *Amcogreen VR* (1982), *American Climbing* (1987), *Amira II* (2004), *Aodai* (1991), *Ashley Improved* (1994), *Beit Alpha* (2004), *Beit Alpha Saria* (1994), *Big Daddy* (1998), *Bountiful Bush* (1998), *Bourbonne Improved* (1991), *Brocade* (1994), *Burpee's Sunnybrook* (1987), *Burpless Improved* (1991), *Bush Whopper* (2004), *Carolina 84* (1998), *Chipper* (1998), *Climbing* (1991), *Colorado* (1987), *Conqueror* (1991), *Coolgreen MR* (1987), *Cornichon, Small Paris* (2004), *Crackerlee* (1998), *Crackerlee Improved* (1994), *Crystal Salad* (1987), *Darkie* (1984), *Darkie Improved* (1984), *Dekah* (2004), *Earliest of All* (1982), *Early Kaga* (1983), *Early Quebec* (2004), *Early Slicers* (1998), *English* (1994), *Fletcher* (1987), *Galaxy* (1991), *Gele Tros* (2004), *Gherkins Pariser* (1994), *Green King* (1991), *Green Ridge* (1994), *Heinz Special* (1998), *High Moor (Scab Resistant)* (1984), *Highmoor* (1984), *Hokus* (1994), *Hokus Original* (1987), *Japanese Burpless Cucumber* (1987), *Japanese Pickling* (2004), *Japanese Red* (2004), *Jersey Pickle* (2004), *Jet* (1991), *Jumbo* (1991), *Kirby (Original Strain)* (1987), *Le Generaux* (1987), *Liberator* (1987), *Long English Greenhouse* (1994), *Long Green Black Spine* (1991), *Long Green Ridge* (1987), *Long Green White Spine* (1998), *Long White Italian Import* (1991), *Long White Salad* (1991), *Marketer, Long* (2004), *Marketsett* (1994), *Medalist* (2004), *Melange Supreme* (1998), *Mideast Peace* (1998), *Model* (1987), *Model* (1987), *Morden Midget* (1994), *MR 17* (1991), *Nejinsky* (2004), *Ohio MR 17* (1998), *Pacer* (1991), *Palmetto* (1987), *Palomar* (1991), *Pixie MR* (1991), *Poinmarket* (1994), *Poinsett Improved* (1984), *Polaris* (1987), *Precoce Grosso Bianco Crema* (2004), *Producer* (1994), *Rhinish Pickle, Ridu* (2004), *Riesenschal* (2004), *Sanjiaku Kiuri* (2004), *Seedless Supreme* (1991), *Sena* (1998), *Setter* (1987), *Siberian Pickling* (1998), *Slendersweet* (1998), *SMR 18* (1998), *SMR 58, RX* (1994), *Spartan Salad* (1991), *Spartan Valour* (1987), *Stono* (1991), *Summer Slicer* (1987), *Super Bush Slicers* (2004), *Tablegreen* (1994), *Tablegreen 65* (1991), *Tableslice* (1991), *Tasty Green (Oriental)* (2004), *Taxpayer* (1991), *Telegraph* (2004), *Telegraph Reselected* (1991), *Telegraph, Rollinson's* (2004), *Tex Long* (1991), *Tiny Dill* (1984), *Tokyo Green* (1991), *Top Green* (1987), *Toska* (1984), *Triumph* (1987), *Venlo'er Export* (1987), *White Spine Improved, Early* (1998), *Windermoor Wonder* (2004), *Wisconsin SMR 15* (1991), *Yellow Giant* (1987), *York State* (1991).

Eggplant
Solanum melongena var. esculentum

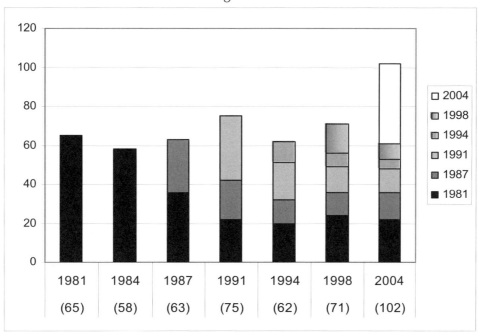

Number of varieties listed in 1981 and still available in 2004: 65 to 22 (34%).
Overall change in number of varieties from 1981 to 2004: 65 to 102 (157%).
Number of 2004 varieties available from 1 or 2 sources: 71 out of 102 (70%).

Almaz - 85 days - Russian var. with dark purple fruits, long fat slicing type. *Source History: 2 in 1998; 1 in 2004.* **Sources: Sa9.**

Antigua - 75-92 days - Tender, 2 x 8 in. long white fruit with lavender-violet stripes, mild flavor, good yields. *Source History: 1 in 1998; 5 in 2004.* **Sources: Ba8, Ho13, Ma18, Sa9, To1.**

Applegreen - 62-70 days - Extra early, apple-green

med-sized non-acid fruits, no need to peel, productive upright plants, sets fruits under adverse Northern conditions, developed by Professor Elwyn Meader, NH/AES, 1964. *Source History: 2 in 1981; 1 in 1984; 1 in 1987; 1 in 1991; 4 in 1998; 18 in 2004.* **Sources: Ag7, Ba8, Bo19, Bu2, Fe5, La1, Ma18, ME9, Pr3, Pr9, Sa9, SE14, Se16, Se26, So1, So25, Ter, To1.**

Aubergine Mohican - 85 days - Bushy dwarf plants bear clusters of long, oval white fruits, ideal for containers. *Source History: 1 in 2004.* **Sources: Tho.**

Baby Marbled - 80 days - Apple-green and white stripes on lavender fruit, 3-4 in. long. *Source History: 1 in 2004.* **Sources: Ma18.**

Bianca Ovale - 75 days - Spineless plants grow 15 in. tall, 3 in., white egg-shaped fruits mature to yellow, very seedy, rare. *Source History: 1 in 1991; 1 in 1998; 2 in 2004.* **Sources: Ba8, Ma18.**

Billeau - 55 days - Short, dark purple fruit, early from transplant. *Source History: 1 in 2004.* **Sources: Sa9.**

Black - 70 days - Early maturing var. with sweet, oblong fruits, 4-6 in. long, 8-10 per plant, does well in northern country. *Source History: 1 in 1998; 2 in 2004.* **Sources: Hi6, Se7.**

Black Beauty - 72-90 days - Bushy spreading 24-30 in. plant with 4-6 purplish-black smooth oval 6.5 x 5 in. dia. fruits, blunt and broad at blossom end, retains color well, 1-3 lbs., holds well, high quality, fine flavor, yields well in the North if the season is long. *Source History: 118 in 1981; 103 in 1984; 105 in 1987; 82 in 1991; 82 in 1994; 81 in 1998; 90 in 2004.* **Sources: Ada, All, Ba8, BAL, Ban, Be4, Bo17, Bo18, Bo19, Bu1, Bu3, Bu8, But, CA25, CH14, CO23, CO30, Co31, Co32, Com, Coo, Cr1, Dam, De6, Dgs, DOR, Ear, Ers, Fo7, Ga1, Ge2, GLO, Gr27, Gr29, GRI, Gur, HA3, He8, He17, Hi13, HPS, Jor, Kil, La1, Lej, Lin, Loc, Ma19, MAY, ME9, Mel, Mey, MIC, Mo13, MOU, OLD, Ont, Or10, PAG, Pe2, Pin, Pr3, Pr9, Ra5, Red, RIS, Roh, Ros, RUP, Sau, SE14, Se25, Se26, Se28, SGF, Sh9, Shu, Si5, Sk2, So1, SOU, TE7, To1, Up2, Vi4, WE10, Wet, Wi2, WI23, Wo8.**

Black Beauty, Imperial - 78-85 days - Excellent strain of this popular variety, plants average 18-24 in. tall and usually bear four large dark-purple, egg-shaped fruit, 1-3 lbs. each, introduced in 1910. *Source History: 1 in 1981; 1 in 1984; 3 in 1987; 2 in 1991; 3 in 1994; 4 in 1998; 3 in 2004.* **Sources: Ri12, Se7, Sto.**

Black Egg - (Early Black Egg) - 63-85 days - Earliest Japanese egg-shaped type, 4 x 2 in., broader toward the blossom end, bushy plants, fruits held well above ground, very prolific, fruits until fall. *Source History: 2 in 1981; 1 in 1984; 1 in 1987; 1 in 1991; 1 in 1994; 6 in 2004.* **Sources: Ba8, La1, Ma18, ME9, Sa9, SE14.**

Black Egg, Early - 65-75 days - Japanese variety, pear-shaped slightly blocky 5-7 in. fruits, 3 ft. bushy plants, unusually tender, fine flavor, sets early and well in cool summer and short season areas, pick when young. *Source History: 6 in 1981; 6 in 1984; 3 in 1987; 3 in 1991; 3 in 1994; 3 in 1998; 1 in 2004.* **Sources: So1.**

Blue Devil - 70 days - Dark purple, 8-10 in. long fruits. *Source History: 1 in 2004.* **Sources: Sa9.**

Bride - *Source History: 1 in 2004.* **Sources: ME9.**

Casper - (Casper White) - 70-90 days - Early, shiny ivory fruits, 6-10 x 2.75 in., mild snow-white flesh, no need to peel if eaten very small, originally from France. *Source History: 2 in 1987; 3 in 1991; 3 in 1994; 5 in 1998; 16 in 2004.* **Sources: Ba8, Bo19, Bou, Coo, Ho13, Hud, La1, ME9, Pe6, Pr9, Sa9, SE14, Se16, Sk2, Twi, WE10.**

Chinese Long - (Chinese Long Purple) - 125 days - Heavy yields, disease tol., uniform long slender 12 x 1.5-2 in. purple fruits, tapers to a point, upright robust plants, good type for the South, from Japan. *Source History: 4 in 1981; 4 in 1984; 3 in 1987; 3 in 1991; 1 in 1994; 1 in 1998; 2 in 2004.* **Sources: Ba8, Kit.**

Chinese White - (Chinese White Sword) - 65 days - White oval fruits with very tender skin, 2 x 7 in., mild sweet taste, turns golden yellow at maturity but should be harvested young when shiny white, grows well in containers. *Source History: 1 in 1991; 1 in 1998; 1 in 2004.* **Sources: Sa9.**

Comprido Verde Claro - 72 days - Tall plant with unusual jagged foliage, 3-4 in. long, tasty colorful fruit, green > orange, good yields, from Brazil. *Source History: 2 in 2004.* **Sources: Ba8, Ma18.**

Dark Long Red - 65 days - Blocky short fruit, deep purple-red skin color. *Source History: 1 in 1994; 1 in 1998; 2 in 2004.* **Sources: Red, Sa9.**

De Barbentane - 75 days - Sturdy vigorous plants, very long cylindrical fruits, shiny purple-black, named after a town in France. *Source History: 1 in 1981; 1 in 1984; 2 in 1987; 2 in 1991; 2 in 1994; 3 in 1998; 2 in 2004.* **Sources: Gou, Lej.**

Diamond - 70 days - 109-149 days from direct seeding, newly bred Ukrainian variety, plants 20-25 in. tall, long, slender, cylindrical fruits, 2-3 in. x 7-9 in., 4-6 oz., dark-violet skin, dense green pulp without bitter mustard taste, recommended for climates where it can be planted and ripened in open soil, not for northern greenhouses, collected from the Ukraine by the Seed Savers Exchange, 1993. *Source History: 1 in 1994; 4 in 1998; 7 in 2004.* **Sources: Ba8, La1, Pr9, Sa9, SE14, Se16, WE10.**

Dourga - (White Dourga) - 65-70 days - Vigorous plant, shiny white cylindrical 2 x 8 in. fruit, excel. flavor and texture, France. *Source History: 1 in 1991; 4 in 1994; 2 in 1998; 7 in 2004.* **Sources: Bo18, Gr29, Ho13, HPS, Ra5, Se28, Wo8.**

Fengyuan Purple - Light lavender-purple fruit to 16 in. long, sweet, tender and mild, good yields, heirloom from Taiwan. *Source History: 2 in 2004.* **Sources: Ami, Ba8.**

Florentine Silk - 90 days - Medium size pink-lavender fruit with white shading on the shoulders, sweet, mild flesh. *Source History: 1 in 2004.* **Sources: Ma18.**

Florida High Bush - (Florida High Bush Select) - 76-87 days - Vigorous upright well-branched plants, large purple egg-shaped fruits held high off ground, broad near blossom end, disease and drought res., hardy, everbearing. *Source History: 13 in 1981; 10 in 1984; 6 in 1987; 4 in 1991; 1 in 1994; 1 in 1998; 2 in 2004.* **Sources: Pr9, Se16.**

Florida Market - (Cook's Strain) - 78-90 days - Vigorous erect 30-38 in. plants, glossy dark purple-black 9.5

x 6.5 in. fruits held well off ground, res. to phomopsis blight and fruit rot, long harvest season, long oval, no neck at stem end, green calyx, suited for deep South, holds up well in the sun, prolific high bush type, fresh market. *Source History: 18 in 1981; 16 in 1984; 10 in 1987; 15 in 1991; 16 in 1994; 17 in 1998; 9 in 2004.* **Sources: Ba8, Ech, Hud, Kil, LO8, ME9, Mey, SE14, WE10.**

Gitana - Long, slim, slightly curved purple eggplant, outstanding flavor, excel. producer. *Source History: 1 in 2004.* **Sources: Se24.**

Green Beauty - One of the most popular eggplants grown in Southeast Asia, produces well in hot weather and rain, sets fruit 30 days after transplant, 10-12 in. long fruit. *Source History: 1 in 2004.* **Sources: Ev2.**

Green Eggs - 80 days - Dwarf plant bears green, 3 in., egg size fruit, seedy, good stuffer, early, originally introduced in the 1860s as Tibet Green. *Source History: 1 in 1998; 1 in 2004.* **Sources: Ma18.**

Green Giant Early - 62 days - Asian type, bell shaped fruit with green calyx, light green skin, firm flesh. *Source History: 1 in 2004.* **Sources: Jo1.**

Hibush Special - 80-82 days - Deep purple-black, oval shape fruit, 8 x 4 in., resists phomopsis fruit rot, grows best in light sandy soil. *Source History: 1 in 1998; 1 in 2004.* **Sources: ME9.**

Highbush Select - 80 days - Teardrop shaped, glossy black fruit holds color well. *Source History: 1 in 1994; 3 in 1998; 2 in 2004.* **Sources: SE4, Sto.**

Italian Pink Bicolor - 70-90 days - Large oval cream and rose-pink fruits mature to pure rose-pink, purple novelty with the European trade, open-pollinated strain that still throws a percentage of bicolored striped fruits. *Source History: 1 in 1991; 3 in 1994; 3 in 1998; 3 in 2004.* **Sources: Ma18, Se7, Sto.**

Italian White - (Melanzana Bianca Ovale) - 72-78 days - Imported from Italy, plump round white fruits, 3-4 in., sweet mushroom-like flavor, best when egg-sized, milder than the purples, never bitter. *Source History: 2 in 1981; 1 in 1984; 1 in 1991; 1 in 1994; 3 in 1998; 4 in 2004.* **Sources: Ho13, Roh, Se7, Se17.**

Jade Sweet - 77 days - Blocky shaped fruit with lime-green skin. *Source History: 1 in 2004.* **Sources: Sa9.**

Japanese Early Purple - An early strain, smaller, slimmer amd more tender than European types. *Source History: 1 in 1981; 1 in 1984; 1 in 1987; 1 in 1991; 1 in 1994; 1 in 1998; 1 in 2004.* **Sources: Se17.**

Japanese Purple Pickling - 75-90 days - Small 18 in. plants bear early, produces masses of small, 6-8 in. fruits, very full- flavored, good for frying and cooking, excel. for pickling, space miser. *Source History: 1 in 1981; 1 in 1984; 1 in 1987; 1 in 1991; 2 in 1994; 1 in 1998; 2 in 2004.* **Sources: Roh, Se7.**

Kurume Long - (Kurume Long Purple, Kurime Long) - 78 days - Slender 10-12 in. fruits, glossy purple-black, can produce 40-50 fruits per plant, res. to FW BW heat and drought, excel. for warm areas, med-late, from Japan. *Source History: 2 in 1981; 2 in 1984; 3 in 1987; 4 in 1991; 5 in 1994; 5 in 1998; 4 in 2004.* **Sources: CO30, Ev2, Kit, Ri2.**

Lao Green Stripe - (Green Tiger Stripe) - 120 days - Plants 3.5 ft. tall, round 2 in. dark cream colored fruits with green tiger stripes on upper half, heirloom from Laos. *Source History: 1 in 1991; 1 in 1998; 1 in 2004.* **Sources: Ho13.**

Lao Lavender - Ornamental plant bears beautiful 3 in. round light purple-lavender fruit, non-bitter skin, no need to peel, from Laos, Asia. *Source History: 2 in 2004.* **Sources: Ami, So25.**

Lebanese Bunching - 80 days - Plant produces clusters of 3-4 fruit per stem, non-bitter, mild flavor. *Source History: 1 in 2004.* **Sources: Ma18.**

Listada de Gandia - 75-90 days - Italian type, white with purple stripes, 6-10 in. long, thin skin, not rec. for northern states, French heirloom, 1850. *Source History: 1 in 1987; 3 in 1991; 3 in 1994; 4 in 1998; 5 in 2004.* **Sources: Ho13, Ma18, Sa9, Se16, So1.**

Little Fingers - 60-75 days - Long slim 4-7 in. fruit in clusters of 5-10, easy to pick, excel. for frying, pickling and Oriental dishes, PVP expired 2000. *Source History: 2 in 1991; 2 in 1994; 4 in 1998; 11 in 2004.* **Sources: BAL, CH14, Ha5, HO2, Ma18, ME9, Orn, OSB, SE4, Ter, To1.**

Little Spooky - 60 days - Plants to 3 ft., cylindrical, 3 x 7 in. white fruit, grown for generations in Hokkaido Island, Japan, where local legend has it that growing these eggplants in the garden will scare away evil night spirits, thus insuring good crops. *Source History: 1 in 2004.* **Sources: He8.**

Long Lavender White - Slender 8 in. long fruit, white skin color matures to white and lavender striped, skin is thin and tender with no bitterness. *Source History: 1 in 2004.* **Sources: Hud.**

Long Purple - (Long Purple Italian) - 70-80 days - Dark-purple productive Italian type, 22-38 in. plants, 8-10 x 2.5 in. dia., 4 or more fruits per plant, long club shape, firm mild flesh, yields well in the north, dates back to the 1850s in America. *Source History: 9 in 1981; 7 in 1984; 11 in 1987; 12 in 1991; 13 in 1994; 16 in 1998; 28 in 2004.* **Sources: Ag7, Ba8, Bo17, Bo19, Bou, Bu3, CO30, Cr1, Dgs, DOR, Fo7, Ga1, He8, He17, La1, LO8, Loc, Na6, Or10, Pe2, Ros, SE14, Se24, Se25, So1, TE7, To1, WE10.**

Long Purple, Early - 70-78 days - Very prolific Italian var., cucumber-shaped, 9-12 x .75 in. dia., young violet skin matures to purple, 20-26 in. plant not heavily branched, slightly bulbous. *Source History: 8 in 1981; 7 in 1984; 7 in 1987; 4 in 1991; 4 in 1994; 4 in 1998; 2 in 2004.* **Sources: Com, Se8.**

Long Purple, Japanese - 70 days - Lavender fruit, 2 x 9 in. *Source History: 1 in 1987; 1 in 1991; 1 in 1998; 1 in 2004.* **Sources: Ho13.**

Long White Sword - 85 days - Meaty fruits 2 x 9 in., mild flavor. *Source History: 1 in 1987; 1 in 1991; 1 in 1994; 1 in 1998; 1 in 2004.* **Sources: Ma18.**

Louisiana Green Oval - 75 days - Sturdy upright 2-3 ft. plants with light-green foliage, large glistening light-green fruits 2-4 lb., high quality, mild flavor, seed cavities confined to the blossom end, best harvested while skin is still glossy. *Source History: 1 in 1987; 1 in 1991; 1 in 1994; 1 in 1998; 2 in 2004.* **Sources: Ho13, Sa9.**

Louisiana Long Green - (Green Banana) - 100 days -

Prolific plants, well-shaped, 3 x 8 in. long light green fruits, mild sweet flavor, excel. quality, much like Louisiana Green Oval but of the Japanese type, rare heirloom from Louisiana. *Source History: 2 in 1991; 3 in 1994; 4 in 1998; 17 in 2004.* **Sources: Ba8, Bu2, Ho13, Ho13, Jo1, La1, ME9, Na6, Orn, Rev, Ri12, Se7, SE14, Se26, Shu, So1, WE10.**

Malaysian Dark Red Long - Medium purple, 2 x 6 in. fruits. *Source History: 1 in 1998; 1 in 2004.* **Sources: Ho13.**

Manjri Gota - Beautiful white fruit with purple stripes, 3 in. wide by 5 in. long. *Source History: 1 in 2004.* **Sources: Red.**

Millionaire - 68 days - Exceptionally handsome Japanese type, upright plants, slender elongated fruits are 10-12 in. long x 1-2 in. dia., extra early. *Source History: 1 in 2004.* **Sources: CO30.**

Morden Midget - 60-80 days - Earliest of all, small bushy sturdy plants, smooth med-sized oval deep-purple fruits, Morden Exp. Farm/Manitoba. *Source History: 3 in 1981; 3 in 1984; 1 in 1987; 1 in 1991; 1 in 1994; 4 in 1998; 5 in 2004.* **Sources: Pe6, Pr3, Sa9, Te4, Tt2.**

New York Improved - 85 days - Compact plants with large black fruit, tasty flesh, good fried or baked, mentioned by Fearing Burr in 1865. *Source History: 1 in 2004.* **Sources: Ba8.**

Oriental Princess - 80 days - Used in Chinese cooking. *Source History: 2 in 2004.* **Sources: CO23, La1.**

Oriental White - Large white pear-shaped fruit, needs 5 mos. warmth (65 to 75 degrees), may start indoors. *Source History: 1 in 1987; 2 in 1991; 1 in 1994; 1 in 2004.* **Sources: Ban.**

Osaka Honnoga - (Burgundy) - Plants to 4 ft., 7.5 in. long purple-black fruits, not bothered by flea beetles, withstands light frost. *Source History: 1 in 1991; 1 in 1994; 2 in 1998; 1 in 2004.* **Sources: Yu2.**

Pandora Striped Rose - Short fruit, cylindrical to tear drop shape, rose color skin with light green calyx, medium vigor, heavy yields. *Source History: 1 in 2004.* **Sources: WE10.**

Paris White - 95 days - Decorative plant bears 2 in. round white fruit. *Source History: 1 in 2004.* **Sources: Ma18.**

Ping Tung Long - 65-90 days - Long purple fruit, 12-18 in. long x 1.25 in. dia., sweet and tender, hardy vigorous disease res. plants, heavy yields, heirloom named for its native town, Ping Tung, Taiwan. *Source History: 2 in 1981; 2 in 1984; 1 in 1991; 3 in 1994; 9 in 1998; 21 in 2004.* **Sources: Ba8, CO30, Coo, Ev2, Ho13, Ki4, Kit, La1, Loc, Na6, Orn, Pin, Pr3, Pr9, Se16, Se17, Se26, So1, To1, Tu6, WE10.**

Ping Tung Long Improved - More upright plant bears up to 20 fruit, 1-2 in. across by 8-10 in. long, very early, very productive. *Source History: 1 in 1991; 1 in 1994; 1 in 1998; 1 in 2004.* **Sources: Red.**

Prospera - 70-78 days - Italian heirloom, round fruit, deep violet skin, mild tasting white flesh, originated in Tuscany. *Source History: 2 in 2004.* **Sources: Se24, To1.**

Purple Ball - Tiny purple fruit, one bite size, tender, excel. for boiling and pickling. *Source History: 1 in 2004.* **Sources: Ev2.**

Purple Dragon - 72 days - Plant grows to 34 in., 9 in. long black fruits with purple calyx. *Source History: 2 in 1998; 1 in 2004.* **Sources: Sk2.**

Purple Long Slender - 65 days - Dark purple eggplant used in Oriental dishes, 2 in. wide by 8 in. long. *Source History: 1 in 2004.* **Sources: Red.**

Pusa Purple Cluster - 65 days - Small 2 in. round purple fruit, from India. *Source History: 1 in 2004.* **Sources: Red.**

Red Egg - 92 days - Decorative plant suitable for pot culture, flavor of orange-red fruit is better than mature red fruit. *Source History: 2 in 2004.* **Sources: Hi13, Tho.**

Red Ruffles - (Bitter Red, Hmong Red, Small Ruffled Red) - Plants 20 in. tall, orange-red 2 in. fruits borne in clusters, bitter fruit may not be palatable to some but is considered a delicacy in Asian cuisine, originally introduced as an ornamental in the 19th century. *Source History: 1 in 1991; 1 in 2004.* **Sources: Ea4.**

Redonda - Tall, strong plant bears huge, teardrop-shaped fruit, some deeply ribbed, good quality firm thick flesh, very rare. *Source History: 1 in 2004.* **Sources: Ba8.**

Rosa Bianca - (Rosa Bianco) - 70-90 days - Italian heirloom, stunning light pink-lavender fruits with occasional creamy white shading, meaty, mild flavor with no bitterness. *Source History: 1 in 1987; 3 in 1991; 4 in 1994; 11 in 1998; 19 in 2004.* **Sources: Ami, Ba8, Bo19, Fe5, Ho13, Ki4, La1, ME9, Na6, Orn, Roh, Sa9, Se7, Se8, SE14, So1, To1, Tu6, Und.**

Rosita - 70-85 days - Beautiful rose-lavender skin, teardrop shape, 6-8 in. long, bitter-free, mild sweet flavor, white flesh, heirloom from Puerto Rico. *Source History: 1 in 1994; 8 in 2004.* **Sources: Ba8, He17, La1, Ma18, Pr9, Sa9, Se16, To1.**

Rotonda Bianca Sfumata di Rosa - 120 days - Italian heirloom, round fruit, white skin color is shaded with rose-pink, mild, delicious flavor, striking appearance is appealing to market growers. *Source History: 2 in 2004.* **Sources: Ba8, La1.**

Round Mauve - Tennis ball size fruit, pretty color, China. *Source History: 2 in 2004.* **Sources: Sk2, WE10.**

Slim Jim - 60-80 days - Long tender 4-6 in. light-purple fruits in clusters of 3 to 5, attractive violet foliage, small 15-18 in. plants are ideal for pots, good pickled when young, unique. *Source History: 7 in 1981; 4 in 1984; 3 in 1987; 1 in 1991; 1 in 1994; 1 in 1998; 3 in 2004.* **Sources: Coo, Hud, Pin.**

Snowy - 60-65 days - Uniform, ivory-white, 7 in. long fruit with mild flavor, no bitterness, good for short growing season areas. *Source History: 5 in 2004.* **Sources: Al25, BAL, Jo1, Na6, OSB.**

Taiwan Long - 70 days - Early maturing, long tender purple fruits, should be picked when young, high yields, continuous producer, disease resistant. *Source History: 1 in 1981; 1 in 1984; 1 in 1987; 2 in 1991; 2 in 1994; 2 in 1998; 1 in 2004.* **Sources: Ev2.**

Thai Dark Round Green - Bushy plants with dark green, 2.5 in. fruits striped with white, excellent for stir fry, used extensively in Thai cuisine, easy to grow, from

Thailand. *Source History: 1 in 2004.* **Sources: Ba8.**

Thai Green - (Thai Long Green) - 70-80 days - Heirloom from Thailand, short 2-3 ft. plants, beautiful light-green 8-12 in. long fruits, tender flesh with superb flavor, withstands light frosts, very popular in Thai cuisine, rarely grown in the U.S. *Source History: 1 in 1987; 3 in 1991; 2 in 1994; 3 in 1998; 10 in 2004.* **Sources: Ba8, Coo, He8, Ho13, La1, Pr9, Se16, So1, So25, WE10.**

Thai Green Pea - 70-90 days - Tiny fruit looks like green peas, strong rich flavor, used for stir fry, soup and curry, heavy yields, heirloom from Siam. *Source History: 2 in 2004.* **Sources: Ba8, La1.**

Thai Light Round Green - Light green with white stripes, used extensively in Thailand. *Source History: 1 in 2004.* **Sources: Ba8.**

Thai Long Purple - Uniform, long thin fruit, dark purple skin color, mild sweet flavor, does well in hot humid climate, from Thailand. *Source History: 1 in 2004.* **Sources: Ba8.**

Thai Oblong Purple - Dark purple fruit with the traditional "Japanese" cylindrical shape, mild flavor, good quality, from Thailand. *Source History: 1 in 2004.* **Sources: Ba8.**

Thai Round - (Brinjal) - 80 days - Attractive 3 ft. bushy plants some with thorns, small round fruits of lavender or green and white turning gold at full maturity, may be seedy or bitter but provide authentic fruits for Southeast Asian cuisine, requires warm moist conditions. *Source History: 1 in 1987; 2 in 1991; 1 in 1994; 1 in 1998; 5 in 2004.* **Sources: Ev2, Ra5, Sa9, Vi5, Wo8.**

Thai Round Purple - Round to slightly oblong fruit, bright purple skin, good for baking and frying, not as popular in Thailand as the green fruited varieties. *Source History: 1 in 2004.* **Sources: Ba8.**

Thai White - (Thai White Ribbed) - 65 days - White, flat, deeply lobed fruit, 6-7 in. dia., excel. flavor and texture, good producer. *Source History: 1 in 1994; 2 in 1998; 1 in 2004.* **Sources: Sa9.**

Thai Yellow Egg - Egg-size fruit, bright golden yellow when ripe, used in many Thai dishes and as garnish, a must for specialty growers. *Source History: 1 in 2004.* **Sources: Ba8.**

Tycoon Oriental - 54-61 days - Elegant 3 ft. plant, slender cylindrical fruit 7-8 x 1.25 in., shiny purple-black skin, purple calyx, very productive. *Source History: 1 in 1987; 2 in 1991; 1 in 1994; 1 in 2004.* **Sources: Bo19.**

Udumalapet - (Udmalbet) - 51-90 days - Rare eggplant from India, colorful egg-shaped fruit, light green streaked with purple, 3 in. long, matures to purple with gold streaks, used in chutney and curry, from a Tamil village. *Source History: 6 in 2004.* **Sources: Ag7, Ami, Ba8, La1, Sa9, Se16.**

Ukrainian Beauty - 75 days - Excellent yields of dark purple fruit with slight green cast. *Source History: 1 in 2004.* **Sources: Sa9.**

Violetta Longue - (Violetta Lunga Precoce) - 70 days - Early variety very popular in the south of France, long bright-purple fruits. *Source History: 1 in 1981; 1 in 1984; 3 in 1987; 2 in 1991; 2 in 1994; 3 in 1998; 3 in 2004.* **Sources: Ber, Gou, Ho13.**

Violetta Siciliana - Medium size fruit, purple and cream skin, white flesh. *Source History: 1 in 1987; 1 in 1991; 1 in 1994; 1 in 1998; 2 in 2004.* **Sources: Ber, So25.**

Violetta Tonda - Large oval fruit, purple skin, white flesh. *Source History: 1 in 1987; 1 in 1991; 1 in 1994; 1 in 1998; 1 in 2004.* **Sources: Ber.**

Violette di Firenze - 60-85 days - Beautiful and unusual lavender fruit sometimes striped with white, large stocky and grooved-- resembles squash, needs plenty of heat--cloches and black plastic required to mature the crop in the North. *Source History: 2 in 1987; 3 in 1991; 3 in 1994; 4 in 1998; 4 in 2004.* **Sources: Coo, Ma18, Pin, Sa9.**

Vitorria - (Long Italian) - 61 days - Plants to 3 ft., long, cylindrical fruits are good for slicing, mild flavor, TMV res. *Source History: 1 in 1998; 3 in 2004.* **Sources: La1, MIC, Ont.**

Waimanalo Long - 80 days - Dark purple fruit, 13-18 in. long, few seeds. *Source History: 1 in 1987; 2 in 1991; 1 in 1994; 3 in 1998; 3 in 2004.* **Sources: Ho13, Ma18, Uni.**

White - 65 days - Plants to 3 ft., egg shaped fruit, white skin, disease res., grown at Monticello by Thomas Jefferson. *Source History: 1 in 1991; 1 in 1994; 2 in 1998; 4 in 2004.* **Sources: Ge2, Or10, Shu, Th3.**

White Beauty - 70 days - Early and flavorful, snow-white med-sized fruits, held well off the ground, free of blemishes, does well in hot humid areas. *Source History: 2 in 1981; 2 in 1984; 1 in 1987; 1 in 1991; 2 in 1998; 2 in 2004.* **Sources: Ho13, So1.**

White Egg - (Japanese White Egg) - 69-75 days - Small white oval 2-3 in. fruits ripen to yellow, more spicy flavor, sets fruits early and continuously throughout season, winter pot plant. *Source History: 5 in 1981; 5 in 1984; 3 in 1987; 2 in 1991; 1 in 1994; 1 in 1998; 4 in 2004.* **Sources: Ba8, La1, ME9, SE14.**

White Sword - 75 days - Cylindrical white fruit with streaks of lavender, mild excel. flavor. *Source History: 1 in 1991; 1 in 1998; 1 in 2004.* **Sources: Se17.**

ZaHara - 70 days - Early maturing eggplant, 2 x 6 in., dark purple skin, mild flesh. *Source History: 1 in 2004.* **Sources: Ma18.**

Varieties dropped since 1981 (and the year dropped):

Agora (1987), *Alba* (2004), *Baby White* (1991), *Baby White Tiger* (1994), *Blacknite* (1982), *Bonica* (1998), *Chieh Tzu* (1982), *Chinese Long White* (1994), *Claresse* (1998), *Cook's Strain* (2004), *Dwarf Black Beauty* (1984), *Dwarf Egg Tree* (1987), *Dwarf Golden* (1987), *Early Black Beauty* (1991), *Early Purple Egg* (1984), *Easter Egg* (2004), *Elephant Tusk* (1994), *Emerald Pearl* (1994), *Farmers Long* (1998), *Florentine* (1998), *Florida Market No. 10* (1994), *Foo Chow Round* (2004), *Fort Meyers Market* (1987), *Gator* (1998), *Golden Yellow* (1987), *Green Grape* (2004), *Green Tomato Eggplant* (1994), *Harris Special Hibush* (1991), *Hito-Kuchi Nasu* (2004), *Improved Purple Thornless* (1982), *Italian Long Purple* (1991), *Jersey King* (1982), *Kemer* (1994),

Kitsuta Chunaga (1987), *Long Purple, Oriental* (1991), *Long Violet* (1987), *Melanzana* (1987), *Melongena Ornamental Mixed* (1987), *Negrita* (1994), *New Orleans Market* (1987), *New York Purple* (1987), *New York Purple Improved* (1991), *New York Round Purple* (1994), *New York Spineless Improved* (1991), *Ornamental* (1991), *Pala* (1994), *Pallida Romanesca* (1994), *Pick Me Quick* (1991), *Pink Bicolor* (2004), *Pinky* (1998), *Pompano Market* (2004), *Pompano Pride* (1998), *Prelane* (1991), *Purple Pickling* (1991), *Romanesca* (1994), *Ronde de Valence* (1991), *Rosa Bianco* (1991), *Small Round Italian* (1987), *Small Ruffled Red* (1994), *Snake Eggplant* (1987), *Snake Eye* (1994), *Soshun Long* (1984), *Sweet Red* (2004), *Taiwanaga* (2004), *Thomas Jefferson* (2004), *Tokyo Black* (1987), *Valance* (1987), *Violetta* (1987), *Violetta Lunghissima* (1987), *Viserba* (2004), *White Italian* (1987), *White Knight* (1998), *White Oval* (1991), *White, Bush* (1998).

Endive (Escarole)
Cichorium endivia

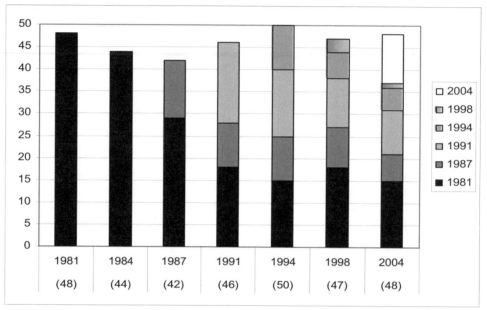

Number of varieties listed in 1981 and still available in 2004: 48 to 15 (31%).
Overall change in number of varieties from 1981 to 2004: 48 to 48 (100%).
Number of 2004 varieties available from 1 or 2 sources: 33 out of 48 (69%).

Batavian, Broad Leaved - (Broad Batavian, Broad Leaved Escarole, Escarole, Batavian Full Heart, Batavian Green) - 85-90 days - Large broad slightly-twisted lettuce-like leaves form round 12-16 in. tight-packed heads, well-blanched creamy-white very deep hearts, thick bulky texture, from the 1860s. *Source History: 38 in 1981; 35 in 1984; 33 in 1987; 27 in 1991; 25 in 1994; 29 in 1998; 31 in 2004.* **Sources: Ag7, All, Bo19, Bu2, CH14, CO23, Cr1, DOR, Fe5, Fo7, HA3, He8, He17, Hi13, Hum, Ma19, Mey, Mo13, Ni1, Ont, Pe2, Ri12, Roh, Sau, Se8, Sh9, Shu, Si5, TE7, Wet, WI23.**

Batavian, Broad Leaved No. 5 - 85 days - Dark-green Full Heart strain, forms an extra large well-filled head which readily blanches itself to a clear-yellow inside. *Source History: 1 in 1981; 1 in 1984; 1 in 1987; 1 in 1991; 1 in 1994; 1 in 1998; 1 in 2004.* **Sources: Dam.**

Batavian, Full Heart - (Improved Batavian, Florida Full Heart, Escarole) - 80-90 days - AAS/1934, deep-green 10-12 in. dia. head, broad thick twisted ruffled outer leaves, erect thick inner leaves blanch the deep full heart to creamy-white, spreading, thick nearly white midribs, slightly frost susc. *Source History: 35 in 1981; 32 in 1984; 29 in 1987; 28 in 1991; 25 in 1994; 20 in 1998; 24 in 2004.* **Sources: Ag7, Ba8, Com, Coo, Dgs, Dom, Ers, Ga1, Ge2, HO2, Kil, La1, Loc, ME9, Mel, OLD, Orn, PAG, RUP, So1, Sto, Twi, Vi4, WE10.**

Belgian - 110 days - Grown for the root. *Source History: 2 in 1991; 1 in 1994; 1 in 1998; 1 in 2004.* **Sources: Lej.**

Bellesque - Frisee type in the Neos class, produces in spring and fall under cool conditions and also resists tipburn when grown in hot summer weather, resists TB and sclerotinia. *Source History: 1 in 2004.* **Sources: Se27.**

Bianca Riccia - 35 days - Fringed light green leaves with pink petiole, heat and cold tolerant, good for year-round growing. *Source History: 1 in 2004.* **Sources: Jo1.**

Bianca Salad - 40 days - Finely cut and fringed baby leaves, matures to light green tinged with pink, does well under adverse weather conditions, easy blanching. *Source History: 1 in 2004.* **Sources: Ge2.**

Bionda a Cuor Pieno - (Blond Full Heart) - 60-65 days - Long, thick, light green crumpled leaves, does best in fall for full size head, grow any time for salad. *Source History: 1 in 1987; 1 in 1991; 1 in 1994; 1 in 1998; 3 in 2004.* **Sources: Ber, Se24, Se26.**

Bossa - 75-90 days - Broad, slightly upright leaves, creamy white midrib, suitable for summer and greenhouse production. *Source History: 1 in 1991; 2 in 1994; 2 in 1998; 1 in 2004.* **Sources: SE4.**

Bubikopf 3 - 55-60 days - Dark green serrated leaves, full head, great taste. *Source History: 1 in 2004.* **Sources: Se24.**

Coquette - (Tres Fine Maraichere) - 90 days - Very finely cut curled leaves. *Source History: 1 in 1991; 1 in 1994; 1 in 1998; 1 in 2004.* **Sources: Tho.**

Cornet d'Anjou - 80 days - Large, broad leaved field endive from France, escarole, best when blanched, named after its shape which resembles the mouth of a horn and its home, Anjou France. *Source History: 1 in 1981; 1 in 1984; 2 in 1987; 2 in 1991; 2 in 1994; 2 in 1998; 1 in 2004.* **Sources: Lej.**

Cornet de Bordeaux - (Chicoree Scarole en Cornet de Bordeaux) - Upright habit, long-lobed white-ribbed outer leaves, not deeply cut, easily tied to blanch heart, excel. flavor. *Source History: 1 in 1981; 1 in 1984; 3 in 1987; 2 in 1991; 1 in 1994; 1 in 1998; 2 in 2004.* **Sources: Lej, Se24.**

Eros - 59 days - Erect, broad green leaves, white midrib, 12-14 in. head size, excellent standability, TB tolerant, productive. *Source History: 1 in 2004.* **Sources: BE6.**

Fine Curled - (Frisee) - 50 days - Finely cut leaves add beauty and body to salads, blanch by tying the heads with a broad rubber band, known to specialty growers and chefs as Frisee. *Source History: 2 in 1991; 2 in 1994; 1 in 1998; 1 in 2004.* **Sources: Ber.**

Fine Curled Louviers - (Chicoree Frissee Fine de Louviers, Fine Louviers) - 68-90 days - Withstands more heat and cold than lettuce, finely curled leaf edges, thin white ribs, compact curled heart, almost identical to Tres Fine Maraichere. *Source History: 2 in 1981; 2 in 1984; 2 in 1987; 2 in 1991; 2 in 2004.* **Sources: Com, Orn.**

Florida Deep Heart - 85-90 days - Southern broad leaf type, upright, 12-15 in., white midribs, thick broad crumpled dark- green leaves, deeper fuller heart than Batavian, blanches creamy-white. *Source History: 15 in 1981; 12 in 1984; 9 in 1987; 12 in 1991; 10 in 1994; 9 in 1998; 4 in 2004.* **Sources: Ha5, La1, ME9, Sh9.**

Frisan - 98 days - Good late fall endive, heads average 17 in. across, long dark-green outer leaves, well filled well blanched centers, high yielding, tolerant to bad weather and low temps. *Source History: 2 in 1987; 2 in 1991; 2 in 1994; 1 in 1998; 1 in 2004.* **Sources: Sto.**

Full Heart - (Florida Deep Heart, Full Heart NR 65) - 65-90 days - Widely used broad leaf or escarole type, low spreading plant, well-blanched and very crumpled inner leaves, dense heart, TB and bolt res., 1960. *Source History: 4 in 1981; 4 in 1984; 4 in 1987; 2 in 1991; 2 in 1994; 8 in 1998; 4 in 2004.* **Sources: Ag7, Ha5, ME9, SE4.**

Galia - 45-88 days - Vilmorin variety, resembles Fine Curled but is more petite with more finely cut leaves, small heads, good self-blanching, bolt res. *Source History: 1 in 1994; 3 in 1998; 4 in 2004.* **Sources: CO30, Coo, OSB, VI3.**

Gloire de l'Exposition - Notched bright leaves surround semi-erect light yellow hearts, a good Italian variety. *Source History: 1 in 2004.* **Sources: Ba8.**

Grado Pancalieri - 80 days - Moderately fine leaves, open head blanches well, Wallone type, good TB tol., slow bolting. *Source History: 2 in 2004.* **Sources: Coo, Jo1.**

Green Curled - (Early Green Curled, Giant Fringed Oyster, Ruffec, Rubble, Fine Green Curled) - 75-98 days - Finely cut dark- green curled leaves, leafy full hearts blanch creamy-white, low-lying spreading 12-18 in. dia. plants, green midribs, gourmet fall salad greens, referred to as "Chicoree Fine d'Ete" by Vilmorin-Andrieux (1885). *Source History: 57 in 1981; 53 in 1984; 49 in 1987; 41 in 1991; 34 in 1994; 31 in 1998; 23 in 2004.* **Sources: All, Be4, Bu2, Cr1, Dam, Ear, HA3, Hal, Ho12, Jun, La1, Lin, LO8, MAY, Mel, Mey, MOU, PAG, Roh, RUP, Sa9, Sau, Si5.**

Green Curled Pancalier - 95 days - Large plant that blanches well with close planting, pink ribbed. *Source History: 2 in 1981; 2 in 1984; 1 in 1987; 1 in 2004.* **Sources: Se26.**

Green Curled Ruffec - (Curled Ruffec, Green Ribbed Ruffec, Green Curled Ruffec Green Ribbed, De Ruffec, Ruffec) - 75-100 days - Deeply cut dark-green leaves blanch easily to creamy-white inside thick deep heart, prostrate habit, thick pale greenish- white 1 in. wide midribs, grows thick, very hardy and res. to cold wet weather, sow summer or autumn, over 100 years old. *Source History: 29 in 1981; 28 in 1984; 25 in 1987; 18 in 1991; 19 in 1994; 22 in 1998; 20 in 2004.* **Sources: Alb, Ba8, CH14, CO23, DOR, Ers, He8, Jor, Kil, Loc, Ni1, OLD, Ri12, Se26, Sh9, Twi, Vi4, WE10, Wet, WI23.**

Grosse Bouclee - (Grosse Boucle) - 50-90 days - French variety with very tight heart, long curled leaves, easy to grow, bolt res., early. *Source History: 5 in 1991; 12 in 1994; 12 in 1998; 9 in 2004.* **Sources: Bo17, CO30, Hig, Na6, Pin, RUP, Se26, VI3, Wi2.**

Large Green Curled White Rib - (Large Green Curled) - 80-92 days - Long well fringed bright-green leaves, 14-18 in. large- framed plants, excel. heart, blanches well, white midrib, most popular in the South. *Source History: 4 in 1981; 2 in 1984; 3 in 1987; 2 in 1991; 1 in 1994; 1 in 1998; 2 in 2004.* **Sources: Bu8, But.**

Lorca - 60-90 days - Large, deep thick heads, fine, very curled, early, TB res., recommended for fall and early winter harvest. *Source History: 6 in 1994; 6 in 1998; 5 in 2004.* **Sources: CO30, Fe5, OSB, RUP, VI3.**

Markant - *Source History: 1 in 1998; 1 in 2004.* **Sources: ME9.**

Moss Curled - (Green Moss Curled, Early Green Moss Curled) - 90-95 days - Feathery dark-green curled finely divided fringed leaves, plants 15 in. in dia. or more, heavy growth, good quality, more heat tolerant than lettuce, faster growth in cold weather. *Source History: 5 in 1981; 4 in 1984; 5 in 1987; 4 in 1991; 2 in 1994; 3 in 1998; 1 in 2004.* **Sources: Ont.**

Nataly - 48 days - Pretty, medium-large heads, cream colored, partially blanched hearts surrounded by green outer leaves, mildly bitter flavor, excellent bolt, TB and bottom rot tolerance, salads, sow spring, summer and fall. *Source History: 2 in 2004.* **Sources: Eo2, Jo1.**

Neos - 45-90 days - Finely curled type with heavy well-blanched heads 11-12 in. across, pale-green midribs, self-blanches to a creamy white center, delicious crisp mildly bitter-sweet taste, slow to bottom rot, for mid-spring to early fall crops. *Source History: 1 in 1991; 3 in 1994; 3 in 1998; 8 in 2004.* **Sources: BE6, Dam, Ga22, Jo1, SE4, St18, Ter, We19.**

Perfect - 80 days - Med-green heads form a rosette of ruffled leaves whose broad stems form a balanced center, very mild and tasty, hardy to -7 deg. F. *Source History: 1 in 1991; 1 in 1994; 1 in 1998; 1 in 2004.* **Sources: Ter.**

Plantation - More uniform Salad King type is larger and darker with more vigor. *Source History: 1 in 2004.* **Sources: ME9.**

Red Endive - Rootstocks produce compact rounded heads of purple-red foliage when being forced after fall harvest of roots, crisp crunchy texture, slightly bitter taste. *Source History: 1 in 1987; 1 in 1991; 1 in 1994; 1 in 1998; 1 in 2004.* **Sources: Ri2.**

Rhodos - (Bejo 1412) - 53 days - Tres Fine Maraichere sel., compact heads are dense and well blanched, fine, curled leaves, can be produced year-round in mild climates. *Source History: 1 in 1994; 1 in 1998; 2 in 2004.* **Sources: BE6, Dam.**

Riccia a Cuor d'Oro - Long, deeply indented green leaves. *Source History: 1 in 1987; 1 in 1991; 1 in 1994; 1 in 1998; 1 in 2004.* **Sources: Ber.**

Riccia Pancalieri - 65 days - Self-blanching, very curled leaves with rose-tinted white midribs, large white heart, sow summer, autumn or winter. *Source History: 1 in 1994; 1 in 1998; 2 in 2004.* **Sources: Bou, Se24.**

Riccia Romanesco da Taglio - 40-45 days - Cutting endive grows in a small upright bunch, thin serrated dark green leaves with classic endive taste and texture, easy to grow, sow anytime, good bolt res., can be cooked. *Source History: 1 in 2004.* **Sources: Se24.**

Riccia Verde - Thick, deeply indented, white ribbed leaves. *Source History: 1 in 1987; 1 in 1991; 1 in 1994; 1 in 1998; 1 in 2004.* **Sources: Ber.**

Salad King - (Salad King Green Curled) - 50-100 days - Giant of curled endives, large dark-green finely cut leaves, vigorous, non-bolt, TB and frost res., tie outer leaves to blanch 22-24 in. heart, white ribs, crops summer and fall, performs well when stressed, adapted to West, full heart, 1957. *Source History: 31 in 1981; 25 in 1984; 26 in 1987; 27 in 1991; 20 in 1994; 24 in 1998; 15 in 2004.* **Sources: Bo19, Bu3, CA25, Com, Fo7, Goo, Ha5, He8, HO2, La1, ME9, Se26, Sh9, Sto, Vi4.**

Salanca - 90 days - Large heads, good res. to cold, TB and bolting. *Source History: 1 in 1991; 2 in 1994; 1 in 1998; 2 in 2004.* **Sources: CO30, VI3.**

Samantha - 85 days - Medium curled leaves, compact, tight, heavy, well-filled, self-blanching heart, strong bolt res. *Source History: 1 in 2004.* **Sources: HO2.**

Tasos - 68-90 days - Wallone variety with strongly curled dark-green outer leaves, for summer and fall crops, very cold tolerant. *Source History: 1 in 1994; 2 in 1998; 2 in 2004.* **Sources: BE6, SE4.**

Tosca - 60-85 days - Med-green leaves, white blanched hearts, delicious mild flavor, very large early Tres Fine Maraichere type, TB tol., very bolt resistant. *Source History: 2 in 1991; 2 in 1994; 7 in 1998; 7 in 2004.* **Sources: CO30, Orn, OSB, RUP, Se8, Tho, VI3.**

Traviata - (Scala) - 90 days - Upright growth, deep self-blanching head with extra-curled toothed leaves, crispy flavor, tipburn sensitive, from France. *Source History: 3 in 1991; 3 in 1994; 3 in 1998; 4 in 2004.* **Sources: CO30, OSB, Pin, VI3.**

Tres Fine Maraichere - 45-70 days - Quick growing miniature French endive, self-blanching, narrow finely curled leaves, cream colored hearts, plant 6 in. each way, delicate flavor, sweeter than Neros, individual servings. *Source History: 1 in 1981; 1 in 1984; 5 in 1991; 10 in 1994; 16 in 1998; 21 in 2004.* **Sources: Bo17, CO30, Fe5, Ga1, Hig, Ki4, Loc, ME9, Na6, Orn, OSB, Pe2, Ri2, Roh, RUP, Se7, Se26, TE7, Ver, VI3, We19.**

Verde a Cuor Pieno - (Green Full Heart) - 60-70 days - Thick, crumpled green leaves, full head, crisp white stems, very large plant, best grown in fall for full size head, cooked, stuffed, fresh salads. *Source History: 1 in 1987; 1 in 1991; 1 in 1994; 1 in 1998; 2 in 2004.* **Sources: Ber, Se24.**

Varieties dropped since 1981 (and the year dropped):

Big Curled Savoy (1987), *Broadleaved Fullheart Winter* (1998), *Bubikopf* (1994), *Coral* (2004), *Cuore Pieno* (1991), *Curled Large Head* (1991), *Curled Wallone* (1987), *Deep Heart* (1987), *Deep Heart (NR 65)* (1987), *Deep Heart Fringed* (1991), *Elna* (1987), *Elodie* (2004), *Festo* (1983), *Fine Maraichere* (1991), *FM 10* (1998), *Giant Market* (1987), *Giant Samy* (1998), *Gloria dell Esposizione* (1998), *Grosse Pommant Seule* (1987), *Grower's Giant* (1994), *Heading Endive* (2004), *Large Green Curled Pink Ribbed* (1987), *Large Green Curled White TBR* (1991), *Leafy Endive* (2004), *Loire Cornet Escarole* (1987), *Maral* (1991), *New Giant* (1991), *Nina* (1998), *NR65* (1991), *Nufema* (1987), *Nuvol* (1998), *Pavia* (1987), *PAX 1003* (2004), *Pink Star* (1991), *Premier Green Curled* (1984), *President* (2004), *Ruffle Leaf* (1998), *Salad King Improved* (2004), *Sanda* (1987), *Sinco* (2004), *Solera* (1984), *St. Laurent* (1987), *Taglio* (2004), *Valdena* (1994), *Wallone* (2004), *White Curled* (2004), *Wivol* (2004).

Fava Bean
Vicia faba

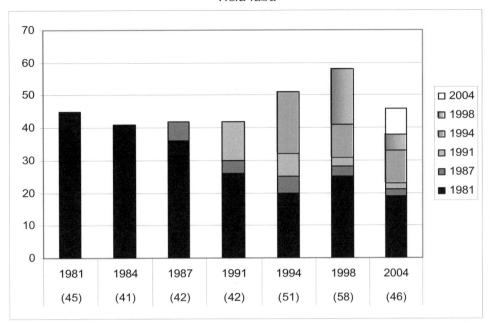

Number of varieties listed in 1981 and still available in 2004: 45 to 19 (42%).
Overall change in number of varieties from 1981 to 2004: 45 to 46 (102%).
Number of 2004 varieties available from 1 or 2 sources: 33 out of 46 (72%).

164 MB 5 Red - Medium-large, dusky red seeds, good producer, rare. *Source History: 1 in 2004.* **Sources: Pr3.**

Aguadulce - (D'Aguadulce a Tres Longue Cosse) - 75 days - Plants grow to 3 ft., 12-16 in. long well-filled pods, nourishing, cooked seeds have a warm nutty flavor, meatier than our lima beans, very disease res., extremely cold tolerant, quite productive, listed in 1885 as 'Agua-Dulce Long Podded' by Vilmorin-Andrieux, French seedhouse. *Source History: 2 in 1981; 2 in 1984; 3 in 1987; 2 in 1991; 2 in 1994; 1 in 1998; 6 in 2004.* **Sources: Bou, In8, La1, ME9, Ri12, We19.**

Aprovecho Select - 75-85 days - Unusually large seeds, sweet flavor, excel. for table use, dev. in Oregon, hardy to 20 degrees F. *Source History: 4 in 1991; 5 in 1994; 3 in 1998; 3 in 2004.* **Sources: Pr3, Se7, Se17.**

Aquadulce Claudia - (Aquadulce) - 75-90 days - Plants 36-40 in. tall, 15 in. pods, white seeds, primary var. for really early crop, can be planted in January or in the autumn, can withstand very cold conditions. *Source History: 1 in 1981; 1 in 1984; 2 in 1987; 11 in 1991; 7 in 1994; 9 in 1998; 4 in 2004.* **Sources: OSB, Ter, Tho, WE10.**

Aztec Yellow - Early variety, pods contain 1-3 large seeds. *Source History: 1 in 1994; 1 in 1998; 1 in 2004.* **Sources: Pr3.**

Banner - (Winter Bean) - 240 days - Sturdy self-supporting bushes, hardy to 10 deg. F., blocky greenish-brown seeds, cooks quickly into thick brown sauce, excel. green manure crop. *Source History: 1 in 1981; 1 in 1984; 2 in 1987; 4 in 1991; 5 in 1994; 5 in 1998; 5 in 2004.* **Sources: Bou, OSB, Se7, So12, Ter.**

Bell - (Peanut, Tic) - 60-75 days - Unusual small-seeded type, grows well in Pacific Northwest where it survives winters, small dark reddish-brown seeds resemble peanuts, good cool weather crop, high yields. *Source History: 2 in 1981; 2 in 1984; 2 in 1987; 4 in 1991; 2 in 1994; 3 in 2004.* **Sources: Pe2, Pr3, Sa5.**

Broad Improved Long Pod - 85 days - Upright plants bear 7 in. pods containing 5 or 6 large flat oblong beans, used as green shell or dry, unique flavor, high yielding med-tall str. with uniform pods. *Source History: 1 in 1981; 1 in 1984; 1 in 1987; 1 in 1991; 1 in 1994; 2 in 1998; 2 in 2004.* **Sources: Ha5, Mey.**

Broad Windsor - (Broad Pod Windsor, Broad Windsor, Exhibition Long Pod, English Windsor Broad, Horse Bean, Conqueror) - 65-85 days - Plant same time as peas, 24-48 in. upright non-branching plants, 5-8 in. x 1 in. glossy-green pods contain 4 to 7 oblong flat light-green shell beans, high protein, pinch tops of plants to set pods after first 4 or 5 flower clusters appear. *Source History: 17 in 1981; 17 in 1984; 15 in 1987; 15 in 1991; 17 in 1994; 23 in 1998; 31 in 2004.* **Sources: Alb, Ba8, Bo17, Bu3, De6, DOR, Ga22, HA3, He17, La1, Lin, Loc, Ma19, ME9, Mo13, MOU, OLD, Ont, OSB, PAG, Pin, Pp2, Pr3, Ra5, Ri12, Roh, Sto, Ter, Ver, Vi4, We19.**

Cairo Market - Early variety with short plants, pods contain 1-5 fairly small seeds. *Source History: 1 in 1994; 1 in 1998; 1 in 2004.* **Sources: Pr3.**

Chak'rusga - (Chakrusga) - 110-140 days - One of the favorite varieties of the Aymara natives in northern Bolivia, multi- branched plant grows 3-4 ft. tall and yields 6-10 pods per stalk, prolific and drought tolerant, good soil builder, hardy to 15 degrees F. *Source History: 1 in 1994; 1 in 1998; 4 in 2004.* **Sources: Ho13, Ri12, Se7, Se17.**

Con Amore - 70 days - Early, rather short sturdy plant, very long 8 in. pods, high yields of oval med-sized seeds, colored flowers, from Holland. *Source History: 1 in 1981; 1 in 1991; 1 in 1994; 1 in 1998; 3 in 2004.* **Sources: Se17, Ves, VI3.**

Conqueror - (Longpod Conqueror) - 78 days - Extra long pods with 8 to 10 large seeds, better yielding variety than Seville for the home garden or market grower. *Source History: 1 in 1981; 1 in 1984; 1 in 1987; 2 in 1991; 1 in 1994; 1 in 1998; 1 in 2004.* **Sources: Hal.**

Copper - Medium size, deep copper colored seeds. *Source History: 1 in 1998; 1 in 2004.* **Sources: Pr3.**

Crimson Flowered - Rare heirloom from England, bright crimson flowered stalks, dates back to 1778. *Source History: 2 in 2004.* **Sources: Ga22, Se17.**

English Long Pod - Distinctive from Windsor types, smaller seed, plants grow vigorously, very productive, plant as early as possible in the spring. *Source History: 1 in 1981; 1 in 1984; 1 in 1987; 1 in 1991; 1 in 1994; 1 in 1998; 1 in 2004.* **Sources: Ont.**

Espanola Valley - Large brown to yellow seeds, grown in New Mexico since the 1800s. *Source History: 1 in 1998; 1 in 2004.* **Sources: Ho13.**

Exhibition Long Pod - 70 days - An improved type, far superior, huge pods often contain 5 very large flat pale-green to cream seeds, 5 to 7 beans per inedible pod, 3.5-5 ft. plants, fine flavor. *Source History: 4 in 1981; 4 in 1984; 4 in 1987; 1 in 1991; 2 in 1994; 2 in 1998; 1 in 2004.* **Sources: Ear.**

Express - 71-78 days - Fast-maturing large-seeded broad bean, from an early spring sowing will outyield all others, up to 34 good pods per plant, winter hardy, outstanding for freezing, does not discolor when frozen. *Source History: 2 in 1981; 2 in 1984; 2 in 1987; 2 in 1991; 2 in 1994; 2 in 1998; 2 in 2004.* **Sources: Ho13, Tho.**

Fava - (Horse Bean, Broad Bean, Windsor Bean, English Broad Bean) - 52-85 days - Favorite throughout Mediterranean, 20- 26 in. plants, huge beans in 7 in. med-green round/oval spongy pods, similar to limas, high protein, plant early, sensitive to heat, inedible pods, contain 5-8 white seeds, frost res., green shell or dry, 20-26 in. plants. *Source History: 23 in 1981; 19 in 1984; 15 in 1987; 13 in 1991; 11 in 1994; 16 in 1998; 11 in 2004.* **Sources: All, Fo7, Ge2, HO2, Mel, Na6, Ni1, SE4, Se26, Sh9, Shu.**

Foul Muddamma - (Egyptian) - 77 days - Suitable for a late spring and summer sowing, good root system, heat tol., plant on 6 in. centers, from Egypt. *Source History: 1 in 1987; 1 in 1991; 1 in 1994; 2 in 1998; 1 in 2004.* **Sources: Ho13.**

Fredericks - Overwintering fava, hardy to at least 15 degrees F., tall stalks to over 6 ft., huge amount of biomass, twice the nitrogen of clovers, good silage for cattle. *Source History: 1 in 1981; 1 in 1984; 1 in 1998; 1 in 2004.* **Sources: Ter.**

Frog Island Nation - 65-75 days - Large purple seeds, selected by Dougo at OAEC from Aprovecho favas, grown out several generations at Abundant Life farm. *Source History: 1 in 2004.* **Sources: In8.**

Guatemalan Purple - 80-90 days (spring planted) 160-200 (fall planted) days, Medium size purple seed, occasional black, from Guatemala by seed collector Ianto Evans. *Source History: 1 in 2004.* **Sources: Se7.**

Habas Jergonas - *Source History: 1 in 2004.* **Sources: Se17.**

Hava - Ancient old world bean, widely distributed by the Spanish, this str. collected from Tierra Amarilla in northern New Mexico, prolific, small and tasty. *Source History: 1 in 1981; 1 in 1984; 1 in 1987; 1 in 1991; 1 in 1994; 1 in 1998; 1 in 2004.* **Sources: Pla.**

Histal - Forms a long, wide, hanging pod. *Source History: 1 in 1994; 1 in 1998; 1 in 2004.* **Sources: Ber.**

Iant's Yellow - Stocky plants produce an abundance of pods with 3 large seeds per pod, from Ianto Evans. *Source History: 1 in 1994; 1 in 1998; 1 in 2004.* **Sources: Pe1.**

Imperial Green Longpod - (Imperial Long Pod) - 84-100 days - Twenty years of research has dev. a plant capable of continually producing 15-20 in. pods with up to 9 large beans, maximum productivity. *Source History: 2 in 1981; 2 in 1984; 2 in 1987; 2 in 1991; 2 in 1994; 4 in 1998; 1 in 2004.* **Sources: Tho.**

Jubilee Hysor - 95 days - Improved variety, well-filled pods with 6-8 beans, superb flavor, heavy yields. *Source History: 1 in 2004.* **Sources: Tho.**

Jumbo - Japanese type, pods and beans are brighter green than western types, 3 large seeds per pod, fine flavor in fresh shell stage, mid-early. *Source History: 1 in 2004.* **Sources: SAK.**

Long Pod - (English Broadbean, Windsor) - 83-90 days - Good substitute for limas or soybeans in short season Northern areas, sow early, susceptible to summer heat, 7 in. pods hold 5 to 7 big red/brown seeds, use like limas. *Source History: 11 in 1981; 8 in 1984; 9 in 1987; 9 in 1991; 4 in 1994; 6 in 1998; 4 in 2004.* **Sources: Com, Hum, Loc, Sol.**

Medes - 70-90 days - Short plant bears many pods with 5-6 small white seeds, excel. fresh, frozen or canned. *Source History: 1 in 1994; 1 in 1998; 2 in 2004.* **Sources: Hi13, Tho.**

Mr. Barton's - Large green seeded variety with fine flavor and texture, used fresh or dried, from a Mr. Barton, Victoria, B.C. *Source History: 1 in 1998; 1 in 2004.* **Sources: Ga22.**

Nintoku Giant - 70-90 days - Vigorous heavy yielding var. with uniform growth habit, 3 large seeds per pod, for fresh market and home use. *Source History: 1 in 1994; 3 in 1998; 3 in 2004.* **Sources: Ev2, Kit, Tho.**

Ojito - Brown speckled seeds .5 x .75 in., raised at Ojito, south of the Spanish village of Chamisa, NM, 7800 ft. *Source History: 2 in 1998; 2 in 2004.* **Sources: Ho13, Na2.**

Purple - (Guatemalan Purple) - Bushes 3-4 ft. tall, medium size, purple seeds with occasional black, originally from Guatemala, from Ianto Evans. *Source History: 2 in 1994; 7 in 1998; 2 in 2004.* **Sources: Ho13, Sa5.**

Sanuki Kotsuba - Rare Japanese str. with small, mahogany-red seeds. *Source History: 1 in 1998; 1 in 2004.*

Sources: Ho13.

Sweet Lorane - 90-100 days - Plants grow to 6 ft. and produce an abundance of small good tasting favas of the cover crop kind, cold hardy. *Source History: 2 in 1994; 4 in 1998; 4 in 2004.* **Sources: Ba8, Ho13, Ter, We19.**

Tarahumara Habas - Green to beige seed with black hilum, summer crop in Sierra Madres, frost hardy, ground and made into thick tortillas or soup, introduced by the Spanish, then grown by the Tarahumara Indians. *Source History: 1 in 1987; 1 in 1991; 2 in 1998; 2 in 2004.* **Sources: Ho13, Na2.**

The Sutton - 95 days - Dwarf 12 in. plants, 5-6 in. pods with 5 small tender beans, excel. for early plastic tunnels or successive sowings or small gardens, excel. for freezing, prolific. *Source History: 2 in 1981; 2 in 1984; 1 in 1987; 1 in 1991; 1 in 1994; 2 in 2004.* **Sources: Pr3, Tho.**

Threefold White - Long slim pods, white flowers, large white seeds retain color when cooked, mild flavor, dates to late 1800s. *Source History: 1 in 1991; 2 in 1994; 1 in 1998; 1 in 2004.* **Sources: Pr3.**

Vroma - 65 days - Flat pods grow 3-4 in. long, easy to pick, good Italian bean flavor, use fresh, canned or frozen. *Source History: 1 in 1994; 1 in 1998; 1 in 2004.* **Sources: Mcf.**

Walter Krivda's - Small seeded, high yields, good fresh or dried, freezes well, sow in November or February for May/June harvest. *Source History: 1 in 2004.* **Sources: Ga22.**

Windsor - (Giant, Four Seeded White, Windsor Long Pod) - 65-75 days - The Soybean of the North, popular in cool humid areas, plant late autumn on the Northwest coast, stores well, sometimes grown for cover crop. *Source History: 12 in 1981; 12 in 1984; 11 in 1987; 9 in 1991; 10 in 1994; 9 in 1998; 11 in 2004.* **Sources: Bou, Dgs, Fe5, Ga1, Goo, He8, Hig, Jo1, Pe2, So12, Tt2.**

Witkiem Major - 60-85 days - Broad Windsor type produces huge crops of thick 10 in. pods, extremely fast growing, crops as quickly as autumn vars. and gives heavier yields, green seeds, strong 36 in. plants, from Holland. *Source History: 2 in 1981; 2 in 1984; 1 in 1987; 1 in 1991; 2 in 1994; 1 in 1998; 1 in 2004.* **Sources: Dam.**

Varieties dropped since 1981 (and the year dropped):

Albinette (1998), *Black* (2004), *Bonnie Lad* (1998), *Broad Windsor Improved* (1991), *Broad Windsor Long Pod* (2004), *Broad Windsor Small Pod* (1987), *Brunette* (1994), *Bunyard's Exhibition* (2004), *Cambridge Scarlet* (2004), *Castillo Franco* (1998), *Cavalier* (1987), *Colossal* (1994), *Crimson Flower* (2004), *Dow Fu* (1994), *Dwarf Broad* (1994), *Equina* (2004), *Felix's Fava* (1998), *Fidrim* (1984), *Frostproof* (2004), *Giant English* (1991), *Habas Chiquitas* (2004), *Habas del Norte* (2004), *Ipro* (1998), *Italian* (2004), *Ite* (1991), *Jade* (2004), *La Mesilla* (2004), *Large Seeded* (1991), *Long Pod, Early Improved* (2004), *Loreta* (1998), *Magdalena* (1998), *Masterpiece* (2004), *Metissa* (1984), *Minica* (1991), *Optica* (1983), *Polar* (2004), *Portuguese* (1998), *Primo* (2004), *Prizetaker Exhibition Long* (1985), *Red Cheek* (1998), *Red Eye* (1998), *Relon* (1991), *San Ildefonso* (1994), *San Luis* (2004), *Statissa* (1998), *Superfin* (1987), *Suprifin* (1994), *Swiss* (2004), *Tezieriviera White Seeded* (1991), *Tezieroma* (1991), *Toto* (2004), *Vadito* (2004), *Verdy* (2004), *Windsor Jubilee* (2004).

Florence Fennel
Foeniculum vulgare

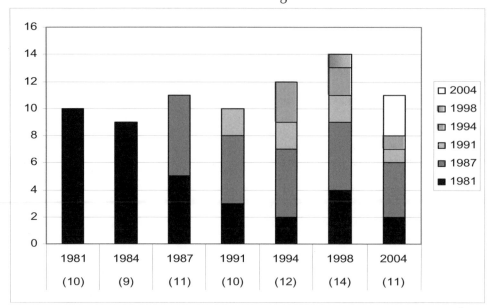

Number of varieties listed in 1981 and still available in 2004: 10 to 2 (20%).
Overall change in number of varieties from 1981 to 2004: 10 to 11 (110%).
Number of 2004 varieties available from 1 or 2 sources: 6 out of 11 (55%).

Autumn Giant - Medium tall plants produce mammoth size pure white heads with fleshy but very tender texture, medium-late maturity. *Source History: 1 in 1994; 1 in 1998; 1 in 2004.* **Sources: GLO.**

Fino - 65-95 days - Italian strain, egg-shaped bulb, crunchy texture, nutty mild celery-anise flavor, stalks and lacy leaves make a wonderful seasoning bed when cooking fresh fish, bolt res. *Source History: 1 in 1987; 1 in 1991; 2 in 1994; 4 in 1998; 3 in 2004.* **Sources: BE6, Ki4, OSB.**

Florence Fennel - (Florence Sweet Fennel, Finocchio, Italian) - 65-100 days - Finocchio in Italy, broad overlapping leaf bases form a large bulb-like enlargement at base of stem, 3 ft. celery-like stalks, annual. *Source History: 39 in 1981; 34 in 1984; 51 in 1987; 65 in 1991; 72 in 1994; 80 in 1998; 76 in 2004.* **Sources: Alb, All, Ba8, Ban, Bo17, Bo19, Bou, Bu2, CA25, CO23, Com, Dam, Dgs, DOR, Eo2, Ers, Fi1, Fo7, Ga21, Ge2, GRI, HA3, Hal, Hi6, HO2, Ho8, Hum, Jor, Jun, Kil, La1, La8, Lej, Lin, Loc, ME9, Mey, Mo13, MOU, Na6, Ol2, Ont, Orn, OSB, Par, Pe2, Pep, Pin, Re8, Red, Ren, Ri2, RUP, Sa6, Sa9, SE4, SE14, Se16, SGF, Shu, So1, So12, Sto, TE7, Ter, Th3, Tho, Thy, Twi, Ves, Vi4, Vi5, WE10, We20, Wi15, WI23.**

Herald - 75 days - Plump bulbs with sweet anise-like flavor, better bolt res. than other varieties. *Source History: 1 in 1991; 1 in 1994; 1 in 1998; 1 in 2004.* **Sources: Tho.**

Montebianco - Solid stalks, mid size round head, very white bulb, tasty, does best in spring and fall but will take some heat. *Source History: 1 in 2004.* **Sources: Se24.**

Montovano - Deep green stalks, large white bulb, nice sweet flavor. *Source History: 1 in 2004.* **Sources: Se24.**

Napoletano - Round, firm, crisp white head. *Source History: 1 in 1987; 1 in 1991; 1 in 1994; 1 in 1998; 1 in 2004.* **Sources: Ber.**

Parma Sel Prado - Head is smaller than Perfezione sel Fano, great taste and texture, prefers light fertile soil. *Source History: 1 in 2004.* **Sources: Se24.**

Perfection - (White Perfection, Perfezione Sel Fano) - 75 days - Northern European strain, that matures more quickly in cool northern climates, medium size bulbs by end of July, bolt res. *Source History: 2 in 1981; 2 in 1984; 1 in 1987; 4 in 2004.* **Sources: Se24, Sh12, Ter, Tu6.**

Romanesco - Classic fennel from Rome, round, firm, crisp white head is used raw or marinated and cooked, delicious, fern-like tops. *Source History:1 in 1987; 1 in 1991; 1 in 1994; 1 in 1998; 3 in 2004.* **Sources: Ber, Sa9, Se24.**

Zefa Fino - (Finocchio) - 65-90 days - Forms a robust tender bulb, slow-bolting, can be spring sown with success while other varieties go to seed before forming a bulb, Swiss Federal Research Station. *Source History: 1 in 1987; 9 in 1991; 12 in 1994; 16 in 1998; 9 in 2004.* **Sources: CO30, Coo, Fe5, Jo1, Ni1, Pe2, Se7, Syn, Tu6.**

Varieties dropped since 1981 (and the year dropped):

Cristal (1991), *Di Napoli* (1983), *Domino* (1987), *Early* (2004), *Mammoth* (2004), *Romano Precoce* (1987), *Romy* (2004), *Sicilian* (1994), *Superwadenromen* (2004), *Sweet Florence* (1998), *Wadenromen Grosso* (2004), *White Mountain* (2004).

Garlic
Allium sativum

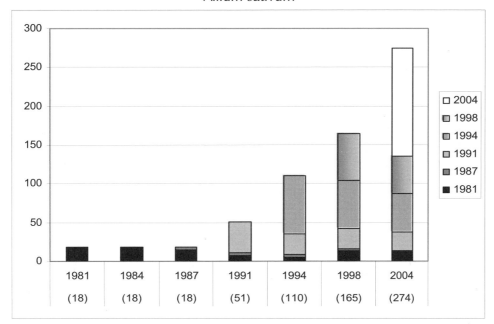

Number of varieties listed in 1981 and still available in 2004: 18 to 12 (67%).
Overall change in number of varieties from 1981 to 2004: 18 to 274 (1522%).
Number of 2004 varieties available from 1 or 2 sources: 195 out of 274 (71%).

Achatami - Softneck, large bulbs, cream colored cloves with white wrappers, mild flavor starts with a bite and mellows, from the Republic of Georgia. *Source History: 1 in 2004.* **Sources: Ga17.**

Acropolis - Softneck artichoke with rich earthy flavor, midseason, originated in Greece. *Source History: 1 in 2004.* **Sources: Bo20.**

Ail de Pays Gers - Artichoke, very large white bulbs, 8-12 large cloves, tan with pink and purple stripes, robust, piquant flavor, from Bergiers in the south of France. *Source History: 1 in 2004.* **Sources: Ga17.**

Ajo Rojo - Similar to Creole Red, a genetic softneck Silverskin, bolts weakly, behaves like *ophio* garlic, beautiful bulbs with striking clove colors, from Spain via virus-free program in Nevada, early 1990s. *Source History: 1 in 1994; 1 in 1998; 3 in 2004.* **Sources: Bi3, Fi6, Go6.**

Alan - Emerges early, large bulbs, from Dr. Boris Andrst, NY. *Source History: 1 in 2004.* **Sources: Fi6.**

Alexandria - (Rocambole Alexandria) - Large-bulbed garlic discovered in an Alexandria, Virginia market, very productive plants with large leaves, emerges early, doubled cloves are common, 8-12 cloves per bulb, light-brown clove skins with thin faint brown lines top to bottom, sometimes darker lines and dark-brown blush near top of clove. *Source History: 1 in 1991; 1 in 1994; 1 in 1998; 1 in 2004.* **Sources: Ga21.**

Armenian - Hardneck/Porcelain, excel. raw flavor heats up slowly and leaves a pleasant lingering aftertaste, delicate creamy texture, flavor is mild but full, excel. baked, from Hadrut Karabach near Azerbaijan border. *Source History: 1 in 1998; 2 in 2004.* **Sources: Bi3, Fi6.**

Asian Tempest - (Seoul Sister) - Beak on bulbil capsule can reach 18 in., bulbs often finely striped, nice clove color and shape, 5-7 cloves per head, 45-50 cloves/6-8 bulbs per pound, long storage, productive in both cold winter and wet mild winter climates, matures earlier than standard Artichokes, harvest as soon as leaf browning begins or bulb wrappers will split open, from South Korea. *Source History: 2 in 1994; 3 in 1998; 7 in 2004.* **Sources: Bi3, Fi6, Fi7, Ga1, Ga17, Ga19, Se16.**

Baba Franchuk's - Hardneck, skin of bulb is white with purple blush, glossy, dark purple clove skin color, 7 cloves per bulb, sharp but smooth flavor resembling German Red, oily texture, from Winnipeg, Canada, rare. *Source History: 1 in 1998; 2 in 2004.* **Sources: Bo21, Sa5.**

Bai Pi Suan - Hardneck, purple stripe variety, marbled, plump cloves, 4-6 per bulb, raw flavor has gentle heat that builds. *Source History: 1 in 1998; 2 in 2004.* **Sources: Bi3, Fi6.**

Bavarian - (Bavarian Purple) - Hardneck, Rocambole, plants and bulbs are larger than those of common varieties, purple skinned cloves, white outer skin, strong flavor, heat diminishes quickly, retains flavor in storage, very good keeper, requires a rich well drained soil and no fresh manure, 7-8 large cloves, 35-40 cloves per pound, stores 9-12 months, mid-season maturity. *Source History: 1 in 1981; 1 in 1984; 1 in 1987; 2 in 1998; 5 in 2004.* **Sources: Bi3, Ni1, Shu, Ter, Ver.**

Beauty - Purple striped variety. *Source History: 1 in 2004.* **Sources: Gl4.**

Beijing - Large bulbs with purple pinstripes, 6-7 huge round beige cloves, excel. flavor. *Source History: 1 in 2004.* **Sources: Ga17.**

Belarus - One of the first Purple Stripes to emerge, foliage is dark green. *Source History: 2 in 2004.* **Sources: Bi3, Fi6.**

Belgian Red - Raw flavor is smooth to very hot, baked flavor is sweet with a tangy aftertaste. *Source History: 1 in 2004.* **Sources: Fi6.**

Blanak - Hardneck, Purple Stripe, glazed, from the Czech collection of Dr. Boris Andrst, a New York grower. *Source History: 2 in 2004.* **Sources: Bi3, Fi6.**

Blazer - (PI 250662 (USSR)) - *A. longicuspis*, hardneck/porcelain, very tall plants, few cloves, heavily striped cloves with some light pink blush, nice garlic flavor when baked, hot raw flavor. *Source History: 1 in 1994; 1 in 1998; 2 in 2004.* **Sources: Bi3, Fi6.**

Blossom - Turban type, large bulbs with dark stripes, smooth and mild baked, crunchy raw with lingering heat. *Source History: 2 in 2004.* **Sources: Bi3, Fi6.**

Bogatyr - (Gatersleben #7204) - Marbled purple stripe, dark purple-brown clove color, few cloves per head, 6-8 cloves per bulb, 35-40 cloves/6-8 bulbs per pound, flower stalks sometimes tinged with red, long storing, Moscow. *Source History: 1 in 1994; 1 in 1998; 8 in 2004.* **Sources: Bi3, Fi6, Ga17, Ga19, Gl4, Go6, Ho13, Se16.**

Broadleaf Czech - (Gatersleben #146) - Purple stripe variety, raw flavor is hot to very hot, tasty, mild, full flavor when baked. *Source History: 1 in 1998; 2 in 2004.* **Sources: Bi3, Fi6.**

Brown Rose - (R/AL/28, Dervish Delight) - *A. longicuspis*, very fat cloves, rose-brown blush and light streaks over buff brown, 5-8 cloves per bulb, 13-16 bulbs per pound, medium to hot flavor. *Source History: 1 in 1994; 1 in 1998; 3 in 2004.* **Sources: Bi3, Fi6, Se16.**

Brown Saxon - (Gatersleben #7116) - Unique Georgian Rocambole with large brown cloves, great storage, double cloves are rare, late harvest. *Source History: 1 in 1998; 2 in 2004.* **Sources: Ch15, Fi6.**

Brown Tempest - *A. longicuspis*, hardneck, bulbs splotched with purple, brown cloves with hint of rose blush, no stripes, 6 cloves per bulb, 16-20 cloves per pound, nice shape and size, initial fiery raw taste mellows to a pleasing garlic flavor, original R/AL/18 PI 493098 had the largest bulbs of any wild garlics collected by Swenson and Simon in the late 1980s. *Source History: 1 in 1994; 4 in 1998; 10 in 2004.* **Sources: Ba8, Bi3, Ch15, Da6, Fi6, Ga1, Ga19, Ga21, Ho13, So1.**

Brown Vesper - Marbled variety, hot and spicy raw, when baked the mild sweet flavor is almost fruity, old R81 originally from USSR by Maria Jenderek. *Source History: 1 in 1998; 2 in 2004.* **Sources: Bi3, Fi6.**

Burgundy - Darker colored Creole str., striking, deep solid burgundy cloves, 8-12 per bulb, medium to large bulb size, clove tips not elongated like other Creole strains, higher sugar content than most, from U.C. Davis via Horace Shaw in Oregon. *Source History: 2 in 1994; 2 in 1998; 3 in 2004.* **Sources: Fi6, Go6, Ho13.**

Bzenc - Hardneck/Purple Stripe, raw flavor is very spicy with the heat building quickly, garlic flavor is retained when baked, from the Czech collection of Dr. Boris Andrst, New York. *Source History: 2 in 2004.* **Sources: Bi3, Fi6.**

California Early White - (California Early, California White) - Softneck, well-filled tight-skinned large slightly flat round bulbs, typically rough looking from large outer cloves, some purple blotch in bulb wrappers, mostly four clove layers with 10-21 cloves, clove color is light-tan or off-white often with some pink blush, not many smallish cloves, better adapted to heat than California Late or Susanville, very vigorous and productive, tolerates hot climates, not a good keeper, widely grown commercially in California where it originated. *Source History: 3 in 1981; 3 in 1984; 4 in 1987; 10 in 1991; 9 in 1994; 8 in 1998; 11 in 2004.* **Sources: Bi3, Bl7, Fi1, Fi6, Ga19, Gur, Ho13, Jun, Pe2, Shu, Ver.**

California Giant - Extremely large bulbs with very thin skin, very mild flavor. *Source History: 2 in 1998; 1 in 2004.* **Sources: Par.**

California Late White - Smooth bulbs, cloves are pink or pink-brown, best keeper of the white garlics, can be stored for 6-8 months in proper conditions, excel. for braiding, matures 2-3 weeks later than California Early White. *Source History: 2 in 1991; 2 in 1994; 3 in 1998; 2 in 2004.* **Sources: Fi6, Pe2.**

Carpathian - Large uniform off-white bulbs with thin copper veins, doubled cloves rare even in big bulbs, large brown cloves, mostly 6-10 cloves per bulb, excel. fiery flavor, vigorous deep-green plants, bulbs mature about one week after Spanish Roja and most other rocamboles, from Carpathian Mountains of southwest Poland. *Source History: 2 in 1991; 2 in 1994; 2 in 1998; 9 in 2004.* **Sources: Ch15, Fi6, Ga19, Ga21, Go6, Ho13, La8, Rev, Sw8.**

Celaya Purple - Hardneck, fiery raw taste with nice finish, midseason, from Mexico. *Source History: 1 in 2004.* **Sources: Bo20.**

Chamiskuri - (Gatersleben #6035) - Softneck, artichoke, consistently productive strain, large bulbs, good storage quality, from Republic of Georgia. *Source History: 1 in 1994; 1 in 1998; 2 in 2004.* **Sources: Bi3, Fi6.**

Chengdu - Turban type, robust flavor with hint of heat, purchased at Chengdu, China market. *Source History: 3 in 2004.* **Sources: Bi3, Fi6, Ho13.**

Chesnok Red - (Shvelisi, Gatersleben 6811) - Hardneck, Purple Stripe, nicely colored, 9-11 cloves per bulb/54-60 per pound, cloves peel easily, good aroma, good lingering flavor, late season, 4-6 month storage, from near Shvelisi, Republic of Georgia. *Source History: 1 in 1994; 3 in 1998; 24 in 2004.* **Sources: Ba8, Bi3, Bl7, Bo20, Bo21, Ch15, Da6, Fi6, Ga1, Ga19, Ga21, Gl4, Go6, Ho13, La1, Ni1, Pe2, Rev, Ri12, Ron, Se7, So1, Sw8, Ter.**

Chet's - (Chet's Italian Purple, Chet's Italian Red) - Large gnarly bulbs, consistently outproduces other garlics, mostly 4 clove layers and 10-20 large cloves total, clove skins are mild-white or yellowish with faint brownish pink blush at base and-or tip and some purple streaking, 6-8 bulbs per pound, noted for its mild flavor, originated with the late Chet Stevenson of Tonasket, WA who found this garlic growing wild in a long abandoned garden in the 1960s and reselected it for the next 25 years. *Source History: 1 in 1981; 1 in 1984; 1 in 1987; 3 in 1991; 3 in 1994; 2 in 1998; 5 in 2004.* **Sources: Bi3, Fi6, Ga19, Go6, Se16.**

Chilean Silver - Silverskin, large white bulbs, 15-18 cloves per bulb, robust spicy flavor, good keeper. *Source History: 1 in 2004.* **Sources: Se7.**

Chinese - Hardneck, Turban Group, brown clove skin color, purple-brown bulb color, 11 cloves per bulb, good keeping quality, good strength, sharp but smooth flavor, oily texture. *Source History: 1 in 1994; 2 in 1998; 1 in 2004.* **Sources: Bi3.**

Chinese Pink - Softneck, extra early maturity, can be harvested late May to early June, 9-15 large cloves arranged in two layers, white outer skins, pink-purple inner skins, pink clove wrappers, mellow flavor, stores 4-5 months, 36-60 cloves per pound. *Source History: 4 in 2004.* **Sources: Bi3, Bl7, Ga17, Ter.**

Chinese Purple - Turban type with turban-shaped capsule striped with red, deep purple bulbs, relatively good storage. *Source History: 2 in 2004.* **Sources: Bi3, Fi6.**

Chinese Red and White - Early maturing garlic with large uniform cloves, red and white stripes, end of the clove has a little tip which looks like a firecracker fuse. *Source History: 2 in 1994; 2 in 1998; 2 in 2004.* **Sources: Shu, Ver.**

Chinese Sativum - Unclassified artichoke, 8 cloves per bulb in a circular configuration, somewhat elongated shape, purple striped bulb wrappers, clove covers are a golden tan with tiny pink veins, fairly hot. *Source History: 1 in 2004.* **Sources: Fi6.**

Chinese Stripe - Turban type, bulbs have delicate purple stripes, mild flavor, Beijing market. *Source History: 2 in 2004.* **Sources: Bi3, Fi6.**

Choparsky - Marbled Group, 3-5 large fat cloves, hot raw flavor, mild garlic flavor when baked, from Siberian Botanical Garden. *Source History: 1 in 1998; 3 in 2004.* **Sources: Bi3, Fi6, Gl4.**

Chrysalis Purple - Hardneck, large heads, cloves peel easily, excel. flavor, 6-10 cloves per bulb, 5-6 bulbs per pound, from Viroqua, WI. *Source History: 1 in 2004.* **Sources: Se16.**

Colorado Black - Produces large bulbs when grown in poor soil, needs lots of water, produces a serpent coil followed by very large bulbils in the scape, heirloom from Fort Collins, CO. *Source History: 2 in 2004.* **Sources: Ga19, Ho13.**

Colorado Purple - Medium-large white bulbs, 8-10 large tan and purple cloves, warm spicy flavor, vigorous, heirloom from Vineland, Colorado. *Source History: 2 in 2004.* **Sources: Ga17, Ga19.**

Creole Red - True stiff stem red garlic, one of the original strains of red garlic grown in the U.S., excel. garlic flavor is stronger than whites, true gourmet garlic, matures early, keeps only 3-6 months, loose cloves are easily peeled, not a good braider. *Source History: 1 in 1991; 3 in 1994; 2 in 1998; 2 in 2004.* **Sources: Bi3, Fi6.**

Cuban Purple - (Rojo de Castro) - Silverskin var. from Spain, medium-large white bulb blushed with purple, 6-8 large deep purple cloves, hot, sweet flavor. *Source History: 1 in 1998; 3 in 2004.* **Sources: Bi3, Fi6, Ga17.**

Czech Red - Obtained from Wayne's Organic Garden. *Source History: 2 in 2004.* **Sources: Bo20, Fi6.**

Czechland Race - Rocambole, raw flavor is described as hot with a sharp zap and lingering flavor, mild when baked with an assertive but pleasant taste. *Source History: 1 in 2004.* **Sources: Fi6.**

De Vivo - Dependable topsetting garlic, very large cloves peel easily, 4-6 per bulb, spicy flavor with lingering aftertaste, heirloom from Dr. Gilbert McCollum of USDA, CV So1 introduction, 1991. *Source History: 1 in 1991; 1 in 1998; 3 in 2004.* **Sources: Fi6, Ga21, So1.**

Dominic's - New York and New England favorite, thick wrappers, large cloves, easy to peel, very tasty, good cold weather bulb. *Source History: 1 in 2004.* **Sources: Ga19.**

Duganskij - Marbled, fat cloves, raw flavor starts smooth and finishes with a bite, sweet baked flavor with pleasant aftertaste, from Kazahkistan via Czechoslovakia. *Source History: 1 in 1998; 3 in 2004.* **Sources: Bi3, Fi6, Gl4.**

Dushanbe - Turban type, light red stripes on bulbil capsule, large plump rose-brown to mahogany cloves cause a hot bite on the tip of the tongue, very early maturity, from Tadzhikistan, former Soviet Republic. *Source History: 2 in 2004.* **Sources: Bi3, Fi6.**

Early Red - *Allium sativum* var. *ophioscorodon*, vigorous large braiding var., rich flavor, does well in the Mid-Atlantic, productive and hardy. *Source History: 1 in 1991; 1 in 1994; 1 in 2004.* **Sources: Ga21.**

Early Red Italian - Medium to large bulbs are purple splotched, mostly 4 clove layers and 11-21 cloves, very fat-round cloves are milk-white with some pink blush at base and/or tip, strain sel. in southern Oregon at Telowa Farms for early maturity although it has not proven so in all locations, very productive and long storing. *Source History: 1 in 1991; 1 in 1994; 1 in 1998; 3 in 2004.* **Sources: Ch15, Fi6, Ga19.**

Enid's Vertalia - Rocambole, large slightly flattened white bulbs, 10-15 cloves, good storage, Vermont. *Source History: 1 in 2004.* **Sources: Hi6.**

F 1 - Bolting type, large cloves, easy to peel, often produces multiple cloves under single skin, weighs up to 6 oz. *Source History: 1 in 1994; 1 in 1998; 1 in 2004.* **Sources: Fi7.**

F 2 - Resembles F1, from Austria. *Source History: 1 in 1994; 1 in 1998; 1 in 2004.* **Sources: Fi7.**

F 3 - Bolting type with 4-8 large cloves in a bulb, up to 5 oz., tiny bulbils can produce mature bulbs in 3-4 years, very strong taste. *Source History: 1 in 1994; 1 in 1998; 1 in 2004.* **Sources: Fi7.**

F 4 - Nonbolting Italian type, large cloves, bulbs up to 5 oz., strong taste, has been grown in New York state for over 35 years. *Source History: 1 in 1994; 1 in 1998; 1 in 2004.* **Sources: Fi7.**

F18 - Rocambole, unknown origin. *Source History: 1 in 2004.* **Sources: Fi7.**

F21 - Rocambole, averages 6 cloves per bulb, from Poland. *Source History: 1 in 2004.* **Sources: Fi7.**

F22 - Bolting type from China. *Source History: 1 in*

1994; 1 in 1998; 1 in 2004. **Sources: Fi7.**

F31 - Grown from F3 bulbils with smaller bulbs and cloves, otherwise similar. *Source History: 1 in 2004.* **Sources: Fi7.**

F32 - Grown from F3 bulbils but is similar to F1. *Source History: 1 in 2004.* **Sources: Fi7.**

F34 - F3 variety, striving for a repeat of 4 cloves per bulb. *Source History: 1 in 2004.* **Sources: Fi7.**

F40 - Softneck, many small cloves per bulb, strong taste. *Source History: 1 in 2004.* **Sources: Fi7.**

F40x - This is F40 which became a hardneck, which is what happens with some varieties of garlic. *Source History: 1 in 2004.* **Sources: Fi7.**

F41 - Similar to F4 but with smaller bulbs and cloves, better yields. *Source History: 1 in 2004.* **Sources: Fi7.**

F44 - Softneck, 8-9 white cloves per bulb, from Austria. *Source History: 1 in 2004.* **Sources: Fi7.**

F45 - Softneck. *Source History: 1 in 2004.* **Sources: Fi7.**

Fantasy - Purple Stripe, medium-large bulb size. *Source History: 1 in 2004.* **Sources: Gl4.**

FC - Hardneck, bulb weighs one-half ounce, 19 cloves, from Cuba. *Source History: 1 in 2004.* **Sources: Fi7.**

Ferganskij - Purple stripe hardneck, good bulb size and color, collected by John Swenson in Samarkand, Uzbekistan Bazaar. *Source History: 1 in 1998; 2 in 2004.* **Sources: Bi3, Fi6.**

Fish Lake 3 - Hardneck with purplish cloves, 6 per bulb, very strong, pleasant flavor, good keeper, oily, crunchy hard texture, from Fish Lake. *Source History: 1 in 1998; 1 in 2004.* **Sources: Sa5.**

Floha - (Gatersleben #264) - Hardneck/Porcelain, raw taste is fiery hot and long lasting, 4-10 cloves per bulb, collected near Floha, Germany. *Source History: 1 in 1994; 1 in 1998; 2 in 2004.* **Sources: Bi3, Fi6.**

French Germinador - Hardneck, attractive large bulbs, violet and cream skin, 8-12 rose-purple cloves, rated among Chester Aaron's top five for superior flavor, from France. *Source History: 2 in 2004.* **Sources: Ga17, Ga19.**

French Purple - Artichoke, beautiful garlic with medium zip. *Source History: 1 in 2004.* **Sources: Gl4.**

French Red - (Hannan) - Bulb wrappers with moderate to heavy purple streaks and blotches, clove skins are light-brown with med-dark reddish brown stripes on top half of clove and burnt brown at base, average of 8 cloves per bulb, slightly tallish clove shape, bulbil capsules somewhat turban-shaped, probably the most highly colored of several strains of French Red in North America. *Source History: 1 in 1991; 1 in 1994; 1 in 1998; 1 in 2004.* **Sources: Bi3.**

French Red Asian - Mutation selected from French Red Hannan (Rocambole) strain in 1990, lively sweet lingering baked flavor, raw flavor starts sweet and becomes extremely hot. *Source History: 2 in 2004.* **Sources: Bi3, Fi6.**

Garlic - (White Garlic, Garlic Bulbs) - Bulbs separated into cloves (5-7) for planting, plant in early spring or fall, grown and stored just like onions, used for flavoring, also in sprays for insect control, keep a braid of dried garlic hanging in the kitchen. *Source History: 22 in 1981; 20 in 1984; 32 in 1987; 26 in 1991; 27 in 1994; 27 in 1998; 17 in 2004.* **Sources: Alb, Dam, Dom, Ers, Fa1, Hal, La1, Lej, Mcf, Mel, Mey, Mo13, Pin, Ri2, Roh, Sh9, Tt2.**

Gem - Marbled group, old RAL27 from M. Jenderek via Phil Simon, mild flavor quickly heats up. *Source History: 1 in 1998; 1 in 2004.* **Sources: Bi3.**

Georgian Crystal - (Cichisdzhvari #1, Gatersleben #6819) - Large white bulbs, hardneck/porcelain, 4-7 flat cloves per bulb, 6-8 bulbs per pound, very mild raw flavor, long storage, from the Republic of Georgia. *Source History: 1 in 1994; 2 in 1998; 13 in 2004.* **Sources: Ba8, Bi3, Ch15, Fi6, Ga17, Ga19, Gl4, Go6, Ho13, Ni1, Se7, Se16, Sw8.**

Georgian Fire - (Cichisdzhvari #4, Gatersleben #6822) - Hardneck, similar to Georgian Crystal, 5-9 cloves per bulb, 30-45 cloves per pound/6-8 bulbs per pound, strong raw taste with nice hotness but not unpleasant, good salsa garlic, very late season maturity, originated in the Republic of Georgia. *Source History: 1 in 1994; 1 in 1998; 10 in 2004.* **Sources: Bi3, Bl7, Bo20, Bo21, Fi6, Ga19, Go6, Sa5, Se16, Ter.**

German (Red) - Fi6 puts the (Red) in parentheses because this is not the same German Red with bright clove colors offered by others, very vigorous plant with deep-purple at base end, large doubled cloves common and easy to peel, often 10-15 cloves per bulb, ideal for sauteeing in butter, originated with old-time gardeners of German descent in Idaho. *Source History: 1 in 1991; 1 in 1994; 1 in 1998; 1 in 2004.* **Sources: Fi6.**

German Brown - Rocambole, different from German Red, huge bulbs, hot spicy flavor, clove color is a distinctive brown rather than red, large scapes, not bothered by cold winters. *Source History: 3 in 2004.* **Sources: Da6, Fi6, Ga19.**

German Extra Hardy - Stiffneck type, vigorous grower with long roots which enables it to overwinter without heaving out of the ground, good flavor, stores well, outside skin color is white, cream clove skin color striped with burgundy, 4-7 cloves per bulb, 4-6 bulbs per pound. *Source History: 1 in 1994; 1 in 1998; 7 in 2004.* **Sources: Bi3, Ga17, Ga19, Go6, Hi6, Jo1, Se16.**

German Porcelain - Hardneck, Porcelain type with delicate purple stripes on white wrapper, 4-7 large cloves, easy peeler, good flavor, good keeper, does well in any climate. *Source History: 6 in 2004.* **Sources: Ba8, Ga1, Ga19, Ho13, Pe2, Rev.**

German Red - (Red German) - *A. sativum* var. *ophioscorodon*, large bright purple heads, yellow fleshed cloves, 8-12 cloves per head, 50-75 cloves per pound, difficult to grow, longest growing and poorest keeper, but most flavorful by far, can be preserved by pureeing or dehydrating or placing peeled cloves in oil, can be grown in mild climates but does best where winters are cold, produces topsets the size of pine nuts. *Source History: 1 in 1981; 1 in 1984; 1 in 1987; 3 in 1991; 9 in 1994; 15 in 1998; 15 in 2004.* **Sources: Ch15, Ec2, Fi1, Ga1, Ga17, Ga19, Ga21, Gl4, Go6, Ho13, Ki4, Ron, Shu, Sw8, Ver.**

German White - Porcelain hardneck, 6-10 cloves per head, 6-8 bulbs per pound, strong flavor, excel. for roasting, highest allicin content, stores 4-6 months, from Germany. *Source History: 1 in 1998; 10 in 2004.* **Sources: Bi3, Bl7, Bo20, Ga19, Gl4, Hi6, Ri12, Ri12, Se7, We20.**

Germany - Flavorful rocambole, fairly good storage. *Source History: 1 in 2004.* **Sources: Fi6.**

Germidour - Rare Artichoke variety, large cloves, 7-8 per bulb, hot, from France. *Source History: 1 in 2004.* **Sources: Bi3.**

Gigantus - Bulb size compares to German Red with milder flavor, 10-15 tan cloves per bulb, double cloves are common. *Source History: 1 in 2004.* **Sources: Ch15.**

Glacier Porcelain - Good looks along with good taste, dev. in Montana. *Source History: 1 in 2004.* **Sources: Gl4.**

Gregory's China Rose - Artichoke, Asiatic or Turban Group, produces a seed stalk, excel. raw flavor, grows almost anywhere, does well in warm winter areas, early harvest, keeps about 6 months. *Source History: 1 in 2004.* **Sources: Go6.**

Grizzly - Large bulbs with many large cloves, often double. *Source History: 1 in 2004.* **Sources: Gl4.**

Guatemalan - Silverskin, medium-large, white bulbs splashed with purple, 6-8 large brown and burgundy cloves, mild, almost nutty flavor, from Ikeda family in the village of Aguacatan near Huehuetenango, Guatemala via Chester Aaron. *Source History: 1 in 2004.* **Sources: Ga17.**

Gypsy Red - Porcelain var. from Romania, tall white bulbs, 4-6 very large purple and tan cloves, hot, strong lingering flavor mellows with cooking. *Source History: 1 in 2004.* **Sources: Ga17.**

Himalayan Red - Unique Artichoke with only 2-3 cloves per bulb. *Source History: 1 in 2004.* **Sources: Fi6.**

Hokkaido Zai Tai - Hardneck, very large bulbs, white wrappers with purple blush, 6-10 purple-skinned cloves, deep, mellow, lingering flavor, grown in Shikoku. *Source History: 2 in 2004.* **Sources: Ga17, Ho13.**

Hungarian Purple Stripe - Purple stripe type, uniform bulbs with purple streaks, 8-12 white cloves per bulb, emerges late in the spring. *Source History: 1 in 2004.* **Sources: Hi6.**

Idaho Silver - Cream-silver bulbs, red-pink cloves, good bulb size and character, mild and sweet baked, when eaten raw, heat starts slowly and increases to very hot, well adapted to northern interior climates with cold winters. *Source History: 1 in 1994; 1 in 1998; 1 in 2004.* **Sources: Fi6.**

Idita Red - Very red, hot Mediterranean type. *Source History: 1 in 1998; 1 in 2004.* **Sources: Ho13.**

Inchelium Red - *A. sativum* var. *sativum*, artichoke type, light-purple blotching on very large bulbs, 4-5 clove layers and 8-22 cloves per bulb, 45-65 cloves/6-8 bulbs per pound, white or yellowish skin with faint purple at the base, bulbs over 8 in. dia. are possible, mild true garlic flavor, higher soluble solids, heavier bulbs, mid-season maturity, stores 6-9 months, discovered on the Colville Indian Reservation in Inchelium, WA, original source unknown. *Source History: 2 in 1991; 5 in 1994; 5 in 1998; 24 in 2004.* **Sources: Bi3, Bl7, Bo20, Bo21, Ch15, Fi6, Ga1, Ga17, Ga19, Ga21, Gl4, Go6, Goo, Hi6, Ho13, Rev, Ri12, Ron, Se7, Se16, So1, Sw8, Ter, We20.**

Iowa - Large bulbs, hot raw flavor, mild rich baked flavor, relatively long storage. *Source History: 1 in 2004.* **Sources: Fi6.**

Israeli - Large bulbs with dark thin red veins and light-purple blotches, clove color varies from light-brown with purple blush at base to mahogany almost completely overlaid with purple blush, doubled cloves are rare, 4-9 cloves per bulb. *Source History: 1 in 1991; 2 in 1994; 2 in 1998; 1 in 2004.* **Sources: Ho13.**

Italian - Softneck garlic with fantastic pungent flavor, far superior to the garlic sold in many stores, good keeper, from a strain originally grown by a certified organic grower in Washington State. *Source History: 1 in 1981; 1 in 1984; 2 in 1987; 2 in 1991; 4 in 1994; 2 in 1998; 1 in 2004.* **Sources: Gl4.**

Italian Early - Softneck variety with larger clove size, white skin, better adapted to summer heat than Italian Late, good braider. *Source History: 1 in 1998; 2 in 2004.* **Sources: Go6, Ron.**

Italian Late - Softneck late season variety with all the best qualities of garlic bred into it, flavorful, pungent cooking type, stores for up to 6-9 months after harvest, good braiding type, 18-22 cloves per head, 110-130 cloves per pound. *Source History: 3 in 1998; 6 in 2004.* **Sources: Ga1, Jun, Ri12, Shu, Ter, Ver.**

Italian Purple - Heirloom artichoke garlic, mild and flavorful, 8-10 pink to red cloves in a rosette, keeps for 7 to 12 months, grown in many climate zones. *Source History: 2 in 1987; 1 in 1991; 1 in 1994; 2 in 1998; 7 in 2004.* **Sources: Bo20, Ga1, Ga19, Gl4, Gur, Ki4, Pe1.**

Jain Shang Dong - True Turban type, red headed scape develops early. *Source History: 1 in 2004.* **Sources: Bi3.**

Japanese - (Sakura) - Produces a flower stalk that does not need to be removed, large white bulbs with 6-8 pale yellow cloves, delicious pungent garlic flavor with pleasant lingering aftertaste. *Source History: 2 in 2004.* **Sources: Bi3, Fi6.**

Japo - From Czech collection of Dr. Boris Andrst, NY. *Source History: 1 in 2004.* **Sources: Fi6.**

Jovak - Hardneck, purple stripe, from the Czech collection of New York grower, Dr. Boris Andrst. *Source History: 2 in 2004.* **Sources: Bi3, Fi6.**

Kaskaskia Red - Red-purple streaks on white bulbs, 5-8 cloves per bulb, flavor described as "hot and zesty", fine southern Illinois var. that may prove to be a good garlic for Zone 7 as it has done well in Virginia, obtained from a local farmer and named by breeders Merlyn and Mary Ann Niedens after the nearby Kaskaskia River. *Source History: 1 in 2004.* **Sources: Rev.**

Kazahkistan - Extremely rare variety. *Source History: 1 in 2004.* **Sources: Gl4.**

Keeper - Hardneck garlic that stores well, up to one year, red striped cloves surrounded by dusky tan wrappers, compelling taste, may be related to the Southern Continental/Creole variety. *Source History: 1 in 2004.* **Sources: Ga19.**

Kettle River Giant - Artichoke, very large, somewhat flattened bulbs can reach 4 in. across, 8-15 cloves per bulb, 35-55 per pound, rich hot flavor, stores well, heirloom from Stevens County, Washington, grown by the Kettle River for over 100 years. *Source History: 5 in 2004.* **Sources: Ga17, Ga19, Go6, Ho13, Se7.**

Khabar - Marbled, mild and creamy baked, good for adding to mashed potatoes, heat eases into a burst of flavor when eaten raw, collected from Khabarofsk, Siberia by Alaskan grower Bob Ellis. *Source History: 2 in 2004.* **Sources: Bi3, Fi6.**

Kiev - Rocambole, broad leaved, large Roja type, crunchy cloves, hot. *Source History: 1 in 2004.* **Sources: Sa5.**

Killarney Red - Hardneck, Rocambole type, pink skinned cloves, 8-10 per bulb, 40-50 per pound, easy to peel, lingering spicy hot flavor, resembles German Red and Spanish Roja, produces better than most under wet conditions, late season. *Source History: 2 in 1998; 12 in 2004.* **Sources: Bi3, Ch15, Da6, Fi6, Ga17, Ga19, Gl4, Go6, Ni1, Ron, Sw8, Ter.**

Korean Red - Hardneck, very hot, large bulbs contain 6-10 very large purple cloves, wrappers are heavily striped with purple, heat mellows to a sweetness with great garlic flavor, easy to peel, good keeper. *Source History: 1 in 1998; 7 in 2004.* **Sources: Bl7, Ga1, Ga17, Ga19, Ni1, Pe2, Ron.**

Korean Rocambole - (Korean Red) - Originally from South Korea, hardneck, medium to large bulbs with moderate purple streaks, cloves have a pink-brown background with a few dull-purple streaks that stand out, often burnt-brown color at base, contrast of colors is more striking than in most rocamboles, 60-65 cloves per pound, winter hardiness uncertain. *Source History: 1 in 1991; 1 in 1994; 2 in 1998; 4 in 2004.* **Sources: Bi3, Fi6, Ho13, Sa5.**

Kyjev - Porcelain hardneck, from the Czech collection of New York grower, Dr. Boris Andrst. *Source History: 2 in 2004.* **Sources: Bi3, Fi6.**

Labera Purple - Silverskin var. from Andalusia district of Spain. *Source History: 1 in 1998; 2 in 2004.* **Sources: Bi3, Fi6.**

Lapanantkari - Hardneck, Porcelain type, faint brown-purple stripes on white bulbs, 4-6 very fat cloves per bulb, purple tinge on brown and cream colored bulbs with thin black stripes, collected by John Swenson from the Republic of Georgia. *Source History: 3 in 2004.* **Sources: Bi3, Fi6, Rev.**

Leah 99 - Porcelain hardneck, big, bold tasting cloves, from Dawn and Ian Glasser, excel. garlic growers, named after their daughter. *Source History: 1 in 2004.* **Sources: Ga19.**

Legacy - Hardneck, large bulbs, 4-5 streaked cloves, moderate heat, complex flavor, very good storage, from Canada. *Source History: 2 in 2004.* **Sources: Ga1, Ga19.**

Leningrad - (Gatersleben #684) - Hardneck, Porcelain type, tall, dark green plant can grow to 4 ft., large paper-white heads, 4-6 purple-flecked cloves per head, fairly hot, stores well, 35-40 cloves per pound, European origin. *Source History: 3 in 1998; 5 in 2004.* **Sources: Bi3, Bo21, Fi6, Ga19, Ga21.**

Locati - Columbia Basin heirloom, 10-12 cloves per bulb, well adapted to northern latitude, winter cold and summer heat, originally from Milan, Italy in the 1920s. *Source History: 1 in 1994; 1 in 1998; 1 in 2004.* **Sources: Go6.**

Lorz Italian - Artichoke type with very large very flat round bulbs, 3-5 clove layers with 12-19 total cloves, not many smallish interior cloves but tend to be squarish due to bulb shape, cloves are milk-white to yellowish with subtle pink blush, some years very hot tasting, 6-8 bulbs per pound, a Northwest heirloom brought to Washington's Columbia Basin from Italy by the Lorz Family before 1900. *Source History: 1 in 1991; 2 in 1994; 1 in 1998; 10 in 2004.* **Sources: Ba8, Bi3, Bl7, Fi6, Ga19, Go6, Hi6, Ho13, Ni1, Se16.**

Lotus - (Kowloon #1) - Turban type, hot when raw, retains some heat baked, from southeast China market. *Source History: 2 in 2004.* **Sources: Bi3, Fi6.**

Lukak - Artichoke, large bulbs, flavor is rich but mild with no bite, very productive, from Czech collection of New York grower, Dr. Boris Andrst. *Source History: 1 in 2004.* **Sources: Fi6.**

Machashi - (Gatersleben 5873) - Softneck, clove colors are variable from nearly solid grey-purple to bright-white with thin purple streaks on the top half of the clove and a light-brown base, inner cloves generally lighter colored and well streaked while outer cloves often blushed brown-purple or grey-purple enough to obscure the streaks, cloves have short stout tails that occasionally elongate to 1.5 in., collected in Georgian SSR and imported from Gatersleben in Eastern Germany. *Source History: 1 in 1991; 1 in 1994; 2 in 1998; 4 in 2004.* **Sources: Bi3, Bl7, Fi6, Ga21.**

Madrid - Mild flavor with only slight heat, 8-12 plump cloves, bought in a Madrid marketplace by a California grower. *Source History: 1 in 2004.* **Sources: Fi6.**

Mahogany - (F/AL/11 (PI 493097)) - Hardneck, rocambole, firm bulbs, deep brown cloves, good flavor with creeping bite, originally from the USSR, 1986. *Source History: 1 in 1994; 1 in 1998; 1 in 2004.* **Sources: Bi3.**

Maiskij - (Maiskii) - Artichoke type, deep yellow bulbs streaked with purple, outer layers of bulbs are purple spattered with large white spots and purple striping toward the top, 6-8 clovers per bulb, collected by John Swenson at Ashkhabad bazaar in Turkmenistan. *Source History: 3 in 2004.* **Sources: Bi3, Fi6, Rev.**

Maitake - Asiatic or Turban var., often develops a scape with a seed head or spathe similar to a hardneck, but the stalk is not thick, making the harvested bulb look like a softneck, beautiful bulb with deep purple stripes, thin clove skins, fair amount of heat with powerful garlic flavor, very early maturing, good choice for mild winter areas. *Source History: 1 in 2004.* **Sources: Ga19.**

Marie's Special - Topsetting garlic that produces numerous small bulblets at the tops of the stalks in August, which can be planted in the fall and transplanted the following spring, bulbs are small but flavorful, will be sent as young plants in the spring, plant 3 inches apart. *Source History: 1 in 2004.* **Sources: Sa9.**

Marino - (Marino Rocambole) - Hardneck, Rocambole, large white bulbs with occasional purple streaks, 6-8 large cloves, strong zesty flavor, exceptional storage for a hardneck, originally from a Mr. Marino of New York state. *Source History: 4 in 2004.* **Sources: Ga19, Ga21, Hi6, Rev.**

Maxatawny - Hardneck, bronze bulbs with 10-15 cloves, a blush of lavender at the base which fades to a rich cream color, mild but pleasant flavor lingers, found among the Mennonites of eastern Pennsylvania by W. Weaver, originally from Silesia (now part of southern Poland) in the 1740s. *Source History: 2 in 2004.* **Sources: Ho13, Rev.**

Metechi - (Gatersleben #7081) - Marbled variety, hardneck, upright, broad leafed plant, large bulbils and bulbil capsule, firm nicely colored bulbs, large fat cloves with thick blushed skins, 4-7 per bulb, 40-45 per pound, fiery taste with a smooth finish, very late season maturity. *Source History: 1 in 1994; 2 in 1998; 11 in 2004.* **Sources: Bi3, Da6, Fi6, Ga1, Ga19, Gl4, Go6, Pe2, Ron, Sw8, Ter.**

Mexican Red - Purple stripe variety, easy to peel, well suited for southern gardens, origin may trace back to Taiwan. *Source History: 1 in 2004.* **Sources: Ga19.**

Mexican Red Silver - Silverskin, raw flavor is hot and strong, light with low heat when baked, from Mexico via John Swenson. *Source History: 1 in 1998; 1 in 2004.* **Sources: Fi6.**

Mexicano-B - Hardneck, 5-6 cloves per bulb. *Source History: 1 in 2004.* **Sources: Bi3.**

Mikulovsky - Productive Artichoke variety, from Czech collection of New York grower, Dr. Boris Andrst. *Source History: 1 in 2004.* **Sources: Fi6.**

Mild French - (Mild French Silverskin) - Large cloved Silverskin var., mostly 4 clove layers and 13-16 cloves total, cloves have a striking red-pink to red blush overlaid with dark thin red lines on a yellow-white background, red colors sometimes fade in highly fertile soil but still beautiful, better adapted to hot dry climates, grows taller and produces larger bulbs and matures slightly earlier than other Silverskin strains. *Source History: 3 in 1991; 3 in 1994; 2 in 1998; 3 in 2004.* **Sources: Fi6, Ga19, So1.**

Monshanskij - (Gatersleben #131) - Czech strain, marbled group, good color. *Source History: 1 in 1998; 2 in 2004.* **Sources: Bi3, Fi6.**

Montana Giant - Rocambole with mild garlic flavor with no aftertaste, all-purpose, 2.5-3 in. heads, light to med. purple colored wrappers, 8-10 large cloves, 40-50 per pound, stores for 6-8 months, mid-season maturity, bred in Flathead Valley of Montana. *Source History: 4 in 2004.* **Sources: Da6, Gl4, Ron, Ter.**

Montana Roja - *A. sativum* var. *ophioscorodon*, an improved Spanish Roja, gourmet flavored rocambole, light to med. purple colored wrapper, 8-14 brownish cloves per bulb, easy peeling, for spring planting. *Source History: 1 in 1994; 4 in 2004.* **Sources: Bi3, Bl7, Da6, Gl4.**

Morado Gigante - Hardneck, large uniform bulbs, 6-8 large cloves, deep burgundy bulb wrappers and clove skins, mellow flavor with no bite, from Chile. *Source History: 2 in 2004.* **Sources: Ga17, Ga19.**

Morado Red - Subtle garlic taste. *Source History: 1 in 2004.* **Sources: Mcf.**

Mother of Pearl - Softneck with more cloves per bulb than others, pungent flavor. *Source History: 1 in 1998; 1 in 2004.* **Sources: Se7.**

Mountain Rocambole - Strong, spicy flavored red-brown cloves. *Source History: 1 in 2004.* **Sources: Da6.**

Mucdi - (Gatersleben #5876) - Late maturing Artichoke garlic, fat wide outer cloves, light tan with light stripes, 12 per bulb, excel. character and storage, from Republic of Georgia. *Source History: 1 in 1998; 2 in 2004.* **Sources: Bi3, Fi6.**

Music - (Musik, Music Pink) - Hardneck, Porcelain type, white bulb wrapper, pink clove skin color, 8 cloves per bulb, 25-30 per pound/4-5 bulbs per pound, medium strength, fine flavor, soft and oily texture, excellent storage, mid-season, pioneered by Canadian Al Music, seems identical to Susan Delafield's, from Ontario, Canada. *Source History: 1 in 1998; 17 in 2004.* **Sources: Bi3, Bl7, Bo20, Da6, Fi6, Ga1, Ga19, Gl4, Go6, Ho13, Pe2, Rev, Ri2, Ron, Se7, Se16, Ter.**

Native Creole - Brilliant color, from CV Seeds of Change, Arizona. *Source History: 2 in 2004.* **Sources: Bi3, Fi6.**

Nichols Silverskin - Developed over many years of selection for strong flavor and size and easy-peeling qualities, garlic helps to lower blood pressure. *Source History: 2 in 1981; 2 in 1984; 1 in 1987; 1 in 1991; 1 in 1994; 1 in 1998; 1 in 2004.* **Sources: Ni1.**

Nootka Rose - Highly colored Silverskin strain, Northwest heirloom from the San Juan Islands off northern Washington coast, mostly 5 clove layers and 15-24 smallish cloves, clove color varies a bit but is usually heavily streaked red on a mahogany background with dark solid bright-red clove tips, cloves often have a long paper tail, very nice for braiding as the red clove color stands out when one or two bulb wrappers are removed. *Source History: 1 in 1991; 1 in 1994; 2 in 1998; 7 in 2004.* **Sources: Ch15, Fi6, Ga1, Ga17, Ga19, Go6, Ho13.**

Northe - Marbled group, early maturity, produces seed stalk early, from Beltsville Ag. Station. *Source History: 1 in 2004.* **Sources: Bi3.**

Northe No. 3 - *A. longicuspis*, tall, early maturing var., often with double umbels on the flower stalk, from Dr. Gil McCollum, Beltsville Res. Sta. *Source History: 1 in 1994; 1 in 1998; 1 in 2004.* **Sources: Fi6.**

Northern Quebec - Hardneck with strong, robust flavor, pale purple clove skin color, 7 cloves per bulb, crunchy texture, one of the first to head, hardy. *Source History: 1 in 1994; 1 in 1998; 1 in 2004.* **Sources: Sa5.**

Northern White - (German Porcelain, Premium Northern White) - Stiffneck, porcelain, bulbs can reach half the size of Elephant garlic, large red-tinged cloves average 6-8 per head, easy to peel, very pungent, excel. for baking, exceptionally hardy in the far north. *Source History: 3 in 1998; 6 in 2004.* **Sources: Bi3, Ga19, Ga21, Ni1, So1, We14.**

Okrent - High yielding Artichoke, long storage. *Source History: 1 in 2004.* **Sources: Fi6.**

Old Yellow - Artichoke, large bulb with unique shape, California. *Source History: 1 in 2004.* **Sources: Gl4.**

Olomuk - Consistently large bulbs, from Dr. Boris Andrst, NY. *Source History: 1 in 2004.* **Sources: Fi6.**

Oregon Blue - Artichoke type, dark green leaves with

purple cast, large flat-round bulbs, 4-5 clove layers with 11-20 cloves, 55-70 per pound, clove skin is off-white, nice hot flavor, very vigorous and productive, good storage, does not produce any stem bulbils which contributes to its productivity, mid-season, maritime Northwest heirloom. *Source History: 1 in 1991; 2 in 1994; 1 in 1998; 4 in 2004.* **Sources: Bi3, Fi6, Ni1, Ter.**

Osage - Stiffneck, redomesticated variety from a remnant colony at one of the principal town sites of the Osage Indians in central Missouri about nine generations ago. *Source History: 1 in 2004.* **Sources: Mo13.**

Ozark - White bulbs over 2 in., grown in Arkansas mountains since the turn of the century. *Source History: 1 in 1998; 1 in 2004.* **Sources: Ho13.**

Pacer - Marbled group, from France as K/S/10. *Source History: 1 in 1998; 2 in 2004.* **Sources: Bi3, Fi6.**

Pearly Red - Fairly hot with lingering raw flavor, spicy sweet flavor when baked, smooth and creamy. *Source History: 1 in 2004.* **Sources: Fi6.**

Pennsylvania Dutch - Hardneck, large bulbs streaked with red, 4-5 cloves, hot. *Source History: 1 in 2004.* **Sources: Ga1.**

Pescadero Red - Creole strain. *Source History: 1 in 2004.* **Sources: Fi6.**

Phillips - Rocambole, large bulbs with large cloves, peel easily, does well in cold areas, introduced to the U.S. from Italy in the 1820s, from Phillips, Maine. *Source History: 1 in 2004.* **Sources: Und.**

PI 250662 (USSR) - Tasty Russian var., 4 large cloves per bulb. *Source History: 1 in 2004.* **Sources: Ho13.**

Pioneer - Artichoke type with excel. size and flavor, 10-20 cloves per bulb, best for braiding. *Source History: 2 in 1998; 1 in 2004.* **Sources: Gur.**

Pitarelli - Stores better than most Rocamboles, heavily striped bulb, dark brown/red cloves, strong, originally from Czechoslovakia in the 1920s. *Source History: 1 in 1994; 2 in 1998; 3 in 2004.* **Sources: Fi6, Gl4, Ho13.**

Polish Hardneck - Porcelain, full-bodied garlic flavor, 4-6 cloves per bulb/40-50 per pound, late season, from Ontario, Canada. *Source History: 1 in 1998; 7 in 2004.* **Sources: Bi3, Ch15, Fi6, Ga19, Gl4, Go6, Ter.**

Polish Jenn - Hardneck, Porcelain type, 6-8 large cloves, 30-45 per pound, white skin, flavorful, stores well, high yields, mid to late season, brought into Canada from Poland in the 1980s. *Source History: 4 in 2004.* **Sources: Bi3, Bl7, Ga1, Ter.**

Polish Softneck - Large softneck, 11-13 cloves per bulb, 65-75 per pound, hot flavor is retained after roasting, raw flavor compares to heat of a hot pepper, good braider, extremely winter hardy, mid-season, introduced to North America around 1900. *Source History: 1 in 2004.* **Sources: Ter.**

Polish White - (New York White) - Softneck type, vigorous plant with deeper green leaves than most, usually 3 or 4 clove layers with average 11 cloves per bulb and no small cloves, skins off-white with some purple blush, more productive and disease tol. in the North and East. *Source History: 1 in 1991; 3 in 1994; 3 in 1998; 7 in 2004.* **Sources: Bi3, Bl7, Fi6, Ga21, Jo1, Ri2, So1.**

Portugese - Turban top, medium large bulbs, purple and white stripes, 6-8 large creamy brown cloves, gourmet flavor with a nice bit of heat, from a farmer's market just outside Lisbon. *Source History: 1 in 2004.* **Sources: Ga17.**

Premium White - (Premium Northern White) - Extra select quality garlic can be directly traced to Northern Germany, hardneck, 5-7 cloves per bulb, large clove size, easy to peel, all-purpose, especially good baker, possibly the most cold-hardy variety known, 16-20 cloves per pound, late season. *Source History: 1 in 1981; 1 in 1984; 1 in 1987; 1 in 1998; 1 in 2004.* **Sources: Ter.**

Prim - Medium-size bulbs with 6-10 cloves, hardy, from Czech collection of Dr. Boris Andrst, NY. *Source History: 1 in 1991; 1 in 2004.* **Sources: Fi6.**

Pskem - Hardneck, Purple Stripe, purple striped white bulbs with faint purple wash, 2-5 huge cloves per bulb, creamy yellow cloves blushed with mauve at the base, collected by John Swenson from Pskem River Valley, Uzbekistan in 1989. *Source History: 1 in 1998; 4 in 2004.* **Sources: Bi3, Fi6, Gl4, Rev.**

Purple Cauldron - Vigorous, large bulbed Artichoke strain, large dark purple stem bulbils and bulbs, from France. *Source History: 1 in 1994; 2 in 1998; 3 in 2004.* **Sources: Bi3, Fi6, Ho13.**

Purple Glazer - (Gatersleben #7092) - Hardneck, Rocambole x Purple Stripe, originally Mchadidzhvari #1 (Mchadijvari #1), smooth bulb and clove wrappers, fat cloves, strong flavor but not hot, no aftertaste, similar to Red Rezan, from the Republic of Georgia. *Source History: 1 in 1994; 1 in 1998; 9 in 2004.* **Sources: Bi3, Bl7, Bo21, Fi6, Ga19, Gl4, Ni1, Rev, Sw8.**

Purple Italian Easy Peel - Zesty mid-season hardneck with a clean, sweet aftertaste, peels quite easily, excellent storage ability, 7-10 cloves per head, 60-80 cloves per pound. *Source History: 3 in 1998; 3 in 2004.* **Sources: Pe2, Se7, Ter.**

Purple Miranda - Large bulbs contain 7-12 cloves, some of the bulbs produce an S-curved stalk and purple topsets resembling rocambole, very early maturing and hardy. *Source History: 1 in 1991; 1 in 2004.* **Sources: Gl4.**

Purple Queen - Softneck type, purple-skinned bulbs, many small cloves, wonderful garlic flavor, from heirloom Hispanic stock. *Source History: 1 in 1994; 1 in 1998; 1 in 2004.* **Sources: Ho13.**

Purple Shandong - Purple clove color, medium hot, from China. *Source History: 1 in 1998; 1 in 2004.* **Sources: Ho13.**

Puslinch - Hardneck, large and small cloves with purplish skin, good keeping quality, excel. strength, sweet, aromatic, full flavor, very oily and crunchy texture, 12 cloves per bulb, from Ontario, Canada. *Source History: 1 in 1998; 1 in 2004.* **Sources: Bo21.**

Pyongyang - (Gatersleben #7041) - Hardneck, Artichoke type, brown cloves with purple blush, 8-10 per bulb, elongated tips, raw texture resembles a green apple, builds to med. heat, light pleasant flavor when baked, longer storage than most Asiatics, from near the North Korean capital of Pyongyang via Gatersleben seed bank. *Source History: 4 in 2004.* **Sources: Bi3, Fi6, Ho13, Rev.**

Red Czar - (Gatersleben #6017) - Hardneck, purple stripe, good bulb coloring, brown cloves, originally from Verchnyava Mcara in the Republic of Georgia. *Source History: 1 in 1994; 1 in 1998; 2 in 2004.* **Sources: Bi3, Fi6.**

Red Janice - Turban type with slightly later maturity and longer storage, bulbs have heavy solid stripes over purple blush, extremely fragrant, sweet spicy baked flavor, raw flavor starts hot and heat continues to build, originally from Nmarazeni in Republic of Georgia. *Source History: 2 in 2004.* **Sources: Bi3, Fi6.**

Red Rezan - (Gatersleben #7205) - Rocambole x Purple Stripe, bulb color dark glazed purple tinged with gold, strong, lasting flavor but not hot, no aftertaste, from south Moscow (Republic of Russia). *Source History: 1 in 1994; 2 in 1998; 4 in 2004.* **Sources: Bi3, Fi6, Ga19, Ho13.**

Red Toch - (Tochliavri) - Artichoke, red and pink striped cloves, 10-18 cloves per bulb, 7-10 bulbs per pound, raw taste described as perfect garlic flavor, from the village of Tochliavri, Republic of Georgia in the former USSR, 1988. *Source History: 1 in 1994; 1 in 1998; 10 in 2004.* **Sources: Ba8, Bi3, Bo20, Ga17, Ga19, Ga21, Go6, Ho13, Se16, So1.**

Redgrain - (Cichisdzhvari #2, Gatersleben #6820) - Hardneck Purple Stripe with fewer cloves per bulb, less purple coloring, from near Cichisdzhvari in Republic of Georgia. *Source History: 1 in 1994; 1 in 1998; 2 in 2004.* **Sources: Bi3, Fi6.**

Rocambole - (Purple Skin Garlic, Top Set Garlic, Rocambole Bavarian, Spanish Garlic, Italian, Serpent, Sand Leek, Spanish Shallot) - *Allium sativum* var. *ophioscorodon*, mild garlic, produces small topsets that are used for flavoring, also segmented cloves and some round bulbs form underground, forms a perennial patch, stems grow in curious curls and knots. *Source History: 4 in 1981; 4 in 1984; 5 in 1987; 5 in 1991; 5 in 1994; 8 in 1998; 4 in 2004.* **Sources: Ho13, Lej, Par, We19.**

Rocambole, French - Hardneck, smooth sweet baked flavor, mild to moderate raw flavor. *Source History: 2 in 2004.* **Sources: Fi6, Gl4.**

Rocambole, Roggero - Hot raw flavor, mild and sweet when baked. *Source History: 1 in 2004.* **Sources: Fi6.**

Romanian Red - (Gypsy Red) - Porcelain var., purple blotched bulb color in poor soil, white color in rich soil, streaked and lined cloves on brown background, 4-5 cloves per bulb, 35-40 cloves per pound, very good keeper, hot pungent taste with long-lasting bite, late season maturity, from Romania where it is called Red Elephant Garlic. *Source History: 2 in 1994; 4 in 1998; 16 in 2004.* **Sources: Ba8, Bi3, Bo20, Ch15, Da6, Fi6, Ga1, Ga19, Ga21, Gl4, Go6, Ho13, La8, Ron, Sw8, Ter.**

Rose de Lautrec - Hardneck from the Tarn region of southern France, large, white skinned bulbs, 6-10 rosy purple skinned cloves per bulb, pungent raw flavor mellows into a delightfully spicy flavor when cooked. *Source History: 1 in 2004.* **Sources: Ga17.**

Rose du Var - Often 12-15 cloves in 5 layers, usually only 2 or 3 large cloves per layer, very uniform, clove skins blushed solid pink or red-pink and often with bright-red patches, originally imported from France by David Cavagnaro. *Source History: 1 in 1991; 1 in 1994; 1 in 1998; 1 in 2004.* **Sources: Fi6.**

Rosewood - (PI 493099 (USSR), Siberian Rose) - *A. longicuspis*, hardneck/porcelain, softly colored large fat cloves, 3-4 per bulb, long storage, withstands cold winters. *Source History: 1 in 1994; 2 in 1998; 7 in 2004.* **Sources: Bi3, Bo20, Ch15, Fi6, Ga19, Gl4, Ho13.**

Russian - *Allium sativum* var. *ophioscorodon*, large to very large bulbs contain an average of 10 brown-red cloves which are easy to peel, excel. keeper, fine flavor, crunchy oily texture. *Source History: 1 in 1991; 1 in 1994; 2 in 1998; 2 in 2004.* **Sources: Gl4, Sa5.**

Russian Giant - Porcelain var. with large, purple streaked bulbs, 8-10 giant tan cloves. *Source History: 2 in 2004.* **Sources: Ga17, Ga19.**

Russian Hardneck - Porcelain type with strong flavor. *Source History: 1 in 2004.* **Sources: Sa5.**

Russian Inferno - Asiatic, softneck, possibly named by Horace Shaw, from Spokane, WA. *Source History: 2 in 2004.* **Sources: Bi3, Fi6.**

Russian Red - Hardneck with broad green leaves, large to very large bulbs, off-white wrappers with copper veins, large easily peeled cloves are all brown, 8-13 cloves per bulb, doubled cloves are common, stores 3-6 months, Pacific Northwest heirloom brought to the interior of British Columbia by Dukhobor immigrants from Russia in the early 1900s. *Source History: 1 in 1991; 1 in 1994; 2 in 1998; 10 in 2004.* **Sources: Bi3, Fi6, Ga19, Hi6, Ho13, Jo1, Rev, Shu, Ver, We20.**

Russian Redstreak - Unique strain dev. from a single bulb mutation from Russian Red in 1988, plants produce a few large bulbils out the side of the false stem close to the top foliage leaf rather than forming typical topsets on a coiled stalk, bulbs have heavy purple blotching and very fat round cloves, large-sized bulbs despite not removing the flower stalk, average of 8 cloves per bulb, occasional double cloves, color varies from light to dark-brown background overlaid with thin brown or red lines plus a dark red-brown to occasionally bright-pink blush, very firm bulbs are excel. for long storage. *Source History: 2 in 1994; 1 in 1998; 2 in 2004.* **Sources: Bi3, Fi6.**

Russian Rose - Purple striped, from Moscow market. *Source History: 1 in 2004.* **Sources: Gl4.**

S and H Silver - Softneck, large cloves, 15-20 per bulb, off white to tan with pink blush at tip, bottom half often brownish, large bulbs, mild and sweet taste builds in hotness and lingers, from S and H Organic Acres. *Source History: 1 in 1998; 1 in 2004.* **Sources: Fi6.**

Sakura - Unusual softneck from Japan, large white bulbs, 6-8 large tan cloves, outstanding flavor with lingering pleasant sweet heat. *Source History: 1 in 2004.* **Sources: Ga17.**

Sam Loiacono - (Loiacono) - Artichoke growth form, purple streaked, large cloves, 3-6 per bulb, excellent flavor, excellent keeper, peels easily, grown by Sam Loiacono and his family of Canandaigua, New York for over 40 years. *Source History: 2 in 2004.* **Sources: Ga21, Und.**

Samarkand - (Persian Star) - *A. longicuspis*, purple stripe, 6-9 cloves per bulb, 35-40 cloves/6-8 bulbs per pound, outer bulb wrapper sometimes smooth white, inner wrappers purple streaked, vivid clove colors, red-tipped cloves with marbled streaks on white or yellow-brown background, pleasant flavor with a mild spicy zing, late season, from Samarkand, Uzbekistan via John Swenson, late 1980s. *Source History: 1 in 1994; 2 in 1998; 13 in 2004.* **Sources: Bi3, Bo21, Ch15, Fi6, Ga17, Ga19, Gl4, Go6, Ho13, Sa5, Se16, Sw8, Ter.**

Shan Tung Purple - Chinese softneck, compact round bulbs, dark blue-purple skin, 6-8 large cream and rose-beige cloves, strong earthy flavor, hot when raw. *Source History: 3 in 2004.* **Sources: Ga17, Ga19, Se7.**

Shandong - Turban type, raw flavor is fiery hot, maintains good garlic flavor when baked, from Shandong Province, China. *Source History: 2 in 2004.* **Sources: Bi3, Fi6.**

Shatili - (Gatersleben #7108) - Hardneck, purple striped cloves, good bulb size, retains flavor when cooked, from Republic of Georgia. *Source History: 1 in 1994; 1 in 1998; 3 in 2004.* **Sources: Bi3, Fi6, Ga19.**

Shvelisi - Purple stripe, large magenta striped bulbs, 8-12 burgundy cloves per head, 7-10 bulbs per pound, medium heat, distinctive aroma, outstanding flavor, from Shvelisi in the Republic of Georgia. *Source History: 2 in 2004.* **Sources: Ga17, Se16.**

Siberian - Purple stripe var., very large bulbs due to weak flower stalks at U.S. latitudes, 7-8 fat, dark brown cloves, 35-40 per pound, mid-season maturity, secured by fishermen trading green leafy vegetables with poor peasants who grew only root crops. *Source History: 1 in 1998; 13 in 2004.* **Sources: Bi3, Bl7, Ch15, Fi6, Ga1, Ga17, Ga19, Gl4, Go6, Ri2, Se16, Sw8, Ter.**

Sicilian - Non-bolting softneck garlic, large white skinned bulbs, excel. flavor with moderate heat, abundant yields, long storage, vigorous, Sicilian heirloom. *Source History: 1 in 1994; 2 in 2004.* **Sources: Ga17, Ga19.**

Sicilian Artichoke - Good producer, long storage. *Source History: 2 in 2004.* **Sources: Bi3, Fi6.**

Sicilian Gold - Heirloom garlic from Sicily, robust flavor yet not too hot, slightly flattened head is protected by golden brown sheaths, nut-brown clove skins, 6-10 cloves per bulb, 40-65 per pound, mid-season. *Source History: 1 in 2004.* **Sources: Ter.**

Sicilian Silver - Not to be confused with Sicilian Artichoke, brown-red cloves, large bulbs if grown in rich soil, good character, raw flavor starts out mild and explodes. *Source History: 1 in 1994; 1 in 1998; 2 in 2004.* **Sources: Fi6, Ron.**

Silver Rose - Softneck, silverskin, one of the longest storing garlics, pink to rose striped clove skin color, small cloves on the inside are surrounded by 7-10 larger cloves, sharp garlic flavor, good for braiding, 12-15 cloves per head, 90-120 cloves per pound. *Source History: 2 in 1998; 3 in 2004.* **Sources: Bi3, Ga1, Jun.**

Silverskin - Beautiful large-cloved Silverskin, white bulb wrappers with faint brown lines that become heavier in each succeeding wrapper, 15-20 good-sized cloves usually in 5 clove layers, clove skins have white to tan background and varying amounts of pink blush especially at top and sides, bottom half of clove is often brownish to pink-brown, even outer cloves are tall and concave, strong distinct flavor, good yields, good keeper. *Source History: 1 in 1981; 1 in 1984; 1 in 1987; 5 in 1991; 2 in 1994; 3 in 1998; 4 in 2004.* **Sources: Bl7, Ga19, Ga21, We19.**

Silverwhite - Resembles Mild French, softneck, productive large bulbed garlic does well in both hot interior and humid maritime climates, California strain. *Source History: 1 in 1994; 1 in 1998; 5 in 2004.* **Sources: Ch15, Fi6, Ga19, Go6, Hi6.**

Simoneti - (Gatersleben #5878) - Extra large bulb size when planted in rich soil, good Artichoke character, 12 large cloves per bulb, cream with pink blush, moderate heat, very good quality, stores quite well, collected by Simons in Rep. of Georgia. *Source History: 1 in 1994; 1 in 1998; 3 in 2004.* **Sources: Fi6, Ga17, Go6.**

Single - Asiatic, softneck, from Chengdu Province, south-central China. *Source History: 2 in 2004.* **Sources: Bi3, Fi6.**

Skuri No. 2 - (Gatersleben #6038) - Hardneck, purple stripe, heavily striped bulbs, soft but striking clove colors, 8-12 cloves per bulb, raw flavor good, mild and earthy, from the Republic of Georgia. *Source History: 1 in 1994; 1 in 1998; 2 in 2004.* **Sources: Bi3, Fi6.**

Slovak - Rocambole from Central Europe, easy to peel large cloves, flavorful without too much heat. *Source History: 1 in 2004.* **Sources: Ga19.**

Sonoran - Turban var., softneck or hardneck, dependent on the weather, very large cream colored bulbs, 8-12 large pink and cream cloves, excel. flavor with moderate heat, stores quite well, very early maturing, originally from the state of Sonora in Mexico. *Source History: 2 in 2004.* **Sources: Ga17, Ga19.**

Spanish Benitee - Silverskin var. from Andalusia district of Spain. *Source History: 1 in 1998; 2 in 2004.* **Sources: Bi3, Fi6.**

Spanish Morado - (Morado de Pedronera) - Silverskin, excellent baked flavor, hot and strong raw, from Andalusia district of Spain. *Source History: 1 in 1998; 2 in 2004.* **Sources: Bi3, Fi6.**

Spanish Roja - (Spanish Red, Oregon Blue, Greek, Greek Blue) - Basically a gourmet garlic for the fresh market, bulb wrappers have light purplish blotches or streaks, clove color varies with soil and climate from teak-brown to darker brown almost obscured by red-purple blush, 6-13 cloves per bulb with 8-10 most common, 50-75 cloves per pound, doubled cloves common in large bulbs especially when grown in rich soil or mild winter climates, easy to peel, excel. flavor, 4-6 months storage if well grown but 2-3 months for many growers, mid-season maturity, brought to Portland by Greek immigrants from the East Coast before 1900, still called Greek or Green Blue by Northwest gardeners, often performs poorly in mild winter climates.

Source History: 1 in 1981; 1 in 1984; 1 in 1987; 4 in 1991; 8 in 1994; 12 in 1998; 20 in 2004. **Sources: Bi3, Bo20, Bo21, Ch15, Fi6, Ga1, Ga19, Ga21, Gl4, Go6, Goo, Gur, Ho13, Ma13, Ni1, Par, Pe2, Se7, So1, Ter.**

St. Helen's - Silverskin, subtle nutty flavor when baked, hot raw, heirloom from western WA. *Source History: 1 in 1998; 1 in 2004.* **Sources: Fi6.**

Summit Roja - Hardneck, strain of Spanish Roja with larger, flatter bulbs up to 3 in. dia., 8-12 large cloves, purple streaked bulbs, good Roja flavor, very productive, found in Summit Valley, WA. *Source History: 1 in 2004.* **Sources: Ga17.**

Susan Delafield's - Hardneck with consistently large bulbs, 8 cloves per bulb, white bulb wrapper, clove skin color is light pink, strong fruity flavor, good keeping quality, texture is creamy, oily and crunchy, one of the first to head. *Source History: 1 in 1994; 1 in 1998; 2 in 2004.* **Sources: Bo21, Sa5.**

Susanville - California Early sel. that is bigger and stores longer, mostly four clove layers with 11-15 cloves and not many small cloves, 55-75 cloves per pound, clove color is off-white or yellowish with some pink blush, excel. flavor, mid-season maturity, great keeper, excel. for roasting, popular commercial strain in California but widely adapted. *Source History: 1 in 1981; 1 in 1984; 1 in 1987; 3 in 1991; 2 in 1994; 1 in 1998; 5 in 2004.* **Sources: Bi3, Fi6, Ga19, Gl4, Ter.**

Sweet Haven - Softneck, uniform cloves, pungent flavor. *Source History: 2 in 2004.* **Sources: Bo21, Sa5.**

Tai Cang - Unique Asiatic variety, white wrappers, 3-6 light purple cloves, from National Seed Corporation of China via G. Czarnecki. *Source History: 2 in 2004.* **Sources: Bi3, Fi6.**

Temptress - (GSF #65) - Rocambole, rather large bulbs, 6-7 cloves, pungent flavor mellows out. *Source History: 1 in 2004.* **Sources: Ga19.**

Thai - Hardneck, skin color of cloves is brown-pink, mild, sweet and musky flavor is strong, then mellow, good keeping quality, not crunchy, 9 cloves per bulb. *Source History: 1 in 1998; 1 in 2004.* **Sources: Bo21.**

Thermadrone - Artichoke type, large bulbs, long storage, commercial strain from France. *Source History: 2 in 2004.* **Sources: Bi3, Fi6.**

Tibet - Hardneck, attractive bulbs with good strong taste, growth appears late, matures late. *Source History: 1 in 2004.* **Sources: Sa5.**

Tipatilla - (Tipatilla Silverskin) - Old mid-season variety, great medium-hot flavor, high oil content, best keeping variety, stores until the next harvest. *Source History: 1 in 1991; 1 in 2004.* **Sources: Se7.**

Transylvanian - Artichoke, hot raw flavor, white bulbs, 10-16 beige cloves per bulb, from Chester Aaron, bought in Transylvania by Prof. Feur of Brandeis University. *Source History: 3 in 2004.* **Sources: Fi6, Ga17, Ga19.**

Trueheart - Unique Artichoke usually forms a solid center rather than many small cloves, from Ron Bennett, NY. *Source History: 1 in 2004.* **Sources: Fi6.**

Tzan - Turban, very hot, collected by John Swenson from Shandong Province via Mexico. *Source History: 2 in 2004.* **Sources: Bi3, Fi6.**

Udabmo - Softneck, Artichoke type, tightly wrapped cream colored cloves, 18-20 per bulb, collected in the mountains of the Republic of Georgia in 1988, arriving in the U.S. via the Gatersleben seed bank. *Source History: 1 in 2004.* **Sources: Rev.**

Ukrainian Hot - Large Roja type, hot. *Source History: 1 in 2004.* **Sources: Sa5.**

Ukrainian Topset - Hardneck, rocambole type. *Source History: 1 in 2004.* **Sources: Bi3.**

Uzbek Turban - Turban type, from the former USSR Allium Station via R. Hannan. *Source History: 2 in 2004.* **Sources: Bi3, Fi6.**

Vekak - Purple stripe, glazed, very productive garlic, large purple striped white bulbs, 10-12 large violet cloves, cooking enhances the flavor, from the Czech collection of Dr. Boris Andrst, a New York grower. *Source History: 3 in 2004.* **Sources: Bi3, Fi6, Ga17.**

Vilnius - Good flavor, originally from market in Vilnius, Lithuania. *Source History: 1 in 2004.* **Sources: Ho13.**

Vladivostok II - Purple stripe, large colorful purple bulb, 6-8 cloves. *Source History: 1 in 2004.* **Sources: Bi3.**

Vostani - Vigorous Porcelain, old variety from near the Washington-British Columbia border. *Source History: 2 in 2004.* **Sources: Bi3, Fi6.**

Walla Walla Early - An Italian Purple var., softneck, pink-red cloves, mild, rich flavor, 12-20 cloves per bulb, braiding type. *Source History: 2 in 1998; 2 in 2004.* **Sources: Fi1, Ho13.**

Webber Greek - Hardneck, from Ellensburg, WA. *Source History: 1 in 1998; 1 in 2004.* **Sources: Fi6.**

Wedam - Hot, spicy raw flavor, sweet, smooth and mild when baked. *Source History: 1 in 2004.* **Sources: Fi6.**

Western Rose - Larger relative of Silver Rose, white outer bulb covering, clove wrappers striped with pink and rose, smaller inner cloves are surrounded by 7-10 large cloves, totaling 12-15 cloves per head, 90-120 cloves per pound, sharp garlic flavor, good braider, long storage, late season maturity. *Source History: 1 in 2004.* **Sources: Ter.**

White Rose - *Source History: 1 in 2004.* **Sources: Gl4.**

Wild Buff - (Silk Road) - *A. longicuspis*, porcelain type, few cloves with little striping, buff brown color, nice earthy flavor, from USDA, Beltsville, MD. *Source History: 1 in 1994; 1 in 1998; 3 in 2004.* **Sources: Bi3, Fi6, Ho13.**

Wild Rocambole - (R/AL/159) - *A. longicuspis*, typical light purple blotching on bulb, light brown cloves, does not look wild despite its name, from Poland via Dr. P.W. Simon, 1986. *Source History: 1 in 1994; 1 in 1998; 1 in 2004.* **Sources: Fi6.**

Wildfire - Hardneck, Porcelain Group, large bulbs, 5-8 cloves, strong, hot flavor. *Source History: 2 in 2004.* **Sources: Bi3, Ga1.**

Wonha - Baked flavor is nutty and smooth, raw flavor starts easy with a strong bite that lingers, flavorful aftertaste, from North Korea via John Swenson. *Source History: 2 in 2004.* **Sources: Bi3, Fi6.**

Xi'an - (Xian) - Large flattened bulbs with few large

cloves, red and brown striped wrappers, 8-12 large dark purple cloves, unique rich flavor with a little heat in the middle, a favorite of Chester Aaron who bought this var. from a garlic smuggler in San Francisco's Chinatown, from Beijing, China. *Source History: 5 in 2004.* **Sources: Bi3, Bo20, Fi6, Ga17, Ho13.**

Yampolskij - Hardneck from Ukraine via Czechoslovakia. *Source History: 1 in 1998; 3 in 2004.* **Sources: Bi3, Fi6, Gl4.**

Yugoslavian - *Allium sativum* var. *ophioscorodon*, vigorous plant with double cloves common, deep-green leaves, large to very large bulbs have copper veins with some purple blotching, solid dark-brown clove color, 9-14 cloves per bulb, good strong flavor, stores well, original stock from Dacha Barinka, British Columbia. *Source History: 2 in 1991; 3 in 1994; 4 in 1998; 7 in 2004.* **Sources: Ba8, Bo21, Ch15, Fi6, Ga21, Gl4, Ho13.**

Zaharada - Hardneck/Porcelain, from the Czech collection of Dr. Boris Andrst, New York. *Source History: 2 in 2004.* **Sources: Bi3, Fi6.**

Zahorsky - From the Czech collection of Dr. Boris Andrst, NY. *Source History: 1 in 2004.* **Sources: Fi6.**

Zemo - (Gatersleben #6307) - Hardneck, porcelain type, large bulbs, few cloves which are well streaked on back and soft brown-pink lines on inside, 25-35 per pound, strong raw taste is pleasant with no aftertaste, late season maturity, from Republic of Georgia. *Source History: 1 in 1994; 1 in 1998; 6 in 2004.* **Sources: Bi3, Fi6, Ga19, Go6, Goo, Ter.**

Varieties dropped since 1981 (and the year dropped):

Ail Rose de Lautrec (2004), *Alison's* (2004), *Arguni* (2004), *Arivaca* (1994), *Atkin's Russian* (2004), *Breathless* (2004), *Chrysalis Rose* (2004), *Colorado Purple, Easy Peel* (2004), *Denman Island* (2004), *Dixon Strain* (2004), *Early Spanish Red* (1994), *F 5* (2004), *F 6* (2004), *F 7* (2004), *F 8* (2004), *F 9* (2004), *F11* (2004), *F12* (2004), *F20* (2004), *F23* (2004), *Fauquier* (1998), *Fish Lake* (1998), *Fish Lake 23* (2004), *French* (2004), *Giant Cajun* (1987), *Giant Imported* (1987), *Gilroy* (2004), *Gomecari* (1994), *Hardstalk Italian* (1987), *Himalayan* (1994), *Hornby* (2004), *Jumbo* (1994), *L.K. Mann's C751* (2004), *Lerg* (1994), *Limburg* (2004), *McDougall's* (2004), *Mexican* (1994), *Nichols Top Setting* (1998), *Ontario* (2004), *Printanor* (1991), *Purple* (2004), *Purple Artichoke* (1998), *Purple Max* (2004), *Purple Rocambole* (2004), *Purple Tip* (2004), *Racey* (2004), *Randl Colorado White* (2004), *Red Revel* (2004), *Salt Spring* (2004), *Salt Spring Select* (2004), *Silverskin Select* (1994), *Silverskin, Masha's Select* (1998), *Spanish* (2004), *Specialty Garlic Cultivar List* (2004), *Stein Mountain* (2004), *Tipitilla* (2004), *Ukrainian Danielo* (2004), *Ukrainian Mavniv* (2004), *Western Rojo* (2004), *White Max* (1998).

Gourd
Lagenaria siceraria

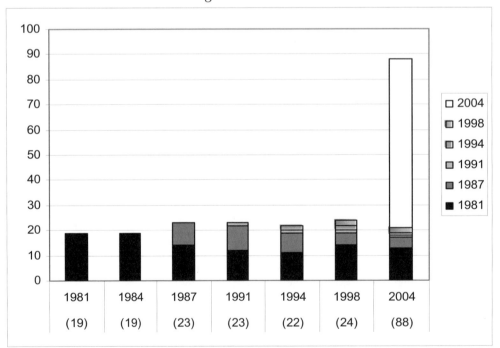

Number of varieties listed in 1981 and still available in 2004: 19 to 13 (68%).
Overall change in number of varieties from 1981 to 2004: 19 to 88 (463%).
Number of 2004 varieties available from 1 or 2 sources: 51 out of 88 (58%).

Acoma Rattle - 100 days - Very flat, bowl shaped gourd, many the size of a dinner plate. *Source History: 1 in 2004.* **Sources: Sa9.**

African Kettle - 120 days - Large gourd averages 8 in. tall by 16 in. diameter. *Source History: 3 in 2004.* **Sources: Go9, Ra5, Sa9.**

African Water Bottle, Giant - 140 days - Large gourd with skinny top, grows 2 ft. long, 9 in. diameter at the base. *Source History: 3 in 2004.* **Sources: Ra5, Ra6, Sa9.**

Apple - (Big Apple, Speckled Apple) - 110 days - Unusual apple shaped gourds with mottled green color, 7 in. tall, 6-8 in. diameter, varying sizes and shapes, 4-7.5 lbs., mottled green when fresh, brown dried color, popular for crafting. *Source History: 36 in 2004.* **Sources: Ba8, Bo19, Bu2, CH14, CO30, Fe5, Go9, Ha5, HO2, Jo1, Jor, Jun, Ma19, ME9, Mel, Ni1, OSB, Pep, Pin, Ra5, Ra6, Rev, Ri12, RIS, Roh, RUP, SE4, SE14, Se16, Se26, Sh9, Sh13, Ter, Und, Vi5, WE10.**

Bali Sugar Trough - 105 days - Oblong shaped gourd grows 18-20 in. long by 8 in. diameter, for long bowls or bottles. *Source History: 2 in 2004.* **Sources: Ho13, Sa9.**

Banana - Grows to 10 in. long, banana shape. *Source History: 2 in 2004.* **Sources: Ra5, Ra6.**

Basket - 120-130 days - Smaller version of Bushel with green weight around 20 pounds, curing time up to one year, fruit set may not occur until 65 days after germ., does best in areas with hot days and warm nights, long trailing vines, low yields, hardy, insect res., green rind cures to buff-tan or light brown color, Hollar introduction. *Source History: 7 in 2004.* **Sources: Bu8, HO2, Ra5, Ra6, RUP, Vi5, Wi2.**

Big African Wine Kettle - Vines can grow 40-60 ft. long, huge kettle gourd with slight taper toward stem end, 18-24 in. diameter. *Source History: 2 in 2004.* **Sources: ME9, Ra6.**

Birdhouse - (Sugar Trough) - 95-110 days - Grown for making inexpensive bird houses, bell shape, smooth hard shell, 10-12 in. wide, 14 in. tall, pale green ripening to straw-beige, use of this type of gourd by the Egyptians dates between 3000 and 4000 B.C. *Source History: 86 in 2004.* **Sources: All, Ba8, Be4, Bo17, Bo19, Bou, Bu1, Bu2, Bu3, Bu8, CA25, CH14, Cl3, Co31, Com, Coo, Dam, De6, DOR, Ers, Fa1, Fi1, Ga1, Go9, Gur, HA3, He8, He17, Hi13, HO1, HO2, Ho8, Ho13, Jo1, Jor, Jun, La1, Lan, LO8, Loc, Ma19, MAY, ME9, Mel, Mey, Mo13, MOU, Ni1, OLD, Par, Pep, Pin, Pr9, Ra5, Ra6, Ri12, RIS, Roh, RUP, Scf, SE4, Se7, Se8, SE14, Se16, Se17, Se26, Se28, Sh9, Shu, Si5, So1, SOU, Sto, Ta5, Ter, Tt2, Twi, Ver, Ves, Vi4, Vi5, WE10, Wet, Wi2, WI23.**

Bottle, Black Seeded - 115 days - Bottle shaped gourd grows 6-8 in. long. *Source History: 1 in 2004.* **Sources: Sa9.**

Bottle, Chinese - 100 days - Small pear shaped gourd, 5 in. tall by 4 in. across, good crafting gourd, flat bottom sits well. *Source History: 2 in 2004.* **Sources: HO2, Sa9.**

Bottle, Dudhi - (Opo Club) - 85-95 days - Prolific, spreading vines bear edible, 24-36 in. long, bat shaped fruit, mild refreshing, juicy flesh, used like summer squash at 8 in. length. *Source History: 2 in 2004.* **Sources: ME9, So12.**

Bottle, Indian - Showy vines with 12 in. leaves, white flowers open in the evening, young gourds are edible. *Source History: 1 in 2004.* **Sources: Ban.**

Bottle, Large - 120 days - Bottle shaped fruit, 8-14 in. tall, 5 in. wide, can be used for birdhouses, pre-1850. *Source History: 15 in 2004.* **Sources: CO30, Ech, Ev2, Fe5, Go9, Hi13, Lan, Na6, Ni1, Pe2, Ra5, RUP, Shu, Sto, Twi.**

Bottle, Mexican - Grows to 12+ in. tall, bottle shape with large bulb at the bottom, nipple on the top of the smaller bulb. *Source History: 2 in 2004.* **Sources: Ra5, Ra6.**

Bottle, Mini - 110-125 days - Miniature version of the birdhouse gourd, 7-8 in. tall, 2 in. dia. at widest point, shell dries to tan with mottled natural brown spots, curing requires several months. *Source History: 26 in 2004.* **Sources: Bo19, CA25, CO30, Eo2, Ers, Ev2, Go9, HO2, Jo1, Jor, ME9, Mel, Ni1, OSB, Pep, Pin, Ra5, Ra6, RIS, Roh, RUP, SE4, SE14, Se26, Sto, Vi5.**

Bottleneck - 100-120 days - Long vines with velvety leaves, white flowers, large pale gourd with long narrow neck, 10 in. across at base, edible when harvested young. *Source History: 1 in 2004.* **Sources: We19.**

Bowl - 110-120 days - Round gourd, flattened top and bottom, can be divided to make two bowls, 8-10 in. wide by 4-5 in. tall. *Source History: 5 in 2004.* **Sources: Pe2, Ra5, Ra6, Ri12, Vi5.**

Bule - (Hardshell Wartie, Peanuts) - 100-120 days - Unique gourd for drying, vigorous vines to 15-35 ft., shape resembles a large apple, 6-8 in. tall, 5-6 in. diameter, hard shell is covered with small warts or tooth-like bumps, France. *Source History: 3 in 2004.* **Sources: ME9, Ri12, Se16.**

Bule, Tarahumara Small - *Source History: 1 in 2004.* **Sources: Se17.**

Bushel - 110-130 days - Huge slate-gray gourd measures 3-5 ft. in diameter, weighs 25-60 pounds, dries to a light weight, used in crafting and as a container, long vines with evening-blooming white flowers that attract moths. *Source History: 39 in 2004.* **Sources: Ba8, Bo19, Bu1, Bu3, Co31, Ech, Ers, Fa1, Fe5, Fi1, Go9, Gur, He17, Hi13, HO2, Ho13, Hud, Jor, La1, Loc, ME9, Mel, Mo13, Ni1, Pep, Ra5, Ra6, Rev, Ri12, RIS, RUP, SE14, Shu, So1, So12, Sto, Vi4, Vi5, WE10.**

Calabash, Long - (Calabash Extra Long) - *Lagenaria siceraria*, Japanese edible gourd, grows up to 36 in. long x 6 in. dia., young fruits are cooked, Dr. Yoo says that sometimes the young plant is used as a rootstock for watermelon. *Source History: 2 in 1981; 2 in 1984; 2 in 1987; 1 in 1991; 1 in 2004.* **Sources: Kit.**

Calabash, Med-Long - Japanese edible gourd, 15 in. long x 4 in. dia., yellow-green skin, white flesh, vigorous grower, heavy producer, can be grown on ground like a pumpkin or along a trellis, used for cooking and soup when young. *Source History: 2 in 1981; 1 in 1984; 2 in 1987; 1 in 1991; 1 in 1994; 1 in 1998; 1 in 2004.* **Sources: Ev2.**

Calabash, Powderhorn - 120 days - Smooth tan gourd with a short curved neck at stem end, 12-15 in. long x 5 in. dia., used as powderhorn in the days of Davie Crockett and for spoons, dippers and containers before that. *Source*

History: 1 in 1987; 1 in 1991; 1 in 1994; 1 in 1998; 5 in 2004. **Sources: Ers, Ra5, Ra6, Vi5, Wi2.**

Calabash, Round - (Penguin) - 120 days - Large bottle or powderhorn type used for making water jugs and crafting, short curved neck at stem end, shaped like a penguin, smooth tan skin, known to grow up to 12 in. dia., 45 lbs., edible when young, easy to grow. *Source History: 2 in 1981; 2 in 1984; 4 in 1987; 3 in 1991; 6 in 1994; 6 in 1998; 32 in 2004.* **Sources: Ba8, Bo19, CA25, Cl3, CO30, Co31, Ev2, Go9, Gur, HO1, HO2, Ho8, Jor, Kit, La1, ME9, Ni1, Par, Pep, Ra5, Ra6, RIS, Roh, RUP, Sa9, SE14, Shu, Sto, Ver, Vi4, Vi5, WE10.**

Cannon Ball - Size and shape of a cannon ball. *Source History: 2 in 2004.* **Sources: Ra5, Ra6.**

Caveman's Club - 120-125 days - Vines to 8 ft., 4-8 gourds per plant, oblong shaped, ridged bulb with long narrow handle, 6 x 11 in. long. *Source History: 11 in 2004.* **Sources: Bo19, Go9, Ha5, ME9, OSB, Ra5, Ra6, Se26, Sto, Ter, Vi5.**

Chinese Fuzzy Gourd - (Winter Melon, Wax Melon) - 85 days - *Benincasa hispida*, derived from a wax gourd, squash-like vegetable gourd, sweet and succulent, grows best in hot weather and rich soil, smaller but heavier yielder. *Source History: 1 in 1981; 1 in 1984; 1 in 1987; 4 in 1991; 5 in 1994; 6 in 1998; 2 in 2004.* **Sources: Ho13, Loc.**

Chinese Wintermelon - (Ash Gourd, Wax Gourd, Chinese Preserving Melon, Mao Gwa) - 120-150 days - *Benincasa hispida*, large green fruit up to 35 lbs., becomes coated with a white waxy substance as it matures, white flesh, tender sweet and succulent, grows best in hot rich well drained soil, stays fresh up to 1 year if properly cured and stored, good in soups, steamed or stir-fried, introduced to the U.S. in 1884 by East Asian immigrants. *Source History: 1 in 1987; 2 in 1991; 2 in 1994; 5 in 1998; 7 in 2004.* **Sources: Com, Ech, Ho8, Ho13, Kit, Loc, Rev.**

Club - Large club shaped gourd to 3 ft. long, some off-types. *Source History: 1 in 2004.* **Sources: Sa9.**

Cork and Twine - 200 days - Bottle shaped gourd suitable for use as a canteen. *Source History: 1 in 2004.* **Sources: Sa9.**

Corsican Flat - (Corsican Bowl, Canteen Gourd, Kalimba Gourd, Sugar Gourd) - 110-130 days - Long vines bloom at night, medium green, flat canteen shaped gourd, up to 12 in. diameter, sometimes has a knob at stem end, medium green when mature, dries to tan, crafted into bowls, CV Sol intro., 1992. *Source History: 22 in 2004.* **Sources: Bo17, Bo19, Fe5, Ga21, Go9, HO1, HO2, Ho8, Jo1, ME9, Ni1, Pep, Ra5, Ra6, Roh, Sa9, Se7, Se17, Se26, Sol, Vi5, WE10.**

Corsican Kettle - *Source History: 1 in 2004.* **Sources: Ra6.**

Cucuzzi - (Italian Edible Gourd, New Guinea Butter Bean or Vine, Hercules Club, Dhudy, Indian Squash) - 55-80 days - Used like summer squash when young, pick when 6 in. long x 1 in. dia., grows up to 48 in. long, smooth pale-green skin, creamy-yellow flesh, hard-shelled gourd with beautiful evening-blooming white flowers. *Source History: 11 in 1981; 10 in 1984; 8 in 1987; 12 in 1991; 12 in 1994; 12 in 1998; 19 in 2004.* **Sources: Ba8, Bu3, CH14, Co31, Ech, Ge2, Hi13, Jor, Kil, Lej, ME9, PAG, Pep, SE4, SE14, Se26, Sto, Vi4, WE10.**

Cucuzzi Caravazzi - 60-65 days - Edible gourd, extra long, light-green, best when grown on a trellis, grows up to 6 ft. in length, used like summer squash when up to 2 ft. long, unique flavor. *Source History: 1 in 1981; 1 in 1984; 1 in 1987; 2 in 1991; 2 in 1994; 5 in 1998; 4 in 2004.* **Sources: Com, HA3, Ho13, Hud.**

Dinosaur Gourd - 115-125 days - Club-shaped spiked gourd, 12-15 in. long, traditionally used to decorate sweat lodges. *Source History: 1 in 1987; 1 in 1991; 2 in 1994; 4 in 1998; 3 in 2004.* **Sources: Ri12, Sa9, Se16.**

Dipper - 120-125 days - Long handle to 2 ft., lower bulbed part makes the dipper. *Source History: 36 in 2004.* **Sources: Bo17, Bo19, Bu8, CA25, CH14, Ers, Go9, Gur, He8, Hi13, HO1, Ho8, Jo1, Jor, La1, Lan, ME9, Mel, Ni1, Par, Ra5, Ra6, Ri12, RIS, Roh, RUP, SE4, SE14, Sh9, Si5, Vi4, Vi5, WE10, Wet, Wi2, WI23.**

Dipper, Apache - Neck handle up to 12 in. long, collected at the San Carlos Reservation in Peridot, AZ. *Source History: 3 in 2004.* **Sources: Na2, Ra5, Ra6.**

Dipper, Extra Long Handled - 120 days - Handle grows 3-4 ft. long with a 5-8 in. bowl, can be used for dippers and crafting, grow on a trellis for straighter, longer handles. *Source History: 19 in 2004.* **Sources: Ba8, Bu3, Cl3, Co31, Com, He17, HO2, Loc, Mo13, OLD, OSB, Pep, Pin, RUP, Sa9, Se8, SE14, Sh13, Shu.**

Dipper, Mini - (Kapok) - 100 days - Popular variety, averages 4 in. long, great for crafting. *Source History: 2 in 2004.* **Sources: Ra6, Sa9.**

Dipper, Short Handled - Thick shelled gourd, deep green skin, 5 in. bowl, 14 in. handle. *Source History: 2 in 2004.* **Sources: Sa9, Se17.**

Dipper, Short Handled Mottled - Thick shelled gourd, green skin mottled with white, 5 in. bowl, 16-18 in. handle. *Source History: 1 in 2004.* **Sources: Sa9.**

Dipper, South Sea Island - 120 days - Grows 16-18 in. long with the dipper section measuring 4-5 inches in diameter. *Source History: 1 in 2004.* **Sources: Sa9.**

Disc Nantes - Hard shelled gourds, used by the Cherokee and Choctaw Indians as a soap dish or for containers. *Source History: 1 in 2004.* **Sources: Sa9.**

Doan Gwa - (Dongua, Tung Gua, Ton Kwa, Tougan, Wax Gourd, White Gourd, Banquet Dish, Chinese Winter Melon) - 120- 150 days - *Benincasa hispida*, Chinese winter melon, 6-12 in. round gourd, sweet tender white flesh, used in soups and as a soup bowl, needs hot soil, keeps well. *Source History: 8 in 1981; 7 in 1984; 6 in 1987; 6 in 1991; 4 in 1994; 2 in 1998; 3 in 2004.* **Sources: Ba8, Ni1, Ri2.**

Gakhaa - 185 days - Large African trough gourd, 18-20 in. long by 7 in. wide at blossom end, used for storage and for fermenting beverages. *Source History: 1 in 2004.* **Sources: Sa9.**

Hawaiian Dance Mask - 140 days - Shaped like a large pear with flat bottom, 10 in. tall, 8 in. diameter at the base. *Source History: 1 in 2004.* **Sources: Sa9.**

Healing Squash - Actually an edible gourd, rampant

grower benefits from trellising, when young fruit is 15-18 in. long, half the fruit can be cut off for eating like squash, while the remainder of the fruit will heal itself and continue to grow, hence the name, mature fruit may reach 4 ft. long, maintained by a family in Italy for almost 200 years. *Source History: 1 in 1998; 1 in 2004.* **Sources:** Scf.

Hopi Rattle - 100-120 days - Grows 4-6 in. long, resembles Acoma Rattle but with a scar on the bottom, used as a ceremonial rattle by the Hopi Indians for centuries, harvest at maturity, seeds will rattle inside when dry. *Source History: 8 in 2004.* **Sources:** Na2, Pe2, Ra5, Ra6, Roh, Sa9, Se7, Se17.

Italian Edible Gourd - (Cucuzzi Long Green, Italian Simeza Teneroni) - 100 days - Grows straight on trellis, edible when small, grows to 48 x 5 in., 15 lbs., hard-shelled at maturity, rampant vines can cover small garden. *Source History: 8 in 1981; 8 in 1984; 8 in 1987; 8 in 1991; 8 in 1994; 10 in 1998; 10 in 2004.* **Sources:** HO2, Jor, Loc, Ra5, Ra6, RIS, RUP, Sa9, SE4, Vi5.

Japanese Long Gourd - 150 days - Green striped skin, grows 24 in. long, 6 in. wide at base, makes an excellent bowl when cut lengthwise. *Source History: 1 in 2004.* **Sources:** Sa9.

Japanese Round - 210 days - Bushel basket type, 18 in. tall by 12-14 in. diameter. *Source History: 1 in 2004.* **Sources:** Sa9.

Lagenaria longissima - 65-97 days - Italian edible gourd, used like summer squash if picked half ripe, does best if staked, Italians say they can build a banquet around a *Lagenaria*. *Source History: 1 in 1981; 1 in 1984; 2 in 1987; 2 in 1991; 2 in 1994; 2 in 1998; 2 in 2004.* **Sources:** Ni1, Sa9.

Legendary Bottle - Large African bottle gourd 15-20 in. long, 10 in. wide at the base, 5 in. wide at the neck, very thick shell, needs leaching for use as a liquid container. *Source History: 3 in 2004.* **Sources:** Hud, Ra6, Sa9.

Little Man - 120 days - Slightly warted round gourd, 4-6 in. diameter. *Source History: 1 in 2004.* **Sources:** Sa9.

Lump in the Neck Bottle Gourd - 115 days - Rounded at the base with a bulbous top. *Source History: 3 in 2004.* **Sources:** Ra5, Ra6, Sa9.

Mao Gwa - (Chinese Fuzzy Gourd, Fuzzy Squash, Small White Gourd, Mao Qua) - 80-100 days - *Benincasa hispida*, pale-green squash-like vegetable gourd, slightly pear-shaped, light-green skin covered with fuzz like a peach, harvest immature fruits at 6 in. long, vines grow best in hot weather and rich soil. *Source History: 9 in 1981; 7 in 1984; 6 in 1987; 6 in 1991; 5 in 1994; 4 in 1998; 4 in 2004.* **Sources:** Ban, Ev2, Sa9, Sto.

Maranka - (Cave Man's Club, Dinosaur Gourd, Swan, Dolphin, Turtle) - 125-130 days - Dark green, 10-16 in. long, 7-14 lbs. shiny fruits, hard shell has knobby protrusions, curved neck tapers at the stem end, will grow straight if supported, known as Dudi in India and Opo in Oriental markets, edible at 8 in. length. *Source History: 1 in 1981; 1 in 1984; 1 in 1987; 1 in 1991; 2 in 1998; 27 in 2004.* **Sources:** Ba8, CA25, Com, Ers, Fe5, He17, HO2, Jo1, Jor, Jun, La1, Mo13, Ni1, Pep, Ra5, Ra6, Rev, RIS, Roh, RUP, SE4, Se7, SE14, Sto, Twi, Vi5, WE10.

Martinhouse - 90-110 days - Purified strain of Birdhouse gourd with slightly longer neck, smooth, tan hard shell, 10 x 15 in. tall, trellis vines for a straight neck. *Source History: 12 in 2004.* **Sources:** Ers, Go9, He8, HO2, ME9, Mo13, OSB, Ra5, Ra6, RUP, SE14, Vi5.

Mauritius Warted - Oblong, 10 in. long gourd is covered with warts, edible when small, dried gourd makes a good guiro, from Mauritius Island in the Indian Ocean. *Source History: 1 in 2004.* **Sources:** Ho13.

Mayo Bilobal - 125 days - Medium size gourd, thick shell, bilobal shape, small upper chamber, from Mayo River, Sonora, Mexico. *Source History: 2 in 2004.* **Sources:** Na2, Sa9.

Mayo Bule, Giant - 140 days - Shape resembles a large straightneck summer squash with slight crook at the top, 17 in. long, 7 in. diameter at the base, more narrow at stem end. *Source History: 1 in 2004.* **Sources:** Sa9.

Mayo Deer Dance Rattle - Fruit shape varies from tear drop to short handled dipper, rattles used for the Deer Dance. *Source History: 1 in 2004.* **Sources:** Na2.

Mayo Gooseneck - 125 days -ND 24 in. long, looks like a goose neck and used for a hunting decoy, will grow straight if grown on a trellis. *Source History: 1 in 2004.* **Sources:** Sa9.

Mayo Warty Bule - 200 days - Covered with knobby growths or warts, 12 in. tall, 10 in. bowls, used for canteens or water jugs, grown in Piedras Verdes, Sonora, Mexico. *Source History: 2 in 2004.* **Sources:** Na2, Sa9.

New Guinea Butter Vine - (Cucuzzi) - 100-115 days - Squash-like long curly fruits, used like summer squash when immature, can be trellised like squash, easily grows to huge size. *Source History: 2 in 1981; 2 in 1984; 2 in 1987; 2 in 1991; 2 in 1994; 1 in 1998; 1 in 2004.* **Sources:** Bu1.

Nigerian Saybo - 110 days - Large bottle gourd with bulbous neck, pointed end. *Source History: 2 in 2004.* **Sources:** Sa9, Se17.

O'odham Dipper - Length varies from 8 to 18 inches long. *Source History: 1 in 2004.* **Sources:** Na2.

Peru Sugar Bowl - 120 days - Gourds measure 6 in. tall by 6 in. wide. *Source History: 1 in 2004.* **Sources:** Sa9.

Peyote Ceremonial - Dipper gourd, 2-4 in. long, 2 lobes, used for crafts and as rattles by Native American Church. *Source History: 3 in 2004.* **Sources:** Na2, Se17, Sh13.

Rattle - 125 days - Prolific vines bear 3-5 in. fruits, ceremonial gourd. *Source History: 1 in 2004.* **Sources:** So12.

San Felipe Pueblo Rattle - Flattened round gourds with some bottle and canteen variations, from San Felipe Pueblo on the Rio Grande. *Source History: 1 in 2004.* **Sources:** Na2.

Siphon - (Japanese Siphon, Syphon) - 110-140 days - Very long night blooming vines, medium green, 5-6 in. round bulb, 2.5 in. neck, will grow straight if trellised, used to siphon wine from barrels. *Source History: 10 in 2004.* **Sources:** Fe5, ME9, Ni1, Pep, Ra5, Ra6, Sa9, Sto, Vi5, WE10.

Snake, New - 180 days - Thin shelled gourd grows 18

in. long by 3 in. diameter, used for rattles, rain sticks and fake snakes. *Source History: 1 in 2004.* **Sources: Sa9.**

Speckled Swan - (Goose Neck) - 90-130 days - Smooth dark green gourd with white speckles, long neck resembles a swan's neck and head, 12-18 in. tall, base measures 6-8 in. diameter, 9-15 lbs., great for crafting, unique. *Source History: 55 in 2004.* **Sources: Ba8, Bo19, Bu2, CA25, CH14, Com, Coo, De6, Dom, Ear, Ers, Fe5, Go9, Ha5, He8, He17, HO1, HO2, Ho13, Hud, Jo1, Jor, Jun, La1, Ma19, ME9, Mel, Mo13, Ni1, OSB, Par, Pe2, Pep, Pin, Ra5, Ra6, Re6, Rev, RIS, Roh, RUP, SE4, SE14, Se26, Sh9, Shu, Sto, TE7, Ter, Twi, Ver, Vi5, WE10, Wi2, WI23.**

SSE Snake - 200 days - Edible when small, grows to 20 in. long by 3 in. wide at blossom end, thin shell is sturdy, good for rattles and snake scare. *Source History: 1 in 2004.* **Sources: Sa9.**

Tarahumara Canteen - 180 days - Quart size, 10 in. tall by 6 in. diameter, thick shelled, slightly pear shaped gourds still used by the Indians of the Copper Canyon in Chihuahua, Mexico. *Source History: 3 in 2004.* **Sources: Ho13, Na2, Sa9.**

Tobacco Box - (Sugar Bowl Gourd) - 200 days - Neckless gourd, 6 in. tall by 3 in. diameter, resembles Corsican Flat but with a high top. *Source History: 5 in 2004.* **Sources: Hud, Ra5, Ra6, Sa9, Vi5.**

Tong Qwa Oblong - 120 days - *Benincasa hispida*, very large 30 x 18 in. gourd, can weigh up to 50 lbs., thick white flesh used for cooking or making Winter Melon Soup, a popular soup found in most Chinese restaurants. *Source History: 1 in 1987; 1 in 1991; 1 in 1994; 1 in 1998; 1 in 2004.* **Sources: Ev2.**

Tong Qwa Round - *Benincasa hispida*, medium size wax gourd, round shape, 15-20 lbs., very productive, popular in Japan and in Oriental supermarkets on the West Coast. *Source History: 1 in 1994; 1 in 1998; 1 in 2004.* **Sources: Ev2.**

Volleyball - *Source History: 1 in 2004.* **Sources: Se17.**

Water Jug - 120 days - Grows to 22 in. long, 9 in. diameter bowl and 2 in. across at stem end. *Source History: 5 in 2004.* **Sources: Ra5, Ra6, Sa9, Se17, Vi5.**

Water Ladle, Giant - 125 days - Grows larger and faster than most gourds of this type, 10-12 in. fruits. *Source History: 5 in 2004.* **Sources: Eo2, Gr29, Re8, Se28, Wo8.**

Wax Gourd - (Winter Melon, Tong Gwa, Doan Gwa, White Gourd) - *Benincasa hispida*, large very round fruits, grown for its thick tender flesh, white powder in the rind indicates maturity, used for Chinese style cooking. *Source History: 2 in 1991; 7 in 1994; 4 in 1998; 5 in 2004.* **Sources: CO30, Jor, Kit, RUP, WE10.**

Wax Gourd, Oblong - Thick fleshed gourd used for soup, 12 by 24 in. long, up to 20 lbs. *Source History: 1 in 1998; 1 in 2004.* **Sources: Jor.**

Zaire Round - Round, flat green gourd, 14 in. wide, Africa. *Source History: 1 in 2004.* **Sources: Hud.**

Zucca Melon Gourd - (Sheep's Nose) - 150 days - Huge gourd, 36 in. long by 12 in. diameter, thin shell, green rind is candied for use in Panatone. *Source History: 1 in 2004.* **Sources: Sa9.**

Zucca Sicillano - (Serpente di Sicilia, Serpent of Sicily) - 65 days - Pale light-green fruits, 24-36 in. long x 3 in. dia., eaten when immature or used for gourd craft decorations when mature and hard shelled. *Source History: 1 in 1981; 1 in 1984; 1 in 1987; 1 in 1991; 1 in 1994; 2 in 1998; 1 in 2004.* **Sources: Se24.**

Varieties dropped since 1981 (and the year dropped):

Early Long Green (2004), *Green Long White Gourd* (1998), *Large Long* (1994), *Large Rounded* (2004), *Round Mao Gwa* (1991), *Round Striped Skinned* (1987), *Wax Gourd, Small* (2004).

Ground Cherry
Physalis spp.

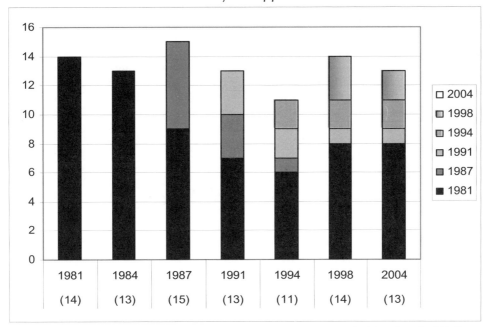

Number of varieties listed in 1981 and still available in 2004: 14 to 8 (57%).
Overall change in number of varieties from 1981 to 2004: 14 to 13 (93%).
Number of 2004 varieties available from 1 or 2 sources: 5 out of 13 (38%).

Aunt Molly's - 65-70 days - Fruit is enclosed in the same kind of papery husk as tomatillos, golden orange when ripe, drops to the ground, stores up to 3 months in the husks, fruit salads and jams, originated in Poland, Territorial Seed Co. intro. *Source History: 1 in 1994; 2 in 1998; 6 in 2004.* **Sources: Ga22, Ho13, Se16, St18, Ter, We19.**

Cape Gooseberry - (Poha) - 75 days - *P. peruviana*, fruits packaged by nature in small paper wrapper that looks like a Chinese lantern, harvest summer and fall, for old-fashioned preserves and Mexican recipes. *Source History: 1 in 1981; 1 in 1984; 5 in 1987; 3 in 1991; 5 in 1994; 5 in 1998; 5 in 2004.* **Sources: CO30, Com, Ga7, So25, WE10.**

Cossack Pineapple - 60 days - *P. pruinosa*, Eastern European relative of the Mexican tomatillo, bushy spreading plants grow 12-18 in. tall, marble-sized pineapple-flavored fruits encased in husks, good for jams and pies. *Source History: 2 in 1991; 4 in 1994; 5 in 1998; 8 in 2004.* **Sources: Ba8, Ho13, Pin, Pr3, So1, Syn, Te4, Vi4.**

Eden - Yellow husked fruit, yellow when ripe, from the Netherlands. *Source History: 1 in 1998; 1 in 2004.* **Sources: Pr3.**

Golden Berry - (Poha Berry) - 200 days - *P. peruviana*, orange semi-tropical fruits, relative of tomatillo, about 6 grams each, 1.5-6 ft. plants, deseeded fruit juice is of similar color and intensity of taste as orange juice. *Source History: 1 in 1981; 1 in 1984; 2 in 1987; 2 in 1991; 3 in 1994; 3 in 1998; 5 in 2004.* **Sources: Ban, Hud, Pr3, So12, Tho.**

Goldie - 75 days - *P. pruinosa*, selected strain, clean sweet flavor when ripe, a treat for hot desserts and preserves .75 in. dia. berries in easily removed papery husks, bears until frost. *Source History: 1 in 1981; 1 in 1984; 2 in 1987; 3 in 1991; 4 in 1994; 3 in 1998; 2 in 2004.* **Sources: So1, Syn.**

Ground Cherry - (Husk Tomato, Golden Husk, Strawberry Husk, Chinese Lantern, Poha, Dwarf Cape Gooseberry, Goldenberry, Winter Cherry) - Common name for a number of species of the genus *Physalis*, native to Europe, North America, Japan and the tropics, the Pilgrims considered it a delicacy, cherry-size reddish-yellow fruits, use raw or dried or in pies or preserves, stores well in husks. *Source History: 9 in 1981; 9 in 1984; 11 in 1987; 10 in 1991; 14 in 1994; 17 in 1998; 23 in 2004.* **Sources: Al25, Ami, Ba8, Be4, Bu1, Eo2, Ers, Ge2, Gr27, Gr28, He17, Hi13, Hum, Jor, Jun, Ma19, MAY, Mel, Mo13, OLD, Pr3, Ri12, SE14.**

Ground Cherry, Giant - 75 days - *P. peruviana*, orange semi-tropical fruits, weighs about 10 grams, 4-6 ft. plants whose roots sometimes overwinter if well mulched, excel. orange colored juice. *Source History: 1 in 1981; 1 in 1984; 2 in 1987; 1 in 1991; 1 in 1994; 2 in 1998; 2 in 2004.* **Sources: Kit, Pe1.**

Hanover - 70-80 days - Old Pennsylvania Dutch strain, sprawling bush, yellow fruit, tastes like a lemony tomatillo. *Source History: 1 in 1998; 3 in 2004.* **Sources: Ea4, Ho13, Vi4.**

Huberschmidt - *P. pruinosa*, fruit enclosed in husks which turn brown and papery as the fruit ripens, tomato-like flavor with sweet aftertaste, popular among Pennsylvania Germans for pies and preserves. *Source History: 1 in 1994; 2 in 1998; 3 in 2004.* **Sources: Ho13, Lan, Roh.**

Husk Tomato - (Strawberry Tomato, Green Sauce, Husk Cherry, Winter Cherry) - 70-86 days - Ancient, small round yellow- gold fruits in brown papery husks, grows just like tomatoes, dozens per 12 in. vine. *Source History: 9 in 1981; 8 in 1984; 10 in 1987; 9 in 1991; 10 in 1994; 8 in 1998; 6 in 2004.* **Sources: Hud, Ni1, RUP, Shu, So25, Und.**

Strawberry - (Yellow Husk, Strawberry Husk, Winter Cherry, Dwarf Cape Gooseberry) - 70-90 days - Small round fruits borne singly, enclosed in papery husk, rich sweet flavor, cherry-like, yellow, used for preserving and pies. *Source History: 2 in 1981; 1 in 1984; 2 in 1987; 2 in 1991; 2 in 1994; 2 in 1998; 2 in 2004.* **Sources: Hud, To3.**

Yellow Husk - (Ground Cherry, Strawberry Tomato, Golden Husk) - 70-90 days - Golden-yellow cherry-size round seedy fruits, borne singly in thin paper husks, heat and drought res., mildly acid. *Source History: 8 in 1981; 8 in 1984; 7 in 1987; 5 in 1991; 5 in 1994; 3 in 1998; 2 in 2004.* **Sources: Dam, To1.**

Varieties dropped since 1981 (and the year dropped):

Chinese Lantern Plant (1987), *Giant Yellow* (1994), *Golden Express Tomatillo* (1991), *Jamberry* (1987), *Local Dwarf* (1994), *Long Ashton Golden Berry* (1991), *Miltomate* (1983), *Miltomate Loco* (1994), *Pineapple Cherry* (1991), *Rendidora Tomatillo* (1991), *Sugar Cherry Tomatillo* (1991), *Sweet Amber* (2004), *Wild Ground Cherry* (1987), *Yellow Improved* (1994).

Jerusalem Artichoke
Helianthus tuberosus

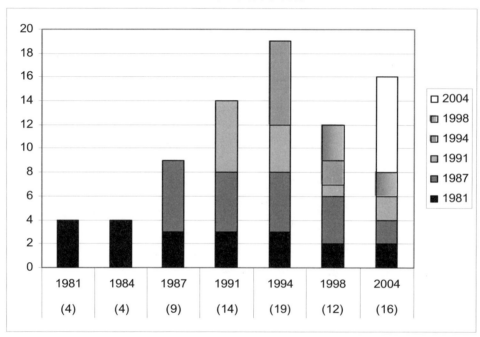

Number of varieties listed in 1981 and still available in 2004: 4 to 2 (50%).
Overall change in number of varieties from 1981 to 2004: 4 to 16 (400%).
Number of 2004 varieties available from 1 or 2 sources: 15 out of 16 (94%).

Challenger - Few blooms, highly productive. *Source History: 1 in 2004.* **Sources: Ma13.**

Clearwater - Smooth tubers, white skin, crisp, tasty white flesh, from Will Bonsall in Maine. *Source History: 1 in 2004.* **Sources: Moo.**

French Mammoth White - Large, knobby white tubers. *Source History: 1 in 1987; 1 in 1991; 1 in 1994; 1 in 1998; 1 in 2004.* **Sources: Ho13.**

Fuseau - Yam type, long straight, knob-free tubers are easy to clean, good flavor, early harvest. *Source History: 1 in 1987; 1 in 1991; 5 in 1994; 4 in 1998; 2 in 2004.* **Sources: Ga21, Sa9.**

Jack's Copperclad - Plants to 10 ft., plump, knobby tubers are copper and rose colored, good yields. *Source History: 1 in 1998; 1 in 2004.* **Sources: Ho13.**

Jerusalem Artichoke - (Sunchoke, Native Sunflower, Sunroot) - Perennial, sunflower-like 6-8 ft. plants, 4 in. daisy-like flowers, crunchy tubers grow like potatoes, spreads rampantly so plant away from garden, native to eastern U.S., use like potatoes or slice raw, tastes like water chestnuts, good for dieters and diabetics, carbohydrates in form of insulin instead of starch. *Source History: 21 in 1981; 19 in 1984; 23 in 1987; 23 in 1991; 17 in 1994; 20 in 1998; 17 in 2004.* **Sources: Bu1, Fa1, Fi1, Gur, Jun, Lej, Mcf, Mel, Moo, Ri2, Sa5, Se7, Shu, So12, Tho, Ver, Yu2.**

Long Red - Sleek red skin. *Source History: 1 in 1991; 2 in 1994; 2 in 1998; 1 in 2004.* **Sources: Sa9.**

Magenta Purple - Attractive tubers are dark purple

with dark pink splotches, prolific. *Source History: 1 in 1991; 1 in 1994; 1 in 1998; 1 in 2004.* **Sources: Ho13.**

Mark's Chokes - Tan skin, white flesh, not excessively knobby, generic var. grown in Maine. *Source History: 1 in 2004.* **Sources: Moo.**

Nakhodka - Early maturing var., light bulb-shaped tubers, many eyes, white skin, white flesh, the result of Soviet breeding efforts, originated in and named after a city in eastern Siberia. *Source History: 1 in 2004.* **Sources: Moo.**

Purple, Giant - Large, sweet tubers, purple skin. *Source History: 1 in 2004.* **Sources: Ho13.**

Roter Topinambur - German var., dark purple with copper highlights. *Source History: 1 in 2004.* **Sources: Ho13.**

Skorospelka - Huge, fat, knobby tubers, cream and lavender color, good yield, from Russia. *Source History: 1 in 1998; 1 in 2004.* **Sources: Ho13.**

Stampede - Perennial, special, high yielding, extra early str., 5-7 ft. plants, flowers in July, matures a month before common vars., winter hardy in severe cold, some large white tubers over 8 oz. *Source History: 1 in 1981; 1 in 1984; 1 in 1987; 1 in 1991; 4 in 1994; 4 in 1998; 2 in 2004.* **Sources: Jo1, Ma13.**

Vadim - Large round irregular tubers, purple skin, from Russia. *Source History: 1 in 2004.* **Sources: Ho13.**

Volgo 2 - Large, smooth egg-shaped tubers, few knobs. *Source History: 1 in 2004.* **Sources: Ma13.**

Varieties dropped since 1981 (and the year dropped):

Brazilian (2004), *Dwarf Sunray* (1998), *Fuseau, Red* (2004), *Garnet* (1998), *Golden Nugget* (2004), *Large Tuber* (1998), *Maine Giant* (1998), *No. 1351* (1998), *Nova Scotia Redskin* (1998), *Red* (1987), *Silver Skin* (1991), *Smooth Garnet* (2004), *Sugarball* (1994), *Usborne's* (1998), *Waldoboro Gold* (1998), *White Mammoth* (1998).

Kale
Brassica oleracea (Acephala Group)

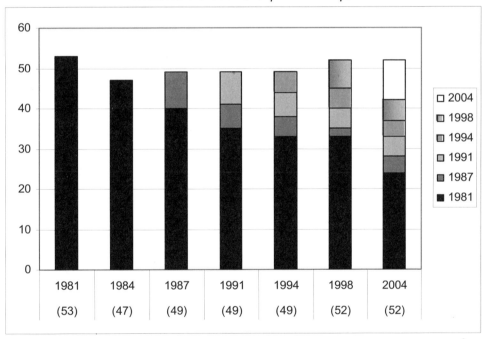

Number of varieties listed in 1981 and still available in 2004: 53 to 24 (45%).
Overall change in number of varieties from 1981 to 2004: 53 to 52 (98%).
Number of 2004 varieties available from 1 or 2 sources: 28 out of 52 (54%).

Asparagus - Violet tinged leaves and stems on 2 ft. plants, good for greens, sprouts eaten like asparagus. *Source History: 1 in 1998; 1 in 2004.* **Sources: Ho13.**

Black - Long savoyed curled leaves are dark green on the top and very dark to black on the underside, delicious. *Source History: 1 in 1994; 1 in 1998; 1 in 2004.* **Sources: So9.**

Black Sea Man Turkish - Dark green smooth leaves grow 3 ft. tall, hardy to 0 degrees F. *Source History: 1 in 2004.* **Sources: Ho13.**

Bona - Baby greens in 30 days - mature greens 65 days - dark-green, 24-30 in. tall plant with medium curled outer leaves, fresh market, Denmark. *Source History: 1 in 1994; 1 in 1998; 1 in 2004.* **Sources: Se8.**

Caulet de Flandre - Old French var. with large smooth green leaves, 3 ft. tall, flavorful, high yields, not cold hardy. *Source History: 1 in 2004.* **Sources: Ho13.**

Crimson Garden - *Source History: 1 in 2004.* **Sources: CO23.**

Curly Green - Tasty even when flowering, good market

garden variety, overwinters well. *Source History: 1 in 1991; 1 in 1998; 1 in 2004.* **Sources: Te4.**

Delaway - Loose headed plant, green and purplish leaves and stems, very tasty, seed received as a cabbage but looks more like kale. *Source History: 2 in 2004.* **Sources: In8, Se17.**

Dwarf Blue Curled - (Norfolk) - 55-90 days - Finely curled bluish-green plume-like leaves, 12-15 in. tall plant with spread of 30 in., stands well, maintains its color, hardy, uniform, low-growing. *Source History: 8 in 1981; 7 in 1984; 8 in 1987; 10 in 1991; 12 in 1994; 10 in 1998; 13 in 2004.* **Sources: Bo17, Bu3, CO23, De6, DOR, Fis, Fo7, He8, Hi13, Ma19, Pin, Se25, Vi4.**

Dwarf Curled - 60 days - Hardy vigorous spreading plants, large coarse leaves have plain centers and cut and frilled edges, deep bluish-green, vitamin rich, light frost improves flavor. *Source History: 3 in 1981; 3 in 1984; 1 in 1987; 1 in 1991; 1 in 1994; 1 in 1998; 1 in 2004.* **Sources: Roh.**

Dwarf Green Curled - (Jamaica Kale, Bloomsdale Kale, Dwarf Curled Scotch) - 50-65 days - Dwarf compact 12-15 in. plants with 24-28 in. spread, deep yellowish-green wrinkled and curly foliage, most palatable when well cooked, immense yields, very hardy, good as a spring or fall crop, very winter hardy, good market variety. *Source History: 26 in 1981; 24 in 1984; 17 in 1987; 12 in 1991; 9 in 1994; 7 in 1998; 5 in 2004.* **Sources: Ear, Ge2, Ho13, Tho, Vi4.**

Flowering Kale, Miniature - (Salad Savoy) - 65 days - Ornamental kale, brilliant range of colors, makes beautiful pot plants for the patio, delicious eating. *Source History: 1 in 1981; 1 in 1984; 1 in 1987; 1 in 1991; 1 in 1994; 1 in 1998; 1 in 2004.* **Sources: Ni1.**

Frilly - Sport of Russian Kale, one month-old seedlings produce leaves with frilled edges resembling curled parsley. *Source History: 1 in 2004.* **Sources: Sa5.**

Gai Lohn - (Guy Lon, Gailan, Chi-e-lan, Chinese Kale, Chinese Broccoli, Gai-Lon) - 70-75 days - *Brassica oleracea* (Alboglabra Group), known as either Chinese kale or Chinese broccoli, 12-14 in., tender stems, flat leaves like mustard greens, open non-heading buds, white or yellow flowers, leaves stems flowers and buds are edible, tangy broccoli taste, sweet hearts, cool weather plants. *Source History: 9 in 1981; 8 in 1984; 9 in 1987; 5 in 1991; 5 in 1994; 7 in 1998; 5 in 2004.* **Sources: Ban, CO23, Ni1, Ri2, Ri12.**

Grun Kohl Lerchenzungen - Flavorful old German str., 3 ft., curled, ruffled green leaves, cold hardy. *Source History: 1 in 2004.* **Sources: Ho13.**

Hanover Salad - (Hanover, Meyer's Brand Spring, Spring, Smooth) - 30 days - *Brassica napus* (Pabularia Group), Siberian kale type that is actually a rape, hardy, smooth-leaved, fast-growing, slow-seeding, long season, best when young and tender. *Source History: 4 in 1981; 4 in 1984; 5 in 1987; 4 in 1991; 4 in 1994; 4 in 1998; 6 in 2004.* **Sources: Cl3, Ho13, ME9, So1, SOU, Wet.**

Harvester - 68 days - Short-stemmed spreading plant, well-curled leaves, does not yellow after heavy frost, smaller yields but shortness and large leaves protect stem from freezing. *Source History: 4 in 1981; 4 in 1984; 2 in 1987; 1 in 1991; 1 in 1994; 1 in 1998; 1 in 2004.* **Sources: Ga1.**

Judy's - 60-65 days - Outstanding spring regrowth, broad, tall foliage, med. curled in fall, well curled in spring, good spring regrowth where it can overwinter. *Source History: 1 in 1994; 1 in 1998; 3 in 2004.* **Sources: Ho13, Pe2, Tu6.**

Kailaan White Flowered - (Gai Lohn) - *Brassica oleracea* (Alboglabra Group), known as either Chinese kale or Chinese broccoli, quick growing with large smooth dark-green leaves and thick tender stalks, flowering stalks and young leaves and buds are all delicious. *Source History: 1 in 1981; 1 in 1984; 2 in 1987; 4 in 1991; 8 in 1994; 10 in 1998; 11 in 2004.* **Sources: Ba8, Ev2, Jor, Kit, Loc, ME9, OSB, Red, RUP, Sto, TE7.**

Konserva - 50-60 days - Broad 24-30 in. tall plants, moss-green foliage, medium curled at first, becomes more curled with cool weather, tremendous yields, first class Danish strain. *Source History: 3 in 1981; 3 in 1984; 4 in 1987; 6 in 1991; 7 in 1994; 7 in 1998; 6 in 2004.* **Sources: CO30, Fe5, Fo7, Ga1, Pin, Se25.**

Lacinato - (Cavolo Palmizio, Nero di Toscana, Dinosaur Kale, Black Cabbage, Black Palm, Tuscan Black Palm Cabbage) - 60-90 days - A rather primitive open kale traced back to the 18th century Tuscany region of Italy, strap-like blue-green leaves 3 in. wide x 10 in. long, ornamental as well as edible, flavor is enhanced by frost, extremely winter hardy, CV Sh12 intro. *Source History: 1 in 1987; 1 in 1991; 1 in 1994; 3 in 1998; 38 in 2004.* **Sources: Ag7, Ami, Bo17, Bou, Co32, Coo, Dam, Fe5, Ga1, Ga22, Ha5, Hi6, In8, Jo1, Ki4, Na6, Ni1, Orn, Pe2, Pin, Pr3, Re6, Ri2, Ri12, Roh, Sa5, Sa9, Se7, Se8, Se16, Se24, Se26, Se27, Sh12, TE7, Ter, Tu6, We19.**

Lacinato Rainbow - Lacinato x Redbor (hybrid) cross, diverse population with leaf qualities of Lacinato, leaves are overlain with hues of red, purple and blue-green, more vigorous and cold hardy than Lacinato, sel. by CV Sh12. *Source History: 1 in 2004.* **Sources: Sh12.**

Lavo - (Lav opret voksende) - Dwarf variety with medium curled outer leaves, sweet, mild-flavored cooked greens, young leaves good in salads, good for mechanical harvest, regrows quickly. *Source History: 2 in 1994; 2 in 1998; 1 in 2004.* **Sources: CO30.**

Maris Kestrel - Good forage variety, shorter and thicker in the stem than Marrow Stem, less prone to lodging, uniform plants with high dry matter and good palatability. *Source History: 1 in 1987; 1 in 2004.* **Sources: Ada.**

Marrowstem - (Green Marrowstem) - Edible swollen stalk, 24-48 in. tall x 4 in. dia., superb vegetable when young, harvest whole plant in fall and store, not affected by light frost, milk producing, very nutritious, abundant leaves, forage or cover type, good autumn livestock feed. *Source History: 6 in 1981; 5 in 1984; 7 in 1987; 3 in 1991; 1 in 1994; 1 in 1998; 2 in 2004.* **Sources: Ho13, Pe1.**

Nero de Philo - New population of black kales created by crossing a very refined heirloom strain with low vigor with a more vigorous but primitive Lacinato, a work in progress by CV Sh12. *Source History: 1 in 2004.* **Sources:**

Sh12.

Ozark Black - Experimental var. selected by Horus Botanicals, 3.5-4 ft. tall with large dark purple to maroon-black leaves, broad and slightly wavy, tasty, hardy to 10 degrees F. *Source History: 1 in 2004.* **Sources: Ho13.**

Palm Tree Cabbage - (Dinosaur Kale) - 100 days - Loose leaf heads, dark "black" green heavily savoyed leaves up to 24 in. long, delicious flavor, excel. in soups and stews, Italian heirloom dating back to the early 1800s. *Source History: 1 in 1991; 1 in 1994; 1 in 1998; 2 in 2004.* **Sources: La1, Red.**

Pentland Brig - 55-75 days - Thousand Headed x Curly Kale, very hardy, high yields of finely curled leaves from early winter through April, then broccoli-like spring sprouts, England. *Source History: 4 in 1981; 3 in 1984; 1 in 1987; 1 in 1991; 1 in 1994; 1 in 1998; 1 in 2004.* **Sources: Bou.**

Premier - (Early Hanover) - 60-68 days - Produces heavy deep-green leaves with slightly scalloped margins more serrated than those of Smooth Long Standing, short main stems develop many growing points, stands 3-4 weeks longer than Smooth Long Standing, higher yielding when seeded in the fall for spring harvest. *Source History: 1 in 1981; 1 in 1984; 3 in 1987; 8 in 1991; 5 in 1994; 10 in 1998; 6 in 2004.* **Sources: Cl3, ME9, Mey, So1, Wet, WI23.**

Purple Napini - Purple stemmed, smooth leaves, grown over winter, in spring the deep purple shoots are harvested, 12-18 in. long, with a texture like asparagus, later shoots are smaller. *Source History: 2 in 1998; 3 in 2004.* **Sources: Ho13, Hud, Sh12.**

Ragged Jack - (Russian Kale, Russian Red, Rugged Jack) - 50-60 days - Ancient variety with beautiful red oak type leaves, considered to be the hardiest and most deliate kale, from Canada, 1885. *Source History: 1 in 1987; 2 in 1991; 4 in 1994; 3 in 1998; 4 in 2004.* **Sources: Bou, He17, Lan, Ri12.**

Russian Red - (Canadian Broccoli, Ragged Jack, Buda) - 50-65 days - Rare strain with purple veination, red frilly tender leaves, wavy margins resemble an oak leaf, excellent flavor, attractive, very frost tolerant, originally from Siberia, brought into Canada by Russian traders about 1885. *Source History: 2 in 1981; 2 in 1984; 3 in 1987; 16 in 1991; 19 in 1994; 38 in 1998; 73 in 2004.* **Sources: Ag7, Ami, Au2, Ba8, Bo18, Bo19, Bu2, CO23, CO30, Co32, Com, Dam, DOR, Ers, Fe5, Ga1, Ga22, Goo, Gr29, Ha5, He8, He17, Hi6, Ho12, Ho13, Hud, Hum, Jo1, Jor, Ki4, La1, Lin, Loc, ME9, Mo13, Na6, Ni1, OLD, Or10, Orn, OSB, Pe1, Pe2, Pin, Pr3, Ra5, Ri2, Roh, RUP, Sa5, Sa9, Se7, Se8, SE14, Se16, Se17, Se26, Se28, Shu, So1, So9, So12, St18, Syn, TE7, Tu6, Twi, Vi4, WE10, We19, Wi2, WI23, Wo8.**

Russian White - 50-60 days - Sweet, mild tasting kale, leaf color similar to Red Russian but with white midrib and veins, cold tol., semi-savoy, hardy. *Source History: 3 in 1998; 6 in 2004.* **Sources: Fe5, Ga1, Ho13, Ki4, Sh12, Ter.**

Savoy - 55 days - Dwarf Scotch Kale x Black Tuscan Kale cross by Dr. Alan Kapuler, slightly ruffled, dark green plants blushed with purple, continuous harvest encourages regrowth, excellent table quality, hardy. *Source History: 1 in 2004.* **Sources: Ni1.**

Scotch, Dwarf Blue Curled - (Dwarf Blue Scotch, Blue Curled Scotch, Blue Scotch, Vates, Norfolk) - 53-65 days - Finely curled blue-green leaves, compact low-growing plant, 12-15 in. tall and 20-35 in. spread, very hardy, overwinter with mulch, does not yellow even in severe cold, light frost improves flavor and adds sweetness, high in vitamin A. *Source History: 17 in 1981; 15 in 1984; 14 in 1987; 18 in 1991; 18 in 1994; 21 in 1998; 22 in 2004.* **Sources: Bu8, Co31, Co32, Cr1, Ea4, Goo, Hum, LO8, Mel, MOU, Ni1, Orn, Ri12, Ros, Se16, Sh9, Si5, SOU, Und, Ves, WE10, Wi2.**

Scotch, Dwarf Green Curled - (Dwarf Green Scotch, Green Curled Scotch, Dwarf Curlie) - 55-65 days - Spreading low- growing plant, 15 in. tall, large finely curled dark-green parsley-like leaves, very hardy, withstands severe frost, sow early spring to August, slight frost improves flavor, can overwinter with mulch, cabbage-like flavor. *Source History: 18 in 1981; 15 in 1984; 13 in 1987; 12 in 1991; 10 in 1994; 10 in 1998; 7 in 2004.* **Sources: All, Dam, Hal, Ho12, Lin, Ont, Sau.**

Siberian - (Early Siberian, Siberian Curled, Early Curled Siberian) - 60-70 days - *B. napus* (Pabularia Group), known as Siberian kale but is actually a rape, extremely hardy, rapid growing, huge blue-green slightly curled feather-shaped leaves, non- heading 12-16 in. spreading plants, sow spring or fall, popular in the South, light frost improves tenderness and flavor, often used for stock feed. *Source History: 24 in 1981; 21 in 1984; 19 in 1987; 22 in 1991; 21 in 1994; 18 in 1998; 23 in 2004.* **Sources: Alb, Bu8, CO23, Co31, Ers, He8, Hi6, Ho13, ME9, Mey, Pe1, Pe2, Ri12, RUP, Se7, Se26, Sh9, Shu, SOU, TE7, Ter, Th3, Wet.**

Siberian Improved - 50-70 days - Dwarf plants 12-15 in. tall, large, thick, blue-green leaves with frilled edges on long, succulent, sweet stalks. *Source History: 13 in 1981; 11 in 1984; 10 in 1987; 10 in 1991; 9 in 1994; 8 in 1998; 12 in 2004.* **Sources: Ada, CH14, DOR, Ga22, Mo13, OLD, Se25, Sk2, Sto, WE10, We19, Wi2.**

Siberian, Dwarf - (German Sprouts) - 60-70 days - *B. napus* (Pabularia Group), extremely hardy, broad thick blue-gray- green plume-like leaves, slightly frilled edges, plants 12-16 in. tall x 24-36 in. dia., not nearly as curled as Scotch types. *Source History: 21 in 1981; 15 in 1984; 12 in 1987; 12 in 1991; 9 in 1994; 9 in 1998; 6 in 2004.* **Sources: ABB, Bo19, Hig, La1, Sau, WI23.**

Spring - (Early Hanover Salad, Smooth, Plain, Spring Sprouts, Spring Smooth Leaf) - 30-35 days - Quick growing thick smooth round blue-green leaves, usually sown in early spring but can be overwintered, hardy. *Source History: 6 in 1981; 6 in 1984; 4 in 1987; 2 in 1991; 1 in 1994; 1 in 1998; 1 in 2004.* **Sources: Wet.**

Squire - (Slow Bolting Vates) - 55 days - Finely curled blue-green foliage, heavy yields, very uniform short stemmed strain of Vates with better tol. to bolting. *Source History: 8 in 1991; 11 in 1994; 8 in 1998; 2 in 2004.* **Sources: Cl3, ME9.**

Tall Green Curled - Deeply cut leaves are curled at the edges, 3-4 ft. plants, med-green, also for winter animal feed. *Source History: 2 in 1981; 1 in 1984; 4 in 1987; 3 in 1991; 2 in 1994; 1 in 1998; 1 in 2004.* **Sources: Pe1.**

Thousand Headed - (Branching Borecole) - 110 days - Large branching European kale, plants 3-4 ft. tall, produces numerous branches of large dark-green tender smooth leaves, suitable for ensilage, very productive, vigorous, excellent chicken and stock feed, still used in Europe for fodder, cold sensitive according to some sources, cold hardy according to others. *Source History: 5 in 1981; 5 in 1984; 5 in 1987; 4 in 1991; 3 in 1994; 4 in 1998; 1 in 2004.* **Sources: Ho13.**

Ursa Red - 59-70 days - Red Russian x Siberian Kale cross, deep purple-red leaves, mild flavor in the summer with increased sweetness after frost, grows back well after repeated cuttings, from Frank Morton. *Source History: 3 in 1998; 6 in 2004.* **Sources: Ga1, In8, Pe1, Se7, Sh12, So9.**

Vates - (Dwarf Blue Vates, Dwarf Blue Scotch, Dwarf Curled Scotch, Dwarf Blue Curled) - 55 days - Improved strain of Dwarf Blue Scotch, more upright, sturdy compact 12-14 in. x 20 in. plants, res. yellowing in cold weather, slow bolt, very hardy, deep blue-green finely curled leaves, good for wintering over, VA/AES. *Source History: 16 in 1981; 15 in 1984; 15 in 1987; 14 in 1991; 17 in 1994; 20 in 1998; 15 in 2004.* **Sources: Bo19, But, CO30, He17, Hig, HO2, ME9, MIC, Mo13, Or10, Sh9, So1, Sto, Twi, WI23.**

Vates, Dwarf Blue Curled - (Blue Curled Vates, Dwarf Curled Vates, Vates Dwarf Blue Curled Scotch, Dwarf Blue Scotch Vates,) - 50-80 days - Slow-bolting Scotch type, finely curled blue-green leaves, low-growing, 12-16 in. tall x 24 in. dia., short stem, hardy, especially tender after light frost, does not yellow from frost or heat, sow early spring or July, non-heading, VA/AES. *Source History: 47 in 1981; 45 in 1984; 41 in 1987; 37 in 1991; 34 in 1994; 37 in 1998; 27 in 2004.* **Sources: ABB, Bu2, CH14, Cl3, Com, Ers, Fi1, Gr27, HA3, Ha5, Jor, Jun, Kil, La1, Loc, MAY, Mey, OLD, OSB, PAG, RIS, RUP, SE4, Se25, Se26, Shu, Wet.**

Walking Stick - (Giant Walking Stick) - 180-300 days - *B. oleracea longata*, grown for nearly two centuries in the Channel Islands where it is a tourist attraction, from an early spring sowing will grow 5-7 ft. in most soils and be topped by a leafy cabbage head, stems can be cut and dried for walking sticks in the fall or left for another year when they may run to seed or continue growing 10-20 ft. tall. *Source History: 2 in 1987; 4 in 1991; 5 in 1994; 5 in 1998; 7 in 2004.* **Sources: Ga22, Hi13, Hud, Ni1, Pe1, Ter, Tho.**

Westlandse, Semi Dwarf - Fine-curled dark-green variety, excel. flavor. *Source History: 2 in 1981; 2 in 1984; 2 in 1987; 1 in 1991; 1 in 1994; 1 in 1998; 1 in 2004.* **Sources: Dam.**

Wild Red - 55 days - Variation on Red Russian, 2 ft. plants with silver-green foliage overlaid with bright red on the stems and leaf joints, extremely hardy and productive. *Source History: 2 in 1998; 3 in 2004.* **Sources: Ho13, Ni1, Sh12.**

Willy's - 55 days - Edible ornamental, centers of deep-red fringed by curled green, use in salads or in place of cabbage in soups and stir fry, light frost improves flavor. *Source History: 1 in 1991; 1 in 1994; 2 in 1998; 2 in 2004.* **Sources: Ho13, Pla.**

Winter Red - 50 days - Dark-green oak leaf cut leaves, may be richer in vitamins and minerals than other greens, red and purple hues intensify after fall frosts, give way to tender and sweet rich dark-green kale when cooked, also good raw, very disease res., dev. by Tim Peters. *Source History: 1 in 1991; 2 in 1994; 3 in 1998; 8 in 2004.* **Sources: Bo17, He8, Pe2, SE14, Se26, Sh12, TE7, Ter.**

Varieties dropped since 1981 (and the year dropped):

Cavalier Forage Kale (Green) (1987), *Cavalier Forage Kale (Red)* (1987), *Coral Prince White Flowering* (1984), *Coral Queen Red Flowering* (1984), *Cottagers* (1994), *Cow Kale* (1987), *Dwarf Green Curled Vates* (1991), *Flowering Kale* (1998), *Grand Frise d'Ecosse* (1991), *Greenpeace Kale* (2004), *Hungry Gap* (1991), *Long Seasons* (2004), *Long Standing Siberian* (1991), *Long Standing Slow Seeding* (1991), *Marrowstem Green Improved* (1987), *Miniature Japanese Ornamental* (1994), *Ornamental* (2004), *Rema* (1983), *Salad King* (1984), *Scope* (2004), *Scotch, Tall Green Curled* (1998), *Sekito* (1998), *Siberian, Blue* (2004), *Siberian, Tall* (1994), *Smooth* (2004), *Smooth Long Standing* (2004), *Spring Sweet* (2004), *Spurt* (1998), *Tall* (1984), *Verdura* (2004), *Westland Winter* (1998), *Westlandse Winterharde* (1987), *White Flowered Kale* (2004), *Xmas Fringed White Flowering* (1998), *Yellow Flowered Kale* (2004), *Zallah* (1984).

Kohlrabi
Brassica oleracea (Gongylodes Group)

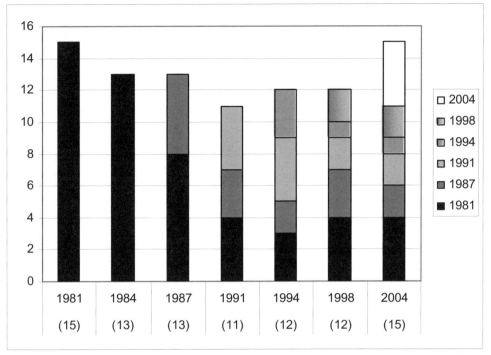

Number of varieties listed in 1981 and still available in 2004: 15 to 4 (27%).
Overall change in number of varieties from 1981 to 2004: 15 to 15 (100%).
Number of 2004 varieties available from 1 or 2 sources: 7 out of 15 (47%).

Azur Star - 50 days - Sweetest and crispest purple type, leaves and stems are purple, bulbs have purple skin and sweet white flesh, 2-3 in. across, edible purple leaves, bolt res., very rare, from Germany. *Source History: 1 in 1981; 1 in 1984; 3 in 2004.* **Sources: Coo, Ho13, Tu6.**

Blaro - (Giant Purple) - 40-50 days - Very refined European purple kohlrabi, less foliage than the Vienna types, fewer stems so bulb is smoother and more tender, purple skin, crisp white flesh. *Source History: 1 in 1981; 1 in 1991; 2 in 1994; 2 in 1998; 1 in 2004.* **Sources: Mcf.**

Blusta - 65 days - Intense purple-blue color, sweet nutty flavor, does not become woody. *Source History: 1 in 2004.* **Sources: Tho.**

Delicatesse, Blue - 60 days - Slightly more uniformity and disease resistance than Purple Vienna, which it resembles. *Source History: 1 in 1994; 1 in 1998; 3 in 2004.* **Sources: Bo19, Vi4, WE10.**

Delicatesse, White - 60 days - Slightly more uniformity and disease resistance than White Vienna, which it is similar to. *Source History: 1 in 1998; 4 in 2004.* **Sources: Bo19, Sk2, Vi4, WE10.**

Dyna - 60 days - Compares to Superschmelz only in purple. *Source History: 1 in 2004.* **Sources: Ter.**

Early Purple Vienna - (Purple Vienna, Early Purple, Purple, Di Vienna Violetto) - 55-69 days - Standard home and market var., purple-skinned bulbs, greenish-white flesh, small tops, slightly larger and later than Early White Vienna, best at 2.5 in. dia. *Source History: 75 in 1981; 67 in 1984; 71 in 1987; 75 in 1991; 72 in 1994; 70 in 1998; 61 in 2004.* **Sources: Alb, Ba8, Be4, Ber, Bo19, Bu8, But, CH14, CO23, Com, Coo, Dam, Dgs, DOR, Ear, Ers, Fi1, Ga1, Ge2, Gr27, HA3, Hal, He8, He17, HO2, Hud, Hum, Jor, La1, Lej, Lin, LO8, Loc, Ma19, MAY, ME9, Mo13, MOU, OLD, Ont, Or10, Orn, Ra5, Ri12, Roh, Ros, RUP, Sa9, Sau, SE4, Se25, Sh9, Shu, So1, Sto, Up2, Ver, Ves, Vi4, WE10, WI23.**

Early White Delicacy - 50-65 days - Harvest the swollen stem just above soil level when tennis ball size, flavor is between a cabbage and turnip. *Source History: 1 in 2004.* **Sources: Bou.**

Early White Vienna - (White Vienna, Large White, Early White, White, Di Vienna Bianco) - 50-65 days - Early dwarf variety, pale-green smooth flattened globe bulb forms above ground, 10-12 in. plants with small leaves, white flesh, best when 2.5 in. dia., unique flavor, successive sowings, good raw in salads or boiled or creamed or frozen, pre-1860. *Source History: 112 in 1981; 102 in 1984; 108 in 1987; 93 in 1991; 89 in 1994; 93 in 1998; 75 in 2004.* **Sources: Alb, All, Be4, Ber, Bu2, Bu3, Bu8, CA25, CH14, CO23, Coo, Cr1, Dam, De6, Dgs, DOR, Ear, Ers, Fe5, Fis, Fo7, Ga1, Ge2, Gr27, Gur, HA3, Ha5, Hal, He8, He17, Hi13, HO2, Hum, Jor, Jun, Kil, La1, Lej, Lin, LO8, Loc, Ma19, MAY, ME9, Mel, Mey, MIC, Mo13, MOU, OLD, Ont, PAG, Pin, Pr3, Ri12, RIS, Roh, RUP, Sa9, Sau, SE4, Se25, Se26, Sh9, Shu, Si5, Sk2, Sto, Tt2, Ves, Vi4, WE10, Wet, Wi2, WI23.**

Gigante - (Gigant Winter) - 130 days - Czechoslovakian heirloom, huge 10 in. dia., weight often exceeds 10 lbs., extraordinary quality, crisp white flesh is tender and mild-flavored, no tough or woody fiber, greens can be prepared like collards or kale, stores well, central Europeans make their own version of sauerkraut using Gigante, reselected by E. M. Meader, U of NH, intro. by CV So1, 1989. *Source History: 3 in 1991; 4 in 1994; 8 in 1998; 11 in 2004.* **Sources: Coo, Fa1, Fe5, Ga1, Ho13, Ni1, Pe6, Pp2, Se25, So1, Tt2.**

Granlibakken - 45-60 days - Tender bulb with sweet flavor and excel. texture, holds up well in the garden, leaves make delicious greens, rich in vitamins and minerals. *Source History: 8 in 1998; 14 in 2004.* **Sources: Com, Dam, Ers, Fi1, Ge2, Gr27, HO2, Ho13, Jun, Lin, Ni1, OLD, Pin, Ver.**

Green - (Green Globe) - Hardy 2 ft. plants, good raw or cooked. *Source History: 1 in 1987; 1 in 1998; 1 in 2004.* **Sources: Ev2.**

Logo - 45 days - Improved white forcing type used for baby vegetable production in Europe, early. *Source History: 2 in 2004.* **Sources: Coo, Tu6.**

Peking - 55 days - Med-size pale-green bulbs, pure white fine quality flesh, Chinese var. *Source History: 1 in 1987; 1 in 1991; 1 in 1994; 2 in 1998; 1 in 2004.* **Sources: Ev2.**

Super Schmelz - (Giant White, Superschmeltz) - 54-70 days - Yields up to 22 lbs. per plant, extremely tender at any size, no bolting or woodiness, from Switzerland. *Source History: 1 in 1991; 3 in 1994; 3 in 1998; 3 in 2004.* **Sources: Mcf, Ter, We19.**

Varieties dropped since 1981 (and the year dropped):

Blauer Speck, Original (1998), *Blue Danish* (1998), *Giant Green Czechoslovakian* (1998), *Kin Men Early White* (1991), *Landro* (1984), *Large Green Short Topped* (1991), *Lauko* (1991), *Prague Early Forcing* (1991), *Purple Delicacy* (1991), *Purple Speck* (2004), *Rapid* (2004), *White Danish* (1987), *White Triumph* (1987), *Wiener Weisser Glas* (1987).

Leek
Allium ampeloprasum

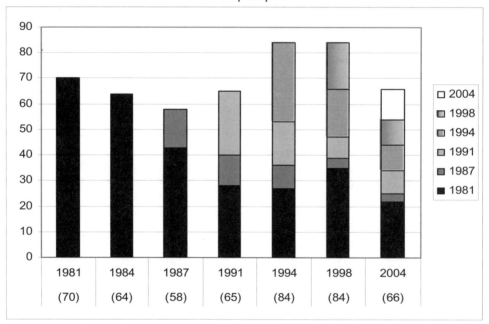

Number of varieties listed in 1981 and still available in 2004: 70 to 22 (31%).
Overall change in number of varieties from 1981 to 2004: 70 to 66 (94%).
Number of 2004 varieties available from 1 or 2 sources: 44 out of 66 (67%).

Albinstar - 100 days - Tightly wrapped leaves allow for easier cleaning, harvest at about 8 in., retains flavor when allowed to mature, very frost hardy, bred in Holland for the gourmet baby vegetable market. *Source History: 3 in 2004.* **Sources: In8, Ki4, Se17.**

American Broad Flag - 130 days - Large-stalked leek, tall and hardy, very strong growing and productive, long thick blanched necks. *Source History: 3 in 1981; 3 in 1984; 2 in 1987; 2 in 1991; 2 in 1994; 2 in 1998; 2 in 2004.* **Sources: All, Com.**

American Flag - (Giant Musselburgh) - 120-155 days - Hardy 15-18 in. plants, blue-green leaves, long thick 7-10 x 1.5 in. well blanched white stalks, standard var. for home gardens and market, good for fall and winter, grows very quickly, uniform. *Source History: 35 in 1981; 29 in 1984; 27 in 1987; 25 in 1991; 29 in 1994; 31 in 1998; 39 in 2004.* **Sources: Bo18, Bo19, Bu8, CA25, CH14, CO23, Eo2, Fi1, Gr29, Hal, He8, He17, HO2, Hud, Hum, Jor, LO8, Ma19, MAY, Mey, Mo13, MOU, Na6, Or10, Orn, PAG, Pe2, Ra5, Ri2, Roh, Ros, Se28, Sh9, TE7, Vi5, WE10, Wet, WI23, Wo8.**

American Flag, Large - (American, Large American, Large Flag, London Flag, Broad London) - 85-150 days - Early home garden var., large extra-long thick 7.5-10 x 1.5 in. dia. stems, smooth clear pearl-white bulb blanching far up, med-dark green broad leaves, hardy, extra large, mild delicate onion flavor, tender, use boiled or in soups and stews. *Source History: 31 in 1981; 29 in 1984; 27 in 1987; 28 in 1991; 25 in 1994; 25 in 1998; 24 in 2004.* **Sources: Alb, Be4, Bu3, De6, DOR, Ear, Ers, Fo7, GRI, HA3, Jun, Lej, Loc, Mel, OLD, Pin, RUP, Sau, Scf, Se26, Shu, So1, Tt2, Ver.**

Ardea - 180 days - Autumn leek, upright dark green foliage, minimal bulbing, excellent flavor, good rust tolerance, winter hardy. *Source History: 1 in 2004.* **Sources: Tho.**

Arena - 105 days - Overwintering type, blue-green leaves, upright, long white 7 in. shafts, non-bulbing, cleans easily, tolerates light frost. *Source History: 1 in 1998; 4 in 2004.* **Sources: ME9, OSB, Sto, Ves.**

Arkansas - 108-150 days - An improved Selecta, more erect growth habit and longer shank than Alaska, storage type, suitable for overwintering. *Source History: 1 in 1991; 4 in 1994; 8 in 1998; 5 in 2004.* **Sources: HO2, ME9, OSB, SE4, Sto.**

Autumn Giant 2 - Porvite - 170 days - Uniform, non-bulbing shafts resist bolting, rust tolerant foliage, stands well. *Source History: 1 in 2004.* **Sources: Tho.**

Autumn Giant Triumphator - (Autumn Giant) - 85-105 days - Top quality leek, extremely long shaft, almost no bulbing, tol. to YSV, for autumn or late autumn or early winter harvest, outstanding in European trials, rarely available. *Source History: 1 in 1981; 1 in 1984; 1 in 1991; 2 in 1994; 2 in 1998; 1 in 2004.* **Sources: Dam.**

Babbington - Cold hardy variety that perennializes thus providing a sustainable food source, heirloom. *Source History: 1 in 2004.* **Sources: Pe1.**

Bandit - 120 days - Dark blue-green foliage, thick, med-long shaft, late maturity, good cold tol. *Source History: 2 in 1998; 1 in 2004.* **Sources: OSB.**

Blue Solaize - (Blue de Solaize, Blue Solaise) - 100-120 days - Truly blue leaves turn violet after a cold period, big med- long shaft, extremely cold res. and hardy, for short season areas and winter harvest, excel. for soup, 19th century heirloom. *Source History: 3 in 1981; 3 in 1984; 4 in 1987; 3 in 1991; 3 in 1994; 5 in 1998; 7 in 2004.* **Sources: Ba8, Co32, Coo, Gou, Ho13, Lej, Se16.**

Broad London - (Large American Flag) - 80-130 days - Long thick blanched necks can be used alone or as a thick puree of green peas and leeks, sweet mild flavor, 6 x 3 in. dia. bulbs. *Source History: 4 in 1981; 3 in 1984; 2 in 1987; 1 in 1991; 1 in 1994; 1 in 1998; 3 in 2004.* **Sources: Hi13, Lin, Vi4.**

Bulgarian Giant - Light-green leaves, thin shaft, erect growth, exceptionally good quality, autumn harvest. *Source History: 1 in 1994; 1 in 1998; 1 in 2004.* **Sources: Ba8.**

Carentan - (Carentan Winter, Improved Swiss Giant) - 95-110 days - Med-dark green leaves, tender 8.5 x 2 in. white stems, ample base, very productive, hardy, fall and winter use, old European variety mentioned by Vilmorin in 1885. *Source History: 9 in 1981; 8 in 1984; 4 in 1987; 4 in 1991; 5 in 1994; 5 in 1998; 10 in 2004.* **Sources: Au2, Ba8, Bo19, Ers, He8, Pe2, Pr3, Te4, Vi4, WE10.**

Carentan, Giant - (Poireau Monstrueux de Carentan, Large Carentian) - 130 days - Very popular imported French leek, cold res., huge thick solid smooth white mild stems, delicate green tops. *Source History: 7 in 1981; 5 in 1984; 3 in 1987; 3 in 1991; 2 in 1994; 2 in 1998; 1 in 2004.* **Sources: Tu6.**

Cisco - Very long shank, easy peel, good bolt tol., early, suitable for fresh and processing, Nickerson-Zwaan var. *Source History: 1 in 2004.* **Sources: VI3.**

Columbus - 80 days - Blue-green foliage, heavy, thick, long shaft, matures early. *Source History: 2 in 1998; 1 in 2004.* **Sources: OSB.**

Dawn Giant - 98 days - Easy to grow, can be direct seeded in early spring for harvest midsummer into fall, 15 in. shaft. *Source History: 1 in 1998; 1 in 2004.* **Sources: Bu2.**

Durabel - 120-150 days - Winter leek from Denmark, small 1-1.5 in. dia. stalks, dark green leaves with erect growth, sweet mild flavor, perfect for winter salads and other raw uses, commercially grown for the gourmet trade. *Source History: 1 in 1981; 1 in 1984; 2 in 1987; 1 in 1991; 3 in 1994; 3 in 1998; 1 in 2004.* **Sources: St18.**

Electra - 125-150 days - French, very adaptable, fast grower, harvest very early or grow for fall and winter, cold res., long keeper, dark blue-green tops, long thick white shanks. *Source History: 4 in 1981; 4 in 1984; 3 in 1987; 1 in 1991; 1 in 1994; 1 in 1998; 1 in 2004.* **Sources: Ha5.**

Elephant Garlic - (Giant French Mild, Jumbo Elephant, Giant Elephant) - *A. ampeloprasum*, actually a giant leek, some over 1 lb., huge 4 oz. cloves, sweet and so extremely mild that it can be sliced into salads or cooked and served buttered, hardy but needs winter protection in the North. *Source History: 18 in 1981; 18 in 1984; 27 in 1987; 28 in 1991; 25 in 1994; 25 in 1998; 26 in 2004.* **Sources: Bo20, Bu1, Fa1, Fi1, Fi7, Fo4, Ga1, Ga17, Ga19, Go6, Gur, Ho13, Jo1, Jun, Mcf, Mel, Ni1, Par, Pe2, Ri2, Ri12, Shu, So1, Ter, Ver, We19.**

Elephant Leek - 85-150 days - Very large extra thick stalks, flavor and texture equal to Musselburgh, slightly earlier, thicker stem, heavier crop, late summer and early fall harvest, winter hardy, can store. *Source History: 7 in 1981; 6 in 1984; 4 in 1987; 5 in 1991; 4 in 1994; 4 in 1998; 1 in 2004.* **Sources: We14.**

Ester Cook - 120 days - Hungarian variety, extra tall, thin white leek, good winter hardiness. *Source History: 2 in 2004.* **Sources: Ho13, Sa9.**

Falltime - 80-90 days - Summer leek with shanks to 3 ft. long, delicious, tender. *Source History: 2 in 1998; 1 in 2004.* **Sources: Se7.**

Furor - 110-140 days - Gennevillier type, dark-green foliage, long thick stems, pungent flavor, can be harvested Dec. to March in mild areas. *Source History: 1 in 1991; 1 in 1994; 1 in 1998; 2 in 2004.* **Sources: VI3, Wi2.**

Gennevilliers - (De Gennevilliers) - 130-150 days - European leek, long med-slender shaft, dark blue-green upright foliage, medium hardy, good in mild climates, can overwinter if temps stay above 30 degrees F. *Source History: 1 in 1981; 1 in 1984; 2 in 1987; 1 in 1991; 1 in 1994; 1 in 1998; 1 in 2004.* **Sources: Lej.**

Giant Musselburgh - (Large Musselburgh, American Flag, Musselburgh, Selected Musselburgh, Scotch Flag) - 80-150 days - One of the best, extremely hardy, enormous size, 9-15 x 2-3 in. dia. tender white stalks, med-dark green fan-shaped leaves, mild flavor, stands winter well, good buncher, for home gardens or local market or shipping. *Source History: 35 in 1981; 30 in 1984; 24 in 1987; 19 in 1991; 14 in 1994; 10 in 1998; 17 in 2004.* **Sources: Ba8, Bou, Bu2, Ers, He8, Hi13, Kil, La1, Ni1, Ont, Ri12, Se7, Se16, Shu, Up2, Ves, WE10.**

Giant Musselburgh Improved - 90 days - Very hardy and exceptionally fine strain of this popular variety. *Source History: 1 in 1981; 1 in 1984; 1 in 1987; 1 in 1991; 1 in 1994; 1 in 2004.* **Sources: Tho.**

Hilari - Dark green shaft, good flavor, heirloom. *Source History: 1 in 2004.* **Sources: So9.**

Imperial - (Copenhagen Market Imperial) - 85 days - Stout, bulbless shaft with good length makes for a large edible area on each plant, medium-early, autumn harvest, Danish origin. *Source History: 2 in 1994; 2 in 1998; 1 in 2004.* **Sources: Fe5.**

Inegol - Long, narrow, tender variety from Turkey, grows 12 in. long and only .6 in. across. *Source History: 1 in 1991; 1 in 2004.* **Sources: Red.**

Jersey - 100 days - Imp. Nebraska type with longer shafts, more upright plants and strong bulbing tolerance, for late summer and early fall harvests. *Source History: 2 in 1998; 3 in 2004.* **Sources: Jor, RUP, Sto.**

Kajak - Good flavor, hardy, heirloom. *Source History: 1 in 2004.* **Sources: So9.**

Kilima - 80-130 days - Very rapid grower, free from bulbing, excel. for summer and autumn crops, slender pure-white tender 12.5-15 in. stems, mild onion like flavor, blue-green foliage. *Source History: 3 in 1981; 3 in 1984; 3 in 1987; 3 in 1991; 4 in 1994; 3 in 1998; 5 in 2004.* **Sources: BAL, Ni1, OLD, OSB, RIS.**

King Richard - 75 days - Fast growing leek, under optimal conditions the white stem can reach 12 in. in length, light-green upright leaves, mild flavor, susc. to heavy frosts. *Source History: 3 in 1981; 3 in 1984; 6 in 1987; 12 in 1991; 14 in 1994; 19 in 1998; 10 in 2004.* **Sources: Coo, Fe5, Gou, Hig, Pin, Se8, Syn, Tho, Tu6, We19.**

Laura - 110-140 days - Winter var. with dark-green upright leaves, thick heavy 10 in. shafts, excel. frost tol., hardy to 12 degrees F, yellow stripe virus tol. *Source History: 1 in 1991; 2 in 1994; 2 in 1998; 4 in 2004.* **Sources: BE6, Dam, Jo1, Ter.**

Leefall - 108-140 days - Upright, dark blue-green foliage, long white shank, 8-9.5 in., good foliage disease tolerance, for fall and early winter harvest, performs well in NJ, CA and other areas. *Source History: 2 in 1998; 2 in 2004.* **Sources: ME9, Sto.**

Leek - 120-130 days - Sweet and most delicate flavor of the onion family, large white mild stalks often 12 x 2-3 in. dia., can transplant from spring until late fall, a gourmet flavoring. *Source History: 6 in 1981; 6 in 1984; 4 in 1987; 4 in 1991; 2 in 1994; 2 in 1998; 3 in 2004.* **Sources: Bo17, Ga1, Gur.**

Leekool - 140 days - Dark green leaves, 8-9 in. white shank, widely adapted. *Source History: 4 in 1998; 3 in 2004.* **Sources: HA20, ME9, Par.**

Lincoln - 50-75 days - Bulgarian Giant type with extra long shaft for the processing industry. *Source History: 1 in 1998; 2 in 2004.* **Sources: Fe5, Jo1.**

London Flag - (London Broad Flag, Long Flag, Broad Scotch, American Flag) - 130-145 days - Long broad white stems, 7- 10 x 1.5 in. dia., large green leaves, very productive and early but sensitive to cold, milder flavored. *Source History: 3 in 1981; 2 in 1984; 2 in 1987; 1 in 1991; 1 in 1994; 1 in 1998; 1 in 2004.* **Sources: Ge2.**

Long d'Hiver de Paris - 150 days - Blue-green leaves, winter hardy, from France. *Source History: 1 in 1987; 1 in 1991; 1 in 1994; 1 in 2004.* **Sources: Lej.**

Long Fall - 80-90 days - Summer leek with shanks up to 3 ft. long, tender and delicious. *Source History: 1 in 1994; 1 in 2004.* **Sources: Tu6.**

Lyon (Prizetaker) - (The Lyon, The Lyon Prizetaker) - 110-135 days - Probably the finest variety in cultivation, very hardy, 36 in. tall plants, produces tender solid long white stems with fine flavor, extremely cold hardy, England, 1886. *Source History: 2 in 1981; 2 in 1984; 1 in 1987; 1 in 1991; 3 in 1994; 2 in 1998; 6 in 2004.* **Sources: Bou, Pe2, Se16, Se26, Tho, Tu6.**

Pancho - 80-100 days - Heavy thick white stems, resists bulbing, tol. to foliage diseases, summer and fall harvest, one of the best early leeks in Europe. *Source History: 1 in 1991; 4 in 1994; 2 in 1998; 1 in 2004.* **Sources: Se7.**

Primor - 135-180 days - Beautiful long-stemmed heirloom var., delicate flavor, very early, somewhat hardy, can be overwintered with good mulch around the base. *Source History: 1 in 1991; 5 in 1994; 5 in 1998; 8 in 2004.* **Sources: CO30, HO2, Orn, Re6, Roh, Se8, Twi, VI3.**

Purple - (Purple Tinged) - 145 days - Decorative foliage, blue-green with purple tinge which deepens to violet in cold weather, excellent quality and flavor, hardy, productive, traditional French cultivar. *Source History: 2 in 2004.* **Sources: Bou, Ho13.**

Rikor - Long shafts, white to pale green stems, green leaves, excellent sweet flavor. *Source History: 1 in 2004.* **Sources: Jo1.**

Rival - Medium green, 34-36 in. tall plant, harvest when stem diameter is between 1-2 in. *Source History: 1 in 2004.* **Sources: OSB.**

Scotland - 85-90 days - Heirloom, short shanks, excel. flavor and texture, good overwintering var. *Source History: 1 in 1994; 2 in 1998; 2 in 2004.* **Sources: Ri12, Se7.**

Sherwood - 75 days - Summer and fall leek with white stems which can grow to 1 ft. before leafing under the right conditions, mild, sweet flavor, high yields, can withstand frost but will not overwinter. *Source History: 1 in 1994; 1 in 1998; 1 in 2004.* **Sources: Se7.**

Siegfried - (Siegfried Frost) - 120-125 days - Winter

leek, sturdy, thick stems, tender texture, extremely bolt res., good resistance to winter rotting, will hold in the garden into April. *Source History: 1 in 1994; 1 in 1998; 4 in 2004.* **Sources: Fe5, Ga22, In8, We19.**

St. Victor - 145 days - Reselected strain of Blue Solaise, late harvest leek with tender fleshed, thick shanks, blue-green leaves deepen to a beautiful deep-violet tint after frost, heavy yields. *Source History: 1 in 1994; 6 in 1998; 3 in 2004.* **Sources: In8, Se17, So12.**

Swiss Giant Coloma - Early variety with long blanched cylindrical bulbs, decorative flower heads, good producer. *Source History: 1 in 1994; 1 in 1998; 1 in 2004.* **Sources: Lan.**

Tadorna - 100-108 days - Blue-green flags, 7-8 in. long shaft, easy to clean, frost tol., matures October-November, holds well in the garden, hardy to 20 degrees F, good leaf spot and leaf stripe virus tol. *Source History: 2 in 1998; 5 in 2004.* **Sources: ME9, RIS, RUP, Sto, Ter.**

Tenor - 105-145 days - Long straight thick shanks, dark blue-green foliage, very cold hardy, suitable for mechanical harvesting, well suited for direct sowing, for fall and early winter harvest. *Source History: 1 in 1991; 4 in 1994; 1 in 2004.* **Sources: CO30.**

Titan - (Titan Summer) - 70 days - Extra long vigorous early type, dark-green leaves, white 6-10 x 1 in. dia. shafts, for early summer or summer planting only, not winter hardy, from Denmark. *Source History: 3 in 1981; 3 in 1984; 4 in 1987; 6 in 1991; 6 in 1994; 7 in 1998; 2 in 2004.* **Sources: Bu2, Ear.**

Toledo - 135 days - Large stem is almost twice the length of some, winter hardiness for cropping from late December to May. *Source History: 1 in 1994; 1 in 2004.* **Sources: Tho.**

Varna - 85-105 days - Top quality leek bred for late fall or early winter harvest, long shaft up to 24 in. above soil line, excellent yields, not winter hardy. *Source History: 1 in 1987; 5 in 1991; 2 in 1994; 3 in 1998; 1 in 2004.* **Sources: Ho13.**

Vernor - Improved Blue Solaise type, blue-green color, tender and aromatic, plant in August for overwintering. *Source History: 1 in 1991; 1 in 1994; 1 in 2004.* **Sources: CO30.**

White - Can be harvested in fall if spring planted, ready in spring if planted in August. *Source History: 1 in 1991; 1 in 1994; 1 in 2004.* **Sources: So9.**

Wild - Unlike seeds of cultivated forms these seeds will germinate irregularly over at least a year with some coming up the first fall, some the next spring, and still others the following fall, keep seed flat in a moist shady spot. *Source History: 1 in 1987; 2 in 1991; 1 in 1994; 2 in 1998; 2 in 2004.* **Sources: Pe8, Ri2.**

Winora - 117 days - Blue-green, upright foliage with medium shaft length, very little bulbing, easy to clean, early winter harvest. *Source History: 1 in 1994; 3 in 1998; 1 in 2004.* **Sources: Tho.**

Winter - 75-85 days - Hardy sel. of composite population, medium-long, broad shanks and leaves. *Source History: 1 in 2004.* **Sources: So12.**

Winter Giant - (Winter Giant-Alaska) - 95 days - Very hardy leek for overwintering or winter storage, thick med-long pure-white stems, good market variety. *Source History: 2 in 1981; 2 in 1984; 2 in 1987; 3 in 1991; 5 in 1994; 3 in 1998; 3 in 2004.* **Sources: In8, Pe1, Se7.**

Varieties dropped since 1981 (and the year dropped):

90-0255 (2004), *A.F.W. No. 6* (1987), *Alabama* (1998), *Alaska* (2004), *Albana* (2004), *Alberta* (1998), *Arcona* (1994), *Argenta* (2004), *Artaban* (1998), *Artico* (1998), *Autumn Goliath* (1994), *Autumn Mammoth* (2004), *Bakker* (1994), *Blue Solaise Sel. Isjbeer* (1987), *Bluvetia* (1991), *Bombarde* (1987), *Broadleaf Chinese* (1994), *Buffalo* (1991), *Bulgarian Triumph* (2004), *Carina* (2004), *Catalina* (1991), *Colonna* (1991), *Conora* (2004), *Conqueror* (1987), *Copenhagen Market Konta* (2004), *Cortina* (1998), *Cylindra* (1994), *Danora* (2004), *De Gennevilliers Furor* (1987), *Derrick* (1998), *Early Bonanza* (1991), *Eskimo Winter Giant* (1987), *Everest* (1984), *Firena* (2004), *French Summer* (1998), *Frost* (2004), *Gabilan* (2004), *Gavia* (1998), *Giant* (1987), *Giant Flag* (2004), *Giant Special* (1991), *Giant Winter Wila* (1987), *Giant Winter-Royal Favorite* (1987), *Goliath* (1998), *Helvetia* (1998), *Hivor* (2004), *Inverno* (2004), *Italian Giant* (1998), *Kalem Kartal* (1998), *Lavi* (2004), *Leader* (1991), *Leegloo* (2004), *Leekwik* (2004), *Ligina* (1987), *Long de Mezieres* (2004), *Longina* (2004), *Mammoth Otina* (2004), *Marble Pillar* (1983), *Molos* (1991), *Nebraska* (1998), *Odin* (1987), *Orato* (2004), *Otina* (2004), *Pandora* (2004), *Pinola* (1994), *Platina* (1987), *Prenora* (2004), *Rese* (2004), *Selandia* (1991), *September Giant* (2004), *SG 178* (1983), *Snow White* (1991), *Snowstar* (2004), *Splendid* (2004), *Stora* (2004), *Strata* (1994), *Summer Early Large* (1983), *Sweet* (1991), *Taurus* (1991), *Thor* (2004), *Tivi* (2004), *Unique* (2004), *Verina* (1987), *Winta* (1987), *Winteruizen* (1991), *Yates Empire* (1983).

Lettuce / Head
Lactuca sativa

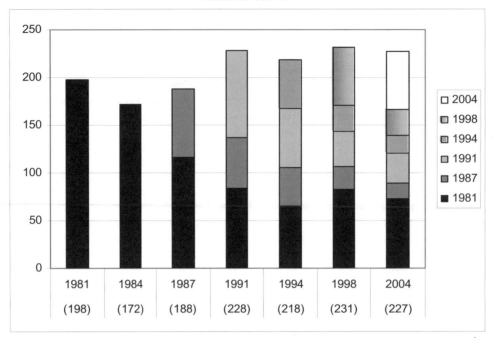

Number of varieties listed in 1981 and still available in 2004: 198 to 73 (37%).
Overall change in number of varieties from 1981 to 2004: 198 to 227 (115%).
Number of 2004 varieties available from 1 or 2 sources: 157 out of 227 (69%).

Akcel - 60-75 days - Good French forcing butterhead, compact heavy deep-green heads, very early, should do well for early baby lettuce market. *Source History: 1 in 1987; 1 in 1991; 1 in 1994; 1 in 1998; 1 in 2004.* **Sources: Tu6.**

All the Year Round - (All Year Round) - 64-73 days - Med-sized med-green head, solid even in hot weather, slow bolt, very hardy, good for far North, sow in spring or late summer, black seed, England. *Source History: 10 in 1981; 8 in 1984; 7 in 1987; 11 in 1991; 8 in 1994; 9 in 1998; 8 in 2004.* **Sources: Fa1, Hum, La1, Se26, Shu, Sk2, Tho, WE10.**

Alpha DMR - Main season iceberg with med. to large, dark green heads, uniform size, sure heading, DM res., PVP 1991 HA20. *Source History: 1 in 2004.* **Sources: HA20.**

Amazon - (HMX7550) - 60-65 days - Medium-large, dark green heads, flat butt, smooth leaf texture, uniform maturity, performs well in both warm and cool conditions, good TB tol. *Source History: 2 in 2004.* **Sources: ME9, OSB.**

Anuenue - 50-54 days - Anuenue (ah-new-ee-new-ee) means 'rainbow', thick crisp bright-green outer leaves surround a large well-packed heart, mild and juicy, resembles a small iceberg lettuce but more heat res. and easier to grow, spring summer and fall crop, U of Hawaii, introduced in 1987. *Source History: 2 in 1987; 7 in 1991; 7 in 1994; 9 in 1998; 8 in 2004.* **Sources: Ech, Eo2, Fe5, Ho13, Jo1, So1, Tu6, Uni.**

Apollo - Early med-size butterhead type for winter culture, sow August to January in greenhouse and coldframe or outdoors in milder regions, short day. *Source History: 1 in 1987; 3 in 1991; 2 in 1994; 2 in 1998; 1 in 2004.* **Sources: Dam.**

Aquarius - Very early, semi-heading Dark Green Boston type with darker color, good TB tol., PVP 1993 Sakata Seed America, Inc. *Source History: 1 in 1998; 1 in 2004.* **Sources: SAK.**

Arctic King - 80 days - Hardy var. for autumn and early winter harvest, does not stand heavy frosts, light-green crinkled leaves, small firm heads, quick to mature, white seed, spring sow also. *Source History: 4 in 1981; 3 in 1984; 2 in 1987; 1 in 1991; 1 in 1994; 3 in 1998; 3 in 2004.* **Sources: Coo, Ter, We19.**

Arizona RZ - *Source History: 1 in 2004.* **Sources: ME9.**

Attrazionne - 50-60 days - Blonde Boston variety, med. round head in the center of thick yet delicate leaves, buttery smooth texture, prefers cooler temps, tolerates some heat. *Source History: 1 in 2004.* **Sources: Ag7.**

Avondefiance - 75 days - Solid heavy med-sized heads, res. to downy mildew so an excellent choice for fall harvest, from England. *Source History: 1 in 1981; 1 in 1984; 2 in 1987; 1 in 1991; 2 in 1994; 1 in 1998; 1 in 2004.* **Sources: Tho.**

Bacardi - 45-55 days - Brilliant red butterhead, added to Mesclun at miniature head size, leaves are slightly savoyed at full size, earlier and more compact than Marvel of Four Seasons types, disease res. *Source History: 3 in 1998; 1 in 2004.* **Sources: Ta5.**

Baja - 63 days - Large butterhead with dark green leaves, loose head, slightly larger than Genecorp Boston, good TB tol. *Source History: 2 in 2004.* **Sources: ME9, Sto.**

Ballade - 80 days - Crisphead, firm med-size globular head, rich bright-green color, slow bolting, adapted for closer plantings, heat res. *Source History: 1 in 1987; 1 in 1991; 1 in 2004.* **Sources: CO30.**

Batavia Blonde - 65 days - Crisphead type with yellow-green leaves, slow bolting. *Source History: 1 in 1994; 1 in 1998; 2 in 2004.* **Sources: Bo17, Se26.**

Batavia Bord Rouge - (Batavia Rouge) - 59 days - Very attractive lettuce, large rosettes of crumpled dark-green leaves fringed with deep-red, flavor initially seems bitter but mellows to a nut-like fineness, stands about 45 days without bolting, France. *Source History: 2 in 1987; 2 in 1991; 2 in 1994; 1 in 1998; 2 in 2004.* **Sources: Lej, Ta5.**

Batavia Laura - (Batavian Loura, Blond Batavian) - 45-55 days - French var. with crisp green savoyed leaves tinged with red at the edges, forms crisp heads, 12-14 in. dia., bolt res. *Source History: 1 in 1987; 3 in 1991; 2 in 1994; 3 in 1998; 3 in 2004.* **Sources: Ho13, Pe1, Se7.**

Batavia, Cheena - Thick green leaves. *Source History: 1 in 2004.* **Sources: Pe1.**

Batavian, Long Standing - 50-60 days - Excellent quality large upright crispheads, thick leaves store water during heat, exceptional bolt res. *Source History: 1 in 1994; 1 in 1998; 2 in 2004.* **Sources: Pe2, Se7.**

Bayview - 70-75 days - Salinas type with large frame, round, dark green heads, tolerates cool, wet early season, bolt tolerant. *Source History: 1 in 1998; 1 in 2004.* **Sources: OSB.**

Bazooka - 70-75 days - Large, firm, compact, domed heads, small core, flavorful, good summer production, uniform yields. *Source History: 1 in 2004.* **Sources: Ves.**

Beatrice - 75 days - Bright-green, solid, crunchy heads, short internal stalk, excel. mildew and root aphid res. *Source History: 1 in 1994; 1 in 1998; 1 in 2004.* **Sources: Tho.**

Ben Shemen - 60-70 days - Summer butterhead type, bred for intense heat and slow bolting, large dark-green compact sweet crisp heads, for late spring planting and Southern gardens, Israel. *Source History: 2 in 1981; 2 in 1984; 1 in 1987; 1 in 1991; 1 in 1994; 1 in 1998; 1 in 2004.* **Sources: Ho13.**

Bibb - (Limestone, Limestone Bibb, Kentucky Bibb) - 54-75 days - AAS, loosely folded 3.5 in. heads, dark-green thick smooth brown-tinged leaves, inside blanches golden-yellow, does best if planted early, bolts in hot weather, less tipburn than most butterheads, bolts in hot weather, black seed. *Source History: 71 in 1981; 64 in 1984; 56 in 1987; 50 in 1991; 44 in 1994; 41 in 1998; 36 in 2004.* **Sources: Ada, All, Bu8, CH14, CO23, Coo, Cr1, Dgs, DOR, Ear, Ers, Gr27, HA3, He17, HO2, Jor, Kil, La1, Lin, LO8, Mel, Mey, Mo13, Ni1, Ont, PAG, Pe2, RUP, Sau, Se17, Se26, Sh9, Te4, Wet, Wi2, WI23.**

Bibb Forcing - 60 days - Butterhead, dark-green leaves, firm and tender, fine flavor, good for forcing. *Source History: 1 in 1981; 1 in 1984; 2 in 1987; 1 in 1991; 1 in 1994; 1 in 1998; 1 in 2004.* **Sources: Ge2.**

Bibb, Burpee's - 75 days - Superior to regular bibb, small loosely folded heads, dark-green outer leaves sometimes tinged brown, inside blanches to golden-yellow, tipburn res., slow bolting. *Source History: 3 in 1981; 3 in 1984; 1 in 1987; 1 in 1991; 1 in 1994; 1 in 1998; 2 in 2004.* **Sources: Bu2, Hi13.**

Bibb, Slow Bolting - 75 days - Dark-green, can withstand hot weather without bolting. *Source History: 3 in 1981; 3 in 1984; 2 in 1987; 2 in 1991; 1 in 1994; 1 in 1998; 1 in 2004.* **Sources: Com.**

Bibb, Summer - (Summer Baby Bibb, Esmeralda) - 60-65 days - Summer str. of Kentucky Bibb, slower bolting, better adapted to Southern heat, used mainly on muck, med-green med-rosette head, deep waxy leaves, firm interior, vigorous, stands 2 to 3 weeks longer than original bibb, recommended for hotbed or greenhouse, black seed, Dr. Raleigh/Cornell, 1963. *Source History: 19 in 1981; 17 in 1984; 16 in 1987; 14 in 1991; 15 in 1994; 9 in 1998; 8 in 2004.* **Sources: Be4, ME9, Na6, Or10, Roh, SE14, TE7, WE10.**

Bibb, Susan's Red - 60 days - Large ruffled tender leaves with red edges, no bitterness, very attractive addition to salads, heirloom. *Source History: 1 in 1994; 1 in 1998; 4 in 2004.* **Sources: Se16, Se17, So1, Und.**

Blonde de Paris - Crisphead, similar to Iceberg. *Source History: 2 in 1991; 2 in 1994; 1 in 1998; 2 in 2004.* **Sources: Sk2, WE10.**

Blush - 65-100 days - Miniature Iceberg, identical to Mini Green but with deep green and red tinged leaves, slow to bolt. *Source History: 4 in 1998; 2 in 2004.* **Sources: Mcf, Tho.**

Blushed Butter Oak - 45-50 days - Compact oakleaf butterhead with pink and green colors, buttery taste, cold hardy, butterhead type x oak leaf cross dev. by Frank Morton of Shoulder to Shoulder Farm and Wild Garden Seeds. *Source History: 2 in 1998; 9 in 2004.* **Sources: Ag7, Fe5, Ki4, Sa5, Sh12, So12, Syn, Ta5, Tu6.**

Bon Jardinier - (Du Bon Jardinier) - 55-60 days - Med-large head, med-green with red tinge. *Source History: 1 in 1987; 2 in 1991; 1 in 1994; 1 in 1998; 1 in 2004.* **Sources: Lej.**

BOS 9032 - Iceberg with large frame, compares with Desert Queen. *Source History: 1 in 2004.* **Sources: ME9.**

BOS 9091 - Iceberg with large frame, compares with Desert Queen. *Source History: 1 in 2004.* **Sources: ME9.**

Boston, Big - 66-77 days - Firm round heads to 12 in. dia., med-green crumpled leaves with reddish-brown tinged edges, forces well, stands cold without injury, good shipper, long standing, buttery-yellow heart, good home garden var. for summer and fall, pre-1894. *Source History: 25 in 1981; 21 in 1984; 14 in 1987; 11 in 1991; 6 in 1994; 8 in 1998; 5 in 2004.* **Sources: Ba8, Com, He8, Or10, Sa9.**

Boston, Burgundy - Bright color throughout, good bolt tol. *Source History: 2 in 1991; 2 in 1994; 2 in 1998; 1 in 2004.* **Sources: WE10.**

Boston, Dark Green - 68-80 days - Med-green med-sized butterhead type, smooth outer leaves with wavy edges, susc. to heat damage so no mid-summer cropping, white seed, darker green than White Boston, produces quality heads under adverse conditions, resembles Cobham Green,

Ferry-Morse, 1957. *Source History: 30 in 1981; 27 in 1984; 23 in 1987; 23 in 1991; 21 in 1994; 12 in 1998; 5 in 2004.* **Sources: DOR, Fo7, HO2, Loc, ME9.**

Boston, White - (Summer Unrivalled, White Big Boston, Special White Boston) - 55-76 days - An improved Big Boston selected under Florida conditions, lighter green, no red tinge, grows higher off ground so no bottom rot, TB res., will not scale in heat or turn red from cold, white seed, good for forcing, home garden or market, excel. on black muck. *Source History: 24 in 1981; 21 in 1984; 17 in 1987; 19 in 1991; 22 in 1994; 18 in 1998; 11 in 2004.* **Sources: Ber, Cr1, Dom, DOR, Ers, Fo7, He8, La1, Ont, PAG, WE10.**

Brauner Trotzkopf - Loose head is translucent green in the middle, shading to light ochre-red at the tips, tender, very hardy. *Source History: 2 in 1998; 1 in 2004.* **Sources: Ho13.**

Brune d'Hiver - (Brown Winter) - 55-65 days - Red butterhead, 8-10 in. head, wrinkled green leaves tinged on edges with amber-red, very cold hardy, French heirloom, 1855. *Source History: 1 in 1981; 1 in 1984; 3 in 1987; 8 in 1991; 10 in 1994; 5 in 1998; 6 in 2004.* **Sources: Ag7, Coo, Fe5, Ri12, Se7, Ta5.**

Bunte Forellenschuss - 40-55 days - Maroon splashed, sweet, apple-green leaves, forms 8-10 in. loose head, "bunte" in German means "colorful", given to SSE in 1973. *Source History: 2 in 2004.* **Sources: Hud, Se16.**

Burgundy Ice - 77 days - Medium size heads, bright green inside, burgundy outer leaves. *Source History: 1 in 2004.* **Sources: Bu2.**

Butter Beauty - Pale green, oval leaves accented with light red and brown, good quality, the first to go to seed. *Source History: 1 in 1998; 1 in 2004.* **Sources: Sa5.**

Butter King - (Buttercrunch) - 60-80 days - AAS/1966, slow to bolt or turn bitter, light-green crisp 12-13 oz. butterheads, does well in Midwest heat, Boston type but nearly twice as large and more tender, grows vigorously, good flavor, disease res., Central Exp. Sta. in Ottawa, Canada. *Source History: 17 in 1981; 15 in 1984; 10 in 1987; 6 in 1991; 4 in 1994; 3 in 1998; 3 in 2004.* **Sources: Se7, Shu, So9.**

Buttercrunch - (Buttercrunch Bibb, Butter King) - 50-75 days - AAS/1963, bibb type with larger more compact yellow-white heart, less bolting, long lasting, heat tol., dark-green reddish-tinged 4.5 in. rosette heads, white seed, 12 in. plant, thick blistered leaves, mosaic free, 456 x Bibb, Dr. Raleigh at Cornell, 1963. *Source History: 136 in 1981; 130 in 1984; 123 in 1987; 117 in 1991; 112 in 1994; 116 in 1998; 111 in 2004.* **Sources: Ada, Ag7, Alb, All, BAL, Be4, Bo19, Bou, Bu2, Bu3, Bu8, But, CA25, CH14, CO23, Co31, Com, Coo, Cr1, Dam, De6, DOR, Ear, Ers, Fa1, Fe5, Fi1, Fis, Fo7, Ga1, Ga7, Ge2, GLO, Goo, Gr27, GRI, Gur, HA3, Ha5, Hal, He8, He17, Hi6, Hig, HO2, Hum, Jo1, Jor, Jun, Kil, La1, Lin, LO8, Loc, Ma19, MAY, Me7, ME9, Mel, Mey, MIC, Mo13, MOU, Ni1, OLD, Ont, Or10, Orn, PAG, Par, Pe1, Pe2, Pe6, Pin, Pr3, Ri12, RIS, Roh, Ros, RUP, Sau, Scf, SE4, Se7, SE14, Se16, Se25, Se26, SGF, Sh9, Si5, Sk2, So1, So12, SOU, Sto, Sw9, Te4, TE7, Ter, Tho, Tt2, Und, Up2, Ver, Vi4, WE10, We19, Wet, Wi2, WI23.**

Butterhead, Green - 58 days - Heirloom from Marburg, Yugoslavia, good flavor and texture with low bitterness, best grown in northern areas. *Source History: 2 in 1991; 1 in 2004.* **Sources: Bo17.**

Butterhead, Red - (Yugoslavian Red Butterhead) - 55-60 days - Tender, red-tinged leaves form large loose heads 10-12 in. wide, white center is sometimes streaked with red, mild flavor, crisp texture, excellent producer, heirloom from peasant family in Marburg, Yugoslavia. *Source History: 1 in 1987; 1 in 1991; 1 in 1994; 1 in 1998; 8 in 2004.* **Sources: Hud, Mcf, Pr3, Se16, So1, St18, Te4, Und.**

Capitan - (Capitane) - 62-65 days - Large vigorous plants, tight med-green heads, excel. shape and uniformity, LMV and bremia res., Boston type butterhead, can be used in greenhouses. *Source History: 5 in 1981; 5 in 1984; 6 in 1987; 5 in 1991; 6 in 1994; 1 in 1998; 2 in 2004.* **Sources: Se7, So1.**

Cardinale - 48-60 days - Red French leaf variety with thick, dark, brilliant red leaves, upright growth means less trouble with bottom rot, TB and bolt res., formerly Vilmorin #602. *Source History: 4 in 1998; 6 in 2004.* **Sources: Ki4, SE4, Sw9, Ter, VI3, We19.**

Carmona - 55 days - Red butterhead with excellent size, color, disease tolerance and holding ability, mild flavor holds well, tolerant to bottom rot, tipburn and virus. *Source History: 1 in 1994; 1 in 2004.* **Sources: Syn.**

Cassandra - 65 days - Pale green butterhead, excellent flavor and quality, resists most races of downy mildew and LMV, good summer and fall crop. *Source History: 1 in 2004.* **Sources: Tho.**

Cavolo di Napoli - Med. size head with bright green outer leaves, inner leaves white, mild flavor with crunchy texture, flavor is best when harvested in cool weather. *Source History: 1 in 2004.* **Sources: CA25.**

Cerise - (Cerize) - 48-64 days - Med-green outer leaves heavily overlain with warm shiny red, med-sized round well formed heads, many blanched interior leaves and mild crisp quality typical of Iceberg types. *Source History: 1 in 1991; 1 in 1994; 2 in 1998; 1 in 2004.* **Sources: Ga7.**

Cinnamon Red - 55-60 days - Beautiful 'true' red butterhead, smooth, thick red leaves, home garden and gourmet markets. *Source History: 1 in 2004.* **Sources: HO2.**

Climax - (Climax M.T.) - 90 days - Large med-dark green solid heads, good butt appearance, tipburn tol., res. to rib discoloration, for winter harvest in Imperial Valley, USDA/Beltsville. *Source History: 5 in 1981; 5 in 1984; 4 in 1987; 5 in 1991; 6 in 1994; 5 in 1998; 3 in 2004.* **Sources: Be4, SE14, WE10.**

Cobham Green - (Green Cobham) - 62 days - Very dark-green for a butterhead, for spring or summer planting, large solid heads, leaves crackling crisp, from England. *Source History: 2 in 1981; 1 in 1984; 1 in 1987; 1 in 1991; 1 in 1994; 1 in 1998; 1 in 2004.* **Sources: Bou.**

Continuity - (Crisp As Ice, Merveille des 4 Saisons) - 70-75 days - Extra early iceberg type, firm solid crisp and fine flavored, reliable header, some tendency to bolt in hot weather, thick dark-green leaves with bronze-red outer leaves. *Source History: 6 in 1981; 6 in 1984; 6 in 1987; 8 in 1991; 9 in 1994; 7 in 1998; 8 in 2004.* **Sources: Bou, Ga22,**

RUP, Sa5, Se7, Ter, WE10, We19.

Coolgreen - Large heads, good color throughout, excel. uniformity, great for fall planting, DM3 res, TB tol., PVP 1998 HA20. *Source History: 1 in 1998; 1 in 2004.* **Sources: HA20.**

Court - 80 days - Iceberg type, dark green, round smooth heads, tight leaf wrapper, ideal for long season production and organic growing, good DM and bolt res. *Source History: 1 in 2004.* **Sources: Tho.**

Craquerelle du Midi - 60 days - Heat tolerant French butterhead, open heart form, very slow to bolt. *Source History: 1 in 1981; 1 in 1984; 1 in 1987; 1 in 1991; 1 in 1994; 1 in 1998; 3 in 2004.* **Sources: Coo, Par, Sa5.**

Crisp As Ice - (Hartford Bronzehead) - 65-74 days - Good hot weather garden var., slow bolt, thick crumpled deep-green leaves with bronze cast, butter-yellow hearts, medium size, fine quality. *Source History: 6 in 1981; 5 in 1984; 4 in 1987; 4 in 1991; 1 in 1994; 1 in 2004.* **Sources: Se17.**

Crispino - 53-57 days - Glossy-green large firm heads, early like Ithaca but larger, less susceptible to twisted leaves and has a higher percentage of good heads, white interior, juicy and mild, Holland. *Source History: 2 in 1991; 1 in 1994; 1 in 1998; 2 in 2004.* **Sources: Jo1, Se26.**

Crispy Frills - 80 days/head, 50 days/leaf, ruffled, frilly, fan shaped leaves, crisp, no bitterness, bolt resistant. *Source History: 2 in 2004.* **Sources: Bu2, Hi13.**

Czechoslovakian - Loose heads with bronze outer leaves, flavorful, brought to New York over 65 years ago. *Source History: 1 in 1998; 1 in 2004.* **Sources: Ho13.**

Density - *Source History: 1 in 1991; 2 in 1994; 1 in 1998; 1 in 2004.* **Sources: WE10.**

Desert Queen - 88 days - Imp. Empire type, medium size head, bolt tolerant, PVP 1988 Seminis Vegetable Seeds, Inc. *Source History: 1 in 2004.* **Sources: Sto.**

Desert Storm - Large framed iceberg, large dark green head, consistent high yields, DM and LMV res., good tolerance to Big Vein, PVP 1997 HA20. *Source History: 1 in 2004.* **Sources: HA20.**

Diamond Gem - Small green heads, firm upright habit, good for individual salads and sandwiches. *Source History: 1 in 1994; 2 in 1998; 1 in 2004.* **Sources: Syn.**

Divina - 59-80 days - Thick and shiny dark-green leaves, large heads weigh over 1 lb., fairly tight for a butterhead type, slow bolting, TB res., from France. *Source History: 3 in 1991; 9 in 1994; 9 in 1998; 8 in 2004.* **Sources: CO30, Dam, Ki4, Pin, SE4, Ta5, VI3, Wi2.**

Du Bon Jardiniere - Light green med. size heads with a rose tinge, French. *Source History: 1 in 1998; 1 in 2004.* **Sources: Ho13.**

Edox - 64 days - Medium size Red Boston, imp. Manto type, deep red, firm leaves, good shelf life, tolerates poor weather and NL1-16 diseases. *Source History: 1 in 2004.* **Sources: Sto.**

El Dorado - 65 days - Compact head with short core, rich green interior, dark green leaves, PVP 1993 Royal Sluis B. V. *Source History: 1 in 1998; 1 in 2004.* **Sources: OSB.**

Empire - (Empire M.T.) - 72-90 days - Medium size solid med-green head, medium size short core, res. to TB and RD, used in warmer areas, very slow bolting, extremely high quality, USDA/1957. *Source History: 8 in 1981; 6 in 1984; 5 in 1987; 8 in 1991; 6 in 1994; 7 in 1998; 2 in 2004.* **Sources: Mey, WE10.**

Encanto - Large, dark green Boston type with uniform maturity, size and color, excel. TB tol. *Source History: 1 in 2004.* **Sources: ME9.**

Ermosa - 48-52 days - Kagran Summer/Mantilia type with deeper green color like Dark Green Boston, tipburn and bolt res., LMV res., plant in spring, summer and fall, intro. in 1993. *Source History: 2 in 1994; 2 in 1998; 2 in 2004.* **Sources: Jo1, Tu6.**

Esmeralda - 53-68 days - Buttery lime-green heads, 1 lb., sweet taste, slow bolting, tender texture, resistant to most lettuce diseases, good yields, PVP 1991 Royal Sluis B. V. *Source History: 3 in 1991; 4 in 1994; 13 in 1998; 19 in 2004.* **Sources: BAL, CA25, Ers, Fe5, GLO, Jun, Lin, Mcf, ME9, Ni1, OSB, RIS, SE4, Sto, Ter, Tt2, Twi, Ver, We19.**

Fallgreen - (FM-1638) - Vanguard leaf type, large, firm head, med. size core, sood shipper, TB res., PVP 1992 HA20. *Source History: 1 in 1991; 1 in 1994; 1 in 1998; 1 in 2004.* **Sources: HA20.**

Fatima - Large, dense butterhead, resists TB, Bremia NL 1-16, LMV. *Source History: 1 in 2004.* **Sources: Tho.**

Flashy Trout Back - Shoulder to Shoulder Farm sel. from Forellenschuss, uniform heads spattered with bright red, resists TB, sclerotinia bottom rot, withstands close spacing, upright growth, crisp texture. *Source History: 3 in 2004.* **Sources: Ga1, Ki4, Sh12.**

Floresta - 65 days - Medium size heads, light green, smooth leaves, tender, excellent flavor. *Source History: 1 in 2004.* **Sources: La1.**

Formidana - 60 days - Green crisphead type, forms 5-7 in. heads on very small plants, continually growing solid heads, recent dev. from Dutch breeding so that plants the size of a silver dollar start to make heads, as the plants grow bigger so do the heads. *Source History: 2 in 1991; 2 in 1994; 1 in 1998; 2 in 2004.* **Sources: Se7, So12.**

Gemini - 65-85 days - Rich deep-green color, uniform compact heads, good reliability under temperature and moisture fluctuation, ideal for fresh food market, PVP 1992 Sakata Seed America, Inc. *Source History: 1 in 1991; 3 in 1994; 3 in 1998; 1 in 2004.* **Sources: SAK.**

Grandpa Admire's - 60 days - Large uniform loose heads are light-green with bronze edges, mild flavor, slow to bolt, stays tender longer than most, white seed, Grandpa George Admire was a Civil War veteran born in 1822, seed given to SSE in 1977 by 90 year-old Cloe Lowrey, Grandpa Admire's granddaughter. *Source History: 2 in 1991; 3 in 1994; 3 in 1998; 3 in 2004.* **Sources: He8, Hud, Se16.**

Great Lakes - (Premier, Premier Great Lakes) - 65-100 days - AAS/1944, iceberg type, firm crisp large bright-green heads, res. to tipburn heat sunburn and rain, widely adapted, slow to bolt, white seed, for Midwest summers, good on muck and upland soils, MIC/AES. *Source History: 81 in 1981; 71 in 1984; 64 in 1987; 46 in 1991; 36 in 1994; 33 in 1998; 25 in 2004.* **Sources: Ada, Alb, BAL, Bu2, Bu8, CA25,**

De6, Ear, Fo7, Ge2, Gr27, He8, Hum, La1, Lin, Ma19, Mey, Ont, Ri12, Roh, SGF, Sh9, Si5, SOU, Wet.

Great Lakes 118 - (Great Lakes 118 M.T.) - 63-90 days - Large cabbage-like heads, thick broad slightly crumpled glossy leaves, scald and TB res., sure header, widely adapted, very hardy, overwinter in mild areas, 1948. *Source History: 13 in 1981; 12 in 1984; 13 in 1987; 11 in 1991; 13 in 1994; 11 in 1998; 11 in 2004.* **Sources: CH14, DOR, GLO, HA3, Jor, LO8, MOU, SE14, TE7, WE10, WI23.**

Great Lakes 366 - (Great Lakes 366 M.T.) - Med-late, large dark-green solid head, acceptable butt appearance, large core, res. to tipburn, popular for early and late harvest, 1954. *Source History: 3 in 1981; 3 in 1984; 5 in 1987; 5 in 1991; 5 in 1994; 2 in 1998; 2 in 2004.* **Sources: DOR, WE10.**

Great Lakes 407 - (Great Lakes 407 M.T.) - 90-100 days - Standard str. for Southwest winter growing areas, large bright-green wrapper leaves protect head, sunburn rain and TB res., withstands adverse weather, slow bolt, 1945. *Source History: 3 in 1981; 3 in 1984; 4 in 1987; 4 in 1991; 4 in 1994; 3 in 1998; 2 in 2004.* **Sources: SE14, WE10.**

Great Lakes 659 - (Great Lakes 659 M.T.) - 80-88 days - Most TB res. and versatile Great Lakes str., popular both North and South for producing solid med-large dark-green heads in cool or warm weather, medium core, white seed, slow bolt, does best in warm weather, good on heavy soils, mosaic free, 1944. *Source History: 25 in 1981; 24 in 1984; 17 in 1987; 19 in 1991; 20 in 1994; 14 in 1998; 15 in 2004.* **Sources: Bu3, Com, Cr1, Dam, Ers, GRI, ME9, MIC, PAG, RIS, Ros, SE14, Sh14, WE10, WI23.**

Great Lakes 659-G - 90 days - Large round solid dark-green heads, small core, res. to tipburn sunburn and cold, white seed, early September plantings produce in 115 days. *Source History: 2 in 1981; 2 in 1984; 1 in 1987; 1 in 1991; 1 in 1994; 1 in 1998; 1 in 2004.* **Sources: Wi2.**

Great Lakes 6238 - 83-94 days - Vigorous, distinctive deep grassy-green, large compact solid heads, for late sowing, stands high temps, SR TB and bottom rot tol., slow bolt, ready all at once. *Source History: 5 in 1981; 5 in 1984; 6 in 1987; 3 in 1991; 2 in 1994; 1 in 1998; 1 in 2004.* **Sources: Kil.**

Great Lakes Mesa 659 - (Mesa 659, Great Lakes Mensa 659, Great Lakes Mesa 659 M.T.) - 83-90 days - Great Lakes 659 str., larger and more uniform and better color, dark-green solid heads, small core, very slow to bolt, heat and TB res., for the East in the fall. *Source History: 15 in 1981; 14 in 1984; 17 in 1987; 17 in 1991; 15 in 1994; 10 in 1998; 7 in 2004.* **Sources: Be4, DOR, Loc, Or10, RUP, SE14, WE10.**

Great Lakes, Early M.T. - Large dark-green head, extra good quality, TB and heat res. *Source History: 2 in 1981; 1 in 1984; 1 in 1987; 1 in 1991; 1 in 1994; 1 in 1998; 1 in 2004.* **Sources: Fis.**

Greenday - 45 days - Dark green, long, heavy leaves, PVP 2000 Seminis Vegetable Seeds, Inc. *Source History: 1 in 2004.* **Sources: RIS.**

Grosse Blonde Paresseuse - 60 days - Butterhead type, well formed head with tightly folded green outer leaves and light golden heart, sweet and tender, small cabbage-like heads, France. *Source History: 1 in 1981; 1 in 1984; 1 in 1987; 2 in 1991; 2 in 1994; 1 in 2004.* **Sources: Ta5.**

Habana - 75 days - Dark leaf butterhead, pale yellow hearts, resists DM and LMV, good for summer and fall production. *Source History: 1 in 2004.* **Sources: Tho.**

Hanson - (Nonpariel, Improved Hanson) - 65-80 days - Very large yellowish-green head, widely adapted, stands heat well, sure header, frilled leaves, white heart, tipburn tol., not for forcing or wintering over, intro. by Henry A. Dreer Company in 1871. *Source History: 13 in 1981; 12 in 1984; 11 in 1987; 9 in 1991; 9 in 1994; 7 in 1998; 8 in 2004.* **Sources: Alb, Bo19, DOR, He8, Se16, Vi4, WE10, Wet.**

Harmony - (Anthem) - 68 days - Uniform, large heavy head size, excel. internal quality, resists DM races I, IIA IIB, III, IV, V. *Source History: 2 in 2004.* **Sources: ME9, RIS.**

Hilde - 63-79 days - Butterhead, Reine de Mai x Unrivalled, vigorous, rapid grower, very promising, for spring and autumn sowing, early frame and cloche culture, France. *Source History: 5 in 1981; 3 in 1984; 2 in 1987; 2 in 1991; 2 in 1994; 1 in 1998; 2 in 2004.* **Sources: Dam, Sa5.**

Hubbard's Market - 65 days - Medium size globular plants, crumpled dark-green leaves with straight edges, buttery heart, fine for cold frame culture, white seed. *Source History: 1 in 1981; 1 in 1984; 1 in 2004.* **Sources: Se17.**

Hyper Red Rumple Waved - 50 days - Plants spread to 12 in., grow to only 5-6 in. tall, deeply savoyed, ruffled dark plum-red leaves, good flavor, pleasing texture, good cold tolerance, selected by Frank Morton from a cross between Valeria and Wavy Red Cos. *Source History: 4 in 2004.* **Sources: Fe5, Sh12, So9, Ter.**

Ice Cube - 85-100 days - Miniature Iceberg about the size of an orange, leaves are slightly less swirled than Mini Green, adapted to West coast conditions. *Source History: 1 in 1998; 1 in 2004.* **Sources: Mcf.**

Ice Queen - 80-85 days - Medium-large heads, bright green outer leaves, adaptable to most climates, good bolt res. *Source History: 3 in 2004.* **Sources: Ag7, Dam, Te4.**

Iceberg - (Black Seeded Iceberg, Giant Crystal Head) - 50-85 days - Compact med-large heads, crisp hearts, light-green leaves with waxy bronze fringe, TB and heat res., good in mid-summer, does well in the East and in mountain areas, not suited for shipping, introduced in 1894. *Source History: 83 in 1981; 74 in 1984; 72 in 1987; 59 in 1991; 49 in 1994; 52 in 1998; 48 in 2004.* **Sources: Ada, Alb, All, Ba8, Be4, Bu3, Bu8, CA25, CH14, Co31, Cr1, DOR, Ear, Ers, Fo7, Gr27, HA3, He8, He17, Hi13, Jor, La1, Lej, Lin, LO8, ME9, Mel, Mey, MIC, Mo13, MOU, OLD, Ont, Or10, PAG, Ros, RUP, Sau, SE14, Se25, Se26, Sh9, SOU, Vi4, WE10, Wet, Wi2, WI23.**

Iceberg A - 85 days - Outer leaves are fringed and savoyed, silvery white inner leaves, firm heads, slow to wilt, widely adapted, Burpee introduction dating back to 1894. *Source History: 1 in 2004.* **Sources: Bu2.**

Iceberg, Red - 60-80 days - Red to purple and green outer wrapper leaves, 8-10 in. fairly tight head, sweet flavor, heat tolerant. *Source History: 1 in 1994; 4 in 2004.* **Sources: Ag7, Roh, Se7, Se16.**

Imperial Winter - 90 days - Very hardy, one of the largest and best varieties to plant in the open in autumn, not recommended for spring or summer sowing. *Source History: 1 in 1981; 1 in 1987; 1 in 1991; 2 in 1994; 2 in 1998; 1 in 2004.* **Sources: Ho13.**

Ithaca - (Ithaca M.T., Ithica M.I., Improved Iceberg) - 53-85 days - Dev. by Dr. Minotti at Cornell for muck or upland, med-dark green, 5 x 5.5 in. dia. and well wrapped, TB and BR res. *Source History: 42 in 1981; 38 in 1984; 33 in 1987; 34 in 1991; 27 in 1994; 28 in 1998; 24 in 2004.* **Sources: Be4, Ers, Fi1, Ga1, Gr27, Gur, Ha5, He17, Hig, HO2, La1, ME9, MIC, Mo13, RIS, Roh, RUP, SE14, Se25, Sh9, Si5, Sto, Twi, WE10.**

Ithaca 989 - 70-85 days - Asgrow selection of original Ithaca, slightly earlier, upright frilled leaves are well whorled around firm attractive heads, med-size, TB tol., for spring and late summer in Eastern U.S. and Canada. *Source History: 1 in 1987; 1 in 1991; 2 in 1994; 1 in 2004.* **Sources: Sto.**

Juliet - 61 days - Fancy medium-size butterhead with thick green leaves overlaid with bronze-red, smooth and buttery tasting, DM tol., MTO tested, from France. *Source History: 1 in 1991; 1 in 1994; 1 in 1998; 1 in 2004.* **Sources: Ta5.**

Kagran Summer - 54 days - Very bolt res. bibb for early-mid summer, large light-green heads, large well-folded firm hearts, res. to TB and hot weather, harvest when mature or gets bitter. *Source History: 2 in 1981; 2 in 1984; 4 in 1987; 6 in 1991; 4 in 1994; 4 in 1998; 2 in 2004.* **Sources: He8, Pla.**

Kagraner Sommer - (Kagraner, Orfeo Kagraner Summer, Butter Bow Head) - 58 days - French butterhead, originally from Germany, slow to bolt in summer heat, soft well-formed medium size light-green heads, midseason. *Source History: 5 in 1981; 5 in 1984; 8 in 1987; 10 in 1991; 10 in 1994; 8 in 1998; 9 in 2004.* **Sources: Alb, DOR, Fe5, Fo7, Lej, Ri2, Se26, Te4, WE10.**

Key Lime - 63 days - Lime-green butterhead type, sweet buttery flavor, large heads can reach 1 lb., slow bolting, originally from southern Florida, thus the name. *Source History: 1 in 2004.* **Sources: He8.**

Kinemontpas - French lettuce, heat res., very slow bolting, large green leaves, delicate gourmet flavor, translation: very slow to bolt. *Source History: 2 in 1981; 1 in 1984; 2 in 1987; 6 in 1991; 2 in 1994; 2 in 1998; 3 in 2004.* **Sources: Coo, Lej, Syn.**

King - 70 days - Heads measure up to 10 in. across, crisp light green leaves are held off the ground, fall and spring planting, from Switzerland. *Source History: 1 in 2004.* **Sources: Ter.**

Krolowa Majowych - *Source History: 1 in 2004.* **Sources: Se17.**

Kwiek - 50-60 days - Large green butterhead, delicious sweet flavor, for late fall holdover or early spring forcing under cover, dev. in Switzerland. *Source History: 1 in 1981; 1 in 1984; 1 in 1987; 1 in 1991; 2 in 1994; 2 in 1998; 1 in 2004.* **Sources: Ho13.**

La Brillante - 60-75 days - Summer lettuce, combines the shiny leaf of a crisphead with butterhead tenderness, resistant to difficult conditions. *Source History: 1 in 1987; 1 in 1991; 1 in 1994; 1 in 1998; 2 in 2004.* **Sources: Pe1, Syn.**

Lakeland - Heavy, dark green tight heads, few outer leaves. *Source History: 1 in 2004.* **Sources: Se26.**

Landis Winter - 45 days - Loose butterhead, matures to 11-12 in. diameter, tender green heads, sweet flavor, resists hard frosts, will survive in cold frame if well established by mid-October, listed in 1878 seed list and 1785 agricultural treatise by Abbe Rozier. *Source History: 1 in 2004.* **Sources: Und.**

Lattughino - (Red Tinged Winter) - Plant in fall for early spring heads, hardy to -5 degrees F, summer heat causes bitterness. *Source History: 1 in 2004.* **Sources: Tu6.**

Legacy - *Source History: 1 in 2004.* **Sources: CO30.**

Lentissima a Montare 3 - Cylindrical, upright heads with tightly packed, smooth leaves, light green color, crisp and thick with sweet mild flavor. *Source History: 1 in 2004.* **Sources: CA25.**

Limestone Bibb - 60-70 days - Small compact rosette heads, thick smooth waxy deep-green leaves, yellow-green hearts, very pleasant, creamy flavor, black seed, first offered in 1850 in Kentucky. *Source History: 1 in 1981; 1 in 1984; 2 in 1994; 2 in 1998; 2 in 2004.* **Sources: Fo7, Ri12.**

Little Gem - (Sucrine) - 45-70 days - Refined Dark Boston type with midget 5 in. heads, tightly wrapped, bright green leaves, pale yellow heart, sweet and crunchy, superb flavor, heat tol., can be served whole. *Source History: 3 in 1991; 8 in 1994; 13 in 1998; 22 in 2004.* **Sources: Ba8, CO23, CO30, DOR, Ga7, He8, Hi13, Ki4, La1, ME9, OSB, Par, Pe2, Pe6, Roh, Sa5, Se7, SE14, So9, Te4, Vi4, WE10.**

Loma - (Little Loma) - 46-55 days - French lettuce with thick, firm, dark green leaves that are slightly frilled around the edge, great for fresh cut salad or harvested whole for baby lettuce sales, tolerates heat, tipburn and mildew, formerly Vilmorin #3042. *Source History: 4 in 1998; 8 in 2004.* **Sources: CA25, Jo1, Na6, Orn, Sw9, Ter, VI3, We19.**

Magenta - 48 days - Improved Sierra with conical head, crispy green heart, darker red color, resists bolting, TB and bottom rot, good taste, spring and summer planting. *Source History: 1 in 2004.* **Sources: Jo1.**

Magnet - 60-75 days - Excellent forcing butterhead, large tender pale-green head, fine flavor, grows quickly, withstands summer heat, res. to adverse conditions. *Source History: 1 in 1987; 1 in 1991; 2 in 2004.* **Sources: In8, Se17.**

Margarita - 68 days - Dark green Boston type, large dark green heads, smooth rounded leaves, DM, LMV and TB res., PVP 2000 Enza Zaden de Enkhuizer Zaadhandel B.V. *Source History: 2 in 2004.* **Sources: ME9, SE4.**

Marvel of 4 Season, Winter Sun Str. - 60 days - Very cold hardy var. can withstand 20 degree F temperature in a well-sealed double-walled greenhouse. *Source History: 1 in 1998; 1 in 2004.* **Sources: Bou.**

Maule's Philadelphia Butter - Most popular salad lettuce in the 19th century, waxy leaves, compact soft head, light green with red flecks, 10 in. diameter, sweet nutty flavor, introduced by Maule in 1870. *Source History: 2 in 2004.* **Sources: In8, Se17.**

Maverick - 88 days - Empire type, med-large head size, medium dark green, good solidity, very good TB and bolting tol., does well in warmer conditions, PVP 1988 Genecorp. *Source History: 1 in 1994; 1 in 2004.* **Sources: Sto.**

May King - (Mai Konig) - 60-65 days - Med-large compact head, light-green red-tipped leaves, small creamy heart, fine quality, greenhouse forcing, scorches in heat, long harvest, white seed. *Source History: 9 in 1981; 7 in 1984; 9 in 1987; 8 in 1991; 8 in 1994; 8 in 1998; 5 in 2004.* **Sources: Coo, DOR, Te4, Tu6, WE10.**

May Queen - (Reine de Mai, Rainha de Mai, Regina di Maggio) - 45-60 days - Earliest butterhead for early and coldframe sowings, medium size pale-green heads tinged with brown, small firm round creamy-yellow hearts, also good for summer, 19th century heirloom. *Source History: 2 in 1981; 1 in 1984; 1 in 1987; 3 in 1991; 5 in 1994; 4 in 1998; 7 in 2004.* **Sources: Ba8, DOR, He8, Mcf, Red, Se24, WE10.**

Melnicky Maj - 50 days - Czech heirloom, green with some light red tinge on some heads. *Source History: 1 in 1998; 1 in 2004.* **Sources: Se17.**

Merveille des Quatre Saisons - (Four Seasons, Meraviglia Delle Quattro Stagioni, Continuity, Marvel of Four Seasons) - 45- 70 days - Unusual French Bibb type, reddish leaves with cranberry-red tips, 8-12 in., pale blond-green tight heart, 12 to 16 in. across, excellent flavor, in hot weather holds flavor but bolts, intro. before 1885. *Source History: 9 in 1981; 8 in 1984; 12 in 1987; 22 in 1991; 26 in 1994; 34 in 1998; 45 in 2004.* **Sources: Au2, Ba8, Ber, Bu2, CO23, CO30, Dam, DOR, Ers, Ga1, Ga22, Goo, He8, Hi13, Hig, Ho13, Ki4, La1, Lej, Loc, ME9, Na6, Orn, Pe2, Pin, Red, Ri12, RUP, Se7, Se8, SE14, Se16, Se17, Se24, Se26, Sk2, So9, St18, Ta5, Te4, TE7, Tu6, VI3, Vi4, WE10.**

Mescher - (Mescher Bibb Schweitzer Strain, Schweitzer's Mescher Bibb) - 50 days - Small tight crisp heads of green leaves ringed with red, excellent flavor and appearance, best grown in cool weather--has survived to 28 deg. F, dates back to the 1700s, this strain brought to the U.S. from Austria in early 1900s and has been maintained as a Schweitzer family heirloom. *Source History: 2 in 1981; 2 in 1984; 4 in 1987; 4 in 1991; 3 in 1994; 4 in 1998; 4 in 2004.* **Sources: Ea4, Ho13, Lan, So1.**

Michelle - Red Batavian type, very dark green during summer heat, shows more red color in cooler temps of spring and fall, excel. bolt and TB tol., compares to Sierra. *Source History: 1 in 2004.* **Sources: ME9.**

Mignonette Bronze - (Early Surehead, Mignonette Bronze Rosette) - 55-67 days - Small crumpled green-on-bronze globular heads, frilled leaves, heart blanches creamy-yellow, fine for tropical or hot weather, slow to bolt, introduced in 1898. *Source History: 12 in 1981; 10 in 1984; 7 in 1987; 8 in 1991; 11 in 1994; 14 in 1998; 18 in 2004.* **Sources: Ba8, CO23, DOR, Fe5, Fo7, He8, Ho13, La1, Lej, Ni1, Orn, Se7, Se8, Sk2, Syn, Vi4, WE10, WI23.**

Mignonette Green - (Manoa, Mignonette Green Rosette) - 65-67 days - Small compact plants, med-deep green crumpled and frilled leaves, globular white heart, slow-bolting, stands heat well, often grown in tropical countries, black seed. *Source History: 9 in 1981; 8 in 1984; 8 in 1987; 7 in 1991; 9 in 1994; 7 in 1998; 2 in 2004.* **Sources: Fo7, He8.**

Mikola - Red butterhead, slightly ruffled, large rounded succulent leaves, withstands summer heat. *Source History: 1 in 2004.* **Sources: Tu6.**

Miluna - Iceberg with crunchy dense hearts, withstands summer heat, resists Bremina NL 1-16. *Source History: 1 in 2004.* **Sources: Tho.**

Minetto - (Minetto 16, Minetto M.T.) - 71-80 days - Small compact med-green solid heads, heat res. iceberg type, for summer crops, best on Florida muck, smaller on sandy soils, slow bolt, dependable header, Cornell. *Source History: 11 in 1981; 9 in 1984; 7 in 1987; 4 in 1991; 3 in 1994; 3 in 1998; 1 in 2004.* **Sources: Dam.**

Mini-Green - 65-75 days - Grapefruit size heads, creamy yellow interior, small core, heat tolerant. *Source History: 1 in 1994; 18 in 1998; 3 in 2004.* **Sources: LO8, Tho, Twi.**

Miura - 75 days - Uniform, dense heads, resists bolting, LMV and mildew res. *Source History: 1 in 1998; 1 in 2004.* **Sources: Tho.**

Montello - 70-88 days - Good sized well covered iceberg type heads, glossy dark-green leaves, smooth midribs, res. to bolting and CRR in hot weather, WI/AES. *Source History: 6 in 1981; 5 in 1984; 15 in 1987; 12 in 1991; 11 in 1994; 3 in 1998; 2 in 2004.* **Sources: Ech, SE14.**

Morgana - Heavy thick-leaved dark-green heads, spring or fall or winter crops, plant Aug. through May, harvest Sept. through June. *Source History: 1 in 1991; 1 in 2004.* **Sources: Pe1.**

Morges - Tender, green heads tinged with red, sow spring and fall. *Source History: 1 in 2004.* **Sources: Tu6.**

Nancy - 58-66 days - Sturdy Boston type longer keeping than others due to unusually thick glossy outer leaves, earlier darker green and slightly smaller than Patty but still large for a butterhead, large well packed heart, excel. quality, res. to rot LMV and 9 races of mildew, spring and fall crops. *Source History: 1 in 1987; 2 in 1991; 4 in 1994; 3 in 1998; 1 in 2004.* **Sources: Jo1.**

Nevada - 58-75 days - Attractive French variety with large foliage and small, tall open heads with thick, mint green leaves, crunchy texture with a nutty flavor, does well in stressful conditions, matures early, bolt and tipburn res. *Source History: 7 in 1994; 18 in 1998; 23 in 2004.* **Sources: Bo17, Bu2, CO30, Dam, Ech, Jo1, Jun, Ki4, ME9, Na6, Orn, Pin, Roh, RUP, Sa9, Se26, So1, Sw9, Ta5, Ter, Ver, Ves, VI3.**

New York - (Wonderful, Nonpariel, Los Angeles, New York Iceberg) - 78-85 days - Huge 3-4 lb. head, almost as solid as cabbage, well blanched interior, heat res., fine quality, pre-1906. *Source History: 11 in 1981; 8 in 1984; 5 in 1987; 4 in 1991; 5 in 1994; 6 in 1998; 3 in 2004.* **Sources: Lin, Ri12, Se7.**

New York 12 - (Wonderful) - 60-90 days - Large sure heading New York str., med-green solid cabbage-like head, often 3 lbs., TB and heat res., excel. for fall crops and shipping, home or market, good in hot weather, poor on low-lying muck, needs well drained soil, distinct flavor. *Source History: 29 in 1981; 23 in 1984; 17 in 1987; 12 in 1991; 11 in 1994; 6 in 1998; 5 in 2004.* **Sources: DOR, He8, MOU, SE14, WE10.**

North Pole - 50-55 days - Compact light-green butterheads, extremely cold res., sow or set out 2-3 weeks before first frost date, harvest through winter and early spring, bolts quite easily. *Source History: 1 in 1981; 1 in 1984; 2 in 1987; 2 in 1991; 1 in 1994; 4 in 1998; 5 in 2004.* **Sources: Coo, Pe2, Roh, Se7, Tu6.**

Okayama Salad - 55 days - Japanese var., small heads with soft, deep green leaves, delicate flavor, med. bolt res. *Source History: 2 in 2004.* **Sources: CO30, Kit.**

Optima - 54-70 days - Boston type, large framed lettuce with thick, dark green leaves with excellent heat tolerance, resists bolting, tipburn and races 1-4 of mildew, LMV tolerant. *Source History: 5 in 1998; 9 in 2004.* **Sources: CO30, ME9, Orn, OSB, SE4, Sto, Ter, Ves, VI3.**

Pablo - (Pablo Batavian) - 60-82 days - A red Iceberg, beautiful plant, deep-red outer leaves contrast with the green inner leaves, crisp and succulent, sweet mild flavor, well-blanched centers, extremely heat res. *Source History: 1 in 1991; 1 in 1994; 5 in 2004.* **Sources: Fe5, Ho13, Pr3, Se16, So1.**

Paradise - Salinas type, beautiful green uniform heads, TB tol. *Source History: 1 in 2004.* **Sources: SAK.**

Parella Green - (Parella Verde) - 50 days - One of the smallest heading lettuces with a small open soft head, flavorful. *Source History: 1 in 1991; 2 in 1994; 2 in 1998; 2 in 2004.* **Sources: CO30, Orn.**

Parella Red - (Parella Rossa) - 52 days - Old Italian heirloom, loose half-size butterhead, sweet leaves tinged with dark ruby-bronze, prefers cool weather. *Source History: 3 in 1991; 3 in 1994; 5 in 1998; 3 in 2004.* **Sources: CO30, Ho13, Orn.**

Patriot - 65-75 days - Salinas type with excellent heading and holding ability, good for warm growing conditions, DM tolerance. *Source History: 1 in 1994; 1 in 1998; 1 in 2004.* **Sources: OSB.**

Pirat - (Sprenkel, Brauner Trotzkopf, Hardhead, Webbers Brownhead) - 50-63 days - Butterhead type, brown-speckled green leaves, creamy heart, extra crisp, round compact head, outstanding flavor, an improved Sprenkel lettuce from Germany, not rec. for the South, 1875. *Source History: 1 in 1981; 1 in 1984; 4 in 1987; 10 in 1991; 5 in 1994; 6 in 1998; 10 in 2004.* **Sources: Ag7, Eco, Fe5, Pe1, Ri12, Se16, Sh12, So9, Sw9, Tu6.**

Pirat, Red - 50-55 days - Reddish bronze slightly crinkled outer leaves, creamy yellow heart, slow to bolt, rot res., improved Merveille des Quatre Saisons. *Source History: 3 in 1991; 1 in 1994; 3 in 1998; 2 in 2004.* **Sources: Pr3, Se7.**

Pro 917 - 92 days - Very large Raleigh type, used in Florida for mid-winter harvest, CRR tolerant, PVP 1992 Seminis Vegetable Seeds, Inc. *Source History: 1 in 2004.* **Sources: Sto.**

Rapsody - Large framed lettuce with a strong well closed base, TB res., resists Bremia NL 1-16, recommended for autumn and winter crops under low light conditions. *Source History: 1 in 1998; 2 in 2004.* **Sources: ME9, Orn.**

Red Butter - 55-70 days - Beautiful outer leaves are crimson and dark green, bright-green interior, shows some res. to TB and bolting. *Source History: 1 in 1994; 2 in 1998; 1 in 2004.* **Sources: OSB.**

Red Butterworth - 40-70 days - Beautiful large red heads with sweet mild flavor, lightly savoyed red-bronze outer leaves, heads mature all at the same time, TB res., withstands high summer temperatures, can be sown spring, summer or fall. *Source History: 9 in 2004.* **Sources: Coo, Ers, Ga22, GLO, Mcf, Ni1, St18, Up2, We19.**

Red Grenoble - (Grenoblaise, Red Grenobloise, Rouge Grenobloise) - 55 days - Red-tinted Batavian or crisphead lettuce, color of plants varies from light-green to magenta, excellent flavor, vigorous grower, very cold hardy, slow to bolt during summer heat, can be cut young as a looseleaf or left to head up, from France. *Source History: 2 in 1981; 2 in 1984; 3 in 1987; 7 in 1991; 6 in 1994; 6 in 1998; 2 in 2004.* **Sources: Ga1, Orn.**

Red Montpelier - (Rougette du Midi, Rougette de Montpelleir) - 60-75 days - Small red butterhead, great for "baby lettuce" or off season cold frame crops, not for summer planting, popular in Northern Italy. *Source History: 1 in 1981; 1 in 1984; 1 in 1987; 1 in 1991; 1 in 1994; 1 in 1998; 2 in 2004.* **Sources: Coo, Se24.**

Red Ridinghood - 60-65 days - Red Boston type similar to Four Seasons but better color and bolt res., heads get quite large, suggested harvesting at one-half to three-fourths pound. *Source History: 1 in 1991; 3 in 1994; 5 in 1998; 4 in 2004.* **Sources: Co32, Fo7, Roh, Se7.**

Red Vogue - 60 days - Flavorful red butterhead type with loose head of smooth dark-red leaves, long harvest with little bitterness, fine flavor, TB tol., PVP 1992 Sakata Seed America, Inc. *Source History: 2 in 1994; 1 in 1998; 3 in 2004.* **Sources: ME9, SAK, Ta5.**

Regina dei Ghiacci - (Estivea Regina de Ghiacci) - Crisp iceberg type heads, buttery texture, dark green spiked leaves, excel. flavor. *Source History: 1 in 1987; 1 in 1991; 1 in 1994; 1 in 1998; 2 in 2004.* **Sources: Ber, Ho13.**

Regina di Maggio - 60 days - Butterhead type, medium size blond-green heads, creamy-yellow interior, firm heart, outer green leaves have a slight rosy tint, Italy. *Source History: 1 in 1981; 1 in 1984; 1 in 1987; 1 in 1991; 1 in 1994; 1 in 1998; 1 in 2004.* **Sources: Ber.**

Reine de Mai - (Pleine Terre) - Old French var., large, tender, pale green heads, creamy yellow heart, early, best for spring planting. *Source History: 1 in 1991; 1 in 1998; 1 in 2004.* **Sources: Ho13.**

Reine des Glaces - (Ice Queen) - 60-65 days - Small slow bolting French variety for summer, dark-green deeply cut pointed almost lacy leaves, crisp on hot days, use as leaf lettuce after heads cut. *Source History: 3 in 1981; 2 in 1984; 6 in 1987; 10 in 1991; 6 in 1994; 4 in 1998; 12 in 2004.* **Sources: Co32, Coo, Fe5, Lej, Pe2, Ri12, Sa5, Se7, Se8, Se16, Se17, Syn.**

Rodan Bronze - 55-60 days - Compact reddish bibb, resembles Deer Tongue, red-bronze tender pointed leaves, a favorite of Alan Chadwick. *Source History: 1 in 1994; 1 in 1998; 3 in 2004.* **Sources: Ho13, Pe2, Tu6.**

Roger - 65 days - Batavian type, 10-12 in. heads with deep brilliant red color, thick sweet leaves, resists Bremia NL 1-16 and LMV. *Source History: 1 in 2004.* **Sources: Ter.**

Rosa - Iceberg type, green leaves with red/burgundy tips, tight center, slow to bolt, bred by Peter's Seed and Research. *Source History: 1 in 2004.* **Sources: Tu6.**

Rossimo - (Rossima) - 52-65 days - Slow bolting variety with large open heads, heavily textured, large bright-red leaves with light green backs, upright, frilled, twisted, blistered and heavily textured, sweet mild flavor. *Source History: 1 in 1994; 2 in 1998; 3 in 2004.* **Sources: Fe5, La1, Se16.**

Rouge Grenoblaise - 45-55 days - Rare red butterhead, red tinged leaves, heat tol., does not get bitter in the heat, tolerates light frosts, French heirloom. *Source History: 4 in 1998; 8 in 2004.* **Sources: CO30, Coo, Ki4, Se7, Se8, So9, So12, VI3.**

Salinas - (Salina, Salinas M.T., Saladin) - 70-90 days - Solid med-large heads, slightly dull-green, flat smooth butt, med- large core, res. to DM TB and big vein, standard shipper for coastal growing areas, black seed, USDA. *Source History: 7 in 1981; 6 in 1984; 9 in 1987; 8 in 1991; 7 in 1994; 12 in 1998; 6 in 2004.* **Sources: CO30, MOU, OSB, Se26, WE10, We19.**

Salinas 88 Supreme - 50 days - Solid round heads up to 2.25 lbs., dark green outer leaves, creamy yellow interior, LMV res., does well in California and Mexico. *Source History: 2 in 2004.* **Sources: SAK, Ter.**

Sandrina Winter - 60 days - Fine butterhead lettuce from Vilmorin, flavorful yellow-green leaves, good choice for fall planting. *Source History: 1 in 1987; 3 in 1991; 2 in 1994; 3 in 2004.* **Sources: Pe1, Ri12, Se7.**

Sangria - 55-72 days - Fancy thick-leaved butterhead prominently tinted with warm rosy red, med-size head with somewhat wavy leaves free of savoy, interior heart leaves blanch pale yellow, slow to bolt and tipburn, dev. by Vilmorin in France, intro. by Johnny's, PVP 1992 Vilmorin, S.A. *Source History: 3 in 1991; 10 in 1994; 12 in 1998; 7 in 2004.* **Sources: CO30, Dam, Na6, Se17, Ves, VI3, Wi2.**

Santa Fe - 55 days - Small blanched heads surrounded by green outer leaves, tinged bronze-red, extra crunchy. *Source History: 1 in 2004.* **Sources: Bu2.**

Sharpshooter - 70 days - Iceberg, attractive, medium size round head, clean interior, tolerates DM-1 and 3, TB, corky root, TB., PVP 2000 Seminis Vegetable Seeds, Inc. *Source History: 1 in 2004.* **Sources: OSB.**

Sherwood - 72 days - Excellent flavor, mildew res. *Source History: 1 in 1998; 1 in 2004.* **Sources: Tho.**

Sierra - (Sierra Batavian) - 45-65 days - Intermediate butterhead type, open head, steely green with bronze tinge at the leaf margins, can be used as a leaf lettuce, excellent heat tolerance, mosaic indexed, dev. by Vilmorin in France, PVP 1992 Vilmorin, S.A. *Source History: 1 in 1981; 1 in 1984; 6 in 1991; 17 in 1994; 23 in 1998; 23 in 2004.* **Sources: Be4, Bu2, CO30, Coo, Dam, Fe5, Ga1, Hig, Jun, ME9, Na6, Orn, Pin, RUP, So1, Sw9, Syn, Ta5, Ver, Ves, VI3, We19, Wi2.**

Skyline - 85 days - Ithaca type, medium dark green, extra solid head, good bolt tolerance, summer/fall harvest. *Source History: 1 in 1998; 1 in 2004.* **Sources: Sto.**

SLE 4105 - Salinas type, vigorous plants, large, uniform heads, TB tol., heavy yields. *Source History: 1 in 2004.* **Sources: SAK.**

Soissons - French heirloom, large heads tinged with red, gourmet flavor. *Source History: 1 in 2004.* **Sources: Syn.**

South Bay - 70-90 days - Firm dark-green heads, 6 in. dia., extra wrapper leaves, good holding ability, free of internal disorders, CRR TB and rib cracking tol., white seed, widely adapted, performs especially well on muck soils, Dr. Guzman/U of Florida. *Source History: 4 in 1987; 3 in 1991; 1 in 1994; 3 in 1998; 1 in 2004.* **Sources: SE4.**

Speckled - 40-55 days - Loose head type can become quite heavy, sweet, apple-green leaves splashed with maroon, may originally have come from Germany or Holland (1660-1700), similar to French heirloom "Sanguine Ameliore," seed brought from Lancaster County, PA to Ontario, Canada by covered wagon by Mennonites in 1799, introduced into commerce as Golden Spotted in 1880 according to William Weaver. *Source History: 1 in 1994; 3 in 1998; 9 in 2004.* **Sources: Ba8, Fe5, Ga1, Ho13, Se16, Se27, Sh12, So1, Syn.**

SSC 1501 - Iceberg type with large frame, similar to Desert Queen. *Source History: 1 in 2004.* **Sources: ME9.**

Sucrine - (Little Gem) - 60 days - French butterhead reminiscent of Buttercrunch but superior in flavor, medium length green leaves, slightly loose head, crisp and sweet, plant in fall and winter, cold res., slow bolt, popular among home gardeners and market growers in Europe, well received in many California specialty restaurants. *Source History: 3 in 1981; 3 in 1984; 3 in 1987; 7 in 1991; 9 in 1994; 6 in 1998; 6 in 2004.* **Sources: Ba8, Com, Lej, Orn, Se7, Tu6.**

Summer Fury M.T. - *Source History: 1 in 1991; 1 in 1994; 1 in 1998; 1 in 2004.* **Sources: WE10.**

Summer Green - 45 days - Butterhead type, almost round medium size glossy med-green leaves, thick fleshed, stands heat well, also cold tol., slow bolt, grow year round, disease res., 1983. *Source History: 1 in 1981; 1 in 1984; 1 in 1991; 1 in 1998; 1 in 2004.* **Sources: Ev2.**

Summer Season M.T. - *Source History: 1 in 1991; 1 in 1994; 1 in 1998; 1 in 2004.* **Sources: WE10.**

Summertime - 62-82 days - Ithaca x Salinas, compact plant habit, solid heads of good flavor, sure-heading, never bitter, very slow bolting in hot weather, tol. to rib discoloration, excel. for home gardens, bred by Dr. Baggett at Oregon State, OR/WA AES introduction, 1989. *Source History: 3 in 1991; 38 in 1994; 47 in 1998; 49 in 2004.* **Sources: Be4, Bo17, Bu2, CH14, Com, Coo, Dam, Ers, Fa1, Fe5, GLO, Gr27, HA3, Ha5, Hal, HO2, Jun, Ki4, La1, Lin, MAY, ME9, Mo13, Ni1, OLD, OSB, PAG, Par, Pe2, Pin, Roh, RUP, SE4, Se8, SE14, Se26, Sh9, Shu, So1, Sw9, Ter, Tt2, Tu6, Twi, Up2, Ver, Ves, WE10, We19.**

Sun Devil - 60 days - Uniform head size, slow bolting, PVP application withdrawn. *Source History: 1 in 2004.* **Sources: RIS.**

Sweet Red - 54 days - Bright burgundy outer leaves, color fades to creamy green near the center of the compact, tender 7 in. heads. *Source History: 1 in 1998; 2 in 2004.* **Sources: Bu2, Hi13.**

Tania - 55-65 days - Uniform tender heads, soft texture, excel. flavor, res. to 4 strains of mildew, will not thrive in midsummer heat, grow for cool harvest, England, PVP expired 1996. *Source History: 2 in 1981; 2 in 1984; 1 in 1987; 1 in 1991; 1 in 1994; 3 in 1998; 1 in 2004.* **Sources: Ha5.**

Target - 65-80 days - Medium size Salinas type, good uniformity, excel. color and adaptability, PVP 1992 Arthur Yates and Co. *Source History: 1 in 1991; 1 in 1998; 1 in 2004.* **Sources: OSB.**

Tennis Ball - (Loos Tennis Ball) - 45-55 days - Small tight rosettes of light-green leaves, yellow-green at the base, gourmet lettuce, grown by Thomas Jefferson at Monticello, pre-1804, very rare. *Source History: 4 in 1991; 2 in 1994; 3 in 1998; 5 in 2004.* **Sources: Ol2, Se16, So1, Te4, Th3.**

Tiber - 65-75 days - Large firm heads, well wrapped, TB tolerant. *Source History: 1 in 1998; 1 in 2004.* **Sources: Ves.**

Tom Thumb - 50-70 days - Miniature butterhead, tennis ball size, creamy-yellow center, small compact plants used whole for salads, med-green slightly crumpled leaves, home gardens and containers, grows well in small pots indoors or out, a miniature Limestone Bibb, oldest American lettuce still growing, dating back to 1830. *Source History: 30 in 1981; 25 in 1984; 33 in 1987; 45 in 1991; 40 in 1994; 38 in 1998; 48 in 2004.* **Sources: Ba8, Be4, Bo17, Bo19, Co32, Com, DOR, Ear, Ers, Fe5, Fis, Fo7, Gr27, He8, Hi13, Hig, HO2, Ho12, Ho13, La1, Lin, ME9, Na6, Ni1, Or10, Orn, PAG, Pe2, Pin, Ra5, Re8, Ri12, Roh, RUP, Sa9, Se7, SE14, Se26, Sh9, So1, Ta5, Ter, Tho, Und, Vi4, Vi5, WE10, WI23.**

Torenia - Type of Kagraner Sommer lettuce with large, tightly packed heads, smooth dark green leaves, excellent bolt resistance. *Source History: 1 in 2004.* **Sources: Coo.**

Triumph - 80 days - New Saladin type of Iceberg, dense round heads, crunchy, flavorful, good root aphid tolerance, spring and summer sowing. *Source History: 2 in 2004.* **Sources: Tho, WE10.**

Val d'Orge - 120 days - Winter lettuce, start in September, transplant late October, mulch over winter, harvest April-May, large heads, blond-green leaves, strong, hardy, Gonthier of Belgium. *Source History: 4 in 1981; 4 in 1984; 4 in 1987; 2 in 1991; 2 in 1994; 2 in 1998; 2 in 2004.* **Sources: Roh, Se7.**

Valdor - 60-75 days - Excellent winter lettuce, large firm deep-green heads, used mostly in greenhouses but can be sown outdoors in the fall or very early spring for an early spring cutting, very cold hardy, good res. to botrytis (gray mold). *Source History: 3 in 1987; 2 in 1991; 1 in 1994; 2 in 1998; 2 in 2004.* **Sources: Bou, Ho13.**

Valleygreen - Uniform, dark green Iceberg type, large frame, TB tol., excel. heat tol., available to commercial lettuce growers in CA and AZ desert areas, PVP application abandoned. *Source History: 1 in 2004.* **Sources: HA20.**

Vanity - 48 days - Large head of slightly savoyed and frilled, light green leaves, retains mildness, slow to bolt and TB. *Source History: 2 in 1998; 1 in 2004.* **Sources: So1.**

Vanmax - Dull olive-green large solid heads, smooth base, medium size core, tipburn res., earlier larger heavier more uniform than Vanguard, med-late, PVP expired 1990. *Source History: 1 in 1981; 1 in 1984; 1 in 1987; 1 in 1991; 1 in 1994; 2 in 1998; 1 in 2004.* **Sources: HA20.**

Victoria - 45-60 days - Large French Canasta type, heavy upright open head of crisp leaves with rich deep-green color, 10 in. dia., crisp juicy and notably sweet, long harvest period, bolt and bottom rot res. *Source History: 1 in 1991; 2 in 1994; 2 in 1998; 4 in 2004.* **Sources: Pe2, Syn, Ter, Tu6.**

Victory - *Source History: 1 in 1998; 1 in 2004.* **Sources: WE10.**

Vista - 50 days - Large firm heads, disease and TB res., semi-early, slow to bolt, mosaic virus indexed, dev. for hot dry areas where summer cultivation of lettuce is difficult. *Source History: 5 in 1991; 4 in 1994; 1 in 1998; 1 in 2004.* **Sources: CO30.**

Wakefield Crunch - 75 days - From the southern regions of Virginia, small green compact heads with crisp mild flavor, withstands summer heat. *Source History: 1 in 1994; 1 in 1998; 1 in 2004.* **Sources: He8.**

Wayahead - (Way-A-Head) - 65 days - Early home garden variety, medium size compact plants, thick crumpled med-bright green leaves, creamy-yellow heart, white seed. *Source History: 1 in 1981; 1 in 1984; 2 in 1987; 1 in 1994; 1 in 1998; 1 in 2004.* **Sources: DOR.**

Webb's Wonderful - 64-87 days - English crisp heart type, crumpled, very large and robust, fine distinct flavor, stands heat well, good for the South, slow to bolt, use successive sowings. *Source History: 6 in 1981; 4 in 1984; 4 in 1987; 3 in 1991; 4 in 1994; 5 in 1998; 6 in 2004.* **Sources: Bou, Fe5, Se16, Se26, Tho, WE10.**

Wendel - Low light, large butterhead with a well closed base, darker green, faster and heavier than Rachael, suitable for short day planting, resists Bremia NL 1-16. *Source History: 1 in 1998; 1 in 2004.* **Sources: Orn.**

Winter Density - 70 days - Mosaic tested, intermediate between butterhead and romaine, crisp sweet succulent and flavorful, best romaine in tests, excellent results in summer also, England. *Source History: 1 in 1981; 1 in 1984; 1 in 1987; 2 in 1991; 2 in 1994; 2 in 1998; 2 in 2004.* **Sources: Coo, WE10.**

Winter Harvest M.T. - *Source History: 1 in 1991; 1 in 1994; 1 in 1998; 1 in 2004.* **Sources: WE10.**

Winter Marvel - (Merveille d'Hiver) - 50 days - French lettuce, large heads, smooth green leaves, excel. market variety, can be autumn sown in certain areas, hardy without cover to 18 degrees F., from the 1850s. *Source History: 3 in 1981; 2 in 1984; 5 in 1987; 4 in 1991; 5 in 1994; 5 in 1998; 4 in 2004.* **Sources: Bou, Coo, DOR, Fe5.**

Wintergreen M.T. - *Source History: 1 in 1991; 1 in 1994; 1 in 1998; 1 in 2004.* **Sources: WE10.**

Winterhaven - Mosaic indexed, PVP expired 1991. *Source History: 1 in 1981; 1 in 1984; 1 in 1987; 1 in 1998; 1 in 2004.* **Sources: SE14.**

Yucaipa - 70 days - Iceberg type with uniform, compact heads, crisp green leaves, non-bitter flavor, slow bolting, tolerates heat, tipburn and rib discoloration, well suited for commercial production. *Source History: 1 in 2004.* **Sources: Ves.**

Yuma - Iceberg type, large dark green heads, moderate cold tolerance, does well in AZ and CA desert areas, PVP expired 2003. *Source History: 1 in 2004.* **Sources: HA20.**

Zimni - 45-50 days - Czechoslovakian forcing lettuce that forms small loose heads with light-green leaves, sweet when picked early. *Source History: 1 in 1991; 1 in 1998; 1 in 2004.* **Sources: So12.**

Varieties dropped since 1981 (and the year dropped):

Ace of Hearts (2004), *Action* (2004), *America* (1998), *Antina* (1991), *Ascona* (2004), *Astra* (1987), *Astral* (1987), *Attraction* (2004), *Attractive* (2004), *Audran* (2004), *Augusta* (1994), *Avoncrisp* (1991), *Babylon* (1998), *Bella* (1991), *Bella Green* (2004), *Benita* (1991), *Bibb, Green* (2004), *Bibb, Red* (1998), *Bibb, Salad* (2004), *Big Ben* (2004), *Big Red Head* (1987), *Bix* (2004), *Bogata* (1998), *Bon Jardinier Select* (1991), *Boston Rose* (2004), *Boston, Canada* (1994), *Boston, Green M.T.* (1998), *Boston, Nova* (1998), *Boston, Red* (2004), *Boston, Summer* (1994), *Bounty* (2004), *Brazil 48* (1991), *Brown German* (1994), *Bullseye* (1994), *Burpee's Iceberg* (1983), *Calmar* (2004), *Calona (L 9752)* (1987), *Campan* (1987), *Canasta* (2004), *Cannon* (1998), *Capo* (1998), *Cappuccio Regina* (1994), *Ceremony I* (2004), *Chaparral* (1998), *Charlene* (1994), *Chessib Bibb* (1991), *Chieftain* (1998), *Chou de Naples Bresi* (2004), *Cindy* (1998), *Citation* (2004), *Climax 84* (1994), *Coach Supreme* (2004), *Commander* (1998), *Coolguard* (1994), *Couve de Napoles* (1998), *Creamy Heart* (1991), *Crestana* (1994), *Crisp and Sweet* (1991), *Crispy* (1987), *Cybele Batavian* (1994), *Dandie* (1991), *Dapple* (1998), *Darka* (1983), *Debby* (1998), *Deci-Minor* (1991), *Del Rey* (2004), *Delta* (2004), *Desertgreen* (2004), *Diamante* (1991), *Diamond Head* (2004), *Diana* (1991), *Dolly* (2004), *Doree Printemps* (1991), *Duchesse* (2004), *Dynasty* (1994), *E 4226* (1994), *Early Great Lakes* (1991), *El Toro* (1998), *Eline* (1998), *Emerald Gem* (1994), *Emperor* (2004), *Empire No. 4164* (1987), *Empress* (1991), *Encore* (1998), *Escort* (2004), *Estelle* (1987), *Etna* (1994), *Etus* (1984), *Excellence* (1987), *Exp E.C. 47* (1987), *Fairton* (1991), *Fariton* (1984), *Fivia* (1998), *Floribibb* (2004), *Florida Bibb* (1991), *FM-8248* (1994), *Fordhook* (1984), *French Crisp* (1998), *Frostproof Latehead* (1994), *Frosty* (1998), *Fulton* (1982), *Gallega* (1991), *Gilaben* (2004), *Gloire de Dauphine* (1994), *Golden State D* (1998), *Goliath* (2004), *Grande Buerre* (1987), *Grazer* (2004), *Great Lakes 54* (1987), *Great Lakes 366-A* (1987), *Great Lakes 407-P* (2004), *Great Lakes 428* (1991), *Great Lakes 118-A* (2004), *Great Lakes B Strain* (1984), *Great Lakes R-200* (2004), *Great Lakes Select* (1998), *Great Lakes, Premier* (1994), *Great Lakes, Special B Str.* (1998), *Green Head* (2004), *Green Lake* (1998), *Hartford Bronze Head* (1987), *Hjerter Es* (2004), *Hot Weather* (1984), *Hudson* (1994), *Iceberg Improved* (1998), *Impact* (2004), *Imperial 44* (1991), *Imperial 456* (1991), *Imperial 847* (1994), *Irma* (1983), *Jacqueline* (2004), *Jessy* (1987), *Kazan Summer* (1994), *King Crown* (2004), *Kloek* (1991), *Kral Maj* (2004), *Kristia* (1994), *L 8785* (1984), *L 9752* (1984), *Lednicky* (2004), *Lianne* (2004), *Lovina* (2004), *Luro* (1987), *Luxor* (1998), *Magnum* (2004), *Maikonig* (1987), *Malika (L 8785)* (1987), *Mantilia* (2004), *Marmer* (1994), *Marvel* (2004), *Marvilha de Inverno* (1994), *Media* (1991), *Meridian* (1987), *Merit* (2004), *Minilake* (1994), *Mirena* (1991), *Mission* (1998), *Mondian* (1987), *Montemar* (2004), *Murietta* (2004), *Musca* (2004), *Musette* (2004), *Nanda (L 0192)* (1987), *Neckarriesen Virus Free* (1984), *New York 515* (1994), *New York 515 Improved* (1991), *New York Mammoth* (1983), *New York Wonderful* (1987), *North Pole Bibb* (1998), *Ocean Side M.T.* (1998), *Oleta* (1984), *Open-Heart* (2004), *Orba* (1987), *Oresto* (1994), *Orfeo* (1998), *Ostinata* (2004), *Oswego* (1994), *Ovation* (2004), *Overture* (1998), *Pacific* (1998), *Palmyran* (1987), *Pandora (L 1149)* (1987), *Panlight (L 9154)* (1987), *Panvit* (1987), *Patience* (1983), *Patty* (1991), *Pavana* (1987), *Pax 1012* (1998), *Pennlake* (1994), *Pennlake M.T.* (1983), *Prado* (2004), *Premier Great Lakes 659* (1987), *Premier M.T. Great Lakes* (1983), *Prima* (1991), *PYB 0136 M.T.* (2004), *PYB 2144 M.T.* (2004), *PYBAS No. 250 M.T.* (2004), *PYBAS No. 251* (2004), *Queen Crown* (2004), *Queen of Malta* (1994), *Queen of May* (1991), *Radian* (1991), *Raleigh* (2004), *Ramcos* (1983), *Ranger* (2004), *Red Bowl* (1998), *Redcap* (1998), *Resistant* (1987), *Reskia* (1991), *Ribb Lettuce* (1983), *Rigoletto* (2004), *Rosanna* (1998), *Rose* (1998), *Rosy* (2004), *Rouge Crisp* (1994), *Rougette du Midi* (2004), *Rychlini* (1994), *Saint Anne's Slow Bolting* (2004), *Salad Crisp* (1998), *Saladin* (1998), *Salinas Sun 2* (1994), *Salverde* (1998), *Sea Green* (2004), *Sea Side M.T.* (1994), *Seagreen* (2004), *Selma Wisa* (2004), *Shogun* (1991), *Shumway's Bibb* (1991), *Sitonia* (1994), *Snowbird* (2004), *St. Anne's Slow Bolting* (2004), *Sudel* (1984), *Summer White* (1994), *Summerlong* (1984), *Super 59* (2004), *Tete de Glace* (1987), *Thai Salad* (2004), *Thialf* (1987), *Timpurie de Mai* (2004), *Titania* (1998), *Tunska Edra* (2004), *Unicum* (1994), *Unrivalled* (1991), *Val Rio* (1987), *Val Tex 41* (1998), *Valprize* (1998), *Valverde* (1987), *Vango* (1998), *Vanguard* (2004), *Vanguard 30* (1994), *Vanguard 75* (2004), *Vasco* (1991), *Velvet Bibb* (1991), *Verano* (1998), *Verian* (2004), *Vicky Bing* (2004), *Voluma* (1991), *Warpath* (2004), *Wayahead, Black Leaved* (1998), *Windermere* (1983), *Winter Crunch* (1994), *Winter Pick M.T.* (2004), *Winter Pride M.T.* (2004), *Winterset* (2004), *Wonderful Head* (1984), *Zielgler Heirloom Bibb* (1998), *Zita* (1984), *Zomerkoning* (1987), *Zwart Duits* (1987).

Lettuce / Leaf
Lactuca sativa

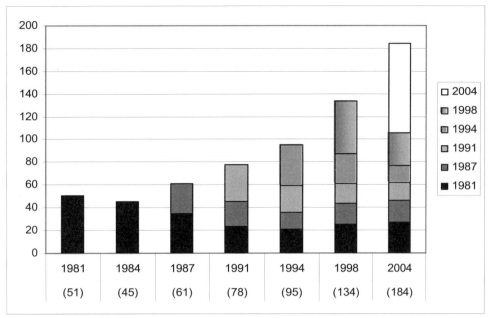

Number of varieties listed in 1981 and still available in 2004: 51 to 27 (53%).
Overall change in number of varieties from 1981 to 2004: 51 to 184 (361%).
Number of 2004 varieties available from 1 or 2 sources: 121 out of 184 (66%).

<u>504 Green</u> - 53 days - Medium-dark green, blistered leaves, excellent bolt tolerance, TB and bolt res., PVP 2000 Seminis Vegetable Seeds, Inc. *Source History: 2 in 2004.* **Sources: SE4, Sto.**

<u>Alpine</u> - 55 days - Large, heavily savoyed leaves with serrated leaf margins, open plant habit, TB, bolt and heat tolerant, PVP application pending. *Source History: 1 in 2004.* **Sources: OSB.**

<u>Antares</u> - 45-50 days - Ornamental, frilled oak leaf, bright red leaves with shades of pink, bronze and chartreuse, colorful and tender, no bitterness in hot weather, Salad Bowl x Rouge d'Hiver cross bred by Frank Morton of Wild Garden Seed in Oregon. *Source History: 4 in 2004.* **Sources: Ag7, Fe5, Sh12, So9.**

<u>Aragon Red</u> - 56 days - Green leaves with red-bronze margins, color intensifies later in the season, nice pale yellow to light green interior, great field holding ability, PVP 1992 Central Valley Seeds, Inc. *Source History: 1 in 1998; 1 in 2004.* **Sources: OSB.**

<u>Aruba</u> - 53 days - Dark red oak leaf type with red-bronze color at baby leaf stage, lobed leaves are slightly ruffled, mild taste, harvest at 28 days for baby leaf. *Source History: 1 in 2004.* **Sources: Ves.**

<u>Asian Green</u> - *Source History: 1 in 1998; 1 in 2004.* **Sources: WE10.**

<u>Asian Red</u> - Bright red leaves with frilled edges, from Asia. *Source History: 1 in 1998; 2 in 2004.* **Sources: Ba8, WE10.**

<u>Atoll</u> - 56 days - Batavia type produces crisp heads, tasty leaves with excellent flavor, does well in the heat, resists many DM strains. *Source History: 1 in 2004.* **Sources: Par.**

<u>Australian Yellow</u> - (Australian Yellowleaf) - 50-54 days - Very early tender yellow-green leaves grow into large crinkled leaves, tender texture, 12 to 16 in. dia., slightly sweet flavor, moderate bolt res., Australian heirloom. *Source History: 1 in 1991; 2 in 1994; 2 in 1998; 4 in 2004.* **Sources: Hud, La1, Se16, So1.**

<u>Austrian Greenleaf</u> - 50 days - Frilly Oak Leaf type with light green leaves, fancy looseleaf. *Source History: 1 in 1998; 1 in 2004.* **Sources: Tu6.**

<u>Bellisimo</u> - Red Lollo Rossa type, retains intense dark red color under short day conditions, higher bolting tolerance than Impuls, Bremia NL 4-16 res. *Source History: 1 in 1998; 1 in 2004.* **Sources: Orn.**

<u>Belowa</u> - 52 days - Upright plants, dark green oak leaf shaped leaves, slow to bolt, well suited for baby leaf production, 25- 30 days for baby leaf, 52 days for full size. *Source History: 1 in 2004.* **Sources: Ves.**

<u>Berenice</u> - 50 days - Oakleaf type with more compact heads, long, narrow, dark green leaves, tender with crispy ribs, slow bolting, LMV tolerant, spring, summer and fall production. *Source History: 1 in 2004.* **Sources: Jo1.**

<u>Bionda Ricciolini</u> - 35 days - Light green, frilly leaves, sow spring, summer or fall. *Source History: 1 in 2004.* **Sources: Se24.**

<u>Biondo a Foglie Lisce</u> - (Bionda Liscia) - 40-60 days - Tender smooth green leaves, quick-growing, can be cut several times beginning a month after sowing, retains sweetness over a long period, sow spring, summer or fall. *Source History: 1 in 1987; 1 in 1991; 1 in 1994; 1 in 1998; 3*

in 2004. **Sources: Coo, Sa5, Se24.**

Biscia Rossa - Crumpled red-green leaves, sow spring, summer, fall. *Source History: 1 in 1987; 1 in 1991; 1 in 1994; 1 in 1998; 2 in 2004.* **Sources: Ber, Se24.**

Black Seeded Simpson - (Best of All, Simpson) - 40-65 days - Large loose light-green lightly crumpled leaves, very early and adaptable, inner leaves blanch almost white, large upright plants, stands heat and drought well, withstands some frost, for spring or early summer, slow bolt, 1850. *Source History: 121 in 1981; 114 in 1984; 112 in 1987; 113 in 1991; 108 in 1994; 104 in 1998; 99 in 2004.* **Sources: All, Au2, Ba8, Be4, Bo17, Bo19, Bu1, Bu2, Bu3, Bu8, But, CA25, CH14, CO23, Co31, Co32, Com, Coo, Cr1, Dam, De6, DOR, Dow, Ear, Eco, Ers, Fe5, Fi1, Fis, Fo7, Ga1, Ge2, Goo, Gr27, Gur, HA3, Ha5, He8, He17, Hi6, Hi13, Hig, HO2, Jo1, Jor, Jun, Kil, La1, Lan, Lej, LO8, Ma19, MAY, Me7, ME9, Mel, Mey, MIC, Mo13, MOU, Na6, Ni1, OLD, Or10, PAG, Pe1, Pe2, Pin, Pla, Pr3, Ri12, RIS, Roh, Ros, RUP, Sa9, Sau, Scf, SE4, Se7, Se8, SE14, Se26, Sh9, Shu, Si5, Sk2, So1, So9, SOU, St18, TE7, Ver, Ves, Vi4, WE10, Wet, Wi2, WI23.**

Blackjack - 60-62 days - Dark red, puckered leaves with ribbed margins, clean interior, entire plant is dark burgundy at baby stage, excellent for salad mix, PVP application pending. *Source History: 1 in 2004.* **Sources: OSB.**

Blushed Icy Oak - 50-60 days - Blushed Butter Oak parentage sel. for crisp leaves and more flavor, 3-lobed leaves blushed with bright pink, green-white self-blanching center, pink to white midribs, cold tolerant, sow in late winter for long harvest, dev. by Frank Morton. *Source History: 3 in 2004.* **Sources: Ag7, Ni1, Sh12.**

Blushed Icy Oak Gene Pool - 49 days - Green leaves tinged with red, juicy sweet flavor, crisp iceberg texture, bitterness only in very hot weather, fairly slow bolting, selected by Frank Morton out of the Blushed butter Oak gene pool. *Source History: 1 in 2004.* **Sources: Fe5.**

Bolzano - Improved form of Lollo Bionda and pale green cousin of Lollo Rossa, slow to bolt. *Source History: 1 in 2004.* **Sources: Coo.**

BOS 9079 - Large frame, good bolt and TB tolerance. *Source History: 1 in 2004.* **Sources: ME9.**

Bronze Arrow - (Bronze Arrowhead) - 40-60 days - AAS/1947, California heirloom, midseason, massive dark-green leaves with bronze trim, delicious, remains fresh and tasty for up to 3 weeks before bolting, hardy and adaptable, intro. as Bronze Beauty by Germania Seed and Plant Co. *Source History: 1 in 1987; 4 in 1991; 2 in 1994; 4 in 1998; 9 in 2004.* **Sources: Bou, Co32, Ho13, In8, Pr3, Se7, Se16, Se17, Te4.**

Bronze Guard - Deep-red Oak Leaf type, full flavor, long standing, well adapted to many climates and seasons. *Source History: 1 in 1994; 1 in 1998; 2 in 2004.* **Sources: Sa5, WE10.**

Brunia - (Brunia Bronze Oak Leaf, Red Oak Leaf) - 50-74 days - Forms very tall rosettes, more delicate in flavor than many butterheads, slender green red-tipped leaves are tender and heavily notched, sweet tangy flavor, ideal for spring or fall, France. *Source History: 4 in 1991; 13 in 1994; 20 in 1998; 10 in 2004.* **Sources: Coo, Fe5, Ki4, Orn, Sa5, Se8, Sw9, Ter, Ves, VI3.**

Carnival - 60 days - Batavia type, delicious dark green, red tinged leaves, good shelf life, DM and bolt tol. *Source History: 1 in 1994; 1 in 2004.* **Sources: OSB.**

Chadwick's Rodan - 58 days - European med-size med-heading leaf lettuce selected by Alan Chadwick, similar to Deer Tongue but with finer tapering bronze-tipped leaves, tender juicy and crunchy, forms a tight blanched heart, produces very little seed. *Source History: 3 in 1991; 3 in 1998; 5 in 2004.* **Sources: Au2, Bou, Ho13, In8, Se17.**

Chicken or Rabbit Lettuce - (Everbearing) - 46 days - Large coarse abundant med-green wavy leaves, 3-4 ft. tall, green feed for chickens or rabbits, keeps growing after cutting or stripping, heavy yields. *Source History: 6 in 1981; 4 in 1984; 1 in 1987; 3 in 1998; 2 in 2004.* **Sources: Ho13, Shu.**

Claret Symphony - Mosaic indexed. *Source History: 2 in 2004.* **Sources: CO23, DOR.**

Cocarde - 49 days - Vigorous giant red Oak Leaf type, large heavy upright trumpet-shaped heads, elongated delicately lobed dark-green leaves with rusty red overlay, excel. cut and come again lettuce with an abundance of delicate but crunchy leaves, will grow more than 12 in. across if given room, slow bolting, unusual French leaf lettuce. *Source History: 4 in 1991; 5 in 1994; 3 in 1998; 4 in 2004.* **Sources: Jo1, Pe2, Se26, Tu6.**

Concept - (BOS9052) - 51 days - Leaves have characteristics of romaine and leaf lettuce, thick, juicy leaves, flavorful. *Source History: 2 in 2004.* **Sources: Jo1, ME9.**

Cracoviensis - 45-65 days - Beautiful long, wavy, twisted lime-green leaves are dusted with red-purple or lavender, retains buttery, non-bitter flavor even after bolting, becomes sweeter as temperatures fall, from Eastern Europe, pre-1885. *Source History: 3 in 1998; 12 in 2004.* **Sources: Ag7, Fe5, Ho13, Ni1, Sa5, Se17, Sh12, Sk2, Syn, Ta5, Ter, Tu6.**

Crissy - 55 days - Red Tango type, dark red, heavily frilled leaves, DM, bolt tol., PVP application pending. *Source History: 1 in 2004.* **Sources: OSB.**

Dano - 58 days - Deep, dark red oak leaf with creamy light green interior, uniform size and maturity, good holding ability, DM tol. *Source History: 1 in 2004.* **Sources: OSB.**

Deep Red - 48 days - Loose clusters of red-tinged frilled curled crisp tender green leaves, upright and vigorous, not long- standing in summer heat, good cool crop, intense colors, PVP expired 1996. *Source History: 1 in 1981; 1 in 1984; 1 in 1987; 1 in 1994; 2 in 1998; 2 in 2004.* **Sources: Ha5, ME9.**

Deer Tongue - (Matchless, Rodin) - 45-80 days - Compact upright 7-8 in. loose heads, triangular round-tipped green leaves, thick midribs, rich nutty flavor with no bitterness, slightly savoyed, stands heat, slow to bolt, excel. quality, was very popular among pioneer families because of its ruggedness and large production of tasty leaves, dates back to the 1740s. *Source History: 6 in 1981; 3 in 1984; 3 in 1987; 8 in 1991; 11 in 1994; 17 in 1998; 28 in 2004.* **Sources: Ag7, Al25, Be4, Bu2, Coo, Eco, Fe5, Ga1, Ge2, Goo, Hi6, Ho13, Hud, Jo1, Lan, Ni1, Pe1, Roh, Se8, Se27,**

Sh12, Sk2, So1, So9, St18, Syn, Ter, We19.

Deer Tongue, Amish - 45-55 days - Old Amish favorite, medium green triangular leaves form a loose upright head, pleasantly sharp flavor, slot to bolt, dependable, high yielder, popular with heirloom growers, 1840. *Source History: 4 in 2004.* **Sources: Ba8, He8, Pr3, Se16.**

Deer Tongue, Red - 45-80 days - Late bolting var. very similar to Deer Tongue, triangular shaped, pointed leaves with red edges, rich nutty flavor, was popular among pioneer families for its productivity and ruggedness. *Source History: 3 in 1994; 9 in 1998; 23 in 2004.* **Sources: Ag7, Ba8, Ea4, Ers, Ga1, He8, Hi6, Ho13, La1, ME9, Orn, Pe2, Pin, Ri12, Roh, RUP, Se7, SE14, Shu, So1, So9, So12, TE7.**

Devil's Tongue - 55 days - Cos-like leaves in loose heads, new growth is a beautiful deep red, turns from light green to green overlaid with red, tender, buttery texture, new and variable strain, black and white seeds are possible. *Source History: 2 in 1998; 4 in 2004.* **Sources: Ho13, Hud, Sh12, Ter.**

Drunken Woman Fringed-Headed - 55 days - Semi-heading, tender, light to bright green leaves with red fringe, sweet, slow to bolt. *Source History: 1 in 1994; 2 in 1998; 4 in 2004.* **Sources: Ga22, Sa5, St18, Und.**

Dunsel - 35 days - Early European round-leaf type, yellowish-green, very tender, pick leaves 3 to 5 weeks after sowing. *Source History: 1 in 1981; 1 in 1984; 1 in 1987; 1 in 1991; 1 in 1994; 1 in 2004.* **Sources: Ta5.**

Early Curled Simpson - (Black Seeded Curled Simpson, Silesia, Silica Curled, Simpson's Curled) - 43-50 days - An old-timer long favored for its sweet taste, pale yellowish-green crinkly leaves form tight center clusters, hardy and quick growing, slow-bolting, widely used by home gardeners as a looseleaf or for bunching. *Source History: 18 in 1981; 17 in 1984; 16 in 1987; 10 in 1991; 9 in 1994; 6 in 1998; 1 in 2004.* **Sources: MAY.**

Emerald Oak - Blushed Butter Oak x Deer Tongue cross, compact green oakleaf lettuce that matures into a dense buttery hearted head, bred by CV Sh12. *Source History: 1 in 2004.* **Sources: Sh12.**

Envy - 54 days - Tender, crisp, dark green leaves, ruffled and frilled, resists TB, head stays open, good holding ability, slowest bolting, PVP application pending. *Source History: 1 in 2004.* **Sources: Jo1.**

Fanfare - (FM-1530) - 48 days - Waldmann's Grand Rapids type, tipburn tol., PVP 1989 HA20. *Source History: 1 in 1987; 2 in 1991; 2 in 1994; 2 in 1998; 1 in 2004.* **Sources: Sh14.**

Ferrari - Upright growth habit makes for easy harvest, excellent shelf life, resists disease. *Source History: 1 in 2004.* **Sources: Orn.**

Feuille de Chene Blonde - (Foglie di Quercia, Curly Oakleaf) - 47 days - True cutting lettuce, slow bolt, heat res., deeply indented light-green tender leaves, sown and cut from early spring right through summer heat into autumn, France. *Source History: 2 in 1981; 2 in 1984; 1 in 1987; 1 in 1991; 1 in 1994; 1 in 1998; 4 in 2004.* **Sources: Ag7, Coo, Dom, Se17.**

Fiere - Retains its blood-red color all season long, disease tol. *Source History: 1 in 2004.* **Sources: Dam.**

Fire Mountain - 58 days - Well closed 24 in. plants, deep burgundy oak leaf shape leaves, sweet, juicy flavor, does not become bitter in the heat, uniform, vigorous grower with good bolt resistance. *Source History: 1 in 1998; 1 in 2004.* **Sources: Tu6.**

Flame - 45-60 days - Prizehead selection with a distinctively bright-red color throughout the leaves and stem, broad frilled leaves hold color well, early maturity, slow to bolt., Harris Moran Seed Co. introduction, 1988. *Source History: 1 in 1987; 2 in 1991; 1 in 1998; 10 in 2004.* **Sources: Ag7, Ba8, Ga1, He8, ME9, Orn, Pe2, SE14, Se16, TE7.**

Four Seasons - Large ruby-red tipped leaves, crisp and delicious. *Source History: 1 in 1991; 2 in 1994; 1 in 1998; 1 in 2004.* **Sources: Coo.**

Funly - 60 days - Lettuce x Batavian endive cross with better taste and higher sugar levels, retains crispness after washing, excellent disease resistance. *Source History: 1 in 2004.* **Sources: Tho.**

Galactic Red - (Galactic) - 58 days - Extremely dark, wine-red shiny broad ruffled leaves, the reddest of the red leaf lettuces, adds color contrast to salads, excellent quality. *Source History: 3 in 2004.* **Sources: Bou, Orn, Pe6.**

Garnet - (FM-1567) - 42-46 days - Improved Prizehead or red Grand Rapids type, more red coloration and holds nicer bunch without over-sizing or folding at the heart, bright-green with prominent red at the tips and where exposed, PVP 1989 HA20. *Source History: 1 in 1987; 4 in 1991; 3 in 1994; 3 in 1998; 1 in 2004.* **Sources: HO2.**

Genecorp Green - 55-58 days - Smooth dark green leaves, improved for size weight and color, some TB tolerance, PVP 1986 Seminis Veg. Seeds, Inc. *Source History: 1 in 1994; 1 in 1998; 1 in 2004.* **Sources: Sto.**

German - (German Leaf) - Old World heirloom, closely folded pale-green leaves with bronze-red edges, tender and flavorful. *Source History: 2 in 1991; 2 in 2004.* **Sources: Me7, Se17.**

Gold Rush - 50-60 days - Frilled, curly, crinkled plant, lime-green loose, thin leaves with deeply cut margins, mild, clean flavor. *Source History: 1 in 2004.* **Sources: Se16.**

Goose - 50 days - Arkansas heirloom, semi-bibb type, hardy savoyed leaf, dark green with bronze overlay, summer and fall, very hardy. *Source History: 1 in 1987; 1 in 1991; 1 in 1994; 2 in 1998; 2 in 2004.* **Sources: Ho13, Sa9.**

Grand Rapids - (Grand Rapids B5, Grand Rapids Black Seeded) - 42-65 days - Large erect bright light-green heavily frilled and curled leaves, for greenhouse or field culture, early, holds well, slow bolting, TB disease and rot res., for home gardens or greenhouses, Michigan State University. *Source History: 82 in 1981; 76 in 1984; 69 in 1987; 52 in 1991; 46 in 1994; 42 in 1998; 33 in 2004.* **Sources: Alb, Be4, Bu8, CA25, CH14, CO30, Cr1, Dam, De6, Dom, Ear, Ers, Fo7, GLO, Gr27, Hal, La1, Lin, Ma19, MAY, Mey, MIC, OLD, Ont, PAG, Roh, RUP, Sau, Se17, Sh9, SOU, Te4, Wet.**

Grand Rapids Forcing - 43-45 days - Popular strain for winter greenhouses, light-green frilled leaves, rapid upright grower, disease res., also for spring outdoor sowing, use successive plantings. *Source History: 5 in 1981; 4 in 1984; 6 in 1987; 2 in 1991; 2 in 1994; 2 in 1998; 1 in 2004.* **Sources: Ge2.**

Grand Rapids Tipburn Res. - 40-65 days - Wavy frilled deeply cut finely blistered light-green leaves, dev. for more uniformity and greenhouse forcing, med-large, more compact than regular Grand Rapids, TB and heat res., black seed, mainly home garden, OH/AES. *Source History: 21 in 1981; 21 in 1984; 16 in 1987; 19 in 1991; 20 in 1994; 18 in 1998; 13 in 2004.* **Sources: CO23, DOR, HO2, Hum, Jor, ME9, Pe6, RIS, SE14, Sto, WE10, We19, WI23.**

Grand Rapids, Seafresh - *Source History: 1 in 2004.* **Sources: DOR.**

Grand Rapids, Waldmann's - (Waldmann's Dark Green) - 43-50 days - Long finely frilled leaves larger darker green later maturing and more heat resistant than regular Grand Rapids, large upright plant, good weight, black seed, attractive variety for home or market, greenhouse or outdoor production. *Source History: 14 in 1981; 14 in 1984; 9 in 1987; 5 in 1991; 8 in 1994; 7 in 1998; 3 in 2004.* **Sources: HO2, Jor, Twi.**

Grandpa's - 50-55 days - Old-timer from the Civil War era, lovely ruffled blush-red leaves, semi-heading, sweet mild flavor, spring, very rare. *Source History: 1 in 1987; 2 in 1991; 1 in 1994; 2 in 1998; 3 in 2004.* **Sources: Hi6, Ho13, Te4.**

Grappa - 55-65 days - Soft leaves surround a blanched creamy heart, buttery flavor, disease res., tenderness does not allow it to be shipped, PVP application abandoned. *Source History: 1 in 2004.* **Sources: Ki4.**

Green Ace - Green leaf color. *Source History: 1 in 2004.* **Sources: CO30.**

Green Ice - (Burpee's Green Ice) - 45 days - Glossy dark-green blistered savoyed leaves, waxy fringed leaf margins, very slow to seed, commercial and home garden use, has the dubious distinction of being first plant ever patented, PVP expired 1990. *Source History: 8 in 1981; 8 in 1984; 13 in 1987; 21 in 1991; 25 in 1994; 29 in 1998; 34 in 2004.* **Sources: Be4, Bu2, CA25, Com, Coo, Ers, Fe5, Gr27, Gur, He17, Hi13, HO2, Jun, La1, Lin, LO8, MAY, Mcf, Mel, Mo13, Orn, Par, Pe2, Pin, Roh, RUP, SE4, Se8, Shu, Sk2, Tt2, Tu6, Twi, WE10.**

Green Valley - Large loose leaf type, dark green color, uniform maturity, texture, color, size and weight, excel. bolt tol., good TB tol. *Source History: 1 in 2004.* **Sources: ME9.**

Green Vision - 57 days - Thick leaves with nicely savoyed texture, good bolt and TB tol., head stays open, very slow bolting, non-suckering, PVP 1999 Central Valley Seeds. *Source History: 1 in 1998; 1 in 2004.* **Sources: Jo1.**

Green Wave - 45 days - Selection from Grand Rapids type, larger longer deeper frilled green leaves, vigorous, disease and heat res., produces all summer long. *Source History: 1 in 1987; 2 in 1991; 2 in 1994; 3 in 1998; 1 in 2004.* **Sources: Ev2.**

GX 514 - *Source History: 1 in 2004.* **Sources: Se17.**

Hungarian Winter - Heirloom, cool season cutting lettuce, hardy. *Source History: 1 in 2004.* **Sources: Syn.**

Ibis - 41 days - Deepest red lettuce on the market, broad, upright leaves are strongly savoyed, holds its rich sweet flavor longer into warm weather, TB tolerant, good for short season areas. *Source History: 1 in 1994; 2 in 1998; 6 in 2004.* **Sources: Al25, In8, Sa5, Se17, Sto, Ta5.**

Italienischer - 55 days - Tall, sturdy, upright plants to 18 in., bright green leaves are sweet and crisp even when mature, plant from spring to fall. *Source History: 1 in 2004.* **Sources: Ter.**

Jeanne - French variety dating back to the 1700s, broad smooth leaves, fine flavor. *Source History: 1 in 2004.* **Sources: In8.**

Jebousic - Vibrant green spiraling arrow tipped leaves, sweet and crunchy, selected against TB, old Czech variety. *Source History: 1 in 2004.* **Sources: In8.**

Lasting Green - 42 days - Dark green color, slow bolting, corky root and DM tol. *Source History: 2 in 2004.* **Sources: ME9, RIS.**

Lau's Pointed - Star shaped plant with long, pointed, thick leaves growing from the center, good for stir fry, from Malaysian highlands. *Source History: 1 in 1998; 1 in 2004.* **Sources: Ho13.**

Leopard - Attractive, pale green leaves speckled and splattered with red, flavorful. *Source History: 1 in 2004.* **Sources: Sa5.**

Les Orielles du Diable - 70 days - Name translates "ears of the devil", glossy, large, tapered burgundy leaves form a starlike rosette, oily, nutty flavor, slow to bolt. *Source History: 2 in 2004.* **Sources: Ga22, Ho13.**

Lingue de Canarino - (Royal Oak Leaf) - 45-60 days - Pale-green leaves resemble round oak leaves and form a tight bunch, very mild and delicious in the spring, continues to be quite edible right through the summer though the taste changes somewhat, does not bolt, name means "canary tongues" or "canary-colored tongues". *Source History: 3 in 1991; 1 in 1994; 1 in 2004.* **Sources: Tu6.**

Lolla Rossa, Selway - *Source History: 1 in 2004.* **Sources: DOR.**

Lollo Bianco - 55 days - Light green, beautiful loosely frilled heads, tender and delicious. *Source History: 1 in 1991; 2 in 1994; 2 in 1998; 2 in 2004.* **Sources: Ba8, WE10.**

Lollo Biondo - 45-70 days - Pale-green cousin of Lollo Rossa, very ruffled leaves, cut leaf by leaf rather than using entire head, sharp, tangy flavor, from Italy. *Source History: 2 in 1987; 2 in 1991; 3 in 1994; 5 in 1998; 6 in 2004.* **Sources: CO30, Dam, Ki4, Orn, Syn, Tho.**

Lollo Rossa Atsina - 55 days - Italian looseleaf, mounded globes of tightly frilled reddish leaves taper down to a light green center, mild flavor, slow to bolt, can be cut repeatedly or harvested as a loose head, specialty market and home garden. *Source History: 6 in 1998; 4 in 2004.* **Sources: Dam, Ers, Ni1, Twi.**

Lollo Rosso - (Lolla Rossa) - 45-75 days - One of the most deeply curled looseleaf letttuces, beautiful magenta leaves with light-green bases, mild flavor, cut and come again, sow spring or fall. *Source History: 5 in 1987; 16 in 1991; 18 in 1994; 26 in 1998; 41 in 2004.* **Sources: Ba8, Bo17, Bo19, CA25, CO30, Coo, Dam, Fe5, Fo7, Ga1, He8,**

Hi6, HO2, Jo1, Ki4, La1, Loc, ME9, Na6, Orn, OSB, Pe2, Pin, Ri2, Ri12, Roh, RUP, Se8, SE14, Se16, Se24, Se26, Sto, Sw9, Syn, Ta5, TE7, Tho, Vi4, WE10, WI23.

Madera - Dark Lolla Rossa type with very uniform, dark red leaves, good texture, outstanding babyleaf production, DM tol. *Source History: 1 in 2004.* **Sources: ME9.**

Mamba - 50 days - Glossy savoyed, deeply lobed oak leaf, firmer than most oak leaf varieties, yellow interior, tolerates TB and bolting, holds well in the field. *Source History: 1 in 2004.* **Sources: OSB.**

Manoa - 50 days - Compact, semi-heading type, delicate light green, crunchy leaves, tolerates heat, popular in Hawaii. *Source History: 2 in 2004.* **Sources: Ev2, Kit.**

Maravilla der Verano Vonny - (Maravilla der Verano, Maravilla Red-Green) - 62 days - Very slow bolting, excellent for summer crops, small heads, deep-green undulated deeply indented leaves, cutting lettuce, crisp even on hot summer days, Holland. *Source History: 1 in 1981; 1 in 1984; 1 in 1991; 1 in 1994; 2 in 1998; 3 in 2004.* **Sources: Ag7, Bou, Dom.**

Mariachi - (SXP0066) - Red Salad Bowl type with very dark red, solid texture, DM tol. *Source History: 1 in 2004.* **Sources: ME9.**

Mariah - Green leaf, corky root and DM tol., good bolt tol. *Source History: 1 in 2004.* **Sources: ME9.**

Marimba - (SXP0071) - Uniform, dark red loose leaf with good texture for babyleaf production. *Source History: 1 in 2004.* **Sources: ME9.**

Marin - Tipburn and bolt tolerance compares to Two Star, slightly heavier than Two Star, especially in cool weather, PVP application pending. *Source History: 1 in 2004.* **Sources: ME9.**

Mascara - 60-65 days - Curly, frilled oak leaf shaped leaves retain dark-red color in hot weather, mild non-bitter flavor, very bolt res., Dutch seed. *Source History: 1 in 1991; 2 in 1994; 5 in 1998; 4 in 2004.* **Sources: Hud, Se16, Sh12, Tho.**

Maserati - Medium size red oakleaf, excellent mildew tolerance. *Source History: 1 in 2004.* **Sources: Coo.**

Mercury - 55 days - Red Lollo type, frilly, tight leaves with striking red color, sweet flavor, delicate texture. *Source History: 1 in 2004.* **Sources: Ves.**

Merlot - (Galactic) - 60 days - Looseleaf type with intense deep burgundy colored leaves, slow bolting, ideal for cut-and- come again culture and baby salad mix. *Source History: 1 in 1998; 5 in 2004.* **Sources: Coo, Fe5, Ho13, In8, Sh12.**

Merveille de Mai - 50 days - Upright, lime-green, rounded leaves, 8 in. tall, leaf formation resembles a butterhead, great texture and flavor, for spring and fall planting. *Source History: 1 in 2004.* **Sources: Ter.**

Mighty Red Oak - 50 days - Finely cut leaves with red-bronze tint, spreads 12-16 in. across, mild flavor, slow to bolt. *Source History: 1 in 1998; 1 in 2004.* **Sources: Bu2.**

Misticanza - Mix of green and red leaves. *Source History: 1 in 1987; 1 in 1991; 1 in 1994; 1 in 1998; 1 in 2004.* **Sources: Ber.**

Monet - 53 days - Deeply cut and curled, dark green leaves, mild flavor, crisp texture, dev. for both greenhouse and outdoor use. *Source History: 4 in 1998; 4 in 2004.* **Sources: Ers, Fe5, Ra5, Ves.**

Natividad - 55 days - Dark red Lolla Rossa type with deeper burgundy red color and smoother leaves. *Source History: 2 in 1998; 1 in 2004.* **Sources: CO30.**

New Red Fire - 43-65 days - The standard for a dark red leaf lettuce, ruffled leaves similar to Red Sails with a lighter color, semi-heading, slow to bolt and holds its color well under difficult conditions, specialty market and home garden. *Source History: 13 in 1994; 17 in 1998; 21 in 2004.* **Sources: CA25, CO30, Fe5, Ga1, Ha5, HO2, Loc, ME9, Orn, OSB, Pe2, Pin, RIS, RUP, SE4, Sto, Ta5, Ter, Tho, Tu6, Twi.**

Oak Leaf - (Green Oak Leaf) - 38-60 days - At one time known as American Oak Leaved, tight rosettes of med-dark green deeply lobed oak leaf shaped leaves, very res. to hot weather, long-standing, never bitter, still fine quality late in summer, upright plant, Vilmorin introduction, 1771, known as Baltimore or Philadelphia Oakleaf in the 1880s. *Source History: 82 in 1981; 75 in 1984; 81 in 1987; 77 in 1991; 70 in 1994; 73 in 1998; 67 in 2004.* **Sources: All, Ba8, Be4, Bo17, Bo19, Bu2, Bu8, But, CH14, CO30, Co32, Com, De6, DOR, Ers, Fis, Ge2, Gr27, HA3, He8, He17, Hig, HO2, Hud, Jor, Ki4, Kil, La1, Ma19, ME9, Mel, Mey, Mo13, MOU, Ni1, OLD, Ont, Orn, PAG, Pin, Pla, Red, Ri2, Ri12, Roh, RUP, Sa5, Sau, SE4, SE14, Se16, Se25, Sh9, Shu, Si5, Sk2, So1, So12, St18, Tho, Tu6, Und, Ver, Vi4, WE10, Wet, WI23.**

Oak Leaf, Baby - 50 days - More compact version of Green Oak Leaf, medium green, oak leaf shaped leaves with rounded lobes, holds well without bolting. *Source History: 1 in 2004.* **Sources: Se16.**

Oak Leaf, Purple - 40-65 days - Vigorous plants, deeply lobed, purple leaves, tasty, beautiful, heat res., slow to bolt. *Source History: 3 in 1998; 6 in 2004.* **Sources: Fo13, Ho13, In8, Pe1, Se17, Se26.**

Oak Leaf, Red - 50-68 days - Same as green Oak Leaf except for color, leaves turn a deep burgundy as they mature, grow in full sun for deep-red color and best growth, shows some heat resistance, unappealing to slugs and snails. *Source History: 1 in 1987; 3 in 1991; 7 in 1994; 10 in 1998; 10 in 2004.* **Sources: Co32, Eo2, Hi6, Ho13, Orn, Pe2, Ri2, Roh, Se7, TE7.**

Oak Leaf, Redder Ruffled - Brick-red leaves with crinkled edges, savoyed in the middle, excellent early spring lettuce, dev. by Frank Morton. *Source History: 3 in 2004.* **Sources: Au2, Sh12, St18.**

Oak Leaf, Royal - 48-50 days - Very heat res. long-standing plants, large rosettes (indistinct heads) of oak leaf shaped dark- green leaves with thick midribs, never bitter when hot, PVP expired 1995. *Source History: 6 in 1981; 5 in 1984; 5 in 1987; 7 in 1991; 10 in 1994; 16 in 1998; 23 in 2004.* **Sources: Bu2, CO30, Coo, Dam, DOR, Ear, Ers, Fe5, Ga1, Ha5, Jun, La1, LO8, ME9, Orn, OSB, Pe2, SE4, Se8, SE14, TE7, Tu6, WE10.**

Oaky Red Splash - 60 days - Bred by Shoulder to Shoulder Farm, oak leaf shaped leaves are red and copper tinged and splashed with dark-red splotches, somewhat upright plant growth forms a blanched heart when mature, thick, juicy midribs sometimes tinged pink, excel. salad var. for continuous harvest, especially succulent and well colored in cold spring or autumn weather. *Source History: 1 in 1994; 1 in 1998; 2 in 2004.* **Sources: Sh12, Ter.**

Paramix - 47 days - Deeply lobed, bright green leaves, excellent for salad mix and baby use, forms an attractive head, tolerates DM, LMV. *Source History: 1 in 2004.* **Sources: OSB.**

Pentared - 75 days - Beautiful dark burgundy slender leaves. *Source History: 1 in 2004.* **Sources: Tho.**

Pink Tip - Med. green leaves with pink overlay. *Source History: 1 in 2004.* **Sources: Ho13.**

Prizehead - (All Cream, Prizehead Leaf, Brown Edged Leaf, Red Leaf, Bronze Prizehead, Red Fringe, Ruby) - 45-55 days - Non-heading despite its name, upright broad deeply curled light-green leaves with bronze-red leaf margins, fast growing, beautiful, med-slow bolting, white seed, never bitter, good var. for home gardens or local market. *Source History: 83 in 1981; 81 in 1984; 74 in 1987; 71 in 1991; 72 in 1994; 62 in 1998; 48 in 2004.* **Sources: Alb, Be4, Bo19, Bu3, Bu8, CH14, Cr1, DOR, Ear, Ers, Fe5, Fis, Fo7, Ge2, Gr27, Hal, He8, He17, HO2, Jor, Jun, Kil, La1, Lin, LO8, Loc, ME9, Mo13, MOU, OLD, Ont, PAG, Pe6, Pr3, Red, RIS, RUP, Sa9, Sau, SE4, SE14, Se26, Sh9, Shu, So9, Vi4, WE10, WI23.**

Prizehead, Early - 47 days - Frilly leaves tinged reddish-brown, tender and sweet, good early lettuce for home gardens, very quick growing, stays tender a long time. *Source History: 2 in 1981; 2 in 1984; 7 in 1987; 3 in 1991; 2 in 1994; 1 in 1998; 1 in 2004.* **Sources: HA3.**

Prizeleaf - 48 days - Tender, green and maroon leaves, excel. flavor, slow bolting. *Source History: 1 in 1998; 1 in 2004.* **Sources: Bu2.**

Quedlinburger Dickkopf - *Source History: 1 in 2004.* **Sources: Ta5.**

Queensland - Australian variety with large leaves, resembles a cos type, attractive yellowish hue, extremely bolt resistant. *Source History: 1 in 1994; 1 in 1998; 1 in 2004.* **Sources: Ech.**

Raisa - 48 days - Dutch variety, oak leaf shaped leaves, more red coloring than Red Oak and more cold tolerance, TB res., PVP 1992 Rijk Zwaan Zaadteelt en Zaadhandel B.V. *Source History: 1 in 1994; 2 in 1998; 2 in 2004.* **Sources: CO30, Ha5.**

Red - Cool weather reddish bibb, does not tolerate heat, rare. *Source History: 1 in 2004.* **Sources: Ea4.**

Red Coral - 55 days - Ruffled red-pink leaves, sweet flavor, long harvest, heirloom. *Source History: 3 in 2004.* **Sources: Co32, He8, Se16.**

Red Curl - Red-purple, wide crumpled leaves. *Source History: 1 in 2004.* **Sources: Ho13.**

Red Embers - Red leaf type, compares to Red Fire. *Source History: 1 in 2004.* **Sources: ME9.**

Red Fire - 45 days - Large savoyed leaves frilled at the edges, intense-red color, crisp and tender, easy to grow, withstands heat as well as cold, very uniform and productive. *Source History: 1 in 1987; 8 in 1991; 6 in 1994; 2 in 1998; 1 in 2004.* **Sources: ME9.**

Red Rage - 45-50 days - Loose, broadleaf type with intense red leaf color and apple-green midribs, the red color does not bleed into the midribs, darker color than New Red Fire, TB tol., PVP 2003 Pybas Vegetable Seed Company, Inc. *Source History: 2 in 1998; 2 in 2004.* **Sources: OSB, Sto.**

Red Rapids - 50 days - Resembles Red Sails but smaller and more savoyed, turns from red to green as leaves become overly mature. *Source History: 1 in 1991; 1 in 2004.* **Sources: Se16.**

Red Ruffled Oaks - 50 days - Open, ruffled, 12 in. heads, oak leaf shaped leaves, red on green, tender, succulent mid-ribs, productive during cool spring nights. *Source History: 1 in 2004.* **Sources: Ter.**

Red Sails - 40-66 days - AAS/1985, radiant bronze-red leaves form a full rosette, delicately ruffled and deeply lobed, mild bitter-free flavor, ready to cut early, holds salad quality a long time, slow to bolt, 6 times the vitamin A and 3 times the vitamin C as supermarket lettuce, most spectacular in cool weather but grows well spring summer and fall, PVP 1986 Seminis Vegetable Seeds, Inc. *Source History: 65 in 1987; 72 in 1991; 68 in 1994; 62 in 1998; 59 in 2004.* **Sources: All, BAL, Be4, Bo17, Bu3, Bu8, CA25, Co31, Com, Coo, Dam, Ers, Fe5, Fi1, Fo7, Ge2, GLO, Goo, Gur, HA3, HO2, Hum, Jo1, Jor, Kil, La1, Lin, Loc, ME9, Mey, MIC, MOU, Ni1, Or10, Par, Pin, RIS, Roh, RUP, Se25, Se26, SGF, Sh9, Si5, Sk2, So1, SOU, St18, Sw9, Ta5, Te4, Ter, Tho, Tt2, Und, Up2, Ver, We19, Wet.**

Red Velvet - 55 days - Very dark lettuce, red-maroon colored leaves, backs of leaves are green tinged with maroon, chewy texture, good choice for mixed greens. *Source History: 3 in 2004.* **Sources: He8, La1, Se16.**

Red Wave - Prizehead type, red leaves, slow to bolt, grows well in warm or cool conditions. *Source History: 1 in 1994; 1 in 1998; 1 in 2004.* **Sources: Ev2.**

Redina - 38-59 days - One of the reddest varieties of this type with deep burgundy leaves with speckled green throats, slow to TB, for spring and fall plantings. *Source History: 6 in 1998; 5 in 2004.* **Sources: CO30, Orn, OSB, Pin, Ta5.**

Redprize - (FM-1559) - 50-70 days - Red leaf type, larger than Prizehead and slower to bolt, PVP 1989 HA20. *Source History: 1 in 1987; 1 in 1991; 2 in 1994; 4 in 1998; 1 in 2004.* **Sources: Sh14.**

Revolution - 38-70 days - Lollo Rossa type, 10-12 in., frizzy edged leaves with intense red, almost black coloring, good flavor, suitable for winter greenhouse, cold tolerant, excel. bolt res. *Source History: 4 in 2004.* **Sources: Coo, Ter, Tho, We19.**

Ricciolina da Taglio - 30-40 days - Wavy light green leaves. *Source History: 1 in 1987; 1 in 1991; 1 in 1994; 1 in 1998; 1 in 2004.* **Sources: Ber.**

Robles Royale - Good size leaf with exel. dark green color, good texture and bolt tol., great improvement over standard Oak Leaf. *Source History: 1 in 2004.* **Sources: ME9.**

Rossa d' Amerique - Pale-green leaves tipped with sparkling rosy red, used as a cutting lettuce (harvested at 4-6 in.) but will form a loose head if thinned. *Source History: 1 in 1991; 2 in 1994; 3 in 1998; 1 in 2004.* **Sources: Coo.**

Rossa di Trento - 45-60 days - A broad savoyed red-tipped cutting lettuce from Milan, Italy, can be grown nearly year round in mild climates. *Source History: 1 in 1987; 2 in 1991; 2 in 1994; 1 in 1998; 4 in 2004.* **Sources: Coo, Se16, Se24, Ta5.**

Royal Green - 45-55 days - Darker green version of Grand Rapids, medium green, smooth leaves, very uniform, TB tol., PVP 1988 Royal Sluis B. V. *Source History: 4 in 1991; 4 in 1994; 4 in 1998; 3 in 2004.* **Sources: ME9, SE4, Sto.**

Royal Red - 45-52 days - Mildly ruffled leaves form an open head, heavily colored with bright cherry red, green leaf bases and undersides, slower to size than other red lettuces but stands well, good bolt and TB res., sweet taste, soft leaf texture, PVP 1988 Royal Sluis B. V. *Source History: 4 in 1991; 3 in 1994; 7 in 1998; 12 in 2004.* **Sources: CO30, Dam, Ers, Fa1, HO2, Jun, ME9, MIC, Ni1, OLD, Ont, PAG.**

Rubin - 50-60 days - Beautiful deep maroon, looseleaf heads with frilled edges, brighter colors in cooler temperatures, hardy, great market variety. *Source History: 6 in 2004.* **Sources: Ag7, Ba8, Pe6, SE14, Ta5, TE7.**

Ruby - 40-65 days - AAS/1958, beautiful bright-green frilled savoyed leaves shaded with intense-red, sel. from Prizehead for more vivid color, heat res., long-standing, slow bolting, white seed, USDA/Beltsville, 1957. *Source History: 42 in 1981; 35 in 1984; 30 in 1987; 21 in 1991; 13 in 1994; 14 in 1998; 13 in 2004.* **Sources: CH14, CO30, DOR, La1, Lej, Ma19, Mo13, OLD, Orn, Sau, St18, Ta5, WI23.**

Ruby Red - (Ruby Red Salad Bowl) - 40-53 days - Glossy bright-green frilled leaves with heavy intense-red shading, large spring plant, does not fade in summer heat, black seed, USDA. *Source History: 25 in 1981; 22 in 1984; 23 in 1987; 15 in 1991; 9 in 1994; 13 in 1998; 11 in 2004.* **Sources: De6, HA3, He17, Hud, MAY, Ont, PAG, Sa9, Sh9, So25, Wi2.**

Ruby Ruffles - 55-60 days - Uniform plant, crisp red leaves with good flavor, good field holding ability, slow to bolt, tolerates DM and LMV, PVP 2002 HA20. *Source History: 2 in 2004.* **Sources: Ha5, HO2.**

Rustica - 60 days - Deepest darkest maroon, frilly outer leaves, bright green center, no bleeding into veins, suitable for baby mix at 27 days - DM tol. *Source History: 2 in 2004.* **Sources: OSB, Ves.**

Salad Bowl - (Green Salad Bowl, Green Salad) - 45-68 days - AAS/1952, light-green long wavy deep-notched leaves, large fast-growing rosette, TB and heat res., will not get bitter in hot weather, very slow bolt, USDA/Beltsville, 1952. *Source History: 129 in 1981; 122 in 1984; 112 in 1987; 104 in 1991; 95 in 1994; 94 in 1998; 82 in 2004.* **Sources: Ag7, Alb, All, Be4, Bo19, Bou, Bu2, Bu3, Bu8, But, CH14, CO23, CO30, Co31, Com, Coo, Cr1, Dam, De6, Dom, DOR, Ear, Ers, Fa1, Fe5, Fi1, Fis, Fo7, Ge2, GLO, Goo, Gr27, Gur, HA3, Ha5, He8, Hi6, Hig, HO2, Hum, Jor, Jun, Kil, La1, Lin, LO8, Loc, MAY, ME9, Mel, Mey, Mo13, MOU, Na6, Ni1, OLD, Ont, Or10, PAG, Pe2, Roh, RUP, Sau, Scf, SE4, SE14, Se26, Sh9, Sh14, Sk2, So1, SOU, Sto, TE7, Ter, Tho, Und, Vi4, WE10, Wet, Wi2, WI23.**

Salad Bowl, Compact - 28-46 days - Smaller, more dense heads than Salad Bowl, leaves are more lobed. *Source History: 1 in 2004.* **Sources: Jo1.**

Salad Bowl, Red - (Red Salad) - 45-50 days - Large upright plants, long deeply lobed bronze-red leaves, blanched inner leaves with a bronze glow, good keeper, decorative, just like Salad Bowl except for color, black seed, very slow bolting for long harvest, 1955. *Source History: 20 in 1981; 18 in 1984; 21 in 1987; 24 in 1991; 31 in 1994; 46 in 1998; 55 in 2004.* **Sources: Ag7, All, Bo19, Bu2, Bu3, But, CA25, CH14, CO23, CO30, Coo, Dam, Dom, DOR, Ers, Fe5, Fis, Ga1, Gr27, Ha5, Hi6, Hig, HO2, Jo1, Ki4, La1, Loc, ME9, Mel, MIC, Mo13, Na6, Orn, OSB, Pe2, Pe6, Roh, Ros, RUP, Se8, SE14, Se16, Se24, Se26, Sh9, Shu, Sto, Sw9, Syn, TE7, Tu6, Ves, Vi4, WE10, WI23.**

Salad Trim - 55 days - Dark purplish-red crisp crinkly leaves, nice cut-and-come-again lettuce, somewhat heat tol., unattractive to slugs and snails. *Source History: 2 in 1981; 1 in 1984; 1 in 1987; 2 in 1991; 1 in 1994; 1 in 1998; 2 in 2004.* **Sources: Se16, So9.**

Salade de Russie - *Source History: 1 in 2004.* **Sources: Ta5.**

Sanguine - Triple red Lollo Rossa with blood-red color throughout the leaves, excellent quality all season. *Source History: 1 in 2004.* **Sources: Dam.**

Sanguine Ameliore - (Strawberry Cabbage Lettuce) - 45 days - Old French var. with deep red-brown mottling clustered toward the pink center of each tongue shaped leaf, retains color, tender texture, excel. quality, intro. to the U.S. in 1906 as Strawberry Cabbage Lettuce by C. C. Morse and Co. *Source History: 2 in 1998; 4 in 2004.* **Sources: Fo7, Ho13, Se16, Ter.**

Sesam - (Dark Lollo Rossa) - 45-55 days - Lolla lettuce with intense cranberry coloring, compact plant habit, holds color well for greenhouse production, slightly pungent flavor is a great addition to salads, Italy. *Source History: 1 in 1994; 2 in 1998; 5 in 2004.* **Sources: Ag7, Ba8, CO30, Ha5, Orn.**

Shining Star - 55 days - Imp. bolting tolerance over Genecorp Green, Waldmann type heads, deep green color, TB tol., PVP 2001 Seminis Vegetable Seeds, Inc. *Source History: 1 in 1998; 2 in 2004.* **Sources: Dam, OSB.**

Simpson Elite - 40-53 days - Imp. Black Seeded Simpson type, holds delicate flavor longer without bitterness, medium light-green, broad, crumpled leaves with curled outer leaf edges, bolts 30 to 40 days later, PVP 1999 Seminis Vegetable Seeds, Inc. *Source History: 33 in 1994; 48 in 1998; 45 in 2004.* **Sources: All, BAL, Be4, Bu2, Bu3, CA25, Co31, Dam, Ers, Fa1, Fe5, Fi1, Ge2, Gr27, Hal, HO2, HPS, Jo1, Jor, Jun, Mcf, MIC, Mo13, Ni1, OLD, Par, Pin, Roh, Ros, RUP, Se25, SGF, Sh9, Shu, So1, SOU, Sto, Sw9, Ter, Tho, Tt2, Twi, Ver, Ves, We19.**

Slobolt - (Longstanding, Harris' Slobolt) - 45-65 days - Large compact plants produce during entire summer, very slow to bolt, thick clusters of light-green crumpled frilled leaves, Grand Rapids type but 2 to 3 weeks longer without bolting, hot weather does not change its growth habit, USDA, 1946. *Source History: 24 in 1981; 21 in 1984; 20 in 1987; 17 in 1991; 12 in 1994; 13 in 1998; 10 in 2004.* **Sources: Ha5, HA20, HO2, RIS, RUP, SE4, Se16, So1, Sto, Ter.**

Slogun - Semi-heading lettuce resembles Sierra but with more bolt resistance and more of a tendency to form a head, always the last to shoot up a seed stalk, less resistance to bottom rot, from PSR Breeding Program. *Source History: 1 in 1998; 1 in 2004.* **Sources: Tu6.**

Smile - New compact oakleaf, tolerates tipburn and mildew and resists aphids. *Source History: 2 in 2004.* **Sources: Coo, Tho.**

Sparta - Batavian type, deep red leaves can be harvested young as leaf lettuce, will form a loose head, resists most strains of Bremia, LMV tolerant. *Source History: 1 in 2004.* **Sources: Coo.**

Speckles - 45-55 days - Apple-green leaves flecked with red-brown polka dots, forms a dense bibb-like head with a blanched heart, sweet buttery flavor, Amish heirloom. *Source History: 1 in 1991; 1 in 1994; 3 in 1998; 11 in 2004.* **Sources: Ers, La1, ME9, Orn, Pe2, Ri2, RUP, SE14, Se26, TE7, Ter.**

Spring Feast - Does well in early spring, soft, crunchy texture. *Source History: 1 in 1998; 1 in 2004.* **Sources: Bou.**

Strela Green - Large plant, long narrow leaves, from SSE member Mary Schultz, pre-1500. *Source History: 1 in 2004.* **Sources: Pe1.**

Summer Ace - *Source History: 1 in 2004.* **Sources: CO30.**

Sunfire - 60 days - Large red oak leaf, firm, narrow leaves, deeper red color toward center of head, holds well, DM tol., PVP application pending. *Source History: 1 in 2004.* **Sources: OSB.**

Sunset - 60 days - AAS/1987, one of the favorites selected in 1996 from Heritage Farm's lettuce collection of 850 varieties, although a spectacular variety, it has little chance of ever becoming commercially available due to the fact that it produces relatively few seeds, so it is up to the home gardener to preserve this variety. *Source History: 1 in 1998; 11 in 2004.* **Sources: Ag7, Ba8, Ers, Ga1, He8, ME9, Orn, Pe2, RUP, Se16, TE7.**

Sunshine - 50-60 days - Attractive cool weather variety with green and bronze-red leaves, delicious mild taste and firm, crisp texture, wilt resistant. *Source History: 1 in 1998; 1 in 2004.* **Sources: CO30.**

Sweetie - 40-70 days - Vigorous large upright plants, large broad flat smooth rich green leaves with delicate tinge near edge, unusually sweet, tender, productive, black seed, cos-like. *Source History: 6 in 1981; 5 in 1984; 7 in 1987; 1 in 1991; 2 in 2004.* **Sources: In8, Se17.**

Tahoe - 60 days - Batavia type, deep red, heavily textured leaves with light green center, does well in cool wet areas. *Source History: 1 in 2004.* **Sources: OSB.**

Tango - 40-52 days - Uniform attractive plants form tight erect rosettes, deeply cut pointed leaves resemble endive in appearance, darker green than most varieties, tender, tangy flavor, vitamin rich, can become bitter in warmer weather. *Source History: 1 in 1987; 4 in 1991; 3 in 1994; 10 in 1998; 21 in 2004.* **Sources: Ag7, Ba8, CO23, CO30, DOR, Ers, Ga1, He8, HO2, Ki4, Loc, ME9, Ni1, Orn, OSB, Pe2, SE4, SE14, Se16, TE7, WE10.**

Thai Green - 55 days - Yellow-green looseleaf lettuce for hot climates, slow to bolt. *Source History: 1 in 1994; 1 in 1998; 1 in 2004.* **Sources: Se7.**

Thai Oakleaf 88 - (Thai 88) - 39 days - Resembles Oak Leaf but with a larger plant and more tender texture, more heat tol. and slower to bolt, dev. at lettuce breeding program in Thailand, intro. in 1988 by CV So1. *Source History: 1 in 1991; 1 in 1994; 1 in 1998; 1 in 2004.* **Sources: So1.**

The Drunkard - (Ubriacona) - 60 days - Crisp leaves matures to loose head, outer leaves red, inner green leaves, grow any time for cutting lettuce or baby heads, sow spring/fall for full size heads, colors best in fall. *Source History: 1 in 2004.* **Sources: Se24.**

Tiara - 45-45 days - Widely adapted variety with blistered and frilled dark green leaves, good tolerance to tipburn and bolting, PVP 1991 Seminis Vegetable Seeds, Inc. *Source History: 2 in 1994; 2 in 1998; 1 in 2004.* **Sources: SE4.**

Tomahawk - Thick, ruffled green leaves with pink and bronze, delicate taste. *Source History: 1 in 1998; 1 in 2004.* **Sources: Se17.**

Triple Red - Small, thick textured loose leaf lettuce, very dark red to burgundy color, most suitable for use in spring mix prepared salads. *Source History: 2 in 2004.* **Sources: ME9, Orn.**

Trocadero - (Laithue Lorthois) - 60 days - Improved butterhead type, light-green slightly wavy leaves with tinge of red at edge, thrives in wet cool weather, plant late summer or early fall, early. *Source History: 2 in 1981; 1 in 1984; 1 in 1987; 1 in 1998; 1 in 2004.* **Sources: Se17.**

Two Star - 50-60 days - Medium size Waldmann type, dark green frilled leaves, high bolt tol., uniform maturity, holds well in the field, PVP 1993 Orsetti Seed Company, Inc. *Source History: 4 in 1998; 7 in 2004.* **Sources: CA25, CO30, Dam, ME9, OSB, SE4, Sto.**

Valeria - 45 days - Lolla Rossa type, intense red leaves with frizzy edges, also suitable for winter greenhouses. *Source History: 2 in 1991; 3 in 1994; 3 in 1998; 1 in 2004.* **Sources: Syn.**

Vulcan - 52 days - Vivid candy-apple red over pale-green background, slightly frilled leaves form large full head that stays open and fresh looking, very slow to bolt or tipburn, PVP 1992 Sakata Seed America, Inc. *Source History: 3 in 1991; 6 in 1994; 2 in 1998; 3 in 2004.* **Sources: SAK, Ta5, Ves.**

Waldmann's Dark Green - (Waldmann's Dark-Green Grand Rapids) - 50 days - Deeper green and larger leaves than other Grand Rapids types, fringed leaves, vigorous, uniform, productive. *Source History: 1 in 1981; 1 in 1984; 6 in 1987; 10 in 1991; 5 in 1994; 4 in 1998; 3 in 2004.* **Sources: Fe5, Sto, Tu6.**

Waldmann's Green - 43-58 days - Large wavy frilled light-green leaves, Grand Rapids type with heavier slightly longer leaves, greenhouse or outdoors, uniform, productive, 1958. *Source History: 4 in 1981; 3 in 1984; 5 in 1987; 8 in 1991; 8 in 1994; 12 in 1998; 10 in 2004.* **Sources: CO23, HA20, La1, Loc, ME9, OSB, RUP, SE4, Vi4, WE10.**

Western Red - *Source History: 1 in 2004.* **Sources: CO30.**

Winterwunder - 75 days - Light green, 9-11 in. heads, extremely winter hardy, bred in northern Europe. *Source History: 1 in 2004.* **Sources: Ter.**

Xena - 55-60 days - Frilled, large, thick dark green leaves, uniform maturity, fresh market, processing and whole leaf pack, long shelf life, good shipper, TB tol. *Source History: 4 in 2004.* **Sources: Ha5, HA20, ME9, OSB.**

Varieties dropped since 1981 (and the year dropped):

American Gathering (2004), *Australian* (1991), *Australian Gathering* (2004), *Baby Green* (2004), *Baby Oak* (1994), *Batavia Blonde No. 6* (2004), *Big Red* (2004), *Bionda Liscia* (2004), *Blonde Feuille de Chene* (1991), *Bolchoi* (2004), *Bologna* (2004), *Boston, Rosanna* (1998), *Bronc Rodon* (2004), *Bronze* (1998), *Bronze Arrow Variant* (2004), *Burpee's Greenhart* (1998), *Centennial* (2004), *Ceremony* (2004), *Ceremony II* (2004), *Chapeau Rouge* (1998), *Chinese Loose Leaf* (1994), *Cireo* (2004), *Cressonnette Marocaine* (2004), *Crispy Sweet* (1991), *De Cortar* (1994), *Early Curled Silesia* (1984), *European Lettuces* (1987), *French Red* (1994), *Frisby* (2004), *Glossy Green* (2004), *GML-7* (2004), *Grand Rapids (U.S. No. 1)* (1991), *Grand Rapids Greenhart* (1998), *Grand Rapids H5-4 TBR* (1991), *Grand Rapids Improved* (1983), *Grand Rapids Old Strain* (1991), *Grand Rapids Special* (1998), *Grand Rapids Washington Str.* (1987), *Grand Rapids, Dark Green* (1994), *Green Oak* (1998), *Impuls* (2004), *Krizet* (2004), *Li'l Sweetie* (1991), *Matchless* (1987), *Mondian* (1987), *Nimes No-Bolt Thick-Leaved* (1987), *Oak Leaf, Piroga* (1998), *Pom Pom* (2004), *Prizehead (R. Strain)* (1991), *PS 64289* (2004), *PS 21192* (2004), *Purple Oak* (1998), *Rebosa* (2004), *Red Fox* (2004), *Red Sceptre* (1998), *Red Tipped* (1991), *Redhead* (1998), *Ritsa* (2004), *Rolina* (2004), *Rosa PSR* (2004), *Royalty* (2004), *Rubra* (1994), *Ruby Be Mine* (1998), *Salad Bowl, Australian* (1994), *Salad Bowl, Red Rebosa* (1994), *Samantha* (2004), *Selma Lollo* (1998), *Sheng Tsai* (1982), *Summer Queen* (1991), *Sunglow* (1987), *Super Prize* (2004), *Taiwan White Leaf* (1983), *Tangent* (2004), *Tigris* (2004).

Lettuce / Romaine
Lactuca sativa

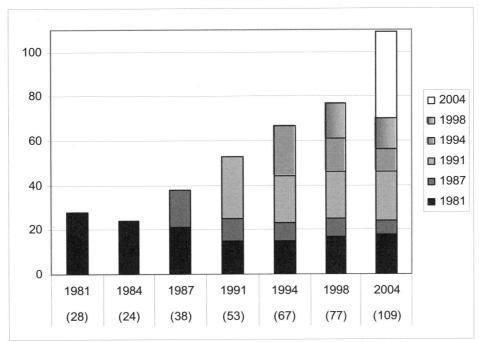

Number of varieties listed in 1981 and still available in 2004: 28 to 18 (64%).
Overall change in number of varieties from 1981 to 2004: 28 to 109 (389%).
Number of 2004 varieties available from 1 or 2 sources: 72 out of 109 (66%).

Apache - 80 days - Cream and pink and green heart, crisp and sweet with bronze-red outer leaves. *Source History: 1 in 1991; 1 in 2004.* **Sources: Ho13.**

Balady - *Source History: 1 in 2004.* **Sources: Se17.**

Ballon - (Balloon) - 66 days - Large French romaine lettuce, enormous heart, yet it is tender and perfectly formed, heat res., seldom runs to seed, excellent for exhibition or general use, 1885. *Source History: 3 in 1981; 2 in 1984; 2 in 1987; 3 in 1991; 3 in 1994; 3 in 1998; 3 in 2004.* **Sources: Coo, Ho13, Syn.**

Bambi - Semi-romaine type with dark green, thick, sweet leaves, excellent for mini head production. *Source History: 1 in 1998; 1 in 2004.* **Sources: Orn.**

Barcarolle - 70-80 days - Tight med-large upright heads, excel. deep-green color, good flavor and texture, finer leaf structure than others, LMV res., crunchy juicy midribs, high quality. *Source History: 5 in 1981; 5 in 1984; 2 in 1987; 3 in 1991; 1 in 1994; 2 in 1998; 2 in 2004.* **Sources: MOU, Se7.**

Bath Cos - Lighter head, upright dark-green leaves, 12 in. dia., spring and fall, pre-1800. *Source History: 1 in 1987; 1 in 1998; 2 in 2004.* **Sources: Ho13, Th3.**

Beretta - Tall framed, dark green romaine, darker green than Ideal Cos with more blistered leaves and similar height, less prone to twisting and cupping, more attractive with a classier look than Ideal Cos, PVP application pending. *Source History: 1 in 2004.* **Sources: ME9.**

Big Country - Tall dark green romaine, tolerates corky root, bolt and burn, slightly darker green than Beretta, PVP application pending. *Source History: 1 in 2004.* **Sources: ME9.**

Big Green - (SXP9810) - Tall heavily textured romaine with dark green lightly savoyed leaves, good weight, good TB and bolt tol. *Source History: 1 in 2004.* **Sources: ME9.**

Bionda Lentissima - 55-65 days - Large, light green heads, sow spring/fall. *Source History: 1 in 2004.* **Sources: Se24.**

Bionda Ortolani - 55-65 days - Classic romaine lettuce, large, dark green head, spring/fall. *Source History: 2 in 2004.* **Sources: Ri12, Se24.**

Blonde Maraichere - 45-55 days - Sturdy elongated heads, well veined outer leaves protect creamy center, subtle delicate flavor not often found in romaine, good for salad blends, Gonthier Seedhouse of Belgium. *Source History: 1 in 1981; 1 in 1984; 1 in 1987; 2 in 1991; 1 in 1994; 3 in 2004.* **Sources: Ag7, Dam, Dom.**

Blushed Butter Cos - 50-60 days - Combination butterhead/romaine lettuce with ruffled savoyed leaves in shades of reds and greens, 10 in. heads, crisp, buttery taste, spring and fall culture, CV Fedco introduction. *Source History: 2 in 1998; 5 in 2004.* **Sources: Fe5, Ki4, Sh12, Syn, Ter.**

BOS 9003 - Short plant frame resembles a leaf type, dark green leaves with a nice texture, sel. for more open prostrate growth habit. *Source History: 1 in 2004.* **Sources: ME9.**

BOS 9114 - Tall, large plant with more open habit, dark green, blistered leaves tinged with red. *Source History: 1 in 2004.* **Sources: ME9.**

BOS 61603 - Tall, upright romaine with very dark green leaves tinged with red. *Source History: 1 in 2004.* **Sources: ME9.**

BOS 61612 - Upright, med. height, dark green leaves tinged with red. *Source History: 1 in 2004.* **Sources: ME9.**

Broad Sword - 60-66 days - Small heads, sweet, dark green, slightly savoyed leaves, mild flavor for a Cos type, very slow to bolt. *Source History: 3 in 1991; 1 in 1994; 1 in 1998; 1 in 2004.* **Sources: Fo7.**

Brown Golding - 55-70 days - Heirloom cos, red-tinged leaves, medium heads, crunchy texture, sweet flavor, high vitamin C content, does best in cool weather. *Source History: 1 in 1981; 1 in 1984; 1 in 1987; 2 in 1991; 3 in 1994; 3 in 1998; 2 in 2004.* **Sources: Ho13, So12.**

Brown Winter - 70 days - Fairly large flat broad leaves which vary from green to bronze shades, smoother and thinner leaves than most romaines, quick-growing, heat and cold res. *Source History: 1 in 1991; 1 in 2004.* **Sources: Se17.**

Bubbles - 65 days - Little Gem type with small head, crisp crinkled leaves, superb flavor, summer-autumn harvest. *Source History: 1 in 1994; 1 in 1998; 1 in 2004.* **Sources: Tho.**

Bunyard's Matchless - Open-hearted romaine with bronze-green center, sweet, juicy ribs. *Source History: 1 in 2004.* **Sources: Sa5.**

Cherry Crisp - Thick, semi-savoy, extremely dark red leaf, sweet flavor, crisp texture, longer shelf life than other baby leaf cos types, does not stick to cutting or cleaning equipment, colorful addition to Mesclun mix. *Source History: 1 in 2004.* **Sources: Orn.**

Chief - Tall dark green romaine, good bolt, burn and corky root tolerance, rounded butt may allow for heavier crates. *Source History: 1 in 2004.* **Sources: ME9.**

Cimmaron - (Deep Red, Little Leprechaun, Rouge d'Hiver) - 60-70 days - Dates back to the 18th century, 10-12 in. deep-red head, creamy yellow-bronze center, good flavor, crisp tender texture, virtually impervious to bolting. *Source History: 3 in 1991; 11 in 1994; 18 in 1998; 25 in 2004.* **Sources: Al25, Bo17, Bo19, Com, Dam, DOR, Fo7, Ga1, He8, Hig, Jor, Jun, La1, Lin, Ni1, OLD, Pe2, Ri12, RUP, Sa5, Se17, Se26, Ta5, Tu6, Yu2.**

Claremont - 46 days - Compact, upright heads, 9 in. tall, ribs are open at the top, resists downy mildew races. *Source History: 1 in 2004.* **Sources: Jo1.**

Corsair - 57-76 days - Sturdy plant with excel. standing ability, solid heavy heads, thick, crunchy leaves, yellow heart, good bolt res., quality is retained late in the season, summer-autumn harvest, good LMV resistance. *Source History: 1 in 1994; 1 in 1998; 2 in 2004.* **Sources: So1, Tho.**

Cosmo - 55-65 days - Bright-green 11 in. tall plants, savoyed leaves with wavy margins which hold firmly into blanched hearts, susc. to tipburn in mid-summer, for early or late crops, contains three times as much vitamin C and six times as much vitamin A as crisphead lettuce. *Source History: 5 in 1981; 3 in 1984; 2 in 1987; 6 in 1991; 6 in 1994; 7 in 1998; 3 in 2004.* **Sources: Bou, Se7, So1.**

Craquante d'Avignon - (Craquerelle du Midi) - 65-70 days - Classic half-size romaine similar to Winter Density in size and shape, 5-7 in. leaves, 6-8 in. heads, deep-green crisp crunchy leaves, cold res., fine flavor and texture, can plant all season. *Source History: 1 in 1981; 1 in 1984; 1 in 1987; 1 in 1991; 1 in 1994; 1 in 1998; 1 in 2004.* **Sources: Se7.**

Crisp Mint - (Erthel) - 60-70 days - Selected as one of the best green leaf varieties in the 1996 Lettuce Growout of at Seed Savers' Heritage Farm, tight compact heads grow upright to a height of 10 in., excel. crisp sweet flavor, some tipburn in July heat. *Source History: 1 in 1998; 4 in 2004.* **Sources: Ag7, Fe5, Sa5, Se16.**

Dark Green Cos - 65-70 days - Med-large self-closing type, large thick dark-green slightly crumpled leaves, self-blanching, upright 8-10 in. plants, finely savoyed, home garden and local market. *Source History: 10 in 1981; 8 in 1984; 8 in 1987; 7 in 1991; 7 in 1994; 6 in 1998; 4 in 2004.* **Sources: All, La1, LO8, WE10.**

Darkland Cos - 50-80 days - One of the leading green romaine varieties in the U.S., produces compact "hearts of romaine" when planted 6-8 in. apart, traditional spacing produces full-size heads, PVP 1992 Central Valley Seeds, Inc. *Source History: 2 in 1998; 3 in 2004.* **Sources: Orn, OSB, RIS.**

De Morges Braun - 65 days - Upright growth habit with centers forming a green contrast to the coppery pink outer leaves, 8- 10 in. tall, smooth, buttery flavor, no bitterness, grows slowly, one of the last to bolt, excellent. *Source History: 1 in 1987; 4 in 2004.* **Sources: Fe5, Ho13, Se17, Ter.**

Delle 7 Lune - 50-60 days - Loose head, deep green leaves with pretty red flecks, spring/fall. *Source History: 1 in 2004.* **Sources: Se24.**

Eruption - 50 days - Bibb-Romaine cross, dark red glossy leaves, blanched interior, 8 in. tall, growth habit resembles Winter Density, also good for baby salad mix, resists DM races 1-16, 19 and 21, LMV tolerant, PVP application pending. *Source History: 2 in 2004.* **Sources: Jo1, Ves.**

Eva's Burgundy Winter Lettuce - Extremely rare heirloom, green leaves tipped with burgundy, shows more red in cooler weather, crisp type, sweet crunchy leaves, resists bolting, grows to 6 ft. after bolting, winter hardy, best planted in the fall when it grows a little and overwinters easily, original plants from 87 year-old Amish lady, Eva, who sold this as cut lettuce at the local farmers' market for over 50 years. *Source History: 1 in 2004.* **Sources: Ami.**

Evergreen - (SXP0137) - Lightly savoyed leaves with heavy texture, excel. color, good size and weight, DM tol., good frost tol. *Source History: 1 in 2004.* **Sources: ME9.**

Fever - Bronze-red leaf color. *Source History: 1 in 1998; 1 in 2004.* **Sources: Pe1.**

Floricos - 60-72 days - Slow bolting, open, savoyed, spring-green leaves, medium size head. *Source History: 2 in 1994; 3 in 1998; 2 in 2004.* **Sources: SE4, Syn.**

Forellenschuss - 50-68 days - Austrian heirloom, name means 'Trout's Back' or 'Speckled Like a Trout' in German, bright apple-green leaves with maroon-scarlet blotches, excel. flavor, somewhat heat tol., but likes mid-day shade. *Source History: 4 in 1998; 21 in 2004.* **Sources: Ag7, Al25, Ba8, Bou, Coo, Fe5, Fo13, Hi6, Ho13, Pe2, Pr3, Ri12, Se7, Se16, Se17, So1, So12, Syn, Ter, Tu6, Yu2.**

Freckles - (Trout Back, Freckles Forellenschuss) - 55-70 days - Beautiful tender romaine, green leaves splashed with red, long standing, resembles Speckles butterhead but with more substantial leaves, heirloom. *Source History: 2 in 1994; 7 in 1998; 26 in 2004.* **Sources: Be4, Ers, Ga1, He8, HO2, Jo1, La1, ME9, Na6, Ni1, Orn, Pe2, Pe6, Pin, RUP, SE14, Se26, Shu, Sk2, St18, Sw9, Ta5, TE7, Tho, Ves, WE10.**

Fresh Heart - (BOS9021G) - Large dark green romaine with good tolerance to cupping and twisting. *Source History: 1 in 2004.* **Sources: ME9.**

Giant Caesar - 70 days - Huge heads, 6 in. wide by 16 in. long, crisp, dark green leaves. *Source History: 1 in 2004.* **Sources: Bu2.**

Green Forest - 56-70 days - Dark green heads to 12 in. tall, tender, partially savoy leaves with smoother ribs which allows for less damage in packing and handling, midrib is crisp and juicy, strong TB and bolt tolerance, PVP application pending. *Source History: 3 in 2004.* **Sources: CA25, RIS, Ves.**

Green Towers - 68-74 days - Improvement over traditional Parris Island types, earlier, taller, large full-bodied heads, slightly savoyed leaves, dull gray-green color, large butt core, concentrated maturity, good shipper, PVP 1987 HA20. *Source History: 1 in 1987; 3 in 1991; 3 in 1994; 4 in 1998; 4 in 2004.* **Sources: Ha5, HA20, HO2, ME9.**

Grenadier - (SXP0059) - Med. height red romaine with dark bronze to red color, ruffled leaf margins, suitable for full size harvest. *Source History: 1 in 2004.* **Sources: ME9.**

Ideal Cos - 73-75 days - Slightly savoyed gray-green leaves, 8-10 in., slow to bolt, widely adapted, mosaic tolerant, PVP 1991 Genecorp. *Source History: 1 in 1994; 3 in 1998; 4 in 2004.* **Sources: ME9, RUP, SE4, Sto.**

Jericho - 60 days - Bred in Israel, sword shaped upright leaves, stays sweet and crispy, even in hot weather. *Source History: 1 in 1991; 2 in 1994; 2 in 1998; 4 in 2004.* **Sources: Ki4, Ri12, Se7, Tu6.**

Kalura - 60-65 days - Large green cos type with high heat tolerance, great taste. *Source History: 1 in 1998; 2 in 2004.* **Sources: Pe2, Tu6.**

Kostadinov - *Source History: 1 in 2004.* **Sources: Se17.**

Larissa - Medium green romaine type with excellent bolt resistance. *Source History: 1 in 2004.* **Sources: Coo.**

Lion's Tongue - 50 days - Loose headed cos type, leaves resemble Deer Tongue but are larger, not prone to slug damage in dry weather. *Source History: 1 in 1998; 2 in 2004.* **Sources: Fo7, Se17.**

Little Caesar - 70 days - Green outer leaves, golden blanched inner leaves, one head feeds two people. *Source History: 1 in 1998; 1 in 2004.* **Sources: Bu2.**

Little Gem Cos - (Sugar Cos, Sucrine) - 56-80 days - Head 6 in. tall x 4 in. dia., thick tight blanched heart, fine flavor, no waste, maximum yield even planted 6.5 x 6.5 in., holds in field a few weeks, heat tol., England. *Source History: 9 in 1981; 8 in 1984; 7 in 1987; 11 in 1991; 8 in 1994; 9 in 1998; 12 in 2004.* **Sources: Ag7, Bou, Coo, Ga1, Na6, Orn, Se8, Se26, TE7, Ter, Tho, Ves.**

Little Leprechaun - 75 days - Compact semi-cos type resembles Winter Density but with leaf colors in shades of mahogany, red and forest green, sweet, crunchy tight core. *Source History: 1 in 1987; 1 in 1991; 1 in 1994; 1 in 1998; 1 in 2004.* **Sources: Ter.**

Lobjoit's Green Cos - (Lobjoit's Green Romaine) - 55-75 days - Large compact well-blanched dark-green firm 12 in. heads with rounded tops, strictly self-closing, very bolt res. in hot dry weather, excel. quality, from England. *Source History: 6 in 1981; 4 in 1984; 5 in 1987; 6 in 1991; 4 in 1994; 2 in 1998; 3 in 2004.* **Sources: Ho13, Se26, So9.**

Majestic Red - (Red Majestic) - 55-75 days - Tall cylindrical heads with tender green leaves tinged with dark-red, sweet and crisp, slow bolting, sel. strain of Rouge d'Hiver. *Source History: 10 in 1991; 4 in 1994; 4 in 1998; 3 in 2004.* **Sources: CO30, Fe5, ME9.**

Major Cos - 68-75 days - Large upright plant, dark green color, early maturing, excellent performer, PVP 1991 Seminis Vegetable Seeds, Inc. *Source History: 1 in 1994; 1 in 1998; 1 in 2004.* **Sources: Sto.**

Medallion - 60 days - Dark green 11-12 in. head resembles Parris Island but stays open and fresh looking much longer, slow bolting, TB tol., resists CRR and LMV, PVP application withdrawn. *Source History: 1 in 1998; 1 in 2004.* **Sources: Jo1.**

Olga - 50-75 days - Cup shaped, glossy green leaves, grows 8 in. tall, flavorful, slow to bolt, sweeter than other Romaine types, cold tolerant. *Source History: 1 in 1994; 14 in 1998; 10 in 2004.* **Sources: CO30, Dam, He8, Ho13, Kil, Lin, Ni1, Pin, Sa5, Ta5.**

Outredgeous - 65-75 days - One of the reddest romaines available, thick, slightly ruffled, vibrant red leaves form a loose romaine head, harvested as baby lettuce or at 10 in. head size, Frank Morton introduction. *Source History: 4 in 2004.* **Sources: Ni1, Sh12, St18, Ter.**

Panther - 75 days - Tall, smooth leaf, broad base, short core, multi-purpose, excellent fresh market heads, withstands heat better than other romaines, from Seminis plant breeders. *Source History: 2 in 2004.* **Sources: Dam, Sto.**

Paris Green Cos - (True Romaine) - 66 days - Large heads of dark-green leaves, sweet and crisp, white seed. *Source History: 2 in 1981; 1 in 1984; 1 in 1987; 1 in 1991; 1 in 1994; 1 in 1998; 1 in 2004.* **Sources: Red.**

Paris White Cos - (White Paris Self-Folding Cos, Trianon, Paris White Romaine) - 50-83 days - Large cylindrical upright plant, tight 8-10 in. tall self-folding conical head, light-green outer leaves, strong midribs, white blanched heart, great flavor, slightly savoyed, for North, home or market, pre-1868. *Source History: 42 in 1981; 37 in 1984; 28 in 1987; 22 in 1991; 18 in 1994; 16 in 1998; 21 in 2004.* **Sources: Bu2, CO23, DOR, Fo7, Fo13, Gur, He8, He17, Hi13, Ho12, La1, Mo13, Ont, PAG, Ri12, Se8, SE14, Shu, Th3, WE10, Wet.**

Parris Island 318 - 75 days - Similar to Green Towers but with less tipburn and slightly better weight, tall narrow head, crisp sweet leaves, mosaic res., dev. by Royal Sluis, PVP application abandoned. *Source History: 3 in 1991; 1 in 1994; 3 in 1998; 4 in 2004.* **Sources: ME9, SE4, Sto, Twi.**

Parris Island Cos - 50-80 days - Upright dark gray-green 8-12 in. head slightly and neatly folded and slightly savoyed, creamy white heart, vigorous, very uniform, med-slow bolt, white seed, TB and mosaic tol., dev. to replace Dark Green Cos, named after Parris Island off South Carolina, Clemson/AES and USDA, 1952. *Source History: 39 in 1981; 38 in 1984; 42 in 1987; 61 in 1991; 62 in 1994; 68 in 1998; 63 in 2004.* **Sources: Alb, Ba8, BAL, Be4, Bo17, Bu2, Bu3, Bu8, CH14, CO23, CO30, Com, Cr1, Dam, De6, Dgs, DOR, Ers, Fe5, Ga1, Ga7, Ge2, Gr27, HA20, Hal, Hi6, HO2, Hum, Jor, Kil, La1, LO8, Loc, Ma19, MAY, Mel, Mey, OLD, Or10, Orn, OSB, PAG, Pe2, Ri2, RIS, RUP, Sa6, Scf, SE4, SE14, Se25, Se26, Sh9, Si5, Sk2, So1, TE7, Tu6, Ves, Vi4, WE10, Wi2, WI23.**

Parris Island Cos 454 - 68 days - Medium green leaves, pale yellow interior, used for hearts, salad packs and conventional packs, superior uniformity, consistent performer. *Source History: 1 in 2004.* **Sources: OSB.**

Petite Rouge - True red baby romaine, easy to grow, well adapted to many climates. *Source History: 1 in 1994; 1 in 1998; 2 in 2004.* **Sources: Ba8, WE10.**

Plato - 72-80 days - Uniform attractive dark-green cos, open style head, excel. flavor, slow bolting, MTO and TB res., intro. 1993. *Source History: 3 in 1991; 12 in 1994; 7 in 1998; 2 in 2004.* **Sources: GLO, Lin.**

Plato II - 60-70 days - Improved Plato with glossy medium-green, meaty leaves with sweet flavor, no bitterness, slow to bolt, virus and TB res. *Source History: 2 in 1994; 5 in 1998; 3 in 2004.* **Sources: Fe5, Pin, Ta5.**

Red Cos - 45-50 days - Large leaves with crimson tinge, old European heirloom. *Source History: 1 in 1998; 2 in 2004.* **Sources: ME9, So12.**

Red Eye Cos - 76 days - Medium size romaine, upright plant habit, rounded base, smooth, bronze-red leaves, green underside, PVP 1992 Seminis Vegetable Seeds, Inc. *Source History: 1 in 2004.* **Sources: Sto.**

Red Leprechaun - 60 days - Plants 8-12 in. tall with succulent puckered leaves, pale-pink on cream in the blanched hearts and burgundy on exposed leaf surfaces, slightly bitter flavor, mature heads can weigh more than one pound, recommended harvest time at three-fourths pound. *Source History: 4 in 1991; 1 in 1994; 2 in 1998; 3 in 2004.* **Sources: He8, Ho13, Se16.**

Red Romana - *Source History: 1 in 1991; 1 in 1994; 1 in 1998; 1 in 2004.* **Sources: WE10.**

Red Ruffles - Vigorous plant to 15 in., deep-burgundy savoyed leaves. *Source History: 1 in 1991; 1 in 1994; 1 in 1998; 1 in 2004.* **Sources: WE10.**

Red Splash Cos - Long, oval, smooth, lime-green leaves with red tips and bold red splashes, forms an open upright head, buttery sweet flavor like Red Cos. *Source History: 1 in 1994; 2 in 1998; 1 in 2004.* **Sources: Hud.**

Remington - 70 days - Tall, medium green plants with slightly blistered leaf, sturdy mid rib, head size averages 13 in., 1.5-2 lbs., strong TB tolerance. *Source History: 2 in 2004.* **Sources: ME9, Ves.**

Remus - Excellent Parris Island type, bright green leaves are slightly more savoyed, more res. to bolting and TB, also resists mildew. *Source History: 1 in 2004.* **Sources: Coo.**

Romaine - (Cos, White Cos) - 66-83 days - Large size but not coarse, med-green 8-10 in. upright tightly folded oval head, greenish-white interior, mild endive-like flavor, pre-1900. *Source History: 33 in 1981; 28 in 1984; 28 in 1987; 20 in 1991; 16 in 1994; 15 in 1998; 16 in 2004.* **Sources: But, Ear, Fis, GLO, Goo, HA3, Lin, Mcf, MIC, Pla, Roh, Ros, Sa9, Sau, SGF, St18.**

Romaine, Blonde - Summer variety from France. *Source History: 1 in 1987; 1 in 1991; 1 in 1994; 1 in 1998; 2 in 2004.* **Sources: Ber, Lej.**

Romaine, Dark Green - 65-70 days - Large dark-green semi-heading type, stands light frost, heirloom. *Source History: 2 in 1981; 1 in 1984; 1 in 1987; 1 in 1991; 1 in 1994; 1 in 1998; 2 in 2004.* **Sources: Fo13, Pr3.**

Romaine, Green - Exceptionally crisp sweet variety, excellent for spring and fall crops. *Source History: 1 in 1987; 1 in 1991; 1 in 1994; 1 in 1998; 1 in 2004.* **Sources: GRI.**

Romaine, Red - 66-70 days - Useful and colorful tart addition to any salad, color varies from green to deep red to bronze, dark color develops best in cool weather, gourmet variety. *Source History: 2 in 1991; 1 in 1994; 4 in 1998; 8 in 2004.* **Sources: Ba8, He17, HO2, La1, Mo13, SE14, Se16, WI23.**

Romana Bionda Degli Ortolani - Huge lettuce with light green leaves and nice heart, very good flavor, name means "Light Green Romaine Sold by Green Grocers", from Italy. *Source History: 1 in 1998; 1 in 2004.* **Sources: Ho13.**

Romance - 50-65 days - Early maturing cos type, excellent uniform compact shape, tender smooth leaves are medium green outside and blonde at the center, crisp but not tough, LMV res. *Source History: 3 in 1981; 3 in 1984; 1 in 1987; 1 in 1991; 1 in 1994; 4 in 1998; 1 in 2004.* **Sources: Ta5.**

Romany - 75 days - Heavy, dark green, upright crisp hearts, flavorful, resists TB and mildew. *Source History: 1 in 2004.* **Sources: Tho.**

Romulus - 59-75 days - Heads average 11-12 in., similar to Parris Island Cos but stays open and fresh looking much longer, sweet delicious flavor, slow bolting and TB tol., PVP 1992 Seminis Vegetable Seed, Inc. *Source History: 3 in 1991; 4 in 1994; 2 in 1998; 9 in 2004.* **Sources: CA25, Ers, GLO, HO2, Jun, Lin, So1, Tt2, Ver.**

Rosalita - 55-65 days - Medium-large upright plants, red Salad Bowl type with shorter thicker leaves that stay crisper longer after picking, formerly Karibu, PVP 1989 Johnny's Selected Seeds and Brinker Orsetti Seed Co., Inc. *Source History: 6 in 1991; 6 in 1994; 7 in 1998; 7 in 2004.* **Sources: CO30, Coo, Jo1, Na6, Orn, OSB, Sw9.**

Rouge d'Hiver - (Red Winter, Cimmaron) - 55-65 days - Extremely beautiful European heirloom from the 1800s, large flat broad leaves are sweet with a buttery texture, color varies from green to bronze to deep-red, quick growing, very heat res. if kept watered, cold res. *Source History: 1 in 1981; 1 in 1984; 6 in 1987; 13 in 1991; 16 in 1994; 21 in 1998; 35 in 2004.* **Sources: Ag7, Au2, Ba8, Bu2, Co32, Coo, Ers, Fe5, Ga1, He8, Hi6, Hig, HO2, Ho13, Ki4, ME9, Na6, Orn, Pe1, Pe2, Red, Sa5, Se7, Se8, SE14, Se16, Se17, Sk2, So9, So12, TE7, Tu6, Up2, Vi4, WE10.**

Ruben's Red - (Rubens Dwarf) - 50-60 days - Semi-dwarf romaine, very full 12 in. heads of open upright leaves, outer leaves are deep cranberry-red fading to cool lime-green hearts, juicy, crisp and flavorful. *Source History: 4 in 1991; 5 in 1994; 8 in 1998; 9 in 2004.* **Sources: Ag7, Ki4, Pe1, Pe2, Roh, Se7, Se16, So9, Tu6.**

Rusty - Cos type, bronze tinged leaves, forms a delicious heart if leaves have not been harvested. *Source History: 1 in 2004.* **Sources: Ga7.**

Seville - 58-77 days - Crisp leaves with smooth texture, sweet flavor, over half of the leaf is covered with dark red color, bolt tol., PVP 2003 Seminis Vegetable Seeds, Inc. *Source History: 2 in 2004.* **Sources: OSB, Tho.**

Shanghai 20 - Medium green leaves with bronze-red-lavender overlay, slightly savoyed, elongated, rounded, long, twisted shape, from China. *Source History: 1 in 1998; 1 in 2004.* **Sources: Ho13.**

Spotted Aleppo - Ornamental green leaves with red-brown speckles, delicious, 1804. *Source History: 1 in 2004.* **Sources: Th3.**

SSC 1128 - Tall med. green romaine, good tolerance to twisting and cupping, corky root tol. *Source History: 1 in 2004.* **Sources: ME9.**

St. Blaise - Small upright light bright-green heads, few outer leaves, adapted to early spring planting under row covers. *Source History: 1 in 1991; 1 in 1994; 2 in 1998; 3 in 2004.* **Sources: Coo, Syn, Tu6.**

Sudia - (Sudia MTO) - 55 days - Thick dimpled dark-green leaves form 14 oz. heads, buttery crisp flavor, resists bottom rot and tipburn, grows well in summer heat, slow to bolt, PVP 1992 Vilmorin S.A. *Source History: 4 in 1991; 5 in 1994; 2 in 1998; 2 in 2004.* **Sources: CO30, Hig.**

Sweet Valentine - 55-76 days - Brilliant red on both sides of the leaves, lower half of each leaf is apple-green, starts out as a head lettuce before forming a loose leaf romaine, excel. uniformity, slow to bolt. *Source History: 1 in 1991; 3 in 1994; 7 in 1998; 5 in 2004.* **Sources: CO30, Fe5, Roh, Se7, So1.**

Tall Guzmaine - 65-72 days - Slow bolting, savoyed leaves, CR, LMV, TBR, U of Florida rel. *Source History: 3 in 1991; 2 in 1994; 2 in 1998; 1 in 2004.* **Sources: RIS.**

Triton - 75 days - Tall plant, large frame and base which adds to the weight, heat tolerant, multi-purpose, produces hearts, fresh market heads and good packaged material. *Source History: 2 in 2004.* **Sources: HA20, Sto.**

Valmaine - (Paris White, Valmaine Savoy) - 68-75 days - Somewhat more open-headed than Parris Island but darker green and hardier in adverse weather, slightly savoyed, 10-11 in. long x 3-4 in. dia., heart blanches to creamy-yellow, white seed, USDA and TX/AES, 1963. *Source History: 31 in 1981; 25 in 1984; 24 in 1987; 22 in 1991; 15 in 1994; 7 in 1998; 7 in 2004.* **Sources: Bou, Ri12, Se8, So12, Ter, Vi4, WE10.**

Vaux Self-Folding - The first self-folding cos, firm erect heads, good flavor, 1873. *Source History: 1 in 1998; 2 in 2004.* **Sources: Ho13, Pr3.**

Verte Mar - 60-65 days - Excellent traditional French romaine, tall, upright tight heads, smooth, medium dark-green leaves, colorful, tender and sweet. *Source History: 1 in 1991; 5 in 1994; 4 in 1998; 6 in 2004.* **Sources: CO30, Ga1, Pin, Ta5, VI3, Wi2.**

Verte Maraichere - 65-75 days - French variety with relatively large uniform heads surrounded by tasty crisp slightly crinkled leaves which are sweeter than American counterparts. *Source History: 3 in 1991; 2 in 1994; 3 in*

1998; 1 in 2004. **Sources: Orn.**

<u>Weatherby</u> - Med. green slightly blistered leaves, shorter than Beretta, good tol. to twisting and cupping, outstanding bolt tol., may lack height for markets requiring a tall romaine. *Source History: 1 in 2004.* **Sources: ME9.**

<u>Winchester</u> - Attractive dark green savoyed romaine, med. height may lack height for markets requiring a tall romaine. *Source History: 1 in 2004.* **Sources: ME9.**

<u>Winter Density</u> - 55-65 days - Large heavy compact very dark-green 9-10 in. heads, slow bolt, tol. light frost, for summer or winter use, tightly-folded round-topped heads, white seed, sweet flavor. *Source History: 10 in 1981; 9 in 1984; 9 in 1987; 12 in 1991; 11 in 1994; 16 in 1998; 20 in 2004.* **Sources: Bo19, Bou, CO30, Com, Eo2, Fe5, Fo7, Hig, Jo1, Na6, Orn, Se7, Se26, So1, So9, Sw9, Syn, Ter, Tu6, Vi4.**

<u>Winter Wonderland</u> - *Source History: 1 in 1991; 1 in 1998; 1 in 2004.* **Sources: WE10.**

<u>Yedi Kule Cinsi</u> - (Yedikule) - 65 days - Turkish heirloom, dark green, waxy large leaves form a loose head, good flavor, heat res. *Source History: 4 in 2004.* **Sources: In8, Red, Se17, Sk2.**

Varieties dropped since 1981 (and the year dropped):

Andros (2004), *Angela* (1994), *Augustus* (2004), *Bautista* (1998), *Bionda Degli Ortolani* (2004), *Capri* (2004), *Clemente* (2004), *Corsica* (1994), *De Frontignan* (1991), *Dwarf Romaine* (1987), *Erthel* (1998), *Floricos 83* (2004), *Guzmaine* (2004), *H.P.S. Winterweight* (1991), *Lente a Monter* (2004), *Light Green Cos* (1998), *Lobjoit's Crisp Mint* (1998), *Lobjoit's White Cos* (1984), *Paris Dark Green* (1991), *Paris White Self-Folding Cos* (1991), *Pyramid Cos* (1998), *Redcurl* (2004), *Romaine Improved* (1991), *Roman Queen* (2004), *Rubra Cos* (1987), *Signal* (1991), *Silver Cos* (1983), *St. Albans Allheart* (1991), *Sugar Cos* (1998), *Sweet Midget Cos* (1984), *Tetue de Nimes* (1994), *Toledo* (1998), *Verde Degli Ortolani* (2004), *Wallop* (1991), *Wavy Red Cos* (2004).

Lima Bean / Bush
Phaseolus lunatus

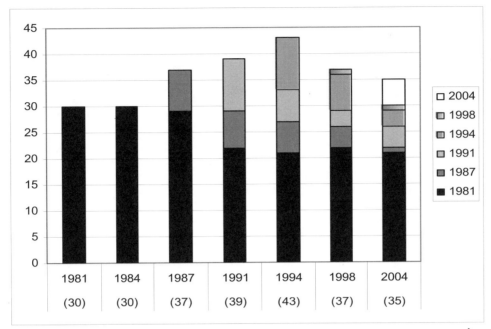

Number of varieties listed in 1981 and still available in 2004: 30 to 21 (70%).
Overall change in number of varieties from 1981 to 2004: 30 to 35 (117%).
Number of 2004 varieties available from 1 or 2 sources: 21 out of 35 (60%).

Baby Bush - 70 days - White beans turn gray-green when cooked, 2.5 in. pods contain four plump beans half as thick as they are long, ripen early and over a long season, freezing or baking, popular since Civil War times. *Source History: 1 in 1981; 1 in 1984; 2 in 1987; 1 in 1991; 3 in 1994; 2 in 1998; 5 in 2004.* **Sources: Bo17, Na6, Pe2, Ri2, TE7.**

Baby Lima 184-85 - 69 days - Compact bush, light green, flat, 3.5 in. pods, green seed, resists A,B,C and D str. of DM, dev. for east coat growing and mechanical harvest. *Source History: 1 in 2004.* **Sources: SE4.**

Baby Lima 878 - 80 days - Dark green 4 in. flat pods, green seeds, upright plants suitable for machine harvest, resists DM strains A,B,C and D of DM. *Source History: 1 in 2004.* **Sources: SE4.**

Bridgeton - 65-76 days - Flattened 3 in. pods form low on compact prolific bushes, 4 small thick green seeds per pod, resists A and D str. of DM, widely adapted, heavy yields, from USDA and DE/AES and NJ/AES. *Source History: 7 in 1981; 6 in 1984; 3 in 1987; 1 in 1991; 2 in 1998; 2 in 2004.* **Sources: Ers, SE4.**

Burpee's Improved - (Burpee's Improved Bush, Wonder Bush Lima) - 70-76 days - Pods borne in clusters of 5 or 6 at center of plant, 4.5-5.5 x 1-1.25 in., contain 3-6 pale-green wrinkled seeds, home market and dry beans, fatter than Fordhook. *Source History: 44 in 1981; 41 in 1984; 38 in 1987; 37 in 1991; 35 in 1994; 33 in 1998; 28 in 2004.* **Sources: All, Be4, Bu2, Bu8, CH14, Co31, De6, Ers, Ge2, Gr27, HA3, Hi13, Jun, La1, ME9, Mel, Mey, Mo13, PAG, Roh, RUP, Sau, Se26, Sh9, Shu, SOU, Ver, Wet.**

Bush Lima, Burpee's - 75 days - A large-seeded lima that's easy to shell, clusters of 5-6 pods 4-5 in. long, each pod contains 4-5 large white flat-oval beans. *Source History: 4 in 1981; 4 in 1984; 5 in 1987; 3 in 1991; 3 in 1994; 2 in 1998; 1 in 2004.* **Sources: HO2.**

Bush Lima, Giant Improved - 75 days - Erect vigorous bush plants, slightly curved 5-6 in. pods with 4 or 5 large beans, high quality, tender, rich buttery flavor, easy to shell, bears until frost. *Source History: 1 in 1981; 1 in 1984; 1 in 1987; 1 in 1991; 1 in 1994; 1 in 2004.* **Sources: MAY.**

Bush Lima, Improved - (Improved Bush) - 75 days - Easy to shell, flatter than Fordhook, 4.5-5.5 x .75-1.25 in. pods with 4-5 pale flat/oval seeds, clusters of 5-6 pods, older large seeded var. *Source History: 4 in 1981; 4 in 1984; 2 in 1987; 1 in 1991; 1 in 1994; 1 in 1998; 1 in 2004.* **Sources: La1.**

Buttergreen - (Bush Butterbean) - 65 days - Matures early, 6 in. pods, 7-9 seeds. *Source History: 2 in 2004.* **Sources: Bu2, Roh.**

Butterpea - *Source History: 1 in 1994; 1 in 1998; 1 in 2004.* **Sources: Sh13.**

Cangreen - (Clark's Bush, Early Thorogreen) - 65-68 days - For canning or fresh, holds green color and buttery flavor after processing, Henderson type but stronger grower, larger, more prolific. *Source History: 7 in 1981; 7 in 1984; 7 in 1987; 5 in 1991; 5 in 1994; 4 in 1998; 2 in 2004.* **Sources: Ada, Co31.**

Dixie Butterpea Speckled - (Butterpea Speckled) - 75-76 days - Vigorous bushy 16-21 in. med-dark green plants, slightly curved 3-3.5 x .5 in. pods, 3 or 4 small nearly round brownish-red seeds speckled with darker-brown, sets well in hot weather, very productive under hot dry conditions, good in drought areas, Southern adapted. *Source History: 26 in 1981; 23 in 1984; 21 in 1987; 16 in 1991; 14 in 1994; 19 in 1998; 20 in 2004.* **Sources: Ada, Al25, Ba8, Co31, Fo7, He17, HO2, Ho13, Kil, La1, ME9, Mo13, OLD, Ra5, RUP, Se26, Sh9, Ver, Vi4, Wi2.**

Dixie Butterpea White - (Butterpea White) - 70-76 days - White baby lima, sets pods in high temps, vigorous bushy 16-23 in. plants loaded with broad oval 3-4 x .5 in. pods, produces until frost, med-dark green plants, med-dark green pods, for home garden or fresh market, white seed, extremely prolific, originally selected by Don Hastings Sr. *Source History: 23 in 1981; 20 in 1984; 19 in 1987; 15 in 1991; 13 in 1994; 20 in 1998; 18 in 2004.* **Sources: Ada, Co31, Ers, Fo7, He17, HO2, Ho13, Ma19, ME9, Mey, Mo13, OLD, RUP, Se26, Shu, SOU, Ver, Wet.**

Early Bush Lima - 65 days - Very early and productive, requires a warm location, good quality either fresh or frozen. *Source History: 1 in 1981; 1 in 1984; 2 in 1987; 1 in 1991; 1 in 1994; 1 in 1998; 1 in 2004.* **Sources: Lin.**

Early Giant Bush Lima - 65 days - Produces 5-6 in. long pods from July until frost, very rugged and upright in growth, 4 or 5 flat oval beans per pod, highest quality, good freezer. *Source History: 1 in 1981; 1 in 1984; 1 in 1987; 1 in 1991; 1 in 1994; 1 in 1998; 1 in 2004.* **Sources: Shu.**

Eastland - 60-76 days - Baby lima, vigorous upright 20 in. plants, med-green 3-4 in. semi-flat pods, 3 small greenish-white seeds per pod, superior quality, DM res., heavy yields, continuous harvest, dependable. *Source History: 4 in 1987; 5 in 1991; 4 in 1994; 5 in 1998; 7 in 2004.* **Sources: Ers, Ha5, Par, Sa9, SOU, Sto, Tho.**

Excel - 65-72 days - Selection from Fordhook 242 by Elwyn Meader, large, spreading plants, 20-24 in. tall, broad flat pods, 3-4 plump green-white seeds per pod, plant after the first week of June or when soil is thoroughly warm, not recommended for planting in Virginia southward since it does not set pods well under high heat, but does well in northern areas. *Source History: 2 in 1994; 2 in 1998; 1 in 2004.* **Sources: So1.**

Fordhook - (Burpee's Fordhook, The Potato Lima) - 70-75 days - Large all-purpose potato-type bush lima, bushy upright plants are 20 in. high and 24 in. across, pods are 4-4.5 in. long x 1 in. wide x .75 in. thick, contain 3 to 4 seeds, large thick oval plump beans, use fresh or dry, good for either freezing or market, from Burpee in 1917. *Source History: 24 in 1981; 19 in 1984; 18 in 1987; 11 in 1991; 13 in 1994; 8 in 1998; 9 in 2004.* **Sources: Co31, Cr1, Dam, He8, He17, LO8, Ont, Ros, Shu.**

Fordhook 90-1 - 80 days - Medium green, 4 in. pods, green seed, tolerates heat and drought, resists A,B,C,D str. of DM. *Source History: 1 in 2004.* **Sources: SE4.**

Fordhook 242 - (Mammoth Wonder Bush, Fordhook Improved, Potato Lima) - 70-85 days - AAS/1945, shorter fatter pods and more heat and drought res. and easier to shell than regular Fordhook, USDA/Beltsville, broad thick 3.5-4 x 1-1.25 in. pod, nutty flavor, 16-20 in. plants, 3 to 5 large flat greenish-white seeds per pod, good for the North or maritime conditions, 1945. *Source History: 90 in 1981; 83 in 1984; 78 in 1987; 71 in 1991; 63 in 1994; 70 in 1998; 57 in 2004.* **Sources: Ada, Be4, Bu1, Bu2, Bu3, Bu8, CH14, Cl3, Com, De6, Dgs, DOR, Ers, Fa1, Fi1, Gr27, GRI, Gur, HA3, Ha5, Hal, HO2, Jo1, Jor, Jun, Kil, La1, Loc, Ma19, MAY, ME9, Mel, Mey, Mo13, OLD, Or10, PAG, Par, RIS, Roh, RUP, Sau, Scf, SE4, Se26, Sh9, Si5, Sk2, So1, SOU, Sto, Twi, Ver, Ves, Vi4, Wet, Wi2.**

Fordhook 1072 - (Green Seeded Fordhook) - 70 days - Dependable, heavy producer, green seeds, otherwise same as Fordhook 242, processing and home garden. *Source History: 1 in 1991; 1 in 1994; 2 in 2004.* **Sources: Eo2, HO2.**

Fordhook, Baby - 70-75 days - Thick seeded small lima, 2.75 x .75 in. pods are slightly curved, each contain 3 to 4 bright- green seeds, 14-16 in. true bush plants, potato type, USDA/Beltsville/1940. *Source History: 8 in 1981; 6 in 1984; 7 in 1987; 6 in 1991; 7 in 1994; 9 in 1998; 8 in 2004.* **Sources: Bu2, Ers, Fo7, Gr27, Hi13, Mey, Roh, Wet.**

Henderson Bush Lima - (Henderson Baby Bush, Henderson Baby Lima, Henderson's Dwarf, Earliest Bush Lima) - 60-75 days - Old favorite, buttery baby lima, slightly curved 2.75-3.5 x .8 in. dark green pods, 3-4 small green seeds that dry creamy- white, sets pods reliably, bears until frost, popular for its flavor, drought res., suited for mech. harvest, found growing along a Virginia roadside by a soldier returning home from the Civil War, introduced in 1888 by Peter Henderson and Co. *Source History: 82 in 1981; 77 in 1984; 74 in 1987; 73 in 1991; 65 in 1994; 62 in 1998; 58 in 2004.* **Sources: Ada, All, Ba8, Be4, Bo19, Bu8, But, CH14, Cl3, Co31, Cr1, De6, DOR, Ear, Ers, Fi1, Fo7, Gr27, HA3, He8, He17, HO2, Ho12, Jor, Kil, La1, LO8, Loc, Ma19, MAY, ME9, Mel, Mey, Mo13, MOU, OLD, Ont, PAG, Pin, Ri12, RIS, Roh, Ros, RUP, Sa9, Sau, Se16, Se26, Sh9, Shu, Si5, SOU, Twi, Und, Ver, Vi4, Wet, Wi2.**

Henderson Improved - 65 days - Glossy dark-green compact 14-16 in. plants, fine leaves, medium-green pods are 3 x .75 in. with 3-4 small seeds, found growing along Virginia roadsides just after the end of the Civil War. *Source History: 2 in 1981; 2 in 1984; 2 in 1987; 2 in 1991; 2 in 1994; 1 in 1998; 1 in 2004.* **Sources: Red.**

Henderson Red - Selected out of Henderson White, 2 ft. bushes, seeds are two-tone red, very productive. *Source History: 1 in 1998; 1 in 2004.* **Sources: Ho13.**

Hopi Lima Red - 110 days - Prolific, dark crimson seed is streaked with black, adapted to the high cool desert, for spring planting, dev. by a Hopi Indian. *Source History: 1 in 1981; 1 in 1984; 1 in 1987; 1 in 1991; 1 in 1994; 1 in 1998; 2 in 2004.* **Sources: Pla, So12.**

Hopi Orange - 90 days - Bush plants 5-6 ft. tall, beautiful orange and black mottled seed, good dried or shelled fresh, very drought tol., preserved by the Hopi Native American Indian, who ground them and used as flour, dates back to 1727-1817 when Nicholas Joseph de Jacquin found this bean growing in the Indian Ocean on the island of Bourbon, possible South American origin, extremely drought and heat res. *Source History: 1 in 1991; 1 in 1994; 1 in 2004.* **Sources: Ami.**

Hopi Tan - 90-110 days - Medium size bush, tan and

black mottled bean with variations, good fresh or dried, very drought tolerant. *Source History: 1 in 1991; 1 in 1994; 2 in 1998; 1 in 2004.* **Sources: So12.**

Hopi Yellow - (Sikyahatiko) - 90 days - Medium-large flattened true lima raised at Hopi, warm "old-gold" color with some streaking, deeply rooted plant produces weak runners. *Source History: 2 in 1991; 1 in 1994; 1 in 1998; 2 in 2004.* **Sources: Na2, Se17.**

Jackson Wonder - (Old Florida Bush Speckled Butter Bean, Calico, Speckled Bush Lima) - 65-76 days - Vigorous erect prolific 13-20 in. plants, broad flat slightly curved 3-3.5 x .9 in. pods, 3-5 buff seeds splashed with purple-black, sets well in hot weather, originated with Thomas Jackson, a farmer near Atlanta, Georgia, late 1800s. *Source History: 46 in 1981; 37 in 1984; 38 in 1987; 38 in 1991; 34 in 1994; 38 in 1998; 35 in 2004.* **Sources: Ada, Ba8, Bu8, CH14, Cl3, Co31, Com, De6, DOR, Fe5, Fo7, HA3, He8, He17, Kil, LO8, Ma19, ME9, Mey, Mo13, OLD, Or10, RIS, Ros, RUP, SE4, Se26, Sh9, Shu, So1, SOU, Te4, Ter, Vi4, Wi2.**

Nemagreen - 68 days - Small flat pale-green seeds, 3 x .75 in. pods, nematode resistant, 1958. *Source History: 3 in 1981; 3 in 1984; 4 in 1987; 3 in 1991; 2 in 1994; 2 in 1998; 3 in 2004.* **Sources: Ers, ME9, Scf.**

Thorogreen - (Green Seeded Henderson, Early Thorogreen, Cangreen, Allgreen) - 65-72 days - AAS, imp. green-seeded baby lima, strong 12-20 in. plant, 3 x .75 in. pods, 3-4 bright-green seeds at processing stage, vigorous despite hot weather, excel. quality, good yields, excel. for canning and freezing. *Source History: 36 in 1981; 30 in 1984; 24 in 1987; 23 in 1991; 22 in 1994; 24 in 1998; 21 in 2004.* **Sources: Be4, Bu8, Cl3, Com, Ers, Fa1, Fi1, Gr27, Gur, He17, HO2, Jun, La1, ME9, Mey, Mo13, Ni1, RUP, Se26, Shu, Ver.**

Thorogreen, Early - (Green Seeded Henderson) - 65-66 days - Small flat rich-green baby lima, heavily productive, sets clusters of 2.75 x .75 in. pods throughout plant, very adaptable and vigorous despite hot weather, 1946. *Source History: 9 in 1981; 9 in 1984; 8 in 1987; 6 in 1991; 5 in 1994; 5 in 1998; 3 in 2004.* **Sources: Bou, SOU, Wi2.**

Verna Shirk - Baby lima, maroon seed with pink marbling. *Source History: 1 in 2004.* **Sources: Lan.**

Wood's Prolific - 65-71 days - Very similar to Henderson Bush Lima but more vigorous, a little larger and somewhat longer in season, 3 or 4 beans per pod, an excel. sheller. *Source History: 5 in 1981; 4 in 1984; 2 in 1987; 2 in 1991; 2 in 1994; 2 in 1998; 3 in 2004.* **Sources: Mey, SOU, Wet.**

Varieties dropped since 1981 (and the year dropped):

Green (1991), *Baby Potato Bush Lima* (1994), *Blue Lake Bush Lima* (1991), *Durango Patol Blanco* (1987), *Eastland Fresh* (1998), *Geneva* (2004), *Hopi Lima Grey* (2004), *Hopi Lima White* (1998), *Indian Red Bush* (1994), *Jackson Wonder* (1991), *King* (2004), *Kingston* (1998), *Lee* (2004), *Mescla Vine* (1998), *Packer DM* (2004), *Packers* (2004), *Pima Beige* (2004), *Pima Lima* (1994), *Prolific Bush Lima* (1991), *Simmon's Red Streak* (2004), *Superba Giant Podded* (2004), *Thaxter* (1998), *UC 92* (2004), *White Butterpea* (1991), *White Light* (1994), *Wild Horse* (1991).

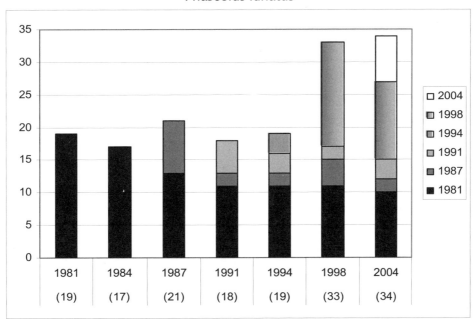

Lima Bean / Pole
Phaseolus lunatus

Number of varieties listed in 1981 and still available in 2004: 19 to 10 (53%).
Overall change in number of varieties from 1981 to 2004: 19 to 34 (179%).
Number of 2004 varieties available from 1 or 2 sources: 24 out of 34 (71%)

Alma's Limas (Alma's PA Dutch Purple Burgundy Lima) - Rare heirloom from Lancaster County, PA, believed to have been brought here from Germany by the family's great grandfather, small, deep purple-black seed, excel. flavor, continued production until frost. *Source History: 1 in 1998; 3 in 2004.* **Sources: Ami, Lan, Roh.**

Big Frosty - Vines grow to 8 ft., very large, white flat seeds with lavender frost. *Source History: 1 in 1998; 1 in 2004.* **Sources: Ho13.**

Black - Med. size purple seeds dry black, productive. *Source History: 1 in 1998; 1 in 2004.* **Sources: Ho13.**

Black Star - Small white seeds with black lines radiating from the eye, does well in dry weather. *Source History: 1 in 1998; 1 in 2004.* **Sources: Ho13.**

Burpee's Best - 92 days - Best features of Fordhook Bush and more productive, pods are 4-4.75 in. long x 1.2 in. wide x .5-.75 in. thick, climbs strongly to 12 ft., good fresh or frozen, potato type. *Source History: 4 in 1981; 3 in 1984; 2 in 1987; 2 in 1991; 2 in 1994; 3 in 1998; 2 in 2004.* **Sources: Bu2, ME9.**

Calico - (Speckled Calico, Florida Butter, Florida Speckled) - 78 days - Prolific, vigorous, heat res., bears until frost, 3-4 in. pods, 3 to 4 large flat speckled seeds, good flavor. *Source History: 2 in 1981; 2 in 1984; 2 in 1987; 1 in 1991; 1 in 1994; 2 in 1998; 2 in 2004.* **Sources: Fi1, Sh13.**

Carolina - (Sieva, Carolina Sieva) - 70-78 days - Quick bearing, dark-green 3-3.5 in. pods, 3 to 4 small flat med-green seeds dry to white, sets reliably, stands more cold than most. *Source History: 7 in 1981; 6 in 1984; 6 in 1987; 3 in 1991; 1 in 1994; 4 in 1998; 4 in 2004.* **Sources: Cl3, Fo7, Ra5, Se26.**

Christmas - (Large Speckled Calico, Giant Calico/Florida Pole, Giant Butter/Florida Speckled Giant Speckled) - 75-100 days - Vigorously climbing 84-120 in. vines, 4.5-5.5 x 1 in. pods, quarter-sized flat white seeds with maroon spots, rich flavor, heavy yields, bears during extreme heat, first cultivated in the U.S. around 1840. *Source History: 35 in 1981; 30 in 1984; 36 in 1987; 33 in 1991; 36 in 1994; 33 in 1998; 35 in 2004.* **Sources: Ada, Ba8, Bu8, CH14, Co31, De6, DOR, Ers, Fo7, Gur, He8, He17, HO2, Ho13, Hud, Kil, La1, Loc, ME9, Mey, Mo13, OLD, RIS, Roh, Ros, Sau, Se16, Shu, Sk2, So1, SOU, Ta5, Ver, Vi4, Wi2.**

Cliff Dweller - 80-90 days - Buff colored seeds with black or dark-wine at the edges, 3-4 seeds per pod, heat and drought res., believed to originate from Apache Indians in the Southwest. *Source History: 1 in 1991; 1 in 1994; 1 in 2004.* **Sources: Sa9.**

Dr. Martin - 90 days - Probably the largest lima, 2-3 times the size of Fordhook 242, fine quality, usually 3 to 4 beans per pod, climbs to about 7 ft., hand-selected seed available. *Source History: 1 in 1981; 1 in 1984; 1 in 1987; 1 in 1991; 2 in 1994; 3 in 1998; 2 in 2004.* **Sources: Lan, Roh.**

Florida Butter - (Speckled Pole, Florida Speckled, Calico Pole, Florida Butter Speckled) - 68-90 days - Plants 8-10 ft. tall, dark-green 3-3.5 in. pods in clusters, 3-4 buff seeds splashed with dark-purple, heavy yields, good tol. to hot humid weather, dependable, used either as green shells or dry. *Source History: 19 in 1981; 17 in 1984; 16 in 1987; 15 in 1991; 7 in 1994; 5 in 1998; 4 in 2004.* **Sources: Bu8, De6, Mey, So1.**

Florida Speckled - (Florida Speckled Butter, Red Speckled, Oval Seeded, Cutshort Cornfield) - 85-90 days - Vigorous 96-120 in. plants, 3.25-3.5 x 9 in. pods in clusters, buff seed with maroon speckles, long continuous productive season, heat res., used in the South for planting between rows of corn. *Source History: 15 in 1981; 13 in 1984; 12 in 1987; 12 in 1991; 11 in 1994; 15 in 1998; 18 in 2004.* **Sources: Ada, Bu2, Co31, Ers, He8, Ho13, Kil, La1, LO8, ME9, OLD, RUP, Sau, Se26, Sh9, Shu, SOU, Ver.**

Ganymede - Beautiful Tennessee heirloom, red black and white med-size bean, 8 ft. twining vines bear prolifically right up to frost, good for eating or display--let them mature on the vine and put them in a jar to admire. *Source History: 1 in 1987; 1 in 1994; 1 in 1998; 2 in 2004.* **Sources: Ho13, Se17.**

Incan Giant White - (Pallares) - Vines grow to 5-6 ft., 1 in. long seeds swell to 2 in. long when cooked, twice the size of the largest limas in America, from Peru. *Source History: 1 in 2004.* **Sources: Red.**

Indian Red Pole - *Source History: 1 in 2004.* **Sources: Se17.**

Johnny's Red - Sieva pole, small dark red seeds mottled with black, pruning runners later in the season stimulates greater pod production, thrives with low fertilization, produces until frost with adequate moisture during summer heat, originated in the Carolinas and Virginia. *Source History: 1 in 2004.* **Sources: Scf.**

King of the Garden - (Henderson's Leviathan, Large White Lima, Garden King) - 85-93 days - Vigorous 8-10 ft. plant, flat med-green 4-6 x 1.25 in. pods, 4 to 6 large creamy-white seeds, huge crops harvested over an extended period, old home garden favorite, introduced in 1883. *Source History: 71 in 1981; 63 in 1984; 59 in 1987; 49 in 1991; 42 in 1994; 48 in 1998; 42 in 2004.* **Sources: Ba8, Be4, Bu8, CA25, CH14, Co31, Com, Ers, Fi1, Gur, HA3, Ha5, He8, He17, HO2, Jun, Kil, La1, MAY, ME9, Mel, Mey, Mo13, OLD, Red, Ri12, Roh, RUP, Sau, SE4, Se25, Se26, Sh9, Shu, Si5, So1, SOU, Sto, Twi, Ver, Vi4, Wet.**

Lima o del Papa - Large, flat white and red seed. *Source History: 1 in 1987; 1 in 1991; 1 in 1994; 1 in 1998; 1 in 2004.* **Sources: Ber.**

Loudermilk - Black and white seed, produces until frost, pod set increased by pruning new growth runners later in the season. *Source History: 1 in 2004.* **Sources: Scf.**

Lynch Butterbean Collection - Beautiful seeds show a vast array of colored patterns, eaten fresh cooked from the garden, blanched and frozen, dried seed can be cooked as dried beans after soaking. *Source History: 1 in 1998; 2 in 2004.* **Sources: Ho13, Scf.**

North Pole - *Source History: 1 in 2004.* **Sources: Se17.**

Pennsylvania German Red - *Source History: 1 in 1998; 1 in 2004.* **Sources: Lan.**

Purple Eye - Vigorous vines to 10 ft., med. white seed with pink or purple eye ring. *Source History: 1 in 1998; 1 in 2004.* **Sources: Ho13.**

Red Calico - 90 days - Bright dark-red seeds splashed

with black streaks and speckles, heirloom originally from South Carolina, grown by the family of T. B. Thweat in Tennessee since 1790. *Source History: 1 in 1991; 1 in 1998; 3 in 2004.* **Sources: Ho13, Scf, Th3.**

Red Saba - Tall vines, med. size dark red-purple seeds, pre-1820. *Source History: 1 in 2004.* **Sources: Ho13.**

Shantyboat - White seed with heavy red markings, grown during the Great Depression by shantyboat dwellers in the mid- country river basins. *Source History: 2 in 1998; 3 in 2004.* **Sources: Ho13, Scf, Sh13.**

Sieva - (Carolina, Small Sieva, Southern Pole Butterbean, Southern Running Butterbean) - 70-89 days - Dark-green 9-10 ft. vines, broad flat med-green 3-4 in. pods, small flat med-green butterbeans dry white, high yields, green shells or dry, excel. for home gardens, also used for fresh market, dates back to 1880. *Source History: 28 in 1981; 22 in 1984; 18 in 1987; 15 in 1991; 14 in 1994; 19 in 1998; 20 in 2004.* **Sources: Ada, Bu8, Co31, Ers, HO2, Ho13, Ki4, Kil, La1, LO8, ME9, Mey, OLD, Pin, Ri12, Ros, Se16, SOU, Th3, Wi2.**

Smith Butterbean - Small seeds are cream with red, tan or black speckles, heavy yield. *Source History: 1 in 1998; 2 in 2004.* **Sources: Ho13, Sh13.**

Snow on the Mountain - *Source History: 1 in 1998; 2 in 2004.* **Sources: Se17, Sh13.**

White Christmas - 80 days - Result of an accidental cross of Christmas and Sieva, beautiful, large ivory-white seed blushed with purple on one corner, high yields, shells easily, produces reliably in hot, humid areas, CV So1 introduction, 2000, original seed from Brian Heatherington, GA. *Source History: 1 in 2004.* **Sources: So1.**

Willow Leaf Colored - Long vine with narrow leaves, colors include tan, black, red and speckled, withstands heat and drought. *Source History: 1 in 1998; 2 in 2004.* **Sources: Ho13, Scf.**

Willow Leaf White - (Willow Leaf Pole) - 65-90 days - Grows 8-10 ft., very narrow dark-green leaves that resemble those of a willow, 3 or 4 white seeds per pod, narrow leaf thought to contribute some drought and heat tolerance, introduced in 1891 by W. Atlee Burpee of Philadelphia as a sport of the Carolina Lima, exceedingly rare. *Source History: 3 in 1981; 2 in 1984; 5 in 1987; 3 in 1991; 4 in 1994; 8 in 1998; 9 in 2004.* **Sources: Ada, Ami, Ho13, Ra5, Scf, Se26, Sh13, So1, Ver.**

Winfield - *Source History: 1 in 1998; 1 in 2004.* **Sources: Sh13.**

Worchester Indian Red Pole - Extremely hardy, heat and drought res. plants, seeds range from dull red to dull maroon-red, still exhibits a wild trait of pod shattering when completely dry, reportedly of Native American origin, pre-1868, introduced in 1990 by CV So1. *Source History: 1 in 1991; 1 in 1994; 1 in 1998; 1 in 2004.* **Sources: So1.**

Varieties dropped since 1981 (and the year dropped):

Aubrey Deane (1998), *Bandy* (1991), *Black-Eyed Butterbeans* (1991), *Butterman* (2004), *Cape May Giant* (2004), *Challenger* (1982), *Civil War* (1998), *Climbing Baby Lima* (1991), *Dreer's Improved* (1987), *Dreer's Improved Challenger* (1987), *Hopi Mottled* (1984), *Illinois Giant* (2004), *Indian Red* (1998), *Jackson Wonder* (1991), *Large Speckled* (1991), *Nugget 55-88* (1987), *Nugget Jr. 79-6* (1987), *Oahu Island Common Lima* (1994), *Pima Orange* (2004), *Prizetaker* (2004), *Red-Speckled Pole* (2004), *Simmons Red Streak Butterbean* (1991), *Speckled Butterpea* (1991).

Melon / Honeydew
Cucumis melo

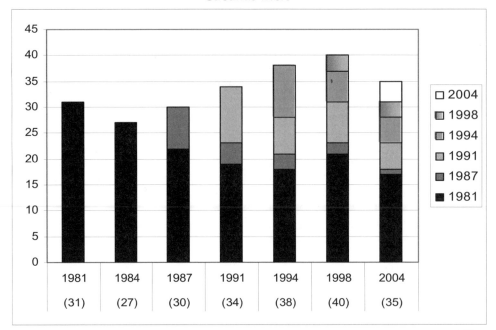

Number of varieties listed in 1981 and still available in 2004: 31 to 17 (55%).
Overall change in number of varieties from 1981 to 2004: 31 to 35 (113%).
Number of 2004 varieties available from 1 or 2 sources: 15 out of 35 (43%).

Amarillo Oro - 100 days - Winter type melon, golden yellow oblong fruit, sweet creamy white flesh, up to 15 lbs., good shipper, pre-1870 European heirloom. *Source History: 5 in 2004.* **Sources: Ba8, He8, Se16, Sk2, WE10.**

Canary Yellow - (French Canary, Italian Yellow Canary, Jaune des Canaries, Jaune de Canary, Jaune Canari) - 88-110 days - Vigorous plant, tough bright light-yellow skin slightly wrinkled, pear-shaped, 7 x 6 in., 4-6 lbs., 2 in. thick pale-green flesh, musky aroma, very sweet flavor, CB res., triangular cavity, keeps well into winter, popular in Europe. *Source History: 7 in 1981; 6 in 1984; 6 in 1987; 12 in 1991; 16 in 1994; 9 in 1998 7 in 2004.* **Sources: Bu3, Gr29, He8, Loc, Syn, WE10, Wi2.**

Casaba - (Canary Cantaloupe) - Golden wrinkled rind, 6 in. long, sweet white flesh, long keeper. *Source History: 4 in 1981; 3 in 1984; 2 in 1987; 2 in 1991; 1 in 1994; 3 in 1998; 2 in 2004.* **Sources: Gr29, Se8.**

Casaba, Bidwell - 90-95 days - Huge melon measures 12-15 in. long by 9 in. wide, 12-16 lbs., sweet orange flesh, from Chico, CA, grown by John Bidwell (1819-1900), a Civil War General and U.S. Senator who procured his stock seed from the USDA in 1869. *Source History: 1 in 2004.* **Sources: Se16.**

Casaba, Golden Beauty - 104-120 days - Tough wrinkled rind matures to golden, 8 x 7 in., 7-8 lbs., thick white aromatic spicy flesh, good for home gardens, crown blight res., excel. shipper, will keep for months, adapted to hot dry climates, especially Arizona and southern California, dates back to the 1920s. *Source History: 24 in 1981; 22 in 1984; 17 in 1987; 18 in 1991; 16 in 1994; 13 in 1998; 14 in 2004.* **Sources: Ba8, Bu3, CO30, He8, Jor, La1, Loc, MOU, Ros, SE14, Se17, Se26, Sk2, WE10.**

Casaba, Santa Claus - (Christmas Melon, Pimonet Piel de Sapo) - 108-110 days - Wrinkled gold and green mottled skin, 12 x 6 in. dia., 7-9 lbs., firm pale-green flesh, sweet flavor, very hard rind, for long warm dry season, fine keeper, crown blight res. *Source History: 4 in 1981; 4 in 1984; 4 in 1987; 3 in 1991; 4 in 1994; 6 in 1998; 4 in 2004.* **Sources: Ba8, Bu3, CO30, Sk2.**

Casaba, Santo Domingo - 100 days - Aromatic pale green flesh is sweet and juicy, 5-6 fruits per vine, heirloom from northern New Mexico Pueblos. *Source History: 1 in 1998; 2 in 2004.* **Sources: So12, Syn.**

Casaba, Santo Domingo Mixed - Honeydew type, basketball size, varied orange and white flesh, prolific, adapted to high cool desert, for spring planting. *Source History: 1 in 1981; 1 in 1984; 1 in 1987; 1 in 1991; 1 in 1994; 1 in 1998; 1 in 2004.* **Sources: Na2.**

Casaba, Sungold - 85-95 days - Full-size Casaba that ripens a month earlier than others, shorter vines, impressive yields, very sweet flavor, tender flesh, dev. by Prof. E. M. Meader at Univ. of NH, rescued from oblivion by Dr. Jeff McCormack. *Source History: 2 in 1981; 2 in 1991; 3 in 1994; 2 in 1998; 3 in 2004.* **Sources: Roh, Sa9, Se7.**

Crenshaw - 90-115 days - Vigorous productive plants, smooth skin corrugated green to yellow, pear-shaped, 8-10 x 6-8 in. dia., 6-10 lbs., salmon-pink flesh, grows best in warmer areas of the southern California coast, needs special packing for shipping. *Source History: 29 in 1981; 26 in 1984; 27 in 1987; 21 in 1991; 20 in 1994; 19 in 1998; 19 in 2004.* **Sources: Ba8, Bu3, CH14, CO30, Gr29, He8, He17, Jor, La1, Loc, Mo13, MOU, RIS, Ros, SE14, Sh9, Shu, WE10, WI23.**

Crenshaw, Bella Dulce - White fleshed, oval fruit, netted skin, ripe when yellow-orange with some green mottling, stabilized and selected by Horse Creek Seed Sanctuary, Horse Creek, CA. *Source History: 1 in 2004.* **Sources: Tu6.**

Crenshaw, White - 100 days - An improved Crenshaw, excel. foliage, creamy white flesh when ripe, excel. for long distance shipping and specialty markets and home gardens. *Source History: 1 in 1991; 2 in 1998; 2 in 2004.* **Sources: He8, SE14.**

Escondido Gold - (Tibetan Gold) - Vigorous vines with huge leaves, round fruit, up to 23 lbs., once popular in California market gardens, heirloom, rare. *Source History: 1 in 1994; 1 in 1998; 1 in 2004.* **Sources: Syn.**

Golden Honeymoon - (Gold Rind Honeymoon) - 92 days - Gold rind, thick bright-green flesh, unique flavor, not too sweet, smaller and two weeks earlier than regular Honeydew, prolific, sunburn res., good keeper, rare. *Source History: 2 in 1981; 2 in 1984; 2 in 1987; 2 in 1991; 2 in 1994; 4 in 1998; 5 in 2004.* **Sources: Ba8, Bo19, La1, Se26, Sk2.**

Green Climbing - (Vert Grimpant) - Small, oval fruit with dark green skin with lighter spots, 4-5 in. long, 1-1.5 lbs., tender green flesh, juicy and sweet with a delightful scent, ripens best when allowed to climb, rare French heirloom. *Source History: 1 in 1998; 3 in 2004.* **Sources: Ba8, Co31, Hud.**

HoneyDew - (Honeydew Green Fleshed) - 95-115 days - Broad oval 7-8 x 7 in. dia. melons, 6 lbs. average, smooth ivory skin, no netting, light-green very sweet juicy flesh, small seed cavity, hard rind, prefers a warm dry climate, good shipping and storage, introduced in 1915. *Source History: 34 in 1981; 30 in 1984; 26 in 1987; 23 in 1991; 26 in 1994; 24 in 1998; 16 in 2004.* **Sources: Ada, Be4, Bu2, Ear, Ers, Fo7, Ge2, Gr29, He17, Kil, La1, Lej, Mey, Mo13, PAG, Wet.**

Honeydew, Early - 70 days - Early maturing honeydew for Northern gardeners, smooth green-gold skin, lime-green flesh, 2.5-3 lb. melons, disease res. *Source History: 3 in 1981; 3 in 1984; 1 in 1987; 1 in 1991; 1 in 1994; 1 in 1998; 1 in 2004.* **Sources: Mcf.**

Honeydew, Gold Rind - (Golden Honeymoon) - 90-115 days - Smooth ivory skin turns light-golden as it ripens, light greenish- white flesh, 7 x 6.5 in., 5 lbs., stems will not slip when ripe, vigorous, high sugar, good shipper. *Source History: 5 in 1981; 5 in 1984; 4 in 1987; 4 in 1991; 5 in 1994; 3 in 1998; 3 in 2004.* **Sources: Bu3, MAY, Ros.**

Honeydew, Green Fleshed - 105-115 days - Smooth hard creamy-white rind without netting or sutures, 7.5-8 x 6.5-7 in. dia., 5.5-6 lbs., ripens to light-gold, thick lime-green flesh, high sugar, well adapted to the Pacific Coast, good for storage and shipping, small cavity. *Source History: 29 in 1981; 26 in 1984; 24 in 1987; 23 in 1991; 24 in 1994; 25 in 1998; 30 in 2004.* **Sources: ABB, Bou, Bu3, Bu8, CH14, CO23, CO30, De6, DOR, He8, Jor, LO8, Loc, Ma19, MIC, MOU, Na6, OLD, Pr9, Ros, RUP, SE14, Se26, Sh9, Shu, So25, TE7, WE10, Wi2, WI23.**

Honeydew, Italian - *Source History: 1 in 2004.* **Sources: Se17.**

Honeydew, Orange Fleshed - 90-110 days - Vigorous, creamy-white skin when ripe, 6.5 x 6 in., 5 lbs., thick firm light-orange flesh, flavor between honeydew and crenshaw, can ship even at half to full-slip. *Source History: 7 in 1981; 6 in 1984; 8 in 1987; 11 in 1991; 12 in 1994; 12 in 1998; 19 in 2004.* **Sources: Ba8, Bo19, Bu3, CO30, Ers, Gr27, He8, He17, La1, La1, Loc, Ma19, Mo13, Ros, SE14, Se26, TE7, WE10, Wi2.**

Honeydew, Sharlyn - 100-105 days - Vigorous disease res. vines produce 6 in. oval fruits, 4-6 lbs., thin green rind matures to yellow or salmon, no ribbing, slight net, creamy white and gold, sweet soft flesh is aromatic and juicy, not recommended for northern growing areas. *Source History: 1 in 1994; 4 in 1998; 6 in 2004.* **Sources: Ba8, CO30, Pe2, Se7, SE14, So12.**

Jaune de Canary - 90-110 days - Heirloom, football shaped fruit, 4-5 lbs., wrinkled yellow rind, white-green flesh is sugar sweet, keeps well, full sun, good yields. *Source History: 4 in 1998; 5 in 2004.* **Sources: Ba8, CO30, Eo2, Fo7, So12.**

Kazakh - 65-70 days - Early, 1-3 lb., softball size fruit, white skin turns bright-yellow when ripe, sweet white flesh, good climber, not a good keeper, rare. *Source History: 1 in 1987; 2 in 1991; 1 in 1994; 3 in 1998; 5 in 2004.* **Sources: Pr3, Ri12, Sa9, Se17, Sk2.**

Marygold - 80-92 days - Bright-yellow 3-4 lb. slightly wrinkled casaba melon with white flesh, much earlier, size better suited for market, cut from the vine, res. to race 2 of fusarium wilt, from the University of Maryland. *Source History: 2 in 1991; 5 in 1994; 3 in 1998; 3 in 2004.* **Sources: Pe6, RUP, SOU.**

O'odham Ke:li Ba:so - Casaba type fruits, light-green flesh, favorite of Tohono O'odham and Pima farmers of the low desert. *Source History: 1 in 1991; 1 in 1994; 1 in 1998; 1 in 2004.* **Sources: Na2.**

Stutz Supreme - 95-105 days - Large 5-10 lb. melon, smooth tan skin, delicious orange flesh, mid-season, bears well until frost, requires heat to dev. full flavor, originally from Joe Stutz of California. *Source History: 1 in 1981; 1 in 1984; 1 in 1987; 2 in 1991; 2 in 1994; 3 in 1998; 3 in 2004.* **Sources: Pe1, Ri12, Se7.**

Swan Lake - 85 days - White fleshed 2-3 lb. fruits, occasional white swirled with orange, exceptional flavor, good open- pollinated line from the hybrid. *Source History: 2 in 1991; 2 in 1994; 3 in 1998; 2 in 2004.* **Sources: Pe2, Se7.**

Sweet Delight - (Delight) - 75-95 days - Honeydew type with more uniformity and sweeter flesh, 7.5 x 7 in., no net, light green flesh, 6 lbs., PVP 1997 Hollar Seeds, Inc. *Source History: 4 in 1994; 6 in 1998; 4 in 2004.* **Sources: Bo17, HO1, OLD, RUP.**

Sweet Freckles - 100 days - Crenshaw type with unusual dark-green "freckles" dotting the light-green skin, about one-half to two-thirds the size of normal crenshaws, sweet, juicy, tender, aromatic, light-orange flesh, tough rind and nose means it will handle better than other crenshaws, ready for picking when the freckles on the belly turn yellow-orange, Farthest North x Casaba crosses, PSR Breeding Program. *Source History: 1 in 1994; 1 in 1998; 1 in 2004.* **Sources: Pe6.**

Tam Dew - 90-110 days - Vigorous plant, smooth creamy-white oval fruits, 7.5 x 6.5 in. dia., 5.5 lbs., firm thick light-green flesh, ripens to the rind, classic honeydew flavor, storage and shipping in warmer areas. *Source History: 8 in 1981; 6 in 1984; 3 in 1987; 3 in 1991; 3 in 1994; 5 in 1998; 7 in 2004.* **Sources: Ba8, HO1, La1, LO8, Scf, Sk2, Wi2.**

Tam Dew Improved - 95-100 days - Round, ribless, 5-6.5 in. dia., 4-5.5 lbs, uniform green flesh, small seed cavity, creamy- white when ripe, hard rind, DM PM1 and PM2 res., excel. shipper, TX/AES. *Source History: 9 in 1981; 8 in 1984; 9 in 1987; 11 in 1991; 9 in 1994; 5 in 1998; 5 in 2004.* **Sources: ABB, CO30, Jor, Sk2, WE10.**

Valencia - (Spanish Melon, Valencia Rocket Melon) - 110-115 days - Slightly elongated dark-green melon, 8 x 6 in. dia., tough wrinkled skin, sweet white flesh, cavity lined with light-orange, keeps up to 4 months, common in Spain, listed by American seedsmen in the 1830s, believed to have come from Italy. *Source History: 4 in 1981; 3 in 1984; 5 in 1987; 1 in 1994; 1 in 1998; 2 in 2004.* **Sources: Ba8, Syn.**

Valencia, Early - *Source History: 1 in 1994; 1 in 1998; 1 in 2004.* **Sources: WE10.**

Valencia, Late - *Source History: 1 in 1991; 1 in 1994; 1 in 1998; 1 in 2004.* **Sources: WE10.**

Varieties dropped since 1981 (and the year dropped):

Afghani Honey Dew (1991), *Amarelo* (2004), *Baby Slip Honeydew* (1991), *Cameo* (1991), *Canary 96* (2004), *Casaba Black* (1991), *Casaba, Sunglo* (1994), *Christmas Long Keeper* (2004), *Cora's Melon* (1998), *Crenshaw, Blanco* (2004), *Crenshaw, Golden* (1998), *Di Brindisi* (1987), *Dwarf Bush Honeydew* (1987), *Escondido Gold Melon* (1991), *Honeyloupe* (2004), *Mayan Sweet* (1994), *Minnesota Honeymist* (1983), *Morgan* (1982), *NV 990* (1994), *Oliver's Pearl Cluster* (1991), *Pinyonet de Valencia* (2004), *San Juan* (2004), *Spanish Winter Valencia* (1987), *Tam Canary* (1998), *Tam Mayan Sweet* (2004), *Valencia Early* (1983), *Valencia Espanhol* (2004).

Melon / Muskmelon
Cucumis melo

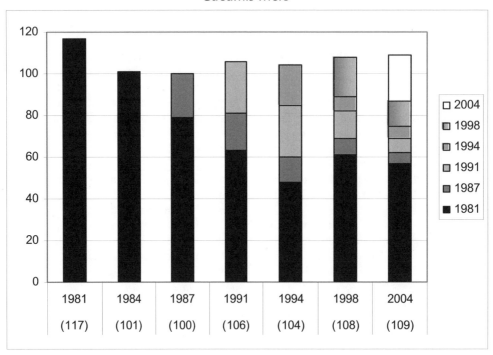

Number of varieties listed in 1981 and still available in 2004: 117 to 57 (49%).
Overall change in number of varieties from 1981 to 2004: 117 to 109 (93%).
Number of 2004 varieties available from 1 or 2 sources: 65 out of 109 (60%).

Albers - Oval fruits are ribbed and netted, firm flavorful orange flesh, ripens about two weeks later than Boughem, from Ontario. *Source History: 1 in 1991; 1 in 1994; 1 in 1998; 1 in 2004.* **Sources: Pr3.**

Altai - Large, almost round fruits, from the former Soviet Union. *Source History: 1 in 1998; 1 in 2004.* **Sources: Pr3.**

Amish - 80-90 days - Fairly short, productive vines, well suited for small gardens, oval, 6-9 in., slightly ribbed fruits with some netting, 6-9 lbs., rather thick rind, sweet, orange, very juicy flesh, excellent full flavor, FW res., tolerates heat, drought, cold and wet. *Source History: 1 in 1994; 2 in 1998; 9 in 2004.* **Sources: Ho13, Pe2, Ri12, Se16, Sh13, Shu, So12, Tu6, Up2.**

Ananas - (Israeli, Sharlyn, Ananas da America, Pineapple Melon) - 86-110 days - Large vigorous productive vines, oblong fruits, 5 lb. average, green rind matures yellow-orange blushed with salmon green, no ribs or sutures, slight net, flesh is creamy yellow with a hint of salmon at full maturity, soft juicy sweet and very aromatic, home garden and local market, PM res., CB tol., heirloom from the 1800s. *Source History: 4 in 1981; 4 in 1984; 6 in 1987; 8 in 1991; 11 in 1994; 15 in 1998; 14 in 2004.* **Sources: Ba8, CO23, DOR, Fo7, He8, HO1, La1, Red, Se24, So9, So25, Syn, WE10, Wi2.**

Ananas d'Amerique a Chair Rouge - Very rare variety that was illustrated in France in 1885 by Vilmorin, netted skin, bright orange flesh, sweet, highly perfumed, dates to the early 1800s, rare. *Source History: 1 in 2004.* **Sources: Ba8.**

Ananas d'Amerique a Chair Verte - (Green Fleshed Pineapple) - 90 days - Historical melon with netted skin, light green firm flesh is sweet and highly perfumed, grown by Thomas Jefferson in 1794, offered commercially in the U.S. in 1824, illustrated in color in France in the Vilmorin Album in 1854. *Source History: 3 in 2004.* **Sources: Ba8, Ea4, La1.**

Anne Arundel - 83 days - Bright green flesh with a tinge of orange in the middle, 3-4 lb. fruit, pre-1800 Maryland heirloom grown by farmers in Anne Arundel County, Maryland, introduced commercially in 1890. *Source History: 1 in 1998; 5 in 2004.* **Sources: Ba8, Ea4, Sa9, Sh13, Th3.**

Bender's Surprise - 90-92 days - Coarse net, prominent ribs, 8 x 7 in. oval melons, skin gray-green > yellow, bright salmon-pink flesh, 6-8 lbs., flavor improves 5-6 days after picking, fine quality. *Source History: 10 in 1981; 6 in 1984; 4 in 1987; 2 in 1991; 1 in 1994; 1 in 1998; 1 in 2004.* **Sources: Sh9.**

Blenheim Orange - 80-90 days - From the Henry Doubleday Res. Inst. in England, finely netted skin, 1-2 lb. fruit with aromatic, orange flesh, for northern, short season areas, 1887. *Source History: 2 in 1994; 2 in 1998; 3 in 2004.* **Sources: Pe2, Ri12, Se7.**

Boughem - Large mild-flavored fruits of various sizes, matures 7-10 days earlier than Far North, seed originally from Russia, has been in Ternier family for over 40 years. *Source History: 1 in 1991; 1 in 1994; 1 in 1998; 1 in 2004.* **Sources: Pr3.**

Burrell's Jumbo - 80-82 days - Improved selection of Hale's Best by the D. V. Burrell Seed Co., very sweet, salmon-orange flesh, up to 5 lbs., ideal for home or market. *Source History: 4 in 2004.* **Sources: Ba8, Bu3, Se16, Vi4.**

Canoe Creek Colossal - 85-90 days - Large, deeply ribbed, football shaped fruits, dark green skin changes to orange when ripe, average 8-15 lbs. but can reach 20 lbs. with regular watering, great flavor, harvest just as they begin to slip. *Source History: 2 in 2004.* **Sources: Ag7, Se16.**

Castella - (Formerly Amber Nectar) - 86 days - Fruits average 3-3.5 lbs., high sugar content, tender, juicy and sweet, freezes well, bred to ripen in northern European conditions. *Source History: 1 in 1994; 1 in 1998; 1 in 2004.* **Sources: Tho.**

Cavaillon Espagnol - 80-100 days - Large oblong heavily netted fruits, 8-10 in. long, up to 6 lbs., green flesh with salmon center, sweet and juicy, yellow skin when ripe, Spanish heirloom that dates back to the 1800s. *Source History: 2 in 1991; 1 in 1994; 1 in 1998; 6 in 2004.* **Sources: Ba8, He8, Se16, Sk2, Vi4, WE10.**

Chimayo - Sweet, orange fleshed oval fruits, Spanish heirloom from northern New Mexico. *Source History: 1 in 1994; 1 in 2004.* **Sources: Na2.**

Collective Farm Woman - 80-85 days - Yellow-gold rind, extra sweet white flesh, 7-10 in., ripens early, heirloom from the Ukraine, very popular on the Island of Krim in the Black Sea, first offered to American gardeners by SSE in 1993. *Source History: 6 in 2004.* **Sources: Ba8, He8, Hi6, Hud, Se16, Sk2.**

Crane - (China) - 75-85 days - Crenshaw type, speckled skin, minimal netting, extremely fragrant, delicious sweet flavor, smooth texture, intro. in the 1920s by Oliver Crane whose family farmed for six generations near Santa Rosa, CA. *Source History: 4 in 2004.* **Sources: Na6, Sa9, Se16, Syn.**

Delicious - 83 days - Slightly oval 7 in. long fruits, 3-4 lbs., thick orange-yellow flesh, preferred by market gardeners because it handles well, matures to light-yellow, strong vines. *Source History: 6 in 1981; 6 in 1984; 4 in 1987; 5 in 1991; 3 in 1994; 5 in 1998; 3 in 2004.* **Sources: BAL, Pr3, So25.**

Delicious 51 - (Cornell Delicious #51, Golden Delicious 51) - 68-95 days - Early str. of Bender's Surprise, 6-7 x 5-5.5 in., 4-5 lbs., round-oval, well netted, slight ribs, salmon-orange flesh, FW res., matures some fruit ahead of most hybrids, home garden, not a shipper, Dr. Munger/Cornell. *Source History: 37 in 1981; 32 in 1984; 31 in 1987; 29 in 1991; 26 in 1994; 35 in 1998; 35 in 2004.* **Sources: All, Ba8, Be4, Bo19, Bu3, CH14, Com, Dam, Ear, Ers, Fe5, HA3, He8, Hi6, HO1, HO2, Hud, Jun, La1, MAY, Mo13, OLD, Ont, Roh, RUP, Sau, Scf, SE4, SE14, Se25, Se26, Sh9, So1, Tu6, WE10.**

Delicious 51, Early - (Early Delicious) - 80-84 days - Slightly oval, 6-7 in., 5.5-6 lbs., creamy-green mature rind, coarse netted, medium ribs, FW res., prolific, not a shipper, Dr. Munger/Cornell. *Source History: 6 in 1981; 4 in 1984; 2 in 1987; 3 in 1991; 1 in 1994; 1 in 1998; 1 in 2004.* **Sources: GRI.**

Delicious, Golden - 83 days - Deep-salmon flesh, 6 in. dia., FW res., rich sweet flavor, disease res. *Source History: 3 in 1981; 2 in 1984; 8 in 1987; 5 in 1991; 4 in 1994; 7 in 1998; 4 in 2004.* **Sources: La1, PAG, Sau, SE14.**

Dr. Jaegar's Mildew Res. - 80 days - A selection from Hales Best, prolific yields of good sized melons, salmon flesh, mildew and wilt res., excel. quality, fairly early. *Source History: 2 in 1981; 1 in 1984; 1 in 1987; 1 in 1991; 1 in 1994; 1 in 1998; 1 in 2004.* **Sources: Shu.**

Early Hanover - 70-85 days - Extra early home-garden variety, 2-3 lbs., 6 in. wide by 5 in. long, globe-shaped, salmon-colored flesh, sweet sugary flavor, harvest slightly prior to full slip for best flavor, almost extinct commercially, intro. in 1895 by T. W. Woods and Sons of Richmond, VA. *Source History: 1 in 1981; 1 in 1987; 1 in 1991; 2 in 1994; 2 in 1998; 2 in 2004.* **Sources: Ri12, Se16.**

Edisto - 75-95 days - Oval 6.5-7 x 5.5-6 in. dia., 4-4.5 lbs., thick orange flesh, small cavity, fine net, light ribs, firm tough rind, PM DM ALS res., excel. for hot humid conditions, fine flavor, excel. shipper, good home garden, Clemson SC/AES, 1957. *Source History: 15 in 1981; 13 in 1984; 12 in 1987; 7 in 1991; 6 in 1994; 5 in 1998; 7 in 2004.* **Sources: Ada, Bo19, Bu3, Scf, Se26, Sk2, Vi4.**

Edisto 47 - 88-95 days - Larger Edisto type, 7.5 x 6.5 in. dia., 4.5 lbs., deep-salmon flesh, fine flavor, hard rind, medium net, slight suture, more vigorous vines, higher DM and PM res., same ALS res., humidity tol., excel. shipper, Dr. Hughes/Clemson SC/AES, 1965. *Source History: 21 in 1981; 20 in 1984; 13 in 1987; 11 in 1991; 10 in 1994; 11 in 1998; 7 in 2004.* **Sources: Ba8, CO30, Kil, SE14, So1, WE10, WI23.**

Eel River - 90-100 days - Delicious orange flesh with a creamy peach flavor, 3-10 lbs., grown by Bear Jones, Humbolt County, CA, seed brought from Japan by his cousin after WW II. *Source History: 3 in 2004.* **Sources: Co32, Ri12, Se7.**

Emerald Gem - 70-90 days - Early, green skin, sweet yellow-orange flesh, sweet and somewhat spicy flavor, fruits up to 3 lbs., Burpee introduction, 1886. *Source History: 2 in 1981; 2 in 1984; 2 in 1987; 4 in 2004.* **Sources: Ba8, Hud, Se16, Te4.**

Emerald Green - 70-90 days - Sweet, juicy pale orange flesh with rich, musky flavor, 2 lbs., high yields, considered the finest tasting melon in 1886. *Source History: 1 in 2004.* **Sources: He8.**

Far North - 65-70 days - Hearts of Gold type and quality, 4-5 in. dia., thick firm salmon flesh, small seed cavity, medium to sparse net, slight suture, small compact vine, grows well in any part of Canada, 14%+ sugar content, good flavor, dev. in Saskatchewan, improved at Morden Manitoba Exp. Farm and Minnesota Hort. Dept., 1950. *Source History: 16 in 1981; 14 in 1984; 10 in 1987; 6 in 1991; 3 in 1994; 2 in 1998; 2 in 2004.* **Sources: Pr3, Sa9.**

Giant Perfection - (Mammoth, Field's Giant, Hollybrook, Shumway's Giant Perfection) - 95-97 days - Very large, round, 8 in. dia., 14-18 lbs., selected from old-fashioned Perfection, very fragrant, deep-orange, firm, high ribs. *Source History: 4 in 1981; 3 in 1984; 5 in 1987; 3 in 1991; 1 in 1994; 2 in 1998; 1 in 2004.* **Sources: Ho13.**

Golden Champlain - 75 days - Earliest quality melon, medium size, nearly round, slight ribs, coarse open net, golden-yellow flesh, skin shows tinge of orange when ripe, short season areas, yields well under adverse weather conditions. *Source History: 2 in 1981; 2 in 1984; 2 in 1987; 2 in 1991; 1 in 1994; 1 in 1998; 1 in 2004.* **Sources: Shu.**

Golden Gopher - 80-87 days - Heavily ribbed fruit, 6 by 5 in., deep orange flesh, FW res., heirloom from the 1930s. *Source History: 1 in 1987; 2 in 1991; 2 in 1994; 4 in 1998; 3 in 2004.* **Sources: Fe5, Sa9, Tu6.**

Golden Jenny - 85 days - More vigorous and compact than Jenny Lind with better insect res., sweet orange flesh, fruits weigh up to .75 lb., dev. by Merlyn Niedens, CV So1 introduction, 1997. *Source History: 1 in 1998; 1 in 2004.* **Sources: So1.**

Goldmaster - 87 days - Fully netted fruit with small seed cavity, non-sutured type, sulfur tol., does well in many areas, dev. by Sunblest. *Source History: 1 in 1991; 2 in 1994; 1 in 1998; 1 in 2004.* **Sources: Loc.**

Granite State - 83-90 days - Early high-quality cantaloupe, bred especially for Northern climates, 4 lb., lightly netted fruits with deep orange flesh, very high sugar content. *Source History: 1 in 1981; 1 in 1998; 1 in 2004.* **Sources: Sa9.**

Green Bush - Flavorful heirloom from Missouri, heavily netted 3 in. fruits on small bushy plant, green flesh with some orange. *Source History: 1 in 2004.* **Sources: Ho13.**

Green Machine - 85 days - Jenny Lind x Kansas cross, so named because of its high productivity, compact vines produce green fleshed, sweet fruit, dev. by Merlyn Niedens, introduced in 1998 by CV So1. *Source History: 1 in 1998; 1 in 2004.* **Sources: So1.**

Green Nutmeg - (Early Green Flesh, Nutmeg) - 61-89 days - Extra early, slightly oval, ribbed, heavy netting, light-green flesh with salmon center, sweet with a unique spiciness, 2-3 lbs., harvest at half slip, home gardens or nearby markets, noted by Fearing Burr in 1863. *Source History: 2 in 1981; 2 in 1984; 3 in 1987; 4 in 1991; 5 in 1994; 4 in 1998; 5 in 2004.* **Sources: Ba8, Ea4, Fe5, La1, Sa9.**

Hale's Best - (Dixie Jumbo) - 75-90 days - Heavily netted, indistinct sutures, thick solid deep-salmon flesh, 6.5 x 5.75 in. dia., popular early shipper, res. to drought and mildew, fine quality, produces heavily, uniform oval shape. *Source History: 22 in 1981; 15 in 1984; 19 in 1987; 16 in 1991; 16 in 1994; 14 in 1998; 15 in 2004.* **Sources: All, Bo19, CH14, Fi1, Gur, Hi13, Hum, Lej, Lin, LO8, Ont, PAG, Pr9, Se25, Wet.**

Hale's Best 36 - (Hale's Best Jumbo 36) - 80-90 days - Popular older shipper, very uniform strain, 6 x 5.5 in., 3-3.5 lb., very sweet thick salmon-pink flesh, small seed cavity, well netted, slight suture, vigorous vine, Imperial Valley favorite. *Source History: 27 in 1981; 22 in 1984; 16 in 1987; 16 in 1991; 10 in 1994; 9 in 1998; 4 in 2004.* **Sources: La1, Sh9, WE10, Wi2.**

Hale's Best 45 - (Imperial, Sweet Eating) - 85-88 days - Almost 100% PM res., can pick at full slip and ship across the country, improves a few days after picked, 6.5 x 5.5 in., 4 lbs., heavy net, developed by a Japanese market grower in California around 1920. *Source History: 10 in 1981; 8 in 1984; 6 in 1987; 4 in 1991; 3 in 1994; 2 in 1998; 2 in 2004.* **Sources: Ba8, Ear.**

Hale's Best Jumbo - (Hale's Best Original, Hale's Best Jumbo #936, Hale's Jumbo) - 68-90 days - Standard early shipper in South and Southeast, 6.5-7.5 x 5.5 in. dia., 3.5-4.5 lbs, heavy coarse net, light rib, firm salmon flesh, good flavor, drought res., introduced in 1923. *Source History: 66 in 1981; 60 in 1984; 53 in 1987; 48 in 1991; 50 in 1994; 50 in 1998; 51 in 2004.* **Sources: Ada, Be4, Ber, Bo17, Bo18, Bu1, Bu2, Bu8, But, CO23, CO30, Coo, Cr1, De6, Dgs, DOR, Ers, Fo7, Ge2, Gr27, HA3, He17, HO1, HO2, HPS, Jor, Kil, La1, Loc, MAY, Mey, MIC, Mo13, Mo20, MOU, OLD, Ri12, Ros, RUP, Sau, Se7, SE14, Se26, Shu, So1, SOU, Vi4, WE10, Wi2, WI23, Wo8.**

Haogen - (Israel) - 80-85 days - Translates in Hebrew as 'the anchor,' thick med-green flesh tinged salmon around seed cavity, unique flavor, vigorous vines bear 10 or more fruit, 3-5 lb., round melons with smooth skin and no netting, spicy, freezes well, considered Israeli but actually an Israeli adaptation of an ancient Southwest Indian melon. *Source History: 1 in 1981; 1 in 1984; 1 in 1987; 3 in 1991; 5 in 1994; 6 in 1998; 8 in 2004.* **Sources: Bou, Pe1, Pe2, Sa9, Se7, Se16, So9, So12.**

Harvest Queen - (Pride of Wisconsin) - 82-95 days - Heavily netted, faintly ribbed, deep-orange firm thick flesh, small cavity, 7 x 6 in. dia., 3-4 lbs., tough rind, ripens to gold, FW res., fine quality, standard shipping melon in Northeast and Mid-Atlantic and Midwest, keeps for 5-6 days, 1954. *Source History: 31 in 1981; 26 in 1984; 20 in 1987; 20 in 1991; 20 in 1994; 9 in 1998; 4 in 2004.* **Sources: Pe2, Roh, Se7, So12.**

Hearts of Gold - (Hoodoo) - 70-97 days - Aromatic sweet deep-orange flesh very thick and firm, small cavity, nearly round 6.5 x 5.5 in. dia., 3.5-4.5 lbs., heavily netted, medium ribs, quite vigorous, very productive, blight res., ships well for moderate distances, dev. by Roland Morrill, intro. about 1895, granted a trademark on December 15, 1914. *Source History: 40 in 1981; 37 in 1984; 39 in 1987; 32 in 1991; 30 in 1994; 27 in 1998; 36 in 2004.* **Sources: Ag7, Ba8, Be4, Bo17, Bu8, CH14, Co31, Co32, DOR, Ers, Ga1, Gr27, He8, He17, La1, Loc, MAY, Mey, MOU, Na6, Pe2, Red, Ri2, Roh, RUP, Sau, SE14, Se16, Se25, Se26, Sh9, Shu, TE7, WE10, Wet, WI23.**

Hearts of Gold Improved - 90 days - Vigorous plants, round-ribbed deep-green fruits, gray net, 6 x 6 in. dia., 3.5 lbs., very sweet tender deep-salmon flesh, small cavity, home and local market. *Source History: 3 in 1981; 2 in 1984; 1 in 1987; 3 in 1991; 2 in 1994; 2 in 1998; 1 in 2004.* **Sources: SOU.**

Hollybrook Luscious - 85-110 days - Deep ribs, dark green rind, heavy net, 8 x 7 in. dia., 8-10 lbs., salmon flesh, good flavor, pick slightly prior to full slip for best flavor, bruises easily, for home gardens and roadside stands, introduced by T. W. Woods and Sons of Richmond, Virginia, 1905, rare. *Source History: 1 in 1981; 1 in 1984; 2 in 2004.* **Sources: Ba8, Se16.**

Honey Rock - (Sugar Rock) - 74-88 days - AAS/1933, 6 x 5.5 in. dia., 3-4 lb., tough gray-green skin, coarse open net, ribbed, thick sweet firm deep-salmon flesh, vigorous, res. to FW, 5-7 melons per plant, for home garden and local market in northeast U.S. and southern Canada, Michigan State University. *Source History: 40 in 1981; 38 in 1984; 29 in 1987; 22 in 1991; 22 in 1994; 21 in 1998; 22 in 2004.* **Sources: Ba8, Bu8, Cr1, De6, Ers, Fi1, He8, HO2, Jor, La1, Ma19, Mel, OLD, RUP, SE14, Se25, Sh9, Sk2, TE7, Vi4, WE10, Wet.**

Imperial - 90-100 days - Oval, heavily netted fruit, 3-5 lbs., delicious flavor, excel. quality, small seed cavity. *Source History: 1 in 1998; 1 in 2004.* **Sources: Se7.**

Imperial 4-50 - 90-95 days - Med-heavy net, medium sutures, 6.5 x 6 in. dia., salmon flesh, res. to some PMs, excel. shipper, best for Texas conditions, large #45 type, holds quality longer. *Source History: 3 in 1981; 3 in 1984; 2 in 1987; 2 in 1991; 2 in 1994; 1 in 1998; 2 in 2004.* **Sources: Bu3, Wi2.**

Imperial 45 - (Hales Best No. 45, Imperial 45 Shipper) - 85-95 days - Basic southern California market melon, 6.5 x 5.5 in., 3.5 lbs., fine solid net, slight ribs, orange flesh, PM res., holds well, shipper. *Source History: 12 in 1981; 9 in 1984; 9 in 1987; 7 in 1991; 7 in 1994; 6 in 1998; 4 in 2004.* **Sources: CH14, CO23, DOR, SE14.**

Imperial 45-ECS (PMR) - 83 days - Well netted, minimal ribs, 6 x 5 in. dia., deep-orange flesh, small seed cavity, res. to some PMs, sets fruits on nodes 3, 4 and 5, other PMR types are later and set on nodes 5 and beyond. *Source History: 2 in 1981; 2 in 1984; 1 in 1987; 2 in 1991; 1 in 1994; 1 in 1998; 2 in 2004.* **Sources: DOR, WE10.**

Iroquois - (Iroquois Extra Select, Giant Early Wonder) - 75-99 days - Coarse heavy net, deep ribs, 7 x 6 in. dia., 5-7 lbs., thick deep-orange flesh, hard gray-green shell, FW res., especially good in midwestern and northeastern U.S. and eastern Canada, Dr. Munger/Cornell, 1944. *Source History: 42 in 1981; 38 in 1984; 31 in 1987; 37 in 1991; 32 in 1994; 35 in 1998; 27 in 2004.* **Sources: Ba8, Be4, Bou, CH14, Com, De6, Dgs, Eco, Eo2, Ers, Gr27, He17, HO2, Jor, La1, Ma19, Mel, Mo13, RIS, RUP, Sau, SE14, Sh9, Shu, So9, WE10, Wi2.**

Israel - (Ogen, Ogden, Haogen, Israeli Ha-Ogen) - 70-85 days - Aromatic sweet pale-green flesh, small cavity, smooth pale- yellow rind, green sutures, each vine produces 10 or more small fruits, grows well anywhere in the U.S. if watered during dry spells, can be frozen, widely coveted, Israel. *Source History: 4 in 1981; 4 in 1984; 5 in 1987; 6 in 1991; 6 in 1994; 4 in 1998; 2 in 2004.* **Sources: Loc, Wi2.**

Israeli - (Old Original, Ananas) - 90-95 days - Vigorous productive vines, large oval 7-8 lb. fruits, no ribs or sutures, sparse netting, yellow-orange rind, creamy-white flesh, very aromatic, unique flavor, fine for home gardens and local markets. *Source History: 1 in 1981; 1 in 1984; 2 in 1987; 3 in 1991; 3 in 1994; 4 in 1998; 6 in 2004.* **Sources: Bo17, He8, Ho13, Se26, Syn, Wi2.**

Itsy Bitsy Sweetheart - 95 days - Bushy non-vining plants, small round, thin rind, bright-orange flesh, heat and drought and mildew res., fine flavor, good for small gardens. *Source History: 2 in 1981; 2 in 1984; 4 in 1987; 6 in 1991; 3 in 1994; 1 in 1998; 1 in 2004.* **Sources: Syn.**

Jake's - 90-110 days - Thin tan rind spotted with orange and gray, orange-yellow flesh, Pueblo tribe members say it looks like Snake in the Shed, an old lost American Indian melon. *Source History: 2 in 2004.* **Sources: Ho13, Se17.**

Jenny Lind - (Jenny Lynn) - 70-85 days - Old heirloom favorite back by popular demand, 5 ft. vines, flat 1-3 lb. fruits, many with knobs or turbans on the blossom end, lime-green flesh, very sweet and aromatic, prolific, matures midseason, known in the Philadelphia markets before 1840, named after a popular singer of that era, introduced in 1846. *Source History: 5 in 1987; 9 in 1991; 16 in 1994; 17 in 1998; 17 in 2004.* **Sources: Ba8, Bu2, Dow, Fe5, Fo7, He8, Ho13, Lan, Pin, Ri12, Roh, SE14, Se16, Sh13, Shu, Te4, Ver.**

Kansas - 90 days - Rare heirloom from Kansas, orange flesh, excellent flavor, fine texture, moderate netting, ridged, oval fruit, 4 lbs., hardy, good res. to sap beetle. *Source History: 1 in 1991; 1 in 1994; 1 in 1998; 2 in 2004.* **Sources: Ba8, So1.**

Kin Makuwa - Japanese heirloom, attractive fruit, bland flesh, likely a pickling melon, grown for centuries in northern Japan, kin means gold. *Source History: 2 in 2004.* **Sources: Ba8, Yu2.**

Mainstream - 85-90 days - High yielding, nearly round 2.75 lb. fruits, firm orange flesh, small cavity, PM and DM res., adapted to the Southeast and elsewhere, long-distance shipper. *Source History: 3 in 1981; 3 in 1984; 6 in 1987; 2 in 1991; 3 in 1994; 3 in 1998; 1 in 2004.* **Sources: Wi2.**

Melon de Castillo - Pale yellow, smooth skin, sweet flesh, from the Sierra Madre in Mexico. *Source History: 1 in 1998; 1 in 2004.* **Sources: Na2.**

Minnesota Midget - 60-100 days - Compact, 3 ft. vines, large crops of 4 in. melons, thick meaty gold-yellow flesh, unique flavor, high sugar content, edible to the rind, high quality, fine dense netting, early, res. to FW, U of MN, 1948. *Source History: 15 in 1981; 15 in 1984; 15 in 1987; 14 in 1991; 15 in 1994; 13 in 1998; 21 in 2004.* **Sources: Ba8, Bo19, Ers, Ga1, He8, Hi6, Hig, Ho12, Jor, La1, Pin, RUP, SE14, Se16, Se25, Sk2, So25, Syn, Te4, Up2, Yu2.**

Montana Gold - 60 days - Extremely early, bred by Fisher's to ripen in their short growing season in Belgrade, MT, fruit starts to set with first blossoms, netted, flavorful. *Source History: 1 in 1981; 1 in 1984; 1 in 1987; 1 in 1991; 1 in 1994; 1 in 1998; 1 in 2004.* **Sources: Fis.**

Montreal - Delicious, sweet, spicy pale green flesh, needs an early start and heat to reach its potential size of 20 lbs., heirloom. *Source History: 2 in 2004.* **Sources: So25, Te4.**

Musketeer - (Bush Musketeer) - 90 days - Greatly improved bush type, round 5.5-6 in. dia., 3.5-4.5 lbs., very sweet deep- salmon flesh, heavy net, 3-5 fruits per plant, gourmet quality, good container plant, outstanding performer, widely adapted, PVP expired 2001. *Source History: 27 in 1981; 26 in 1984; 30 in 1987; 18 in 1991; 14 in 1994; 7 in 1998; 2 in 2004.* **Sources: Kil, Wet.**

North Carolina - 110 days - Giant cantaloupe, 20-30 lbs., as sweet and juicy as ordinary size cantaloupes, easy to grow. *Source History: 1 in 1987; 1 in 1991; 1 in 1994; 3 in 1998; 4 in 2004.* **Sources: Bu1, Fa1, Pp2, Tt2.**

Northern Arizona - Very sweet, early var., 4-6 in. fruit, smooth yellow skin with orange and green markings, pink-orange flesh, sweet cantaloupe flavor with perfumey overtones. *Source History: 1 in 1998; 1 in 2004.* **Sources: Bou.**

Nutmeg - (Eden Gem, Rocky Ford, Netted Gem) - 80-87 days - In 1863 Fearing Burr, Jr. described 12 varieties suitable for the garden, and Nutmeg was ranked as one of "the very best", greenish-orange flesh, 2 lbs., fine aroma, good flavor, midseason, grows well in northern climates *Source History: 1 in 1987; 3 in 1991; 4 in 1994; 6 in 1998; 6 in 2004.* **Sources: Co32, Fo7, Ho13, SE14, Se16, Syn.**

Nutmeg, Extra Early - 60-75 days - Netted skin, green flesh, mildly spicy, tolerates cool conditions, pre-1835. *Source History: 2 in 1998; 2 in 2004.* **Sources: Hi6, Ri12.**

Ogen - 70-95 days - Actually an Israeli adaptation of an ancient SW Indian melon, each vigorous vine produces 10 or more small fruits, 3 lbs., smooth pale-yellow rind with green sutures, thick medium-green flesh tinged salmon around the seed cavity, freezes well, grows anywhere in mainland U.S. *Source History: 3 in 1991; 3 in 1994; 6 in 1998; 5 in 2004.* **Sources: Ba8, Bo17, Se26, So25, WE10.**

Oka - Bizard Island strain, cross of the green fleshed Montreal Market x Banana, large netted melon, flattened shape, orange, aromatic flesh with rich muskmelon flavor, an heirloom that had disappeared for years before being rediscovered on the Island of Bizard, Quebec, Canada, bred around 1912 by Father Athanase of the Trappist Monastery at La Trappe, Quebec, Canada. *Source History: 1 in 1998; 5 in 2004.* **Sources: Co32, Se16, Te4, Up2, Yu2.**

Old Time Tennessee Muskmelon - 90-110 days - Huge, deeply creased rind, convoluted shape, delicious when fully ripe, do not pick too soon, fragile garden-to-table melon, use at once for best flavor and texture. *Source History: 5 in 1981; 4 in 1984; 4 in 1987; 5 in 1991; 4 in 1994; 2 in 1998; 2 in 2004.* **Sources: Ho13, So1.**

Old West Virginia Heirloom - 85 days - Medium size, 3-5 lb. fruit, light netting, productive. *Source History: 1 in 2004.* **Sources: Sa9.**

Oran's Melon - 80-90 days - Large round fruit, netted and ribbed rind, 6-8 lbs., sweet orange flesh, from the Oran Ball family in the Ozarks. *Source History: 2 in 2004.* **Sources: Co32, Se17.**

Pear - 85-90 days - Football shape fruits, yellow when ripe, sweet, orange flesh, family heirloom over 100 years old. *Source History: 3 in 1998; 1 in 2004.* **Sources: Se17.**

Perlita - (Tam Perlita) - 80-94 days - Well netted, faint ribs, 6 x 5.5-6 in. dia., 2.5-3 lbs., salmon-orange flesh, high yielding shipper, used mainly in the lower Rio Grande valley, TX/AES. *Source History: 14 in 1981; 13 in 1984; 13 in 1987; 9 in 1991; 5 in 1994; 2 in 1998; 1 in 2004.* **Sources: LO8.**

Persian - 95-115 days - Dark-green firm rind, no ribs, fairly heavy netting, bright-orange flesh, large fruits to 6 lbs., distinctive sweet flavor, does not do well in cool areas. *Source History: 6 in 1981; 5 in 1984; 5 in 1987; 3 in 1991; 2 in 1994; 2 in 1998; 4 in 2004.* **Sources: CO30, Gr29, He8, La1.**

Persian Small - 110-115 days - Dark-green and ribless with fine net, bright-orange flesh, 8 x 6 in. dia., 6-7 lbs., small cavity, adapted to California, deep root system requires less moisture, rare old-time variety. *Source History: 3 in*

1981; 3 in 1984; 2 in 1987; 1 in 1991; 1 in 1994; 2 in 1998; 5 in 2004. **Sources: Ba8, Loc, Se26, Sk2, Vi4.**

Petit Gris de Rennes - 80-85 days - Excellent French variety, grey-green rind, flavorful sweet orange flesh, 2-3 lbs., well adapted to cool climates, early maturity. *Source History: 2 in 2004.* **Sources: Ba8, So12.**

Planters Jumbo - 85-91 days - Firm rind, light rib, smooth heavy netting, thick deep-orange flesh, 7 x 6 in. dia. and 4-4.5 lbs., PM and DM res., another sel. from same cross that produced Gulfstream, vigorous vines do well in drought or high rainfall, for home or local market or bulk shipping in South, USDA and Clemson/SC, 1954. *Source History: 21 in 1981; 17 in 1984; 17 in 1987; 14 in 1991; 11 in 1994; 14 in 1998; 13 in 2004.* **Sources: Ba8, Bu3, CH14, Gr27, HO1, Kil, La1, Scf, SE14, Se26, Vi4, WE10, Wi2.**

PMR 45 - (Mildew Resistant No. 45, Resistant No. 45, No. 45 SJ) - 85-95 days - Close high net, faint ribs, 5-6 x 4.5-5.5 in. dia., 2-3 lbs., immune to PM1, highly disease tol., thick firm orange flesh, small cavity, benchmark for comparison, hard rind, long-distance West Coast shipper, pick at full slip, vigorous, heavy yields, USDA. *Source History: 10 in 1981; 9 in 1984; 11 in 1987; 8 in 1991; 6 in 1994; 4 in 1998; 6 in 2004.* **Sources: Bu3, CO30, HO2, Ros, Se26, Wi2.**

Pride of Wisconsin - (Queen of Colorado, Harvest Queen) - 85-100 days - Round-oval, 7 x 6.5 in. melons, faint ribs, coarse net, thick orange flesh, gray skin, very hard shell, holds well after fully ripe, grows well in heavy soils, FW res., for home garden and local market. *Source History: 21 in 1981; 16 in 1984; 11 in 1987; 3 in 1991; 2 in 1994; 1 in 2004.* **Sources: Se16.**

Rampicante Zuccherino - 80 days - Early maturing melon, very sweet salmon-orange flesh, 2-2.5 lbs., eaten with Parma or other hams in Italy. *Source History: 1 in 2004.* **Sources: Se24.**

Retato Degli Ortolani - 80 days - Orange flesh, netted skin, 3-4 lbs., said to be the melon to eat with Parma ham. *Source History: 1 in 1987; 1 in 1991; 1 in 1994; 1 in 1998; 1 in 2004.* **Sources: Se24.**

Rocky Ford - 85-95 days - Well netted skin, slight rib, firm, sweet salmon-colored flesh, PM res., good shipper. *Source History: 2 in 2004.* **Sources: Mey, Se7.**

Rocky Ford, Green Fleshed - (Eden Gem, Netted Gem, Gold Lined Rocky Ford, Nut Meg, Improved Rocky Ford, Genuine Rocky Ford) - 84-96 days - Derived from Netted Gem which developed from a chance seedling of Nutmeg, deep fine-grained green flesh, 5.5 x 4.5-5 in. dia., 2-3 lbs., heavy net, slightly ribbed, very rust res., prolific, for home gardens and local markets, also a good shipper, dev. by J. W. Eastwood of Colorado in 1881. *Source History: 40 in 1981; 36 in 1984; 36 in 1987; 25 in 1991; 23 in 1994; 24 in 1998; 36 in 2004.* **Sources: Ada, Ba8, Bu2, Bu3, Bu8, CH14, DOR, Ea4, Ga1, Gr27, He8, He17, Hi6, Hi13, HO1, La1, Lan, Mo13, Pe2, Ros, Sa9, Se8, SE14, Se26, Shu, Sk2, So1, So25, SOU, Syn, TE7, Vi4, WE10, Wet, Wi2, WI23.**

Schoon's Hard Shell - (New Yorker, Illinois Hardshell) - 88-95 days - Very hard shell, first rate shipper, stands well in field or market, 7-8 x 6-7 in. dia., 5-8 lbs., slow to ripen, thick red-salmon flesh, very highly flavored, shape and appearance like Bender's Surprise, gray-green skin > yellow at maturity, coarse gray-yellow rope-like netting, heirloom from New York. *Source History: 15 in 1981; 13 in 1984; 12 in 1987; 12 in 1991; 10 in 1994; 8 in 1998; 8 in 2004.* **Sources: Ba8, Bo19, Bu3, Ri12, Se16, Se26, Sh9, Vi4.**

Shumway's Giant Muskmelon - (Tip-Top) - 81-90 days - Giant size slightly flattened round fruits, deeply ribbed, lightly netted, deep salmon-orange flesh, 8-10 lbs., high sugar content, excel. eating quality. *Source History: 1 in 1981; 2 in 1998; 2 in 2004.* **Sources: Sa9, Shu.**

Sierra Gold - 85-90 days - Well netted, indistinct ribs, 6-6.5 in. dia., 2.5-3.5 lbs., salmon-colored flesh, small cavity, heavy foliage, excel. flavor, PM1 res., sulphur tol., 1956. *Source History: 8 in 1981; 8 in 1984; 10 in 1987; 10 in 1991; 8 in 1994; 4 in 1998; 4 in 2004.* **Sources: Ba8, Sau, SE14, Se26.**

Sleeping Beauty - 85 days - Compact vines produce uniform fruit in a small area, smooth, round fruit with white-green appearance before ripening, orange flesh .05-.75 lb., the name refers to the tendency of the fruits to lie together in groups, dev. by Merlyn Niedens, introduced by CV So1, 1997. *Source History: 1 in 1998; 1 in 2004.* **Sources: So1.**

Spanish Moon - Heirloom. *Source History: 1 in 2004.* **Sources: Bu2.**

Spear - 85 days - Football-shaped melon, gray-green with slight netting, orange flesh, excellent flavor, weighs 3-6 lbs., produces good size fruits under cool growing conditions, longtime favorite in western Oregon. *Source History: 1 in 1981; 1 in 1984; 1 in 1987; 1 in 1991; 2 in 1994; 4 in 1998; 1 in 2004.* **Sources: Sa9.**

Sugar Rock - (Honey Rock) - 74-88 days - AAS/1933, almost round, 5.5-5 in., 4 lbs., coarse heavy netting, not ribbed, very tough rind, very sweet thick firm deep-orange flesh, gray-green skin, does well in northern U.S. and Canada. *Source History: 6 in 1981; 4 in 1984; 4 in 1987; 3 in 1991; 2 in 1994; 2 in 1998; 1 in 2004.* **Sources: Und.**

Sweet Granite - 65-80 days - Sweet aromatic bright-orange flesh, med-net, football-shaped, 2.5-3.5 lbs., dev. for direct seeding in Northern cool summer areas, does not keep for more than 1 to 2 days after slipping, Prof. Meader/U of NH, 1966. *Source History: 9 in 1981; 6 in 1984; 7 in 1987; 9 in 1991; 7 in 1994; 6 in 1998; 3 in 2004.* **Sources: Goo, Jo1, So9.**

Sweet Passion - 85 days - Sweet orange flesh, 3-4 lbs., small seed cavity, drought res., wilt tol., it is rumored to cause a state of passion if eaten straight from the garden on a moon-lit night, from the 1920s. *Source History: 1 in 1998; 1 in 2004.* **Sources: He8.**

Sweetheart - 65 days - Globe shaped, 4 in. fruit, smooth green-white skin, sweet, thick, salmon-red flesh, does well in short season areas. *Source History: 1 in 1994; 1 in 1998; 1 in 2004.* **Sources: Ver.**

Tam Uvalde - 85-90 days - Well netted, indistinct ribs, 5 x 6 in. dia., 3-4.5 lbs., deep-orange flesh, concentrated set, no crown set, DM PM1 and PM2 res., good Rio Grande shipper, TX/AES. *Source History: 12 in 1981; 11 in 1984; 10 in 1987; 9 in 1991; 7 in 1994; 3 in 1998; 3 in 2004.*

Sources: CO30, LO8, Wi2.

Tendral Verde Tardif - Oblong fruit, dark green rind, sweet white flesh, very late, popular winter melon in Europe, rare. *Source History: 1 in 1991; 1 in 1994; 1 in 1998; 2 in 2004.* **Sources: Ba8, WE10.**

Thai Long - Medium size oblong fruit, green rind, sweet orange flesh, does well in humid areas and the tropics, popular in Thailand. *Source History: 1 in 2004.* **Sources: Ba8.**

Thayer - 95 days - Kansas heirloom, vigorous plants, slight ribbing, some netting, orange flesh, exceptional flavor, up to 10 lbs., excellent market melon, very rare. *Source History: 1 in 2004.* **Sources: Ba8.**

Tip-Top - 80-90 days - Large slightly flattened fruits, moderate net, sweet juicy firm bright-salmon flesh, fine flavor, for local markets, Livingston Seed Co., intro. in 1892. *Source History: 3 in 1981; 1 in 1984; 2 in 1987; 2 in 1991; 1 in 1994; 1 in 1998; 2 in 2004.* **Sources: Ba8, Ho13.**

Top Mark - 85-95 days - Vigorous dark-green plants, 7 x 6 in. dia., well netted, very slight ribs, thick deep-salmon flesh, DM PM CB and scab res., sulphur tol., Southwest shipper, 1963. *Source History: 11 in 1981; 11 in 1984; 14 in 1987; 14 in 1991; 17 in 1994; 14 in 1998; 13 in 2004.* **Sources: Bu3, CH14, CO23, CO30, DOR, Fo7, He8, La1, Ros, SE14, Se26, WE10, Wi2.**

Top Net SR - 86 days - Slightly oval 3-4 lb. fruit, medium full net, thick salmon flesh, superb firmness makes it an outstanding shipping melon, crown blight and sulfur tol., PVP expired 2002. *Source History: 1 in 1994; 1 in 1998; 1 in 2004.* **Sources: Loc.**

Turkey - 90 days - Very old Southern var. with long oval fruits, 12-16 in. long, slight net, light-green sutures, 8-18 lb., thick firm salmon flesh, better keeper than Old Time Tennessee, home and local market. *Source History: 1 in 1981; 1 in 1984; 1 in 1987; 1 in 1991; 1 in 1994; 1 in 1998; 1 in 2004.* **Sources: Sa9.**

Uncle E - 85-90 days - Lightly netted, roundish football shape fruit, golden when ripe, sweet, cream colored flesh. *Source History: 2 in 1998; 1 in 2004.* **Sources: Se17.**

Vandehay Family - *Source History: 1 in 2004.* **Sources: Se17.**

Weeks' North Carolina Giant - 85 days - Developed by North Carolina farmer Ed Weeks, 20-25 lbs. is common on good soil with lots of humus and water during normal season, sweet juicy flesh, his record: 39 lbs., 1977. *Source History: 4 in 1981; 2 in 1984; 1 in 1987; 1 in 2004.* **Sources: Hi13.**

Zatta - 80-85 days - Ribbed fruit, green and yellow skin, very sweet orange flesh, 4 lbs. *Source History: 1 in 2004.* **Sources: Se24.**

Varieties dropped since 1981 (and the year dropped):

Amarillo Liso (2004), *Armstrong Extra Early* (1987), *Aurora* (2004), *Autumn Paradise* (1994), *Berthold* (1991), *Big Daddy* (1994), *Bit O'Honey* (1987), *Burpee's Fordhook Gem* (1987), *Bush Midget* (2004), *Chilton* (1994), *Chinese Melon* (1994), *Cinco* (1991), *Colossal Furrowed* (1998), *Cossack* (1998), *Crane Melon* (1998), *Cum Laude* (1983), *Dessert Sun* (1987), *Dulce* (1998), *Early Chaca* (1994), *Early Green Flesh* (1991), *Early Large Prescott* (2004), *Early May* (1984), *Early Rocky Ford* (1987), *Eureka* (1987), *Extra Early Hackensack* (1991), *Gold Lined Rocky Ford* (1991), *Golden Delight* (1984), *Golden Honey Delicious 51* (1994), *Golden Perfection* (1987), *Greeley Wonder* (1984), *Gulf Coast* (1991), *Gulf Stream* (1998), *Gulfstream* (2004), *Gusto 45* (1998), *Hackensack* (1987), *Hale's Best 36 Improved* (1991), *Hale's Best Jumbo Improved* (1991), *Hearts of Gold, Morrill's* (2004), *Hicross Gulf Sweet* (1998), *Hiline* (1998), *Honey Combed* (2004), *Honey Rock Improved* (1987), *Honeybush* (2004), *Imperial 45 PMR Crownset* (1998), *Imperial 45-S12* (1998), *Jumbo* (1998), *Kangold* (2004), *King Henry* (1991), *Knight's Early Green Flesh* (1982), *Large White Prescott* (2004), *Magdalena* (2004), *Marathon* (1991), *Market Pride* (2004), *Market Star* (1998), *Melba* (2004), *Melon Ball* (1987), *Midget Muskmelon* (1994), *Milwaukee Market* (1982), *Mini* (1987), *Minnesota Honey* (1991), *Mountain Magic* (2004), *Mr. Ugly* (1998), *Muskotaly* (1998), *Navajo Mix* (2004), *Navajo Yellow* (2004), *Northwest* (1983), *Oregon Delicious* (2004), *Osage* (1984), *Pecos* (1984), *Pennsweet* (1987), *Perfection* (1994), *Perlita 45, Tam* (1994), *Persian Medium* (1998), *Pike* (2004), *PMR 45 K Strain* (1994), *Queen of Colorado* (1987), *Resistant 4-50* (2004), *Resistant 45* (2004), *Rochet* (1998), *Rocket de Valencia* (1994), *Rocky Ford Pollock 10-25* (1987), *Rocky Ford, Green Flesh Imp.* (1994), *Shipmaster* (1998), *Short 'n Sweet* (1991), *SJ-45* (1998), *Skagit Golden* (1984), *Smith's Perfect* (2004), *Southland* (1987), *Spanish Espanhol* (1998), *Spartan Rock* (2004), *SR 59* (1987), *SR 91* (1987), *Sugar Salmon* (1991), *Sunrise Melon* (1994), *Tendral Verde Tardio* (2004), *Tennessee* (1994), *Texas Resistant No. 1* (1982), *Top Mark Crownset* (2004), *Unwins' No Name* (1983).

Melon / Other
Cucumis melo

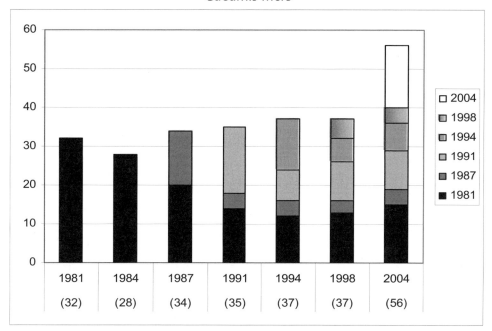

Number of varieties listed in 1981 and still available in 2004: 32 to 15 (47%).
Overall change in number of varieties from 1981 to 2004: 32 to 56 (175%).
Number of 2004 varieties available from 1 or 2 sources: 42 out of 56 (75%).

Ao Uri - Japanese pickling melon, 10-12 x 4.5 in. dia., dark-green rind, thick white flesh, used to make many types of Oriental pickles but excellent for any rind pickle recipes. *Source History: 1 in 1981; 1 in 1984; 1 in 1987; 1 in 1991; 1 in 1994; 1 in 1998; 1 in 2004.* **Sources: Kit.**

Armenian - (Armenian Burpless Cucumber, Yard Long, Guta, Snake or Serpent Cucumber) - 50-75 days - *Cucumis melo*, (Flexuosus Group), light-green, heavily ribbed, 24-36 x 3-4 in. dia., best at 12-18 in., mild flavor, easily digested skin and all, fluted slices, prolific, twists and turns on ground, hangs fairly straight from trellis. *Source History: 33 in 1981; 30 in 1984; 29 in 1987; 31 in 1991; 30 in 1994; 31 in 1998; 41 in 2004.* **Sources: Bo17, Bo19, Bou, Bu2, Bu3, Co32, Com, Eo2, Ge2, Goo, He8, HO1, Ho13, Hud, Jo1, Kit, La1, Lej, Loc, MOU, Na6, Ni1, Or10, Orn, Pe2, Pe6, Pla, Ra5, Red, Rev, Ri12, Ros, Se7, SE14, Se26, Shu, Sw9, Syn, Ter, We19, Wi2.**

Banana - 80-100 days - Long tapering banana-shaped melon, 16-24 x 4 in. dia., 5-8 lbs., smooth yellow skin, salmon-pink flesh, sweet spicy flavor, good late-maturing variety that can stand sun, dates back to late 1800s. *Source History: 40 in 1981; 35 in 1984; 31 in 1987; 26 in 1991; 23 in 1994; 18 in 1998; 28 in 2004.* **Sources: Ag7, Ba8, Bo19, Bu1, Bu8, CH14, Co31, Fi1, Gr29, He8, Hi13, HO1, Ho12, La1, Mey, Ra5, Ri12, Sau, SE14, Se16, Se28, Sh9, Shu, SOU, WE10, Wi2, WI23, Wo8.**

Blanco - (White, Navajo) - 85-110 days - Improved crenshaw type, oval fruit, creamy white rind, salmon-pink flesh, extra sweet flavor, 5 lbs., great for specialty markets and shipping. *Source History: 1 in 1991; 1 in 1998; 1 in 2004.* **Sources: Ba8.**

Boule d'Or - 120 days - Taste almost as legendary in Provence as that of the Charentais--but a very different type of melon, hard yellowish lightly netted skin, pale-green flesh, ripens in late Sept. when direct seeded in late May, winter keeper, store in a cool dry place, listed by Vilmorin in 1885, very rare. *Source History: 1 in 1987; 1 in 2004.* **Sources: Ba8.**

Carosello Mezzo Lungo Barese - Football shaped, corrugated fruit, soft thin skin covered with soft fuzz, very tasty, cucumber flavor with a distinct melon overtone, typical of Bari region. *Source History: 1 in 1987; 1 in 1991; 1 in 1994; 1 in 1998; 2 in 2004.* **Sources: Ber, So25.**

Charentais - (Igor, French Gourmet) - 75-90 days - True cantaloupe, smooth light-green striped thin rind matures creamy yellow, salmon flesh with superb flavor, grapefruit size, about 2 lbs., extremely fragrant when ripe, needs some extra attention. *Source History: 10 in 1981; 8 in 1984; 7 in 1987; 10 in 1991; 6 in 1994; 10 in 1998; 20 in 2004.* **Sources: Ba8, Bo18, Bou, Coo, Gou, He8, Ho13, La1, Lej, Mcf, Na6, Ni1, Ri12, Roh, Se7, Se8, Se16, Se24, So25, WE10.**

Charlynne - 95-100 days - Aromatic European type, large productive vines, 6-8 in. oval fruit, 4.5-6.5 lbs., very sweet white flesh, slight netting, popular for roadside stands and home gardens. *Source History: 1 in 1991; 1 in 1994; 1 in 1998; 1 in 2004.* **Sources: Loc.**

China Long - (China Long Green, China, Chinese Long Jumbo, Lungo di Cina) - 65-75 days - *Cucumis melo*,

(Flexuosus Group), vigorous productive vines, very dark-green smooth skin, 20-36 x 2-3 in., crisp firm mild flesh, almost seedless, not suited for pickles, very mosaic tol., produces over the entire season, very dependable even in adverse conditions, straighter fruits if hanging from a trellis. *Source History: 14 in 1981; 13 in 1984; 10 in 1987; 8 in 1991; 5 in 1994; 4 in 1998; 2 in 2004.* **Sources: Ba8, Ear.**

Chinese Snake - *Source History: 1 in 1991; 1 in 2004.* **Sources: WE10.**

Cob Melon - Seeds on a cob like corn, sweet creamy-white flesh slightly grainy, med-large fruits mottled light and dark-green, common a century ago in gardens throughout West. *Source History: 2 in 1981; 2 in 1984; 2 in 1987; 2 in 1991; 1 in 1994; 1 in 1998; 1 in 2004.* **Sources: Ho13.**

D'Algiers - Compact vines, ribbed fruit, dark green rind looks black with silver splashes, ripens to yellow with red splashes, smooth, creamy flesh is highly perfumed, ancient variety from France, possible African origin. *Source History: 2 in 2004.* **Sources: Ba8, Hud.**

Early Black Rock - French heirloom with slightly flattened, ribbed, smooth fruit, dark green skin is almost black, orange when ripe, dark orange flesh, fragrant and sweet, 2-3.5 lbs., once popular but now rare. *Source History: 1 in 1998; 1 in 2004.* **Sources: Hud.**

Early Silverline - 75-80 days - Crisp, white, refreshingly sweet flesh, 1-2 lbs., very productive. *Source History: 1 in 2004.* **Sources: Se16.**

Giallo Canary Rugoso - Type uncertain. *Source History: 1 in 1991; 1 in 1994; 1 in 1998; 1 in 2004.* **Sources: WE10.**

Giallo da Inverno 3 - 90-95 days - Yellow Winter Melon, white fleshed cooking melon, yellow skin, sweet flesh, 4-5 lbs., productive, vigorous. *Source History: 1 in 2004.* **Sources: Se24.**

Ginger's Pride - Huge oblong fruit weighs 14-18 lbs., green rind ripens to yellow-orange at maturity, superior quality, sweet melting flesh, from Kentucky. *Source History: 1 in 2004.* **Sources: Ba8.**

Golden Crispy - (Golden Sweet, Japanese Golden Crispy) - 80 days - Japanese melon eaten like an apple, golden yellow edible skin, white crisp flesh, flavor resembles pears, prolific, fruits develop 30 days after flowering. *Source History: 2 in 1994; 3 in 1998; 4 in 2004.* **Sources: Ba8, Ban, Sa9, So25.**

Half Long of Puglia - Green cucumber/melon type, early and productive, for fresh eating and pickling. *Source History: 1 in 2004.* **Sources: Se24.**

Honey Gold No. 9 - (Honey Gold Sweet) - 85-86 days - Egg-shaped, 10 oz., smooth golden-yellow skin, crisp apple-like white flesh, eat skin and all, honey flavored, aromatic, does not slip, trellis, keeps well, Japan. *Source History: 4 in 1981; 4 in 1984; 2 in 1987; 1 in 1991; 1 in 1998; 1 in 2004.* **Sources: Ho13.**

Isleta Pueblo - Orange and green flesh, ribbed, heat adapted, collected near Albuquerque, NM. *Source History: 1 in 1991; 1 in 1994; 1 in 1998; 1 in 2004.* **Sources: Na2.**

Italian Cucumber, Fuzzy White - 50 days - Rare Italian heirloom, vines bear numerous small white cucumbers with fuzzy skin, crisp flesh with slightly sweet flavor, frequent picking increases fruit production. *Source History: 1 in 2004.* **Sources: Und.**

Jharbezeh Mashadi (Persia) - 85 days - Ancient melon, considered one of the best melons in the world by Marco Polo over 700 years ago. *Source History: 1 in 1981; 1 in 1984; 1 in 2004.* **Sources: Red.**

Kroumir - French heirloom, dark green skin, fragrant, thick juicy flesh, 2 lbs. *Source History: 1 in 2004.* **Sources: Hud.**

Kyoto Three Feet - (Kyoto) - 62 days - *Cucumis melo*, (Flexuosus Group), light-green smooth skin, 25-36 x 1.5 in., straight if trellised, seeds in blossom end, 75% seedless, sow early summer for fall harvest, Japan. *Source History: 9 in 1981; 7 in 1984; 5 in 1987; 3 in 1991; 4 in 1994; 5 in 1998; 4 in 2004.* **Sources: Kit, Ni1, Tho, Yu2.**

Mango Melon - (Vine Peach) - 80-90 days - Native American variety used for pies and preserves and eaten raw, 3 in. fruit, yellow rind, bland white flesh, resembles flavor and texture of a mango, vines resist pests, documented in this country in the 1850s. *Source History: 1 in 1994; 1 in 1998; 3 in 2004.* **Sources: Ba8, So25, Und.**

Metki Dark Green - Ancient variety, used like a cucumber, mild and tasty when 18 in. or less, can grow over 3 ft. long, introduced from Armenia into Italy as early as the 1400s *Source History: 1 in 1991; 1 in 1994; 2 in 1998; 2 in 2004.* **Sources: Ba8, WE10.**

Metki Painted Serpent Melon - 65 days - Rare Armenian type cucumber that is botanically a melon, very long fruit, dark green and pale green stripes, excellent flavor, popular at farmers markets, ancient Armenian heirloom which dates as far back as the 1400s. *Source History: 4 in 2004.* **Sources: Ba8, He8, La1, WE10.**

Metki White - (Metky White Trailing) - Armenian cucumber, botanically a melon, used as a cucumber when under 18 in. for best taste, easy to grow. *Source History: 2 in 1991; 1 in 1994; 2 in 1998; 2 in 2004.* **Sources: Ba8, WE10.**

Million Dollar - Soft flesh with sweet flavor, perfumes the garden air as it ripens, from the Cambridge steamship wreck along the coast of Maine in 1886. *Source History: 1 in 1998; 1 in 2004.* **Sources: Ho13.**

New Mexico Melon - Ribbed fruits with green/yellow skin, sweet and juicy orange white green or yellow flesh, from Alameda, New Mexico. *Source History: 1 in 1994; 1 in 1998; 1 in 2004.* **Sources: Na2.**

Noir des Carmes - 75 days - Translates "Early Black Rock", slightly flattened, ribbed smooth fruits, very dark green, almost black skin starts turning orange at maturity, 2-3 lbs., dark orange flesh, French heirloom from before 1880. *Source History: 1 in 1987; 1 in 1991; 1 in 1994; 2 in 1998; 5 in 2004.* **Sources: Co31, Ho13, Pr3, Se16, Syn.**

Ojo Caliente - Oval smooth-skinned fruits can weigh 5-7 lbs., pale-green flesh with a tinge of orange, sweet and juicy, originally from a farmer in northern NM. *Source History: 1 in 1991; 1 in 1998; 1 in 2004.* **Sources: Na2.**

Oriental Pickling Melon - (Shimauri Stripe) - 73 days - *C. melo* var. *conomon*, dark green with slender white stripes, smooth rind, 12 in. long, 3 in. dia. at widest point, used for various Oriental pickles and for cooking. *Source History: 1 in 1981; 1 in 1984; 1 in 1987; 3 in 1991; 2 in 1994; 3 in 1998; 2 in 2004.* **Sources: Ev2, So25.**

Oshiro Uri - (Numane) - 70 days - Japanese pickling melon, long oval light-green fruits turn white when mature, med-thick crisp flesh, excellent for any rind pickle recipes. *Source History: 1 in 1981; 1 in 1984; 1 in 1987; 1 in 1991; 1 in 1994; 1 in 1998; 1 in 2004.* **Sources: Kit.**

Pickling Melon, Numame - Fruits resemble a large pale cucumber, 10-11 in. long, firm thick flesh makes good Oriental or watermelon pickles. *Source History: 1 in 1994; 1 in 1998; 1 in 2004.* **Sources: Ev2.**

Piel de Sapo - (Toad Skin, Frog Skin, Pinyonet de Valencia) - 90-110 days - Rare melon from Spain, oval fruit, mottled green and yellow skin, 7-9 lbs., 8-12 in. long, sweet pale green to white flesh, very hard rind, fine keeper, common in European and Spanish markets. *Source History: 5 in 2004.* **Sources: Ag7, Ba8, CO30, He8, Se16.**

Plum Grannie - (Queen Anne's Pocket Melon, Dwarf Pomegranate, Vine Pomegranate, Plum Granny) - 75 days - Tiny melons with sweet fruit-like aroma, orange skin with darker striping, edible but smells better than it tastes, carried to mask body odors in ancient times. *Source History: 2 in 1991; 1 in 1994; 2 in 1998; 7 in 2004.* **Sources: Ba8, Co31, Ri12, Sa9, Scf, Se16, Und.**

Prescott Fond Blanc - 85-95 days - Flattened, ribbed French cantaloupe with warts and bumps, 4-9 lbs., grey-green rind ripens to straw color, salmon-orange flesh, excellent flavor and fragrance, good yield, drought and wilt res., mentioned in the 1860s but is likely much older. *Source History: 1 in 1998; 4 in 2004.* **Sources: Ba8, Fe5, La1, Se16.**

Quito - Small, bright yellow, smooth skinned melon, tart flesh, dates to the 1880s, France. *Source History: 1 in 2004.* **Sources: Hud.**

Ribertejo Blanco - *Source History: 1 in 1994; 1 in 1998; 1 in 2004.* **Sources: WE10.**

Sakata's Sweet - 85-90 days - Rounded 10-12 oz., baseball size melons, gray-green skin ripens yellow-green, soft green flesh, highly aromatic, eat skin and all, does not slip, can trellis, good keeper, grown in the East for centuries, now making an appearance in American markets. *Source History: 3 in 1981; 3 in 1984; 1 in 1987; 1 in 1991; 1 in 2004.* **Sources: Se16.**

Savor - 78 days - Improved French Charentais type with excel. flavor, sweet orange flesh, faintly ribbed, smooth gray-green skin with dark green sutures, 2-2.5 lbs., resists Fusarium oxysporum races 0, 1 and 2, PM res. *Source History: 1 in 1994; 1 in 2004.* **Sources: VI3.**

Serpent Cucumber - 65 days - *Cucumis melo*, (Flexuosus Group), long slim light-green ribbed 30-75 in. fruits, used for slicing up to 10 in., curls into realistic snake-like shapes, mild, burpless. *Source History: 3 in 1981; 3 in 1984; 2 in 1987; 2 in 1991; 2 in 1994; 2 in 1998; 3 in 2004.* **Sources: Bu1, Ra5, Vi5.**

Snap Melon, Kachra - *Cucumis melo* var. *momordica*, unusual var. from India, so named since the mature fruits explode, scattering the seeds, young tender fruits are eaten raw or cooked, ripe fruits eaten as a desert. *Source History: 1 in 2004.* **Sources: Hud.**

Souhela - *Source History: 1 in 1994; 1 in 1998; 1 in 2004.* **Sources: WE10.**

Spanish Melon - 140 days - Dark-green wrinkled skin, 8-12 lbs., green-gold flesh, sweet flavor, will keep up to 4 mos. if kept cool and dry. *Source History: 1 in 1991; 1 in 1994; 1 in 1998; 1 in 2004.* **Sources: Pla.**

Thai Round Green - Round, medium size fruit, pale green flesh, grows well in hot humid areas and the tropics, popular in Thailand, rare in the U.S. *Source History: 1 in 2004.* **Sources: Ba8.**

The Duke - 70 days - Used as a cucumber when immature, mild flavor, no bitterness, popular in the Mideast. *Source History: 1 in 2004.* **Sources: Sa9.**

Tigger - Armenian heirloom, vibrant yellow rind with fire-red zigzag stripes, extra sweet fragrance, sweet white flesh with smooth flavor, up to 1 lb., tolerates dry conditions, excellent specialty market variety. *Source History: 1 in 2004.* **Sources: Ba8.**

Tondo Liscia Manduria - 60 days - Cucumber/Melon type, round fruit, good flavor, productive. *Source History: 2 in 2004.* **Sources: Se24, So25.**

Tortarello Abruozzese - Cuke/Melon type from Abruzzo, 10-11 in. long, light green, ribbed, fresh eating. *Source History: 1 in 2004.* **Sources: Se24.**

Tortarello Barese - 60 days - Long, corrugated green fruit covered with soft hairs, typical of Bari region. *Source History: 1 in 1987; 1 in 1991; 1 in 1994; 1 in 1998; 2 in 2004.* **Sources: Ber, So25.**

Vedrantais - (Verdrantais) - 92 days - Charentais type, smooth and round, light gray-green, slight ribbing, very sweet firm dark-orange flesh, 1-2 lb., will not slip so grow on trellis or in greenhouse, FW tol., Europe. *Source History: 1 in 1981; 1 in 1987; 2 in 1991; 2 in 1994; 3 in 1998; 5 in 2004.* **Sources: Bu2, Hi13, Pe2, So12, Tu6.**

Vine Peach - (Mango Melon, Vegetable Peach) - 80-90 days - *Cucumis melo*, (Chito Group), native American annual, peach size and color, flavor and texture much like mango, easy to grow, very productive, vigorous spreading vines may be staked, ready when stem slips freely, peel and cut in half and remove seeds, best when cooked, for canning and preserves and pies and pickling whole (pick while still green for pickling). *Source History: 8 in 1981; 7 in 1984; 6 in 1987; 5 in 1991; 7 in 1994; 6 in 1998; 3 in 2004.* **Sources: Bu1, La1, Shu.**

White Striped Snake - Resembles Armenian but with white stripes along the sides, very rare. *Source History: 1 in 1998; 1 in 2004.* **Sources: Ho13.**

Varieties dropped since 1981 (and the year dropped):

Ao-Shima (1987), *Bananalope* (1991), *Cantalun* (1998), *Cavaillon* (1998), *Charentais Improved* (1991), *Charentais: Ido* (1983), *China-Schlangen* (1991), *Chun Gwa* (1994), *Donkey Melon* (1994), *Dudaim Fragrant* (1994), *Early Silver Line* (1998), *Giallo de Napoli* (1998), *Ginsen* (1991), *Golden Pear Giant* (1991), *Hami Sweet* (1998), *Hopi Melon* (1991), *Ito Uri* (1982), *Jade* (1987), *Kanro* (1987), *Katsura* (2004),

Kuromon (1987), *Kyoto Two Feet* (1991), *Lemon Sweet* (1991), *Mayo* (1991), *Nara No. 1* (1991), *Narita Green Melon* (1994), *Numame Early* (1987), *Oriental Pear* (1991), *Pickling Melon, Tokyo Early White* (1998), *Printadou* (2004), *Santo Domingo Orange Meated* (1991), *Sweet Gold* (1983), *Tardif Espagnol* (2004), *Tiger* (2004), *Tokyo Early* (1987), *Tokyo Giant* (1987), *Turner Cantaloupe* (1991), *Verte de Treste* (1991), *Yuki Large* (1998).

Mushroom
Various fungi

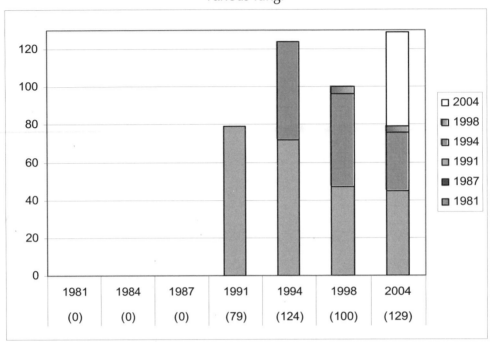

Number of varieties listed in 1991 and still available in 2004: 79 to 45 (57%).
Overall change in number of varieties from 1991 to 2004: 79 to 129 (163%).
Number of 2004 varieties available from 1 or 2 sources: 115 out of 129 (89%).

Agaricus abruptibulbus - *Source History: 1 in 1994; 1 in 1998; 1 in 2004.* **Sources: Fl3.**

Agaricus arvensis - *Source History: 1 in 1994; 1 in 1998; 1 in 2004.* **Sources: Fl3.**

Agaricus bernardii - *Source History: 1 in 2004.* **Sources: Fl3.**

Agaricus bisporus - *Source History: 1 in 1994; 1 in 1998; 1 in 2004.* **Sources: Fl3.**

Agaricus bitorquis (rodmani) - *Source History: 1 in 1998; 1 in 2004.* **Sources: Fl3.**

Agaricus campestris - *Source History: 1 in 1994; 1 in 1998; 1 in 2004.* **Sources: Fl3.**

Agaricus silvaticus - *Source History: 1 in 1994; 1 in 1998; 1 in 2004.* **Sources: Fl3.**

Almond - (Almond Agaricus) - *Agaricus subrufescens*, long forgotten cousin to the commercially grown Champignon, delightful almond scent, at one time cultivated commercially on a large scale in greenhouses. *Source History: 2 in 1991; 3 in 1994; 2 in 1998; 3 in 2004.* **Sources: Fi2, Ra3, We2.**

Angel's Wings - *Pleurocybella porrigens. Source History: 1 in 2004.* **Sources: Fl3.**

Armillariella mellea - (Honey Mush) - *Source History: 1 in 1994; 1 in 1998; 1 in 2004.* **Sources: Fl3.**

Auricularia auricula - *Source History: 1 in 1994; 1 in 1998; 1 in 2004.* **Sources: Fl3.**

Boletus bicolor - *Source History: 1 in 1994; 1 in 1998; 1 in 2004.* **Sources: Fl3.**

Boletus chrysenteron - *Source History: 1 in 1994; 1 in 1998; 1 in 2004.* **Sources: Fl3.**

Boletus edulis - *Source History: 1 in 1994; 1 in 1998; 1 in 2004.* **Sources: Fl3.**

Boletus frostii - *Source History: 1 in 2004.* **Sources: Fl3.**

Boletus mirabilis - *Source History: 1 in 1994; 1 in 1998; 1 in 2004.* **Sources: Fl3.**

Boletus pinophilus - *Source History: 1 in 2004.* **Sources: Fl3.**

Button - (Champignon) - *Agaricus brunnescens. Source History: 5 in 1991; 4 in 1994; 3 in 1998; 5 in 2004.* **Sources: Mus, Par, Se24, Ver, We2.**

Calvatia cyathiformis - *Source History: 1 in 2004.* **Sources: Fl3.**

Calvatia gigantia - *Source History: 1 in 2004.* **Sources: Fl3.**

Cantharellus cibarius - *Source History: 1 in 1994; 1 in 1998; 1 in 2004.* **Sources: Fl3.**

Cantharellus cinnabarinus - *Source History: 1 in*

1994; 1 in 1998; 1 in 2004. **Sources: Fl3.**

Cantharellus lateritius - *Source History: 1 in 1994; 1 in 1998; 1 in 2004.* **Sources: Fl3.**

Cantharellus minor - *Source History: 1 in 1994; 1 in 1998; 1 in 2004.* **Sources: Fl3.**

Cantharellus subalbidus - *Source History: 1 in 1994; 1 in 1998; 1 in 2004.* **Sources: Fl3.**

Cantharellus tubaeformis - *Source History: 1 in 1994; 1 in 1998; 1 in 2004.* **Sources: Fl3.**

Cantharellus umbonatus - *Source History: 1 in 2004.* **Sources: Fl3.**

Champignon - *Agaricus bisporus*, small to medium-large mushroom with smooth chocolate-brown cap, pure-white stem, solid meaty mushroom with firm white flesh, excel. flavor both raw and cooked, grown on straw based compost, most widely cultivated mushroom in the world and a mainstay in French cooking, well adapted to indoor and outdoor "natural culture" cultivation. *Source History: 1 in 1991; 1 in 1994; 1 in 1998; 2 in 2004.* **Sources: Mcf, Ra3.**

Chicken of the Woods - (Sulfer Shelf) - *Polyporus sulphuraceus*, inhabits a wide variety of hardwoods and softwoods throughout North America, very vigorous strain, flavor of white chicken meat, considered a favorite edible among many mycophiles. *Source History: 4 in 1991; 4 in 1994; 1 in 1998; 3 in 2004.* **Sources: Fl3, Fun, Ra6.**

Clavicorona pyxidata - *Source History: 1 in 2004.* **Sources: Fl3.**

Clitocybe nuda - *Source History: 1 in 2004.* **Sources: Fl3.**

Clitocybe odora - *Source History: 1 in 2004.* **Sources: Fl3.**

Collybia dryophila - *Source History: 1 in 2004.* **Sources: Fl3.**

Conifer Coral - *Hericium abietis*, grows exclusively on conifers (not pines) and is native to western North America, beautiful mushroom with cascading white icicle-like spines, delicate and pleasant flavor, stumps or logs buried in sawdust can be inoculated. *Source History: 1 in 1991; 1 in 2004.* **Sources: Fun.**

Conocybe lactea - *Source History: 1 in 1994; 1 in 1998; 1 in 2004.* **Sources: Fl3.**

Corprinus atramentarius - *Source History: 1 in 2004.* **Sources: Fl3.**

Corprinus micaceus - *Source History: 1 in 2004.* **Sources: Fl3.**

Corprinus radians - *Source History: 1 in 2004.* **Sources: Fl3.**

Cortinarius alboviolaceus - *Source History: 1 in 2004.* **Sources: Fl3.**

Craterellus fallax - *Source History: 1 in 1994; 1 in 1998; 1 in 2004.* **Sources: Fl3.**

Enoki - (Enokitake, Velvet Stem Enoki, Winter Mushroom, Golden Needle, Snow Puff) - *Flammulina velutipes*, small yellowish-orange cap, tough velvety brown stem, commonly grows in large families that have dozens of mushrooms clustered together, found on dead conifers and their stumps in the Pacific Northwest, grows at low temps, delicate flavor, pleasantly chewy texture, eaten raw in salads or added to soups and stir fry in the last minute of cooking. *Source History: 4 in 1991; 3 in 1994; 2 in 1998; 3 in 2004.* **Sources: Fun, Ra3, We2.**

Flammulina velutipes - *Source History: 1 in 1994; 1 in 1998; 1 in 2004.* **Sources: Fl3.**

Garden Giant - (Giant Stropharia, King Stropharia) - *Stropharia rugoso-annulata*, large mushroom with thick white stem and wine-red cap, can weigh as much as 7 lbs., ideal for growing in the shade of garden vegetables. *Source History: 1 in 1991; 2 in 1994; 1 in 1998; 1 in 2004.* **Sources: Ra3.**

Gomphus clavatus - *Source History: 1 in 1994; 1 in 1998; 1 in 2004.* **Sources: Fl3.**

Hen of the Woods - (Maitake) - *Grifola frondosa*, popular woodland mushroom prefers hardwood stumps, available as grain and sawdust spawn. *Source History: 1 in 1991; 2 in 1994; 1 in 1998; 2 in 2004.* **Sources: Fl3, Fun.**

King Pleurotus - *Pleurotus ergyngii*, medium to large mushroom with reddish-brown to gray-brown cap and central off-white stem, native of the Southern European steppes and sub-tropical areas of North Africa, in nature it grows on roots of umbelliferous plants (carrot or parsnip family), in Europe it is cultivated on chopped straw, robust fleshy mushroom, sweet flavor, meaty texture. *Source History: 2 in 1991; 2 in 1994; 1 in 1998; 1 in 2004.* **Sources: Ra3.**

Laccaria amethystina - *Source History: 1 in 2004.* **Sources: Fl3.**

Laccaria laccata - *Source History: 1 in 2004.* **Sources: Fl3.**

Laccaria ochropurpurea - *Source History: 1 in 2004.* **Sources: Fl3.**

Laccaria trallisata - *Source History: 1 in 2004.* **Sources: Fl3.**

Lactarius hygrophoroides - *Source History: 1 in 2004.* **Sources: Fl3.**

Lactarius paradoxus - *Source History: 1 in 2004.* **Sources: Fl3.**

Leccinum aurantiacum - *Source History: 1 in 2004.* **Sources: Fl3.**

Leccinum scabrum - *Source History: 1 in 2004.* **Sources: Fl3.**

Lepiota americana - *Source History: 1 in 1994; 1 in 1998; 1 in 2004.* **Sources: Fl3.**

Lepiota naucina - *Source History: 1 in 2004.* **Sources: Fl3.**

Lepiota procera - *Source History: 1 in 2004.* **Sources: Fl3.**

Lion's Mane - *Hericium erinaceus*, sister species to *Hericium abietis*, differs in its preference for hardwoods, this species particularly loves oaks and maples, stumps or partially buried logs are recommended sites for its cultivation. *Source History: 1 in 1991; 1 in 1994; 2 in 1998; 2 in 2004.* **Sources: Fun, Mus.**

Maitake - (Hen of the Woods) - *Grifola frondosa*, large polypore composed of multiple and clustered grey-brown caps with a branched and compound stem, in nature it grows on oak and other hardwood stumps, delicious, highly prized in the Orient, growing methods are similar to those used for

Shiitake since it is a wood decomposer, believed to contain medicinal properties. *Source History: 3 in 1991; 2 in 1994; 3 in 1998; 4 in 2004.* **Sources: Fi2, Mus, Ra3, We2.**

<u>Marasmius oreades</u> - *Source History: 1 in 2004.* **Sources: Fl3.**

<u>Matsutake</u> - (Pine Mushroom) - *Armillaria ponderosa*, this mushroom is mycorrhizal which means it grows in symbiotic association with the living roots of specific green plants, in this case, Pine, Douglas Fir and Hemlock trees. *Source History: 1 in 1991; 1 in 1994; 1 in 1998; 1 in 2004.* **Sources: We2.**

<u>Morchella elata</u> - *Source History: 1 in 2004.* **Sources: Fl3.**

<u>Morel</u> - *Morchella esculenta. Source History: 4 in 1991; 5 in 1994; 4 in 1998; 6 in 2004.* **Sources: Fi2, Fl3, Fun, Mus, Ra3, We2.**

<u>Morel, Black</u> - *Morchella angusticeps*, brown conical cap that is composed of ridges and pits, creamy white stem, grows primarily in soil and often in areas where the soil has been disturbed, also commonly found in old apple orchards and in forests the year following a forest fire. *Source History: 4 in 1991; 3 in 1994; 1 in 1998; 1 in 2004.* **Sources: Fun.**

<u>Naematoloma sublateritium</u> - *Source History: 1 in 2004.* **Sources: Fl3.**

<u>Nameko</u> - *Pholiota nameko*, flavor resembles cashews, chestnut-brown, slimy caps. *Source History: 3 in 1991; 2 in 1994; 1 in 1998; 2 in 2004.* **Sources: Fun, We2.**

<u>Oyster</u> - *Pleurotus ostreatus*, easily grown mushroom thrives on hardwoods throughout North America, white to pale gray or brown color, strong, delicious flavor. *Source History: 5 in 1991; 5 in 1994; 4 in 1998; 5 in 2004.* **Sources: Fl3, Fun, Ra6, Rai, Se24.**

<u>Oyster, Blue Capped Strain</u> - *Pleurotus columbinus. Source History: 1 in 1991; 1 in 1994; 2 in 1998; 1 in 2004.* **Sources: We2.**

<u>Oyster, Blue Dolphin</u> - *Pleurotus ostreatus*, very pretty strain, requires a cool temperature to develop pewter color and produce high yields, good choice for cool fruiting rooms. *Source History: 1 in 2004.* **Sources: Fi2.**

<u>Oyster, Florida Strain</u> - *Pleurotus sapidus. Source History: 1 in 1991; 3 in 1994; 1 in 1998; 1 in 2004.* **Sources: Fl3.**

<u>Oyster, Golden</u> - *Pleurotus cornucopiae*, small to medium-large mushroom, white stem, lemon-yellow cap, commonly grows in tight floral clusters from a bulbous base, has slightly thinner flesh then the other *pleurotus* and is therefore more delicate, very aromatic. *Source History: 2 in 1991; 2 in 1994; 3 in 1998; 3 in 2004.* **Sources: Fi2, Ra3, We2.**

<u>Oyster, Grey Dove</u> - *Pleurotus ostreatus*, clusters of silver colored mushrooms, produces well on straw, sawdust and natural log cultivation. *Source History: 1 in 2004.* **Sources: Fi2.**

<u>Oyster, Italian</u> - *Pleurotus pulmonarius*, gorgeous clusters of thick stemmed, sturdy mushrooms, mature to a parchment color, mild flavor, vigorous producer. *Source History: 1 in 2004.* **Sources: Fi2.**

<u>Oyster, King</u> - *Pleurotus eryngii*, stout, sturdy mushroom with an earthy flavor and almost crunchy texture, the cap resembles brushed suede, lower yields than other Oyster mushrooms, good shipper, longer shelf life. *Source History: 1 in 2004.* **Sources: Fi2.**

<u>Oyster, Phoenix</u> - *Pleurotus sajor-caju*, medium to large mushroom with white stem and dark brown cap which becomes lighter brown with maturity, solid and fleshy when young, originally found in India growing on succulents in the Himalayan foothills, rapid growing cycle, harvest mushrooms 14-20 days after spawning. *Source History: 1 in 1991; 1 in 1994; 1 in 1998; 3 in 2004.* **Sources: Fun, Mus, Ra3.**

<u>Oyster, Pink</u> - *Pleurotus djamor*, brilliant pink when young, tart when cooked, for summer culture only, cannot tolerate cool temps. *Source History: 1 in 1998; 2 in 2004.* **Sources: Fi2, Fun.**

<u>Oyster, Strawberry</u> - *Pleurotus flabellatus*, bright pink mushrooms. *Source History: 1 in 1991; 1 in 1994; 1 in 1998; 1 in 2004.* **Sources: We2.**

<u>Oyster, Tree</u> - *Pleurotus ostreatus*, dark-violet colored oyster mushrooms in early spring and again in late fall, grown primarily outdoors on hardwood bolts (usually larger aspen logs) cut into one foot lengths. *Source History: 3 in 1991; 3 in 1994; 5 in 1998; 5 in 2004.* **Sources: Fi2, Fun, Mus, Ra3, We2.**

<u>Oyster, White</u> - *Pleurotus ostreatus*, white trumpet shaped caps, look more fragile than other strains. *Source History: 1 in 1994; 1 in 1998; 1 in 2004.* **Sources: Dom.**

<u>Oyster, Yellow</u> - (Yellow Trumpet, Golden Mushroom) - *Pleurotus cornucopiae. Source History: 2 in 1991; 1 in 1994; 1 in 2004.* **Sources: Se24.**

<u>Paddy Straw</u> - (Fukurotake) - *Volvariella volvacea*, grey-brown cap occasionally streaked with black fibrils, off-white to dull-brown stem, widely grown in the tropical areas of SE Asia on waste rice straw, hence the name, needs high temperatures and humidities to grow well, harvested in the immature "egg" stage and must be eaten or marketed within days or they will spoil, very pleasant texture and flavor, Western Biologicals (We2) offers a Phillipines strain and a Laotian strain. *Source History: 3 in 1991; 3 in 1994; 2 in 1998; 2 in 2004.* **Sources: Ra3, We2.**

<u>Panellus serotinus</u> - *Source History: 1 in 2004.* **Sources: Fl3.**

<u>Pholiota aegerita</u> - Known in Italy as Funghi Pioppino, pioppe translates poplar, usually found under poplar trees, hence the name, sow onto straw or logs, prefers poplar but any hardwood will work. *Source History: 1 in 2004.* **Sources: Se24.**

<u>Phylloporus rhodoxanthus</u> - *Source History: 1 in 1994; 1 in 1998; 1 in 2004.* **Sources: Fl3.**

<u>Pluteus cervinus</u> - *Source History: 1 in 2004.* **Sources: Fl3.**

<u>Pom Pom</u> - (Pom Pom Blanc, Monkey Head, Lion's Mane, Beard Tooth) - *Hericium erinaceus*, grown indoors under cool, humid conditions of 60 degrees F or less with plenty of fresh air, the aromatic, slightly sweet mushrooms grow in frilly white clusters, valued by chefs for retaining

their white color and ability to stay crisp after sauteing. *Source History: 2 in 1991; 3 in 1994; 2 in 1998; 2 in 2004.* **Sources: Fi2, We2.**

Portobello - (Cremini, Brown Button) - *Agaricus bisporus*, more flavorful than the white variety, cultivation similar to Brown Button. *Source History: 1 in 1998; 3 in 2004.* **Sources: Fl3, Mus, We2.**

Prince - (The Prince) - *Agaricus augustus*. *Source History: 3 in 1991; 3 in 1994; 1 in 1998; 1 in 2004.* **Sources: Fl3.**

Psathyrella canodolleana - *Source History: 1 in 2004.* **Sources: Fl3.**

Reishi - (Ling Zhi Mushroom, Mannentake) - *Ganoderma lucidum*, very aggressive species, quickly assaults both conifer and hardwood stumps, produces a conk-like mushroom similar to the famed Artist Conk but with a smooth polished surface, deep reddish brown color on top and white pored underlayer, used in teas to impart a subdued non-narcotic peaceful feeling, heralded by Chinese and Japanese for its health stimulating properties, produces bonsai-like formations as it matures. *Source History: 4 in 1991; 4 in 1994; 4 in 1998; 4 in 2004.* **Sources: Fun, Mus, Ra3, We2.**

Reishi, Red Dragon - *Ganoderma lucidum*, red strain, reliable fruiter. *Source History: 1 in 2004.* **Sources: Fi2.**

Russula aeruginea - *Source History: 1 in 1994; 1 in 1998; 1 in 2004.* **Sources: Fl3.**

Russula brevipes - *Source History: 1 in 2004.* **Sources: Fl3.**

Russula crustosa - *Source History: 1 in 1994; 1 in 1998; 1 in 2004.* **Sources: Fl3.**

Russula lutea - *Source History: 1 in 2004.* **Sources: Fl3.**

Russula mariae - *Source History: 1 in 1994; 1 in 1998; 1 in 2004.* **Sources: Fl3.**

Russula virescens - *Source History: 1 in 2004.* **Sources: Fl3.**

Russula xerampelina - *Source History: 1 in 1994; 1 in 1998; 1 in 2004.* **Sources: Fl3.**

Shaggy Mane - *Corprinus comatus*, medium to large mushroom with tall white cap and white stem, name comes from the shaggy tufts or scales adorning the cap, pleasing delicate flavor and texture. *Source History: 5 in 1991; 4 in 1994; 2 in 1998; 3 in 2004.* **Sources: Fun, Ra3, We2.**

Shiitake - (Donku, Shiangku, Black Forest, Chinese Mushroom) - *Lentinula edodes*, small to medium size mushroom with light-brown fuzzy stem and a pale to dark reddish brown cap, dense solid flesh, very strong pleasant flavor, grown on oak or chestnut or alder logs, high on the list of gourmet mushrooms, has proven medicinal properties. *Source History: 8 in 1991; 6 in 1994; 6 in 1998; 7 in 2004.* **Sources: Fl3, Fun, Ha14, Mus, Ra6, Rai, We2.**

Shiitake, 26 - Producees very high quality mushrooms in 16-20 weeks, benefits from a longer spawn run time, most consistent of strains on second flush, cure in and out of the bag, for use in composite substrates indoors. *Source History: 1 in 2004.* **Sources: Fi2.**

Shiitake, 730 - Large mushrooms produced in 16 weeks, requires careful management during incubation, cures well once removed from the bag, for use in composite substrates indoors. *Source History: 1 in 2004.* **Sources: Fi2.**

Shiitake, 77A - Warm weather strain. *Source History: 1 in 1991; 1 in 1994; 1 in 1998; 1 in 2004.* **Sources: Ra3.**

Shiitake, Bolshoi Breeze - Produces when spring temperatures consistently reach 50 degrees F and again in the fall when temps fall to 50 degrees F, easy harvest, fruits indoors in winter when placed in a cool room after a cold water soak. *Source History: 1 in 2004.* **Sources: Fi2.**

Shiitake, CS- 11 - Cold-weather strain, robust, thick fleshed mushrooms with heavy white fringe, short stout stems, top quality. *Source History: 1 in 1991; 1 in 1994; 1 in 1998; 1 in 2004.* **Sources: No12.**

Shiitake, CS- 15 - All-season rapid fruiting strain, thick caps with heavy white fringe, thick stems in good proportion to cap, mushrooms often produced in clumps. *Source History: 1 in 1991; 1 in 1994; 1 in 1998; 1 in 2004.* **Sources: No12.**

Shiitake, CS- 16 - Cool-cold weather strain, robust dense thick fleshed caps with heavy white fringe, high quality mushroom. *Source History: 1 in 1991; 1 in 1994; 1 in 1998; 1 in 2004.* **Sources: No12.**

Shiitake, CS- 24 - Warm-weather rapid fruiting strain, robust, thick-fleshed caps with a thick white fringe, top quality, stems are thick and well proportioned. *Source History: 1 in 1991; 1 in 1994; 1 in 1998; 1 in 2004.* **Sources: No12.**

Shiitake, CS- 41 - All-season rapid fruiting strain, med-thick to thick caps with minimal white fringe. *Source History: 1 in 1991; 1 in 1994; 1 in 1998; 1 in 2004.* **Sources: No12.**

Shiitake, CS-118 - Cool-weather rapid fruiting strain, medium thick-fleshed cap with dense white fringe, retains incurved margin longer than other strains, thin stems, densely fringed and well proportioned. *Source History: 1 in 1991; 1 in 1994; 1 in 1998; 1 in 2004.* **Sources: No12.**

Shiitake, CS-125 - Warm-weather rapid fruiting strain, medium to thick fleshed mushrooms with moderate white fringe, produced at high temps, heavy fruitings with small to medium mushrooms. *Source History: 1 in 1991; 1 in 1994; 1 in 1998; 1 in 2004.* **Sources: No12.**

Shiitake, CW 25 - Produces a medium-thick fleshed mushroom, smooth cap with lacy ornamentation in its rim, very popular strain among indoor growers. *Source History: 1 in 1991; 1 in 1994; 1 in 1998; 1 in 2004.* **Sources: Fi2.**

Shiitake, East Wind - Productive during the warm summer months, uniform size caps, Japanese strain. *Source History: 1 in 2004.* **Sources: Fi2.**

Shiitake, Felt Hat - More fleshy but less dense than other warm weather strains, used both in and outdoors, pick as soon as the cap separates from the stem for best quality. *Source History: 1 in 1994; 1 in 1998; 1 in 2004.* **Sources: Fi2.**

Shiitake, JC Star - Ideal summer fruiting strain that will fruit naturally when others are dormant, vigorous growth of medium to large size mushrooms, very responsive to temperature changes. *Source History: 1 in 2004.* **Sources:**

Shiitake, Night Velvet - Produces heavily on logs and blocks, umbrella shape, strong production during hot summer months, especially in the South. *Source History: 1 in 2004.* **Sources: Fi2.**

Shiitake, QR - Nice quality mushrooms in 12-16 weeks, will benefit from a longer spawn run time, cures in and out of the bag, fast spawn run, for use in composite substrates indoors. *Source History: 1 in 2004.* **Sources: Fi2.**

Shiitake, Snowcap - Produces a thick-fleshed uniform mushroom with abundant veil remnant tufts, long natural fruiting season with the heaviest production during spring and fall, use this strain on your largest logs as it does not respond well to force fruiting via soaking, primarily an outdoor strain. *Source History: 1 in 1991; 1 in 1994; 1 in 1998; 1 in 2004.* **Sources: Fi2.**

Shiitake, South Wind - Ornate mushrooms with fast spawn run, very sensitive to fruiting stimuli, whether it be temperature change or log movement, best strain to use in rain rooms to initiate fruiting, holds up well to July and August heat under protected conditions. *Source History: 1 in 1994; 1 in 2004.* **Sources: Fi2.**

Shiitake, West Wind - Wide range strain with very fast spawn run, consistent high cumulative yield due to long fruiting period, produces good quality mushrooms for the life of the log under both warm and cool conditions, recommended for beginners and commercial growers alike. *Source History: 1 in 1991; 1 in 1994; 1 in 2004.* **Sources: Fi2.**

Shiitake, WR 46 - Fast spawn run and relative ease of fruiting characterize this strain, ideal choice for indoor and outdoor cultivation, first flushes are high yielding and produce a lovely mushroom with cool temperatures. *Source History: 1 in 1991; 1 in 1994; 1 in 2004.* **Sources: Fi2.**

Shiitake, WW 44 - Produces a medium-thick fleshed mushroom with a high cap to stem ratio, lacy ornamented cap skirt and consistently high quality production make this a popular warm weather strain among indoor year round growers. *Source History: 1 in 1991; 1 in 1994; 1 in 1998; 1 in 2004.* **Sources: Fi2.**

Shimeiji - (Japanese Oyster Mushroom) - *Lyophyllum descastes*, firm, tender flesh, excel. flavor. *Source History: 1 in 1991; 1 in 1994; 1 in 1998; 1 in 2004.* **Sources: We2.**

Stropharia rugoso-annulata - *Source History: 2 in 1991; 2 in 1994; 2 in 1998; 2 in 2004.* **Sources: Fl3, We2.**

Stropharia, The King - (The Garden Giant) - *Stropharia rugoso-annulata*, huge burgundy colored mushroom, not recommended for areas where pine, cedar or other aromatic trees are growing. *Source History: 1 in 1991; 1 in 1994; 2 in 1998; 1 in 2004.* **Sources: Fun.**

Stropharia, Wine-Red - *Stropharia rugoso-annulata*. *Source History: 2 in 1991; 1 in 1994; 1 in 1998; 2 in 2004.* **Sources: Fi2, Rai.**

Tricholoma flavovirens - *Source History: 1 in 2004.* **Sources: Fl3.**

Tricholoma portentosum - *Source History: 1 in 2004.* **Sources: Fl3.**

Volvariella bombycina - *Source History: 1 in 2004.* **Sources: Fl3.**

Varieties dropped since 1981 (and the year dropped):

Agaricus subrufescens (2004), *Agrocybe agaerita* (2004), *Amanita muscaria* (1994), *Blewit* (2004), *Boletus ornatipes* (2004), *Boletus zelleri* (2004), *Brick Cap* (1994), *Brown Button* (1998), *Button, Light Cream* (1998), *Button, Off-White* (1998), *Button, White* (1998), *Cauliflower Mushroom* (1998), *Coral Hericium* (1998), *Elm Oyster* (2004), *Fairy Ring* (1994), *Kuehneromyces mutabilis* (1998), *Kuritake* (1998), *Old Fashion Dark Brown* (1998), *Oyster, Abalone* (1998), *Oyster, Bikini* (1998), *Oyster, Brown* (2004), *Oyster, Brown Capped Strain* (1998), *Oyster, Cool Weather Fruiting* (2004), *Oyster, Grey* (2004), *Oyster, Grey-Brown* (2004), *Oyster, Grey-Purple* (2004), *Oyster, Indian* (2004), *Oyster, Low Temperature Strain* (1998), *Oyster, Native B.C. Strain* (1998), *Oyster, Regal* (1998), *Oyster, Salmon* (2004), *Oyster, Sporeless Strain* (1998), *Oyster, Strain #21* (2004), *Oyster, Strain #23* (2004), *Oyster, Warm Weather Fruiting* (2004), *Pleurotus citrino-pileatus* (2004), *Pleurotus cystidiosus* (1994), *Pleurotus pulmonarius* (1998), *Rhizopus oligosporus* (1998), *Shiitake, 77B* (2004), *Shiitake, CW 12* (2004), *Shiitake, Flying Goose* (1998), *Shiitake, Jupiter* (2004), *Shiitake, Mars* (2004), *Shiitake, Mercury* (2004), *Shiitake, Neptune* (2004), *Shiitake, Pluto* (2004), *Shiitake, Saturn* (2004), *Shiitake, Twice Flowering* (1994), *Shiitake, V-3* (1998), *Shiitake, Venus* (2004), *Shiitake, WR 85* (1998), *Shiitake, WW 70* (1998), *Vita-Flake* (1998), *Wood Ear* (1998)

Mustard Greens
Brassica juncea

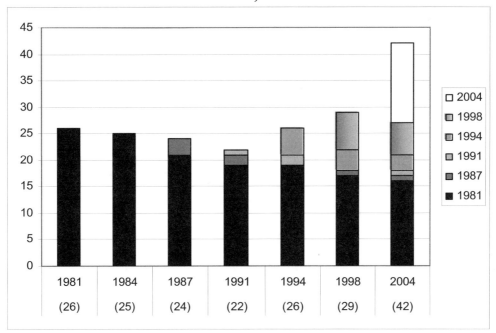

Number of varieties listed in 1981 and still available in 2004: 26 to 16 (62%).
Overall change in number of varieties from 1981 to 2004: 26 to 42 (162%).
Number of 2004 varieties available from 1 or 2 sources: 31 out of 42 (74%).

Aka Takana - India mustard, large dark purplish-red pungent leaves with wide white ribs, fast grower. *Source History: 1 in 1981; 1 in 1984; 1 in 1987; 1 in 1991; 1 in 1994; 1 in 1998; 1 in 2004.* **Sources: Kit.**

Ao Takana - India mustard, coarsely savoyed, large bright-green pungent leaves, winter hardy, slow bolter. *Source History: 1 in 1981; 1 in 1984; 1 in 1987; 1 in 1991; 1 in 1994; 1 in 1998; 1 in 2004.* **Sources: Kit.**

Black Winter - 45 days - Small seeded mustard with med. size, dark green leaves, dark brown-bronze tinge on outer leaf edges, mildly hot raw leaves with sweet quality, cooked leaves are darker green, spring/fall planting, pre-1890 Virginia heirloom from the Alford family in Morrison, TN. *Source History: 1 in 2004.* **Sources: Ga21.**

Broad Leaved - Chinese strain with huge plants to 3 ft. across, wide, flat leaves, med. green, hot flavor in warm weather which gets milder in the fall. *Source History: 1 in 1998; 1 in 2004.* **Sources: Ho13.**

Cherokee Blue - Broad, flat, blue-purple leaves with white ribs, plant grows 2 ft. across, mild flavor. *Source History: 1 in 1998; 1 in 2004.* **Sources: Ho13.**

China Takana - Vigorous semi-heading type with bright green frilled leaves, mustard pungency, good raw or steamed. *Source History: 1 in 2004.* **Sources: Ho13.**

Chinese - Large leaves with sharp, peppery flavor which mellows after cooking, does best in cool weather, heirloom from Thailand. *Source History: 2 in 2004.* **Sources: Ba8, So25.**

Chinese Nan-Fong - Looseleaf plant with large savoy leaves, deep purple-red with white midrib, nice mustard pungency, good for cooking and pickling. *Source History: 1 in 1998; 1 in 2004.* **Sources: Ev2.**

Florida Broad Leaf - (Large Smooth Leaf) - 40-60 days - Large semi-upright 16-22 in. spreading leafy plant, round/oval serrated dark-green leaves, 8-10 in. long x 8 in. wide, tender cream-colored midribs, grows vigorously, popular in the South. *Source History: 63 in 1981; 56 in 1984; 59 in 1987; 57 in 1991; 53 in 1994; 54 in 1998; 46 in 2004.* **Sources: Ada, Bu2, Bu3, Bu8, CA25, CH14, Cl3, CO23, CO30, Co31, Com, Cr1, De6, Dgs, DOR, Ers, HA3, He17, Hi13, HO2, Hud, Jor, Kil, La1, LO8, Loc, Ma19, ME9, Mey, Mo13, OLD, Or10, PAG, RIS, Ros, RUP, Sau, Se26, Sh9, SOU, Sto, Twi, Vi4, Wet, Wi2, WI23.**

Florida Giant - *Source History: 1 in 1991; 1 in 2004.* **Sources: WE10.**

Fordhook Fancy - (Fordhook, Burpee's Fordhook Fancy) - 35-50 days - Fringed deeply curled dark-green leaves curl backwards, 15-18 in. spread, slow to bolt, mild flavor, ostrich plume type. *Source History: 9 in 1981; 6 in 1984; 7 in 1987; 6 in 1991; 5 in 1994; 3 in 1998; 1 in 2004.* **Sources: Pe6.**

Gai Choi - (Chinese Mustard Cabbage, India Mustard, Kai Choi, Nam Fong Loose Leaf) - 35-60 days - Chinese mustard spinach or India mustard, 6-8 in. tall, mild clear mustard flavor but not pungent, easy to grow, spring or fall, much lighter than Dai Gai Choi. *Source History: 4 in 1981; 4 in 1984; 3 in 1987; 4 in 1991; 2 in 1994; 2 in 1998; 2 in 2004.* **Sources: Kit, Loc.**

Giant Curled - 50 days - Deep-green leaves, deeply crimped and frilled at edges, upright slightly spreading plants, vigorous grower, mild, hardy, for the South. *Source History: 2 in 1981; 2 in 1984; 1 in 1987; 3 in 1991; 2 in 1994; 1 in 1998; 1 in 2004.* **Sources: But.**

Giant Red - (Giant Japanese, Miike Purple) - 40-65 days - *B. juncea* var. *rugosa*, Japanese mustard, large thick tender deep purpish-red savoyed leaves with white midribs, strong mustard flavor, good pickled, very slow to bolt. *Source History: 2 in 1981; 2 in 1984; 2 in 1987; 3 in 1991; 10 in 1994; 16 in 1998; 40 in 2004.* **Sources: Ba8, Bo17, Bou, CO23, Com, Dam, Gou, Gr28, He17, Jor, Ki4, Kit, La1, ME9, Mo13, Na6, Orn, Pe2, Pe6, Pin, Pr3, Rev, Ri12, Roh, Sa9, Se7, SE14, Se26, So1, St18, Sto, TE7, Ter, Tho, Tu6, Twi, Ver, WE10, We19, WI23.**

Giant Southern Curled - (Giant Southern Curled Long Standing) - 45-70 days - Old southern favorite, large wide long oval bright-green leaves, curled and fringed on the edges, slow to bolt, cold tol., pre-1880. *Source History: 5 in 1981; 5 in 1984; 5 in 1987; 4 in 1991; 3 in 1994; 3 in 1998; 7 in 2004.* **Sources: CO30, Co31, OLD, Ri2, So1, SOU, Vi4.**

Green in Snow - (Hsueh Li Hung) - 50 days - *B. juncea* var. *multiceps*, leaf mustard from northern China, 20 in. plants, about 20 main leaves, mild slight mustardy flavor, for winter greenhouse greens or sow early spring or late fall. *Source History: 7 in 1981; 7 in 1984; 6 in 1987; 9 in 1991; 5 in 1994; 4 in 1998; 2 in 2004.* **Sources: Kit, Ni1.**

Green Wave - (Green Wave Long Standing, Yellow/Green Curly, Chinese Green Wave) - 40-60 days - AAS/1957, large upright plant, Southern Curled type but more deeply frilled and finely cut leaves, spineless, darkest green of any curly mustard, spicy hot flavor, popular with commercial growers, slow to bolt, stands 2-4 weeks longer than others of its type. *Source History: 40 in 1981; 31 in 1984; 34 in 1987; 36 in 1991; 34 in 1994; 30 in 1998; 27 in 2004.* **Sources: CH14, Com, Dam, DOR, Fe5, Ge2, He8, He17, HO2, Jo1, Jor, La1, MAY, ME9, Mey, Mo13, Ni1, Pin, RIS, RUP, Se7, Sh9, Sh12, So9, So12, Ter, WI23.**

Horned Mustard - 40-60 days - Bright-green, indented, frilled leaves, enlarged, curved petioles, semi-heading, nice mustard pungency, bud shoots used for pickling by the Japanese. *Source History: 2 in 1994; 3 in 1998; 4 in 2004.* **Sources: Kit, ME9, Sh12, Ter.**

Hsueh Li Hung - (Snow Mustard) - 42 days - Grows about 20 in. tall, large dark-green leaves with narrow lobes, non-heading, mild, slightly mustard flavor, very hardy, often picked when the leaves are growing up through the snow, from northern China. *Source History: 1 in 1981; 1 in 1984; 1 in 1987; 1 in 1991; 1 in 1994; 2 in 1998; 1 in 2004.* **Sources: Ho13.**

India Mustard - 65 days - Mustard with semi-closed head, large leaves on broad thick stems, more mustardy taste but still mild, may be eaten as a vegetable or made into sour-salt mustard. *Source History: 1 in 1981; 1 in 1984; 1 in 1987; 1 in 1991; 3 in 1994; 3 in 1998; 1 in 2004.* **Sources: Tho.**

India Mustard, Red - Large, thick, tender, red-purple leaves, white midrib, hot peppery taste raw, milder if cooked, tolerates mild frosts, sow early spring or fall. *Source History: 1 in 1998; 1 in 2004.* **Sources: Ho13.**

Kekkyu Takana - 55 days - Chinese mustard, forms a small head wrapped with wide, hairless tender leaves, mild mustard pungency increases as it matures, good for pickling. *Source History: 1 in 2004.* **Sources: Kit.**

Leaf Heading Mustard - 43 days - Chinese leaf mustard, slightly whorled large light-green leaves, 20 in. wide by 12 in. high, sweet and tender with mild pungency, forms loose heads in cold fall weather, good for kimchee. *Source History: 1 in 1981; 1 in 1984; 1 in 1987; 1 in 1991; 1 in 1994; 2 in 1998; 1 in 2004.* **Sources: Ho13.**

Magma - Pretty, ruffled leaves are deep purple on top with green undersides, used as salad green, peppery flavor. *Source History: 2 in 1998; 3 in 2004.* **Sources: Ho13, Hud, Sh12.**

Miike Giant - 40 days - Late giant Japanese Tendergreen type, thick dark-green leaves, very crumpled and frilled, strong and cold res., slow to bolt, pungent, popular in the South. *Source History: 5 in 1981; 3 in 1984; 3 in 1987; 2 in 1991; 3 in 1994; 3 in 1998; 6 in 2004.* **Sources: Dam, Ho13, Kit, ME9, Orn, SE4.**

Miike Long Shoot - (Miike Long Shoot) - Reselection of Miike Giant, red veined, sweetly pungent leaves with broad, often undulant, midrib, young plants used in salads, long flower stem to 18 in., good in stir fry, best planted for fall harvest as cold weather enhances sweetness. *Source History: 1 in 1994; 1 in 1998; 1 in 2004.* **Sources: Sh12.**

Miike Purple - (Red Giant Indian Mustard, Giant Japanese) - A Japanese mustard for cutting at 4-6 in., deep purple, clear peppery taste, can also be grown out for boiling greens, ornamental. *Source History: 1 in 1987; 2 in 1991; 2 in 1994; 2 in 1998; 1 in 2004.* **Sources: Ev2.**

Oak Fire - Red and purple oak leaf types with a spicy touch of fire, used young for salads or later as cooked greens, more winter hardy, better disease res. than other red or purple hot mustard types, PSR Breeding Program. *Source History: 1 in 2004.* **Sources: Pe6.**

Old Fashion - (Old Fashion Ragged Edge Mustard, Hen Pecked) - 40-42 days - Plants to 2 ft., long ruffled leaves, no finer mustard for salad greens, bolts early but has the best flavor, fine quality accounts for its long lasting popularity, used widely in Virginia and the Carolinas. *Source History: 6 in 1981; 6 in 1984; 6 in 1987; 4 in 1991; 4 in 1994; 8 in 1998; 6 in 2004.* **Sources: Cl3, DOR, Ers, Ho13, SOU, Wet.**

Osaka Purple - 40-80 days - Plants grow 12-14 in. tall, mild, flavorful, tender green leaves tinged with purple-red, best flavor when young, pungent and sharp when mature, Japan. *Source History: 10 in 2004.* **Sources: Fo13, Kit, La1, ME9, Pe2, Roh, Se8, Ter, We19, Yu2.**

Peacock Tail - 40 days - Leafy mustard, harvested at 12 in. when tender and tasty, 18 in. tall when mature, stir fry and pickling, fast grower. *Source History: 2 in 1998; 1 in 2004.* **Sources: Ho13.**

Purple Wave - 70-80 days - Stabilized open-pollinated cultivar, Osaka Purple x Green Wave Mustard cross, light purple leaves edged with green, hot and spicy, developed by Alan Kapuler. *Source History: 1 in 2004.*

Sources: Se7.

Red Leaf - Med. green, slightly crumpled leaves, long stalks, mild flavor, Taiwan. *Source History: 2 in 2004.* **Sources: CO30, Ho13.**

San Ho Giant - Large Chinese str., light green leaves, wide white stalks, semi-heading in cool weather. *Source History: 1 in 2004.* **Sources: Ho13.**

Sawtooth - Teardrop shaped spineless leaves, very curly with extremely frilled edges, dark green upright plant, spicy flavor, babyleaf harvest in 21 days. *Source History: 3 in 2004.* **Sources: CO23, DOR, ME9.**

South African Spinach Mustard - Plants grow 12 in. tall, purple stems, small hairy green leaves with serrated edges, mild flavor, bolts in warm weather. *Source History: 1 in 2004.* **Sources: Ho13.**

Southern Giant Curled - (Southern Giant Curled Longstanding, Southern Curled) - 35-60 days - Large upright vigorous hardy plants, early curled type, spread of 18-24 in., large bright-green leaves with crumpled frilled edges, mild but mustardy flavor, cold res., slow to bolt, can be sown spring or fall, for home or market gardens. *Source History: 80 in 1981; 68 in 1984; 64 in 1987; 61 in 1991; 56 in 1994; 63 in 1998; 62 in 2004.* **Sources: ABB, Ada, Ag7, Ba8, Be4, Bo17, Bo19, Bou, Bu3, Bu8, CH14, Cl3, CO23, Com, Coo, Cr1, De6, DOR, Ers, Fi1, Fo7, HA3, Ha5, He8, He17, HO2, Jor, Jun, Kil, La1, LO8, Loc, Ma19, ME9, Mel, Mey, Mo13, Ont, Or10, Orn, PAG, Ri12, RIS, Roh, Ros, RUP, Sa9, Sau, Scf, SE4, Se26, Sh9, Shu, Sto, TE7, Twi, WE10, We19, Wet, Wi2, WI23, Yu2.**

Spicy Curls - Gorgeous array of spicy warm baby-leaf head mustards, curly frilled reds or purples, sel. for better than average winter hardiness and disease res., PSR Breeding Program. *Source History: 1 in 2004.* **Sources: Pe6.**

Spinach Mustard - 35 days - Large thick oblong leaves, mild mustard-spinach flavor, heat and drought res. *Source History: 1 in 1994; 1 in 2004.* **Sources: Ho13.**

Tsoisim - Vietnamese type. *Source History: 1 in 2004.* **Sources: Red.**

Yanagawa Takana - 40 days - Broadleaf mustard with bright green aromatic leaves with mildly pungent flavor, med. ribbed stems, can be used fresh in a salad or in soup or stirfry or pickled, all plant parts are edible. *Source History: 1 in 2004.* **Sources: Kit.**

Yukina Savoy - 21-45 days - Longer leaves than Tatsoi at the baby leaf stage, 10-12 in. tall, savoyed leaves, heat and cold tolerant. *Source History: 2 in 2004.* **Sources: Ho13, Jo1.**

Varieties dropped since 1981 (and the year dropped):

Chinese Kai Choi (1987), *Common Leaved Mustard* (2004), *Dai Gai Choi* (1998), *Giant Curled Chinese* (1998), *Hakarashina* (1991), *Kai Choi* (1987), *Mayo Quelite* (2004), *Mostaza Roja* (2004), *Slobolt* (1998), *Softleaf* (1987), *Swollen Stem* (1991), *Takana* (1987), *Wild, Tarahumara Espinaca* (2004), *Yellow/Green Curly* (1991), *Zero* (1983).

Okra
Abelmoschus esculentus

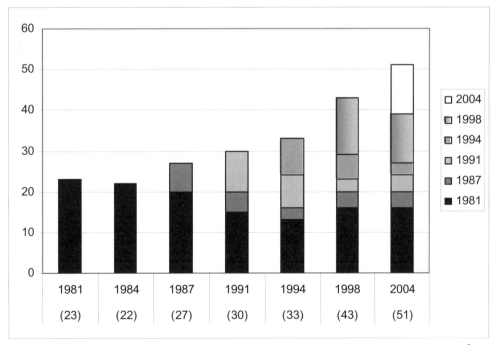

Number of varieties listed in 1981 and still available in 2004: 23 to 16 (70%).
Overall change in number of varieties from 1981 to 2004: 23 to 51 (222%).
Number of 2004 varieties available from 1 or 2 sources: 35 out of 51 (69%).

African - Tender large pods, production continues when days are short, from Africa. *Source History: 1 in 2004.* **Sources: Ech.**

African Red - Tall plants grow 6-8 ft., red stems and red tinted leaves, long red pods to 8 in. *Source History: 1 in 1998; 2 in 2004.* **Sources: Ho13, Se17.**

African Striped - Bushy spreading plant to 4 ft., med. size fruit tinged with red, long season, withstands drought. *Source History: 1 in 2004.* **Sources: Ho13.**

Alabama Red - 70 days - Fat red pods, great fried, heirloom from Alabama, rare. *Source History: 1 in 2004.* **Sources: Ba8.**

Alice Elliot - Plants to 4 ft., 6 in. green pods, does well in heat and drought, brought to Oklahoma from Missouri during the Land Rush of 1889, grown locally for over 100 years. *Source History: 1 in 1998; 1 in 2004.* **Sources: Ho13.**

Artist - 55 days - High yielding 4 ft. plants, pentagonal 3 in. purple-red pods turn dark green when boiled. *Source History: 1 in 1991; 2 in 1994; 1 in 2004.* **Sources: OLD.**

Beck's Gardenville Giant Pod - Texas heirloom from Malcomb Beck in San Antonio, 6 in. pods, vigorous, productive, drought tolerant plant, good summer producer, brought to Texas in the 1940s by German immigrants. *Source History: 1 in 1994; 2 in 1998; 1 in 2004.* **Sources: Ho13.**

Bengali - *Source History: 1 in 2004.* **Sources: Se17.**

Blondy - (White Blondy, Blondy Dwarf Spineless) - 48-50 days - AAS/1986, dwarf plant 3 ft. high, 2 ft. wide, spineless ribbed creamy-lime pods, pick when about 3 in., very productive, seed can be sown directly in the garden when soil reaches 70 deg. F., ideal for the North. *Source History: 15 in 1987; 15 in 1991; 11 in 1994; 9 in 1998; 2 in 2004.* **Sources: Ho13, Lej.**

Borneo - Attractive, long, slender, slightly curved pods, retain tenderness longer than most others. *Source History: 1 in 2004.* **Sources: Ech.**

Bowling Red - Tall 8 ft. plants with red stems, 7 in. red pods, heavy yields. *Source History: 1 in 1998; 1 in 2004.* **Sources: Ho13.**

Burgundy - (Red Burgundy) - 49-65 days - AAS/1988, attractive 4-5 ft. plant with green leaves and burgundy stems branches and leaf ribs, tender 6-8 in. burgundy pods, color not retained when cooked, good yields, bred by Leon Robbins, Clemson U. *Source History: 1 in 1987; 19 in 1991; 25 in 1994; 34 in 1998; 51 in 2004.* **Sources: Ada, Ag7, Ba8, Bo19, Bou, CO23, CO30, Ech, Gou, He8, He17, HO2, Ho13, Jor, Kil, La1, Lej, Ma19, Me7, ME9, Mel, MIC, Mo13, Ni1, OLD, Pe2, Pin, Red, Rev, Ri2, Ri12, Ros, RUP, Sa9, Scf, SE14, Se16, Se25, Se26, Sh9, Sh13, Shu, Sk2, So1, So25, TE7, Und, Vi4, WE10, Wet, WI23.**

Burmese - 58 days - Heirloom from Burma, plants begin to bear when 18 in. tall and continue until frost, spineless pods range from 9-12 in. long, mature from light green to creamy yellow-green, less gooey than other okra which makes this variety more appealing. *Source History: 1 in 1998; 1 in 2004.* **Sources: So1.**

Cajun Jewel - 53 days - Spineless plants 3-4 ft. tall, pods up to 7 in. long, excel. flavor, introduced by CV So1 in 1989, from Louisiana, 1950s. *Source History: 1 in 1991; 1 in 1994; 2 in 1998; 2 in 2004.* **Sources: Ho13, So1.**

Cherokee Long Pod - Prolific plants, 6 ft. tall, light green, 15 in. pods, good slicing type. *Source History: 1 in 1998; 1 in 2004.* **Sources: Ho13.**

Choppee - Heirloom from the Jacobs family since the mid-1800s, grown in the Choppee area of South Carolina near Georgetown which is named after Native Americans indigenous to this area. *Source History: 1 in 2004.* **Sources: Fe5.**

Clemson 80 - (Clemson Spineless 80) - 52-55 days - Straight plants 4 to 5 ft. tall, spineless med-green pods, early and high yielding, slightly taller and more open than Clemson Spineless, Clemson University. *Source History: 2 in 1981; 2 in 1984; 8 in 1987; 12 in 1991; 20 in 1994; 21 in 1998; 22 in 2004.* **Sources: ABB, Bo17, CO23, Cr1, DOR, Fi1, Jor, Kil, MAY, ME9, Pe2, RIS, Scf, SE4, SE14, Sto, TE7, Twi, Vi4, WE10, Wi2, WI23.**

Clemson Spineless - 50-64 days - AAS/1939, most popular okra, 3.5-5 ft. plants, med-foliage, straight deep-green spineless ribbed pods 6.5-9 x 1.5 in. (best at 3-3.5 in.), prolific, fine quality, for the Cotton Belt, Clemson SC/AES. *Source History: 94 in 1981; 86 in 1984; 87 in 1987; 79 in 1991; 68 in 1994; 72 in 1998; 70 in 2004.* **Sources: Ada, Ag7, All, Ba8, Be4, Bou, Bu1, Bu2, Bu3, Bu8, CA25, CH14, Cl3, CO30, Co31, Com, De6, Dgs, Dow, Ers, Fo7, Gr27, Gur, HA3, Ha5, He8, He17, Hi13, HO2, HPS, Hud, Hum, Kil, La1, Lej, Lin, LO8, Loc, Ma19, Mel, Mey, MIC, Mo13, MOU, OLD, Or10, PAG, Par, Pin, Red, Ri12, Roh, Ros, RUP, Sa9, Sau, Se16, Se25, Se26, SGF, Sh9, Shu, Si5, So1, SOU, Tho, Vi4, WE10, Wet, Wi2.**

Cow Horn - 55-60 days - Plants 6-7 ft. tall, long slender ribbed and spined 10-12 in. med-green pods, for home gardens and fresh market, pre-1865. *Source History: 2 in 1981; 2 in 1984; 2 in 1987; 2 in 1991; 1 in 1994; 3 in 1998; 9 in 2004.* **Sources: Ba8, He8, He17, HO2, Ho13, Hud, SE14, Shu, So1.**

Dwarf Green Long Pod - (Dwarf Long Pod, Prolific Long Pod Green, Dwarf Stalk Long Green, Dwarf Green, Early Dwarf Green) - 50-64 days - Fleshy tapered dark-green lightly ribbed 7-8 x 1-1.5 in. pods, 2.5-3 ft. sturdy plants, lobed leaves, spineless, heavy yields, fully formed seeds can be cooked like peas, fine for the North too. *Source History: 64 in 1981; 56 in 1984; 53 in 1987; 45 in 1991; 40 in 1994; 37 in 1998; 29 in 2004.* **Sources: Ada, Bo18, Bu1, Bu3, CH14, CO23, Co31, Cr1, DOR, Ear, Ers, Ge2, HA3, He17, LO8, MAY, Mo13, OLD, Ros, RUP, Sau, Se25, Se26, Se28, Sh9, SOU, Vi4, Wet, Wi2.**

Emerald - (Velvet, Emerald Green Velvet Spineless, Emerald Green) - 55-58 days - Spineless Velvet type, 6-9 ft. tall plants, round smooth dark-green 7-9 in. pods, perfectly round cross-section, green-gray lightly lobed leaves, high quality and yields even in North, Campbell Soup Co., 1950. *Source History: 36 in 1981; 31 in 1984; 26 in 1987; 17 in 1991; 14 in 1994; 13 in 1998; 19 in 2004.* **Sources: Ba8, Bu8, CO23, CO30, DOR, Eo2, Gr27, HO2, La1, LO8, Loc, ME9, Mi12, SE14, Se26, Sh9, Und, Wi2, WI23.**

Evertender - 50-65 days - Imported from India, 4-7 ft.

plants, extra long 5-7 in. green pods, tender until almost mature, disease res. *Source History: 1 in 1981; 1 in 1984; 2 in 1987; 2 in 1991; 2 in 1994; 2 in 1998; 3 in 2004.* **Sources: Ho13, Sa9, So1.**

Ganawia - Tunisian variety, grows 3-4 ft. tall, 4 in. pods, extremely slimey when cooked, harvest young. *Source History: 1 in 2004.* **Sources: Ho13.**

Grandpa's - Prolific, 5 ft. plants, rare. *Source History: 2 in 1998; 1 in 2004.* **Sources: Se17.**

Green Velvet - (Spineless Green Velvet) - 60 days - Excel. variety with round smooth light-green tapered spineless 7-8 in. pods, very prolific, 5-6 ft. plant, should keep picked, retains form and color during canning, becoming hard to find. *Source History: 5 in 1981; 4 in 1984; 3 in 1987; 2 in 1991; 2 in 1994; 3 in 1998; 3 in 2004.* **Sources: CH14, He17, LO8.**

Hill Country Red - 70 days - Beautiful 5-6 ft. plant with bronze-red fruit, does well in summer heat, slender pods to 5 in., can be eaten raw in salads, rare heirloom from south Texas. *Source History: 1 in 1994; 1 in 1998; 3 in 2004.* **Sources: Ba8, Ho13, Na2.**

Jade - 55 days - Plants average 4.5 ft. tall, straight dark-green pods stay tender to 6 in., high yields, recommended for home gardens and farmer's market, U of Arkansas, 1991, CV So1 introduction. *Source History: 1 in 1994; 1 in 1998; 2 in 2004.* **Sources: Sh13, So1.**

James Hopper - 70 days - Tasty, 3-5 in. pods on medium height plants. *Source History: 1 in 2004.* **Sources: Sa9.**

Lee - (Lee Dwarf) - 50-53 days - New, 2.5-3 ft., spineless dark-green 6-7 in. tapered pods with slight ridges, no large lateral branches, higher yields, longer harvest, University of Arkansas. *Source History: 8 in 1981; 8 in 1984; 6 in 1987; 7 in 1991; 9 in 1994; 5 in 1998; 5 in 2004.* **Sources: CO23, DOR, Fi1, Par, Wi2.**

Long Pod - Plant grows 2.5 ft. tall, 7-8 in. long green pods, used boiled or fried or in soups, will store frozen. *Source History: 2 in 1981; 2 in 1984; 1 in 1998; 3 in 2004.* **Sources: Bu8, De6, WI23.**

Longhorn - Grown locally by Cajuns in Louisiana for at least 75 years, 8-10 ft. tall plants bear very large pods, over 12 in. long, best gumbo strain. *Source History: 1 in 1998; 1 in 2004.* **Sources: Ho13.**

Louisiana Green Velvet - 57-65 days - Large vigorous 6 ft. plants, heavy producer, round slender smooth, light-green 6-7 in. spineless pods, produces all season, pods retain color when processed, 1940. *Source History: 8 in 1981; 6 in 1984; 8 in 1987; 9 in 1991; 9 in 1994; 7 in 1998; 5 in 2004.* **Sources: Ag7, Ho13, Ri12, Se7, Wi2.**

Louisiana Red - Old str. from Louisiana with red tinted stalks and leaves, 5-6 ft. tall plant, dark red pods, 6-7 in. long. *Source History: 1 in 1998; 1 in 2004.* **Sources: Ho13.**

Louisiana Short - 85 days - Stocky branching plants with shorter, 6 in. fat pods, good for gumbo, heirloom. *Source History: 1 in 1998; 2 in 2004.* **Sources: Ho13, Sa9.**

Mammoth Spineless Long Pod - 55-65 days - Plants 4-6 ft. tall, intense green 6-8 in. pods with tapered spine, slight rib, excel. quality, stays tender a long time. *Source History: 1 in 1991; 1 in 1994; 2 in 1998; 1 in 2004.* **Sources: Se7.**

My Joannie - Mississippi heirloom with pinkish blue, 6-8 in. pods on 5 ft. plant with red stalks. *Source History: 1 in 2004.* **Sources: Ho13.**

Penta Green - One of the best o.p. okra for cooler, short season areas, excel. Japanese var. that disappeared commercially with the advent of hybrids. *Source History: 1 in 1998; 1 in 2004.* **Sources: Pe6.**

Pentagreen - 70 days - Early high-yielding variety, sturdy plants, glossy green 5-angled upright pods, pick when forefinger size, rare var. from Japan. *Source History: 3 in 1981; 2 in 1984; 1 in 1987; 2 in 1991; 1 in 1994; 2 in 2004.* **Sources: Ho13, Sa9.**

Perkins Dwarf, Sun - 50-55 days - Improved str., plants grow 5 ft. tall, 9 x 1.25 in. pods are straight ridged and tapered near the tip, not prickly. *Source History: 2 in 1991; 2 in 1994; 1 in 1998; 1 in 2004.* **Sources: CO30.**

Perkins Mammoth - (Perkins Mammoth Green, Perkins Mammoth Long Pod, Perkins Mammoth Spineless) - 50-67 days - Intense green 6-9 in. pods, thick tapered spined and slightly ribbed, 4-6 ft. plants, strong grower, high yields, excel. quality, stays tender a long time, pick often to extend production. *Source History: 16 in 1981; 13 in 1984; 10 in 1987; 12 in 1991; 8 in 1994; 7 in 1998; 8 in 2004.* **Sources: HA3, He8, La1, ME9, Mey, Se25, Shu, SOU.**

Perkins Spineless - (Perkins Dwarf Spineless, Perkins Dwarf Green Spineless, Perkins Dwarf Long Green, Perkins Large Pod) - 50-62 days - Campbell Long Green x Clemson Spineless, early, semi-dwarf plants, deep-cut leaves, spineless deep-green pods, Campbell Soup Co., 1944. *Source History: 26 in 1981; 21 in 1984; 17 in 1987; 15 in 1991; 13 in 1994; 10 in 1998; 7 in 2004.* **Sources: But, Co31, Kil, SE14, Sk2, WE10, Wet.**

Pinky - Heirloom from Mississippi, 6-7 ft. tall plants, ridged, pink pods up to 8 in. long. *Source History: 1 in 1998; 2 in 2004.* **Sources: Ho13, Se17.**

Red Germany - Multi-branched 4 ft. plants bear short, fat pods with a red tint on the tip. *Source History: 1 in 1998; 1 in 2004.* **Sources: Ho13.**

Red Okra - (Purple Okra) - 55-65 days - Red pods, 3-3.5 ft. bushy plants also reddish-green, heavy yields, 6-7 in. pods hold color when processed, taller than Clemson, like the green types except for color. *Source History: 14 in 1981; 12 in 1984; 14 in 1987; 8 in 1991; 7 in 1994; 4 in 1998; 2 in 2004.* **Sources: CO30, Ki4.**

Red Velvet - 55-70 days - Reddish pods leaves and stems, tender slightly ribbed 6-7 in. pods hold their color when cooked, 4-5 ft. plants, fine quality, good producer, use fresh or frozen. *Source History: 3 in 1981; 3 in 1984; 6 in 1987; 5 in 1991; 1 in 1994; 3 in 1998; 4 in 2004.* **Sources: Ho13, Roh, Se7, So12.**

Sarajevo Bamija - Extremely spiney 5 ft. plants, 3-4 in. pods, red ribs, good flavor, very productive. *Source History: 1 in 2004.* **Sources: Ho13.**

Silver Queen - 80 days - Vigorous branched 6 ft. tall plants, white-green 7 in. long pods, tender when young, good flavor, productive. *Source History: 2 in 2004.* **Sources:**

Ho13, Se16.

Star of David - (Old Fashioned Okra) - 60-75 days - Israeli variety, unbranched stalks grow 8-10 ft., purple coloration on top of leaf petioles and major leaf veins, 5-9 x 1.25-1.5 in. pods with medium spines if left to mature, best when small, keep well- picked, cannot be shipped to AZ, CA or WA, intro. in 1987 by CV So1. *Source History: 2 in 1987; 3 in 1991; 4 in 1994; 3 in 1998; 7 in 2004.* **Sources: Ba8, He8, Hud, Sa9, Se7, Se16, So1.**

UGA Red - Attractive reddish semi-dwarf plants, 1 x 8 in. scarlet pods turn green when cooked, heavy yields, U of Georgia AES. *Source History: 1 in 1987; 1 in 1998; 1 in 2004.* **Sources: Ho13.**

White Velvet - (Creole) - 55-60 days - Plants grow 3-5 ft. tall, 6-7 in. green-white pods round thick and velvety smooth, meaty and tender when young, no ridges, quite productive, introduced by Peter Henderson in 1890. *Source History: 18 in 1981; 14 in 1984; 10 in 1987; 7 in 1991; 4 in 1994; 3 in 1998; 2 in 2004.* **Sources: Ada, Ba8.**

Yuma Red - Old Arizona desert str. with 7 ft. plants, 4-5 in. maroon pods, good production until frost. *Source History: 1 in 2004.* **Sources: Ho13.**

Varieties dropped since 1981 (and the year dropped):

Big 'un (2004), *Brazil* (2004), *Cajun Queen* (1994), *Candelabra Branching, Park's* (1994), *Clemson 800* (1991), *Dink Hendricks* (1998), *Eastern Texas* (1998), *Emerald, Split Leaf* (1994), *Gold Coast* (2004), *Guarijio "Nescafe"* (2004), *Guarijio Nescafe* (1991), *Little Egypt* (1994), *Louisiana Queen Velvet* (1991), *Mammoth Long* (1991), *Prelude* (1994), *Red River* (1991), *Red Wonder* (1998), *Tenderpod* (1998), *Weldon Prolific* (1987), *White* (2004).

Onion / Bunching
Allium fistulosum

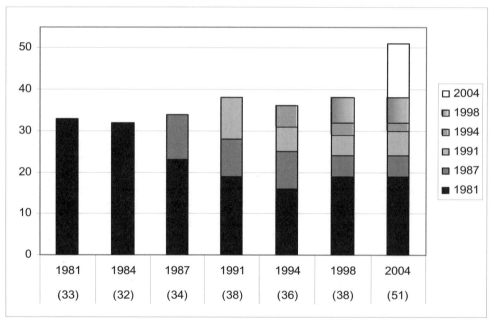

Number of varieties listed in 1981 and still available in 2004: 33 to 19 (58%).
Overall change in number of varieties from 1981 to 2004: 33 to 51 (155%).
Number of 2004 varieties available from 1 or 2 sources: 37 out of 51 (73%).

Annual Bunching - (Annual Green Bunching) - 105 days - Selected for long thick white single stems, clear white sweet mild flesh, good keeper, for green bunching or white sets or pickling. *Source History: 4 in 1981; 4 in 1984; 3 in 1987; 2 in 1991; 3 in 1994; 1 in 1998; 1 in 2004.* **Sources: Ear.**

Beltsville Bunching - 60-65 days - White Portugal x Nebuka, bred to resist bulbing, bulbs swell slightly, crisp mild flesh of fine eating quality, light seeder, vigorous, withstands hot dry weather better than any other bunching variety, winter hardy, PR and yellow dwarf and smut and thrips res., widely adapted, Dr. Jones at USDA/Beltsville, 1951. *Source History: 14 in 1981; 11 in 1984; 10 in 1987; 9 in 1991; 7 in 1994; 4 in 1998; 2 in 2004.* **Sources: DOR, ME9.**

Bunching - Non-bulbing onion for use as scallions. *Source History: 1 in 1991; 2 in 1994; 2 in 1998; 1 in 2004.* **Sources: Bo17.**

Crimson Forest - 60 days - Bulb-type with multiple leaves, dark-red to purple color fades toward the center, pungent. *Source History: 1 in 1991; 2 in 1994; 1 in 1998; 3 in 2004.* **Sources: Ba8, Pin, WE10.**

Deep Purple - 60 days - Red buncher that retains color at any temperature or age, spring or summer sowing. *Source*

History: 3 in 1998; 2 in 2004. **Sources: Jo1, So1.**

Emerald Isle - (XP 3619) - 65 days - Long white bulbless shank with strong straight top, erect foliage, pink root res., PVP 1991 Seminis Vegetable Seeds, Inc. *Source History: 1 in 1991; 3 in 1994; 1 in 2004.* **Sources: Sto.**

Evergreen White Bunching - (Sakatas Evergreen Hardy Long White Bunching Onion, Japanese Bunching, Nebuka, Evergreen Bunching) - 60-120 days - Clusters of 4-9 long slender silvery shanks, non-bulbing, green bunching or scallions, hardy, slow bolt, winters well, PR and thrips and smut res., leek-like stalks divide continuously, white skin so no stripping for market, plant spring or fall. *Source History: 42 in 1981; 38 in 1984; 38 in 1987; 46 in 1991; 49 in 1994; 44 in 1998; 53 in 2004.* **Sources: Bo19, Bu2, CO23, Dam, DOR, Eo2, Ers, Ev2, Fe5, Fis, Ga1, Ga22, Ge2, HA3, He8, He17, Hi6, Hi13, Hig, Hud, Hum, In8, Jo1, Jor, Jun, Kil, Kit, ME9, Mo13, Ni1, OLD, Par, Pe2, Pr3, Ri2, Roh, Ron, Ros, RUP, SE14, Se26, Shu, So1, Syn, TE7, Tt2, Tu6, Twi, Vi4, Vi5, WE10, Wi2, WI23.**

Evergreen, Select Dividers - From selected dividing plants, for rapid propagation. *Source History: 1 in 2004.* **Sources: Syn.**

Feast - 68 days - Imp. Tokyo Long White type, upright plant habit, 15-20 in., DM AB tol., good heat tol. *Source History: 2 in 1998; 2 in 2004.* **Sources: SE4, Sto.**

Fukagawa - 60-70 days - Pleasant sweet flavor, non-bulbing, popular with Japanese cooks. *Source History: 1 in 2004.* **Sources: Ki4.**

Get Set Red - (Welsh Get Set Red) - Perennial bunching onion, multiplies at the base, 18 in. tall with 2 in. red color at the base, overwinters and sets seed. *Source History: 1 in 1998; 2 in 2004.* **Sources: In8, Se17.**

Green Bunching, Edo - Versatile Japanese splitting type with white stalks, mild flavor, tender leaves used fresh or cooked, sow seeds in spring or fall. *Source History: 2 in 2004.* **Sources: DOR, Kit.**

Green Bunching, Koba Strain - *Source History: 1 in 1981; 1 in 1984; 1 in 1987; 2 in 1991; 1 in 1994; 1 in 1998; 1 in 2004.* **Sources: Uni.**

Guardsman - 50 days - Salad onion with strong, upright foliage and a vigorous root system, emerald green foliage, 22 in. plants, small white bulbs, from England. *Source History: 1 in 1998; 1 in 2004.* **Sources: Ter.**

Hardy White Bunching - 70 days - Improved strain of Japanese bunching He-Shi-Ko, hardy non-bulbing type, long slim pure- white stems, used for summer crops or wintering over for next spring. *Source History: 2 in 1981; 1 in 1984; 2 in 1987; 2 in 1991; 2 in 1994; 2 in 1998; 2 in 2004.* **Sources: Lin, Sto.**

He-Shi-Ko Bunching - (He-Shi-Ko Long White Bunching, Evergreen Bunching, Scallions, White Welsh Bunching) - 60-80 days - Perennial bunching onion, non-bulbing, 3 to 5 slim tender 12-14 in. silvery-white stalks grow and divide from base, white pungent flesh, PR and thrips and smut and yellow dwarf res., will overwinter. *Source History: 25 in 1981; 23 in 1984; 17 in 1987; 19 in 1991; 17 in 1994; 17 in 1998; 9 in 2004.* **Sources: Alb, Com, DOR, Gur, Jor, Ni1, Pin, Ver, WE10.**

Ishikura Improved - Improved version of Ishikura, grows to 2.5 ft. tall, long attractive white stems, deep green leaves, delicious. *Source History: 2 in 2004.* **Sources: Ev2, ME9.**

Ishikuri - (Ishikura, Ishikura Long White, Ishikuro) - 65-75 days - Long day, Japanese bunching onion, blue-green leaves, long single leek-like stalks, 12-24 x .75-1 in. dia., no splitting, winter hardy, summer or fall. *Source History: 7 in 1981; 7 in 1984; 6 in 1987; 8 in 1991; 14 in 1994; 16 in 1998; 14 in 2004.* **Sources: CH14, Hi13, HO2, Jor, Kit, ME9, Orn, OSB, Pe6, RIS, RUP, Se26, Sto, Tho.**

Japanese Bunching - (Japanese White Bunching, He-Shi-Ko, White Spanish) - 60 days - Non-bulbing hardy perennial, long white shank, can summer plant, needs decent weather, tends to double, dig when pencil to carrot size. *Source History: 5 in 1981; 4 in 1984; 4 in 1987; 1 in 1991; 4 in 2004.* **Sources: Ra5, Se28, Wet, Wo8.**

K-99 Bunching - Improved Southport type. *Source History: 2 in 1991; 3 in 1994; 5 in 1998; 3 in 2004.* **Sources: CO30, DOR, WE10.**

Kincho - 75 days - Tokyo Long White type, long slim pure-white stems 18-22 in. tall, better uniformity than Hardy White or Ishikura, late summer fall or overwintering, winter hardy in southern Ontario, New York, New Jersey and Ohio. *Source History: 1 in 1987; 3 in 1991; 5 in 1994; 7 in 1998; 6 in 2004.* **Sources: Kit, ME9, Mey, OSB, Sto, We19.**

Kujo Multistalk - (Kujo Green Multistalk) - *Allium fistulosum*, perennial, stalks divide at base producing clusters of 5-6 scallions, usually in the fall of 1st or 2nd year clumps are divided, can be sown spring through late summer. *Source History: 3 in 1981; 2 in 1984; 2 in 1987; 4 in 1998; 1 in 2004.* **Sources: Ev2.**

Kuronobori - 60 days - Japanese green bunching onion, single stalk type, 16 in. white stalk plus 14 in. dark-green leaves, mildly pungent, plant all year round. *Source History: 1 in 1987; 1 in 1991; 1 in 1994; 1 in 1998; 1 in 2004.* **Sources: Ev2.**

Long White Bunching - (Nebuka Long White Bunching, Long White Summer Bunching) - 120 days - Long-stemmed onion, does not form bulbs, summer production or overwinter to produce early scallions for spring bunching. *Source History: 4 in 1981; 3 in 1984; 5 in 1987; 5 in 1991; 4 in 1994; 4 in 1998; 4 in 2004.* **Sources: Ho13, SE4, Sto, WE10.**

Long White Sakata - 75 days - White, non-bulbing type, outstanding yields. *Source History: 1 in 1998; 2 in 2004.* **Sources: ME9, SAK.**

Long White Tokyo - 68 days - Dark-green leaves, single white stalks grow 20-22 in. long, highly res. to hot weather, also fairly cold res., excel. for fall and summer bunching, not winter. *Source History: 1 in 1981; 1 in 1984; 1 in 1994; 1 in 1998; 2 in 2004.* **Sources: CO30, Ves.**

Multi-Stalk 9 - (Iwatsuki) - 80 days - Thick rough soft sweet 12 in. green leaves, white 7 in. stalks in clusters of 8-10, suited for seeding all year round. *Source History: 1 in 1981; 1 in 1984; 1 in 1987; 1 in 1991; 1 in 1994; 1 in 1998; 1 in 2004.* **Sources: Ho13.**

Pacific 22 - 65 days - Mild flavor, resistant to many

foliar disesases due to vigorous root system, little bulbs form when not harvested early which make great pickling onions. *Source History: 1 in 2004.* **Sources: We19.**

Pacific Pearl - 50 days - Pearl onion developed in Oregon, day neutral so it can be planted almost any time of the year and still produce bulbs, harvest at .25-.50 in. dia. size for salads, garnish or pickling. *Source History: 1 in 1998; 1 in 2004.* **Sources: Ter.**

Parade - 70-75 days - Vigorous upright grower with no branching, waxy dark green stalks. *Source History: 2 in 1994; 4 in 1998; 4 in 2004.* **Sources: BE6, Ga1, SE4, Ves.**

Perennial Bunching - 70 days - Long white shanks, mild delicate flavor, will not form a bulb, seed. *Source History: 2 in 1981; 2 in 1984; 1 in 1987; 1 in 1991; 1 in 1994; 1 in 1998; 1 in 2004.* **Sources: Ear.**

Ramrod - White Lisbon type, sweet, mild flavor, withstands winter cold, harvest in spring from June-July sowing. *Source History: 1 in 2004.* **Sources: Ter.**

Red Beard - 60 days - Japanese red scallion, red coloration on the stem's wrapper leaf develops in cool weather, best planted for late summer or fall harvest, fast-germinating, vigorous, best when bulbs are just beginning to swell, mild crisp flesh, excellent raw or grilled with oil. *Source History: 1 in 1987; 7 in 1991; 6 in 1994; 8 in 1998; 6 in 2004.* **Sources: Gou, Ho13, Kit, So12, Ter, Ves.**

Red Bunching - (Toga) - 60 days - Mild flavor, single stem up to 12 in., plant spring or summer. *Source History: 1 in 1991; 1 in 2004.* **Sources: Ga1.**

Red Goddess - Deep red-purple bulb end, cold tolerant. *Source History: 1 in 2004.* **Sources: OSB.**

Red Stalk - Perennial bunching onion, green onions with some red on the stalk. *Source History:1 in 2004.* **Sources: Pr3.**

Santa Clause - 56 days - Ishikuro type, harvest as a spring onion 6-8 weeks after sowing, still tasty and tender 2-3 months later when the size of a leek, red color at the base improves with earthing up and intensifies with the onset of cooler temperatures. *Source History: 2 in 1994; 1 in 1998; 1 in 2004.* **Sources: Tho.**

Snowstorm - Small flat bulb with mild flavor. *Source History: 1 in 2004.* **Sources: OSB.**

Sperling's Toga - Produces shallot-size bulbs, very early green onions, rare. *Source History: 1 in 2004.* **Sources: Sa5.**

Summer Bunching - 75 days - Can be direct seeded or sown 6 to a cell in plug flats and transplanted on 8 in. centers, hill the bunches like leeks to produce the Japanese specialty Nebuka. *Source History: 1 in 1987; 1 in 1991; 1 in 1994; 1 in 1998; 2 in 2004.* **Sources: Coo, Ev2.**

Summer Isle - Low pungency, sweet flavor, great for salads and stir fry, Japan. *Source History: 1 in 2004.* **Sources: Tho.**

Tokyo Bunching - 110 days - Resembles Japanese Bunching, stiffer more upright tops, mostly single stems, almost no bulbing especially in fall, will not winter over, good for summer or fall. *Source History: 2 in 1981; 2 in 1984; 2 in 1987; 3 in 1991; 3 in 1994; 3 in 1998; 2 in 2004.* **Sources: He8, Twi.**

Tokyo Natsuguro - Similar to Kuronobori, dark-green leaves up to 3 ft. with long white stalk which is tender and mildly pungent, favorite in Japan. *Source History: 1 in 1991; 1 in 1994; 1 in 1998; 1 in 2004.* **Sources: Ev2.**

Tokyo, Long White - (Japanese Bunching, Tokyo Long Bunching) - 65-95 days - Looks like a long slim leek, non-bulbing, white-skinned sweet mild stalks, can grow to 30 inches, upright stiff blue-green top, not hardy. *Source History: 8 in 1981; 7 in 1984; 15 in 1987; 15 in 1991; 15 in 1994; 19 in 1998; 14 in 2004.* **Sources: Ba8, CO23, CO30, DOR, Hal, Jor, Kit, ME9, MOU, RIS, SE14, Sto, WE10, WI23.**

Welsh Onion - (White Welsh, Early White Welsh, Japanese Bunching, Nebuka) - Hardy bulbless Siberian perennial, similar to chives but stronger, 24 in. plants, dark-green hollow 12-14 in. leaves, hill up soil to blanch stems, overwinters easily, mild, excel. quality, intro. into England in 1629. *Source History: 9 in 1981; 7 in 1984; 7 in 1987; 10 in 1991; 11 in 1994; 12 in 1998; 12 in 2004.* **Sources: Bou, Com, Ho13, Kit, La8, Lej, Re8, Ri2, Sa5, So12, Te4, Vi5.**

Welsh Onion, Red - Perennial bunching onion with red stalks to 14 in., spreads by offsets. *Source History: 1 in 2004.* **Sources: Hud.**

White Bunching - (White Lisbon, Hardy White Bunching, White Sweet Bunching) - 40-120 days - Quick growing, 14-18 in. silvery-white stalks, stands heat well, holds in bunching condition, not a good keeper, plant in fall for early spring harvest. *Source History: 10 in 1981; 8 in 1984; 11 in 1987; 11 in 1991; 13 in 1994; 15 in 1998; 12 in 2004.* **Sources: All, Bu2, CO30, Fi1, He8, Hi13, Ma19, MAY, RUP, SE14, Tho, WE10.**

White Spear Bunching - 60-65 days - Earlier, taller and more upright than Nebuka, blue-green shanks are 5-6 in. long, very attractive, can be overwintered if mulched. *Source History: 1 in 1981; 1 in 1984; 5 in 1987; 10 in 1991; 11 in 1994; 14 in 1998; 10 in 2004.* **Sources: CH14, Cl3, Eo2, ME9, Pe2, RIS, Se8, Sh14, So1, Tu6.**

Winter White Bunching - 60-120 days - Dual purpose onion that resists bulbing, longer harvest period than regular varieties, sow either late August for late May pulling or in spring and summer for summer/fall harvest, 24 in. tall plant. *Source History: 1 in 1987; 1 in 1991; 1 in 1994; 1 in 2004.* **Sources: Ter.**

Yakko Summer - 100 days - *Source History: 2 in 1981; 2 in 1984; 1 in 1987; 1 in 1991; 1 in 1994; 1 in 1998; 1 in 2004.* **Sources: Red.**

Varieties dropped since 1981 (and the year dropped):

Alpine (1994), *Asagi Bunching* (2004), *Chung* (1998), *Evergreen Bunching Tokyo* (1991), *Green Bunching* (1991), *Iwatsuki* (1987), *Javelot* (2004), *K-99 Bunching Select* (2004), *Kincho III* (1994), *Kuiyo Regular Strain* (1994), *Kujohoso* (1987), *Kujyo Slender Strain* (1987), *Kyoto Market* (1987), *Mt. Blanc* (1991), *Multi-Stalk 2* (1987), *Multi-Stalk 5* (1991), *New Wonder Bunch* (1983), *Prosperity Green Onion* (1987), *Red Welsh Bunching* (1998), *Staysgreen Bunching* (1994), *Stokes Early Mild Bunching* (1987), *Tsukuba Long White* (1991), *White Knight* (1998), *White Salad* (2004).

Onion / Common / Red-Purple
Allium cepa (Cepa Group)

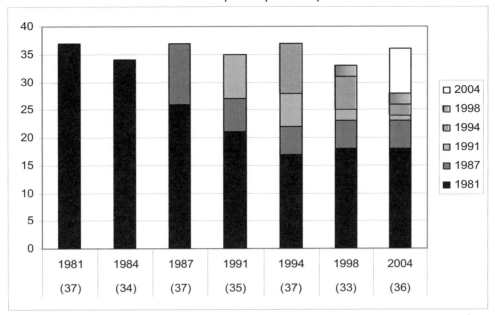

Number of varieties listed in 1981 and still available in 2004: 37 to 18 (49%).
Overall change in number of varieties from 1981 to 2004: 37 to 36 (97%).
Number of 2004 varieties available from 1 or 2 sources: 21 out of 36 (58%).

Amposta - *Source History: 1 in 1994; 1 in 2004.* **Sources: WE10.**

Benny's Red - (Bennies Red) - 108-112 days - Spanish type, improved over Southport Red Globe, bright-red skin, large deep- globe bulbs, soft pinkish-white flesh, long day, fair to poor keeper, home garden. *Source History: 8 in 1981; 7 in 1984; 11 in 1987; 19 in 1991; 11 in 1994; 8 in 1998; 11 in 2004.* **Sources: Bu3, Dam, DOR, Fe5, Ga1, HO2, Pe2, RUP, Sh9, Ter, Tu6.**

Burgundy - (Red Hamburger, Red Burgundy, Hamburger Queen) - 95-165 days - Southern short day type, large flattened globes 3-4 in. dia., dark-red skin, soft white flesh with red rings, very mild and sweet, short storage, may be picked young for bunching, botrytis squamosa and PRR res. *Source History: 24 in 1981; 21 in 1984; 24 in 1987; 20 in 1991; 23 in 1994; 15 in 1998; 16 in 2004.* **Sources: CO23, Dgs, Fa1, He8, La1, LO8, MIC, MOU, Sau, SE14, Se26, SGF, Vi4, WE10, Wi2, WI23.**

California Early Red - Late maturity, large red flattened globe, soft mild white sweet flesh, very good color, slow-bolting, intermediate day, short storage period. *Source History: 3 in 1981; 3 in 1984; 3 in 1987; 3 in 1991; 3 in 1994; 2 in 1998; 1 in 2004.* **Sources: WE10.**

Early Red Burger - 172 days - Early, fresh market, good color, thick and flat, med-large, dark-red skin, soft mild white flesh with light-red ringing, short keeper, PR tol., non-bolt, California. *Source History: 2 in 1981; 2 in 1984; 1 in 1987; 1 in 1991; 3 in 1994; 3 in 1998; 2 in 2004.* **Sources: CH14, Loc.**

Flat of Italy - Old Italian red cipollini type, bright red flat onions, fresh eating or cooking, good for fresh markets, early, mentioned by Vilmorin in 1885. *Source History: 1 in 2004.* **Sources: Ba8.**

Genovese - Mild red cipollo type, flatter shape than Savona, great in salads, harvest young for skewers. *Source History: 1 in 2004.* **Sources: Se24.**

Italian Blood Red Bottle - 120 days - Large beautiful bottle-shaped onion, blood-red color, spicy tangy flavor. *Source History: 3 in 1981; 2 in 1984; 1 in 1987; 1 in 1991; 1 in 1994; 2 in 1998; 2 in 2004.* **Sources: Ho13, Ni1.**

Italian Red Early - *Source History: 1 in 1981; 1 in 1984; 1 in 2004.* **Sources: Bo17.**

Mora di Bassano o di Genova - Large red onion. *Source History: 1 in 1987; 1 in 1991; 1 in 1994; 1 in 1998; 1 in 2004.* **Sources: Ber.**

Purplette - (Baby Purplette) - 60-65 days - The first purple-red skinned mini onion, early maturing, delicate mild flavor, attractive either topped or bunched, can be harvested very young as baby bunching onions with purple pearl ends, turn pastel pink when cooked or pickled. *Source History: 2 in 1987; 6 in 1991; 6 in 1994; 3 in 1998; 4 in 2004.* **Sources: Hig, Jo1, Ron, Ter.**

Red Bermuda - 93 days - Early, large and flat, 4 in. globes, red skin, solid fine-grained pinkish flesh, sweet mild flavor, short day, similar to Yellow Bermuda and Crystal Wax except for color. *Source History: 4 in 1981; 2 in 1984; 3 in 1987; 2 in 1991; 5 in 1994; 3 in 1998; 1 in 2004.* **Sources: Fo7.**

Red Creole - 85-190 days - Hard thick flat reddish-buff bulbs, pungent red-purple flesh, good storage for tropics if dry and ventilated, med-to-small, PR res., short day for South, PVP expired 1992. *Source History: 10 in 1981; 8 in 1984; 6 in 1987; 11 in 1991; 15 in 1994; 13 in 1998; 11 in 2004.* **Sources: CO23, CO30, DOR, Fo7, He8, Kil, LO8, Loc, Ros, SE14, Sh14.**

Red Creole C-5 - 160 days - Small to medium size thick flat bulbs, reddish buff skin, red pungent flesh, fair storage, adapted to 24-28 deg. latitude. *Source History: 1 in 1987; 2 in 1991; 2 in 1994; 2 in 1998; 2 in 2004.* **Sources: WE10, WI23.**

Red Eyes - 100 days - Good yields of mild tasting, firm, red globes, med. storage. *Source History: 2 in 1998; 1 in 2004.* **Sources: Bu3.**

Red Globe - 100-110 days - Late, med-large globe, red-purple skin, white flesh tinged with pink, strong flavor, very productive, good keeper. *Source History: 2 in 1981; 2 in 1984; 2 in 1987; 4 in 1991; 3 in 1994; 3 in 1998; 4 in 2004.* **Sources: Ban, Hum, Ont, PAG.**

Red Grano - (Red Grano PRR) - Medium maturity, large red top-shaped onion, soft mild flesh, PR res., short day, short to medium storage period. *Source History: 2 in 1981; 2 in 1984; 4 in 1987; 5 in 1991; 6 in 1994; 3 in 1998; 5 in 2004.* **Sources: Bo19, CO23, Fis, SE14, WE10.**

Red Hamburger - (Red Italian, Red Burgundy) - 100-180 days - Flat but deep, 4 in. dia., purple-red skin, white flesh with pinkish-tinge near skin darkens to deep red at center, rather pungent, good keeper. *Source History: 10 in 1981; 9 in 1984; 10 in 1987; 7 in 1991; 4 in 1994; 4 in 1998; 4 in 2004.* **Sources: Fi1, Ge2, Gur, MAY.**

Red Long of Tropea - 90 days - Specialty variety traditionally grown in Mediterranean Italy and France, not a storage onion *Source History: 1 in 2004.* **Sources: Jo1.**

Red Simiane - 105 days - Elongated, cylindrical shaped purple-red bulb, average length 4 in., 2 in. dia., mild sweet flavor, from France. *Source History: 2 in 1994; 1 in 1998; 1 in 2004.* **Sources: Gou.**

Red Torpedo - (Italian Red Torpedo, Early Red Torpedo, Torpedo Onion) - 95-200 days - Long torpedo or spindle-shaped onion, 1 lb. ave., purple-red skin, soft light-red flesh, mild sweet flavor, intermediate day, med-short keeper, mainly used for fresh market, heirloom. *Source History: 12 in 1981; 11 in 1984; 10 in 1987; 5 in 1991; 6 in 1994; 8 in 1998; 9 in 2004.* **Sources: Bou, Coo, Loc, Na6, Orn, Pe2, Red, Ri12, Tu6.**

Red Valley - Mini onion with red-purple skin, dark green leaves, harvest when the size of a golf ball, can be grown as full size onions in the Deep South. *Source History: 1 in 1998; 1 in 2004.* **Sources: Dam.**

Red Wethersfield - (Hamburger, Large Red Wethersfield, Red Weathersfield) - 100-115 days - Large flattened globe, deep purple-red skin, fairly firm pink-tinged white flesh with red circles, fine strong flavor, long day, vigorous, productive, adaptable, medium keeper, popular with home gardeners but mainly used to produce sets, first listed in 1834 by Hovey and Co., dev. by growers in Wethersfield, CT. *Source History: 31 in 1981; 28 in 1984; 30 in 1987; 23 in 1991; 22 in 1994; 18 in 1998; 19 in 2004.* **Sources: CH14, CO23, Co32, Com, DOR, Dow, Fo7, Fo13, HA3, He17, Hud, Ol2, Ri12, Sa9, Se7, Se16, Se25, Th3, Wet.**

Redman - (4PHDR) - 100-175 days - Red long day Bennie's Red type, good interior color, medium pungency, good yields, medium storage, fresh market. *Source History: 1 in 1987; 3 in 1991; 11 in 1994; 7 in 1998; 1 in 2004.* **Sources: Me7.**

Rossa di Bassano o di Genova - 75 days - Flattened shape, dark red skin, rose colored flesh, 5 in., wonderful flavor, not a storage type, traditionally braised and served whole in Italy. *Source History: 1 in 2004.* **Sources: Pin.**

Rossa di Milano - 110-120 days - Italian onion with a flat top that tapers to a barrel-shaped bottom, mildly hot, long to intermediate day type, stores well. *Source History: 1 in 1991; 1 in 1994; 2 in 1998; 3 in 2004.* **Sources: Coo, Ri12, Se7.**

Rossa Lunga di Firenze - 100 days - Early intermediate, long, light red onion, good flavor, 18 in. stalks, looks like bunching onion but has much longer storage. *Source History: 3 in 2004.* **Sources: Ga1, Orn, Se26.**

Rouge de Florence - (Florence Red, Red of Florence) - Bottle-shaped red onion from France, mild flavor, may be sown spring or fall, rare Italian heirloom. *Source History: 1 in 1981; 1 in 1984; 1 in 1987; 1 in 1991; 1 in 1994; 1 in 1998; 4 in 2004.* **Sources: Ba8, Bu2, Lej, Se24.**

Rouge Pale de Niort - *Source History: 1 in 2004.* **Sources: Se17.**

Ruby - 100-105 days - Deep globe med-large deep-red onion, med-firm pungent flesh, tight heavy deep-red scales, heavy yields, med-long day, excel. keeper when properly cured, 1965. *Source History: 5 in 1981; 4 in 1984; 4 in 1987; 4 in 1991; 3 in 1994; 4 in 1998; 3 in 2004.* **Sources: Loc, Se25, Sh9.**

Savona - Flattened red bulb, mild flavor, great in salads, use when young for skewers. *Source History: 1 in 2004.* **Sources: Se24.**

Shonan Red - 160 days - Medium day, reddish colored bulb, excel. flavor with sweet, mild pungency, best in salads and sandwiches, dev. by the Tokyo AES. *Source History: 1 in 2004.* **Sources: Kit.**

Southport Red Globe - 100-120 days - Large deep globe, dark glossy purple-red skin, mild firm white pink-tinged flesh, pungent flavor, heavy yields, long day, for early market and medium storage, mainly used in the North as a dry onion and for sets, intro. to American gardeners in 1873. *Source History: 61 in 1981; 56 in 1984; 38 in 1987; 33 in 1991; 32 in 1994; 24 in 1998; 9 in 2004.* **Sources: Bou, Cr1, Ear, Fo13, GLO, Me7, Se7, Se16, Tu6.**

Stockton Early Red - 180 days - Large thick flat onion, thin dark-red skin, mild white flesh with light-red ringing, very productive and vigorous, highly non-bolting and PR tol., short keeper. *Source History: 3 in 1981; 2 in 1984; 2 in 1987; 5 in 1991; 3 in 1994; 4 in 1998; 2 in 2004.* **Sources: Loc, Syn.**

Stockton Red - (Stockton California Red) - 180-200 days - Selected from Stockton Early Red, 7-10 days later, more globe-shaped, slightly smaller, intermediate day type, well adapted to central California. *Source History: 2 in 1981; 2 in 1984; 2 in 1987; 1 in 1991; 3 in 1994; 3 in 1998; 3 in*

2004. **Sources: Coo, Loc, Se7.**

Sweet Flat Red - Sweet Spanish variety, produces 3-4 in. semi-flat red bulbs by late July, white flesh with a little red in each ring, as mild as Bermudas but better adapted to the Midwest, great for summer salads and hamburgers, not a good keeper. *Source History: 2 in 1987; 2 in 1991; 2 in 1994; 1 in 1998; 1 in 2004.* **Sources: Ga1.**

Varieties dropped since 1981 (and the year dropped):

A and C Big Red (1991), *A and C No. 192* (1987), *Benizome Red* (1994), *Big Red* (1984), *Braunschweiger* (1994), *Brunswick Blood Red* (1983), *Brunswick Rodo* (2004), *California Red* (1991), *Calred* (1987), *Cherokee Red* (2004), *Colorado Big Red* (1987), *Creole C-5* (2004), *Eglisau Red* (2004), *French Purple Epicure* (1991), *Genova Red* (1984), *Giza 6* (2004), *Long Red Italian* (1994), *Milanese Red* (1998), *Mountain Red* (1994), *Red Amposta* (1998), *Red Brunswick* (1991), *Red Creole Cal-Select* (1998), *Red Dutch* (1991), *Red Florence* (1998), *Red Giant* (1998), *Red Mac* (1994), *Red Regal* (1994), *Red Salad* (1987), *Red Sunset* (1991), *Redmate* (2004), *Regal* (1994), *Round Purple* (1991), *Simione Red Bottle* (2004), *Spanish Red* (1998), *Stockton Red Medium* (1991).

Onion / Common / White
Allium cepa (Cepa Group)

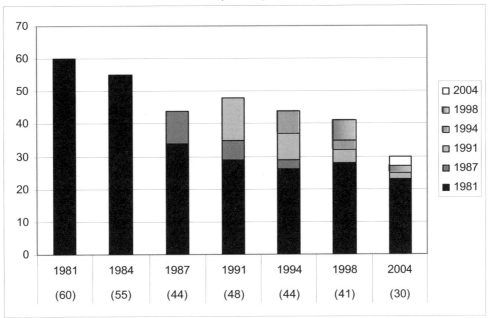

Number of varieties listed in 1981 and still available in 2004: 60 to 23 (38%).
Overall change in number of varieties from 1981 to 2004: 60 to 30 (50%).
Number of 2004 varieties available from 1 or 2 sources: 11 out of 30 (37%).

Awahia - Pungent flavor. *Source History: 2 in 1991; 1 in 1994; 1 in 1998; 1 in 2004.* **Sources: Uni.**

Barletta - (St. Jans, Early White Queen Pickling, White Barletta, White Pearl) - 70-100 days - Round silver-white pickling onion, soft mild flesh, sow thick early in spring, England. *Source History: 11 in 1981; 9 in 1984; 13 in 1987; 13 in 1991; 12 in 1994; 6 in 1998; 9 in 2004.* **Sources: Bo19, Dam, Jor, Ki4, Lej, Pin, Se24, Sto, WE10.**

Bianca di Maggio - 80-110 days - Cipollini variety, long day, medium to small, flat white onions, yellow-tan skin, delicate sweet flavor, used in Italy for pickling, grilling and in salads, limited storage. *Source History: 1 in 1998; 2 in 2004.* **Sources: Ba8, Na6.**

Blanc Hatif de Paris - Short day, flat white bulb, looks similar to Borettana, matures early, short storage. *Source History: 1 in 2004.* **Sources: Ga1.**

Blanco Duro - 120-130 days - Medium maturity, large high globe, hard pungent flesh, not prone to greening of fleshy scales, long day, long storage, pink root tolerant, PVP expired 1997. *Source History: 1 in 1981; 1 in 1984; 1 in 1987; 5 in 1991; 5 in 1994; 6 in 1998; 3 in 2004.* **Sources: Bu3, Loc, Tu6.**

Borettano, Bianca - Flattened small white onion, extra sweet flavor, 2.5 in. wide by 1 in. high, good for specialty grower and home gardener. *Source History: 1 in 2004.* **Sources: Orn.**

Crystal Wax - (Crystal White Wax, Crystal Wax Pickling PRR, Crystal Wax Bermuda, White Bermuda, White Pickling) - 90- 185 days - White Bermuda type, flat, almost no neck, med-large (.5-.75 in. dia. after 60 days - 1 in. after 90), waxy white coarse mild flesh, short day, PR res., adapted to the South, especially Texas, good for green bunching or pearl onions for pickling, not a good keeper, PVP expired 2001. *Source History: 38 in 1981; 29 in 1984;*

34 in 1987; 30 in 1991; 28 in 1994; 22 in 1998; 19 in 2004. **Sources: Alb, CH14, CO23, DOR, Ear, Ers, Fe5, Fi1, He8, Kil, LO8, MOU, Ont, Ros, SE14, Se26, Twi, Ves, WE10.**

Eclipse L303 - (White Eclipse PRR, L303 Early White Eclipse PRR) - 80-175 days - Med-size, flat thick, pencil thin neck, almost no skin, crisp mild flesh, PR and doubling res., non-bolt, short day, short keeper, USDA and TX/AES. *Source History: 5 in 1981; 5 in 1984; 5 in 1987; 4 in 1991; 4 in 1994; 4 in 1998; 4 in 2004.* **Sources: LO8, OSB, Tt2, WE10.**

Kelsae Sweet Giant - (Giant Magnificent) - 110 days - Gigantic top-shaped onions up to 5 lbs., dense solid flesh, mild sweet flavor, excel. for storage. *Source History: 4 in 1991; 4 in 1994; 6 in 1998; 2 in 2004.* **Sources: Lin, Tt2.**

Numex Luna - Short day, PVP 1997 New Mexico State University AES. *Source History: 1 in 2004.* **Sources: CO23.**

Pompeii - 65-100 days - Med-size, flat, shiny skin, mild flesh, perfect pickling onion, short day for areas of 24-28 degrees latitude, poor keeper. *Source History: 1 in 1981; 1 in 1984; 1 in 1987; 2 in 1991; 5 in 1994; 3 in 1998; 2 in 2004.* **Sources: DOR, Hig.**

Ringmaster - (White Sweet Spanish Ringmaster, Ringmaster P.R.R. Dry) - 110-130 days - Improved White Utah with single centers, huge, globe to high globe, small necks, clear white shiny scales adhere well for handling and storage, firm sweet and mild flesh, PR res., high yields, long day type for the North, keeps well if stored well, excellent for rings as you might guess. *Source History: 19 in 1981; 18 in 1984; 11 in 1987; 10 in 1991; 10 in 1994; 9 in 1998; 4 in 2004.* **Sources: Bu3, CO23, DOR, Vi4.**

Southern White Globe - 112 days - *Source History: 1 in 1981; 1 in 1984; 2 in 1991; 2 in 1994; 1 in 1998; 1 in 2004.* **Sources: WE10.**

Southport White Bunching - (Southport White Globe/Green Bunching) - 65-110 days - Slow-bulbing strain of Southport White Globe, widely used buncher, long white shank, deep bluish-green top, med-pungent, not as mild as some but more flavor, grows quickly. *Source History: 8 in 1981; 6 in 1984; 7 in 1987; 7 in 1991; 5 in 1994; 4 in 1998; 5 in 2004.* **Sources: Ca26, Ha5, HO2, PAG, Sto.**

Southport White Globe - 65-120 days - Standard processing type, med-size high globe, very firm pungent flesh, long day for the North, used for bunching or fall onions, good market variety if harvested promptly and dried in shade to preserve whiteness, may be best white keeper. *Source History: 52 in 1981; 46 in 1984; 31 in 1987; 21 in 1991; 21 in 1994; 19 in 1998; 14 in 2004.* **Sources: Bou, Bu8, He17, In8, La1, ME9, Mey, Ont, RIS, Se7, SE14, Sh9, So9, Sto.**

Southport White Globe 404 - 110 days - Bunching globe type, dark green foliage, long white stalks with little bulbing. *Source History: 4 in 1998; 6 in 2004.* **Sources: CH14, CO23, DOR, Loc, WE10, WI23.**

White Bermuda - (Crystal Wax, Eclipse 1303) - 90-180 days - Med-large, flat, crisp mild solid flesh, slow bolt, good slicer and pickler, good keeper, for southern latitudes. *Source History: 11 in 1981; 10 in 1984; 10 in 1987; 14 in 1991; 8 in 1994; 6 in 1998; 5 in 2004.* **Sources: CO23, Roh, RUP, Sau, Wi2.**

White Ebenezer - 100-105 days - Med-size thick flat, clear white skin, solid fine-grained flesh, long day, excel. keeper, widely grown for white onion sets, also for pickling. *Source History: 9 in 1981; 6 in 1984; 9 in 1987; 9 in 1991; 7 in 1994; 9 in 1998; 5 in 2004.* **Sources: Com, DOR, Fo7, Ga1, Pin.**

White Grano - 95-110 days - Medium maturity, large, top-shape, soft mild flesh, short day, short storage. *Source History: 3 in 1981; 3 in 1984; 1 in 1987; 1 in 1991; 2 in 1994; 2 in 1998; 3 in 2004.* **Sources: CO23, Pe2, SE14.**

White Grano PRR - (Early White Grano PRR) - Medium maturity, large, rounded top-shape, soft mild flesh, productive, PR res., short day, short storage. *Source History: 3 in 1981; 2 in 1984; 2 in 1987; 4 in 1991; 6 in 1994; 3 in 1998; 1 in 2004.* **Sources: WE10.**

White Lisbon - (Brooks PRR) - 60 days - Bred especially for use as a scallion, juicy green tops, long white shafts, mild sweet flavor, crisp texture, does not keep well, pre-1865. *Source History: 6 in 1981; 5 in 1984; 9 in 1987; 6 in 1991; 7 in 1994; 3 in 1998; 5 in 2004.* **Sources: Bou, CA25, CO23, Jor, Vi4.**

White Lisbon Bunching - (White Bunch, White Lisbon) - 60-110 days - Vigorous and early, widely used for greenhouse bunching from seed, strong bright-green tops hold well after harvest, long clear-white mild sweet shanks, small-bulbing, best for spring but also good winter and fall, heat and cold res., fine quality. *Source History: 30 in 1981; 25 in 1984; 21 in 1987; 17 in 1991; 15 in 1994; 14 in 1998; 9 in 2004.* **Sources: CH14, Cr1, Dam, De6, DOR, La1, PAG, SGF, Sto.**

White Pearl - (Barletta, White Pearl Pickling, Early White Pearl, Crystal Wax) - 60-75 days - Small silver-white pickling onion, fine flavor and texture, sow seed thickly as with bunching onions to obtain even-size bulbs, good yields. *Source History: 7 in 1981; 7 in 1984; 6 in 1987; 4 in 1991; 2 in 1994; 2 in 1998; 2 in 2004.* **Sources: Fis, Lin.**

White Portugal - (Silverskin, American Silverskin, Early Silverskin Pickling, Portuguese Silverskin) - 96-150 days - Large, thick flat, clear silver-white skin, firm fine grained flesh, mild and sweet, short day but grown in the North too, good keeper - retains flavor well, used for slicing pickling bunching and sets, pre-1800. *Source History: 48 in 1981; 41 in 1984; 30 in 1987; 20 in 1991; 12 in 1994; 7 in 1998; 8 in 2004.* **Sources: DOR, Fo7, HA3, Hud, Ol2, Roh, Sa9, Se17.**

White Queen - Very early, pure-white bulbs, 1.5 in. dia., very mild flavor, unsurpassed for very early table use, excellent for small pickles. *Source History: 1 in 1981; 1 in 2004.* **Sources: Se26.**

White Sweet Spanish - (Giant White Sweet Spanish, White Spanish, Extra Early White Spanish) - 65-130 days - Large, globe, shiny skin, mild firm flesh, dark-green tops, for green bunching or grow to 5.5 in. dia. and 2+ lbs., long day, not a good keeper, 1961. *Source History: 56 in 1981; 51 in 1984; 57 in 1987; 51 in 1991; 44 in 1994; 45 in 1998; 42 in 2004.* **Sources: BAL, Bo18, Bu8, Dam, De6, Ear, Ers, Fa1, Fis, Ga1, Ge2, Gr27, Gr29, Gur, HA3, Ha5, He8, He17, La1,**

Lin, ME9, MIC, Mo13, MOU, OLD, Pin, Ra5, Roh, RUP, Sau, Se25, Se28, SGF, Sh9, Shu, Si5, Sto, Ver, WE10, Wet, Wi2, Wo8.

<u>White Sweet Spanish Bunching</u> - (White Spanish Bunching, Extra Early White Sweet Spanish) - 65-75 days - Special strain for bunching or stripping, slender blue-green tops, long thick mild white stems, slow to bulb but will reach 5.5 in. dia., vigorous. *Source History: 12 in 1981; 10 in 1984; 7 in 1987; 5 in 1991; 3 in 1994; 3 in 1998; 3 in 2004.* **Sources: Bu3, HO2, Se26.**

<u>White Sweet Spanish Jumbo</u> - (White Sweet Spanish Utah Jumbo) - 120-130 days - Selected from White Sweet Spanish Utah for huge size, globe, soft to fairly firm flesh, mild flavor, heavy yields, thrips res., long day, fair keeper, for fall shipping and storage, Utah. *Source History: 14 in 1981; 14 in 1984; 12 in 1987; 6 in 1991; 5 in 1994; 2 in 1998; 3 in 2004.* **Sources: Jor, Loc, Ma19.**

<u>White Sweet Spanish Utah</u> - (Utah Jumbo, White Spanish Jumbo, White Sweet Spanish Utah Bunching) - 110-130 days - Late, large (3.5-4 in.), globe, soft to fairly firm flesh, mild flavor, light-green foliage, thrips tol., heavy yields, med-short storage, UT/AES. *Source History: 17 in 1981; 14 in 1984; 12 in 1987; 8 in 1991; 11 in 1994; 11 in 1998; 15 in 2004.* **Sources: CA25, CH14, DOR, Ga1, GLO, Hal, HO2, Loc, Ni1, Ont, RIS, Ros, SE14, Vi4, WI23.**

<u>White Sweet Spanish Valencia</u> - (White Bunching Sweet Spanish Valencia) - 105-120 days - Slightly earlier than Utah strain, large, globe, fairly firm mild flesh, long day, med-short keeper, mainly for green bunching trade. *Source History: 7 in 1981; 7 in 1984; 3 in 1987; 6 in 1991; 8 in 1994; 7 in 1998; 4 in 2004.* **Sources: Fo7, He8, SE14, WE10.**

Varieties dropped since 1981 (and the year dropped):

Autumn No. 1 (1987), *Barletta Barla* (1983), *Bianca di Guigno* (1987), *Brooks* (1991), *Creoso* (1987), *Dehyso* (1987), *Dr.'s Early White* (1987), *Early Aviv* (2004), *Early Pompei* (1987), *El Toro* (2004), *Everest* (2004), *Fresno White* (2004), *Gladstone* (2004), *HM 1* (1987), *Hysol* (1987), *Ingegnoli White Giant* (1984), *Jane Lot* (1987), *Knob Hill* (1991), *Lisbon Winter Bunching* (1991), *Long White Globe* (1983), *Owa* (2004), *Paradise* (1994), *Paris Cocktail* (1994), *Perfecto Blanco* (1991), *Pompeii Perla Prima* (1998), *Primero* (1998), *Queen Silvato* (2004), *Quicksilver* (1994), *Reina Rondella* (1991), *Ringmaster Bunching* (1994), *Rio Blanco Grande* (2004), *Robusta* (2004), *Shamrock* (1987), *Silver Queen* (1998), *Silver Spring* (1994), *Silverskin* (1998), *Snow Drop* (1994), *Snow White* (1994), *South Australian White Globe* (1987), *Tampico* (1987), *Temprana* (1991), *Texas Grano 1025Y* (1998), *Texas White Grano* (1991), *Valenciana* (2004), *White Creole* (1998), *White Dutch* (1994), *White Early Paris* (1983), *White Globe* (1998), *White Grano New Mexico Strain* (2004), *White Hawk* (2004), *White Keeper* (1994), *White Spanish Green Bunching* (2004), *White Sweet* (1987), *White Sweet Slicer* (1998), *White Sweet Spanish Utah Dry* (1994), *White Velvet* (1987), *Wonder of Pompeii* (2004).

Onion / Common / Yellow-Brown
Allium cepa (Cepa Group)

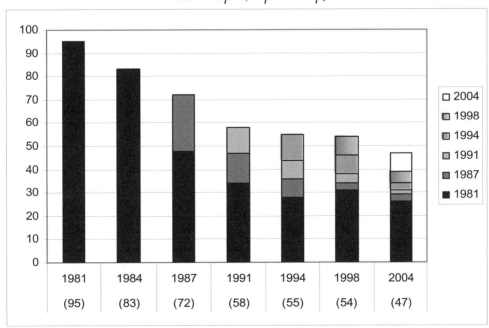

Number of varieties listed in 1981 and still available in 2004: 95 to 26 (27%).
Overall change in number of varieties from 1981 to 2004: 95 to 47 (49%).
Number of 2004 varieties available from 1 or 2 sources: 31 out of 47 (66%).

Ailsa Craig - (Ailsa Craig Exhibition) - 110-140 days - Named after a small island off the coast of England that is round and solid rock, Danver's Yellow x Cranston's Excelsior cross, huge straw-yellow globe, some over 2 lbs., table quality, perfect for exhibition, firm sweet mild flesh, long keeper, day-neutral, Sweet Spanish type, intro. in 1887 by David Murray, a gardener for the Marquis of Ailsa. *Source History: 6 in 1981; 5 in 1984; 5 in 1987; 4 in 1991; 7 in 1994; 9 in 1998; 15 in 2004.* **Sources: Bou, Ear, Fe5, Ho13, Jo1, Jun, Pin, Ri12, Se16, Se25, Sto, Ter, Tu6, Und, We19.**

Australian Brown - 100-250 days - Medium size hard flattened globe, thick dark reddish-brown skin, firm lemon-yellow flesh, extremely pungent, long keeper, central California coast, storage and sets, in 1894 C. C. Morse and Co. obtained 5 pounds of Brown Spanish seed from Australia and sold it to W. Atlee Burpee in 1897 who changed the name to Australian Brown. *Source History: 10 in 1981; 8 in 1984; 1 in 1987; 3 in 1991; 1 in 1994; 2 in 1998; 2 in 2004.* **Sources: Ba8, Se16.**

Bedfordshire Champion - 100-110 days - Special selection from the extensive gardens at Culzean Castle in Scotland, good size globes, brown skin, solid white flesh, mildly pungent, keeps well into winter, England. *Source History: 4 in 1981; 1 in 1984; 2 in 1987; 2 in 1991; 1 in 1994; 1 in 1998; 1 in 2004.* **Sources: St18.**

Borettana - (Italian Button Onion, Cipollini) - 110-120 days - Italian heirloom, long day onion, 1.5-2 in. dia., .75 in. long, rose-bronze skin color, firm flesh, traditionally served whole in sweet-sour sauce or pickled, boiled or for kebobs. *Source History: 1 in 1994; 2 in 1998; 11 in 2004.* **Sources: Coo, Fe5, Ga1, Ki4, Ni1, Pe2, Re6, Ron, Ron, Se16, Se24.**

Borettano, Red Striped - 100 days - Cippolini type, golden yellow skin striped with red, flattened, saucer-shaped Italian onion, sweet flavor, 2-2.5 in. wide by 1 in. tall, heirloom. *Source History: 2 in 2004.* **Sources: Ho13, Orn.**

Borettano, Yellow - Cippolini type, larger size than most pearl onions, unique flattened shape, 2-3 in. wide by 1 in. thick, golden skin, delicate sweet flavor, very popular in Italian cuisine. *Source History: 2 in 2004.* **Sources: Ho13, Orn.**

Cipolle di Rovato - Classic flat Italian yellow onion with flying saucer shape, used for pickling and slicing. *Source History: 1 in 2004.* **Sources: Red.**

Clear Dawn - 104 days - High quality, uniform yellow onion, excellent storage quality. *Source History: 1 in 1998; 3 in 2004.* **Sources: Bou, Pe2, Tu6.**

Downing Yellow Globe - 105-112 days - Improved Southport type for Northern muck, large deep globes, dark gold-yellow skin, firm creamy-white flesh, pungent, long day, good skin retention, long keeper. *Source History: 15 in 1981; 12 in 1984; 9 in 1987; 9 in 1991; 6 in 1994; 4 in 1998; 1 in 2004.* **Sources: Cr1.**

Early Yellow Globe - (Extra Early Yellow Globe) - 90-110 days - Standard early var. for muck soils, med-size med-hard globe, deep bronze-yellow skin, med-soft med-pungent white flesh, long day, fairly good keeper, early shipping, used mainly in the Northeast. *Source History: 40 in 1981; 34 in 1984; 32 in 1987; 30 in 1991; 21 in 1994; 7 in 1998; 7 in 2004.* **Sources: Dam, Fo13, HA3, Hum, Me7, Ont, Up2.**

Excel 986 - (Bermuda 986) - 120-180 days - Sweet, mild, flat but deep, gold skin, short day, PR res., uniform maturity, no doubles or splits, medium keeper, for the South, USDA and TX/AES, 1946. *Source History: 6 in 1981; 4 in 1984; 2 in 1987; 3 in 1991; 4 in 1994; 3 in 1998; 1 in 2004.* **Sources: Kil.**

Exhibition - 100-110 days - Giant Sweet Spanish type, extra sweet, mild flavor, short storage, compares to Walla Walla. *Source History: 2 in 2004.* **Sources: OSB, Twi.**

Giallo di Milano - 110 days - Golden yellow form of Rossa di Milano with the same characteristics, braids well. *Source History: 1 in 1991; 1 in 1994; 1 in 1998; 1 in 2004.* **Sources: Coo.**

Giant Zittau - 110 days - Possibly the best keeping onion available, golden-brown, 5 in. semi-globular bulbs, spring or autumn sowing, productive, good for pickles, excel. keeper, pre-1885. *Source History: 1 in 1981; 1 in 1984; 1 in 1987; 1 in 1991; 1 in 1994; 2 in 1998; 1 in 2004.* **Sources: Bou.**

Gold Coin - 80 days - Flat yellow onion, 1.5-3 in. diameter, stores well, pungent and sweet, nice for braiding, recommended for latitudes 38-60 degrees, necks too thick at lower altitudes. *Source History: 1 in 1998; 1 in 2004.* **Sources: Jo1.**

Golden Globe - 110 days - Oblong globe to bottleneck, med-size, rich golden-yellow skin, med-firm flesh, fairly pungent, long day, used chiefly to produce sets. *Source History: 2 in 1981; 2 in 1984; 2 in 1987; 5 in 1991; 1 in 1994; 2 in 1998; 1 in 2004.* **Sources: DOR.**

Imai Extra Early Yellow - (Imai Early Yellow) - 150 days - Sel. from Senshyu Yellow, 8-12 oz. yellow globes, mildly pungent pure-white flesh, scallions March-May, matures mid-May in normal years, keeps until fall. *Source History: 1 in 1981; 1 in 1987; 1 in 1991; 1 in 2004.* **Sources: Kit.**

New York Early - 94-100 days - An elite o.p. strain of Early Yellow Globe sel. from stock seed from Orange County growers, long day Northern type, globe shape, very hard bulbs, sweet enough for sandwiches and salads, shows some tolerance to pink rot, good keeper. *Source History: 1 in 1987; 5 in 1991; 7 in 1994; 8 in 1998; 5 in 2004.* **Sources: Ga1, Jo1, Me7, Sto, Tu6.**

Newburg - 110-120 days - Long to intermediate day type, globe shaped 3-4 in. bulbs with amber-brown wrapper leaves, crisp white flesh, small necks, exceptional keeper, Seeds of Change original. *Source History: 1 in 1998; 2 in 2004.* **Sources: Se7, So9.**

Numex Starlite - Short day, sweet bulb onion, grows very large in southern areas, PVP 1995 New Mexico AES. *Source History: 1 in 1998; 2 in 2004.* **Sources: CO23, Enc.**

Pearls of Heaven - Silky golden skin on small globes, sweet mild flavor, long keeper. *Source History: 2 in 2004.* **Sources: In8, Se17.**

Piatta of Bergamo - Yellow Cipolla type, very flat shape. *Source History: 1 in 2004.* **Sources: Se24.**

Pukekohe Longkeeper - (M and R Pukekohe Longkeeper) - 100-120 days - Med-large globe, amber to light-brown skin, firm pungent white flesh, 5-12 oz., 3-4 in. across, extremely productive and long-keeping, from original Turbot Strain of Pukekohe Longkeeper. *Source History: 3 in 1981; 3 in 1984; 2 in 1987; 2 in 2004.* **Sources: In8, Se17.**

Ringer - (Ringer Grano) - Early, top shape, very large if thinned to 4-5 in. on rich soil, single centered, very thick rings, PR res., short day, short storage, mainly for onion rings. *Source History: 4 in 1981; 4 in 1984; 3 in 1987; 3 in 1991; 3 in 1994; 3 in 1998; 1 in 2004.* **Sources: CO23.**

Riverside Sweet Spanish - (Yellow Sweet Spanish Riverside) - 115 days - Large globe, small neck, amber-orange skin, mild and sweet, high yields, thrips and blight res., short keeper, dev. for starting in greenhouse for transplant. *Source History: 3 in 1981; 3 in 1984; 2 in 1987; 2 in 1991; 2 in 1994; 2 in 1998; 1 in 2004.* **Sources: Sto.**

Sherkston Sweet Giant - *Source History: 1 in 1998; 1 in 2004.* **Sources: Pp2.**

Stockton Early Yellow - 184 days - Med-large flattened globes, thin light brownish-yellow skin, soft white slightly pungent flesh, high yield, non-bolt, PR tol., short keeper, California fresh market. *Source History: 2 in 1981; 1 in 1984; 1 in 1987; 1 in 1991; 1 in 1994; 1 in 1998; 1 in 2004.* **Sources: Loc.**

Stuttgarter - (Stutgarten Riesen, Stuttgart, Stuttgarts Yellow, Stuttgarter Giant) - 120 days - Med-large flat thick onion, light yellow-brown skin, long day, productive in sandy soils, poor in deep muck, used to produce yellow onion sets which keep without splitting or sprouting prematurely during warm spring weather, sow in July for sets. *Source History: 11 in 1981; 9 in 1984; 17 in 1987; 15 in 1991; 12 in 1994; 16 in 1998; 9 in 2004.* **Sources: Bu1, Dam, DOR, Ha5, Moo, Pin, Roh, Tu6, Up2.**

Texas Early Grano 502 - (Texas Early Grano 502 PRR, Texas Grano 502 PRR, Texas Grano 502, Texas Yellow Grano 502 PRR) - 168-179 days - Vidalia type, large uniform globe/top shape, straw-colored skin, thin scales, soft white flesh, sweet mild flavor, short day, split and bolt and PR res., limited keeper, widely used in southern latitudes for winter production, TX/AES, 1944. *Source History: 20 in 1981; 17 in 1984; 15 in 1987; 17 in 1991; 19 in 1994; 18 in 1998; 13 in 2004.* **Sources: Ba8, CO23, CO30, Dgs, DOR, Kil, La1, LO8, Ri12, SE14, TE7, WE10, WI23.**

Texas Grano - 90-120 days - Sweet yellow-skinned onion, short-day, preferred for fall planting but may be planted in the spring for fall harvest, good keeper. *Source History: 2 in 1994; 3 in 1998; 3 in 2004.* **Sources: Fo7, Ge2, MIC.**

Texas Grano 1015Y - (Texas Grano Spring Sweet, Texas Supersweet) - 175 days - Large 1 lb. straw-yellow deep globe, very sweet mild flesh, short day, PRR, heavy yields, fair storage, suited for the South, PVP expired 2001. *Source History: 6 in 1987; 11 in 1991; 14 in 1994; 16 in 1998; 17 in 2004.* **Sources: Bu3, CH14, CO23, Ers, Fi1, Ga1, Gur, Jun, LO8, ME9, Pe2, Ros, SE14, Ter, WE10, Wi2, WI23.**

Texas Yellow Grano - Large top-shaped bulbs, thin skin, soft white flesh, sweet mild flavor, short keeper, shows good res. to splitting and bolting, pink root tol. *Source History: 1 in 1994; 2 in 1998; 2 in 2004.* **Sources: Bu2, Hi13.**

Valencia - (Utah Jumbo) - 105-130 days - Utah str. of Yellow Sweet Spanish, 4-6 in., large globes up to 1 lb., mild flavor, thick necks must dry well for storage, does not tol. rough harvesting or bulk storage. *Source History: 2 in 1991; 1 in 1994; 2 in 1998; 3 in 2004.* **Sources: DOR, Ri12, Se7.**

Valencia, Burrell's - 115-120 days - AAS, largest and heaviest yielding strain of Sweet Spanish so far, full globe, deep bronze skin, hardy, withstands thrips, keeps well, ships well. *Source History: 1 in 1981; 1 in 1984; 1 in 1987; 1 in 1991; 2 in 1994; 2 in 1998; 1 in 2004.* **Sources: Bu3.**

Walla Walla Sweet Yellow - (Walla Walla, Sweet Walla Walla, Walla Walla Yellow Sweet Spanish, Giant Walla Walla) - 100-150 days - Ten days earlier than other Spanish types, light-brown skin, white flesh, very mild and sweet, summer ripening for fresh use, very cold hardy, not a keeper, spring sown, does well in the Northwest. *Source History: 15 in 1981; 15 in 1984; 28 in 1987; 36 in 1991; 36 in 1994; 47 in 1998; 57 in 2004.* **Sources: BAL, Be4, Bo17, Bou, Bu1, Bu2, Co32, Com, Coo, Ear, Eco, Ers, Fa1, Fe5, Fi1, Fis, Ga1, Ge2, Gou, Gr27, Gur, Ha5, Hi6, HO2, Hum, Jo1, Jor, Jun, La1, Loc, ME9, Mo13, MOU, Na6, Ni1, Or10, OSB, Par, Pe2, Pin, Ri2, Ri12, Roh, RUP, Se7, SE14, Se26, SGF, So9, TE7, Ter, Tt2, Twi, Und, Ver, Ves, We19.**

Yellow Bermuda - 90-185 days - Early shipping type, med-size deep flat bulbs, small necks, loose straw-yellow skin, mild white coarse flesh, short day, PR res., not for areas further north than Virginia. *Source History: 21 in 1981; 16 in 1984; 15 in 1987; 11 in 1991; 9 in 1994; 6 in 1998; 4 in 2004.* **Sources: Fo7, La1, LO8, Sau.**

Yellow Dutch - Good flavor and texture, produces green onions soon after planting or will grow into large mature onions for winter storage, only sets are available. *Source History: 1 in 1981; 1 in 1984; 8 in 1987; 5 in 1991; 1 in 1994; 2 in 1998; 2 in 2004.* **Sources: Alb, Dam.**

Yellow Ebenezer - (Japanese, Japanese Ebenezer, Pungent) - 100-140 days - Flattened 3 in. bulbs, thick brownish-yellow skin, firm yellowish-white flesh, quite pungent when mature, long day, heavy yields, stands handling well, fair keeper, extensively used to produce sets in the North, 1906. *Source History: 33 in 1981; 27 in 1984; 18 in 1987; 15 in 1991; 10 in 1994; 12 in 1998; 10 in 2004.* **Sources: Bu8, DOR, Fo7, Gur, La1, ME9, Mey, Shu, So1, Wet.**

Yellow Globe - 85-110 days - Medium to large flattened globes, tan skin, pungent flavor, keeps well, good for home gardens. *Source History: 1 in 1987; 6 in 1991; 4 in 1994; 1 in 1998; 1 in 2004.* **Sources: Te4.**

Yellow Globe Danvers - (Danvers Yellow Globe) - 100-114 days - Med-large slightly flattened globes, no thick necks, dark golden-brown skin, med-heavy scales, firm white fine-grained flesh, med-strong flavor, long day, heavy yields, keeps well, often grown for sets, standard for garden or market. *Source History: 41 in 1981; 36 in 1984; 25 in 1987; 13 in 1991; 8 in 1994; 5 in 1998; 3 in 2004.* **Sources: Ear, Lin, Roh.**

Yellow Grano New Mexico Str. - 167-185 days -

Early market type, large top-shape, light-yellow skin, soft mild white flesh, short day, PR res., tops somewhat res. to thrips, short keeper, NM/AES, PVP expired 1997. *Source History: 9 in 1981; 9 in 1984; 6 in 1987; 6 in 1991; 4 in 1994; 3 in 1998; 1 in 2004.* **Sources: Ros.**

Yellow Keeper - *Source History: 1 in 2004.* **Sources: Se17.**

Yellow of Parma - Large, oblong-globe shaped golden onions, excellent keeper, late, rare Italian variety. *Source History: 1 in 2004.* **Sources: Ba8.**

Yellow Rock - 100 days - Extensively used for sets, flattened, bronze-yellow skin, heavy yielder, excel. keeper. *Source History: 1 in 1981; 1 in 1984; 1 in 1987; 3 in 1991; 2 in 1994; 3 in 1998; 2 in 2004.* **Sources: DOR, Ga1.**

Yellow Sweet Spanish - (Giant Yellow Sweet Spanish, Yellow Sweet Spanish Jumbo) - 105-130 days - Large (4-6 in., 1+ lb.) globe, yellow-brown heavy skin, mild creamy-white flesh, long day, used for bunching when young, heavy thick necks must dry well, good keeper and shipper. *Source History: 61 in 1981; 58 in 1984; 57 in 1987; 52 in 1991; 48 in 1994; 47 in 1998; 47 in 2004.* **Sources: All, BAL, Ber, Bo18, CA25, Com, De6, Ers, Fa1, Fis, Gr27, Gr29, HA3, Ha5, He8, He17, La1, Lin, Loc, Ma19, MAY, Mey, MIC, Mo13, OLD, Or10, PAG, Ra5, RIS, Roh, RUP, Sau, Se25, Se26, Se28, SGF, Shu, Si5, So1, So25, Twi, Ver, Vi4, Vi5, Wet, Wi2, Wo8.**

Yellow Sweet Spanish Colo. 6 - 110-135 days - Dev. for Arkansas Valley of Colorado, large (3.5-4 in.) deep globe, yellow- brown skin, huge tonnage per acre, thrip and mildew res., handles well, excel. keeper, CO/AES. *Source History: 5 in 1981; 4 in 1984; 2 in 1987; 2 in 1991; 1 in 1994; 1 in 1998; 1 in 2004.* **Sources: Bu3.**

Yellow Sweet Spanish Utah - (Spanish Utah Strain, Yellow Sweet Spanish Utah Jumbo, Utah Jumbo, Yellow Valencia Sweet) - 110-130 days - Deep globe, up to 6 in. dia. and 2 lb. on rich soil, small neck, shiny straw-brown skin, white med-firm flesh, good mild flavor, light-green foliage has some res. to thrips, long day, heavy yielder, fair keeper, UT/AES. *Source History: 51 in 1981; 48 in 1984; 45 in 1987; 33 in 1991; 25 in 1994; 21 in 1998; 20 in 2004.* **Sources: Alb, Bo19, CH14, CO23, Cr1, Dam, Dgs, DOR, Ear, Ge2, GLO, HO2, Jor, MOU, Ont, Ros, SE14, Sh9, WE10, WI23.**

Varieties dropped since 1981 (and the year dropped):

A and C No. 193 (1983), *A and C No. 195* (1983), *Alix (MA 380)* (1987), *Apollo* (1987), *Apollo II* (1987), *Aragon* (2004), *Autumn Spice* (1983), *Autumn Splendor* (1982), *Ben Shemen* (2004), *Bingo* (1987), *Bolstar PV* (1984), *Bottle Onion* (1987), *Breakthru* (2004), *Brigham Yellow Globe* (1991), *Canada No. 1* (1991), *Cipolle Gialla Gigante* (1987), *Colossal* (2004), *Contessa* (2004), *Copper Coast* (1987), *Dia Maru* (1998), *Dorata di Parma* (2004), *Downing Yellow Globe Trapps* (1991), *Downings Early Globe* (1991), *Early Grand* (1987), *Early Sweet* (1994), *Early Yellow Sweet* (1991), *Giant Spanish Valencia* (1991), *Glory* (1987), *Golden Ball* (1994), *Golden Delight* (1987), *Golden Mosque* (1998), *Granex 33* (2004), *Granex 429* (2004), *Gustado* (1987), *Indian Princess* (1991), *Indian Queen* (1991), *Iowa 44* (1987), *Jaune Paille des Vertus* (1987), *Jumbo* (1987), *Kaizuka Extra Early* (1987), *Karbo* (1987), *Lancastrian* (2004), *Late Sprouting Yellow* (1987), *Longkeeper* (1987), *Maui Onion* (1991), *Mayon* (1987), *Michigan Yellow Globe* (1983), *No Tears* (1991), *Numex BR 1* (1998), *NuMex Sunlite* (1998), *NuMex Suntop* (1998), *Odorless Lime* (1991), *Oregon Yellow Danvers* (1984), *Owa Slicing* (1994), *Owy Iti* (1991), *Owy No. 171 (Winter Onion)* (1984), *Perfect Slice* (1984), *Polar* (1987), *Radar* (1994), *Reliance* (2004), *Rijnsburger* (1991), *Rijnsburger Wijbo* (1983), *Ringer CAL 160* (2004), *Riverside (49 Strain)* (1991), *Riverside (Utah Strain)* (2004), *Riverside Strain Valencia* (1987), *Round Yellow* (1991), *San Joaquin* (1987), *Senshyu Yellow Globe* (1991), *Slicing Onion* (1987), *Southport Yellow Globe* (2004), *Stockton Yellow* (1998), *Sturon* (1987), *Stuttgarter Diskos* (2004), *Sweet Spanish Las Animas Str.* (1987), *Sweet Winter* (1994), *Tearless* (1987), *Texas Grano 438* (2004), *Texas Grano 1015* (1998), *Texas Grano 1030Y* (1994), *Texas Grano Valley Sweet* (1998), *Texspan* (1998), *Turbo (SG 913)* (1994), *Valencia, Yellow Sweet Spanish* (1994), *Valenciana Bonanza* (1991), *Willamette Sweet* (2004), *Yellow Bermuda 986 Excel PRR* (1991), *Yellow Creole* (1994), *Yellow Early Globe H36* (1991), *Yellow Globe Mountain Danvers* (1991), *Yellow Spanish Utah Jumbo* (1998), *Yellow Sweet Spanish Peckham* (1994), *Yellow Sweet Spanish PRR* (1987), *Yellow Sweet Spanish Utah Dry* (2004), *Yellow Tampico* (1987), *Yellow Vertus* (1991), *Yellowstone* (1994), *Zittau Giant Yellow Zirius* (2004), *Zittauer Gelbe* (1987).

Onion / Multiplier / Root
Allium cepa (Aggregatum Group)

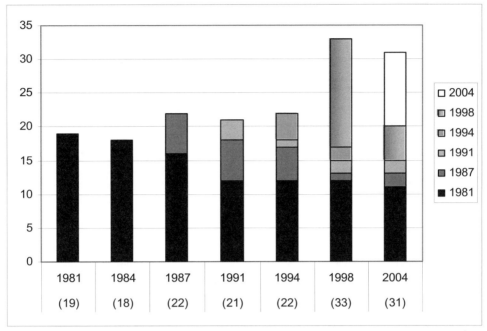

Number of varieties listed in 1981 and still available in 2004: 19 to 11 (58%).
Overall change in number of varieties from 1981 to 2004: 19 to 31 (163%).
Number of 2004 varieties available from 1 or 2 sources: 18 out of 31 (58%).

Eveready - Productive green harvest, small brown bulbs, hardy, said to have been grown by London gardeners during the Blitz. *Source History: 1 in 2004.* **Sources: Ho13.**

Greeley Bunching - Med. size brown bulbs with spicy flavor, very productive, brought to Kansas in a covered wagon many years ago. *Source History: 1 in 2004.* **Sources: Ho13.**

Multiplier Onion - Perennial, Grows to golf ball size, great flavor, multiply year round. *Source History: 1 in 1981; 1 in 1984; 5 in 1987; 5 in 1991; 6 in 1994; 6 in 1998; 5 in 2004.* **Sources: Alb, Dom, Jun, Mcf, Ont.**

Multiplier Onion, Yellow - (Shallots) - Perennial, Dig in fall or overwinter for early green onions, often 10 bulbs per cluster, mild-flavored, hardy everywhere, good keepers. *Source History: 4 in 1981; 4 in 1984; 3 in 1987; 2 in 1991; 2 in 1994; 3 in 1998; 4 in 2004.* **Sources: Bi3, Dam, Fa1, Ga1.**

Pink Salad - Productive shallot type, small brown-pink bulbs, good raw flavor in salads. *Source History: 1 in 2004.* **Sources: Ho13.**

Potato Onion - (Multipliers, Shallots) - Perennial, Multiplier onion, each bulb divides underground forming a clump of 6-8 green onions, mild in flavor, used in salads, no color given. *Source History: 2 in 1981; 2 in 1984; 3 in 1987; 1 in 1991; 1 in 1994; 4 in 1998; 3 in 2004.* **Sources: Lin, Ma13, So12.**

Potato Onion, Brown - Medium-large brown bulbs, good flavor, requires well drained soil, not a keeper. *Source History: 1 in 2004.* **Sources: Ho13.**

Potato Onion, Red - Family heirloom from central Virginia, bronze-red skin, pink flesh, color appears to be affected somewhat by growing conditions, slight vertical ridges, similar in size and flavor to Yellow Potato Onion, may be hardier and a better keeper. *Source History: 2 in 1987; 2 in 1991; 1 in 1994; 1 in 2004.* **Sources: Ga21.**

Potato Onion, Yellow - (Hill Onion, Mother Onion, Pregnant Onion) - 250 days - Divides underground to form clusters of up to 15, for green bunching onions in spring or store 3 in. dia. bulbs up to 12 months, hardy to -25 deg. F, rare, dates back to the late 1800s. *Source History: 1 in 1981; 1 in 1984; 1 in 1987; 1 in 1991; 2 in 1994; 4 in 1998; 4 in 2004.* **Sources: Ga21, Ho13, Ron, Und.**

Shallot - (Multipliers, French Shallots) - Small tender mild-flavored species that multiplies underground, mostly used as green onions but does form a slight bulb, one bulb will produce up to 30, does best in soil that is not too fertile, used in many French recipes. *Source History: 22 in 1981; 19 in 1984; 28 in 1987; 24 in 1991; 23 in 1994; 20 in 1998; 16 in 2004.* **Sources: Bu1, Bu2, Coo, Ers, Fi1, Gur, Hal, Ho13, Jun, Lej, Lin, Mcf, Mey, Pin, Roh, Ver.**

Shallot, Drittler White Nest - Brought to the U.S. by German pioneers who settled in north central Arkansas on Colony Mountain, nest onion with large variable size bulbs, Drittler family heirloom since 1885. *Source History: 1 in 1991; 1 in 1998; 1 in 2004.* **Sources: Ho13.**

Shallot, Dutch - (Yellow Dutch) - Perennial, Multiplier, orange-yellow skin, white flesh, excel. onion flavor, good keeper. *Source History: 1 in 1981; 1 in 1984; 4 in 1987; 4 in 1991; 5 in 1994; 6 in 1998; 6 in 2004.* **Sources:**

Bi3, Ga1, Moo, Pe2, Ri2, Ter.

Shallot, Dutch Red - Grow to 3 in. diameter, excel. flavor, will rot in cold wet springs. *Source History: 1 in 2004.* **Sources: Ho13.**

Shallot, French - Perennial, Dark orange-brown skin with pink cast, purplish-white flesh, flavor of garlic, each division grows into cluster by fall, excel. green onions all summer. *Source History: 4 in 1981; 4 in 1984; 3 in 1987; 3 in 1991; 6 in 1994; 6 in 1998; 6 in 2004.* **Sources: Dam, Dom, Gou, Ri2, Ter, Wo6.**

Shallot, French Epicurean - Perennial, Mild delectable onion-flavored bulb, can be fall or spring planted, harvest in fall and dry thoroughly and store in cool dry room, or leave in ground over winter. *Source History: 1 in 1981; 1 in 1984; 1 in 1987; 1 in 1991; 1 in 1994; 1 in 1998; 1 in 2004.* **Sources: Ni1.**

Shallot, French Red - 130 days - Reddish pink bulb scales, pale purple-pink flesh, mature bulb size ranges from 1-2 in. dia., greatly valued in gourmet cooking, widely adaptable as a perennial for delicious green tops. *Source History: 1 in 1998; 3 in 2004.* **Sources: Ga21, Ho13, Ki4.**

Shallot, Frog's Legs - Perennial, Larger than chicken eggs, elongated like frog legs, dark orange-brown skin, mildest sweet purplish-white flesh, each division produces cluster of 15 or more. *Source History: 2 in 1981; 2 in 1984; 1 in 1987; 1 in 1991; 2 in 1994; 2 in 1998; 3 in 2004.* **Sources: Bi3, Lej, Ri2.**

Shallot, Golden Snow Shoe - Gourmet grade shallot from an East German family's heirloom collection in 1986, survives temperatures of -28 degrees F with snow cover, yields about 5-6 times the amount originally planted, 20-28 per pound. *Source History: 1 in 2004.* **Sources: Bi3.**

Shallot, Grandma Featherston's - Elongated, brown skinned bulbs with good mild flavor, storage not as long as others, brought to Arkansas from Alabama around 1883. *Source History: 1 in 1998; 1 in 2004.* **Sources: Ho13.**

Shallot, Gray - Perennial, In France gray shallots are considered best, esteemed for strong distinctive flavor, grayish- orange or dull-brown skins, purplish-white flesh, short storage. *Source History: 3 in 1981; 3 in 1984; 3 in 1987; 3 in 1991; 3 in 1994; 8 in 1998; 7 in 2004.* **Sources: Bi3, Ga1, Lej, Ma13, Ri2, Ter, We14.**

Shallot, Griselle - French gourmet favorite, elongated curved bulb with hard gray skin, pink flesh, good keeper, rapid spring growth, shy fall growth. *Source History: 1 in 2004.* **Sources: Ho13.**

Shallot, Holland Red - Short, flat round bulbs with coppery red outer skin, red-purple flesh, milder flavor than Brittany, superior storage qualities, increases ten from one. *Source History: 1 in 1998; 3 in 2004.* **Sources: Ga1, Jun, Ver.**

Shallot, Joe's - Tall greens, long white bulbs, very productive, Louisiana heirloom. *Source History: 1 in 2004.* **Sources: Ho13.**

Shallot, Ken's - Gourmet-grade shallot, large squat cloves with excellent flavor, good sheathing, yields increase 5-6 times original planting, hardy and dependable, stores well, from Massachusetts. *Source History: 1 in 2004.* **Sources: Und.**

Shallot, Odetta's White - (Odetta's Onion) - Delicate flavor, uses include scallions and green tops and pearl white bulbs, good for pickling, family heirloom from Kansas prior to 1900. *Source History: 1 in 1991; 1 in 1994; 1 in 1998; 1 in 2004.* **Sources: So1.**

Shallot, Old German - Small roundish pointed shape, rich brown skin color, good flavor, keeps until June. *Source History: 1 in 2004.* **Sources: Ron.**

Shallot, Prince de Bretagne - Pink-brown skinned, mildly flavored shallot, good producer and keeper, local favorite in Brittany. *Source History: 1 in 1998; 1 in 2004.* **Sources: Ho13.**

Shallot, Red - (French Red) - Perennial, Originally from France, one bulb will produce up to 30 bulbs, its unique flavor is a gourmet treat, extensively used in sauces in French restaurants. *Source History: 1 in 1981; 1 in 1984; 3 in 1987; 4 in 1991; 4 in 1994; 3 in 1998; 3 in 2004.* **Sources: Jo1, Pe2, So1.**

Shallot, Red Sun - Clusters of 8-10 flattened bulbs, bronze-red skin, light purple flesh, excellent for spring planting. *Source History: 1 in 2004.* **Sources: Moo.**

Shallot, Sante - Extra large, productive French shallot, for spring or fall planting. *Source History: 2 in 1998; 3 in 2004.* **Sources: Bi3, Ga7, Ma13.**

Tohono O'odham I'itoi's - Light purple shallot, copper colored skin, multiplies vigorously to 20-30 cloves per bulb, flavor can be strong or delicate, said to be brought from sacred desert mountains, from the Tohono O'odham Indians of southern Arizona and adjacent portions of Mexico. *Source History: 1 in 1987; 1 in 1991; 1 in 1994; 2 in 1998; 2 in 2004.* **Sources: Ho13, Rev.**

Varieties dropped since 1981 (and the year dropped):

Brown Shallot (1991), *Dutch Brown Shallot* (1991), *Echalotes Poires* (1983), *European Golden* (2004), *German Giant* (2004), *Heko* (2004), *Pink Skinned Shallot* (1991), *Potato Onion, White* (1994), *Potato Onion, Yellow (1886)* (2004), *Round Shallot* (1987), *Shallot, Asian Purple* (2004), *Shallot, Atlantic* (2004), *Shallot, Atlas* (2004), *Shallot, Brittany Red* (2004), *Shallot, Chinese* (2004), *Shallot, Giant Red* (1998), *Shallot, Jersey* (1998), *Shallot, Pear* (1998), *Shallot, Sanders' White* (1994), *Shallot, Success* (1998), *Welsh* (2004), *White Multiplier Onion* (1998), *Yellow Shallot* (1991).

Onion / Multiplier / Topset
Allium cepa (Proliferum Group)

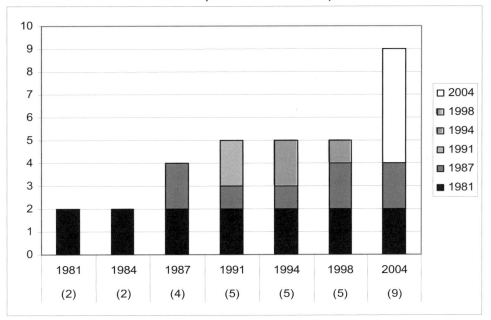

Number of varieties listed in 1981 and still available in 2004: 2 to 2 (100%).
Overall change in number of varieties from 1981 to 2004: 2 to 9 (450%).
Number of 2004 varieties available from 1 or 2 sources: 7 out of 9 (78%).

Catawissa Onion - (Tree Onion, Walking Onion, Winter Onion, Egyptian Top Onion) - Perennial, Extremely hardy Egyptian Onion from Canada, forms cluster of sets on tip of stalk, also divides underground to form permanent clump, the hollow green part is excellent stuffed with cream cheese or other spreads, produces a crop in early spring and another in the fall, not recommended for the extreme south (Zones 9-10). *Source History: 1 in 1981; 1 in 1984; 1 in 1987; 1 in 1991; 2 in 1994; 3 in 1998; 3 in 2004.* **Sources: Ga1, Ni1, Ter.**

Catawissa, Red - Red topsets, dev. by F. F. Merceron in Pennsylvania in the mid-1850s, rediscovered by W. Weaver. *Source History: 1 in 2004.* **Sources: Ho13.**

Egyptian Onion - (Tree Onion, Walking Onion, Top Onion, Topper, Everlasting, Egyptian Multiplier, Winter Onion) - Perennial, *Allium cepa* var. *proliferum*, hardy shallot-like bulbs, forms cluster of sets on tip of 24-36 in. seedstalk, early sprouts for greens, topsets for little onions, self-sows as stalks fall, great flavor, divide mother plants every few years. *Source History: 12 in 1981; 11 in 1984; 12 in 1987; 9 in 1991; 9 in 1994; 17 in 1998; 15 in 2004.* **Sources: Ga1, Ho13, Lej, Ma13, Ni1, Par, Pin, Ri2, Ron, Sa6, Sa9, Shu, So9, Ter, Yu2.**

McCullar's White Topset - Heirloom from Missouri, white-fleshed counterpart to the yellow-fleshed Pran, highly esteemed cultivar from Pakistan and Kasmir, produces a number of 1 in. or larger white bulbs below ground level (for eating) plus pea-size bulbils at the top of the flower stalk (for replanting), prolific, fast top growth, very cold hardy, excel. keeper, used primarily for greens when other onions are dormant. *Source History: 1 in 1987; 1 in 1991; 1 in 1998; 1 in 2004.* **Sources: Ho13.**

Moritz Egyptian - Deep green leaves capped by dark maroon topsets, similar to the Egyptian Onion except the bulbs are a deeper color (red-purple) and topsets are slightly larger, may produce additional sets in the middle of the stalk, Missouri heirloom grown by several families in a one-block area since 1940. *Source History: 1 in 1987; 1 in 1991; 1 in 1994; 1 in 1998; 1 in 2004.* **Sources: Ga21.**

Mrs. Cox's Topset - Nice topsets, purplish bulbs, heirloom from 96 year-old Mrs. Cox. *Source History: 1 in 2004.* **Sources: Ho13.**

Purple Topset - Purple-brown skin, very hardy. *Source History: 1 in 2004.* **Sources: Ho13.**

Volga German Luftzwiebel - Brown-red bulbs, large stalks with large spicy topsets, from Steinwenden, Germany, originally from Khasakstan by a family of Volga Germans. *Source History: 1 in 2004.* **Sources: Ho13.**

White Topset - Tall greens, white bulbs and topsets. *Source History: 1 in 2004.* **Sources: Ho13.**

Varieties dropped since 1981 (and the year dropped):

Egyptian Onion, Red (1998), *Heritage Sweet White Topset* (2004), *Norris Egyptian* (1994).

Onion / Other Species
Allium spp.

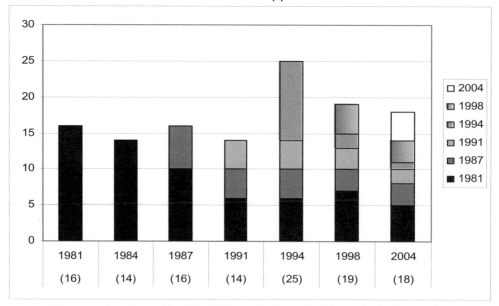

Number of varieties listed in 1981 and still available in 2004: 16 to 5 (31%).
Overall change in number of varieties from 1981 to 2004: 16 to 18 (113%).
Number of 2004 varieties available from 1 or 2 sources: 13 out of 18 (72%).

Allium nutans - *Source History: 1 in 2004.* **Sources: Se17.**

Broad Leaf Chinese Leek - (Chinese Chives) - Perennial, *Allium senescens*, flat dark-green leaves, strong flavor, white flowers, heat and cold res., harvest 3-4 times per year, grows almost anywhere. *Source History: 2 in 1981; 2 in 1984; 1 in 1987; 1 in 1991; 2 in 1994; 1 in 1998; 1 in 2004.* **Sources: Ev2.**

Chinese Leek Flower - (Chinese Chives) - *A. odorum*, young stalks and flowers can be eaten, unusual leek-like flavor, excellent for soups or frying or pickling. *Source History: 1 in 1987; 1 in 1991; 2 in 1994; 1 in 1998; 1 in 2004.* **Sources: Kit.**

Chinese Leek, Hiro Haba - Leaves up to 14 in. long, edible leaves, buds and stems, used in Chinese and Japanese cuisine. *Source History: 1 in 2004.* **Sources: Kit.**

Chives - (Fine Leaved Chives, Schnittlauch, Grass Onion, Ciboulette) - *Allium schoenoprasum*, hardy perennial, heavy clusters, 6-12 in. tall clumps, cut and come again, lavender flowers, delicate onion flavor, once established will last many years, good for pot culture, finely chopped tops used as a flavoring. *Source History: 88 in 1981; 82 in 1984; 119 in 1987; 121 in 1991; 121 in 1994; 116 in 1998; 128 in 2004.* **Sources: ABB, Ag7, Alb, All, Ba8, Ban, Be4, Ber, Bo17, Bo19, Bou, Bu2, Bu3, CA25, CH14, CO30, Com, Coo, Cr1, Dam, De6, Dgs, DOR, Ear, Eco, Eo2, Ers, Ev2, Fa1, Fe5, Fi1, Fis, Fo7, Fo13, Ga1, Ga21, Ge2, GLO, Goo, Gr28, GRI, Gur, HA3, Ha5, Hal, He8, Hi6, Hig, HO2, Ho13, HPS, Hud, Hum, Jo1, Jor, Jun, Ki4, Kil, La1, Lan, Lej, Lin, LO8, Loc, Ma19, MAY, Me7, ME9, Mel, Mey, Mo13, MOU, Na6, Ni1, Ol2, Ont, Orn, OSB, PAG, Par, Pe2, Pep, Pin, Pla, Pr3, Re8, Ren, Ri2, Ri12, RIS, Roh, Ros, RUP, Sa12, Sau, Scf, SE4, Se7, Se8, SE14, Se24, Se26, SGF, Sh9, Sh14, Shu, Sk2, So1, So25, Sto, Te4, TE7, Ter, Tho, Thy, Tt2, Tu6, Twi, Ver, Ves, Vi4, WE10, We19, We20, Wet, Wi2, Wi15, WI23.**

Chives, American (Garlic) - Old perennial variety, white flowers. *Source History: 1 in 1991; 1 in 1994; 1 in 1998; 1 in 2004.* **Sources: So9.**

Chives, Chinese - (Garlic Chives) - *Allium tuberosum*, hardy perennial from Siberia and east Asia, forms no bulb, grown for its flat ashy-grey leaves that are .25 in. wide x 6-14 in. long, slight garlic flavor, dense clumps. *Source History: 3 in 1981; 3 in 1984; 3 in 1987; 25 in 1991; 22 in 1994; 20 in 1998; 16 in 2004.* **Sources: Com, DOR, Fe5, Ga1, GLO, Jor, Jun, OSB, Par, Roh, Se7, Sh14, Sto, Ves, WE10, Wet.**

Chives, Curly - *A. senescens*, low-growing flat leaves grow in circular pattern, rosy flowers, for rock gardens, 8 in. high. *Source History: 1 in 1998; 1 in 2004.* **Sources: Sa6.**

Chives, Fine-Leaf - (Fine Chives) - Same chives flavor but with fine leaves that require less chopping and provide a different texture in the garden. *Source History: 2 in 1998; 3 in 2004.* **Sources: Jo1, La8, Re6.**

Chives, Forsgate - Grows 18-24 in., edible rose-pink flowers. *Source History: 1 in 1994; 1 in 1998; 1 in 2004.* **Sources: Ni1.**

Chives, Grande - *A. schoenoprasum*, large leaves up to 5 mm across, suitable for fresh or frozen commercial use. *Source History: 1 in 2004.* **Sources: Ri2.**

Chives, Grolau - (Windowsill Chives) - Especially developed for forcing in greenhouses, medium thick foliage, good strong flavor, good yield when continuously cut, less susceptible to turn yellow or leggy. *Source History: 1 in 1987; 2 in 1991; 1 in 1994; 2 in 1998; 3 in 2004.* **Sources: Ni1, Ri2, Ter.**

Chives, Large - (Ciboule) - *A. schoenoprasum*, large purple flowers, lots of leaves. *Source History: 1 in 1987; 1 in 1991; 1 in 1994; 1 in 1998; 1 in 2004.* **Sources: La8.**

Chives, Mauve Garlic - *A. tuberosum*, edible, attractive mauve flowers, twisted flat leaves with good garlic-shallot flavor, 12 in. tall. *Source History: 1 in 1991; 1 in 1994; 2 in 1998; 2 in 2004.* **Sources: Ri2, Thy.**

Chives, Staro - Thicker leaf for freezing, drying or fresh use. *Source History: 1 in 1998; 1 in 2004.* **Sources: Jo1.**

Chives, Wilau - *A. schoenoprasum* 'Wilau', erect, blue-green, medium coarse stems, excellent flavor, good for commercial field production for the fresh and dried market and for late season forcing in pots. *Source History: 1 in 2004.* **Sources: Ri2.**

Flowering Leek Tenderpole - (Chinese Leek Flower) - *Allium odorum*, young stalks and flower buds can be eaten, unusual leek-like flavor, excel. for soups and frying. *Source History: 1 in 1981; 1 in 1987; 1 in 1991; 1 in 1994; 1 in 1998; 1 in 2004.* **Sources: Ev2.**

Garlic Chives - (Broad Leaved Garlic Chives, Chinese Chives, Oriental Chives, Chinese Leek) - Perennial, *Allium tuberosum*, heavy cluster of chives, distinct garlic flavor, dark grey-green 16 in. leaves are flatter and slightly larger, multiplies rapidly, hardy but must be protected during severe winters. *Source History: 26 in 1981; 24 in 1984; 46 in 1987; 39 in 1991; 61 in 1994; 70 in 1998; 84 in 2004.* **Sources: Ag7, Alb, Ba8, Bo17, Bo19, Bou, Bu2, Bu3, CA25, CO30, Co32, Coo, Dam, Dgs, Ear, Eo2, Ers, Ev2, Fi1, Fo7, Ga21, Ga22, Ge2, Goo, Gr28, GRI, Gur, HA3, He8, Hi6, Hig, Ho13, HPS, Hud, Hum, Jo1, Jor, La8, Lan, Lej, Lin, ME9, Mo13, MOU, Na6, Ni1, Ol2, Ont, Pe2, Pe6, Pep, Pin, Pr3, Re6, Re8, Ren, Ri2, Ri12, RUP, Sa6, Sa12, SE4, SE14, Se26, SGF, So1, So12, So25, Syn, Te4, TE7, Ter, Tho, Thy, Tt2, Tu6, Twi, Ver, Vi4, Vi5, We19, We20, WI23, Yu2.**

Varieties dropped since 1981 (and the year dropped):

Baia do Cedo (1991), *Baia Peiforme* (1991), *Bermuda Excel 986* (1991), *Bonanza Georo* (1991), *Borettana per Sotto Aceti* (1991), *Chinese Leek* (2004), *Chives Improved* (2004), *Chives, Amel Priest* (1998), *Chives, Blue Spear* (1998), *Chives, Bode Pastel* (1998), *Chives, Curly Mauve* (1998), *Chives, Forescate Progeny* (1998), *Chives, Marsha* (1998), *Chives, Marsha Sibling* (1998), *Chives, Nira* (2004), *Chives, Snowcap* (1998), *Chives, Stella* (1998), *Chives, Yellow Chinese* (1998), *Ciboulette Civette* (1987), *Garlic Chives, Vietnamese* (2004), *Gow Choy* (2004), *James Long Keeping* (1991), *New Belt* (1987), *Nira* (1984), *Nira Broad Leaved Chives* (1987), *Papago Onions* (1987).

Parsley
Petroselinum crispum

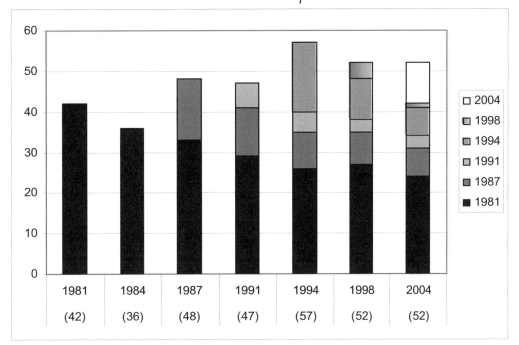

Number of varieties listed in 1981 and still available in 2004: 42 to 24 (57%).
Overall change in number of varieties from 1981 to 2004: 42 to 52 (124%).
Number of 2004 varieties available from 1 or 2 sources: 30 out of 52 (58%).

Afro - 75 days, Bred in Denmark, fine extremely curled frilly green leaves on long stems, holds well in the garden, will not get moldy or gray in cold fall weather, sweet and crisp. *Source History: 1 in 1981; 1 in 1984; 2 in 1987; 2 in 1991; 1 in 1994; 1 in 1998; 1 in 2004.* **Sources: Tho.**

Aphrodite - Strong stems, extra tightly curled leaves, vigorous, home garden and commercial pot plant production. *Source History: 1 in 2004.* **Sources: Ri2.**

Banquet - 76-80 days, Dev. in Denmark, cool and cold weather tol., tightly curled, deep-green color especially in cooler seasons, good to overwinter, stiff erect stems, long season. *Source History: 2 in 1981; 1 in 1984; 3 in 1987; 2 in 1991; 2 in 1994; 3 in 1998; 3 in 2004.* **Sources: GRI, Ha5, Pep.**

Bartowich Long - 95 days, Old World type, slender 7-8 in. long roots store like carrots, can overwinter if mulched, exceptional flavor, roots used raw in salads or cooked, flat leaves used for flavoring. *Source History: 1 in 1994; 1 in 1998; 1 in 2004.* **Sources: Jo1.**

Budapest Market - Hungarian winter hardy variety, flat leaf, sweet and delicious. *Source History: 2 in 2004.* **Sources: In8, Se17.**

Catalogno - 75 days, True Italian flat leaf parsley, stronger flavor than curly leaf types, best for salads, soups, stews and drying. *Source History: 1 in 1987; 1 in 1991; 1 in 1994; 1 in 1998; 1 in 2004.* **Sources: Coo.**

Champion Moss Curled - 70-85 days, Dark-green tightly curled finely cut frilly leaves, dwarf compact 8-15 in. plants, very robust and quick growing, best var. for growing under glass, beautiful, used either green or dried. *Source History: 27 in 1981; 22 in 1984; 19 in 1987; 9 in 1991; 7 in 1994; 5 in 1998; 4 in 2004.* **Sources: Gur, Ho12, Sau, Sto.**

Clivi - (Persil) - 65-68 days, Entire plant intensely fragrant, curled dented leaves, very dwarf, rapid grower, continuous cutting all summer, base leaves do not yellow, from Belgium. *Source History: 2 in 1981; 2 in 1984; 1 in 1987; 1 in 1991; 2 in 1994; 1 in 1998; 1 in 2004.* **Sources: Tho.**

Comune o Aromatico Bolognese - Smooth, deep green leaves, aromatic. *Source History: 1 in 1987; 1 in 1991; 1 in 1994; 1 in 1998; 1 in 2004.* **Sources: Ber.**

Crispum - Type of plain leaf Italian, very strong flavor, good for fresh use and drying. *Source History: 1 in 1994; 1 in 2004.* **Sources: Pep.**

Curly Parsley - (Curled) - Tightly curled leaves, good salad and seasoning herb, one of the best sources of iron and vitamins A and C. *Source History: 1 in 1981; 1 in 1984; 3 in 1987; 8 in 1991; 8 in 1994; 9 in 1998; 12 in 2004.* **Sources: Dow, Fo7, Gr28, GRI, He8, Lej, Ren, Sa5, So12, Syn, Thy, Wi15.**

Dark Green Curled - 78-85 days, French Perfection type, finely cut tightly curled dark-green leaves, rapid recovery, continuous harvests until fall, will not yellow if hot and humid, rugged stem. *Source History: 2 in 1981; 2 in 1984; 2 in 1987; 2 in 1991; 2 in 1994; 1 in 1998; 1 in 2004.* **Sources: Red.**

Dark Green Italian - (Plain Italian Dark Green, Italian Dark Flat-Leaf) - 72-80 days, Shiny dark-green flat leaves, med-deep cut but not curled, similar to Plain but has wider leaves, larger plant and stronger flavor than the curled types, erect vigorous growing celery leaf type, excel. variety for seasoning. *Source History: 14 in 1981; 14 in 1984; 16 in 1987; 13 in 1991; 13 in 1994; 16 in 1998; 22 in 2004.* **Sources: CA25, CH14, CO30, Dom, Ga22, Ha5, He8, HPS, In8, Jun, La1, Lin, Ma19, ME9, PAG, Pe6, SE4, Se25, Sh9, Sto, WE10, We19.**

Darki - 70-80 days, Dark-green intensely curled heavy leaves, short internodes, excel. cold tol., vigorous and adaptable, holds longer when cut, European award winner. *Source History: 6 in 1981; 5 in 1984; 4 in 1987; 2 in 1991; 6 in 1994; 8 in 1998; 3 in 2004.* **Sources: Bou, Dom, Sto.**

Decora - (Decora Triple Curled) - 70-80 days, Selected from Paramount, larger thick stout dark-green leaves, remains well curled in hot weather, thicker stems, compact, vigorous, wilts less after cut, PVP expired 1996. *Source History: 2 in 1981; 2 in 1984; 3 in 1987; 2 in 1991; 4 in 1994; 4 in 1998; 2 in 2004.* **Sources: SAK, Twi.**

Delikat Original - 71 days, Large fine curled dark-green leaves on long stiff stems, good winter hardiness, easy to cut and bunch, from seedsmen at L. Daehnfelt in Denmark. *Source History: 2 in 1981; 2 in 1984; 1 in 1998; 1 in 2004.* **Sources: In8.**

Double Curled - (Double Green Curled, Evergreen Double Curled, Evergreen, Persil Frisee Double) - 70-76 days, Dense dark-green finely crumpled closely curled leaves, vigorous compact 12-14 in. tall plants, frost tol., home or market gardeners, for garnishing and flavoring. *Source History: 16 in 1981; 12 in 1984; 9 in 1987; 6 in 1991; 6 in 1994; 7 in 1998; 3 in 2004.* **Sources: All, CO30, Sa9.**

Envy - 90 days, Deep green, curled variety. *Source History: 1 in 1994; 1 in 1998; 1 in 2004.* **Sources: Tho.**

Evergreen - 70-80 days, Vigorous growing, large dense coarse-cut deep-curled dark-green compound leaves, more frost res. than others, uniform, home or market gardens, biennial, FER/1940. *Source History: 5 in 1981; 5 in 1984; 2 in 1987; 4 in 1991; 5 in 1994; 4 in 1998; 3 in 2004.* **Sources: Bu3, Se26, Vi4.**

Extra Curled Dwarf - (Emerald, Champion Moss Curled) - 85 days, Compact plants, fine-cut curled dark-green moss-like leaves. *Source History: 5 in 1981; 5 in 1984; 4 in 1987; 5 in 1991; 4 in 1994; 3 in 1998; 3 in 2004.* **Sources: Bu2, Ge2, Ni1.**

Extra Triple Curled - (Moss Curled, Sherwood) - 70-75 days, Finely cut and closely curled, used fresh for garnishing, dried for winter use or potted and grown inside, 8-12 in. and dark-green, compact and productive. *Source History: 11 in 1981; 11 in 1984; 11 in 1987; 9 in 1991; 6 in 1994; 5 in 1998; 5 in 2004.* **Sources: Alb, BAL, Kil, Se26, Sh14.**

Fakir - (Hamburg Turnip Rooted, Turnip Rooted Parsley) - 85 days, Coarse, dark green plain leaves resemble flat leaf parsley, grown for parsnip-shaped roots, white skin and flesh, eaten raw, steamed or cooked in soups or stews, rich distinctive flavor resembles carrots and celeriac combined with parsley flavor. *Source History: 1 in 1991; 1 in 1994; 1 in 1998; 5 in 2004.* **Sources: BE6, In8, Pin, SE4, Se17.**

Favorit - 75 days, Improved moss curled type with very dark-green, densely curled leaf, strong upright plant to 13 in., leaves held high off the ground, good for fresh market and drying. *Source History: 2 in 1994; 2 in 1998; 2 in 2004.* **Sources: BE6, Ter.**

Forest Green - (Moss Curled, Double Curled) - 70-78 days, Finely cut deeply curled leaves, dark-green all summer, does not yellow or fade in adverse weather, long-stemmed, early vigorous erect plants, triple curled, 10-12 in. plants, best curled variety for flavoring and garnishing. *Source History: 23 in 1981; 21 in 1984; 27 in 1987; 33 in 1991; 36 in 1994; 36 in 1998; 40 in 2004.* **Sources: ABB, Bo19, CA25, CH14, Co31, Fe5, Ga21, Hig, HO2, HPS, Hum, Jo1, La8, Lin, LO8, MAY, Mey, Mi12, OLD, Ont, OSB, Pe1, Pe2, Pla, Ri2, Ri12, RIS, Roh, SE4, Se7, Sh9, Sh12, So1, So9, SOU, Sto, Twi, Ves, We19, WI23.**

Giant Catalogna - (Italian Flat Leaf) - Hearty Italian strain, large plants to 24 in. tall, flat dark-green leaves, exceptionally aromatic, prolific, requires moist partially shaded location, biennial. *Source History: 1 in 1987; 1 in 1994; 2 in 1998; 3 in 2004.* **Sources: CA25, Jo1, Orn.**

Giant Italian - (Gigante d'Italia, Prezzemolo, Celery Parsley, Giant Plain, Italian Giant Dark Green) - 85-90 days, Bushy massive thick-stalked 2-3 ft. plants, large quantities of plain flat leaves with strong pungent parsley flavor, prized by Italian cooks for seasoning, very spicy and flavorful, can be eaten like celery. *Source History: 5 in 1981; 5 in 1984; 3 in 1987; 7 in 1991; 9 in 1994; 9 in 1998; 11 in 2004.* **Sources: CO23, Coo, Fe5, Jo6, Ki4, Ni1, Pe2, Pin, Re6, Red, Se16.**

Gigante di Napoli - Smooth, deep green leaves, long stems, aromatic. *Source History: 1 in 1987; 1 in 1991; 1 in 1994; 1 in 1998; 2 in 2004.* **Sources: Ber, Se24.**

Green Curled - Finely cut, deeply curled parsley for home or commercial use. *Source History: 2 in 1987; 1 in 1991; 1 in 1994; 1 in 1998; 1 in 2004.* **Sources: Ers.**

Green River - 70-75 days, Superior double curled type, 10 in. tall, plants can withstand up to five cuttings without bolting, good disease res., tolerates heat, good for bunching. *Source History: 1 in 1991; 9 in 1994; 12 in 1998; 13 in 2004.* **Sources: Bo17, CO30, Dam, Ge2, GLO, Jor, Loc, Ni1, Ri2, RUP, Se26, Si5, St18.**

Greenstar - Compact plants, strong stems, dark green curled leaves, good for greenhouse, cold frame and outdoor planting. *Source History: 1 in 2004.* **Sources: Tu6.**

Hamburg - (Hamburg Long, Hamburg Turnip/Parsnip/Stump/Thick Rooted, Root) - 85-95 days, *P. hortense*, fleshy white tapered parsnip-like roots cooked as vegetable, 8-10 x 2 in. dia., dark-green flat plain deeply cut leaves used like ordinary parsley, med-long stems, stores well, resembles slender parsnips, pre-1600s. *Source History: 71 in 1981; 66 in 1984; 60 in 1987; 54 in 1991; 48 in 1994; 51 in 1998; 45 in 2004.* **Sources: ABB, Alb, Ba8, Bo19, Bu2, CA25, Com, Dam, Dom, Ear, Ers, Fe5, Fo7, Ge2, GRI, HA3, HO2, Hud, Jun, La1, Lan, Lej, Lin, ME9, Mel, Mey, Ni1, Ol2, OLD, Ont, Pep, Ri2, RIS, Roh, RUP, SE14, Sh9, So12, Sto, Ter, Tt2, Ver, Vi4, WE10, WI23.**

Hardy Italian - Overwintering flat leaf strain survives temperatures to below 10 degrees F., pest and disease res., from Shoulder to Shoulder Farm. *Source History: 1 in 1994; 1 in 1998; 1 in 2004.* **Sources: Sh12.**

Hosszu - Large flat leaves with sweet flavor, tolerates partial shade and drought, Hungary. *Source History: 1 in 2004.* **Sources: In8.**

Italian Parsley - (Italian Single Leaf, Flat-Leaved Italian, Italian Dark Green Plain Leaf) - 78 days, *P. neapolitanum*, strong 8-16 in. plants with flat leaves, stronger flavored than the curled types, excel. for seasoning, parsley leaves turn bitter when plant starts to flower. *Source History: 2 in 1981; 2 in 1984; 12 in 1987; 16 in 1991; 22 in 1994; 28 in 1998; 38 in 2004.* **Sources: Ag7, Au2, But, Dgs, Gur, Hal, Hi6, La8, Mi12, Na6, OSB, Pe1, Pe2, Pin, Re8, Red, Ren, Ri2, Roh, RUP, Sa9, Scf, Se7, SE14, Se26, Si5, Sk2, So9, So12, Syn, TE7, Th3, Thy, Tt2, Ves, Vi4, We20, Wi15.**

Italian Sweet - (JPY 194) - 75 days, Flat leaf Italian type with a sweet parsley taste. *Source History: 1 in 1998; 1 in 2004.* **Sources: Jo1.**

Krausa - (Krausa Market) - 68-75 days, Rich green leaves, very densely curled, plants are exceptionally even and compact, strong stems, fast regrowth, reselected strain, reliable. *Source History: 2 in 1981; 2 in 1991; 2 in 1994; 2 in 1998; 5 in 2004.* **Sources: BE6, Coo, Dam, OSB, Par.**

Moss Curled - (Fine/Double/Triple/Dark Moss Curled, Paramount, Forest Green) - 70-85 days, Vigorous productive compact 12 in. tall plant, very dark-green finely cut deeply curled leaves, grows so thickly curled that the plant resembles a bunch of moss, high yielding strain, very uniform, subtly flavored seasoning, 1865. *Source History: 29 in 1981; 24 in 1984; 51 in 1987; 55 in 1991; 54 in 1994; 49 in 1998; 55 in 2004.* **Sources: Ag7, Be4, Bo17, Bo19, Bu3, But, CA25, Com, Cr1, Dam, De6, Dgs, Ear, Eo2, Fi1, Fis, Ga1, Ga22, Goo, HA3, Hal, Hi6, Jor, Jun, La1, Lin, Loc, Ma19, Mel, Mey, Na6, PAG, Pe2, Pr3, Ri2, Roh, Ros, RUP, Sa12, Scf, Se7, SE14, Se26, Shu, Si5, Te4, TE7, Ter, Tt2, Tu6, Ver, Vi4, WE10, We20, WI23.**

Pagoda - 72-85 days, Long stems, 10-12 in., triple curled dark-green leaves, fast regrowth after cutting. *Source History: 1 in 1991; 2 in 1994; 2 in 1998; 1 in 2004.* **Sources: Shu.**

Paramount - 75-85 days, Finely-cut, triple curled leaves, long stems are extremely suitable for bunching. *Source History: 1 in 1981; 1 in 1984; 1 in 1987; 8 in 1991; 6 in 1994; 7 in 1998; 7 in 2004.* **Sources: Bu8, CO23, DOR, Ers, Fa1, HO2, PAG.**

Petra - 80-90 days, Moss curled type with long, strong stems, good regrowth, recommended for bunching, from Holland. *Source History: 1 in 1994; 2 in 1998; 1 in 2004.* **Sources: Ki4.**

Plain - (Italian Dark Green Plain/Single/Flat/Broad Leaf, Celery Leaf Parsley, Common, Simple) - 60-85 days, Vigorous 12 in. erect plants, large plain flat glossy dark-green leaves, long stems, more flavor than the curled types, the preferred parsley in Europe, 1806. *Source History: 82 in 1981; 72 in 1984; 75 in 1987; 77 in 1991; 79 in 1994; 81 in 1998; 74 in 2004.* **Sources: ABB, All, BAL, Bo19, Bou,**

Bu2, Bu3, CH14, CO23, CO30, Com, Cr1, Dam, DOR, Dow, Ear, Ers, Fe5, Fis, Fo7, Ga1, Ga21, Ge2, GLO, Gr28, GRI, HA3, Hig, HO2, Ho8, Hud, Hum, Jor, Kil, La1, Lej, LO8, Loc, ME9, Mel, Mey, Mo13, MOU, Ni1, Ol2, OLD, Ont, Pla, Pr3, Ri2, RIS, RUP, Sa6, Sau, SE4, Se8, SE14, Se17, SGF, Sh9, Sh12, So1, Sto, TE7, Ter, Tho, Tu6, Twi, Und, Ver, Vi4, WE10, Wet, WI23.

Prezzemolo Comune 2 - 75-80 days, Flat leaf type, excel. flavor, smaller leaves than Gigante di Napoli, works well from transplants. *Source History: 1 in 2004.* **Sources: Se24.**

Rina - Double Moss Curled type with long stems. *Source History: 1 in 2004.* **Sources: Coo.**

Rosette - 90 days, New triple curled type, compact plants, tighter curl than Moss Curled, good color retention over long period. *Source History: 1 in 2004.* **Sources: Tho.**

Sherwood - 75 days, Moderately coarse double-to-triple curled leaves, excellent dark-green color, 10-12 in. plants, very uniform, vigorous, holds color well even after multiple harvests, home garden and fresh market. *Source History: 3 in 1987; 14 in 1991; 5 in 1994; 2 in 1998; 1 in 2004.* **Sources: Bou.**

Survivor - 72 days, Selection from Dark Green Italian, flat leaf, very cold hardy and disease resistant. *Source History: 1 in 2004.* **Sources: Ni1.**

Sweet Curly - Double curled, green leaves, sweet, nutty tasting flavor without harsh metallic overtones, vigorous grower. *Source History: 2 in 2004.* **Sources: Pe2, Re6.**

Thai - Used in Thai and Oriental dishes, rare in the U.S., seed from Bangkok. *Source History: 1 in 2004.* **Sources: Ba8.**

Triple Curled - 68-75 days, Closely curled dark-green leaves, nutritious, actually a biennial but grown as an annual for the foliage, slow to germinate, fast-growing, uniform strain. *Source History: 18 in 1981; 14 in 1984; 14 in 1987; 14 in 1991; 17 in 1994; 19 in 1998; 22 in 2004.* **Sources: CH14, CO23, CO30, DOR, Ers, Fo7, Gr27, Kit, La1, ME9, Mo13, MOU, OLD, Orn, Pin, SE14, Se16, Twi, WE10, Wet, Wi2, WI23.**

Turkish - Flat-leaved parsley with mild, sweet flavor, originally from Istanbul. *Source History: 1 in 1994; 1 in 1998; 3 in 2004.* **Sources: Pe1, Pr3, Te4.**

Unicurl - 73 days, Finely curled dark-green leaves which curl in instead of out, foliage shakes clean of dust and dirt, easier to clean than any other strain. *Source History: 2 in 1981; 2 in 1984; 1 in 1987; 1 in 1991; 1 in 1994; 1 in 1998; 1 in 2004.* **Sources: Sto.**

Vaughan's XXX - 75 days, Excellent strain of triple curled parsley, compact, dark-green finely cut and curled leaves. *Source History: 1 in 1987; 1 in 1991; 1 in 1994; 1 in 1998; 1 in 2004.* **Sources: SGF.**

Varieties dropped since 1981 (and the year dropped):

Bravour (2004), Compact Curled (1991), Curlina (2004), Curly-Q (1994), Dark Green Winter (1983), Decorator (1994), Deep Green (1998), Dwarf Triple Curled (1983), Early Sugar (2004), Emerald Dark Green (1984), Esquire (2004), Exotica (1998), Frisca (2004), Frison (1998), Garland (2004), German Parsley (1994), Green Carpet (1998), Green Velvet (1998), Green Winter (1991), H.P.S. Supercurled (1991), Halblange Wurzelpetersilie (1987), Italian Rock (1998), Market Gardener's, Improved (1998), Minncurl (1984), Moskrul Fonvert (1987), Multikrul (1982), New Dark Green (1991), Omega Root Parsley (1998), Optima (2004), Parus (2004), Perfection Curled, Forest Green (2004), Picnic (2004), Short Sugar (1998), Titan (2004), Toso (1994), Wild (1991), Yuen Sai (1991).

Parsnip
Pastinaca sativa

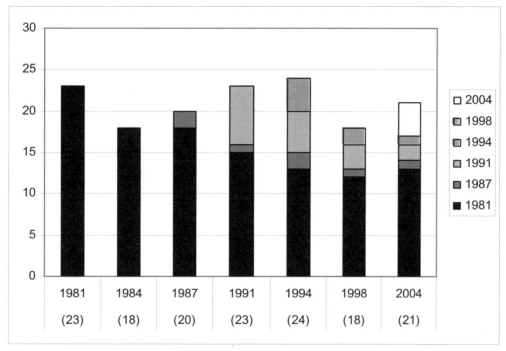

Number of varieties listed in 1981 and still available in 2004: 23 to 13 (57%).
Overall change in number of varieties from 1981 to 2004: 23 to 21 (91%).
Number of 2004 varieties available from 1 or 2 sources: 12 out of 21 (57%).

All American - (All America, All American Hollow Crown) - 95-145 days, Tapering 10-12 in. long x 3 in. dia. roots, hollow crown, white skin, white fine-textured flesh, free from side roots and fiber, attains thickness fairly early, small core, high sugar, excel. quality, stores well in a fruit cellar or in the ground. *Source History: 29 in 1981; 26 in 1984; 24 in 1987; 16 in 1991; 16 in 1994; 15 in 1998; 24 in 2004.* **Sources: Bo18, CH14, Dgs, DOR, Ers, Gr29, Gur, He17, La1, Loc, MAY, ME9, Mo13, Ni1, Re8, Ri12, SE4, Se28, So12, St18, Vi5, WE10, WI23, Wo8.**

Andover - 110-120 days, Similar to Harris Model but with more vigorous foliage, good storage variety, bright white roots grow 10-12 in. long, res. to canker in storage, PVP 1992 Regents of U of MN. *Source History: 1 in 1991; 3 in 1994; 12 in 1998; 9 in 2004.* **Sources: Alb, Fe5, HO2, RIS, RUP, SE4, Se26, Sto, Ves.**

Arrow - 110 days, Clean white root, good canker tolerance, high yields, stores well. *Source History: 2 in 2004.* **Sources: OSB, SE4.**

Avonresister - 95-110 days, By far the most resistant variety to canker, small roots, thin to only 3 in., more uniform in size and shape than most other varieties. *Source History: 2 in 1981; 1 in 1984; 1 in 1987; 1 in 1991; 1 in 1994; 1 in 2004.* **Sources: Tho.**

Cobham Improved Marrow - 120 days, English variety, half-long, 8 in., smooth tapered white roots, very high sugar content, outstanding flavor and appearance, selected for resistance to canker, high germinating seed. *Source History: 1 in 1987; 3 in 1991; 3 in 1994; 3 in 1998; 6 in 2004.* **Sources: Me7, Sa9, Se7, Syn, Ter, Tu6.**

Dlouhy Bily - *Source History: 1 in 2004.* **Sources: Se17.**

Guernsey - (Demi-Long de Guernesey) - 95-120 days, Med-long broad tapered roots, smooth white skin, productive, harvest after heavy frost which changes starch to sugar, store in moist sand or dig when convenient. *Source History: 3 in 1981; 2 in 1984; 2 in 1987; 3 in 1991; 2 in 1994; 2 in 1998; 2 in 2004.* **Sources: Gou, Lej.**

Harris Model - (All America, Early Short, Jung's White Sugar) - 100-130 days, Moderately tapered, 10-12 x 3.5 in., skin is whiter than others, almost no hollow crown, no side roots, very tender white flesh, good flavor, fair sized core, clean and refined appearance, grows well on marshland, preferred for packaging. *Source History: 36 in 1981; 33 in 1984; 36 in 1987; 40 in 1991; 40 in 1994; 37 in 1998; 38 in 2004.* **Sources: Alb, Bo19, CH14, CO23, CO30, Cr1, Dgs, DOR, Ear, Ers, Fe5, Ga1, Ha5, Hig, HO2, Hum, Jor, Jun, ME9, MOU, Na6, Pe2, Pin, RIS, RUP, Se25, Se26, Sh9, Shu, St18, Sto, Tu6, Up2, Ves, WE10, We19, Wet, WI23.**

Harris Model, Early - 80-100 days, Very heavy yielder, 10-12 in. long x 3.5 in. dia., less taper, almost no hollow crown and no side roots, skin and interior white, tender, excel. quality, pulls quite easily. *Source History: 6 in 1981; 6 in 1984; 6 in 1987; 4 in 1991; 4 in 1994; 4 in 1998; 5 in 2004.* **Sources: Bu3, De6, Ma19, Sk2, Vi4.**

Hollow Crown - (Guernsey, Sugar, Large Sugar, Long

Smooth, Hollow Crown Thick Shoulder) - 65-135 days, Long smooth white roots, 10-15 in. long x 2.75-3 in., uniformly tapered, sweet white fine-flavored flesh, no side roots, heavy yields when grown in deeply prepared soils, good winter keeper, pre-1852. *Source History: 71 in 1981; 66 in 1984; 63 in 1987; 53 in 1991; 54 in 1994; 55 in 1998; 47 in 2004.* **Sources: Alb, All, Ba8, Bo19, Bu2, Bu8, But, CH14, CO23, Co31, Com, DOR, Ear, Ers, Fis, Fo7, Ge2, Gr27, HA3, He8, He17, Hi13, Jun, La1, Lin, Mcf, ME9, Mel, Mo13, Ol2, OLD, Or10, PAG, Pe1, Roh, RUP, Sa9, Sau, Se25, Shu, Si5, So1, SOU, Ver, Vi4, WE10, WI23.**

Hollow Crown Improved - 95-120 days, Selected for whiter skin color and slightly longer root, 10-12 in. long x 3 in. dia. at the shoulder, large smooth medium tapered, very few side roots, tender fine-grained white flesh, short top, good in deeply prepared soils, good winter keeper, an old standby. *Source History: 18 in 1981; 13 in 1984; 13 in 1987; 10 in 1991; 10 in 1994; 9 in 1998; 6 in 2004.* **Sources: Be4, Hal, Jor, Mey, Ont, Sh9.**

Kral Russian - Short round roots, good for poor shallow soil, nice mild flavor, very hardy. *Source History: 2 in 2004.* **Sources: In8, Se17.**

Lancer - (Whitespear) - 110-120 days, Refined Harris Model type, very sweet, easy to lift, excel. res. to canker. *Source History: 1 in 1991; 4 in 1994; 4 in 1998; 1 in 2004.* **Sources: Jo1.**

Model Hollow Crown - 95-110 days, Med-long smooth roots, fine-grained white flesh, excel. flavor, good keeper, leave in ground as late as frost permits in the fall, freezing improves flavor. *Source History: 2 in 1981; 2 in 1984; 1 in 1987; 1 in 1991; 1 in 1994; 1 in 1998; 1 in 2004.* **Sources: Tt2.**

Offenham - 100 days, Suited for a wide range of soils, thick fleshy med-length root with broad, thick shoulders, tender flesh, sweet flavor, early cropping, good freezer. *Source History: 1 in 1981; 1 in 1987; 3 in 1998; 2 in 2004.* **Sources: Ho13, So12.**

Short Thick - (Short and Thick) - 110 days, Excellent on shallow or heavy or clay soils, similar to Hollow Crown except root is shorter, excel. quality, early, very tender and sweet. *Source History: 3 in 1981; 3 in 1984; 2 in 1987; 2 in 1991; 2 in 1994; 1 in 2004.* **Sources: Pr3.**

Survivor - Sweet, tender roots can withstand less than ideal growing conditions, PSR Breeding Program. *Source History: 1 in 2004.* **Sources: Pe6.**

Tender and True - 102 days, Reputed to be the longest parsnip with the finest flavor, 3 in. dia., almost coreless, good resistance to canker. *Source History: 1 in 1981; 1 in 1984; 2 in 1987; 2 in 1991; 3 in 1994; 3 in 1998; 3 in 2004.* **Sources: Bou, So12, Tho.**

The Student - 95-125 days, Med-long tapering root to 30 in., depending on soil, fairly narrow shoulders, heavy crops, one of the best for flavor, possibly still available from Unwins in England, introduced in 1860. *Source History: 2 in 1981; 2 in 1994; 3 in 1998; 2 in 2004.* **Sources: Bou, Ho13.**

Turga - 100 days, Med. long, stout roots, sweet flavor with a hint of coconut, excellent hardiness and germination rate, Hungary. *Source History: 1 in 1994; 3 in 1998; 5 in 2004.* **Sources: Ho13, In8, Se17, So12, Tu6.**

White Gem - 120 days, Bred for increased canker res., clean very white skin, rounded broad shoulders, short to med. length, fine flavor and quality, good home garden variety, England. *Source History: 2 in 1981; 2 in 1984; 1 in 1987; 1 in 1991; 1 in 1994; 1 in 1998; 1 in 2004.* **Sources: Tho.**

Varieties dropped since 1981 (and the year dropped):

Alba (1994), *Cobham* (1994), *Eden Valley Hollow Crown* (1998), *Evesham* (1983), *Exhibition Long-Rooted* (1983), *Fullback* (1998), *Giant* (1998), *Hollow Crown, Gaze's Improved* (2004), *Ideal Hollow Crown* (1991), *Long White* (1994), *Oatney* (2004), *Short Thick Improved* (1991), *Stump Rooted Improved* (1998), *White King* (2004), *White Model* (1998), *White Wonder* (1984).

Pea / Garden
Pisum sativum var. sativum

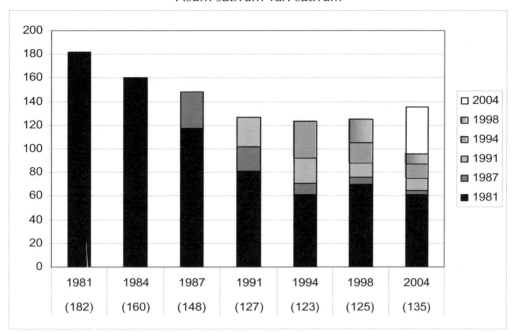

Number of varieties listed in 1981 and still available in 2004: 182 to 61 (34%).
Overall change in number of varieties from 1981 to 2004: 182 to 135 (74%).
Number of 2004 varieties available from 1 or 2 sources: 87 out of 135 (64%).

Alaska - (Earliest of All) - 50-60 days, Light-green 26-36 in. plants, 2.5-3 x .5 in. round straight pale-green pods with 5-8 small smooth-skinned peas, grows well in cool soils, FW res., heavy yields, for early first peas or dried split peas in short season areas, dry in 80 days, dates back to 1880. *Source History: 56 in 1981; 45 in 1984; 40 in 1987; 38 in 1991; 31 in 1994; 32 in 1998; 30 in 2004.* **Sources: Ba8, Be4, Bo19, CH14, Co31, CON, Cr1, DOR, Ers, Fo7, HA3, He8, He17, Ho12, Hum, Jor, La1, Lin, LO8, Mcf, ME9, Mo13, OLD, Ont, Ri12, Sa9, SOU, Tho, Vi4, Wi2.**

Alaska, Early - 52-58 days, Smooth seeded, hardy, for earliest planting, 2.5 in. pods contain 5 to 8 peas, 28 in. light-green slender plants mature uniformly, heat and drought res., for fresh use, canning, sprouting, split pea soup and cover cropping, 1881. *Source History: 15 in 1981; 12 in 1984; 11 in 1987; 7 in 1991; 6 in 1994; 8 in 1998; 9 in 2004.* **Sources: Ada, But, Cl3, De6, Fa1, Fo13, Gr27, Ma19, Wet.**

Alaska, Extra Early - (Extra Early Wilt Resistant) - 52-58 days, Improved wilt res. strain, very early and hardy, med-sized pods with 5 to 6 smooth blue-green peas, 30 in. plants, heavy concentrated crop. *Source History: 12 in 1981; 12 in 1984; 10 in 1987; 9 in 1991; 11 in 1994; 12 in 1998; 9 in 2004.* **Sources: Bu8, MAY, Mey, RUP, Sau, Se26, Sh9, Shu, Sk2.**

Alderman - (Tall Telephone, Improved Telephone, Telephone Dark Podded, Rondo o Alderman) - 70-78 days, Sel. from Duke of Albany, large dark-green vigorous indet. 5-6 ft. vines, large 4.5-5 x .75 in. plump straight pointed pods with 8-10 very large dark-green peas, holds well, RW res., climber, needs support, for canning and quick freezing, intro. in 1891. *Source History: 39 in 1981; 33 in 1984; 28 in 1987; 29 in 1991; 25 in 1994; 23 in 1998; 18 in 2004.* **Sources: Ber, Bou, CON, DOR, Fe5, HA3, Ha5, He8, Ho13, Loc, OLD, Ont, OSB, Pe1, Sa9, SE4, Ter, We19.**

Allsweet III - 60 days, Plants 22 in. tall, 7-9 peas per pod, very sweet, high yields, canner pea holds its color in water, excellent disease resistance. *Source History: 2 in 2004.* **Sources: Ers, Fi1.**

Ambassador - 90 days, Robust, 30 in. plants bear dark green pods, 3 in. long, 8-10 peas per pod, PEMV PM FW res. *Source History: 1 in 1998; 2 in 2004.* **Sources: Bou, Ga7.**

American Wonder - 60-65 days, Dark-green 12-18 in. plants, straight firm plump 3 in. pods contain 5 to 7 very sweet peas, good main crop mid-season pea, very res. to drought, extra dwarf and early, first referenced in 1898. *Source History: 10 in 1981; 6 in 1984; 4 in 1987; 2 in 1991; 1 in 1994; 1 in 2004.* **Sources: Ea4.**

Annonay - 60 days, Straight green pods contain 6-8 peas, popular for home gardeners in the North of France. *Source History: 1 in 1998; 1 in 2004.* **Sources: Gou.**

Argenteum - (Silver Leaf Pea) - 75 days, Silvery leaves on 2 ft. plants, two-tone purple flowers, beautiful. *Source History: 2 in 2004.* **Sources: Ho13, Ta5.**

Arise - 59 days, Leafless plant grows to 16 in. tall, very large, dark green peas, easy to pick, PVP 1992 Seminis Vegetable Seeds, Inc. *Source History: 1 in 1998; 1 in 2004.* **Sources: Bu2.**

Attar - Mixed Ethiopian landrace, tall vines, various small soup types. *Source History: 1 in 2004.* **Sources: Ho13.**

Balmoral - 80 days, Plants grow 28-30 in., pods average 7 peas, sow May-June for Sept-Oct. harvest, PW resistant, good DM tolerance. *Source History: 1 in 2004.* **Sources: Tho.**

Blue Bantam - (Peter Pan, Hundredfold, Laxtonian) - 60-64 days, Early, 14-18 in. vines, quite wide 4-4.5 in. dark-green pods with 7-9 very large deep bluish-green peas, high yields, long harvest, for freezing and cooking, 1902. *Source History: 10 in 1981; 10 in 1984; 9 in 1987; 7 in 1991; 3 in 1994; 4 in 1998; 2 in 2004.* **Sources: Fo7, Ho13.**

Bolero - 65-69 days, Full season maturity freezer type, combines double and triple pod set with improved plant type and high yields, 7-9 peas per pod, 26-29 in. plants, FW res., PVP expired 1994. *Source History: 2 in 1981; 2 in 1984; 2 in 1987; 2 in 1991; 3 in 1994; 6 in 1998; 7 in 2004.* **Sources: Gr27, HO2, Jor, ME9, RUP, SE4, Sto.**

Bounty - 61-68 days, Vines 24-30 in. long, sets very heavily with 3.5 in. blunt pods filled with about 8 med-size dark-green peas, resists wilt and PM, PVP 1986 Novartis Seeds, Inc. *Source History: 2 in 1991; 6 in 1994; 4 in 1998; 2 in 2004.* **Sources: Pr3, Sh9.**

British Wonder - 50-55 days, Vines grow 2-3 ft. tall, require trellising, excel. yields of sweet green peas, introduced in England by Taber and Cullen in 1890, sent to the USDA for trial in 1903, introduced by W. Atlee Burpee in 1904. *Source History: 1 in 2004.* **Sources: Se16.**

Burpeeana Early - 63 days, All purpose, straight 3 in. pods with 8 to 10 med-large peas, 18-24 in. vines, retains color and flavor during freezing, many doubles, good in the North, prolific, 1949. *Source History: 4 in 1981; 2 in 1984; 3 in 1987; 4 in 1991; 4 in 1994; 2 in 1998; 2 in 2004.* **Sources: Bu2, Hi13.**

Canoe - 70 days, So named because of the long, slightly curved pointed pods which resemble a canoe, semi-leafless plants, 28 in., up to 11-12 peas per pod, good freezer. *Source History: 1 in 2004.* **Sources: Tho.**

Capucijners - Wrinkled soup peas look like brown-green pebbles, staking not necessary but helpful, fragrant maroon and white flowers, fresh peas are not sweet but are edible, best when dried and cooked whole into a rich brown gravy. *Source History: 1 in 1991; 1 in 1994; 3 in 1998; 1 in 2004.* **Sources: So12.**

Caseload - 57 days, Self-supporting 2.5 ft. plants, 2.5-3 in. pods, 6-7 med. size peas, sweet flavor, holds flavor longer than most while still on the vine. *Source History: 1 in 2004.* **Sources: Na6.**

Champion of England - 67-75 days, Older midseason variety, blunt 3 in. pods, 6-10 peas per pod, said to be the oldest wrinkled pea in cultivation, introduced to England by William Fairbeard in 1843. *Source History: 1 in 1994; 2 in 1998; 2 in 2004.* **Sources: Pr3, St18.**

Chicharo Criollo - Vines grow 5 ft., produce many small pods with tasty small peas, resists aphids, from Zapotecs in Mexico. *Source History: 2 in 1998; 1 in 2004.* **Sources: Ho13.**

Citadel - (French Citadel) - 60 days, Tiny French peas, sweet at any stage, 3-4 in. plump pods contain 6-8 peas which are about half the size of American shelling peas, vigorous, disease resistant plants to 26 in., exceptional yields, PVP expired 2004. *Source History: 3 in 2004.* **Sources: Lin, Pin, Re6.**

Coral - 50-60 days, Vines to 2 ft., early pod set is low on the plant, good yields, very good quality, good production in a small space, suitable for container growing, does well in dry conditions, resists FW and PLRV. *Source History: 1 in 1994; 2 in 1998; 3 in 2004.* **Sources: Fe5, Me7, Tu6.**

Dakota - 52-60 days, Vines to 36 in. tall, best when trellised, double pods at each node, 8-9 peas per pod, sweet flavor, resists PM and FW1, PVP 2002 Syngenta Seeds, Inc. *Source History: 10 in 2004.* **Sources: Coo, Ers, Jo1, Jor, Ki4, ME9, RUP, Ter, Twi, Ves.**

Dark Pod Telephone - (Alderman) - 76 days, *Source History: 1 in 1981; 1 in 1984; 1 in 1987; 1 in 1998; 1 in 2004.* **Sources: PAG.**

Daybreak - 52-65 days, Plant grows to 20 in., 3 in. pods with 6-7 peas, very early, easy to shell, FW res., PVP 1986 Novartis Seeds, Inc. *Source History: 6 in 1991; 12 in 1994; 3 in 1998; 1 in 2004.* **Sources: Lin.**

Dual - 71 days, Full season variety, 30 in. vines, 4 in. pods consistently borne as doubles, 10-14 med-size med-dark peas per pod, high yielding, excellent for eating fresh or freezing, PVP expired 2000. *Source History: 1 in 1987; 2 in 1991; 2 in 1994; 5 in 1998; 11 in 2004.* **Sources: CH14, Ers, Gur, Jor, Jun, ME9, MOU, SE4, Shu, Ver, Ves.**

Early Bird - (Laxton's Superb) - 60-61 days, Light-green 18-24 in. vines, loaded with curved pointed dark-green 4-4.5 in. pods with 7 to 10 peas, very res. to cold wet weather, excel. flavor. *Source History: 9 in 1981; 9 in 1984; 9 in 1987; 8 in 1991; 7 in 1994; 7 in 1998; 1 in 2004.* **Sources: Alb.**

Early Freezer 680 - 58-63 days, Double podded variety, good color and quality and high yields, good canning and freezing var., PVP expired 2001. *Source History: 2 in 1991; 3 in 1994; 6 in 1998; 5 in 2004.* **Sources: Bu2, Dam, Ers, HO2, Roh.**

Early Frosty - 60-68 days, High yielding freezer pea, 28-30 in. vines, 3.5-4 in. blunt double pods contain 7 to 8 uniform dark-green med-sized peas, excel. quality and yields and vigor, wilt res., for freezing or fresh use or fresh market, good home garden variety. *Source History: 26 in 1981; 23 in 1984; 24 in 1987; 30 in 1991; 32 in 1994; 34 in 1998; 23 in 2004.* **Sources: All, But, CH14, CON, Ers, Fe5, Gr27, Hal, He8, He17, HO2, Jor, Jun, La1, Mo13, MOU, PAG, Roh, RUP, SE4, Shu, Si5, Wet.**

Eclipse - 65 days, Grows up to 3 ft., 7-8 peas per pod, yields well, produces a sweeter pea than standard garden varieties, sweetness holds longer before being converted to starch, this was achieved through conventional breeding, not through genetic engineering, PVP 2001 Seminis Vegetable Seeds, Inc. *Source History: 3 in 2004.* **Sources: Hum, OSB, SE4.**

Elf - Abundant production of green pods on 2.5 ft. vines, small sweet green peas good for "petit pois". *Source History: 2 in 2004.* **Sources: In8, Se17.**

Endeavour - 80 days, Early maincrop pea, 26-28 in., semi-leafless vines nearly self-supporting if grown in a block, long double pods with 8 peas, excellent flavor, PW resistant, good DM tol. *Source History: 1 in 2004.* **Sources: Tho.**

Fairbeard's Nonpareil - Older midseason variety with 3 ft. vines, rather small peas, English type. *Source History: 1 in 1994; 2 in 1998; 2 in 2004.* **Sources: Ho13, Pr3.**

Feltham First - 45-58 days, Very early, 18 in. plants, produces heavy crops of round-seeded deep-green peas in large pointed 3.5 in. pods, ideal for autumn or March sowing, excel. freezer. *Source History: 1 in 1981; 1 in 1984; 2 in 1987; 2 in 1991; 2 in 1994; 1 in 1998; 1 in 2004.* **Sources: So12.**

Fortune - 80 days, Curved, pointed 3-4 in. pods, superior flavor, autumn sown. *Source History: 1 in 2004.* **Sources: Tho.**

Freezer - (Berthold Freezer) - 62 days, Dark-green pods filled with large peas of excel. quality, suitable for quick freezing, productive. *Source History: 1 in 1981; 1 in 1984; 2 in 1987; 1 in 1991; 1 in 2004.* **Sources: Mcf.**

Freezer 69 - 69 days, An improved Homesteader dev. for the freezing industry, 24-30 in. dwarf plant, 6 in. pods with 6-7 peas, heavy yields, stands well in warm weather, good canner, ASG, 1960. *Source History: 6 in 1981; 5 in 1984; 3 in 1987; 3 in 1991; 3 in 1994; 2 in 1998; 2 in 2004.* **Sources: Ear, Lin.**

Freezonian - (Thomas Laxton W-R) - 60-63 days, AAS/1948, Thomas Laxton x Wilt Res. World's Record, 30-40 in. vigorous vines, blunt 3-3.5 x .6 in. pods with 7 to 8 dark-green wrinkled peas, high sugar content, FW res., RR tol., trellis for heavy early yield, can freeze or market. *Source History: 40 in 1981; 36 in 1984; 36 in 1987; 27 in 1991; 20 in 1994; 19 in 1998; 14 in 2004.* **Sources: All, Be4, Bu3, CON, Fa1, HA3, HO2, Mel, Roh, RUP, SE4, Sh9, Shu, Vi4.**

Fristo - 57 days, Introduced as a replacement for and improvement of Citadel with the same sweet gourmet flavor, good disease resistance, PVP 1997 Seminis Vegetable Seeds, Inc. *Source History: 1 in 2004.* **Sources: Ers.**

Frosty - 60-67 days, Early freezer, strong vigorous 28 in. vines, large blunt 3.5 in. pods, 7-9 dark-green olive-shaped peas, sweet, always double, FW res., heavy yields, widely adapted, Novartis Seeds, Inc., 1960. *Source History: 16 in 1981; 15 in 1984; 13 in 1987; 10 in 1991; 8 in 1994; 11 in 1998; 9 in 2004.* **Sources: Be4, HA3, Ha5, MAY, ME9, Mey, Sh9, SOU, Ves.**

Galena - (FP2237) - 61 days, Heavy yielder with most pods borne in pairs, 8-10 peas per pod, flavorful, tolerates F1 and 2 and PLRV, PVP 2002 Syngenta Seeds, Inc. *Source History: 2 in 2004.* **Sources: Ers, We19.**

Garden Sweet - 70 days, Retains garden-fresh flavor for up to two days, 25% more sugar than other peas, 3-4 ft. vines require support, 3-4 in. pods with 9-10 seeds, large crops. *Source History: 1 in 2004.* **Sources: Bu2.**

Gloire de Quimper - *Source History: 1 in 2004.* **Sources: Se17.**

Green Arrow - (Green Shaft) - 62-70 days, English bred main crop var., 24-28 in. vines, 4-5 in. slim pointed pods with 8 to 11 small deep-green peas, pods borne double and set high, res. to DM FW RR and LCV, wrinkled seed. *Source History: 73 in 1981; 69 in 1984; 76 in 1987; 76 in 1991; 73 in 1994; 82 in 1998; 80 in 2004.* **Sources: Ada, Alb, All, Ami, Be4, Bo17, Bou, Bu1, Bu2, Bu3, CA25, CH14, Cl3, Co31, Com, CON, Coo, Dam, De6, Dgs, DOR, Ers, Fa1, Fe5, Fi1, Fis, Ga1, Goo, Gr27, Gur, HA3, Ha5, He8, He17, Hi6, HO2, Hud, Hum, Jor, Jun, Lin, Ma19, MAY, Mcf, ME9, Mel, Mey, Mo13, MOU, Ni1, OLD, Ont, Or10, OSB, PAG, Par, Pin, Pr3, Ra5, Ri12, RIS, Roh, RUP, Sa9, Sau, Se16, Se26, Sh9, Shu, Si5, SOU, St18, Tho, Tt2, Tu6, Twi, Up2, Ver, Ves, Wet.**

Greensage - 70 days, Greenshaft x semi-leafless variety, self-supporting vines if grown in a block, 3.5 in. curved pods, contain 8-11 peas, sweet flavor. *Source History: 2 in 2004.* **Sources: Hi13, Tho.**

Hatif de Anonay - French var., 2 ft. vines, small pods with small tasty sweet peas. *Source History: 1 in 1998; 1 in 2004.* **Sources: Ho13.**

Homesteader - (Lincoln) - 64-67 days, Vigorous productive 18-30 in. plants, 4 in. rather slender pointed slightly curved pods, 7 to 10 dark-green peas, heat res., best pea for Canadian prairies. *Source History: 9 in 1981; 8 in 1984; 8 in 1987; 8 in 1991; 8 in 1994; 8 in 1998; 8 in 2004.* **Sources: Alb, Ear, He8, Lin, Ont, Pr3, Te4, Tt2.**

Hurst Green Shaft - 70 days, Outstanding pod length, 4-4.5 in. with 9-11 peas per pod, borne double in the tops of 28 in. plants, wrinkled seed, maincrop var., excel. flavor, good freezer. *Source History: 1 in 1981; 1 in 1984; 1 in 1987; 1 in 1991; 3 in 1994; 1 in 1998; 1 in 2004.* **Sources: Tho.**

Hylite - 68 days, Dark-seeded Perfection, 30 in. vines, 3.5 in. pods contain 7 to 9 med-sized dark-green peas, excel. for freezing. *Source History: 1 in 1981; 1 in 1984; 1 in 1987; 1 in 1991; 1 in 1994; 1 in 1998; 1 in 2004.* **Sources: Fis.**

Jaguar - 80 days, Plants to 26-28 in. tall, double pods, 6-7 peas per pod, sow March-June for harvest from June-September, PW resistant, good DM tolerance. *Source History: 1 in 2004.* **Sources: Tho.**

Kelvedon Wonder - (Kelvedon) - 60-75 days, Dark-green pointed well-filled 3 in. pods, 18 in. plants, very sweet, very prolific, excel. for early spring and successive sowings and quick freezing, wrinkled seed. *Source History: 4 in 1981; 3 in 1984; 4 in 1987; 2 in 1991; 2 in 1994; 1 in 1998; 2 in 2004.* **Sources: Bou, Tho.**

Knight - 56-62 days, Progress strain, extra large 3.5-4 in. pointed well-filled pods contain 9 to 10 bright-green peas, vigorous productive 20-21 in. plants, PEMV BYMV PM and CW res., NY/AES. *Source History: 14 in 1981; 13 in 1984; 20 in 1987; 25 in 1991; 33 in 1994; 37 in 1998; 29 in 2004.* **Sources: All, Be4, CH14, Cl3, CON, Ga1, Gr27, Ha5, Hal, HO2, Jo1, Jor, ME9, Mey, Mo13, OLD, Pin, RIS, Roh, SE4, Sh9, SOU, Sto, Sw9, Tho, Tt2, Ves, We19, Wet.**

Kurume Hi-Crop - 70 days, Japanese pole garden pea, white flowers, 8 seeds per pod, good yields, high tolerance to cold temperatures. *Source History: 1 in 2004.* **Sources: Kit.**

Lacy Lady - (Novella Lacy Lady) - 55-64 days, No

leaves, just tendrils, no staking, plant two rows 12 in. apart and just climbs itself, 2.5-3 in. double pods near tops of sturdy 18 in. true bush plants, disease and wilt res., PVP expired 2001. *Source History: 16 in 1981; 16 in 1984; 18 in 1987; 10 in 1991; 4 in 1994; 1 in 1998; 1 in 2004.* **Sources: Ver.**

Laxton's Progress - 58-64 days, Largest podded early dwarf type, med-dark green 10-20 in. vines with heavy foliage, dark-green plump 4.5 x .9 in. pod with 8-10 wrinkled apple-green seeds, very productive, does well in cooler climates, good freezer, used extensively by shippers. *Source History: 34 in 1981; 30 in 1984; 29 in 1987; 15 in 1991; 13 in 1994; 8 in 1998; 5 in 2004.* **Sources: Bu3, Bu8, Ear, Hal, Lin.**

Laxton's Progress Improved - 55-62 days, Earlier, vigorous 16-18 in. dwarf plants, 7 to 9 large peas of good quality, excel. early main crop variety. *Source History: 3 in 1981; 3 in 1984; 2 in 1987; 2 in 1991; 2 in 1994; 2 in 1998; 1 in 2004.* **Sources: Sto.**

Laxton's Progress No. 9 - (Progress No. 9, Greater Progress, Blue Bantam) - 58-65 days, Leading large podded home garden pea, earlier taller hardier and larger than Laxton's Progress, vigorous dark-green 15-20 in. plants, 4-5 in. x .9 in. pods with 6 to 9 large wrinkled dark-green seeds, fine flavor and quality, some splitting, FW res., consistent heavy crops, dev. from British lines. *Source History: 61 in 1981; 56 in 1984; 56 in 1987; 54 in 1991; 50 in 1994; 57 in 1998; 51 in 2004.* **Sources: Ag7, Alb, All, Be4, Bo17, CA25, CH14, Cl3, CON, Dam, De6, Dom, DOR, Ers, Fis, Gr27, HA3, He8, HO2, Jor, Jun, La1, Loc, Ma19, MAY, Mcf, ME9, Mey, MOU, Ont, OSB, PAG, Pe2, Roh, RUP, Sau, SE4, Se26, Sh9, Shu, Si5, SOU, Sto, TE7, Twi, Ver, Ves, Vi4, WE10, Wet, Wi2.**

Lincoln - (Homesteader) - 60-70 days, Older type for home and market gardeners, wilt res. 18-30 in. vines, 3-3.5 in. tightly filled pods mostly in pairs, 6 to 9 small peas per pod, very productive, easy to shell, excel. for growing in hot weather, does well in the North, intro. to American gardeners in 1908. *Source History: 42 in 1981; 37 in 1984; 41 in 1987; 37 in 1991; 35 in 1994; 42 in 1998; 41 in 2004.* **Sources: Al6, All, Be4, Bu3, CH14, CON, Coo, Dam, DOR, Dow, Ers, Fe5, Fis, Ga1, Gr27, Gur, HA3, Ha5, He17, Hi6, HO2, Jor, La1, Mcf, ME9, Mo13, MOU, PAG, Pin, Ri12, Roh, RUP, Se8, Se16, Shu, Sk2, Sto, Und, Ver, Ves, Vi4.**

Little Marvel - (Improved American Wonder, Extra Early Little Marvel) - 58-64 days, Vigorous bushy 15-20 in. dwarf plant, 3-4 x .6 in. square-ended tightly packed pods with 6-7 med-sized dark-green peas, mostly borne double, fine quality, FW res., heavy yields, extended season, old dependable var. introduced as Sutton's Little Marvel in 1900. *Source History: 121 in 1981; 109 in 1984; 100 in 1987; 94 in 1991; 80 in 1994; 74 in 1998; 61 in 2004.* **Sources: Alb, All, Ba8, Be4, Bu3, CH14, Co32, CON, Cr1, De6, Dom, DOR, Ear, Ers, Fa1, Fe5, Fi1, Ge2, Gr27, Gur, HA3, Ha5, Hal, He17, HO2, Ho12, Hum, Jor, Jun, La1, Lej, Lin, LO8, Loc, Ma19, MAY, ME9, Mel, Mey, Mo13, MOU, OLD, Ont, PAG, Ri12, Roh, RUP, Sau, SE4, Se8, Se16, Sh9, Shu, Si5, So1, Tho, Tt2, Ver, Vi4, Wet, Wi2.**

Little Marvel Improved - 63-65 days, Blunt-ended 3 in. pods well filled with 6-8 dark-green peas, vigorous 18 in. vines, no staking, fine quality, early, high yields, excel. on Florida muck soils. *Source History: 3 in 1981; 3 in 1984; 2 in 1987; 2 in 1991; 1 in 1994; 1 in 1998; 1 in 2004.* **Sources: Kil.**

Maestro - (Improved Alaska) - 55-61 days, Nearly immune to PM, tol. to PEM BYM and CW, heavy crops of 4-4.5 in. pods, 9 to 12 dark-green med-sized peas, 24-27 in. plants, Geneva NY/AES. *Source History: 15 in 1981; 14 in 1984; 23 in 1987; 28 in 1991; 29 in 1994; 42 in 1998; 25 in 2004.* **Sources: Be4, Bu2, DOR, Fi1, Ga1, Gur, He17, Hig, HO2, Jor, Jun, ME9, Mo13, Or10, Par, Pr3, Roh, RUP, Sh9, Shu, Si5, St18, Ter, Tu6, Wet.**

Manitoba - Vines are self-supporting if mass planted, good cover crop, the large tendrils are an excellent addition to salads, pods contain 8-10 seeds. *Source History: 3 in 1998; 1 in 2004.* **Sources: Sa5.**

Maxigolt - 62 days, Vines grow 3.5-5 ft., broad 3.5 in. pods, sweet large dark green peas, wilt res. *Source History: 1 in 1998; 1 in 2004.* **Sources: Jo1.**

Mayfair - 72-75 days, Long 40 in. vines but tend not to tangle, 4-4.5 in. double pods, high sugar, easy harvest, holds well, heat and PM tol., much tested in Northeast, NY/AES. *Source History: 1 in 1981; 1 in 1984; 2 in 1987; 5 in 1991; 3 in 1994; 3 in 1998; 3 in 2004.* **Sources: Alb, Fe5, Tu6.**

Mennonite - Garden pea, can be dried for soup, creamy color seed, rare. *Source History: 1 in 2004.* **Sources: Ta5.**

Mesa - 64 days, Excellent canner variety, 17 in. semi-leafless plants, 2-3 pods per node, 8-9 peas per pod, resists PM, FW, LCV. *Source History: 1 in 2004.* **Sources: MOU.**

Miragreen - 70 days, Large very sweet peas, 48 in. vines, very hardy, unusually high sugar content, produces an abundance of med-sized pods containing 8 to 10 peas, FER, 1953. *Source History: 2 in 1981; 1 in 1984; 3 in 1987; 2 in 1991; 1 in 1994; 2 in 1998; 2 in 2004.* **Sources: Fe5, Ho13.**

Monico - 57 days, Compact 20-24 in. plants produce 2-3 pods per node, 8 seeds per pod, sweet, juicy flavor, huge yields, strong disease tol., PVP 2001 Seminis Vegetable Seeds, Inc. *Source History: 1 in 2004.* **Sources: Ver.**

Montana Marvel - 64 days, Pods 3.5 in. long, filled with 8 or 9 med-sized peas, retains color when frozen, 18 in. vines. *Source History: 1 in 1981; 1 in 1984; 2 in 1987; 2 in 1991; 2 in 1994; 2 in 1998; 2 in 2004.* **Sources: Fis, Hig.**

Morse's Progress No. 9 - 64 days, Pointed 4.5 in. dark-green pods with 8 to 10 dark-green peas, 20 in. plants, res. to FW, wrinkled seed, popular early variety for home gardens and early market. *Source History: 2 in 1981; 2 in 1984; 2 in 1987; 2 in 1991; 1 in 1994; 1 in 1998; 1 in 2004.* **Sources: Kil.**

Mr. Big - 58-72 days, AAS/2000, vines to 3-4 ft. tall, large pods measure 4.5 in. long by .5 in. diameter, 9-10 large peas per pod, retains color and sweet flavor for extended time, retains quality when frozen, high percentage of double pod set, resists FW1 and PM, PVP 2002 Seminis Vegetable Seeds, Inc. *Source History: 35 in 2004.* **Sources: Ag7, Bu1, Bu2, CA25, Dam, Ear, Ers, Fa1, Ge2, Ha5, Hal, Hi13, HO2, Ho13, Jor, Jun, Lin, MAY, Mcf, ME9, MOU, Ni1, OLD, Ont, OSB, Par, Pin, RUP, SE4, Shu, Sto, Tt2, Twi,**

Ver, Ves.

Mrs. Van's - 70-90 days, High quality staking pea for fresh eating and freezing, from the Heritage Seed Program. *Source History: 2 in 1994; 2 in 1998; 1 in 2004.* **Sources: Fo13.**

New Century - 58 days, Compact 30-36 in. vines bear 6 in. pods with 8-9 peas, matures all at once, resists PM and PEMV. *Source History: 1 in 2004.* **Sources: Ter.**

Northern Sweet - 70 days, Largest podded pea, sweet and productive, 24 in. vines, produces over a long period. *Source History: 2 in 1981; 1 in 1984; 1 in 1987; 1 in 1991; 1 in 1994; 2 in 1998; 2 in 2004.* **Sources: Ho13, Pr3.**

Novella - (Novella Leafless, Bikini) - 64-70 days, Unique, strong tendrils instead of leaves, 20-28 in. plants, med-green 3 in. pods mostly in pairs, 9 small to med-sized peas, long harvest, heavy yields, needs no staking, tendrils interlock keeping plants off ground, PVP expired 1996. *Source History: 41 in 1981; 39 in 1984; 37 in 1987; 32 in 1991; 24 in 1994; 15 in 1998; 7 in 2004.* **Sources: Be4, Ear, Fis, MAY, Mcf, Pr3, Shu.**

Novella II - 65-68 days, Almost leafless 28 in. vines, 3-3.25 in. pods with 8 or 9 peas, usually 2 pods per cluster, needs no staking, res. to common wilt, novelty, PVP expired 2000. *Source History: 1 in 1981; 1 in 1984; 4 in 1987; 8 in 1991; 10 in 1994; 13 in 1998; 9 in 2004.* **Sources: Dam, Ga1, HO2, Jor, Mel, RUP, Se25, Ves, We19.**

Olympia - (Early Olympia) - 62 days, Improved Laxton's Progress type, vigorous dark-green 16-18 in. vines, 4-4.5 in. pods with 9 peas, many doubles, heavy yields, concentrated set, holds well after harvest, PEMV BYMV and PM res., NY/AES. *Source History: 6 in 1981; 6 in 1984; 12 in 1987; 15 in 1991; 16 in 1994; 7 in 1998; 9 in 2004.* **Sources: CON, HO2, Hum, ME9, Mo13, Ont, RUP, Sto, We19.**

Onward - 70-72 days, Main crop variety, 24 in. tall vines, well-filled blunt-ended 3.5 in. pods with 6 to 8 large tender peas, pods borne in pairs, old English favorite. *Source History: 3 in 1981; 1 in 1984; 2 in 1987; 1 in 1991; 3 in 1994; 2 in 1998; 2 in 2004.* **Sources: DOR, Tho.**

Oregon Pioneer - (OSU 700) - 61-102 days, Det. plant habit, 24 in. tall, well-filled straight pods borne as doubles, earliness and good flavor make this a standard for the early garden, PEMV PW and FW1 res., dev. by Dr. James Bagget at OSU. *Source History: 2 in 1991; 2 in 1994; 2 in 1998; 3 in 2004.* **Sources: Ni1, Ter, We19.**

Oregon Trail - (OSU 695) - 55-71 days, Bush plant bears until midsummer, twin podded, up to 9 peas per pod, suitable for mech. harv., prefers cool moist weather, multiple disease res., Dr. Jim Baggett, Oregon State Univ. *Source History: 2 in 1991; 4 in 1994; 7 in 1998; 12 in 2004.* **Sources: DOR, HO2, Jor, ME9, Ni1, OSB, RUP, Se7, Se17, So12, Ter, We19.**

Paladio - 62 days, Compact 25 in. plant, 8 large seeds per pod, for fresh market, canning, freezing, suitable for mechanical harvest, FW tolerant, PVP 1997 Seminis Veg. Seeds, Inc. *Source History: 1 in 1998; 3 in 2004.* **Sources: OSB, SE4, Sto.**

Parsley - (Prairie Parsley) - Unusual var. with edible tendrils that resemble curled parsley, tasty. *Source History: 1 in 1991; 1 in 1994; 2 in 1998; 1 in 2004.* **Sources: Pr3.**

Patriot - 65-72 days, An improved Prairie var., 20-24 in. plants, loose 3.5-4 in. pods with 9 to 10 dark-green peas, double at joints, easy to shell, retains good flavor and texture when frozen. *Source History: 12 in 1981; 9 in 1984; 7 in 1987; 9 in 1991; 8 in 1994; 8 in 1998; 1 in 2004.* **Sources: Fe5.**

Payload - 56 days, Plants 20-24 in. tall, deep green peas, sweet flavor, resists F1 and 2, extra early maturity, PVP expired 2003. *Source History: 3 in 2004.* **Sources: Ers, Fi1, Sk2.**

Peewee - *Source History: 1 in 2004.* **Sources: Se17.**

Perfection, Dark Green - (Dark Skinned Perfection, Perfection Dark Seeded) - 65-70 days, Dark-green 3.5-4 in. x .6 in. straight blunt pods in clusters, 7-9 dark-green peas, 26-30 in. vines need staking, FW and heat res., can be planted late, excellent canner, holds color during processing, also a good table pea, old New England favorite but good in any climate. *Source History: 6 in 1981; 6 in 1984; 5 in 1987; 6 in 1991; 5 in 1994; 9 in 1998; 4 in 2004.* **Sources: CON, DOR, He8, MOU.**

Perfection, Early - 66 days, Similar to Perfection but sturdier and more drought tolerant, vines grow 26-30 in., 3.5 in. straight blunt pods. *Source History: 1 in 1981; 1 in 1984; 2 in 1987; 2 in 1991; 1 in 1994; 3 in 1998; 1 in 2004.* **Sources: Dgs.**

Petit Pois - 58-59 days, Very tender small-seeded French peas, small pods, prolific 20-30 in. vines, very hardy, plant early, withstood frosts that froze 1 in. of soil, commercial or home. *Source History: 2 in 1981; 2 in 1984; 2 in 1987; 1 in 1991; 1 in 1998; 2 in 2004.* **Sources: Se8, Ta5.**

Petit Provencal - 60 days, Early and very productive, probably the most popular pea in France, 16 in. plants, should be planted early. *Source History: 3 in 1981; 2 in 1984; 1 in 1987; 2 in 1991; 1 in 1994; 3 in 1998; 3 in 2004.* **Sources: Gou, Ho13, Pr3.**

Picolo Provenzale - 55-60 days, Dwarf bush to 36 in. tall, rustic 'country style' pea. *Source History: 1 in 2004.* **Sources: Se24.**

Pioneer - 55-70 days, Bush plant grows 20-24 in., 3 in. pods, good yields, bred by Jim Baggett of Oregon State University. *Source History: 1 in 1994; 2 in 1998; 2 in 2004.* **Sources: Pe2, Se7.**

Pisello Nano Sole di Sicilia - Translates "Sicilian Sunshine", short plant produces sweet peas all summer with sufficient moisture. *Source History: 1 in 2004.* **Sources: Ta5.**

Plein de Panier - French translation is "full basket", intro. in 1872 by Thomas Laxton, an English seedsman. *Source History: 1 in 2004.* **Sources: Ta5.**

Prince Albert - Most popular variety of English pea in America during the mid-19th century, grown in England before 1837, introduced to the U.S. in 1845. *Source History: 1 in 1994; 2 in 2004.* **Sources: Ea4, Th3.**

Puget - 70-75 days, Heavy cropper with up to five good sized peas per pod, high sugar content is retained over a long cropping period, excel. freezer. *Source History: 1 in 1994; 2 in 2004.* **Sources: In8, Se17.**

Quantum - 69 days, Sweet large peas, excellent freezer, resists FW1, tolerates PM and PLRV, PVP 1992 Seminis Vegetable Seeds, Inc. *Source History: 1 in 2004.* **Sources: Ers.**

Rally - 61-67 days, Plants to 20 in., 2 pods per node, 8-10 small sweet peas per pod, FW and PM tolerant/resistant, PVP expired 1993. *Source History: 3 in 2004.* **Sources: Ers, Jun, PAG.**

Record - 60 days, Vines 3 ft. tall, 2.5 in. pods, very sweet, does well in summer. *Source History: 1 in 1998; 1 in 2004.* **Sources: Dam.**

Rondo - 74-77 days, Stocky vigorous det. 28-30 in. wilt res. plants, dark-green 4.5 x .75 in. pods contain 8 to 10 dark-green peas, some pods borne in pairs, FW res., late maturing. *Source History: 3 in 1981; 3 in 1984; 3 in 1987; 3 in 1991; 2 in 1994; 3 in 1998; 6 in 2004.* **Sources: DOR, HO2, ME9, SE4, Sto, Tho.**

Round Green - 55 days, Round green peas for dry use, especially for pea soup, also for fresh use but not as sweet as the wrinkled varieties, strong with 24 in. vines. *Source History: 1 in 1981; 1 in 1984; 1 in 1987; 1 in 1991; 1 in 1994; 1 in 1998; 1 in 2004.* **Sources: Dam.**

Saturn - 80 days, Huge crops of 3 in. log pods with 9 sweet juicy peas, dark green foliage, 30 in. tall. *Source History: 1 in 2004.* **Sources: Tho.**

Serpette - From France. *Source History: 1 in 1987; 1 in 1991; 1 in 1994; 1 in 1998; 1 in 2004.* **Sources: Lej.**

Shoshone - Nice garden/table pea. *Source History: 1 in 2004.* **Sources: Ta5.**

Sickle - (Risser Sickle Pea) - Sickle-shaped pods, slightly square seeds, grow with support, from the Risser family in Lancaster County, pre-1900. *Source History: 1 in 1994; 1 in 1998; 1 in 2004.* **Sources: Lan.**

Sierra Madre del Sur - Smooth-seeded pea, small pods, white flowers, very sweet and flavorful, especially good raw, can also be used as dry peas, possibly aphid res., Zapotec Indian seed. *Source History: 2 in 1981; 2 in 1984; 1 in 1987; 1 in 2004.* **Sources: Se17.**

Spanish Skyscraper - Vines grow 5-6 ft., short pods contain large seeds, heavy yields, dev. by Ken Allan, plant breeder from Kingston, Ontario, Canada. *Source History: 3 in 1994; 3 in 1998; 3 in 2004.* **Sources: Al6, Pr3, Te4.**

Sparkle - 55-60 days, Extra early freezer, compact 15-18 in. vines, blunt 2.5-3.25 x .6 in. pods contain 6 to 8 small peas, concentrated set, CW res., good percentage of doubles gives high yields. *Source History: 16 in 1981; 15 in 1984; 19 in 1987; 20 in 1991; 19 in 1994; 8 in 1998; 5 in 2004.* **Sources: DOR, Ers, Ho13, ME9, RUP.**

Spring - 52-57 days, Fast-growing 18-22 in. plant sets 5 to 7 blunt dark-green 3 in. pods with 5 to 7 med-sized peas, dev. for commercial freezing, excel. quality, very early, PVP expired 1991. *Source History: 3 in 1981; 3 in 1984; 4 in 1987; 8 in 1991; 10 in 1994; 18 in 1998; 18 in 2004.* **Sources: Ag7, Be4, Bu2, CH14, Ear, Ers, Gr27, HO2, Jor, Mcf, ME9, Mo13, Roh, RUP, SE4, Sto, Tt2, Ves.**

Straight Arrow - Main season, fresh market var., 4 in. long slender dark green pods, 10-12 peas per pod, do not get bitter or starchy in the heat, very productive. *Source History: 1 in 2004.* **Sources: Dam.**

Stratagem - 70-72 days, Stout vigorous stems on strong 2 ft. plant, 8 to 10 large peas per 4 in. pod, late, dev. in 1879 by Carter Seeds in England, introduced to North America in 1883, received honor awards from Royal Hort. Soc. *Source History: 2 in 1981; 2 in 1984; 2 in 1987; 1 in 1991; 2 in 1994; 1 in 1998; 2 in 2004.* **Sources: Ho12, Lin.**

Summer Pea - Fresh green shelling pea stands up to summer heat better than any others, vines grow 4-5 ft. tall, best grown with support. *Source History: 1 in 1991; 1 in 1994; 1 in 1998; 1 in 2004.* **Sources: Sa5.**

Survivor - 70 days, Leafless vines to 24 in. with intertwining tendrils which require less support, 8 seeds per pod, excellent disease resistance, PVP 2002 Seminis Vegetable Seeds, Inc. *Source History: 1 in 2004.* **Sources: Bu2.**

Sutton's Harbinger - 52-60 days, Received an Award of Merit from the Royal Hort. Soc. in 1901, 28-32 in. tall plants, 6-8 peas per pod, heavy crops, excel. quality, dev. in England in 1898. *Source History: 2 in 2004.* **Sources: Ag7, Se16.**

Tacoma - 63 days, Plant has few leaves but many tendrils for support, pods are held high on the plant for easy harvest, good choice for wide row intensive planting, disease res., PVP 1997 Seminis Vegetable Seeds, Inc. *Source History: 1 in 2004.* **Sources: Hum.**

Tall Arrow - Early variety with quality equal to that of Lincoln, dev. by Ken Allan, plant breeder from Kingston, Ontario, Canada. *Source History: 2 in 1994; 1 in 1998; 1 in 2004.* **Sources: Al6.**

Tarahumara - Dry farmed, from Sierra Madres of Mexico, not heat adapted, plant in spring in cooler climates. *Source History: 1 in 1987; 1 in 1991; 1 in 2004.* **Sources: Na2.**

Telefono - Long green pods with large round seeds. *Source History: 1 in 1987; 1 in 1991; 1 in 1994; 1 in 1998; 1 in 2004.* **Sources: Ber.**

Telephone, Dwarf - (Little Gem, Daisy, Dwarf Alderman) - 70-78 days, Late var., extends season, light-green 24 in. plants, 4.5 in. broad straight pointed pods with 8 to 9 peas, heavy crops, wilt res., some doubles, 1888. *Source History: 13 in 1981; 10 in 1984; 8 in 1987; 7 in 1991; 3 in 1994; 6 in 1998; 3 in 2004.* **Sources: CON, Fo7, HA3.**

Telephone, Tall - (Alderman, Merrimack Improved Telephone) - 68-78 days, Climbs 4-6 ft. with support, large pods, easy to pick, good late main crop variety, 4.5-6 in. oval dark-green pods with 8 or 9 peas, res. to FW, very productive, for home gardens or market or freezing or shipping, 1878. *Source History: 39 in 1981; 37 in 1984; 36 in 1987; 31 in 1991; 26 in 1994; 24 in 1998; 20 in 2004.* **Sources: All, Bo19, Co32, Com, Dam, Ear, Fa1, Fo7, Hal, Hum, Lin, Or10, Pin, Se17, Se24, Sh13, Shu, SOU, Ver, Ves.**

The Pilot - Vines up to 5 ft. long, large pointed deep-green pods, for spring or autumn sowing, extremely hardy, from England. *Source History: 1 in 1981; 1 in 1984; 1 in 1987; 1 in 1991; 1 in 1994; 3 in 1998; 3 in 2004.* **Sources:**

Bou, Ho13, Pr3.

<u>Thomas Laxton</u> - (Freezonian) - 55-65 days, All purpose, 30-36 in. vines, dark-green square-ended straight 3.5-4.5 x .6 in. single pods with 7-9 large dark-green peas, very FW res., holds well, heavy yields, good in maritime conditions, heirloom. *Source History: 59 in 1981; 50 in 1984; 48 in 1987; 42 in 1991; 32 in 1994; 27 in 1998; 27 in 2004.* **Sources: Ada, Bu2, Bu8, CON, Cr1, De6, DOR, Fo7, HA3, He8, He17, Hi13, Hud, Jor, La1, Ma19, Mey, Ont, PAG, Ri12, RUP, Sau, Se26, Shu, Ves, Vi4, Wi2.**

<u>Thomas Laxton Improved</u> - 62 days, Vines 36 in., 3.5 in. pods, wilt res., unsurpassed freezing quality. *Source History: 2 in 1981; 2 in 1984; 2 in 1987; 3 in 1991; 3 in 1994; 2 in 1998; 1 in 2004.* **Sources: SOU.**

<u>Tom Thumb</u> - (Meteor) - 50-58 days, Rescued by Thelma Sanders of Missouri who has grown this var. since 1920, dwarf 6- 8 in. vines, perfect for indoor or outdoor pot culture, good yields, able to withstand temps down to 20 degrees F, pre-1855. *Source History: 1 in 1991; 1 in 1998; 5 in 2004.* **Sources: Co32, Ho13, Se16, So12, Ta5.**

<u>Top Pod</u> - 64-72 days, Sets huge 5-6 in. pods near top of 2 ft. plant, 8-10 fat peas per pod, heat resistance similar to Wando but with higher yields, good flavor, PM res., PVP 1992 Novartis Seeds, Inc. *Source History: 7 in 1994; 7 in 1998; 7 in 2004.* **Sources: CH14, Gr27, HO2, Jun, Lin, ME9, Ver.**

<u>Triple Treat</u> - (Tri-Pod Cluster) - 65 days, Triple pods in fertile soils, higher yields, 3-3.5 in. pods with 7 to 9 dark-green peas, 30 in. vines, res. to PW and mosaic, excel. quality, home use or market, new. *Source History: 2 in 1981; 2 in 1984; 4 in 1987; 3 in 1991; 3 in 1994; 2 in 1998; 1 in 2004.* **Sources: PAG.**

<u>Twinkle</u> - 75 days, Plants grow 18-22 in., many small pods with 6 juicy sweet peas, sow in February under cloches for May harvest, direct sown in March-April for June-July harvest, PW res. *Source History: 1 in 2004.* **Sources: Tho.**

<u>Utrillo</u> - 71 days, High quality fall harvest type, straight 5 in. pods on 30 in. plant, 8-10 seeds per pod, PVP expired 2004. *Source History: 2 in 1991; 3 in 1994; 4 in 1998; 6 in 2004.* **Sources: Fi1, HO2, ME9, SE4, Sto, Ter.**

<u>Victory Freezer</u> - 65-74 days, AAS/1948, med-heavy dark-green vigorous 30 in. vines, 3.25 in. pods, large dark-green peas, FW res., double pods, bred for freezing but also good fresh or canned. *Source History: 8 in 1981; 7 in 1984; 5 in 1987; 3 in 1991; 4 in 1994; 6 in 1998; 2 in 2004.* **Sources: Alb, Fi1.**

<u>Wando</u> - 64-72 days, Laxton's Progress x Perfection, for late sowings because of its heat res., also pollinates well under cold conditions, sturdy upright 24-30 in. plants, 3-3.5 in. straight blunt pods, 7-9 peas per pod, high yielding, Dr. Wade/Southeast Veg. Breeding Lab (SC), 1943. *Source History: 94 in 1981; 84 in 1984; 86 in 1987; 77 in 1991; 61 in 1994; 64 in 1998; 61 in 2004.* **Sources: Alb, All, Be4, Bo19, Bu1, Bu2, Bu8, CH14, Cl3, Co31, Com, CON, Cr1, De6, Dgs, Ers, Fa1, Fo7, Ga1, Gr27, Gur, HA3, Ha5, He17, Hi13, HO2, Jor, Jun, La1, LO8, Ma19, MAY, ME9, Mel, Mey, Mo13, MOU, Or10, PAG, Pla, Roh, Ros, RUP, Sa9, Sau, Scf, SE4, Se8, Se25, Se26, Sh9, Shu, Si5, So1, SOU, Twi, Ver, Ves, Vi4, Wet, Wi2.**

<u>Waverex</u> - 65 days, Petit pois, very productive 15-20 in. plant, 7-8 very sweet small peas per pod, heavy yields, grows well in any cool climate, dev. in West Germany, fine freezer. *Source History: 4 in 1981; 4 in 1984; 3 in 1987; 3 in 1991; 2 in 1994; 3 in 1998; 3 in 2004.* **Sources: Bou, Coo, Ter.**

<u>Willet Wonder</u> - 70 days, Grown primarily in the deep south for fresh peas, freezes well. *Source History: 3 in 1981; 3 in 1984; 3 in 1987; 2 in 1991; 2 in 1994; 3 in 1998; 1 in 2004.* **Sources: Ada.**

<u>World's Record</u> - 57-61 days, Improved strain of Gradus, extra early, 26-36 in. plants, large pointed 3-4 in. med-dark green pods with 7 or 8 peas, wrinkled seed, good freezer, excel. quality. *Source History: 5 in 1981; 4 in 1984; 3 in 1987; 3 in 1991; 1 in 1994; 2 in 1998; 2 in 2004.* **Sources: All, HA3.**

Varieties dropped since 1981 (and the year dropped):

Alaska 14-A (WR) (1991), *Alaska 28-57 WR* (1994), *Alaska 423* (1991), *Alaska M-163 Small Sieve* (1991), *Aldot* (1983), *Almota* (1998), *Alpine* (1994), *Anoka* (2004), *Argona* (1998), *Aristagreen* (2004), *Asgrow No. 40* (1982), *Aspen* (1987), *Aurora* (1994), *Banquet* (2004), *Beagle* (2004), *Big Sky* (2004), *Bikini* (1998), *Blue Pod* (1991), *Burpee's Blue Bantam* (1987), *Cabree* (2004), *Califa* (1987), *Cameo* (1984), *Canadian* (1983), *Caractacus* (2004), *Caribou* (2004), *Century (Yellow)* (1987), *Cera Sierra* (1991), *Chinook* (1982), *Code 1* (1994), *Columbia* (1991), *Coronado* (1987), *Coronet* (1987), *Corvalette* (2004), *Corvallis* (1994), *Cukor Borsi* (2004), *Curly* (1998), *Dane* (1998), *Dark Podded Thomas Laxton* (1991), *Dawn* (1983), *Deep Freeze* (1987), *Dessert Frosty* (1987), *Duet* (1994), *Dwarf Progress* (1987), *Early Abundant* (1983), *Early All Sweet* (1987), *Early Market* (1991), *Early May* (1998), *Early Onward* (2004), *Early Patio* (1991), *Early Perfection 15* (1991), *Early Perfection 11* (1991), *Early Perfection 3019* (1991), *Early Sweet* (1998), *Early Sweet 5* (1991), *Early Wonder* (2004), *Eldo* (1998), *Fabina* (1991), *Fonado* (1987), *Fordhook Wonder* (1987), *Freezer 37* (1987), *Freezer 133* (1987), *Freezer 692* (1987), *Freezer Gene* (1991), *Frescoroy* (1994), *Fridget* (1991), *Frizette* (1991), *Frizzy* (2004), *Frontier* (1987), *Frost Bite* (1984), *Frostiroy* (1994), *Giant Stride* (2004), *Gigante a Fiore Violetto* (2004), *Gloriosa* (1987), *Golden Sweet Vine* (2004), *Gradus* (1991), *Greater Progress* (1994), *Green Isle* (1998), *Grenadier* (1991), *Holiday* (1991), *Home Freezer* (2004), *Hundredfold* (2004), *Hurst Green Arrow* (1987), *Hustler* (1987), *Icer No. 95* (1998), *Jof* (1987), *Karina* (1998), *Karisma* (2004), *Kickham* (2004), *Kleine Rheinlanderin* (1987), *Kosta* (1998), *Laxtonian* (2004), *Laxton's Progress K Strain* (1987), *Laxton's Superb* (2004), *Le Bon Petit* (1994), *Lenca* (1994), *Little Gem* (1991), *Little Marvel No. 30* (1991), *Lotus* (1998), *Mammoth Early Canner* (1987), *Markado* (1987), *Markana* (2004), *Mars* (1994), *Mendota* (1994), *Meteor Round Seeded* (2004), *Midseason Freezer* (1987), *Mighty Midget* (1984), *Morse's No. 60* (1991), *Multistar* (2004), *Nain Tres Hatif Annonay* (1984), *New Era* (1983), *No. 40* (1994), *Nott's*

Excelsior (1987), *O'odham--Adapted Import* (2004), *Orcado* (1987), *Pacemaker* (1994), *Papago Peas* (1994), *Parlay* (1991), *Pegado* (1987), *Perfected Freezer* (1994), *Perfection* (1998), *Perfection (Freezer)* (1991), *Perfection (Wilt Resistant)* (1991), *Petit Pois, Darvon* (1998), *Petit Pois, Giroy* (2004), *Polaris* (1994), *Poppet* (1994), *Potlatch* (1983), *Prairie Sweet* (1991), *Precoce du Midi* (1998), *Precovelle* (1998), *Premier* (1982), *Premium Gem* (1983), *Premium Little Gem* (1991), *Primdor* (1998), *Proval* (1994), *PW 606* (1994), *Recette* (1987), *Relavil* (1983), *Roi des Conserves* (1998), *Roi des Fins Vert* (1984), *Rurik* (1987), *Samish* (2004), *Scout* (1983), *Semi-Leafless* (2004), *Senatore* (1994), *Sima* (2004), *Skinado* (1987), *Sounder* (1991), *Spanish Arrow* (2004), *Spanish Lincoln* (2004), *Sprite* (2004), *Stampede* (1998), *Stratagem Improved* (2004), *Sunset Wonder* (1994), *Super Sweet* (1982), *Superfection* (1991), *T and T Sweet Parsley Pea* (1994), *Taos Pueblo Peas* (1987), *Tasty Giant* (1987), *Telephone, Tall Improved* (1994), *Three Kings* (1991), *Tiny Tim* (1984), *Tonka* (1994), *Trio* (1998), *Triplet* (1994), *Tripod* (1998), *Twiggy* (1998), *Two Hundredfold* (1987), *Unwin's Early* (1984), *Vedette* (1991), *Venus* (1994), *Vitalis* (1987), *Western Giant* (1982), *Winfrida* (1991).

Pea / Edible Pod
Pisum sativum var. *sativum*

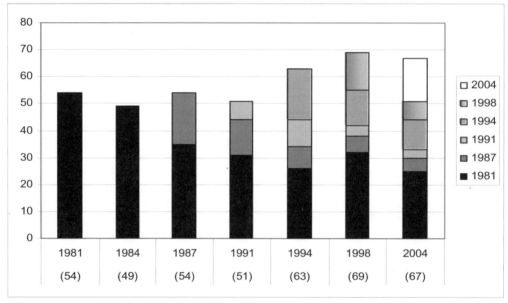

Number of varieties listed in 1981 and still available in 2004: 54 to 25 (46%).
Overall change in number of varieties from 1981 to 2004: 54 to 67 (124%).
Number of 2004 varieties available from 1 or 2 sources: 32 out of 67 (48%).

Amish Snap - 60-70 days, Vines 4-6 ft. tall, moderately sweet peas in translucent pods, yields over a six-week period if kept well picked, Amish heirloom from Lancaster County, PA. *Source History: 4 in 1998; 5 in 2004.* **Sources: Ami, Ho13, In8, Se16, Und.**

Blizzard - 63 days, High quality determinate Snow Pea, 30 in. plants produce two 3 in. pods at each node, tender, high yields in small space, excellent for stir fry and Oriental recipes, PVP 1987 Novartis Seeds, Inc. *Source History: 2 in 1987; 4 in 1991; 1 in 1994; 1 in 1998; 1 in 2004.* **Sources: Fe5.**

Bon Appetite II - 60 days, Crisp, stringless snap pea, 2.5-3 in. pods, excellent raw, steamed or frozen, plants need support, PM res. *Source History: 1 in 1994; 1 in 1998; 1 in 2004.* **Sources: Eco.**

Brazilian - (Brazilian Snow Pea) - 100 days, Robust vines to 5 ft., pink flowers, large tender edible pods, good soup pea when left to mature in the pods. *Source History: 1 in 1998; 2 in 2004.* **Sources: Ga22, Sa5.**

Carouby de Maussane - (Mangetout Carouby) - 65-70 days, Plant grows to 4-5 ft., early edible-podded sugar pea, very sweet and tender, pick the pods when the peas are barely visible, pois mangetout from France. *Source History: 1 in 1981; 1 in 1984; 2 in 1987; 3 in 1991; 3 in 1994; 5 in 1998; 11 in 2004.* **Sources: Al25, Ba8, Coo, Gou, Ho13, Lej, Se17, Sk2, So12, Vi4, WE10.**

Cascadia - 58-70 days, Vines to 3 ft., fiberless 3 in. sweet pods and peas, good yields, multiple disease res., matures one week later than Sugar Ann, Dr. James Baggett, OSU. *Source History: 8 in 1994; 25 in 1998; 24 in 2004.* **Sources: Bo17, Coo, DOR, Ers, Fe5, Hig, HO2, Jor, Loc, ME9, Mel, Ni1, Pe6, Pr3, Ros, RUP, Sau, Se7, Se26, Sto, Sw9, Ter, Tu6, WE10.**

Chinese Giant Snow - Older variety, larger pods than Oregon Giant, pick pods when still flat. *Source History: 1 in 1998; 2 in 2004.* **Sources: Pr3, Ra5.**

Chinese Snow - (Dwarf White Sugar, China Snow) - 65 days, Small slender 26 in. plants with white blossoms, pick before seeds get plump. *Source History: 5 in 1981; 2 in 1984; 4 in 1987; 4 in 1991; 5 in 1994; 1 in 1998; 1 in 2004.* **Sources: In8.**

Chinese Snow, Tall - Tall 6 ft. vines, large, light green, flattened pods, pick just when the seeds are forming. *Source History: 1 in 1998; 1 in 2004.* **Sources: Ho13.**

Dwarf Gray Sugar - (Dwarf Gray Sugar Cabbage Pea) - 57-75 days, Earliest and most dwarf edible podded pea, prolific bushy 24-30 in. vines need no staking, flat fleshy curved semi-pointed light-green 2.5-3 x .5 in. pods, borne in clusters at tops of plants, FW res., pre-1773, intro. in 1892 by D. M. Ferry and Company. *Source History: 81 in 1981; 70 in 1984; 69 in 1987; 55 in 1991; 48 in 1994; 50 in 1998; 60 in 2004.* **Sources: Ag7, Alb, Bo19, Bou, Bu3, Bu8, CH14, Cl3, Co31, Co32, Com, CON, Cr1, De6, Ear, Eo2, Ers, Ev2, Fa1, Fe5, Fi1, Fis, Fo13, Goo, Gur, HA3, He8, He17, Hi6, Hi13, Ho12, Hud, Hum, Jor, Kit, La1, Lej, LO8, MAY, ME9, Mo13, MOU, Ol2, OLD, Ont, Or10, PAG, Red, RUP, Sau, Se16, Se25, Se26, Sh9, Sk2, SOU, Twi, Ver, Vi4, Wet.**

Dwarf Sugar - 65-68 days, Light-green 2.5-3 in. pods, curved and indented between peas, borne in clusters at tops of vines, plump and fleshy, stringless and brittle when young, 26-30 in. vines. *Source History: 6 in 1981; 4 in 1984; 3 in 1987; 3 in 1991; 3 in 1994; 4 in 1998; 3 in 2004.* **Sources: Bou, Hal, Tt2.**

Dwarf White Blossom - 50-64 days, Edible podded pea for home garden or market, 36 in. vines, 2.5-3 in. pods, very tender, must be picked when peas just start to form for best quality. *Source History: 1 in 1981; 1 in 1984; 2 in 1987; 2 in 1991; 2 in 1994; 1 in 1998; 1 in 2004.* **Sources: Roh.**

Dwarf White Sugar - 50-65 days, Vigorous 24-30 in. vines, 2-2.5 in. sweet stringless pods, harvest when peas begin to dev., keep picked for maximum yields, produces abundantly until hot weather. *Source History: 9 in 1981; 7 in 1984; 5 in 1987; 7 in 1991; 8 in 1994; 9 in 1998; 9 in 2004.* **Sources: Fe5, Hi6, HO2, La1, ME9, PAG, Par, SE4, Si5.**

Early Snap - 58-60 days, Quality and yields like Sugar Snap but 10-14 days earlier, 3.25 in. thick-walled fleshy pods, meaty, 18-22 in. vines, little or no staking, BYM res., NY/AES. *Source History: 14 in 1981; 13 in 1984; 9 in 1987; 7 in 1991; 11 in 1994; 7 in 1998; 5 in 2004.* **Sources: Bu1, CON, Fa1, Sh9, Wi2.**

Golden Sweet - 55-80 days, Grown from seeds imported from India, the only yellow-colored edible podded pea in SSE's collection of over 1,200 varieties of peas, 6-8 ft. long vines, pale-yellow pods are edible but not very sweet, small dry peas are good in soup, beautiful bicolor purple flowers, heat tol., rare. *Source History: 2 in 1991; 3 in 1994; 6 in 1998; 8 in 2004.* **Sources: Ag7, Co32, Ho13, Pe1, Pr3, Se16, Ta5, Te4.**

Goliath - 68 days, Plants to 4 ft., abundant yields of large green stringless edible pods, 4.5 in., remain tender on the vine, long shelf life, continuous production throughout the season, resists F1 and PM. *Source History: 7 in 2004.* **Sources: Bu1, Fi1, Jor, Jun, Shu, Twi, Ver.**

Grandma Hershey's Sugar - Grow with support, dev. by the Isaac Hershey family, the developer of Lancaster Surecrop Corn, who suggest letting the seed develop in the pod slightly longer for a sweeter flavor rather than eating in the flat pod stage. *Source History: 1 in 1994; 1 in 1998; 1 in 2004.* **Sources: Lan.**

Green Beauty Vine - Vigorous plants to 8 ft., edible 8 in. pods. *Source History: 1 in 2004.* **Sources: Pe1.**

Ho Lohn Dow - (Snow Pea) - Small-podded snow pea, bush strain which needs no staking, sow in March for harvest in May or in August for October, used raw in salads or stir fried. *Source History: 1 in 1981; 1 in 1984; 3 in 1987; 1 in 1991; 1 in 1994; 3 in 1998; 4 in 2004.* **Sources: Ban, Sa5, Sto, Yu2.**

Honey Pod - 70 days, Vines 20 in. tall, very crisp sweet fleshy med-green 3 in. pods, uniform in shape and appearance, ready to eat at .4 in. dia., high yields, usually 2 pods each node, PVP expired 2002. *Source History: 3 in 1981; 3 in 1984; 4 in 1987; 6 in 1991; 8 in 1994; 4 in 1998; 1 in 2004.* **Sources: Tho.**

Jade - Semi-leafless type, 3 ft. tall, from Saskatchewan. *Source History: 1 in 1981; 1 in 1984; 2 in 1998; 1 in 2004.* **Sources: Pr3.**

Kawana - 60 days, Japanese snow pea, dark red flowers produce large flat pods, excel. in stir fry or salad, sow spring or fall. *Source History: 2 in 2004.* **Sources: Kit, SAK.**

Little Sweetie - 60 days, Early dwarf variety, med-sized 2.5 in. pods filled tightly with peas, produces heavily, excel. flavor. *Source History: 1 in 1981; 1 in 1984; 1 in 1994; 1 in 1998; 1 in 2004.* **Sources: Ear, Sto.**

Mammoth Melting Sugar - 65-75 days, Wilt res. 4-5 ft. vines should be trellised, thick stringless 4-5 x .75 in. pods used like snap beans, likes cool weather, high yielding, early and uniform, productive, creamy-white seed, sweeter and more tender than Dwarf Gray Sugar. *Source History: 50 in 1981; 45 in 1984; 39 in 1987; 41 in 1991; 42 in 1994; 43 in 1998; 50 in 2004.* **Sources: Ag7, Ba8, Be4, Bo19, Bu2, Bu8, CH14, CO30, Co32, Com, CON, DOR, Ech, Ev2, Fe5, Fi1, Fo7, HA3, Hal, He8, He17, Hi6, Hi13, Hig, HO2, Jor, Jun, Kit, Loc, Mcf, ME9, Mel, Mo13, Orn, Pe2, Ra5, Red, Ri2, Ri12, Roh, RUP, Sa9, Sau, Shu, So1, TE7, Ver, Vi4, WE10, Wi2.**

Mangia Tutto - (Rampicante Mangiatutto) - *Source History: 1 in 1987; 1 in 1991; 2 in 1998; 2 in 2004.* **Sources: Ber, Se17.**

Manoa Sugar - (Chinese Pea) - PM res. *Source History: 1 in 1987; 2 in 1991; 1 in 1994; 1 in 1998; 1 in 2004.* **Sources: Uni.**

Mega - 90-98 days, Enation res. var. that produces in both cool and warm weather, vigorous 2.5-3 ft. plants, light-green 4 in. pods, crisp, juicy, sweet. *Source History: 1 in 1994; 3 in 1998; 3 in 2004.* **Sources: Pr3, Ter, We19.**

Melting Sugar - (Snow Pea) - 72-74 days, Late tall type, 48-54 in. vines, broad flat 4-5 in. pale-green pods, 8-10 large round creamy-white peas per pod, excel. for home or market. *Source History: 9 in 1981; 9 in 1984; 8 in 1987; 4 in 1991; 5 in 1994; 6 in 1998; 3 in 2004.* **Sources: GRI, Kil, Twi.**

Norli - 50-58 days, Early, white-flowered, 18 in. vines,

2.5 in. pods, one of the best sugar peas for taste and yield, plant early, very sweet, for salads, stir fry and freezing. *Source History: 3 in 1981; 3 in 1984; 4 in 1987; 4 in 1991; 6 in 1994; 4 in 1998; 2 in 2004.* **Sources: Ag7, Dam.**

Opal Creek Golden Snap - Vines grow 5-6 ft. long, bicolor purple flowers, yellow snap pods, tasty and sweet, unique, bred by Alan Kapuler of Peace Seeds, named in commemoration of the struggle to maintain our old growth forests. *Source History: 2 in 2004.* **Sources: In8, Se17.**

Oregon Giant - 60-74 days, Disease res. variety dev. by Dr. James Bagget from OSU, 30-36 in. bush produces high yields of 4.5-5 in. flat pods which are larger and sweeter than Oregon Sugar Pod, starts production early and continues over a long period, PW FW PEMV res. *Source History: 17 in 1994; 32 in 1998; 31 in 2004.* **Sources: CO30, Coo, De6, DOR, Ers, Fe5, Ga1, HO2, Jo1, Jor, Jun, La1, ME9, OLD, OSB, Pe2, Pe6, Pr3, Re6, Ri12, Roh, RUP, Scf, Se7, Sto, Sw9, Ter, Und, Ves, WE10, We19.**

Oregon Giant Sugar Pod - (OSU 706) - 70 days, Bush 30-36 in. plants, large wide 5.5 x 1 in. pods, sweet and succulent, semi-wrinkled seeds lack bitterness associated with some edible podded vars., PEMV PW FW1 res., OSU release, introduced in 1991. *Source History: 3 in 1991; 4 in 1994; 3 in 1998; 3 in 2004.* **Sources: Ev2, Na6, Ni1.**

Oregon Sugar Pod - 58-70 days, Productive 24-30 in. dwarf vines, 4-4.5 x .75 in. smooth edible pods with 6-8 large light- green peas, pods borne double, FW PEM and PS res., white flowers, mild flavor, fine quality, Dr. Baggett/Oregon State. *Source History: 53 in 1981; 46 in 1984; 40 in 1987; 36 in 1991; 21 in 1994; 20 in 1998; 24 in 2004.* **Sources: Alb, Bo17, Bou, CA25, Ear, Ers, HA3, He8, He17, Hi6, La1, ME9, Mey, Mo13, Pe2, Ri2, Ri12, Roh, Ros, Sa9, TE7, Tho, Up2, Vi4.**

Oregon Sugar Pod II - 60-68 days, Adds mildew res. to permit fall harvest, 28-30 in. vines, smooth light-green pods still stringless at 4 in., heavy yields, good freezer, PM FW PEMV res., Dr. Baggett/Oregon State. *Source History: 5 in 1981; 5 in 1984; 22 in 1987; 33 in 1991; 42 in 1994; 46 in 1998; 45 in 2004.* **Sources: Bu1, Bu2, Bu3, Bu8, CH14, Cl3, CO30, CON, Dam, De6, Dgs, DOR, Ev2, Fa1, Fe5, Fi1, Ga1, Gur, Ha5, Hig, HO2, Hum, Jor, Loc, Ma19, ME9, MOU, Ni1, Ont, OSB, PAG, Pin, RIS, RUP, SE4, Se8, Se25, Se26, Shu, Si5, Ter, Twi, Ver, Ves, WE10.**

Osaya Endo - 72 days, Pole type strain of Mammoth melting Sugar, large sweet tender bright yellow pods, white-flowering, for eating raw or stir fry, sow spring/fall. *Source History: 1 in 1981; 1 in 1984; 1 in 1987; 1 in 1991; 2 in 1994; 2 in 1998; 5 in 2004.* **Sources: Fe5, Ho13, Se17, So12, Tu6.**

Purple Podded - Vines grow to 6 ft., beautiful lilac flowers, purple pods, harvest when young, large pods get tough and stringy. *Source History: 1 in 1994; 2 in 1998; 1 in 2004.* **Sources: Ma13.**

Risser Early Sugar - 60-70 days, Productive tall vines, Lancaster County heirloom grown by several generations of the Risser family. *Source History: 1 in 1994; 2 in 1998; 1 in 2004.* **Sources: Lan.**

Sandy - 75 days, Plants grow 3.5-4 ft. tall, produce many edible tendrils, pods are most flavorful at 4 in. length. *Source History: 1 in 2004.* **Sources: Ter.**

Schweizer Riesen - 60-65 days, Translates "Swiss Giant", 5 ft. vines require support, large, tender snow pea, 1 x 5 in. pods with 9 seeds, retains sweetness even when mature, old Swiss heirloom. *Source History: 1 in 1998; 4 in 2004.* **Sources: Pe2, Se17, So12, Tu6.**

Snow Green - 56-70 days, Vines to 28 in., pods to 3.5 in., can be shelled, crisp and flavorful, shoots can be used in salads, stir fry or as garnish, PM and PLRV res., PVP 2000 Novartis Seeds, Inc. *Source History: 5 in 2004.* **Sources: Ers, Jo1, Jor, Ki4, OSB.**

Snow Pea - (Sugar Pea, Chinese Snow Pea, Ho Lohn Dow, Tall Sugar Pea, Pois Mangetout Grace) - 60-70 days, Chinese edible podded 3 in. pea, stops producing if allowed to seed, pick when peas become visible, easy to grow. *Source History: 12 in 1981; 10 in 1984; 10 in 1987; 7 in 1991; 2 in 1994; 4 in 1998; 5 in 2004.* **Sources: All, Gr29, Se28, Vi5, Wo8.**

Snow Pea Shoots - 65 days, Considered a delicacy, sold at a premium in only a few Oriental supermarkets on the West Coast, young tips, branches and leaves used, tender shoots are excel. for stir-fry and salad. *Source History: 1 in 1994; 2 in 1998; 2 in 2004.* **Sources: Ev2, Kit.**

Snow Wind - 70 days, Snow pea, semi-leafless plants to 30 in. tall, flat, dark green 3.5 in. pods, smaller than Oregon Giant, PLRV, PM tolerance, resists Bean Leaf Roll, PVP 2000 Novartis Seeds, Inc. *Source History: 4 in 2004.* **Sources: Ers, OSB, Par, We19.**

Snowbird - 58 days, Very early snow pea, erect dwarf 16-18 in. plants need no support, huge yields of 3 in. pods borne in double and triple clusters, good for short seasons and short space. *Source History: 2 in 1981; 2 in 1984; 1 in 1987; 1 in 1991; 1 in 1994; 1 in 1998; 3 in 2004.* **Sources: Bu2, Hi13, Se17.**

Snowflake - 58-72 days, Imp. short-vined Oriental snowpea, dark-green 4 x .6-1 in. flat pods stay straight, 24-30 in. vines grown with or without support, high-yielding, a definite refinement. *Source History: 4 in 1981; 4 in 1984; 8 in 1987; 12 in 1991; 13 in 1994; 15 in 1998; 9 in 2004.* **Sources: DOR, Fo7, Ho13, Jor, ME9, Pla, RUP, Sh9, Wet.**

Sugar Ann - (Dwarf Sugar Ann, Sugar Anne) - 52-75 days, AAS/1984, 24-30 in. vines need no support, med-green 2.5-3 in. round blunt pods with 7 peas, Sugar Snap quality but 14 days earlier and bush, sweet pods freeze well, no trellising, PVP expired 2000. *Source History: 56 in 1981; 56 in 1984; 65 in 1987; 63 in 1991; 59 in 1994; 72 in 1998; 40 in 2004.* **Sources: Bu3, CA25, CH14, Com, Dam, De6, DOR, Ear, Ers, Fe5, Fi1, Fis, Fo7, Ga1, Gou, Gr27, Gur, HA3, Hal, Hig, HO2, Jo1, Jor, Kil, La1, Me7, ME9, Mey, Na6, Or10, PAG, Pe6, Pr3, RUP, SE4, Se25, Sh9, Si5, Tho, Wet.**

Sugar Bon - 54-69 days, Early short-vined Sugar Snap type pea, med-green 2.5-3 in. blunt pods contain 7 peas, 18-24 in. compact bushy plants need no support, PM FW1 res., 2 weeks earlier than Sugar Snap, perfect for small gardens and even for container culture, PVP expired 2000. *Source History: 33 in 1981; 23 in 1984; 22 in 1987; 13 in 1991; 16* in 1994; 15 in 1998; 8 in 2004. **Sources: Bu2, Fo7, Hi13, Jor, La1, Mey, Pr3, So1.**

Sugar Daddy - 62-75 days, First stringless snap pea, dev. after 25 years of research and over 100 second-generation crosses, dwarf 24-30 in. vines need little support, 2.5-3.5 in. pods stringless and thick-fleshed, double pods at each node at top of plant for easy picking, 6-7 med-green peas per pod, full season--3 pickings, heavy yields, PM and Pea Leafroll res., FW1 susc., Dr. Calvin Lamborn/Gallatin Valley, PVP expired 2003. *Source History: 37 in 1987; 39 in 1991; 36 in 1994; 38 in 1998; 19 in 2004.* **Sources: All, Bou, CH14, DOR, Fa1, Fo7, HA3, He8, Hig, Jor, Jun, ME9, Mey, Pr3, RUP, SE4, Sh9, So1, SOU.**

Sugar Gem - 60 days, Plant to 26 in., completely stringless, sweet, crisp pods, PM tol., not suited to wet areas or seasons, can be sown in successive plantings throughout the summer. *Source History: 1 in 1994; 2 in 1998; 1 in 2004.* **Sources: Tho.**

Sugar Lace - (SP 550) - 68 days, Semi-leafless 30 in. plant, no need to stake, stringless, tender green pods, good for commercial and home garden, PM and LCV res., PVP 2001 Novartis Seeds, Inc. *Source History: 1 in 1998; 11 in 2004.* **Sources: Cl3, De6, Fi1, Ga1, Hum, Jor, Lin, Ma19, Mcf, Tt2, We19.**

Sugar Lace II - 68 days, First semi-leafless sugar snap type, 24 in. tall plant requires no support, good yields of 3.5 in. pods with sweet flavor, PM and PEMV resistant, PVP 2002 Novartis Seeds, Inc. *Source History: 6 in 2004.* **Sources: Bu3, Ers, HO2, OLD, Par, Pin.**

Sugar Lode - 60-70 days, Plants 3 ft. tall, pods sweeten slowly, good pea flavor becomes sugary, recommended for second early crop after Sugar Ann and for late crop after Sugar Snap, heat res. *Source History: 1 in 1991; 1 in 1994; 1 in 1998; 4 in 2004.* **Sources: Fe5, Hi6, Pe2, Tu6.**

Sugar Lord - R.H.S. Award of Garden Merit winner, super sweet flavor, abundant producer. *Source History: 1 in 2004.* **Sources: Ga7.**

Sugar Pod 2 - 60-70 days, Bush plant bears 3-4 in. juicy pods, two per node, prolific, Baggett/OSU. *Source History: 1 in 1994; 3 in 1998; 3 in 2004.* **Sources: Pe2, Se7, So9.**

Sugar Snap - (New Sugar Snap, Cascadia) - 53-72 days, AAS/1979, thick round meaty 2.5-3.5 in. pods, can be snapped like green beans or left on vine until 5-7 med-green peas mature, dev. strings at full maturity, 48-72 in. vines need support, frost res., PW tol., freezes well, cannot stand the heat of canning, PVP expired 1994. *Source History: 142 in 1981; 124 in 1984; 126 in 1987; 106 in 1991; 90 in 1994; 77 in 1998; 64 in 2004.* **Sources: Ag7, Alb, All, Be4, Bu1, Bu2, Bu3, Bu8, But, CH14, Co32, Com, Coo, Dam, De6, Dgs, DOR, Ear, Ers, Fa1, Fe5, Ga1, Ge2, Goo, Gur, HA3, Hal, He8, He17, Hi6, Hi13, Hig, HO2, Hum, Jo1, Jun, Kil, La1, Lin, Loc, MAY, ME9, Mel, Mey, Mo13, Ont, PAG, Pla, Pr3, Red, Roh, RUP, Sau, SE4, Se7, Se25, Se26, Sk2, So1, Te4, Ter, Tho, We19, Wet.**

Sugar Snap, Dwarf - Early short-vined snap pea with edible pods that are very similar to Sugar Snap, 18-24 in. vines, 8 peas per pod. *Source History: 3 in 1981; 2 in 1984; 1 in 1987; 2 in 1991; 2 in 1994; 2 in 1998; 1 in 2004.* **Sources: Ada.**

Sugar Sprint - 58-62 days, Sweet, tender 3 in. stringless snap pods, sturdy 20-30 in. plants, tolerant to PEMV and PM, fresh market, freezing and shipping, PVP 2002 Syngenta Seeds, Inc. *Source History: 36 in 2004.* **Sources: Be4, Bu3, Bu8, CH14, Cl3, Coo, Dam, Ers, Ha5, He8, HO2, Hum, Jo1, Jun, La1, MAY, Mcf, ME9, Mel, Mey, Mo13, MOU, Ni1, OLD, OSB, PAG, Par, Ra5, RIS, Roh, RUP, SE4, Sto, Twi, Ver, Ves.**

Sugar Star - 70-90 days, Early Sugar Daddy type, vigorous 24-30 in. plants, fleshy tender stringless round sugar snap pods, 2.5-3 in., two per node, fresh, canning or freezing, earlier disease tolerance to PLRV and PM. *Source History: 10 in 2004.* **Sources: CH14, Ers, Loc, ME9, Par, Roh, RUP, Shu, Ter, Ver.**

Sugaree Snap - 53-72 days, Strong vines grow 6-10 ft. tall, 3.5-4 in. sweet pods, o.p. snap pea bred by Kusra Kapuler of Peace Seeds. *Source History: 2 in 2004.* **Sources: In8, Pe1.**

Super Snappy - 65 days, Vigorous vines grow 28-32 in., need no support, large pods 5-5.75 in. long .75 in. wide, 8-10 peas per pod, sweet and crisp, PM tol. *Source History: 2 in 1994; 2 in 1998; 4 in 2004.* **Sources: Bu2, Hi13, Roh, SE4.**

Super Sugar Mel - 68-80 days, Tall bush plant, large 4 in. pods with 2 pods per node, tol. to PLRV, PVP expired 2002. *Source History: 4 in 1987; 12 in 1991; 17 in 1994; 18 in 1998; 2 in 2004.* **Sources: Se8, Wet.**

Super Sugar Snap - 62-70 days, Vine grows 5 ft., 3-3.5 in. light green pods, used both as sugar pea and shelled, good disease res., requires a trellis, bred by Dr. Calvin Lamborn, PVP 2000 Novartis Seeds, Inc. *Source History: 27 in 1998; 35 in 2004.* **Sources: Bou, Bu2, Bu3, Cl3, Co31, Dam, Dgs, Fi1, Ha5, Hig, HO2, Hum, Jor, Jun, Ki4, Lin, MOU, Na6, Ni1, OLD, OSB, Par, Pe2, Pin, Re6, RIS, RUP, Sh9, Shu, SOU, St18, Tt2, Twi, Ver, Ves.**

Sweet Snap - 60-66 days, Prolific snap pea, 2-3 in. med-green crisp tender fiber-free pods with 6 or 7 peas, vigorous 34-36 in. vines need staking, tol. to mildew and legume yellow virus, not for canning, PVP expired 2001. *Source History: 21 in 1981; 20 in 1984; 22 in 1987; 17 in 1991; 13 in 1994; 6 in 1998; 2 in 2004.* **Sources: Cr1, Mcf.**

Taichung No. 11 - 70 days, Chinese snow pea, pink flowers, tender, flat, 3-4 in. long pods, sow spring or fall. *Source History: 2 in 2004.* **Sources: DOR, Kit.**

Taichung No. 13 - 70 days, Chinese edible pod sugar pea, white flowers form sweet, round, plump pods, sow spring or fall. *Source History: 2 in 2004.* **Sources: DOR, Kit.**

Taiwan Sugar - Large sweet pods, excellent for stir fry and salads, originated in Taiwan. *Source History: 2 in 2004.* **Sources: Ba8, TE7.**

Yokomo Giant - Productive Japanese var. with large flat edible pods, tall vines. *Source History: 1 in 2004.* **Sources: Ho13.**

Varieties dropped since 1981 (and the year dropped):

A74-128 (1991), *Agio* (1994), *Bamby* (1991), *Banana Nano* (2004), *Bohlen's Swedish Snow Pea* (2004), *Bush Snapper* (1998), *Caroubel* (1991), *Corgi* (2004), *Cukrowy Ilowiecki* (2004), *De Grace* (1994), *Dwarf Melting Sugar* (1987), *Early Sugar Pea* (1982), *Edible Podded Pea* (1991), *Edula* (1987), *Giant Dwarf Sugar* (1994), *Giant Melting Sugar* (1991), *Golden Edible Pod* (2004), *Golden Snow* (1998), *Green Sugar Pods* (1994), *Hendricks* (1998), *Hungarian* (2004), *Hyogo Sugar* (2004), *Mangetout Geant* (1998), *Melting Sugar, Tall* (1994), *Nippon Kinusaya* (1991), *Nordland* (2004), *Pois Mangetout Carouby* (1987), *Rembrandt* (1994), *Russian Sugar* (2004), *Sapporo Express* (2004), *Seedling Sugar* (2004), *Short 'N Sweet* (1998), *Small Podded China Pea* (2004), *Snappy* (2004), *Snow Dwarf* (1987), *Snow Pea, Dwarf* (1998), *Snowbiz* (1983), *Snowhite* (2004), *Sugar Bush* (1998), *Sugar Dwarf Sweetgreen* (1983), *Sugar Mel* (2004), *Sugar Pop* (2004), *Sugar Rae* (2004), *Sugar Snap Stringless* (1991), *Super Sugar Pod* (1998), *Super Sweetpod* (1984), *Sweet and Snappy* (1985), *Sweet Pod, Burpee's* (1998), *Sweetgreen* (1987), *Tall Melting Pod* (1987), *Tezieravenir* (1983), *Titania* (1991), *Tokyo Sugar Red* (1998).

Pea / Soup
Pisum sativum var. sativum

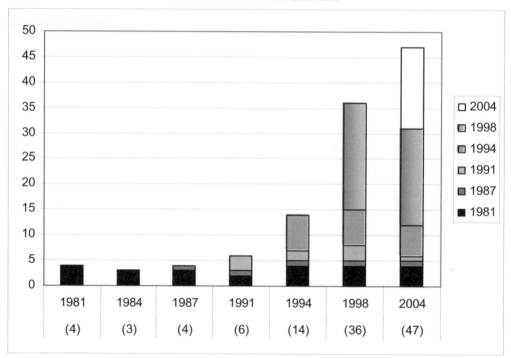

Number of varieties listed in 1981 and still available in 2004: 4 to 4 (100%).
Overall change in number of varieties from 1981 to 2004: 4 to 47 (1175%).
Number of 2004 varieties available from 1 or 2 sources: 41 out of 47 (87%).

<u>Abyssinicum</u> - *Source History: 1 in 2004.* **Sources: Se17.**

<u>Alverjon Temporal</u> - Dry farm pea, grown in the Manzano Mountains in northern New Mexico at 7500 ft., introduced by the Spanish. *Source History: 1 in 2004.* **Sources: Na2.**

<u>Amplissimo Victoria</u> - 90 days, Smooth, large soup peas in long wide pods, plant requires staking, cooked pea has a garbanzo flavor, good for making hummus. *Source History: 1 in 1994; 2 in 1998; 3 in 2004.* **Sources: Fe5, Ho13, So12.**

<u>Australian</u> - High yields of brown speckled, rectangular seeds, 5 ft. vines, from Australia. *Source History: 1 in 1998; 1 in 2004.* **Sources: Ho13.**

<u>Austrian Winter</u> - (Austrian Field Pea) - Hardy overwintering pea, strictly for stock feed, grows on poor soil, sown in fall to prevent winter wind and water erosion, plow under in spring. *Source History: 3 in 1981; 3 in 1984; 4 in 1987; 8 in 1991; 10 in 1994; 8 in 1998; 7 in 2004.* **Sources: Ada, Bou, Ga1, Hi6, Ni1, Pe2, Ron.**

<u>Bill Jump's Soup Pea</u> - 80 days, Old strain from Washington, 6 ft. vines, short slim pods with 6 small round seeds, green or tan speckled with purple. *Source History: 2 in 1994; 3 in 1998; 1 in 2004.* **Sources: Ho13.**

<u>Bisilia Market</u> - Productive medium tall vines, seed colors include tan, brown, and green with purple spots, mixed market sample from Turkey. *Source History: 1 in 2004.* **Sources: Ho13.**

<u>Capucijners, Blue Pod</u> - (Capucijners Purple Pod, Dutch Grey) - 65-75 days, Early blue-podded var., heavy yielding 4 in. vines, greyish-brown peas, mainly for dry use, plant early, harvest when vines are dry, also pick green and

can and freeze, dev. by the Capuchin Monks in Europe in the 1500s. *Source History: 1 in 1981; 1 in 1984; 1 in 1987; 2 in 1991; 2 in 1994; 5 in 1998; 11 in 2004.* **Sources: Ami, Dam, Fe5, Fo7, Ho13, In8, Pr3, Rev, Se17, Ta5, Te4.**

Capucijners, Dwarf French - Vines grow 2 ft., large, grey-brown seeds, very productive. *Source History: 1 in 1998; 1 in 2004.* **Sources: Ho13.**

Capucijners, Hala - Soup pea from the Netherlands, large brown seed, productive. *Source History: 1 in 1998; 1 in 2004.* **Sources: Ho13.**

Capucijners, Holland - (Grey Pea) - 85 days, Traditional pea unique to the farm communities of north Holland, name derives from its markings which resemble the cowls of Capuchin friars, compact plants bloom profusely with fragrant violet-red and pink flowers and produce large wrinkled brown-gray peas, not sweet but edible fresh, best cooked whole, make a rich brown gravy. *Source History: 1 in 1987; 3 in 1991; 1 in 1994; 2 in 1998; 1 in 2004.* **Sources: Ho13, Syn.**

Capucijners, Raisin - Dwarf type with 24 in. vines, large brownish peas for winter use, mainly grown for dry use, excel. flavor, also can pick green like normal peas. *Source History: 1 in 1981; 1 in 1984; 1 in 1987; 3 in 1991; 3 in 1994; 4 in 1998; 3 in 2004.* **Sources: Dam, Ho13, Pr3.**

Carlin - Tall, heavy yielding plants, medium size round seed is heavily spotted with golden brown, dates back to Elizabethan England. *Source History: 2 in 1998; 1 in 2004.* **Sources: Pr3.**

Carling - Old English soup pea, small brown seeds produced on tall vines. *Source History: 1 in 2004.* **Sources: Ho13.**

Century - Smooth round soup pea, heirloom from the 1940s originally distributed as Creamette. *Source History: 1 in 1994; 2 in 1998; 1 in 2004.* **Sources: Pr3.**

Corne de Belier - Tall vines produce round white soup type seeds. *Source History: 1 in 2004.* **Sources: Ho13.**

Creamette - Vines to 5 ft., large, smooth, round cream colored soup pea, high yields. *Source History: 1 in 1998; 1 in 2004.* **Sources: Ho13.**

Darlaine - Yellow soup pea, plant can be self-supporting. *Source History: 1 in 2004.* **Sources: Sa5.**

Dutch - Capucijners type with 2 ft. vines, large brown seeds, good soup pea. *Source History: 1 in 1998; 1 in 2004.* **Sources: Ho13.**

Eroica - Capucijners type from the Netherlands, 3 ft. plants, two-tone flowers, large wrinkled brown seed, makes a tasty soup. *Source History: 1 in 1998; 1 in 2004.* **Sources: Ho13.**

Ervilha Torta Flor Roxa - Name means "red flowering pea", 6-8 peas per 3 in. pod, nice for soup but not great fresh, from Brazil. *Source History: 2 in 2004.* **Sources: Se17, Ta5.**

Gruno Rosyn - Capucijners type from Holland, 3 ft. plants, large brown-buff mottled seeds, good soup pea. *Source History: 1 in 1998; 1 in 2004.* **Sources: Ho13.**

Kazankij - Large, smooth white-tan seed, 4 ft. vines, old Russian soup pea. *Source History: 1 in 1994; 2 in 1998; 1 in 2004.* **Sources: Ho13.**

King Tut - Purple pods on 4 ft. vines, pink and purple flowers, brown seeds, pods edible when young, good soup type, reportedly found in an Egyptian tomb. *Source History: 1 in 1998; 2 in 2004.* **Sources: Ho13, Se17.**

Margaret McKee's Baking Pea - Productive capucijners type, 4 ft. vines, large, dark brown speckled seeds. *Source History: 1 in 1998; 1 in 2004.* **Sources: Ho13.**

Marrowfat - 60-72 days, Old English soup pea dating to the 1700s, large, wrinkled, green-white seeds. *Source History: 1 in 1998; 3 in 2004.* **Sources: Ho13, Ta5, Th3.**

Mexican Soup - Tall 6 ft. vines, productive soup pea, large, smooth, light tan seeds, good for winter drying. *Source History: 1 in 1998; 1 in 2004.* **Sources: Ho13.**

Monk Pea - 90 days, Heirloom pea, larger size than a normal pea, grown and preserved for centuries by monks in the Netherlands, brown with a rough finish resembling a nut. *Source History: 1 in 1994; 2 in 1998; 1 in 2004.* **Sources: Se25.**

Ojito - Tan to yellow seeds, from Ojito, NM. *Source History: 1 in 2004.* **Sources: Na2.**

Old Scotch - Vines 5 ft., smooth green seeds with white blotch, soup pea. *Source History: 1 in 1998; 1 in 2004.* **Sources: Ho13.**

O'odham - Cultivated for more than 300 years as a desert winter crop by O'odham farmers, planted in rotation with cotton in commercial fields to reduce Texas root rot, alkali tol., hardy. *Source History: 1 in 2004.* **Sources: Na2.**

PI 120630 (Turkey) - Vines 4-5 ft. tall, red flowers, med. size purple-brown peas, good in soup, from Turkey, USDA. *Source History: 1 in 1998; 1 in 2004.* **Sources: Ho13.**

Prussian Blue - 75-95 days, Smooth-seeded, blue-green, soup pea, white flowers, pods have blue cast, heat tol., from Europe, 1806. *Source History: 1 in 1991; 2 in 1994; 2 in 1998; 3 in 2004.* **Sources: Ho13, Lan, Th3.**

Purple Passion - 80 days, Tall 5 ft. vines, purple flowers, purple pods and seeds, good soup pea, not the same as Capucijners Purple pea, extremely rare. *Source History: 1 in 1998; 2 in 2004.* **Sources: Ho13, Ta5.**

Russian - *Source History: 1 in 2004.* **Sources: Se17.**

San Luis - Smooth skinned seeds, tan to light yellow. *Source History: 1 in 2004.* **Sources: Na2.**

St. Hubert - Medium-tall plants, round, smooth green seeds, brought to Quebec by French settlers. *Source History: 1 in 1998; 2 in 2004.* **Sources: Ho13, Pr3.**

Subfrufrum - Vines 6 ft. tall, mahogany-purple seeds, round and dimpled, soup pea from Germany. *Source History: 1 in 1998; 1 in 2004.* **Sources: Ho13.**

Taos - large, smooth, tan to light green seeds, good soup pea, grown in Taos Pueblo, NM. *Source History: 1 in 2004.* **Sources: Na2.**

Tepehuan - Tan seeds, originally from the Sierra Madre in southern Chihuahua, Mexico. *Source History: 1 in 2004.* **Sources: Na2.**

Trapper - (Wild Pea) - *P. sativum* var. *arvense*, field type pea used as a shelled pea when ripe or during the winter for soups or boiled as a vegetable, young shoots are steamed or used in stir fry, good soil builder, cold tolerant, wild ancestor of the garden pea *Source History: 1 in 1981; 1 in*

1984; 2 in 1994; 2 in 1998; 2 in 2004. **Sources: Ho13, Und.**

Truchas Alverjon - Smooth soup pea, tan and green color, introduced by the Spanish. *Source History: 1 in 2004.* **Sources: Na2.**

Vadito - Smooth, tan and green seeds, soup pea, from Vadito, NM. *Source History: 1 in 2004.* **Sources: Na2.**

Yoquivo del Sur - Smooth beige seed, from remote region near Yoquivo del Sur. *Source History: 1 in 2004.* **Sources: Na2.**

Zapotec Chicharo Criollo - Oaxacan dry soup pea, midseason maturity. *Source History: 1 in 1998; 1 in 2004.* **Sources: So12.**

Zeiner's Gold - Productive, 5-8 ft. vines, double pods contain 5 round, orange peas. *Source History: 1 in 1998; 1 in 2004.* **Sources: Ho13.**

Varieties dropped since 1981 (and the year dropped):

Blue Pod Desiree (2004), *Capuchin* (2004), *Capucijners, Tall* (2004), *Field Pea* (1991), *Holland Brown* (2004), *Kimberly* (2004), *Mellow Yellow* (2004).

Peanut
Arachis hypogaea

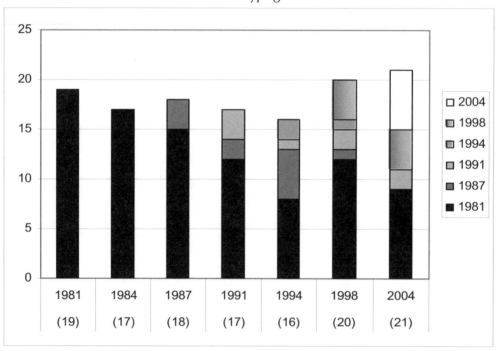

Number of varieties listed in 1981 and still available in 2004: 19 to 9 (47%).
Overall change in number of varieties from 1981 to 2004: 19 to 21 (111%).
Number of 2004 varieties available from 1 or 2 sources: 16 out of 21 (76%).

Carolina Black - 110 days, Rare heirloom from North Carolina, slightly larger than Spanish peanuts, black skin, sweet flavor, 2 nuts per shell, CV So1 introduction, 1999. *Source History: 1 in 2004.* **Sources: So1.**

Carwile's Virginia - 130-140 days, Grown by Frank Carwile for over 75 years since he was given this peanut by a traveler when he was 8 years old, he tried other varieties after that but found none as flavorful, 2-4 nuts per pod, average disease res., excel. drought res., 1910. *Source History: 1 in 1991; 1 in 1998; 1 in 2004.* **Sources: So1.**

Ecuadorian Purple - Productive 2 ft. bushes, yellow flowers spiral down into the ground after fertilization to form peanuts at the end of "pegs", dark purple skin, requires 5 month growing season. *Source History: 1 in 2004.* **Sources: Ho13.**

Georgia Green - 140 days, Highly productive runner type, small seeds, resists tomato-spotted wilt virus, PVP 1996 University of Georgia Res. Fdn., Inc. *Source History: 1 in 2004.* **Sources: Scf.**

Gregory - Virginia type with intermediate growth habit between bunch and runner, very large seeds, pink seed coat, high percentage of extra large kernels, high calcium, PVP 2002 NC ARS. *Source History: 1 in 2004.* **Sources: Scf.**

Jamaican Red - Short plants, 3-4 red nuts per pod, matures early, Jamaica. *Source History: 1 in 2004.* **Sources: Ho13.**

Jumbo - (Mammoth Jumbo) - Huge peanuts, about twice as large as normal, excel. flavor, easy to grow and harvest and shell. *Source History: 3 in 1981; 3 in 1984; 5 in 1987; 3 in 1991; 1 in 1994; 1 in 1998; 1 in 2004.* **Sources: CH14.**

Jumbo Improved - 145 days, *Source History: 1 in 1981; 1 in 1984; 1 in 1987; 1 in 1991; 1 in 1994; 1 in 1998;*

1 in 2004. **Sources: Mey.**

NC-V11 - 150 days, Runner plant habit, PVP 1991 NC ARS. *Source History: 1 in 2004.* **Sources: Bu8.**

NC-7, Carolina Jumbo - 125 days, Plants 16 in. tall with 3 ft. runners, 2-3 peanuts per pod, 560 peanuts per lb., can be grown in the North with lighter yields than in North Carolina, PVP expired 1997. *Source History: 1 in 1991; 1 in 1994; 3 in 1998; 3 in 2004.* **Sources: OLD, Roh, Scf.**

Pre-Civil War - Light buff colored kernels, from Mississippi, pre-1860. *Source History: 1 in 1998; 2 in 2004.* **Sources: Ho13, Se17.**

Purple Eagle - Medium-tall plants, deep violet skins, 2-3 kernels per pod. *Source History: 1 in 1998; 1 in 2004.* **Sources: Ho13.**

Spanish - (Sweet Spanish) - 100-120 days, Early small sweet peanut, 2-3 kernels per pod, dwarf plants grow close, easy to cultivate and gather, grows quick for North, needs lots of calcium. *Source History: 10 in 1981; 5 in 1984; 7 in 1987; 5 in 1991; 3 in 1994; 3 in 1998; 2 in 2004.* **Sources: Ho13, Shu.**

Spanish, Early - (Valencia) - 100-110 days, Early and prolific, small pods with 2 or 3 small kernels, dwarf erect bushes like warm sandy soil with southern exposure, good crops grown in most areas of Canada. *Source History: 12 in 1981; 10 in 1984; 7 in 1987; 7 in 1991; 6 in 1994; 6 in 1998; 3 in 2004.* **Sources: Gr27, Gur, Jun.**

Suffolk Black - Tall plants, dark purple skin, 3-4 kernels per pod. *Source History: 1 in 1998; 1 in 2004.* **Sources: Ho13.**

Tennessee Red - (Valencia Tennessee Red) - 120-140 days, One of the finest large varieties, each shell contains 3 or 4 red-skinned peanuts, mild sweet flavor, extremely productive. *Source History: 6 in 1981; 5 in 1984; 6 in 1987; 7 in 1991; 7 in 1994; 8 in 1998; 6 in 2004.* **Sources: CH14, He17, Mo13, Par, Red, Und.**

Texas Red and White - Plants to 12 in. tall, bright red skins with white streaks, sweet flavor, productive, early. *Source History: 1 in 1998; 2 in 2004.* **Sources: Ho13, Se17.**

Valencia - (Valencia Spanish, Valencia Early Spanish) - 110-150 days, Large Spanish type adapted to variety of soils, 3 small sweet kernels per pod, heavy bearing, does well in Ontario, rich-flavored favorite. *Source History: 7 in 1981; 6 in 1984; 9 in 1987; 6 in 1991; 5 in 1994; 5 in 1998; 4 in 2004.* **Sources: Co31, Ear, Ont, So12.**

Virginia - (Virginia Bunch, Early Bunch Virginia, Large Virginia Bunch) - 120 days, Largest kerneled of the bunch peanuts, 1-2 per pod, week earlier than running var. Virginia Jumbo, prefers sandy soil. *Source History: 4 in 1981; 3 in 1984; 4 in 1987; 5 in 1991; 3 in 1994; 3 in 1998; 1 in 2004.* **Sources: Sau.**

Virginia Improved - 120 days, Highly productive large-podded var., compact 18 in. tall bush, heavy erect stalks and large leaves, adapted for growing in the North where it does well. *Source History: 3 in 1981; 3 in 1984; 1 in 1987; 1 in 1991; 1 in 1994; 1 in 1998; 1 in 2004.* **Sources: Fi1.**

Virginia, Jumbo - 120 days, Heavy yields, 2 or 3 extra large rich-flavored kernels per pod, grows well in loose or light sandy soils, makes valuable fodder for stock, running variety, grows quite rapidly, large pods and nuts, prefers warm dry sandy soil. *Source History: 16 in 1981; 12 in 1984; 10 in 1987; 12 in 1991; 11 in 1994; 14 in 1998; 14 in 2004.* **Sources: Be4, Bu2, Ers, Gur, He17, Ho13, La1, Mel, Mo13, Par, Ri12, Sh9, Shu, So1.**

Varieties dropped since 1981 (and the year dropped):

Arizona Native (2000), *Early Prolific* (1987), *Florigiant* (1998), *Florunner* (2004), *Garoy* (2004), *Mammoth Virginia* (1991), *Park's Whopper* (2004), *Pronto* (2004), *Spanish Improved* (2004), *Spanish Tamnut* (1994), *Spanish, Star* (1994), *Sunrunner* (1991), *Tolson* (1983), *Weeks N.C. Giant* (1987).

Pepper / Hot
Capsicum annuum

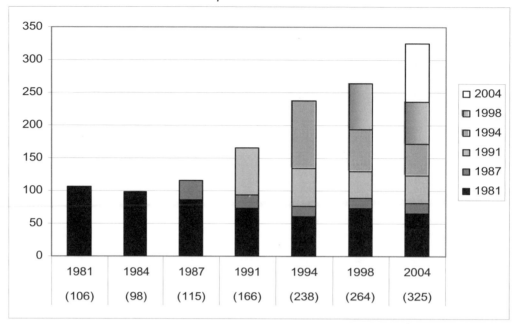

Number of varieties listed in 1981 and still available in 2004: 106 to 66 (62%).
Overall change in number of varieties from 1981 to 2004: 106 to 325 (307%).
Number of 2004 varieties available from 1 or 2 sources: 194 out of 325 (60%).

Abeytas - 90 days, Moderately hot chile, 2.5 x 4.5 in., from Rio Grande Valley community. *Source History: 1 in 1998; 1 in 2004.* **Sources: So12.**

Achar - Mildly pungent, 3-4 in. long fruit, deep red, excel. for pickles, India. *Source History: 1 in 1998; 2 in 2004.* **Sources: Gr29, Pep.**

Aci Sivri - 75-90 days, Plant grows 36-48 in., up to 50 or more 7-10 in. long fruit per plant, excellent sweet-hot taste, heat varies from mild to burning hot, does well in cool weather, rare Turkish heirloom. *Source History: 3 in 1991; 3 in 1994; 4 in 1998; 9 in 2004.* **Sources: Bo9, Gr29, Ni1, Pe1, Pep, Red, Ri12, Se7, So12.**

Afghan Long - 75-90 days, Twisted tapering fruit up to 3 in. long, green > orange > red, heat increases at color stages. *Source History: 1 in 1991; 1 in 1994; 1 in 1998 1 in 2004.* **Sources: Pep.**

Afghan Short - *Source History: 1 in 1991; 1 in 1994; 1 in 1998 1 in 2004.* **Sources: Gr29.**

Agua Blanca - 75 days, Mild green chili up to 10 in. long, thicker flesh than the average green chili, makes good mild chili rellenos. *Source History: 2 in 1998; 2 in 2004.* **Sources: Gr29, Pep.**

Alcalde - Fruit matures red, 4 in., mild to medium heat with slightly sweet complex flavor, New Mexico. *Source History: 1 in 2004.* **Sources: Na2.**

Almapaprika - 70-80 days, Popular Hungarian pickling pepper, mushroom-shaped 1 x 2 in. fruit, white > orange > red, thick crunchy medium hot flesh. *Source History: 1 in 1991; 1 in 1994; 1 in 1998; 2 in 2004.* **Sources: Gr29, Se16.**

Amando - 80 days, Plant grows to 3 ft. tall, upright, pendant, Serrano type fruit, .6 x 3.5 in., Netherlands. *Source History: 1 in 1991; 1 in 1994; 1 in 1998; 2 in 2004.* **Sources: Gr29, Pep.**

Ammazzo - (Joe's Round) - 65 days, Italian cherry pepper, name translates as 'nosegay' or 'small bunch of flowers,' ornamental plant bears clusters of small hot peppers that resemble red marbles in a green bowl .05 x .75 in., 8-15 per cluster. *Source History: 1 in 1998; 1 in 2004.* **Sources: So1.**

Anaheim - (Anaheim Chili, Long Green Chili) - 70-90 days, Vigorous bushy upright 24-30 in. plants, tapered pointed 6-8 x 1.5 in. two-celled med-hot fruits, pendant habit, med-thick dark-green > red flesh, continuous bearing, good cover, widely grown in the South and California, good canned or dried or fried. *Source History: 43 in 1981; 37 in 1984; 42 in 1987; 35 in 1991; 38 in 1994; 44 in 1998; 59 in 2004.* **Sources: Ba8, Bo9, Bo19, Bu2, CH14, CO23, Coo, Dgs, DOR, Ers, Fo7, Ge2, Gr27, Gr29, He8, He17, HPS, Hud, Jo6, Ki4, Lej, LO8, Ma18, MAY, ME9, Mo13, MOU, Na6, Ont, OSB, PAG, Pe7, Pep, Pin, Re8, Red, Ri2, Ri12, Roh, Se7, Se8, SE14, Se25, Se26, SGF, Shu, Sk2, So1, So12, Te8, Tho, Thy, To3, To10, Tu6, Vi4, WE10, Wi2, WI23.**

Anaheim M - (Anaheim Mild) - 75-80 days, Milder than Anaheim TMR 23, tapered med-walled 7 x 1.75 in. peppers, dark- green > red, mostly 2-celled and flat, fairly mild, upright 24 in. plants, pendant, bears continuously. *Source History: 11 in 1981; 10 in 1984; 8 in 1987; 6 in 1991; 7 in 1994; 9 in 1998; 6 in 2004.* **Sources: CO30, Enc, Fe5, Gr29, Pep, Pla.**

Anaheim TMR - (Anaheim 23 TMV, California Green Chili) - 74-79 days, Longer thicker-fleshed pods, slightly less pungent than Anaheim Chili, TM res, 7-8 x 1.5-2 in., green > red, pendant, tapered, everbearing, 2 or 3 lobes, med-hot. *Source History: 4 in 1981; 3 in 1984; 5 in 1987; 7 in 1991; 9 in 1994; 10 in 1998; 9 in 2004.* **Sources: Com, Ga1, GLO, HO2, MIC, Or10, RUP, To1, WE10.**

Anaheim TMR 23 - 74-80 days, Late var. planted extensively in the South and California, sturdy erect 24-34 in. plants, 7-8 in. x 2 in. tapered fruits, mildly hot, TM res., deep-green > red, freeze or can or dry. *Source History: 8 in 1981; 8 in 1984; 7 in 1987; 10 in 1991; 14 in 1994; 16 in 1998; 9 in 2004.* **Sources: ABB, Bu3, CA25, Jun, Loc, ME9, Ni1, RIS, Twi.**

Ancho - (Poblano (so called when fresh), Ancho San Luis) - 88-150 days, Fat heart-shaped 4 x 3 in. relleno chili, black- green > red-brown, mildly hot distinct flavor, very wrinkled when dry, 36 in. bushy plant, everbearing, pendant, Mexico. *Source History: 10 in 1981; 10 in 1984; 7 in 1987; 12 in 1991; 17 in 1994; 20 in 1998; 36 in 2004.* **Sources: Bo19, Bu2, CO30, Com, Enc, Fo7, Goo, Gr28, Gr29, He8, He17, Ho13, Jo6, Jun, Ki4, ME9, Mi12, Pe7, Pep, Pin, Pla, Red, Ri2, Ri12, Sa9, Scf, Se7, Se8, SE14, Te8, Thy, To1, To10, Vi4, WE10, We19.**

Ancho 101 - 68-85 days, Heart-shaped stuffing pepper, 4 x 2.5 in. at shoulder, med-thick walls, black-green > red, rich flavor, concentrated set, pendant habit, 30-36 in. plants. *Source History: 6 in 1981; 5 in 1984; 4 in 1987; 11 in 1991; 14 in 1994; 17 in 1998; 19 in 2004.* **Sources: Bo18, CA25, CH14, Fe5, Gou, HPS, Jor, Mo13, MOU, Orn, Ra5, Ra6, Ros, RUP, Se26, Ter, To1, To3, Wo8.**

Ancho Costeno - 80 days, Uniform, 3 x 6 in. fruits, dark green > dark chocolate-brown, mild, for chile rellenos or used dry in soups and sauces. *Source History: 1 in 2004.* **Sources: Loc.**

Ancho Gigantea - 90 days, Larger Ancho type that holds its size better in warm weather, medium thick fruit, dark green > brown, 1000-1500 Scoville units, fresh market and home gardens. *Source History: 4 in 1998; 6 in 2004.* **Sources: HO2, Pep, RUP, SE14, Se16, Sk2.**

Ancho San Luis - 76-80 days, Medium upright plant produces high quality 5-6 in. fruit, dark green > red > mahogany, moderately pungent, excel. for fresh market, processing or drying. *Source History: 1 in 1981; 1 in 1984; 2 in 1987; 2 in 1991; 4 in 1994; 7 in 1998; 6 in 2004.* **Sources: Bu3, HO2, Hud, Loc, ME9, To1.**

Apaseo - Thin, cylindrical pods, 6-12 in. long, mild flavor, brown at maturity. *Source History: 1 in 1998; 1 in 2004.* **Sources: Enc.**

Arizona Toothpick - 70 days, Bushy, 2.5 ft. tall plant with smooth leaves, straight upright fruit .25 x 2.5 in. long. *Source History: 1 in 1991; 1 in 1994; 1 in 1998; 2 in 2004.* **Sources: Gr29, Pep.**

Aurora - 60-75 days, Plant to 2 ft., purple foliage, small lavender pepper ripens orange and finally red, 1.5 in. long, extremely hot. *Source History: 1 in 1987; 1 in 1991; 3 in 1994; 2 in 1998; 6 in 2004.* **Sources: Co32, Pep, Pr9, Se16, So1, Ta5.**

Autopick - 65-75 days, The result of more than 10 years of breeding, stem stays on the fruit at harvest, 2.5 ft. plant produces dozens of 3-4 in. cone shaped fruit, matures red, hot, full-bodied flavor, ideal for salsa and grilling, PVP 1999 Drs. Fred C. and Nancy A. Elliott. *Source History: 1 in 2004.* **Sources: Ter.**

Bacio di Satana - 90 days, Hot round cherry pepper, poor soil fertility produces hotter fruit, name translates Satan's Kiss. *Source History: 2 in 2004.* **Sources: Ri12, Se24.**

Bahamian Chile - 67 days, Small leaved, 2 ft. plants bear upright clusters of .25 x 1.5 in. fruit, green > red. *Source History: 1 in 1998; 1 in 2004.* **Sources: Gr29.**

Balloon - 90-100 days, Tall branching 3-4 ft. plants, small leaves, strange unique bell-shaped fruit about 3 in. across with 3 or 4 flat square-tipped wings, thin crunchy flesh, wings are sweet but the seeds and placenta are extrmely hot. *Source History: 6 in 2004.* **Sources: He8, Hud, Pr9, Ri12, Se16, WE10.**

Barker - 80 days, Very hot variety of New Mexico Long Green Chili, lacks uniformity and yield of improved varieties but has remained popular due to its pungency, 5-7 in. long, eaten at red or green stage. *Source History: 2 in 1994; 4 in 1998; 3 in 2004.* **Sources: Enc, Pep, Ros.**

Beaver Dam - 80 days, Hungarian heirloom brought to Beaver Dam, Wisconsin in 1929 by the Joe Hussli family, crunchy flesh, mildly hot when seeded, excel. flavor, lime-green > red. *Source History: 1 in 2004.* **Sources: Se16.**

Beijing - 65 days, Plant to 20 in., pleasantly hot orange-red fruit up to 4 in. long. *Source History: 1 in 1991; 1 in 1994; 1 in 2004.* **Sources: Pep.**

Bellingrath Gardens - 80-85 days, Sub-variety of Hot Ecuador, 24-30 in. plant, green and purple and white variegated leaves .5-1 in. pods, purple > green > red, very hot, recommended for ornamental gardens, colorful hot pepper vinegars and spicy cuisine, named in honor of famous gardens in Mobile, AL. *Source History: 1 in 1991; 1 in 1994; 3 in 1998; 5 in 2004.* **Sources: Ho13, Ra6, Re8, So1, To3.**

Berbere - (Short Wrinkled Berbere) - 85-95 days, Re-introduction from Ethiopia, fiery 4-8 in. wrinkled pods turn bright-red when ripe, dried pods used in stir fry or ground into pepper, 30,000-100,000 Scoville heat units. *Source History: 4 in 1991; 1 in 1994; 1 in 1998; 1 in 2004.* **Sources: Bo9.**

Black Cluster - Attractive plant with dark-purple foliage and stems .5 in., deep-red pods. *Source History: 1 in 1994; 1 in 1998;; 1 in 2004.* **Sources: We11.**

Black Dallas - 80 days, Large 3 ft. tall plants, 5 in. long thin green and black fruit ripens red, hot, abundant yields. *Source History: 2 in 1987; 2 in 1991; 2 in 1994; 3 in 1998; 3 in 2004.* **Sources: Gr29, Ho13, Pep.**

Black Plum - 75 days, Thin walled, pendant fruit, 1 x 1.5 in., green > red. *Source History: 1 in 1991; 1 in 1994; 2 in 1998; 2 in 2004.* **Sources: Gr29, Pep.**

Black Prince – Erect .5 in. long purple fruits. *Source History: 1 in 1981; 1 in 1991; 1 in 2004.* **Sources: We11.**

Bouquet - 75 days, Ornamental 2.5 ft. tall plant with

purple leaves and flowers, edible fruit, white > red. *Source History: 1 in 1994; 1 in 1998; 2 in 2004.* **Sources: Gr29, Pep.**

Buenos Carbide - Pendant fruit, light yellow > orange. *Source History: 1 in 2004.* **Sources: Te8.**

Bulgarian Carrot - (Shipkas) - 65-70 days, Florescent-orange, thin walled fruit borne in large clusters close the the main stem of the 18 in. plant, 1.5-3.5 in. long, hot but fruity, Bulgarian heirloom. *Source History: 7 in 1998; 18 in 2004.* **Sources: Fe5, He8, Hud, Ma18, ME9, Par, Pep, Pla, Pr9, Ra6, Re8, SE14, Se16, Ta5, Te8, To1, To3, Ver.**

Burning Bush - Bush plant is covered with 2 in. long, pencil-thin fruit, matures to red, very productive. *Source History: 1 in 2004.* **Sources: Te8.**

Cabeza de Lagarto - Small leaves on med. size plant, 1 in. fruit, name translates Head of the Lizard, which the fruit resembles, ripens from green > brown-red > red, from Nicaragua. *Source History: 1 in 2004.* **Sources: Ba16.**

Calistan - *Source History: 2 in 1991; 2 in 1994; 1 in 1998; 1 in 2004.* **Sources: WE10.**

Caloro - (Caloro Yellow Wax Hot) - 75-80 days, Strong 25 in. plants, good foliage, sets continuously, 3.5 x 1.5 in., bright- yellow > orange-red, med-thick flesh, pointed, pendant, everbearing, fresh market, ideal for pickling, TM res. *Source History: 10 in 1981; 8 in 1984; 8 in 1987; 7 in 1991; 11 in 1994; 12 in 1998; 11 in 2004.* **Sources: Enc, Fo7, Gr29, Ho13, ME9, Pep, Red, Ros, SE14, WE10, WI23.**

Cambuci - 80-90 days, Ornamental, fruit looks like Bishop's Crown but with less heat, often mistaken for flowers, only center part of fruit is hot. *Source History: 1 in 2004.* **Sources: So25.**

Candlelight - 80-100 days, Ornamental plants to 12-16 in., clusters of thin tapered 1 in. fruit, green > yellow > orange > red, mild heat. *Source History: 4 in 2004.* **Sources: He8, Pep, Se16, Sk2.**

Cantina Yellow - Productive plant bears many fruit, 1 x 2.75 in. *Source History: 1 in 2004.* **Sources: Te8.**

Caribe - Medium heat increases to hot, 2-3 in. chile, southern Chihuahua, Mexico. *Source History: 1 in 2004.* **Sources: Na2.**

Casados Native - Spanish heirloom from Casados Farms in El Guique, NM, 5500 ft., 3-5 in. long, med. to hot flavor. *Source History: 1 in 1998; 1 in 2004.* **Sources: Na2.**

Cascabel - (Cascabelle) - 75 days, Mirasol type, spicy conical bells, 1.5-1.75 x .75 in. dia., dark-green > red > red brown, mildly hot, bushy plant, dried or in sauces and gravies, Mexico, name means Little Bell in Spanish. *Source History: 5 in 1981; 4 in 1984; 3 in 1987; 5 in 1991; 10 in 1994; 7 in 1998; 6 in 2004.* **Sources: Enc, Gr29, ME9, Pep, SE14, To1.**

Cascabella - 65-80 days, Plant to 28 in., conical tapered 1.5 in. long fruit, light-yellow > red, mildly pungent, great for hot pickled chilis. *Source History: 2 in 1987; 2 in 1991; 4 in 1994; 11 in 1998; 8 in 2004.* **Sources: Enc, Gr29, Ho13, ME9, Pep, SE14, Se26, To1.**

Cayenne - 70-75 days, Hot 5-6 x .75 in. thin-walled pods, 2 cells, pendant, concentrated fruit set, dark-green > red, used fresh or dried (crushed or powdered), for home or market. *Source History: 19 in 1981; 15 in 1984; 14 in 1987; 19 in 1991; 17 in 1994; 15 in 1998; 20 in 2004.* **Sources: Ban, Bo9, Bo17, Bo19, CO30, Coo, Goo, Lin, Ol2, Pla, Pr3, Ri2, Ri12, Sa6, So9, So12, Te8, Th3, Thy, Tu6.**

Cayenne, Buist's Yellow - Plants 2 ft. tall, 3-5-5 in. long, pendant fruit with somewhat twisted shape, slightly curled at the tip, light green > lemon yellow > dark golden yellow-orange, seed from collection of William Woys Weaver, maintained by his father since before 1945, description matches that in the Buist's seed catalog from the 1870s. *Source History: 1 in 1998; 2 in 2004.* **Sources: Ea4, Ho13.**

Cayenne, Carolina - 70 days, Strong plants from 24-36 in. tall, pungent fruit matures to deep red, pods borne well above ground level, can be dried on the bush, 80,000-100,000 Scoville heat units, good RKN res., Clemson U and USDA. *Source History: 1 in 1994; 5 in 1998; 5 in 2004.* **Sources: Enc, Gr29, Pep, Scf, We11.**

Cayenne, Charleston Hot - 65-75 days, USDA introduction known as the hottest cayenne type, sturdy 18-24 in. plant, normal foliage color is yellow-green, not to be mistaken for a nitrogen deficiency, fruit size is 5 in. long by .75 in. wide, light green > orange > deep red, fine flavor, about 20 times hotter than the typical cayenne, 100,000 Scoville heat units, excel. producer, res. to many types of nematodes. *Source History: 7 in 1994; 12 in 1998; 17 in 2004.* **Sources: Enc, Gr29, HPS, Jo6, Kil, ME9, Pe7, Pep, Ra6, Re8, Scf, SE14, Se26, Te8, To1, To3, We11.**

Cayenne, Golden - 60-72 days, Larger hotter version of Ultra Cayenne, pencil thin 6 in. fruit ripens to crimson, used in place of Jalapeno, Serrano or Habanero, 1000-1500 Scoville heat units. *Source History: 2 in 1991; 4 in 1994; 6 in 1998; 11 in 2004.* **Sources: Gr29, Ho13, Ma18, ME9, Mo20, Pep, SE14, Sto, TE7, To1, WE10.**

Cayenne, Island Delight - Cayenne chile from Jamaica. *Source History: 1 in 2004.* **Sources: Te8.**

Cayenne, Joe's Long - 60-85 days, Thin fleshed, tapering fruit, 8-10 in. long, good for ristras and hot pepper flakes, from Joe Sestito of Troy, NY via his brother, seeds from Italy. *Source History: 1 in 2004.* **Sources: Jo1.**

Cayenne, Large Red Thick - (Large Thick, Large Long Red) - 64-80 days, Flesh twice as thick, 5-7 x 1.5 in., large shoulders, wrinkled tapered and curved, concentrated set and harvest, pendant, strong upright plants, dark-green > scarlet. *Source History: 17 in 1981; 14 in 1984; 14 in 1987; 14 in 1991; 16 in 1994; 21 in 1998; 14 in 2004.* **Sources: CA25, Enc, HO2, Jor, MIC, Ra6, Re8, RIS, RUP, Sa9, Se26, To1, To3, WE10.**

Cayenne, Long Red - (Long Cayenne) - 70-75 days, Fiery hot even when small, 4-6 x .5-1 in. wrinkled pods, waxy dark-green > crimson, often curled and twisted, tapers to point, 2 cells, med-thick flesh, 20-30 in. plants with good cover, 1828. *Source History: 68 in 1981; 62 in 1984; 55 in 1987; 43 in 1991; 42 in 1994; 39 in 1998; 40 in 2004.* **Sources: Ada, All, Ber, Bou, Bu2, Bu8, Co31, Co32, Com, De6, Fo13, Ge2, He8, He17, HO2, La1, Lej, LO8, Mel, Mey, MIC, Mo13, Mo20, MOU, Ont, Or10, PAG, Pin,**

Roh, Sau, Se8, Sh9, So1, SOU, Sto, Sw9, To10, Up2, Vi4, Wet.

Cayenne, Long Red Thick - (Finger, Long Thick Cayenne) - 70-76 days, Larger dia. and longer, blunt-ended, 6-7 x 1.25 in., 2 cells, 22-24 in. plant, med-thick flesh, meatier but not as hot as Slim Cayenne. *Source History: 9 in 1981; 9 in 1984; 9 in 1987; 8 in 1991; 7 in 1994; 8 in 1998; 6 in 2004.* **Sources: CH14, HA3, ME9, Pep, SE14, Twi.**

Cayenne, Long Slim - (Long Red Slim Cayenne, Slim Cayenne, Long Narrow Cayenne, Long Red Narrow) - 70-75 days, Long slender wrinkled very hot peppers, 5-6 x .75 in., thin walls, 2 cells, dark-green > bright-red, tapered, easy to dry, large vigorous 22-30 in. spreading plants, pendant, bountiful harvests, for pickles and canning and drying. *Source History: 38 in 1981; 35 in 1984; 37 in 1987; 32 in 1991; 34 in 1994; 38 in 1998; 38 in 2004.* **Sources: Ag7, BAL, CA25, CH14, Cr1, Dam, Ear, Enc, Ers, Fe5, Fo7, Ga1, GLO, Jo6, Jor, Loc, MAY, ME9, Ni1, OLD, OSB, Pe2, Pep, Re8, Red, RIS, Ros, RUP, Se25, SGF, Shu, Si5, TE7, To1, To3, WE10, Wi2, WI23.**

Cayenne, Long Thin - (Long Red Thin Cayenne) - 67-80 days, Very hot thin-walled curled twisted pointed fruits, 4-6 x .75 in., pendant, 2 cell, pungency develops early, used from 1 in. on up, slender. *Source History: 8 in 1981; 7 in 1984; 8 in 1987; 7 in 1991; 9 in 1994; 9 in 1998; 12 in 2004.* **Sources: CO23, DOR, GRI, HPS, Kil, Ra5, Ra6, SE14, Se26, Sk2, Ter, We19.**

Cayenne, Purple - 70-85 days, Plants 2 ft. tall, covered with dozens of purple blossoms followed by dark purple, thin fruit, 5- 7 in. long, very hot. *Source History: 1 in 1998; 7 in 2004.* **Sources: Ma18, ME9, Pe2, Pin, SE14, TE7, WE10.**

Cayenne, Thomas Jefferson - Med. hot fruit, planted by Thomas Jefferson at his birthplace in Shadwell in 1767, just before his 24th birthday. *Source History: 1 in 2004.* **Sources: Ami.**

Cayenne, Ultra - (Slim Cayenne) - 78 days, Tall bushy plant, large straight fruit. *Source History: 2 in 1991; 1 in 1994; 1 in 1998; 1 in 2004.* **Sources: Pep.**

Chaco - (El Chaco) - 75 days, Plants grow 18 in. tall, slightly curved, 4 in. long fruits, green > yellow > deep orange-red, hotter than Jalapeno, Cayenne, Tabasco or Thai Hot, excel. for drying, preserving or fresh use, named after the vast plains next to the Paraguay River in South America. *Source History: 1 in 1987; 1 in 1991; 1 in 1998; 1 in 2004.* **Sources: He8.**

Chahua - Hot pepper from Mexico. *Source History: 1 in 2004.* **Sources: Te8.**

Chamborate - 80 days, Plant grows 3 ft. tall, pointed fruit, .6 x 3.5 in., medium green > red. *Source History: 1 in 1994; 1 in 1998; 1 in 2004.* **Sources: Gr29.**

Chayusa Ruso - Many pods per plant, light yellow > orange-red, from Brazil. *Source History: 1 in 2004.* **Sources: Te8.**

Cherry - Very pretty 2 ft. plant with pods growing upwards, cream > yellow > red. *Source History: 1 in 1991; 1 in 2004.* **Sources: LO8.**

Cherry, Hungarian - Cherry pepper on compact plant. *Source History: 1 in 2004.* **Sources: Te8.**

Cherry, Large Red - (Large Hot Cherry, Cherry Hot) - 69-85 days, Very hot 1 x 1.5 in. flattened-globe, med-dark green > cherry- red, upright habit, everbearing 24 in. plants, rather late but very productive, great for small pickles, good for canning, easily dried. *Source History: 44 in 1981; 42 in 1984; 38 in 1987; 38 in 1991; 39 in 1994; 36 in 1998; 34 in 2004.* **Sources: BAL, Bu2, Bu3, CH14, Com, Enc, Ge2, GLO, Gr28, GRI, He8, HO2, Jo6, Jor, ME9, MIC, MOU, Pep, Pla, Ra6, Re8, Red, RIS, Ros, RUP, SE4, SE14, Se26, Sto, To1, To3, To10, Twi, WE10.**

Cherry, Red - 75-82 days, Small prolific upright plants, 1-1.25 x 1-1.5 in. dia., slightly flattened cherry-shaped med-walled fruits, deep-green > dark-red, used mostly for pickling, hot. *Source History: 13 in 1981; 11 in 1984; 15 in 1987; 13 in 1991; 8 in 1994; 9 in 1998; 11 in 2004.* **Sources: Bo9, Bo19, Ga1, Gr29, HA3, Mey, Mo20, Pe7, SE14, Sh9, Si5.**

Cherry, Small Red - (Small Hot Cherry) - 75-80 days, Quite hot .75 x 1.25 in. dia. fruits, med-thick walls, med-green > red, 18-22 in. bushy upright plants, bears continuously, mainly for pickling. *Source History: 10 in 1981; 10 in 1984; 4 in 1987; 6 in 1991; 3 in 1994; 2 in 1998; 1 in 2004.* **Sources: Sh9.**

Chi-Chien - 80 days, Upright, very hot thin-skinned red pepper, 2-3 in., easy to grow, popular in the Orient. *Source History: 1 in 1991; 2 in 1994; 3 in 1998; 5 in 2004.* **Sources: Ev2, Gr29, Ho13, Jo6, Pep.**

Chilaca - Pasilla type, medium hot fruits mature to dark brown, dried pods used in sauces such as moles and adobados, fruit is called Chilaca when eaten fresh, from Mexico. *Source History: 1 in 1991; 2 in 1994; 3 in 1998; 2 in 2004.* **Sources: Ho13, Pep.**

Chilaucle Negro - Medium hot, round, muffin shape fruit, 2-3 in. dia., dark green > dark brown-black, 2-3 ft. plants. *Source History: 1 in 1998; 1 in 2004.* **Sources: Ho13.**

Chilcostle - Thin fleshed pods, 1.5 in. wide by 5-6 in. long, green > brown > red, medium heat, used to make dark sauces and mole, great for ristras, from Oaxaca, Mexico. *Source History: 1 in 2004.* **Sources: Ba16.**

Chile Colorado - Long poblano-shaped chile, 3.5-4 in. long, mild heat, southern Chihuahua, Mexico. *Source History: 1 in 2004.* **Sources: Na2.**

Chile Lombak - 90 days, Heirloom from Lombak, Indonesia, excel. hot drying pepper with great taste and rich color, 5-6 in. long fruit, the lower third of the fruit is without heat and may be eaten like a sweet pepper, green > orange > red. *Source History: 1 in 1998; 1 in 2004.* **Sources: So1.**

Chili Grande - 60 days, Large tapered fruit, 3 in. long, very hot. *Source History: 2 in 1998; 2 in 2004.* **Sources: Pep, Sto.**

Chimayo - 95 days, Plants 24-30 in. tall, slightly curved fruits, thin flesh, mild when green, turns quite hot when red, legendary chili dev. by Spanish people in village of Chimayo. *Source History: 2 in 1981; 2 in 1984; 3 in 1987; 2 in 1991; 3 in 1994; 7 in 1998; 7 in 2004.* **Sources: Enc, Gr29, Na2, Pep, Pla, Red, Se8.**

Chinese Multicolor - (Chinese Five Color, Joe McCarthy Pepper) - Dark-green 3-4 ft. plant, beautiful 1 in.

cone-shaped fruits, purple > cream > yellow > orange > red, very hot. *Source History: 1 in 1991; 1 in 1994; 2 in 1998; 3 in 2004.* **Sources: So1, Te8, We11.**

Coban - Bushy plants spread to 2 ft., extremely hot triangular fruits are less than 1 in. long, traditionally cured by smoking over wood fires, collected near Chichicastenanejo in the Guatemalan highlands. *Source History: 2 in 1991; 2 in 1994; 2 in 1998; 1 in 2004.* **Sources: Ho13.**

Cochiti - 130 days, Hot cone-shaped fruit, 3.5 x 1 in., medium to very hot, green > red, collected at Cochiti Pueblo. *Source History: 3 in 1994; 1 in 1998; 3 in 2004.* **Sources: Gr29, Na2, Pep.**

College 64 L - (Anaheim Chili College 64L) - 74-76 days, Mildly hot thick-walled 2.5 x 6-8 in. long peppers, med-green > red, excel. for chile rellenos, used fresh (roasted and peeled) or dried and powdered, good for mild salsa and guacamole. *Source History: 5 in 1981; 4 in 1984; 3 in 1987; 2 in 1991; 1 in 1994; 2 in 1998; 1 in 2004.* **Sources: Ter.**

Congo de Nicaragua - Very hot, small piquin type pods held upright on the med. size plant, green > orange > red, Nicaragua. *Source History: 1 in 2004.* **Sources: Ba16.**

Country Girl - 80 days, Experimental variety resulting from a cross in H.W. Alfrey's garden, very large fruit, ranges from mild to hot. *Source History: 1 in 1987; 1 in 1991; 1 in 1994; 1 in 1998; 1 in 2004.* **Sources: Pep.**

Cow Horn - 70-75 days, Green 3 ft. plant, 6-10 in. slender curved red pods resemble cow's horns, dark green > red, great for sauces. *Source History: 1 in 1984; 1 in 1987; 2 in 1991; 2 in 1994; 2 in 1998; 3 in 2004.* **Sources: CA25, Gr28, Pep.**

Crimson Hot - 60 days, Similar to Hot Portugal but not quite as hot, long tapered 6-6.5 in. fruits, deep waxy crimson, med- thick flesh, good yields, recommended for ketchup and pickling. *Source History: 1 in 1981; 2 in 1984; 2 in 1987; 2 in 1991; 2 in 1994; 2 in 1998; 4 in 2004.* **Sources: Gr29, Pep, Ra6, Sto.**

Criolla de Tres Cantos - Prolific hot pepper, tapered shape, thick walls. *Source History: 1 in 2004.* **Sources: Ta5.**

Criolla Sella - 90-120 days, *C. baccatum*, Bolivian pepper, 2 ft. plant, 1 x 3 in. fruit, green > yellow, tasty fresh or dried. *Source History: 1 in 1998; 3 in 2004.* **Sources: Ba16, Se7, Se26.**

Curry - 75 days, Upright plants produce 6 in. long fruits with the distinctive flavor found in Indian curry dishes, from India. *Source History: 1 in 1998; 2 in 2004.* **Sources: CO30, To3.**

Cyklon - 80-100 days, Tapered fruits are 2 in. at the shoulder by 4-5 in. long, just the right amount of heat for cooking without burning your mouth, used extensively by the spice industry in Poland for drying due to its rather thin flesh. *Source History: 1 in 1998; 5 in 2004.* **Sources: Pep, Pr3, Pr9, Se16, Ta5.**

Czechoslovakian Black - (Czech Black) - 65-70 days, Heirloom from Czechoslovakia, productive 2.5 ft. bushes with dark- green leaves, purple veins, 2 in. fruit has shape of a jalapeno, dark-green > red, dark-green color looks black, mildly hot. *Source History: 3 in 1994; 5 in 1998; 7 in 2004.* **Sources: Ag7, Fe5, Fo7, Ho13, Ma18, Ra6, So1.**

Da Nang Market Chili - Tall bush bears abundant yields of cayenne type fruit. *Source History: 1 in 2004.* **Sources: Te8.**

Datil - 80 days, Habanero type native to St. Augustine, Florida where it has been grown for about 300 years, very difficult to grow out of its native habitat, thin walled fruit, 2 in., green > yellow-orange, extremely hot, unique smoky fruity flavor. *Source History: 2 in 1998; 4 in 2004.* **Sources: Ba16, Gr29, Ho13, Te8.**

De Arbol - (Tree Chili, Pico de Pajaro, Cola de Rata) - 80-90 days, Tall perennial chile bush with little spear-shaped red arrows .5 x 2 in., one of the hottest, resembles a little Cayenne pepper, productive, from Chihuahua, Mexico *Source History: 2 in 1981; 2 in 1984; 2 in 1987; 5 in 1991; 12 in 1994; 15 in 1998; 17 in 2004.* **Sources: Enc, Gr28, Gr29, Ho13, Jo6, Loc, ME9, Na2, Pep, Pla, Red, Ri2, Se8, SE14, Thy, To1, We11.**

De Arbol Purple - Ornamental plant, foliage colors include green, white, purple and lavender, thin fleshed pointed pods, 2-3 in. long, very hot, purple > green > red, from Mexico. *Source History: 1 in 2004.* **Sources: Ba16.**

De Comida - Glossy red peppers, 4-6 in. long, rather hot, usually used dried in making 'mole'--a pre-Columbian gravy-like sauce, Zapotec Indian seed from Oaxaca, Mexico. *Source History: 3 in 1981; 2 in 1984; 3 in 1987; 3 in 1991; 2 in 1994; 1 in 1998; 2 in 2004.* **Sources: Sk2, Te8.**

De Oaxaca - Distinctive, mild flavor, brown when mature. *Source History: 1 in 1998; 1 in 2004.* **Sources: Enc.**

De Rata - Plants to 4 ft. tall, 2 ft. across, 4-5 in. long slender pods, very prolific, up to 100 fruit per plant, Mexico. *Source History: 1 in 2004.* **Sources: Red.**

Del Arbol de Baja California Sur - Thin, 4 in. bright red fruit, medium hot. *Source History: 1 in 2004.* **Sources: Na2.**

D'Espelette - Deep red fruit, fairly hot, popular in the Pyrennees region of southern France, near Spain. *Source History: 1 in 2004.* **Sources: Hud.**

Diablo Grande - 65 days, Tapered fruit, 2 x 7 in., light green > dark red, fiery hot. *Source History: 1 in 1998; 1 in 2004.* **Sources: Pep.**

Dragon's Claw - 85 days, Beautiful red fruit, tapered and curved at the tip, 10 in. long, relatively tame, heat is in the ribs and seeds. *Source History: 1 in 1998; 1 in 2004.* **Sources: Fo7.**

Dulce Medeterraneo, Yellow - Huge 10+ in., thick fleshed pods, green > yellow, otherwise same as red version. *Source History: 1 in 2004.* **Sources: Ba16.**

Dunso - Tapered fruit with fairly hot but distinct flavor. *Source History: 1 in 1994; 1 in 1998; 1 in 2004.* **Sources: Pep.**

Dutch Hot - Plants bear many long twisted cylindrical fruit, good for container culture, dates back to the Dutch occupation of Indonesia. *Source History: 1 in 2004.* **Sources: Te8.**

Eastern Rocket - 65-68 days, Heavy yields of flattened thick-fleshed fruits, 3-6 x 1-2 in. at shoulder, tapers to blunt point, light-green > glossy bright-red, med-hot to hot, productive. *Source History: 3 in 1981; 3 in 1984; 4 in 1987;*

4 in 1991; 3 in 1994; 1 in 1998; 2 in 2004. **Sources: Gr29, Pep.**

Ecuador - 70 days, Plant has beautiful dark-green and purple leaves, purple flowers, 1 in. pods, green > purple > red, very hot. *Source History: 2 in 1991; 2 in 1994; 1 in 1998; 5 in 2004.* **Sources: Ho13, Pep, Se17, Te8, We11.**

Embers - Striking plant to 18 in., bears numerous small fruit, green > orange > red. *Source History: 1 in 2004.* **Sources: Te8.**

Escondida - Medium heat develops slowly in the mouth, New Mexico. *Source History: 1 in 2004.* **Sources: Na2.**

Espanola - 65-70 days, Vigorous plants to 2 ft., medium-hot, thin walled, red fruit dries easily, 6-7 in., makes excel. chile powder and salsa. *Source History: 1 in 1994; 1 in 1998; 3 in 2004.* **Sources: Gr28, Gr29, Red.**

Espanola Improved - 65-75 days, Special short season chili resulting from a cross between Sandia and a Northern native type, pods similar to Sandia except slightly thinner wall, 5-6 in. long, will turn red even in cool Northern climates, uniform heavy producer, extremely hot, New Mexico, 1983. *Source History: 2 in 1987; 5 in 1991; 6 in 1994; 8 in 1998; 9 in 2004.* **Sources: Bu3, Enc, Jo6, Pep, Pla, Ros, Se7, Se26, To1.**

Ethiopian Brown - Thin fleshed, 4 in. pods, green > reddish brown > deep brown, low heat, dries well, from Ethiopia. *Source History: 1 in 2004.* **Sources: Ba16.**

Fiesta - 90 days, Compact plant, slim erect 2 in. rocket-like fruit appears above the foliage, white > yellow > bright red. *Source History: 1 in 1994; 2 in 1998; 3 in 2004.* **Sources: So25, Und, We11.**

Fiji Embers - Abundant production of medium size heart shaped fruit. *Source History: 1 in 2004.* **Sources: Te8.**

Filius Blue - 75-90 days, Ornamental plants with purplish foliage speckled with white, small round-oval, upright purple-blue fruit, red when mature, very hot when young. *Source History: 3 in 1998; 10 in 2004.* **Sources: Ami, Ma18, ME9, Ra6, Re8, SE14, TE7, Te8, To1, To3.**

Fips - 90 days, Small 8 in. plant covered with conical, 1 in., red and yellow, edible fruit. *Source History: 1 in 1994; 2 in 1998; 4 in 2004.* **Sources: ME9, Pep, Pin, SE14.**

Fire - 75 days, Incredibly hot 2 x .5 in. peppers, med-green > red, lacy foliage is quite ornamental. *Source History: 3 in 1981; 2 in 1984; 1 in 1987; 3 in 1991; 3 in 1994; 2 in 1998; 1 in 2004.* **Sources: Pep.**

Firecracker - 66-75 days, Ornamental multi-branched, 3 ft. plant with purple blossoms, pointed, glossy 1 in. fruit, purple > brilliant red, fiery hot. *Source History: 2 in 1994; 5 in 1998; 6 in 2004.* **Sources: Bo19, Eo2, He8, Pe7, Tho, WE10.**

Fireworks - 55-66 days, Attractive plant with dark foliage, 8-10 in. tall, 1 in. pods, purple > cream > yellow > orange > red, fiery hot. *Source History: 1 in 1994; 1 in 1998; 4 in 2004.* **Sources: Gr29, Pep, Roh, Se26.**

Fish Pepper - 80-90 days, Multi-branched, 2.5 ft. plants, white and light green mottled leaves with curled edges, pendant, 2.5 in. fruit, striped light and dark green > striped orange and red > solid red, very hot, African-American heirloom from the Philadelphia/Baltimore area, pre-1947. *Source History: 1 in 1998; 13 in 2004.* **Sources: Ami, Ba16, Co32, Ea4, Fe5, Ho13, Ma18, Mcf, Se16, So25, To1, To3, Und.**

Floral Gem - 75-80 days, Erect 22-26 in. plants, med-size foliage, conical tapered 3-lobed fruits, 2-2.5 x 1-1.5 in. dia., very hot thin flesh, waxy-yellow > deep-red, continuous, can or pickle. *Source History: 9 in 1981; 9 in 1984; 8 in 1987; 4 in 1991; 3 in 1994; 5 in 1998; 4 in 2004.* **Sources: ME9, Pep, SE14, TE7.**

Flourescent Purple - Beautiful plant with fluorescent purple and white foliage .75-1 in. hot fruit, green > purple > red, dries well, easy to germinate, transplant and grow. *Source History: 1 in 1994; 1 in 1998; 2 in 2004.* **Sources: Ba16, Pe7.**

Fresno Chili - 75-80 days, Very hot, 2-3 x 1-1.25 in. dia., tapers to point, vigorous 24-30 in. plants, dense dark-green foliage, TM res., smooth, good pickler, market at mature green stage, from Mexico. *Source History: 8 in 1981; 8 in 1984; 5 in 1987; 4 in 1991; 5 in 1994; 5 in 1998; 11 in 2004.* **Sources: Bu3, CO30, Gr29, ME9, Pep, Red, Se7, SE14, Se26, Sk2, WE10.**

Fresno Chili Grande - (Fresno Hot) - 75-78 days, Small very hot pepper, 2-3 x 1.5 in. dia., tapers to a point, med-thick flesh, bright med-green > red, prolific, used fresh in salads and sauces or pickled. *Source History: 5 in 1981; 5 in 1984; 3 in 1987; 3 in 1991; 2 in 1994; 2 in 1998; 1 in 2004.* **Sources: Loc.**

Georgia Flame - 90-120 days, Plants to 24 in. tall, fruit is 2 in. at shoulder by 6-8 in. long, crunchy flesh is not tough, produces heavily over a long period, good salsa pepper with just the right amount of heat, from the Republic of Georgia. *Source History: 6 in 2004.* **Sources: He8, Pep, Pr9, Se16, Ta5, Ter.**

Georgia White - Cayenne shaped fruit, mild flavor becomes more pungent as it matures from white > orange > dark red. *Source History: 1 in 2004.* **Sources: Fe5.**

Goat Horn - (Sweet Spanish Long, Italian Green Frying) - 70 days, Plants grow 4-5 ft. tall, deep-green > cherry-red, 5 in. long x about 1 in. thick, often curled and twisted, easily dried. *Source History: 1 in 1981; 1 in 1984; 2 in 1987; 3 in 1991; 3 in 1994; 4 in 1998; 6 in 2004.* **Sources: CO30, Ev2, Pep, Ra6, Re8, Se24.**

Grandpa's Favorite - 90 days, Resembles a small, meaty, juicy jalapeno, medium hot. *Source History: 2 in 1998; 2 in 2004.* **Sources: Pep, We11.**

Grandpa's Home Pepper - 70 days, Small plant produces 50 or more small brilliant red semi-hot peppers, derives its name from the fact that many Siberian homes have this pepper plant growing in a pot in a window, produces all winter long in low- light conditions. *Source History: 1 in 1991; 1 in 1994; 1 in 1998; 1 in 2004.* **Sources: Hig.**

Guajillo - (Guajillo Costeno) - 70-100 days, Shiny, 4 in. long, bright-red fruit, hot but tolerable, annual, good for drying, one of the most common chiles grown in Mexico. *Source History: 2 in 1981; 2 in 1984; 2 in 1987; 2 in 1991; 4 in 1994; 9 in 1998; 7 in 2004.* **Sources: Enc, Ho13, Jo6, Pep, Pla, Red, To1.**

Guam Boonies - Hardy 4-4.5 ft. plant, slender 1 in. pods yellow > orange > red, very hot, excellent dried and crushed, needs an early start in the North. *Source History: 1 in 1991; 1 in 1994; 1 in 1998; 1 in 2004.* **Sources: We11.**

Guilin - Prolific, compact plant, 2-3 in. long tapered fruit turns red very quickly, thin flesh, medium heat. *Source History: 1 in 1994; 1 in 1998; 1 in 2004.* **Sources: Al6.**

Gumdrop - 95-100 days, Compact plants to 18 in., fruits are the size, shape and color of gumdrops, colors range from tangerine > lemon > grape > orange > red with all colors on the plant at one time, very hot. *Source History: 1 in 1998; 2 in 2004.* **Sources: Fo7, He8.**

Hahong Koch'o - (Takanotsume, Japanese Hot Claw, Rooster Spur) - 100 days, Upright clusters of bright red fruit resembles claws, .25 x 2-3 in., dries well, does well in wet climates, from Korea. *Source History: 1 in 1994; 1 in 1998; 1 in 2004.* **Sources: Pep.**

Haitian - 67-85 days, Plant grows 1-3 ft., thin skinned fruits dry quickly, 3 in. long, green > bright red, prolific. *Source History: 1 in 1998; 1 in 2004.* **Sources: Pep.**

Hanoi Market Yellow - Market chile from Vietnam. *Source History: 1 in 2004.* **Sources: Te8.**

Hanoi Red - Vietnamese Oriental pepper, ripens red. *Source History: 1 in 2004.* **Sources: Te8.**

Hawaiian Sweet Hot - Slightly smaller fruit than Jalapeno with more rounded shoulders, sweet heat, good in salads, pickling, fresh and salsa, from Hawaii. *Source History: 2 in 2004.* **Sources: Ba16, Red.**

Hinkelhatz - 105 days, Rare Pennsylvania Dutch heirloom dating back to before 1880, pendant fruits the size and shape of chicken's hearts, which is the translation, 12 in. plant grows 1.5 ft. wide, good drying pepper, very hot. *Source History: 1 in 1998; 5 in 2004.* **Sources: Ami, Ea4, Fo7, Pe7, Ta5.**

Hinkelhatz, Yellow - Selected from the red Hinkle Hatz, may get some red fruit. *Source History: 1 in 2004.* **Sources: Ea4.**

Holiday Cheer - 90 days, Round ball-like fruit, cream > orange > bright red, can be very hot. *Source History: 3 in 1994; 3 in 1998; 5 in 2004.* **Sources: Enc, ME9, OSB, Pep, SE14.**

Holiday Time - 90 days, AAS/1980, ornamental plant grows 6-8 in., green purple and red fruit, Honduras. *Source History: 3 in 1994; 2 in 1998; 3 in 2004.* **Sources: ME9, OSB, We11.**

Holland Hot Finger, Red - Semi-fleshy pods grow 6-7 in. long by .5 in. wide, green > bright red, hot, used in salsa, stews and chile. *Source History: 1 in 2004.* **Sources: Ba16.**

Holland Hot Finger, Yellow - Semi-fleshy, 6-7 in. long pods, green > beautiful yellow, used in salsa, stews and chile. *Source History: 1 in 2004.* **Sources: Ba16.**

Hot Apple - (Hungarian Hot Apple, Almapaprika) - 55-90 days, Upright plant growth, butter-yellow globe cheese type ripens to scarlet-orange, 2.5 in. across x 1.5 in. deep, semi-hot, suitable for eating raw or lightly cooked, most widely grown pepper in Hungary. *Source History: 1 in 1987; 3 in 1991; 2 in 1994; 4 in 1998; 2 in 2004.* **Sources: Pep, We11.**

Hot Portugal - 60-65 days, Sturdy, upright plants, fiery hot, large smooth glossy bright-scarlet fruits, 6-7 in. long, pointed shape, heavy yields, twice as large as old Giant Cayenne, offered by Joseph Harris and Company of Coldwater, New York in 1935. *Source History: 3 in 1981; 4 in 1984; 6 in 1987; 4 in 1991; 3 in 1994; 6 in 1998; 12 in 2004.* **Sources: Fe5, Ha5, ME9, Pep, Pr9, Ra6, Red, RUP, Se16, Sto, Ta5, To3.**

Huasteco - 78-85 days, Serrano type, upright, 29-31 in. plant, 2-3 in. long fruit, 2500-4000 Scoville heat units, dark green > red, PVY TEV and PMV res., the name refers to the indigenous people of important serrano producing areas in Mexico. *Source History: 6 in 1998; 5 in 2004.* **Sources: Gr29, ME9, Pep, Ra6, RUP.**

Hungarian Rainbow Wax Block - 60-62 days, Only semi-hot, excel. for flavoring or relishes or packing by themselves, blocky shape, often 3-4 lobes, yellow fruits mature through rainbow of colors to red. *Source History: 1 in 1981; 1 in 1984; 1 in 1987; 2 in 1991; 1 in 1994; 1 in 1998; 1 in 2004.* **Sources: Sh9.**

Hungarian Rainbow Wax Short - 60-80 days, Shorter blocky type, medium hot, very attractive fruits, yellow later turning red. *Source History: 1 in 1981; 1 in 1984; 1 in 1987; 1 in 1991; 1 in 2004.* **Sources: Pep.**

Hungarian Semi-Hot - 80 days, Resembles Szentesi but larger and color is more yellow with slight green tinge, blocky, 5.5 x 3.5 in. fruit, good stuffer. *Source History: 1 in 2004.* **Sources: Pep.**

Hungarian Yellow Wax Hot - (Bulgarian, Hungarian Wax, Hungarian Wax Hot, Hungarian Long Wax, Hot Banana Hungarian Wax No. 2) - 58-85 days, Med-hot med-walled 5-8 x 1-2 in. peppers, tapers to a point, waxy yellow > crimson, both upright and pendant, strong upright 16-24 in. plant, everbearing, pickle or can, best hot pepper for cool areas, matures in Nova Scotia in sheltered exposures, fresh home use or market or processing. *Source History: 116 in 1981; 107 in 1984; 106 in 1987; 97 in 1991; 94 in 1994; 99 in 1998; 101 in 2004.* **Sources: ABB, Ada, Ag7, Alb, BAL, Be4, Bo17, Bo19, Bou, Bu2, Bu3, Bu8, CA25, CH14, CO23, Co31, Co32, Com, Coo, Cr1, Dam, De6, Dgs, DOR, Ear, Enc, Ers, Fe5, Fo7, Ga1, Ge2, GLO, Gr28, Gr29, GRI, HA3, Ha5, He8, He17, Hig, HO2, HPS, Hud, Hum, Jo1, Jo6, Jor, Jun, Kil, La1, Lej, LO8, Ma19, ME9, Mel, Mey, Mi12, MIC, Mo13, Mo20, MOU, Ni1, OLD, Ont, Or10, OSB, Pep, Pin, Ra6, Re8, Red, Ri12, RIS, Roh, Ros, RUP, Sa9, Sau, Se7, SE14, Se26, SGF, Sh9, Shu, Si5, So1, SOU, Sto, Sw9, Te8, To1, To3, To10, Twi, Ver, Ves, Vi4, WE10, Wet, Wi2, WI23.**

Hungarian, Black - 70-80 days, Ornamental Hungarian heirloom, green foliage is highlighted by purple veins and beautiful purple flowers, plants grow 30-36 in. tall and stand very well on their own, 2-3 in. fruits resemble the shape of a Jalapeno, black > red, mildly hot with good flavor. *Source History: 2 in 1998; 14 in 2004.* **Sources: Ba8, Bo19, Co31, He8, La1, ME9, Pep, Pr9, Ri12, SE14, Se16, Se26, TE7, To3.**

Inca Glow - Vigorous plant bears large fruit, ripens to

red, from Peru. *Source History: 1 in 2004.* **Sources: Te8.**

Inca Hot - Peruvian variety with lemon-yellow pods, slightly wrinkled, sweet flesh with very hot interior, color change from green > yellow with the tip of the fruit changing to yellow first. *Source History: 1 in 2004.* **Sources: Te8.**

Inca Red Drop - Bushy plant bears many fruit which look like bright red drops, good for container culture, Peru. *Source History: 1 in 2004.* **Sources: Te8.**

India Chile - 70 days, Tall plant to 3 ft., upright .25 x 2.5 in. fruit, light yellow > green, productive. *Source History: 1 in 2004.* **Sources: Pep.**

Isleta - Medium hot chiles traditionally strung into ristras, then ground into chile powder, grown in the Isleta Pueblo near Albuquerque, New Mexico. *Source History: 1 in 1991; 1 in 1994; 1 in 1998; 1 in 2004.* **Sources: Na2.**

Italian White Wax - 70-75 days, Popular home garden variety for pickling, 34-40 in. upright plant with spreading lacy light-green foliage, 2-3 x 1/2 in. fruit tapering to a point, thin pale-yellow walls turn pale-red at maturity, pendant, continuous set, snappy mild flavor when young, quite pungent when mature. *Source History: 1 in 1987; 1 in 1991; 3 in 1994; 2 in 1998; 1 in 2004.* **Sources: Pla.**

Jalapeno - (Green Chili Jalapeno, Chile Chipotle) - 65-80 days, Very hot sausage-shaped fruits, med-thick walls, 3-3.5 x 1 in., tapers to a blunt rounded point, dark-green > red, pendant, productive 24-36 in. plants, long harvest, pickled in Mexico and Southwest, named after the city of Xalapa in Veracruz, Mexico where it is no longer grown, 2,500-10,000 Scoville heat units, large dried mesquite-smoked red jalapenos are known as chipotle chilis which are used extensively in Southwestern and Mexican cooking. *Source History: 67 in 1981; 61 in 1984; 62 in 1987; 59 in 1991; 60 in 1994; 57 in 1998; 57 in 2004.* **Sources: Ada, Ag7, Al25, Alb, All, Ban, Be4, Bo9, Bo19, Bu2, But, Com, Coo, Cr1, De6, Dgs, Enc, Fi1, Ge2, Goo, Gr29, GRI, Gur, HA3, He8, He17, Hi13, Kil, Lej, LO8, Ma19, ME9, Mey, Mi12, MIC, Mo20, MOU, Ont, OSB, Pep, Pla, Red, Ri2, Ri12, Ros, Sa9, Sau, Se7, Se8, SGF, Shu, So1, Sto, Te8, Thy, To10, Vi4.**

Jalapeno M - 72-75 days, Sausage-shaped blunt-ended thick-walled fruits, 3.5 x 1.5 in., dark-green > red, very pungent, upright 26-36 in. dark-green plants, good cover, pendant habit, used for market gardens and processing, very prolific, continuous production. *Source History: 22 in 1981; 19 in 1984; 25 in 1987; 28 in 1991; 35 in 1994; 41 in 1998; 38 in 2004.* **Sources: ABB, BAL, Bu3, CH14, CO23, CO30, Co31, DOR, Ers, GLO, Gr27, Ha5, Jor, Lin, ME9, Mel, Mo13, OLD, OSB, PAG, Par, Ra6, Red, RIS, Roh, RUP, SE4, SE14, Se26, Si5, Sk2, SOU, To1, Twi, WE10, Wet, Wi2, WI23.**

Jalapeno, Early - 60-68 days, Just like Jalapeno but earlier and better adapted to cool coastal conditions and concentrated set, thick-walled 3 in. cone-shaped fruits ripen to red, hottest when fully ripe, compact non-brittle bushes grow to 24 in. tall. *Source History: 20 in 1981; 20 in 1984; 24 in 1987; 33 in 1991; 36 in 1994; 36 in 1998; 42 in 2004.* **Sources: Bo17, Bu3, CA25, CH14, Dam, Ear, Ers, Fe5, Fis, Ga1, Gr29, Hi6, Hig, HO2, HPS, Jo1, Jor, Jun, La1, MAY, ME9, Na6, Ni1, Pe2, Pe7, Pep, Pin, Ra6, Re8, Red, RUP, SE14, Se25, Se26, Sh9, TE7, Ter, To3, Tu6, WE10, We19, WI23.**

Jalapeno, Gigantia - (Giant Jalapeno) - Large red fruit is 5 in. long, good for stuffing, thick walls, med. heat, fleshier than regular Jalapeno. *Source History: 2 in 1998; 3 in 2004.* **Sources: Ho13, ME9, SE14.**

Jalapeno, Golden - Very hot, 2 in. fruits, bright orange > red. *Source History: 1 in 1998; 1 in 2004.* **Sources: Ho13.**

Jalapeno, Purple - 75 days, Larger fruit size than regular Jalapeno, green fruit changes to dark purple, holds that color a long time before ripening to red, hot, attractive addition to pickled peppers. *Source History: 1 in 1998; 3 in 2004.* **Sources: ME9, SE14, To1.**

Jalapeno, Tam Mild - (Tam Jalapeno No. 1 Mild) - 67-73 days, The gringo pepper, 1/4-1/2 as hot, more productive, 31 in. dark-green plant, virus res., Jalapeno size shape and flavor, freezes and cans well, good for pickling, prolific producer, excel. yields, can be eaten raw or roasted or pickled, Benigno Villalon/Texas A and M. *Source History: 15 in 1981; 15 in 1984; 13 in 1987; 24 in 1991; 32 in 1994; 34 in 1998; 31 in 2004.* **Sources: Ba8, Be4, Bo19, Bu3, CH14, Enc, Fi1, Fo7, Ga1, Gr29, Jor, LO8, ME9, MIC, MOU, Pep, Ra6, Re8, Ros, RUP, Scf, SE14, Se26, SGF, Sw9, To1, To3, Twi, Up2, WE10, Wi2.**

Jalapeno, Tam Vera Cruz - 60-64 days, Slightly tapered cylindrical Jalapeno, dark-green > red, very hot, used fresh, in sauces or pickled, PVP 1992 TX AES. *Source History: 1 in 1991; 3 in 1994; 5 in 1998; 5 in 2004.* **Sources: Enc, Ra6, Re8, Scf, To3.**

Jaloro - (Yellow Jalapeno) - 70-75 days, Compact plant, high yields, 20,000-25,000 SHU, multiple virus res., used at yellow and red stages, high in ascorbic acid and flavinoids, Texas A and M University. *Source History: 1 in 1994; 4 in 1998; 9 in 2004.* **Sources: Eo2, Gr28, Gr29, ME9, Pep, Scf, SE14, Te8, To1.**

Jarales - Fruit size and shape varies, somewhat hot, slightly sweet chile, from Jarales, NM. *Source History: 1 in 1998; 1 in 2004.* **Sources: Na2.**

Jemez - 90 days, Medium-hot chiles average 3 in. long, grown in Jemez Pueblo in northern New Mexico along the Rio Jemez, a tributary to the Rio Grande. *Source History: 1 in 1991; 2 in 1994; 2 in 1998; 2 in 2004.* **Sources: Na2, So12.**

Jerusalem - 75 days, Pendant, cone shaped fruit .5 x 1.5 in., light green > white. *Source History: 1 in 2004.* **Sources: Pep.**

Jigsaw - 90 days, Compact plant with variegated foliage, small fruit points upward, purple, green and white turning to red. *Source History: 1 in 1998; 2 in 2004.* **Sources: Enc, Pep.**

Jose Luis Majorca - Fairly hot drying pepper, 4 in. long thin red fruit, easy to grow, from Majorca, an island off the coast of Spain. *Source History: 1 in 2004.* **Sources: Sk2.**

Kalia - 100 days, Deep red, pointed, very hot fruit, 1 in. long, from India. *Source History: 1 in 1994; 1 in 1998; 1 in 2004.* **Sources: Pep.**

Karlo - 50-80 days, Semi-hot or semi-sweet Roumanian

type, short very compact plant, top-shaped peppers dev. very early, 3-4 x 2.5-3 in. at shoulder, med-thick flesh, yellow > red, old heirloom from Europe. *Source History: 3 in 1981; 3 in 1984; 3 in 1987; 3 in 1991; 3 in 1994; 1 in 1998; 1 in 2004.* **Sources: Fo7.**

Korean Dark Green - Korean heirloom, 2 ft. plants, dark green foliage, 3-4 in. long fruit, dark green > red, spicy hot flavor, great for kimchee. *Source History: 1 in 2004.* **Sources: Ra6.**

Korean Hot - Round white pods with purple marbling .5 in., white > yellow > orange > red, very hot. *Source History: 1 in 1994; 1 in 1998; 1 in 2004.* **Sources: Ea4.**

Kori Sitakame - Med. size fruit, sweet and hot, will produce in the low desert with shade but does better the second year if overwintered, collected in Norogachi, a Tarahumara pueblo in the mountains of Chihuahua, Mexico. *Source History: 1 in 1998; 1 in 2004.* **Sources: Na2.**

Kung Pao - 85-90 days, Large 30 in. plants, thin walled cayenne type, 4.5 in. fruit dries fast, 10,000 Scoville heat units. *Source History: 4 in 2004.* **Sources: Ge2, GLO, Pep, Te8.**

Largo Purple - Ornamental plant with white, dark green and violet foliage, grows 2.5-3 ft. tall, conical fruit matures to deep red, hot. *Source History: 1 in 2004.* **Sources: To1.**

Laungi - 90 days, Long fruit, hot, from India. *Source History: 1 in 1998; 1 in 2004.* **Sources: Pep.**

Little Elf - Ornamental. *Source History: 3 in 2004.* **Sources: ME9, SE14, TE7.**

Little Nubian - 95 days, Beautiful ornamental 2 ft. plants, purple-black foliage, lavender flowers, 1.5 in. long bell-shaped fruit, glossy black > garnet red, hot crunchy flesh with unique flavor, seed from the personal collection of Will Weaver's grandfather. *Source History: 1 in 2004.* **Sources: Und.**

Long Green Buddha - 60-70 days, Bushy 2 ft. plant, long conical shape, pungent flavor with seeds, without seeds has a strong green pepper taste, yields 12-20 fruits per plant. *Source History: 1 in 1991; 2 in 1994; 1 in 1998; 2 in 2004.* **Sources: Pep, Sk2.**

Long Thick Red - (Chile Grande) - 60 days, Really hot, small 3-4 in. long fruits, smooth and very uniform, tapered slightly but not twisted, dark-green > red early in season, good chili type. *Source History: 1 in 1981; 1 in 1984; 1 in 1987; 2 in 1991; 2 in 1994; 1 in 1998; 1 in 2004.* **Sources: Pep.**

Luteum - Bushy, 4-6 ft. plant with small leaves, very high yields of thin, 1-2 in. long fruit, green > red, very hot. *Source History: 1 in 1998; 1 in 2004.* **Sources: Ho13.**

Macskapiros, Cat Red - Med. size plant bears 2-3 in. long pods that grow in all directions on the plant, green > dark red > red, good drying pepper, makes a good hot Hungarian paprika. *Source History: 1 in 2004.* **Sources: Ba16.**

Macskasarga, Cat Yellow - Hungarian pepper with slightly fruity flavor, thin fleshed, 2-3 in. long pods, green > yellow, for salsas and hot sauce, also good smoked or dried. *Source History: 1 in 2004.* **Sources: Ba16.**

Magenta - Purple stemmed plant with green and white foliage, .6 in. fruit, resembles Varingata. *Source History: 1 in 2004.* **Sources: Pep.**

Marbles - 55-60 days, Ornamental edible, 12-18 in. tall plants produce marble-size fruits borne on the top of the plant, yellow > purple > orange > flame-red, from Dr. Baggett, OSU. *Source History: 2 in 1994; 7 in 1998; 8 in 2004.* **Sources: Fis, Par, Pep, Ra6, Re8, Ter, To1, WE10.**

Marseillais, Petit - Wrinkled yellow fruit, 3-4 in. long, mildly hot, French heirloom. *Source History: 1 in 2004.* **Sources: Hud.**

Martin's Carrot - Bright orange, pointed fruits, 1 x 2 in., dev. in the 1800s by a Mennonite horticulturist. *Source History: 1 in 1998; 1 in 2004.* **Sources: Ho13.**

Masquerade - 100 days, Ornamental plant to 10 in. tall, erect fruit, purple > buff > orange > red. *Source History: 1 in 1994; 2 in 2004.* **Sources: BAL, Pep.**

Matchbox - 70 days, An open-pollinated selection of the hybrid Super Chili, the pepper is in its sixth year but is not completely settled down, 1-1.5 ft. plants bear heavy yields of upright, pointy, tapered 1.5-2 in. fruits, ripening green > scarlet, thin walled fruit is fiery hot, excel. fresh or dried, yields well in hot and dry or cold conditions, early midseason maturity. *Source History: 1 in 1998; 2 in 2004.* **Sources: Fe5, Hi6.**

Maule's Red Hot - (Maule's New Red Hot, Ladyfinger) - Tapered, long cayenne type fruit, good flavor, high yielding, produces well in northern areas, rare. *Source History: 1 in 2004.* **Sources: Ea4.**

Mayan Cobanero Love - One of the world's hottest peppers, originally grown by the Mayans of Guatemala origins, bright red, heart-shaped fruit .5 in. long. *Source History: 1 in 2004.* **Sources: Red.**

Mesa - Good powder production variety, tall upright plants, 7 in. bright-red fruit, mildly pungent. *Source History: 1 in 1994; 1 in 1998; 1 in 2004.* **Sources: WE10.**

Mexican Negro - (Pasilla, Mulato) - Long thin black pods, 5-6 in. long, rich, unique flavor. *Source History: 1 in 1981; 1 in 1984; 1 in 1987; 1 in 1991; 1 in 1994; 2 in 1998; 1 in 2004.* **Sources: Red.**

Mexican Red Hot - 68-85 days, Small chili peppers .5 x 2.5 in., thin walls, used for flavoring Mexican and Oriental cooking. *Source History: 1 in 1987; 2 in 1991; 3 in 1994; 1 in 1998; 1 in 2004.* **Sources: Pep.**

Mira Sol - (Mirasol, Mirasol Guajillo) - 76-100 days, Popular hot pepper in Mexico, pods borne erect from the top of stems, 3- 4 in. long, very hot, use small amount of dried pods for flavoring, use green in salsa and guacamole, name means "look at the sun." *Source History: 1 in 1981; 1 in 1984; 2 in 1987; 5 in 1991; 4 in 1994; 7 in 1998; 9 in 2004.* **Sources: Enc, Gr29, Ho13, Mi12, Pep, Pla, Se17, So12, To1.**

Mulato - (Mulato Isleno) - 70-90 days, Excel. for stuffing as rellenos, very fat, 6 x 3 in. dia. at stem, mild and chewy, distinctive flavor, dark-green > red > red-black, 1000-1500 Scoville units, from Mexico, use fresh or dry. *Source History: 6 in 1981; 5 in 1984; 3 in 1987; 1 in 1991; 9 in 1994; 30 in 1998; 23 in 2004.* **Sources: Bo17, Com, Enc, Fo7, Ge2, Gr29, Ho13, Jor, Lin, MIC, Pep, Pla, Ra6, Re8,**

Red, Ri2, RUP, Se8, Se17, Ter, Thy, To1, To3.

Navaho - Thick-walled, 2-lobed, red fruit, 6-8 in. long, mild flavor with very slight pungency, upright plant lends itself well for hand picking or mechanical harvest, good for drying and powder production. *Source History: 1 in 1994; 1 in 1998; 3 in 2004.* **Sources: Red, Sk2, WE10.**

Negro - 77 days, Long slender cayenne type black fruits, moderately hot, great for drying. *Source History: 1 in 1987; 2 in 1991; 1 in 1994; 4 in 2004.* **Sources: Ho13, Pep, Sa9, Se17.**

Negro de Valle - Medium hot, dark brown, 6 in. chile. *Source History: 1 in 2004.* **Sources: Na2.**

Nepali Orange - (Polo Pipiki) - Extremely rare pepper brought to Fiji by migrants from India over 100 years ago, 3 ft. plant bears 400-500 fruit, upright or hanging, 1.5 in. long by .25 in. across, beautiful color change from yellow > orange > red with the heat going from medium hot to very hot when red. *Source History: 1 in 2004.* **Sources: Red.**

New Mex Joe Parker - 65 days, High yielding 2.5 ft. plant, tapered 6-7 in. pods with mild lively flavor. *Source History: 1 in 1998; 3 in 2004.* **Sources: Gr29, Lej, SE14.**

New Mexico Extra Hot - 90 days, Fruit ripens green > red, 6.5-7 in. long. *Source History: 1 in 1981; 1 in 1987; 1 in 1991; 1 in 1994; 1 in 1998; 1 in 2004.* **Sources: Pep.**

New Mexico Improved - (Mira Sol) - 75 days, Thick-walled, 4.5-5 x 1-1.25 in. at shoulder, tapers to a point, green > red, upright 27 in. plants, good for canning and freezing and drying. *Source History: 2 in 1981; 1 in 1984; 1 in 1987; 1 in 1991; 1 in 1994; 1 in 1998; 4 in 2004.* **Sources: Bu3, CH14, Red, Se26.**

New Mexico No. 6 - 75 days, Medium hot, 30 in. plants, 6-6.5 x 1.5 in. fruits, med-green > red, thick walls, for canning and freezing. *Source History: 2 in 1981; 2 in 1984; 1 in 1987; 1 in 1991; 3 in 1994; 3 in 1998; 6 in 2004.* **Sources: CH14, Dgs, Gr29, Pep, Ra6, Ros.**

New Mexico No. 6 Mild - 80 days, Mild all-purpose chili, best when fresh but can be dried or powdered. *Source History: 1 in 1981; 1 in 1984; 1 in 1987; 1 in 1991; 1 in 1994; 1 in 1998; 1 in 2004.* **Sources: Pla.**

New Mexico No. 64 - 70-75 days, Thin, 2-celled pods, 4.5-5.5 in. by 2.5 in. wide, dark green > red, mild pungency, drying and chili powder. *Source History: 3 in 1994; 2 in 1998; 1 in 2004.* **Sources: Pep.**

New Mexico No. 6-4L - New Mexican type, commercial var. for fresh green or dried red chile, 5-7 in. long, medium heat. *Source History: 1 in 1994; 1 in 1998; 3 in 2004.* **Sources: Enc, Red, WE10.**

Nosegay - 70-80 days, Tiny ornamental 6 in. plants bear a rainbow of colors with ripening fruit, marble size, white to orange to deep red, hot and edible, plant leaves resemble bay leaves. *Source History: 4 in 2004.* **Sources: Ami, He8, Pr9, Se16.**

NuMex Centennial - Medium, compact plant with purple leaves and flowers, small fruits point upward on the plant, purple > cream > orange > red. *Source History: 3 in 1998; 2 in 2004.* **Sources: Enc, To1.**

NuMex Eclipse - 120 days, New Mexican type, 4-6 in. long fruit, chocolate-brown at maturity. *Source History: 3 in 1994; 4 in 1998; 4 in 2004.* **Sources: Enc, Gr29, Pep, Pla.**

NuMex Joe E. Parker - 65-95 days, Version of NuMex 64, 2.5 ft. plant, tapered 6-7 in. fruit, thick flesh, green > red, mild to medium flavor, high yields. *Source History: 2 in 1994; 8 in 1998; 16 in 2004.* **Sources: Bo17, Bo19, Bu3, Enc, Jo1, ME9, Na6, Pep, Red, RUP, Scf, Se26, Sk2, Sw9, TE7, To1.**

Numex Six - (Numex 6) - 78 days, Mild to medium hot 7 x 1.5 in. fruits, 2-3 cells. *Source History: 1 in 1987; 3 in 1998; 1 in 2004.* **Sources: PAG.**

NuMex Sunburst - 120 days, Flavorful, 2-3 in. upright fruit, bright orange when mature, 500-1000 Scoville heat units. *Source History: 5 in 1998; 3 in 2004.* **Sources: Enc, Pep, To1.**

NuMex Sunflare - 120 days, DeArbol type, 2-3 in. long, matures to a bright red. *Source History: 4 in 1998; 3 in 2004.* **Sources: Enc, Pep, To1.**

NuMex Sunglow - 75-120 days, Flavorful Arbol type, 2-3 in. upright fruit, bright yellow when mature, 500-1000 Scoville heat units. *Source History: 3 in 1998; 3 in 2004.* **Sources: Enc, Pep, To1.**

NuMex Sunrise - 75-120 days, First chile pepper that turns bright yellow when mature, 4-6 in. long, excel. for drying, typical chile pepper flavor, rel. by NM/AES. *Source History: 1 in 1991; 3 in 1994; 7 in 1998; 7 in 2004.* **Sources: Enc, Gr29, Ma18, Pep, Pla, TE7, To1.**

NuMex Sunset - 75-120 days, Smooth fruits turn a beautiful glowing orange at maturity, 4-6 in. long, typical chile pepper flavor, good ristra chile, rel. by NM/AES. *Source History: 1 in 1991; 3 in 1994; 7 in 1998; 8 in 2004.* **Sources: Enc, Gr29, Ma18, Pe2, Pep, Pla, TE7, To1.**

NuMex Sweet - 74-78 days, Mildest Anaheim, strong plant, 2 x 7 in. long fruit. *Source History: 2 in 1998; 1 in 2004.* **Sources: Enc.**

NuMex Twilight - 70-90 days, Piquin type, ornamental 1-2 ft. tall plant with many small fruit in all color shades, very hot, bred by Paul Bosland, Jaime Iglesias and Max Gonzalez at NM State U from a semi-wild plant found growing in Jalisco, Mexico. *Source History: 2 in 1994; 6 in 1998; 14 in 2004.* **Sources: Enc, Jo1, Jo6, ME9, Orn, Pep, Pla, Ra6, Re8, Scf, SE14, Te8, To1, WE10.**

NuMex, Big Jim - (Big Jim, New Mexico Big Jim) - 75-80 days, Med-hot and fleshy, tapered pods average 7.68 x 1.89 in. at shoulder, up to 4 oz., 24 to 40 pods on 16-24 in. plants, flowers when hot and dry, ripens all at once, NM/AES. *Source History: 12 in 1981; 12 in 1984; 11 in 1987; 11 in 1991; 13 in 1994; 17 in 1998; 27 in 2004.* **Sources: Bu3, CH14, Enc, Gr29, He17, Ho13, HPS, La1, Loc, Ma18, ME9, MOU, Pe7, Pep, Pla, Ra6, Re8, Ros, Scf, Se8, Se26, Te8, To1, To3, To10, WE10, We11.**

NuMex, R. Naky - (New Mexico R. Naky Chili) - 80 days, Sturdy very prolific plants, heavy foliage for sunburn protection, large long wide smooth pods, very low level of pungency, may soon replace other mild chilis, named after chile researcher Dr. Nakyama of New Mexico State Univ. *Source History: 1 in 1981; 1 in 1984; 3 in 1987; 5 in 1991; 5 in 1994; 9 in 1998; 5 in 2004.* **Sources: Enc, Ho13, Pep, Pla, WE10.**

Ole - (FM-297, Jumbo Jalapeno) - 60-68 days, Looks like a jumbo jalapeno, 3.5 in. long dark green fruit, good for fresh market or processing, resists tobacco etch virus, dev. by TX A and M, PVP 1993 HA20. *Source History: 8 in 1994; 13 in 1998; 6 in 2004.* **Sources: Ga1, Gr29, Ma18, Ni1, Pep, Sh14.**

Orange Blossom - 70 days, Bushy plant to 2 ft., upright .5 in. fruit with blunt nose, green > orange. *Source History: 1 in 2004.* **Sources: Pep.**

Oranger Ungerischer Pfefferoni - (Orange Hungarian) - Pointed, orange fruit, 3 in. long, good for drying, from Budapest. *Source History: 1 in 1998; 1 in 2004.* **Sources: Ho13.**

Ordono - 85-90 days, Ornamental pepper produces green, yellow, orange, purple and red 1 in. upright fruits, hot and edible, collected in Batopilas Canyon, Chihuahua, Mexico. *Source History: 2 in 1991; 2 in 1994; 2 in 1998; 4 in 2004.* **Sources: Ho13, Na2, Se17, So12.**

Oriental Hot - Attractive plant bears upright 3 in. fruit, dark-green > deep crimson-red. *Source History: 1 in 1994; 1 in 1998; 1 in 2004.* **Sources: WE10.**

Orozco - 90 days, Ornamental plant, very hot purple fruit. *Source History: 1 in 2004.* **Sources: Pep.**

Orozco, Golden Orange - 70-90 days, Extremely hot orange fruits resemble carrots in appearance, straight and tapered, dev. by John Adams from several hot Eastern European varieties. *Source History: 1 in 1994; 3 in 1998; 2 in 2004.* **Sources: Fo7, So1.**

Ortega - 70-green/95-red days, Mildly hot Anaheim chile, 2 in. across by 6-7 in. long, slightly flattened with a blunt point, green > red, does very well in the Southwest, variably in the North, special stock of Anaheim TMR 23 sel. by the Ortega Chile Company. *Source History: 5 in 2004.* **Sources: Bo18, Gr29, Ra5, Vi5, Wo8.**

Ot-Chi-Thien - 75-85 days, Plant to 2.5 ft., long pointed red fruit, extremely hot, produces until frost, Thailand. *Source History: 1 in 2004.* **Sources: Ta5.**

Paprika - (Culinary Paprika, California Paprika, Mild California, Hungarian Paprika) - 80-120 days, Upright spreading 30-36 in. plant, flattened med-wrinkled 6 x 1.6 in. fruits, 4 lobes, generally quite mild, spicy, dried when red and ground for use as a seasoning, thin walls, Hungarian paprika is supposedly the most flavorful. *Source History: 6 in 1981; 6 in 1984; 7 in 1987; 8 in 1991; 6 in 1994; 6 in 1998; 5 in 2004.* **Sources: Fo7, Ra6, Re8, Se26, Te8.**

Paprika, Csereszyne - Hungarian hot cherry pepper, med. size plants bear 1 in. hot cherry size fruit, green > dark red > red, good for pickling and pizza topping. *Source History: 1 in 2004.* **Sources: Ba16.**

Pasilla - (Pasilla Bajio, Chilaca) - 75-80 days, Mildly hot and slightly sweet, slender 8-14 in. peppers, dark-green > red > brown, unique rich flavor, Pasilla translates "little raisin", used in "mole", a delicious paste used in many Mexican dishes. *Source History: 4 in 1981; 4 in 1984; 2 in 1987; 2 in 1991; 13 in 1994; 19 in 1998; 23 in 2004.* **Sources: Ba8, Ers, Gr29, He8, Ho13, ME9, Pe7, Pep, Pin, Pla, Ra6, Re8, Ri2, RIS, RUP, Se8, SE14, Se26, Sw9, Thy, To1, To3, WE10.**

Patagonia - Medium hot, cone-shaped upright fruit, yellow mottled with purple > orange > red, heirloom from Patagonia, AZ. *Source History: 1 in 2004.* **Sources: Na2.**

Pepperoncini - (Italian Pepperoncini) - 62-75 days, Mildly warm, conical 3-5 x .75-1.5 in., often picked green at 2-3 in. and pickled or used fresh, thin walls, pendant, fine flavor, small bushes, from southern Italy. *Source History: 16 in 1981; 16 in 1984; 16 in 1987; 14 in 1991; 11 in 1994; 16 in 1998; 33 in 2004.* **Sources: Ba8, Bo19, CO30, Ge2, Gr28, Gr29, He8, HO2, HPS, Jo6, Ma18, ME9, Ni1, Pe7, Pep, Ra5, Ra6, Ri12, RUP, SE14, Se26, Se28, So25, Sw9, Ta5, TE7, Te8, Ter, To1, To3, Vi5, We19, Wo8.**

Pepperoncini, Golden Greek - 65 days, Greek variety, fruit is shorter than the Italian strain, green > yellow, mild flavor, good for pickling, processing market. *Source History: 3 in 1998; 10 in 2004.* **Sources: Ba8, Co31, HO2, ME9, Ni1, Ra6, RUP, SE14, TE7, To1.**

Pequeno - 60 days, Plant grows 16 in. tall, upright fruit .5 x 1.5 in., light green > red. *Source History: 1 in 2004.* **Sources: Pep.**

Peter Pepper - 90 days, Rather dense 20-30 in. plants, penis-shaped fruits, 3.5-4 x 1-1.5 in. dia., very hot, excel. flavored chili powder, a real conversation piece from your garden. *Source History: 1 in 1981; 1 in 1984; 4 in 1987; 4 in 1991; 6 in 1994; 8 in 1998; 15 in 2004.* **Sources: Enc, Eo2, Fo7, Gr29, Jo6, Pe7, Pep, Pla, Ra6, Se28, Shu, Te8, To3, Und, We11.**

Peter Pepper, Orange - Medium-large plant bears uniquely shaped pendant pods, 3-5 in., ripens green > orange, excellent flavor, can be dried, smoked or pickled. *Source History: 1 in 2004.* **Sources: Ba16.**

Peter Pepper, Yellow - 90 days, Stubby yellow variety of the red Peter Pepper, 2-3 in. long, hot. *Source History: 1 in 1994; 4 in 1998; 3 in 2004.* **Sources: ME9, SE14, Te8.**

Pico de Gallo - (Rooster's Beak) - Plants grows to 3 ft., bears dozens of small thin curved 3 in. fruits, very hot, the heat builds in the mouth gradually, Sonora, Mexico. *Source History: 2 in 1987; 2 in 1991; 1 in 1994; 2 in 1998; 2 in 2004.* **Sources: Ho13, Na2.**

Pico de Pajaro - (Bird's Beak) - Knobby, curved fruit, 5-5.5 in. long, mild, from Yecora, Sonora. *Source History: 1 in 1991; 1 in 1994; 1 in 2004.* **Sources: Na2.**

Pili Pili - 80 days, Plant grows to 2.5 ft., pendant fruit, 1 x 3 in., green > red, good for chile powder, East Africa. *Source History: 1 in 1991; 1 in 1994; 1 in 1998; 1 in 2004.* **Sources: Pep.**

Pimenta di Chiero - (Smell Pepper) - So named because of its unique smell, almost round, yellow-white fruit, hot, South America. *Source History: 1 in 1998; 1 in 2004.* **Sources: We11.**

Pinguita de Mono - Peruvian variety, translates Little Monkey Dick, bushy plants bear 1 in. pods, hot with a delayed burn, unusual pleasant flavor enhances stews, meats, salsas and mole. *Source History: 1 in 2004.* **Sources: Ba16.**

Pizza Pepper - 59-80 days, Plant grows 14-18 in. tall, pendulous 3-4 in., cone shaped fruit is flavorful when green and mildest when red, early and prolific, very thick walls, good pepper flavor without all the heat, released by Dr.

James Baggett of OSU. *Source History: 2 in 1998; 2 in 2004.* **Sources: Ra6, Ter.**

Poblano - (Ancho) - 75 days, Heart-shaped pepper called Ancho in its dried state in which it lends its distinctive aroma and red-brown color to chili powder, called Poblano in its fresh green state in which it may be peeled and stuffed for chili rellenos. *Source History: 2 in 1987; 2 in 1991; 2 in 1994; 5 in 1998; 10 in 2004.* **Sources: Ban, Ge2, Gr29, ME9, Na6, Pep, Ra6, Re8, Sa5, Und.**

Poinsettia - 65-90 days, Upright clusters of 2-3 in. hot peppers, matures red, compact plant to 16 in., beautiful foliage, bears abundantly. *Source History: 1 in 1981; 1 in 1984; 1 in 1987; 2 in 1991; 4 in 1994; 5 in 1998; 6 in 2004.* **Sources: Fo7, ME9, Pep, SE14, TE7, To1.**

Portugese Hot - *Source History: 1 in 1998; 2 in 2004.* **Sources: SE14, WE10.**

Prairie Fire - 80-90 days, Bushy plants bear hundreds of small, very hot fruit, yellow-green > red, foliage changes from chartreuse > cream > yellow > orange > red, PVP 2000 Hollar and Company, Inc. *Source History: 1 in 1998; 6 in 2004.* **Sources: HO1, Jo1, Pep, Ra6, Re8, Tho.**

Pretty in Purple - 60 days, Ornamental, pungent fruits also used as a hot pepper seasoning, short spreading upright plants with purple or variegated purple and green foliage, purple flowers .75 in. round glossy fruit, deep-purple > scarlet. *Source History: 1 in 1994; 2 in 1998; 7 in 2004.* **Sources: Fe5, He8, Jo1, Pr3, Ra6, Re8, So25.**

Pretty Purple - 65-90 days, Amazingly beautiful, leaves stems and fruits are all deep-purple, holds color from seedling to full-grown 12-18 in. plant, conical fruits ripen to red, extremely hot, 5,000 Scoville units. *Source History: 1 in 1981; 1 in 1984; 1 in 1987; 2 in 1991; 8 in 1994; 10 in 1998; 13 in 2004.* **Sources: Com, Ha5, ME9, Orn, Ra6, Re8, SE14, So1, TE7, Te8, To1, To3, WE10.**

Pricky Nu - *Source History: 1 in 2004.* **Sources: WE10.**

Pueblo - 75-85 days, Improved Anaheim type, 6 in. long tapered fruit dries to rich red, fruits grow in upright clumps, excel. flavor with mild pungency, fresh market or processing. *Source History: 2 in 1994; 2 in 1998; 4 in 2004.* **Sources: Ers, Red, Te8, WE10.**

Pulla - (Puya) - 90 days, Fruit is 6 in. long, brown-red when mature, fruity flavor with a hint of licorice. *Source History: 2 in 1998; 3 in 2004.* **Sources: Enc, Pep, Red.**

Purple Delight - 60 days, Pretty plant with purple and green leaves, small purple flowers, glossy fruit ripens scarlet. *Source History: 1 in 1994; 1 in 1998; 1 in 2004.* **Sources: Pep.**

Purple Glow in the Dark - Incredibly beautiful plant with leaves mottled in a fluorescent purple and white, making it look like it glows in the dark, deep purple stems and branches, tiny fruit, green > purple > deep red. *Source History: 1 in 2004.* **Sources: Ami.**

Purple Prince - Attractive plant with deep-purple foliage, shiny, purple, 1 in. fruits, very hot. *Source History: 1 in 1994; 1 in 1998; 1 in 2004.* **Sources: WE10.**

Purple Tiger - 70-80 days, Ornamental plant with foliage colors including variegated green, white and purple, teardrop- shaped fruit, deep-purple > red, extremely pungent, makes excellent seasoning, 15,000-30,000 Scoville heat units. *Source History: 1 in 1994; 1 in 1998; 3 in 2004.* **Sources: He8, Sk2, WE10.**

Quatro Milpas - Mildly flavored, dark brown-red fruits, 7-8 in. long, grown in Quatro Milpas, Sonora. *Source History: 2 in 1998; 2 in 2004.* **Sources: Ho13, Na2.**

Ram's Horn - Productive, 3 ft. plants, 6-8 in. long, tapered, deep red fruit. *Source History: 1 in 1998; 1 in 2004.* **Sources: Ho13.**

Red Torpedo - Bright red, 4 in. long fruit, flavorful, medium hot, family heirloom. *Source History: 1 in 1998; 1 in 2004.* **Sources: Pe7.**

Relleno - 65-75 days, Green tapered 6.5 in. pods ripen fiery red, mildly hot, favorite with big canning companies, excellent in sauces, good for stuffing. *Source History: 1 in 1987; 2 in 1991; 2 in 1994; 2 in 1998; 6 in 2004.* **Sources: Gr28, Pe2, Ri12, Sa5, Se7, Se26.**

Ring of Fire - (Ring-O-Fire Cayenne) - 60-85 days, Extra early, short smooth pencil-thin Cayenne type, 5-6 in. long x .4 in. dia., very hot, concentrated heavy yields from 16 in. plants. *Source History: 1 in 1981; 2 in 1984; 3 in 1987; 3 in 1991; 5 in 1994; 8 in 1998; 9 in 2004.* **Sources: Eo2, Gr28, Gr29, Pe2, Pep, Se7, Se26, Sto, To1.**

Rio Grande Hot - 80-90 days, Medium-hot fruits are slightly larger than Jalapeno but milder, 1.75 x 6 in., excel. for processing. *Source History: 1 in 1994; 2 in 1998; 8 in 2004.* **Sources: Ba16, Enc, Jo6, Pe2, Pep, Se7, Se17, Se26.**

Riot - 60-90 days, Compact 10-12 in. mound-like plants covered with hot red chili-like peppers, light-green > yellow and purple > orange > red, ornamental, Baggett/OSU. *Source History: 2 in 1994; 3 in 1998; 4 in 2004.* **Sources: Jo1, Ra6, Se26, Ter.**

Rojo Morado - Purple stemmed plant bears upright tiny purple fruit. *Source History: 1 in 2004.* **Sources: Pep.**

Rooster Comb - 65 days, Fruit grows in clusters on 16 in. plant, .5 x .25 in., light green > red. *Source History: 1 in 1991; 1 in 2004.* **Sources: Pep.**

Rooster Spur - (Takanotsume, Hahong Koch'o) - 70 days, Late, very small thin fruits, 1-1.5 in. long x .5 in. dia., tapered tips, fruits point upwards, very hot, excel. dried and ground for chili powder. *Source History: 1 in 1981; 1 in 1984; 2 in 1987; 3 in 1991; 4 in 1994; 4 in 1998; 4 in 2004.* **Sources: Gr29, Pep, Te8, We11.**

Rouge de la Bresse - Compact plants bear slender, shiny red fruit, 2-3 in. long, med. hot flavor, French var. from the Rhone area. *Source History: 1 in 2004.* **Sources: Hud.**

Roumanian Hot - (Roumanian Wax, Roumanian Medium Yellow) - 65-75 days, Blocky 4 x 2.5 in. med-walled fruits, tapers to blunt point, yellow > red, med-hot, upright, 14 in. plants, bears until frost, for home or processing. *Source History: 12 in 1981; 9 in 1984; 10 in 1987; 7 in 1991; 5 in 1994; 6 in 1998; 10 in 2004.* **Sources: Bu2, CO30, He8, ME9, Pep, Ra6, RUP, SE14, Sk2, WE10.**

Round Chilly - Very popular Indian pepper, round cherry-red fruit, 1.5 in. across, thin walls, sweet spicy flavor. *Source History: 1 in 2004.* **Sources: Red.**

Royal Black - 88 days, Well-branched 2.5 ft. plant with black leaves, purple flowers, variable size, bullet shaped fruit, black > red, hot, original seed from Carolyn Male, CV So1 introduction, 1995. *Source History: 1 in 1994; 2 in 1998; 6 in 2004.* **Sources: Ho13, ME9, Pep, SE14, So1, TE7.**

Russian - 75 days, Smaller plant to 24 in., fruit with shape similar to Pequin but slightly smaller and thicker, .5 x 2 in., green > red. *Source History: 1 in 1991; 1 in 1994; 1 in 1998; 1 in 2004.* **Sources: Pep.**

San Felipe - 65 days, Med-hot fruits average 5 in. long, planted mid-May in San Felipe Pueblo, grown with irrigation along the Rio Grande in northern New Mexico. *Source History: 2 in 1991; 2 in 1994; 1 in 1998; 1 in 2004.* **Sources: Na2.**

San Juan Tsile - 75-80 days, Plants to 2 ft., 5 in. long, flavorful pods, med. hot, slightly curved cowhorn shape, thin walls, excel. drying pepper, traditional chile still grown by elderly farmers in the pueblo north of Espanola, New Mexico. *Source History: 1 in 1994; 1 in 1998; 2 in 2004.* **Sources: Ag7, Na2.**

Sandia - (Anaheim Chili Sandia, Hot Anaheim) - 77-80 days, Upright 24 in. plants, good foliage cover, flat and tapering, 2 cells, 6-7 x 1.5 in., very pungent med-thick flesh, dark-green > bright-scarlet, pendant, everbearing. *Source History: 3 in 1981; 3 in 1984; 7 in 1987; 6 in 1991; 9 in 1994; 14 in 1998; 14 in 2004.* **Sources: Bu3, CH14, Enc, Gr29, He8, Ho13, Pep, Pla, Red, Ros, Se17, Se26, To1, WE10.**

Santa Fe Grande - (Caribe) - 75-100 days, Hot conical blunt 3.5-5 x 1.5 in. med-walled fruits, yellow > orange-red, upright 24-28 in. plants, TM res., pendant, continuous heavy yields, for home or market., intro. by Petoseed Co., 1965. *Source History: 13 in 1981; 13 in 1984; 15 in 1987; 20 in 1991; 19 in 1994; 24 in 1998; 28 in 2004.* **Sources: Bo19, Bu3, CH14, CO30, Enc, Ers, Gr29, Jo6, ME9, MOU, OSB, Pep, Pla, Pr9, Ra6, Re8, Red, Ros, RUP, Se8, SE14, Se16, Se26, Sk2, To10, WE10, Wi2, WI23.**

Santaka - (Hontaka, Chili Japones) - 70-80 days, Very hot, deep-scarlet, tapered 2.5-3 in. fruits, compact plants, good dried as chili powder, its heat will not cook away on the stove, from Japan. *Source History: 2 in 1981; 2 in 1984; 1 in 1987; 1 in 1991; 2 in 1994; 2 in 1998; 5 in 2004.* **Sources: Bo17, Enc, Pep, Se8, We11.**

Santo Domingo - (Santo Domingo Pueblo) - Hot chiles average 3 in. long, traditionally strung into ristras for drying then crushed into powder or rehydrated, collected from the Santo Domingo Pueblo in northern New Mexico. *Source History: 2 in 1991; 3 in 1994; 3 in 1998; 2 in 2004.* **Sources: Na2, Pep.**

Sapporo - 60-75 days, Thick skinned fruit, 4-5 in. long, green > red, young fruits used fresh for cooking, mature fruits are best when dried, resembles Fushimi but hotter. *Source History: 1 in 1998; 2 in 2004.* **Sources: Kit, Pep.**

Sebes, Orange - (Czech Orange) - 75 days, Plant to 36 in., requires staking or caging, hot banana type fruit with smooth medium thick walls, matures to bright orange, good for stuffing and frying, from Czechoslovakia. *Source History: 1 in 1994; 2 in 1998; 2 in 2004.* **Sources: Fo7, Ho13.**

Serrano - (Serrano Hidalgo) - 75-90 days, Very hot, 1-3 x .5 in. dia., candle flame shaped fruit tapers to blunt point, glossy green > orange-red, med-thin walls, pendant fruit position, vigorous and everbearing, erect branching 26-34 in. plants produce up to 50 pods each, good cover, prolific, distinctive flavor is described as clean with no-lingering heat, 10,000-20,000 Scoville heat units, used for fresh salsa and guacamole, popular along the Mexican border, first grown in the mountains of Northern Puebla and Hidalgo in Mexico. *Source History: 24 in 1981; 23 in 1984; 27 in 1987; 42 in 1991; 46 in 1994; 48 in 1998; 63 in 2004.* **Sources: BAL, Ban, Bo9, Bo17, Bo19, Bu2, Bu3, CH14, CO23, CO30, Coo, Dgs, DOR, Enc, Ers, Fo7, Ga1, Ge2, Gr28, Gr29, GRI, He8, He17, Jo6, Jor, Ki4, LO8, Loc, ME9, Mi12, MIC, Mo13, Mo20, MOU, Na6, Ni1, Pep, Pin, Pla, Ra6, Re8, Red, Ri2, Ri12, RIS, Ros, RUP, Se7, Se17, Se25, Se26, Sk2, So1, So12, Te8, Thy, To1, To3, To10, Vi4, We11, Wi2, WI23.**

Serrano Hidalgo, Tam - Mild Serrano-type, pendant fruits, green > red. *Source History: 1 in 1987; 3 in 1991; 6 in 1994; 4 in 1998; 1 in 2004.* **Sources: WE10.**

Serrano Tampequino - 75-80 days, Name means "from the mountains", first grown in the mountains of northern Puebla and Hidalgo in Mexico, large plant bears club shaped, smooth fruit, 1.5 in. long, medium thick walls, green > red, very pungent, fresh market and pickling, 3,000 Scoville heat units. *Source History: 1 in 1987; 2 in 1991; 1 in 1994; 8 in 1998; 13 in 2004.* **Sources: Ba8, Co31, Com, HO2, Jo6, Kil, ME9, Ra6, Re8, RUP, Se8, SE14, WE10.**

Serrano, Purple - 85 days, Very hot, deep purple fruit, 2.25 in. long, ideal for fresh and salsa, slightly larger than Serrano. *Source History: 2 in 1998; 7 in 2004.* **Sources: Jo6, ME9, Ra6, Re8, SE14, To1, To3.**

Sinahuisa - Fleshy pods are 1.5-2 in. long, seeds are hot, from communal farm in Sonora, Mexico. *Source History: 1 in 1991; 1 in 1994; 1 in 1998; 1 in 2004.* **Sources: Na2.**

Sirichi Thai - Plants 3-4 ft. tall with small purple and green leaves with some white variegation, .5 in, pendant fruit, dark purple > red, very hot. *Source History: 1 in 1998; 1 in 2004.* **Sources: Ho13.**

Slim Jim - (Foot Long) - Very large chili peppers, up to 12 in. long, even the smallest are over 7 in., green > crimson, semi- hot, use all summer and then dry for winter use. *Source History: 1 in 1981; 1 in 1984; 1 in 1987; 2 in 1991; 3 in 1994; 2 in 1998; 2 in 2004.* **Sources: Bu1, Pep.**

Small Chili - 75-82 days, Pointed 5.5 x 2.5 in. cone shaped, bright red fruits, med-thick walls, good for pepper sauce, pickling or drying. *Source History: 2 in 1981; 2 in 1984; 3 in 1987; 2 in 1991; 3 in 1994; 2 in 1998; 5 in 2004.* **Sources: ME9, Mey, Re8, SE14, Te8.**

Sonora Anaheim - 75-77 days, Two-celled fruit, 8 in. long, slightly longer than Anaheim TMR 23, dark green > dark red, mildly pungent, good for fresh market and processing, PVP 1992 Seminis Vegetable Seeds, Inc. *Source History: 2 in 1994; 3 in 1998; 4 in 2004.* **Sources: CH14, Loc, Ros, To1.**

St. Helena Acorn - Plant grows 4-5 ft. tall, produces many small hot fruit the shape and size of an acorn, good clean pepper flavor, purple blushed green fruit ripens to bright red in areas with longer growing seasons, St. Helena is an island in the South Atlantic which was made famous by Napoleon. *Source History: 1 in 2004.* **Sources: Al6.**

Sue Senger's Chili - Bushy 2-3 ft. plants bear big spicy hot tapered fruit, ripens red, high yields. *Source History: 1 in 2004.* **Sources: Sa5.**

Szentesi Semi-Hot - 60 days, An elite Romanian type, unusual tapered pendant fruit 4.5 in. long, ripens lime-green to lime-yellow to orange, sold at lime-yellow stage, indoor or outdoor production, slightly pungent. *Source History: 1 in 1987; 2 in 1991; 2 in 1994; 2 in 1998; 4 in 2004.* **Sources: Pep, Se17, Sto, Ta5.**

Tears of Fire - 65 days, Thick fleshed, 1-2 in. teardrop cherry type, used at green, brown or red stages. *Source History: 1 in 2004.* **Sources: Eo2.**

Tennessee Spice - 82 days, Pointed fruit matures to red, shape and size resembles a Tabasco, very hot. *Source History: 1 in 2004.* **Sources: Sa9.**

Tennessee Teardrops - 80 days, Plant to 24 in. tall, 2 in. tapered fruits are borne upright, thick flesh, black > red, occasionally yellow > red, heirloom. *Source History: 1 in 1998; 1 in 2004.* **Sources: Fo7.**

Texas Chile - 75 days, Medium green, 3 in. fruit on 2.5 ft. plant. *Source History: 1 in 2004.* **Sources: Pep.**

Thai Bird - (Thai Red Chili) - Plants 3-4 ft. tall, 1-1.5 in. thin pods grow upward, green > red, extremely hot, useful for making hot pepper vinegar, heirloom from Thailand. *Source History: 1 in 1991; 1 in 1994; 1 in 1998; 3 in 2004.* **Sources: Ba8, Enc, Te8.**

Thai Orange - (Thai Orange-Red) - 80-90 days, Productive, 4 ft. plants, tapered, pendant, 2.5 in. fruit, dark green > bright orange, very hot, great for drying. *Source History: 1 in 1991; 1 in 1998; 3 in 2004.* **Sources: Ho13, Se16, Ta5.**

Thai Yellow Chili - Golden-yellow version of Thai Red, very hot and flavorful, from Thailand. *Source History: 1 in 2004.* **Sources: Ba8.**

Tomato Pepper - 80 days, Sturdy, productive plants, flattened, globe-shaped fruits with sweet thick flesh, mildly hot, good for eating raw, pickling or making salsas and sauces, Hungarian heirloom acquired by the Joe Hussli family of Beaver Dam, Wisconsin, 1929. *Source History: 1 in 2004.* **Sources: Und.**

Treasures Red - 90 days, Short plant to 8 in., tapered 2 in. fruit, cream > orange > bright red, ornamental. *Source History: 2 in 2004.* **Sources: BAL, Pep.**

Tri-Fetti - (Variegata, Tricolor Variegata) - 72-80 days, Ornamental plant with variegated green, white and purple foliage, small purple fruit turns red when ripe. *Source History: 2 in 1998; 6 in 2004.* **Sources: Ami, Pin, Se26, Shu, TE7, To1.**

Vallero - Plants to 3 ft., horn-like, dark-red pods, 5 in. long, mature to dark chocolate-brown color, medium in heat and size, used dried for red chile sauces and soups, collected in the Buena Ventura Valley in Chihuahua, Mexico. *Source History: 1 in 1991; 4 in 1994; 5 in 1998; 3 in 2004.* **Sources: Ho13, Na2, Pla.**

Varingata - 50-75 days, Ornamental plant grows 8-12 in. tall with green white and purple variegated foliage, purple and white blooms, small green and purple eggplant-like fruits mature to red. *Source History: 1 in 1994; 4 in 1998; 9 in 2004.* **Sources: Bo17, Jo1, ME9, OSB, Pep, Roh, SE14, To3, Ver.**

Velarde - Medium to mild heat, 3.5-4 in. long, from Velarde, NM. *Source History: 1 in 2004.* **Sources: Na2.**

Venezuela - Hotter than a Habanero, small 1.5-2 ft. plant is a dark iridescent purple color, bears drooping clusters of small pods, cream > orange > red. *Source History: 2 in 1994; 1 in 1998; 1 in 2004.* **Sources: We11.**

Vietnamese Pepper - (Oriental Toothpick) - Small red peppers, edible but very hot, fast grower, very ornamental. *Source History: 1 in 1981; 1 in 1984; 1 in 1987; 2 in 1991; 2 in 1994; 1 in 1998; 2 in 2004.* **Sources: Ban, Vi5.**

Waialua - Bacterial wilt res. *Source History: 1 in 1991; 1 in 1994; 1 in 1998; 1 in 2004.* **Sources: Uni.**

Wenk's Yellow Hots - Medium to hot fruit with thick, waxy yellow walls, bright orange > red, 2-4 in. long, tapers to a blunt point, grown by the late Erris Wenk, one of the last large local truck farmers in Albuquerque's South Valley. *Source History: 1 in 1998; 1 in 2004.* **Sources: Na2.**

West Virginia Pea - Upright clusters of 1 in. cherry peppers on bushy erect plant, green > maroon > red, good dried, pickled or fresh, West Virginia heirloom. *Source History: 1 in 2004.* **Sources: Ba16.**

White Hot - Branched 2 ft. plants bear upright clusters of 1 in., creamy white fruit, red at maturity, very hot. *Source History: 1 in 1998; 1 in 2004.* **Sources: Ho13.**

White Wax - 75 days, Plants up to 40 in., colorful fruit, mauve > bright red, good for pickling. *Source History: 1 in 1991; 1 in 1994; 1 in 1998; 1 in 2004.* **Sources: Pep.**

Yunan - 80 days, Clusters of 3 in. fruit on 2 ft. tall plant. *Source History: 1 in 1998; 1 in 2004.* **Sources: Gr29.**

Yung Ko - 80 days, Hot long curved pepper, dark-green > red, used fresh or dried (crushed or powdered), similar to Cayenne but with less heat, imported from Taiwan. *Source History: 1 in 1981; 2 in 1984; 2 in 1987; 2 in 1991; 2 in 1994; 2 in 1998; 2 in 2004.* **Sources: Gr29, Pep.**

Zia Pueblo - Mild flavor, sweet when red, 3 in. long. *Source History: 1 in 2004.* **Sources: Na2.**

Varieties dropped since 1981 (and the year dropped):

4 Inch Thick Chili (1987), *Acorn* (1998), *Agni Sikha II* (2004), *Anaheim Chili Improved* (1998), *Anaheim G Strain Mild* (1994), *Ancho Improved* (1994), *Angelo Grande* (1998), *Bahru* (1998), *Barney* (2004), *Bermuda* (1998), *Black and Green Ripens Red* (1991), *Black Pungent* (1998), *Caienna Hot Red* (1987), *Caliente* (1998), *California* (1984), *Cayenne, Early* (1998), *Cayenne, Iberian Long* (2004), *Cayenne, Southern Heat* (2004), *Cayenne, Turkish* (2004), *Cayman Hotter* (1998), *Cherokee* (2004), *Cherry (India)* (1994), *Cherry (Northern Peru)* (1994), *Cherry (Russia)* (1998), *Chilacate* (2004), *Chili Red Pickling* (1991), *China*

Express (2004), *Chinese Ornamental* (2004), *College No. 64* (1983), *Colorado, Hot* (2004), *Colorado, Mild* (2004), *Craig's Chili* (2004), *Czechoslovakian Black Sweet* (1998), *De Agua* (1991), *Di Napoli* (1987), *Dog Tooth* (1991), *Durca* (1998), *Early Bonanza Jalapeno* (1991), *East Indian Hot* (1998), *Ecuadorean All-Purple Dwarf* (1998), *El Paso* (2004), *Ethiopian* (1998), *Fiesta Grande* (1998), *Florida Gem* (1987), *Gambia* (2004), *Garden Nugget* (2004), *Geno's Giant* (1998), *Gold Spike* (1983), *Gorbaci* (1994), *Guatamala* (1998), *Hades Hot* (1998), *Horn Hot Wax* (1994), *Hot Claw* (1998), *Hungarian Block Type* (1991), *Inayague* (2004), *Indian* (1998), *Jalapeno Improved* (1994), *Jalapeno L* (1994), *Jalapeno, Mexican* (1994), *Jalapeno, Tam Mild 2* (2004), *Kahari Mirch* (2004), *Kashmiri Degi Mirch* (1994), *Korean* (1998), *Kreta* (1983), *Lago Pickling Wax Pepper* (1994), *Laotian* (2004), *Little Boy* (1998), *Long Hot Chili* (2004), *Long Red Marconi* (1991), *Long Red Market Growers'* (1994), *Long Thick Yellow Cayenne* (1991), *Malahat* (1998), *Malibu Purple* (2004), *Manbila* (1998), *Mexican Chili Improved* (2004), *Mexican Hat* (1998), *Midnight Special* (1994), *Mini Bell Hot* (2004), *Miracle Willie* (1998), *Monkey Face* (1998), *Moravian* (2004), *Nambe* (1998), *New Mexico* (2004), *New Mexico 64* (2004), *Oaxaqueno* (1982), *Ot Khol* (1998), *Pasilla (Ancho Mejorado)* (1987), *Phillipino* (1998), *Picante* (1998), *Presov* (1998), *Pungent Pride* (2004), *Purple Blossom* (1994), *Purple Mystery* (1998), *Purple Pepper* (2004), *Purple Variegated* (1998), *R. J. Reynolds* (1998), *Rainbow Wax Semi-Hot* (1998), *Rat Tail* (2004), *Rat Turd Pepper* (2004), *Rio Grande 21* (1987), *Rooster Beak* (1994), *Rosso Piccante di Cayenna* (1998), *Rouge Long* (1991), *Roumanian Hot (Bullnose Type)* (1991), *Roumanian Semi-Hot* (2004), *Sandia A* (1994), *Serrano 63585* (1998), *Serrano Tampequino 74* (2004), *Serrano, Purple Brazilian* (1998), *Seven Year South American* (1998), *Siling Pari* (1998), *Simla Mirch* (2004), *Stripe* (2004), *Suave Rojo* (1998), *Sucette de Provence* (1991), *Sylvan's Select* (2004), *Tabiche* (1994), *Taco* (1984), *Tatong* (1994), *Tennessee Firecracker* (1998), *Texas* (1998), *Thai Dark Red* (1994), *Tuste Verde* (1998), *Vinni's Hot Pepper* (2004), *Vurino* (1987), *Weeks Hot-Hot Midget* (1987), *White Fire* (2004), *Wild Grove* (1998), *Yellow Globe* (1991), *Yellow Hot Apple Pepper* (1991).

Pepper / Sweet / Bell
Capsicum annuum

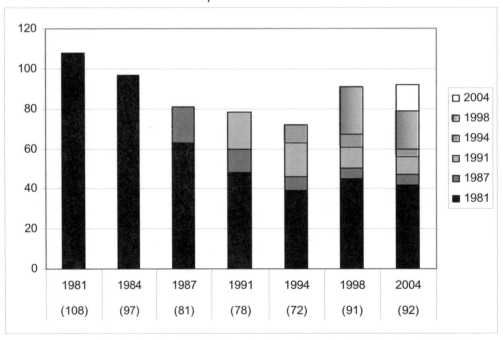

Number of varieties listed in 1981 and still available in 2004: 108 to 42 (39%).
Overall change in number of varieties from 1981 to 2004: 108 to 92 (85%).
Number of 2004 varieties available from 1 or 2 sources: 47 out of 92 (51%).

Albino - (Albino Bullnose, White Bullnose) - 75-80 days, Very early sweet bell pepper, stays white a long time before finally turning red, dwarf bush, suitable for far Northern regions, new color and taste for salads, offered commercially since the turn of the last century. *Source History: 1 in 1981; 2 in 1984; 3 in 1987; 2 in 1991; 3 in 1994; 3 in 1998; 13 in 2004.* **Sources: Ba8, Bo18, Gr29, Ho13, Ma18, Pep, Ra5, Ra6, Sa9, Se28, To3, Vi5, Wo8.**

Ariane - 70-75 days, First true orange bell pepper, medium-green > vibrant orange, sweet and flavorful in green or orange stages, early and very productive, dev. to honor the House of Orange, the Royal family of Holland. *Source History: 1 in 1981; 1 in 1984; 1 in 1987; 3 in 1991; 3 in 1994; 7 in 1998; 7 in 2004.* **Sources: Gr29, Ho13, Ki4, Pep,**

Pr3, Sa5, To1.

Baby Belle - 68 days, Tiny fruit, 2 x 2.5 in., most 4-lobed, emerald-green > red, TMV res. *Source History: 2 in 1994; 1 in 1998; 5 in 2004.* **Sources: Ra6, Re8, Se28, To3, WE10.**

Big Dipper - 73 days, Plants to 24 in. tall, large, blocky fruits, 4.5 x 4.5 in., 4-lobed, excellent flavor. *Source History: 1 in 2004.* **Sources: Bu2.**

Big Red - 75 days, Very sweet thick flesh, green > red, 3-4 lobes, high yields, excel. for fresh market and home garden. *Source History: 1 in 1991; 3 in 1994; 2 in 1998; 8 in 2004.* **Sources: Be4, Bo19, Loc, ME9, Ra6, RUP, Se25, WE10.**

Bull Nose - (Sweet Bullnose, Large Bell, Sweet Mountain, Large Sweet Spanish) - 55-80 days, Very early bell pepper, both hot and sweet, ribs are quite pungent but the rest is mild, square 4 x 3.5 in. dia. fruits, dark-green > scarlet, thick heavy walls, uniform in size and shape and earliness, prolific, recently becoming quite scarce, from India, 1759. *Source History: 20 in 1981; 17 in 1984; 11 in 1987; 7 in 1991; 6 in 1994; 3 in 1998; 5 in 2004.* **Sources: He8, La1, Ri12, So1, Wet.**

California Wonder - (Burpee's California Wonder) - 68-89 days, Large blocky thick-walled stuffing pepper, 4-4.5 x 3.5-4 in. dia. and 6 oz., glossy deep-green > red, 3 or 4 thick lobes, upright prolific everbearing 24-30 in. plants, leading market and shipping pepper, also good home garden var. and freezer, introduced in 1928. *Source History: 108 in 1981; 94 in 1984; 83 in 1987; 66 in 1991; 67 in 1994; 63 in 1998; 74 in 2004.* **Sources: Ada, Ag7, All, Ba8, Be4, Bo17, Bo19, Bou, Bu2, Bu8, CH14, Co31, Com, Cr1, Dam, De6, DOR, Ear, Ers, Fa1, Fo7, Ga1, Ge2, Gr27, Gr28, Gr29, GRI, HA3, He8, He17, Hi6, Hi13, HO2, HPS, Jo6, La1, Lej, Lin, LO8, Ma19, MAY, Me7, Mel, Mey, Mo13, MOU, Ni1, Pe2, Pep, Pin, Pr9, Ra6, Re8, Red, Ri12, RIS, Roh, Ros, Sau, Se26, SGF, Sh9, Shu, Si5, Sk2, So1, SOU, Sto, To3, Tu6, Ves, Vi4, Wet, Wi2.**

California Wonder 300 - (California Wonder 300 TMR) - 73-76 days, Heavy smooth thick-walled blocky 4.75 x 4 in. fruits, mostly 4-lobed, green > red, pendant, vigorous 24-28 in. plants, TM res., good for cool short seasons, prolific, Asgrow, 1965. *Source History: 20 in 1981; 19 in 1984; 19 in 1987; 18 in 1991; 22 in 1994; 20 in 1998; 18 in 2004.* **Sources: Bu3, CA25, CH14, CO23, Fis, Jor, Loc, ME9, MIC, OLD, RUP, SE14, Se25, Sh14, Ter, Twi, WE10, WI23.**

California Wonder Purple - 72 days, Resembles California Wonder except fruits mature to dark purple, thick walls, 6 oz. *Source History: 1 in 1998; 1 in 2004.* **Sources: He8.**

California Wonder Select - 72 days, Blocky fruit with extra thick walls, 4 x 4 in., good for stuffing. *Source History: 2 in 1998; 1 in 2004.* **Sources: BAL.**

California Wonder, Early - 66 days, Early heavy yields, thick blocky fruits, deep-green > crimson, 4 lobes, popular with Canadian growers, sets well during cool nights, sel. from Early Calwonder. *Source History: 5 in 1981; 3 in 1984; 3 in 1987; 5 in 1991; 5 in 1994; 4 in 1998; 6 in 2004.* **Sources: Bo19, Ge2, MOU, Pep, Ra6, Re8.**

California Wonder, Golden - 72-78 days, Blocky thick-fleshed 3.75 x 3.5 in. dia. fruits, med-green > golden-yellow > deep- orange, very sweet, 4-lobed, sturdy upright 22-30 in. plants, upright and pendant, everbearing. *Source History: 15 in 1981; 12 in 1984; 6 in 1987; 7 in 1991; 6 in 1994; 7 in 1998; 15 in 2004.* **Sources: Ba8, Bo19, Co31, Ear, Ge2, GLO, He8, HO2, OLD, Or10, Pe2, Pep, Shu, Vi4.**

Calwonder - (Calwonder TMR) - 65-75 days, Improved California Wonder, earlier, 4.5 x 3-4 in. dia., thick walls, square blocky shape, dark-green > rich-red, sweet mild and meaty, susc. to mosaic. *Source History: 8 in 1981; 7 in 1984; 6 in 1987; 8 in 1991; 5 in 1994; 4 in 1998; 6 in 2004.* **Sources: Enc, Ri2, Se7, So9, TE7, WE10.**

Calwonder, Early - 63-80 days, Popular Southern market and processing type but affected by hot dry weather, also good for short season areas, 4-4.5 x 3.5-4 in., 3 or 4 lobes, dark-green > red, vigorous upright bushy everbearing 24-28 in. plant, upright through pendant habit, very productive--often 12 peppers at one time. *Source History: 36 in 1981; 31 in 1984; 31 in 1987; 25 in 1991; 20 in 1994; 15 in 1998; 8 in 2004.* **Sources: All, Gr29, Hal, Kil, Ont, PAG, SGF, To1.**

Calwonder, Golden - 62-73 days, Smooth blocky and blunt-ended, 5 x 4 in. dia., 4-lobed, light-green > golden-yellow, thick- walled, very sweet and mild, 22-26 in. sturdy upright plants, sets continuously. *Source History: 18 in 1981; 15 in 1984; 17 in 1987; 17 in 1991; 20 in 1994; 26 in 1998; 31 in 2004.* **Sources: CA25, CH14, Com, De6, Gr27, Gr29, He17, HPS, Kil, La1, Loc, Mey, Mo13, MOU, Ont, Ra6, Re8, Ri2, Roh, RUP, SE14, Se26, Sh9, Sk2, So1, TE7, To1, To3, WE10, Wet, WI23.**

Capistrano - 70-76 days, Upright slightly spreading plant, 20-24 in., very large 3-4 lobed fruits, 4 in. long and 3.5 in. wide, dark-green > red, TMV res., commercial market and processing var., PVP 1994 Seminis Vegetable Seeds, Inc. *Source History: 5 in 1991; 11 in 1994; 11 in 1998; 9 in 2004.* **Sources: Eo2, Gr29, Jor, Loc, ME9, Pep, Ra6, RUP, Si5.**

Carmagnola Ross - Large red bell, good producer. *Source History: 1 in 2004.* **Sources: Se24.**

Carolina Wonder - 75 days, Nearly the same as Yolo Wonder B except it is highly resistant to southern rootknot nematode, large bells ripen to red, less prone to developing fungus in the seed cavity, dev. by US Veg Lab. *Source History: 1 in 1998; 1 in 2004.* **Sources: So1.**

Charleston Belle - 67-85 days, Resembles Keystone Resistant Giant, 3-4 in. long bells ripen to red, sweet flavor, highly resistant to southern rootknot nematode, dev. by the U.S. Veg. Lab. *Source History: 2 in 1998; 2 in 2004.* **Sources: So1, Ter.**

Chinese Giant - 73-90 days, Extremely large almost square fruits, 6 x 4-5 in. dia., med-thick sweet mild flesh, rich dark- green > bright cherry-red, produces early, ideal for home gardens, introduced in 1900 by W. Atlee Burpee. *Source History: 2 in 1981; 3 in 1984; 2 in 1987; 2 in 1991; 4 in 1994; 6 in 1998; 21 in 2004.* **Sources: Ba8, Bo18, Bu2,**

Co31, Ea4, Ers, Gr29, Hi13, HPS, Hud, Pep, Ra5, Ra6, SE14, Se26, Shu, TE7, To1, To3, Vi5, Wo8.

Chocolate (Prof. Meader's) - 75 days, Pointed fruit, 2 x 4 in., matures to chocolate-brown skin color, brick-red flesh, sweet flavor, dev. by E. M. Meader, U of NH. *Source History: 1 in 1998; 3 in 2004.* **Sources: Ho13, Ma18, Scf.**

Chocolate Baby - (Chocolate Cherry) - 60 days, Thick fleshed cherry type, green > red > chocolate, great for stuffing and baby vegetable market. *Source History: 4 in 1998; 3 in 2004.* **Sources: Ho13, Pep, So25.**

Corona - 66 days, Compact plants, med-size blocky fruits, green > orange, sweet flavor, the var. used to produce the stunning orange Holland imports, TMV res., intro. in 1991. *Source History: 1 in 1991; 2 in 1994; 2 in 1998; 1 in 2004.* **Sources: So1.**

Criolla - 80 days, Richly flavored, thick walled bells with a rich, robust flavor, green > scarlet. *Source History: 1 in 2004.* **Sources: Ma18.**

Diamond - (Diamond White) - 60-80 days, Medium thick fleshed, 3-4 lobed bell, translucent white > pale yellow > scarlet red. *Source History: 3 in 1998; 8 in 2004.* **Sources: CA25, HO2, Pe2, Pep, RUP, SE14, TE7, Twi.**

Doe Hill Golden Bell - Matures earlier than regular size bells, 24 in. plants produce flat shaped bells, 1 in. wide by 2.25 in. tall, 4-6 lobes, green > orange, sweet flavor with a fruity, multi-dimensional quality, nice salad pepper, widely adapted, disease res., early maturity, pre-1900 family heirloom from the Doe Hill area in Highland County, VA. *Source History: 2 in 2004.* **Sources: Ga21, So1.**

Earliest Red Sweet - 55-70 days, Earliest bell pepper in Canada, small plants with 5 or 6 fruits, 4 x 3.25 in. dia., 2 or 3 lobes, bright-red, not uniform in shape, will mature in Nova Scotia. *Source History: 3 in 1981; 3 in 1984; 3 in 1987; 6 in 1991; 7 in 1994; 4 in 1998; 3 in 2004.* **Sources: Gr28, Hig, Tu6.**

Earlired - 70 days, Extremely early bell type, thin walls, ripens in problem areas where others will not, Plant Science Dept./University of Manitoba. *Source History: 1 in 1981; 1 in 1984; 1 in 1987; 1 in 1991; 1 in 1994; 2 in 1998; 2 in 2004.* **Sources: Pe2, Sa9.**

Early Mountain Wonder - 70 days, High Altitude Gardens selection from California Wonder, 24-48 in. plants, blocky green fruit, 4 x 4 in., thick walls, matures to red. *Source History: 1 in 2004.* **Sources: Hig.**

Early Niagara Giant - (Early Niagara) - 65 days, Square 4.5 in. green peppers, 3-4 lobes, thick flesh, 24 in. plants set well during cool summers, 6-8 fruits per plant, ripen to red, good tol. to TMV, home or market gardens. *Source History: 2 in 1981; 1 in 1984; 3 in 1987; 4 in 1991; 2 in 1994; 2 in 1998; 7 in 2004.* **Sources: Gr29, Ho13, Pep, Ra6, Sa9, So25, Sto.**

Emerald Giant - (Bull Nose) - 72-80 days, Improved Keystone Res. Giant type, selected for erect habit and heavy yields, blocky 4-4.5 x 3.5-3.75 in. dia., thick flesh, dark-green > red, 4-lobed, vigorous productive 26-30 in. upright plant, everbearing, pendant, TM res., excel. for the South, 1963. *Source History: 38 in 1981; 33 in 1984; 28 in 1987; 22 in 1991; 20 in 1994; 16 in 1998; 18 in 2004.* **Sources: Ba8, Bo17, Bu3, CH14, Ers, Gr29, GRI, HO2, Jor, Loc, ME9, Pep, Ra6, RIS, SE14, Se25, Se26, WE10.**

Emerald Giant Resistant - 74 days, Med-thick walls, 4.5 x 3.75 in. dia., 4 lobes. *Source History: 1 in 1981; 1 in 1984; 1 in 1987; 1 in 1991; 1 in 1994; 1 in 1998; 1 in 2004.* **Sources: Mey.**

Frank's - 56 days, Extremely productive, fast maturing pepper, compact plants bear medium size, elongated bells turn red quickly, very sweet. *Source History: 1 in 2004.* **Sources: Sa9.**

Garden Sunshine - 80-100 days, Bushy plants to 12-16 in., large bell shaped fruits hold for weeks at the yellow-green stage before turning to orange then red, great extended harvest period, best used when yellow or orange, excel. quality, very juicy, mild flavor. *Source History: 1 in 1981; 1 in 1984; 1 in 1987; 1 in 1991; 1 in 1994; 1 in 2004.* **Sources: Se16.**

Golden Bells - 90 days, Vigorous, productive plants bear large green bells that mature to a rich gold. *Source History: 2 in 1998; 1 in 2004.* **Sources: Enc.**

Golden Queen - 65 days, British heirloom, excellent flavor, in Victorian times these were grown in stovehouses and kitchen garden borders. *Source History: 1 in 1998; 1 in 2004.* **Sources: Se26.**

Golden Summit - 65-75 days, Exceptionally attractive pepper from an heirloom strain developed under 200 years of intensive Yugoslavian market gardening, golden-green fruits ripen to a bright-gold and then to a bright deep-red, thick sweet juicy walls, bears early even if the nights are cool. *Source History: 1 in 1987; 2 in 1991; 2 in 1994; 2 in 1998; 2 in 2004.* **Sources: He8, Pep.**

Golden Wonder - 70 days, California Wonder type, upright sturdy plant, sweet thick-fleshed blocky fruits, green > golden-yellow > deep-orange, Burgess, 1943. *Source History: 2 in 1981; 2 in 1984; 2 in 1987; 2 in 1991; 2 in 1994; 1 in 1998; 1 in 2004.* **Sources: MIC.**

Goldie - 60 days, Vigorous 16 in. plants produce 12-14 small golden bell-shaped fruits, 2.25 x 2.25 in., thick .25 in. sweet flesh, 3-4 lobes, some tapered types, pale-green > yellow > red. *Source History: 1 in 1981; 2 in 1984; 2 in 1987; 2 in 1991; 1 in 1994; 2 in 1998; 1 in 2004.* **Sources: Gr28.**

Gourmet - 85 days, Swiss var., 4-lobed blocky bell with heavy, thick orange walls, very sweet flavor, 4-5 fruit per plant, TMV res. *Source History: 1 in 2004.* **Sources: Ter.**

Granny Smith - 57 days, Med. size fruit, 3-4 in. across, green > dark red, mild and juicy. *Source History: 2 in 1998; 1 in 2004.* **Sources: To3.**

Gusto - (FM 284) - 75 days, Semi-erect, 24-36 in. plant, slightly elongated, pendant fruit, 4 x 5 in., med. dark green, TMV res., high yield potential, PVP 2000 HA20. *Source History: 2 in 1998; 1 in 2004.* **Sources: HA20.**

Hercules Yellow Golden - 70 days, Large ornamental, bell type, yellow fruits with mild delicate flavor. *Source History: 1 in 1981; 1 in 1984; 2 in 1987; 2 in 1991; 1 in 1994; 1 in 1998; 1 in 2004.* **Sources: Ho13.**

Hungarian Yellow Stuffing - 70 days, Vigorous 24 in. plants provide an excellent foliage cover, huge 4-lobed fruits mature to a deep-red, very sweet, heavy yields, perfect for stuffing. *Source History: 1 in 1987; 1 in 1991; 1 in 1994; 1 in 2004.* **Sources: Gr29.**

Jupiter - 66-74 days, Stocky 30 in. plant, dense leaf canopy affords fruit protection from sunburn, large blocky 4-lobed fruit 4.5 x 4.5 in., thick green walls ripen red, sweet, high-yielding, widely adapted, TMV tol., PVP expired 2003. *Source History: 12 in 1987; 23 in 1991; 27 in 1994; 30 in 1998; 30 in 2004.* **Sources: Be4, Bu8, CH14, Ers, Gr29, GRI, Ha5, HO2, HPS, Jor, Loc, ME9, Mey, MIC, Mo13, MOU, Pep, Ra6, Ri12, RUP, Se26, SGF, Sh9, SOU, Sto, To1, To3, Twi, Und, We19.**

Kaala - Bacterial wilt res. var. developed for subtropic areas, 3 in. long fruit, U of Hawaii. *Source History: 1 in 1991; 1 in 1994; 2 in 1998; 1 in 2004.* **Sources: Uni.**

Kandil - (Kandil Dolma) - Classic bell pepper, 4 in. tall by 6 in. across, translucent. *Source History: 1 in 1998; 1 in 2004.* **Sources: Red.**

Keystone Resistant - 76-80 days, California Wonder type, big blocky pendant fruits, mosaic res., heavy foliage against sunscald, green > red, widely adapted, good for Mid-Atlantic region. *Source History: 1 in 1981; 2 in 1984; 3 in 1987; 5 in 1991; 5 in 1994; 3 in 1998; 4 in 2004.* **Sources: Fo7, HPS, Mel, Pep.**

Keystone Resistant Giant - 72-80 days, Florida Giant type, large fancy quality 4-5 x 3.75 in. fruits, mostly 4-lobed, thick dark- green flesh, pendant habit, vigorous upright 28-30 in. plants with large dark-green leaves, thick stems resist breaking despite continuous harvest, heavy yields, TMV res. *Source History: 30 in 1981; 27 in 1984; 29 in 1987; 19 in 1991; 19 in 1994; 18 in 1998; 16 in 2004.* **Sources: BAL, CH14, CO23, Ge2, Gr29, GRI, He8, HO2, MIC, MOU, Sau, SGF, Sh9, So1, Und, WE10.**

Keystone Resistant Giant No. 3 - 72-80 days, Blocky 4-4.5 x 3.5-3.75 in. dia. fruits, very thick walls, mostly 4-lobed, dark- green > red, pendant, vigorous erect 24-36 in. plants with thick stems, everbearing plants hold up well, TM res., heavy yielder, widely adapted, developed for shipping and processing. *Source History: 30 in 1981; 26 in 1984; 20 in 1987; 16 in 1991; 14 in 1994; 16 in 1998; 11 in 2004.* **Sources: Bu3, CA25, Enc, GLO, Kil, La1, Mey, RUP, SE14, To3, Wet.**

Keystone Resistant Giant No. 4 - (Keystone Giant No. 4, Resistant Giant No. 4) - 73 days, Sweet variety, mostly 4-lobed, dark-green to red, thick walls, 4 x 3.75 in., TMV tol., for commercial growers. *Source History: 1 in 1981; 1 in 1984; 2 in 1987; 2 in 1991; 1 in 1994; 1 in 1998; 1 in 2004.* **Sources: ME9.**

King of the North - 57-68 days, Uniform Bull Nose type for short season areas, 5-6 x 3-4 in. dia., medium flesh, sweet taste, slight taper, 3 lobes, dark-green > deep-red, prolific cold tol. spreading plant. *Source History: 12 in 1981; 10 in 1984; 5 in 1987; 5 in 1991; 4 in 1994; 8 in 1998; 18 in 2004.* **Sources: Com, Ers, Fe5, Fi1, Gr29, Gur, He8, He17, Hi6, HPS, Pe2, Pep, Ra6, Re8, Se25, Se26, To3, Tu6.**

Lamayo White - Plants grow 3 ft. tall, sweet, blocky, 5 in. long fruit, some are pointed, thick walls, light green > orange > red. *Source History: 1 in 1998; 1 in 2004.* **Sources: Ho13.**

Midnight Dream - Blocky, thick-walled purple bell, emerald-green > brown > purple. *Source History: 1 in 1994; 1 in 1998; 1 in 2004.* **Sources: WE10.**

Midway - 67-80 days, Reselected Staddon's Select, blocky 4.5 x 4-4.5 in. dia., mostly 4-lobed, bright med-green > red, thick walls, pendant habit, mild sweet flavor, tall open 18-26 in. plants, TMV res., continuous harvest, heavy yields, excellent early variety in short season areas, for home or market. *Source History: 23 in 1981; 19 in 1984; 21 in 1987; 16 in 1991; 16 in 1994; 13 in 1998; 12 in 2004.* **Sources: Alb, Bu1, CA25, Com, Ear, GLO, HO2, MIC, Ra6, Re8, Sh9, Sto.**

Miniature Chocolate Bell - 55-70 days, Small 2 in. long bell, green > chocolate-brown, stuffing, pickling and salads, family heirloom from SSE member Lucina Cress, Ohio. *Source History: 7 in 2004.* **Sources: Hud, Ma18, Pr9, Sa5, Sa9, Se16, Ter.**

Montana Wonder - 56-68 days, Small plants bear sweet, med. size blocky fruit with thick flesh, ripens to red quickly. *Source History: 1 in 1998; 1 in 2004.* **Sources: Sa9.**

Morgold - Very early large 3-4 in. fruits, ripens to glossy golden-yellow, very sweet, dwarf 12 in. plant, dev. for southern Canadian prairie by Morden Exp. Station in Manitoba, 1952. *Source History: 2 in 1981; 1 in 1984; 1 in 2004.* **Sources: Ga21.**

Napolean Sweet - 70-90 days, Plants to 2 ft., 8 in long fruit with 4.5 in. circumference, borne upright on the plant, very sweet, best used in green stage before it matures to red, offered by the L. L. Olds Seed Company in 1923. *Source History: 2 in 2004.* **Sources: Co32, Se16.**

Olena Red - 80 days, Medium size bells, thin walls, crunchy flesh, delicious strong pepper flavor, green > red, produces until frost. *Source History: 2 in 2004.* **Sources: Ho13, Ma18.**

Orange Bell - 80-100 days, Large, blocky bells, 3.5 x 4 in. long, thick walls, 3-4 lobes, excel. sweet flavor, dark green > orange, lower yields but high quality fruit. *Source History: 1 in 1998; 7 in 2004.* **Sources: Co32, Ho13, Ma18, Ri12, Sa9, Se16, Ta5.**

Orange Sun - 75-80 days, Thick flesh, 3-4 lobes, 4-5 in. long, green > orange, excel. flavor and quality. *Source History: 4 in 1998; 11 in 2004.* **Sources: Ba8, Bo19, Com, HO2, ME9, Pe6, RUP, SE14, TE7, To1, Twi.**

Ozark Giant - 68-75 days, Very large long smooth thick-fleshed fruits, 4-lobed, shiny dark-green, weighs up to 1 lb., early. *Source History: 2 in 1981; 2 in 1984; 1 in 1987; 1 in 2004.* **Sources: Sa9.**

Permagreen - 73-79 days, Blocky and very early, retains its deep dark-green color when fully mature, does not turn red, excel. yield, good keeper, adapted to the Northern states and Canada, Dr. Meader, NH/AES. *Source History: 2 in 1981; 2 in 1984; 2 in 1987; 2 in 1991; 1 in 1994; 1 in 1998; 1 in 2004.* **Sources: Sa9.**

Pimlico - Sturdy plant provides good leaf cover, large blocky fruit, thick walls, resist bruising, 3-4 lobes, green > red, more disease resistance than commercially available o.p. peppers, CV Harris Moran intro., PVP application withdrawn

1997. *Source History: 1 in 2004.* **Sources: HA20.**

Purple Beauty - 70-75 days, Compact plants with thick protective foliage, thick-walled meaty 3-4-lobed fruit, tender-crisp texture, mild sweet flavor, dev. from hybrid Purple Belle. *Source History: 32 in 1991; 34 in 1994; 37 in 1998; 49 in 2004.* **Sources: BAL, Bo17, Bo19, CA25, CH14, Com, Dam, Fe5, Ga1, Ge2, GLO, Gr29, Ha5, He17, HO2, Ho13, HPS, Jor, Kil, Loc, MIC, Mo13, Na6, Pe2, Pe6, Pep, Pin, Ra6, Re8, Ri12, Roh, RUP, Sa9, SE14, Se25, Se26, Se28, Shu, Sw9, Te4, TE7, Ter, To1, To3, Twi, Vi4, WE10, We19, WI23.**

Quadrato Asti Giallo - (Asti Square Yellow, Yellow from Asti) - 70-80 days, Large blocky 3 or 4-lobed Italian bell, golden- yellow when ripe, crisp thick flesh with delicious sweet spicy flavor, most famous pepper in Italy. *Source History: 1 in 1981; 1 in 1984; 2 in 1991; 1 in 1994; 3 in 1998; 11 in 2004.* **Sources: Ba8, Ber, Co31, Pep, Pr9, Ra6, Red, Se16, Se24, So25, Te4.**

Quadrato d'Asti Giallo Rossa - 80-90 days, Deep-gold and red sweet bell types mixed, mostly 4-lobed, grown together in Italy as a single crop, freezes well, good roasted. *Source History: 1 in 1981; 1 in 1984; 1 in 1987; 1 in 1998; 2 in 2004.* **Sources: Ba8, Se17.**

Quadrato d'Asti Rosso - 80-85 days, Classic blocky bell pepper from Asti, thick red walls, sweet flesh, large productive plant. *Source History: 4 in 2004.* **Sources: Co31, Ra6, Red, Se24.**

Red Giant - 74 days, *Source History: 2 in 1998; 1 in 2004.* **Sources: SE14.**

Red Miniature - (Mini Bell, Mini Red) - 55-90 days, Full-size plants with loads of tiny sweet bell-shaped peppers, excel. for stuffing, family heirloom given to the Seed Savers Exchange by Lucina Cress, Ohio SSE member. *Source History: 4 in 1991; 4 in 1994; 5 in 1998; 12 in 2004.* **Sources: Fo7, Gr28, Ho13, Jor, ME9, OSB, Pep, Pr9, SE14, Se16, Ter, Vi5.**

Resistant Giant No. 4 - 72-80 days, Blocky 4.5 x 4.5 in. fruits, 4-lobed, med-dark green thick flesh, 28 in. plants with good cover, widely adapted, dependable under a wide range of conditions. *Source History: 3 in 1981; 3 in 1984; 5 in 1987; 5 in 1991; 5 in 1994; 4 in 1998; 2 in 2004.* **Sources: RIS, Twi.**

Ruby Giant - 70-72 days, Vigorous upright 28 in. plant with large dark-green foliage, 4-lobed dark-green 5 x 4 in. fruits ripen red, thick sweet flesh. *Source History: 2 in 1987; 2 in 1991; 1 in 1994; 1 in 1998; 2 in 2004.* **Sources: Gr29, Pep.**

Ruby King - 64-69 days, Early, 24 in. plant, 4-6 x 3 in. dia., dark-green > bright ruby-red, 3 lobes, medium thick sweet extremely mild flesh, thin skin, excel. for frying. *Source History: 4 in 1981; 3 in 1984; 3 in 1987; 1 in 2004.* **Sources: Se26.**

Staddon's Select - (Missile) - 60-72 days, Slightly variable 4 x 4 in. med-fleshed fruits, 3 or 4 lobes, TM res., everbearing, pendant, vigorous branching 26 in. plants, large heavy market type, excel. color, slightly rough, produces continuously if kept well picked, adapted to the East and the North, heavy crops even in hot dry weather, 1963. *Source History: 21 in 1981; 18 in 1984; 19 in 1987; 16 in 1991; 12 in 1994; 11 in 1998; 12 in 2004.* **Sources: Fe5, Gr29, GRI, HA3, Hig, Pep, Pla, SE4, Se26, Ter, Ves, We19.**

Sunbright - 75 days, Yellow sister to Big Red, sweet thick flesh, 5 x 7 in. dia. fruits, green > yellow. *Source History: 1 in 1991; 3 in 1994; 4 in 1998; 6 in 2004.* **Sources: He8, Pe6, Ra6, RUP, SE14, WE10.**

Sunrise - (Sunrise Orange) - 60-75 days, Vigorous bushy plants produce beautiful yellow bells, 4.5 x 3.5 in. fruits turn bright-red at maturity, medium-thick walls, early and productive, market and home gardens, TM res., one of the earliest and most productive varieties to grow in the North. *Source History: 4 in 1991; 1 in 1994; 4 in 1998; 6 in 2004.* **Sources: Eco, Ga1, Gr28, Pep, Ri12, Se7.**

Super Stuff - 67 days, Deep tapered bell fruits 6 in. deep x 3.75 in. across, 3-4 lobes, .25 in. thick flesh, ripens from pale- green to butter-yellow to orange to red, very sweet, better stuffing type than Romanian. *Source History: 1 in 1987; 1 in 1991; 2 in 1994; 1 in 1998; 2 in 2004.* **Sources: Pep, Sto.**

Sweet Chocolate - (Choco) - 58-86 days, Early, med-sized, tomato-shaped, glossy-green > chocolate-brown, thick flesh, good frozen whole, North and Canada, new color for salads, rare, NH/AES, 1965. *Source History: 5 in 1981; 4 in 1984; 8 in 1987; 9 in 1991; 7 in 1994; 9 in 1998; 24 in 2004.* **Sources: Al25, Bou, Co32, Com, Coo, Gr28, He8, Hi6, Hig, Ho13, Jo1, Orn, Pe2, Pep, Ra6, Re8, Ri2, Sa9, SE14, Se16, Se28, So1, TE7, Und.**

Sweet Red - Open-pollinated selection from a supermarket hybrid dev. by Ken Allan, Ontario, Canada, blocky fruit ripens to red by late August when planted in soil that has been warmed under clear plastic, as is done for growing sweet potatoes in northern areas. *Source History: 1 in 2004.* **Sources: Al6.**

Taurus - 76 days, Large branched plant produces fewer misshapen fruit, 4-lobed, 5 x 6 in., TMV TEV res., PVP 1999 Novartis Seeds, Inc. *Source History: 2 in 1998; 3 in 2004.* **Sources: Ra6, RIS, RUP.**

Turkish 11-B-14 - Sweet bell, 5 in. wide by 4 in. tall. *Source History: 1 in 1994; 1 in 2004.* **Sources: Red.**

Ultra Stuff - 74-80 days, Large 2-3 lobed fruit, 3.75 x 6 in., thick flesh, light green > yellow > orange > red. *Source History: 1 in 1998; 2 in 2004.* **Sources: Pep, Sto.**

Wisconsin Lakes - 84 days, Early sweet pepper, med-large and thick fleshed, ripens to bright red even in the North, adapted to the Midwest, and excel. for Northern areas, U of WI. *Source History: 1 in 1981; 1 in 2004.* **Sources: Sa9.**

World Beater - (Ruby Giant, Giant World Beater) - 73-74 days, Dev. from a cross of Chinese Giant x Ruby King, huge blocky non-tapering fruits, 5 x 3.5 in. dia., heavy crops, some plants produce 32 marketable fruits, thick flesh, keeps well in prime condition, introduced before 1912. *Source History: 8 in 1981; 8 in 1984; 8 in 1987; 4 in 1991; 7 in 1994; 4 in 1998; 3 in 2004.* **Sources: Pep, So1, Tu6.**

Yankee Bell - 60-80 days, Northstar/Lady Bell type developed for northern growers, fruits are blocky, 3-4 lobed, holds well into the sweet red stage, strongly branched plants with good leaf cover. *Source History: 1 in 1994; 1 in 1998; 3 in 2004.* **Sources: Jo1, Ra6, Re8.**

Yellow Belle - 65 days, California Wonder type, extremely heavy yields, 3.5 x 3 in. dia. fruits, mild thick flesh, 3-4 lobed, medium green > golden yellow > red, very early. *Source History: 3 in 1981; 3 in 1984; 4 in 1987; 5 in 1991; 6 in 1994; 5 in 1998; 6 in 2004.* **Sources: Gr29, Pe2, Pep, Ri12, So1, Sto.**

Yellow Miniature - (Mini Yellow) - 70-85 days, Light celery-green > golden yellow fruit, 2 in. long, sweet flavor, stuffing, pickling and salads, family heirloom from SSE member Lucina Cress, Ohio. *Source History: 1 in 1991; 3 in 1998; 13 in 2004.* **Sources: Hud, Jor, Ma18, ME9, OSB, Pep, Pr9, Sa9, SE14, Se16, Ta5, TE7, Ter.**

Yolo Wonder - 70-80 days, Improved strain of California Wonder, larger and more mosaic res., 4-4.5 x 3.75-4 in. dia., thick flesh, 3-4 lobes, dark-green > red, pendant, compact spreading 24-28 in. plant, dense foliage protects against sunscald. *Source History: 55 in 1981; 43 in 1984; 31 in 1987; 22 in 1991; 25 in 1994; 24 in 1998; 27 in 2004.* **Sources: BAL, Bou, CA25, Com, Fo7, Ge2, Gr29, GRI, He8, He17, Jor, La1, Lej, LO8, ME9, Mel, Mo13, Or10, Pep, Ra6, Re8, RIS, SE14, Se25, Se26, SGF, Sw9.**

Yolo Wonder B - 72-80 days, Sturdy upright 18-24 in. plants, large dark-green foliage, smooth blocky fruits, 4 x 3.5 in. dia., mostly 4-lobed, firm thick flesh, dark-green > red, pendant. *Source History: 6 in 1981; 6 in 1984; 10 in 1987; 11 in 1991; 11 in 1994; 9 in 1998; 5 in 2004.* **Sources: CH14, CO23, DOR, HO2, WE10.**

Yolo Wonder B Improved TMR - 77 days, Uniform sel. by Ferry-Morse from original Yolo Wonder dev. by Campbell Soup Co., TMV res., 4 x 4 in., 4 lobes, thicker walls, sweet, pendant, well shaped. *Source History: 3 in 1981; 3 in 1984; 3 in 1987; 2 in 1991; 2 in 1994; 2 in 1998; 1 in 2004.* **Sources: Wi2.**

Yolo Wonder L - 70-78 days, Vigorous more upright 18-32 in. plants, larger 4.25 x 3.75 in. fruits, 3-4 lobes, thick-walled, very mild and sweet flesh, good cover against sunburn, TM res., longer picking period, very adaptable. *Source History: 29 in 1981; 28 in 1984; 27 in 1987; 23 in 1991; 21 in 1994; 13 in 1998; 8 in 2004.* **Sources: Bu3, Kil, Loc, MIC, MOU, Sh9, WE10, Wet.**

Varieties dropped since 1981 (and the year dropped):

16 to 1 (1987), *All Big* (1984), *Amethyst* (2004), *Argo* (1987), *Big Belle* (1991), *Big Green M.R.* (1987), *Big Pack* (1987), *Burpee's Fordhook* (1987), *Butter Belle* (2004), *Calcom "A"* (1994), *California Wonder A* (1994), *California Wonder Special* (1987), *California Wonder TMR* (1998), *Calwonder Resistant* (1998), *Carnosissimo di Cuneo* (1998), *Chinese Emerald Giant* (1982), *Delaware Bell* (1982), *Delray Bell* (1987), *Earliest* (1987), *Early Canada Bell* (1994), *Early M.R. Keystone Giant* (1987), *Early Set* (1991), *Early Thick Meat* (1987), *Florida Resistant Giant* (1991), *Florida VR-2* (1994), *Gedeon* (1987), *Georgia Bell* (2004), *Giant Resistant* (1982), *Giant Sweet Green* (2004), *Gold Star* (1991), *Gold Topaz* (1987), *Golden Ball* (1991), *Golden Dwarf* (1987), *Grande Rio 66* (1998), *Hercules Sweet Red* (1987), *Jung's Yellow Belle* (2004), *Jupiter Elite* (2004), *Jupiter Sterling* (2004), *Kansas Wonder* (1982), *King* (1998), *Lamuyo* (1987), *Liberty Bell* (1987), *Lincoln Bell* (1998), *Mammoth Ruby King* (1987), *Manda, Yellow Bell* (2004), *Manto* (1982), *Market Master* (1987), *Martindale* (1991), *Merced TMR* (2004), *Mercury* (2004), *Merrimack Wonder* (1994), *Michigan Wonder (Burgess)* (1991), *Miss Belle* (1987), *Missouri Wonder M.R.* (1984), *Oakview Wonder* (1991), *Pennbell* (1987), *Pip* (2004), *Piperade* (1987), *Ponderosa (TMR)* (1991), *Prima Belle* (2004), *Purple Lobe* (1998), *Quadrato d'Asti* (1991), *Red Baby* (2004), *Reus Dulce Espanol* (1991), *Roma Yellow Belle* (2004), *Ruby King Improved* (1982), *Saitama Early* (1987), *Spartan Emerald* (1983), *Sultan* (1994), *Sweet Cream* (1994), *Sweet Spanish Green Carre* (1984), *Ta Tong* (2004), *Tambel-2* (1998), *Tarog* (2004), *Tisana* (1987), *Titan* (1994), *Vinedale Early Red* (1994), *VR-2* (1987), *Yellona* (1998), *Yellow Baby* (2004), *Yolo Wonder A* (2004), *Yolo Wonder B Improved* (2004), *Yolo Wonder L Improved TMR* (1991), *Yolo Wonder TMR* (1991), *Yolo Wonder Y* (1991), *Yolo Wonder 43* (1987), *Yolo Wonder Improved* (1994).

Pepper / Sweet / Non-Bell
Capsicum annuum

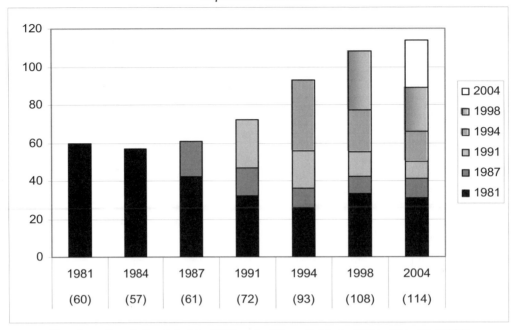

Number of varieties listed in 1981 and still available in 2004: 60 to 31 (52%).
Overall change in number of varieties from 1981 to 2004: 60 to 114 (190%).
Number of 2004 varieties available from 1 or 2 sources: 62 out of 114 (54%).

Aconcagua - (Giant Aconcagua) - 70-75 days, Long frying pepper named for Mt. Aconcagua in Argentina, Cubanelle- shaped but much larger, tapered 7-11 x 2.25-2.5 in., yellow/green > red, very sweet when light-green, medium flesh, pendant, 28-30 in. plants, dark-green foliage gives fruits excellent coverage, continuous fruiting, use fresh or roast and peel. *Source History: 10 in 1981; 3 in 1984; 4 in 1987; 5 in 1991; 5 in 1994; 15 in 1998; 28 in 2004.* **Sources: Ba8, Ers, Fo7, Fo13, Gr29, HO2, Ho13, HPS, Ma18, Pep, Pla, Ra5, Ra6, Re8, Ri12, RUP, Se7, SE14, Se17, Se26, Se28, Sh9, Shu, To1, To3, Vi5, WE10, Wo8.**

Alba Regia - Conical, white fruit with red blush at maturity, 3 x 5 in., often held upright on the 3 ft. plant, sweet and crunchy, Italian. *Source History: 1 in 1998; 1 in 2004.* **Sources: Ho13.**

Antohi Romanian - 53-78 days, Upright plant growth, pointed fruits are smooth, 4 in. long, 2 in. dia., pale-yellow > red, yields early and heavily, seeds brought to this country by Jan Antohi who was a touring acrobat who defected to the U.S., returned from Romania after a visit in 1991 with seeds of this heirloom pepper. *Source History: 1 in 1994; 2 in 1998; 9 in 2004.* **Sources: Goo, Jo1, Mi12, Pep, Ra6, Sw9, Tu6, Up2, Yu2.**

Apple - 57-77 days, Med. size plants yield well in diverse climates, 3-4 in. heart shaped fruit with mild, sweet, fruity taste, thick tender flesh. *Source History: 3 in 1998; 7 in 2004.* **Sources: Gou, Jo1, Lej, Ra5, Ra6, Re8, Sa5.**

Blanco del Pais - 75 days, Plants to 3 ft., 4.5 in. light green fruit. *Source History: 1 in 2004.* **Sources: Pep.**

Boldog Hungarian Spice - 51-80 days, Compact plants yield deep crimson fruit, 6-7 in. long, makes sweetly spicy paprika, rich flavor with just a touch of heat, from central Hungary. *Source History: 5 in 1998; 13 in 2004.* **Sources: Fe5, Goo, Gr29, Hi6, Jo1, Jo6, Pep, Pr3, Ri2, Se25, Se26, Sw9, Twi.**

Bujan - 80 days, Sturdy 2 ft. plants each bear an average of five thick walled fruit, dark green > scarlet, some weigh over 2 lbs., four-lobed at the shoulder and tapered to a rounded point, pungent, sweet flavor. *Source History: 1 in 1994; 1 in 1998; 1 in 2004.* **Sources: Ho13.**

Bulgarian - 75 days, Plants to 36 in., open foliage, somewhat variably shaped tapered fruit, 1-1.25 in. wide by 4 in. long, abundant yields, good keeper, excel. for drying, sunscald and disease res., ethnic var. from Bulgaria. *Source History: 1 in 2004.* **Sources: Ga21.**

Bulgarian No. 5 - Fairly long pointed fruit, sweet flavor, high yields, productive in northern areas, from Bulgarian market, rare. *Source History: 1 in 2004.* **Sources: Ea4.**

Bulgarian Sweet - 60 days, Semiblocky sweet pepper, 6-7 in. long, green > red, good for roasting. *Source History: 1 in 1991; 1 in 2004.* **Sources: Cl3.**

Buran - 90-100 days, Polish heirloom, 18-24 in. plant, 3-lobed fruit, 4 in. long, 3 in. at shoulder, very sweet flavor, green > red, sweet flavor at green or red stage. *Source History: 6 in 2004.* **Sources: Co32, Pep, Pr9, Ri12, Se16, Ter.**

California Rose - *Source History: 1 in 1994; 1 in 1998;*

1 in 2004. **Sources: WE10.**

Calrose - 85 days, Pimento pepper with flattened shape, some are slightly elongated. *Source History: 1 in 2004.* **Sources: Sa9.**

Centinel - (Tatli Sivri) - Resembles Corbaci but even sweeter, 5 in. long curved pods .5-.75 in. across, light green > red, good for eating fresh, pickling or stir fry, can produce 785 fruit from only 10 plants. *Source History: 2 in 1991; 2 in 1994; 2 in 1998; 1 in 2004.* **Sources: Red.**

Cherry Sweet - (Large Cherry, Red Cherry, Small Sweet Red Cherry) - 70-80 days, Pickling pepper, strong upright bushy everbearing 20 in. plant, cherry-shaped 1 in. x 1.5 in. dia. fruits, med-thin crisp shell, dark-green > deep-crimson, pendant, good foliage cover prevents sunburn, productive, for bite-size pickles or canning or stuffing, pre-1860. *Source History: 60 in 1981; 58 in 1984; 49 in 1987; 48 in 1991; 42 in 1994; 34 in 1998; 36 in 2004.* **Sources: Bu3, CH14, Cl3, Enc, Ge2, GLO, Gr29, He17, HO2, Ho13, LO8, Loc, Ma18, MAY, ME9, MIC, Mo13, MOU, Ni1, Pep, Ra6, Re8, RIS, Ros, RUP, Se26, Se28, SGF, Sh9, So1, So25, Sto, To1, Vi4, WE10, Yu2.**

Cherry, Super Sweet - 75 days, Dark-green easy-to-pick fruit by late July, large 1.75 in. dark cherry-red fruit by mid-August, TMV and crack tol., very heavy early yields. *Source History: 1 in 1987; 2 in 1991; 2 in 1994; 2 in 1998; 3 in 2004.* **Sources: Gr29, Pep, Sto.**

Chervena Chujski - (Chervena Chushka) - 85 days, Healthy plants produce 2 x 6 in. long fruits with bright red flesh which is very sweet, almost candy-like, green > brown > bright red, Bulgarian heirloom. *Source History: 1 in 1998; 8 in 2004.* **Sources: Co32, He8, La1, Pep, Pr9, Ri12, Sa5, Se16.**

Chile Negro - Black sweet med-size fruits, 5-7 in. long, dries well, Parral, Chihuahua, Mexico. *Source History: 1 in 1987; 1 in 1991; 1 in 1994; 2 in 1998; 1 in 2004.* **Sources: Na2.**

Conquistador - (Nu Mex Conquistador) - 72-75 days, Plants 16 in. tall, mild sweet paprika pepper without the heat, used at the green stage for mild chiles, ripened to red for a premium paprika, 4-4.5 in. long, thick walls, NM State U. *Source History: 1 in 1991; 2 in 1994; 5 in 1998; 6 in 2004.* **Sources: Bu3, Enc, Gr29, Pep, Scf, Se26.**

Corbaci - (Sari Tatli Sivri) - Strong plant to 3 ft., thin, bright red fruit to 10 in. long, sweet, great for fresh eating, from Turkey. *Source History: 1 in 1991; 2 in 1994; 3 in 1998; 4 in 2004.* **Sources: Ho13, Pep, Red, So25.**

Corno di Toro Giallo - (Corno di Toro Yellow, Yellow Bull Horn) - 90-100 days, Horn of the Bull, 8 in. long x 1.5-2 in. dia. at shoulder, yellow > deep-golden, curved like the horn of a bull, peppery flavor is neither hot nor very sweet, from Italy. *Source History: 2 in 1981; 2 in 1984; 3 in 1987; 6 in 1991; 9 in 1994; 7 in 1998; 13 in 2004.* **Sources: Ba8, Ber, Gr29, Ho13, Ma18, ME9, Orn, Pep, Ra6, Red, Se28, To3, Ver.**

Corno di Toro Rosso - (Red Bull Horn, Corno di Toro Rouge) - 75-100 days, Very attractive Italian heirloom, 8-10 in. long, curved fruit in the shape of a bull's horn, 18 in. plant, can be harvested at green or red stage, popular in Italy and Spain. *Source History: 1 in 1987; 5 in 1991; 9 in 1994; 10 in 1998; 25 in 2004.* **Sources: Ag7, Ba8, Bou, Dom, Gr28, Gr29, Ho13, HPS, Ma18, ME9, Na6, Orn, Pep, Pin, Ra5, Ra6, Red, Se24, Se26, Se28, Ter, To3, Ver, Vi5, We19.**

Corno di Toro, Red and Yellow - 70 days, Tall branching plants, 8-10 in. long curved tapered pointed shiny fruits, both bright-yellow or deep-red cultivars included. *Source History: 2 in 1991; 2 in 1994; 5 in 1998; 5 in 2004.* **Sources: Hi13, Ki4, Se7, So25, To1.**

Crimson Sweet - Fleshy fruit, 6-7 in. long, good for frying or roasting. *Source History: 1 in 1998; 1 in 2004.* **Sources: Pe7.**

Cuban - 60-65 days, Irregular, 2 x 6 in. long fruit, 2-3 lobes, green > yellow. *Source History: 1 in 1998; 1 in 2004.* **Sources: Ra5.**

Cubanelle - (Cubanella, Cubanella Long Sweet/Red Sweet Long/Sweet Frying, Cubanelle Italian, Cuban) - 62-70 days, Sweet frying pepper, 4.5-6 x 2-2.5 in. dia., 3 lobes, tapers to a blunt end, slightly irregular and roughened, med-thick waxy flesh, yellow-green > red, distinct flavor, pendant, 28-30 in. bushy plant, everbearing. *Source History: 50 in 1981; 45 in 1984; 48 in 1987; 57 in 1991; 55 in 1994; 54 in 1998; 54 in 2004.* **Sources: BAL, Bo19, Bu2, Bu3, CH14, CO23, CO30, Com, Cr1, DOR, Enc, Fe5, Fo7, Fo13, Ge2, GLO, Gr29, GRI, HA3, Ha5, He8, HO2, Hud, Jo6, Jor, Kil, La1, ME9, Mey, MIC, Ont, Or10, PAG, Pep, Pin, Ra6, Re8, Red, Ri12, RIS, RUP, Sa9, SE14, Se25, Se26, Se28, SGF, Sh9, Sto, To1, To3, WE10, Wi2, WI23.**

Cuneo - 80 days, Name derives from the town of Cuneo in the Piedmont region of northern Italy where it is the primary market grower's pepper, clear bright-yellow softball-size fruit with a pointed tip and barely discernible lobes, sweet. *Source History: 1 in 1987; 1 in 2004.* **Sources: Ma18.**

Dainty Sweet - 80 days, Ornamental compact bush, 2 in. long fruit held above foliage, light-yellow > light-purple > bright orange-red. *Source History: 1 in 1994; 3 in 2004.* **Sources: Sa5, WE10, We19.**

Datil, Sweet - 90 days, Plants have large, pale green leaves, 1.5 in. fruit, pale green > orange > red, same flavor and smoky aroma as regular Datil but without the heat. *Source History: 1 in 1998; 1 in 2004.* **Sources: Fo7.**

Demre - Similar to Centinel and Corbaci, 6-7 in. long pods, 1 in. across, Turkey. *Source History: 1 in 1994; 1 in 1998; 1 in 2004.* **Sources: Red.**

Dulce Medeterraneo, Red - Huge 10+ in., thick fleshed pods, green to red, very sweet, size and growth habit resembles Big Jim and Anaheim, but no heat, good for roasting and stuffing. *Source History: 1 in 2004.* **Sources: Ba16.**

Elephant Trunk - One of the longest mild sweet peppers, 8-10 in. long, grows thickly set on vigorous plants, sweet flesh, dark-green > scarlet-red, for home gardens, excel. pickler. *Source History: 1 in 1981; 1 in 1984; 1 in 1987; 1 in 1991; 1 in 1994; 1 in 1998; 1 in 2004.* **Sources: Red.**

Feherozon - (Feher Ozon Paprika) - 90 days, Excellent

paprika pepper, extremely productive dwarf plants grow 12-15 in. tall, produce 3 x 4-5 in. long fruits, creamy white > orange > red, exceptionally sweet flesh, dried fruits are ground to make fresh paprika, CV So1 introduction, 1999. *Source History: 1 in 1991; 1 in 1994; 1 in 1998; 2 in 2004.* **Sources: Fe5, So1.**

Friariello - (Italianelle, Sweet Italian Frying Pepper, Italianelle Sweet Frying) - 60 days, Frying pepper, also used for fresh eating and pickling, Italy. *Source History: 1 in 1994; 1 in 1998; 1 in 2004.* **Sources: So25.**

Frigitello - 70 days, Vigorous 3-4 ft. plants, 3-4 in. long tapered red fruit, sweet flavor, used fresh or dried or pickled, Italian heirloom. *Source History: 1 in 2004.* **Sources: Coo.**

Fushimi, Long Green - 65-80 days, Famous old Japanese var., many long slender mild thin-walled 6 in. pods, bright-green glossy skin is wrinkled when mature, vigorous rather tall plant, easy to grow. *Source History: 2 in 1981; 2 in 1984; 2 in 1987; 2 in 1991; 1 in 1994; 3 in 1998; 7 in 2004.* **Sources: CO30, Ev2, Ho13, Jo6, Kit, Ma18, Pep.**

Gayle (Chocolate) - 80 days, Tapered fruit, 6-8 in. long, thick sweet flesh with slightly spicy flavor, green > brown, burgundy- red inside, delicious when fried, prolific. *Source History: 1 in 1998; 4 in 2004.* **Sources: Fe5, Gr29, Ho13, La1.**

Georgescu - 75 days, Elongated bell, flesh resembles Cubanelle, sweet, celery-green > yellow > red, good production. *Source History: 2 in 2004.* **Sources: Ma18, Se17.**

Georgescu Chocolate - 80 days, Oblong fruit, ripens to chocolate-brown, suitable alternative for Sweet Chocolate, from Romania. *Source History: 1 in 2004.* **Sources: Sa9.**

Goccia d'Oro - Excellent frying pepper, long fruit, thin skin, sweet flavor, green, yellow or red-yellow when ripe. *Source History: 1 in 2004.* **Sources: Se24.**

Golden Treasure (Italian) - 80-100 days, Italian pepper with sweet med-thick flesh and thin skin, 1.75 x 8-9 in. at shoulder, 2 lobes, green > yellow, excel. for frying, roasting and fresh eating. *Source History: 3 in 1991; 3 in 1994; 2 in 1998; 8 in 2004.* **Sources: Ba8, Gr29, He8, Pep, Pr9, Se16, Sk2, Ta5.**

Healthy - 90 days, Early maturing 30 in. plants, attractive, sweet, wedge shaped, 3-celled fruits, 2.5 in. at triangular shoulder by 4 in. long, yellow > orange > red, does not need much sun, will ripen during cloudy summers, resistant to disease and rotting, developed at the Institute of Vegetable Breeding and Seed Production on the west edge of Moscow. *Source History: 1 in 1994; 2 in 1998; 7 in 2004.* **Sources: In8, Pep, Pr9, Se16, Ta5, To3, Und.**

Italia - 55-75 days, Long Italian Sweet type similar to Corno di Toro, long green 8 x 2.5 in. fruits ripen early to a dark crimson, sweet full pepper flavor expresses well in sauces and fried preparations, easy to grow. *Source History: 1 in 1987; 2 in 1991; 1 in 1994; 1 in 1998; 2 in 2004.* **Sources: Jo1, Sw9.**

Italian Frying - 50 days, Early maturing, long, shiny sweet pepper matures to red, looks like a hot pepper. *Source History: 2 in 2004.* **Sources: Ra6, Sa9.**

Italian Giant - 78 days, Large frying pepper, 7-9 in. long x 2 in. dia. *Source History: 1 in 2004.* **Sources: He8.**

Italian Green Frying - (Ram's Horn, Goat Horn) - 78 days, Umbrella-shaped 25 in. tall plants, med-sized foliage, flat 6 in. long fruits, dark-green > very dark-red, 2-celled, thin flesh, very mild, pendant. *Source History: 1 in 1981; 1 in 1984; 2 in 1987; 2 in 1991; 3 in 1994; 2 in 1998; 3 in 2004.* **Sources: Gr29, GRI, Pep.**

Italian Sweet - (Long John, Italian Long Sweet) - 58-75 days, For the Northeast, great flavor when fried, 5.5-6.5 x 1.5-2 in., 3- lobed, tapered, med-thick mild flesh, thin skin, pendant, everbearing, TM res., highly flavored. *Source History: 14 in 1981; 13 in 1984; 13 in 1987; 15 in 1991; 8 in 1994; 10 in 1998; 7 in 2004.* **Sources: BAL, Ga1, Gr29, Loc, Pep, Sa5, Se7.**

Kapya - (Yag Biberi) - Unusual horn-shaped red fruit, 7 in. long by 3 in. across, thin, crunchy flesh, unique flavor. *Source History: 1 in 2004.* **Sources: Red.**

Klari, Baby Cheese - (Golden Delicious Apple Pepper) - 55-65 days, Sweet version of Almapaprika, small, flat, white flattened round fruit, 1 x 2 in., 4 oz., white > yellow > red, great for pickling whole, so named because of its resemblance to an old-fashioned wheel of cheese and Klari was the name of the woman who maintained the seed, from Hungary. *Source History: 4 in 1998; 3 in 2004.* **Sources: Ami, Fe5, Sa5.**

Le Rouge Royal - 75-80 days, Tall plants to 46 in., large, sweet fruit, 2-3 lobes, 6 x 8 in., green > red, crisp sweet fruit, huge yields. *Source History: 1 in 1998; 4 in 2004.* **Sources: Ho13, Sa9, So25, To3.**

Lemme's Italian Sweet - 74-81 days, Productive 36 in. plants benefit from staking or caging, slightly curved, tapered, 6-8 in., thin-walled fruits are suitable for drying even in humid areas, distinctive, sweet flavor. *Source History: 1 in 1998; 2 in 2004.* **Sources: Ma18, Sa9.**

Lipstick - 53 days, Shiny smooth 5 in. long cone-shaped dark-green fruits ripen to a glossy rich red, thick juicy sweet flesh, dependably early, heavy yields even in a cool summer, salads cooking or roasting. *Source History: 2 in 1987; 4 in 1991; 2 in 1994; 4 in 1998; 5 in 2004.* **Sources: Ba8, Jo1, Pep, Sw9, Und.**

Lombardo - Early maturing, thin skinned, long fruit, green > red, good frying pepper, excel. for pickling. *Source History: 1 in 2004.* **Sources: Se24.**

Marconi, Golden - (Marconi Yellow) - 80-90 days, Sweet Italian pepper, medium thick sweet flesh, thin skin, 8-12 in. long by 3 in. across, 2-3 lobes, green > yellow, used for salads and drying. *Source History: 6 in 1998; 9 in 2004.* **Sources: Gr29, Ho13, Ma18, ME9, Pep, SE14, Se26, To1, WE10.**

Marconi, Purple - 70-90 days, Abundant yields of 4-6 in. long tapered fruit with blunt end, sweet flavor, ripens to purple, traditionally used for frying, also eaten fresh. *Source History: 2 in 1998; 6 in 2004.* **Sources: ME9, OSB, Pe2, SE14, TE7, To1.**

Marconi, Red - (Marconi, Red Marconi, Italian Red Marconi, Marconni Rosso) - 70-85 days, Somewhat elongated fruit up to 12 in. long x 3 in. across at the shoulders, ripens to red, early and sweet, Italy. *Source*

History: 2 in 1981; 2 in 1984; 4 in 1987; 6 in 1991; 11 in 1994; 17 in 1998; 35 in 2004. **Sources: Ag7, Al25, Ba8, Bo19, CA25, Com, Coo, Dom, Gr28, Gr29, HA3, La1, Ma18, ME9, Mel, Ont, Pe2, Pep, Pr9, Ra6, Re8, Red, Ri12, RUP, Se8, SE14, Se16, Se26, Se28, So25, Ta5, TE7, To1, To3, WE10.**

Nardello - (Jimmy Nardello, Jimmy Nardello's Sweet Italian Frying Pepper) - 65-75 days, Productive, low growing, 24 in. plants, thin, tapered, 8-10 in. long, crinkly fruits ripen rapidly to bright red, slightly spicy, smoky flavor, freezes and dries well, Italian heirloom from the Nardello family. *Source History: 1 in 1981; 1 in 1984; 2 in 1987; 1 in 1991; 3 in 1994; 7 in 1998; 18 in 2004.* **Sources: Ag7, Fe5, Ho13, Ma18, Pe2, Pep, Pr3, Pr9, Ri12, Sa5, Sa9, Se7, Se16, So1, So9, Te4, To1, Up2.**

Neapolitan - 80 days, Large sweet bell, green > red, excel. producer. *Source History: 2 in 2004.* **Sources: Ea4, Ta5.**

Odessa Market - 77 days, Top-shaped fruit, thick sweet walls, 4-6 in. long, ripens lime-green > orange > red, from John Trumpeter, Clinton, Iowa. *Source History: 1 in 2004.* **Sources: Sa9.**

Padrone - Thin-skinned sweet frying pepper, 2 x 3 in. green fruit, later fruits may become spicy, from Galacia in Spain. *Source History: 1 in 2004.* **Sources: Se24.**

Papri Mild - 80-120 days, Large upright spreading med-green 30-36 in. plants, good cover, tapering flat 6 x 1.5 in. fruit, deep-green > dark-red, thin flesh, high flavor but low pungency. *Source History: 2 in 1981; 2 in 1984; 1 in 1987; 1 in 1991; 5 in 1994; 4 in 1998; 1 in 2004.* **Sources: Se8.**

Paprika (Spain) - Productive 2.5 ft. plant, thin walled fruit, 6 in. long, matures to deep red, dried and ground for paprika. *Source History: 1 in 1998; 1 in 2004.* **Sources: Ho13.**

Paprika, Andrea - Straggly 3 ft. plant, semi-blocky, 3-4 in., orange-red fruit. *Source History: 1 in 1994; 1 in 1998; 1 in 2004.* **Sources: We11.**

Paprika, Frutka - Flavorful fruit with the shape of a small tomato, outstanding flavor. *Source History: 1 in 1994; 1 in 1998; 1 in 2004.* **Sources: We11.**

Paprika, Hungarian - 70-77 days, Plants to 18 in. tall, 1 x 5 in., sweet, spicy flavored red fruits dry to a leathery texture and are then dried and ground into an excellent paprika, not for fresh eating. *Source History: 3 in 2004.* **Sources: Ga21, Sa9, So1.**

Paprika, Kalosca Sweet Spice - 70 days, Slightly ruffled, ornamental foliage, produces an abundance of thin walled, sweet fruit, 1 x 4 in., green > scarlet. *Source History: 2 in 2004.* **Sources: Ma18, Ta5.**

Paprika, PCR - Large, tapered, semi-blocky, 4 in. fruit, orange-red, prolific. *Source History: 1 in 1994; 1 in 1998; 1 in 2004.* **Sources: We11.**

Piement des Landes - High yields of 3 x 9 in., large yellow fruit, France. *Source History: 1 in 1998; 1 in 2004.* **Sources: Ho13.**

Pimento - (Pimiento, Red Pimento, Red Heart Pimiento) - 65-90 days, Upright prolific 26-30 in. plant, smooth heart-shaped fruits, 3-3.5 x 2.5 in. dia., extremely thick walls, very sweet at maturity, dark-green > deep bright-red, grown extensively in the Southern states for canning, heavy yields. *Source History: 39 in 1981; 36 in 1984; 37 in 1987; 27 in 1991; 30 in 1994; 25 in 1998; 41 in 2004.* **Sources: Ada, Bo19, CA25, Co31, Co32, Com, Enc, Fo7, Gr29, HA3, He8, He17, HO2, Jo6, Jor, LO8, ME9, Mey, MOU, Na6, Ont, Or10, Orn, Pin, Pla, Ra5, Ra6, Re8, Ri12, Ros, Sau, SE14, Se28, So25, SOU, Te8, Vi4, Vi5, We11, Wet, Wo8.**

Pimento L - 74-100 days, Large heart-shaped 4.5 x 3.5 in. fruits, very thick walls, dark-green > bright-red, TM res., upright 18-24 in. plants, proven Southern producer, processing or canning. *Source History: 7 in 1981; 7 in 1984; 9 in 1987; 11 in 1991; 14 in 1994; 24 in 1998; 23 in 2004.* **Sources: CH14, Eo2, HPS, Kil, La1, Loc, MIC, MOU, Pep, Ra6, Re8, RIS, Roh, RUP, Se7, Se26, Se28, Sh9, Si5, To1, To3, WE10, Wi2.**

Pimento, Amish - 85-90 days, Extra sweet pepper, 2 x 4 in. squat ribbed fruit on compact 1-2 ft. plant, thick sweet flesh with fruity taste, ripens bright red, for fresh eating, stuffing, roasting or pickling, rare Amish heirloom. *Source History: 3 in 2004.* **Sources: Ag7, Ami, Fe5.**

Pimento, Canada Cheese - 75 days, Large Sweet Cherry type, 1.25 x 2 in. dia., small fruits turn red at maturity, excel. for pickling. *Source History: 1 in 1981; 2 in 1984; 2 in 1987; 2 in 1991; 2 in 1994; 2 in 1998; 2 in 2004.* **Sources: Gr29, Pep.**

Pimento, Choco - (Choco Pepper) - 60-80 days, Pimento type, beautiful shiny chocolate-brown fruits, tapered shape, 3-4 in. long, sweet mild flavor, hearty plants produce abundantly, excel. for stuffing or salads. *Source History: 1 in 1981; 2 in 1984; 2 in 1987; 4 in 1991; 4 in 1994; 2 in 1998; 1 in 2004.* **Sources: Fo7.**

Pimento, Gambo - 90 days, Popular tomato-shaped pimento, flat shiny deep-red fruits, 2.5 in. deep x 4 in. dia., mostly 4- lobed, crisp flesh, very juicy and sweet, cooking enhances the sweetness, med-tall vigorous plants. *Source History: 1 in 1981; 1 in 1984; 2 in 1987; 2 in 1991; 3 in 1994; 2 in 1998; 1 in 2004.* **Sources: So1.**

Pimento, Morrow - (Calwonder) - *Source History: 1 in 1987; 1 in 1991; 1 in 1994; 1 in 1998; 1 in 2004.* **Sources: Pep.**

Pimento, Perfection - (True Heart) - 72-80 days, Very sweet and mild, heart-shaped, 2-2.5 x 3 in. dia., very thick smooth walls, dark-green > dark-red, upright and pendant, dark-green 28 in. plants, good canner. *Source History: 19 in 1981; 19 in 1984; 20 in 1987; 13 in 1991; 8 in 1994; 5 in 1998; 3 in 2004.* **Sources: Ge2, SGF, So1.**

Pimento, Red Cheese - 75-82 days, Huge flat squash-shaped fruits, 2.25 x 4 in. dia., thick fleshed, excel. for stuffing, green > red. *Source History: 1 in 1981; 1 in 1984; 3 in 1998; 4 in 2004.* **Sources: He8, ME9, Orn, SE14.**

Pimento, Red Ruffled - 70-85 days, The best of the pimento types, 3 x 2.5 in., thick-walled fruits mature to a dark-red, 8-10 fruits per plant. *Source History: 1 in 1994; 3 in 1998; 3 in 2004.* **Sources: Pe2, Se7, So12.**

Pimento, Select - 72-75 days, Thick-walled heart-shaped 3.5 x 2.5 in. smooth fruits, bright-green > deep-red, mild, upright 33 in. plants, concentrated set, excel. freezer, for home or market. *Source History: 13 in 1981; 11 in 1984; 8 in 1987; 6 in 1991; 3 in 1994; 4 in 1998; 1 in 2004.* **Sources: Hud.**

Pimento, Sheepnose - 70 days, Plants 22 in. tall by 16 in. wide, sparse foliage, cheese pimento shaped red fruits, 3 in. deep x 4 in. dia., very meaty and good for canning, cold tolerant, Ohio heirloom. *Source History: 1 in 1994; 2 in 1998;; 4 in 2004.* **Sources: He8, Pr9, Se16, Ter.**

Pimento, Sunnybrook - (Sweet Cheese, Burpee's Sunnybrook) - 65-73 days, Tomato-shaped pepper, 2-2.5 x 3 in. dia., smooth sweet and very mild, deep-green > scarlet, upright dark-green 23-28 in. plants. *Source History: 9 in 1981; 4 in 1984; 5 in 1987; 5 in 1991; 6 in 1994; 5 in 1998; 5 in 2004.* **Sources: Gr29, Ra6, Re8, Se28, To3.**

Pimento, Super Red - 70 days, Squash or flattened pimento type, earlier larger TMV tol. version of the original, huge 5.75 in. wide x 3.25 in. deep fruit ripens green to red, sweet .5 in. thick flesh. *Source History: 1 in 1987; 4 in 1991; 4 in 1994; 3 in 1998; 4 in 2004.* **Sources: Pep, Pr3, Sto, To1.**

Pimento, Yellow Cheese - 73-90 days, Large squash-shaped fruits, 25% larger than Sunnybrook Pimento, contrasts well with red or green cheese types in baskets or relish, green > yellow and orange. *Source History: 1 in 1981; 2 in 1984; 4 in 1987; 7 in 1991; 5 in 1994; 6 in 1998; 7 in 2004.* **Sources: He8, Ho13, ME9, Pep, Ra6, SE14, Sto.**

Pimiento, Ashe County Heirloom - 70 days, Bright red, thick fleshed squat fruit, ideal for stuffing, sweet flavor, great raw, retains texture and flavor when cooked, roasted or canned, can tolerate cool evenings and morning dew, drought tolerant, from a seed saver in North Carolina. *Source History: 1 in 2004.* **Sources: Rev.**

Rotunda - Flattened, round, dark red fruit, 3-4 in., from Russia. *Source History: 1 in 1998; 1 in 2004.* **Sources: Ho13.**

Roumanian Rainbow - 60 days, Short, compact plants, fruit starts out ivory > persimmon orange > red, all three colors on the plant at one time, 4-5 in. long, sweet flavor. *Source History: 1 in 1998; 2 in 2004.* **Sources: Ta5, To1.**

Roumanian Sweet - 60-80 days, Med-walls, 4-4.5 x 2-2.5 in., tapers to blunt point, smooth, yellow > bright-red, ribs sometimes slightly pungent, upright 22-24 in. plants, conc. set, good pickler. *Source History: 15 in 1981; 16 in 1984; 10 in 1987; 12 in 1991; 9 in 1994; 9 in 1998; 12 in 2004.* **Sources: CO30, Ge2, Gr29, La1, ME9, Pep, Ra6, Re8, RUP, SE14, Sh9, WE10.**

Round of Hungary - 55-75 days, Pimento cheese pepper with sweet thick flesh, ribbed, flattened shape, stuffing, cooking and salads, from Switzerland. *Source History: 2 in 2004.* **Sources: Jo1, Sw9.**

Shishito - (Shisatou) - Slightly wrinkled, cylindrical med-green fruits, 7 in. long x 2 in. dia., mild flavor, productive spreading plants, vigorous under most soil conditions, heat res., for home gardens, from Japan. *Source History: 1 in 1981; 1 in 1984; 2 in 1994; 2 in 1998; 3 in 2004.* **Sources: Ev2, Ho13, Jo6.**

Shishitou - 70-80 days, Japanese sweet pepper, smooth crispy texture, somewhat milder flavor than the Korean Hot pepper, wrinkled 3 x .5 in. dia. fruits, early. *Source History: 1 in 1981; 2 in 1984; 2 in 1987; 1 in 1991; 2 in 1994; 2 in 1998; 4 in 2004.* **Sources: Gr28, Gr29, Kit, Pep.**

Sigaretta per Sotto Aceti - *Source History: 1 in 1987; 1 in 1991; 1 in 1994; 1 in 1998; 1 in 2004.* **Sources: Ber.**

Spicy Puerto Rican No-Burn - Plant bears an abundance of round red pods which resemble Christmas tree ornaments, neither spicy nor very sweet. *Source History: 1 in 1994; 1 in 1998; 1 in 2004.* **Sources: We11.**

Super Shepherd - 68 days, Most popular Italian sweet pepper, dark-red 7.5 in. fruits, tapers to blunt point, thick sweet juicy flesh, distinct flavor, perfect size to process or fry. *Source History: 3 in 1981; 3 in 1984; 5 in 1987; 4 in 1991; 4 in 1994; 5 in 1998; 11 in 2004.* **Sources: Ber, Ga21, Gr29, Pep, Pr3, Ra6, Sa5, Se17, So1, Sto, Up2.**

Swallow - Rounded fruit tapers to a point, 3 x 4 in. thick walls, light green > red, Russia. *Source History: 1 in 1998; 1 in 2004.* **Sources: Ho13.**

Sweet Banana - (Yellow Banana, Long Sweet Hungarian, Hungarian Wax Sweet, Hungarian Banana, etc.) - 58-75 days, Catch-all group for Hungarian Yellow Wax Sweet types, known by a dozen slightly varying names, upright prolific 16-24 in. plant, 6 x 1.5-2 in. dia., tapers to blunt point, thick sweet mild waxy flesh, pale-green > yellow > orange > red, pendant fruiting habit, from Hungary, 1941. *Source History: 110 in 1981; 97 in 1984; 102 in 1987; 89 in 1991; 85 in 1994; 88 in 1998; 93 in 2004.* **Sources: ABB, Ada, Alb, BAL, Be4, Bo19, Bu2, Bu3, CA25, CH14, CO23, CO30, Co31, Co32, Com, Cr1, Dam, De6, Dom, DOR, Ear, Enc, Ers, Fi1, Fo7, Fo13, Ge2, GLO, Gr27, Gr29, GRI, Gur, HA3, Ha5, He8, He17, HO2, HPS, Hum, Jor, Kil, La1, LO8, Loc, MAY, ME9, Mel, Mey, MIC, Mo13, Mo20, MOU, Ni1, OLD, Ont, Or10, OSB, PAG, Pep, Pin, Ra6, Re8, Red, RIS, Roh, Ros, RUP, Sa5, Sau, Se8, SE14, Se25, Se26, Se28, SGF, Sh9, Sh14, Shu, Si5, Sk2, So1, So25, SOU, Sto, Te8, To1, To3, Twi, Vi4, WE10, Wet, Wi2, WI23.**

Sweet Banana, Early - 60-65 days, Dwarf Sweet Banana str., compact bushy plant, earlier, much more productive, 6 x 1.75 in. at stem end, tapers to point, pale-yellow > red, everbearing, pendant. *Source History: 3 in 1981; 2 in 1984; 1 in 1987; 2 in 1991; 1 in 1994; 1 in 1998; 1 in 2004.* **Sources: Pep.**

Sweet Banana, Giant Yellow - 60 days, Unique large sweet banana type from Hungary, purple stemmed plants loaded with 7 in. pendant fruits, thick flesh, easy to pick. *Source History: 2 in 1991; 2 in 1994; 1 in 1998; 1 in 2004.* **Sources: Pep.**

Sweet Cayenne - 75 days, Thin walled, long, cayenne shaped fruit grows to 1 ft. long, crimson red when ripe, sweet crunchy flesh, same uses as a frying pepper. *Source History: 1 in 1998; 5 in 2004.* **Sources: Ma18, ME9, SE14, To1, WE10.**

Sweet Melrose - (Melrose, Italianelle) - 60-70 days, Sweet Italian frying pepper, 4 in. long x 1-1.5 in. at shoulder, tapers to a point, thin walls, crinkled flesh, dark-green > red. *Source History: 3 in 1981; 3 in 1984; 3 in 1987; 3 in 1991; 2 in 1994; 2 in 1998; 5 in 2004.* **Sources: Co31, Ge2, Re8, Se28, Und.**

Sweet Melrose, Improved - (Italianelle Sweet Frying) - 60 days, Excellent frying pepper, tapered fruits 3-4 in. long, 1 in. dia., thin crinkled flesh, dark green > deep red, heavy yields. *Source History: 1 in 1994; 1 in 1998; 1 in 2004.* **Sources: Ra6.**

Sweet Pickle - 65-85 days, Ornamental bedding plant, 12-15 in. tall, covered with clusters of 2 in. oval fruits, yellow > orange > red > purple, sweet thick walls, for pickles or salad use. *Source History: 2 in 1981; 1 in 1984; 2 in 1987; 4 in 1991; 5 in 1994; 5 in 1998; 10 in 2004.* **Sources: Gr28, Gr29, Ho13, ME9, OSB, Par, Pep, Sa9, SE14, To1.**

Sweet Spanish - (Doux Espagne) - Long sweet pepper, very popular in France. *Source History: 1 in 1981; 1 in 1984; 2 in 1987; 1 in 1991; 1 in 1994; 1 in 1998; 1 in 2004.* **Sources: WE10.**

Sweet Wrinkled Old Man - (Petit Star) - Wrinkled green fruit, 2-3 in. long .5 in. across, very popular in Japanese and Caribbean cooking. *Source History: 1 in 2004.* **Sources: Red.**

Szegedi - (Giant Szegedi, Giant Szegedi Hungarian Yellow) - 70-75 days, Paprika pepper, larger than Roumanian, better foliage than Dutch Treat, up to 10 in. long x 3 in. across at shoulder, pale yellow > deep orange, tapers slightly, wilt res., dev. in Szeged, Hungary. *Source History: 1 in 1981; 1 in 1984; 1 in 1987; 5 in 1991; 7 in 1994; 7 in 1998; 8 in 2004.* **Sources: Gr29, Ho13, Pep, Ra6, Se7, Sto, To1, We11.**

Tangerine - 70-85 days, Selected from Yellow Cheese Pimento, 3 in., round and flattened, green > bright orange fruit, thick walls. *Source History: 3 in 1994; 2 in 1998; 3 in 2004.* **Sources: Ho13, Ma18, Se7.**

Taormina - Italian heirloom, 4-5 ft. plants, very sweet, tapered red fruit, 3 x 8 in. *Source History: 1 in 1998; 1 in 2004.* **Sources: Ho13.**

Tennessee Cheese - 75-82 days, Pimento type, round 3 in. red fruit with a flattened base, thick skin, unique flavor, heirloom from Spain where it is still used to make high quality paprika. *Source History: 3 in 1998; 9 in 2004.* **Sources: Ho13, Ma18, OSB, Pep, Ra6, Re8, Ri12, To3, Ver.**

Tequila Sunrise - 77-100 days, Ornamental but delicious pepper, 12-14 in. tall x 12 in. wide plants with upright habit, deep- green 4-5 in. long fruits ripen to golden-orange, firm crunchy thin flesh with sweet, slightly sharp flavor when ripe, good for adding texture and color to salsa, abundant yields. *Source History: 1 in 1984; 3 in 1987; 3 in 1991; 2 in 1994; 1 in 1998; 8 in 2004.* **Sources: He8, Pep, Pr9, Sa5, Se16, So1, Ta5, Ter.**

Tollies Sweet Italian - 75-85 days, Good abundance of 4-5 in. long tapered fruit, green > deep glossy red, fresh eating and canning, sweet flavor, Italian heirloom. *Source History: 1 in 2004.* **Sources: Se16.**

Topepo Rosso - Flattened round red fruit, 2 x 2 in., sweet flavor, eaten fresh, roasted or pickled. *Source History: 4 in 2004.* **Sources: Red, Se24, Se26, So25.**

Topito Cheese - (Tondo Topito) - 70-78 days, Heirloom from Johnson County, North Carolina, great stuffing pepper with smooth, uniform fruits, apple shape with slightly flattened bottoms, extremely thick walls, sweet, crisp flesh. *Source History: 3 in 1994; 2 in 1998; 1 in 2004.* **Sources: Fo7.**

Weaver's Mennonite Stuffing - Old Pennsylvania heirloom via William Woys Weaver. *Source History: 1 in 2004.* **Sources: Ta5.**

White Cheese, Giant - 60-80 days, Early cheese pepper, white fruit, 1 x 2.25 in. *Source History: 1 in 1998; 1 in 2004.* **Sources: Sa9.**

Yesil Tatli - Early maturing fruit resembles a cayenne .5 in. wide x 6 in. long, yellow-green > vibrant red, mild flavor, originally from Istanbul. *Source History: 1 in 1998; 2 in 2004.* **Sources: Ho13, Se17.**

Varieties dropped since 1981 (and the year dropped):

Agronomico (1987), *Andean* (2004), *Belconi* (1998), *Burpee's Early Pimento* (1991), *Calahorra (Morron de Espana)* (1991), *Carliston Sweet* (2004), *Carnosissimo di Cuneo* (1998), *Carousel* (2004), *Casca Dura Ikeda* (1998), *Cherrytime* (2004), *Cubanelle Improved* (1994), *Douce de Espagne* (2004), *Dulce de Espana* (1994), *Dutch Treat* (1991), *Dwarf Romanian* (1987), *Equadorean Indian Spherical* (1987), *Espanto* (1983), *Gift from Moldova* (2004), *Greygo Baby* (2004), *Horizon* (1998), *Iberian* (2004), *Ise* (1991), *Italian Gold* (1998), *Jericho* (1998), *Kashmir, Giant Long* (2004), *Kishinev* (2004), *Kisser Pepper* (2004), *KRG No. 4* (1998), *Little Dickens* (2004), *Long Red Mild* (2004), *Long Thick Red Sweet* (1984), *Long Yellow Sweet* (1998), *Maor* (1994), *Mayata* (1998), *Mild California* (1987), *Mild California* (1987), *Mole Sweet* (2004), *Montego* (1998), *Naples* (1987), *Old Italian* (1994), *Orange Sweetie* (1998), *Papri Mild II* (2004), *Papri Sweet* (1998), *Paradicsom Alaku Sarga Szentesi* (1998), *Paradicsom Alaku Zold Szentesi* (2004), *Peperone Sigaretta* (1994), *Petite Yellow* (1994), *Pimento (Sunnybrook #4)* (1991), *Pimento Grande* (2004), *Pimento, de Cocinar* (1998), *Pimento, Figaro* (2004), *Pimento, Large Red* (1994), *Pimento, Round Tomato-Shaped* (2004), *Pinocchio* (1984), *Pusztagold* (1987), *Red Belgian* (2004), *Rio Grande 66* (1994), *Romanian White Gypsy Sweet* (1998), *Roumanian Sweet, Bullnose* (1998), *Sari Sivri Sweet* (1994), *Sari Tatli Sweet* (1994), *Shepherd* (1991), *Sicilian Giant Yellow* (1987), *Sovereign* (1998), *Spartan Garnet Pimento* (1987), *Sweet Petite* (1987), *Tomato Pimento* (1987), *Topaz, Yellow* (2004), *Topito* (1998), *Triton* (1991), *Trottolino Amoroso* (1998), *Twiggy* (1991), *Viejo Arruga Dulce* (2004), *Vinedale* (1991), *W H B 83* (1987), *White Cloud* (1998), *Yellow Wax Jugo* (2004), *Zapotek Paprika* (1994).

Pepper / Other Species
Capsicum spp.

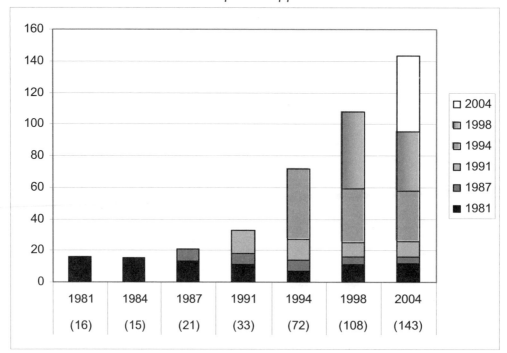

Number of varieties listed in 1981 and still available in 2004: 16 to 12 (75%).
Overall change in number of varieties from 1981 to 2004: 16 to 143 (894%).
Number of 2004 varieties available from 1 or 2 sources: 98 out of 143 (69%).

<u>African Surprise</u> - Ornamental plant with colorful, hot fruit, from Gabon, Africa. *Source History: 1 in 2004.* **Sources: Fe5.**

<u>Aji Amarillo</u> - (Aji Orange) - 80-90 days, *C. baccatum* var. *pendulum*, 3-5 in. cylindrical orange fruits, medium to hot, spicy smoky flavor, eaten fresh or dried and ground into a powder, South America. *Source History: 1 in 1994; 3 in 1998; 7 in 2004.* **Sources: Ba16, Enc, Ri12, Se7, So25, Te8, We11.**

<u>Aji Andean</u> - *Source History: 1 in 2004.* **Sources: Se17.**

<u>Aji Benito</u> - Hot pepper from Bolivia, robust plant. *Source History: 1 in 2004.* **Sources: Te8.**

<u>Aji Brown</u> - *C. baccatum*, sweet, mild flavor, 5-7 in. long, brown at maturity. *Source History: 1 in 1994; 4 in 1998; 2 in 2004.* **Sources: Enc, Ho13.**

<u>Aji Cito</u> - Good production of hot peppers, from South America. *Source History: 1 in 2004.* **Sources: Te8.**

<u>Aji Colorado</u> - 75-85 days, *C. baccatum*, sprawling plant to 2 ft., elongated, orange-red, 3-5 in. fruit, pungent flavor, from the Andes. *Source History: 1 in 1994; 1 in 1998; 3 in 2004.* **Sources: Ba16, Ri12, Se7.**

<u>Aji Cristal</u> - 90-100 days, *C. baccatum*, prolific set of 3.5 in. long tapered fruit, light cream-green > orange-red, best flavor when immature, originated in Curico, Chile. *Source History: 6 in 2004.* **Sources: Ba16, Pr9, Se16, Se17, Se26, To3.**

<u>Aji Dulce</u> - 110 days, *C. chinense*, 2 ft. plants, pendant, muffin shaped fruit, 1 in. wide x 2 in. long, green > red, same fruity flavor and aroma of Habanero with little or no heat, good flavor, from Venezuela. *Source History: 2 in 1998; 5 in 2004.* **Sources: Ho13, Pep, So1, So25, Und.**

<u>Aji Habanero</u> - *C. baccatum*, pendant 2-3 in. pods with variable shape, fruity aroma, hot, matures to golden yellow, from Chile. *Source History: 1 in 2004.* **Sources: Ba16.**

<u>Aji Omnicolor</u> - *C. baccatum*, low sprawling plant bears 2-3 in. flat fruit, teardrop shape, ripens from yellow-white with slight purple tinge to deep coral-orange red, crisp flesh, hot, fruity flavor, good for hot sauce, salsa and salads, Peru. *Source History: 1 in 2004.* **Sources: Ba16.**

<u>Aji Panca (Peru)</u> - Tapered, mild fruit, 1 in. wide by 5 in. long, red at maturity. *Source History: 1 in 1994; 1 in 1998; 1 in 2004.* **Sources: Gr29.**

<u>Aji Peruvian</u> - *C. baccatum*, tall branching plant to 5 ft., 2-3 in. long pods, green > deep red, hot medium thin flesh with a slow lingering heat, good fresh or dried, Peru. *Source History: 1 in 2004.* **Sources: Ba16.**

<u>Aji Red</u> - (Aji Rojo) - 85-90 days, *C. baccatum*, perennial in some areas, wrinkled fruit, 3-5 in. long, orange-red peppers are generally dried into powder and used in sauces and stews, often pickled at green stage when mildly hot, hotter when red, good in salsa with lime, 30,000-50,000 Scoville units, tolerates light frosts, Peruvian pepper from the Inca Empire, still cultivated in the foothills of the Andean Sierras. *Source History: 1 in 1998; 3 in 2004.* **Sources: Se17, So25, To1.**

<u>Aji Umba</u> - *C. chinense*, light lime-green > yellow-

orange, heat, flavor and aroma resembles that of Habanero, fruit shape is more box-like, similar uses including hot sauces, salsas, drying and smoking, Suriname. *Source History: 1 in 2004.* **Sources: Ba16.**

Aji Verde - *C. baccatum*, med. size branching plant, long fruit to 4-5 in., green > red, nice heat with slightly fruity taste, salsa, stuffing, roasting and smoking, Peru. *Source History: 1 in 1998; 2 in 2004.* **Sources: Ba16, Pep.**

Aji Yellow - (Kellu Uchu) - *C. baccatum*, 2-3 in. long fruit, fairly mild when picked green, very hot when red, 30,000-50,000 Scoville heat units, somewhat frost res., favorite pepper of the Incas. *Source History: 1 in 1994; 4 in 1998; 5 in 2004.* **Sources: Gr29, Ho13, Pep, Red, To1.**

Aji Yuquitania - (Amazonian Chile) - Perennial 4 ft. bush in the tropics, more adapted to heavy rain and poor soil than most chiles, grown since ancient times for use as a spice by the Cubeo natives of the Rio Vaupes, Columbia, hot peppers were fire- roasted then slowly dried for months in the rafters above the cook fires, result is a smoke-cured flavor which is aromatic and savory even above the pungency, intro. by Andrew Weill in late 1970s as powdered spice to benefit the tribe, so hot it could not be measured on the Scoville Scale for chile hotness, Cubeo people have since become extinct. *Source History: 1 in 1991; 1 in 1994; 1 in 2004.* **Sources: We11.**

Amarillear - 100-120 days, *C. baccatum*, conical, yellow-orange, very hot fruit grows upright on 3 ft. plant, from Bolivia. *Source History: 1 in 1998; 2 in 2004.* **Sources: Gr29, Pep.**

Api - (Cabai Api) - *C. frutescens*, 4 ft. tall plant, 2 ft. across, produces an abundance of upright pods, sometimes purple- tinged, 2.5 in. long x .25 in. wide, hot, used at green or red-ripe stage in Chinese cooking to give dishes a fragrant piquancy, originally from China. *Source History: 1 in 1998; 1 in 2004.* **Sources: Red.**

Aribibi Gusano - Bolivian "Caterpillar Pepper". *Source History: 1 in 2004.* **Sources: Te8.**

Assam - *C. frutescens*, one of the world's hottest peppers, very thin walled orange-red fruit, 3-6 in. long .75 in. across, from India. *Source History: 1 in 1998; 1 in 2004.* **Sources: Red.**

Azr - *C. frutescens*, 4 ft. tall plants with attractive fiery purple, ridged pods, red when ripe, 2-3 in. long x .25 in. across, used in India's hot dishes, as hot as Habanero. *Source History: 1 in 1998; 1 in 2004.* **Sources: Red.**

Bailey Chile Piquin - Imp. piquin var. with distinctive taste, small fruit, less than .5 in. wide x 1 in. long, very hot. *Source History: 1 in 1998; 1 in 2004.* **Sources: Enc.**

Balada - Abundant production of 4-5 in. long tapered fruit, hot. *Source History: 1 in 2004.* **Sources: Te8.**

Bangalore Torpedo - *C. frutescens*, narrow fruit, 5 in. long, light green > red, fruits dry nicely on the plant. *Source History: 1 in 1998; 1 in 2004.* **Sources: Red.**

Barbados - (Bishop's Crown, Monks Cap, Orchid,) - 100 days, *C. baccatum*, large plant bears 3-sided pods, green > orange > red, med-hot, late maturing, fruit shape resembles a bishop's crown. *Source History: 1 in 1994; 2 in 1998; 5 in 2004.* **Sources: Ba16, Pep, So25, Te8, We11.**

Beni Highlands - Abundant production of lemon-yellow fruit, Bolivia. *Source History: 1 in 2004.* **Sources: Te8.**

Bird Aji - *C. baccatum*, pea size pods with slightly fruity flavor, green > orange > orange-red, resembles Tepin in size but is more juicy, liked by birds, from the wilds of Bolivia. *Source History: 1 in 2004.* **Sources: Ba16.**

Bird's Eye - (Mexican Peanuts, Trinidad Bird, Tepin, Chiltepin) - Extremely hot, tiny .25 in. bead-like peppers, grows wild in Texas and the Southwest, reportedly so hot that its juice will blister your fingers, difficult to germinate, propagated by birds. *Source History: 2 in 1981; 2 in 1984; 1 in 1987; 1 in 1991; 1 in 1994; 2 in 1998; 4 in 2004.* **Sources: Gr29, Pep, Te8, WE10.**

Birgit's Locoto - *C. baccatum*, 2-3 in. pods, lime-green > orange > orange-red, crisp fruity flavor, excel. for fresh use, hot sauce or smoked, Bolivia. *Source History: 1 in 2004.* **Sources: Ba16.**

Bode - *C. chinense*, round .5 in. fruit, very mild flavor, ripens red, brought to the U.S. by a Brazilian exchange student. *Source History: 1 in 1998; 1 in 2004.* **Sources: Ho13.**

Bolivian Rainbow - 75-95 days, *C. frutescens*, beautiful everbearing plants with purple foliage and flowers, 2-4 ft. tall, tiny pointed purple fruits, rainbow multi-colored > purple, very hot, grown in Bolivia. *Source History: 1 in 1994; 8 in 1998; 9 in 2004.* **Sources: Ba16, Fo7, Ho13, Pe7, Ri12, Se7, So25, To1, Tu6.**

Bradley's Bahamian - *C. frutescens*, large branching plant bears 1.25 in. pods, thin juicy flesh, similar to Tabasco with less heat, lime yellow-green > orange > red, makes an excellent hot sauce, Bahamas. *Source History: 1 in 2004.* **Sources: Ba16.**

Brazil - 80 days, Tall thin plant to 36 in., thin walled 2 in. fruit on long upright stems. *Source History: 1 in 2004.* **Sources: Pep.**

Brazilian Malagueta - *C. frutescens*, South American form of Tabasco, 4 ft. tall plants, 2 ft. across, upright 1-2 in. fruit, dark green > red. *Source History: 1 in 2004.* **Sources: Red.**

Brazilian Starfish - *C. baccatum*, pods shaped like a spaceship, green > orange > orange-red, fruity flavor, variable heat, Brazil. *Source History: 1 in 2004.* **Sources: Ba16.**

Brown Congo - *C. chinense*, green > chocolate-brown fruit, super hot, originated in Trinidad. *Source History: 1 in 1994; 1 in 1998; 2 in 2004.* **Sources: Pe7, Te8.**

C. baccatum var. pendulum aji - Very hot, red fruit, 3 x 1 in., perennial in warm climates. *Source History: 1 in 1998; 1 in 2004.* **Sources: Ho13.**

C. baccatum var. valentine - Prolific plant bears hundreds of small round pods, yellow > orange > red, beautiful edible ornamental. *Source History: 1 in 1994; 1 in 1998; 1 in 2004.* **Sources: We11.**

C. chacoense - Growth habit of plant resembles an elm tree, very hot, small fruits, CV Ho13 offers a red and yellow variety. *Source History: 1 in 1994; 1 in 1998; 1 in 2004.* **Sources: Ho13.**

Cabai Burong - (Cabai Barong) - Long red, hot fruit, from Malaysia. *Source History: 1 in 1994; 2 in 1998; 2 in 2004.* **Sources: Gr29, Pep.**

Cachucha - *C. chinense*, 1.25 x 1-2 long in. pods, light green > pale yellow-orange, mild flavor, resembles Habanero but without the heat, similar uses, Mexico. *Source History: 1 in 2004.* **Sources: Ba16.**

Calusa Indian Mound - *Source History: 1 in 2004.* **Sources: Se17.**

Cayenne, French - (Piccante di Cajenne) - *C. frutescens*, 4 in. long fruit .5 in. across, thick flesh. *Source History: 1 in 2004.* **Sources: Red.**

Cay-Viet - Hot pepper from Vietnam, 1.5 in. *Source History: 1 in 2004.* **Sources: Te8.**

Centennial - 130 days, Tiny pods, purple > white > red, all colors on the plant at one time, edible at all stages, named for the colors of the flag, good houseplant. *Source History: 3 in 2004.* **Sources: Gr29, Pep, Pla.**

Chandigarh - *C. frutescens*, East Indian pepper, plants grow 3 ft. tall by 2 ft. across, pendant pods, 3 in. long, heavy producer. *Source History: 1 in 2004.* **Sources: Red.**

Chiltepin - 80-95 days, Very hot miniature chili pepper usually harvested in the wild in the southwestern U.S., oval fruit, green > red, good houseplant, native Sonoran perennial. *Source History: 7 in 1998; 12 in 2004.* **Sources: Bu2, Enc, He8, Hi13, Ho13, ME9, Pla, SE14, Se17, Te8, To1, To3.**

Chiltepines, Hermosillo Select - Large fruited variety from Sonora, Mexico, plants grows 3-4 ft., .75 in. red fruit, extremely hot. *Source History: 1 in 1994; 1 in 1998; 1 in 2004.* **Sources: Ho13.**

Chiltepines: Sinaloa - (Tepin) - 95 days, Wild fruits from Sinaloa, Mexico, bushy plants produce round pods .25-.5 in., grows in partial shade but prefers full sun, has an intense burst of blistering heat that dissipates quickly, 50,000-100,000 Scoville heat units, ripe pods are dried and ground into powder, will grow in partial shade, does well in containers, sold by Southern Mayos at the railroad crossing between San Blas and El Fuerte in Sinaloa, Mexico. *Source History: 1 in 1994; 1 in 1998; 1 in 2004.* **Sources: Bo9.**

Christmas - 90 days, *C. baccatum*, compact 8 in. plant, cone-shaped, small fruit, green > purple > red. *Source History: 2 in 1994; 2 in 1998; 1 in 2004.* **Sources: Ho13.**

Cleo's Dragon - *C. chinense*, similar to a large red habanero, very hot. *Source History: 1 in 1998; 1 in 2004.* **Sources: We11.**

Cobincho - *C. exile*, plants 2 ft. tall, purple stems, bullet shaped, red fruit, very hot. *Source History: 2 in 1998; 1 in 2004.* **Sources: Ho13.**

Cobra - *C. frutescens*, very hot pepper from India, plants 4 ft. tall, 2 ft. across, leaves protect fruit from sunburn, hanging pods, 3-4 in. long, very productive, pods dry nicely on plants. *Source History: 1 in 1998; 1 in 2004.* **Sources: Red.**

Congo Trinidad - *C. chinense*, very hot pepper from Trinidad, 2 in. pendant pods ripen green > yellow > orange > red with all colors on the plant at the same time, fruity flavor, same uses as Habanero. *Source History: 1 in 2004.* **Sources: Ba16.**

Cumari - *C. chinense*, branching plant, very hot pea-size pods ripen yellow, aromatic, fruity flavor, Brazil. *Source History: 1 in 2004.* **Sources: Ba16.**

Dagger Pod - *C. frutescens*, very thin, slightly curved fruit, 2-3 in. long x .25 in. across, hot. *Source History: 1 in 1998; 1 in 2004.* **Sources: Red.**

Deco de Moca - *C. baccatum*, name translates to Little Girl's Finger, branched plants bear 3 in. pods with a smoky fruity flavor, excel. for salsa, sauce and smoked, Brazil. *Source History: 1 in 2004.* **Sources: Ba16.**

Dong Xuan Market - Vietnamese pepper with classic Oriental pungency. *Source History: 1 in 2004.* **Sources: Te8.**

Fatalii - 90 days, *C. chinense*, 24-30 in. plant, bright yellow shrivelled fruit with a distinctly pointed end, 2-3 in. long, excel. citrus flavor but extremely hot, good for container culture, from the central African Republic. *Source History: 1 in 1998; 3 in 2004.* **Sources: Pe7, Se16, To1.**

Flame Fountain - *C. frutescens*, long, pointed chile with very hot flesh, ripens deep red, India. *Source History: 1 in 1994; 2 in 1998; 1 in 2004.* **Sources: Red.**

Fountain Red - *C. frutescens*, 4 ft. tall plant, 3 ft. across, variable fruit size and hotness. *Source History: 1 in 2004.* **Sources: Red.**

Goat's Weed - Plants to 4.5 ft., leaves with fuzzy pubescence, upright, tapered, 1.5 in., pointed fruits, deep biting heat, green > black > red, dries well, from Venezuela. *Source History: 1 in 1998; 2 in 2004.* **Sources: Ba16, Ho13.**

Grove - *C. frutescens*, plant is multi-limbed at the base, light-green > orange, fruit shape similar to Pequin. *Source History: 1 in 1994; 1 in 2004.* **Sources: Pep.**

Habanero - 90-100 days, *C. chinense*, native to the Yucatan, 36 in. plants, tapered lantern-shaped 1 x 1.5 in. pods, thin wrinkled light-green flesh ripens to a lovely golden-orange, reportedly 1,000 times hotter than Jalapeno, 200,000 Scoville heat units, slow to germinate, must be grown in warm moist conditions, a Caribbean favorite used in sauces. *Source History: 6 in 1987; 17 in 1991; 55 in 1994; 82 in 1998; 97 in 2004.* **Sources: Ba16, BAL, Ban, Bo9, Bo18, Bo19, Bu1, Bu2, Bu3, Bu8, CA25, CH14, Co31, Com, Coo, Enc, Eo2, Ers, Fe5, Fi1, Fo7, Fo13, GLO, Goo, Gr29, GRI, Gur, HA3, Ha5, He8, He17, Ho13, HPS, Hud, Jo1, Jo6, Jor, Jun, Ki4, Kil, La1, Lej, LO8, Ma18, MAY, Mel, Mey, Mi12, MIC, Mo13, MOU, Na6, Ni1, OLD, Or10, Orn, OSB, PAG, Pe2, Pep, Pin, Pla, Ra5, Ra6, Red, Ri2, Ri12, RIS, Roh, Ros, RUP, Sa9, Scf, SE4, Se7, Se8, SE14, Se16, Se25, Se26, SGF, Shu, So1, Sto, Sw9, Te8, Ter, Tho, Thy, To1, To3, Twi, Vi4, Vi5, WE10, Wi2, Wo8.**

Habanero, Brown - *Source History: 4 in 2004.* **Sources: Bo19, ME9, SE14, TE7.**

Habanero, Caribbean Red - 90-110 days, *C. chinense*, compact 30 in. plant, 1 x 2 in. wrinkled fruit is nearly twice as hot as most commercial habanero vars., 445,000 Scoville heat units, lime-green > red. *Source History: 14 in 1998; 34 in 2004.* **Sources: BAL, Bu2, CA25, CH14, Dam, Ers, Gur, HO2, HPS, Jo1, Jo6, Jun, Ki4, Kil, La1, Ma18, MAY, Mo20, Ni1, Pe7, Pep, Ra6, Re8, RIS, Scf, Se26, Shu, Te8, To1, To3, Tt2, Ver, WE10, We19.**

Habanero, Chocolate Congo - 90 days, *C. chinense*, spreading, 3-4 ft. plant, rounded, convoluted, somewhat squat, 2 in. fruit, orange-chocolate color, very hot, from Trinidad. *Source History: 1 in 1998; 3 in 2004.* **Sources: Ho13, Pep, Red.**

Habanero, Golden - 90-100 days, *C. chinense*, incredibly hot pepper grows on a large plant, tapered fruit, 1.5 x 3 in., deep gold when ripe, use rubber gloves when handling, 40 times hotter than Jalapeno. *Source History: 1 in 1994; 5 in 1998; 9 in 2004.* **Sources: Ami, Ga1, HO2, Ho13, Jor, ME9, Pe7, SE14, Sk2.**

Habanero, Orange - (Scotch Bonnet) - 90-100 days, *C. chinense*, Caribbean variety with fiery hot fruits, beautiful 3-4 ft. plant, late. *Source History: 1 in 1994; 3 in 1998; 7 in 2004.* **Sources: Ba16, ME9, Mi12, Mo20, Pe7, Te8, We11.**

Habanero, Red - (Lucifer's Dream) - 90-100 days, *C. chinense*, an even hotter str. of Habanero, large bright red fruit, 300,000+ Scoville heat units, do not grow where children may play. *Source History: 4 in 1998; 12 in 2004.* **Sources: Ba16, Co31, Enc, Ga1, HA3, Ho13, ME9, Pr9, RUP, Te8, To1, Vi4.**

Habanero, Tazmanian - *C. chinense*, shorter, more bushy plant produces ultra-hot fruit, Pepper Joe original. *Source History: 1 in 2004.* **Sources: Pe7.**

Habanero, White - *C. chinense*, 3.5 ft. branched plant, creamy white, 1-2 in. pods, very hot, late, rare. *Source History: 1 in 1994; 2 in 1998; 8 in 2004.* **Sources: Ba16, Bo19, Ho13, ME9, Pep, Ra6, SE14, We11.**

Habanero, White Bullet - *C. chinense*, most rare, hottest and earliest maturing of the Habanero group, white fruit, 1 in. long by .5 in. wide, can produce up to 1,000 fruit per plant for a total of five pounds, not available for shipment out of the U.S., trademarked. *Source History: 1 in 2004.* **Sources: Red.**

Hawaiian - *C. frutescens*, grown throughout the islands and in parts of U.S., sometimes called The Tabasco Pepper, hot and spicy, 4 ft. bush, bears nearly year-round after it matures at 8 months. *Source History: 1 in 1981; 1 in 1984; 1 in 1998; 1 in 2004.* **Sources: Eo2.**

Hot Paper Lantern - 70-90 days, *C. chinense*, elongated version of habanero with same blistering heat, wrinkled lantern shaped fruit, 3-4 in. long, fruits mature from lime-green to shades of orange and finally scarlet-red, better production in the North. *Source History: 2 in 2004.* **Sources: Jo1, Jo6.**

Indian P-C 1 - *C. frutescens*, one of the world's hottest peppers from East India, 3-4 ft. tall plant, 2-3 in. long by .25 in. wide fruit, orange-red with a curve at the end, grows horizontally on the plant. *Source History: 1 in 1994; 1 in 1998; 1 in 2004.* **Sources: Red.**

Jamaica Gold - 75 days, *C. chinense*, thin walled fruit, green > bright gold, 150,000 Scoville units, originated in Jamaica. *Source History: 1 in 1998; 1 in 2004.* **Sources: TE7.**

Jamaican - (Jamaica Red) - 75 days, *C. chinense*, plants can grow to 5 ft., lantern-like fruits, deep orange > fiery red, 150,000 Scoville heat units, resembles Habanero but with larger yields and no bitter aftertaste, freezes well, Jamaican origin. *Source History: 1 in 1994; 5 in 1998; 6 in 2004.* **Sources: ME9, Pe7, Pep, SE14, Te8, To1.**

Jamaican Scotch Bonnet - 70-120 days, Small red or yellow fruit, smoky flavor, unbelievably hot. *Source History: 1 in 1998; 4 in 2004.* **Sources: Ra6, Re8, Te8, To3.**

Japones - 75 days, *C. frutescens*, thick stemmed, 2 ft. tall plant bears clusters of upright .25 x 1.5 in. fruit. *Source History: 1 in 1998; 3 in 2004.* **Sources: Gr28, Pep, Red.**

Jellybean - Unusual fruit looks like a large, orange-red jellybean, extremely hot, flowery hot scent is released when fruits are cut, fruits produce only a few seeds, evolved from a wild chile pepper. *Source History: 1 in 1994; 1 in 2004.* **Sources: Pe7.**

Joker's Hat - Shape resembles Bishop's Crown. *Source History: 1 in 2004.* **Sources: Te8.**

Jwala - *C. frutescens*, wrinkled fruit, 4 in. long, green > red, pungent mature fruit, excel. for fresh market and dried, the most popular hot pepper grown and used in India. *Source History: 1 in 1994; 2 in 1998; 2 in 2004.* **Sources: Ev2, Red.**

Kellu Uchu - *C. baccatum*, upright, pendant fruit, 1.5 in. dia., green with purple > bright yellow, 3 ft. plant, very hot, from Peru. *Source History: 1 in 1998; 1 in 2004.* **Sources: Ho13.**

Kovinchu - 115 days, *C. baccatum*, bright red, oblong fruit .5 x 5 in., sweet, spicy flavor that is not hot. *Source History: 2 in 1998; 2 in 2004.* **Sources: Ho13, So1.**

Lemon - 70-95 days, *C. chinense*, Ecuadorian heirloom, slender, 4-5 in., tender skinned fruit, lemon-yellow color, refreshing pungent aroma, best for fresh use, also sauces. *Source History: 3 in 2004.* **Sources: Bu2, Pe7, Ta5.**

Lemon Drop - 100 days, *C. baccatum*, plant is covered with many 2 in. long, wrinkled, bright yellow, conical fruit, nice citrus flavor with intense heat, few seeds, 15 seeds or less per fruit, from Peru. *Source History: 1 in 1998; 5 in 2004.* **Sources: Ba16, Ho13, Pr9, Se16, Ta5.**

Limo - Squash shaped yellow fruit, good for container culture, from Peru. *Source History: 1 in 2004.* **Sources: Te8.**

Limon Chile - 90 days, *C. chinense*, bushy 2 ft. plants, fat, pointed fruit, 1.5 in., delicious smoky-citrus flavor, dark green > yellow, from Peru. *Source History: 2 in 1998; 3 in 2004.* **Sources: Ba16, Ho13, Ma18.**

Louisiana Hot - (Louisiana Arledge Hot) - 70-75 days, Heirloom Tabasco pepper preserved by the Arledge family from Louisiana, very hot 4 in. tapered fruit, green > red, prolific bearer. *Source History: 1 in 1981; 1 in 1987; 6 in 1991; 6 in 1994; 5 in 1998; 5 in 2004.* **Sources: Ba16, Ga1, Pep, Pr3, So1.**

Madras - (LCA 305) - *C. frutescens*, vigorous 3 ft. tall plants grow 3 ft. across, bright red fruit, 3 in. long by .5 in. across, East India. *Source History: 1 in 2004.* **Sources: Red.**

Manzano Amarillo - (Tree Chilipepper, Rojo, Canario) - 90 days, *C. pubescens*, sprawling small tree, blocky yellow 5 in. fruits with black seeds, rather hot when turns light-green, can bear 15 years, 8 in. trunk, best cool weather chili, hardy. *Source History: 3 in 1981; 2 in 1984; 3 in 1987; 4 in 1991; 3 in 1994; 4 in 1998; 4 in 2004.* **Sources: Ba16, Ho13, Pep, Red.**

Manzano Rojo - (Red Apple Chile) - Like the yellow-fruited Manzano Amarillo, this is a perennial chile with distinctive downy leaves and purple flowers and black seeds, med-sized blocky fruits ripen a deep glossy red, milder and more bell pepper- flavored than the Amarillos when the core and inner membranes are removed, sweeter flesh and hotter core than Amarillo, which gives quite a range of flavors. *Source History: 1 in 1991; 1 in 1994; 1 in 1998; 1 in 2004.* **Sources: Red.**

McMahon's Texas Bird Pepper - *C. annuum glabriusculum*, native to southwest Texas, decorative 8-12 in. plant is covered with tiny, edible, sparkling red fruits, ideal for pot culture. *Source History: 1 in 1998; 2 in 2004.* **Sources: Mi12, Th3.**

Merah - (Tempaton) - *C. frutescens*, red fruit dries nicely, from Malaysia. *Source History: 1 in 1991; 2 in 1994; 2 in 1998; 2 in 2004.* **Sources: Red, Te8.**

Mushroom - (Squash Pepper, Cheese Pepper) - 75-90 days, Plant grows 24-30 in., very hot 2 x 2 in. fruit, green > red, hot with a fruity aftertaste, good pickled or dried for seasoning. *Source History: 1 in 1994; 4 in 1998; 7 in 2004.* **Sources: Ho13, Pe7, Ra6, Re8, Sk2, To1, To3.**

Pea - *C. baccatum*, fruit shaped like a pea, green > red, hot. *Source History: 1 in 2004.* **Sources: Te8.**

Pequin - (Piquin, Chili Piguin, NuMex Bailey Piquin Chile) - 80-150 days, Wild ancestor to the domesticated *Capsicums*, no longer than 3 in., fruits grow erect from the top of the stems, very hot, ornamental, prolific, indoor or outdoor. *Source History: 1 in 1981; 1 in 1984; 3 in 1987; 4 in 1991; 4 in 1994; 8 in 1998; 8 in 2004.* **Sources: Hud, Pep, Pla, Ra6, Re8, Red, To1, To3.**

Pequin, African - Bushy, multi-branched plant, productive, from Sudan region of Africa. *Source History: 1 in 2004.* **Sources: Te8.**

Pequin, Large Black - Plant to 3 ft., purple stems, purple overtones on leaves .5 in. long fruit, purple-black > red, very hot. *Source History: 1 in 2004.* **Sources: Ho13.**

Peru Yellow - (Aji Limon) - 80 days, *C. baccatum*, gangly 3 ft. tall plant, flavorful lemon-yellow fruit, up to 4.5 in. long, thin walls, shaped like a large Tabasco, hot. *Source History: 1 in 1987; 1 in 1991; 1 in 1994; 2 in 1998; 1 in 2004.* **Sources: Pep.**

Peruvian Brown - (PI 315010) - Tall 4-5 ft. plants benefit from staking, 4 in. long brown fruit, very hot, USDA. *Source History: 1 in 1998; 1 in 2004.* **Sources: Ho13.**

Peruvian Purple - 90 days, *C. frutescens*, beautiful purple plant, 2 ft. tall, perennial in warmer climates, stubby upright 1 in. fruits, deep-purple > red, mildly hot. *Source History: 1 in 1994; 1 in 1998; 2 in 2004.* **Sources: Ri12, Se7.**

Petine - Beautiful plant with dark green leaves with purple cast, bears many .75 in. thin green pods, red when ripe, very hot. *Source History: 1 in 1994; 1 in 2004.* **Sources: Te8.**

PI 260574 (Bolivia) - *C. baccatum*, erect branching plant bears small heart shaped fruit, heat strikes at the back of the throat, fruity, from Bolivia. *Source History: 1 in 2004.* **Sources: Ba16.**

Piment bec d'Oiseau Noir - French name translates "Beak of the Black Bird Pepper", 3 ft. plant with purple stems and leaves, bright green new leaves .5 in. fruit, dark purple > red, very hot. *Source History: 1 in 2004.* **Sources: Ho13.**

Puca Uchu - (Cuero de Ora (Horn of Plenty), Aji Rojo) - *C. baccatum* var. *pendulum*. *Source History: 2 in 1994; 2 in 1998; 1 in 2004.* **Sources: Red.**

Pungent Red - *C. frutescens*, plants 4 ft. tall, 2-3 ft. across, variable size and hotness of fruit. *Source History: 1 in 2004.* **Sources: Red.**

Punjab Small Hot - *C. frutescens*, 4 ft. tall plant, beautiful dark red pendant fruit, 2 in. long x .25 in. wide, purple > red, twice the heat as habanero. *Source History: 1 in 1998; 1 in 2004.* **Sources: Red.**

Purira - 68-72 days, *C. frutescens*, bushy plants, 2 in. fruits borne upright, yellow and purple blotches turn orange-red when ripe, so intensely hot that they could not be measured on the Scoville scale, disease res. *Source History: 3 in 1998; 6 in 2004.* **Sources: Ba16, Eo2, Ho13, Pe2, Se7, Se26.**

Pusa Jwala - 62-85 days, *C. frutescens*, plants grow 3 ft. tall, slender 3 in. elongated cones, purple > red, very hot, for drying, India. *Source History: 1 in 1991; 3 in 1994; 2 in 1998; 2 in 2004.* **Sources: Se17, Ta5.**

Quintisho - Tiny round fruit looks like a small cherry tomato, yellow-orange > red, from Bolivia. *Source History: 1 in 2004.* **Sources: Te8.**

Rainbow Bird - 100 days, Very hot, tapered fruit, 1 x 1.5 in., green > purple > orange > red. *Source History: 1 in 1991; 1 in 1994; 2 in 1998; 1 in 2004.* **Sources: Fo7.**

Raja's Pride - *C. frutescens*, 4 ft. tall plant, 2 ft. across, some with purple stems, 3-4 in. long fruit .5 in. wide, very productive, fruit dries nicely on the plant, East India. *Source History: 1 in 2004.* **Sources: Red.**

Red Chili - (Small Red Chili, Long Red Chili, Finger Cayenne) - 80-85 days, *C. frutescens*, hot conical 1.75-2.5 x .5 in. fruits, tapers to a blunt point, thin walls, upright, pale yellow-green > bright-red, 18-20 in. erect bushy plants, concentrated set, heavy yields, very hot, for pickling or drying or sauce, 70,000 Scoville units. *Source History: 54 in 1981; 46 in 1984; 43 in 1987; 34 in 1991; 28 in 1994; 23 in 1998; 13 in 2004.* **Sources: Com, Gr28, Gr29, He8, Pep, Ra6, Re8, Red, RUP, SGF, Shu, To1, WE10.**

Rocoto - (Tree Pepper, Manzano) - 90-110 days, *C. pubescens*, Ecuadorian var. similar to Zapotec Chile Manzano, fuzzy- leaved plant with purple flowers that can live 10 years in mild climates, fruits resemble 2 in. wide bell peppers but are very hot, prized by the Incas for its unique flavor, stands more cold than most peppers. *Source History: 2 in 1991; 3 in 1994; 5 in 1998; 6 in 2004.* **Sources: Enc, Gr29, Pep, Red, Se8, To1.**

Scotch Bonnet, Chocolate - (Jamaican Hot Chocolate) - 120 days, *C. chinense*, 3 ft. plants, 1.5 in., round, convoluted fruit, green > dark brown, very hot, from the Caribbean. *Source History: 1 in 1994; 2 in 1998; 4 in 2004.* **Sources: Gr29, Pep, Se17, To1.**

Scotch Bonnet, Orange - (Orange Bumpy Scotch Bonnet) - 75-80 days, *C. chinense*, fruit is rippled with bumps which gives it a bright lantern-like appearance, hot juicy flesh, tasty, good in salsa, stir fry or dried, rare. *Source History: 3 in 2004.* **Sources: HPS, ME9, Se8.**

Scotch Bonnet, Red - (Jamaica Red) - 75-100 days, *C. chinense*, Jamaican favorite, mushroom shaped, bright red fruit on 24 in. plant, fruity aroma with blistering heat, 200,000 Scoville units. *Source History: 1 in 1991; 3 in 1994; 7 in 1998; 13 in 2004.* **Sources: Bo17, Bo18, Com, Enc, ME9, Orn, Ra5, Red, Se8, SE14, Se26, To1, Wo8.**

Scotch Bonnet, Yellow - (Jamaica Yellow, Yellow Mushroom) - 70-100 days, *C. chinense*, yellow version of the Jamaican Hot Red, 200,000 Scoville heat units, green > yellow. *Source History: 1 in 1991; 3 in 1994; 6 in 1998; 13 in 2004.* **Sources: Bo19, Enc, Eo2, He8, Jo6, ME9, Ni1, Orn, Pep, Red, Ri2, SE14, WE10.**

Squash, Red - (Mushroom Pepper, Rocotillo) - 90 days, *C. chinense*, plants are 24-30 in. tall, benefit from staking, abundance of mushroom-shaped 2 x 2 in. fruits, thin skin, green > orange > scarlet > red, easy to dry, much more rare than Yellow Squash. *Source History: 3 in 1991; 5 in 1994; 2 in 1998; 5 in 2004.* **Sources: Ba16, Mi12, Pep, WE10, We11.**

Squash, Yellow - (Yellow Mushroom) - 90-100 days, *C. chinense*, flat disk-shaped pepper .5-1 in. deep x 2-2.5 in. dia., light-green > yellow, 18-24 in. bushes, does not store whole well, must cut open to dry, red-ripe strain exists. *Source History: 2 in 1981; 2 in 1984; 4 in 1987; 5 in 1991; 5 in 1994; 7 in 1998; 11 in 2004.* **Sources: Ami, Ba16, Ho13, Pep, Ra6, Re8, Sk2, To1, To3, WE10, We11.**

Suryamukhi Cluster - (Surajmukhi) - *C. frutescens*, 2 ft. tall plant, small dark red fruit, 3-4 in. long, fruit held upright on the plant in clusters of up to 12 pods, from India. *Source History: 2 in 1998; 2 in 2004.* **Sources: Pep, Red.**

Tabasco - 80 days, *C. frutescens*, tall plant bears up to 100 upright pointed 1.5 in. fruit, light yellow-green > red, thin flesh, flavor is sharp with biting heat, 30,000 to 50,000 Scoville heat units, used in hot sauces, vinegars and curries, best in the South and East, known to have been cultivated near Tabasco, Mexico, in the early 1840s and imported to Louisiana in 1848. *Source History: 7 in 1981; 5 in 1984; 4 in 1987; 8 in 1991; 13 in 1994; 20 in 1998; 46 in 2004.* **Sources: Ba8, Ba16, BAL, Bo9, Bo18, Bo19, Bu3, Co31, Enc, Fo7, Ge2, Gr29, Gur, He8, He17, Ho8, Ho13, HPS, Jo6, Ki4, La1, Lej, Ma18, ME9, Na2, Pe7, Pep, Pla, Ra5, Ra6, Red, Ri12, RUP, Se8, SE14, Se26, Shu, Te8, Tho, To1, To3, Twi, Vi5, WE10, We11, Wo8.**

Tabasco, Costa Rican - Slightly curved fruit borne on compact plants. *Source History: 1 in 2004.* **Sources: Te8.**

Tabasco, Dr. Greenleaf's - 120 days, *C. frutescens*, very hot, virus res., heavy yields of red pods for spicy Louisiana pepper sauce or fiery pickles, for the South, Auburn University. *Source History: 1 in 1981; 1 in 1984; 1 in 1987; 2 in 1991; 1 in 1994; 3 in 1998; 3 in 2004.* **Sources: Enc, Gr29, Pep.**

Tabasco, Short Yellow - 75 days, *C. frutescens*, plant grows 3 ft. tall, small leaves, small fruit, yellow-green > yellow-orange. *Source History: 1 in 1994; 1 in 1998; 2 in 2004.* **Sources: Gr29, Pep.**

Takanotsume - (Rooster Spur, Japanese Hot Claw, Hahong Koch'o) - 68 days, *C. frutescens*, very hot flame-shaped pepper, 1.5 x .5 in. dia., dark-green > red, takes a long time to mature, used fresh or dried (crushed or powdered), imported from Japan. *Source History: 2 in 1981; 3 in 1984; 3 in 1987; 3 in 1991; 3 in 1994; 2 in 1998; 2 in 2004.* **Sources: Kit, Pep.**

Tepin - (Chiltecpin, Bird Pepper) - 90 days, *C. anuum* var. *aviculare*, wild pepper native from Southwest U.S. to South America, red pea-sized hot fruit is used fresh or dried, tender perennial (marginally hardy in Zone 8), known as "bird pepper" perhaps because mockingbirds are fond of the fruit. *Source History: 1 in 1981; 1 in 1984; 1 in 1987; 1 in 1991; 3 in 1994; 6 in 1998; 10 in 2004.* **Sources: Bo19, Gr29, He8, Ho8, ME9, Pep, Red, SE14, So25, To1.**

Thai Hot - (La Chaio, Duey Kai) - 68-75 days, *C. frutescens*, from Thailand, mound-shaped 8 in. plants, covered with extremely hot 1 in. green and red peppers held upright, 80,000 Scoville heat units, widely used in Oriental dishes, makes a good ornamental. *Source History: 1 in 1981; 1 in 1984; 6 in 1987; 9 in 1991; 22 in 1994; 28 in 1998; 28 in 2004.* **Sources: CA25, Com, Coo, Enc, Eo2, Ev2, Fe5, Fo7, Ge2, Gr29, Ha5, He8, Hig, Ho13, Jor, Loc, ME9, Pep, Pla, Red, Ri2, SE14, Ta5, Te4, To1, To3, WE10, We11.**

Thai Hot, Large - 42 days, Produce twice the amount of flesh per pod as Thai Hot, just as spicy, deep green > deep red, dev. by R.E. Bernstrom, Richmond, KY. *Source History: 1 in 1994; 1 in 1998; 1 in 2004.* **Sources: Par.**

Thai Sun - Small plant grows 10-12 in. tall and 1-1.5 ft. wide, 1 in. fruits face upward, each plant produces hundreds of fruit, easily grown in containers, ripens early, produces fruit all season, pods contain few seeds. *Source History: 1 in 1994; 1 in 1998; 1 in 2004.* **Sources: Pe7.**

Trinidad - (Trinidad Seasoning) - 75-90 days, *C. chinense*, seasoning pepper with mild, fruity flavor, golden yellow at maturity, small, wrinkled globes, 1-2 in., heirloom from Trinidad. *Source History: 4 in 1998; 7 in 2004.* **Sources: Ba8, Ho13, ME9, Pep, Ra6, Sa9, SE14.**

Trupti - *C. frutescens*, mildly hot fruit, slender, 4-5 in. long, usually curved like a fishhook, good for pickling like pepperoncini, ripens red, East India. *Source History: 1 in 1998; 1 in 2004.* **Sources: Red.**

USDA GRIF9221 - *C. baccatum*, rampant plant bears waffled fruit, fruity flavor and aroma, deep green > orange > red, Colombia. *Source History: 1 in 2004.* **Sources: Ba16.**

Venezuelan Purple - Ornamental plant grows 3 ft. tall with deep purple leaves and stems, .25 in. dia. fruits ripen purple > deep red. *Source History: 1 in 1987; 1 in 1998; 1 in 2004.* **Sources: Fo7.**

Willing's Barbados - 100 days, Ornamental plant to 3 ft. tall, leaves resemble boxwood leaves, tiny fruits look like barberries, maintained in the Philadelphia area since the late 1700s, brought there by Charles Willing of Barbados, West Indies. *Source History: 1 in 1998; 1 in 2004.* **Sources: Fo7.**

Wirri Wirri - Woody 3 ft. shrubs, red, rounded, upright .5 in. fruit, very hot, from the Caribbean. *Source History: 1 in 1998; 1 in 2004.* **Sources: Ho13.**

Xigole - (Bachelor, Soltero) - Very small extremely hot peppers grow upward, bright-red when mature, long-lived perennial in cool Oaxacan Mountains, needs more warmth than Manzano to thrive. *Source History: 1 in 1991; 1 in 1998; 1 in 2004.* **Sources: Ho13.**

Yatsafusa - 80 days, *C. frutescens*, leading hot pepper in Japan where it is grown extensively for the spice trade, 20 in. plants, 3 in. thin-walled tapered pods ripen red, fiery hot, productive. *Source History: 3 in 1987; 4 in 1991; 4 in 1994; 3 in 1998; 5 in 2004.* **Sources: Gr29, Kit, Pep, Red, Se17.**

Zimbabwe Bird - (River Pepper, Chile Tepin, Turkey Pepper) - *C. frutescens*, small round red fruit, extremely hot, grows wild in southwestern United States. *Source History: 2 in 1994; 1 in 2004.* **Sources: Red.**

Varieties dropped since 1981 (and the year dropped): *baccatum* (2004), *C. baccatum* var. *baccatum* (1998), *C. baccatum* var. *praetermissum* (2004), *Calcutta Long* (2004), *Capsicum Pubescens* (1987), *Chiltepin, Purple* (2004), *Chiltepin, Puya* (2004), *Chiltepines, Tarahumara* (2004), *Chiltepines: Guarijio* (1998), *Chiltepines: Mayo* (2004), *Chiltepines: Pima Bajo Indian* (1991), *Chiltepines: Tarahumara Indian* (1994), *Chiltepines: Texas* (2004), *Chiltepines: Tohono O'odham* (1998), *Chiltepines: Warihio Indian* (1991), *Cuerno de Oro* (1998), *Desi Teekhi* (2004), *Guachinango* (2004), *Habanero, Red Savina* (2004), *Honduran Wild* (1998), *Japanese Type* (1994), *Mauritius* (1998), *Mexican Tree* (2004), *PA-353* (2004), *PA-398* (2004), *Papago Indian Chiltepines* (1987), *Pequento* (1998), *Rocotillo* (2004), *Scotch Bonnet, Early* (2004), *South African No. 1* (1998), *South African No. 2* (1998), *Tiny Samoa* (1998), *Tuste Blanco* (2004), *Ulupica* (2004), *Una de Pavo* (1998), *Wild Dynamo* (2004), *Yellow Bumpy* (1998).

Potato
Solanum tuberosum

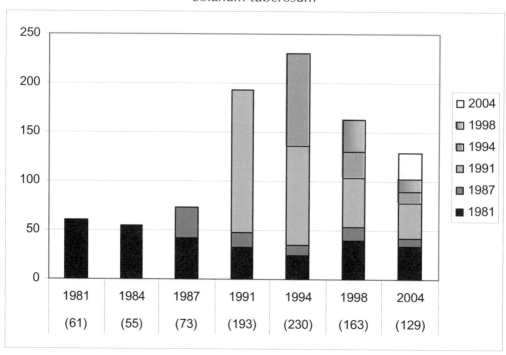

Number of varieties listed in 1981 and still available in 2004: 61 to 34 (56%).
Overall change in number of varieties from 1981 to 2004: 61 to 129 (211%).
Number of 2004 varieties available from 1 or 2 sources: 79 out of 129 (61%).

A. C. Ptarmigan - Good early fresh market variety, oval tubers, shallow eyes, buff skin, cream/light yellow flesh, good for chips and boiling, no darkening after cooking. *Source History: 1 in 2004.* **Sources: Ear.**

A79543-4R - Red skin, matures early. *Source History: 1 in 1998; 1 in 2004.* **Sources: Lin.**

Adora - 55+ days, Oblong tubers, fairly smooth skin, medium shallow eyes, light yellow flesh, uniform size, good keeper, maintains color after cooking, good disease resistance, bred by A. D. Mulder in the Netherlands, PVP 2000 HZPC Holland B.V. *Source History: 1 in 2004.* **Sources: Ron.**

Alaska Frostless - 75+ days, Small to medium tubers, white as snow, very hardy, needs good fertile soil to produce, vines able to withstand frost to 27 degrees F., midseason. *Source History: 1 in 1991; 1 in 1994; 2 in 2004.* **Sources: Bec, Ga1.**

Alaska Sweet Heart - Red skin, red flesh, pretty, shows great promise for uniqueness and fresh market, midseason, from an Alaskan breeding program. *Source*

History: 1 in 1991; 1 in 1994; 4 in 1998; 1 in 2004. **Sources: Dam.**

Alby's Gold - 80+ days, Large attractive round to oval tubers, smooth thin yellow skin, deep yellow flesh, shallow eyes with slightly prominent eyebrows, good yields, excellent all-purpose potato, moderate scab resistance, Dutch introduction. *Source History: 1 in 2004.* **Sources: Ron.**

All Blue - (Purple Marker) - 100-135 days, Deep-blue skin, flesh is blue to the center, very novel but fine flavored, yields well, midseason, good keeper, attractive in salads, eye-catching no matter how you serve it. *Source History: 1 in 1981; 1 in 1984; 2 in 1987; 9 in 1991; 14 in 1994; 20 in 1998; 26 in 2004.* **Sources: Bu1, Bu2, Dam, De6, Ear, Ers, Fa1, Fi1, Ga1, Gur, Hi6, Jo1, Jun, Ki4, Mi8, Moo, Pe2, Pin, Roh, Ron, Se7, Se16, Ter, Tt2, Ver, Wo6.**

All Red - (Cranberry Red) - 70-90 days, Round-oblong tubers, red skin and light-red flesh, firm, good taste, color retained after cooking, low starch content makes this var. a good boiling potato for salads or any dish that requires potatoes to retain their shape, mid to late maturity. *Source History: 1 in 1991; 3 in 1994; 5 in 1998; 10 in 2004.* **Sources: Ga1, Hi6, Jo1, Mi8, Moo, Pin, Ron, Se16, Ter, Wo6.**

Anna Cheeka's Ozette - (Haida, Kasaan) - Historic heirloom said to be brought from Peru in the late 1700s by Spanish explorers and traded with the Makah-Ozette Indian tribe and maintained ever since, 2-8 in. long tubers, thin skin, flaky creamy yellow flesh, many deeply set eyes arranged in a spiral pattern around the tuber, earthy taste. *Source History: 1 in 1991; 1 in 1994; 2 in 1998; 1 in 2004.* **Sources: Ron.**

Anoka - Earliest of the white vars., shows scab res., excel. yields, bred in Minnesota in 1958 by Prof. Orrin C. Turnquist. *Source History: 2 in 1991; 3 in 1994; 1 in 1998; 2 in 2004.* **Sources: Jun, Ron.**

Arran Pilot - (Salt Water) - Round to oblong tubers, white skin and flesh, midseason, first grown on the island of Arran and probably on land that was sea washed during winter storms. *Source History: 1 in 1991; 1 in 1994; 1 in 1998; 1 in 2004.* **Sources: Ron.**

Atlantic - 90-110 days, Large, upright plants make for easy hoeing, round, medium size white tubers, smooth lightly netted skin, good boiler and baker, dependable high yields, medium maturity, USDA, introduced in 1978. *Source History: 1 in 1981; 1 in 1984; 2 in 1987; 3 in 1991; 1 in 1994; 2 in 1998; 2 in 2004.* **Sources: Ga1, Mi8.**

Austrian Crescent - Prolific yields of 4-8 oz. fingerlings to 10 in., yellow-tan skin, light-yellow flesh, midseason. *Source History: 2 in 1991; 3 in 1994; 5 in 1998; 3 in 2004.* **Sources: Mi8, Moo, Ron.**

Bake King - 90+ days, Mid-season maturity, oval to oblong tubers, white russet skin, generally shallow eyes, good yields, outstanding baker, very mealy, high solid and low water content, rel. by Cornell University, 1967. *Source History: 1 in 1981; 1 in 1991; 1 in 1994; 2 in 2004.* **Sources: Ga1, Ron.**

Banana - (Russian Banana) - 95-100 days, Waxy yellow fingerling 1-8 in. long, pale-yellow flesh, stays firm when boiled, scab res., late maturity. *Source History: 4 in 1991; 5 in 1994; 7 in 1998; 7 in 2004.* **Sources: Dam, Ear, Lin, Mi8, Moo, Ri2, Se7.**

Banana, Russian - 90-105 days, Small banana-shaped fingerling tubers, smooth buff-yellow skin, light-yellow flesh, appealing waxy cooked quality, nice in potato salads and mixed vegetable dishes, tubers store well, scab res., dev. in the Baltic region of Europe-Asia, very popular across Canada as a market potato. *Source History: 4 in 1991; 5 in 1994; 6 in 1998; 9 in 2004.* **Sources: Ga1, Hi6, Jo1, Ki4, Pe2, Ron, Se16, Tt2, Wo6.**

Beltsville - Big round tasty white tubers, crisp texture, withstands common potato diseases. *Source History: 1 in 1991; 2 in 1994; 4 in 1998; 2 in 2004.* **Sources: Shu, Ver.**

Bintje - (Yellow Finnish) - 110-120 days, Late maturing variety with gold skin, yellow flesh, oblong, shallow eyes, heavy yields, excel. texture and flavor, drought res., been around since 1911, from the Netherlands. *Source History: 1 in 1981; 1 in 1984; 5 in 1987; 7 in 1991; 6 in 1994; 10 in 1998; 7 in 2004.* **Sources: Ga1, Lin, Mi8, Moo, Ri2, Ron, Se7.**

Bison - Early, med-sized round baking potato, deep-red skin, smooth shallow eyes, highly res. to late blight and scab, excel. for baking. *Source History: 1 in 1981; 1 in 1984; 1 in 1987; 2 in 1991; 1 in 1994; 2 in 1998; 1 in 2004.* **Sources: Ron.**

Bliss Triumph - (Red Bliss) - Light-red skin, white flesh, good yielder, early maturing, old timer released in 1878. *Source History: 1 in 1991; 2 in 1994; 1 in 1998; 1 in 2004.* **Sources: Ron.**

Blossom - Striking pink skin, pink flesh, pink flower blossoms, mild but earthy flavor, excel. mashed or baked, midseason, bred by Ewald Eliason. *Source History: 2 in 1991; 2 in 1994; 3 in 1998; 2 in 2004.* **Sources: Mi8, Moo.**

Blue Eyed Russian - Pretty tubers with yellow skin with a waxy surface, striking blue eyes, rare. *Source History: 1 in 1998; 1 in 2004.* **Sources: Ho13.**

Brigus - 75+ days, Round-oblong, uniform tubers with bluish purple skin and creamy yellow flesh, shallow eyes, excel. boiled, wart and late blight res., high yields. *Source History: 2 in 1991; 2 in 1994; 1 in 1998; 2 in 2004.* **Sources: Bec, Dam.**

Buffalo - (Red Ruby) - From a cross made with Bison with a more oblong shape, brilliant red skin, very sweet flavor, good keeper, high res. to scab VW LB and HH, Lauer/U of MN. *Source History: 1 in 1998; 3 in 2004.* **Sources: Ga1, Pe2, Ron.**

Butte - 90-135 days, Long uniform russet baker, up to 7% higher yield, 25% more No. 1's, 58% higher vitamin C, 20% more protein, shallow eyes, white flesh, disease res., late maturity, good keeper, U of Idaho. *Source History: 2 in 1981; 2 in 1984; 2 in 1987; 6 in 1991; 6 in 1994; 8 in 1998; 7 in 2004.* **Sources: Coo, Hi6, Moo, Ron, Se16, Ter, Wo6.**

Butterfinger - (Swedish Peanut) - Fingerling with a relatively high starch content, russetted thin skin, nutty tasting yellow flesh, retains firmness when cooked. *Source History: 1 in 1998; 2 in 2004.* **Sources: Ga1, Ron.**

Cal Red - 65+ days, Small to medium round tubers, red

skin, shallow eyes, cream colored flesh, great cooking potato, moderate scab resistance. *Source History: 2 in 2004.* **Sources: Ga1, Ron.**

Cal White - Large upright plants, oblong tubers, white skin, shallow eyes, smooth creamy flavor, moderately high vitamin C content, resists growth cracks, second growth, hollow heart and net necrosis. *Source History: 1 in 2004.* **Sources: Moo.**

Candy Cane - 80+ days, Medium size tubers, oblong shape, bright red skin, creamy yellow interior, mottled ring of red just under skin surface, sweet flavor. *Source History: 1 in 1994; 2 in 2004.* **Sources: Ga1, Ron.**

Caribe - (Purple Caribe) - 70-90 days, Blue-skinned white-fleshed smooth uniform tuber, rich flavor, fairly starchy, a bit waxy, excellent yielder, early, named not for its origin but because this var. was once widely grown in New England for export to the Caribbean. *Source History: 1 in 1981; 1 in 1984; 2 in 1987; 10 in 1991; 15 in 1994; 16 in 1998; 13 in 2004.* **Sources: Bec, Ear, Ga1, Hi6, Lin, Mi8, Moo, Par, Ron, Se16, Ter, Tt2, Wo6.**

Carola - (Carole) - 90-110 days, German-bred potato, oval buff-skinned tubers, flavorful yellow flesh, moist creamy texture similar to fingerlings, maintains new potato qualities for months in root cellar, midseason, scab and blight res., German introduction, 1979. *Source History: 3 in 1991; 5 in 1994; 8 in 1998; 9 in 2004.* **Sources: Ga1, Hi6, Mi8, Moo, Pin, Ron, Se16, Ter, Wo6.**

Chaleur - Canadian variety, large tubers are excellent for fries, boiling and baking, average keeper, shows improved tolerance to hot dry conditions, early. *Source History: 1 in 2004.* **Sources: Dom.**

Charlotte - 55+ days, Golden yellow skin and flesh, shallow eyes, extra early maturity, excel. fresh from the garden, very good producer, excellent storage. *Source History: 1 in 1991; 1 in 1994; 1 in 2004.* **Sources: Ron.**

Cherokee - 80+ days, Round-flat tubers with very white skin, good yielder and boiler, stores well, late maturity, good res. to scab and LB, adapted to muck and clay soils, bred in Maine in 1941, introduced in 1954. *Source History: 1 in 1981; 1 in 1984; 1 in 1987; 2 in 1991; 4 in 1994; 1 in 1998; 1 in 2004.* **Sources: Ron.**

Cherry Red - Uniform, red skinned tubers, small to medium size, white flesh, excel. disease res., good boiled and baked, tubers are tightly clustered under the plant, unknown origin. *Source History: 1 in 1994; 5 in 1998; 2 in 2004.* **Sources: Ga1, Mi8.**

Chieftain - Early, medium-spreading plant with light violet blossoms, shallow to medium eyes, round to oblong tubers with attractive red skin, greater yields than Norland, good results when grown in clay soils. *Source History: 1 in 1981; 1 in 1984; 3 in 1987; 2 in 1991; 4 in 1994; 6 in 1998; 5 in 2004.* **Sources: Bec, Dam, Dom, Ga1, Moo.**

Chippewa - 110 days, Smooth white flattened oval potatoes, shallow eyes, heavy yields, good quality and appearance, succeeds under almost all conditions of soil and weather, disease res., introduced in 1933. *Source History: 1 in 1981; 1 in 1984; 2 in 1987; 2 in 1991; 3 in 1994; 2 in 1998; 2 in 2004.* **Sources: Mey, Ron.**

Coastal Russet - Russet Burbank parentage, medium size tubers, russet skin, white flesh, good storage, resists net necrosis and hollow heart, tolerates common scab, purple flowers with cream tips held on long peduncles held high above the foliage. *Source History: 1 in 2004.* **Sources: Moo.**

Conestoga - Newly licensed, blocky flattened tubers buff skin, white flesh, good for steaming and roasting, excellent yielder, medium maturity, resists scab, early blight, net necrosis, leaf roll. *Source History: 1 in 1981; 1 in 1984; 1 in 1987; 1 in 1991; 1 in 1994; 1 in 2004.* **Sources: Moo.**

Daisy Gold - 80 days, Large, long tubers, shallow eyes, smooth yellow skin, deep yellow, moist flesh with flaky texture, boiling and baking, nematode resistant. *Source History: 2 in 2004.* **Sources: Bu2, Ga1.**

Denali - 90+ days, Oval to oblong tubers, tough skin, shallow eyes, dry and mealy texture, very good baked or mashed, tolerates heat and frost, moderately resistant to hollow heart, good storage, midseason, developed in Alaska, 1978. *Source History: 1 in 1991; 1 in 1994; 1 in 2004.* **Sources: Ron.**

Desiree - 95-100 days, Popular red-skinned midseason variety, deep golden flesh with moist creamy texture, delicate flavor, good for growing from seed, crops marginally later than tubers, from the Netherlands, 1962. *Source History: 1 in 1987; 4 in 1991; 9 in 1994; 7 in 1998; 8 in 2004.* **Sources: Bec, Dam, Ga1, Jun, Mi8, Moo, Ron, Se7.**

Durango Red - (Durango) - 90-100 days, Small to medium upright plants, red-purple flowers, good yields of round to slightly oval, dark red-skinned tubers, smooth white flesh, good all-purpose potato. *Source History: 2 in 2004.* **Sources: Mi8, Moo.**

Early Ohio - Early, white, fine flavor, excel. for frying, good keeper, probably bred from Early Rose, once a favorite with older gardeners in northern Midwest and Ohio Valley, it is dying out with them, early 1900s. *Source History: 3 in 1981; 3 in 1984; 1 in 1987; 3 in 1991; 2 in 1994; 2 in 1998; 4 in 2004.* **Sources: Fi1, Ga1, Gur, Ron.**

Elba - Versatile round white tuber, slightly dry texture, good baker, res. to early-late blight and VW and golden nematode. *Source History: 2 in 1991; 1 in 1994; 1 in 1998; 2 in 2004.* **Sources: Moo, Wo6.**

Epicure - Popular since 1879, slightly irregular shape with deep eyes, thin white skin, delicious creamy white flesh, best eaten boiled. *Source History: 3 in 1998; 1 in 2004.* **Sources: Ga1.**

Eramosa - New Canadian variety from New Brunswick, earliest in most area trials, smooth round white potato, does not tend to oversize, res. to wilt and fusarium and rhizoctonia. *Source History: 1 in 1991; 3 in 1994; 3 in 1998; 1 in 2004.* **Sources: Dom.**

Erik - 60+ days, Early red, excel. boiled, high res. to late blight, some scab res., tol. to wet soil, named after Erik the Red who was a legendary Scandinavian explorer, introduced in 1983. *Source History: 2 in 1991; 1 in 1994; 1 in 1998; 2 in 2004.* **Sources: Bec, Dam.**

Fingerling - (Yellow Fingerling) - Little yellow-fleshed salad potato, from early German settlers, 2-4 in. x 1 in. dia., yellow skin and flesh, delicious fried or boiled in jackets and cut up for salad. *Source History: 1 in 1981; 1 in 1984; 1 in 1987; 2 in 1991; 2 in 1994; 2 in 1998; 4 in 2004.* **Sources: Bec, Jun, Shu, Ver.**

Frontier - Improved Butte, oblong tubers, firm white flesh, contains 20% more protein than others, developed at the U of Idaho. *Source History: 2 in 1994; 3 in 1998; 1 in 2004.* **Sources: Fi1.**

Frontier Russet - 85 days, Early maturing, easily grown russet with dry white flesh, large tubers store well, resists scab, fusarium dry rot, verticilium, bruising, knobbiness and growth cracks. *Source History: 3 in 1994; 6 in 1998; 1 in 2004.* **Sources: Moo.**

Garnet Chile - 90-100 days, Heirloom parent of most of the modern red potatoes, lumpy, thick red skin with deep eyes, good keeper, late, introduced in 1853. *Source History: 1 in 1991; 2 in 1994; 2 in 1998; 2 in 2004.* **Sources: Mi8, Moo.**

Gem Russet - 75+ days, Long white tubers, light netted skin, good potato taste, excellent boiler and baker, high yields, good storage, good disease res., PVP application pending. *Source History: 2 in 2004.* **Sources: Bec, Mcf.**

German Butterball - Good all-purpose potato, small to medium size oblong tubers with thick, netted golden yellow skin, yellow flesh, remarkable flavor, stores well, late. *Source History: 1 in 1991; 1 in 1994; 5 in 1998; 10 in 2004.* **Sources: Bec, Dam, De6, Ga1, Jun, Mcf, Mi8, Moo, Ron, Se7.**

Gold Nugget - Medium to large russet shaped tubers, netted skin, yellow flesh, rich buttery flavor, midseason. *Source History: 1 in 1991; 1 in 1994; 1 in 1998; 1 in 2004.* **Sources: Ron.**

Goldrush - Parentage includes Lemhi Russet, Norgold and Russet Burbank, flavorful midseason russet, vigorous vines with upright growth habit, scab res., bred by Robert Johansen, North Dakota State University, 1993, PVP. *Source History: 3 in 1994; 6 in 1998; 4 in 2004.* **Sources: Bu1, Fa1, Jo1, Jun.**

Green Mountain - Medium size tubers, light tan skin is half russet and half white, mealy flesh, excel. flavor retained over long storage period, tolerates a wide variety of soils, excel. producer, mid to late maturity, originated in Vermont, introduced in 1885. *Source History: 2 in 1981; 2 in 1984; 5 in 1987; 8 in 1991; 9 in 1994; 7 in 1998; 4 in 2004.* **Sources: Bec, Moo, Pin, Ron.**

Huckleberry - 80+ days, Maroon skin, red flesh with some white marbling, typical potato shape, medium-large size, midseason, named after the legendary wild huckleberry. *Source History: 1 in 1991; 1 in 1994; 2 in 1998; 3 in 2004.* **Sources: Mi8, Moo, Ron.**

IdaRose - 75+ days, Uniform, round tubers, bright red skin, white flesh, high yields, good disease tolerance, from western Canada, PVP application pending. *Source History: 2 in 2004.* **Sources: Bec, Mcf.**

Inca Gold - Round dumpling type potato, golden skin splashed with purple, yellow flesh with somewhat nutty taste, vigorous root system, good storage. *Source History: 1 in 2004.* **Sources: Ron.**

Irish Cobbler - (Cobbler) - 100 days, Said to be named for an Irish shoemaker in New Jersey who grew it from tubers selected from a mixture of Early Rose seed, early, round to oblong, white smooth skin and white flesh, strong med-deep eyes, high yields, very widely adapted all-purpose potato, good for boiling and baking, short storage life, released in 1876. *Source History: 10 in 1981; 9 in 1984; 9 in 1987; 12 in 1991; 11 in 1994; 12 in 1998; 11 in 2004.* **Sources: Bec, Dam, De6, Ers, Ga1, Mey, Moo, Roh, Ron, Sau, Tt2.**

Island Sunshine - Private patented variety with very dark yellow skin, excel. table quality, late maturity, moderate res. to A2, total resistance to A1 late blight, PVP 2000 Loo Brothers. *Source History: 6 in 1998; 3 in 2004.* **Sources: Hi6, Moo, Wo6.**

Katahdin - 110 days, Glossy thin white skin, shallow eyes, oval, good baked fried mashed or boiled, slightly earlier than Green Mountain, dev. by USDA, good for spring or fall plantings, good quality, disease res., old standard since 1932. *Source History: 4 in 1981; 4 in 1984; 6 in 1987; 5 in 1991; 5 in 1994; 7 in 1998; 8 in 2004.* **Sources: Ers, Ga1, Mey, Moo, Pin, Roh, Ron, Sau.**

Kennebec - (White Kennebec) - 80+ days, Med-late, smooth white skin, smooth-textured white flesh, shallow eyes, res. to mosaic and late blight and net necrosis, oval to oblong, heavy mid-season yields, 4 to 6 large white tubers, heavy yields on most soils, thin-skinned, good-keeping all-purpose potato, excellent for boiling, adapted to North, 1948. *Source History: 24 in 1981; 22 in 1984; 22 in 1987; 27 in 1991; 26 in 1994; 30 in 1998; 29 in 2004.* **Sources: Be4, Bec, Bu1, Bu2, Co31, Dam, De6, Dom, Ear, Ers, Fa1, Fi1, Ga1, Gur, Jo1, Jun, Lin, Mcf, Mey, Moo, Pe2, Pin, Roh, Ron, Sau, Shu, Tt2, Ver, Wet.**

Kerr's Pink - 90-100 days, Scottish variety, round tubers, light-pink skin, fine-grained white flesh, red eyes, excel. flavor, midseason, introduced in Ireland in 1917. *Source History: 1 in 1994; 1 in 1998; 4 in 2004.* **Sources: Ga1, Mi8, Moo, Ron.**

Keystone Russet - 90-100 days, More blocky shape than Silverton, good for baking and frying, PVP application pending. *Source History: 1 in 2004.* **Sources: Mi8.**

King Edward - 80+ days, Long oval tubers, yellow and red skin, shallow eyes, creamy white floury flesh, rarely discolors with cooking, high quality, mid to late season. *Source History: 1 in 1994; 1 in 2004.* **Sources: Ron.**

Kipfel - Golden skin, yellow flesh, crescent shape, res. to scab and blight. *Source History: 1 in 1991; 3 in 1998; 1 in 2004.* **Sources: Mi8.**

Krantz - Interesting round russet which produces medium to large bakers, also good for fries and mashing, extreme res. to hollow heart, scab and late blight, susc. to early blight, early, from a Canadian grower. *Source History: 1 in 1991; 2 in 1994; 4 in 1998; 3 in 2004.* **Sources: Ga1, Moo, Ron.**

La Ratte - 100-120 days, Fingerling type with yellow skin color, smooth, buttery texture, unique nutty flavor, good for boiling and roasting. *Source History: 2 in 2004.* **Sources: Mi8, Moo.**

La Rouge - Red skinned tuber, white flesh, fair boiling

quality, good baking quality, early sizing, good for new potatoes, holds up well in hot weather, loses color in storage, scab res., Louisiana release, 1962. *Source History: 2 in 1991; 1 in 1994; 3 in 1998; 1 in 2004.* **Sources: Moo.**

Magic Molly - (Black Beauty) - Seedling from Red Beauty, produces many medium oval tubers, upright plant with huge stolons, fibrous root system, produced by Bill Campbell, Alaska Research Station. *Source History: 1 in 2004.* **Sources: Ron.**

Maris Piper - Well known UK variety with creamy white skin and flesh, great for mashed potatoes, high yielder. *Source History: 3 in 1998; 2 in 2004.* **Sources: Ga1, Mi8.**

Maroon Bells - Medium to large, round tubers, maroon-red skin and flesh, floury texture when cooked. *Source History: 1 in 2004.* **Sources: Ron.**

Morning Gold - Good yields of medium size brown to gold skinned oblong tubers, medium yellow flesh, baking, salads, steaming or boiling, PVP 2000 HZPC Holland B.V. *Source History: 1 in 1998; 2 in 2004.* **Sources: Ga1, Tt2.**

NDC4069-4 - 90-100 days, Darker red skin and flesh than other red varieties, bred in North Dakota, selected by Dr. David G. Holm in Colorado. *Source History: 1 in 2004.* **Sources: Mi8.**

Netted Gem - (Russet Burbank) - Baking potato, white russet skin, oblong with generally shallow eyes, flavor improves in storage, best selling main crop variety, excel. keeper, late. *Source History: 5 in 1981; 5 in 1984; 5 in 1987; 3 in 1991; 2 in 1994; 1 in 1998; 1 in 2004.* **Sources: Alb.**

Nicola - 90-110 days, Waxy type, long oval tubers, yellow skin, yellow flesh, med. size spreading plant, white flowers, used for boiling, mashing, roasting and potato salad, excel. disease res. *Source History: 2 in 2004.* **Sources: Mi8, Moo.**

Nooksack - Broad leaved plants, well adapted white potato with larger more round shape than Russet, good uniformity and disease res. and yields, late, excellent storage, does well in coastal and wetter regions. *Source History: 2 in 1991; 1 in 1994; 3 in 1998; 2 in 2004.* **Sources: Ga1, Ron.**

Norgold M - Same Norgold Russet flavor, good for dry and drought type soils, keeps on growing when the weather gets hot, early, released in early 1980s for southern states. *Source History: 1 in 1991; 1 in 1994; 1 in 1998; 2 in 2004.* **Sources: Ga1, Ron.**

Norgold Russet - Med-early, long and smooth, well netted, brown russet skin, scab res., quality and maturity and yields like Irish Cobbler, space 10 in. to avoid hollow heart, North Dakota State U, 1964. *Source History: 8 in 1981; 7 in 1984; 8 in 1987; 10 in 1991; 8 in 1994; 4 in 1998; 3 in 2004.* **Sources: Bu1, Fi1, Gur.**

Norkotah Russet - 65+ days, Smooth white tubers with shallow eyes and russet skin, matures early, bears heavy crops, res. to scab and hollow heart, dev. by Dr. Robert Johnson of North Dakota, 1987. *Source History: 7 in 1991; 7 in 1994; 11 in 1998; 14 in 2004.* **Sources: De6, Ear, Ers, Fis, Ga1, Jun, Lin, Mi8, Pe2, Pin, Ron, Shu, Tt2, Ver.**

Norland - (Red Norland, Early Bird) - Extra early, smooth, shallow eyes, nearly round, thin red skin, white fine quality flesh, moderately res. to scab, med-large spreading vines with med-large slightly closed leaves, purple blooms, heavy yields, excel. cooking quality, hardy, dev. in North Dakota in 1957 for Northern areas. *Source History: 19 in 1981; 17 in 1984; 18 in 1987; 20 in 1991; 18 in 1994; 19 in 1998; 19 in 2004.* **Sources: Alb, Bec, Bu1, Dam, Ear, Ers, Fi1, Ga1, Gur, Jun, Mcf, Mey, Na6, Pin, Ri2, Roh, Ron, Shu, Wet.**

Norland, Dark Red - Almost burgundy skin, white flesh, good keeper, matures two weeks later than its parent Red Norland, skin fades to pink in storage, good disease res., moderate scab res. *Source History: 1 in 1991; 5 in 1994; 9 in 1998; 7 in 2004.* **Sources: De6, Fa1, Fis, Ga1, Jo1, Moo, Tt2.**

Nosebag - (French Fingerling) - Purple-pink fingerling, yellow flesh, one story is that it arrived in a horse's feedbag. *Source History: 1 in 1991; 1 in 1994; 5 in 1998; 6 in 2004.* **Sources: Bu2, Ga1, Ho13, Mi8, Moo, Ron.**

OAC Ruby Gold - Patented variety from the University of Guelph breeding program, red skinned tubers with yellow flesh, excellent cooking quality similar to Yukon Gold, high yields. *Source History: 1 in 1998; 1 in 2004.* **Sources: Bec.**

Onaway - Katahdin is one parent, early-maturing, round, all white, high yielder, late blight and scab res., good for winter storage, released in Michigan in 1956. *Source History: 1 in 1987; 3 in 1991; 3 in 1994; 4 in 1998; 3 in 2004.* **Sources: De6, Moo, Wo6.**

Ozette - 100-120 days, Yellow fleshed tuber with excel. nutty flavor, dry texture, roasting or baking, ancient heirloom variety grown by the Makah people of northwest Washington. *Source History: 1 in 1994; 1 in 1998; 3 in 2004.* **Sources: Ga1, Mi8, Moo.**

Peanut - (Swedish Peanut, Mandelpotatis, Almond) - 105-135 days, Peanut shaped tubers with old-time potato taste, yellow flesh, brown almost netted skin, hardy, good keepers, drier and flakier texture than most fingerlings, mod. blight and scab tol., about 50 years old, Sweden. *Source History: 1 in 1991; 2 in 1994; 6 in 1998; 5 in 2004.* **Sources: Hi6, Mi8, Moo, Ter, Wo6.**

Pink Pearl - 80+ days, Late-maturing, long oval, pink skin, shallow eyes, white flesh, good cooking quality, high yields, good storage, wart and blight res., very attractive, bred in Canada, 1962. *Source History: 1 in 1987; 2 in 1991; 1 in 1994; 1 in 1998; 2 in 2004.* **Sources: Ga1, Ron.**

Pontiac - (Red Pontiac, Improved Red Bliss, Red Chieftain, Dakota Chief) - 100 days, Round tubers, thin reddish skin, many shallow eyes, crisp white flesh, does well in heavy muck soils, heavy yields in most soils, good keeper, all purpose potato, popular with commercial growers and home gardeners, 1939. *Source History: 19 in 1981; 16 in 1984; 21 in 1987; 24 in 1991; 23 in 1994; 25 in 1998; 24 in 2004.* **Sources: Be4, Bec, Bu1, Bu2, Co31, De6, Ear, Ers, Fa1, Fi1, Ga1, Gur, Jun, Lin, Mcf, Mey, Moo, Roh, Ron, Sau, Shu, Tt2, Ver, Wet.**

Prince Hairy - (NYL 235-4) - Medium-large round tubers, white skin, white flesh, highly res. to potato beetles, potato leaf hoppers and flea beetles, not genetically engineered. *Source History: 2 in 2004.* **Sources: Hi6, Wo6.**

Princesse (TM) Laratte - 90-100 days, Resembles

Russian Banana with its yellow skin and flesh, rich, nutty flavor, purees nicely yet maintains a firm texture when cooked, mid to late season, specialty gourmet potato from France, trademarked. *Source History: 3 in 2004.* **Sources: Ga1, Ki4, Ron.**

Ptarmigan - Slight oval shaped tubers, light yellow skin, yellow flesh, shallow eyes, extra early maturity, dev. in 1992, Canada. *Source History: 1 in 2004.* **Sources: Ga1.**

Purple Peruvian - (Peruvian Blue) - 85-100 days, The only purple fingerling, medium to large tubers are purple throughout, may be higher in protein than current cultivars, requires rich soil and ample water supply, South America. *Source History: 1 in 1987; 5 in 1991; 3 in 1994; 3 in 1998; 5 in 2004.* **Sources: Ga1, Ho13, Mi8, Moo, Ron.**

Red Cloud - 90-110 days, Flavorful all-purpose red-skinned tuber, shallow eyes, retains good skin color during long storage life, bred by Drs. Robert O'Keefe and Alexander Pavlista, U of Nebraska. *Source History: 3 in 1994; 6 in 1998; 8 in 2004.* **Sources: Coo, Dom, Hi6, Mi8, Moo, Par, Ter, Wo6.**

Red Dale - (Reddale) - Chieftain x Erik, red tubers, white flesh, good eating quality, great for baking, scab and wilt res., good keeper, early, Minnesota release, 1984. *Source History: 1 in 1981; 1 in 1984; 1 in 1987; 3 in 1991; 6 in 1994; 4 in 1998; 5 in 2004.* **Sources: Hi6, Moo, Ron, Ter, Wo6.**

Red Gold - 80+ days, Canadian variety, pale orange-red skin is lightly netted, delicate yellow flesh, good taste, excellent boiled, high-yielding under most conditions, appears to have some disease resistance, midseason, Canadian introduction, 1987. *Source History: 1 in 1987; 5 in 1991; 5 in 1994; 9 in 1998; 7 in 2004.* **Sources: Dam, Ga1, Mcf, Mi8, Moo, Na6, Ron.**

Red La Soda - (Red Lasota) - 80+ days, Bright-red 1-2 lb. tubers, slightly waxy white flesh, med-to-late, heavy yields under adverse conditions, adapted to wide range of soils and climates, home garden, used in the South for the first early market potatoes, introduced in 1954. *Source History: 1 in 1981; 1 in 1984; 1 in 1987; 4 in 1991; 5 in 1994; 9 in 1998; 6 in 2004.* **Sources: De6, Ers, Ga1, Gur, Moo, Ron.**

Red Thumb - 75-90 days, Rare fingerling with brilliant red skin and unusual red-pink flesh, medium-oblong tubers with very shallow eyes, uniform shape, prolific, well suited to the restaurant trade. *Source History: 1 in 1994; 1 in 1998; 4 in 2004.* **Sources: Ga1, Mi8, Moo, Ron.**

Reda - Certified seed from experimental station in Alaska, red skin, creamy white flesh, dumpling shape, extremely hardy, old-time European variety. *Source History: 1 in 2004.* **Sources: Ron.**

Rose Finn Apple - 105-135 days, Medium size tubers, rose buff skin with deep-yellow flesh blushed with red, waxy texture, mid to late maturity, best results when planted 10-12 in. apart in the row, overcrowding may result in tubers growing as a connected cluster. *Source History: 1 in 1991; 4 in 1994; 6 in 1998; 9 in 2004.* **Sources: Coo, Ga1, Hi6, Ho13, Mi8, Moo, Par, Ter, Wo6.**

Rose Gold - 90-110 days, Smooth red skin, golden yellow dry flesh, good baked or steamed, fair keeper, good disease res., midseason. *Source History: 2 in 1994; 3 in 1998; 6 in 2004.* **Sources: Ga1, Hi6, Mi8, Se16, Ter, Wo6.**

Rote Erstling - 100-120 days, Swedish cultivar with large tubers, skin color is maroon with plum blush, yellow flesh, skin resists flaking at harvest and darkens during storage, early, name means "early red." *Source History: 2 in 1994; 3 in 1998; 2 in 2004.* **Sources: Mi8, Ri2.**

Ruby Crescent - (Rose Finn Apple) - 85-90 days, Large fingerling type, deep-pink skin, yellow flesh, superb flavor, excellent steamed, baked, fried or used for salad, possibly the best tasting fingerling, late. *Source History: 1 in 1987; 5 in 1991; 3 in 1994; 2 in 1998; 2 in 2004.* **Sources: Pe2, Ron.**

Ruby Red - 65 days, Disease resistant variety from Minnesota, high yields of red skinned tubers with very white flesh. *Source History: 2 in 1998; 1 in 2004.* **Sources: Ki4.**

Russet Burbank - (Netted Gem, Idaho Netted Gem, The Russet) - 120-140 days, Idaho baking type, late, heavy brown russet skin, shallow eyes, oblong and well netted, does well on some muck soils, often sold as Idaho Russet, white flesh, bred by Luther Burbank from a seedball on an Early Rose plant, 1874. *Source History: 7 in 1981; 7 in 1984; 9 in 1987; 13 in 1991; 14 in 1994; 14 in 1998; 10 in 2004.* **Sources: Bec, De6, Ear, Fi1, Ga1, Lin, Mcf, Moo, Ron, Tt2.**

Russet Nugget - 90+ days, Popular russet for fresh eating or storage. *Source History: 2 in 1998; 2 in 2004.* **Sources: Mi8, Se7.**

Russian Blue - (All Blue) - Unique blue color throughout, color retained after cooking, excellent yields, good disease tolerance, midseason. *Source History: 1 in 1998; 5 in 2004.* **Sources: Bec, Dom, Lin, Ri2, Shu.**

Saginaw Gold - Oval tuber with pale-yellow flesh, very good for boiling and baking and French fries, high res. to virus and hollow heart, excel. yields, joint release by U.S. and Canada. *Source History: 3 in 1991; 3 in 1994; 2 in 1998; 1 in 2004.* **Sources: Moo.**

Sangre - (Red Sangre) - 80+ days, Early, round 2.5-3 in. tubers, deep wine-red skin, white flesh, better color, shape, uniformity and keeping qualities than other red potatoes, good boiler and baker, scab and hollow heart res., no darkening after cooking, name means "blood" in Spanish, introduced in 1984. *Source History: 1 in 1981; 1 in 1984; 1 in 1987; 4 in 1991; 5 in 1994; 6 in 1998; 6 in 2004.* **Sources: Ear, Ga1, Mi8, Moo, Ron, Se7.**

Sebago - Chippewa x Katahdin, late maturity, round tubers, smooth white skin, shallow eyes, good cooking quality, res. to late blight and common scab and potato wart and canker, USDA, 1938. *Source History: 3 in 1981; 3 in 1984; 3 in 1987; 4 in 1991; 2 in 1994; 3 in 1998; 1 in 2004.* **Sources: De6.**

Shepody - Newly licensed, long and white, high yielder, excellent French fries, late maturity, susceptible to early and late blight, Canadian release, 1980. *Source History: 1 in 1981; 1 in 1984; 2 in 1987; 3 in 1991; 7 in 1994; 7 in 1998; 6 in 2004.* **Sources: Bec, Dom, Ga1, Pin, Ron, Tt2.**

Sieglinde - Oblong shape, thin yellow skin, yellow flesh, good flavor, high yields, German heirloom. *Source History:*

1 in 2004. **Sources: Ron.**

Silverton Russet - 90-110 days, New variety from Colorado that does well in hot climates, general use potato, good for baking and frying, PVP application pending. *Source History: 1 in 2004.* **Sources: Mi8.**

Snowden - 100-120 days, Uniform medium round tubers, smooth white skin, white flesh, excellent chipper, also good for boiling and baking, matures late, stores well, from Wisconsin. *Source History: 2 in 2004.* **Sources: Ga1, Mi8.**

Superior - 105 days, Smooth oval white tubers with few shallow eyes, good size and uniform, early, excel. quality, heat drought and disease res., spring or fall crops, University of Wisconsin, 1961. *Source History: 8 in 1981; 7 in 1984; 7 in 1987; 5 in 1991; 5 in 1994; 11 in 1998; 7 in 2004.* **Sources: De6, Ers, Jo1, Jun, Mey, Roh, Shu.**

Tobique - Medium-maturing, tan skin with pink splashes, good potato taste, excellent boiler and baker, Canadian release, 1976. *Source History: 1 in 1987; 1 in 1991; 1 in 1994; 1 in 1998; 1 in 2004.* **Sources: Moo.**

Tolaas - Compact plant, high yields of large, smooth skinned tubers, white flesh, well adapted to heavy dry soil, good storage, midseason, makes long French fries, good for baking and boiling. *Source History: 1 in 2004.* **Sources: Dom.**

Tundra - 65+ days, White skin and flesh, good for baking and chips, developed by Dr. Dearborn in Alaska. *Source History: 1 in 2004.* **Sources: Ron.**

Viking - (Purple Viking) - 80-100 days, Very early tuber set, 3-4 in. potatoes in 70 days, matures to jumbos, shallow eyes, smooth red skin, white flesh, drought res., high yields, fine quality, University of North Dakota, 1962. *Source History: 4 in 1981; 4 in 1984; 5 in 1987; 6 in 1991; 5 in 1994; 8 in 1998; 8 in 2004.* **Sources: Bec, Dam, Ga1, Mcf, Mi8, Moo, Ron, Se7.**

Viking, Red - 85 days, Smooth red skin, white flesh, good yield, scab res., drought tol., midseason, released in North Dakota, 1963. *Source History: 3 in 1991; 1 in 1994; 2 in 1998; 2 in 2004.* **Sources: Ear, Tt2.**

Warba - (Early Warba, White Warba) - Widely grown early var. for short season areas, smooth, white skin, high yields. *Source History: 6 in 1981; 5 in 1984; 4 in 1987; 4 in 1991; 3 in 1994; 2 in 1998; 1 in 2004.* **Sources: Lin.**

Warba, Pink Eye - Very unique potato, golden skin with reddish-pink splashed around and in the pronounced eyes, very flavorful, uniform size, scab res., early, great keeper. *Source History: 1 in 1991; 1 in 1994; 1 in 1998; 1 in 2004.* **Sources: Ga1.**

Warba, Red - 80+ days, Old standard that has been around for over 50 years, oval shape, smooth red skin, white flesh, pronounced eyes, good boiler and baker, early, released in Minnesota, 1939. *Source History: 1 in 1981; 1 in 1984; 1 in 1987; 2 in 1991; 2 in 1994; 1 in 1998; 1 in 2004.* **Sources: Ron.**

White Cobbler - Popular early white baking potato, rather coarse texture, does not store as well as Kennebec or Red Pontiac. *Source History: 2 in 1981; 2 in 1984; 4 in 1987; 2 in 1991; 2 in 1994; 2 in 1998; 1 in 2004.* **Sources: Fi1.**

White Rose - Fairly early, white skin, deep pink eyes, long flattened shape, mealy texture, excel. for boiling and salad, does not store well, from New York, 1893. *Source History: 2 in 1981; 2 in 1984; 3 in 1987; 2 in 1991; 1 in 1994; 4 in 1998; 1 in 2004.* **Sources: Ga1.**

Yellow Finn - (True Yellow Finn) - 95-100 days, Single drop tubers, round pancake or dumpling shape, yellow flesh, delicious, good for boiling or baking, spreads in a wide area underground, long bloom time, moisture stress results in small tuber formation. *Source History: 2 in 1987; 11 in 1991; 10 in 1994; 12 in 1998; 9 in 2004.* **Sources: Ga1, Jun, Ki4, Mi8, Moo, Na6, Pe2, Ron, Se7.**

Yukon Gold - 85-95 days, Yellow flesh, excellent yielder, storage potato, early maturity, shallow pink eyes, mod. susceptible to scab, res. to leafroll, viruses A and X, Canadian release, 1980. *Source History: 1 in 1981; 1 in 1984; 5 in 1987; 18 in 1991; 24 in 1994; 34 in 1998; 39 in 2004.* **Sources: Be4, Bec, Bu1, Bu2, Co31, Coo, Dam, De6, Dom, Ear, Ers, Fa1, Fi1, Fis, Ga1, Gur, Hi6, Jo1, Jun, Lin, Mcf, Mey, Mi8, Moo, Na6, Par, Pe2, Pin, Ri2, Roh, Ron, Se7, Se16, Shu, Ter, Tt2, Ver, Wet, Wo6.**

Varieties dropped since 1981 (and the year dropped):

Acadia Russet (2004), *Ake Truedsson Blue* (1998), *Alaska Red* (1998), *All Pink* (2004), *Allegany* (2004), *Antigo Gem* (1998), *Asparagus* (1998), *Augsburg Gold* (2004), *Australian Crawlers* (2004), *Aylesbury Gold* (1998), *Banana True Seed* (1994), *Banana, White* (2004), *Batoche* (1991), *Beauty of Hebron* (1998), *Belle de Fontenay* (1998), *Belrus* (2004), *Berita* (1994), *Bevelander* (1998), *Black Joe's Toes* (1998), *Black Man's Toes* (1994), *Black Russian* (1994), *Blaze* (2004), *Blue Christy* (1998), *Blue Cloud* (2004), *Blue Goose* (1994), *Blue Mac* (2004), *Blue Mountain* (1998), *Blue Shetland* (1994), *Blue Tom Cat* (1998), *Blue Victor* (2004), *Blushing* (2004), *Brador* (1994), *Buckskin* (1994), *Candy Stripe* (2004), *Caribe Sport* (2004), *Cariboo* (1998), *Carlotta* (2004), *Carlton* (2004), *Cascade* (2004), *Castile* (2004), *Catriona* (2004), *Centennial Russet* (2004), *Century Russet* (2004), *Champion* (1998), *Cherries Jubilee* (1998), *Chile Ancud* (1998), *Chinook* (1994), *Chipeta* (2004), *Cinderella* (2004), *Cinnabar* (2004), *Cow Horn* (2004), *Cream Puff* (1998), *Crystal* (1994), *Daku Round Purple* (1998), *Danish Pepper* (1994), *Davis Purple* (1998), *Dazoc* (2004), *Deep Red* (2004), *Delta Gold* (1994), *Donna* (1994), *Dorita* (1994), *Dublin* (1994), *Duke of York* (1998), *Early Bangor* (1998), *Early Blue* (2004), *Early Epicure* (1991), *Early Gem* (1998), *Early Red, Fisher's* (1998), *Early Rose* (2004), *Early Vermont* (1998), *Epicure Red Banana* (1998), *Eureka Purple* (1998), *Explorer* (1994), *Favorite Red* (1998), *Feldeslohn* (1998), *Fenton Blue* (1994), *Fina* (1998), *Flava* (2004), *French Yellow* (1998), *Fruhmolle* (1998), *Funday* (1994), *German Finger* (2004), *German Yellow* (2004), *Glacier* (1998), *Gladstone* (2004), *Gold Coin* (1998), *Gold Rae* (2004), *Gold Russet* (1994), *Great Northern* (2004), *Greta* (1998), *Haida* (1998), *Hampton* (1991), *Hansa* (1998), *Harmony Beauty* (1994), *Heidzel Blue* (1994), *Highlite Russet* (2004), *Hilite* (1994), *Hindenberg* (1998),*

Hudson (1998), Humboldt Red (1998), Huron (1998), Indian Pit (2004), Irene (1998), Irish Treasure (1998), Islander (1994), Isles of North Germany (1998), Itasca (2004), Jemseg (2004), Johnny Gunther (1998), Kanona (2004), Kasaan (2004), Kerry Blue (1998), Keswick (2004), Laaland Pride (1998), LaBelle (1998), LaChipper (2004), Lady Finger (1998), Langlade (1998), Larota (2004), Lavender (2004), Lemhi Russet (2004), Levitt's Pink, Dark (1998), Levitt's Pink-Fleshed (1998), Little Boy Blue (1998), Long Blue Andean (1998), Longlac (2004), Mandel (1998), Mandel, Red (1998), Marygold (1998), Matsuyama (1998), Mayfair (1998), McIntyre (1998), McNeilly Everbearing (1994), Menominee (1998), Minnesota Rose (1994), Minnesota Russet 1087-4 (1998), Mystery (1998), Newfoundland Blue (1994), Newfoundland Jumbo (1998), Norchip (2004), NorDonna (2004), Norway (1994), Norway Blue (1984), Norwegian (1998), OAC Royal Gold (2004), Oberarnmacher Fruhe (1998), Ogonok (1998), Old Spanish (2004), Ontario (1998), Pawnee (1998), Peach Blow (1998), Pimpernel (2004), Pink Champaign (2004), Pink Peach (1994), Pinto (1998), Pioneer (1994), Poorlander (2004), Pressmar's German (1998), Purple (1994), Purple Baker (1998), Purple Chief (2004), Purple Marker (2004), Purple Mountain (1994), Ranger Russet (2004), Ratte (1998), Red Beauty (1998), Red Dutch (1998), Red Eye (1998), Red McClure (1984), Red Peruvian (1998), Red Sun (2004), Redsen (2004), Rhinered (1998), Rideau (2004), Rodbrokig Svensk (1998), Rojo (2004), Romeo (1998), Rosa (1994), Rosa Yellow Pear (1998), Rose Queen (1994), Rosie (2004), Round Blue Andean (1998), Rural New Yorker (1998), Russet (1991), Russet Centennial (1991), Russet Sebago (1994), Russette (1991), Sable (1991), Sante (2004), Satin Princess (1998), Scotia Blue (1994), Scotland (1994), Seneca Horn (2004), Sequoia (2004), Sharon's Blue (1994), Shaw No. 7 (1994), Sheffield Yellow (1998), Siberian (1994), Sierra (2004), Simcoe (1991), Snowchip (1994), Snowflake 1873 (1998), Steuben (1998), Strawberry (1994), Sunrise (1998), Sunset (2004), Super 777 (1994), Swartz Potatais (1998), Sweet Yellow Dumpling (1991), Trent (1994), Trina (1998), True Potato Seed - Mixed (1987), True Potato Seed - Red (1987), True Potato Seed - Russet (1987), True Potato Seed - White (1987), Urgenta (2004), Ute Russet (1998), Ware's Pride (1998), Waseca (1998), White Bliss (1998), White Dumpling (1998), White Elephant (1998), Yellow Danish (1994), Yellow Dutch (1994), Yellow Rose (1994), Yukon Purple (1998).

Quinoa
Chenopodium quinoa

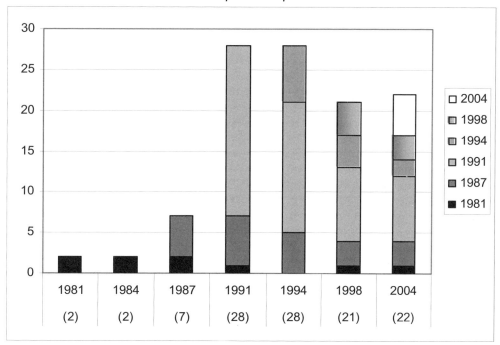

Number of varieties listed in 1981 and still available in 2004: 2 to 1 (50%).
Overall change in number of varieties from 1981 to 2004: 2 to 22 (1100%).
Number of 2004 varieties available from 1 or 2 sources: 17 out of 22 (77%).

<u>407</u> - Golden seedheads, excel. flavor. *Source History: 1 in 1998; 1 in 2004.* **Sources: So12.**

<u>407 Black</u> - Ebony sel. from 407, unusual nutty flavor. *Source History: 1 in 1998; 1 in 2004.* **Sources: So12.**

<u>Apelawa</u> - Plants 4-5 ft. tall, green and red seedheads, mid-season var., from southern Bolivia and northern Chile. *Source History: 1 in 1991; 2 in 1998; 1 in 2004.* **Sources: So12.**

Cahuil - (Cuyu) - Early maturing, 42 in. plants, medium size light-green seed with some variation, suitable for lower elevations. *Source History: 1 in 1987; 2 in 1991; 1 in 1994; 2 in 1998; 1 in 2004.* **Sources: So12.**

Cherry Vanilla - Clean, white seed, if whiteness of the seed relates to less saponin on the seedcoat, strains like this could lead to broader acceptance of quinoa as an alternative to rice, since saponin must be rinsed from the grain before cooking to avoid its soapy bitterness. *Source History: 1 in 1998; 2 in 2004.* **Sources: Hud, Sh12.**

Dave - Large seed, intense orange color, high yields. *Source History: 1 in 1994; 2 in 2004.* **Sources: Se17, Syn.**

Dave No. 407 - 90-100 days, Medium sized yellow-gold seedheads with yellow-brown seeds on 5-6 ft. tall plant, probably the most widely grown var. in western Canada, very short season, named after Dave Cusack who was instrumental in introducing quinoa into North American agriculture. *Source History: 3 in 1991; 2 in 1994; 4 in 1998; 2 in 2004.* **Sources: In8, Se7.**

Faro - (Farro) - South Chilean variety, 48 in. plants, light-green foliage, small white seeds, long season, high yielding, adaptable, performs well in Oregon. *Source History: 2 in 1987; 5 in 1991; 3 in 1994; 1 in 1998; 3 in 2004.* **Sources: Se7, Se17, So12.**

Faro Red - Vigorous 4 ft. plants, large red heads which contain hundreds of small white seeds, does best in cool weather. *Source History: 1 in 1991; 2 in 1994; 3 in 1998; 4 in 2004.* **Sources: Ho13, In8, Se17, Syn.**

Isluga Yellow - 90-110 days, Early-maturing high-yielding tall variety, beautiful golden-yellow or pink heads, med-sized yellow seeds, has grown consistently well in a variety of Western mountain and coastal sites, from northern Chile. *Source History: 3 in 1991; 3 in 1994; 3 in 1998; 2 in 2004.* **Sources: Se7, So12.**

Kaslala - 100-120 days, One of the main varieties still cultivated by the Aymara and Quichua Indians, 5-6 ft. tall, pink and gold seed head, from the central and north altiplano of Bolivia. *Source History: 1 in 1991; 1 in 1994; 2 in 1998; 3 in 2004.* **Sources: In8, Se17, So12.**

Kilo - Can grow 6-7 ft. tall, huge seed heads, resembles Lambs Quarters, not day length sensitive. *Source History: 2 in 2004.* **Sources: Sa9, Syn.**

Melang - Beautiful plant. *Source History: 1 in 1991; 1 in 1994; 1 in 1998; 1 in 2004.* **Sources: Bou.**

Multi-hued - Plants grow 5-6 ft. tall, flowers range from red through orange and yellow to purple and mauve, very productive in northern latitudes. *Source History: 2 in 1991; 2 in 1994; 2 in 1998; 2 in 2004.* **Sources: Bou, Ho13.**

Orange Head - Tall, productive plants, orange seed heads, selected by Shoulder to Shoulder Farm. *Source History: 1 in 2004.* **Sources: Hud.**

Pison - Plants grow 4-5.5 ft. tall, medium to large yellow and dark-white seeds, mid-season, from southern Bolivia. *Source History: 1 in 1991; 1 in 1994; 2 in 1998; 3 in 2004.* **Sources: In8, Se17, So12.**

Quinoa - (Petty Rice, Mother Grain) - 150 days, A Chilean cultivar that was fertile in Oregon, related to amaranth, 5 ft. tall plants, thick tops, light colored seeds. *Source History: 1 in 1981; 1 in 1984; 5 in 1987; 3 in 1991; 3 in 1994; 4 in 1998; 1 in 2004.* **Sources: Pla.**

Rainbow - Multicolored seedheads, edible seeds, young flowers and leaves and stem tips are cooked like spinach, mild flavor. *Source History: 1 in 2004.* **Sources: Hud.**

Redhead - Sel. from No. 409, brilliant red-pink seedheads, high yields, yellow-white seed, excel. thick-leaved salad green. *Source History: 1 in 1994; 1 in 1998; 1 in 2004.* **Sources: Sh12.**

Temuco - 90-130 days, Midseason, 48 in. plants, yellow-green heads with some golden, white seeds, originally from 39 deg. lat. in Chile, has performed well in Washington, California and New Mexico, often called Mother Grain of the Incas. *Source History: 1 in 1987; 8 in 1991; 9 in 1994; 7 in 1998; 3 in 2004.* **Sources: Bou, Goo, Se7.**

Wild Black - Tricolor seedheads, rose, violet and cream. *Source History: 1 in 2004.* **Sources: Syn.**

Wild Black Bolivian - 100-105 days, Plants to 3.5-4 ft. tall, shiny black seeds, smaller than common quinoa, adds crunchy texture to muffins and cookies. *Source History: 1 in 2004.* **Sources: So12.**

Varieties dropped since 1981 (and the year dropped):

Ajencha (1994), *Chesan* (2004), *Cochabamba 250* (1998), *Colorado 407* (1998), *Dulce Saj* (1994), *Early Orange No. 77* (1998), *Ecuador* (2004), *Glorieta* (1994), *Gossi* (2004), *Isluga* (1998), *Kcoito* (2004), *Lake Titicaca* (2004), *Linares Black/Red* (1998), *Linares No. 407* (1998), *Linares No. 509* (1994), *Lipez* (1998), *Llico* (1998), *Millahue* (1994), *No. 77* (1998), *No. 409* (1998), *Tunari* (1998).

Radish
Raphanus sativus

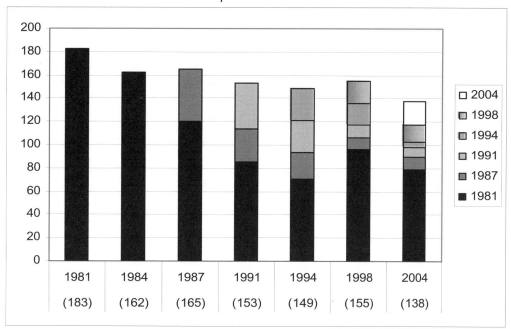

Number of varieties listed in 1981 and still available in 2004: 183 to 79 (43%).
Overall change in number of varieties from 1981 to 2004: 183 to 138 (75%).
Number of 2004 varieties available from 1 or 2 sources: 77 out of 138 (56%).

All Seasons - (Tokinashi, All Seasons White) - 45 days, White, tapered, use until 12 x 2.5 in., grows to 24 in., holds 6 weeks before pithy or strong, slow-bolting, sow anytime, good keeper. *Source History: 13 in 1981; 13 in 1984; 13 in 1987; 11 in 1991; 9 in 1994; 5 in 1998; 5 in 2004.* **Sources: Bu2, Hi13, Kit, Loc, OSB.**

Altabelle - 28 days, Uniform, deep red roots, slow to become pithy, disease res., trademarked. *Source History: 2 in 1998; 5 in 2004.* **Sources: Lin, ME9, OSB, RUP, Sto.**

Altaglobe - 28 days, Large, cherry-red globe, retains color, crisp flesh, resists pithiness, tolerates DM and Rhizoctonia, resists FW. *Source History: 1 in 1998; 5 in 2004.* **Sources: Ers, HO2, Sk2, Ter, We19.**

Bartender Red Mammoth - 35 days, Bright deep-red long tapered roots, 9 x 1.25 in. dia., firm crisp pungent pink flesh, med- tall vigorous tops, heavy yielder under semi-tropical conditions, rare heirloom. *Source History: 1 in 1981; 1 in 1984; 1 in 1987; 2 in 1991; 4 in 1994; 4 in 1998; 7 in 2004.* **Sources: Ba8, Fo7, He8, Hud, Ri12, Sk2, WE10.**

Beer Garden - (German Beer) - White radish from Germany that grows as big as a turnip, traditionally served sliced and either dipped in sugar, or salted and washed down with beer, produces edible seed pods if sown early. *Source History: 1 in 1981; 1 in 1984; 1 in 1987; 1 in 1991; 2 in 1994; 2 in 1998; 2 in 2004.* **Sources: Gr28, Lej.**

Black Spanish, Long - (Noir Gros Long d'Hiver) - 55-70 days, Black outer skin, good-keeping winter var. with cylindrical roots, 7-10 x 2-2.5 in. dia. at shoulder, pure-white crispy flesh, pungent flavor, roots store well, strong tops, pre-1828. *Source History: 31 in 1981; 28 in 1984; 26 in 1987; 18 in 1991; 13 in 1994; 12 in 1998; 5 in 2004.* **Sources: Ba8, DOR, He17, Ho13, La1.**

Black Spanish, Round - (Noir Gros Rond d'Hiver) - 53-80 days, Large turnip-shaped 3-4 in. dia. globes, deep-black skin, solid crisp pungent pure-white flesh, tall tops, primarily for winter storage, sow in July or August, will keep all winter stored in moist sand, intro. before 1824. *Source History: 49 in 1981; 43 in 1984; 43 in 1987; 42 in 1991; 43 in 1994; 55 in 1998; 59 in 2004.* **Sources: Ba8, Bo17, CO23, Co32, Com, Dam, DOR, Ear, Ers, Fe5, Fis, Fo7, Gr27, HA3, He8, Hi13, HO2, Ho12, Ho13, Hud, Jor, Jun, La1, Lin, ME9, Mel, Mey, Mo13, Ol2, OLD, Ont, Or10, PAG, Ri12, RIS, Roh, RUP, Sa5, Sa9, Sau, SE4, Se7, Se8, Se16, Se25, Se26, Sh9, Shu, Si5, Sk2, So1, So25, Sto, Syn, Ter, Up2, Vi4, WE10, We19.**

Black Winter Round - 55 days, Imported from Italy, large smooth and round, deep-black skin, solid white flesh, sow July or August for fall or winter use, keeps all winter stored in moist sand. *Source History: 1 in 1981; 1 in 1984; 1 in 1987; 1 in 1991; 1 in 1994; 1 in 1998; 1 in 2004.* **Sources: Tu6.**

Blanc de Sezanne - Beautiful med. size oval roots, white with a light magenta top, tasty, very vigorous. *Source History: 2 in 2004.* **Sources: In8, Se17.**

Brightest Breakfast - 20 days, Oblong scarlet red radish with white tip, 1.5 in. long, mild and sweet, introduced in the 1870s. *Source History: 1 in 1991; 1 in 1994; 1 in 1998; 1 in 2004.* **Sources: He8.**

Burpee White - (Burpee's Round White, Burpee White Globe, Snoball) - 25 days, Nearly round, pure-white skin and flesh, med-tall strap-leaved foliage, tender and mild, best at .75-1 in. dia., holds longer than most without getting soft and pithy. *Source History: 14 in 1981; 12 in 1984; 9 in 1987; 7 in 1991; 3 in 1994; 3 in 1998; 3 in 2004.* **Sources: Alb, Bu2, Ear.**

California Mammoth White - (California White Mammoth) - 50-60 days, Winter radish with large long pure-white 8-12 x 2-2.5 in. cylindrical roots, fine quality, mild-flavored solid crisp white flesh, dates back to the 1800s. *Source History: 8 in 1981; 8 in 1984; 5 in 1987; 4 in 1991; 4 in 1994; 3 in 1998; 4 in 2004.* **Sources: Ba8, DOR, Ers, MAY.**

Candela di Ghiaccio - Italian, 8-10 in. long white roots, very uniform, crisp texture, mild pungency. *Source History: 1 in 1987; 2 in 2004.* **Sources: Se24, Sk2.**

Champion - 20-30 days, AAS/1957, for upland soils and home garden, bright-red 1-1.5 in. globes, 3.5-4.5 in. tops, firm mild flesh, grows large but not pithy, holds well, can stand cold, large Cherry Belle type, sow either spring or fall. *Source History: 84 in 1981; 80 in 1984; 76 in 1987; 72 in 1991; 73 in 1994; 72 in 1998; 61 in 2004.* **Sources: ABB, Alb, Be4, Bu1, Bu2, Bu8, CA25, CH14, Cl3, CO23, CO30, Com, Dgs, DOR, Ear, Ers, Fe5, Fi1, Fis, Fo7, Ga1, Gr27, Gur, HA3, Ha5, He17, Hig, HO2, Hum, Jor, Jun, Kit, La1, Lin, MAY, ME9, Mey, Mo13, MOU, Ni1, OLD, Or10, PAG, Pe6, Roh, RUP, Sa9, Sau, SE4, Se7, SE14, Se26, Sh9, Shu, Si5, Sto, Tt2, WE10, Wet, Wi2, WI23.**

Cherry Belle - 20-30 days, AAS/1949, round, bright cherry-red .75-1.5 in. dia., 2.5-3 in. tops, crisp firm-white flesh, from Holland for indoor forcing or on muck or mineral soils, good keeper, can plant all summer long, res. pithiness, earlier than Comet, not recommended for sowing after May 10th. *Source History: 132 in 1981; 123 in 1984; 115 in 1987; 103 in 1991; 98 in 1994; 100 in 1998; 99 in 2004.* **Sources: Ag7, Alb, All, Be4, Ber, Bo17, Bou, Bu2, Bu3, Bu8, CA25, CH14, Cl3, CO23, CO30, Co31, Cr1, Dam, De6, Dgs, DOR, Ear, Eco, Ers, Fa1, Fe5, Fi1, Fis, Ga1, Ge2, Goo, Gr27, Gur, HA3, Ha5, Hal, He8, He17, Hi6, Hi13, HO2, Hum, Jor, Jun, La1, Lej, Lin, LO8, Loc, Ma19, MAY, Mcf, ME9, Mey, Mo13, MOU, Na6, Ni1, OLD, Ont, OSB, PAG, Pe2, Pin, Pr3, Pr9, Ri2, Ri12, Roh, RUP, Sa9, Sau, Scf, SE4, Se7, Se8, SE14, Se26, Sh9, Shu, Si5, Sk2, So1, So9, SOU, Sw9, Te4, TE7, Ter, Tho, Tt2, Twi, Ver, Ves, Vi4, WE10, Wet, Wi2, WI23.**

China Rose - (Rose of China, Scarlet China, China Rose Winter, Chinese Red Winter, Winter Rose) - 27-55 days, Chinese winter radish brought to Europe from China by Jesuit missionaries, smooth rose-colored skin, 6-8 x 2 in. dia., becomes larger towards tip, tall tops, excel. for sprouting, good keeper, sow spring or fall, 1850. *Source History: 56 in 1981; 49 in 1984; 54 in 1987; 48 in 1991; 46 in 1994; 59 in 1998; 57 in 2004.* **Sources: Alb, Ba8, Be4, Bo18, Bu8, CH14, Com, Dam, DOR, Ear, Eo2, Ers, Ev2, Fo7, Ga1, Gr27, Gr29, HA3, He8, He17, HO2, Hud, Jor, Jun, Kit, La1, Lin, LO8, Ma19, MAY, Mey, Mo13, Ol2, OLD, Ont, Or10, PAG, Pe2, Pr3, Ra5, Roh, RUP, Sa9, Sau, Se16, Se25, Se26, Se28, Shu, Si5, So1, So12, Sto, Vi4, WE10, Wet, Wo8.**

China White - (China White Winter) - 55-60 days, Least pungent winter radish, pickling or winter storage, 5-8 x 2.5 in. dia., plant mid to late August, withstands light frost, stores crisp and mild all winter. *Source History: 8 in 1981; 8 in 1984; 15 in 1987; 18 in 1991; 17 in 1994; 18 in 1998; 21 in 2004.* **Sources: Bu8, CH14, DOR, Ers, Gr27, He17, HO2, Lej, LO8, ME9, Mo13, OLD, Ont, PAG, Pr9, Ros, RUP, Sa9, SE4, Sh9, So12.**

Chinese - 55 days, Roots are tender and crisp with almost no pungency, 6 in. long x 3 in. dia., white with green neck above ground, pickling or raw eating. *Source History: 1 in 1981; 1 in 1984; 2 in 1987; 2 in 1991; 1 in 1994; 1 in 1998; 1 in 2004.* **Sources: So25.**

Chinese White Celestial - (Celestial White Winter, Celestial Mammoth White Stump Root, Calif. Mammoth White) - 55-60 days, Cylindrical and stump-rooted, 6-9 x 2.5-3 in. dia., pure-white skin, large coarse tops, solid crisp very mild white flesh, good for pickling, excel. keeper. *Source History: 28 in 1981; 24 in 1984; 21 in 1987; 17 in 1991; 15 in 1994; 15 in 1998; 11 in 2004.* **Sources: Co31, La1, Mey, Ni1, Roh, Se26, Shu, SOU, Sto, Wet, Wi2.**

Chinese, Green Goddess - 80 days, Pear shaped roots, green skin, pale green crisp flesh with mild flavor, stands well in the ground through Christmas, bring inside for storage, sow June-July to avoid bolting. *Source History: 1 in 2004.* **Sources: Tho.**

Comet - 23-26 days, AAS/1936, deep-globe to round, best at .75-1 in., bright-red skin, white flesh, 4 in. tops, tangy flavor, not pithy even when good-sized, suited for mid-season crops, remains firm longer than other early globe types, ideal for cello-pak, good variety for either forcing or bunching. *Source History: 37 in 1981; 31 in 1984; 33 in 1987; 22 in 1991; 18 in 1994; 15 in 1998; 12 in 2004.* **Sources: Alb, Bu8, CO30, DOR, Ear, Gr27, He8, Loc, Sh14, Sk2, Sto, WI23.**

Crimson Giant - (Early Crimson Giant, Silver Dollar) - 25-30 days, Largest early turnip-shaped type, deep-crimson skin, mild white flesh, 4-5 in. top, heat res., grows apple size if given room but stays tender and sweet, always solid, never gets pithy, vigorous. *Source History: 51 in 1981; 47 in 1984; 53 in 1987; 46 in 1991; 42 in 1994; 43 in 1998; 33 in 2004.* **Sources: All, Be4, Bo17, Bo19, Bu2, Bu8, CH14, CO30, De6, DOR, Ers, Fi1, Gr27, He8, Hi13, Jor, La1, Lej, LO8, Loc, MOU, OLD, Red, Roh, Ros, RUP, Sau, Se26, Sh9, Sh14, Shu, WE10, WI23.**

Daikon - (Daikon Long White, Giant Oriental Fall Radish) - 65 days, Daikon is Japanese for radish, fine for spring or fall planting, pure-white long pointed root, 12-18 in. long x 2-3 in. dia., crisp firm flesh. *Source History: 7 in 1981; 7 in 1984; 8 in 1987; 16 in 1991; 12 in 1994; 12 in 1998; 17 in 2004.* **Sources: Ban, Bo19, Com, Dgs, Ga1, HA3, He17, Kil, Ma19, Mcf, Mo13, OLD, Pr3, Ri2, Se26, Shu, Te4.**

Daikon, Early 40 Days, 38-45 days, Early, sweet slender 12-20 in. white roots, small erect tops, greenish-white stems, likes mild weather, developed for sprouting, seeds sprouted like bean sprouts. *Source History: 3 in 1981; 3 in 1984; 2 in 1987; 1 in 1991; 1 in 1994; 1 in 1998; 1 in 2004.* **Sources: Ev2.**

Daikon, Mino Early - (Mino Early Sakata Improved, Mino Early No. 1, Chinese Radish Mino Early, Lo Bok) - 45-60 days, Japanese, white, 16-20 x 3 in., 1-6 lbs., mildly pungent, brittle, vigorous and fast growing, heat and mosaic res. *Source History: 3 in 1981; 3 in 1984; 6 in 1987; 4 in 1991; 4 in 1994; 4 in 1998; 3 in 2004.* **Sources: Ev2, Jor, Loc.**

Daikon, Sprouting - *Source History: 1 in 1981; 1 in 1984; 1 in 1987; 4 in 1991; 4 in 1994; 4 in 1998; 3 in 2004.* **Sources: CO23, DOR, La1.**

Daikon, Tokinashi - (All Seasons) - 65-70 days, Long white Japanese radish, mildly pungent, good for pickling, for early autumn and spring sowings. *Source History: 1 in 1981; 1 in 1984; 1 in 1987; 4 in 1991; 2 in 1994; 3 in 1998; 1 in 2004.* **Sources: So12.**

Daikon, Winter - *Source History: 1 in 1991; 1 in 1994; 1 in 1998; 1 in 2004.* **Sources: But.**

D'Avignon - 21-25 days, Longest French variety, slender tapered root, 3-4 in. long, 66-75% red with a white tip, good pithiness tolerance, from southern France. *Source History: 1 in 1987; 1 in 1991; 2 in 1994; 2 in 1998; 7 in 2004.* **Sources: Ag7, Coo, Jo1, Se8, Sw9, VI3, Wi2.**

Dessert Cherry - 22 days, Smooth cherry-red globes, crisp tender white flesh, small tops are excellent for bunching, holds quality longer in hot weather than other varieties. *Source History: 2 in 1981; 2 in 1984; 1 in 1987; 1 in 1991; 1 in 1994; 1 in 1998; 1 in 2004.* **Sources: Lin.**

Duro - Very large roots, red skin, white center, 4 in. diameter, retains firmness, spicy. *Source History: 1 in 2004.* **Sources: Tho.**

Edible Longpod - 55-60 days, Grown for the 10-14 in. edible immature seed pods, used like snap beans when immature, adds a nice pungent zing to dishes, worthy "new" old crop. *Source History: 1 in 1991; 1 in 1998; 1 in 2004.* **Sources: So12.**

Eighteen Days, (Demi-Longue de 18 Jours) - 18 days, The quickest, eighteen days from seed to the table, old o.p. var. with red roots and white tips. *Source History: 1 in 1981; 1 in 1984; 3 in 1987; 1 in 1991; 1 in 1998; 1 in 2004.* **Sources: Gou.**

Fancy Red - 25 days, Med-sized uniform round bright red roots, very fine taproots, res. to FW, high tol. to black root and RS, fancy, Dr. Humaydan and Dr. Williams at University of Wisconsin, PVP expired 1995. *Source History: 1 in 1981; 1 in 1984; 1 in 1987; 3 in 2004.* **Sources: HA20, Jor, RIS.**

Flamboyant 3 - Long red root to 4 in., white tip, pretty and tasty. *Source History: 1 in 2004.* **Sources: Se24.**

Flamboyant Sabina - 28 days, Half long root, bright red with 1/5 white tip, early. *Source History: 1 in 1994; 2 in 1998; 1 in 2004.* **Sources: Tho.**

Flamivil - 25 days, Semi-long, very scarlet with small white tip, resists pithiness, heirloom breakfast radish from France. *Source History: 1 in 1981; 1 in 1984; 1 in 1987; 3 in 1991; 4 in 1994; 2 in 1998; 3 in 2004.* **Sources: Ki4, Orn, VI3.**

Four Seasons - Grown for its leaves, not the roots, leaves have a mild flavor, grows well in early spring and autumn. *Source History: 1 in 1998; 1 in 2004.* **Sources: Te4.**

French Breakfast - (French Breakfast Early) - 20-30 days, Oblong and blunt, 1.5-2 x .75 in. dia., rose-scarlet with white tip, 5-6 in. top, crisp white flesh, small slender taproot, distinct mildly pungent flavor, top quality, grown since the 1880s. *Source History: 89 in 1981; 81 in 1984; 84 in 1987; 82 in 1991; 80 in 1994; 84 in 1998; 97 in 2004.* **Sources: Ag7, Alb, All, Ba8, Be4, Bo17, Bo19, Bou, Bu2, Bu8, CH14, CO30, Co32, Com, Dam, De6, Dgs, DOR, Ear, Ers, Fa1, Fe5, Fi1, Fis, Fo7, Fo13, Ga1, Ga22, Gr27, Gur, HA3, Ha5, Hal, He8, He17, Hi6, Hi13, Hig, HO2, Hum, Jor, Jun, Kil, Kit, La1, Lej, Lin, LO8, Loc, Ma19, MAY, Mcf, Mey, Mo13, MOU, Na6, Ni1, OLD, Ont, Or10, Orn, OSB, PAG, Pe2, Pin, Pr3, Ra5, Ri2, Ri12, Roh, Ros, RUP, Sa9, Sau, Scf, Se7, Se8, SE14, Se16, Se26, Sh9, Shu, Sk2, So9, Sto, TE7, Tho, Tt2, Tu6, Up2, Ver, Ves, Vi4, WE10, We19, Wet, WI23.**

French Breakfast 25 L D Orig. - Large top strain producing a high percentage of first quality roots, cylindrical carmine-red roots with white tips, blunt bottoms, grows well spring through summer. *Source History: 1 in 1981; 1 in 1984; 1 in 1987; 1 in 1991; 1 in 1994; 1 in 1998; 1 in 2004.* **Sources: Ter.**

French Dressing - 21 days, Uniform, medium size root .75 x 2 in. long, bright red with white tip *Source History: 1 in 2004.* **Sources: Bu2.**

French Golden - 32 days, Light golden brown skin, 4-5 in. long roots, crisp icy white flesh, distinctive piquant flavor, mid- season, adds a beautiful color to salads. *Source History: 3 in 1981; 2 in 1984; 1 in 1987; 1 in 1998; 1 in 2004.* **Sources: Ho13.**

Fuego - 24-25 days, Bright scarlet-red, slightly deeper than round, 2.5-2.75 in. top, FW ABR and RS res., crisp white flesh, newly dev. by Univ. of WI and MN and FL, bunching and cello-pak, introduced by Rogers NK. *Source History: 8 in 1981; 5 in 1984; 7 in 1987; 12 in 1991; 14 in 1994; 11 in 1998; 6 in 2004.* **Sources: CH14, DOR, HO2, Jor, Ki4, RUP.**

Galahad - 20 days, Medium top version of Saxafire, same tol. to club root, dark-green tops are excel. for bunching. *Source History: 1 in 1981; 1 in 1984; 2 in 1987; 2 in 1991; 2 in 1994; 1 in 1998; 1 in 2004.* **Sources: Sto.**

Gaudry ll - Improved Sparkler type, two-tone color tops, bi-color roots, half red and half white, does not split or get woody, excel. fresh market var. *Source History: 1 in 2004.* **Sources: Dam.**

German Giant - (Parat) - 29 days, Very large round red radish from Germany, scarlet-red skin, crisp white slightly pungent flesh, harvest from marble to baseball size, will not get woody or spongy. *Source History: 1 in 1981; 1 in 1984; 1 in 1987; 2 in 1991; 3 in 1994; 4 in 1998; 15 in 2004.* **Sources: Be4, Bu1, Ers, Fa1, He8, He17, HO2, Jor, Jun, Mo13, SE14, TE7, Ver, Vi4, WE10.**

Gournay Violet - (De Gournay, Violet de Gournay) - 65-70 days, French purple-skinned white-fleshed cylindrical storage radish, 8-10 x 1.5 in. dia., tapers towards tip, very sweet mild flavor, grows rapidly. *Source History: 4 in 1981; 4 in 1984; 1 in 1987; 1 in 1991; 1 in 1994; 1 in 1998; 1 in 2004.* **Sources: Lej.**

Green Meat - (Continental Fancy) - Unique and beautiful with dark-green neck with green flesh, white tip, 3 x 10 in. long, for pickling, salads or cooking. *Source History: 1 in 1987; 2 in 1991; 2 in 1994; 4 in 1998; 4 in 2004.* **Sources: Dam, Ev2, OSB, Pin.**

Green Skin and Red Flesh - 55 days, Unusual fall or winter radish, green skin, reddish-purple flesh, 4-5 in. long x 4 in. dia., 1.5 lbs., eaten raw or overnight pickling or in salads, popular in Peking. *Source History: 1 in 1981; 1 in 1984; 1 in 1987; 1 in 1991; 3 in 1994; 3 in 1998; 1 in 2004.* **Sources: So12.**

GV - Long, pointed white root, 12-18 in. long by 2-3 in. diameter, firm crisp flesh, baby leaves used in salad mixes. *Source History: 1 in 2004.* **Sources: Orn.**

Hailstone - (White Globe Hailstone, White Globe, Round White, White Button) - 23-30 days, Pure-white skin, firm flesh stays crisp for a long time, dark-green strong med-sized tops, larger than most early globe types, plant either spring or fall. *Source History: 19 in 1981; 16 in 1984; 20 in 1987; 17 in 1991; 12 in 1994; 18 in 1998; 26 in 2004.* **Sources: Ba8, Bu3, CO23, DOR, Ers, Fe5, Fis, Gr27, He8, HO2, Hum, Jor, Ki4, Ma19, MOU, OLD, RUP, SE14, Se26, Sh9, Sk2, Sto, Twi, Vi4, WE10, WI23.**

Hild's Blue - 30 days, Bright blue-purple 4 in. long tapered roots, white flesh, zesty flavor. *Source History: 1 in 2004.* **Sources: OSB.**

Icicle - 27-30 days, Straight white long oval 5-6 x 1 in. root with tapered tip, med-small tops, clear-white brittle mild flesh, for home garden or market, plant spring or fall. *Source History: 22 in 1981; 20 in 1984; 18 in 1987; 16 in 1991; 22 in 1994; 21 in 1998; 10 in 2004.* **Sources: Ada, Bu3, Bu8, Hum, Mey, Mo13, PAG, SE4, Se7, Tu6.**

Icicle, Long White - 27-30 days, Early long white 4.5-6 in. radish, slender, tapers at tip to a point, clear-white mild flesh, grows quick, does not get pithy, small tops, can force in greenhouse. *Source History: 9 in 1981; 9 in 1984; 10 in 1987; 7 in 1991; 7 in 1994; 8 in 1998; 9 in 2004.* **Sources: Com, Dam, Fi1, HA3, Hal, Hud, Lej, Or10, Wet.**

Icicle, Short Top - 23-30 days, Short tops, cylindrial 4-6 in. nearly straight roots, clear-white throughout, very smooth, tender and mild, grows rapidly, earliest long white radish, uniform. *Source History: 7 in 1981; 4 in 1984; 5 in 1987; 5 in 1991; 6 in 1994; 6 in 1998; 6 in 2004.* **Sources: Ha5, Kil, ME9, Roh, Ter, We19.**

Icicle, White - (Lady Finger, White Finger) - 27-32 days, White uniform slim 4-6 x .5-1 in. roots, tapers to a point, very slender taproot, crisp mild white flesh, 5-6 in. tops good to eat, easy to force, fine for outdoors, holds its quality well, pre-1865. *Source History: 90 in 1981; 82 in 1984; 79 in 1987; 73 in 1991; 62 in 1994; 63 in 1998; 66 in 2004.* **Sources: ABB, Alb, All, Be4, Bo17, Bu2, CH14, CO30, Co31, Co32, Cr1, De6, Dgs, DOR, Ear, Ers, Fa1, Fis, Fo7, Gr27, Gur, He8, He17, Hi13, Jo1, Jor, Jun, Kit, La1, Lin, LO8, Loc, Ma19, MAY, Me7, Mel, MOU, Ni1, OLD, Ont, Pe2, Pin, Ros, RUP, Sa9, Sau, Se8, SE14, Se16, Se17, Se26, Sh9, Shu, Si5, Sk2, So1, So9, SOU, Sto, TE7, Twi, Ves, Vi4, WE10, Wi2, WI23.**

Icicle, White (Short Top) - 27-30 days, Short top str. of White Icicle for forcing or for muck or upland soils, pure-white slender 5-6 in. roots, crisp and mild, 3.75-4.5 in. tops ideal for bunching. *Source History: 9 in 1981; 9 in 1984; 9 in 1987; 8 in 1991; 6 in 1994; 6 in 1998; 3 in 2004.* **Sources: Ga1, HO2, RIS.**

Indian White - Long root to 2.5 ft., firm white, crisp flesh, mild flavor. *Source History: 1 in 1994; 1 in 2004.* **Sources: Ban.**

Japanese Ball - Round, white, 3-4 in. root, sweet if grown in the fall, pungent in spring with warmer weather. *Source History: 1 in 1998; 1 in 2004.* **Sources: Ho13.**

Jumbo - 24 days, European radish with apple-sweet taste, 3 in. dia., bright red round roots, no pithiness during hot weather, Stokes introduction. *Source History: 1 in 1994; 1 in 1998; 1 in 2004.* **Sources: Ear.**

Kim Giant - Root remains crisp and tender up to the size of a baseball if well fed and watered, edible pods are good in salads, stir fry or pickled. *Source History: 1 in 2004.* **Sources: Sa5.**

Korea Green - 55 days, Large Korean radish with green skin and flesh, 10 in. long, good for fresh use, pickled or added to salads. *Source History: 1 in 1991; 2 in 1994; 2 in 1998; 1 in 2004.* **Sources: Ho13.**

Lady Slipper - 27-32 days, Round/oblong soft pink roots, white flesh, retains crispness, holds up to one month under refrigeration. *Source History: 2 in 2004.* **Sources: Ers, OLD.**

Lo Bok - 45-60 days, Chinese radish, pure-white roots, 5.5 x 1.75 in. dia., large 12 in. tops can be cooked with roots, crisp white flesh, matures in 45 days in spring, 60 days in fall, there are many varieties of Lo Bok which may have white or pink or black skins. *Source History: 3 in 1981; 3 in 1984; 5 in 1987; 3 in 1991; 2 in 1994; 2 in 1998; 1 in 2004.* **Sources: Ban.**

Long Black - (Noir Gros Long d'Hiver) - Thin and cylindrical, 10-12 x 1.5 in. dia., black skin, white flesh, pungent taste, sow July or August, excel. keeper. *Source History: 2 in 1981; 2 in 1984; 3 in 1987; 1 in 1991; 1 in 1994; 1 in 1998; 1 in 2004.* **Sources: Lej.**

Long Scarlet - (Cincinnati, Cincinnati Market, Long Bright Scarlet, Long Scarlet Cincinnati) - 25-30 days, Long tapered deep-red roots, 6 x .75 in. roots, crisp tender white flesh, medium tops, for bunching and home gardens, pre-1870s heirloom. *Source History: 13 in 1981; 12 in 1984; 10 in 1987; 9 in 1991; 8 in 1994; 12 in 1998; 8 in 2004.* **Sources: Ba8, DOR, Ers, Sa9, SE14, Sk2, Ter, WE10.**

Long Scarlet Short Top - (Cincinnati Market, Long Red Scarlet Short Top) - 27-30 days, Tapered 5-6 in. bright-scarlet roots, mild white flesh, 5 in. light-green tops, older long var. for home and marketing, introduced before 1800. *Source History: 11 in 1981; 8 in 1984; 9 in 1987; 4 in 1991; 3 in 1994; 2 in 1998; 1 in 2004.* **Sources: Ho12.**

Madras Sweet Edible Pod - 30 days, Seed stalk grows 3-4 ft. tall, sweet, mildly hot, 2-3 in. pods are delicious raw or cooked, attracts beneficial insects, root is not edible, heirloom from 1885. *Source History: 3 in 1998; 3 in 2004.* **Sources: Ho13, In8, Se17.**

Mammoth White - Elongated shape. *Source History: 1 in 1981; 1 in 1984; 2 in 1987; 1 in 1991; 1 in 1994; 2 in 1998; 1 in 2004.* **Sources: Pp2.**

Marabelle - 21-23 days, Extra short tops, perfectly round, med-large, bright-red, unbelievably uniform, successive outdoor crops all season, greenhouse too, A.R. Zwaan of Holland, 1982. *Source History: 1 in 1981; 1 in 1984; 1 in 1987; 3 in 1991; 4 in 1994; 2 in 1998; 1 in 2004.* **Sources: Se8.**

Minowase - (Minowase Dark-Green Leaved, Minowase Long, Minowase Early, Minowase Cross No. 1, Minowashi) - 45-58 days, Pure-white, 12-24 x 3 in. dia., very smooth, tender, little pungency, sow summer or fall, early, often seen in Chinatown markets, disease res, unique taste, uniform, brittle juicy flesh, from Japan. *Source History: 6 in 1981; 6 in 1984; 8 in 1987; 7 in 1991; 8 in 1994; 14 in 1998; 12 in 2004.* **Sources: Ba8, DOR, Ers, Gur, Ho13, Kit, Ont, Roh, So12, Und, WE10, WI23.**

Mirabeau - 30 days, Pale red cylindrical roots, white tips, good for bunching. *Source History: 1 in 2004.* **Sources: Tho.**

Misato Green - Misato Red counterpart, bright green flesh, up to 10 in. long, crisp flesh, easy to grow, from Japan. *Source History: 2 in 1994; 3 in 1998; 2 in 2004.* **Sources: Ba8, Kit.**

Misato Green Flesh - 60 days, Green color all through the root, 2 x 10 in., fine grained and sweet. *Source History: 2 in 1998; 2 in 2004.* **Sources: Ni1, RUP.**

Misato Red - (Misato Rose) - 60 days, Large round red Chinese radish, 5 in. dia., solid white dry flesh, not for salad use, cut leaves, petioles tinged with red, late in becoming pithy, excellent keeper, fall/winter type. *Source History: 1 in 1981; 1 in 1984; 1 in 1987; 1 in 1991; 2 in 1994; 3 in 1998; 2 in 2004.* **Sources: Ba8, RUP.**

Misato Rose Flesh - 60-65 days, Imported from China, round white roots with green shoulders, 4 in. dia., red flesh, not for spring sowing, stores well, good in salads and cooked. *Source History: 1 in 1987; 5 in 1991; 5 in 1994; 7 in 1998; 8 in 2004.* **Sources: Com, Coo, Fe5, Kit, Ni1, Pin, RUP, Ver.**

Miyashige - (White Myashige) - 50-78 days, Popular fall and winter Japanese radish, 12-15 x 2-3 in. dia., 2-6 lbs., pure-white, stump-rooted, crisp, pale-green band at neck, excel. quality, pickling and storage. *Source History: 9 in 1981; 9 in 1984; 7 in 1987; 8 in 1991; 7 in 1994; 11 in 1998; 12 in 2004.* **Sources: Ev2, Hig, Ho13, Jo1, Kit, Ni1, Pe2, Ri12, Se7, So9, So12, Syn.**

Miyashige Long, White Neck - (Pointed Rooted) - A fall and winter type, 14-16 in. long x 2.5-3 in. dia., pure-white skin all the way to the crown, flesh of very fine texture, mildly pungent, excellent quality. *Source History: 1 in 1981; 1 in 1984; 1 in 1987; 1 in 1991; 1 in 1998; 1 in 2004.* **Sources: Ev2.**

Munchen Bier - (Munich Bier, Madras, German Beer) - 52-67 days, Unique German var., produces masses of tender juicy stringless sweet 2-3 in. seedpods at top of 24 in. stem, whole or chopped in salads, boiled or steamed or stir-fried, 1885. *Source History: 1 in 1981; 1 in 1984; 1 in 1987; 2 in 1991; 4 in 1994; 5 in 1998; 10 in 2004.* **Sources: Bou, Coo, Ho13, Jun, Lan, Ri12, Se16, Se26, Shu, Ver.**

Munchener Bier - 55-60 days, Famous German Oktoberfest radish, eaten sliced on black bread or with pretzels and washed down with fine German beer, 3.5-4 in. globes, firm white pungent flesh. *Source History: 2 in 1981; 2 in 1984; 1 in 1998; 1 in 2004.* **Sources: Fe5.**

Nerima - (Nerima Long Takakura, Nerima Longest, Nerima Long Neck, Japanese Nerima) - 62-65 days, Japanese, 20-30 x 3 in. dia., both ends taper, crisp mild white flesh, mosaic res., late, slow-bolting. *Source History: 8 in 1981; 8 in 1984; 6 in 1987; 4 in 1991; 2 in 1994; 2 in 1998; 3 in 2004.* **Sources: Dam, Kit, OSB.**

Novired - 21 days, Round roots, medium tops, color does not bleed when washed, imported from Holland. *Source History: 2 in 1991; 2 in 1994; 1 in 1998; 1 in 2004.* **Sources: Gou.**

Ohkura - (Okhura, Winter Queen, Stump Rooted) - 65-80 days, Japanese, fall and winter, stump-rooted, 12-20 x 3-4 in. dia., sow 75-80 days before first frost, pure-white flesh, sprawling leaves, excel. keeper and shipper. *Source History: 7 in 1981; 6 in 1984; 4 in 1987; 2 in 1991; 1 in 1994; 3 in 1998; 2 in 2004.* **Sources: Ho13, Se7.**

Parat - (German Giant, Parat Large Round Red, Parat Sperling) - 29-40 days, Large round radish imported from Germany, used from marble to baseball size, will not get woody or spongy or split, scarlet skin, white flesh, slightly pungent. *Source History: 1 in 1981; 1 in 1984; 3 in 1987; 7 in 1991; 6 in 1994; 10 in 1998; 11 in 2004.* **Sources: Fi1, Gr27, Gur, Ho13, Lej, Pin, Ri12, Se7, Se26, Shu, So9.**

Philadelphia White Box - Small round white root, good for cold frame culture, dev. in the 1890s. *Source History: 1 in 1981; 1 in 1984; 1 in 1987; 1 in 2004.* **Sources: Ho13.**

Pink Beauty - 27 days, Rose-pink color, round and smooth, small taproot, unique flavor, crisp white flesh, grows large but stays mild and not pithy, adds a new color to relish trays. *Source History: 2 in 1981; 2 in 1984; 4 in 1987; 2 in 1991; 3 in 1994; 4 in 1998; 7 in 2004.* **Sources: Ba8, Be4, Gr27, HO2, ME9, SE14, TE7.**

Pink Celebration - 30 days, Round root, 1-1.25 in., pink skin, mild white flesh, holds well in storage. *Source History: 3 in 2004.* **Sources: CO23, Ers, He17.**

Plum Purple - (Purple Plum) - 25-30 days, Bright-purple skin, firm white flesh, crisp, sweet and mild all season, never pithy or hot, hardy and adaptable, PVP expired 2003. *Source History: 2 in 1987; 14 in 1991; 12 in 1994; 8 in 1998; 15 in 2004.* **Sources: Ag7, Bo19, Co31, Coo, Ers, Fe5, HO2, La1, Lin, Loc, Roh, Sa9, SE14, Sk2, We19.**

Purple - Light purple skin color, 3 in. long, crisp, pungent flesh. *Source History: 1 in 1998; 1 in 2004.* **Sources: Ho13.**

Purple Olive-Shaped - 35-40 days, Size varies from olive to small plum, beautiful violet skin, complex flavor changes with the weather, hot flavor when grown in warm weather, flavor sweetens in cool weather, a pre-1760 French variety from William Woys Weaver who inherited it from his great grandfather. *Source History: 3 in 2004.* **Sources: Ea4, Ho13, In8.**

Rapid Red Sanova - Round root, intense red color, nice sweet taste. *Source History: 1 in 2004.* **Sources: Se24.**

Rat's Tail - (Oriental Seed Pod Radish) - 50 days, Developed in South Asia, brought to India from Java by the British, grown solely for its slender dark-purple 10-12 in. long edible seedpods, no root worth mentioning, pick before pods develop fiber, eaten raw or cooked or pickled, slightly hot radish taste, briefly grown in U.S. kitchen gardens in the 1860s. *Source History: 1 in 1981; 1 in 1984; 1 in 1987; 1 in 1991; 2 in 1994; 6 in 1998; 17 in 2004.* **Sources: Bou, Fe5, Gr28, Ho13, Hud, Ki4, Mcf, Pin, Pr3, Ri12, Sa5, Se8, Se16, Se17, Syn, Ter, Tho.**

Raxe - Large German bunching type, resists pithiness, resists bolting when planted from spring to fall, excel. market item. *Source History: 1 in 2004.* **Sources: Dam.**

Rebel - 25-30 days, Medium tops, bright red roots, good for bunching, resists pithiness, good shelf life, early or late seeding. *Source History: 4 in 1998; 5 in 2004.* **Sources: BE6, Ki4, OSB, SE4, Ves.**

Red Boy - 22-23 days, Round to slightly oval bulb, very bright-red, very short 2.25-2.5 in. tops, fine taproot, holds for a longer period than most varieties, high quality, spicy, does well on muck or high nitrogen soils or greenhouse, for home or market use, excel. for bunching or poly bags. *Source History: 21 in 1981; 13 in 1984; 12 in 1987; 4 in 1991; 3 in 1994; 5 in 1998; 2 in 2004.* **Sources: DOR, Sto.**

Red Devil - 21-24 days, Intense bright-red, pure-white crisp flesh, round to olive-shaped, can eat when dime size, grows larger than a quarter, bunching or cello-pak, ships well. *Source History: 5 in 1981; 5 in 1984; 1 in 1987; 1 in 2004.* **Sources: CO30.**

Red Flame - 24 days, French Breakfast type, 3 in. long, cylindrical shape, bright red with white tip, good for slicing. *Source History: 1 in 1998; 1 in 2004.* **Sources: Par.**

Red Fortress - 25 days, Bright red globes, crisp white flesh, little or no pithing, disease res. *Source History: 1 in 2004.* **Sources: HA20.**

Red King - 25 days, Firm med-size globular bright-red roots, mild flavor, med-size tops, holds very well in the field, first radish on the market to combine resistances to Fusarium yellows and club root (race 6), Harris Moran, PVP expired 2001. *Source History: 1 in 1987; 2 in 1991; 2 in 1994; 2 in 1998; 1 in 2004.* **Sources: Tu6.**

Red Meat - (Continental Fancy, Watermelon Radish) - 50-60 days, Intriguing gourmet radish with bright-red heart, white globes are tender and sweet, 3.25 in. dia., best as late fall crop, colorful in salads. *Source History: 1 in 1987; 3 in 1991; 3 in 1994; 5 in 1998; 7 in 2004.* **Sources: Dam, Jo1, Jor, Lej, Orn, OSB, Ves.**

Red Silk - 26-30 days, Healthy short tops, smooth-textured 1 in. roots, crunchy white flesh, excel. quality, resists pithing, bunching, PVP 1999 HA20. *Source History: 1 in 2004.* **Sources: ME9.**

Red Skoaring - Round red roots. *Source History: 1 in 1998; 1 in 2004.* **Sources: Tu6.**

Reggae - 26-28 days, Tall tops, egg shaped red roots, vigorous, very high crack res., resistant to flea beetles and root maggots, slow bolting. *Source History: 1 in 1998; 3 in 2004.* **Sources: Coo, Pin, SE4.**

Riso - 85 days, Winter type, similar to Nerima but earlier maturing, slender roots up to 18 in. long, pickling and cooking. *Source History: 1 in 1981; 1 in 1984; 1 in 1987; 1 in 2004.* **Sources: Kit.**

Rudolf - (Bejo 1837) - 25 days, Short tops, bright red roots, crisp and tasty, excellent tolerance to cracking and pith development, early maturity. *Source History: 1 in 1998; 1 in 2004.* **Sources: Ves.**

Sakurajima - (Mammoth White Globe, Sakurajima Mammoth, Sakurajima Giant) - 60-150 days, Japanese, slow-growing, summer sowings in Florida produce roots to 14 in. dia. and 65 lbs., 10 lb. roots from mid-August sowings in Maryland, pure-white tender flesh, no pungency, vigorous, stored like carrots for winter use, good fodder or livestock feed. *Source History: 17 in 1981; 15 in 1984; 9 in 1987; 5 in 1991; 5 in 1994; 7 in 1998; 6 in 2004.* **Sources: Bu1, Fa1, Ho13, Kit, Ni1, Syn.**

Salad Rose - 35 days, Beautiful rose-pink root, 8 in. long, young leaves good in salad or stir fry, good fall crop, from Russia. *Source History: 1 in 1998; 2 in 2004.* **Sources: Bu2, Hi13.**

Saxa - (Saxa Short Top) - 18-25 days, Deep-red globes with very small tops, pure-white flesh, plant spring or fall, ideal for forcing or garden, holds well in good condition, fully dev. in 3 weeks. *Source History: 6 in 1981; 4 in 1984; 5 in 1987; 4 in 1991; 2 in 1994; 3 in 1998; 5 in 2004.* **Sources: Be4, Coo, Dam, Ers, Fi1.**

Scarlet Globe - 23-25 days, Bright-scarlet olive-shaped 1 in. dia. roots, sweet brittle white flesh, very small tops, thread-like taproot, holds well in good condition, plant spring or fall, good either for forcing or outdoor culture. *Source History: 29 in 1981; 23 in 1984; 16 in 1987; 14 in 1991; 11 in 1994; 9 in 1998; 9 in 2004.* **Sources: Ada, Ca26, Co31, Ers, Gr27, Hal, Mel, Se26, SOU.**

Scarlet Globe Special - 23 days, Finest market garden strain of Scarlet Globe, perfectly globe-shaped, thin taproot, uniform bright scarlet color. *Source History: 1 in 1981; 1 in 1984; 1 in 1987; 1 in 1991; 1 in 1994; 1 in 1998; 1 in 2004.* **Sources: Sto.**

Scarlet Globe, Early - (Early Globe) - 20-28 days, Deep-globe, bright-red skin, white flesh, bright-green 2.5-3 in. tops, for frame or greenhouse forcing or for peat or muck soils during summer, market gardener str., adapted to mech. harv., good on upland or sandy soils, excel. buncher. *Source History: 55 in 1981; 49 in 1984; 50 in 1987; 47 in 1991; 48 in 1994; 52 in 1998; 40 in 2004.* **Sources: Ag7, Alb, All, Ba8, Be4, Bu1, Bu3, Bu8, But, CH14, Cr1, De6, Dgs, Dom, Fo7, Ge2, HA3, He8, Hum, Jor, Jun, La1, LO8, Ma19, MAY, Mey, Ont, Pla, Roh, RUP, Sau, SE4, SE14, Se16, Sh9, Shu, Vi4, WE10, Wet, Wi2.**

Scarlet Globe, Early (Short) - (Short Top Scarlet Globe) - 20-25 days, Short Top str. bred for greenhouse forcing and growing on peat or muck, 2.5-3 in. tops, bright-scarlet skin, pure-white flesh, strong short tops in cool weather, valuable for 2nd and 3rd spring crops, also for fall planting. *Source History: 17 in 1981; 15 in 1984; 11 in 1987; 6 in 1991; 6 in 1994; 4 in 1998; 4 in 2004.* **Sources: HO2, Kil, PAG, Ros.**

Scarlet Globe, Medium Top - 23 days, *Source History: 3 in 1991; 3 in 1994; 3 in 1998; 2 in 2004.* **Sources: DOR, WI23.**

Scarlet Knight - 22-23 days, Smooth and round, bright-scarlet skin, crisp white flesh, light-green 3-3.25 in. tops for easy bunching, fusarium wilt res., good for home garden or market, hardy. *Source History: 14 in 1981; 9 in 1984; 9 in 1987; 4 in 1991; 2 in 1994; 1 in 1998; 1 in 2004.* **Sources: HO2.**

Scarlet Long - 26-30 days, Short green tops, tapering 5-6 in. long roots, deep-scarlet skin, firm white flesh, stays mild well into summer. *Source History: 1 in 1981; 1 in 1984; 2 in 1987; 1 in 1991; 2 in 2004.* **Sources: Kit, LO8.**

Scarlet Turnip White Tip - (Early Scarlet Turnip White Tip, Early Scarlet White Tipped, Scarlet White Tip, Sparkler) - 24-26 days, Nearly round turnip-shaped bright-scarlet root with small white tip, sweet flesh stays fresh and crisp a long time, 1-1.5 in. dia., gets huge in good conditions, does not get hollow or pithy, intro. before 1859. *Source History: 21 in 1981; 18 in 1984; 14 in 1987; 12 in 1991; 12 in 1994; 11 in 1998; 7 in 2004.* **Sources: Ba8, HA3, Ho12, Hud, Lin, Pr3, SOU.**

Sezanne - Round white root with pink-red on top, Europe. *Source History: 1 in 2004.* **Sources: Sk2.**

Shangtung Green Skin - 55-60 days, Chinese fall and winter radish, grows partly above ground, green above and white below, 12 x 2 in. dia., 1 lb., sweetness increases during storage, can overwinter. *Source History: 1 in 1981; 1 in 1984; 1 in 1987; 1 in 1991; 2 in 1998; 2 in 2004.* **Sources: Ho13, So12.**

Shinrimei - 50-60 days, Unique Chinese radish, ball shaped root with deep green shoulder, crisp, bright red flesh, mild, sweet flavor, name translates "beauty heart" in Chinese. *Source History: 1 in 1994; 3 in 1998; 5 in 2004.* **Sources: Ev2, Ho13, In8, Ki4, Se17.**

Shogoin - (Shogoin Round Giant, Shogoin Large Round, Shogoin Giant Fall) - 50-80 days, Japanese white winter radish, turnip-shaped, 5.5 x 6 in. dia., sweet and mild, sow July or August, very productive. *Source History: 5 in 1981; 5 in 1984; 4 in 1987; 4 in 1991; 4 in 1994; 4 in 1998; 4 in 2004.* **Sources: Ev2, Kit, Ni1, Pe6.**

Shunkyoh Semi-Long - 30-35 days, Imported from northern China, red 5 in. long pink-red roots, crisp white flesh, edible long pink stems are pickled and used in stir fry, sow early spring or fall. *Source History: 1 in 1987; 1 in 1991; 1 in 1994; 1 in 1998; 6 in 2004.* **Sources: Jo1, Kit, Ra5, Se28, Vi5, Wo8.**

Silver Dollar - (Giant White, Hailstone) - 30 days, Deep-crimson globe-shaped roots, tender white flesh, will grow as big as silver dollars, yet stay crisp and mild. *Source History: 2 in 1981; 2 in 1984; 2 in 1987; 2 in 1991; 2 in 1994; 1 in 1998; 1 in 2004.* **Sources: MAY.**

Snow Belle - (Burpee Snow Bell) - 21-30 days, Smooth pure-white skin, globe-shaped, firm white flesh, tangy flavor, med- dark green 3.75 in. tops, very early and uniform, good for bunching, attractive, PVP expired 2000. *Source History: 6 in 1981; 5 in 1984; 15 in 1987; 19 in 1991; 19 in 1994; 18 in 1998; 6 in 2004.* **Sources: Fi1, HO2, Lin, LO8, RUP, Wet.**

Sora - 26 days, Large smooth bright-red roots, solid texture with high tol. to sponginess, good tol. to hot weather, suitable for spring or summer or fall crops. *Source History: 1 in 1991; 1 in 1994; 1 in 1998; 3 in 2004.* **Sources: OSB, Pe2, Tu6.**

Sparkler - (Scarlet Sparkler White Tip, Early Scarlet Turnip White Tip, Brightest White Tip) - 24-28 days, Bright-scarlet skin with white on lower 1/3, round to round-oval, 1-1.25 in. dia., sweet juicy white flesh, medium tops, for home or market gardens, snappy flavor. *Source History: 83 in 1981; 76 in 1984; 80 in 1987; 71 in 1991; 64 in 1994; 65 in 1998; 57 in 2004.* **Sources: Ada, Alb, All, Be4, Bu3, Bu8, CH14, CO30, Co31, Cr1, De6, Dgs, DOR, Ear, Ers, Fi1, Fis, Fo7, Ga1, Hal, He17, HO2, Hum, Jor, Jun, Kil, La1, Lej, LO8, Loc, Ma19, MAY, ME9, Mey, Mo13, Ni1, OLD, Ont, PAG, Par, Ri2, Roh, Ros, RUP, Sau, SE14, Se26, Sh9, Si5, So1, Sto, TE7, Vi4, WE10, Wet, Wi2, WI23.**

Sprouts - Specially bred for sprouting, normally takes 2-4 days to reach .5-1 in. long when they are at their best, hot flavor, use sparingly. *Source History: 2 in 1991; 1 in 1994; 4 in 1998; 2 in 2004.* **Sources: Kit, WE10.**

Tai-Mei-Hwa - Popular in San Francisco's Chinatown, 6 x 3 in. root, blunt end, white, no pungency. *Source History: 1 in 1998; 1 in 2004.* **Sources: Ho13.**

Todo Red - (Todo Red Sprouting) - Planted for microgreens, red root. *Source History: 2 in 2004.* **Sources: CO23, ME9.**

Tokinashi - (All Seasons) - 55-85 days, Japanese, tapered, white, 12-14 x 2-3 in., 1-2 lb., harvest young roots in 30 days, mildly pungent brittle flesh, good for spring sowing, slow bolt. *Source History: 6 in 1981; 6 in 1984; 7 in 1987; 4 in 1991; 5 in 1994; 6 in 1998; 5 in 2004.* **Sources: Bou, In8, Ni1, Pe6, Se17.**

Tondo Rosso a Punta Bianca - 35-45 days, Half red and half white roots. *Source History: 1 in 1987; 1 in 1991; 1 in 1994; 1 in 1998; 1 in 2004.* **Sources: Ber.**

Tricolour - Older Chinese variety, three colors on the root, green on top, white on the bottom and pink flesh, late season. *Source History: 1 in 2004.* **Sources: Pr3.**

Valentine - 25 days, A novelty, round green and white radish for salads, red flesh at maturity. *Source History: 2 in 1987; 2 in 1991; 2 in 1994; 2 in 1998; 1 in 2004.* **Sources: Sto.**

Wakayama White - 70 days, Crisp, tender, long white root, harvest when 12-14 in. long, stir fry or pickled, sow late summer to early fall. *Source History: 1 in 2004.* **Sources: Kit.**

White Globe - (Snow Belle, Hailstone) - 25 days, Early white variety, round to flattened globe in shape, white skin, firm white flesh, attractive. *Source History: 5 in 1981; 4 in 1984; 6 in 1987; 7 in 1991; 9 in 1994; 8 in 1998; 9 in 2004.*

Sources: CH14, Fo7, Ga1, He17, La1, Mo13, Pe2, Ri12, Sau.

White Prince, The Big - 26 days, Largest size of any early round white radish, most are size of golf balls, holds its mildness in spite of large size, crisp white flesh. *Source History: 2 in 1981; 2 in 1984; 1 in 1991; 1 in 1998; 1 in 2004.* **Sources: Pr3.**

White Strassburg - 35-40 days, Excel. heat res. summer var., Icicle shaped but not as large or brittle, mild and sweet even in mid summer, white skin and flesh, grows quick, holds very well, old variety. *Source History: 3 in 1981; 2 in 1984; 1 in 1998; 1 in 2004.* **Sources: Sa9.**

White Tipped Scarlet Globe - 45 days, Roots are the size of a golfball, pungent and crisp, sweet when cooked. *Source History: 1 in 1981; 1 in 1984; 1 in 1998; 1 in 2004.* **Sources: So12.**

White Winter - White winter radish, 7-8 in. long, blunt tip, crisp, solid flesh, good cooked, tasty in cool weather. *Source History: 1 in 1987; 1 in 1991; 1 in 1994; 2 in 1998; 2 in 2004.* **Sources: Ba8, Sau.**

Varieties dropped since 1981 (and the year dropped):

Akamaru (1987), *Andino* (1987), *Aomarukoshin* (1987), *Arabic Red* (2004), *Awa Pickling* (1991), *Belle Glade* (1998), *Benary's Reform* (1987), *Bisai* (1998), *Black* (1991), *Black Globe* (1991), *Black Luxury* (1991), *Brightest Long Scarlet* (1994), *Burgess Special* (1994), *Burpee White Round* (1982), *Burpee's All Season* (1982), *Cavalier* (2004), *Cavalrondo* (2004), *Cedar* (1987), *Cerise* (1998), *Cherry Belle 6D* (1991), *Cherry Belle Extra Short Top* (2004), *Cherry Bomb* (2004), *Chinese Greenskin* (1998), *Chinese, Par Chi* (1994), *Chinese, Santung* (1998), *Chinese, Ta Mei Hwa* (1998), *Copenhagen Market Dema* (2004), *Crimson and Long* (1994), *Crystal White* (1991), *Daiko* (1994), *Daikon, Round* (1994), *Dandy* (1998), *Danra* (2004), *Delikat* (1987), *Delongpont* (1984), *Delta Red* (1998), *Earliest* (1991), *Early Delta* (1987), *Early Lo Bok* (1991), *Easy Red* (1998), *Edible Long Purple Podded* (1998), *Fancy Red II* (1994), *Far Red* (1991), *Faribo White Snoball* (1987), *Fire Candle* (2004), *Firecracker* (1984), *Flambo* (1991), *Flamboyant* (2004), *Flamboyant 25* (2004), *Flamboyant Vidan* (2004), *Fluo* (2004), *Fruhrot* (1994), *Gala* (2004), *Gaudo* (1994), *Giant Radish* (1991), *Global* (2004), *Golden* (1994), *Green Meat Chinese* (1994), *Green Skin* (2004), *Half Long Scarlet White Tip 35* (2004), *Half Long Scarlet White Tip 393* (2004), *Half Long Scarlet White Tip 875* (2004), *Half Long Scarlet White Tip Fota* (2004), *Hankook Kim-jJng* (1991), *Herbstrettich Munchner Bier* (1987), *Hsiao Hung Lopo* (1982), *Icicle Medium Top* (1991), *Icicle Syla* (2004), *Icicle, Early* (2004), *Ideal White* (1987), *Inca* (1994), *Jaba* (1998), *Japanese* (2004), *Japanese Long White* (1987), *Jumbo Scarlet* (1991), *Kaiware Sprouting* (1987), *Lady Finger* (1991), *Langer Schwarzer Winter* (1987), *Large Round Red* (1987), *Late Lo Bok* (1991), *Long Red* (2004), *Long Red Bartender* (1987), *Long Red Italian* (1998), *Long Scarlet Amiens* (1991), *Long Scarlet Icicle* (1983), *Long White* (2004), *Marketgardener* (1984), *Martian* (1991), *Mexican Bartender* (1998), *Mikado* (2004), *Minokunichi Improved* (1991), *Minowase Summer Cross No. 1* (1998), *Minowase Summer Cross No. 3* (1994), *Mr. Big* (1991), *Munchner Weisser Treib* (1987), *National* (2004), *Ne Plus Ultra (Forcing)* (1983), *New White Chinese* (1991), *Non Plus Ultra* (1987), *Non Plus Ultra 406* (2004), *Novitas* (1987), *NS 586* (1994), *NV 3258* (1991), *Oliva* (1991), *Oregon Scarlet Globe* (1987), *Oriental* (1994), *Osaka Shijunichi* (1991), *Par Chi Lo Bok* (1991), *Perfection Extra Early* (1991), *Pointed Root Miyashige Long* (1987), *Poker* (2004), *Pontvil* (1998), *Prinz Rotin* (1998), *Radar* (1987), *Rapid Red* (1991), *Rave d'Amiens* (1991), *Rave Longue Saumonee* (1998), *Red Ball* (2004), *Red Baron* (2004), *Red Beret* (2004), *Red Boy II* (1998), *Red Cherry* (2004), *Red Devil B* (2004), *Red Fall* (1994), *Red Meat Chinese* (1991), *Red Pak* (1998), *Red Prince* (2004), *Red Prince Improved* (2004), *Red Prince K Strain* (1994), *Red Rose* (1991), *Revosa* (2004), *Ribella* (1998), *Rosie* (1983), *Rosy Gem* (1982), *Rota* (1998), *Rovi* (2004), *Saxa Korto* (2004), *Saxafire* (1994), *Saxa-Nova* (1987), *Scarlet Champion* (1994), *Scarlet Globe Forcing* (1982), *Scarlet Globe Greenhouse* (1987), *Seoul Spring Radish* (1991), *Sezanne Type* (1991), *Shijunichi* (1987), *Southern Market Globe* (1991), *Sparkler, Cooper's* (1994), *Stop Lite* (1991), *Stump Root Miyashige Long* (1987), *Sunyoh Semi-Long* (1994), *Sure Shot* (1991), *Ta May Hua* (1987), *Ta Pai Lopo* (1982), *Takuan* (1998), *Tan Kua Pai* (1991), *Tendersweet* (1983), *Terao Spring White* (1987), *Uzbekistan* (2004), *Vintage* (2004), *White and Long* (1998), *White Ball* (2004), *White Giant Globe* (2004), *White Giant Radish* (1994), *White Mammoth* (1983), *White Prince* (1998), *White Round* (2004), *White Snowball* (1987), *White Turnip* (1983), *Winter King Miyako* (2004), *Winter Queen* (2004), *Winter Red* (1987), *Winterretch* (1987), *Zwaans Champion* (1984).

Runner Bean
Phaseolus coccineus

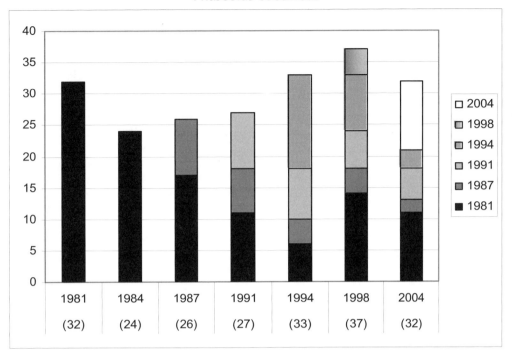

Number of varieties listed in 1981 and still available in 2004: 32 to 11 (34%).
Overall change in number of varieties from 1981 to 2004: 32 to 32 (100%).
Number of 2004 varieties available from 1 or 2 sources: 23 out of 32 (72%).

Achievement Merit - 85 days, British Selection of Achievement bean, red flowers, tender sweet long pods. *Source History: 1 in 2004.* **Sources: Tho.**

Aleppo Arab - 90-100 days, White flowers, large white seed used as dry bean, very young pods used as string beans, from Aleppo, Syria market, thought to be similar to runner beans documented in Aleppo, 1575. *Source History: 1 in 2004.* **Sources: Se17.**

Apricot Runner - 100 days, Beautiful apricot flowers on rambling vines, unusual. *Source History: 1 in 2004.* **Sources: Bu2.**

Aztec Dwarf White - (Potato Bean, Aztec Half Runner) - 55-75 days, Archaeological finds of this bean indicate a history of cultivation stretching back to the Anasazi and perhaps the Aztecs, bush plants have short runners to 3 ft. that do not need staking, white flowers, white seeds the size of plump lima beans, drought tol. and hardy, stand up in hot days and cool nights. *Source History: 1 in 1987; 5 in 1991; 2 in 1994; 3 in 1998; 4 in 2004.* **Sources: Ho13, Pla, Se7, Tu6.**

Black - 105 days, Variant of Scarlet Runner, can be used as a snap but most often used dry, excellent flavor, holds shape and color well when cooked. *Source History: 2 in 1994; 1 in 1998; 1 in 2004.* **Sources: Ho13.**

Blackcoat - 95-105 days, Vines to 7 ft., orange-red flowers, black seeds, excellent in soups and salads, used as snaps in spite of strings and as a green shelly, recorded as rare in 1654. *Source History: 1 in 1991; 1 in 1994; 1 in 1998; 2 in 2004.* **Sources: Se7, Se17.**

Butler - (The Butler) - 65 days, Excel. runner bean, stringless fiberless pods average 12 in., grows quickly, huge crops, excel. flavor, sets well in hot weather, quite ornamental. *Source History: 1 in 1981; 1 in 1984; 1 in 1987; 2 in 1991; 2 in 1994; 1 in 1998; 1 in 2004.* **Sources: Bou.**

Desiree - 65-85 days, Entirely stringless, white flowers, huge crops, often 40 fleshy 10-12 in. pods per plant, shy seeder, yields well when dry, BCMV and SS res., excel. flavor, Holland. *Source History: 2 in 1981; 1 in 1984; 2 in 1987; 2 in 1991; 3 in 1994; 2 in 1998; 1 in 2004.* **Sources: Tho.**

Droitwich Champion - *Source History: 1 in 1991; 1 in 2004.* **Sources: WE10.**

Emergo - (White Runner Bean, Snij-Pronker Emergo, White Emergo) - 80 days, White-flowered Scarlet Runner type, strong growing plants withstand poor weather, high yields, fine quality but pick young. *Source History: 3 in 1981; 3 in 1984; 1 in 1987; 3 in 1991; 2 in 1994; 3 in 1998; 3 in 2004.* **Sources: Com, Fo7, Ho13.**

Goliath - (Prizetaker) - 65 days, Extremely productive, produces pods up to 22 in. long, fine texture, heavy crops, reliable and uniform, excel. quality, black/purple seeds, England. *Source History: 2 in 1981; 2 in 1984; 1 in 1987; 2 in 1991; 1 in 1994; 1 in 1998; 1 in 2004.* **Sources: Ver.**

Granny's Scarlet Runner - Profuse vibrant scarlet blossoms, large seeds are purple on black, lavender with black flecks and all black, grow in clusters, good snaps when young, also makes a fine dry bean. *Source History: 2 in 2004.* **Sources: In8, Se17.**

Hammond Scarlet Runner Bush - (Hammond's Dwarf Red Flowered) - 55-60 days, Plants grow 16-18 in. tall, produce beautiful ornamental red-orange blossoms throughout summer which attract hummingbirds, flat 6-7 in. pods, good snap beans when young, later use as a shell bean. *Source History: 2 in 1987; 2 in 1991; 3 in 1994; 2 in 1998; 6 in 2004.* **Sources: Ba8, Bou, Roh, Sk2, Tho, Ver.**

Hestia - 70 days, Dwarf plants produce long, straight, stringless pods held high off the ground, white and vermilion flowers, early. *Source History: 2 in 2004.* **Sources: Par, Tho.**

Jack and the Beanstalk - 80-85 days, Polish heirloom, to 20 ft., best used when immature in the green shelling stage, large white seeds good for soup, rare. *Source History: 2 in 2004.* **Sources: Pe1, Se7.**

Lady Di - 65 days, High yields of 12 in. fleshy pods, longer harvest season due to slow seed development, good quality. *Source History: 1 in 1994; 1 in 1998; 1 in 2004.* **Sources: Tho.**

Painted Lady - (Painted Runner) - 68-90 days, Beautiful red and white blossoms, pods 9-12 in. long, seeds are pink-brown with dark streaks, edible flowers and beans, can be used to cover a porch or fence, good for freezing, from the Archives, 1855. *Source History: 1 in 1981; 1 in 1984; 1 in 1987; 4 in 1991; 3 in 1994; 5 in 1998; 19 in 2004.* **Sources: Ba8, Bou, Bu2, Co32, Coo, Ga7, Jo1, Ni1, Pe2, Pr3, Re6, Rev, Ri12, Se16, Sw9, Ter, Th3, Ver, We19.**

Prizewinner - 70-75 days, Improved European strain of Scarlet Runner, large fleshy slightly fuzzy 12-16 x .5 in. pods, large violet seeds, fine flavor, showy scarlet flowers, holds well on vine. *Source History: 4 in 1981; 3 in 1984; 2 in 1987; 3 in 1991; 4 in 1994; 2 in 1998; 2 in 2004.* **Sources: Sk2, WE10.**

Red Rum - 60-85 days, Sparse foliage results in high yields of 6-8 in. succulent pods, HB tolerant. *Source History: 1 in 2004.* **Sources: Tho.**

Royal Standard - 65 days, Smooth bright-green 20 in. pods are completely stringless, good pod set under adverse conditions, very heavy yields. *Source History: 1 in 1991; 1 in 1994; 1 in 1998; 1 in 2004.* **Sources: Tho.**

Sadie's Horse Bean - 75-80 days, Vines to 8 ft., scarlet or white flowers enjoyed by hummingbirds, huge maroon and black seed, also tan and brown, white, mottled striped and speckled, fun to shell, heirloom from a North Carolina mountain family over 100 years ago. *Source History: 2 in 2004.* **Sources: Ri12, Se17.**

Salt and Pepper - *Source History: 1 in 2004.* **Sources: Se17.**

Scarlet Emperor - 60-90 days, Fine heavy-cropping variety with smooth-textured 12-15 in. dark-green pods, beautiful purple and black mottled seeds up to 1 in. long, excel. for kitchen or exhibition work, showy scarlet-orange flowers, ceremonial bean used by SW native and Mexican indigenous cultures, 1906. *Source History: 1 in 1981; 1 in 1984; 2 in 1987; 6 in 1991; 9 in 1994; 13 in 1998; 11 in 2004.* **Sources: Coo, Ga1, Ni1, Pe1, Pr3, Se7, Se26, Sw9, Ter, Tho, We19.**

Scarlet Runner - (Best of All, Scarlet Emperor) - 65-100 days, Vines 6-18 ft., enjoys cool weather, half-hardy roots, brilliant scarlet-orange flowers, beautiful mauve on black seeds, stringless med-green pod, green shells or snaps, good in English climate, attracts hummingbirds, discovered in South America in the 1600s, brought to England in the 1700s, intro. to North America in the 1800s. *Source History: 49 in 1981; 46 in 1984; 52 in 1987; 49 in 1991; 53 in 1994; 50 in 1998; 62 in 2004.* **Sources: Alb, All, Ba8, Bo17, Bu2, Co32, Com, Cr1, De6, Dow, Ear, Ers, Fe5, Fis, Fo7, Fo13, Ge2, Goo, HA3, Hal, He8, Hi13, Ho13, Hum, Jo1, La1, Lan, Lin, LO8, Mcf, Mi12, Na6, Ol2, OLD, Ont, Pe2, Pin, Pla, Ra5, Re6, Red, Rev, Ri12, Roh, Scf, SE4, Se16, Shu, So1, So12, Sto, Te4, Th3, Tt2, Tu6, Und, Up2, Ver, Ves, Vi4, WE10, Wet.**

Scarlet Runner, Dwarf Bees - 65-80 days, Bush runner bean, 20-24 in., bright scarlet blossoms produce edible long flat snap beans when immature, "shelley" beans and dry beans for cooking when left to dry on the vine, germinates in cooler soil than most beans, will cross with runner beans, attractive to hummingbirds, rare heirloom from 1853. *Source History: 2 in 1991; 5 in 1994; 5 in 1998; 6 in 2004.* **Sources: Bou, Fo13, Pr3, Ter, Und, We19.**

Scarlet Runner, Tarahumara - 90 days, Large deep-green plants, bright scarlet flowers, fuzzy 7 in. pods, purple seeds flecked with black, dried beans are edible but not as tasty as other dry bush bean varieties, very attractive to hummingbirds and beneficial insects. *Source History: 1 in 1991; 1 in 1994; 1 in 1998; 1 in 2004.* **Sources: Se8.**

Streamline - 65-75 days, Fine flavored pods in clusters all along vine, pods up to 12-16 in., pick young, early, excel. freezer, an improved Scarlet Emperor, highly commended by Royal Hort. Society in 1970. *Source History: 2 in 1981; 2 in 1984; 1 in 1987; 3 in 1991; 3 in 1994; 3 in 1998; 5 in 2004.* **Sources: Com, Roh, Sk2, Tho, WE10.**

Sun Bright - 90 days, Gold tinged foliage and red flowers provides a nice color contrast, bears a good crop until frost, slightly less vigor than green leaf varieties. *Source History: 2 in 2004.* **Sources: Hi13, Tho.**

Sunset - 60-75 days, Decorative salmon-pink flowers, continues to crop throughout the season, distinctive runner bean flavor, excel. for freezing. *Source History: 1 in 1981; 1 in 1984; 1 in 1987; 4 in 2004.* **Sources: Co32, Coo, Se16, Tho.**

Tocamres Chocolate - Large brown runner bean with a rich, starchy flavor. *Source History: 1 in 1994; 1 in 1998; 1 in 2004.* **Sources: Ho13.**

White Dutch - 88 days, Very large pure-white seed, plants grow over 10 ft., blossoms attract hummingbirds, very hardy, highest quality, for small snaps or sliced larger pods or shell, pre-1825. *Source History: 4 in 1981; 4 in 1984; 3 in 1987; 3 in 1991; 3 in 1994; 5 in 1998; 2 in 2004.* **Sources: Ho13, Ver.**

White Lady - 85 days, Clusters of medium green smooth, stringless pods, white seeds, white flowers not as attractive to birds, sets pods in high temperatures. *Source History: 2 in 2004.* **Sources: Hi13, Tho.**

Varieties dropped since 1981 (and the year dropped):

Achievement (1987), *Ayocote* (1982), *Aztec White* (1987), *Best Of All* (1994), *Betty* (2004), *Black Knight* (2004), *Brown and Pink Seeded* (1987), *Buckskin* (1983), *Buton* (2004), *Butterfly* (1987), *Champion Scarlet Runner* (1983), *Chiapas Black* (1982), *Crusader* (1987), *Czar* (2004), *Donna* (2004), *Emergo Stringless* (1991), *Emperor* (2004), *Enorma* (2004), *Erecta* (2004), *Goodrich's White Soup* (1991), *Grammy Tilley* (2004), *Gulliver* (2004), *Hija* (1998), *Ivanhoe* (2004), *Jerusalem* (2004), *Kelvedon Marvel* (2004), *Mergoles* (1983), *Mexican Rainbow Beans* (1987), *Oregon Lima* (1994), *Paleface* (1983), *Pickwick* (2004), *Polish Giant Lima* (1987), *Red Knight* (2004), *Scarlet Runner, Wild* (1994), *Souper Bean* (1991), *Tarahumara Bordal* (2004), *Tarahumara Tekomari* (2004), *Whatcom Lima* (1987), *White Knight* (1994), *White Stringless* (1998), *Zembylas* (2004).

Rutabaga (Swede Turnip)
Brassica napus (Napobrassica Group)

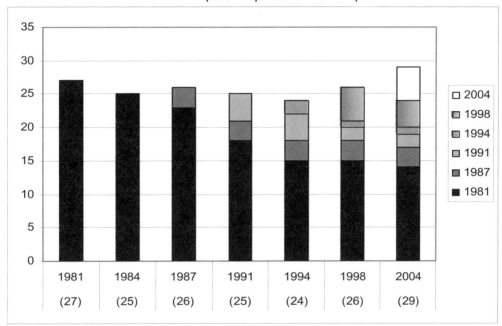

Number of varieties listed in 1981 and still available in 2004: 27 to 14 (52%).
Overall change in number of varieties from 1981 to 2004: 27 to 29 (107%).
Number of 2004 varieties available from 1 or 2 sources: 24 out of 29 (83%).

Altasweet - (Stoke's Altasweet) - 90-92 days, Shape and color of Laurentian but milder sweeter more tender and less woody, smooth thin skin, deep-yellow flesh, keeps well in the ground all winter, heavy yields, first real improvement in years, dev. by Mr. Simonet from a cross with a sweet summer turnip. *Source History: 5 in 1981; 3 in 1984; 6 in 1987; 5 in 1991; 2 in 1994; 2 in 1998; 1 in 2004.* **Sources: Me7.**

American Purple Top - (American Purple Top Yellow, American Yellow) - 80-120 days, The standard for both home and market, nearly globe, 4-6 in. dia., 16-20 in. blue-green cut-leaf top, deep-buff to light-yellow skin with purple on top, light-yellow flesh, slight taproot, firm fine-grained yellow flesh, sel. for sweetness and small neck, milder and sweeter when grown in mountain districts, introduced before 1920 as an improved str. of Purple Top Yellow. *Source History: 72 in 1981; 67 in 1984; 60 in 1987; 58 in 1991; 57 in 1994; 64 in 1998; 54 in 2004.* **Sources: ABB, All, Ba8, Bo19, Bu3, Bu8, CH14, Cl3, CO30, Cr1, De6, DOR, Ers, Fis, Fo7, Ga1, Gr27, Gur, HA3, Ha5, He8, HO2, Hum, Jor, Jun, Kil, La1, Lin, LO8, Loc, Ma19, MAY, ME9, Mey, Mo13, MOU, Ol2, OLD, PAG, Roh, Ros, RUP, Sau, Se16, Se25, Sh9, So1, SOU, Twi, Ver, Vi4, WE10, Wi2, WI23.**

American Purple Top Improved - (Laurentian) - 85-92 days, Round buff 5-7 in. roots purplish-red above ground, light-yellow fiberless flesh cooks to orange, best flavor cooked with skins on, small necks. *Source History: 5 in 1981; 5 in 1984; 6 in 1987; 6 in 1991; 5 in 1994; 3 in 1998; 2 in 2004.* **Sources: Fa1, Wet.**

Aspas - *Source History: 1 in 2004.* **Sources: Se17.**

Best of All - 90 days, Home garden gourmet var. from England, purple topped rutabaga, narrow 18 in. cut leaves, creamy- yellow mild flesh, fine in-the-ground keeper, up to 6 in. dia. *Source History: 2 in 1981; 1 in 1984; 1 in 1987; 1 in 2004.* **Sources: Tho.**

Bristol White - Rare English str. with white flesh, 6 in. roots are white with rose colored tops, sweeter than yellow fleshed strains. *Source History: 1 in 1998; 1 in 2004.* **Sources: Ho13.**

Canadian Gem - 120 days, Similar to Purple King, excel. Swede turnip for fall and winter use, firm sweet fine-flavored fine- grained light-yellow flesh, high quality, good keeper. *Source History: 2 in 1981; 2 in 1984; 2 in 1987; 2 in 1991; 1 in 1994; 2 in 1998; 1 in 2004.* **Sources: Alb.**

Doon - 87 days, Major type, purple topped, large roots, excellent flavor, maintain quality in storage. *Source History: 2 in 2004.* **Sources: Ho13, Pin.**

Fortin's Family - 100-120 days, Quebec heirloom, purple roots with yellow flesh, excellent flavor, appears to have some insect and disease res., good winter keeper. *Source History: 1 in 1998; 1 in 2004.* **Sources: Pr3.**

Golden Neckless - (American Purple Top) - 90-95 days, Yellow skin with purple top, yellow flesh, excel. shape, very hardy and heavy, excel. color and flavor, good keeper, very productive. *Source History: 3 in 1981; 3 in 1984; 3 in 1987; 1 in 1991; 1 in 1994; 1 in 1998; 2 in 2004.* **Sources: Se26, Shu.**

Joan - 90-100 days, Uniform purple topped, round roots, yellow flesh, sweeter and milder than other yellow vars., hard frost improves flavor. *Source History: 1 in 1998; 3 in 2004.* **Sources: Dam, Jo1, Se7.**

Laurentian - (Laurentian Purple Top, Laurentian Golden, Laurentian Swede, Perfect Model) - 90-122 days, An improved American Purple Top, creamy-yellow globe roots are deep purpish-red above ground, 4-6 in. dia., pale-yellow firm mild flesh, good flavor, no side roots, med-short top, almost neckless, uniform, for fall and winter, storage and shipping, Canada. *Source History: 45 in 1981; 41 in 1984; 41 in 1987; 31 in 1991; 30 in 1994; 31 in 1998; 28 in 2004.* **Sources: Alb, Bo19, CH14, Cl3, Dam, Ear, Fe5, He8, HO2, Jor, Lin, Mcf, Me7, ME9, OSB, PAG, Pin, Pr3, RUP, SE4, Se16, Se25, Sto, Syn, Tt2, Up2, Vi4, WE10.**

Laurentian, Certified No. 1 - 90 days, The strain recommended for use by commercial growers, grown and packaged in 1 lb. packages only under the Canadian Dept. of Agriculture's supervision, seed is sized in half 64's for precision seeding. *Source History: 1 in 1987; 3 in 1991; 2 in 1994; 2 in 1998; 2 in 2004.* **Sources: Ont, Sto.**

Laurentian, Neckless - 90 days, Purple top globe with white bottom, tender sweet fine-grained rich yellow flesh, smooth, free from excess roots, good keeper. *Source History: 2 in 1981; 2 in 1984; 1 in 1987; 1 in 1991; 1 in 1994; 1 in 1998; 1 in 2004.* **Sources: Fa1.**

Laurentian, OCS - Seed specially grown for Ontario Seed Company from sel. certified stock, produces purple-top yellow rutabaga, excel. winter keeper. *Source History: 1 in 1991; 1 in 1994; 1 in 1998; 1 in 2004.* **Sources: Ont.**

Laurentian, Thomson - 120 days, Improved strain of original Laurentian, smooth globe-shaped roots with deep-purple shoulders, fine-grained yellow flesh, especially recommended for commercial growers but also well suited for the home garden. *Source History: 1 in 1991; 2 in 1994; 3 in 1998; 1 in 2004.* **Sources: Ves.**

Macomber - (Sweet German) - 80-92 days, Round, 5-6 in., 4-5 lbs., smooth white skin with greenish-purple top, almost no neck, sweet white flesh, not woody or pithy when big, good keeper, dates back to 1863. *Source History: 9 in 1981; 8 in 1984; 1 in 1987; 1 in 1991; 1 in 1994; 1 in 1998; 2 in 2004.* **Sources: Ho13, Sa9.**

Marian - (Marian Swede) - 76-95 days, Mildew and club root res., purple globes, good quality, fine texture, high yields, very hardy, harvest when tennis ball size, dev. in Wales, widely grown in Europe. *Source History: 3 in 1981; 2 in 1984; 1 in 1987; 3 in 1991; 4 in 1994; 3 in 1998; 3 in 2004.* **Sources: Ga22, Ter, We19.**

Mary - 70 days, Uniform roots are slow to oversize, bred for faster root dev. and good eating quality, excel. ability to overwinter in the field, PSR Breeding Program. *Source History: 1 in 2004.* **Sources: Pe6.**

Monarch - *Source History: 1 in 1998; 1 in 2004.* **Sources: Pp2.**

Old Jake - Mellow roots with excellent flavor, very hardy, found in the remains of a shipwreck on North Haven Island, Maine. *Source History: 1 in 2004.* **Sources: In8.**

Pike - 100 days, Maine purple top variety similar to Laurentian, roots grow to 6 in., somewhat larger roots and tops, later maturity, same storage quality, flavorful yellow flesh. *Source History: 1 in 1987; 1 in 1991; 1 in 1994; 2 in 1998; 1 in 2004.* **Sources: Ho13.**

Purple Top Yellow - (Purple Top) - 90 days, Smooth and globe-shaped, deep purplish-red above ground and light-yellow below, fine-grained yellow flesh cooks to bright-orange, for fall and winter use, stores well. *Source History: 11 in 1981; 6 in 1984; 8 in 1987; 6 in 1991; 5 in 1994; 5 in 1998; 6 in 2004.* **Sources: Bu2, But, Co31, Mel, Ni1, Ri12.**

Purple Top Yellow Improved - (Purple Top Improved) - 90 days, Short neck str., solid yellow flesh, sweeter than ordinary turnips, smooth, not too large, good keeper when full grown. *Source History: 4 in 1981; 3 in 1984; 2 in 1987; 2 in 1991; 2 in 1994; 1 in 1998; 1 in 2004.* **Sources: Com.**

Purple Top, Swede - (Acme) - 90 days, Excellent stock, does best on moist rich soils, may grow woody or pungent if soil is too dry, easily grown, long-standing, dependable winter storage. *Source History: 1 in 1987; 1 in 1991; 1 in 1994; 1 in 1998; 1 in 2004.* **Sources: Bou.**

Tona a Colletto Viola - Round white root with violet neck. *Source History: 1 in 1994; 1 in 1998; 1 in 2004.* **Sources: Ber.**

Virtue - 130 days, Red skin, fine grained sweet yellow flesh, winter hardy. *Source History: 1 in 2004.* **Sources: Tho.**

Wilhelmsburger Gelbe - (Pandur) - 85-95 days, Very early, very cold hardy, perfect for short season areas, nicely flavored root with greenish top, ability to store in a cool place for several months, no hint of fiber. *Source History: 1 in 1981; 1 in 1984; 1 in 1987; 1 in 1991; 1 in 1994; 1 in 1998; 2 in 2004.* **Sources: Ho13, Sa9.**

York - (York Purple Top Swede, York Swede Certified) - 115-120 days, Identical to Laurentian except res. to 2 of the 3 club root strains, dev. at Charlottetown Exp. Station, sparse leaves, small neck. *Source History: 8 in 1981; 8 in 1984; 6 in 1987; 5 in 1991; 4 in 1994; 4 in 1998; 2 in 2004.* **Sources: Hal, Ves.**

Varieties dropped since 1981 (and the year dropped):

American Yellow Improved (1987), *Bucks County* (1991), *Champion Purple Top* (1994), *Eastern Laurentian* (1984), *Elephant (or Monarch)* (1991), *Fortune* (1991), *Laurentian, Canada No. 1* (1998), *Laurentian, Thompson Strain* (1998), *Lizzy* (2004), *Long Island Improved* (1994), *Long Island Purple Top* (1991), *Purple King* (1998), *Purple Top, Gaze's Special* (2004), *Red Chief* (1984), *Superba* (1987), *Thompson* (1994), *Wilhelmsburger Green Top* (2004).

Salsify
Tragopogon spp.

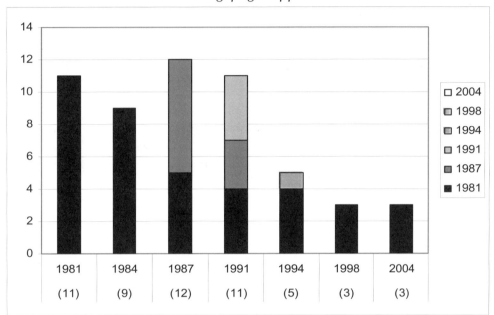

Number of varieties listed in 1981 and still available in 2004: 11 to 3 (27%).
Overall change in number of varieties from 1981 to 2004: 11 to 3 (27%).
Number of 2004 varieties available from 1 or 2 sources: 1 out of 3 (33%).

Mammoth - (Mammoth White) - 150 days, Creamy-white roots, oyster-like flavor, dig after frost which improves texture and flavor, store most for winter in sand in cellar, mulch some for spring harvest. *Source History: 2 in 1981; 2 in 1984; 1 in 1987; 1 in 1991; 3 in 1994; 4 in 1998; 2 in 2004.* **Sources: HA3, Si5.**

Mammoth Sandwich Island - (Sandwich Island) - 110-180 days, Dull-white roots resemble slender parsnips, tapers to a point, 8-9 x 1.5-2 in. at shoulder, creamy-white flesh, oyster-like flavor, better if dug after frost, overwinter or store like carrots, excel. winter keeper, baked or creamed, soups or stews, pre-1900. *Source History: 86 in 1981; 79 in 1984; 75 in 1987; 64 in 1991; 59 in 1994; 58 in 1998; 58 in 2004.* **Sources: Alb, All, Ba8, Be4, Bo18, Bo19, Bu2, Bu3, Bu8, CH14, CO23, CO30, Com, Dam, Dgs, Eo2, Ers, Fis, Fo7, Gr27, Gr29, He8, He17, Hi13, HO2, Jo1, Jun, La1, Ma19, MAY, Me7, ME9, Mey, Mo13, Ni1, OLD, Ont, Pin, Ra5, Roh, RUP, Sa9, Sau, Se26, Se28, Sh9, Shu, So1, So12, SOU, Sto, Tho, Vi4, Vi5, WE10, Wet, WI23, Wo8.**

Salsify - (Oyster Plant, Hawwerwurtzel) - 90-120 days, Parsnip-like 6-8 x 1-1.5 in. tapered roots, creamy-white flesh, needs a deep soil, best when dug after frost, hardy, overwinter for spring, or cool damp storage, pre-1800. *Source History: 9 in 1981; 8 in 1984; 6 in 1987; 5 in 1991; 4 in 1994; 9 in 1998; 4 in 2004.* **Sources: Ami, Gr28, Me7, Ol2.**

Varieties dropped since 1981 (and the year dropped):

Black Giant Russian (1991), *Blanc Ameliore* (1994), *Creamy White* (1991), *Eenjarige Nietschieters* (1987), *European* (1994), *Flandria Scorzonera* (1994), *French Blue Flowered* (1991), *Geante Noire de Russie Donia* (1987), *Gigantia* (1994), *Gigantia Scorzonera* (1994), *Madagascar Long Black* (1987), *Mammoth Long Island* (1998), *Mammoth Sandwich Island Imp.* (1991), *Pronora Verbeterde* (1987), *Vegetable Oyster* (1991), *White* (1998), *White French* (1994), *White Mammoth* (1983).

Scorzonera (Black Salsify)
Scorzonera hispanica

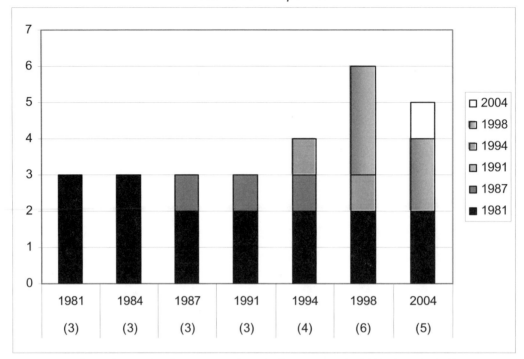

Number of varieties listed in 1981 and still available in 2004: 3 to 2 (67%).
Overall change in number of varieties from 1981 to 2004: 3 to 5 (167%).
Number of 2004 varieties available from 1 or 2 sources: 2 out of 5 (40%).

<u>Belstar Super</u> - 80 days, Black skinned, long straight roots, 9-11 in., white flesh. *Source History: 1 in 2004.* **Sources: Jo1.**

<u>Duplex</u> - Black roots, 8-10 in. long, creamy white flesh, young leaves, flower buds and flower petals all edible. *Source History: 1 in 1998; 2 in 2004.* **Sources: Com, Dam.**

<u>Geante Noire de Russie</u> - 100-135 days, Extra long var. of black salsify favored by European cooks, black-skinned white roots, very long, very cylindrical, unique oyster-like flavor, limited storage, from Belgium. *Source History: 1 in 1981; 1 in 1984; 1 in 1987; 1 in 1991; 1 in 1994; 2 in 1998; 4 in 2004.* **Sources: Ami, Fe5, Sa9, So12.**

<u>Giant Russian</u> - 120 days, Long black roots, smooth, firm texture, good onion-garlic flavor, widely grown in France. *Source History: 1 in 1998; 1 in 2004.* **Sources: Lej.**

<u>Scorzonera</u> - (Black Salsify, Long Black, Black-Skinned) - 95-120 days, Black skin, dull-white flesh, excel. flavor, many think it is better than white salsify, needs loose soil, repels carrot flies. *Source History: 7 in 1981; 6 in 1984; 9 in 1987; 8 in 1991; 8 in 1994; 9 in 1998; 5 in 2004.* **Sources: Bou, Eo2, Lan, Ri12, Tho.**

Varieties dropped since 1981 (and the year dropped):

Annual Blackrooted (1998), *Hoffman's Lange Pfahl* (2004), *Lange Jan* (2004).

Sorghum
Sorghum bicolor

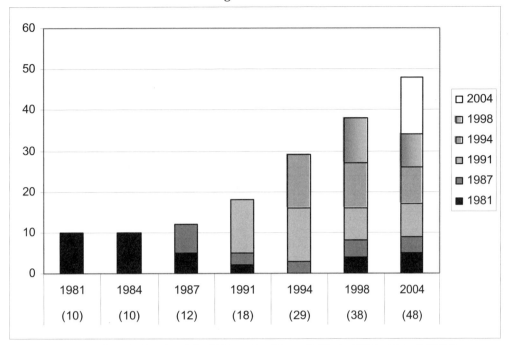

Number of varieties listed in 1981 and still available in 2004: 10 to 5 (50%).
Overall change in number of varieties from 1981 to 2004: 10 to 48 (480%).
Number of 2004 varieties available from 1 or 2 sources: 36 out of 48 (75%).

Amber - 95-110 days, Molasses sorghum, stalks grow 6-12 ft. tall, harvest when amber seeds are covered with shiny black coating, also used for molasses. *Source History: 1 in 1987; 3 in 1991; 2 in 1994; 9 in 1998; 12 in 2004.* **Sources: Bo18, GLO, Gr29, HO2, Ho13, Jor, Roh, Se28, So1, So12, Vi5, Wo8.**

Amber, White Seeded - 120 days, Purplish seed heads on 5 ft. stalks, white seeds blushed red-amber, used as a cooking grain, flour or cereal, ornamental, requires warmth to germinate. *Source History: 1 in 2004.* **Sources: Tu6.**

Apache Red - 120 days, Red seedheads attract birds, stalk chewed like candy when seeds have turned red, from San Carlos Reservation, Arizona, probably introduced in 19th Century. *Source History: 1 in 1987; 2 in 1991; 2 in 1994; 3 in 1998; 3 in 2004.* **Sources: Na2, So12, Syn.**

Black - 120 days, Flour and seed variety, black hulls, light colored grain, decorative plant, compact club-shaped seedhead, productive, stands well until frost. *Source History: 1 in 2004.* **Sources: Bou.**

Black Kaffir - 105-120 days, Plants grow 7-12 ft. tall, compact head, white seed, flour grain for seed and feed. *Source History: 1 in 1991; 1 in 1994; 6 in 1998; 6 in 2004.* **Sources: Bo18, Gr29, RUP, Se28, Vi5, Wo8.**

Black Milo - Plant grows 6-8 ft., glossy black seedheads, ornamental. *Source History: 1 in 2004.* **Sources: Syn.**

Brandes - Sweet syrup sorghum, dev. by Dr. Bill Knight at MS State Univ., tall 9 ft. talks, large seed heads with white seed, long season. *Source History: 1 in 1981; 1 in 1984; 1 in 1998; 1 in 2004.* **Sources: Ho13.**

Broomcorn - 110-120 days, Not a true corn, plant looks and grows like corn, straw of mature plants is used for brooms. *Source History: 16 in 1994; 23 in 1998; 48 in 2004.* **Sources: Be4, Bou, CA25, Coo, De6, Ea4, Ech, Fi1, Ge2, GLO, Gr28, Gr29, Gur, Ha5, He8, He17, HO2, Jor, Jun, Ma19, MAY, ME9, Mel, Mo13, Ni1, Ol2, OLD, Pin, Ren, RIS, RUP, SE4, SE14, Se16, Se28, Sh9, Shu, Sto, Ter, Tu6, Twi, Und, Ver, Ves, Vi4, Vi5, WE10, Wo8.**

Broomcorn, Black - 115 days, Tall stalks to 7 ft., 7 in. heads with black seeds, good bird seed producer and very ornamental, from Africa. *Source History: 2 in 1994; 1 in 1998; 2 in 2004.* **Sources: Fe5, Sa9.**

Broomcorn, Hungarian Black Seeded - 115 days, Heirloom traditionally used for making brooms, 8 to 12 ft. stalk produces long seed heads loaded with shiny black seeds. *Source History: 2 in 1994; 2 in 1998; 2 in 2004.* **Sources: Sa9, So1.**

Broomcorn, Hungarian Red - 100-110 days, Corn-like stalks produce red seedhead, edible, drought tolerant, good for making brooms, good bird forage. *Source History: 5 in 1994; 6 in 1998; 4 in 2004.* **Sources: Pe2, Se7, Se17, Syn.**

Broomcorn, Iowa Red - Stalks grow 8 ft. tall with red heads. *Source History: 1 in 1994; 1 in 1998; 1 in 2004.* **Sources: Sa9.**

Broomcorn, Iowa Sweet - 125 days, Sweet sugar cane type with thick juicy stalks, 9 ft. tall. *Source History: 1 in 1998; 1 in 2004.* **Sources: Sa9.**

Broomcorn, Mayo - Tassel spikes are used to make brooms, Tarahumara sometimes use seed as aid in fermenting tesuigno (beer), stalks and leaves can be used for fodder. *Source History: 1 in 1987; 1 in 1991; 1 in 1994; 1 in 1998; 1 in 2004.* **Sources: Na2.**

Broomcorn, Red - 105-120 days, Unusual strain produces bright red brooms instead of the standard yellow. *Source History: 4 in 1994; 8 in 1998; 13 in 2004.* **Sources: Bo18, Fe5, Fo7, Ge2, GLO, Gr29, HO2, Jor, ME9, Roh, Se28, Vi5, Wo8.**

Broomcorn, Red Deer - 120 days, The original sorghum used for broom making. *Source History: 1 in 2004.* **Sources: RUP.**

Broomcorn, Shaker Standard - Huge panicles of edible grain, ornamental, great for brooms. *Source History: 1 in 1994; 1 in 1998; 1 in 2004.* **Sources: Syn.**

Broomcorn, Yellow - 105 days, Produces a spray of seed heads 24-36 in. long which grow from the top of a corn-like plant where a tassel would normally be. *Source History: 1 in 1994; 1 in 1998; 1 in 2004.* **Sources: Fo7.**

Cana Ganchado - Sweet canes can be 6 ft. tall, dark-maroon seeds, grown by the Guarijio in Sonora, Mexico. *Source History: 1 in 1991; 1 in 1994; 2 in 1998; 2 in 2004.* **Sources: Ho13, Na2.**

Cane Sorghum - Old-time cane sorghum, produces large amount of forage and high amount of protein for cattle, also used to make sorghum molasses for home use. *Source History: 1 in 1981; 1 in 1984; 1 in 1987; 1 in 2004.* **Sources: Ech.**

Crookneck Milo - 115 days, Plants to 5 ft. tall with large seed head that tips over. *Source History: 1 in 2004.* **Sources: Sa9.**

Dale, John Coffer Strain - Extra early maturity, very sweet 6-7.5 ft. canes, also good grain production, dev. by John Coffer in the mountainous area of New York. *Source History: 1 in 1998; 2 in 2004.* **Sources: Bou, Pe6.**

Dwarf Grain - Plants grow 3 ft. tall, thick seedheads, large seeds. *Source History: 1 in 1991; 1 in 1994; 1 in 1998; 1 in 2004.* **Sources: Syn.**

Early Hegari - 95-105 days, White grained seed heads on 3.5-4 ft. plants. *Source History: 1 in 2004.* **Sources: So12.**

Egyptian Wheat Sorghum - Fluffy loose seed heads, 10-12 in. long, straw-colored seeds, harvest for fresh use at any stage of development. *Source History: 1 in 2004.* **Sources: Ge2.**

Golden Shallu - Stalks to 5 ft., light amber grain, med. size, high yields of grain with low tannin content which tastes the best, rare. *Source History: 2 in 1998; 1 in 2004.* **Sources: Ho13.**

Ho-K - Sweet sorghum, easily grown, same culture as corn, chunks of stalk chewed like candy, can extract juice and boil into syrup, seeds used for bird feed. *Source History: 1 in 1981; 1 in 1984; 1 in 1987; 2 in 1991; 1 in 1994; 1 in 1998; 1 in 2004.* **Sources: So1.**

Honey Drip - 105-110 days, Red hulled seeds, medium tall, 8-10 ft. stalks, a very old sweet sorghum variety that has become quite rare. *Source History: 1 in 1981; 1 in 1984; 1 in 1998; 3 in 2004.* **Sources: Ba8, Ho13, So12.**

Mennonite - Red-hulled, tall very thick stalks, 7-9 ft., from a Mennonite who used to have a sorghum mill near Jamesport, Missouri, makes a light-colored syrup, an old-time cane. *Source History: 1 in 1981; 1 in 1984; 1 in 1987; 1 in 1991; 2 in 1994; 4 in 1998; 3 in 2004.* **Sources: Ho13, Se17, So1.**

Northern Sugar Cane - Produces extra sweet, thick, 8-10 ft. tall canes even in the north. *Source History: 1 in 1998; 1 in 2004.* **Sources: Pe6.**

Onavas Red - Sweet juicy stalks produce many tillers, burgundy-red seeds, from the Pima Bajo village on the Rio Yaqui, Sonora, Mexico. *Source History: 1 in 1991; 1 in 1994; 1 in 1998; 1 in 2004.* **Sources: Na2.**

Perennial, M6-1 - Grows 2.5-4 ft. tall, light medium-brown seed, matures fast, sets and matures seeds in cooler weather, perennial where the ground does not freeze, good for northern, coastal and mountain gardeners, from Peters Seed and Research *Source History: 1 in 2004.* **Sources: Bou.**

Popping - 88-120 days, Makes a tasty popcorn alternative, pops up small with a delicious but different taste, white seed. *Source History: 1 in 1998; 4 in 2004.* **Sources: Bou, Ge2, Pe6, Shu.**

Rio - White grain sorghum, 7-8 ft. canes, for cane sorghum syrup, dev. by USDA, reselected by PSR. *Source History: 1 in 2004.* **Sources: Pe6.**

Rox Orange - 100-130 days, Good sorghum for table use and forage, 7-8 ft., grows well in any area that corn can be grown. *Source History: 4 in 1994; 3 in 1998; 5 in 2004.* **Sources: Ba8, Ho13, Ra5, Sa9, Shu.**

Salts Red - 105 days, Sweet sorghum from Missouri, loose grain heads, red seeds. *Source History: 1 in 2004.* **Sources: Sa9.**

San Felipe Pueblo - Plants to 5 ft. tall, black seeds, stalks chewed. *Source History: 1 in 2004.* **Sources: Na2.**

Sand Mountain - High sugar content at maturity. *Source History: 1 in 2004.* **Sources: Se17.**

Santa Fe Red - Slender heads, red seeds, cut stalks chewed, raised at the Santo Domingo Pueblo. *Source History: 1 in 2004.* **Sources: Na2.**

Tarahumara Popping - White seed popped by Indians like miniature popcorn, collected from Batopilas Canyon in southern Chihuahua, Mexico. *Source History: 1 in 1991; 1 in 1994; 2 in 1998; 2 in 2004.* **Sources: Ho13, Na2.**

Tennessee Tall Girls - Old strain from Texas with thin stalks, 8-9 ft. tall, makes a very good light syrup. *Source History: 1 in 1998; 2 in 2004.* **Sources: Ho13, Se17.**

Tepehuan Popping - (White Popping) - 110 days, Slender plants with many tillers, white seeds used like popcorn, from Mexico. *Source History: 1 in 1991; 1 in 1994; 4 in 1998; 3 in 2004.* **Sources: Jor, Pin, So12.**

Texacoa - 115 days, Grows 4-5 ft. tall, white grain, used for porridge. *Source History: 2 in 1998; 2 in 2004.* **Sources: So12, Syn.**

Texas Black Amber Molasses - 110-120 days, Semi-loose heads with plump black seeds used for molasses, silage and floral arrangements, heirloom from Waco, Texas. *Source History: 1 in 1991; 2 in 1994; 3 in 1998; 4 in 2004.* **Sources:**

Ge2, Na2, Pin, RUP.

Texas Long Sweet - Thick, 8 ft. stalks, makes a light colored syrup, old str. from Texas. *Source History: 1 in 1998; 1 in 2004.* **Sources: Ho13.**

Tohono O'odham - Perennial in mild winter areas, black-hulled seeds, stalk chewed, once used for molasses, cultivated by the O'odham (Gila River Pima) of southwestern Arizona. *Source History: 1 in 1987; 2 in 1991; 2 in 1994; 3 in 1998; 2 in 2004.* **Sources: Na2, Se17.**

White African - Stalks to 8 ft., white seed with black hull, makes dark syrup. *Source History: 1 in 1991; 1 in 1994; 1 in 1998; 2 in 2004.* **Sources: Hud, Se17.**

White Mountain Apache - Red seeds, from Cibeque, AZ. *Source History: 1 in 2004.* **Sources: Na2.**

Varieties dropped since 1981 (and the year dropped):

Ames Amber (1991), *Black African* (1998), *Black Strap* (1991), *Broomcorn, Hadley Deer* (2004), *Broomcorn, Hadley T115* (2004), *Broomcorn, Satties Museum* (1998), *Eddie Coleman's* (1987), *Gooseneck* (2004), *Gooseneck Brown Durra* (1998), *Jawar* (2004), *Lesotho* (2004), *Red Kafir* (1994), *Sart* (1991), *Sorgum* (1991), *Sugar Drip* (1994), *Wacona Orange* (1987), *White Pearl* (1998).

Soybean
Glycine max

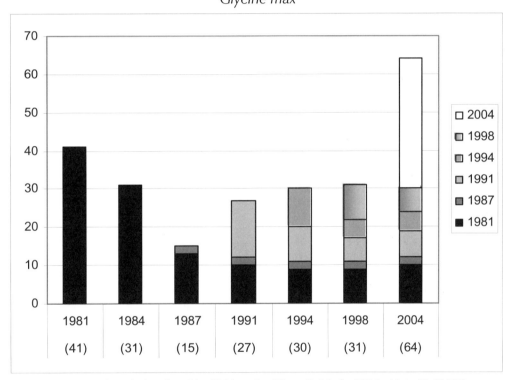

Number of varieties listed in 1981 and still available in 2004: 41 to 10 (24%).
Overall change in number of varieties from 1981 to 2004: 41 to 64 (156%).
Number of 2004 varieties available from 1 or 2 sources: 41 out of 64 (64%).

Agate - 65-70 days, Early variety with beautiful olive green-brown seeds, 1 ft. plant, good eating quality, grown in New Mexico, 1929. *Source History: 1 in 1987; 2 in 1991; 3 in 1994; 3 in 1998; 5 in 2004.* **Sources: Ho13, Pe1, Sa5, Se16, Syn.**

Altona - 100 days, Popular early soybean for the North, dev. in Manitoba, shiny yellow with black hilum, slightly larger than Fiskeby V, good yields and quality, dependably early. *Source History: 1 in 1981; 1 in 1984; 1 in 2004.* **Sources: Syn.**

Amish Green - *Source History: 1 in 2004.* **Sources: Se17.**

B 006 Brun Hatif Rouest - *Source History: 1 in 2004.* **Sources: Se17.**

Beer Friend - 70-85 days, Japanese type, bushy 2.5-3 ft. plant, white flowers, high yields of dark green pods, 2-3 seeds per pod, long shelf life, tasty snack food when dried and salted, rich in protein, calcium and vitamins A and B. *Source History: 4 in 2004.* **Sources: Ev2, Kit, Sto, Ter.**

Beijing Black - High yields of small, kidney shaped seeds. *Source History: 1 in 2004.* **Sources: Syn.**

Black Eyebrow - (Hei Mei Tou) - Plants to 3 ft., brown seed with black saddle over eye, from Manchuria in 1911. *Source History: 1 in 1998; 3 in 2004.* **Sources: Ho13, Se17, Syn.**

Black Jet - 90-110 days, Early maturing black soybean for short season areas, prolific 24 in. plants, thin skinned, good flavor, cooks more quickly than other blacks, dev. by

Johnny's Selected Seeds. *Source History: 1 in 1981; 1 in 1984; 1 in 1987; 3 in 1991; 5 in 1994; 8 in 1998; 9 in 2004.* **Sources: Bou, Ho13, Jo1, Ma13, Pr3, Sa5, Syn, Te4, Tu6.**

Black Pearl - 85 days, Vigorous 2.5-3 ft. tall plants, .25 in. black pearl-size seeds resemble lentils, rich flavor. *Source History: 1 in 2004.* **Sources: Ter.**

Butterbeans - 90 days, Bright-green soybean dev. for eating fresh, well-branched stocky 24 in. plants, res. to lodging, very prolific pod set, many with 3 large beans, can or freeze. *Source History: 3 in 1981; 3 in 1984; 4 in 1987; 6 in 1991; 3 in 1994; 4 in 1998; 8 in 2004.* **Sources: Be4, Bou, Hi6, Jo1, Na6, Sa9, Syn, Te4.**

Canatto - *Source History: 1 in 2004.* **Sources: Se17.**

Chico - 95-105 days, Prolific bushy plants, small yellow seeds, more digestible than larger strains, clear hilum, resembles Natto but is earlier. *Source History: 1 in 1998; 4 in 2004.* **Sources: Ho13, Se17, So12, Syn.**

Crest - *Source History: 1 in 2004.* **Sources: Se17.**

Dieckman Black - Glossy black, large, round seed, very productive, medium early. *Source History: 2 in 2004.* **Sources: Pe1, Syn.**

Edible Soybean - (Edamame) - 85-100 days, Bushy erect plants are loaded with pods, each contains 2 to 3 oval bright- green nearly round seeds, use green or dried, cook like limas, 50 pods per plant. *Source History: 13 in 1981; 13 in 1984; 11 in 1987; 7 in 1991; 6 in 1994; 6 in 1998; 21 in 2004.* **Sources: Bu1, Co31, De6, Eo2, Gr29, He8, La1, Ma19, Mey, OLD, Pe2, Ra5, Red, Ri12, Roh, RUP, Se28, Sh9, Si5, Ter, Wo8.**

Envy - 75-115 days, Upright 24 in. plants, seeds green inside and out, high yields, exceptional quality, early ripening, popular in the Northern states and Canada, dev. by Prof. Meader at U of NH. *Source History: 4 in 1981; 3 in 1984; 3 in 1987; 3 in 1991; 1 in 1994; 5 in 1998; 12 in 2004.* **Sources: Bou, Goo, Ho13, Jo1, Na6, Sa9, Scf, Se16, Se26, Syn, Te4, Tho.**

Evan's Mutation - Productive 3 ft. plant, yellow seed with faint gray saddle. *Source History: 1 in 1998; 1 in 2004.* **Sources: Ho13.**

Grand Forks - Very large yellow seeds partly overlaid with brown, sweet buttery flavor, easily digestible, dev. in North Dakota for early maturity in northern climates. *Source History: 2 in 1991; 2 in 1994; 4 in 1998; 5 in 2004.* **Sources: Ho13, Pe1, Pr3, Sa5, Syn.**

Green Legend - 75 days, Dark green pods with white hairs, tasty, very productive. *Source History: 1 in 2004.* **Sources: Ev2.**

Hakucho Early - 65-80 days, Early bush Japanese soybean, 3 small yellow/green seeds per pod, dwarf 1 ft. plant, early enough for the North, very concentrated pod set, quick heavy yields, eaten in the green stage, also good when dried or made into soy milk. *Source History: 6 in 1981; 6 in 1984; 4 in 1987; 2 in 1991; 1 in 1994; 4 in 1998; 7 in 2004.* **Sources: Coo, Ev2, OSB, Par, Syn, We19, Wi2.**

Hedge - Tall plant, continuous pod set, variably round yellow seeds. *Source History: 1 in 2004.* **Sources: Syn.**

Hokkaido Black - 105-110 days, Bushy plants, tasty small seeds, originated in northern Japan. *Source History: 2 in 1998; 6 in 2004.* **Sources: Ho13, Pe1, Sa5, Se17, So12, Syn.**

Jacques Black - Moderate yield of large round black seeds. *Source History: 1 in 2004.* **Sources: Syn.**

Jacques Brown - 95-100 days, Plants to 2.5-3 ft. tall, fuzzy pods, brown seeds, from Canada. *Source History: 3 in 2004.* **Sources: Ho13, Se17, Syn.**

Jewel - 95-105 days, Bushes to 2.5 ft., shiny yellow seeds with black saddle, Manchuria. *Source History: 2 in 2004.* **Sources: Se17, Syn.**

Kahala - Nematode res. *Source History: 2 in 1991; 1 in 1994; 1 in 1998; 1 in 2004.* **Sources: Uni.**

Kegon - (83) - Sweet and tender with a mild flavor, highly productive, mid-early maturity, can be frozen. *Source History: 1 in 1994; 1 in 1998; 1 in 2004.* **Sources: SAK.**

Korean - Tall, broad-leaf plant, black seeds, vigorous, cold hardy. *Source History: 1 in 2004.* **Sources: Syn.**

Kura Kake Daizu - 95-100 days, High yielding plants, gold and brown seeds with sweet, buttery taste. *Source History: 1 in 1998; 3 in 2004.* **Sources: Se17, So12, Syn.**

Kuromane - Japanese var., large black seeds used as green shell beans in mature and immature stages. *Source History: 1 in 1994; 1 in 1998; 2 in 2004.* **Sources: Se17, Syn.**

Lammer's Black - 100-120 days, Best black soybean for short season areas, prolific 2 ft. plants with fuzzy foliage, thin skinned, good flavor, black seeded vars. are the best for use in tofu and tempe, heirloom. *Source History: 1 in 1991; 2 in 1994; 3 in 1998; 5 in 2004.* **Sources: Ho13, Roh, Se17, So12, Syn.**

Lanco - (Lanco Green) - 85-90 days, Erect vigorous plants produce abundance of pods containing bright-green seeds at maturity, suitable for canning or freezing, also for roasting when dried. *Source History: 1 in 1981; 1 in 1994; 2 in 1998; 2 in 2004.* **Sources: Roh, Syn.**

Laredo - Historic Dust Bowl cultivar, indet. forage variety, black seeds. *Source History: 1 in 2004.* **Sources: Syn.**

Late Giant Black Seeded - Popular Japanese var., green beans at edible stage, matures to black color, long season extends 5 months, possibly day-length sensitive. *Source History: 1 in 1981; 1 in 1984; 2 in 2004.* **Sources: Ev2, Ho13.**

Manitoba Brown - 80-100 days, Flavorful, round to oval deep brown seeds, well adapted to northern climates. *Source History: 3 in 1998; 5 in 2004.* **Sources: Ho13, Pr3, Sa5, Sa9, So12.**

Maple Arrow - 77-100 days, Very popular early soybean developed in Canada, 30 in. plants produce med-size pods containing 2-3 beans each, good fresh green (prepare like peas) or dry when they are bright-yellow with a brown hilum, rich flavor and high protein content, high yields, dependable for short-season growers, best yield when planted around June 1, good for processed soy products such as tofu and tempeh. *Source History: 2 in 1987; 1 in 1991; 1 in 1994; 2 in 1998; 4 in 2004.* **Sources: Ho13, Me7, Syn, Ver.**

McCall - 106 days, Attractive med-small all-yellow multi-purpose soybean, 24-30 in. plants with excel res. to lodging (falling over), high yields, dev. by MN/AES to replace Wilkin. *Source History: 1 in 1981; 1 in 1984; 1 in 2004.* **Sources: Syn.**

Misono Green - 85 days, Productive early variety, vigorous, 2 ft. plants, 3 in. pods with 2-4 buff colored seeds per pod, does well in the maritime Northwest. *Source History: 1 in 2004.* **Sources: Ter.**

Morlanvia - *Source History: 1 in 2004.* **Sources: Se17.**

N-03 Rouest 13 A1 2 - *Source History: 1 in 2004.* **Sources: Se17.**

Natto - Traditionally used to make Natto miso, very small yellow seeds, high yields, good for sprouting, prone to shattering. *Source History: 1 in 1991; 1 in 1994; 2 in 1998; 2 in 2004.* **Sources: Sa5, Syn.**

Norman - *Source History: 1 in 2004.* **Sources: Se17.**

Ozzie - 110-120 days, Large downy leaves on 3 ft. bushes, small round buff colored seed, productive. *Source History: 1 in 2004.* **Sources: Se17.**

PI 290156 - *Source History: 1 in 2004.* **Sources: Se17.**

PI 458541 (China) - Experimental germplasm from USDA. *Source History: 2 in 2004.* **Sources: Se17, Syn.**

PI 522192 (USSR) - *Source History: 1 in 2004.* **Sources: Se17.**

Prize - 85-120 days, Green shells in 85 days, erect 30 in. bushes, large oval bright-green high protein beans, fresh/freeze/can/sprout, very adaptable, does rather well even in Florida. *Source History: 14 in 1981; 14 in 1984; 11 in 1987; 7 in 1991; 4 in 1994; 2 in 1998; 1 in 2004.* **Sources: Syn.**

Saint Ita - 95 days, Compact upright plants keep pods off the ground, 2-3 seeds per pod, excel. edamame, similar to lima beans but better adapted to Northern conditions, twice as high in protein, can be cooked in the pod at green stage, shelled or left to dry, more easily digested than yellow soybeans. *Source History: 1 in 1991; 1 in 1994; 4 in 2004.* **Sources: Me7, Pe1, Se17, Syn.**

Sapporo Midori - 112 days, Sturdy plants with heavy clusters of bright green 2.5 in. pods covered with white pubescence, 3 beans per pod. *Source History: 1 in 2004.* **Sources: OSB.**

Sayamusume - (Edamame Sayamusume) - 85 days, Consistent heavy yielder, 2 ft. plants, 3-3.5 in. pods, 3-4 light green seeds per pod, buttery sweet flavor, high vitamin content, rich in protein, calcium, potassium and phytoestrogens. *Source History: 2 in 2004.* **Sources: Se7, Ter.**

Shirofumi - 80-90 days, Vigorous vines to 3 ft., harvest plump green pods before any yellowing from late August into September, large pale green seeds, good lima substitute, extremely productive. *Source History: 3 in 2004.* **Sources: Fe5, Hi6, Se16.**

Shironomai - 70-75 days, Green soybean for fresh beans, good for cooking, canning or freezing, good source of protein, used as an alternative to milk and meat in China. *Source History: 2 in 1994; 1 in 1998; 4 in 2004.* **Sources: Par, SAK, Sto, Syn.**

Syncom - Yellow seeds, highly productive. *Source History: 1 in 2004.* **Sources: Syn.**

Taiwame - One of the most popular varieties grown in Taiwan, does very well in sub-tropical climates. *Source History: 1 in 2004.* **Sources: Ev2.**

Tohya - 88 days, Edible green soybean from Japan, vigorous vines, early concentrated sets of light green pods, 3 seeds per pod, easy to pick, good substitute for limas in short season areas. *Source History: 1 in 1994; 4 in 1998; 2 in 2004.* **Sources: Bo17, Syn.**

Ugra Saja - Compact plants produce high yields of large oval dark brown seeds, Sweden. *Source History: 2 in 2004.* **Sources: Se17, Syn.**

Velvet - Plants have whitish appearance because of white silky hairs on the leaves, white flowers, white seeds, good indicator str. for looking at cross-pollination between cultivars. *Source History: 2 in 2004.* **Sources: Pe1, Se17.**

Vinton 81 - 82 days, Plants 36 in. tall, yellow seeds, popular for tofu and soy milk production, great used fresh like a shell bean or eaten raw. *Source History: 1 in 1991; 2 in 1994; 1 in 1998; 6 in 2004.* **Sources: Ba8, Com, Ma13, Or10, Pin, Syn.**

White Lion - 70-111 days, Popular vegetable in Japan, 2 ft. tall plants, large dark green pods contain 3 seeds, recommended for fresh use. *Source History: 1 in 1991; 1 in 1994; 2 in 1998; 3 in 2004.* **Sources: Ev2, Kit, OSB.**

Wisconsin Black - Moderate yield, black seeds. *Source History: 2 in 2004.* **Sources: Se17, Syn.**

Yeda-Mame - Large vigorous plant with good pod set, medium early maturity, good for fresh market and freezing. *Source History: 2 in 1994; 2 in 1998; 2 in 2004.* **Sources: Kit, Ni1.**

Yusuzumi - 150 days, Good yields of large dark green pods, white pubescence. *Source History: 2 in 2004.* **Sources: OSB, SAK.**

Varieties dropped since 1981 (and the year dropped):

Ada (1987), *Ag Canada* (1998), *Blue Ribbon* (2004), *Brown* (1994), *Browncoat* (1987), *Chinese* (1987), *Daizu* (2004), *Disoy* (1984), *Early Akita Yellow* (1987), *Early Green Okuhara* (1984), *Express Japanese Green* (1984), *Extra Early* (1998), *Fiskeby V* (1998), *Frostbeater* (1987), *Giant Green Soybean* (1987), *Green Soy Early* (1984), *Hodgeson* (1991), *Japanese Black* (1984), *Japanese Black Soy* (1982), *Karikachi* (1998), *Kuroname* (1994), *Late Blackcoat* (1991), *Late Whitecoat* (1987), *Light Yellow Dwarf Early* (1984), *Maple Amber* (1994), *Maple Ridge* (1998), *Midnight* (1998), *Mikawashima Green* (1994), *Northern Express* (1987), *Okuhara Early* (1987), *Panda* (1998), *Panther* (2004), *Pickett* (1987), *Rook's* (1994), *Sodefuri Green* (1987), *Tamanishiki* (1982), *Tsurunoko* (1987), *Vegetable Soybean* (1998), *Verde* (2004), *Vinton* (1987), *Wilkin* (1983), *Yellow Soybean* (2004), *Yellowcoat* (1987).

Spinach
Spinacia oleracea

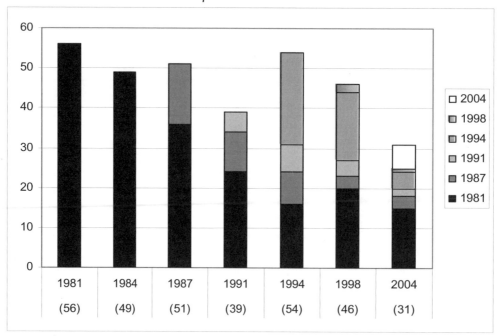

Number of varieties listed in 1981 and still available in 2004: 56 to 15 (27%).
Overall change in number of varieties from 1981 to 2004: 56 to 31 (55%).
Number of 2004 varieties available from 1 or 2 sources: 20 out of 31 (65%).

America - (America Long Standing, American) - 43-55 days, AAS/1952, long-standing Bloomsdale type, 8 in. high plants, thick deep-green savoyed leaves, slow-growing, slow-bolting, heat and drought res., fine quality, heavy yields, well suited for spring sowing in long day areas, home garden, fresh or can or freeze. *Source History: 27 in 1981; 23 in 1984; 18 in 1987; 15 in 1991; 11 in 1994; 7 in 1998; 8 in 2004.* **Sources: CO30, Cr1, He8, Pe2, Se7, Se16, Se24, So9.**

Bergonia - Early maturing winter variety for use under glass, sow at intervals from autumn to spring, some DM resistance. *Source History: 1 in 1994; 1 in 1998; 1 in 2004.* **Sources: Bou.**

Bloomsdale - (Long Standing) - 39-60 days, Sow very early, thick twisted crumpled glossy dark-green leaves, fine quality, quick-growing, heavy yielder, withstands heat or cold, slow to bolt, good long distance shipper, excellent fresh or canned, pre- 1908. *Source History: 23 in 1981; 19 in 1984; 17 in 1987; 17 in 1991; 18 in 1994; 21 in 1998; 36 in 2004.* **Sources: Ag7, Bo17, Bu8, CH14, CO30, De6, Ear, Fo7, Goo, Gr27, Gr28, He8, Lin, LO8, Ma19, MAY, Mo13, OLD, Or10, Pe2, Pr3, Ra5, Roh, Ros, Scf, Se7, SE14, Se16, Se26, Sk2, So9, TE7, Up2, Ves, WE10, We19.**

Bloomsdale Dark Green - (Bloomsdale Savoy Extra Dark Green Strain, Long Standing) - 40-50 days, Res. to premature bolting so widely used for early spring planting or wintering over, vigorous upright plants, large dark-green fleshy deeply crumpled leaves. *Source History: 15 in 1981; 10 in 1984; 8 in 1987; 6 in 1991; 5 in 1994; 4 in 1998; 2 in 2004.* **Sources: Dam, HO2.**

Bloomsdale Long Standing - (Bloomsdale Savoy L.S., Bloomsdale L.S. Dark Green, American Bloomsdale Extra L.S) - 39- 48 days, Stands well in hot weather, shipped extensively from the deep South, dark-green thick glossy savoyed blistered crumpled leaves, vigorous semi-upright leaves, 1925. *Source History: 90 in 1981; 85 in 1984; 83 in 1987; 74 in 1991; 72 in 1994; 79 in 1998; 67 in 2004.* **Sources: Ada, All, Ba8, Be4, Bu2, Bu3, But, CA25, CO23, Co31, Co32, Com, Dgs, DOR, Dow, Ers, Fa1, Fe5, Fi1, Fis, Fo13, Ga1, Ge2, Gur, HA3, Hal, He8, He17, Hi6, Hi13, Hig, Hud, Jor, Jun, La1, Lej, Mel, MOU, Na6, Ni1, Ont, Orn, PAG, Pin, Pla, Red, Ri2, Ri12, RIS, RUP, Sa9, Sau, Se8, Se25, Sh9, Shu, Si5, So1, SOU, Sto, Tho, Tu6, Ver, Vi4, Wet, Wi2, WI23.**

Bloomsdale Savoy Resistant - (Bloomsdale Savoy) - 45-50 days, Heavy yielder, very fast growing, upright plants can be mech. harv., large med-green med-savoyed leaves, highly blight and BM res., fine quality, very cold hardy. *Source History: 4 in 1981; 3 in 1984; 3 in 1987; 2 in 1991; 2 in 1994; 2 in 1998; 3 in 2004.* **Sources: ME9, Mey, Ter.**

Bloomsdale, Winter - (Winter Bloomsdale Long Standing) - 45-50 days, Firm thick dark-green deeply savoyed leaves, spreading and slow-growing, rugged hardy and cold tol., good for summer and fall crops, ideal for wintering over, will not bolt in hot weather, blight and mosaic and BM res., takes temp. extremes better than many hybrids. *Source History: 19 in 1981; 17 in 1984; 17 in 1987; 17 in 1991; 16 in 1994; 14 in 1998; 7 in 2004.* **Sources: Au2, In8, Me7, SE4, Se27, So1, Tu6.**

Broad Leaved Prickly Seeded - (Standwell Broad Leaved Prickly) - 40 days, Thick-stemmed, large leaved hardy strain, very long-standing, recommended for autumn sowing. *Source History: 1 in 1981; 2 in 1987; 1 in 1991; 2 in 1994; 2 in 1998; 3 in 2004.* **Sources: Bou, Lan, Th3.**

Butterflay - 45-55 days, Flavorful, medium dark green savoyed leaves, bolts late, spring or fall crop, from Germany. *Source History: 2 in 2004.* **Sources: Pe2, Tu6.**

Early No. 7 - 37 days, Large vigorous plants, somewhat spreading, large med-dark green, med-savoyed, DM and CMV res., grows fast, heavy yields, for fall and winter harvest in South and West. *Source History: 1 in 1981; 1 in 1984; 1 in 1987; 1 in 1991; 1 in 1994; 1 in 1998; 1 in 2004.* **Sources: Kil.**

Erste Ernte - 30 days, Name translates 'first harvest' in German, upright arrow-shaped leaves held above the ground, harvest when 5 in. long by 4 in. wide, early to mature and early to bolt, retains good flavor after bolting. *Source History: 1 in 2004.* **Sources: Ter.**

F-380 - PVP 2001 Arkansas AES. *Source History: 1 in 2004.* **Sources: ABB.**

Fall Green - (AR 82-5, Improved Green Valley) - 35 days, Probably the best white rust res. spinach for fall planting, semi-savoyed, dev. by the University of Arkansas. *Source History: 4 in 1991; 5 in 1994; 3 in 1998; 1 in 2004.* **Sources: ME9.**

Gamma - Dark green leaves, long standing, resists DM 1 and 2, tolerates race 3, from Switzerland. *Source History: 1 in 2004.* **Sources: Tu6.**

Giant Nobel - (Giant Thick Leaf, Long Standing Gaudry) - 40-56 days, Large spreading plants, huge thick smooth dark-green pointed leaves with round tips, heavy yields, excel. variety for canning, slow to bolt, for late spring. *Source History: 16 in 1981; 14 in 1984; 14 in 1987; 12 in 1991; 12 in 1994; 13 in 1998; 14 in 2004.* **Sources: Bo19, CH14, DOR, Ers, Fi1, Gr27, He8, He17, Hud, Ros, Se26, Syn, Vi4, WE10.**

Giant Winter - (Geant Hiver) - 45 days, Smooth large semi-savoyed med-green leaves, cold hardy, special strain for late summer or fall seeding for crop in early spring, also good for normal seeding. *Source History: 2 in 1981; 2 in 1984; 4 in 1987; 4 in 1991; 5 in 1994; 5 in 1998; 5 in 2004.* **Sources: Bou, Dam, Fe5, In8, Ter.**

Gigante d'Inverno - Large, deep green wavy leaves. *Source History: 1 in 1987; 1 in 1991; 1 in 1994; 1 in 1998; 1 in 2004.* **Sources: Ber.**

Low Acid - (Monnopa) - 60 days, Round-leaved var., very little oxalic acid in the leaves, sweetest of all spinach leaves, very high in vitamins A and C and E. *Source History: 2 in 1991; 4 in 1994; 1 in 1998; 1 in 2004.* **Sources: Bou.**

Medania - (Mediana) - 40-70 days, Smooth-leaved, Viking type with fine full-bodied flavor, sow early spring and late summer, heat and cold res., especially bred to resist bolting, DM1 and 3 res., high-yielding. *Source History: 2 in 1987; 3 in 1991; 3 in 1994; 4 in 1998; 4 in 2004.* **Sources: CO30, Gur, HO2, Se8.**

Merlo Nero - Vigorous Bloomsdale type, dark-green, oval, savoyed leaves, medium early. *Source History: 1 in 1994; 1 in 1998; 1 in 2004.* **Sources: Se24.**

Monstrous Viroflay - (Monstrueux de Viroflay) - 43 days, Large erect smooth dark-green leaves, fast-growing, mainly a fall or winter crop, produces heavy yields. *Source History: 1 in 1981; 1 in 1984; 1 in 1987; 1 in 1991; 1 in 1994; 1 in 1998; 1 in 2004.* **Sources: Lej.**

Nobel - (Giant Thick Leaved, Long Standing Gaudry) - 43-46 days, Huge smooth triangular dark-green leaves with rounded tips, spreading vigorous plants, slow bolt, prolific, canner. *Source History: 9 in 1981; 8 in 1984; 8 in 1987; 5 in 1991; 7 in 1994; 10 in 1998; 6 in 2004.* **Sources: Bu3, La1, Pe6, RUP, Shu, Und.**

Picnic - 39-48 days, Semi-upright, dark green, thick, glossy savoyed leaves, withstands hot weather. *Source History: 1 in 1998; 1 in 2004.* **Sources: Eco.**

Resistoflay - (Viroflay 99 MR) - 37-45 days, Large smooth med-green leaves, large erect plants, fall and winter var. in West, sow summer or fall, res. to DM2 and BM, USDA and TX/AES. *Source History: 6 in 1981; 6 in 1984; 5 in 1987; 7 in 1991; 5 in 1994; 5 in 1998; 2 in 2004.* **Sources: Ers, WE10.**

Round - (Summer, Medania) - 28-42 days, Large round leaves, mild and sweet, plant in succession for a prolific supply throughout the season, should be kept moist, tends to run to seed if it dries out. *Source History: 1 in 1987; 1 in 1991; 1 in 1994; 1 in 1998; 1 in 2004.* **Sources: Bou.**

Spring Giant - Large flat smooth leaves with crisp, firm texture, sweet with no bitterness even in dry conditions. *Source History: 1 in 2004.* **Sources: St18.**

Steadfast - 50 days, Extremely bolt resistant variety with smooth, dark-green leaves, not as productive as some but remains usable for up to two months longer. *Source History: 1 in 1994; 2 in 1998; 1 in 2004.* **Sources: Ter.**

Super Verano - Med-early variety with smooth, oval, green leaves. *Source History: 1 in 1994; 1 in 1998; 1 in 2004.* **Sources: CO30.**

Viking - (Northland, Heavy Pack) - 45-46 days, Large vigorous plant, huge arrow-shaped thick broad dark-green leaves with round tips, med-savoy, slow bolt, leading canning var., 1933. *Source History: 5 in 1981; 4 in 1984; 3 in 1987; 2 in 1991; 3 in 1994; 3 in 1998; 2 in 2004.* **Sources: Ber, Cr1.**

Viroflay - (Monstrueux de Viroflay) - 40-50 days, Plants up to 2 ft. in dia., large dark-green crisp smooth leaves can grow up to 10 in. long and 8 in wide at the base, low acid, 1866. *Source History: 2 in 1981; 2 in 1984; 3 in 1987; 7 in 1991; 7 in 1994; 9 in 1998; 11 in 2004.* **Sources: CO23, CO30, DOR, Fo7, He8, Ho13, La1, Ri12, Se7, Se8, WE10.**

Wintergreen - PVP pending. *Source History: 1 in 2004.* **Sources: ABB.**

Varieties dropped since 1981 (and the year dropped):

All Season (1991), *America Savoy* (1991), *AR354* (1998), *AR380* (1998), *Asian Delight* (1987), *Bloomsdale (Wisconsin)* (1991), *Bounty* (1983), *Bouquet* (1994), *Broad Leaved Summer* (2004), *Canadian Green* (1998), *Cold Resistant Savoy* (1998), *Dixie Market* (1994), *Dominant*

(2004), *Erik (Early Slow Bolting)* (1991), *Estivato* (2004), *Estivato Ete* (1987), *Fabris* (1991), *Fares* (2004), *First Crop* (2004), *Giant Munsterland* (1984), *Giant Viroflay* (1991), *Green Valley II* (1991), *Hiverna* (2004), *Hojo* (1987), *Jiromaru* (1983), *Keystone No. 7241* (1984), *King of Denmark* (2004), *King of Denmark Konsan* (2004), *Lessley Dark Green Savoy* (1987), *Long Standing Savoy* (1984), *Long Standing Savoy No. 653* (1991), *Lorelei* (2004), *Marvelous Long Standing* (1994), *Matador* (2004), *Matador Matarno* (2004), *Matador Nobio* (2004), *Monnopa* (2004), *Munsterlander* (1994), *Munsterlander Prickly* (2004), *Munsterlander Round* (2004), *No. 7241* (1987), *Nordic IV* (2004), *Nores* (1991), *Norfolk* (2004), *Norland Spring* (1998), *Northland* (1994), *Old Dominion* (2004), *Ozarka* (1991), *Ozarka II* (1998), *Popey* (1998), *Popeyes Choice* (1998), *Resistoflay Secundo* (1987), *Savoy Supreme* (1994), *Sohshu* (1998), *Special Summer Savoy* (1987), *Spencer* (1991), *Spinoza* (1991), *Summer Giant L* (2004), *Supergreen* (2004), *Suttons Sigmaleaf* (1991), *Thick Leaf* (1994), *Thick Leaved Improved* (1984), *Ujo* (1991), *Ushiwakamaru* (1991), *Valiant* (1991), *Virginia Savoy Blight Res.* (1998), *Vital-R* (1991), *Winterriesen Wiri* (2004), *Wobli* (1998), *Ziromaru* (1987).

Squash / Maxima
(Cucurbita maxima)

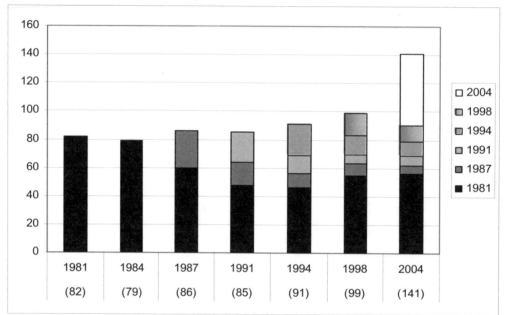

Number of varieties listed in 1981 and still available in 2004: 82 to 56 (68%).
Overall change in number of varieties from 1981 to 2004: 82 to 141 (172%).
Number of 2004 varieties available from 1 or 2 sources: 90 out of 141 (64%).

American Indian - *Source History: 1 in 2004.* **Sources: Se17.**

Amish Pie - 90-105 days, Good processing pumpkin with 5 in. thick flesh, 60-80 lbs., the moist firm flesh is excel. for making pies, heirloom from an Amish gardener in the Maryland mountains. *Source History: 1 in 1998; 3 in 2004.* **Sources: Hud, Pr9, Se16.**

Argentine - 100 days, Derived by successive hand pollinations of Argentina Primitive, grey-green, ribbed, dry, flavorful, good quality flesh, heavy producer. *Source History: 2 in 1994; 2 in 1998; 1 in 2004.* **Sources: Sa9.**

Argentine Primitive - Heirloom, ancestor of the Buttercups, round flat fruits up to 8 lbs., excellent food quality. *Source History: 1 in 1987; 1 in 1991; 1 in 1994; 1 in 1998; 1 in 2004.* **Sources: So9.**

Arikara - (Arikara Long) - 72-110 days, Grown by Arikara tribe who lived along the Missouri River in western North Dakota, extremely heavy producer of pink, slightly warted, Hubbard shaped fruits, up to 15 lbs., thin, green-yellow, somewhat watery flesh, reportedly an excellent keeper, from September harvest until following August, produces well under stress and drought, pre-Columbian. *Source History: 2 in 1987; 4 in 1991; 4 in 1994; 1 in 1998; 1 in 2004.* **Sources: Sa9.**

Atlantic Giant - 110-125 days, Huge upright round orange fruits weighing up to 800 lbs., developed by the Dill family of Nova Scotia from Goderich Giant, Genuine Mammoth and Mammoth Tours seedstock, introduced as a public variety in 1978 by Howard Dill, this variety is often confused with the PVP variety "Dill's Atlantic Giant," which was derived by Howard Dill from the same seed pool but is longer, less tall and generally smaller. *Source History: 6 in 1981; 6 in 1984; 23 in 1987; 30 in 1991; 38 in 1994; 23 in 1998; 34 in 2004.* **Sources: Bo18, Bu3, Cl3, CO30, Com,**

Ear, Ers, Fo7, Ga1, Gr28, Gr29, He8, He17, Hum, In8, Ma19, MAY, ME9, Mo13, Na6, Or10, Pp2, Ra5, Rev, SE4, Se17, Se26, Se28, Tho, Tt2, Ves, Vi4, Vi5, Wo8.

Atlantic Giant, Dill's - 110-125 days, Huge, red-orange skin, orange flesh, reported weights in excess of 600 lbs., dev. by Howard Dill of Nova Scotia, often confused with the non-PVP variety 'Atlantic Giant' which was also developed and released by the Dill family, this variety is longer and not as tall as 'Atlantic Giant,' PVP 1986 Howard Dill Ent. *Source History: 6 in 1981; 4 in 1984; 9 in 1987; 13 in 1991; 7 in 1994; 31 in 1998; 42 in 2004.* **Sources: Bu1, CA25, Ca26, CH14, Dam, De6, Dil, Fa1, Fi1, GRI, Gur, Ha5, Hal, Hi13, HO2, HPS, Jo1, Jor, Jun, La1, Loc, Mcf, ME9, Mel, MOU, Ni1, OLD, Ont, OSB, Pin, Ri12, RIS, Roh, RUP, SE14, Shu, Sto, Sw9, Ter, Ver, WE10, Wi2.**

Australian Butter - 90-100 days, Excel. baker with thick dry orange flesh, hard shell, 12-15 lbs., small seed cavity, great keeper, Diggers Garden Club. *Source History: 1 in 2004.* **Sources: Se16.**

Baby Queen - 100 days, Turban shape tapering toward blossom end, gray-blue skin, about 12 in. diameter, resembles Queensland Blue but smaller. *Source History: 1 in 2004.* **Sources: ME9.**

Banana - 105 days, Winter trailing sort, 24 x 6 in. dia. with pointed end, grey-green and pink types, sweet dry solid fine-grained flesh, good for pies and baking, excel. keeper. *Source History: 5 in 1981; 5 in 1984; 5 in 1987; 3 in 1991; 2 in 1994; 2 in 1998; 1 in 2004.* **Sources: Ros.**

Banana, Blue - (Gray Banana, Green Banana) - 105-120 days, Gray-green skin, 18-30 x 6-8 in., 25-30 lbs., string and fiber-free, some resistance to squash bugs, prolific despite disease or bad weather, developed 100 years ago by Aggeler and Musser Seed in Los Angeles, CA, once popular in the Old West. *Source History: 4 in 1981; 3 in 1984; 5 in 1987; 1 in 1991; 1 in 1994; 2 in 1998; 2 in 2004.* **Sources: Ba8, HO2.**

Banana, Guatemalan Blue - 90-110 days, Hard smooth slate-gray shell, 36 x 8 in., sweet flavor, fiberless, good keeper, freezes and bakes well, grown by South American Indians over 1,000 years before Columbus. *Source History: 2 in 1981; 1 in 1984; 2 in 2004.* **Sources: Sa9, Se16.**

Banana, Orange - 100 days, Short, fat orange skinned banana squash. *Source History: 1 in 2004.* **Sources: Sa9.**

Banana, Pink - 100-120 days, Cylindrical with tapered blossom end, 18-24 x 5.5-7 in. dia., 10 lbs., hard smooth deep-pink skin, thin brittle rind, solid fiberless yellow-orange flesh, for pies, common in American pioneer gardens. *Source History: 15 in 1981; 12 in 1984; 13 in 1987; 13 in 1991; 11 in 1994; 11 in 1998; 14 in 2004.* **Sources: ABB, Bo17, Bo19, Bu2, DOR, Gr27, Ho12, Kil, La1, Pin, Ri12, Sau, Sh9, WI23.**

Banana, Pink Jumbo - 105-115 days, Vining plant, nearly cylindrical fruit tapered at blossom end and slightly curved, 30-48 in. long x 8-12 in. dia., 70-75 lbs., thin brittle smooth skin turns pink-orange at maturity, yellow-orange flesh is thick firm dry sweet and not stringy, pick when less than 30 in., productive, keeps well, good for pies or baking or canning, developed about 100 years ago by Aggeler and Musser Seed, Los Angeles, CA. *Source History: 29 in 1981; 26 in 1984; 25 in 1987; 31 in 1991; 31 in 1994; 29 in 1998; 35 in 2004.* **Sources: Ba8, Bo18, Bu3, Bu8, CH14, CO23, Ers, Fi1, Gr29, Hi13, HO1, HO2, Ho13, Jor, Loc, Ma19, ME9, Mo13, MOU, OLD, OSB, Pp2, Ra5, RIS, RUP, SE14, Se26, Se28, Shu, Ter, Vi4, Vi5, WE10, Wi2, Wo8.**

Berrettina Piacentina - 105 days, Heirloom squash from Northern Italy, round, grey-green squash, yellow-orange flesh, good flavor, 3-5 lbs. *Source History: 1 in 2004.* **Sources: Se24.**

Big Max - (Big Mac, Big Mack) - 110-120 days, Largest pumpkin for faces or novelty, nearly round, 17-18 in. dia., fruits average 100 lbs., slightly rough red-orange skin, bright yellowish-orange 3-4 in. thick flesh, large exhibition type, good for pies or canning. *Source History: 86 in 1981; 79 in 1984; 79 in 1987; 77 in 1991; 71 in 1994; 67 in 1998; 65 in 2004.* **Sources: ABB, Ada, Ag7, All, Ba8, BAL, Be4, Bo19, Bu2, Bu3, Bu8, CA25, CH14, Cl3, CO23, CO30, Com, De6, Dgs, DOR, Ers, Fe5, Fo7, Ga1, Gr27, HA3, He8, HO1, HO2, Jor, La1, Loc, Ma19, MAY, ME9, Mel, Mey, MOU, OLD, Ont, Or10, PAG, Pe2, Pla, Ra5, Ri2, RIS, Roh, RUP, Sau, SE4, Se7, SE14, Se26, Sh9, Shu, So1, TE7, Twi, Vi4, Vi5, WE10, Wet, Wi2, WI23.**

Big Moon - 110-120 days, Extra large true pumpkin, carefully tended fruits average over 100 lbs.--some to 200 lbs., can reach 40 in. dia., slightly rough heavily-ribbed med-orange skin, thick light-orange flesh, large brown seeds, large vines, similar to Mammoth Gold, PVP expired 1996. *Source History: 28 in 1981; 27 in 1984; 31 in 1987; 31 in 1991; 25 in 1994; 23 in 1998; 23 in 2004.* **Sources: Bu1, Bu8, CA25, CH14, Coo, Ge2, GLO, Gr27, HPS, Ki4, Kil, Lin, MOU, OLD, Par, Ra5, RUP, Sh9, Shu, Twi, Vi5, We19, Wet.**

Big Tom - 120 days, Large orange pumpkin weighs 18+ lbs., flat on both ends with smooth hard bright-orange rind, orange-yellow flesh, measures 11 x 13 in. at maturity. *Source History: 1 in 1987; 1 in 1991; 1 in 1994; 1 in 1998; 1 in 2004.* **Sources: Cr1.**

Black Forest - 95 days, Kabocha type but slightly smaller, size of Buttercup without the button, dark-green skin, flat-round shape, deep-orange flesh is medium-dry, rich and sweet, early, stores well, bred by Johnny's Selected Seeds. *Source History: 1 in 1994; 2 in 1998; 4 in 2004.* **Sources: Jo1, Pe2, Pr3, Tu6.**

Blue Ballet - 90-100 days, Smaller version of Blue Hubbard, blue-grey smooth skin, 4-7 lbs., bright orange fiberless flesh, stores well. *Source History: 1 in 1994; 3 in 1998; 4 in 2004.* **Sources: Jo1, Pr3, Se7, Ter.**

Blue Blockers - 90-100 days, Vining habit, mottled light blue fruit with rounded shoulders, 5-6 lbs., good flavor, unknown variety name, seed from South Africa. *Source History: 1 in 2004.* **Sources: Rev.**

Boston Marrow - 90-110 days, Reddish-orange skin, 12-16 x 9-12 in. dia., 10-20 lbs., yellow-orange fine-grained thick flesh, fall and winter var., widely grown, shaped like a rounded Hubbard, introduced in 1831 from seeds obtained from Native Americans in New York State. *Source History:*

6 in 1981; 3 in 1984; 5 in 1987; 4 in 1991; 7 in 1994; 8 in 1998; 3 in 2004. **Sources: Fo7, La1, ME9.**

Bush Baby - 105 days, Vining habit, Jarahdale type, small blue fruit, 6-10 lbs., flattened square shape, smooth, some with slight ribbing, swollen ring at blossom end, delicious flesh, from Australia. *Source History: 2 in 2004.* **Sources: Ba8, Rev.**

Buttercup - (Turk's Cap) - 90-110 days, Turban or drum-shaped, dark-brown/green smooth thick tough skin, faint stripes, 4- 4.5 x 6-8 in. dia., 3.5-4 lbs., thick firm dry golden-yellow flesh, excel. keeper, introduced in 1920. *Source History: 72 in 1981; 69 in 1984; 60 in 1987; 50 in 1991; 43 in 1994; 45 in 1998; 42 in 2004.* **Sources: Alb, All, Ba8, Be4, Bu2, CH14, Co32, Cr1, Dam, Dgs, Ear, Fa1, Fi1, Ga1, Ge2, Goo, Gr27, Gr28, Gur, Ha5, He8, Lej, Lin, Loc, MAY, Me7, Mel, OSB, PAG, Pe2, Pin, Pr3, Ri12, Ros, Sa5, Sau, Se7, So9, Syn, Ter, Tu6, We19.**

Buttercup, Bitterroot - 98-110 days, Reselected buttercup str., uniform shape with square shoulders, small cup size, smooth creamy texture, 2.5-4 lbs., good keeper. *Source History: 1 in 1998; 4 in 2004.* **Sources: ME9, Pe6, SE14, TE7.**

Buttercup, Brown Seeded - Small turban, flavorful, stores well, occasional white seed, from Turtle Tree Seed Farm. *Source History: 1 in 2004.* **Sources: Tu6.**

Buttercup, Burgess Strain - 85-110 days, Flattened turbans, 5-8 in. dia., 3-5 lbs., thin hard dark-green rind, slight ribs, distinctive blossom-end button, very sweet med-orange stringless fiberless flesh, fine quality, small seed cavity, harvest when yellow spot touching ground turns orange, med-keeper, good maritime variety, the most common of the many Buttercup types. *Source History: 26 in 1981; 24 in 1984; 30 in 1987; 44 in 1991; 46 in 1994; 50 in 1998; 53 in 2004.* **Sources: ABB, Ag7, Bo17, Bu3, CH14, Cl3, CO23, Com, De6, Dil, DOR, Ers, Fe5, Fis, Fo7, GRI, HA3, Hal, Hi6, HO1, HO2, Jo1, Jor, Jun, La1, Ma19, ME9, Mo13, MOU, OLD, Ont, Or10, Rev, Ri2, RIS, RUP, SE4, Se8, SE14, Se25, Se26, Sh9, Shu, Si5, So1, Sto, TE7, Twi, Ver, Ves, Vi4, WE10, WI23.**

Buttercup, Bush - (Emerald Strain) - 85-100 days, Upright bushes, small smooth rounded turbans, the color of Blue Hubbard, thick deep sweet dry yellow-orange flesh, 3-5 lbs., high quality, heavy yields, 1952. *Source History: 13 in 1981; 11 in 1984; 12 in 1987; 10 in 1991; 8 in 1994; 14 in 1998; 8 in 2004.* **Sources: Bu1, Fa1, Fe5, Fis, HA3, Jor, Sh9, WE10.**

Buttercup, Discus Bush - 90 days, Bush plant spreads to 3 ft., bears up to 4 full size, 3 lb., dark green fruits, thick golden flesh is moderately dry, cooks up sweet, NDS breeding program. *Source History: 1 in 1998; 1 in 2004.* **Sources: Ter.**

Buttercup, Kindred - (Orange Buttercup, Kindred Turban) - 80-105 days, AAS/1969, ripens to reddish-gold, 3.5-4.5 x 7-8 in. dia., 3-5 lbs., semi-vining, 1 in. fine-grained golden-yellow flesh, keeps well. *Source History: 16 in 1981; 11 in 1984; 3 in 1987; 2 in 1991; 1 in 1994; 1 in 1998; 2 in 2004.* **Sources: Sa9, Tu6.**

Buttercup, Verdas - 100 days, More of a Kabocha type, cupless buttercup. *Source History: 1 in 1998; 1 in 2004.* **Sources: Sa9.**

Calabaza del Norte - Skin color varies from dark green to orange, shape resembles acorn with a small nipple on blossom end, sweet thick orange flesh, originally collected in Vadito, common throughout the northern villages of New Mexico. *Source History: 1 in 2004.* **Sources: Na2.**

Candy Roaster - (Candy Roaster Melon Squash) - 95-115 days, Vining plant bears bumpy, large, teardrop shaped fruit, sea-green skin color, orange tasty flesh, 10-18 lbs., almost extinct, not to be confused with the North Georgia Candy Roaster, traced back to North Carolina. *Source History: 4 in 2004.* **Sources: Ag7, Ba8, Fe5, Rev.**

Chersonskaya - *Source History: 1 in 2004.* **Sources: Se17.**

Courge Olive Verte - 120 days, Fruit looks like huge ripe olives with dark olive to green-black skin color, round shape with tapered ends, 5-8 lbs., fine flavor, good keeper, true 19th-century French heirloom. *Source History: 1 in 2004.* **Sources: Se17.**

Crown Pumpkin - 95 days, Large blue buttercup type, tasty golden flesh, ring at blossom end, from South Africa. *Source History: 1 in 2004.* **Sources: Ba8.**

Delicious, Golden - (Hubbard Golden Delicious) - 100-105 days, Heart or top-shaped fruits, 10-12 x 8 in. dia., 7-9 lbs., reddish-orange skin with green blotches, bright orange-yellow thick fine-grained flesh, vines yield well, higher vitamin C, not watery, dev. especially for baby food, excel. flavor, good keeper, also for canning or freezing. *Source History: 21 in 1981; 18 in 1984; 18 in 1987; 12 in 1991; 13 in 1994; 15 in 1998; 12 in 2004.* **Sources: Ba8, Fe5, Ha5, HO1, HO2, Jor, Lej, ME9, Rev, RUP, SE14, Sto.**

Delicious, Green - 100-105 days, Heart-shaped, smooth hard dark-green skin with lighter blossom end stripes, rounded swelling around the stem, tapers to a point at blossom end, 10 x 12 in., 6-12 lbs., sweet thick dry yellow-orange flesh, good flavor and keeper, vining. *Source History: 13 in 1981; 7 in 1984; 4 in 1987; 3 in 1991; 2 in 1994; 3 in 1998; 1 in 2004.* **Sources: Rev.**

Emerald - (Emerald Bush Buttercup) - 75-90 days, Buttercup type and size, gray-green skin, thick orange high-quality flesh, true bush squash, 3-4 ft. vines, slightly smaller yields, an occasional longer vine may dev. later in the season, keeps well if harvested promptly when mature, ND Univ. *Source History: 5 in 1981; 1 in 1984; 4 in 1987; 6 in 1991; 8 in 1994; 7 in 1998; 1 in 2004.* **Sources: Com.**

Essex Hybrid - Turban x Hubbard cross discovered by Aaron Low of Essex, MA when passing a field of American Turban squash, 8-12 lb. fruit, drum-like shape, deep orange skin with apricot stripes, good quality, intro. in 1883. *Source History: 1 in 2004.* **Sources: Ea4.**

Essex Turban - 105-110 days, Very warty bright orange skin, hard, thick rind, flavorful Hubbard type flesh, 12-15 lbs. *Source History: 1 in 1994; 1 in 2004.* **Sources: RUP.**

Flat White Pumpkin - (Flat White Boer) - 120 days, Pure white skin, 6-8 in. tall by 12 in. diameter, 15-30 lbs., sweet orange flesh, leaves no fiber when flesh is passed through a sieve, high sugar content, makes a lighter colored pie, keeps well, seed from South Africa. *Source History: 2 in 1981; 1 in 1984; 1 in 1987; 1 in 1991; 2 in 1994; 2 in 1998; 2 in 2004.* **Sources: Lej, Rev.**

Fortna White - Prolific vines produce 5-10 fruits each, small to medium size, pear-shaped pumpkin with white skin, creamy yellow flesh, good for pies, occasionally a small-necked pumpkin will appear, save seeds from the pear-shaped fruits only, heirloom from the Fortna family of Pennsylvania for more than 50 years. *Source History: 1 in 1994; 2 in 1998; 2 in 2004.* **Sources: Lan, Roh.**

Galeux d'Eysines - 95-100 days, Raised bumps over the skin surface make this a very unusual pumpkin, pink with beige scars or bumps that resemble peanuts, shaped like a wheel of cheese, unknown in the U.S. before its discovery at the 1996 Pumpkin Fair in Tranzault, France. *Source History: 2 in 2004.* **Sources: Rev, Se16.**

Georgia Roaster - (Georgia Candy Roaster) - 100-110 days, Vining plant, long cylindrical fruit with thin pink skin, 20-30 in. long, up to 60 lbs., delicious orange fiberless flesh, baking and pies, good keeper, very rare. *Source History: 1 in 1998; 6 in 2004.* **Sources: Ba8, HO2, ME9, RUP, SE14, Sk2.**

German - 87 days, Slightly ribbed fruit, pink skin, blocky shape, 10-15 lbs., 1.5 in. thick flesh, productive. *Source History: 1 in 1998; 1 in 2004.* **Sources: Sa9.**

Gold Nugget - (Golden Nugget) - 75-95 days, AAS/1966, early, compact runnerless plant, Buttercup shaped, 1-3 lbs., hard red-orange skin with faint salmon-orange stripes, sweet dense med-thick fiberless dry dark yellow-orange flesh, good table quality and keeper, Dr. Holland at ND/AES. *Source History: 39 in 1981; 26 in 1984; 17 in 1987; 19 in 1991; 15 in 1994; 15 in 1998; 19 in 2004.* **Sources: Alb, CO30, Coo, Fis, Ga1, Hig, Ho13, Jor, ME9, OSB, Pe6, RIS, RUP, Se7, SE14, Ter, Ves, WE10, We19.**

Goldkeeper - 105 days, Vining plant habit, pink to red skin color, variable shape, thick yellow-orange flesh, excellent for pies and squash recipes, most fruits resemble Golden Delicious Hubbard at the stem end but not as pointed at the blossom end, used by the Stokely Canning Company for many years for their canned pumpkin operation. *Source History: 1 in 2004.* **Sources: Rev.**

Hidatsa - Hubbard shaped orange fruit, medium-large size, excellent keeper, reliable producer when stressed, from the Fort Berthold Reservation, North Dakota. *Source History: 1 in 2004.* **Sources: Tu6.**

Hokkaido, Green - 98-105 days, Rounded, slight ribs, dull slate-green skin, sweet fiberless yellow flesh is dry and almost flaky, not particularly high yielding, Japanese var. *Source History: 4 in 1981; 4 in 1984; 2 in 1987; 2 in 1991; 3 in 1994; 4 in 1998; 4 in 2004.* **Sources: Fe5, Ri12, Se7, So12.**

Hokkaido, Orange - Delicious orange-colored 4-5 lb. fruits, tapers to a point at both ends, excel. for pumpkin pies, originally from Japan. *Source History: 1 in 1981; 1 in 1984; 1 in 1998; 1 in 2004.* **Sources: So9.**

Hokkaido, Semi-Bush - 95-105 days, Hokkaido selection with more compact plant habit, slate-green skin, rich orange flesh, 2-4 lbs., rare. *Source History: 1 in 2004.* **Sources: Se7.**

Hopi Orange - 90-110 days, Dark orange fruit, 10-15 lbs., tasty yellow-orange fibrous flesh, good for baking, good keeper, resists squash bugs. *Source History: 2 in 1998; 1 in 2004.* **Sources: Se7.**

Hopi Pale Grey - 100 days, Vigorous vines, light and dark gray skin, 6-10 lbs., ridged cheesebox-round shape, cantaloupe color flesh, bland flavor, good keeper. *Source History: 1 in 1991; 2 in 1994; 2 in 1998; 2 in 2004.* **Sources: Au2, Se17.**

Hopi White - 100 days, White-skinned heirloom, bright orange flesh, round to oblong ridged fruit, 7-9 in. across, 6-12 lbs., acquires pink blush in storage, great keeper. *Source History: 1 in 2004.* **Sources: Se17.**

Hopi White Pumpkin - 110 days, White skin, thick orange flesh, good for pies and ornamental, 30-40 lbs., Native American. *Source History: 1 in 2004.* **Sources: Se8.**

Hubbard - 100-110 days, Coarsely warted hard globular fruits with pointed ends, dark bronze-green shell, 12 x 9-10 in. dia., vigorous vines, home or market, often broken and sold in chunks. *Source History: 6 in 1981; 6 in 1984; 4 in 1987; 4 in 1991; 3 in 1994; 4 in 1998; 1 in 2004.* **Sources: CO30.**

Hubbard, Anna Swartz - 90-100 days, Family heirloom given to Anna Swartz by a friend in the 1950s, flesh color and flavor resembles that of yams, 5-8 lbs., Anna loved this var. because of its extremely hard shell and excel. storing qualities. *Source History: 1 in 1998; 3 in 2004.* **Sources: Pr9, Se16, Te4.**

Hubbard, Azure - 110 days, Blue-green, grey fruit with necks at both ends, hard, slightly ridged shell, 15-20 in. long, 9-12 in. diameter, 14-30 lbs., yellow-orange, thick flesh is dry, fine-grained and sweet, great for pies, boiling or freezing, good keeper. *Source History: 1 in 1994; 1 in 1998; 1 in 2004.* **Sources: Eco.**

Hubbard, Baby Blue - (Kitchenette) - 90-100 days, Blue Hubbard x bush buttercup str., early, high quality, 4-6 lbs., sweet dry thick orange-yellow flesh, light-blue skin, semi-bush, good keeper, U of NH, 1953. *Source History: 4 in 1981; 4 in 1984; 5 in 1987; 9 in 1991; 5 in 1994; 8 in 1998; 11 in 2004.* **Sources: Dam, Ers, Fe5, Hig, HO2, ME9, RUP, SE14, Shu, St18, WE10.**

Hubbard, Baby Green - (Baby Green Kitchenette Hubbard) - 90-105 days, Sweet nutty yellow fine-grained flesh, 7 x 7 in. fruits, tapers towards stem end, all the fine qualities of Green Hubbard, but smaller more popular fruit size, good baked or in pies. *Source History: 4 in 1981; 4 in 1984; 3 in 1987; 2 in 1991; 1 in 1994; 3 in 1998; 6 in 2004.* **Sources: Ers, Fis, ME9, RUP, SE14, Sk2.**

Hubbard, Blue - 100-120 days, Blue-gray-green slightly-ridged fruits, hard coarsely-warted shell, neck at both ends, 15-20 in. long x 9-12 in. dia., 15-30 lbs., thick dry fine-grained yellow-orange flesh, very sweet, very productive, great keeper, for pies baking boiling or freezing, favorite in New England, dev. by Harris Seeds. *Source History: 34 in 1981; 31 in 1984; 36 in 1987; 38 in 1991; 44* in 1994; 48 in 1998; 45 in 2004. **Sources: All, Bo19, Bu2, Bu3, CA25, Com, De6, Ers, Fo7, GRI, HA3, Ha5, He8, HO1, HO2, Ho13, Hud, Hum, Jo1, Jor, La1, Lej, Ma19, ME9, Mo13, Ni1, OLD, Or10, OSB, PAG, Pin, RIS, Roh, RUP, SE4, SE14, Se25, Se26, Sh9, Shu, So9, Ter, Vi4, WE10, WI23.**

Hubbard, Blue (New England Str.) - 110-120 days, Huge Blue Hubbard type, appears to be a cross between the original Hubbard x Marblehead or Middleton Blue, 20-30 x 12-14 in. dia., bumpy and warted, 30-45 lbs., thick orange flesh, fine texture and flavor and quality, excel. keeper, introduced in 1909. *Source History: 10 in 1981; 8 in 1984; 9 in 1987; 9 in 1991; 10 in 1994; 14 in 1998; 12 in 2004.* **Sources: CH14, Fe5, Ga1, Ge2, HO1, HO2, Jor, Loc, ME9, Rev, RUP, Sto.**

Hubbard, Blue (Special) - 110 days, Blue-grey skin, 12 x 7 in., yellow-orange flesh, firm texture. *Source History: 3 in 1991; 2 in 1994; 1 in 1998; 1 in 2004.* **Sources: MOU.**

Hubbard, Chicago Warted - (Green Chicago Warted Hubbard, Hubbard Improved) - 105-115 days, Vigorous vines, dark- green fruits, very hard shell thickly covered with heavy warts, true Hubbard shape, 12-14 x 10 in. dia., average 12-16 lbs., thick dry sweet fine-grained golden-yellow flesh, keeps until late spring, good shipper, for pies baking or freezing, dev. by Budlong Gardens of Chicago, introduced by Vaughans Seed Store of Chicago in 1894. *Source History: 16 in 1981; 12 in 1984; 13 in 1987; 13 in 1991; 14 in 1994; 15 in 1998; 20 in 2004.* **Sources: Ba8, CH14, Ers, HO2, Ho13, Jor, Jun, ME9, Mo13, OLD, Rev, Ri12, RUP, SE14, Se26, Sh9, Shu, Twi, Vi4, WE10.**

Hubbard, Dulong Qhi - 82 days, Deeply ribbed, blue-gray fruit, slightly flattened shape, 8-12 lbs., richly flavored golden yellow flesh is 5-7 in. thick, moderate res. to PM, from Australia. *Source History: 1 in 2004.* **Sources: Ter.**

Hubbard, Golden - (Red Hubbard, Genesee Red Hubbard) - 90-105 days, Hard, moderately-warted, red-orange rind with tan stripes at blossom end, 10-12 x 8-9 in. dia., 8-12 lbs., solid thick sweet dry fine-grained orange-yellow flesh, fine flavor, excellent keeper, can or freeze, popular with home gardeners and commercial growers, introduced in 1898. *Source History: 42 in 1981; 32 in 1984; 38 in 1987; 35 in 1991; 36 in 1994; 34 in 1998; 35 in 2004.* **Sources: CO30, Co32, Dam, De6, Ers, Fi1, Fis, Gr28, HO2, Ho13, Jor, La1, Lin, Loc, Ma19, ME9, Mo13, OLD, Pin, Pr9, Rev, RIS, Roh, RUP, SE4, SE14, Se16, Sh9, Shu, Si5, Sk2, Sto, Twi, Vi4, WE10.**

Hubbard, Green - (True Hubbard, True Green Hubbard, Improved Hubbard) - 100-115 days, Dark bronze-green med-warted thick hard shell, tapered at both ends, 12-16 in. x 9-10 in. dia., 10-15 lbs., sweet fine-grained dry golden-yellow mealy flesh, fine flavor, excel. shipper and keeper, large vines, introduced for sale in the 1840s by seedsman, J. H. Gregory, who named it after Elizabeth Hubbard of Marblehead, MA. *Source History: 43 in 1981; 36 in 1984; 34 in 1987; 33 in 1991; 30 in 1994; 31 in 1998; 23 in 2004.* **Sources: All, Ba8, CH14, Ear, Ers, HO2, Ho13, La1, Lin, Loc, Ma19, ME9, Mey, MOU, Ont, RUP, Se7, Se8, Sh9, Si5, So9, Wet, WI23.**

Hubbard, Green Improved - 95-120 days, Large vining plants, dark bronze-green fruits turn a deeper bronze at maturity, hard tough moderately rough rind, typical Hubbard shape tapered at both ends, 12 x 9 in. dia., 10-12 lbs., very thick sweet dry yellow- orange flesh, exceptional winter keeper, for freezing baking or pies. *Source History: 22 in 1981; 18 in 1984; 14 in 1987; 8 in 1991; 6 in 1994; 8 in 1998; 6 in 2004.* **Sources: Bou, Dam, De6, Dgs, DOR, Ros.**

Hubbard, Little Gem - 70-80 days, Miniature Golden Hubbard type, bright-orange skin, 3-4 lbs., sweet deep-orange dry flesh, excel. for pies or vegetable use, good keeper, early, well named. *Source History: 2 in 1981; 2 in 1984; 2 in 1987; 2 in 1991; 1 in 1994; 1 in 1998; 1 in 2004.* **Sources: Ves.**

Hubbard, Mini Green - 100 days, Smaller, imp. version of Kitchenette Hubbard, med. green, 2.5 lbs. *Source History: 1 in 1998; 1 in 2004.* **Sources: Sto.**

Hubbard, Minnie's Apache - Fruits vary in size and shape, light to dark-orange skin, bright-orange flesh is non-stringy and sweet, white and tan seeds. *Source History: 1 in 1991; 1 in 2004.* **Sources: Na2.**

Hubbard, Navajo - 105 days, Large size, light green-blue to dark green to orange skin, tasty orange flesh, large tan seeds, good for roasting, grows well in cooler weather where season length permits, originally collected at Fort Defiance on the Navajo Nation. *Source History: 2 in 2004.* **Sources: Na2, So1.**

Hubbard, Smooth Green - Differs slightly from the 19th century True Green Hubbard in that it has somewhat smoother skin (sometimes debatedly), long vines, 8-12 lb. green fruit tapers at both ends, thick orange-yellow flesh, long storage, may be from an early selection of True Green or possibly the other way around, intro. by the Gregory Seed Company of Marblehead, MA in 1857, possibly brought to Boston by a sea captain around 1798, rare. *Source History: 1 in 2004.* **Sources: Ea4.**

Hubbard, Sugar - 110 days, Smooth Blue Hubbard str. resulting from a cross between Sweet Meat and True Hubbard with moist golden flesh, blue-gray skin, 15-20 lbs., vining, good keeper, Northwest heirloom developed by the Gill Brothers Seed Company in Gresham, Oregon. *Source History: 1 in 1981; 1 in 1984; 1 in 1987; 1 in 1991; 1 in 1994; 1 in 1998; 1 in 2004.* **Sources: Ter.**

Hubbard, True - (Genuine Hubbard, Improved Hubbard, True Green Hubbard, Green Mountain, Green Hubbard) - 105-115 days, Smooth dark bronze-green skin, few very slight bumps, 12-15 x 10 in., 15-35 lbs., thick fine-grained sweet dry yellow- orange flesh, keeps until spring, becoming scarce, introduced in the 1790s. *Source History: 8 in 1981; 7 in 1984; 6 in 1987; 4 in 1991; 6 in 1994; 8 in 1998; 8 in 2004.* **Sources: Bu2, HA3, Jor, Mo13, OLD, Rev, SE14, WE10.**

Hubbard, Warted - (Warted Green Hubbard) - 110-115 days, Thick heavily warted dark-green hard rind, will not bruise easily, 12-30 lbs., dry deep-orange flesh, some keep up to 9 months, late. *Source History: 12 in 1981; 11 in 1984; 13 in 1987; 8 in 1991; 7 in 1994; 7 in 1998; 4 in 2004.*

Sources: Bu3, RIS, Sa9, Ves.

Hubbard, Warted Improved - 105-110 days, More warted and darker-green, thin very hard skin, 13 x 10 in. dia., 12-14 lbs., very thick orange-yellow dry sweet flesh, for winter storage and market, vining, late. *Source History: 5 in 1981; 3 in 1984; 3 in 1987; 3 in 1991; 1 in 1994; 2 in 1998; 1 in 2004.* **Sources: PAG.**

Indian Pumpkin - Fruit shaped like a top, blue-green skin, almost sringless bright orange flesh, used for custards and pies, from the North Valley of Albuquerque, NM. *Source History: 1 in 2004.* **Sources: Na2.**

Iran - 95 days, Beautiful, large round fruit, shows three colors at maturity, salmon, white and green, up to 25 lbs., good eating qualities, decorative. *Source History: 1 in 2004.* **Sources: Se16.**

Jarrahdale - 95-100 days, Superb eating pumpkin from Australia, similar to Sweet Meat but superior in range of cultivation and flavor, very deeply ribbed, 12-20 lbs., wider than tall shape, hard blue-grey skin with orange-yellow sweet dry stringless flesh, excel. for baking or in pies, extended storage, very rare in the U.S. *Source History: 1 in 1991; 1 in 1994; 5 in 1998; 26 in 2004.* **Sources: Ba8, Bo19, Bou, Bu2, Com, Ers, Fe5, He17, HO2, Jo1, Jor, Lej, ME9, Na6, OSB, Par, Rev, Ri12, RIS, RUP, Sa9, SE4, SE14, Sk2, Und, WE10.**

Jaune Gros de Paris - Large yellow fruit can weigh over 100 lbs., round flattened shape with light ribbing, good in pies, soups and baked, good keeper, still popular in France. *Source History: 1 in 2004.* **Sources: Ba8.**

Kabocha - 110 days, Hokkaido type, grey skinned fruit, 4-5 lbs., excel. quality dry flesh. *Source History: 1 in 1994; 3 in 1998; 3 in 2004.* **Sources: Pr9, So12, Syn.**

Kakai - 100 days, Austrian type, medium-small, black striped pumpkins, 5-8 lbs., large, dark green hulless seeds are delicious when roasted, source of pumpkin seed oil. *Source History: 1 in 2004.* **Sources: Jo1.**

King of Mammoth - (Pot Iron, Mammoth Chili, Golden KOM, King Mam. Gold, Jumbo, Kentucky Large Field) - 105-130 days, Slightly ribbed flattened globe, light-yellow skin mottled with orange, often slightly netted, solid coarse good quality flesh, 15-18 x 18-24 in. dia., 50-75 lbs., some over 100, pre-1824 introduction. *Source History: 22 in 1981; 20 in 1984; 9 in 1987; 5 in 1991; 3 in 1994; 1 in 1998; 1 in 2004.* **Sources: Ba8.**

Kuri, Blue - (Tokyo Squash) - 100 days, Smooth blue-gray-green mature skin, very sweet thick fine-grained fine quality orange-yellow flesh, 4-5 lbs., good for short season, vining, Japanese seed. *Source History: 3 in 1981; 3 in 1984; 1 in 1987; 4 in 1991; 8 in 1994; 7 in 1998; 7 in 2004.* **Sources: CO30, Ev2, Jor, Kit, Loc, ME9, RUP.**

Kuri, Red - (Orange Hokkaido, Baby Red Hubbard) - 90-95 days, Japanese winter squash, bright reddish-orange teardrop-shaped fruits, 5-10 lbs., smooth-textured flesh, high yields, for pies and mashing. *Source History: 3 in 1981; 3 in 1984; 6 in 1987; 10 in 1991; 13 in 1994; 11 in 1998; 20 in 2004.* **Sources: Al25, Ba8, Com, Coo, Ers, Fe5, Hi6, HO2, Jo1, ME9, OSB, Pe2, Pr3, Se7, Se8, SE14, Se25, Sk2, So9, Te4.**

Large Turk's Turban - Large green turban-shaped gourd. *Source History: 1 in 1981; 1 in 1984; 1 in 1987; 1 in 2004.* **Sources: Sto.**

Lower Salmon River - 94 days, Beautiful pink turban type, large tasty fruit, 8-18 lbs., 1.5 in. thick flesh, excellent for pies. *Source History: 1 in 2004.* **Sources: In8.**

Lumina - 80-115 days, Creamy white skinned fruit grow 8-10 in. across, 10-16 lbs., bright orange flesh, makes great pies, skin will turn bluish white under stress, stores well, good for fall decorations, PVP 1994 Hollar Seed Company. *Source History: 33 in 1994; 54 in 1998; 69 in 2004.* **Sources: Bo17, Bo18, Bo19, Bu1, Bu2, Bu3, Bu8, CA25, Ca26, CH14, Cl3, Co31, Com, Dam, De6, Dgs, Ear, Fa1, Fe5, Fi1, Gr27, Gr29, Gur, Ha5, He17, Hi13, HO1, HO2, HPS, Jor, Jun, Lej, Loc, Ma19, MAY, ME9, Mel, Mo13, MOU, Ni1, OLD, Ont, OSB, Pin, Ra5, Ri12, RIS, Roh, Ros, RUP, Sau, SE4, Se26, Se28, Sh9, Shu, Si5, Sk2, SOU, Sto, Sw9, Ter, Twi, Ves, Vi5, We19, Wet, Wi2, Wo8.**

Mammoth Gold - (Jumbo, King of Mammoth, Virginia Mammoth) - 105-120 days, Smooth golden-orange pumpkin, faintly ribbed, irregular globe, 18-24 in. dia., 40-60 lbs., some over 100 lbs., skin mottled with pinkish-orange and yellow, thick pale yellow-orange flesh, edible but coarse, used for pies and novelty. *Source History: 15 in 1981; 14 in 1984; 13 in 1987; 14 in 1991; 15 in 1994; 18 in 1998; 18 in 2004.* **Sources: Ada, Bu8, CH14, Cl3, Co31, Ers, ME9, Ra5, RUP, Sau, SE14, Sk2, SOU, Vi5, WE10, Wet, Wi2, WI23.**

Mammoth King - (Jumbo, Mammoth Chile) - 120 days, Usually slightly flattened salmon-orange globes, 18-24 in. dia., 60-100 lbs., firm coarse good-quality yellow flesh, table use or stock feed, shallow ribs, introduced before 1824, rare. *Source History: 3 in 1981; 2 in 1984; 2 in 1987; 1 in 1991; 1 in 1994; 1 in 1998; 1 in 2004.* **Sources: La1.**

Mammoth Orange Gold - 110 days, Deep-orange rind, 20 x 20 in., 75-80 lbs. or more, thick orange flesh, smooth firm large pumpkin. *Source History: 2 in 1981; 2 in 1984; 1 in 1987; 1 in 1991; 1 in 1994; 1 in 1998; 1 in 2004.* **Sources: Mey.**

Marblehead, Umatilla - 110 days, Vigorous vining Hubbard type, blue-grey-green skin, no warts, slight ridges, nearly round, 16-30 in., thick deep-orange flesh, fine flavor and texture, CTV res. *Source History: 2 in 1981; 2 in 1984; 2 in 1998; 1 in 2004.* **Sources: So12.**

Marblehead, Yakima - Round Blue Hubbard type, sprawling vines, gray-green skin with pink blush, 10 x 13 in., 8-24 lbs., a variant of Umatilla Marblehead, 1896. *Source History: 2 in 1981; 1 in 1987; 1 in 1991; 1 in 1994; 1 in 1998; 2 in 2004.* **Sources: In8, Se17.**

Marina di Chioggia - 80-105 days, Beautiful heirloom, ridges and bumps on the gray-green skin look like ocean waves, 5-10 lbs., stores well, excellent for squash gnocchi, staple in Venetian markets, heirloom Sea pumpkin of Chioggia, on the coast of Italy. *Source History: 1 in 1994; 1 in 1998; 5 in 2004.* **Sources: Ba8, Com, Coo, Hud, Se24.**

Mayo Blusher - 100-120 days, White round to oval fruits, pale blue or white skin with slight indentations, sweet apricot-colored flesh, 7-10 lbs., excellent for pumpkin pie, good keepers, the Mayo or Yoreme are a Native American people of western Mexico. *Source History: 1 in 1987; 1 in 1991; 1 in 1994; 2 in 1998; 3 in 2004.* **Sources: Na2, Rev, Se17.**

Mexigold - 80 days, Glenn Drowns' creation from the late 1970s, very early, sweet, pink skinned Delicious type, not vine borer res., but recovers well from vine borer attacks. *Source History: 1 in 1998; 1 in 2004.* **Sources: Sa9.**

Mooregold - 90-105 days, More vigorous and productive than buttercup, 4-5 x 6-7 in. dia., thick bright-orange flesh, bright-orange skin with distinct salmon stripes, uniform, U of WI. *Source History: 2 in 1981; 2 in 1984; 5 in 1987; 5 in 1991; 5 in 1994; 5 in 1998; 6 in 2004.* **Sources: Ga1, Jor, Jun, ME9, OLD, SE14.**

Mormon - Hubbard with thick orange flesh, blue-green skin, occasional orange skinned fruit, considered to have come from the Mormons. *Source History: 1 in 2004.* **Sources: Na2.**

Mountaineer - 75-90 days, Very early winter squash dev. by Ken Fisher, 5-6 lb. fruits, color varies but selected for golden color and Hubbard shape, thick orange-yellow flesh, bakes dry, keeper. *Source History: 1 in 1981; 1 in 1984; 2 in 1987; 2 in 1991; 2 in 1994; 3 in 1998; 3 in 2004.* **Sources: Fis, Hig, Pe6.**

Muscat de Provence - 80 days, Deeply ridged, flattened fruit with smooth orange skin, 5-10 lbs., sweet flesh, good winter keeper, from the south of France, rare. *Source History: 1 in 2004.* **Sources: Coo.**

Nanicoke Indian - 110 days, Initial fruits set on bush plants and later sprawls, 3-6 lb. turbans, 3-10 in. by 4-7 in. high, good for soup, baking and pies, Nanicoke nation heirloom. *Source History: 2 in 2004.* **Sources: Ri12, Se17.**

Nepalese - 100 days, Vining plant produces variably shaped fruit with strong upright stems, deep brown-grey mottled with blue or blue-gray mottled with light blue or plain pink skin color, 12-25 lbs., firm bright yellow flesh, excellent keeper, tan colored seeds, great for pies. *Source History: 1 in 2004.* **Sources: Rev.**

New Zealand Blue - 110-120 days, Flattened, ribbed, gray-blue squash, deep orange, sweet flesh, 8 lbs., 8 x 7 in., good keeper, makes an interesting ornamental. *Source History: 2 in 2004.* **Sources: ME9, OSB.**

Ol Zeb's - (Old Zeb) - 110 days, Brightest orange skin, long handle, 22-28 lbs., high yields, Rupp introduction, dev. in Canada. *Source History: 5 in 2004.* **Sources: HO2, Jor, ME9, OSB, RUP.**

Old Blue - 95-100 days, Vining plant habit, variably shaped fruit, dark green skin with muted gold spots, 5-10 lbs., orange-yellow flesh with rich strong flavor, almost meaty when cooked. *Source History: 1 in 2004.* **Sources: Rev.**

Padana - 105 days, Pumpkin shape, alternate vertical gray-green and orange ribs, sweet dry orange flesh, for soup, gnocchi, roasting and Halloween pumpkins, good storage, from the Northwest of Italy. *Source History: 2 in 2004.* **Sources: Ri12, Se24.**

Penasco Cheese - Flat, ribbed cheese shaped fruit, gray or pale pink skin, sweet orange flesh, 5-8 lbs., from Penasco, NM. *Source History: 1 in 2004.* **Sources: Na2.**

Queensland Blue - 110-120 days, Striking blue, flattened, ribbed, 10-20 lb. fruits, golden flesh, sweet flavor, good keeper, Australian variety, introduced to the U.S. in 1932. *Source History: 1 in 1987; 2 in 1991; 3 in 1994; 3 in 1998; 12 in 2004.* **Sources: Ba8, Bo19, Ers, He8, HO2, Hud, Lej, ME9, Pr9, SE14, Se16, Sk2.**

Quintale seme Giallo - 105 days, Pumpkin shape, orange skin, sweet dry orange flesh, 5 lbs., for Halloween pumpkin carving, soup, gnocchi and roasting, good storage. *Source History: 1 in 2004.* **Sources: Se24.**

Rainbow - 85 days, Small banana type with pink and blue stripes, 12 in. long, 2-4 lbs., good flavor and keeping qualities. *Source History: 1 in 1998; 2 in 2004.* **Sources: Ho13, Sa9.**

Red Chestnut - (Sweet Chestnut Red) - A smaller Oriental type, bright red skin, chestnut shape, dry sweet flesh, good for cooking and pies, good keeper. *Source History: 2 in 1981; 1 in 1984; 1 in 1987; 1 in 2004.* **Sources: Yu2.**

Red Gold - 92 days, Unique, no hard shell, when boiled after seed is removed everything including the skin can be mashed up together, mottled salmon skin color, 1 in. thick yellow flesh, 6-10 lbs., dev. by Prof. Meader of New Hampshire. *Source History: 1 in 1981; 1 in 1984; 1 in 1998; 1 in 2004.* **Sources: Sa9.**

Red Warty Thing - 100 days, Essex Turban type, round red fruit completely covered with bumps, tasty stringless flesh, great market variety, excellent fall decoration. *Source History: 2 in 2004.* **Sources: Ba8, Ri12.**

Redlands Trailblazer - 90 days, Parentage includes the Australian heirloom Jarrahdale and a wild species squash, 8 ft. vines, slightly ribbed and flattened, gray fruit, 12 in. dia., sweet moist gold flesh, stores well, disease res., bred by Dr. Mark Herrinton at the Redlands Res. Sta. in Queensland, Australia. *Source History: 1 in 2004.* **Sources: Ter.**

Relleke - 120 days, Old-fashioned antique pumpkin, enormous size, excellent flavor and keeping ability. *Source History: 1 in 2004.* **Sources: Und.**

Rio Lucio Pumpkin - Hubbard with medium to large, variably shaped fruit, gray-green, dark green or orange skin, large tan seeds, from Rio Lucio, NM. *Source History: 1 in 2004.* **Sources: Na2.**

Rouge Vif d'Etampes - (Rouge d'Etampes, Red Etampes, Deep Red d'Etampes, Etampes Pumpkin, Cinderella) - 95-150 days, Antique french heirloom, flat shiny red-orange fruits, 6 in. deep x 18 in. dia., narrow deep-ribbed sections, rough bumpy skin, can grow to 40 lbs., beautiful and decorative, popular in the Central Market in Paris in the 1880s, first offered commercially in America by W. Atlee Burpee in 1883. *Source History: 2 in 1981; 4 in 1987; 7 in 1991; 21 in 1994; 35 in 1998; 51 in 2004.* **Sources: Al25, Ba8, Bo19, Bu2, CA25, CO30, Co32, Coo, Dam, Eo2, Fe5, Gou, Gr28, He8, He17, Hi13, Ho13, Hud, Jo1, Jor, Jun, Ki4, Lej, Loc, Me7, Na6, Ni1, OSB, Pe2, Pin, Pr3, Ra5, Re6, Rev, Ri12, RIS, Roh, RUP, Sa9, Se7, Se16, Shu, So1, Ter, Twi, Up2, Ver, VI3, WE10, We19, Wi2.**

Show King - 120 days, Fruits weighing 334 242 224 221 201 and 177 lbs. on one vine, skin colors: orange and yellow and light- blue and blue-green and gray and dark-green, former record of 438.5 lbs., Howard Dill. *Source History: 8 in 1981; 7 in 1984; 6 in 1987; 3 in 1991; 2 in 1998; 2 in 2004.* **Sources: Dil, Pp2.**

Sibley - (Pike's Peak) - 110 days, Hubbard type, short runners, round to elongated, teardrop shaped, slate-blue fruits 5-15 lbs., orange flesh, excel. flavor, introduced by Sibley and Co. in New York, 1887. *Source History: 1 in 1987; 1 in 1991; 3 in 1994; 1 in 1998; 3 in 2004.* **Sources: Hud, Pr9, Se16.**

Sweet Meat - (Sweet Meat Oregon) - 85-110 days, Hard slate-gray skin, thick golden-yellow sweet dry stringless fiberless flesh, 10-20 lbs., flavor sweetens with age, baked or pies, keeps many months, vigorous vines needs lots of room, dev. in Oregon, Northwest favorite, introduced by the Gill Bros. Seed Company of Portland, Oregon. *Source History: 13 in 1981; 12 in 1984; 14 in 1987; 17 in 1991; 18 in 1994; 19 in 1998; 27 in 2004.* **Sources: Ba8, Ers, Fe5, Ga1, Ha5, Hum, Jor, La1, ME9, Mo13, MOU, Ni1, OLD, OSB, Pe1, Pe6, Pin, Rev, RUP, SE14, Se25, So1, Syn, Ter, WE10, We19, WI23.**

Sweetkeeper - 95-105 days, Excel. flavor, seems to be quite high in sugar, like all squash it keeps best when dry and above freezing, warm or cool temps do not matter but it must be dry. *Source History: 1 in 1981; 1 in 1984; 1 in 1987; 2 in 2004.* **Sources: Al25, Co32, Se7.**

Sweetmeat, Katy's Super Sweet - Blue-green fruit, 15-20 lbs., very nice orange flesh, good flavor, sweeter than Sweetmeat with better flavor. *Source History: 1 in 2004.* **Sources: In8.**

Taos - Traditional hubbard type squash from Taos Pueblo, NM. *Source History: 1 in 2004.* **Sources: Na2.**

Taos Pueblo - 90-100 days, Rare squash from the Taos Pueblo, cheese box shaped, gray-green skin, 5-15 lbs., very good keeper, great for pies. *Source History: 1 in 1991; 1 in 2004.* **Sources: Se17.**

Tiny Turk - 100 days, Miniature selection of Turk's Turban, color and shape very similar, varies from half to nearly full size, 1.5-3 lbs. *Source History: 1 in 1991; 1 in 1994; 1 in 1998; 3 in 2004.* **Sources: HO2, Ni1, RUP.**

Triamble - (Triangle, Tristar, Shamrock) - 180 days, Three-cornered shape, large fruits up to 15 lbs., very tough thick slate- grey rind, firm deep orange flesh, quite sweet, excel. keeper, first grown in the U.S. in 1932 with seed from Arthur Yates and Co., Sydney, Australia. *Source History: 1 in 1981; 1 in 1984; 2 in 1987; 2 in 1991; 1 in 1994; 2 in 2004.* **Sources: Hud, Se16.**

Trojan - 110 days, Consistent producer of large pumpkins with huge handles, deep orange skin with sutures, 17 x 15 in., 20-30 lbs. *Source History: 2 in 2004.* **Sources: HO2, SE4.**

Turban, Mini-Red - 80-100 days, Well colored small turban squash, brilliant red-orange cap, 6-8 in. tall, high quality, extremely prolific, may be the variety called Small Chinese Turban described in Vilmorin's "The Vegetable Garden" (1885). *Source History: 7 in 2004.* **Sources: Jor, ME9, Pr9, Ra6, RIS, Se16, Vi5.**

Turban, Red Warren - (Warren) - Discovered in a field of Essex Hybrid, orange, drum-like fruit with somewhat turban shape, prominent pale pink acorn, intro. by James Gregory, Marblehead, MA, 1890. *Source History: 1 in 2004.* **Sources: Ea4.**

Turban, Turk's - (Aladdin's Turban, Turk's Cap Gourd) - 80-125 days, Distinctive cap or turban, orange and red and white on green, 8-12 in. dia., 5-10 lbs., fair table quality, durable if not bruised, decorative gourd for fall displays, resembes a small orange Butternut with a colorful turban striped with green and yellow and red, pre-1800. *Source History: 15 in 1981; 14 in 1984; 37 in 1987; 41 in 1991; 44 in 1994; 53 in 1998; 67 in 2004.* **Sources: ABB, Ba8, Bo19, Bu3, Bu8, CA25, CH14, Cl3, CO30, Com, Dam, De6, Dil, Ers, Fe5, Fo13, Gur, HA3, Ha5, HA20, He17, HO1, HO2, Jo1, Jor, Jun, La1, Loc, Ma19, ME9, Mey, Mo13, MOU, Ni1, Ol2, OLD, OSB, Pep, Pin, Ra5, Ra6, Rev, Ri12, RIS, Roh, Ros, RUP, SE4, SE14, Se16, Se26, Se28, Sh9, Shu, Si5, Sk2, Sto, Tho, Twi, Ver, Vi4, Vi5, WE10, We19, Wet, Wi2, WI23.**

Uchiki Kuri - (Orange Hokkaido) - 95-100 days, Heavy yields of tear drop shaped fruit, orange-red rind, clear yellow flesh, sweet nutty flavor, 10 lbs., improved and selected from hubbard, very good for cool areas, Japan. *Source History: 1 in 1981; 1 in 1984; 3 in 1991; 6 in 1994; 8 in 1998; 8 in 2004.* **Sources: Bou, CO30, Dam, Ev2, Jor, Kit, Loc, RUP.**

Valenciano - 110 days, French heirloom, flattened shape, 11-15 in. diameter by 6-8 in. tall, slightly ribbed, white skin, thick orange flesh suitable for pies, great for painting. *Source History: 1 in 1991; 1 in 1994; 1 in 1998; 4 in 2004.* **Sources: Jo1, Na6, Ri12, RIS.**

Vert D'Espagne - Large pumpkins, maroon-orange skin, flavorful orange flesh, up to 35 lbs. *Source History: 1 in 2004.* **Sources: Hud.**

Weeks North Carolina Giant Pumpkin - 120 days, Grows huge with sufficient moisture and space and fertilizer, 210 lbs., 9 ft. circumference, suitable for pies. *Source History: 1 in 1981; 1 in 1984; 1 in 2004.* **Sources: Hi13.**

White Pumpkin - Small pumpkin with white rind, yellow-orange flesh, good for cooking, heirloom from Hendersonville, NC. *Source History: 1 in 2004.* **Sources: Scf.**

Windsor Cup - 90-100 days, Compact bush plants bear buttercup type fruit with delicious yellow-orange flesh, 4-5 lbs., dark green to gray-green skin color, good storage. *Source History: 2 in 2004.* **Sources: Ha5, ME9.**

Zapallo Plomo - 100 days, Deeply ribbed fruit, beautiful light turquoise-blue skin with attractive netting, dry flesh, flavorful, rare variety from the Southwest, reputed to be the inspiration for topaz colored jewelry for the early Indian people. *Source History: 3 in 2004.* **Sources: Ba8, RUP, Se17.**

Varieties dropped since 1981 (and the year dropped):

Acorn (Maxima) (1991), *Amerindian* (1998), *Arikara Round* (1998), *Autumn Pride* (1994), *Ayote Cascara Dura* (1998),

Baby Hubbard (1991), *Banana, Goldpak* (1998), *Banana, Hopi Long Blue* (1994), *Blue Hubbard (Colby Strain)* (1991), *Blue Hubbard (Select)* (1991), *Bohemian* (1991), *Boston Marrow Necky* (1991), *Boston Marrow Special* (1991), *Buttercup (Perfection)* (1991), *Buttercup, Golden Bush* (1998), *Buttercup, Red Splash* (1998), *Buttercup, Semi-Bush* (1994), *Chestnut* (2004), *Chioga* (1991), *Chuska Mt. Navajo* (1994), *Doe* (1991), *Forragero* (1994), *Giant Pumpkin, Burgess* (2004), *Gold Mountain* (2004), *Golden Turban* (1998), *Great Pumpkin, The* (2004), *Greengold* (2004), *Hubbard Rugueuse* (1998), *Hubbard, Black* (1998), *Hubbard, Golden Delicious* (1998), *Hubbard, Gray* (1998), *Hubbard, NK580* (1994), *Hubbard, Sibley* (1994), *Hundredweight* (1987), *Hungarian Mammoth* (1998), *King of Giants* (1991), *La Calabasa* (1994), *Large Mammoth* (2004), *Leningrad Giant Early Pumpkin* (1987), *Long Napoli* (1994), *Manitoba Miracle* (1987), *Marblehead* (2004), *Mayo White Giant* (1987), *My Best Pie* (2004), *Navaho Ni-es-pah* (1987), *Old Humboldt* (1994), *Old Winter* (1998), *Oregold* (1987), *Quality Winter Squash* (2004), *Silver Bell* (2004), *Slate Banana* (1987), *Spanish* (1994), *Stambolka* (1991), *Tokyo Earliest* (1987), *Turban, Lakota Sioux* (2004), *Turban, Windsor Black* (2004), *Verruquex du Portugal* (2004), *Virginia Mammoth* (1987), *Vounicheo* (1991), *Warted Hubbard (Denning Strain)* (1987), *Whangaparoa Crown Pumpkin* (2004), *White Cheesequake Pumpkin* (1991), *Wickersham Sweet Potato* (2004), *Zapallo Macre* (2004), *Zipinki Campana* (2004), *Zucca Marina di Chioggia* (1994).

Squash / Mixta
(Cucurbita mixta)

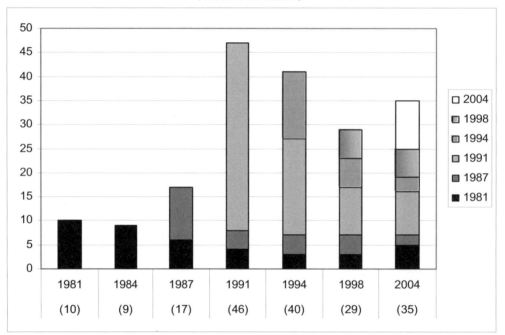

Number of varieties listed in 1981 and still available in 2004: 10 to 5 (50%).
Overall change in number of varieties from 1981 to 2004: 10 to 35 (350%).
Number of 2004 varieties available from 1 or 2 sources: 26 out of 35 (74%).

<u>Calabaza de las Aguas</u> - Common squash which is boiled and sweetened with brown sugar or eaten young as calabacitas, from Ures, Sonora, Mexico. *Source History: 1 in 1991; 1 in 1998; 1 in 2004.* **Sources: Na2.**

<u>Campeche</u> - 100-110 days, Silver seeded type, 3-8 lbs., variable shapes range from spherical to oval to light bulb-like shape, Campeche is a state in southern Mexico. *Source History: 1 in 1994; 1 in 1998; 2 in 2004.* **Sources: Rev, Sa9.**

<u>Cochiti Pueblo</u> - 110-140 days, Variable shapes, squat and oblong, light orange flesh, 1.5-2.5 lbs., good for baking, good storage. *Source History: 2 in 2004.* **Sources: Ri12, Se17.**

<u>Cushaw, Gila</u> - 110 days, Large fruit, yellow with green stripes, U to club shape, 10-12 lbs., large edible seeds, long keeper. *Source History: 1 in 1998; 1 in 2004.* **Sources: So12.**

<u>Cushaw, Gila Cliff Dweller</u> - Cushaw grown by the Gila Indians, white skin, 6-12 lbs., late bloomer, good keeper. *Source History: 1 in 1991; 1 in 1994; 1 in 1998; 2 in 2004.* **Sources: Se17, Syn.**

<u>Cushaw, Gold Striped</u> - 110 days, Sport/mutant of Green Striped Cushaw, bright lemon-yellow rind with darker yellow stripes, 15-20 lbs., attractive light and dark-green variegated foliage. *Source History: 1 in 1981; 4 in 2004.* **Sources: Gr27, HO2, Ri12, Scf.**

<u>Cushaw, Golden Striped</u> - 110 days, Long neck may be straight or crooked, white with green stripes, gold stripes on the sunny side, smooth sweet golden flesh, 25-30 lbs., good keeper, rare. *Source History: 2 in 1991; 2 in 1998; 3 in

2004. **Sources: Ho13, RUP, Sa9.**

Cushaw, Green Striped - (Striped Crookneck, Striped Cushaw) - 75-115 days, Formerly known as Improved Cushaw, hard thin smooth rind, green mottled stripes over creamy-white, 16-20 x 8-10 in. dia. bowl, 12-16 lbs., pear-shaped with crookneck, prolific, old favorite, thick fine-grained sweet pale-yellow flesh, canning boiling baking pies or stock feed, from the Caribbean, pre-1893. *Source History: 49 in 1981; 41 in 1984; 41 in 1987; 34 in 1991; 38 in 1994; 41 in 1998; 46 in 2004.* **Sources: Be4, Bu3, Bu8, CA25, CH14, Cl3, Co31, Com, DOR, Ers, Gr27, He17, HO1, HO2, Jor, Kil, La1, LO8, ME9, Mey, Mo13, Ol2, OLD, OSB, Ra5, Ri12, RIS, Roh, Ros, RUP, Sa9, Sau, SE4, Se8, SE14, Se26, Sh9, Shu, Sk2, So1, SOU, Vi5, WE10, Wet, Wi2, WI23.**

Cushaw, Hopi - 110-120 days, Mottled green and yellow skin with yellow stripes, yellow flesh, 10-20 lbs., water jug shape, excel. storage, squash bug res., pre-Columbian. *Source History: 1 in 1981; 1 in 1984; 1 in 1991; 4 in 2004.* **Sources: Rev, Sa9, Se17, So12.**

Cushaw, Hopi Black Green - Pear-shaped cushaw, black-green with yellow spotted skin, yellow flesh, 40-60 lbs., excel. keeper, good for pies. *Source History: 1 in 1991; 1 in 1994; 1 in 1998; 2 in 2004.* **Sources: Ho13, Se17.**

Cushaw, Hopi Vatnga - Crookneck fruits can be striped or solid green, yellow flesh, 15-25 lbs., fine quality winter squash, thick hard shells are sometimes made into musical instruments, seeds are ground for Kachina face paint. *Source History: 2 in 1991; 2 in 1994; 1 in 2004.* **Sources: Na2.**

Cushaw, Illinois - *Source History: 1 in 2004.* **Sources: Ea4.**

Cushaw, Papalote Ranch - Pear-shaped fruits, solid green skin, yellow flesh, 10 lbs., good keeper, rare variety from Mexico. *Source History: 3 in 1991; 2 in 1994; 1 in 1998; 1 in 2004.* **Sources: Na2.**

Cushaw, San Juan Pueblo - Light cream and green striped, pear shaped fruit, up to 15 lbs., tan seeds. *Source History: 1 in 2004.* **Sources: Na2.**

Cushaw, Santo Domingo - 100-110 days, Rare long season variety, pear shaped, 12-15 lbs., dark-green and med-green striped skin, cream colored flesh, tan seed, cook like winter squash, can be eaten like summer squash if picked young, edible seeds, excel. keeper. *Source History: 1 in 1987; 2 in 1991; 1 in 1994; 3 in 1998; 2 in 2004.* **Sources: Ho13, Se17.**

Cushaw, Tricolor - 95-110 days, Beautiful large squash with crooked neck, skin is streaked with white, green and orange, 12-15 lbs., specialty market growers, CV Rev suggests this variety will lose its vigor and bright color contrasts if seed is not saved from the dark green striped and all-white fruit that grow along the "Tricolors proper". *Source History: 3 in 2004.* **Sources: HO2, ME9, Rev.**

Cushaw, White - (White Cushaw, White Crookneck, Jonathan, Jonathan Pumpkin) - 85-115 days, White skin with pale-yellow background, fruits 20-45 lbs., taste differs from other squashes, similar if not identical to Salem White, may have been commercially available before 1860. *Source History: 2 in 1991; 3 in 1994; 7 in 1998; 16 in 2004.* **Sources: Ba8, Ea4, Ers, Gr27, HO2, Ho13, La1, ME9, Rev, RIS, RUP, Sa9, Scf, SE14, Shu, WE10.**

Hopi Taos - 110-120 days, Pear-shaped fruit, green and yellow skin, yellow flesh, 20-30 lbs., excel. keeper, Hopi, rare. *Source History: 1 in 1991; 2 in 2004.* **Sources: Ri12, Se17.**

Japanese Pie - 110 days, Pear-shaped, crookneck, med-sized, 12 lbs., dark-green skin with white stripes, fine-grained orange-yellow flesh, small cavity, for baking and pies, keeps well. *Source History: 3 in 1981; 2 in 1984; 1 in 1987; 1 in 2004.* **Sources: Se17.**

Magdalena Striped Cushaw - Eaten young or mature, keeps well, from Sonora, for planting during the summer rains in low hot desert regions. *Source History: 1 in 1981; 1 in 1984; 1 in 1987; 1 in 1991; 1 in 1994; 1 in 1998; 1 in 2004.* **Sources: Na2.**

Mexican Xtop - Round cushaw type with bowl shape, green and white striped skin, yellow flesh, mild flavor, huge silver-edged seeds, from Mexico. *Source History: 1 in 2004.* **Sources: Ba8.**

Papalote - 110-115 days, Short-neck bottle shape, 6-8 lbs., netted patterned skin, sweet. *Source History: 1 in 1998; 1 in 2004.* **Sources: So12.**

Salem White - Cream colored skin, crooked neck, bulbous blossom end, ornamental, similar if not identical to White Cushaw or Jonathan. *Source History: 1 in 2004.* **Sources: Ea4.**

San Pedro Ha:l - Yellow skin with some green mottling, hard shell, originally from the Tohono O'odham village of San Pedro. *Source History: 1 in 2004.* **Sources: Na2.**

Silver Edged - A Mesoamerican land race, round 12 in. fruit, grown for large tasty spectacular seeds, donated by Anita Williams. *Source History: 1 in 1987; 1 in 1991; 1 in 1994; 2 in 2004.* **Sources: Ho13, Na2.**

Silver Seed Gourd Thick Margin - 120 days, Large seeds with silver margin, grown for the seeds which are roasted. *Source History: 1 in 1994; 1 in 1998; 1 in 2004.* **Sources: Sa9.**

Tamala - 114 days, Green and white striped, bowl shaped fruit, heavy producer of silver-rimmed seeds. *Source History: 1 in 1998; 1 in 2004.* **Sources: Sa9.**

Tennessee Sweet Potato - 85-100 days, Tennessee heirloom from the 1800s with white, pear shaped fruits, light yellow flesh, 10-15 lbs., good quality, excel. keeper. *Source History: 2 in 1994; 3 in 1998; 4 in 2004.* **Sources: Ho13, Sa9, Se16, Se17.**

Tequila Black - Black skinned, bell shaped fruit with very large stems, 10 lbs., good for pies and pumpkin seeds, from Tequila, Mexico. *Source History: 1 in 2004.* **Sources: Ba8.**

Tohono O'odham Ha:l - Heat tolerant, short season crop, prized for the immature fruits, mature fruit has light orange flesh, stores well. *Source History: 1 in 2004.* **Sources: Na2.**

Veracruz Pepita - (Pepita) - Round flattened fruits, white with green mottled stripes, grown for the long narrow seeds which are toasted for snacks or ground to prepare pipian sauces, originally collected in Veracruz, southern Mexico. *Source History: 1 in 1991; 1 in 1994; 2 in 2004.* **Sources: Na2, Syn.**

Winter Vining - 105-109 days, Yellow fleshed, green striped cushaw with thick mottled stripes, slight crookneck, 12-15 lbs., close relative of the Green Striped Cushaw. *Source History: 1 in 1998; 3 in 2004.* **Sources: Ea4, Rev, Sa9.**

Yoeme (Yaqui) Kama - Size, shape and color of fruit varies, flavorful yellow flesh, from Ures, Sonora. *Source History: 1 in 2004.* **Sources: Na2.**

Zapotec - Small, flattened, pumpkin shape squash, 10 in. dia., white with green to gold stripes, sweet, dry flesh, stores well. *Source History: 1 in 1998; 2 in 2004.* **Sources: Ho13, Se17.**

Zebra Mystery - 120 days, Sprawling plants with huge leaves with silver spots, flattened fruits, deep yellow flesh, nutty and sweet flavor, 8 in. dia., white with faint green stripes, 5-10 lbs., resists borers, good storage, possibly a Southwest Indian type. *Source History: 1 in 1998; 3 in 2004.* **Sources: Ho13, Se17, So1.**

Varieties dropped since 1981 (and the year dropped): *Apache Giant* (1991), *Calabaza Taos* (1998), *Calabaza, Mexican White* (1994), *Chompa* (2004), *Cushaw, Albino Pepita* (2004), *Cushaw, Cochiti Pueblo* (2004), *Cushaw, Green and Yellow Striped* (1994), *Cushaw, Hermington Striped* (1994), *Cushaw, Hopi Taos* (1998), *Cushaw, Hopi Teardrop* (1998), *Cushaw, Hopi White Albino* (1998), *Cushaw, Mexican Solid Green* (1994), *Cushaw, Mexican Wild* (1998), *Cushaw, Papillon Nicaraguan* (1994), *Cushaw, Parral* (2004), *Cushaw, Pepita* (1998), *Cushaw, Santo Domingo No. 2* (1994), *Cushaw, Santo Domingo White* (1994), *Cushaw, Striped Cochiti Pueblo* (1998), *Douglas Striped Cushaw* (1987), *Elfrida Taos* (1998), *Gila Pima Ha:l* (1998), *Hard Shelled* (1998), *Hindu* (1994), *Hopi* (1998), *Hopi No. 4* (1998), *Hopi Teardrop* (2004), *Japanese* (1994), *Mayo Arrote* (2004), *Mexican Wild White* (1998), *Moctezuma Giant* (1991), *O'odham Papago* (1994), *Papago* (1991), *Papago Striped Cushaw* (1987), *Pepinas* (1994), *Pueblo Indian* (1998), *San Bernarda Feral* (1998), *San Bernardo* (2004), *Silver Edged Mexican Cashaw* (1994), *Silver Edged Mexican Pepita* (1998), *Striped Belize* (2004), *Tamala de Carne* (2004), *Tamala de Hueso* (2004), *Tarahumara Calabaza* (1991), *Tennessee Sweet Potato Squash* (1991), *Tohona O'odham* (2004), *Wild Mexican* (1994).

Squash / Moschata
(Cucurbita moschata)

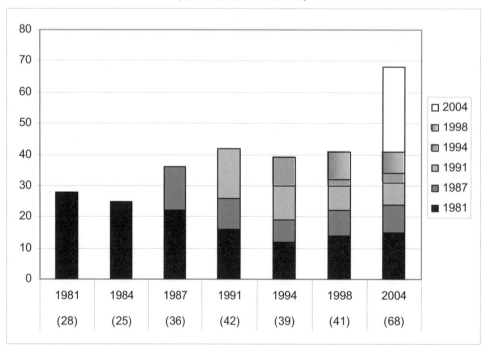

Number of varieties listed in 1981 and still available in 2004: 28 to 15 (54%).
Overall change in number of varieties from 1981 to 2004: 28 to 68 (243%).
Number of 2004 varieties available from 1 or 2 sources: 49 out of 68 (72%).

African - 95-105 days, Ribbed, butternut shaped fruit, buff colored skin, sweet, moist orange flesh, 6-8 lbs., 4-5 fruit per plant, keeps for over one year. *Source History: 1 in 2004.* **Sources: So12.**

African Winter - Heirloom from Zaire, 15-18 in. long fruit with long neck, small cavity the size of a golf ball in the bulbous bottom portion, firm flesh, vines will root into the ground along the runner stems, allowing the plant to take up nutrients and water despite stem damage in other areas, making it especially tolerant of the squash vine borer, long storage life, hardy and disease res. *Source History: 1 in 2004.* **Sources: Scf.**

Bugle - 90 days, Med. size butternut, vining plant, high yields, crack res., PM tolerant, PVP 2001 Cornell University Exp. Station. *Source History: 3 in 2004.* **Sources: HO2, Pe2, TE7.**

Butterbush, Burpee's - (Butterbush) - 75 days, First bush butternut, 3-4 ft. wide plants each bear 4 to 6 smaller 1.75 lb. fruits, red-orange flesh bakes faster than regular butternut, keeps until mid-winter. *Source History: 11 in 1981; 9 in 1984; 12 in 1987; 9 in 1991; 10 in 1994; 9 in 1998; 8 in 2004.* **Sources: Be4, Bu2, Ers, Hi13, Pe6, Pin, Roh, Sh9.**

Butternut - 75-110 days, Light-tan bottle-shaped fruits with thick solid necks, rich golden-orange thick dry fine-grained sweet flesh, good flavor and texture, 8 x 5 in. bowl, 3-5 lbs., thin hard skin, good keeper. *Source History: 68 in 1981; 58 in 1984; 46 in 1987; 34 in 1991; 31 in 1994; 29 in 1998; 23 in 2004.* **Sources: Alb, All, But, Co31, Dgs, Ear, Ge2, Goo, He8, Hum, Lej, LO8, Mey, Pe2, Pr9, Ri12, Roh, Sau, Sh9, SOU, Te4, Tu6, Wet.**

Butternut, Baby - 85-105 days, Extra early, good for the North, 6-8 in., solid meat except egg-sized seed cavity, deep-yellow fine-textured sweet dry flesh, excel. table quality, no crooknecks. *Source History: 3 in 1981; 3 in 1984; 3 in 1987; 1 in 1991; 1 in 1998; 1 in 2004.* **Sources: Sa9.**

Butternut, Nicklow's Delight - 105 days, Fruit to 8 in., gummy stem and black rot tolerant. *Source History: 1 in 2004.* **Sources: Jor.**

Butternut, Ponca - 80-100 days, Earlier, compact vigorous vines, smaller fruit, 6-8 in., 1.5-2.5 lbs., hard smooth rind, creamy-tan skin, thick necks, most fruits nearly cylindrical, some slightly bulbous at flower end, high quality light-orange flesh, heavy yields, good keeper, U of NE. *Source History: 25 in 1981; 20 in 1984; 17 in 1987; 19 in 1991; 15 in 1994; 7 in 1998; 5 in 2004.* **Sources: Fe5, Gr28, Hi6, HO2, Und.**

Butternut, Puritan - 100-110 days, Vigorous and uniform, 5 x 10 in., blocky smooth cylindrical fruits, no crooknecks, just like Waltham for flesh quality and yields and storage. *Source History: 3 in 1981; 2 in 1984; 2 in 1987; 2 in 1991; 4 in 1994; 2 in 1998; 3 in 2004.* **Sources: CA25, SE4, Se26.**

Butternut, Waltham - 83-115 days, AAS/1970, most popular butternut, more uniform shape and size, fewer crooknecks, better interior texture and color, 8-12 x 3-5 in., 3-6 lbs., smooth light-tan skin, thicker cylindrical necks, small seed cavity, rich dry yellowish-orange flesh, nutty flavor, vigorous vines, higher yields, stores better, dev. and intro. by Bob Young of Waltham, MA, Waltham MA/AES. *Source History: 74 in 1981; 68 in 1984; 75 in 1987; 76 in 1991; 75 in 1994; 84 in 1998; 92 in 2004.* **Sources: ABB, Ada, Ag7, Ba8, BAL, Be4, Bo17, Bo19, Bou, Bu2, Bu3, Bu8, CA25, CH14, Cl3, CO23, CO30, Co32, Com, Cr1, Dam, De6, DOR, Dow, Ers, Fa1, Fe5, Fi1, Fo7, Fo13, Gou, Gr27, Gr28, Gur, HA3, Ha5, HA20, Hal, Hi6, Hi13, HO1, HO2, Jo1, Jor, Jun, Kil, La1, Loc, Ma19, MAY, Mel, Mo13, MOU, Na6, OLD, Ont, Or10, OSB, PAG, Pin, Pr3, Pr9, RIS, Ros, RUP, Sa5, Sa9, Scf, SE4, Se7, SE14, Se16, Se25, Se26, SGF, Sh9, Shu, Sk2, So1, So9, Sto, Syn, Te4, TE7, Ter, Twi, Up2, Ver, Vi4, WE10, Wi2, WI23.**

Calabaza Segualca - Sweet fleshed fruit, tan skin color, shapes include oblongs, flats and crooknecks, does well in the heat, collected in north and central Mexico by Baker Creek Heirloom Seeds. *Source History: 1 in 2004.* **Sources: Ba8.**

Calabaza, Cuban - *Source History: 1 in 2004.* **Sources: Se17.**

Calabaza, United Nations Plaza - 125 days, Vining habit, deeply ribbed, somewhat flattened, flared out from the stem and tapered in toward the blossom end, 4-6 lbs., green and white over buff colored skin, Calabaza is Spanish for squash, variety name unknown, from farmer's market at United Nations Plaza in San Francisco, CA. *Source History: 1 in 2004.* **Sources: Rev.**

Canadai Mezoides - 85-120 days, Heavy producing butternut type that ripens to tan in storage, fine grained, yellow-orange flesh, excellent for pie, borer res., small seeds. *Source History: 1 in 1994; 2 in 2004.* **Sources: Sa9, Se17.**

Carrizo - Formerly listed as Sonora/Sinaloa Border, small orange tear drops and necked, good for soup and puree, local market variety. *Source History: 1 in 1987; 1 in 1991; 1 in 1998; 1 in 2004.* **Sources: Na2.**

Cheese - (Kentucky Field, Long Island Cheese) - 100-150 days, Flattened pumpkins, slate-gray or blue or tan skins, usually slightly ribbed, orange flesh is coarse and fibrous yet sweet, good for pies, succeeds best in areas with long growing seasons, introduced commercially by McMahon in 1815. *Source History: 2 in 1981; 2 in 1984; 1 in 1987; 1 in 1998; 2 in 2004.* **Sources: Fe5, Sa9.**

Chirimen - (Chu Chirimen, Bizen Chirimen) - 100-125 days, Japanese pumpkin, warted flattened globes with distinct ribs, dark-green skin matures to buff-orange, thick sweet yellow-orange flesh, 7-10 lbs., prolific, first offered by the Aggeler and Musser Seed Company of Los Angeles in 1922. *Source History: 2 in 1981; 2 in 1984; 3 in 1987; 1 in 1991; 3 in 1994; 3 in 1998; 3 in 2004.* **Sources: Kit, ME9, Pr9.**

Choctaw Sweet Potato - 100-120 days, Vines to 20 ft., deep orange oval fruit, flavor of sweet potatoes, makes great pies, good storage, heirloom from the Creek and Choctaw nations, rare. *Source History: 2 in 2004.* **Sources: Ri12, Se17.**

Creole - Oversized fat butternut shape, some fruits oval, tan skin, orange flesh, 40-50 lbs., good keeper, excel. African- American pie pumpkin, been around for 300 years, very rare. *Source History: 2 in 1991; 1 in 1994; 1 in 1998; 1 in 2004.* **Sources: Ho13.**

Crookneck, Canada - Small fruit, 10-12 in long, thin neck grows straight to crooked, 5 in. dia. at the seed cavity, edible skin, delicious red-orange flesh, rare, first listed by Hovey in 1834, Thorburn in 1840. *Source History: 1 in 2004.*

Sources: Ea4.

Cuban - 123 days, Blocky butternut type, shape varies slightly. *Source History: 1 in 1998; 1 in 2004.* **Sources: Sa9.**

Cushaw, Golden - (Mammoth Golden Cushaw, Golden Crookneck, Golden Cushaw Pumpkin) - 100-120 days, Smooth hard golden yellowish-orange rind, crookneck, 18-20 x 9 in. dia. bowl, 10-12 lbs, fine-flavored sweet solid dry flesh, small seed cavity, good keeper, for baking or boiling or canning or pies, dairymen use them for stock feed. *Source History: 15 in 1981; 12 in 1984; 5 in 1987; 2 in 1991; 5 in 1994; 3 in 1998; 2 in 2004.* **Sources: OSB, Shu.**

Cushaw, Orange - 110-115 days, Bright orange skin with some white markings, 10-20 lbs., excel. for baking and canning. *Source History: 3 in 1998; 5 in 2004.* **Sources: Ba8, Ers, SE14, Shu, WE10.**

Cushaw, Orange Striped - 110 days, Resembles Green Striped Cushaw, pale orange skin with dark orange stripes, 5-10 lbs., edible flesh is excellent for pies, also ornamental. *Source History: 2 in 1998; 5 in 2004.* **Sources: Ers, La1, ME9, SE4, SE14.**

Dickinson - (Dickinson Field) - 115 days, Nearly round to elongated fruits, 18 in. long x 14 in. dia., up to 40 lbs., slightly furrowed but smooth buff-colored rind, sweet orange high quality flesh, for canning and pies. *Source History: 1 in 1987; 1 in 1991; 2 in 1994; 1 in 1998; 8 in 2004.* **Sources: Ba8, HO2, LO8, ME9, RUP, SE14, Se17, TE7.**

Flat Tan Field - *Source History: 1 in 2004.* **Sources: Se17.**

Futtsu Early Black - (Futtsu, Futtsu Black) - 100-120 days, Flattened globes with deep ribs and heavily warted shells, deep green to nearly black skin turns dark orange with silver dusting in storage, 7 in. wide, 3-8 lbs., deep orange-yellow flesh, fine grained, sweet nutty flavor. *Source History: 2 in 1981; 2 in 1987; 1 in 1991; 5 in 2004.* **Sources: Ba8, Red, Sa9, Se17, Ter.**

German Above Ground Sweet Potato - Winter squash with shape resembling a sweet potato, grows 18 in. long with 6 in. diameter and tapered at both ends, can be trellised or allowed to run on long vines, good keeper. *Source History: 1 in 2004.* **Sources: Scf.**

Gray - 92 days, Early cheese type, blocky ribbed fruit, 1.5 in. thick medium orange flesh, 10-15 lbs. *Source History: 1 in 2004.* **Sources: Sa9.**

Guarijio Segualca - Variable size and shape of fruit, stores well, from San Bernardo, Sonora. *Source History: 1 in 2004.* **Sources: Na2.**

Jap - 110 days, Lovely, medium size pumpkin, green skin with cream stripes, rich orange flesh is dry, good for frying and baking, North American heirloom from the 1860s. *Source History: 1 in 2004.* **Sources: Te4.**

Kentucky Field - (Dickinson) - 115-120 days, Variable, ribbed, dull-orange, flat to almost round, 10-12 x 12-15 in. dia., 10-15 lbs., thick yellow-orange flesh, prolific, South, can or stock feed. *Source History: 10 in 1981; 7 in 1984; 4 in 1987; 3 in 1991; 3 in 1994; 4 in 2004.* **Sources: Bu8, HO2, ME9, SE14.**

Kikuza Early White - (Kikuza Winter) - 90-100 days, Rare winter squash from southern China, ribbed fruits, 4-7 lbs., firm sweet fine-textured flesh, excel. eating quality, heat res., vigorous. *Source History: 2 in 1981; 2 in 1984; 1 in 1987; 1 in 1991; 1 in 1994; 1 in 2004.* **Sources: Se16.**

La Primera - 115 days, Tropical Calabaza squash, green fruit with lighter green markings turning tan, flavorful orange flesh, 15-30 lbs. *Source History: 1 in 1991; 1 in 2004.* **Sources: Ech.**

Long Island Cheese - 71-108 days, Flattened, deeply ribbed, buff colored fruits, deep orange, sweet flesh, excellent in pies, 6-10 lbs., stores well, named for its resemblance to a wheel of cheese, heirloom long remembered as a great pie squash by people in the New York and New Jersey area. *Source History: 3 in 1998; 19 in 2004.* **Sources: Ba8, Ers, He8, HO2, Jo1, ME9, Ol2, OSB, Pr3, Pr9, Ri12, RIS, RUP, SE14, Se16, Se26, Sk2, So12, Vi4.**

Long Island Milk Pumpkin - Minute traces of white on the skin makes this squash different from other tan cheese squashes, 10-20 lbs., tan round flattened shape like old-time rounds of cheese, excel. flavor, from eastern Long Island, NY, cheese squashes were first depicted in a painting by Lucas Van Valkenborch, circa 1530-1597, and first listed in America in 1815. *Source History: 1 in 2004.* **Sources: Ea4.**

Long of Naples - Large fruit with oblong-butternut shape, flavorful, rich sweet bright orange flesh, 20-35 lbs., good for areas with long warm growing season, good for home or market grower, Italian heirloom, listed in America by Fearing Burr in 1863, very rare in the U.S. *Source History: 2 in 2004.* **Sources: Ba8, CA25.**

Magdalena Big Cheese - 105 days, Buff/orange, fluted, squat pumpkin, from Sonora, for planting during the summer rains in low hot desert regions. *Source History: 1 in 1981; 1 in 1984; 2 in 1987; 4 in 1991; 2 in 1994; 1 in 1998; 1 in 2004.* **Sources: Na2.**

Mayo Segualca - Mottled green fruits, slightly ribbed with butternut-like necks, sweet flavorful flesh, Guarijio var. collected in San Bernardo, Sonora, Mexico. *Source History: 1 in 1991; 1 in 1994; 1 in 1998; 1 in 2004.* **Sources: Na2.**

Middle Rio Conchos - Mixed shapes and sizes, originally collected from a market in Saucillo, Chihuahua, Mexico. *Source History: 1 in 2004.* **Sources: Na2.**

Milk - 95-105 days, Old-timer, flattened, ribbed cheese pumpkin, 15-20 lbs., light-tan color, pale, creamy orange flesh. *Source History: 1 in 1987; 1 in 2004.* **Sources: Sa9.**

Mrs. Amerson's Pumpkin - 122 days, Two shapes, either a flattened cheese or a tall blocky cheese, 1.5 in. thick, bright orange flesh, excellent insect and disease resistance, high yields. *Source History: 1 in 1998; 1 in 2004.* **Sources: Sa9.**

Neck Pumpkin - (Cornfield Crookneck) - 110-120 days, Old type squash related to butternut squash, vining plant, 18-30 in. long fruits with 4-5 in. necks and 9 in. dia. bulbs, green fruit turns pale yellow when mature, thick solid dry sweet orange flesh, excel. for pumpkin pie filling, stores well. *Source History: 1 in 1981; 1 in 1984; 2 in 1987; 3 in 1991; 6 in 1994; 10 in 1998; 12 in 2004.* **Sources: Bu2, Bu8, Fe5, Ha5, Ho13, ME9, Mo13, Ol2, Roh, RUP, SE4, Se26.**

North Falkland Island - 100 days, Vining plants

produce beautiful egg-shaped fruit, either all white or white with dark green stripes from end to end or white with jagged bands of golden yellow, presumably grown for the seeds, although the name implies its origin is the Falkland Islands, which are the northern islands off the southern coast of Argentina, the foggy and windy growing conditions there would seem inhospitable to a heat-loving *Cucurbita*. *Source History: 2 in 2004.* **Sources: Rev, Se17.**

Old-Fashioned Tennessee Vining - Tennessee heirloom, oval-shaped fruits average 15-30 lbs., tan skin, deep-orange flesh, res. to squash vine borer, CV Sol introduction, 1988. *Source History: 1 in 1991; 1 in 1994; 2 in 1998; 3 in 2004.* **Sources: Ba8, Se17, So1.**

Papaya Pumpkin - 90 days, Novel Asian winter squash, 2 lb. fruits strongly resemble papayas, very sweet thick flesh, nutty flavor, trailing vines. *Source History: 1 in 1987; 1 in 1991; 1 in 1994; 1 in 1998; 1 in 2004.* **Sources: Ev2.**

Piedras Verdes Segualca - From Piedras Verdes, Sonora, Mexico. *Source History: 1 in 2004.* **Sources: Na2.**

Pima Bajo - Small fruits with narrow necks, striped green and white, collected near Onavas, Sonora. *Source History: 1 in 1991; 1 in 1994; 1 in 1998; 1 in 2004.* **Sources: Na2.**

Quaker Pie Squash - Light tan fruit, light orange flesh resembles butternut, 6-8 lbs., original seeds obtained from a Quaker family, W. Atlee Burpee introduction, 1888. *Source History: 2 in 2004.* **Sources: Ba8, Se17.**

Rio Fuerte Mayo Arrote - Largish seed, *C. mixta* introgression. *Source History: 1 in 1987; 1 in 1991; 1 in 1994; 1 in 1998; 1 in 2004.* **Sources: Na2.**

Rio Fuerte Mayo Segualca - Planted in fall in Sinaloa, round, fluted large cheese type fruit, flavorful orange flesh. *Source History: 1 in 1987; 1 in 1991; 1 in 1994; 1 in 1998; 1 in 2004.* **Sources: Na2.**

Rio Mayo Segualca - Large round fluted cheese type, flavorful orange flesh. *Source History: 1 in 2004.* **Sources: Na2.**

Seminole Pumpkin - 95-115 days, Hard-shelled winter squash perfectly adapted for the South, found growing wild in the Everglades by early explorers and settlers, vining habit, buff-colored 6-8 in. fruits, tolerant to both dry and rainy extremes, vigorous, productive, delicious baked, boiled or steamed. *Source History: 1 in 1987; 1 in 1991; 1 in 1994; 2 in 1998; 5 in 2004.* **Sources: Ba8, Fe5, Ho13, Rev, So1.**

Sicilian - 123 days, Tan, blocky, flat cheese, 1 in. thick, med. orange flesh. *Source History: 1 in 1998; 1 in 2004.* **Sources: Sa9.**

Sucrine du Berry - Oblong, bell-shaped fruit, green-tan skin color, 2 lbs., sweet flesh, used in jams and soups, heirloom from the center of France. *Source History: 1 in 2004.* **Sources: Ba8.**

Tahitian Melon Squash - (Tahitian Squash) - 160-200 days, Huge club-shaped fruits, looks like a giant smooth Butternut, 8 to 40 lbs., tender deep golden-orange flesh, highest sugar content of any winter squash, gets sweeter during storage, 100 lbs. or more per vigorous mildew res. vine, needs long warm growing season, brought into the U.S. by Steve Spangler of Exotica Seeds. *Source History: 9 in 1981; 9 in 1984; 13 in 1987; 13 in 1991; 12 in 1994; 8 in 1998; 12 in 2004.* **Sources: Bo18, Bu1, Ech, Gr29, Ho13, Ra5, Se7, Se17, Se28, So12, Vi5, Wo8.**

Tan Cheese - 110-125 days, One of the oldest cultivated squash varieties, cheese box shaped fruit, 6-8 in. high, smooth tan skin, moderately deep ribs, sweet dark-orange, 1.5 in. flesh, 6-12 lbs., good keeper, pre-1824. *Source History: 1 in 1994; 1 in 1998; 2 in 2004.* **Sources: Sa9, So1.**

Texas Indian Moschata - 100-110 days, Large leaved rambling vines, large orange fruit with flattened pumpkin shape, ripens to mellow tan-orange skin color, 10-15 lbs., extremely long storage, from Curtis Showell, rare. *Source History: 1 in 1994; 2 in 2004.* **Sources: Ri12, Se7.**

Thai Large Pumpkin - Flattened fruit with ribbed, textured black skin which changes to tan in storage, thick sweet yellow flesh, 10 lbs., heavy yields, suitable for hot humid climates, from Thailand. *Source History: 1 in 2004.* **Sources: Ba8.**

Thai Small Pumpkin - Small fruit, flattened shape, ribbed rind is green to tan with spots, thick good quality flesh, good yields, does well in hot humid climates, from Thailand. *Source History: 1 in 2004.* **Sources: Ba8.**

Tromboncino - (Tromboncino Rampicante) - 80-100 days, Long thin fruit curves to a bell at the flower end like a trombone or question mark, fine sweet flavor, harvestable at several stages, for best quality grow on a trellis and harvest at 8-18 in. *Source History: 1 in 1987; 2 in 1991; 2 in 1994; 2 in 1998; 8 in 2004.* **Sources: Coo, Ki4, Pe2, Re6, Sa9, So25, Ter, We19.**

Trombone - 110 days, Long necked butternut type, can produce two shapes, sweet orange flesh, closely related to Tahitian Melon. *Source History: 1 in 1998; 1 in 2004.* **Sources: Sa9.**

White Rind Sugar - Old rare strain with very solid, orange flesh, variable shapes, hard shell with buff colored skin, deep ridges, 15 lbs., great for baking and pies. *Source History: 1 in 1991; 1 in 1994; 2 in 1998; 1 in 2004.* **Sources: Ho13.**

Xhosa - 120 days, Vining habit, matures to light yellow tan with slight green markings, 10-12 lbs., makes tasty although watery pie, variable shape, traditional variety of the Xhosa people, one of the major ethnic groups in eastern South Africa. *Source History: 1 in 2004.* **Sources: Rev.**

Yoeme Segualca - Large fluted fruit with flattened shape, muted orange skin, excellent flavor, from the Yoeme village of Vicam, Sonora. *Source History: 1 in 2004.* **Sources: Na2.**

Yokohama - Flat, ribbed fruit with dark green to tan skin color, dry orange flesh is fine-grained and sweet, introduced to America about 1860, listed by Burr in 1863 and Vilmorin in 1885, from Japan. *Source History: 1 in 2004.* **Sources: Ba8.**

Zucca Piena di Napoli - 100 days, Butternut shape, gray-green skin, yellow flesh with very delicate flavor, 4-5 lbs., stores well, great for gnocchi. *Source History: 1 in 1987; 1 in 1991; 1 in 1998; 2 in 2004.* **Sources: Red, Se24.**

Zucchetta, Rampicante - (Tromboncino, Zucca d'Albenga, Tromba d'Albenga) - 60 days, Plants produce 5 ft. runners, harvest fruits at 10 in. long, light yellow-green skin, slender with bulbous end, will grow into a twisted 3 ft. long squash if left on the vine, firmer than zucchini, mild delicious flavor, used in Italy for stuffing in gnocchi and ravioli, great for baking and pies. *Source History: 2 in 1991; 3 in 1994; 2 in 1998; 4 in 2004.* **Sources: Ba8, Com, Pin, Se24.**

Varieties dropped since 1981 (and the year dropped):

(1982), *Badger's Heirloom Pumpkin* (1987), *Big Cheese, Rio Mayo* (1994), *Butternut Improved* (2004), *Butternut, Bush* (2004), *Butternut, Hercules* (1998), *Butternut, Peacenut* (1994), *Butternut, World Feeder* (1998), *Calabasa de Castilla* (1987), *Calabasa, Tan* (1994), *Calabaza, Marian Van Atta* (1998), *Calabaza: La Primera* (1994), *Crooked Neck* (1998), *Cutchogue Cheese* (1994), *Early Butternut* (1991), *Eastern Butternut* (1982), *Guaymas Arrote* (1991), *Hopi, Tan* (2004), *Hyuga Early Black* (1987), *Landreth Cheese* (2004), *Large Cheese* (1991), *Long Cheese* (1998), *Long Keeper* (1991), *Mayo Kama* (1998), *Musquee de Provence* (1991), *Pat* (1994), *Patriot Butternut* (1991), *Paw Paw* (2004), *Seminole Acorn* (2004), *Sweet Potato, Shumway's* (2004), *Upper Ground Sweet Potato* (1998), *Wisconsin Cheese* (1998), *Wyoming Crookneck* (1994).

Squash / Pepo
(Cucurbita pepo)

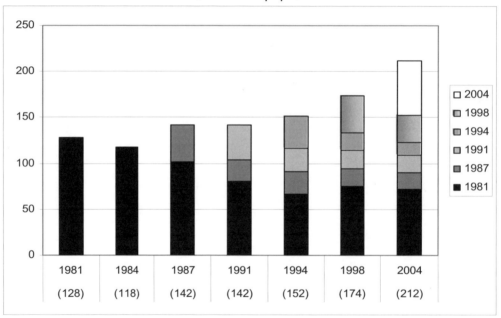

Number of varieties listed in 1981 and still available in 2004: 128 to 72 (56%).
Overall change in number of varieties from 1981 to 2004: 128 to 212 (166%).
Number of 2004 varieties available from 1 or 2 sources: 112 out of 212 (53%).

Acoma Pumpkin - 90-100 days, Vines grow 8 ft. long, blue fruits with thick flesh, used as winter squash or for pies, good keeper, collected at Sky City, the oldest occupied pueblo in the Southwest, grown for over 1,000 years. *Source History: 1 in 1991; 2 in 1994; 1 in 1998; 1 in 2004.* **Sources: Na2.**

Acorn - (Table Queen) - 80-85 days, Vigorous vines, med-sized dark-green deeply ribbed fruits, sweet dry golden-yellow flesh, excel. quality for baking, used as either summer or winter squash, thin hard shell when mature, fine flavor, good keeper, heavily productive, pre-1835. *Source History: 15 in 1981; 15 in 1984; 9 in 1987; 6 in 1991; 5 in 1994; 6 in 1998; 4 in 2004.* **Sources: Alb, Gr28, Pe2, Shu.**

Acorn, Autumn Prince - 70 days, Nicely ribbed green-black acorn, 1.5-2 lbs., good yield potential. *Source History: 1 in 2004.* **Sources: ME9.**

Acorn, Ebony - (Table Queen Ebony Strain, Improved Table Queen, Munger Strain, Ebony) - 80-100 days, Prolific Ebony str., rounder, deeper green-black, shallower ridges, slightly larger, 6 x 5 in. dia., 1.5-2 lbs., excel. yellow flesh, productive vigorous vines, Dr. Munger/Cornell. *Source History: 34 in 1981; 30 in 1984; 35 in 1987; 35 in 1991; 37 in 1994; 15 in 1998; 9 in 2004.* **Sources: Bou, But, Hig, HO2, Lej, ME9, SE14, Shu, WE10.**

Acorn, Fisher's - (Fisher's Early Acorn) - 85 days, Early acorn strain dev. by Fishers, compact vines. 7-8 in. long x 6 in. dia., distinctive sharp ridges, dark-green color, earlier and larger than Table Queen. *Source History: 1 in 1981; 1 in 1984; 1 in 1987; 1 in 1991; 1 in 1994; 1 in 2004.* **Sources: Fis.**

Acorn, Fordhook - 56 days, Long vines, oblong acorn-shaped fruit, tan skin, used green as summer squash after 56 days or mature for baking after 85 days, W. Atlee Burpee introduction in 1890, named after their Fordhook Farm in Pennsylvania, popular in the 1920s. *Source History: 1 in 1991; 2 in 1998; 4 in 2004.* **Sources: Ba8, Pr9, Se16, TE7.**

Acorn, Paydon Heirloom - Extremely productive rambling vines, beige skinned acorn squash, matures to orange in storage, traveled from France to Louisiana, then to Basco, IL where it was grown since the 1860s, finally to Iowa, extremely rare and endangered. *Source History: 3 in 2004.* **Sources: Ea4, In8, Se17.**

Acorn, Royal - (Royal Table Queen, Mammoth Table Queen, Royal Mammoth) - 80-90 days, Larger and more prolific than Table Queen, 7.5-8 x 6-6.5 in., 3-4 lbs., dull dark-green turning dull-orange in storage, sweet dry fine-flavored 1.5 in. thick light-yellow flesh, excel. baker, high yields, use as summer squash or store when mature. *Source History: 30 in 1981; 24 in 1984; 24 in 1987; 21 in 1991; 13 in 1994; 10 in 1998; 8 in 2004.* **Sources: Be4, Cr1, La1, Rev, Se7, Se25, Sh9, SOU.**

Acorn, Table Gold - 50-90 days, Productive compact dark-green bush plants, good fruit set, 50 days for light-yellow nutty-tasting 3 x 2 in. fruits, 82 days for 5 in. golden-yellow winter fruits, 1-2 lb. fruit, from Dr. Oved Shifriss. *Source History: 2 in 1981; 3 in 1991; 17 in 1994; 19 in 1998; 17 in 2004.* **Sources: Ba8, Dam, Ers, Ga1, HO2, Jor, Loc, ME9, Pe2, Pe6, RIS, RUP, SE4, SE14, TE7, Twi, WE10.**

Acorn, Table King - (Table King Bush) - 70-90 days, AAS/1974, vigorous compact bush plants, 2 ft. tall with 4 in. spread, 6-7 x 5 in. dia., 2 lb., hard glossy dk. gray-green ribbed fruits, thick pale-orange flesh, excel. flavor imp. in storage, 5-8 fruits per plant, PVP expired 1995. *Source History: 50 in 1981; 45 in 1984; 36 in 1987; 35 in 1991; 31 in 1994; 30 in 1998; 27 in 2004.* **Sources: Bo17, Bu2, De6, Dgs, Ers, Fa1, Ga1, Hi13, HO2, Hum, Jor, Ma19, MAY, ME9, MOU, OLD, PAG, RUP, SE14, Se26, Sh14, Tt2, Twi, Ves, WE10, We19, WI23.**

Acorn, Table Queen - (Danish, Pepper Squash, Des Moines) - 58-90 days, Very early for a running squash, 5-6 x 4-4.5 in., 1.25-2 lbs., hard dark-green thin rind, med-deep ribs, light-yellow to deep-orange sweet dry thick flesh, small seed cavity, 6-8 ft. trailing vines, very vigorous and prolific, can withstand poor soil conditions, good keeper, bakes well, introduced by the Iowa Seed Company of Des Moines, IA in 1913. *Source History: 81 in 1981; 69 in 1984; 68 in 1987; 61 in 1991; 56 in 1994; 64 in 1998; 64 in 2004.* **Sources: Ag7, All, Ba8, Bu2, Bu3, Bu8, CA25, CH14, CO23, Com, Dam, De6, Dgs, DOR, Dow, Ear, Fe5, Fi1, Fo7, Ge2, Gr27, HA3, Hal, He8, Hi6, HO1, Jor, Jun, Kil, La1, Lin, LO8, Loc, Ma19, MAY, Mel, Mey, Mo13, Ni1, OLD, Ont, PAG, Ri12, RIS, Roh, Ros, Sau, SE4, Se8, SE14, Se16, Se26, Sh9, Si5, Sk2, So1, Te4, Up2, Vi4, WE10, We19, Wet, Wi2, WI23.**

Acorn, Table Queen Bush - (Burpees Bush Table Queen) - 80-95 days, Runnerless true bush plants 36 in. dia., fruits similar in all respects to Table Queen, 3-8 fruits per plant, 5 x 4 in., very productive in a small space, fair keeper, widely adapted, orange flesh cooks dry and sweet, 1948. *Source History: 20 in 1981; 18 in 1984; 16 in 1987; 12 in 1991; 13 in 1994; 15 in 1998; 13 in 2004.* **Sources: Bu2, Bu3, CO30, Ers, Jor, Jun, Kil, RUP, SE14, Sh9, So1, TE7, Vi4.**

Acorn, Table Queen Mammoth - (Royal Acorn) - 80-85 days, Similar to Des Moines Table Queen but much larger, 8 x 6 in. and 3-4 lbs., 2 in. thick pale-orange dry flesh, tough dark greenish-black deep-grooved rind. *Source History: 9 in 1981; 8 in 1984; 5 in 1987; 4 in 1991; 5 in 1994; 7 in 1998; 2 in 2004.* **Sources: HO2, Twi.**

Acorn, Tuffy - 90 days, Flesh is thicker, sweeter, drier than other acorns, 2 lbs., 5-6 fruits per plant, flavor sweetens in storage. *Source History: 1 in 1998; 2 in 2004.* **Sources: Jo1, Se7.**

African Warty - (Verruqueuse Africaine) - 130 days, Unusual looking gourd is covered with warts, fairly long shape, 6 in. diameter, rare. *Source History: 5 in 2004.* **Sources: Ba8, Ra5, Ra6, Sa9, Vi5.**

Ancestors - 100 days, Warty ribbed pumpkin weighing 10-12 lbs. *Source History: 1 in 1994; 1 in 1998; 1 in 2004.* **Sources: Sa9.**

Apple, Small - Small apple shaped gourd, popular for crafting. *Source History: 1 in 2004.* **Sources: Ba8.**

Autumn Wings - 100 days, Unusual gourd with 5 or 6 double fins or wings, small, medium and large sizes, 2.5-3 in. dia. bulbs, Small Autumn Wings neck size is 4-5 in., Medium Autumn Wings neck size is 6 in., Large or Swan Autumn Wings neck size is 8-12 in. long, colors range from creams and yellows to whites and greens, need to be planted early and allowed to cure before harvest, wings break off easily if picked when immature. *Source History: 8 in 2004.* **Sources: Ers, HO2, Jor, Mo13, OSB, Pin, RUP, SE4.**

Baby Bear - 105 days, AAS/1993, dev. by Johnny's from a cross of New England Pie and a small naked seeded pumpkin, deep orange fruits 5-6 in. wide by 3.5-4 in. high, 1.5-2 lbs., strong stem is easy to grip, fine grained flesh excel. for pies, semi-hulless seeds for toasting, blight and frost tol., PVP 1996 Johnny's Selected Seeds. *Source History: 42 in 1994; 36 in 1998; 27 in 2004.* **Sources: BAL, Be4, Bu2, CH14, Cl3, HA3, Hal, He17, HO2, Jo1, Jor, La1, Lin, ME9, Mey, Mo13, MOU, Ni1, Pin, Pr3, RIS, Roh, RUP, Se8, So1, Twi, Wet.**

Baby Boo - 95-110 days, Miniature white pumpkin weighs 3-4 oz., 2-3 in. wide, edible white flesh, plant later than Jack-B-Quick types since it turns pale-yellow at full maturity, PVP 1997 John Jaunsem. *Source History: 12 in 1994; 28 in 1998; 36 in 2004.* **Sources: Bu1, Bu2, CH14, De6, Ers, Fi1, Ge2, Gur, He17, Hi13, HO2, Jor, Jun, Loc, Ma19, ME9, Mel, Mo13, OLD, Orn, OSB, Pin, Ra5, Re8, RIS, Roh, Ros, RUP, Sau, SE4, Sh9, Si5, Sto, Ter, Vi5, Wi2.**

Baby Pam - 100-105 days, Very uniform 5 x 5.5 in. dia. fruits, deep-orange flesh excellent for pies, tightly secured stems, vigorous 10-12 ft. vines, heavy yields, smooth skin, indistinct ribs. *Source History: 1 in 1981; 1 in 1984; 7 in 1987; 11 in 1991; 12 in 1994; 14 in 1998; 19 in 2004.*

Sources: CA25, CH14, Com, Fe5, Ha5, HO2, Jo1, Jor, Loc, ME9, Ra5, Re8, RIS, RUP, SE4, Si5, Tu6, Vi5, Wet.

Bakery's - Green and orange, oblong fruit with good flavor, 2-3 lbs., old str. from New York. *Source History: 1 in 1998; 1 in 2004.* **Sources: Ho13.**

Big Red California Sugar - 82 days, Deep orange skin, 1 in. yellow flesh, 3-4 lbs. *Source History: 1 in 1998; 1 in 2004.* **Sources: Sa9.**

Black Sweet Potato - 120 days, More blocky version of Green Striped Cushaw with light color. *Source History: 1 in 2004.* **Sources: Sa9.**

Bushkin - 95 days, Short-vine plants spread only 5-6 ft., 1-3 fruits per plant, bright golden orange pumpkins weigh 8-10 lbs., thick delicious light-yellow flesh, edible seeds, good for pies and carving and seed snacks, PVP expired 2003. *Source History: 2 in 1991; 1 in 1994; 1 in 1998; 1 in 2004.* **Sources: Bu2.**

Calabaza Mercado Verde - Round fruit, green striped skin, 2 lbs., used for baking or as summer squash when picked young, collected at market in central Mexico by Baker Creek Heirloom Seeds. *Source History: 1 in 2004.* **Sources: Ba8.**

Caserta - 50-57 days, AAS/1949, early bush cocozelle, spiny semi-open cut-leaf plants can set 30 cylindrical slightly tapered fruits, thin light-green skin with dark-green stripes, firm creamy-white flesh, pick when 4-6 x 2-3.5 in, U of CT. *Source History: 15 in 1981; 12 in 1984; 13 in 1987; 12 in 1991; 9 in 1994; 11 in 1998; 12 in 2004.* **Sources: Bo19, CO23, HA20, Loc, Roh, Ros, Sa9, Se7, SE14, Sh9, Sk2, WE10.**

Casper - 90-105 days, White skin with blue cast, 8-12 in. dia., 10-20 lbs., good flavor, use like squash or for pies, great for Halloween. *Source History: 3 in 1994; 6 in 1998; 14 in 2004.* **Sources: CO30, Ers, He8, HO2, Jor, ME9, Mel, Ri2, RIS, RUP, SE14, Sto, TE7, Twi.**

Cheyenne Bush - 80-90 days, Extremely early bush pie pumpkin adapted to great plains, compact bush plant, small bright-orange fruits, 5-8 lbs., fine grained deep golden-yellow solid flesh, fair table quality, dev. from a cross of Cocozelle x New England Pie, WY/AES. *Source History: 1 in 1981; 2 in 2004.* **Sources: Pr9, Se16.**

Chinese Miniature - Flattened 6 in. orange ornamental pumpkin. *Source History: 1 in 1987; 1 in 1991; 1 in 1994; 1 in 1998; 1 in 2004.* **Sources: Ba8.**

Chinese Miniature, White - Tiny flat white pumpkin, 3-4 in., edible, easy to grow, high yields, great for fall decorating, compares to Baby Boo, from China. *Source History: 1 in 2004.* **Sources: Ba8.**

Cinderella - (Rouge Vif D'Etampes) - 84-115 days, Round 10 in. pumpkins on bush vines like summer squash, needs only 6 sq. ft. of well drained soil, uniform globes, 20-25 lbs., bright-orange skin, excel. flavor, stores well for winter use. *Source History: 7 in 1981; 4 in 1984; 3 in 1987; 2 in 1991; 6 in 1994; 16 in 1998; 27 in 2004.* **Sources: Bo17, Bo19, CH14, Cl3, CO30, Dil, Ers, Fa1, Fi1, Ha5, HO2, La1, Ma19, ME9, Mel, Mo13, OLD, Pe2, Re6, Ros, SE4, Se8, SE14, Sk2, TE7, Vi4, WI23.**

Cocozelle - (Italian Cocozelle, Zucchini Cocozelle Bush, Zucchini, Italian Vegetable Marrow, Cocozella di Napoli) - 42-120 days, Bush type plant, long cylindrical bush zucchini, dark-green with light-green stripes, edible in 42-45 days, ripens to yellow, very firm greenish-white flesh, 10-12 in. young fruits, good market strain, intro. before 1885. *Source History: 23 in 1981; 19 in 1984; 18 in 1987; 28 in 1991; 25 in 1994; 25 in 1998; 35 in 2004.* **Sources: Ba8, Bo19, Bu2, Co31, Com, Cr1, Dow, Fe5, Ga1, Ge2, Goo, HA3, Hig, HO2, Ho13, Kil, LO8, Na6, Ni1, Or10, Pe1, Ri12, Sau, Se7, Se8, SE14, Se17, Se26, Sh9, So1, So9, Syn, WE10, Wet, WI23.**

Cocozelle (Green Bush) - 55 days, Bush type plant, long cylindrical fruit with marbled light and dark green skin, 4-8 in. long. *Source History: 1 in 1981; 1 in 1984; 1 in 1987; 1 in 1998; 1 in 2004.* **Sources: Gou.**

Cocozelle, Green Striped - 50-60 days, Deep-green when young but yellow-striped later, smooth cylindrical fruits, 12 x 3 in. dia., thick tasty green flesh. *Source History: 2 in 1981; 2 in 1984; 2 in 1987; 2 in 1991; 2 in 1994; 2 in 1998; 1 in 2004.* **Sources: La1.**

Cocozelle, Long - (Long Type Cocozelle, Open Bush Long Cocozelle) - 44-63 days, Long smooth narrow cylindrical 7-9 in. fruits with ridges, dark-green with light stripes, harvest at 6-8 in., firm pale-green flesh, cut-leaf semi-open bush plant, popular zucchini type for home or market gardens and shipping. *Source History: 7 in 1981; 7 in 1984; 4 in 1987; 4 in 1991; 3 in 1994; 4 in 1998; 1 in 2004.* **Sources: PAG.**

Connecticut Field - (Southern Field, Big Tom, Large Yellow, Holloween, Jaune des Champs, Yankee Cow Pumpkin) - 100- 120 days, Large globes flattened at ends, 10-14 x 12-15 in. dia., 15-25 lbs., hard slightly-ribbed thin bright yellow-orange rind, thick coarse deep-yellow sweet flesh, some up to 20 in. dia., cornfield type for canning or stock feed, Native American origin, pre-1700. *Source History: 104 in 1981; 96 in 1984; 94 in 1987; 97 in 1991; 91 in 1994; 85 in 1998; 85 in 2004.* **Sources: ABB, Ada, Ag7, Alb, All, Ba8, Ber, Bo18, Bo19, Bu2, Bu3, But, CA25, Ca26, CH14, Co32, Com, Dam, De6, Dil, DOR, Dow, Ers, Fa1, Fe5, Fi1, Ge2, Goo, Gr27, Gr29, Gur, HA3, Hal, He8, He17, Hi13, HO1, HO2, Hum, Jor, Jun, Kil, La1, Lin, Loc, Ma19, Me7, ME9, Mel, Mey, Mo13, MOU, Ol2, OLD, Ont, Or10, PAG, Pr9, Ra5, Rev, Ri12, Roh, Ros, RUP, Sa9, Sau, Se26, Sh9, Shu, Si5, So1, So9, SOU, Sto, Th3, Twi, Up2, Ves, Vi4, Vi5, WE10, Wet, Wi2, WI23, Wo8.**

Connecticut Sweet Pie - 95 days, Tall pie-type pumpkin weighs 6-8 lbs. *Source History: 1 in 1994; 1 in 1998; 1 in 2004.* **Sources: Sa9.**

Cornfield Pumpkin - 90 days, Traditionally grown by farmers as a dual crop with their corn, 12-16 in. wide by 10-12 in. tall perfectly shaped fruit, very sturdy stems, remain in excel. condition until Halloween and beyond, carving and fall decorations *Source History: 1 in 2004.* **Sources: Se16.**

Cow Pumpkin - 110 days, Old-time variety brings premium prices at the Indianapolis Farmers Market, med-size globular-to- elongated fruit, tannish-orange skin, thick sweet fine-grained light-orange flesh, delicious as winter squash or in pumpkin pies, makes an excellent jack-o'-lantern. *Source History: 1 in 1987; 1 in 1994; 1 in 1998; 1 in 2004.* **Sources: Sa9.**

Crookneck, Early Yellow Summer - (Early Golden Summer Crookneck, Early Prolific Crookneck, Dwarf Summer Crookneck, Yellow Crookneck) - 42-60 days, Semi-open plant, smooth light-yellow fruits with curved neck, mature orange and warted, 8-10 in. with 3-4 in. bowl, pick at 5-6 in., creamy-white sweet mild flesh, excel. flavor, dates back to pre-Columbus era. *Source History: 98 in 1981; 85 in 1984; 90 in 1987; 83 in 1991; 78 in 1994; 86 in 1998; 91 in 2004.* **Sources: Ada, Ag7, All, Ba8, Bo17, Bou, Bu2, Bu3, Bu8, But, CA25, CH14, CO23, Co31, Co32, Com, Cr1, Dam, De6, DOR, Dow, Eo2, Ers, Fe5, Fi1, Ga1, Goo, Gr27, Gr28, GRI, HA3, He8, He17, Hi6, Hi13, Hig, HO1, Ho12, In8, Jo1, Jor, La1, LO8, Loc, Ma19, MAY, ME9, Mel, Mey, MIC, Mo13, MOU, Na6, Ol2, Or10, Orn, PAG, Pe2, Pe6, Pr3, Red, Ri2, Ri12, RIS, Roh, Ros, RUP, Sau, Scf, Se7, SE14, Se16, Se17, Se25, Sh9, Shu, Sk2, So1, So9, SOU, Sw9, TE7, Ter, Tu6, Ver, Vi4, WE10, We19, Wet, Wi2, WI23.**

Crookneck, Goldarch - 42 days, Butter-yellow, 3 to 5 days earlier than Golden Summer Crookneck, longer and more cylindrical, arched neck, almost no warts so acceptable longer, uniform, bush. *Source History: 1 in 1981; 1 in 1984; 1 in 1987; 1 in 1991; 1 in 1994; 2 in 2004.* **Sources: OLD, PAG.**

Crookneck, Yellow Summer Imp. - 50-55 days, Light lemon-yellow strain with open plant habit, small smooth fruits with curved necks, very early and prolific, uniform, bright color gives good market appearance. *Source History: 3 in 1981; 3 in 1984; 2 in 1987; 2 in 1991; 3 in 1994; 3 in 1998; 2 in 2004.* **Sources: Kil, Sa9.**

Crown of Thorns - 100 days, Creamy white to green and yellow striped, crown shaped gourd with ten points or thorns. *Source History: 21 in 2004.* **Sources: Bo17, Bo19, Bu8, Cl3, Ers, HA3, Hi13, HO1, HO2, Jor, ME9, Pep, Pin, Ra6, RIS, Se28, Sh9, Sto, Tho, Vi5, WE10.**

Dancing Gourd - (Tennessee Dancing Gourd, Spinning Gourd) - 90-100 days, Tiny bottle-shaped gourds, 2 in. long, hard green and white striped shell dries to tan, spins like a top, used in crafting, heirloom from Tennessee. *Source History: 6 in 2004.* **Sources: Ba8, He17, Pr9, Ra5, Ra6, Se16.**

Delicata - (Delicatessen, Sweet Potato Squash, Delicata Sweet Potato, Peanut) - 95-100 days, Ivory-cream skin with dark- green stripes, 7-9 x 3 in. dia., rich sweet potato-like flavor, short prolific vines, good keeper, introduced in 1894. *Source History: 10 in 1981; 7 in 1984; 14 in 1987; 24 in 1991; 31 in 1994; 46 in 1998; 58 in 2004.* **Sources: All, Ba8, Be4, Bo19, Bou, Bu2, Bu3, CH14, CO30, Coo, De6, Ers, Ga1, Ge2, Goo, Gr27, GRI, Ha5, Hi6, HO2, Hum, Jor, Jun, La1, Loc, Ma19, ME9, Mel, Mo13, Na6, OLD, Ont, Orn, Pe1, Pe2, Pin, Pla, Rev, Ri12, RIS, Roh, RUP, SE4, Se7, Se8, Shu, So1, So9, So12, Sto, Sw9, Tt2, Tu6, Twi, Ver, WE10, We19, WI23.**

Delicata JS - 100 days, Distinct strain from Johnny's Seeds, short vines, 7-9 in. long, 3 in. wide, 1.5-2 lbs., cream skin with dark green stripes and flecks, very sweet orange flesh with superior taste, stores well, excellent for stuffing and baking, no curing required. *Source History: 1 in 2004.* **Sources: Jo1.**

Delicata, Cornell's Bush - 80-100 days, AAS/2002, healthy semi-bush 3-4 ft. vines, dark green striped fruit, whiter and more plump than Delicata JS, similar sweet cooked quality, PM tolerant, Dr. Molly Jahn and George Moriarty, Cornell U, PVP 2002 Cornell University Exp. Station. *Source History: 22 in 2004.* **Sources: Bu2, Dam, Fe5, Ge2, GLO, Gur, Hal, HO2, HPS, Hum, Jo1, Jor, Jun, MAY, Ni1, OSB, Par, Se26, Ter, Ver, Ves, We19.**

Delicata, Zeppelin - 95-100 days, Long tapered fruit with excellent Delicata taste, 6-9 in. long, up to 3 in. diameter, ivory skin with dark green stripes which turn a delicate orange in storage, 1 lb. average, light orange flesh, good storage, dev. by Frank Morton. *Source History: 4 in 2004.* **Sources: Fe5, Ki4, Sh12, St18.**

Dinosaur Egg - (Sacred Indian Rattle) - 90 days, Long vines, oval to egg-shaped fruit, 3-8 in. long, can be dried to make rattles, children like to grow them as 'dinosaur eggs.' *Source History: 1 in 2004.* **Sources: Red.**

Fairytale - 100-110 days, Bright orange pumpkin with unique shape, 8 in. deep by 12-18 in. wide, 4 in. thick flesh, deeply serrated orange-tan exterior, 10-30 lbs., dense flesh, delicious, cooks and tastes like a winter squash, shape resembles Cinderella but with deeper ridges, carries the lineage of 3 heirloom lines. *Source History: 5 in 1998; 10 in 2004.* **Sources: Cl3, Ha5, HA20, HO2, Jor, ME9, Orn, Ri12, Sk2, Ter.**

Fat Boy - 90-115 days, Blocky, round, dark orange pumpkin with heavy, black stem, 15-25 lbs. *Source History: 3 in 1998; 1 in 2004.* **Sources: HO2.**

Field Pumpkin - *Source History: 1 in 1981; 1 in 1984; 1 in 2004.* **Sources: Eco.**

Ghost Rider - 110-115 days, Long dark-green tightly held stem, slightly ridged deep-orange fruits average 15-20 lbs., orange flesh, very uniform attractive Halloween pumpkin, suitable for either carving or processing, PVP abandoned 1990. *Source History: 21 in 1991; 20 in 1994; 21 in 1998; 18 in 2004.* **Sources: All, Bu2, CA25, CH14, CO30, De6, Dil, HO2, Jor, Jun, MAY, ME9, Ra5, RIS, RUP, SE4, Sh9, Vi5.**

Gold Rush - 120 days, Large round fruit, dark orange, thick flesh, large thick handles are 4-5 in long and 3 in. thick, attached firmly, 20-40 lbs., canning, carving or stock feed. *Source History: 1 in 1998; 5 in 2004.* **Sources: HO2, Mo13, RUP, SE4, Si5.**

Golden Scallopini Bush - 55-65 days, Patty pan type, excel. quality, 3-5 in. golden fruit with scalloped edges, pick when young and tender, heirloom. *Source History: 1 in 1991; 1 in 1994; 2 in 1998; 4 in 2004.* **Sources: Gr28, Pe2, Se7, So9.**

Half Moon - (Half Moon Large Yellow) - 100-110 days, Dark-orange, narrow ribs, 14 in. dia., 14-18 lbs., dark-yellow thick coarse flesh, quite uniform, larger well-attached stem-handle, tall Connecticut Field type. *Source History: 11 in 1981; 11 in 1984; 13 in 1987; 9 in 1991; 6 in 1994; 6 in

1998; 5 in 2004. **Sources: CH14, Ers, Jor, RUP, Sh9.**

Halloween - 100 days, Halloween is usually thought of as a synonym for Jack-o'-Lantern. *Source History: 1 in 1981; 1 in 1984; 3 in 1987; 1 in 1991; 1 in 1994; 3 in 1998; 5 in 2004.* **Sources: Bo18, Bu8, HO1, OLD, Vi5.**

Happy Jack - 105-110 days, Large vigorous vines, Howden-type fruits are dark orange with strong dark green stems, 16-22 lbs., used for baking and processing, attractive pumpkin, PVP abandoned 1993. *Source History: 4 in 1991; 20 in 1994; 3 in 1998; 1 in 2004.* **Sources: MIC.**

Honeyboat - 105 days, Delicata type selected for a more typical and elongated Delicata shape, high quality sister line to Sugar Loaf with the same rusty tan and green striped coloration, 2-3 in. wide by 6-9 in. long, excellent eating quality, good storage. *Source History: 1 in 1998; 2 in 2004.* **Sources: In8, Pe6.**

Hopi Pumpkin - Round or elongated fruit with either striped or solid green skin which yellows as it matures, from Arizona. *Source History: 1 in 1994; 1 in 2004.* **Sources: Na2.**

Howden - (Howden Field) - 105-115 days, Imp. Connecticut Field type, larger, more uniform and symmetrical, rich-orange hard rind, thick flesh, 20-25 lbs., large spreading vines, good keeper, BR tol., PVP expired 1994. *Source History: 5 in 1981; 5 in 1984; 6 in 1987; 17 in 1991; 22 in 1994; 56 in 1998; 75 in 2004.* **Sources: ABB, Ag7, All, BAL, Be4, Bo19, Bu1, Bu3, Bu8, CA25, Ca26, CH14, Cl3, Com, Dam, De6, Dil, Ers, Fa1, Fe5, Fi1, Ga1, Ge2, Gr27, GRI, HA3, Ha5, HA20, Hal, He17, Hi6, HO1, HO2, Hum, Jor, Jun, La1, Loc, Ma19, MAY, ME9, Mel, Mey, MIC, Mo13, MOU, Na6, OLD, Ont, OSB, PAG, Par, Pe2, Pin, Ra5, Ri12, RIS, Roh, Ros, RUP, SE4, Se7, Sh9, Si5, SOU, Sto, TE7, Ter, Twi, Ver, Vi5, WE10, We19, Wi2, WI23.**

Howden Biggie - 110-115 days, Large version of Howden with dark orange skin, globe to tall globe shape, strong dark stems, 30-60 lbs., consistent even shape over its range of sizes allows for more marketable fruit, PVP 1996 Howden Farm. *Source History: 6 in 1998; 11 in 2004.* **Sources: CH14, Cl3, Ha5, HA20, HO2, Jor, Loc, MAY, ME9, Mo13, OSB.**

Idaho Gem - 80 days, Pie type weighing 4-6 lbs., tolerates cool weather, dev. by Glenn Drowns. *Source History: 1 in 1987; 1 in 1994; 1 in 1998; 1 in 2004.* **Sources: Sa9.**

Jack Be Little - (Little Sweetie) - 85-110 days, Delightful miniature pumpkin, small vigorous vines produce 8-20 flattened deeply ribbed fruits, 2 in. tall x 3-6 in. dia., 3-8 oz., deep-orange rind and flesh, edible but most appealing as a fall decoration, shelf life up to 12 mo. if cured fully on the vine. *Source History: 36 in 1987; 79 in 1991; 79 in 1994; 87 in 1998; 85 in 2004.* **Sources: ABB, All, BAL, Be4, Bo17, Bo19, Bu1, Bu2, Bu3, Bu8, CA25, CH14, Cl3, CO30, Coo, Dam, De6, Dgs, Dil, Ear, Eo2, Ers, Fa1, Fe5, Fi1, Fo7, Ga1, GLO, Gr27, Gur, HA3, He8, He17, Hi13, HO1, HO2, HPS, Hum, Jo1, Jor, Jun, La1, Lej, LO8, Loc, Ma19, MAY, Me7, ME9, Mel, Mey, Mo13, MOU, Na6, Ni1, OLD, Ont, Orn, OSB, PAG, Pin, Ri12, RIS, Roh, Ros, RUP, Sau, SE4, Se8, SE14, Se26, Sh9, Si5, Sk2, So25, SOU, Sto, Ter, Twi, Ves, WE10, We19, Wet, Wi2, WI23.**

Jack-B-Quik - 95 days, More ribbed selection of Jack-B-Little with fewer off-type fruits, med-size plant, orange 2 x 3 in. dia. fruits weigh 1/4 lb. *Source History: 1 in 1991; 2 in 1994; 3 in 1998; 3 in 2004.* **Sources: HO2, RUP, Sau.**

Jack-O-Lantern - (Halloween, Large Connecticut Yellow Field) - 75-115 days, Smooth med-orange skin, shallow ribs, 9-10 x 7-10 in. dia., 10-18 lbs., med-thick sweet fine-grained pale-orange flesh, good cooking quality, stores well, sel. from cross of Connecticut Field x Golden Oblong to be the size of a human head. *Source History: 98 in 1981; 86 in 1984; 87 in 1987; 80 in 1991; 71 in 1994; 67 in 1998; 63 in 2004.* **Sources: ABB, Ada, Alb, All, Be4, Bo17, Bu2, Bu3, CA25, CH14, CO23, CO30, Co31, Com, Cr1, De6, Dil, DOR, Ear, Ers, Fa1, Fis, Gr29, Gur, HA3, Hal, He17, Hi13, HO2, Hum, Jor, La1, Lej, Loc, Ma19, MAY, Mel, Mey, Mo13, MOU, Ont, Or10, PAG, Ra5, Ri12, Ros, RUP, Sau, SE14, Se26, Se28, Sh9, Shu, Sk2, SOU, Twi, Vi4, Vi5, WE10, Wet, Wi2, WI23, Wo8.**

Jack-o-Lite - 90 days, Classic Halloween pumpkin used for carving and baking, 12-15 lbs. *Source History: 1 in 1994; 2 in 1998; 2 in 2004.* **Sources: Se7, So9.**

Jumpin Jack (May) - 110-120 days, Very vigorous vines, attractive large dark-orange fruits ranging from 20-60 lbs., solid black-green handle ranges from 4-10 in. length, fruit shape similar to Howden only larger with thick dense deep-orange flesh, very good eating quality, dev. in Michigan after 30 years of growing and selection, PVP. *Source History: 1 in 1991; 2 in 1994; 2 in 1998; 1 in 2004.* **Sources: RUP.**

Kojac - 75 days, Short upright plant with open growth habit, spineless, good yields of dark green fruit, from the UK. *Source History: 1 in 2004.* **Sources: Tho.**

Lady Godiva - (Hulless) - 100-110 days, Grown only for its naked hulless greenish seeds, nutritious and rich in protein, great roasted or raw, 8-12 lbs., 12-15 pumpkins per plant, inedible flesh, 1 lb. of seeds in each fruit. *Source History: 15 in 1981; 10 in 1984; 5 in 1987; 4 in 1991; 2 in 1994; 5 in 1998; 5 in 2004.* **Sources: Ho13, Pr3, Sa9, Se7, Se17.**

Lady Godiva Bush - 100 days, Recently developed compact bush form of Lady Godiva, plants can be grown very close together, round fruits larger than the vining variety, productive, for cooking or jack-o'-lanterns although not richly colored, naked seeds. *Source History: 1 in 1987; 1 in 1998; 1 in 2004.* **Sources: Syn.**

Lakota - 85-105 days, Fine grained, orange flesh with sweet, nutty flavor, good for baking, beautiful orange and green skin color, 5-7 lbs., developed from Lakota Sioux squash, rare. *Source History: 1 in 1998; 5 in 2004.* **Sources: Bu2, Hi13, Ri12, SE4, Se7.**

Lebanese - 78 days, Nearly oval fruit, yellow-orange skin. *Source History: 1 in 1998; 2 in 2004.* **Sources: Bo17, Sa9.**

Lebanese Light Green - 40-50 days, Summer squash that is very popular in the Middle East, sweet fruits, excel. flavor, keep fruits picked small but still fine when up to 2 lbs., capable of withstanding tough climatic conditions. *Source History: 1 in 1994; 2 in 1998; 1 in 2004.* **Sources: Bou.**

Lemon - 60 days, Large bush plant produces fruit the color, size and shape of a lemon, good flavor, great fried, excellent pest resistance, heat tol., heirloom. *Source History: 1 in 1998; 3 in 2004.* **Sources: Ba8, Ho13, Sk2.**

Little Gem - 65 days, Small decorative pumpkin, vining type, edible when apple size and green, turns yellow and then orange as it matures and hardens, when baked whole cut small steam release hole, sells well as miniature for Halloween, seed from Africa. *Source History: 2 in 1981; 2 in 1984; 3 in 1987; 3 in 1991; 8 in 1994; 7 in 1998; 4 in 2004.* **Sources: Fo7, Ho13, Lej, We19.**

Little October - 95-100 days, Round to semi-flat, dark orange skin, dark green stem, 3-4 in. diameter, 8-12 oz., ideal for painting for Halloween, PVP 2001 Willhite Seed Inc. *Source History: 4 in 2004.* **Sources: HO1, RUP, SE4, Wi2.**

Long Pie Pumpkin - (Indian, Oblong, Nantucket) - 95-110 days, Heirloom from Maine, 3-6 lbs., shape and color resembles an overgrown zucchini at maturity, ready for harvest as long as there is an orange spot where the fruit touches the ground, turns a brilliant orange in storage, supposedly originated on the Isle of St. George in the Azores, brought to Nantucket on a whaling ship in 1832, where it was first known as the Nantucket pumpkin, very similar to Golden Oblong, a Burpee intro. in 1889, but this var. is bigger, may be a selection from it. *Source History: 1 in 1998; 11 in 2004.* **Sources: Ag7, Ea4, Fe5, Ga1, Hi6, Ho13, In8, Ol2, Rev, Sa9, Se17.**

Lubnani - (Vegetable Marrow White, Lebanese White Bush Squash, Lubnani Summer) - 52 days, Bush summer squash very popular in the Middle East, sweet excellent flavor, fruit looks like a small honeydew melon and is sweet so we call it the honeydew of squashes, rind cooks up well and is easily eaten, fruits weigh 1-2 lbs., quick maturing, withstands variable climatic conditions. *Source History: 1 in 1987; 3 in 1991; 1 in 1994; 3 in 1998; 2 in 2004.* **Sources: Sk2, Te4.**

Luxury Pie Pumpkin - (Winter Luxury) - 95-100 days, Extremely productive, nearly globe, 10 in. dia., 6-8 lbs., finely netted golden russet skin which is a sign of finest quality, excel. keeper for winter use, ripens early, a lighter yellow version was originally introduced in 1893 by Johnson and Stokes, the present, more orange colored strain was introduced by Gill Brothers of Portland, Oregon, after 1920. *Source History: 1 in 1981; 1 in 1984; 2 in 1987; 1 in 1991; 1 in 2004.* **Sources: Rev.**

Mandan - 47-56 days, Grown for generations by the Mandan tribe, bush plants until first fruits mature then runners grow to produce a second crop, white skin with green stripes and green mottling between stripes, pale-green flesh, excel. flavor, intro. in 1912 by the Oscar Will Seed Company. *Source History: 1 in 1991; 1 in 1998; 2 in 2004.* **Sources: Pr3, Sa9.**

Mandan Strain No. 1 - White with green and yellow stripes, tasty as summer squash and good for winter storage, from Oscar Will Seeds of Bismark, ND via Mandan tribe, 1912. *Source History: 2 in 1998; 1 in 2004.* **Sources: Se17.**

Mandan Yellow - 83 days, Blocky, flattened shape, pale yellow with green stripes. *Source History: 1 in 1998; 1 in 2004.* **Sources: Sa9.**

Marrow, Long Yellow - 55 days, Bush type, creamy white fruit ripens to pale yellow. *Source History: 1 in 1998; 1 in 2004.* **Sources: Sa9.**

Marrow, Tender and True - 60 days, Productive bush type with round grey-green pumpkin shape fruit, excellent for frying, England. *Source History: 1 in 1998; 1 in 2004.* **Sources: Sa9.**

Marrow, Trailing Green - 70 days, Vigorous vines, striped fruit remains edible even when quite large, late for a summer squash. *Source History: 1 in 1998; 1 in 2004.* **Sources: Sa9.**

Mayeras - 92 days, Round fruit, creamy grey > creamy tan, 4-6 lbs., used as a summer squash. *Source History: 1 in 1998; 1 in 2004.* **Sources: Sa9.**

Melonnette de Vendee - Round to oval, ribbed yellow-orange fruit, sweet orange flesh, 2-4 lbs., for baking and pies, good keeper, rare French variety. *Source History: 1 in 2004.* **Sources: Ba8.**

Menominee - *Source History: 1 in 2004.* **Sources: Se17.**

Miniature Ball - 90 days, Round gourd, baseball size and shape, green and white stripes, 3.5 in. *Source History: 3 in 2004.* **Sources: ME9, Pep, Sto.**

Montana Frost - 90 days, White skin, orange flesh, good keeper, good for carving. *Source History: 1 in 2004.* **Sources: Fis.**

Montana Jack - 90-95 days, Sel. for early maturity in cool northern summers, vining habit, thick flesh, 8-15 lbs., sturdy green stem, great for pies and carving, bred by John Navazio. *Source History: 1 in 1998; 6 in 2004.* **Sources: Ga1, HO2, ME9, OSB, Rev, SE14.**

Morgan - *Source History: 1 in 2004.* **Sources: Se17.**

Mt. Pima Vavuli - Pumpkin-shaped fruits, green and yellow striped skin, orange flesh, 10 lbs., edible seeds, excel. keeper, good for pies, obtained from the Pimas, rare. *Source History: 1 in 1987; 2 in 1991; 2 in 1994; 1 in 2004.* **Sources: Na2.**

Munchkin - 85-110 days, Baby pumpkin, flattened scalloped fruits 3-4 in. dia., 1/4 lb., bright-orange skin, med-thick orange-yellow flesh, good sweet flavor, small seed cavity, high yielding, for baking carving or decorating, holds firmness a long time. *Source History: 3 in 1987; 6 in 1991; 7 in 1994; 7 in 1998; 11 in 2004.* **Sources: CH14, Cl3, Ha5, HA20, Jor, Loc, Mo13, OSB, Ra5, Vi5, Wi2.**

Nest Egg - (Japanese Nest Egg, White Egg) - 90-100 days, Hard, white fruit the size and shape of a hen's egg, used as nest eggs, often found growing wild in the Ozarks, popular in the 1800s. *Source History: 30 in 2004.* **Sources: Ba8, Bo17, Bo19, Bu3, Cl3, Co31, Gur, HO1, HO2, Ho8, La1, ME9, Ni1, Pep, Pin, Ra5, Ra6, Ri12, RIS, Roh, RUP, Sa9, Scf, Se26, Sh9, So12, Sto, Twi, Vi5, WE10.**

New England Pie - (Small Sugar) - 100-105 days, New England Pie is usually thought of as a synonym for Small Sugar, bright orange pie pumpkin, 5-8 lbs., superb flavor.

Source History: 2 in 1981; 2 in 1984; 3 in 1987; 6 in 1991; 6 in 1994; 5 in 1998; 5 in 2004. **Sources: Dow, Fe5, He8, Hi6, Se26.**

Northern Bush - 90 days, Bush pumpkin dev. by Fishers Garden Store from Cheyenne Bush x Orange Winter Luxury, 5-8 lb. fruits, rich-orange skin, fine-grained golden-yellow thick flesh. *Source History: 1 in 1981; 1 in 1984; 1 in 1987; 1 in 1991; 2 in 1994; 2 in 1998; 2 in 2004.* **Sources: Fis, Hig.**

Northern Gold - 90 days, Fruit turns gold color weeks before harvest, extra sweet, smooth, orange flesh, 7-10 lbs. *Source History: 1 in 1998; 1 in 2004.* **Sources: Fis.**

Odessa - 65 days, Vigorous, spreading plant bears oblong, 8-12 in. long beige/pale green fruit, delicious, from Russia. *Source History: 1 in 1998; 3 in 2004.* **Sources: In8, Sa9, Se17.**

Old Timey Flat Pumpkin - *Source History: 1 in 2004.* **Sources: Se17.**

Omaha - (Omaha Indian) - 80 days, Tall orange 3-4 lb. fruits, med. orange, size variable, prolific, Native American type, original seed collected from the Omaha Indians by Dr. Melvin Gilmore, intro. by the Oscar Will Company in 1924. *Source History: 1 in 1987; 1 in 1991; 1 in 1994; 1 in 1998; 2 in 2004.* **Sources: Ea4, Sa9.**

Orange Gourd - 85 days, Round smooth squash looks just like a navel orange, mentioned in Vilmorin's "The Vegetable Garden" in 1885. *Source History: 1 in 2004.* **Sources: Ba8.**

Orange Warted - 90-100 days, Productive vines, round to slightly flattened 4-6 in. orange gourd, warted skin, excel. for fall decorations and arrangements. *Source History: 7 in 2004.* **Sources: ME9, Pep, Ra5, Ra6, Se16, Sto, Vi5.**

Pacheco Pumpkin - Round to elongated shape, bright yellow skin, flavorful cream colored flesh, Sonora. *Source History: 1 in 2004.* **Sources: Na2.**

Pankow's Field - 100-120 days, Unique big thick sturdy stems, smoother than Howden and somewhat more variable in color and shape, deep round to tall, 20-30 lbs., extra thick flesh keeps them symmetrical. *Source History: 1 in 1981; 1 in 1984; 2 in 1987; 5 in 1991; 6 in 1994; 6 in 1998; 7 in 2004.* **Sources: Ha5, HO2, Jor, MOU, Ra5, RIS, Vi5.**

Patissons Golden Marbre - French heirloom, rare scallop type, tall bush plants, tender golden flesh with orange marbling, excellent flavor, used as summer or winter squash, very rare. *Source History: 1 in 2004.* **Sources: Ba8.**

Patissons Panache Blanc et Vert - 60-70 days, White bush scallop with dark green radial streaks, delicious fried or baked, pre-1885 French heirloom. *Source History: 2 in 2004.* **Sources: Ba8, Se16.**

Patissons Panache Jaune et Vert - 55-70 days, Beautiful French heirloom, scalloped fruit, cream-yellow skin with contrasting green radial streaking, flavorful summer squash or used as winter squash for baking. *Source History: 2 in 2004.* **Sources: Ba8, Se16.**

Pear Bi-Color - 90 days, Smooth, pear shape, yellow stripes on green bottom half, yellow to buff upper part, 3-5 in., used for old-fashioned darning ball. *Source History: 27 in 2004.* **Sources: Bo17, Bu8, CH14, Cl3, Ers, Go9, He17, HO1, HO2, Jor, ME9, Mo13, Ni1, OLD, Pep, Ra5, Ra6, RIS, Roh, RUP, SE4, Se26, Sh9, Sto, Twi, Vi5, WE10.**

Piena di Napoli - (Carpet Bag, Bedouin) - 120-150 days, Italian winter squash, fruits grow 20-28 in. long with a diameter of more than 7 in., fragrant yellow-orange flesh, sweet flavor, 20-35 lbs., stores quite well. *Source History: 1 in 1994; 1 in 1998; 1 in 2004.* **Sources: Sa9.**

Redondo di Tronco - 100 days, Italian winter squash, compact growth habit, fruits average 1 lb., dark-green skin, orange flesh. *Source History: 3 in 1991; 3 in 1994; 1 in 1998; 3 in 2004.* **Sources: HA20, Sk2, WE10.**

Rondini - 65 days, Vining plants produce 10 dark green fruit, tennis ball size, used like zucchini when immature, develops a hard shell at maturity, stores well. *Source History: 1 in 2004.* **Sources: Ter.**

Rondo de Nice - 52 days, Small round zucchini type, smooth med-green skin mottled yellow, space saver, 10-12 in. between plants, prolific, harvest when golf ball size, fine flavor, cook as a courgette or try it boiled in lightly salted water for 10- 15 min., good freezer. *Source History: 1 in 1987; 2 in 1998; 2 in 2004.* **Sources: Loc, Tho.**

Salvadorean Ayote - Grown in Central America for its 1 in. long seeds which have a very thin skin, can be eaten raw or toasted or seeds are ground and added to chicken, tamales and stews, adds a distinctive delicious flavor and aroma, long vines, fruit is the shape of a calabash gourd, 8 in. tall, light yellow when ripe. *Source History: 1 in 2004.* **Sources: Red.**

Scallop, Benning's Green Tint - (Green Bush Scallop, Green Tinted Bush, Green Tinted Patty Pan, Patty Pan, Bennings Green Summer) - 47-56 days, Pale-green pie-shaped fruits with scalloped edges, pale-green fine-textured flesh, 2.5 in. deep x 3.5 in. dia., small blossom scar, semi-open bush, heavy yields, improved Patty Pan type, better yields than older bush scallops, introduced about 1914. *Source History: 26 in 1981; 21 in 1984; 20 in 1987; 19 in 1991; 19 in 1994; 21 in 1998; 25 in 2004.* **Sources: Al25, Ba8, Cl3, Fe5, Fo7, HO2, Ho13, Jor, La1, Loc, ME9, Mey, Orn, Pe6, Pr3, Sa9, SE14, Se25, So1, Ter, Und, WE10, We19, Wet, WI23.**

Scallop, Early White Bush - (White Patty Pan) - 46-60 days, Very productive closed bush plants, round flat scalloped greenish-white fruits, white when mature, 2.5-3 in. deep x 4-9 in. dia., 2.5 lbs., best at 4 in. dia., firm mild sweet greenish-white flesh, popular in East for home garden or market, pre-1722. *Source History: 38 in 1981; 34 in 1984; 33 in 1987; 34 in 1991; 29 in 1994; 32 in 1998; 38 in 2004.* **Sources: ABB, Bo19, Bu3, Bu8, CH14, Cl3, CO23, Co31, Com, Cr1, De6, DOR, Gr27, He17, HO1, Kil, La1, LO8, Loc, ME9, Mey, Mo13, Ol2, OLD, Or10, RIS, Roh, Ros, SE4, Se8, SE14, Shu, Sk2, So1, SOU, Vi4, WE10, WI23.**

Scallop, Golden Bush - 60-68 days, Open bushes, round flat golden-yellow disks with scalloped edges, 2-3 in. dia., mild firm pale-green flesh, very prolific plant. *Source History: 1 in 1981; 1 in 1984; 2 in 1987; 5 in 1991; 4 in 1994; 7 in 1998; 5 in 2004.* **Sources: He8, Lej, Sa9, Shu, WE10.**

Scallop, Patty Pan - (Patty Pan White Bush Scallop,

Early White Bush, Custard Marrow) - 54-56 days, Flattened fruits with scalloped edges, pale greenish-white when young, matures to white, fine texture, bush. *Source History: 7 in 1981; 6 in 1984; 4 in 1987; 5 in 1991; 5 in 1994; 6 in 1998; 4 in 2004.* **Sources: Bo19, Bou, Co31, He8.**

Scallop, Sunburst - 50 days, Compact, open plant habit, almost entirely golden colored fruits, sel. from a hybrid. *Source History: 1 in 1991; 2 in 2004.* **Sources: Cl3, Na6.**

Scallop, White Bush - (White Patty Pan, Custard, Early White Bush) - 47-65 days, Standard Patty Pan type, very early, pale greenish-white pie-shaped scalloped fruits, fine-grained fine-flavored white flesh, 3 in. deep x 5-7 in. dia., smooth, up to 2.5 lbs., ancient Native American variety which was grown by the Northern Indians for hundreds of years, this type was depicted by Europeans back to 1591. *Source History: 32 in 1981; 30 in 1984; 28 in 1987; 31 in 1991; 26 in 1994; 30 in 1998; 19 in 2004.* **Sources: Ada, Ba8, Ear, Ers, Fi1, Fis, HA3, HO2, Ho13, Hud, Hum, Lej, Ma19, PAG, RUP, Sau, Sh9, Wet, Wi2.**

Scallop, Yellow Bush - 49-54 days, Vigorous bush plants, large golden-yellow 7 in. dia. scalloped fruits, distinctive flavor, studies at Iowa State show it is the least attractive to squash bugs. *Source History: 8 in 1981; 6 in 1984; 7 in 1987; 10 in 1991; 8 in 1994; 7 in 1998; 10 in 2004.* **Sources: De6, HO2, Ho13, La1, LO8, ME9, RUP, SE14, Sh9, WI23.**

Scallop, Yellow Custard - 65 days, Old-timer, bush plants, med-size orange fruits, 5-6 in. flattened patty pan, good flavor, texture and color, constant fruit set when picked regularly, overly mature fruit makes a nice ornamental, rare native American squash. *Source History: 1 in 1987; 2 in 1991; 2 in 1994; 2 in 1998; 4 in 2004.* **Sources: Ba8, Gr28, In8, So9.**

Scarchuk's Supreme - (Formerly Nine-One) - 83 days, Resembles Delicata color pattern, cream with green stripes, does better than most Acorn types, dev. by Dr. Scarchuk at U of CT. *Source History: 1 in 1998; 1 in 2004.* **Sources: Sa9.**

Schooltime - 95 days, Deep orange fruit, 8-10 lbs., strong dark green handles make it ideal for school tours of PYO. *Source History: 1 in 2004.* **Sources: RIS.**

Shenot Crown of Thorns - (Ghost Gourd) - 95-100 days, Improved Crown of Thorns type, pronounced thorny warts, many patterns in white, green and orange. *Source History: 31 in 2004.* **Sources: ABB, Bu2, Bu3, CH14, Fe5, Gur, HA3, He17, HO1, HO2, Jo1, Jor, Jun, ME9, Mo13, Ni1, OLD, OSB, Ra5, Ra6, RIS, Roh, RUP, SE4, Se26, Sto, Twi, Ver, Ves, Vi5, WE10.**

Small Flat Striped - 80-90 days, Small green fruit, flattened shape, 2 in. long, dark green stripes on white and light yellow. *Source History: 12 in 2004.* **Sources: Cl3, HO1, HO2, Jor, Pep, Ra5, Ra6, RIS, RUP, Sh9, Twi, Vi5.**

Small Orange Ball - 80-100 days, Small dark orange round gourd, 2-3 in., smooth skin, tiny black handle, suitable for varnishing. *Source History: 21 in 2004.* **Sources: CH14, Cl3, Dil, Go9, HO1, HO2, Jor, ME9, Ni1, Pep, Pr9, Ra5, Ra6, RIS, RUP, SE4, Se16, Sto, Twi, Vi5, WE10.**

Small Spoon Gourd - 80-90 days, Huge yields of small, spoon shaped, orange and green striped gourds, hollow out to use like a spoon, excellent for fall decorating. *Source History: 29 in 2004.* **Sources: Ba8, Bo19, CA25, CH14, Cl3, Ers, Go9, HA3, HO1, HO2, Jor, ME9, Ni1, Pep, Pin, Ra5, Ra6, RIS, RUP, SE4, Se28, Sh9, Shu, So1, Sto, Twi, Vi5, WE10, Wi2.**

Small Sugar - (New England Pie Pumpkin, Northeast Pie, Boston Pie, Sugar Pie, Early Sugar Pie, Sweet, Sugar) - 80-118 days, Classic pie pumpkin, small round slightly flattened light-ribbed orange globes, 7-9 x 8-12 in. dia., 5-8 lbs., fine-grained stringless sweet thick yellow flesh, widely believed to be the best for canning or pies, 1863. *Source History: 99 in 1981; 88 in 1984; 79 in 1987; 79 in 1991; 73 in 1994; 76 in 1998; 81 in 2004.* **Sources: ABB, Ag7, Alb, All, Be4, Bu2, Bu3, Bu8, CA25, Ca26, CH14, Cl3, CO23, CO30, Co32, Com, Coo, Dam, Dgs, Ear, Ers, Ga1, Ge2, Goo, Gr27, Gr28, Gur, HA3, He17, Hi13, HO1, HO2, Hud, Hum, Jo1, Jor, La1, Loc, MAY, ME9, Mel, Mo13, Na6, Ni1, OLD, Ont, Or10, PAG, Pe2, Pr3, Ra5, Rev, Ri2, Ros, RUP, Sa5, Sau, SE4, Se7, Se8, SE14, Se25, Sh9, Shu, Sk2, So1, So12, SOU, Sto, Syn, TE7, Ter, Tu6, Und, Ves, Vi4, Vi5, WE10, We19, Wet, Wi2.**

Small Sugar, Asgrow Strain - 100 days, Selection of Small Sugar with a superior handle, med-size vines, 6 x 7 in. dia. fruits weigh 5-6 lbs. *Source History: 1 in 1991; 1 in 1994; 1 in 1998; 1 in 2004.* **Sources: RUP.**

Sonoran Hulless - 95 days, Moderate vining habit, tasty olive-green seeds. *Source History: 1 in 1998; 1 in 2004.* **Sources: So12.**

Southern Miner - 124 days, Ridged hard shell, mottled green to yellow, 1 in. yellow flesh. *Source History: 1 in 1998; 1 in 2004.* **Sources: Sa9.**

Spookie - (Spookie Pie, Deep Sugar Pie, Improved Sugar Pie) - 90-110 days, Sugar Pie x Jack-o'-Lantern, heavier yield, deeper fruit shape, more uniform, 6-7 x 5-6 in. dia., 6-10 lbs., smooth hard red-orange skin, excel. quality, fine-textured sweet thick yellow-orange flesh, great for pies. *Source History: 18 in 1981; 16 in 1984; 12 in 1987; 7 in 1991; 11 in 1994; 13 in 1998; 11 in 2004.* **Sources: CH14, CO30, HO2, Jor, ME9, Mey, RIS, Sa9, SE14, Tt2, WE10.**

Storage - 90 days, Small sugar pie type. *Source History: 1 in 1994; 1 in 1998; 1 in 2004.* **Sources: Sa9.**

Straightneck - (Early Straightneck, Yellow Straightneck, Early Yellow Straightneck) - 46-53 days, Heavy-yielding bush plants, creamy-yellow fruits, edible from 4-10 in. long, excel. flavor and quality, pre-1938. *Source History: 11 in 1981; 7 in 1984; 6 in 1987; 7 in 1991; 5 in 1994; 6 in 1998; 6 in 2004.* **Sources: BAL, Bo17, But, HA3, Sa9, So1.**

Straightneck, Early Prolific - (Prolific Straightneck, Yellow Straightneck, Golden Straightneck, Burpee Early Prolific Straightneck) - 42-56 days, AAS/1938, uniform lemon-yellow lightly-warted club-shaped fruits, 10-14 x 2.5-3.5 in. when mature, best at 4-7 in., firm fine-grained thick flesh, excel. quality, popular in the Northeast and Northern areas. *Source History: 79 in 1981; 73 in 1984; 72 in 1987; 61 in 1991; 60 in 1994; 62 in 1998; 58 in 2004.* **Sources: Ada, All, Ba8, Be4, Bo19, Bu2, Bu3, Bu8, CA25, CH14,**

CO23, Co31, Co32, Com, Cr1, De6, DOR, Ers, Fi1, Fis, Ga1, Ge2, Gur, He17, Hi6, HO1, HPS, Jor, Kil, La1, LO8, Loc, Ma19, ME9, Mel, Mey, MIC, Mo13, MOU, Or10, PAG, RIS, Roh, Ros, RUP, Sau, Scf, SE4, SE14, Se25, Sh9, Shu, SOU, TE7, Vi4, WE10, Wet, WI23.

Straightneck, Early Prolific Improved - Unusually attractive straight fruits, slightly smaller at stem end, light-cream, usually eaten when 4-12 in. long, bush plants can be closely spaced, heavy yields. *Source History: 2 in 1981; 2 in 1984; 2 in 1987; 1 in 1991; 1 in 1994; 1 in 1998; 1 in 2004.* **Sources: Wi2.**

Straightneck, Saffron Prolific - 43 days, Sturdy, open plant bears uniform, tapered, smooth fruit, good for commercial trade and a great improvement for the home garden, PVP application abandoned. *Source History: 1 in 1998; 4 in 2004.* **Sources: Bu2, HO1, OLD, Wi2.**

Streaker Jack - 90 days, Orange and green striped fruit contain very large, dark green naked seeds that are roasted, popped or crushed for oil, research shows that pumpkin seeds are rich in vitamin E, zinc, carotenoids and photosterines, CV Ter sel. from seed received from one of the world's leading producers of Styrian pumpkin seed oil in Austria. *Source History: 1 in 2004.* **Sources: Ter.**

Striped Pear - 80-90 days, Prolific 8-12' vines, green and white stripes run lengthwise along the pear shaped gourd. *Source History: 6 in 2004.* **Sources: ME9, Ra5, Ra6, Se16, Sto, Vi5.**

Styrian Hulless - (Naked Seeded) - 90-120 days, Vining habit, large green pumpkins up to 10-20 lbs., seed is delicious eaten raw or toasted, primarily pressed for high-quality oil, unique variety from Styria, a region in Austria. *Source History: 2 in 1991; 2 in 1994; 2 in 1998; 9 in 2004.* **Sources: Ho13, Hud, Rev, Ri2, Sa5, Sa9, Se17, So12, Und.**

Sugar Loaf - (Tan Delicata) - 100 days, Selection from Delicata with dark-green and tan stripes, trailing plants run about 12 ft., yellow-orange fine-textured med-dry flesh, mild sweet flavor, stored quality remains high, for home garden and commercial use, dev. by Dr. James R. Baggett of Oregon State Univ., 1990. *Source History: 5 in 1991; 14 in 1994; 11 in 1998; 13 in 2004.* **Sources: Ba8, Bu3, He8, HO2, Jor, Ni1, OSB, Pe6, RUP, SE4, Se26, Sk2, We19.**

Sugar Pie - (New England Pie, Early Sugar, Early Sweet, Early Sweet Sugar, Small Sugar, Small Sweet) - 90-115 days, Small round pumpkins, flattened at ends, very finest quality, 6 x 7 in. dia., 6-8 lbs., thick sweet orange-yellow fine-grained flesh, best var. for general use, excel. for pies and canning, good keeper, described by Fearing Burr in 1863. *Source History: 29 in 1981; 26 in 1984; 29 in 1987; 26 in 1991; 29 in 1994; 32 in 1998; 25 in 2004.* **Sources: Ba8, Bo17, Bo19, Cr1, De6, Dil, DOR, Fo7, Gou, Hal, Jun, Kil, Lej, Lin, LO8, Ma19, Mey, MOU, Red, Ri12, RIS, Roh, Sa9, Se26, WI23.**

Sweet Dumpling - (Vegetable Gourd) - 83-100 days, Oriental variety, compact 5-6 ft. plant, teacup-shaped ivory-colored fruits striped and mottled dark-green, 3-4 in. dia., 7 oz., tender sweet orange flesh, 10 fruits per plant, conc. set, suitable for stuffing, requires no curing, stores 3-4 months. *Source History: 13 in 1981; 11 in 1984; 13 in 1987; 29 in 1991; 34 in 1994; 34 in 1998; 49 in 2004.* **Sources: Ba8, Be4, Bo19, Bu2, Bu8, CH14, CO30, De6, Ers, Fe5, Fi1, Fo13, Ge2, Gr28, Ha5, HO2, Hum, Jo1, Jor, Lej, Loc, Ma19, ME9, Mo13, OSB, Pe6, Pin, Re8, Rev, Ri2, Ros, RUP, SAK, SE4, Se7, Se8, SE14, Sh9, Shu, So9, Sto, Sw9, Syn, Ter, Ver, Ves, WE10, We19, WI23.**

Sweet Potato - (Delicata) - 100 days, Long keeping small variety with cream colored skin with dark green stripes, sweet flavor. *Source History: 1 in 1981; 1 in 1984; 1 in 1987; 4 in 1991; 3 in 1994; 2 in 1998; 1 in 2004.* **Sources: He8.**

Sweetie Pie - 110 days, Miniature pumpkin grown from stock seed imported from China where it is considered a delicacy, flattened deeply ribbed fruits, 1.75 in. deep x 3 in. dia., 5 oz., edible and decorative. *Source History: 2 in 1987; 2 in 1991; 1 in 1994; 1 in 1998; 2 in 2004.* **Sources: Ca26, Sto.**

Table Dainty - 65-70 days, Vining type, blocky green and yellow striped zucchini type fruit, 6 in. long, productive, produces many male flowers which are good for frying or eating fresh, dev. in 1909. *Source History: 1 in 1998; 1 in 2004.* **Sources: Sa9.**

Tarahumara - Pumpkin-shaped fruits, green with beige ribs, from LaBufa, deep in Batopilas Canyon, Chihuahua, Mexico. *Source History: 1 in 1991; 1 in 1994; 1 in 1998; 1 in 2004.* **Sources: Na2.**

Tarahumara Pumpkin - Round to slightly elongated fruit with protruding ribs at stem end, matures to bright yellow skin color, from Dr. Robert Bye of the Jardin Botanico in Mexico City. *Source History: 1 in 2004.* **Sources: Na2.**

Tatume - 50-61 days, Summer squash popular in Mexico and Texas, round/oval med-dark green fruits with faint stripes, 5-7 x 2-2.5 in. dia., firm, fine flavor, vigorous spreading vines. *Source History: 6 in 1981; 6 in 1984; 9 in 1987; 9 in 1991; 6 in 1994; 7 in 1998; 8 in 2004.* **Sources: ABB, Bu3, LO8, Sa9, Se8, Syn, Te4, Vi4.**

Ten Commandments - 95 days, Softball size fruit with five pairs of protruding prongs pointing toward the blossom end, nice mix of bright striped, mottled, multicolored fruit. *Source History: 1 in 2004.* **Sources: Se16.**

Tepehuan - Fruits come in a variety of shapes and colors from dark green to stripes, long season, may not produce seed in northern climates. *Source History: 1 in 1991; 1 in 1994; 1 in 1998; 1 in 2004.* **Sources: Na2.**

Thelma Sanders' Sweet Potato - 85-96 days, Acorn-shaped fruits up to 6 in. long, ripens cream > light gold, golden yellow flesh, superb flavor, family heirloom from Thelma Sanders in Adair County, Missouri. *Source History: 1 in 1991; 1 in 1994; 1 in 1998; 2 in 2004.* **Sources: Se16, So1.**

Tom Fox - 110 days, Long vines produce fruits in a variety of shapes, well-ribbed fruit, deep-orange skin, very large dark-green strong handles, developed by New Hampshire farmer Tom Fox, PVP 1997 Jo1. *Source History: 1 in 1994; 1 in 1998; 3 in 2004.* **Sources: Jo1, Jor, RIS.**

Tonda di Piacenza - (Round of Parma) - 55 days, Round zucchini from northern Italy, dark green skin, harvest at baseball size or smaller, freezes well. *Source History: 2 in 2004.* **Sources: CA25, Se24.**

Triple Treat - 110 days, Round and bright-orange, 10 in. dia., 6-8 lbs., good carving size, thick deep-orange flesh for pies and puddings, hulless seeds for roasting or raw, keeps well, PVP expired 1994. *Source History: 5 in 1981; 5 in 1984; 6 in 1987; 8 in 1991; 8 in 1994; 7 in 1998; 6 in 2004.* **Sources: Bu2, Ers, Hi13, HO2, Jor, RUP.**

Turner Family - 95-103 days, Ribbed, firm shell, deep orange skin, strong stem, 10-15 lbs. *Source History: 2 in 1998; 2 in 2004.* **Sources: Rev, Sa9.**

Turner's Select - 95 days, Prolific vines bear 5-6 round and barrel shaped fruits each, 6+ lbs., sweet moist flesh with flavor that compares to winter squash, stores well. *Source History: 1 in 2004.* **Sources: So12.**

Tweet-ee-oo Bakers - *Source History: 1 in 2004.* **Sources: Se17.**

Uconn - 86 days, Bush plant, fist size acorn fruit. *Source History: 1 in 1998; 1 in 2004.* **Sources: Sa9.**

Veg. Marrow, Bush - (White Bush Vegetable Marrow) - 60-75 days, Snow-white summer squash, grows to about 12 in. long, harvest at 8 in., early maturity, a great favorite with the English. *Source History: 4 in 1981; 4 in 1984; 4 in 1987; 6 in 1991; 5 in 1994; 7 in 1998; 4 in 2004.* **Sources: Dam, Sto, WE10, We19.**

Veg. Marrow, Green Bush - (Long Green Bush Vegetable Marrow) - 65 days, Dark and light-green striped skin, pale greenish- white flesh, 15-16 x 5-6 in. dia., mild flavor, good market variety. *Source History: 6 in 1981; 4 in 1984; 3 in 1987; 3 in 1991; 5 in 1994; 3 in 1998; 2 in 2004.* **Sources: Tho, WE10.**

Veg. Marrow, Italian - (Cocozelle) - 57-60 days, Trailing vines, cylindrical dark-green 3 lb. fruits, best used at 8 in. length, greenish-white flesh, excel. quality, old variety from Italy. *Source History: 1 in 1981; 1 in 1984; 1 in 1987; 2 in 1991; 2 in 1994; 2 in 1998; 2 in 2004.* **Sources: All, He8.**

Veg. Marrow, Lebanese - (Lebanese White Bush, White Bush Vegetable Marrow, Cousa, Lebanese Zucchini) - 45-50 days, Rare Lebanese var., oblong fruits, pale-green to cream-white skin, larger at blossom end, harvest when 7 in. long, tasty fried or baked, good for fresh market. *Source History: 3 in 1981; 3 in 1984; 3 in 1987; 3 in 1991; 4 in 1994; 6 in 1998; 5 in 2004.* **Sources: Ba8, Fe5, Sh9, Syn, WE10.**

Veg. Marrow, Long White Bush - (Long White Marrow) - 43-60 days, Prolific bush plants, tapered creamy-white 6-18 in. fruits, creamy-white fine-textured flesh, white zucchini type, very abundant, popular in Italy and England. *Source History: 17 in 1981; 16 in 1984; 12 in 1987; 8 in 1991; 7 in 1994; 6 in 1998; 4 in 2004.* **Sources: Alb, Hal, Hud, Ont.**

Vegetable Marrow - 43-65 days, Bush zucchini type, fast-growing, very prolific, produces fruit on nearly every leaf node, several different types: pale-green, creamy-white, green and yellow striped. *Source History: 7 in 1981; 7 in 1984; 7 in 1987; 4 in 1991; 5 in 1994; 4 in 1998; 1 in 2004.* **Sources: Lin.**

Vegetable Spaghetti - (Spaghetti Squash) - 70-115 days, Vigorous spreading vines, rounded cylindrical buff-tan fruits, 8-12 in. long x 5-6 in. dia., 5-7 per plant, ready to harvest when skin color changes from cream to buff, keeps up to six months, boil whole or bake, flesh peels out in strands of low calorie spaghetti, Manchuria, 1890. *Source History: 128 in 1981; 119 in 1984; 121 in 1987; 118 in 1991; 108 in 1994; 109 in 1998; 105 in 2004.* **Sources: ABB, Alb, All, Ba8, Be4, Bo17, Bo19, Bou, Bu2, Bu3, Bu8, But, CA25, CH14, Cl3, CO23, CO30, Com, Dam, De6, Dgs, DOR, Ear, Ers, Fe5, Fil, Fis, Fo7, Fo13, Ge2, GLO, Goo, Gr27, Gr29, GRI, Gur, HA3, Ha5, Hal, He8, Hig, HO1, HO2, Hum, Jo1, Jor, Kil, Kit, La1, Lej, Lin, LO8, Loc, Ma19, MAY, ME9, Mel, Mey, Mo13, MOU, Ni1, OLD, Ont, OSB, PAG, Pin, Pr3, Pr9, Ra5, Red, Rev, Ri2, Ri12, RIS, Roh, Ros, RUP, SAK, Sau, SE4, Se7, SE14, Se25, Se26, Se28, Sh9, Shu, Si5, Sk2, So1, So25, Sto, Te4, Ter, Tu6, Twi, Up2, Ves, Vi4, WE10, We19, Wet, Wi2, WI23, Wo8.**

Wee-B-Little - (Wee Be Little) - 85-110 days, AAS/1999, vines 6-8 ft., small 1 lb. fruit, 3 x 3.5 in. globe shape, a true miniature pumpkin, spineless, sturdy stems, smooth skin, yellow > orange, decorative, PVP 1999 Novartis Seeds, Inc. *Source History: 40 in 2004.* **Sources: BAL, Bu3, CA25, CH14, Cl3, Co31, De6, Dgs, Ers, Fe5, Ga1, HO2, Jo1, Jor, Jun, Ki4, Lin, Ma19, MAY, ME9, Mey, MIC, MOU, OLD, OSB, PAG, Par, Pin, RIS, Roh, RUP, Scf, SE4, SE14, SGF, SOU, Sto, Tt2, Twi, Ves.**

White Scallop Rattle, Sacred - 95 days, Mature 7 in. gourds resemble white sunbursts, traditionally carved or painted for Native American ceremonial rattles, extremely rare. *Source History: 3 in 2004.* **Sources: Ra5, Red, Und.**

Winter Luxury - (Winter Luxury Pie, Queen Pumpkin) - 85-100 days, Heavily netted golden russet pumpkin, 10 in. dia., 8 lbs., excel. keeper, 1893. *Source History: 1 in 1981; 1 in 1984; 1 in 1987; 2 in 1991; 2 in 1994; 5 in 1998; 4 in 2004.* **Sources: In8, Jun, Sa9, Se17.**

Wood's Earliest Prolific - (Wood's Early Prolific) - 50 days, Smaller fruit than standard bush scallop squash with thicker flesh and fewer scallops, produces fruit right up until frost with continuous picking, heirloom first offered by T. W. Woods and Sons, Richmond, Virginia, 1899. *Source History: 1 in 1987; 4 in 1991; 3 in 1994; 2 in 1998; 5 in 2004.* **Sources: Co32, Gr28, Pr9, Se16, So9.**

Xochitian Pueblo - 83 days, Hard shell, 1 in. yellow flesh, high yields, Mexico. *Source History: 1 in 1998; 1 in 2004.* **Sources: Sa9.**

Yellow Large Paris - (Jaune Gros de Paris) - 110 days, Large yellow fruits up to 25 lbs., used in France to make pumpkin soup, very good for pies. *Source History: 1 in 1981; 1 in 1987; 1 in 1991; 1 in 1994; 1 in 2004.* **Sources: ME9.**

Young's Beauty - 95-120 days, Large New England Pie type, 7.5 x 8 in. dia., 10-12 lbs., hard med-ribbed dark-orange skin, thick yellow-orange flesh, fine for Halloween, high yields, uniform. *Source History: 13 in 1981; 7 in 1984; 6 in 1987; 11 in 1991; 12 in 1994; 6 in 1998; 9 in 2004.* **Sources: Ba8, Fe5, HO2, ME9, Pe2, Pr3, Rev, Sa9, SE14.**

Yugoslavian Finger Fruit - 100 days, Bush habit, ten finger-like points on large, fluted acorn type fruits, 4-5 lbs., cream colored skin, used when young as summer squash, tasty as winter type and ornamental, very rare. *Source History: 2 in 2004.* **Sources: Ba8, Hud, Rev.**

Zapallito del Tronco - (Redondo, South American Round, Redondo de Tronco Verde, Calavacita) - 60 days, Bushy upright plant, acorn squash type 2.5-3.5 in. dia. fruit are a wonderful cross between winter and summer squash, edible rind, seed cavity can be eaten just like cucumber, excel. keeper and shipper. *Source History: 1 in 1981; 1 in 1984; 1 in 1987; 2 in 1991; 4 in 1994; 1 in 1998; 3 in 2004.* **Sources: Bou, Ga1, Se8.**

Zucchini - (Italian, Italian Marrow, Cocozelle, Bush Marrow) - 50-65 days, Dark-green smooth cylindrical firm tender fruits, highly productive bush plants, harvest at 6-8 in., for home and market gardens. *Source History: 25 in 1981; 24 in 1984; 18 in 1987; 8 in 1991; 8 in 1994; 10 in 1998; 6 in 2004.* **Sources: Ada, Be4, But, Co31, Ont, Sau.**

Zucchini, Alberello Di Sarzana - (Little Tree of Sarzano) - 50-55 days, Slim, speckled, medium green fruit, slight ribbing, tender, delicious, delicate flavor, PM res., rarely available in the U.S. *Source History: 2 in 2004.* **Sources: Hud, Se24.**

Zucchini, Black - (Black Italian Marrow, Black Shell Zucchini, Dark Zucchini) - 44-64 days, Closed spiny everbearing bush plant, dark green-black long straight cylindrical fruits, slight ridges, best when 6-8 in., crisp fine-textured greenish-white flesh, vigorous, first offered by the Jerome B. Rice Seed Company of Cambridge, NY in 1931. *Source History: 61 in 1981; 49 in 1984; 53 in 1987; 39 in 1991; 31 in 1994; 33 in 1998; 31 in 2004.* **Sources: All, Bu3, CA25, DOR, Ear, Fe5, Fi1, Fo7, He8, He17, HPS, La1, Lin, Ma13, Mel, Mey, Mo13, Pin, Ri12, RIS, RUP, SE14, Se25, Se26, Sh9, Shu, Sk2, Syn, Vi4, WE10, Wet.**

Zucchini, Black Beauty - (Black Zucchini) - 60-63 days, Semi-upright open plants, blocky glossy black-green fruits, 6-8 in. long x 2 in. dia., very tender firm creamy-white flesh, fine flavor, very early and productive, dark-green at usable stage, almost black at maturity, excellent freezer, introduced to U.S. markets in the 1920s, offered commercially in the 1930s. *Source History: 18 in 1981; 16 in 1984; 16 in 1987; 25 in 1991; 42 in 1994; 40 in 1998; 59 in 2004.* **Sources: Ag7, Ba8, Bo17, Bou, Bu2, Bu8, CH14, CO23, Co32, Com, Coo, Cr1, De6, Dom, DOR, Ers, Ga1, Ga7, Ge2, Gr27, Hal, Hi6, Hig, HO1, Hum, Jor, Jun, La1, Loc, Ma19, MAY, Me7, ME9, MOU, OLD, Or10, PAG, Pe2, Roh, Scf, SE4, Se7, SE14, Se25, SGF, So1, SOU, Te4, TE7, Ter, Und, Up2, Ves, Vi4, WE10, We19, Wet, Wi2, WI23.**

Zucchini, Black Magic - 44-62 days, Semi-spineless sturdy compact plants, 1.5-2 ft. tall x 3-3.5 ft. wide, very uniform straight smooth glossy black-green fruits, very small seed cavity, high yields, PVP expired 1996. *Source History: 13 in 1981; 13 in 1984; 14 in 1987; 10 in 1991; 5 in 1994; 3 in 1998; 1 in 2004.* **Sources: Gur.**

Zucchini, Black Shelled - Older running variety, fruits still good when quite large. *Source History: 1 in 1991; 1 in 1994; 1 in 1998; 1 in 2004.* **Sources: So9.**

Zucchini, Bolognese - 55-60 days, Harvest at 3-4 in. size, short thick fruit, med. green skin with light speckles, no ribs, same great taste and texture as Italian zucchini. *Source History: 1 in 2004.* **Sources: Se24.**

Zucchini, Burpee's Fordhook - 57 days, AAS, vigorous prolific bush plants, deep blackish-green cylindrical long slender smooth fruits, white flesh that freezes well, best at 8-12 in., slightly curved. *Source History: 4 in 1981; 3 in 1984; 3 in 1987; 3 in 1991; 2 in 1994; 2 in 1998; 2 in 2004.* **Sources: Bu2, Hi13.**

Zucchini, Burpee's Golden - 54 days, Glossy golden-yellow med-long slender cylindrical fruits, vigorous compact bush, dev. by Burpee's breeders from material from Dr. Shifress of Rutgers, PVP expired 1994. *Source History: 4 in 1981; 4 in 1984; 4 in 1987; 1 in 1991; 1 in 1994; 1 in 1998; 3 in 2004.* **Sources: Bu2, Hi13, HO2.**

Zucchini, Costata Romanesca - (Costa Romanesco, Roman Ridged Zucchini) - 60 days, Unusual pale-green fruits with slight ridges, name means "Ribbed Roman", a favorite variety in Italy where it is picked 6 in. long or less with flowers still on and fried whole, sweet, mildly nutty flavor, longer vines than other summer squash varieties. *Source History: 2 in 1987; 2 in 1991; 2 in 1994; 6 in 1998; 22 in 2004.* **Sources: Ba8, Fe5, Hi6, Ho13, Jo1, Na6, Orn, Pe1, Pe2, Pr3, Re6, Red, Rev, Sa9, SE14, Se24, Se26, So1, So12, So25, Tu6, WE10.**

Zucchini, Cristoforo - 54 days, Large, open plants bear cylindrical light green fruits mottled with darker stripes, high yields, selection of Genovese squash, Italian heirloom. *Source History: 1 in 2004.* **Sources: Coo.**

Zucchini, Dark Green - (Green Zucchini) - 45-65 days, Early vigorous bush plants, straight smooth dark-green mottled fruits, best at 6-8 in., pale greenish-white firm flesh, prolific, heavy yields, fine flavor--stronger than Grey Zucchini. *Source History: 26 in 1981; 23 in 1984; 23 in 1987; 30 in 1991; 30 in 1994; 33 in 1998; 30 in 2004.* **Sources: Alb, Bo19, Bu2, CO23, Dam, DOR, Ga1, HA3, He8, Hi6, HO2, HPS, Hud, Kil, La1, Loc, Mcf, MIC, Pe2, Pla, Red, Ri2, Ros, RUP, SE14, Shu, So1, TE7, WE10, WI23.**

Zucchini, French White Bush - 50-55 days, Vigorous white bush zucchini from France, nice mild flavor, firm meat, few seeds, at peak flavor when 6-10 in. long, versatile, good customer reaction. *Source History: 3 in 1981; 2 in 1984; 2 in 1987; 2 in 1991; 4 in 1994; 2 in 1998; 2 in 2004.* **Sources: Ho13, Ni1.**

Zucchini, Genovese - 50-55 days, Large bushy plants, pale yellow-green fruit, no ribbing, very nice flavor, early and productive, many blossoms for cooking, rare. *Source History: 4 in 2004.* **Sources: Hud, SE14, Se24, WE10.**

Zucchini, Gold Satin - 60 days, First relatively spineless, open-pollinated, golden zucchini with rich flavor, from PSR Breeding Program. *Source History: 1 in 1994; 1 in 1998; 1 in 2004.* **Sources: Pe6.**

Zucchini, Golden - (Golden Bush) - 50-55 days, Bush plants, bright-golden cylindrical fruits, medium long and slender, best when picked at 8-10 in. length, excellent flavor, not as productive as the greens, Seeds of Change selection. *Source History: 2 in 1981; 2 in 1984; 2 in 1987; 3 in 1991; 2 in 1994; 7 in 1998; 16 in 2004.* **Sources: Ba8, De6, Ers, Fis, He8, Ho13, PAG, Red, RUP, Se7, SE14, Shu, Sk2, So9, WE10, WI23.**

Zucchini, Green Satin - 60 days, Relatively spineless, open-pollinated, green zucchini with a sweeter flavor than some, home garden variety, from the PSR Breeding Program. *Source History: 1 in 1994; 1 in 1998; 1 in 2004.* **Sources: Pe6.**

Zucchini, Grey - 42-65 days, Med-green skin mottled and flecked with grey, almost cylindrical, tapers slightly to stem end, pick at 6-8 in., small seeds, open semi-spiny bush plant, heavy early yields, no conc. fruit set so better home garden var. than the hybrids, popular in Southwest and Mexico, 1957. *Source History: 39 in 1981; 35 in 1984; 31 in 1987; 26 in 1991; 25 in 1994; 28 in 1998; 23 in 2004.* **Sources: ABB, Ba8, Bu3, CH14, CO23, DOR, Ers, HA20, He8, Hig, HO1, Jor, Kil, LO8, OLD, Ros, SE14, Se26, Shu, So1, Vi4, WE10, Wi2.**

Zucchini, Grise de Algiers - 56 days, Light grey-green fruit, delicious flavor, best when used at 4-6 in. stage, popular in French Africa. *Source History: 1 in 2004.* **Sources: Coo.**

Zucchini, Lungo Bianco - Bush plants, light green-cream skin, mild sweet flesh, from Italy. *Source History: 3 in 2004.* **Sources: Ba8, Red, SE14.**

Zucchini, Midnight - 54 days, Vigorous, everbearing bush plant, long, straight, cylindrical dark-green fruit with slight ridge, 6-8 in. long. *Source History: 1 in 1994; 1 in 1998; 1 in 2004.* **Sources: Eco.**

Zucchini, Milano Black - 40-50 days, Traditional Italian zucchini, dwarf bushes, black skinned fruit, sweet and most flavorful when picked when no longer than 8 in. long. *Source History: 1 in 2004.* **Sources: Ki4.**

Zucchini, Nano Verde di Milano - Long fruit, deep green skin, green-white flesh. *Source History: 1 in 1987; 1 in 1991; 1 in 1994; 1 in 1998; 2 in 2004.* **Sources: Ber, CA25.**

Zucchini, Nimba - 50 days, Early, hardy variety from Poland, bush plant, medium green skin, older fruits have green striped skin, tasty fried, frequent pickings increases yields, good for cool short season areas. *Source History: 6 in 2004.* **Sources: Ba8, He8, Hud, Pr3, Pr9, Se16.**

Zucchini, Ronde de Nice - (Round French Zucchini, Tondo Nizza) - 45-50 days, Round zucchini, exceptional flavor, delicate skin bruises easily, vigorous fast-growing plant, serve steamed at 1 in. dia. or stuffed at 4 in., sought after in France. *Source History: 2 in 1981; 2 in 1984; 2 in 1987; 4 in 1991; 13 in 1994; 12 in 1998; 26 in 2004.* **Sources: Ba8, Bou, Coo, DOR, Gou, Gr28, HO2, Ho13, Ki4, Lej, Orn, Pe2, Re6, Sa9, Se7, Se8, SE14, Se24, Se25, So25, Syn, Tu6, Ver, VI3, WE10, We19.**

Zucchini, Round - (Round Bush Zucchini) - 45 days, Bush, mottled green deeply cut leaves, 2.5 in. dia. greenish-grey globes, small blossom scar, creamy-yellow solid flesh, productive, for home gardens. *Source History: 4 in 1981; 3 in 1984; 6 in 1987; 8 in 1991; 9 in 1994; 15 in 1998; 22 in 2004.* **Sources: Bo17, Bu3, CO23, Com, Ers, Fis, He8, HO1, Hud, Hum, Jor, LO8, Or10, Pr3, RIS, Roh, SE14, Se26, Shu, Vi4, WE10, WI23.**

Zucchini, Sakiz - 50 days, Heavy yields of light-green fruit, tender and delicious, very good flavor, tolerates hot and dry conditions, rare. *Source History: 1 in 1991; 1 in 1994; 1 in 2004.* **Sources: Sk2.**

Zucchini, Siciliano - Very long, light green squash. *Source History: 1 in 1987; 1 in 1991; 1 in 1994; 1 in 1998; 1 in 2004.* **Sources: Ber.**

Zucchini, Small Green Algerian - Light green skin striped with darker green. *Source History: 1 in 2004.* **Sources: Pr3.**

Zucchini, Striato d'Italia - (Italian Striped) - 50-60 days, Large, bushy, open plants, many flowers, light and dark green striped fruit, 8-9 in. long, slightly thicker at blossom end, harvest when small, superb flavor and texture, long season, from Italy. *Source History: 1 in 1987; 1 in 1991; 1 in 1994; 2 in 1998; 12 in 2004.* **Sources: Ag7, Ba8, Ber, Bo19, Dom, Ki4, Mcf, Pin, Red, SE14, Se24, WE10.**

Zucchini, Verde d'Italia - (Green of Italy) - 50-60 days, Large plant, many flowers, med. green skin speckled with white, ribbed fruit, good taste, produces over a long season. *Source History: 1 in 2004.* **Sources: Se24.**

Zucchini, White Egyptian - 50-55 days, Bush plant, oblong, pale green to white fruit, excellent culinary choice for home gardens. *Source History: 1 in 1987; 1 in 1991; 2 in 1994; 1 in 2004.* **Sources: Se7.**

Zucchini, Yellow - 55 days, Bush plant, bright yellow skin, creamy flesh, best when picked at 7 in. *Source History: 1 in 1998; 1 in 2004.* **Sources: SE14.**

Varieties dropped since 1981 (and the year dropped):

Abobrinha Redonda de Tronco (1991), *Acoma* (2004), *Acorn (Bush)* (1991), *Acorn (Ebony Bush)* (1991), *Acorn (Royal Bush)* (1991), *Acorn, Chestnut* (1994), *Acorn, Des Moines* (1994), *Acorn, Golden* (2004), *Acorn, Jersey Golden* (2004), *Acorn, Showell White Bush* (1994), *Acorn, Snow White* (2004), *Acorn, Swan White* (2004), *Argentina Summer* (1994), *Arlesa* (2004), *Austrian Bush* (2004), *Baby Straightneck* (1991), *Benning's Small Scar Scallop* (1991), *Caserta Improved* (1987), *Cocozelle (Striped Bush)* (1991), *Cocozelle, Baby* (1994), *Cocozelle, Green* (1998), *Cocozelle, Long Dark* (1998), *Crookneck, Confederate Gold* (1994), *Crookneck, Dixie* (2004), *Crookneck, Giant Summer* (1994), *Cupid* (1984), *Early Northern Pie* (1994), *Early Summer K Str. Crookneck* (1987), *Early Sweet Sugar Pumpkin* (1994), *Eat-All* (2004), *Gem (Rolet Strain)* (1991), *Gill's Golden Pippin* (2004), *Golden Acorn* (1991), *Gourmet Globe* (1991), *Green Bush Imp. Veg. Marrow* (1982), *Green Striped Veg. Marrow* (1991), *Green Trailing Veg. Marrow* (1987), *Hallo-Queen* (1998), *Hopi* (1998), *Hopi No. 1* (1998), *Hopi No. 2* (1994), *Hopi No. 5* (1998), *Huicha* (2004), *Ichabod* (2004), *Idaho* (1994), *Ingot* (1998), *Jack-O'-Lantern, Hopi White* (2004), *Jackpot (O.P.)* (1994), *Kumi Kumi* (2004), *Kuta* (1991), *Large Table Queen Acorn* (1987), *Lime Green Zucchini* (1987), *Little Boo* (1998), *Little Gem A* (1998), *Little Lantern* (2004), *Long Striped Cocozelle* (1987), *Long*

Wht. Trailing Veg. Marrow (1983), *Lunghisimo Bianco di Palermo* (1987), *Mammoth White Bush Scallop* (2004), *Manteca Large White* (2004), *Marrow, Warted* (1998), *Mini-Jack Pumpkin* (2004), *Minnesota Sweet* (2004), *Naked Seeded* (1998), *Potimarron* (2004), *Precoce Maraiche Re* (1998), *Pugliese* (1998), *Ranger Crookneck* (1991), *Rocky Mountain Pie* (2004), *Rolet* (1998), *Royal Knight* (1994), *Seneca Gourmet Zucchini* (1987), *Shawnee* (1998), *Small Green Algerian* (1991), *Small Green Algerian* (1991), *Small Table Queen Acorn* (1991), *Snack 'R Jack* (2004), *Southern Chihuahua Arrota* (1991), *Storr's Green* (1987), *Straightneck, Connecticut* (1994), *Straightneck, Giant Summer* (1994), *Straito d'Italia* (1987), *Streaker* (1987), *Sugar Baby* (1998), *Summer Zucchini* (1991), *Super Acorn* (1991), *Super Select Zucchini* (1991), *Tahitian Small* (2004), *Tallman* (2004), *Thomas Halloween* (2004), *Tours Pumpkin* (1991), *Tremendous Green Veg. Marrow* (1983), *Tricky Jack* (1987), *Veg. Marrow, Danish Strain* (2004), *Veg. Marrow, English* (2004), *Wee Willie* (1991), *White Lebanese Zucchini* (1991), *White Zucchini* (1987), *Winter Acorn* (1994), *Winter Queen* (1998), *Worcester Indian* (1998), *Zapallo Italiano* (1991), *Zucchini (Brunino)* (1991), *Zucchini (Select)* (1991), *Zucchini, Cereberus Bush* (2004), *Zucchini, Gourmet Globe* (1994), *Zucchini, Odessa* (2004), *Zucchini, True French* (1998), *Zucchini, Wes Brown's Vining* (1994), *Zucchini, White* (2004), *Zucchini, Zuboda* (2004).

Squash / Other Species
Cucurbita spp.

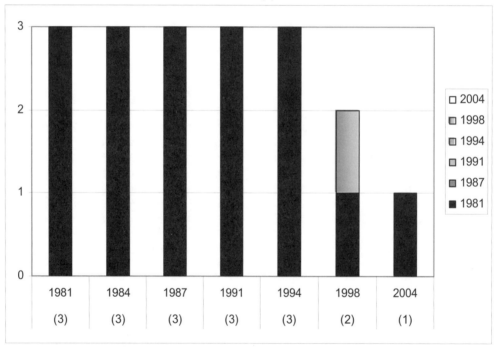

Number of varieties listed in 1981 and still available in 2004: 3 to 1 (33%).
Overall change in number of varieties from 1981 to 2004: 3 to 1 (-66%).
Number of 2004 varieties available from 1 or 2 sources: 1 out of 1 (100%).

<u>Buffalo Gourd</u> - (Mock Orange, Calabazilla, Missouri Fetid Gourd) - *Cucurbita foetidissima*, vigorous perennial wild *cucurbit*, yellow round hard-shelled 3 in. fruits, flat white seeds are rich in oil and protein, plant has a bad smell, highly drought tolerant, huge starch root can grow 15 ft. deep and 1 ft. across, tapered with some smaller side roots, grows wild in Southwest and Mexico, possible new oil-seed crop. *Source History: 1 in 1981; 1 in 1984; 3 in 1987; 3 in 1991; 2 in 1994; 1 in 2004.* **Sources: Rev.**

Varieties dropped since 1981 (and the year dropped):

Chichicoyota (1998), *Chilacayote* (2004), *Mexican Wild Pumpkin* (2004).

Sunflower / Ornamental
Helianthus spp.

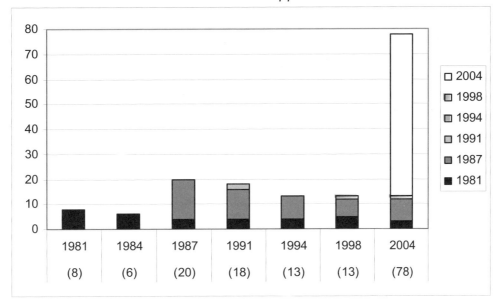

Number of varieties listed in 1981 and still available in 2004: 8 to 3 (37%).
Overall change in number of varieties from 1981 to 2004: 8 to 78 (975%).
Number of 2004 varieties available from 1 or 2 sources: 46 out of 78 (59%).

Amber Eyes - Small flowers, amber rings. *Source History: 1 in 2004.* **Sources: So9.**

Apache Brown Striped - Med-size heads, does fine in Tucson, from San Carlos Reservation, Arizona. *Source History: 1 in 1987; 1 in 1991; 1 in 1994; 1 in 1998; 1 in 2004.* **Sources: Na2.**

Autumn Beauty - 90-110 days, Novel multiflorous type, 7 ft. plants, single flowers 4-8 in. dia., bloom over a long season, sulphur-yellow, copper, bronze and purple shades, seed is very attractive to birds, annual. *Source History: 1 in 1987; 1 in 1991; 50 in 2004.* **Sources: All, Ba8, Bo17, CA25, Ca26, Co32, Com, De6, Dgs, Dow, Ers, Fe5, Ga1, Gr28, Ha5, He8, Hi13, Hud, Hum, Jo1, Jor, Jun, Lej, LO8, Loc, MAY, ME9, Mey, Mo20, MOU, OLD, OSB, PAG, Red, Roh, SE14, Se16, Se26, Sh14, Sk2, So1, Und, Ver, Ves, Vi4, Vi5, Wet, Wi2, Wi15, WI23.**

Autumn Time - Multi-branching plant, 4-5 ft., shades of red, rust and tan, ideal border plant. *Source History: 1 in 2004.* **Sources: Tho.**

Bellezza d' Autuno - Slender 6 ft. stalks bear multiple 5 in. blooms, colors range from light cream to deep mahogany and bi- colors, blooms from July through autumn. *Source History: 1 in 2004.* **Sources: Pin.**

Black Market - Short stalks, med. fat head. *Source History: 1 in 1998; 1 in 2004.* **Sources: Syn.**

Cinnamon Sun - Gorgeous new variety with 5-6 in. flowers with glowing cinnamon-bronze petals, chocolate centers. *Source History: 2 in 2004.* **Sources: Pe2, Re6.**

Cucumber Leaved - (Cucumber Leaf) - *H. debilis*, bushy plant with branching stems, 3-5 ft. tall, 4 in. bronze-yellow flowers, good cut flower. *Source History: 7 in 2004.* **Sources: Bo18, Com, Gr29, Hud, Se28, Vi5, Wo8.**

Cutting Gold - Yellow petals, black centers, for mass production of branching cut flowers, 5 ft. tall. *Source History: 2 in 2004.* **Sources: Ca26, Sto.**

Discovery Mix - 80-100 days, Stalks 6-9 ft. tall with 6-15 branches each bearing 6 in. flowers, beautiful mix dev. from several sunflower families, from the Seeds of Change breeding program. *Source History: 2 in 2004.* **Sources: Pe2, Se7.**

Dwarf Yellow Spray - Compact plant to 18-24 in., bright yellow flowers, beds or borders. *Source History: 2 in 2004.* **Sources: Roh, Tho.**

Earth Walker - Tall, multi-headed stems, 6-9 ft., 8-10 in. blooms, beautiful earth tones, diverse color mix including browns and oranges, long flowering period, borders and cutting. *Source History: 5 in 2004.* **Sources: GLO, HPS, Jun, Pin, Tho.**

Elf - Small plants, 14-16 in. tall, 4 in. flower heads. *Source History: 1 in 2004.* **Sources: Bu2.**

Endurance - 85-95 days, *H. argophyllus x annuus*, result of a spontaneous cross of Israeli x Silverleaf sunflowers, up to 50 branches per 6-9 ft. plant, each ending in a cluster of 2-5 flowers, Seeds of Change original. *Source History: 1 in 2004.* **Sources: Se7.**

Evening Sun - (Abendsonne) - 100 days, Plants grow 6-8 ft. tall, yellow, red and bronze flower petals surround a red-black center, 8-10 in. flowers. *Source History: 14 in 2004.* **Sources: HA3, Hi13, Hud, MAY, Pe2, Red, Roh, Se7, Se16, Se17, Sh13, So1, Tu6, We19.**

Few-Leaved Sunflower - *H. occidentalis*, 3-6 ft. tall plant, few leaves, 2.5 in. bright orange-yellow flowers, blooms August- September. *Source History: 1 in 2004.* **Sources: Hud.**

Florenza - Well-branched plant grows 4-4.5 ft. tall, 5 in. blooms with large mahogany-red central ring, pale-yellow petal tips, bedding or container plant, attractive cut flower. *Source History: 4 in 2004.* **Sources: Ge2, GLO, Ha5, OSB.**

Floristan - 70-90 days, Bushy plant to 48 in., 5-6 in. blooms, dark russet at base of petals with light yellow tip, excellent for cutting. *Source History: 8 in 2004.* **Sources: GLO, Gur, Jo1, Mcf, Pin, Ter, Twi, Ves.**

Gelber Diskus - Height of 4-5 ft. *Source History: 1 in 2004.* **Sources: Wi15.**

Gloriosa Polyheaded - 80-90 days, Beautiful cut flower, single petalled, 4-6 in. poly-headed flowers, yellow, orange, orange-yellow with red markings, 10-40 branches per plant, rare. *Source History: 3 in 2004.* **Sources: In8, Pe2, Se7.**

Happy Face - Green eyes surrounded by golden yellow petals, 5 in. diameter, plant grows 2.5-4 ft. tall. *Source History: 1 in 2004.* **Sources: Bu2.**

Havasupai Striped - 85 days, From the bottom of the Grand Canyon, up to 10 ft. tall, long narrow seeds, not good in Tucson. *Source History: 1 in 1987; 1 in 1991; 1 in 1994; 1 in 1998; 2 in 2004.* **Sources: Na2, Se17.**

Henry Wilde - Single golden yellow with black centers, flowers 12 in. across, plants 9 ft. tall. *Source History: 1 in 2004.* **Sources: Vi5.**

Holiday - 90-110 days, Multi-branching, 60-84 in. plant bears many 3-5 in. blooms, golden yellow petals, dark brown disks, classic. *Source History: 1 in 2004.* **Sources: Jo1.**

Hopi Black Dye - (Hopi Dye) - 75-90 days, Diverse range of flower types but all with golden yellow single petals and dark blue-black centers, black shells used as a dye for baskets, seeds also edible, mound dirt around aerial roots to anchor 6-9 ft. plants, traditionally spring-planted in the high cool desert, Hopi Indian seed. *Source History: 1 in 1981; 1 in 1984; 3 in 1987; 5 in 1991; 3 in 1994; 5 in 1998; 9 in 2004.* **Sources: Ba8, Ho13, In8, Na2, Pla, Sa9, Se7, So1, So12.**

Ikarus - 85-90 days, Sister to Soraya, grows to 48 in., stems 30 in., 4-6 in flowers, brilliant light yellow petals, sturdy side stems, good cut flower. *Source History: 4 in 2004.* **Sources: Bo18, Ge2, Hal, Vi5.**

Incredible - (Incredible Dwarf) - 60-70 days, Plant grows 18 in. tall, single yellow 10 in. flowers, does well in containers. *Source History: 4 in 2004.* **Sources: GRI, Hum, Jor, ME9.**

Indian Blanket - 75-90 days, Resembles Autumn Beauty with darker colors, multi-heading, 5-7 ft. plant, 4-5 in. blooms. *Source History: 4 in 2004.* **Sources: Ba8, Jor, ME9, SE14.**

Irish Eyes - Short plants to 24-30 in. tall, loaded with multiple blooms, green center surrounded by pointed golden petals, one of the best for pot culture and cutting. *Source History: 1 in 2004.* **Sources: Se16.**

Italian White - 75-85 days, Highly-branched 4-6 ft. plants bear long-stemmed 4 in. flowers, creamy white petals surrounding a dark disk, tiny seed, annual, does not last as a cut flower, best enjoyed on the plant. *Source History: 3 in 1987; 1 in 1991; 31 in 2004.* **Sources: Ag7, Ba8, Bo18, CA25, Coo, De6, Ge2, Gr29, GRI, He8, Hi13, HPS, Hum, Jor, Lej, ME9, Ont, OSB, Par, SE14, Se16, Se28, Sk2, So1, So25, Ter, Tho, Ves, Vi5, We19, Wo8.**

Jerusalem Dwarf - 50-60 days, True dwarf sunflower, single 4-5 in. flowers, good container plant. *Source History: 1 in 2004.* **Sources: Se7.**

Large Flowering - Single large flowered sunflower for early planting, giant heads with a profusion of seeds, 6-8 ft. tall, very showy, raised for bird seed, fairly heavy feeders, likes an alkaline soil. *Source History: 1 in 1987; 1 in 1991; 1 in 1994; 2 in 1998; 1 in 2004.* **Sources: Sh9.**

Lemon Queen - 70-100 days, Tall 7-8 ft. plants, 5 in., bright lemon-yellow blooms with chocolate colored centers. *Source History: 17 in 2004.* **Sources: Ba8, De6, Ers, Fe5, Ga1, Gur, He8, Jor, ME9, Ont, OSB, Roh, RUP, SE14, Se16, Wet, Wi15.**

Lemonhead - The result of a sunflower breeding project by Glenn Drowns over the past 12 years, cross of a wild Polish yellow x orange sunflower, mix of single and double flowers with 95% bright lemon yellow color. *Source History: 1 in 2004.* **Sources: Sa9.**

Lyng's California Greystripe - 120 days, Improved Mammoth Russian with more consistent seed color, bright golden yellow flowers up to 10 in. across, one flower per plant, attracts birds, PVP 1996 Ed J. Lyng Company. *Source History: 1 in 2004.* **Sources: Jo1.**

Maya - Grows 3-4 ft. tall, yellow petals. *Source History: 1 in 2004.* **Sources: Wi15.**

Moonwalker - Sturdy 8-12 ft. tall plants bear multiple 6-8 in. blooms, dark chocolate-brown centers, pale yellow ray petals, Dutch variety. *Source History: 7 in 2004.* **Sources: Bu2, Coo, GLO, Jun, Se16, Sey, Tho.**

Multiflora Single Yellow - Single yellow flowers on tall plants. *Source History: 1 in 2004.* **Sources: Sa9.**

Music Box - Multi-branching dwarf plants to 24-30 in., colors range from yellows and creams to mahogany bi-colors, brown centers. *Source History: 10 in 2004.* **Sources: CA25, Com, GRI, Hal, Jun, Pin, Pr3, Re6, Se16, We19.**

Orange Sun - Plants grow 6-8 ft. tall, bright calendula-orange flowers explode with a flash of color, the double flowers have a unique outer row of single petals surrounding large 5-6 in. heads. *Source History: 5 in 2004.* **Sources: Coo, GLO, MAY, Roh, Se16.**

Pacino - 70-84 days, Miniature plants to 12-16 in., 4-5 in. blooms, yellow petals, yellow centers. *Source History: 11 in 2004.* **Sources: Bo18, Dom, Ge2, GLO, GRI, Jo1, Pin, Sto, Ter, Ves, Vi5.**

Pan - Multi-branched plants, small golden yellow flowers with contrasting dark center. *Source History: 1 in 2004.* **Sources: GLO.**

Peace and Freedom - Polyhead, Autumn Beauty type. *Source History: 1 in 2004.* **Sources: Syn.**

Primrose Yellow - Well-branched plant grows 8 ft. tall, lemon-yellow flowers with chocolate-brown center. *Source History: 1 in 2004.* **Sources: Coo.**

Red Menace - Special select polyhead, red-gold

shades. *Source History: 1 in 2004.* **Sources: Syn.**

Red Sun - 85-95 days, Multiflora type to 5-6 ft. tall, bears medium size 5-6 in. blooms in shades of deep dark orange-red with a hint of yellow around the brown center, extended flowering season. *Source History: 19 in 2004.* **Sources: Bo18, Gr28, Gr29, HA3, He8, HPS, Hud, Jor, La1, ME9, Mi12, Roh, SE14, Se28, Sey, Sh9, Ter, Vi5, Wo8.**

Ring of Fire - 120 days, AAS/2001, heavily branched plant is 5-6 ft. tall with 4-5 in. flowers, petals are dark red at the base with golden yellow tips, excellent cut flower. *Source History: 25 in 2004.* **Sources: Bo18, Ca26, Dom, Ers, Ge2, GLO, GRI, Gur, Ha5, HPS, Jun, MAY, Mcf, MOU, OSB, Par, Ri12, Se16, Sey, Sto, Ter, Twi, Ver, Ves, Vi5.**

Santa Fe - Grows 5' tall, double orange-yellow flowers. *Source History: 2 in 2004.* **Sources: Ni1, Wi15.**

Sara's - Plants grow 8-10 ft. tall with many large heads, several seed colors, original seed from a now deceased Seed Savers Exchange member in Mexico. *Source History: 1 in 2004.* **Sources: Sa9.**

Schnittgold - Single-flowered 5 ft. plants, deep yellow petals, brown ring surrounds a flat black center which gets hollow in the center, perfect for making sunflower bird wreaths, 'Schnitt' means 'cut' in German. *Source History: 1 in 2004.* **Sources: Fe5.**

Selma Suns - Sturdy plants bear from 6 to 30, 2-6 in. heads, earth tone colors include red, orange, brown, green and yellow, beautiful cut flower. *Source History: 1 in 2004.* **Sources: So1.**

Short Stuff - (Kid Stuff) - Short plants to 3 ft., 10 in. blooms. *Source History: 2 in 2004.* **Sources: Ba8, Hi13.**

Silverleaf - (Gold and Silver) - 100-110 days, *H. argophyllus*, stalks to 5 ft., multi-flowering, fuzzy, silvery foliage, 4 in. yellow blooms with dark centers, south Texas species, rare. *Source History: 4 in 2004.* **Sources: Ri12, Roh, Se7, Tho.**

Small Black - 100 days, Peredovik type, grows 4-5 ft. tall, excellent bird food, sow May-August. *Source History: 1 in 1987; 2 in 1991; 2 in 1994; 1 in 1998; 1 in 2004.* **Sources: Ada.**

Sole d' Oro - Sturdy stalks to 3-4 ft., large, golden yellow, double blooms, blooms from July through the fall, good cut flower. *Source History: 1 in 2004.* **Sources: Pin.**

Solina - Short plant to 12-16 in. in pots and 20 in. tall in the garden, large yellow 8 in. flowers. *Source History: 1 in 2004.* **Sources: GLO.**

Sonja - 90-110 days, Plant grows 40-45 in. tall, strong straight stems, 4 in. blooms, bright golden orange petals, good cut flower, trademarked. *Source History: 17 in 2004.* **Sources: Bo18, CA25, Ers, Ge2, GLO, GRI, Gur, Hi6, Hud, Jo1, Pin, So1, Tt2, Twi, Ves, Vi5, We19.**

Soraya - 77-91 days, AAS/2000, branching 72 in. plants bear 4-6 in. golden orange flowers with dark center, up to 20-25 stems per plant, stems average 20 in. long, first sunflower to win an All-America Selections award. *Source History: 12 in 2004.* **Sources: Bo18, Ca26, Dom, Fe5, Ge2, Hal, HPS, MAY, Par, Sey, Ter, Twi.**

Sundance Kid - 60 days, Well-branched 15-17 in. plants, large semi-double flowers, 4-6 in., resemble a lion's mane, colors range in shades from bronze to yellow, long vase life. *Source History: 7 in 2004.* **Sources: GLO, GRI, Ha5, Jo1, MOU, Par, Ter.**

Sungold - (Sungold Double) - 90-110 days, Large, 5-8 in. golden flowers borne freely on 6 ft. stalks, makes a good screen or hedge. *Source History: 1 in 1981; 11 in 1987; 6 in 1991; 4 in 1994; 4 in 1998; 13 in 2004.* **Sources: Bo18, Cr1, Ear, Ers, Ge2, Gr29, Jo1, Sa9, Se28, Sh9, Ter, Vi5, Wo8.**

Sungold, Dwarf - Plants 2 ft. tall with dwarf branching habit, 4-6 in. flowers are like Tall Sungold. *Source History: 1 in 1987; 2 in 1991; 2 in 1994; 1 in 1998; 5 in 2004.* **Sources: Fe5, Hum, PAG, Wet, WI23.**

Sungold, Tall - 70-100 days, Plants 5-6 ft. tall, large double blooms, 5-6 in., taller version of Teddy Bear. *Source History: 3 in 1987; 3 in 1991; 1 in 1994; 1 in 1998; 5 in 2004.* **Sources: Ont, Ri12, Sk2, Ves, WI23.**

Sunrays - 42 days, Earliest blooming sunflower, green-gold disc surrounded by bright golden yellow petals, requires extra light for longer stems for greenhouse production in winter. *Source History: 1 in 2004.* **Sources: GLO.**

Sunrise - (Jerusalem Sunrise Lemon) - 70-80 days, Multi-branched plants to 5 ft., 4-6 in. lemon-yellow flowers, good cut flower *Source History: 5 in 2004.* **Sources: Roh, Se7, Se17, Sh13, So9.**

Sunset - Mahogany petals tipped with gold, dark center, 6 in. diameter blooms, 3 ft. plants. *Source History: 1 in 2004.* **Sources: Bu2.**

Sunshine - 80-90 days, Plants to 6-8 ft. tall, 4-8 in. blooms, semi-double golden petals, two distinct flower types, some with gold centers, others with green centers, Seeds of Change original. *Source History: 1 in 2004.* **Sources: Se7.**

Sunsplash - Height to 5 ft., yellow petals tipped with gold surround a red halo around a dark disc. *Source History: 1 in 2004.* **Sources: GLO.**

Supermane - 100-110 days, Large, thick, polypetalled sunflower, great cut flowers, selected from an old Turkish variety, Seeds of Change original. *Source History: 2 in 2004.* **Sources: Pe2, Se7.**

Supreme Mix - Ongoing sel. from years of public domain sunflower breeding by CV Pe1, single and poly headed, early and late flowering, single, double and tiger's eye petal morphs, colors include bronze, amber, red, gloriosa, yellow and lemon, includes crosses with the Texas endemic *H. argophyllus*. *Source History: 1 in 2004.* **Sources: Pe1.**

Taiyo - Favorite old Japanese variety, stalk height varies from 10-15 ft., 5 in. flowers, yellow-gold petals contrast with the brown center, last a long time after cutting. *Source History: 1 in 1987; 7 in 2004.* **Sources: Ag7, Com, Coo, GRI, Se16, Tho, Vi5.**

Tangina - Grows 3 ft. tall, orange petals, dark centers, 4 in. diameter flowers. *Source History: 1 in 2004.* **Sources: Bu2.**

Teddy Bear - 50-65 days, Small bushy sunflower grows to 18-36 in., double yellow blooms 3-6 in., black seeds, good cut flower. *Source History: 1 in 1981; 1 in 1984; 2 in 1987; 3 in 1991; 10 in 1998; 50 in 2004.* **Sources: All, Ba8, Bo17,**

Bo18, Bu2, CA25, Ca26, De6, Ear, Ers, Fis, Ga1, Ge2, Ge2, Gr28, Gr29, GRI, Ha5, He8, Hi13, Jo1, Jor, Jun, La1, LO8, MAY, ME9, Mi12, Mo20, OLD, Ont, OSB, Ri12, Roh, SE14, Se16, Se26, Se28, Sh9, So1, So25, Sto, Syn, Ter, Tho, Ves, Vi5, We19, Wi15, Wo8.

<u>Tiger's Eye Mix</u> - 85-100 days, From Gloriosa x Lion's Mane cross, a beautiful blend of multi-colored, double and single petalled flowers, 4-8 in. blooms, 10-30 branches per 6-8 ft. plant, good for wild adaptation, Seeds of Change original. *Source History: 2 in 2004.* **Sources: Pe2, Se7.**

<u>Ukrainian Ornamental</u> - *Source History: 1 in 2004.* **Sources: Se17.**

<u>Valentine</u> - Plant reaches 5 ft. tall, 5-6 in., pale lemon-yellow flowers have dark disk in the center, excellent cut flower, bred by Dr. Kovacs of Budapest. *Source History: 28 in 2004.* **Sources: Ag7, Bo18, CA25, Coo, Dom, GLO, Gr29, GRI, Ha5, HPS, Hud, Jo1, Jun, Na6, OSB, Pe2, Re6, Roh, Se16, Se28, Sey, Ter, Tho, Ver, Ves, Vi5, We19, Wo8.**

<u>Vanilla Ice</u> - 90-100 days, *H. Debilis* or *H. Cucumerifolius*, the cucumber-leaved sunflower, 5 ft. multi-branching plants, single pale lemon-yellow flowers with brown center, good for cutting. *Source History: 22 in 2004.* **Sources: Bo17, Bo18, Bu2, Ers, Fe5, Ge2, GLO, Gr29, Hi13, Jor, La1, ME9, OSB, Pin, Ri12, Roh, SE14, Se26, Se28, Vi5, Wi15, Wo8.**

<u>Velvet Queen</u> - 70-110 days, Strong, well-branched 5-7 ft. plants, velvety, dark mahogany-red petals surround an almost black center, great for cutting, birds enjoy the seeds. *Source History: 30 in 2004.* **Sources: Ba8, Com, Coo, De6, Fi1, Ge2, Gur, He8, Jo1, Jor, Lej, MAY, Mcf, ME9, Ont, Orn, PAG, Ri12, Roh, RUP, SE14, Se16, Sk2, So1, Tho, Ves, We19, Wet, Wi2, Wi15.**

<u>Yellow Disc</u> - (Yellow Disk) - Med. height stalks bear 4-5 in. blooms, concave mahogany disk when flowers start to open, pulls itself out into a ring of yellow pollen against the mahogany as it opens, attractive. *Source History: 2 in 2004.* **Sources: Hi13, Pin.**

Varieties dropped since 1981 (and the year dropped):

Arrowhead (1998), *Double Sun Gold* (1987), *Fireworks 1976* (1984), *Hopi Striped* (1991), *Jumbo* (2004), *Piccolo* (1991), *Sunburst* (1998), *Texas Wild* (1994), *The Sun* (1998), *Turkish Sun* (1991), *White Seeded* (2004), *Wild, California Serpentine* (1994), *Zebulon* (1994).

Sunflower / Edible
Helianthus annuus

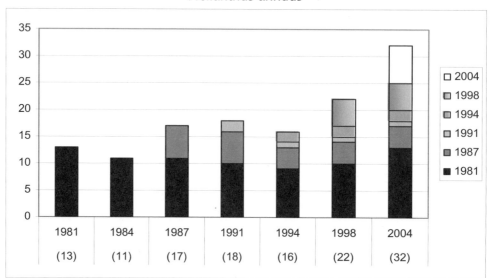

Number of varieties listed in 1981 and still available in 2004: 13 to 13 (100%).
Overall change in number of varieties from 1981 to 2004: 13 to 32 (246%).
Number of 2004 varieties available from 1 or 2 sources: 18 out of 32 (56%).

<u>Arikara</u> - 90-95 days, Sturdy plants up to 12 ft. tall, edible, multi-patterned seeds are black, white and streaked variations, heads 12-16 in. across, from the Arikara tribe in North Dakota. *Source History: 1 in 1998; 5 in 2004.* **Sources: Ag7, Hud, In8, Se16, So12.**

<u>Big Smile</u> - 50-55 days, Plants to 2 ft., 10 in. blooms, edible seeds, good border or container plant. *Source History: 1 in 1998; 14 in 2004.* **Sources: Bo18, Dom, Ge2, GLO, GRI, Ha5, HPS, Jo1, Mcf, MIC, OSB, Roh, Ter, We19.**

<u>Black Russian</u> - 90-95 days, Plants up to 12 ft. tall, 15-18 in. heads, black seeds, leaves - cattle feed; stem fiber - making paper; seeds - raw, oil, breads, poultry feed, roasted coffee substitute; pith - life preservers; dried stems - firewood. *Source History: 1 in 1981; 1 in 1984; 2 in 1987; 2 in 1991; 5 in 2004.* **Sources: Ba8, Bu8, Jor, La1, ME9.**

<u>Black Seeded</u> - 100 days, Tall upright stalks bear huge, sunshine-yellow flowers, large meaty seeds, flavorful when roasted, also enjoyed by the birds. *Source History: 2 in 1981;

1 in 1987; 2 in 1998; 1 in 2004. **Sources: Gur.**

Black Stripe - 70 days, Smaller than Gray Stripe but highly productive. *Source History: 1 in 1987; 2 in 1991; 2 in 1994; 2 in 1998; 2 in 2004.* **Sources: But, Red.**

California Valley Yokut Indian - Beautiful, multi-flowered plants grow 6 ft. tall by 6 ft. across, 2-3 in. flowers, 100-150 per plant, harvested for their tiny edible seeds by the Yokuts of California's San Joaquin Valley, hand-collected from Yokuts ancient village sites by CV Red, where these flowers still grow, germination may take several weeks, seed needs cold nights to germinate, start seeds in pots, best along a sunny wall or fence, rare. *Source History: 1 in 1994; 1 in 1998; 2 in 2004.* **Sources: Red, Se17.**

Dwarf - 90-95 days, Grows to 4-5 ft. tall, 6-8 in. heads, black shells, some with white stripes. *Source History: 1 in 2004.* **Sources: So12.**

Giant - (Giganteus, Gigante) - 80-95 days, Stalks grow to 12-14 ft. or more, heads can reach 15 in. dia., prolific producer of plump thin-shelled striped seeds rich in protein, plant in full sun after last frost date. *Source History: 1 in 1987; 2 in 1991; 4 in 1994; 6 in 1998; 12 in 2004.* **Sources: Gr28, Gr29, In8, Par, Pla, Se28, Ter, Tt2, Tu6, Twi, Ves, Wo8.**

Giant Single - (Giganteus) - Single flowering, golden-yellow, 8 ft. stalks, 12 in. heads, heat tolerant, high protein content, excel. flavor, easy to grow. *Source History: 1 in 1981; 1 in 1984; 1 in 1987; 1 in 1991; 1 in 1994; 1 in 2004.* **Sources: We19.**

Grey Stripe - (Mammoth Grey Stripe, Giant Gray Stripe) - 75-90 days, Heads up to 20 in. across, huge crops, large plump grey-striped seeds, sturdy 8-12 ft. stalks, good roasted, high protein, chicken or bird feed. *Source History: 5 in 1981; 4 in 1984; 14 in 1987; 21 in 1991; 16 in 1994; 20 in 1998; 22 in 2004.* **Sources: Bu1, Ca26, Coo, Dgs, Ear, Eo2, Ers, Fa1, Fe5, Fi1, Gr27, Gur, He8, Jun, OLD, PAG, Pin, Pp2, Ri12, RUP, SE4, Vi4.**

Hidatsa No. 1 - 100 days, Stalks to 8 ft. tall, yellow flowers to 8 in., striped seed is attractive to birds. *Source History: 2 in 2004.* **Sources: In8, Se17.**

Israeli - 90-100 days, Single, 10-14 in. golden flower on stocky, 6-8 ft. plant, black and white edible seeds. *Source History: 1 in 1998; 2 in 2004.* **Sources: Pe2, Se7.**

Large Grey Striped - (Mammoth Gray Striped, Large Gray, Large Striped) - 75-90 days, Huge 12 in. dia. or larger heads, 6-10 ft. tall, good windbreak or screen, plant on east side of garden so they will not shade it out. *Source History: 3 in 1981; 3 in 1984; 1 in 1987; 3 in 1991; 8 in 1994; 7 in 1998; 11 in 2004.* **Sources: CA25, De6, GRI, He8, Loc, MAY, ME9, Mel, Se17, So1, Wi15.**

Large Seeded Tall - Large heads on tall 5-7 ft. stalks, large seeds with excel. flavor. *Source History: 2 in 1981; 2 in 1984; 1 in 1987; 1 in 1998; 1 in 2004.* **Sources: Lin.**

Mammoth - (Mammoth Gray Stripe) - 80 days, Large thin-shelled striped seeds, plump meaty and high protein, for roasting or poultry/bird feed, 6-12 ft. stalks make good screens, productive even on poor soils. *Source History: 11 in 1981; 11 in 1984; 17 in 1987; 20 in 1991; 19 in 1994; 21 in 1998; 25 in 2004.* **Sources: Be4, Bo17, Bo18, Bou, Bu2, Bu3, CH14, Ge2, HA3, Ha5, Hi6, Hi13, Hum, Jor, La1, Me7, Mo13, MOU, Ni1, Pe2, Ri2, Roh, Ros, TE7, WI23.**

Mammoth Russian - (Commander Strain, Russian, Large Russian, Early Russian, Russian Giant, Geant de Russie) - 75-90 days, Heads often 12 in. across, 8-14 ft. plants, 23.5 ft. record, large striped seeds, rich in oil and protein, eaten dried and/or roasted, hardy and disease res., 1200-1500 lbs. per acre, burned stalks produce a high-phosphorus fertilizer, introduced in 1888. *Source History: 17 in 1981; 14 in 1984; 29 in 1987; 31 in 1991; 32 in 1994; 28 in 1998; 33 in 2004.* **Sources: Ag7, Alb, All, Bo18, Com, Cr1, Dam, Ech, Fis, Fo7, Goo, Hal, Mey, Mo20, Na6, Ont, OSB, Pe2, Pr3, Sau, Se7, Se28, Shu, Si5, So1, So25, Sto, Syn, Tho, Tt2, Ves, Wet, Wo8.**

Mammoth Russian, White Seeded - Large heads, white seeds. *Source History: 1 in 2004.* **Sources: Syn.**

Maximilian - *H. maximilianii*, tuberous roots, grows 7-8 ft. tall, native to U.S. prairie states, used by Sioux and other tribes, not Jerusalem artichoke, good wildlife food plant. *Source History: 1 in 1981; 1 in 1984; 1 in 1991; 6 in 1994; 12 in 1998; 20 in 2004.* **Sources: Ada, Bo9, Bo18, Fi1, Ge2, Ho8, Hud, Jo1, Mcf, Par, Pe2, Re6, Se7, Se8, Se28, Sh7, Und, Vi4, We19, Wo8.**

Miriam Edible - 60-70 (flower) 100-110 (mature seed) days, Plants to 5-6 ft., large seed heads with 1 in. gray seeds with white stripes, single yellow petals, yellow centers. *Source History: 1 in 2004.* **Sources: Se7.**

Oilseed - 95 days, Small black-shelled narrow seeds used to press quality sunflower oil for salads and cooking, oil rich - over 40%, good small stock feed, selected early commercial var. *Source History: 1 in 1981; 1 in 1984; 2 in 2004.* **Sources: De6, So12.**

Peredovik - Russian strain grown commercially for the pressing of quality sunflower oil for cooking, small black-shelled seeds make good food for birds and poultry and rabbits. *Source History: 1 in 1981; 1 in 1987; 2 in 1991; 4 in 1994; 3 in 1998; 2 in 2004.* **Sources: Ech, Wi1.**

Purple Seeded - Large size, extraordinary color, rare. *Source History: 1 in 2004.* **Sources: Syn.**

Rostov - True Russian black-seeded type, seed heads average 11 in. dia., from the Ukraine. *Source History: 1 in 1981; 1 in 1984; 1 in 1987; 2 in 1991; 2 in 1994; 5 in 1998; 4 in 2004.* **Sources: Sa9, Se16, Se17, Syn.**

Russian Mammoth, Diane's Strain - 120 days, Plants to 10 ft., large heads average 20 in. diameter, may benefit from staking due to the heavy heads, late maturity. *Source History: 1 in 1994; 1 in 1998; 2 in 2004.* **Sources: Pe6, Pp2.**

Seneca - 90 days, Plants to 4-8 ft., mostly single, 6-10 in. heads with purple, white and grey seeds, from Seneca tribe in NY, grown for oil that was used for ceremonial masks. *Source History: 2 in 1998; 2 in 2004.* **Sources: In8, Se17.**

Skyscraper - 75-85 days, Strong stalks grow 12 ft. tall with bright yellow flowers with 14 in. dia., heat and drought tol., large edible seed. *Source History: 1 in 1998; 12 in 2004.* **Sources: Ech, Ers, Gur, He8, Jor, La1, ME9, RIS, SE14, Sh9, TE7, Wi2.**

Sundak - (Early Sundak) - 65-80 days, Earlier, shorter version of Russian Mammoth, one med-sized well-filled head per heavy, 5-6 ft. stalk, large grey and black striped seeds, rust res., good feed for wild birds, poultry and parrots, confectionary variety. *Source History: 7 in 1981; 6 in 1984; 9 in 1987; 6 in 1991; 4 in 1994; 5 in 1998; 4 in 2004.* **Sources: Eco, Ga1, Pe6, Tt2.**

Sunspot - (Sunspot Dwarf) - 80 days, Compact bushes just 2 ft. tall, 10-12 in. flower heads filled with edible seeds, blooms 9 weeks from sowing, PVP 1992 K. Sahin, Zaden B.V, Dr. I.E. Sahin. *Source History: 5 in 1991; 12 in 1994; 18 in 1998; 33 in 2004.* **Sources: All, Bu2, CA25, De6, Ech, Ers, Fe5, Fi1, GLO, Gur, HA3, Hal, He8, Hi13, Jor, La1, MAY, ME9, Mey, Ni1, OLD, Ont, OSB, Roh, RUP, SE14, Se24, So1, Sto, Tho, Twi, Wi2, Wi15.**

Sunzilla - Tall branching stalks to 14 ft., gold flower petals, edible seeds. *Source History: 1 in 2004.* **Sources: Re6.**

Tarahumara White - 100 days, Adapted 30 year introduction, probably Mennonite origin, adopted by the Tarahumara Indians, 7-10 ft., solid gold flower, 8-10 in. heads, ivory color seed, some stripes, excellent producer, large heads in Tucson with irrigation, wind and heat resistant, unusual and rare. *Source History: 1 in 1987; 2 in 1991; 4 in 1994; 10 in 1998; 14 in 2004.* **Sources: Ag7, Co32, Ho13, Hud, Mi12, Pe2, Ri12, Sa9, Se7, Se16, So9, So12, Syn, Tu6.**

Titan - Plants up to 12 ft. tall, large yellow heads up to 18-24 in. across, extremely large seeds, one of the tallest sunflowers with the largest head and seed size available. *Source History: 1 in 2004.* **Sources: Se16.**

Wild - Classic bright-yellow 3-4 in. sunflowers bloom in late summer, petals enhance salads, small edible seeds, easy to grow, likes full sun and well drained even disturbed soil, drought tol. once established, sow in fall or early spring. *Source History: 1 in 1987; 1 in 1991; 1 in 2004.* **Sources: Hig.**

Varieties dropped since 1981 (and the year dropped):

Black Stripe, Large (1994), *Dwarf Russian* (2004), *Progress* (1994).

Sweet Potato
Ipomoea batatas

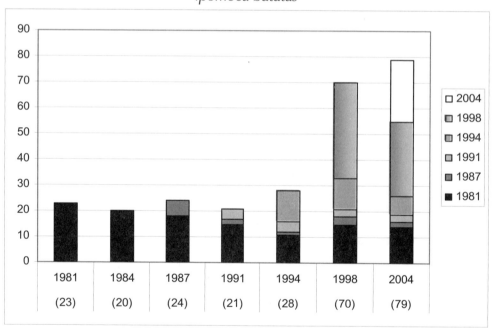

Number of varieties listed in 1981 and still available in 2004: 23 to 14 (61%).
Overall change in number of varieties from 1981 to 2004: 23 to 79 (343%).
Number of 2004 varieties available from 1 or 2 sources: 64 out of 79 (81%).

B-18 - Long vines produce heavy yields of small to medium size tubers, orange skin and flesh, excel. flavor, only surviving product of a Canadian breeding program at Simcoe in the 1960s. *Source History: 1 in 1994; 1 in 1998; 1 in 2004.* **Sources: Al6.**

Batas - 120 days, White skin, white flesh. *Source History: 1 in 2004.* **Sources: Sa9.**

Beauregard - 90-105 days, Red-orange skin, orange flesh, extremely high yields with little cracking, matures quickly, does best with adequate nutrition and plastic row cover, Louisiana AES. *Source History: 2 in 1991; 6 in 1994; 11 in 1998; 15 in 2004.* **Sources: Al6, Br12, Bu2, Ers, Fi1, Fre, Ga1, Gur, Jun, Mel, Par, Ri12, Sa9, Scf, Ste.**

Bermuda Pink - Vining plant, round, somewhat lumpy tubers, white with pink splotches, cream flesh, very sweet. *Source History: 1 in 1998; 1 in 2004.* **Sources: Ho13.**

Bermuda White - 120 days, Vigorous vines, white skin, cream flesh, average yield, very late maturity. *Source History: 1 in 1998; 1 in 2004.* **Sources: Sa9.**

Betty's - 100 days, Semi-bush plant, yellow flesh tubers, red skin. *Source History: 1 in 2004.* **Sources: Sa9.**

Brazilian - 130 days, Vigorous vines produce excel. yields of white skinned tubers with white flesh, midseason. *Source History: 2 in 1998; 2 in 2004.* **Sources: Ho13, Sa9.**

Brinkley White - 105 days, Vining plant produces excel. yields of white skinned tubers with white flesh, midseason. *Source History: 1 in 1998; 1 in 2004.* **Sources: Sa9.**

Carolina Bunch - Bunch plant habit requires less garden space, uniform, smooth, light copper skin, deep orange flesh, disease res., USDA/ARS and Clemson U. *Source History: 1 in 1998; 1 in 2004.* **Sources: Scf.**

Carolina Rose - Attractive, rose colored skin, deep orange flesh, blocky tubers, FW res., NC State U. *Source History: 1 in 1998; 1 in 2004.* **Sources: Scf.**

Carver - 100 days, Semi-bush plant with ivy leaves, pink-red skin, orange flesh, above average yields, midseason. *Source History: 1 in 1981; 1 in 1984; 2 in 1998; 2 in 2004.* **Sources: Ma13, Sa9.**

Centennial - 90-150 days, Leading U.S. variety, vigorous vines can reach 20 ft., bright copper-orange skin, deep-orange fine-grained soft flesh, med-large cylindrical tapered roots, yields its large crop early, good-sized sprouts but few in number, good quality, widely adapted, good in short season areas and in the South, keeps well, LA/AES, 1957. *Source History: 13 in 1981; 12 in 1984; 14 in 1987; 13 in 1991; 10 in 1994; 13 in 1998; 16 in 2004.* **Sources: Br12, Bu2, Ers, Fa1, Fi1, Fre, Ga1, Gur, Jun, Mel, Par, Ri12, Sa9, Shu, Ste, Ver.**

Continental Red - 100 days, Semi-bush plant, ivy leaves, pink-red skin, orange flesh, average yields, midseason. *Source History: 1 in 1998; 1 in 2004.* **Sources: Sa9.**

Copper Jewel - 115 days, Vigorous vines, deep copper colored, small uniform roots. *Source History: 1 in 2004.* **Sources: Sa9.**

Crystal White - 110 days, Semi-bush plant with dark green normal leaves, cream colored skin, white flesh, excel. yields, midseason. *Source History: 1 in 1998; 1 in 2004.* **Sources: Sa9.**

Cuban - 100 days, Large tubers, pink-purple skin, white flesh. *Source History: 1 in 2004.* **Sources: Sa9.**

Darby - 110 days, Uniform tubers, dark red skin, deep orange flesh, good eating quality when cured, baked flesh is soft and juicy. *Source History: 2 in 2004.* **Sources: Sa9, Ste.**

Edna Evans - 120 days, Vining plant, normal leaf shape, creamy pink skinned tubers with yellow flesh, average yields, very late maturity. *Source History: 1 in 1998; 1 in 2004.* **Sources: Sa9.**

Excel - 110 days, Vigorous vines with good ground cover, attractive light copper skin, orange flesh, early, excel. for baking or making fried chips, disease and insect res. similar to Regal, USDA/ARS and Clemson U. *Source History: 2 in 1994; 4 in 1998; 3 in 2004.* **Sources: Al6, Sa9, Scf.**

Frazier White - 105 days, Vining plant, pale green colored leaves, uniform tubers with white skin and flesh, excel. yields, early. *Source History: 2 in 1998; 2 in 2004.* **Sources: Ma13, Sa9.**

Garnet Red - 115 days, Very vigorous vines, deep red skin, orange flesh. *Source History: 1 in 1998; 1 in 2004.* **Sources: Sa9.**

Georgia Jet - (Jet) - 80-100 days, Fast-growing variety, semi-bush, uniform No. 1 size tubers, dark-red skin, moist orange flesh, delicious flavor, high vitamin A content, very productive, very hardy, thrives both North and South. *Source History: 9 in 1987; 9 in 1991; 9 in 1994; 13 in 1998; 19 in 2004.* **Sources: Br12, Bu1, Bu2, Ers, Fa1, Fi1, Fre, Ga1, Gur, Jun, Ma13, Mel, Par, Ri12, Sa9, Shu, Ste, Ter, Ver.**

Georgia Jet Semi-Bush - Sport of Georgia Jet with vines about two-thirds the length, same size roots, flesh is more moist, less prone to cracking, more productive. *Source History: 1 in 1998; 1 in 2004.* **Sources: Al6.**

Georgia Red - 120 days, Vining, copper/red skin, resembles Porto Rico but more productive, fine eating and keeping qualities, heavy yields, midseason, dev. by GA/AES especially for Georgia growers. *Source History: 1 in 1981; 1 in 1998; 1 in 2004.* **Sources: Sa9.**

Georgia Yam - 90 days, Large tubers, orange skin, deep orange flesh, heavy yields. *Source History: 1 in 2004.* **Sources: Sa9.**

Ginseng Red - 105 days, Semi-bush, ivy leaves, pink-red skin, orange flesh, excel. yields, early. *Source History: 3 in 1998; 3 in 2004.* **Sources: Ho13, Ma13, Sa9.**

Gold Nugget - 120 days, Semi-bush, ivy leaves, pink-red roots, above average yields, midseason. *Source History: 1 in 1998; 1 in 2004.* **Sources: Sa9.**

Gold Rush - 105 days, Light-copper skin, deep-orange fleshy roots, purple stems, res. to FW, resembles Porto Rico, good yields, dev. by LA/AES for growers in the South. *Source History: 1 in 1981; 1 in 1987; 1 in 2004.* **Sources: Sa9.**

Golden Slipper - 120 days, Vining plant, orange-pink skin, orange flesh, average yields, very late. *Source History: 1 in 1998; 1 in 2004.* **Sources: Sa9.**

Goldmar - *Source History: 1 in 1998; 1 in 2004.* **Sources: Mey.**

Goldstar - 100 days, Ivy leaf, orange fleshed, pink-red roots. *Source History: 1 in 2004.* **Sources: Sa9.**

Hayman White - 120-130 days, Very old white variety, white skin and flesh, very sweet, average yields, late. *Source History: 3 in 1998; 2 in 2004.* **Sources: Mey, Sa9.**

Hernandez - 100-125 days, Deep orange flesh, copper colored skin, good yields, to attain good size, excel. baking and processing qualities, moderate disease res., not recommended for zones 3 and 4, dev. by Louisiana AES. *Source History: 3 in 2004.* **Sources: Mel, Scf, Ste.**

Ivis White Cream - 95 days, Uniform roots with cream skin and flesh, excel. yields, midseason. *Source History: 1 in 1998; 1 in 2004.* **Sources: Sa9.**

Japanese - 105 days, Semi-bush, ivy leaves, pink-red skin, pale orange flesh, excel. yields, midseason. *Source*

History: 1 in 1998; 1 in 2004. **Sources: Sa9.**

Jersey Yellow - 120 days, Vining type, roots with cream colored skin, white flesh, above average yields, very late. *Source History: 1 in 1981; 1 in 1984; 1 in 1987; 1 in 1998; 1 in 2004.* **Sources: Sa9.**

Jewel - (New Jewell, New Golden Jewell) - 100-135 days, Vining, bright-copper skin, deep-orange flesh, rich flavor, soft texture, bakes quickly, well shaped, high yields, keeps well and retains weight, home or market, early-maturing, fine-grained moist flesh, excel. quality, dis. res., NC State U. *Source History: 9 in 1981; 9 in 1984; 11 in 1987; 10 in 1991; 10 in 1994; 10 in 1998; 11 in 2004.* **Sources: Al6, Ers, Fre, Jun, Mel, Par, Sa9, Scf, Shu, Ste, Ver.**

Jumbo - 90 days, Semi-bush, orange-red flesh tubers, excel. yields, early. *Source History: 1 in 1998; 1 in 2004.* **Sources: Sa9.**

Korean Purple - 100 days, Vining plant, excel. yields of purple skinned tubers with white flesh, midseason. *Source History: 2 in 1998; 4 in 2004.* **Sources: Al6, Ho13, Ma13, Sa9.**

Kumara - (Owairaka Red) - Possibly from the original stock of sweet potatoes brought to New Zealand by the Maoris about 1500 years ago, known to them as kumara, very rough, purple skinned tubers with pink-white flesh. *Source History: 1 in 1994; 1 in 1998; 1 in 2004.* **Sources: Al6.**

Laceleaf - 100 days, Semi-bush plant with green-purple ivy leaves, pink skin, pale orange flesh, excel. yields, midseason. *Source History: 1 in 1998; 1 in 2004.* **Sources: Sa9.**

Mahon - Rich orange flesh, rose colored skin, some variation in flesh color but always sweet and creamy with few strings, not attractive to white tail deer, Dr. David Bradshaw. *Source History: 1 in 2004.* **Sources: Scf.**

Martins - 100 days, Orange fleshed tuber with pinkish skin. *Source History: 1 in 2004.* **Sources: Sa9.**

Maryland 810 - 110 days, Vining plant, red skin, deep orange flesh. *Source History: 1 in 2004.* **Sources: Sa9.**

Memphis Pride - 100 days, Vining plant, golden orange flesh. *Source History: 1 in 2004.* **Sources: Sa9.**

Nancy Hall - 110-120 days, Light-yellow skin, sweet juicy waxy deep-yellow flesh, excel. baker, great taste instead of beauty, produces abundantly, lush green vining foliage, not for eating fresh out of the field but is sweet and tasty after curing in storage, an old favorite. *Source History: 5 in 1981; 2 in 1984; 4 in 1987; 3 in 1991; 2 in 1994; 10 in 1998; 8 in 2004.* **Sources: Al6, Ers, Ho13, Mel, Sa9, Shu, Ste, Ver.**

O Henry - 115 days, White skin, white flesh. *Source History: 1 in 2004.* **Sources: Sa9.**

Oakleaf - 105 days, Semi-bush, green-purple ivy leaves, pink skin, excel. yields, midseason. *Source History: 1 in 1998; 1 in 2004.* **Sources: Sa9.**

Oklahoma Red - 120 days, Semi-bush plant, regular leaf, red skin, orange flesh. *Source History: 1 in 1991; 1 in 1994; 1 in 1998; 1 in 2004.* **Sources: Sa9.**

Old Orange - 120 days, Vining plant, large, uniform roots with orange skin and flesh, excel. yields, midseason, tough to get to sprout. *Source History: 1 in 1998; 1 in 2004.* **Sources: Sa9.**

Painter - 125 days, Deep orange skin and flesh, requires hot weather to do well. *Source History: 1 in 2004.* **Sources: Sa9.**

Paramutai - *Source History: 1 in 2004.* **Sources: Al6.**

Poplar Root - 120 days, Very vigorous vines, large slim roots, white skin and flesh. *Source History: 1 in 2004.* **Sources: Sa9.**

Porto Rico, Bush - (Porto Rican, Vineless P.R., Bunch P.R., New Bunch P.R., Light Yellow Yams) - 110-150 days, Semi- upright bush type plants, easy to grow, 12-30 in. tall, copper skin, deep-orange sweet flesh, old-fashioned flavor, excel. baker, needs to cure before it tastes good, GA/AES, South. *Source History: 12 in 1981; 11 in 1984; 13 in 1987; 12 in 1991; 10 in 1994; 13 in 1998; 13 in 2004.* **Sources: Bu2, Ers, Fa1, Fi1, Fre, Ga1, Mel, Mey, Par, Sa9, Shu, Ste, Ver.**

Red Ivy Leaf - 105 days, Semi-bush, ivy leaves, red skin, orange flesh, above average yields, early. *Source History: 1 in 1998; 1 in 2004.* **Sources: Sa9.**

Red Jewel - 150 days, Red skin, deep-orange flesh, bakes quickly with soft texture, good yields, easy to grow. *Source History: 1 in 1987; 2 in 1991; 2 in 1994; 3 in 1998; 2 in 2004.* **Sources: Mey, Ste.**

Red Yams - (Red Nuggets) - 105 days, Looks like a yam, not a baker but a frying type sweet potato, red skin, flesh color inside with an orange milky ring around the deep part of the flesh, bears heavily, good keeper, vine type plant. *Source History: 2 in 1981; 2 in 1984; 2 in 1987; 1 in 1991; 1 in 1994; 2 in 1998; 2 in 2004.* **Sources: Fre, Sa9.**

Redmar - *Source History: 1 in 1998; 1 in 2004.* **Sources: Mey.**

Regal - 110 days, Purple-red skin, orange flesh, excel. baking quality, stores well but not as long as Jewel, USDA/ARS, Clemson U and Texas A and M. *Source History: 1 in 1994; 2 in 1998; 4 in 2004.* **Sources: Al6, Ma13, Sa9, Scf.**

Shoregold - 120 days, Orange skin, yellow-orange flesh. *Source History: 1 in 2004.* **Sources: Sa9.**

Southern Delight - 110 days, Regular leaves, reddish skin, orange flesh. *Source History: 1 in 2004.* **Sources: Sa9.**

Southern Queen - 120 days, Vining plant, white skin, white flesh. *Source History: 1 in 1981; 1 in 2004.* **Sources: Sa9.**

Spanish White - Medium production of white skinned tubers, light salmon flesh, mild flavor. *Source History: 1 in 1994; 1 in 1998; 1 in 2004.* **Sources: Al6.**

Sumor - 90-115 days, Unusual sweet potato with white to yellow flesh, smooth texture, can be used in Irish potato recipes and potato salad, can be used right after digging, becomes sweeter and retains quality after curing and extended storage, res. to heat drought disease and many insects, Clemson U, SC AES and USDA/ARS. *Source History: 2 in 1994; 3 in 1998; 3 in 2004.* **Sources: Al6, Sa9, Scf.**

Sunnyside - 105 days, Cream-yellow skin and flesh,

SWEET POTATO

very productive. *Source History: 1 in 2004.* **Sources: Sa9.**

Superior - Experimental var. MD-305 from Maryland, ivy-like leaves, medium production. *Source History: 1 in 1994; 2 in 1998; 1 in 2004.* **Sources: Ma13.**

Tainung 65 - Large tuber, light pink to red skin, slightly dry, creamy flesh, productive, keeps in storage up to 12 months, experimental variety from Taiwan. *Source History: 2 in 2004.* **Sources: Al6, Ma13.**

Tennessee Top Mark - 110 days, Vining plant, golden orange flesh. *Source History: 1 in 2004.* **Sources: Sa9.**

Toka Gold - (Golden Kumara, Toka Toka Gold) - Yellow skinned tubers, deep yellow flesh streaked with orange, good flavor, texture is more dry than commercial North American varieties, from New Zealand. *Source History: 1 in 1998; 1 in 2004.* **Sources: Al6.**

Vardaman - 90-110 days, Bunch or vineless var., golden-yellow skin darkens soon after digging, bright deep red-orange flesh, purple plants mature to dark-green, heavy yields, USDA, 1983. *Source History: 3 in 1981; 3 in 1984; 11 in 1987; 10 in 1991; 9 in 1994; 12 in 1998; 14 in 2004.* **Sources: Br12, Bu2, Ers, Fi1, Fre, Ga1, Gur, Jun, Mel, Par, Sa9, Shu, Ste, Ver.**

Violetta - 100 days, Vining plant, bright purple skin, very sweet white flesh. *Source History: 1 in 2004.* **Sources: Sa9.**

W-374 - 115 days, Experimental type, shows potential. *Source History: 1 in 2004.* **Sources: Sa9.**

White Bunch - (Triumph, Southern Queen, Poplar Root, Choker) - 120 days, White inside and out, very sweet, dry flesh, unusual. *Source History: 1 in 1981; 1 in 1984; 1 in 1987; 1 in 1991; 2 in 1998; 2 in 2004.* **Sources: Shu, Ver.**

White Crystal - *Source History: 1 in 1991; 1 in 1994; 1 in 1998; 1 in 2004.* **Sources: Fre.**

White Delite - 120 days, Vining plant, pale bright purple skin, white flesh, average yields, late. *Source History: 1 in 1998; 1 in 2004.* **Sources: Sa9.**

White Queen - 105 days, Vining plant, white skin and flesh, below average yields, very late. *Source History: 1 in 1998; 1 in 2004.* **Sources: Sa9.**

White Star - 105 days, Vigorous vines, uniform white roots. *Source History: 1 in 2004.* **Sources: Sa9.**

White Triumph - (White Yams) - 90-120 days, Vine type plant, small roots, smooth white skin and snow-white flesh, cooks very well. *Source History: 2 in 1981; 2 in 1984; 2 in 1987; 1 in 1991; 1 in 1994; 2 in 1998; 3 in 2004.* **Sources: Fre, Ga1, Sa9.**

White Yam - (White Triumph, Southern Queen, Poplar Root, Choker, White Bunch) - 105 days, Cotton-white inside and out, very sweet and dry, pink-red ivy leaves, one of the oldest in America, unusual. *Source History: 2 in 1981; 1 in 1984; 2 in 1987; 3 in 1991; 2 in 1994; 4 in 1998; 5 in 2004.* **Sources: Ers, Mel, Par, Sa9, Ste.**

Varieties dropped since 1981 (and the year dropped):

Gold (2004), *Amish Pink* (2004), *Amish Red* (2004), *Arcadian* (1991), *Burgandy* (1991), *Carter* (2004), *Copperskin* (1994), *Cordner* (2004), *Jamaican Red* (2004), *Jasper* (1991), *Nugget* (1998), *Okinawa* (2004), *Old Kentucky* (2004), *Porto Rico, Running* (2004), *Purple* (2004), *Purple Yam* (2004), *Red Wine Velvet* (2004), *Resisto* (2004), *Rose Centennial* (2004), *Southern Delite* (2004), *Stoker's Red* (1994), *Toka* (2004), *Travis* (1998), *Tuskegee* (2004), *White Nancy Hall* (1987), *Yellow Yam* (1994).

Swiss Chard
Beta vulgaris (Cicla Group)

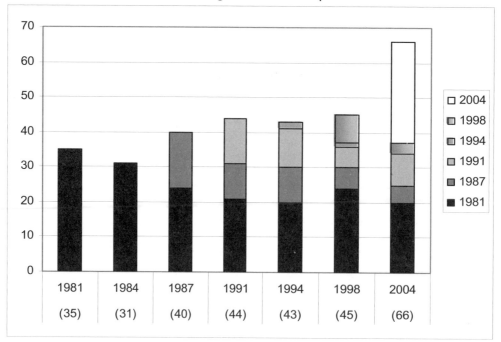

Number of varieties listed in 1981 and still available in 2004: 35 to 20 (57%).
Overall change in number of varieties from 1981 to 2004: 35 to 66 (189%).
Number of 2004 varieties available from 1 or 2 sources: 44 out of 66 (67%).

Acelga - 60 days, Smooth leaves on white ribs, very sweet, productive, Uruguay. *Source History: 1 in 1991; 1 in 1994; 2 in 2004.* **Sources: In8, Se17.**

Argentata - (Bionda a Costa) - 55-60 days, Italian heirloom, vigorous 2-3 ft. plants with silvery-white crispy midribs and savoyed deep-green leaves, mild, clean, sweet flavor, cold tolerant. *Source History: 1 in 1991; 2 in 1994; 2 in 1998; 4 in 2004.* **Sources: Fe5, Ki4, Na6, Orn.**

Barese - 26-28 days, Babyleaf market var., quick grower, smooth dark glossy green leaves with slightly curled edges. *Source History: 2 in 2004.* **Sources: CO23, ME9.**

Bietola da Coste Bionda de Lione - Huge leaves with thick stems form large heads, outstanding flavor. *Source History: 1 in 2004.* **Sources: Se24.**

Breeders Rainbow 2000 - Mix of the brightest, boldest colors, stem and leaf veining in shades of red, magenta, gold, orange and white. *Source History: 1 in 2004.* **Sources: Pe6.**

Bright Lights - 50-62 days, AAS/1998, 20 in. plants, stalks are gold, yellow, green, orange, pink, violet, striped, red and white, green and bronze foliage, beautiful. *Source History: 31 in 1998; 56 in 2004.* **Sources: All, BAL, Be4, Bu2, Com, Coo, Dam, Ear, Eo2, Fa1, Fe5, Fi1, Ga1, Ge2, GLO, Gur, HA3, Ha5, Hal, Hi13, HO2, Hum, Jo1, Jor, Jun, Ki4, Lin, MAY, Mcf, ME9, Mel, MIC, MOU, Na6, Ni1, OLD, Ont, Or10, OSB, Par, Pe2, Pin, Re6, Roh, Se25, Se26, Shu, Sto, Sw9, Ter, Tho, Tt2, Twi, Und, Ver, We19.**

Bright Yellow - 30-57 days, Deep green leaves with yellow stems and leaf veins, sweet mild taste, young leaves added to salad, mature stalks eaten raw or steamed, PVP 2002 Jo1. *Source History: 4 in 2004.* **Sources: Hi13, Jo1, Roh, Tho.**

Broadstem Green - 50-60 days, Large green leaves and broad white stems make an excel. steamed vegetable, grows to 24 in., dev. by Alan Kapuler. *Source History: 2 in 1991; 2 in 1994; 1 in 1998; 1 in 2004.* **Sources: Se7.**

Burgundy Giant - 55 days, Distinct type of Rhubarb Chard, deep burgundy-crimson, dark-green leaves are slightly crumpled and veined with crimson, ribs are narrow but fleshy, decorative. *Source History: 1 in 1981; 1 in 1984; 1 in 1987; 1 in 1991; 1 in 1994; 1 in 1998; 1 in 2004.* **Sources: Ear.**

Canary - Rich luminous yellow stems. *Source History: 2 in 2004.* **Sources: ME9, Pe6.**

Cardinal - Bright red stems. *Source History: 2 in 2004.* **Sources: ME9, Pe6.**

Chadwick's Choice - White stems, wide green leaves, naturalized for more than 20 years at Mariposa Ranch in California, originally from Alan Chadwick. *Source History: 1 in 2004.* **Sources: Tu6.**

Charlotte - (Scarlet Charlotte) - 50-59 days, Bright-scarlet colored veins and stalk, cold weather brings out the dark purple-red color of the heavily savoyed leaves, better than common rhubarb chard. *Source History: 1 in 1991; 1 in 1994; 4 in 1998; 7 in 2004.* **Sources: CO23, Coo, Fe5,**

Ho13, Pe2, Pin, Re6.

Chismahoo - Thin white stem ribs, light green leaves, clear clean flavor, cold res. *Source History: 2 in 2004.* **Sources: In8, Se17.**

Costa Argentata - 60-90 days, Long, wide green leaves, flat white ribs. *Source History: 1 in 1987; 1 in 1991; 1 in 1994; 1 in 1998; 1 in 2004.* **Sources: Ber.**

Dark Green White Ribbed - (Dark Green, White Ribbed, White Silver Ribbed, Smooth Leaf) - 50-60 days, Upright 20-26 in. plant, large med-dark green slightly crumpled leaves, broad white midribs, heavy yields. *Source History: 8 in 1981; 8 in 1984; 6 in 1987; 3 in 1991; 3 in 1994; 2 in 1998; 3 in 2004.* **Sources: Bo17, Loc, ME9.**

Dorat - (Yellow Dorat) - 60 days, Imp. on Giant Lucullus, milder flavor, better uniformity and vigor, very tender, pale yellowish-green leaves on thick wide creamy-white stalks, heat res., prepare stalks like asparagus, gourmet, Denmark. *Source History: 2 in 1981; 2 in 1984; 2 in 1987; 2 in 1991; 3 in 1994; 3 in 1998; 3 in 2004.* **Sources: CO30, Gou, Ho13.**

Erbette Leaf - (Erbette) - 45 days, Refined variety of chard grown for its fine-textured and fine flavored tops, keep cut for continued harvest, does not bolt in hot weather, drought res., winter hardy to zone 5 in cold frame. *Source History: 1 in 1987; 1 in 1991; 1 in 1994; 2 in 2004.* **Sources: Coo, Und.**

Feurio - Beautiful red chard. *Source History: 1 in 1998; 1 in 2004.* **Sources: Tu6.**

Flamingo - Luminous pink stems. *Source History: 2 in 2004.* **Sources: ME9, Pe6.**

Fordhook Giant - (Burpees Fordhook Giant, Fordhook Large Ribbed Dark Green, Fordhook) - 50-60 days, Broad dark-green heavily crumpled leaves with white veins, 24-28 in. thick white 2.5 in. wide stalks, abundant crops all season and even after the first light frosts, excel. for greens, also good for poultry feed, from the 1750s. *Source History: 77 in 1981; 72 in 1984; 72 in 1987; 64 in 1991; 59 in 1994; 64 in 1998; 60 in 2004.* **Sources: Alb, All, Ba8, Be4, Bo17, Bu2, But, CH14, CO23, Com, Cr1, De6, DOR, Ear, Ers, Fe5, Fis, Fo7, Ga1, Ga22, He8, Hi13, Hig, HO2, Jo1, Jor, La1, Lej, Lin, Ma19, Me7, ME9, Mey, OLD, Ont, Or10, Orn, PAG, Pe2, Pe6, Pin, Ros, RUP, SE4, SE14, Se16, Se25, Se26, Sh9, Shu, TE7, Ter, Twi, Ver, Ves, Vi4, WE10, We19, Wi2, WI23.**

French - 60 days, Very tender thick green leaves on large white stalks, heavy yielder, can be sown twice a year in milder climates, heat res., improvement over common Swiss chard. *Source History: 2 in 1981; 1 in 1984; 2 in 1987; 1 in 1991; 1 in 1994; 2 in 1998; 1 in 2004.* **Sources: Ni1.**

French Green - 50-60 days, Well developed white midribs, savoyed thick green leaves, vigorous, adapts to both heat and cold. *Source History: 1 in 1981; 1 in 1984; 1 in 1987; 1 in 1991; 1 in 1994; 1 in 1998; 4 in 2004.* **Sources: Fo13, In8, Pr3, Se17.**

Glatter Silber - Bright green leaves, broad silvery white stems, vigorous and productive. *Source History: 1 in 2004.* **Sources: Tu6.**

Gold Glebe - 65 days, Smooth, ribbed leaves in shades of yellow, mild flavor, productive. *Source History: 1 in 2004.* **Sources: Sa9.**

Golden - 55-60 days, Brilliant golden stalks and midribs, tasty, heat tolerant, also withstands light frost, less prone to fungal and bacterial leaf spotting than other specialty chards, French heirloom from the 1830s. *Source History: 1 in 1998; 4 in 2004.* **Sources: In8, Se7, Se27, Sh12.**

Golden Sunrise - 53-59 days, Grows 23-26 in., stalks, midribs and veins are deep golden color, leaves are lightly crumpled, medium green, holds color well, never tough or bitter, very productive, good frost tol., PVP applied for. *Source History: 7 in 2004.* **Sources: CO23, Dam, DOR, He17, ME9, Ni1, Wi2.**

Green - Smooth-leaved green variety, popular in Europe, small stalks, excel. cooked green, spinach beet. *Source History: 1 in 1981; 1 in 1984; 1 in 1987; 3 in 1991; 2 in 1994; 2 in 1998; 2 in 2004.* **Sources: Dam, So9.**

Italian Silver Rib - 50 days, Crinkled, broad, deep green leaves with wide, silver midribs, clean mellow flavor, good spinach alternative. *Source History: 2 in 2004.* **Sources: Pe2, Re6.**

Japanese - (Nihon) - Broad deep-green leaves and thick light-green stalks, looks like a round-leaved spinach. *Source History: 1 in 1991; 1 in 1994; 2 in 1998; 1 in 2004.* **Sources: Ho13.**

Large Smooth White Ribbed - (Large White Ribbed) - 55-60 days, Large broad creamy-white ribs, smooth dark-green 20-26 in. leaves, vitamin rich, very productive, for home garden or market, easy to grow. *Source History: 5 in 1981; 5 in 1984; 8 in 1987; 9 in 1991; 10 in 1994; 12 in 1998; 12 in 2004.* **Sources: ABB, CO23, DOR, He8, Hum, Me7, Sa9, SE4, Se26, Sto, Twi, WI23.**

Large White Broad Ribbed - (Large Ribbed Whited, Large White Rib) - 55-60 days, Broad thick smooth white tender midribs, thick tender foliage, excel. flavor cooked, good variety for commercial growers. *Source History: 5 in 1981; 5 in 1984; 3 in 1987; 7 in 1991; 9 in 1994; 11 in 1998; 6 in 2004.* **Sources: CH14, Ha5, La1, Mey, RIS, WE10.**

Large White Ribbed Dark Green - (Large Ribbed Dark Green) - 55-68 days, Very broad thick white stems, some 8 in. wide, creamed like asparagus, dark-green almost smooth leaves cooked like spinach, productive. *Source History: 13 in 1981; 11 in 1984; 8 in 1987; 7 in 1991; 7 in 1994; 6 in 1998; 5 in 2004.* **Sources: Bu3, Hud, Kil, Roh, SE14.**

Lucullus - (Swiss Lucullus, Green Lucullus, Lucullus Light Green, Green White) - 45-60 days, Light yellowish-green large thick heavily-crumpled white-veined leaves, white 12 x 2.5 in. rounded midribs, strong vigorous 24-30 in. plants, stands hot weather well, grows until frost, fine flavor, very heavy yields, named after the Roman General Lucius Lucullus who was famous for his banquets, 1914. *Source History: 71 in 1981; 63 in 1984; 61 in 1987; 57 in 1991; 55 in 1994; 50 in 1998; 47 in 2004.* **Sources: All, Bo19, Bou, CH14, CO30, Co31, Co32, Dam, DOR, Ear, Ers, Fi1, Fis, Ge2, Gr27, Gur, HA3, Hal, He8, He17, Jun, La1, Lej, Lin, LO8, MAY, ME9, Mel, Mey, Mo13, MOU, PAG,**

RUP, Sa9, Sau, Se8, SE14, Se26, Sh9, Sol, SOU, Tho, Ves, Vi4, WE10, Wet, WI23.

Lucullus, Giant - 40-55 days, Light yellowish-green heavily crumpled leaves, cream-colored 6 in. wide stalks, 24-28 in. upright plants, produces over a long season when only outer leaves picked. *Source History: 8 in 1981; 6 in 1984; 4 in 1987; 2 in 1991; 2 in 1994; 2 in 1998; 1 in 2004.* **Sources: Com.**

Magenta Sunset - 55 days, PVP applied for. *Source History: 3 in 2004.* **Sources: CO23, DOR, Se26.**

Monstruoso - Select Italian strain, visually striking plant, broad white petioles are great in stir fry or raw, very productive. *Source History: 1 in 1991; 1 in 1994; 2 in 1998; 2 in 2004.* **Sources: Coo, Ho13.**

Neon Lights - 60 days, Beautiful mix of brilliant colors, brighter than Bright Lights, bright green leaves used as spinach. *Source History: 1 in 2004.* **Sources: Bu2.**

Orange Fantasia - 65 days, Dark green shiny leaves with bright orange stalks and veins, excellent as small leaf salad green, non-fading color, good bolt tolerance, PVP pending. *Source History: 2 in 2004.* **Sources: Ves, WE10.**

Orea - 48 days, Light green leaves, bright white stalks, stir fry, salads, steamed. *Source History: 1 in 2004.* **Sources: Bu2.**

Oriole - Tangerine-orange stems. *Source History: 2 in 2004.* **Sources: ME9, Pe6.**

Paris, Green White Ribs - From France. *Source History: 1 in 1987; 1 in 1991; 1 in 1994; 1 in 1998; 1 in 2004.* **Sources: Lej.**

Parrot - *Source History: 1 in 2004.* **Sources: ME9.**

Perpetual - (Perpetual Spinach, Spinach Beet, Leaf Beet) - 50-60 days, Fine old European str., smooth dark-green leaves, small green midribs, drought and frost and bolt res., long perpetual harvest, introduced in 1869. *Source History: 4 in 1981; 3 in 1984; 8 in 1987; 4 in 1991; 3 in 1994; 13 in 1998; 20 in 2004.* **Sources: Bou, Com, Ear, Ers, Fe5, Ga22, Hal, Hi13, HO2, Ho13, Hum, Ki4, Lin, Ni1, Or10, Ra5, Ter, Tho, Ves, We19.**

Pink Lipstick - Savoyed blistered leaves, magenta-pink stems and petioles, excellent for salad packs. *Source History: 1 in 2004.* **Sources: WE10.**

Rainbow Chard - (Five Color Chard, Five Color Silverbeet) - 60 days, Australian heirloom, beautiful ornamental technicolor mixture of Swiss chards, shades of red orange purple yellow and white, tender and tasty. *Source History: 1 in 1981; 1 in 1984; 1 in 1987; 9 in 1998; 40 in 2004.* **Sources: Ag7, Ba8, Bo19, Bu3, CO30, Co32, Com, Ers, GRI, He8, He17, Hi6, HO2, Ho13, Hud, In8, La1, LO8, Loc, ME9, Mo13, Orn, Pe2, Red, Ri2, Ri12, RUP, Sa5, Se8, SE14, Se16, Se17, Se26, Sol, So12, Te4, TE7, Tu6, Vi4, WE10.**

Red Stalk - Green leaves with red veins, crisp red stalks. *Source History: 1 in 1991; 1 in 1994; 1 in 2004.* **Sources: So9.**

Rex Wide Ribbed - 55 days, Vigorous med-tall plants, large glossy dark-green savoyed leaves, fleshy crisp silvery-white 2 in. wide stalks, heat and cold res., sow spring or summer, Denmark. *Source History: 3 in 1981; 3 in 1984; 2 in 1987; 1 in 1991; 1 in 1994; 1 in 1998; 1 in 2004.* **Sources: CO30.**

Rhubarb Chard - (Burpee's Rhubarb Chard, Crimson Rhubarb Chard, Rhubarb/Ruby Red, Ruby, Red, Strawberry) - 50-60 days, Deep-crimson stalks and leaf veins, contrasting dark-green heavily crumpled leaves, cook stalks like asparagus and leaves like spinach, both also used raw, unique flavor, very ornamental, 1857. *Source History: 64 in 1981; 58 in 1984; 64 in 1987; 57 in 1991; 50 in 1994; 62 in 1998; 63 in 2004.* **Sources: ABB, All, Bou, Bu2, CH14, CO30, Co31, Co32, Com, Dam, Dgs, Fis, Fo7, Goo, Gou, HA3, Ha5, Hal, Hi13, HO2, Hud, Hum, Jo1, Jor, Jun, La1, Lej, Lin, LO8, Loc, ME9, Mel, Mo13, MOU, Na6, OLD, Or10, OSB, PAG, Pe2, Pin, Pr3, RIS, RUP, Sa9, Sau, Se7, SE14, Se26, Sh9, Sh12, SOU, St18, Sto, TE7, Ter, Tho, Twi, Ver, Ves, WE10, We19.**

Rhubarb Supreme - European heirloom dating back to 1857, superior to Ruby Red or Rhubarb in that it has a consistently higher percentage of red petioles which are wider with less white and pink stripes, appreciable levels of Horizontal res. to DM and Cercospera leaf spot, more bolt tolerance, selected for superior seedling vigor. *Source History: 2 in 2004.* **Sources: In8, Se27.**

Ruby - (Ruby Red, Rhubarb Chard) - 55-60 days, Tender sweet rhubarb-like ruby-red stalks, color extends into leaf veins, dark-green heavily crumpled leaves, 18-24 in. plants, yields all summer and into the fall, used fresh or frozen, very ornamental. *Source History: 20 in 1981; 14 in 1984; 18 in 1987; 18 in 1991; 22 in 1994; 26 in 1998; 38 in 2004.* **Sources: Ag7, Alb, Au2, Ba8, Bo17, Bo19, Bu3, CO23, DOR, Ers, Fa1, Fe5, Ga1, He8, He17, Hi6, Ki4, Kil, ME9, Mey, Ni1, OLD, Orn, Pla, Ri12, Roh, Ros, SE4, SE14, Se16, Sk2, Sol, TE7, Vi4, WE10, Wet, Wi2, WI23.**

San Francisco Wild - 60 days, Smooth, light green leaves to 9 in., tender stems, an Italian chard that went wild in the San Francisco area, excel. flavor, no bitterness. *Source History: 1 in 1981; 1 in 1984; 1 in 1987; 1 in 1991; 1 in 1994; 2 in 1998; 1 in 2004.* **Sources: Red.**

Seafoam - *Source History: 1 in 2004.* **Sources: DOR.**

Silverado - 55-60 days, Refined Lucullus, 14-16 in. plant, white stems, dark-green savoyed leaves, fine taste, slow bolting. *Source History: 6 in 1991; 7 in 1994; 10 in 1998; 11 in 2004.* **Sources: ABB, Co31, ME9, Mey, OSB, OSB, RUP, SE4, SOU, Sto, Ves.**

Snow Queen - Moderately early-maturing, short plants, dark emerald-green leaves on thick broad pure-white ribs, beautiful and flavorful. *Source History: 1 in 1987; 1 in 1991; 1 in 1994; 1 in 1998; 1 in 2004.* **Sources: GLO.**

Spinach-Beet Greens - 60 days, spinach substitute, bushy 2.5 ft. plants with large soft green leaves useful for constant supply of tasty greens from spring to fall, used raw in salads, cooked like spinach or stir fry, withstands heat, keeps producing until heavy frost, heirloom. *Source History: 1 in 2004.* **Sources: Und.**

Star 1801 - *Source History: 1 in 2004.* **Sources: CO23.**

Verde da Costa - (Verde a Costa Blanca) - 55-60 days, Superior Italian chard, large tender slightly crumpled leaves, medium silver-white rib, cook leaves like spinach, flavorful.

Source History: 1 in 1991; 2 in 2004. **Sources: CA25, Se24.**

Verde da Taglio - 55-60 days, Green, small ribbed leaves with mild spinach taste, cook like spinach, thin stems are very sweet and tender. *Source History: 1 in 1987; 1 in 1991; 1 in 1994; 1 in 1998; 2 in 2004.* **Sources: Ber, Se24.**

Verte a Carde Blanche - 54 days, Dark-green leaves with large white ribs, early and cold res., popular French var. *Source History: 1 in 1981; 2 in 1998; 2 in 2004.* **Sources: Ho13, Pin.**

Virgo - Monstruoso type with broad white stems, excellent hot weather substitute for bok choy, use instead of celery in stir fry, lightly savoyed green foliage, vigorous. *Source History: 1 in 2004.* **Sources: Coo.**

White Cloud - 35-60 days, Plant grows 21-26 in. tall, green leaf, white stalk and ribs, tasty. *Source History: 3 in 2004.* **Sources: CO23, DOR, He17.**

White King - 55 days, Uniform sel. from Lucullus, large thick white ribs, extra dark-green heavily savoyed leaves, very upright plant, stalks appear almost celery-like. *Source History: 2 in 1981; 2 in 1984; 1 in 1987; 1 in 1991; 1 in 1994; 1 in 1998; 1 in 2004.* **Sources: Sto.**

Witerbi Mangold - 60-70 days, Short broad stems, large savoyed leaves with white veins *Source History: 1 in 2004.* **Sources: Se7.**

Yellow - Brilliant yellow stems, delicious picked in the baby stage for salads, steamed when mature. *Source History: 4 in 2004.* **Sources: Ba8, ME9, SE14, TE7.**

Varieties dropped since 1981 (and the year dropped):

Ampuis (1987), *Bietola da Taglio* (1987), *Blonde a Carde Blanche* (1994), *Breeders Rainbow 95* (2004), *Broad White Stem* (1991), *Burgundy* (1998), *Candy Stripe* (2004), *Common Green* (1998), *Da Taglio Verde Scuro* (1987), *De Languedoc* (1991), *De Nice* (1991), *French Prize* (1987), *Greek Spinach* (1998), *Lyon, Blonde White Ribs* (1998), *Magenta Mix* (2004), *Nihon* (1994), *Palak Durga* (2004), *Paros* (2004), *Pavich* (2004), *Red Rib* (2004), *Redstem* (1991), *Shirokuki Ohba* (1987), *Silver Giant* (2004), *Silver Lyons* (1984), *Silver Vein* (1987), *Spinach Beet* (1991), *Swiss* (2004), *Swiss Chard of Geneva* (2004), *Swiss Gold* (2004), *Vulcan* (2004), *Wintergold* (1998).

Tepary Bean
Phaseolus acutifolius (Latifolius Group)

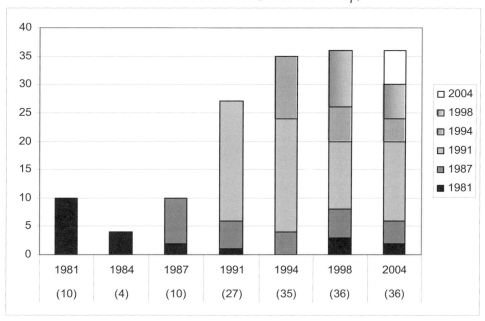

Number of varieties listed in 1981 and still available in 2004: 10 to 2 (20%).
Overall change in number of varieties from 1981 to 2004: 10 to 36 (360%).
Number of 2004 varieties available from 1 or 2 sources: 32 out of 36 (89%).

Big Fields Brown - Typical brown tepary, traditionally planted in August, harvested in November. *Source History: 1 in 2004.* **Sources: Na2.**

Big Fields White - White var. rarely grown anymore, from the Tohono O'odham village of Big Fields. *Source History: 1 in 1998; 1 in 2004.* **Sources: Na2.**

Black - Sel. from white tepary seed purchased many years ago in a Tucson Mexican market, similar to historic Tohono O'odham and Yuma vars., late season maturity. *Source History: 1 in 1991; 1 in 1994; 3 in 1998; 2 in 2004.* **Sources: Ho13, Na2.**

Blue Speckled - 85-98 days, Pods to 2-3 in., plump seeds, pretty, prolific, thrives on heat, smooth flavor, traditional native staple in the Southwest, high protein, pre-Columbian. *Source History: 2 in 1981; 1 in 1984; 2 in 1987; 3 in 1991; 3 in 1994; 5 in 1998; 6 in 2004.* **Sources: Ag7,**

Ho13, Na2, Se7, So12, Syn.

Blue-Grey Speckled - Early ripening variety. *Source History: 1 in 1994; 1 in 1998; 1 in 2004.* **Sources: Pr3.**

Brown Speckled - 96 days, Seeds are beige speckled on light gray, originally separated out of Blue Speckled. *Source History: 2 in 1991; 2 in 1994; 2 in 1998; 2 in 2004.* **Sources: Ho13, Na2.**

Cocopah Brown - Golden brown seeds with some mottling, from the Cocopah who live and farm along the Lower Colorado River. *Source History: 2 in 1998; 2 in 2004.* **Sources: Ho13, Na2.**

Cocopah White - (Frijoles Cucupa) - Grown by the Chupa (River People) or Cocopah living in Sonora, Mexico. *Source History: 1 in 1994; 1 in 1998; 1 in 2004.* **Sources: Na2.**

Colonia Morelos Speckled - Medium size, flat seeds, colors include brown, blue, beige, yellow and tan speckles on light tan background and gray-black speckles, matures early, originally from Colonia Morelos, Sonora. *Source History: 1 in 2004.* **Sources: Na2.**

Guarijio White - 92 days, Formerly Warihio White, medium-large somewhat flattened white tepary, prolific even in hot dry weather, native staple in the Sonoran regions of Mexico. *Source History: 1 in 1991; 2 in 1994; 1 in 1998; 2 in 2004.* **Sources: Na2, Syn.**

Hopi White - Large tepary, dry farmed by Hopi farmers in northeastern Arizona. *Source History: 1 in 1994; 1 in 2004.* **Sources: Na2.**

Kickapoo White - Native Mexican bean adopted by the Kickapoos a transplanted Algonquian tribe, Rio Bavispe, Sonora. *Source History: 1 in 1987; 1 in 1991; 1 in 1994; 1 in 2004.* **Sources: Na2.**

Light-Brown Seeded - *Source History: 1 in 1991; 1 in 1994; 1 in 1998; 1 in 2004.* **Sources: Red.**

Little Tucson Brown - Round, burnt-orange seeds, originally from Tohono O'odham reservation, Arizona. *Source History: 1 in 2004.* **Sources: Na2.**

Mayo White - Small white and green bean, dry farm staple. *Source History: 1 in 1987; 1 in 1991; 1 in 1994; 1 in 1998; 1 in 2004.* **Sources: Na2.**

Mitla Black - 75-90 days, Slightly smaller and rounder than Sonoran tepary beans, same good flavor and nutrition, less day-length sensitive, grow better in northern latitudes than the Sonoran types, collected in Oaxaca, Mexico. *Source History: 1 in 1991; 3 in 1994; 6 in 1998; 8 in 2004.* **Sources: Fe5, Pe1, Pr3, Sa5, Se7, Se8, So12, Syn.**

O'odham Brown - 60-90 days, Adapted to the desert climate, requires very little water, most often planted just before monsoon season in Tucson, flourishes with only the summer rains, produces pods very quickly. *Source History: 1 in 1987; 2 in 1991; 1 in 1994; 1 in 1998; 1 in 2004.* **Sources: So12.**

Paiute Mixed - Bean colors include chocolate brown, speckled light tan and burnt orange, from Shivwits Paiute Res. in Utah. *Source History: 2 in 1998; 2 in 2004.* **Sources: Ho13, Na2.**

Paiute Yellow - Ochre colored traditional favorite from the Kaibab Indian Reservation in southern Utah. *Source History: 1 in 1991; 1 in 1994; 1 in 2004.* **Sources: Na2.**

Papago Brown - Light brown pods produced quickly, desert adapted, most often planted just before summer rains. *Source History: 1 in 1998; 1 in 2004.* **Sources: Ho13.**

Pima Beige and Brown - Seeds are mixed shades of beige, gold and tan, originally collected at Santan, Arizona on the Gila River Indian Reservation. *Source History: 1 in 1991; 1 in 1994; 2 in 1998; 2 in 2004.* **Sources: Ho13, Na2.**

Pinacate - 90-95 days, Tan slightly mottled seeds, originally obtained from an extremely arid runoff farm in Mexico in the Sierra El Pinacate Protected Zone. *Source History: 1 in 1991; 1 in 1994; 2 in 1998; 2 in 2004.* **Sources: Na2, So12.**

Pinacate, Black - 90-95 days, Selected from Pinacate mother strain, moderate yields of dappled ebony seeds. *Source History: 1 in 2004.* **Sources: So12.**

Sacaton Brown - Medium size, orange-tan seeds, once commercially cultivated by the Gila River Pima near Sacaton, Arizona. *Source History: 1 in 1998; 1 in 2004.* **Sources: Na2.**

San Felipe Pueblo White - Formerly San Felipe Pueblo, large white seeds grown along the Rio Grande River in northern New Mexico. *Source History: 1 in 1991; 1 in 1998; 1 in 2004.* **Sources: Na2.**

San Ignacio - White seeds, collected on the Rio Magdalena in northern Sonora. *Source History: 1 in 2004.* **Sources: Na2.**

San Pablo Balleza - Black tepary from 5,000 ft. elevation in the Chihuahua desert of Mexico. *Source History: 1 in 1991; 1 in 1994; 1 in 1998; 1 in 2004.* **Sources: Na2.**

Sonoran - 85-110 days, Rambling 2-3 ft. plants produce 3-4 in. pods, small beans with over 30% crude protein, under native desert conditions will mature in 70 days, longer growing season required with cooler temps, native to the Sonoran Desert. *Source History: 1 in 1987; 2 in 1991; 2 in 1994; 3 in 1998; 4 in 2004.* **Sources: Ho13, Pla, Se7, So12.**

Sonoran Gold - Colorful heirloom from the Southwest, golden brown, flattened seed, drought res. *Source History: 1 in 1994; 6 in 1998; 3 in 2004.* **Sources: Ho13, Pr3, Syn.**

Sonoran White - Small to medium size seed, matures early. *Source History: 1 in 2004.* **Sources: Na2.**

Tepary - (Tepari) - 90 days, *Phaseolus acutifolius*, drought res. species for hot arid climates, grown by Mexican Indians near Tehuacan 5,000 years ago, colors from white to brown. *Source History: 3 in 1981; 3 in 1984; 4 in 1987; 1 in 1994; 1 in 1998; 1 in 2004.* **Sources: Ech.**

Tohono O'odham Brown - (Menager's Dam Brown) - Red-brown bean from Menager's Dam near the Mexican border. *Source History: 1 in 1991; 1 in 1994; 1 in 1998; 1 in 2004.* **Sources: Na2.**

Tohono O'odham White - Medium size, white seeds, several elderly traditional O'odham tell us they prefer the taste of these white teparies. *Source History: 1 in 1991; 1 in 1994; 2 in 2004.* **Sources: Ho13, Na2.**

Virus-Free Yellow - Ochre color seeds, USDA selected and grown out in Tucson, do not infect by growing near other teparies as others may carry BCMV. *Source History: 1 in 1991; 1 in 1994; 1 in 1998; 1 in 2004.* **Sources: Na2.**

Yaqui - Beige bean from a traditional Yaqui village on the southern coastal plain of Sonora. *Source History: 1 in 1991; 1 in 1994; 1 in 1998; 1 in 2004.* **Sources: Ho13.**

Yoeme Brown - Formerly Yoeme, colorful mix of tan-brown and pink-brown seeds, from a traditional Yoeme village on southern Sonora's coastal plain. *Source History: 1 in 1998; 1 in 2004.* **Sources: Na2.**

Varieties dropped since 1981 (and the year dropped): *Speckled* (1982), *Brown* (1998), *Colonia Morelos White* (2004), *Cumpas* (2004), *Durango* (2004), *Durango Aluvia* (1991), *Light Brown* (1984), *Light Brown Mottled* (1998), *Light-Green Seeded* (1994), *Mitla Speckled* (1998), *Namaquipe White* (2004), *Paiute White* (1998), *Pale Green* (1984), *Sacaton White* (2004), *San Pedro* (2004), *Santa Rosa* (2004), *Speckled* (2004), *Tiburon Island* (1998), *Ures White* (1998), *Virus-Free White* (1998), *White* (2004), *Wild Broad Leafed* (1987), *Wild Narrow Leafed* (1987), *Wild Willow-leaf* (2004).

Tomatillo
Physalis ixocarpa

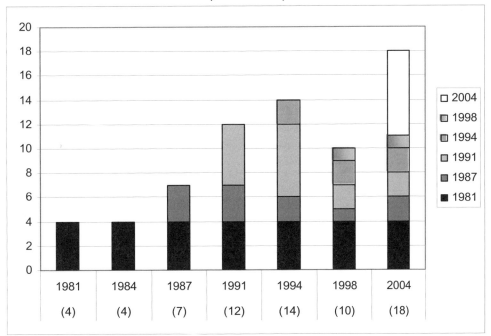

Number of varieties listed in 1981 and still available in 2004: 4 to 4 (100%).
Overall change in number of varieties from 1981 to 2004: 4 to 18 (450%).
Number of 2004 varieties available from 1 or 2 sources: 7 out of 18 (39%).

Cisineros - 75 days, Giant strain, green fruits with diameter up to 6 cm, lime taste, harvest when still green. *Source History: 1 in 2004.* **Sources: So25.**

Giant Yellow - 70-80 days, Vigorous indet., rambling vines covered with 1.25 in. sweet fruit. *Source History: 2 in 1998; 3 in 2004.* **Sources: Sa9, Se17, So25.**

Grande Rio Verde - 74 days, Medium det. plant, 3-3.5 oz. globe shaped fruit, sweet-tart taste, good yields. *Source History: 4 in 2004.* **Sources: ME9, OSB, RIS, SE14.**

Indian Strain - 55 days, Smaller, faster maturing var., adds a sweet-sour flavor to Mexican dishes, good in salsas and salad dressings, Territorial intro. *Source History: 2 in 1994; 4 in 1998; 3 in 2004.* **Sources: Ho13, Hud, Ma13.**

Indian Summer - Later maturity than De Milpa with higher fruit production. *Source History: 1 in 2004.* **Sources: St18.**

Large Green - 75 days, Husked fruit, 2-3 in. dia., does not keep as long as purple variety. *Source History: 1 in 1981; 1 in 1984; 1 in 1987; 1 in 1991; 2 in 1994; 3 in 1998; 6 in 2004.* **Sources: Bo19, Ho13, Hud, MOU, Red, To1.**

Merida Market - Smaller fruits than Purple but more dependable yields, bothered less by husk worms. *Source History: 1 in 2004.* **Sources: Sa9.**

Mexican Strain - 65 days, Larger and more flavorful var., 2 in. dia., used in Mexican dishes and spaghetti sauce, high yields, Territorial intro. *Source History: 2 in 1994; 2 in 1998; 1 in 2004.* **Sources: Ter.**

Miltomate - (Miltomato Loco) - 90 days, Small plants bear many tiny, pea-size purple fruit used in Mexico to make salsa. *Source History: 3 in 2004.* **Sources: Se17, Se26, So25.**

Miltomato Vallisto - Larger plants and fruit than Miltomato loco and Sunberry, dark purple berries, naturally sweet fruity flavor, makes good jam or pie filling or can be eaten fresh, productive, requires full sun. *Source History: 1 in 2004.* **Sources: So25.**

Pineapple - 75 days, Short, spreading plants, fruity taste

resembles pineapple, good for fruit salsa. *Source History: 3 in 2004.* **Sources: CO30, SE14, To1.**

Purple - (Purple Husk Tomatillo) - 68 days, Rare heirloom variety, smaller 1-1.5 in. fruit has sweeter flavor than green tomatillo, makes attractive purple salsa, noticeably less sticky than most tomatillos, easy to grow. *Source History: 1 in 1987; 5 in 1991; 6 in 1994; 9 in 1998; 16 in 2004.* **Sources: Ba8, Gr28, Jo1, Jo6, Lej, ME9, Ni1, Pep, Pin, Pr3, Sa9, SE14, Se16, TE7, Ter, To1.**

Purple de Milpa - 70-90 days, The small purple-tinged variety which grows wild in cornfields, 3-4 ft. tall plant, nickel-size fruit in papery husks which sometimes completely encloses the fruit, making it harder to harvest than large-fruited vars., sharp flavor preferred by some cooks. *Source History: 1 in 1981; 1 in 1984; 2 in 1987; 3 in 1991; 3 in 1994; 9 in 1998; 13 in 2004.* **Sources: Goo, Hi13, Ho13, Jo1, Pe2, Pr3, Ri12, Roh, Se7, Se26, So1, Syn, Und.**

Tepehuan - *P. philadelphica*, small sweet fruit, common condiment ingredient, may not produce until September. *Source History: 1 in 1987; 1 in 1991; 1 in 1994; 1 in 2004.* **Sources: Na2.**

Tomate Verde - 60-95 days, Vines grow 8 ft. or longer, 2-3 in. fruit, adds unique sweet-sour flavor to Mexican dishes, green fruit is enclosed in a papery husk. *Source History: 1 in 1991; 8 in 1994; 27 in 1998; 45 in 2004.* **Sources: Ba8, Bo17, Com, Coo, Dam, Ers, Fe5, Gr29, GRI, He8, Hi6, Hi13, Jo1, Jo6, Jor, Jun, Ki4, ME9, Na6, Or10, OSB, Par, Pe2, Ra5, Ra6, Re6, RIS, Roh, Scf, Se17, Se25, Se26, Se28, So1, So25, Sto, Syn, To1, Twi, Ves, Vi4, Vi5, We19, WI23, Wo8.**

Tomatillo - (Green Tomatillo, Green Husk Tomatillo, Mexican Ground Cherry) - 70-80 days, Mexican husk tomato, bushy plants reach 3-4 ft. across and almost as tall, fruit is ripe when it bursts through the husk, green sticky and sweet, blended with hot peppers to make Mexican green sauce, eaten off the vine, used in pies, jams, sauces and salsas, very easy to grow. *Source History: 12 in 1981; 10 in 1984; 15 in 1987; 29 in 1991; 33 in 1994; 44 in 1998; 42 in 2004.* **Sources: Ami, Bo9, Bu2, CA25, CO23, Coo, Enc, Fo7, Ga22, Ge2, Gr28, He17, Jor, La1, Lej, LO8, Loc, Mel, MIC, Ni1, Pep, Pin, Pla, Ra5, Ra6, Re8, Ren, Ri2, RUP, Sa5, Sa9, SE4, Se8, SE14, Se16, So12, Te4, TE7, Tho, Vi5, WE10, Wi2.**

Verde Puebla - (Puebla Verde) - 60-90 days, Det. plant, 1.25 in. dia., globe shaped fruit in a papery husk, rich flavor, more sprawling vines than tomatoes, for homemade Mexican green sauce. *Source History: 3 in 1991; 5 in 1994; 10 in 1998; 10 in 2004.* **Sources: CO30, Ga1, Hum, Lin, MAY, Orn, Ri12, RUP, Se7, To1.**

Zuni - *P. philadelphica*, predates tomato, small, sweet, husk, prolific, used in salsa, spring planted, high cool desert, Zuni Indian seed. *Source History: 1 in 1981; 1 in 1984; 1 in 1987; 2 in 1991; 1 in 1994; 2 in 1998; 1 in 2004.* **Sources: Ho13.**

Varieties dropped since 1981 (and the year dropped):

Giant Green (1998), *Medium Purple* (1998), *Medium Yellow* (1998), *Tarahumara* (1998).

Tomato / Orange-Yellow
Lycopersicon lycopersicum

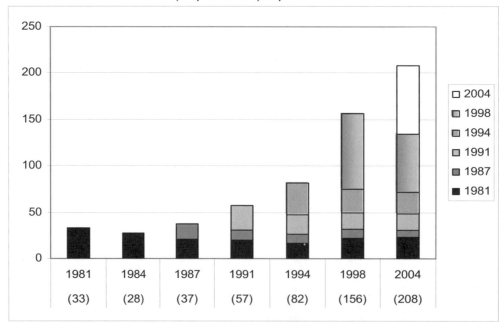

Number of varieties listed in 1981 and still available in 2004: 33 to 23 (70%).
Overall change in number of varieties from 1981 to 2004: 33 to 208 (630%).
Number of 2004 varieties available from 1 or 2 sources: 119 out of 208 (57%).

Amana Orange - 80-90 days, Indet., glowing orange beefsteak type, up to 5 in. dia. fruit, 1-2 lb., mild flavor, late season, heirloom from Amana, Iowa. *Source History: 8 in 1998; 23 in 2004.* **Sources: Ba8, CO30, Coo, Ga1, Go8, He17, Ho13, Ma18, ME9, Mel, Pe2, Ra5, Ra6, Re8, RUP, Sa9, SE14, Ta5, TE7, To1, To3, To9, WE10.**

Amber-Colored - 68-75 days, Dwarf, upright, 12 in. bush plants with rugose leaves, heavy crops of round, 2 in., amber-yellow fruits, from Russia. *Source History: 1 in 1994; 3 in 1998; 2 in 2004.* **Sources: Ma18, Syn.**

Amish, Large Yellow - 78 days, Indet., large 4 in. fruit, yellow-orange skin, apricot color flesh with unique sweet flavor, heirloom. *Source History: 1 in 2004.* **Sources: To9.**

Anna Hermann - Rare Russian tomato named after a singer who liked this variety, med. size yellow fruit, heart shape, mild flavor, resembles Lemon or Wonder Lights. *Source History: 1 in 2004.* **Sources: So25.**

Aranyalma - 75 days, Hungarian variety, name translates Golden Apple, indet. plant with few disease problems, 3-4 oz. bright yellow fruit, fruity, sweet tomato flavor. *Source History: 1 in 2004.* **Sources: Go8.**

Aunt Gertie's Gold - 75-85 days, Large indet. potato leaf plant, golden fruit up to 1 lb., some blemishes but the taste is excellent, original seed from Virginia. *Source History: 1 in 2004.* **Sources: He18.**

Azoychka - 81-85 days, Russian heirloom, productive, det., 4 ft. plants require some support, round, slightly flattened, 3-3.5 in., yellow fruits, best described as a citrus taste, "Azoychka" is a Russian woman's name. *Source History: 1 in 1994; 5 in 1998; 12 in 2004.* **Sources: CO30, He18, Ma18, Ra5, Ra6, Re8, RUP, Ta5, To1, To3, To9, To10.**

Balconi Yellow - 77 days, Clusters of delicious bright yellow fruit, same culture as Balconi Red, ideal for containers. *Source History: 1 in 2004.* **Sources: Tho.**

Banana - (Orange Banana) - 85 days, Indet., beautiful, persimmon colored paste tomato, 2 x 4 in. long, orange skin with thick orange flesh, fruity sweet flavor, dries very well, excel. yield. *Source History: 2 in 1998; 2 in 2004.* **Sources: Ag7, Sa9.**

Banana Legs - (South American Banana) - 72-90 days, Indet., lacy foliage, 4 in. long, yellow, paste-like fruit resembles a banana, 8-10 per lb., heirloom. *Source History: 3 in 1994; 12 in 1998; 30 in 2004.* **Sources: Bo19, CO30, Fo7, Ga1, Gr28, Gr29, He8, He18, Ho13, Loc, Ma18, ME9, Orn, Pe2, Ra5, Ra6, Ri2, SE14, Se17, Se28, Sh13, So25, Ta5, TE7, To1, To3, To9, Vi4, Vi5, Wo8.**

Basinga - 80 days, Indet. with sparse foliage, may benefit from some shade, large, pale lemon-yellow globes, 4 x 5 in., mild flavor, cooks into a sweet delicate sauce. *Source History: 1 in 1998; 5 in 2004.* **Sources: Ap6, Ma18, Se16, Ta5, To9.**

Beam's Yellow Pear - 70-80 days, Indet., endless supply of 1.5 in. pear tomatoes with great taste, ideal for salads. *Source History: 2 in 2004.* **Sources: Pr9, Se16.**

Beefsteak, Chuck's Yellow - 80 days, Indet., excellent yellow tomato with rich full flavor into the cold weather of fall, creamy texture resembles that of Yellow Brandywine. *Source History: 1 in 2004.* **Sources: Ma18.**

Beefsteak, Orange - 80 days, Indet., yellow-gold 4 in. fruit, sweet mild flavor, high yields. *Source History: 1 in 2004.* **Sources: To9.**

Big Golden Delicious - 85 days, Indet., 1+ lb. beefsteak fruit, deep golden color, great producer. *Source History: 1 in 2004.* **Sources: Ma18.**

Big Orange - 75 days, Indet., orange beefsteak fruit, 1 lb., good sweet flavor. *Source History: 1 in 2004.* **Sources: To9.**

Big Yellow - 80 days, Indet., large flattened, yellow-orange fruit, some green shoulders, 8-10 oz., nice flavor. *Source History: 1 in 1998; 1 in 2004.* **Sources: To9.**

Blondkopfchen - 70-75 days, Indet., .5-1 in. yellow cherries borne in clusters of 20-30, fruity flavor, very prolific, name translates Little Blond Girl, from Helliner, France via Norbert Parreira. *Source History: 1 in 1998; 7 in 2004.* **Sources: Ea4, Ma13, Ma18, Se16, So25, Ta5, To9.**

Brandywine, Platfoot Yellow - 85-90 days, Improved version of Yellow Brandywine with increased productivity and smoother shape, indet. vines, potato leaf foliage, deep golden yellow fruit, 1-2 lbs., rich flavor, from Gary Platfoot of Ohio. *Source History: 3 in 2004.* **Sources: Ap6, To1, To9.**

Brandywine, Yellow - 76-100 days, Indet., fuzzy potato leaf vines, large fruit can weigh up to 2 lbs., excellent flavor for a yellow variety, delicious creamy texture, exceptional quality, temperature extremes may affect fruit shape, heirloom from SSE member Barbara Lund, OH, via Charlie Knoy, Bloomfield, IN. *Source History: 6 in 1994; 22 in 1998; 48 in 2004.* **Sources: Al25, Ba8, Bo19, CA25, CO30, Co32, Ers, Fe5, Ga1, Ge2, Gr28, He8, He18, HO2, Ho13, Jo1, Jun, Lan, Loc, Ma18, ME9, MIC, Or10, Orn, Pe7, Ra5, Ra6, Re8, Ri12, RUP, Scf, SE14, Se26, Shu, So1, So9, So25, Sw9, Te4, TE7, Ter, To1, To3, To10, Up2, Vi4, WE10, We19.**

Brimmer, Yellow - 79 days, Indet., large yellow-orange beefsteak fruit, mild luscious flavor. *Source History: 1 in 2004.* **Sources: To9.**

Brown's Yellow Giant - 90 days, Indet., regular leaf, 3 x 5 in. fruit, 10-16 oz., mild fruity flavor, late season. *Source History: 1 in 2004.* **Sources: To9.**

Caro Rich - 72-80 days, Indet., does not require trellising, old variety, large 5-8 oz. beefsteak type fruits, rich deep-golden color, high carotene content, very low acid, too late for far Northern areas. *Source History: 1 in 1981; 1 in 1984; 4 in 1987; 9 in 1991; 10 in 1994; 15 in 1998; 20 in 2004.* **Sources: Bo9, Bou, Ga22, Gr28, Ma18, Ni1, Pe2, Ra5, Ra6, Re8, Se7, Se8, Se17, So12, Syn, Te4, To1, To3, To10, WE10.**

Cerise Orange - 75-80 days, Indet., bright orange-yellow cherries with subtle striping, fruity sweet flavor, from Norbert Parreira, Helliner France, 1992. *Source History: 3 in 2004.* **Sources: Ma18, To1, To9.**

Ceses Russian - Indet., baseball size yellow fruit, good taste, productive. *Source History: 1 in 2004.* **Sources: Up2.**

Chello - 55-80 days, Early yellow cherry on compact semi-bush plant, excellent sweet tomato flavor, 1 in. golden yellow fruits, bears all summer. *Source History: 1 in 1994; 2 in 1998; 3 in 2004.* **Sources: Ra5, Ra6, So25.**

Cherry, Amish - 75 days, Large, vigorous plants produce an abundance of orange cherry fruits covered with tiny golden hairs, unusual flavor resembles cranberries, grown by the Amish community in the U.S. *Source History: 1 in 2004.* **Sources: So25.**

Cherry, Aunt Ruby's Yellow - 75 days, Vigorous indet. Vines .75-1 in. round fruit, fruity tart flavor, very productive, heirloom. *Source History: 1 in 1998; 2 in 2004.* **Sources: Gr28, To9.**

Cherry, Esther's - Indet., regular leaf, very productive yellow cherry with good taste. *Source History: 1 in 2004.* **Sources: Up2.**

Cherry, Galina's - 59-75 days, Indet. plant produces long clusters of round yellow 1 in. fruit, weigh about 1 oz., 6-8 per cluster, very flavorful, good sweet/acid balance, meaty enough for salsa and sauce, from Siberia. *Source History: 1 in 1991; 3 in 1994; 4 in 1998; 8 in 2004.* **Sources: Hig, Ma18, Sa5, Se17, Ta5, Te4, To1, To9.**

Cherry, Lollipop - 70-80 days, Indet., clusters of light lemon-yellow cherry tomatoes, sweet, tangy flavor with a touch of citrus, good yields, good foliage disease res., bears until frost. *Source History: 2 in 1998; 4 in 2004.* **Sources: Ba8, So1, So25, Und.**

Cherry, Porters Pride - 70 days, Indet., round yellow cherry type fruit. *Source History: 1 in 2004.* **Sources: Pe2.**

Cherry, Tennessee Yellow - 72-85 days, Indet., high yields of .75 in. yellow fruit in clusters of 6-8, delicious fruity flavor. *Source History: 1 in 1998; 1 in 2004.* **Sources: To9.**

Cherry, Una's Yellow - 75-80 days, Larger size, flavorful yellow cherry, high yields. *Source History: 1 in 1994; 1 in 1998; 2 in 2004.* **Sources: Sa9, Ta5.**

Cherry, Warren's Yellow - 75-80 days, Vigorous, med. size indet. plants bear yellow cherry fruit. *Source History: 1 in 1998; 8 in 2004.* **Sources: CA25, He18, ME9, Ra5, Ra6, Re8, SE14, TE7.**

Cherry, Yellow - 70-95 days, Good-sized vigorous indet. plant, yellow cherry-shaped fruits .75-1 in. dia., hardy. *Source History: 3 in 1981; 2 in 1984; 5 in 1987; 5 in 1991; 6 in 1994; 4 in 1998; 7 in 2004.* **Sources: Ga1, Gr28, Ma18, Sa9, Se26, Te4, To3.**

Coeur de Pigeon Jaune - 75 days, Indet., bright yellow plums with fruity flavor, good for snacking and salad, prolific, from France. *Source History: 2 in 2004.* **Sources: Ma18, To9.**

Colossal Golden, Burgess - 90 days, Large spreading well-branched plants, heavy protective foliage, deep flat thick-fleshed meaty fruits average 14 oz. but have grown to 6 in. across and 2.5 lbs., 1948. *Source History: 1 in 1981; 1 in 1984; 1 in 1987; 1 in 1991; 2 in 1998; 2 in 2004.* **Sources: CO30, To1.**

Czech's Excellent Yellow - 70-75 days, Indet., 3 in. round, orange-yellow fruit, 3-4 oz., surprisingly strong flavor for a yellow tomato, great for salads, great var. from Ben Quisenberry. *Source History: 4 in 2004.* **Sources: Pr9, Se16, St18, Ta5.**

Dad's Sunset - 75-95 days, Meaty paste tomato with tangy, yellow flesh, 4-12 oz., thick yet tender walls, excellent flavor, slightly tart, good keeper, good insect res., ripens to a golden orange resembling the setting sun. *Source History: 1 in 1994; 4 in 1998; 4 in 2004.* **Sources: Ba8, So1, Ta5, To9.**

Dakota Gold - 80 days, Det., large, pale-yellow fruit, 3 in. dia., 10 oz., low acid. *Source History: 1 in 1994; 3 in 1998; 3 in 2004.* **Sources: Fo7, Sa9, Ta5.**

D'anjou Pear - Indet., fern-like foliage, fruity sweet taste, most flavorful of the yellow pears. *Source History: 1 in 2004.* **Sources: Se27.**

De Barao Gold - 96 days, Indet., orange-yellow, elongated globe shaped fruit, 2 oz. *Source History: 1 in 1998; 1 in 2004.* **Sources: Sa9.**

Dicoff's Yellow - 80 days, Indet., good set of 10-12 oz. solid fruit, flavorful keeper, late season. *Source History: 1 in 2004.* **Sources: To9.**

Dixie Golden Giant - 85-100 days, Indet., giant golden-yellow tomato, sweet mild taste, good for slicing, vigorous grower, some fruits weigh 2-2.5 lbs., grown by an Amish family since the 1930s. *Source History: 1 in 1981; 3 in 1998; 6 in 2004.* **Sources: Ra5, Ra6, Re8, To1, To3, To9.**

Djena Lee's Golden Girl - 80 days, Family heirloom since 1929, won 1st prize at the Chicago Fair 10 years in a row, indet., yellow-orange fruits, richly balanced sweet tangy flavor, grown by Djena Lee of Minnesota, who was part Indian and granddaughter of Minnesota financier, Jim Lee, after moving to Illinois in 1929, Djena gave some plants to Rev. Morrow who was 15 years old at the time, pronounced 'Zshena.' *Source History: 1 in 1987; 1 in 1991; 2 in 1994; 6 in 1998; 10 in 2004.* **Sources: Bou, Go8, He18, ME9, Pe2, SE14, So1, Te4, To9, Tu6.**

Dr. Wyche's Yellow - 75-90 days, Indet., sparse foliage, slightly flattened, golden yellow fruit with some green shoulders, 8- 20 oz., meaty flesh with few seeds, rich flavor, heirloom named after Dr. Wyche. *Source History: 1 in 1998; 12 in 2004.* **Sources: Ap6, He18, Ma18, Pr9, Sa9, Se16, Se26, So25, Ta5, To1, To3, To9.**

Earl of Edgecombe - 73-85 days, Indet., sparse foliage, 3 in., round orange fruit, free of cracks and blemishes, 6-12 oz., firm, meaty flesh, good flavor, resists cracking, blossom end rot and fruit diseases, New Zealand heirloom named after the heir to the title of the 7th Earl of Edgecombe in the 1960s. *Source History: 2 in 1998; 10 in 2004.* **Sources: Ap6, He18, Ma18, Sa5, Sa9, So1, So25, Ta5, To9, To10.**

Elbe - 89 days, Indet. plants benefit from caging, beefsteak type fruit has been a favorite since 1889, unique flavor is sweet and tart at the same time, 8-12 oz., deep yellow to orange, named after the Elbe River in Germany. *Source History: 2 in 1998; 5 in 2004.* **Sources: He8, So25, Ta5, To9, Yu2.**

Elfie - 85-90 days, Indet., bright deep orange beefsteaks, slight green cast on shoulders, 8-12 oz., juicy flesh is sweet and sugary, few seeds, no cracking. *Source History: 2 in 2004.* **Sources: Ma18, Ta5.**

Esther Hess Yellow - 68-80 days, Rougher version of

Golden Sunrise, sprawling indet. vines, 1-2 in. seedy yellow fruits, tangy flavor, heirloom via SSE. *Source History: 1 in 1998; 3 in 2004.* **Sources: Ea4, Sa9, Ta5.**

Extra Eros Zlatolaska - 86 days, Det. plants, orange-red, 3 oz. plums. *Source History: 1 in 1998; 2 in 2004.* **Sources: Sa9, Ta5.**

Fablonelystynj - 70 days, Indet., tiny yellow beefsteaks, some irregular shapes, tasty. *Source History: 1 in 2004.* **Sources: Ma18.**

Fargo - 80 days, Det. vines bear an abundance of 2.5 in. yellow fruit, same good quality flavor as Yellow Pear but with shorter vines. *Source History: 1 in 2004.* **Sources: To9.**

Faribo Goldheart - 81 days, Indet., abundant production of 3 in. orange globes, delicious, late season. *Source History: 1 in 2004.* **Sources: To9.**

Farnsworth - 91 days, Indet., very large oxheart shaped fruit, yellow flesh. *Source History: 1 in 2004.* **Sources: Sa9.**

Flamme - (Jaune Flamme, Flammee, Jaune Flammee) - 70-90 days, Indet., beautiful salad tomato, apricot shaped fruit, lovely persimmon-orange color inside and out, sweet intense flavor with fruity overtones, 2-3 oz. fruit borne in clusters of six, 1.5 in., French heirloom. *Source History: 1 in 1998; 17 in 2004.* **Sources: Ap6, Bu2, Ga1, Go8, He8, He18, Hud, Ma18, Pr9, Sa9, SE14, Se16, Ta5, To1, To3, To9, Und.**

Floragold Basket - 85 days, Dwarf det. plant produces many golden yellow cherry tomatoes, meaty flesh, ornamental. *Source History: 1 in 1991; 1 in 1994; 1 in 1998; 1 in 2004.* **Sources: So25.**

Fruity Orange - High yields of tasty orange cherries. *Source History: 1 in 2004.* **Sources: Ea4.**

Galina Grande - 75 days, Indet., round, creamy yellow cherry, 1.5 in., flavor has good sweet/citrus balance, midseason. *Source History: 1 in 2004.* **Sources: To9.**

German Golden Ribbed - Indet., regular leaf, heavily ribbed golden fruit, mild flavor, good yields, midseason. *Source History: 1 in 2004.* **Sources: Up2.**

German Large - 87 days, Large plants, deep orange, round fruit, semi-sweet flavor. *Source History: 1 in 1998; 2 in 2004.* **Sources: Sa9, Se17.**

German Orange Strawberry - 90 days, Indet., large orange fruit, oxheart shape, paste type, heirloom. *Source History: 1 in 1998; 3 in 2004.* **Sources: Ga1, ME9, SE14.**

Giant Belgium, Yellow - (Goldie) - 85-95 days, Robust indet., globe shaped fruit, golden yellow skin, smooth blossom ends, tasty mild flesh, low acid flavor, 150 year-old heirloom dating back to the pioneer days. *Source History: 1 in 1998; 6 in 2004.* **Sources: Ers, ME9, Ra5, Ra6, Re8, SE14.**

Glory of Moldova - 98 days, Indet., oval orange fruits, 4 oz., orange flesh, mild flavor, good for juice, late. *Source History: 3 in 1998; 4 in 2004.* **Sources: Sa9, Sh13, Ta5, To9.**

Gold Dust - 60-65 days, Extra early orange skinned and fleshed smooth fruits on semi-det. plants, firm, crack resistant and blemish free, uniform ripening, dev. at the Univ. of NH. *Source History: 1 in 1981; 1 in 1984; 1 in 1987; 1 in 1991; 2 in 1994; 3 in 1998; 7 in 2004.* **Sources: Fis, Pr3, Ra5, Ra6, Re8, Se26, Ta5.**

Gold Nugget - 55-70 days, Early, det., uniform compact plants 24 in. tall average, bright golden fruits ping pong ball size, many seedless fruits, mild flavor, prolific, Dr. James Baggett/OSU. *Source History: 1 in 1981; 1 in 1984; 5 in 1987; 14 in 1991; 20 in 1994; 20 in 1998; 29 in 2004.* **Sources: Bo19, CO30, Ea4, Ers, Ga1, Ga22, Gr28, Hig, Hum, Jo1, ME9, Ni1, Pin, Ra5, Ra6, Re8, RUP, Se26, Sk2, So25, Syn, Ter, Tho, To1, To3, Up2, Vi4, WE10, We19.**

Gold Rush - Bushy plants, extra early yellow paste tomato, meaty flesh. *Source History: 1 in 1998; 1 in 2004.* **Sources: Pr3.**

Golden Delight - 57-79 days, Extremely early, compact det. plants, butter-yellow 3-8 oz. globe-shaped fruits, heavy yields, low acid but excel. flavor, dev. at South Dakota State University. *Source History: 6 in 1981; 6 in 1984; 8 in 1987; 6 in 1991; 5 in 1994; 2 in 1998; 4 in 2004.* **Sources: Ap6, Gr28, Sa9, Tu6.**

Golden Dwarf Champion - 83 days, Compact, rugose plants, medium size, bright butter-yellow version of Pink Champion, an old Shumway variety. *Source History: 1 in 1981; 1 in 1984; 1 in 1987; 1 in 1991; 1 in 2004.* **Sources: Sa9.**

Golden Egg - 68-80 days, Indet., 1.5 in. egg shaped fruit in clusters of 6-10, gold skin, firm yellow flesh, semi-sweet flavor. *Source History: 3 in 2004.* **Sources: Sa9, To10, Up2.**

Golden Grape - (Yellow Grape) - 85 days, Indet., fruit clusters measure 9 in. across, grape size, yellow cherry, intense flavor. *Source History: 1 in 1998; 2 in 2004.* **Sources: Gr28, To9.**

Golden Monarch - 80 days, Indet., good production of 10-14 oz., pale yellow fruit, flattened shape, 3 in., listed by Buist in 1946. *Source History: 1 in 2004.* **Sources: To9.**

Golden Pearl - 60-67 days, Old-time cherry the size of a thumbnail, indet. vines bear dozens of grape-like clusters of golden yellow fruits, sweet fruity flavor, crack res. *Source History: 1 in 1994; 1 in 1998; 1 in 2004.* **Sources: Ma18.**

Golden Queen - (Jubilee) - 65-85 days, Indet., large smooth waxy yellow-orange fruits, 8-12 oz., meaty, few seeds, sweet mild flavor, low acid, dates back to 1882, Livingston introduction. *Source History: 5 in 1981; 3 in 1984; 6 in 1987; 7 in 1991; 7 in 1994; 7 in 1998; 11 in 2004.* **Sources: Coo, He8, Lan, Ma18, Ont, Pin, So25, Ta5, To9, Up2, Vi4.**

Golden Queen Sun Ray - Indet., lower acid orange-yellow fruit, meaty flesh, milder flavor, good leaf cover and disease res. *Source History: 1 in 1991; 1 in 1994; 1 in 1998; 1 in 2004.* **Sources: Lin.**

Golden Roma - 60 days, Indet., bright lemon-yellow fruit, long shape, 1.5 in. wide x 3 in. long, sweet meaty flesh, good processing tomato. *Source History: 7 in 2004.* **Sources: Ba8, Go8, ME9, Pe2, Ra6, SE14, TE7.**

Golden Russian Ball - (Zolotoi Char) - Med. size, yellow-orange, round fruit, nice mild flavor, Russian. *Source History: 2 in 1998; 2 in 2004.* **Sources: CO30, Ho13.**

Golden Sunburst - (Golden Moneymaker) - 80 days, Indet. vine, excellent quality 3 in. golden fruit, suited for fresh market or home garden, Royal Horticultural Show (England) winner. *Source History: 1 in 1991; 2 in 1994; 3 in 1998; 6 in 2004.* **Sources: CO30, Ga1, ME9, SE14, Ta5, WE10.**

Golden Sunrise - 75-78 days, Indet. plant, smooth med-sized yellow fruits, tangy distinctive flavor especially good for a yellow type, average disease tol., England. *Source History: 3 in 1981; 3 in 1984; 4 in 1987; 4 in 1991; 3 in 1994; 2 in 1998; 2 in 2004.* **Sources: Se26, Tho.**

Goldene Konigen - 75-80 days, Indet., pale yellow, 2 in. round fruit, flavorful, better adapted to cool summers than most yellow varieties, German heirloom. *Source History: 1 in 1994; 1 in 1998; 1 in 2004.* **Sources: Syn.**

Goldie - (Dixie Golden Giant, Yellow Giant Belgium) - 85-100 days, Very vigorous indet. vines, huge golden slightly flattened globes, smooth blossom ends, mild fine flavor, mostly giants, possibly a yellow version of Giant Belgium, 150 year old heirloom from pioneer gardens. *Source History: 2 in 1981; 2 in 1984; 4 in 1987; 4 in 1991; 6 in 1994; 9 in 1998; 16 in 2004.* **Sources: Bo9, CO30, Fe5, Ho13, ME9, Red, Sa9, SE14, Se17, Se26, So25, Syn, Ta5, Te4, To1, To3.**

Goldie Boy - Indet., regular leaf, yellow fruit, mild yet full flavor, early. *Source History: 1 in 2004.* **Sources: Up2.**

Goldmine - 55-60 days, Productive, det. vines, extra early, golden yellow Roma type fruit, sweet flavor, PSR. *Source History: 1 in 1998; 1 in 2004.* **Sources: Pe6.**

Gramma Climen Hagen - 78 days, Indet., 3 in. round bright orange fruit, 6-10 oz., meaty flesh, good citrus/acid flavor, good yields, heirloom from T. McIntee's grandmother. *Source History: 1 in 2004.* **Sources: To9.**

Grape Tress - Gold fruit the size of a cherry in clusters of 30-50. *Source History: 1 in 2004.* **Sources: Pe1.**

Green Gage - 75 days, Indet., clusters of pale yellow fruit, 1.5 in. dia., strong flavor, green in the name possibly refers to the green gel around the seeds, according to William Woys Weaver, this var. was mentioned as early as 1867 by the name Yellow Plum, rare. *Source History: 4 in 2004.* **Sources: Ea4, Ga1, Sa9, To9.**

Hahnstown Yellow - Translates hentown in Pennsylvania Dutch, det. plant bears many round, blemish-free, pale yellow fruit, mild sweet flavor. *Source History: 1 in 2004.* **Sources: Ami.**

Hartman's Yellow Gooseberry - 75 days, Indet. plant produces 1 in. golden yellow fruits which grow in clusters, mildly acidic flavor with full texture. *Source History: 1 in 1998; 3 in 2004.* **Sources: Lan, To9, To10.**

Hawaiian Pineapple - 90 days, Indet., large golden orange beefsteak weighs up to 2 lbs., when fruit is very ripe the flavor resembles that of a ripe pineapple. *Source History: 3 in 1998; 7 in 2004.* **Sources: Ami, Ea4, Ma18, So25, To1, To9, To10.**

Hazel Gold - 80 days, Indet., 4 in. golden fruit with excellent flavor, good yields, heirloom. *Source History: 1 in 2004.* **Sources: To9.**

Herman's Yellow - 80 days, Indet., droopy foliage, heart shaped, golden orange fruit. *Source History: 1 in 2004.* **Sources: Sa9.**

Hopkins - 75 days, Indet., 2-4 oz., translucent yellow fruit, very sweet flavor, high yields, from the gardens of the Edgar Allen Poe estate. *Source History: 1 in 2004.* **Sources: Ma18.**

Hourma - 101 days, Heavy yields of peach colored globes, 6 oz. *Source History: 1 in 1998; 1 in 2004.* **Sources: Sa9.**

Hssaio Hungshih - Bright, oval to almost pear shaped fruit, prolific, presumed to be from China, very rare. *Source History: 1 in 2004.* **Sources: Ea4.**

Hughs - 90 days, Indet. vines, meaty pale-yellow fruits, touch of red on blossom end, up to 2.5 lbs., flavor compares to that of the best heirloom red tomatoes, heirloom from Madison County, Indiana since 1940. *Source History: 1 in 1991; 2 in 1994; 6 in 1998; 7 in 2004.* **Sources: Ap6, Ho13, Sa9, So1, Ta5, To1, To9.**

Ida Gold - 55-59 days, Extra early, branching det. plant, small orange 2-3 oz. fruits, mild low-acid flavor, produces over an extended period, from U of Idaho for cold Northern areas. *Source History: 4 in 1981; 3 in 1984; 6 in 1987; 7 in 1991; 5 in 1994; 6 in 1998; 8 in 2004.* **Sources: Fe5, Ga1, Hig, Pe2, Pr3, Sa9, So25, Syn.**

Ildi - 68-77 days, Det., 5-6 ft. tall, small yellow pears in trusses of 80 fruit, 3-4 trusses per plant, well balanced flavor, can be hung in a cool garage, fruit keeps for weeks without dropping. *Source History: 3 in 2004.* **Sources: So25, Tho, To1.**

Ilse's Yellow Latvian - 75-80 days, Indet., bright yellow-gold, 3 in., round-oblong fruit, 8 oz., tasty yellow, good processor, heirloom. *Source History: 1 in 1998; 2 in 2004.* **Sources: Ma18, To9.**

Indian Moon - 75-80 days, Navajo heirloom, indet., med. size golden globes, meaty flesh, hardly any seed cavities, low acid flavor with complex acid-sweet balance, excellent saucer/canner, possibly the best tasting yellow tomato, rare. *Source History: 1 in 1998; 4 in 2004.* **Sources: Ea4, Ga22, Ma18, To9.**

Italian Gold - 75-80 days, Roma style tomato, compact, det. plant, 6 oz., makes rich, thick tomato sauce, resists V F1 and F2. *Source History: 1 in 1998; 5 in 2004.* **Sources: Ra5, Ra6, Re8, Se17, To9.**

Italian Quadratic Yellow - 94 days, Indet. plants, bright yellow fruits with a distinctive nipple at the blossom end, 4 oz., peach size. *Source History: 1 in 1998; 1 in 2004.* **Sources: Sa9.**

Jaffa - 75 days, Indet., fruit resembles an orange, dev. by German tomato specialist, Manfred Hahm-Hartmann, heirloom. *Source History: 1 in 2004.* **Sources: To9.**

Jaune Negib - 75 days, Short, indet. vines, yellow, 2-3 in., slightly flattened, round fruit, mildly flavored white flesh, nice yield, heirloom. *Source History: 1 in 1998; 1 in 2004.* **Sources: To9.**

Jubilee - (Golden Jubilee, Orange or Yellow Jubilee, Burpee's Golden Jubilee, Sunray VF) - 72-85 days, AAS/1943, med- sized indet. plants, narrow foliage and fair cover, golden-orange slightly flattened globes, 2.25-2.75 x 2.5-3.5 in. dia., 6-7 oz., meaty thick walls, solid, few seeds, mild but good tomato flavor, heavy yields, home garden and local market, not for the far North, grow on the ground or use short stakes, bred by W. Atlee Burpee Seeds, the result of a six-generation sel. from a Tangerine x Rutgers cross. *Source History: 59 in 1981; 51 in 1984; 55 in 1987; 51 in 1991; 53 in 1994; 52 in 1998; 60 in 2004.* **Sources: All, Ba8, Bo9, Bu2, Bu8, CA25, CH14, Co31, Cr1, Dgs, DOR, Ear, Ers, Fe5, Fo7, Ga1, Ge2, GLO, Gr27, HA3, Ha5, He8, He17, He18, Hi13, HO2, HPS, Kil, La1, Loc, MAY, Mel, Mey, MIC, Mo13, Orn, Pep, Ra5, Ra6, Re8, Ri12, RIS, RUP, Sau, SE14, Se26, Sh9, Shu, Sk2, So25, SOU, Ta5, To1, To3, Up2, Vi4, WE10, Wet, Wi2, WI23.**

Jumbo Jim Orange - 75 days, Indet., perfect orange, 10-20 oz. globes, heat and disease resistance. *Source History: 2 in 2004.* **Sources: Ma18, Ta5.**

Kaki Coing - 84 days, Semi-det. plant produces oblong, thick skinned orange fruit, 2 x 3 in., paste-like dry flesh, mild sweet flavor. *Source History: 1 in 2004.* **Sources: To9.**

Kellogg's Breakfast - 80-90 days, Heirloom from West Virginia via Darrell Kellogg, indet. vines, large pale orange beefsteak, thin skin, 1-2 lbs., nice rich flavor. *Source History: 5 in 1998; 20 in 2004.* **Sources: Ba8, Ga1, He8, He18, Loc, Ma18, ME9, Orn, Pin, Ra5, Ra6, Re8, Sa9, SE14, Se16, So25, To1, To3, To9, To10.**

Kimberton Hills Yellow - 75-80 days, Indet., sturdy vines, medium size yellow fruit, grown and selected for many years by Hubert Zipperlen, of Camphill Village, Kimberton Hills, who died at age 90 in 2002. *Source History: 1 in 1994; 1 in 1998; 2 in 2004.* **Sources: Pe2, Tu6.**

Lemon Boy - 75-80 days, Bushy plants, true lemon-yellow 3-4 in. fruits, hardy prolific bearer, flavor similar to red slicing tomatoes, introduced as a hybrid but now has become a stable open-pollinated variety. *Source History: 1 in 1991; 3 in 1994; 3 in 1998; 4 in 2004.* **Sources: Bu2, CA25, Gr27, Gr28.**

Lillian's Yellow Heirloom - (Lillian's Yellow) - 73-98 days, Indet., potato leaf, pale yellow, globe shaped fruits up to 16 oz., green shoulders, superb rich flavor, meaty flesh, few seeds, low yields, heirloom from Manchester, Tennessee. *Source History: 4 in 1998; 16 in 2004.* **Sources: Fe5, He8, He17, Ho13, Ma18, Ra5, Ra6, Re8, Sa9, Sh13, So25, Ta5, To1, To3, To9, To10.**

Limmony - (Lemony) - 75-80 days, Russian heirloom, healthy indet. plants, lemon-yellow beefsteak, 8-10 oz., 4-5 in. dia., solid meaty flesh, delicious tangy lemon flavor that sets it apart from other yellow tomatoes, good yields. *Source History: 7 in 2004.* **Sources: Ea4, He18, Ho13, Ma18, So25, To1, To9.**

Lincoln Gold - Huge beefsteak, excel. flavor. *Source History: 1 in 1994; 1 in 2004.* **Sources: Syn.**

Little Yellow Pot Tomato - Compact upright plant, yellow cherry fruit. *Source History: 1 in 2004.* **Sources: Pr3.**

Livingston's Gold Ball - 65-73 days, Indet. vine bears clusters of deep yellow, golf ball size fruit, tasty, some blight tolerance, 1892. *Source History: 1 in 1998; 2 in 2004.* **Sources: Sa9, Vi4.**

Mandarin Cross - 75-80 days, Vigorous grower produces heavy crops of golden orange, 8 oz. fruits in clusters of 4-5, meaty flesh, luscious acid balance. *Source History: 3 in 1998; 2 in 2004.* **Sources: Gr28, Sa9.**

Manyel - 70-91 days, Indet. vines provide light foliage cover, lemon-yellow, flattened globes, 6-16 oz., outstanding rich lemony taste, translates "Many Moons", heirloom reportedly of recent Native American origin. *Source History: 11 in 1998; 16 in 2004.* **Sources: Ap6, CO30, Fo7, He18, Ho13, Ma18, ME9, Sa9, SE14, Sh13, So1, So25, Ta5, TE7, To1, To9.**

Mary Reynolds - 75-80 days, Indet., large dusky orange beefsteaks, productive, rare heirloom from Virginia. *Source History: 2 in 2004.* **Sources: Ma18, Ta5.**

Mennonite Orange - 90-95 days, Indet., excellent beefsteak type slicing tomato, meaty orange flesh with very small seed cavities, sweet flavor, midseason maturity, rare Mennonite heirloom dates back to 1852. *Source History: 2 in 2004.* **Sources: Fo13, Te4.**

Mexican Yellow - 85 days, Indet., highly productive plants bear 12-16 oz. yellow beefsteak fruit, mild sweet flavor. *Source History: 2 in 2004.* **Sources: Ho13, Ma18.**

Miniature Orange - (Mini Orange) - 65-85 days, Indet., small smooth round bright-orange fruits about the size of a golfball, grows in clusters of 4-6, fine flavor, green > yellow > orange, fruit set continues when night temps are above 70 degrees F, recommended for hot southern areas. *Source History: 1 in 1991; 4 in 1994; 5 in 1998; 15 in 2004.* **Sources: CO30, Ga1, He8, Ma18, ME9, OSB, Pe2, Ra5, Ra6, Re8, Sa9, SE14, So1, Ta5, TE7.**

Mirabell - 67-74 days, Indet., very tiny yellow cherry, clusters of 6-8, tart fruity flavor, high yields, from Germany. *Source History: 4 in 1998; 4 in 2004.* **Sources: CO30, Ma18, Sa9, To9.**

Moon Glow - 85 days, Indet., med-sized blunt-pointed globes, 6-8 oz., orange fruits, mild flavor, excel. keeper. *Source History: 1 in 1981; 1 in 1984; 2 in 1987; 2 in 1991; 1 in 1994; 2 in 1998; 1 in 2004.* **Sources: To9.**

Morden - (Morden Yellow) - Outstanding early tomato, ripens to orange/yellow, dark green shoulders darken to light brown, det. vines, dev. by Dr. Walkof at Morden Exp. Farm in Manitoba, adapted to short season areas, 1950. *Source History: 1 in 1981; 1 in 1998; 2 in 2004.* **Sources: Gr28, Yu2.**

Mortgage Lifter, Yellow - 80-95 days, Indet., 4 in., round, flattened fruit, yellow with orange blush, pink in the center, rich, fruity flavor, from a sport of Mortgage Lifter, introduced by Underwood Gardens, Ltd. *Source History: 2 in 1998; 2 in 2004.* **Sources: To9, Und.**

Mountain Gold - 72-80 days, Det., deep yellow-tangerine 3.5 in. fruits, mild low acid flavor, superior in both appearance and holding quality, F F1 and F2 res., Gardner/NC State U, PVP 1992 NC Ag. Res. Ser. *Source History: 1 in 1991; 15 in 1994; 19 in 1998; 15 in 2004.* **Sources: Bu3, CA25, HO2, Jor, MIC, Mo13, Orn, Ra5, Ra6, Re8, Scf, Se26, To1, To3, Vi4.**

Native Sun - 50 days, Det. vines bear large early yellow fruits, flavorful, 6-10 oz., PSR Breeding Program. *Source History: 1 in 1998; 1 in 2004.* **Sources: Pe6.**

Nebraska Wedding - 90-105 days, Det. vines, attractive, 3-4 in. globes in clusters, orange skin and flesh, wonderfully balanced sweet/acid flavor, good for commercial production, old Great Plains heirloom. *Source History: 2 in 1994; 10 in 1998; 26 in 2004.* **Sources: CO30, Ers, Fo7, Go8, He17, He18, Hud, Ma18, Me7, ME9, Pr9, Ra5, Ra6, Re8, Ri12, RUP, Sa9, SE14, Se16, Se26, Sk2, Ta5, Te4, To3, To9, To10.**

New Moon - 81 days, Indet., pale yellow 2.5 in. long fruit, oval shape, firm flesh, flavor is mild with a hint of acid. *Source History: 1 in 2004.* **Sources: To9.**

New Sun - 60 days, Det., bright yellow, 8-16 oz. fruit, flavorful, compares to Native Sun but with larger plant and fruit size, firmer flesh with better flavor, disease resistant, from Peters Seed and Research. *Source History: 2 in 2004.* **Sources: Bou, Pe6.**

Northern Sun - 58-65 days, Compact plants, med-size yellow fruits, delicious flavor, one of the earliest. *Source History: 1 in 1991; 1 in 1994; 1 in 1998; 1 in 2004.* **Sources: Me7.**

Old Colussus - 110 days, Indet, huge fruit weigh 2 lbs. or more, heirloom from the hills of Arkansas. *Source History: 1 in 2004.* **Sources: Re8.**

Old Ivory Egg - 80 days, Indet., pale yellow, 1 x 2 in. fruit, ripens to creamy yellow, mild sweet flavor, productive, heirloom. *Source History: 1 in 1998; 2 in 2004.* **Sources: To1, To9.**

Old Wyandotte - 70-90 days, Indet., productive early yellow tomato, fruit is somewhat lobed and flattened, 8-10 oz., excel. fruity flavor, disease res. *Source History: 2 in 1994; 3 in 1998; 2 in 2004.* **Sources: Ta5, To9.**

Olga's Yellow Round Chicken - 75-80 days, Compact indet. plants to 3-4 ft. produce an abundance of 5-10 oz. bright orange fruits, round shape, flavor is a nice balance of sweet and slightly acid, Siberia. *Source History: 1 in 1994; 1 in 1998; 3 in 2004.* **Sources: Ma18, Ta5, To9.**

Olivette Jaune - 65-81 days, Indet., large plum to olive shaped, yellow cherry tomatoes, from France. *Source History: 1 in 1998; 3 in 2004.* **Sources: Ma18, Sa9, Ta5.**

Orange - 75-78 days, Old traditional Russian variety, leafy, productive, 3 ft. bush plant, good yields of lobed, flattened, 3.5- 4.5 in. fruits with true orange color, needs warm temperatures to develop well, originally obtained from Vavilov Institute. *Source History: 1 in 1994; 1 in 1998; 1 in 2004.* **Sources: Sa9.**

Orange Banana - (Liane Orange) - 70-95 days, Indet., 3-4 in. long paste tomatoes with pointed ends, 4-5 oz., unique orange color, sweet flavor, excel. for drying, salads, salsas and sauces. *Source History: 1 in 1998; 20 in 2004.* **Sources: Ba8, Coo, Ea4, Fe5, Go8, La1, Ma18, ME9, Pr9, Ra6, Ri12, SE14, Se16, Se26, Sh13, So25, TE7, To1, To3, To9.**

Orange Cherry - 65 days, Indet., slightly elongated globes, excel. flavor. *Source History: 1 in 1987; 1 in 1998; 1 in 2004.* **Sources: Ma18.**

Orange Cherry, Potato Leaf - *Source History: 2 in 2004.* **Sources: ME9, SE14.**

Orange Crimea - 107 days, Semi-det. plant, peach shaped fruit, orange skin, 8 oz. *Source History: 1 in 1998; 1 in 2004.* **Sources: Sa9.**

Orange Heart - 87 days, Indet., 4-6 oz. orange globes, orange-red flesh, semi-sweet, fruity flavor. *Source History: 1 in 1998; 3 in 2004.* **Sources: Ma18, Sa9, Up2.**

Orange King - 55 days, Home garden variety with bright orange fruits weighing up to .75 lb., almost twice the size of Orange Queen with equally good flavor, compact det. plant, uniform ripening, sets fruit under high or low temperature conditions, PSR Breeding Program. *Source History: 1 in 1994; 2 in 1998; 2 in 2004.* **Sources: Bou, Pe6.**

Orange Oxheart - 70-90 days, Excel. orange version of Pink Oxheart, large vines are best when caged to protect fruit from sunscald, heart-shaped fruit, 10-12 oz., firm solid flesh, few seeds, family heirloom originating in Virginia. *Source History: 2 in 1991; 4 in 1994; 6 in 1998; 18 in 2004.* **Sources: Bo19, Ga1, Ma18, ME9, Pe2, Ra5, Ra6, Re8, Ri12, Roh, Sa9, SE14, TE7, Ter, To1, To3, To10, WE10.**

Orange Queen - 60-90 days, The standard for early orange slicers, open det. vines, bright-orange beefsteak fruits, 4-6 oz., mild flavor, early-maturing. *Source History: 2 in 1981; 2 in 1984; 4 in 1987; 6 in 1991; 6 in 1994; 4 in 1998; 4 in 2004.* **Sources: Pr3, Ra6, Se7, Se26.**

Orange Roma - 69 days, Early maturing variety, indet., flavorful orange fruit. *Source History: 1 in 2004.* **Sources: To9.**

Orange Strawberry - 80-90 days, Indet., wispy foliage, beautiful, bright orange, 3 in., strawberry shape fruit, 8-16 oz., sweet rich taste, low acid, few seeds. *Source History: 3 in 1998; 9 in 2004.* **Sources: Ho13, Ma18, Na6, Shu, So25, Sw9, To1, To3, To9.**

Orange-Glow - Sweet mild flavor, low acid, high beta carotene content. *Source History: 1 in 2004.* **Sources: Pe7.**

Paragon Yellow - 85 days, Indet., 4-6 oz. orange-yellow globes, free of cracks and blemishes, nice fruity flavor. *Source History: 1 in 1998; 3 in 2004.* **Sources: Ho13, Ma18, Sa9.**

Patio Orange - 50 days, Slightly flattened globes, deep-orange, good disease res. *Source History: 1 in 1994; 1 in 1998; 2 in 2004.* **Sources: Bo19, WE10.**

Peach, Yellow - (Peche Jaune, Sorbet de Citron, Garden Peach) - 67-80 days, Midseason, productive indet., tasty yellow 1 oz. fruits that sometimes have a slightly fuzzy skin, soft, seedy flesh, intense fruity sweet flavor improves with age, Massachusetts heirloom dating back to 1862. *Source History: 1 in 1987; 2 in 1991; 1 in 1994; 1 in 1998; 5 in 2004.* **Sources: Ami, Ea4, Ma18, ME9, To9.**

Pear, Yellow - 70-80 days, Med-large open indet. vines, clusters of small waxy-yellow pear-shaped fruits with definite necks, 1.5-2 in. x .75-1 in. dia., mild flavor, bears all summer long, ASC and FW1 res., heat resistant, home gardens, for eating whole or preserves or pickles, first described by Persoon in 1805. *Source History: 57 in 1981; 49 in 1984; 59 in 1987; 58 in 1991; 62 in 1994; 73 in 1998; 94 in 2004.* **Sources: Ag7, All, Ap6, BAL, Be4, Bo9, Bo17, Bo19, Bu1, Bu2, CA25, CH14, CO30, Co32, Com, Coo, Dam, Eo2, Ers, Fi1, Fis, Fo7, Ga1, Ga22, Ge2, Gr27, Gr28, He8, He17, He18, Hi6, Hi13, HO2, Hum, Jo1, Jo6, Jor, Jun, La1, LO8, Loc, ME9, Mey, Mi12, MIC, Mo13, MOU, Na6, Ni1, Ol2, Or10, Orn, Pe6, Pin, Pla, Pr3, Ra5, Ra6, Red, Ri12, RIS, Roh, Ros, RUP, Sa9, Se7, Se8, SE14,**

Se17, Se26, SGF, Sh9, Shu, So1, So9, So12, So25, Sw9, Syn, Ta5, Te4, Ter, To1, To3, To9, To10, Up2, Vi4, WE10, We19, Wet, Wi2, WI23, Yu2.

Peg's Round Orange - 80 days, Indet., deep orange 2.5 in. fruit, sweet flavor, low acid, named after seed donor, grown on a Cox's Creek Kentucky farm for over 20 years. *Source History: 1 in 2004.* **Sources: To9.**

Persimmon - 80-96 days, Heirloom dating back to the 1800s, yellow beefsteak type, indet., very large, persimmon-colored fruit is uniform and blemish free, 12-32 oz., low acid, small seed cavity, very flavorful, not a good keeper. *Source History: 1 in 1981; 1 in 1984; 1 in 1987; 1 in 1991; 5 in 1994; 18 in 1998; 37 in 2004.* **Sources: Al25, Ap6, Be4, CO30, Coo, Ers, Fo7, He17, He18, Ho13, Ma18, ME9, Mi12, Na6, Pe2, Ra5, Ra6, Re8, Ri12, RUP, Sa9, Se7, SE14, Se17, Sk2, So1, So25, Sw9, Te4, Ter, To1, To3, To9, To10, Tu6, Up2, We19.**

Pirkstine Orange - 80 days, Indet., bright orange fruits shaped like a short banana, great yields, from Latvia. *Source History: 1 in 2004.* **Sources: Ta5.**

Pirkstine Yellow - 80 days, Indet., good yields of yellow fruit, looks like a short banana, from Latvia. *Source History: 1 in 2004.* **Sources: Ta5.**

Plum Lemon - 75-95 days, Indet., 4-5 ft. plants, bright yellow, 3 x 2 in. dia. fruits with two distinct chambers, plum or lemon shaped with a point on the tip, "Sleevuhvidnaya" means "plum-shaped" in Russian, collected by Seed Savers' Kent Whealy from an elderly seedsman at the Moscow Bird Market in 1991, originates in the St. Petersburg region of Russia. *Source History: 1 in 1994; 8 in 1998; 22 in 2004.* **Sources: Ba8, Bo19, CO30, Com, Go8, He8, He18, Ma18, Mcf, ME9, Pr9, Ra5, Ra6, Re8, RUP, SE14, Se16, So25, Ta5, TE7, To1, To9.**

Plum, Yellow - 70-85 days, Large open indet. plants, lemon-yellow skin, oval plum-shape, 1.5-1.75 in. long x 1 in. across, firm flesh, sweet mild flavor, grows in clusters, very productive, bears until frost, ASC res., home garden var., good in salads or preserves. *Source History: 29 in 1981; 28 in 1984; 28 in 1987; 20 in 1991; 22 in 1994; 18 in 1998; 19 in 2004.* **Sources: Fe5, Fo7, Ga1, Ge2, He17, Jor, La1, Ma13, ME9, Ri12, SE14, Se17, Se26, Sh9, Sh13, To1, To3, Vi4, WE10.**

Polonaise Jaune - 85 days, Indet., 6-8 oz., golden yellow fruit, meaty flesh, superior flavor. *Source History: 1 in 2004.* **Sources: Ma18.**

Ponderosa, Golden - (Railroad Strain) - 78-90 days, West Virginia heirloom, appears to be a variant of Golden Ponderosa, propagated by employees of the C and O Railroad in West Virginia prior to 1940, indet., somewhat rough large-cored yellow-gold fruits weigh over a pound, shows well until late summer, not very tolerant of foliage diseases, best suited for cage culture. *Source History: 1 in 1987; 2 in 1991; 3 in 1994; 5 in 1998; 10 in 2004.* **Sources: Ba8, Ho13, Ra5, Ra6, Re8, Sa9, Shu, So1, Ta5, To3.**

Potato Leaf Yellow - 85 days, Indet., potato leaf plant, golden oblate fruit up to 2 lbs., sweet. *Source History: 1 in 1998; 1 in 2004.* **Sources: Ma18.**

Powers Heirloom - 80 days, Indet., 3-5 oz. fruits measure 3.5 in. long, oval shape, clear yellow color, sweet, juicy flavor with a fruity overtone, great paste tomato. *Source History: 1 in 1998; 4 in 2004.* **Sources: So25, Ta5, To1, To9.**

Prize of the Trials - 75-80 days, Indet. vines, 1-2 oz., orange, apricot-size fruits, clusters of 6-8, good flavor, high yields, crack res. *Source History: 1 in 1994; 1 in 1998; 2 in 2004.* **Sources: Se7, To10.**

Qiyanai-Huang - Compact plants bear med. size orange fruit with pointed end, from China, very rare. *Source History: 1 in 2004.* **Sources: So25.**

Reinhard's Tomate - 76 days, Large indet. vines produce many golden yellow 1 in. cherries, thin skin, sweet flavor, German variety from Reinhard Kraft. *Source History: 2 in 2004.* **Sources: Ba8, To9.**

Roman Candle - 75-90 days, Indet., yellow, banana shaped fruit, 2 in. wide x 4 in. long, solid yellow flesh, excellent flavor, originated as a sport from Speckled Roman. *Source History: 6 in 2004.* **Sources: Ag7, Ba8, Go8, Ma18, Pr9, Se16.**

Roughwood Golden Plum - 76-85 days, Indet., Yellow Brandywine x San Marzano cross, med. size orange paste, beautiful intense deep golden color, very dry flesh, mild flavor, resists drought and white flies. *Source History: 1 in 1998; 5 in 2004.* **Sources: Ami, Ma18, Sa9, So25, To9.**

Ruby Rakes Yellow - 70-85 days, Indet., 2.25 x 1.75 in. globes, lemon-yellow skin, yellow flesh, very high yields. *Source History: 2 in 1998; 2 in 2004.* **Sources: Sa9, Se17.**

Ruffled Yellow - (Ruffled Golden, Ruffled) - 80 days, Semi-det., yellow fruits, formed like an accordian, makes fancy dessert containers when cut in half and hollowed out, decorative in salads, excel. mild flavor. *Source History: 3 in 1981; 2 in 1984; 5 in 1987; 6 in 1991; 5 in 1994; 6 in 1998; 4 in 2004.* **Sources: Gr28, Se7, To1, To9.**

Russian Lime - 75 days, Indet., clusters of bell shaped, lemon-yellow fruit, pointed end, 2 x 4 in., dry flesh, tasty, makes a flavorful sauce. *Source History: 1 in 1998; 3 in 2004.* **Sources: Ma18, Ta5, To9.**

Russian Persimmon - 80 days, Det., 42 in. plants, sweet, round, 2.5-3.5 in., orange fruits, outstanding flavor, Russian heirloom. *Source History: 1 in 1994; 4 in 1998; 5 in 2004.* **Sources: Co32, Ga1, Se16, Sh13, To1.**

Russian Yellow - 78 days, Indet., flattened yellow 3 in. globes, mild sweet flavor. *Source History: 1 in 1994; 1 in 2004.* **Sources: To9.**

Scotland Yellow - 70-73 days, Indet., 3-4 oz. bright yellow fruit, delicious sweet tangy flavor, healthy vines, good keeper. *Source History: 1 in 2004.* **Sources: Go8.**

Sibirishe Orange - 85 days, Indet., beautiful blocky fruit, waxy skin, meaty flesh, excel. flavor, good production. *Source History: 2 in 2004.* **Sources: So25, Ta5.**

Small Lap - 85 days, Semi-det., slightly flattened, orange globe. *Source History: 1 in 1998; 1 in 2004.* **Sources: Sa9.**

Spark's Yellow - 86 days, Indet., potato leaf plant, 3 in., round yellow-orange beefsteak fruit, juicy flesh, delicious, seeds obtained by Craig LeHoullier from Don Sparks of Kentucky. *Source History: 1 in 2004.* **Sources: To9.**

Stor Gul - 90 days, Indet., med. yellow-orange fruit, 12 oz., good flavor. *Source History: 1 in 1998; 1 in 2004.* **Sources: To9.**

Sundrop - (Burpee's Sundrop) - 70-76 days, Indet. small globular deep-orange fruits, 2 oz., 1.75 in. dia., firm meaty and sweet, resist cracking or bursting even late in the season. *Source History: 3 in 1987; 8 in 1991; 4 in 1994; 3 in 1998; 1 in 2004.* **Sources: Ma18.**

Sungold Select - Non-hybrid version of Sungold .5-1 oz. gold-orange fruit, tender, sweet flesh, flavorful, occasional red fruit, bred by Reinhard Kraft, Germany. *Source History: 2 in 2004.* **Sources: Ba8, So25.**

Sunray - (Golden Sunray, Jubilee) - 72-91 days, Vigorous large sturdy indet. vines, heavy foliage, mild but rich flavor, same size and color and quality as Jubilee plus highly res. to FW, also ASC res., flavor rivals the best reds, usually staked, widely adapted, USDA/Beltsville, 1950. *Source History: 47 in 1981; 36 in 1984; 36 in 1987; 20 in 1991; 10 in 1994; 8 in 1998; 17 in 2004.* **Sources: Coo, Ga1, Gr28, Pr9, Ra6, Sa9, SE14, Se16, Se26, Syn, Ta5, TE7, To1, To3, Vi4, WE10, WI23.**

Sunshine - 70 days, Vigorous det. plants produce heavy yields of sweet yellow cherries, low acid, uniform size and shape, good dryer, crack resistant, PSR. *Source History: 1 in 1994; 1 in 1998; 2 in 2004.* **Sources: Pe6, Sa5.**

Sweet Orange - (Sweet Orange Roma) - 76-80 days, Productive, leggy indet. plant bears pleasantly sweet and tart orange fruit with a nice crispness, high yields, PSR. *Source History: 2 in 1998; 3 in 2004.* **Sources: Bou, Ma18, To9.**

Sweet Orange II - 65-70 days, Indet., orange cherry, outstanding fruity taste, good crack resistance, vigorous, developed by PSR. *Source History: 2 in 2004.* **Sources: Bou, Pe6.**

Tangella - 65-80 days, Indet., 2 in., round, 2-3 oz. bright orange fruit in clusters of 5-6, fruity flavor with a good acid tang, sibling of Tigerella, originated in England. *Source History: 1 in 1998; 6 in 2004.* **Sources: Ap6, He18, Sa9, So25, To1, To9.**

Tangerine - 80-85 days, Flavorful enough to live up to its tart name, indet., deep yellow-orange beefsteak-type fruit with tangerine shape, 5-12 oz., meaty flesh, slightly tart flavor, high yields. *Source History: 1 in 1991; 3 in 1994; 5 in 1998; 19 in 2004.* **Sources: Ag7, CO30, He8, He18, Loc, ME9, Pe2, Ra5, Ra6, Re8, Sa5, SE14, Sw9, Ta5, Te4, TE7, To1, To3, To10.**

Taxi - 65-70 days, Excellent early yellow tomato, compact determinate vines need no staking, firm meaty bright-yellow baseball-size fruits, 4-6 oz., round to slightly oblate, smooth and blemish free, small stem scar, flavorful, easy to grow. *Source History: 2 in 1987; 8 in 1991; 9 in 1994; 8 in 1998; 10 in 2004.* **Sources: Hig, Jo1, Pr3, Ri12, Sa5, Se26, Syn, Ta5, Ter, We19.**

Tiger Paw - 77 days, Indet., thin skin, pale yellow blushed with pink, slightly lobed, flattened fruit, tropical sweet flavor with hint of citrus, slight indentations have the look of a cat's paw, exerted stigma so keep distance from other varieties, Tennessee heirloom. *Source History: 1 in 2004.* **Sources: To9.**

Tommy Toe Yellow - 75 days, Indet., slightly larger fruit than the red version, deep yellow, fine flavor. *Source History: 1 in 2004.* **Sources: Ma18.**

Valencia - 76 days, Johnny's selection of a Maine family heirloom with a Spanish accent, midseason, indeterminate, round smooth bright-orange fruits average 8-16 oz., meaty interiors with few seeds, full tomato flavor. *Source History: 1 in 1987; 3 in 1991; 3 in 1994; 4 in 1998; 6 in 2004.* **Sources: Ap6, Ma18, Se8, Ta5, To1, To9.**

Verna Orange - 84 days, Indiana heirloom, indet., huge orange oxheart-shaped fruits, up to 3 lbs., superb flavor, semi- hollow seed cavities in hot climates, otherwise flesh is very meaty with few seeds, heavy yielder despite late season susceptibility to foliage diseases. *Source History: 1 in 1991; 1 in 1994; 6 in 1998; 6 in 2004.* **Sources: Ho13, Ma18, Se17, So1, Ta5, To1.**

Wapsipinicon Peach - 80-86 days, Indet., 2 in. yellow fruit, fuzzy skin, peach size, excel. taste, smooth texture, named after the Wapsipinicon River in northeast Iowa, rare. *Source History: 1 in 1998; 7 in 2004.* **Sources: Ag7, Ami, Ea4, He18, Pr9, Sa9, Se16.**

Wendy - (Yellow Wendy) - 65-80 days, Semi-det. plants produce many round yellow 2 in. fruits, sweet flavor, from Australia. *Source History: 1 in 1994; 2 in 1998; 2 in 2004.* **Sources: Fo7, Hig.**

West Virginia Yellow - 85 days, Indet., dark golden yellow fruit, 1 lb., very good sweet flavor. *Source History: 1 in 2004.* **Sources: To9.**

Wonder Light - (Lemon Vine, Limon-Liana) - 78 days, Indet., yellow fruits resemble lemons in size, shape and color, good taste, original seed to Johnny's Seeds from Kate Rogers Gessert of Albany, OR, who obtained from a gardener in Irkutsk, Siberia, collected in Moscow by Kent Whealy in 1991. *Source History: 1 in 1998; 8 in 2004.* **Sources: CO30, Ga22, Ge2, Gr28, Jo1, Ta5, Te4, To9.**

Woodle Orange - 75 days, Indet., perfect large orange globes, full bodied flavor, high quality, excellent for market and home growers, excellent production late into the season, rare. *Source History: 2 in 2004.* **Sources: Ma18, Sa5.**

Yellow Belgium - (Goldie) - 85-95 days, Heirloom, indet. vines, yellow to orange med-size fruits, round to heart-shaped, 12- 16 oz., good solid flesh, fine flavor. *Source History: 4 in 1991; 3 in 1994; 3 in 1998; 6 in 2004.* **Sources: Ami, Ho13, Ma18, Sa5, Sa9, Se17.**

Yellow Bell - 60-92 days, Rare family heirloom from Tennessee, indet. pear-shaped fruits like Roma, 4-5 in. long x 1.5 in. dia., ripens from green to creamy-yellow to yellow, 5 fruits per cluster average--12 fruits max., excellent paste tomato. *Source History: 1 in 1987; 1 in 1991; 1 in 1994; 6 in 1998; 8 in 2004.* **Sources: ME9, Sa9, SE14, So1, So25, Ta5, Te4, To1.**

Yellow Canary - 55-100 days, Bright-yellow companion to Red Robin, compact branching plant, 1 in. fruit, sweet and tangy flavor, excellent for patio or indoor culture. *Source History: 4 in 1991; 6 in 1994; 4 in 1998; 5 in 2004.* **Sources: Ge2, GLO, OSB, To1, Twi.**

Yellow Currant - 69 days, Indet., .5 in. globes, yellow skin and flesh, not a true currant. *Source History: 1 in 1998;*

1 in 2004. **Sources: Sa9.**

Yellow Krim - 80-85 days, Good for short growing season, yellow-orange skin, pale yellow flesh, very sweet flavor, from the Ukraine, heirloom. *Source History: 1 in 2004.* **Sources: Ri12.**

Yellow Oxheart - 80-95 days, Indet., pale-yellow, almost white version of Orange Oxheart that is milder and more seedy, blocky, heart shaped fruit, golden yellow flesh, rich, full, well balanced flavor, low disease and drought tol. and poor seedling vigor, otherwise a superb tomato, family heirloom from Willis, Virginia, 1915. *Source History: 3 in 1991; 3 in 1994; 5 in 1998; 6 in 2004.* **Sources: CO30, Ga1, Ma18, SE14, So1, To1.**

Yellow Peach - Indet., yellow, 1.5 in., round fruit with fuzzy skin like a peach, mild fruity flavor, productive. *Source History: 2 in 1998; 2 in 2004.* **Sources: CO30, SE14.**

Yellow Perfection - 70-75 days, Indet., potato leaf, sweet yellow fruits, prone to cracking and shoulder scratches in wet weather, rare English heirloom. *Source History: 1 in 1991; 1 in 1994; 6 in 1998; 9 in 2004.* **Sources: Ap6, Co32, Hi6, Ri12, Ta5, Te4, TE7, To9, To10.**

Yellow Perfection, Unwins - 75 days, Indet., brilliant, thin skinned, yellow fruit, 1-2 oz., originally from a famous old British seed company. *Source History: 1 in 1981; 1 in 1987; 1 in 1991; 1 in 1994; 2 in 1998; 1 in 2004.* **Sources: Se7.**

Yellow Ping Pong - 75 days, Yellow-fruited version of Pink Ping Pong, indet. plant bears clusters of golden yellow 2 in. fruit, sweet and juicy. *Source History: 1 in 1981; 1 in 1984; 1 in 1987; 1 in 1991; 1 in 1994; 2 in 1998; 1 in 2004.* **Sources: To1.**

Yellow Pygmy - 65 days, Compact plants work well for container culture. *Source History: 1 in 1998; 1 in 2004.* **Sources: Sa9.**

Yellow Ruffles - 86-90 days, Indet., 3 in., hollow, ruffled fruits are excel. for salads and stuffing. *Source History: 5 in 1998; 4 in 2004.* **Sources: Ma18, ME9, Sa9, SE14.**

Yellow Stuffer - 76-90 days, Indet. plants, 3 in. fruit, resembles a golden pepper with seeds clustered near the top, insides of fruit removes easily for stuffing. *Source History: 23 in 1991; 19 in 1994; 19 in 1998; 25 in 2004.* **Sources: Bo19, CO30, Ea4, Ers, He18, HO2, Ho13, LO8, Ma18, ME9, Pe2, Pin, Ra5, Ra6, Re8, Sh9, Sk2, So25, Te4, TE7, To1, To3, To10, Vi4, WE10.**

Varieties dropped since 1981 (and the year dropped):

Belgium Orange (1994), *Bursztyn* (2004), *Carobetta* (2004), *Cherry Gold* (2004), *Cherry, Golden Baby* (1998), *Colossal Yellow, Burgess* (1994), *Dixie Gold* (1987), *Doni* (2004), *Dwarf Yellow Champion* (2004), *Fireworks Yellow* (2004), *Garden Peach Fuzzy* (2004), *Gift of Moldova* (1998), *Gold Medal Yellow* (2004), *Golden Giant* (1987), *Golden Pygmy* (1991), *Grimmes Yellow* (1998), *Jersey, Golden* (2004), *Jubilee Treasure* (2004), *K-1* (1987), *Kentucky Volunteer* (1994), *Late Orange* (1987), *Lemon Bush* (1998), *Lemon Gold* (1994), *Lemon Yellow* (1991), *Mandarin* (2004), *Mrs. Lindsey's KY Heirloom* (1994), *Mystery Gold* (1998), *Orange King 94* (2004), *Paskal Yellow* (2004), *Peace Yellow Paste* (1994), *Peach, Rev. Morrow's* (1994), *Perfection* (1984), *Plum, Australian Yellow* (1994), *Ponderosa, Yellow* (1998), *PSR-941S* (2004), *PSR-941SS* (2004), *Rosalie's Sweet Horizon* (2004), *Sugar Pear* (2004), *Sugar Yellow* (1987), *Sun Belle* (2004), *Sun Cheers MR* (2004), *Sun Free* (2004), *Sweet Gold* (2004), *Sweet Gold MR* (2004), *Yellow Brandywine, Platfoot* (2004), *Yellow Candle* (1994), *Yellow Delicious* (2004), *Yellow Gold* (1987), *Yellow Marble* (2004), *Yellow Mortgage Lifter* (1991), *Yellow Paste* (1998), *Yellow Pigmy* (1998).

Tomato / Pink-Purple
Lycopersicon lycopersicum

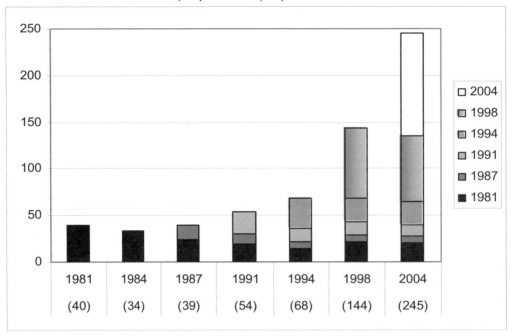

Number of varieties listed in 1981 and still available in 2004: 40 to 20 (50%).
Overall change in number of varieties from 1981 to 2004: 40 to 245 (613%).
Number of 2004 varieties available from 1 or 2 sources: 155 out of 245 (63%).

Acme - 82 days, Indet., potato leaf plant, pink globes, 14 oz. *Source History: 1 in 2004.* **Sources: Sa9.**

Afghanistan - 84 days, Indet., flattened, 6 oz. fruit, 3.5 x 2.5 in., pink flesh, slight ribbing. *Source History: 1 in 1998; 1 in 2004.* **Sources: Sa9.**

Airyleaf - 75 days, Indet., fine leaf foliage, 2 x 4 in. sweet pink plums, from the Black Forest region of Germany. *Source History: 2 in 2004.* **Sources: Ma18, To9.**

Aker's Plum Pink - 83 days, Paste tomato, thick, broad pear shape. *Source History: 1 in 2004.* **Sources: Sa9.**

Amana Pink - 85 days, Indet., dusty rose-pink, heart shaped fruit, 1-2 lbs., dense flesh with small seed cavities, rich, sweet flavor. *Source History: 1 in 2004.* **Sources: Ma18.**

Amish Salad - 70-75 days, Indet., sprawling plants, pink, oval shaped cherry tomatoes, 2 oz., firm flesh, excellent flavor, Amish heirloom. *Source History: 3 in 1998; 15 in 2004.* **Sources: Ba8, CO30, Ga1, In8, La1, Ma18, ME9, Ra6, Sa9, SE14, Se17, Se26, Shu, To3, To9.**

Anderson - 80 days, Indet., abundant yields of large pink beefsteak fruit, from Beaver Falls, PA. *Source History: 1 in 2004.* **Sources: Ta5.**

Anna Russian - 65-70 days, Indet. plant with small foliage and wispy vines, pink-red, heart shaped fruit weighs about 1 lb., juicy flesh with superb flavor, originally from a Russian immigrant. *Source History: 1 in 1994; 4 in 1998; 22 in 2004.* **Sources: Ap6, Ers, Ga1, He8, He18, Ho13, Ma18, ME9, Ra5, Ra6, Re8, SE14, Se26, Sh13, So25, Sw9, Ta5, To1, To3, To9, To10, Vi4.**

Argentina - 80 days, Indet., regular leaf, 3-4 in. pink beefsteak, excellent sweet/acid flavor balance. *Source History: 2 in 2004.* **Sources: He18, To9.**

Arkansas Traveler - (Traveler) - 80-90 days, Southern heirloom, indet., smooth rose colored fruit, excel. shape, mild flavor, juicy, tolerant to heat and humidity, crack and disease resistant, old reliable variety dev. at the U of Arkansas, pre-1900. *Source History: 2 in 1991; 9 in 1994; 14 in 1998; 41 in 2004.* **Sources: Ap6, Ba8, Bo9, CA25, CO30, Co32, Ers, Fi1, Ga1, Ga21, Ge2, Gr29, He17, He18, HO2, HPS, ME9, Pe2, Ra5, Ra6, Ri12, Roh, RUP, Sa9, Se7, SE14, Se26, Se28, Sh13, So1, Sw9, Ta5, To1, To3, To9, To10, Up2, Vi4, Vi5, WE10, Wo8.**

Arkansas Traveler 76 - (Arkansas Pink, Traveler 76) - 76 days, Med-sized indet. vines, good cover, smooth pink 6-8 oz. flattened globes, medium maturity, for home or market in humid areas, crack res., excel. disease res., improved, dev. for the South, University of Arkansas. *Source History: 4 in 1981; 3 in 1984; 4 in 1987; 1 in 1991; 3 in 1994; 3 in 1998; 5 in 2004.* **Sources: Jo1, ME9, SE14, Vi4, WE10.**

Aunt Ginny's Purple - 75-85 days, Indet., potato leaf plant, smooth beefsteak, little cracking, 12-16 oz., excel. flavor resembles Brandywine, German heirloom. *Source History: 5 in 1998; 11 in 2004.* **Sources: He18, Ho13, Ma18, Sa9, Se26, Sh13, So25, Sw9, Ta5, To1, To9.**

Bali - 75 days, Large det. plants, 2-3 in., ribbed fruit, raspberry-pink color, sweet spicy flavor, heavy yields, good for home and market gardeners, from the Island of Bali, Indonesia. *Source History: 3 in 2004.* **Sources: Ami, Ba8,**

So25.

<u>Ballad</u> - Large Ukraine/Russian pink tomato, 7+ ft. plants bear boat-shaped pink fruit, over 1 lb., sweet well balanced flavor, early, cold tolerant, seed from Sergey who lived in the capital of Ukraine for many years. *Source History: 1 in 2004.* **Sources: Ami.**

<u>Bear Claw</u> - 85 days, Indet., large beefsteak with variable shapes, outstanding flavor, from Ben Quisenberry collection. *Source History: 2 in 2004.* **Sources: He18, ME9.**

<u>Beefsteak, African</u> - (German Beefsteak) - 85 days, Very large heavy thick-meated fruits, deep-crimson pink, non-acid, almost seedless. *Source History: 1 in 1991; 1 in 1994; 2 in 1998; 1 in 2004.* **Sources: So25.**

<u>Beefsteak, Watermelon</u> - 75-95 days, Indet. vines, extremely meaty, pink skin, purplish-red mild flesh, some weigh 2 lbs. and more, large oblong fruits give it its name, up to 50 lbs. of fruit per vine, an old West Virginia strain that dates back a century. *Source History: 1 in 1981; 1 in 1984; 2 in 1987; 3 in 1991; 5 in 1994; 10 in 1998; 25 in 2004.* **Sources: Ba8, Bu2, CA25, CO30, Ers, Fo7, He8, He18, Hi13, Ho13, Ma18, ME9, Ra5, Ra6, Re8, Sa9, SE14, Sh9, Shu, Sk2, So9, So25, To1, To3, To10.**

<u>Belgian Beauty</u> - Indet., low acid pink flesh, 1-2 lbs., sweet flavor, pre-1900. *Source History: 1 in 2004.* **Sources: Lan.**

<u>Black Mountain Pink</u> - 80 days, Indet., pink fruit up to 3 lbs., deep rose colored flesh with garnet streaks, outstanding flavor, from an abandoned homestead in the Black Mountain area of Harlan County, Kentucky, 1933. *Source History: 1 in 2004.* **Sources: Ma18.**

<u>Blue River</u> - 80 days, Indet., huge pink beefsteaks, flavorful meaty flesh. *Source History: 1 in 2004.* **Sources: Ta5.**

<u>Boondocks</u> - 70-80 days, Indet., potato leaf plants, pink-purple fruit, green shoulders, up to 1 lb., rich flavor, high in both sugar and acid, from Joe Bratka. *Source History: 1 in 1998; 6 in 2004.* **Sources: Ho13, Ma18, Sh13, Ta5, To1, To9.**

<u>Bradley</u> - 75-85 days, Large indet. plant, fairly good cover, smooth mild pink 6 oz. slightly flattened globes, FW ASC and ST res., University of Arizona for home and market gardeners in humid areas. *Source History: 7 in 1981; 6 in 1984; 7 in 1987; 5 in 1991; 9 in 1994; 9 in 1998; 19 in 2004.* **Sources: CA25, CH14, He17, He18, Ma13, ME9, MIC, Ra5, Ra6, Re8, Sa9, SE14, Se26, Ta5, To1, To3, Vi4, WE10, WI23.**

<u>Brandywine</u> - 75-100 days, Probably from the Ben Quisenberry collection, an old-timer prized for its incredible flavor, medium to large size, 8-16 oz fruit, potato leaf indet. plant, matures very late in northern areas, named after Brandywine Creek in Chester County, PA, Amish heirloom since 1885, introduced by the Philadelphia seed firm, Johnson and Stokes, in 1889. *Source History: 1 in 1981; 1 in 1984; 5 in 1987; 14 in 1991; 27 in 1994; 63 in 1998; 105 in 2004.* **Sources: Al6, Al25, All, Ami, Ba8, Be4, Bo9, Bo17, Bo19, Bou, Bu1, Bu2, CA25, CO30, Co31, Co32, Com, Coo, Dam, Ers, Fa1, Fe5, Fi1, Fo7, Fo13, Ga1, Ge2, Goo, Gr28, GRI, Gur, He8, He17, Hi6, Hi13, HO2, Ho13, HPS, Hud, Jo1, Jun, Ki4, La1, Lan, Lin, Loc, Ma18, ME9, Mel, Mi12, MIC, Na6, Ni1, Ol2, Ont, Or10, Orn, Par, Pe2, Pe6, Pep, Pin, Pla, Pr3, Pr9, Ra5, Ra6, Re8, Red, Ri2, Ri12, Roh, RUP, Sa9, Scf, Se7, Se8, SE14, Se17, Se26, Sh9, Sh13, Shu, So1, So9, So12, So25, Sto, Sw9, Ta5, Te4, TE7, Ter, Th3, Tho, To1, To3, To9, To10, Und, Ver, Ves, Vi4, WE10, We19.**

<u>Brandywine Sport</u> - 75 days, Indet., potato leaf plant, 4 oz. pink plum, sweet pink-red flesh, very flavorful, seed from Robert Sala. *Source History: 1 in 2004.* **Sources: Ma18.**

<u>Brandywine, Cherry</u> - 70 days, Indet., miniature Brandywine with the same fabulous flavor, 1.5 oz. *Source History: 1 in 2004.* **Sources: Ma18.**

<u>Brandywine, Joyce's</u> - 80 days, Indet., potato leaf foliage, flat, pink-red fruit. *Source History: 1 in 2004.* **Sources: Sa9.**

<u>Brandywine, Pawer's</u> - 85 days, Another flavorful Brandywine strain, indet. plant. *Source History: 1 in 2004.* **Sources: Ma18.**

<u>Brandywine, Purple</u> - Indet., potato leaf, dark pink, round, irregular fruit in clusters of 3-5, 4 x 5 in., productive, great acid flavor. *Source History: 1 in 1998; 1 in 2004.* **Sources: Sa5.**

<u>Brandywine, Sudduth's</u> - 90 days, Indet., large fruits can weigh up to 2 lbs., never to be surpassed for full-tomato flavor, this str. was grown by the Sudduth family for nearly 100 years, they obtained the original seed from Johnson and Stokes, who introduced Brandywine in 1889, Ben Quisenberry, who operated Big Tomato Gardens until his death in 1986, obtained his seed from the Sudduths. *Source History: 1 in 1998; 6 in 2004.* **Sources: Ap6, He8, He18, Sa9, Se16, To9.**

<u>Brianna</u> - 80-85 days, Indet., potato leaf, huge blossoms, 4-5 in. pink fruit, up to 1 lb., mild, low acid flavor, rather late ripening, Celtic origin. *Source History: 1 in 1998; 4 in 2004.* **Sources: Ap6, ME9, Sa9, To9.**

<u>Brilliant Pink</u> - Large 4 in. fruit, semi-sweet flavor. *Source History: 1 in 2004.* **Sources: Sa5.**

<u>Brimmer, Pink</u> - 83 days, Deep purple-pink fruits up to 2.5 lbs., nearly all meat, low acid, high sugar content, few seeds and no core, must be staked, ideal for slicing, possibly derived from Ponderosa around 1889, from Virginia. *Source History: 1 in 1981; 1 in 1987; 1 in 1991; 2 in 1994; 2 in 1998; 2 in 2004.* **Sources: Ho13, So1.**

<u>Bull's Heart</u> - 87 days, Very old Russian variety, indet. 6 ft. plants, large pink oxhearts average 4 in., can grow to 3 lbs., only a few large fruits per plant, fruit will rot at stem if temperature gets cold, difficult to grow, seed originally obtained from the Vavilov Institute. *Source History: 1 in 1994; 6 in 1998; 12 in 2004.* **Sources: Co32, Ga1, He18, Ho13, Re8, Ri12, Se26, Sh13, Sw9, Ta5, To1, To3.**

<u>Burcham's New Generation</u> - 100 days, Indet., huge flattened deep pink flattened fruit, meaty flesh, good old-time flavor, productive. *Source History: 1 in 2004.* **Sources: He18.**

<u>Bushy Chabaroysky</u> - 78 days, Bushy det. plant to 1.5 ft., potato leaf foliage, dusty pink salad size fruit, good flavor. *Source History: 1 in 2004.* **Sources: Ta5.**

Butter and Bull Heart - 75 days, Indet., oxheart type fruit, 8-16 oz., meaty, seedless flesh, high yields, good sauce tomato. *Source History: 1 in 2004.* **Sources: Ma18.**

Butterworth's Potato Leaf - 80 days, Indet., potato leaf, 1 lb. pink beefsteak, flavorful. *Source History: 1 in 2004.* **Sources: Ma18.**

Canabec Rose - 78 days, Det., medium size, 8 oz., red-pink fruit, tasty, midseason, from Quebec. *Source History: 1 in 1994; 2 in 1998; 2 in 2004.* **Sources: Pr3, Sa9.**

Caspian Pink - 75-80 days, Indet., good yield of 12-16 oz. pink fruit, juicy sweet flesh, originated in Russia, in an area between the Caspian and Black Seas. *Source History: 31 in 2004.* **Sources: Al25, CA25, CO30, Coo, Ers, Ge2, GLO, HPS, Jun, Ma18, Ni1, Ra5, Ra6, Re8, Ri12, Se8, Se26, Shu, Sk2, So25, Sw9, Ta5, To1, To3, To9, To10, Tt2, Ver, Ves, Vi4, Yu2.**

Century - *Source History: 1 in 2004.* **Sources: Se17.**

Cherokee Purple - (Cherokee) - 72-90 days, Relatively short, indet. vines, purple-pink-brown flesh color with green gel when not fully ripe, flattened globes, 10-12 oz., flavorful, from tomato grower Craig LeHoullier via a J.D. Green of Tennessee who claimed it to be more than 100 years old and originally from the Cherokee people. *Source History: 5 in 1994; 25 in 1998; 60 in 2004.* **Sources: Al25, Ami, Ap6, Ba8, Bo17, Bo19, Bou, Bu2, CA25, CO30, Co32, Com, Coo, Fe5, Ga1, Go8, He8, He17, He18, Hi13, Ho13, Jo1, La1, Loc, Ma18, ME9, Mel, Mi12, Mo13, Ni1, Orn, Pe2, Pr9, Ra5, Ra6, Re8, Ri12, Roh, RUP, Sa9, SE14, Se16, Se17, Se26, Sh9, Sh13, Shu, So1, So25, Sw9, Ta5, Te4, TE7, Ter, To1, To3, To9, To10, Up2, Vi4.**

Cherokee Purple, Potato Leaf - 80 days, Indet., large potato shaped leaves offer good cover for the fruit, 10-12 oz., brown to purple color, sweet zingy flavor, dev. by Baker Creek Heirloom Seeds. *Source History: 1 in 1998; 1 in 2004.* **Sources: Sa9.**

Cherry, Pearly Pink - 65-77 days, Indet. vines bear clusters of .5-1 in. dark pink, flavorful fruits. *Source History: 2 in 1994; 3 in 1998; 4 in 2004.* **Sources: Ho13, OSB, Sa9, Se17.**

Cherry, Pink - 70-85 days, Early indet., pale-skinned salad tomato .75 in. dia., heirloom. *Source History: 2 in 1981; 1 in 1984; 2 in 1987; 3 in 1991; 3 in 1994; 4 in 1998; 3 in 2004.* **Sources: Gr28, Sto, To1.**

Cherry, Porter's Dark - 74 days, Indet., small pink oval fruits in clusters of 6-10, rich tomato flavor, long season, heirloom. *Source History: 1 in 2004.* **Sources: To9.**

Cherry, Sara Pink - Flavorful, large pink cherry, from Germany, rare. *Source History: 1 in 2004.* **Sources: Ea4.**

Cherry, Valley Girl - 90 days, Indet., large purple-pink cherries, many look like small beefsteaks, intense sweet flavor when picked fully ripe. *Source History: 2 in 2004.* **Sources: Gr28, Ma18.**

Chris Ukrainian - 70-90 days, Indet., light pink, flattened beefsteak, 1-2 lbs., meaty flesh with balanced sweet taste, Ukrainian family heirloom. *Source History: 3 in 2004.* **Sources: Ap6, He18, Sa9.**

Church - 85-90 days, Indet., large creamy pink fruits weigh over 2 lbs., solid flesh with tiny seed cavities, excellent sweet flavor, heirloom from the Church Family in Hot Springs, Virginia. *Source History: 2 in 2004.* **Sources: Ma18, To9.**

Clear Pink Early - 58-85 days, Old Russian variety, very productive, 30 in. plants, round, 2.5-3.5 in., 4-5 oz., dark-pink, pulpy fruit with few seeds and little juice, good slicing tomato. *Source History: 1 in 1994; 2 in 1998; 3 in 2004.* **Sources: Ra6, Se7, To1.**

Cleota Pink - 70 days, Tennessee heirloom, flavor compares or exceeds that of Brandywine, large beefsteaks with deep rose, fragrant flesh, early maturity, indet. vines, saved by a woman named Cleota. *Source History: 1 in 2004.* **Sources: Ma18.**

Cooper's Special - 75 days, Det. plant bears smooth pink globes, 2-3 oz., from Mr. Cooper in Florida. *Source History: 1 in 2004.* **Sources: Ma18.**

County Agent - 65 days, Indet., large pink fruit, 24 oz., juicy flesh. *Source History: 1 in 2004.* **Sources: Sa9.**

Crimean - 100 days, Det. plants with thick foliage, 4 oz., rosy colored fruits. *Source History: 1 in 1998; 1 in 2004.* **Sources: Sa9.**

Crimson Jack - 85 days, Family heirloom from Hot Springs, VA, indet., pink fruit, 12 oz., excellent rich tomato flavor, late season. *Source History: 2 in 2004.* **Sources: He18, To9.**

Crnkovic Yugoslavian - 67-90 days, Indet. vines bear beautiful pink beefsteaks that weigh 1+ lbs., perfect shoulders rarely crack, outstanding full tomato flavor, brought to the U.S. by Yasha Crnkovic. *Source History: 1 in 1998; 7 in 2004.* **Sources: He18, Ho13, Sa9, Sh13, Ta5, To1, To9.**

Dr. Buresh Pink Italian - 90 days, Indet., potato leaf, 1-1.5 lb. beefsteaks, flavorful, sweet flesh. *Source History: 1 in 2004.* **Sources: Ma18.**

Dr. Lyle - 77-80 days, Large indet. plant, gray-green fuzzy leaves, 1.5-2 lb., oblate-lobed pink fruits with outstanding flavor, blossom end scars, heirloom. *Source History: 2 in 1998; 7 in 2004.* **Sources: Fo7, Ho13, Ma18, Sa9, So25, To1, To9.**

Dr. Neal - (Lambert's General Grant) - 85 days, Very tall, indet. vines, heavy foliage, 1-2 lb., 4 in., round, flattened, pink-red slicer, meaty flesh, mildly acid, rich flavor, according to William Woys Weaver, this tomato was released under the name General Grant by a gardener by the name of Lambert in Bellefonte, PA, his name was added to it since another General Grant tomato already existed, thought to be a cross or sport from a now extinct tomato variety, Boston Market. *Source History: 1 in 1998; 6 in 2004.* **Sources: Ea4, He18, Ma18, Sa9, To1, To9.**

Dufresne - Quebec heirloom, large pink fruit, was grown by Fr. Savignac in the Joliette region. *Source History: 1 in 2004.* **Sources: So25.**

Dufresne No. 2 - Large spreading plant to 5-6 ft., pink, 3-4 in. fruit with tender skin, excellent taste, late, dev. by Quebec plant breeder. *Source History: 2 in 1998; 1 in 2004.* **Sources: Te4.**

Dutchman - (Der Dutchman) - 80-110 days, Indet., large flat purple-pink fruits, up to 3 lbs., mild flavor, almost solid flesh with few seeds, mid-season, an old strain from 1920 or earlier. *Source History: 1 in 1981; 1 in 1984; 2 in 1987; 3 in 1991; 5 in 1994; 6 in 1998; 17 in 2004.* **Sources: Ba8, Bo9, CO30, Ers, He18, Loc, ME9, Ra5, Ra6, Re8, Ri12, SE14, Sk2, TE7, To1, To9, To10.**

Early Wonder - 55 days, Det., round, dark pink fruit, 6 oz., excellent tomato flavor, good for short season areas. *Source History: 1 in 2004.* **Sources: To1.**

Eckert Polish - 84 days, Heavy yields of pink, semi-pear shape fruits. *Source History: 2 in 1998; 1 in 2004.* **Sources: Sa9.**

Eschelman's Pride - 80 days, Indet., nice pink fruit. *Source History: 1 in 2004.* **Sources: Ta5.**

Eva Purple Ball - 70-78 days, Heirloom from the Black Forest region of Germany from Joe Bratka's grandfather, smooth round fruits weigh 4-5 oz., blemish-free, tender texture, peels easily, does very well in hot humid areas, excel. resistance to foliar and fruit diseases, from the late 1800s. *Source History: 1 in 1994; 9 in 1998; 27 in 2004.* **Sources: Ap6, Ba8, CO30, Ea4, Ga1, Ga21, He8, He18, Loc, Ma18, ME9, Pr3, Ra5, Ra6, Re8, Sa9, SE14, Se17, Sh13, Sk2, So1, So25, Ta5, To1, To3, To9, To10.**

Florida Pink - 85-90 days, Indet. vines bear huge pink fruit, up to 3 lbs. or more, crack-free, smooth skin, juicy, delicious sweet flavor, few seeds, good fruit set in summer heat, late. *Source History: 1 in 1998; 3 in 2004.* **Sources: Ma18, To1, To9.**

Genuine Italian - (Geniune Italian Potato Leaf) - 82 days, Indet., 5.25 x 3 in., flattened pink globes, non-acid, average yield. *Source History: 1 in 1981; 1 in 1984; 1 in 1987; 1 in 1998; 1 in 2004.* **Sources: Sa9.**

Gerig - 89 days, Central Iowa heirloom, indet., flattened globe, high acid. *Source History: 1 in 2004.* **Sources: Sa9.**

German Giant - 77-82 days, Potato leaf indet. plant, smooth flattened globes, 5 x 3.5 in., pink skin and flesh, 2 lbs. or more, sweet flavor. *Source History: 1 in 1998; 2 in 2004.* **Sources: Sa9, To1.**

German Head - 80-90 days, Indet., large dark-pink beefsteak-type fruits, 1-2 lb., smooth and well shaped, excel. flavor, crack res., high yields, old variety. *Source History: 2 in 1991; 4 in 1994; 5 in 1998; 7 in 2004.* **Sources: CO30, He18, Ho13, Ma18, ME9, To1, To3.**

German Heart - 61 days, Indet., large flat pink fruit. *Source History: 1 in 2004.* **Sources: Sa9.**

German Johnson - (German Johnson Pink) - 76-80 days, Heirloom from Virginia and North Carolina, one of the 4 parent lines of Mortgage Lifter, indet. large rough fruits averaging .75-1.5 lbs., pink skin with yellow shoulders, very mild, low acid, very meaty, few seeds, heavy yields, good slicing or canning, fair disease res. *Source History: 2 in 1987; 1 in 1991; 8 in 1994; 17 in 1998; 32 in 2004.* **Sources: Ba8, Bo9, CA25, CO30, Co31, Ers, Fe5, Ge2, Gr27, He17, He18, HO2, Ho13, HPS, Ma18, ME9, MIC, Ra5, Ra6, RUP, SE14, Se25, Se26, Sh9, Shu, So1, Sw9, To1, To3, To9, To10, WE10.**

German Pink - 78-85 days, Indet., potato leaf plant, 10-16 oz., deep pink fruit with excellent, sweet flavor, meaty flesh, few seeds, excel. canner, one of the two original Bavarian varieties that started SSE. *Source History: 3 in 1998; 10 in 2004.* **Sources: Coo, Gr28, Gr29, Pr9, Se16, Se26, Se28, Ta5, To1, Wo8.**

German Queen - 80 days, Indet., potato leaf plant, slightly flattened, pinkish beefsteak fruit, 1 lb., meaty flesh, incredible flavor, heirloom. *Source History: 1 in 1994; 2 in 1998; 4 in 2004.* **Sources: Ho13, Se7, Ta5, To1.**

Giant Belgium - (Giant Belgian) - 82-90 days, Indet., huge, dark-pink, 1.5-3 lb. average, some almost 5 lbs., smooth blossom end, mild flavor, solid meat, mid-season, so sweet that some folks make wine from it. *Source History: 2 in 1981; 2 in 1984; 3 in 1987; 3 in 1991; 5 in 1994; 11 in 1998; 32 in 2004.* **Sources: Ba8, Bu2, CA25, CO30, Ers, Gr29, He8, He17, He18, Ho13, HPS, Ma18, ME9, Ra5, Ra6, Ri12, SE14, Se17, Se26, Se28, Shu, Sk2, So25, Sw9, Ta5, To1, To3, To9, To10, Vi4, Vi5, Wo8.**

Giant Rosy - 74-84 days, Indet., 12-16 oz., flat globe fruits with pink skin. *Source History: 1 in 1998; 1 in 2004.* **Sources: Sa9.**

Gigantesque - Tall heavy bushes bear absolute huge fruits, orange-red skin, meaty pink flesh with very little gel, few seeds, super taste, over 2 lbs., more than 5 in. across, disease res., from Sergey who says tomatoes from the south of Ukraine are the most tasty. *Source History: 1 in 2004.* **Sources: Ami.**

Gillogly Pink - 80-85 days, Indet., lobed pink fruit, 4-5 in. diameter, up to 1 lb., meaty flesh, flavorful, has been in the Gillogly family of Ohio for 85 years. *Source History: 3 in 2004.* **Sources: He18, Se17, Ta5.**

Giuseppe's Big Boy - 80 days, Indet., potato leaf plant, oblate pink, 8-16 oz. fruit, sweet flavor. *Source History: 1 in 2004.* **Sources: Ma18.**

Glazier's Giant - 80 days, Originally known as the Archer tomato, indet., regular leaf, 3-4 in. pink fruit, grown for many years by a Mr. Glazier. *Source History: 1 in 2004.* **Sources: He18.**

Glesener - 96 days, Indet., potato leaf foliage, semi-sweet fruit, family heirloom. *Source History: 1 in 1998; 1 in 2004.* **Sources: Sa9.**

God Love - 75 days, Indet., good production of 1 x 2 in. pink plums, excellent robust flavor, good for salad and snacking, heirloom. *Source History: 1 in 2004.* **Sources: To9.**

Gogosha - 75-85 days, Indet., potato leaf, excellent yields of large pink fruit, 1-2 lbs., average yield, good disease tolerance, from Tarnipal, Ukraine, late 1800s. *Source History: 3 in 2004.* **Sources: Ap6, Sa9, Ta5.**

Grace Lahman's Pink - 80 days, Excellent main crop tomato produces rose-pink globes until frost, mild, sweet, old-fashioned flavor, indet. vines. *Source History: 2 in 2004.* **Sources: Ma18, To10.**

Grape, Pink - Pink fruit the size of a grape, midseason maturity. *Source History: 1 in 2004.* **Sources: Gr28.**

Gregori's Altai - 62-67 days, Indet., large 4 in. purplish red fruits weigh 8-12 oz., dense flesh, long harvest season, originated in the Altai Mountains on the border of China, from Novosibirsk, Siberia, rare. *Source History: 1 in 1991; 3 in 1994; 3 in 1998; 6 in 2004.* **Sources: He18, Hig, Sa5,**

So25, Ta5, To1.

Grosse Cotelee - 80 days, Indet., 3 in., round pink fruit, 8-12 oz., excel. rich flavor. *Source History: 1 in 1998; 1 in 2004.* **Sources: To9.**

Gulf State Market - (Gulf State) - 77-80 days, Strong med-sized blight-res. vines withstand bad weather, good cover, large firm smooth purplish-pink globes ripen uniformly in clusters of 5-7, meaty, few seeds, some tol. to cracking, for home gardens and fresh market, best main crop pink. *Source History: 5 in 1981; 3 in 1984; 1 in 1987; 1 in 1991; 1 in 1994; 3 in 1998; 7 in 2004.* **Sources: CO30, He18, Ra6, Re8, Ta5, To1, To3.**

Happy Jack - 85 days, Indet., heart shaped fruit, 12-16 oz., blossom end has a flattened point, originated in Belgium. *Source History: 1 in 1998; 1 in 2004.* **Sources: Fo7.**

Heart - 90 days, Heirloom from the Vavilov Institute, indet., pink, oxheart shaped fruit, 1-2 lbs., few seeds, good flavor. *Source History: 1 in 2004.* **Sources: To9.**

Heart of the Bull - 80-90 days, Indet., heart shaped, meaty fruit, few seeds, not the same as Bull's Heart, heirloom from Italy. *Source History: 1 in 1998; 3 in 2004.* **Sources: Ma18, Se17, Ta5.**

Henderson - Huge beefsteak, 2+ lbs., pink-red meaty flesh, excellent canner, Peter Henderson introduction, 1891. *Source History: 1 in 2004.* **Sources: Co31.**

Honey - 75-85 days, Indet., med. size pink fruit, very sweet flavor, few seeds, poor keeper, from Latvia. *Source History: 2 in 2004.* **Sources: Ta5, To9.**

Hungarian - 85 days, Indet. vines do best when staked and pruned to one single stem, large beefsteak-type from Hungary, ribbed irregular-shaped bright-pink fruits often weigh 1 lb. or more, good flavor. *Source History: 1 in 1991; 1 in 1994; 1 in 1998; 3 in 2004.* **Sources: He8, He18, Ta5.**

Hungarian Heart - 85 days, Indet., huge pink oxheart with no cracking, weighs 1 lb. or more, very productive, seeds were originally brought to America in 1901 from a small village 20 miles from Budapest. *Source History: 2 in 1998; 3 in 2004.* **Sources: Ma18, Se16, To1.**

Hungarian Oval - 72-90 days, Indet., potato leaf, large, 5 in., 8-16 oz., slightly oval, pink-purple fruit, good flavor, mildly acid, meaty flesh, few seeds. *Source History: 1 in 1998; 3 in 2004.* **Sources: Sa9, Ta5, To9.**

Ispolin - 70 days, Indet. vines can reach 7 ft., large round pink 1-2 lb. fruit, excellent flavor with an earthy mild sweetness, very productive, name means 'giant' in Russian. *Source History: 1 in 1994; 1 in 1998; 2 in 2004.* **Sources: Hig, To9.**

Italian Purple - 85 days, Indet., potato leaf, large, 4-5 in., pink-purple fruit, green shoulders, good, mildly acid, old-time flavor, heirloom. *Source History: 1 in 1998; 1 in 2004.* **Sources: To9.**

Italian Sweet - 90 days, Indet., potato leaf plant, dark rose-pink beefsteak fruit, excellent eating quality, very productive. *Source History: 1 in 2004.* **Sources: Ma18.**

Jahamato Pink Paste - 85 days, Variety from New Zealand, indet., wispy vines produce 6 oz. pink fruit with pear shape, meaty flesh, excellent flavor. *Source History: 1 in 2004.* **Sources: To9.**

Japanese Oxheart - 90 days, Indet. plant, fern-like foliage, 10-16 oz., pink fruit is shaped like a heart, juicy, flavorful flesh with few seeds. *Source History: 1 in 1998; 3 in 2004.* **Sources: Ho13, So25, To1.**

Jeff Davis - 80-90 days, Indet. vines, oval, 10-16 oz. purple-pink fruit with a small core and few seeds, low acid, good disease res., from Alabama in the 1890s where it became popular along the path of the old B and O Railroad. *Source History: 3 in 1994; 3 in 1998; 8 in 2004.* **Sources: Ga1, He8, Ma18, Ra5, Ra6, Re8, To3, To9.**

Jefferson Giant - 85-90 days, Indet. vines, meaty fruits with reddish pink flesh, few seeds, 1-2 lbs., exceptional flavor, 1880s. *Source History: 1 in 1991; 1 in 1994; 2 in 1998; 1 in 2004.* **Sources: He8.**

Julia Child - 78 days, Indet., potato leaf, lightly fluted, deep pink beefsteak fruit, 4 in., firm juicy flesh, robust tomato flavor, good production. *Source History: 1 in 2004.* **Sources: To9.**

June Pink - (Pink Earliana) - 68-74 days, Identical to Earliana except for color and longer production, large rosy-pink globes borne in clusters of 6 to 10, very solid, free of cracks around the stem, med-heavy indet. vines, very productive, best-selling pink in markets, heirloom from New Jersey, introduced in 1900, released to the commercial trade in 1906. *Source History: 5 in 1981; 3 in 1984; 2 in 1987; 2 in 1991; 2 in 1994; 4 in 1998; 10 in 2004.* **Sources: Ea4, Ma18, Pr3, Ra5, Ra6, Re8, Sa9, So1, Syn, To3.**

Kalman's Hungarian Pink - 85-100 days, Indet., 2 x 4 in., oval plums, 6-8 oz., extra sweet flesh, one of the best tasting pink tomatoes, brought to America in 1967 by Kalman Lajvort, Edison, NJ. *Source History: 5 in 2004.* **Sources: Ami, Ma18, So25, Ta5, To9.**

Kelly's Pink - *Source History: 1 in 2004.* **Sources: Se17.**

Khabarovsky - 78 days, Det. stocky plant, pink-red salad size fruit. *Source History: 1 in 2004.* **Sources: Ta5.**

Khaborovsky 308 - 75 days, Indet. plants, 6 oz., slightly oblong pink fruits, great taste. *Source History: 1 in 1998; 2 in 2004.* **Sources: Sa9, Ta5.**

Kornesevsije - 85-93 days, Indet., 10-16 oz., bright pink, flattened fruits, good producer. *Source History: 1 in 1998; 4 in 2004.* **Sources: He18, Sa9, Ta5, To9.**

Laketa - 77 days, Long pointed pear type, solid like lemons, non-acid and sweet. *Source History: 1 in 1981; 1 in 1984; 2 in 1987; 3 in 1991; 1 in 1998; 1 in 2004.* **Sources: Sa9.**

Large Pink - 107 days, Indet., flattened, 4 x 2.5 in. fruit, pink-orange skin, pink flesh, average yield. *Source History: 1 in 1998; 1 in 2004.* **Sources: Sa9.**

Large Pink Bulgarian - 85 days, Indet., dark pink, large, round, flattened fruit, 1-3 lbs., solid flesh with few seeds, mild sweet flavor, low acid. *Source History: 2 in 1998; 4 in 2004.* **Sources: Ap6, Ma18, To1, To9.**

Livingston's Beauty - 75-85 days, Indet., 4-5 oz. round pink fruit, from Livingston Heritage Collection, 1886. *Source History: 1 in 1998; 3 in 2004.* **Sources: Ea4, Sa9, Vi4.**

Livingston's Globe - (Globe) - 85 days, Cross between Livingston's New Stone x Ponderosa, one of the parents of Marglobe, productive indet. plant, rose-red fruit, 7 oz., up to

13 oz., good mild flavor, rel. in 1905, rare. *Source History: 4 in 2004.* **Sources: Ea4, He18, Ta5, Vi4.**

Livingston's Main Crop Pink - 78-90 days, Vigorous indet. plant, large, smooth, pink globes, 4-6 oz., very flavorful, heirloom. *Source History: 1 in 1998; 1 in 2004.* **Sources: Sa9.**

Louisiana Pink - 75-80 days, Indet., 3-4 in. red-pink salad type globes, delicious, very good yields. *Source History: 2 in 2004.* **Sources: He18, To9.**

Mac Pink - 62 days, Large det., open plant, 5 oz., excel. fruit quality, extremely early, dev. at MacDonald Campus of McGill Univ. by E. Gyapay and B. Bible and C. Chong, 1973. *Source History: 1 in 1981; 1 in 1984; 1 in 1987; 3 in 1991; 1 in 2004.* **Sources: Pr3.**

Madagascar - 75 days, Sprawling vines, pretty rose-pink plums blushed with magenta, sweet flavor, high yields. *Source History: 1 in 2004.* **Sources: Ma18.**

Magnus - (Livingston's Magnus) - 80-90 days, Indet., potato leaf plant, broad foliage, 6-8 oz. oblate pink fruit, firm flesh with excellent flavor, heavy cropper, no cracking, released by Livingston Seed Co. in 1900. *Source History: 5 in 2004.* **Sources: Ag7, Ea4, He18, To9, Vi4.**

Mammoth Sandwich - 90 days, Indet. vines, meaty, deep pink beefsteaks, up to 1.5 lbs. *Source History: 1 in 2004.* **Sources: Ma18.**

Marianna's Peace - 85 days, Indet., potato leaf, beautiful 1-2 lb. pink beefsteak fruit, creamy dense flesh, rich sweet old- fashioned tomato flavor, seeds brought to America during WW II, originally from Czechoslovakia. *Source History: 2 in 2004.* **Sources: So25, To9.**

Maritime Pink - 69 days, Indet., deep rose-pink, 5-7 oz. fruits, flavor starts lemony and ends sweet, from Sakhalin Island, Siberia. *Source History: 1 in 1994; 1 in 1998; 1 in 2004.* **Sources: Ta5.**

Marizol Bratka - (Purple Brandy) - 85-90 days, Indet., potato leaf, Brandywine x Marizol Purple cross by Joe Bratka, red-pink beefsteak fruit to 5 in., 10-16 oz., juicy, low acid flesh, mild flavor, late ripening. *Source History: 3 in 1998; 5 in 2004.* **Sources: Ea4, Ma18, To1, To9, Yu2.**

Marizol Pink - 85 days, Indet., shy producer of 6-12 oz. fruit, deep rose flesh, wonderful complex flavor. *Source History: 1 in 2004.* **Sources: Ma18.**

Marizol Purple - 80-90 days, Old German var. from the Black Forest, indet., potato leaf plant, oblate purple-pink fruit, 8-16 oz., good sweet/acid flavor balance, good disease res., dates back to the 1800s. *Source History: 1 in 1994; 8 in 1998; 14 in 2004.* **Sources: Ba8, CO30, He18, Ho13, Ma18, Ra5, Ra6, Re8, Sa9, Se26, Ta5, To1, To3, To9.**

Marlowe Charleston - 80-88 days, Indet., potato leaf, pink, 3 in., round fruit, 10-12 oz., unique acidic flavor with a mild finish, good disease res., small yield. *Source History: 1 in 1998; 5 in 2004.* **Sources: He18, Ho13, ME9, Sa9, To9.**

Mary Ann - 78 days, Indet., 1 lb. fruits are deep pink if grown with lots of foliage cover and orange-red when grown in full sun, sweet old-fashioned flavor, good disease res. *Source History: 1 in 1998; 1 in 2004.* **Sources: Ap6.**

Maryland Pink - 75 days, Huge indet. plant bears 1+ lb. pink beefsteak fruit, excellent flavor. *Source History: 1 in 2004.* **Sources: He18.**

Mexican - (Mexican Potato Leaf) - 85 days, Indet., potato leaf, abundance of 8 oz., 4-6 in. flat pink fruit, tolerates heat and drought. *Source History: 2 in 2004.* **Sources: He18, Ta5.**

Mexico - 75-80 days, Indet., 1 lb. pink fruit, outstanding flavor and size maintained through the growing season, brought into the U.S. by a Mexican family. *Source History: 2 in 1998; 7 in 2004.* **Sources: He18, Ma18, Sh13, So25, Ta5, To1, To9.**

Micado Violettor - (Mikado Violet) - 80-105 days, Old Australian cultivar, indet., potato leaf foliage, crack-free pink fruit of excel. quality and flavor, 4-6 oz. *Source History: 2 in 1994; 2 in 1998; 7 in 2004.* **Sources: He18, Ra6, Sa9, Se7, Se26, Ta5, To1.**

Mike's Italian - 80 days, Indet., flavorful, pink heart shaped plums, rich flavor, good for fresh eating. *Source History: 1 in 2004.* **Sources: Ma18.**

Milgrene Rose - 80 days, Very productive, small pink fruit, from France. *Source History: 1 in 2004.* **Sources: Ta5.**

Mission Dyke - 65-78 days, Indet., med-pink globes in clusters of 3-6, all are large, perfect blossom ends, smooth shoulders, mild, disease and drought res., indet., tested well in tropics of Puerto Rico, vigorous. *Source History: 1 in 1981; 1 in 1984; 2 in 1987; 3 in 1991; 2 in 1994; 2 in 2004.* **Sources: ME9, Ta5.**

Missouri Pink Love Apple - 80 days, Indet., potato leaf plant, pink beefsteak fruit, 6-18 oz., complex sweet flavor rivals that of Brandywine and Cherokee Purple, excellent main crop tomato, grown by a Mr. Barnes during the Civil War era, who thought tomatoes were not edible so he grew them for scenery. *Source History: 5 in 2004.* **Sources: He18, Ma18, Ta5, To9, To10.**

Monfavet - 70 days, French var., large fruit, intense red color, disease res. *Source History: 1 in 1987; 1 in 1991; 1 in 1994; 1 in 1998; 1 in 2004.* **Sources: Gou.**

Mortgage Lifter, Estler's - 85 days, Estler family heirloom dating back to 1922, indet., most prolific of the Mortgage Lifter strains with the largest fruit, 1+ lb., pink-red fruit, excel. flavor, said to be older than Radiator Charlie's, original seed was sent to Australia in the 1930s by Mr. Estler, possibly the ancestor of the Australian variety of Mortgage Lifter. *Source History: 2 in 2004.* **Sources: Ap6, He18.**

Mortgage Lifter, McGarity's - 85 days, Indet., good production of classic Mortgage Lifter type fruit. *Source History: 1 in 2004.* **Sources: He18.**

Mortgage Lifter, Pink - 83-90 days, Old-timer still in demand, indet., large well shaped fruits, 1-2 lb., dark-pink skin, very meaty, few seeds, much like Giant Belgium but not quite as big, named by a man who sold this tomato to pay off his farm which he was about to lose, dates to the 1930s. *Source History: 2 in 1987; 6 in 1991; 7 in 1994; 15 in 1998; 30 in 2004.* **Sources: Ba8, Be4, Bo9, Bu2, Fi1, Fo7, Ga1, Ge2, Gr27, Gr28, Gur, He8, Loc, Ma18, ME9, Ni1, Ra5, Ra6, Re8, RUP, Sa5, Se26, Sh13, Shu, Sk2, Sw9, Ter, To1, To10, WE10.**

Mortgage Lifter, Radiator Charlie's - 79 days, Dev.

in the 1930s by M. C. Byles in Logan WV, where Radiator Charlie owned a radiator repair business. He had no formal education or plant breeding experience, yet he created this tomato by cross-breeding German Johnson, Beefsteak and an Italian and English variety, Radiator Charlie sold the plants for $1 each. After 6 years he paid off the $6,000 mortgage on his house. Fruits may weigh up to 4 lbs., few seeds, indet., released to Southern Exposure Seed Exchange as an exclusive variety in 1985 with all rights reserved. *Source History: 1 in 1991; 3 in 1994; 3 in 1998; 17 in 2004.* **Sources: Ami, Ap6, CA25, Co32, Eo2, Fo13, Ge2, He17, Hi13, Ho13, Jun, Ra6, Ri12, So1, To3, To9, Ver.**

Mortgage Lifter, Reiger's - 80 days, Indet., regular leaf, similar to other types of Mortgage Lifter. *Source History: 1 in 2004.* **Sources: He18.**

Mother Russia - 72-80 days, Indet., 10-12 oz., pink fruit in the shape of a perfect heart, dense meaty flavorful flesh. *Source History: 1 in 1994; 2 in 1998; 2 in 2004.* **Sources: Hig, Ta5.**

Mr. Hawkins - 70 days, Indet., potato leaf and regular leaf, pink-red, flattened, 4 x 5 in., round fruit with fruity flavor, seed given to Kent Whealy by a former school teacher and named after him. *Source History: 1 in 1998; 1 in 2004.* **Sources: To9.**

Mr. Underwood's Pink German Giant - 90 days, Indet., huge, ribbed, oblong beefsteaks up to 3 lbs., dense meaty flesh, few seeds. *Source History: 1 in 2004.* **Sources: Ma18.**

Mrs. Benson - 70 days, Indet., potato leaf, beautiful fruit with no cracks or blemishes, 1 lb., few seeds, old-time flavor, good acid balance, seed from Mrs. Benson, Oswego, IL, whose family had grown it for many years, introduced by Underwood Gardens, Ltd. *Source History: 1 in 2004.* **Sources: Und.**

Nectar Rose - 75 days, Indet., 4-6 oz. rose-pink salad tomato, nice flavor, good yields. *Source History: 1 in 2004.* **Sources: Ta5.**

Nectarine - 75 days, Vigorous indet. vines, regular leaf, grows in clusters of 3-5 fruit, red-yellow glossy, blemish-free skin with green stripes that changes to salmon-pink when mature, slight oblate shape, 4-6 oz., meaty flesh, very unusual. *Source History: 5 in 2004.* **Sources: He8, He18, ME9, SE14, So25.**

New Zealand Paste - 80 days, Heat and drought res. plants, large bell shaped pink fruit, meaty flesh, excel. flavor, heirloom. *Source History: 2 in 1994; 1 in 1998; 1 in 2004.* **Sources: Ta5.**

New Zealand Pear - (New Zealand Pink Paste) - 90 days, Paste variety, indet., size and shape of Bartlett pear, 6-12 oz., pink with green shoulders, nearly seedless meaty flesh, needs staking, great saucer and slicer, heirloom from New Zealand. *Source History: 1 in 1987; 1 in 1994; 4 in 1998; 4 in 2004.* **Sources: Ma18, Se17, Se26, To9.**

Newtown Italian Plum - Oxheart type fruit, large to huge size, very meaty pink flesh with few seeds, from a caretaker at Fairfield Hills, a former state mental hospital in Newtown, CT, endangered. *Source History: 1 in 2004.* **Sources: Ea4.**

Nicky Crain - 88 days, Fine foliage, oxheart shaped fruit, up to 1 lb. *Source History: 1 in 1998; 2 in 2004.* **Sources: Sa9, Sh13.**

Nina's Heirloom - 85 days, Indet., beautiful large pink beefsteak. *Source History: 1 in 2004.* **Sources: Ta5.**

Noire des Cosebeuf - 87 days, Wrinkled, dark pink-purple fruit. *Source History: 1 in 1998; 1 in 2004.* **Sources: Sa9.**

Oaxacan - (Oaxacan Pink) - 75-85 days, Indet., med. size, oblate, attractive pink fruit, juicy and seedy, sweet flavor, marbled flesh, 10 per lb., delicious, Mexico. *Source History: 1 in 1994; 3 in 1998; 3 in 2004.* **Sources: Ri12, Se7, Ta5.**

Olena Ukrainian - 75-85 days, Indet., potato leaf plant, blemish-free pink globes, 1-2 lbs., excellent sweet complex flavor, heirloom from Olena Warshona, Odessa, Ukraine. *Source History: 1 in 1998; 6 in 2004.* **Sources: Ap6, He18, Ma18, Sa9, Se26, To1.**

Oli Rose de St. Dominique - Medium size fruit with distinctive oval shape, pink flesh with sweet taste, prolific, flavor seems better in less than ideal growing seasons, collected by CV Ea4 in a market in Arles, France, rare. *Source History: 1 in 2004.* **Sources: Ea4.**

Omar's Lebanese - 80 days, Indet., large, pink, irregular, flattened fruit to 5 in., 1-1.5 lbs., sweet flavor seems to be expressed best in northern areas while the quality is more variable in southern areas, heirloom from Lebanon brought to this country by a Lebanese college student who got the seed from farmers living in the Lebanese hills. *Source History: 4 in 1998; 15 in 2004.* **Sources: Ami, Ba8, Co31, Ers, Ma18, ME9, SE14, Sh13, So1, So25, Sw9, To1, To3, To9, To10.**

Opal Essence - 85-90 days, Indet., medium size round fruit with iridescent pink skin, very tasty. *Source History: 2 in 1998; 4 in 2004.* **Sources: Ami, Se17, So25, Ta5.**

Ororiko - Indet., large round pink fruit, 2.5-3 in., sweet taste, prolific. *Source History: 1 in 2004.* **Sources: Se17.**

Oxheart - 80-95 days, Large open indet. plants, fair cover, deep rosy-pink firm heart-shaped shallowly furrowed 7 oz. fruits, thick meaty mild walls, few seeds, clusters of 2 to 7, ASC res. BER susc., old favorite with home gardeners, adapted to high humidity, oxheart shape is the result of a single gene mutation, 1925. *Source History: 53 in 1981; 49 in 1984; 45 in 1987; 38 in 1991; 40 in 1994; 33 in 1998; 38 in 2004.* **Sources: Ba8, Ber, Bo9, Bo19, Bu1, CA25, CO30, Com, Ers, Ga1, Ge2, Gr28, He17, He18, HO2, Jo6, La1, Lan, Lej, ME9, Na6, Or10, Pe7, Pr3, Ra5, Ra6, SE14, Se24, Sh9, Sh13, So1, Te4, To1, Tt2, Up2, Vi4, WE10, Wet.**

Oxheart, Giant - (Oxheart Giantissimo, Jung's Giant Oxheart) - 87-95 days, Distinct shape like a big oxheart or a delicious apple, smooth rosy-pink fruits up to 2 lbs., firm flesh, few seeds, pleasing taste, heavy yielder, unique, 1925. *Source History: 1 in 1981; 1 in 1984; 3 in 1987; 3 in 1991; 5 in 1994; 8 in 1998; 9 in 2004.* **Sources: Fo7, Ni1, Ra5, Ra6, Sa9, Se26, Shu, To3, To10.**

Oxheart, Light Pink - 75 days, Indet., heart shaped fruit, 12 oz., meaty and flavorful. *Source History: 2 in 1998;*

5 in 2004. **Sources: Ra5, Ra6, Re8, Se26, To3.**

Ozark Pink - 65-80 days, Indet. plants grow 5 ft. tall, pink 7 oz. flattened globes, FW1 res., recommended for hot humid disease-prone areas, included in its parentage are Bradley, Arkansas Traveler 76, Heinz 1439 and Campbell 1327. *Source History: 1 in 1994; 3 in 1998; 2 in 2004.* **Sources: Sh13, To9.**

Ozark Pink VF - 65 days, Productive, indet. vines are excellent for staking, large, 6-10 oz., beautiful globe shaped deep pink fruits, uniform ripening, not overly firm but do resist cracking, midseason, recommended for hot, humid, disease-prone areas, dev. from a complex pedigree including Bradley, Arkansas Traveler 76, Heinz 1439 and Campbell 1327, U of AR. *Source History: 1 in 1991; 1 in 1994; 2 in 1998; 3 in 2004.* **Sources: Ga21, Pe6, So1.**

Palestinian - 80 days, Indet., unique, delicious heart-shaped pink fruits weighing 1 lb. each, few seeds. *Source History: 1 in 1991; 1 in 1994; 3 in 1998; 1 in 2004.* **Sources: Ta5.**

Peach, Pink - Tall plants bear rosy pink fruit, textured skin feels tacky because of small soft hairs, speckled with tiny pale spots, tender juicy flesh, said to have been selected in France in the 1880s. *Source History: 1 in 2004.* **Sources: So25.**

Pear, Pink - 90 days, Large indet. plant, small pear-shaped fruit, slightly larger than Red Pear and Yellow Pear, used like cherry tomatoes. *Source History: 1 in 1994; 1 in 1998; 1 in 2004.* **Sources: So25.**

Pear, Ukrainian - 80 days, Vigorous indet. vines, pink-purple, pear shaped fruit is 3-4 in. long, 6 oz., delicious sweet flavor, heirloom brought from Yalta, Ukraine, by a Peace Corps volunteer. *Source History: 1 in 2004.* **Sources: To1.**

Peru - 89 days, Indet., average yields of 2.5 x 1.75 in., pink globes. *Source History: 1 in 1998; 1 in 2004.* **Sources: Sa9.**

Ping Pong - 70-85 days, Compact dwarf plants, distinctive rosy-pink dime-sized fruits, very sweet, tender skin, same size fruits as Tiny Tim but plants not quite as dwarf. *Source History: 2 in 1981; 1 in 1984; 1 in 1987; 4 in 1998; 5 in 2004.* **Sources: Bo19, CO30, Ga1, Ra5, Ra6.**

Pink - 79-90 days, Indet., 4 x 3 in., flattened globes, pink skin and flesh, average yield. *Source History: 1 in 1998; 2 in 2004.* **Sources: Sa9, Ta5.**

Pink Accordion - Attractive, large pink fruit, ruffled like an accordion, sweet mild flavor, semi-hollow, excellent for stuffing. *Source History: 2 in 2004.* **Sources: Ba8, Ra6.**

Pink Climber - 115 days, Vigorous, indet. plant, potato leaf foliage, 12-16 oz. flat globes. *Source History: 1 in 1998; 1 in 2004.* **Sources: Sa9.**

Pink Ice - 70-80 days, Indet., good foliage cover, oblong, pale pink large size cherry, mild flavor. *Source History: 3 in 2004.* **Sources: Ma18, Sa9, Ta5.**

Pink Monserrat - 76 days, Indet., beautiful plink slicer, 3 in., sweet mild flavor, low acid, heirloom. *Source History: 1 in 2004.* **Sources: To9.**

Pink Niblets - 70 days, Indet., square-oval pink cherry, wonderful flavor, huge yields. *Source History: 1 in 2004.* **Sources: Ta5.**

Pink Petticoat - 70-76 days, Indet., semi-hollow, ruffled fruit, 4 oz., mild sweet flavor, good stuffer. *Source History: 4 in 2004.* **Sources: Ma18, Ra5, Ra6, Re8.**

Pink Ping Pong - 75 days, Indet., 2 x 2 in. round pink fruit the size of a ping pong ball, great flavor, highly productive. *Source History: 9 in 2004.* **Sources: Ea4, Ma18, Pe2, Sa9, TE7, To1, To3, To9, Tu6.**

Pink Plum - 75-90 days, Large 1-2 in. oval light-pink fruits, meaty, mild flavor. *Source History: 1 in 1991; 1 in 1994; 1 in 1998; 1 in 2004.* **Sources: Sa9.**

Pink Potato Top - 107 days, Indet., 5.25 x 3.5 in., flattened globes, pink flesh and skin, average yield. *Source History: 1 in 1998; 1 in 2004.* **Sources: Sa9.**

Pink Russian - 94 days, Indet., very large, flattened, pink fruits, 10-16 oz. *Source History: 1 in 1998; 1 in 2004.* **Sources: Sa9.**

Pink Russian 117 - 90 days, Pink sport of Russian 117, med. size, pink, heart shaped fruit, 10-14 oz., meaty, low acid, very sweet. *Source History: 1 in 1998; 1 in 2004.* **Sources: To9.**

Pink Stuffer - 80 days, Indet., long scalloped fruit, excellent for stuffing. *Source History: 3 in 2004.* **Sources: ME9, Pin, SE14.**

Pink Sweet - 80 days, Indet., perfect, flattened, pink slicers to 1 lb., excel. tomato flavor, originally called No Name, but those who grew it thought it deserved better, re-named by Craig LeHoullier, respected tomato authority. *Source History: 1 in 1998; 2 in 2004.* **Sources: Sa9, To9.**

Polish C - 67 days, Indet., potato leaf, flat pink fruit with green shoulders, pink flesh. *Source History: 1 in 2004.* **Sources: Sa9.**

Polish Heart - 80 days, Indet., large pink heart shaped fruit, good taste, from Poland. *Source History: 1 in 2004.* **Sources: Ta5.**

Polish Non-Acid - 82 days, Indet., potato leaf, large flattened pink fruit, sweet, non-acid flesh. *Source History: 1 in 2004.* **Sources: Sa9.**

Pomadora, Pink - 75 days, Indet., pink version of Pomadora, not as productive, from Italy. *Source History: 1 in 2004.* **Sources: Ma18.**

Pomme d'Amore - 75-95 days, Indet., prolific large pink 2-3 in. cherry, delicious firm flesh, bears early and throughout the season, isolated for a century on the Canary Islands, a relic of the first introduced tomatoes, translates "apple of love". *Source History: 2 in 1998; 3 in 2004.* **Sources: Bo9, Co32, Se26.**

Ponderosa - (Ponderosa Pink, Pink Beefsteak, Henderson #190, Henderson #400, Tall Pond. Pink) - 80-100 days, Giant home garden strain for slicing and canning in humid areas, large vigorous open spreading indet. vines, fair to poor cover, does best when staked, dark purplish-pink 12 oz. flat fruits, some 2 lbs., clusters of 3 to 5, rough irregular shape, small cavities, few seeds, sweet and mild, firm and very meaty, will crack, late-maturing, heavy-yielding, intro. by seedsman Peter Henderson, 1891. *Source History: 63 in 1981; 52 in 1984; 44 in 1987; 35 in 1991; 29 in 1994; 32 in 1998; 33 in 2004.* **Sources: Ba8, Bo9, CA25, CH14, Dam, Ea4, Eo2, Ers, Ge2, Gr27, Gr28, He8, He17, HO2, LO8,**

Ont, Ra5, Ra6, RIS, RUP, SE14, Se26, Sh9, Shu, SOU, Ta5, To1, To3, Up2, Vi4, WE10, Wet, Wi2.

Potato Leaf Pink - 79-90 days, Indet., 5 x 3 in., flattened fruit, pink skin and flesh, excel. yield. *Source History: 1 in 1998; 1 in 2004.* **Sources: Sa9.**

Prude - 80 days, Indet., large purple-pink beefsteak fruit, very nice looking, good yields. *Source History: 1 in 2004.* **Sources: Ta5.**

Pruden's Purple - (Prudence) - 65-85 days, Considered to be one of the great re-discoveries in heirloom tomatoes, indet. potato leaf plants, 10-16 oz. fruits with distinctive ridges, dark-pink skin color, firm meaty crimson flesh, no cracking, few seeds, widely adapted. *Source History: 1 in 1991; 5 in 1994; 15 in 1998; 39 in 2004.* **Sources: Ag7, Ba8, CO30, Ea4, Fe5, Fo7, Ga1, Ge2, Go8, Gr28, He8, He18, Ho13, Jo1, Ma18, ME9, Na6, Pe2, Pin, Ra5, Ra6, Ri12, RUP, Sa5, Se8, SE14, Se26, Sh13, So25, Sw9, Ta5, Te4, TE7, To1, To3, To9, To10, Up2, Vi4.**

Pumpkin - Vines require staking and pruning, large pink fruit with unusual ribbed shape and pointed end, up to 1 lb., good flavor. *Source History: 1 in 1994; 2 in 2004.* **Sources: Te4, Up2.**

Purple - 95 days, Deep pink fruit. *Source History: 1 in 1998; 2 in 2004.* **Sources: Mi12, Sa9.**

Purple Calabash - 80-90 days, Novelty, prolific, drought tolerant plants, small ruffled purple-pink fruits blushed bronze or chocolate-brown when ripe, identical to ones figured in 16th century herbals, 5-6 oz. average weight, intense flavor, fairly crack resistant, stores well. *Source History: 3 in 1987; 5 in 1991; 7 in 1994; 12 in 1998; 28 in 2004.* **Sources: Ag7, Ami, Ba8, Bo9, CO30, Co32, Ea4, Ga1, Go8, Gr28, Ho13, Hud, ME9, Pr3, Sa9, SE14, Se17, Se26, So25, Ta5, Te4, TE7, Ter, Th3, To1, To9, To10, Up2.**

Purple Perfect - (Pale Purple Perfect, Perfect Purple) - 85 days, Indet., potato leaf plant produces dusky pink fruits that ripen to a pale rose, 6-8 oz., distinct sweet-tart flavor. *Source History: 1 in 1998; 5 in 2004.* **Sources: He18, Ma18, Sa9, To1, To9.**

Purple Price - 70-80 days, Indet., potato leaf, medium-large, pink-purple fruit, green shoulders, excel. flavor. *Source History: 1 in 1998; 2 in 2004.* **Sources: He18, Ma18.**

Purple Prince - 75 days, Resembles Cherokee Purple but matures earlier, indet., potato leaf, med-large pink-purple fruit, 10 oz., excellent flavor, heirloom. *Source History: 4 in 2004.* **Sources: Al25, Co32, So25, To9.**

Purple Smudge - 76 days, Extremely productive plants bear purplish colored fruit from the start, ripening to a pink-red color with a purple cast which is most visible in cold areas, low acid, high yielder. *Source History: 1 in 1998; 2 in 2004.* **Sources: Sa5, Sa9.**

Red Butter Heart - 91 days, Indet., large, pink, heart shape fruit in clusters of 5-6, 16 oz., 5 x 4 in. *Source History: 1 in 1998; 2 in 2004.* **Sources: Ta5, To9.**

Redfield Beauty - 80-98 days, Heirloom tomato sel. from Livingston's Beauty in 1885, indet., flat pink fruit, hard flesh, excel. full flavor, off-types possible (CV Ea4). *Source History: 3 in 2004.* **Sources: Ea4, Sa9, Ta5.**

Richardson - 85 days, Indet., gigantic, juicy beefsteak. *Source History: 1 in 2004.* **Sources: Ma18.**

Rosalie's Big Rosy - 80 days, Indet. plants produce many huge fruit, many over 3 lbs., excel. flavor, smooth blossom end, holds well on the plant and after picking, family heirloom, CV Underwood introduction. *Source History: 1 in 1998; 1 in 2004.* **Sources: Und.**

Rose of Berne - (Rose de berne) - 75 days, Sturdy indet. plants, clusters of tasty pink-red fruit with nearly translucent skin, 6-8 oz., spicy rose-pink flesh, sweet and juicy, good yields, late, Swiss heirloom. *Source History: 1 in 1991; 1 in 1994; 9 in 2004.* **Sources: Coo, Ea4, Fe5, So25, Syn, Ta5, To1, To10, Und.**

Rose Quartz - 65-75 days, Indet., large pink cherry tomatoes borne in clusters, excellent flavor, from Japan. *Source History: 2 in 2004.* **Sources: So25, To9.**

Royal Hillbilly - 78-85 days, Indet., thin-walled, purple-pink fruit, 12-16 oz., exceptional flavor, good yields. *Source History: 4 in 2004.* **Sources: Ho13, Ma18, Sa9, Und.**

Roza - 63-91 days, Med-sized det. plants, excel. foliage cover, med-sized firm pink-red globes, holds well on vine, produces throughout season, VW FW and CTV res., ripens uniformly, dev. in Prosser, WA. *Source History: 1 in 1981; 1 in 1991; 1 in 1994; 3 in 1998; 1 in 2004.* **Sources: Sa9.**

Russian Heart - 85 days, Indet., wispy foliage, pink-red, oxheart shaped fruit, some with tapered ends and others round, 12- 16 oz., flavor is a wonderful blend of sweet yet acid. *Source History: 1 in 1998; 1 in 2004.* **Sources: To1.**

Russian Rose - 78-84 days, Indet. plants bear large rose-pink fruit, 6-12 oz., meaty flesh, from Russia. *Source History: 1 in 1998; 4 in 2004.* **Sources: Sa9, Ta5, To1, To9.**

Sabre - 80-90 days, Indet., potato leaf, huge pink beefsteak fruit, 2 lbs., tart but tasty. *Source History: 2 in 1998; 2 in 2004.* **Sources: He18, Sa9.**

Sandul Moldovan - 73-85 days, Indet., large beefsteak fruit, pink with green shoulders, 1 lb., good high sugar/high acid balance, heirloom from the Sandul family of Moldova in the Moldovan Region. *Source History: 4 in 1998; 7 in 2004.* **Sources: Ami, Ap6, Ho13, Sa9, Ta5, To1, To9.**

Schwarze Sarah - 80 days, Indet., tasty pink-purple fruit, dark green shoulders, grey-purple streaks throughout the flesh, German heirloom. *Source History: 1 in 2004.* **Sources: To9.**

Segler - 93 days, Indet., blocky, 3 oz., bright pink fruits. *Source History: 1 in 1998; 1 in 2004.* **Sources: Sa9.**

Sheboygan - 80 days, Indet., pretty pink paste with pointed shape, good keeper, grown by Lithuanian immigrants in Sheboygan, WI since the early 1900s. *Source History: 2 in 2004.* **Sources: Sa5, Ta5.**

Shirley S. - 80 days, Indet., potato leaf plant bears huge beefsteaks up to 2.5 lbs., dense, sweet flesh with small seed cavities, very productive. *Source History: 2 in 2004.* **Sources: Ma18, Ta5.**

Shirvey - 82 days, Indet., large pink 4 in. globes, rich tomato flavor, late. *Source History: 1 in 2004.* **Sources: To9.**

Siberian Pink - 69 days, Early, small pink fruit, 6-10 per cluster, very high yields, selected by Glenn Drowns. *Source History: 1 in 1998; 1 in 2004.* **Sources: Sa9.**

Sochulak - 70 days, Italian heirloom, indet., oxheart type, medium to very large, oblate pink fruit, meaty texture suitable for fresh use or tomato paste. *Source History: 2 in 1994; 2 in 1998; 1 in 2004.* **Sources: Fe5, Ta5.**

Soldacki - 75 days, Indet., dark pink, flattened fruits, 1 lb., thin skin, firm flesh, low acid, brought to Ohio from Krakow, Poland, 1900. *Source History: 8 in 1998; 21 in 2004.* **Sources: CO30, Ers, Fe5, He8, He18, Ma18, ME9, Ra5, Ra6, Re8, Ri12, Sa9, SE14, Se16, Sh13, So25, Ta5, To1, To3, To9, To10.**

Speckled Peach - 80 days, Indet., dark pink 2 in. fruit, slightly oval shape, slightly fuzzy skin with a few tiny speckles, from Reinhart Kraft, Germany. *Source History: 1 in 2004.* **Sources: To9.**

Stump of the World - (Big Ben) - 105 days, Indet., potato leaf, 6 ft. plants, smooth, pink, 4-5 in. fruits, some weighing up to 2 lbs., small seed cavity with few seeds, outstanding flavor, crosses easily, isolate plants kept for seed, from Ben Quisenberry. *Source History: 1 in 1994; 2 in 1998; 5 in 2004.* **Sources: Ho13, Ma18, Ta5, To9, To10.**

Tappy's Finest - 77-90 days, West Virginia heirloom, indet., large slightly flattened pink globes, very small core, few seeds, very meaty and flavorful, excel. home garden var. for slicing and canning, does best where summers are moderate to cool, originally from Italian seed stock, pre-1948. *Source History: 1 in 1981; 1 in 1984; 1 in 1987; 2 in 1991; 1 in 1994; 4 in 1998; 6 in 2004.* **Sources: Ap6, Sa9, So1, Ta5, To1, To9.**

Thai Pink - 75 days, Indet., pear shaped fruit, 1-3 oz., ruby pink skin with slight iridescent luster, smooth, sweet flesh, prolific, heirloom from Thailand. *Source History: 3 in 1998; 5 in 2004.* **Sources: Ap6, Ba8, Ni1, So25, Ta5.**

The 1884 Tomato - 85-90 days, Indet. vines, dates back to the Ohio Valley flood of 1884, original was found growing in a pile of debris beside the river at Friendly, WV by a Mr. Williamson who was so impressed that he saved the seed, it survives today as a local favorite that is not widely shared, good slicer and canner, few seeds, sweet flavor, late. *Source History: 1 in 1991; 2 in 1998; 8 in 2004.* **Sources: Fo7, He18, Ho13, ME9, Sa9, Ta5, To1, To9.**

Tiffen Mennonite - 75-85 days, Indet., potato leaf plant, large, dark pink fruit with outstanding flavor and smooth texture, up to 4 in. across, heirloom. *Source History: 1 in 1994; 8 in 1998; 16 in 2004.* **Sources: Ami, CO30, Ers, Fe5, Ga1, He8, He18, HO2, Ho13, Ma18, Ra5, Ra6, Re8, Sa9, Ta5, To1.**

Togo Trefle - Indet., deeply ribbed 2 in. pink fruit, excellent rich sweet-tart taste, extremely hardy and very tolerant to drought and heat, rare heirloom from the West African nation of Togo. *Source History: 1 in 2004.* **Sources: Ami.**

Trucker's Favorite - 75 days, Uniform, 2 in. pink globes, excel. flavor, good yields, introduced around 1912. *Source History: 2 in 2004.* **Sources: Ea4, To9.**

Ukrainian Heart - 70 days, Semi-indet., large pink, 5 in., round fruit, high shoulders, flat sides, mild flavor, to 1.5 lbs. *Source History: 1 in 1998; 1 in 2004.* **Sources: Sa9.**

Una Hartsock's Beefsteak - Indet., elongated shape, violet skin, delicious, productive. *Source History: 1 in 2004.* **Sources: Ma13.**

Urbanite - 90 days, Vigorous indet. vines produce pink, grapefuit size fruits, Danish tomato from the 1930s. *Source History: 1 in 1998; 2 in 2004.* **Sources: Fo7, He18.**

Vermillion - Indet., large pink fruit, 8-10 oz., fine flavor. *Source History: 1 in 1998; 1 in 2004.* **Sources: Abu.**

Violaceum Krypni-Rozo - 75-85 days, Very rare French heirloom, indet. plants, good yields of pink fruit, all shapes and sizes, up to 2 lbs., sharp-sweet taste. *Source History: 3 in 1998; 3 in 2004.* **Sources: He18, Ho13, To1.**

Wagner's Italian - 80 days, Indet., productive plants, plum to heart shaped dusky pink fruit, delicious, sweet flavor, very rare. *Source History: 1 in 2004.* **Sources: Ma18.**

West Virginia Straw - 82-85 days, Indet., potato leaf, high yields of sweet pink, beefsteak fruit weighing up to 1 lb. each. *Source History: 2 in 2004.* **Sources: He18, Ma18.**

Wheatly's Frost Resistant - 66-79 days, Indet., excel. yields of .75 x 1.5 in. plum shape fruit, pink-red skin and flesh. *Source History: 1 in 1998; 1 in 2004.* **Sources: Sa9.**

Wins All - 80 days, Improved Ponderosa type, flavorful old variety with potato leaf foliage, indet., large pink, slightly flattened fruits, 18-24 oz., few seeds, low yields but worth growing, introduced by Peter Henderson and Co. in 1924 as "Number 400", renamed in 1925 following a contest for a name. *Source History: 1 in 1994; 3 in 1998; 11 in 2004.* **Sources: Ba8, Ea4, Ga1, He8, He18, ME9, Ra6, Sa9, So1, To1, To9.**

Winsted Oxheart - Fair production of fleshy pink fruit, good flavor, dates back to the first half of the 20th century, from Winsted, CT, endangered. *Source History: 1 in 2004.* **Sources: Ea4.**

Wolford's Wonder - 90 days, Indet. plants grow 5 ft. or more, firm solid heart-shaped fruits average 1.5-3 lbs., thick pink skins, exceptionally juicy, small seed cavity, original seed from Max Wolford, a big tomato contest winner. *Source History: 1 in 1987; 1 in 1998; 2 in 2004.* **Sources: Hud, To9.**

Yasha Yugoslavian - Vigorous plant, large pink oxheart fruit, 1-1.5 lbs., excel. flavor, consistent high yields. *Source History: 1 in 2004.* **Sources: Ho13.**

Zapotec Ribbed - (Zapotec Pleated, Zapotec Pink Ribbed) - 80-90 days, Indet., hollow ruffled pink fruit with with yellow blush, bell pepper shape, 4 in., up to 1 lb., can be stuffed and baked, few cracks, from Zapotec Indians of Oaxaca, Mexico. *Source History: 1 in 1987; 3 in 1991; 2 in 1994; 3 in 1998; 13 in 2004.* **Sources: Ami, Ea4, He18, Hud, Jo6, Ma18, Na6, Pe2, Ri12, Se7, To1, To9, To10.**

Zaryanka - 65 days, Semi-det. plants, pink-red fruit, 3-6 oz., early, name means "sunrise" in Russian. *Source History: 1 in 1994; 1 in 1998; 1 in 2004.* **Sources: Hig.**

Varieties dropped since 1981 (and the year dropped):

Beauty Special (1987), *Bellerose* (1994), *Blue Ridge* (1994), *Cherry, Old Fashioned Pink* (1994), *Dwarf Champion* (1994), *Early Detroit* (2004), *Enrollado* (2004), *Everbearing* (1994), *Festive Pink* (2004), *Holmes Mexican Tomato*

(1991), *Hungarian* (2004), *Jan's Pink* (2004), *Jumbo* (2004), *Jung's Giant Oxheart* (1987), *Large Ribbed Pink Zapotec* (1991), *Mexican Ribbed* (2004), *Ohio MR13 Forcing (TMV)* (1991), *Olympic* (1994), *Olympic Pink* (1994), *Omhiya* (1998), *Pink Champion* (1994), *Pink Delight* (1998), *Pink Droplet* (1998), *Pink Giant Subacid* (1987), *Pink Globe* (1994), *Pink Persimmon* (1984), *Pink Tree* (1987), *Pomme d'Amour* (1998), *Ponder-Heart* (2004), *Porter (Small Fruited Strain)* (1994), *Porter Pink* (1998), *Purple King* (1983), *Raspberry Colored* (2004), *Ruby Rakes* (2004), *Ruffled Pink* (2004), *San Marzano, Pink* (1994), *Sasha's Pink* (2004), *Spartan Pink 10* (1994), *Subalpine Plenty* (1991), *Super Purple King* (1983), *Supreme Gulf State* (1983), *Tennessee Peach Fuzz* (2004), *Tomboy* (1994), *Traveler* (1987), *Upright* (1987), *Vermillion* (99999), *VF Pink* (1994).

Tomato / Red
Lycopersicon lycopersicum

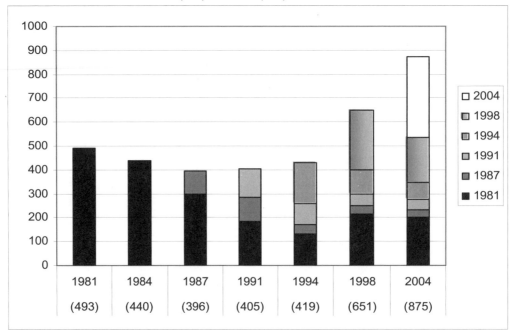

Number of varieties listed in 1981 and still available in 2004: 493 to 203 (41%).
Overall change in number of varieties from 1981 to 2004: 493 to 875 (177%).
Number of 2004 varieties available from 1 or 2 sources: 629 out of 875 (72%).

<u>1 x 6</u> - 80 days, Indet., thick walled, red banana shaped fruit, 1 x 6 in., few seeds. *Source History: 1 in 2004.* **Sources: Ma18.**

<u>506 Bush</u> - 62 days, Det., vines to 18 in., medium-large fruit, mild flavor with a hint of sweetness, blemish-free, tolerates drought, good yields. *Source History: 2 in 2004.* **Sources: He18, Ra6.**

<u>ABC Potato Leaf</u> - 62-70 days, Indet., potato leaf, very small cherry, cracks during wet spells, sweet fruity flavor. *Source History: 1 in 1998; 2 in 2004.* **Sources: Sa9, So25.**

<u>Abraham Lincoln</u> - 77-87 days, Old H.R. Shumway variety with dark-red meaty fruits, smooth sweet solid and juicy, some to 3 lbs., often 9 fruits per cluster totaling 7 lbs., no cracks or seams, few seeds, indet. vines, bronze-green foliage. *Source History: 1 in 1981; 1 in 1984; 5 in 1987; 2 in 1991; 2 in 1994; 3 in 1998; 11 in 2004.* **Sources: Fo7, Gr29, He17, Ho13, ME9, SE14, Se26, Se28, To9, To10, Vi5.**

<u>Abraham Lincoln, Original</u> - (Buckbee's Abraham Lincoln) - 87 days, Rebred by Shumway in 1986 to improve purity--other varieties that had crossed with the original over the years have now been completely bred out, slightly later and larger than the Regular Abraham Lincoln, 6-10 oz., may not ripen in the Upper Midwest or northern New England, originally released by Buckbee Seed Company in Illinois around 1923. *Source History: 1 in 1987; 2 in 1991; 3 in 1994; 5 in 1998; 10 in 2004.* **Sources: Ea4, Ers, Ma18, Ra5, Ra6, Shu, To1, To3, Vi4, Wo8.**

<u>Abraham Lincoln, Regular</u> - (Early Abraham Lincoln) - 70-77 days, Not as pure as Original Abraham Lincoln, indet., somewhat smaller fruits and earlier maturing, distinctive slightly acid flavor, excel. foliage disease res. *Source History: 1 in 1987; 3 in 1991; 4 in 1994; 3 in 1998; 3 in 2004.* **Sources: Shu, So1, To1.**

<u>Ace</u> - 75-90 days, Med-large det. vines, good cover, large smooth firm flattened-globes, F2 and ASC res., good color, fresh market or canning, thick walls, Campbell Soup Co., 1953. *Source History: 13 in 1981; 12 in 1984; 9 in 1987; 6 in 1991; 4 in 1994; 3 in 1998; 3 in 2004.* **Sources: Bo17, He8, Sau.**

<u>Ace 55 VF</u> - 75-85 days, Vigorous med-large det.

plants, good foliage cover, red thick-walled flattened globes, 5 x 6 in. dia., 7 oz., res. to VW FW1 ACS and cracking, commercial processing or shipping, because of its low acidity this tomato must be pressure-canned or canned with lemon juice to avoid the risk of botulism, dev. by Asgrow. *Source History: 18 in 1981; 13 in 1984; 18 in 1987; 18 in 1991; 24 in 1994; 23 in 1998; 30 in 2004.* **Sources: Ber, Bo19, CH14, Dam, Dgs, DOR, Ga1, He17, He18, HO2, Loc, ME9, Mel, MIC, MOU, Ra5, Ra6, Re8, RIS, RUP, SE14, Se26, Syn, To1, To3, Up2, Vi4, WE10, Wi2, WI23.**

Ace 55 VF Improved - 90 days, Med-large det. vines, large heavy firm red flattened-globes, med-late, res. to VW and FW1, some res. to cracking, for processing or fresh market, excel. shipper. *Source History: 1 in 1981; 1 in 1984; 2 in 1987; 1 in 1991; 2 in 1994; 2 in 1998; 1 in 2004.* **Sources: Ros.**

Adelia - Productive old var., med. size red fruits, slightly flattened shape. *Source History: 1 in 2004.* **Sources: Ho13.**

AH 3 - 87 days, Indet., semi-sweet globes, 8 oz., average yields. *Source History: 1 in 1998; 2 in 2004.* **Sources: Sa9, Ta5.**

AHLO - 89 days, Indet. plants, mildly acid globes, 8 oz. *Source History: 1 in 1998; 1 in 2004.* **Sources: Sa9.**

Ailsa Craig - 70-75 days, Scottish heirloom, excel. forcing variety for greenhouse or garden, med-sized uniform fruits, deep- red, fine flavor, vining, very early, good shape and size, fine cropper. *Source History: 4 in 1981; 2 in 1984; 2 in 1987; 2 in 1991; 3 in 1994; 4 in 1998; 6 in 2004.* **Sources: Bou, Sa9, St18, Ta5, Te4, Up2.**

Aker's Plum Red - 75 days, Red, pear shaped fruit, family heirloom of Carl Aker of Pennsylvania. *Source History: 2 in 2004.* **Sources: Sa9, To9.**

Aker's West Virginia - 70-85 days, Large indet., smooth red fruit, 10-20 oz., little cracking, delicious flavor, midseason, West Virginia family heirloom. *Source History: 1 in 1998; 7 in 2004.* **Sources: Ap6, Ma18, Sa9, Sh13, To1, To9, To10.**

Alaska - 63 days, Dev. by Edward Lowden from Farthest North, indet., large, round, red salad type cherry with good flavor, heirloom. *Source History: 1 in 1981; 1 in 2004.* **Sources: To9.**

Alaskan Fancy - 55 days, Det., early yet flavorful oval red plums, 1.5 x 2 in., 2 oz. *Source History: 5 in 2004.* **Sources: Ga1, Ra5, Ra6, Re8, To3.**

Alberta Peach - (Elberta Peach) - *Source History: 1 in 1998; 2 in 2004.* **Sources: ME9, SE14.**

Alicante - 68-80 days, English favorite broiled for breakfast, will not wilt when baked or broiled, emerges firm and full bodied, uniform and solid, greenback-free, heavy crops, excel. old-time tomato flavor. *Source History: 2 in 1981; 2 in 1984; 2 in 1987; 3 in 1991; 6 in 1994; 4 in 1998; 14 in 2004.* **Sources: Ers, Ga7, Ga22, Ra5, Ra6, Re8, Sa9, Sk2, Ta5, Tho, To9, Up2, WE10, We19.**

Alice Roosevelt - 80 days, Large indet. plants, 6-8 oz., red fruit, old-fashioned flavor. *Source History: 1 in 2004.* **Sources: Ma18.**

All Meat - 78 days, Indet., regular leaf, 1 x 1.5 in. fruit, juicy, meaty red flesh, midseason. *Source History: 1 in 2004.* **Sources: To9.**

Allerbest - 80 days, Heirloom from Munich, Germany, indet., 4-5 oz. globe shaped, bright red fruit, firm meaty flesh, few seeds, old-fashioned tomato flavor, dates back to 1916. *Source History: 1 in 2004.* **Sources: He8.**

Almetia - 70 days, Red, pear shape fruit, 1-2 oz., resists blight and blossom end rot, from Oklahoma. *Source History: 1 in 1998; 1 in 2004.* **Sources: Se17.**

Alpha - Very early, square, 2-lobed fruit. *Source History: 1 in 1994; 2 in 1998; 1 in 2004.* **Sources: Pr3.**

Alpine - 83 days, Indet., above average yields of flattened, 2.5 x 2 in. globes, dev. before 1950, Cheyenne, Wyoming. *Source History: 1 in 1998; 1 in 2004.* **Sources: Sa9.**

Amateur's Dream - 65 days, Indet. vines reach 5 ft., round, red large fruits, flavorful crimson flesh. *Source History: 1 in 1994; 2 in 1998; 1 in 2004.* **Sources: Hig.**

American Beauty - Indet., med. size round red fruit, good yields, dates back to the first half of the 19th century, rare. *Source History: 1 in 2004.* **Sources: Ea4.**

Amish Oxheart - Possibly a sub-variety of Amish Paste with more double fruits and juicier flesh, 8 oz. fruit, oxheart shape, paste type with rich sweet flavor for fresh eating, excellent for canning and freezing, Amish origin. *Source History: 1 in 2004.* **Sources: Ami.**

Amish Paste - 75-85 days, Indet., acorn-shaped deep-red fruits, 8-12 oz., thick flesh, few seeds, needs staking, Amish heirloom from Wisconsin, originally from Lancaster, PA. *Source History: 1 in 1991; 3 in 1994; 19 in 1998; 61 in 2004.* **Sources: Ag7, Al25, Ap6, Au2, Be4, Bo17, Bou, Bu2, CO30, Co32, Coo, Ers, Fe5, Fo13, Go8, Gr28, Gr29, He8, He17, He18, Hi6, HO2, Ho13, HPS, Hud, Jun, Lan, Ma18, ME9, Mel, Mo13, Pe2, Pr9, Ra5, Ra6, Ri2, Ri12, Roh, RUP, SE14, Se16, Se17, Se25, Se26, Se28, Sh13, Shu, So1, So25, Ta5, Te4, TE7, To1, To3, To10, Und, Up2, Ver, Vi4, Wo8, Yu2.**

Amish Red - 75 days, Red counterpart to Amish Gold, indet plant, midseason. *Source History: 1 in 1998; 1 in 2004.* **Sources: To9.**

Amy's Sugar Gem - Red Cherry x Tappy's Finest cross dev. by Dr. Jeff McCormack, indet., vigorous vines, small-med. fruit, 2 oz., 1.5 in. wide x 1.75 in. high, meaty flesh with small core, sweet, full flavor, home gardeners and fresh market growers, named for Amy Boor Hereford whose grandmother 'Tappy' introduced Dr. McCormack to heirloom tomatoes in 1982, 'Sugar Gem' part of the name relates to the sweet flavor and tiny light gold sparkles in the red skin. *Source History: 2 in 2004.* **Sources: Ga21, So1.**

Anahui - 90 days, Det. plant, round red fruit, 2-3 in., sweet flavor, nematode res., does well in hot humid conditions. *Source History: 2 in 1991; 1 in 1994; 1 in 1998; 2 in 2004.* **Sources: To9, Uni.**

Andes - 75 days, Indet., red, pepper shape fruit, 2 x 5 in., meaty flesh with few seeds, rich flavor. *Source History: 1 in 1998; 1 in 2004.* **Sources: To9.**

Andrew Rahart's Jumbo Red - 82-95 days, Indet., flattened, 24-32 oz. scarlet fruits, semi-sweet, old-time tomato flavor, good disease res., Andrew Rahart lived north of New York City where he collected seeds from local immigrants. *Source History: 4 in 1998; 13 in 2004.* **Sources: Ap6, Fo7, He18, Ma18, Ra5, Ra6, Re8, Sa9, Ta5, To1, To3, To9, To10.**

Angelina's Italian - 75 days, Indet., med. size square paste, nice taste, brought from Italy by Angelina Genovese in 1931. *Source History: 1 in 2004.* **Sources: Ta5.**

Angelo's Italian Red - 80 days, Indet., deep red, boat shaped beefsteak, 2+ lbs., meaty flesh, real old-time tomato flavor, from Italy. *Source History: 1 in 2004.* **Sources: Ma18.**

Angelo's Red - 75 days, Indet., regular leaf, 12-14 oz. red beefsteak fruit, good yields. *Source History: 1 in 2004.* **Sources: Ta5.**

Angora - 62-100 days, Brilliant-red tomatoes appear through indet. foliage with grayish-white fuzz on leaves and stems, med-sized solid and smooth, rarely cracks, rogue out the few plain plants. *Source History: 2 in 1981; 2 in 1984; 3 in 1987; 4 in 1991; 5 in 1994; 6 in 1998; 10 in 2004.* **Sources: CO30, Ga1, ME9, Ra5, Ra6, Re8, Sa9, SE14, Te4, WE10.**

Angora Supersweet - 75-80 days, Indet., unique silvery gray foliage, slightly fuzzy 1 in. cherry tomatoes, excel. sweet flavor, dev. by Joe Bratka. *Source History: 1 in 2004.* **Sources: Se16.**

Anna Aasa - 75 days, Indet., intensely flavored cherry tomato borne in long clusters, productive. *Source History: 1 in 2004.* **Sources: Ma18.**

Arctic Blum - 79 days, Semi-det., 8 oz. fruit, semi-sweet. *Source History: 1 in 1998; 1 in 2004.* **Sources: Sa9.**

Ardwyna - 80 days, Large slightly ribbed fruit, meaty flesh, few seeds, sauce and canning, great fresh flavor, very high yields. *Source History: 1 in 2004.* **Sources: Ga22.**

Atkinson - 70 days, Vigorous, indet., heavy foliage, firm smooth light-red flattened globe, 6 oz. to over 1 lb., FW1 and RKN res., prolific, meaty, for humid areas, Auburn University, 1966. *Source History: 7 in 1981; 5 in 1984; 3 in 1987; 1 in 1991; 3 in 1994; 7 in 1998; 3 in 2004.* **Sources: ME9, SE14, WE10.**

Auld Sod - 75 days, Indet., huge crops of bright red cherry fruit with true tomato flavor, from the Sod in Ireland. *Source History: 1 in 1998; 1 in 2004.* **Sources: Ta5.**

Aunt Lucy's Italian Paste - 85 days, Indet., typical Italian paste with meaty red flesh, few seeds, originally from Italy. *Source History: 1 in 2004.* **Sources: To9.**

Aunt Mary's Paste - 55-60 days, Heirloom from Wisconsin, det., elongated fruit is 2.5-3.5 in. long, 4-6 oz., sweet. *Source History: 2 in 1998; 1 in 2004.* **Sources: Hi6.**

Auriga - 72 days, Indet., small oblate, orange-red fruits, 2 oz., have the highest carotene content of any tomato. *Source History: 1 in 1998; 1 in 2004.* **Sources: Gr28.**

Aurora - 65 days, Indet., round red 5-8 oz. fruits, exceptional flavor, from Krasnoyarsk, Siberia. *Source History: 2 in 1994; 2 in 1998; 3 in 2004.* **Sources: Hig, Te4, To10.**

Aussie - 85 days, Australian heirloom, large, vigorous indet. plant bears large 1-2 lb. beefsteak fruit, scarlet skin, pinkish flesh, good sugar/acid flavor balance. *Source History: 4 in 2004.* **Sources: He18, Sa5, Se26, To1.**

Austin's Red Pear - 80 days, Indet., 2 in. red fruit, excel. flavor, very productive, occasional yellow fruit. *Source History: 1 in 2004.* **Sources: Se16.**

Australia - 75 days, Indet., large red beefsteak to 5 in. across, soft texture, mild flavor, late maturity, heirloom. *Source History: 1 in 1998; 1 in 2004.* **Sources: To9.**

AYWC - 93 days, Indet plants., 6 oz. red globes. *Source History: 1 in 1998; 1 in 2004.* **Sources: Sa9.**

Azteca B - 95 days, Semi-det. plants, 6 oz. globes. *Source History: 1 in 1998; 1 in 2004.* **Sources: Sa9.**

Azure - 75 days, Extremely prolific, 3 ft. plants, commercial variety with oval, plum shaped, 2 in., red fruits, dense flesh, good for canning and pickling whole, oval shape fits in jar better, skin does not break during processing, conflicting information about whether it was developed in the Ukraine or Moldova. *Source History: 1 in 1994; 1 in 1998; 2 in 2004.* **Sources: Se17, Ta5.**

Baccone - 75-80 days, Indet., round, smooth red fruit, 4 oz., flavorful, good producer, from Bolognese, Italy. *Source History: 2 in 2004.* **Sources: Ma18, Ta5.**

Balconi Red - 77 days, Compact, bushy dark green foliage, clusters of bright red fruit, ideal for containers. *Source History: 1 in 2004.* **Sources: Tho.**

Baller - 71 days, Indet., Roma type fruit is oblong with swollen end, 3 in., mild flavor. *Source History: 1 in 1998; 1 in 2004.* **Sources: To9.**

Basket Pak - 76 days, Prolific indet. vines, clusters of 5-9 rich-red fruits, 1.5-1.75 in. dia., everbearing, fine flavor, widely adapted, good long distance pink shipper, 1963. *Source History: 3 in 1981; 2 in 1984; 3 in 1987; 3 in 1991; 2 in 1994; 1 in 1998; 1 in 2004.* **Sources: Syn.**

Basketvee - 70 days, Green-pick basket shipper, det. vines, easy peel sister line of Veebright, thick-fleshed 9 oz. solid fruits, VF CF and crack tol., widely grown in south Ontario and NY state. *Source History: 2 in 1981; 2 in 1984; 6 in 1987; 5 in 1991; 2 in 1994; 4 in 1998; 7 in 2004.* **Sources: Pep, Ra5, Ra6, Re8, Se26, Sto, Ves.**

Baylor Paste - 75-78 days, Family heirloom from the Baylor family of Texas, indet., plum-shaped, thick-walled, bright red fruits average 4-5 oz., no cracks or blemishes. *Source History: 1 in 1994; 1 in 1998; 3 in 2004.* **Sources: Fo7, Ma18, Ta5.**

BC- 2 - 77-87 days, Bush plant to 3 ft., round, red 2 in. salad size fruit, good flavor, heavy crop, from Bulgaria. *Source History: 1 in 2004.* **Sources: Ta5.**

Beach Boy - 80 days, Indet., orange-red fruit grows 4-7 in. long, paste type with few seeds, abundant producer. *Source History: 1 in 2004.* **Sources: Ma18.**

Bearo - Large plants grow over 15 ft., oblong red fruit, acid flavor, disease tolerant, rare. *Source History: 3 in 2004.* **Sources: Hud, Sa5, Sk2.**

Beefsteak - (Scarlet Beefsteak, Red Beefsteak, Red Ponderosa, Crimson Cushion) - 62-115 days, Scarlet sport of Ponderosa, vigorous indet. vines, should stake, fair cover, flat solid meaty juicy and bright-red, 10 oz., some to 2 lbs., excel. slicer, rich sub-acid flavor, rather rough and somewhat ribbed, for home gardeners. *Source History: 84 in 1981; 68 in 1984; 65 in 1987; 51 in 1991; 50 in 1994; 51 in 1998; 61 in 2004.* **Sources: All, Au2, Ba8, Ber, Bo9, Bo17, Bo19, Bu3, CA25, CH14, Cr1, De6, Dgs, Ers, Fi1, Gr29, GRI, HA3, He8, He17, He18, HO2, HPS, Jor, La1, Lej, ME9, Mel, Mo13, MOU, OLD, Ont, Or10, PAG, Pe2, Ra5, Ra6, Ri2, Ri12, Ros, RUP, Sa9, Sau, SE14, Se25, Se26, Se28, SGF, Sh9, Sk2, TE7, To1, To3, To9, Up2, Ves, Vi4, WE10, Wet, WI23, Wo8.**

Beefsteak, Buckbee's - Beautiful 19th century variety with medium-large fruit, red meaty flesh, delicious, rare. *Source History: 1 in 2004.* **Sources: Ea4.**

Beefsteak, Bush - (Early Bush Beefsteak) - 60-65 days, Vigorous dwarf vines, solid meaty and crack res., rich red inside, very few seeds, 8 oz., abundant yields, excel. fine-flavored mild eating tomato, 25-30 lbs. per plant, dev. at Morden Exp. Farm in Manitoba, ideal for the Canadian prairies. *Source History: 16 in 1981; 16 in 1984; 15 in 1987; 12 in 1991; 11 in 1994; 10 in 1998; 12 in 2004.* **Sources: Alb, Ear, Ga1, Lin, MAY, Ra6, Re8, Se26, Sto, To1, To3, Tt2.**

Beefsteak, Garvey's - Indet., regular leaf, red fruit, 5 in. across, from a Mr. Garvey of Philadelphia. *Source History: 1 in 2004.* **Sources: He18.**

Beefsteak, Giant - (Red Beefsteak) - 100 days, Improved selection of Beefsteak, indet., very large brilliant scarlet fruits, 12-32 oz., mild flavor. *Source History: 1 in 1981; 1 in 1984; 2 in 1987; 2 in 1991; 1 in 1994; 2 in 1998; 2 in 2004.* **Sources: GLO, He8.**

Beefsteak, Sugar - 74 days, Indet., potato leaf plant, 3.5 x 2 in., flattened fruit, 8 oz., red flesh, above average yield. *Source History: 1 in 1998; 1 in 2004.* **Sources: Sa9.**

Beefsteak, Super VFN - 80-97 days, An improved beefsteak, fruits average 12-17 oz., meaty with smaller blossom end scar and smoother shoulders, prolific vigorous indet. plants, FW VW and nem. res., PVP expired 2000. *Source History: 1 in 1981; 1 in 1984; 5 in 1987; 10 in 1991; 10 in 1994; 9 in 1998; 11 in 2004.* **Sources: Bu2, Gur, He18, HO2, MAY, Ra5, Ra6, Roh, Sa9, To1, To3.**

Beefsteak, Urban - 63-75 days, Indet., flattened red globes, one crop, high yields. *Source History: 1 in 1998; 1 in 2004.* **Sources: Sa9.**

Beefsteak, Yates - 85 days, Superb Australian canning type. *Source History: 1 in 1998; 1 in 2004.* **Sources: Sa9.**

Beginner - 50-55 days, Det. vines, 3-6 oz. fruit, has shown good res. to blossom end rot even under drought conditions, experimental variety, PSR. *Source History: 1 in 1998; 1 in 2004.* **Sources: Pe6.**

Believe It Or Not - 85-90 days, Indet. plant bears an abundance of smooth red fruit that can weigh up to 2 lbs., excellent sweet flavor, great sandwich tomato, heirloom. *Source History: 4 in 2004.* **Sources: Te4, To1, To9, Up2.**

Bellstar - (Bellestar) - 68-80 days, Early, large 4-6 oz. plum fruits, compact det. plants, excel. for all types of processing, good fresh flavor for salads, dev. by Dr. Metcalf in Ontario, 1984. *Source History: 2 in 1981; 2 in 1984; 5 in 1987; 7 in 1991; 8 in 1994; 8 in 1998; 14 in 2004.* **Sources: Fe5, Ga1, Hig, Jo1, Pr3, Ra5, Ra6, Re8, Se26, Sto, Ta5, Ter, To1, Up2.**

Belyj Naliv - 55 days, Short indet. plant to 42 in., fruit size and flavor somewhat variable, 1.5 in. tall x 2 in. wide, sweet tangy flavor, resist cracking, Russia. *Source History: 1 in 2004.* **Sources: So1.**

Benewah - 60-70 days, Sets fruits from earliest blossoms in cool weather, det., firm bright-red 2.5-3 in. fruits, good yields, for cooler mountain areas, from N. Idaho Seed Foundation, 1970s. *Source History: 3 in 1981; 2 in 1984; 3 in 1987; 2 in 1991; 1 in 1998; 1 in 2004.* **Sources: Sa9.**

Berkshire Oxheart - Not a true oxheart but a good quality old tomato brought from Vermont to the Berkshires (Stockbridge) in the 1930s, endangered. *Source History: 1 in 2004.* **Sources: Ea4.**

Berkshire Polish - Excellent all-purpose tomato, originating in the Americas, then on to Poland and settling in Berkshire County for a generation. *Source History: 1 in 2004.* **Sources: Ea4.**

Berwick German - 85 days, Indet., round to oblong, 2 x 4 in. meaty fruit, few seeds, good flavor. *Source History: 1 in 1998; 2 in 2004.* **Sources: Ho13, To9.**

Besser - 75 days, Red cherry tomato produces .75 in. dia. fruit in clusters of up to 12, very sweet, great in salads, from the Freiburg region in southern Germany, dates back to the late 1800s. *Source History: 1 in 1998; 2 in 2004.* **Sources: He8, Ta5.**

Big Boy - 82 days, Indet., 3 x 2.75 in. flattened gobes, red skin and flesh, acid flavor. *Source History: 1 in 1998; 1 in 2004.* **Sources: Sa9.**

Big Italian Plum - 75-80 days, Indet., regular leaf, look like fat red banana peppers, rich flavor, great for paste and canning, excellent producer. *Source History: 2 in 2004.* **Sources: Ma18, To9.**

Big Red - 85 days, Large indet. vines, red fruit, 16-24 oz., good old-fashioned tomato taste, from southern Illinois. *Source History: 1 in 2004.* **Sources: He8.**

Big's Red Siberian - 72 days, Large indet. plant does not require staking, large round red fruit, 3-4 in., 8-12 oz., flavorful, named after "Big" from Wyoming, High Altitude Gardens introduction. *Source History: 1 in 1994; 1 in 1998; 1 in 2004.* **Sources: Hig.**

Bijskij Zeltyi - 70 days, Disease resistant German variety, 3-4 oz. reddish pink fruit, plum shape with slightly pointed end, sweet intense tomato flavor, very productive, good dried, canned or fresh. *Source History: 1 in 2004.* **Sources: Go8.**

Bisignano - 80 days, Indet., sparse foliage, large heart shaped fruit, 3-4 in., occasional globe shapes, flavorful. *Source History: 2 in 2004.* **Sources: Sh13, To9.**

Bisignano No. 2 - 70 days, Indet., Italian plum, produces four different fruit shapes, oval, globe, plum and large heart shapes, thick meaty deep orange-red flesh with rich tomato flavor, excellent for processing, produces over a long season, from a Mr. Bisignano. *Source History: 1 in 1998; 2 in 2004.* **Sources: Ma18, Ta5.**

Bison - 65-77 days, Old Oscar Will variety from 1937, det., very early, 3 oz., deep red globes, produces deep-red fruits in 65 days on a dwarf plant without any pruning.

Source History: 1 in 1981; 1 in 1984; 1 in 1998; 2 in 2004. **Sources: Sa9, Ta5.**

Bitchyeh Gertzeh - 75 days, Det., 4 ft. plants, clusters of round red globes, 12 oz., from Moscow, Russia. *Source History: 1 in 2004.* **Sources: Ta5.**

Blocked - 74 days, Vigorous indet., 1 in. red fruit in clusters of 3-4, delicious, heirloom. *Source History: 1 in 2004.* **Sources: To9.**

Bloody Butcher - 55-65 days, Indet., tall plants bear clusters of 5-9 red fruits, 2 in. diameter, 3-4 oz., very seedy, early maturity, heirloom. *Source History: 5 in 2004.* **Sources: Bu2, Ga7, Hi13, Roh, To10.**

Blue Beech - 90 days, Indet., Roma type, 6-8 oz. red fruit, green shoulders, few seeds, makes a flavorful sauce, well adapted to cold climate, original seed brought to Vermont from Italy during World War II, Fedco introduction. *Source History: 1 in 2004.* **Sources: Fe5.**

Blue Ridge Mountain - 75-85 days, Indet., potato leaf plant, 14-16 oz., flavorful fruit with fine texture and old-fashioned tomato flavor, heirloom. *Source History: 2 in 1994; 2 in 1998; 5 in 2004.* **Sources: Ap6, Ho13, ME9, Ta5, To9.**

Bobbie - 75 days, Indet., red-pink, heart shape fruit with meaty flesh, few seeds, fruity flavor, heirloom. *Source History: 1 in 1998; 1 in 2004.* **Sources: To9.**

Bonny Best - (John Baer, Super Standard Bonny Best, Chalk's Early Jewel) - 66-85 days, Old favorite, large open indet. plant, poor cover, 6-10 oz. smooth bright-scarlet globes, solid and meaty, variable, either flat culture or staking, widely adapted, especially good in Northern areas, popular for home or market gardens and growing under glass, prone to catfacing, dev. at the Bonny Plant Farm in Union Springs, AL, introduced about 1908. *Source History: 36 in 1981; 33 in 1984; 39 in 1987; 31 in 1991; 29 in 1994; 24 in 1998; 29 in 2004.* **Sources: All, Bo9, Co32, Ers, Fe5, Ga1, Ga22, Gr28, He8, He18, ME9, MIC, Ont, Pla, Pr3, Ra5, Ra6, Re8, Ri12, RUP, Sa9, SE14, Se26, Sk2, Ta5, Te4, To1, To3, Up2.**

Bonny Best Improved - 70 days, Med. early, heavy crops of smooth, bright red, flat globes. *Source History: 1 in 1994; 3 in 1998; 2 in 2004.* **Sources: Dam, La1.**

Bounty - 65 days, Compact self-pruning det. plant, med-sized scarlet globes, good size and quality, few seeds, good for dry sections of the Canadian prairies, ND/AES, 1941. *Source History: 7 in 1981; 6 in 1984; 6 in 1987; 1 in 1994; 1 in 2004.* **Sources: Mo20.**

Box Car Willie - 80 days, Indet., abundant yields of round red fruits, 12-18 oz., excel. sweet-tart flavor, crack free, disease resistant, old timer from Joe Bratka. *Source History: 3 in 1998; 24 in 2004.* **Sources: Ap6, Ers, Fo7, Ga1, Ge2, He8, He17, He18, Ho13, Ma18, Ra5, Ra6, Re8, Se26, Sh13, Shu, Sk2, So25, Sw9, Ta5, To1, To3, To9, To10.**

Bradley, Red - 65-70 days, Short stocky det. plants, abundant production of egg shaped, med. size fruit, red skin, juicy red flesh, resists cracking, holds well on the plant, excel. slicer, originally dev. in New Brunswick for growing along Canada's eastern coast. *Source History: 1 in 2004.* **Sources: Ag7.**

Brandywine, Glick's Strain - Potato leaf, large fruit with rich sweet flavor, originally from the Glick's Seed Company, Lancaster, PA, one of the distributors who introduced the Brandywine tomato. *Source History: 2 in 2004.* **Sources: Ea4, Se17.**

Brandywine, Heart-Shaped - 70 days, Resembles Red Brandywine with a brighter red color, indet. plant, high yields of brilliant red fruits, some are heart shaped and weigh 4-6 oz., others weigh 2 lbs. *Source History: 1 in 2004.* **Sources: Ma18.**

Brandywine, OTV - 72 days, Indet., potato leaf vines, 3 in., round, flat fruit, rich red color with orange undertones, sweet flavor, the most productive, heat tolerant Brandywine strain from a natural cross between Yellow Brandywine x unknown red tomato, "OTV" stands for "Off the Vine," a newsletter once published by Carolyn Male and Craig LeHoullier. *Source History: 3 in 1998; 8 in 2004.* **Sources: Ami, He18, Ma18, Sa9, Sh13, So1, To1, To9.**

Brandywine, Red - 80-100 days, Amish heirloom dating back to 1885, named after Brandywine Creek in Chester County PA, large indet. vines, deep-red fruits, 8-12 oz., excel. old-fashioned tomato flavor. *Source History: 2 in 1991; 4 in 1994; 22 in 1998; 39 in 2004.* **Sources: Ap6, Bo19, CA25, CO30, Ers, Ga1, Gr28, He8, He17, He18, Ho13, Jun, Ki4, Lan, Loc, Ma18, ME9, MIC, Mo13, Or10, Orn, Ra5, Re8, Roh, Sa9, SE14, Se17, Se26, Shu, Sk2, Syn, Te4, TE7, To1, To3, To9, To10, Up2, WE10.**

Brandywine, Red Landis - Round, very red fruit, ripens early, productive. *Source History: 1 in 2004.* **Sources: Sa5.**

Brazil - 95 days, Indet. plants, flattened, 5 oz., semi-acid fruits. *Source History: 1 in 1998; 1 in 2004.* **Sources: Sa9.**

Break o'Day - 63-78 days, Open med-size spreading plants, smooth firm 6 oz. orange-scarlet globes, thick and meaty, high yields, wilt and cold res., popular with market gardeners, good shipper. *Source History: 8 in 1981; 4 in 1984; 4 in 1987; 1 in 1991; 1 in 1998; 3 in 2004.* **Sources: Ap6, Sa9, Ta5.**

Brookpact - Early, det., 7-8 oz, flattened globe, smooth bright-red skin, res. to cold sterility and cracking, reliable, dev. at Brooks Hort. Station for Canadian prairies. *Source History: 2 in 1981; 2 in 1984; 1 in 1987; 1 in 1991; 4 in 1994; 3 in 1998; 3 in 2004.* **Sources: Pr3, Te4, Up2.**

Buckbee's New Fifty Day - Old 19th century variety with that good old-fashioned tomato flavor, 6-10 oz., productive, rare. *Source History: 1 in 2004.* **Sources: Ea4.**

Bulgarian Dix - 76 days, Indet., round red fruit, 1.5 in., good acid flavor, midseason maturity. *Source History: 1 in 2004.* **Sources: To9.**

Bulgarian No. 7 - Indet., regular leaf, red fruit, 2-4 oz., strong sweet flavor. *Source History: 1 in 2004.* **Sources: He18.**

Bulgarian Triumph - 80 days, Indet., 2 in. red fruits become sweeter as they ripen on the vine, 3-4 oz., grow in clusters of 4-6, blemish-free, very productive, Bulgarian heirloom. *Source History: 1 in 1994; 2 in 1998; 8 in 2004.* **Sources: Ap6, Ra5, Ra6, Re8, Te4, To1, To9, To10.**

Bull Sac - (Sausage) - 80-90 days, Large indet. plant, 3 x 6 in. long fruit with a distinctive pointed tip, mild taste, few seeds, very productive. *Source History: 1 in 1987; 1 in 1991; 1 in 1994; 1 in 1998; 2 in 2004.* **Sources: Ma18, Ta5.**

Bull's Heart Select - Huge heart shaped fruit with fresh flavor, good production, ripens early. *Source History: 1 in 2004.* **Sources: Sa5.**

Burbank - (Burbank Red Slicing) - 70 days, Heirloom from the master, stocky det. 18-36 in. bushes, 6-8 oz., red fruits, very productive, highest total free amino acids of all tomatoes tested, does well in drier climates, dev. around 1915 by Luther Burbank. *Source History: 1 in 1987; 6 in 1991; 4 in 1994; 6 in 1998; 13 in 2004.* **Sources: Au2, Co32, He18, In8, Pe2, Se7, Se17, So9, Syn, Te4, Ter, To9, To10.**

Burgundy Traveler - 80 days, Indet., very similar to Brandywine. *Source History: 3 in 1994; 2 in 1998; 3 in 2004.* **Sources: He18, Ma18, WE10.**

Burkina Faso - 75 days, Det. plant benefits from staking because of the weight of the huge crops, red fruit, small pointed end, nice flavor, West African origin. *Source History: 1 in 2004.* **Sources: Ta5.**

Burrell's Special - *Source History: 1 in 2004.* **Sources: Bu3.**

Bush Whopper - Indet., uniform dark red fruit, excellent flavor, ripens midseason, very high yields. *Source History: 1 in 2004.* **Sources: Up2.**

Cabot - 58-80 days, Sel. from Scotia, just slightly earlier, compact det. plants more open and easier to pick, 2 in. red fruits, no green shoulders, very firm, does well in cool damp climates, Kentville. *Source History: 3 in 1981; 3 in 1984; 4 in 1987; 4 in 1991; 2 in 1994; 3 in 1998; 3 in 2004.* **Sources: Sa9, Ta5, Ves.**

Cal Ace - (Cal Ace VF, Cal Ace 55 VF) - 75-90 days, Earlier than Ace, thicker walls, smaller blossom end scar, med-large semi-det. vines, heavy cover, bush or pole production, large smooth 8 oz. flattened globes, many over 1 lb., meaty and sweet, FW VW and ASC res., sets well in bad weather, uniform ripening, good in California, fresh market. *Source History: 19 in 1981; 14 in 1984; 15 in 1987; 10 in 1991; 9 in 1994; 9 in 1998; 7 in 2004.* **Sources: Bu3, Ers, He18, ME9, RUP, Se26, WE10.**

Cal Culterator - 80 days, Indet., 6 oz. globes, firm flesh, good shipper. *Source History: 1 in 1998; 1 in 2004.* **Sources: Sa9.**

Cal Lake #4 - 80 days, Indet., flattened 3.5 x 2.5 in. red fruit, 6 oz., average yield. *Source History: 1 in 1998; 1 in 2004.* **Sources: Sa9.**

Calypso - 75-80 days, Bred for humid tropical areas, det., good cover, short pole or bush production, 7 oz. slightly flattened globes, res. to GLS AB VW FW1 FW2 and ASC, fresh market. *Source History: 7 in 1981; 3 in 1984; 7 in 1987; 5 in 1991; 4 in 1994; 5 in 1998; 4 in 2004.* **Sources: He18, Lej, Lin, WE10.**

Campbell 33 - Deep-oblate fruit, multi-lobed, very early, VW res. *Source History: 1 in 1991; 2 in 1994; 4 in 1998; 1 in 2004.* **Sources: WE10.**

Campbell 146 - (KC 146) - 78-81 days, Slightly flattened 7 oz. red globes, res. to FW TMV GLS and cracking, uniform ripening, intro. by Campbell Soup Co. *Source History: 3 in 1981; 2 in 1984; 1 in 1998; 1 in 2004.* **Sources: Sa9.**

Campbell 1327 - 69-120 days, Vigorous med-sized det. plants, good cover, firm smooth 7 oz. flattened globes, res. to VW FW1 and cracking, sets fruits well under unfavorable conditions, fairly conc. set, fairly crack res., good flavor, large firm main crop tomato for processing and fresh market, Campbell Soup Co. *Source History: 24 in 1981; 23 in 1984; 26 in 1987; 20 in 1991; 18 in 1994; 17 in 1998; 19 in 2004.* **Sources: Bu3, Bu8, CH14, Ge2, He17, He18, Jor, ME9, Mey, MOU, Ra5, Ra6, Re8, RIS, RUP, Se26, To1, To3, Wi2.**

Campbell's - 69 days, Semi-det., bright-red fruit with light-green shoulders, oblate, smooth and firm, excel. flavor, dev. by the soup company. *Source History: 2 in 1994; 2 in 2004.* **Sources: Ta5, To10.**

Canabec Rouge - 70 days, Det., 4-6 oz. red globes, high yields, for cool climates. *Source History: 1 in 1998; 2 in 2004.* **Sources: Sa9, Up2.**

Canabec Super - Short vines produce attractive pink, 6 oz. fruit, good fruit set in cool weather, early maturity. *Source History: 1 in 1998; 1 in 2004.* **Sources: Te4.**

Canadian Stuffing - 70-75 days, Indet., hollow core, firm walls, good stuffer, makes a thick sauce with little cooking down, resembles a blocky red bell pepper. *Source History: 2 in 2004.* **Sources: Pe2, Tu6.**

Canary Salad - Indet., large red cherry, golf ball size, tasty, early, prolific. *Source History: 1 in 2004.* **Sources: Up2.**

Canestrino I - 75 days, Indet., plum fruit with unusual shape that resembles a fat chocolate kiss, excellent flavor and production, name means "little basket" in Italian, grown in Tuscany. *Source History: 2 in 2004.* **Sources: Ma18, So25.**

Canestrino II - Indet., slightly smaller that Canestrino I without the fluting, excellent flavor, good production, from Italy. *Source History: 1 in 2004.* **Sources: Ma18.**

Cannonball - 66-75 days, Vining plants, globe-shaped red fruits average 6.3 oz., uniform color, main season variety, intro. by North Dakota State University in 1973. *Source History: 2 in 1981; 2 in 1984; 1 in 1991; 1 in 1994; 2 in 1998; 2 in 2004.* **Sources: Sa9, Ta5.**

Cardinal - (The Cardinal) - 80 days, Found growing in a field of Acme tomato, medium indet. vines, 3 in. dia. fruit, delicious, productive, widely adapted from North to South, pre-1879, Burpee introduction, 1884. *Source History: 1 in 1987; 1 in 2004.* **Sources: Ea4.**

Caribe - 70-80 days, Compact det. plants, good cover, smooth jointless 8 oz. red flattened globes, firm, VW FW1 FW2 ASC and GLS res., med-early, humid adapted, shipping and market. *Source History: 2 in 1981; 1 in 1984; 1 in 1987; 1 in 1991; 1 in 1994; 1 in 1998; 2 in 2004.* **Sources: He18, WE10.**

Carmello - 70-75 days, French tomato, indet., fine flavor with a perfect balance of sweet and acid, thin skin, good for fresh eating, sauteed and salads, no cracking, disease res. *Source History: 3 in 1994; 2 in 1998; 3 in 2004.*

Sources: Na6, Te4, To9.

Carol Chyko's Big Paste - 80 days, Large healthy indet. vines, orange-red beefsteak fruit, solid juicy flesh, few seeds, crack res., good yields. *Source History: 2 in 2004.* Sources: Ho13, Ma18.

Caro-Red - 78-86 days, Indet., deep carrot-orange flesh, 6 oz., semi-acid, distinct flavor, contains 10 times more vitamin A than other tomatoes, one fruit is 150% of adult requirement, resembles Rutgers, from Purdue. *Source History: 2 in 1981; 1 in 1984; 1 in 1987; 2 in 1991; 2 in 1998; 1 in 2004.* Sources: Sa9.

Castlerock - 128 days, Excel. vine storage, medium compact plant, blocky fruit, PVP expired 2000. *Source History: 1 in 1991; 2 in 1994; 2 in 1998; 1 in 2004.* Sources: WE10.

Ceylon - 50-80 days, Indet., multi-ribbed, flat fruit, red with some orange shoulders, 3.5 oz., good tart flavor, heirloom. *Source History: 2 in 1998; 3 in 2004.* Sources: Hi6, Sa9, To9.

Chalks Early Jewel - 70-80 days, Similar to June Pink except for larger size and red color, indet., round, red 2 in. globes, good sweet/acid flavor balance, not recommended for the South where diseases are a problem, developed by James Chalk of Norristown, PA in 1899, introduced in 1910. *Source History: 1 in 1981; 1 in 1991; 1 in 1998; 2 in 2004.* Sources: Sa9, To9.

Champ - 80 days, Indet., round red fruit, 3-4 in., excellent sweet, plum-like flavor, heirloom. *Source History: 1 in 2004.* Sources: To9.

Chapman - 80 days, Indet., large red fruit, 1-2 lbs., good flavor. *Source History: 1 in 1998; 2 in 2004.* Sources: Ma18, To9.

Charlie Chaplin - 75-78 days, Indet., flavorful 6-12 oz., nearly hollow fruit, resembles a bell pepper when sliced, good stuffer. *Source History: 6 in 2004.* Sources: Loc, Ma18, ME9, Ra5, Ra6, Re8.

Cheerio - 55 days, Very early, branching but determinate compact plants, med-large bright-red cherry fruits, delicious full tomato flavor, heavy yields, tolerates cold and adverse weather conditions, stake or ground culture, suitable for commercial production and home gardens, old New England heirloom. *Source History: 2 in 1987; 4 in 1991; 1 in 1994; 2 in 1998; 2 in 2004.* Sources: Ma18, Syn.

Cheetam's Potato Leaf - Indet., large potato leaf foliage, early high yields of medium size red fruit, sweet tangy flavor with excellent tomato taste. *Source History: 1 in 2004.* Sources: Go8.

Cherokee - 78-84 days, Vigorous indet. plant, slightly-flat 7 oz. red fruits, firm meaty interior, high quality, long harvest, VW FW1 FW2 ASC ST res., home and local market, PVP expired 2001. *Source History: 3 in 1981; 3 in 1984; 2 in 1987; 1 in 1991; 4 in 1994; 4 in 1998; 7 in 2004.* Sources: Ra5, Ra6, Re8, Sa9, Ta5, Vi4, WE10.

Cherry Roma - 75-80 days, Indet., very heavy set of 1 in. long plum shaped fruit, addictive, sweet spicy flavor, great fresh or dried. *Source History: 4 in 2004.* Sources: Pr9, Sa5, Se16, Up2.

Cherry Salad - 58 days, Fisher's extra early cherry tomato, self-pruning compact plants loaded with scarlet-red 1 in. fruits. *Source History: 1 in 1981; 1 in 1984; 1 in 1987; 1 in 1991; 1 in 1994; 1 in 1998; 1 in 2004.* Sources: Fis.

Cherry, Baxter's Early Bush - 72 days, Det. plant needs no staking or caging, 7-10 days earlier than most open-pollinated cherries, tasty firm fruit, resistant to splitting, outstanding vigor and productivity with fruit set even under adverse temps, exceptional keeper, PVP expired 2003. *Source History: 1 in 1981; 1 in 1984; 5 in 1987; 8 in 1991; 7 in 1994; 4 in 1998; 9 in 2004.* Sources: Bo19, Lin, Ra5, Ra6, Re8, Se26, Te4, To1, WE10.

Cherry, Borgo Cellano - 75 days, Abundant production of oval fruit with a pointed tip, good sauce and salad tomato, Italy. *Source History: 1 in 2004.* Sources: To10.

Cherry, Camp Joy - 65-80 days, Indet., large cherry fruits, fine blend of sweet and acid, indet. plant, bears until frost, from Santa Cruz, dev. by Alan Chadwick, originator of bio-intensive gardening. *Source History: 2 in 1991; 3 in 1994; 4 in 1998; 4 in 2004.* Sources: Ta5, Te4, Up2, Yu2.

Cherry, Chadwick's - 70-90 days, Possibly the best overall cherry tomato, selected by English horticultural genius Alan Chadwick, indet. vines to 5 ft., unforgettable flavor, highly productive, rare. *Source History: 1 in 1991; 1 in 1994; 5 in 1998; 16 in 2004.* Sources: Bou, Com, Eco, Ga22, Pe2, Ra5, Ra6, Re8, Sa5, Se7, Se17, Se26, So9, Te4, To9, To10.

Cherry, Early - 56 days, Compact upright plants, bright-red round mild 1.5 in. fruits, bears heavily for about 30 days, as early as the Sub-Arctics with excel. flavor, Cornell. *Source History: 2 in 1981; 2 in 1984; 3 in 1987; 2 in 1991; 3 in 1994; 3 in 1998; 1 in 2004.* Sources: Ter.

Cherry, Fence Row - 74-85 days, Indet., large size cherry tomatoes in clusters of 5-8, found growing wild in a fence row near Paris, IL in 1975. *Source History: 4 in 1998; 4 in 2004.* Sources: Fo7, He18, Sa9, To9.

Cherry, Fox - 80-90 days, Indet., potato leaf, red fruits, 30 per lb., rich tomato flavor, prolific, heirloom. *Source History: 1 in 1987; 2 in 1991; 3 in 1994; 3 in 1998; 4 in 2004.* Sources: Ra6, Ri12, Se7, Se26.

Cherry, Koralik - 61 days, Det., heavy yields of 1 in., bright red fruit in clusters of 6-8 fruit, great flavor, ripens all at once, from Russia. *Source History: 1 in 2004.* Sources: Ter.

Cherry, Large Red - 70-80 days, Vigorous indet. vines, half-dollar size red fruits, 1.25-1.5 in. dia., productive, needs staking, sweet juicy flavor, widely used in markets and for berry boxes, res. to ASC, green fruits may be pickled, ripe fruits often used in preserves. *Source History: 32 in 1981; 27 in 1984; 33 in 1987; 36 in 1991; 38 in 1994; 35 in 1998; 39 in 2004.* Sources: Bu3, But, CO23, Com, Ers, Ga1, GRI, HA3, He8, He17, HO2, Jor, Kil, Loc, Ma19, ME9, MIC, Mo13, MOU, Or10, Pep, Pr9, Ra5, Ra6, Re8, RIS, Ros, RUP, SE14, Se16, Se26, Sh9, Shu, Sk2, To1, To3, WE10, Wi2, WI23.

Cherry, Mildred Cowger's Belgian - 75-80 days, Indet., large cherry, dev. and saved by gardener Mildred Cowger. *Source History: 2 in 2004.* Sources: Pe2, Tu6.

Cherry, MS-5 - 80 days, Indet., pink-red cherry, intense flavor, very productive. *Source History: 1 in 2004.* **Sources: To9.**

Cherry, Old Fashioned Red - 72 days, Vigorous vines, small red marble-size ridged fruits, heat res., dates to the 1800s. *Source History: 2 in 1991; 1 in 2004.* **Sources: He18.**

Cherry, Oregon - 60 days, One of the oldest extra early cherry tomatoes, compact det. plants, small to med. size round red fruit, thin skin, sweet flavor, Oregon State University. *Source History: 1 in 1981; 1 in 1984; 1 in 1987; 1 in 1991; 1 in 2004.* **Sources: Ter.**

Cherry, Peacevine - 75 days, So named because of the high amino acid content which has a calming effect on the body, very high in vitamin C, indet. vines need trellising, high yields of clusters of small red fruit, has currant tomato (*L. pimpinellifolium*) in its ancestry, selected by Peace Seeds from Sweet 100 cherry tomato. *Source History: 3 in 1994; 8 in 1998; 15 in 2004.* **Sources: Ap6, Fe5, Fo13, Gr28, Hi6, Ho13, Pe1, Pe2, Se7, Se17, So9, St18, Ta5, To9, To10.**

Cherry, Red - 70-80 days, Strong vigorous indet. plants, small round red fruits, 1 in. dia. average, borne in large clusters, bears all summer for a long harvest period, widely used shipper, heavy yields, fine flavor, raw or in salads, intro. before 1840. *Source History: 64 in 1981; 56 in 1984; 56 in 1987; 34 in 1991; 36 in 1994; 31 in 1998; 26 in 2004.* **Sources: Alb, Be4, Bo19, Com, Cr1, De6, Ev2, Fi1, Fis, Ge2, Gr27, Gr28, GRI, Hal, La1, Mey, PAG, Ra5, Ra6, Red, Roh, Sau, Sh9, So25, We19, Wet.**

Cherry, Small Red - (Old-Fashioned Red Cherry) - 75 days, Large open indet. plants, smooth nickel-size globes .5-.75 in., volunteers readily each spring from fruits left in garden, ASC res., good in salads or for pickles or preserves, pre-1840. *Source History: 11 in 1981; 10 in 1984; 10 in 1987; 15 in 1991; 13 in 1994; 11 in 1998; 19 in 2004.* **Sources: All, Bo9, Bu8, CA25, CH14, Ers, Ga1, GLO, He18, LO8, ME9, MIC, Ra5, Ra6, Re8, SE14, So1, Vi4, WE10.**

Cherry, Stan's Best - Large red-orange cherry, sweet flavor, highly productive, midseason. *Source History: 1 in 2004.* **Sources: Te4.**

Cherry, Sugar - 63-76 days, Currant type tomato with .75 in. orange-red fruits borne in clusters of 12, intensely sweet flavor, indet. vines with small leaves, sparse foliage cover, CV So1 introduction, 1994. *Source History: 1 in 1994; 2 in 1998; 5 in 2004.* **Sources: Eco, Ga1, Gr28, Pe2, So1.**

Cherry, Swander - 78 days, Bushy indet. vines produce red .5 in. fruits in clusters of 8, very sweet flavor, heirloom. *Source History: 1 in 2004.* **Sources: To9.**

Cherry, Washington - 60 days, Globe-shaped 1-1.25 oz. fruits (no ovals), almost identical to Large Red Cherry but with deeper red color and better quality, requires no staking, thick walls, meaty and flavorful, good keeping quality on or off the vine, dev. by Washington State Univ. *Source History: 2 in 1991; 2 in 1994; 2 in 1998; 2 in 2004.* **Sources: Ga1, Syn.**

Cherry, Wickline - 75-85 days, Indet., well-branched plant, pink-red, egg-shaped fruits borne in clusters of four, tolerates cool wet conditions, heirloom from Wickline, Pennsylvania, CV So1 introduction, 1991. *Source History: 1 in 1991; 1 in 1994; 1 in 1998; 2 in 2004.* **Sources: Sa9, So1.**

Chesapeake - 70-75 days, Large indet. plant, 10-12 oz. red-orange globes, old-fashioned tart tomato flavor, reportedly a very popular tomato in the Baltimore area years ago. *Source History: 2 in 2004.* **Sources: He18, Ta5.**

Chianti Rose - 80 days, Cross between the famous pink Brandywine and a treasured family variety from Italy, vigorous potato leaf vines, large, thin skinned fruit, rosy red, smooth creamy flesh is juicy and flavorful, resists cracking, high yields. *Source History: 2 in 2004.* **Sources: Pe2, Re6.**

Chibikko - Plants just over 12 in. tall, produces many 1 in., tangy, red fruits, from Japan. *Source History: 1 in 1998; 1 in 2004.* **Sources: Ho13.**

Chico - 70-80 days, Indet., clusters of small pink-red plums with wonderful sweet flavor, juicy flesh, excellent all-purpose tomato. *Source History: 3 in 2004.* **Sources: LO8, Ma18, Ta5.**

Chico III - (Improved Chico Grande) - 70-98 days, Mech. harv., med-sized compact det. plant, firm pear-shaped paste tomato, 2.5-3 in. dia., 2-3 oz., FW1 GLS ASC AT and ST res., uniform and productive, very conc. set, sets fruits well during high temps, from TX/AES for the Southwest. *Source History: 15 in 1981; 12 in 1984; 8 in 1987; 12 in 1991; 10 in 1994; 4 in 1998; 2 in 2004.* **Sources: Syn, WE10.**

Chinese Red - 85-90 days, Indet., long red paste type fruit, 2 x 7 in., thick dry flesh, produces over a long season. *Source History: 2 in 2004.* **Sources: Ma18, To9.**

Chipollino - 88 days, Indet., round red globes, 3-5 oz., mildly acid flavor. *Source History: 1 in 1998; 2 in 2004.* **Sources: Sa9, Ta5.**

Chuck's - 85 days, Indet., red beefsteak fruit, 12-20 oz. *Source History: 1 in 2004.* **Sources: Ma18.**

Climbing Trip-L-Crop - (Italian Tree) - 80-90 days, Extremely vigorous 10-20 ft. potato-leaf vines, solid meaty bright-red flat fruits, 2.75-3.5 x 5 in. dia., up to 2 lbs., often yields 2-3 bushels per vine, good flavor, small seed cells near its center, excel. slicer, grow on a trellis. *Source History: 16 in 1981; 13 in 1984; 15 in 1987; 8 in 1991; 5 in 1994; 6 in 1998; 17 in 2004.* **Sources: Cr1, Gr29, Ho13, HPS, ME9, Ra5, Ra6, Sa9, Se26, Se28, Shu, So1, TE7, To3, To9, To10, Wo8.**

Clustermato - 87 days, Indet., above average yields of 3 x 3 in., 6 oz., round fruit, red flesh, dev. prior to 1950. *Source History: 1 in 1998; 1 in 2004.* **Sources: Sa9.**

CN415 - Det., small firm fruit slightly larger than Sweet Million, beefsteak shape, good flavor, very high vitamin C content. *Source History: 1 in 1994; 1 in 1998; 1 in 2004.* **Sources: Al6.**

Cobourg - Indet., regular leaf, meaty flesh, good for sauce, midseason. *Source History: 1 in 2004.* **Sources: Up2.**

Cold Set - 60-74 days, Seedlings withstood 18 degree F. spring frosts in Canadian trials, for direct seeding, germinates in cool soils and cold weather, compact bushy det. self-pruning plants, round deep-red meaty 4-6 oz. fruits, will not tolerate continuous wet weather, grown successfully as far south as Texas, Ontario Agriculture College, 1963. *Source History: 11 in 1981; 8 in 1984; 10 in 1987; 9 in 1991; 6 in 1994; 6 in 1998; 3 in 2004.* **Sources: Fi1, Gur, Sa9.**

Colorado Red - 80 days, Indet., flavorful round red fruit. *Source History: 1 in 2004.* **Sources: Ta5.**

Columbia - 80-85 days, Det., 4-6 oz., pointed fruit, nice slicer and canner, dev. at WSU-USDA Exp. Sta., Prosser, WA. *Source History: 2 in 1998; 2 in 2004.* **Sources: Sa9, Ta5.**

Comstock's Sauce and Slice - 92 days, Indet., excellent paste with meaty flesh, 1 lb., few seeds, sweet flavor, good for sauce and eating fresh, Comstock, Ferre and Co. introduction. *Source History: 1 in 2004.* **Sources: Com.**

Coop Red - 62 days, Indet., 5-7 oz. red globes. *Source History: 1 in 2004.* **Sources: Sa9.**

Coral - 78 days, Det. plant to 3 ft., red fruit, 10-12 oz., meaty flesh with 3 seed cavities, from Moldova. *Source History: 1 in 2004.* **Sources: Ta5.**

Corne de Bouc - (Goat's Horn) - 75-80 days, Indet., 5 in long pointed fruit, good fresh and dried, French heirloom from Quebec, Canada. *Source History: 2 in 2004.* **Sources: Pe2, Tu6.**

Cosmonaut Volkov Red - 72-75 days, Indet., 5 ft. plants, round, slightly flattened, 4 in. red fruits, Russian gardeners grow this variety for prize winning fruits, up to a kilogram (2.2 lbs.), named for famous Russian cosmonaut who was killed while landing. *Source History: 1 in 1994; 4 in 1998; 7 in 2004.* **Sources: Fe5, Hi6, So25, Syn, Ta5, To9, To10.**

Costa Rica - 75 days, Indet., 1-1.5 in., round red fruit, 4 oz., good tomato flavor, prolific. *Source History: 1 in 1998; 2 in 2004.* **Sources: Sa9, To9.**

Costoluto Catanese - Prior to the introduction of this ancient variety in the 1600s, Europe had no tomatoes, red, med. size, ridged fruit, 3.5 in., excel. flavor. *Source History: 1 in 2004.* **Sources: Red.**

Costoluto Fiorentino - 70-90 days, Indet., large beefsteak, 6-8 oz., excellent flavor is high in both sugar and acid, good canner, slight late blight res., Italian heirloom from the Tuscany region. *Source History: 1 in 1991; 1 in 1998; 7 in 2004.* **Sources: Ga22, He18, Se24, Se26, So25, To1, To10.**

Costoluto Genovese - (Italian Costoluto Genovese) - 80-90 days, Old unimproved Italian variety, indet. vines, lumpy flattish convoluted fruits, 6-8 oz., absolutely delicious, slightly tart flavor, does well in hot weather but continues to produce when the weather turns cool. *Source History: 2 in 1987; 3 in 1991; 6 in 1994; 8 in 1998; 27 in 2004.* **Sources: Ap6, Coo, Ga1, Gr28, He8, He18, Ki4, Ma18, Na6, Orn, Ra5, Ra6, Re8, Red, Se7, Se17, So1, Sw9, Syn, Ta5, Te4, Ter, Th3, To1, To3, To9, To10.**

Cougar Red - 65 days, Vigorous, semi-det. plants, medium size fruit, meaty flesh, good sugar/acid flavor balance, produces in cool temps in areas with short growing season, ripens until frost. *Source History: 1 in 2004.* **Sources: Ga1.**

Coursen Roy's Stuffing - 80 days, Indet., bell pepper shaped fruit, thick walls, hollow interior makes a good stuffing tomato. *Source History: 1 in 2004.* **Sources: Ma18.**

Coustralee - 72-82 days, Indet., meaty, orange-red, smooth fruit, 4 in. across, 1-2 lbs. *Source History: 2 in 1998; 8 in 2004.* **Sources: CO30, Gr28, ME9, Ra6, Re8, Sa9, SE14, To1.**

CPC-2 - Vigorous productive det. plants, fair to good cover, red 8 oz. flattened-globes, some blossom end scar, FW and VW res., med-early, semi-arid, market and processing. *Source History: 1 in 1981; 1 in 1984; 1 in 2004.* **Sources: WE10.**

Crackproof, Burgess - 80 days, Indet., resists cracking. *Source History: 1 in 2004.* **Sources: He18.**

Crazy - 60 days, Indet., red .75 in. round cherries borne in clusters of 6, sweet flavor with good acid balance. *Source History: 1 in 1998; 1 in 2004.* **Sources: To9.**

Creole - 72-78 days, Vining, smooth med-large red fruits, good-textured firm flesh, good tomato flavor, high quality, high yields, res. to FW and BER, dev. at Louisiana State University for hot, humid climates. *Source History: 2 in 1981; 1 in 1984; 1 in 1987; 1 in 1994; 5 in 1998; 6 in 2004.* **Sources: He18, Ho13, Sh13, Ta5, To1, To9.**

Crimson Cushion - (Beefsteak, Red Ponderosa) - 80-95 days, Very old beefsteak variety, late season, vigorous indeterminate vines, huge ribbed irregular fruits weigh 1.5-2 lbs., rich tomato flavor, introduced in 1892 by Peter Henderson. *Source History: 1 in 1987; 2 in 1991; 2 in 1994; 3 in 1998; 7 in 2004.* **Sources: Ba8, Ge2, He8, Ra6, Re8, Ta5, Vi4.**

Crimson Sprinter - Largest tomato to abort main stem like Beaverlodge Sub-Arctics, wide spreading vines, med-sized unblemished mild high-crimson fruits, 7-9 oz., ripens quite early. *Source History: 1 in 1981; 1 in 1984; 1 in 2004.* **Sources: Se27.**

Culstrage - 87 days, Indet., flattened red fruit. *Source History: 1 in 2004.* **Sources: Sa9.**

Cuor Di Bue - 65-70 days, Name translates Bull's Heart, large, heart shaped red fruit, 6-12 oz., flavorful meaty flesh, large indet. plant, popular in Italy. *Source History: 4 in 2004.* **Sources: Hud, Se24, So25, To10.**

Cuore de Toro - 89 days, Italian heirloom, det., oxheart shaped red fruit up to 2 lbs., very sweet flavor, excel. producer. *Source History: 3 in 2004.* **Sources: Ra5, Ra6, To10.**

Cuostralee - 75-85 days, Indet., large, flavorful, red beefsteak, 1-3 lbs., meaty flesh, few seeds, intense balanced flavor, French heirloom. *Source History: 4 in 1998; 11 in 2004.* **Sources: Ap6, Ba8, He8, He18, Ho13, Ma18, Ra5, So25, TE7, To9, To10.**

Cup of Moldova - 80 days, Round red fruit, from Moldova. *Source History: 1 in 1998; 1 in 2004.* **Sources: Sa9.**

Currant, Hawaiian - 75-80 days, Indet., clusters of red, pea size fruit, holds on the plant until all in the cluster are ripe, very sweet, late. *Source History: 2 in 2004.* **Sources: Ga22, To9.**

Currier - 82 days, Indet., 3.5 x 3 in. globes, 8 oz., red flesh, excellent yield. *Source History: 1 in 1998; 1 in 2004.*

Sources: Sa9.

Czech Plum - 80 days, Indet., 6-10 oz. plums with excellent flavor, productive, from Czechoslovakia. *Source History: 2 in 2004.* **Sources: Ma18, Ta5.**

Czech Select - 65 days, Large red fruit, excel. flavor, good yield, from Milan Sodomka. *Source History: 1 in 1998; 2 in 2004.* **Sources: Co32, Se26.**

Czech's Bush - 70 days, Another great tasting tomato from the Ben Quisenberry collection, stakeless, strong, low growing plants produce heavy yields of round, red, 4 oz. fruit, attractive rugosa foliage, original seed sent to Mr. Quisenberry in 1976 by Milan Sodomka of Czechoslovakia. *Source History: 1 in 1981; 1 in 1984; 1 in 1998; 2 in 2004.* **Sources: Pr9, Se16.**

Czech's Excellent - Light pink gourmet fruit, rare. *Source History: 1 in 1994; 1 in 1998; 1 in 2004.* **Sources: Syn.**

Dad's Barber Paste - 86 days, Indet., meaty red sauce tomato with a nipple on one end, 4 in. long, very sweet flavor, late season. *Source History: 1 in 2004.* **Sources: To9.**

Dad's Mug - 80-95 days, Indet., large mug-shaped deep-scarlet fruit, 8-12 oz., thick walls, nearly solid center with few seeds, very flavorful for slicing or eating out-of-hand but outstanding as a stuffing tomato, heirloom. *Source History: 1 in 1987; 4 in 1991; 5 in 1994; 7 in 1998; 5 in 2004.* **Sources: Fo7, Ma18, Sh13, Ta5, To1.**

Dana - 90 days, Det. plants with bright red, 6 oz. fruits, acid, Russian variety. *Source History: 1 in 1998; 2 in 2004.* **Sources: Sa9, Ta5.**

Dansk Export - 65 days, Indet., medium sized red salad fruit, 2 oz. *Source History: 1 in 1994; 1 in 1998; 1 in 2004.* **Sources: Sa9.**

D'Australie - (Erica d'Australie) - 85-90 days, Indet., red, meaty, 4 in. round fruit, 1-2 lbs., good flavor, nice acid balance, good fruit set in temperatures over 90 degrees, originally from Norbert Parreira, Helliner, France. *Source History: 1 in 1998; 3 in 2004.* **Sources: He18, Sh13, To9.**

Davington's Epicure - Indet. plants bear huge crops of 2-4 oz. red fruit in clusters of 4-6, firm flesh, excellent sweet taste, only minimal signs of late blight at the end of the season, good salsa and canning tomato, heirloom. *Source History: 1 in 2004.* **Sources: Go8.**

Daydream - Large flattened red globes, late season maturity. *Source History: 1 in 2004.* **Sources: He18.**

De Barao Red - 70-95 days, Indet., 3-4 oz., non-juicy globe shaped fruits, good for sauce and canning, from Irkutsk, Siberia. *Source History: 1 in 1998; 8 in 2004.* **Sources: Goo, Ma18, Ra5, Ra6, Re8, Ri12, Sa9, Ta5.**

De Barrao II - 72 days, Compact semi-det. plant, 3 in. long 6-8 oz. pear-shaped Roma-type fruits, stores for months if kept cool, name comes from the hat that the Russian calvary once wore. *Source History: 1 in 1991; 1 in 1994; 3 in 1998; 2 in 2004.* **Sources: Hig, Jo1.**

Debbie - 75 days, Beefsteak type, red-orange, flattened, 12 oz. fruit. *Source History: 1 in 2004.* **Sources: Sa9.**

Delicious - (Burpee's Delicious, Improved Beefsteak) - 77-89 days, Indet., excel. slicer, most over 1 lb., many 2-3 lbs., world record of 7 lbs. 9 oz., smooth and solid, seldom cracks, small cavities, almost solid meat, excel. flavor, dev. by W. Atlee Burpee from Beefsteak after thirteen years of breeding and selection. *Source History: 21 in 1981; 19 in 1984; 25 in 1987; 24 in 1991; 31 in 1994; 37 in 1998; 47 in 2004.* **Sources: Be4, Bo9, Bu1, Bu2, CA25, Dgs, Ear, Ers, Fi1, Fo7, Ga1, Ge2, Gr27, GRI, Gur, He8, He17, He18, HO2, HPS, Jor, La1, Loc, ME9, Mey, MIC, PAG, Ra5, Ra6, Re8, Roh, RUP, Sa9, SE14, Se25, Se26, Sh9, Shu, So1, Ter, To1, To3, To9, Und, Up2, Vi4, WE10.**

Delmar - 75 days, Indet., heavy producer of 6-8 oz. red fruit, once a major part of the tomato industry in the Delaware/Maryland area. *Source History: 1 in 1998; 1 in 2004.* **Sources: He18.**

Denali - 65 days, Early var. from Alaska, plum size and shape, rich, tangy flavor, splits in wet conditions. *Source History: 1 in 1998; 1 in 2004.* **Sources: Fo7.**

Deutscher Fleiss - Potato leaf foliage, 2-4 oz. red fruit, sweet flavor, retains firmness when fully ripe, resists cracking, early maturity, German heirloom. *Source History: 1 in 2004.* **Sources: Te4.**

Diener - 75 days, Indet., bred from Santa Clara Canner with darker crimson color, large size, meaty flesh, superb flavor, USDA. *Source History: 2 in 2004.* **Sources: He18, To9.**

Ding Wall Scotty - 75 days, Narrow, upright, indet. plants, 2 in. round, red fruit with pointed ends, borne in clusters of 3-4, mildly acid, rich flavor, high yields. *Source History: 1 in 1998; 2 in 2004.* **Sources: Ta5, To9.**

Dinner Plate - 75-100 days, Indet., heart-shaped fruits so large that one slice can fill a dinner plate, 24-30 oz., fine quality, delicious flavor, heirloom. *Source History: 3 in 1991; 4 in 1994; 5 in 1998; 11 in 2004.* **Sources: Fo7, Gr29, He18, Ma18, Ra5, Ra6, Sa9, Se28, To1, To3, Wo8.**

Disarsa - 70 days, Indet., 3 in. round, red, flattened fruit, mild flavor. *Source History: 1 in 1998; 1 in 2004.* **Sources: To9.**

Ditmarsher - Small, sprawling 12 in. plant works well in a hanging basket, oval pink cherry size fruit, early. *Source History: 2 in 2004.* **Sources: Gr28, So25.**

Dix Doights de Naples - 70 days, Indet., elongated fruit is borne in clusters of three or more, wonderful flavor, name means "Ten Fingers of Naples", from Italy. *Source History: 1 in 2004.* **Sources: Ma18.**

Dominick's Paste - 85 days, Italian heirloom, shaped somewhat like a Roma with more juice, excellent flavor and texture, multi-purpose, indet., introduced by Peters Seed and Research. *Source History: 1 in 1994; 1 in 1998; 1 in 2004.* **Sources: Ma18.**

Dona - 75 days, Productive indet. plants, 3-6 oz., deep red-orange fruit, meaty flesh, few seeds, good disease res., open-pollinated version of the French variety. *Source History: 4 in 2004.* **Sources: He18, Na6, Se17, To9.**

Doublerich - 60-80 days, Indet., solid meaty med-sized deep-red globes, 50-60 units of vitamin C, 12-25 normally, few seeds, for short season areas, good for canning and juices, ND/AES and NH/AES. *Source History: 3 in 1981; 3 in 1984; 4 in 1987; 5 in 1991; 5 in 1994; 5 in 1998; 7 in 2004.* **Sources: Bou, Ea4, Se7, So9, So12, Te4, Ter.**

Doucet's Quebec Early Market - Det., regular leaf, 2-3 in. fruits, tasty, very early. *Source History: 1 in 2004.* **Sources: Up2.**

Druzba - 75-95 days, Indet., name means friendship in Bulgarian, fruits are 4 in. round and slightly flattened, 8-16 oz., bright red with pink-gold shoulders, blemish-free, very good flavor is sweet yet tart, high acid, heavy set continues throughout the season, Bulgarian heirloom via Norbert Parreira, Helliner, France. *Source History: 7 in 1998; 22 in 2004.* **Sources: Ap6, Bu2, CO30, He8, He18, Hud, Ma18, Pr9, Ra5, Ra6, Re8, Sa9, Se16, Sh13, So1, So25, St18, Ta5, To1, To3, To9, To10.**

Dunkin's Delight - Det., 6-8 oz. fruit, crack res., robust flavor, slicing and canning, very productive. *Source History: 1 in 1998; 1 in 2004.* **Sources: Se26.**

E. H. House - 72 days, Bush type plant to 3 ft., 2.5 in., round or slightly oval fruit. *Source History: 1 in 1998; 1 in 2004.* **Sources: Sa9.**

Earliana - (Sparks Earliana) - 58-70 days, Standard early staking variety, small indet. open spreading vigorous vines, small foliage, clusters of 6-10 red flattened globes averaging 4.5-5 oz., solid firm flesh, very productive, good in short season areas, especially popular in the East, used extensively by truckers, 1900. *Source History: 37 in 1981; 31 in 1984; 31 in 1987; 18 in 1991; 10 in 1994; 5 in 1998; 11 in 2004.* **Sources: Au2, Cr1, He8, He18, Ra5, Ra6, Re8, Sa9, Syn, Up2, Vi4.**

Earliana Sunnybrook - (Burpee Sunnybrook Earliana) - 73 days, Det. plant, 6 oz. flattened fruit. *Source History: 1 in 1981; 1 in 1998; 1 in 2004.* **Sources: Sa9.**

Earlibell - 86 days, Det., 4 x 3.75 in. red canning tomato, good flavor. *Source History: 1 in 1998; 2 in 2004.* **Sources: Sa9, Ta5.**

Earliest Paste - 60 days, Det., pear shaped, meaty, thick fleshed fruit, good for tomato sauce and paste. *Source History: 1 in 1998; 1 in 2004.* **Sources: Fis.**

Earlirouge - 63-70 days, Moira and Starfire are in the parentage, det., firm 6 oz. slightly flattened globes, high crimson inside and out, dev. in Canada for northern growers, excel. for maritime conditions, crack-free, VW tol. *Source History: 10 in 1981; 8 in 1984; 10 in 1987; 12 in 1991; 11 in 1994; 4 in 1998; 1 in 2004.* **Sources: Pr3.**

Early A Pac - 88 days, Indet., flattened 4.25 x 2.75 in. red fruit. *Source History: 1 in 1998; 1 in 2004.* **Sources: Sa9.**

Early Annie - (Early Annie) - 60-65 days, Det., slightly flattened 3-4 in., bright red smooth fruits, meaty with few seeds, flavorful, great for canning. *Source History: 2 in 1994; 3 in 1998; 3 in 2004.* **Sources: Fo7, Me7, To9.**

Early Chatham - 61-65 days, Self-pruning small det. vines, can withstand cold nights, bright-red med-large smooth fruits, uniform color, adapted to the northern tier of states, 1941. *Source History: 7 in 1981; 4 in 1984; 3 in 1987; 2 in 1998; 1 in 2004.* **Sources: Pr3.**

Early Fruit - 75 days, Det. plants to 3 ft., produce heavy crop of round red fruit with nipple on blossom end, 2 in., from Russia. *Source History: 1 in 2004.* **Sources: Ta5.**

Early Glee - 80 days, Semi-det., 16 oz. fruit, high yields. *Source History: 1 in 1998; 1 in 2004.* **Sources: Sa9.**

Early Large Red - 75 days, Good all-purpose tomato, med. size red fruit, known in France in the 18th century, America in the 1840s. *Source History: 1 in 1981; 1 in 1984; 2 in 2004.* **Sources: Ea4, He18.**

Early Red Chief - 65-70 days, Bush plants bear huge crops of 10 oz. red fruits over a long picking season, early tomatoes shipped when market prices high, later pickings for canning trade. *Source History: 4 in 1981; 2 in 1984; 1 in 1987; 1 in 1991; 1 in 1998; 3 in 2004.* **Sources: In8, Se7, Se17.**

Early Rouge - 68-75 days, Compact indet. vines, 6-8 oz., 4 x 3.5 in. flattened pink-red globe, semi-sweet, bright red flesh, good flavor, crack res. slicing tomato. *Source History: 2 in 1998; 3 in 2004.* **Sources: Sa9, Ta5, Up2.**

Early Summer Sunrise - 60 days, Indet. vines, 5-8 oz. fruit with good old-fashioned tomato juiciness, flavor and texture, old var. that disappeared 10-15 years ago, originally sold by Stokes Seeds. *Source History: 1 in 1998; 2 in 2004.* **Sources: Pe2, Tu6.**

Ecuador - 85 days, Indet., above average yields of flattened, ribbed red fruit, 3 x 2 in. *Source History: 1 in 1998; 1 in 2004.* **Sources: Sa9.**

Edelrot - 74 days, Det., semi-acid, 2-3 oz. fruit. *Source History: 1 in 1998; 1 in 2004.* **Sources: Sa9.**

Ed's Fat Plum - 80 days, Large red plum type selected for fatness by a gardener in Connecticut for over 20 years, great sauce and drying tomato, fat tapered fruit, occasional round variant, slightly juicy, meaty flesh, tangy acid taste. *Source History: 2 in 2004.* **Sources: Ea4, To9.**

Egg - 75-80 days, Red, egg shaped fruit is the size and shape of a chicken egg, very smooth solid red flesh, a firm tomato that keeps well, mild flavor, unblemished fruit even in poor growing seasons, 1950s. *Source History: 1 in 1981; 1 in 1984; 2 in 1987; 1 in 1991; 4 in 1994; 3 in 1998; 4 in 2004.* **Sources: Ba8, Ma18, Syn, WE10.**

Egyptian - (Nile River Egyptian) - 85 days, Indet. plant, small leaves, 3 x 3 in. red fruit, round shape tapers to a point, good for canning, supposedly a descendent from seeds found in a 4000 year-old tomb. *Source History: 3 in 2004.* **Sources: Gr28, So25, To9.**

Eleanor - 85 days, Indet., 2 x 4 in. red paste tomato with pointed end, meaty flesh, unique sweet flavor, late season. *Source History: 1 in 2004.* **Sources: To9.**

Elfin - 55 days, Short det. plants bear clusters of grape-shaped cherry fruit, sweet flavor. *Source History: 1 in 2004.* **Sources: To1.**

Elim - 75 days, Det., 2 ft. plant, med. size, round red fruit, blemish-free, good taste, high yields, from Russia. *Source History: 1 in 2004.* **Sources: Ta5.**

Elizabeth Crook's - Italian heirloom from Elizabeth Crook, heavy indet. vines, red fruit, 3 x 4 in., thin skin. *Source History: 1 in 2004.* **Sources: Se17.**

Enormous - Maule Seed Company introduction, 1899, sent out in trial seed packets with the name "Eight to the Yard", bred by a Mr. Miesse, not as large as Maule claims or it may have gotten smaller over the last 100 years, 8-10 oz. fruit, good old-fashioned tomato flavor, dependable producer, very rare. *Source History: 1 in 2004.* **Sources: Ea4.**

Enormous Plum - 85 days, Indet., large heart shape fruits, many over 1 lb., solid flesh, rich flavor, excellent for canning or fresh use. *Source History: 2 in 2004.* **Sources: Ma18, To9.**

Enterprise - (Enterprize) - 79 days, Indet., high yields of uniform size, round, red fruit. *Source History: 1 in 1998; 1 in 2004.* **Sources: Sa9.**

Epoch - 80 days, Very compact bush plants, rugose foliage, can be set 2 ft. apart each way, large smooth tomatoes, borne in centers of plants, excel. quality, crack-free, Purdue, 1960. *Source History: 1 in 1981; 1 in 1984; 2 in 1987; 1 in 1991; 1 in 2004.* **Sources: Sa9.**

Ermak - 78 days, Bushy det. plant to 3 ft., blocky oblong paste type fruit, meaty flesh with few seeds, no juice, from Russia. *Source History: 1 in 2004.* **Sources: Ta5.**

Ernie's Plump - 80 days, Indet., uniquely shaped fruit, grows as a plump double pear with a tiny blossom scar, 8-12 oz., extremely rich flavor makes the best rich red sauce, also excellent fresh, from Italy. *Source History: 1 in 2004.* **Sources: Ma18.**

Ernie's Pointed - 75 days, Indet., icicle shaped fruit, 8 oz., meaty flesh, good sauce and canning tomato, from Italy. *Source History: 1 in 2004.* **Sources: Ma18.**

Eros - 74-84 days, Received in 1976 from Milan Sodomka, Czechoslovakian tomato breeder, det., early, med-round fruits, good for salads, rare East German heirloom. *Source History: 1 in 1981; 1 in 1984; 1 in 1987; 1 in 1991; 3 in 1994; 2 in 1998; 4 in 2004.* **Sources: Co32, Fo13, Sa9, Se17.**

Ethel Watkins Best - Bright red fruit, seed from Steve Parton of Melbourne, Australia. *Source History: 1 in 2004.* **Sources: He18.**

Ethiopia-Roi Humbort - 85 days, Indet., high yields of 2 in. red pear shaped paste fruit, flavorful. *Source History: 1 in 2004.* **Sources: To9.**

Eye - 79 days, Semi-det. plants, 8 oz. fruit. *Source History: 1 in 1998; 1 in 2004.* **Sources: Sa9.**

F. H. Crow - 79 days, Indet., potato leaf plants bear round, red fruits. *Source History: 1 in 1998; 1 in 2004.* **Sources: Sa9.**

Fa - 90 days, Indet., medium sized red fruit, Hungarian origin. *Source History: 1 in 1994; 1 in 1998; 1 in 2004.* **Sources: Sa9.**

Fan-3 - 73-83 days, Indet. potato leaf plant, flattened 4 x 2.5 in. red fruit, high yields. *Source History: 1 in 1998; 1 in 2004.* **Sources: Sa9.**

Federle - 85 days, Indet., excellent salsa tomato, 6-7 in. long paste, very few seeds, rich full flavor unlike most other banana pepper-shaped tomatoes, good processor. *Source History: 3 in 2004.* **Sources: Pr9, Se16, To10.**

Fireball - 60-80 days, Compact det. vines for close planting, good foliage cover, early firm smooth 4 oz. scarlet globes, excel. interior color and yields, free from cracks, for market, a sensation when intro. in 1952, it ended the commercial growing of stake tomatoes in Ontario, adapted to East and North, Canada, northern Europe. *Source History: 26 in 1981; 17 in 1984; 21 in 1987; 12 in 1991; 8 in 1994; 5 in 1998; 3 in 2004.* **Sources: PAG, Sau, Yu2.**

Firesteel - 68 days, Vigorous fairly-open indet. vines, medium foliage, smooth meaty 12 oz. bright-scarlet globes, FW and crack res., sets on well during dry spells, slicing and juice, bred by Clare Barber of Mitchell, SD, 1940. *Source History: 2 in 1981; 1 in 1991; 2 in 1994; 2 in 2004.* **Sources: He18, Pr3.**

Fireworks - 50-60 days, The largest of the super-early red tomatoes with excellent flavor, 6-12 oz. fruits on healthy indet. vines, round globes with pointed end, from PSR. *Source History: 4 in 1998; 3 in 2004.* **Sources: To1, To9, Tu6.**

Fireworks II - 45-50 days, One of the earliest, large red tomatoes, ripens consistently earlier than Fireworks with the same good flavor, no yellow-fruited (off-type) plants, compact det. plants, 4-8 oz. fruit, PSR. *Source History: 1 in 1998; 1 in 2004.* **Sources: Pe6.**

First Pick - 60 days, Det., deep red, globe shaped fruit, 4-5 oz., sets fruit in cold weather. *Source History: 1 in 2004.* **Sources: He8.**

Flavour Steak - 75 days, Indet., large, uniform-ripening, bright-red, 1-1.5 lb. fruit, tender flesh, sweet, rich tomato flavor, good crack resistance, from PSR Breeding Program. *Source History: 1 in 1994; 1 in 1998; 2 in 2004.* **Sources: He18, Ma18.**

Floradade - 72-83 days, Large det. plant, good cover, firm smooth 5-7 oz. slightly deep globes, red with green shoulder, VW FW1 FW2 GLS ASC BER CF CS and ST res., jointless, good yields, bears less on sandy soil, fresh market, dev. by U of FL for VW-infected calcareous soils of Dade County, FL. *Source History: 18 in 1981; 17 in 1984; 18 in 1987; 21 in 1991; 22 in 1994; 21 in 1998; 21 in 2004.* **Sources: Bu3, CO23, DOR, Ers, Ga1, HA20, He18, Hig, HO2, Jor, Kil, ME9, Ra5, Ra6, Re8, RUP, SE14, Se26, Vi4, WE10, WI23.**

Floradel - 75-85 days, Indet., for fresh market pole culture in humid areas, medium indet. plant, large leaves, firm fleshy 6 oz. globes, res. to FW1 GLS LM GW AB and cracking, U of FL. *Source History: 16 in 1981; 12 in 1984; 12 in 1987; 9 in 1991; 7 in 1994; 7 in 1998; 3 in 2004.* **Sources: Cr1, Kil, WE10.**

Florida Basket - 56-80 days, Excellent basket variety, larger and slightly huskier strain of Florida Petite, very sweet thin-skinned 1-2 in. dia. fruits, excellent growth habit for hanging baskets, 4 plants per 10 in. basket, open-pollinated status uncertain--sources disagree, U of Florida Research Station. *Source History: 5 in 1981; 5 in 1984; 5 in 1987; 1 in 1991; 1 in 1994; 1 in 1998; 1 in 2004.* **Sources: To1.**

Florida Petite - 40 days, Dwarf plants grow 8-9 in. tall and wide, deep red, round fruit, 1.25 in. dia., resists gray leaf spot. *Source History: 1 in 1998; 1 in 2004.* **Sources: To1.**

Fond Red Mini Plum - 75 days, Indet., excellent flavor, long shape resembles grapes. *Source History: 1 in 1998; 2 in 2004.* **Sources: Ra6, Re8.**

Forest Fire - 50-55 days, Matures as fast as Prairie Fire but is much firmer and crack res., det. vines, 3 in. fruit, better flavor than Sub-Arctic Maxi, excel. concentrated yields, good multi-purpose tomato for coastal, high mountain or first-of-season situations, PSR. *Source History: 1 in 1998; 1 in 2004.* **Sources: Pe6.**

Frisco - Med-large vine, good foliage cover, large fruit with elongated round shape, very high yields. *Source History: 1 in 2004.* **Sources: HA20.**

Frosty F. House - 67 days, Semi-det., 8 oz., flattened globes with red flesh, above average yield. *Source History: 1 in 1998; 1 in 2004.* **Sources: Sa9.**

Fuzzy Bomb - 86 days, Plant appearance and fruit resembles Angora, indet., potato leaf plant is covered with white fuzz, 12-16 oz. red fruit, good flavor, late, dev. by Tadd Smith, Franklin, NC. *Source History: 1 in 2004.* **Sources: To9.**

Gala - Jointless det. plant, large uniform red fruit, blocky shape, fruit distribution throughout the plant allows for improved mechanical harvest, high yields, V FW 1 and 2 res., good for "Saladette" markets, CV Harris Moran intro. *Source History: 1 in 2004.* **Sources: HA20.**

Gallo Plum - 75 days, Indet., dense paste type with firm flesh, 2 x 4-5 in., productive, excel. for all types of processing and fresh sauce. *Source History: 1 in 1998; 1 in 2004.* **Sources: Ma18.**

Gandia Lyc '77 - Small blocky red fruit. *Source History: 1 in 2004.* **Sources: Ea4.**

Garden Peach - (Yellow Peach) - 75-90 days, The original Longkeeper, indet. vines, lovely foliage, large red fruits with fuzzy skin and a peach blush, 1-2 in. dia., mild rich flavor that improves with age, low acid, keeps up to 4 months, old-time favorite introduced in America by James J. H. Gregory in 1862. *Source History: 1 in 1987; 2 in 1991; 3 in 1994; 11 in 1998; 26 in 2004.* **Sources: Ap6, CO30, Co32, Coo, Ea4, Fe5, Fo7, Ga1, He8, Ho13, ME9, Pe2, RUP, Sa9, SE14, So25, Syn, TE7, Ter, To1, To3, To9, Tu6, Und, Ver, Vi4.**

Garden Peach, Red - 85 days, The same as Garden Peach but with red skin. *Source History: 2 in 2004.* **Sources: Ra6, Ter.**

Gardener's Delight - (Sugar Lump) - 50-80 days, Vigorous indet. plants, best if staked, clusters of 6-12, bright-red fruit .75- 1.5 in., tendency to crack, sweet flavor, bears until frost, German-bred. *Source History: 12 in 1981; 10 in 1984; 12 in 1987; 10 in 1991; 17 in 1994; 19 in 1998; 33 in 2004.* **Sources: Bo17, Bou, Bu2, Coo, Ers, Fe5, Ga1, Ga7, Ga22, Ge2, Gr28, ME9, Pe2, Pin, Pr3, Ra5, Ra6, Re8, Sa5, SE14, Se26, Sk2, St18, Ta5, Te4, TE7, Tho, To1, To3, To9, Up2, WE10, We19.**

Gary's Late Canner - 80-85 days, Indet., produces large scarlet plums all summer long and into the fall. *Source History: 2 in 2004.* **Sources: Ma18, Ta5.**

Gem State - 58-68 days, Dev. for northern Rocky Mts. at U of Idaho, small 1.5-2 oz. red fruits, compact det. thick-stemmed potato-leaf plants do not sprawl, excel. for containers. *Source History: 3 in 1981; 3 in 1984; 6 in 1987; 6 in 1991; 4 in 1994; 3 in 1998; 3 in 2004.* **Sources: Hig, Sa9, Se8.**

Germaid Red - 76 days, Indet., large red beefsteak type fruit, 10-16 oz., large yields over a long season. *Source History: 2 in 2004.* **Sources: Ma18, Sa9.**

German - (German Bush) - 72-80 days, Old-fashioned slicing tomato, flat-round to round shaped fruits on compact 4-5 ft. vines, crack res., early, delicious. *Source History: 1 in 1994; 5 in 1998; 3 in 2004.* **Sources: He18, To1, WE10.**

German Bushy - 64 days, Det. plants to 2 ft., rugose foliage, 2 in. round red fruit in clusters of 4, from Russia. *Source History: 1 in 2004.* **Sources: Ta5.**

German Red - 80 days, Amish tomato, indet. plant produces high yields of large, smooth red fruit averaging 1.5-2 lbs., solid meaty flesh, excel. flavor, good slicer and canner. *Source History: 1 in 1994; 1 in 1998; 1 in 2004.* **Sources: Fo7.**

German Red Strawberry - 80-85 days, Indet. plant with open growth habit, light foliage cover, color and shape resembles strawberries, 3 in. wide by 3.5 in. long, 10 oz., meaty flesh with little juice, few seeds, old-fashioned tomato flavor with a lingering sweetness, German heirloom, pre-1900. *Source History: 9 in 1998; 23 in 2004.* **Sources: Ap6, Ers, Ga1, He8, He18, Ho13, Hud, Lan, Ma18, ME9, Ra5, Ra6, Ri12, SE14, Sh13, Shu, So1, So25, Ta5, To1, To3, To9, Up2.**

German Roper - Plants produce abundance of medium size cherry fruit, from Germany. *Source History: 1 in 2004.* **Sources: Ba8.**

Giant 11 - 89 days, Indet., potato leaf vines, red-pink, slightly flattened beefsteak, 4-5 in., mild fruity flavor, low acid. *Source History: 1 in 1998; 1 in 2004.* **Sources: To9.**

Giant Colossal - Heavy producer of 10-24 oz. red fruit, good flavor, heirloom. *Source History: 1 in 2004.* **Sources: To10.**

Giant Italian Plum - (Giant Italian Paste) - 85-100 days, Plum-shaped 1 lb. fruits, thick flesh, few seeds, ideal for making paste or thick sauces or ketchup, good stuffer, imported from Italy in the 1920s. *Source History: 2 in 1991; 1 in 1994; 2 in 1998; 1 in 2004.* **Sources: Fo7.**

Giant Paste - 72-85 days, Medium tall indet., multi-use smooth plum type fruits average 6-10 oz., good yield. *Source History: 1 in 1994; 1 in 1998; 2 in 2004.* **Sources: Ma18, To1.**

Giant Pepperview - 85 days, Indet., pepper shaped fruit, paste type with dry meaty flesh, fine flavor. *Source History: 1 in 2004.* **Sources: Ma18.**

Giant Syrian - 75-80 days, Indet., large, heart shaped fruit, 1 lb., meaty flesh, few seeds, excellent flavor. *Source History: 1 in 1998; 4 in 2004.* **Sources: He18, Pr9, Se16, To9.**

Gilbert Italian Plum - 85 days, Indet., abundant yields of large red plum fruit, good flavor. *Source History: 1 in 2004.* **Sources: Ta5.**

Gilbertie Paste - 80-85 days, Indet., red fruit up to 7 in. long, often with a crook at the tip, sweet, dry flesh, heirloom from Connecticut. *Source History: 1 in 1998; 1 in 2004.* **Sources: Hi6.**

Giles Mullis Plum - 85 days, Indet., large, deep red plums, good flavor, very prolific. *Source History: 1 in 2004.* **Sources: Ta5.**

Giuseppi's Big Boy - 85 days, Highly flavored

beefsteak, rare. *Source History: 1 in 1994; 1 in 1998; 1 in 2004.* **Sources: Syn.**

Glacier - 55-90 days, Spreading det. plant, potato leaf, 2.5-3 in. crack-free round red fruits ripen evenly, very sweet and flavorful, may appear weed-like at first, does well in cool climates, distributed for the first time in 1985, Sweden. *Source History: 1 in 1981; 1 in 1984; 5 in 1987; 9 in 1991; 7 in 1994; 10 in 1998; 20 in 2004.* **Sources: CO30, Fe5, Ga1, Hi6, Me7, ME9, Pe2, Pla, Ra5, Ra6, Sa5, SE14, So1, So25, Ta5, Ter, To1, To3, To9, Up2.**

Glamour - 72-82 days, Semi-upright spreading indet. plants, fair cover, smooth bright-red thick-walled firm mild meaty flattened globes, bright-red flesh with small core, fine flavor, crack res., sets first clusters heavily, good for hot dry Northeast and Midwest areas, home or market, good canner, similar to Sioux, 1957. *Source History: 30 in 1981; 25 in 1984; 30 in 1987; 28 in 1991; 26 in 1994; 21 in 1998; 21 in 2004.* **Sources: All, CH14, Dam, Ge2, GRI, He17, HO2, MIC, Ont, Ra5, Ra6, Re8, RUP, SE4, Sh9, Sto, Ta5, Te4, To1, Up2, Vi4.**

Glasnost - 59-75 days, Indet. vines do not require staking, 6-10 oz., 3 in. fruits, shiny smooth deep reddish-orange skin, dense meaty flesh, excel. flavor, from Siberia. *Source History: 1 in 1991; 3 in 1994; 3 in 1998; 3 in 2004.* **Sources: Hig, Ta5, To9.**

Gloriana, Burpee - 83 days, Det. plants, 6 oz. globes. *Source History: 1 in 1998; 1 in 2004.* **Sources: Sa9.**

Goliath - 65-85 days, Indet. vines, light-red fruits with irregular shape, excel. flavor, solid flesh, few seeds, extremely large fruits can grow to 3 lbs., first grown in late 1800s. *Source History: 1 in 1991; 2 in 1994; 4 in 1998; 7 in 2004.* **Sources: Gr27, He8, He18, Ho13, Se26, So25, To1.**

Good Old Fashioned Red - 75 days, Indet., flattened red smooth beefsteak, 12-16 oz., intense flavor, highly productive. *Source History: 2 in 2004.* **Sources: He18, To9.**

Graham's Good Keeper - 70 days, Indet. vines produce an abundance of medium-size, 3 in. oval, sweet fruit, good fresh, frozen or in sauces, good keeper, midseason. *Source History: 1 in 1994; 1 in 1998; 3 in 2004.* **Sources: Ga22, Sa5, Up2.**

Grandfather Barnini's Oxheart - Oxheart shaped fruit from Pittsfield, MA, originally from Italy. *Source History: 1 in 2004.* **Sources: Ea4.**

Grandma Mary - 70 days, Indet., high yields of 3 in. long red fruit, 1.5 oz., heirloom. *Source History: 1 in 2004.* **Sources: To9.**

Grandma Mary's Paste - 68-70 days, Heirloom, indet. plant, 6-10 oz. fruit, useful for any type of sauce, early, prolific, introduced by CV Fedco in 1992. *Source History: 3 in 1994; 3 in 1998; 2 in 2004.* **Sources: Fe5, To1.**

Grandpa's Cocks Plume - 70 days, Indet., 4-6 in. light-red fruits, 14-20 oz., juicy fine-textured deep-red interior, originally from the Altai Mountains between Siberia and Mongolia. *Source History: 1 in 1991; 2 in 1998; 2 in 2004.* **Sources: Hig, Ta5.**

Grandpa's Golf Balls - 80 days, Indet., large size red cherry, very flavorful, abundant yields. *Source History: 1 in 2004.* **Sources: Ta5.**

Grandpa's Minnesota - (Grandpa's Minnesota Hardy) - 70-75 days, Indet., round, 1 in. red cherry, prolific, mild sweet flavor. *Source History: 1 in 1998; 2 in 2004.* **Sources: Ma18, To9.**

Grape - (Red Grape, Christmas Grape) - 55-70 days, Indet. plant bears long grape-like clusters of elongated red cherry tomatoes, sweet complex flavor, resists cracking, heat tolerant, disease resistant, has become a popular supermarket item. *Source History: 18 in 2004.* **Sources: Ea4, Eo2, Ga7, Gr29, He17, He18, Hi13, Ma18, Na6, Ont, Ra6, Se28, So25, Te4, To1, Vi4, Vi5, Wo8.**

Grappoli d'Inverno - 75 days, Indet., large clusters of plum shaped fruit, excellent fresh and dried, rare. *Source History: 4 in 2004.* **Sources: CA25, Ea4, Ma18, So25.**

Great Divide - 80 days, Indet., nice old-time, flavorful large red beefsteak, good yields. *Source History: 1 in 2004.* **Sources: Ta5.**

Greater Baltimore - 74-81 days, Red main crop var., indet., clusters of heavy meaty fine-flavored fruits, 6 oz., good shipper, productive, blight res., bred by the Birds-Eye Hort. Res. Lab., Albion, New York, originally marketed by Joseph Harris Co., Rochester, NY, released in 1957. *Source History: 1 in 1981; 2 in 2004.* **Sources: He18, Vi4.**

Greek Asemina - Round red fruit with piercing flavor like a cherry tomato, ripens early. *Source History: 1 in 2004.* **Sources: Sa5.**

Greek Domata - 75 days, Indet., 8-12 oz. red slicer, excellent flavor, productive, long season, from Dionysiou Monastery, Athos, Greece. *Source History: 2 in 2004.* **Sources: Ta5, To9.**

Grightmire's Pride - 65 days, Det. prolific plants, large early pinkish-red heart-shaped var. originally from Yugoslavia, Fred Grightmire of Dundas, Ontario has refined this var. for many years into a very meaty low acid fruit with few seeds and good taste. *Source History: 1 in 1991; 2 in 1994; 2 in 1998; 4 in 2004.* **Sources: Dam, Pr3, Ta5, Up2.**

Grosse Lisse - 80 days, Indet., 3 in. round fruit, meaty flesh, nice flavor, main crop var., original seed from New Zealand. *Source History: 1 in 1981; 1 in 1984; 1 in 1987; 1 in 1991; 1 in 2004.* **Sources: He18.**

Grosse Ronde - 75 days, Old French variety, semi-det., clusters of med. size, slightly flat angular 3 in. red fruit, meaty with plenty of juice. *Source History: 1 in 2004.* **Sources: To9.**

Grosso Rosso Nostral - Traditional tomato, good eating and cooking, slightly flattened, ruffled red fruit, slight acid, possibly dates to the early 19th century or before. *Source History: 1 in 2004.* **Sources: Ea4.**

Grozney 91 - 80 days, Long pointed fruit, from the town of Grozney on the Caspian Sea. *Source History: 2 in 2004.* **Sources: He18, Ta5.**

Grushovka - 62-70 days, Siberian variety, det. 2.5 ft. plant produces 30-40 slightly oblong fruits, 3 in. long, 6-8 oz., flavorful. *Source History: 1 in 1991; 3 in 1994; 2 in 1998; 6 in 2004.* **Sources: He18, Hig, So25, Ta5, To1, To9.**

Ham Green Favorite - 82 days, Indet., uniform flattened fruit. *Source History: 1 in 2004.* **Sources: Sa9.**

Hanky Red - 75 days, Det. plant, 3 x 2.5 in., round fruit,

red flesh, 6 oz., slightly acid, excel. yield. *Source History: 1 in 1998; 2 in 2004.* **Sources: Sa9, Ta5.**

Harbinger - 70-80 days, Older flavorful English variety, early indet., ripens quickly, med-sized fruits, thin skin, fine flavor, good yields, bears until frost, introduced in early 1900s. *Source History: 1 in 1981; 1 in 1984; 2 in 1987; 2 in 1991; 1 in 1994; 1 in 1998; 3 in 2004.* **Sources: Bou, Ga22, Te4.**

Harry's Italian Plum - 85 days, Indet., classic sauce tomato, 6-8 in. long red Italian paste. *Source History: 1 in 2004.* **Sources: Ma18.**

Harzfeuer - German translation is Resin Fire, orange-red, med. size fruit, 6-10 oz., juicy flesh, slight acid taste, abundant yields. *Source History: 1 in 2004.* **Sources: Ami.**

Hasting's Everbearing Scarlet Globe - *Source History: 1 in 2004.* **Sources: Ea4.**

Hawaiian - 66 days, Indet., thick foliage, tiny red cherry tomato, super sweet flavor, disease res., excel. production. *Source History: 1 in 1994; 1 in 1998; 3 in 2004.* **Sources: Co31, Ea4, Sa9.**

Hayslip - 75-89 days, Med-large det. plant, 6 oz. flattened globe with green shoulders, FW1 FW2 V2 GLS ASC and ST res., BER CF cracking and black shoulder tol., jointless, fresh market. *Source History: 4 in 1981; 4 in 1984; 9 in 1987; 12 in 1991; 11 in 1994; 10 in 1998; 7 in 2004.* **Sources: Fo7, He18, Kil, ME9, SE14, Ta5, WE10.**

Healani - 82-90 days, Bred in Hawaii for res. to hot weather diseases, det. plant, 2.5 in. red globe fruit, 4-8 oz., sweet flavor, late. *Source History: 3 in 1991; 2 in 1994; 1 in 1998; 2 in 2004.* **Sources: To9, Uni.**

Heidi - 75-80 days, Med. indet. plant bears 2.5 in., 3-4 oz., pear shape, red paste fruit, rich tomato flavor, high yields, heat and drought resistant, from Africa. *Source History: 1 in 1998; 4 in 2004.* **Sources: Ma18, So25, To9, To10.**

Heinz 1350 - 72-75 days, Med-sized leafy semi-det. plant, good cover, 6-8 oz. red slightly flattened globes, FW1 VW ASC and crack res., prolific main crop var., fairly conc. set, dev. by the H. J. Heinz Co., popular with canneries and home gardeners east of the Rockies, 1963. *Source History: 48 in 1981; 45 in 1984; 32 in 1987; 16 in 1991; 7 in 1994; 4 in 1998; 1 in 2004.* **Sources: So1.**

Heinz 1439 - 70-75 days, Vigorous med-sized det. plant, good cover, smooth slightly flattened globes, 6 oz., VW FW1 and crack res., fairly conc. set, good processor, for general climates. *Source History: 12 in 1981; 9 in 1984; 13 in 1987; 20 in 1991; 13 in 1994; 12 in 1998; 13 in 2004.* **Sources: Bu3, Dam, HO2, Jor, Lin, Ra5, Ra6, Se26, Sh9, To1, To3, WE10, Wi2.**

Heinz 2274 - Processing and fresh market. *Source History: 1 in 1981; 2 in 1987; 1 in 1998; 1 in 2004.* **Sources: WE10.**

Heinz 2653 - (Romanelle) - 68 days, New standard of earliness for processors, ripens entire crop on the vine even in cold or coastal areas, very firm 2.5 oz. fruits, compact plants, res. to VW and FW. *Source History: 2 in 1981; 2 in 1984; 3 in 1987; 9 in 1991; 2 in 1994; 2 in 1998; 1 in 2004.* **Sources: Fe5.**

Heinz VF - 75 days, Det., bright red fruit, resists cracking., F and V resistant. *Source History: 1 in 2004.* **Sources: Vi4.**

Hellfrucht - (Sweet 'n' Bright) - 72-85 days, Very productive plant, produces a heavy crop of 3-4 oz., bright-red, firm-fleshed fruit, from Germany. *Source History: 1 in 1981; 1 in 1984; 1 in 1998; 1 in 2004.* **Sources: Pe2.**

High Carotene - 76 days, Large indet. plants require staking, 2-3 in. red globes with a hint of orange, meaty and flavorful, 2-3 times more beta-carotene than other varieties, high acid content, good canner. *Source History: 2 in 2004.* **Sources: Ra6, Ter.**

High Country - 60-70 days, Indet. plants, med-size scarlet pear-shaped fruits, 3-4 oz., thick meaty flesh, good for making paste and sauce, a single plant can produce over 100 fruit, from Montana. *Source History: 1 in 1991; 2 in 1994; 1 in 1998; 2 in 2004.* **Sources: Hig, To10.**

Hilltop - 75 days, Sturdy det. plant, 10-12 oz., firm red fruit, flavor has a good balance of sugar and acid, good producer *Source History: 1 in 1998; 3 in 2004.* **Sources: He18, Ta5, To1.**

Himmelssturmer - 90 days, Indet., wispy vines require 8 ft. tall stakes, name translates "sky climber", pear shaped red fruit, from Germany. *Source History: 1 in 2004.* **Sources: Ta5.**

Hires Rootstock - *Source History: 1 in 1981; 1 in 1984; 1 in 2004.* **Sources: He18.**

Hog Heart - 86 days, Indet., scarlet-orange fruit, 6-8 in. elongated shape, dry flesh, few seeds, sweet, excel. flavor, peels easily, original seed from a woman in Massachusetts who got it from an Italian immigrant between 1910 and 1920. *Source History: 1 in 1998; 3 in 2004.* **Sources: Fe5, To9, Up2.**

Holland - 75-80 days, Indet., smooth, round red fruit, 8-12 oz., abundant production, juicy sweet flesh, excellent flavor with slight acid undertones, adaptation of a greenhouse variety done by a New Jersey gardener. *Source History: 3 in 2004.* **Sources: He18, To1, To9.**

Holmes Mexican - 85 days, Indet., large pink flattened, ribbed fruit, good taste and yield. *Source History: 1 in 1994; 1 in 1998; 2 in 2004.* **Sources: Sa9, Ta5.**

Homestead - 80-83 days, Strong semi-det. plant, big leaves, firm meaty bright-red med-sized globes, FW and crack res., local market or can or pink ship in the South, especially Florida, developed in the 1950s. *Source History: 7 in 1981; 4 in 1984; 4 in 1987; 7 in 1991; 12 in 1994; 10 in 1998; 14 in 2004.* **Sources: Ada, Ba8, Bo19, Fo7, Ge2, He17, Lej, LO8, ME9, Sa9, Shu, SOU, WE10, WI23.**

Homestead 24 - 70-85 days, Large semi-det. plant, dense foliage, smooth firm dark-red 7-8 oz. flattened globes, FW1 GLS ASC CF and crack res., sets well in hot weather, good in humid areas, well suited to all Florida soils, productive, pink ship or green wrap or local market or home garden, 1966. *Source History: 34 in 1981; 25 in 1984; 23 in 1987; 15 in 1991; 13 in 1994; 10 in 1998; 17 in 2004.* **Sources: CA25, Eo2, Ga1, He8, He18, Kil, Ra5, Ra6, Re8, SE14, Se26, So1, Ta5, To1, To3, Vi4, Wi2.**

Homestead 500 - 75-85 days, Slightly-larger med-sized det. plant, good cover, smoother red firm slightly flattened 8 oz. globes, FW and ASC res., green wrap or shipping or local market. *Source History: 9 in 1981; 8 in 1984; 9 in 1987; 8 in 1991; 5 in 1994; 3 in 1998; 2 in 2004.* **Sources: To1, Wi2.**

Homestead Improved VF - *Source History: 1 in 1981; 1 in 1984; 1 in 1987; 1 in 1991; 3 in 1994; 3 in 1998; 1 in 2004.* **Sources: DOR.**

Homesweet F2 - 70 days, Det., stabilized o.p. variety produces an abundance of 9 oz. bright red fruit, well balanced flavor, resists VW, AB, two races of FW and 75% resistant to root knot nematodes, widely adaptable, second generation removed from a hybrid. *Source History: 2 in 2004.* **Sources: Kil, To1.**

Horvath Plum - 85 days, Indet., red, pepper shape paste type, 4 in. long, seed from the Adriadic Coast of Italy. *Source History: 2 in 2004.* **Sources: Ma18, Ta5.**

Howard German - 80 days, Indet. vines need staking, variably shaped fruit, from elongated and narrow to elongated and plump, shape resembles a large poblano pepper, red skin, meaty red flesh, few seeds, excel. mild sweet flavor, good resistance to disease and insects, pre-1900. *Source History: 1 in 1994; 3 in 1998; 8 in 2004.* **Sources: Ami, He18, Lan, Ma18, Roh, So25, Ta5, To1.**

Hubie - Indet., 6 oz., red fruit. *Source History: 1 in 1998; 1 in 2004.* **Sources: Se17.**

Hungarian Italian - 79-90 days, Healthy indet. vines, med-size paste tomato, 2-3 oz. fruits borne in clusters of 4, good holding quality and disease resistance, very prolific, bears until frost. *Source History: 1 in 1987; 1 in 1991; 2 in 1994; 3 in 1998; 5 in 2004.* **Sources: Ma18, Sa9, So1, Ta5, To1.**

Illini Star - 65 days, Indet., 6-8 oz. deep red fruit, excel. flavor, strong disease and split res., good choice for market growers and home gardeners, dev. by Merlyn Niedens. *Source History: 1 in 2004.* **Sources: So1.**

IM 83-1 Tall - 72 days, Indet., round flattened, tiny cherry, above average yield. *Source History: 1 in 1998; 1 in 2004.* **Sources: Sa9.**

Immune - (Irion) - 55 days, Eastern European type, fruit and plant size similar to Bush Cherry, formerly offered as Irion. *Source History: 1 in 1987; 1 in 1991; 2 in 1994; 1 in 1998; 1 in 2004.* **Sources: Co32.**

Imperial - 85 days, Indet., 4 in. round red fruit in clusters of 3-4, rich acid flavor, heirloom. *Source History: 1 in 2004.* **Sources: To9.**

Impostor - 85 days, Indet., large fruit with great flavor, did well in spite of heat and drought, received as Sojourner South American but did not fit that description, thus renamed. *Source History: 1 in 2004.* **Sources: Ta5.**

Imur Prior Beta - (IPB) - 60 days, Dev. in Norway, grown in short season high mountain valleys of southern Chili, reselected by Edward Lowden, seems adapted to high altitudes in Washington and Oregon, indet. potato-leaf vines need staking, almost 2 oz., slightly tough skin, tolerant of cool growing conditions, came back after top-killed by frost that killed Sub-Arctics. *Source History: 5 in 1981; 5 in 1984; 5 in 1987; 5 in 1991; 2 in 1994; 7 in 1998; 4 in 2004.* **Sources: Hum, In8, Pr3, Sa9.**

India - 86 days, Indet., 3.25 x 2.5 in., flattened red fruit, above average yields. *Source History: 1 in 1998; 1 in 2004.* **Sources: Sa9.**

Indian River - 72-85 days, Large vigorous indet. plants, smooth firm thick-walled deep-scarlet globes, 3 in., res. to FW GLS and AB, tol. to GW LM BER and CS, heavy crops in warm humid Southern areas. *Source History: 7 in 1981; 6 in 1984; 7 in 1987; 4 in 1991; 1 in 2004.* **Sources: To9.**

Ingegnoli Giant - (Pomodoro Ingegnoli Gigante) - Ingegnoli seedhouse in Italy reports a 3.5 lb. fruit, large and smooth, few seeds, highly disease res. *Source History: 1 in 1981; 1 in 1984; 1 in 2004.* **Sources: Se17.**

IPK T 120 (Italy) - Med. size oblong paste type, strong tomato flavor, good for sauce. *Source History: 1 in 2004.* **Sources: Ea4.**

IPK T 912 (Italy) - Long round paste type with small point on end, slightly acid flavor. *Source History: 1 in 2004.* **Sources: Ea4.**

IPK T 955 (Italy) - Old traditional type, ruffled red fruit, thin walls, 4-6 oz., low acid. *Source History: 1 in 2004.* **Sources: Ea4.**

IPK T 983 (Italy) - Large red fruit, plum shape. *Source History: 1 in 2004.* **Sources: Ea4.**

IPK T 985 (Italy) - Small, square shouldered fruit, medium-high acid. *Source History: 1 in 2004.* **Sources: Ea4.**

IPK T 1118 (Italy) - Very good sauce tomato, small oblong red fruit, uncooked flavor is good for a paste type. *Source History: 1 in 2004.* **Sources: Ea4.**

Irish - 75 days, Each plant produces dozens of tennisball-sized fruits over long season, seed from Ireland. *Source History: 1 in 1991; 1 in 1994; 1 in 1998; 1 in 2004.* **Sources: Fo7.**

Italabec - 71 days, Det., 4 oz., red fruit, meaty flesh. *Source History: 1 in 2004.* **Sources: Sa9.**

Italian - (Italian Sweet) - 85 days, Indet., potato leaf plant, flattened, globe shape, red-orange fruits, 14 oz., sweet, juicy, old-fashioned tomato flavor. *Source History: 2 in 1998; 2 in 2004.* **Sources: Sa9, To1.**

Italian Giant - 70-100 days, Italian heirloom, indet., deep red, flattened fruits average 1 lb., meaty and delicious, dry flesh, nearly seedless, from the 1920s. *Source History: 4 in 1998; 9 in 2004.* **Sources: He8, Ma18, Pr9, Ra5, Ra6, Re8, Sa9, Se16, To3.**

Italian Paste - 90 days, Indet., abundance of 3-5 oz. red fruit, meaty flesh, few seeds, good flavor, seeds originally brought from Italy during WW II. *Source History: 1 in 2004.* **Sources: To9.**

Italian Pear - 75 days, Indet., bright red, torpedo shaped fruit, pointed ends, meaty flesh, flavorful. *Source History: 1 in 2004.* **Sources: Ma18.**

Italian Pepper Shaped - 85 days, Indet., red paste type, resembles a banana pepper, 4-6 in. long, meaty flesh with few seeds, makes a flavorful sauce. *Source History: 1 in 2004.* **Sources: Ma18.**

Italian Winter - 75-85 days, Large red cherry size fruit, whole plants traditionally uprooted and hung over winter for fresh use as needed. *Source History: 1 in 1998; 1 in 2004.*

Sources: So12.

Jaga - 66 days, Det. plant to 2 ft., round red fruit, from Poland. *Source History: 1 in 2004.* **Sources: Ta5.**

Jawor - 70 days, Bushy 2 ft. det. plant, rugose foliage, 2 in. red fruit, golden shoulders, from Poland. *Source History: 1 in 2004.* **Sources: Ta5.**

Jersey Devil - 90-100 days, Long tapering pointed fruits look like very large frying peppers, 4-6 in., rich red color, good full tomato flavor, deliciously sweet, few seeds, immature green fruit make excellent pickles, large indet. plants need staking or caging, home gardens. *Source History: 1 in 1987; 1 in 1991; 2 in 1998; 3 in 2004.* **Sources: Ap6, Ta5, To1.**

Jersey Giant - Early, indet. Jersey Devil type but larger and less slender, red 4-12 oz. fruits, irregular in shape but smooth and crack-free as Jersey Devil, good flavor. *Source History: 1 in 1987; 1 in 1991; 1 in 2004.* **Sources: Se17.**

Jewish - Indet., regular leaf, large heart shaped fruit, 1-2 lbs., excel. flavor, good yields. *Source History: 1 in 2004.* **Sources: Up2.**

Jitomate Bulito - 80 days, Indet., box shaped, slightly pointed red fruit, seeds are in a pocket in the center of the thick red firm flesh, delicious, Zapotec Mexican cultivar. *Source History: 2 in 2004.* **Sources: Ag7, Sa5.**

Joe's Favorite - 80 days, Indet., clusters of very nice small fruit, paste type. *Source History: 1 in 2004.* **Sources: Ta5.**

Joe's Plum - 85 days, Indet., productive plants, huge red plums, 8-12 oz., solid flesh, few seeds, rich flavor, from Italy. *Source History: 1 in 2004.* **Sources: Ma18.**

John Baer - (Bonny Best) - 70 days, Indet., med-sized red globes, meaty and tasty, good for eating fresh or canning, good producer. *Source History: 2 in 1981; 2 in 1984; 1 in 1987; 1 in 1991; 1 in 1994; 2 in 1998; 3 in 2004.* **Sources: Sh13, Ta5, Up2.**

John Lossaso's Low Acid Ruby - 80 days, Indet., regular leaf, 8-10 oz. red fruit, different but good flavor. *Source History: 1 in 2004.* **Sources: He18.**

Jubileum - Rare Hungarian variety, deep red, medium size fruit with pointed ends. *Source History: 1 in 2004.* **Sources: So25.**

Jungle Salad - 74 days, Indet., high yielding, med. size red cherry, semi-sweet flavor. *Source History: 1 in 1994; 1 in 1998; 1 in 2004.* **Sources: Sa9.**

Justine - 85-110 days, Indet., scarlet orange-red fruit with no yellow shoulders, unusual flat heart shape, blemish-free skin, solid red flesh with excellent flavor, rare. *Source History: 2 in 2004.* **Sources: Ma18, So25.**

Kanner Hoell - (Canner Howle) - 85 days, Indet., 4.5 x 2.75 in. flattened globes, red flesh, perfect size for canning whole, an heirloom from the collection of Rev. C. Frank Morrow. *Source History: 1 in 1981; 2 in 1998; 2 in 2004.* **Sources: Sa9, Ta5.**

Karzetek Tallinski - 75 days, Bushy det. 3 ft. plant, med. size, round, red-orange fruit, from Russia. *Source History: 1 in 2004.* **Sources: Ta5.**

Katrina - 65 days, Det. 3 ft. plant, round red fruit, 3 seed cavities, Russia. *Source History: 1 in 2004.* **Sources: Ta5.**

Kecskemeti 509 - 70 days, Indet., med. size, round red fruit, sweet flavor, Hungary. *Source History: 1 in 2004.* **Sources: Ta5.**

Kecskemeti 886 - 78 days, Det. 2 ft. plants, round red fruit, salad size, Hungary. *Source History: 1 in 2004.* **Sources: Ta5.**

Kennington's Big Red - 80 days, Indet., flavorful red fruit, family heirloom from Kennington family of Oregon. *Source History: 1 in 2004.* **Sources: To9.**

Kentucky Beefsteak - 75-115 days, Indet., large beefsteak, globe to oval shape, deep yellow-orange color, up to 2 lbs., some fruit shows pink tinge on blossom end, heirloom found in the hills of eastern Kentucky. *Source History: 4 in 1998; 17 in 2004.* **Sources: Ba8, CO30, Ers, He18, ME9, Ra5, Ra6, Re8, SE14, Se26, Shu, Sk2, Ta5, To1, To3, To9, To10.**

Kewalo - 80 days, Det., round red fruit, well suited for the tropics, BW and nem. res., also does well in more temperate areas. *Source History: 1 in 1987; 3 in 1991; 3 in 1994; 2 in 1998; 3 in 2004.* **Sources: Ech, To1, Uni.**

Kille No. 7 - 74-96 days, Med-sized plants with good foliage cover, red globe-shaped fruits, medium to large, good second early variety, for either home gardens or market. *Source History: 2 in 1981; 2 in 1984; 3 in 1987; 1 in 1991; 1 in 1998; 1 in 2004.* **Sources: Sa9.**

King Humbert - (King Umberto) - 89 days, One of the oldest named tomato varieties, 2 oz., red pear shaped fruit, paste type, good for drying, named after the king of Italy, pre-1800. *Source History: 2 in 2004.* **Sources: Ea4, Sa9.**

Kootenai - 70-75 days, Early, compact vines with large juicy, round red fruits, 2-3 in. dia., fine flavor, heavy yields, low det. habit perfect for cloche or poly-tunnel culture, ripens mid-July in Oregon, University of Idaho. *Source History: 2 in 1981; 2 in 1984; 4 in 1987; 5 in 1991; 3 in 1994; 6 in 1998; 6 in 2004.* **Sources: Fo7, In8, Pla, Sa9, Ter, We19.**

Kora - 68 days, Indet., potato leaf, oblate, multi-lobed variety, well adapted to mech. harvest, VW res., Poland. *Source History: 1 in 1994; 1 in 1998; 1 in 2004.* **Sources: Ta5.**

Kotlas - (Sprint) - 60 days, Formerly offered as Sprint, indet. plant which should be staked and only minimally pruned, small round 2 oz. red fruits, sweet flavor, Kotlas in the Soviet Union is the sister city to Waterville, (the closest city to Johnny's), citizens of Kotlas visited Waterville in the spring of 1990, as a token of friendship Johnny's Selected Seeds re-named Sprint tomato Kotlas. *Source History: 1 in 1991; 2 in 1994; 2 in 1998; 2 in 2004.* **Sources: Hig, So25.**

Krakowsie Wczesne - 78 days, Indet., round, med. size red fruit, orange-yellow shoulders, Poland. *Source History: 1 in 2004.* **Sources: Ta5.**

Kujawski - 66 days, Bushy plant to 3 ft., med. size fruit, orange-red skin, red flesh, good taste, good producer, Poland. *Source History: 1 in 2004.* **Sources: Ta5.**

La Roma - 80-90 days, Compact det., extremely productive, very hard fruits, o.p. version of La Roma hybrid. *Source History: 1 in 1998; 5 in 2004.* **Sources: He18, Ra5, Ra6, Re8, Sa9.**

Lafayette - 77 days, Indet., 3-5 oz. red globes, high

yields. *Source History: 1 in 2004.* **Sources: Sa9.**

Landino di Panachia - (Ladino di Panocchio) - 75 days, Indet. vines, flattened, ruffled, medium size red globes, 3-8 oz., pleasant flavor. *Source History: 2 in 2004.* **Sources: Ea4, Ma18.**

Landis - 60-65 days, Formerly known as Landis Frost Resistant, round tennis ball size fruit, excellent flavor, no cracks, very hardy. *Source History: 2 in 2004.* **Sources: Pe2, Tu6.**

Landreth, The - Very rare variety, red fruit with good old-time taste, productive. *Source History: 1 in 2004.* **Sources: Ea4.**

Landry's Russian - 90 days, Exceptional heirloom brought to the prairie provinces of Canada by Russian immigrants many years ago, indeterminate, round uniform 8 oz. fruit, deliciously sweet, consistently high-yielding. *Source History: 1 in 1987; 1 in 1991; 2 in 1994; 2 in 1998; 2 in 2004.* **Sources: Ta5, Te4.**

Langada - 100 days, Indet., 4 x 2.5 in., flattened gobes, red flesh, average yield, from Greece prior to 1950. *Source History: 1 in 1998; 1 in 2004.* **Sources: Sa9.**

Large - 110 days, Indet. vine, huge fruits are free of cracks, big stem and blossom scars, sel. from an heirloom variety bred by Tim Peters, not recommended for north of Longview, WA. *Source History: 1 in 1991; 1 in 1994; 1 in 1998; 1 in 2004.* **Sources: Ter.**

Large Red - 85 days, Indet. vines provide med. foliage cover, fruits measure 2 x 4 in. wide, sweet flavor with a bit of tang, listed in the 1843 Shaker seed catalog at New Lebanon, NY, noted by Fearing Burr in his 1865 book. *Source History: 1 in 1981; 1 in 1984; 1 in 1987; 1 in 1998; 3 in 2004.* **Sources: He18, So1, Up2.**

Large Red, Brown's - Long, rangy vines prefer staking, thin skinned red fruit, weighs up to 1.5 lbs., good taste, late maturity, good for fried green tomatoes. *Source History: 1 in 1998; 1 in 2004.* **Sources: Te4.**

Larissa - 75 days, Compact plant produces a continuous crop of red, med. size fruit, fine flavor with a touch of acidity, from Larissa Avrorina in Russia. *Source History: 1 in 1998; 1 in 2004.* **Sources: Bou.**

Latah - 60 days, Early, large det. vines, 2-3 in. fruits, very heavy producer, University of Idaho. *Source History: 1 in 1981; 1 in 1984; 2 in 1987; 1 in 1991; 1 in 1994; 2 in 1998; 3 in 2004.* **Sources: Ma13, Sa9, St18.**

Lebanon - 85 days, Indet., flat, med. size red fruit. *Source History: 1 in 1998; 1 in 2004.* **Sources: Sa9.**

Lebyajinsky - 78 days, Det., 2 ft. tall bushes, pointed red, slightly heart shaped fruit, Russia. *Source History: 1 in 2004.* **Sources: Ta5.**

Lee Williams Red - 85 days, Red-orange, large size cherry. *Source History: 1 in 2004.* **Sources: Sa9.**

Legend - 70 days, Det., 3 ft. plant, no staking required, uniform, oblate, seedless fruit, 8-16 oz., nice blend of sugars and acids, seeded fruit are produced at the end of the growing season, does well in cool summers, resists late blight fungus, introduced by J. R. Baggett, OSU. *Source History: 7 in 2004.* **Sources: Hi13, Hum, Ni1, Pin, Pr3, Ter, To1.**

Lerica - 75 days, Indet., abundant production of 1.5 x 2.5 in. red paste type fruit, meaty firm red flesh. *Source History: 1 in 2004.* **Sources: To9.**

Liberator - 72-84 days, New jointless variety, determinate, firm med-size, 8 oz. fruits, concentrated set, uniform ripening, heavy yields, widely adapted, withstands stress well, strong disease res., handles and ships well, PVP expired 2000. *Source History: 3 in 1987; 2 in 1991; 1 in 1998; 2 in 2004.* **Sources: Sa9, Ta5.**

Lickurich - 57-77 days, Det., 2 ft. tall plant, med. size, round red fruit, salad size, good flavor, Russia. *Source History: 1 in 2004.* **Sources: Ta5.**

Lida Ukrainian - 84 days, Short det. vines, clusters of 2.5 in., round, red fruit, 4-6 oz., good flavor, pre-1920s heirloom brought to America by Russian immigrants. *Source History: 1 in 1998; 2 in 2004.* **Sources: Sa9, To9.**

Liliana - 80 days, Looks a small peach with a small nipple at the blossom end. *Source History: 1 in 2004.* **Sources: Sa9.**

Lillian's Red Kansas Paste - 70-80 days, Indet. plants, 12 oz., red fruit, not a paste type, solid red flesh, outstanding flavor. *Source History: 1 in 1998; 2 in 2004.* **Sources: He18, Ma18.**

Lima - 65 days, Det. plant grows 3 ft. tall, oval red fruit, 3 oz., excel. flavor, from market place in small Polish town. *Source History: 1 in 2004.* **Sources: Ta5.**

Line 125 - 65 days, Det. plant to 3 ft., rugosa foliage, round red fruit with good flavor, Moscow. *Source History: 1 in 2004.* **Sources: Ta5.**

Lisa King - 75-100 days, Det. vines, 8-10 oz. fruits with very small blossom end scars, crack and VFN res., bred by Tim Peters. *Source History: 1 in 1991; 1 in 1994; 5 in 1998; 3 in 2004.* **Sources: Ga1, Pe6, To1.**

Little Julia - 75 days, Indet., high yields of marble-size, bright red fruit with thin skin, does well in wet or dry conditions. *Source History: 1 in 2004.* **Sources: Und.**

Livingston's Favorite - 76-90 days, Indet., 6-10 oz., smooth red fruit, exceptional flavor and color, originally bred for canneries, withstands long distance shipping, Livingston introduction, 1883, rare. *Source History: 2 in 1998; 6 in 2004.* **Sources: Ea4, Ri12, Sa9, Se17, Sh13, Vi4.**

Livingston's Marvelous - High yielding old-time tomato with good flavor, very rare. *Source History: 1 in 2004.* **Sources: Ea4.**

Livingston's Perfection - 95 days, Indet., smooth, bright red, tough skin, excel. quality, retains good size to season's end, good market or home use tomato, introduced in 1880. *Source History: 1 in 2004.* **Sources: Vi4.**

Long Tom - 85-90 days, Indet., good canning tomato, 2 x 5 in. long, 8 oz., ideal for salads, few seeds, listed by Ben Quisenberry in 1981. *Source History: 1 in 1981; 1 in 1984; 1 in 1987; 1 in 1998; 5 in 2004.* **Sources: Ag7, Coo, Pr9, Se16, Ta5.**

Lucky Leprechaun - 75 days, Det. plant grows 2-3 ft. tall, bears dozens of ping pong-ball size, bright red fruit, rich tomato flavor, Irish heirloom dating back to the early 1900s. *Source History: 1 in 2004.* **Sources: He8.**

Lukullus - 78 days, Indet., 4 ft. plant, round red fruit, good flavor, from a small Polish town. *Source History: 1 in*

2004. **Sources: Ta5.**

Lunch Bucket - Huge plants grow to 8 ft., red fruit, spicy, salty flavor. *Source History: 1 in 1998; 1 in 2004.* **Sources: Sa5.**

Lutescent - (Livingston's Honor Bright) - 75-80 days, Indet. with pale yellow-green, curled foliage, white flowers, small to med. round fruit turns from pale green > white > yellow > orange > red, unusual, Livingston, 1897. *Source History: 1 in 1998; 6 in 2004.* **Sources: Ea4, Ri12, Sa9, So25, To9, Vi4.**

Lynnwood - 75 days, Indet., regular leaf, red fruit, excellent flavor, abundant production over a long season, from the village of Lynnwood, VA located where Madison Run joins the Shenandoah River. *Source History: 1 in 2004.* **Sources: He18.**

Macero II - 76 days, Det., large firm elongated pear with good deep red color, excel. flavor, good for salads and sauce, disease res. *Source History: 1 in 1991; 2 in 1994; 1 in 1998; 1 in 2004.* **Sources: Ha5.**

Magnum Beefsteak - 80 days, Hardy, productive indet. plants produce gigantic red beefsteaks with superb flavor, excellent disease res., selected by Chuck Wyatt. *Source History: 2 in 2004.* **Sources: He18, Ma18.**

Malaysian - Det., large yields of red fruit, good fresh eating variety. *Source History: 1 in 2004.* **Sources: Sa5.**

Mama Mia's - 70 days, Det., large red Italian roma type, withstood heat and drought, from Am. Hort. Society, 1994. *Source History: 1 in 2004.* **Sources: Ta5.**

Manalucie - 75-95 days, Large vigorous indet. plants, needs staking, good cover prevents sunspot, smooth deep-scarlet 6- 8 oz. deep globes, res. to FW1 GLS LM BER and cracks, AG and BW tol., large-fruited market type for pole production in humid areas, picked pink, ripens slowly, U of FL. *Source History: 24 in 1981; 21 in 1984; 17 in 1987; 9 in 1991; 5 in 1994; 7 in 1998; 10 in 2004.* **Sources: Ers, He18, Ra5, Ra6, Sa9, Se26, Sk2, Ta5, To3, WE10.**

Manapal - 80 days, Large indet. vines, heavy cover, large thick-walled smooth firm deep-red globes, clusters of 3 to 5, pole culture in home gardens or winter greenhouses, res. to FW GLS LM AB and GW, tol. to BER and CF, almost free from culls, for humid areas, picked pink, for shipping and local market, FL/AES. *Source History: 13 in 1981; 10 in 1984; 9 in 1987; 5 in 1991; 2 in 1994; 1 in 1998; 2 in 2004.* **Sources: He18, WE10.**

Manitoba - (Manitoba Bush) - 58-80 days, Non-staking bush type, med-large smooth firm bright-red fruits, 6 oz., very productive, F1 and V res., dev. at Morden Exp. Farm in Manitoba for southern Canadian prairies, 1956. *Source History: 10 in 1981; 10 in 1984; 9 in 1987; 8 in 1991; 7 in 1994; 15 in 1998; 23 in 2004.* **Sources: CO30, Ear, Ers, Ga1, He8, He18, ME9, Pr3, Ra5, Ra6, Re8, Sa9, SE14, Se26, Ta5, Te4, Ter, To1, To3, To9, Tt2, Up2, WE10.**

Manitoba Improved - 60 days, Det. non-staking plant, bright-red 4 oz. fruits, early and reliable enough to be a good choice for prairie or northern gardens. *Source History: 1 in 1991; 2 in 1994; 1 in 2004.* **Sources: Lin.**

Mankin Plum - Oblong, orange-red paste tomato, very meaty flesh with few seeds, mild sweet flavor, best for cooking, very rare. *Source History: 1 in 2004.* **Sources: Ea4.**

Manzana - 91 days, Indet., very flattened red fruit, 3 x 1.5 in., average yield. *Source History: 1 in 1998; 1 in 2004.* **Sources: Sa9.**

Marazzi's Giant - 85 days, Indet., large red beefsteak, good flavor, good yields. *Source History: 1 in 2004.* **Sources: Ta5.**

Marglobe - 75-80 days, Heavy vigorous det. vines, dense foliage, smooth thick-walled 6-10 oz. meaty scarlet flattened globes, FW1 GLS VW LM tol., NHR RES., home garden or can or ship, USDA, 1927. *Source History: 73 in 1981; 62 in 1984; 55 in 1987; 43 in 1991; 41 in 1994; 43 in 1998; 37 in 2004.* **Sources: Ada, All, Bu8, CH14, Co31, Com, Cr1, Eo2, Ers, Fo7, Ge2, HA3, He8, He18, Hi13, HO2, La1, Lej, LO8, Mel, Mey, MIC, Mo13, PAG, Ri12, Roh, RUP, Sa9, Sau, Se24, Sh9, Shu, SOU, To1, Up2, Vi4, Wi2.**

Marglobe Improved - (Marglobe VF) - 70-80 days, Select strain of Marglobe, shorter more det. plants, adds res. to FW VW and rust, smooth firm crack-res. red globes, 5-8 oz., thick walls and small cells, holds well in white ripe stage for shipping. *Source History: 7 in 1981; 7 in 1984; 9 in 1987; 8 in 1991; 10 in 1994; 10 in 1998; 8 in 2004.* **Sources: DOR, GLO, ME9, Ra5, Ros, So1, To1, Wet.**

Marglobe PS - 72-78 days, Outstanding uniform strain, med-large det. plant, heavily productive, smooth solid 6 oz. flat fruits, res. to VW FW ASC and NHR. *Source History: 5 in 1981; 5 in 1984; 4 in 1987; 2 in 1991; 2 in 1994; 3 in 1998; 1 in 2004.* **Sources: GRI.**

Marglobe Select - (Marglobe S) - 72-77 days, Old favorite, med-sized firm smooth globes, res. to FW, med-sized indet. vines, good cover, widely adapted, for stake or cage production. *Source History: 5 in 1981; 5 in 1984; 7 in 1987; 5 in 1991; 4 in 1994; 4 in 1998; 7 in 2004.* **Sources: HPS, OLD, Ra5, Ra6, Se26, Ta5, To3.**

Marglobe Supreme - (Supreme Marglobe Improved) - 73-77 days, Better size and uniformity, less cracking and better color around stem, deep-scarlet smooth globes, for humid climates, home or market or can or ship, FW and blight and rust res. *Source History: 5 in 1981; 4 in 1984; 5 in 1987; 4 in 1991; 4 in 1994; 5 in 1998; 12 in 2004.* **Sources: Ba8, Bo19, CA25, He17, He18, Jo6, Or10, Ra5, Ra6, SE14, WE10, WI23.**

Maria - Rare Hungarian heirloom, meaty flesh, good flavor. *Source History: 1 in 2004.* **Sources: Sa5, Sa9.**

Maria Augustina - Red paste type with long pointed shape, best for cooking and drying, high yielding, originally from Italy, obtained by CV Ea4 from a local radio broadcaster's mother-in-law. *Source History: 1 in 2004.* **Sources: Ea4.**

Marianne's - 75-86 days, Indet., 1.75 x 6 in., pale red paste tomato, 4 oz., excellent flavor balance, above average yield, good main crop tomato, from France. *Source History: 2 in 1998; 2 in 2004.* **Sources: Ma18, Sa9.**

Marion - (Marion F) - 70-78 days, Rutgers type but earlier, larger and more disease free, FW GLS ST and crack res., some tol. to AB, med-large strong indet. vines, good cover, med-large smooth dark-scarlet slightly deep globes, small blossom end scar, high yields, adapted to the Southeast, home or market, Clemson SC/AES. *Source History: 13 in 1981; 10 in 1984; 12 in 1987; 6 in 1991; 6 in 1994; 6 in 1998; 10 in 2004.* **Sources: Ada, DOR, He8, ME9, Scf, SE14, Ta5, To1, To3, WE10.**

Marizol Red - 80-85 days, Indet., large 14 oz. red fruit, outstanding flavor, possibly the best tasting of the Marizol varieties, originally from the Black Forest region of Germany. *Source History: 2 in 2004.* **Sources: To1, To9.**

Mark Twain - 75-85 days, Large beefsteak with old-time flavor, indet., 2 lb. average, fruit has a striped appearance as it approaches maturity, ripens to a bright red about a week later, retains a slightly flattened and ribbed shape. *Source History: 1 in 1994; 1 in 1998; 2 in 2004.* **Sources: Fe5, Fo7.**

Market at Huachinango Puebla Mexico - Unusual blue-green foliage, slightly flattened pink-red fruit, small but larger than cherry size, thick walls, good flavor with fruity undertones, traditional Mexican variety. *Source History: 1 in 2004.* **Sources: Ea4.**

Market King - 75 days, Indet., 1 x 2.5 in. red fruit, clusters of 8-12 fruit, up to 20 clusters per plant, once a common commercial var. in England but discontinued due to Common Market agreement, very similar to Moneymaker. *Source History: 1 in 2004.* **Sources: Ta5.**

Market Miracle - 65-80 days, Compact semi-det. plants, 8-12 oz., round red fruits, good flavor. *Source History: 1 in 1994; 1 in 1998; 2 in 2004.* **Sources: Hig, To10.**

Marmande - (De Marmande VF) - 60-75 days, Med-sized semi-det. plants, clusters of slightly ribbed 4-6 oz. flat fruits, FW1 and VW res., ships well, excel. color and taste, French heirloom developed by Vilmorin Seed Co., 1897. *Source History: 9 in 1981; 9 in 1984; 8 in 1987; 13 in 1991; 18 in 1994; 14 in 1998; 17 in 2004.* **Sources: Ba8, Dom, DOR, Fo7, Gou, Lej, Ma18, Ra5, Ra6, Re8, Ri12, Se24, So25, Ta5, To1, To3, To9.**

Marmande, Super - 72 days, Large slightly flat fruits, some slightly misshapen, fleshy and flavorful, prolific det. plants often bear 10 small grapefruit size fruits, aborts in high temps, excel. flavor, France. *Source History: 3 in 1981; 2 in 1984; 4 in 1987; 4 in 1991; 7 in 1994; 8 in 1998; 9 in 2004.* **Sources: Bou, Hi13, Hud, Sa9, Se7, Se26, Syn, Tho, WE10.**

Martian Giant - 90-100 days, Beefsteak size, bright red fruit, 10-16 oz., good acid/sweet flavor balance, Seeds of Change original. *Source History: 1 in 2004.* **Sources: Se7.**

Martino's Roma - 75-90 days, Det. plants, rugose foliage, paste tomato, 2-3 oz., 1.5 x 2.5 in., dry meaty flesh, few seeds, blossom end rot res., Italy. *Source History: 7 in 1998; 14 in 2004.* **Sources: Ap6, CO30, He8, He18, Ma18, ME9, Sa9, SE14, Se16, So25, Ta5, To1, To9, To10.**

Maryland Large Red - 80-90 days, Indet., huge beefsteak type, 10-24 oz., good flavor. *Source History: 1 in 1998; 2 in 2004.* **Sources: He18, Sa9.**

Maskabec - 56 days, Det., 4-6 oz. red globes, high yields. *Source History: 1 in 2004.* **Sources: Sa9.**

Maslov's Giant - 78 days, Indet., 4 ft. vines, large red flattened beefsteak fruit, Russia. *Source History: 1 in 2004.* **Sources: Ta5.**

Matchless - Nice tomato but not the original, some lobed fruits. *Source History: 1 in 2004.* **Sources: Ea4.**

Matchless, Austin - 80-90 days, Selected from the original Matchless for better flavor, relatively short plants with deep green rugose foliage, flavorful red-orange fruit, dates to 1908. *Source History: 2 in 2004.* **Sources: Ea4, He18.**

Matchless, Burpee's - 79-90 days, Det., rugose foliage, flattened 3.25 x 2.5 in. fruit, red flesh, high yield, Burpee introduction, 1894. *Source History: 2 in 1998; 2 in 2004.* **Sources: Ea4, Sa9.**

Matina - 58-75 days, Large indet., potato leaf foliage, clusters of round red fruit, 2-4 oz., no green shoulders, flavor equals that of a large beefsteak, good balance of sweet and slightly tart flavor, disease resistant, from Germany. *Source History: 9 in 2004.* **Sources: Ap6, Coo, Dam, He18, Sa9, Se7, So25, Ter, To1.**

Maule's Earliest of All Field 43 - Very early rare tomato, listed in the 1912 Maule Seed Book. *Source History: 1 in 2004.* **Sources: Ea4.**

Maule's Success - Heirloom with the classic old-time taste, 1912 or before, rare. *Source History: 1 in 2004.* **Sources: Ea4.**

Medford - 83-90 days, Semi-det., good sized midseason tomato, 5-8 oz. fruit with good flavor, requires plentiful calcium and adequate water to prevent blossom-end rot, originated in southern Oregon but will ripen further north, heirloom. *Source History: 1 in 1981; 1 in 1984; 1 in 1994; 3 in 1998; 2 in 2004.* **Sources: Au2, Sa9.**

Mega Tom - 79 days, Indet., huge 2 lb. red fruit, great flavor. *Source History: 1 in 2004.* **Sources: Se26.**

Memorial Polish - 80 days, Bushy plants bear high yields of large fruit, slightly ribbed, meaty flesh, few seeds, sauce and fresh eating. *Source History: 1 in 2004.* **Sources: Ga22.**

Memory to Vavilov - 67 days, Det. 3 ft. plants, round red fruit, salad size, named for Dr. Vavilov, heroic Russian plant scientist. *Source History: 1 in 2004.* **Sources: Ta5.**

Merveille des Marches - Commercial variety from France, indet., large fruit, firm flesh with good color, dates back to the 1880s or before. *Source History: 1 in 2004.* **Sources: Pr3.**

Mexican Beefsteak - 78 days, Det., 4 x 2.5 in., flattened globes, red flesh, average yield. *Source History: 1 in 1998; 1 in 2004.* **Sources: Sa9.**

Mexican Paste - 80 days, Det., large oval fruits with blemish-free skin, good canner. *Source History: 1 in 2004.* **Sources: Ma18.**

Mexican Salad - 86 days, Det., 3.5 x 2.5 in., red globes, above average yield. *Source History: 1 in 1998; 1 in 2004.* **Sources: Sa9.**

Mexico Midget - 60 days, Indet., 6 tomato plants produced 5 gallons of fruit from one picking, round, dark crimson .5-1 in. fruit with an incredible flash of tomato flavor. *Source History: 1 in 1998; 1 in 2004.* **Sources: Se16.**

Micarda - Semi-det. vines, elongated fruit with point on

one end, midseason. *Source History: 1 in 2004.* **Sources: Goo.**

Micro-Tom - 63-88 days, World's smallest tomato, compact vine for pot culture, tasty miniature fruit, does best in lightly shaded area, FW1 BER and crack res., PVP 1992 FL AES. *Source History: 4 in 1998; 6 in 2004.* **Sources: GRI, MIC, Ra5, Ra6, Re8, So25.**

Middle Tennessee Low Acid - 90 days, Large indet., regular leaf, 1-2 lb. fruit, sweet flavor, low acid. *Source History: 1 in 2004.* **Sources: He18.**

Mikado Ecarlate - 69 days, Indet., 6-8 oz., red globes, very good flavor, pre-1930s French heirloom. *Source History: 2 in 2004.* **Sources: He18, Sa9.**

Mikado Scarlet - 83 days, Indet., deep red, flattened fruits. *Source History: 1 in 2004.* **Sources: Sa9.**

Mikarda Sweet - 67 days, Indet., clear-red elongated pear-shaped, 3-4 in., 5-7 oz. fruit, thick-textured flesh, sweet flavor combined with slight acid bite, from Siberia. *Source History: 1 in 1991; 1 in 1994; 2 in 1998; 1 in 2004.* **Sources: Goo.**

Minelli Plum - 75 days, Indet. vining plant, red plum type makes great paste and sauce. *Source History: 1 in 2004.* **Sources: Ma18.**

Miracle of World - 104 days, Indet., 4 oz., hard fleshed paste. *Source History: 1 in 1998; 1 in 2004.* **Sources: Sa9.**

Mishca - Excellent producer, processing quickly for salsa or paste, local Salt Spring variety selected by Mishca Goldberg. *Source History: 1 in 2004.* **Sources: Sa5.**

Moby Grape - Productive grape tomato is a new step in breeding for genetic elasticity as it is a "multi-line" variety that is made up of several phenotypically similar but genetically unique sister lines, excel. sweet-tart flavor ratio, better yields in the cool coastal Northwest than Red Grape. *Source History: 1 in 2004.* **Sources: Se27.**

Moira - 66-76 days, Bush beefsteak type, uniform round 6 oz. fruit, firm crimson flesh, compact healthy det. plants, BER and crack res., stays blemish-free in adverse weather, dev. by the U of Toronto. *Source History: 6 in 1981; 6 in 1984; 7 in 1987; 5 in 1991; 2 in 1994; 2 in 2004.* **Sources: Ea4, Up2.**

Moneymaker - 75-80 days, Old English greenhouse variety from Bristol, England, sets freely in any weather, smooth 4 oz. red deep globes, clusters of 6-10 fruit, heavy yields, med-early, adapted to high humidity, vigorous indet. open plants, poor cover, very similar to if not the same as Market King. *Source History: 7 in 1981; 6 in 1984; 8 in 1987; 6 in 1991; 8 in 1994; 10 in 1998; 19 in 2004.* **Sources: Ba8, Bo19, Bou, Fo7, Fo13, Ga22, He18, Ra5, Ra6, Se7, Se26, Ta5, Tho, To3, To10, Up2, Vi4, WE10, We19.**

Mong - 90 days, Indet., large 2-3 lb. fruit, meaty flesh, mellow, sweet fruity flavor, small seed cavity, unbelievable size with no special treatment, CV Und obtained seeds from Marvin Reiter, Palatine, IL whose dad's 90 year-old neighbor in Iowa, Mr. Mong, grew it, originally from Europe. *Source History: 1 in 2004.* **Sources: Und.**

Monster - Indet., very large late tomato. *Source History: 1 in 1991; 1 in 1998; 1 in 2004.* **Sources: Pp2.**

Monte Verde - 85 days, Large det., non-curled foliage, firm, green shouldered fruit, deep oblate to flattened globe shape, crack res., excel. flavor, for fresh market, roadside, local market and home gardeners, VFF res., NC State U, PVP 1997 NC Ag. Res. Service. *Source History: 1 in 1998; 2 in 2004.* **Sources: Ra6, Se26.**

Montreal Tasty - Compact vines bear large red slicing tomatoes, midseason maturity. *Source History: 1 in 2004.* **Sources: Te4.**

Morena - 95 days, Semi-det., 2 oz., pink-red, thick paste fruits. *Source History: 1 in 1998; 1 in 2004.* **Sources: Sa9.**

Mormon World's Earliest - 95 days, Indet., large red slicer. *Source History: 1 in 1994; 1 in 1998; 1 in 2004.* **Sources: Sa9.**

Mortgage Lifter, Australian - 85 days, Huge indet. plant, regular leaf, 1+ lb. fruit, said to be a descendant of the Estler's version of Mortgage Lifter which was sent to Australia in the 1930s. *Source History: 1 in 2004.* **Sources: He18.**

Mortgage Lifter, Red - 83 days, An improved version of Radiator Charlie's Mortgage Lifter with added disease res. and higher yields, indet., 10-14 oz. fruit, red rather than pink-red color, bears until frost, dev. by Jeff McCormack. *Source History: 1 in 1991; 3 in 1994; 5 in 1998; 13 in 2004.* **Sources: Bo19, CO30, Ers, He18, HO2, SE14, So1, Syn, Te4, Th3, Und, Up2, Vi4.**

Moscow - 70 days, Good foliage cover, moderately large smooth fruit, heavy yields, oustanding canner. *Source History: 1 in 1991; 2 in 1994; 2 in 1998; 1 in 2004.* **Sources: Hig.**

Moskvich - 60-75 days, Indet., early, smooth globe shaped red fruit with small stem scar, 4-6 oz., smooth texture, rich taste, disease res., from Eastern Siberia. *Source History: 1 in 1994; 2 in 1998; 18 in 2004.* **Sources: Ge2, Go8, Goo, He18, Jo1, Mi12, Ra5, Ra6, Re8, Ri12, Sa9, Scf, So25, Ta5, Te4, To10, Tu6, Vi4.**

Mountain Boy - 60 days, Det., 3-3.5 in. red fruit. *Source History: 1 in 1998; 1 in 2004.* **Sources: Fis.**

Mountain Princess - 45-68 days, Det., bright red 8 oz. fruit, mild flavor, early ripening, grown for generations in the Monongahela National Forest region of West Virginia. *Source History: 1 in 1994; 2 in 1998; 5 in 2004.* **Sources: He8, Hi6, Ma13, ME9, Ta5.**

Msia - Malaysian var. with wonderful tomato flavor, 3 in. round fruit with pointed end, thin skin, good for fresh eating. *Source History: 1 in 2004.* **Sources: Red.**

Mt. Athos - 70-90 days, Unusual Greek strain, large blocky fruits, thick meaty flesh, good for slicing and cooking, dev. by Alan Kapuler. *Source History: 1 in 1994; 1 in 1998; 1 in 2004.* **Sources: Ta5.**

Mt. Roma - 68 days, Cold hardy paste tomato, compact det. plant, firm, 2-3 oz. fruit. *Source History: 1 in 1994; 1 in 1998; 1 in 2004.* **Sources: Hig.**

Muchamiel - 75 days, Indet., slightly flattened red fruit, some ribbing, yellow-green shoulders, good taste, name translates "much honey" in Spanish, from Spain. *Source History: 1 in 1998; 1 in 2004.* **Sources: Ta5.**

Mule Team - 75-90 days, Indet. plant, uniform 3.5 in. red globes, 8-12 oz., sweet flavor with a slight amount of tang, heat and disease res. vines bear continuously until frost, heirloom from Joe Bratka. *Source History: 5 in 1998; 19 in 2004.* **Sources: Ap6, CO30, Ers, Fo7, Go8, He8, He18, Ra5, Ra6, Re8, Sa9, Sk2, So1, Sw9, Ta5, To1, To3, To9, To10.**

Myona - 90-95 days, Indet., large plants, round paste, 3-4 oz., good flavor, excel. yields, originally from an Italian market gardener in the 1960s, who when asked what variety it was said, 'It's-a my own'a.' *Source History: 2 in 1998; 2 in 2004.* **Sources: To9, Tu6.**

Napoli - (Napoli VF) - 75-125 days, Italian paste type, med-sized compact det. plant, 3 oz., bright-red pear-shaped solid fruits, VW FW1 and ASC res., very concentrated set, mech. harv. or home garden. *Source History: 12 in 1981; 10 in 1984; 9 in 1987; 9 in 1991; 12 in 1994; 11 in 1998; 12 in 2004.* **Sources: DOR, He8, He18, Ma18, ME9, Sa9, SE14, So25, Syn, Ta5, Vi4, WE10.**

Nepal - 75-85 days, Indet. vines, 10-12 oz., very deep-red color, disease and heat res., high yields, good quality, intense flavor, resists cracking, good keeper when picked late in the season and wrapped in paper, from the Himalayan Mountains. *Source History: 2 in 1981; 2 in 1984; 2 in 1987; 7 in 1991; 9 in 1994; 9 in 1998; 14 in 2004.* **Sources: Al6, Ap6, He18, Ho13, Pe2, Sa9, Sh13, So25, St18, Sw9, Te4, To1, To9, Tu6.**

Neptune - 67 days, Det. vines grow 2.5-3 ft. tall, provide good cover for 4 oz. red fruit, borne in clusters of 2-4, specifically bred for heat tolerance and bacterial wilt resistance, recommended for gardeners and growers in hot, humid, rainy growing areas where it is difficult to grow tomatoes, dev. by Dr. J. W. Scott at the Gulf Coast Res. and Ed. Center, U of FL, released to CV So1, 1999. *Source History: 1 in 2004.* **Sources: So1.**

New Hampshire Surecrop - Indet., regular leaf, uniform fruit, excel. taste, productive, from Plant Gene Resources. *Source History: 1 in 2004.* **Sources: Up2.**

New Jersey Championship - Indet., regular leaf, uniform fruit with good taste, high yields, midseason, supposedly won the tomato trials at the New Jersey State Fair years ago. *Source History: 1 in 2004.* **Sources: Up2.**

New Round Paste - 82 days, Indet., flavorful orange-red, 3 in. fruit, round to slightly oval shape. *Source History: 1 in 2004.* **Sources: To9.**

New Yorker - (New Yorker 5, New Yorker V) - 60-75 days, Bush Beefsteak type, vigorous med-small compact det. plant, smooth meaty 4-6 oz. scarlet globes, well colored, sets fruit under cool conditions, res. to VW ASC and late blight, early, uniform concentrate ripening, adapted to Northern short season areas, Dr. Robinson/Geneva NY/AES. *Source History: 17 in 1981; 15 in 1984; 16 in 1987; 10 in 1991; 8 in 1994; 7 in 1998; 18 in 2004.* **Sources: Cr1, Ers, Ga1, He18, HO2, ME9, Ra5, Ra6, Re8, RUP, SE14, Se26, Sk2, Syn, Ta5, To1, To3, WE10.**

Nick's - 85 days, Indet., Italian paste type fruit, produced all summer, makes a rich sauce or paste. *Source History: 1 in 2004.* **Sources: Ma18.**

No. 506 - 62-80 days, Rugged upright 18 in. plants, potato leaf, brilliant-red fruits, mild and slightly sweet, defies drought, yields well on poor soil, spot and cracking and sunburn res., never watery. *Source History: 1 in 1994; 3 in 1998; 1 in 2004.* **Sources: To1.**

Northampton Italian Plum - Red paste tomato, long tapered shape, great for sauce and drying, from Northampton, MA, rare. *Source History: 1 in 2004.* **Sources: Ea4.**

Northern Delight - 65-74 days, Det. plants bear high yields of 2 in. red fruit with good flavor for an early tomato, cold tolerant plants, NDS. *Source History: 1 in 1994; 3 in 1998; 4 in 2004.* **Sources: Ma13, Pr3, Sa9, Ter.**

Northern Exposure - 67-75 days, Indet., medium-large red globes, from Alaska. *Source History: 1 in 1998; 1 in 2004.* **Sources: Bu2.**

Northern Light - 55 days, Smooth bright-red 2-3 in. fruits, determinate 14-18 in. plants with potato-like leaves, good for pots and limited space. *Source History: 1 in 1981; 1 in 1984; 1 in 1987; 2 in 1991; 2 in 1994; 2 in 1998; 2 in 2004.* **Sources: Fis, Hig.**

Northern Ontario Roadside - 75 days, Large indet., red fruit up to 4 in. dia., red flesh with mellow rich flavor, few seeds, does well in most climates, common roadside market variety. *Source History: 1 in 2004.* **Sources: Ag7.**

Nova - 65-75 days, Small early Italian paste type, New Yorker x Roma, 2 oz. elongated red pear, thick meaty deep-red flesh, 1.25 x 3 in., VW res., some tol. to FW, commercially scarce, originally dev. at the NY AES, introduced years ago by Stokes Seeds. *Source History: 10 in 1981; 10 in 1984; 9 in 1987; 10 in 1991; 8 in 1994; 5 in 1998; 3 in 2004.* **Sources: Ag7, Me7, Pr3.**

Novinka Kubani - 77 days, Det. 3 ft. plant, perfect round red fruit, blemish-free, good flavor, Russia. *Source History: 1 in 2004.* **Sources: Ta5.**

Novogogoshary - 80 days, Semi-det. plants, hollow red fruits, huge yields. *Source History: 1 in 1998; 1 in 2004.* **Sources: Sa9.**

Odessa - 85 days, Det. plant, 4-6 oz. red fruit, heavy production, from Odessa, Ukraine. *Source History: 1 in 2004.* **Sources: He18.**

Odessa No. 2 - 58 days, Ukrainian variety with det. vines, thick meaty med-size globes that ripen early, should do well in northern and short-season areas. *Source History: 1 in 1991; 3 in 1994; 3 in 1998; 1 in 2004.* **Sources: Hig.**

Old '49er - 80-85 days, Vigorous indet., med. size red fruit, excel. flavor. *Source History: 1 in 2004.* **Sources: Ta5.**

Old Brooks - 70-85 days, Indet., large smooth blemish-free fruits, 1 lb., no cracking, small core, few seeds, somewhat acidic, early and productive, good canner, splendid sandwich tomato, some disease res., heirloom. *Source History: 4 in 1991; 7 in 1994; 9 in 1998; 17 in 2004.* **Sources: CO30, Ers, Fo7, He8, He18, Ma18, Ra5, Ra6, Re8, Sa9, Se17, Ta5, Te4, To1, To3, To9, Up2.**

Old Ferry Morse Beefsteak - 80-85 days, Indet. vine, flattened beefsteak type. *Source History: 1 in 1998; 1 in 2004.* **Sources: Sa9.**

Old Virginia - 80-90 days, Indet., regular leaf, old-time red beefsteak, over 1 lb., meaty flesh, few seeds, sweet, intense flavor, heirloom from the Giltner family of Virginia. *Source History: 6 in 2004.* **Sources: He18, Ho13, ME9,**

Ta5, To9, To10.

Olga's Biggest - 65 days, Indet. plants, softball-size wide-ribbed bright-red fruits, 4-6 in. across, some weigh 1.5 lbs., good tomato flavor. *Source History: 1 in 1991; 1 in 1998; 1 in 2004.* **Sources: Hig.**

Olkhon Island - Productive, bright red, 2 in. fruit, mid to late tomato with better early blight res. than most, from an island in Lake Baikal, Siberia. *Source History: 1 in 1998; 1 in 2004.* **Sources: Tu6.**

Olomovic - 75-85 days, Received in 1976 from Milan Sodomka, Czechoslovakian tomato breeder, semi-det. vines, 5 oz. flattened globes, excel. flavor, midseason, prolific, does well in northern areas. *Source History: 2 in 1981; 1 in 1984; 1 in 1987; 1 in 1991; 3 in 1994; 3 in 1998; 5 in 2004.* **Sources: In8, Sa5, Sa9, Sk2, Ta5.**

Onyx - 64 days, Det. 3.5 ft. plant, 2 in. red fruit, Russia. *Source History: 1 in 2004.* **Sources: Ta5.**

Opalka - (Polish Torpedo) - 85 days, Polish heirloom, indet., wispy foliage, solid, meaty fruit, 9-11 oz., 4-6 in. long, looks like a long red pepper, few seeds, good for sauce and fresh use, from Poland to Amsterdam, NY about 1900. *Source History: 1 in 1994; 5 in 1998; 15 in 2004.* **Sources: Ap6, Ea4, He18, Ma18, ME9, Pin, Ri12, SE14, Se16, Se26, Ta5, Te4, To1, To3, To9.**

Optimus - 80-85 days, Indet. plant bears a good set of 3 in. red fruit with excel. old-time tomato taste, resists cracking, Ferry-Morse, 1885. *Source History: 1 in 1998; 3 in 2004.* **Sources: Ag7, Ea4, To9.**

Oregon 11 - 55-65 days, Det. 36 in. dia. bushes, small round red fruits, juicy and tart, tends to crack, very early, early fruit sets are seedless or parthenocarpic, Oregon State U. *Source History: 2 in 1981; 2 in 1984; 2 in 1987; 3 in 1991; 2 in 1994; 1 in 1998; 1 in 2004.* **Sources: We19.**

Oregon Spring - 55-80 days, Russian Severianin x Starshot, very early, det., luscious, nearly seedless fruits, 4-6 per lb., uniform size, very good flavor, home garden and fresh market, Dr. Baggett/Oregon State U. *Source History: 3 in 1987; 15 in 1991; 23 in 1994; 30 in 1998; 40 in 2004.* **Sources: Bou, Com, Ear, Fe5, Fis, Ga1, Goo, He17, Hi6, Hig, HO2, Hum, Jo1, Me7, ME9, MOU, Ni1, Pe2, Pe6, Pin, Pr3, Ra5, Ra6, Re8, RUP, Se7, Se8, SE14, Se25, Se26, Se28, Ta5, Ter, To1, To3, To9, Tu6, Up2, WE10, We19.**

Oregon Star - 80-85 days, Sister to Oregon Pride, det., irregularly shaped, 1-2 lb., large seedless fruits, late fruits are smaller and contain seeds, good for slicing and home processing, Dr. James R. Baggett/OSU. *Source History: 3 in 1994; 3 in 1998; 2 in 2004.* **Sources: Ni1, Pr3.**

Orenburg Giant - 75 days, Indet., beautiful round crimson fruit weighs up to 3 lbs., very flavorful, originally from European Russia but tested in Siberia by Dr. Andriev. *Source History: 1 in 1994; 1 in 1998; 1 in 2004.* **Sources: Hig.**

Oroma - (Oregon Roma) - 71-90 days, Result of a cross with Oregon Spring as a parent, det. vine, red pear shaped fruit borne in clusters of 4-7 fruits, 6 per lb., semi-seedless, peels easily, makes thick sauce and paste, good fruit set at high or low temperatures, V res., Baggett/OSU. *Source History: 3 in 1994; 4 in 1998; 1 in 2004.* **Sources: Ni1.**

Otraolny - 70 days, Indet., red, slightly oblong, 3-4 oz. fruits, smooth flavor starts sweet and ends with a slightly acid bite, from Siberia. *Source History: 1 in 1994; 1 in 1998; 1 in 2004.* **Sources: Hig.**

Oxheart, Red - *Source History: 4 in 2004.* **Sources: CO30, ME9, SE14, TE7.**

Ozark Champion - 75 days, Indet. plant with unusual prostrate growth habit, med. size round red fruit, old-timey flavor, prolific, received as Missouri Pink Love Apple. *Source History: 1 in 2004.* **Sources: Ta5.**

Packhouse - 84 days, Indet., large size red cherry fruits. *Source History: 1 in 1998; 1 in 2004.* **Sources: Sa9.**

Page German - 80 days, Indet., large red beefsteak fruit, very good taste, from the foothills of the Blue Ridge Mountains. *Source History: 1 in 2004.* **Sources: Ap6.**

Painted Ukrainian - 79 days, Indet., heart shaped fruit, 2.5 x 3 in., some subtle vertical striping, delicious, heirloom. *Source History: 1 in 2004.* **Sources: To9.**

Pakmor VF - (Packmore) - Compact det. plant, fair cover, red 7 oz. flattened globes, some shoulder roughness, VW and FW res., early bush production in semi-arid areas for fresh market. *Source History: 2 in 1981; 1 in 1984; 1 in 1987; 2 in 1991; 2 in 1994; 2 in 1998; 1 in 2004.* **Sources: WE10.**

Pantano Romanesco - 70-80 days, Vigorous indet. plants produce an abundance of deep red, scalloped red fruit, some green shoulders, 8-12 oz., meaty flesh, excellent rich tomato flavor, rare Italian heirloom from the former marshes near Rome, Italy. *Source History: 7 in 2004.* **Sources: Co31, Ma18, Ri12, Se24, Ta5, To1, To10.**

Paragon Livingston - 75-90 days, From an imp. stock grown in 1848 by tomato canner, Harrison W. Crosby, indet. vines with light foliage, clusters of med. size, red flavorful fruit, excel. old-time taste, can be disease sensitive, one of the prominent canning tomatoes in New Jersey, introduced by Livingston, 1870. *Source History: 4 in 1998; 9 in 2004.* **Sources: CO30, Ea4, Ho13, Ri12, Sa9, Sh13, Ta5, To9, Vi4.**

Patio King - 75 days, Ornamental 30 in. plants usually need no staking, sturdy dark-green thick leaves, 4-5 oz. fruits, mild flavor, tropical appearance, small gardens and larger pots. *Source History: 1 in 1981; 1 in 1984; 1 in 1998; 2 in 2004.* **Sources: Sa9, Vi5.**

Peach, Red - 70-80 days, Nineteenth-century Russian variety, indet. tasty red 1 oz. fruits that sometimes have a slightly fuzzy skin, looks like a tiny peach, juicy flesh, mild flavor, does not keep well, great for snacks. *Source History: 1 in 1987; 2 in 1991; 2 in 1994; 2 in 1998; 4 in 2004.* **Sources: CO30, Go8, SE14, To9.**

Pear Shaped Red - Sprawling plants, .5 oz., red pear shaped fruit, very productive. *Source History: 1 in 2004.* **Sources: Ea4.**

Pear, Red - (Italian Red Pear, Red Pomodoro) - 70-78 days, Vigorous med-large indet. plant, fair cover, clusters of sweet scarlet-red pear-shaped fruits, 1.75-2 x 1 in. dia., flavorful salad tomato that bears all summer, home garden favorite for eating fresh or canning or preserves. *Source History: 31 in 1981; 26 in 1984; 29 in 1987; 27 in 1991; 23* in 1994; 24 in 1998; 45 in 2004. **Sources: Ag7, Bo9, Bo19, Bu1, CA25, CO30, Co32, Coo, Ers, Fi1, Fo13, Ge2, Gr28, He8, He18, Hi6, Ho12, Jo1, La1, Loc, ME9, Mi12, MOU, Or10, Orn, PAG, Pin, Ra5, Ra6, Ri12, RUP, Sa9, SE14, Se24, Se26, Shu, So25, Ta5, Te4, To1, To3, To10, Up2, Vi4, WE10.**

Pearson - 80-90 days, Old-fashioned acid flavor, excel. size and yields, det. plant, large fruits, some green shoulders even when ripe, excel. for canning, tolerates semi-arid growing conditions, 1910. *Source History: 2 in 1981; 1 in 1984; 2 in 1987; 4 in 1991; 3 in 1994; 6 in 1998; 5 in 2004.* **Sources: Ba8, Loc, Ra6, Sk2, WE10.**

Pearson Improved - (Pearson Improved VF, Pearson A-1 Improved) - 78-93 days, Large vigorous slightly open med-det. plant, good cover, smooth 7 oz. flattened globes, excel. flavor, VW and FW res., adapted to California and semi-arid regions, high quality, for green wrap or processing or shipping, UC/Davis. *Source History: 17 in 1981; 15 in 1984; 11 in 1987; 7 in 1991; 5 in 1994; 4 in 1998; 6 in 2004.* **Sources: Bu3, He8, Ros, Sa9, Se26, Syn.**

Peasant - 67 days, Siberian variety, det., great paste tomato, 3-4 oz., thick-walled fruits, Roma shape. *Source History: 1 in 1994; 1 in 1998; 1 in 2004.* **Sources: Hig.**

Pepper Tomato - Indet., stuffing tomato, looks like a red bell pepper, acidic flavor, pre-1900. *Source History: 2 in 2004.* **Sources: Gr28, Lan.**

Peremoga - 70 days, Indet., 3-6 oz. red fruit, sweet juicy flesh with superb acid/sugar ratio, excellent yields, from Estonia. *Source History: 1 in 2004.* **Sources: Go8.**

Perestroika - 62-67 days, Indet. plants, 8-10 oz. red-orange fruits, very meaty, good flavor, few seeds, Siberia. *Source History: 1 in 1991; 2 in 1994; 1 in 1998; 2 in 2004.* **Sources: He18, Hig.**

Peron - (Peron Sprayless, Great Peron) - 68 days, Sprayless tomato, disease and crack res., outperforms hybrids, vitamin rich, deep-red 10 oz. globes, 3.5 in. across, fine quality, midseason, stake, Greece, 1951. *Source History: 3 in 1981; 3 in 1984; 5 in 1987; 7 in 1991; 9 in 1994; 13 in 1998; 19 in 2004.* **Sources: CO30, Eco, Ers, Fo7, Ga1, Ho13, Ra5, Ra6, Re8, Se7, Se17, So1, So9, Syn, Te4, Ter, To1, To3, WE10.**

Persey - 64 days, Det. 3 ft. plant, 2.5 in. red fruit, from Semco, Moldova. *Source History: 1 in 2004.* **Sources: Ta5.**

Pete's Italian Plum - 80-90 days, Indet., meaty wide red plums, 6-8 oz., 6 x 2 in., from Italy. *Source History: 2 in 2004.* **Sources: Ma18, Ta5.**

Petitbec - 61 days, Det., normal leaf, large red cherry, huge yields. *Source History: 1 in 2004.* **Sources: Sa9.**

Peto 76 - Large det. plants, very firm 5 oz. square-round fruits, med-late, general adaptation, VW and FW res., excel. for peeling whole, suited for mechanical harvest. *Source History: 2 in 1981; 2 in 1984; 1 in 1987; 2 in 1991; 2 in 1994; 1 in 1998; 1 in 2004.* **Sources: ME9.**

Petomech - (Redstone) - 80-90 days, Det., med. size, red paste tomato with squarish shape. *Source History: 1 in 1981; 1 in 1984; 1 in 1987; 1 in 1991; 1 in 1994; 1 in 1998; 1 in 2004.* **Sources: So25.**

Phil's Fantastic - Seed sent to P and P Seeds from Phil Lillie. *Source History: 1 in 2004.* **Sources: Pp2.**

Photon - 70 days, Det., 3 ft. plant, salad size, round red fruit, good flavor, Russian. *Source History: 1 in 2004.* **Sources: Ta5.**

Picardy - 76 days, Dates back to the 1890s, large indet. vines need support, slightly flattened globe shape, red fruit, 5 oz., produces large initial set of fruit and continues to bear until frost, canning or fresh, France. *Source History: 1 in 1998; 2 in 2004.* **Sources: He8, Ta5.**

Piedmont - 75 days, Det., thick-walled fruits average 7-8 oz., fresh market, disease res., rel. by NC State. *Source History: 1 in 1987; 3 in 1994; 3 in 1998; 5 in 2004.* **Sources: He18, Ra5, Ra6, Ta5, WE10.**

Pioneer II - 77-90 days, Received in 1976 from Milan Sodomka, Czechoslovakian tomato breeder, semi-indet. vines, med. size fruit, delicate flavor, stores better than most if picked in dry weather before killing frost. *Source History: 1 in 1981; 1 in 1984; 1 in 1987; 2 in 1998; 1 in 2004.* **Sources: Sa9.**

Pittman Valley Plum - 88 days, Elongated pink-red paste, 4-5 in. long, 3-4 oz., few seeds, excel. sauce, some ripe fruit drop but with little damage, can hold for up to 2 weeks, German heirloom from Pittman Valley, PA. *Source History: 1 in 1998; 1 in 2004.* **Sources: So1.**

Plainsman - 65-79 days, Small compact plants, good cover, smooth 5-6 oz. bright-red globes, space close for heavy yields, good processor, performs well on high dry Texas plains, TX/AES. *Source History: 3 in 1981; 2 in 1984; 2 in 1987; 1 in 1991; 1 in 1998; 2 in 2004.* **Sources: Sa9, Ta5.**

Plum, Red - 73-78 days, Large indet. plants, fair cover, deep-red oval plum-shaped fruits, 1.75 in. long, borne in clusters, very prolific yielder, unusually sweet, rich flavor, res. to ASC, excel. home garden var. for eating raw or canning or in salads. *Source History: 16 in 1981; 13 in 1984; 15 in 1987; 15 in 1991; 16 in 1994; 6 in 1998; 6 in 2004.* **Sources: La1, ME9, SE14, Ta5, To1, WE10.**

Plumpton King - Indet., 4-6 oz. bright red fruit, sweet and tart flavor, true old-fashioned tomato taste, retains flavor in the canning process. *Source History: 1 in 2004.* **Sources: Go8.**

Polar Baby - 55-60 days, Very early variety bred in Alaska, det. plants, 2.5-4 oz. fruits with remarkably sweet flavor, crack res. *Source History: 6 in 1998; 3 in 2004.* **Sources: Pe6, Pr3, To1.**

Polar Beauty - 63 days, Bred in Alaska for extremely rapid growth and extra early maturity, second to ripen in the Polar Series, short, bushy plants, 2.5-4 oz., crack-free fruit with good sweet flavor. *Source History: 4 in 1998; 2 in 2004.* **Sources: Sa9, To1.**

Polar Gem - 56 days, Latest to ripen in the Polar Series but still ahead of other earlies, bred in Alaska, det. plant is similar to Oregon 11 in growth habit but with improved leaf cover, 2.5-4 oz. fruits with sizes being larger in the far north, good, sweet tomato flavor. *Source History: 1 in 1998; 2 in 2004.* **Sources: Pe6, Sa9.**

Polar Star - 65 days, Det., 3-4 oz., red fruit with the pattern of a star on the blossom end, round to slightly flattened shape, sweet yet tangy taste, Alaska. *Source History: 3 in 1998; 1 in 2004.* **Sources: To1.**

Polish - 80-90 days, Vigorous indet., potato leaf plants set fruit well in cool weather, brick-red fruits up to 1 lb., exquisite flavor, smuggled out of Poland on the back of a postage stamp. *Source History: 2 in 1998; 6 in 2004.* **Sources: CO30, Fo13, He18, Se17, Sw9, To1.**

Polish Egg - Indet., egg shaped, red fruit, 3 oz., very tasty, extremely productive, early, rare. *Source History: 2 in 2004.* **Sources: So25, Up2.**

Polish Giant - 82-90 days, Indet., elongated paste up to 2 lbs., 4-6 in. long, purple-red skin, smooth pointed blossom end, Polish heirloom. *Source History: 1 in 1994; 2 in 1998; 8 in 2004.* **Sources: Ra5, Ra6, Re8, Se26, Ta5, To3, To10, Ver.**

Polish Linguisa - 73 days, Vigorous indet., 3-4 in. pointed fruit, 8-10 oz., excellent paste tomato, also good for slicing, high in vitamins A and C, holds well on the plant, bears until frost, family heirloom from New York, dates back to the 1800s, Poland. *Source History: 12 in 2004.* **Sources: Coo, Ers, Ge2, Go8, Ni1, Ra6, Se26, Sh9, Shu, Sk2, To1, To3.**

Polish Paste - 85-110 days, Indet., huge 5.5-6.5 in. pear-shaped fruits weigh up to 1 lb., ribbed and irregular, exquisite flavor, Poland. *Source History: 2 in 1991; 4 in 1994; 6 in 1998; 4 in 2004.* **Sources: He8, ME9, SE14, TE7.**

Pollock - Selected from Bonny Best by Andy Pollock, selected for earliness and productivity, rich flavor. *Source History: 1 in 2004.* **Sources: Sa5.**

Polo - *Source History: 1 in 2004.* **Sources: Se17.**

Pomadora - 75 days, Indet., oval to heart shaped plump fruit, 4-8 oz., meaty flesh, few seeds, sweet rich flavor, withstands hot dry summer weather, bears a nice fall crop, Italy. *Source History: 2 in 2004.* **Sources: Ma18, Up2.**

Ponder Heart - 70-90 days, Ponderosa x Oxheart cross, indet., medium-large size red fruit with pointed blossom end, excellent flavor, late, Japan. *Source History: 2 in 2004.* **Sources: He18, Ma18.**

Ponderosa Scarlet - (Ponderosa Red) - 80 days, Beefsteak type, large indet. open plants benefit from support, fair to poor cover, flat rough meaty 10-24 oz. deep-red fruits, mild and sweet, for humid areas, grown in the U.S. since 1891. *Source History: 11 in 1981; 10 in 1984; 12 in 1987; 11 in 1991; 14 in 1994; 15 in 1998; 17 in 2004.* **Sources: Ea4, Ge2, Gr28, HA3, He17, Hud, Loc, Ra5, Ra6, Re8, RIS, Se26, Sh9, Ta5, To3, Up2, WE10.**

Pop-In - 60-65 days, Indet. plant does well in containers and hanging baskets, bears tiny red tear-drop shape fruit, flavor is sweet with a slight bite. *Source History: 1 in 2004.* **Sources: Go8.**

Popovich - 65-72 days, Red Siberian tomato, 8-12 oz., tasty. *Source History: 1 in 1994; 1 in 1998; 2 in 2004.* **Sources: Goo, He18.**

Porter - 65-78 days, Large indet. plants, fair cover, med-small red deep-globes, mid-season, sunburn and crack res., heavy yields on all soils even when hot and dry, bears until frost. *Source History: 10 in 1981; 9 in 1984; 9 in 1987; 9 in 1991; 6 in 1994; 9 in 1998; 13 in 2004.* **Sources: Fo7, He8, He17, LO8, ME9, Ra5, Ra6, Re8, Sa9, SE14, Ta5, To1, To3.**

Porter Improved - (Porters Pride) - 65-78 days, Heavy yields in high temp. and low humidity, large indet. plant, 4 oz. deep-red globes larger than Porter but same setting ability, ASC res., excel. keeping quality, V and F1 resistant. *Source History: 9 in 1981; 8 in 1984; 6 in 1987; 5 in 1991; 2 in 1994; 3 in 1998; 7 in 2004.* **Sources: Ga1, ME9, Se7, SE14, TE7, Ter, To1.**

Potentate - (Harrison's English Strain) - 75-84 days, Indet., main crop, 2-3 in. fruit, nice acidic flavor, disease res., widely grown in New Zealand. *Source History: 1 in 1981; 1 in 1984; 1 in 1987; 1 in 1991; 1 in 1998; 1 in 2004.* **Sources: Sa9.**

Pozdnyakov's - 75 days, Det. 3 ft. plants produce an abundance of round red fruit, good flavor, from Eastern Bloc. *Source History: 1 in 2004.* **Sources: Ta5.**

Prairie Fire - 50-65 days, Dev. by Fishers from Benewah which was dev. by the U of Idaho from a cross with a Sub-Arctic type, small dwarf vines produce large 1.5-3 in. extra-red fruits, good for canning slicing or general use. *Source History: 1 in 1981; 1 in 1984; 2 in 1987; 2 in 1991; 6 in 1994; 8 in 1998; 8 in 2004.* **Sources: Fis, Fo7, Ga1, Hig, Pe6, Sa9, Ter, To1.**

Prairie Pride - 55 days, Very early dwarf compact plants, med-sized deep thick-walled fruits, extended keeping quality, low acid, superior flavor, U of Manitoba. *Source History: 1 in 1981; 1 in 1984; 1 in 1987; 2 in 1991; 4 in 1994; 2 in 1998; 1 in 2004.* **Sources: Pr3.**

Precocibec - 70 days, Det., 3-4 oz. red fruit, early maturity. *Source History: 1 in 1998; 2 in 2004.* **Sources: Pr3, Sa9.**

Prelude - 67 days, Bushy plant to 4 ft., round red salad size fruit, good flavor, Russia. *Source History: 1 in 2004.* **Sources: Ta5.**

Principe Borghese - 70-75 days, Tomato used for sun-drying in Italy, det. vines need support, 1-2 oz. plum-shaped fruits with pointy ends borne in clusters, very meaty with little juice and few seeds, good processing variety, in dry areas branches of these tomatoes can be hung up to dry until leathery. *Source History: 1 in 1987; 10 in 1991; 14 in 1994; 25 in 1998; 54 in 2004.* **Sources: Al25, Ap6, Bo19, Bu2, CO30, Co32, Com, Coo, Dom, Ers, Fe5, Ga22, Gr28, He8, He17, He18, Hi6, Ho13, Hum, In8, Ma18, ME9, Pe2, Pin, Pr3, Pr9, Ra5, Ra6, Red, Ri12, Sa5, Se8, SE14, Se16, Se17, Se24, Se25, Se26, Sh13, Sk2, So1, So12, So25, Sw9, Syn, Ta5, Te4, Ter, To1, To3, Tu6, Up2, Vi4, We19.**

Pritchard's Scarlet Topper - (Pritchard, Scarlet Topper) - 66-90 days, Large globes, brilliant-red inside and out, rugged vines with abundant foliage, highly res. to FW, very productive, bears over a long season, for home or market. *Source History: 1 in 1981; 1 in 1984; 1 in 1987; 2 in 1991; 3 in 1994; 4 in 1998; 3 in 2004.* **Sources: Ma18, Sa9, To9.**

Productiva - 70-80 days, Bulgarian variety, indet., medium-large fruit averages 3 in. diameter, red fruit, minimal blemishing, good yields. *Source History: 1 in 2004.* **Sources:**

Hi6.

Prometeo - 90 days, Strong, 4 ft. bush plants, blocky, oval, 1.5 x 2 in. red fruits, newly bred variety preferred by Russians for canning whole, name refers to the Greek hero, Prometheus. *Source History: 1 in 1994; 1 in 1998; 1 in 2004.* **Sources: Ta5.**

PSR 37 - 60 days, Bred from the older, more aromatic strain of Early Girl with is no longer available, indet. vines produce an abundance of firm, crack res., 4-6 oz. fruits, excellent multi-purpose tomato from the PSR Breeding Program. *Source History: 1 in 1994; 1 in 1998; 1 in 2004.* **Sources: Pe6.**

Puck - 60 days, Small plant, salad size fruit, dev. in England prior to 1950. *Source History: 1 in 1998; 1 in 2004.* **Sources: Sa9.**

Puebla - 76 days, Med. size red pear shaped fruit, 3 oz., firm flesh, excellent flavor, disease res. *Source History: 3 in 2004.* **Sources: Ra5, Ra6, Re8.**

Pusa Ruby - 60-80 days, Early var. from India, abundant crops of uniformly deep-red med-sized fruits, very seedy, tart flavor, indet., very disease res., good keeper, popular with home and market growers in India. *Source History: 2 in 1981; 2 in 1984; 1 in 1987; 1 in 1991; 3 in 1994; 4 in 1998; 3 in 2004.* **Sources: Sa9, Se17, Ta5.**

Quebec 59 - Abundant yields of slightly flattened red fruit, strong flavor, once very popular but now is hard to find. *Source History: 1 in 2004.* **Sources: Sa5.**

Quebec 314 - Medium size round red fruit, slightly sweeter than Quebec 59. *Source History: 1 in 2004.* **Sources: Sa5.**

Quebec 1121 - (Square) - Early square tomato, used mainly for paste and puree, very firm and meaty, dark-crimson flesh, dev. by R. Doucet at Agri. Res. Sta. in St. Hyacinthe. *Source History: 1 in 1981; 1 in 1984; 1 in 1987; 2 in 1991; 1 in 1994; 1 in 1998; 1 in 2004.* **Sources: Pr3.**

Quedlinburger Frue Liebe - 70 days, Spindly, indet., potato leaf vines, red, 1.5 in. round, 4-lobed fruit borne in clusters of 4, good acid flavor, dev. for cool rainy nights, productive in colder summers, old German variety, means "Early love of Qued Linburg". *Source History: 1 in 1998; 1 in 2004.* **Sources: To9.**

Quigley's Italian Paste - 90 days, Indet., late red paste, heavy producer, Mr. Quigley obtained seed from his Italian neighbor, originally from Italy. *Source History: 1 in 2004.* **Sources: Ta5.**

Quinte - (Easy Peel) - 70 days, Smithfield Exp. Farm (1976) for Bay of Quinte canning district, compact plants, firm 7 oz. high crimson fruits, VW res., peels like a peach without scalding, cooler climates. *Source History: 2 in 1981; 2 in 1984; 3 in 1987; 3 in 1991; 1 in 2004.* **Sources: Up2.**

Readheart - 78 days, Indet. plant bears large, flattened globes, average yield. *Source History: 1 in 1998; 1 in 2004.* **Sources: Sa9.**

Red Alert - 50-55 days, Bush plant, may crop a week earlier and yield more heavily than most early varieties, bushy open habit, 20-24 fruits per lb., 4-5 lbs. per bush. *Source History: 1 in 1987; 1 in 1991; 1 in 1994; 1 in 1998; 2 in 2004.* **Sources: Tho, Ves.**

Red Beauty - 80 days, Det. plant, uniform 3.5 by 2.5 in. fruit, 4-5 oz., flavorful, good yield. *Source History: 1 in 1998; 2 in 2004.* **Sources: Sa9, Up2.**

Red Calabash - 75-85 days, Indet., ruffled shape, very sweet thick-skinned fruits, 2-3 oz., may have come from Chiapas State in Mexico. *Source History: 2 in 1991; 1 in 1994; 1 in 1998; 9 in 2004.* **Sources: Co32, Ea4, Ma18, Ra6, Ri12, Sa5, Se7, Se26, To10.**

Red Chief - 80 days, Indet., smooth med. size red globes, good yields. *Source History: 1 in 1981; 1 in 2004.* **Sources: Ta5.**

Red Cloud - 83-90 days, Vigorous det. plant needs staking, clusters of slightly pointed fruit, 2.5 x 2 in., 6-8 oz., good for canning, adapted to high elevations and the North, FW and VW res., does best in warm not hot weather, heavy yields. *Source History: 3 in 1981; 3 in 1984; 2 in 1987; 1 in 1991; 1 in 1998; 2 in 2004.* **Sources: Sa9, Ta5.**

Red Cup - 75-90 days, Heirloom stuffing tomato, hollow teacup-size fruits with 3 lobes, good flavor, indet. *Source History: 3 in 1991; 3 in 1994; 5 in 1998; 3 in 2004.* **Sources: Ma18, Ter, To1.**

Red Fig - (Pear Shaped) - 70-85 days, Indet., sweet tasty fruit, 1.5 in. pear shape with long slender neck, look like tiny bowling pins, heavy yields, used as fig substitute by gardeners years ago who packed away crates of dried preserved tomatoes for winter use, dates back to the 1700s. *Source History: 5 in 2004.* **Sources: Ba8, Co31, Ma18, Ra6, Se16.**

Red House Free Standing - 75-85 days, Stakeless bushy plants with dense foliage, red, round 6-8 oz. fruits, tolerates heat, wilt and crack res., does well in areas with hot summers. *Source History: 1 in 1994; 1 in 1998; 1 in 2004.* **Sources: Se7.**

Red King - 65-73 days, Compact det. vines, smooth 6 oz. deep globes, crack-free, tol. to FW GW AB ASC TMV BER and soft rot, extremely solid, ships better, remains in better condition. *Source History: 1 in 1981; 1 in 1984; 3 in 1987; 1 in 1991; 1 in 2004.* **Sources: To9.**

Red October - 68 days, Indet., storage tomato, 8 oz. fruit retains good flavor on the vine and after harvest, will keep 3-4 weeks longer than other varieties if harvested when fully ripe, good disease res. *Source History: 1 in 2004.* **Sources: Hi13.**

Red Rasp - (Rasp Large Red) - 92 days, Semi-det. spindly plant, large, orange-red fruit, 12-16 oz., excellent texture and flavor, dev. by T. Rasp of Cheektowaga, NY, heirloom. *Source History: 1 in 1998; 3 in 2004.* **Sources: Ap6, Sa9, To9.**

Red Robin - 55-100 days, Pot type variety, dwarf compact 8-12 in. plants, small round orange-scarlet fruits 1-1.5 in. dia., sweet with a mild acid flavor, highly ornamental. *Source History: 3 in 1987; 11 in 1991; 9 in 1994; 6 in 1998; 13 in 2004.* **Sources: Ge2, GLO, He18, Ma18, MIC, OSB, Ra5, Ra6, Re8, To1, To3, To10, Und.**

Red Rock - (Early Red Rock) - 80 days, Med-sized det. plant, slightly open habit, smooth thick-walled 3 oz. red globes, VW FW GLS and crack res., jointless, for mech. harv. in humid or general areas. *Source History: 6 in 1981; 4 in 1984; 3 in 1987; 5 in 1991; 2 in 1994; 2 in 1998; 1 in 2004.* **Sources: So9.**

Red Rocket - 60 days, Det., bushy plant, heavy crop of bright red fruit, 8-10 oz., blemish-free, good blend of sugar and acid, heavy yields. *Source History: 1 in 2004.* **Sources: To1.**

Red Rose - 80-90 days, Brandywine x Rutgers, vigorous indet. vines, large pink-red fruit, 6-10 oz., flavor and texture resembles that of Brandywine, crack and disease resistance is inherited from Rutgers. *Source History: 3 in 1994; 4 in 1998; 3 in 2004.* **Sources: Ga21, Sa9, To1.**

Red Sausage - (Sausage) - 75-85 days, Mid-season indet. vines, paste tomato up to 6 in. long, few seeds, fine for sauces or paste, very prolific, unusual, heirloom. *Source History: 2 in 1991; 5 in 1994; 8 in 1998; 21 in 2004.* **Sources: CO30, Ers, Gr28, Gr29, He8, He18, HO2, HPS, ME9, Ra5, Ra6, SE14, Se26, Se28, Sk2, So25, To1, To3, To10, Vi4, Wo8.**

Red Star - Six-lobed fruit, flavorful. *Source History: 1 in 2004.* **Sources: Ga7.**

Red Supreme - (Edelrot) - 80 days, Obtained in 1976 from Milan Sodomka, Czechoslovakian tomato breeder, bush habit, med-size fruits, med-early, does not keep well in continued wet weather, formerly offered as Edelrot. *Source History: 1 in 1981; 1 in 1984; 1 in 1987; 1 in 1991; 2 in 1994; 1 in 1998; 1 in 2004.* **Sources: Co32.**

Redskin - 63 days, Sprawling plants, 4-6 oz., solid flesh with few seeds, looks like Oxheart. *Source History: 1 in 1994; 1 in 1998; 2 in 2004.* **Sources: Pr3, Sa9.**

Reif Italian Heart - 75-85 days, Tall, spindly, indet. plant, large, red, heart shaped fruit, 8-18 oz., meaty flesh with few seeds, full flavor, Italian origin. *Source History: 1 in 1998; 7 in 2004.* **Sources: He18, Jo1, Ma18, Ri12, Sa9, Se26, Ta5.**

Reigart - Indet., small plum shaped fruit, excellent flavor, slightly acid, high yields, German heirloom, pre-1900. *Source History: 1 in 2004.* **Sources: Lan.**

Rentita - Hungarian variety, beautiful rugose foliage, medium size red fruit. *Source History: 1 in 2004.* **Sources: He18.**

Rideau Sweet - 75 days, Indet., exceptionally sweet, the most flavorful regular size red cherry, resulted from a volunteer seedling of either Sweet 100 or Sweet Million. *Source History: 1 in 1998; 3 in 2004.* **Sources: Al6, Ma18, Te4.**

Rief's Red - *Source History: 1 in 1998; 1 in 2004.* **Sources: Sh13.**

Riesentraube - 76-85 days, German heirloom, name translates "giant bunch of grapes", compact indet. plant with good foliage cover, up to 350 flowers per cluster which produces bunches of 20-40 1.5 x 1.25 in. red fruits, flavor resembles a beefsteak tomato, may have been grown by the Pennsylvania Dutch as early as 1855. *Source History: 1 in 1994; 15 in 1998; 36 in 2004.* **Sources: Ap6, CO30, Co32, Coo, Ea4, Ers, Fo7, He8, He17, He18, Ho13, Lan, Ma18, ME9, Pr9, Ra5, Ra6, Re8, Ri12, Sa9, Se8, SE14, Se16, Sh13, Sk2, Sol, So25, Te4, TE7, Ter, To1, To3, To9, To10, Und, Yu2.**

Rio Fuego - 125 days, Det. vines, blocky red fruit, for fresh market or processing. *Source History: 1 in 1991; 3 in 1994; 4 in 1998; 1 in 2004.* **Sources: WE10.**

Rio Grande - 75-85 days, Large determinate plants, rugose foliage, 8-12 oz., fruits can weigh up to 5 lbs., high viscosity, well adapted to hot days and cold nights, withstands extremes in temp., res. to ASC FW1, FW2 and VW. *Source History: 1 in 1981; 1 in 1984; 4 in 1987; 7 in 1991; 6 in 1994; 9 in 1998; 11 in 2004.* **Sources: CA25, CO23, DOR, HA20, He18, Hud, Loc, Ma18, Ta5, To1, WE10.**

Robert Wolfe's - 75 days, Indet., 6-10 oz. red fruit, good flavor, good production through heat and drought. *Source History: 1 in 2004.* **Sources: Ta5.**

Rocket - 50-64 days, Older var., extremely early, fruits average 3-4 oz., color and flavor superior to most early varieties, recommended by U of Manitoba. *Source History: 3 in 1981; 3 in 1984; 7 in 1987; 3 in 1991; 1 in 1994; 2 in 1998; 3 in 2004.* **Sources: Eco, Ga1, Sa9.**

Rockingham - 75-80 days, Indet. plants, potato leaf foliage, firm red 6-8 oz. almost-round fruits, LB PI and TR1 res., for home gardens or market, adapted to Northern states, U of NH/1962. *Source History: 3 in 1981; 1 in 2004.* **Sources: He18.**

Rocky - 75-105 days, Re-introduction from Italy, very productive large red paste tomato, 1 lb., one of the best flavored sauce tomatoes, very rare, 1916. *Source History: 1 in 1991; 1 in 1994; 1 in 2004.* **Sources: To9.**

Roma - (Italian Roma) - 73-92 days, Italian paste and canning tomato, strong vigorous compact det. vines, firm meaty 1.5 x 3 in. long pear-shaped fruits, up to 200 per plant, few seeds, highly wilt res., USDA/Beltsville. *Source History: 33 in 1981; 30 in 1984; 31 in 1987; 22 in 1991; 21 in 1994; 23 in 1998; 32 in 2004.* **Sources: Ag7, Bo17, CA25, Ear, Fo7, He8, He17, He18, Hi13, HO2, Jo6, MAY, Mel, MIC, MOU, PAG, Pe2, Pin, Ra5, Ra6, Re8, Ri2, Sa5, Sa9, Se7, Se26, Shu, SOU, Sw9, TE7, Up2, WI23.**

Roma Long - 76 days, Det., 5 oz. pear-shaped fruits, VF res. *Source History: 1 in 1991; 1 in 1994; 2 in 1998; 1 in 2004.* **Sources: Jor.**

Roma VF - (Large Red Plum) - 70-78 days, Vigorous strong med-large det. plant, heavy cover, mild solid thick-walled red pear-to-plum fruits, few seeds, 2.5-3.5 x 1.5 in., 2-3 oz., AB VW FW1 and ASC res., solid meat, process or fresh market, fairly conc. set, popular. *Source History: 58 in 1981; 54 in 1984; 56 in 1987; 57 in 1991; 59 in 1994; 66 in 1998; 54 in 2004.* **Sources: All, Ba8, BAL, Be4, Bo19, Bu2, Bu3, CH14, Com, Cr1, Dam, De6, Ers, Fi1, Ge2, Gr27, GRI, Gur, HA3, Ha5, Hal, HPS, Kil, La1, Lej, Lin, LO8, Loc, Ma19, Mey, Mo13, Ni1, Ont, Pep, RIS, Roh, Ros, RUP, Sau, SE14, Se24, SGF, Sh9, Si5, So1, Sto, Syn, To1, To3, Ves, Vi4, WE10, Wet, Wi2.**

Roma VFN - 76 days, Main tomato grown primarily for paste and puree or adding body to juice, ample vine foliage to protect huge crops of bright-red plum-shaped fruits in large clusters, det., fruits are about 3 x 1.5 in. with red smooth tough skin that is easily peeled, meaty solid flesh with few seeds, excel. disease resistance, grows well caged.

Source History: 1 in 1981; 1 in 1984; 1 in 1987; 3 in 1991; 2 in 1994; 2 in 1998; 3 in 2004. **Sources: Bou, Bu8, ME9.**

Roma, Jumbo - 90 days, Indet., extra large meaty red fruit up to 1 lb., few seeds, productive. *Source History: 2 in 2004.* **Sources: Ma18, To9.**

Roma, Super VF - 70-90 days, Bred for making ketchup or soup or tomato juice, thick and juicy, nearly seedless, plum shape, widely accepted as a considerable improvement on Roma. *Source History: 1 in 1981; 1 in 1984; 1 in 1987; 1 in 1991; 1 in 1994; 1 in 1998; 2 in 2004.* **Sources: Loc, Tho.**

Romeo - The biggest Roma, up to 2 lbs., surpasses all other large roma varieties for earliness, uniform size, kitchen value and disease res., healthy vigorous indet. vines, PSR Breeding Program. *Source History: 1 in 2004.* **Sources: Pe6.**

Romeo and Juliette - Cross between Roma x hybrid Juliette that occurred spontaneously in a backyard garden, saved and stabilized, large indet., loaded with many small red fruit, Roma shape, excellent flavor. *Source History: 1 in 2004.* **Sources: Bou.**

Ropreco - (Ropreco Paste) - 70-85 days, Det. 4 ft. dia. vines need no staking or training, heavy yields of bright red fruit, cooks down well into sauce or paste, earlier than San Marzano with more manageable vines, excel. disease res. *Source History: 1 in 1981; 1 in 1984; 1 in 1987; 5 in 1991; 4 in 1994; 5 in 1998; 9 in 2004.* **Sources: He18, Pe2, Ra6, Se7, Se8, Se26, So9, Syn, Ta5.**

Rosalie's Large Paste - 80-90 days, Indet., red, 6-8 oz. fruit, 3 x 3 in., paste tomato that is juicy enough to make a good slicer, great flavor, family heirloom, CV Underwood Gardens introduction. *Source History: 1 in 1998; 4 in 2004.* **Sources: Ma18, Ta5, To9, Und.**

Rose - 78 days, Indet., large rose-red fruit, meaty flesh, rivals Brandywine for taste, Amish origin, original seed to Johnny's Seeds from John David Helsel in 1991 via his sister, Dr. Grace Kaiser, a physician to the Amish in New Holland, PA, given seeds in 1960 by Hannah Lapp. *Source History: 1 in 1998; 4 in 2004.* **Sources: Ami, Jo1, Scf, Ta5.**

Rosies - Indet., regular leaf, small fruit, tasty, early maturity. *Source History: 1 in 2004.* **Sources: Up2.**

Rossol - (Rossol VFN) - 70 days, Det., attractive large paste tomato resembles Roma. *Source History: 1 in 1998; 1 in 2004.* **Sources: He18.**

Rousich - 70 days, Det., potato leaf plant, red fruit. *Source History: 1 in 1998; 1 in 2004.* **Sources: Sa9.**

Row Pac - 80 days, Det., red salad tomato, excel. canner, dev. by WSU/USDA Exp. Sta. *Source History: 3 in 1998; 1 in 2004.* **Sources: Sa9.**

Royal Chico - (Royal Chico VFN) - 72-125 days, Productive Roma type, larger heavier more uniform crop, mech. harvest, vigorous compact med-large det. plant, good cover, firm bright-red meaty pear-shaped fruits, 2.8-3.5 oz., FW VW AB ASC RKN and nem. res., fairly conc. set, does well in humid areas, excel. for home gardeners and commercial processors, paste or can. *Source History: 18 in 1981; 17 in 1984; 11 in 1987; 9 in 1991; 4 in 1994; 5 in 1998; 5 in 2004.* **Sources: Ba8, HO2, Ra6, Sk2, WE10.**

RP 23-2 - 84 days, Indet., 1.75 in. globe, red skin and flesh, above average yield. *Source History: 1 in 1998; 1 in 2004.* **Sources: Sa9.**

Ruffled Red - 86 days, Ruffled hollow red fruits, 3 in. across, excel. for salads and stuffing. *Source History: 1 in 1998; 2 in 2004.* **Sources: Gr28, Sa9.**

Russian 117 - 74-90 days, Indet., wispy foliage, unusual double heart shape, red fruit, 4 x 5 in., 1-2 lbs., meaty flesh with few seeds, real tomato flavor. *Source History: 2 in 1998; 6 in 2004.* **Sources: Ap6, Ho13, Sa9, To1, To9, To10.**

Russian Big Roma - 85 days, Indet., 2 x 4 in. red paste, 10 oz., juicy, good sweet flavor, heirloom. *Source History: 1 in 1998; 2 in 2004.* **Sources: Ma18, To9.**

Russian Currant - 70-77 days, Not a true currant but a small bright red cherry borne in cascading clusters of 10-12 fruits, mild flavor, heavy yields. *Source History: 1 in 1998; 1 in 2004.* **Sources: Sa9.**

Russian Giant - 70 days, Indet., slightly oblong, pointed red fruit, very productive. *Source History: 1 in 2004.* **Sources: Ta5.**

Russian No. 33 - 70 days, Det. plant with fine, feather-like foliage, 2 oz., flattened fruit. *Source History: 1 in 1998; 1 in 2004.* **Sources: Sa9.**

Russian Red - 70-79 days, Med-tall upright bushes, thick heavy leaves and foliage, med-sized red globes, tolerates low temperatures, imported from New Zealand. *Source History: 4 in 1981; 4 in 1984; 3 in 1987; 3 in 1991; 3 in 1994; 3 in 1998; 3 in 2004.* **Sources: Pr3, Sa9, Tu6.**

Russo Sicilian - (Russo Sicilian Toggeta) - 70 days, Det. vines, red stuffing tomato with thick fleshy walls, hollow interior, 6 oz., intensely fragrant, from Sicily. *Source History: 2 in 2004.* **Sources: Ma18, To9.**

Rutgers - (Jersey) - 60-100 days, Large vigorous productive thick-stemmed det. vines, heavy foliage, bright blood-red 6-8 oz. globes, old-time flavor, deep color throughout, free from cracks, widely adapted, dev. by Campbell's in 1928 from a cross of Marglobe and J.T.D., later refined by the NJ/AES, released in 1940. *Source History: 68 in 1981; 57 in 1984; 66 in 1987; 56 in 1991; 57 in 1994; 52 in 1998; 65 in 2004.* **Sources: Ada, Ba8, Be4, Bo9, Bo19, Bou, Bu2, Bu8, CA25, CH14, Com, Cr1, De6, DOR, Ers, Fe5, Fi1, Fo7, Gr28, He8, He17, He18, Hi13, HO2, Hud, La1, Lej, LO8, Ma19, MAY, ME9, Mel, Mey, MIC, Mo13, MOU, Or10, PAG, Pe2, Ra5, Ra6, Re8, Ri12, Roh, Ros, RUP, Sau, Scf, SE14, Se26, SGF, Sh9, Shu, Sk2, SOU, Ta5, To1, To3, To10, Tu6, Up2, Vi4, WE10, Wi2, WI23.**

Rutgers Improved VF - 73-86 days, Adds FW and VW res. to old Rutgers lines, more compact, strong vigorous det. vines, generally upgraded large fruits, excel. early variety for canning and juice, Rutgers Univ., 1943. *Source History: 9 in 1981; 9 in 1984; 9 in 1987; 7 in 1991; 6 in 1994; 6 in 1998; 4 in 2004.* **Sources: BAL, Ge2, So1, Wet.**

Rutgers PS - (Rutgers PS-R) - 75-82 days, Same fruit as Rutgers, higher yields, less subject to cracking FW VW and ASC res., med-large det. plant, firm meaty fruits, excel. flavor. *Source History: 5 in 1981; 5 in 1984; 3 in 1987; 3 in 1991; 2 in 1994; 4 in 1998; 1 in 2004.* **Sources: GRI.**

Rutgers Select - (Rutgers S) - 73-82 days, Vigorous med-large indet. plants, thick stems, good cover, smooth firm red globes, 8 oz., thick walls, small cells, canner or green wrap shipper or general market. *Source History: 6 in 1981; 6 in 1984; 9 in 1987; 9 in 1991; 7 in 1994; 9 in 1998; 5 in 2004.* **Sources: HPS, Ra5, Ra6, RIS, To1.**

Ruth's Perfect - 80 days, Indet., beautiful red fruit, 7-8 oz., delicious, bio-dynamic origin. *Source History: 1 in 1994; 1 in 1998; 3 in 2004.* **Sources: Ta5, To9, Tu6.**

Sainte Lucie - 85 days, French variety, indet. plant bears large harvests of 1 lb. fruit, blemish-free, flavorful, solid meaty flesh, from Norbert Parreira, France. *Source History: 2 in 2004.* **Sources: To1, To9.**

Saladmaster - 64-75 days, Semi-det., 4-6 oz., flattened red globes, mild flavor, high yields, dev. at WSU-USDA Exp. Station. *Source History: 2 in 1998; 2 in 2004.* **Sources: Sa9, Ta5.**

Sallisaw Café - 75 days, Indet., high yields of bright red 1 in. cherries, super sweet flavor. *Source History: 1 in 2004.* **Sources: Ma18.**

Saltspring Sunrise - (Salt Spring Sunrise) - 70-80 days, Very early, small to medium sized red fruits, det. plants, dev. by the late J. James of Salt Spring Island off British Columbia. *Source History: 2 in 1981; 1 in 1984; 2 in 1987; 1 in 1991; 1 in 1998; 2 in 2004.* **Sources: Sa5, Ta5.**

San Francisco Fog - 70-75 days, Indet. vines, red 2 in. fruits with tough skin, mild flavor, adapted to cool wet California coast. *Source History: 1 in 1981; 1 in 1991; 1 in 1994; 5 in 1998; 4 in 2004.* **Sources: Ho13, Ta5, To1, To9.**

San Marzano - (Italian Canner, La Padrino) - 70-90 days, Large indet. vines, good cover, rectangular flat-sided intensely red fruits, 1.5 x 3.5-4 in. long, borne in clusters, holds well on vine or in storage, crack res., excel. for paste or puree or canning, processor favorite due to high solids, standard pear, Italy. *Source History: 45 in 1981; 38 in 1984; 36 in 1987; 32 in 1991; 25 in 1994; 28 in 1998; 40 in 2004.* **Sources: Ag7, Al25, Ber, Bo9, Bo19, But, Coo, Dam, Dom, Gr28, He8, Hi13, LO8, ME9, Mo20, Na6, PAG, Ra5, Ra6, Re8, Roh, Ros, Sa9, Se7, SE14, Se24, Se25, Se26, Sh9, So9, Sto, Sw9, Te4, TE7, To1, To3, To10, Tu6, Up2, WE10.**

San Marzano Lampadino - (San Marzano La Padrino) - 75-82 days, Improved San Marzano from Italy, vigorous indet., lobed fruits look like long bell peppers, excel. flavor, good disease res. *Source History: 1 in 1994; 1 in 1998; 2 in 2004.* **Sources: Ma18, Pin.**

San Marzano Redorta - 78-85 days, Heirloom from Tuscany, indet., larger fruit than San Marzano, 9-10 oz. average, fresh use and cooking, named for a mountain in Bergamo, Italy. *Source History: 2 in 2004.* **Sources: Se24, To1.**

San Marzano, Super - 70 days, Improved var. of San Marzano that has been dehybridized over a few years, 1 x 5 in. red paste, good yields. *Source History: 1 in 2004.* **Sources: To9.**

Sandia Gem - Medium-large red fruit, good flavor, midseason, seeds found in the Sandia mountain area in Peru in 1985, leather pouch holding the seeds carbon dated from early 1800s, 3 seeds germinated. *Source History: 1 in 2004.* **Sources: Gr28.**

Sandpoint - 58-70 days, Early, compact vines, 2-3 oz. fruits, good in the garden or in pots, U of Idaho. *Source History: 1 in 1981; 1 in 1984; 3 in 1987; 1 in 1991; 1 in 1998; 1 in 2004.* **Sources: Sa9.**

Santa - 58-63 days, Named for the town of Santa, Idaho, heavy fruit set, ripens uniformly, bright-red 1.5 oz. fruit, extremely fine flavor, distinct imp. over other super-early types, cool tol. *Source History: 2 in 1981; 1 in 1984; 2 in 1987; 1 in 1991; 1 in 1994; 1 in 1998; 2 in 2004.* **Sources: Roh, Ta5.**

Santa Clara Canner - (NSL 34243) - 80-90 days, Indet., 8-10 oz. red fruit, rich complex flavor, its history suggests it originated in Italy and was used in the California canning industry. *Source History: 2 in 1998; 4 in 2004.* **Sources: Ap6, He18, Ta5, To1.**

Santa Cruz Kada - 85 days, Midseason, indeterminate, very vigorous plant with good cover, firm red plum shaped fruit, good shipper, does well in foggy, cool climate and coastal areas. *Source History: 1 in 1981; 1 in 1984; 2 in 1987; 2 in 1991; 4 in 1994; 3 in 1998; 1 in 2004.* **Sources: Hud.**

Santiam - 58-75 days, Extra early, determinate 30 in. plants, mostly seedless 4-5 oz. fruits, good quality, mild, slightly acid, sweet and juicy, abundant yields, good home garden and fresh market, Dr. Baggett/Oregon State U. *Source History: 4 in 1987; 7 in 1991; 6 in 1994; 5 in 1998; 6 in 2004.* **Sources: Ga1, ME9, Se7, SE14, Up2, WE10.**

Santorini Paste - 75 days, Indet., abundant yields of 4 oz. red, flat-ribbed fruit, seedy flesh, great flavor, from the Island of Santorini in Greece, rare. *Source History: 1 in 2004.* **Sources: Ta5.**

Sasha's Altai - (Sasha's Altai Pride, Sasha's Pride) - 55-60 days, Indet. vines, med-size slightly flattened bright-red 3 in. fruits, 4-8 oz., juicy, very sweet flavor, Sasha traveled over 80 kilometers into the Altai Mountains to get what he believes is Siberia's best tomato. *Source History: 1 in 1991; 2 in 1994; 2 in 1998; 6 in 2004.* **Sources: Hig, Ra5, Ra6, Re8, Sa5, Ta5.**

Saucey - (Saucy Paste) - 65-85 days, Compact det., blocky red, paste type fruits, borne in clusters of 5-10 fruit, 2-3 oz., good flavor both fresh and processed, stores well for a couple of weeks on or off the plant, sets fruit easily under adverse weather conditions, Baggett/OSU. *Source History: 3 in 1994; 5 in 1998; 5 in 2004.* **Sources: Eco, Ga1, Ni1, Pe2, Ter.**

Sausalito Cocktail - 68 days, Semi-indet., clusters of 5-6 bright red .75 in. fruit, high shoulders rounding to a slight nipple on the bottom, slightly acidic, bold rich flavor, FW1 and V res., released by Dr. Martha Mutschler of Cornell University. *Source History: 1 in 2004.* **Sources: Ter.**

Sauvignac - Indet., 3 in. fruit, good yield, midseason, from Quebec. *Source History: 1 in 2004.* **Sources: Pr3.**

Schellenberg's Favorite - 110 days, Heirloom var. from the Schellenberg family near Mannheim, Germany, huge beefsteak fruit, 1-2 lbs., oval shape, red-orange color, crack res., large indet. vines. *Source History: 1 in 1998; 1 in*

2004. **Sources: He8.**

Scotia - 60-72 days, Det., deep-red 4-9 oz. fruits, sets well in cool weather, excel. for maritime, smooth and firm, good flavor, reliable, Kentville/Nova Scotia. *Source History: 6 in 1981; 6 in 1984; 8 in 1987; 9 in 1991; 7 in 1994; 11 in 1998; 8 in 2004.* **Sources: Dam, Hal, Hi6, Sa9, Sto, Ta5, Up2, Ves.**

Seache's Italian - 80 days, Indet., long red Italian paste, 2 x 6 in., orange-red, meaty flesh is nearly seedless, from Mr. Seache of Wisconsin. *Source History: 1 in 2004.* **Sources: Ma18.**

Sebastopol - 85 days, Vigorous large indet. vines, large red fruit, good flavor, grown for over 70 years by an elderly lady from Sebastopol. *Source History: 1 in 2004.* **Sources: To9.**

Semenarma Ljubljana - 75-80 days, Heavy crops of small to med. size bright red-orange beefsteak fruit, solid flesh, tasty sharp flavor when not fully ripe which develops into a sweet rich tomato taste. *Source History: 1 in 2004.* **Sources: Go8.**

Sementi - Indet., regular leaf, large fruit with mellow flavor, moderately productive, late. *Source History: 1 in 2004.* **Sources: Up2.**

September Dawn - 76-80 days, Det., 3.5 x 2.25 in. red fruit, good flavor, produces well, good fall market variety. *Source History: 1 in 1987; 3 in 1991; 1 in 1994; 1 in 1998; 1 in 2004.* **Sources: Ta5.**

Sequoia Alpine - 80-87 days, Semi-det. plants with 6 oz., globe shaped, tart red fruits, average yields. *Source History: 1 in 1998; 2 in 2004.* **Sources: Sa9, Ta5.**

Sergey's Ukrainian Red - Det. bush plant bears small round red-pink fruit, nice clean flavor, another of Sergey's family tomatoes, Ukraine. *Source History: 1 in 2004.* **Sources: Ami.**

Sheriff - Medium vine with good foliage cover, large fruit with square shape, high yields. *Source History: 1 in 2004.* **Sources: HA20.**

Sheyenne - 60-90 days, Det. plant, fair to good cover, medium to large red globes, extra early, general purpose, used primarily by home gardeners, ND/AES, 1960. *Source History: 3 in 1981; 2 in 1984; 1 in 1987; 1 in 1991; 1 in 1998; 2 in 2004.* **Sources: Sa9, Ta5.**

Shilling Giant - 90 days, Indet., huge oxheart type to 2 lbs., red flesh, few seeds, rich flavor. *Source History: 1 in 1998; 2 in 2004.* **Sources: Ho13, Ma18.**

Shirley Amish Red - 76 days, Indet., potato leaf, round, slightly flattened, 4 in., 12-16 oz., red-pink fruit, very good flavor. *Source History: 1 in 1998; 1 in 2004.* **Sources: To9.**

Shoshone - 49-64 days, Small det. plants, edible orange-ripe fruits at about 38 days, not uniform ripening, red med-thick skinned 1 oz. fruits in trusses, yield exceeds Sub-Arctics, flavor not intense. *Source History: 2 in 1981; 1 in 1984; 2 in 1987; 1 in 1991; 1 in 1998; 1 in 2004.* **Sources: Sa9.**

Shriver - 85 days, Indet., large pink-red fruit, 10-16 oz., sweet mild flavor, in a Morgantown, PA family for 80 years. *Source History: 1 in 2004.* **Sources: To9.**

Siberia - 40-75 days, Smuggled out in 1975 and given to the owner of Bjorkman's Greenhouses in Sundre, Alberta by a Russian traveler who said it was being grown experimentally in Siberia, seems to be able to set fruit at 38 deg. F., sturdy dark-green plants, frost susc., 40-60 meaty fruits, some 7 oz. if pruned to 15-20, does well in shade, good container plant, valuable for Alaska or Yukon or mountain areas. *Source History: 1 in 1981; 1 in 1984; 7 in 1987; 6 in 1991; 7 in 1994; 16 in 1998; 21 in 2004.* **Sources: CA25, Eco, Ers, Fo7, Ga1, Goo, Gr27, Ho13, HPS, Hum, Pe2, Ra5, Ra6, Re8, Sa9, Shu, So25, Ta5, To3, Tt2, Vi4.**

Siberian - 40-70 days, Russian, dwarf sprawling plant, early, egg-shaped, high color, heavy crops of 1.5 oz. fruit with good strong flavor, great success at Churchill, Manitoba and even on Baffin Island. *Source History: 1 in 1981; 1 in 1984; 6 in 1987; 7 in 1991; 6 in 1994; 11 in 1998; 14 in 2004.* **Sources: Ba8, CO30, Hig, La1, ME9, Pr3, Sa9, Se8, SE14, Se16, Se26, Sk2, To1, WE10.**

Siberian, Early - 57-60 days, Indet., 3-4 oz. round red fruit, from Siberia. *Source History: 1 in 1994; 1 in 1998; 2 in 2004.* **Sources: He18, Hig.**

Sicilian Saucer - Italian heirloom, very large flattened beefsteak fruit, paste type, not pretty but makes a great sauce. *Source History: 1 in 2004.* **Sources: Dam.**

Sierra Sweet - 80 days, Heirloom. *Source History: 1 in 1998; 1 in 2004.* **Sources: Ga1.**

Sigma Bush - *Source History: 1 in 1998; 1 in 2004.* **Sources: WE10.**

Siletz - 52-75 days, Improved Oregon Spring type with a somewhat more vigorous plant and earlier ripening fruit, 8-16 oz., blemish free, det. vines, resistant to V and F1, better resistance to cracking and cat facing, from Dr. James Baggett/OSU. *Source History: 8 in 1998; 14 in 2004.* **Sources: Eco, Ga1, Goo, Gr28, He18, Hig, Hum, Ni1, Pe2, Pr3, St18, Ter, To1, We19.**

Silvery Fir Tree - (Carrot Top Tomato) - 58 days, Unique, 24 in. plant with unusual dissected foliage that looks like a carrot top, heavy crops of round, flattened 2-3.5 in. red fruit, Russian origin. *Source History: 1 in 1994; 12 in 1998; 23 in 2004.* **Sources: Ami, CO30, Co32, Ga1, Ga22, Go8, Hi6, Hig, Pr3, Ra5, Ra6, Re8, Ri12, Sa5, Se7, Se16, Se26, Sh13, Ta5, To1, To3, Up2, Yu2.**

Sioux - 70-80 days, Strong med-large indet. vines, med-heavy open foliage, solid meaty dark-red med-sized smooth globes, 4-6 oz., crack res., very productive, ripens uniformly, good eating, dependable fruit set in hot weather, for home gardens or early fresh market in the Midwest and Upper Mississipi Valley, Allred x Stokesdale, NE/AES, 1944. *Source History: 7 in 1981; 5 in 1984; 5 in 1987; 3 in 1991; 1 in 1994; 1 in 1998; 6 in 2004.* **Sources: He18, Ma18, Sa5, Sa9, To1, Und.**

Sioux, Super - (Improved Sioux, Super Lakota) - 70 days, Rather open med-sized indet. plant, fair cover, smooth thick-walled fleshy 4 oz. red globes, ASC res., sets well during high temps, good for hot dry areas, old-time acid flavor, All Red x Stokesdale developed at Nebraska Exp. Sta. *Source History: 18 in 1981; 18 in 1984; 16 in 1987; 9 in 1991; 9 in 1994; 7 in 1998; 23 in 2004.* **Sources: Ap6, Bu3, CO30, Ea4, Ers, He8, He17, He18, ME9, Pe2, Ra5, Ra6,**

Re8, Se7, SE14, Se26, Sh13, Sk2, So9, Ta5, To1, To3, To10.

Skorospelka - Russian variety, stocky bushes, 2-3 oz. fruits borne in clusters of 6, high quality, early maturity. *Source History: 1 in 2004.* **Sources: Pe1.**

Slava - 65 days, Indet., potato leaf, round, 1-2 oz. fruit, pointed end, some late blight res., heavy producer, from Czech Republic, Slava translates "Glory". *Source History: 1 in 1998; 4 in 2004.* **Sources: Co32, In8, Se26, So25.**

Sojourner South American - 85-95 days, Indet., regular leaf, bright red, large blocky plum shaped fruit, sweet flavor, good fruit set in heat and drought. *Source History: 1 in 2004.* **Sources: He18.**

Sophie's Choice - 55-70 days, Det. plant, 18-24 in. tall, extra early ripening, bears large, 6-12 oz. fruit, orange-red skin, deep red flesh, does not handle heat or drought well, heirloom from Edmonton, Canada. *Source History: 1 in 1998; 2 in 2004.* **Sources: Ap6, So1.**

Spanish Sun - 80 days, Indet., medium size fruit, red skin with slight orange cast, 8 oz., nice flavor and appearance. *Source History: 1 in 2004.* **Sources: Ta5.**

Spitze - 75 days, Indet., long, pointed fruit resembles a banana pepper, 4-5 in. long and 2 in. across, paste tomato that is excel. for sauce as well as fresh eating, Romania. *Source History: 2 in 1998; 1 in 2004.* **Sources: Se16.**

Spoon - 65-90 days, Very small fruit about half the size of currant tomatoes, high sugar content, crack res., excel. dried for sauces and Italian dishes. *Source History: 3 in 1998; 2 in 2004.* **Sources: Ma18, To10.**

Spring King - 60 days, Compact det. vines with excellent leaf cover, firm, globe to slightly oblate fruits are crack and blemish free, 10-16 oz., tart flavor, good disease res., large early yields, from PSR Breeding Program. *Source History: 1 in 1994; 1 in 1998; 2 in 2004.* **Sources: Bou, Pe6.**

Sprint - (Kotlas) - 59 days, Extra early, indeterminate, somewhat sparse foliage, round red 2 oz. fruits with green shoulders, sweet delicious flavor, tremendous staying power due to its good early blight resistance, should be staked and only minimally pruned. *Source History: 2 in 1987; 3 in 1991; 3 in 1994; 3 in 1998; 3 in 2004.* **Sources: Se8, Tu6, Up2.**

Square Paste - 75 days, Firm red fruit with slightly angular shape, 2 in., earliest of the California mechanical harvest processing types. *Source History: 2 in 2004.* **Sources: Ra5, Ra6.**

St. Pierre - 74-88 days, Indet. vines, tender, thin-skinned 2 in. red fruit, green shoulders, sweet rich tomato flavor, good producer under adverse weather conditions, heirloom. *Source History: 1 in 1991; 4 in 1994; 7 in 1998; 12 in 2004.* **Sources: Ba8, Ers, Gou, He18, Ra5, Re8, Sa9, Se7, Sk2, Ta5, To9, WE10.**

Stakebreaker - 85 days, Indet., large yields of 4 in. round red fruit in clusters of 3-4, excellent flavor, late. *Source History: 1 in 2004.* **Sources: To9.**

Stakeless - (Red House) - 75-85 days, Needs no staking, 18-24 in. bush plants hold fruits off ground, dense potato-leaf foliage, firm meaty 6-8 oz. red globes, FW AB and crack res., for hot summer areas, Delaware AES. *Source History: 9 in 1981; 9 in 1984; 13 in 1987; 11 in 1991; 7 in 1994; 5 in 1998; 2 in 2004.* **Sources: MIC, Sk2.**

Starfire - 55-65 days, Bushy det. plants with strong branches, no staking required, solid meaty 5-8 oz. bright-red flat fruits, good interior color, few seeds, not inclined to sunscald, sets fruits in cold conditions, good on sandy soils, uniform, dev. at Morden Exp. Farm in Manitoba for the Canadian great plains and areas farther east, 1963. *Source History: 20 in 1981; 15 in 1984; 13 in 1987; 13 in 1991; 6 in 1994; 3 in 1998; 2 in 2004.* **Sources: Tt2, Yu2.**

Starling - 81 days, Indet., globe shaped fruit, solid red flesh, 12 oz., crack resistant. *Source History: 1 in 2004.* **Sources: Sa9.**

Stone - 70-90 days, Old reliable canner, large vigorous indet. vines, smooth solid bright deep-scarlet oval fruits, 10 oz., heavy yields, uniform ripening, wilt res., remarkable holding qualities, introduced by the Livingston Seed Company of Columbus OH, late 1800s, original plant was found by Mr. Nichols between rows of Favorite and Beauty tomatoes. *Source History: 2 in 1981; 2 in 1984; 5 in 1987; 5 in 1991; 8 in 1994; 7 in 1998; 12 in 2004.* **Sources: Bo9, Fo7, He8, Ma18, Ra5, Ra6, Re8, Sa9, So1, Ta5, To3, To9.**

Stone, Dwarf - 85 days, Compact plants with sparse rugose foliage, 6-8 oz. fruits are slightly smaller than standard Stone, smooth skin, prolific, uniform ripening, released by Livingston in 1902. *Source History: 2 in 2004.* **Sources: Ea4, Vi4.**

Stuffing - (Pepper Tomato, Hollow Tomato, Vera Tomato Pepper) - 78-90 days, Productive indet. plants, bright-red firm hollow fruits, 2.75 x 3 in. dia., matures hollow like bell peppers, mild flavor, for stuffing with cold salads or slicing. *Source History: 8 in 1981; 3 in 1984; 6 in 1987; 2 in 1991; 3 in 1994; 2 in 1998; 4 in 2004.* **Sources: Ra6, Se26, To3, Vi4.**

Stuffing, Burgess - 78 days, Red tomato bred specifically for stuffing, core is almost completely hollow, 2.5 x 3+ in. long, decorative vertical exterior ribs, can be stuffed and baked or served raw, indet. *Source History: 2 in 1981; 2 in 1984; 4 in 1987; 4 in 1991; 2 in 1994; 3 in 1998; 2 in 2004.* **Sources: Fo7, To1.**

Stupice - 52-85 days, Received in 1976 from Milan Sodomka, Czechoslovakian tomato breeder, indet. potato leaf vines to 4 ft., small to med-sized, 3-6 oz. fruit, very early, exceptional flavor, good yielder, good in cool weather. *Source History: 1 in 1981; 1 in 1984; 4 in 1987; 4 in 1991; 11 in 1994; 21 in 1998; 49 in 2004.* **Sources: Al6, Ap6, Bou, Co32, Coo, Ea4, Ear, Fo7, Ga1, Ga22, Go8, He8, He18, Hi6, Hig, Ho13, In8, Loc, Ma18, ME9, Na6, Pe1, Pe2, Pe6, Pr9, Ra5, Ra6, Re8, Ri12, Sa9, Se7, SE14, Se16, Se17, So1, St18, Te4, TE7, Ter, To1, To3, To9, To10, Tt2, Tu6, Up2, Vi4, We19, Yu2.**

Stuse - 85 days, Indet., potato leaf plant, 3.5 x 1.75 in., flattened globes, red flesh, excel. yield. *Source History: 1 in 1998; 1 in 2004.* **Sources: Sa9.**

Sub-Arctic Cherry - 40-45 days, Ripens huge clusters of .5 in. cherry tomatoes on small compact plants, just under .5 oz., very early, Beaverlodge Research Station, Alberta. *Source History: 3 in 1981; 2 in 1984; 1 in 1987; 1 in 1991; 1 in 1994; 1 in 1998; 1 in 2004.* **Sources: Gr28.**

Sub-Arctic Early - 53 days, Det., small 1.4 in. dia. cherry tomatoes, 1 oz., ripens almost the entire crop extra early, develops several trusses, crack res., acceptable eating quality. *Source History: 4 in 1981; 2 in 1984; 1 in 2004.* **Sources: Ga1.**

Sub-Arctic Maxi - 48-64 days, Early dwarf bushes, concentrated clusters of deep-red 2.5-3 oz. fruits, very susc. to early blight, largest and latest Sub-Arctic type, very heavy producer, the Beaverlodge Sub-Arctics all have a unique growth habit, they abort the main stem and then produce quickly and heavily on fast-growing lateral branches. *Source History: 17 in 1981; 10 in 1984; 16 in 1987; 14 in 1991; 5 in 1994; 5 in 1998; 8 in 2004.* **Sources: Alb, Pr3, Ra5, Ra6, Re8, Sto, To1, To3.**

Sub-Arctic Plenty - (World's Earliest) - 40-59 days, Very early, upright det. plants, upright stems packed with 1.6-1.9 in. fruits, 1.5-2 oz., 75% ripen at once, excel. cold set ability, heavy yields almost anywhere, if started indoors can succeed in southern Yukon, BER res., Dr. Harris, Beaverlodge Research Sta., Alberta. *Source History: 15 in 1981; 14 in 1984; 9 in 1987; 10 in 1991; 11 in 1994; 11 in 1998; 12 in 2004.* **Sources: Alb, Dam, Ers, Fi1, Gr27, Gur, Hig, ME9, Pr3, SE14, Tho, Tt2.**

Sugar Beefsteak, Potato Leaf - 70 days, Indet., sweet beefsteak fruit. *Source History: 1 in 2004.* **Sources: Ma18.**

Sugar Lump - (Extra Early Sugar Lump, Jung's Sugar Lump) - 65-80 days, Extra early, heavy yields until frost, smooth, sweet, deep-red fruits, 1.5-2 in. dia., clusters of 6-12, excel. quality and flavor, for canning whole or salads or juice, heirloom from the 1800s. *Source History: 6 in 1981; 5 in 1984; 8 in 1987; 5 in 1991; 5 in 1994; 7 in 1998; 12 in 2004.* **Sources: CO30, Ers, Ga1, He8, ME9, Ra5, Ra6, Re8, SE14, Shu, Sk2, To3.**

Sunset's Red Horizon - 70-80 days, Heirloom native to Southern Russia, indet., wispy vines resemble that of Anna Russian, large red fruit, 4-5.5 in., meaty flesh, excellent flavor, resists blossom end rot and cracking, bears into November in Oregon, named after Sunset magazine, brought to the U.S. in 1999 by Russian immigrant, Nik Peplenov, from the Rostov Don region of Russia. *Source History: 1 in 2004.* **Sources: To9.**

Super Italian Paste - 73-97 days, Italian heirloom, indet. plant, 10 oz., 6 in. long paste tomatoes, plant tends to wilt. *Source History: 1 in 1981; 1 in 1984; 1 in 1987; 2 in 1991; 1 in 1994; 6 in 1998; 8 in 2004.* **Sources: Bu2, He18, Hi13, Ma18, ME9, SE14, Sh13, To1.**

Superfantastic - 70 days, Open-pollinated version of hybrid Fantastic with same excellent flavor, 4-6 oz., red fruit, 2 ft. det. vines. *Source History: 1 in 1998; 2 in 2004.* **Sources: Ag7, Te4.**

Sweet 100 OP - (Sweet Cherry OP) - 75 days, Hybrid dev. at least 20 years ago, we have had an open-pollinated line for many years now, good sweet-acid balance, highest vitamin C content of any cherry tomato, indet. *Source History: 2 in 1991; 1 in 2004.* **Sources: Ma13.**

Sweet Million - Open-pollinated cherry type with wonderful tangy flavor, large spreading plant with wispy foliage, high vitamin C content, early. *Source History: 1 in 1994; 1 in 2004.* **Sources: Up2.**

Sweet n' Bright - Indet., red, 3 oz., perfect fruit, disease resistant, midseason, bio-dynamic origin. *Source History: 1 in 1994; 1 in 1998; 1 in 2004.* **Sources: Tu6.**

Sweet Olive - 57 days, Indet., grape type fruit with unique olive shape, red fruit borne in cascading trusses, no splitting. *Source History: 4 in 2004.* **Sources: Hi13, Ra5, Ra6, Re8.**

Sweetie - 50-79 days, Large indet. vines, very sweet 1-1.5 in. dia. globes, 12-14% sugar content, use for juice or preserves with no added sugar, clusters of 15-20, everbearing, ASC res., less than 1 oz., for home gardeners. *Source History: 17 in 1981; 17 in 1984; 20 in 1987; 28 in 1991; 26 in 1994; 25 in 1998; 27 in 2004.* **Sources: Dam, Ga1, Gr28, Gur, HO2, Hum, Kil, Lin, Me7, ME9, Ni1, Ont, Pe2, Ra5, Ra6, Ri2, RUP, Se8, SE14, Se26, Sh9, SOU, Ta5, TE7, Ter, To3, Tt2.**

Swift - 54 days, Det. vines, no staking, smooth, 3 oz., ripens uniformly, high yields, for high altitude home gardens in Canadian prairies and northern B.C., Swift Current Exp. Sta. *Source History: 4 in 1981; 4 in 1984; 6 in 1987; 6 in 1991; 4 in 1994; 1 in 1998; 1 in 2004.* **Sources: Pr3.**

Swiss Alpine - 80 days, Old Swiss variety, productive indet. vines, 4-5 oz. fruit, red flesh with true tomato flavor, sets fruit in cool weather, from the 1800s, CV Heirloom Seeds introduction. *Source History: 1 in 1994; 1 in 1998; 2 in 2004.* **Sources: He8, Ta5.**

Sybirski Wczesny - 77 days, Det. 2 ft. plants, round red salad size fruit, from Russia. *Source History: 1 in 2004.* **Sources: Ta5.**

Sylvan Gaume - 80 days, Indet., huge, heart-shaped red fruit up to 3 lbs., dense meaty flesh, few seeds. *Source History: 1 in 2004.* **Sources: Ma18.**

Sztambowyj Karlikowy - 75 days, Det., bushy 3 ft. plant, salad size fruit, red-orange skin, red flesh, good flavor, from Russia. *Source History: 1 in 2004.* **Sources: Ta5.**

Table Talk, Burpee's - 78 days, Det. plants, 3 in. globe fruit. *Source History: 1 in 1998; 1 in 2004.* **Sources: Sa9.**

Tadesse - 77 days, Indet., 2-3 in., round, red-pink fruit, nice acid flavor, from Ethiopia. *Source History: 1 in 1998; 1 in 2004.* **Sources: To9.**

Tanana - (Early Tanana) - 60-80 days, Very early det. var., cold res. plants hold their foliage well, uniform red 3.5-4 oz. fruits will not sunscald, University of Alaska. *Source History: 2 in 1981; 1 in 1984; 2 in 1987; 2 in 1991; 1 in 1998; 1 in 2004.* **Sources: Sa9.**

Taos - 65 days, Early indet., round red fruit, 3 in., meaty flesh with mild flavor, very productive. *Source History: 1 in 2004.* **Sources: To9.**

Tappy's Heritage - 85 days, Indet., selected from a cross of Tappy's Finest and a red-fruited tomato, 6 oz. fruits ripen from pink > orange-red > deep red on the bottom with orange-red shoulders, borne in clusters of 2-4, sweet full flavor, meaty flesh, above average insect and disease res. *Source History: 1 in 1998; 2 in 2004.* **Sources: Ba8, So1.**

Taps - 80-85 days, Robust, healthy indet. plants produce large, firm pink-red fruits, 1-2 lbs., good fresh or cooked.

Source History: 1 in 1994; 1 in 1998; 3 in 2004. **Sources: He18, Pe7, To9.**

Tartar from Mongolstan - Large red fruit, delicious, very rare. *Source History: 1 in 2004.* **Sources: Co31.**

Teardrop - Indet., shape resembles a tear drop, nice firm flesh, no cracks or splits. *Source History: 1 in 2004.* **Sources: Ma13.**

Teepee - 75 days, Indet., sprawling vines, 6-8 oz., scalloped fruit, excel. flavor, from Navajo Indian Tribe. *Source History: 1 in 1998; 1 in 2004.* **Sources: Ma18.**

Teton de Venus - 96 days, Indet. vine with finely cut sparse foliage, heart-shaped fruit with nippled ends borne in pairs, thick, pink-red flesh, very sweet flavor. *Source History: 3 in 2004.* **Sources: Ma18, Sa9, Yu2.**

The President - 67 days, Flavorful Russian variety, det., 3 ft. bushy plant, round red fruit, good salad size. *Source History: 1 in 2004.* **Sources: Ta5.**

Thessaloniki - 66-80 days, Developed in Greece, indet. plant needs staking, uniform baseball-sized smooth solid red fruits, 4-6 oz., res. to sunburn and cracks and spots, perfect blossom ends, mild, will not rot when ripe, introduced to the U.S. by Glecklers Seedsmen of Ohio in the 1950s. *Source History: 1 in 1981; 1 in 1984; 2 in 1987; 2 in 1991; 3 in 1994; 9 in 1998; 27 in 2004.* **Sources: Ba8, CO30, Ers, Ga1, Gr28, He18, HO2, Hud, Ma18, ME9, Na6, Pep, Ra5, Ra6, Re8, Ri12, RUP, Se7, SE14, Se17, Se26, So25, Ta5, To1, To3, To10, WE10.**

Three Sisters - 75-85 days, Seeds of Change o.p. breeding project, includes three distinct forms, single-pleated salad size, a Roma type and a pleated flattened globe type, indet., potato leaf foliage. *Source History: 1 in 2004.* **Sources: Se7.**

Tibet - Attractive red-purple fruit, 6 oz. *Source History: 1 in 1998; 1 in 2004.* **Sources: Sa5.**

Tibet-Appel - 80 days, Bushy det. plant bears med. size round pink-red fruit, thin skin, great taste, originated in Tibet. *Source History: 1 in 2004.* **Sources: Go8.**

Tiny Tim - 45-75 days, Small tree-like det. plants, 8-16 in. tall x 6-12 in. dia., .75-1.5 in. scarlet fruits, used primarily for container culture, matures fruits when planted in a 3 in. pot, ASC and ST res., rugose leaf, ornamental and productive, for pots and small spaces, U of NH, 1945. *Source History: 78 in 1981; 75 in 1984; 62 in 1987; 47 in 1991; 37 in 1994; 31 in 1998; 34 in 2004.* **Sources: Alb, All, Bo19, Bu1, Cr1, Dam, Ear, Ers, Fo7, Fo13, Ge2, Hal, He8, Lin, ME9, MIC, Ont, Pin, Ra5, Ra6, Re8, SE14, Se26, So1, So25, Sto, Ta5, To1, To3, Tt2, Up2, Ves, Vi4, WE10.**

Tip-Top - 74 days, Determinate plants require no staking, 3 oz. fruits almost perfectly round above a distinctive tip, borne in large clusters but much larger and sweeter than cherry or small paste types, thick walls and low water content, excellent for salads and also useful for paste. *Source History: 1 in 1987; 1 in 1991; 1 in 1994; 3 in 1998; 2 in 2004.* **Sources: Fo7, Pin.**

Tlacolula Ribbed - 85 days, Indet., flattened, deeply ribbed red-pink fruit, 5 oz., occasional light green striping, moderately juicy, firm flesh is slightly hollow, Mexican heirloom. *Source History: 2 in 2004.* **Sources: So25, To9.**

Togorific - 83 days, Indet., small, ruffled red fruit, very productive, from Iran-Iraq area. *Source History: 1 in 2004.* **Sources: Sa9.**

Tolli Roma - 94 days, Semi-det., 3 oz. paste tomato. *Source History: 1 in 1998; 1 in 2004.* **Sources: Sa9.**

Tom Patti's Italian Paste - 85 days, Indet., resembles Amish Paste but has better flavor, solid red heart shaped fruit, high yields, seed brought from Italy by Tom Patti in the early 1960s. *Source History: 2 in 2004.* **Sources: Ta5, To9.**

Tommy Toe - 70-80 days, Indet., hundreds of apricot-size fruits produced over an extended period, excellent flavor, salads and juice, resists early and late blight, heirloom from the Ozark Mountain region. *Source History: 2 in 1998; 21 in 2004.* **Sources: Coo, Ga1, Go8, He18, HO2, Ma18, ME9, Pr9, Ra5, Ra6, Re8, Ri12, Sa9, SE14, Se16, Ta5, Ter, To1, To3, To9, To10.**

Tonadose des Conores - Bright red cherry fruit, excellent sweet flavor, high yields, French heirloom. *Source History: 2 in 2004.* **Sources: Ba8, Ho13.**

Tondino di Manduria - 80 days, Semi-det., 2-2.5 in. small red fruit, oblong shape, pointed blossom end, flesh has more juice than Roma, originally from Italy. *Source History: 1 in 2004.* **Sources: To9.**

Tony's Italian - 80 days, Indet., solid fleshed, 4-6 oz. scarlet plums, few seeds, brought to New Jersey in the early 1920s by a Sicilian family. *Source History: 1 in 2004.* **Sources: Ma18.**

Trauffaut Precoce - Small round red fruit on plants that lack vigor but the sweet taste makes it worth growing, from a French gardener via J. L. Hudson. *Source History: 1 in 2004.* **Sources: Hud.**

Trimson - 75-77 days, Sister line to Moira, larger 6-6.5 oz. fruits, mild sweet flavor, not as uniform in poor weather, very deep-red flesh, vigorous, det., cool climate canner for eastern Ontario. *Source History: 2 in 1981; 1 in 1984; 1 in 2004.* **Sources: He18.**

Trip-L-Crop, Burgess - (Climbing Trip-L-Crop) - 90 days, Has been known to climb 25 ft., large crimson fruits grow to 6 in. across, meaty, good for slicing and canning, very productive, up to 2 bushels per plant, not disease res. *Source History: 1 in 1981; 1 in 1984; 2 in 1987; 3 in 1991; 2 in 1994; 3 in 1998; 4 in 2004.* **Sources: Bu1, SE14, To1, Vi5.**

Trophy - 78-80 days, Indet., clusters of med. size red globes, 5-7 oz., sweet mild flavor, early, introduced in 1870 by Colonel George Waring of Newport, RI, who sold them for $5 per packet, purchased by many in pursuit of the $100 reward for largest specimen tomato, dev. by Dr. Hand of Baltimore, MD. *Source History: 5 in 2004.* **Sources: Co32, Ea4, Sa9, Se16, Up2.**

Tropic - (Tropic VFN) - 80-85 days, Large strong indet. vines, good cover, solid smooth 8 oz. flattened globes, GLS FW1 VW LM GW ST ASC and TMV1 res., TMV4 tol., market pole type for fall and spring greenhouse production or outdoor growing in the South or Southwest, also used for hydroponic crops, pink harvest, exel. shipper, Florida State U. *Source History: 26 in 1981; 24 in 1984; 24 in 1987; 12 in 1991; 9 in 1994; 10 in 1998; 11 in 2004.* **Sources: Ech,**

He18, Kil, La1, ME9, Ra5, Sa9, SE14, So1, To1, WE10.

TSAC - 73-80 days, Medium size red cherry, high yields. *Source History: 1 in 1998; 1 in 2004.* **Sources: Sa9.**

Tuscanini - 85 days, Indet., large flavorful paste type, 12 oz., meaty flesh, few seeds. *Source History: 1 in 2004.* **Sources: Ta5.**

UC 82 B - 80 days, Med-sized det. plants, good cover, med-small red square-round fruits, res. to VW FW and ASC, mid- season, very firm, excel. holding ability, UC/Davis for West Coast. *Source History: 3 in 1981; 2 in 1984; 3 in 1987; 3 in 1991; 3 in 1994; 3 in 1998; 1 in 2004.* **Sources: WE10.**

Uncle Ike's Big Red - 90 days, Indet., large round red fruits. *Source History: 1 in 1998; 1 in 2004.* **Sources: Se17.**

Urbikany - (Czech Select) - 55-70 days, Received in 1976 from Milan Sodomka, Czechoslovakian tomato breeder, early, good quality, prolific very compact det. plant, semi-flat mild red 3 oz. fruits. *Source History: 2 in 1981; 2 in 1984; 2 in 1987; 3 in 1991; 3 in 1994; 3 in 1998; 3 in 2004.* **Sources: He8, Hig, Sa9.**

V. R. Moscow - 70-95 days, Outstanding canner dev. at Utah State for intermountain area, med-large smooth deep-red fruits, thick walls, fine quality, VW res., prolific in foliage and fruit. *Source History: 1 in 1981; 1 in 1984; 1 in 1987; 1 in 1998; 2 in 2004.* **Sources: Sa9, Ta5.**

Valiant - 65-80 days, Fairly open spreading indet. vines, fair cover, very smooth solid meaty 15 oz. globes ripen dark-red all at once, produces heavily even in hot dry weather, good color, mild flavor, for juice or slicing or canning or shipping, adapted to the North and Canada, also popular for home and market in the East, introduced in 1937. *Source History: 25 in 1981; 19 in 1984; 13 in 1987; 6 in 1991; 4 in 1994; 1 in 1998; 2 in 2004.* **Sources: Sa9, Ta5.**

Valle Nacional Oaxaca - Traditional Mexican variety, large, thick walled red cherry, flesh is slightly sweet with acid undertones. *Source History: 1 in 2004.* **Sources: Ea4.**

Valley Girl - 80-90 days, Productive bush type produces flavorful, medium-large red, firm fruit, smooth skin, crack resistant, sets in hot and cool conditions, midseason. *Source History: 1 in 1994; 2 in 2004.* **Sources: Ma18, Ta5.**

Varibest - Indet., med. size fruit, great for fresh eating or canning, highly productive. *Source History: 1 in 2004.* **Sources: Up2.**

Vatan - 77 days, Indet., round red fruit, salad size, from Russia. *Source History: 1 in 2004.* **Sources: Ta5.**

Veeroma - 72 days, Early Roma type but better-yielding, det., crack res., medium-red, adapted to mech. harv. for ketchup or paste, also whole pak or pick your own, FW and VW tol. *Source History: 2 in 1981; 2 in 1984; 3 in 1987; 4 in 1991; 2 in 1994; 2 in 1998; 1 in 2004.* **Sources: Sto.**

Vendor - 68-83 days, Semi-compact indet., greenhouse type, very firm, 6-10 oz., bright-red, egg-shaped maturing to globe, tol. to TM and most leaf molds, crack-free, for fall production, stake, uniform ripening. *Source History: 10 in 1981; 8 in 1984; 13 in 1987; 9 in 1991; 6 in 1994; 4 in 1998; 4 in 2004.* **Sources: Ear, Gr28, Hig, Ont.**

Vendor VFT - 76 days, Dr. Munger of Cornell has bred VW and FW res. into Vendor, some tol. to TM and LM, smooth bright-red 4-6 oz. fruits, small stem scar, excel. eating quality, indet. *Source History: 1 in 1981; 1 in 1984; 2 in 1987; 2 in 1991; 1 in 1998; 1 in 2004.* **Sources: Sto.**

Ventura - (Ventura Paste) - 78 days, Compact det. plants, fairly good cover, firm pear-shaped red fruits, FW and GLS res., very concentrated set, good color, processor, earliest mech. harv. pear, from Bulgaria. *Source History: 3 in 1981; 2 in 1984; 2 in 1987; 1 in 1991; 1 in 2004.* **Sources: Ta5.**

VF 134-1-2 - (UC 134 VF) - 128 days, Vigorous med-size det. plant, good cover, solid smooth square 3 oz. fruits, med- maturity, VW FW1 ASC ST res., good vine storage and bulk handling, widely adapted. *Source History: 4 in 1981; 4 in 1984; 5 in 1987; 2 in 1991; 1 in 1994; 2 in 1998; 1 in 2004.* **Sources: WE10.**

VF 145B-7879 - (Strain B) - Highly det. small vines, good cover, firm 5-6 oz. deep globes, FW and VW res., for processing and mech. harv., can be direct seeded, fine flavor, popular in California. *Source History: 7 in 1981; 6 in 1984; 5 in 1987; 3 in 1991; 1 in 1994; 1 in 1998; 1 in 2004.* **Sources: WE10.**

VF 6203 - Mech. harv., med-sized det. plant, red square-round 2.5-3 oz. fruits, med-early FW1 VW ASC res., excel. yields even after rains, very thick walls, conc. harvest. *Source History: 2 in 1981; 2 in 1984; 4 in 1987; 5 in 1991; 2 in 1994; 2 in 1998; 1 in 2004.* **Sources: WE10.**

Victor - 65-70 days, AAS/1941, small det. plants, fair cover, med-small deep-scarlet flattened globes, rather rough, uniform ripening, high yields, for home gardens in short season areas. *Source History: 3 in 1981; 2 in 1984; 1 in 1987; 1 in 1991; 1 in 1994; 1 in 1998; 2 in 2004.* **Sources: Pr3, Sa9.**

Victoria - 79 days, Highest yielding of the potato leaf plants, flat, red, 7 oz. fruits. *Source History: 1 in 1998; 2 in 2004.* **Sources: Sa9, Ta5.**

Victory - 72 days, Indet. vines, red-orange fruit, 3 in. dia., 6-8 oz., slightly flattened shape, surprising flavor with good blend of acidity and sweetness, from Sakhalin Island, USSR. *Source History: 2 in 1994; 1 in 1998; 1 in 2004.* **Sources: Vi4.**

Viktorina - 60 days, Early Russian variety for home use, med. size red fruit, thick skin, flavor is a combination of sweet and tang, fruit holds well due to thicker skin and resistance to fruit diseases, produces fruit longer than other Russian varieties. *Source History: 1 in 2004.* **Sources: So1.**

Vilms - 85-95 days, Extremely productive plant, small paste tomatoes, 2-3 in. long. *Source History: 1 in 1998; 1 in 2004.* **Sources: Sa9.**

Viola Italian - 80 days, Indet., large solid plums, 4-8 oz., juicy enough for slicing. *Source History: 1 in 2004.* **Sources: Ma18.**

Visitation Valley - 70-90 days, Indet., small, round, red fruit, 7 oz., aromatic, intensely sweet, dense flesh, named for San Francisco fog belt where it originated, rare. *Source History: 1 in 1998; 2 in 2004.* **Sources: In8, Ta5.**

Vita 29 - 78 days, Red globes, 3 x 2 in., good canner. *Source History: 1 in 1998; 1 in 2004.* **Sources: Sa9.**

Vivid - Det., regular leaf, flavorful fruit, good yields, early. *Source History: 1 in 2004.* **Sources: Up2.**

Vogliotti - 80 days, Beefsteak x Camalia Paste, semi-det. vines are vigorous and productive, deep-red 3-4 in. fruits, tasty, good on heavier soils. *Source History: 1 in 1991; 1 in 2004.* **Sources: Ta5.**

Wagner - 80-90 days, Med. size red fruit, classic tomato taste, more heat tolerance than most, late maturity, heirloom. *Source History: 1 in 2004.* **Sources: Sk2.**

Walter - 75-80 days, Large det. plant, good cover, smooth firm 7 oz. deep globes, res. to GW GLS ST FW1 FW2 ASC BER CF CS and cracking, can be grown on the ground or with short stakes, for the South, fresh market, Dr. Bradenton/FL/AES. *Source History: 25 in 1981; 23 in 1984; 22 in 1987; 13 in 1991; 6 in 1994; 4 in 1998; 3 in 2004.* **Sources: He18, Kil, WE10.**

Walter Villemaire - 75-79 days, Improved sel. of Walter dev. at Villemaire Farm in Florida, stronger large det. plant, larger 8 oz. deep globes, more production, otherwise similar, disease and crack res. *Source History: 4 in 1981; 4 in 1984; 4 in 1987; 5 in 1991; 5 in 1994; 4 in 1998; 5 in 2004.* **Sources: Ra5, Ra6, Re8, Se26, Ta5.**

Wanda's Potato Top - 70-95 days, Very early, large indet., potato leaf, 12-18 oz. pink-red fruit, old-fashioned tomato flavor. *Source History: 1 in 1994; 4 in 1998; 4 in 2004.* **Sources: He18, Ma18, ME9, Se17.**

Wayahead, Jung's Improved - (Wayahead) - 63-80 days, Good-sized smooth bright-scarlet fruits, almost round, solid flesh, indet. vines bear a very long time, fine flavor, excel. for early market. *Source History: 3 in 1981; 3 in 1984; 3 in 1987; 4 in 1991; 2 in 1994; 3 in 1998; 3 in 2004.* **Sources: Jun, Sa9, Ta5.**

Wayahead, Jung's Super Select - 63 days, Seed saved from the most select plants and first or crownset fruits. *Source History: 1 in 1991; 1 in 1994; 1 in 2004.* **Sources: Jun.**

Whippersnapper - 45-62 days, Compact det. plants with numerous side branches, quick-ripening cherry tomato, oval 1 in. long fruits, excel. flavor, pinkish-red skin, loaded with fruits. *Source History: 5 in 1981; 3 in 1984; 4 in 1987; 7 in 1991; 8 in 1994; 8 in 1998; 7 in 2004.* **Sources: Fe5, Hig, Ma18, Pr3, Sa9, Syn, Te4.**

Wilford - 70 days, Small plants with less leaf cover than most, ping pong ball size fruit. *Source History: 1 in 1998; 1 in 2004.* **Sources: Sa9.**

Willamette - 70-100 days, Bushy upright det. plant, good cover, very smooth mild meaty red 3 in. slightly deep globe, uniform good color, small stem and blossom scars, very popular in the Pacific Northwest, adapted to Western valleys of the U.S. and Canada, also performs well in New York and Michigan, Oregon State U. *Source History: 7 in 1981; 4 in 1984; 4 in 1987; 4 in 1991; 5 in 1994; 5 in 1998; 6 in 2004.* **Sources: Ni1, Pe1, Sa9, So9, Ta5, Vi4.**

Willow Pond Big Red Beefsteak - Large round red beefsteak, prolific, quite early. *Source History: 1 in 2004.* **Sources: Sa5.**

Window Box - (Long Island Dwarf) - 60-75 days, Very early, very dwarf det. plants 8 in tall, oval, 3 oz, red fruit, excel. for window boxes, dev. at U of NH. *Source History: 2 in 1981; 2 in 1984; 1 in 1987; 2 in 1991; 2 in 1994; 6 in 1998; 2 in 2004.* **Sources: He18, Sa9.**

Wisconsin 55 - 72-80 days, Large smooth and deep-red, BER and crack res., ripens evenly, 8 oz., uniform size, strong skin, solid walls, excel. shipper, tol. to AB and GLS, for good soils only, becoming hard to find, remembered as one of the best home and market tomatoes in the Madison area, U of WI, 1949. *Source History: 6 in 1981; 6 in 1984; 7 in 1987; 6 in 1991; 4 in 1994; 4 in 1998; 7 in 2004.* **Sources: Ge2, He18, ME9, Pr9, Se16, Se27, To1.**

World Record - Det., regular leaf, 2-3 in. fruit with excellent taste, very good yields, midseason. *Source History: 1 in 2004.* **Sources: Up2.**

Wuhib - 83 days, Plum type, small red fruit, almost paste tomato. *Source History: 1 in 2004.* **Sources: Sa9.**

Yucutan - Red fruit with flattened shoulders, acid taste with fruity undertones. *Source History: 1 in 2004.* **Sources: Ea4.**

Yugoslavian - 79 days, Indet., very large blocky pear shape fruit with thick, meaty flesh. *Source History: 1 in 1998; 1 in 2004.* **Sources: Sa9.**

Yukon - Indet., 2-3 in. round red fruit, early. *Source History: 1 in 2004.* **Sources: Gr28.**

Yuliana - 86 days, Semi-det., 6 oz., red globes. *Source History: 1 in 1998; 1 in 2004.* **Sources: Sa9.**

Zarnitsa - 60 days, Short indet. vines suited for staking or ground culture, name translates 'summer lightning', red fruits average 2.5 in. wide x 2 in. tall, excel. flavor is well balanced, sweet, buttery and smooth, crack res., Russia. *Source History: 1 in 2004.* **Sources: So1.**

Zieglers Fleisch - 90 days, German translation is "Brickmaker's Flesh," 3-5 oz. red fruit, flavor resembles Brandywine, good for containers, European heirloom variety. *Source History: 1 in 1994; 2 in 1998; 2 in 2004.* **Sources: He18, Te4.**

Zogola - 77-85 days, Indet., huge, flat, ribbed fruit, 1-2 lbs., slice easily covers a slice of bread, juicy flesh, balanced sweet flavor, from Poland. *Source History: 4 in 1998; 5 in 2004.* **Sources: Ap6, Sa9, To1, To9, To10.**

Varieties dropped since 1981 (and the year dropped):

Ace, Royal (1998), Adoration (2004), Advantage (1991), Alta Terra (1994), Altajskij Urozajnij (2004), Andino VF (1994), Aneta (2004), Apex 1000 (1994), Arctic Sweet (1991), Arla (1984), Azteca (1994), Beaut (2004), Beefsteak Improved (2004), Beefsteak, Burgess Giant (2004), Beefsteak, Mr. Underwood's (1998), Best of All (1998), Best of Bonn (1991), Bielo Russian (2004), Big A Sandy (2004), Big Brother (2004), Big Italian Paste (2004), Big Ray's Argentina Paste (2004), Big Set (2004), Bigro (1991), Blazer (1998), Bo-Kay (1984), Bonanza (1994), Bonner (1994), Bonner Idaho (1991), Bonnyvee (1994), Brag (2004), Brookpact, Porter's Early (2004), Bruinima Produckt (2004), BSI 80 VF (1991), BSI 804 (1991), Budenovka (2004), Burgess Early Salad (1987), Burgess Everbearing (1987), Burgis (1987), Burnley Bounty (1987), Burpee's Cherry (1991), Butte (1994), C-17 (1991), Cal Ace TM VF (1987), Cal J (2004), Cal Supreme (1987), Camarillo (1991), Campbell 17 (1994), Campbell 19

(1994), *Campbell 28* (1998), *Campbell 32* (1994), *Campbell 34* (1987), *Campbell 37* (1994), *Campbell 38* (1991), *Campbell 39* (1987), *Campbell 37A* (1991), *Canabec 564* (1998), *Canada Hanging Basket* (1987), *Canner's Delight* (1998), *Cannery Row* (1998), *Carrot* (2004), *Castleace VF* (1987), *Castleblock* (1987), *Castlemart II* (1994), *Castlemor F* (1984), *Castlemor Improved VF* (1994), *Castlepear* (1987), *Castlepeel II* (1998), *Castlered* (1994), *Castlerock II* (1998), *Castlerock N* (1998), *Castleroyal* (1994), *Castlestar E-77* (1991), *Castlestar EHV* (1994), *Castlex 499* (1987), *Castlex 622* (1984), *Castlong* (1994), *Centennial Rocket* (2004), *Champs of New Jersey* (1994), *Charlie's Red Staker* (2004), *Cheers* (2004), *Cheery Delight* (1998), *Cherriette of Fire* (2004), *Cherriettes* (1998), *Cherry Delight* (2004), *Cherry Delight "E"* (2004), *Cherry Delight "R"* (2004), *Cherry, Cornell Early* (1994), *Cherry, Dr. Meek's Self Sow* (1994), *Cherry, Everglades* (1994), *Cherry, Loomis Potato Leaf* (2004), *Cherry, Sun* (1998), *Cherry, Sweet Cheree* (1998), *Chico 111F* (1987), *Chico Paste* (1994), *Chinese* (2004), *Chinese Long* (1987), *Chonto* (1991), *Chu Yu* (1983), *Climbing* (1994), *Clover* (1994), *CN420* (1998), *CN421* (1998), *Cocktail* (1987), *Colossal Crimson, Burgess* (1994), *Colossal Red, Burgess* (1994), *Colusa* (1994), *Coracio di Boi* (1994), *Cover Up* (2004), *Crack-Proof* (1991), *Crimson Pasta* (2004), *Crimson Trellis* (1994), *Crimsonvee* (1998), *CS - 290* (1987), *Culiacan* (1987), *De la Plata* (1991), *Del Oro* (1994), *Dorchester* (1984), *Doukhobor* (1998), *Droplet* (2004), *Dwarf Champion* (1994), *Dwarf Early 506* (2004), *DX 52-12* (2004), *E 3202* (1991), *E 6203* (1991), *Earliana 498* (1987), *Earliana, Spark's Improved* (2004), *Earlibright* (2004), *Earliest and Best* (1991), *Early Bird* (1991), *Early Castlemech* (1984), *Early Castlepeel* (1984), *Early Cluster* (1987), *Early Dwarf Cherry* (1987), *Early Giant* (1987), *Early Hi-Crimson* (1994), *Early Pak No. 7* (2004), *Early Red Bush No. 506* (1984), *Early Red Rock* (1991), *Early Rose* (1984), *Early Stone* (2004), *Early Swedish* (2004), *Early Temptation* (1998), *Early Wonder, Burgess* (1994), *Earlypak* (2004), *Early-Set* (1991), *Eilon (Hason) 228VF* (1987), *Enduro* (1998), *Ensenada* (1994), *ES-58* (1987), *Eureka* (2004), *Eurofresh* (2004), *Everbearing Scarlet Globe* (1982), *Everlasting* (1991), *Fakel* (1998), *Far North* (1994), *Faribo Springtime* (1987), *Fern* (1998), *Field's Red Bird* (1987), *Fire Cheers* (2004), *Fire Drill* (2004), *Floralou* (1994), *Florida Lanai* (1991), *Florida MH-1* (2004), *Forme de Coeur* (2004), *Frost Resistor* (1998), *Full Flavor Paste* (1998), *Ganti* (2004), *Garden State* (2004), *Gardener VF* (1998), *German Beefsteak* (1987), *German Dwarf Bush* (1998), *German Giant Red* (2004), *German Salad* (1987), *Giant Climbing, Jung's* (1994), *Giant Italian Potato Leaf* (2004), *Gito* (1994), *Gladness of Garden* (2004), *Gourmet VF* (2004), *Gran Marzano* (1991), *Grand Pak* (1982), *Grosse Cerise Q* (1998), *Harvestvee* (2004), *Heinz 1370* (2004), *Heinz 1409* (1994), *Heinz 1706* (1994), *Heinz 1765* (2004), *Heinz J 30* (1987), *Highlander* (1994), *Holland Red* (2004), *Homestead 61* (1987), *Homestead Elite* (1991), *Horizon* (1994), *Hungarian* (2004), *Hunt 100* (2004), *Hunter (FM-60431)* (1991), *Imperial Improved* (1984), *Impulse* (1998), *Irion* (1984), *Italian Banana* (2004), *Italian Canner* (1998), *Italian Plum* (1982), *Italian Stallion* (2004), *J. Moran* (1994), *J.S.S. No. 3570* (1987), *Jefferson* (1984), *Jersey* (2004), *Jolimac* (1987), *Jumbo Jim* (1991), *Jungle Island* (1994), *K-2* (1987), *K-4* (1987), *Kanatto* (1987), *Kenearly* (1983), *Large Italian* (1987), *Large Red Plum* (1991), *Lillian* (1987), *Lorissa* (1998), *M 82-1-8* (1991), *Mammoth Wonder* (1994), *Marbi* (1994), *Marbon* (1994), *Marglobe, Meaty* (1994), *Mark Allen* (1998), *Marmande Extra* (1994), *Marmande Improved* (1987), *Mars* (2004), *Marvel* (1991), *Marzano Big Red* (1994), *Marzano Lampadina Extra* (1987), *Memoirs* (2004), *MH-1* (1991), *MH-6203 VF* (2004), *MH-9209 VF* (1994), *Michigan State Forcing* (1987), *Mike's Bride* (2004), *Minibel* (1991), *Monte Grande* (1987), *Monterey* (1991), *Moto-Red Greenhouse* (1991), *Mountain Girl VF* (2004), *Mountain One* (1998), *Murrieta* (1994), *Mustang* (2004), *Nemared* (1987), *Nematex* (1984), *New Sidor* (1987), *New Yorker (Transplant Str.)* (2004), *Nigeria* (2004), *No. 80* (1984), *No. 95* (1994), *No. 670* (1998), *No. 3032* (1987), *No. 7870 (Ohio)* (1991), *No. 9208 (Ferry Morse)* (1991), *Norcal* (1994), *Nota* (2004), *Ohio 7870* (2004), *Old Timey Beefsteak* (2004), *Old's Old* (2004), *Omnipak (FM-91304)* (1994), *Opal's Homestead* (2004), *Oregon Pride* (2004), *Ottawa 39* (1987), *Ottawa 78* (1984), *Outdoor Girl* (2004), *Pacesetter 490* (1994), *Pacesetter 502 VF* (1991), *Pacesetter 882* (1987), *Pakmor B* (2004), *Palermo* (1994), *Party* (1998), *Patio Prize* (1994), *Paul Bunyan* (1994), *Payette* (2004), *Pearmech* (1987), *Pearpeel* (1991), *Pearson ET* (1987), *Peelmech* (1994), *Peelo* (2004), *Pelican* (2004), *Peto 13* (1987), *Peto 77* (1987), *Peto 80* (1987), *Peto 81* (1998), *Peto 86* (2004), *Peto 94-C* (1991), *Peto 95-43* (1994), *Peto 98* (1998), *Peto 102* (1987), *Peto 111* (1994), *Peto 343* (1994), *Peto 460* (1987), *Petoearly* (1987), *Petomech II* (2004), *Pierrette* (1994), *Pipo* (1991), *Pixie* (1987), *Plum 198* (1987), *Plum 317* (1987), *Plum Fryer* (1987), *Plum Seedless* (2004), *Pole Boy* (1987), *Pole Boy 83* (1994), *Pole Boy No. 83* (1987), *Polish Potato Leaf* (1994), *Portuguese Grande* (1994), *Potomac* (1987), *Pritchard* (1998), *Processor 40* (1987), *PSR 100-2975* (2004), *PSR 286* (2004), *PSR 546* (2004), *PSR 951* (2004), *PSR 952 A* (2004), *PSR 952 B* (2004), *PSR 953* (2004), *PSR 954* (2004), *PSR 955* (2004), *PSR 956* (2004), *PSR 957* (2004), *PSR 958* (2004), *PSR 959* (2004), *PSR 1071* (2004), *PSR V-931* (1998), *PSR V-932* (1998), *PSR V-933* (2004), *Quebec 5* (1994), *Quebec 13* (1994), *Quebec Select* (1982), *Ramapo OP* (1994), *Red Bird* (1994), *Red Cloud Improved VF* (1991), *Red Cloud VF* (1991), *Red Dawn* (2004), *Red Emperor* (2004), *Red Hunter* (1998), *Red Lightning* (2004), *Red Top* (1984), *Redheart* (2004), *Rehort* (1987), *Reine de Ste. Marthe* (1994), *Rio Chico* (1987), *Rio Grande Cal Supreme* (2004), *Rio Magic* (1994), *Rodade* (2004), *Roforto* (1991), *Roma 198* (1987), *Roma F-021* (1991), *Roma Gigante VFN* (1991), *Roma Puree* (2004), *Romance* (2004), *Rosabec* (2004), *Rossol VFN* (2004), *Royal Improved Ace* (1991), *Ruby* (1998), *Ruby Gold Sport* (2004), *Rutgers 39 VF* (1994), *Rutgers 8828* (1987), *Rutgers California Supreme* (2004), *Saint Pierre* (2004), *Saladette* (1994), *Salt Spring Island* (1982), *Salt Spring Sunrise* (1998), *San Marzano Large Fruited* (1994), *San Pablo* (1998), *Santa Cruz* (1984),

Santa Cruz 22 (1991), *Santa Cruz Angela* (2004), *Santa Cruz Kada Gigante* (1991), *Santa Cruz Yokota* (1991), *Santiam PL* (2004), *Saturn* (1987), *Sausage Tomato* (1987), *Scarlet Egg* (1994), *Scarlet Heirloom* (1998), *Scoresby Dwarf* (1987), *Scoursby's Dwarf* (1994), *Seattle Best of All* (2004), *Selandia* (1994), *Shia Kuan* (1987), *Short-N-Sweet* (2004), *Showell's Red* (1998), *Sicilian Plum* (1998), *Slumac* (1987), *Small Wonder* (2004), *Smoky Mountain* (2004), *Southland W.R.* (1991), *Soviet Elite* (1998), *Space Tomato* (1994), *Spanish Paste* (2004), *Special Back* (1987), *Spectrum* (1984), *Springtime* (1984), *Square* (2004), *ST81* (1987), *Starfire Improved* (1994), *Starshot* (1998), *Start Z* (2004), *Stockton 81* (1987), *Stokesalaska* (2004), *Stokesdale* (1994), *Stone Improved* (1991), *Stupendous* (2004), *Sub-Arctic 25* (2004), *Sub-Arctic Delight* (1991), *Sub-Arctic Midi* (1987), *Sub-Arctic Paste* (2004), *Sugar Red* (1987), *Summer Sun* (1994), *Summertime* (1994), *Summertime Improved* (1994), *Summit* (1998), *SUN 499* (1994), *SUN 1641* (1994), *SUN 1642* (2004), *SUN 1643* (1998), *SUN 1648* (1994), *SUN 5715* (2004), *SUN 6000* (1994), *SUN 6002* (2004), *SUN 6040* (1998), *SUN 6095* (1998), *Suncoast VFF* (2004), *Sunlight* (1991), *Sunrise I* (2004), *Sunrise II* (1998), *Sunrise III* (2004), *Sunrise V* (2004), *Sunrise VII* (2004), *Super Break o'Day* (1984), *Super Stokesdale* (1991), *Superbec* (1998), *Sureset* (1994), *Sweet Billion* (2004), *Sweetie Supreme* (2004), *Tamiami* (1994), *Teresa* (2004), *TH318* (2004), *The Amateur* (2004), *The Prince* (2004), *Titano M* (1991), *Tomato Russe* (1994), *Tracy* (1994), *Triumph* (1994), *Tropi-Gro* (1987), *Tyboroski Giant Plum* (2004), *UC 82* (1998), *UC 82 A* (1991), *UC 82 C* (1991), *UC 82 L* (2004), *UC 97-3* (1998), *UC 105-J* (1984), *UC 134-1-2* (1987), *UC 204-9 or 9b* (1991), *UC 204-B* (1991), *UC 204C* (1998), *UC 204-C* (1991), *UC 211-52-3* (1987), *UC 211-58-1* (1991), *UC M75-28-2-3* (1991), *Unwins Amature* (1994), *Unwin's Histon Early* (1983), *Upline* (2004), *Urbana* (2004), *Usabec* (1994), *Valina* (1998), *Vancouver Island* (1994), *Veebright* (1987), *Veemore* (1984), *Veepick* (1998), *Veepro* (1991), *Veeset* (1987), *Vesuvius* (1991), *VF 10* (2004), *VF 13-1-34* (1984), *VF 13-l* (1987), *VF 13-L-34* (1991), *VF 65* (1991), *VF 65-433* (1987), *VF 134 E* (1987), *VF 145-21-4* (1991), *VF 145-21-4 Select* (1987), *VF 145-513* (1987), *VF 145-7879* (1987), *VF 198* (1994), *VF 270* (1991), *VF 270-C* (1987), *VF 315* (1987), *VF 317* (1987), *VF 590* (1984), *VF 590-5-8* (1991), *VF 3032* (1991), *VF 3202* (1987), *VF 6201* (1987), *VF 7879/FAR* (1987), *VF 9209* (1991), *VFN 8* (1994), *VFN Bush* (1987), *VFN Bush No. 3* (1991), *Vision* (1987), *Vladivostok* (1998), *Volgogradskiye 323* (2004), *Walter Improved* (2004), *Walter Improved F1-2* (1987), *Walter PF* (1987), *Warm Fuzzies* (2004), *West Virginia '63* (1998), *Western Processor 92* (1998), *Western Processor 94* (1998), *Westover* (1994), *Willamette Early Cherry* (1987), *Wondervee* (1987), *World's Largest* (1987), *Yolla* (1987), *You-Go* (2004), *Yuba* (1998), *Yugoslavian Homegrown Tomato* (1987), *Zagola Prizetaker* (1994).

Tomato / Other Colors
Lycopersicon lycopersicum

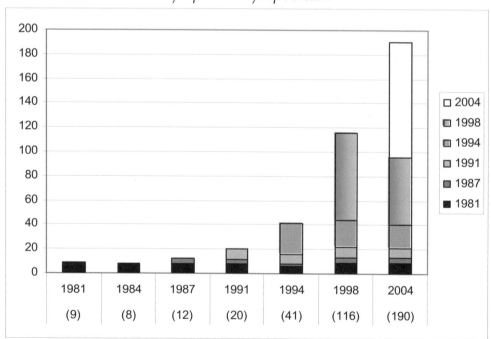

Number of varieties listed in 1981 and still available in 2004: 9 to 9 (100%).
Overall change in number of varieties from 1981 to 2004: 9 to 190 (2111%).
Number of 2004 varieties available from 1 or 2 sources: 105 out of 190 (55%).

<u>Alleghany Sunset</u> - Indet., yellow-orange fruit, red marbled flesh, flavorful, very pretty when sliced. *Source*

History: 1 in 2004. **Sources: Up2.**

Amazon Chocolate - Ukrainian/Russian tomato that is almost twice the size of other black tomatoes, deep purple-brown color, 1 lb. or more, deep smoky rich flavor is excellent, from Sergey who lived in the Ukraine for almost 40 years and moved to the U.S. in 1993. *Source History: 1 in 2004.* **Sources: Ami.**

Ananas - 85 days, Indet., pale yellow fruit blushed with red on blossom end, 3-4 in. across, mildly acid, sweet flavor, heirloom. *Source History: 1 in 2004.* **Sources: To9.**

Arkansas Marvel - 100 days, Indet., yellow-orange flesh with red marbling, up to 1 lb., mild, peach-like flavor, heirloom. *Source History: 1 in 2004.* **Sources: To9.**

Armenian - 85-90 days, Indet., beefsteak type, flattened and somewhat scalloped, orange-yellow flesh mottled with red, 1 lb., very good flavor, juicy. *Source History: 3 in 2004.* **Sources: Ho13, Ta5, To9.**

Aunt Ruby's German Green - 80-95 days, Indet., regular leaf foliage, large green beefsteak with delicious, sweet, spicy flavor, 12-16 oz., light green when ripe with a tint of yellow and a pink blush underneath, heirloom from Ruby Arnold of Greenville, TN. *Source History: 9 in 1998; 41 in 2004.* **Sources: Ap6, Ba8, Bo17, CO30, Coo, Ea4, Fe5, Ga1, Go8, He8, He18, Ho13, Hud, Jun, Ma18, ME9, Pe2, Pin, Pr9, Ra5, Ra6, Re8, Ri12, Sa9, SE14, Se16, Se26, Sh13, So25, Sw9, Ta5, Te4, TE7, To1, To3, To9, To10, Up2, Ver, Vi4, Yu2.**

Aviuri - 85-90 days, Orange striped globes, 6 oz. *Source History: 1 in 1998; 1 in 2004.* **Sources: Sa9.**

Beaute Blanche du Canada - 86 days, Indet. plants to 3 ft., white to pale cream, 5 oz., flattened fruits, very flavorful. *Source History: 1 in 1998; 3 in 2004.* **Sources: Gr28, Sa9, Up2.**

Beefsteak, White - (Giant White Beefsteak) - 80-90 days, Indet., boat-shaped, 10 oz. fruit, ripens pale yellow-white, very sweet, good texture and quality. *Source History: 2 in 1994; 4 in 1998; 6 in 2004.* **Sources: Bo19, Ers, Ra5, Sk2, To3, WE10.**

Bianca - Rare European variety, sweet small white cherries blushed with pink at blossom end. *Source History: 2 in 2004.* **Sources: Ba8, Pe7.**

Big Rainbow - 80-102 days, Heirloom from Polk County, Minnesota, indet., 6 ft. plants, huge rib-shouldered golden fruits with streaks of red running through the flesh, fully ripened fruits are gold on stem end, red on blossom end, some weigh up to 2 lbs., subject to cracks and catfacing but has superb flavor, CV So1 introduction, 1990. *Source History: 2 in 1991; 5 in 1994; 11 in 1998; 29 in 2004.* **Sources: Al25, Ami, Ba8, Bou, Bu2, CO30, Coo, Ers, Ga1, He8, He18, Ho13, La1, Loc, Ma18, ME9, Ra6, Re8, Sa9, SE14, Se26, Sh13, So1, Te4, To1, To3, To9, Up2, Vi4.**

Big Tiger - 85 days, Indet., yellow-red fruit, 8 oz., good flavor, good yields. *Source History: 1 in 2004.* **Sources: To9.**

Big White - 81 days, Indet., flattened, 10 oz. fruit. *Source History: 1 in 1998; 1 in 2004.* **Sources: Sa9.**

Big White Pink Stripes - 71-95 days, Indet., globe shaped fruit, 12-16 oz., pale peach skin with cream flesh, some will turn pinkish, sweet-tart flavor. *Source History: 3 in 1998; 3 in 2004.* **Sources: Ma18, Sa9, To9.**

Black - 75-85 days, Russian heirloom, very early indet., 4-16 oz., fruit color is such a dark red it is considered black, thin skin, soft texture, flavorful, grows under adverse conditions. *Source History: 1 in 1991; 2 in 1994; 5 in 1998; 9 in 2004.* **Sources: Ea4, Ga22, Gr28, Ma18, Sa9, So25, Ter, To1, To9.**

Black Aisberg - 80 days, Indet., 4-8 oz. beefsteak fruit with attractive brown-red skin, gray-olive shoulders, rich deep flavor with spicy overtones, good producer. *Source History: 2 in 2004.* **Sources: Ea4, Ma18.**

Black Early - 70 days, Related to Cherokee Purple and Black Krim, 8-10 oz., purple-brown fruit. *Source History: 1 in 2004.* **Sources: To10.**

Black Ethiopian - 69-81 days, Indet., beautiful bronze to brown-red, 4-6 oz. plums, rich sweet flavor, for sauce and salad, probably originated in the Ukraine, rare. *Source History: 4 in 2004.* **Sources: Ea4, Ma18, Ta5, To9.**

Black from Tula - 75-85 days, Indet., largest of the black tomatoes, 4-5 in., flattened fruit, 8-12 oz., dark brown to purple with green shoulders, rich smoky flavor, sets fruit well in hot weather, rare Russian heirloom. *Source History: 4 in 1998; 24 in 2004.* **Sources: Ami, Ap6, Ba8, Co32, Ea4, Ga1, Go8, He8, He18, Ma18, ME9, Na6, Pe2, Pr9, Ri12, Sa9, SE14, Se16, Se26, TE7, To1, To3, To9, Und.**

Black Krim - 69-90 days, Indet. plants must be staked or caged, flattened, globe shaped, 8-12 oz., dark red-purple fruit with delicate skin, green shoulders, green tinted flesh, full flavor with slight saltiness that enhances the taste, very juicy, always places at the top in tomato taste trials, originated in Krymsk on the Black Sea in Russia. *Source History: 1 in 1994; 16 in 1998; 54 in 2004.* **Sources: Al25, Ap6, Ba8, Bo17, Bo19, Bou, Bu2, CO30, Co31, Co32, Coo, Ea4, Ers, Fe5, Fo7, Ga1, Go8, Gr29, He8, He17, He18, Ho13, Hud, Loc, Ma18, ME9, Na6, Orn, Pr3, Ra5, Ra6, Ri12, SE14, Se16, Se26, Se28, Sh13, Shu, Sk2, Sw9, Syn, Ta5, Te4, To1, To3, To9, To10, Tu6, Up2, Vi4, Vi5, WE10, We19, Wo8.**

Black Pear - 75-82 days, Indet., potato leaf foliage, 6-10 oz., grey-brown fruits with green shoulders, flavorful, productive. *Source History: 1 in 1998; 5 in 2004.* **Sources: Ea4, Gr28, Ma18, Sa9, Up2.**

Black Plum - 75-82 days, Russian commercial variety, better color than other "black" tomatoes, indet., 6 ft. plants, oval or plum shaped, 2 in. fruits, deep mahogany to brown, produces steadily and consistently, black color develops best when hot and sunny, very little if dark and rainy. *Source History: 1 in 1994; 10 in 1998; 29 in 2004.* **Sources: Ea4, Ga1, Go8, Gr28, He18, Ho13, In8, Ma13, Ma18, Pr3, Pr9, Ra6, Sa9, Se7, Se16, Se17, Se26, Sh13, So9, So25, St18, Sw9, Ta5, Ter, To1, To3, To9, To10, Und.**

Black Prince - 70-90 days, Indet. vines, smooth round 5 oz. fruits, deep-garnet skin color, dark-red to brown flesh color, tender, juicy and flavorful, harvest when shoulders are dark and still show a trace of green, Russian heirloom. *Source History: 13 in 1998; 28 in 2004.* **Sources: Ami, CO30, Co32, Ea4, Ga1, He8, He18, HO2, Loc, ME9, Ni1,**

Orn, Pe2, Ra5, Ra6, Re8, RUP, Sa9, SE14, Se26, So1, Sw9, Ta5, TE7, To1, To3, To9, Up2.

Black Russian - 75-80 days, Semi-det., thin brownish skin, red-pink, purple flesh, meaty, juicy flavor resembles that of Brandywine, dark purple stripes radiate from blossom end to stem, green shoulders, thrives under cool growing conditions, heirloom. *Source History: 3 in 1998; 11 in 2004.* **Sources: Ami, Go8, He8, Ki4, Orn, Pr3, Ra5, Ra6, Re8, Ri12, Tho.**

Black Sea Man - 75-80 days, Very early str. from Russia, det. plant to 3 ft., pink-brown fruits with olive-green shading, 6-8 oz., good spicy flavor, does not like summer heat. *Source History: 3 in 1998; 6 in 2004.* **Sources: Ho13, Ma18, Pr3, Pr9, Ta5, To10.**

Black Zebra - 80-85 days, Beautiful bi-color fruit striped with brown and green, 1.5 in. fruit, sweet complex flavor with hint of smoke, light yields, created by Jeff Dawson. *Source History: 3 in 2004.* **Sources: Ea4, So25, To9.**

Blackstar - 80 days, Indet., purple-pink-brown skin with olive-green shoulders, 16+ oz., earthy spicy flavor, productive. *Source History: 1 in 2004.* **Sources: Ma18.**

Blue - 78-80 days, Indet., potato leaf, medium size "black" tomato, maroon with purple-gray shoulders, 8 oz., sweet flavor, rare. *Source History: 4 in 2004.* **Sources: Ma18, Ta5, To1, To9.**

Bordo - 75-80 days, Indet. plants, 8 oz. globes, red-brown on green background, productive. *Source History: 1 in 1998; 2 in 2004.* **Sources: Ma18, Sa9.**

Brandywine, Black - 80-120 days, Indet., potato leaf, 12-16 oz., purple-maroon, oval fruit, deep fruity flavor, pre-1900. *Source History: 5 in 1998; 20 in 2004.* **Sources: Ami, Ga1, He8, He18, Jun, Lan, Loc, Ma18, ME9, Or10, Ra5, Ra6, Re8, Sa5, SE14, Se26, Shu, To1, To3, To10.**

Brown Flesh - 70-75 days, Indet., hollow stuffing type, 3 lobes, red-mahogany with green stripes. *Source History: 3 in 2004.* **Sources: Gr28, Ma18, To9.**

Burracker's Favorite - 82 days, Indet., huge red-orange bicolor fruit, 1-2 lbs., nice flavor, from the territory of Burracker's Hollow in the foothills of the Blue Ridge Mountains. *Source History: 2 in 2004.* **Sources: Ta5, To9.**

Candy's Old Yellow - 75-80 days, Indet., medium size beefsteak, yellow flesh blushed with pink, mild flavor is sweet and fruity. *Source History: 2 in 2004.* **Sources: Ma18, To9.**

Carbon - 76-85 days, One of the darkest "black" tomatoes, indet., 8-12 oz. fruit, flattened round shape, no cracks or blemishes, dark purple-brown skin, dark red flesh mixed with clay and olive, rich sweet flavor. *Source History: 4 in 2004.* **Sources: Ma18, To1, To9, Up2.**

Cherokee Chocolate - 77 days, Indet., flattened fruit with beautiful mahogany-brown color. *Source History: 1 in 2004.* **Sources: Sa9.**

Cherry, Aunt Ruby's German - Medium green skin, tinged with yellow when ripe and blushed with amber on blossom end, green flesh, superb spicy sweet flavor, 12-16 oz., unique, attractive, heirloom from Ruby Arnold, Greeneville, TN. *Source History: 1 in 2004.* **Sources: Ea4.**

Cherry, Aunt Ruby's German Black - Small black cherry found among Aunt Ruby's German Green, not fully stable, expect some off-types. *Source History: 1 in 2004.* **Sources: Ea4.**

Cherry, Bi-Color - 78 days, Indet., small round cherry type, heirloom. *Source History: 2 in 2004.* **Sources: Ga1, ME9.**

Cherry, Black - 65 days, Indet. plants produce an abundance of round red-black cherries with sweet complex flavor, introduced by Tomato Growers Supply Company. *Source History: 2 in 2004.* **Sources: So25, To1.**

Cherry, Ghost - 75-85 days, Indet., 2 oz., cream colored skin with yellow-peach blush, slightly fuzzy skin, almost square shape, very sweet. *Source History: 8 in 2004.* **Sources: Ga1, Gr28, Ho13, Ma18, Pr3, To1, To9, Up2.**

Cherry, Isis Candy - 67-90 days, Indet. plants bear throughout a long season, 2 in. yellow-gold cherry tomatoes with red marbling, sweet, rich, fruity flavor. *Source History: 1 in 1998; 12 in 2004.* **Sources: Ea4, Gr28, Ho13, In8, Se16, Se17, So25, Sw9, Ta5, Ter, To1, To9.**

Cherry, Sarah Goldstar - 65 days, Indet., productive cherry tomato, yellow blushed with red, very sweet taste, Germany, rare. *Source History: 3 in 2004.* **Sources: Ea4, Ho13, So25.**

Cherry, Snow White - 68-75 days, Indet., ivory .5 in. fruits ripen to pale yellow, mild flavor, developed by Joe Bratka. *Source History: 4 in 1998; 12 in 2004.* **Sources: Ba8, Go8, Ma18, Ra6, Sa5, Sa9, Syn, Ter, To1, To9, To10, Up2.**

Cherry, White - Indet., pale yellow-ivory fruit, 1 oz., huge yield. *Source History: 3 in 1998; 6 in 2004.* **Sources: Bo19, CO30, ME9, SE14, Se26, TE7.**

Copia - 75 days, Indet., yellow-red beefsteak, 3-4 in. across, Marvel Stripe x Green Zebra cross from Sharon and Jeff Dawson, named after the American Center for Wine, Food and the Arts in Napa, CA (COPIA). *Source History: 1 in 2004.* **Sources: To9.**

Cuban Black - 80-88 days, Semi-det., 4-6 oz., flattened red-green fruit, green shoulders, flavor is intensely sweet with earthy tones, makes unique fresh sauce, prolific. *Source History: 1 in 1998; 4 in 2004.* **Sources: Ea4, Ma18, Sa9, Ta5.**

De Barao Black - 75-88 days, Indet., gray-black, semi-sweet, 3 oz. fruit, nice spicy sweet flavor, Moldova, rare. *Source History: 1 in 1998; 3 in 2004.* **Sources: Ma18, Sa9, Ta5.**

Dorothy's Green - 80 days, Indet., med. size green fruit with amber tint, 1 lb., sweet, spicy taste, heirloom. *Source History: 2 in 1998; 4 in 2004.* **Sources: Ho13, Ma18, Ta5, To1.**

Dr. Carolyn - 65-75 days, Indet., light yellow, 1 in. round cherry, excel. balance of sweet and tart flavor, 1.25 x 1.25 in. fruits borne in clusters of six, sport of Galina named after Dr. Carolyn Male. *Source History: 2 in 1998; 8 in 2004.* **Sources: Ap6, Ea4, Ga21, He18, Ma18, So1, To1, Up2.**

Duggin White - 80 days, Indet., 8-12 oz. fruit, cream color blushed with yellow, clean, crisp flavor. *Source History: 1 in 2004.* **Sources: Ma18.**

Early Yellow Stripe - 80 days, Indet., semi-hollow fruit, 2 in., pale red with yellow stripes, highly productive. *Source History: 1 in 1998; 1 in 2004.* **Sources: To9.**

Elberta Girl - Indet. vine with blue-green fuzzy leaves, 5-6 ft. tall, red and green striped fruit, 2.5 in. dia., good acidic tomato flavor, heirloom. *Source History: 2 in 1994; 3 in 1998; 9 in 2004.* **Sources: Ami, CO30, Ers, Pr3, Ra5, Ra6, Sa9, To1, To3.**

Esikos Botermo - 84 days, Indet., 2.5 in. orange-red globes striped with light orange. *Source History: 1 in 2004.* **Sources: To9.**

Eva's Amish Stripe - Amish heirloom, huge yellow fruit with yellow and red marbling throughout the flesh and red on the blossom end, 1-3 lbs., low acid with sweet flavor, resembles Old German but ripens much later, grown by 87 year-old Eva for over 40 years, original seed from an Amish lady her father sold cattle to in Belleville, PA. *Source History: 1 in 2004.* **Sources: Ami.**

Evergreen - (Emerald Evergreen, Tasty Evergreen) - 70-80 days, Vigorous indet. vines, flesh and gel remain green when ripe, very solid med-large fruits, 10-16 oz., skin ripens from green to light yellow-brown, delicious mild flavor, for slicing or frying or pickling or making conserves, Glecklers Seedsmen introduction, 1950. *Source History: 7 in 1981; 5 in 1984; 4 in 1987; 5 in 1991; 5 in 1994; 14 in 1998; 37 in 2004.* **Sources: Ami, Ba8, Bou, CO30, Coo, Ers, Gal, Gr27, Gr28, Gr29, He18, Ho13, Loc, Ma18, ME9, Mel, Na6, Ni1, Orn, Pe2, Ra5, Ra6, RUP, Sa9, SE14, Se28, Sk2, Ta5, Te4, TE7, To1, To3, To9, Up2, Vi5, WE10, Wo8.**

Flame - (Old Flame, German Johnson, Rainbow, Hillbilly, Mr. Stripey) - 80-85 days, Indet., beefsteak size fruit with red streaks throughout rose-gold flesh, rich tangy flavor, crack res., beautiful when sliced, from West Virginia. *Source History: 8 in 1994; 10 in 1998; 14 in 2004.* **Sources: Ba8, CO30, Ers, Gr27, He18, HO2, Loc, ME9, Orn, SE14, Sh9, Sk2, Up2, WE10.**

Fuzzy - 80 days, Unusual tomato from Mexico, indet., dusty rose-pink, 2-4 oz. fruit is covered with silver fuzz, sweet and juicy. *Source History: 1 in 2004.* **Sources: Ma18.**

Garden Lime - 81-86 days, Indet., flat, green-gold fruits, 12 oz., semi-sweet. *Source History: 1 in 1998; 1 in 2004.* **Sources: Sa9.**

Georgia Streak - 85-90 days, Indet. vines, yellow flesh blushed with red, red core on blossom end, 1-2 lbs., heirloom from Georgia. *Source History: 1 in 1991; 1 in 1994; 5 in 1998; 8 in 2004.* **Sources: Fo7, Ra5, Ra6, Re8, So1, Ta5, To1, To3.**

German Black - 80 days, Indet., fluted, purple-gray fruit, green shoulders, rose flesh with olive tones, nice rich sweet flavor, great producer, from Tennessee. *Source History: 1 in 2004.* **Sources: Ma18.**

German Gold - 85-88 days, Indet., large, 1-2 lb. golden fruit with red blossom end and red streaks radiating up the sides, sweet, mild flavor, Germany, 1890. *Source History: 4 in 1998; 8 in 2004.* **Sources: Co32, Fo13, Gr28, He8, Ho13, Se17, Se26, Ta5.**

German Striped Stuffer - 78 days, Indet., flavorful stuffing tomato, red skin striped with yellow. *Source History: 2 in 2004.* **Sources: Al25, To9.**

Giant Fiolet - 71 days, Semi-det. plants, 8-12 oz., flattened, grey-brown fruit, unique flavor. *Source History: 1 in 1998; 2 in 2004.* **Sources: Ho13, Sa9.**

Gold Medal - 75-90 days, Indet., large yellow fruit, big blossom end, yellow flesh blushed with red, sweet, low acid flavor, 1- 2 lbs., grows well in cool overnight temperatures, Quisenberry variety. *Source History: 5 in 2004.* **Sources: Go8, Pr9, Se16, Ter, To9.**

Gold 'n Green - 75 days, Sturdy plant, small fruit, bronze on the outside and bright-green on the inside, fine tasting novelty, heavily concentrated fruit set. *Source History: 1 in 1991; 1 in 1998; 2 in 2004.* **Sources: Ta5, To9.**

Golden Queen, USDA Strain - 75 days, The original Golden Queen tomato as described by Livingston in 1882, indet., yellow fruit with pink blush on the blossom end that radiates to the stem end, superior flavor sets it apart from the solid yellow Golden Queen. *Source History: 2 in 2004.* **Sources: So25, To1.**

Golden Treasure - 70 days, First long-keeping tomato with gold skin color, ripens from green to golden in about 1-1.5 months, can keep for up to 4 months, vigorous indet. vines, 2.5 in. dia., 2.5-4 oz., acidic tomato flavor becomes more mild in storage, PSR Breeding Program. *Source History: 2 in 1994; 2 in 1998; 1 in 2004.* **Sources: To10.**

Golden Treasure UR - Similar to Golden Treasure except fruits average 20-30% larger and are uniform ripening without green shoulders, crisp texture, good tangy flavor, long-keeping, PSR Breeding Program. *Source History: 1 in 1994; 1 in 1998; 1 in 2004.* **Sources: Pe6.**

Grandma Oliver's Green - 80 days, Indet., bright green flesh, skin tinged with amber when ripe, 8-12 oz., beautiful, flavorful and unusual, ripens late, huge yields. *Source History: 1 in 1991; 2 in 2004.* **Sources: Ta5, To9.**

Granny Smith - 80 days, Indet., 8 oz. fruit, ripens to rich green color, rich tomato taste. *Source History: 1 in 2004.* **Sources: Ri12.**

Great White - (White Beefsteak) - 80-85 days, Vigorous heavy indet. foliage, large white fruits, 14-16 oz., white flesh throughout, mild, low acid flavor, no catfacing or cracking, few seeds, ripens midseason, heirloom. *Source History: 2 in 1991; 2 in 1994; 12 in 1998; 26 in 2004.* **Sources: Ami, Ba8, CO30, Co31, Gal, Go8, He18, HO2, Ho13, La1, Ma18, ME9, Mel, MIC, Na6, Pep, Ra6, Re8, RUP, SE14, Se17, Se26, So25, To1, To9, Up2.**

Green - 70-80 days, Indet., medium-large fruit, olive skin with gold stripes, dark green shoulders, emerald-green flesh, excel. yield, fresh taste. *Source History: 1 in 1998; 3 in 2004.* **Sources: Gr28, He18, Ma18.**

Green and Yellow - 78 days, Indet., stuffing type, resembles Green Bell Pepper tomato but with larger fruits and more hardy plants, may set an occasional solid yellow fruit. *Source History: 1 in 2004.* **Sources: Ma18.**

Green Bell Pepper - 75-90 days, Indet., pepper shaped, hollow stuffing type, green flesh, yellow and green striped skin, from Tater Mater Seed. *Source History: 2 in 1998; 2 in 2004.* **Sources: Ma18, To1.**

Green Cage - 75 days, Indet., 2-3 oz. golden fruits, nice fruity flavor, high yields, so named because of the green gel around the seeds, pre-1900 Old English variety. *Source History: 1 in 2004.* **Sources: Ma18.**

Green Grape - 70-87 days, Old-fashioned bush tomato, Yellow Pear x Evergreen, distinctive small green-yellow fruits borne in clusters of 4-12, resemble large muscat grapes, green flesh and juice, easily cultivated, dev. and rel. by Tater Mater Seed, 1986. *Source History: 1 in 1987; 5 in 1991; 9 in 1994; 20 in 1998; 43 in 2004.* **Sources: Ami, Ba8, Bo19, Bou, CO30, Coo, Ga1, Gr28, He8, He18, La1, Loc, Ma18, ME9, Na6, Ni1, Orn, Pin, Ra5, Ra6, Re8, RUP, Sa9, SE14, Se16, Se17, Se26, Shu, So1, So25, Sw9, Syn, Ta5, Te4, Tho, To1, To3, To9, To10, Und, Up2, Vi4, WE10.**

Green Pineapple - 80 days, Indet., small beefsteak, smooth skin, khaki-green with yellow blossom end, chartreuse-green flesh with fruity aroma, complex flavor with bit of spice. *Source History: 3 in 2004.* **Sources: Ma18, So25, To1.**

Green Sausage - 78-100 days, Indet., bushy plants, elongated fruit shape resembles sausage, green skin with faint yellow stripes, delicious green flesh, makes a tasty fresh green salsa. *Source History: 2 in 2004.* **Sources: Ma18, Tho.**

Green Velvet - 80 days, Indet., small flattened green beefsteak with gold blush when ripe, 4 oz., chartreuse flesh is extremely sweet and flavorful. *Source History: 5 in 2004.* **Sources: Ea4, Ma18, So25, Ta5, Up2.**

Green Zebra - 75-90 days, Indet., yellow-green, 1.5-2.5 in. fruits with dark green vertical stripes, emerald-green flesh with mild but not bland flavor, very productive, vines show some res. to septoria leaf spot, dev. in 1985 by heirloom tomato breeder Tom Wagner of Tater Mater Seeds. *Source History: 7 in 1994; 21 in 1998; 55 in 2004.* **Sources: Ag7, Al25, Ap6, Ba8, Bo17, Bu2, CO30, Co31, Co32, Coo, Ea4, Fe5, Fo7, Fo13, Ga1, Ga22, Go8, Gr28, He8, He18, Hud, Jo1, Jo6, Ki4, Loc, Ma18, ME9, Na6, Orn, Pe2, Pr9, Ra5, Ra6, Re8, Ri12, Roh, RUP, Sa9, SE14, Se16, Se17, Se26, So1, So25, Sw9, Syn, Te4, TE7, To1, To3, To9, To10, Tu6, Up2, WE10.**

Greenwich - 80 days, Med. size fruit, burnt orange skin, lime-green flesh, excellent flavor. *Source History: 1 in 1998; 1 in 2004.* **Sources: Sa5.**

Halfmoon China - 75-80 days, Indet., beefsteak fruit, 3 in., 8-10 oz., pale yellow to white fruit with a hint of melon flavor. *Source History: 4 in 2004.* **Sources: Ma18, Ta5, To9, To10.**

Hazel Mae - 85 days, Indet., large beefsteak up to 1 lb., yellow skin striped with dark red, some red marbling on blossom end, juicy flavorful flesh, good yields. *Source History: 2 in 2004.* **Sources: To9, To10.**

Hess - 85 days, Indet., large yellow fruit with red marbling, 14 oz., mild fruity flavor, German heirloom. *Source History: 2 in 1998; 2 in 2004.* **Sources: Ho13, To9.**

Hillbilly - (West Virginia Hillbilly) - 85-94 days, Indet., smooth, ribbed, 4-5 in. beefsteak fruits, 1-2 lbs., yellow-orange skin mottled with red, flesh is streaked with red, low acid, Ohio heirloom originally from West Virginia. *Source History: 2 in 1994; 14 in 1998; 34 in 2004.* **Sources: Be4, CA25, CO30, Ers, Fo7, Gr29, He8, He17, He18, HO2, Ho13, HPS, Hud, Loc, ME9, Mel, Ra5, Ra6, Ri12, RUP, Se17, Se26, Se28, Shu, Sk2, Sw9, Ta5, To1, To3, To9, To10, Vi4, Vi5, Wo8.**

Hillbilly Potato Leaf - 85 days, Indet., gorgeous slicing tomato, sweet juicy yellow flesh is streaked with red, 4-6 in. flattened fruit, 1 lb., heavy producer, Ohio. *Source History: 2 in 2004.* **Sources: Pr9, Se16.**

Holy Land, Yellow Strain - Flavorful large fruit, yellow flesh with some red marbling, from Israel. *Source History: 1 in 2004.* **Sources: Ho13.**

Indian - 90 days, Indet., 8 oz., reddish green or bronze fruit with green shoulders, acid taste. *Source History: 1 in 1998; 2 in 2004.* **Sources: Ho13, Sa9.**

Indische Fleisch - 80 days, Indet., beautiful brown-purple fruits, dark olive shoulders, 6-16 oz., mahogany-red flesh is juicy with rich, spicy, earthy flavor, productive, from Germany. *Source History: 2 in 2004.* **Sources: Ma18, To9.**

Isis - 70-84 days, Indet., creamy yellow, ribbed beefsteaks, 4 oz., fruity sweet flesh, low acid. *Source History: 3 in 2004.* **Sources: Ma18, Sa5, Sa9.**

Ivory Egg - 70-75 days, Indet., cream colored fruits, egg size, good for salad and sauce, prolific. *Source History: 3 in 2004.* **Sources: Ma18, Ta5, Up2.**

Japanese Trifele Black - 80 days, Indet., potato leaf, beautiful pear shaped fruit, 5-8 oz., mahogany and black skin with green shoulders, flavorful. *Source History: 4 in 2004.* **Sources: Ea4, Ho13, Ma18, So25.**

La Carotina - 75 days, Indet., vigorous productive plants, 2 in., blood-orange fruits, 7 oz., high carotene content. *Source History: 2 in 2004.* **Sources: Ma18, To9.**

Lenny and Gracies Kentucky Heirloom - 80-85 days, Indet., light yellow 4 in. beefsteak blushed with light pink on the bottom, 6-8 oz., sweet, old-fashioned flavor. *Source History: 3 in 2004.* **Sources: Ho13, Ma18, To9.**

Liberty Bell - 80 days, Indet., 4-5 oz., red to green stuffing tomato, same shape as bell peppers, thick shell wall, mild flavor, small seed core that looks like a strawberry is easily removed. *Source History: 1 in 1981; 1 in 1984; 2 in 1987; 2 in 1991; 2 in 1994; 1 in 1998; 6 in 2004.* **Sources: Ma18, Ra5, Ra6, Re8, Ta5, To9.**

Liberty Tree - Indet., 4-5 in. fruit, pick when almost red, ripens indoors, does not tolerate heat, long season. *Source History: 1 in 2004.* **Sources: Sk2.**

Lime Green Salad - 58-85 days, Compact, rugose plant bears juicy, 2.5-3 oz. fruits, beautiful lime-green flesh with tangy flavor, skin turns from apple-green to yellow-green when ripe, early, from Tom Wagner, independent breeder. *Source History: 1 in 1994; 2 in 1998; 5 in 2004.* **Sources: Ga1, Sa9, So25, To1, To9.**

Little White - Plant bears many pale yellow fruits. *Source History: 1 in 1998; 1 in 2004.* **Sources: Se17.**

Livingston's Golden Queen - Med. size yellow fruit, blushed with red at base, mild flavor with very good sweet/acid tomato taste, good yields, listed in Livingston's 1887 Seed Annual. *Source History: 1 in 2004.* **Sources: Ea4.**

Long-Keeper, Burpee's - 78-90 days, Semi-det., light

golden-orange to red skin, high acid med-red flesh, resists fruit rot, gather unblemished fruits before frost, if properly stored will keep in good condition for up to 12 weeks. *Source History: 11 in 1981; 10 in 1984; 16 in 1987; 33 in 1991; 42 in 1994; 35 in 1998; 40 in 2004.* **Sources: Bou, Bu1, Bu2, CA25, Com, Dam, Ers, Fi1, Fo7, Ga1, Ga22, Ge2, Gr27, Gur, He8, He17, Hi13, Hig, HO2, Lin, Mo13, Pin, Ra5, Ra6, Re8, RUP, Se25, Se26, Sk2, So1, So25, St18, Ter, To1, To3, Tt2, Up2, Vi4, We19, Wet.**

Mammoth German Gold - (German Gold) - 85 days, Late season, indet. large golden fruits with red streaks throughout, 2+ lbs., looks like fresh peaches when cut up, introduced in late 1800s. *Source History: 1 in 1987; 1 in 1991; 2 in 1994; 6 in 1998; 6 in 2004.* **Sources: He18, Lan, Ma18, Roh, To1, Up2.**

Marizol Black - Round, 4 in. fruit, purple-bronze skin with green shoulders, productive, Germany. *Source History: 1 in 2004.* **Sources: Ho13.**

Marizol Gold - 85-90 days, Indet., large 1-2 lb. yellow fruit with red blush on the bottom, sweet taste, resists cracking, from the Black Forest region of Germany, dates to the 1800s. *Source History: 5 in 1998; 8 in 2004.* **Sources: CO30, He18, Ho13, La1, Ma18, Ta5, To1, To9.**

Marizol Majic - 70 days, Cherry to salad size, multi-colored fruit with fruity flavor, rare. *Source History: 1 in 2004.* **Sources: So25.**

Marvel Striped - 85-110 days, Vigorous indet. vines bear large heart-shaped orange and yellow fruits with red splashy stripes, 12-16 oz., very sweet, prolific, withstand heat and drought, 5th generation seeds from Oaxaca, Mexico, heirloom from the Zapotec people of southern Mexico. *Source History: 1 in 1987; 2 in 1991; 5 in 1994; 5 in 1998; 14 in 2004.* **Sources: Ami, CO30, Co32, Ma18, Na6, Orn, Ra6, Ri12, Se7, Se26, So25, Syn, To1, To9.**

Mary Robinson's German Bi-Color - 80 days, Indet., gold flesh with red streaks, slightly pointed at blossom end, 1+ lb., excellent taste, no blemishes, productive. *Source History: 1 in 1998; 4 in 2004.* **Sources: Ea4, Ho13, Ma18, Ta5.**

Mirabelle Blanche - 75-80 days, Indet., round, creamy-white cherry tomato with pink tinge on the bottom, sweet flavor, from France. *Source History: 3 in 1998; 3 in 2004.* **Sources: Coo, Ho13, Ma18.**

Moldovan Green - 80 days, Semi-det., semi-flattened, green-gold fruits, 10 oz., very rare, from Moldova. *Source History: 2 in 1998; 1 in 2004.* **Sources: Sa9.**

Mortgage Lifter Bicolor Strain - 80 days, Bicolor strain of Mortgage Lifter, huge indet. vines, 1 lb. beefsteaks, golden yellow flesh marbled with red, very good tomato flavor, heirloom. *Source History: 2 in 2004.* **Sources: To1, To9.**

Mr. Brown - 80 days, Indet., beautiful slicer with chocolate color flesh and green seed gel, rich sweet flavor, dev. by Jeff Dawson. *Source History: 1 in 2004.* **Sources: To9.**

Mr. Stripey - 80 days, Highly productive English heirloom, indet., golf ball size fruits, low acid, yellow flesh with pink stripes, pink heart, mild flavor. *Source History: 6 in 1994; 12 in 1998; 29 in 2004.* **Sources: Bo19, CA25, CO30, Ers, Ge2, Gr27, He17, He18, Ho13, HPS, Jun, La1, Loc, MIC, Mo13, Pe7, Ra5, Ra6, Re8, Ri12, SE14, Se25, Se26, Sh9, Shu, To3, To10, Vi4, WE10.**

Mystery Keeper - Ripens from the inside out, harvest just after green color lightens before frost, slight acid flavor, requires no special care after harvest. *Source History: 1 in 2004.* **Sources: Ma13.**

Noir de Crimee - 75-80 days, Indet., 4-8 oz. fruit, deep purple-black with green shoulders at maturity, rich dusky sweet flavor with smoky overtone, strain of Black Krim which it resembles, Russian heirloom. *Source History: 5 in 2004.* **Sources: Ap6, He18, ME9, Sa9, To10.**

Noire Charbonneuse - 70-90 days, Indet., purple-brown fruit, green shoulders, 8-12 oz., superb flavor, wonderful tomato smell, translates "black like coal." *Source History: 4 in 2004.* **Sources: Ho13, Ma18, So25, Ta5.**

Northern Lights - 55-75 days, Indet., yellow-orange fruit with red blush on bottom, red center, 8-12 oz., 4 in., intense flavor. *Source History: 1 in 1998; 2 in 2004.* **Sources: To1, To9.**

Nyagous - 80 days, Indet., 6 oz., flattened, gray-brown globes, blemish-free, baseball size, clusters of up to 6 fruit, excel. full flavor, very productive, great for markets, from Germany via Reinhard Kraft. *Source History: 1 in 1998; 2 in 2004.* **Sources: Sa9, Se16.**

Oaxacan Jewel - 75-85 days, Indet., potato leaf, beautiful golden orange flesh with deep red streaks throughout, sweet, fruity flavor, from the mountains of the state of Oaxaca, Mexico. *Source History: 4 in 2004.* **Sources: Ho13, Ma18, So25, Ta5.**

Old Flame - 80-90 days, Huge beefsteak type with fruits up to 2 lbs., pink-red, meaty flesh with a yellow streak radiating through, rich flavor, very productive. *Source History: 2 in 1998; 4 in 2004.* **Sources: Eco, Gr28, Se17, Te4.**

Old German - 75 days, Best of several varieties obtained from the Virginia Mennonite community, indet., yellow fruits with a red center visible on the surface and throughout the core, unusual boat shape, often weigh over a pound, outstanding flavor, few seeds, not a heavy producer, not drought tolerant but flavor and color make it worth growing, CV So1 introduction, 1985. *Source History: 1 in 1987; 1 in 1991; 2 in 1994; 11 in 1998; 33 in 2004.* **Sources: Ami, Ba8, Bo19, CA25, CO30, Ers, Ga1, Gr29, He18, Ho13, HPS, Loc, ME9, Pe2, Pep, Ra5, Ra6, Ri12, RUP, SE14, Se26, Se28, Shu, So1, Sw9, Ta5, TE7, To1, To3, To10, Ver, Vi4, Wo8.**

Old Yellow Candystripe - 77-83 days, Indet., large, flattened, bi-color fruits, pale yellow flesh with semi-pink center, sweet flavor with good acid balance, productive. *Source History: 1 in 1998; 2 in 2004.* **Sources: Sa9, So25.**

Oliver's German White - Vigorous vines grow 4.5 ft. tall, lower branches grow sideways 7-8 ft., white skinned fruit is absolutely pure white throughout, delicate mild flavor, known as Weisse (White) in Germany. *Source History: 1 in 2004.* **Sources: Ami.**

Orange Russian 117 - 83 days, Russian 117 x Georgia

Streak cross, large, 1-2 lb., heart shape fruit, yellow-orange skin with pink blush, meaty flesh with red center, flavor is a perfect balance of fruit and acid, dev. by Jeff Dawson. *Source History: 1 in 1998; 1 in 2004.* **Sources: To9.**

Paul Robeson - 74 days, Russian heirloom, indet., round, 4 in., slightly flattened fruit with brown to dusky dark red skin, brown center, dark green shoulders, earthy flavor, original seed from Marina Danilenko, a Moscow seedswoman, named after Paul Robeson, operatic vocal artist who was an advocate of equal rights for Blacks. *Source History: 2 in 1998; 2 in 2004.* **Sources: So1, To9.**

Peach Blow Sutton - 70-92 days, Indet., attractive med. size fruit, blend of red, yellow and orange flesh colors, mild flavor, skin is sticky as fruit ripens. *Source History: 4 in 2004.* **Sources: Ga1, He18, Sa9, TE7.**

Peppermint - 81 days, Indet., med. size fruit with red-pink skin with yellow stripes before ripe, 8-12 oz., meaty flesh, sweet. *Source History: 1 in 1998; 1 in 2004.* **Sources: To9.**

Petite Pomme Blanche - Flavorful white cherry tinted with yellow, very productive, midseason. *Source History: 2 in 2004.* **Sources: Ea4, Gr28.**

Pineapple - 75-95 days, Indet. vines, large beefsteak fruit up to 2 lbs., yellow skin color streaked with red at blossom end, pink streaks radiate throughout the yellow flesh, few seeds, pretty sliced, excellent mild flavor. *Source History: 3 in 1991; 7 in 1994; 19 in 1998; 44 in 2004.* **Sources: Ba8, Bo9, Bo19, CA25, CO30, Ers, Fo7, Ga1, Go8, Gr27, Gr28, Gr29, He8, He18, HO2, Ho13, HPS, Hud, Jun, Loc, Ma18, ME9, Ni1, Pin, Ra5, Ra6, Ri12, Sa5, SE14, Se17, Se26, Se28, Sh13, Shu, Sw9, Ter, To1, To3, To9, Ver, Vi4, Vi5, WE10, Wo8.**

Pineapple/Fog - Natural cross of Pineapple x San Francisco Fog, large cherry fruit with flavor of Pineapple and body of San Francisco Fog. *Source History: 1 in 2004.* **Sources: Sa5.**

Pink Grapefruit - 65-70 days, Vines to 3.5-4 ft., med-sized fruits, yellow skin and pink flesh like a pink grapefruit, 4-6 oz., delicious mild flavor, low acid content, dates back to 1900, located again with the help of the SSE. *Source History: 1 in 1981; 1 in 1984; 2 in 1987; 3 in 1991; 3 in 1994; 6 in 1998; 5 in 2004.* **Sources: Ami, Ho13, Lan, Ma18, Roh.**

Pink Lemon - 75 days, Indet., large, 5 in., 8-10 oz., round, flattened, yellow-orange flesh with red burst in the center, rich fruity flavor. *Source History: 1 in 1998; 1 in 2004.* **Sources: To9.**

Pink-White - Light pink-white fruit, 1-2 lbs., nice flavor. *Source History: 1 in 2004.* **Sources: Ho13.**

Pixie Striped - 85 days, Dwarf plant produces 2 in. red fruit striped with yellow, hollow seed cavity, average flavor, ideal for containers, late season. *Source History: 1 in 2004.* **Sources: To9.**

Poll Robson Angolan - 65 days, Indet. plants produce greenish red-brown globes, about 6 oz. *Source History: 1 in 1998; 1 in 2004.* **Sources: Sa9.**

Potato Leaf White - 80-85 days, Novelty, indet. potato leaf plants, creamy white round fruits with sweet mild tasting flesh. *Source History: 4 in 1994; 4 in 1998; 3 in 2004.* **Sources: Re8, Sa9, To3.**

Purple Russian - 80-86 days, Indet., pink-purple sauce tomato, short sausage shape with pointed ends, 6 oz., excel. sweet flavor, heavy yields, good cold tol., no cracks or bad shoulders, Ukraine. *Source History: 4 in 2004.* **Sources: Ma18, Se16, So25, To9.**

Red Belly - Large fruit up to 2 lbs., yellow flesh with red blossom end and red stripes radiating up toward the stem, mild non-acid flavor, not prone to rot, late. *Source History: 1 in 2004.* **Sources: Te4.**

Red Georgia - 90 days, Very productive, indet. vines, sel. from Georgia Streak x Stupice, 2.5 in., round, slightly flattened fruit, pink-red skin with very pale yellow stripes, sweet fruity flavor, may also produce orange fruit. *Source History: 1 in 1998; 1 in 2004.* **Sources: To9.**

Red Gold Stripe - 75-80 days, British var., indet., red skin with golden streaks, 8 oz., better flavor than Tigerella. *Source History: 3 in 1998; 6 in 2004.* **Sources: CO30, Ga1, He18, Loc, Ra6, RUP.**

Red Marbled Yellow Bell - 76 days, Indet., bell-shaped, oblong fruit, yellow blushed with red, fruity flavor. *Source History: 1 in 2004.* **Sources: To9.**

Red Streak - 86 days, Indet., 12 oz., bi-color beefsteak fruit. *Source History: 1 in 2004.* **Sources: Sa9.**

Red Yellow Cap - 78 days, Indet., yellow-orange fruit blushed with red, 3-4 in., fruity flavor, heirloom. *Source History: 1 in 2004.* **Sources: To9.**

Red Zebra - Yellow-red striped fruit. *Source History: 1 in 2004.* **Sources: Ga7.**

Regina's Yellow - (Regina's Bicolor) - 80-90 days, Indet., regular leaf, 8 oz. fruit, red with yellow streaks, sweet fruity flavor, from Regina Yanci, Ohio, rare. *Source History: 5 in 2004.* **Sources: Ea4, He18, Sa9, Ta5, To9.**

Rose Beauty - Indet., regular leaf, pale yellow fruit with rose streaks on blossom end, good flavor, moderately productive. *Source History: 1 in 2004.* **Sources: Up2.**

Ruby Gold - (Georgia Streak) - 85 days, West Virginia heirloom from the Ben Quisenberry collection, indet., large yellow fruits streaked with red, pink or red blush at blossom end, sweet flavor, meaty flesh. *Source History: 1 in 1981; 1 in 1984; 1 in 1987; 2 in 1991; 3 in 1998; 1 in 2004.* **Sources: Fe5.**

Ruby Treasure - 78-85 days, Firm, 6-8 oz. red fruit, good storage variety, excellent flavor and aroma increases as it ripens, will keep in storage for 2-3 months when harvested green, PSR. *Source History: 1 in 1994; 3 in 1998; 3 in 2004.* **Sources: Bou, Pe6, Pr3.**

Sara Black - Indet., very tasty, med. size green-black fruit, heirloom from Germany. *Source History: 1 in 2004.* **Sources: Ea4.**

Schimmeig Creg - 75 days, Indet., 4-6 oz. red fruit with yellow stripes. *Source History: 2 in 2004.* **Sources: Se17, To10.**

Schimmeig Striped Hollow - 80-85 days, Indet., bell pepper shaped, hollow fruit with red and orange stripes, 2.5-5 oz., high yields, from Germany. *Source History: 3 in 1998; 7 in 2004.* **Sources: Ami, Fe5, Ho13, Ma18, Ra6, Ri12, Ter.**

Shah - (White Brandywine, Mikado White) - 80 days,

Indet., potato leaf, 8-12 oz. flattened globes, yellow-white flesh, superb flavor is full-bodied and sweet with a hint of pear, concentrated flavor when dried, high yields, may be a sport of Turner's Hybrid which was sold by W. Atlee Burpee in 1886, name is taken from the 1800s Gilbert and Sullivan Operetta. *Source History: 4 in 2004.* **Sources: Ami, Ea4, So25, Und.**

Siniy - 70 days, Det., potato leaf, pink-black, slightly ribbed oblate beefsteak fruit, 8-12 oz., from Estonia. *Source History: 1 in 2004.* **Sources: Ta5.**

Slovenian Black - 80 days, Indet., round, garnet-brown fruit, large yields. *Source History: 1 in 2004.* **Sources: Ta5.**

Snow White, Super - 75 days, Larger than Snow White but otherwise similar, ivory fruit ripens to pale yellow, good size for salads, indeterminate. *Source History: 2 in 1998; 3 in 2004.* **Sources: CO30, To1, To9.**

Snowball - (White Beauty) - 78 days, Indet., 5 ft. plants, round, flattened, 3-3.5 in. white fruits, flesh is almost white, few seeds, very mild, makes a beautiful white sauce. *Source History: 2 in 1981; 1 in 1984; 1 in 1987; 1 in 1991; 2 in 1994; 3 in 1998; 2 in 2004.* **Sources: Lej, Shu.**

Snowstorm - 70 days, Indet., excellent production of 3 in. round fruit, cream color blushed with golden to peach when ripe, sweet fruity flavor, long keeper, disease resistant. *Source History: 2 in 2004.* **Sources: Ma18, So25.**

Southern Night - 80-84 days, Very rare, old traditional Russian variety, leafy, 40 in., potato leaf plants, round, slightly flattened, 3.5 in., black-red fruits, difficult to grow, name refers to pitch-black nights in southern territories of the former Soviet Union, unlike the white nights in Moscow. *Source History: 1 in 1994; 5 in 1998; 8 in 2004.* **Sources: Fo13, Ho13, Ma18, Ra5, Re8, Se17, Ta5, To1.**

Speckled Roman - 85 days, Dev. by SSE member John Swenson as a result of a stabilized cross of Antique Roman x Banana Legs, very productive indet., 3 in. wide by 5 in. long fruit with jagged orange and yellow striped skin, meaty flesh, few seeds, great tomato taste, one of the best for processing. *Source History: 11 in 2004.* **Sources: Al25, Ami, Ba8, Co31, Ho13, Ma18, Mcf, Se16, So25, Und, Up2.**

Striped Cavern - 80-100 days, Indet., firm blocky thick-walled fruits, very meaty, core is near the stem, average 8 oz. at maturity, abundant yields, keeps 4 weeks when harvested ripe--longer if semi-ripe, good home-grown flavor for eating fresh, superb for stuffing, firm walls do not split or crumble. *Source History: 1 in 1987; 2 in 1991; 1 in 1994; 3 in 1998; 9 in 2004.* **Sources: Ba8, He8, Pr3, Pr9, Ra6, Sa9, Se16, Tho, To1.**

Striped German - 78-90 days, Medium tall indet. plants, red and yellow skin and flesh color, huge fruits up to 2.5 lbs., fruity flavor, beautiful sliced, ripens late August and September in the North, from Hampshire County, WV. *Source History: 1 in 1994; 4 in 1998; 10 in 2004.* **Sources: Ap6, CO30, He18, Loc, Ma18, ME9, Ra5, Ra6, Re8, Te4.**

Striped Roman Yellow - Indet., yellow with red streaks, the reverse of Striped Roman, early. *Source History: 1 in 2004.* **Sources: Up2.**

Striped Stuffer - 85-90 days, Stuffing tomato with strong walls, 6 oz. fruit, red striped with orange. *Source History: 1 in 2004.* **Sources: Hi13.**

Sun and Snow - Indet. vines produce beautifully shaped, 4-8 oz. fruits with iridescent yellow skin, flesh is snow-white at early ripening stages, when fully ripe both skin and flesh may be yellow, pleasant flavor, heavy yields, resists some common diseases, PSR Breeding Program. *Source History: 1 in 1994; 1 in 1998; 1 in 2004.* **Sources: Pe6.**

Sutton - (Sutton White) - 75-83 days, Indet., cream colored fruit, 8 oz., blushed with yellow when ripe, all shapes and sizes, sweet-tart flavor, very rare. *Source History: 2 in 1998; 3 in 2004.* **Sources: Ma18, Sa9, To9.**

Tiger - Fruit matures from light green with dark green stripes to red with orange stripes, 2 in. diameter, tangy flavor, high yields. *Source History: 1 in 2004.* **Sources: Ho13.**

Tiger Stripe - Small to medium size fruit, red-orange with yellow streaks, tender skin, very sweet. *Source History: 1 in 1994; 2 in 1998; 1 in 2004.* **Sources: Sa5.**

Tiger Tom - 70-75 days, Productive, 6 ft. plants, round, 2-3 in., red fruits with jagged golden stripes, delightful tart flavor, believed to have been developed in Slovakia (formerly Czechoslovakia), from Ben Quisenberry. *Source History: 2 in 1994; 3 in 1998; 5 in 2004.* **Sources: Go8, Pr9, Sa9, Se16, Vi4.**

Tigerella - 55-75 days, Indet. plants, huge crops of red fruits with orange stripes, 1.5-2 in., immature fruits light green with dark-green stripes, tangy tomato flavor. *Source History: 3 in 1981; 1 in 1984; 1 in 1987; 6 in 1991; 8 in 1994; 8 in 1998; 25 in 2004.* **Sources: Al25, CO30, Ers, Ga1, Gr28, He8, He18, Hud, ME9, Mi12, Pr3, Ra5, Ra6, Re8, Se7, SE14, So25, Syn, Tho, To1, To3, To9, Up2, Ver, WE10.**

Tiger-Like - 72 days, The tastiest striped tomato, det., 2 in. fruit, orange-red skin striped with yellow, sweet, complex flavor, from Russia. *Source History: 1 in 2004.* **Sources: Ter.**

Tigerly - 78 days, Moldovan var. with large striped red fruit. *Source History: 1 in 1998; 1 in 2004.* **Sources: Sa9.**

Tomalillo - Tiny light green cherry tomato, good flavor, great for garnish, will crack if picked too late, very rare. *Source History: 1 in 2004.* **Sources: Ea4.**

Tomango - 88 days, Stuffing type, red with yellow streaks. *Source History: 1 in 1998; 1 in 2004.* **Sources: Sa9.**

Tonnelet - 75 days, Indet., unique tomato from Belgium, beautiful, egg shaped fruit, red and yellow striped skin, thick red flesh, delicious flavor, yields until frost. *Source History: 2 in 1998; 1 in 2004.* **Sources: To9.**

Transparent - 75 days, Resembles Lutescent, flattened globes, clear color, mild, juicy flesh, blight tolerant, skin has the feel of "gummy bears". *Source History: 1 in 2004.* **Sources: Sa9.**

Turkish Striped Monastery - Indet., high yields of 2 in. red fruits striped with gold, possibly the best tasting striped tomato, long season, very rare to endangered, seed collected by CV Ea4 from a monastery garden outside Istanbul. *Source History: 1 in 2004.* **Sources: Ea4.**

Wagon Wheel - 90 days, Indet., large 1 lb. fruit with yellow and red stripes, flavorful. *Source History: 1 in 1998;*

1 in 2004. **Sources: Se17.**

White - 86-95 days, Indet., fruit ripens to yellowish white throughout at maturity, delicate flavor. *Source History: 1 in 1991; 2 in 1994; 3 in 1998; 2 in 2004.* **Sources: Ho13, Sa9.**

White Beauty - (Snowball) - 76-85 days, Semi-indet., 8 oz. green fruits ripen to ivory-white, firm flesh is almost snow-white, delicious mild flavor, tastes red, high sugar content, crack resistant, tolerates shade, does well in hot, humid areas, acquired by Ben Quisenberry in 1982 from John Parker of Toledo, OH, dates back to the 1920s. *Source History: 8 in 1981; 6 in 1984; 10 in 1987; 10 in 1991; 10 in 1994; 8 in 1998; 8 in 2004.* **Sources: Bo9, Bu1, Co32, Ea4, Se16, Se26, Syn, To1.**

White Bush - 75-79 days, Det., compact plant requires no staking, med. size 2 in. fruit with clear skin, great mild fruity flavor, low acid, good slicer, crack resistant. *Source History: 1 in 1998; 3 in 2004.* **Sources: Sa5, To1, To9.**

White Heirloom - 75 days, Indet., abundant yields of cream colored beefsteaks, 4-12 oz., flavorful. *Source History: 1 in 2004.* **Sources: Ma18.**

White Potato Leaf - 75-85 days, Indet. with potato leaf foliage, flattened 4-5 in. fruits, creamy white to pale yellow skin color, pure white flesh, juicy and mild with an acidic yet creamy flavor. *Source History: 3 in 2004.* **Sources: Ami, Ra5, Ra6.**

White Princess - (Jumbo White) - 85 days, The best of the "whites" with very large, 1-2 lbs., beautiful cream-white fruits, very good flavor. *Source History: 1 in 1998; 1 in 2004.* **Sources: Ho13.**

White Queen - 85 days, Indet., 3 in., round, flat, white fruit, 8-12 oz., ribbed shoulders, some pink blush or streaks on blossom end, no cracks, flavorful, heirloom. *Source History: 2 in 1998; 6 in 2004.* **Sources: Ap6, Ea4, Sa9, To1, To9, Up2.**

White Rabbit - 76-95 days, large indet., sprawling plant bears clusters of tiny pea-size fruit with a sweet tropical flavor, creamy white color, dev. by Joe Bratka. *Source History: 4 in 2004.* **Sources: Gr28, Ho13, Ma18, To9.**

White Wonder - 80-90 days, Indet., 4 oz., large firm and creamy-white fruit, high sugar content (studies show that all tomatoes are acidic, light-colored ones are not actually sub-acid, they just have more sugar). *Source History: 1 in 1981; 1 in 1984; 2 in 1987; 2 in 1991; 7 in 1994; 8 in 1998; 26 in 2004.* **Sources: Ag7, Al25, Ba8, Bo19, Bu2, CO30, Coo, Ge2, Gr29, He8, Jun, La1, ME9, Pe6, Ra5, Ra6, Ri12, Sa9, SE14, Se28, So1, To3, To10, Vi4, Vi5, Wo8.**

Winter Gold - Largest, most rot resistant longkeeping tomato, productive plants with rugose foliage, 4-8 oz. fruits are a rich gold inside and out, crisp texture is similar to a pepper or apple, tangy to tart flavor, uniform ripening, begins to soften after 2 or 3 months in storage, stores better than the well-known Longkeeper, PSR. *Source History: 1 in 1994; 1 in 1998; 1 in 2004.* **Sources: Pr3.**

Winterkeeper - 9291 days, Indet., 10 oz. fruit, ripens in storage, green color turns pale yellow outside and red inside. *Source History: 1 in 2004.* **Sources: Sa9.**

Yellow Heartstock - Indet., regular leaf, yellow fruit with reddish tint on the blossom end, good flavor, late. *Source History: 1 in 2004.* **Sources: Up2.**

Yellow Out Red In - 85-99 days, Semi-det., solid, 6 oz. globes, yellow skin, red flesh, acid, good keeper, very tart. *Source History: 1 in 1998; 2 in 2004.* **Sources: Ma18, Sa9.**

Yellow Stone - 80 days, Prolific indet. plant bears round orange-yellow fruit with red blush on blossom end, 12 oz., mild fruity taste. *Source History: 1 in 2004.* **Sources: To9.**

Zigan - (Gipsy) - Rare Russian variety, dark red-purple skin streaked with green and greenish shoulders, best flavor before fully ripe, otherwise bland. *Source History: 1 in 2004.* **Sources: So25.**

Varieties dropped since 1981 (and the year dropped):

Amish Gold (2004), *Arctic Lights* (2004), *Blushing* (1984), *Centerpiece* (2004), *Cherry, Tigerette* (2004), *Deweese Streaked* (2004), *Dourne d'Hivre* (2004), *Dwarf Gold Treasure* (2004), *Early Stripe* (1984), *Etoile Blanche d'Anvers* (2004), *Evergreen Improved* (2004), *Forever Green* (2004), *Green Apple* (2004), *Green Thumb* (2004), *GT 12-22-41* (1998), *Heavy Stripes* (2004), *K-3* (1987), *K-5* (1987), *KT-3* (1987), *KT-4* (1987), *L-1* (1987), *Malaysian M'sia* (1998), *Missouri* (1998), *Neverwill* (2004), *Olympic Flame* (2004), *P-1* (1987), *P-2* (1987), *P-3* (1987), *P-4* (1987), *P-5* (1987), *Peach* (1987), *Ponderosa Sunrise* (2004), *Potato Red Skins* (2004), *Strain B* (1994), *Sunset* (2004), *T-1* (1987), *T-2* (1987), *T-3* (1987), *T-4* (1987), *T-5* (1987), *Thompson Seedless Grape* (2004), *Tiger Fuzz* (2004), *Tiny Tiger* (2004).

Tomato / Other Species
Lycopersicon spp.

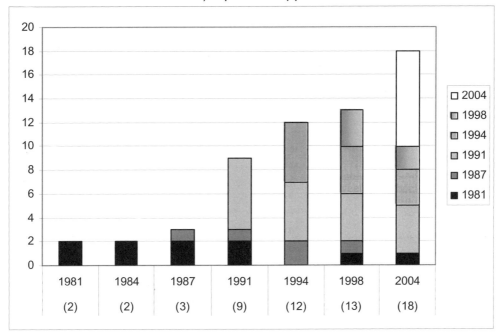

Number of varieties listed in 1981 and still available in 2004: 2 to 1 (50%).
Overall change in number of varieties from 1981 to 2004: 2 to 18 (900%).
Number of 2004 varieties available from 1 or 2 sources: 11 out of 18 (61%).

Broad Ripple Yellow Currant - 75 days, Tall, vigorous, indet. plants, incredibly prolific all season, tiny .5 in., round yellow fruits, found growing in a street crack in Indianapolis, IN, early 1900s. *Source History: 1 in 1994; 2 in 1998; 5 in 2004.* **Sources: Co32, Pr9, Ri12, Se16, Yu2.**

Cherry, Matt's Wild - 45-60 days, Indet., richly flavored, small fruit, high sugar content, blight res., these grow wild in Mexico, seed from Hidalgo, Mexico via Teresa Arellanos de Mena, friend of former U of ME ag faculty members, Drs. Laura Merrick and Matt Liebman. *Source History: 4 in 1998; 14 in 2004.* **Sources: Ami, Co31, Go8, Gr28, He8, Hi6, Ho13, Jo1, Ma18, Mi12, Se7, St18, Te4, To10.**

Cherry, Wild - (Currant Tomato) - 60 days, *L. pimpinellifolium*, red dime-size wild tomato from Mexico, rampant grower that could cover an arbor or trellis. *Source History: 1 in 1991; 1 in 1994; 1 in 2004.* **Sources: To1.**

Ciudad Victoria - 64 days, *L. esculentum* var. *cerasiforme*, small round sweet fruits borne in clusters of 9-13, late maturing, never cracks, even after a heavy rain, semi-cultivated tomato from dooryard gardens in Ciudad Victoria, Tamaulipas, Mexico. *Source History: 1 in 1991; 1 in 1994; 1 in 1998; 5 in 2004.* **Sources: Ba8, Ea4, Fe5, Ho13, Na2.**

Coyote - 60-75 days, Indet., clusters of pea size, cream-yellow fruit, very sweet flavor, said to be wild, from Vera Cruz, Mexico. *Source History: 6 in 2004.* **Sources: Ga22, Gr28, Ho13, Ma13, Ma18, To9.**

Cuatomate - 60-70 days, Semi-det., continuous production of small round .75 in. fruit, orange > red, sweet-tart taste, germination can take up to 2 weeks, originally from the Zapotecs in southern Oaxaca, Mexico. *Source History: 1 in 2004.* **Sources: Ag7.**

Currant Tomato (Holds Fruit) - (Non-Shattering, Red Currant, Raisin Tomato) - 70 days, *L. pimpinellifolium*, midseason, long trusses of tiny red pea-size tomatoes on indeterminate vines, this variety holds its fruit until picked, can be crossed with regular tomatoes even though they are different species. *Source History: 1 in 1981; 1 in 1984; 5 in 1987; 6 in 1991; 2 in 1994; 2 in 1998; 2 in 2004.* **Sources: Hud, Sa9.**

Currant, Gold Rush - 75-80 days, *L. pimpinellifolium*, indet., .25 in. fruits borne in trusses of 10-12, chosen for its manageable growth habit and excel. fruit set, hundreds of fruits per plant, excel. clean flavor, fruit holds on the vine, Dutch seedsman selection. *Source History: 1 in 2004.* **Sources: Se16.**

Currant, Lemon Drop - 68 days, *L. pimpinellifolium*, strong vigorous indet. vines produce high yields of round, lemon- yellow .4 in. fruits borne in clusters, sweet yellow flesh, adds color to salads. *Source History: 1 in 1994; 3 in 1998; 4 in 2004.* **Sources: Fo7, Ga1, Pe6, Se16.**

Currant, Red - 65-80 days, *L. pimpinellifolium*, indet., clustered currant-sized fruits, intensely flavored, packets of red currant seeds may produce 10% yellow currant plants. *Source History: 3 in 1991; 9 in 1994; 17 in 1998; 39 in 2004.* **Sources: Bo17, Bo19, CO30, Co32, Coo, Fo13, Ga1, Ge2, Hi6, Ho13, Ki4, Loc, ME9, Mi12, Na6, Ni1, Pin, Pr3, Ra5, Ra6, Re8, Ri2, RUP, Se7, Se8, SE14, Se26, So1, Sw9,**

Te4, TE7, To1, To3, Twi, Up2, Vi4, WE10, We19, Yu2.

Currant, Sweet Pea - 75-80 days, *L. pimpinellifolium*, indet., great companion for Gold Rush, hundreds of fruits per plant borne in trusses of 10-12, holds on the vine, excel. clean flavor. *Source History: 1 in 2004.* **Sources: Se16.**

Currant, White - Large vining plants bear clusters of tiny fruit, cream-white with yellow tint, superb sweet flavor, huge yields, self-seeding, wild type from Mexico. *Source History: 2 in 2004.* **Sources: Ami, Ba8.**

Currant, Yellow - (Gold Currant) - 70-83 days, *L. pimpinellifolium*, indet. vines with heavy foliage, pale yellow, currant- sized fruit grows in clusters, intensely flavored, disease resistant, rare. *Source History: 4 in 1991; 12 in 1994; 11 in 1998; 28 in 2004.* **Sources: Bo17, Bo19, Bou, CO30, Coo, Ge2, Gr28, Ho13, Ki4, ME9, Mi12, Na6, Orn, Pin, Pr3, Ri2, RUP, Se7, Se8, SE14, Se26, Sw9, Syn, Te4, TE7, To1, To9, WE10.**

L. cheesemanii f. minor - Small, bright orange, pear-oval shaped fruit with a piercing sweetness, used in plant breeding because of its high Beta Carotene content, very rare. *Source History: 2 in 2004.* **Sources: Ba8, Sa5.**

L. Columbianum - 76 days, Indet., wild cherry tomato with intense tomato taste, very good disease resistance, from Columbia. *Source History: 1 in 2004.* **Sources: To9.**

Punta Banda - *L. esculentum*, plants produce hundreds of red, meaty, thick skinned fruits despite less than ideal conditions, great paste tomato, collected on the Punta Banda Peninsula in Baja, CA. *Source History: 1 in 1998; 2 in 2004.* **Sources: Hud, Na2.**

Texas Wild - 63-73 days, Low sprawling indet. plants, sweet red currant tomato with fruits from .25-.5 in., bears through summer heat until frost. *Source History: 1 in 1994; 1 in 1998; 1 in 2004.* **Sources: To9.**

Zlutakytice - Indet., tiny yellow fruit with pointed end, gooseberry size, midseason, from Australia. *Source History: 2 in 2004.* **Sources: Gr28, So25.**

Varieties dropped since 1981 (and the year dropped):

Chiapas Wild (2004), *Currant HF* (2004), *Currant Tomato (Drops Fruit)* (2004), *Currant, Puerto Escondido Red* (1994), *Giant Tree* (2004).

Turnip
Brassica rapa (Rapifera Group)

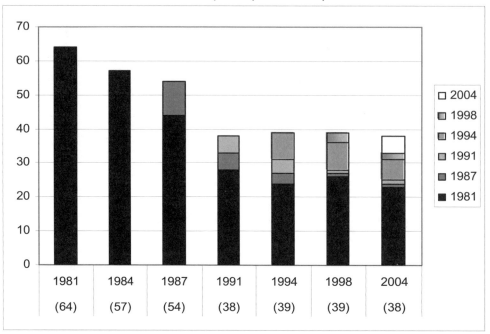

Number of varieties listed in 1981 and still available in 2004: 64 to 23 (36%).
Overall change in number of varieties from 1981 to 2004: 64 to 38 (59%).
Number of 2004 varieties available from 1 or 2 sources: 25 out of 38 (66%).

Amber Globe - (Yellow Globe) - 63-80 days, Smooth yellow 6 in. dia. globes, sweet pale-yellow flesh, fine grained, 18-20 in. overall height, for fall plantings, table use or stock feed, pre-1888 heirloom. *Source History: 13 in 1981; 10 in 1984; 13 in 1987; 10 in 1991; 7 in 1994; 8 in 1998; 8 in 2004.* **Sources: Ba8, Co31, DOR, La1, Mey, So1, SOU, Wet.**

Croissy - Elongated creamy white root from 6-9 in., looks like a Daikon, radish-like flavor is bitter but pleasant, very disease res. *Source History: 3 in 1994; 3 in 1998; 1 in 2004.* **Sources: Ho13.**

De Berlin - White turnip, shred and add to salad. *Source History: 1 in 1991; 1 in 1994; 1 in 1998; 1 in 2004.* **Sources: Lej.**

De Milan - (De Milan Rouge) - 35 days, Flat French turnip for spring planting, rose-colored shoulder and white bottom, fine tip, rapid producer, thrives in very cool spring weather, can pick young. *Source History: 1 in 1981; 1 in 1984; 2 in 1987; 2 in 1991; 3 in 1994; 4 in 1998; 2 in 2004.* **Sources: Ho13, Lej.**

Des Vertus Marteau - (Navet) - 50-60 days, Most popular turnip in Paris for over 100 years, 6 in. long roots, thicker at bottom, sweet tender white flesh, best when 2/3 grown, no turnip bite, popular with French market growers in the late 1800s. *Source History: 2 in 1981; 2 in 1984; 2 in 1987; 2 in 1991; 1 in 1994; 1 in 1998; 2 in 2004.* **Sources: Ba8, Lej.**

Early Italian White Red Top - Spring turnip or meirapen, plant in early spring for early mild turnips, pick when 3-4 in. wide. *Source History: 1 in 1981; 1 in 1984; 1 in 1987; 1 in 1991; 1 in 1994; 1 in 1998; 1 in 2004.* **Sources: Dam.**

Early Snowball - (Early Six Weeks) - 40-45 days, Perfectly round smooth white roots, firm tender fine-grained white flesh, grows rapidly, adapted to early spring plantings, excel. quality. *Source History: 6 in 1981; 5 in 1984; 3 in 1987; 1 in 1991; 1 in 1994; 1 in 1998; 1 in 2004.* **Sources: Ear.**

Fortress - 55 days, Greens used only, roots are not edible, upright plant habit enables clean harvest, deep green, thick, strap leaves, mild flavor, short stems allow for higher quality bunched and processed product, high tolerance to White Spot, cold and bolting. *Source History: 1 in 2004.* **Sources: HO2.**

Gilfeather® - 75-85 days, Unusually sweet, delicately mild, creamy-white, egg-shaped, 3 in. dia. average, but grows much larger without getting strong flavored, stores best in the ground, also stores well in cool fruit cellar, Vermont heirloom dev. by John Gilfeather in the late 1800s, trademarked. *Source History: 1 in 1981; 1 in 1984; 3 in 1987; 5 in 1991; 6 in 1994; 10 in 1998; 8 in 2004.* **Sources: Coo, Du7, Fe5, Jo1, Lan, Pr3, Roh, Ver.**

Golden Ball - (Gold Ball, Orange Jelly) - 38-70 days, Yellow 3-4 in. dia. globes, sweet tender mild deep golden-yellow flesh, very fine flavor, medium-sized cutleaf tops, keeps well, nice to cook for mashing, tastes more like a rutabaga than an early summer turnip, excel. for home gardens, 1859. *Source History: 15 in 1981; 13 in 1984; 14 in 1987; 12 in 1991; 14 in 1994; 18 in 1998; 19 in 2004.* **Sources: Ba8, CO30, Dam, Ear, Fe5, He8, Ho13, La1, ME9, Pin, Pr3, Ri12, Sa9, Se7, Se16, So12, Syn, TE7, WE10.**

Hinona Kabu Pickling - 40 days, Bright-purple and white elongated root, grows larger than average but also makes a great baby turnip for salads when just 1.5 in. long, good cooked or pickled as well. *Source History: 1 in 1987; 1 in 2004.* **Sources: Kit.**

Ivory - 35 days, Small cylindrical roots, ivory-white skin, mild taste, ideal for succession sowing. *Source History: 1 in 2004.* **Sources: Tho.**

Large Yellow Globe - 70 days, Light-yellow with green top, grows to large size, reliable winter keeper, good quality, popular var. for table use. *Source History: 1 in 1981; 1 in 1984; 1 in 1987; 1 in 1991; 1 in 1994; 1 in 1998; 1 in 2004.* **Sources: Com.**

Milan Purple Top - 30-35 days, Flat, round, white root with purple top, white flesh, good market variety, 19th century heirloom from Italy. *Source History: 1 in 1994; 1 in 1998; 1 in 2004.* **Sources: Ba8.**

Ochsenhorner - Rare old Swedish str., name translates "oxhorn", long twisted root grows out of the ground, white with purple top, strong turnip flavor, not as cold hardy as the round strains, long storage. *Source History: 1 in 2004.* **Sources: Ho13.**

Ohno Scarlet - 55 days, Red flattened globes, white flesh, red-veined greens, when pickled the red color spreads over the white making a fine red-fleshed turnip pickle, heirloom from Japan. *Source History: 4 in 1981; 3 in 1984; 2 in 1987; 2 in 1991; 2 in 1994; 2 in 1998; 2 in 2004.* **Sources: Ho13, Syn.**

Orange Jelly - (Golden Ball) - Old Scottish str., maincrop, 3-4 in. deep-golden yellow roots cook well for mashing, fine flavor, good keeper, can also be overwintered and used as winter greens. *Source History: 1 in 1981; 1 in 1984; 2 in 1987; 1 in 1991; 1 in 1994; 2 in 1998; 3 in 2004.* **Sources: Bou, Ho13, Ol2.**

Petrowski - (Petrowski de Berlin, Petrowski Yellow) - 33-38 days, Slightly flattened golden globe roots, med. size, heirloom grown by early Russian immigrants to Alaska when it was owned by Russia. *Source History: 2 in 1994; 2 in 1998; 1 in 2004.* **Sources: Ho13.**

Purple Top Globe - (Purple Top) - 55-58 days, Large round and very smooth, purplish-red above ground and white below, crisp tender sweet white flesh, fine quality, good keeper, home or shipping. *Source History: 5 in 1981; 4 in 1984; 4 in 1987; 4 in 1991; 4 in 1994; 5 in 1998; 6 in 2004.* **Sources: Hum, Pe6, Roh, Sa9, Si5, So9.**

Purple Top Strap Leaf - (Purple Top Flat Strap Leaf, Early Purple Top Strap Leaf) - 45-60 days, Popular older home garden variety, 14-16 in. plants, med-sized tops, med-green strap leaves, flattened globe, 4-5 in. dia., purplish-red above ground and white below, sweet tender fine-textured white flesh, very early and productive, for table use and stock feed, pre-1845. *Source History: 19 in 1981; 18 in 1984; 14 in 1987; 7 in 1991; 5 in 1994; 4 in 1998; 2 in 2004.* **Sources: Sa9, Shu.**

Purple Top White Globe - (Red Top White Globe, Early Purple Top White Globe, Mammoth Purple Top White Globe) - 45- 65 days, Med-green cut-leaf 14-22 in. tops, uniform 6 in. smooth white globes, best at 3-4 in., sweet mild fine-grained white flesh, purplish-red above ground and white below, widely used for market and home, large tops make good greens, bears longer than most, good condition until quite large, stores well, pre-1880. *Source History: 134 in 1981; 122 in 1984; 122 in 1987; 109 in 1991; 108 in 1994; 112 in 1998; 100 in 2004.* **Sources: ABB, Ada, Alb, All, Ba8, Bo17, Bu1, Bu2, Bu3, Bu8, But, CA25, CH14, Cl3, CO23, Co31, Com, Cr1, Dam, De6, Dgs, DOR, Ear, Ers, Fe5, Fi1, Fis, Fo7, Ga1, Ge2, Gr27, Gur, HA3, Ha5, Hal, He8, He17, Hi13, Hig, HO2, Jo1, Jor, Jun, Kil, Kit,**

La1, Lej, Lin, LO8, Loc, Ma19, MAY, Me7, ME9, Mel, Mey, Mo13, MOU, Na6, Ni1, OLD, Ont, Or10, Orn, OSB, PAG, Pe2, Pin, Pr3, Pr9, Ri12, RIS, Ros, RUP, Sau, Scf, SE4, Se7, SE14, Se16, Se26, Sh9, Shu, Sk2, So1, SOU, Sto, Syn, Te4, TE7, Ter, Twi, Ver, Ves, Vi4, WE10, We19, Wet, Wi2, WI23.

Red Globe - 65 days, Red skinned, globe-shaped roots look like a large radish, 3 in. dia., crisp, tender white flesh, excel. keeper *Source History: 1 in 1994; 1 in 1998; 1 in 2004.* **Sources: Fo7.**

Red Milan - (De Milan Rouge) - 35 days, Recommended for growing as a baby vegetable, round with bright red blushed shoulders, white bottoms, matures early. *Source History: 2 in 1994; 2 in 1998; 2 in 2004.* **Sources: Gou, Ki4.**

Red Round - 55 days, Nearly identical to Ohno Scarlet turnip with a rounder shape, bright red skin, white flesh with a variable rose blush. *Source History: 1 in 2004.* **Sources: Fe5.**

Scarlet Ball - Imported from Japan, semi-globe, looks like a table beet, deep scarlet-red skin, white flesh, red-veined foliage, red stems, skin colors the flesh when cooked. *Source History: 1 in 1981; 1 in 1984; 2 in 1987; 2 in 1991; 3 in 1994; 4 in 1998; 2 in 2004.* **Sources: Ho13, Ni1.**

Scarlet Ohno Revival - Diverse Revival version of Scarlet Ohno, round to flattened, bright scarlet skinned roots, smooth strap-leaf greens with some scarlet-red and a few purple, ongoing project in the garden of Frank Morton. *Source History: 2 in 2004.* **Sources: In8, Sh12.**

Seven Top - (Seven Top Strap-Leaf, Southern Prize, Foliage Turnip, Winter Greens, Seven Top Green) - 45-50 days, Used only for greens, tough woody inedible roots, dark-green (yellowish?) cutleaf 16-22 in. tops, usually grown as a winter annual for spring greens and sprouts, very popular variety in the South, pre-1880. *Source History: 66 in 1981; 60 in 1984; 54 in 1987; 48 in 1991; 47 in 1994; 53 in 1998; 53 in 2004.* **Sources: ABB, Ada, Bo17, Bu8, CH14, Cl3, CO30, Co31, Com, Cr1, De6, DOR, Ers, Fo7, HA3, He8, He17, Hi13, HO2, Hud, Jor, La1, LO8, Loc, Ma19, ME9, Mel, Mey, OLD, Or10, PAG, Pe2, Ri12, RIS, Roh, RUP, Sau, Scf, SE4, SE14, Se26, Sh9, Shu, So1, SOU, Sto, TE7, Twi, Vi4, WE10, Wet, Wi2, WI23.**

Shogoin - (Japanese Foliage, Foliage, Shogoin Round) - 30-70 days, Dual-purpose, 30 days for greens, 70 days for roots, top-shaped pure-white 2-4 in. roots, very tender mild white flesh, quick-growing, thrives in hot dry weather, erect 20 in. strap-leaf light-green tops, tol. to aphid damage and lice. *Source History: 50 in 1981; 41 in 1984; 43 in 1987; 34 in 1991; 33 in 1994; 36 in 1998; 28 in 2004.* **Sources: Ada, Bu8, CH14, DOR, Ers, Ev2, Fo7, He8, Jor, Kil, Kit, La1, Lej, LO8, Ni1, OLD, Roh, Ros, RUP, Sau, Se26, Sh9, Shu, Ter, Vi4, We19, Wi2, WI23.**

Snowball - 30-45 days, Excel. small round pure-white summer turnip, crisp sweet tender white flesh, 3-4 in. diameter, formerly called Early Six Weeks. *Source History: 6 in 1981; 5 in 1984; 7 in 1987; 6 in 1991; 6 in 1994; 3 in 1998; 4 in 2004.* **Sources: Bou, Fis, Ho13, Tho.**

Tokyo Market - (White Tokyo) - 20-56 days, Very early, smooth slightly flattened globes, pure-white skin and flesh, very mild delicate flavor, solid crisp and uniform, slow bolt, spring or fall, from Japan. *Source History: 10 in 1981; 7 in 1984; 5 in 1987; 5 in 1991; 8 in 1994; 7 in 1998; 6 in 2004.* **Sources: Ev2, Hig, Kit, Loc, MIC, RUP.**

Turnip Top Greens - Smooth green leaves, cut at 6 in. stage, will grow back for later cuttings, for quick early cooking greens in spring, cook like spinach, distinctive taste. *Source History: 1 in 1981; 1 in 1984; 1 in 1987; 1 in 1991; 2 in 1994; 2 in 1998; 2 in 2004.* **Sources: Dam, Hal.**

Turnip Top Greens, Nozawana - 40-65 days, Long dark green leaves and petioles are used for pickling and stir fry, edible 2 in. purple-topped roots, resistant to cold and heat, more tender and sweeter after a frost, from Japan. *Source History: 1 in 1998; 4 in 2004.* **Sources: Ev2, Ho13, Kit, Ni1.**

Veitch's Red Globe - Resembles French Breakfast radish, roots have a white base, red top, smooth skin, flavorful white flesh, grows fast. *Source History: 1 in 1998; 1 in 2004.* **Sources: Bou.**

Vertus - (Navet des Vertus Marteau) - 35-60 days, Very early, pure-white and half-long, 4-5 in. long x 1 in. dia., tapers at both ends, grows fast, best when young, can sow all season, French. *Source History: 2 in 1981; 2 in 1984; 1 in 1987; 2 in 1991; 2 in 1994; 3 in 1998; 1 in 2004.* **Sources: So12.**

Waldoboro Green Neck - (Cambridge) - 50 days, Old Maine heirloom from the 1780s, possible French origin, legend has it that the original seeds came from the shipwrecked "Cambridge" from England, which ran aground in the 1800s, large root with green shoulder. *Source History: 1 in 1994; 2 in 1998; 1 in 2004.* **Sources: Ho13.**

White Egg - (Early White Egg, Snowball) - 45-65 days, Med-green cutleaf 17 in. tops, fast-growing egg-shaped 3-4 x 2-3 in. dia. white root, pure-white firm fine-grained sweet tender mild flesh, grows partly above ground, crown shows green tint, garden and market var., popular in the South. *Source History: 25 in 1981; 21 in 1984; 20 in 1987; 17 in 1991; 19 in 1994; 29 in 1998; 24 in 2004.* **Sources: Ada, Ba8, Com, DOR, Ers, Fe5, He8, He17, HO2, Ho13, Hud, Jor, Kil, Lej, Loc, Mo13, Or10, RUP, Sa9, Sh9, So1, Vi4, WE10, WI23.**

Yellow Globe - 73 days, Globe-shaped, yellow, smooth, fine-grained fruit with fine flavor. *Source History: 2 in 1981; 2 in 1984; 1 in 1987; 2 in 1991; 1 in 1994; 2 in 1998; 4 in 2004.* **Sources: HA3, HO2, Hud, SE14.**

Yorii Spring - (Yorri Spring) - 38 days, Very early, small slightly flattened sweet white roots, slow to bolt, best in cool areas, large yields of greens, from Japan. *Source History: 3 in 1981; 3 in 1984; 3 in 1987; 2 in 1991; 2 in 1994; 4 in 1998; 1 in 2004.* **Sources: Ho13.**

Varieties dropped since 1981 (and the year dropped):

All Seasons (2004), *Alsacean Round* (1984), *Amberglow* (1987), *Barive (Stock Turnip)* (1987), *Bency* (1987), *Centoventina* (1994), *Charlestowne* (1991), *Cocktail* (1991), *Cow Horn* (1991), *Crawford* (2004), *De Nancy* (1994), *Early*

Red (1998), *Early White Ball* (1991), *Early White Flat Dutch* (1987), *Early White Globe* (1991), *Early White Milan Red Top* (1987), *Extra Early White Ball* (1982), *Fall Foliage Turnip* (1998), *German Sweet* (1991), *Golden Amber* (1987), *Greenneck Salad* (1987), *Gros Longue Alsace* (1991), *Herbst Rube* (1994), *Hinona* (1991), *Japanese* (1994), *Kanamachi* (1987), *Large White Norfolk* (1991), *Longue de Caluire* (1991), *Manchester* (1984), *Manchester Market* (1983), *Milan Early Red Top* (1991), *Milan White* (2004), *Namenia* (1983), *Novantina* (1994), *Ping Pong* (1991), *Pomeranian White Globe* (1998), *Presto* (1998), *Purp. Top Wh. Globe Short Top* (1987), *Purple Top Milan* (1991), *Purple Top White Globe Smith Str.* (1994), *Purple-Top Yellow Aberdeen* (1987), *Quarantina* (2004), *Red Top Globe* (1984), *Rese Purple Top White Globe* (1987), *Roots* (1991), *Sakata's Foliage Turnip* (1991), *Sessantina* (2004), *Someya Kanamach* (1991), *Spring* (1983), *Tamahikari* (1994), *Vermont* (2004), *White Globe* (2004), *White Stone* (1991), *Yellow Purple Top Aberdeen* (1991).

Watermelon
Citrullus lanatus

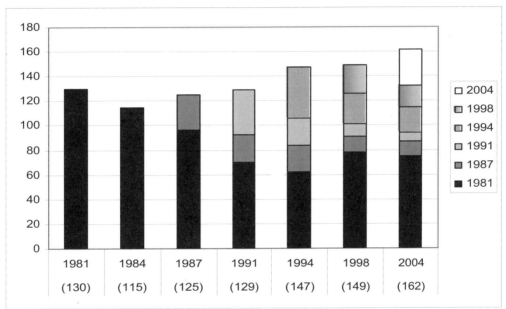

Number of varieties listed in 1981 and still available in 2004: 130 to 75 (58%).
Overall change in number of varieties from 1981 to 2004: 130 to 162 (125%).
Number of 2004 varieties available from 1 or 2 sources: 96 out of 162 (59%).

Acoma - Dark green skin, red flesh, round to slightly elongated shape, believed to have been grown by the ancient ancestors of Acoma Pueblo. *Source History: 1 in 2004.* **Sources: Na2.**

Ali Baba - Excellent variety with extremely sweet flavor, crisp texture, hard gray-green rind, 16-25 lbs., excellent shipper, does well under dry conditions, from Iraq, introduced by Baker Creek Heirloom Seeds via Azia Nail, a French seed saver. *Source History: 2 in 2004.* **Sources: Ba8, Ho13.**

Allsweet - 90-104 days, Dev. from a sister line of Crimson Sweet, similar except for elongated shape, 17-19 in. long x 7 in. dia., averages 25-28 lbs., tough .75 in. yellowish green rind with dark-green broken stripes, res. to FW AN1 and AN3, stands long in field without becoming overripe, few small seeds, excel. shipper, Dr. Hall, Kansas State U. *Source History: 24 in 1981; 22 in 1984; 23 in 1987; 22 in 1991; 28 in 1994; 30 in 1998; 32 in 2004.* **Sources: ABB, Bu3, CH14, Cl3, CO23, CO30, DOR, Ers, Fi1, Ga1, Gr27, HO1, HO2, Jor, La1, Me7, ME9, OLD, Pe2, Pe6, Ra5, Roh, Ros, RUP, SE14, Se26, SOU, TE7, Twi, WE10, Wi2, WI23.**

Arikara - Heirloom variety from the Arikara Indians, dark green rind, pink to yellow seedy flesh, high yields. *Source History: 2 in 1994; 2 in 1998; 1 in 2004.* **Sources: Syn.**

Astrakhanski - 75-85 days, Round, 10-12 in. fruits, dark green rind with light mottled stripes, delicate, pink flesh, great flavor, early ripening, named after the city of Astrakhan which is located on the Volga River near the Caspian Sea. *Source History: 1 in 1994; 1 in 1998; 1 in 2004.* **Sources: Se17.**

Au-Golden Producer - 86-90 days, High quality golden fleshed melon, blocky shape, 20-25 lbs., med-green with dark striped skin, superior flavor, firm texture, PVP 1993 AL/AES. *Source History: 4 in 1994; 5 in 1998; 6 in 2004.* **Sources: Jor, Pe6, RIS, RUP, Se25, Wi2.**

Au-Jubilant - 90-95 days, Jubilee type, imp. red flesh color, sweeter juice, slightly larger, DM and AN1 tol., some tol. to GSB AN2 and FW, large black stippled seeds, PVP

1987 AL AES. *Source History: 3 in 1981; 3 in 1984; 6 in 1987; 9 in 1991; 6 in 1994; 4 in 1998; 1 in 2004.* **Sources: Wi2.**

Au-Producer - 85-90 days, Vigorous vines, size and appearance of Crimson Sweet, better flesh color, 15-25 lbs., sweeter, very high yields, FW DM and AN1 tol., some tol. to GSB and AN2, small dark-brown seeds, PVP 1987 AL AES. *Source History: 4 in 1981; 4 in 1984; 6 in 1987; 10 in 1991; 11 in 1994; 5 in 1998; 5 in 2004.* **Sources: CH14, HO1, Jor, Se25, Wi2.**

Au-Sweet Scarlet - 80-85 days, Globular shaped fruit, 20-25 lbs., crimson flesh, excel. quality, Auburn Univ., PVP 1993 AL/AES. *Source History: 6 in 1994; 4 in 1998; 3 in 2004.* **Sources: RIS, RUP, Wi2.**

Big Crimson - 90 days, Almost identical to Crimson Sweet but larger and several days later, 35-40 lbs., tremendous yields, very good shipper, disease tol. unknown, PVP 1986 Coffey Seed Co. *Source History: 1 in 1981; 1 in 1984; 1 in 1987; 1 in 1991; 1 in 1994; 1 in 1998; 3 in 2004.* **Sources: Ers, Ra5, Wi2.**

Black Diamond - (Diamond, Cannonball, Black Cannonball, Florida Giant, Clara Lee) - 75-95 days, Nearly round, 14-16 x 12 in. dia., 30-50 lbs., some up to 125 lbs., dark blue-green thick tough rind, sweet firm coarse bright-red flesh, tough vigorous vines, widely used throughout the South for long-distance shipping. *Source History: 40 in 1981; 35 in 1984; 39 in 1987; 36 in 1991; 38 in 1994; 30 in 1998; 28 in 2004.* **Sources: Be4, Bu1, Bu3, Bu8, CA25, CH14, DOR, Ers, Fi1, Fo7, Gr27, Gur, He8, He17, HPS, Jor, Mo13, OLD, Pr9, Ra5, RIS, Sau, SE14, Se26, Shu, WE10, Wi2, WI23.**

Black Diamond Yellow Belly - (Superior Black Diamond) - 90-95 days, Tough blue-black rind, spot where melon touches ground turns yellow when ripe, oblong, prominent creases, 19 x 17 in., 30-50 lbs., red flesh, stores well. *Source History: 17 in 1981; 14 in 1984; 10 in 1987; 9 in 1991; 7 in 1994; 7 in 1998; 7 in 2004.* **Sources: Bu3, Ho13, MAY, Ra5, SE14, Vi4, Wi2.**

Black Diamond Yellow Flesh - (Yellow Meated Black Diamond) - 90 days, Black Diamond shape and size with tender yellow flesh, mid-season, 60-70 lbs., thin tough dark-green rind, fine shipper. *Source History: 2 in 1981; 2 in 1984; 2 in 1987; 2 in 1991; 3 in 1994; 3 in 1998; 5 in 2004.* **Sources: Ba8, Ra5, Ros, Se26, Wi2.**

Blackstone - 85-91 days, Tough hard rind, very dark-green, earlier smaller Black Diamond type, 17 x 15 in. dia., 20-25 lbs., AN1 HH and WH res., FW tol., prolific high-quality shipper, 1956. *Source History: 9 in 1981; 9 in 1984; 7 in 1987; 3 in 1991; 1 in 1994; 1 in 1998; 1 in 2004.* **Sources: Sa9.**

Blacktail Mountain - 65-76 days, Developed by Glenn Drowns for short season areas, round fruit, dark green rind with very faint stripes, orange-red flesh, 6-10 lbs., small brown seeds have a black tail, thrives in northern gardens or in hot, humid southern areas, 1977. *Source History: 1 in 1994; 3 in 1998; 7 in 2004.* **Sources: Ba8, Sa9, Se16, Sh13, So1, Syn, Tu6.**

Calsweet - (Calsweet Supreme) - 90-92 days, Blocky, 21 x 18 in., 20-30 lbs., dark-green with light-green stripes, bright-red flesh, FW res., popular shipper especially in Southwest, less breakage, PVP expired 1993. *Source History: 9 in 1981; 9 in 1984; 8 in 1987; 8 in 1991; 9 in 1994; 11 in 1998; 14 in 2004.* **Sources: ABB, CO23, DOR, Ers, HO1, Loc, MOU, Ra5, RUP, SE14, Twi, WE10, Wi2, WI23.**

Calsweet Classic - 92 days, Blocky, oblong shape, firm dark flesh, brown seeds, some disease res. *Source History: 1 in 1994; 2 in 1998; 2 in 2004.* **Sources: ABB, CH14.**

Cannonball - 90 days, Round melon with solid dark green rind, red flesh. *Source History: 1 in 2004.* **Sources: Sa9.**

Carolina Cross - 90 days, Long, light-striped, holds current world record of 255 lbs., germinates slowly in cool soils, won Hope Arkansas Watermelon Contest 7 out of the last 8 years, two consistent variations appear, a long type with few seeds that averages 40 in. and a blunt type with many seeds that averages 35 in. and puts on more good shapes. *Source History: 1 in 1981; 1 in 1984; 1 in 1987; 4 in 1991; 5 in 1994; 5 in 1998; 5 in 2004.* **Sources: Ge2, He17, Mo13, Pp2, Shu.**

Carolina Cross No. 183 - 90-100 days, One of the world's largest watermelons, thick rind, very sweet fine grained flesh, can grow to 200 lbs., needs a long hot summer. *Source History: 13 in 1994; 15 in 1998; 12 in 2004.* **Sources: Bu1, Bu2, Ers, Fa1, Fi1, Gur, Hi13, Jun, Kil, MAY, Ter, Wi2.**

Charleston Elite - 83 days, Consistently larger and more uniform shape than other Charleston Gray types, 25-35 lbs., long cylinder with blunt ends, gray-green skin, bright-red flesh, excel. quality, excel. disease res., PVP 1986 Novartis Seeds, Inc. *Source History: 1 in 1987; 4 in 1991; 8 in 1994; 5 in 1998; 4 in 2004.* **Sources: Be4, CH14, HO2, RUP.**

Charleston Gray - 85-100 days, Tough med-thick hard rind, gray-green with fine darker-green veins, 30-40 lbs., 20-24 x 10 in. dia., red fiberless flesh, USDA Southeast Vegetable Breeding Lab in South Carolina, FW AN1 HH WH GN sunburn and heat res., 1954. *Source History: 64 in 1981; 58 in 1984; 52 in 1987; 47 in 1991; 45 in 1994; 40 in 1998; 38 in 2004.* **Sources: Ba8, Bu2, Bu8, CA25, CH14, CO23, CO30, Co31, De6, Dgs, DOR, Ers, Fi1, Fo7, Gr27, HA3, He8, He17, Hi13, HO1, HO2, La1, Loc, Ma19, ME9, Mey, MIC, Mo13, OLD, Roh, Ros, Sau, SE14, Se26, Sh9, Shu, Si5, WI23.**

Charleston Gray No. 5 - 90 days, Increased vigor, faster starting young plants, 10-20% increase in yields over regular Charleston Gray, similar in all other respects, good in North or South. *Source History: 5 in 1981; 5 in 1984; 10 in 1987; 7 in 1991; 1 in 1994; 1 in 1998; 2 in 2004.* **Sources: Bu1, HO1.**

Charleston Gray No. 133 - 85-90 days, Exactly like Charleston Gray but greater FW res. for Southern areas, oblong with rounded ends, 24 x 10 in., 20-40 lbs., thin tough light-green rind with darker veins, FW AN1 and AN3 res., dark-red fine-quality flesh, large brown seeds, Purdue. *Source History: 22 in 1981; 19 in 1984; 21 in 1987; 22 in*

1991; 22 in 1994; 14 in 1998; 10 in 2004. **Sources: Bu3, HPS, Jor, Kil, Ra5, Sa9, SOU, WE10, Wet, Wi2.**

Chelsea - 90-100 days, Sweet pink flesh, white seeds, 15-20 lbs., keeps for several weeks after picking, from Chelsea, Iowa, where melons were grown on the sandy hills north of town in the early 1900s, farmers would fill their horse-drawn "triple box" wagons in the field, haul them to town and sell the melons right from the wagons *Source History: 1 in 2004.* **Sources: Se16.**

Chilean Black Seeded - 85-95 days, Round dark-green striped fruits, 10 in. dia., 18-20 lbs., crisp bright red-orange fine- grained flesh, prolific, tough rind, good shipper, large black seeds, very rare, pre-1910 heirloom that is nearly extinct. *Source History: 5 in 1981; 4 in 1984; 3 in 1987; 5 in 1991; 3 in 1994; 2 in 1998; 2 in 2004.* **Sources: Ba8, Se25.**

Chris Cross - 85-90 days, Good yields of 15-20 lb. fruit, good drought tol., result of a cross between Hawksbury x Dixie Queen made by Chris Christenson in Montrose, Iowa, in 1950. *Source History: 1 in 2004.* **Sources: Se16.**

Chubby Gray - 90 days, Short chubby rectangular melon, 28-30 lbs., gray rind, firm red flesh, black seed, good shipper, long shelf life, high yields, disease tol., PVP expired 2000. *Source History: 1 in 1981; 1 in 1984; 1 in 1987; 1 in 1991; 1 in 1994; 1 in 1998; 2 in 2004.* **Sources: Ra5, Wi2.**

Citron, Green Seeded - 95-98 days, *Citrullus lanatus* var. *citroides*, round and smooth, greenish-white pulp, bright-green seeds, 6-10 in. dia., productive, keeps for months, larger than Red Seeded. *Source History: 6 in 1981; 5 in 1984; 3 in 1987; 2 in 1991; 1 in 1994; 4 in 1998; 4 in 2004.* **Sources: Lan, Lin, Pr3, Sa9.**

Citron, Red Seeded - (Colorado Preserving Melon, Red Seeded Round) - 80-100 days, *Citrullus lanatus* var. *citroides*, hardy prolific vines, smooth nearly-round fruits, 6-8 in., 10-12 lbs., pale-green rind striped and spotted with darker-green, clear- white solid inedible flesh, bright red seeds, tasteless raw but delicious as preserves or sweet pickles or candied fruit. *Source History: 30 in 1981; 26 in 1984; 26 in 1987; 10 in 1991; 7 in 1994; 9 in 1998; 10 in 2004.* **Sources: Ho13, Lin, Ol2, Ont, Pr3, Sa9, Se16, Shu, Te4, We19.**

Citron, Red Shine - 85 days, Widely grown in the 1930s and 1940s as a pickling or preserving melon, 6-7 in. across, the entire melon is cut into chunks and pickled, almost all candied fruit is made from citron. *Source History: 1 in 1998; 1 in 2004.* **Sources: Ter.**

Cobb Gem - (Giant Cobb Gem, Kolb) - 100 days, Giant melons, some over 130 lbs., nearly round, dark and light-green stripes, red flesh, grayish-black med-size seeds, fair cutting quality, primarily for show. *Source History: 6 in 1981; 6 in 1984; 7 in 1987; 6 in 1991; 8 in 1994; 7 in 1998; 8 in 2004.* **Sources: Ers, He17, Mo13, Ra5, RUP, Shu, Syn, Wi2.**

Colocynth Melon - 90-95 days, *Citrullus colocynthis*, round fruits, 4-6 in. dia., light green rind with dark green splashes, bitter flesh is pickled, edible seeds are toasted, grown to stablilize sand dunes. *Source History: 2 in 2004.* **Sources: Gr29, So12.**

Colorado Striped Tarahumara - 95 days, Oblong melon, broad dark green stripes on green rind, orange-pink flesh color varies with the closeness to the rind, seedy flesh. *Source History: 1 in 2004.* **Sources: Sa9.**

Congo - 85-99 days, AAS/1950, med-thick tough rind, med-green with dark-green stripes, cylindrical and blunt-ended, 27 x 12 in., 30-50 lbs., firm med-red high-quality flesh, tests show 9.4% sugar, AN1 AN2 AN3 res., long-bearing, shipper for Florida and the Southeast, USDA/SC, pre-1910 heirloom. *Source History: 44 in 1981; 39 in 1984; 38 in 1987; 27 in 1991; 22 in 1994; 18 in 1998; 15 in 2004.* **Sources: Ba8, CA25, Co31, Ers, Fi1, He8, Kil, La1, Mey, SE14, SOU, Vi4, WE10, Wet, WI23.**

Cream of Saskatchewan - 80-85 days, Excellent 14 in. fruits, pale-green skin with dark stripes, thin .25 in. rind, sweet white flesh, 8-10 lbs., black seeds, does well in cool northern climates, thin rind makes it suitable for home gardeners and local markets, not a shipper, probably brought to Saskatchewan by Russian immigrants. *Source History: 1 in 1987; 1 in 1994; 7 in 1998; 16 in 2004.* **Sources: Ba8, Co31, Fe5, Ga1, He8, Ho13, La1, ME9, Pe6, Pr3, Sa9, Se8, SE14, Se16, So25, Te4.**

Crimson Sweet - (Green and White Ruby) - 80-97 days, AAS/1964, tough med-thick hard rind, light-green with small dark- green stripes, blocky, 12 x 10 in., 25 lbs., bright deep-red flesh, high sugar content, few small seeds, white > cream on bottom when ripe, FW1 FW2 AN1 and AN3 res., shipping and local market, Dr. Hall, KS/AES. *Source History: 85 in 1981; 81 in 1984; 82 in 1987; 83 in 1991; 82 in 1994; 82 in 1998; 80 in 2004.* **Sources: Ba8, Be4, Ber, Bo17, Bou, Bu1, Bu2, Bu3, Bu8, CA25, CH14, Cl3, CO23, Com, De6, DOR, Ear, Ers, Fa1, Fi1, Ge2, Gr27, Gur, Ha5, He17, Hi13, HO1, HO2, HPS, Hum, Jor, Jun, Kil, La1, Loc, Ma19, MAY, ME9, Mel, Mey, MIC, Mo13, MOU, OLD, Ont, Or10, Pe2, Pe6, Ra5, Ri2, RIS, Roh, Ros, RUP, Sau, SE4, Se7, Se8, SE14, Se25, Se26, SGF, Sh9, Shu, Si5, So1, So9, So25, SOU, Sto, Syn, TE7, Ter, Tu6, Twi, Ver, WE10, Wet, Wi2, WI23.**

Crimson Treat - (Early Crimson Treat) - 85 days, Early, small, Crimson Sweet icebox type, blocky fruit, light and dark stripes, bright red crisp flesh, 8-10 lbs. *Source History: 1 in 1998; 3 in 2004.* **Sources: Ba8, RUP, Vi4.**

Curtis Showell White Flesh - Extremely rare, large round oblong fruit, black rind, cream-peach colored flesh is very sweet, 20 lbs., possibly one of the only large light colored strains left, most of this type have become extinct. *Source History: 1 in 2004.* **Sources: Ba8.**

Desert King, Yellow Flesh - 85 days, Light-green, round to slightly oblong, very sweet deep-yellow flesh, med-thin tough rind, 8-10 ft. vine, holds quality on vine a month after ripe and does not sunburn, can ship. *Source History: 4 in 1981; 4 in 1984; 4 in 1987; 10 in 1991; 10 in 1994; 10 in 1998; 9 in 2004.* **Sources: Ers, Gr27, Gur, He17, Mo13, Ros, RUP, Se7, Wi2.**

Dixie Lee - 90-92 days, Smooth tough .5 in. rind, light-green with narrow dark-green stripes, 11 x 10 in. dia., 20-30 lbs., dark intense-red flesh, res. to FW and AN1, good shipper, FL/AES, PVP. *Source History: 9 in 1981; 8 in 1984; 8 in 1987; 8 in 1991; 5 in 1994; 1 in 1998; 2 in 2004.*

Sources: Ers, Wi2.

Dixie Queen - 70-90 days, Thin tough rind, light-green striped with dark-green, 15-16 x 12-13 in. dia., 30-50 lbs., firm bright-scarlet flesh, small white seeds, almost seedless, res. to FW, very vigorous and productive, uniform size and appearance, good shipper, pre-1935. *Source History: 37 in 1981; 33 in 1984; 32 in 1987; 28 in 1991; 25 in 1994; 21 in 1998; 11 in 2004.* **Sources: Ba8, Bu8, CH14, Ear, Ers, Gr27, Ra5, SE14, Sh9, WE10, Wet.**

Early Arizona - 83 days, Dark green rind, very sweet, pink-red flesh. *Source History: 1 in 2004.* **Sources: Sa9.**

Early Midget - 68-80 days, Excel. melon for small gardens or short season areas, extremely early, excel. quality and flavor, few seeds, very thin, faintly striped rind, rich-red very sweet flesh, 7 in. long. *Source History: 1 in 1981; 1 in 1984; 1 in 1987; 2 in 1991; 2 in 1994; 1 in 2004.* **Sources: Sa9.**

Early Moonbeam - 80 days, Open pollinated sel. from Yellow Doll hybrid that will breed true, green and white striped skin, yellow flesh, flavorful, 5-8 lbs., excel. market crop, good for northern climates, from plant breeder Alan Kapular. *Source History: 2 in 1994; 3 in 1998; 5 in 2004.* **Sources: Fe5, Pe2, Se7, So9, Tu6.**

Early Yates - 83-100 days, Round-oblong fruit with light-green rind, 8-12 lbs., pink-red flesh. *Source History: 1 in 1994; 1 in 1998; 1 in 2004.* **Sources: Sa9.**

Fairfax - 85-90 days, Oblong fruits striped light and dark-green, 25 x 9 in., 35-40 lbs., very sweet bright-red flesh, good texture, FW and AN res., thin tough rind, USDA and Clemson, 1952. *Source History: 4 in 1981; 4 in 1984; 5 in 1987; 2 in 1991; 3 in 1994; 5 in 1998; 4 in 2004.* **Sources: Ba8, He8, Sa9, WE10.**

Florida Giant - (Black Diamond, Improved Black Diamond, Cannonball, State Fair) - 90-95 days, Smooth thick tough hard rind, very dark-green with bluish cast, nearly round, 15 in. dia. or larger, 30-60 lbs., firm sweet bright-red flesh, good shipper, vigorous prolific vines, Southern favorite for home and market, old-fashioned quality. *Source History: 25 in 1981; 23 in 1984; 21 in 1987; 15 in 1991; 16 in 1994; 16 in 1998; 16 in 2004.* **Sources: Bu8, CA25, Cr1, DOR, Fo7, Kil, La1, Mey, Ra5, Ros, SE14, Sh9, SOU, WE10, Wet, WI23.**

Garden Baby - 68-78 days, Round 8 in. icebox melon, 5 lbs., dark-green and med-green stripes, red flesh, very productive, extremely compact vines, very early, reliable in cooler areas. *Source History: 3 in 1981; 3 in 1984; 3 in 1987; 4 in 1991; 1 in 1994; 1 in 1998; 1 in 2004.* **Sources: Fa1.**

Garrisonian - (Rattlesnake, Georgia Rattlesnake) - 85-95 days, Hard tough rind, light-green skin with dark-green Rattlesnake markings, 22-24 x 10-12 in., 35-45 lbs., firm sweet bright-rose flesh, excel. quality, AN res., wilt susc., popular attractive market melon in the South and East, USDA Southeast Vegetable Breeding Lab/SC, 1957. *Source History: 20 in 1981; 17 in 1984; 13 in 1987; 8 in 1991; 6 in 1994; 4 in 1998; 5 in 2004.* **Sources: Bu8, He17, Ra5, SE14, WE10.**

Georgia Giant, Striped - 100 days, Pink-red flesh, 20-40 lbs., striped rind. *Source History: 1 in 1994; 1 in 1998; 1 in 2004.* **Sources: Sa9.**

Georgia Rattlesnake - 90 days, Old Southern favorite prized for its size productiveness and eating quality, 22 x 10 in. dia., 25-30 lbs., thin very tough rind, light-green with irregularly mottled dark-green Rattlesnake-like stripes, firm sweet bright-rose flesh, dull-white seeds with black tips, excellent shipper, developed in Georgia in the 1830s. *Source History: 6 in 1981; 3 in 1984; 3 in 1987; 1 in 1991; 1 in 1994; 2 in 1998; 10 in 2004.* **Sources: Ba8, Bu2, Co31, He8, La1, ME9, Ra5, SE14, Shu, WE10.**

Giza - (Giza No. 1) - Very large seeds are used like sunflower seeds in the Mideast. *Source History: 2 in 1991; 3 in 1994; 2 in 1998; 2 in 2004.* **Sources: Ba8, Vi4.**

Gold Baby - Very sweet creamy-yellow flesh, 5-8 lb. fruits, deep-green rind with dark-green stripes, easy to grow, quite uniform and prolific. *Source History: 1 in 1981; 1 in 1984; 1 in 1987; 1 in 1991; 1 in 1994; 1 in 1998; 1 in 2004.* **Sources: Syn.**

Golden Honey - (Golden Honey Round) - 85-88 days, Dark-green rind irregularly striped with darker-green, 12 x 10 in. dia., 20-30 lb., sweet sugary crisp bright golden-yellow flesh, vigorous productive vines, Tendersweet sometimes listed as a synonym, but this is questionable. *Source History: 16 in 1981; 14 in 1984; 12 in 1987; 12 in 1991; 9 in 1994; 5 in 1998; 1 in 2004.* **Sources: Se7.**

Golden Midget - (Early Golden Midget) - 65-82 days, Thin tough green rind turns golden-orange when mature, compact vines need only a small space, 7-8 in. oval melons, very sweet rich-red flesh, for short season areas, dark-colored seeds, widely adapted, excel. for either small garden areas or container culture, dev. by Elwyn Meader and Albert Yaeger at the U/NH in 1959. *Source History: 19 in 1981; 18 in 1984; 8 in 1987; 1 in 1991; 1 in 1998; 5 in 2004.* **Sources: Ba8, Fe5, Ho13, Se16, So25.**

Gourd Tinda - (Punjabi Tinda, Akra Tinda, Baby Indian Pumpkin, Apple Gourd) - 45-55 days, *C. lanatus* var. *fistulosus*, tender tasty apple size fruits on 3-4 ft. tall vines, light to dark green color with soft hairs on the skin, 2-3 oz., flesh resembles zucchini, heavy yields, can be harvested every one to two days. *Source History: 1 in 1994; 1 in 1998; 3 in 2004.* **Sources: CO23, So25, Wi2.**

Graybelle - (Greybelle) - 80 days, Nearly round fruit resembles Early Canada, light grey-green rind with dark veins, dark pink flesh, 15 lbs., resists sunburn. *Source History: 2 in 1981; 2 in 1984; 2 in 1987; 1 in 1991; 3 in 1994; 3 in 1998; 1 in 2004.* **Sources: WE10.**

Hopi Red - Small to medium size fruit to 10 lbs., green rind, red flesh, intro. by the Spanish, the Hopis have selected this str. for generations to be drought tolerant and to keep in storage, often up to 6 months. *Source History: 2 in 1998; 2 in 2004.* **Sources: Ho13, Se17.**

Hopi Red Meated - (Kawayvatnga) - 80-90 days, Small, tasty, prolific, some bonus yellow-meated genes, does well in Tucson. *Source History: 1 in 1987; 1 in 1991; 2 in 1994; 2 in 1998; 1 in 2004.* **Sources: Na2.**

Hopi Yellow Meated - (Sikyatko) - 100 days, Small round melon 3-4 lb. average, crisp and sweet, sand dune dry farmed, adapted to high cool desert regions, crushed seeds are used to grease stone on which the traditional bread piki is baked, Hopi Indian seed. *Source History: 1 in 1981; 1 in 1984; 2 in 1987; 1 in 1991; 2 in 1994; 3 in 1998; 3 in 2004.* **Sources: Ho13, Na2, Se17.**

Iopride - (Pride of Iowa) - 85-105 days, Dark-green with darker-green stripes, oblong and blocky, 20-40 lbs., sweet bright-red flesh, high sugar content, excel. quality, holds long at maturity, top yielder in tests, especially good on upland soils, dev. for high tol. to FW and AN1, U of IA. *Source History: 6 in 1981; 4 in 1984; 5 in 1987; 3 in 1991; 1 in 1994; 1 in 1998; 1 in 2004.* **Sources: Sa9.**

Irish Grey - 85 days, Greyish-green rind, 22 x 12 in. dia., 30 lb. average, bright-red flesh, high sugar content, good shipper, popular in the 1920s, found in an unoccupied feeding pen in Georgia in 1913. *Source History: 1 in 1981; 1 in 2004.* **Sources: Ho13.**

Japanese Cream Fleshed Suika - Rare Japanese variety, green striped rind, crisp cream colored flesh, excellent refreshing taste with mild flavor, 8 lbs. *Source History: 1 in 2004.* **Sources: Ba8.**

Jemez - Skin color varies from pale to dark green, considered a "native" watermelon which is still rare in any pueblo. *Source History: 1 in 2004.* **Sources: Na2.**

Jubilee - (Wilt Resistant Jubilee, Black Seeded Garrisonian) - 85-95 days, Med-thick tough rind, greenish-white with dark- green stripes, 22-25 x 11-13 in. dia., 25-40 lbs., firm sweet fine-textured bright-red flesh, excellent large late shipper, long season areas, FW AN1 res. *Source History: 47 in 1981; 44 in 1984; 46 in 1987; 38 in 1991; 35 in 1994; 31 in 1998; 27 in 2004.* **Sources: Ba8, CA25, CH14, Cl3, CO30, Ers, Fi1, Gr27, HO2, HPS, Jor, Kil, Loc, ME9, Mey, MIC, OLD, Ra5, RIS, Ros, RUP, Sau, SE14, SOU, WE10, Wet, Wi2.**

Jubilee F.F.S.S. - 95 days, Tough striped rind, elongated melons, 28 x 11.5 in. dia., 25-30 lbs., bright-red flesh, res. to FW1, large mottled seed, good shipper. *Source History: 1 in 1981; 1 in 1984; 1 in 1991; 1 in 1998; 1 in 2004.* **Sources: Sh9.**

Jubilee II - PVP 1992 FL AES. *Source History: 1 in 1991; 4 in 1994; 4 in 1998; 2 in 2004.* **Sources: CH14, Wi2.**

Jubilee Improved - (Special Jubilee, Jubilee II) - 90-95 days, Jubilee strain from University of Florida with much greater fusarium wilt res., selected to survive in fields which now cause severe loss. *Source History: 3 in 1981; 2 in 1984; 3 in 1987; 4 in 1991; 5 in 1994; 3 in 1998; 4 in 2004.* **Sources: Bu3, CO23, DOR, WI23.**

Jubilee, Bush - 90-100 days, Bush plants spread only 3-5 ft. but produce heavily, 10-20 lb. fruit smaller version of original Jubilee, higher sugar content, FW and AN res., PVP expired 2001. *Source History: 21 in 1987; 21 in 1991; 15 in 1994; 9 in 1998; 6 in 2004.* **Sources: Be4, Co31, Mey, OLD, SOU, Wet.**

Jubilee, Registered - Special class of seed for growers who want the highest possible disease tolerance with the Jubilee var. *Source History: 2 in 1994; 2 in 1998; 2 in 2004.* **Sources: ABB, Bu3.**

Juliett - (NV 4317) - 90 days, Jubilee type with greater FW tolerance, 25-32 lbs., elongated shape with blunt ends, medium green skin with dark green stripes, PVP 1992 Novartis Seeds, Inc. *Source History: 2 in 1994; 1 in 1998; 3 in 2004.* **Sources: CH14, Cl3, SE4.**

Jumanos - Pale to dark green skin, small round size, from Redford, TX, known as Sandia Tuliza across the border in Chihuahua, Mexico. *Source History: 1 in 2004.* **Sources: Na2.**

Karpas Greek - 110 days, Semi-oblong shape, rind is a blend of medium green with dark green stripes, deep red flesh. *Source History: 1 in 2004.* **Sources: Sa9.**

King and Queen - (Christmas, Winterkeeper, Winter King and Queen) - 80-90 days, Russian, light-green with thin dark-green stripes, smooth and round, 10 in. dia., soft pinkish-red flesh, vigorous spreading vines, high yields, excel. shipper, will keep for months unless it is picked too soon. *Source History: 9 in 1981; 8 in 1984; 9 in 1987; 8 in 1991; 9 in 1994; 7 in 1998; 6 in 2004.* **Sources: CH14, HO1, Jor, MOU, Red, Se25.**

King Winter - 90 days, Pale green rind with faint stripes, sweet, dark red flesh, small seeds, 10-15 lbs., good keeper. *Source History: 1 in 2004.* **Sources: Syn.**

Kleckley Resistant No. 6 - (Kleckley No. 6, Kleckley Sweet Wilt Resistant No. 6, Kleckley Sweet Improved No. 6) - 85-88 days, Adds wilt res. and higher sugar content to Kleckley Sweet, very sweet deep-red flesh, thin tough brittle dark-green rind, FW res., 30-35 lbs., no bottle-necks, stands handling and shipping well, IA/AES. *Source History: 15 in 1981; 11 in 1984; 12 in 1987; 9 in 1991; 9 in 1994; 5 in 1998; 3 in 2004.* **Sources: Fo7, Sau, Wet.**

Kleckley Sweet - (Monte Cristo, Wonder Melon) - 80-90 days, Glossy dark-green rind, too thin to ship, oblong fruit with square ends, 25-40 lbs., very sweet deep-red flesh, broad stringless heart, good home garden var., large white seeds, introduced by W. Atlee Burpee in 1897. *Source History: 20 in 1981; 16 in 1984; 16 in 1987; 15 in 1991; 12 in 1994; 12 in 1998; 12 in 2004.* **Sources: Ba8, Ers, Gr27, He8, He17, Ra5, SE14, Se25, Sh9, Shu, Vi4, WE10.**

Kleckley Sweet Improved - (Wonder Melon) - 85 days, Very thin dark-green rind with faint ribbing, long with rounded ends, bright-red sweet flesh, juicy to the rind, not a shipper, for home gardens and local markets. *Source History: 6 in 1981; 6 in 1984; 6 in 1987; 5 in 1991; 4 in 1994; 3 in 1998; 3 in 2004.* **Sources: Com, La1, SOU.**

Klondike - (Klondyke) - 78-85 days, Med-thin dark-green rind, blocky fruits, 14 x 10 in. dia., 20-25 lbs., very sweet fine- textured deep-red flesh, for home gardens and shipping in Pacific Coast areas. *Source History: 7 in 1981; 7 in 1984; 7 in 1987; 4 in 1991; 6 in 1994; 3 in 1998; 1 in 2004.* **Sources: Sau.**

Klondike Peacock Improved - 80-90 days, Deep-oval fruits, blossom end taper, 15 x 10 in., 20-25 lbs., bright-red flesh, thin tough dark-green rind, good shipper, FW res. standard in Mexico and the Southwest. *Source History: 2 in 1981; 2 in 1984; 1 in 1987; 1 in 1991; 1 in 1994; 1 in 2004.* **Sources: Ros.**

Klondike R-7 - 82-85 days, Oval, 15 x 10 in., 20-25 lbs., thin tough rind, dark-green with shallow creases, slight taper to blossom end, deep-red flesh, FW res., good shipper, very sweet. *Source History: 4 in 1981; 4 in 1984; 3 in 1987; 1 in 1991; 1 in 1994; 1 in 1998; 1 in 2004.* **Sources: WE10.**

Klondike RS 57 - (Klondike Striped RS 57) - 85-90 days, Cylindrical, blockier ends, 16 x 10 in., 30 lbs., deep-red flesh, FW res., tough med-thick striped rind, arid irrigated areas, UC/Davis. *Source History: 3 in 1981; 3 in 1984; 1 in 1987; 1 in 1991; 2 in 1994; 2 in 1998; 1 in 2004.* **Sources: WE10.**

Klondike Striped Blue Ribbon - (Klondike Blue Ribbon, Klondike Striped, Wilt Resistant Klondike, Klondike Sugar) - 80-90 days, Thin hard tough rind, light-green with dark-green stripes, 15-16 x 10-11 in. dia., 20-30 lbs., very sweet stringless flesh, scarlet to the rind, FW res., AN tol., shows no sunburn, high quality, med-distance shipper, resembles Blue Ribbon. *Source History: 22 in 1981; 17 in 1984; 15 in 1987; 11 in 1991; 10 in 1994; 9 in 1998; 8 in 2004.* **Sources: Ba8, He8, He17, Loc, MOU, SE14, Vi4, WE10.**

Kolb's Gem - (American Champion) - 100 days, Huge melons can grow up to 130 lbs., nearly round, sweet red flesh, dates back to the 1880s. *Source History: 2 in 2004.* **Sources: Ba8, Co31.**

Legacy - 85 days, Allsweet type with dark green striped rind, oblong shape, 22-25 lbs., deep red, firm flesh, flavorful, no splits or cracks in the heart, wilt tol., some ability to resist sunburn, PVP 2000 Wi2. *Source History: 1 in 1998; 4 in 2004.* **Sources: Bu3, Cl3, He17, Wi2.**

Long Crimson - 80-95 days, Thick, blocky, rectangular, 25-8 lbs., rind like Crimson Sweet , brighter sweet very firm flesh, late, black seeds, disease tolerance yet unknown, tough rind, PVP 1986 Coffey Seed Co. *Source History: 1 in 1981; 1 in 1984; 1 in 1987; 2 in 1991; 1 in 1994; 1 in 1998; 3 in 2004.* **Sources: Ers, Ra5, Wi2.**

Louisiana Sweet - 90 days, Med-green rind with dark-green stripes, round oblong shape, crisp sweet red flesh, high sugar level, fruits average 20-30 lbs., FW and AN tol., PVP 1987 LA AES. *Source History: 2 in 1991; 2 in 1994; 3 in 1998; 6 in 2004.* **Sources: Ers, He8, ME9, Ra5, SE14, Wi2.**

Luscious Golden - 85 days, Long fruit with deep green rind, orange flesh, very good quality. *Source History: 1 in 2004.* **Sources: Sa9.**

Malali - (Malali Israeli) - 90-100 days, Light red flesh, tan seeds outlined in black, grows to 10 lbs., unique flavor, appearance similar to Crimson Sweet, originated in Israel. *Source History: 2 in 1991; 1 in 1994; 2 in 1998; 1 in 2004.* **Sources: Se7.**

Mayo - Prolific heat tolerant vines produce small fruit of various colors, from Mayo farmers in Sinaloa, Mexico. *Source History: 1 in 1994; 1 in 1998; 1 in 2004.* **Sources: Na2.**

Mayo Sandia - Small, red fleshed melons from Piedras Verdes, Sonora, Mexico. *Source History: 1 in 1998; 1 in 2004.* **Sources: Na2.**

Melitopolisky - 100 days, Early ripening var., round 10-12 in. fruits, alternating black and green stripes, sweet red flesh, from the Volga River region of Russia, an area famous for its melons. *Source History: 1 in 1998; 1 in 2004.* **Sources: Se16.**

Mickylee - 72-82 days, New icebox type, round to oval 7-15 lb. fruit, tough light gray-green rind with faint net, bright-red flesh, small black seed, superior fruit-holding ability, FW and AN1 res., long shelf life, fresh market shipper, PVP 1987 FL AES. *Source History: 6 in 1987; 18 in 1991; 18 in 1994; 11 in 1998; 10 in 2004.* **Sources: CH14, CO30, HO1, HO2, Jor, Kil, Loc, RUP, Se26, Wi2.**

Moon and Stars - (Van Doren's Moon and Stars) - 95-100 days, Legendary variety rediscovered in rural Missouri, large oval 40 lb. fruits, dark-green skin speckled with bright-yellow splashes ranging from star to moon size, leaves are also speckled, very sweet bright-red flesh, obtained from Merle Van Doren near Macon, Missouri, originally introduced by Peter Henderson and Co. in 1926. *Source History: 2 in 1987; 7 in 1991; 12 in 1994; 20 in 1998; 47 in 2004.* **Sources: Ba8, Be4, Bo17, Bou, Bu2, Coo, Dow, Eco, Ers, Fe5, Fi1, Ga1, He8, He17, Hi6, Hi13, HO2, Ho13, La1, Lan, Mcf, ME9, Par, Pe2, Pin, Ra5, Red, Roh, Se7, SE14, Se16, Se17, Se25, Se26, Sh13, Shu, Sk2, So12, So25, Sw9, TE7, Th3, Tu6, Und, Vi4, WE10, WI23.**

Moon and Stars, Long Milky Way - 95 days, Most flavorful of the Moon and Stars melons, elongated fruit with a large yellow moon, speckled with pinpoint stars which resembles a panorama of the Milky Way, 35 lbs., dev. by breeder Glenn Drowns. *Source History: 1 in 1994; 2 in 1998; 2 in 2004.* **Sources: Ba8, So1.**

Moon and Stars, Pink Flesh Amish - Amish heirloom, 25-30 lb. fruits, smooth rind unlike the rougher-skinned slightly ridged rinds characteristic of other Moon and Stars strains, outstanding sweet flavor, brown seed slightly mottled with beige, possible variant of 'Sun, Moon, and Stars' which was introduced by Peter Henderson and Co. in 1920. *Source History: 1 in 1987; 1 in 1991; 1 in 1994; 1 in 1998; 1 in 2004.* **Sources: So1.**

Moon and Stars, Yellow Flesh - 90-95 days, Family heirloom from Georgia, rare white-seeded yellow-fleshed variety, years ago a melon of this description was routinely shipped from Bermuda to some Southern states around Christmas time, 18-24 in. long, 20-25 lbs., excellent flavor, not quite as sweet as pink-fleshed variety, some disease and drought tol., 1900. *Source History: 1 in 1987; 1 in 1991; 4 in 1994; 3 in 1998; 6 in 2004.* **Sources: Ba8, Hud, Sa9, Se16, So1, Syn.**

Moon and Stars, Cherokee - 90 days, Yellow spotted foliage and fruit, dark green rind, nice pink sweet flesh, crisp and tasty, black seeds, 20-24 in. long, 30-40 lbs. *Source History: 1 in 1998; 2 in 2004.* **Sources: Sa9, Se16.**

Mountain Hoosier - 85 days, Extremely productive, very tasty, slightly oblong, dark-green rind, very sweet crisp deep-red flesh, grows up to 80 lbs., white seeds with black rims and tips, pre-1937. *Source History: 1 in 1981; 1 in 1984; 2 in 1987; 2 in 1991; 2 in 1994; 2 in 1998; 2 in 2004.* **Sources: Ba8, Ers.**

Mountain Sweet Yellow - 95-100 days, Yellow fleshed version of Mountain Sweet which was popular in the 1840s in markets throughout New Jersey, Pennsylvania and New York, large striped fruit with deep yellow flesh, high sugar content, sweet and tasty, 20-35 lbs. *Source History: 3 in 2004.* **Sources: Ba8, Hud, Se16.**

Navajo Red Seeded - Light red sweet flesh, red seeds. *Source History: 1 in 2004.* **Sources: Na2.**

Navajo Sweet - 85-90 days, Round 10-20 lb. fruits, small dark-green striping on pale-green background, firm red flesh, unusually sweet flavor, very good shelf life, PVP abandoned 1999. *Source History: 1 in 1991; 3 in 1994; 5 in 1998; 6 in 2004.* **Sources: Bu3, He8, Red, Se8, Se17, Se26.**

Navajo Winter - Red fleshed melon to 15 lbs., black seeds, good keeper. *Source History: 1 in 1994; 2 in 1998; 1 in 2004.* **Sources: Na2.**

New Hampshire Midget - 65-82 days, AAS/1951, very early, small vigorous prolific vines, thin brittle gray-green rind striped and mottled with dark-green, 6-7 x 5.5-6 in. dia., 4-6 lbs., high quality, strawberry-red flesh, black seeds, heavy yields, U of NH, 1951. *Source History: 41 in 1981; 37 in 1984; 24 in 1987; 11 in 1991; 5 in 1994; 4 in 1998; 2 in 2004.* **Sources: Sa9, Up2.**

North Star - 85-90 days, Selection from Moon and Stars with many stars and Milky Way pattern on deep green skin, bright pink crisp flesh, sweet flavor, round shape, up to 40 lbs., requires fairly long season. *Source History: 1 in 2004.* **Sources: Und.**

Odem - Small oval melons, Israeli heirloom via seedsman Alan Kapuler, rare. *Source History: 1 in 2004.* **Sources: Syn.**

Orangeglo - 85-100 days, Rank vigorous vines bear quite heavily, very thin light-green rind with darker stripes, oblong melons to 50 lb., crisp sweet orange flesh, home or local market. *Source History: 2 in 1981; 1 in 1984; 1 in 1987; 1 in 1991; 2 in 1994; 3 in 1998; 8 in 2004.* **Sources: Ba8, Fe5, Ra5, Sa9, Se16, So25, Syn, Wi2.**

Osh Kirgizia - 90 days, Productive var. with sweet, pink-red flesh, 10-15 lbs., light green skin with jagged green stripes, performs very well in northern climates, introduced to the U.S. by the Seed Savers Exchange in 1992, from Russia. *Source History: 1 in 1998; 2 in 2004.* **Sources: Ba8, Se16.**

Peacock Improved - (Peacock) - 85-90 days, Tougher med-thick dark-green rind, slight furrowing, 15 x 12 in. dia., 20-25 lbs., firm sweet orange-red flesh, FW susc., widely used shipper in California and the Southwest. *Source History: 8 in 1981; 8 in 1984; 7 in 1987; 5 in 1991; 5 in 1994; 6 in 1998; 4 in 2004.* **Sources: CO23, CO30, DOR, WE10.**

Peacock Striped - Dark green striped rind, bright red flesh, sweet flavor, 25 lbs., good shipper, once common in the Southwest, the Peacock varieties originated from a Klondike cross by Bob Peacock around 1935. *Source History: 2 in 1981; 2 in 1984; 2 in 1987; 1 in 1998; 2 in 2004.* **Sources: Ba8, Hud.**

Peacock WR 60 - 85-88 days, Thin tough rind, dark-green with faint darker-green stripes, 15 x 10 in., 20-25 lb, sweet fine-textured bright-red flesh, tol. to FW, used as a shipping variety in California and Arizona and Mexico. *Source History: 8 in 1981; 7 in 1984; 6 in 1987; 5 in 1991; 3 in 1994; 2 in 1998; 2 in 2004.* **Sources: Loc, WE10.**

Petite Sweet - 72-75 days, Slightly oval, blocky, 5-12 lbs., very sweet firm pinkish-red flesh, few small seeds, FW AN1 and AN3 res., light and dark-green stripes, very early, Kansas State U. *Source History: 10 in 1981; 8 in 1984; 3 in 1987; 1 in 2004.* **Sources: Sa9.**

Petite Yellow - 80 days, Excellent quality, sweet bright yellow flesh, 6-10 lbs., good market variety. *Source History: 4 in 2004.* **Sources: Ba8, ME9, Sa9, SE14.**

Picnic - 90-100 days, Peacock type, more uniform size, oblong, blocky, thin dark-green med-tough rind, 15 x 10 in., 18-20 lbs., very sweet red-orange flesh, small black seeds, for semi-arid regions, FW res. *Source History: 1 in 1981; 1 in 1984; 1 in 1987; 1 in 1991; 1 in 2004.* **Sources: Se16.**

Quetzali - 85 days, Dark green rind is splotched with lime-green sponge prints, dense pink flesh with few seeds, 7-13 lbs., 12% sugar content, PVP 2000 Novartis Seeds, Inc. *Source History: 1 in 1998; 6 in 2004.* **Sources: Bo17, Fe5, Ki4, Se26, Sh13, So1.**

Rattlesnake Melon - (Garrisonian) - 90 days, Extremely sweet firm bright-rose flesh, about 22 in. long x 10 in. dia., light-green skin with dark rattlesnake markings, old Southern favorite, becoming quite rare. *Source History: 1 in 1981; 1 in 1987; 3 in 1991; 2 in 1994; 2 in 1998; 3 in 2004.* **Sources: Fo7, Gr27, Gur.**

Red Cloud - Dark green rind, sweet red flesh, red-pink seeds, 8-10 lbs., sometimes grows with a handle on one end of the fruit, thought by the U of Arizona to possibly be a different species than the *Citrullus vulgaris* that was introduced by the Spanish in the 1500s, seeds were discovered in 1928 in a cave in Arizona, maintained by Art Combe since then. *Source History: 1 in 1998; 1 in 2004.* **Sources: Ho13.**

Rio Mayo Sakobari - Small red-fleshed fruits, dry farmed, dry tropics. *Source History: 1 in 1994; 1 in 1998; 1 in 2004.* **Sources: Na2.**

Rio San Miguel - Grown for the edible seeds which are black, red and mottled, small round green fruit, flavorless flesh, originally from near Polanco, Chihuahua, Mexico. *Source History: 1 in 2004.* **Sources: Na2.**

Russian - Red fleshed Russian variety, oblong shape, 5-12 lbs., fairly early, from Russia. *Source History: 1 in 2004.* **Sources: Ea4.**

San Juan - Light to dark green, solid or striped skin, yellow to red flesh, white to black seeds, sweet and productive, seed collected from an elder in San Juan Pueblo. *Source History: 1 in 2004.* **Sources: Na2.**

Santo Domingo Winter - Round fruit, pale green to nearly white with faint stripes, muted red flesh, good keeper. *Source History: 1 in 2004.* **Sources: Na2.**

Scaley Bark - 98 days, Tough skin resembles the bark of a tree with the light and dark variegations, round fruit, 15-20 lbs., sweet crimson flesh, excel. for preserves, rare old strain. *Source History: 2 in 1998; 2 in 2004.* **Sources: Ho13, Shu.**

Scaley Bark, White - 98 days, White skinned version of Scaley Bark, tough thin rind, red flesh, flavorful. *Source History: 1 in 1998; 1 in 2004.* **Sources: Shu.**

Small Shining Light - 80-90 days, Old traditional Russian variety, round, 10-12 in. fruits with dark green rind and sweet, red flesh, will ripen in northern U.S. *Source History: 1 in 1994; 1 in 1998; 3 in 2004.* **Sources: Ea4, Hud, Se16.**

Snakeskin, Bush - 100-110 days, Space saving 3-5 ft. bush plants, 25-45 lb. oval fruits, sweet flavor. *Source History: 1 in 1994; 1 in 2004.* **Sources: Se7.**

Stone Mountain - (Dixie Belle) - 85-95 days, Very thick tough rind, dark-green with ribs shading to darker-green, 17 x 14 in., 35-40 lbs., deep-red flesh, few large black seeds with white tips, almost all solid heart, sweet to the rind, vigorous and prolific, fine home garden variety, not a long distance shipper, named after Stone Mountain Confederate Memorial in Georgia where it was found. *Source History: 26 in 1981; 21 in 1984; 19 in 1987; 14 in 1991; 13 in 1994; 9 in 1998; 1 in 2004.* **Sources: Sau.**

Stone Mountain Improved No. 5 - 85 days, Excellent melon, wilt resistant. *Source History: 1 in 2004.* **Sources: Wet.**

Strawberry - 85 days, Fruits from 15-25 lbs., dark green rind with darker green stripes, 8 x 20 in., delicate texture, distinctive sweet flavor, sel. from a Florida heirloom by Walt Childs, CV So1 introduction, 1989. *Source History: 2 in 1991; 1 in 1994; 1 in 1998; 1 in 2004.* **Sources: So1.**

Sugar Baby - (Icebox, Tough Sweets, Icebox Midget) - 68-86 days, Superior ice box type, very early, thin hard tough rind, distinct stripes when immature becoming almost black when ripe, round, 7-8.5 in. dia., 6-12 lbs., firm sweet red-orange flesh, small dark-brown apple-like seeds, drought res., bulk shipping and local markets, 1959. *Source History: 130 in 1981; 118 in 1984; 116 in 1987; 114 in 1991; 110 in 1994; 103 in 1998; 103 in 2004.* **Sources: Alb, All, Ba8, Be4, Ber, Bou, Bu1, Bu2, Bu3, Bu8, But, CA25, CH14, CO23, CO30, Co31, Com, Coo, Cr1, Dam, De6, DOR, Ear, Ers, Fe5, Fi1, Fo7, Ga1, Ge2, Gr27, GRI, Gur, HA3, Ha5, Hal, He8, He17, Hi6, Hi13, Hig, HO1, HO2, HPS, Jo1, Jor, Jun, Kil, La1, Lin, Loc, Ma19, MAY, Me7, ME9, Mel, Mey, MIC, Mo13, Mo20, MOU, Na6, OLD, Ont, Or10, Par, Pin, Pla, Pr3, Pr9, Ra5, Red, RIS, Roh, Ros, RUP, Sau, SE4, Se7, Se8, SE14, Se25, Se26, Sh9, Shu, Si5, So1, So9, So25, SOU, Sto, Syn, Te4, Ter, Tt2, Twi, Und, Ver, Ves, Vi4, WE10, Wet, Wi2, WI23.**

Sugar Baby, Bush - 75-80 days, Bush type, dwarf vines grow only 3-3.5 ft. long, most plants bear 2 round-oval 8.5-10 in. fruits weighing 8-12 lbs., dark-green rind, juicy sweet bright-scarlet flesh, firm texture, med-small seeds. *Source History: 1 in 1987; 1 in 1991; 3 in 1994; 2 in 1998; 3 in 2004.* **Sources: Bu2, MAY, Roh.**

Sugar Lump, Yellow - 85 days, Thin rind with small stripes, pale lemon-yellow flesh. *Source History: 1 in 2004.* **Sources: Sa9.**

Sugarlee - 90 days, Moderately vigorous and prolific, round striped med-sized melons, high quality firm sweet red flesh, med- large black stippled seed, early, good yields, shipper. *Source History: 1 in 1981; 1 in 1984; 1 in 1987; 1 in 1991; 1 in 1994; 1 in 1998; 2 in 2004.* **Sources: Ra5, Wi2.**

Sun Sweet - (Sunsweet, Sun Sugar) - 85 days, Dark green with light green stripes, sweet, bright-red flesh, 18-22 lbs., PVP 1986 NO1. *Source History: 1 in 1987; 7 in 1991; 4 in 1994; 4 in 1998; 4 in 2004.* **Sources: Bu3, CH14, Fe5, Twi.**

Sunsugar - 85 days, Allsweet type, blocky oval fruit, rind is dark green with light green stripes, bright red flesh, 18-21 lbs., productive, PVP 1986 Novartis Seeds, Inc. *Source History: 1 in 2004.* **Sources: Cl3.**

Super Sweet - 90-104 days, Thick very tough rind, med-green with dark stripes, round, 8 in., 10-15 lbs., FW AN1 and AN3 res., another selection from the same 3-way cross that produced Crimson Sweet, Kansas State U. *Source History: 5 in 1981; 4 in 1984; 2 in 1987; 1 in 1998; 1 in 2004.* **Sources: He8.**

Sweet Princess - 85-96 days, Thin tough rind, yellow-green with med-green stripes, 15 x 8 in., 20-30 lbs., sweet crisp deep-pink flesh, FW AN1 and AN3, very small seed, good shipper, NC/AES. *Source History: 11 in 1981; 11 in 1984; 8 in 1987; 4 in 1991; 4 in 1994; 4 in 1998; 1 in 2004.* **Sources: WE10.**

Sweet Siberian - 75-80 days, Light green rind, sweet orange flesh, 8 lbs., good for northern growing areas, from Russia. *Source History: 2 in 1998; 4 in 2004.* **Sources: Ho13, Sa9, Se16, So25.**

Tastigold - 80 days, Round melon with gray rind, averages 22-24 lbs., yellow flesh with small black seeds, good wilt tol., superior to Tendergold in taste, PVP 1992 Coffey Seed Co. *Source History: 1 in 1991; 1 in 1994; 1 in 1998; 1 in 2004.* **Sources: Wi2.**

Tendergold - 80 days, Oval fruit, dark green rind with mottled stripes of lighter green, creamy yellow, firm flesh turns orange at maturity, dark seeds, 25-30 lbs., good keeper and shipper. *Source History: 1 in 1994; 2 in 1998; 3 in 2004.* **Sources: Ros, Si5, Syn.**

Tendergold, Willhite - 80 days, Dark-green with mottled lighter stripes, much tougher rind, good shipper, sweet yellow flesh matures more orange, taste improves with time, 22-28 lbs., some FW tol. *Source History: 2 in 1981; 2 in 1984; 2 in 1987; 2 in 1991; 2 in 1994; 1 in 1998; 1 in 2004.* **Sources: Wi2.**

Tendersweet - 75-90 days, Med-thick tough rind, dark-green with light-green stripes, 12-18 x 8-12 in. dia., 25-50 lbs., very sweet tender yellow-orange flesh, high sugar content, for home gardens or shipping or market. *Source History: 15 in 1981; 13 in 1984; 10 in 1987; 9 in 1991; 6 in 1994; 4 in 1998; 4 in 2004.* **Sources: DOR, Ho13, Mo13, Pin.**

Tendersweet Orange Flesh - 90 days, Oblong 18 x 12 in. dia. Fruits, med-thick light-green rind with dark-green stripes, 35- 40 lbs., deep-orange flesh, high sugar, white seed with black markings and tips, great market variety. *Source History: 8 in 1981; 7 in 1984; 10 in 1987; 8 in 1991; 8 in 1994; 9 in 1998; 12 in 2004.* **Sources: Ba8, Bu3, CH14, Ers, He8, ME9, SE14, Se26, Vi4, WE10, Wi2, WI23.**

Tendersweet Yellow Flesh - 80-90 days, Oblong 20 x 12 in. dia. melon, med-thick rind, light-green with dark-green stripes, 30-35 lbs., sweet tender yellow flesh, same seeds, has been grown to 90 lbs. *Source History: 8 in 1981; 7 in 1984; 8 in 1987; 2 in 1991; 2 in 1994; 1 in 1998; 1 in 2004.* **Sources: Shu.**

Texas Giant - 95 days, Improved strain of Black Diamond, large round melon, 40-50 lbs., glossy black-green rind, very sweet tender blood-red flesh, no strings, can grow over 75 lbs. *Source History: 2 in 1981; 1 in 1984; 1 in 1987; 1 in 1998; 1 in 2004.* **Sources: Sa9.**

Texas Pink - 107 days, Medium green skin color, pink flesh, blocky shape, 10 lbs. *Source History: 1 in 1994; 1 in 1998; 1 in 2004.* **Sources: Sa9.**

Thai Black - Small to medium size melon, dark green rind, deep red flesh, sweet and crisp, does well in humid climates, introduced by Baker Creek Heirloom Seeds, from Thailand. *Source History: 1 in 2004.* **Sources: Ba8.**

Tohono O'odham Yellow Meated - 90 days, Sweet, crisp, fast-growing, high-yielding, up to 40 lbs., produces well until frost, rare. *Source History: 1 in 1987; 1 in 1991; 2 in 1994; 2 in 2004.* **Sources: Na2, Sh13.**

Tom Watson - 80-95 days, Tough elastic vines, thin tough rind, dark-green with darker-green veining, 22-24 x 10-12 in., 25- 40 lbs., sweet firm crisp coarse dark-red flesh, large brown seeds spotted with white, prolific, widely used home garden variety, also an ideal shipping melon, best adapted to Canada, pre-1900. *Source History: 23 in 1981; 18 in 1984; 15 in 1987; 12 in 1991; 7 in 1994; 6 in 1998; 5 in 2004.* **Sources: Ba8, He8, Shu, Vi4, WE10.**

Tom Watson, White Seeded - 90-95 days, Vigorous rank vines, large leaves help prevent sunburn, shiny blue-green rind, red flesh, with proper pruning and favorable conditions grows well over 100 lbs., pre-1900 heirloom. *Source History: 1 in 1981; 1 in 1987; 2 in 1991; 3 in 1994; 1 in 1998; 1 in 2004.* **Sources: Fo7.**

Uncle E - 90-100 days, Round green fruit, sweet red flesh, up to 30 lbs., Mississippi heirloom. *Source History: 2 in 1998; 1 in 2004.* **Sources: Se17.**

Verona - 70-86 days, Excel. Black Diamond type, very uniform and early, productive, very tough thin dark-green rind, firm red flesh, excel. flavor, 10-30 lbs., AN and FW tol., 30-40 lbs., 1965. *Source History: 2 in 1981; 1 in 1984; 1 in 1987; 1 in 1991; 1 in 1994; 3 in 1998; 4 in 2004.* **Sources: Fe5, Se7, So12, Wi2.**

Weeks' North Carolina Giant - (Giant Watermelon) - 100 days, Dev. by NC farmer Ed Weeks after 12 years of work, held former world record of 197 lbs. in 1975, oblong and striped, thick rind, can gain 3-4 lbs. per day. *Source History: 4 in 1981; 2 in 1984; 2 in 1987; 2 in 1991; 1 in 2004.* **Sources: Hi13.**

White Meated - Round fruit, light green rind with stripes, juicy white flesh with delicious fruity flavor, 2-3 lbs., white fleshed melons were common in the 1800s. *Source History: 1 in 2004.* **Sources: Ba8.**

White Wonder - 80 days, White flesh is almost transparent, very sweet and crisp, green rind with dark-green stripes, slightly smaller than Sugar Baby, unique variety for home gardens, from Japan, very rare. *Source History: 1 in 1981; 1 in 1984; 1 in 2004.* **Sources: Ba8.**

Will's Sugar - 90 days, An old Oscar Will var., round, med. green skin, pale red flesh, 10-14 lbs. *Source History: 1 in 1998; 1 in 2004.* **Sources: Sa9.**

Winter Melon - (King Winter Melon, Winter, King and Queen) - 78-80 days, Almost round 8-12 in. very solid fruits, almost as hard as a Citron, about 10 lbs., thin hard light-green rind ripens pale-yellow, very sweet bright-red flesh, very prolific, keeps 2 months if properly stored, longer if dipped in wax. *Source History: 3 in 1981; 3 in 1984; 6 in 1987; 5 in 1991; 2 in 1994; 2 in 1998; 1 in 2004.* **Sources: Shu.**

Winter Queen - (Black Seeded Winter Queen, King and Queen, Winter Melon) - 85-90 days, Excel. keeper, grown mostly for storage, tough rubbery rind, yellow-green with faint pale-green stripes, ripens to pale-yellow, 9.5 x 9 in. dia., 20-25 lbs., deep-red flesh, harvest before frost, edible until Christmas if stored properly. *Source History: 6 in 1981; 4 in 1984; 4 in 1987; 2 in 1991; 2 in 1994; 2 in 1998; 2 in 2004.* **Sources: Ni1, Pla.**

Wondermelon - 93 days, Oblong, dark green melon, 25-40 lbs., white seeds, pink-red flesh. *Source History: 1 in 1998; 1 in 2004.* **Sources: Sa9.**

Yellow Crimson - 80 days, Round green and light-green striped melon, very similar to Crimson Sweet in size and appearance, bright-yellow flesh, black seed, very good taste, PVP 1986 Coffey Seed Co. *Source History: 1 in 1987; 1 in 1991; 2 in 1994; 3 in 1998; 3 in 2004.* **Sources: Ers, So25, Wi2.**

Yellow Shipper - (Daisy) - 85 days, Tough rind is dark green with light green stripes, oblong fruit, 13-20 lbs., ideal for shipping in bins, crisp sweet, yellow flesh, some tol. to wilt and AN, found growing in a field of watermelon by a Burrell Seed Co. customer. *Source History: 2 in 1998; 3 in 2004.* **Sources: Ba8, Bu3, La1.**

Varieties dropped since 1981 (and the year dropped):

Art Combe Red Seeded (2004), *Asahi Sugar* (1991), *Berthold Sweet Jumbo* (1991), *Big Bully* (1998), *Blacklee* (1994), *Blue Rind* (1998), *Calhoun Gray* (1998), *Candy Red* (2004), *Carolina Cross No. 200* (2004), *Charlee* (1998), *Charleston Certified* (1998), *Charleston Gray No. 6* (1991), *Charleston Gray, Bush* (2004), *Charleston Sweet* (1983), *Citron, Black Seeded* (1998), *Citron, California* (2004), *Citron, California Wild* (1998), *Cole's Early* (1998), *Crimson Diamond* (1991), *Cutter 55* (1994), *Dixie Queen, Wilt Resistant* (2004), *Early Canada* (2004), *Early Gray* (1987), *Early Kansas* (1991), *Family Fun* (1983), *Far North* (2004), *Florida Favorite* (1994), *Florida Giant Yellow* (1998), *Fourth of July* (2004), *Halbert's Honey* (1987), *Harris' Early* (1987), *Honey Island* (1987), *Ice Box* (2004), *Ice Cream* (2004), *Ice Cream, Black Seeded* (2004), *Jubilee, Golden* (2004), *Jumbo State Fair* (1991), *Kaho* (1998), *Kansas King Jubilee* (1987), *Kengarden* (1991), *Kenya* (2004), *Klondike 155-88* (1991), *Klondike Black Seeded No. 3* (1984), *Klondike II* (2004), *Klondike Peacock* (1994), *Klondike Striped No. 11* (1991), *Klondike Wilt Res. No. 7* (1984), *Klondyke Wilt Resistant* (1991), *Kodama Yellow* (1984), *Ledmon* (1998), *Louisiana Queen* (2004), *Market Midget* (1987), *Mayo Red Meated* (1987), *Midget* (1987), *Minilee* (2004), *Moon and Stars No. 2* (1994), *Moon and Stars, Hopi* (1994), *Moon and Stars, Japan* (1994), *Moon and Stars, Sugar Baby* (1994), *Moon and Stars, White Seeded* (1998), *Mountain Sweet* (1983), *Nancy* (1998), *Northern Sweet* (1998), *Otume* (1998), *Peacock 500* (1991), *Peacock Special Shipper* (1998), *Peerless* (1987), *Piedras Verdes Mayo* (1994), *Rainbow* (2004), *Red 'n Sweet* (1991), *Red Russian* (1998), *Red Sugar Lump* (1987), *Rio Grey - Round Charleston* (1987), *Rio Mayo Miguel* (2004), *Ruby Diamond* (1998), *Ruby Gem* (1998), *Sandia Temporal* (1998), *Shin Yamato* (1991), *Skagit Gem* (1984), *Smokeylee* (1994), *Stars and Stripes* (2004), *Stone Mountain No. 5* (1994), *Sugar Bush, Burpee's* (2004), *Sugar Doll* (1987), *Sugar Loaf* (1994), *Sugar Loaf* (1994), *Sugarcreek* (2004), *Summit* (1982), *Sunny Boy* (1987), *Sunshade* (1998), *Super Tresor* (1994), *Sweet Northern Prize* (2004), *Sweet South* (1994), *Sweet Treat* (1994), *Swift Waters* (2004), *Texas Tendersweet* (1987), *Warpaint* (2004), *Yellow Doll (O.P.)* (1998), *Yellow Gem Sweet* (1998), *Yugoslavian* (2004).

Miscellaneous

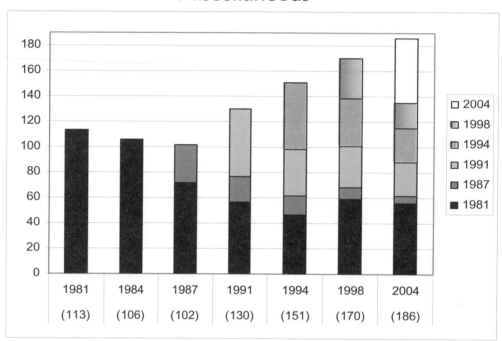

Number of varieties listed in 1981 and still available in 2004: 113 to 56 (50%).
Overall change in number of varieties from 1981 to 2004: 113 to 186 (165%).
Number of 2004 varieties available from 1 or 2 sources: 112 out of 186 (60%).

Arugula, Greek - Greek variety, grows quickly, will self-seed. *Source History: 1 in 1998; 1 in 2004.* **Sources: Te4.**

Arugula, Italian - *Eruca sativa*, easy to grow addition to salads, tangy leaves with nutty zing. *Source History: 2 in 2004.* **Sources: Pe2, Re6.**

Arugula, Sputnik - *Eruca sativa*, variable leaf shape ranges from deeply lobed like wild arugula all the way to strap leaves, no bitterness even when the weather gets hot, vigor and leaf size comes from the genetic diversity derived from the French and Italian arugula varieties that are in its parentage. *Source History: 2 in 2004.* **Sources: In8, Se27.**

Arugula, Sylvetta - (Rucola Selvatica) - 50 days, *Eruca selvatica*, shorter and smaller than regular Arugula, with more deeply lobed leaves, white flowers, and is slower to bolt. *Source History: 6 in 1998; 10 in 2004.* **Sources: Coo, Fe5, Ga22, Jo1, Orn, Pe2, Pin, Sa5, Se7, Se16.**

Arugula, Wild - (Rustic) - 55-60 days, *Diplotaxis muralis*, grows more slowly than the regular cultivar, the finely cut, deeply indented, dark green leaves are followed by pretty little yellow flowers that serve well as edible garnish, tangy, spicy flavor. *Source History: 4 in 1998; 12 in 2004.* **Sources: Bo17, Bu2, CO23, Ki4, Ni1, Re6, Ri2, SE14, Se24, TE7, Tho, WE10.**

Aztec Red Spinach - 60-80 days, *Chenopodium berlanderi*, bright red color of lower leaves retained when cooked quickly for 30-60 seconds, seedheads used in stir fry, seeds for red tortillas or as red sprouts, summer heat causes more leaf production. *Source History: 1 in 1994; 1 in 1998; 4 in 2004.* **Sources: Ni1, Red, Sh12, Und.**

Beetberry - (Strawberry Blite, Strawberry Spinach) - 90 days, *Chenopodium capitatum*, bush is 1.5-2 ft. tall, red .5 in., mulberry-like berries with strawberry flavor, leaves and shoots make good greens, berries can be dried or frozen. *Source History: 1 in 1981; 1 in 1984; 2 in 1987; 3 in 1991; 4 in 1994; 6 in 1998; 7 in 2004.* **Sources: Pe6, Roh, Sa5, Sa9, Se7, Syn, Tu6.**

Bietina - 60 days, *Beta*, 18 in. tall Italian green, cut-and-come-again, buttery sweet spinach taste, full flat green leaves used in salads, matures leaves used as spinach substitute, harvest from spring to late fall. *Source History: 1 in 2004.* **Sources: Und.**

Black Pepper - *Piper nigrum*, black and white pepper is obtained from the mature and immature fruits of the female plants, respectively, attractive tropical vine with male and female plants, makes a nice houseplant. *Source History: 1 in 1981; 1 in 1984; 1 in 1987; 1 in 1991; 3 in 1994; 1 in 1998; 1 in 2004.* **Sources: Ban.**

Brazilian Pepper - *Schinus terebinthifolius*, source of red peppercorns, bright-red color, unique flavor, tropical tree will grow as a potplant but may not produce peppercorns indoors, can cause rashes to sensitive skin. *Source History: 1 in 1991; 1 in 1994; 1 in 1998; 1 in 2004.* **Sources: Ri2.**

Buckwheat - 60-80 days, *Fagopyrum esculentum*, very short season grain, often used as an emergency crop when others fail, valuable for grain and honey or green manure or cover crops, staggered maturity and ripening, combine when most seed is ripe and shatters least, can be ground into flour for buckwheat pancakes, brown dye from flowers. *Source History: 4 in 1981; 4 in 1984; 27 in 1987; 31 in 1991; 30 in 1994; 26 in 1998; 34 in 2004.* **Sources: Ada, All, Bou, Coo, Dam, ECK, Fi6, Ga1, Hi6, Jo1, Kes, Lan, Loc, Mel, Mey, Mo13, OSB, Pe2, Pin, Pla, Pr3, Ri2, Roh, Ron, Sa12, Shu, So1, St18, Ter, Thy, Tt2, We19, We20, Wi1.**

Buckwheat, Gornosorskaya - Russian heirloom. *Source History: 1 in 1998; 1 in 2004.* **Sources: Syn.**

Buckwheat, Kaneko Soba - 70 days, Hulless var. from northern Japan, vigorous plants to 3 ft., fairly drought tol., produces seed used to make a whole grain cereal called Kasha. *Source History: 1 in 1991; 1 in 1994; 2 in 1998; 2 in 2004.* **Sources: So12, Syn.**

Buckwheat, Kinnauri Faffra - Used for gruel and thick flatbreads, landrace from the Himalayan region of Himachal Pradesh, adjacent to Tibet, grown between 9,000-10,000 ft. *Source History: 1 in 2004.* **Sources: So12.**

Buckwheat, Medawaska - 100-110 days, *Fagopyrum tartaricum*, old Asian crop originally from the altiplano of the Himalayas, especially Nepal, 4-6 ft. plants, semi-hulless seeds, more drought res. variety. *Source History: 1 in 1994; 4 in 1998; 2 in 2004.* **Sources: Ho13, Syn.**

Buckwheat, Rani Dev Faffra - Known as the Goddess Queen of the Himalayas, a Kinnauri Himalayan folkrace, 4-5 ft. tall, looks like a mass of gently wafting clouds with a shimmering mass of pink flowers that float ethereally through the garden. *Source History: 1 in 2004.* **Sources: So12.**

Buckwheat, Sarasin - *Fagopyrum tartaricum*. *Source History: 1 in 2004.* **Sources: Syn.**

Buckwheat, Silverhull - 85-90 days, Prolific heirloom, harvest whole plants, cure in sun for 7-10 days until 85% of seeds show a silvery sheen. *Source History: 1 in 2004.* **Sources: So12.**

Buckwheat, Sperli's - Unique seed shape and color, Germany. *Source History: 1 in 1998; 1 in 2004.* **Sources: Syn.**

Buckwheat, Sprouting - Cover crop and sprouting. *Source History: 1 in 1994; 1 in 1998; 1 in 2004.* **Sources: Syn.**

Buckwheat, Tartary - *Fagopyrum tartaricum*, folkrace buckwheat from the Kinnauri Himalayas, adjacent to Tibet, slender, 3- 4 ft. plant, small green-yellow flowers, for short season areas and poor soils, toasted and ground for porridge. *Source History: 1 in 1998; 1 in 2004.* **Sources: So12.**

Burdock, Takinogawa Long - 105 days, *Arctium lappa*, long thin roots used in Japanese dishes, hardy, harvest in fall, winter and spring. *Source History: 3 in 2004.* **Sources: Loc, Sa9, Tu6.**

Burdock, Wantanabe Early - 150 days, Long brown-skinned roots rich in fiber, tasty in tempura or sauteed with other vegetables or baked with beans, young leaves can be eaten fried when 12 in. tall. *Source History: 1 in 1987; 2 in 1991; 1 in 1994; 2 in 1998; 1 in 2004.* **Sources: Ter.**

Caper Bush - *Capparis spinosa inermis*, capers are made from the unopened flower buds which are pickled to develop their pungent flavor, straggly, low growing shrub found in rocky areas of the Mediterranean and North Africa, slow germination, typically less than 30%, not recommended for the novice gardener. *Source History: 1 in 1994; 3 in 1998; 4 in 2004.* **Sources: Hud, Ri2, Se24, Sh14.**

Celtuce - (Chinese Lettuce, Burpee's Strain, T and M Celtuce, Asparagus Lettuce, Stem Lettuce, Wo Sun) - 80-90 days, Actually a stem lettuce from China, 2-4 times higher in vitamin C than lettuce, leaves used for salad greens or boiled, then let plants bolt and cut seedstalks before flowers appear, stalks and heart of central stem eaten raw or cooked like celery, dates back to the 1840s. *Source History: 19 in 1981; 18 in 1984; 24 in 1987; 27 in 1991; 22 in 1994; 20 in 1998; 14 in 2004.* **Sources: Ba8, Bo19, CO30, Com, DOR, Ear, Ev2, Fo7, Kit, Ni1, Ra5, Sa5, Tho, WE10.**

Chia - 120 days, *Salvia columbariae*, California native plant, used to make sprouts, a refreshing drink, also ground into meal for delicious muffins, blue flowers go well with California poppies. *Source History: 1 in 1987; 2 in 1991; 5 in 1994; 9 in 1998; 4 in 2004.* **Sources: Ho8, Hud, Pla, So12.**

Chia Roja - *Chenopodium berlandieri* var. *nuttaliae*, seed is toasted and used as pinole or atole, ground seed is also added to corn nixtamal to make red tamales (chapatas), leaves also edible, grown and harvested by Tarascan Indian farmers in Michoacan, southern Mexico. *Source History: 2 in 1991; 1 in 1994; 1 in 1998; 1 in 2004.* **Sources: Ho13.**

Chia, Tarahumara - 110-120 days, *Salvia tiliafolia*, vibrant 6 ft. bushes with blue flowers that yield many highly nutritious seeds that still are eaten by the Tarahumara and other Indian tribes of the Madrean region, the seeds are roasted and ground and added to water to form a gel, one tablespoon reputedly can sustain a person for 24 hours, good for long distance running. *Source History: 3 in 1991; 3 in 1994; 6 in 1998; 8 in 2004.* **Sources: Ga21, Hud, Na2, Ri12, Se7, Se17, So12, Syn.**

Chinese Artichoke - (Mint Spiral Roots) - *Stachys affinis*, very hardy ancient Chinese perennial, tuberous rooted member of mint family, intro. in 1880s to France where it is called Crosnes, then spread to Belgium, called Chorogi in Japan, grows as a low many branched bushy plant which produces 40 or 50 white lumpy 2 x .5 in. dia. tubers, dries out quickly, needs heavy compost. *Source History: 1 in 1981; 1 in 1984; 3 in 1987; 5 in 1991; 1 in 1994; 4 in 1998; 1 in

2004. **Sources: Ho13.**

Chinese Parsley - (Coriander, Yuen Sai, Cilantro) - 35 days, *Coriandrum sativum*, special flavor, young 6-8 leaf seedlings are harvested and used in salads or soups or Chinese dishes, citrus flavor, seeds used as condiment. *Source History: 2 in 1981; 2 in 1984; 3 in 1987; 59 in 1991; 45 in 1994; 44 in 1998; 39 in 2004.* **Sources: ABB, Alb, All, Ban, Coo, Dgs, Ear, Ev2, Fe5, Fo7, Ge2, GRI, He8, HPS, Ki4, Lin, MAY, Mey, MOU, Ont, PAG, Pe1, Pin, Pr3, Re8, Ren, Ri2, Roh, RUP, SGF, Sh9, Shu, So1, Sto, Te4, Tho, Ves, Vi5, Wet.**

Chufa (Ground Almond) - *Cyperus esculentus*, 5 vars. in U.S., tasty peanut-sized tubers, 30 lbs. of seed per acre yields 3,000 lbs., grown in food plots for wild turkeys from Canada to Florida, before 1940 was grown as a snack food and for hog feed, does best in sandy loam at 6.8 pH, eat raw or sun-dried or pressure-cooked, declared a weed in California--Yellow Nutsedge. *Source History: 1 in 1981; 1 in 1984; 8 in 1987; 5 in 1991; 4 in 1994; 7 in 1998; 8 in 2004.* **Sources: Ada, Gl2, Ho13, Hud, Kes, Ma13, Shu, Wi1.**

Cilantro - (Mexican Parsley, Chinese Parsley, Coriander) - Young plants (cilantro-30 days) and seeds (coriander-70 days) widely used in Mexican and Oriental cooking, parsley-like slightly peppery flavor. *Source History: 3 in 1981; 3 in 1984; 5 in 1987; 46 in 1991; 65 in 1994; 77 in 1998; 90 in 2004.* **Sources: ABB, Au2, Ba8, Bu2, Bu3, But, CA25, Cl3, Com, Com, Cr1, Dam, De6, Dow, Eo2, Ers, Fa1, Fi1, Fis, Ga1, Ga21, Ga22, Ge2, GLO, Gr28, Gur, HA3, Hal, Hi6, Hig, Ho8, Ho13, HPS, Hud, Hum, In8, Jo1, Jor, Jun, Kil, Kit, La8, Lan, Lej, Lin, Ma19, Me7, Mel, Mi12, Mo13, Na2, Ni1, Ol2, OLD, Orn, OSB, Par, Pe2, Pep, Pin, Pla, Red, Ros, Sa6, Sa12, Scf, SE4, Se7, Se8, Se16, Se24, Se25, Se26, Sh14, Sk2, So9, So25, SOU, Syn, TE7, Thy, Tt2, Tu6, Twi, Und, Vi4, WE10, We19, We20, Wi15.**

Cilantro, Cal Long Standing - *Source History: 1 in 2004.* **Sources: Bo17.**

Cilantro, Yen Sai - Sweetest of all the cilantros with thicker leaves, slightly longer production. *Source History: 1 in 1998; 1 in 2004.* **Sources: Red.**

Cinnamon Vine - (Chinese Yam, Jinengo Potato) - *Dioscorea japonica*, vining plant with shiny leaves, tiny fragrant flowers, tiny potato-like tubers grow above ground on the vines, this plant is actually a true yam. *Source History: 1 in 1981; 1 in 1987; 2 in 1991; 2 in 1994; 3 in 1998; 2 in 2004.* **Sources: Ga21, Ri2.**

Colbaga - Dev. by Prof. Meader of New Hampshire, color and quality of a rutabaga but grows above ground like a kohlrabi, unique delicate flavor, used like turnips, good keeper. *Source History: 3 in 1981; 2 in 1984; 2 in 1987; 1 in 1991; 1 in 1994; 1 in 1998; 2 in 2004.* **Sources: Fis, Ho13.**

Comfrey - (Russian Comfrey, Russian Broadleaf Comfrey, Quaker Comfrey, Prickly Comfrey) - Perennial, *Symphytum* spp., known for centuries as the healing herb, greens used like spinach, dried for tea, roots used like parsnips, also for cattle and goat feed, high in protein, prefers some shade, harvest at 12 x 12 in. dia., grows 3-6 ft. tall, some vague warnings of toxicity problems several years ago from Australia. *Source History: 19 in 1981; 15 in 1984; 17 in 1987; 10 in 1991; 6 in 1994; 5 in 1998; 4 in 2004.* **Sources: Fa1, Mel, Ni1, Ri2.**

Coriander, Indian - (Dhani-ya) - East Indian type, grown mainly for the oblong seeds, crushed leaves are more aromatic and have superior flavor to commercial coriander, seeds are sweeter with fuller flavor, eaten like candy when coated with sugar, used in curries. *Source History: 1 in 1991; 1 in 1994; 1 in 1998; 1 in 2004.* **Sources: Red.**

Coriander, Leisure - (Long Standing) - Tender, round leaves are delicious in salsa, dried seeds used commercially in candy, sauces and soups, slow bolting. *Source History: 6 in 1998; 9 in 2004.* **Sources: CH14, CO23, DOR, HO2, Jor, Pe6, Ri2, RUP, Wi2.**

Coriander, Santo - *Coriandrum sativum*, much improved strain for production of parsley-like leaves, slow bolting and frost tol. *Source History: 2 in 1991; 10 in 1994; 17 in 1998; 14 in 2004.* **Sources: Ag7, CH14, Com, Ha5, HO2, ME9, Mo20, Par, RIS, SE4, Sto, Syn, Ter, Ver.**

Coriander, Slo-bolt - 45 days, *Coriandrum sativum*, heat res. stain that holds at the broadleaf stage much longer. *Source History: 1 in 1991; 6 in 1994; 15 in 1998; 21 in 2004.* **Sources: Bo17, CO30, Com, Coo, DOR, Hud, La1, Loc, Na6, Pe2, Re6, Red, Sa5, SE4, Se7, SE14, So12, Thy, Twi, WE10, WI23.**

Coriander, Standby - 40-50 days, Bred for fall sowing in August to Sept. and harvesting winter through early spring, super slow bolting, extra winter hardy, PSR. *Source History: 1 in 2004.* **Sources: Pe6.**

Coriander, Vietnamese - Perennial, *Polygonium odoratum*, excel. flavor, may be grown indoors in good light. *Source History: 1 in 1991; 1 in 1994; 1 in 1998; 1 in 2004.* **Sources: Ri2.**

Cotton - 150 days, *Gossypium hirsutum*, annual, the cotton grown today originated in the Americas, 2-4 ft. tall plants with white flowers which form seedpods which spit open at maturity, cotton fiber is actually hairs attached to the seed. *Source History: 1 in 2004.* **Sources: Red.**

Cotton, Erlene's Green - Light olive-green fibers can be spun off the seed, harvest bolls soon after opening to prevent fading in the sunlight, turns a yellow-green color after spinning and washing, family heirloom from Erlene Melancon, TX. *Source History: 2 in 2004.* **Sources: Ga21, So1.**

Cotton, Green - *Gossypium sp.*, old str. from Texas, light green fibers, attractive plant with bronze-red leaves and stems. *Source History: 1 in 1998; 1 in 2004.* **Sources: Ho13.**

Cotton, Honey Gold - Light tan-gold fibers. *Source History: 1 in 2004.* **Sources: Ho13.**

Cotton, Hopi Short Staple - 100 days, *Gossypium hirsutum* var. *puntatum*, short stapled fiber and oilseed cotton, large white blooms, short growing season, harvest pods as they mature, originated in Central America and traded north, prehistorically grown by the Hopi. *Source History: 3 in 2004.* **Sources: Ho13, Na2, So12.**

Cotton, Nankeen - Pre-1860 heirloom, naturally brown cotton with non-fading attractive dark copper colored lint which is retained by the seed, thus easier to spin, brightens with washing, plant has longer, thinner branches with more finely divided leaves than other cotton, long bloom cycle, white and pink-red blooms from mid-summer until late fall, does well in poor dry soil, slightly better insect res., planted as an ornamental in some parts of Louisiana, CV So1 introduction, 1997. *Source History: 3 in 2004.* **Sources: Ga21, Ho13, So1.**

Cotton, Peruvian Brown - *Gossypium hirsutum*, long growing season, soft, short staple cotton, colors range from red- brown to olive-green, yellow flowers, natural colored cotton fibers have been grown by native peoples from Mexico to South America for over 2000 years. *Source History: 2 in 2004.* **Sources: Ho13, Na2.**

Cotton, Sacaton Aboriginal - *Gossypium hirsutum* var. *punctatum*, related to Hopi cotton, maintained by the Field Station in Sacaton, AZ for many years by the name Sacaton Aboriginal, grown until 1900 by the Pimans for food and fiber. *Source History: 1 in 2004.* **Sources: Na2.**

Cotton, Sea Island - *Gossypium barbadense*, 3 ft. shrub, yellow flowers, light-tan color, from South America where its main use is for cotton seed oil. *Source History: 1 in 2004.* **Sources: Ho13.**

Curled Mallow - (Malva Crispa, Curly Mallow) - 50-57 days, *Malva verticillata*, old-time salad plant now rarely grown, ornamental cousin of hibiscus, tender glossy lobed 4 in. leaves, elegantly frilled edges, delicate almost bland flavor, never bitter throughout summer, small edible flowers appear only on the central stalk, interesting addition to any salad, excellent garnish, not for cooking, reseeds dependably. *Source History: 4 in 1987; 6 in 1991; 2 in 1994; 5 in 1998; 6 in 2004.* **Sources: Gr28, Ni1, Pe1, Ri2, Sa12, Syn.**

Dandelion - 90-95 days, *Taraxacum officinale*, plant spreads to 2 ft. across, cabbage-like, vitamin rich leaves can be eaten raw or stir fried, roots are roasted and made into a coffee-like beverage. *Source History: 7 in 1994; 22 in 1998; 18 in 2004.* **Sources: Bou, Ers, Ga21, He8, He17, Ho8, Ho13, Jo1, La1, Ni1, Orn, Ri2, Sa6, Se24, Sto, TE7, Tu6, WE10.**

Dandelion, French - (Vert de Montmogny, Montmagny) - 75-90 days, *Taraxacum officinale*, narrow, deeply cut, 8-10 in. leaves are eaten raw or stir fried, taste like mild chicory, high in minerals and magnesium, harvest when very young and tender, older leaves are steamed or sauteed like spinach. *Source History: 3 in 1994; 5 in 1998; 3 in 2004.* **Sources: Coo, Ho13, Pe2.**

Dandelion, Improved - (Fullheart Improved) - 48 days, *Taraxacum officinale sativum*, tender fleshy leaves resemble domestic var., they are more savoyed and nearly 12 in. long, ideal for salads or cooked like spinach, very high iron content. *Source History: 1 in 1994; 2 in 1998; 2 in 2004.* **Sources: Pin, Ri2.**

Devil's Claw - (Unicorn Plant) - *Proboscidea parviflora*, low spreading annual up to 3 ft. tall and wide, abundant foliage and large orchid-like blooms, unique decorative seed pods can be eaten when young or pickled like cucumbers, edible seeds, dried black claws (seed pods) are split and woven to make the black designs on Pima and Papago baskets. *Source History: 1 in 1991; 1 in 1994; 3 in 1998; 2 in 2004.* **Sources: Ho13, So12.**

Devil's Claw, Domesticated - (Tohono O'odham White Seeded Domesticate) - *P. parviflora* var. *hohokamiana*, domesticated by Sonoran tribes, long black fibers used in basket weaving, white seed eaten or pressed for oil, 15 in. claws. *Source History: 2 in 1981; 2 in 1984; 2 in 1987; 2 in 1991; 2 in 1994; 2 in 1998; 1 in 2004.* **Sources: Na2.**

Devil's Claw, Eagle Creek - *P. parviflora* var. *parviflora*, medium length claws, white seeds. *Source History: 1 in 1994; 1 in 1998; 1 in 2004.* **Sources: Na2.**

Devil's Claw, Hopi - *P. parviflora* var. *hohokamiana*, long claws, pale pink flowers, white seeds, from Hopi Reservation. *Source History: 1 in 2004.* **Sources: Na2.**

Devil's Claw, Morelos - *Proboscidea parviflora* var. *hohokamiana*, wild annual with 3-4 in. claws, fragrant violet flowers, black seeds, from Morelos in southern Mexico. *Source History: 1 in 2004.* **Sources: Na2.**

Devil's Claw, O'odham - Long spines up to 12 in. long, colorful plants, their stickiness traps some insects, dried seed pod used for crafting. *Source History: 1 in 1998; 1 in 2004.* **Sources: Sa9.**

Devil's Claw, Paiute - *P. parviflora* var. *hohokamiana*, domesticated variety with white seeds, grown in southwest Utah. *Source History: 1 in 1994; 1 in 1998; 1 in 2004.* **Sources: Na2.**

Devil's Claw, Po'onomp - *Proboscidea parviflora* var. *hohokamiana*, domesticated var., white seeds, original seed from the Supai via a Chemehuevi basketmaker in 1979. *Source History: 1 in 2004.* **Sources: Na2.**

Devil's Claw, San Carlos Apache Domesticated - *P. parviflora* var. *hohokamiana*, claws used in basketry, white seeds, collected from plants growing in a field of blue corn in 1978. *Source History: 1 in 2004.* **Sources: Na2.**

Devil's Claw, Wild Multiclawed - *P. parviflora* var. *parviflora*, short-clawed var. with black seeds, pink or purple flowers. *Source History: 1 in 1994; 1 in 1998; 1 in 2004.* **Sources: Na2.**

Duck Potato - *Saggitaria latifolia*, cultivated in Asia, tasty, egg size tubers, does best in 6 in. of water, forms colonies. *Source History: 1 in 1998; 1 in 2004.* **Sources: Pe8.**

Earth Chestnut - (Tuberous-Rooted Caraway) - *Bunium bulbocastaneum*, perennial, low-growing crown resembles parsley, produces masses of sweet crunchy .75 in. tubers, unique, from Eurasia. *Source History: 2 in 1994; 3 in 1998; 3 in 2004.* **Sources: Ho13, Ni1, So12.**

Egusi - *Citrullus* spp., grown in Africa for the seeds which contain 50% oil and 30% protein, eaten when under 14 in. long, can be roasted, mashed into peanut butter consistency and used in soups, taste is similar to zucchini, seeds from larger size fruit make an almost nutty addition to stir fry, close relative of watermelon which the 6 in. fruits resemble, white flesh is inedible. *Source History: 1 in 1994; 1 in 1998; 2 in 2004.* **Sources: Ech, Ho13.**

Ethiopian Cabbage - (Abyssinian Mustard, Karate) - *Brassica carinata*, growth habit resembles collards, 3 ft. tall plant with med. green, wide leaves, young tender leaves are eaten raw, older leaves are cooked, flowers are used like broccoli, seeds produce edible oil, heat res., from North Africa. *Source History: 1 in 1998; 1 in 2004.* **Sources: Ho13.**

Ginger Roots - 240 days, Knobby root is grated for flavoring, sweetly pungent flavor, used in Oriental cooking. *Source History: 2 in 1981; 2 in 1984; 4 in 1987; 3 in 1991; 3 in 1994; 1 in 1998; 1 in 2004.* **Sources: Gur.**

Ginseng - (American Ginseng) - *Panax quinquifolium*, a poorly managed planting should average about 2,000 lbs. per acre, one person can care for about 2 acres of ginseng grown from seed which takes a year or more to germinate, plant seeds anytime from September until the ground freezes in a well-drained moist location, seedlings will appear the following spring and will mature in 5-7 years. *Source History: 5 in 1991; 9 in 1994; 11 in 1998; 12 in 2004.* **Sources: Ba6, Ban, Bi3, Ga21, Ho8, Jo1, Lej, Mel, Ri2, Sh15, So1, So12.**

Ginseng, Siberian - *Eleutherococcus senticosus*, controversial hardy prickly shrub, its restorative and strength and stamina properties have reportedly been documented scientifically, unlike true ginseng it has calming relaxant effect and does not require shade, hardy in Ontario, potential economic crop. *Source History: 1 in 1981; 1 in 1984; 2 in 1987; 2 in 1991; 1 in 1994; 3 in 1998; 3 in 2004.* **Sources: Ho8, Ho13, Ri2.**

Gobo - (Edible Burdock, Burdock) - 120 days, *Arctium lappa*, edible burdock, dev. for centuries in Japan, tender roots grow quickly to 40 x 1 in. dia., tapers near end, snaps like a carrot if bent, fiberless, biennial, large leaves the first year, roots stay fiberless and continue to grow, likes deep fertile sandy soils, 6 ft. tall plants, flowering stems edible, seeds can be sprouted, tender young roots eaten like celery, stalks also steamed. *Source History: 6 in 1981; 4 in 1984; 6 in 1987; 7 in 1991; 6 in 1994; 9 in 1998; 7 in 2004.* **Sources: Bou, Ga21, Ho8, Kit, Pe2, Ri2, We20.**

Gobo, Takinogawa Long - 110-130 days, *Arctium lappa*, edible burdock, widely grown around Tokyo, red stalked, smooth slender 18-50 in. roots, excel. flavor and texture, strong aroma, serve boiled, Japan. *Source History: 6 in 1981; 4 in 1984; 4 in 1987; 9 in 1991; 11 in 1994; 15 in 1998; 10 in 2004.* **Sources: Ev2, Ho13, Jo1, Kit, Ni1, OSB, Ri2, Roh, Se7, So12.**

Gobo, Watanabe Early - 110 days, Slender roots about 40 in. long, very rapid grower, fine textured flesh, good flavor. *Source History: 1 in 1981; 1 in 1984; 3 in 1998; 4 in 2004.* **Sources: Ev2, Kit, OSB, So12.**

Gobo, Yanagawa - 120 days, Popular in Japanese cuisine, used like Salsify, strong aroma, 36-42 in. long. *Source History: 1 in 2004.* **Sources: OSB.**

Good King Henry - (Mercury, Lincolnshire Spinach) - *Chenopodium bonus-henricus*, long, arrow-shaped leaves cooked and eaten much like asparagus and used in salads, rich in vitamins A and C and calcium, hardy self-seeding perennial, German name is Der Gut Heinrich. *Source History: 1 in 1987; 1 in 1991; 10 in 1994; 10 in 1998; 7 in 2004.* **Sources: Bou, Coo, Ga22, Lan, Ni1, Pe8, Ri2.**

Gopher Purge - (Mole Plant, Gopher Plant, Gopher Spurge) - *Euphorbia lathyris*, unusual Mediterranean native, tall fast growing ornamental, sap causes skin rash so warn children, sap in roots repels gophers and moles, decorative biennial plant with strange greenish flowers, prefers light sandy soil and full sunlight. *Source History: 5 in 1981; 5 in 1984; 10 in 1987; 9 in 1991; 8 in 1994; 6 in 1998; 2 in 2004.* **Sources: Lan, Ri2.**

Groundnut - (American Groundnut) - *Apios americana*, hardy vine with clusters of fragrant brown-pink flowers, edible sweet starchy tubers, 17% protein, staple fare for the Indians of the eastern U.S., tubers sent. *Source History: 1 in 1991; 1 in 1998; 2 in 2004.* **Sources: Ho13, Pe1.**

Horseradish - Perennial, *Armoracia rusticana*, large very pungent crisp white roots, long-lasting perennial, multiplied with root cuttings, finely grated and used in sauces or with meats. *Source History: 13 in 1981; 11 in 1984; 12 in 1987; 13 in 1991; 13 in 1994; 16 in 1998; 9 in 2004.* **Sources: Fa1, Ho8, Jo1, Ma13, Mel, Par, Pin, Ri2, Ter.**

Horseradish, Bohemian - (Maliner Kreb, Maliner Kren, Bohemian Giant) - Perennial, Very large white-fleshed high quality roots, not bitter, superior to and earlier than ordinary horseradish, very hot but sweet agreeable taste, hardy everywhere. *Source History: 10 in 1981; 9 in 1984; 8 in 1987; 9 in 1991; 9 in 1994; 7 in 1998; 5 in 2004.* **Sources: Bu1, DAl, De6, Jun, Ni1.**

Horseradish, Maliner Kren - Perenniel days, *Armoracia rusticana*, true Bohemian horseradish, standard variety for condiment use, vigorous grower, produces large white roots, excel. quality, quite pungent, fine flavor. *Source History: 3 in 1981; 3 in 1984; 4 in 1987; 3 in 1991; 2 in 1994; 5 in 1998; 4 in 2004.* **Sources: La1, Mey, Shu, Ver.**

Horseradish, Variegated - *Armoracia rusticana* 'Variegata', less aggressive than the green leaved variety, white develops in leaves the second year, condiment made from the roots. *Source History: 1 in 2004.* **Sources: Ri2.**

Huazontle - (Aztec Red Spinach) - 60 days, *Chenopodium berlandieri*, young leaves and stems are eaten as greens, seed stalks are blanched and fried in egg and cheese batter or stripped from the stem and made into patties. *Source History: 1 in 1994; 6 in 1998; 2 in 2004.* **Sources: Hud, Ri2.**

Jicama - (Mexican Turnip, Sa Gord, Yam Bean) - 120-270 days, *Pachyrrhizus erosus*, very long vines, culture like sweet potatoes, resembles flat brown turnips, 6 in. dia. if blossoms and seedheads kept pinched off, 1-6 lbs., crisp white flesh, texture like Jerusalem artichoke, tastes like water chestnuts, needs sun and a rich light soil, drought res., keeps well, eat raw or steam or bake. *Source History: 18 in 1981; 14 in 1984; 17 in 1987; 16 in 1991; 16 in 1994; 18 in 1998; 19 in 2004.* **Sources: Ban, Bo18, CO30, Ech, Gr29, Gur, Hi13, Hud, Kit, LO8, Loc, Pin, Ra5, Ri2, Ri12, Sa12, Se28, Shu, Wo8.**

Job's Tears - *Coix lachryma-jobi*, edible, not ornamental type, Japanese buff-colored shells bear rouge-brown seeds on loose panicles like millet with plant type like maize or sorghum, 4.5-5.5 ft. tall. *Source History: 1 in 2004.* **Sources: So12.**

Lambs Quarters - (White Goosefoot) - *Chenopodium album*, entire plant is edible, leaves used in salads, leaves and stems can be steamed or boiled, seeds are ground for meal or flour, easy to grow. *Source History: 1 in 1981; 1 in 1984; 4 in 1987; 7 in 1991; 5 in 1994; 7 in 1998; 8 in 2004.* **Sources: Bou, Ho8, Lej, Ri2, Sa6, So9, So12, We20.**

Lambs Quarters, Edulis - *Chenopodium alba*, stable true-breeding bushy form of the common wilding, excellent flavor, leaves and tender stem tips are used in salads, high in protein, potassium, calcium and vitamin A, no oxalic aftertaste. *Source History: 1 in 1994; 1 in 1998; 2 in 2004.* **Sources: Ni1, Sh12.**

Lambs Quarters, Goosefoot - 45 days, *Chenopodium giganteum*, beautiful leaves are green with magenta pink center, adds color to salads, good cooking green. *Source History: 1 in 1994; 4 in 1998; 4 in 2004.* **Sources: Hud, Pe6, So12, So25.**

Lambs Quarters, Indian Greens - *Chenopodium murale*, Southwest desert native grows 6 ft. tall, slender plant with heavy seed set. *Source History: 1 in 2004.* **Sources: Syn.**

Lambs Quarters, Magentaspreen - (Purple Goose-foot) - 65 days, *Chenopodium giganteum*, European species, grows to 8 ft., leaves like lambs quarters, magenta-colored young leaves, delicate flavor, readily self-seeds, CV Peace Seeds intro., 1983. *Source History: 2 in 1991; 7 in 1994; 7 in 1998; 6 in 2004.* **Sources: Ga22, Gr28, Ho13, Se7, Sh12, Syn.**

Malabar Spinach - (Poi Sag, Ceylon Spinach, Malabar Nightshade, Malabar Climbing Spinach, Land Kelp) - 70 days, *Basella alba*, excel. spinach substitute, big glossy bright-green leaves, tender .5 in. stems, will climb 6-10 ft., cooked leaves have the taste and texture of kelp, used as a thickening agent in the Orient, used in place of okra in the South. *Source History: 8 in 1981; 5 in 1984; 5 in 1987; 11 in 1991; 12 in 1994; 10 in 1998; 17 in 2004.* **Sources: Ban, Bou, CO30, Com, Ech, Eo2, Ev2, Jo1, Kit, ME9, Ni1, OSB, Red, Ri12, Sa9, Ter, Und.**

Malabar Spinach, Red Strain - (Basella Rubra, Basella Malabar Red Stem) - 110 days, *Basella rubra*, heat loving perennial from India, red leaf veins and stems, free branching climber, cut sprouts to eat, regrows rapidly, for food and dye and jelly. *Source History: 2 in 1981; 2 in 1984; 2 in 1987; 5 in 1991; 8 in 1994; 6 in 1998; 9 in 2004.* **Sources: CO30, Ev2, Hud, Jo1, Kit, Par, Pin, Sa9, So25.**

Melo Khiya - *Corcorus olitorius*, green leafy vegetable, 3.5 ft. tall, lanceolate leaf, pick frequently, no oxalic acid, from Egypt. *Source History: 1 in 2004.* **Sources: Sa9.**

Millet, Candlestick - Heads are borne on tall stalks and shaped like candlesticks, many tillers, grown as food grain on 50 million acres in India and Africa where many delicious recipes have developed. *Source History: 1 in 1991; 1 in 1994; 1 in 1998; 2 in 2004.* **Sources: Kus, Syn.**

Millet, Chiwapa Barnyard - 110 days, *Echinochloa frumentacea*, 3.5-4 ft. tall, selected str. of minor millet crop of India. *Source History: 1 in 2004.* **Sources: So12.**

Millet, Dragon's Claw - (Birdsfoot Millet, Finger Millet) - 120 days, *Eleusine coracana*, staple crop of arid and semi-arid regions of Asia and Africa, 3 ft. plants, dark crimson seeds borne on seed stalks like a bird's foot or clasping hand, rich, nutty flavor, highest iodine content among cereal grains, very high iron content, drought and alkaline tol., abundant tillering, prolific yields, known as 'ragi' in India, used for chapattis, can be stored for 50 years. *Source History: 2 in 2004.* **Sources: So1, So12.**

Miner's Lettuce - (Claytonia, Winter Purslane, Cuban Spinach) - 40 days, *Montia perfoliata* or *Claytonia perfoliata*, used as a salad green by the 49ers during the California gold rush, unusual dark-green plant, succulent, likes rich moist soil, will self seed. *Source History: 1 in 1981; 1 in 1984; 3 in 1987; 3 in 1991; 8 in 1994; 14 in 1998; 12 in 2004.* **Sources: Bou, CO30, Com, Coo, Dam, Fe5, Jo1, Ni1, Pe8, Red, Syn, Tu6.**

Minutina - (Erba Stella) - 50 days, *Plantago corynopsus*, cold hardy salad plant for fall, winter and spring production, edible flower buds, regrows after cutting. *Source History: 2 in 1998; 4 in 2004.* **Sources: Coo, Fe5, Jo1, Orn.**

Mitsuba - (Japanese Celery, Japanese Parsley) - 60-70 days, *Cryptotaenia japonica*, Japanese celery-parsley, thin 36 in. leafstalks with 3 heart-shaped leaves, unique flavor, for soup flavoring or cooked greens or bean sprouts. *Source History: 13 in 1981; 12 in 1984; 10 in 1987; 13 in 1991; 11 in 1994; 10 in 1998; 16 in 2004.* **Sources: Bo17, Bu2, Ev2, Kit, Lej, ME9, Ni1, Ren, Ri2, Roh, Sa12, Se26, So12, So25, Sto, We19.**

Mustard, Black - 60 days, *Brassica nigra*, grows to 3-10 ft. tall, young leaves used as greens and in salads, seeds mixed with vinegar and used as a flavoring, ground seeds used to make table mustard. *Source History: 2 in 1981; 1 in 1984; 3 in 1987; 6 in 1991; 9 in 1994; 8 in 1998; 8 in 2004.* **Sources: Fo7, Ga21, Gou, Hud, Re8, Red, So1, Vi5.**

Mustard, Brown - *Brassica juncea*, plant to 4 ft., dark seeds, least pungent, young thinnings make tasty mustard greens in the spring. *Source History: 2 in 1994; 1 in 1998; 2 in 2004.* **Sources: He8, Sa12.**

Mustard, Burgonde - 85 days, *Brassica hirta (Sinapis alba)*, standard brown mustard seed variety, used for the many different hot brown stone-ground and French mustards, small brown seed, 9,900 per oz. average. *Source History: 2 in 1987; 1 in 1991; 1 in 1998; 1 in 2004.* **Sources: Te4.**

Mustard, Fine White - *Brassica hirta (Sinapis alba)*, to 4 ft., used to produce seeds for condiments, milder than black var., finest strain for salad purposes. *Source History: 1 in 1981; 1 in 2004.* **Sources: Ni1.**

Mustard, White - 10-16 days, *Brassica hirta*, mature seeds are ground to make mustard condiment or sprouted, spicy new leaves in salads or sandwiches, grown in Europe for over 400 years. *Source History: 4 in 1981; 4 in 1984; 3 in 1987; 3 in 1991; 2 in 1994; 2 in 1998; 6 in 2004.* **Sources: Com, Red, Ri2, Sa12, So12, Syn.**

Mustard, Wild - (Tarahumara Mostaza, Mocoasali) - *Brassica campestris*, Old World introduction to Mexico, commonly found in fields, slightly spicy leaves are harvested young, eaten in salads or cooked. *Source History: 1 in 1991; 1 in 1994; 1 in 1998; 1 in 2004.* **Sources: Na2.**

Mustard, Yellow/White - *Brassica hirta (B. alba)*,

native to Europe, cultivated as an annual, 12-24 in. tall, common in cooking, milder than Black Mustard and far more widely used. *Source History: 1 in 1981; 1 in 1984; 3 in 1987; 2 in 1991; 4 in 1994; 4 in 1998; 1 in 2004.* **Sources: Thy.**

New Zealand Spinach - (Everbearing, Everlasting, Tetragone, Della Nuova Zelanda, Perpetual Spinach, Sea Spinach) - 50- 70 days, *Tetragonia expansa*, not a true spinach but similar in flavor and use, slow to germinate, soak for 24 hours, plant 24 in. apart, large strong spreading plants branch freely, good for summer greens, thrives in hot weather, won't bolt or get bitter, small brittle fleshy deep-green leaves, pick only the 4 in. tips of branches all summer and fall, disease and insect res., suited for home gardeners only, this New Zealand native was first brought to Europe by Captain Cook in the 1770s after he discovered it to be a valuable source of vitamin C. *Source History: 108 in 1981; 103 in 1984; 98 in 1987; 85 in 1991; 84 in 1994; 85 in 1998; 72 in 2004.* **Sources: Ag7, All, Ba8, Be4, Ber, Bo17, Bo19, Bou, Bu2, But, CH14, CO30, Com, Coo, Dam, De6, Ear, Eo2, Ers, Fe5, Fo7, Gr27, HA3, Hal, He8, He17, Hud, Hum, Jo1, Jun, Kil, La1, Lej, Lin, Ma19, MAY, ME9, Mel, Mey, Mo13, Ni1, Ol2, Ont, Pin, Ra5, Red, Ri2, Roh, Ros, RUP, Sa6, Sa9, Sau, Se8, Se16, Se25, Se26, Sh9, Shu, Sk2, So1, Sto, Te4, Ter, Tho, Tt2, Ver, Vi4, WE10, We19, Wet, WI23.**

Orach - (Green Orach, Giant Lambs Quarters, Mountain Spinach, Belledame, Butter Leaves, Arroche, Wild Spinach) - 37-50 days, *Atriplex hortensis*, hardy branching annual that is used like spinach, large tender broad triangular soft leaves, one of the oldest cultivated plants, was grown by the Romans, widely grown in France, 4-6 ft. tall green plants. *Source History: 7 in 1981; 5 in 1984; 10 in 1987; 10 in 1991; 13 in 1994; 12 in 1998; 13 in 2004.* **Sources: Bou, Com, Ga1, Ga21, Gr28, Hud, Ni1, Pe6, Pla, Pr3, Ri2, Sa12, Th3.**

Orach, Aurora Mix - 38 days, Mix of gold, red, green, pink, carmine and purple leaves, good taste, culture similar to spinach, bred by Frank Morton. *Source History: 3 in 2004.* **Sources: Ag7, Fe5, Sh12.**

Orach, Fire Red - Greatly imp. red orach, brilliant crimson-purple even at sprout and micro greens stages, retains color as plant matures, eating quality compares to spinach, micro greens harvest in 14 days, babyleaf at 22 days. *Source History: 2 in 2004.* **Sources: CO23, ME9.**

Orach, Golden - 75-80 days, *Atriplex hortensis*, grows to 5 ft., used in salads, good bee plant. *Source History: 1 in 2004.* **Sources: Ter.**

Orach, Green Velvet - Very tall, large leaves, green with pinkish stems and seed heads, vigorous. *Source History: 1 in 2004.* **Sources: Sh12.**

Orach, Magenta Magic - Darkest red of all the orach varieties, upright plant habit, magenta-red color, slightly spicy flavor holds well through summer's heat. *Source History: 1 in 2004.* **Sources: So1.**

Orach, Maroon - *Atriplex hortensis*, plants grow 3-6 ft., dark brown-purple leaves. *Source History: 2 in 1994; 3 in 1998; 1 in 2004.* **Sources: Orn.**

Orach, Oracle - 40 days, *Atriplex hortensis*, deep purple leaves are 4-5 shades darker than Red Orach, harvest leaves when plant is 1 ft. tall. *Source History: 2 in 2004.* **Sources: Jo1, ME9.**

Orach, Purple - 30 days, *Chenopodium hortensis*, purple leaves are almost black under some growing conditions, used in salad mixes and as a colorful cooked green, PSR sel. *Source History: 1 in 1998; 9 in 2004.* **Sources: Ami, Ba8, Pe6, Ri2, SE14, So9, TE7, Ter, WE10.**

Orach, Purple Savoyed - *Atriplex hortensis*, landscape and salad plant with especially thick, rumpled leaves. *Source History: 1 in 1994; 2 in 1998; 2 in 2004.* **Sources: Hud, Sh12.**

Orach, Red - (Purple Orach, Ruby Orach Mountain, Crimson Plume, Purple Passion) - 37-60 days, *Atriplex hortensis*, beautiful red variant, plants grow 4-6 ft. tall, can be used when young (6-8 in. tall) or leaves can be harvested from the branches of large plants, hot weather spinach substitute. *Source History: 4 in 1981; 3 in 1984; 7 in 1987; 13 in 1991; 16 in 1994; 21 in 1998; 29 in 2004.* **Sources: Au2, Ba8, Bo17, Bou, CO23, Coo, Dam, Fe5, Ga1, Goo, Gr28, Ho13, Hud, Lej, ME9, Ni1, Orn, OSB, Pe2, Roh, Sa5, Sa9, Se7, Se26, So12, TE7, Th3, Tu6, WE10.**

Orach, Triple Purple - Same rumpled texture as Purple Savoy with darker purple color, occasional green and light green plants appear. *Source History: 1 in 2004.* **Sources: Sh12.**

Orach, Yellow - (Golden) - 37 days, *Atriplex hortensis*, plant grows 5-6 ft., leaves not as large as Green Orach, used as cooked greens in France, immature seedheads make a nice winter decoration. *Source History: 1 in 1987; 3 in 1991; 3 in 1994; 2 in 1998; 2 in 2004.* **Sources: Ni1, Sh12.**

Perilla, Green - (Ao Shiso, Beefsteak Plant) - 55 days, *P. frutescens*, ruffled green leaves, subdued color, strong, exotic flavor, use sparingly. *Source History: 3 in 1981; 3 in 1984; 15 in 1987; 16 in 1991; 18 in 1994; 19 in 1998; 26 in 2004.* **Sources: Bo17, CO30, Dam, Ev2, Fo7, Ga21, GRI, Ho8, Ho13, Jo1, Kit, Lej, ME9, Na6, Ni1, Orn, Re8, Ren, Ri2, Sa6, Sa9, Se7, Se26, Thy, Vi5, We19.**

Perilla, Hojiso - 60 days, Japanese specialty, grown for leaves and flower buds, green leaf tops, red underneath, sow late spring to early summer. *Source History: 1 in 2004.* **Sources: Kit.**

Perilla, Purple - (Akashisho, Purple Shiso, Beefsteak Plant) - *Perilla frutescens*, 3 ft. tall purple plant, dark-purple crinkled aromatic leaves, distinct strong peppery flavor, chosen var. for pickling and seafood, heat tol. but needs constant moisture. *Source History: 5 in 1991; 8 in 1994; 11 in 1998; 11 in 2004.* **Sources: Ho13, La8, Ni1, Ren, Ri12, Sa6, Se7, Se17, So12, So25, Thy.**

Perilla, Red - (Red Shiso) - 55-60 days, *Perilla frutescens*, 3 ft. plant with deeply serrated leaves, dark purple shading to rich purple-bronze, sharp, pungent flavor. *Source History: 3 in 1981; 3 in 1984; 16 in 1987; 20 in 1991; 17 in 1994; 25 in 1998; 27 in 2004.* **Sources: Bo17, CO30, Coo, Dam, Ev2, Fe5, Fo7, Ga21, Ge2, GRI, Ho8, Jo1, Kit, Lej, ME9, Orn, Pin, Re8, Ri2, Roh, Sa5, Sa9, Se26, So1, Sto, Vi5, We20.**

Pigeon Pea - Perennial, *Cajanus cajon*, widely grown in the tropics, perennial legume, one plant will supply a family, likes bright warm well-drained soil, keep moist until 12 in. tall. *Source History: 1 in 1981; 1 in 1984; 2 in 1987; 2 in 1991; 4 in 1994; 3 in 1998; 2 in 2004.* **Sources: Ban, Ech.**

Pleurisy Root - (Butterfly Weed, Indian Paintbrush) - Perennial, *Asclepias tuberosa*, beautiful native American plant, 24 in. tall, yellow and orange flowers, tasty young pods are boiled, young shoots in salads or like asparagus. *Source History: 2 in 1981; 2 in 1984; 3 in 1987; 3 in 1991; 1 in 1994; 1 in 1998; 1 in 2004.* **Sources: Ri2.**

Poppy, Breadseed - (Hungarian Blue) - 85 days, *Papaver*, plants grow tall, beautiful lavender flowers form seed pods, seeds used in breads and rolls, heirloom. *Source History: 5 in 1991; 9 in 1994; 12 in 1998; 10 in 2004.* **Sources: Au2, Ga21, Ho13, In8, Lej, Pe2, Re6, Ri2, So12, Tt2.**

Poppy, Czechoslovakian White - Plants to 4 ft., pale-mauve flowers, large seed pods, white seeds used as a walnut substitute for baking in Eastern Europe. *Source History: 1 in 1994; 4 in 1998; 1 in 2004.* **Sources: Ri2.**

Poppy, Hutterite Breadseed - Plants grow to height of 4 ft. under ideal conditions, pale-mauve flowers with deep purple throats. *Source History: 1 in 1991; 2 in 1994; 3 in 1998; 3 in 2004.* **Sources: Ho13, Se8, Te4.**

Poppy, Przemko Breadseed - *Papaver*, large plants produce 4 in. blooms in shades of deep red to purple, seed pods produce large amount of poppy seed for use in baking, prevent seed loss by staking seed stalks, does best in cool areas. *Source History: 1 in 1998; 2 in 2004.* **Sources: Ga21, So1.**

Poppy, White Breadseed - 85-90 days, Blooms in shades of pink, white seeds. *Source History: 1 in 2004.* **Sources: So12.**

Prairie Turnip - *Psoralea esculenta*, tubers were an important staple for Native Americans of the prairie region, high in protein and starch, bland taste, prefers full sun, nitrogen fixer. *Source History: 1 in 1998; 1 in 2004.* **Sources: Pe8.**

Prickly Pear - *Opuntia compressa*, 8-10 in. tall x 4-7 in. long pads, yellow 2-3 in. flowers, purplish edible 2 in. fruits, hardy in Zones 4 and 5. *Source History: 1 in 1991; 1 in 1994; 1 in 1998; 1 in 2004.* **Sources: So12.**

Purslane - 50 days, *Portulaca oleracea*, considered a weed in the U.S., a favorite in China, the Mideast, Europe and Mexico, delicious sharp flavor due to high vitamin C, eaten raw or as cooked greens. *Source History: 2 in 1981; 1 in 1984; 9 in 1987; 4 in 1991; 6 in 1994; 13 in 1998; 15 in 2004.* **Sources: Bou, CO30, Com, Ge2, Ho8, Hud, Lej, Lin, Roh, So12, Tho, Thy, WE10, We19, We20.**

Purslane, Erect Largeleaf - 60-70 days, *Portulaca oleracea*, French variety with upright stems and leaves with tart, lemony flavor, more refined and delicate flavor than common variety. *Source History: 1 in 1991; 2 in 1994; 2 in 1998; 1 in 2004.* **Sources: Se7.**

Purslane, Garden - (Green Purslane) - 60 days, Popular Dutch var., upright growth habit keeps leaves clean, cut when 4-6 in. tall, for salads or like spinach, 4 times size of wild purslane. *Source History: 3 in 1981; 2 in 1984; 3 in 1987; 3 in 1991; 6 in 1994; 6 in 1998; 4 in 2004.* **Sources: Dam, Ri2, Sh12, Ter.**

Purslane, Golden - 50-70 days, *Portulaca oleracea*, 12-16 in. tall, semi-upright spreading growth, golden color, flavor milder than Jade Green, excellent for salads, high in vitamin C and omega-3 fatty acid. *Source History: 7 in 1998; 13 in 2004.* **Sources: Dam, Fe5, Gou, Hud, Orn, Ri12, Se7, Sh12, So9, So12, Syn, Ter, WE10.**

Purslane, Goldgelber - 50 days, *Portulaca oleracea sativa*, similar to regular cultivated green purslane but more upright and larger leaves which have a golden cast, succulent texture, mildly ascerbic flavor. *Source History: 1 in 1991; 3 in 1994; 4 in 1998; 6 in 2004.* **Sources: Coo, Fe5, Ga21, Na6, Pin, So1.**

Purslane, Large Leaf Golden - (Pourpier Dore) - *Portulaca oleracea* var. *sativa*, large broad juicy leaves, popular in France for salads or cooked greens or in soups, grows well in hot weather. *Source History: 2 in 1981; 1 in 1984; 1 in 1987; 2 in 1991; 4 in 1994; 4 in 1998; 2 in 2004.* **Sources: Ni1, Ri2.**

Quail Grass - *Celosia argentia*, grows to 5-6 ft., not a grass but an African vegetable used as a spinach substitute during hot weather when spinach will not grow, excel. taste and color, black leaves change to green when immersed in the cooking water (will not change when steamed), African relative of *celosia* or cockscomb. *Source History: 1 in 1994; 1 in 1998; 1 in 2004.* **Sources: Ech.**

Quelite - *Chenopodium berlandieri*, flavorful stems and leaves are prepared like spinach, for spring and summer planting, cultivated race originally from Atlixco, Mexico. *Source History: 1 in 1994; 2 in 1998; 1 in 2004.* **Sources: Na2.**

Rape - (Rape Salad, Canola) - 21 days, *Brassica napa*, very easy to grow, just broadcast seed once a month for fall and winter and spring greens, hardy, fine flavored and tender, used as an oilseed for Canola oil and sprouting seed. *Source History: 3 in 1981; 3 in 1984; 5 in 1987; 5 in 1991; 3 in 1994; 3 in 1998; 5 in 2004.* **Sources: Ga1, La1, Pe2, Ron, Sh9.**

Rape, Dwarf Essex - 40-60 days, *Brassica napus*, leaves, flower buds and seeds edible, 20 days/greens, 40-60 days/seeds, vigorous, cold tolerant, grows in all soil types, leaves used in salads or like mustard greens, unopened flower buds eaten like broccoli, seeds for oil, excellent cover crop. *Source History: 7 in 1981; 5 in 1984; 15 in 1987; 17 in 1991; 18 in 1994; 11 in 1998; 12 in 2004.* **Sources: ABB, Ada, Bu8, Co31, Com, Ho13, Kes, ME9, Mey, Ra5, Sau, Wi1.**

Rape, Yu Choy - 40-60 days, a.k.a. Yu Tsai Sum, tender young leaves and flower stalks are used extensively in stir fry cooking, tolerates heat but not cold, plant in early summer, up to 6 weeks before frost, early spring plantings may bolt. *Source History: 1 in 1998; 3 in 2004.* **Sources: Ho13, Kit, We19.**

Rape, Yu Choy (50 days) - Fast growing, quickly goes to flower, harvest the plant for vegetable use when bolting. *Source History: 1 in 1991; 1 in 1994; 1 in 1998; 1 in*

2004. **Sources: Ev2.**

Rape, Yu Choy (60 days) - Slightly slower in bolting than Yu Choy (80 days), equally good in quality and taste. *Source History: 1 in 1991; 1 in 1994; 1 in 1998; 1 in 2004.* **Sources: Ev2.**

Rape, Yu Choy (80 days) - Slow bolting var. grown mainly for tender delicious leaves and stems. *Source History: 1 in 1991; 1 in 1994; 1 in 1998; 1 in 2004.* **Sources: Ev2.**

Rocket - (Rocquette, Roquette, Roqueta, Rogula, Rauke, Rucola, Arugula, Italian Cress) - 30-50 days, *Eruca sativa*, ancient potherb and spicy salad plant, strong peppery flavor, deeply cut leaves and stems and flowers are eaten, use leaves when 2-3 in. long, quick to bolt, erect hardy annual, likes cool weather, edible oil from seeds, grows wild in southern France and Italy. *Source History: 16 in 1981; 15 in 1984; 25 in 1987; 50 in 1991; 65 in 1994; 89 in 1998; 113 in 2004.* **Sources: Ag7, All, Ami, Au2, Ba8, Ber, Bo17, Bo19, Bou, But, CA25, Cl3, CO23, CO30, Coo, Cr1, Dam, DOR, Ear, Eo2, Ers, Fe5, Fo7, Ga1, Ga7, Ge2, GLO, Goo, Gr28, GRI, Gur, HA3, Ha5, Hal, He8, He17, Hi6, Hi13, HO2, Ho8, Hud, Hum, In8, Jo1, Jor, Jun, Ki4, Kil, La1, La8, Lin, Loc, MAY, Mey, Mo13, Mo20, MOU, Na6, Ni1, Ont, Or10, Orn, OSB, PAG, Pe1, Pe2, Pep, Pin, Pla, Pr3, Re8, Red, Ren, Ri2, Ri12, RIS, Roh, RUP, Sa5, Sa6, Scf, SE4, Se7, Se8, SE14, Se16, Se17, Se24, Se26, Sh9, Sh14, So1, So9, So12, St18, Sto, TE7, Ter, Th3, Thy, Tt2, Tu6, Twi, Ver, Ves, Vi4, Vi5, WE10, We19, We20, Wet, Wi2, WI23.**

Rocket, Slow Bolt - Higher yields, holds in the field without bolting an extra ten days. *Source History: 1 in 2004.* **Sources: Twi.**

Rocket, Turkish - (Hill Mustard) - *Bunias orientalis*, perennial salad plant from Poland and some parts of Russia, young leaves and spreens are a steamed vegetable, rare. *Source History: 1 in 1994; 2 in 1998; 3 in 2004.* **Sources: Ga22, Ho13, Sa5.**

Rocket, Wild - *Diplotaxis arucoides*, deep green, finely cut leaves, edible yellow flowers, harvest leaves when plant is no more than 8-10 in. tall and leaves are 4-5 in. long, more pungent taste than regular arugula, spring/fall sowing, Italian favorite. *Source History: 4 in 2004.* **Sources: Ba8, DOR, So12, Und.**

Roquette: Arugula - 45-55 days, Unique pungent flavor similar to watercress, gather arrow-shaped leaves before flowering, sow outdoors in late spring, ready in 6-7 weeks, easy to grow. *Source History: 1 in 1981; 1 in 1984; 16 in 1987; 11 in 1991; 12 in 1994; 10 in 1998; 7 in 2004.* **Sources: ABB, Alb, Bu2, Com, Hig, Lej, Tho.**

Roselle, Thai Red - *Hibiscus sabdariffa*, entire plant is red, citrus-flavored flowers, leaves and calyces are used to make cranberry flavored red beverage, sauce, jelly, pie and tea, from Thailand. *Source History: 1 in 2004.* **Sources: Ba8.**

Sea Kale - Perennial, *Crambe maritima*, stout branching 20-24 in. stem, broad thick fringed leaves, peculiar green color, leafstalks blanched in darkness, tender slightly bitter shoots, native to Europe. *Source History: 2 in 1981; 2 in 1984; 1 in 1987; 2 in 1991; 3 in 1994; 3 in 1998; 1 in 2004.* **Sources: Pe8.**

Sea Kale: Couve Tronchuda - *Brassica oleracea* var. *tronchuda*, Portuguese or Sea Kale Cabbage, six varieties: Manteiga, Tronchuda Gloria de Portugal, Tronchuda Portuguesa, Tronchuda de Valhascos, Pencuda da Povoa, Pencuda Espanhola; see catalog for descriptions. *Source History: 1 in 1981; 1 in 1984; 1 in 1987; 1 in 1991; 1 in 1994; 1 in 1998; 1 in 2004.* **Sources: Red.**

Sea Kale: Lily White - Perennial, *Crambe maritima*, small crisp white heads that snap off lightly when bent, spring sowing provides harvest the following late winter and spring and for many years after, very rare and hard to find. *Source History: 2 in 1981; 2 in 1984; 2 in 1987; 2 in 1991; 2 in 1994; 2 in 1998; 2 in 2004.* **Sources: Bou, Tho.**

Serifon - (Hsueh-Li-Hung) - 37 days, *Cernua* Group, deep-green leaves with slight pungency, easy grower with cold weather hardiness, used for pickling or boiling or stir-fry. *Source History: 1 in 1991; 2 in 1994; 1 in 2004.* **Sources: ME9.**

Sesame - (White Sesame, Benne) - *Sesamum indicum*, 18-48 in. tender annual from India and the Orient, many pods, seeds used for oil or sprouts, toasted seeds decorate breads and cheeses. *Source History: 12 in 1981; 10 in 1984; 15 in 1987; 18 in 1991; 19 in 1994; 19 in 1998; 16 in 2004.* **Sources: Ada, Ev2, Fo7, Ge2, Kit, La1, Lej, Ni1, Par, Re8, Ri2, Sa6, Th3, Thy, Twi, Vi5.**

Sesame, Afghani - 110-120 days, *Sesamum indicum*, dependable early maturing variety with tan seed, 3 ft. tall, likes hot weather. *Source History: 1 in 1994; 3 in 1998; 3 in 2004.* **Sources: Ho13, Roh, Se7.**

Sesame, Black - Taller and more irregular than Line 31 with seeds in shades of dark brown to black. *Source History: 2 in 2004.* **Sources: Ev2, Sa9.**

Sesame, Black Thai - 120 days, Plants grow to 5 ft., white/pink flowers, pods contain black seeds which are more flavorful than the tan, matures late, used in South Asian cookery. *Source History: 1 in 1994; 3 in 1998; 2 in 2004.* **Sources: Ho13, So12.**

Sesame, Brown Turkey - 115 days, Easy to grow, cut and dry as pods begin to shatter, 5 ft. tall, dark brown seeds, Eurasian oilseed. *Source History: 1 in 1994; 2 in 1998; 3 in 2004.* **Sources: Ho13, So12, Syn.**

Sesame, Golden Tohum - 110 days, Plants grow 3.5-4 ft. tall, produces moderate yields of golden yellow seed which makes an excel. tahini, a lightly toasted sesame butter. *Source History: 2 in 2004.* **Sources: Ho13, So12.**

Sesame, Kingoma - 85 days, Tan seeds, young leaves used in salads or cooked with soy sauce and sesame oil, rich in calcium, vitamins C, E and F and lecithin. *Source History: 1 in 2004.* **Sources: Kit.**

Sesame, Kurogoma - 85 days, Tropical plant does best in hot weather, black seeds are more pungent than white var., leaves are eaten in Korea, sow late spring to early summer. *Source History: 1 in 2004.* **Sources: Kit.**

Sesame, Landrace Mix - 95-100 days, Breeding pool of strains from Turkey, Iran, Afghanistan and Thailand, colors include dark brown, tan, black and off-white seeds. *Source History: 1 in 2004.* **Sources: So12.**

Sesame, Line 31 - Plants to 4 ft. tall, white seeds. *Source History: 3 in 2004.* **Sources: Ho13, Sa9, Syn.**

Sesame, Shirogoma - 85 days, Tropical plant likes hot summer weather, seeds are delicious roasted, sow late spring to early summer. *Source History: 1 in 2004.* **Sources: Kit.**

Sesame, Tan Anatolia - 115 days, Heirloom from Anatolia (Asia Minor, now extreme East Turkey), 3 ft. plants, light brown seeds, very tasty. *Source History: 1 in 1994; 2 in 1998; 3 in 2004.* **Sources: Ho13, So12, Syn.**

Shungiku - (Med-Leaved, Garland/Edible Chrysanthemum, Chopsuey Green, Tanghoe Garland, Tang-Hoe, Tung Hao, Chuba) - 35-80 days, *Chrysanthemum coronarium*, edible leaved chrysanthemum, grow and use like spinach, 3.5 ft. plant, pick individual leaves or harvest whole plant when only 4-5 in. tall, aromatic, neutralizes fish odors, enhances flavors in general, Mediterranean native taken to the Orient in ancient times, now coming to the U.S. from Japan. *Source History: 22 in 1981; 21 in 1984; 22 in 1987; 22 in 1991; 27 in 1994; 30 in 1998; 35 in 2004.* **Sources: Au2, Ba8, Ban, Bou, CO23, CO30, Com, Coo, Dam, Dgs, Eo2, Goo, Gr28, Ho13, Kit, La1, Lej, ME9, Ni1, Orn, Pe2, Ren, Ri2, Sa5, Sa6, Sa9, Se7, Se8, So12, Sto, Syn, Ter, Thy, WE10, We19.**

Shungiku, Large Leaf - (Large Leaved Garland, Tanghoe) - 75-80 days, Large smooth oblong broad thick leaves, slightly indented, popular Far Eastern veg., special flavor for cooking or soup greens. *Source History: 4 in 1981; 2 in 1984; 3 in 1987; 4 in 1991; 4 in 1994; 6 in 1998; 3 in 2004.* **Sources: CO23, Ev2, OSB.**

Shungiku, Maiko Garland - 40 days, Dark-green vigorous plants with aromatic leaves, prefers cool weather conditions. *Source History: 1 in 1991; 1 in 2004.* **Sources: Kit.**

Shungiku, Small Serrated - (Chopsuey Green, Serrate Leaf Shungiku, Small Leaf Shungiku, Small Leaved Garland) - 30-40 days, *Chrysanthemum coronarium*, small dark-green serrated leaves, mild but aromatic flavor, edible flowers. *Source History: 4 in 1981; 3 in 1984; 4 in 1987; 5 in 1991; 5 in 1994; 7 in 1998; 5 in 2004.* **Sources: Ev2, Fe5, Jor, Kit, RUP.**

Skirret - *Sium sisarum*, hardy east Asian perennial, carrot family, grown for clustered white slender roots, pleasant flavor when cooked, likes abundant moisture and good soil. *Source History: 1 in 1981; 1 in 1984; 1 in 1987; 1 in 1991; 3 in 1998; 2 in 2004.* **Sources: Fe5, Ho13.**

Strawberry Spinach - *Blitum foliosum*, old-fashioned vegetable from 1600 or earlier, neither spinach nor strawberry plant, small plants with edible foliage used raw in salads or as a spinach substitute, produces small sweet strawberry-like berries which are attached to the stem, found in southern Europe in a monastery garden. *Source History: 1 in 1994; 5 in 1998; 11 in 2004.* **Sources: Ami, Bou, Co32, Com, Hi13, Ni1, Pr3, Re8, Se16, So25, Und.**

Teff - *Eragrostis abyssinica*, grows to 2 ft., seeds used in Ethiopia for making bread and beer, high in calcium and iron, teff seeds were found in the Dassur Pyramid in Egypt, carbon dated back to 3359 B.C. *Source History: 1 in 1991; 2 in 1994; 3 in 1998; 2 in 2004.* **Sources: In8, Red.**

Teff, Ivory - 90-100 days, *Eragrostis Tef*, white colored seed. *Source History: 1 in 2004.* **Sources: So12.**

Teff, Reddish Brown - 90-100 days, *Eragrostis Tef*, darker grains with more earthy flavor. *Source History: 1 in 2004.* **Sources: So12.**

Ung Choi - *Ipomoea aquatica*, grown around or in rice paddies in Asia, entire plant is edible, known as "water spinach", boiled or stir-fried. *Source History: 1 in 1998; 1 in 2004.* **Sources: Ho13.**

Vetch, Winter Tare - *Vicia sativa*, short spreading plants, buff-grey angular seeds, used for flour/bread, from Nepal. *Source History: 1 in 2004.* **Sources: So12.**

Vitamin Green - 45 days, *Brassica rapa narinosa*, Japanese greens, rich in vitamins A and C, thick narrow stems are white inside and green outside, broad curled leaves, can be grown almost year round, vigorous. *Source History: 1 in 1981; 1 in 1984; 1 in 1987; 2 in 2004.* **Sources: Eo2, Jo1.**

Water Lotus - *Nelumbo nucifera*, cultivated in Asia for its edible root (lotus root), edible young leaves, and large seeds which are eaten fresh or dried, taste like chestnuts, grows in 1-4 ft. of water, unusual seed pods make a great dried flower. *Source History: 1 in 1998; 1 in 2004.* **Sources: Pe8.**

Wild Rice - (Giant Wild Rice) - Perennial, *Zizania aquatica*, Native American delicacy, the Sioux and Chippewa fought wars over it, needs slow moving water from 6-36 in. deep, soft mud bottom is best but can thrive on sand, prolonged flooding drowns it out, does poorly in landlocked ponds or waters salty to taste or on white marl bottom, best in Northern states, ducks love it. *Source History: 2 in 1981; 2 in 1984; 3 in 1987; 3 in 1991; 3 in 1994; 5 in 1998; 5 in 2004.* **Sources: Ada, Kes, Ri2, So12, Wi1.**

Varieties dropped since 1981 (and the year dropped):

Achocha (1987), *Blonde Belle-Dame Orach* (1987), *Buckwheat, Japanese* (1998), *Buckwheat, Mancan* (2004), *Burdock, Edible* (2004), *Burdock, Sakigake* (2004), *Butterfly Pea* (1987), *Canna, Edible* (1998), *Celtuce (Narrow Leaved)* (1987), *Celtuce (Round Leaved)* (1987), *Chapil* (2004), *Chia, Blue Seeded* (1994), *Chia, Desert* (2004), *Chia, Golden* (1998), *Chilacayote White Seeded* (1983), *Chual* (2004), *Claytonia, Emerald Green* (1998), *Coriander, Ho San* (1994), *Coriander, Moroccan* (2004), *Coriander, Tian Chin* (2004), *Corn Rocket* (2004), *Cuachipil* (1994), *Devil's Claw, Chihuahua Wild* (2004), *Devil's Claw, Pima Bajo* (2004), *Devil's Claw, Pima Black Seed* (1994), *English Giant Rape* (1991), *French Brown* (1998), *German Greens* (1991), *German Spinach* (2004), *Ginger, Chinese* (1994), *Gobo, Wild* (1998), *Gobo: Nakanomiya Long* (1987), *Gobo: Sakigake* (1991), *Gobo: Tokiwa* (1987), *Gobo: Tokiwa Long* (1987), *Granada* (1991), *Green Boy Mustard* (1998), *Green Engtsai (or Kancon)* (1991), *Green Magic Mexican Tea* (1994), *Grun Rheinische Kopf Orach* (1987), *Guaco* (2004), *Gunga Pidgeon Pea* (1987), *Hawaiian Pidgeon Pea* (1987), *Hoja del Pescado* (1994), *Horseradish Tree* (1998), *Huauzontli* (1991), *Hulless Buckwheat* (1987), *Iceplant* (1987), *Indian Greens* (1998), *Japanese Parsley* (2004),

Jicama Montes (2004), *Kankon* (1998), *Kudzu Vine* (1991), *Lambs Quarters, California Yokuts* (2004), *Lambs Quarters, Gabrielino So. California* (2004), *Lambs Quarters, Morton's Select* (2004), *Licorice* (1987), *Madeira Vine* (1987), *Mao Kua* (1982), *Miner's Lettuce, Emerald Green* (2004), *Miner's Lettuce, Miner's Pick* (2004), *Moringa Tree* (2004), *Mrs. Hawkin's Spinach* (2004), *Mustard, English White* (1994), *Naranjilla* (1998), *Narovit* (1994), *Ong Choi* (1994), *Ong Choy: Green Stem China* (1991), *Ong Choy: Green Stem Taiwan* (1991), *Ong Choy: White Stem China* (1991), *Ong Choy: White Stem Taiwan* (1991), *O'odham Onk I:waki* (2004), *Orach, Giant Calite* (1998), *Papago Sugar Cane* (1987), *Para Cress* (1991), *Perilla, Cumin Scented Green* (2004), *Perilla, Flatleaf Pickling* (1998), *Perilla, Purple Curly* (1994), *Perilla, Purple Extra Curly* (1994), *Pitaya Cactus* (1987), *Poppy, Carnation* (2004), *Poppy, Mauve Flowered* (2004), *Poppy, White Persian Breadseed* (2004), *Prickly Pear, Major* (1994), *Purslane, Golden Tall* (2004), *Purslane, Jade Green* (2004), *Purslane, Matweed* (2004), *Rampion* (1994), *Rape, Westar* (1994), *Rocket, Enrico Rao* (1998), *Roselle* (2004), *Schwarzwurzein* (2004), *Sesame, Anatolian Tan* (1994), *Sesame, Thai Black* (2004), *Sesame, Turkish Brown* (2004), *Shungiku, Fine Leaf* (2004), *Silky Squash* (1982), *Sinaloan Wild Devil's Claw* (1991), *Six Foot Luffa (Cee Gwa)* (1991), *Taro* (1987), *Teff, A.L. White* (2004), *Tel-Tex Greens* (1991), *Tetragone Cornue* (1991), *Tilney* (1994), *Tree Tomato: Golden Peach* (1982), *Tree Tomato: Orange Ecuador* (1987), *Tree Tomato: Red Tamarillo* (1998), *Tree Tomato: Ruby Red* (1987), *Tree Tomato: Standard Yellow* (1998), *Turkey Red Wheat* (1987), *Unicorn Plant* (1994), *Vine Spinach* (1991), *Vista Spineless Cactus* (1987), *Warihio Panic Grass* (1991), *Water Spinach* (2004), *Water Spinach (Gr. Leaf and Cane* (1991), *Water Spinach (No. 2)* (1987), *Water Spinach, Large Leaf* (1994), *Zucchini-Melon* (1987).

Miscellaneous Cucurbitaceae

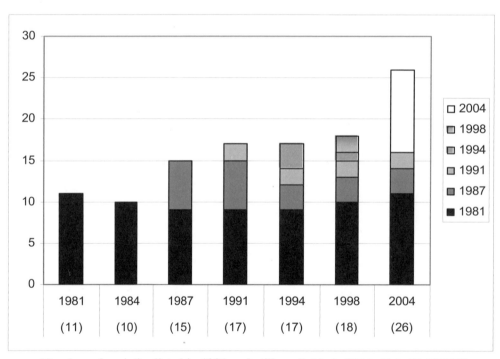

Number of varieties listed in 1981 and still available in 2004: 11 to 11 (100%).
Overall change in number of varieties from 1981 to 2004: 11 to 26 (236%).
Number of 2004 varieties available from 1 or 2 sources: 18 out of 26 (69%).

<u>African Horned Cucumber</u> - (Jelly Melon, Kiwano) - *Cucumis metuliferus*, climbing vine, thorny oval fruit filled with greenish-gold gel and lots of seeds, flavor reminiscent of pomegranate and citrus, good on ice cream, requires warm growing conditions. *Source History: 2 in 1987; 2 in 1991; 1 in 1994; 3 in 1998; 2 in 2004.* **Sources: So25, WE10.**

<u>Balsam Apple</u> - *Momordica balsamina*, quick growing annual vine requires staking, showy 3 in. long orange fruits, Bitter Melon prized in the Orient, grown for ornament and screens. *Source History: 1 in 1981; 1 in 1984; 1 in 1987; 1 in 1991; 2 in 1994; 3 in 1998; 1 in 2004.* **Sources: Th3.**

<u>Balsam Pear</u> - (Bitter Melon, Bitter Gourd, Foo Gwa) - 80 days, *Momordica charantia*, cucumber-like vine, heavily warted deep-green fruits, 8-10 x 3-4 in., used like cucumbers when young, common in tropics, bursts open to show seeds and red pulp, immature pulp cooked to make a bitter garnish for meats, many medicinal uses. *Source History: 12 in 1981; 11 in 1984; 12 in 1987; 10 in 1991; 13 in 1994; 13 in 1998;*

16 in 2004. **Sources: Ba8, Bo17, CO30, Ge2, Hud, Jor, Kit, Loc, Mi12, Ri2, Ri12, Roh, Se26, Sto, Th3, Wi2.**

Balsam Pear, Foo Gwa - (Fu Kua, Ku Kua, Carella, Reishi, Chinese Bitter Melon, Long Bitter, Med-Long Bitter, Bitter Cucumber) - 80 days, *Momordica charantia*, strains from 5-12 in. long and 2.5-4 in. dia., light-green skin turns silvery-white, must stake vines, slightly bitter. *Source History: 7 in 1981; 6 in 1984; 5 in 1987; 5 in 1991; 2 in 1994; 1 in 2004.* **Sources: Ban.**

Bitter Gourd, India Long Green - Long fruit to 12 in., rough toothed green skin, does well in warm climates, produces heavily over a long period. *Source History: 1 in 2004.* **Sources: Ev2.**

Bitter Gourd, India Long White - Knobby white skin resembles India Long Green, 8-12 in long, benefits from support, does well in warm climates. *Source History: 1 in 2004.* **Sources: Ev2.**

Bitter Melon, Hong Kong - (Balsam Pear) - 80 days, *Momordica charantia*, short warty tapered fruit, about 5 in. long, dark-green skin, light-green flesh, requires staking. *Source History: 1 in 1981; 1 in 1984; 2 in 1987; 2 in 1991; 2 in 1994; 1 in 1998; 2 in 2004.* **Sources: Ev2, Ni1.**

Bitter Melon, Large - (Balsam Pear, Bitter Gourd Taiwan Large) - 80 days, *Momordica charantia*, warty fruit tapers toward blossom end, 12 in., long x 4 in. dia., light-green skin matures to silver-white, requires staking. *Source History: 1 in 1981; 1 in 1984; 2 in 1987; 4 in 1991; 4 in 1994; 6 in 1998; 6 in 2004.* **Sources: CO23, Ev2, Ga21, Ho8, Jor, RUP.**

Bitter Melon, Taiwan Large - Large fruit to 10-12 in. long, 1 lb., green skin, white flesh, popular in Taiwan and subtropical areas. *Source History: 1 in 2004.* **Sources: Ev2.**

Bitter Melon, Thailand - 85 days, *Momordica charantia*, very popular variety in Thailand and Vietnam, small fruit, 3-4 in. long, deep green skin with blistered bumps, smaller than regular bitter melon, very productive. *Source History: 1 in 1991; 1 in 1994; 1 in 1998; 3 in 2004.* **Sources: Ba8, Ev2, Sa9.**

Chayote - *Sechium edule*, pear-shaped fruit, pale-green firm flesh, only youngest fruits are eaten raw, sweet fresh flavor with a hint of zucchini squash, more mature ones can be baked or sauteed. *Source History: 2 in 1991; 2 in 1994; 3 in 1998; 1 in 2004.* **Sources: Ed3.**

Diplocyclos palmatus - Climbing vine, pale yellow flowers, 3-5-lobed 5 in. leaves and young fruits are boiled and eaten, native to tropical Africa to Asia. *Source History: 1 in 2004.* **Sources: Hud.**

Fuzzy Gourd, Japanese - *Benincasa hispida*, smaller str. grows 6-8 in. long, oblong shape, productive, good storage. *Source History: 1 in 2004.* **Sources: Ho13.**

Fuzzy Gourd, Large Long - *Benincasa hispida*, measures 8 x 36 in., 20 lbs. *Source History: 1 in 2004.* **Sources: Ho13.**

Gherkin - (Small Gherkin) - 60 days, Introduced in 1793 from Africa, pale-green 1.5-2.5 in. oval cucumbers with prickly spines, prolific vines, produces 40 or more fruits per plant. *Source History: 2 in 1981; 2 in 1984; 5 in 1987; 2 in 1991; 1 in 1994; 1 in 1998; 1 in 2004.* **Sources: Roh.**

Gherkin, Mexican Sour - 60-70 days, Abundant crops of 1-2 in. fruit that look like miniature watermelons, ripened fruit falls from the vines, sweet cucumber flavor contrasted by a surprising tartness, tasting like they were already pickled, great for trellising. *Source History: 1 in 2004.* **Sources: Se16.**

Jamaican - 65 days, Gherkin type, resembles West India Gherkin but is earlier and more productive. *Source History: 1 in 2004.* **Sources: Sa9.**

Jelly Melon - (African Horned Melon, Kiwano Horned Cucumber, Kiwano, Karoo Cucumber) - 110-140 days, *Cucumis metuliferus*, exotic horned fruit turns orange when ripe, lime-green 'jelly sacks' inside, faintly banana-like flavor, or pomegranate and citrus, grow like cucumbers, vigorous climber, does well if trellised, keeps up to 6 months. *Source History: 1 in 1987; 1 in 1991; 3 in 1994; 3 in 1998; 8 in 2004.* **Sources: Ba8, Co31, Ho13, Rev, Ri2, Sa9, Se17, Syn.**

Luffa Sponge - (Vine Okra, Chinese Okra, Dishrag Gourd, Vegetable Sponge, Oh-Hechima, Cee Gwa, Northern Sponge Gourd) - 100-150 days, *Luffa aegyptiaca*, climbing vine, many 12-15 in. pods. some to 30 in., when under 6 in. is used like zucchini, when large and dry is crushed and peeled for fibrous sponge, yellow flowers, needs hot long season and well manured sunny soil. *Source History: 27 in 1981; 24 in 1984; 52 in 1987; 52 in 1991; 56 in 1994; 51 in 1998; 74 in 2004.* **Sources: Ba8, Ban, Bo17, Bo19, Bu2, Bu3, Bu8, CA25, CO23, CO30, Co31, Com, Dam, Dgs, DOR, Ear, Ech, Eo2, Ers, Ev2, Fe5, Go9, Gr29, HA3, He8, He17, Hi13, Ho8, Hud, Jor, Kil, La1, LO8, Loc, MAY, ME9, Mel, Mo13, MOU, Ni1, Par, Pep, Pin, Ra5, Ra6, Re8, Red, Ri2, Ri12, Roh, Ros, RUP, Sa9, Sa12, Scf, SE14, Se25, Se26, Se28, Sh9, Sh14, Shu, So1, Te4, Tho, Thy, Twi, Ver, Vi4, Vi5, WE10, Wet, WI23, Wo8.**

Luffa, Angled - 85 days, Best when grown on a trellis, used as summer squash substitute when young. *Source History: 1 in 2004.* **Sources: Sa9.**

Luffa, Cee Gwa - (Chinese Okra, Luffa Gourd, Gee Gwa) - 90 days, *Luffa aegyptiaca*, nature's sponge, it looks like okra, young fruits eaten raw or cooked, ridges removed, mature fruits used for scrubbers. *Source History: 9 in 1981; 8 in 1984; 9 in 1987; 6 in 1991; 4 in 1994; 5 in 1998; 5 in 2004.* **Sources: Ban, CO30, Hud, Loc, Sto.**

Luffa, Ridged - 90 days, *Luffa acutangula*, long thin fruits with 8-10 ridges running from end to end, tender very crisp flesh when young, up to 24 in. long, requires staking, scrubbers. *Source History: 1 in 1981; 1 in 1984; 2 in 1987; 3 in 1991; 3 in 1994; 2 in 1998; 2 in 2004.* **Sources: Ev2, Kit.**

Malabar Gourd - *Cucurbita ficifolia*, hard-shelled perennial short day gourd, 8 lbs., white flesh resembles rice noodles, used in fruit salads. *Source History: 1 in 1987; 2 in 1991; 1 in 2004.* **Sources: Pe1.**

Sikkim - *C. sativus* var. *sikkimensis*, short fat fruit with maroon netting on skin. *Source History: 1 in 2004.* **Sources: Hud.**

Snake Gourd - (Serpent Gourd) - *Trichosanthes anguina*, ornamental vine with 4-5 ft. long fruits, often grown on a fence or trellis, fruits curl into curious shapes, used extensively in India and southeast Asia, rare. *Source History: 2 in 1981; 1 in 1984; 3 in 1994; 4 in 1998; 16 in 2004.* **Sources: Ba8, Ban, Ech, Go9, HO2, Ho8, La1, Pin, Ra5, Ra6, Ri2, Ri12, RUP, SE14, Shu, Vi5.**

West India Gherkin - (West Indian Gherkin, Burr, Maxixe**)** - 60-65 days, *C. anguria*, large vines with leaves that look like watermelon leaves, large crops of oval burr-like fruits, 2-3 x 1-1.5 in. dia., fleshy prickles, black spines, distinct flavor, rather elastic, small pickles or relish, dates back to 1793. *Source History: 26 in 1981; 26 in 1984; 19 in 1987; 16 in 1991; 14 in 1994; 9 in 1998; 27 in 2004.*
Sources: Ami, Ba8, Bo18, Bo19, Com, Fo7, Gr29, He8, He17, La1, Lan, Ol2, Or10, Pin, Pr3, Ra5, Sa9, Se8, Se28, Sh9, Shu, So25, Und, Vi4, Vi5, WE10, Wo8.

Varieties dropped since 1981 (and the year dropped):

Bitter Gourd, Batu Improved (2004), *Bitter Gourd, Fontana* (2004), *Calabazilla* (2004), *Fig Leaf Gourd* (1994), *Luffa, Smooth* (2004), *Luffa, Wild* (2004), *Vegetable Pear* (2004).

Miscellaneous Leguminosae

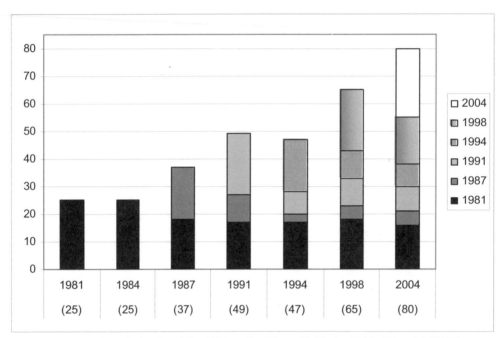

Number of varieties listed in 1981 and still available in 2004: 25 to 16 (64%).
Overall change in number of varieties from 1981 to 2004: 25 to 80 (320%).
Number of 2004 varieties available from 1 or 2 sources: 63 out of 80 (79%).

Adzuki - (Japanese Red, Chinese Red**)** - 90-118 days, *Vigna angularis*, colorful red Japanese bean, long narrow pods contain 7-10 small dark-red seeds, tender and delicious, high in protein, small bushes, young pods are eaten like snow peas, has been sprouted for centuries in the Orient. *Source History: 5 in 1981; 5 in 1984; 8 in 1987; 9 in 1991; 11 in 1994; 9 in 1998; 12 in 2004.* **Sources: Ev2, Hud, MOU, Sa9, Se17, Se25, So12, Syn, Ter, Tho, Ver, Yu2.**

Adzuki Mame, Late Tamba - 90 days, Dwarf plant, rose colored flowers, dark red bean used for culinary, pastries and confections, sprouts used in salads. *Source History: 1 in 1998; 1 in 2004.* **Sources: Kit.**

Adzuki, Buff - 100 days, *Vigna angularis*, rare str. with buff colored seeds, slightly sweet, pods 4-5 in., shells easily, needs a warm growing season. *Source History: 1 in 1998; 2 in 2004.* **Sources: Ho13, Yu2.**

Adzuki, Grey Mottled - *Vigna angularis*, grey-speckled seeds on short plants, productive. *Source History: 1 in 2004.* **Sources: Ho13.**

Adzuki, Red/White - *Vigna angularis*, red and white seed, productive. *Source History: 1 in 2004.* **Sources: Ho13.**

Asparagus Bean - (Yard Long Bean, Dow Gauk, Kaoh Siung**)** - 70-90 days, *Vigna* spp., thin round 20-24 in. pods in clusters, best for table use when less than 15 in., light string, climber, black or brown or red or red/tan mottled seed, use like snap beans, asparagus-like flavor, vigorous vines. *Source History: 16 in 1981; 12 in 1984; 16 in 1987; 14 in 1991; 14 in 1994; 11 in 1998; 12 in 2004.* **Sources: Ba8, Bou, Com, Coo, DOR, Hi13, Kit, Ol2, Pp2, Roh, Shu, Th3.**

Asparagus Bean, Black Seeded - 75 days, Vigorous climbing vines, 15 x .5 in. green pods, slightly bent at tip, thick crisp skin, black seeds when mature, harvest pods when young. *Source History: 3 in 1981; 3 in 1984; 2 in 1987; 2 in 1991; 3 in 1994; 4 in 1998; 2 in 2004.* **Sources: Ev2, Ho13.**

Asparagus Bean, Black Stripe Seed - New variety from Taiwan, crisp delicious pods, heavy yields. *Source*

History: 1 in 1994; 2 in 1998; 2 in 2004. **Sources: Ev2, Ho13.**

Asparagus Bean, Dow Gauk - (Asparagus Bean, Yard Long Bean, Indian Jori) - 65-75 days, *Vigna* spp., needs support and rich warm soil, 24 in. pods grow in pairs, has a light taste that is different from other green beans. *Source History: 8 in 1981; 7 in 1984; 7 in 1987; 4 in 1991; 2 in 1994; 2 in 1998; 1 in 2004.* **Sources: Ri2.**

Asparagus Bean, Kaohsiung - Dark green, thick pods, black seed, named after place of origin, Kaohsiung, Taiwan. *Source History: 2 in 1998; 2 in 2004.* **Sources: Ev2, Ho13.**

Asparagus Bean, Purple Pod - 92 days, Vigorous vines, purple pods usually borne in pairs, can grow to over 2 ft. long, used like snaps when young and tender, few seeds per pod, very rare in the U.S. *Source History: 1 in 1991; 1 in 1998; 1 in 2004.* **Sources: Ho13.**

Asparagus Bean, Red Seeded - 75 days, Light-green pods, smaller dia. than black-seeded type, red seeds when mature, popular in Southern Asia and Japan. *Source History: 3 in 1981; 3 in 1984; 2 in 1987; 2 in 1991; 3 in 1994; 5 in 1998; 5 in 2004.* **Sources: Ev2, Ho13, Red, So1, Wi2.**

Asparagus Bean, Two-Tone - 88 days, Thick green pods, bicolored white and red-brown seeds, more productive than other vars. but require longer growing season. *Source History: 1 in 1991; 1 in 1998; 1 in 2004.* **Sources: Ev2.**

Asparagus Bean, White Seed - Long pods with light green skin, crisp and sweet, very popular in Taiwan. *Source History: 2 in 1998; 2 in 2004.* **Sources: Ev2, Ho13.**

Asparagus Pea - (Winged Pea) - 50-65 days, *Lotus tetragonolobus*, curious rectangular flanged fleshy green pods look like miniature Winged Beans, not related to either asparagus or peas, cooked whole when young, unusual delicate flavor, mentioned as early as 1734 by Philip Miller, one of the earliest garden writers, probably from northwest Africa. *Source History:2 in 1981; 2 in 1984; 3 in 1987; 1 in 1991; 1 in 1994; 1 in 1998; 4 in 2004.* **Sources: Hi13, Pin, Se16, Tho.**

Erythrina berteroana - Large trifoliate leaves on 20 ft. tall tropical tree, clusters of 5 flowers, 18 in. long, later form 11 in. twisted pods, scarlet seeds, tree bears flowers when leafless, tender leaves and flowers eaten as a vegetable. *Source History: 1 in 2004.* **Sources: Hud.**

Fagiolino Dolico Veneto - 65 days, Pods are borne upright in pairs like "antlers" on the plant, sweet nutty flavor, green-black pods, pick while thin. *Source History: 1 in 1998; 1 in 2004.* **Sources: Und.**

Garbanzo - (Chick Pea, Chestnut Bean) - 65-105 days, Small pods usually contain one round light-brown curiously bumpy bean, boiled or roasted when dry, also used as green shells, drought res., distinctive flavor, used in salads or roasted or creamed for dips. *Source History: 20 in 1981; 15 in 1984; 8 in 1987; 5 in 1991; 7 in 1994; 11 in 1998; 9 in 2004.* **Sources: Bou, Bu2, La1, Loc, Se24, Se25, Se26, Ter, Ver.**

Garbanzo, Black Seeded - Warm weather crop being developed for larger-scale production in the interior, can mature west of but Cascades but not recommended there, Ethiopian origin. *Source History:1 in 1981; 1 in 1984; 1 in 1987; 2 in 1991; 2 in 1994; 1 in 1998; 1 in 2004.* **Sources: Pr3.**

Garbanzo, Blond - Mounded plants to 1 ft. high, small balloon-like pods contain 1-2 seeds, does not like cold soil. *Source History: 1 in 1991; 1 in 2004.* **Sources: Pr3.**

Garbanzo, Brown - *Source History: 1 in 1991; 1 in 1998; 1 in 2004.* **Sources: Pr3.**

Garbanzo, Kabuli - (Kabouli Black) - 95-115 days, *Cicer arietinum*, 2 ft. plants, 2-seeded pods, solid black med-sized beans, unique, somewhat tol. of cold soils, dev. at Washington State University at Pullman, experimental var. from Kabul, Afghanistan, about 95 seeds per oz. *Source History: 2 in 1981; 2 in 1984; 3 in 1987; 5 in 1991; 4 in 1994; 8 in 1998; 11 in 2004.* **Sources: Ag7, Fe5, Ga1, Ho13, Pe2, Ri12, Roh, Sa5, Se7, So12, Syn.**

Garbanzo, Kala Channa - *Cicer arietinum*, small brown seeded chickpea from India, eaten fresh or dried, good in dry, short season areas, not for cool coastal areas. *Source History: 1 in 1998; 1 in 2004.* **Sources: Hud.**

Garbanzo, Mayo Winter - Plump beige beans, introduced staple, dry farmed in Sonora with winter rains. *Source History: 1 in 1987; 1 in 1991; 1 in 1998; 1 in 2004.* **Sources: Na2.**

Garbanzo, Myles - 80-90 days, Rare brown chickpea, for use in the interior NW and other short season dry areas, not for coastal fog belts, asochyta blight res., native to India. *Source History: 2 in 1998; 2 in 2004.* **Sources: Ho13, So9.**

Garbanzo, Red Channa - 105-110 days, Desi type, pleasant earthy flavor, commonly eaten as a dry toasted snack, also eaten like green shell peas, longer season than black, from India. *Source History: 1 in 1998; 1 in 2004.* **Sources: So12.**

Garbanzo, Tarahumara - 105-110 days, *Cicer arietinum*, kabuli type, small dark-brown seeded peas, very insect res., adaptable to growing at different elevations, sea level to 10,000 ft., very drought tol., fall crop from the bottom of Copper Canyon, Mexico. *Source History: 1 in 1987; 2 in 1991; 1 in 1994; 2 in 1998; 3 in 2004.* **Sources: Na2, Se17, So12.**

Gram, Black - 80 days, Mung bean eaten as a green vegetable when young, the green pods are cut into small chunks, 3-4 ft. tall plants, very small seeds, popular in Indonesian cuisine. *Source History: 1 in 2004.* **Sources: Sa9.**

Gram, Green - 80 days, Tan podded gram bean. *Source History: 1 in 2004.* **Sources: Sa9.**

Gram, Horse - *Macrotylomum uniflorum*, tiny brown seeds produced on tall vines, used in curries and boiled, Australia. *Source History: 1 in 2004.* **Sources: Ho13.**

Guar Bean - (Cluster Bean) - 90-100 days, *Cyamopsis tetragonalobus*, plant to 3 ft., protein rich guar gum used world-wide as a thickener in various foods, young pods eaten as a vegetable in India. *Source History: 3 in 1994; 3 in 1998; 1 in 2004.* **Sources: Ban.**

Hog Peanut - *Amphicarpa bracteata*, vining legume to 10 ft., underground sweet seeds, delicious raw, also used in soup, formerly cultivated in the southern U.S. by Eastern Indians and early settlers. *Source History: 1 in 2004.* **Sources: Ho13.**

Hyacinth Bean - Perennial, *Dolichos lablab*, tender plant grows 10-20 ft. tall, showy large purple flowers in upright 6 in. spikes followed by 2.5-5 in. pods filled with dark seeds, ripe seeds are boiled for food (water discarded repeatedly) or sprouted like mung beans, young green seeds can be eaten, immature pods are excel. table vegetable, young leaves and flowers are boiled and eaten like spinach, grown by Thomas Jefferson, pre-1802. *Source History: 4 in 1991; 12 in 1994; 16 in 1998; 26 in 2004.* **Sources: Ba8, Ban, CA25, Com, Dam, Ech, Ev2, Ge2, GRI, Ho13, Hud, Mel, Mey, Mi12, Par, Pe2, Pin, Ri2, Roh, Ros, Sa6, Sh13, So1, Sto, Th3, Tt2.**

Hyacinth Bean, Akahana Fujimame - 100 days, Japanese var. with red flowers, flat, thick curved pods, young pods are cooked or used in stir fry. *Source History: 1 in 2004.* **Sources: Kit.**

Hyacinth Bean, Murasakiirohana Fujimame - 100 days, Purple flowers, green leaves with purplish stems and veins, thick, flat, curved red pods, young beans are cooked or used in stir fry. *Source History: 1 in 2004.* **Sources: Kit.**

Hyacinth Bean, Shirohana Fujimame - 100 days, Fast growing ornamental vine with white flowers, thick, flat, curved edible pods are used in stir fry or cooked. *Source History: 1 in 2004.* **Sources: Kit.**

Hyacinth Bean, White Flowered - *Dolichos lablab*, lush, light green vines, white flowers are lightly stained with pale lilac- pink on the upper lobes, white seeds. *Source History: 2 in 1998; 3 in 2004.* **Sources: Ev2, Hud, Pin.**

Jackbean - (Chickasaw Lima Bean) - *Canavallia ensiformis*, 2-3 ft. plant, ovate leaflets are dull-green and veined, purple flowers, pendant 18 in. pods, white seeds, ripe seeds used as coffee substitute after extensive boiling and peeling to remove the toxic alkaloid, West Indies. *Source History: 1 in 1981; 1 in 1984; 1 in 1987; 1 in 1991; 1 in 1994; 2 in 1998; 2 in 2004.* **Sources: Ban, Ho13.**

Lentil - (Italian Green) - 90 days, Ancient annual semi-climbing legume, extensively grown abroad, small green disk- shaped beans, seeds store best if left in pods, excel. in soups. *Source History: 5 in 1981; 4 in 1984; 5 in 1987; 3 in 1991; 4 in 1994; 4 in 1998; 7 in 2004.* **Sources: Bou, Ha5, Kus, MOU, Pin, Pr3, Ter.**

Lentil, Beluga - (Black Beluga) - Firm flesh, small black seeds, appearance resembles caviar, thus the name, from the Middle East, PVP 2003 MI State University. *Source History: 1 in 1998; 2 in 2004.* **Sources: Ho13, So12.**

Lentil, Chilean - (Green Seeded Lentil) - Small seeded green lentil, short plants with attractive fern-like leaves, improves soil nitrogen reserves, can be grown in the winter in areas where winter vegetables are grown. *Source History: 2 in 1991; 2 in 1994; 1 in 1998; 1 in 2004.* **Sources: Kus.**

Lentil, Crimson - *Lens culinaris*, pods contain 1-2 tiny oval light brown seeds, bright orange interior, originated in Giza, Egypt. *Source History: 1 in 1994; 2 in 1998; 1 in 2004.* **Sources: Kus.**

Lentil, Du Puy - *Lens culinaris*. *Source History: 1 in 2004.* **Sources: Se17.**

Lentil, French Green - Green seeds speckled with purple. *Source History: 1 in 2004.* **Sources: Pr3.**

Lentil, Horse - Attractive plant with finely divided leaves, blue flowers, high yields of flat pods with grey-green angular seeds, used fresh and dried. *Source History: 1 in 1994; 1 in 1998; 1 in 2004.* **Sources: Ho13.**

Lentil, Masoor - (Red Lentil) - Widely used in India and the Middle East, small seeds with dark skin, salmon-orange inside, quick cooking, turn yellow when cooked, sweet, mild flavor. *Source History: 1 in 1998; 1 in 2004.* **Sources: Hud.**

Lentil, Petite Crimson - Original str. from Giza, Egypt, popular gourmet market var., small light brown seed with red-orange cotyledon. *Source History: 1 in 2004.* **Sources: Ho13.**

Lentil, Red Chief - *Lens culinaris*, heavy bearing, semi-upright plants, large beige seed with bright red interior, dev. at WA State U, 1980. *Source History: 1 in 1994; 1 in 1998; 2 in 2004.* **Sources: Ho13, Kus.**

Lentil, Red Seeded - Tiny, disk shaped legume. *Source History: 1 in 1994; 2 in 1998; 2 in 2004.* **Sources: Be4, Pr3.**

Lentil, Spanish Pardina - *Lens culinaris*, Spanish favorite, small green-brown seed with orange cotyledon, firm texture when cooked, nutty flavor. *Source History: 1 in 2004.* **Sources: Ho13.**

Lentil, Syrian Red - 90-95 days, Cold hardy plant grows 1-1.5 ft. tall. *Source History: 1 in 1998; 1 in 2004.* **Sources: So12.**

Lentil, Tarahumara Pinks - Small mottled seed, acclimatized Old World introduction, from the Sierra Madre in Mexico, for desert winter planting, Tarahumara Indian seed. *Source History: 1 in 1981; 1 in 1984; 1 in 1987; 1 in 1991; 1 in 1994; 1 in 1998; 1 in 2004.* **Sources: Na2.**

Lentil, Verte du Puy - 90-95 days, Airy looking plants, leaves are green, speckled with purple, stands heat and drought well, from France. *Source History: 2 in 1998; 2 in 2004.* **Sources: Ho13, So12.**

Lupin, Kiev Sweet Edible - 85-90 days, *Lupinus albus*, 3 ft. plants with creamy violet flowers, sets spiny clustered pods, nutritious seeds high in digestible protein (40%), does best planted in early spring. *Source History: 1 in 1991; 1 in 1994; 1 in 1998; 1 in 2004.* **Sources: Ho13.**

Lupin, Lebanese - Plants to 2 ft., .75 in., flat pale tan seeds. *Source History: 1 in 2004.* **Sources: Ho13.**

Monkeytail Bean - Twiner with fat zipper pods, unique. *Source History: 1 in 1994; 1 in 1998; 1 in 2004.* **Sources: Syn.**

Moth Bean - (Mother Bean, Mat Bean) - 95 days, *Vigna aconitifolia*, sprawling mat-forming plants do well in very hot weather (100-120 deg. F), young tender pods eaten in India as a vegetable, dry beans used like lentils in "dahl" or fried in oil and salted as a snack called "bhujia", high in digestible protein (22-24%). *Source History: 1 in 1987; 2 in 1991; 2 in 1994; 2 in 1998; 7 in 2004.* **Sources: Ami, Bou, Ho13, Hud, Sa9, Se7, So12.**

Mung Bean - (Look Dow, Mungo Bean, Oriental Mung Bean) - 75-130 days, *Vigna radiata*, small olive-green beans, Chinese delicacy, enjoys hot weather, 15-24 in. plants, 3-4 in. pods contain the small green okra-like seeds, seeds sprout in 2- 5 days, nutritious sprouts used extensively in Oriental cooking. *Source History: 33 in 1981; 28 in 1984; 26 in 1987; 20 in 1991; 21 in 1994; 28 in 1998; 20 in 2004.* **Sources: Ada, Be4, Bu2, Dam, ECK, Ers, Ev2, Ha5, Kit, La1, Loc, MOU, Ont, Pin, Se7, Se17, Se25, Syn, Ter, Ver.**

Mung, Black - Naturalized in north California. *Source History: 1 in 1994; 1 in 1998; 1 in 2004.* **Sources: Syn.**

Mung, Black/Kali Gram - Plants to 2.5 ft. tall, yellow flowers, thick upright clusters of small hairy pods bear small black seeds, esteemed in India for fermented bean dishes, beans have glutinous or sticky starch quality. *Source History: 1 in 2004.* **Sources: So12.**

Rice Bean, Red - Long vines, yellow flowers, seeds are red with white hilum, the size of a large rice grain, likes hot weather, grown in villages in India as a protein source, steamed young leaves and green pods are eaten. *Source History: 1 in 1998; 2 in 2004.* **Sources: Ho13, Hud.**

Rice Bean, White - Species uncertain, bush type similar to Mung Beans, 2 ft. tall, white seed resembles rice grains, drought tol., from Germany, 1860. *Source History: 1 in 1998; 1 in 2004.* **Sources: Ho13.**

Siberian Pea Tree - *Caragana arborescens*, small shrub-like tree, fast-growing to 12-20 ft., produces loads of long edible green pea pods, thick fern-like foliage and bright-yellow flowers spring through summer, very hardy. *Source History: 1 in 1987; 1 in 1991; 1 in 2004.* **Sources: Hud.**

Su Li Bean - High yielding kidney x soy cross from China. *Source History: 1 in 2004.* **Sources: Syn.**

Sword Bean, Akanata Mame –70 days, *Canavalia gladiata*, Pink-red flowers and pods with ridged edge, harvest immature pods at 4 in. length, prepare same as snap beans, from Japan. *Source History: 1 in 2004.* **Sources: Kit.**

Sword Bean, Shironata Mame -70 days, *Canavalia gladiata*, White flowers, green pods with ridged edge, harvest when immature at 4 in. length, prepare like snap beans. *Source History: 1 in 2004.* **Sources: Kit.**

Urd Bean - 90 days, *Phaseolus mungo*, plants grow much like mung beans but mature earlier and are more prolific, small reticulated grey seeds, bean beetle and drought resistant. *Source History: 1 in 1991; 1 in 1994; 1 in 1998; 3 in 2004.* **Sources: Ho13, Se7, Syn.**

Velvet Bean - *Mucuna pruriens*, vigorous climber produces long bunches of purple flowers, short thick pods, coffee-like substitute is made from the ground and roasted seeds, also quite ornamental. *Source History: 1 in 1991; 2 in 1994; 1 in 1998; 1 in 2004.* **Sources: Ban.**

Velvet Bean: Florida Speckled - 120 days, A prolific green snap bean for warm climates, large vines, purple blooms, black gray and mottled seeds, mature seeds must be well cooked to be eaten, frequently grown as winter forage for milk cows swine or deer. *Source History: 2 in 1987; 1 in 1991; 1 in 1994; 2 in 1998; 1 in 2004.* **Sources: Gl2.**

Winged Bean - *Psophocarpus tetragonolobus*, tropical perennial, tuberous roots pods and leaves edible, 4-sided pods to 24 in., high protein seeds and roots, will not flower in the U.S., climber. *Source History: 8 in 1981; 6 in 1984; 3 in 1987; 2 in 1991; 3 in 1994; 3 in 1998; 6 in 2004.* **Sources: Ban, Ech, Ev2, Ho13, Pr3, Tho.**

Winged Bean Day Neutral - *Psophocarpus tetragonolobus*, tropical perennial, usually this plant cannot be grown north of southern Florida because of day length restrictions, but this var. blooms regardless of day length, leaves, flowers and roots are edible, light blue flowers are used for food coloring, winged, 4-sided pods are eaten raw or cooked like snap beans, seeds are rich in protein. *Source History: 1 in 2004.* **Sources: Hud.**

Yard Long - (Asparagus Bean, Dow Gauk, Chinese Yard Long, Orient Wonder) - 70-80 days, Imported from China, 24-36 in. bright-green pods are thin as pencils, good as snap beans when young, widely grown in Japan and China and Africa, good yields, vigorous growing vines, not good for the northern United States. *Source History: 19 in 1981; 14 in 1984; 20 in 1987; 24 in 1991; 27 in 1994; 19 in 1998; 20 in 2004.* **Sources: Alb, Bo17, Bo19, Bu2, Ech, HA3, Jor, Kil, Loc, MAY, Ni1, Ont, Or10, Pin, Ra5, Re8, Scf, Se24, So12, WE10.**

Yard Long, Akasanjaku - 75 days, Stringless green pods grow 12-16 in. long, red seeds, same use as string beans, cut in 2 in. lengths, fast cooking, popular in Szechwan dish called dry-fried beans, from Asia. *Source History: 1 in 2004.* **Sources: Kit.**

Yard Long, Dolique - 80 days, Pods grow to 3 ft. long. *Source History: 1 in 2004.* **Sources: Lej.**

Yard Long, Extra Long Black - (Extra Long Black Seeded) - *Vigna sesquipedalis*, tall climbing cousin of cowpeas, longer and more slender than red-seeded type, deep-green 24-30 in. pods, for snaps at 15-18 in., drought and heat tol. *Source History: 1 in 1981; 1 in 1984; 1 in 1987; 2 in 1998; 2 in 2004.* **Sources: Ev2, Red.**

Yard Long, Kurojuroku - 75 days, Dark green, slender, stringless 14-18 in. long pods, each with 10-20 black seeds, thrives in the heat, high yields, highly prized in Asia for the sweet crunchy pods. *Source History: 1 in 2004.* **Sources: Kit.**

Yard Long, Liana - 80 days, Vines grow over 8 ft., long tender, dark green pods can grow up to 30 in., steamed or used in stir-fry, does best in warm soil and warm growing conditions. *Source History: 5 in 1998; 5 in 2004.* **Sources: Dam, Eo2, Jo1, Ter, Ver.**

Yard Long, Long White Snake - 80 days, Over 33 in. long, nearly 10 in. longer than other vars., light green pods are wrinkled like a snake skin, unusual white seeds, favorite heirloom in Malaysia. *Source History: 1 in 1998; 1 in 2004.* **Sources: Red.**

Yard Long, Orient Wonder - 70 days, Bright green, stringless, slender long pods, brown-red seeds, same use as string beans, cut into 2 in. sections and deep fry, stir fry, steam or cook, used in the popular Szechwan dish called dry-fried beans, tolerates heat but sets better in cool or drier weather, from Asia. *Source History: 3 in 2004.* **Sources: Ev2, Kit, SAK.**

Yard Long, White Seeded - 60 days, Firm but tender light green pods to 24 in. long, white seeds, uses include deep fry, stir fry, steaming or cooking, heat tolerant. *Source History: 1 in 2004.* **Sources: Kit.**

Varieties dropped since 1981 (and the year dropped):

Adzuki Express (2004), *Adzuki Tix-Su-Dow* (1994), *Adzuki, Violet Speckled* (1998), *Alphatoco* (1994), *Ayalet Garbanzo* (1987), *Charlotte* (1998), *Chick Pea, Bush* (1998), *Chickpea (Garbanzo) UC-5* (1987), *Chinese Long Bean/Cowpea Cross* (1998), *Cluster Bean Barasati* (1998), *Egyptian Fava* (1987), *Garbanzo, Brown Seeded* (1998), *Garbanzo, Dolores de Hidalgo* (2004), *Garbanzo, Gene Pool Mix* (2004), *Garbanzo, Green Seeded* (1994), *Garbanzo, Hannan* (2004), *Garbanzo, Indian Red 'Desi'* (1998), *Garbanzo, Sarah* (2004), *Garbanzo, Sonoran Brown* (1998), *Garbanzos del Norte* (2004), *Guaje* (1994), *Japanese* (1994), *Jumbo Mung* (1987), *Lentil, Brewer* (2004), *Lentil, O'odham* (1998), *Lentil, Pink* (1998), *Mifuki Asparagus Bean* (1987), *Mother Beans* (1994), *Mung Bean: Berken* (2004), *Mung: Urd Bean* (2004), *Peruano* (1991), *Rice Pea* (1991), *Snowbean* (1994), *Southern Garbanzo* (1991), *Spaghetti* (2004), *Swordbean* (1987), *Velvet Bean: 90 Day* (1991), *Velvet Bean: 120 Day* (1998), *Wild Bean* (1987), *Wild Black Jones Bean* (1991), *Wild Cocolmeca* (1998), *Wild Common Bean* (1994), *Wild Frijolillo* (1994), *Wild Mexican Pole* (1994), *Wild Mitten Leaf Desert Bean* (1991), *Winged Bean, Thai Bin Dow* (1994), *Yard Long, Extra Long Red Seed* (1994), *Yard Long, Purple* (2004), *Yard Long, Sabah Snake* (2004).

Miscellaneous Solanaceae

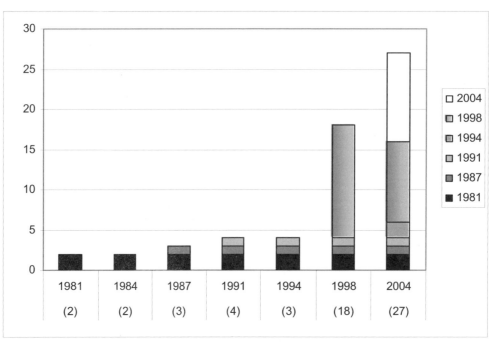

Number of varieties listed in 1981 and still available in 2004: 2 to 2 (100%).
Overall change in number of varieties from 1981 to 2004: 2 to 27 (1350%).
Number of 2004 varieties available from 1 or 2 sources: 24 out of 27 (89%).

Cannibal Tomato - (Poro Poro) - *S. uporo*, plants 2-3 ft. tall, 2 in. fruit, red when mature, used like tomatoes to make sauces, leaves are used in cooking. *Source History: 1 in 1998; 2 in 2004.* **Sources: Ho13, Hud.**

Chichiquilite - 70 days, *S. nigrum*, small berries resemble nightshade, prepare same as Garden Huckeberry. *Source History: 1 in 2004.* **Sources: Sa9.**

Egg Tree - 90 days, *Solanum melongena*, three months from seed produces small white eggs, needs good light and well drained soil and warmth, interesting windowsill plant, grows quickly. *Source History: 2 in 1981; 2 in 1984; 3 in 1987; 1 in 1991; 1 in 1998; 2 in 2004.* **Sources: Ban, Bu1.**

Eggplant Collards, Gbogname - *S. macrocarpon*, 2.5-3 in. fruits are cooked like eggplant, young leaves are cooked and eaten like collards, perennial in mild climates, withstands heat, ancient vegetable from Togo, West Africa. *Source History: 1 in 1998; 2 in 2004.* **Sources: Ba8, Hud.**

Eggplant, African Egg - 75 days, *S. aethiopicum*, rare African eggplant, duck egg-shaped fruit, white > orange, tender and mild, resemble miniature pumpkins, excel. yields. *Source History:1 in 1998; 1 in 2004.* **Sources: Sa9.**

Eggplant, Black Stem - *S. gilo*, tall shrub with downy leaves, round to spindle-shaped edible fruits, 1-3 in., green > red, eaten raw or cooked, said to taste like carrots or green beans, Africa and Brazil. *Source History: 1 in 2004.* **Sources: Hud.**

Eggplant, Brazil - Small orange oval shaped fruit. *Source History: 1 in 2004.* **Sources: WE10.**

Eggplant, Cookstown Orange - 69 days, Round smooth fruit, green with yellow stripes, red at maturity, sweet, somewhat bitter taste, good yields, rare. *Source History: 2 in 1998; 2 in 2004.* **Sources: Ba8, Sa9.**

Eggplant, Gboma - *S. macrocarpon*, short plant, 1 in. green fruit, eaten as a cooked vegetable, leaves are steamed, from tropical Africa. *Source History: 1 in 1998; 1 in 2004.* **Sources: Ho13.**

Eggplant, Goyo Kumba - *S. aethiopicum*, bright red fruit, 2-3 in., slightly flattened, from Africa. *Source History: 1 in 2004.* **Sources: Ho13.**

Eggplant, Paris White/Amboni - *S. aethiopicum*, 4 ft. tall plants, egg shaped fruit is mild and tasty, white > red, from Ghana, West Africa, very rare. *Source History: 1 in 1998; 1 in 2004.* **Sources: Ba8.**

Eggplant, Red Burkina Faso - *S. aethiopicum*, beautiful large plant to 6 ft. with dark purple stems and large green leaves, slightly flattened red-orange fruit is 5 in. across, deeply ribbed and wrinkled, from western African nation of Burkina Faso. *Source History: 1 in 2004.* **Sources: Ami.**

Eggplant, Red China - 50 days, Beautiful spiny plant grows 3-4 ft. tall, flat, ribbed fruit is 2.5 x .75 in., green > orange, look like tiny pumpkins, best eaten in green stage. *Source History: 1 in 1998; 2 in 2004.* **Sources: Ba8, Sa9.**

Eggplant, Ruffled Red - (Hmong Eggplant) - *S. integrifolium*, flattened, ribbed, deep red-orange fruit, 2-3 in. across, ornamental plants to 2-3 ft. tall with dark purple stems, pleasantly bitter flavor, used in Asian stir fry, referred to as mini- pumpkins by florists. *Source History: 1 in 1994; 1 in 1998; 1 in 2004.* **Sources: Hud.**

Eggplant, Turkish Italian Orange - *S. integrifolium*, spineless plants grow 4 ft. tall, bite-sized orange-red fruits, heirloom originally from Turkey, more recently from Italy, intro. by CV So1, 1990. *Source History: 1 in 1991; 1 in 1994; 1 in 1998; 1 in 2004.* **Sources: So1.**

Eggplant, Turkish Orange - 80-85 days, *S. integrifolium*, 2 ft. plants with many 2 in. orange fruit that look like tiny pumpkins, strong flavor, brought to the U.S. from Italy and Turkey in the Civil war era to use as ornamental plants. *Source History: 1 in 1998; 9 in 2004.* **Sources: Ami, Ba8, Bo19, Hi13, Ho13, La1, Ma18, ME9, Se7, SE14, Syn.**

Garden Huckleberry - (Wonderberry, Sunberry) - 75-80 days, *Solanum melanocerasum*, plants resemble tomatoes, 3 ft. tall, hundreds of round shiny purple-black .5-.75 in. fruits in clusters, pick when glossy black color turns to dull-black, hardy and prolific, thrives anywhere, tasteless when raw and unsweetened, makes delicious pies and preserves, good for freezing and canning. *Source History: 19 in 1981; 16 in 1984; 17 in 1987; 12 in 1991; 14 in 1994; 17 in 1998; 16 in 2004.* **Sources: Ba8, Bu1, Fa1, Gr27, He17, Hi13, Ho13, Lej, Mo13, Roh, Sa9, Se16, So12, So25, Sto, Vi4.**

Garden Huckleberry, Fruity - 130 days from seed, fully ripened fruit is sweet and edible right off the plant, fruit cap turns yellow-brown when mature, no bitterness, fruit is 2-3 times the size of Mrs. B's and less mushy, but not quite as sweet, excellent for pies, sauces, jams and jellies, freezes well, from PSR Breeding Program. *Source History: 1 in 2004.* **Sources: Pr3.**

Garden Huckleberry, Mrs. B's - 55 days, Mrs. B's Non-Bitter Garden Huckleberry, a nightshade, 3 ft. plants, sweet 3/8 in. fruits borne in clusters, slightly smaller and more intense flavor than regular Garden Huckleberry. *Source History: 2 in 1987; 2 in 1991; 3 in 1994; 2 in 1998; 1 in 2004.* **Sources: Pr3.**

Kangaroo Apple - (Poro Poro) - *S. aviculare*, 3-5 ft. plants, deeply lobed leaves, 1 in. oval fruit, harvest mature orange fruit after it falls off the plant, loses acidity at maturity, eaten raw, boiled or baked, from Australia. *Source History: 1 in 2004.* **Sources: Ho13.**

Litchi Tomato - *S. sisymbriifolium*, 4-5 ft., thorny plant bears 1 in., bright red fruit in clusters, seedy, sweet, undomesticated species from South America. *Source History: 1 in 1998; 1 in 2004.* **Sources: Ho13.**

Ogomo - *S. melanocerasum*, African variety of garden huckleberry with large purple-black berries, edible when ripe, cooked. *Source History: 1 in 2004.* **Sources: Hud.**

S. aethiop - (Jaxatu Soxna) - African variety, leaves eaten as spinach, 4 in. ribbed fruit, edible, said to be sweet, slightly bitter. *Source History: 1 in 2004.* **Sources: Hud.**

S. sisymbriifolium - White or pale blue 1.5 in. flowers, bright red berries .75-1 in., edible. *Source History: 1 in 2004.* **Sources: Hud.**

S. maglia - Wild potato species from South America, small tubers, blue skin, blue-white flesh, stores well, withstands temps to -20 degrees F, sent in Fall only. *Source History: 1 in 2004.* **Sources: Ho13.**

Sunberry - 75 days, *S. Burbankii*, Luther Burbank selection, compact plants, purple-blue fruit with white bloom, slightly sweet, used in pies, preserves and wines, said by admirers to rival and even surpass blueberries, bred in the early 1900s by Luther Burbank who called it Sunberry and lamented its being renamed Wonderberry by the dealer who purchased and introduced it, critics immediately claimed Burbank had simply reintroduced *S. nigrum*, Garden Huckleberry, as a new plant, Burbank said this was the result of many years of crossing of *Solanum guinense*, a species native to Africa, and *S. villosum*, indigenous to Europe. *Source History: 3 in 1998; 5 in 2004.* **Sources: Ho13, Ma13, Se16, So25, Und.**

Tree Tomato - (Ruby Red Tamarillo) - Perennial, *Cyphomandra betaceae*, fast growing tree to 10 ft., oblong purple- red fruits taste almost like tomatoes, whitish flesh, dark-red seeds, very easy to grow, for greenhouse or 5 gallon container, must protect. *Source History: 2 in 1981; 2 in 1984; 4 in 1987; 5 in 1991; 5 in 1994; 5 in 1998; 3 in 2004.* **Sources: Ban, Ho13, Hud.**

Varieties dropped since 1981 (and the year dropped):

Chice de Cierva (2004), Eggplant African (2004), Eggplant Aubergine du Mali (2004).

Disease Abbreviations

AB	Alternaria Blight (Early Blight)
ABR	Aphanomyces Black Root
ALS	Alternaria Leaf Spot
AN	Anthracnose
AN1	Anthracnose (Race 1)
AN2	Anthracnose (Race 2)
AN3	Anthracnose (Race 3)
ASC	Alternaria Stem Canker
BB	Brown Blight
BC	Brown Check
BCMV	Bean Common Mosaic Virus
BER	Blossom End Rot
BM	Blue Mold
BR	Black Rot (or Brown Rot)
BSR	Bacterial Soft Rot
BS	Brown Spot
BW	Bacterial Wilt
BYM	Bean Yellow Mosaic
BYMV	Bean Yellow Mosaic Virus
CB	Crown Blight
CF	Catface
CMV	Cucumber Mosaic Virus (Spinach Blight)
CRR	Corky Root Rot
CS	Crease Stem
CT	Curly Top
CTV	Curly Top Virus
CW	Common Wilt
DM	Downy Mildew
FW	Fusarium Wilt
FW1	Fusarium Wilt (Race 1)
FW2	Fusarium Wilt (Race 2)
GLS	Gray Leaf Spot (Stemphylium)
GN	Ground Neck
GP	Greasy Pod
GSB	Gummy Stem Blight
GW	Gray Wall
HB	Halo Blight
HH	Hollow Heart
IC	Internal Cork
LB	Late Blight
LCV	Leaf Curl Virus
LMV	Lettuce Mosaic Virus
LM	Leaf Mold (Cladosporium)
NHR	Nail Head Rust
NY15	New York 15 Strain (of Bean Virus 1)
nem.	Nematodes
PEMV	Pea Enation Mosaic Virus
PI	Phytopithora Infestans
PM	Powdery Mildew
PMV	Pod Mottle Virus
PPMV	Pepper Mottle Virus
PR	Pink Root
PS	Pea Streak
PVY	Potato Virus Y
PW	Pea Wilt
RB	Rib Blight
RCVM	Red Clover Vein Mosaic
RD	Rib Discoloration
RKN	Root Knot Nematode
RR	Root Rot
RS	Rhizoctonia Scurf
SB	Spanish Blight
SBW	Southern Bacterial Wilt
SM	Squash Mosaic
SR	Slime Rot (or Soft Rot)
SS	Stipple Streak
ST	Stemphylium
TB	Tipburn
TEV	Tobacco Etch Virus
TLS	Target Leaf Spot
TMV	Tobacco Mosaic Virus
TMV1	Tobacco Mosaic Virus Race 1
TMV2	Tobacco Mosaic Virus Race 2
TMV3	Tobacco Mosaic Virus Race 3
TMV4	Tobacco Mosaic Virus Race 4
TR1	Tomato Race 1
TY	Top Yellows
VW	Verticillium Wilt
WH	White Heart
WCM	Western Celery Mosaic
WMV	Watermelon Mosaic Virus
YSV	Yellow Streak Virus

Symbols and Other Abbreviations

/	per
>	changing to
#	number
AAS	All American Selection
AES	Agricultural Experiment Station
B. S.	black seed
conc.	concentrated
cont.	continued
det.	determinate
dev.	developed
dia.	diameter
dwf.	dwarf
E.	East
excel.	excellent
exp.	experiment
F.	Fahrenheit
ft.	foot (30.48 cm)
ger.	germinate or germination
gyn.	gynoecious
hort.	horticultural
imp.	improved
in.	inch (2.54 cm)
indet.	indeterminate (vining)
intro.	introduced or introduction
lb.	pound (453.59 grams)
mech. harv.	mechanical harvest
med.	medium
mo.	month
N. E.	Northeast
N.	North
no.	number
o.p.	open-pollinated
oz.	ounce (28.35 grams)
peren.	perennial
PVP	Plant Variety Protection
reg.	regular
res.	resistant
rtl.	retail
SASE	self-addressed stamped envelope
sel.	selected or selection
St.	State
sta.	station
str.	strain
susc.	susceptible
S.	South
S. E.	Southeast
S. W.	Southwest
temp.	temperature
tol.	tolerant
W. S.	white seed
whsl.	wholesale
U	University
USDA	U.S. Dept. of Agriculture
veg.	vegetable
var.	variety

DISCARD

Hartness Library
Vermont Technical College
One Main St.
Randolph Center, VT 05061